A TEXTBOOK OF

MACHINE DESIGN

DR. P.C. SHARMA
DR. D.K. AGGARWAL

A TEXTBOOK OF
MACHINE DESIGN

(MECHANICAL ENGINEERING DESIGN)

(IN S.I. UNITS)

For

B.Sc. (Engineering); A.M.I. E. (Section B);
Students and Engineers in the Field

By

DR. P.C. SHARMA
B.Sc. Engg. (Mech.) (Hons.)
M.Sc. Engg. (Mech.) (Distinction)
Ph.D. (I.I.T. Delhi); M.I.S.M.E. (India)
Ex Principal SUS College of Engg. & Tech.
Tangori (Mohali) (Punjab)
Formerly of
Punjab Engineering College Chandigarh

DR. D.K. AGGARWAL
B.Sc. Engg. (Mech.) (Hons.)
M.Sc. Engg. (Mech.) (Distinction)
Ph.D. (Pb) F.I.E. (India)
Prof. & Head Deptt. of Mechanical Engg.
IEET
Baddi (H.P.)

S.K. KATARIA & SONS®

Publisher of Engineering Books
4885/109, Prakash Mahal, Dr. Subhash Bhargav Lane,
Opposite Delhi Medical Association, Daryaganj, New Delhi-110002
Phone: +91-11-43551243; Mobile: +91-9871775858, 7678425209
e-mail: katariabook@gmail.com, katariabooks@yahoo.com
Website: www.skkatariaandsons.com
(e-books, subscriptions and bulk buying also available on our website)

Published by:

S.K. Kataria & Sons®
4885/109, Prakash Mahal, Dr Subhash Bhargav Lane,
Daryaganj, New Delhi-110002
Phone: +91-11-43551243; Mobile: +91-9871775858, 7678425209
e-mail: katariabook@gmail.com, katariabooks@yahoo.com

Ninth Edition : 1999; Thirteenth Edition: 2017, Reprint: 2026

***Price :* ₹ 895/-**

Laser Typesetting by:

SARA ASSIGNMENTS, Patparganj, Delhi-110092.

Printed by:

A.S. Offset Printers, Delhi-110051.

Preface To The Thirteenth Edition

This standard treatise on "Machine Design" has remained very popular with the readers for the last 40 years (The book was first published in Jan. 1976).

The publication of the latest edition has provided an opportunity to the authors to go through the whole text and incorporate changes in the text and as far as possible rectify all the errors. Lot of changes have been made in chapters 8 and 9, that is, "Design of Welds" and "Design of Threaded Joints". The whole text has been recomposed on computer and all the figures haven been redrawn on computer. This has given an entirely new outlook to the text.

Any suggestions to further enhance the utility of the book will be most welcome and thankfully acknowledged.

Authors

ACKNOWLEDGEMENT

Authors express their sincere thanks to the Director General, Indian Standards Institute for granting permission to produce certain extracts from the following Indian Standards:

(1) IS 1560-1978 (Pt. I)
(2) IS 1570-1979 (Pt. II)
(3) IS 2524-1963
(4) IS 1076-1967
(5) IS 1076-1967
(6) IS 2709-1964
(7) IS 1928-1961
(8) IS 1929-1961
(9) IS 813-1961
(10) IS 4211-1971 (Pt. V)
(11) IS 2585-1961
(12) IS 1363-1967
(13) IS 2393-1980
(14) IS 6688-1981
(15) IS 2610-1964
(16) IS 2048-1975
(17) IS 2293-1974
(18) IS 2291-1981
(19) IS 2291-1980
(20) IS 2292-1974
(21) IS 2643-1975 (Pt. I)
(22) IS 3823-1966 (Pt. I & II)
(23) IS 1624-1966 (Pt. II)
(24) IS 1991-1980
(25) IS 2122-1973 (Pt. I)
(26) IS 1691-1980
(27) IS 2494-1974
(28) IS 3153-1965
(29) IS 1804-1976
(30) IS 2403-1975
(31) IS 2535-1968
(32) IS 4694-1968
(33) IS 7008-1973 (Pt. II & III)

For full details, the readers may refer to the above Indian Standards. These standards are available from Indian Standards Institution, New Delhi, and its Branch and Regional offices at Ahmedabad, Bangalore, Bombay, Calcutta, Chandigarh, Hyderabad, Kanpur, Madras and Patna.

P.C. Sharma
D.K. Aggarwal

CONTENTS

SECTION—IV PRESSURE VESSELS

SECTION—V I.C. ENGINE PARTS

SECTION—VI MISCELLANEOUS TOPICS

Section I

GENERAL

CHAPTER 1 Introduction

1.1 ENGINEERING DESIGN

It may be defined as the iterative decision making activity to create the plans by which resources are converted, preferably optimally, into systems, processes or devices to perform the desired functions and to meet human needs.

The main purpose of courses in engineering design is to help the student inculcate the skill in applying the knowledge he has acquired in his science, engineering science, social studies and humanities courses, to the solution of practical engineering problems. The object is not to teach a new engineering course but to help the students learn how to use purposefully and effectively what is already known to him.

"Machine Design" or **"Mechanical Design"** is one of the branches of Engineering Design. It concerns primarily the systems by which energy is converted into useful mechanical forms, and of mechanisms required to convert the output of the machines to the desired form. A course on **"Machine Design"** acquaints the reader with the fundamentals of designing the most commonly used parts, elements and units of the various machines. The design may lead to an entirely new machine or an improvement on an existing machine. Thus Machine Design is the creation of the right combination of correctly proportioned moving and stationary components so constructed and joined as to enable the liberation, transformation, and utilisation of energy. Machine design is an iterative process in which we proceed through several steps, evaluate the results, and then return to an earlier phase of the procedure. The aim of the iterative process is to optimize the design. The course on "**Machine Design**" helps in inculcating in the reader the habits of thought in selecting the material and best suited shapes, computing the dimensions on the basis of strength, and specifying the manufacturing requirements, i.e., surface finish, accuracy, etc. "*Machine Design*" is nothing but the application of:

Mathematics
Kinematics
Statics
Dynamics
Mechanics of Materials
Engineering Materials
Mechanical technology of metals and
Engineering Drawing

It may also involve the application of other subjects, such as electrical theory, thermodynamics and hydraulics as in the design of electric motors and generators, I.C. Engines and steam engines and turbines, gas turbines, water pumps, water turbines and other hydraulic machines, etc.

1.1.1 Some Design Definitions

1. Empirical Design: The design based on past experience and existing practice is known as the "Empirical Design". In this, various dimensions are set out as a proportion of certain main dimension, e.g., the various dimensions of a flange coupling are given as proportions of the shaft diameter. The conditions under which certain machine elements serve are many and varied and, therefore, are not subject to accurate and detailed evaluation. Because of this, machine designers widely use approximate formulas and empirical relations. The centre distance of belt drives is taken in terms of diameters of the two pulleys, and so on.

2. Rational Design: It is purely a mathematical design and it is based upon principles of mechanics, etc. In this design, the different relations developed/studied in the mechanics of solids are used to decide the different dimensions of a product.

3. Combined Rational and Empirical Design: As the heading indicates, it is a combination of rational and empirical approach for doing a design. In this, the design of a component is based on solid mechanics, but in no case the individual dimension is to be less than the existing or past practice. For example, the various dimensions of a knuckle joint are initially taken in proportion to the diameter of the rods to be joined. These dimensions are then checked for strength with the help of appropriate relations.

4. Design by Evolution: In the past, the technology development was slow. The society was not eager or impatient for new engineering designs. The designs tended to evolve over a long span of time. The designing was thought to be an activity of grey haired persons who will obediently translate practical needs into drawings of products. Devices or technical systems changed gradually, each change making a small improvement on the previous model. The designs were without adequate reasons. The leisurely pace of technological changes reduced the risk of making major errors. This was design by evolution.

5. Design by Innovation: These days the competition has widened. The slow pace of technological changes is not the need of the day. Present circumstances require bolder, faster improvements. A new design is based on ideas and concepts not tried earlier. So the technical risks which the designers must face today are very great. The need of evaluating information in the face of uncertainty and to analyse complicated interactions of components is immense. So the products are to be designed de-novo. This is design by innovation.

6. New Design: These designs are created from a scratch through the application of scientific laws, technical ability and creative thinking. In order to venture successfully into new design fields, personal qualities of high order are needed in the designer. Locomotive, aeroplanes, steam engines, motor car, etc. are some of the examples of the same. It is of course quite difficult to make a new design. New design is also called as "Inventive or creative design". A designer's work may border on invention. If a problem arises that suggests a machine for the solution and no suitable machine exists, a designer will have to create it. He is, at this stage, more, interested in how the job is to be done than what materials he should use and how the parts should be proportioned. Here, we give the example of a machine which was invented to separate fertile seed from infertile seed. Since a fertile seed is of the same size as an infertile one, screening will not make the separation. A dead seed, however, is lighter than a live one. So, a new and specialised machine was invented which would detect this weight difference. The well-known phenomenon of gravity was applied to create this machine.

7. Adaptive Design or Redesign: There are many fields where development has practically ceased, e.g., bicycles, I.C. engines, etc., so that there is hardly anything left for the designer to do except make minor modifications usually in the dimensions of the product. The design activity can be achieved with ordinary technical training and it demands no special knowledge or skill. This work is concerned with adaptation of existing designs. The final outcome does not differ much from the initial product. Much of designer's work consists of redesign. A press is to be designed for more capacity, a turret lathe is to be redesigned to accommodate large stock or an automobile engine is to be redesigned to develop more power. Here, the mechanisms are already established and the designer does not have to devise different ones. The same linkage is to be used in the press, the same spindle speeds are to be used on the lathe, and the automobile engine is to have the same number of cylinders. The designer's problem is to decide what parts are to be changed and what parts are to remain the same. In making these decisions, he must never loose sight of the fact that changes are costly and that his redesign must produce the desired result with the minimum of alteration.

8. Developed Design: Development design is also started from an existing design. The final outcome may differ quite markedly from the initial product. Here, considerably more design abilities and scientific training are needed.

9. Functional Design: The most important criterion to be satisfied by the design of a machine component or a machine is that it should work. Such a design is called as "Functional design". Sound functional design assures that the product will operate satisfactorily for a specified period of time in the manner intended. This design takes into account all those details that affect the operation of a product, i.e., physical characteristics of the materials (strength, stiffness, weight, etc.) and the various service characteristics (corrosion resistance, conductivity, etc.). So functional design is the "strength design". It itself does not take into consideration ease of manufacture or maintenance and it frequently neglects the appearance and human engineering.

10. Initial and final design: In the present day methods of designing a machine part, two distinct but closely related divisions of the problem present themselves:

(i) First the machine part is tentatively proportioned so as to suit the relationships associated with the chosen criterion of serviceability (strength, stiffness, wear resistance, etc.). This stage of design widely draws upon various assumptions and may be called "initial or preliminary design".

(ii) Second, the chosen critical criterion of serviceability is used to determine the actual stresses and the factor of safety of the part at hand and to compare them with the maximum and minimum safe values. In other words, the proposed part is checked for compliance with a relevant condition for strength. This stage may be called "final design".

Obviously, the objective of final design is to refine the results presented by preliminary design. It is carried out when the parts geometry has already been outlined in the course of initial design or adopted from structural considerations. It also serves to determine loads on existing machines or machine parts.

As noted above, design is an iterative process in which we proceed through several steps, evaluate the results, and then return to an earlier phase of the procedure. But under no circumstances does it reduce to substituting initial data into ready-made formulas.

11. Design for Realization: It consists of transforming the ideas or concepts into reality. Here each detail of the design concept is developed in relation to the useful functions required and also such technical functions as of strength, etc. Here the design is not only meant to ascertain their primary functions but also to maintain required degree of accuracy, efficiency, reliability, safety, etc.

12. Design for Profits: It consists of analysing those factors or circumstances which affect the value of product in relation to its cost.

13. Process Design: It is usually limited to the design of tools, equipment, components, apparatus, etc. for mass production systems. Design features selected by other groups for the product parts, that are being produced by the production processes, are directly or indirectly related to developments in production processes.

14. Industrial Design: It is an independent design effort that is not part of a particular product producing organization. Individuals interested in this type of design activity develop combined abilities in styling, engineering, industrial processes and product design to become consultants for the manufacturing type of organization. Further specialization of design efforts is sometimes done, such as chassis design, engine design, tool design, die design, body design, etc. in an automobile industry.

15. Design by Experience: Occasionally, there are times when a design decision can be based on experience and excessive analysis is unjustified. Let us take a very simple example. Suppose, you are asked to devise a means of fastening an automobile license plate to its bracket. Let us assume that you decide that a bolt in a hole is as simple a scheme as you can contrive. Now, you have to decide about the diameter of the bolt and its material. You would certainly not take much time to arrive at the correct design. Your decision would be based on your previous experience with bolts. Obviously, a very small bolt would hold the plate in place but very small bolts are hard to handle. Surely you would specify a standard bolt that could be purchased at any hardware store and you would select a size large enough to handle conveniently, yet not so large as to look awkward.

Now about the bolt material. Probably, the choice would be between steel, which rusts, and brass or aluminium which does not. The choice will depend upon your personal opinion. If you prefer to use the same bolts over again you would probably select brass or aluminium. But if you treat the rusting steel as insurance against the bolt working loose, you might prefer to cut the bolts off at the end of the year and put in new ones. So, as a designer you must make up your mind. Hence for such simple design problems, detailed analysis is unjustified from the cost point of view. The decisions can be based on past experience.

16. Optimum Design: An engineer's objective is always to get maximum efficiency and economy out of men, money and materials to meet the identified needs in the best possible manner. In this direction optimization is an important concept in engineering.

The iterative nature of design provides an opportunity to improve the design on the basis of preceding outcome. In this manner, we can arrive at a design which provides the best possible technical conditions, e.g., maximum on best performance of the product with the least total cost per unit (material cost, manufacturing cost and maintenance cost) consistent with weight, size, strength and similar considerations. Such a design is called "optimum design." Many techniques for optimizing a design have been developed. Optimizing a design means maximizing the desirable effects and minimizing the undesired ones. However, the various optimizing methods often have limited application in a complex design situation. In the usual situation, the actual design parameters chosen by the designer are a compromise among several alternatives. In a problem, there may be too many design variables. It may not be possible to include all of them in the optimization due to considerations like available time and other constraints. The parameters chosen for the design are then close to but not at optimum values. These are referred to as "optima values", i.e., the best that can be achieved within the total constraints of the system. So, by the term "optimal design" we mean the best of all feasible designs.

1.1.2 Machines and Machine Elements

A machine may be defined as a combination of stationary and moving parts constructed for the useful purpose of generating, transforming, or utilizing mechanical energy. Machines may be classified as below:

1. Machines for generating mechanical energy, also known as prime movers. These machines convert some form of energy (electrical, heat, hydraulic, etc.) into mechanical work, e.g., steam engines, steam turbines, I.C. engines, gas turbines, water turbines, etc.

2. Machines for transforming mechanical energy, known as converting machines. These types of machines transform mechanical energy into another form of energy, e.g., electric generators, air or hydraulic pumps, etc.

3. Machines for utilizing mechanical energy. These machines receive mechanical energy and deliver and utilize it as such in the performance of useful work, e.g., lathe and other machine tools, etc.

All machines are constituted of elements or parts and units. A machine element or part is a separate part of machine, either integral or consisting of several small pieces which are rigidly joined together by rivetting, welding, etc. When many individual joined machine parts are grouped together and work together these are known as machine units. The machine elements may be classified in two groups as below:

1. General purpose elements: These machine elements are common to machines of various types and applications, e.g., nuts, bolts, keys, axles, shafts, couplings, bearings, etc.

2. Special purpose elements: These machine elements are employed only with a particular type of machine, or machines, e.g., pistons, connecting rods, cam shafts, push rods, turbine blades, etc.

The general purpose machine elements can be further sub-divided into the following groups:

1. Fasteners: These are the machine elements which connect or join the parts of a machine. These joints are of two types; "permanent joints" and "temporary joints". The permanent joints are those which can not be disassembled into separate elements without damaging or destroying them. The examples are: riveted joints, welded joints, etc. On the other hand, the temporary joints are those which can be easily disassembled without damaging or destroying the joints at all. The examples are nut and bolt joints, cotter joints, etc.

2. Elements of Rotary motion drive: These elements belong to the units or machines which transmit power. The examples are belt, rope, chain, gear and worm drives, shafts, axles, couplings, bearings, etc.

1.2 BASIC REQUIREMENTS FOR MACHINE ELEMENTS AND MACHINES

While designing a machine element or a complete machine, the following things must be considered carefully and adequately by the designer:

- Cost
- High output and Efficiency
- Strength
- Stiffness or rigidity
- Wear resistance
- Light weight and minimum dimensions
- Reliability
- Durability
- Economy of Performance
- Accessibility
- Processability
- Compliance with state standards.
- Lubrication
- Operation safety
- Ease of assembly
- Ease and simplicity of disassembly
- Ease and simplicity of servicing and control
- Economy of repairs and maintenance
- Use of standard parts
- Use of easily available materials
- Appearance
- Number of machines to be built.

Cost is the most important item in the design of machine elements of machines. For a given machine, the best design is, that will do the required job with the least over all cost per unit, consistent with weight, size, probability and similar considerations.

The machine element must be sufficiently strong so that it is not damaged or permanently deformed during the specified life of the machine. Also, the machine element must be sufficiently rigid so that the elastic deformation under the applied forces does not exceed a specified limit. Wear of the machine elements should be minimised by proper lubrication, by increasing the strength and hardness of the working surfaces or by increasing the contact surface so that during the service life of the machines, its strength is not unduly reduced or the mating of the machine element is not affected.

The machine element should be sufficiently strong, rigid and wear resistant with minimum weight and least dimensions. This can be achieved by the wide use of light weight rolled sections; by employing the latest methods of surface hardening i.e., shot peening, case hardening, nitriding, porous chrome plate high frequency current hardening, by using high strength grades of cast iron and light alloys and if possible by improving the design of the machine elements.

Reliability, durability are very important for the function of the machine. Lubrication is also very important and adequate lubrication with minimum wastage should be provided.

For the safety of the machine and the operator the hazard producing things should be eliminated and the design should conform to safety code.

The designer must see that the machine can be assembled with a minimum of difficulty. If the machine is very complex such as automobiles, typewriters, sewing machines, etc. and if there is to be mass production of these, then the use of so called "**unit assembly**" is quite common. For the case of repair of individual elements or "unit assemblies" the machine should be easily accessible.

For the economy of maintenance and repairs, the wear should be kept to a minimum and cheap and simple replacement should be available.

For the economy of performance, the correct mechanical, hydraulic, thermodynamic and other principles must be applied.

The design can be modified by the possibility of using standard parts and the material on **hand**. The shape and the material of the machine should be selected so that it can be produced with minimum charges on labour and processing methods. The complex and costly methods of production should be avoided. Appearance is also an important factor, especially for the product which has competition in the market. Lastly, the machine element or the machine must comply with requirements of the standards set up by the State.

1.3 DESIGN PROCEDURE

When a new machine or its elements are to be designed, no rigid rules can be specified for a designer to follow. The problem can be solved in many ways. However, a designer may proceed as follows:

1. Make a written statement of the problem, as completely and as clearly as possible, and also of the purpose for which the machine is to be designed.
2. Make selection of the possible mechanisms which will give the desired motion or group of motions.
3. Determine the forces acting on and the energy transmitted by each element of the machine.
4. Select the material best suited for each element of the machine.
5. Determine the allowable or design stress considering all the factors that affect the strength of the machine part.
6. Determine the size of each member, with a view to preventing undue distortion or breakage under the applied loads.
7. Modify the machine elements in parts to agree with the past experience and judgement and to facilitate manufacture.
8. Make assembly and detail drawings of the machine with complete specification for the materials and the manufacturing methods, i.e., accuracy, surface finish, etc.

So the design of a new machine or its elements may follow a tentative plan shown in Fig. 1.1. As is shown by broken arrows, the design plan is not a rigid one way process but there is continuous feed back. After the general specifications of the machine or a machine member have been set, the kinematic scheme is decided. This is followed by a force analysis. This force analysis at this stage will be rather incomplete for designs where dynamics is of importance, since masses of the moving parts are not yet known. With this preliminary information, a tentative design of the components is done because the forces are not known exactly. After this, a more exact analysis of forces is made and the designs refined. There are many factors other than strength and rigidity which might affect and modify the final decisions. These factors are: availability of materials and manufacturing techniques, weight and space limitations, appearance, etc.

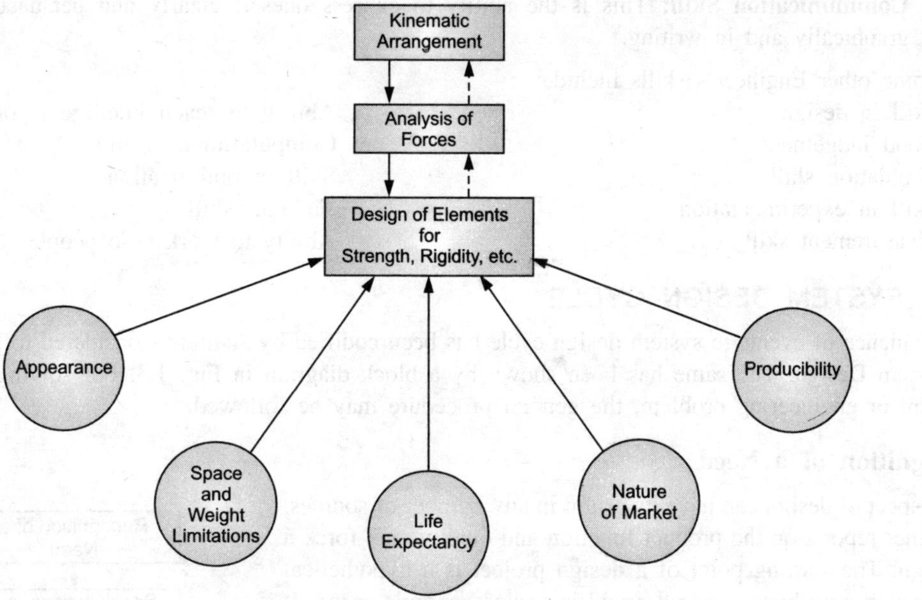

Fig. 1.1 *Design plan*

Figure 1.2 depicts the evolution of Machine Design.

It is a wrong procedure to design a machine without the benefit of the past knowledge and experience about that machine. The design of machine is based on the past experience, factual knowledge and Engineer's skill. Factual knowledge means that the engineer or designer must know the theory and the other knowledge in the appropriate branch of engineering. For example, for machine design, the engineer must have a knowledge of kinematics, statics, dynamics, mechanical technology of metals, mechanics of materials, etc. The factual knowledge and the experience are then combined with creativity of engineer's skill to evolve a new machine. Engineer's skill includes:

1. **Inventiveness:** This is the ability to think of or discover valuable and useful ideas or concepts for things or processes to achieve given objectives.
2. **Engineering Analysis:** This is the ability to analyse a given component, system, or process using engineering or scientific principles in order to arrive quickly at meaningful answer.

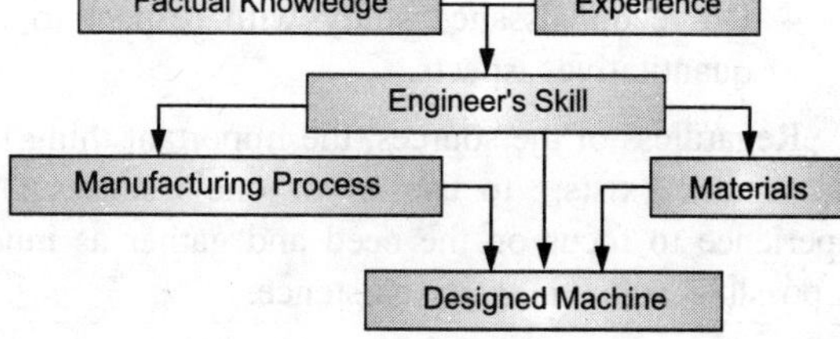

Fig. 1.2 *Evolution of machine design*

3. **Engineering Science:** This means the thorough knowledge and training in depth in an engineering science speciality.
4. **Interdisciplinary Ability:** The engineer should have the ability to deal competently and confidently with basic problems or ideas from disciplines outside one's speciality.
5. **Mathematical Skill:** This means thorough knowledge of mathematics.
6. **Decision Making:** This is the ability to make decisions in the event of uncertainty but with full and balanced grasp of all the factors involved.
7. **Manufacturing Processes:** The engineer should have a knowledge of and an appreciation for the potential and limitations of both old and new manufacturing processes.
8. **Communication Skill:** This is the ability to express oneself clearly and persuasively, orally, graphically and in writing.

Some other Engineer's skills include:

Skill in design	Ability to reach intelligent conclusions
Good judgement	Computational ability
Simulation skill	Skill in optimization
Skill in experimentation	Thought skill
Measurement skill	Ability to work with people.

1.4 SYSTEM DESIGN CYCLE

The sequence of events in system design cycle has been codified by Asimow considered to be the father of System Design. The same has been shown by a block diagram in Fig. 1.3. For solving any design problem or engineering problem, the general procedure may be followed.

Recognition of a Need

This aspect of design can have its origin in any number of sources. Customer reports on the product function and quality may force a redesign. The starting point of a design project is a hypothetical need which may have been observed in socio-economic scene. It may be phrased in the form of a primitive statement based on either market or consumer studies or on untested observations. The need may be existing or may not yet exist, but the evidence may be existing that the need is latent, i.e., the need can be evoked when economic means for its satisfaction becomes available. In what ever way the need may be there, in starting a project its economic existence must be established with sufficient confidence.

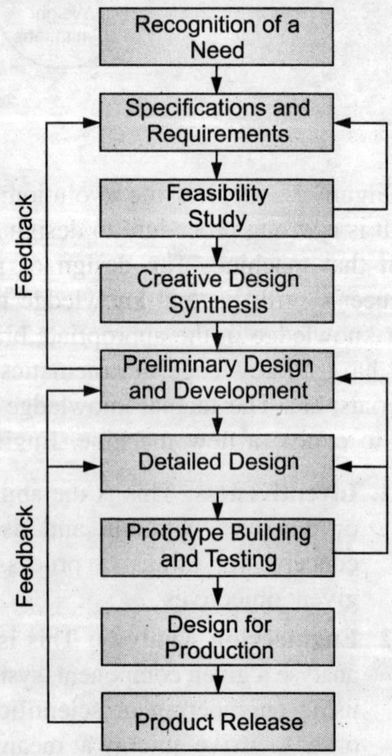

Fig. 1.3 *Design cycle*

The following points will help at this stage:

1. Study the nature of the need.
2. Establish the need reasonably well to the extent possible.
3. Make primitive statement of need.
4. Do reconnaissance study with respect to qualitative and quantitative aspects.

Regardless of the sources, the important thing is to recognize that a need exists, to use all of one's senses and background experience to focus on the need and gather as much information as possible concerning its existence.

Specifications and Requirements

Once a need has been established, its requirements must be carefully spelled out. Very often the specifications are stated in such general terms as to indicate that the consumer has only a vague idea of what he wants. On the other hand the military, the atomic energy commission and a variety of sophisticated engineering clients submit specifications that are written in great details. In these requirements design and performance requirements are carefully stated. The health, safety, or any other legal requirements will be part of the original specifications. In addition government, commercial and industrial standards are made part of specification. Also, the specifications are prepared keeping in mind the production competence and capability of the company.

Feasibility Study

Once the problem's need has been established and the specifications have been prepared, accepted and submitted, the feasibility study is done. The purpose of this study is to check the possibility of success or failure of the proposed project both from technical and economic stand point. Feasibility can cut short fruitless design effort. In this study, various questions are to be answered.

1. Is any natural law being defied?
2. Are some of the specifications beyond what is technically available at present?
3. Is there any dependence on scarce materials?
4. Will the cost of the end product be too high?

The persons doing the feasibility study will be the engineers with strong design background, knowledge of engineering sciences, a good knowledge of material usage, knowledge of production methods and requirements of the sales department. In fact, it is likely that the men responsible for the feasibility study will be same engineers who will ultimately have the design responsibility for the over all project.

Quite often, as a result of the feasibility study, some changes are to be made in the specifications and requirements in order to provide a greater likelihood of project success. This action is indicated as a feed back loop in Fig. 1.3.

Creative Design Synthesis

Once the feasibility of the design is established, the creative design synthesis has to be done. This phase is the most challenging and interesting part of design. Here, the designer can act as engineer, inventor, and artist all in one, because now he is called upon to create.

By creativity is meant simply the demonstrable ability to get new and useful ideas for meeting engineering goals. The qualities of the new design will be:

1. The quality of newness or uniqueness.
2. Things which are either useful (as with a product) or appreciated (as with an art work) are called creative. Creative things are sometimes either purposeful or beautiful or both.
3. The third quality in the design solution will be that it will have simplicity. A solution having the qualities of newness and usefulness are appreciated but complex would not be known as a creative solution.

Thus inventive things are called creative or inventive when they have the above mentioned three qualities. Educators have not, as yet, devised a method for teaching creativity, although methods to stimulate it have been tried. Creativity is an aspect of human behaviour that is still being investigated by psychologists, and an in-depth understanding is still lacking. It is generally agreed that all of us have the ability to create at various competence levels. Creativity involves thinking, and certainly we all think, to a greater or lesser extent. Proper educational preparation should enhance creative process. It must be remembered that psychological handicaps tend to lessen the capacity for logical thought and creative effort.

The reader will agree that the above given definition of creativity has a qualitative nature. It remains sometimes a matter of opinion whether a particular thing is creative or not. But the above given definition is useful because most people will agree most of the time about the particular use of the word. Familiar examples which can be observed as creative are: the transistor, the zipper fastener, the cherry shoe polish container, nylon, the engine, the heat sealed plastic bag, fibreglass, textiles, plywood, ball point pen, vacuum cleaners, extruded wire insulation and so on.

Preliminary Design and Development

After the process of creativity design is complete, there will be one or more possible designs that satisfy the given set of specifications and requirements. It then becomes necessary to decide which of the solutions to choose for the preliminary design and development stage. The bases for making decisions are many and varied. Once the choice has been made, the layout drawings showing machine are made to determine the overall configuration and to establish functional relations between various parts of machine or system. These layout drawings will contain important dimensions and notations as well as auxiliary and sectional views. Note, this stage is primarily concerned with checking the validity of the functional and overall size requirements of the specification.

Rarely the specifications and requirements are achieved during this phase. Therefore, referring to Fig. 1.3, we see a feed back from this phase to specifications and requirements. This indicates that some specifications are to be modified in order to complete the design. The stages of the preliminary design can be summarized as below:

1. Selection of the most useful solution from the several suggested solutions.
2. Formulation of useful model preferably mathematical for this selected solution.
3. Analysis of this model.
4. Formation of criterion in order to determine and measure which values we are attempting to maximise or minimise.
5. Optimization so as to best meet the objective as expressed by the criterion.
6. Prediction of performance.
7. Preparing layout of the selected solution, making a check for its function.

Detailed Design

Many of the activities of the detail design phase of the design process are often missed by many designers, particularly for smaller jobs; because so much of the detailing in the sense of component design is already done in the preliminary design phase. The designer tends to finish off his design by making a complete assembly drawing of the "final design" in the preliminary design phase itself. Particularly, this happens when the outcome is from an optimization model of the product or a pilot plant of the process. But this could not really be the end of a designer's responsibility. We must know the distinction between the processes of obtaining a final design and detailing it. Irrespective of mass manufactured components or individual fabricated component, a complete set of detail and assembly drawings is needed for successful manufacture, fabrication, production, etc. On these drawings full dimensions and tolerances with dimensions of the parts and components, specifications of the material and any other conceivable data the production shops may need to know for producing the product are to be mentioned. This phase of the design work consists of two major activities:

1. Detailing the parts, component and their assembly.
2. Going into sufficient details of fabrication/manufacture in implementing the design.

Both these activities call for a good knowledge on the part of the designer of existing and available technical knowledge. If he lacks this knowledge he should collect it or consult proper authorities before his detailed design. Thus we see that the detailed design is the stage at which design idea, i.e., the concepts are transformed to reality. In this stage each detail of the design concept is developed in relation to the use, functions as of strength, mobility, conductivity, etc. The specialized knowledge of strength of materials, theory of machines, thermodynamics, fluid mechanics, electromagnetics, etc. is used for detailing each individual part of the designed product. Further, it is not only to ascertain their primary functions, but also to maintain efficiency, required degree of accuracy, reliability, safety margins, quality standards, appearance, features and others. Many times, a good design has to be abandoned for problems of materials, procurements, material conversion, fabrication and assembly. Problems involved in tooling, machining, casting, moulding, welding, testing and other manufacturing processes are too many to enumerate.

The second phase as indicated earlier requires a good knowledge of manufacturing processes. The manufacturing method is strongly influenced by the material selected as well as the quantity desired. For minimum manufacturing cost to be there, the method of manufacturing should be such that the following are avoided.

1. Heavy weight of material
2. Reclamping of work
3. Auxiliary equipment
4. Special purpose tools
5. Change of machine
6. Expensive machinery operations
7. Finishing operations
8. Rejects
9. Difficult assembly

Prototype Building and Testings

After completing all the details, the sub-assemblies and assembly drawings including the materials and parts list, the completed design is sent to the prototype or model shop for fabrication. At this juncture the parts are fabricated, commercial components are purchased, and the machine or system, after having been assembled, is ready for evolution and testing. This testing can help in:

1. Producing acceptable performance
2. Physical realizability of the system and the components
3. To verify design hypotheses
4. To generate new design information
5. To reveal flaws
6. To develop improved design concepts
7. To expose difficulties overlooked in the theoretical design
8. To provide convincing evidence
9. To increase validity of the results.

The output of this testing may or may not yet yield data that requires changes or modifications in the preliminary design or detailed design area. This is shown in the Fig. 1.3. After the changes or/and modifications are made, when necessary, the new components are put into the prototype assembly for continued testing and evaluation. The procedure of making continuous revisions and improvements to the design is repeated until the design engineer is satisfied that the performance specifications are met. At this point, all the drawings, parts and materials lists are sent to the production engineering department where the drawings are modified so that the completed project can be designed for production.

Design for Production

In addition to being functionally sound, a product must have sales appeal and must be competitive in price. In order for a product to be made economically, it must be designed so that the most appropriate material and processes will be utilized. This is called "Design for production". For this consideration is given to design changes in the functional design that would be compatible with the most economical methods of production. In modern day nomenclature, this is sometimes called *value analysis* and it is becoming an important consideration in design. For example, a production engineer may consider a part suitable for manufacture by stamping, casting, or perhaps forging. For large scale manufacturing, any of these processes may be more economical than individual part machining. Of course, in making decision, consideration must be given to necessary tooling costs. Another example of what the production engineer may look for is the possibility of combining several fabricated parts into one or replacing some parts with commercially available equivalents. Still another consideration of the production engineer is the possibility of replacing some of the materials with equivalent satisfactory but less expensive materials. When the production drawings have been completed, they are sent to the production department for product release.

Product Release

Production prototype are usually made and tested, and any drawback that can be easily corrected is generally referred back to preliminary design and development or to detailed design for alteration as shown in Fig. 1.3 a feed back loop.

By no means is the above description complete in all its detail nor is it the only path that is followed in the design of a product, device, or a system. However, to appreciate fully the aforementioned description of the design cycle, it is necessary to be involved with the procedure on a daily basis.

1.5 HUMAN FACTORS ENGINEERING

Technology ultimately serves humanity. Hence, the well being of human beings as operators of machinery and as users of products, is of great importance in any manufacturing activity. For this, it is essential to understand the interaction of a human being with machines and the workplace environment for the proper design of the machinery and the development of safe and efficient working conditions in manufacturing plants. This field concerning all aspects of human-machine interaction is known as: "Human factors engineering" or "Human Engineering". This area of study is also called "Ergonomics". This word is from Greek; 'Ergo' means work and 'nomos' means rule, law; 'nomics' means management.

Human factors engineering is an inter-disciplinary subject. It deals with applying the knowledge gained from anatomy, applied psychology (industrial psychology), industrial engineering, medicine, climatology, physiology, bio-mechanics, bio-engineering and anthropometry (the science of dimensions of humans). The study concerns the characteristics and capabilities of human body and mind. These include such factors as: Height, Weight, Vision, Hearing, Posture, Strength, Age, Intelligence, Educational level, dexterity and reaction time. Numerous statistical data pertaining to these characteristics are available.

Goals of Human-Factors Engineering

1. To design a product that is pleasing and easy to use.
2. To maximize human values such as safety, health and comfort and satisfaction of operators of machines, and to minimize fatigue and stress. All these factors lead to increase in the productivity of the human beings.
3. To maximize the quality and efficiency of work.

Ergonomics has received a big impetus due to Space programme and Automobile industry.

The measurements (body dimensions) of average-size adult males and females, as determined from studies made by different authors may be used for most designs requiring close adaptation to human physical characteristics. Since, many designs involve both foot and hand movements, these average body dimensions, as tabulated, include arms, hands, legs and feet along with other parts of the human body. As noted above, data relating to body proportions and dimensions are known as "anthropometric data". Other factors that must be dealt with in design include: motor activities and body orientation; the five human senses (sight, hearing, touch, smell and taste); atmospheric environment, temperature, humidity and light and finally accelerative forces if they are exceptional and are likely to cause undue physical discomfort.

The study of body motion deals primarily with the effective range of operation of parts of the body, usually the arms and legs and the amount of body force a human being may reasonably be expected to exert in the performance of an assigned task. For example, the normal working area for a person in seated position extends about 40 cm from the body. The greatest work rate that an average healthy male can provide on a sustained basis is about 500 W.

Vision is an important factor in all designs where visual gauges or coloured lights on control panels are a part of a control system that involves manual operation. Control panels for such equipment must be designed to be within the visual range of the average person and distinctive colours must be selected and used for the coloured lights. The colours selected must be easily recognizable and must be capable of quickly attracting the operator's attention. Vision studies have produced much new information of use in highway design.

While designing a large industrial plant or a particular work area in the plant, the cockpit of an aeroplane or the cabin of a space vehicle, special attention has to be given to working environment, that is, temperature, humidity, lighting, colour schemes and sound, etc.

Human factors engineering is applied to a wide range of consumers goods such as : automobiles, refrigerators, furniture, office equipment, lawn movers, hand tools and so on.

1.5.1 Aesthetics or Visual Design

The ultimate objective of any design and manufacturing activity is to sell the designed and manufactured product and earn profit. In to-days' highly competitive market, performance of the product alone may no longer be sufficient to sell a product. The product should have the eye-pleasing appearance for customer acceptance. Without this, there will not be sufficient sales and hence profit, even though the product might otherwise have been carefully engineered. A pleasing design appearance is a way of adding perceived value to a product. This is known as "Aesthetics" or "Visual design".

Visual design includes the use of line, form proportion, texture and colour to produce the eye-pleasing appearance. This part of design which is concerned with appearance is properly called "Styling". Big companies have their own "Styling". "Divisions". Small companies hire trained stylists for a product.

1.6 PRINCIPLES OF COMPUTER AIDED DESIGN (CAD)

Design is a key feature of any production system and its importance is growing. It is becoming a major determinant of competitiveness due to emphasis on design and product innovation stressing novel features, frequent changes, greater response to customer requirements and more rapid new product development cycles. All this necessitates the use of not only new design tools such as Computer Aided Design (CAD), but also approaches to integrate design with manufacturing.

Definition of CAD

Computer-aided design can be simply defined as: Using a Computer in the design process.

The more comprehensive definitions can be:

"By CAD, we mean the development and use of special Computer programmes (Software) to help the designer in carrying out routine computations for the design of products and processes".

"CAD can be defined as the use of Computer systems to assist in the development, analysis, modification, and optimization of an engineering design. It also includes storing and communication of design information".

CAD helps in getting the analytical results very quickly. This will enable the designer to evaluate more than one design alternatives which otherwise is not possible. Also, optimum design solutions can be obtained by using sophisticated programs. This will result in significant savings in unit costs.

All the above has been made possible by two significant developments:

1. **NC Drafting Machines.** These machines enable part and assembly drawings to be produced, based on the computational results of CAD.

2. **Computer Graphics.** (CG) With this, the designer is able to interact with the computer using a probe or a light pen on a graphics terminal. The analytical design is also displayed on the same terminal, by the computer. If the designer wants a change in the computed solution, he can do so by sketching the desired modification on the graphics terminal with his light pen. The modified design is analysed by the computer and results are again displayed on the graphics terminal.

When the designer is finally satisfied with the design, it is assigned a classification number. All the data concerning the product, namely, the component and part numbers, detail and assembly drawing numbers, material for each part, dimensions, tolerances and surface finish requirements are stored in a master file.

1.6.1 The Components of a CAD System

A typical CAD system consists of three main parts:

1. **Hardware:** It consists of the Computer and the Input-Output (I/O) devices. Input devices are generally used to transfer information from a human or storage medium to the computer. A keyboard is the standard input device used to transmit alphanumeric data to the system. The standard output device for CAD is a CRT (Cathode Ray Tube).

2. **Operating System Software:** It is the interface between the CAD application software and the hardware.

3. **Application Software:** It is the heart of a CAD system. It consists of programs that do: 2-D, 2.5-D or 3-D Geometric modelling; engineering analysis; and drafting.

(a) **Geometric Modeling:** Geometric modeling is an important part in CAD. It allows the creation of a geometric model to represent the size and shape of the component. It corresponds to the synthesis phase of the design process.

For geometric modeling, the designer gives the following commands to the computer:

(i) Commands for generating basic geometric elements such as points, lines and circles, etc.
(ii) Commands for scaling, rotation or other transformations of the geometric elements.
(iii) Commands which join the various elements into the desired shape of the part being created.

The Computer converts these commands into a mathematical model, stores it in the computer data file and displays it as an image on the CRT screen.

The geometric models may be:

(i) 2-D for flat objects.
(ii) 2.5-D for parts with constant section and having no side-wall details.
(iii) 3-D for generalized part shape.

The wire frame geometric modeling is the simplest and most common. It is easy to create and requires less computer time and memory. However, it is difficult to interpret the image. The perception of wire frame modeling can be enhanced by: Colour graphics, Dashed lines to portray invisible edges, removal of hidden lines and surface models. Surface models define part geometry precisely and can be used to produce NC machining instructions automatically. However, these models don't give information about the part being solid or hollow and about the internal features of the part. The most advanced technique of geometric modeling is 3-D Solid modeling.

(b) **Engineering Analysis:** All the engineering design problems require some type of analysis such as:

(i) Stress-strain calculations
(ii) Heat transfer computations.
(iii) Fluid flow problems
(iv) Lubrication problems
(v) Static and dynamic analysis of complex structures such as aircraft, cars, dams, bridges, etc. and so on.

The Finite Element Method (FEM) is probably the most powerful and widely used tool in CAD.

(c) **Design Review and Evaluation :** It has already been discussed in Art. 1.6 under "Computer Graphics".

(d) **Automated Drafting :** See Art. 1.6. Automated drafting increases productivity four times over manual drafting. The quality of the drawings is improved and consistent. In addition, automated drafting allows automatic dimensioning, generation of cross hatched areas, sectional views, scaling and enlarged views of particular part details.

Other Benefits of the System are

(i) Drawings can be stored and recalled easily.
(ii) By means of a coding system, similar part designs can be grouped into classes and the similarities related. By this, the designer can recall a particular design and modify it to suit a different application. This will be much more convenient and cheaper than always designing new parts. This feature is very useful in linking the CAD system with the CAM (Computer aided Manufacturing) system.

Another powerful feature of CAD systems is "Simulation". Basically, the principle behind simulation is that it is cheaper, faster and easier to explore the behaviour of products and systems before they are made or used, through simulation in a CAD system. A variety of software techniques and mathematical models are used for this purpose. By FE analysis and simulation techniques, the final product design can be progressively improved.

For many companies, a key design problem is the storage and retrieval of information to wide and old product ranges which are still active and for which orders may be expected. CAD systems allow the use of computer storage to hold electronic information and the use of database management techniques to radically improve the speed and efficiency of access. Another advantage is that where a new order is only slightly different to a part already held in the library, the library drawing can be modified quickly to create the new design. This cuts the overall response time.

Thus, the CAD system performs the following two activities:

1. Product design.
2. Production preparation, that is, part drawings, assembly drawings, bill of materials, etc.

EXERCISES

1. Define ''Engineering Design".
2. What is the objective of teaching "Engineering Design"?
3. What is the scope of "Machine Design"?
4. Define a Machine.
5. How the machines are classified?
6. How the machine elements are classified?
7. Discuss the basic requirements for machine elements and machines.

8. Write a tentative design procedure.
9. Draw a tentative design plan.
10. With the help of a sketch, depict the evolution of machine design.
11. Discuss the qualities of an engineer.
12. Discuss the system design cycle in brief.
13. Differentiate between empirical design and rational design.
14. Discuss combined rational and empirical design.
15. Discuss the following: (a) Design by evolution, (b) Design by innovation, (c) New design.
16. It is said that to make a new design is difficult, comment.
17. Discuss in detail about adaptive design or redesign.
18. Discuss the following: (a) Developed design, (b) Functional design, (c) Initial and final design.
19. Briefly discuss : Design for realization, Design for profits, Process design and Industrial design.
20. Design by experience is important, comment.
21. Define and discuss optimum design.
22. Iterative nature of design activity improves the design, comment.
23. What is "optimal design"?
24. Define CAD.
25. List the components of a typical CAD system.
26. Write a short note on 'Geometric modeling'.
27. What is 'Computer Graphics'?
28. Write the benefits of CAD.
29. Write a short note on "Human factors" Engineering.
30. Write the goals of "Human factors" Engineering.
31. What is "Aesthetics or Visual Design"?
32. What is "Styling"?
33. What is the height of letters/numbers on the indicator of visual indicators?

    Ans. The height of letter/numbers on the indicator, should be as follows:

$$\text{Height of letter/numbers on the indicator} \geq \frac{\text{Reading distance}}{200}.$$

34. What is the importance of "Colours" in the "Aesthetic design" of a product?

    Selection of a proper colour for the product is an important factor in its "Aesthetic design". The following recommendations are due to Morgan:

Colour	**Meaning**
Red	Danger, Hazard, Hot
Orange	Possible danger
Yellow	Caution
Green	Safety
Blue	Caution, cold
Grey	Dull

35. What is Sequential Design Process?

    It can be depicted as follows:

36. What is Simultaneous Design Process?

It can be Depicted as follows:

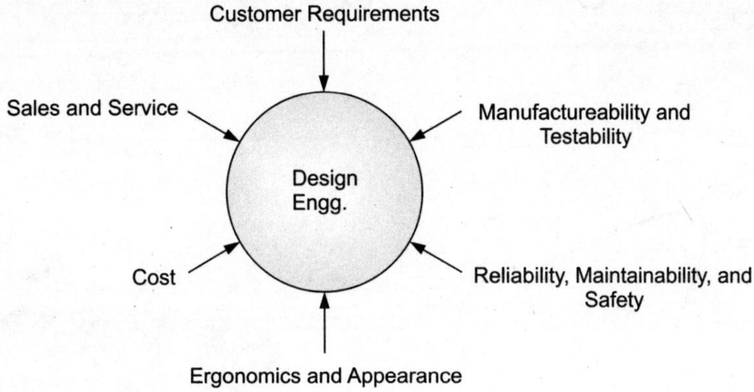

❑❑❑

CHAPTER 2

Engineering Materials

2.1 GENERAL DISCUSSION

After the general layout of the machine has been made and the mechanisms to be used have been decided upon, it is necessary to select the proper material for each machine member. This involves the consideration of many factors such as the engineering properties of the available materials, weight, size, shape of the member, material cost, fabricating cost, overhead charges and any properties peculiar to the use of which the member is to be put. A few examples of the engineering properties to be considered are strength, stiffness, toughness, ductility, fatigue resistance, resilience, shock resistance, corrosion resistance, wear resistance, hardness, frictional qualities, machinability, casting properties, forging properties, temperature properties and electrical properties.

The selection of a material is one of the decisions the designer is supposed to make and the selection of the proper material has always been a difficult problem and is one that requires an increasing amount of experience as the variety of materials available is constantly increasing. Previously the designers were handicapped by the lack of ferrous and non-ferrous materials suitable for their particular needs but now the problem is of selecting the most suitable one from the hundreds of materials available. The proper material to be used is one that meets the engineering requirements of the designers, the production and heat-treating requirements of the production department and the cost requirement of the purchasing agent.

It is important that materials be used in such a way as to take full advantage of their natural characteristics. Thus it is usually desirable to think of design change when selecting materials. For example, a moulded plastic part may take advantage of the natural colour of the material and of insulating properties in a design which could not be employed if sheet metal were used. In the application of the

Table 2.1 Average properties of pure metals

Material	ρ kg/m^3	E GN/m^2	C GN/m^2	$\frac{1}{m}$	f_{yp} MN/m^2	f_{ut} MN/m^2	δ*	HB	t_m °C	α μm/m°C	K* W/m°C
Aluminium	2720	70	26	0.33	24.5	77.0	25	23	650±8	23.6	22.1
Berylium	1830	230	—	—	385.0	630.0	5	—	1283±39	12.4	169.5
Cobalt	8850	210	—	—	308.0	700.0	13	—	1490	12.23	69.2
Columbium	8600	159	—	—	168.0	273.0	49	—	2470	7.2	—
Copper	8900	119	45	0.33	70.0	224.0	45	42	1083	16.7	338.5
Hafuium	11.4×10^3	140	—	—	154.0	413.0	35	—	2095±110	6.12	—
Lead	11.35×10^3	14	49	0.43	14.0	17.5	50	5	327	29.5	34.5
Molybdenum	10.2×10^3	294	—	—	525.0	700.0	30	250	2605±50	4.8	13.0
Nickel	8900	210	77	—	140.0	490.0	40	100	1440±6	13.2	60.5
Tantalum	16.6×10^3	210	—	—	700.0	700.0	3	123	3000±50	6.58	53.5
Tungston	19.3×10^3	350	—	—	—	2100.0	0	—	337	4.32	200.3
Uranium	18.7×10^3	210	—	—	175.0	630.0	13	—	1130	—	25.9
Vanadium	6000	150	—	—	385.0	475.0	34	—	1750±50	7.75	—
Ziroconium	67.9×10^3	84	—	—	112.0	452.0	31	77	1860	5.22	—

ρ = mass density, E = Modulus of Elasticity, C = Modulus of Rigidity, $\frac{1}{m}$ = Poisson's ratio

δ* percentage elongation in a gauze length of 50 mm

K* thermal conductivity in W/m°C

HB = Brinell Hardness Number

α = Coefficient of Expansion

f_{yp} = yield point stress

f_{ut} = ultimate tensile stress

t_m = melting temperature

Table 2.2 Average properties of alloys

Material	ρ kg/m^3	E GN/m^2	C GN/m^2	$\frac{1}{m}$	f_{yp} MN/m^2	f_{ut} MN/m^2	δ*	HB	t_m °C	α μm/m°C	K* W/m°C
Aluminium (1) 2024—T_4	2270	74	28	0.33	308.0	420.0	12	—	580±78	23.2	121
Aluminium (2) 7076—T_6	7800	73	27.3	0.33	490.0	560.0	10	150	556±11	23.6	129.7
Brass yellow hard (3)	8460	1050	39.2	0.35	420.0	518.0	10	180	932	18.9	119.3
Cast iron grey	7200	91	98.5	0.20	—	150.0T 750.0	—	180	1178	12.05	46.5
Cast iron ductile	7200	175	65	—	476.0	630.0	7	235	1150	13.5	32.8
Phosphor Bronze	8800	112	42	—	455.0	560.0	8	160	1050	16.9	69.2
Steel 0.2% C—HR	7830	210	84	0.27	280.0	490.0	35	125	1516	12.05	51.9
Steel 0.2% C—CR	7830	210	84	0.27	455.0	560.5	20	160	1516	12.05	51.9
Steel 1.0% C—HR	7830	210	84	0.27	588.0	100.0	8	293	—	13.1	51.9
Steel 1% C Hardened and tempered at 430° C	7830	210	84	0.27	965.0	1400.0	12	388	—	—	51.9
Steel 0.38 (4) to 0.43% C Hardened and tempered at 430°C	7750	203	77	0.30	1330.0	1410.0	15	388	—	—	51.9
Stainless steel	7910	203	74.2	0.30	700.0	980.0	30	280	1415	15.8	16.3

1. Wrought copper—Aluminium alloy solution treated followed by natural aging.
2. Wrought Zinc—Aluminium alloy solution treated followed by artificial aging.
3. Cast iron —T—Tension, C—compression, HR—Hot rolled, CR—Cold rolled.
4. Nickel—molybdenum steel with a carbon content of 0.38 to 0.43% C.

moulded plastics or metal castings, on the other hand, it is sometimes necessary to assemble a part from a number of pieces. The right material may not be chosen unless it is used in a design that takes full advantage of its useful characteristics while avoiding less favourable properties.

2.2 MATERIALS OF CONSTRUCTION CAN BE CLASSIFIED INTO TWO GROUPS

1. Metallic materials.
2. Non-metallic materials.

Metallic materials can be further split into ferrous and non-ferrous materials.

Ferrous materials consist mainly of iron with comparatively small addition of other materials. Ferrous materials are iron and its alloys such as cast iron, grey cast iron, malleable cast iron, wrought iron, and steels of low carbon content and high carbon content.

Non-ferrous materials contain little or no iron. The materials are copper, zinc, tin, lead and aluminium. The alloys produced by combining materials are brass, bronze, gun-metal, duralumin, etc.

Non-metallic materials include plastics, rubber, leather, carbon, wood, glass, etc.

2.3 PROPERTIES OF CONSTRUCTION MATERIALS

The economical use of construction materials in mechanical design requires considerable knowledge of their mechanical properties. These properties of the materials are those properties which describe the behaviour of the material under mechanical usage. Tables 2.1 to 2.4 of the most important properties of materials commonly used in machine design, are included for reference. The properties of materials vary with the mechanical and thermal treatment received during manufacture, the tabulated values must be considered as average design values only. Higher values may be obtained by careful manufacture and heat treatment.

2.4 DEFINITIONS

Homogeneity: A material that exhibits the same properties throughout is said to be homogeneous. Homogeneity is an ideal state that is not achieved by real materials, particularly metals. However, the variation in properties is so small that calculations for stress and deflection assume that a material is homogeneous throughout.

Isotropy: A material that displays the same elastic properties in all loading directions is said to be isotropic. The equations of elasticity and strength of materials are based upon this assumption.

Plasticity: When a material is subjected to an external load of such magnitude that deformation continues with no apparent further increase in load, the material is said to have become plastic. Once in this region, the material will have experienced a permanent set or deformation and will not return to its original size and shape when the load is removed. Thus, plasticity can be considered as the opposite of elasticity. This definition is for a theoretically perfect material. Actually a material such as metal will continue to deform with only very small further increase of load.

Proportional limit: It is defined as that point on a stress-strain curve beyond which the stress is no longer proportional to strain. The actual value of the proportional limit is quite difficult to establish because it greatly depends on the quality of measuring instruments.

Proof stress: It is that stress which will cause a small but permanent strain in a material. This permanent strain is specified as 0.01 per cent in 50 mm or called the 0.01 per cent offset.

Modulus of Elasticity: It is the ratio of stress to strain within the elastic limit. Then it is really a measure of the stiffness (for tension and compression) or rigidity for shear of material. For ductile materials the modulus of elasticity is same for both tension and compression and it is represented by E. In particular, it is called “Young’s modulus”. For brittle materials such as cast iron, this modulus is not the same for tension and compression. For the case of shear or torsion, this modulus is called the shear modulus or modulus of rigidity or transverse modulus and it is represented by C.

Hook's Law: It states that stress is proportional to strain upto the proportional limit of the material.

Elastic limit (proportional unit): The greatest stress a metal can withstand without permanent elongation **i.e.** when the load is released, the sample will return to its original length.

Yield point (yield strength): The stress at which appreciable elongation occurs without increase in stress.

Ultimate strength: Maximum stress required to break a specimen.

Modulus of elasticity: It is the ratio of stress to strain within the elastic limit. It is a measure of stiffness.

Elongation: The ratio of the increase in gauge length to the original gauge length, expressed in percentage.

Table 2.3 Average properties of ceramics

Material	ρ kg/m³	E GN/m²	$\frac{1}{m}$	f_{ut} MN/m²	f_{uc} MN/m²	t_m °C	α μm/m°C	K* W/m°C
Polycrystalline glass	2500	88	0.25	140.0	—	1250	11.8	1.95
Fused silica glass	2200	73.5	0.17	—	—	1585	0.54	1.38
Tungston carbide	15300	700	0.24	—	4200.00	—	3.96	120.9
Moulded graphite	1550	5.6	—	420.0	24.0	—	2.16	138.4
Alumina ceramics	3700	315	—	280.0	2450.0	2015	7.2	—
Concrete medium strength	2500	21–28	10–20	2.1–2.8	21.0–28	—	10.8	—

Reduction in area: The decrease in the X-sectional area of the test specimen, after fracture, to the original area expressed in percentage.

Ultimate strain: The unit elongation (elongation per unit of length) at the specimen breaking point. It is a measure of ductility.

Ductility and Brittleness: Ductility of a metal is to deform plastically without fracturing. In general, it means deformation under low stressing instead of sudden impact. Although there are other measures, ductility is most commonly measured by means of elongation and reduction of area in the tensile test

$$\%\text{ elongation} = \frac{\text{Final gauge length} - \text{original gauge length}}{\text{Original gauge length}} \times 100$$

$$\%\text{ reduction of area} = \frac{\text{Original area} - \text{final area}}{\text{Original area}} \times 100$$

A material is generally classified as brittle if the percentage elongation is less than 5 in a gauge length of 50 mm.

Stiffness: It is the ability of a material to resist deformation, that is stretching, Bending and Twisting etc. It is defined by the modulus of elasticity, that is, only within the elastic range. A material with a higher value of '*E*' is stiffer than other.

Indentation hardness: In metal working, hardness generally implies resistance to penetration. It may, however, include resistance to scratching, abrasion or cutting, Indentation hardness is probably the most widely used mechanical testing procedure. It is a non-destructive test, relatively inexpensive, and can be performed by semi skilled operators. The Brinell hardness test measures the diameter of impression of a hard ball under compressive load on the flat surface of the metal sample (units N/mm²). The Rockwell hardness test measures the depth of penetration of diamond point (Rockwell C) of steel ball (Rockwell B) into the sample.

Table 2.4 Average properties of organic materials

Material	ρ kg/m³	$\frac{1}{m}$	E (1) GN/m²	f_{ypt} or f_{ut} MN/m²	f_{uc} or f_{ypc} MN/m²	δ*	HB	t_m°C	α μm/m°C	K* W/m°C
Thermoplastics										
Acrylic plastics, cast	1130	0.4	T–28 × 10³	56.0	84.0	4	100	77	90	0.207
Cellulose acetate moulded	1300	—	B–17.5 × 10³	35.0	140.0	2	100	—	126	0.140
Teflon	2300	0.4	T-3500 C–5600	14.0	7.0	235	85	260	54	0.256
Nylon 66, general purpose (polyamide)	1140	0.4	T–29 × 10³	56.0	90.0	60	118	121	99	0.244
Polystyrene, general purpose	1040	—	T–35 × 10³	49.0	100.0	2	80	66	126	—
Polyethlene, medium density	930	0.45	T–4200	14.0		350	—	81	225	0.349
Thermosets										
Epoxy, general purpose	1200	—	T–45.5×10³	49.0	210.0	4	100	80	60	0.779
Phenolic, moulded type, general purpose	1440	—	T–70×10³	52.0	210.0	0.6	115	163	38	0.235
Silicon Mineral filled, moulded	1900	—	B–80.5×10³	28.0	125.0	—	88	288	54	0.207
Rubber natural moulded	920	0.5	—	21.0	—	800	—	66	62	—
Wood										
Douglas fir, air dry (parallel to grain)	480	—	B–133×10³	58.0	52.0	—	—	—	5	1.337
White oak, air dry (parallel to grain)	670	—	B–113.4×10³	55.0	50.0	—	—	—	—	—

1. Modulus of elasticity measured in tension (T), compression (C) or bending (B)
f_{ypt}—yield strength in tension, f_{ypc}—yield strength in compression.

Fatigue: A description of the behaviour of metals under the action of alternating (cyclic) loads as distinguished from their behaviour under steady loads. Cyclic stressing can occur in bending, tension, compression, torsion, or in combinations of these, and may be found in such diverse applications as axles, connecting rods, springs, aircraft landing gears and ship's hulls.

Impact strength: The ability of a material to withstand shock loading.

Transverse shear strength: The value obtained by dividing the breaking load by the transverse shear area. It is needed in the design of rivets, bolts, pins and similar parts.

Toughness: It is the amount of energy required upto fracture. It is the ability of a material to resist suddenly applied load. It depends both on the strength and ductility of a material. It is measured by the total area under stress-strain curve upto fracture.

Malleability: It is the property of a material which permits it to be drawn in the form of sheets in the cold state. Gold, copper and aluminium have good malleability.

Machinability: Property which refers to the relative ease with which a material can be cut. For example brass can be easily machined as compared to steel.

Damping Capacity: Energy dissipated as heat by a unit volume of the material during a completely reversed cycle of stress. High damping capacity is desirable to decrease vibration and to decrease chatter in machine tools. Cast iron has a high damping capacity.

Creep: At high temperatures, such as are encountered in steam boilers, turbines, and piping, the deformation of materials ceases to be elastic and becomes plastic with a continuous increase under a constant load. The equilibrium between stress and load is not established, even after a very long time. The material under tensile stress continues to stretch or creep. Creep is measured in terms of plastic deformation during a certain time. The limiting creep stress for a certain temperature is the maximum stress under which the material will not fail during a prescribed length of time.

2.5 FERROUS MATERIALS

Ferrous materials are composed of iron and carbon plus a number of other elements which are present either (1) impurity elements carried over from the raw materials or unavoidable introduced during the manufacturing processes or (2) intentionally added alloying elements.

The general classes of ferrous metal products differ from one another principally in the amount of carbon content in each. Commercially pure iron contains upto about 0.01% carbon, steels from 0.8 to 2.0% carbon and cast iron over 2.0% carbon.

2.6 CAST IRON

It is cheapest of the cast materials and in the plain or alloyed state it can be cast almost to any size and form. Cast iron without the addition of alloying elements is weak in tension and shear, strong in compression and has low resistance to impact. The stress-strain diagram is curved all the way from the zero to the rupture stress. There is no well defined proportional limit, the ratio of stress to strain is not constant, and the modulus of elasticity decreases as the stress increases. The damping capacity of cast iron is much greater than that of steel.

Grey cast iron has the carbon present in the free or graphitic state and is soft, easily machined, and only moderately brittle. It is not as abrasive wear resistant as chilled iron, it will resist galling wear as in the case of brake shoes, cylinders and pistons. In general it may be assumed that the lowest tensile strength is associated with greatest amount of graphite, and the higher tensile strength with the smallest amount of graphite. The modulus of elasticity of grey cast iron is somewhat smaller than that of steels. Grey cast iron has high damping capacity.

White cast iron is formed when the casting is rapidly cooled and it has most of the carbon in the combined state. Therefore, it is hard and brittle and cannot be machined with ordinary cutting tools, but requires grinding as a shaping process. This material is seldom used for complete casting, but is formed on wearing surfaces by the use of chill moulds. White cast iron exhibits white fracture.

Chilled cast iron is white cast iron composing parts of castings that are cooled rapidly to provide hard wear-resistant surfaces. The depth of the chill can be controlled by composition and heat treatment. Chilled cast iron is used in such members as dies, car wheels and cam followers. In general chilled cast iron is used for a limited number of engineering applications in which extreme hardness and wear resistance properties are required. It has low impact resistance.

Malleable cast iron is made by heating white cast iron castings for a period of several days in air tight pots filled with an oxide of iron and other packing materials, such as slag, sand, silica, gravel and pebbles. The action of heat and iron oxide partially removes the carbon and reduces the remainder from combined state to a globular form of free carbon so that after a slow cooling a strong, soft and somewhat ductile casting is obtained. Since the carbon change reaches only to a depth of about 10 mm, this process is not suitable for heavy castings. Castings upto 175 kg have been made, however. The difference between malleable cast iron and grey cast iron is the form in which the free carbon occurs. In grey cast iron the free carbon occurs in the form of flat or plate like particles, whereas in malleable cast iron the carbon is in the form of irregularly shaped spherical particles. Malleable cast iron is tough and ductile, resistant to impact and easily machinable. The higher strength and ductility of malleable cast iron are attributed to the shape of the carbon particles. The application of malleable cast iron is considerably more limited than that of grey cast iron because it is more expensive to produce and the better mechanical properties are not required in most instances. The use of malleable cast iron usually involves parts of complex shape that often need considerable machining to meet the specification such as agricultural implements, automobiles, motor trucks, locomotives, rail-road cars, pipe fittings, flanges, valves, chains, etc.

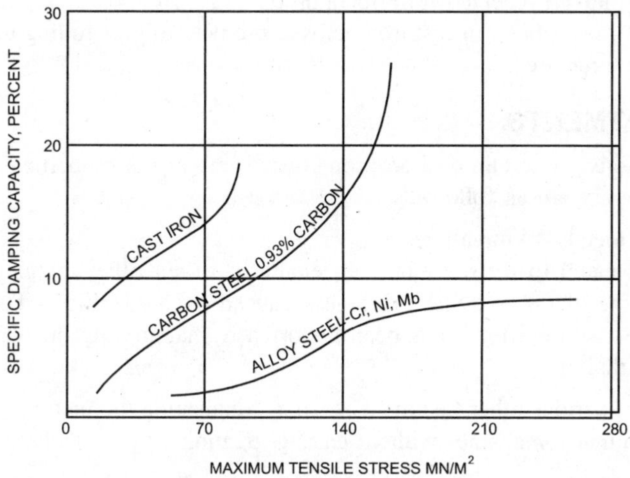

Fig. 2.1 *Comparison of damping characteristics for cast iron and two steels*

In general the damping capacity of cast iron is high, as shown in Fig. 2.1. This property makes it suitable for beds, bases and frames of machinery that are subject to vibration and for crankshafts and cam-shafts. The effect of mass cooling rates on the structure and properties of casting should be considered.

2.7 ALLOY CAST IRON

Alloying elements are added to cast iron to improve the mechanical properties or to influence certain characteristics, such as machinability, heat or corrosion resistance. Alloying elements are seldom added to cast irons unless special properties are required which cannot be otherwise attained. The more commonly used alloying elements are copper, chromium, molybdenum, nickel, titanium, vanadium. These alloying elements are added singly or in various combinations.

Nickel is used in amount of 0.25% to 5.0% in low-alloy cast irons. This promotes machinability of cast iron. In heat and corrosion cast irons as much as 35% nickel is used. "NI-Resist" a typical heat and corrosion resistant cast iron has a composition of nickel 14%, copper 6%, chromium 4%, silicon 2%, total carbon 3%. "NI-Resist" can withstand a temperature of 800°C and is used in a wide variety of media such as hydraulic, sulphuric, acetic, citric and formic acids, chlorides, caustic, sea water, etc.

Chromium in cast iron promotes hardness. Tensile strength and wear resistance are also increased by moderate chromium additions. Chromium content tends to lower machinability and increase the heat resistance property of cast iron.

Molybdenum is added to cast iron in amounts ranging from 0.25% to 1.25%. The principal function is an appreciable increase in tensile strength, transverse strength, deflection and hardness. Molybdenum is frequently used in cast iron in combination with nickel or chromium or nickel and chromium.

Vanadium is added to cast iron in amounts of 0.10 to 0.50%. Vanadium additions increase tensile strength, transverse strength and hardness.

Copper is added to cast iron in amounts upto about 1.0%. Its functions are to increase fluidity for improved mould filling ability, impart corrosion resistance, and improve mechanical properties, notably toughness and hardness. Machinability of cast iron is also slightly improved by copper additions.

Silicon in the range upto about 2.5% promotes resistance to elevated temperatures. Silicon in amounts of 11% to 17% is used in an important class of cast irons. Remarkable resistance to all types of acids, except hydrofluoric and not hydrochloric, is outstanding characteristic of these alloys.

Until quite recently **boron** received little recognition as an addition to regular grey cast iron .05% boron, 3.5% carbon and 1.0% silicon in cast iron helps to promote roll for rolling mills, to increase surface hardness and refine the structure.

2.8 HEAT TREATMENTS

Metal is heat treated to give it certain desired properties. Some of the properties which may be required and the treatment necessary are as follows:

(a) To soften the metal – **Annealing**.
(b) To harden the metal to resist wear or to enable it to cut other metals – **Hardening**.
(c) To remove some of the extreme brittleness caused by hardening – **Tempering**.
(d) To refine the structure after it has been distorted by hammering or working when in the cold state – **Normalising**.

In addition there are various other treatments such as toughening the metal to better withstand shock, toughening soft steel so that it machines without tearing, treating special steels to increase their strength.

2.9 HEAT TREATMENT OF CAST IRON

Several types of heat treatments are used to alter or to exchange some properties of cast iron and thus to increase their usefulness.

Aging: It is applied to relieve the casting stresses without materially affecting physical properties. It is carried on during 1 to 5 hours in the temperature range of 450°C to 550°C.

Annealing: It is carried out for 1 to 5 hours at 660°C to 870°C, the temperature depends upon the size of the part. It is intended to reduce hardness and to facilitate machining. However, annealing is done at the expense of some strength.

Baking: It is applied to castings that have been pickled in acid to remove sand and scale. Pickling makes casting brittle but baking for a few hours at 150°C removes this brittleness.

Quenching: Cast iron in oil or water after it has been heated above the critical range increases its hardness and also its brittleness.

Table 2.5 Gives the properties of the ferrous cast alloys.

Table 2.5 Mechanical properties of ferrous cast alloys

Material name	f_{ut} MN/m^2	Elastic limit Average			Modulus of elasticiy		Endurance limit MN/m^2	Brinell hardness	f_{uc} MN/m^2
		Tension MN/m^2	Compression MN/m^2	Shear MN/m^2	E GN/m^2	C GN/m^2			
Cast iron	140.0	43.5	350.0	43.5	70	29.4	45.0	163–230	900
	– 350.0	– 210.0	– 840.0	– 210.0	– 140	– 56	– 126.0		
Malleable iron	350.0	230.0	230.0	160.0	175	70	168.0	120–190	390
	– 490.0	– 335.0	– 335.0	– 245.0	– 190	– 77	– 210.0		
Iron copper alloy	400.0	210.0	280.0	168.0	168	70	168.0	280	350
Machanite	315.0	130.0	475.0	125.0	133	52.5	135.0	196–250	800
	– 525.0	– 350.0	– 700.0	– 140.0	– 196	– 77	– 195.0		
Ni—Cast iron	245.0	119.0	630.0	105.0	112	45.5	91.0	180	800
Ni, 1.25 to 1.75	– 310.0	– 126.0	– 700.0	– 112.0	– 126	– 52.5	– 112.0		
Ni—Test Cast iron	315.0	70.0	665.0	175.0	161	56	140.0	220	780
Ni 1.8 Mo 0.3									
Ni Resist Cast iron	175.0	56.0	630.0	56.0	84	35	56.0	120	750
Nitralloy Cast iron	455.0	245.0	1225.0	245.0	161	66.5	210.0	1050	1550
Ductile iron	420.0	315.0	—	280.0	175	65.1	168.0	140–230	—
	– 630.0	– 455.0		– 420.0			– 280.0		

Mass density of all cast irons is about 7210 kg/m^3

Ultimate shear strength values of representative materials:

Material	f_{us}/f_{ut}
Aluminium	0.6
Cast iron	1.3
Copper	0.9
Malleable iron	0.9
Steel	0.75

Drawing or tempering: This is done by reheating the quenched metal to a temperature below the critical temperature. It reduces the brittleness but still leaves an increase of hardness. By such a treatment a Brinell hardness number from 200 to 400 can be attained, the value depending on the quenching and tempering temperature.

2.10 CAST STEELS

A plain carbon steel is an alloy of iron and carbon varying up to 1.7%. The plain or alloyed cast steel is used where castings of improved properties are required over iron or malleable castings. Cast steel is tougher and stronger, weighs less for the same strength of stiffness and has a higher endurance limit than cast iron. They are used for machine members of intricate shape that require high strength and impact resistance, such as locomotive frames, large internal combustion engine frames, gears, wheels and many small and intricate highly stressed machine parts. Steel castings are more difficult to produce than iron castings and are more expensive. Corrosion resistance can be improved by the use of alloys or protective coatings. Alloy-steel castings have also been developed to meet the demands of industry for greater strength and reliability of cast machine members.

2.11 WROUGHT STEEL

The carbon limits for the three classes of carbon steel are low carbon steel, below 0.2% carbon, medium carbon steel, 0.20% to 0.50% carbon and high carbon steel, above 0.5%. The plain carbon steel is the most common material found in machine members, except cast iron. The mechanical working in the manufacturing process refines the structure and produces a more uniform steel having greater strength, greater toughness, and more durability than is obtained in castings. The properties of the steels vary greatly with the carbon content and with the form in which the carbon occurs.

Low carbon steels, containing upto 0.2% carbon, are soft, very ductile, easily machined and easily welded by any process. Low carbon steels are unresponsive to heat treatment and possess low strength. Medium carbon steels, containing from 0.2 to 0.5 per cent carbon are stronger and tougher than the low carbon steels, machine well and respond to heat treatment.

High carbon steels, containing over 0.5% carbon, respond readily to heat treatment. In the heated state, they may have very high strengths combined with hardness, but are not so ductile as the medium carbon steels. In the higher carbon ranges, the extreme hardness is accompanied by excessive brittleness. The higher the carbon content, the more difficult it is to weld these steels.

2.12 ALLOY STEELS

While a plain carbon steel is an alloy of iron and carbon with small amounts of manganese, silicon, sulphur and phosphorous, the term "alloy-steel" is applied when one or more elements other than carbon are introduced in sufficient quantities to modify its properties substantially. The principal alloying elements used in steel are chromium, nickel, manganese, silicon, molybdenum, vanadium, tungsten and to a lesser extent copper, cobalt, beryllium, boron and silver.

Chromium: The addition of chromium results in the formation of various carbides of chromium which are very hard, yet the resulting steel is more ductile than a steel of the same hardness produced by a simple increase in carbon content. Chromium also refines the grain structure so that these two combined effects result in both increased toughness and increased hardness. The addition of chromium increases the critical range of temperatures and raises the strength at high temperature. Alloy of chromium resists abrasion and wear.

Nickel: It also increases the critical range of temperature. Nickel is soluble in ferrite and does not form carbides of oxides, and thus increases the strength without decreasing the ductility. Case hardening of nickel steel results in a better core that can be obtained with plain carbon steels. Chromium is frequently used with nickel to obtain the toughness and ductility provided by the nickel and the wear resistance and hardness contributed by the chromium.

Manganese: It is added to all steels as deoxidizing and desulphurizing agent, but if the sulphur content is low and the manganese content is high (over one per cent), then it is classified as a manganese alloy. It lowers the critical range of temperatures. It increases the time required for transformation, so that oil quenching becomes practicable.

Silicon: It is added to all steels as a deoxidizing agent. When added to very-low-carbon steels it produces a brittle and a high magnetic permeability. The principal use of silicon is with other alloying elements, such as manganese, chromium, and vanadium, to stabilize the carbides.

Molybdenum: It acts very much like chromium but is more powerful in action. It also increases the depth of hardening after heat-treatment. Molybdenum finds its greatest use when combined with other alloying elements such as nickel, chromium or both. Nickel-molybdenum and nickel-chromium-molydbenum steels retain the good features of the nickel chromium steels and in addition have better machining qualities. Molybdenum increases the critical range of temperature. Except for carbon, it has the greatest hardening effect and results in the retention of a great deal of toughness.

Vanadium: It is used to toughen and strengthen the steel to reduce the grain size and to act as a cleaner and designer. It has the desirable effect of increasing the life of tools, springs and other members subjected to high temperatures. As vanadium has a very strong tendency to form carbides, hence it is used only in small amounts.

Tungston: It is widely used in tools steel because the tool maintains its hardness even at red heat. Tungston produces a fine dense structure and adds both toughness and hardness. Its effect is similar to molybdenum except that greater quantities must be added.

The steels may be forged, swaged, rolled, drawn and welded and it is especially useful to machining and finishing. It finds in-numerable applications in members subjected to high stresses due to external loads either static or cyclic, to shock loading, or to stress concentration, and in parts subjected to wear, high or low temperature, of corrosion. It is also used in applications where special properties, electrical or thermal requirements, are involved and where space and weight must be conserved. Some applications of steel in machinery part are in crankshafts, connecting rods, piston rods, keys, pins, rivets and bolts, ball and roller bearings, springs, shafting, gears, valves, frames of heavy stationary and transport equipment, tubes, levers, arms, dies and rolls. Table 2.6 gives the mechanical properties of steel and wrought iron.

2.13 HEAT TREATMENT OF STEELS

The elastic limit, ultimate strength and hardness of carbon steel are changed by heating the steel and then allowing it to cool rapidly or slowly. The higher carbon steels and alloy steels are especially responsive to heat-treatment. Heat-treatment of various steel alloys may be found from Indian Standards and other engineering hand books and will not be included here. The various processes that may be used are described below:

Annealing: The purpose of annealing is to soften the steel so that it may be more easily machined and to relieve that internal stress which may have been caused by working the metal or by unequal contraction in casting. The process of annealing involves, heating the metal slowly to the required temperature, then holding at that temperature for long enough to enable the internal changes to take place, and finally cooling slowly. The annealing temperatures for steels of different carbon contents are given below. Annealing reduces the hardness, increases the ductility, and usually reduces the strength.

Carbon content %	Annealing temperature °C
less than 0.12 (dead mild)	875–925
0.12 to 0.25 (mild)	840–970
0.3 to 0.5	815–840
0.5 to 0.9	780–810
0.9 to 1.3 (tool steels)	760–780

Table 2.6 Mechanical properties of steels and wrought Iron

Material Name	ρ kg/m^3	f_{ut} MN/m^2	Elastic limit			Modulus of Elasticity		Endurance limit bending MN/m^2	Elongation in 50 mm (%)	Brinell hardness
			Tension MN/m^2	Compr. MN/m^2	Shear MN/m^2	E GN/m^2	C GN/m^2			
Steel Casting 0.2%C	7750	420.0	175.0	230.0	105.0	203	78.4	170.0	30	120
Steel Casting 0.3%	7750	500.0	210.0	273.0	120.0	203	78.4	200.0	27	140
Steel Casting 0.4%	7750	560.0	225.0	300.0	140.0	203	81.2	225.0	22	160
Alloy Steel Casting	7750	630.0	420.0	420.0	250.0	203	79.1	—	20	187
Stainless steel C = 0.1, Cr = 12 N*i* = 1.0	7800	630.0 –1330.0	385.0 –910.0	385.0 –910.0	230.0 –560.0	203	78.4	280.0 –490.0	—	80–380
Stainless Steel C = 0.1, Mn = 0.4, S*i* = .35, Cr = 12, N*i* = 0.6	7850	735.0	420.0	420.0	250.0	210	84	280.0	—	200
Carbon steel	7830	380.0 –1050.0	217.0 –560.0	217.0 –560.0	140.0 –350.0	203 –210	78.4	170.0 –420.0	22–30	110–300
Nickel steel	7830	490.0 –840.0	315.0 –670.0	415.0 –700.0	190.0 –420.0	210	84	280.0 –420.0	22–29	140–263
Cr-Ni-steel	7830	600.0 –1120.0	420.0 –840.0	420.0 –980.0	250.0 –500.0	210	87.5	350.0 –560.0	17–26	117–325
Cr-V-steel	7830	1400.0	1190.0	1330.0	700.0	217	91	560.0	13	390
Cr-Ni V-steel	7830	1120.0	910.0	910.0	560.0	214	87.5	4.9	—	340
Wrought iron	7680	330.0	180.0	170.0	110.0	189	70	140.0	25	100
Wrought iron Fe 99.94%	7950	308.0	175.0	175.0	105.0	220	84	168.0	30	100

Normalising: When steel is cold worked (hammered, rolled, bent, etc., in the cold condition) the crystal structure is distorted, and the metal may be brittle and unreliable. To refine structure of steel and remove strains which may have been caused by cold working, normalising is done of steel. Normalising is special kind of annealing in which the material is heated slowly to its annealing temperature below that range, in still air at room temperature.

Quenching is rapid cooling from above the critical temperature by emission in cold water or other cooling medium. It is a hardening treatment. The degree of hardness depends on the amount of carbon present, and on the rate of cooling, which can be varied by using such cooling mediums as ice water, cool water, oil, hot oil, molten lead, etc. The oil quench is quite slow but prevents quenching cracks caused by rapid contraction of the object being treated. Quenching in water is used for carbon steels and for medium carbon low alloy steels. Some alloy steels, notably those containing tungston, harden when slowly cooled in air. The quenching treatment raises the strength of the metal and increases the wear resistance, but makes the metal brittle and low ductility.

Tempering or Drawing: When a steel specimen has been fully hardened it is very hard and brittle and has high residual stresses. The steel is unstable and tends to contract on aging. These internal stresses can be relieved by an additional heating process which softens the material and toughens it. This heating process is called tempering or drawing. After the specimen has been fully hardened by being quenched from above the critical temperature, it is reheated to some temperature below the critical temperature for a certain period of time and then allowed to cool in still air. The temperature to which it is reheated depends upon the composition and the degree of hardness or toughness desired. Tempering reduces the elastic limit and ultimate strength, slightly, but they are still higher than they were before drawing.

Case Hardening or Carburizing: It is a process of hardening the outer portion of the metal by prolonged heating free from contact with air while packed in carbon in the form of bone char, leather scraps or charcoal or barium carbonate. The outer material absorbs carbon depending upon the composition of part, time and temperature. When the hot metal is quenched this high-carbon steel hardens, whereas the low-carbon steel of the core remains soft and ductile. Case hardening is applied to soft steels with a carbon content of 0.2 per cent or less which cannot be hardened by simple heating and quenching.

Cyaniding: For the rapid surface hardening of small parts such as screws, clamps, washers, etc., a form of surface hardening may be used. This is done by cyanide salt and may be carried out either by placing the article in the fused salt, holding it at the quenching temperature for a few minutes and then quenching out, or by sprinkling the red hot article with powdered salt, heating up again quenching. Cyaniding produces a thin but very hard case in a very short time.

Nitriding: Nitriding is surface hardening accompanied by heating certain steel alloys called Nitralloys. The treatment consists in heating the part to a temperature of about 560°C inside a chamber through which a stream of ammonia is passed. The part is rough machined before being nitrided, and is then heated to the nitriding temperature without ammonia, in order to produce whatever distortion may occur. After that the part is finish-machined and nitrided. The nitrided surface has hardness from 730 to 1100 BHN and resists corrosion. Nitriding also raises the endurance limit of the steel.

Flame Hardening: It is a process in which a steel part is heated locally above the critical temperature and then quenched. The depth of the flame hardened layer can be varied from 1.5 mm to about 3.0 mm, the exact depth depending on the service requirements. After being quenched the piece must be stress relieved by tempering, a temperature of 200°C is usually sufficient. Plain carbon steels with a carbon content of 0.30 to 0.70 per cent are best suited for this method. It is often applied to gear teeth, cams, castings, forgings and rolled sections irrespective of their size.

2.14 DESIGNATION OF STEEL : INDIAN STANDARDS (I.S. 1570–1978)

Two systems of notation are recommended by I.S.

(a) based on ultimate tensile strength

(b) based on chemical composition

(a) **Based on ultimate tensile strength:** This is applicable to carbon and low alloy steels. The symbol consists of the letters Fe followed by the number representing the ultimate tensile strength in N/mm^2, e.g., Fe 410. Table 2.7 gives steels specified by tensile strength but without detailed chemical composition.

Table 2.7 Tensile properties and applications

Designation	f_{ut} N/mm^2	δ*	Typical uses
Fe 360	360	25	Structural steel for chemical pressure vessels and other general engineering purposes.
Fe 410	410	23	Structural steel for bridges and building construction, railway railing stock, screw spikes, oil well casing, tube piles, and other general engineering purposes.
Fe 490	490	21	Structural steel for mines, forgings for marine engines, sheet, pilling and machine parts.
Fe 620	620	15	High tensile steel for tramway axles and seamless tubes.

(b) **Based on chemical composition:** This type of designation is preferred if subsequent heat treatment is carried out.

1. **Plain carbon steels** are designated by C followed by number representing average percentage carbon content in hundredths of a percent, e.g., C-14-carbon steel with 0.14% carbon, (Old System). Table 2.8 to 2.9 gives carbon steels specified by chemical composition, as per New System. First figure indicates 100 times the average percentage of carbon content, then letter C and the last figure indicates 10 times the average percentage of Mn content rounded off to the nearest integer.

Table 2.8 Specified ultimate tensile limits for plates, sections, bars, billets and forgings in the hot-rolled or normalized conditions

Designation	C Present	Mn Present	f_{ut} N/mm^2	δ*
10 C 4	0.15 max.	0.30–0.60	340–420	26
15 C 4	0.20 max.	0.30–0.60	370–490	25
15 C 8	0.10–0.20	0.60–0.90	420–500	25
30 C 8	0.25–0.35	0.60–0.90	500–600	21
40 C 8	0.35–0.45	0.60–0.90	580–680	18
45 C 8	0.40–0.50	0.60–0.90	630–710	15
50 C 4	0.45–0.55	0.60–0.90	660–780	13
55 C 8	0.50–0.60	0.60–0.90	720	13

Table 2.9 Specified tensile limits for cold drawn bars

Designation	f_{ut} N/mm^2				Percentage elongation			
	upto 20 mm	20–40 mm	40–63 mm	over 63mm	upto 20mm	20–40 mm	40–63 mm	over 63mm
10 C 4	490	450	410	360	11	13	15	18
15 C 8	540	510	470	430	11	13	15	17
20 C 8	540	510	470	430	10	11	13	15
30 C 8	610	570	530	490	8	10	11	13
40 C 8	610	610	570	540	8	8	10	11
50 C 4	670	630	610	590	8	8	10	10
55 C 8	730	690	670	630	8	8	10	10

Table 2.10 indicates the typical applications of carbon steels.

Table 2.10 Typical applications

Steel designation	Typical applications
4 C 4	Dead soft steel generally used in electrical industry.
10 C 4	Case hardening steel used for making camshafts, cams, light duty gears, worms, gudgeon pins, selector forks, spindles, bowls, ratchets, chain wheels, tappets, etc.
15 C 4	Used for lightly stressed parts. The material, although easily machineable, is not designated specifically for rapid cutting, but is suitable where cold web, such as bending and riveting may be necessary.
30 C 8	Used for making certain types of cold, formed parts, such as shift and brake levers. After suitable case hardening or hardening and tempering, this steel is also made use of in making parts, such as socket, tierod, yolk adjustable control lever cable, shaft fork and rear hub, 2-wheeler and 3-wheeler lambretta parts, such as sprocket, lever, hubs for forks, cams, rocker arms and bushes are made of this steel. Tubes for aircraft, automobiles, bicycle and furniture are also made of this steel.
40 C 8	Steel for crankshafts, shafts, spindles, automobile axle beams, push rods, studs, connecting rods, bolts, lightly stressed gears, etc.
45 C 8	Steel for spindles of machine tools, bigger gears, bolts and shafts.
50 C 4	Steels for making keys, shafts, cylinders, machine components requiring moderate wear resistance. In surface hardened condition it is also suitable for large pitch worms and gears.
55 C 4 and 55 C 8	Steel used for making gears, cylinders, cams, keys, crank shafts, sprockets, and machine parts requiring moderate wear resistance for which toughness is not of primary importance.

2. **Alloy Steels:** Here the letter C is omitted and the number representing the carbon content are followed by chemical symbols for one or more of the predominant alloying elements followed by numbers indicating their average contents. For average alloy content upto 1 percent, alloy index number will be: Average alloy content upto 2 decimal places underlined by a bar. For average alloy content 1 percent and over, Alloy index number will be: Rounded to the nearest whole number. Upto 0.5 rounded down, 0.5 and above rounded up.

e.g., 15 Cr $\underline{65}$ — Chromium steel with 0.15% carbon and 0.65% Cr.
40 Cr $\underline{14}$ — Chromium steel with 0.4% carbon and 0.14% Cr.
20 Cr 18 Ni 2 — Nickel-Chromium steel with 0.2% carbon, 18% chromium and 2% nickel.

Table 2.11 gives the properties of alloy steels (other than stainless and heat resisting steels).

Table 2.11 Specified mechanical properties for bars and forging in the hardened and tempered conditions–oil hardened otherwise stated

Designation	f_{ut} N/mm^2	f_{yp} mini. N/mm^2	Percentage elongation minimum	Brinell hardness
20 Mn 2	600–750	440	18	170–217
	700–850	500	16	201–248
27 Mn 2	600–750	440	18	170–217
	700–850	500	16	201–248
37 Mn 2	600–750	440	18	170–217
	700–850	540	18	201–248
	800–950	600	16	229–277
	900–1050	700	15	255–311
35 Mn 2	800–950	600	16	229–277
Mo 45	900–1050	700	15	255–311
	1000–1150	800	13	285–341
40 Ni 2 Cr 1	800–950	600	16	229–277
Mo 28	900–1050	700	15	255–311
	1000–1150	800	13	285–311
	1100–1250	880	11	311–363
	1200–1350	1000	10	341–401
	1550 mini.	1300	6	444 mini.

Table 2.12 Specified mechanical properties for case hardening steel in the refined and quenched condition

Designation	f_{ut} mini. N/mm^2	Percentage elongation minimum
11 Mn 2	600	17
15 Cr 65	600	13
17 Mn 1Cr 95	800	10
16 Ni 1Cr 80	850	12
20 Mn 1 Cr 1	1000	8

Table 2.13 gives the typical applications of alloy steels.

Table 2.13 Typical applications

Designation	Typical uses
20 Mn 2 and 27 Mn 2	Used for welded structures, crank shafts, steering levers, shafting spindles, etc.
35 Mn 2–Mo 28	Used for making general engineering components, such as crank shafts, connecting rods, axle shafts, levers, bolts, etc.
15 Cr 65	Used for roller bearings, measuring instruments, piston pins, and differential pipes.
17 Mn 1 Cr 95	Used for small gear wheels and shafts, piston joints and steering regulators.
20 Mn 1 Cr 1	Used for medium size gear wheels, and shafts of vehicles.
40 Cr 1 Mo 28	Used for making axle shafts, crankshafts, connecting rods, gears, high tensile bolts and studs, propeller shafts, joints, etc.
16 Ni 1 Cr 80	Used for gear box and transmission components, cams, camshafts, etc.
40 Ni 2 Cr 1 – Mo 28	Used for high strength machine parts, collets, spindles, screws, high tensile bolts and studs, gears, pinions, axle shafts, tappets, crankshafts, connecting rods, boring bars, arbours, etc.

Many more Indian Standards are available for different ferrous materials.

2.15 MATERIALS FOR HIGH TEMPERATURE SERVICE

Metal parts for steam engines, I.C. engines, valves, superheated steam equipment, and similar service are stressed at temperatures ranging from 100°C to 1000°C. The metals used for such parts must be specially selected from those materials which retain a large percentage of their strength at high temperatures and which do not creep. High temperature or heat resisting alloys can be principally of three types:

(*a*) Alloys of the first group contain over 50% iron with additions of 10 to 30 percent chromium and varying amount of copper, nickel, tungston and silicon.

(*b*) Alloys of second group contain upto 25% iron, 50 to 60% chromium and nickel and small addition of maganese, nickel, tungston and silicon.

(*c*) Alloys of the third group contain from 0.5% to 0.6% iron as an impurity, over 80% nickel, considerable chromium and small amounts of manganese and silicon. The alloys of this group also have excellent resistance to corrosion and temperatures upto 1000°C.

2.16 NON-FERROUS MATERIALS

Aluminium and aluminium alloys: Aluminium having a specific gravity of only 2.7 is approximately two-fifths as heavy as iron and one-third as copper. Pure aluminium resists oxidation, is highly ductile and has good forming properties, but it has poor casting characteristics and machining properties. The average properties of pure metals and alloys are given in Tables 2.1 and 2.2. Pure aluminium has a tensile strength of about 77 MN/m^2, but this can be increased considerably by cold working and also by alloying with other materials. The modulus of elasticity of aluminium and its alloys is 70 GN/m^2 which means that it has about one-third the stiffness of steel.

The most useful alloying elements for aluminium are copper, silicon, manganese, magnesium and iron. The effects of copper as an alloying element are to raise the ultimate strength and endurance limit and to improve the casting characteristics and machinability, but to lower its resistance to corrosion. Aluminium copper alloys are used in such applications as crank-cases, transmission housings and fittings. The aluminium silicon alloys have better corrosion resistance and mechanical properties but poorer machinability than aluminium copper alloys. They are suitable for marine castings, water jackets, housings and castings, where a minimum of machining is required. The (0.4% Cu, 5% Mg, 0.5%Mn) aluminium magnesium manganese alloy known as "duralumin" is suitable for parts that require severe working and it has good corrosion resistance and strength. It is obtainable in sheets, plates, tubes, rods, wires, extruded sections, bolts and rivets. The plates can be spot welded and are widely used in aircraft construction.

Aluminium alloys castings made in sand moulds, permanent metal moulds, and die casting machines are generally stronger than the poorer grades of cast iron. Better surface finish, closer dimensional tolerances, and better mechanical properties, together with the savings in machining and finishing costs, make the use of permanent moulds and die castings desirable when the quantity justifies the extra cost of equipment.

Table 2.14 gives the properties and uses of some aluminium alloys.

Table 2.14 Characteristics of some aluminium-alloys

S.No.	Normal composition % remainder aluminium	Outstanding characteristics	Typical uses
1.	5.0 Si	Excellent casting characteristcs and pressure tightness, good weldability and very good corrosion resistance.	Carburettor bodies, refrigerator shelf brackets and fittings, cooking utensils, general purpose castings of thin sections.
2.	4.5 Cu-5.5 Si	Good casting characteristics and pressure tightness, good weldability.	Ornamental grills. reflectors and general purpose castings, used for castings requiring leak tightness and moderate strength.
3.	7.0 Cu-1.7 Zn	Good casting characteristics and machining property.	Vacuum cleaner housings, general purpose castings.
4.	7.0 Cu-2.0 Si 1.7 Cu and 7.0 Cu-3.5 Si	Provides better pressure tightness, modification alloy-3.	Washing machine agitators, hydraulic brake piston, diesel cylinder, head covers, general purpose castings with pressure requirements.
5.	10.0 Cu-0.2 Mg	Retains strength at elevated temperatures, good hardness, machinability and wear resistance.	Meter parts, bushings, bearing caps, automotive pistons.
6.	0.8 Cu-0.8 Fe-12.0 Si-1.0 Mg-2.5 Ni	Good high temperature properties, low coefficient of expansion and machinability and weldability.	Pulleys and sheaves, automotive and diesel pistons.
7.	3.5 Cu, 9.0 Si-0.8 Mg-0.8 Ni	Similar to 6 but with improved casting characteristics.	Pulleys and sheaves, automotive and diesel pistons.
8.	10.0 Cu-0.2 Mg-0.8 Si	Hard in cast condition, good casting and machinability.	Sole plates for electric hand irons.
9.	4.0 Cu-1.5 Mg-2.0 Ni	Good strength at elevated temperature.	Aircraft generator housings, motor cycle, diesel and aircraft pistons, air cooled heads.
10.	7.0 Cu-5.5 Si-0.3 Mg	Good casting characteristics with good machinability	Pistons
11.	4.5 Cu-2.5 Si	Good casting characteristics with good machinability and weldability.	Aircraft fittings, aircraft gun control parts, aircraft wheels, railroad car seat frames, compressor connecting rods, fuel pump bodies, aircraft gear housing.
12.	3.8 Mg-1.8 Zn	Good resistance to corrosion and tarnishing.	Cooking utensils.
13.	3.8 Mg-1.8 Si	Improved foundary characteristics but reduced mechanical properties, excellent resistance to corrosion.	Cooking utensils and pipe fittings, for marine as well as general use.
14.	3.5 Cu-6.3 Si.	Excellent casting characteristics with good mechanical properties.	Typewriters, frames, engine parts, water cooled cylinder heads.
15.	3.5 Cu-9.0 Si.	Low coefficient of thermal expansion, therefore, suitable for high temperature applications.	Gas meters and regular parts internal combustion engine cylinder heads.

S.No.	Normal composi tion% remainder aluminium	Outstanding characteristics	Typical uses
16.	5.0 Si-0.5 Mg 1.3 Cu.	Very good casting characteristics and pressure tightness.	Aircrafts super charge cover, fuel pump bodies, aircraft fittings, supercharger impellers, gears for liquid cooled aircraft cylinder heads.
17.	7.0 Si-0.5 Mg.	Good casting and weldability, good pressure tightness and resistance to corrosion.	Machine parts, handwheels, aircraft pump parts.
18.	6.5 Si-1.0 Cu-1.0-Ni	Excellent bearing qualities with good compressive yield strength.	Bearings, bushings, and other castings requiring good bearing characteristics.

Composition and properties of aluminium alloys are given in Table 2.15.

Standard diameters of aluminium wires in millimetre. 0.32, 0.36, 0.40, 0.45, 0.50, 0.56, 0.68, 0.71, 0.80, 0.90, 1.00, 1.12, 1.25, 1.40, 1.60, 1.80, 2.00, 2.24, 2.50, 2.80, 3.18, 3.55, 4.00, 4.50, 5.00, 5.60, 6.30, 7.10, 8.00 and 9.00

Tolerances on the diameter of aluminium wires are:

Diameter in mm.	Tolerances
0.32–0.63	± 0.015
0.71–1.12	± 0.025
1.25–9.00	± 0.050 or ± 1% of dia. whichever is maximum

2.17 COPPER

Copper is easily distinguished from all other metals on account of its red colour. Commercial copper is a tough, ductile, and malleable metal containing less than 5 percent of such impurities as tin, lead, nickel, bismuth, arsenic and antimony. Because of its high electrical conductivity it is used extensively for wire and cable, and all parts of electrical apparatus which must conduct the current. Copper is also a good conductor of heat and is highly resistance to corrosion by liquids. For this reason it is used in locomotive fire boxes, water heating apparatus, water pipes, vessels in brewery and chemical plants and condenser tubes. For its high heat conduction it is used for soldering iron bits.

The alloys of copper form an important group of materials with a wide variety of properties. Some have high strengths, some are excellent bearing materials, some retain their strength at high temperature, and others are valuable for their corrosion resistance. Three groups of copper alloys used in machine design are the **brasses**, the **bronzes** and **monelmetal**. Copper is also used as an alloy in steel to increase its resistance to corrosion.

Alloys of Copper and Zinc-Brass

An important use of zinc is for alloying with copper to give the various classes of brass. The alloys are of importance in view of the great variety of mechanical properties that they may be obtained, the wide range of production process to which they lend themselves and their resistance to atmospheric effects and corrosion. Suitable types of brass lend themselves to casting, hot forging, cold rolling into sheets, drawing into wire and being extruded through dies to give special shaped bars.

By adding small quantities of other elements (aluminium, iron, manganese and tin) the strength of the bars may be greatly increased from its normal strength 300–375 MN/m^2 and a range of "high tensile" brasses is available having ultimate strengths as high as 600 MN/m^2.

Table 2.15 Composition and properties of aluminium alloys

Percent of alloy					kg/m^3 mass density	f_u		Yield point		Endurance limit bending MN/m^2	Elongation in 50 mm in percent	General information
Cu	Si	Mn	Mg	Ni		f_{ut} MN/m^2	f_{us} MN/m^2	Tension MN/m^2	Comp. MN/m^2			
4.5	1.5	0.3	—	—	2800	220.0 280.0	168.0– 210.0	112.0 –210.0	112.0– 266.0	42.0– 50.0	5.00	Structural casting requiring high strength and shock resistance.
4	0.7	0.3	1.5	2.0	2720	220.0 –280.0	190.0	200.0 –210.0	240.0 –320.0	56.0– 67.0	0.5	Air cooled cylinder heads, pistons of high accuracy I.C. engines.
1.3	5.0	0.5	0.5	—	2630	220.0 –260.0	210.0	175.0 –190.0	200.0 –180.0	60.0 –65.0	1.75	General use for high strength and pressure tightness.
0.6	5.0	0.3	—	—	2680	–135.0 170.0	98.0– 125.0	63.0	70.0 –63.0	45.5	7.5	Intricate casting with thin sections, corrosion resistance.
0.9	12.0	0.1	1.1	2.5	2650	220.0	170.0	195.0	210.0	—	0.5	Piston in I.C. engines low heat expansion.
4.5	2.5	—	0.3	0.3	2800	280.0 –270.0	210.0 –225.0	155.0 –140.0	155.0 –230.0	66.0 –77.0	5.5	Structural casting requiring high strength and shock resistance, modified for use in permanent moulds.
0.6	12.0	0.3	—	—	2650	260.0	—	126.0	—	105.0	1.8	Excellent casting characteristics.
3.5	8.5	0.5	0.1	0.5	2750	315.0	—	175.0	—	—	2.0	Good mechanical properties.
1.0	—	Sn 6.5	—	1.0	2880	140.0	98.0	60.0	60.0	63.0	10	Bearing alloy.
4.5	0.5	0.6	1.5	—	2780	190.0 –470.0	125.0 –290.0	77.0 –320.0	—	84.0 –125.0	20	Sheets, plates, tubes, rivets with high strength.
4.4	0.8	0.8	0.4	—	2800	190.0 –490.0	125.0 –320.0	98.0 –420.0	—	77.0– 125.0	20	Highly stressed forgings.
4.5	1.0	—	0.6	—	2680	310.0	225.0	240.0	180.0	77.0	12	Complicated shapes.
0.9	12.5	0.2	1.1	0.9	2680	360.0	265.0	295.0	—	110.0	5	Press forged pistons I.C. engines.
1.0	1.5	—	—	0.5	2820	150.0	—	105.0	105.0	—	15	Bearing alloy.

$E = 79$ GN/m^2, $C = 27$ GN/m^2

The melting point of brass varies according to its composition but most of the brass in the common range liquify between temperatures of 850°C and 960°C. Hard brass may be softened by heating to about 750°C

To improve the machining quality of brass 1 or 2% of lead is often added, as in its ordinary condition the metal is soft and ductile and tends to drag under the tool. Small amounts of tin are sometimes added to brass to increase its hardness and to increase its resistance against the corrosive action of sea water. Some of the chief brasses together with their applications are given in Table 2.16.

Table 2.16 Properties and uses of brasses

Copper	zinc	Properties and uses
85	15	Gilding metal, cheap jewellery.
75	25	Brazing brass. Used where parts must be brazed or silver soldered. Steam condenser tubes.
70	30	Great ductility and strength. Drawing into tubes and cartridge cases. Drawing into wire. Cold rolling into strip, cold pressing.
66	34	English standard brass. Casts well and may be rolled and hammered. Little lead added to improve machining.
60	40	Muntz metal. A yellow brass can be easily hot worked. Used for general range of articles, e.g., water fittings, house hold articles, sheeting for wooden ships, marine condenser tubes and heat exchanger parts.
58.5	40	Free cutting. Does not bend well. High speed turning and screwing.
50 58.5	50 36	Low melting point. Does not work well. Brazing spelter.
Aluminium Manganese Iron	3 1.5 1.0	High tensile brass, ultimate strength 900 MN/m^2.

Copper tin alloys: (The bronzes)

When copper alloyed with tin forms a set of alloys called bronzes, which are important metals in engineering practices.

Gunmetal contains 88% copper, 10% tin and 2% zinc. The zinc is added to increase its fluidity. The metal is used chiefly for casting which must be strong and resistant to corrosion by water and the atmosphere. Gunmetal is not suitable for cold working but may be forged at about 600ºC, (Just below red heat). Another common bronze is made of 5% tin, 5% zinc, 5% lead and 85% copper.

The bronzes are more costly than the brasses and hence are used only when the cheaper alloys do not prove to be satisfactory. They are used in the highest quality thermostatic bellows and other parts that require resistance to severe stretching together with good tensile and elastic strength.

Copper and tin alone do not readily alloy and small quantities of phosphorous are generally added, giving rise to common trade name phosphorous bronze. The bronze are particularly useful because of corrosion resistance, good fatigue properties and the useful spring characteristics, the latter being due to high elastic limit and creep resistance at room temperatures. As the tin increases upto 8%, the strength and ductility increase rapidly. More than 8% tin increases the strength, but the ductility drops off.

Copper-Nickel Alloys : Monel Metal

Alloys containing less than 50% nickel are called cupro nickels or nickel silvers. These alloys also contain a small percentage of zinc. These alloys are silver in appearance and tarnish very little under atmospheric conditions. They are malleable and can be worked without annealing.

Monel metals are the most important nickel copper alloys, having high strength and toughness and excellent resistance to salt water, brine, caustic soda, hydrochloric acid, sulphuric acid and other corrosive agents.

Table 2.17 Composition and mechanical properties of copper alloys

Name	Percent of metal						Mass density kg/m^3	f_{ut} MN/m^2	Elastic limit			Modulus of elasticity		Endurance limit bending MN/m^2	Elongation in 50 mm. (%)
	Cu	Zn	Sn	Pb	Al	Ni			Tension MN/m^2	Comp. MN/m^2	Shear MN/m^2	E GN/m^2	C GN/m^2		
Admiral metal	71	28	1.0	—	—	—	8850	210.0 –633.0	75.0	—	—	—	—	—	65– 14
Aluminium brass	76	22	—	—	2.0	—	8500	280.0 –350.0	225.0	—	—	91	35	—	17–52
Aluminium bronze	79	—	—	5.0 (Fe)	11.0	5.0	7750	450.0	98.0	98.0	84.0	119	45.5	98.0	20
Ambrac	65	5	—	—	—	3.0	8840	385.0 –560.0	140.0 –315.0	—	—	133	45	—	—
Brass red	84	10.0	3.0	3.0	—	—	8450	200.0	40.0	63.0	63.0	91	35	42.0	16
Brass Yellow	62	35	1.0	2.0	—	—	8480	180.0 –530.0	70.0 –220.0	63.0 –140.0	63.0 –175.0	87.5 –112	31.5 –42	49.0 –140.0	20
Beryllium copper	98	—	—	—	—	2.0 beryllium	8730	420.0 –1084.0	49.0 –420.0	—	—	133	56	80.0	—
Bronze superstrong	86.5	—	—	3.3 (Fe)	10.2	—	7500	810.0	230.0	210.0	—	105	40	160.0	—
Gear Bronze	86.5	—	—	3.3 (Fe)	10.2	—	8730	245.0	125.0	140.0	91.0	94.5	37	84.0	10
Gun metal	88	2.0	10.0	—	—	—	8800	280.0	84.0	70.0	70.0	182	66.5	—	—
Manganese Bronze	90	5.8	—	2.0	2.0 (Fe)	—	7750	455.0	175.0	—	155.0	112	—	112.0	25
Monel metal	30	—	—	—	—	67	8850	505.0 –840.0	175.0 –490.0	154.0 –420.0	160.0 –280.0	175	63	140.0 –350.0	20
K-monel	29.0	—	—	—	2.75	66	8600	1050.0	630.0	560.0	385.0	178	64	420.0	30
Muntz metal	60.0	40.0	—	—	—	—	8400	315.0 –490.0	154.0 –315.0	—	—	90	35	105.0 –175.0	30
Phosphor bronze	80 –90	—	20– 10	1.0 (P)	—	—	8870	385.0	125.0	110.0	105.0	109	43	84.0	—

Monel metal contains 67% nickel and 30% copper with a small amount of iron, manganese, silicon and carbon. It is a general purpose material. It is used under corrosive temperatures upto 260ºC and can be used for high temperatures if no sulphur is present. It holds its strength at elevated temperature better than bronzes and some steels and is used for turbine blades, high temperature valves, and for springs subjected to temperatures upto 200ºC. Unlike plain steel, it increases in strength, hardens, and fatigue strength as the temperature decreases to as low as—150ºC, the elongation and impact strength remaining practically constant.

Copper-aluminium Alloy (Aluminium Bronze)

Aluminium bronze is copper with 2% to 11% aluminium. Its tensile strength increases, as the aluminium content becomes greater. A 10% aluminium bronze is used in cast and hot worked condition for parts requiring high tensile strength and resistance to corrosion, wear and alternating stress. Iron upto 0.5% increases the strength and hardness without reducing the ductility appreciably. An alloy containing 11 per cent aluminium, 5 percent nickel, 5% iron, 79% copper has extreme hardness and wear resistance at high temperatures, a quality essential in aircraft engine valve seats. In general the aluminium bronzes are much used in aircraft for bushings, gears, bearings, valve guides, shock absorber, pistons and similar parts.

Other Copper Alloys: The number of copper alloys adopted to special purposes is very great. One of the latest development is beryllium copper. Alloys containing from 1.5 to 2.5 percent beryllium takes an exceptionally high characteristics of strength, endurance and hardness through cold working and heat treatment. The main use of this alloy in machinery is for springs, particularly in internal combustion engines, high duty gears, valve sleeves and valve seats.

The composition and mechanical properties of most frequently used copper alloys are given in Table 2.17.

Bearing Metal

It has been found by experience that to give an efficient bearing combination, the following conditions are necessary:

1. That the shaft and bearing be dissimilar in their natures with bearing softer than the shaft.
2. That the most efficient bearing metal is one consisting of small pieces of comparatively hard metal embedded in the softer body of another metal.

In general, a metal consisting of one uniform constituent does not serve well as bearing. The "White" metals all contain tin alloyed with various other metals, of which antimony is always present. It is thought that in practice the soft constituent wear away slightly and the shaft is carried on "high spots", the lower level of the remainder of the bearing to maintain the film of oil necessary for efficient

Table 2.18 White bearing metals composition and uses.

Composition				Uses
Tin	Antimony	Copper	Lead	
93	3.5	3.5	—	Big end bearings, motor and aero engines.
86	10.5	3.5	—	Main bearings, motor and aero engines.
80	11	3	6	Bearing for heavy loads and high speeds.
60	10	1.5	28.5	Bearings for engines and electric machines, railways and tramways.
40	10	1.5	48.5	Heavy pressure and bearing speed.
20	15	1.5	63.5	Medium pressure and speed.
5	15	—	80	Long bearings with medium load.

Table 2.19 Properties of bearing metals

Metal	Percentage of Ingredients				Compressive elastic limit MN/m^2	Brinell Hardness at 27.5ºC	Class of use
	Copper	Lead	Tin	Miscell-aneous			
Bronze	91–92	—	8–9	P = 0.5	260.0	150–200	High shock load
Bronze	91–92	—	8–9	P = 0.5	105.0	80–90	High shock load
Bronze	86–89	0.2	9–11	Zn = 1.3	84.0	60	Shaft bearings
Phosphor Bronze	80–82	9–10	9–11	P = 0.7	105.0	55–80	Shock load
Aluminium Nickel Bronze	80	Al = 10	5	Ni = 5.0	315.0	180	High shock load
Aluminium alloy	1.0	Al = 91.0	6.5	Ni = 1.5	59.5	35–50	Severe loads
Aluminium alloy	1.0	Al = 90.5	6.5	Ni = 1.5 Si = 1.5	105.0	40–55	General automotive
Plastic Bronze	65	30	5	—	24.5	30	Crank Pin bearings
Allan red metal	50	50	—	—	17.5	13.5	Crank Pin bearings
Babbit metal	4.5	0.35	90–92	Sb = 4.5	9.1	17	Light loads
Babbit metal	2.75 –3.25	25	60–62	Sb = 10.5	9.1	22	Moderate loads
Babbit metal	—	75	10	Sb = 15	84.0	23	Crank Pin bearings
Parson's white brass	2.5–5	0.2	65–65	Zn = 33	105.0	18	Light load
Lumen brass	10	—	Zn = 86	Al = 4	260.0	116	Light load, high speed

lubrication. The white metals all have a low melting point which facilitates the formation of bearing by casting the metal from a laddle.

The reader will notice from Table 2.18, that some of the metals are composed mainly of tin whilst in other lead is the predominant constituent. These are named as tin base and lead base metals respectively and those with a tin base are often called "Babbit metals".

The composition and mechanical properties of Bearing metals are given in Table 2.19.

2.18 STELLITE

Stellite is cobalt base metal alloyed with various proportion of chromium (35), cobalt (45) and tungston (15) with iron, manganese and silicon present only as impurities. The most notable property of these alloys is high red hardness. They have high abrasive resistance, non-tarnishing, corrosion resistance and nonmagnetic properties. These properties make it excellent cutting tool for various machine operations on cast iron and malleable iron and on some steels. Because of the cost of stellite, the larger tools are made of steel with stellite tips welded on it. Coating of stellite, applied by welding, are used for surface protection against wear and abrasion on such items as oil-well pits, cement-mill grinding rigs, and for some bearing surfaces whose lubrication is impossible or unreliable. The coefficient of friction on dry metals varies from 0.15 to 0.24 with an average of 0.18. Mechanical properties of various stellite materials are given in Table 2.20.

Table 2.20 Mechanical properties of stellite

Material	Ultimate strength			Brinell hardness	Properties and uses
	Tension MN/m^2	Comp. MN/m^2	Tension at 500°C MN/m^2		
Welded	265.0	1792.0	168.0	512	Not machinable, grind
Cast	265.0	2177.0	175.0	600	—
Cast	600.0	1540.0	126.0	402	—
Forged	938.0	1800.0	164.0	—	Can be rolled, forged and punched at 100°C.
Welded	462.0	1350.0	133.0	444	Hard-surfacing weld rod, not used for cutting tools.

2.19 NON-METALLIC MATERIALS

The commonly used non-metallic materials are wood, glass, rubber, leather, carbon and plastics.

Wood: Wood has some uses in machine members. It is used where light weight parts subjected to moderate shock loading are required, e.g. in circuit breaker operating rods where non-metallic bearing material is desirable.

Glass: It is usually thought of as a weak material. The greatly improved mechanical properties of the newer glass composition justify their consideration in design problems. Glass parts can be moulded by heat and pressure or by heating and blowing. They can be finished and cut by grinding and can be machined by carbide tools, but the surface will be too rough for ordinary uses. Glass is used for all parts of centrifugal pumps for acids, piping for mechanical processes, pipe fittings, exchangers and lining for tanks and fittings.

The tensile strength of glass ranges from 42.0 to 84.0 MN/m^2 with small surface scratches reducing the strength by 50 percent. Compressive strength of glass is over 700 MN/m^2. The modulus of elasticity varies from 45 GN/m^2 to 70 GN/m^2. Allowable design stresses range from 3.5 to 7.0 MN/m^2.

Rubber: Rubber and similar synthetic materials such as Neoprene have a variety of application in machinery. Rubber should be protected from high temperature, oil and sunlight. It is an excellent material for seats and diaphragms, for water lubricated bearings, for parts subjected to vibrations (such as vibration mountings, flexible coupling and flexible bearing) and for tubes and hose. In industry, hard rubber is used for electric insulation, switch handles, bearings, etc. Table 2.21 gives the different properties of rubber.

Leather: Leather is very flexible and will stand considerable wear under suitable conditions. Modulus of elasticity varies according to load. It is used in belt drives and as a packing or as washers.

Table 2.21 Properties of rubber and rubber like materials

Material	Form	Compressive strength MN/m^2	Tensile Strength MN/m^2	Shear Strength MN/m^2	Max. temp. for use °C	Effect of heat
Duprene	—	—	1.4–28.0	—	150	Stifens slightly
Koro seal	Hard	—	14.0–63.0	—	100	Softens
Koro seal	Soft	—	3.5–17.5	—	85	Softens
Pilo form	Plastic	60.0–77.0	28.0–35.0	49.0—63.0	70–120	Softens
Rubber	Hard	14.0–105.0	7.0–70.0	63.0—105.0	55–70	Softens
Rubber	Soft	—	3.5–4.2	—	60–90	Softens
Rubber	Linings	—	—	—	85	Softens

Carbon: Since long carbon has been used in electrical insulations and it has not been generally used in mechanical insulations. Use of carbon gives low friction losses, low wear rates when operating against metals and is used in chemical handling equipment. Pump rotors, vanes and gears of carbon have been used with good results to replace similar parts of bronze and laminated plastics. Clutch plates and rings of metal and cork have been replaced by carbon parts. The properties of carbon are given in Table 2.22. Graphite, a form of carbon, has long been used in bronze bearings as a lubricant, but now bearings made entirely of carbon are being used with operating pressure as high as 7.0 MN/m^2 at low speed.

Table 2.22 Properties of carbon

Strength MN/m^2	
Compression	14.0–175.0
Tension	2.8–26.5
Shear	22.4
Modulus of elasticity	9.31–17 GN/m^2
Maximum operating temperature 450°C.	
Coefficient of friction	
Carbon on hard steel, unlubricated	0.18–0.3
Well lubricated	0.04–0.08
Carbon on glass, unlubricated	0.17

Plastics: These are difficult to define because any acceptable definition usually gives too many exceptions. Common usage has applied the term to a class of materials familiar to everyone in such products as buttons, fountain pens, telephones, dials, knobs, etc. The word plastic was used originally to indicate a material that could be formed or moulded by pressure at moderately elevated temperature. Some of the plastics are available as thin sheets, foams, coatings, petroleum laminates and filament for weaving. Raw materials are mainly derived from petroleum and agriculture. Plastic molding is one of the high volume, low cost production methods with products that are replacing those formerly made from wood, glass or metal.

Plastic materials when used have the following advantages:

1. Good accuracy, light, smooth surfaces, low thermal conductivity, pleasing appearance, self lubrication characteristics, resistance to corrosion, and good dielectric strength.
2. Compared to metal, plastics have better resistance to shock and vibration and higher abrasion and wear resistance.
3. They can be either transparent or opaque and can be made in desired colour.
4. No finishing is required after removal from the mould.
5. Although the moulds are expensive, but are long lasting and capable of producing many thousands of parts.

The use of most of the plastics is limited due to the following reasons:

1. Low strength and rigidity, low heat resistance, and sometimes low dimensional stability.
2. Small loads at room temperatures will induce a continuous type of creep behaviour. This effect is intensified at higher temperatures.
3. The thermal expansion of plastic usually runs from five to ten times that of metal.
4. Embrittlement with age is another disadvantage.
5. The cost of the plastic materials is high, but has been decreasing while other materials have been increasing in price.

Types of Plastics: Plastic materials can be broadly classified as: (1) Thermoplasts or thermoplastic materials, (2) Thermosets or thermosetting materials.

Thermoplastics, when moulded, undergo no chemical change. Although rigid at room temperature, heating beyond softening point will cause the material to assume as viscous or liquid state. Thermoplastics are sold under such trade name as Celluloid, Nitron, Pyralin, Bakelite AC, Fibestos, Nixonite and Tenite.

Thermosets, undergo a chemical change when heated. This change is permanent and the material cannot be softened by heating. Thermosets cannot be extruded or calendered. Thermosetting plastics are sold under trade names such a Bakelite, Durite, Textolite, Bakelite uses, Beetle and Plaskon.

Average properties to some of the plastics are given in Table 2.4. The chemistry of plastic materials is very involved. Many hundreds of such materials are available and the number is continuously increasing. The most widely used moulding plastics are given in Tables 2.23 and 2.24. The table also indicates the common trade names, characteristic properties and typical applications.

Table 2.23 Names, characteristics and uses of thermoplastics

Chemical Classification	Trade Name	Characteristics	Typical Applications
1. Cellulose acetate	Tenite 1 plastacele Fibestos Lumarith	Nonflammable, good dielectric, good toughness, high impact strength, moulding and fabricating versatility, available in wide range of colours, Dimensional stability may not be good.	Machine guards and covers, tool and cutlery handles, toys and knobs. Brush backs, Jewellery, Electrical insulation.
2. Cellulose	Celluloid pyralin Nitron Pyroxylin	Flammable, but otherwise same as cellulose acetate. Oldest commercial plastic.	Toilet articles, Movie film, Drawing instruments, Piano and typewriter keys.
3. Methyl	Lucite Plaxiglass Crystalite	Lightweight, weather resistant, Best of plastic for light piping, transmission and edge illumination effects.	Aircraft cockpits and windows, outdoor signs, Wall panels, Name plates, Illuminating accessories, Housing and containers.

Chemical Classification	Trade Name	Characteristics	Typical Applications
4. Polyamide	Nylon zytel	Low coefficient of friction. Dimensional stability. Good electrical properties. Not affected by gasoline or hydraulic fluids, bearings can operate without lubricant. Abrasion resistance.	Small moulded gear and bushings, brush bristles, fuel container, coatings.
5. Polyethylene	Alathon Polythene	Flexible and tough. Excellent dielectric properties. Flammable, but slow burning. Dimensional variations difficult to predict.	Automotive parts. House wares, containers, electrical cable jacketing, tubing, squeeze bottles.
6. Polystyrene	Lystrex Styron Cerex	Resistant to acids and alkalies. Good dimensional stability. Good electrical properties.	Electrical insulations, containers and closures. Instrument panels, knobs, wall tile, radio cabinets.
7. Polytera-flouro Ethylene (TFE)	Teflon	Low coefficient of friction. Resistance to chemical attack. Tough at low temperature. Bushing can operate without lubricant in presence of abrasives.	Gaskets, packings, seals, bushings, lining for pipe and vessels. Not stick surfacing. Electrical insulation.
8. Polyvinyl Chloride	Vinylite Q Koroseal Tygon Geen	Flame resistant, Resistant to chemicals, oils, and solvents. Abrasion resistant. Sound and vibration deadening.	Automotive panels, parts for vacuum cleaners and refrigerator, floor covering, pipe fittings, tanks, balls and floats.

Table 2.24 Names, characteristics and uses of thermosets

Chemical Classification	Trade Name	Characteristics	Typical Applications
1. Epoxy	Araldite Oxiron	Good toughness. Resistant to acids, alkaline and solvents. Excellent adhesion to metal, glass and wood.	Adhesive and coatings, tools and dies, filament wound vessels, laminates for aircraft, patching compound for metal and plastics.
2. Melamine formaldehyde	Melmae Resimeve	Good for application requiring cycling between wet and dry conditions. Hard and abrasion resistant. Good dielectric. Does not impart taste to food.	Table-ware, electric insulation, automotive ignition parts, cutlery handles, jars and bowles.
3. Phenol-formaldehyde	Bakelite Marblette Durez Catalin Resinox	Good dimensional stability. Excellent insulating qualities. Inert to most solvents and weak acids. Good strength around inserts.	Industrial electrical parts automotive electrical components, paper impregnated battery separators.
4. Phenol furfural	Durite	Similar to Phenolformaldehyde.	Electrical insulation. Mechanical parts. Housings and containers.
5. Alkyd (Modified polyester)	Glyptal Duraplex Backosol Rezyl	Can be made flexible, resilient or rigid. Can resist acids but not alkalies, with glass fibre reinforcement resist salt water and fungus growth.	Boats, Tanks, Bins, Housings, Truck, Trailer and tractor components. Ducts shrouds. Vaulting poles, Athletic goods.

2.20 SOME OTHER MATERIALS

1. **Powdered Materials:** Parts made of powdered materials are manufactured by the methods of powder metallurgy. The parts are fabricated by compacting, binding, and sintering powdered metallic and non-metallic materials. Parts need no machining and have better strength, friction and anti-friction properties, and hardness than ordinary cast alloys. However, the part size and weight are limited and intricate parts can not be fabricated. Powdered materials are used to advantage for friction-drive wheels, brake disks, bushes, washers, rollers, light-duty gears, self-lubricated bearings and many auto-mobile parts.

2. **Composite Materials:** Composite materials can be defined as the structures made up of two or more distinct starting materials. The starting materials can be organic, metals or ceramics. The components of a composite material do not occur naturally as an alloy, but are separately manufactured before these are combined together mechanically or metallurgically. Due to this, they maintain their identities even after a composite material is fully formed. However, the starting materials combine to rectify a weakness in one material by strength in another. Hence a composite material exhibits properties distinctly different from those of the individual materials used to make the composite. Thus, a composite material or structure possesses a unique combination of properties such as stiffness, strength, hardness, weight, conductivity, corrosion resistance and high temperature performance, etc, that is not possible by the individual materials.

From the machine design point of view, the various composite materials are : particulate composites, reinforced composites and laminates.

Particulate composites consist of discrete particles of one material surrounded by a matrix of another material. These composite structures are fabricated by the powder metallurgy technique. The examples are: grinding wheels, sintered carbides, ceramics, and other cutting tool materials, electrical contact points, electrical bushes for motors, electric resistance welding electrodes and so on.

Reinforced materials form the biggest and most important group of composite materials. The purpose of reinforcing is always to improve the strength properties. Reinforcement may involve the use of a dispersed phase or strong fibre, thread or rod. The various fibre reinforced composites are:

(a) **Metal-Matrix Composites:** The reinforced fibres are of patented steel, stainless steel, tungston, molybdenum, boron, graphite, Al_2O_3, SiC, Si_3N_4 etc. and the matrix materials can be aluminium, titanium, nickle, and cobalt. These composites are used at very high temperatures and where extreme strength is desired. Being comparatively light, these materials are far stronger and wear resistant than steels and high alloys.

(b) **Glass-fibre reinforced Plastics:** Used for fabricating boat hulls, car bodies, truck bodies and air-craft fittings, automotive parts and sports equipment etc.

(c) **Organic or Resin Matrix Composites:** Graphite reinforced epoxy is being used to make many parts of a fighter plane.

(d) **C-C Composites:** These materials have graphite fibres in a carbon matrix. This material is being used to fabricate parts of missiles and space shuttles, aerospace turbine and jet engine composites and rocket nozzles.

3. **Ceramics:** Ceramics are inorganic and non-metallic materials, which are processed and used at high temperatures. Most of the raw materials for ceramics are of earthy nature. Ceramic materials are used in industry for many purposes. They account for nearly 7% of the tonnage of major engineering materials, compared to about 11% of that of steel. Ceramic articles of industry are: dinnerware, electrical and chemical porcelain, refractory bricks and tiles, glass, porcelain enamels, abrasives, cutting tools (WC, Al_2O_3, SiC, Si_3N_4), bricks and tiles, cements and concretes, whitewares, mineral ores, slags and fluxes and insulators, etc.

Because of their high melting point and relative resistance to corrosion, ceramics make good candidates for high temperature application. Many ceramics retain their strength to much higher temperatures than metals. But, due to their brittleness, many ceramics are susceptible to thermal shock. Porous ceramics (for thermal insulation) are resistant to thermal shock, whereas the same ceramic in the dense form for structural use may be susceptible.

Since ceramics are brittle materials their use in the past has severely been limited in structural applications. They have no yield strength and fail when the local stress exceeds the materials fracture strength. Due to this, they are very sensitive to the distribution of flaws. However, with the development of higher-purity ceramic materials with improved high temperature strength, such as Si_3N_4 and SiC and with the development of processing methods to produce raw materials with low porosity and more consistent properties, the building of high temperature aircraft structure components from ceramic materials is being seriously considered. Ceramics are much stronger under compressive loading.

2.21 SELECTION OF MATERIALS IN MECHANICAL DESIGN—A SYSTEMATIC APPROACH

Introduction

In mechanical engineering design practice, selection of materials is considered to be more of an art than a science. Conventional practices and individual experiences determine the type of the material to be used in the application. This practice has the advantage that it saves the design time. It is also considered to be a safe practice. But at times this practice is uneconomical. It is so because the material technology is advancing fast. Secondary conventional practices overlook objectives of the product for which the material/materials are being selected.

In Authors opinion methodology Systems Design can be very well adopted for purposes of economical selection of materials. It is expected to be so because of orderly, systematic and scientific nature of the methodology. In the following paragraphs step wise details of this systematic approach are explained. Application of the approach is illustrated with the help of a case study.

- Economical Product
  - Cost
    - Initial cost
    - Processing cost
    - Maintenance cost
  - **User's Requirements**
    - Efficiency
    - Mass production
    - Accuracy
    - Aesthetics
    - Noise
    - Safety
    - Functional requirement for reliable performance.
    - Weight
    - Other objectives
  - **Time**
    - Delivery time
    - Life

Stepwise Details of the Approach

First step: Define the material Requirements.

Material requirements can be defined by keeping in view overall objectives of the product for which the selected material is to be used. The main objective in Mechanical Design is an economical product. Analysis of the term economical reveals objectives of the product. An economical product implies meeting the user's requirements at a reasonable cost within desired time. User's Requirements, cost and time can be further split to give product objectives as shown above.

In general one or more of these objectives are to be fulfilled in product design. Very often these objectives do not complement each other, for example incorporation of high efficiency is bound to increase the cost of the product. This often leads to erroneous subjective decisions. One of the methods to make the decisions scientific is by quantifying relative importance of each objective. This method is explained in the case study.

Keeping in view the material properties and product objectives, material requirements can be listed. For example the objective of light weight product can be met through material requirements namely low specific weight and high strength. Thus each objective can be reduced to a set of material requirements. This relationship for other objectives in terms of some material requirement is given in Table 2.25.

Table 2.25

Product objectives	Material requirement
1. Initial cost	Market rate of material, transportation cost, strength and others.
2. Processing cost	Machinability, forgability, castability, capacity for being hot rolled/cold rolled, weldability, brazing characteristics, finishing characteristics, scrap characteristics/ blank size, melting point, heat treatment characteristics and others.
3. Maintenance cost	Wear and failure characteristics, lubrication properties, corrosion, and others.
4. Efficiency	Surface properties like coefficient of friction, surface finish obtainable, corrosion characteristics, hardness, thermal characteristics and others.
5. Mass production	Castability, machinability, forgebility and other processing characteristics and others.
6. Accuracy	Wear characteristics, surface finish obtainable, heat treatment characteristics, hardness, stiffness and others.
7. Aesthetics	Natural appearance, finishing characteristics, adaptability for surface coatings and others.
8. Noise	Damping characteristics, lubrication properties, stiffness and others.
9. Safety	Chemical properties, failure characteristics and others.
10. Functional requirements for reliable performance	Strength (tensile, compressive, shear, bending, torsional, buckling, etc.), endurance limit, yield characteristics, hardness. Impact resistance, creep, specific heat, thermal and electrical conductivity, coefficient of thermal expansion, intensity of magnetisation and others.
11. Weight	Specific weight, strength and others.
12. Delivery time	Processing characteristics and others.
13. Life	Wear resistance, corrosion resistance, failure characteristics and others.

Find Possible Materials

Second Step: Each material requirement can be met through a number of materials. For example low specific weight requirement can be fulfilled by materials namely Aluminium, and Nylon. Thus on the basis of the material requirements a number of materials can be thought of. But, out of these materials a number of materials may not be feasible choices because of influencing factors. These influencing factors are imposed due to limitations of resources, restriction imposed by the product user and certain environmental conditions. A number of such influencing factors are given in Table 2.26.

Table 2.26

Class of influencing factors	Influencing factors
1. Resources limitations	Availability of materials. Manufacturing process available. Available modes of transportation and others.
2. User's Restrictions	Maximum permissible cost Maximum permissible noise level Maximum acceptable weight Minimum acceptable life Desired quantity Desired delivery date Inhibitions for aesthetics Maintenance level acceptable and others.
3. Environmental conditions	Restriction on utilization of indigenous resources Environmental conditions for service life. Restriction on scrap utilization. Standardisation requirement Social, economic and political factors and others.

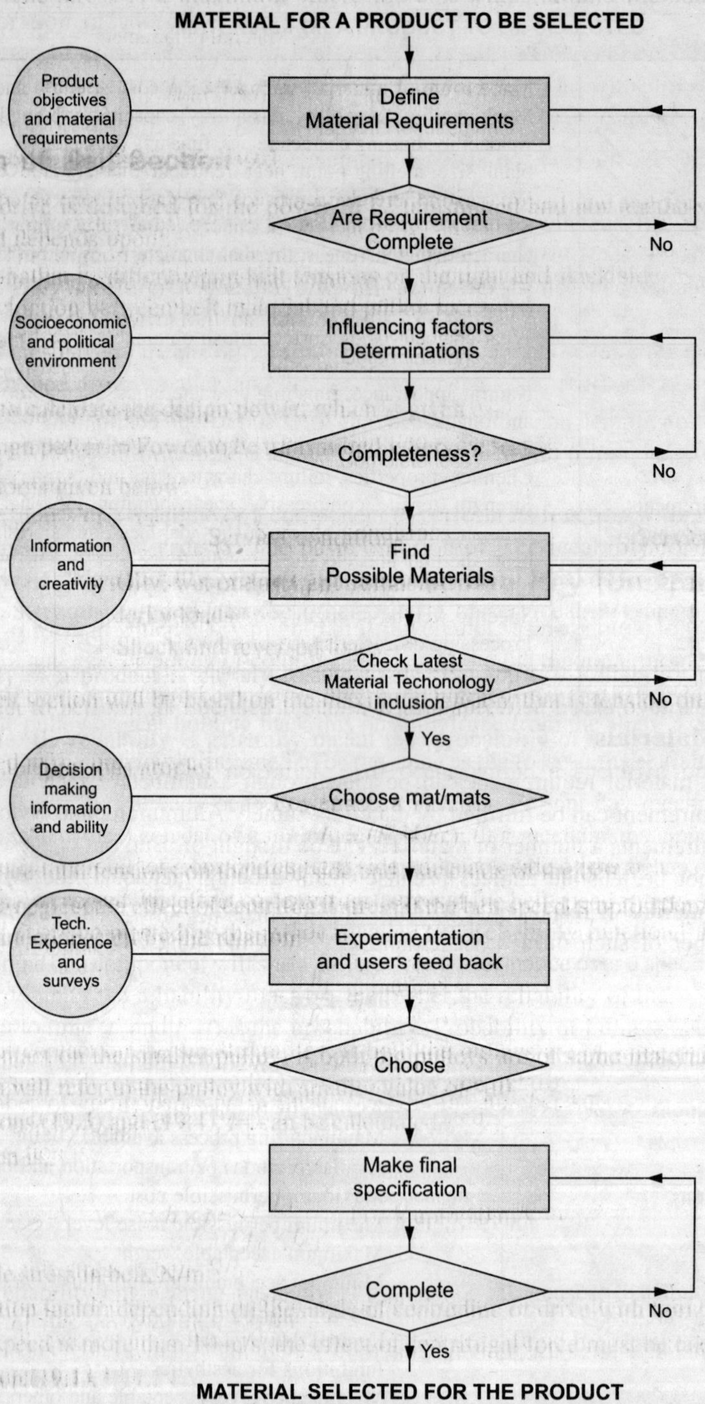

Fig. 2.2 *Material selected for the product*

Keeping in view these influencing factors a number of materials can be rejected resulting into a set of possible materials.

Make a Choice

Third Step: The possible materials chosen on the basis of step two can now be compared for the optimum choice. The method of making this optimum choice in the light of product objectives is illustrated in the case study. In case more than one material appears to be suitable, the final choice should be determined through experimentation and user's feed back. The selected material/materials should be specified in terms of local/National/International standards.

The approach is shown graphically in Fig. 2.2.

Case Study

Selection of material for gears used in paper Industry.

First Step:

Product objectives and, therefore, requirements in this case are observed to be as follows:

Product objectives	Requirements
Low maintenance	Low wear, low lubrication requirements, low rate of failures and high corrosion resistance.
Maximum life	-do-
Minimum noise	High damping characteristics.
Low initial cost	Low market rate and high strength.
Maximum Safety	High chemical resistance and low failure rate.

Second Step: The above listed material requirements can be fulfilled by large number of materials.

Requirements	Materials
Low wear	Cast iron, cast steel, alloy steel, nylon, aluminium alloy, gun metal, high carbon steel, medium carbon steel, phospher bronze.
Low lubrication	Cast iron, nylon
Low rate of failures	Nylon, alloy steel.
High corrosion resistance	Nylon, gun metal, alloy steel, phospher bronze.
High damping characteristics	Nylon, cast iron, wood, aluminium alloy.
Low market rate	Cast iron, wood, medium carbon steel.
High chemical resistance	Nylon, alloy steel, phospher-bronze.
High strength	Nylon, alloy steel.

Out of the above mentioned materials, some of the materials may not be feasible due to following influencing factors.

1. Noise level should not be higher than 80 dB.
2. Lubrication requirement should not be more than four times a year.
3. Manufacturing technology for making accurate gears out of wood in medium lot production is not available.

Because of these influencing factors alloy steel, wood, phospher-bronze and gun metal are not feasible.

Third step: Now the possible materials as decided from step two namely cast iron, cast steel, nylon, aluminium alloy, high carbon steel and medium carbon steel should be compared to make the final choice.

This comparison can be done by quantifying relative importance of product objectives and by quantifying the extent to which each possible material meets the product objectives. For the study under discussion the quantified importance is shown in Table 2.27.

Table 2.27

Product objective	Relative importance	Explanation
Low maintenance Maximum life Minimum noise Low initial cost Maximum safety	8 8 6 8 6	The figures under the column of relative importance are given by converting the feeling of relative importance into a corresponding figure as mentioned below. Rel. Importance felt — Figure Insignificant — 2 Important — 4 Very important — 6 Exceptionally Important — 8 Top priority — 10

Similarly the extent to which each material meets the product objective can be quantified. For this example such weightages are shown in Table. 2.28.

Table 2.28

Product objectives	Relative importance	Cast iron (a)	Cast iron (b)	Cast steel (a)	Cast steel (b)	Nylon (a)	Nylon (b)	Al alloy (a)	Al alloy (b)	H.C.S. (a)	H.C.S. (b)	M.C.S. (a)	M.C.S. (b)
Low maintenance	8	6	48	4	32	10	80	6	48	6	48	4	32
Max. life	8	6	48	4	32	10	80	6	48	6	48	4	32
Minimum noise	6	8	48	4	24	10	60	6	36	4	24	4	24
Low initial cost	8	6	48	6	48	2	16	4	32	6	48	6	48
Maximum Safety	6	4	24	4	24	10	60	8	48	6	36	4	24
Total of figures in (b)			216		160		296		212		204		160

Explanation of Table No. 2.28.

Figures in column marked (a) have been obtained on the basis of following scale.

Extent to which the material fulfils the product objective	Figure
Idealy suited	10
Exceptionally suited	8
Well suited	6
Suited average	4
Not significantly suitable	2

Figures in column (b) have been obtained by multiplying the figure in the "Relative importance" column with the corresponding figure in column (a).

Total of figures in column (b) give a comparative picture of suitability of material under question. Thus from the Table 2.28 Nylon appears to be most suitable.

By referring to relevant literature the Nylon to be used should be type 6/6. This material is commercially available under the trade name of "Nylatron' with the POLYMER Corporation, Pennsylvania, U.S.A.

Concluding Remarks

Authors feel the material technology in the country is not according to the position our country possesses on the world of Industrial map. One major reason for the same appears to be due to lack of pressure on the demand of advanced material technology, which in turn is caused because of acceptance and usage of conventional material technology. It is hoped that teaching of the suggested approach in technical Institutions (the approach is believed to be equally applicable to other disciplines) and it's practice in the field shall generate the mentioned pressure and thus help in the advancement of material technology.

Some other examples are given below

1. Pulley 100 mm in diameter running at 5000 rev/min can be made from cast iron or aluminium. Which material should be chosen for this? For such a high speed of rotation the pulley should be finished all over to balance it correctly. Since the speed of cutting aluminium is eight to ten times greater than that of cast iron, the reduction in the cost of machining and expenditures involved not only doubly compensates for the higher cost of material but also reduces the cost of the aluminium pulley as compared with a cast iron one by 25 percent. In this case, therefore, the pulley should be made from aluminium.

2. The blades of hydraulic turbines should possess adequate strength and resistance to corrosion. Until recently these blades were made from expensive stainless steel which is difficult to machine. In actual fact only the surface of the blades is required to resist corrosion, therefore, the blades need not be made entirely from stainless steel. This can be made from carbon or less expansive alloy steel and then faced with thin coating of stainless steel. This material has been used by one of the Russian manufacturers of blades, which required half the usual time for manufacture and cost 30 percent less.

3. The materials of the friction surfaces of friction clutches should have the properties of high coefficient of friction, removal of heat effectively and should protect the surface against scoring. There is no homogeneous material which can satisfy them all. Powder-metal materials are most effective in this respect. The composition of the charge of copper base powder-metal friction materials is copper 60-70 percent, iron 5-10 percent, lead 5-15 percent, tin 5-15 percent, carborundum, quartz and other abrasives 2.5-7 percent.

4. In recent years, the high alloy steels used in the manufacture of gears have been replaced in many cases by carbon or alloy steels. The gear disc, containing the main mass of metal, experiences only negligible stresses, the teeth are the elements which carry the heaviest load. They should be highly wear resistant, have a high strength of working surfaces and a high limit of endurance in bending. The first is obtained by induction hardening of the active profile of the teeth and the second by strengthening the surface at the tooth root.

5. A link in a mechanism has to be designed. The following requirements have to be met.

Function: To connect two other links through pins of 20 mm diameter.

Mechanical environmental Conditions............Length of link : 150 mm, Mechanical loading : Relatively light loading

Chemical environment Conditions............Indoor location, humid atmosphere.

Service life............10 years.

Maintenance............Minimum (not critical)

Appearance............Shape free from sharp corners.
Weight............Not critical.
Quantity required 20
Delivery............4 weeks.
Manufacturing cost............Minimum.

Materials: While selecting the most suitable material, material and manufacturing method have to be considered together.

(a) Steel
(b) Aluminium alloy
(c) Cast iron

1. Casting
   (d) Malleable iron
   (e) Grey iron
   (f) Brass
   (g) Bronze
2. Mechanical from bar stock
3. Forging
   (a) Hand forged
   (b) Drop forged
4. Welded construction.

Mainly from cost consideration the materials, cast steel, brass, bronze, drop forged steel are rejected. A decision Table given below is used to evaluate the other materials.

Requirements	Weightage	Materials											
		Malleable iron		Cast iron		Alum. casting		M.S. (Machined)		M.S. (Welded)		M.S. (Hand-forged)	
	a_1	b_1	$a_1 \times b_1$	b_2	$a_1 \times b_2$	b_3	$a_1 \times b_3$	b_4	$a_1 \times b_4$	b_5	$a_1 \times b_5$	b_6	$a_1 \times b_6$
Mechanical loading	7	7	49	7	49	5	35	10	70	10	70	10	70
Climatic effects	5	6	30	6	30	8	40	4	20	4	20	4	20
Appearance	3	10	30	10	30	10	30	8	24	6	18	7	21
Manufacturing cost	10	4	40	5	50	4	40	5	50	10	100	5	50
			149		159		145		164		208		161

From this table it is clearly seen that mild steel with welded construction is the best choice. It may be noted that the decision table is most useful in rejecting the undesirable materials. Generally more detailed analysis is required to choose the best materials from a list of two or three materials with total points $(a_i \times b_i)$ close to each other. Even if the preliminary selection based on the decision table suggests steel as the most suitable material, the designer has still to decide which grade of the steel is optimum one. For this work information on the past experience on the use of materials for various applications is of great help.

For the rational designing of machine elements the above factors affecting the choice of materials should be thoroughly mastered.

2.21.1 'Local Quality' Principle of Material Selection

As discussed above, a crucial step in the design of a machine part, governing the quality of that part and the machine as a whole, is the correct choice of materials of construction. A detailed study of the many materials that are applicable to a given design is in itself a speciality.

When specifying the material for a given part, the designer must ensure that the recommended material:

1. suits the chosen critical criterion of serviceability, or in other words, is adequately strong, stiff, and wear resistant.
2. is optimal for the given size and guarantees the minimum weight of the part and the machine as a whole.
3. suits the purpose and operating conditions of the part.
4. has an adequate corrosion resistance, enhances or minimizes friction (whichever is desired), provides adequate insulation from heat and electricity, etc.
5. suits best the shape and the chosen manner of manufacture, that is, can readily be stamped, forged, welded, cast, or machined.
6. is inexpensive and easily available.

In many cases, it is hard to find the material that would satisfy the above requirements all at the same time. Moreover, if such a material can be found, it will not be cheap at all. On the face of it, the situation could readily be tackled by way of compromise, However, a more reasonable approach will be to employ what we call the "local quality" principle. A few examples have already been given under Art. 2.20. Other examples are:

(i) In a gear wheel, most of the metal is spent to form hub and web which experience insignificant stresses in service. The elements that carry the lion's share of the imposed load are the teeth. Accordingly, they must offer substantial resistance to wear and contact fatigue for one, and show a high endurance limit in bending, for another. The former requirement is met by heat treating the working profile of the teeth (by induction hardening); the latter requirement is satisfied by strain-hardening the roots of the teeth (the idea is to induce residual compressive stresses in the metal). Thus the desired effect is seen to be achieved by modifying locally the properties and state of the part's material.

(ii) Large gears more than 600 mm in diameter are generally assembled from two parts, namely a gear ring (rim) made of high-grade steel and a hub made solid with a web of ordinary steel.

(iii) Locomotive and car wheels are similar in construction and are said to be tyred.

(iv) Worms are likewise built of two materials differing in properties and cost; the gear rim is made of bronze and the rest of the worm body, of cast iron or steel.

(v) The same applies to belt pulleys, chain sprockets, carriers in epicyclic gearing, flexible spines, bearing inserts and cases, and so on.

The "local quality" principle is also embodied in compound material, such as composites, clad metals, and reinforced materials, which are finding ever increasing application in mechanical engineering.

EXERCISES

1. Classify the materials of construction.
2. What do you mean by the term "Mechanical properties of material"?
3. What is malleable iron? What advantage has this form over white or grey cast iron?
4. Why are metals in their pure form unsuitable for industrial use?
5. How does carbon content affect cast iron, wrought iron, and steel with reference to hardness and toughness?

6. What is the chief reason for the use of alloy steels in machine parts?
7. What is meant by the term indentation hardness, fatigue, machinability and damping capacity?
8. Explain the difference between malleable cast iron and grey cast-iron?
9. Differentiate between hot and cold working.
10. Under what circumstances alloys of cast iron are used?
11. Explain in detail the affects of nickel, copper, chromium and molybdenum as alloying elements in cast iron.
12. When the cast steels are preferred over cast iron?
13. Enlist the properties of 0.2% carbon steel.
14. Name the steel alloys used for high temperature service.
15. List the alloying materials of aluminium.
16. What alloy steel is suitable for springs?
17. Name the process by which steel is commonly produced.
18. What are the copper alloys?
19. List the advantages and disadvantages of bronze over brass for industrial uses.
20. What is the composition of red brass and yellow brass and where these are commonly used?
21. Differentiate between the tin base and lead base alloys.
22. What are the constituents of Muntz metal and monel metal?
23. What are the constituents of Babbit metal?
24. What are the constituents of stellite and where it is commonly used?
25. List the important non-metallic materials of construction in machine design.
26. Explain clearly the use of wood and glass in place of steel in machine design.
27. Enumerate the advantages and disadvantages of plastic materials over metallic materials.
28. Give the trade names of plastic items used in daily life.
29. Give the materials which are commonly used for the following parts of a steam engine.

    (a) Cylinder head and cylinder (b) Connecting rod
    (c) D-slide valve (d) Crankshaft
    (e) Piston (f) Piston rings.

30. What are the cupronickles or nickel silver alloys?
31. (a) Enumerate the factors and properties of materials a designer is required to consider while designing a machine part.
    (b) While looking at factors and properties in (a) discuss material for the following:
    (i) Nib of pen (ii) Shaving blade (iii) spring in cursor of slide rule (iv) drawing board pins (v) tooth brush (vi) soap box (vii) pressure cooker.
32. Discuss the materials of the following :
    (a) rivets used in a boiler (b) gears of a lathe
    (c) connecting rod of an I.C. engine (d) valve of a safety valve
    (e) body of a safety valve (f) steam pipe
    (g) water pipe.
33. Give two typical applications of each of the following alloys :
    (a) 70/30 copper zinc alloy (b) monel metal
    (c) babbit metal (d) copper-aluminium alloy
    (e) 92/8 copper lead alloy.
34. Write short notes on:
    (a) powdered materials (b) composite materials
    (c) ceramics.
35. Discuss the principle of "Local quality" for selecting the materials.
36. Discuss the "systematic Approach" for the selection of materials in Machine Design.
37. Name the materials which are commonly used for the following mechanical parts and justify your choice:-
    (a) Taps, dies and drills (b) Axles
    (c) Gasket (d) Hand wheel
    (e) Flat belt (f) Valve spring

□□□

CHAPTER 3

Stress Analysis

3.1 STRAIN

When an external force or load is applied to a body, the body will move in the direction of the applied force unless that force is balanced by an opposite force called reaction. Such forces always act in pairs, although mention of the reaction is often omitted. The action of this pair of forces produces a deformation of the body, also called a *strain*. These deformations can be longitudinal (lengthening or shortening of the body) and angular, a change of angle between the faces.

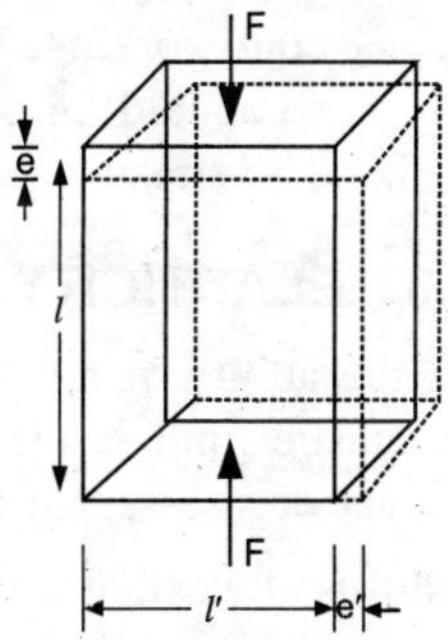

Fig. 3.1

Unit strain. The deformation per unit length in the direction of a load is called unit strain, or deformation. It is symbolised by S and it is a dimensionless number. Thus

$$S = \frac{e}{l} \qquad ...(3.1)$$

Unit deformation θ produced by angular distortion is called *shear strain* and is always measured in radians (Fig. 3.2).

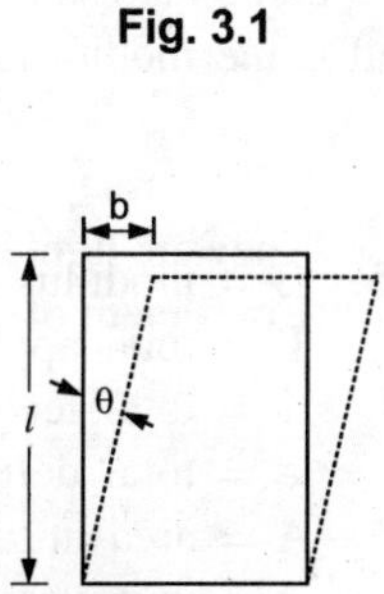

Fig. 3.2

Because tan θ = θ, as the angle is very small, therefore, the shear strain may be given by the relation,

$$\theta = \frac{b}{l}$$

Poisson's ratio $\left(\frac{1}{m}\right)$: Accompanying a longitudinal deformation e is a lateral deflection e' (Fig. 3.1). The deformation e and e' have opposite signs.

Thus, longitudinal strain $S = \frac{e}{l}$, transverse strain $S' = \frac{e'}{l'}$

The ratio of transverse strain to longitudinal strain is called poisson's ratio and is represented by $1/m$. Values of $1/m$ for various materials are given in Tables 2.1 to 2.4.

3.2 STRESS

When any solid body is subjected to external forces, (Fig. 3.1) there are set up within the body resisting forces called stresses. These forces per unit area are called the unit stresses and the units are N/m^2. Unit stress (f) is given by the relation.

$$f = \frac{F}{A} \qquad ...(3.2)$$

where f is unit stress in N/m^2

A = total area in m^2

F is the total force acting through the centroidal axis of the body whose total area is A, in N.

Stress (f) may be assumed to be the unit stress distributed over the entire area. In case of eccentric loading, abrupt changes in section, non-homogeneity of material and other forms of stress raisers, the stress is an average stress over the cross-sectional area and is not uniform.

A stress may be normal, tangential or oblique to the plane on which it acts. A normal stress or the normal component of an oblique stress is referred to as a *direct stress*. Such a stress may be a *tensile stress* (f_t) or a *compressive stress* (f_c). A tangential stress is called *shear stress* (f_s). It may be produced either by transverse load or by torsion.

In the design of various machine members the stress situation is rarely as simple as expressed by relation 3.2. For this reason it is necessary to study the basic assumptions which were used in deriving this relation.

1. The load is steady or static. When the load is moving, the inertia of the body will affect the magnitude of stresses.
2. The body is free from initial stresses. The initial stresses which are built into the material by cold working, heat treatment, or assembly operations are frequently of great importance and so should be considered in addition to the stress obtained from relation 3.2.
3. The stress is uniformly distributed over the cross-section of the bar.

3.3 ELASTICITY

It is the ability of a material to return to its original dimensions after the removal of external forces. The elastic limit is the limit of stress within which the deformation completely disappears after the removal of stress, that is no set remains.

Modulus of elasticity: Fortunately, for most materials the deformation is proportional to the applied loads until a certain stress is reached. The ratio of the unit stress to the unit strain within the proportional limit is called the modulus of elasticity of the material as is given by the relation.

$$E = \frac{f}{s} = \frac{F/A}{e/l} = \frac{Fl}{Ae} \qquad ...(3.3)$$

where E = modulus of elasticity in N/m^2

F = total applied load in N,

l = total length in m,

e = total deformation,

A = area normal to the applied load in m^2.

f = unit stress in compression or tension.

The modulus of elasticity defined above applies to tensile and compressive loads. It is also known as *Young's modulus.*

Modulus of Elasticity in shear or coefficient of rigidity: When the external forces cause adjacent planes to slide relative to each other, the deformations are shearing deformations (Fig. 3.2.) and the modulus of elasticity in shear is given by the relation.

$$C = \frac{f_s \times l}{\theta \times r} \quad ...(3.4)$$

where C = modulus of elasticity in shear or coefficient of rigidity in N/m^2.

l = length in m

f_s = unit shear stress in N/m^2.

θ = twist expressed in radians.

r = distance from the surface to the centre line in m.

Bulk modulus of elasticity: It is the ratio of normal stress, applied to all six faces of a cube, to the change of volume. It is usually represented by *K*.

The following relations exist between the modulus of elasticity in tension or compression E, modulus of elasticity in shear *C*, bulk modulus of elasticity *K* and Poisson's ratio $1/m$.

$$E = 2C\left(1 + \frac{1}{m}\right) \quad ...(3.5)$$

$$K = \frac{E}{3}\left(1 - 2 \times \frac{1}{m}\right) \quad ...(3.6)$$

Values of *C* computed by equation 3.5 are slightly smaller than those found from actual tests, but they may be used in the absence of information about *C*.

3.4 STATIC LOAD

Static loads are those which are applied to a structure very gradually, increasing from zero to their final value and then either do not change their magnitude, direction or point of application with time at all or change them very slightly, so that accelerations occuring in this case can be neglected.

3.5 SIMPLE STRESSES

Tension, compression and shear are called simple stresses when they can be considered single. A simple stress is considered to be distributed uniformly over the cross-section of the part to which the force is applied.

Tensile stress: The stress f_1 in simple tension in machine member is equal to the external force *F* in *N* divided by the *X*-sectional area *A* of the member in square *m*. Thus

$$f_t = \frac{F}{A} \quad ...(3.7)$$

Compressive stress: For a straight short compression member in which the resultant load *F* acts through the centre of gravity of every cross-section, the stress f_c is given by the relation

$$f_c = \frac{F}{A} \quad ...(3.8)$$

If the length of the member under compression is 20 times the least radius of gyration of its normal cross-section, the member must be treated as a column, and the stresses are determined by one of the column formulas.

Shear stress: It exists between two parts of a body when, over the common surface each part exerts over the other, an equal and opposite force, tangential to the surface of contact as indicated in Fig. 3.3. Ordinarily it is assumed that the shear stress is distributed uniformly over the cross-section, normal

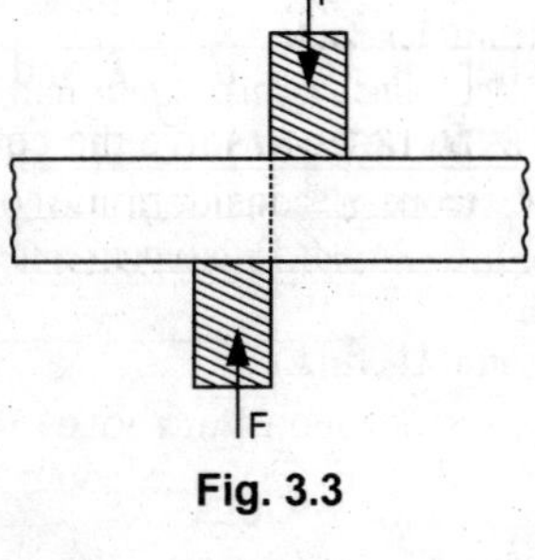

Fig. 3.3

$$f_s = \frac{F}{A} \quad ...(3.9)$$

to its axis, in which case the actual conditions are different. The stresses are greatest near the point of application of load. The assumption of uniform distribution of shear stress should never be made where shear is due to bending.

3.6 TORSION

Stress produced by torsion and bending are termed ***compound stresses,*** because torsion and bending produce normal and tangential stresses in the same plane simultaneously.

Torsional stress: When a solid round bar is subjected to the action of a torsional moment M_t, shearing stresses of magnitude varying linearly from zero at the centre to a maximum at the surface will be induced in the bar. The stress at the surface is

$$f_s = \frac{M_t Y}{J} \quad ...(3.10)$$

where M_t = torsional moment in Nm
r = radius of bar in m,
Y = Distance from the axis of rotation or neutral axis to the stressed fibre. It will be equal to the radius of a circular shaft.
J = polar moment of inertia in m^4.
f_s = Shearing stress in N/m^2

Polar moment of inertia, $J = \frac{\pi d^4}{32}$, where d is the diameter of the round bar in m. For a hollow round circular bar, $J = \frac{\pi}{32}(d_o^4 - d_i^4)$, where d_o and d_i are the outside and inside diameters respectively in m. Relation 3.10 can also be written as

$$f_s = \frac{M_t}{Z_p} \quad ...(3.11)$$

where Z_p = polar section modulus in m^3.

Torque and deflection relations for some sections are given in Table 3.1.

It is important to note that relation (3.10) is valid only for circular bars and that it gives the stress in the outer fibres of the shaft. This is the maximum shearing stress and is the one ordinarily used in design.

The relation between the angular deflection in radians and the torsional moment M_t can be given as

$$\left.\begin{aligned} \theta &= \frac{M_t l}{JC} \\ \text{and} \quad \theta &= \frac{2lf_s}{dC} \text{ in terms of shear stress} \end{aligned}\right\} \quad ...(3.12)$$

where θ, M_t, l, d, J, f_s and $\boldsymbol{C}$ have the usual meanings.

In order to solve the equations for torsional shearing stress, it is often necessary to obtain the torques M_t from a consideration of the horse power and speed of the rotating shaft. The relation is given below:

$$\text{Power} = \frac{2\pi n M_t}{60} = F \times V, \text{ watts} \quad ...(3.13)$$

where M_t = torsional moment in *Nm*
n = Shaft speed in rev/min
F = Force in *N*
V = Velocity in m/s

3.7 BENDING

It is difficult to conceive of a machine or a structure in which there do not exist at least a few elements subjected to bending forces. When a load is applied to a machine member transversely (Fig. 3.4.) it produces stresses of several kinds.

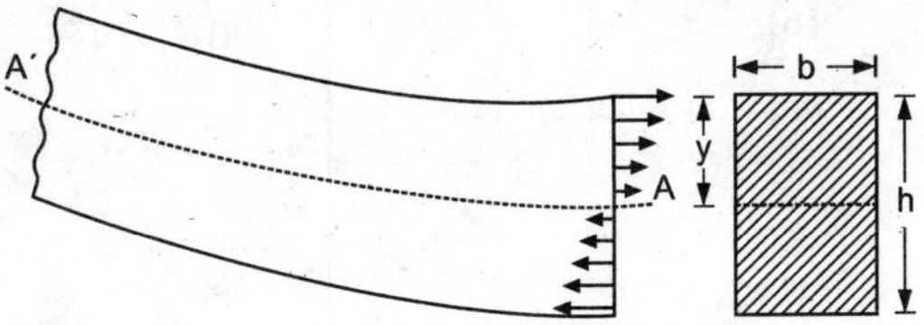

Fig. 3.4 Assumed distribution of normal stresses in a beam

Tensile and compressive stress: A bent beam deforms in such a way that the fibres on the convex side are elongated, that is in tension, while the fibres on the concave side are shortened, that is, in compression. Between these fibres there is a neutral surface which is neither elongated nor compressed and, therefore, does not have any tension or compression, bending stress. This surface in straight beam passes through the centre of area of cross-section of the beam, if the beam is made of homogeneous material. The stress in the beam is proportional to the distance of the fibre from the neutral surface provided that no fibre is stressed beyond the proportional limit. The direct tension or compression stress due to bending is given by the relation:

$$f_b = \frac{M_b Y}{I} \qquad \text{...(3.14)}$$

or

$$f_b = \frac{M_b}{Z}$$

where f_b = bending stress in N/m^2
M_b = bending moment in Nm
Y = distance from the neutral surface to the fibre where the stress is (f_b) in m.
I = rectangular moment of inertia of the cross-sectional area about the neutral axis in m^4.
Z = section of modulus in m^3.

Table 3.1 Torsion of shafts of various cross-sections

Cross-section	Torsional resisting Moment	Angular twist	
		In terms of torsional moment	In terms of max. shear
1. Circular section 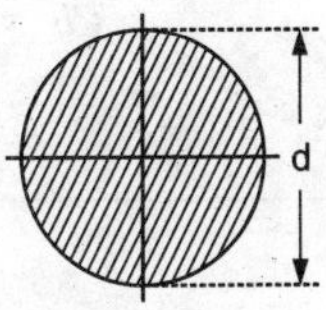 d	$\frac{\pi}{16} d^3 f_s$ f_s max. at circumference	$\frac{32}{\pi d^4}\frac{M_t l}{C}$	$\frac{2 f_s l}{dC}$

Cross-section	Torsional resisting Moment	Angular twist	
		In terms of torsional moment	In terms of max. shear
2. Hollow circular section – 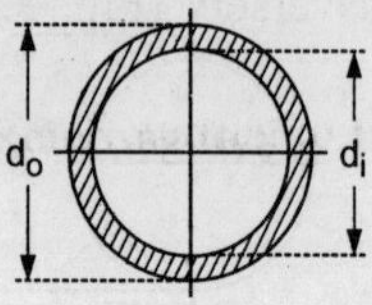	$\frac{\pi}{16}\left(\frac{d_o^4 - d_i^4}{d_o}\right)f_s$ f_s max. at outer circumference	$\frac{32}{\pi\left(d_o^4 - d_i^4\right)} \times \frac{M_t l}{C}$	$\frac{2 f_s l}{d_o C}$
3. Solid elliptical section 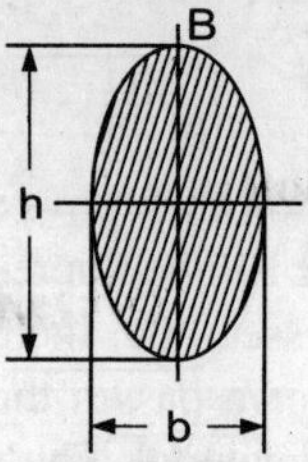	$\frac{\pi}{16} b^2 h f_s f_s$ max. at A $\frac{\pi}{16} b h^2 f_s$, f_s max. at B ($h>b$)	$\frac{16}{\pi}\frac{(b^2 + h^2) M_t l}{b^3 h^3 C}$	$\frac{f_s (b^2 + h^2) l}{b h^2 C}$
4. Solid rectangular section 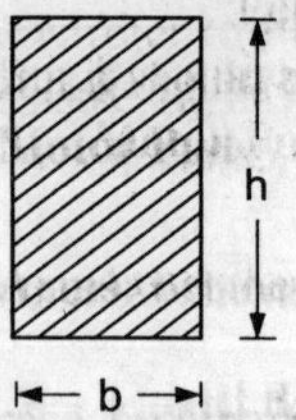	$\frac{2}{9} b^2 h f_s$, f_s max. at middle of side h. f_s max. at middle of b, $\frac{2}{9} b h^2 f_s$ ($h>b$)	$3.6 \frac{(b^2 + h^2)}{b^3 h^3} \times \frac{M_t l}{C}$	$0.8 \frac{f_s (b^2 + h^2) l}{C b h^2}$
5. Solid square section 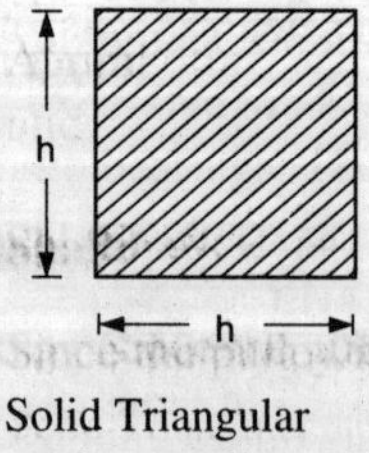	$\frac{2}{9} h^3 f_s$ f_s max. at middle of side	$7.2 \frac{1}{h^4} \frac{M_t l}{C}$	$1.6 \frac{f_s}{C} \frac{l}{h}$
6. Solid Triangular section (equilateral) 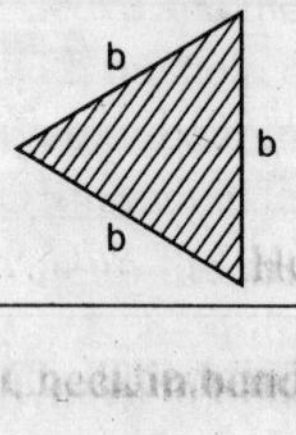	$\frac{b^3}{20} f_s$ f_s max at middle of each side	$46.2 \frac{1}{b^4} \frac{M_t l}{C}$	$2.31 \frac{f_s l}{C b}$

Cross-section	Torsional resisting Moment	Angular twist	
		In terms of torsional moment	In terms of max. shear
7. Solid hexagonal 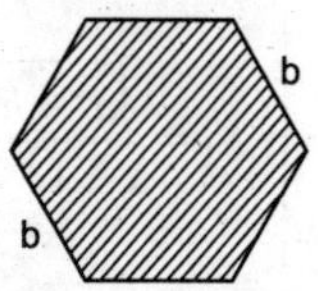	$\frac{b^3}{1.09} f_s$	$0.967 \frac{1}{b^4} \frac{M_t l}{C}$	$0.9 \frac{f_s l}{Cb}$

Properties of various cross-sections are given in Table. 3.2.

Table 3.2 Properties of various cross-sections

Form of section	Area A	Distance from centroid to extremities of section y_1, y_2	Moments of Inertia I_1 and I_2 about principal central axis 1 & 2	Radii of gyration K_1 & K_2 about principal central axis 1 and 2
1. Square 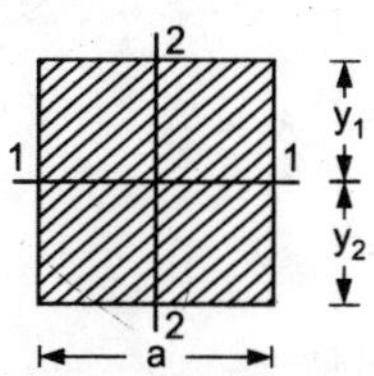	$A = a^2$	$y_1 = y_2 = \frac{1}{2} a$	$I_1 = I_2 = \frac{1}{12} a^4$	$K_1 = K_2 = 0.289\ a$
2. Rectangular 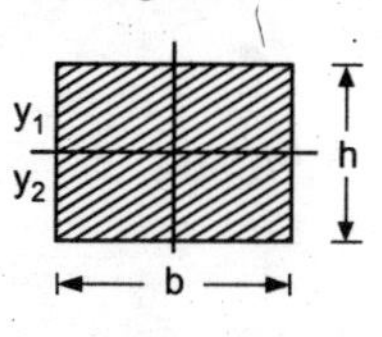	$A = bh$	$y_1 = y_2 = \frac{1}{2} h$	$I_1 = \frac{1}{12} bh^3$	$K_1 = 0.289\ h$
3. Triangular 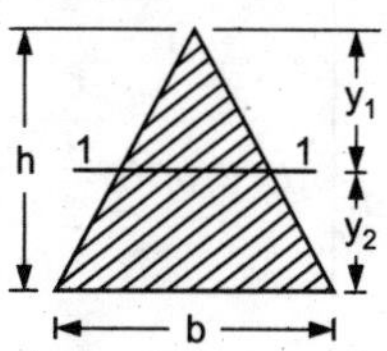	$A = \frac{1}{2} bh$	$y_1 = \frac{2}{3} h,\ y_2 = \frac{1}{3} h$	$I_1 = \frac{1}{36} bh^3$	$K_1 = 0.2358\ h$
4. Hollow rectangular 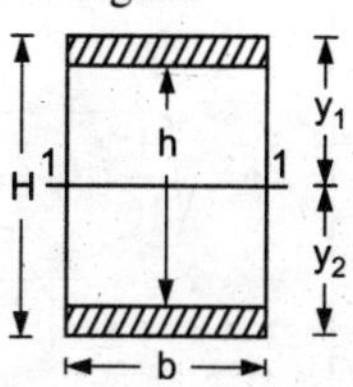	$A = Hb - hb$	$y_1 = \frac{1}{2} H = y_2$	$I_1 = \frac{b}{12}(H^3 - h^3)$	$K_1 = \sqrt{\frac{H^3 - h^3}{12\,(H - h)}}$

Form of section	Area A	Distance from centroid to extremities of section y_1, y_2	Moments of Inertia I_1 and I_2 about principal central axis 1 & 2	Radii of gyration K_1 & K_2 about principal central axis 1 and 2
5. Hollow square 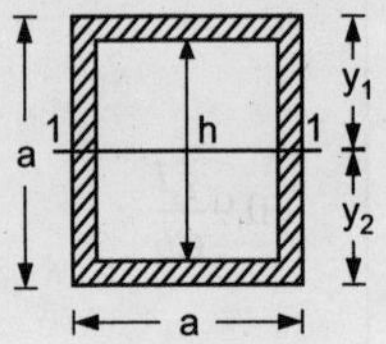	$A = a^2 - h^2$	$y_1 = y_2 = \frac{1}{2}a$	$I_1 = \frac{1}{12}[a^4 - h^4]$	$K_1 = \sqrt{\frac{a^2 + h^2}{12}}$
6. Trapezoidal 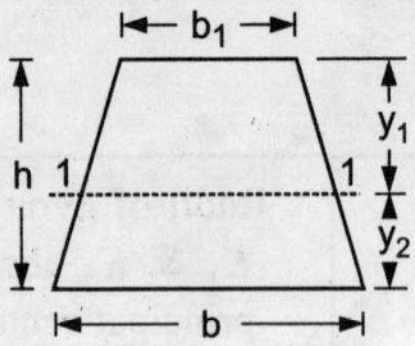	A = $\frac{1}{2}(b+b_1)h$	$y_1 = \frac{2b+b_1}{3(b+b_1)} \times h$ $y_2 = \frac{b+2b_1}{3(b+b_1)} \times h$	$I_1 = \frac{h^3(b^2+4bb_1+b_1^2)}{36(b+b_1)}$	$K_1 = \frac{h\sqrt{2(b^2+4bb_1+b_1^2)}}{6(b+b_1)}$
7. Solid circle 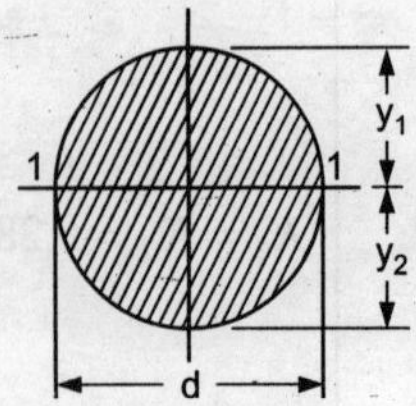	$A = \frac{\pi}{4}d^2$	$y_1 = y_2 = \frac{d}{2}$	$I_1 = \frac{\pi}{64}d^4$ $= 0.05d^4$ (approx.)	$K_1 = \frac{d}{4}$
8. Regular hexagon 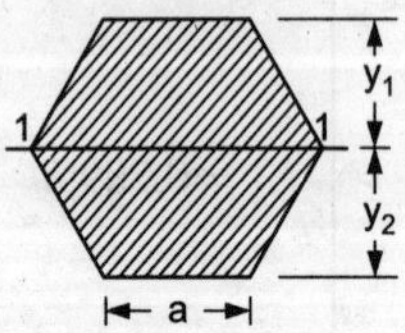	$A = \frac{3\sqrt{3}}{2}a^2$	$y_1 = y_2 = \frac{\sqrt{3}}{2}a$	$I_1 = \frac{5}{16}\sqrt{3}\,a^4$	$K_1 = \sqrt{\frac{5}{24}}\,a$
9. Very thin annulus 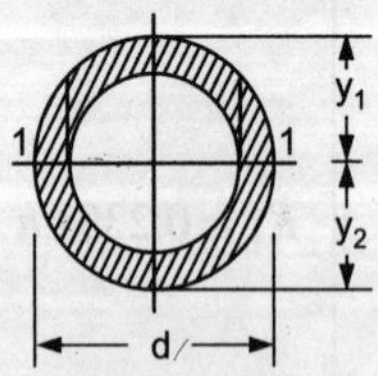	$A = \pi dt$	$y_1 = y_2 = \frac{1}{2}d$	$I_1 = \frac{\pi}{8}d^3t$	$K_1 = 0.707\frac{d}{2}$
10. Hollow circle 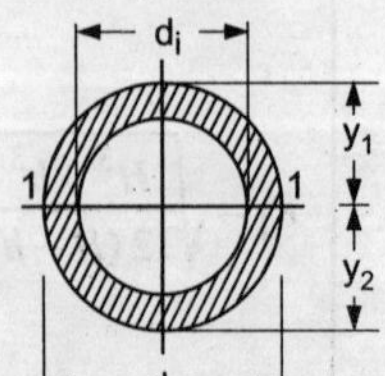	$A = \frac{\pi}{4}$ $(d_o^2 - d_i^2)$	$y_1 = y_2 = \frac{d_o}{2}$	$I_1 = I_2 = \frac{\pi}{64}(d_o^4 - d_i^4)$	$K_1 = \sqrt{\frac{1}{16}(d_o^2 + d_i^2)}$

Form of section	Area A	Distance from centroid to extremities of section y_1, y_2	Moments of Inertia I_1 and I_2 about principal central axis 1 & 2	Radii of gyration K_1 & K_2 about principal central axis 1 and 2
11. Elliptical 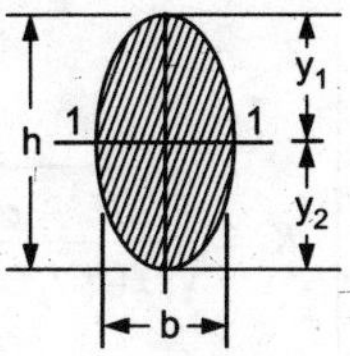	$A = \frac{1}{4}\pi bh$	$y_1 = y_2 = \frac{1}{2}h$	$I_1 = \frac{\pi}{64}bh^3$ $I_2 = \frac{\pi}{64}hb^3$	$K_1 = \frac{1}{4}h$ $K_2 = \frac{1}{4}b$
12. Hollow Elliptical 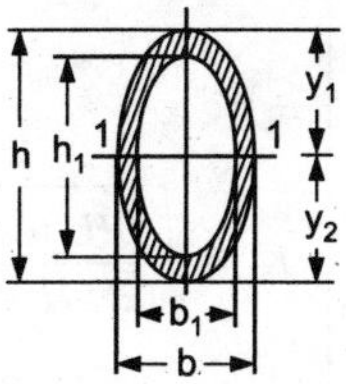	$A = \frac{1}{4}\pi(bh - b_1h_1)$	$y_1 = y_2 = \frac{1}{2}h$	$I_1 = \frac{1}{64}\pi(bh^3 - b_1h_1^3)$	$K_1 = \frac{1}{4} \times \sqrt{\frac{bh^3 - b_1h_1^3}{bh - b_1h_1}}$
13. I-section 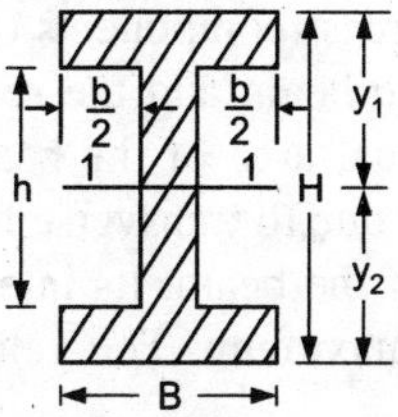	$A = HB - h \times \frac{b}{2} \times 2$	$y_1 = y_2 = \frac{H}{2}$	$I_1 = \frac{BH^3 - bh^3}{12}$	$K_1 = \sqrt{\frac{BH^3 - bh^3}{12(BH - bh)}}$
14. T-section 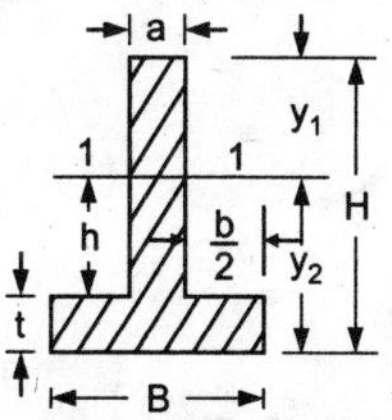	$A = Bt + (H - t)a$	$y_1 = H - y_2$ $y_2 = \frac{aH^2 + bt^2}{2(aH + bt)}$	$I_1 = \frac{By_2^3 - bh^3 + ay_1^3}{3}$ $h = y_2 - t$	$K_1 = \sqrt{\frac{I_1}{Bt + a(H - t)}}$
15. Channel section 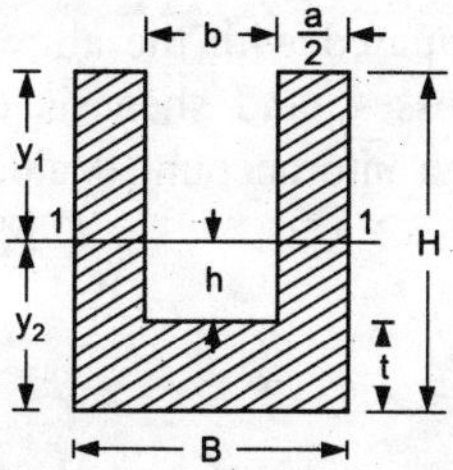	$A = Bt + (H - t)a$	$y_1 = H - y_2$ $y_2 = \frac{aH^2 + bt^2}{2(aH + bt)}$	$I_1 = \frac{By_2^3 - bh^3 + ay_1^3}{3}$ $h = y_2 - t$	$K_1 = \sqrt{\frac{I_1}{Bt + a(H - t)}}$

Form of section	Area A	Distance from centroid to extremities of section y_1, y_2	Moments of Inertia I_1 and I_2 about principal central axis 1 & 2	Radii of gyration K_1 & K_2 about principal central axis 1 and 2
16. H-section 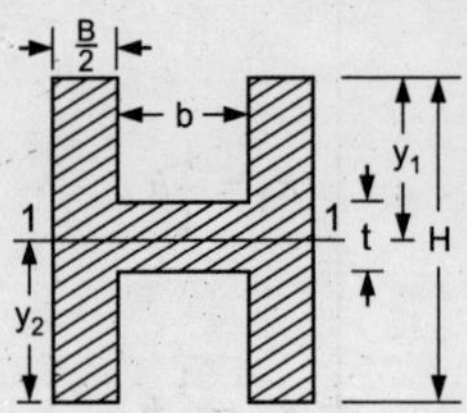	$bt + H \times$ $B = A$	$y_1 = y_2 = \dfrac{H}{2}$	$I_1 = \dfrac{BH^3 + bt^3}{12}$	$K_1 = \sqrt{\dfrac{BH^3 + bt^3}{12(BH + bt)}}$
17. Cross-section 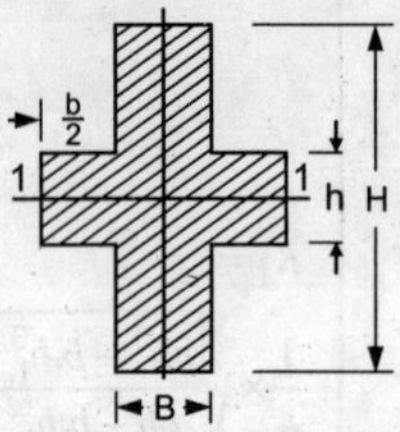	$bh + BH = A$	$y_1 = y_2 = \dfrac{H}{2}$	$I_1 = \dfrac{BH^3 + bh^3}{12}$	$K_1 = \sqrt{\dfrac{BH^3 + bh^3}{12(BH + bh)}}$

Shear stress in beams: Vertical shear in a beam is the algebraic sum of all the forces (components of the loads and reactions), normal to the neutral surface, acting between one end of the beam and the section at which the shear is desired. The total vertical shear divided by the cross-sectional area of the beam is the average shear at the section but not the maximum shear stress. This shear stress due to transverse force, unlike a simple shear stress, is not distributed uniformly over the cross-sections of the beam. Its intensity is zero at the outer fibres and increases towards the neutral plane, where it is a maximum. The value of the shear stress is given by the relation:

$$f_s = \left(\frac{F}{bI}\right) AY \qquad ...(3.15)$$

where f_s = shear stress at a distance from the neutral surface in N/m^2.
F = total vertical force on the section in N.
b = width of the beam at distance a from the neutral surface in m.
I = rectangular M.O.I. of the cross-sectional area of the beam in m^4.
A = cross-sectional area included from a to the outer surface of the beam in m^2.
Y = distance from the neutral surface to the centre of gravity of the area A in m.

In relatively long metal beams the actual shear stress is always small compared with the allowable value when the beam can safely withstand the tensile and compressive stresses, and shear is often neglected. In short beams, however, the shear stress be large enough to be taken into account. Values of shear stress in beams caused by bending are given in Table 3.3.

Table 3.3 Shear stress in beams, caused by bending

Type	Section	Shear stress at a distance y from the neutral axis (N/m^2)	Maximum shear stress N/m^2
(*a*)	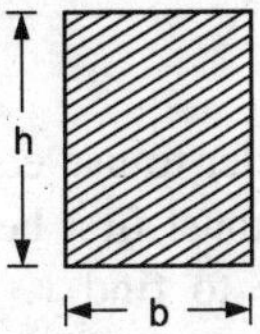	$\frac{3F}{2bh}\left[1-\left(\frac{2y}{r}\right)^2\right]$	$\frac{3F}{2bh}=1.5\,\frac{F}{A}$ (for $y=0$)
(*b*)	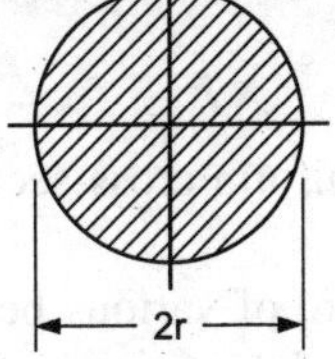	$\frac{4F}{3\pi r^2}\left[1-\left(\frac{y}{r}\right)^2\right]$	$\frac{4F}{3\pi r^2}=1.33\,\frac{F}{A}$ (for $y=0$)
(*c*)	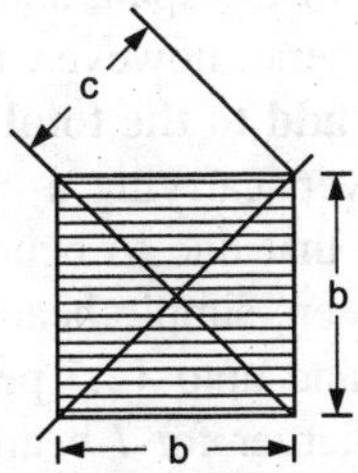	$\frac{F\sqrt{2}}{b^2}\left[1+\frac{y\sqrt{2}}{b}-4\left(\frac{y}{b}\right)^2\right]$	$0.591\,\frac{F}{A}$ (for $y=\frac{c}{4}$)
(*d*)	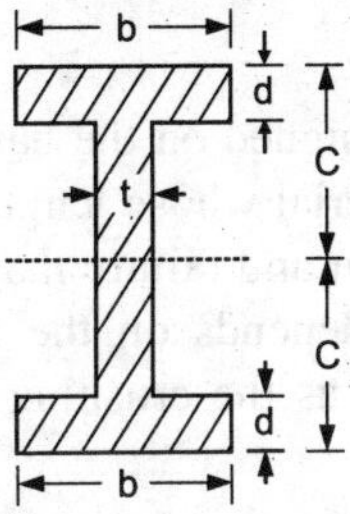	---------------------	$\frac{3F}{4t}\left[\frac{bc^2-(b-t)d^2}{bc^2-(b-t)d^3}\right]$ (for $y=0$)

3.8 DEFLECTION OF BEAMS

When a beam is subjected to bending, the fibres on one side elongate, while the fibres on the other side shorten (Fig. 3.4). These changes in length cause the beam to deflect. All points on the beam except those directly over the support fall below their original position. The distance that a point on the beam moves is called the deflection of the beam at the point. The rigidity of a beam is measured by its deflection.

The **elastic curve** is the curve taken by its neutral axis. The radius of curvature R at any point is given by the relation

$$R=\frac{EI}{M_b} \qquad ...(3.16)$$

where E, I and M_b have the usual meanings.

A beam bent to a circular curve of constant radius has a constant bending moment. Replacing R in the equation by its approximate geometrical value, $\frac{1}{R} = \frac{d^2y}{dx^2}$, the fundamental equation from which the elastic curve of a bent beam can be developed and the deflection of any beam obtained is

$$\frac{d^2y}{dx^2} = \frac{M_b}{EI} \quad ...(3.17)$$

where y is the deflection of the beam and x is the distance from the support. In order to use equation 3.17, an expression for the bending moment M_b in terms of x for the case considered must first be established and substituted in that equation, the resulting equation is then integrated twice to find y.

In general the deflection relation may be expressed by the relation

$$y = \frac{1}{m}\frac{FL^3}{EI} \quad ...(3.18)$$

Where m is a coefficient. See Table 3.4 for values of y for beams of uniform cross-section, loaded transversely.

Since I varies as the cube of the depth, the **stiffness,** or inverse deflection, of various beams varies, other factors remaining constant, inversely as the load, inversely as the cube of the span, and directly as the cube of the depth. This deflection is due to bending moment only. In general, however, the bending of beams involves transverse shearing stresses which cause shearing and thus **add to the total deflection.** Experiments indicate that in the case of a deep, rectangular section cantilever, carrying a concentrated load at the free end, the deflection due to shear may be upto 3.1 percent of that due to bending. If this beam supports a uniformly distributed load, it may be upto 4.1 percent. A deep, simple beam deflection may increase upto 15.6 percent when carrying a uniformly distributed load and upto 12.5 percent when the load is concentrated at mid-span. These added deflections can be even larger for I beams.

3.9 COLUMNS

Short compression members subjected to centrally applied loads may be designed on the basis of direct compression. If, however, a cast iron compression member of a ductile material whose length is 8 times the least dimension perpendicular to its axis, it is classed as a column or strut and failure may be caused by buckling. The failure occurs as a result of lateral deflection, which depends on the modulus of elasticity of the material and the slenderness ratio of the member, as well as the crushing strength of the material.

Table 3.4 Beams of uniform cross-section, loaded transversely

Beam Case	Maximum bending moment M_b	Maximum Shear force F_s	Dangerous section	Maximum deflection y	Reaction	
					R_1	R_2
(a)	Fl	$-F$	At support	$-\frac{Fl^3}{3EI}$	F	.....
(b)	$\frac{wl^2}{2}$	$-wl = -F$	At support	$-\frac{Fl^3}{8EI}$	F	.....
(c)	$\frac{Fl}{4}$	$\pm\frac{F}{2}$	At centre	$-\frac{Fl^3}{48EI}$	$\frac{F}{2}$	$\frac{F}{2}$

Beam Case	Maximum bending moment M_b	Maximum Shear force F_s	Dangerous section	Maximum deflection y	Reaction	
					R_1	R_2
(d)	$\frac{F \times l_1 \times l_2}{l}$	$-\frac{F \times l_1}{l}$	At load	$= \frac{F \times l_2}{3EIl}\left[\frac{l_1(l_1+l_2)^{3/2}}{2}\right]$	$\frac{Fl_2}{l}$	$\frac{F \times l_1}{l}$
(e)	$F \times l_1$	$\pm F$	Between the load	$-\frac{F \times l_1}{24EI}[3l^2 + 4l_1^2)$	F	F
(f)	$\frac{wl^2}{8}$	$\pm\frac{wl}{2} = \pm\frac{F}{2}$	At centre	$-\frac{5Fl^3}{384EI}$	$\frac{F}{2}$	$\frac{F}{2}$
(g)	$-\frac{3}{16}Fl$	$\frac{-11}{16}F$	At fixed support	$\frac{-7}{768}\frac{Fl^3}{EI}$	$\frac{5}{16}F$	$\frac{11}{6}F$
(h)	$+\frac{Fl}{8}$	$\frac{F}{2}$	At support and at centre	$-\frac{Fl^3}{192EI}$	$\frac{F}{2}$	$\frac{F}{2}$
(i)	$-\frac{wl^2}{8}$	$\pm\frac{wl}{2} = \pm\frac{F}{2}$	At support	$-\frac{Fl^3}{384EI}$	$\frac{F}{2}$	$\frac{F}{2}$
(j)	$\frac{wl^2}{8}$	$\frac{5}{8}wl = \frac{5}{8}F$	At fixed support	$-\frac{Fl^3}{185EI}$	$\frac{3}{8}F$	$\frac{5}{8}F$
(k)	$-Fl_1$	$\pm F$	At support points	$f_1 = \frac{wl_1l^2}{8EI}$ $f_2 = \frac{wl_1^2}{3EI}\left[l_1 + \frac{3}{2}l\right]$	F	F

Notation. F = load, w = load per unit length. M_b is positive when clockwise, y is positive when upward, loads and reactions are positive when acting as shown, shear force is positive when upward.

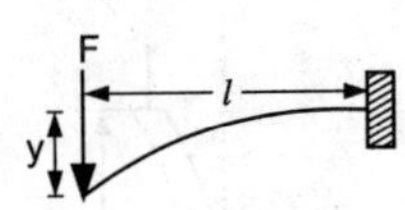

Cantilever end load

(a)

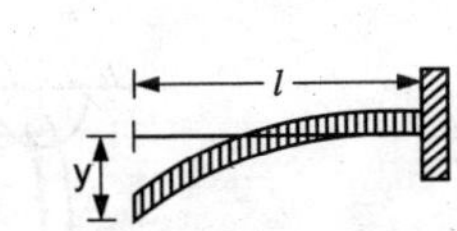

Cantilever uniform load

(b)

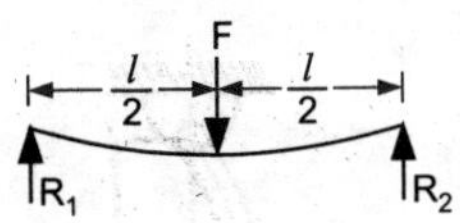

End supports centre load

(c)

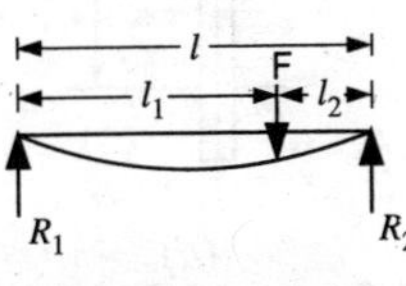

End supports intermediate load

(d)

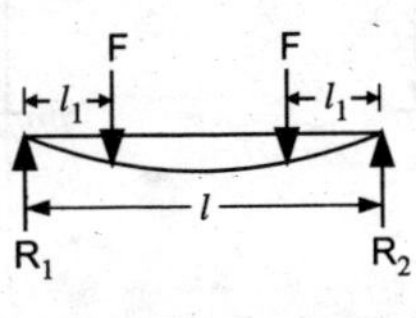

End supports with two loads

(e)

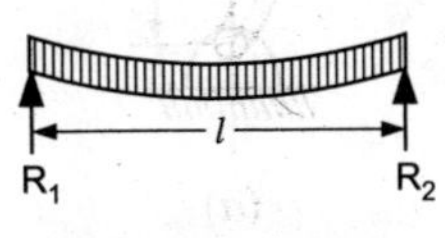

End supports uniform load

(f)

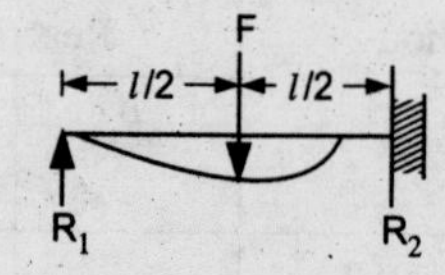

One end fixed other supported centre load
(*g*)

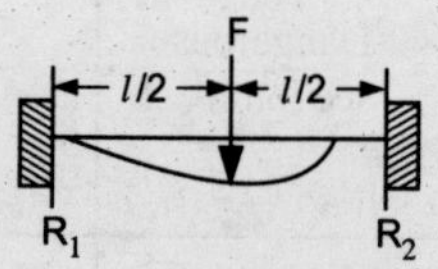

Both ends fixed centre load
(*h*)

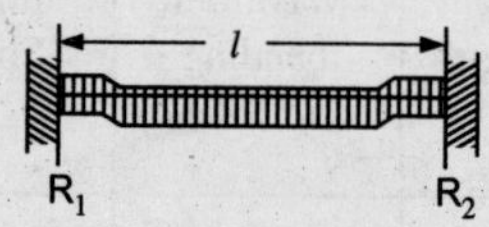

Both ends fixed uniform load
(*i*)

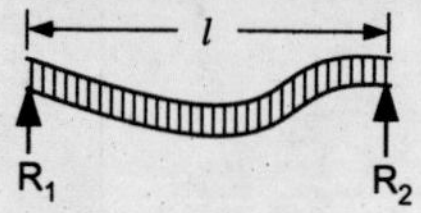

One end fixed other supported uniform load
(*j*)

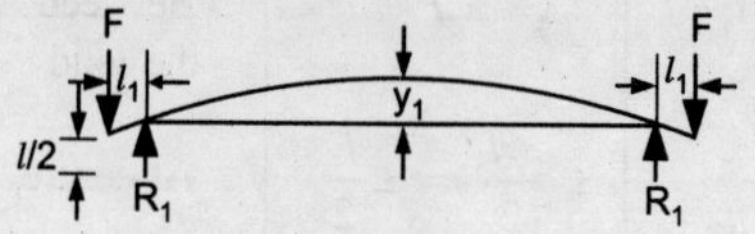

Simple supports over hang load
(*k*)

Slenderness ratio. The ratio of the length of the column to the least radius of gyration of the cross-section is the slenderness ratio. The slenderness ratio divides the column into two groups i.e., **short column and long columns.** The radius of gyration (k) of any section with regard to any axis through the centre of gravity of the section can be found by the relation:

$$k = \sqrt{\frac{I}{A}} \quad \text{...(3.19)}$$

where I is the moment of inertia of the section with regard to the same axis and A is the area of section. Expressions for the radius of gyration for various cross-sections commonly used as column are given in Table 3.2.

Cast iron column with a slenderness ratio not greater than 80 and column of steel and other ductile materials for which slenderness ratio is less then 100, are considered **short columns** and slenderness ratio above these values give **long columns**. When the slenderness ratio is less than 30 the effect of lateral deflection is negligible, and the member may be considered to be in simple compression. In machine design most of the members tend to be short columns.

End conditions. The critical loads for columns with different end conditions can be obtained by comparison. In Fig. 3.5 the different end conditions of columns are shown.

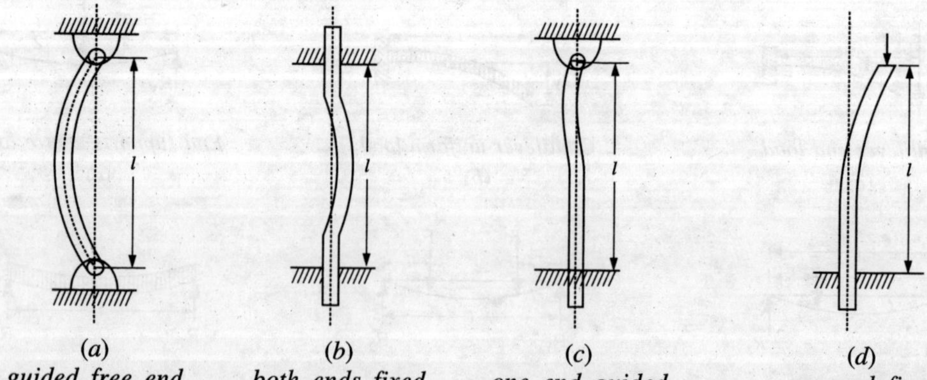

(*a*) *guided free end* $c = 1$

(*b*) *both ends fixed* $c = 4$

(*c*) *one end guided and other fixed* $c = 2$

(*d*) *one end fixed and other free* $c = 0.25$

Fig. 3.5 *End fixity coefficients for columns*

The determination of an end fixity coefficient to be used in a particular design will require the exercise of some judgement by the designer. In very less occasions the end conditions for the columns are the same as shown in Fig. 3.5. For example, it is almost impossible to achieve complete fixity for the ends since some deflection and rotation will occur in the members to which the ends of the columns are attached. Thus the highest value of c that should ever be used might be 3 or possibly 3.5 in some cases.

Design of columns: Several formulas have been proposed for determining the relation between the external load F and the stress in columns.

Euler's Column Formula: Euler's theory is the foundation of all column theories. The crushing loads determined by this theory are given by

$$F = \frac{c\pi^2 EI}{l^2} \qquad ...(3.20)$$

where F = Crushing load causing failure in N

I = least moment of inertia in m^4.

l = length of column in m

E = modulus of elasticity in N/m^2.

c = end fixing coefficient. Values of c for various conditions are given in Fig. 3.5.

Denoting the cross-sectional area of the rod by A, the above relation gives.

$$\frac{F}{A} = \frac{c\pi^2 EAk^2}{Al^2} = \frac{c\pi^2 E}{(l/k)^2} \qquad ...(3.21)$$

It can be seen from the relation that it does not hold if F/A exceeds the stress at the limit of proportionality. Hence l/k must exceed $\sqrt{\dfrac{c\pi^2 E}{F/A}}$ which implies that the formula is only applicable to *long columns*. Most columns encountered in machine design, however, are of such proportions that Euler's formula is not applicable.

Rankine formula: If the ultimate load for a very short member ($l/k < 30$) in compression is $F_1 = f_c \times A$ and that for a long column

$$F_2 = \frac{c\pi^2 EI}{(l)^2}$$

The equation $\dfrac{1}{F} = \dfrac{1}{F_1} + \dfrac{1}{F_2}$ is approximately true in each case.

Therefore,

$$\frac{1}{F} = \frac{1}{f_c A} + \frac{l^2}{c\pi^2 EI}$$

$$F = \frac{f_c A}{1 + fA\dfrac{l^2}{c\pi^2 EI}} = \frac{f_c A}{1 + a\dfrac{l^2}{k^2}} \qquad ...(3.22)$$

where F = safe load in N

f_c = safe compressive stress in N/m^2

a = a constant, the values of which are given in Table 3.5.

In this formula, the use of f_{uc}, the ultimate compressive stress in N/m^2, will cause F to become the ultimate or breaking load in N. This formula is only applicable to short columns.

The **Rankine-Gordan formula** is an empirical expression and modified form of formula 3.22.

$$F = \frac{f_c A}{1 + \alpha(l/d)^2} \qquad ...(3.23)$$

Where d is the diameter or least width of the section and α, another constant.

The results obtained by this formula agree with values obtained by experiments.

Table 3.5

Material	Ultimate compressive stress f_{uc} MN/m^2	The constant (a)
Cast iron	570.0	$\frac{1}{1500}$
Wrought iron	342.0	$\frac{1}{9000}$
Medium steel	420.0	$\frac{1}{7500}$
Hard steel	535.0	$\frac{1}{5000}$
Timber	17.5-28.0	$\frac{1}{750}$

The Ritter Formula: Which is modification of the Rankine Gordan formula, has the advantage that it can be applied for any material for which the elastic limit f_{yp} and the modulus of elasticity E are known. Its induced stress is

$$= \frac{F}{A}\left[1+\left(\frac{1}{k}\right)\frac{f_{yp}}{c\pi^2 E}\right] \qquad ...(3.24)$$

where designations are already mentioned.

J.B. Johnson Formula: Of the many formulas available, one of the most widely used is called the parabolic, or J.B. Johnson formula. It is completely empirical and is an attempt to overcome deficiencies of the Euler equation. This formula is extensively used by machine designers for determining the critical load of steel columns. It may be expressed as follows:

$$F_{cr} = Af_{yp}\left[1-\frac{f_{yp}}{4c\pi^2 E}\left(\frac{l}{k}\right)^2\right] \qquad ...(3.25)$$

where f_{yp} is the yield point stress of the material and other symbols have the same meaning as before.

The parabola of equation 3.25 is tangent to the Euler curve at $F_{cr}/A = \frac{f_{yp}}{2}$. The slenderness ratio at which this occurs thus depends upon the yield strength of the material chosen for the column and upon the end condition constant. To the left of this tangency point Johnson's equation applies as shown in Fig. 3.6 and to the right Euler's equation applies. The value of l/k at the tangent point is

$$\left(\frac{l}{k}\right)t = \sqrt{\frac{2\pi^2 cE}{f_{yp}}} \qquad ...(3.26)$$

The value of l/k at the tangent point is the highest for which the parabolic formula should be used and the lowest for which the Euler's formula should be used. Johnson's formula produces results which compare very favourably with experiment.

$$f_{yp} = 250.0 \text{ MN/m}^2,\ E = 210 \text{ GN/m}^2,\ c = 1$$

Straight line column formula: For short columns of the materials like heat treated aluminium alloys it has been found that the critical load is predicted very closely by the following formula

$$F_{er} = A\left[f_{yp} - \frac{2f_{yp}}{3\pi}\sqrt{\frac{f_{yp}}{3cE}}\,\frac{l}{k}\right] \qquad ...(3.27)$$

where $\dfrac{2f_{yp}}{3\pi}\sqrt{\dfrac{f_{yp}}{3cE}}$ = a constant

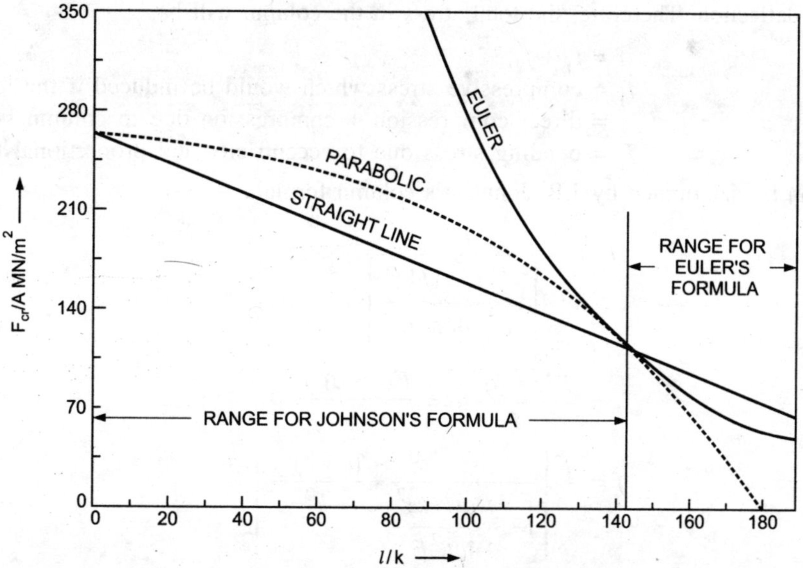

Fig. 3.6 *Comparison of critical loads F_{cr}/A as given by the common column formulas*

The l/k value at the point of tangency between the straight line column formula and Euler's formula is as follows:

$$\left(\frac{l}{k}\right)_t = \pi\sqrt{\frac{3cE}{f_{yp}}} \qquad ...(3.28)$$

The results obtained by this relation are in fair agreement with results from tests on columns made from brittle materials.

3.10 ECCENTRIC LOADING

When short column (l/k less than 25 or 30) are loaded eccentrically in compression or in tension, i.e., not through centre of gravity, a combination of axial and bending stress results. The maximum stress is the algebraic sum of these two stresses. In Fig. 3.7 a load F acts in a line of symmetry at the distance e from centre of gravity. The stresses produced by this force in the body are:

(1) Compressive stress due to F as if it acted through centre of gravity (2) bending stress due to the bending moment of Fe. Thus the combined stress f at any point y is

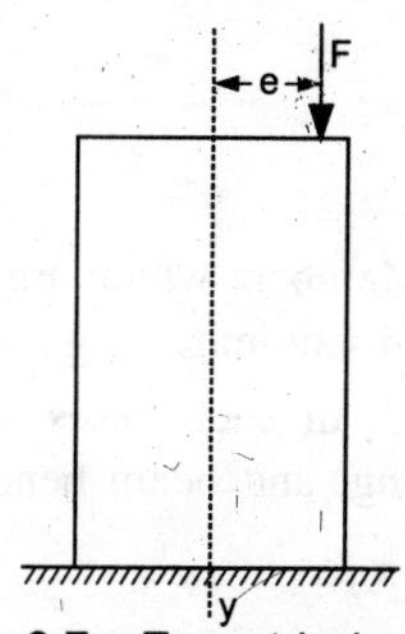

Fig. 3.7 *Eccentric loading of column*

$$f = \frac{F}{A} + \frac{Fey}{I} \qquad ...(3.29)$$

y is positive for points on the same side of centre of gravity as F, and negative on the opposite side. This relation can also be used for tensile loading.

The column formulae given in Article 3.9. are all developed on the assumption that the compression load is applied at the neutral axis of the cross-section and that the column is initially straight. If the column slenderness ratio is less than 30, fair results will be obtained from the use of equation 3.29. For greater slenderness ratios the column bending can not neglected. Also, due to the bending moment, $F.e$, due to be eccentricity of the load, the column will deflect somewhat more than it would if the load were applied axially. This will have the effect of increasing the original eccentricity e, by an additional amount, δ , due to this deflection. Therefore, the total stress in the column will be,

$$f = f_1 + f_2,$$

where f_1 = compressive stress which would be induced if the load were axial = direct compression + compression due to column bending

and f_2 = bending stress due to eccentricity and proportional to $F(e+ \delta)$

Now, f_1 can be determined by J.B. Johnson's column formula:

$$f_1 = \frac{F}{A\left[1 - \frac{f_{yp}(l/k)^2}{4c\pi^2 E}\right]}$$

and,

$$f_2 = \frac{F(e+\delta).y}{I} = \frac{F(e+\delta).y}{Ak^2}$$

$$\therefore \quad f = \frac{F}{A}\left[\frac{1}{1 - \frac{f_{yp}(l/k)^2}{4c\pi^2 E}} + \frac{(e+\delta).y}{k^2}\right] \qquad ...(3.30)$$

For columns with l/k ratio greater than 100, and for whom the stresses are normal, the deflection, δ, may be neglected. However, for longer columns and where the stresses are high, the value of 'δ' may be determined by the theory of elasticity (Refer to any standard book on 'strength of Materials'). For ordinary problems in Machine Design, this refinement may be neglected, and the equation (3.30) will get reduced to,

$$f = \frac{F}{A}\left[\frac{1}{1 - \frac{f_{yp}(l/k)^2}{4c\pi^2 E}} + \frac{e.y}{k^2}\right] \qquad ...(3.30\ a)$$

The critical load will, thus, be given as,

$$F_{cr} = \frac{f_{yp}.A}{\frac{1}{1 - \frac{f_{yp}(l/k)^2}{4c\pi^2 E}} + \frac{e.y}{k^2}} \qquad ...(3.30\ b)$$

Members which are required to withstand axial loads and transverse loads simultaneously are called 'Beam Columns', e.g., eccentrically loaded strut, Fig. 3.8(*a*), and the beam, Fig. 3.8 (*b*).

In such cases also, the three factors should be considered, i.e., direct compression, column bending, and beam bending. The total stress will be given as,

$$f = \frac{F}{A} + \frac{M_b.y}{I} = \frac{F}{A} + \frac{M_b.y}{Ak^2}$$

where M_b = maximum bending moment due to axial and transverse loads acting simultaneously.

It is usually difficult to determine M_b analytically. For case (Fig. 3.8 *a*), M_b is given as,

$$M_b = F.e.sec\frac{l}{2k}\sqrt{\frac{F}{EA}}$$

$$\therefore \qquad f = \frac{F}{A}\left[1+\frac{e.y}{k^2}.\sec\frac{l}{2k}\sqrt{\frac{F}{EA}}\right] \qquad ...(3.31)$$

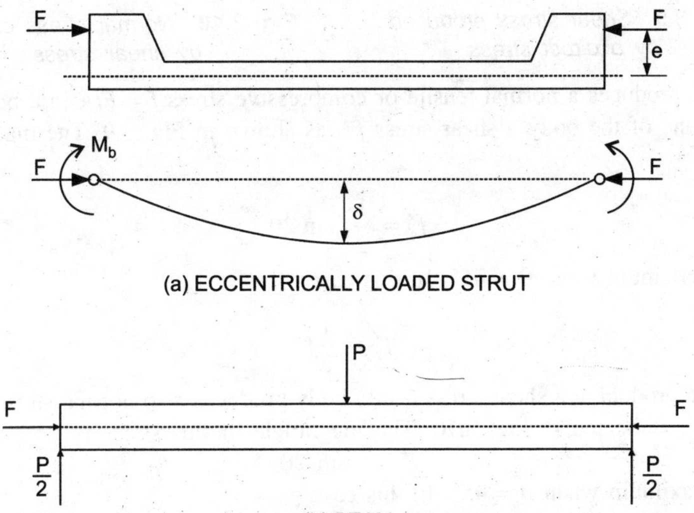

(a) ECCENTRICALLY LOADED STRUT

(b) BEAM

Fig. 3.8 *Beam columns*

For the case, Fig. 3.8 (*b*),

$$M_b = \frac{Pl}{4U}\tan U$$

where $\quad U = \frac{\pi}{2}.\sqrt{\frac{F}{F_{cr}}}$

where F_{cr} = critical load from Euler's equation,

$$\therefore \qquad f = \frac{F}{A} + \frac{P.l.y}{4IU}.\tan U \qquad ...(3.32)$$

3.11 COMBINED STRESSES

The various cases of loading, that is, tension, bending, direct shear and torsion, as discussed, induce two fundamental kinds of stress, which are (1) normal stress, as in compression, tension, or bending, where the stress is normal to the area on which it acts and (2) direct shear stress, or torsion, where the stress acts along the area. Whatever the condition of stress at a point may be, it can always be expressed in terms of two components, the normal component and the shear component.

Most design problems deal with systems in which several types of loading exist simultaneously such as bending and torsion in shafts. It is necessary to determine the maximum value of normal stress and shear stress, so that they may be compared with the allowable values of the stresses for the material.

Induced stresses: Because of the deformation produced in a machine part by the applied loads, other stresses known as induced stresses or secondary stresses are created.

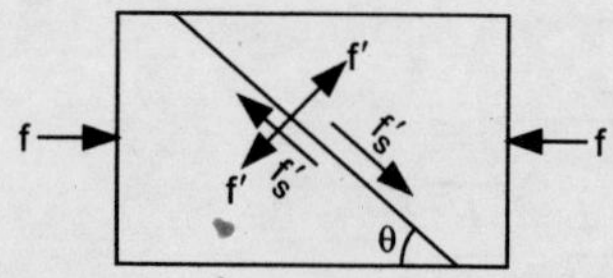

Fig. 3.9 *Shear stress produced by a direct stress*

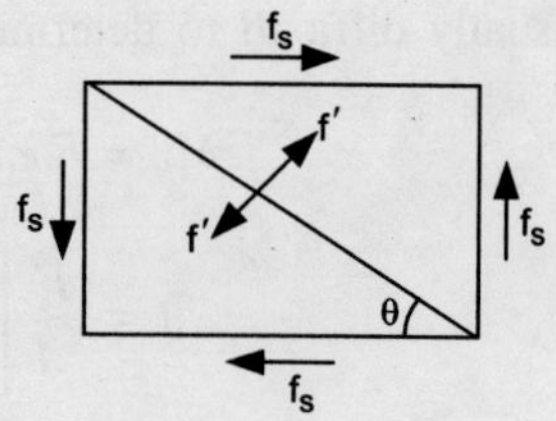

Fig. 3.10 *Normal stress caused by shear stress*

When a force F produces a normal tensile or compressive stress $f = F/A$ in a body, it also produces on every oblique plane of the body a shear stress fs', as shown in Fig. 3.9. The magnitude of this shear stress is

$$f_s' = \frac{1}{2} f \sin 2\theta \qquad ...(3.33)$$

This stress is maximum when $\theta = 45°$. In this case

$$f_s' = \frac{1}{2} f \qquad ...(3.34)$$

Similarly a force producing a shear stress f_s in a body produces also normal stress f' on every oblique plane of the body, as indicated in Fig. 3.10. The magnitude of this stress is

$$f' = f_s \sin 2\theta \qquad ...(3.35)$$

This stress is maximum when $\theta = 45°$. In this case

$$f' = f_s \qquad ...(3.36)$$

Lateral stress: When a force produces a normal stress f, it also causes a lateral strain, which may be considered as caused by a stress f' normal to f. The magnitude of this fictitious stress f' is

$$f' = -\frac{1}{m} f \qquad ...(3.37)$$

where negative sign shows that the nature of the stress f' is opposite to that of f. If f is a tensile stress, f' is a compressive one and vice versa.

Normal and shear stresses:

Consider an elemental block $ABCD$ at a point where the normal stresses f_x and f_y are acting as shown in Fig. 3.11(a). A shear stress f_s also acts on planes BC and AD along with an associated complementary shear stress on planes AB and DC. The stress condition in Fig. 3.11(a) is the general plane stress condition. We are interested to find the normal and shear stress on a plane AM inclined at an angle θ with the direction of f_y. The free body ABM is shown in Fig. 3.11(b). It may be shown that the principal stresses in the material are given by

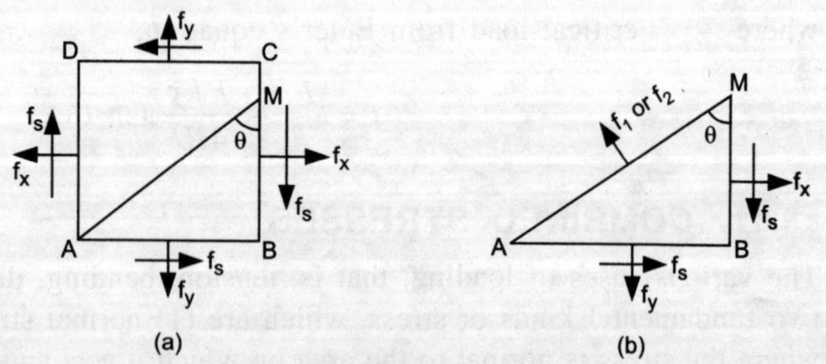

Fig. 3.11

$$= \frac{1}{2}(f_x + f_y) \pm \frac{1}{2}\sqrt{(f_x - f_y)^2 + 4 f_s^2} \qquad ...(3.38)$$

and they act on planes given by the relation of

$$\tan 2\theta = \frac{2 f_s}{f_x - f_y} \qquad ...(3.39)$$

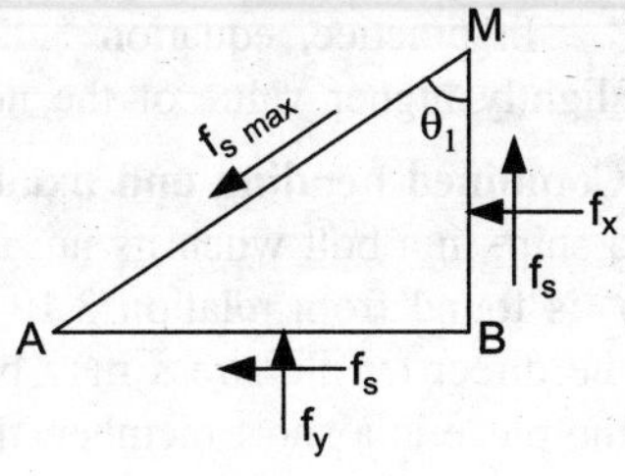

Fig. 3.12

In equation 3.38 the positive sign gives the major stress (maximum principal stress, f_1) and the negative sign the minor stress (minimum principal stress, f_2) acting on the planes given by two solutions of equation (3.39) respectively at right angles. Thus the expressions for f_1 and f_2 are

$$f_1 = \frac{1}{2}(f_x + f_y) + \frac{1}{2}\sqrt{(f_x - f_y)^2 + 4f_s^2}$$

$$f_2 = \frac{1}{2}(f_x + f_y) - \frac{1}{2}\sqrt{(f_x - f_y)^2 + 4f_s^2}$$

Similarly the maximum principal shear stress, f_s max., is given by (Fig. 3.12).

$$f_{s\,\max.} = \pm\frac{1}{2}\sqrt{(f_x - f_y)^2 + 4f_s^2} \qquad ...(3.40)$$

In this the signs denote the direction. The planes on which these stresses act are given by the two solutions of

$$\tan 2\theta_1' = \frac{f_x - f_y}{2f_s} \qquad ...(3.41)$$

This gives the planes inclined at 45° to those given by equation (3.39).

The maximum principal shear stress in the material will be half the difference between the greatest and the least principal stress, due notice being taken of the sign of the smallest stress when it differs from that of the largest stress. In the majority of the cases dealt, the value of f_y is zero, then, if f_x is greater than f_y and of the same sign, the maximum principal shear stress in the material is

$$f_{s\,\max} = \frac{f_x - 0}{2} = \frac{1}{2}f_x$$

when f_x and f_y are of different signs the maximum principal shear stress in the material is

$$f_{s\,\max} = \frac{f_x - f_y}{2}$$

Many cases also occur in practice in which one of the normal stresses in the above equations is zero, so taking $f_y = 0$, above equations become

$$\left.\begin{aligned} f_1 &= \frac{1}{2}(f_x) + \frac{1}{2}\sqrt{f_x^2 + 4f_s^2} \\ f_2 &= \frac{1}{2}(f_x) - \frac{1}{2}\sqrt{f_x^2 + 4f_s^2} \end{aligned}\right\} \qquad ...(3.42)$$

$$f_{s\,\max} = \frac{1}{2}\sqrt{f_x^2 + 4f_s^2} \qquad ...(3.43)$$

In all the above equations, the normal stresses are assumed to be tensile (positive) and f_x is assumed greater than f_y. If these stresses become compressive (negative) their signs must be changed in the above equations.

These equations apply to all cases of normal stress combined with shear, as when a beam is under both shear and tension or compression caused by bending, or when a bolt is under tension in the above equations.

According to maximum-strain theory of failure the design should be governed by the stress that corresponds to the maximum strain, which takes into account the lateral deflection. The normal stress (f_t) which produces the maximum strain is given by the relation

$$f_t = \left(1 - \frac{1}{m}\right)\frac{1}{2}f_x + \frac{1}{2}\left(1 + \frac{1}{m}\right)\sqrt{(f_x^2) + 4f_s^2} \qquad ...(3.44)$$

In practice, equation 3.42, is usually preferred to equation 3.44, although equation 3.44 gives a slightly higher value of the normal stress, specially when the shear stress f_s is relatively high.

Combined bending and axial load: A combination of these loadings may be seen in propeller shaft of a ship, in a bolt when its nut is tightened or in a shaft for driving a worm gear. The primary shear stress f_s is found from relation 3.10 for the torsional moment involved. If the axial load is tension as in bolt, the direct tensile stress may be found by relation 3.7, on the other hand if the load is compressive and the piece is a short member, the compressive stress is found by relation 3.8, if the piece must be treated as a column, relation for column may be used. the combined normal stress is found by relation 3.42, and the combined shear stress by relation 3.43.

Combined torsion and bending: In general, the forces acting on a shaft do not reduce to an axial couple, so that their effect in this case is not simple torsion. But torsion and bending are combined in all shafts used for transmission of power by means of pulleys, sprockets, or gears, etc.

The maximum bending moment due to a number of forces may be found from the bending moment diagram. If the forces are not parallel, this maximum bending moment may be obtained by drawing the diagram for each of the planes in which bending is produced by the separate forces. Then, a resultant bending moment diagram may be obtained, of which the ordinate at any point is the vector sum of the ordinates at the point of the separate diagrams.

Having determined the maximum torsional moment M_t, and the maximum bending moment M_b, the problem is treated as one of complex stress. The maximum shear stress is found from relation 3.10 and the bending moment causes a direct tension or compression, which is found by relation 3.14. The combined stresses are found in the same way as in the case of normal stress and shear stress by using relations 3.42 and 3.43.

3.12 DYNAMIC LOADS

A very few machine parts are subjected to constant, or static loads. The loads on most parts vary, sometimes in magnitude, sometimes in direction, and frequently both in magnitude and direction. These variable loads are known as *dynamic loads and the stresses* produced as *dynamic stresses.* In machine design, dynamic stresses are of far importance than static stresses. The dynamic loads which produce such stresses may be divided into two groups:

1. The first type consists of variable outside loads.
2. Second those resulting from the inertia of the mass of the member itself when it moves with a variable speed.

Loads produced by variable outside forces are called *live loads*, as contrasted with *static loads* or *dead loads.*

The *live loads* can be further split into two categories:

(a) a live load results when the gradual change of force occurs of already applied loads.

(b) the sudden application of an outside force.

Impact load: Shock or impact load is defined as the very rapid application of an outside force to a structure of machine part which result in stresses and waves and vibration in the material.

The practical application where dynamic loads are encountered are hoist of a crane drive shaft, fly-wheel, pulley, etc. Also large structures, such as buildings, bridges, and the like are subjected to variable loads, but proportionally to a lesser extent. In buildings, for example, much of the loading on the beams, girders and columns is dead load, that is, the weight of the structure itself.

Inertia stresses: Inertia loads are caused by acceleration that acts when a change of velocity, either linear or angular takes place. The acceleration may be due to change in the direction of the motion (as of a rotating crank), a change in the magnitude of the velocity (as of a locomotive coupling rod), or a change in both direction and magnitude as of a connecting rod.

The general expression for the centrifugal force (F_c) created in a mass m is given as

$$F_c = \frac{mv^2}{r} \quad \text{...(3.45)}$$

where v is the circumferential velocity in m/s, and r is radius of curvature, e.g., radius of pulley or flywheel in metres.

Centrifugal force can create tensile stresses, as in the case of the rim of a rotating pulley, or it can create bending stress, as in the case of a locomotive coupling rod or a connecting rod.

3.13 IMPACT STRESSES

The phrases shock loads, impact loads, and transient forces, all have roughly the same meaning when applied to the design of machines. We usually think of impact as the collision of two bodies in which at least one of the two has an appreciable velocity. This would also constitute a shock load. But a shock load produced by an explosion is not an impact because only one body is involved. A transient force is any kind of a force which endures for a short period of time.

In the past it has been the custom of many designers to simply double the factors of safety used, in order to allow for shock or impact on a machine. Yet there is a considerable evidence that the strength of the material used actually increases when the load is applied quite rapidly. This would seem to justify the use of a smaller factor of safety, rather than larger one.

Impact energy: If the weight of the moving body is P in N and its velocity is v m/s, its kinetic energy in Nm is

$$E_i = \frac{Pv^2}{2g} \quad \text{...(3.46)}$$

For stress calculations it is convenient to express the impact energy as if it were produced by falling body. The height of fall, in metre, that would develop the velocity v in m/s is

$$h = \frac{v^2}{2g} \quad \text{...(3.47)}$$

hence, the impact energy of a body falling from a height h is

$$E_i = Ph \quad \text{...(3.48)}$$

where E_i refers to the instant when the falling-body strikes the other body.

The total work done by a falling body can also be computed by equation 3.48 if h is understood to include both the height of the fall before contact and the deformation of the body that is struck.

Impact stress: The method to be described is a short cut method. It has been used for a great many years and gives results which are on the conservative side. Following are the assumptions made in this method:

1. The inertia of the system is to be neglected.
2. The deflection of the system is directly proportional to the force.
3. The materials obey Hook's Law.
4. There is no loss in energy, that is, the support is unyielding.

To investigate the effect of impact, an elastic system loaded by a falling weight P, through a height h, as shown in Fig. 3.13, will be used as an example. When the weight P strikes the rod, the rod will be elongated a distance e'. The external work performed by the weight will be

$$U = P(h + e') \quad \text{...(3.49)}$$

The stress in the rod was zero before the weight struck it and after it had been elongated a distance e', the stress becomes f_t'. The internal work of the variable tensile force is

$$U = \frac{1}{2} A . f_t' e' \qquad ...(3.50)$$

where A is the cross-sectional area of the rod. Equating the internal work to the external work and solving for f_t', the result is

$$f_t' = 2P\left(\frac{h + e'}{Ae'}\right) \qquad ...(3.51)$$

but $\frac{P}{A} = f_t$ the static stress, and $e' = \frac{lf_t'}{E}$.

Putting this in equation 3.51 and solving for f_t', we get

$$f_t' = \frac{P}{A} + \frac{P}{A}\sqrt{1 + \frac{2hEA}{Pl}} \qquad ...(3.52)$$

$$f_t' = \frac{P}{A} + \frac{P}{A}\sqrt{1 + \frac{2h}{e}} \qquad ...(3.53)$$

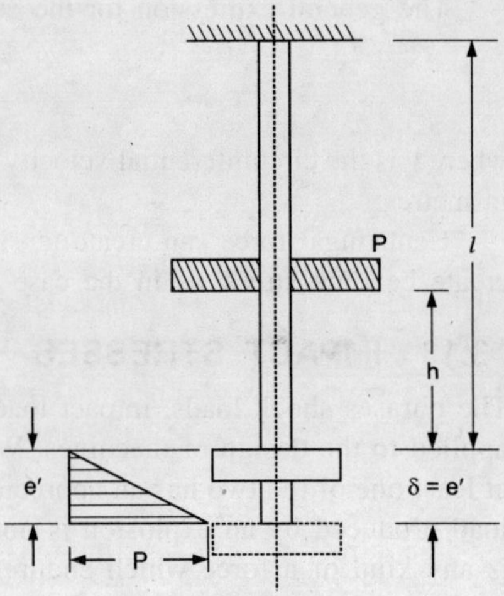

Fig. 3.13

Equation 3.53 is general, and f_t' is the impact stress due to the P force which, when applied as static load, would give the deformation e.

Since the stresses and deformations are proportional, and general equation for the deformation under impact action is

$$e' = e\left(1 + \sqrt{1 + \frac{2h}{3}}\right) \qquad ...(3.54)$$

The equation 3.52 applies for a direct load.

When a beam is subjected to impact, the deformation e in the relation 3.54 means the deflection y, which is computed for the type of the beam and the loading in the particular case. Therefore, for *impact and bending*, equations 3.53 and 3.54 apply without any other changes.

Impact and Torsion: Equations 3.53 and 3.54 also can be used when a force causing torsion is applied with impact. In this case, the deformation e means the angular deflection θ in radians, and the travel h of the force must also be expressed in radians.

Sudden load: The first term in the relation 3.52 is the stress which would exist if the load were gradually applied. It is interesting to see that, if the load were applied without initial velocity, that is, h in the relation is zero, then the equation would reduce to

$$f_t' = 2\frac{P}{A} \qquad ...(3.55)$$

and the deformation from relation 3.54 is

$$e' = 2e \qquad ...(3.56)$$

Relation 3.55 indicates that the effect of applying a load suddenly is twice as severe as applying it statically. This is known as suddenly applied load without velocity approach. An example of this type of loading is the pressure caused by the explosion of gases in an internal combustion engine cylinder which is caused by a rapid building of pressure in a time of the order of 0.001 second.

The following are the three important points while designing member for impact loads:

1. Parts for impact load should be designed to have maximum volume of materials stressed to the highest existing stress.

2. Stress concentrations, that cannot be relieved by a small amount of plastic flow, must be eliminated.
3. If plastic flow is tolerable and advantage is to be taken of its energy-absorbing potential, then ductile materials having a considerable potential for work hardening should be employed. The work hardening will cause strain to be distributed more generally throughout the material.

3.14 RESILIENCE

The resilience of a material is its capacity to absorb potential energy within the elastic range. Since resilience is measured by the amount of deformation, or strain, it is also known as strain energy. Resilience is a function of internal stress, regardless of whether they are created by static loads or dynamic loads, but resilience is important mainly in case of dynamic loads.

The amount of resilience is equal to the work required to deform the body from zero stress to stress f, when f does not exceed the elastic limit. For normal stress,

$$\text{Resilience} = \text{Work of deformation}$$

$$U = \text{Average force} \times \text{deformation} = 1/2\ F \times e \qquad \text{...(3.57)}$$

where U is the resilience in Nm for F to be the applied force in N and e is the deformation in m.

If a load P is applied with impact, the force F may be taken as

$$F = \frac{P \times e'}{e} = \frac{P \times f'}{f} \qquad \text{...(3.58)}$$

where e' and f' are deformation and stress under impact loading respectively, f is the direct stress.

Modulus of resilience. U_p, or *unit resilience* is the elastic energy stored up in a cubic metre of the material at the elastic limit. It is usually expressed in Nm/m^3.

By substituting for F equal to Af and e equal to f_1/E in relation 3.57. We obtain

$$U = \frac{1}{2}\frac{f^2 Al}{E} = \frac{1}{2}\frac{f^2 V}{E} \qquad \text{...(3.59)}$$

where V is the volume of the body in m^3.

Substituting $f_{yp} = f$, for the stress at the elastic limit for unit volume

$$U_p = \frac{f_{yp}^{\ 2}}{2E} \qquad \text{...(3.60)}$$

The meaning of the modulus is shown graphically on the stress-strain (Fig. 3.14). Point A represents the stress intensity at the proportional limit. The line AB is drawn vertical. The shaded area then represents the amount of energy which can be stored in a unit volume of material without exceeding the proportional limit. It can be seen from the diagram that if a material is to be capable of absorbing large amount of energy, that is, if it is to have a high modulus, the area OAB must be large. This area is large if (*a*) the proportional limit is high (*b*) the modulus of elasticity is low.

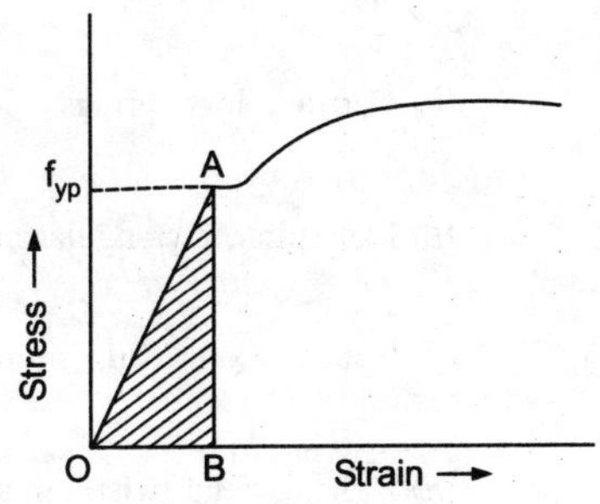

Fig. 3.14

The modulus of elasticity of a given metal is practically constant. The elastic limit, on the other hand, depends to a great extent upon the method of shaping the piece, which may be casting, forging, rolling. Also the elastic limit depends upon the addition of certain chemical elements, such as carbon, chromium, nickel and vanadium and finally upon the subsequent heat treatment. By proper heat treatment the elastic limit of certain alloy steels can be made 2.4 times as great as that of untreated metal and the resilience can be made 5.7 times as great.

The unit resilience or modulus of resilience for any other kind of stress, as shearing, bending, torsion, is a constant times one half the square of the stress divided by the appropriate modulus of elasticity. For values see Table 3.6.

Table 3.6 Maximum resilience per unit of Volume U_p

Types of loading	Modulus of resilience N/m^2 per m^3
1. Tension or compression	$\frac{1}{2} \times \frac{f_{yp}^2}{E}$
2. Shear, simple transverse	$\frac{1}{2} \frac{f_s^2}{C}$
3. Bending in beams (free ends).	
(*a*) Rectangular section, bent in arc of circle, no shear.	$\frac{f_{yp}^2}{6E}$
(*b*) Circular section, bent in arc of circle, no shear.	$\frac{f_{yp}^2}{8E}$
(*c*) Concentrated centre load, rectangular cross-section.	$\frac{f_{yp}^2}{18E}$
(*d*) Concentrated centre load, circular cross-section.	$\frac{f_{yp}^2}{24E}$
(*e*) Uniform load, rectangular cross-section.	$\frac{5}{36} \frac{f_{yp}^2}{E}$
(*f*) Concentrated centre load, I-beam section.	$\frac{3}{32} \frac{f_{yp}^2}{E}$
4. Torsion	
(*a*) Solid circular.	$\frac{f_s^2}{4C}$
(*b*) Hollow round bar $D_1 > D_2$.	$\left[1+\left(\frac{D_2}{D_1}\right)^2\right]\frac{f_s^2}{C}$
5. Springs.	
(*a*) Carriage leaf springs.	$\frac{f_{yp}^2}{6E}$
(*b*) Flat spiral, rectangular cross-section.	$\frac{f_{yp}^2}{24E}$
(*c*) Helical, axial load, circular wire.	$\frac{f_s^2}{4C}$
(*d*) Helical, axial twist, circular wire.	$\frac{f_{yp}^2}{8E}$
(*e*) Helical, axial twist, rectangular wire.	$\frac{f_{yp}^2}{6E}$

f_{yp} = longitudinal elastic stress,
f_s = shearing stress,
E = modulus of elasticity,
C = shearing modulus of elasticity.

Example 3.1. A machine member 0.05 m diameter and 0.25 m long is supported at one end as a cantilever. The arrangement of load on the member is shown on each figure. Determine the tensile, compressive and shear stress in each case.

Solution: (*a*) **Bending and torsion:** Points *A* and *B* are critical as shown in the figure. Load of 2750 N will cause bending and the torque 250 Nm will cause the torsional shear stress.

By using relation 3.14 bending stress

$$= \frac{M_b Y}{I} = \frac{2750 \times 0.25 \times 0.025}{\frac{\pi}{64} \times (0.05)^4} = 56 \text{ MN/m}^2$$

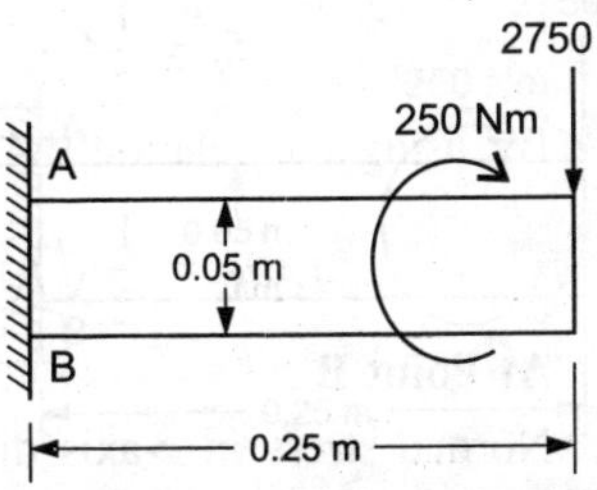

$\therefore$ $f_x = 56$ MN/m^2 at point *A*

$f_x = -56$ MN/m^2 at point *B*

By using relation 3.10

$$\text{Torsional shear stress} = \frac{M_t Y}{J} = \frac{250 \times 0.025}{\frac{\pi}{32}(.05)^4} = 10.2 \text{ MN/m}^2$$

$\therefore$ $f_s = 10.2$ MN/m^2 at points *A* and *B*

Now the maximum and minimum principal stresses can be obtained at the points *A* and *B* by the relation 3.42.

Taking f_x as positive (tensile) at point *A*

$$f_1 = \frac{1}{2} f_x + \frac{1}{2}\sqrt{f_x^2 + 4 f_s^2} = \frac{56}{2} + \frac{1}{2}\sqrt{56.0^2 + 4 \times 10.2^2}$$

$$= 28.0 + 29.8 = 57.8 \text{ MN/m}^2 \quad \text{(tension at point } A\text{)}$$

$$f_2 = \frac{56.0}{2} - 29.8 = -1.8 \text{ MN/m}^2 \quad \text{(compression at point } A\text{)}$$

Taking f_x as negative (compressive) at point *B*

$$f_1 = -28.0 + 29.8 = 1.8 \text{ MN/m}^2 \quad \text{(tension at point } B\text{)}$$

$$f_2 = -28.0 - 29.8 = -57.8 \text{ MN/m}^2 \quad \text{(compression at point } B\text{)}$$

The maximum principal shear stress can be calculated from relation 3.43.

$$f_{s\,max} = \pm \frac{1}{2}\sqrt{f_x^2 + 4 f_s^2} = \pm 29.8 \text{ MN/m}^2$$

$$= 29.8 \text{ MN/m}^2 \quad \text{(Shear at point } A\text{)}$$

$$= -29.8 \text{ MN/m}^2 \quad \text{(Shear at point } B\text{)}$$

Note that the magnitudes of the stresses at point *A* and *B* are the same. The signs of the maximum normal stresses indicate tension or compression and the sign of the maximum principal shear stress is of no result because the design will be based on the magnitude.

(*b*) **Bending and axial load:** Points *A* and *B* are critical in the figure. Loads of 2750 N will cause bending and load of 13.75 kN will cause axial tension. Shear stress f_s is zero at the critical points *A* and *B*.

At point A

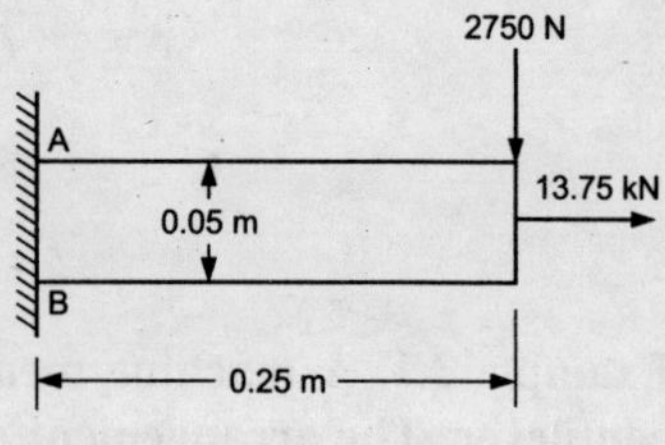

Normal stress in x-axis direction

$$= f_x = \frac{F}{A} + \frac{M_b Y}{I}$$

$$= \frac{13.75\times 10^3}{\frac{\pi}{4}(0.05)^2} + \frac{2750\times 0.25\times 0.025}{\frac{\pi}{64}\times(0.05)^4}$$

$$= 7.0 + 56.0 = 63.0 \text{ MN/m}^2 \text{ (tension)}$$

By using the relation 3.42

$$f_1 = \frac{1}{2}f_x + \frac{1}{2}\sqrt{f_x^2 + 4f_s^2} = \frac{1}{2}f_x + \frac{1}{2}f_x = f_x$$

$$= 63.0 \text{ MN/m}^2 \quad \text{(tension)}$$

$$f_2 = \frac{1}{2}f_x - \frac{1}{2}\sqrt{f_x^2 + 4f_s^2} = \frac{1}{2}f_x - \frac{1}{2}f_x = 0$$

By using the relation 3.43,

$$f_{s\,max} = \frac{1}{2}\sqrt{f_x^2 + 4f_s^2} = \frac{1}{2}f_x = \frac{63.0}{2} = 31.5 \text{ MN/m}^2 \quad \text{(shear)}$$

At Point B

Normal stress in x-axis direction

$$= f_x = \frac{F}{A} - \frac{M_b Y}{I}$$

$$= 7.0 - 56.0 = -49.0 \text{ MN/m}^2 \quad \text{(compression)}$$

By using the relation 3.42

$$f_1 = \frac{1}{2}f_x + \frac{1}{2}\sqrt{f_x^2 + 4f_s^2} = \frac{1}{2}(-49) + \frac{1}{2}\sqrt{(-49.0)^2} = -\frac{49.0}{2} + \frac{1}{2}\times 49.0 = 0$$

$$f_2 = \frac{1}{2}fx - \frac{1}{2}\sqrt{f_x^2 + 4f_s^2} = \frac{1}{2}(-49.0) - \frac{1}{2}\sqrt{(-49.0)^2}$$

$$= -\frac{49.0}{2} - \frac{49.0}{2} = -49.0 \text{ MN/m}^2 \quad \text{(compression)}$$

By using the relation 3.43

$$f_{s\,max} = \frac{1}{2}\sqrt{f_x^2 + 4f_s^2} = \frac{1}{2}\sqrt{(-49.0)^2} = 24.5 \text{ MN/m}^2 \quad \text{(shear)}$$

(*c*) **Torsion and axial load:** The critical points are the points on the outer surface of the member. Load of 13.75 kN will cause the axial tension and the torque 250 Nm will cause torsional shear stress.

$$\text{Direct tensile stress} = f_x = \frac{F}{A} = \frac{13.75\times 10^3}{\frac{\pi}{4}\times 0.05^2} = 7.0 \text{ MN/m}^2$$

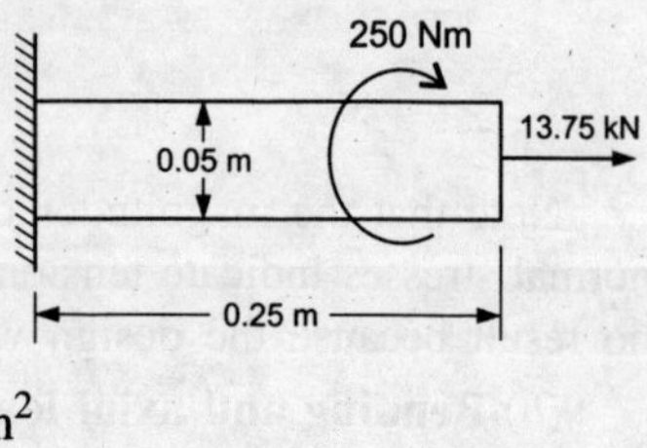

By using the relation 3.10

f_s = torsional shear stress

$$= \frac{M_t \times Y}{J} = \frac{250\times 0.025}{\frac{\pi}{32}\times 0.05^4} = 10.2 \text{ MN/m}^2$$

Now the maximum and minimum principal stresses can be obtained by relation 3.42.

$$f_1 = \frac{1}{2}f_x + \frac{1}{2}\sqrt{f_x^2 + 4f_s^2} = \frac{1}{2} \times 7.0 + \frac{1}{2}\sqrt{7.0^2 + 4\times 10.2^2}$$

$$= 3.5 + \frac{1}{2} \times 21.6$$

$$= 3.5 + 10.8 = 14.3 \text{ MN/m}^2 \quad \text{(tension)}$$

$$f_2 = \frac{1}{2}f_x - \frac{1}{2}\sqrt{f_x^2 + 4f_s^2} = \frac{1}{2} \times 7.0 - \frac{1}{2}\sqrt{7.0^2 + 4\times 10.2^2}$$

$$= 3.5 - 10.8 = -7.3 \text{ MN/m}^2 \quad \text{(compression)}$$

By using the relation 3.43

$$f_{s\,\max} = \frac{1}{2}\sqrt{f_x^2 + 4f_s^2} = \frac{1}{2}\sqrt{7.0^2 + 4\times 10.2^2}$$

$$= 10.8 \text{ MN/m}^2 \quad \text{(shear)}$$

(*d*) **Bending, axial load and torsion:** Maximum stress will occur at points *A* and *B*. Load of 2750 N will cause bending, load of 13.75 kN will cause axial tension and torque 250 Nm will cause torsional shear stress.

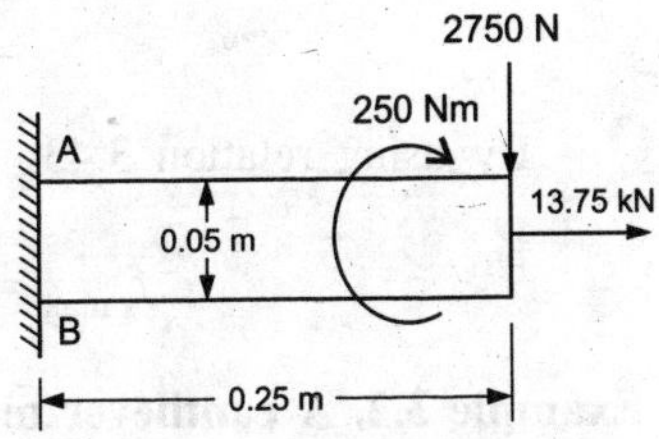

At Point A

Normal stress in *x*-axis direction

$$= f_x = \frac{F}{A} + \frac{M_bY}{I} = \frac{13.75\times 10^3}{\frac{\pi}{4}\times 0.05^2} + \frac{2750\times 0.25\times 0.025}{\frac{\pi}{64}\times (0.05)^4}$$

$$= 63.0 \text{ MN/m}^2$$

By using the relation 3.10

$$\text{Torsional shear stress} = f_s = \frac{M_tY}{J} = \frac{250\times 0.025}{\frac{\pi}{32}\times (0.05)^4} = 10.2 \text{ MN/m}^2$$

By using the relation 3.42

$$f_1 = \frac{1}{2}f_x + \frac{1}{2}\sqrt{f_x^2 + 4f_s^2} = \frac{1}{2} \times 63.0 + \frac{1}{2}\sqrt{63.0^2 + 4\times 10.2^2}$$

$$= \frac{63.0}{2} + \frac{66.2}{2} = 31.5 + 33.1 = 64.6 \text{ MN/m}^2 \quad \text{(tension)}$$

$$f_2 = \frac{1}{2}f_x - \frac{1}{2}\sqrt{f_x^2 + 4f_s^2} = \frac{1}{2} \times 63.0 - \frac{1}{2}\sqrt{63.0^2 + 4\times 10.2^2}$$

$$= 31.5 - 33.1 = -1.6 \text{ MN/m}^2 \quad \text{(compression)}$$

By using the relation 3.43

$$f_{s\,\max} = \frac{1}{2}\sqrt{f_x^2 + 4f_s^2} = \frac{1}{2}\sqrt{63.0^2 + 4\times 10.2^2}$$

$$= \frac{66.2}{2} = 33.1 \text{ MN/m}^2 \quad \text{(shear)}$$

At Point B

$$\text{Normal stress in } x\text{-axis direction} = \frac{F}{A} - \frac{M_b \times Y}{I} = f_x$$

$$= \frac{13.75 \times 10^3}{\frac{\pi}{4} \times 0.05^2} - \frac{2750 \times 0.25 \times 0.025}{\frac{\pi}{64} \times 0.05^4}$$

$$= 7.0 - 56.0 = -49.0 \text{ MN/m}^2 \quad \text{(Compression)}$$

f_s as calculated above = 10.2 MN/m^2

By using the relation 3.42

$$f_1 = \frac{1}{2} f_x + \frac{1}{2}\sqrt{f_x^2 + 4 f_s^2} = \frac{1}{2} \times (-49.0) + \frac{1}{2}\sqrt{(-49.0)^2 + 4 \times 10.2^2}$$

$$= -\frac{49.0}{2} + \frac{1}{2} \times 53.0 = -24.5 + 26.5 = 2.0 \text{ MN/m}^2 \quad \text{(tension)}$$

$$f_2 = \frac{1}{2} f_x - \frac{1}{2}\sqrt{f_x^2 + 4 f_s^2} = \frac{1}{2} \times (-49.0) - \frac{1}{2}\sqrt{(-49)^2 + 4 \times 10.2^2}$$

$$= -\frac{49.0}{2} - \frac{1}{2} \times 53.0 = -24.5 - 26.5 = -51.0 \text{ MN/m}^2 \quad \text{(compression)}$$

By using relation 3.43

$$f_{s\,max} = \frac{1}{2}\sqrt{f_x^2 + 4 f_s^2} = \frac{1}{2}\sqrt{(-49)^2 + 4 \times 10.2^2} = \frac{53.0}{2} = 26.5 \text{ MN/m}^2 \quad \text{(shear)}$$

Example 3.2. A cantilever member 0.1 m long having cross section of 0.05 m × 0.25 m supports a load of 27.5 kN. What is the maximum shear stress and where does it occur?

Solution: The maximum shear stress may occur at points *A-A* due to the bending moment, or it may occur at points along *B-B* due to the transverse shear load.

At points along *A-A*

By using relation 3.14 bending stress

$$= \frac{M_b Y}{I} = \frac{27.5 \times 10^3 \times 0.1 \times 0.125 \times 12}{0.05 \times 0.25^3} = 5.3 \text{ MN/m}^2$$

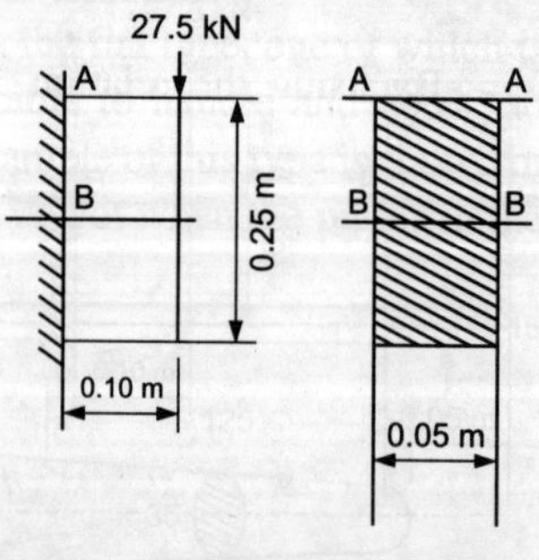

$$\therefore \quad f_x = 5.3 \text{ MN/m}^2, f_s = \text{zero}$$

By using relation 3.43

$$f_{s\,max} = \frac{1}{2}\sqrt{fx^2 + 4 fs^2} = \frac{1}{2}\sqrt{5.3^2 + 4 \times 0^2}$$

$$= \frac{1}{2} \times 5.3 = 2.65 \text{ MN/m}^2 \quad \text{(shear)}$$

At points along *B-B*

Maximum shear stress at the neutral axis due to transverse shear load on a rectangular cross-section is given by the relation

$$f_{s\,max} = \frac{3}{2} \cdot \frac{F}{A} = \frac{3}{2} \times \frac{27.5 \times 10^3}{0.25 \times 0.05} = 3.3 \text{ MN/m}^2 \quad \text{(shear)}$$

Therefore, the maximum shear stress is due to the transverse shear load and it occurs along the neutral axis at *B-B*.

Example 3.3. A critical point in a machine member is subjected to biaxial loading which produces f_x, f_y and f_s stresses as shown in the figure. Determine the maximum and minimum principal stresses and the maximum shear stress. Take the third principal stress zero.

Solution: The maximum and minimum principal stresses can be calculated by relation 3.38.

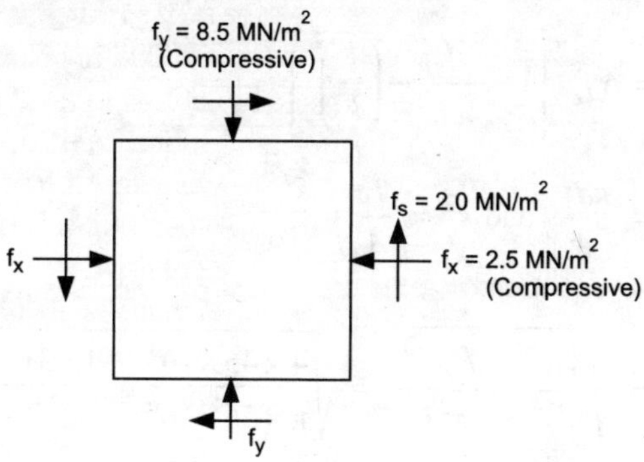

$$f_1 = \frac{1}{2}(f_x + f_y) + \frac{1}{2}\sqrt{(f_x - f_y)^2 + 4f_s^2} = \frac{1}{2}(-2.5 - 8.5) + \frac{1}{2}\sqrt{(-2.5-(-8.5))^2 + 4\times 2.0^2}$$

$$= -\frac{11.0}{2} + \frac{1}{2} \times 7.2 = -5.5 + 3.6 = -1.9 \text{ MN/m}^2 \quad \text{(compression)}$$

$$f_2 = \frac{1}{2}(f_x + f_y) - \frac{1}{2}\sqrt{(f_x - f_y)^2 + 4f_s^2} = \frac{1}{2}(-2.5 - 8.5) - \frac{1}{2}\sqrt{(-2.5-(-8.5))^2 + 4\times 2.0^2}$$

$$= -\frac{11.0}{2} - \frac{7.2}{2} = -5.5 - 3.6 = -9.1 \text{ MN/m}^2 \quad \text{(compression)}$$

Maximum shear stress is given by relation

$$f_{s\,max} = \frac{f_2 - 0}{2} = \frac{-9.1 - 0}{2} = -4.55 \text{ MN/m}^2.$$

Example 3.4. The connecting rod of a steam locomotive is circular in cross-section with a diameter 0.075 m. The length of the rod is 1.5 m. The ends of the rod may be assumed to be pin jointed. Take the safe stress for the material in compression as 63 MN/m², elastic limit as 280 MN/m² and modulus of elasticity as 210 GN/m². Determine the safe load for the rod.

Solution: Johnson's formula is

$$F = A.f_c\left[1 - \frac{f_{yp}}{4c\pi^2 E}\left(\frac{1}{k}\right)^2\right]$$

Now

$$k^2 = \frac{I}{A} = \frac{\pi d^4}{64} \times \frac{4}{\pi d^2} = \frac{d^2}{16} = \frac{(0.075)^2}{16} = 0.0003515 \text{ m}^2$$

$$l = 1.5 \text{ m}, f_c = 63.0 \text{ MN/m}^2, f_{yp} = 280.0 \text{ MN/m}^2$$

$$E = 210 \text{ GN/m}^2, c = 1$$

$$A = \frac{\pi}{4} \times (0.075)^2 = 0.00442 \text{ m}^2$$

$$\therefore \quad F = 0.00442 \times 63 \times 10^6 \left[i - \frac{280 \times 10^6}{4 \times 1 \times \pi^2 \times 210 \times 10^9} \times \frac{1.5 \times 1.5}{0.0003515}\right] = 218 \text{ kN}.$$

Example 3.5. A steel rod, circular in cross-section and 0.66 m long, must support repeated axial reversing loads upto 45 × 10³ N as a part of a large printing press. Calculate the required diameter. Take, f_c = 56 MN/m², f_{yp} = 315 MN/m², E = 210 GN/m². The ends are partially fixed to make the value of c = 2.

Solution: Johnson's equation is:

$$F = A.f_{c'}\left[1-\frac{f_{yp}}{4c\pi^2 E}\left(\frac{1}{k}\right)^2\right]$$

Now $$A = \frac{\pi d^2}{4} \text{ and } k^2 = \frac{I}{A} = \frac{d^2}{16}$$

Putting these in the above equation,

$$d = \sqrt{\frac{4F}{\pi f_c}+\frac{4f_{yp}.l^2}{c\pi^2 E}} = \sqrt{\frac{4\times 45\times 10^3}{\pi\times 56\times 10^6}+\frac{4\times 315\times 10^6\times 0.66\times 0.66}{2\times\pi^2\times 210\times 10^9}}$$

$$= 0.03503 \text{ m, say } 0.0355 \text{ m}$$

Check of l/k ratio:

$$\frac{l}{k} = \frac{0.66\times 4}{0.0355} = 74.3 < 100, \text{ therefore, the use of Johnson's formula is justified.}$$

Example 3.6. A small crane, Fig. 3.15 has a clear swing of 0.7 m. The lower part of the crane is a straight prism. The radius of curvature of the upper part is large compared with the depth of the cross-section, and moment arms on the sections of the curved part are less than that applied to those of the lower part. The Section *X–Y* is shown. Find the load corresponding to a maximum fibre stress (compression) of 63.0 MN/m² at *Y*.

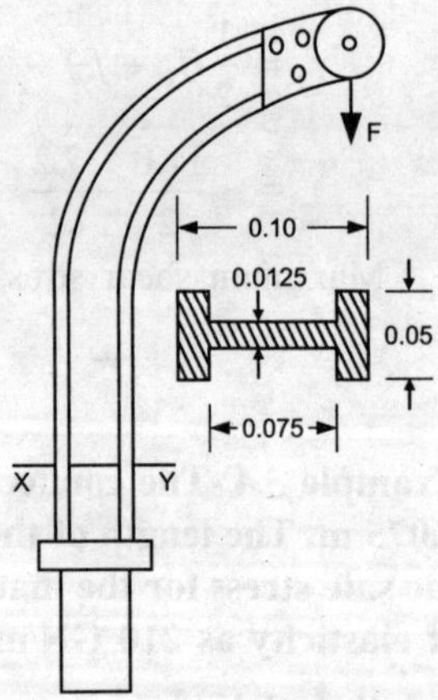

Fig 3.15 *Dimensions in m*

Solution: The following equation can be safely employed:

$$f = \frac{F}{A}+\frac{F.ey}{I}$$

Now A = (0.075 × 0.0125) + 2 × 0.0125 × 0.05 = 0.0022 m²

$$I = \frac{1}{12}[0.05\times 0.10^3 - 0.0375\times(0.075)^3] = 285\times 10^{-8}\ \text{m}^4$$

$$e = 0.70 + 0.05 = 0.75 \text{ m}, \; y = 0.05 \text{ m}$$

Now, from the above equation, $$F = \frac{f\times A\times I}{I + Ae.y}$$

$$= \frac{63\times 10^6\times 0.0022\times 285\times 10^{-8}}{285\times 10^{-8}+0.0022\times 0.75\times 0.05} = 4630 \text{ N}$$

Example 3.7. A circular wooden pole 2.5 meters high is required to carry transformer weighing 11 kN with an eccentricity of 0.25 m. Find the pole diameter at the middle so that the stress due to this load will not exceed 3.5 MN/m². Let f_{yp} = 21.0 MN/m²; E = 10.50 GN/m²; C = 0.25.

Solution: For trial sake, let us assume the diameter of the pole as 0.2 m.

∴ From equation (3.30),

$$f = \frac{F}{A}\left[\frac{1}{1-\frac{f_{yp}\times(l/k)^2}{4c\pi^2 E}} + \frac{(e+\delta).y}{k^2}\right]$$

Now $A = \frac{\pi}{4} \times .04 = 0.0314 \text{ m}^2,\ k^2 = \frac{d^2}{16} = 0.0025 \text{ m}^2$

$e = 0.25$ m; $y = 0.1$ m

Neglecting 'δ',

$$f = \frac{11\times10^3}{0.0314}\left[\frac{1}{1-\frac{21\times10^6\times2.50\times2.50}{4\times0.25\times\pi^2\times10.5\times10^9\times0.0025}} + \frac{0.25\times0.1}{0.0025}\right]$$

$$= 4.08 \text{ MN/m}^2$$

So, the pole can take a stress of 4.08 MN/m^2, therefore, it is O.K.

Example 3.8. A crank shaft 20 mm diameter is mounted on ball bearings and carries a cam midway between the bearings. The distance between the bearings is 150 mm. The follower which weighs 2500 N is out of adjustment so that impact is produced when it is contacted by the cam. The height of fall is 0.25 mm. Determine the maximum bending stress in the shaft.

Solution: Data given

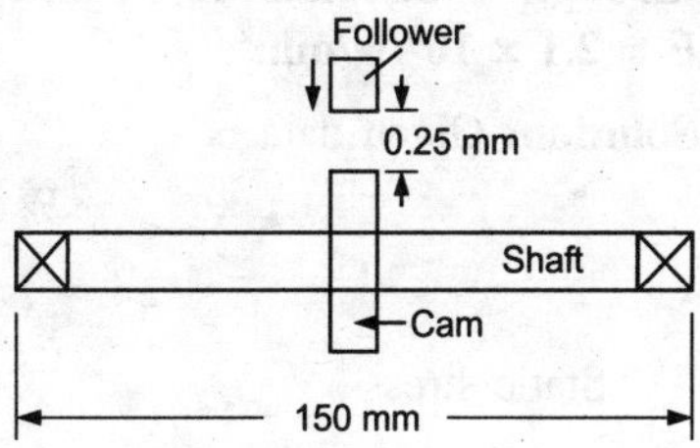

$h = 0.25$ mm, $l = 150$ mm

$w = 2500$ N, $d = 20$ mm

Take $E = 2.1 \times 10^5$ N/mm^2

$$f_b = \frac{M_b}{Z_b} = \frac{2500}{2} \times \frac{32}{\pi(20)^3} \times 75$$

$$= 119.5 \text{ N/mm}^2$$

Static deflection of shaft $= e = \frac{Wl^3}{48EI} = \frac{2500\times(150)^3\times64}{48\times2.1\times10^5\times\pi\times(20)^4} = 0.1065$mm

Impact stress due to bending

$$f_b' = f_b\left[1+\sqrt{1+\frac{2h}{e}}\right] = 119.5\left[1+\sqrt{1+\frac{2\times0.25}{0.1065}}\right] = 404.7 \text{ N/mm}^2.$$

Example 3.9. A cantilever beam of section width 50 mm and depth 150 mm is 1.5 m long. This beam is struck by a load of 1000 N which falls from a height of 10mm at its free end. Determine maximum deflection, maximum stress, impact factor and maximum load intensity. Also determine static deflection and static stress.

Solution: Given data is

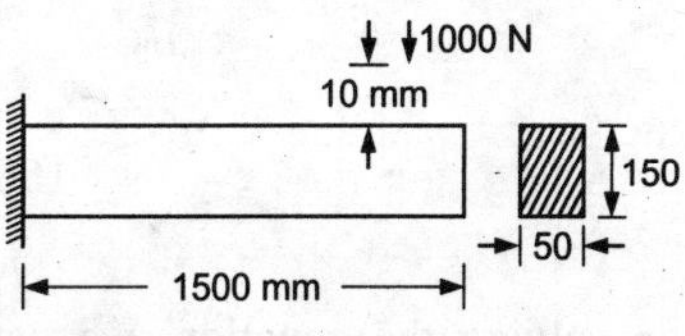

$W = 1000$ N, $h = 10$ mm

$b = 50$ mm, $d = 150$ mm

Take $E = 2 \times 10^5$ N/mm^2

e = static deflection

$$= \frac{1}{3}\frac{Wl^3}{EI} = \frac{1}{3} \times \frac{1000\times(1500)^3\times12}{2.0\times10^5\times50\times150^3} = 0.4 \text{ mm}$$

Instantaneous maximum deflection

$$e' = e\left[1+\sqrt{1+\frac{2h}{e}}\right] = 0.4\left[1+\sqrt{1+\frac{2\times 10}{0.4}}\right] = 3.25 \text{ mm}$$

f_b = static bending stress

$$= \frac{M_b}{Z_b} = \frac{1000\times 1500\times 6}{50\times(150)^2} = 8.0 \text{ N/mm}^2$$

Instantaneous maximum stress in bending

$$f_b' = f_b\left[1+\sqrt{1+\frac{2h}{e}}\right] = 8.0\left[1+\sqrt{1+\frac{2\times 10}{0.4}}\right] = 65.13 \text{ N/mm}^2$$

$$\text{Impact factor} = \frac{f_b'}{f_b} = \frac{\text{maximum bending stress}}{\text{static bending stress}} = \frac{65.13}{8.0} = 8.14$$

Maximum load intensity = impact factor × static load

= 8.14 × 1000 = 8140 N

Example 3.10. Find the diameter of a steel rod 1.5 m long if it resists the impact of 2500 N dropped through a distance of 50 mm along the axis. The maximum stress is limited to 150 N/mm². $E = 2.1 \times 10^5$ N/mm².

Solution: Given data is

$$W = 2500 \text{ N},\ h = 50 \text{ mm},\ E = 2.1\times 10^5 \text{ N/mm}^2$$

$$l = 1500 \text{ mm},\ f'_{max} = 150 \text{ N/mm}^2$$

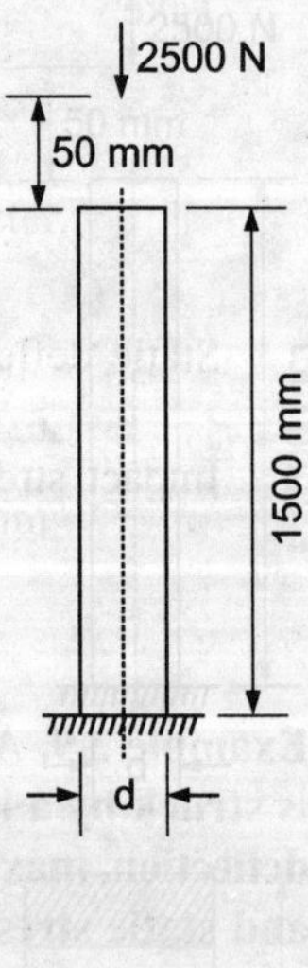

Static stress

$$f = \frac{\text{Force}}{\text{area}} = \frac{2500\times 4}{\pi d^2} = \frac{3183.09}{d^2} \text{ N/mm}^2.$$

Impact stress from a direct load

$$f'_{max} = f\left[1+\sqrt{1+\frac{2h}{e}}\right]$$

where $e = \dfrac{Wl}{AE}$

$$\therefore \quad f'_{max} = f\left[1+\sqrt{1+\frac{2hAE}{Wl}}\right]$$

$$150 = \frac{3183.09}{d^2}\left[1+\sqrt{1+\frac{2\times 50\times \pi d^2\times 2.1\times 10^5}{4\times 2500\times 1500}}\right]$$

on solving the equation, we get

$$d = 70.7 \text{ mm}.$$

Example 3.11. A weight of 2.5 kN falls through a height of 3 mm and the load strikes the round head K of round steel bar 25 mm in diameter and 500 mm long as shown in the figure. Determine:

(*a*) Stress induced in the bar. Take $E = 2 \times 10^5$ N/mm^2. Neglect the inertia of the bar.
(*b*) Compare the stress induced from the same load acting statically.
(*c*) Find how much the length of the rod can be changed before the impact strength exceeds 130 MPa?
(*d*) Determine how much clearance can be increased before the impact strength exceeds 130 MPa?
(*e*) Compare the influence of the inertia of the bar upon the impact stress. Take the density of the bar material as 78×10^{-6} N/mm^3.

Solution: Given data is

$$W = 2.5 \text{ kN}, \quad h = 3 \text{ mm}$$
$$d = 25 \text{ mm}, \quad l = 500 \text{ mm}$$

(*a*) Stress induced in the bar without considering its inertia. Impact energy due to the falling body

$$K_i = Wh = 2.5 \times 10^3 \times 3 = 7500 \text{ Nmm}$$

Resilience is given by

$$U = \frac{f^2 V}{2E} = \frac{f^2 \times \pi(25)^2 \times 500}{2 \times 4 \times 2 \times 10^5} = 0.6135 f^2$$

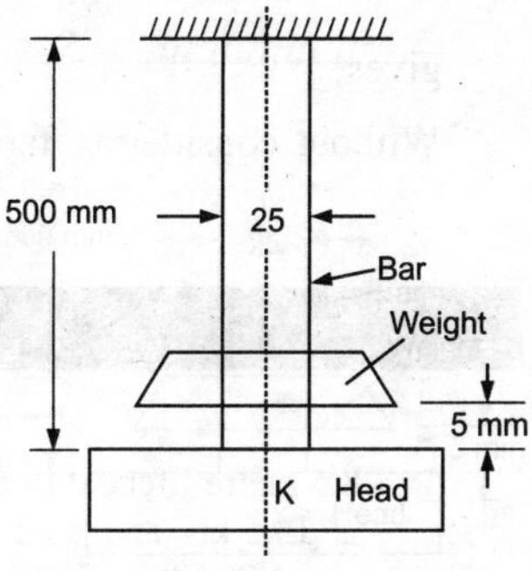

We know that, impact energy = Resilience

$$K_i = U$$
$$7500 = 0.6135\, f^2.$$

gives,

$$f = 110.56 \quad \text{MPa}$$

(*b*) Static stress

$$f = \frac{W}{A} = \frac{2.5 \times 10^3 \times 4}{\pi(25)^2} = 5.09 \text{ MPa}$$

Comparison

$$\frac{\text{Impact stress}}{\text{Static stress}} = \frac{110.56}{5.09} = 21.7$$

(*c*) To find new length of the bar when impact strength is 130 MPa.

Impact energy $K_i = Wh = 7500$ Nmm.

$$\text{Resilience} = U = \frac{f^2 V}{2E} = \frac{(130)^2 \times \pi(25)^2 \times l_1}{4 \times 2 \times 10^5 \times 2}$$
$$= 20.79\, l_1$$

We know that,

$$K_i = U$$
$$7500 = 20.79\, l_1$$

$\therefore$

$$l_1 = 361.63 \text{ mm}$$
$$\text{Change in length} = l - l_1 = 500 - 361.63 = 138.37 \text{ mm}$$

(*d*) To find new height of fall h_1 when $f = 130$ MPa

Using l = 500 mm.

$$K_i = W \times h_1 = 2.5 \times 10^3\ h_1 = 2500\ h_1 \text{ Nmm}$$

equating $$K_i = U$$

$$2500\ h_1 = 10369 = 0.6135 f^2 = 0.6135 \times 130^2$$

gives $$h_1 = 4.14 \text{ mm}$$

Change in clearance required = $h_1 - h = 4.14 - 3 = 1.14$ mm

(*e*) To find impact stress *f*, considering inertia of the bar (self weight).

$$\text{Weight of bar} = \text{Volume} \times \text{Density}$$

$$= \frac{\pi}{4}(25)^2 \times 500 \times 78 \times 10^{-6} = 19.14 \text{ N}$$

$$\text{Effective weight} = 2.5 \times 10^3 + 19.14 = 2519.14\text{N}$$

$$K_i = Wh = 2519.14 \times 3 = 7557.42 \text{ Nmm}$$

$$U = \frac{f^2 \pi (25)^2 \times 500}{4 \times 2 \times 2 \times 10^5} = 0.6135 f^2$$

equating $$K_i = U$$

$$7557.43 = 0.6135 f^2$$

gives, $$f = 110.98 \text{ N/mm}^2$$

Without considering the self weight of the bar, the impact stress is 110.56 N/mm^2

EXERCISES

1. 8.5 metre lifeboat to carry 60 persons is supported from two davits. The weight of the boat fully equipped is 13.5 kN and an estimate of the weight of the occupants is 13.5 kN. Assuming that the overhang, e = 1.9 metre, fig. 3.16 and that davits share the load equally. If the allowable stress is 105.0 MN/m^2 determine the diameter '*d*'.
2. A circular column, the material of which is carbon steel as rolled with ultimate strength of 440.0 MN/m^2 is to have a length of 2.7 m, and is to support an eccentric load of 72 kN at a distance of 0.075 m from the centre line. (*a*) What should be the outside diameter if the column is hollow and inner diameter = 0.75 × outer diameter? (*b*) What should be the diameter if the column is solid?

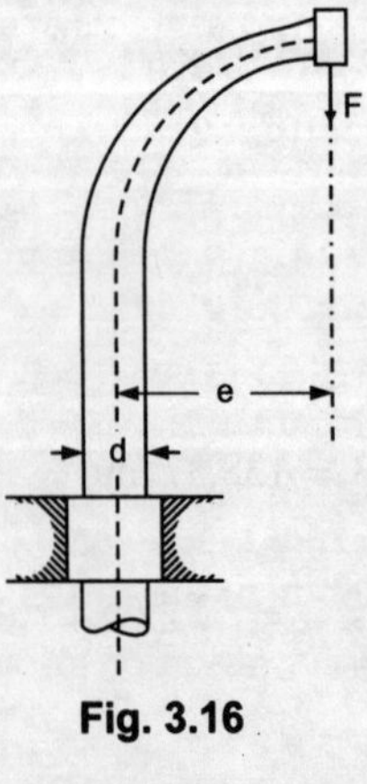

Fig. 3.16

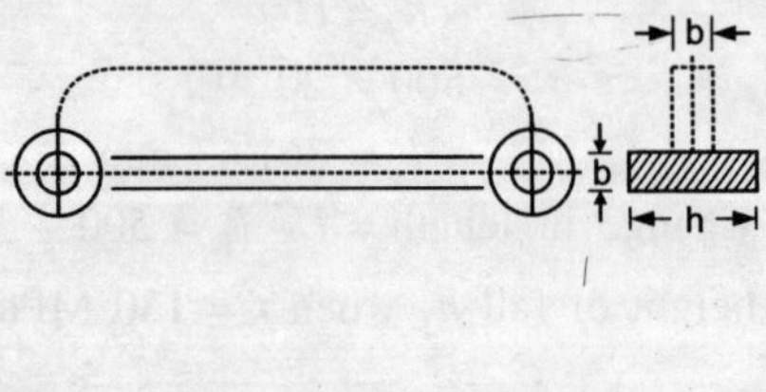

Fig. 3.17

3. The link shown in Fig. 3.17 is made of cast steel and is to support a compressive load. Originally the link was made with rectangular cross-section with L = 0.25 m; h = 0.05 cm; and b = 0.0125 m. In order to strengthen the link, the cross-section was changed to an inverted T as shown by dotted lines with h and b same as before. (*a*) What load could each link carry for factor of safety = 3? (*b*) What was the percentage increase in strength?
4. A pressure 1.4 MN/m^2 on a piston 0.0335 m^2 in area, is transmitted as an axial load to a piston rod, 0.75 m in length. Calculate the required diameter of the rod. Assume a design stress of 63 MN/m^2; and an elastic stress of 420 MN/m^2. The modulus of elasticity is 195 GN/m^2. One end is pinned, and one end is fixed.
5. A steam engine piston rod is to be made of nickel steel. The rod is to be 1.4 m long and the maximum steam load carried by the rod is 200 kN. Assuming that yield point for the steel is 455 MN/m^2; the end-fixity coefficient is 3.5 and the factor of saftey 4, determine the diameter for the rod.
6. The piston rod of a steam engine may be considered to be a column. Determine the recommended diameter of the rod, assuming the following:

Diameter of cylinder	= 0.45 m
Maximum net steam pressure on piston	= 0.6 MN/m^2
Distance from piston to cross-head	= 1.6 m
Material	= C — 45 steel
Factor of saftey	= 4
End-restraint coefficient	= 3.5

7. The piston rod of a steam engine may be considered to be a column with one fixed and one guided end. Determine the size of rod necessary for a 0.45 m cylinder using steam at 2.45 MN/m^2 pressure. The rod is 1m in length and is made of steel having an ultimate strength of 490 MN/m^2 and an elastic limit of 265 MN/m^2. The permissible working stress is 56 MN/m^2.
8. Define the following : Strain, unit strain, shear strain, Poisson's ratio, and stress.
9. What assumptions are made in the definition of stress?
10. Define and differentiate between modulus of elasticity and modulus of rigidity.
11. Define Bulk modulus of elasticity.
12. What assumptions are made in the derivation of torsion and bending relations?
13. What is meant by stiffness?
14. What is meant by: Column, slenderness ratio, end conditions of a column, long column, short column and a strut.
15. Under what conditions the Euler's formula and Rankine's formula, Rankine Gorden Formula and J.B. Johnson formula are applicable?
16. Under what conditions the straight line column formula is used?
17. Explain the term Beam Column.
18. What is meant by: Dynamic loads, static loads or dead loads, live loads, impact loads, inertia loads, and Resilience.
19. What type of stress is produced because of centrifugal force.
20. What assumptions are made for finding out impact stress?
21. A cantilever beam of rectangular cross-section has a span of 800 mm. The rectangular cross-section has a depth of 200 mm. The free end of the beam is subjected to a transverse load of 1000 N that drops on it vertically from a height of 40 mm. The permissible stress in tension is 120 N/mm^2. Determine the width of the rectangular cross-section. Take $E = 2.1 \times 10^5$ N/mm^2.
22. A simply supported steel beam 500 mm long has a rectangular cross-section 25 × 40 mm and a limiting stress of 350 MPa.
    (*a*) Find the maximum load that can be dropped from a height of 120 mm to the centre of the beam with 40 mm side vertical and again with 25 mm side vertical.
    (*b*) Also, find the maximum centre load with 40 mm side vertical and again with 25 mm side vertical. (Hint static load is to be found).
23. Hooke's law is valid upto the
    (a) elastic limit (b) yield point (c) fracture point.
24. What are:-
    (a) Principal planes (b) Principal stresses.
25. Define the Castigliano's theorem.

❑❑❑

CHAPTER 4

Designing for Strength

4.1 GENERAL

The fundamental requirements for the machine elements, on which their efficiency depends, are: strength, stiffness, wear and corrosion resistance. Wear is must when two surfaces of machine elements have relative motion. However, it can be minimised by good design. Selection of good materials and surface protection will help in preventing the corrosion of the machine surfaces. The other factor that can cause failure of the machine part is its excessive deformation. Deformation means the change of shape due to elastic strain, plastic strain, and fracture.

When a machine part or element is subjected to a load, internal stresses are induced in the part by the applied load. These internal stresses resist the deformation or distortion of the part. If the dimensions of the part are based on the elastic deformation so as to prevent excessive elastic deformation, the design is called "rigidity" or "stiffness" design. On the other hand, if the design is based on the induced stress or on plastic deformation, it is said to be the "Strength Design".

Most machine parts are designed on the basis of "strength design" due to the following reasons:

1. It is easier to base the design on strength than on stiffness.
2. The machine parts designed on the basis of strength usually have sufficient elastic rigidity stiffness.
3. For parts which are not simple in form, it is very difficult to determine the deflections than to determine the induced stresses.
4. It is more difficult or impossible to set a limit for permissible elastic deflections than to fix upon the allowable stress.

However, members like springs are designed for both strength and stiffness.

The machine members may be classified as elastically unstable and elastically stable. The members those will fail by buckling or collapsing are said to be elastically unstable, for example, long and slender columns and thin tubes. Elastically stable members fail by breaking, by being permanently deformed or by suffering so much elastic deformation that they will not function properly. Generally speaking, most machine and structural members are elastically stable.

4.2 SOME REMARKS ON THE STRENGTH OF A MATERIAL

After the material for a machine element has been selected, the results of great many strength tests of that material should be available to the designer before he can take up the design of that machine element. For all these tests, the specimens should have the same size, same heat treatment and surface finish as that of the machine element to be designed. Also, the specimens should be tested under the same loading conditions as the machine element will experience during actual service. For example, if a machine element is to be designed for bending it should be tested under a bending load and if the service loading is to be combined bending and torsion, the specimen should be tested under combined bending and torsion. Such tests will provide very useful and precise information to the designer regarding the strength of the material. He will be in a position to ascertain accurately the reliability of the machine part for a given service life and to arrive at a very accurate design.

The cost of getting such detailed results will be very high and can be justified only if there will be a mass production of that part so that the cost of tests per part becomes less and also if the failure of that part may endanger the human life. The examples can be: parts of aeroplanes, automobiles and refrigerators etc. In this book, we shall be mainly concerned with the design of parts made in small to moderate quantities so that the detailed test results are not justified, or there may be the case of improvement in the design of a part. For such cases, the designer will have only the published data about the strength of the material in the form of yield strength, ultimate strength and percentage elongation etc. Such a data is usually obtained from the simple tension test in which the load is applied gradually and at room temperature, With this data, the designer is expected to design machine parts subjected to static, dynamic and variable loads, combined stress states and high and low temperature etc. It means that the basic problem of the designer is to use the simple tension test data and relate it to the strength of the machine part, regardless of the loading condition and state of stress etc. How this is done, we shall study it in this chapter.

The stress-strain relationship from the simple tension test, for some typical material are shown in Fig. 4.1. The typical shape of the stress-strain diagram for a ductile material (low carbon steel or mild steel) is shown in Fig. 4.1(a).

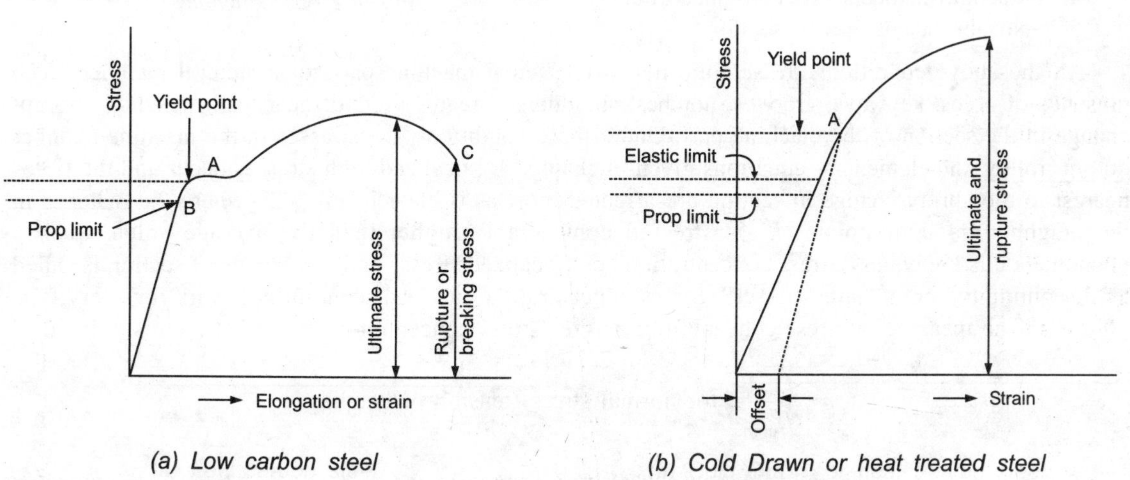

(a) Low carbon steel *(b) Cold Drawn or heat treated steel*

This material has a well defined 'yield point' at which a marked increase in elongation occurs without any increase in load. After the ultimate or maximum stress is reached, there starts a marked reduction in the cross-sectional area of the specimen (the phenomenon being known as 'Necking') and ultimately the specimen breaks. Fig. 4.1(b) shows stress-strain diagram for a material of low ductility, for example, heat treated or cold drawn steel. It is clear that these materials do not exhibit a well defined 'yield point'. For such materials, the yield strength can be taken corresponding to a permanent set of 0.10 to 0.20 per cent. Fig. 4.1(c) shows the stress-strain diagram for a perfectly brittle material, where fracture occurs without any plastic deformation.

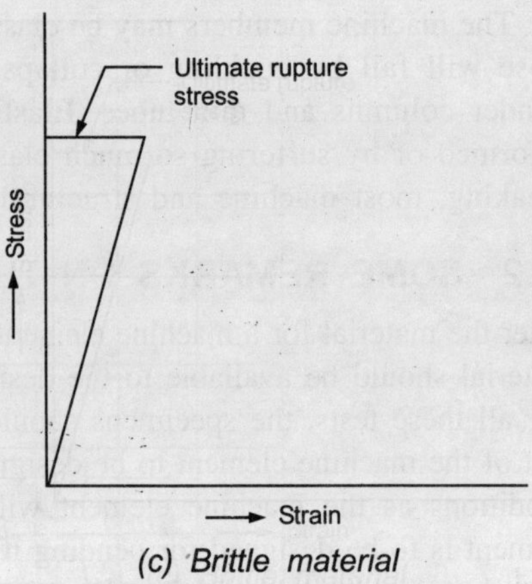

(c) Brittle material

Fig. 4.1 *Stress-strain diagram*

'Ductility' is the property of a material to absorb overloads. If two materials have the same hardness and ultimate strength, the material with greater ductility will absorb more overloads as it will be capable of a relatively large plastic deformation before fracture [Fig. 4.1(a)]. Ductility is measured by the 'percentage elongation', which takes place in a material at fracture. The usual dividing line between ductility and brittleness is 5 per cent elongation. A material having less than 5 per cent elongation is termed as 'Brittle' and the one with more than 5 per cent elongation is termed as 'Ductile'. Ductility can also be measured in terms of percentage reduction in cross-sectional area at fracture. When a machine element is to resist wear, erosion or plastic deformation, hardness is generally the most important property.

4.3 STRESS CONCENTRATION

The derivation and the use of the elementary formulas applying to simple tension, compression, shear, bending and torsion, and to combined stresses, depends upon the following assumptions:

1. Stress is proportional to strain.
2. Modulus of elasticity is the same in tension and in compression.
3. No high intensity contact pressures at the regions of the contact of support and loading.
4. Machine members have constant section.
5. No abrupt changes in section.

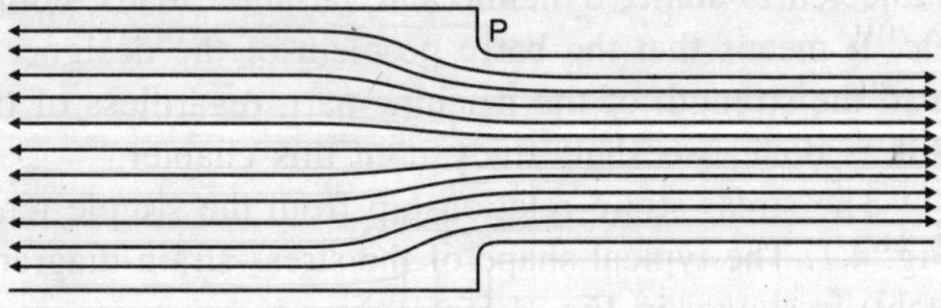

Fig. 4.2 *Discontinuity*

All the above conditions are seldom met in the actual machine parts of structural members. The presence of holes, keyways, grooves, notches, shoulders, threads, roots of gear teeth, results in abrupt change in the section of the machine part. Under these conditions, the stresses in the machine member do not follow the elementary equations given in chapter 3. Localized high stresses occur and the fibres nearest to the abrupt change in section are affected most as is clear in Fig. 4.2, where the material in the neighbourhood of points '*P*' is stressed considerably higher than the average value. Such a phenomenon is known as "Stress concentration" or "localized stress" and the change of section is called a "discontinuity" or a "stress raiser". Stress concentration due to discontinuity occurs for every type of stress. The measure of stress concentration is the "stress concentration factor", which is defined as

$$K_t = \frac{f_{t\,\text{max.}}}{f_{t\,\text{nom.}}} \text{ for normal stress (tension or bending)}$$

$$K_{ts} = \frac{f_{s\,\text{max.}}}{f_{s\,\text{nom.}}} \text{ for shear stress (torsion)}$$

where f_t nom. and f_s nom. are based on the elementary formulas of chapter 3. Maximum stress at the discontinuity will be:

$$f_{t\,max.} = \frac{P}{A}K_t \text{ direct load}$$

$$= \frac{M_b}{Z}.K_t \text{ bending} \qquad ...(4.1)$$

and
$$f_{s\,max.} = \frac{M_t}{Z_p}.K_t \text{ torsion}$$

where A, Z and Z_p are based on net sections. Fig. 4.3.

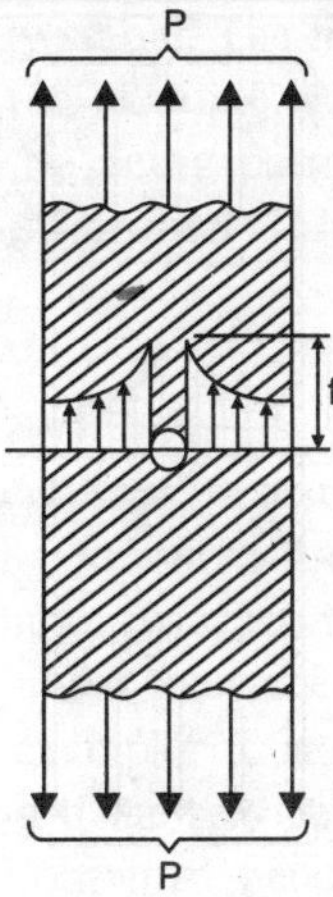

Fig. 4.3

The factor 'K_t' is usually called the theoretical stress concentration factor. It can be determined mathematically or experimentally by the following methods:

1. Photo-elasticity.
2. Plaster model.
3. soap film method
4. precision strain gauge.
5. membrane analogy for torsion.
6. electric analogy for torsion.

It depends only on the type and size of the discontinuity or on its geometric from. Therefore, it is commonly called as "form stress factor". The actual "stress concentration factor" is different from "form stress factor" since it depends on material and type of loading also. The effect of actual "stress concentration" is less than the "form stress factor" since most of the engineering materials are elastic and capable of yielding at the point of excessive localised stress. This will spread the localised stress and reduce its intensity. Some values of "form stress factor" K_t are given in Figs. 4.4 to 4.9 for static loading.

The "form stress factor" K_t and the actual "stress concentration factor" "K" are connected by the following relation:

$$K = 1 + q\,(K_t - 1) \qquad ...(4..2)$$

where q is known as the "Index of sensitivity" of the material to abrupt change of section. From the above equation:

$$q = \frac{K-1}{K_t - 1} \qquad ...(4.3)$$

The value of 'q' depends upon the material and type of loading. It is less for ductile materials and more for brittle materials. It is least for static loads applied steadily and goes on increasing with intensity of shock and is highest for repeated heavy shock conditions.

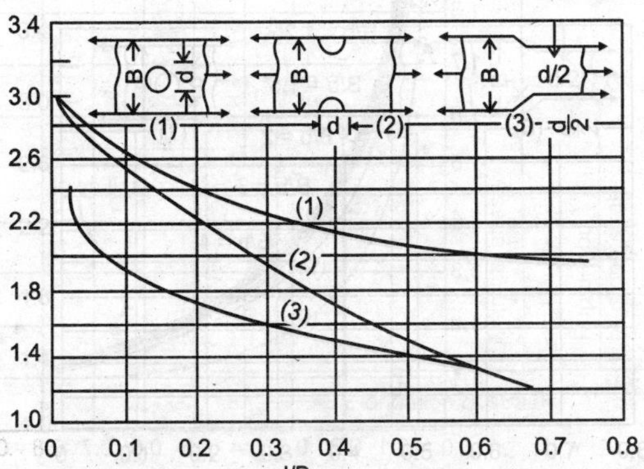

Fig. 4.4 *Tension or compression*

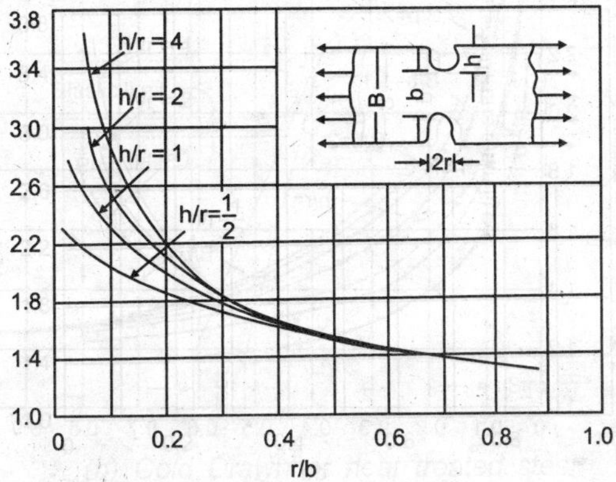

Fig. 4.5 *Tension or compression*

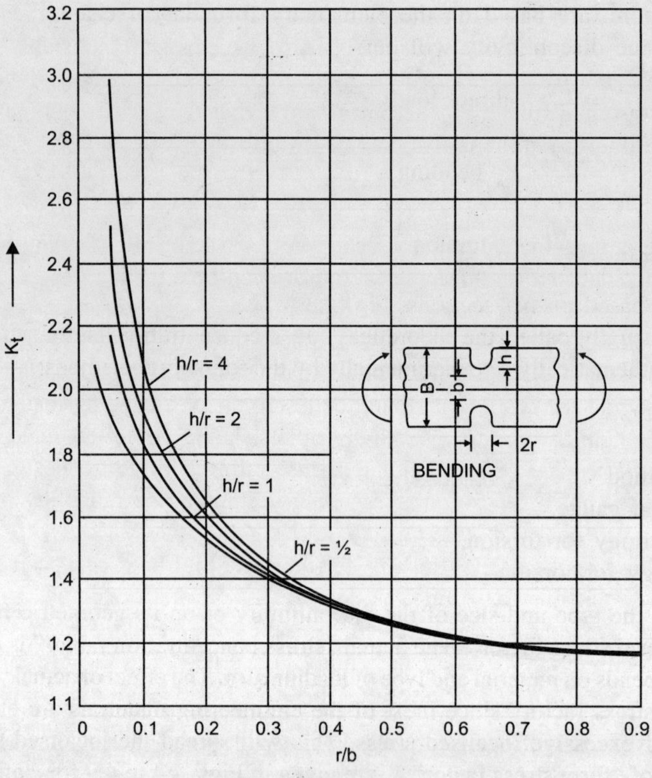

Fig. 4.6 *Bending*

The localized stresses at the concentrations can be greatly reduced by properly forming the part, by employing fillets of sufficient radius, and by avoiding perforations.

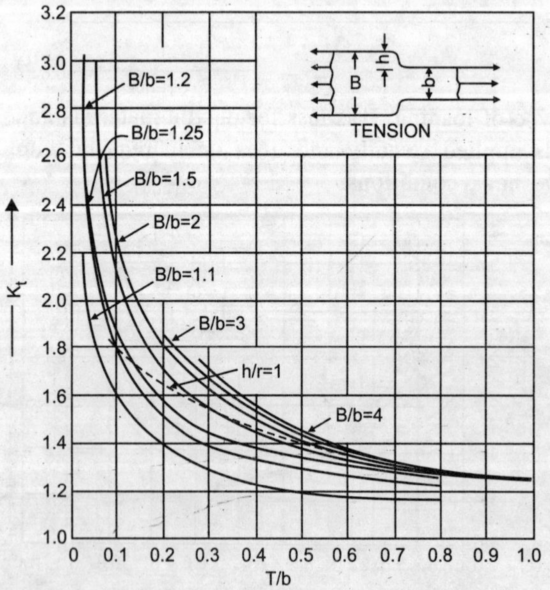

Fig. 4.7 *Tension or compression*

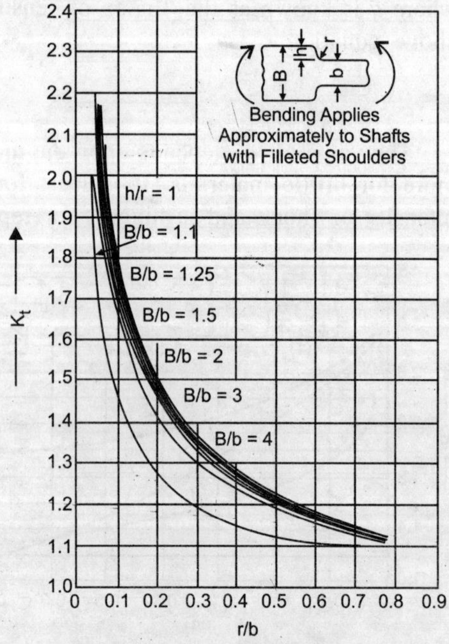

Fig. 4.8 *Bending*

4.3.1 Index of Sensitivity

4.3.1.1 Static Load Applied Steadily

Stress concentration is a highly localised effect. The high stresses actually exist in only a very small region in the vicinity of the discontinuity (Fig. 4.2). In the case of ductile materials, the first load applied to the part will cause yielding at the discontinuity which relieves the stress concentration. Thus when the parts are subjected to static loads and are made of ductile materials, it is not necessary to use a stress concentration factor at all. So, for ductile materials, the value of 'q' *is* zero and the stress concentration factor is 1. Since in the case of cast iron there are so many discontinuities, the value of 'q' may be assumed zero. For other metals, the value of q are as below:

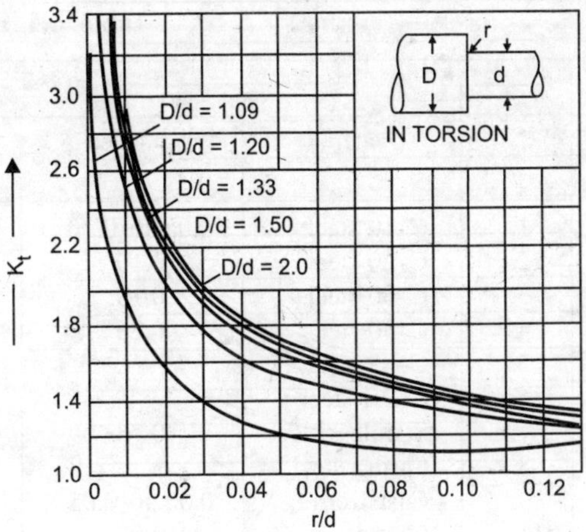

Fig. 4.9 *Shaft in torsion*

= 0.10 for brittle materials, e.g., hardened steel.
= 0.20 for very brittle materials, e.g., steels that are quenched but not drawn.

4.3.1.2 Static Load Applied Suddenly

(shock loading):

q = 0.4 for materials which are ductile and very soft.
= 0.6 for other ductile materials.
= 1.0 for hard and brittle materials.
= 0.5 for cast iron.

4.3.2 Stress Concentration, Variable Loading (Fatigue Loading)

Stress concentration is most important in the case of fatigue loading of the machine parts. Due to the presence of the stress concentrations, the fatigue resistance of the parts is lowered. The effect of such stress raisers may be expressed in the form of a fatigue stress concentration factor, K_f which is defined by the equation

$$K_f = \frac{\text{Endurance limit of notch free specimen}}{\text{Endurance limit of notched specimen}}$$

The factor, K_f is related to the form factor K_t in the following manner:

$$K_f = 1 + q_f (K_t - 1) \quad \text{...(4.4)}$$

So,
$$q_f = \frac{K_f - 1}{K_t - 1} \quad \text{...(4.5)}$$

where q_f is the notch sensitivity for fatigue loading. For various materials, the value of q_f varies from 0 to 0.70. The index is smaller for ductile materials and it increases as the ductility is decreased, as by heat treatment.

The values for q_f are tabulated in Table 4.1.

Table 4.1 Notch sensitivity index, q_f

S.No.	Material	Condition of Specimen		
		Annealed	Heat treated and drawn at 650ºC	Heat treated and drawn at 480ºC
1.	C–10 steel	0.05 to 0.10	—	—
2.	C–20 steel	0.10	—	—
3.	C–30 steel	0.18	0.35	0.45
4.	C–50 steel	0.26	0.40	0.50
5.	C–85 steel	—	0.45	0.57
6.	Cr-Ni steel	—	0.25	0.70
7.	Stainless steel	0.16	—	—
8.	Spring steel	—	0.38	—
9.	Cast Iron	0.00 to 0.05	—	—
10.	Copper	0.07	—	—
11.	Duraluminium	0.05 to 0.13	—	—

The values for notch sensitivity factor, q_f, may also be obtained from the relation:

$$q_f = \frac{1}{1+\left(\frac{a}{r}\right)} \quad ...(4.6)$$

where 'r' is the notch root radius in mm and 'a' is a material constant which depends on the strength and ductility of the material. The values of 'a' are given in Table 4.2. The values of 'a' given in table are for axial and bending loads. For torsional loads, the values of 'a' may be taken as 0.6 times the values given in the Table. 4.2.

Table 4.2 Average values of Material constant 'a'

Material	a, mm
Steel	
(i) Annealed and Normalized	0.25
(ii) Quenched and Tempered	0.06
Aluminium:	
(i) Sheet	0.50 to 1.25
(ii) Bar	0.08 to 0.20

The values of 'q_f' calculated by using the relation (4.6) are plotted in Fig. 4.10.

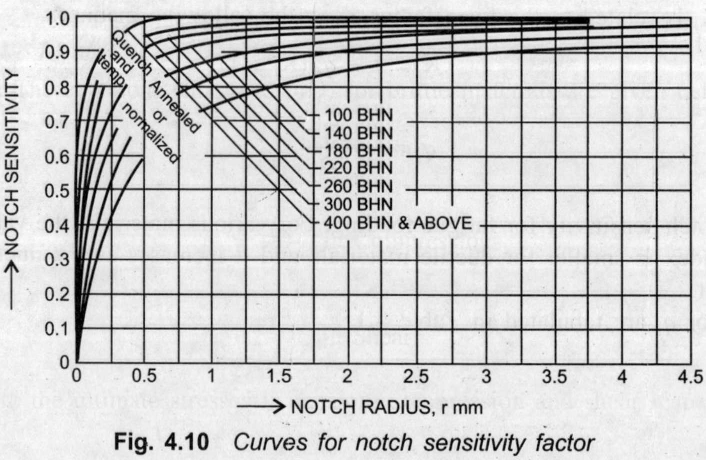

Fig. 4.10 *Curves for notch sensitivity factor*

4.4 STRENGTH DESIGN

As already discussed, strength is a major factor in the design of machine members.

To design a member or part of machine for strength, the two most important considerations are: Selection of material and selection of design stress or allowable stress or permissible stress that will be used to calculate the proportion of the part which will be sufficiently strong. Design stress is the limit which should not be crossed by the maximum induced stress, nominal or actual stress. It is the stress fixed by the designer and used by him in computing the proportions of a part for safe strength in service. The nominal or actual stress is that which is computed by one of the relations of Chapter 3. Working stress is that which actually occurs in the machine part under operating conditions. Working stress should never exceed the design stress.

Selection of material has been discussed in chapter 2. Selection of proper design stress is the most important problem and at the same time, it is the most vague and difficult one. To arrive at a proper value of design stress, the following factors should be considered:

1. Analysis of the service conditions that must be met.
2. Evaluating the strength of the material under service conditions.
3. Deciding on the margin of safety desired between the service conditions and actual failure.

The service conditions that should be fully analysed are:

(i) Magnitude of the loads applied.
(ii) Method of load application i.e., steady, impact or variable.
(iii) Type of stress induced in the part i.e., tensile, compressive, shear or combination of these.
(iv) Temperature
(v) Appearance
(vi) Use of standard parts
(vii) Ease of assembly
(viii) Ease of repair
(ix) Wearing qualities.

The different aspects of strength design will be discussed in this chapter for different types of loading conditions, for example, static loading and fatigue loading.

4.4.1 Static Loading

Static loading may be defined as a loading which is applied to a part very gradually, increasing from zero to its maximum value. The application of load is so slow that the accelerations occurring in this case can be neglected. Also, after the load reaches its final value, it does not change its magnitude, direction or point of application with time.

4.4.1.1 Type of Failure

The main function of a machine element is to resist the applied loading. That is, the part must function satisfactorily throughout its expected service life. When a machine part is no longer able to perform its function satisfactorily, it is considered as failed. When a machine element is subjected to static loading, it may fail in two manners, namely,

(i) Failure due to yielding.
(ii) Failure due to fracture, rupture or breaking.

If we refer to Fig. 4.1, it is clear that for ductile materials [Fig. 4.1(*a*)], large plastic deformations occur after the yield point, resulting in permanent deformation of the machine element which ultimately breaks at point '*C*'. Hence for ductile materials, failure is usually considered to have occurred when yielding, that is plastic deformations reach a limit where the engineering usefulness of the part is destroyed, even though there is no rupture or fracture of the machine part. Thus the yield point stress is the criterion of failure for the ductile materials subjected to static loading. In the case of brittle materials [Fig. 4.1*b*, *c*], the ratio of yield strain to ultimate strain is very nearly equal to unity.

So, the brittle materials are considered to have failed by fracture with little or no plastic deformation. So, the ultimate strength or the breaking strength is the criteria of failure for brittle materials.

4.4.1.2 Prevention of Failure

The failure of a machine element can be prevented if the maximum induced stress due to external loading does not exceed the failure stress for the material. As discussed above, the failure stress for ductile materials is the yield point stress and for brittle materials it is the ultimate strength of the material. Hence, the failure of a machine part can be avoided if the maximum induced stress due to external loading is kept less than the yield point stress for ductile materials and less than the ultimate stress for the brittle materials. This is done by using a factor of safety: F_S, ($F_S > 1$) on the failure stress. The stress which is obtained by using a factor of safety on the failure stress is termed as 'allowable stress', 'design stress', or the 'working stress'.

The failure stress, the allowable stress, and the factor of safety are related as,

$$\text{Allowable stress} = \frac{\text{Failure Stress}}{\text{Factor of Safety}} \qquad ...(4.7)$$

To avoid failure, the following condition then must be satisfied

$$\underset{\text{(Due to loading)}}{\text{Maximum induced stress}} \leq \text{Allowable Stress} \qquad ...(4.8)$$

The allowable stress can be obtained easily if the material properties and the desired factor of safety are known.

This is explained below:

For Ductile Materials: The allowable stress 'f_a' for ductile materials subjected to static loading is obtained as given below:

$$f_{at} = \frac{f_{yp}}{F_S}, \text{ in Tension}$$

$$f_{ac} = \frac{f_{yp}}{F_S}, \text{ in Compression} \qquad ...(4.9)$$

$$f_{as} = \frac{f_{yps}}{F_S}, \text{ in Shear}$$

where f_{at}, f_{ac} and f_{as}, are the allowable stresses in tension, compression and shear respectively, f_{yp} is the yield point stress which is taken to be the same in tension and compression and f_{yps} is the yield point stress in shear.

Note: $f_{yps} = (0.5 \text{ to } 0.6) \times f_{yp}$

For Brittle Materials: The equations similar to (4.9) for brittle materials are given below:

$$f_{at} = \frac{f_{ut}}{F_S}, \text{ in Tension}$$

$$f_{ac} = \frac{f_{uc}}{F_S}, \text{ in Compression} \qquad ...(4.10)$$

$$f_{as} = \frac{f_{us}}{F_S}, \text{ in Shear}$$

where f_{ut}, f_{uc} and f_{us} are the ultimate stresses in tension, compression and shear respectively.

The concepts discussed above are illustrated in Fig. 4.11, with the help of stress-strain curves.

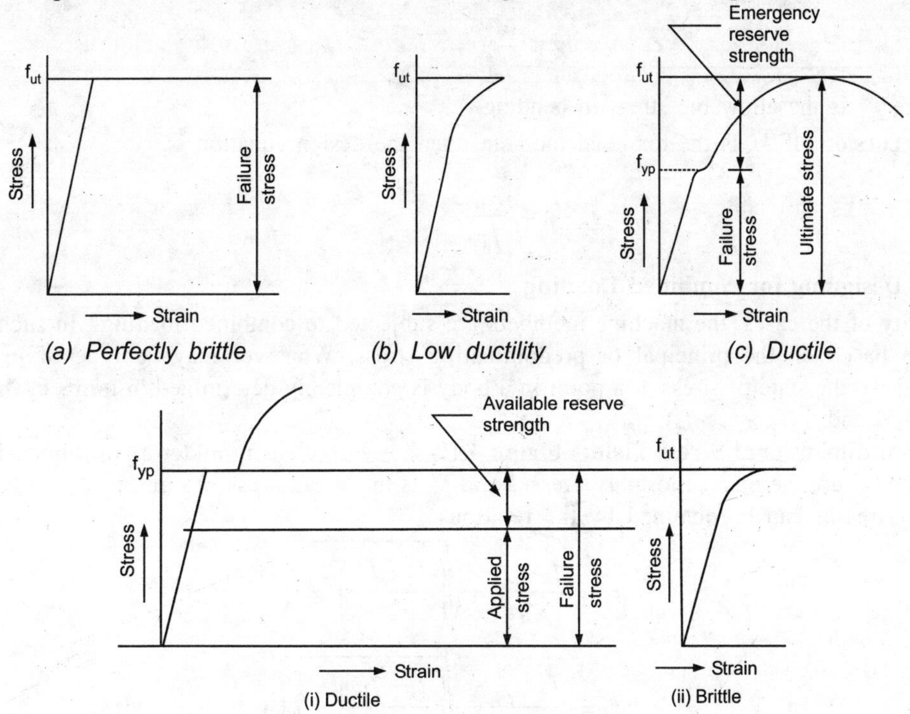

(a) Perfectly brittle (b) Low ductility (c) Ductile

(i) Ductile (ii) Brittle

(d) Concept of factor of safety

Fig. 4.11 Stress strain curves for typical materials

Then once the maximum induced stress due to external loading on machine element is estimated (in terms of its dimension), its dimensions can be determined with the help of equations (4.8, 4.9 and 4.10). This is explained in the next Articles.

4.4.1.3 Designing for Simple Loading

Simple loads are those which can be considered singly, for example, uniaxial tension or compression, bending and torsion. For such cases, the determination of the maximum induced stress due to external loading is quite simple (See equations 3.7 to 3.14, Chapter 3). After that the proportions of the machine element can be determined easily in terms of area of cross-section 'A' for tensile and compressive loads, section modulus 'Z' for bending loads and polar section modulus 'Z_p' for torsional loads, as explained below:

(i) Simple Tension and Compression: If F is the tensile or compressive load, then the maximum induced tensile stress, f_t, *is*

$$f_t = \frac{F}{A}$$

∴ To avoid failure, the design equation is

$$f_t \le f_{at}$$

$$\therefore \qquad \frac{F}{A} \le f_{at} \text{ or } A \ge \frac{F}{f_{at}} \qquad ...(4.11)$$

∴ If F and f_{at} are known, A can be determined.

Similarly for compression, $$A \ge \frac{F}{f_{ac}} \qquad ...(4.12)$$

(ii) Bending: If M_b is the applied bending moment, then the design equation will be,

$$Z \geq \frac{M_b}{f_{ab}} \quad ...(4.13)$$

where f_{ab} is the allowable stress in bending.

(iii) Torsion: If M_t is the torsional moment, then the design equation is,

$$Z_p \geq \frac{M_t}{f_{as}} \quad ...(4.14)$$

4.4.1.4 Designing for combined Loading

In majority of the cases, the machine members are subjected to combined loading. In such cases, the design is based on the principal or predominating stress. Whatever may be type of loading (two dimensional), the state of stress at a point in a body is completely determined in terms of the principal stresses f_1, f_2 and f_3 ($f_1 > f_2 > f_3$).

1. **Two-dimensional Stress Field:** Figure 4.12 shows an element under a two dimensional stress field. f_x and f_y are the direct (normal) stresses and f_{xy} is the shear stress. As discussed in chapter 3, the principal stresses can be obtained by the relations:

$$f_1 = \frac{f_x + f_y}{2} + \sqrt{\left(\frac{f_x - f_y}{2}\right)^2 + f_{xy}^2} \quad ...(4.15)$$

$$f_2 = \frac{f_x + f_y}{2} - \sqrt{\left(\frac{f_x - f_y}{2}\right)^2 + f_{xy}^2} \quad ...(4.16)$$

It is to be noted that the equations (4.15 and 4.16) give algebraic maximum and minimum values of principal stresses, where,

f_x is the direct or normal stress (tension or compression) and may be due to either bending or axial loading or a combination of the two. Tensile stress will be taken as positive and compressive stress as negative.

f_y is the direct or normal stress (tension or compression) and in a direction normal to the f_x stress.

f_{xy} is the shear stress.

f_1 and f_2 are called the max. and min. principal stresses and act on planes that are at 90° to each other, called the principal planes. No shear stress acts on these planes. For a two dimensional case, the principal stress, f_3, is zero.

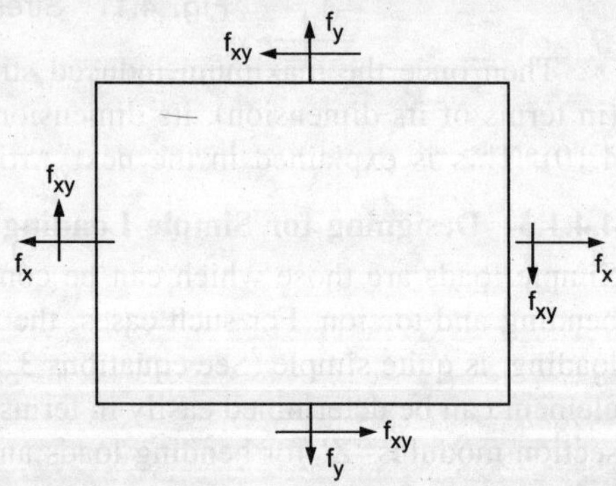

Fig. 4.12 *Two-dimensional stress field*

The maximum shear stress is equal to half of the greatest difference of any two of the three principal stresses (consider also the principal stress which is zero).

Hence

$$f_{s\,(\max)} = \text{the largest of} \begin{cases} \dfrac{|f_1 - f_2|}{2} \\ \dfrac{|f_1|}{2} \\ \dfrac{|f_2|}{2} \end{cases} \quad ...(4.17)$$

Note: If f_{xy} is zero, f_x and f_y are the principal stresses.

2. Three-dimensional Stress Field: In the general three dimensional case, the three principal stresses will be given by the roots of the determinant.

$$\begin{vmatrix} f_x - f & f_{xy} & f_{xz} \\ f_{xy} & f_y - f & f_{yz} \\ f_{xz} & f_{yz} & f_z - f \end{vmatrix} \quad ...(4.18)$$

Maximum shear stress is given as,

$$fs_{(\max)} = \text{the largest of} \begin{cases} \dfrac{|f_1 - f_2|}{2} \\ \dfrac{|f_2 - f_3|}{2} \\ \dfrac{|f_3 - f_1|}{2} \end{cases} \quad ...(4.19)$$

3. Theories of Failure: The purpose of the theories of failure is to predict what combination of principal stresses will result in failure. Many theories of failure have been proposed by different authors, but we shall discuss here only the following three theories which are used most commonly:

(i) **Maximum Normal Stress Theory:** According to the maximum normal stress theory, the yielding at a point occurs only when the maximum principal stress on any plane passing through that point reaches a value equal to the elastic limit as computed in simple tension test, neglecting any other stress that may occur on other planes passing through the point. Thus the criteria for failure is,

$$f_{max} = f_{yp} \quad ...(4.20)$$

(a) If the external static loads produce direct tension or compression stresses f_x, f_y and f_z acting normal to one another, i.e., shear stresses are zero, then, as already discussed, these stresses are the principal stresses. For such a case, failure would occur when any one of these stresses reached the yield stress, the action of the other two stresses having no effect on failure.

(b) In the case of external load producing both direct (tension or compression) and shear stress, the principal stresses can be determined from equations (4.15, 4.16) or equation (4.18) as the case may be.

If we designate the largest of the positive stresses as f_{max} (tensile) and the largest of the negative principal stresses as f_{max} (compressive), then in-corporating a factor of safety, the design criterion (equation 4.20) can be written as:

$$f_{max} \leq \text{Allowable stress}$$

$$\text{i.e.,} \quad f_{max}\ (\text{tensile}) \leq \frac{f_{ypt}}{F_S} \quad ...(4.21)$$

$$\text{and} \quad f_{max}\ (\text{compressive}) \leq \frac{f_{ypc}}{F_S}$$

In the case of static load problem for a material with no distinct yield point, ultimate strength in tension and compression may be used in place of yield point stress. Therefore, the design criterion becomes,

$$f_{max}\ (\text{tensile}) \leq \frac{f_{ut}}{F_S} \quad ...(4.22)$$

$$\text{and} \quad f_{max}\ (\text{compressive}) \leq \frac{f_{uc}}{F_S}$$

(ii) **Maximum Shear Stress Theory:** This theory is based on the assumption that the failure (yielding) occurs when the maximum induced shear stress reaches a value equal to the shear stress at the instant of yielding in a simple tension test. We have already noted that the shear stress at yielding is half the yield stress in tension. Thus, the designing criterion is,

$$f_{s(max)} \le \frac{f_{ypt}}{2F_S} \qquad \text{...(4.23)}$$

It is known that in ductile materials, a shear failure occurs in the standard tension test. If the shear strength of a material exceeds '0.5 f_{ypt}', the maximum shear stress is still limited to (0.5 f_{ypt}/F_S), in order to avoid tensile failure when applying the maximum shear stress theory.

(a) For a three dimensional case, from equation (4.19), the design will be,

$$\text{The largest of} \begin{Bmatrix} |f_1 - f_2| \\ |f_2 - f_3| \\ |f_1 - f_3| \end{Bmatrix} \le \frac{f_{ypt}}{F_S} \qquad \text{...(4.24)}$$

(b) For a two dimensional case, $f_3 = 0$, the design criterion will be,

$$\text{The largest of} \begin{Bmatrix} |f_1| \\ |f_2| \\ |f_1 - f_3| \end{Bmatrix} \le \frac{f_{ypt}}{F_S} \qquad \text{...(4.25)}$$

(b (i)) For the two dimensional case, if the principal stresses f_1 and f_2 are of opposite sign, then the maximum shear stress will lie in the plane of f_1 and f_2. The value of the maximum shear stress will be given as, $1/2(f_1 - f_2)$, so from equations (4.15) and (4.16).

$$f_{s(\max)} = \left[\left(\frac{f_x - f_y}{2}\right)^2 + f_{xy}^2\right]^{1/2} \qquad \text{...(4.26)}$$

(b (ii)) If f_1 and f_2 are of the same sign, then the maximum shear stress will not lie in the plane of f_1 and f_2. Its value will then be given as,

$$f_{s(\max)} = \frac{|f_1|}{2} \text{ or } \frac{|f_2|}{2} \qquad \text{...(4.27)}$$

(b (iii)) If f_y is zero, then the maximum shear is given as,

$$f_{s\,(\max)} = \frac{1}{2}\sqrt{f_x^2 + 4f_{xy}^2} \qquad \text{...(4.28)}$$

(iii) **Distortion Energy Theory:** The difference between the strain energy and the contribution to uniform stress is known as the distortion energy. If it is given as,

$$U_d = \frac{1+m}{mE}\left(f_1^2 + f_2^2 + f_3^2 - f_1 f_2 - f_2 f_3 - f_3 f_1\right) \qquad \text{...(4.29)}$$

where $\frac{1}{m}$ is the Poisson's ratio and E is young's modulus.

The failure criterion based on distortion energy is obtained by comparing distortion energy per unit volume at a point in a machine element with the tension energy per unit volume at failure in the simple tension test. In the tension test.

$$f_1 = f_x = f_{ypt},\ f_2 = f_3 = 0$$

Replacing the yield point stress by the allowable stress, we set $f_1 = (f_{ypt}/F_S)$ in equation (4.29) to obtain the distortion energy for tension. Equating the result to the right hand side of equation (4.29), we obtain the distortion energy theory for three dimensional case as,

$$\left(\frac{f_{ypt}}{F_S}\right)^2 = (f_1^2 + f_2^2 + f_3^2 - f_1f_2 - f_2f_3 - f_3f_1) \quad ...(4.30)$$

For two-dimensional case, $f_3 = 0$, so,

$$\left(\frac{f_{ypt}}{F_S}\right)^2 = f_1^2 + f_2^2 - f_1f_2 \quad ...(4.31)$$

or

$$= f_x^2 + f_y^2 + 3f_{xy}^2 \quad ...(4.32)$$

(iv) **Choice of the Theory of Failure:** Well documented experimental results by various authors on the various theories of failure, indicate that the distortion energy theory predicts results which are always on the safer side. Maximum principal stress theory gives conservative result only if the sign of the two principal stresses is the same (two dimensional case). Therefore, the use of maximum principal stress theory for pure torsion is ruled out where the sign of the two principal stresses are opposite.

When the fracture of a tension specimen loaded upto rupture is examined, it shows that for ductile materials, failure occurs along lines at angles of approximately 45° with the load axis. This indicates a shear failure. Brittle materials on the other hand, rupture on planes normal to the load axis, indicating that the maximum normal stress determines failure.

Because of the above mentioned observation, it is universally accepted that for brittle materials, the maximum normal stress theory is the most suitable. For ductile materials, the maximum shear stress theory as well as the distortion energy theory are applicable since both predict yielding. However, maximum shear stress theory gives conservative results and it is simpler to use as compared to distortion energy theory, so it is universally accepted as the theory of failure for ductile materials. But, where low weight is desired, the distortion energy theory is recommended.

In brief, we can write the design equation for ductile and brittle materials subjected to combined static loads, as given below.

(a) Ductile Materials

(i) Under combined static loading, the machine parts made of ductile material will fail by yielding.

(ii) The working or allowable stress is, therefore, based on the yield point stress.

(iii) The maximum shear stress theory will be used for the design because it is conservative and easy to apply.

(iv) The design equations applicable will be (4.23 to 4.28) as the case may be.

(b) Brittle Materials

(i) Failure in brittle materials, for example, cast iron, takes place by fracture.

(ii) Brittle materials do not have a distinct yield point and so, the ultimate strength is used as the basic for determining the allowable or design stress.

(iii) Separate design equations should be used in tension and compression, since for materials like cast iron, the ultimate compressive strength is considerably greater than the ultimate tensile strength.

(iv) The maximum principal stress theory will be used for the design. Due consideration will be given to the sign of principal stresses.

(v) If both the principal stresses (two dimensional case) are of the same sign, the effect of the smaller stress is neglected and the design equations will be,

$$f_1 \leq \frac{f_{ut}}{F_S}, \text{ in tension} \qquad ...(4.33)$$

and

$$f_1 \leq \frac{f_{uc}}{F_S}, \text{ in compression}$$

(vi) If the two principal stresses are of opposite sign, then the maximum principal stress theory does not give conservative result. In that case, 'Coulomb-Mohr Theory' is used for the design of machine parts made of brittle materials. The design equation will be,

$$\frac{f_1}{f_{ut}} + \frac{f_2}{f_{uc}} = \frac{1}{F_S} \qquad ...(4..34)$$

If f_1 is tensile and f_2 is compressive, then in the above equation, both f_2 and f_{uc} are negative quantities.

4.4.1.5 Effect of Size

As the size of the machine part goes on increasing, its uniformity of structure goes on decreasing. So a machine part with larger section has inferior mechanical properties to those with smaller sections. The values for the elastic limit or yield stress for various materials given in Chapter 2, are obtained from a test specimen of small cross-section, about 12.5 mm in thickness or diameter. With the increase in size, the limit will go on decreasing and to a first approximation, it may be taken as a linear decrease, as shown in Fig. 4.13.

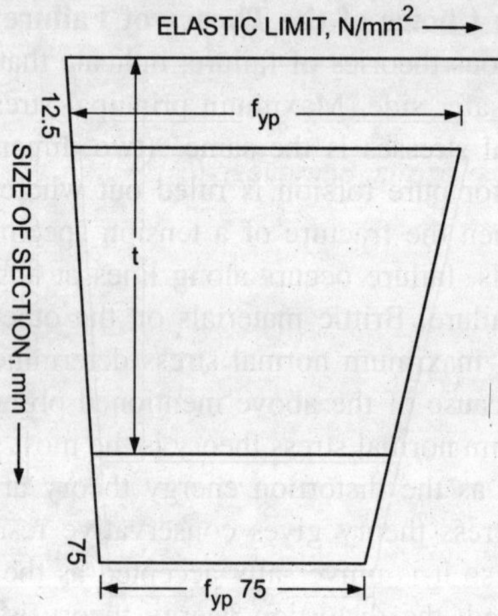

Fig. 4.13 *Change of elastic limit with size of section*

If the yield stress for another section say 75 mm section, is known then the elastic stress for any section 't' between 12.5 mm and 75 mm can be calculated from the relation:

$$f'_{yp} = f_{yp} - \frac{(f_{yp} - f_{yp75})(t - 12.5)}{75 - 12.5}$$

or

$$f'_{yp} = f_{yp} \times E_s$$

where E_s is known as the size coefficient, or factor depending on the section and material of the specimen, and is equal to

$$1 - 0.016\left(1 - \frac{f_{yp}75}{f_{yp}}\right)(t - 12.5)$$

where f'_{yp} = yield stress at any thickness t between 12.5 mm and 75 mm.

f_{yp} = yield stress for 12.5 mm thickness.

and f_{yp} 75 = yield stress for 75 mm thickness.

Based on the tests conducted on specimens of 75 mm diameter of different materials and under different conditions, the size factors are given in Table 4.3.

Table 4.3 Size coefficient (75 mm specimen,) 'E_s'

Material	Coefficient of specimen			
		Annealed and drawn at		
	United	750ºC	575ºC	500ºC
Low carbon steel	0.84			
Medium carbon steel	—	0.85	0.59	0.53
Nickel steel	—	0.86	0.71	
Ni-Cr-Steel	—	0.80	0.70	0.65
Ductile Iron	0.80	0.98	—	—
Wrought Iron	0.55	—	—	—
Cast Iron I.S.15	0.55	—	—	—
Cast Iron I.S. 20	0.73	—	—	—
Cast Iron I.S. 25	0.60	—	—	—
Aluminium	0.93	—	—	—
Bronze	0.90	—	—	—
Monel Metal	0.80	—	—	—

The size coefficient 'E_s' can be related to 'E_s'' as below:

For specimen size of less than 12.5 mm.

$$E_s = 1$$

For specimen size 12.5 mm and 75 mm

$$E_s = 1 - \left[\frac{d-12.5}{75-12.5}(1-E_s')\right], \quad d = \text{Size of the specimen}$$

For $\quad d \geq 75$ mm,

$$E_s = E_s'$$

Beyond 75 mm thickness of section, the decrease in elastic limit is negligible.

Note: The size factor is a refinement which is used only in designing light and heavily stressed machine parts. This refinement is not considered for the design of many ordinary parts.

4.4.2 Impact Loading

Impact or shock loads are those loads which are applied very rapidly to the machine part and which result in stress waves and vibration in the material. It is generally very difficult to determine the exact magnitude of stresses produced by impact or shock. These are usually estimated by multiplying the static stresses with a dynamic or shock factor. The shock factor 'K_i' is defined as,

$$K_i = \frac{\text{Stress produced by shock of suddenly applied load}}{\text{Stress produced by same load applied gradually}}$$

From equation (3.53), Chapter 3, for direct or bending stress, it is given as

$$K_i = 1 + \sqrt{1+\frac{2h}{y}} \qquad \text{...(4.35)}$$

where h = height of free fall to produce the velocity of impact.

y = deformation produced under the same load under static conditions.

For torsional loads, 'h' will be the impact travel and 'y' will be the angular deformation in radians.

For gradually applied loads, 'k_i' = 1.

For suddenly applied load without impact or initial velocity, i.e., when h = 0, the factor k_i = 2.

But since the support is never perfectly rigid, the factor or impact factor may be taken as, equal to 1.5. So the values of shock or impact factor may be taken as,

$$K_i = 1.0, \text{ for gradually applied or steady loads}$$
$$= 1.0 \text{ to } 1.5 \text{ for minor shocks}$$
$$= 1.5 \text{ to } 2.0 \text{ for heavy shocks}$$

Design equation for static loads applied suddenly will be,

$$\text{Dynamic stress} \leq \text{Design stress or Allowable stress}$$

The design or allowable stress will be obtained from the static tests.

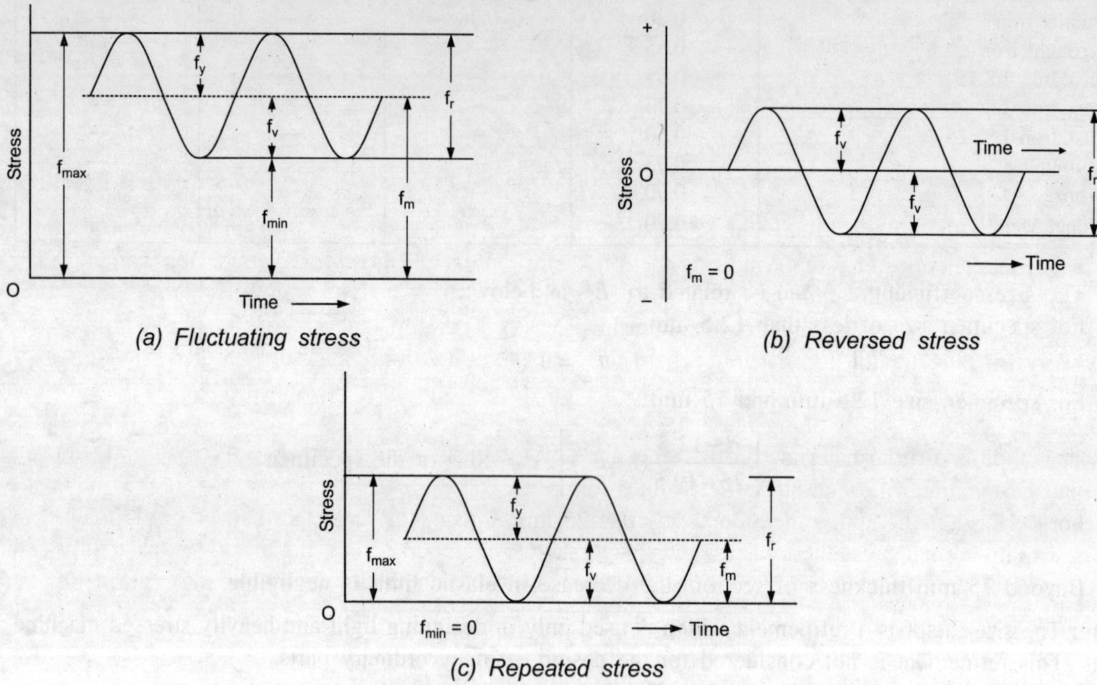

(a) Fluctuating stress (b) Reversed stress (c) Repeated stress

Fig. 4.14 *Types of cyclic loading*

4.4.3 Fatigue Loading

In static loading, the load is applied gradually whose magnitude and direction do not change after the load reaches its maximum value. Also, the static load is applied only once, But in fatigue loading (variable loading or cyclic loading), the load is repeated cyclically. The example can be a rotating shaft subjected to bending loads also in which a fibre on the surface undergoes both tension and compression for each revolution of the shaft.

The various types of cyclic loading are shown in Fig. 4.14. It is clear from the figure that the cyclic variation of the load is above or below a mean or average value. In other words, the loading consists of two components, an average or mean stress and the variable or reversed stress which is superposed upon the former.

Before defining the various types of cyclic loading, we define below the various terms connected with these:

(i) Stress cycle: A stress cycle is the smallest portion of the stress-time plot which is repeated periodically and identically.

(ii) Maximum Stress f_{max}: *It* is the largest algebraic stress in a stress cycle. Tensile stress is taken as positive and compressive stress as negative.

(iii) Minimum Stress, f_{min}: It is the smallest algebraic stress in a stress cycle. Tensile stress is taken as positive and compressive stress as negative.

(iv) Range of stress, f_r: It is the algebraic difference between the maximum and minimum stresses in stress cycle, i.e.,

$$f_r = f_{max} - f_{min} \qquad ...(4.36)$$

(v) Variable Stress, or Stress amplitude or alternating stress, f_v; It is equal to one half of the range of stress, i.e.,

$$f_v = \frac{f_{max} - f_{min}}{2} \qquad ...(4.37)$$

(vi) Mean Stress, f_m: Mean stress or steady component stress is the algebraic mean of the maximum and minimum stress in one stress cycle, i.e.,

$$f_m = \frac{f_{max} + f_{min}}{2} \qquad ...(4.38)$$

(vii) Stress Ratio: It is defined as the algebraic ratio of the minimum stress to the maximum stress in a stress cycle, i.e.,

Stress ratio, $$r = \frac{f_{min}}{f_{max}} \qquad ...(4.39)$$

(viii) Amplitude ratio, $$A = \frac{f_v}{f_m} = \frac{1-r}{1+r} \qquad ...(4.39a)$$

Now, we define below the various types of cyclic or variable loading:

(i) **Fluctuating Loading, Fig. 4.14 (a):** In this type of loading, there are both mean (f_m) and variable (f_v) components of stress. A completely reversed variable stress, f_v, is superimposed on the steady or mean stress, f_m. Figure 4.14 (a) shows a reversed stress superimposed on a tensile stress. There can also be a case of reverse stress component superimposed on compressive stress.

(ii) **Reversed Loading, Fig. 4.14 (b):** In this type of loading, the mean stress, f_m, is Zero. The variable stress reverses completely from tension to an equal value in compression, or from clockwise shear to an equal value of shear in the opposite direction.

(iii) **Repeated Loading, Fig. 4.14 (c):** In this type of loading, $f_{min} = 0$, so that the stress varies from maximum to zero in each cycle. Figure shows the case when the stress is tensile. There can be a similar case when the stress is compressive.

It is well known fact that in the case of cyclic loading, the fracture of the machine part occurs at stresses very much lower than their ultimate statical strength. Also, if the stresses are not merely repeated but reversed, that is, the material is subjected to repeated stresses of opposite signs, the resistance to fracture is less than if the same intensity of only one kind of stress were repeated. In all such cases, the material is often said to have become "fatigued" and this type of failure is known as "fatigue failure".

A fatigue failure usually has its origin in some surface of internal imperfections in the from of minute inclusions. These imperfections can be due to: manufacturing methods, heat treatment, environment, handling, raw material production and surface coating, etc. Localised high stresses are induced in the material because of these discontinuities. Under cyclic loading micro flaws or small cracks begin to appear at these locations. These cracks are continually "closed". These cracks tend to extend across the section if the cyclic loading is repeated a great many times, and a stage will be reached when the net sound section is reduced so much that fracture takes place. So fatigue failure is better known as "gradual fracture" or "progressive fracture". Fatigue failures are quite damaging because such failures occur with no warning.

4.4.3.1 Endurance Strength, f_f

The endurance strength or fatigue strength is the maximum, completely reversed stress under which a material will fail after it has experienced the stress for a specified number of cycles. Thus when a value for the endurance strength of a material is stated, it must be accompanied by the number of stress cycles.

4.4.3.2 Endurance Limit, f_e^*:

The endurance limit or fatigue limit is the maximum completely reversed stress for which it is assumed a material will never fail regardless of the number of stress cycles. Thus, the endurance limit is stated without an associated number of cycles to failure. From experience and for practical reasons, the number of stress cycles may be taken as 10^7 to define the endurance limit for ferrous materials. As will be discussed ahead, there is no endurance limit for non-ferrous materials.

The endurance limit of a material is determined by the rotating beam method, in which the test specimen is rotated while it is loaded in bending. A point on the outer fibre of the specimen is stressed alternately by equal amount of compressive and tension, i.e., a completely reversed stress cycle. In this a constant bending load is applied and the number of revolutions (stress reversals) of the beam required for failure is recorded. The first test is made at a stress which is some what below the ultimate strength of the material. The second test is made with a stress which is less than that used in the first. This process is continued and results plotted on log-log paper with endurance strength or fatigue strength on the vertical ordinate and number of cycles on the horizontal ordinate (S-N curves), Typical S-N curves for a ferrous and a non-ferrous material are shown in Fig. 4.15. In the case of steel, a knee occurs in the graph and beyond this knee, failure will never occur, no matter how great the number of cycles. The endurance strength corresponding to the knee is called the 'Endurance Limit' or 'Fatigue Limit'. For non-ferrous metals such as brass, copper alloys, aluminium and magnesium, the curve becomes horizontal and hence these materials do not have an endurance limit. Therefore, only the endurance strength can be established for these materials.

When no specific endurance limit information is available the following relationships between the endurance limit and tensile strength for untouched and polished specimens may be used:

$f_e^* = 0.5.\ f_{ut}$ for wrought steel where $f_{ut} < 140$ N/mm^2 and BHN < 400.

$= 700$ N/mm^2 for wrought steel, where $f_{ut} > 140$ N/mm^2.

$= 0.4\ f_{ut}$ for cast steel and CI.

$= 0.38\ f_{ut}$ for magnesium cast alloys and magnesium wrought alloys (based on 10^6cycle life).

$= 0.45\ f_{ut}$ for Nickle based alloys and copper based alloys

$= 0.38\ f_{ut}$ for wrought Aluminium alloys upto tensile strength of 280 N/mm^2 (based on 5×10^8 cycle life).

$= 0.16\ f_{ut}$ for cast Aluminium alloys upto a tensile strength of 350 N/mm^2 (based on 5×10^8 cycle life).

When the stress is not completely reversed, but varies from maximum either in tension or compression to lower value in either tension or compression or to a zero stress, the endurance limit will be higher than obtained with complete stress reversals. For any stress cycle, the relationship is,

$$f_e' = \frac{3f_e^*}{2-r}, \text{ for wrought ferrous materials} \qquad ...(4.40)$$

where f_e' gives the endurance limit for any type of stress cycle and r is the stress ratio $\left(\frac{f_{min}}{f_{max}}\right)$.

For cast iron, when the tension is numerically the maximum stress in any stress cycle, the endurance limit may be obtained from equation (4.40). However, when compression is numerically the greater stress, then,

$$f_e' = \frac{6f_e^*}{1-5r} \quad ...(4.41)$$

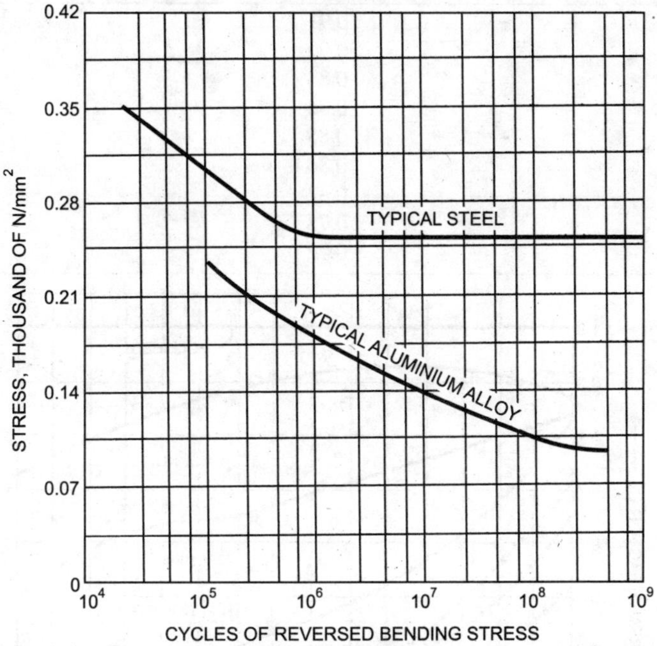

Fig. 4.15 *S–N Curves*

When the torsional stress is not completely reversed, the endurance limit is given by the formula:

$$f_{es}' = \frac{2f^*_{es}}{1-r} \quad ...(4.42).$$

where f_{es}' is the endurance limit for completely reversed torsion stress.

(a) Modifying factors for Endurance Limit:

The endurance limit of a material depends upon many factors, some of these being given below:

1. Manufacturing methods and heat treatment.
2. Surface condition.
3. Size of section
4. Reliability factor
5. Temperature factor
6. Discontinuities in the section.
7. Types of stress,
8. Amplitude of stress variation.
9. Mean value of stress.

The first two factors are inter-related. We shall discuss below, these factors one by one:

1. Surface Condition: The surface of the rotating beam specimen is highly polished. However, most machine elements do not have such a high quality finish. So, the endurance limit found experimentally (rotating beam method) should be modified accordingly. The modification factors 'C_F' are given in Fig. 4.16, for steels only.

The surface finish factors for non-ferrous metals should be taken as unity, because, the tabulated endurance limit of these materials include the surface finish effect. The values for 'C_F' are also given in Table 4.4.

Table 4.4 Values for 'C_F'

f_u MN/m^2	Machined Surface	Hot Rolled Surface
420	0.91	0.72
490	0.90	0.68
560	0.88	0.62
630	0.86	0.58
700	0.85	0.55
770	0.84	0.50
840	0.82	0.48
1050	0.78	0.38
1400	0.72	0.30

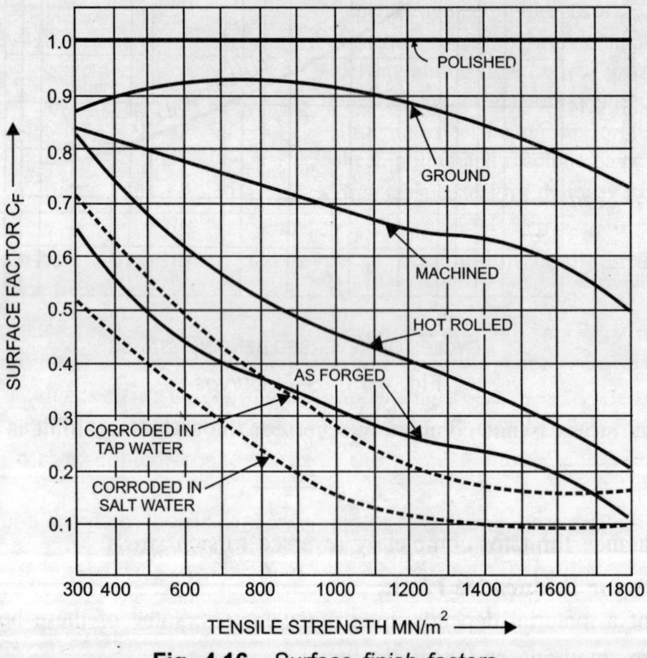

Fig. 4.16 *Surface finish factors*

2. Size Factor: The standard test specimen used in rotating beam method has a nominal diameter of 0.3 inch (about 7.6 mm). Therefore, for machine elements of various sizes, a size correction factor 'C_s' should be applied. The value of factor may be taken as given below:

$$C_s = 1.0,\ d \le 7.6 \text{ mm}$$
$$= 0.85,\ 7.6 \le d \le 50 \text{ mm}$$
$$= 0.75,\ d \ge 50 \text{ mm}$$

The dimension 'd' corresponds to the section depth for non-round sections, in bending.

3. Load Factor: The factor 'C_L' is a correction factor for the type of loading other than reversed bending. The values given in the table for endurance limit are for completely reversed stresses. But many machine members undergo different stress cycles than the completely reversed stress. Therefore, the endurance limit should be modified accordingly by incorporating the load factor 'C_L', whose values may be taken as given below:

$$C_L = 1.0, \text{ Reversed bending load}$$
$$= 0.85, \text{ Reversed axial loading (steels)}$$
$$= (0.70 \text{ to } 1.0), \text{ Reversed axial loading, non-ferrous metals.}$$
$$= 1.0, \text{ Reversed axial loading, C.I.}$$
$$= 0.58, \text{ Reversed torsional (steels)}$$
$$= 0.80, \text{ Reversed torsional (C.I.)}$$
$$= 0.48, \text{ Reversed torsional, copper,}$$

4. Reliability Factor, C_R: It has been mentioned ahead (Art. 4.4.3.5) that the tabulated values of fatigue strengths are at 50% survival or reliability, for which reliability, factor is taken as unity. The reliability factor for various percentages of reliability is given below:

R	C_R
50	1
90	0.897
95	0.868
99	0.814
99.9	0.783

5. Temperature Factor: It has been written ahead (Art. 4.4.3.5) that the fatigue tests are carried out at room temperature for which the value of C_T is taken as unity. The factor, C_T, is taken as:

$$C_T = 1 \quad \text{for } T \le 300°\text{C}$$
$$= 0.5 \quad \text{for } T > 300°\text{C}.$$

6. Working Equation for Endurance Limit: Considering the above three factors (1, 2, 3), the working equation for endurance limit can be written as below:

$$f_e = C_F\, C_s\, C_l f_e^* \qquad \text{...(4.43)}$$

The effect of discontinuities in the section (stress concentration factor), the mean stress and the variable stress will be discussed in the next article.

In equation 4.43, the reliability factor and the temperature factor are both taken as equal to unity. However, for other design condition, the factor C_R and C_T can be incorporated in this equation. The effect of stress-concentration can also be considered by including a factor C_e in equation 4.43, which is $= \dfrac{1}{k_f}$

where k_f = Fatigue stress concentration factor. With all this the equation 4.43, can be written in the generalised form as:

$$f_e = f_e^* . C_F . C_s . C_l . C_R . C_T . C_e \qquad \text{...(4.43}a\text{)}$$

4.4.3.3 Designing for Fatigue Loading

The problem of fatigue loading was discussed under Article 4.4.3. The design of a machine element under fatigue loading will depend upon the type of stress cycle.

(a) **Completely Reversed Stress Cycle, Fig. 4.14b.** Here, the mean stress is zero and endurance limit stress is used for determining the failure criterion. The maximum stress induced in the material will be the variable stress, f_v. However, the calculated variable stress should be increased by the fatigue stress concentration factor, since stress concentration is very important in fatigue loading.

The design equation will be,

$$K_f f_v \le \text{Design stress}$$

or

$$f_v \le \frac{f_e}{K_f \times FS} \qquad \text{...(4.44)}$$

A similar equation can be written for completely reversed shearing stress cycle.

(b) **Stress Cycles other than completely Reversed Stress Cycle:** Here, the mean stress is not zero and both variable stress and the mean stress contribute to failure. The results of variable loading made with different and with various combinations of mean and variable stresses are shown in Fig. 4.17 with the failure points marked thereupon. This is sometimes known as Haig-Soderberg diagram. The parabolic failure curve was proposed by Gerber, while the straight line by Goodman.

Test data generally falls closer to the parabolic curve. However, due to the scatter in the points, the straight line relationship is usually preferred in engineering design. It is apparent that the failure points are not much dependent upon the mean stress when it is compressive, but it is very much related to mean or average stress, when it is tensile. In other words, it means that failure due to fatigue is not so common when the mean stress (static stress) is compressive. So, the more important problem is the combination of steady tensile stress and the reversed or the variable stress.

1. **Soderberg Criterion:** In Fig. 4.17, the failure is not very explicit. However, according to Soderberg criterion, the failure line is taken as the straight line joining the endurance limit, f_e and the yield strength, f_{yp} as shown in the figure. This failure line is termed as 'Soderberg line' and it automatically preclude the possibility of failure by yielding. Here the effect of strain hardening is neglected. A typical Soderberg triangle is shown in Fig. 4.18. Line *AB* connecting f_e and f_{yp} is the failure stress line. By applying a suitable factor of safety to both endurance limit and yield strength, the line *CD* can be drawn parallel to the line *AB*. The line *CD* may be considered as safe stress line. Consider a design point '*E*'. From similar triangles *OAB* and *CEF*,

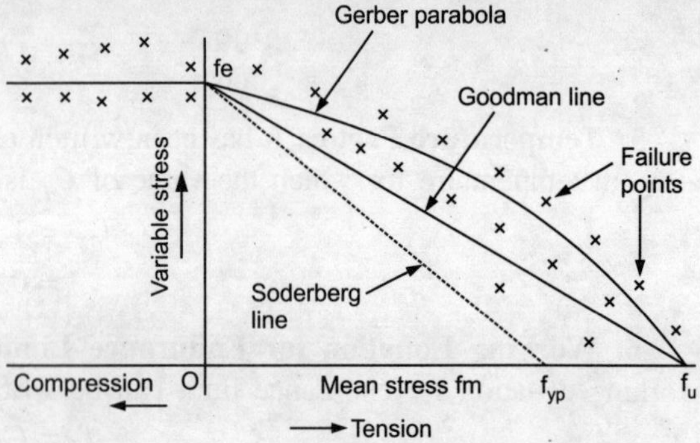

Fig. 4.17 *Results of variable loading*

$$\frac{FE}{FC} = \frac{OB}{OA}$$

or

$$\frac{FE}{OC - OF} = \frac{OB}{OA}$$

$$\therefore \quad \frac{f_v}{\frac{f_{yp}}{FS} - f_m} = \frac{f_e}{f_{yp}}$$

This equation may be written as

$$\frac{1}{F_S} = \frac{f_m}{f_{yp}} + \frac{f_v}{f_e} \quad \text{...(4.45)}$$

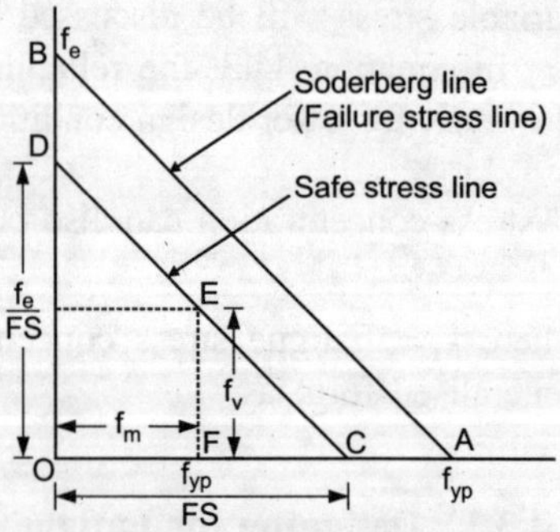

Fig. 4.18 *Soderberg triangle*

As already discussed, suitable stress concentration factor must be applied under fatigue loading. For ductile materials, stress concentration may be ignored under steady loads. Therefore, in problems involving fatigue loading for ductile materials, the stress concentration factor should be applied to only the reversed component of stress (Soderberg criterion is mainly applicable for ductile materials since it precluded the possibility of failure by yielding). Therefore, the equation (4.45) may be modified as:

$$\frac{1}{F_S} = \frac{f_m}{f_{yp}} + \frac{K_f \times f_v}{f_e} \quad \text{...(4.46)}$$

for ductile materials in tension and compression

and

$$\frac{1}{F_S} = \frac{f_{ms}}{f_{yps}} + \frac{K_f \times f_{vs}}{f_e} \quad \text{...(4.47)}$$

for ductile materials in torsion.

where, f_{yp} is the yield point stress in tension or compression and must be given the same sign as the mean stress, f_m

F_S = 1.25 to 3.0 for ordinary design and > 3, if the consequences of failure are serious.

Soderberg criterion is particularly applicable to ductile materials. However, for its application, the yield point stress must be known. If for a given material, there is no distinct yield point, the yield strength based on the offset method be used.

(ii) Goodman Criterion; According to goodman criterion, the failure stress line connects the endurance limit and the ultimate strength, Fig. 4.17. Since the 'Goodman line' connects the endurance limit and the ultimate strength of a material, it can be used for any type of material (Brittle or Ductile). However, for brittle materials, the stress concentration factors should be applied to both mean stress and variable stress, K_t to the mean stress and K_f to the variable stress. Therefore, the design equations, according to Goodman criterion, will be:

Ductile Materials

$$\frac{1}{F_S} = \frac{f_m}{f_u} + \frac{K_f \times f_v}{f_e}, \text{ in tension and compression} \qquad ...(4.48)$$

and

$$\frac{1}{F_S} = \frac{f_{ms}}{f_{us}} + \frac{K_f \times f_{vs}}{f_e}, \text{ in torsion} \qquad ...(4.49)$$

where f_u is the ultimate strength in tension or compression and must be given the same sign as the mean stress, f_m.

Brittle Materials

$$\frac{1}{F_S} = \frac{K_t \times f_m}{f_u} + \frac{K_f \times f_v}{f_e} \text{ in tension or compression} \qquad ...(4.50)$$

and

$$\frac{1}{F_S} = \frac{K_t \times f_{ms}}{f_{u_s}} + \frac{K_f \times f_{vs}}{f_e}, \text{ in torsion} \qquad ...(4.51)$$

Note: It should be noted that in equations (4.48 to 4.51), we have used the ultimate strength instead of yield point stress, therefore, in these equations, the factor of safety should be increased as compared to the value used in equations (4.46 and 4.47).

Note: It should be noted that steady or static stress is not the same as the mean stress. In fact, it may have any value between 'f_{max}' and 'f_{min}'. The steady stress exists because of a fixed load or preload applied to the part, and it is usually independent of the varying portion of the load. A helical compression spring, e.g., is always loaded into a space shorter than the free length of the spring. The stress created by this initial compression is called the steady or static component of stress.

Note: For fatigue loading, brittle materials are usually not used.

4.4.3.4 Theories of Failures Applied to Fatigue Loading

When there are both a variable normal stress and a variable shear stress, the design equations can be derived by the theory of combined stresses. From equation 4.46, we have,

$$\frac{f_{yp}}{F_S} = f_m + K_f f_v \frac{f_{yp}}{f_e} \qquad ...(4.52)$$

The right hand side of equation (4.52) may be considered as the static equivalent of the state of stress, Thus, we define equivalent normal stress, f_{ne} as,

$$f_{ne} = f_m + K_f f_v \frac{f_{yp}}{f_e} \qquad ...(4.53)$$

Similarly, from equation 4.47, we can define the equivalent shear stress as,

$$f_{se} = f_{ms} + K_f f_{vs} \frac{f_{yps}}{f_e} \qquad ...(4.54)$$

In the above equation, the value of the yield point in shear, f_{yps} may be taken as 0.6 times the yield point in tension. This is in close agreement with experimental torsional shear stress.

Now, with the help of equations (4.53 and 4.54), we can apply the theories of failure to fatigue loading.

(a) Maximum Shear Stress Theory: The maximum equivalent shear stress, $f_{se(\max)}$, in designing with ductile materials is (see equations 4.28),

$$f_{se}(\max) = \frac{1}{2}\sqrt{f_{ne}^2 + 4f_{se}^2} \qquad ...(4.55)$$

The design equation will be (see equation 4.23).

$$\frac{f_{yp}}{2F_S} = \frac{1}{2}\sqrt{f_{ne}^2 + 4f_{se}^2}$$

or

$$\frac{f_{yp}}{F_S} = \left[\left(f_m + K_f \times f_v \times \frac{f_{yp}}{f_e}\right)^2 + 4\left(f_{ms} + K_f \times f_{vs} \times \frac{f_{yps}}{f_e}\right)^2\right]^{\frac{1}{2}} \qquad ...(4.56)$$

(b) Maximum Normal Stress Theory: The equivalent maximum normal stress to be used when designing with brittle materials, is given as (see equation 4.15),

$$f_{ne(\max)} = \frac{1}{2}f_{ne} + \frac{1}{2}\sqrt{f_{ne}^2 + 4f_{se}^2}$$

∴ The design equation is,

$$f_{ne(\max)} = \frac{f_{yp}}{F_S} = \frac{1}{2}\left[f_{ne} + \sqrt{f_{ne}^2 + 4f_{se}^2}\right] \qquad ...(4.57)$$

(c) Distortion Energy Theory: For a two dimensional case, the design equation will be (see equation 4.31),

$$\frac{f_{yp}}{F_S} = \sqrt{(f_1^2 + f_2^2 - f_1 f_2)} \qquad ...(4.58)$$

Here, for each principal stress, the equivalent value will be substituted.

For the special case, where $f_y = f_z = f_{xz} = f_{yz} = 0$, We have (from equation 4.32),

$$\frac{f_{yp}}{F_S} = \sqrt{(f_x^2 + 3f_{xy}^2)} \qquad ...(4.59)$$

Substituting equivalent normal stress for 'f_x' and equivalent shear stress for f_{xy}, we have,

$$\frac{f_{yp}}{F_S} = \left[\left(f_m + K_f f_v \frac{f_{yp}}{f_e}\right)^2 + 3\left(f_{ms} + K_f f_{vs} \frac{f_{yps}}{f_e}\right)^2\right]^{\frac{1}{2}} \qquad ...(4.60)$$

4.4.3.5 Note

S–N curve (Fig. 4.15) is a plot of stress versus number of cycles to failure. A log scale is almost always used for number of cycles. The stresses are usually nominal, that is, no adjustment is made for stress concentration. The tests to determine endurance limit/strength are at room temperature and for frequencies in the range 200 to 10,000 cycles per min.

High-cycle fatigue, $N > 10^5$

Low-cycle fatigue, $N < 10^4$ or 10^5.

The scatter of failure points (Fig. 4.17) signifies the statistical nature of fatigue failures. At least 10 samples are tested for fatigue tests on a material. The values of fatigue strengths, f_f at N number of cycles, found in the literature, is at 50% survival at N cycles for completely reversed cycle. Similarly, tabulated values of f_e* are frequently (but not always) values for 50% survival for infinite life, for completely reversed cycle.

Fatigue Design Criteria

1. Infinite Life Design: Fatigue design procedure discussed so far in this chapter is for infinite life. This is the oldest design criteria and is valid when the component is subjected to a very large number of cycles of uniform stress. Here, the failure stress is taken as the "endurance limit" of the material.

2. Finite Life Design or Safe-life Design: The component is initially defect free and it is assumed that the component will develop a critical crack after a finite life. This approach is based on the statistical nature of the fatigue failure. The failure stress is the fatigue strength at the finite life (or at the number of cycles desired), and not the fatigue limit. The mechanical components designed for finite life are: Anti-friction bearings, for which Basic Load rating is specified for Rated Life or Average Life (See Chapter 17); Design of Pressure Vessels; Design of Bridges; Design of Jet Engines; Design of Airforce Aircraft, where the safe life is taken as (1/4) of the life demonstrated in full-scale fatigue tests. This is done to account for material properties variations, variations in as manufactured quality and environmental effects (the stress cycles are normally not uniform. They are random and irregular, e.g., aircraft air wings which are subjected to periodic unpredictable overloads due to gusts).

Design Procedure for Finite Life: Let us do it for N cycles. On a log-log plot of f_f and cycles to failure, join the fatigue strength points at 10^3 cycles and at infinite life, by a straight line. Read the ordinate corresponding to N cycles to give the fatigue strength at N cycles.

Mathematically: f_f at $N = 10^3$ cycles

Type of Load	f_f
Bending	(0.8 to 0.9) fut
Axial	0.75 fut
Torsion	0.72 fut for ferrous materials 0.63 fut for non-ferrous materials

The effect of size and surface are neglected. Also, at $N = 10^3$ cycles, the effect of K_f is neglected.

We know that at $N = 10^6$ (infinite life).

$$f_f = f_e^*$$

Figure 4.19 Shows the fatigue curve. The line joining point A, for

$N = 10^3$ cycles and point B for

$N = 10^6$ cycles will be a straight line on $f - N$ log-log graph.

f_{fN} = Fatigue strength at N cycles.

The equation of the straight line AB can be written as,

$$\log_{10} f_f = b \log_{10} N + c$$

where 'b' and 'c' are constants.

Now at $N = 10^3$, $f_f = 0.85 f_u$ (Average Value), and

at $N = 10^6$, $f_f = f_e$

With these boundary conditions,

$$b = -\frac{1}{3}\log_{10}\left(\frac{0.85\, f_u}{f_e}\right) \quad \text{and} \quad c = \log_{10}\left[\frac{(0.85\, f_u)^2}{f_e}\right]$$

$\therefore$ The equations become,

$$f_f = 10^c \cdot N^b$$

or

$$N = 10^{-c/b} \cdot f_f^{1/b}$$

3. Fail-safe Design: Under this approach, it is presumed that cracks may occur in a component during its service. However, the component is designed so that the cracks do not grow large enough leading to failure, before they are detected and repaired. This approach has been developed in the Aircraft industry, where neither a large F.O.S. can be tolerated due to weight consideration nor a small F.O.S., because of danger to life. Under this approach, multiple-load paths and crack stoppers are built into the structure. At the same time, rigid regulations and criteria for inspection and detection of cracks are strictly followed.

4. Damage Tolerant Approach: This is the latest approach to fatigue design and is an extension of the previous approach. It is certain that cracks will appear in a component during its service. The theories of fracture mechanics are employed to determine whether the cracks will grow large enough to cause failure before they are detected during periodic inspection. The materials employed in the design are with high fracture toughness and slow crack growth. The success of this approach will depend upon a reliable Non-Destructive Evaluation and in being able to identify the damage-critical areas in the design.

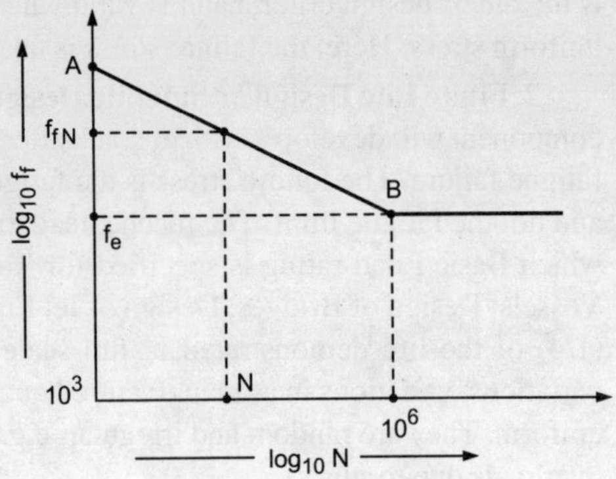

Fig. 4.19 *Fatigue curve*

4.4.3.6 Commulative Fatigue Damage

In the usual fatigue testing methods, the stress cycle is considered to be regular (i.e., varies sinusoidally) and the stress amplitude (f_v) to be constant until the component fails. However, in many practical situations, both these conditions are not met. The stress cycle may be irregular and the stress amplitude is not constant. Again, the component may be subjected to one stress level for a certain number of cycles and to another stress level for a different number of cycles. The damage to the component under the above mentioned types of fatigue loading is termed as "Commulative Fatigue Damage". There have been many approaches to determine the total life of a component subjected to Commulative Fatigue Damage. However, a common approach is "Miner's rule". It is a "linear commulative fatigue damage rule". According to this rule, the total life of a component can be estimated by adding up the percentage of life consumed by each overstress cycle (overstressing is the process of testing a virgin specimen for some number of cycles less than failure at a stress above the endurance limit and subsequently running the specimen to failure at another test stress).

If $Z_1, Z_2, ..., Z_k$ represent the number of cycles of operations at specific overstress levels and $N_1, N_2, ..., N_k$ represent the life (in cycles) at these same overstress levels, then, as per "Miner's rule",

$$\frac{Z_1}{N_1} + \frac{Z_2}{N_2} + ... + \frac{Z_k}{N_k} = 1 \quad \text{or} \quad \sum_{j=1}^{j=k} \frac{Z_j}{N_j} = 1.$$

4.5 FACTOR OF SAFETY

It has been pointed out that to avoid failure of a machine part, the design stress is kept well within the yield stress or the ultimate stress. According to the older definition of factor of safety, it is a totally empirical number used to reduce the ultimate strength or the yield strength of a material so as to obtain a value of design stress which would prevent the failure of the machine part. It used to be a simple number which was varied with application and depended on the experience of part failures for which a clear explanation was not available. So, it used to be called as "Factor of ignorance" also.

It also takes into account the other uncertainties which the designer has to face, e.g., variations in material properties, Effect of size, Type of loading, effect of forming processes, effect of heat treatment on material properties, natural properties, effects of wear, life and reliability requirements and overall concern for human life.

The present definition which is more logical is that the factor of safety is the amount by which the design stress is kept below the limit stress and is expressed by their ratio (see equation 4.7). The limit stress or the failure stress is the maximum stress to which a machine part can be subjected without being fractured or damaged.

For ductile materials subjected to static loading, the limit stress or the failure stress is the yield stress, therefore.

$$\text{Factor of safety} = \frac{\text{yield stress}}{\text{design stress}}$$

or

$$F_S = \frac{f_{yp}}{f_a} \qquad ...(4.61)$$

For brittle materials, the limit stress or the failure stress is the ultimate stress of the material, therefore;

$$\text{Factor of safety} = \frac{\text{ultimate stress}}{\text{design stress}}$$

or

$$F_S = \frac{f_u}{f_a} \qquad ...(4.62)$$

For materials, subjected to varying load, the limit stress or the failure stress is the endurance limit, therefore,

$$\text{Factor or safety} = \frac{\text{Endurance limit}}{\text{design stress}} \text{ for completely reversed stress cycle.}$$

or

$$F_S = \frac{f_e}{f_a} \qquad ...(4.63)$$

For steel, $f_e = (0.8 \text{ to } 0.9) \times f_{yp}$ ...(4.64)

where f_e = endurance limit for completely reversed stress.

However, for stress cycle other than completely reversed cycle, the design will be based either on Soderberg or Goodman criterion as discussed earlier.

It should be noted that for elastically stable machine members, the factor of safety is used to get a value of allowable stress but for elastically unstable members, it is applied to the load. The buckling, critical or collapsing load being the product of the factor of safety and the working load.

The following factors of safety, based on yield strength, are considered to be reasonable for machine design:

(i) F_S = 1.25–1.5, for exceptionally reliable materials used under controlled conditions and subjected to loads and stresses that can be determined certainly.

(ii) F_S = 1.5–2, for well known materials, under reasonable constant environmental conditions, subjected to loads and stresses that can be determined readily.

(iii) F_S = 2–2.5, for average materials operated in ordinary environments and subjected to loads and stresses that can be determined.

(iv) F_S = 2.5–3, for less tried materials under average conditions of environment, load and stress.

(v) F_S = 3–4, for untried materials used under average condition of environment, load and stress.

(vi) F_S = 3–4, for better known materials used under uncertain environments and subjected to uncertain stresses.

The above values for factor of safety are for static loading. For other types of loading, the values may be taken as given below:

(vii) **Impact Loading:** The values given in items (i to vi) are acceptable, but an impact factor must be included.

(viii) **Fatigue Loading:** The values given in items (i to vi) are acceptable, but must be applied to endurance limit rather than the yield strength of the material.

Note: For brittle materials, where ultimate strength is used as the criterion, the factors given in items (i to vi) should be approximately doubled.

4.5.1 Evaluation of Factor of Safety

As already discussed, if the allowable stress is based on the failure or limit stress, then, the factor of safety is defined as the quotient of the failure stress divided by the allowable stress. Such a factor of safety is known as "Real factor of safety". However, if the factor of safety is based on the ultimate strength, then it is known as "Apparent factor of safety" and is equal to ultimate strength, divided by the allowable stress. Even though the "Real factor of safety" is the more logical one, the "Apparent factor of safety" is universally used. It may be evaluated as given below:

(a) **Static loading:**

The factor of safety may be taken as the product of two factors a and R, i.e., $F_S = a \times R$.

Where a = elastic ratio and is used to reduce the allowable stress to a value below the elastic limit, i.e. it is the ratio of ultimate strength to the elastic limit, usually taken as the yield stress.

The factor 'R' is known as the "Reliability factor" to account for the unknown conditions and contingencies under which the machine part must work. Its value is affected by the following considerations and uncertainties:

1. Degree of reliability of the material.
2. Effect of initial stresses due to casting, cold working, welding, heat treating or assembly.
3. Degree of certainty about the magnitude and line of action of the applied forces.
4. Reasonability and safety of the assumptions made in the stress analysis.
5. Overloads, which might be unforeseen, unknown or accidental.
6. Variation of load with wear and deformation.
7. Temperature effects.

To account for the above factors, the "Reliability factor" is used to reduce the test values of yield point, ultimate strength or the endurance limit.

The recommended values of R are:

R = 1, When the forces are determined to the best accuracy and the material is of the best specification and the production is to the close dimensional control and where weight is at a premium, e.g., design of air craft and space vehicles,

= 1.1 to 1.5 for reliable ductile materials of uniform structure, where loads are predictable and low weight is important,

= 1.5 to 2, for the condition same as above but where weight is not critical

= 2 to 3 for ordinary materials, where overloads are possible and reliability is important.

= 3 for C.I. and other brittle materials

= 3 to 4 for timber.

Consideration of Stress Concentration

As already mentioned, under static loads (applied gradually), the failure load is not markedly affected by stress concentration for ductile materials, since the material can yield locally in the overstressed region and the stress is thereby relieved. Therefore, it is customary and permissible to neglect the effect of stress concentration in the design of machine members made of a ductile material and subjected to a gradually applied static load. Brittle materials are unable to yield locally at point of high stress and so the stress concentration is taken into account even when the loading is steady.

(i) Ductile Materials:

$$f_a = \frac{f_u}{a \times R}$$

or $$f_a = \frac{f_{yp}}{R}, \text{ in tension and compression}$$

and $$f_a = \frac{f_{yps}}{R}, \text{ in pure shear}$$

(ii) **Brittle Materials:**

$$f_a = \frac{f_{ut}}{aRK}, \text{ in tension}$$

and $$f_a = \frac{f_{uc}}{RaK}, \text{ in compression}$$

(b) Suddenly applied static load (Impact or shock loading)

$$F_S = a \times R \times K_i \times K$$

where K_i = shock factor or impact factor (see Art. 4.4.2)

So, for ductile materials,

$$f_a = \frac{f_u}{F_S} = \frac{f_u}{aK_iRK}$$

or $$f_a = \frac{f_{yp}}{K_iRK}$$

For brittle materials, $$f_a = \frac{f_u}{aK_iRK}$$

(c) Fatigue Loading:

$$FS = a\ K_i\ R\ K_f$$

For static or steady load conditions, the finish has no or little effect on the strength and so the surface finish factor, C_F, is usually taken as unity. But for fatigue loading, the surface finish is very important since the presence of surface defects such as tool marks, file marks, mill scale notches and grooves greatly affect the fatigue strength of the materials. The values for the factor C_F are given in Fig. 4.16.

$\therefore$ $$f_a = \frac{f_u}{aK_iRK_f}$$

or $$f_a = \frac{f_e}{K_iRK_f}$$

where $$a = \frac{f_u}{f_e} \text{ in fatigue loading}$$

The effect of 'C_F' is included while determining 'f_e' (see equation 4.43).

4.5.2 Simplified Method of Determining Factor of Safety

In practice, the factor of safety is usually determined in the following simplified manner:

$$F_S = a\ K_i \times R$$

The above factors have been explained earlier. The factor 'a' gives the elastic limit or yield point, if the load is steady and the fatigue strength, if the load varies, i.e.,

$$a = \frac{f_u}{f_{yp}} \text{ for steady load } = \frac{f_u}{f_e} \text{ for varying load}$$

(i) For Ductile materials, not treated, the elastic limit is about 50 to 60 per cent of the ultimate strength. Cast iron has no real elastic limit, but the stress strain curve approximates a straight line up to one-third or one-half of the ultimate strength. Therefore, the value of 'a' for steady load may be taken as below:

a = 2 for ductile materials.
= 2 to 3 for C.I. and brittle materials.
= 3 for timber.

For other materials and other kind of steels, the designer should refer to the tables where the elastic limits of such materials are listed.

(ii) For varying load, it may be assumed that the endurance limit is half the ultimate strength, both for stress of one kind and reversed stresses. Therefore, the factor 'a' may be taken as 2. For more conservative results, the values given below may be taken:

Ductile materials: 1.8–2.0 for one directional varying load.
3.0 for reversed loading.
Brittle materials; 2.5 for one directional varying load.
4.0 of reversed loading.

(iii) The factor K_i is already defined and it is known as the shock factor. It depends upon the manner in which the load is applied.

K_i = 1, for steady or very gradually applied load.
= 2, for absolutely suddenly applied load.

The intermediate conditions, e.g., the rise of pressure in engines or surging of a crowd on a bridge may be allowed for according to designer's judgement.

(iv) The factor 'R' is known as reliability factor or the real margin of safety. It provides against such unknown factors, as defective or unreliable materials, poor workmanship, careless maintenance, and unexpected overloads. Values for the factor 'R' have already been listed.

4.5.3 Values for Factor of Safety

The most commonly used values for factor of safety based on the ultimate strength are listed in Table 4.5.

Table 4.5 Factors of safety

Material	Kind of Load				
		Repeated stress in one direction		Repeated stress reversed	
	Steady Stress	Gradually applied	Shock	Gradually applied	Shock
1. Brittle metal C.I., and alloys	4	6	12	8	14
2. Ductile metal, W.I., Mild steel	3	5	10	6	12
3. Cast steel	3	5	10	8	15
4. Hard Steel	4	6	12	9	14
5. Copper and other soft metals and alloys	5	6	10	8	15
6. Timber	5	8	15	10	18
7. Leather	8	10	12	12	15
8. Masonary and Brick Work	10	15	20	20	30

4.6 EFFECT OF HIGH AND LOW TEMPERATURES

The above analysis refers to a treatment of allowable stress and factor of safety at ordinary temperatures. At high temperatures, there is a modification of the strength of the materials. If this high temperature is maintained for a long time, structural changes take place which further affect the strength and other properties of the material. Also at high temperatures, there will be continuous increase in strain or deformation of the material under the action of the applied load. This will result in the change in size and this phenomenon is known as "Creep", which is a function of temperature, stress and time. The design stress should be selected that will keep the creep or deformation within prescribed limits during the life of the machine. Or suitable materials should be selected which give good performance at high temperatures.

At low temperatures, there is embrittlement of the materials although their strength and elasticity are not affected much. Therefore, at low temperatures, the impact properties must be given careful consideration in the selection of the materials and the design stresses. Nickel in steel helps to improve the impact resistance at low temperatures.

4.7 OTHER CRITERIA OF SERVICEABILITY OF MACHINE ELEMENTS

As discussed under Art. 4.1., the main criteria of serviceability of a machine element or a machine are: Strength, stiffness or rigidity, wear and corrosion resistance. The other factors that can influence the design of the element or, perhaps the entire system are: contact fatigue, vibration resistance and heat resistance. The strength design has already been discussed in the previous pages.

4.7.1 Wear Resistance

As already noted, wear is must when two surfaces of machine elements have relative motion. Wear is the process of gradual change in the surface properties of a component due to friction. Wear resistance means the ability of a part to resist superficial deterioration in friction. The service life of many machine elements is reduced by the wear of their working surfaces. Wear could eventually distort the original geometry and surface conditions of a part. As the amount of wear builds up clearances increase resulting in additional loads on, and damage to the machine components.

The operation of a gear becomes jerky and the non-uniform rotation of gear causes an additional dynamic load on the teeth. In this case, their ability to transmit useful load diminishes. The lubrication of the thrust surface of the shaft and bearing deteriorates: the precision of the movement of parts, for example, machine tool spindles, lathe carriages along the bedways, etc. is disturbed; and the airtightness of the joints is destroyed and so on.

Wear being a function of many variables, no well substantiated direct methods of analysis have yet been evolved. Instead, machine parts are usually analysed for wear resistance by indirect methods. The most common analysis procedure is to select contact surface dimensions such that the working unit pressure p or product pv (v is the velocity of sliding) proportional to the work done by the frictional forces would not exceed the maximum permissible values $[p]$ or $[pv]$ established from past experience that is,

$$p \leq [p] \text{ and/or } pv \leq [pv]$$

Calculations for wear are carried out together with calculations for strength. In the case of gears, for example, the formula for strength of gears in bending is supplemented with a correction factor for wear.

To prevent the failure of expensive components, the friction surface of cheaper components are made from a relatively soft material which effectively resists wear and does not allow the rapid wear of the mating surface of the more expensive component. Babbits, bronze, some grades of cast iron and plastics are used as such anti-friction materials.

Wear can be minimized by lubricating contact surfaces, sealing contact areas to protect them against ingress of abrasive particles, reducing the roughness of contact surfaces, using suitable protective coating, and surface hardening.

4.7.2 Contact Fatigue Endurance or Surface Strength

The strength about which we have been talking so far refers to the volume strength of the machine part. Even with adequate volume strength, the operating capacity of many machine elements is restricted by the insufficient strength of the working surfaces.

Contact stresses arise when two members are pressed against each other by certain forces, the contact area being small as compared with the size of the members. Contact-stress problems arise in the contact of a ball and a race in a bearing, a wheel and a rail, two cylinders, mating gear teeth, a cam and its follower, etc.

As the two mating spherical surfaces rotate under load, their constituent points find themselves subjected to zero-to-compression stress cycles. Each point is loaded only when it is located within the contact area and remains unloaded over the rest of the cycle. Variable contact stresses like that bring about fatigue in the upper layers of the mating parts. As the internal damage due to fluctuating stresses builds up, invisible cracks appear at the contact surfaces. As the process is continually repeated, the cracks keep growing until a spall of metal chips off the surface of the driven wheel. This mode of failure has come to be known as "pitting or contact fatigue"

H. Hertz was the first to formulate the solution of the contact-stress problem. The contact stress is given by Hertz's formula as:

$$f_h = 0.418\ \sqrt{qE/r} \qquad ...(4.65)$$

where q = load per unit length of the contact area.

$$= \frac{\text{Compressive normal load}}{\text{Width of the wheel}}$$

E = Reduced modulus of elasticity of the compressed bodies

$$= \frac{2E_1E_2}{E_1 + E_2}$$

and r = Reduced radius of curvature in the area of contact of the compressed bodies.

$$= \frac{r_1 r_2}{r_1 \pm r_2}$$

The minus sign is taken for a case of contact between a concave surface with radius r_1 and a concave surface with radius r_2

It is far more difficult to determine the allowable stresses in surface strength calculations than in volume strength calculations, because the corresponding phenomenon have not yet been sufficiently studied. The contact stress as given by equation (4.65) should not exceed the surface endurance strength of the material (see chapter on Gears).

A number of factors affect the magnitude of the limit of contact endurance, of which the most important are: the properties of the lubricant, the ratio between the hardness of the working surfaces in contact, the quality of machining, etc. Operating experience and special research show that greater oil viscosity increases the limit of contact endurance. Improved hardness (by surface hardening) and machining of the contact surfaces increase their resistance to printing. Other methods are: assembling the units under preload or pretension and reducing the number of joints.

4.7.3 Vibration Resistance

At high speed operation, the machines and their components are subject to vibration. Vibration loads are extremely undesirable since they may cause fatigue failure under certain conditions. When the frequency of load variation approaches the natural free vibration frequency of a machine or its components, resonance sets in. This sharply increases oscillating amplitudes and sometimes leads to failure.

The vibration of spindles of metal-cutting machine tools deteriorates the quality of the machined surface and quickly blunts or even destroys the tool. Oscillations of parts of gears (toothed wheels, shafts, housings of reduction gears, etc.) due to the elasticity of parts and their elements (wheel teeth, walls of housings, etc.), cyclic errors in the working surfaces of teeth and changes in the tooth deformations and friction forces generate noise.

For high speed operation, components are specially designed to resist vibration, they should be stiff enough to avoid resonance or pass it quickly. In many cases, this can be accomplished by changing the dynamic properties of the system, that is, by changing the moment of inertia of the component masses and varying the flexibility of the joints. When such measures prove ineffective, the system is provided with dampers or antivibrators.

4.7.4 Heat Resistance

Heat resistance means the ability of a machine or its components to operate at specific elevated temperatures throughout the expected service life of the system. As the service temperature rises, the mechanical properties of metals and lubricating oils deteriorate, the amount of wear increases, clearances undergo significant changes, and additional dynamic loads are imposed on the machine components. In view of this, many pairs in rubbing contact (for example, worm gear drives and bearings) are analysed for heat balance to ensure that they will operate under normal temperature conditions. If necessary, their construction may be modified or cooling may be employed.

4.7.5 Stiffness or Rigidity Design

Stiffness or rigidity of a machine component is its ability to resist deformations under the action of external load. Alongwith strength, rigidity is also a very important operating property of many machine components, for example, helical and leaf springs, elastic elements in various instruments, shafts, bearings, toothed and worm gears and so on. In many cases, this parameter of operating capacity proves to be most important and to ensure it the dimensions of the part have to be increased to such an extent that the actual induced stresses become much lower than the allowable ones. Property rigidity is also necessary to ensure that the mated parts and the machine as a whole operate effectively.

Forces subject the parts to elastic deformations: shafts are bent and twisted, bolts are stretched, etc. When a shaft is deflected, its journals are misaligned in the bearings thereby causing the uneven wear of the shells, heating and seizure in the sliding bearings. With the deflection of an insufficiently rigid shaft, the operating conditions of antifriction bearings sharply deteriorate if the bearings can not self-aligning. Deflections and angles of turn of shafts at the places where gears are fitted cause the non-uniform load distribution over the length of the teeth. In especially unfavourable cases, this may be the cause of tooth failure.

Rigidity is particularly important for ensuring the adequate accuracy of items produced on machine tools. Calculation for rigidity is obligatory in designing statically indeterminable systems since the bending moments and torque, normal and shear forces can not be determined here only by the equilibrium conditions; an additional requirement is the equalisation of displacement. It is important to evaluate rigidity when designing components loaded with compressive forces (lifting screws, lead screws, springs, etc.) for stability, when the components are designed for operation under dynamic loads. However, in certain cases machine elements should be able to give in to a certain extent. Flexible bolts, teeth and the rims of toothed wheels, etc. have better endurance.

The rigidity of machine elements is found with the help of formulae from the theory of strength of materials. The actual displacements (deflections, angles of turn, angles of twist) should not be more than the allowable (rated) values. As follows from the formulae for deflections (See Arts. 3.6. to 3.9. last chapter) the degree of component rigidity can be actively influenced by a proper choice of materials and of design methods. The most important design methods for increasing the rigidity of machine elements are:

(a) The decrease in the arms of bending and twisting forces.
(b) The incorporation of additional supports.
(c) The application of cross-sections which effectively resist torsion (closed tubular) and bending (in which the cross-sectional area is removed as far as possible from the neutral axis).
(d) The decrease of the length of the parts in tension and the increase of their cross-sectional area.

It is clear from above that the stiffness of a member depends not only on the shape and size of its cross-section, but also on elastic modulus of the material used.

Now the strength characteristics of steel continuously increase, whereas the values of E remain practically the same. Therefore, parts whose dimensions are limited from stiffness considerations need not be made of quality alloy steels provided the use of such steel is not necessitated by some other requirements, that is improved corrosion resistance, etc.

Present-day machines have to meet ever increasing stringent accuracy requirements, operate at greater speeds, and are assembled from smaller and lighter components. So, stiffness is continually gaining importance as a criterion of serviceability.

4.7.6 Reliability

'Serviceability' is the fitness of a machine or a component to perform its function within the specifications. It is estimated in terms of various criteria. The basic criterion of serviceability for most parts is their "Strength", which gives an idea about the ability of a product to resist breakdown or plastic deformation under applied loads. Strength criterion and the other criteria of serviceability have been discussed in Art. 4.1 onwards.

The serviceability of a product is closely related to its 'Reliability'. Reliability may be defined as the ability of a product to perform its intended function within specified limits over a specified operating time or its service life. By reliability is generally meant the 'probability of survival' of a given machine or component, that is, the chances that no failure will occur within a specified operating time. The meaning of the term 'failure' has already been explained in Art. 4.4. Present day mechanical systems (for example, automatic transfer lines) consist of a multitude of interacting mechanisms, apparatus and devices. Failure of any one critical member of a complex system may disturb normal operations of the entire system and result in spoilage, increased idle time, or even an accident.

A very important criterion of reliability is what is called as "Reliability function." It characterizes the probability that a machine or a component will show failure-free performance over a specified operating time. Obviously, the closer to unity the reliability function, the higher the reliability of the system. If $R(t)$ is the reliability with respect to time 't' and $F(t)$ is the unreliability (Probability of failure), then

$$R(t) + F(t) = 1$$

If N_o is the number of components put on test, $N_s(t)$ is the number surviving to or at time 't' and $N_f(t)$ is the number failed, then

$$N_s(t) + N_f(t) = N_o$$

$$R(t) = \frac{N_s(t)}{N_o} = 1 - \frac{N_f(t)}{N_o}$$

The subject of reliability engineering basically is a study of the causes, distribution and prediction of failure and is out of the scope of this book.

CAUSES OF UNRELIABILITY

There are five general causes that account for unreliability of a component, machine or a system:

1. Design Mistakes: These include: incomplete information on loads and working conditions, improper selection of materials, non-consideration of some of the important operating factors and erroneous calculations.

2. Manufacturing defects: For example, poor surface quality leading to fatigue cracks, decarburization or cracks during heat treatment, etc. These defects may arise due to : poor working conditions, unclear instructions, poor supervision, inadequate training and poor motivation, etc.

3. Maintenance: Service life is directly proportional to proper maintenance. A good design strategy is to produce maintenance - free products.

4. Exceeding design limits: If the operating conditions exceed the design limits (of load, temperature, speed, etc.), the component is prone to failure.

5. Environmental factors: Service life is reduced if the component or machine works in conditions for which it was not designed, for example, rain, ice, high humidity, etc.

LONGEVITY

It is the ability of a component or machine to ensure safe operation before a limit state is reached under given conditions. The meaning of "limit state" is that further use of a component or a machine is impossible at all or not advisable for safety reasons. The length of time during which a machine can be operated before a limit state is reached is termed as "service life." The time can be measured from the beginning of the service or resumption of service after a medium repair or an overhaul. It is desirable that in a machine, the service lives of component parts should be the same, or at least chosen as multiples of each other.

MAINTAINABILITY

It is another important criterion of reliability. It implies that a machine or its components should be designed so as to facilitate prevention and/or detection of failures and damage and their rectification by means of maintenance and repair with part replacement whereever necessary.

The reliability of machine parts and machines as a whole can effectively be improved by reducing the stress level or, equivalently, by increasing the factor of safety. However, this normally calls for larger and heavier elements, which practice, as we have learned previously, is to be avoided. So, as a compromise, it will be a good plan to use high-strength materials and to make the most of hardening technologies. If the operating conditions are such that accidental overloads may be expected, provision should be made for suitable safeguards, elastic (flexible) connections, and similar measures.

It is an easy matter to see that the above requirements are closely interrelated. Only if all of them are properly satisfied will the designer be safe to say that the proposed machine conforms to appropriate standards as regards its functionalism and dependability.

The improvement of the reliability and service life of machines and their components is an important task indeed. A better reliability helps to cut down spare parts requirements and idle time, to raise labour productivity, and to keep the cost of maintenance and repair to a minimum. Improved reliability and longer service life may also mean longer replacement periods and higher efficiency of machines and mechanisms.

Thus, reliability may be regarded as the basic criterion of quality. In fact, it gives an idea about the level of design, manufacturing processes employed, operating conditions, etc.

SOLVED EXAMPLES

Example 4.1. A flat bar 32 mm wide and 12 mm thick is loaded by a steady tensile load of 85 kN. The material is mild steel with yield point stress of 315 N/mm². Find the factor of safety based on the yield point.

Solution: Area of cross-section of bar = $0.032 \times 0.012 = 3.84 \times 10^{-4}$ m^2

∴ Tensile stress induced in the bar (which will be principal stress)

$$= \frac{F}{A} = \frac{85 \times 1000}{3.84 \times 10^{-4}} = 221.36 \text{ MN/m}^2$$

∴ Maximum shear stress induced in bar, see equation (4.27) = $(f_s)_{max}$

$$= \frac{221.36}{2} = 110.68 \text{ MN/m}^2 = 110.68 \text{ N/mm}^2$$

Now yield point stress (Tensile) = 315 N/mm²

$$\therefore \quad \text{Factor of safety (equation 4.23). } F_S = \frac{f_{ypt}}{2 \times (f_s)_{max}} = \frac{315}{2 \times 110.68} = 1.42$$

Example 4.2. The stresses at a point in a body are: f_x = 91 MN/m²; f_y = 21 MN/m²; f_{xy} (shear stress) = 84 MN/m²; yield point stress = 280 MN/ m²:
Find the factor of safety:
(a) By the maximum shear stress theory.
(b) By the distortion energy theory.

Solution: The first step is to calculate the principal stresses. Being two dimensional problem equations (4.15 and 4.16), will be used. Therefore,

$$f_1 = \frac{f_x + f_y}{2} + \sqrt{\left(\frac{f_x - f_y}{2}\right)^2 + f_{xy}{}^2}$$

$$= \frac{91 + 21}{2} + \sqrt{(35)^2 + (84)^2}$$

$$= 56 + 91 = 147 \text{ MN/m}^2 \text{ (i.e., Tensile)}$$

and $f_2 = 56 - 91 = -35$ MN /m² (i.e., Compressive)

(a) Maximum shear stress theory

Now, maximum shear stress induced in the material

$$= \text{the largest of } \frac{1}{2}|(f_1 - f_2)|, \frac{1}{2}|f_1|, \frac{1}{2}|f_2| \quad \text{(See equation 4.17)}$$

$$= \text{the largest of } 91, 73.5, 17.5$$

i.e., $(f_s)_{max}$ = 91 MN/m².

$$\therefore \quad F_S = \frac{f_{ypt}}{2.(f_s)_{max}} \text{ (Equation 4.23)}$$

$$= \frac{280}{2 \times 91} = 1.54$$

(b) Distortion energy theory: Equation (4.3)

$$\frac{f_{ypt}}{F_S} = \sqrt{f_1{}^2 + f_2{}^2 - f_1 f_2} = \sqrt{147^2 + (-35)^2 - 147(-35)}$$

$$= 167.3 \text{ MN/m}^2$$

$$\therefore \quad F_S = \frac{280}{167.3} = 1.67$$

Example 4.3. A 50 mm diameter shaft is loaded statically in pure torque at a shear stress of 70 N/mm^2. If the yield point stress is 435 N/mm^2. Find the factor of safety:

(a) By the maximum shear stress theory.

(b) By the distortion energy theory.

Solution: Here, the case is of pure shear, so

$$(f_s)_{max} = f_1 = -f_2 = 70 \text{ N/mm}^2$$

(a) $F_S = \dfrac{f_{ypt}}{2\times(f_s)_{max}} = \dfrac{435}{140} = 3.10$

(b) From equation (4.31),

$$\frac{f_{ypt}}{F_S} = \sqrt{f_1^2 + f_2^2 - f_1 f_2}$$

$$\frac{f_{ypt}}{F_S} = \sqrt{70^2 + (-70)^2 - 70(-70)} = 121.24$$

$$\therefore \quad F_S = \frac{435}{121.24} = 3.58$$

Example 4.4. A long straight tube, 76 mm internal diameter and 2.5 mm thick is subjected to an internal pressure of 5.6 N/mm^2. Consider it as a thin cylinder. If the tube is subjected to a twisting moment of 70 Nm, Elastic limit stress = 282 N/mm^2, determine the factors of safety on the three common theories of failure.

Solution: By Strength of Materials relations:

$$\text{Stress acting in the longitudinal direction} = \frac{pr}{2t} = \frac{5.6\times38}{2\times2.5} = 42.56 \text{ N/mm}^2$$

$$\text{Stress acting in the circumferential direction} = \frac{pr}{t} = 85.12 \text{ N/mm}^2$$

$$\text{Shear stress due to torsion} = \frac{2M_t R}{\pi\left(R^4 - r^4\right)} = \frac{2\times70\times1000\times40.5}{\pi\left(40.5^4 - 38^4\right)}$$

$$= 2.985 \text{ N/mm}^2$$

$$\text{Maximum principal stress } f_1 = \frac{1}{2}\left\{\left(f_x + f_y\right) + \sqrt{\left(f_x - f_y\right)^2 + 4f_s^2}\right\}$$

$$= \frac{1}{2}\left\{(85.12 + 42.56) + \sqrt{42.56^2 + 4(2.985)^2}\right\}$$

$$= \frac{1}{2}\{127.68 + 42.977\} = 85.33 \text{ N/mm}^2$$

$$\text{Minimum principal stress} = \frac{1}{2}(127.68 - 42.977) = 42.35 \text{ N/mm}^2$$

(a) Maximum principal stress theory

$$f_{max} = f_1 = 85.33 \text{ N/mm}^2$$

$\therefore$ From equation (4.51)

$$F_S = \frac{f_{ypt}}{f_{max}} = \frac{282.0}{85.33} = 3.30$$

(b) Maximum shear stress theory

Here the principal stresses are : 85.33, 42.35 and – 5.6

$$\therefore \quad (f_s)_{max} = \frac{1}{2}|f_1 - f_3| = \frac{1}{2}(85.33 + 5.6) = 45.465 \text{ N/mm}^2$$

$$\therefore \quad F_S = \frac{282.0}{2 \times 45.465} = 3.10$$

(c) Distortion energy theory: Equation (4.30)

$$\frac{f_{ypt}}{F_S} = [7281.2089 + 1793.5225 + 31.36 - 3613.7255 + 237.16 + 477.848]^{½}$$

$$= 78.787 \text{ N/mm}^2$$

$$\therefore \quad F_S = \frac{228.0}{78.787} = 3.56$$

Example 4.5. The stresses at a point in a cast iron body are : f_x = 14 N/mm², f_y = – 42 N/mm², f_s = 21 N/mm². The stress concentration factor is equal to 2 for all stresses. f_{uc} = – 560 N/mm². Find factor of safety.

Solution: Now

$$K_{f1} = 2\left[\left(\frac{f_x + f_y}{2}\right) + \sqrt{\left(\frac{f_x - f_y}{2}\right)^2 + f_s^2}\right]$$

$$= 2[-14 + \sqrt{784.00 + 441.00}]$$

$$= 2\,[-14.0 + 35.0] = 42.0 \text{ N/mm}^2$$

$$\therefore \quad K_{f2} = 2[-14 - 35.0] = -98 \text{ N/mm}^2$$

As the two principal stresses are of opposite sign, the design equation will be equation (4.34), i.e.,

$$\frac{1}{F_S} = \frac{K_{f1}}{f_{ut}} + \frac{K_{f2}}{f_{uc}} = \frac{42}{140.0} + \frac{-98.0}{-560.0}$$

$$\therefore \quad F_S = 2.19.$$

Example 4.6. Determine the factor of safety to be used in the design of connecting rod for:
(a) a double acting steam engine, for ordinary general purpose.
(b) a blooming mill reversing engine.
(c) Airplane engine.
The material of connecting rod is ductile, medium carbon steel.

Solution: (a) For double acting steam engine, for ordinary general service, since the load is reversed, it is under varying load. Therefore, the various factors may be taken as below:

$$a = 3$$

K_i = 1.5 , since the rise of pressure is not sudden.

R = 1.75, since the weight is not critical

$$\therefore \quad FS = a \times K_i \times R$$

$$= 3 \times 1.5 \times 1.75 = \text{about } 8$$

(b) In the case of blooming mill, extreme impact and rough handling is possible and breakdown would mean great danger and economic loss. So,

$$a = 3$$
$$K_i = 2$$
$$R = 2 \text{ to } 3, \text{ say } 2.5$$
$$\therefore \quad F_S = 3 \times 2 \times 2.5 = 15$$

(c) In the case of airplane, lightness is indispensable, so

$$a = 2 \text{ to } 3, \text{ say } 2.5$$
$$K_i = 1.5$$
$$R = 1$$
$$F_S = 2.5 \times 1.5 \times 1, \text{ say } 4.$$

Note: In the above problem the values of factor of safety are based on ultimate stress of material.

Example 4.7. Determine the factor of safety for the shaft of a water wheel running under a high head and liable to shock, because of the suddenly varying water pressure. The water wheel is to run 12 hours a day, 300 days in the year, for an expected life of 20 years. The material is to be steel. Find the allowable stress if the ultimate strength of the material is 490 MPa.

Solution: Here the stress is reversed, tension and compression, therefore,

$$a = 3, \text{ Shear may be neglected for simplicity}$$
$$K_i = 1.5$$
$$R = 1.75, \text{ Since the weight is not critical}$$
$$\therefore \quad F_S = 3 \times 1.5 \times 1.75, \text{ say } 8$$
$$\therefore \quad \text{Allowable stress} = \frac{490}{8}, \text{ say } 60 \text{ MN/m}^2.$$

Example 4.8. A cold drawn steel bar is to withstand a tensile preload of 36.3 kN and a fluctuating tensile load varying from 0 to 72.60 kN. The bar has a geometric stress concentration factor of 2.02 corresponding to a fillet whose radius is 4.75 mm. Determine the size of the bar for an infinite life and a factor of safety of 2. The material properties are f_{yp} = 588.0 N/mm², f_{ut} = 700 N/mm².

Solution: The endurance limit, f_e^*, may be taken as,

$$f_e^* = 0.5, f_{ut} = 350.0 \text{ N/mm}^2$$

From Fig. 4.16 the surface finish factor, $C_F = 0.73$

For axial loads, the load factor, $C_L = 0.85$

From equation (4.6), the notch sensitivity index,

$$q_f = \frac{1}{1+\left(\frac{a}{r}\right)}$$

From Table 4.2 let $a = 0.25$

$$\therefore \quad q_f = \frac{1}{1+\frac{0.25}{4.75}} = 0.95$$

$$\therefore \quad K_f = 1 + q_f (K_t - 1) = 1 + 0.95 (2.02 - 1)$$
$$= 1.96$$

Neglecting the size factor, the modified endurance limit from equation (4.43) is,

$$f_e = C_F \cdot C_L \cdot f_e^*$$
$$= 0.73 \times 0.85 \times 350.0 = 217.2 \text{ N/mm}^2$$

Next, we determine the stresses in terms of bar dimensions.

The static stress is, $$f_{st} = \frac{F_{st}}{A} = \frac{36.30\times10^3}{\pi d^2/4} = \frac{46.22}{d^2} \text{ kN/mm}^2$$

The stress range is,

$$f_r = f_{max} - f_{min} \ (f_{min} = 0, \text{ here})$$
$$= \frac{72.6\times10^3}{\pi d^2/4} = \frac{92.44}{d^2} \text{ kN/mm}^2$$

$\therefore$ Variable stress, $f_v = \dfrac{46.22}{d^2}$ kN/mm^2

Mean stress $= \dfrac{f_{max}+f_{min}}{2} = \dfrac{46.22}{d^2}$ kN/mm^2

Total mean stress = Mean stress due to varying load + Static stress due to preload.

$$\therefore \quad f_m = \frac{92.44}{d^2} \text{ kN/mm}^2$$

$\therefore$ From equation (4.46),

$$\frac{1}{F_S} = \frac{f_m}{f_{yp}} + \frac{K_f \times f_v}{f_e}$$

$$\therefore \quad \frac{1}{2} = \frac{92.44\times10^3}{d^2\times588.0} + \frac{1.96\times46.22\times10^3}{d^2\times217.2} = \frac{0.5742\times10^3}{d^2}$$

$$\therefore \quad d^2 = 1.1484 \times 10^3$$

$$\therefore \quad d = 33.89 \text{ mm, say 34 mm.}$$

Example 4.9. A steel connecting rod is to be subjected to a reverse axial loading of 180 kN. Determine the required diameter of the rod, using a factor of safety 2. Assume no column condition. For the material of the rod,

$$f_{ut} = 1050.0 \text{ N/mm}^2 ; f_{yp} = 894.0 \text{ N/mm}^2.$$

Solution: The endurance limit in reversed bending may be taken as half the tensile strength. Therefore,

$$f_e^* = \frac{1}{2} \times 1050.0 = 525.0 \text{ N/mm}^2.$$

Here, there is complete reversal of stress, therefore,

$f_m = 0$, and

$$f_v = \text{Variable stress} \ \frac{F}{A} = \frac{4\times180}{\pi d^2} = \frac{720}{\pi d^2} \text{ kN/mm}^2$$

where d = diameter of rod.

Now $$f_e = C_F \cdot C_s \cdot C_L \cdot f_e^*$$

Value for C_F (Table 4.4) = 0.78

$$C_s = 0.85, \text{ say ; } C_L = 0.7, \text{ for reversed axial loading}$$

$$\therefore \quad f_e = 0.78 \times 0.85 \times 0.7 \times 525.0 = 243.653 \text{ N/mm}^2$$

The design equation is (Equation 4.46),

$$\frac{1}{F_S} = \frac{f_m}{f_{yp}} + \frac{K_f \times f_v}{f_e}$$

let $K_f = 1$

$$\therefore \quad \frac{1}{2} = 0 + \frac{1 \times 720 \times 10^3}{\pi d^2 \times 243.653} = \frac{0.9406 \times 10^3}{d^2}$$

From here, d = 43.4 mm; Take d = 44 mm.

Example 4.10. A hot rolled steel rod is to be subjected to a torsional load that will vary from a – 100.0 Nm to + 400.0 Nm. Determine the diameter of rod using a factor of safety of 1.75 For the material of the rod, take:

$$\mathbf{f_{ut} = 489.0 \text{ MN /m}^2; f_{yp} = 315.0 \text{ MN/m}^2}$$

Solution: Take $f_e^* = \frac{1}{2} \times f_{ut} = 244.5 \text{ MN/m}^2$

and $f_{yps} = 0.6 \times f_{yp} = 189.0 \text{ MN/m}^2$

Now mean torque is, $T_m = \frac{1}{2}[T_{max} + T_{min}]$

$$= \frac{1}{2}(400.0 - 100.0) = 150.0 \text{ Nm}$$

The variable torque is, $T_v = \frac{1}{2}(T_{max} - T_{min}) = 250.0 \text{ Nm}$

$$\therefore \quad f_{ms} = \frac{T_m}{Z_p} = \frac{16 \times T_m}{\pi d^3} = \frac{2400.0}{\pi d^3} \text{ N/m}^2$$

$$f_{vs} = \frac{16 \times T_v}{\pi d^3} = \frac{4000}{\pi d^3} \text{ N/m}^2$$

Now, $f_e = C_F \cdot C_S \cdot C_L f_e^*$

C_F = 0.68 (Table 4.4), C_s = 0.85, say; C_L = 0.58

$$\therefore \quad f_e = 0.68 \times 0.85 \times 0.58 \times 244.5 = 81.966 \text{ MPa (in shear)}$$

∴ From equation (4.47),

$$\frac{1}{F_S} = \frac{f_{ms}}{f_{yps}} + \frac{K_f \times f_{vs}}{f_e}$$

Let $K_f = 1$

$$\therefore \quad \frac{1}{1.75} = \frac{2400.0}{\pi d^3 \times 189 \times 10^6} + \frac{1.0 \times 4000}{\pi d^3 \times 81.966 \times 10^6}$$

From here, d = 0.03215 m, say 32.5 mm.

Example 4.11. A cold drawn steel rod of circular cross-section is subjected to variable bending moment of 500.0 to 1000.0 Nm as the axial load varies from 4.50 kN to 13.50 kN. The maximum bending moment occurs at the same instant that the axial load is maximum. Find the diameter of rod neglecting stress concentration. Take $F_S = 2, f_{ut} = 560.0$ MN/m²; $f_{yp} = 476.0$ MN/m².

Solution: The stress due to bending moment and axial loading should be determined first.

(a) Bending moment:

$$M_m = \frac{1}{2}(500.0 + 1000.0) = 750.0 \text{ Nm}$$

$$M_v = \frac{1}{2}(1000.0 - 500.0) = 250.0 \text{ Nm}$$

$$\therefore \quad f_m = \frac{32.M_m}{\pi d^3} = \frac{32 \times 750.0}{\pi d^3} \text{ N/m}^2$$

$$\therefore \quad f_v = \frac{32.M_v}{\pi d^3} = \frac{32 \times 250.0}{\pi d^3} \text{ N/m}^2$$

Here,
$$f_e = C_F \cdot C_s \cdot C_L \cdot f_e^*$$

$$f_e^* = \frac{1}{2} \times 560.0 = 280.0 \text{ MN/m}^2$$

$$C_F = 0.88 \text{ (Table 4.4)},\ C_s = 0.85, \text{ say},\ C_L = 1.0$$

$$\therefore \quad f_e = 0.88 \times 0.85 \times 1.0 \times 280.0 = 209.44 \text{ MN/m}^2$$

Now equivalent normal stress is, (Equation 4.53),

$$f_{ne} = f_m + K_f\, f_v \,.\, \frac{f_{yp}}{f_e}$$

Taking $K_f = 1.0$

$$\therefore \quad f_{ne} = \frac{32 \times 750.0}{\pi d^3} + \frac{32 \times 250.0 \times 476 \times 10^6}{\pi d^3 \times 209.44 \times 10^6} = \frac{42.215 \times 10^3}{\pi d^3} \text{ N/m}^2$$

(b) Axial loading:

$$F_m = \frac{1}{2}(13.50 + 4.50) = 9.0 \text{ kN}$$

$$F_v = \frac{1}{2}(13.50 - 4.50) = 4.50 \text{ kN}$$

$$\therefore \quad f_m = \frac{9000 \times 4}{\pi d^2}, \text{ and } f_v = \frac{4500 \times 4}{\pi d^2}$$

$$f_e = C_F \cdot C_S \cdot C_L \cdot f_e^*$$

Here,
$$C_F = 0.88,\ C_s = 0.85,\ C_L = 0.70$$

$$\therefore \quad f_e = 146.6 \text{ MN/m}^2$$

∴ Equivalent normal stress due to axial loading,

$$f_{ne} = f_m + K_f \cdot f_v \cdot \frac{f_{yp}}{f_e}$$

$$= \frac{36\times10^3}{\pi d^2} + 1.0\times\frac{18\times10^3}{\pi d^2}\cdot\frac{476.0\times10^6}{146.6\times10^6} = \frac{94.40\times10^3}{\pi d^2} \text{ N/m}^2$$

$\therefore$ Total equivalent normal stress $= \left(\frac{42.215\times10^3}{\pi d^3} + \frac{94.40\times10^3}{\pi d^2}\right)$ N/m^2

This should be equal to $\frac{f_{yp}}{F_S} = \frac{476}{2} = 238$ MN/m^2

$$\therefore \quad \frac{42.215\times10^3}{\pi d^3} + \frac{94.40\times10^3}{\pi d^2} = 238\times10^6$$

By trial and error, $d = 0.0395$ m, say, 40 mm.

Example 4.12. A hot rolled steel shaft is subjected to a torsional load that varies from 300 Nm clockwise to 100 Nm anti-clockwise as an applied bending moment at a critical section varies from +400 Nm to –200 Nm. The shaft is of uniform cross-section and no keyaway is present at the critical section. Determine the required shaft diameter by taking $F_S = 1.5$. For the material, take

$$f_{ut} = 560 \text{ MPa}; f_{yp} = 420 \text{ MPa}$$

$$f_e^* = 0.5 \times f_{ut} = 280 \text{ MPa}.$$

Solution: (a) Determine the equivalent normal stress due to bending moment.

$$M_m = \frac{1}{2}(400.0 - 200.0) = 100 \text{ Nm}$$

$$M_v = \frac{1}{2}(400.0 + 200.0) = 300 \text{ Nm}$$

$$\therefore \quad f_m = \frac{32\times M_m}{\pi d^3} = \frac{1020}{d^3} \text{ N/m}^2$$

$$f_v = \frac{32\times M_v}{\pi d^3} = \frac{3060}{d^3} \text{ N/m}^2$$

Now $\quad f_e = C_F \times C_s \times C_L \times f_e^*$

Here, $\quad C_F = 0.62,\ C_S = 0.85,\ C_L = 1.0$

$\therefore \quad f_e = 0.62 \times 0.85 \times 1.0 \times 280.0 = 147.56$ MPa

The equivalent normal stress is (Equation 4.53),

$$f_{ne} = f_m + K_f \times f_v \times \frac{f_{yp}}{f_e},$$

Taking $K_f = 1.0$

$$f_{ne} = \frac{1020.0}{d^3} + 1.0 \times \frac{3060.0}{d^3} \times \frac{420\times10^6}{147.56\times10^6} = \frac{9730.0}{d^3} \text{ N/m}^2$$

(b) Determine the equivalent shear stress due to torsional loads

$$T_m = \frac{1}{2}(T_{max} + T_{min}) = \frac{1}{2}(300.0 - 100.0) = 100.0 \text{ Nm}$$

$$T_v = \frac{1}{2}(T_{max} - T_{min}) = 200.0 \text{ Nm}$$

$$\therefore \quad f_{ms} = \frac{16 \times T_m}{\pi d^3} = \frac{510.0}{d^3} \text{ N/m}^2$$

$$f_{vs} = \frac{16 \times T_v}{\pi d^3} = \frac{1020.0}{d^3} \text{ N/m}^2$$

Here, $f_e = C_F \times C_S \times C_L \times f_e^*$

$C_F = 0.62,\ C_S = 0.85,\ C_L = 0.58$

$\therefore \quad f_e = 0.62 \times 0.85 \times 0.58 \times 280.0 = 85.585$ MPa

The equivalent shear stress is (Equation 4.54),

$$f_{se} = f_{ms} + K_f \cdot f_{vs} \cdot \frac{f_{yps}}{f_e}$$

Take $K_f = 1.0$ and $f_{yps} = 0.6 \times f_{yp} = 0.6 \times 420.0 = 252$ MPa

$$\therefore \quad f_{se} = \frac{510.0}{d^3} + \frac{1020.0}{d^3} \times \frac{252 \times 10^6}{85.585 \times 10^6} = \frac{3513.33}{d^3}$$

Applying the maximum shear stress theory.

$$f_{se(\max)} = \frac{1}{2}\sqrt{f_{ne}^2 + f_{se}^2} = \frac{1}{2d^3}\sqrt{9730.0^2 + 4 \times 3513.33^2} = \frac{6000.96}{d^3} \text{ N/m}^2$$

Now $$f_{se(\max)} = \frac{f_{yps}}{FS} = \frac{0.5 \times 420.0}{1.5} = 140.0 \text{ MPa}$$

$$\therefore \quad \frac{6000.96}{d^3} = 140 \times 10^6 \quad \therefore \quad d^3 = 42.857 \times 10^{-6} \text{m}^3$$

$$\therefore \quad d = 0.03514 \text{ m, say 36 mm}$$

Example 4.13. A flat plate of width 60 mm has a central hole of 10 mm diameter. This plate is subjected to an axial load of 10 kN (tensile). Determine the thickness of plate. Take f_{yp} = 300 MPa for plate material and FS = 2.5.

Solution: Refer to Fig. 4.4. It is case (1)

$$B = 60 \text{ mm}, \ d = 10 \text{ mm}$$

$$\therefore \quad \frac{d}{B} = \frac{10}{60} = 0.166$$

$\therefore$ From Fig. 4.4 curve (1), $K_f = 2.55$

Now $$K_t = \frac{f_{t\max}}{f_{t\,nom}}$$

$$f_{t\max} = \frac{f_{yp}}{F_S} = \frac{300}{2.5} = 120 \text{ MPa}$$

$$f_{t\,nom} = \frac{P}{A}, \ A \text{ is based on net section, see Equation (4.1)}$$

$$\therefore \quad \frac{10\times10^3}{(B-d)\times t} = \frac{120}{2.55}$$

$$\therefore \quad t = \text{plate thickness} = \frac{10\times10^3\times2.55}{120\times50} = 4.25 \text{ mm.}$$

Example 4.14. Determine the maximum stress produced in a stepped shaft stepped down from 40 mm to 30 mm with a fillet radius of 6 mm. It is subjected to a twisting moment of 100 Nm.

Solution: Refer to Fig. 4.9

$$D = 40 \text{ mm}, d = 30 \text{ mm}, r = 6 \text{ mm}$$

$$\text{Now } K_{ts} = \frac{f_{s\max}}{f_{s\,\text{nom}}}$$

$$\therefore \quad f_{s\max} = K_{ts}\, f_{s\,\text{nom}}$$

Now from Fig. 4.9, $\frac{r}{d} = \frac{6}{30} = 0.2, \frac{D}{d} = \frac{40}{30} = 1.33$

K_{ts} will be ≈ 1.23

Now $f_{s\,\text{nom}} = \frac{Mt}{Zp} = \frac{100\times1000\times16}{\pi\times(30)^3} = 18.86$ MPa

$\therefore \quad f_{s\max} = 1.23 \times 18.86 = 23.20$ MPa.

Example 4.15. A stepped shaft, with the reduction ratio of 1.2 is to have a fillet radius of 10% of the smaller diameter. The material of the shaft has notch sensitivity factor of 0.925; a shear stress of 160 MPa at yield and a shear stress of 120 MPa at endurance limit. Determine the diameter of the shaft at the minimum cross-section to sustain a twisting moment if that fluctuates between 500 Nm to – 800 Nm.

Take; $C_F = 0.95$, $C_S = 0.85$ and $F_S = 2$

Solution: Refer to Fig. 4.9

$$\frac{D}{d} = 1.2, \frac{r}{d} = 0.1$$

$\therefore$ From the Figure $K_t = 1.33$

Now $K_f = 1 + q_f\,(K_t - 1),\ q_f = 0.925$

$\therefore \quad K_f = 1 + 0.925 \times 0.33 = 1.305$

Now $T_{\max} = 500$ Nm and $T_{\min} = -800$ Nm

$$T_m = \frac{1}{2}(T_{\max} + T_{\min}) = \frac{1}{2}(500-800) = -150 \text{ Nm}$$

$$T_v = \frac{1}{2}(T_{\max} - T_{\min}) = 650 \text{ Nm}$$

$$f_{ms} = \frac{16\times T_m}{\pi d^3} = \frac{-2400}{\pi d^3} \text{ N/m}^2$$

$$f_{vs} = \frac{16\times T_v}{\pi d^3} = \frac{10400}{\pi d^3} \text{ N/m}^2$$

Now $f_{yps} = 160$ MPa, $f_e^* = 120$ MPa (in shear)

$\therefore \quad f_e = f_e^* \times C_L \times C_F \times C_s = 120 \times 1.0 \times 0.95 \times 0.85 = 97$ MPa

Now from equation (4.47)

$$\frac{1}{FS} = \frac{f_{ms}}{f_{yps}} + \frac{K_f.f_{vs}}{f_e}$$

$$\frac{1}{2} = -\frac{2400}{\pi d^3 \times 160 \times 10^6} + \frac{1.305 \times 10400}{\pi d^3 \times 97 \times 10^6}$$

From here, $$d^3 = \frac{249.8}{\pi \times 10^6} \text{ m}^3$$

$\therefore$ $$d = 43 \text{ mm.}$$

Example 4.16. A cylindrical rough machined member made from C-50 steel of 50 mm is reduced to 30 mm diameter by a 5 mm fillet. It is subjected to light shock load producing a completely reversed stress in bending and the life is estimated to be 10 million cycles. Find the bending moment.

Take; $FS = 1.5, f_{yp} = 373$ MPa, $f_e^* = 317$ MPa

Solution: Here $$f_m = 0$$

$$M_v = \frac{1}{2}(M_{\max} - M_{\min})$$

Now $$M_{\min} = -M_{\max}$$

$\therefore$ $$M_v = M_{\max} = M_b, \text{ Nm}$$

$$f_v = \frac{32M_b}{\pi d^3} = \frac{32M_b}{\pi(30)^3} = \frac{32M_b}{\pi(0.03)^3} \text{ N/m}^2$$

Now refer to Fig. 4.8,

$B = 50$ mm, $b = 30$ mm, $r = 5$ mm

$$\frac{r}{b} = \frac{1}{6} = 0.167, \quad \frac{B}{b} = \frac{50}{30} = 1.67$$

$$K_t = 1.5$$

Since nothing is mentioned about notch sensitivity, let us take

$$K_f = K_t = 1.5$$

Now Soderberg design equation (4.46) is;

$$\frac{1}{FS} = \frac{f_m}{f_{yp}} + \frac{K_f \times f_v}{f_e}$$

or since the cycle is completely reversed, the design equation becomes,

$$f_v \le \frac{f_e}{K_f \times FS} \quad \text{...equation (4.44)}$$

Now $C_L = 1.0, \quad C_S = 0.85$

Since $f_e^* = 317$ MPa, so let $f_u = 2\times f_e^* = 634$ MPa

$\therefore$ from Table 4.4, $C_F = 0.86$

$\therefore$ $$f_e = 317 \times 1.0 \times 0.85 \times 0.86 = 191.73 \text{ MPa}$$

$\therefore$ $$\frac{32M_b}{\pi(0.03)^3} = \frac{191.73 \times 10^6}{1.5 \times 1.5}$$

From here, $$M_b = 226.2 \text{ Nm.}$$

EXERCISES

1. List the fundamental requirements for machine members for their efficient working.
2. Differentiate between "Strength Design" and "Stiffness Design".
3. What is the criteria of failure for;
   (a) Ductile materials subjected to static load?
   (b) Brittle materials subjected to static load?
   (c) Ductile material subjected to varying load?
4. Define and differentiate between failure stress, design stress and working stress.
5. Define and discuss: Factor of safety, size factor, Index of sensitivity, endurance limit, fatigue strength, Reliability factor?
6. What do you understand by stress concentration? Define and differentiate between form stress factor and stress concentration factor.
7. What is the difference between endurance limit and fatigue strength of a material?
8. What are the various types of varying stresses?
9. A connecting rod made of annealed steel with ultimate strength of 750 MPa in tension, is to be designed for an oil engine. Determine the factor of safety to be used in the design. Also calculate the design stress.
10. What should be factor of safety for the design of crank shaft for:
    (a) A double acting engine?
    (b) An oil engine?
    (c) A Punching press?
11. What should be the factor of safety for:
    (a) Design of coil spring for valve gear of an oil engine?
    (b) Design of coil spring for spring loaded safety valve?
    (c) Design of leaf spring for a Truck?
12. What factor of safety should be taken in the design of a gear tooth in bending for:
    (a) Spur gear?
    (b) Helical gear?
13. What is "Reliability factor"? Discuss it in detail.
14. Write about the various "Theories of failure".
15. Write short notes on:
    (a) Designing for wear resistance.
    (b) Designing for contact strength.
    (c) Designing for Heat resistance.
    (d) Designing for Vibration resistance.
    (e) Stiffness design.
16. How the localised stresses in machine members can be avoided?
17. How is the "factor of safety" evaluated for different types of loading?
18. What is the effect of "High temperature" and "low temperature" on the strength of a material?
19. What maximum axial force can be applied on a plate of width 50 mm and thickness of 10 mm with a central hole of 10 mm diameter without exceeding its yield point stress of 62.5 MPa across its width.
    (Ans. 10 kN)
20. A flat bar (Refer to Fig. 4.5) is subjected to an axial load of 50 kN. Assuming that the stress in the bar is limited to 200 MPa, find the thickness of the bar. Take:
    $B = 150$ mm, $2r = 38$ mm, $h = 19$ mm **(Ans 4.7 mm)**
21. Determine the maximum stress produced in a rectangular plate 50mm wide, 8 mm thick with a central hole of 10 mm. It is loaded in an axial tension of 1 kN. **(Ans. 7.81 MPa)**
22. A connecting rod for a steam engine is subjected to an axial load of 75 kN which is completely reversed. Determine the required diameter of the rod, using a factor of safety 2.5. For the material of the rod:
    $$f_{yp} = 310 \text{ MPa}, f_{ut} = 580 \text{ MPa}.$$
    Take: $C_F = 0.78$ and $C_S = 0.85$ **(Ans. 42 mm)**

23. The bending stress in a machine part fluctuates between a tensile stress of 300 MPa and a compressive stress of 150 MPa. What should be the minimum tensile strength of the material to carry this fluctuation infinitely ? Use factor of safety equal to 2.
24. A cantilever beam of 2000 mm span of rectangular cross-section of depth 80 mm is subjected to transverse load at its free end that fluctuates between 800 N downwards and 2 kN upwards. Determine the width of the beam.
Take: f_{yp} = 300 MPa, f_e = 250 MPa and F_S = 2.25 **(Ans : 32 mm)**
25. Determine the wall thickness of a pressure vessel of 1200 mm mean diameter subjected to an internal pressure that fluctuates between 5 N/mm^2 and 10 N/mm^2
Take: f_{yp} = 200 MPa, f_e = 160 MPa and F_S = 2 **(Ans. 65 mm)**
26. A flat bar with a central hole (see Fig. 4.4) is subjected to a fluctuating tensile load varying from 15000 N to 3750 N. Find the width of the bar, thickness being 6.25 mm.
Take: f_{yp} = 420 MPa, f_e = 175 MPa, F_S = 2
In Fig. (4.4) Case 1, d/B = 0.2 **(Ans. 41 mm)**
27. A steel shaft is subjected to a completely reversed bending moment of 100 kNm. The shaft transmits 735.5 kW at 100 rev/min. The torque varies over a range of ± 40%. Determine the diameter of shaft.

Take: f_{yp} = 400 MPa, f_e^* = 300 MPa, F_S = 2.5.
28. A C-40 steel bar is subjected to a fluctuating load that varies from a maximum of 330 kN tension, to a minimum of 110 kN compression. The mechanical properties of the steel are:
f_u = 1090 MPa ; f_{yp} = 1010 MPa, f_e = 510 MPa
Determine the bar diameter to give infinite fatigue life based on a safety factor of 2.5. **(Ans. 38.7 mm)**
29. Estimate the life of a shaft 35 mmϕ made of alloy steel for which f_{ut} = 550 MPa. The bending stress at the most critical section has been determined to be 180 MPa.

Solution: $f_e = f_e^* \cdot C_F \cdot C_s \cdot C_l \cdot C_R \cdot C_T \cdot C_e$

Now $C_l = C_R = C_T = 1.$ Take $C_F = 0.85,\ C_s = 85,\ C_e = \dfrac{1}{k_f} = \dfrac{1}{1.585}$

Now $f_e^* = 0.5 \times f_{ut} = 275$ MPa.

$\therefore$ $f_e = 275 \times 0.85 \times 0.85 \times 1 \times 1 \times 1 \times \dfrac{1}{1.585} = 125.35$ N/mm^2

Now f_f = 180 MPa which is $> f_e$, $\therefore$ the shaft will have finite life.

Now, $N = 10^{\frac{-c}{b}} \cdot f_f^{\frac{1}{b}}$

$$b = -\frac{1}{3}\log_{10}\left(\frac{0.85 f_u}{f_e}\right) = -0.1906 \text{ and } c = \log_{10}\left[\frac{(0.85 f_u)^2}{f_e}\right] = 3.24$$

$$\therefore \quad N = \frac{3.24}{10^{0.1906}} \times 180^{-\frac{1}{0.1906}} = 147000 \text{ cycles.}$$

30. A rotating beam specimen made of $C-45$ steel with f_{ut} = 630 N/mm^2 is subjected to completely reversed cycle. Determine the endurance strength of the specimen for a life of 90,000 cycles. **(Ans. f_f = 386 MPa)**

CHAPTER 5

Manufacturing Considerations in Design

5.1 STANDARDISATION

The purpose of standardisation is to establish mandatory or obligatory norms or standards to which the different types, grades, parameters (e.g., dimensions), quality characteristics, test methods and rules of marking, packing and storage of finished items, raw materials and semifinished articles should conform. Its aim is to minimise variety so that number of types, dimensions and sizes etc., are limited to a definite number of models. Standardisation is of great importance in Mechanical Engineering in which a large variety of types and sizes of items in single quantities (heavy machine tools, presses, sewing machines etc.) exist, a wide range of materials used and a great number of diverse manufacturing processes adopted. Standardisation helps in the manufacture of required machines comparatively quickly and economically despite this diversity.

There are two stages in standardisation:

(a) The creation of models in which the properties, which a given items should possess are established.

(b) The reduction of the number of these items, processes and methods to a rational minimum.

In mechanical engineering the following items are included in standardisation:

(i) Rules of mechanical drawing, notation of technical units.

(ii) Materials, their chemical composition, basic mechanical properties and heat treatment.

(iii) Shapes and dimensions of most widely used parts and units such as nuts, bolts, studs, screws, washers, cotter pins, rivets, keys, dowels, belts, chains, couplings, rolling contact bearings, etc.

(iv) Structural element of the majority of machine elements such as modules of gears and worm wheels, diameters and widths of pulleys, etc.

(v) Accuracy i.e. fits and surface finish of machine elements.

Two terms closely related to standardisation are "normalisation" and unification.

Normalisation is the standardisation established by separate departments or plants. The aim of unification of the machine elements and units is to minimise the types of the items by eliminating the superfluous diversity of products and production materials, etc. This is achieved by utilizing units and parts common to machines of different dimensions and applications.

Standardisation has got the following uses:

1. It makes the mass production possible, reducing the manufacturing costs and labour requirements.
2. The standardisation of specifications and methods of testing the machine elements helps in improving their quality and hence the service life.
3. The repair and maintenance of the machines is simplified since the worn out or damaged parts can easily be replaced by the standard ones.
4. It reduces the time and efforts needed to create and manufacture new machines since the standard elements and units can be used to assemble a new machine.

5.1.1 Preferred Numbers

The use of preferred numbers is to reduce unnecessary variations in the sizes of an article and thus minimising the variety of production and of the materials. These are specially recommended values and these specify the proportions of items and structures, power, capacities, speed and all other parameters used in production and expressed numerically. Supposing a machine is to be built in several sizes with different capacities of powers. It becomes necessary to decide the capacities which will cover a certain range efficiently and with a minimum number of sizes. Now it has been established by experience that a certain range can be covered efficiently with a minimum number of sizes by the use of geometrical progression with constant ratio. In our decimal system, it is easy and convenient to select a series of numbers from 10 to 100.

Coarse series with the ratio, $r = \sqrt[5]{10} = 1.6$ and written as R_5

the next series with the ratio, $r = \sqrt[10]{10} = 1.15$ and written as R_{10}

the next series with the ratio, $r = \sqrt[20]{10} = 1.12$ and written as R_{20}

and the finest series with the ratio, $r = \sqrt[40]{10} = 1.06$ and written as R_{40}.

These four series are called "Basic Series". The other series known as "Derived Series", can be obtained simply by multiplying the base size by 10, 100, etc., or by dividing the base sizes by 10, 100, etc. The preferred numbers in the series R_5 are 1, 1.6, 2.5, 4.0, 6.3, and 10. The numbers have been rounded. An international standard for preferred numbers is shown in Table 5.1.

Table 5.1

Ordinal numbers of series	Series of preferred numbers			
	R_5	R_{10}	R_{20}	R_{40}
0	1.00	1.00	1.00	1.00
1	—	—	—	1.06
2	—	—	1.12	1.12
3	—	—	—	1.18
4	—	1.25	1.25	1.25
5	—	—	—	1.32
6	—	—	1.40	1.40
7	—	—	—	1.50
8	1.60	1.60	1.60	1.60
9	—	—	—	1.70
10	—	—	1.80	1.80

Ordinal numbers of series	Series of preferred numbers			
	R_5	R_{10}	R_{20}	R_{40}
11	—	—	—	1.90
12	—	2.00	2.00	2.00
13	—	—	—	2.12
14	—	—	2.24	2.24
15	—	—	—	2.36
16	2.50	2.50	2.50	2.50
17	—	—	—	2.65
18	—	—	2.80	2.80
19	—	—	—	3.00
20	—	3.15	3.15	3.15
21	—	—	—	3.35
22	—	—	3.55	3.55
23	—	—	—	3.75
24	4.00	4.00	4.00	4.00
25	—	—	—	4.25
26	—	—	4.50	4.50
27	—	—	—	4.75
28	—	5.00	5.00	5.00
29	—	—	—	5.30
30	—	—	5.60	5.60
31	—	—	—	6.00
32	6.30	6.30	6.30	6.30
33	—	—	—	6.70
34	—	—	7.10	7.10
35	—	—	—	7.50
36	—	8.00	8.00	8.00
37	—	—	—	8.50
38	—	—	9.00	9.00
39	—	—	—	9.50
40	10.00	10.00	10.00	10.00

As an application of preferred numbers, supposing tractors are to be produced in five types to fulfil all the tasks of agriculture. The power needed for the tractors is to be established. Using series R_5, the powers will be 10, 16, 25, 40 and 63 h.p. Using series R_{10}, the power will be 12.5, 20, 31.5, 50 and 80 h.p. According to the preferred numbers, the load carrying capacities of automobiles should be; 16, 25, 40, 63 and 100 kN; and of dump lorries: 40, 63, 100, 160, 250, 400 and 630 kN.

The choice of a particular series will depend upon the quantity of production. The greater the quantity of production, the greater can be the number of typical sizes. For limited quantities, the series R_5 is used and as the production increases, change over is made to the series R_{10}, R_{20} and R_{40}.

Note: For detailed study of "preferred number", refer Indian Standards.

5.2 INTERCHANGEABILITY : TOLERANCES AND FITS

Interchangeability is a major feature of modern mechanical engineering and is indispensable to series and mass production of machine elements and parts and their assemblies. Uniformity is the essence of cheap mass production. With all the human skill it is never possible to make a part exactly to a given dimension due to the inaccuracy of manufacturing methods and other factors. If the dimension is to be maintained to within a very close degree of accuracy, lot of time will be consumed and even then there would be still variation. But for mass production where the work has to be done in a set competitive time, greater variations will result.

This fact is recognised and certain variations are always allowed in the size of the machine elements or parts. When an assembly is made of two parts, one is known as male (enveloped) surface and the other one as female (enveloping) surface. The female surface is referred as "Hole" and the male surface as "Shaft". Now supposing shafts and holes are to be assembled and these are to be manufactured in mass production. As explained above, certain variations in the sizes of the shafts and holes will be allowed during manufacture. If these variations are set in such a manner that any shaft will make an assembly with any hole and vice versa, the method is called an "interchangeable system".

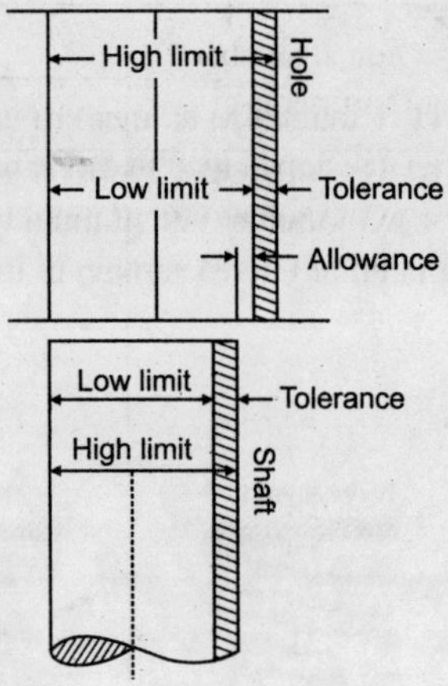

Fig. 5.1 *Limits and tolerances*

The dimension obtained by calculations for strength is called the "Basic dimension" or "Size" or "Nominal dimension", whereas the dimension as measured of a manufactured part is called the "Actual size". It is clear from the above analysis that the actual size will never be equal to the basic size and it is sufficient if it is within predetermined limits. The dimensions within which the actual size of a part may vary are called "Maximum Size" or "High limit", and "Minimum size", or "Low limit".

The difference between the High limit and the Low limit is called the "tolerance". If all the tolerance is allowed on one side of the nominal size, e.g., $2.500^{+0.003}_{-0.000}$, the system is said to be "Unilateral", but if it is divided, some being allowed on either side of the nominal size, e.g., $2.500^{+0.0015}_{-0.0015}$, the system is known as "Bilateral". Fig. 5.1 shows the limit and tolerance for hole and shaft.

Certain terminology connected with limits and tolerances is given below:

Deviation: It is the algebraic difference between a size (actual, maximum etc.) and the corresponding basic size.

Actual deviation: It is the algebraic difference between an actual size and the corresponding basic size.

Upper deviation: It is the algebraic difference between the minimum limit of size and the corresponding basic size.

Lower deviation: It is the algebraic difference between the minimum limit of size and the corresponding basic size.

Zero Line: It is a straight line to which the deviations are referred in a graphical representation of limits and fits.

It is a line of Zero deviation, and represents the basic size. When the Zero line is drawn horizontally, positive deviations are shown above and the negative deviations below.

Fundamental deviation: It is one of the two deviations that is conventionally chosen to define the position of the tolerance zone in relation to the zero line.

Tolerance Zone: It is the zone bounded by two limits of size of a part in the graphical representation of tolerance. It is defined by its magnitude and by its position in relation to the zero line.

The above definitions are explained in Fig. 5.2.

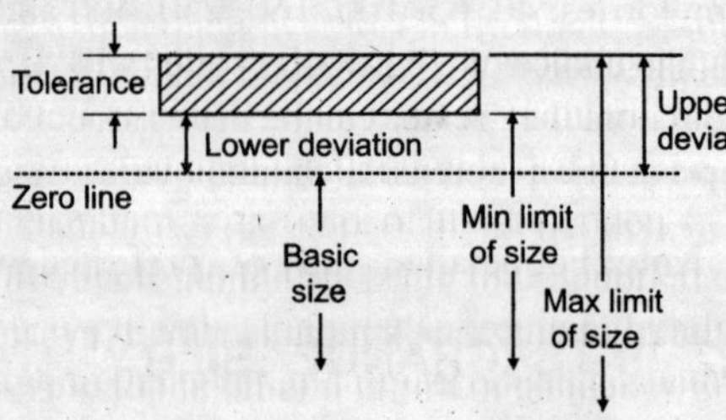

Fig. 5.2 *Terminology*

5.2.1 Fits

When machine elements are assembled to form a machine, their joints may be either movable or permanent depending upon the conditions in which machine elements work in a machine. For example, a shaft rotating in a bush can move in relation to it and so forms a moving joint, whereas a pulley is mounted on the shaft forms a fixed joint. The nature of joint or "fit" is characterised by the presence and size of the 'clearance' (for movable joints) or "interference (for fixed joints). To form a clearance, the shaft diameter should be smaller than the hole diameter while for interference, the shaft diameter is larger than the hole diameter.

Even if the manufacture of the mating parts is perfectly accurate, they should have different dimensions and vary from the nominal size. The difference in their sizes provides the required class of fit. This variation is called the "allowance". The three common types of fits are:

Clearance fit: In this type of fit, a clearance between the shaft and hole is always provided. The tolerance zone of the hole is entirely above that of a shaft.

Interference fit: This fit always provides an interference. The tolerance zone of the hole is entirely below that of a shaft.

Transition fit: This fit may provide either a clearance or an interference fit. The tolerance zones of the hole and shaft overlap.

The various allowances for different fits may be obtained either:

1. by keeping the hole constant and varying the shaft diameter to give the required fit.
2. keeping the shafts constant and varying the hole.
   The first one is known as "Hole Basis" method and the second one as "Shaft basis" method.

The "hole basis" system is most commonly used since it is easier and more convenient to manufacture shafts of varying sizes than holes of varying sizes.

The various fits are shown in Fig. 5.3. It is clear that lower deviation is zero for hole basic system and for shaft basis system, the upper deviation is zero.

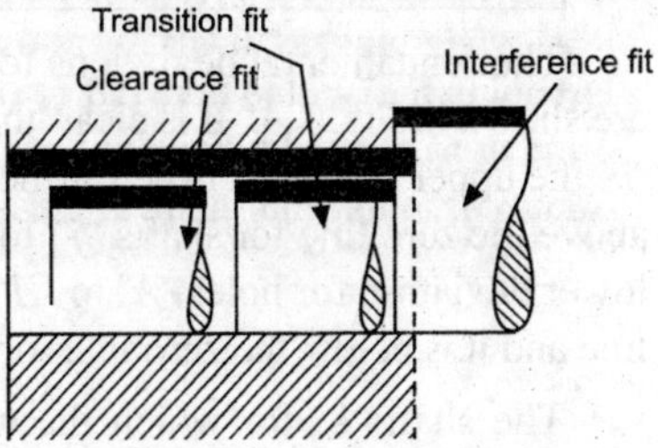

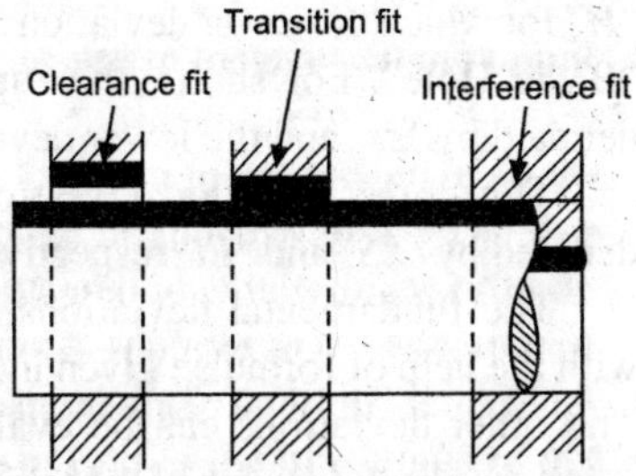

Fig. 5.3 *Fits*

5.2.2 The Indian Standard System

The system of limits and fits recommended in IS–919–1963 comprises 18 grades of fundamental tolerances, i.e., grades of accuracy of manufacture and 25 types of fundamental deviations indicated by letter symbols for both holes and shafts (capital letter *A* to *ZC* for holes and small letters a to *zc* for shafts) in diameter steps upto 500 mm.

Innumerable fits ranging from extreme interference to those of extreme clearance can be obtained by suitable combination of fundamental tolerances and fundamental deviations mentioned above. All but very exceptional engineering requirements are covered from very coarse work to fine gauze manufacture. A unilateral hole basis system is recommended but a full range of holes suitable for use on shaft basis system, unilateral, or bilateral, is also included.

The 18 grades of tolerances are designated as IT01, ITO, IT1 to IT16. These are known as standard tolerances. The value of standard tolerance is calculated as follows:

The value is in microns.

For IT01, value $= 0.3 + 0.008\,D$

For ITO, value $= 0.5 + 0.012\,D$ (D is in mm)

For IT1, value $= 0.8 + 0.020\,D$

The various diameter steps as per I.S.I. are: 1–3, 3–6, 6–10, 10–14, 14–18, 18–24, 24–30, 30–40, 40–50, 50–65, 65–80, 80–100, 100–120, 120–140, 140–160, 160–180, 180–200 mm. D is the geometric mean for a particular range of size.

The values of IT2 to IT4 are regularly scaled approximately geometrically between the values of IT1 to IT5. For grades 5 to 16, the values of tolerance are determined in terms of the standard tolerance unit, i in microns,

where $i = 0.45\sqrt[3]{D + 0.001D}$.

The values are given below:

Grades	IT5	IT6	IT7	IT8	IT9	IT10	IT11	IT12	IT13	IT14	IT15	IT16
Values	$7i$	$10i$	$16i$	$25i$	$40i$	$64i$	$100i$	$160i$	$250i$	$400i$	$640i$	$1000i$

The fundamental deviations for holes and shafts are shown in Fig. 5.4. It is clear that for shafts '*a*' to '*h*' the upper deviation is below the zero line and it is above the zero line for shafts "*j*" to *zc*. Similarly, the lower deviation for holes '*A*' to '*H*' is above the zero line and it is below the zero line for holes '*J*' to '*ZC*'.

The shaft '*h*' for which the upper deviation is zero is called the "Basic Shaft". Similarly the hole "*H*" for which the lower deviation is zero is called the "Basic Hole". For shafts, the upper deviation is denoted by '*es*' and the lower deviation by '*ei*'.

The corresponding deviations for holes are denoted by '*ES*' and '*EI*' respectively.

The fundamental deviations can be calculated with the help of formulae given in the specifications. The other deviations can be evaluated directly by using the absolute value of the tolerance IT. Thus, for shafts,

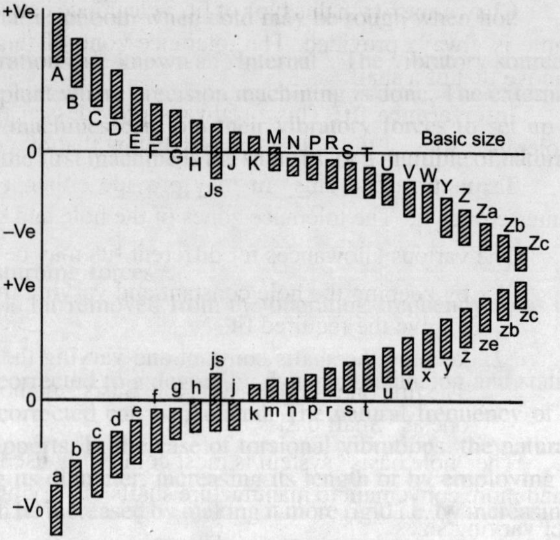

Fig. 5.4 *Fundamental deviations*

$$ei = es - IT \quad \therefore \quad es = ei + IT.$$

Similarly, for holes: $ES = EI + IT$

The values of the deviations for the shafts are given in Table 5.2, for sizes up to 500 mm.

Table 5.2

Upper Deviation (*es*) Lower Deviation (*ei*)

Shaft Designation	*es*, Microns	Shaft Designation	*ei*., Microns
a	$= -(256+1.3D);\ D \le 120$	*j*5 to *j*8	
	$= -3.5D,\ D \ge 120$	*k*4 to *k*7	$+0.6\ \sqrt[3]{D}$
b	$= -(140 + 0.85D);\ D \le 160$	*k* for grades	
	$= -1.8\ D;\ D \ge 160$	≤ 3 and ≥ 8	0
c	$= -\ 52\ D^{0.2};\ D \le 40$	*m*	+ (IT7 – IT6)
	$= -\ (95 + 0.8D);\ D > 40$	*n*	$+5D^{0.34}$
d	$-16D^{0.44}$	*p*	+ IT7+ 0 to 5
e	$-11D^{0.41}$	r	geometric mean of values of *ei* for *p* and *s*.
f	$-5.5D^{0.41}$	*s*	+IT8 + 1 to 4; $D \le 50$
g	$-2.5D^{0.34}$		+IT7 + 0.4*D*; $D > 50$
h	0	*t*	+IT7+ 0.63*D*
		u	+IT7+ *D*
		v	+IT7 + 1.25*D*
		x	+IT7 + 1.6*D*
		y	+IT7 + 2*D*
		z	+IT7 + 2.5*D*
		z_a	+IT8 + 3.15*D*
		z_b	+IT9 + 4*D*
		z_c	+IT10 + 5*D*

For holes, the deviations are derived from the corresponding values of shafts. For holes, the limits are same as the shaft limits of the same symbol i.e. letter and grade, but disposed on the other side of the zero line. Thus,

EI = Upper deviation (es) of shaft of the same letter symbol but of opposite sign

and ES = Lower deviation (ei) of shaft of the same letter symbol but of opposite sign.

The most commonly used shafts and holes are given below in Table 5.4.

Table 5.3 The values of the tolerance grades from IT1 to IT16 for diameter steps up to 500 mm are given below:

Diameter steps	Tolerance grades, microns									(1 Micron = 0.001 mm)						
in mm	IT1	IT2	IT3	IT4	IT5	IT6	IT7	IT8	IT9	IT10	IT11	IT12	IT13	IT14	IT15	IT16
Over 1 To and inc. 3	1.5	2	3	4	5	7	9	14	25	40	60	90	140	250	400	600
Over 3 To and inc. 6	1.5	2	3	4	5	8	12	18	30	48	75	120	180	300	480	750
Over 6 To and inc. 10	1.5	2	3	4	6	9	15	22	36	58	90	150	120	360	580	900
Over 10 To and inc. 18	1.5	2	3	5	8	11	18	27	43	70	110	180	270	430	700	1100
Over 18 To and inc 30	1.5	2	4	6	9	13	21	33	52	84	130	210	330	520	840	1300
Over 30 To and inc. 50	2	3	4	7	11	16	25	39	62	100	160	250	380	620	1000	1600
Over 50 To and inc. 80	2	3	4	5	13	19	30	46	74	120	190	300	460	740	1200	1900
Over 80 To and inc. 120	3	4	6	10	15	22	35	54	87	140	220	350	540	870	1400	2200
Over 120 To and inc. 180	4	5	8	12	18	25	40	63	100	160	250	400	630	1000	1600	2500
Over 180 To and inc. 250	5	7	10	14	20	29	46	72	115	185	290	460	720	1150	1800	2900
Over 250 To and inc. 315	6	8	12	16	23	32	52	81	130	210	320	520	810	1300	2100	3200
Over 315 To and inc 400	7	9	13	18	25	36	57	89	140	230	360	570	890	1400	2300	3600
Over 400 To and inc. 500	8	10	15	20	27	40	63	97	155	250	400	630	970	1550	2500	4000

Table 5.4

SHAFTS		HOLES	
Letter symbol ↓	Grades	Letter Symbol ↓	Grades
a	9,11	*A*	9, 11
b	8, 9, 11	*B*	8, 9, 11
c	8, 9, 11	*C*	8, 9, 11
d	5 to 11	*D*	6 to 11
e	5 to 9	*E*	5 to 10
f	4 to 9	*F*	5 to 9
g	4 to 7	*G*	5 to 7
h	1 to 16	*H*	1 to 16
i	5 to 7	*I*	6 to 8
js	1 to 16	*Js*	1 to 16
k	1 to 7	*K*	5 to 8
m	4 to 7	*M*	5 to 8
n	4 to 7	*N*	5 to 11
p	4 to 7	*P*	5 to 9
r	4 to 7	*R*	5 to 8
s	4 to 7	*S*	5 to 7
t	5 to 7	*T*	5 to 7
u	5 to 8	*U*	6 to 7
v	5 to 7	*V*	6 to 7
x	5 to 8	X	6 to 7
y	6 to 7	*Y*	7
z	6 to 7	*Z*	7 to 8
z_a	6 to 7	Z_A	7 to 8
z_b	7 to 8	Z_B	8 to 9
z_c	7 to 8	Z_C	8 to 9

A fit is designated by the basic size common to both hole and shaft, followed by symbols corresponding to each element, the hole is quoted first. If the basic size is 60 mm, the hole is H_7 and the shaft is f_8, then the fit will be indicated as 60 H_7/f_8.

Note: For thorough study on limits, tolerance and fits, refer: IS-919–1963; IS-2909–1964; IS-2709–1964.

The tolerance has a direct effect on the cost of manufacture of a part. Smaller the tolerance, the more accurately a part or machine element will be finished and the smoother will be the operation of the mechanism being designed. But the machining of part will be more expensive. As an approximation, it may be assumed that the product of the tolerance and cost of machine is constant. The variation of cost with tolerance is shown in Fig. 5.5. Therefore, the designer should analyse and decide the proper tolerance for a part and it should neither be less nor more than what is really necessary.

The most commonly used types of fits are given below in Table 5.5.

Table 5.5

Types of Fit	Class of Fit	With Holes				Remarks
		H_6	H_7	H_8	H_{11}	
CLEARANCE	Shaft a	—	—	—	a_{11}	Large clearance fits result, not widely used.
	b	—	—	—	—	— do —
	c	—	c_8	c_9*	c_{11}	Suitable for slack running fit.
	d	—	d_8	*$d_8 - d_{10}$	d_{11}	Suitable for loose running fits, e.g., loose pulleys, plummer blocks.
	e	e_7	e_8	*$e_8 - e_9$		For loose running or clearance fits, e.g. properly lubricated bearings.
	f	*f_6	*f_7	*f_8	—	Suitable for normal running fits, e.g., bearing] of small electric motors, pumps, etc. gearbox bearings.
	g	*g_5	*g_6	g_7	—	Suitable for close running fits or sliding fit. e.g., piston and slide values spigot or location fits.
	h	*h_5	*h_6	*h_7–h_8	h_{11}	For precision sliding. Used widely for non-running parts.
TRANSITION	j	*j_5	*j_6	h_7	—	Push fit for very accurate location with easy assembly and dismantling e.g., coupling spigots and recesses.
	k	*k_5	*k_6	k_7	—	True transition fit, e.g., keyed shaft, non-running locked pins etc.
	m	*m_5	*m_6	m_7	—	Medium keying fit, e.g. ball bearing races of medium duty.
	n	*n_5	*n_6	n_7	—	Heavy keying fit, for tight assembly for mating parts.
INTERFERENCE	p	p_5	*p_6	—	—	Light press fit with easy dismantling for non-ferrous parts. Standard press fit with easy dismantling for non-ferrous part assembly.
	r	r_5	*r_6	—	—	Medium drive fit with easy dismantling for ferrous parts assembly. Light drive fit with easy dismantling for non-ferrous parts assemlly.
	s	s_5	*s_6	s_7	—	Heavy drive fit ferrous parts permanent or semipermanent assembly. Standard press fit for non-ferrous metals.
	t	t_5	*t_6	t_7	—	Force fit on ferrous parts for permanent assembly.
	u	u_5	u_6	*u_7	—	Heavy force fit or shrink fit.
	v, x, y, and z	—	—	—	—	Very large interference fit. Not recommended for use.

*Fits recommended for common use.

5.3 MANUFACTURING PROCESSES

The manufacturing processes can be conveniently divided into two groups:

1. Processes which are used for the preliminary shaping of the machine parts, mostly by the use of heat.
2. Processes which give final shape of the machine part, by cold machining.

Processes included in the first group are casting, forging, rivetting and welding. Each process has its possibilities and limitations.

Second group includes the following processes: turning, boring, planing, shaping, milling, drilling, reaming, broaching, spot facing, grinding, gear and screw cutting, tapping, honing and polishing.

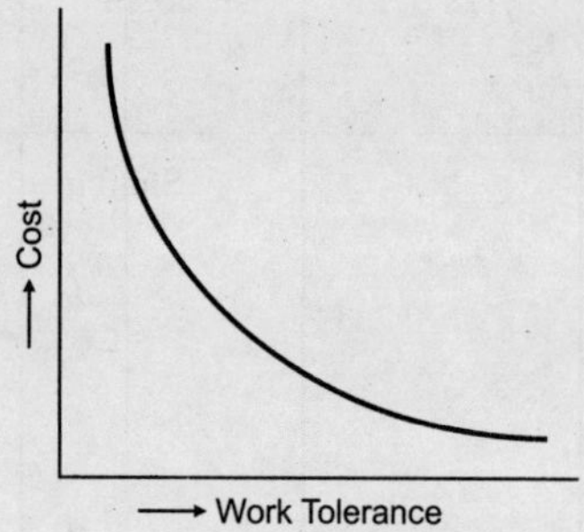

Fig. 5.5 *Variation of cost with tolerance*

5.3.1 Casting

It is one of the most widely used manufacturing methods in mechanical engineering. The following are the most commonly used processes for Iron and Steel Castings.

(i) Sand moulding, (ii) Permanent mould casting, (iii) Shell moulding, (iv) Centrifugal casting.

Another casting process is "Die Casting" which is mostly used for non-ferrous metals. Sand-casting process has the following advantages:

(i) The choice of the metals which can be cast is very wide. Most metals and alloys can be cast.
(ii) Parts of considerable complexity of shape can be manufactured by this process.
(iii) Parts of largest sizes can be made as sand casting.

5.3.1.1 Iron Casting

As already discussed in chapter, cast iron is available in many types. These types and their properties are given in Table 5.6.

From the table, it is clear that Grey cast iron has got excellent cast-ability properties, therefore, it is used most widely in castings.

Table 5.6

Type of Cast Iron	Hardness BHN	Max. Service Temp. °C	Abrasive wear Resistance	Damping Capacity	Corrosion Resistance	Weld-ability	Machin-ability	Cast-ability
Grey	140–300	370	P to F	E	P to F	P	G	E
White	300–575	1100	E	P	F	VP	VP	G
High Alloy:								
Wear Resistant	250–700	650	E	P	F	VP	VP	G
Heat Resistant	100–500	980	P to G	P to G	F to E	P	F to P	G
Corrosion resistant	100–500	810	P to G	P to G	G to E	P	E to P	G
Ductile	140–300	400	P to F	F	P to F	P	E to F	E to G
Malleable:								
Standard	110–156	400	P	F	P to F	P	E	G
Pearlitic	150–300	400	F	F to P	P to F	P	G to F	G

E = Excellent; G = Good; F = Fair; P = poor; VP = Very poor

Table 5.7

Types of Steel	Hardness BHN	Max. Service Temp °C	Abrasive wear Resistance	Damping Capacity	Corrosion Resistance	Weld-ability	Machin-ability	Cast-ability
Low C	108–150	480	P	P	P	E	F	F
Medium C	140–240	480	F	P	P	G	F	F
High C	200–400	480	G	P	P	G	F to P	F
Low alloy	105–500	540	F to G	P	P to F	G to F	F	F
High alloy:								
Heat resistant	130–400	1160	P to G	P	G to F	G to F	F to P	F
Corrosion resistant	140–400	980	P to G	P	G to E	G to F	F to P	F
Wear resistant	180–550	260	F	P	F	F	F	F

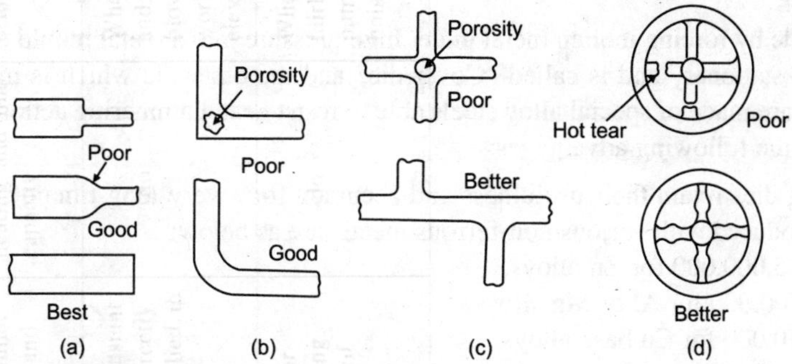

Fig. 5.6

The following basic rules should be followed in the design of castings:

1. If possible, the castings should be designed with uniform section thickness. Where it is not possible, light sections should be blended into heavy ones. An abrupt change in section thickness must always be avoided to avoid shrinkage defects and stresses concentration at the sharp corners. This can be achieved by the use of fillets.
2. The minimum thickness for grey iron casting can be as low as 3 mm, but the average value is 6 mm for parts up to 45 cm in length or diameter and it gradually increases to 18 mm for large and heavy castings. For malleable and ductile iron castings the minimum thickness may be taken as 3 mm but the average minimum thickness for heavier castings should be about 6 mm.
3. Ribs should be used chiefly for static loads. These should be avoided where impact loads are expected since these increase the rigidity of the parts. Wide and low ribs are safer than thin and high ones.
4. Since cast iron is much more stronger in compression than in tension, so, the iron castings should be loaded in compression as far as possible. The tensile and bending stresses can be eliminated or minimised by proper design.
5. Don't use iron casting for impact and shock loading.
6. Don't use cast iron at temperatures above 300ºC since its strength decreases after 400ºC.
7. Avoid concentration of material so that no shrinkage cavities are formed.
8. In the case of Grey Iron castings, the patterns and cores should be simplest and holes should be provided to knock out cores. If possible, the casting should be so designed that no cores are required.
9. Parting lines should be made as even as possible.
10. Patterns should be provided with ample draft for their easy withdrawal from the mould.

Figure 5.6 shows design features to alleviate problems due to:

(a) wall thickness variation
(b) hot spot in corner.
(c) hot spot at cross ribs.
(d) hot tearing.

5.3.1.2 Steel Casting

Steel is not as fluid as cast iron, therefore, complicated shapes, sharp corners and thin sections can be casted. Of the various types of steel, medium carbon steel is most commonly used in steel castings. The various types of steels and their properties are given in Table 5.7.

The minimum thickness of steel castings is usually taken as 6 mm. However, in small casting, this can be equal to 4.5 mm. All steel castings must be annealed to relieve them of internal stresses.

Their strength can be increased by subsequent heat treatment.

5.3.1.3 Die Casting

Die castings are made by forcing molten metal under high pressure into a metal mould called a "die". There are two dies, one is stationary and is called "Cover die" and the other die which is movable is known as "Ejector die". Dies are made of special alloy steels able to resist heat, hammering action, erosion, and wear. Die casting has got the following advantages:

1. Die casting dies retain their usefulness and accuracy for a very long time of production. The die castings produced with various non-ferrous metals are as below:
   500,000 to 5,000,000 for Zn alloys.
   50,000 to 500,000 for Al or Mg alloys.
   20,000 to 40,000 for Cu base alloys.
2. High production rates.
3. Close dimensional tolerances.
4. Thin sections can be cast.
5. Intricate parts can be die cast.
6. Good surface finish and good details on the cast surfaces can be obtained.
7. Walls of varying thickness can be obtained in the same casting.
8. Cored holes down to 0.75 mm diameter are possible at accurate locations.

At the same time, die casting has the following limitations:

1. Only certain non-ferrous alloys can be cast economically, such as Zinc, Aluminium.
2. The maximum size of die casting is limited by the size of the dies and capacities of the die casting machines available. For, Zn, the maximum size is about 90 kg and for Al about 35 kg.
3. Due to high cost of dies and machines, the die casting is economical only for mass production.
4. Due to some entrapped air, the die castings are usually porous resulting in reduced mechanical properties.

In addition to the points for sand casting, the following points should be kept in mind while designing a die casting:

1. As the light alloys are more sensitive to stress concentration, large fillets should be provided, notches should be avoided and the wall thickness should be kept as much uniform as possible. The minimum thickness for brass and bronze is 2.3 mm, and for Al it is 3 mm.
2. The light alloys have high coefficient of linear expansion. They loose their strength at temperatures above 150ºC. In no case, the die casting be used at temperatures above 260ºC.
3. Die castings should be designed so that the cost of flash removal is minimum.
4. Allowance should be made for the surfaces to be machined after die casting.

5.3.2 Forging

A forging process can be defined as the plastic deformation of a metal caused by the stresses imposed on it through hammering or squeezing operations. The forging processes can be classified as below:

1. Smith forging or hand forging.
2. Drop forging.
3. Machine forging or upset forging.
4. Press forging

The forging process has got the following advantages:

(i) Better mechanical properties, especially if the fibre flow lines are properly directed.
(ii) Minimum weight.
(iii) Ability to withstand unpredictable loads.
(iv) Minimum metal removal in machining.
(v) Relatively smooth surfaces of the forging.
(vi) Freedom from internal defects.
(vii) Rapid duplication of the parts.
(viii) Wide range of forgeable metals is available.
(ix) Forgings are readily welded.
(x) Forgings can be held within fairly close dimensional tolerances.

Forging process has got the following limitations:

(i) Due to high temperature of metal, there is a rapid oxidation or scaling of the surface resulting in poor surface finish.
(ii) Tooling and handing costs are high.
(iii) Many intricate and cored shapes possible by casting process cannot be forged.
(iv) Usually forgings cost more than castings.

The following points should be kept in mind while designing a forging:

1. The directions of the fibre flow lines should be kept in mind so that full advantages are taken of them.
2. Adequate draft should be provided to permit easy removal of the forgings from the dies.
3. If possible, the parting line on a forging should be in one plane.
4. Ample fillets and curvatures should be given to the forging.
5. The parting line, if possible, should divide the forging about in half. This will facilitate the forging operation.
6. Ribs should be low and wide.
7. Pockets and recesses should be minimum to avoid increased die wear.
8. Sections should not be so thin as to restrict the flow of metal.

5.3.3 Welding

Welding may be defined as the metallurgical joining together of two metal pieces to produce a single piece of the metal. The most widely used welding processes are arc-welding and gas-welding. Welding is especially suitable for low quantity production to avoid the time and cost necessary to obtain a foundary pattern or a forging die. Most weldments are made of low carbon steel.

The main advantages of welded joints over casting process are:

(i) Lighter weight.
(ii) Lower cost.
(iii) Greater strength.

(iv) Saving of material.
(v) Maximum homogeneity.

The main advantages of welded joints over riveted joints are: lighter weight, greater strength, lower cost and noiselessness.

The following points should be kept in mind while designing a weldment:

1. Weldments should be designed to require a minimum of weld metal.
2. Thermal contraction of metal, which has been heated by welding may cause internal residual stresses and distortion. These can be controlled or reduced by:

   (a) Preheating
   (b) Minimum number of welds
   (c) Smallest size of weld that fulfills requirements.
   (d) Maximum use of intermittent welds.

3. Sharp discontinuities in metal should be kept at a minimum since these cause stress concentrations.
4. An important strength weld should not be located where much of it may be removed later by machining.
5. Welds should be located so that adequate strength will be provided at the proper places on a structure or part.
6. As far as possible, a straight line force pattern should be provided.
7. Laps, straps and stiffening angles should be provided.
8. Where ever possible use butt welds.
9. The ends to be welded together must be of equal thickness.
10. Welds at the vulnerable cross-section should be avoided.
11. The use of welding fixtures should be avoided.
12. Welds should not be subjected to bending.
13. A weld should not be located at the point of maximum deformation.
14. Ribs should be designed correctly and these should be used with care.

5.3.4 Manufacturing Process for Plastics

The various manufacturing processes for plastics are:

(i) Compression moulding.
(ii) Transfer moulding.
(iii) Injection moulding.

While designing the moulded plastic parts, the following design rules should be followed in addition to the general rules already enumerated for sand casting and die casting:

1. Allow for shrinkage after moulding.
2. Allow atleast a minimum draft of 1/2 to 1 degree for easy withdrawal of the parts from the mould.
3. Avoid under cuts since these will need cores or split cavity moulds.
4. If possible, the parting line should be located in one plane.
5. Design corners with ample radii of fillets.
6. Design for uniform section thickness and uniform mass distribution for optimum flow of the plastic in the mould.
7. Avoid sharp discontinuities to eliminate stress concentration.
8. Ribs should be used to increase strength and rigidity and to reduce distortion.
9. Rib height should not be more than its thickness.
10. Specify as shallow a draw as possible.

5.3.5 Comparison of Manufacturing Processes

The various manufacturing processes have been compared in Tables 5.8 to 5.11, on the basic of joinability, metal forming, Design considerations and cost considerations.

5.4 SURFACE ROUGHNESS

Whatever may be manufacturing process, the machine elements or parts retain the surface irregularities left after machining. Surface roughness or surface finish is of great importance in mechanical engineering. Whenever two machine parts are assembled, the type of joint and the class accuracy will depend upon the surface roughness of the mating surfaces. The higher the requirements for the nature of the joint and accuracy of the part, the higher the required class of surface finish, i.e., the small irregularities of the surface.

Selection of the class of accuracy and surface finish is a vital stage in design, since it predetermines the efficiency of the joint operation, the manufacturing cost of the part and the method of machining. The machining operation giving high class of accuracy and surface finish will be more expensive. Therefore, unless it is absolutely necessary, high surface finish is not desired.

Two methods of assessing the surface roughness are: Root Mean Square value (R.M.S. value) and the Centre Line Average value (C.L.A. value). In both the methods, the surface roughness is measured as the average deviation from a nominal surface. The R.M.S. value is the square root of the arithmetic mean of the square of the irregularity heights. The distance above the centre line is taken as positive and below it as

Table 5.8 Manufacturing process: joinability considerations

Material	Welding			Brazing	Soldering	Adhesive bond [TS, TP, EM][1]	Adhesive bond[2]	Thread Fastening	Riveting
	Arc	Oxyacet	Resistance						
Cast Iron	C	R	O	D	O	(TS) C, (TP) O	C	C	R
Carbon steel	R	R	R	R	D	(TS) C, (TP) O	C	R	R
Stainless steel	R	C	R	R	C	(TS) C, (TP) O	C	R	C
Aluminium-Magnesium	C	C	C	C	O	(TS) R, (TP) O	R	C	C
Copper	C	C	C	R	R	(TS) C, (TP) O	C	C	C
Nickel	R	C	R	R	C	(TS) C, (TP) O	C	R	C
Titanium	C	×	C	D	O	(TS) O, (TP) ×	C	C	×
Lead	C	C	×	×	R	R	R	R	R
Zinc	C	C	O	×	R	R	R	R	R
Thermoplastic	R	R	C	×	×	C	O	O	R
Ceramics	×	O	×	×	×	R	R	C	O
Leather	×	×	×	×	×	(EM) R, (TS) C	R	O	R
Fabric	×	×	×	×	×	(EM) R	R	O	R

R : Recommended; C: Common; D = Difficult; O: Seldom used; × : Not used
[TS; Thermoset; TP : Thermoplastic : EM : Elastomeric][1]; [2](Modified Compression Epoxy)

Table 5.9 Manufacturing processes: Metal forming considerations

Manufacturing Process	Irons	Low C-steel alloys	Heat and corrosion resistant alloy	Aluminium alloy	Copper alloy	Lead alloy	Magnesium alloy	Nickel alloy	Precious Metals	Tin alloy	Titanium alloy	Zinc alloy
Sand casting	A	A	A	A	A	B	A	A	—	B	—	B
Shell mould casting	A	B	B	A	A	—	—	B	—	—	—	—
Permenant mould casting	A	B	—	A	B	B	A	B	—	B	—	B
Die casting	—	—	—	A	B	A	A	—	—	B	—	A
Plaster mould casting	—	—	—	A	A	—	—	—	—	—	—	—
Investment casting	—	A	B	A	A	—	A	B	B	—	—	—
Centrifugal casting	A	A	A	B	B	—	—	B	—	—	—	—
Drop forging	B	A	A	B	B	—	B	B	—	—	B	—
Press forging	—	A	A	B	B	—	B	B	—	—	B	—
Upset forging	—	A	A	B	B	—	B	B	—	—	B	—
Cold headed parts	—	A	B	A	A	B	—	B	B	—	—	—
Stampings, Drawing	—	A	B	B	A	—	B	B	B	—	B	B
Spinning	—	A	B	A	A	B	B	A	B	—	B	B
Screw Machining	B	A	B	A	A	—	B	A	B	—	B	B
Roll forming	—	A	—	B	B	—	B	—	—	—	B	B
Extrusion	—	B	—	A	A	B	A	B	—	B	B	—
Powder Metallurgy	A	A	B	B	A	—	—	B	B	—	B	—
Electro forming	A	—	—	B	A	B	—	A	A	B	—	B

A: Materials most Frequently used; B: Also materials currently used.

Table 5.10 Manufacturing Process: Design Considerations

S. No.	Manufacturing Process	Material choice	Complexity of part	Maximum Size	Minimum Size	Mechanical Properties	Precision and Tolerance	Special Structural Properties	Surface Smoothing	Surface Detail	Remarks
1.	Sand casting	Wide; Ferrous Non ferrous Light Metals	Considerable	Largest	3mm	Fair to High	± 0.6 to 0.3mm per mm	Good Bearing structure	Poor	Poor	Usually require some machining before use
2.	Shell Mould Casting	Wide	Moderate	Best for smaller parts	1.5mm	Good	± 0.003 to 0.005 mm per mm	High Quality for cast metals	Good	Good	Best of low costing method
3.	Permanent Mould casting	Restricted: brass, bronze, aluminium	Limited	Moderate	2.5mm	Fair good	± 0.6 mm per mm	None	Good	Good	Economical for mass production
4.	Plaster Mould casting	Narrow: Brass, bronze aluminium	Considerable	Moderate	0.75mm	Fair	± 0.25 to 0.125 mm	None	Good	Good	Little finishing required.
5.	Investment casting	Wide	Considerable	Moderate	0.75mm	Good	± 0.125 mm per mm	None	Excellent	Excellent	Best for too complicated parts.
6.	Die casting	Narrow: Zinc, aluminium, brass, magnesium	Considerable	Large	0.06 mm	Fair to Good	± 0.025 mm per mm	None	Good	Good	Most economical where applicable
7.	Drop Forging	Medium:	Moderate	Large	Small	High	± 0.25 to 0.75 mm	Toughness	Fair	Fair	Used for high strength
8.	Press Forging	Medium: best for non ferrous alloys	Limited	Moderate	Small	High	Medium better than drop forging	Toughness	Fair	Fair	Greater complexity than drop forging.
9.	Upset Forging	Medium: Many ferrous and non-ferrous alloy	Limited	Medium	Moderate	High	Medium	Toughness	Medium	Medium	Best suited to small parts.
10.	Welding	Wide	Considerable	Unlimited	Moderate	Variable	Medium, not highly precise	Toughness	Depends upon components	None	Versatile process.

Table 5.11 Manufacturing processes: cost considerations

S.No.	Manufacturing Process	Raw Material Cost	Tool and Die Costs	Direct Labour Cost	Finishing Cost	Scrap loss
1.	Sand Casting	Low to Medium depending on metal	Low	High	High	Moderate; Scrap can be remelted.
2.	Shell Mould Casting	Low to Medium	Low to Moderate	Low	Low	Low
3.	Permanent Mould Casting	Medium	Medium	Moderate	Low to Moderate	Low
4.	Plaster Mould Casting	Medium	Medium	High	Low	Low
5.	Investment Casting	High: best for Special and costly alloys	Low to Moderate	High	Low	Low: Scrap is remelted.
6.	Die Casting	Medium	High	Low to Medium	Low	Low: Scrap can be remelted.
7.	Drop forging	Low to Moderate	High	Medium	Medium	Moderate
8.	Press forging	Low to Moderate	High	Medium	Medium	Moderate
9.	Upset forging	Low to Moderate	High	Medium	Medium	Medium: Lowest of forging processes
10.	Welding	Low to Moderate	Low to Moderate	Medium to high	Medium	Low: Practically none.

negative. The C.L.A. value is defined as the sum of all the areas between the surface and the centre lines both above and below it and divided by the length chosen. Both the value are measured in micro metres (μm).

1 micro metre = 10^{-6} m.

According to the Indian Standards, the surface roughness is assessed in terms of 'CLA' value denoted as *Ra*. The surface roughness is represented as in Fig. 5.7.

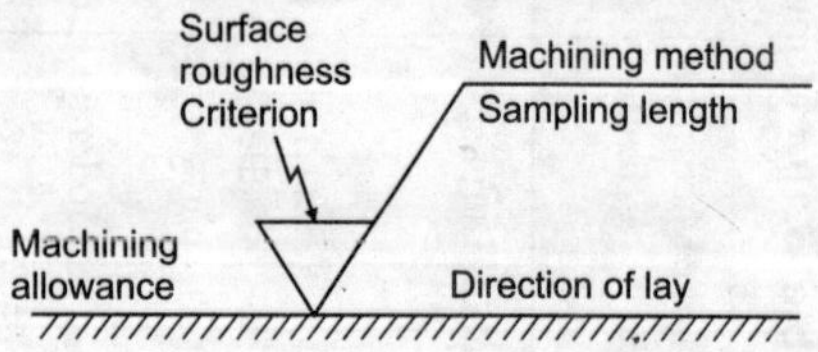

Fig. 5.7 *Representation of surface roughness.*

The direction of lay is represented by the following symbols:

= Lay parallel to the line representing the surface to which the symbol is applied.

⊥ Lay perpendicular to the line representing the surface to which the symbol is applied.

X Lay angular in both directions to line representing the surface to which the symbol is applied.

M Lay multidirectional.

C Lay approximately circular relative to the centre of the surface to which the symbol is applied.

R Lay approximately radial relative to the centre of the surface to which the symbol is applied.

Supposing the value of *Ra* is 0.15 μm with cut off length 0.30 mm and the directions of lay being perpendicular and the method of machining lapping, these things will be symbolically represented as shown in Fig. 5.8. The machining allowance is 0.15 mm.

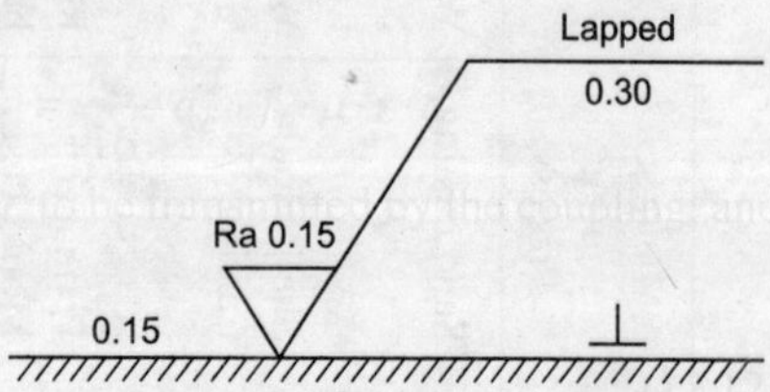

Fig. 5.8

The surface roughness on drawings is represented by the symbols shown below:

Symbol	*Ra* value in μm
∇	8 to 25
∇∇	1.6 to 8
∇∇∇	0.025 to 1.6
∇∇∇∇	< 0.025.

The range of surface roughness that can be produced by various machining operations is given in Table 5.12.

Table 5.12

Machining operation	*Ra* value in μm	Machining operation	*Ra* value in μm
Flame cutting	12.5 to 25	Drilling	0.5 to 6.25
Chipping and Hand sawing	12.5 to 25	Reaming	0.5 to 2.5
Hand grinding	6.25 to 25	Surface grinding	0.5 to 6.25
Disk grinding, Filling	2.5 to 25	Cylindrical grinding	0.25 to 2.5
Turning, Milling, Shaping	1 to 12.5	Honing, Lapping	0.05 to 0.50
Radical cut off saw	1 to 12.5	Polishing, Buffing	0.05 to 0.50
Boring	0.5 to 6.25	Super finish	0.025 to 0.25

SOLVED EXAMPLES

Example 5.1. Find the values of allowance, hole tolerance and shaft tolerance for the following dimensions of mated parts according to basic hole system.

Hole: 37.50 mm **Shaft: 37.47 mm**
37.523 mm **37.445 mm**

Solution: (i) Hole tolerance = high limit – low limit = 37.523 –37.500 = 0.023 mm

(ii) Shaft tolerance = 37.470 –37.445 = 0.025 mm

(iii) Referring to Fig. 5.1. Minimum allowance = Low limit of hole – H.L. of Shaft

= 37.500 – 37.470 = 0.030 mm

Maximum allowance = H.L. of hole – L.L. of shaft = 37.523 – 37.445 = 0.078 mm.

Example 5.2. A 75 mm shaft rotates in a bearing. The tolerance for both shaft and bearing is 0.075 mm and the required allowance is 0.10 mm. Determine the dimensions of the shaft and the bearing bore with the basic hole standard.

Solution: It is clear that,

L.L. of Hole = 75.00 mm

H.L. of Hole = 75.00 + 0.075 mm = 75.075 mm

H.L. of Shaft = L.L. of Hole – allowance

= 75.00 – 0.10 = 74.90 mm

L.L. of Shaft = 74.90 – 0.075 = 74.825 mm.

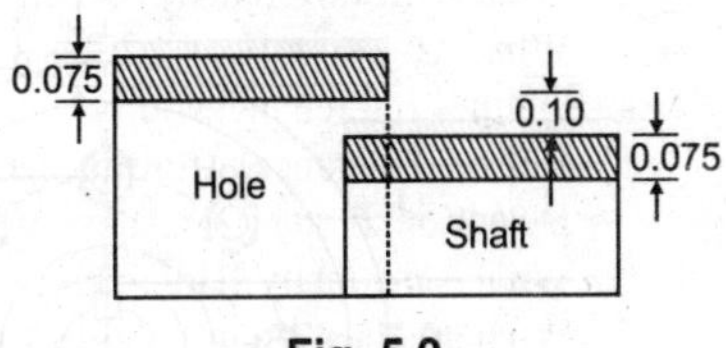

Fig. 5.9

Example 5.3. A medium force 0.225 mm fit on a 75 mm shaft requires a hole tolerance of 0.225 mm, a shaft tolerance of 0.223 mm and an average interference of 0.0375 mm. Determine the proper hole and shaft dimension with the basic hole standard.

Solution: It is clear from the Fig.

L.L. of Hole = 75.00 mm

H.L. of Hole $= 75.00 + 0.225 = 75.225$ mm

L.L. of Shaft $=$ H.L. of Hole + Interference

$= 75.225 + 0.0375$

$= 75.2625$ mm

H.L. of Shaft $= 75.2625 + 0.225$

$= 75.4875$ mm

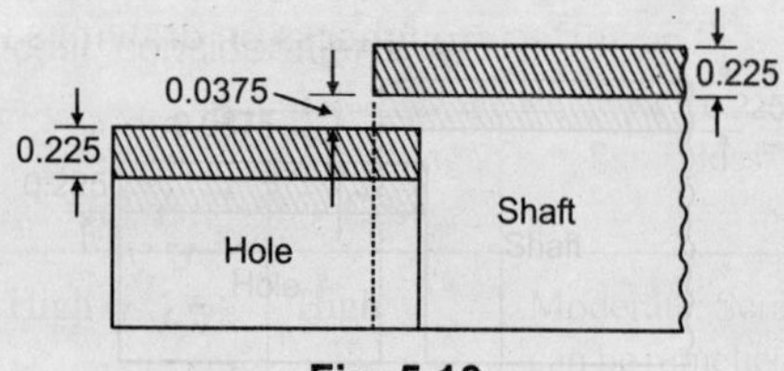

Fig. 5.10

Example 5.4. Calculate the fundamental deviation and tolerances and hence obtain the limits of sizes for the hole and shaft in the following fit: 60 mm H_8f_7. The diameter steps are 50 mm and 80 mm.

Solution: (a) Now tolerance unit, $i = 45\sqrt[3]{D} + 0.001\,D$ microns

D = geometric mean of diameter steps $= \sqrt{50 \times 80} = 63.25$ mm

$\therefore \quad i = 0.45\sqrt[3]{63.25} + 0.001 \times 63.25 = 1.856$ microns $= 0.001856$ mm

Now for hole of grade 8, the fundamental tolerance $= 25i =$ **0.0464 mm**

and for shaft of grade 7, the fundamental tolerance $= 16i =$ **0.0297 mm**

(b) It is clear from Fig. 5.4 and Table 5.2 that the fundamental deviation for hole '*H*' = 0. From Table 5.2, the fundamental deviation for shaft '*f*'

$= -5.5\,D^{0.41}$ microns

$= -5.5\,(63.25)^{0.41}$

$= -30.113$ microns $= 0.0301$ mm

(c) Limits of size

(i) Hole: L.L. of size

$=$ Basic size $= 60$ mm

H.L. of size = L.L. + tolerance

$= 60.0000 + 0.0464 = 60.0464$ mm

(ii) Shaft: H.L. of size = basic size – fundamental deviation

$= 60.0000 - 0.0301 = 59.9699$ mm

L.L. of size = H.L. of size – tolerance

$= 59.9699 - 0.0297 = 59.9402$ mm.

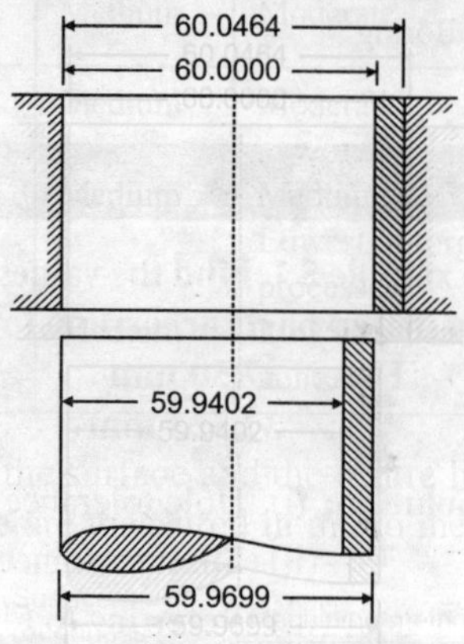

Fig. 5.11

EXERCISES

1. What is the purpose of "standardisation"?
2. What is the difference between "normalisation" and "unification"?
3. What is the significance of preferred Number?
4. Define "Interchangeability" and discuss its importance.
5. Define: Tolerance, Allowance, Low limit of size, High limit of size, Unilateral system and bilateral system tolerance. Deviation, fundamental deviation, Zero line, Tolerance Zone.
6. What is meant by "Hole-Basis system" and "Shaft-Basis system"? Which is better?
7. What are the various types of fits? Explain them with the help of neat diagrams.
8. What is meant by "Basic Hole" and "Basic Shaft"?
9. How will you write the fit: shaft '*m*' of grade 5 and Hole '*H*' of grade 11 and the basic size is 40 mm.
10. What are the advantages and disadvantages of Casting?
11. What precautions should be observed in the case of sand casting?
12. What are the advantages and limitations of Die Casting?
13. Classify the forging process.
14. Enumerate the advantages and limitations of forging process?

15. What precautions should be observed while designing a Forging?
16. What are the advantages of welding over other manufacturing processes?
17. What precautions should be observed while designing a Forging?
18. Discuss the importance of surface finish in Machine Design.
19. How would you assess the surface roughness?
20. What are the units of surface roughness?
21. How the various grades of surface roughness are represented?
22. Determine the values of allowance, hole tolerance, and shaft tolerance for the following dimensions of the mated parts with the basic hole system:

    Hole: 43.75 mm, 43.755 mm — Shaft: 43.725 mm, 43.708 mm

23. A medium force fit on a 75 mm diameter shaft requires a hole tolerance of 0.0225 mm, a shaft tolerance of 0.0225 mm and an average interference of 0.0375 mm. Determine the proper dimensions of the hole and shaft.
24. A hole is dimensioned as $25^{-0.03}_{+0.03}$ mm and the shaft is dimensioned as $25^{-0.02}_{+0.00}$ mm. Determine the hole tolerance and the allowance of the fit. What type of fit will be established?
25. Calculate the tolerance, fundamental deviation and hence obtain the limits of size for the hole and shaft in the following fit: 25 mm $H_8 - d_{10}$. The diameter steps are 18 mm and 30 mm.
26. Recommend the suitable class of fit for the following components:-

    (a) Gear box bearings (b) keyed shaft (c) permanent assembly

27. Preferred numbers are arranged:

    (a) Logarithmically (b) Randomly (c) in G.P.

28. According to BIS, the total number of grades of tolerances are:

    (a) 6 (b) 12 (c) 18 (d) 24

CHAPTER 6

Seals and Gaskets

6.1 DEFINITIONS

(a) Seals: In engineering practice, one comes across various machines, pipe fittings and certain types of expansion pieces, which have moving joints. Due to over pressure, the fluids such as gas, steam or liquids leak through the gaps between the moving and stationary parts. To prevent or reduce these leaks, seals are employed. Seals are equally effective to prevent the flow of a substance between two machine members which do not have relative motion. The basic functions of seals may be summarized as below:

1. To seal in oil, grease or other liquids, e.g., for preventing the leakage of oil from around a rotating shaft or for confining grease inside a ball or roller bearing.
2. To seal out dirt, fluids and contaminants.
3. To maintain applied pressure or vacuum and also to maintain the desired pressure difference between the inner and outer portion of a sealed volume.
4. To allow a controlled atmosphere within a sealed volume, e.g., in refrigeration system and turbines.

(b) Gaskets: A gasket is a relatively elastic or plastic device which is used to create and maintain a barrier against the flow of fluids across mating surface of a mechanical assembly, when the surfaces do not have a relative motion with respect to each other. The common example of the gasket is the member placed between the cylinder head and cylinder block of an engine to make the joint tight. When the gasket is used, the mating surfaces need not be accurately machined as in the case of metal joint.

6.2 SEALS

The design of a sealing device will depend upon the type and properties of the sealed medium, i.e., pressure and temperature, as well as by the speed and direction of movement of the sealed parts. The basic types of seals are given below:

(a) Face seals.
(b) Dynamic shaft seals.
(c) Split ring seals.
(d) Diaphragm seals.
(e) Labyrinth or positive clearance seals:
(f) Compression packings.
(g) Moulded packings.

6.2.1 Face Seals

The function of a mechanical face seal is to effect a pressure tight seal between a rotating shaft and a member through which the shaft passes. Sealing is achieved by continuous contact between two smooth ultra flat radial faces, one attached to the housing and the other to the shaft. There can be different design details for the face seals, but all make use of the following elements:

(a) Rotating seal ring.
(b) Stationary seal ring.
(c) Spring loading devices.
(d) Static seals.

The purpose of the spring is to allow one face to move axially to take up wear or play.

The spring assembly also helps in energizing the end face member axially to keep the faces together during periods of shut down or lack of hydraulic pressure. The static seal insures the sealing of the shaft. The static seal may be in the form of O-ring, V-ring, U-cup, wedge, or bellows.

A basic end face seal is shown in Fig. 6.1. A shaft with an O-ring as its sealing member is provided with a housing that incorporates one of the sealing faces. The housing encloses the O-ring and effects a preload on the shaft, thus insuring its sealing. A stationary member is provided in the end.

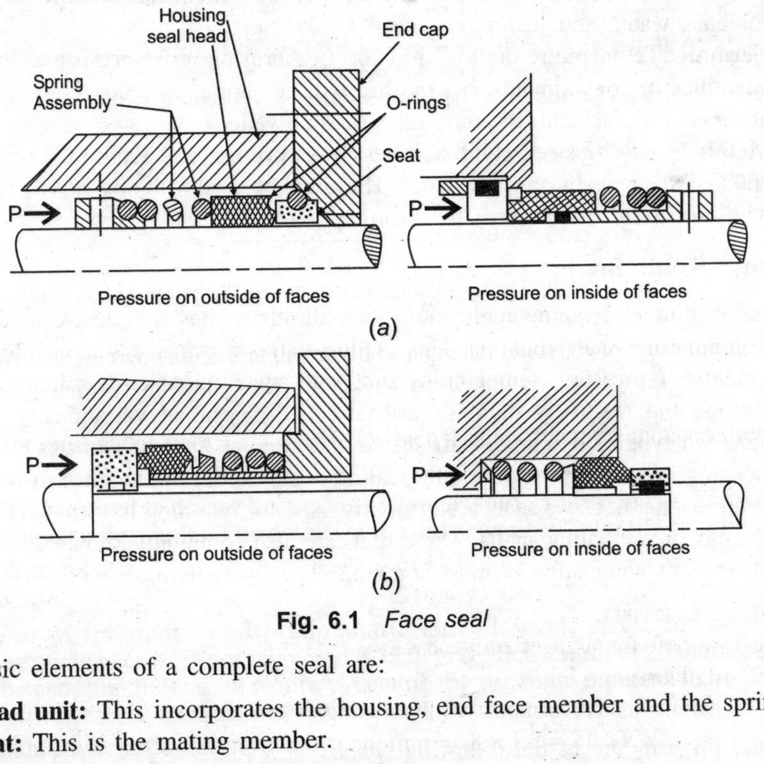

Fig. 6.1 *Face seal*

The two basic elements of a complete seal are:

1. **Seal head unit:** This incorporates the housing, end face member and the spring assembly.
2. **Seal seat:** This is the mating member.

There are two designs for the seal head: (a) Rotating and (b) Stationary. If the head rotates, the seat is stationary and vice versa. Rotating heads are used with shaft made of high grade material and which have been machined to a close tolerance. They are generally assembled on relatively simple sleeves or shafts. Stationary heads are used for comparatively high shaft speeds, because a stationary seal head requires less critical dynamic balancing as compared to rotating head. Where operating conditions are simple, these seals are more practical because of the space available for seal configuration. One drawback of the stationary head is that, close bore tolerance and high grade polish or finish is needed for the material. The design of the seat varies with rotating and stationary seals.

(*a*) **Application of face seals:** Face seal can function positively, both statically, and dynamically, with bidirectional shaft rotation, large pressure reversals, and a wide diversity of mediums, temperatures, pressures and speeds. Even if the shaft bore is slightly misaligned or concentric, these seals will operate satisfactorily and these seals do not cause shaft wear. Also the shaft finish, roundness, hardness and materials are not highly critical.

These seals have been upto pressure of 21 N/mm^2 in heavy duty process pumps. Special design have been developed for pressures upto 70 N/mm^2 for speed exceeding 250 m/s in aircraft applications and for temperatures between –200ºC to +655ºC. High pressures can also be handled by employing tandem seal arrangement.

(*b*) **Materials:** The common materials used for face seals are given below:

(*i*) **Resin:** This may be filled, moulded or reinforced. This can be used for temperatures from –15ºC to 100ºC. The lower limit is the brittle point of the resin. This material has excellent abrasion resistance in absence of foreign solids and it can be used with mineral oils, hydraulic fluids, water and synthetic oils.

(*ii*) **Carbon:** Temperature limit is upto 260ºC. Abrasion resistance is excellent even with clean fluids. It can be used with practically all industrial chemicals and corrosives as well as oils, solvents, water, and steam.

(*iii*) **Ceramic:** Temperature limit is upto 260ºC. Abrasion resistance is excellent even with high quantities of contaminants, but the life will be shortened accordingly. It can be used with most industrial chemicals, oils, fuels, and solvents.

(*iv*) **Metal:** It may be steel, solid stain less steel bronze, and satellite. Temperature limit upto 260ºC. Excellent abrasion resistance. These can be used with oils, fuels, and fluids compatible with corrosion resistance of the metal.

6.2.2 Dynamic Shaft Seals

These seals are used in more dynamic applications than all other types and these can accommodate rather high speed and continuous shaft rotation. This is due to their positive sealing ability, versatility in meeting requirements of pressure, temperature and shaft speed; low cost, tolerance of a variety of lubricating conditions and relatively high insensitivity to dimensional variations.

The most common type of dynamic shaft seal is the "Radial lip seal". In this type, a flange or lip attached to the housing is held against the shaft. Sealing is due to the interference fit between the flexible sealing element and the shaft. These seals are primarily used for retaining lubricants in equipment having rotating, reciprocating, or oscillating shafts. These seals are also commonly known as "oil seals" or "shaft seals". The rotating shaft application is most common. The lip seals have the following advantages;

1. Small space necessary.
2. Relative low cost for high effectiveness.
3. Ease of installation.
4. Ability to handle many variables while sealing.

Types: Radial lip seals are of the following types:

(*a*) **Single lip:** Fig.6.2(*a*). This design is a low cost design for viscous fluid or grease retention at low speeds. The metal case holds the sealing element in position.

(*b*) **Single lip, spring loaded:** Figure 6.2(*b*). This is the improved design of the single lip seal. Misalignment, shaft runout, seal shrinkage, surge pressures, and other cyclic factors tend to lift the lip off the shaft. To avoid this, a finger or garter spring is added to hold the lip against the shaft. This design is used for oil sealing applications or severe grease sealing applications at higher speeds and in clean atmosphere.

(*c*) **Double lip one lip spring loaded, and one not spring loaded:** Figure 6.2(*c*). In this design, a second lip known as "dust lip" is added to exclude dirt and other contaminants. This lip is not spring loaded. This lip can be a separate element or moulded integrally with the primary lip. This type is the general-purpose design for oil sealing application or severe grease sealing applications with dust, dirt and moisture exclusion.

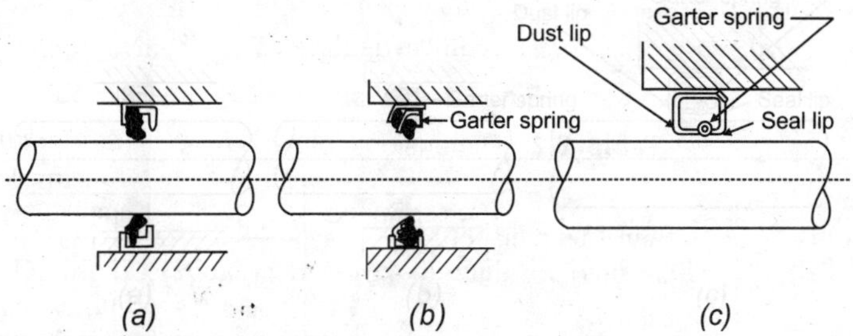

Fig. 6.2 Radial lip seal

Effectiveness and life of the radial seal depend upon the service conditions, most important of which may be listed as below:

(*i*) Shaft speed, alignment, and dynamic runout.
(*ii*) Fluid pressure on the seal.
(*iii*) Temperature at the seal.
(*iv*) Condition of the shaft.
(*v*) Nature of the sealed medium.

Long life of the seal is considered to be more than 1000 hours. Medium life can be defined as leakage beginning at 400 to 600 hours of operation, short life is of about 100 hrs. For increased seal life, the following guidelines may be followed:

1. Lip seals should never run without lubrication.
2. Shafts must have good surface finish.
3. The seals should be undisturbed by "take-part" and end play or eccentricity.
4. Dirt and girt must be kept away from the shaft and seal interface.
5. Lubricants should have low reactivity with sealing elements.
6. Installation and break in must be performed carefully.

Seals are rated in m/s of linear speed. The fluid pressure capabilities of the standard seals are shown in Table 6.1.

Table 6.1 Operating pressure limits

Shaft Speed m/s	Max. permissible pressure, N/mm^2
0 to 5	0.050
> 5, < 10	0.035
> 10	0.020

Materials. The common materials used for shaft seal are:

Nitrile, Polyacrylic, leather, and Silicon. Nitrile has excellent abrasion resistance and it can be used for temperature range of –10 to +77°C for continuous operation and + 16°C for intermittent operation, Leather has got excellent resistance to abrasion and for this, the temperature range is –10°C to +96°C The abrasion resistance of silicon is good and the temperature range is –27°C to 260°C.

The dimensions of the standard oil seals are given in Fig. 6.3 and Table 6.2.

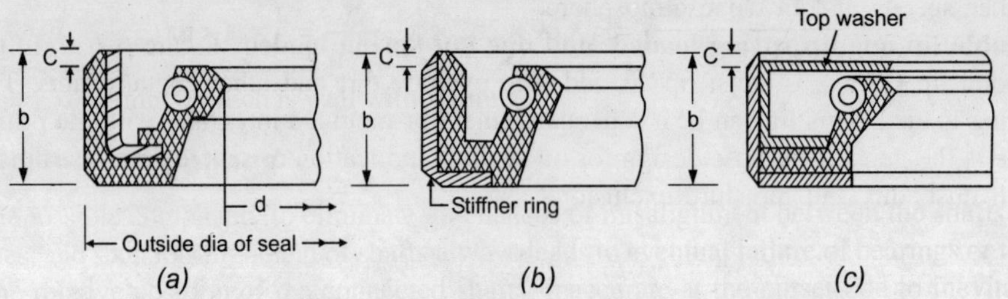

Fig. 6.3 *Oil seals*

Table 6.2 Dimensions of oil seals

Shaft diameter mm	Nominal bore dia. of the Housing, mm	b, mm		c
		Types A and B	Type C	mm
6, 7	16, 22	7	—	0.3
10	19, 22, 24, 26	7	—	0.3
15	24, 26, 30, 32, 35	7	—	0.3
20	30, 32, 35, 40, 47	7	—	0.3
25	35, 40, 42, 47, 52	7	9	0.4
30	40, 42, 47, 52, 62	7	9	0.4
35	47, 50, 52, 62	7	9	0.4
40	52, 55, 62, 72	7	9	0.4
45	60, 62, 65, 72	8	10	0.4
50	65, 68, 72, 80	8	10	0.4
60	75, 80, 85, 90	8	10	0.4
70	90, 100	10	12	0.5
80	100, 110	10	12	0.5
90	110, 120	12	15	0.8
100	120, 125, 130	12	15	0.8
120	150, 160	12	15	1
140	170	15	15	1
160	190	15	15	1
180	210	15	15	1
200 to 250 with step of 10	230 to 280 with step of 10	15	15	1
260 to 500 with step of 20	300 to 530 with step of 20	20	20	1

6.2.3 Split Ring Seals

Expansion split rings (Piston rings) are another common type of seal. These are split to make them resilient and are usually of rectangular cross-section. These rings are used in compressors, pumps, steam engines, internal combustion engines, etc. These are slipped over the piston whose surface has grooves of the same profile as the ring section. In order to ensure the reliability of the seal, more than one ring are employed as shown in Fig. 6.4.

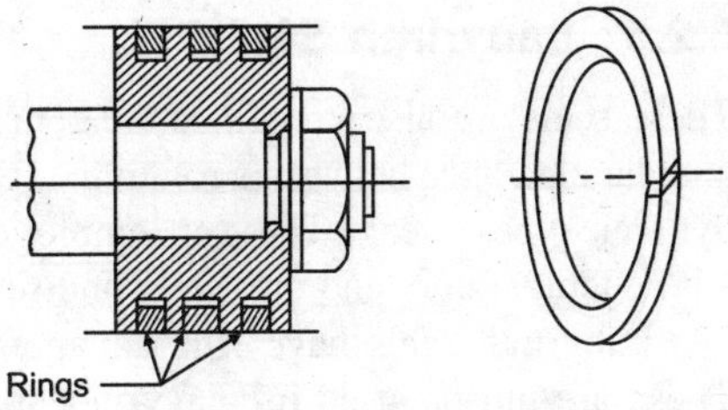

Fig. 6.4 *Split ring seal*

The usual materials for rings are cast iron or forged bronze. The split ring seals are used at high sliding speed. 15 m/s; high temperatures, upto 1200ºC; and high pressures, upto 20.4 N/mm^2.

6.2.4 Diaphragm Seals

A diaphragm is a dividing membrane which spans the gap between a stationary and moving member to prevent transfer of fluid between two separated areas or chambers.

Diaphragms may be grouped under three categories:

(*a*) Those which act only as a separating membrane with no pressure differential existing between the separated zones.

(*b*) Static diaphragms which act as separator between two fluids, and are subjected to no or little motion, generally less than 2.5 mm displacement.

(*c*) Dynamic diaphragms which act as sealing member as well as transmit a force or pressure.

Diaphragms can also be classified as:

(1) Flat diaphragms

(2) Rolling diaphragms.

A flat diaphragm has got no convolutions, or convolutions less than 180º. A rolling diaphragm has full 180º convolutions. During operation, as it moves through the total stroke, it rolls back and forth from the piston to the cylinder wall.

The most used type is the dynamic diaphragm.

"Flat" diaphragms are not necessarily flat. Even if the diaphragm is made flat, convolutions are formed by fluids pressure during the working stroke. A flat diaphragm also gets distorted when there is a relative movement of the diaphragms assembly.

A rolling diaphragm is used as a long travel bellow or radially sealed hydraulic piston with zero leakage. The operation of a rolling diaphragm assembly is shown in Fig. 6.5.

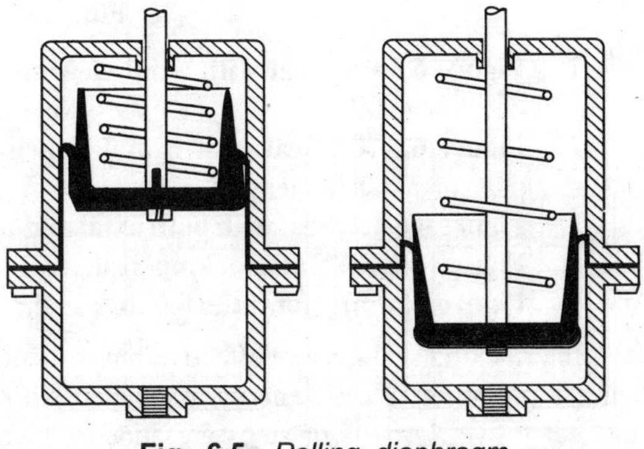

Fig. 6.5 *Rolling diaphragm*

A term used with rolling diaphragms is "Flex Life". In operation, a rolling diaphragm is subjected to a change in a circumference. When any point on the side wall rolls from its moulded diameter to that of the piston cylinder wall, its circumstance changes. The number of strokes before failure is called "flex life". Another term used with diaphragms is the "Burst Strength". This is the pressure at which the diaphragm fails. It is affected by the two properties of the materials; tensile strength and modulus.

Materials. A diaphragm is made of fabric which is impregnated with an elastomer for impermeability.

The most widely used fabrics are ; polyesters, nylon, and nomex. Polyester and nylon have excellent resistance to abrasion and fatigue. Diaphragms made of polyester yarns are quite strong and can operate at higher pressure ratings or longer cycle lives. These can be used in flat diaphragms and temperature of 120ºC. Nomex fabric is very good for high temperature operations and it has replaced glass and fluorocarbon fabrics. It can be used for prolonged exposures at 260ºC.

The common elastomers are;

(i) Nitrile butadiene (ii) Neoprene (iii) Epichlorohydrin
(iv) Silicon (v) Fluorosilicon (vi) Ethylene
(vii) Viton (viii) Polyacrylate (ix) Polyurethane
(x) Butyl (xi) Hyplaon.

6.2.5 Labyrinth Seals

These seals are also known as clearance seals, because these limit the leakage by closely controlling the annular clearance between a rotating or reciprocating shaft and its relatively stationary housing. These are the dynamic type of seals. These are employed when a small loss in efficiency due to leakage can be tolerated and where temperature and pressure conditions, and shaft speed prevent use of positive contact seals.

Labyrinth seals have the advantages of reliability, simplicity, and flexibility in material selection. These are mostly used in heavy industrial power, and aircraft applications where relatively high leakage rates can be tolerated and design simplicity is a necessity. A labyrinth seal consists of one or more thin strips or knives which are attached to either the stationary housing or the rotating shaft. Design clearance between the knives and the housing, or the strips and the shaft, are dependent upon bearing clearances, amplitude of shaft run out, and differential expansion.

The various types of labyrinth seals are shown in Fig. 6.6.

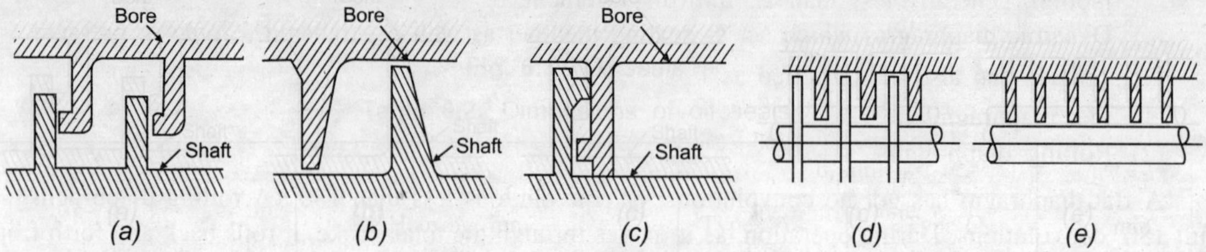

Fig. 6.6 *Labyrinth seals*

1. **Figure 6.6 (a). Seal with axial clearance:** This type is used when radial clearance is impossible due to radial expansion or excessive shaft run out.
2. **Figure 6.6 (b). Seal with radial clearance:** This type is employed when there is excessive end play or axial movement.
3. **Figure 6.6. (c). Seal with both axial and radial clearance:** The design makes the seal more effective.
4. **Figure 6.6 (d). Interlocking seal.**
5. **Figure 6.6 (e). Non-interlocking seal.**

Interlocking seal is more effective but is difficult to assemble. Non-Interlocking seals are less effective since a portion of kinetic energy associated with each stage is carried over to the next stage. To reduce this carry over, groove or steps are added between stages to deflect the expanding fluid into the space between the strips, Fig. 6.7. In Labyrinth seals, the leakage of the fluid is limited as the fluid is made to loose velocity pressure as it is throttled through the annular orifices.

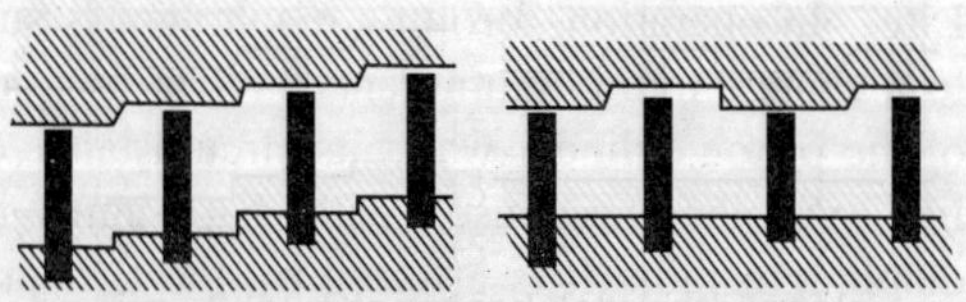

Fig. 6.7 *Staggered labyrinth seals*

6.2.6 Compression Packings

This is the oldest type of seal, and still finds application in pumps, valves and pistons to seal hot water, steam and other fluids. The packing is usually made twisted, woven braided cords or rings of vegetable fibres, asbestos, or soft metals. It is impregnated with a binder such as graphite and lubricant, e.g., grease. The packing is then packed between the shaft and the housing. The seal is produced when packing gets squeezed between the throat of a stuffing box and its gland, Fig. 6.8. The cones in the flange and gland make the packing material to flow inwards the shaft to make a seal against the moving shaft or rod and flow outward to make a seal against the bore of the box. Due to continuous operation, the packing loses its resilience and becomes compacted and a gap appears between the packing and the shaft. To compensate for this wear, the gland is adjusted by means of bolts. The compression packing is used at low speeds upto 3 m/s and in high temperature and pressure conditions.

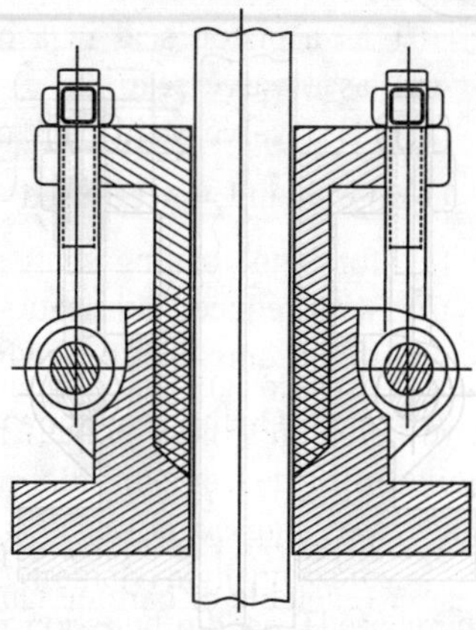

Fig. 6.8 *Compression packing*

Materials: The materials used for compression packing should satisfy the following conditions;

1. Be sufficiently plastic to conform to the shaft and bore under gland pressure.
2. Be able to withstand fluid used for lubrication.
3. Be sufficiently elastic to absorb any shaft gyration.
4. Not corrode or abrade the shaft or rod.
5. Require minimum adjustment and renewal.

The common materials are asbestos and fibres such as jute, flax, linen and duck. Asbestos is good on smooth shafts only. It can be used up to a temperature of 315ºC and for hot water, superheated steam, hot oils, air and other hot gases. Fibres have good abrasion resistance and these can be used for cold and hot water, low pressure steam, oils and ammonia and upto a temperature of 100ºC.

6.2.7 Moulded Packing

These packings are often called automatic hydraulic, or mechanical packings. After installation, these usually do not need gland adjustment. The pressures of the fluid being sealed, seal the packing against the wearing surface. The moulded packings are of two types:

1. **Lip type packings:** This type includes flange, cup, U-cup, U-ring, and V-ring packings.
2. **Squeeze type packings:** This type includes O-ring and other similar forms. Sealing is achieved by the interference built into the ring.

In general, the term U-cup applies to a homogeneous rubber packing of U-shaped cross-section; whereas U-ring refers to leather and fabric-reinforced packing. Of all the types of moulded packings, the V-ring and O-ring packings are used exclusively for dynamic application. They are mostly employed for reciprocating parts.

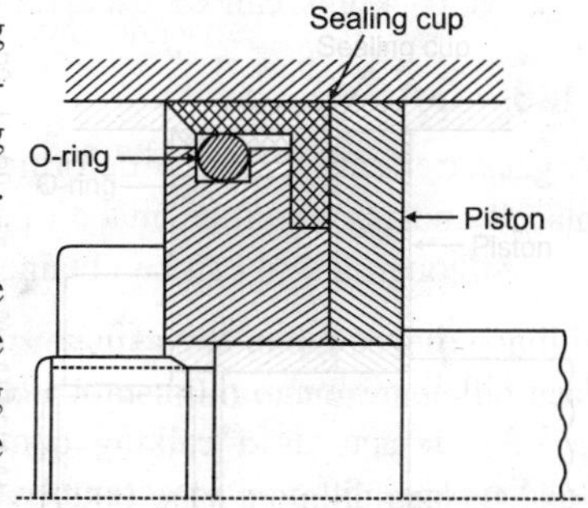

Fig. 6.9 *Cup packing*

A common cup packing design is shown in Fig. 6.9. The pressure of the fluid being sealed acts on the edges of the cup and forces the cup against the cylinder wall. This arrangement is used for low sliding speeds upto 1 m/s. To avoid or reduce the wear of the cup, the surface over which it slides should be of high degree of surface finish.

A typical V-ring packing is shown in Fig. 6.10. The V-rings are assembled in multiples and are usually spring loaded to take up the wear.

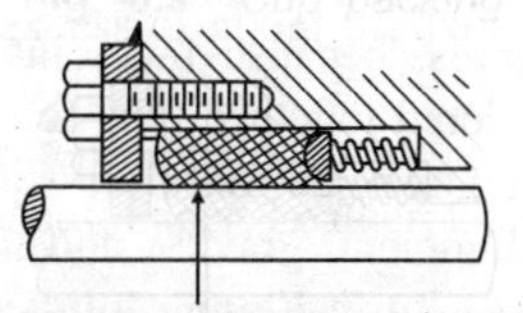

Fig. 6.10 *V-ring packing*

A typical U-ring packing design is shown in Fig. 6.11.

Squeeze type moulded packings have the following advantages:

1. Low initial cost.
2. Adaptability to limited space.
3. Easy to install.
4. High efficiency.
5. Sealing in both directions.
6. No need for adjustment.
7. Tolerance to wide ranges of pressure, temperature and fluids.
8. Relatively low friction.

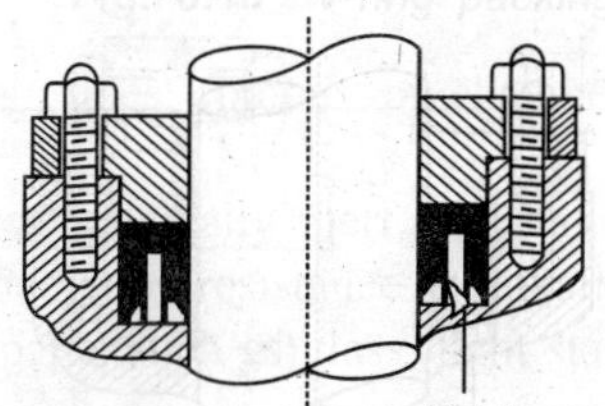

Fig. 6.11 *U-ring packing*

O-ring is the most commonly used type of squeeze type of packing. It finds application in:

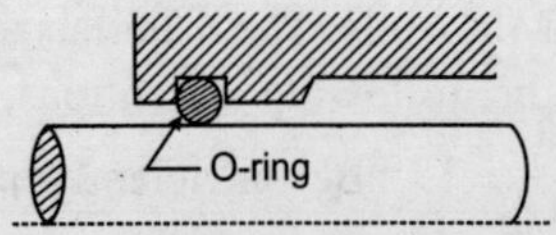

Fig. 6.12 *O-ring seal*

(*i*) as a piston seal in a hydraulic actuating cylinder;
(*ii*) as a valve seat;
(*iii*) as a valve stem packing.

O-ring should not be used:

(*a*) for high rotating shaft speeds.
(*b*) with reduced lubrication since that reduces the ring life.
(*c*) in extremely dirty environment or with a lubricant carrying abrasive dirt particles.

A typical O-ring seal shown is in Fig. 6.12.

Materials

1. *Leather:* For many applications, leather is still the best packing material. It may be either vegetable or chrome tanned. To fill up the voids, it is impregnated with synthetic rubber. Leather is affected by high temperatures, extreme acidity or alkalinity. It has low frictional characteristics, high tensile strength, and resist extrusion. The temperature range is –18ºC to +82ºC. Leather packing should also not be used with pressurised steam.
2. *Fabric-reinforced Rubber:* These packings are made of synthetic rubber and fabric.
   The commonly used fabrics are: duck, nylon, and asbestos. For temperature below 120ºC, duck is used and for temperatures higher than this, asbestos is employed. Nylon is used for high flexibility and greater strength.
3. *Homogeneous Packings:* These packings are made from synthetic rubber and do not contain a fabric reinforcement. These are used for tolerance range of –180ºC to +260ºC.
4. *Metallic Packings*: These packings are made of aluminium, copper, or lead. If fibre filled, these packings can be used for temperatures from –21ºC to +532ºC.

6.3 GASKETS

A gasket effects a seal by deforming and filling surface irregularities of the member between which it is placed and is, therefore, made of relatively soft material compared to flange mating surfaces.

Although a seal can be obtained without the use of gasket, it provides the following extra advantages;

1. It promotes the efficient initial seal.
2. It prolongs the useful life of an assembly.
3. It acts as a calking compound in compensating for errors and irregularities in joined rigid members of an assembly.

The joint should be designed and constructed to make the best use of the properties of the gasket materials, because the joint is only as good as its gasket. Hence, joint and gasket design must be considered together. The forces acting on the gasket are quite complex and variable. Some of these may be summarized as: Bolt squeeze on gasket, Fluid pressure, due to environmental conditions. Also, misalignment, gravity, and thermal movement try to separate faces. The two basic types of joints are:

(*a*) **Basic Flange joints:** Figure 6.13.
(*b*) **Metal to Metal Joints:** Figure 6.14.

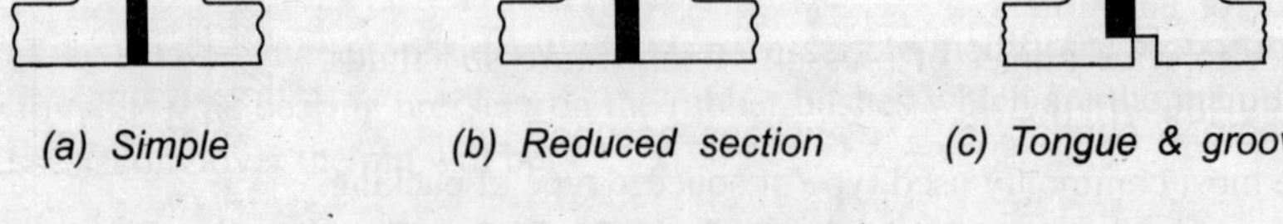

(a) Simple (b) Reduced section (c) Tongue & groove

Fig. 6.13 *Basic flange joint*

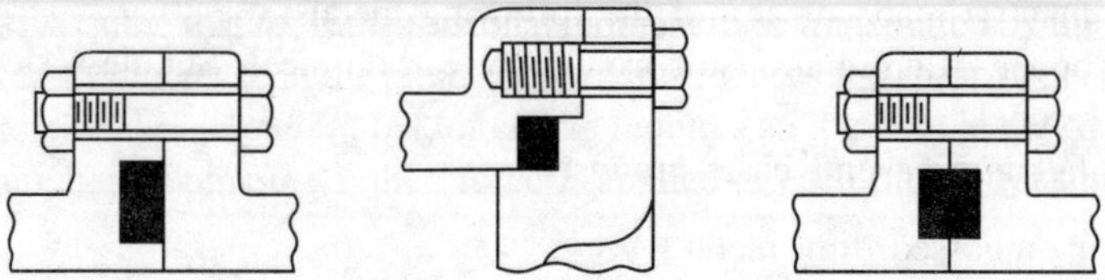

Fig. 6.14 *Metal to metal joint*

The flange joints are suitable for all kinds of flat gaskets. Simple flange joint, Fig. 6.13 *(a)* is suitable for pressures upto 1.5 N/mm^2. For higher fluid pressure, the reduced section joint, Fig. 6.13 *(b)* , and tongue and groove joint, Fig. 6.13 *(c)* are preferred. In reduced section joint, the gasket wall section has been reduced to give a high gasket stress without change in bolt stress. The tongue and groove joint is used for applying extremely high stress to gaskets, and for holding extremely high pressures.

Metal to metal joints are suitable for compressible materials such as cork and rubber. These are useful for limiting the amount of stress on the gasket and are particularly good where it is essential to maintain accurate internal clearance or alignments.

The choice of the gasket material for any application depends upon:

(*i*) operating conditions.
(*ii*) mechanical features of the flanged assembly.
(*iii*) gasket characteristics.

The following requirements must be met by the gasket material:

(*a*) Impermeability.
(*b*) Ability to flow into joint imperfections when compressed.
(*c*) Maintain seal inspite of age, and variations in temperature and pressure.
(*d*) Resistance to environmental deterioration.

Depending upon the materials, the gaskets may be grouped as below:

1. Metallic gaskets.
2. Non-metallic gaskets.

Non-metallic gaskets are used for low pressures whereas for high pressures and under severe condition, metallic or combination gaskets are used. A rough guide as to whether a metallic or non-metallic gasket should be used for given condition is:

If operating pressures in N/mm^2 × operating temperature in ºC is > 900 only metallic gaskets should be used.

Generally, a non-metallic gasket should not be used for pressures in excess of about 8.7 N/mm^2 and temperatures above 450ºC and below –18ºC.

The most common compositions for non-metallic gaskets can be written as below:

Asbestos products; Cork and Rubber; Cork composition; Rubber and plastics; Paper; Leather; Composite and miscellaneous gaskets.

Asbestos products can work efficiently upto a temperature of about 260ºC. These gaskets are tough and durable and dimensionally stable. These are used for heavy duty bolted and threaded joints, as in water and steam pipe fittings, manifold connections, etc.

Cork and rubber gaskets are used for general purposes except steam lines, combustion chambers, etc. these are best used at temperatures below 70ºC.

Cork is light weight, non-capillary, does pot deteriorate with age, practically inert, and has a high co-efficient of friction. It has excellent oil and solvent resistance but poor resistance to alkalies and corrosive acids. These gaskets are used for mating rough or irregular parts, such as glass, light stamping, unfinished casting, etc. Temperature limit is above 70ºC.

Metallic gaskets may be made of any of the following metals:

Lead, Brasses, Copper, Aluminium, Soft iron, Low carbon steel, Stainless steel, Titanium, Silver, Nickels, Monel and Inconel.

Metallic gaskets fall into several basic groups:

1. Corrugated or embossed, thin metal.
2. Metal jacketed, soft filler.
3. Spiral wound.
4. Plain or machined flat metal.
5. Round cross-section, solid metal.
6. Special, heavy cross-section, solid metal.
7. Pressure actuated, light cross-section.

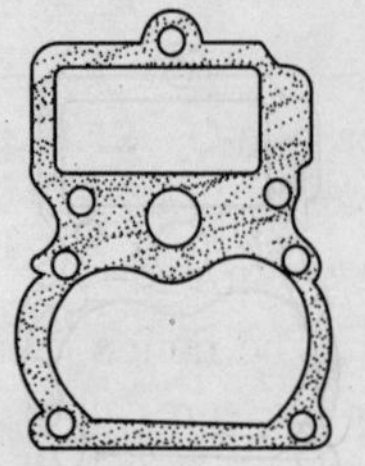

Fig. 6.15 *Gasket*

A typical gasket is shown in Fig. 6.15.

EXERCISES

1. Enumerate the functions of seals and gaskets.
2. How will you classify the seals?
3. Discuss the various types of Labyrinth Seals.
4. What is the function of garter spring in an oil seal?
5. Write about the various types of Labyrinth Seals.
6. Where the split ring seals are used?
7. Where diaphragm seals are used?
8. Define "flex life" and "burst strength" of diaphragms.
9. What is the difference between U-cup and U-ring?
10. What are the requirements of the materials for a gasket?
11. Write the advantages of gaskets.
12. Sketch the Basic Flange joints.
13. Sketch the metal to metal gasket joints.
14. Write a short note on: "Metallic Gaskets" and "Non-Metallic gaskets".
15. What is silicone?

    Ans. It is synthetic rubber. It is used to make seals, gaskets and O-rings.

❑❑❑

Section II

FASTENERS

FASTENERS

Every machine is made up of hundreds or often thousands of parts, for example, an automobile together with the engine has about 16 thousand parts, and a vertical boring mill has over 20 thousand parts. In order to function in machines, parts are joined together to form **sliding** and **fixed joints**. The first group includes joints which link, e.g., a connecting rod and a crank pin, a shaft and shaft bearings. The second group is illustrated by joints fastening the plates of a boiler together or with its dishes, a cover and bearing housing or a rod and piston.

The presence of sliding joints is exclusively dictated by machine kinematics. Fixed joints are employed to split the machine into units and sets and the units into parts and elements to facilitate or make possible the manufacture of the machine, its transportation and maintenance.

Engineering practice qualifies as joints proper only fixed joints. The classification of the joints is shown below in the chart:

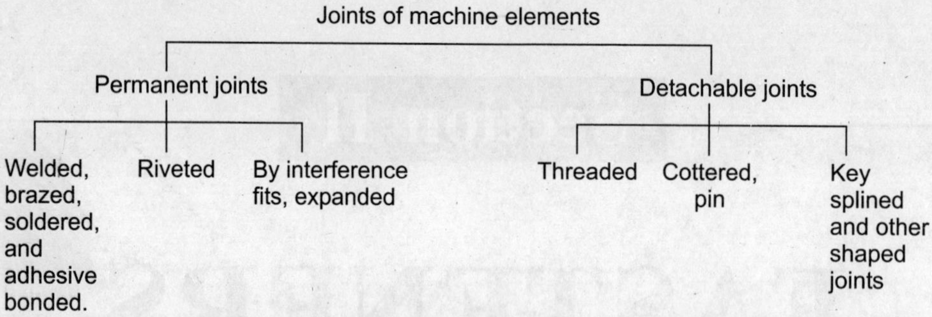

Permanent joints: These joints do not allow a work to be disassembled without destroying the connecting components. Such joints are made in locations where they either do not increase weight of the work at all over the weight of an integral work or increase it negligibly. These joints can be made by mechanical methods, that is, by riveting, expanding, interference fits, welding, brazing, and adhesive bonding.

In modern engineering practice preference is given to the welded, brazed, soldered and adhesive bonded joints. Adhesive bonded joints have been particularly developed in aircraft industry. Adhesives are employed in the automobile industry to fasten friction linings to clutch discs and brake bands.

In general the course of machine design elements examines the joints made by riveting, and interference fits which are particularly widely used.

Detachable joints: These joints allow the disassembly of a unit without damaging the fastened elements and the connecting components. The types of the detachable joints are threaded joints, pin and cotter joints as well as key, splined and others which can be called shaped joints.

The designing of joints is an extremely responsible task, since the majority of failures occur in a machine exactly at the joints. As well as being economical, joints should satisfy, depending on their purpose, the requirement of strength, tightness and stiffness.

CHAPTER 7 Riveted Joints

Riveting is a popular method of fastening and joining because of its simplicity, dependability, and low cost. Its advantages are:

1. Production rates are high.
2. Initial and maintenance costs for equipment are low.
3. Either metallic or nonmetallic materials can be joined.
4. Dissimilar metals and assemblies having a number of parts with non-uniform thickness can readily be fastened.
5. The fastener, being a cold headed part, has the uniform grain structure and inherent strength of a forging.
6. Unskilled labour can be used to operate equipment.
7. The rivet can be made of a variety of materials ranging from monel or inconel to lead or zinc.
8. The rivet can be used not only as a fastener but as a pivot, cam follower, electrical contact, or other functional component.

On the other hand there are some disadvantages associated with the process:

1. Parts once fastened, cannot be easily disassembled.
2. The tensile and fatigue strengths of joints made with some types of rivets are considerably lower than bolted or welded assemblies.
3. Joints normally are neither water tight nor air tight.
4. Protruding rivet heads may be undesirable in food machinery, chemical equipment, or other sanitary assemblies.

Compared with screwed or bolted joints, riveting requires no additional work or supplementary parts such as lock washers, supplementary nuts, or safety wire. Compared with welded joints, riveting makes possible consistently uniform results, without resort to X-ray or other costly methods of inspection.

When riveting is done cold, the strength of the material joined is not affected by heat as in the case of welding, where it is sometimes necessary to use a heavier gage of material to compensate for the loss of strength due to the annealing effect of the heat generated in welding.

Riveted joints may be roughly divided into three categories:

1. Where strength and rigidity are the chief requirements, e.g., coal bunkers, low pressure liquid containers, bins for bulk materials, and ship hulls are included in this class. Ships hulls and liquid containers must also be liquid tight, but since the pressures to be dealt with are low, therefore, it is not difficult to make the joint leak proof.
2. In this type the strength and rigidity are required, but in addition the joint must be proportioned so that leakage will not occur. Boiler drums, high pressure liquid containers, and gas tanks are included in this class.
3. In this case strength and rigidity are the only requirements. Machine frames, building structures, and bridges are included in this class.

7.1 METHODS OF RIVETING

In its simplest form riveting consists of inserting a ductile metal pin through holes in two or more parts and forming over the ends of the metal pin so as to secure the parts firmly together.

Rivets may be driven by hand or by riveting machines. In hand riveting the original rivet head is backed up by a hammer or heavy bar and then the die or set as shown in Fig. 7.1. is placed against the end to be headed and the blows are applied by a hammer. This forms the second head called the **point**. In machine riveting the die is a part of the hammer, which is operated by air, hydraulic, or steam pressure. Whenever possible, rivets are closed red hot under hydraulic pressure, because the steady pressure produced on the end bar of the rivet staves it up better and causes it to fill the hole more completely than is likely to be the case with steam, pneumatic, or hand riveting, where the action is percussive. Cold riveting is used when the steel rivets are upto 12 mm in diameter. For larger diameters, hot riveting is used where the entire length of the rivet is heated to a light red.

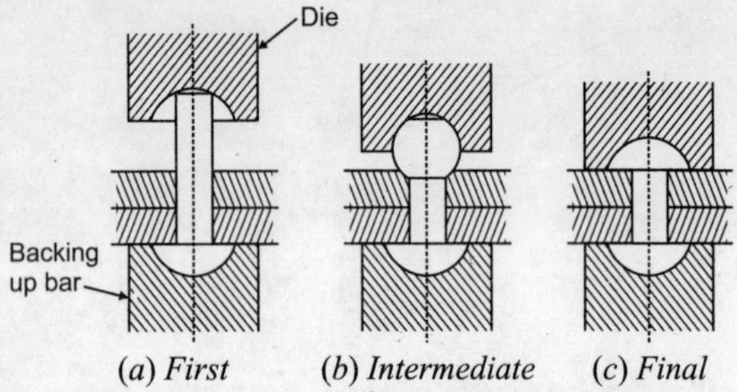

Fig. 7.1 *Progressive stages in riveting*

In structural and pressure-vessel riveting the diameter of rivet hole is usually 1.5 mm larger than the nominal diameter of rivet. After the rivet is driven, it will contract on cooling. The lateral contraction will be slight, but there will be a longitudinal tension introduced in the rivet that will draw the plates tightly together in well driven rivets. This tension will cause frictional resistance between the plates that will resist sliding of the plates under load. This resistance is seldom considered in calculations of the strength of riveted joints, because the coefficient of friction between the plates is difficult to determine, and neglecting this friction is on the side of safety. Also in high temperature installations, relaxation may cause the tension on the rivets to be reduced.

7.2 RIVET HEADS

The forms of heads generally used in boiler joints are shown in Fig. 7.2. More forms of heads are given in I.S. 1928-1961 for boiler joints and I.S. 1929-1961 gives forms of rivet heads for general purpose.

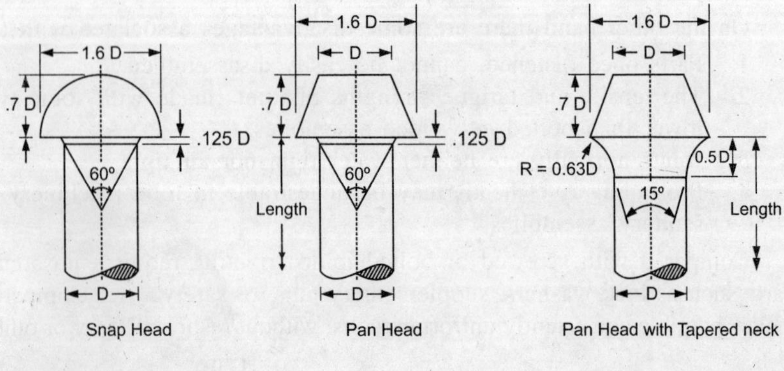

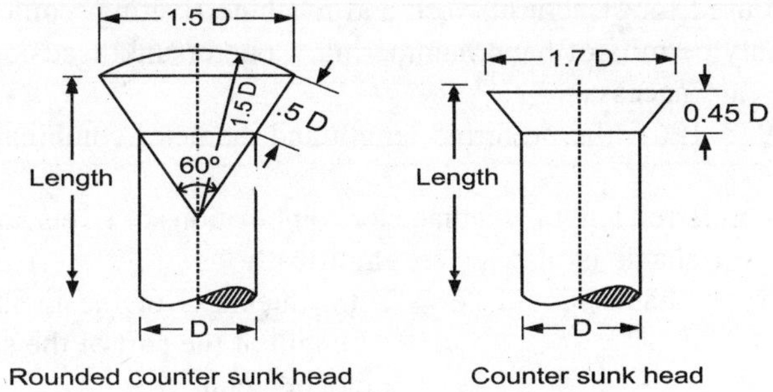

Fig. 7.2 *Rivet heads*

Table 7.1 Preferred length and diameter combinations for rivets used in boilers (All dimensions in mm)

Length	Diameter													
	12	14	16	18	20	22	24	27	30	33	36	39	42	48
28	X	–	–	–	–	–	–	–	–	–	–	–	–	–
31.5	X	X	–	–	–	–	–	–	–	–	–	–	–	–
35.5	X	X	X	–	–	–	–	–	–	–	–	–	–	–
40	X	X	X	X	–	–	–	–	–	–	–	–	–	–
45	X	X	X	X	X	–	–	–	–	–	–	–	–	–
50	X	X	X	X	X	X	–	–	–	–	–	–	–	–
56	X	X	X	X	X	X	X	–	–	–	–	–	–	–
63	X	X	X	X	X	X	X	X	–	–	–	–	–	–
71	X	X	X	X	X	X	X	X	X	–	–	–	–	–
80	X	X	X	X	X	X	X	X	X	–	–	–	–	–
85	–	X	X	X	X	X	X	X	X	X	–	–	–	–
90	–	X	X	X	X	X	X	X	X	X	–	–	–	–
95	–	X	X	X	X	X	X	X	X	X	X	–	–	–
100	–	–	X	X	X	X	X	X	X	X	X	X	–	–
106	–	–	X	X	X	X	X	X	X	X	X	X	–	–
112	–	–	X	X	X	X	X	X	X	X	X	X	–	–
118	–	–	–	X	X	X	X	X	X	X	X	X	X	–
125	–	–	–	–	X	X	X	X	X	X	X	X	X	X
132	–	–	–	–	–	X	X	X	X	X	X	X	X	X
140	–	–	–	–	–	X	X	X	X	X	X	X	X	X
150	–	–	–	–	–	–	X	X	X	X	X	X	X	X
160	–	–	–	–	–	–	X	X	X	X	X	X	X	X
180	–	–	–	–	–	–	–	X	X	X	X	X	X	X
200	–	–	–	–	–	–	–	–	X	X	X	X	X	X
224	–	–	–	–	–	–	–	–	–	X	X	X	X	X
250	–	–	–	–	–	–	–	–	–	–	–	–	X	X

Preferred numbers are indicated by X.

Snap heads are also used for structural work and machine riveting, conical heads are mainly used where restricted space only permits of hand hammering. Countersunk heads are also used in shipbuilding where flush surfaces are necessary.

According to I.S. 1928–1961, the preferred length and diameter combinations of rivets shall be as given in Table 7.1.

For general purposes preferred length and diameter combinations of rivets are given in I.S. 1929–1961.

The length of the rivet shank is also given empirically as:

The length of the rivet shank should be = Total thickness of the members being riveted + Length of the part of the shank required to form the closing end

= Total thickness of the members being riveted + (0.7 to 1.3) D.

Table 7.2 Gives the diameters of rivets and corresponding rivet holes in mm for boiler joints

Nominal diameter of Rivet in mm	Rivet hole diameter in mm	Nominal diameter of Rivet in mm	Rivet hole diameter in mm
12	13.5	27	29
14	15.5	30	32
16	17.5	33	35
18	19.5	36	38
20	21.5	39	41
22	23.5	42	44
24	25.5	48	50

7.3 TESTING OF MANUFACTURED RIVETS (I.S. 1928-1961)

Samples of manufactured rivets are selected from the bulk in such number as specified or approved by the purchaser, and shall withstand the following tests:

Bend Test: The rivet shank shall be bent cold, and hammered until the two parts of the shank touch in the manner as shown in Fig. 7.3 (*a*). There shall be no fracture on the outside of the bend.

Flattening test: The rivet head to be flattened, while hot, in the manner shown in Fig. 7.3 (*b*) without cracking at edges. The head to be flattened until its diameter is 2.5 times the diameter of the shank.

(*a*) *Bend test* (*b*) *Flattening test*

Fig. 7.3

7.4 RIVET MATERIALS

Rivets are made of wrought iron or soft steel for most uses, but where corrosive resistance or light weight is a requirement, rivets of copper or aluminium alloy are used. Rivet materials must he strong and highly ductile. I.S. 1960–1962 gives the specifications for steel rivets and stay bars for boilers. Rivets are used in such a manner that they resist shearing action. However, if the tensile loading is un-avoidable, the safe tension value of a rivet may be taken as one-half the safe single shear value.

The materials to be joined and the material used for the rivet should he carefully considered from the stand point of electrolytic corrosion in the presence of moisture. If desired by the purchaser, the rivet shall be painted with one coat of boiled linseed oil or suitable preventive material to be agreed mutually between the supplier and purchaser.

The chemical composition, tensile strength and percentage elongation of steel for boiler construction, according to I.S. 2100–1962 are given below in Tables 7.3 and 7.4.

Table 7.3 Chemical composition

Grade	Carbon percent	Manganese percent	Silicon percent	Sulphur percent	Phosphorus percent
1.	0.20 max.	0.60 to 0.90	0.10 to 0.35	0.50 max.	0.050 max.
2.	0.16 to 0.24	1.3 to 1.7			

Table 7.4 Mechanical properties

Grade	Tensile strength MN/m^2	Yield strength in percent of minimum tensile strength	Percentage elongation
1.	420.0—500.0	55	25
2.	520.0—620.0	50	20

7.5 KINDS OF RIVETED JOINTS

Joints may be single, double, triple, quadruple riveted in which case one, two, three, or four rows of rivets are used (Fig. 7.4). If the rivets in the rows are in line crosswise, the joint is said to be "chain riveted", while if the rivets in adjacent rows offset by one half the centre to centre distance, the joint is said to be "zig-zag riveted".

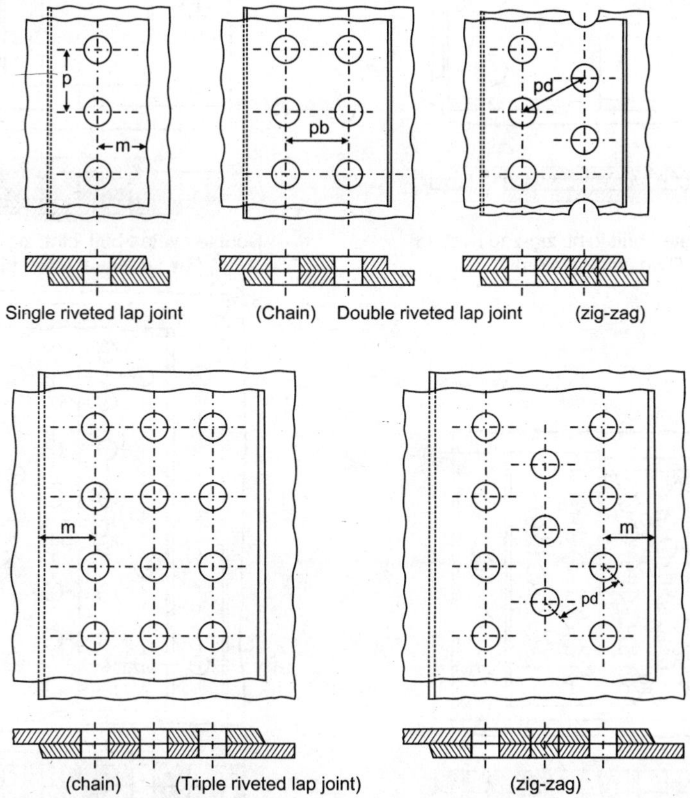

Fig. 7.4 *(a)*

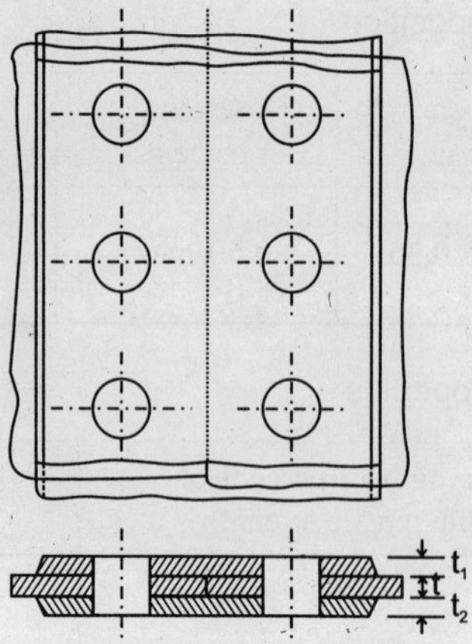

Single riveted butt joint
(Two cover plates)

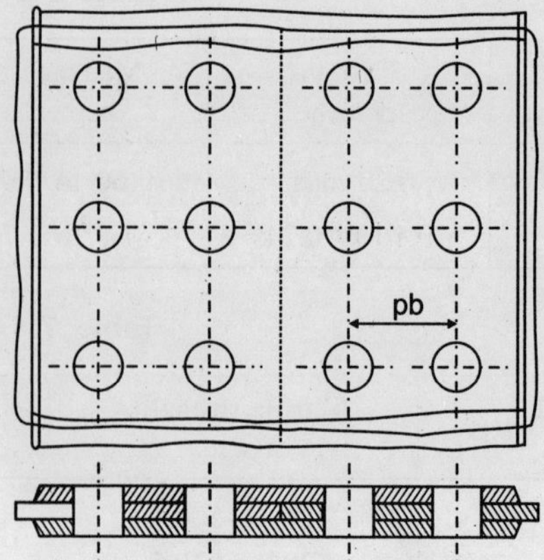

Double riveted butt joint, chain riveting
(Two cover plates)

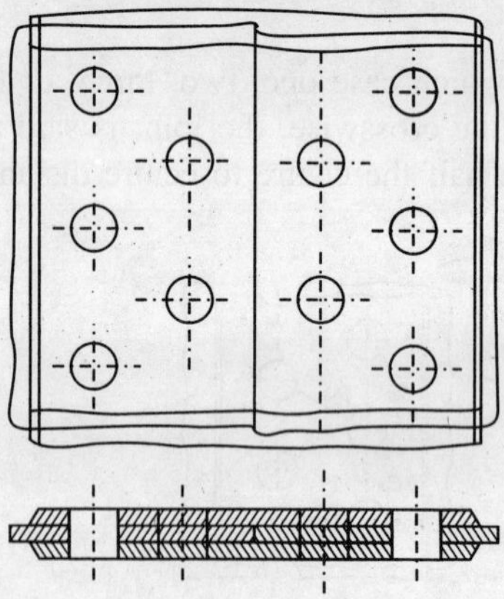

Double riveted butt joint, zig-zag riveting
(Two cover plates)

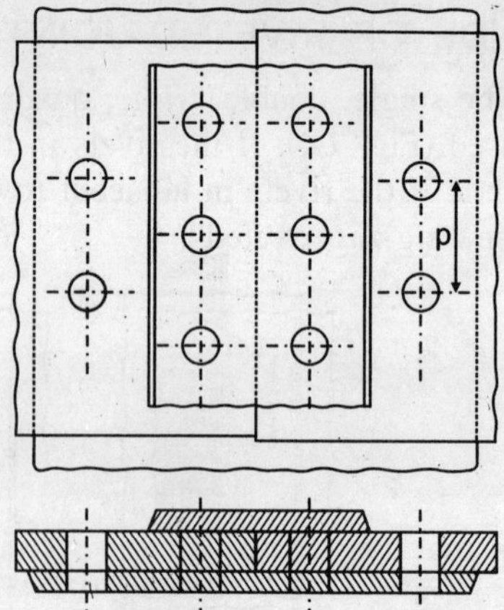

Double riveted butt joint, zig-zag riveting
(Two unequal cover plates)

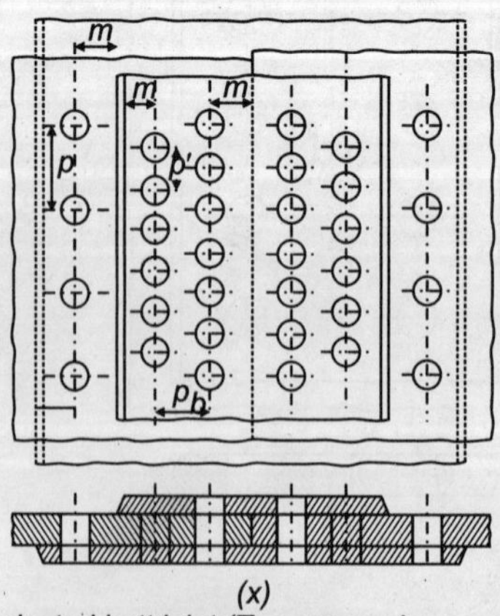

(x)
Triple riveted butt joint (Two unequal cover straps)

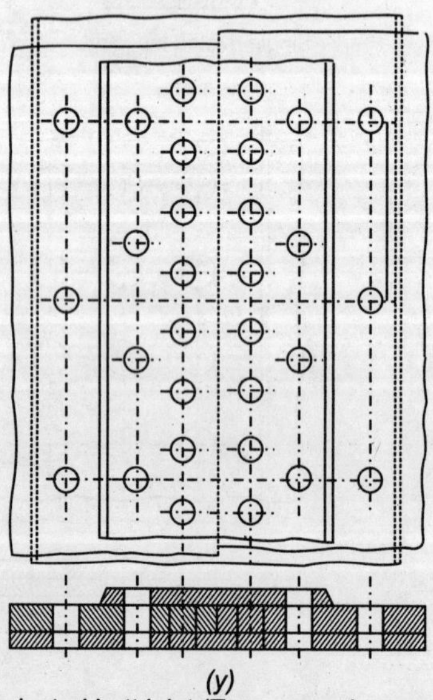

(y)
(Triple riveted butt joint (Two unequal cover straps)

Fig. 7.4 *(b)*

There are two principal kinds of riveted joints, lap joints and butt joints. In the lap joint, plates to be joined overlap each other a sufficient amount for riveting, while in the butt joint, the plates are in the same plane and butt against each other. Butt joints are connected by a short plate called a cover plate or butt strap, either on one side or on both sides of the plates to be joined. Normally two cover plates are used and of equal width, but sometimes the outer strap may be narrower than the inner one.

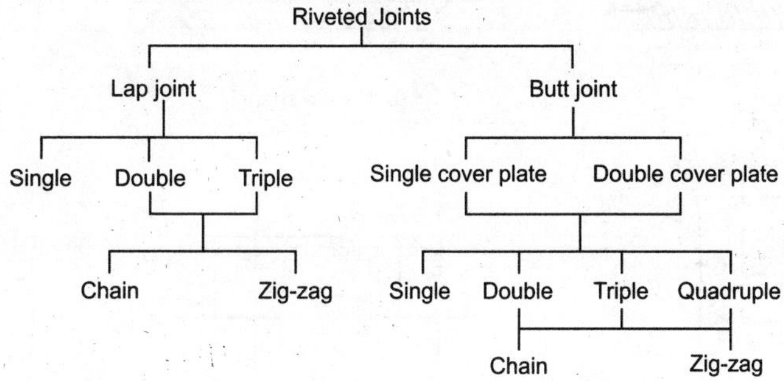

Terminology: The following is the terminology used in riveted joints.

Gage line is the line through the centers of a row of rivets, parallel to the edge of the plate.

Pitch is the distance between the centers of adjacent rivets measured on the gage line. In the book it shall he denoted by p. If the spacing on one gage line is greater than on another, as shown in Fig. 7.4 (*b*), the pitch is the maximum rivet spacing. A **unit strip** or **unit length** is equal in width to the pitch.

Back pitch (Transverse pitch) is the distance between two adjacent gage lines in the same plate as shown in Figs. 7.4 (*a*) and 7.4 (*b*). It is also called transverse pitch and denoted by pb or pt.

Diagonal pitch in zig-zag riveting is the distance between the centers of adjacent rivets on adjacent gage lines. It is represented by pd as shown in Fig. 7.4 (*a*).

Marginal distance is the distance from the gage line to the edge of the plate. It is also called lap distance or margin and is represented by m, as shown in Fig. 7.4 (*a*).

7.6 CAULKING AND FULLERING

Joints in pressure vessels like steam boilers, spheres, air receivers and tanks, etc. are made fluid tight by **Caulking** as shown in Fig. 7.5 (*a*). A narrow blunt chisel like tool (*M*) called a caulking tool, about 5 mm. thick and 38 mm. in breadth, the edge ground to an angle of 80°. It is moved after each blow along the edge of the plate, which is planed to a bevel of 80° to facilitate the forcing down of edge. It can be noted that the tool burrs down the plate at *C* forming a metal to metal joint, care being taken not to damage the plate below the tool or spring the joint open. Generally both edges, *B* and *C* are caulked and the rivet head also as at *A*.

A more satisfactory way of making the joint staunch and tight is known as **fullering** which has largely superseded caulking. The fullering tool having a thickness at the end equal to that of the plate, is used in such a way that the greatest pressure due to the blows occurs at [Fig. 7.5 (*b*)] near the joint, giving a clean finish, with less risk of damaging the plate.

7.7 FAILURE OF RIVETED JOINTS

These joints may fail in a number of ways as shown in Fig. 7.6. The actual stresses set up in a joint are complicated and not subject to exact calculations. The usual methods employed in the design of such joints are based on the apparent direct stress, with the unknown stress provided for in the factor of safety. Common assumptions may be stated as follows:

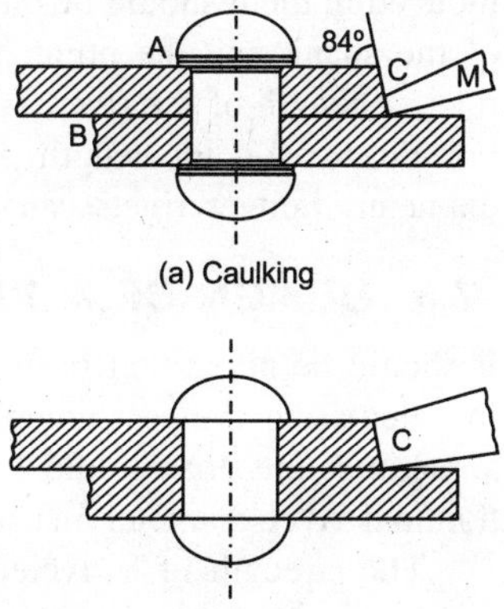

(a) Caulking

(b) Fullering

Fig. 7.5

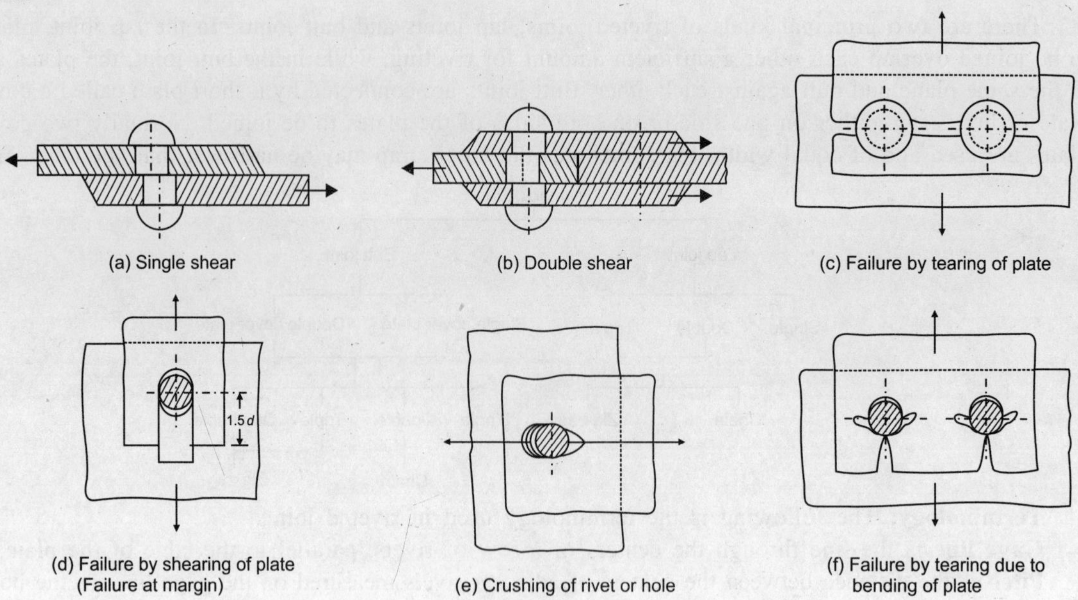

Fig. 7.6

1. The load is distributed among the rivets according to the shear area.
2. There is no bending stress in the rivets.
3. The tensile stress is equally distributed over the section or the metal between the rivets.
4. In a rivet subjected to double shear, the shear is equally distributed between the two areas in shear.
5. The holes into which the rivets are driven do not weaken the member if it is in compression.
6. After driving, the rivets completely fill the holes.
7. Friction between adjacent surfaces does not affect the strength of the joint.

It may be noted that in the first six of the assumptions listed above, the stresses actually developed in the joint are higher than would be indicated by calculations based on these assumptions.

Assumption 7, however decreases the actual stress produced. The exact analysis of the strength of a riveted joint could be determined only with a test. The experience of many years has shown that reasonable results are obtained when a boiler or structural joint is designed as indicated.

Rivets should always be placed at right angles to the acting forces, and the maximum stress induced in them should be either shear or crushing. In a long rivet the initial stress set up at the junction of the shank and the point, when the rivet cools, is dangerous. This initial stress increases with the relative length of the rivet, and in a very long rivet it may cause the head to snap off without any load. For this reason the length of the rivet between the heads should not exceed four or five times its diameter. Longer rivets can be used if the head end is cooled before the rivet is placed in its hole.

7.8 DESIGN OF A RIVETED JOINT

It should be noted that in designing riveted joints of pressure vessels the diameter of the hole is used in calculating the resistance of the rivets both in shear and crushing. In structural joints the rivets sometimes are driven cold and sometimes are replaced by bolts, it is, therefore, customary to use the nominal rivet diameter in calculating the resistance of rivets.

The strength of a riveted joint may be investigated by considering a **pitch length** of each joint.

Let d = diameter of rivet hole.

p = pitch of rivets.

t = thickness of the plates.

f_t = tensile stress, or resistance of plates in tension.
f_s = shearing stress, or resistance of rivets in shear.
f_c = crushing stress, or resistance of plates and rivets to crushing.
m = edge distance or marginal distance.
z = number of rivets in a pitch length.

The different ways in which the joint may fail are:

1. **The plate may split and fracture in front of the rivet** (Fig. 7.6 *f*)
   To resist breaking of the plate in front of the rivet we make the distance from the centre of the rivet to the edge of the plate at least 1.5 *d*. This distance will be safe against both shearing and tearing by the rivet pressure. For the sake of additional safety, in single riveted joints m is given the distance $2d$. On the other hand, in boiler work the margin should not be made larger than necessary, as it is then more difficult to caulk the joint and to make it steam tight.
2. **The plates may tear in the line of minimum section:** (Fig. 7.6 *c*)
   The line of minimum section is the line through the centre of holes, because, due to the drilled holes, the plate length between the holes along the pitch line is considered to be the weakest.

   $$\text{Area of plate along this line} = (p - d)t$$

   $$\text{Plate resistance to tension} = (p - d)t\, f_t \qquad ...(7.1)$$
3. **Plates and rivets may crush or fail in compression:** (Fig. 7.6 *e*)
   This mode of failure will occur when both plate and rivet are stronger in their respective mode of failure, rivet in shear and plate due to tearing.
   This would make the joint loose.

   $$\text{Resistance to crushing} = d \times t \times f_c \times z \qquad ...(7.2)$$
4. **The rivets may shear:** (Figs. 7.6 *a* and 7.6 *b*)
   The shear failure of rivets will occur when the plate is sufficiently strong.

In case of lap joints and butt joints with one strap, the rivets will be in single shear, while in case of two straps, the rivets will be in double shear. In addition the shearing resistance will depend upon that how many rivets fail in shearing in one pitch length.

$$\text{Resistance to single shear} = \pi/4\, d^2 \times f_s \times z \qquad ...(7.3)$$

$$\text{Resistance to double shear} = \pi/4\, d^2 \times f_s \times k \times z \qquad ...(7.4)$$

where $k = 2$ a constant for double shear theoretically. But in actual practice it is usual to use a factor of 1.75 to 1.875. The Indian Boiler Regulations (I.B.R.) specify the factor as 1.875.

Double and triple riveted joints may fail by a combination of shearing, crushing and tearing.

Single riveted lap joint: Resistance to tearing, crushing and shear are given by relations 7.1, 7.2 and 7.3 respectively. In these relations the value of z is one.

Double riveted lap joint: (Fig. 7.4 *a* Zig-Zag)

$$\text{Resistance to shearing two rivets} = 2 \times \pi/4\, d^2 \times f_s \qquad ...(7.5)$$

$$\text{Resistance to tearing between two rivets} = (p - d)\, t \times f_t \qquad ...(7.6)$$

$$\text{Resistance to crushing in front of two rivets} = 2 \times d \times t \times f_c \qquad ...(7.7)$$

$$\text{Strength of solid plate} = pt \times f_t \qquad ...(7.8)$$

Double riveted butt joint (Two unequal cover plates) (Fig. 7.4 *b*)

$$\text{Resistance to tearing at outer row of rivets} = (p - d)\, t \times f_t \qquad ...(7.9)$$

Resistance to shearing one rivet in double shear and one in single shear

$$= k \times \pi/4\, d^2 \times f_s + \pi/4\, d^2 \times f_s \qquad ...(7.10)$$

Resistance to tearing at inner row of rivets and

$$\text{shearing one of the outer row of rivets} = (p - d)\, t \times f_t + \pi/4\, d^2 \times f_s \qquad ...(7.11)$$

$$\text{Resistance of crushing in front of two rivets} = 2\, dt\, f_c \qquad ...(7.12)$$

Crushing in front of one rivet and shearing one rivet $= dt\, f_c + \pi/4\, d^2 \times f_s$ ...(7.13)

Strength of solid plate $= p \times t \times f_t$ ...(7.14)

Triple riveted butt joint: [Fig. 7.4 *b* (*x*)]

Resistance to tearing at outer row of rivets $= (p - d)\, t \times f_t$ ...(7.15)

Resistance to shearing four rivets in double shear and one in single shear

$= k \times (\pi/4\, d^2 \times f_s)\, 4 + \pi/4\, d^2 \times f_s$ ...(7.16)

Resistance to tearing at middle row of rivets and shearing one rivet

$= (p - 2d)\, t \times f_t + \pi/4\, d^2 \times f_s$ ...(7.17)

Resistance to crushing in front of four rivets and shearing one rivet

$= 4\, dt\, f_c + \pi/4\, d^2 \times f_s$ ...(7.18)

Resistance to crushing in front of five rivets $= 5\, dt \times f_c$ ...(7.19)

Strength of solid plate $= pt \times f_t$ ...(7.20)

The above relations help to understand the different relations for failure of riveted joints.

7.9 EFFICIENCY OF RIVETED JOINT

Plates jointed by riveting are never so strong as the original plates. The efficiency of the joint may be defined as the ratio of the load that will produce the allowable tension stress in the unpunched plate. A well designed joint should have the same efficiency in regard to every possibility of failure. In practice no joint is equally strong in every possible method of failure. The efficiency is usually represented as percentage.

$$\text{Tearing efficiency} = \eta_t = \frac{\text{tearing resistance of weakest part of strip}}{\text{tearing resistance of unpunched plate}} = \frac{(p-d)t \times f_t}{pt\, f_t} = \frac{p-d}{p}$$

$$\text{Shearing efficiency} = \eta_s = \frac{\text{resistance to shearing of rivets in a strip}}{\text{tearing resistance of unpunched plate}} = \frac{\pi/4\, d^2\, f_s \times k \times z}{p \times t \times f_t}$$

$$\text{Crushing efficiency} = \eta_c = \frac{\text{resistance to crushing of rivets or holes in a strip}}{\text{tearing resistance of unpunched plate}} = \frac{dt \times f_c \times z}{p \times t \times f_t}$$

The lowest of the three efficiencies is taken as the efficiency of the joint. The average efficiency obtained in commercial boiler joints is given in Table 7.5. The larger value in case being that obtained with the thinner plates.

Table 7.5 Efficiency and maximum pitch of commercial boiler joints

Type of joint	Efficiency percent	Maximum	Maximum pitch in mm
Lap joint			
Single riveted	45–60	63.3	$1.31\, t + 41.28$
Double riveted	63–70	77.5	$2.62\, t + 41.28$
Triple riveted	72–80	86.6	$3.47\, t + 41.28$
Butt joints (2 cover plates)			
Single riveted	55–60	63.3	$1.75\, t + 41.28$
Double riveted	70–83	86.6	$3.50\, t + 41.28$
Triple riveted [5 rivets per pitch, straps of unequal width]	80–90	95.0	$6\, t + 41.28$
Quadruple riveted	85–94	98.1	

I.B.R. 1950 allows a maximum figure of 85% efficiency for the best joint.

7.10 BOILER JOINTS

There are two types of joints needed to fabricate a boiler:

(i) *Longitudinal joint:* This joint is used to form a cylindrical shell by joining the ends of a plate.

(ii) *Circumferential joint:* This joint is used to get the required length of the shell and to close the ends.

The following procedure should be adopted for the design of the joints according to I.B.R.

By using the thin cylinder formula, the thickness of the boiler shell is determined, the formula is

$$t = \frac{PD}{2f_t \times \eta} + 1 \text{ to } 3 \text{ mm (corrosion allowance)}$$

where t = thickness of the boiler shell in mm

P = pressure of steam in N/mm^2 gauge

f_t = permissible tensile stress in N/mm^2

D = internal diameter of boiler shell in mm

η = longitudinal efficiency of joint.

The thickness of the boiler shell should not be less than 7 mm.

7.10.1 Longitudinal Joint

The longitudinal joint in a boiler shell is usually a butt joint with two cover plates. This joint is more efficient than lap joint. It is stiffer also and helps to maintain circularity of the shells at the ends of the plate. The longitudinal joints are staggered, because the shell is made up of a number of rings and this is done for convenience of connecting rings at places where both longitudinal and circumferential joints occur.

The efficiency of the joint chosen can be taken from Table 7.5. According to I.B.R. the factor of safety should never be less than 4. Table 7.6 gives the factor of safety to be used in various kinds of joints in boilers.

Table 7.6

Type of Joint	Hand riveting	Machine riveting
Lap joint	4.75	4.5
Single strap butt joint	4.75	4.5
Single riveted butt joint	4.75	4.5
Double riveted butt joint	4.25	4.0

1. After the thickness of the boiler shell is calculated, rivet diameter can be found. The rivet hole diameter should not be less than the thickness of the plate, otherwise the punch will be crushed earlier. In practice, therefore, the rivet hole diameter is given by Unwin's empirical formula, that is

$$d = 6.05\sqrt{t} \text{ to } 7.05\sqrt{t} \qquad \text{...(7.21)}$$

where t is the thickness of boiler shell in mm. But if the thickness of plate is less than 8 mm, the value of d is usually calculated by equating shearing resistance of rivets to crushing resistance. Table 7.2 gives the size of rivet diameter for rivet hole diameter and Table 7.1 gives the preferred length and diameter combinations.

2. The pitch of the rivets is calculated by considering the tension and shear or, in other words to make the joint as secure against fracture by tearing of the plates as it is against fracture through shearing of rivets. Its minimum value can be $2d$, which spacing is necessary for formation of head. Table 7.5 gives the maximum pitches of boiler joints adopted from I.B.R. The usual value for pressure vessel is $3d$ to make the joint leak proof, upto double riveted joints.

3. According to I.B.R., the thickness (t_1) for the butt straps are the following, but in no case the thickness of the butt strap shall be less than 10 mm.

t_1 = 1.125 t for chain riveted single butt strap

t_1 = 1.125 $t\left(\frac{p-d}{p-2d}\right)$ for single butt straps of every alternate rivet in outer rows being omitted.

t_1 = 0.625 t for double butt straps of equal width having chain riveting, (Fig. 7.4 *b* double riveted butt joint chain riveting).

t_1 = 0.625 $t\left(\frac{p-d}{p-2d}\right)$ for double butt straps of equal width having every alternate rivet in the outer rows being omitted.

For widths of the butt straps unequal, the thickness are given by

t_1 = 0.75 t for wide strap on the inside.

t_2 = 0.625 t for narrow strap on the outside.

4. The distance between the rows denoted by pb, should not be less than 0.6 times pitch for Zig-Zig riveting and 0.8 times pitch for chain riveting. According to I.B.R. the distance between the rows (pb) shall not be less than $0.33\,p + 0.67\,d$ for zig-zag riveting and $2d$ for chain riveting in lap and butt joints and in which there is an equal number of rivets in each row.
5. Again according to I.B.R., in joints in which the number of rivets in the outer rows is one half of the number in each of the inner rows and in which the inner rows are chain riveted, the distance between the outer rows and the next rows shall be not less than $0.33\,p + 0.67\,d$ or $2d$ whichever is greater. The distance between the rows in which there are full number of rivets (chain) shall be not less than $2d$.

   In the above paragraph if all the conditions remaining same but the inner rows are zig-zag riveted the distance between the outer rows and the next rows shall not be less than $0.2\,p + 1.15\,d$. The distance between the rows in which there are full number of rivets (zig-zag) shall be not less than $0.165\,p + 0.67\,d$.

   In all the above distances between the rows p is the pitch of rivets in outer rows.
6. The margin, $m = 1.5\,d$.
7. The diagonal pitch (in the case of zig-zag joint):

$$pd \geq (p + d)/2 \quad \text{or} \quad = (2.25 \text{ to } 2.50)\,d$$

7.10.2 The Procedure Given Below is for Design of Circumferential Lap Joint of a Boiler

The circumferential joint is a lap joint with one ring overlapping the other alternately.

1. The diameter of the rivet and the thickness of the boiler shell will be same as for longitudinal joint.
2. Now the number of rivets required in the seam is determined. As the circumferential joint is a lap joint, the rivets shall be in single shear.

$$\text{Shearing resistance of one rivet} = \pi/4\, d^2 f_s = F'$$

Let F be the total force which acts on the seam $= \pi/4\, D^2\, P = F$

When this force F is applied symmetrically, then the total number of rivets for circumferential joint

$$Z = \frac{F}{F'} = \frac{\pi/4\, D^2 P}{\pi/4\, d^2 f_s} = \left(\frac{D}{d}\right)^2 \frac{P}{f_s}$$

In computing the total number of rivets for crushing forces, the required number of rivets in the seam will be

$$Z = \frac{\pi/4 D^2 P}{d \times t \times f_c}$$

The greater of the two values of Z which are found is taken as the basis for the seam design. If the load coming on the joint is of varying type, and varies between $F_{max.}$ and $F_{min.}$, the allowable stress for a work made from low carbon steel is decreased by multiplying it by the correction factor $\dfrac{1}{1-\dfrac{1}{3}\dfrac{F_{min.}}{F_{max.}}} \leq 1$. The force $F_{max.}$ and $F_{min.}$ appear in the formula each with its sign.

3. It is known that a cylindrical vessel need be only as thick to guard against circumferential break as against a longitudinal break. The circumferential (girth) joints, therefore, need have no more than half the strength of the longitudinal joints. Therefore, once the efficiency of the longitudinal joint is known, the efficiency of the circumferential joint can be calculated. The strength of the circumferential joint in no case shall be less than 42% of the strength of the calculated thickness of the plate.

   Assuming this to be the lowest efficiency of the circumferential lap joint in tearing, the pitch of the rivets can be calculated from relation

$$\eta = \frac{p-d}{p}$$

4. Now the number of rivets in one row of a circumferential lap joint can be calculated.

$$z = \frac{\pi(D+t)}{p}$$

   Once the number of rivets in one row of a circumferential lap joint are known, the number of rows in the joint can be known.

$$\text{Number of rows} = \frac{\text{Total number of rivets (Z)}}{\text{Number of rivets in one row (z)}}$$

   From the number of rows, the type of the joint can be decided. And then the pitch and number of rivets in a row can be readjusted. The pitch for the joint to be leak proof should be checked from the I.B.R.

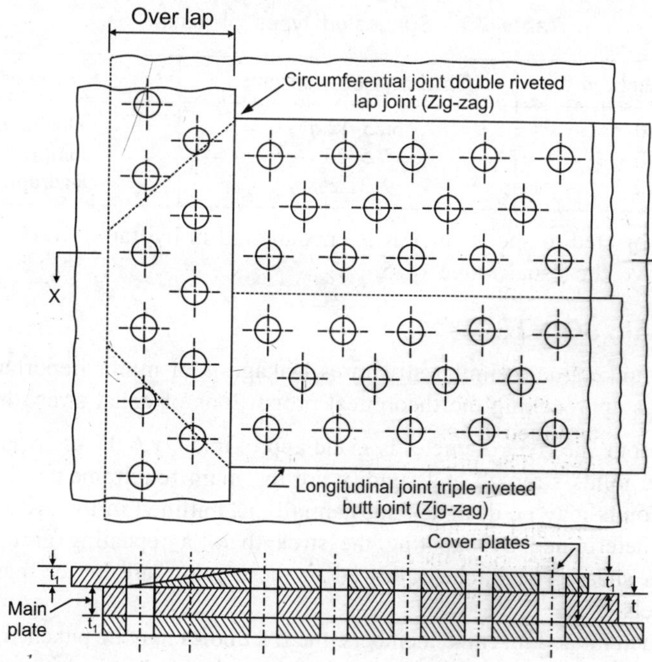

Fig. 7.7 *Junction of four plates*

5. The back pitch is calculated as mentioned earlier.
6. Where the longitudinal joint meets the circumferential joint on steam boiler we have the situation which is met by various methods. One of the methods of joining the longitudinal and circumferential joint is shown in Fig. 7.7. The end of the longitudinal strap *A* fits down underneath the circumferential plate in a recess to receive it.
7. When the distance between the rows of rivets is known, the overlap of the plate can be fixed. Overlap = (number of rows of rivets – 1) $pb + 2m$.

7.11 FRICTIONAL RESISTANCE OF RIVETED JOINTS

Rivets on cooling contract longitudinally and draw the plates together with considerable force. They also contract laterally and, therefore, do not completely fill their holes when cold. Before shearing can take place, it is consequently necessary that the plates shall slip on each other, such slipping, however, being resisted by the friction of the surfaces in contact. According to C. Bach, this frictional resistance, when slipping begins, ranges from 100.0 to 210.0 N/mm^2 of rivet section at each pair of surfaces in contact. As any appreciable slip of a boiler joint will result in leakage.

The tightness of the joint is guaranteed if the following requirement is satisfied $f = \dfrac{F}{\pi/4\, d^2 Z}$, where f is the allowable value of the specific resistance to slipping which depends on the design of the joint, coefficient of friction between the plates and the terminal riveting temperature.

7.12 SOME PRACTICAL CONSIDERATIONS

Tests indicate that when the rivets are placed at least 1.5 *d* from the plate edge there is no danger of the plate shearing or tearing in front of the rivet. Ensuring a leak proof joint by caulking would be difficult when this distance is too large. The A.S.M.E. boiler code requires that the edge distance must be not less than 1.5 *d* or more than 1.75 *d*. The longitudinal joints of shells in excess of 900 mm diameter must be of the double-strap butt construction. Table 7.7 gives the suggested type of joints.

Table 7.7 Suggested types of joints

Diameter of shell, *m*	Thickness of shell mm	Type of joint
0.60–1.80	6.25–12.5	Double riveted
0.90–2.10	7.5–25	Triple riveted
1.50–2.70	9–31.25	Quadruple riveted

When the joint is subjected to shock such as is encountered in hydraulic work, the factor of safety should be increased above the usual value of 5.

7.13 STRUCTURAL JOINTS

In ordinary tanks, coal bunkers, and similar structures leakage is of minor importance, therefore, they may have the proportions approaching the theoretical proportions of equal strengths in all methods of failure. For practical reasons the rivet diameter is made approximately 6.05 $\sqrt{t}$, where t is the thickness of plate in mm. In these joints strength and rigidity are the main requirements, therefore caulking of joints is not done. The joints may be designed substantially as outlined in the preceding article, except that the pitch may be determined by equating the strength of a repeating group in tension to the strength of the rivets in shear. The pitches allowed in these joints are larger than that employed in design of pressure vessels.

In the riveting of structural and machine members, the rivet holes may be punched, drilled and reamed, or some combination of these operations may be used. If the punched holes are without being reamed, as in structural work, it is assumed in calculating the strength of the plate that the punching operation may damage the plate to the extent of 0.08 mm. Therefore, in finding the area of the plate between rivet holes in a tension plate, the rivet hole should be considered as having a diameter that is 0.3 mm greater than the nominal diameter of rivet. In calculating the strength of a structural riveted joint, the nominal diameter of the rivet should be used in determining the strength of the rivet, since it is assumed here that the rivet may not completely fill the hole throughout the length of the rivet. Table 7.8 gives standard size of rivet and rivet holes (I.S. 1929–1961) for general purpose. Single or double riveted lap joints are used for storage tanks. The edges of storage tanks are stiffened by angle irons.

The permissible stresses for the design of riveted connections in bridges, building structures, and machine frames are somewhat higher than those used in pressure vessel design. The common permissible stresses used in such joints are given below, for mild steel structure. The safety factor is lower-about 2.0.

For plates in tension ... ... 140.0 N/mm^2
Shear on power driven rivets ... ... 106.0 N/mm^2
Shear on hand driven rivets ... ... 70.0 N/mm^2
For crushing of rivets and plates on power driven rivets
In single shear ... ... 225.0 N/mm^2
In double shear ... ... 280.0 N/mm^2

Table 7.8 Standard sizes of rivets and rivet holes

Diameter of rivet hole in mm.	13.5	15.5	17.5	19.5	21.5	23.5	25.5	29	32	35	38	41	44	50
Diameter of the shank of rivet in mm.	12	14	16	18	20	22	24	27	30	33	36	39	42	48

Crushing on hand driven rivets,
In single shear ... ... 112.5 N/mm^2
In double shear ... ... 140.0 N/mm^2

In structural riveting, spacing of the rivets should be governed by the stability of the joint, the existence of a margin between the rivet hole and the edge of the plate, and clearance for the die during riveting. Practical rules give for the pitch the limits $16t \geq p \geq 3d$ — where t is the thickness of the thinnest plate used in the design. An excessively large pitch prevents intimate contact between the members and the water may then deteriorate the joint by rusting. Table 7.9 gives the different values of pitch due to Rotscher. In no case the marginal distance (m) should be less than 1.5 d, if the tensile force is acting normal to the edge, the edge distance should be greater than $2d$. Distance between rows of rivets is 2.5 to $3d$.

Table 7.9

Thickness of plate mm	Diameter of rivet hole mm	Diameter of rivet mm	Pitch of rivets $p = 3d + 5$ mm	Marginal pitch m in mm
2	8.4	8	29	16
3	9.5	9	32	17
4	11	10	35	17
5–6	13	12	41	18
6–8	15	14	47	21
8–12	17	16	53	25
11–15	21	20	65	30

Because of the widespread use of aluminium in aircraft, rivets made of this material and its alloys (duralumin) have been the subject of considerable development. The design of aluminium riveted structures is similar to that of steel structures, the main difference being the greater variety of rivet alloys available. The allowable stresses for duralumin are given below:

Shear (rivet)	...	...	...	70.0–90.0 N/mm^2
Tensile (plate)	...	...	...	110.0–135.0 N/mm^2
Crushing (rivet and plate)	...	...	...	165.0–205.0 N/mm^2

Aluminium rivets can be obtained with the same heads as steel rivets. The proportions for aluminium riveting will be almost same as for other structural joints.

The procedure followed for the design of structural joints would be as given below:

(*a*) Determine the load to be transmitted by each member of the structure.
(*b*) Decide the shape and size of the member.
(*c*) Diameter of rivet will be determined from the thickness of the structural member.
(*d*) Shearing or crushing will decide the number of rivets required in each member.
(*e*) The rivets in the joints are spaced and arranged in such a manner as to utilize the material in the most economical way, avoiding eccentric loading as far as possible.
(*f*) Check the joint strength under different ways of failure.

The resistance of a rivet in double shear is taken 75 per cent more than in single shear in order to allow for possible eccentricity of load or defective workmanship. In double strap butt joints, the thickness of the strap is 0.75 times the thickness of the main plates and in single strap butt joints, the thickness of the strap is 1.25 times the thickness of the main plate.

Example 7.1. A single riveted lap joint is to be made of 10 mm plates. Find the diameter of rivets, their pitch and the efficiency of the joint. Take f_s = 64.0 N/mm^2, f_t = 80.0 N/mm^2 and so design the joint that its strength to withstand shear of rivets equals its strength to withstand tearing of the plate across the line of the rivet holes.

Solution: As the thickness of the plate is more than 8 mm, therefore, the diameter of rivet hole can be determined by Unwin's formula,

$$d = 6.05\sqrt{t} = 6.05\sqrt{10} = 19.15 \text{ mm.}$$

Use a standard size of 19.5 mm from Table 7.2. Also from the same table diameter of rivet is 18 mm.

Equating tearing of the plate to shearing of the rivet [Refer Fig. 7.4(*a*)].

$$(p - d)\, t \times f_t = \pi/4\, d^2 f_s$$

$$(p - 19.5) \times 10 \times 80.0 = \pi/4\,(19.5)^2 \times 64.0$$

gives p = 43.4 mm say 43.5 mm.

According to I.B.R. minimum pitch = $2d$ = 2 × 19.5 = 39 mm.

Maximum pitch = 1.31 t + 41.28 = 1.31 × 10 + 41.28 = 54.38 mm.

The value 43.5 mm lies in between the minimum and maximum values.

$$\text{Efficiency} = \frac{(p-d)t \times f_t}{p \times t \times f_t} = \frac{p-d}{p} = \frac{43.5-19.5}{43.5} \times 100 = 52.2\%.$$

Example 7.2. A double riveted lap joint is to be made between 9.5 mm plates. If the safe working stresses are f_t = 80.0 N/mm^2, f_s = 60.0 N/mm^2 and f_c = 1.5 f_t, calculate the rivet pitch, and distance between rows of rivets for the joint. State how the joint will fail.

Solution: Refer Fig. 7.4 *a* for double riveted lap joint chain and zig-zag, diameter of rivet hole by Unwin's formula

$$d = 6.05\sqrt{t} = 6.05\sqrt{9.5} = 19 \text{ mm.}$$

According to I.S., the size of rivet hole from table 7.2 will be 19.5 mm and diameter of rivet 18 mm.

In one pitch length there will be two rivets in single shear. Equating the tearing of the plate to shearing of the rivets.

$$(p - d)t\,f_t = \pi/4\, d^2 f_s \times 2$$

$$(p - 19.5)\,(9.5) \times 80.0 = \pi/4\,(19.5)^2 \times 60.0 \times 2$$

gives, $p = 66.5$ mm.

According to I.B.R. minimum pitch = $2d = 2 \times 19.5 = 39$ mm.

Maximum pitch = $2.62\,t + 41.28 = 2.62 \times 9.5 + 41.28 = 67$ mm.

Adopted pitch = 67 mm.

$$\text{Efficiency} = \frac{p-d}{p} = \frac{67-19.5}{67} = \frac{47.5}{67} \times 100 = 71.8\%$$

According to I.B.R., the distance between the rows for chain riveting $\geq 2 \times d = 2 \times 19.5 = 39$ mm.

Also the distance between the rows $\geq 0.8 \times p$

$$\geq 0.8 \times 67$$

$$\geq 53.6 \text{ mm.}$$

Take distance between the rows as 53.5 mm.

For zig-zag riveting, distance between the rows $\geq 0.33\,p + 0.67\,d$

$$\geq 0.33 \times 67 + 0.67 \times 19.5$$

$$\geq 45.1 \text{ mm.}$$

Also, the distance between the rows $\geq 0.6\,p$

$$\geq 0.6 \times 67$$

$$\geq 40.2 \text{ mm}$$

So take distance between the rows for zig-zag riveting as 45 mm.

Example 7.3. A triple riveted lap joint is to be made between 6.5 mm plates. The allowable stresses $f_t = 35.00$ N/mm², $f_s = 29.00$ mm², $f_c = 52.50$ N/mm². Calculate the rivet diameter, rivet pitch and back pitch, zig-zag riveting is to be used. Indicate how the joint will fail.

Solution: As the thickness of the plate is less than 8.0 mm., therefore, the diameter of rivet hole shall be determined by equating the shearing resistance of rivets to crushing resistance. Refer Fig. 7.4 (*a*) for triple riveted lap joint zig-zag.

There will be three rivets in single shear and three rivets in crushing. Equating the shearing and crushing resistances

$$3\pi/4\, d^2 f_s = 3dt\,f_c$$

$$3 \times \pi/4\, d^2 \times 29.00 = 3 \times d \times 6.5 \times 52.50$$

$$d = 15 \text{ mm.}$$

According to I.S., the size of the rivet hole from table 7.2 will be 15.5 mm. and diameter of rivet 14 mm.

To find out the pitch of rivet, we equate the tearing resistance of plate to shearing resistance of rivets.

$$(p - d)\,t\,f_t = 3\pi/4\, d^2 f_s$$

$$(p - 15.5)6.5 \times 35.00 = 3 \times \pi/4\,(15.5)^2 \times 29.00$$

$$p = 87 \text{ mm.}$$

From table 7.5, maximum pitch = $3.47\,t + 41.28 = 3.47 \times 6.5 + 41.28 = 64$ mm.

Therefore, we take the maximum pitch as 64 mm. Distance between the rows of rivets (back pitch) for zig-zag riveting

$$= 0.6 \times p = 0.6 \times 64 = 38.4 \text{ mm.}$$

According to I.B.R., minimum back pitch = $0.33\,p + 0.67\,d = 0.33 \times 64 + 0.67 \times 15.5 = 31.6$ mm.

Therefore, back pitch = 39 mm.

$$\text{Tearing efficiency } \eta_t = \frac{p-d}{p} = \frac{64-15.5}{64} \times 100 = 76\%$$

$$\text{Crushing efficiency } \eta_c = \frac{3dt \times f_c}{pt\ f_t} = \frac{3 \times 15.5 \times 6.5 \times 52.50}{64 \times 6.5 \times 35.00} \times 100 = 109\%$$

$$\text{Shearing efficiency } \eta_s = \frac{3 \times \pi/4d^2 f_s}{p \times t \times f_t} = \frac{3 \times \pi/4 \times (15.5)^2 \times 29.00}{64 \times 6.5 \times 35.00} \times 100 = 112.5\%$$

which indicates the joint will fail by tearing of the plates.

Example 7.4. Design a double riveted butt joint with two cover plates for the longitudinal seam of a boiler shell, 0.75 m diameter, to carry a maximum steam pressure of 1.05 N/mm². The allowable stresses are $f_t = 35.0$ N/mm², $f_s = 28.0$ N/mm². Assume the efficiency of the joint 75%.

Solution: This thickness of the plate would be obtained from the relation

$$t = \frac{PD}{2f_t\eta} = \frac{1.05 \times 750}{2 \times 35.0 \times 0.75} = 15 \text{ mm.}$$

As t is greater than 8 mm, the diameter of the rivet hole is calculated by the formula $6.05\sqrt{t} = 6.05\sqrt{15}$ = 25 mm. For boiler joints the standard diameter of rivet hole from table 7.2 is 25.5 mm and diameter of rivet is 24 mm.

As the joint is double strap butt joint, the rivets will be in double shear, and there will be two rivets per pitch length.

Strength of two rivets in double shear $= 2 \times \pi/4\ d^2 \times f_s \times 1.875$

According to I.S. the strength of rivets in double shear is taken as 1.875 the strength in single shear.

Shearing resistance $= 2 \times \pi/4\ (25.5)^2 \times 28.0 \times 1.875 = 53.50$ kN.

Strength of two rivets in crushing $= 2 \times d \times t \times f_c$

$$= 2 \times 25.5 \times 15 \times 52.5 = 40.20 \text{ kN.}$$

As the strength in crushing is less, therefore, the strength in crushing will be equated to strength in tearing of plate.

$$40.2 \times 10^3 = (p-d)\ t \times f_t = (p - 25.5)\ 15 \times 35.0$$

$$p = 102 \text{ mm.}$$

Minimum value of pitch $= 2 \times d = 2 \times 25.5 = 51$ mm.

According to I.B.P., maximum pitch for double riveted butt joint with two cover plates

$$= 3.5\ t + 41.28 = 3.5 \times 15 + 41.28 = 94 \text{ mm.}$$

Thus adopt the pitch as 95 mm.

Strength of the solid plate $= p \times t \times f_t = 95 \times 15 \times 35.0 = 50.00$ kN

$$\text{Strength of the joint} = \frac{40.2 \times 10^3}{50 \times 10^3} \times 100 = 80\%$$

which is more than the required efficiency of the joint.

According to I.B.R., the thickness of butt straps of equal width will be

$$t_1 = 0.625\ t = 0.625 \times 15 = 9.4 \text{ mm; say 10 mm.}$$

Example 7.5. A boiler joint is to be designed for a steam pressure of 2.5 N/mm². The inside diameter of the largest course of the drum is 1.35 m. The complete joint is shown in Fig. 7.4 b (*X*). The allowable stresses are $f_t = 78.0$ N/mm², $f_s = 62.5$ N/mm², $f_c = 135.0$ N/mm².

Solution: The thickness of the boiler shell is determined by the relation

$$t = \frac{PD}{2f_t \times \eta} + 1.0$$

1 mm. is taken as the corrosion allowance. As the joint is a triple riveted butt joint with two unequal cover plates (5 rivets per pitch) the efficiency of the joint may be taken as 85% from the Table 7.5.

$$t = \frac{2.5 \times 1350}{2 \times 78.0 \times 0.85} + 1.0 = 26.5 \text{ mm.}$$

According to I.B.R., for widths of butt straps unequal $t_1 = 0.75\,t$ for wide strap on inside.

$$t_1 = 0.75 \times 26.5 = 19.8 \text{ mm. say } 20 \text{ mm.}$$

$$t_2 = 0.625\,t \text{ for narrow strap on the outside.}$$

$$t_2 = 0.625 \times 26.5 = 16.5 \text{ mm.}$$

To calculate the pitch on the outer row, equate the resistance of the plate in tearing to the resistance of shearing of rivets. In one pitch length there are four rivets in double shear and one in single shear. Take strength of rivet in double shear 1.875 times the strength of joint in single shear. As the thickness of plate is more than 8 mm, therefore, the diameter of rivet hole

$$d = 6.05\sqrt{t} = 6.05\sqrt{26.5} = 31.1 \text{ mm.}$$

The standard size of rivet hole is 32 mm. Diameter of rivet will be 30 mm.

Now equating the tearing resistance to shearing resistance

$$(p-d)\,t \times f_t = 4 \times \pi/4\, d^2 \times f_s \times 1.875 + \pi/4\, d^2 \times f_s$$

$$(p-32) \times 26.5 \times 78.0 = (\pi/4 \times (32)^2 \times 62.5)\,(4 \times 1.875 + 1)$$

$$p = 239 \text{ mm say } 240 \text{ mm.}$$

This is the pitch along the outer most row. Pitch along the inner most row will be 120 mm.

The actual efficiency of the joint can now be calculated from the strength of the joint in all the possible modes of failure.

Strength of unpunched plate$= pt \times f_t = 240 \times 26.5 \times 78.0 = 495.00$ kN

The joint may fail by tearing the plate in the first row, then

$$= (p-d)\,t \times f_t = (240-32) \times 26.5 \times 78.0 = 430.00 \text{ kN}$$

This joint may fail by shearing all the rivets, then

$$= 4 \times \pi/4\, d^2 \times f_s \times 1.875 + \pi/4\, d^2 \times f_s$$

$$= \pi/4 \times (32)^2 \times 62.5\,(4 \times 1.875 + 1) = 427.00 \text{ kN}$$

The joint may fail by crushing all rivets, then

$$= (4dt + dt_1)\,f_c = (4 \times 32 \times 26.5 + 32 \times 20)\,135.0 = 545.00 \text{ kN}$$

Since the last two relations show that the joint is stronger in crushing than in shear, it is useless to check any other type of the failure involving crushing of the rivets.

The joint may fail by tearing the plate between the rivets in the second row and shearing the rivet in the outer row, then

$$= (p-2d)\,t \times f_t + \pi/4\, d^2 \times f_s$$

$$= (240 - 2 \times 32)\,26.5 \times 78.0 + \pi/4\,(32)^2 \times 62.5 = 413.30 \text{ kN.}$$

The joint may fail by tearing the plate between the rivets in the third row and shearing two rivets in double shear in second row and shearing the rivet in the outer row, then

$$= (p-2d)\,t \times f_t + 2 \times \pi/4\, d^2 \times f_s \times 1.875 + \pi/4\, d^2 \times f_s$$

$$= (240 - 2 \times 32)\,26.5 \times 78.0 + 2 \times \pi/4\,(32)^2 \times 62.5 \times 1.875 + \pi/4\,(32)^2 \times 62.5$$

$$= 601.30 \text{ kN.}$$

By looking at the various possible modes of failure, we find that the joint will fail by tearing the plate between the rivets in the second row. Therefore, the efficiency of the joint as designed will be

$$= \frac{413.3 \times 10^3}{495.0 \times 10^3} \times 100 = 83.8\%$$

which is very much near the assumed efficiency and may be considered to be satisfactory.

The design is now completed by determining the back pitches and the distance from the rivet holes to the plate edges.

The edges distance from the centre of rivet holes to the plate edges,

$$m = 1.5\,d = 1.5 \times 32 = 48 \text{ mm}$$

The back pitch between the centre of the second and third row for zig-zag riveting is given by

$$= 0.6\,p = 0.6 \times 240 = 144 \text{ mm.}$$

According to I.B.R., *pb* shall not be less than $0.33\,p' + 0.67\,d$ for zig-zag riveting, therefore,

$$pb = 0.33 \times 120 + 0.67 \times 32 = 39.6 + 21.4 = 61 \text{ mm.}$$

Adopt the back pitch as 62.5 mm.

Example 7.6. Determine the main dimensions for longitudinal and circumferential joints for a cornish boiler whose inner diameter is 1.7 *m* and the pressure of steam is 2.05 N/mm^2. The complete joint is shown in Fig. 7.8. The design will be based on a double-strap butt joint quadruple riveted, for the longitudinal seam, with an assumed efficiency of 85 per cent (Taken from table 7.5). This assumed efficiency will be checked later by the actual calculations on the joint. The ultimate stresses are $f_c = 780.0$ N/mm^2, $f_t = 470.0$ N/mm^2, $f_s = 390$ N/mm^2. Take factor of safety as 5. It will be assumed that the rivet in double shear will have an effective resisting area not greater than 87.5% over that in single shear.

Solution: The allowable stresses will be

$$f_c = \frac{780.0}{5} = 156.0 \text{ N/mm}^2$$

$$f_t = \frac{470.0}{5} = 94.0 \text{ N/mm}^2$$

$$f_s = \frac{390.0}{5} = 78.0 \text{ N/mm}^2.$$

The thickness of the boiler shell is determined by the relation

$$t = \frac{PD}{2f_t \times \eta} + 1.0$$

1.0 mm is taken as the corrosion allowance

$$t = \frac{2.05 \times 1700}{2 \times 94.0 \times 0.85} + 1.0 = 22.8 \text{ mm, say 23 mm.}$$

Table 7.7 suggests that the thickness of boiler shell for a quadruple joint can vary from 9.0 to 31.25, for a boiler diameter of more than 1.5 m. Therefore, 23 mm thickness is alright.

According to I.B.R., for unequal width of butt straps

$$t_1 = 0.75\,t \text{ for wide strap on the inside}$$
$$= 0.75 \times 23 = 17.2 \text{ mm say 17.5 mm}$$
$$t_2 = 0.625\,t \text{ for narrow strap on the outside}$$
$$= 0.625 \times 23 = 14.4 \text{ mm say 14.5 mm}$$

Longitudinal joint: Diameter of rivet hole by Unwin's formula

$$d = 6.05\sqrt{t} = 6.05\sqrt{23} = 29 \text{ mm}$$

The standard size of rivet hole from table 7.2 is also 29 mm. Diameter of rivet will be 27 mm.

To calculate the pitch along the row *A–A*, equate the resistance of the plate in tearing to the resistance of shearing of rivets.

In one pitch length there are eight rivets in double shear and three in single shear.

$$(p - d)\,t f_t = \pi/4\, d^2 \times f_s\,(8 \times 1.875 + 3)$$
$$(p - 29)\; 23 \times 94.0 = \pi/4\,(29)^2 \times 78.0\,(15 + 3)$$
$$p = 455 \text{ mm, we adopt 450 mm as the pitch of the longitudinal joint.}$$

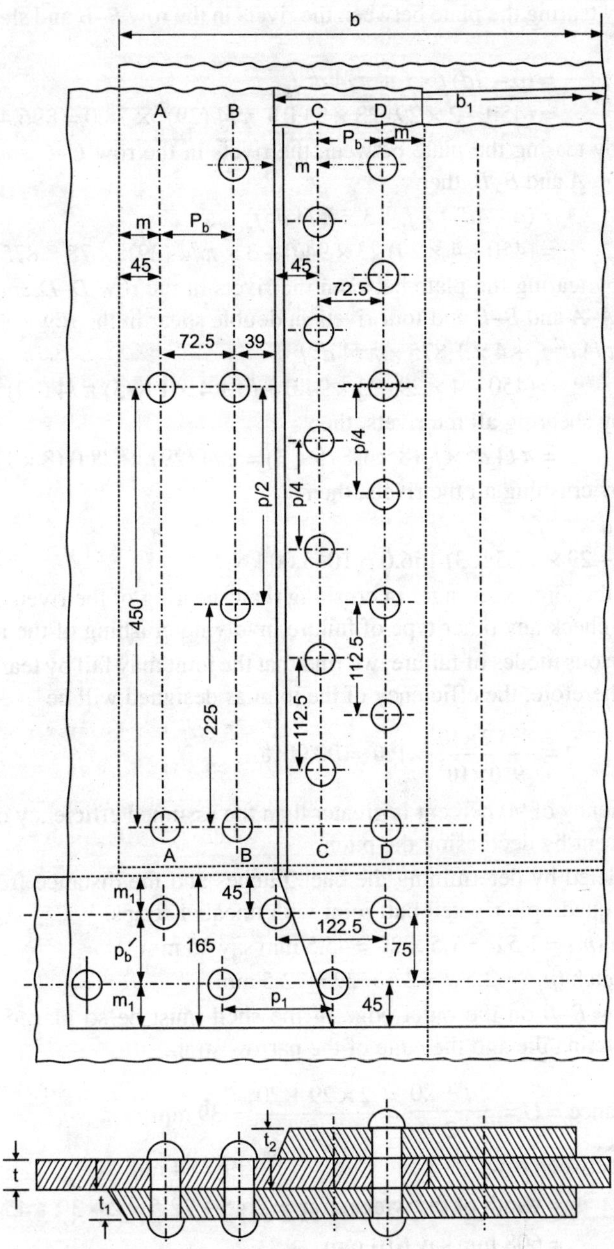

Fig. 7.8

Pitch along row $B–B$ = 225 mm

Pitch along row $C–C$ and $D–D$ = 112.5 mm

The actual efficiency of the joint can now be calculated from the strength of the joint in all the possible modes of failure.

Strength of unpunched plate = $pt \times f_t = 450 \times 23 \times 94.0 = 970.00$ kN

The joint may fail by tearing the plate in the row $A–A$, then

$$= (p - d)\, t \times f_t = (450 - 29)\, 23 \times 94.0 = 907.00 \text{ kN.}$$

The joint may fail by tearing the plate between the rivets in the row *B–B* and shearing the rivet in the row *A–A*, then

$$= (p - 2d)\, t \times f_t + \pi/4\, d^2 f_s$$
$$= (450 - 2 \times 29)\, 23 \times 94.0 + \pi/4\, (29)^2 \times 78.0 = 896.40 \text{ kN.}$$

The joint may fail by tearing the plate between the rivets in the row *C–C* and shearing three rivets in single shear in the rows *A–A* and *B–B,* then

$$= (p - 4d)\, t \times f_t + 3 \times \pi/4\, d^2 f_s$$
$$= (450 - 4 \times 29)\, 23 \times 94.0 + 3 \times \pi/4\, (29)^2 \times 78 = 875.20 \text{ kN.}$$

The joint may fail by tearing the plate between the rivets in the row *D–D* and shearing three rivets in single shear in the rows *A–A* and *B–B* and four rivets in double shear in the row *C–C*, then

$$(p - 4d)\, t \times f_t + 3 \times \pi/4\, d^2 f_s + 4 \times 1.875 \times \pi/4\, d^2 f_s$$
$$= (450 - 4 \times 29)\, 23 \times 94.0 + (3 + 4 \times 1.875)\, \pi/4\, (29)^2 \times 78.0 = 1261.00 \text{ kN.}$$

The joint may fail by shearing all the rivets, then

$$= \pi/4\, d^2 \times f_s\, (8 \times 1.875 + 3) = \pi/4\, (29)^2 \times 78.0\, (8 \times 1.875 + 3) = 925.00 \text{ kN.}$$

The joint may fail by crushing all the rivets, then

$$(8d \times t + 3 \times d \times t_1) f_c$$
$$(8 \times 29 \times 23 + 29 \times 17.5 \times 3)\, 156.0 = 1065.00 \text{ kN.}$$

Since, it shows that the joint is stronger in crushing than shearing of the rivets and tearing of the plates, therefore, it is useless to check any other type of failure involving crushing of the rivets.

By looking at the various modes of failure, we find that the joint may fail by tearing the plate between the rivets in the row *C–C*. Therefore, the efficiency of the joint as designed will be

$$\eta = \frac{875.2 \times 10^3}{970 \times 10^3} \times 100 = 90.00\%.$$

The calculated efficiency of 90 percent is greater than the assumed efficiency of 85%. If so desired, the efficiency can be 85 percent by decreasing the pitch.

The design is completed by determining the back pitches and the distance from the rivet holes to the plate edges. The joint is quadruple riveted butt joint with unequal straps.

Marginal pitch (m) $= 1.5\, d = 1.5 \times 29 = 45.5$ mm say 45 mm

Transverse or back pitch (p_b) $= 2.5\, d = 2.5 \times 29 = 72.5$ mm.

The rivets in the row *B–B* on the outer edge of the shell must be so placed as to provide sufficient clearance between the riveting die and the edge of the narrow strap.

$$\text{The distance} = G = \frac{2d + 20}{2} = \frac{2 \times 29 + 20}{2} = 39 \text{ mm}$$

Therefore, distance between rows *B–B* and *C–C* = 45 + 39 = 84 mm

$$b = \text{width of the wider strap} = 2[45 + 72.5 \times 2 + 39 + 45 + 45]$$
$$= 608 \text{ mm say } 610 \text{ mm}$$
$$b_1 = \text{width of narrow strap} = 2[45 + 72.5 + 45]$$
$$= 325 \text{ mm say } 330 \text{ mm.}$$

Circumferential joint:

The diameter of rivet hole and the thickness of boiler shell will be same. As the joint is a lap joint, therefore, the rivets shall be in single shear.

Shearing resistance of one rivet $= \pi/4\, d^2 \times f_s = \pi/4 \times (29)^2 \times 78.0 = 51.40$ kN.

The force tending to shear all the rivets will be

$$\pi/4\, D^2 P = \pi/4\, (1700)^2 \times 2.05 = 4.65 \text{ MN.}$$

The number of rivets is calculated by equating the strength of rivets in shearing to the bursting force.

$$4.65 \times 10^6 = Z \times \pi/4\,(29)^2 \times 78.0$$

$$Z = 90.3, \text{ say, } 91.$$

The number of rivets can also be calculated by equating the strength of rivets in crushing to the bursting force.

$$4.65 \times 10^6 = Z \times d \times t \times f_c$$

$$Z = \frac{4.65 \times 10^6}{29 \times 23 \times 156.0} = 44.8 \text{ say } 45.$$

Take the number of rivets as 92.

From Table 7.7, the circumferential lap joint will be double riveted, as the boiler shell diameter is 1.7 m.

Therefore, the number of rivets per row $= \frac{92}{2} = 46$

Hence the pitch of rivets $= \frac{\pi(D+t)}{46} = \frac{\pi(1700+23)}{46} = 122$ mm say 122.5 mm

Efficiency of the joint $= \frac{p_1 - d}{p_1} = \frac{122.5 - 29}{122.5} = 76.5\%$

Usually the efficiency of circumferential lap joint is half of the longitudinal joint. But this is much more than that. To reduce the efficiency the pitch of the rivets can be decreased.

As these rivets shall be arranged in zig-zag in two rows, therefore, distance between the rows ($p_b{}'$) will be

$$p_b{}' = 0.33\,p + 0.67\,d = 0.33 \times 122.5 + 0.67 \times 29 = 60 \text{ mm.}$$

But it should not be less than $0.6\,p = 0.6 \times 122.5 = 73.5$ mm.

Take $\qquad p_b{}' = 75$ mm.

Marginal distance $m_1 = 1.5\,d = 1.5 \times 29 = 43.5$ mm. say 45 mm.

Overlap of the plate $= m_1 + m_1 + p_b{}' = 45 + 45 + 75 = 165$ mm.

Example 7.7. Two lengths of a mild steel tie rod (flat plate) 190 mm × 12 mm are to be connected together by means of a lap joint, using only four rivets. Find the diameter of the rivets, when the rods are subjected to tensile loading. Indicate the arrangement of the rivets to give the strongest joint. The allowable stresses are $f_t = 90.0$ N/mm², $f_s = 75.0$ N/mm², $f_c = 150.0$ N/mm².

Solution: The diameter of the rivet can be given by Unwin's formula

$$d = 6.05\sqrt{t} = 6.05\sqrt{12} = 21.1 \text{ mm, say, } 22 \text{ mm.}$$

For lap joints the rivets fail in shear. Therefore, the resistance of the rivet in shearing

$$= \pi/4\,d^2 \times f_s = \pi/4\,(22)^2 \times 75.0 = 29.50 \text{ kN}$$

Resistance in crushing of one rivet $= d \times t \times f_c$

$$= 22 \times 12 \times 150.0 = 39.50 \text{ kN.}$$

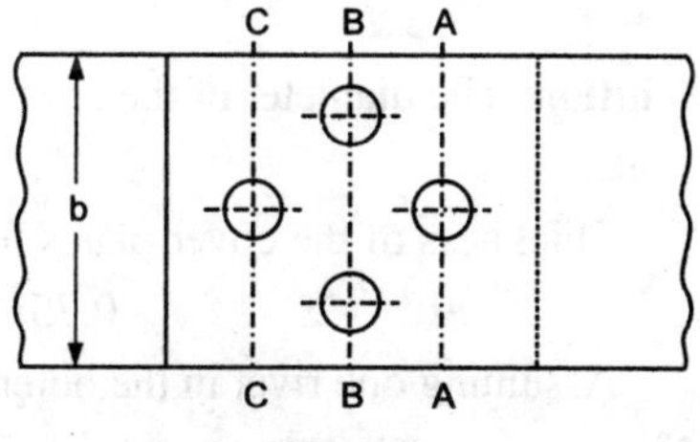

Fig. 7.9

Hence the shearing strength, remains the basis for design. Experience has shown that a joint of diamond form, or sometimes it is called "Lozenge form" gives the greatest possible strength. Therefore, this form of arrangement will be tried in this problem and Fig. 7.9 indicates the arrangement. In this the first row has only one rivet, second row two rivets and in successive rows the rivets are increased. The plate is weakened only by one rivet in the outermost row A–A. As in this problem only 4 rivets are to be used in a width of plate 190 mm, therefore, a bigger diameter of the rivet may be used. This can be found by equating the resistance in tearing of plate in row A–A to shearing of all the rivets.

$$4 \times \pi/4\, d^2 \times f_s = (b-d) \times t \times f_t$$
$$4 \times \pi/4\, d^2 \times 75.0 = (190-d)\, 12 \times 90.0$$
$$d = 28 \text{ mm}$$

Strength in tearing of the joint along A–$A = (b-d) \times t \times f_t = (190-28) \times 12 \times 90.0 = 185.00$ kN.

Shearing strength of all rivets $= 4 \times \pi/4\, (28)^2 \times 75.0 = 181.00$ kN.

Strength at section B–B in tearing when one rivet in A–A will fail in shearing

$$= \pi/4\, d^2 \times f_s + (b-2d) \times t \times f_t$$
$$= \pi/4\, (28)^2 \times 75.0 + (190-56) \times 12 \times 90.0 = 201.00 \text{ kN.}$$

The tearing strength at C–C will be same as at A–A.

Therefore, the maximum load this joint can take in tension will be 181 kN.

Example 7.8. Two mild steel tie bars, for a bridge structure are to be joined by means of a butt joint with double straps. The thickness of the tie bar is 12 mm and carries a tensile load of 400 kN. Design the joint completely taking the allowable stresses as $f_t = 100.0$ N/mm², $f_s = 75.0$ N/mm², $f_c = 150.0$ N/mm².

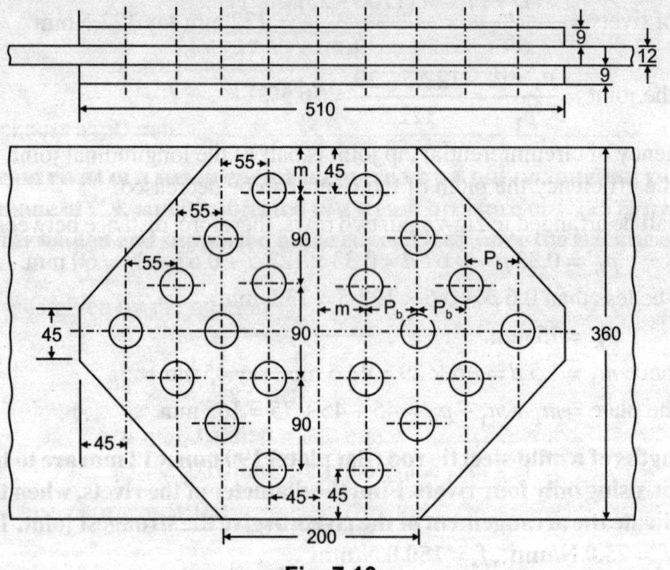

Fig. 7.10

Solution: The diameter of the rivets is determined by the formula

$$d = 6.05\sqrt{t} = 6.05\sqrt{12} = 21.1 \text{ mm say } 22 \text{ mm}$$

Thickness of the cover plates for double strap butt joint will be

$$0.75t = 0.75 \times 12 = 9 \text{ mm}$$

Assuming one rivet in the outer row, resistance to tearing of plate at this row

$$= (b-d) \times t \times f_t$$
$$400 \times 10^3 = (b-22)\, 12 \times 100.0$$
$$b = 355 \text{ mm say } 360 \text{ mm}$$

The joint is a butt joint with two straps, therefore the rivets shall be in double shear.

Shearing resistance of one rivet $= \pi/4\, (d)^2\, f_s \times 1.75 = \pi/4\, (22)^2 \times 75.0 \times 1.75 = 49.90$ kN.

Crushing resistance of one rivet $= d \times t \times f_c = 22 \times 12 \times 150.0 = 40.50$ kN

As the crushing resistance, is less than the shearing resistance, therefore, the number of rivets will be decided on the basis of crushing resistance of rivets.

$$\text{Therefore, number of rivets} = \frac{400 \times 10^3}{40.05 \times 10^3} = 10$$

The placing of the rivets shall be in diamond form as shown in Fig. 7.10. In first row one rivet, second row two rivets, third row three rivets and last row four rivets.

Now the strength of the joint can be checked up

$$\text{Strength of solid plate} = b \times t \times f_t = 360 \times 12 \times 100.0 = 432.00 \text{ kN}$$

$$\text{Resistance of all the rivets in shearing} = Z \times \pi/4\,(d)^2 \times f_s \times 1.75$$

$$= 10 \times \pi/4\,(22)^2 \times 75.0 \times 1.75 = 499 \text{ kN}$$

Resistance of all the rivets in crushing

$$= Z \times d \times t \times f_c = 10 \times 22 \times 12 \times 150.0 = 405 \text{ kN}$$

Resistance of the joint in tearing along the first now

$$= (b - d) \times t \times f_t = (360 - 22) \times 12 \times 100.0 = 400 \text{ kN}$$

Failure of the joint in tearing along the second row will take place by crushing one rivet in the first row.

Resistance of the joint in tearing along the second row

$$= (b - 2 \times d)\, t \times f_t + d \times t \times f_c = (360 - 44) \times 12 \times 100.00 + 22 \times 12 \times 150.0$$

$$= 418 \text{ kN}$$

Resistance of the joint in tearing along third row

$$= (b - 3d) \times t \times f_t + 3 \times d \times t \times f_c$$

$$= (360 - 3 \times 22)\, 12 \times 100.0 + 3 \times 22 \times 12 \times 150.0 = 473 \text{ kN}$$

Resistance of the joint in tearing along the fourth row

$$= (b - 4d) \times t \times f_t + 6 \times d \times t \times f_c$$

$$= (360 - 4 \times 22)\, 12 \times 100.0 + 6 \times 22 \times 12 \times 150.0 = 565 \text{ kN.}$$

The tearing of the plate in the second, third and fourth rows with shearing of rivets in the previous rows need not be checked as the shearing resistance is more than the crushing resistance.

By looking at the different modes of failure, we find that the resistance of the joint along the first row in tearing is the minimum.

$$\therefore \qquad \eta_t = \frac{400 \times 10^3}{432 \times 10^3} = 92.8\%$$

The distance between the rows (p_b) will be taken

$$= 2.5\,d = 2.5 \times 22 = 55 \text{ mm}$$

$$\text{Marginal distance } m = 1.5\,d = 1.5 \times 22$$

$$= 33 \text{ mm. Take it as 45 mm}$$

Pitch of rivets $= 4d = 4 \times 22 = 88$ mm say 90 mm

All the dimensions are shown in Fig. 7.10.

It can be noted that, if instead of diamond riveting, more than one rivet are put on the first row the efficiency of the joint will be

$$= \frac{(b - 2d) \times t \times f_t}{b \times t \times f_t} = \frac{(360 - 2 \times 22)}{360} = 87.9\%.$$

EXERCISES

1. How do you classify the joints in Engineering?
2. In modern engineering which joints are preferred.
3. Differentiate between Interference and adhesive bonded joints.
4. Give advantages and disadvantages of riveted joints.
5. How fastening of boiler plates is done and why?
6. What are the different methods of riveting?
7. What is the method adopted for riveting of boiler plates?
8. Name the rivet heads used in boiler construction.
9. Name the rivet joints used in boiler plates.
10. Why is butt joint preferred over lap joint?
11. Name the rivet heads used in structural work.
12. How is the testing of manufactured rivets done?
13. Explain the terminology gage line, diagonal pitch, marginal distance, transverse pitch, pitch.
14. Which method is adopted to ensure leakproof joint?
15. What are the various modes in which a riveted joint may fail?
16. Name the various assumptions taken in the design of riveted joint.
17. What do you mean by the term efficiency of a riveted joint?
18. According to I.B.R. what is the highest efficiency required of a riveted joint?
19. What formula is used for calculating the diameter of a rivet hole?
20. Design a single riveted lap joint to connect two mild steel plates, 12.5 mm thick, the joint being designed for maximum efficiency. The allowable stresses are f_t = 35.0 N/mm^2, f_s = 28.5 N/mm^2, f_c = 53.0 N/mm^2. Also, calculate the load this joint will carry per pitch length.
21. Determine the efficiency of a single riveted lap joint having a rivet diameter of 29 mm, pitch 80 mm and a plate thickness of 12.5 mm.
22. A double riveted lap joint is made of steel having an ultimate strength of 390.0 N/mm^2 in tension and 675.0 N/mm^2 in crushing. The rivets have a strength of 310.0 N/mm^2 in shear. The plates are 20 mm thick. Rivets 25 mm in diameter are used in 26.5 mm holes spaced 80 mm on centres. Determine the manner in which this joint will fail and at what load.
23. If a double-strap triple riveted joint with equal width straps has a higher efficiency than a similar joint with unequal width straps, why is the latter joint generally used in boiler construction?
24. The longitudinal seam of a cylindrical boiler drum is of the double riveted butt joint type. The shell is 12.5 mm thick, the nominal diameter of rivet is 21 mm, the pitch is 125 mm, the short pitch is 62.5 mm, the transverse pitch is 70 mm, and all margins are 35 mm. The thickness of the inner, wide cover plate is 9 mm. and that of the narrow cover plate is 12.5 mm. Determine the manner of possible failure and the efficiency of this joint using the allowable stresses.
25. Design a double strap, double riveted butt joint for the longitudinal seam of a boiler joint, 0.75 m diameter, to carry a maximum steam pressure of 1.05 N/mm^2 gauze. The allowable stresses are f_t = 35.0 N/mm^2, f_s = 28.5 N/mm^2, f_c = 53.0 N/mm^2.
26. Design the longitudinal joint for a 1.25 m steam boiler, that is to carry a steam pressure of 5.25 N/mm^2. You may assume the ultimate tensile strength of the plate as 420.0 N/mm^2, crushing strength 650.0 N/mm^2 and shearing strength 398.0 N/mm^2. The joint efficiency may be assumed as 85%. Sketch the designed joint with all dimensions.
27. Design the circumferential and longitudinal joints of a 1.2 m diameter horizontal return tabular boiler which must carry a working pressure 1.15 N/mm^2 gauze. The design will be based on a triple riveted, double strap butt joint (unequal straps) for the longitudinal seam, with an assumed efficiency of 85 per cent. This assumed efficiency will be checked later on by the calculations on the joint. Plates are to be made of mild steel having an ultimate strength in tension and shear of 430.0 N/mm^2 and 330.0 N/mm^2 respectively. Factor of safety may be taken as recommended by I.B.R. Assume the rivets to be 87.5% more stronger in double shear than in single shear. Prepare the working drawings of the designed joint.

28. A triple riveted butt joint with double cover plates is used for the longitudinal joint of a lancashire boiler having internal diameter 2.5 m. Working steam pressure 1.1 N/mm^2. The efficiency of the joint may be assumed as 85%. Calculate the plate thickness for mild steel of 459.0 N/mm^2 ultimate tensile strength. The diameter of the rivet hole may be fixed by Unwin's formula. Assume the ratio of tensile to shear stress as 1.23. Design a suitable circumferential joint also. Show the designed joints by a complete drawing, giving at least three rivets in the outer row. Calculate the actual efficiencies of your designed joints.

29. Two steel plates 12.5 by 150 mm are connected by a double strap triple riveted butt joint with 15 mm rivets. The inner row of rivets contains 3 rivets with a pitch of 50 mm. The second row contains two rivets with a pitch of 50 mm. The third row contains one rivet. The back pitch is 37.5 mm and the distance from the outer and the inner rivets to the plate edges is 25 mm. The cover plates are 7.5 mm thick and the corners are cut away at 45 degrees, starting at middle or second row of rivets. When the total load transmitted by the plates is 180 kN determine the following unit stresses: (*a*) Compression in main plate; (*b*) compression in cover plate; (*c*) Tension in main plate at outer, middle and inner rows; (*d*) Tension in cover plate at outer, middle and inner rows; (*e*) Shear in rivets; (*f*) Shear in main plate between inner row rivets and plate edge; (*g*) Shear in cover plate between outer rivet and plate edge.

30. A mild steel tie rod for a bridge structure is 250 mm wide and 18 mm thick. Two such rods are to be connected by a butt joint having two straps. Design this joint completely allowing working stresses as f_t = 100.0 N/mm^2, f_s = 75.0 N/mm^2, f_c = 150.0 N/mm^2. Shearing resistance of rivets in double shear may be taken 75 per cent more than single shear.

31. Two 12 mm thick tie plates are connected by a lap riveted joint. The tie carries a load of 150 kN, calculate
(*a*) width of the plate
(*b*) diameter and number of rivets
(*c*) rows of rivets and arrangement of rivets and
(*d*) efficiency of the joint.
The allowable values of stresses are f_t = 80 N/mm^2, f_s = 60 N/mm^2 and f_c = 120 N/mm^2.

32. Two lengths of flat steel bar 18 mm thick are to be connected by a double cover butt joint to carry a load of 400 kN. Design and draw a neat dimensioned two views sketch of the joint. Assume f_t = 90 N/mm^2, f_s = 75 N/mm^2, f_c = 150 N/mm^2.

33. Determine the number of rivets and the dimensions of the joint for a butt connection of two strips by means of two straps. Given: strip thickness = 12 mm; strip width = 200 mm; strap thickness = 6 mm; Rivet diameter = 17 mm; Tensile load = 280 kN.

34. Give the relation between rivet diameter and the hole diameter into which it will be inserted.

Ans.

Rivet φ, mm	2	2.3	2.6	3	3.5	4	5	6	7	8
Hole φ, mm	2.1	2.4	2.7	3.1	3.6	4.1	5.2	6.2	7.2	8.2

35. When a lap joint is subjected to Tensile load, the stress induced in the rivets is:
(*a*) Shear stress (*b*) Tensile stress
(*c*) Compressive stress (*d*) Bending stress. **[Ans.: (*a*)]**

❑❑❑

CHAPTER 8

Welded Joints

8.1 GENERAL REMARKS

It is the art of joining metals by pressure after heating to a plastic or semimolten state, or of joining the metals by fusion alone. In the past, many designers have avoided welded designs, because of the fear of faulty and weak joints. But now a days, because of the development made in the welding technique, welding materials, and inspection methods the welded joints have become as reliable and free from faults as any riveted or any cast construction. **The welding is used in the following constructions:**

1. The use of welded steel construction for the frames, housing, and connecting members of all types of machinery has reached the point designer must consider the welding process of fabrication when determining the material and construction to be used.
2. For light weight designing welding has definite advantages. Standard rolled shapes, such as channels, *T*-beams and *I*-beams, may often be used to advantage in welded structure. In many cases, the use of welded structures results in lighter, stronger, and better looking construction with a smaller amount of welding.
3. The use of welding in the construction of pressure drums for boilers, refinery stills, and chemical processes has given satisfactory results over many years.

Following are the three reasons for which welding as the fabrication process is selected:

1. Fabrication in which welding is used as an alternate method for casting or forging. In this application, cost of construction and serviceability are important in making the choice between welding and the alternative methods. Cost of construction involves a consideration of the following items along the welding, casting and forging fabrication processes:

Welding	Casting	Forging
(*a*) Cost of materials, as structural steel	(*a*) Cost of materials, as iron and steel	(*a*) Cost of materials
(*b*) Cost of cutting and forming parts.	(*b*) Cost of patterns and moulds.	(b) Cost of dies.
(*c*) Cost of positioning parts, welding and machining.	(*c*) Cost of casting and machining.	(*c*) Cost of forging and machining.

Serviceability takes into account strength and stiffness of the parts and those parts which are subjected to variable loads and/or vibrations, the damping capacity of the structure is also considered. Strength requirements are favoured by the welding of structural steel parts. Deflection requirements are favoured by the larger sections of castings, and damping capacity is met by larger volumes of cast and forged parts. Comparison of fabrication methods bearing support are given in table 8.1 in percentage of number.

Table 8.1

Requirement	Welded	Forged steel	Cast steel	Cast iron
Static strength	100	100	100	100
Deflection	259	259	191	100
Damping capacity	26.1	26.1	27.8	100
Weight	71.1	71.0	76.0	100
Cost	45.0	300	110	100
Thickness of flanges	54.0	54.0	60.0	100

2. Application, in which riveting is replaced by welding. In such an application, welding is favoured from the stand points of strength and saving in weight. The cost of holding the parts during welding and cost of welding itself, including materials, equipments, and operations, may favour riveting in some applications.
3. Welding used as a repair medium. Welding can be easily and economically employed to reunite metal at a crack, to build up a small part that has broken off, such as a gear tooth or to repair a worn surface such as the blade of a hydraulic turbine and the surface of a bearing.

The various advantages of welding mentioned above lead to its wide spread use in all branches of mechanical engineering.

8.2 WELDED JOINTS

A welded joint is a combination of parts fastened together by means of a weld. According to their purpose welded joints are subdivided into strong and tight-strong. Their functions are the same as those of riveted joints, i.e., to transmit load from one member to the other in strong joint and to provide, besides, for fluid and gas impermeability in tight strong joints.

Welded joints are divided into the following five types, i.e., butt, *edge,* lap, corner and Tee joints as shown in Fig. 8.1.

Seams used to join members located in one plane are called butt-jointed. The type of a butt-joined seam is determined by the manner in which the edges of the element to be jointed are prepared.

The various types of Butt joints are: Square, Single-V, double V, Single-U, Double-U, Single-J, Double-J. See Table 8.2. For Edge joint, see Table 8.2 for Tee-joints, see, Fig. 8.2 and Fig. 8.4 for single-fillet joint and Fig. 8.3 and Fig. 8.5 for Double-fillet joint.

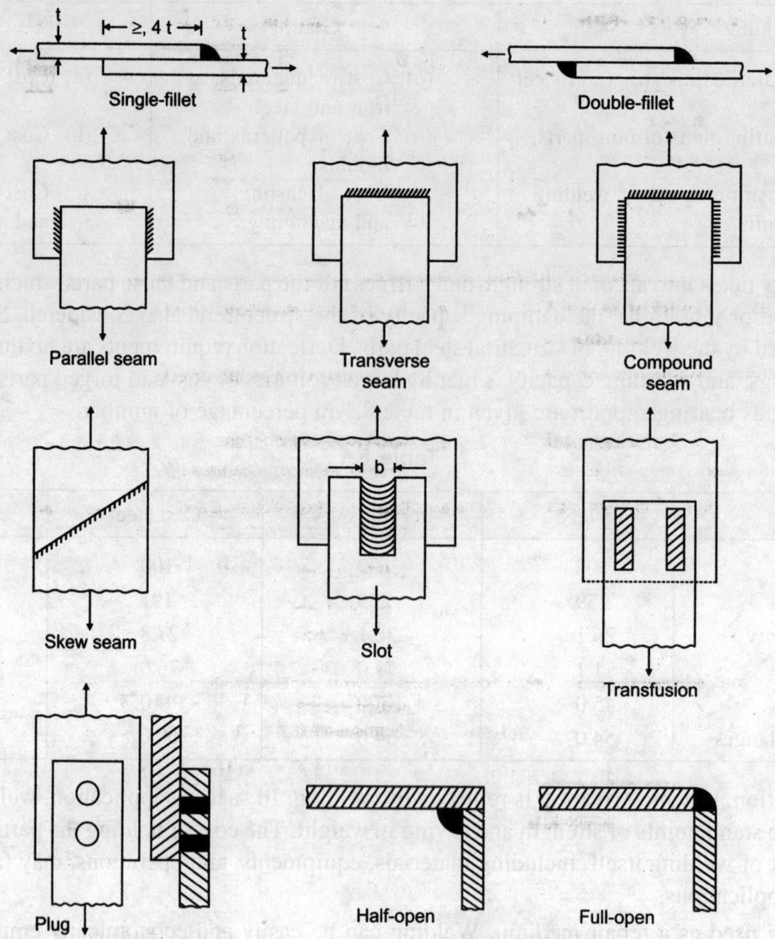

(a) *Lap Joints* (b) *Corner Joint*

Fig. 8.1 *Welded joints*

When comparing *V* and *U*-joints it should be noted that *U*-joints require less electric power and electrode material since the X-section of a *U*-joint for a given plate thickness is less than that of a *V*-joint. Another advantage of *U*-joint lies in the fact that due to the insignificant bevel of the edges the shrinkage of the molten metal on cooling occurs almost uniformly over the entire section and the plates warp less than in case of *V*-joints.

Double *V* and *U*-joints are by no means inferior to single *V*-joints when plates of 12 mm and over are welded because the X-sectional areas of double *V* and *U* joints is 30–40 percent less than that of single *V*-joints at the same bevel angle. The shortcomings of double *V* and *U*-joints accrue from poor penetration liable to occur inside the butt and the greater cost of edge preparation especially in the case of *U*-joints.

The various types of Lap joints and corner joints are shown in Fig. 8.1.

8.3 REPRESENTATION OF WELDS

The scheme adopted in standard I.S. 813–1961 for representation of weld on drawings requires the use of the following elements:

(*a*) A basic symbol used to specify the type of the weld.
(*b*) An arrow and a reference line to indicate the location of the weld in a joint.
(*c*) Supplementary symbols to indicate special instructions, such as finish, weld-all-round, etc.
(*d*) Dimension of the weld in cross-section and in length, where required.

Table 8.2 gives the basic symbols used to specify the type of the weld.

Significance of the position of Arrow Head: In the case of joints in which only one plate is prepared, the arrow shall point towards that plate, the side of the plate on which, the edge preparation is done being indicated by the disposition of the symbol relative to the reference line. For this see Figs. 8.2, 8.3, 8.4 and 8.5.

Table 8.2

From of Weld	Sectional representation	Appropriate symbol
Fillet	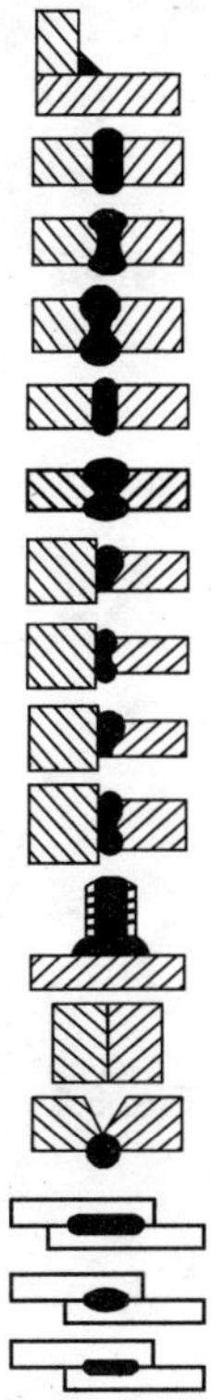	
Square butt		
Single V butt		
Double V butt		
Single U butt		
Double U butt		
Single bevel butt		
Double bevel butt		
Single J butt		
Double J butt		
Stud		
Bead edge or seal		
Sealing run		
Spot		
Seam		
Stitch		

8.4 STRENGTH OF WELDED STEEL JOINTS

The properties of weld metal deposited by the gas or arc process, using shielded arc welding rod, are shown in Table 8.3. With the shielded arc, the weld metal has better properties, as indicated in the table.

Table 8.3 Strength of shielded—Arc flush—Steel welds

Engineering property	Limit Stress MN/m^2			Recommended design stress MN/m^2		
	Base Metal elastic stress	Deposited Metal		Safety factor 2 Static load	Safety factor 2 Load P varies from 0 to P	Safety factor 2.75 load P varies from $+P$ to $-P$
		Elastic Limit	Endurance Limit			
Tension	220.0	280.0	150.0	110.0	110.0	55.0
Compression	250.0	300.0	—	128.0	110.0	55.0
Bending	250.0	300.0	180.0	128.0	110.0	64.0
Shear	140.0	165.0	—	75.0	70.0	35.0
Shear and Tension	—	—	—	75.0	70.0	35.0

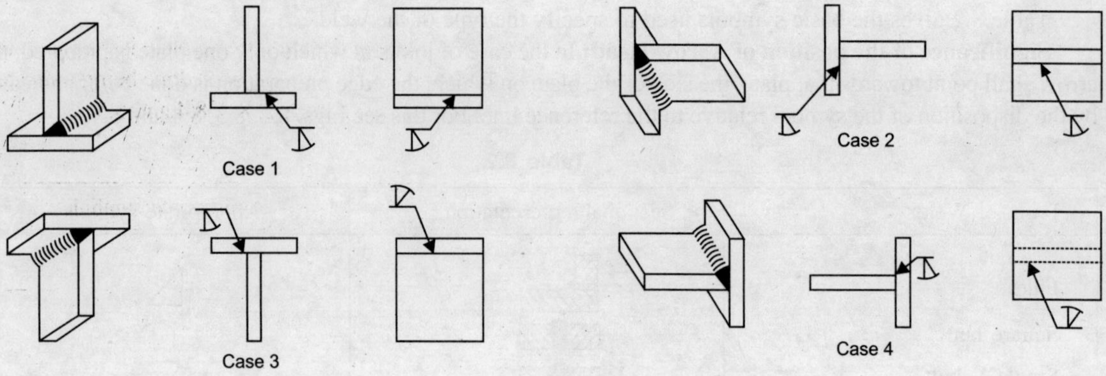

Fig. 8.2

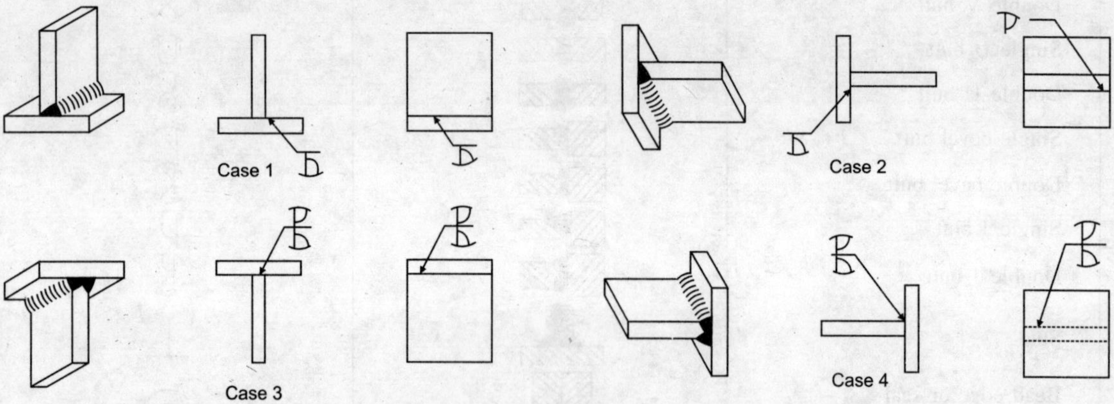

Fig. 8.3

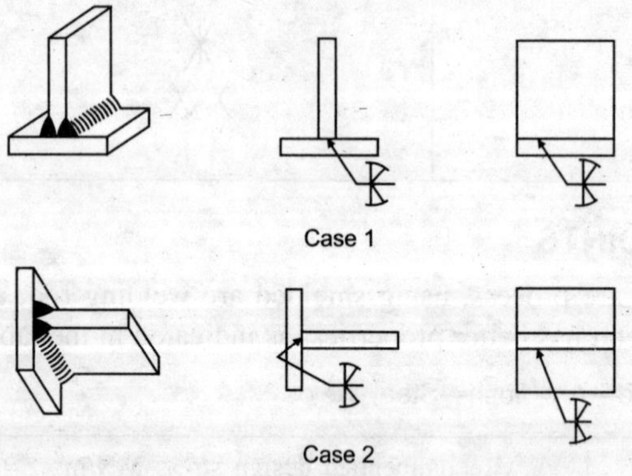

Fig. 8.4

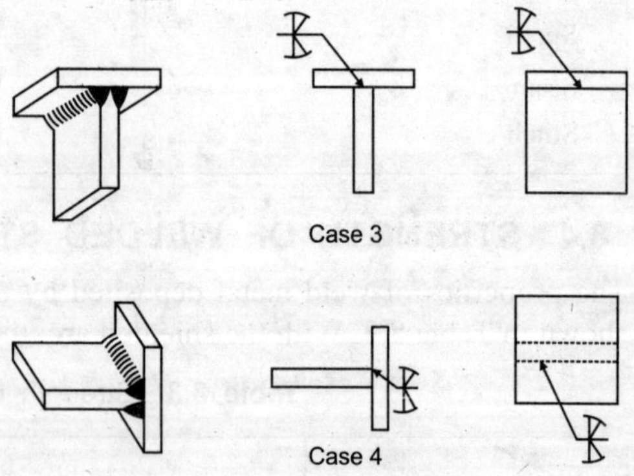

Fig. 8.5

Butt welds may be assumed to have 80 per cent of the strength of the base metal if the welds are flush and 100 per cent if they are reinforced by 15 per cent, or if a reinforcing plate is used. Plug or rivet welds are as strong as the weld metal in shear. The working stresses outlined above are for low and medium carbon steels. Alloy steels, if welded with the proper technique and with shielded rods of the proper composition, will have much higher strengths. The above values of stresses mentioned are for accurate design. These values apply for painstaking workmanship, proper supervision, and careful inspection. With less skill or less care, the recommended design stresses should be lowered. For bare electrode welds, the recommended stresses in the above table be multiplied by 0.8 to 0.85, the factor depending on the skill of the welder.

8.5 DESIGN OF WELDED JOINTS FOR STATIC LOADS

Welds are designed by keeping in mind the following assumptions:

1. Load is distributed uniformly among the entire weld length.
2. Stress is spread uniformly over its effective section.

(*a*) **Strength of butt joints:** Butt jointed seams are designed for tension or compression. The thickness *t* of the thinner of the joined plates being taken as the design throat of the weld as shown in Fig. 8.6. (sketch indicates equal thickness of plates).

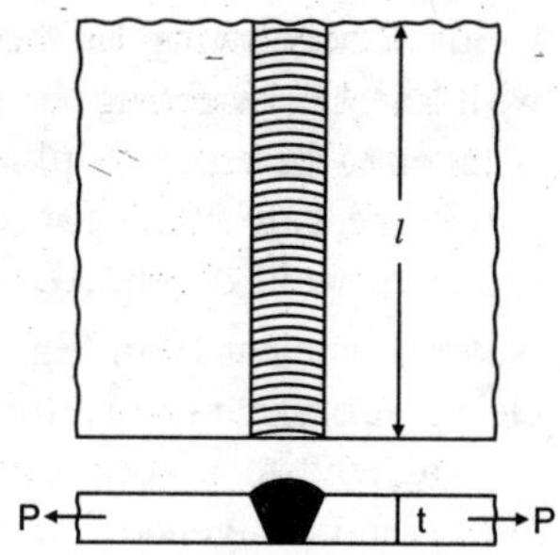

Fig. 8.6

The effective length of a butt jointed seam in joints in tension is found from the relation

$$l = \frac{P}{t \times f_t} \quad ...(8.1)$$

where P = applied load to the weld in tension.

f_t = allowable stress for the weld in tension in N/mm^2.

This stress can be taken from table 8.3 for steels keeping in mind the type of the electrode used.

For situations having compressive load on weld, f_t is replaced by the allowable stress for compression f_c for the given material of electrode.

The calculation of the stress does not consider the reinforcement of the welds. The reinforcement is always provided as shown in Fig. 8.6. Normally the length of the weld will be equal to the width of the plates.

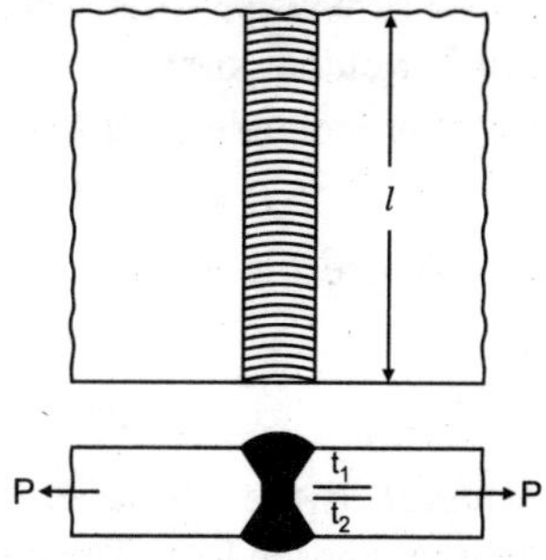

Fig. 8.7

Figure 8.7 indicates the double *V*-welded joint. The joint is subjected to a load *P* in tension.

The effective length of a double *V*-butt joint seam in tension is calculated from the relation

$$l = \frac{P}{(t_1 + t_2)\, f_t} \quad ...(8.2)$$

where t_1 and t_2 = throat thicknesses of the *V* at top and bottom as shown in figure.

In the case of butt joints usually, the minimum area is greater than the area of plate in cross-section and thus, if the weld has strength equal to that of the plate, the joint is stronger than the plate. The table 8.4 gives the thickness of plates for which the different butt joints are used.

Table 8.4

Type of joint	Thickness of plate in mm
Single butt joint	1–6
Double square butt joint	3–8
Single *V*-butt joint	4–26, angle of *V* 55°–70°
Double *V*-butt joint	12–60, angle of *V* 55°–70°
Single *U*-butt joint	20–60
Double *U*-butt joint	30–60

(*b*) **Strength of Lap joints:** Lap joints are made with fillet welds. The overlap of the plates should be $\geq 4t$. The fillet lap joints are of two types:

(*i*) Transverse fillet welded joints, or front fillet welds.

(*ii*) Parallel fillet welded joints, or longitudinal or side fillet welds.

A weld is said to be "transverse fillet weld" if its direction is square to the direction of the acting force (Fig. 8.8) and "Longitudinal fillet weld", if its direction is parallel to the direction of the force, Fig. 8.10.

The normal profile of a fillet weld is an isosceles right angle, Fig. 8.9. Experiment has proved that failure of fillet welds occurs owing to shearing along the section coinciding with line *AE*, bisecting the right angle. Leg of weld or cathetus is taken to be equal to plate thickness, t ($AB = AC = t$). Fillet welds are calculated for shear along this section (*AE*). The effective weld depth, $AD = t \cos 45° = 0.7\,t$.

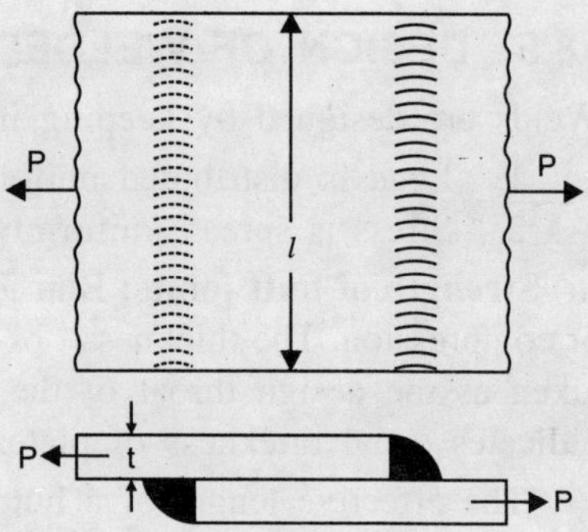

Fig. 8.8 *Double transverse fillet welded joint*

Note: It is clear from Fig. 8.10 that the parallel fillet weld is a clear case of shear stress. However, a transverse fillet weld, Fig. 8.8, is subjected to a combination of normal and shear stresses. However, it is a general practice to design both cases by assuming the failure to occur at the throat, Fig. 8.9, by shearing.

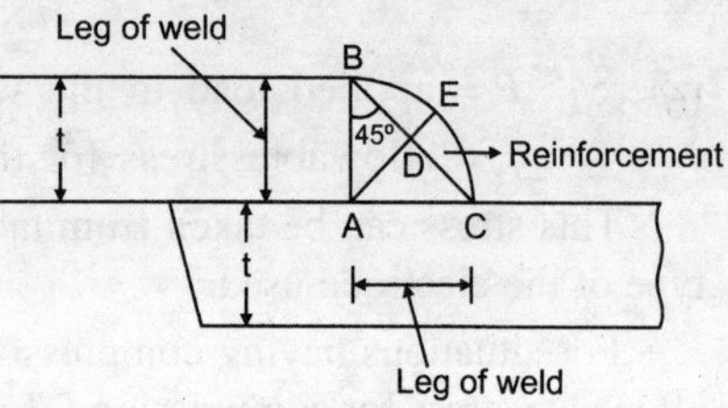

Fig. 8.9

Transverse fillet welded joints: Figure 8.8 shows double fillet welded joint.

For a double fillet lap joint the equation for strength will be

$$P = 2 \times AD \times l \times f_s = 2 \times 0.7t \times l \times f_s$$

Hence the necessary length of the weld will be

$$l = \frac{P}{2 \times 0.7t \times f_s} \text{ mm.} \qquad ...(8.3)$$

where f_s is the allowable stress for the weld metal in shear in N/mm^2 chosen from Table 8.3 for steels.

For single fillet lap joint, the necessary length of the weld will be

$$l = \frac{P}{0.7t \times f_s} \text{ mm} \qquad ...(8.4)$$

Parallel fillet welded joints: Figure 8.10 shows double fillet welded joint.

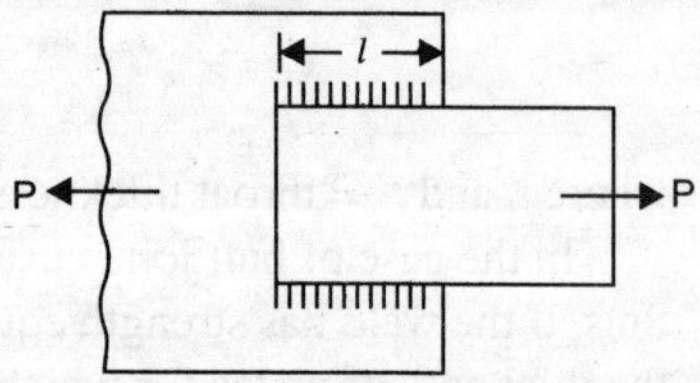

Fig. 8.10 *Double parallel fillet welded joint*

For a double fillet lap joint (parallel) equation for strength will be

$$P = 2 \times 0.7t \times l \times f_s$$

where l is the length of one length of weld.

Hence the necessary length of the weld will be

$$l = \frac{P}{2 \times 0.7 \times t \times f_s} \text{ mm} \qquad ...(8.5)$$

where f_s is the allowable shear stress for the weld metal in N/mm^2 chosen from Table 8.3.

For a single fillet lap joint (parallel), the necessary length of the weld will be

$$l = \frac{P}{0.7t \times f_s} \text{ mm} \qquad ...(8.6)$$

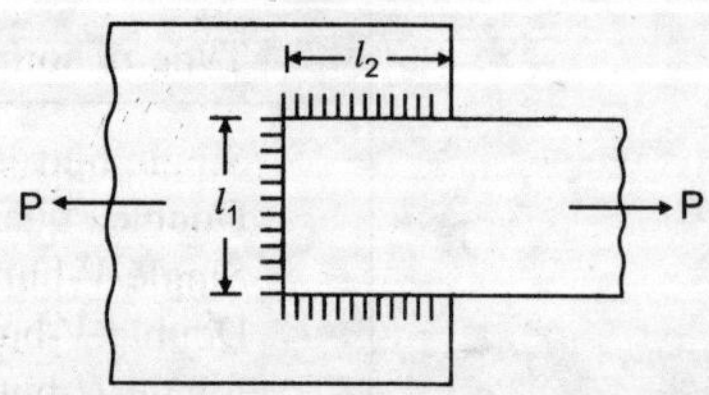

Fig. 8.11 *Compound fillet welded joint*

Compound fillet welded joints: It is a combination of transverse and parallel fillet welded joints as shown in Fig. 8.11. The figure shows a single fillet transverse and double fillet parallel joint. The calculation is based on the principle of independence of force action.

For a compound weld as shown in figure the safe load it can carry is

$$P = P_{\text{transverse}} + P_{\text{parallel}}$$

$$P = 0.7t \times l_1 \times f_s + 2 \times 0.7t \times l_2 \times f_s \qquad ...(8.7)$$

where l_1 is normally the width of the plate. Thus from relation 8.7 the length l_2 of the weld can be calculated.

Normally all the welds should be placed symmetrically about the axis of the welded part unless the loading is unsymmetrical. In determining the length of weld required, 12.5 mm should be added to the length of each weld to allow for starting and stopping of the weld. In computing the area of the X-section of a reinforced fillet, the throat dimension AE in Fig. 8.9 should be taken as $1.2 \times AD$.

The efficiency of a welded joint is the resistance of the joint to the resistance of the weakest section of the base metal. Since the lengths of the welds in many cases are not limited, an efficiency of 100 per cent or even higher can be easily obtained.

Intermittent welds should not be less than 37 mm long spaced not less than 16 times the plate thickness of more than 1.0 m in the clear.

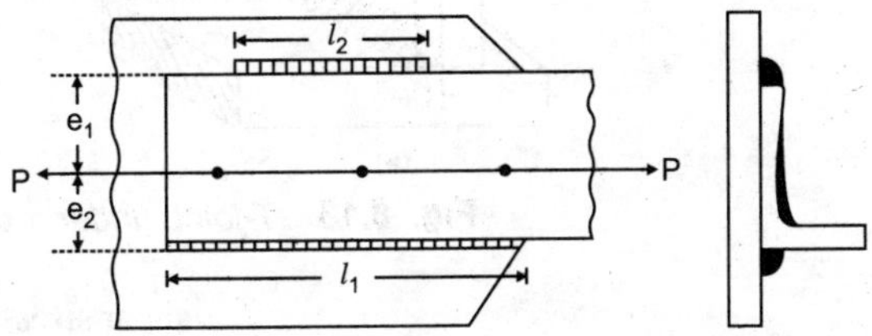

Fig. 8.11 *(a) Unsymmetrical fillet welds*

Unsymmetrical fillet welds: In the case of unsymmetrical parallel fillet welds, Fig. 8.11(a), the total weld length (2 l in Fig. 8.10 as found from equ. 8.5), should be divided into components l_1 and l_2 in inverse proportion to the distance e_1 and e_2 that is,

$$l_1 = \text{Total length of weld} \times \frac{e_1}{e_1 + e_2}$$

and

$$l_2 = \text{Total length of weld} \times \frac{e_2}{e_1 + e_2}$$

(*c*) **Fillet welds in Tee-joints:** A Tee-joint subjected to the action of a tensile force is shown in Fig. 8.12 (*a*). In this the load is perpendicular to the plane of the weld. Its calculations will be in the similar way as done for transverse fillet weld. Fig. 8.12 (*b*) shows a Tee-joint load parallel to the plane of the weld with a small eccentricity. Calculation for this will be the same way as for parallel fillet welding.

A Tee-joint with considerable eccentricity is shown Fig. 8.13 (*a*). The eccentricity of loading from the plane of welding is e. The strength of the welds will be calculated for joint action of bending moment and shear force according to the formula of total stress.

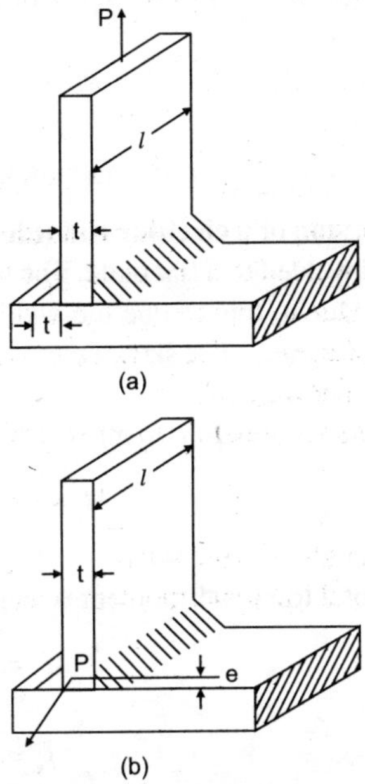

Fig. 8.12 *T-joint with tensile and small eccentric loading*

The stress in the weld due to force P will be shearing, i.e., same as for parallel fillet weld.

$$f_s = \text{shear stress} = \frac{P}{2 \times 0.7t \times l}$$

for weld on both sides of the leg of T-joint.

Bending moment on the weld due to eccentricity $e = P \times e$.

Stress in the weld due to bending moment

$$f_b = \frac{Pe}{0.7 \times 2 \times t \times \frac{l^2}{6}} = \frac{Pe}{2 \times 0.7 \times t \times l^2} \times 6\,;$$

$$\left(Z = \frac{1}{6} l^2 \times \text{throat thickness} = \frac{1}{6} l^2 \times 0.7t \right)$$

The coefficient 0.7 takes into account the possibility of failure at the throat. Now the total stress will be

$$(f_s)_{\text{max.}} = \sqrt{f_s^2 + (f_{b/2})^2} \le |f_s|_{\text{allowable}} \qquad \text{...(8.8)}$$

where $|f_s|_{\text{allowable}}$ is the maximum allowable stress for the weld metal in shear in N/mm^2.

In case the welding is acted upon by a bending moment M_b and force P as shown in Fig. 8.13 (*b*).

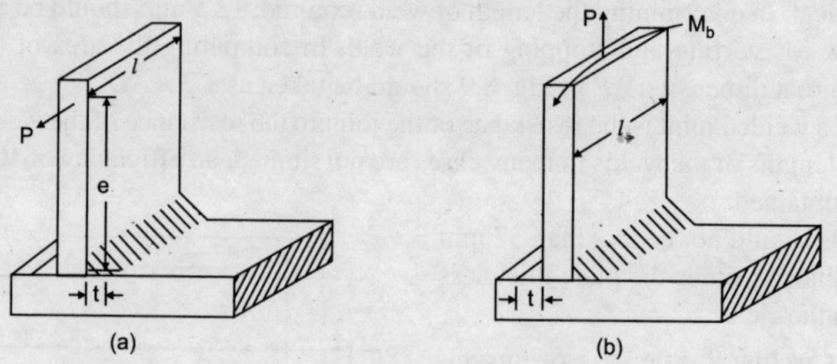

Fig. 8.13 *T-joint under eccentric and bending loading*

$$f_s = \text{shear stress} = \frac{P}{2 \times 0.7 \times t \times l}$$

for weld on both sides of the leg of *T*-joint.

Stress in the weld due to bending moment M_b

$$f_b = \frac{M_b}{0.7 \times 2 \times t \times \frac{l^2}{6}} = \frac{M_b}{2 \times 0.7t \times l^2} \times 6$$

Now the stress equation will take the from

$$(f_s)_{\text{max.}} = \sqrt{\left(\frac{f_b}{2}\right)^2 + f_s^2} \leq |f_s|_{\text{allowable}} \quad ...(8.9)$$

where $|f_s|_{\text{allowable}}$ is the maximum allowable stress for the weld metal in shear in N/mm^2.

(*d*) **Torsion of a circular rod fillet weld:** Figure 8.14 shows a circular rod welded to a flat plate. The torsional moment acting on the rod is M_t. Due to this torque the weld will be under shear. The weld is circumferential. The size of the weld is *t* as shown in figure. Diameter of circular rod is *d*.

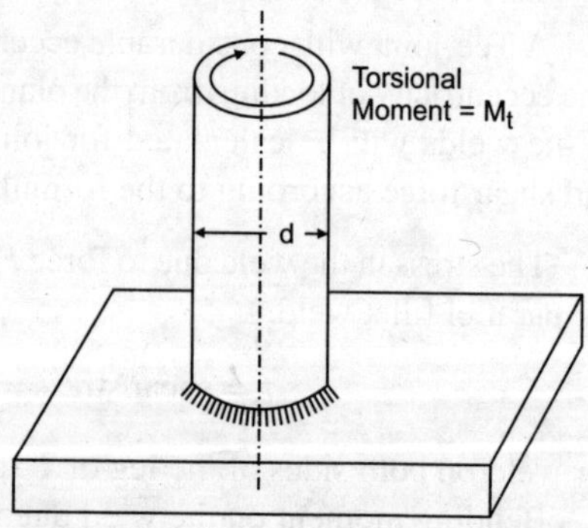

Fig. 8.14 *Circular rod fillet weld subjected to torsional moment M_t*

The torsional moment which a weld of unit length can carry is

$$M_t = l \times 0.7t \times f_s \times \frac{d}{2}$$

Length of weld = πd

Total torsional moment which the weld can carry

$$M_t = \pi d \times 0.7t \times f_s \times \frac{d}{2}$$

$$f_s = \frac{2M_t}{0.7t \times \pi d^2} \quad ...(8.10)$$

where f_s is the allowable shear stress for the weld metal in N/mm^2.

or $$f_s = \frac{M_t}{I_P} \times r_{\text{max.}}, \quad \text{where } I_P = \text{Polar M.O.I.} = \frac{\pi d^3}{4} \times a \; (a = \text{throat thickness} = 0.7t)$$

and $r_{\text{max.}} = d/2$

(*e*) **Bending of a circular rod fillet weld:** In Fig. 8.14, instead of M_t, a bending moment M_b is acting at the top end.

$$f_b = \frac{M_b}{Z}$$

$$Z = \text{Section modulus} = \frac{\pi}{4}d^2 \times a$$

where $a \cong 0.7t$, t = leg of the weld

$$\therefore \quad f_b = \frac{M_b \cdot 4}{\pi d^2 \times 0.7t} \qquad ...(8.11)$$

(f) **Long Tee-fillet weld joint subjected to torsion:** See Fig. 8.15.

$$f_s = \frac{M_t}{I_p} \times r_{max.}$$

$$I_p = 2 \times \frac{al^3}{12}$$

where $a = 0.7t$

$r_{max.} = l/2$

$$\therefore \quad f_s = \frac{3 \cdot M_t}{0.7\, tl^2} \qquad ...(8.12)$$

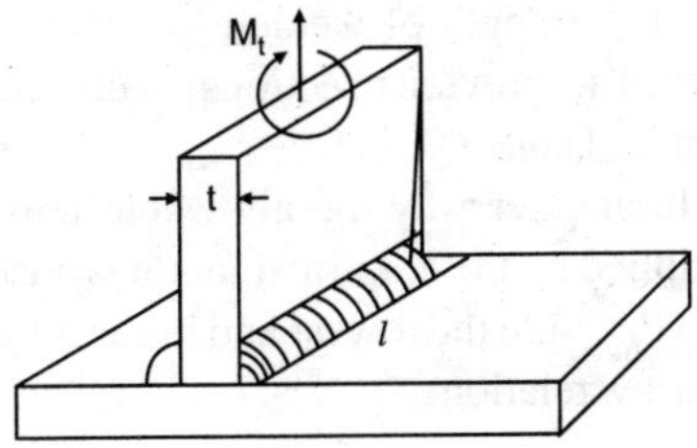

Fig. 8.15

(g) **K-welds in Tee-joints:** Figure 8.16

(i) Subjected to pure tensile load, P. Design will be like Butt Joint

$$l = \frac{P}{f_t \times t}$$

(ii) Combined Tensile force and Bending Moment:

$$f_t = \frac{P}{l \times t} + \frac{6\,M_b}{tl^2}$$

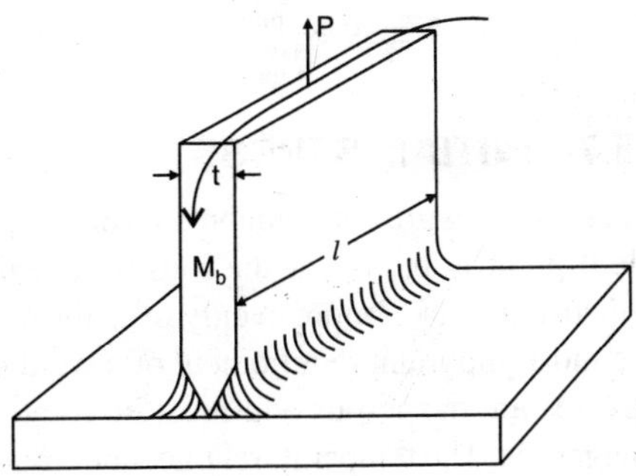

Fig. 8.16

Slot welds: The slot is made parallel to the applied load, width of slot, $b = 2t$

$$\therefore \quad l = \frac{P}{b \times f_s} = \frac{P}{2t \times f_s}$$

Slot length $l \geq 30t$.

Plug welds: are used to increase the contact tightness of the joint elements in combination with other loaded welds.

Plug diameter = $2t$.

Transfusion weld: It is a modification of a slot weld. The elements are jointed by welding through the upper plate which should not be more than 12 mm thick. These welds are made by automatic SAW.

Corner joint: is a special case of T-joint. They are generally confined to structures expected to carry light loads are not checked for strength.

Spot welded joints: Figure 8.17

these joints are designed in shear

$$f_s = \frac{4P}{\pi d^2 \times z}$$

where z = Number of spots

$d = 1.2\,t + 4\text{mm for } t \leq 3 \text{ mm}$

$= 1.5t + 5\text{mm for } t > 3 \text{ mm.}$

$p = 3\,d.$

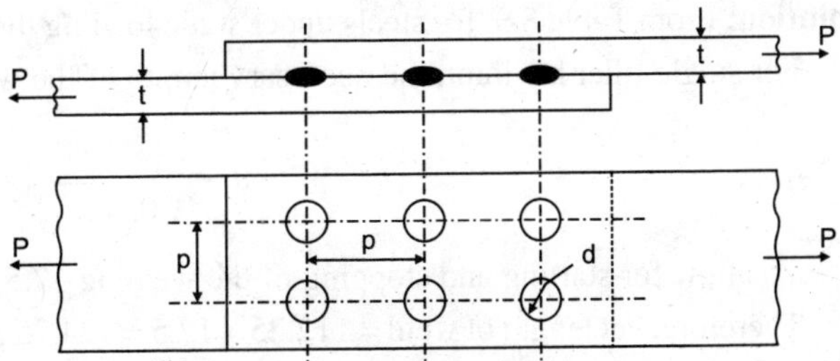

Fig. 8.17

8.6 STRENGTH OF WELDS AT VARYING LOADS

The quality of welding determines the strength of welded seams at varying loads. Small faults in the form or a local poor penetration, undercuts and fractures, being points of stress concentration, reduce the endurance limit of the welded joints as compared to those of the base metal. Automatic welding gives higher strength at varying loads than that of joints welded manually. Joints formed by the resistance butt welding method also have strength at varying load.

The strength of welded joints at varying loads is calculated according to the formulae of static loads (derived in previous sections) with reduced allowable stress. The allowable stresses for varying loads are given in Table 8.3.

In another way the allowable tensile and shear stresses obtained from Table 8.3 at static loading are multiplied by the empirical factor whose value depends on the weld design and the ratio of $P_{min.}/P_{max.}$, where $P_{min.}/P_{max.}$ are the lowest and highest loads. The empirical factor for any number of cycles of varying loads is given by relation

$$\frac{1}{1.3-0.3\dfrac{P_{min.}}{P_{max.}}}, \text{ for fillet, corner and slot welds and } \frac{1}{1-0.3\dfrac{P_{min.}}{P_{max.}}}, \text{ for butt welds}$$

8.7 INITIAL STRESS

These stresses are unavoidable in welds. Initial stresses are not very dangerous for static loading because if the yield point is passed at the most stressed point, the metal starts to creep at this point and the load then redistributes itself more evenly over the cross-section. However, this is only true for ductile materials, hence the most important requirement of a weld material is that it should be ductile. Under varying loads the initial stresses are dangerous and must be removed by annealing. Under impact load the initial stresses are very dangerous. The temperature, rate, and time of heat treatment depend on the size and thickness of the parts and on the plasticizing rate of the material. Highly stressed parts whose failure would involve life hazard or expensive structure damage, such as welded pressure vessels and welded penstocks, should be stress relieved. In welded steam-generator drums, stress relieving is effective in reducing caustic embrittlement. Hand peening of the joint while it is hot, is effective in increasing the strength of the joint, but overpeening should be avoided. Indications are that peening reduces residual stresses in parts of the weld that may be in the plastic state and that it also introduces some cold working.

SOLVED EXAMPLES

Example 8.1. A structural steel plate 75 mm wide and 10 mm thick loaded statically in tension by a load of 55 kN is welded to a plate with normal welds as shown in Fig. 8.1 (lap joint with transverse welding). Determine the size of the weld required.

Solution: From Table 8.3, for steels under static loading the allowable shear stress $f_s = 70$ N/mm².

For single fillet lap joint, the necessary length of the weld will be

$$l = \frac{P}{0.7t \times f_s} = \frac{55000}{0.7 \times 10 \times 70} = 112.25 \text{ mm}$$

To allow for starting and stopping of the welding 12.5 mm should be added.

Therefore, net length of weld = 112.25 + 12.5 = 124.75 mm which is more than the width of the plate, i.e., 75 mm.

Therefore, a double fillet lap weld has to be provided. Hence each welding length will be

$$= \frac{P}{0.7 \times 2 \times t \times f_s} = \frac{55000}{0.7 \times 2 \times 10 \times 70} = 56.125 \text{ mm} = 57 \text{ mm say.}$$

Net length of each weld = 57 + 12.5 = 69.5 mm.

Example 8.2. A 12.5 mm plate 50 mm wide carries a static tensile load of 36.00 kN and is to be welded to a plate by parallel fillet welds, as shown in Fig. 8.10. Determine the length of each weld.

Solution: From table 8.3, the allowable shear stress for static loading is 75.0 N/mm^2. Let l be the length of parallel fillet weld on each side. The size of the weld is 12.5 mm.

Necessary length of the weld will be

$$l = \frac{P}{2 \times 0.7 \times t \times f_s} = \frac{36000}{2 \times 0.7 \times 12.5 \times 75.0} = 27.5 \text{ mm}$$

Net length of each weld = 27.5 + 12.5 = 40 mm.

Example 8.3. A 150 × 100 × 12.5 mm angle is to be welded to a steel plate by fillet welds along the edges of the 150 mm leg. The angle is subjected to a tensile load of 225 kN. Determine the weld lengths required if placed as shown in Figure 8.18. The line of action of the load may be assumed to be the gravity axis of the angle.

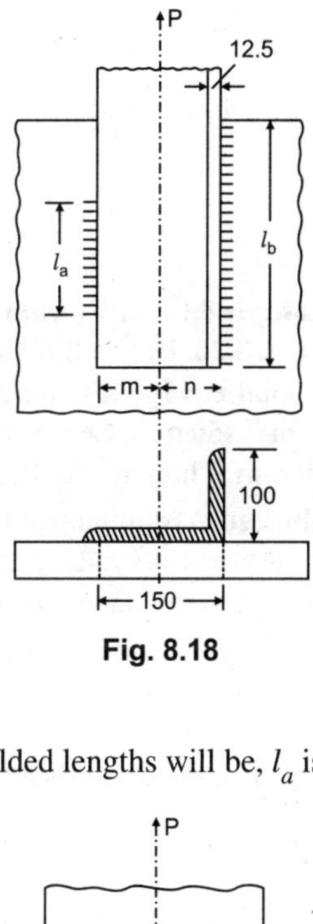

Fig. 8.18

Solution: From steel tables, the distances m and n are 100 mm and 50 mm respectively. It is a case of parallel fillet weld with unequal welded lengths l_a and l_b. Hence the total length of the weld required is

$$l = l_a + l_b = \frac{225000}{0.7 \times 12.5 \times 75.0} = 345 \text{ mm}$$

Here, $f_s = 75 \text{ N/mm}^2$

The individual welded lengths are

$$l_a = \frac{345 \times 50}{150} = 115 \text{ mm}$$

and $$l_b = \frac{345 \times 100}{150} = 230 \text{ mm}$$

Allowing for 12.5 mm for starting and stopping of the weld, the individual welded lengths will be, l_a is 127.5 mm and l_b is 242.5 mm.

Example 8.4. A plate 100 mm wide and 12.5 mm thick is to be welded to another plate by means of single transverse and double parallel fillet welds. Determine the length of the weld run in each case if the joint is subjected to varying loads. The recommended design stress in tension is not to exceed 70.00 N/mm^2 and in shear 56.00 N/mm^2 for static loading.

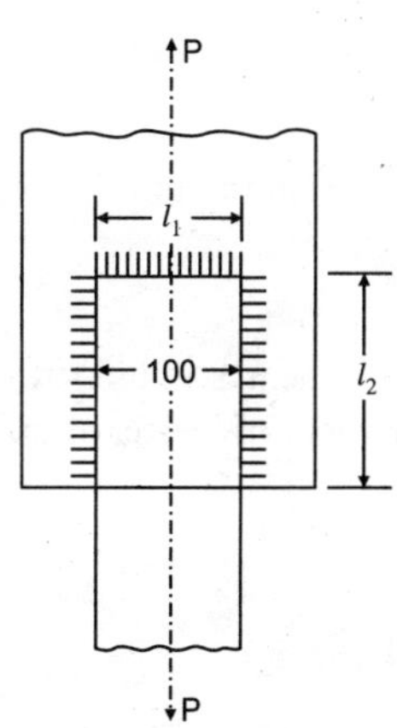

Fig. 8.19

Solution: The joint is shown in Fig. 8.19. Let the maximum force carried by the plate is P.

$$P = 100 \times 12.5 \times 70.00 = 87500 \text{ N}$$

We find from table 8.3 that for varying load from $+P$ to $-P$, the allowable stresses are half than for static loading, i.e., the varying load factor is 2.0.

Therefore, allowable stresses in tension and shear under varying load will be 35.00 N/mm^2 and 28.00 N/mm^2 respectively.

Effective length of the single transverse weld

$$= 100 - 12.5 = 87.5 \text{ mm}$$

For a compound weld as shown in Fig. 8.19, the safe load it can carry is

$$P = P_{transverse} + P_{parallel}$$
$$P = 0.7 \times t \times l_1 \times f_s + 2 \times 0.7t \times l_2 \times f_s$$
$$87500 = 0.7 \times 12.5 \times 87.5 \times 28.00 + 2 \times 0.7 \times 12.5 \times l_2 \times 28.00$$

gives. $l_2 = 135$ mm

Hence net length of parallel weld for each leg = 135 + 12.5 = 147.5 mm.

Example 8.5. Determine the maximum torsional moment that a 50 mm diameter solid steel shaft welded to a flat plate can sustain when the leg of the fillet weld is 12.5 mm. Assume that the stress is applied under static conditions.

Solution: For given conditions from Table 8.3, the allowable stress in shear will be 75.0 N/mm². The maximum shear stress intensity in the weld is given by the relation

$$f_s = \frac{2M_t}{0.7 \times \pi d^2 t}$$

where f_s is the allowable shear stress in the weld material. On putting the values, we get

$$75.0 = \frac{2 \times M_t}{0.7 \times 12.5 \times \pi \times (50)^2}$$

gives M_t = torsional moment = 2570 Nm.

Example 8.6. An *I*-beam is welded all around as shown in Fig. 8.20. Determine the stresses in the fillets of weld for load P = 100 kN. Height of beam 252 mm, length of beam = 150 mm, Leg of the weld = 6 mm, flange width = 180 mm. Thickness of flange and web = 6 mm.

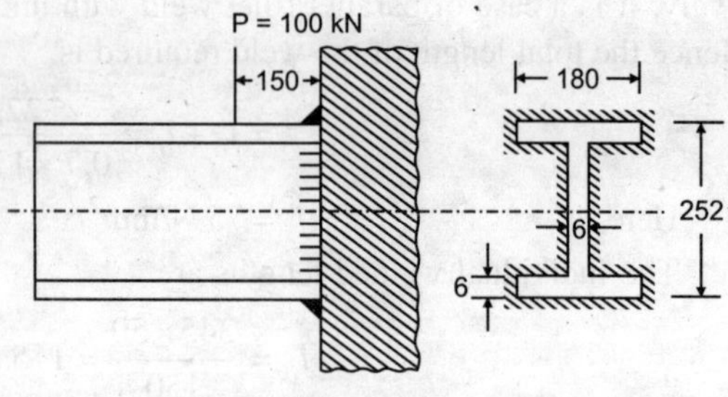

Fig. 8.20

Solution: Assuming that the shearing stress is only taken by the vertical welds of the *I*-section. The short vertical welds with the flange thickness is neglected.

Therefore, the effective length of the weld on each side of web

$$= 252 - 6 \times 2 - 12.5 = 227.5 \text{ mm}$$

Thus, the mean shear stress due to load P

$$f_s = \frac{P}{0.7t \times 2 \times l} = \frac{100 \times 10^3}{0.7 \times 2 \times 6 \times 227.5} = 52.0 \text{ N/mm}^2$$

Bending moment on the weld due to eccentricity of 150 mm

$$= \frac{100 \times 10^3 \times 150}{1000} = 15 \times 10^3 \text{ Nm}$$

The maximum stress from the bending moment 15×10^3 Nm at the farthest edge of the vertical plate at the moment of inertia of the fillets being $I = 80 \times 10^6$ mm⁴.

$$f_b = \frac{15 \times 10^3 \times 10^3}{0.7 \times 80 \times 10^6} \times y_a = \frac{15 \times 10^3 \times 10^3 \times 120}{0.7 \times 80 \times 10^6} = 32.2 \text{ N/mm}^2.$$

$$(f_s)_{max.} = \sqrt{(f_b/2)^2 + f_s^2} = \sqrt{(16.1)^2 + 52^2} = 54.4 \text{ N/mm}^2.$$

This is below the allowable stress which is equal to 70.0 N/mm², and the welded connection can be approved.

Example 8.7. Design the longitudinal joint for a 1.25 m diameter steam boiler carrying a steam pressure of 5 N/mm^2. The allowable stress of the plate material is 80 N/mm^2 intension. Length of the shell of boiler is 5 m. Take allowable stress for shielded arc weld = 95 N/mm^2.

Solution: By using the thin cylinder formula

$$t = \frac{PD}{2f_t} = \frac{5\times125}{2\times80} = 3.90 \text{ cm}$$

Adopt $t = 4.0$ cm.

The longitudinal joint of the boiler shell will always be butt type.

Area of the weld = $0.7t \times l \times 2$

where t is thickness of the cover plate and l is the length of shell.

Tearing strength of weld = $0.7t \times l \times 2 \times$ allowable stress for weld material = $1.4t \times 500 \times 95$

Load acting on the longitudinal joint = $P \times l \times D = 5 \times 500 \times 125$ N

Equate the load to weld strength, $1.4\,t \times 500 \times 95 = 5 \times 500 \times 125$

$$t = 4.70 \text{ cm}$$

Take $t = 5.0$ cm.

Thus the thickness of the cover plate will be 5.0 cm and it is welded all along its length by fillet welds of size 5.0 cm.

EXERCISES

1. Define welding. In what cases welding as fabrication process is preferred over casting and forging?
2. How do you classify welding?
3. Differentiate among forge, pressure, and spot welding.
4. What are the advantages of seam welding over spot welding?
5. Under what special conditions gas welding is preferred over other forms of welding?
6. What are the advantages of cold welding?
7. What are the factors to be kept in mind while selecting a correct method of welding? Explain them in detail.
8. How do you classify the welded joints?
9. What are the assumptions made in design of welded joints and how for one is justified in making these assumptions?
10. Why reinforcement is normally required in welded joints?
11. Two 12.5 mm plates are placed so that they overlap 50 mm and are joined by a single lap weld placed normal to the load. The weld is a 15 mm fillet weld 0.15 m long. Determine the total static load that this joint will sustain.
12. Two 15 mm plates are placed so that they overlap each other and are welded with two 15 mm fillet welds each 0.20 m long and placed normal to the load. Determine the total static load that this joint can sustain.
13. What load can be sustained by the welded joint in problem 12 if the welds are placed parallel to the load line?
14. Two 20 mm plates are 0.3 m wide and are joined by butt welds. Determine the load that can be sustained by the joint if the load is applied with moderate shock.
15. A 10 mm by 0.1 m steel plate is welded to a steel gusset plate by means of parallel fillet welds. Determine the length of welds required if a load of 70×10^3 N is to be transmitted.
16. A 150 × 100 × 12.5 mm angle is welded to a steel plate by two fillet welds along the edges of the 150 mm leg. The angle is subjected to a tension load of 300 kN. Determine the lengths of the welds required if the load is applied with heavy shock.
17. Determine the maximum torsional moment that a 55 mm diameter solid steel shaft welded to a flat plate can sustain when the leg of the fillet weld is 12.5 mm. The allowable shear stress in the weld material under static conditions is not to exceed 75.0 N/mm^2.
18. Determine the stresses in transverse double fillet welds and the plates for a lap joint with applied static load of 3×10^5 N. Take t = 10mm, and length of each weld as 100 mm. (**Ans:** f_s = 71.5 N/mm^2; f_t = 100 N/mm^2).
19. Determine the requisite length of welds joining a steel strip to a plate, by a compound fillet lap joint (See Fig. 8.11). The strip dimensions are: width × thickness: 150mm × 10 mm. load = 240 kN.
20. Design a weld to join an unequal angle to gusset plate (See Fig. 8.11(a). The joint is to support a force of 100 kN. Take f_s = 95 N/mm^2, Angle flange thickness = 6 mm.
    Given, e = 75 mm; e_2 = 20 mm. (**Ans:** l_1 = 183 mm; l_2 = 67 mm.)

❑❑❑

CHAPTER 9

Threaded Joints

9.1 GENERAL REMARKS

Some machine parts must be so constructed that they may be readily connected or disconnected without damage to the machine or the fastening. This requirement may be for the purpose of holding or adjustment in assembly or service, inspection, repair, or replacement, or it may be for manufacturing or assembly reasons. The parts may be rigidly connected, or provision may be made for predetermined relative motion.

A screw fastening is composed of two elements, as illustrated by a bolt and nut. Frequently one of the parts to be connected is so constructed that it becomes one of the elements of the fastening. In fact, both of the parts to be connected may serve as the elements of the fastening as in the case of a handle with a thread turned on one end to engage in a tapped hole in a lever.

A number of advantages offered by threaded joints favour their large scale application in mechanical engineering.

1. Threaded couples are highly reliable in operation.
2. Threaded couples are easy to assemble and disassemble.
3. For various operating conditions, a wide range of threaded couples can be adopted.
4. As all the screws are standardised, therefore, they are relatively cheap to produce, and permit a high degree of interchangeability.
5. Highly efficient methods of manufacturing can be applied.
6. Offer-high load carrying capacity.

Some of the disadvantages offered by the screw fastenings are:

1. The presence of numerous points of stress concentration on the threaded surfaces which makes them to go under loads causing varying stresses and necessitates the employment of special methods to improve their endurance.

2. The processing inadequacy of certain specific types of threaded parts which nevertheless have the highest strength.

Threaded bolts and screws are generally used for the following purposes in engineering:

1. These are used to hold the components together, for example, the heads of cylinders, machine members that must be readily disassembled, and parts of large machines that must be made in small units for case in manufacturing, assembling, or shipping.
2. Screws are also used for the transmission of power, for example, the lead screws on machine tools, screws on presses, and similar devices. See Chapter 23.
3. Screws are also sometimes used as a means of adjustments for obtaining accurate movements in measuring instruments such as micrometers.

9.2 FORMS OF SCREW THREADS

A number of factors are to be considered while selecting a thread profile. These factors are:

1. Easy production of thread profile.
2. The measurement of the thread profile should be straight forward.
3. The thread form should enable a good bearing between the threads on bolt and nut.
4. The thread be as sharp as possible, consistent with required strength. Large thread angles cause large bursting force on the nut.

The purpose for which the screw is to be used will frequently determine the shape or profile of the thread. Threads are standardized to permit inter-changeability of bolt and nut of the same nominal diameter.

The common forms of thread profiles are:

V, Trapezoidal (Acme), Buttress, square and knuckle, etc. Square, Buttress and Trapezoidal threads are mainly used for power screws (See Chapter 23). For fastening purposes, metric Vee threads are mainly used. Its profile is an equilateral triangle with included angle equal to 60°, pitch equal to base of the triangle and crests truncated by 1/8 × height of the triangle in screws and by 1/4 of the height in a nut, See Fig. 9.1. This is done, because the sharp tops and bottoms of V-threads tend to weaken the screw, especially when it is subjected to shock and repeated stresses. It is also difficult to maintain a sharp point on the cutting tool used in forming the thread and on the top of a thread cut in cast iron.

The V threads have their advantages, which are:

1. V-threads are more suitable for fastening purposes as they offer greater frictional resistance to motion than the threads of other profiles.
2. V-threads are stronger than the other threads. They have twice the amount of the internal resisting power, resisting the shear action at the root of the thread than a square thread.
3. V-threads can be cut easily by a tap, die or on machine.
4. V-threads due to high resisting frictional force, which they offer, are not suitable for use where power is to be transmitted.

9.3 THREAD NOMENCLATURE

Refer to Figs. 9.1 to 9.5.

1. **Screw thread:** The ridge produced by forming, on the surface of a cylinder, a continuous helical or spiral groove of uniform section such that the distance measured parallel to the axis between two corresponding points on its contour is proportional to their relative angular displacement about the axis. Such a thread (formed on the surface of a cylinder) is called parallel screw thread. It can be external (Bolt) or internal (Nut) (formed on the internal surface of a hollow cylinder).
2. **Form:** The shape of one complete profile of the thread between corresponding points, at the bottom of adjacent grooves, as shown in an axial plane section.

3. **Basic form:** The theoretical form on which the design forms for both the external and internal threads are based (Fig. 9.1)
4. **Design form:** The form of the external and internal threads in relation to which the limits of tolerance are assigned, Fig. 9.2.
5. **Flanks:** Those parts of the surface, on either side of the thread, the intersection of which with an axial plane are theoretically straight lines. (Fig. 9.5).
6. **Crest:** That part of the surface of a thread which connects adjacent flanks at the top of the ridge. (Figs. 9.3 and 9.4).
7. **Root:** That part of the surface of a thread which connects adjacent flanks at the bottom of the groove, Figs. 9.3 and 9.4.

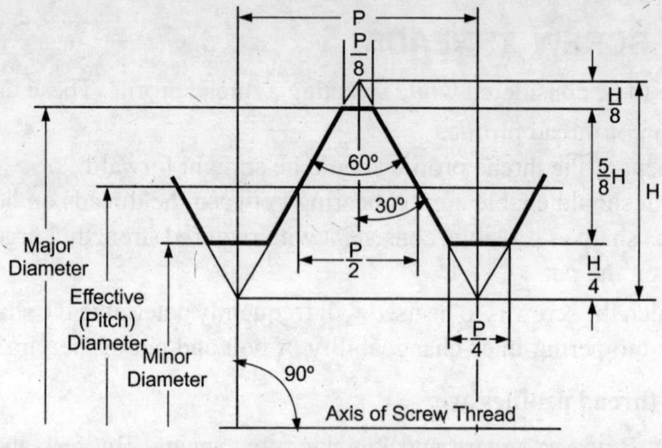

Fig. 9.1 *Basic profile of ISO metric screw thread*

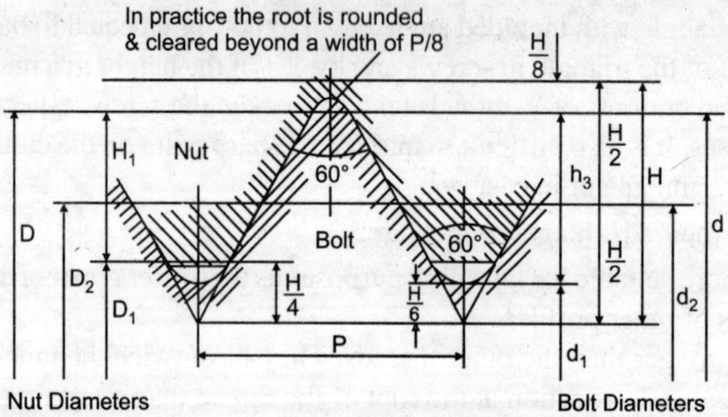

Fig. 9.2 *ISO metric screw thread design profiles of nut and bolt*

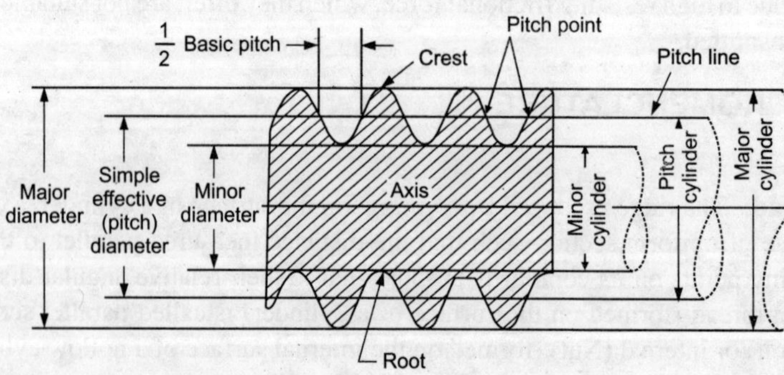

Fig. 9.3 *External parallel screw thread*

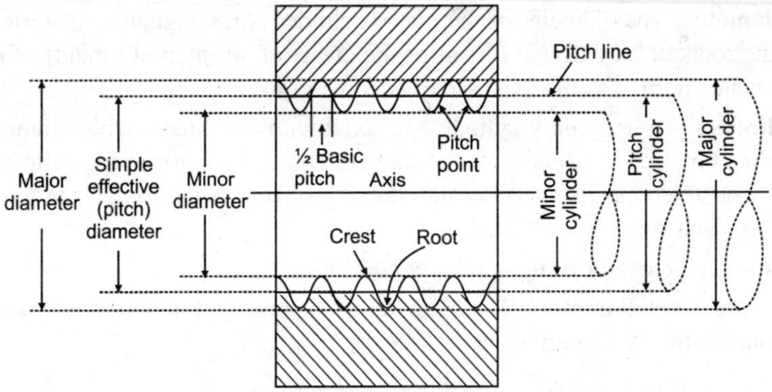

Fig. 9.4 *Internal parallel screw thread*

8. **Included angle of thread:** The angle between the flanks of the thread measured in an axial section, Fig. 9.5.
9. **Apex:** The sharp corner of the fundamental triangle opposite to its base, Fig. 9.5.
10. **Depth of thread:** It is the distance measured normal to the axis, between the crest and the root, see Figs. 9.3 and 9.4.
11. **Pitch:** The distance, measured parallel to the axis, between corresponding points on adjacent thread forms in the same axial plane section and on the same side of the axis.

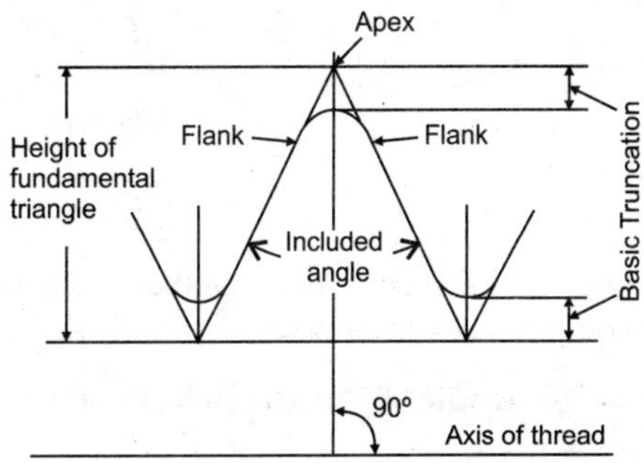

Fig. 9.5 *Fundamental triangle and basic truncation of a parallel screw thread*

12. **Lead:** The distance, measured parallel to the axis, between corresponding points on consecutive contours of the same thread helix in the same axial plane section and on the same side of the helix. This term also denotes the distance that a nut advances along the axis of a fixed screw, if turned through one complete revolution. The relation between the lead, L, and pitch, p, is

$$L = pn$$

where n = number of starts of the thread.

13. **Helix angle, α:** It is the angle made by the helix of the thread (that is, by the tangent to the helix) with the axis at the pitch line.
14. **Lead angle, λ:** It is the angle made by the tangent to the helix with the plane normal to the axis at the pitch line. Thus,

$$\alpha + \lambda = 90°$$

15. **Hand of the thread:** The thread can be R.H. or L.H. It is R.H. when the cylinder is turned clock-wise to engage the second member and when C.C.W., it is L.H.

or

A thread is R.H. if, when viewed axially, it winds in a clock wise and receding direction. And, a thread is a L.H, when viewed axially, it winds in a C.C.W. and receding direction.

or

The helical line ascending from left to right over the visible part of the cylinder surface, its axis being vertical is termed "right-hand" and so on.

16. **Major diameter:** The diameter of the major cylinder (An imaginary cylindrical surface which just touches the crests of an external thread or the roots of an internal thread) of a parallel thread in a specified plane, normal to the axis, See Figs. 9.3 and 9.4. The screw is specified on this diameter.

17. **Minor diameter:** The diameter of the minor cylinder (An imaginary cylindrical surface which just touches the roots of an external thread or the crests of an internal thread) of a parallel thread in a specified plane, normal to the axis, see Figs. 9.3 and 9.4.
18. **Pitch cylinder:** An imaginary cylinder, co-axial with the thread, which intersects the surface of a parallel thread in such a manner that the intercept on a generator of the cylinder between the points where it meets the opposite flanks of the thread groove is equal to half the basis pitch of the thread (See Figs. 9.3 and 9.4).
19. **Pitch line:** The generator of the pitch cylinder or cone.
20. **Pitch (or Effective) diameter:** The diameter of the pitch cylinder of a parallel thread in a specified plane normal to the axis (see Figs. 9.3 and 9.4).

Basic Profile of ISO Metric Screw thread (IS: 1362–1959)

Refer Fig. 9.1.

$$H = 0.866\ 03\ p$$
$$H/8 = 0.108\ 25\ p$$
$$H/4 = 0.216\ 51\ p$$
$$\frac{5}{8}H = 0.541\ 27\ p; \quad H/6 = 0.144\ 34\ p$$

where H = Height or depth of the fundamental triangle. It is the distance, measured perpendicular to the axis from its appex to its base.

Design profile of Nut and Bolt (IS: 1362–1959):

Refer Fig. 9.2.

Capital letters for Nut and Small letters for Bolt.

$$D = d = \text{Major Diameter}$$
$$\text{Pitch diameter, } D_2 = d_2 = d - \frac{3}{4}H = d - 0.6495\ p$$
$$\text{Minor diameter (Nut) } D_1 = d_2 - 2[H/2 - H/4] = d - 1.0842\ p$$
$$\text{Minor (Root, Core) diameter, (Bolt) } d_1 = d_c = d_2 - 2[H/2 - H/6] = d - 1.226\ 8\ p$$
$$H_1 = \frac{D - D_1}{2} = \frac{5}{8}H = 0.541\ 27\ p$$
$$h_3 = \frac{d - d_1}{2} = \frac{17}{24}\text{H} = 0.613\ 4\ p.$$

9.4 DESIGNATION OF INDIAN STANDARD THREAD [I.S. 4218 (PART IV) 1978]

These threads are designated by the letter M followed by diameter and pitch, the two being separated by the sign *X*. Where there is no mention of pitch, it shall mean that a coarse pitch will be used (for example, M16). Metric threads with fine pitches also state the pitch in millimetres (for example, M16 × 1.25). Further, the three classes of tolerances namely fine, medium and coarse shall be denoted by the letters *f*, *m* and *c* respectively. But in the case of medium class of tolerance, the suffix *m* may be dropped. Some of the examples are shown below:

	M 12 × 1.25		M 12 *f* (means M 12 × 1.75 *f*)
	M 12 × 1.25 m		M 12 m (means M 12 × 1.75 m)
or	M 12 × 1.25	or	M 12 or (M 12 × 1.75)
	M 12 × 1.25 *c*		M 12 *c* (means M 12 × 1.75 *c*)

Tables 9.1 and 9.2 give the design dimensions of screw threads for screws, bolts and nuts for coarse and fine series as recommended by I.S. 4218–1978 (Part VI).

Table 9.1 Coarse series

Designation	Pitch p mm.	Minor diameter bolt mm	Minor diameter nut mm.	Stress area mm^2	Designation	Pitch p mm.	Minor diameter Bolt mm	Minor diameter nut mm	Stress area mm^2
M1	0.25	0.693	0.729	0.460	M 7	1	5.773	5.18	28.9
M1.1	0.25	0.793	0.829	0.588	M 8	1.25	6.466	6.647	36.6
M 1.2	0.25	0.893	0.929	0.732	M10	1.5	8.160	8.376	58.0
M1.4	0.30	1.032	1.075	0.983	M 12	1.75	9.853	10.106	84.3
M1.6	0.35	1.171	1.221	1.27	M 14	2	11.546	11.835	115
M1.8	0.35	1.371	1.421	1.70	M 16	2	13.546	13.835	157
M 2	0.40	1.509	1.567	2.07	M 18	2.5	14.933	15.294	192
M 2.2	0.45	1.648	1.713	2.48	M 20	2.5	16.933	17.294	245
M 2.5	0.45	1.948	2.013	3.39	M 22	2.5	18.933	19.294	303
M 3	0.5	2.387	2.459	5.03	M 24	3	20.320	20.752	353
M 3.5	0.6	2.764	2.850	6.78	M 27	3	23.320	23.752	459
M 4	0.7	3.141	3.242	8.78	M 30	3.5	27.706	27.211	561
M 4.5	0.75	3.580	3.688	11.3	M 33	3.5	28.706	29.211	694
M 5	0.8	4.019	4.134	14.2	M 36	4	31.093	31.670	817
M 6	1	4.773	4.918	20.1	M 39	4	34.093	34.670	976

Table 9.2 Fine series

Designation	Pitch p mm	Minor diameter bolt mm.	Minor diameter nut mm.	Stress area mm^2	Designation	Pitch p mm.	Minor diameter bolt mm.	Minor diameter nut mm.	Stress area mm^2
M 8 × 1	1	6.773	6.917	39.2	M 22 × 1.5	1.5	20.160	20.376	333
M 10 × 1.25	1.25	8.466	8.647	61.6	M 24 × 2	2	21.546	21.835	384
M 12 × 1.25	1.25	10.466	10.674	92.1	M 27 × 2	2	24.546	24.835	496
M 14 × 1.5	1.5	12.160	12.376	125	M 30 × 2	2	27.546	27.835	621
M 16 × 1.5	1.5	14.160	14.376	167	M 33 × 2	2	30.546	30.835	761
M 18 × 1.5	1.5	16.160	16.376	216	M 36 × 3	3	32.319	32.752	865
M 20 × 1.5	1.5	18.160	18.376	272	M 39 × 3	3	35.319	35.752	1028

Fine pitch threads are more reliable in respect of self locking than coarse pitch threads. Therefore, fine pitch threads should probably be used where there are alternating loads or vibrations, as well as with thin walled parts.

9.5 DESIGNATION OF BOLTS, SCREWS AND NUTS

(*a*) The square bolts, screws and nuts shall be designated as illustrated in table below as per I.S. 2585–1968. The square bolts may also be used with hexagon nuts conforming to I.S. 1363–1967 in which the designation will be as illustrated below in table.

Designation System for square bolts, screws and nuts.

Description			Designation
Fastener	Thread Size	Length mm.	
Square bolts with a square nut	M 10	30	Square bolt M 10 × 30 N–I.S. 2585
Square bolts only	M 10	30	Square bolt M 10 × 30–I.S. 2585
Square screw	M 10	30	Square screw M 10 × 30–I.S. 2585
Sqare nut	M 10	–	Square nut M 10 I.S. 2585
Square bolt with a hexagon nut Conforming to I.S. 1363–1967	M 10	30	Square bolt M 10 × 30 HN–I.S. 2585

(*b*) The hexagonal bolts, screws and nuts shall be designated as illustrated in table below as per I.S. 1363–1967.

Designation system for Hexagonal bolts, Nuts and lock nuts.

Description			Designation
Fastener	Size	Length	
Bolt with nut and lock nut	M 16	70	Hex. bolt M 16 × 70 NL–I.S. 1363
Bolt with nut	M 16	70	Hex. bolt M 16 × 70 N–I.S. 1363
Bolt only	M 16	70	Hex. bolt M 16 × 70–I.S. 1363
Screw	M 16	70	Hex. Screw M 16 × 70–I.S. 1363
Nut	M 16	—	Hex. nut M 16–I.S. 1363
Lock nut	M 16	—	Hex. Lock nut M 16–I.S. 1363
Double chamfered nut	M 16	—	Hex. nut DC M 16–I.S. 1363

9.6 COMMON SCREW FASTENERS

The choice of type of fastening is very important ; the location or placing is also equally important, Fastenings should be located and placed so that they will be subjected to tensile and for shear loads, and bending of the fastening should be reduced to a minimum. Bending of the fastening due to misalignment, tightening up loads, or external loads has been responsible for many failures.

Some common types of screw fastenings are described as follows:

9.6.1 Bolts, Screws and Studs

Through bolts : A through bolt is shown in Fig. 9.6 (*a*). It is a round bar with a head on one end and thread at the other end to take a nut. It is passed through drilled holes in the two parts to be fastened together and clamps securely to each other as the nut is screwed on to the threaded end. A through bolt is also called simply a bolt. Nuts are made either hexagonal or square. Heads of bolts are also made either hexagonal or square. Square heads of bolts and square nuts are used mostly with rough bolts, as in construction work. Through bolts may or may not have a machine finish. When through bolts are put under tension by a load along its axis, a through bolt should be an easy fit in the holes, thus in this case the shank of the bolt can be rough. If a through bolt is loaded in shear, it carries the load on the shank area, therefore, the bolt shank should be machine finish and the holes should be reamed so that the bolt shank fits snugly therein. However, the threaded end must pass freely through the holes. Through bolts should be used whenever the available space and other conditions permit their use.

Different minor characteristics, which through bolts may possess to fit them to a specific usage, create different names by which they become known, such as machine bolts, carriage bolts, automobile bolts, coupling bolts, etc.

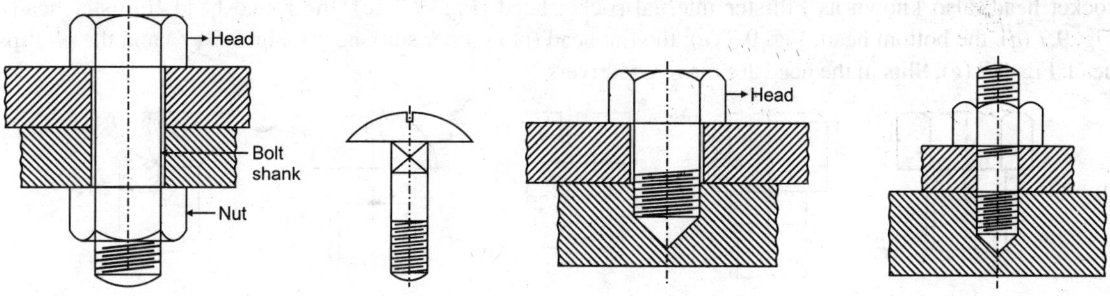

(a) Through bolt (b) Carriage bolt (c) Tap bolt or screw (d) Stud bolt

Fig. 9.6 *Bolts, screws and studs*

A machine bolt has a rough shank, but the head of the bolt and nut may be rough or finished as desired. A coupling bolt is a machine bolt that is finished all over and is made to be fitted into a reamed hole having the same diameter as the bolt. A coupling bolt is under shear.

A carriage bolt is used when the head must rest against wood. The part of the shank near the head is square to prevent the bolt from turning when the nut is tightened, Fig. 9.6 (*b*).

An automobile bolt is finished all over and uses finer pitches. Head of the bolt and nut are hexagonal and are somewhat smaller in size. In order that a screw driver can be used, the head of the bolt is often slotted.

Tap bolt and Cap Screws: When a bolt passes through a hole in one part and is screwed into a tapped hole in the second part, which acts as a nut, thus holding the two parts together, it is called a tap bolt, Fig. 9.6 (*c*). A tap bolt does not use any nut. Its head may be either square or hexagonal, although the latter is preferred. Tap bolts, whose heads are saw-slotted for use with a screw-driver, are known as Cap Screws and are desirable where lack of space or other considerations prevent the use of through bolts, but should not be used when they must be removed frequently as this may ruin the thread in the tapped hole. These bolts are threaded for a length of two diameters plus 6 mm. and should enter the tapped hole at least one diameter in steel, at least 1.5 diameters in cast iron, and three diameters in aluminium. Tap bolts and cap screws are cheaper than studs and hence are often used instead of them. When a tap bolt or cap screw is loaded in shear, it carries the load on the root area, whereas a through bolt carries a shear load on the larger shank area. Long cap screws and tap bolts should not be used, because they twist when they are tightened and later, when they un-twist, this reduces the pressure between the surfaces they hold together.

Machine Screws: These are small cap screws having coarse and fine thread series. Machine screws have a slit at the head for a screw driver. They are used for assembling of small components in a machine.

Stud bolts: A stud bolt is often called simply a stud as shown in Fig. 9.6 (*d*). Studs are threaded on both ends and are used where through bolts are undesirable. The threads are either of coarse or fine series, depending on the material into which the stud is screwed. A stud requires less construction cost as compared to the through bolt and a stud also results in saving of weight and space. The stud has a replacement advantage over the tap bolt in the event of wear of the joints are disassembled. To secure frictional resistance against turning when the nuts are removed, stud should enter the holes at least 1.25 diameters in steel, 1.5 diameters in brittle materials and 3.0 diameters in soft material as aluminium. The holes should be drilled at least 0.5 diameter deeper than the threaded length of the studs. Studs are particularly convenient for positioning the covers of a cylinder head.

9.6.2 Head Forms

Figure 9.7 shows a variety of head forms available on standard bolts, studs and screws. The hexagonal head, Fig. 9.7 (*a*), is the most commonly used form. Other forms are — the square head (Fig. 9.7 (*b*), the hexagon socket head, also known as Fillister internal socket head (Fig. 9.7 (c), the round head (Fillister head), Fig. 9.7 (*d*), the bottom head, Fig. 9.7 (*e*); the flat head (or counter sunk head), Fig. 9.7 (*f*) and the phillips head, Fig. 9.7 (*g*). Slits in the head are for screw drivers.

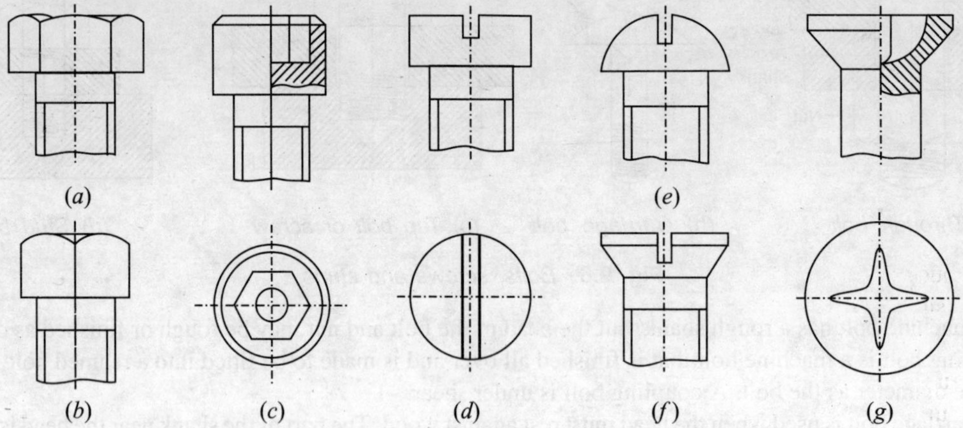

Fig. 9.7 *Head forms*

9.6.3 Set Screws

These are used to prevent relative motion by means of pressure exerted on their points. Fig. 9.8 illustrates set screws with heads and without heads. The latter are preferred. The holding power of a set screw is the frictional resistance set up at the point. An example of such an installation is where a set screw is screwed through the hub of a gear or pulley and by its pressure against the shaft, the gear or pulley is made to rotate with the shaft. According to experiments conducted by Mr. R.H.D. Pinkey, the holding power of set screws is given in Table 9.3.

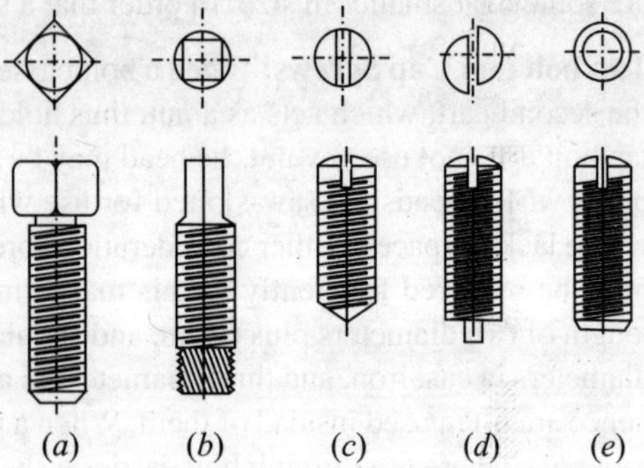

Fig. 9.8 *Set screws*

The proper size of set screws may be determined by the empirical formula

$$d_0 = 0.125\,D + 7.5 \text{ mm.} \qquad \text{...(9.1)}$$

where d_0 = set screw diameter, mm.

D = shaft diameter, mm.

Table 9.3 Safe holding power of set screws

Diameter of set screw mm.	Holding power in N	Diameter of set screw mm.	Holding power in N
6.0	450	15.0	3900
7.5	750	18.5	5850
9.0	1160	21.0	8350
10.5	1650	25.0	11.4×10^3
12.5	2250	28.0	15×10^3
13.5	3000	31.0	19.1×10^3

The frictional resistance which it sets up at the shaft can be known if the torsional moment on the shaft can be calculated. Then by the formula

$$F_t \times R = M_t \quad ...(9.2)$$

where F_t is the tangential force or frictional resistance.

R = Radius of shaft

M_t = Torsional moment on shaft

Once the F_t is known the holding power, *i.e.*, the power which can be transmitted by a set screw can be determined by the formula

$$\text{Power} = \frac{F_t\, 2\pi Rn}{60}, \text{ watts} \quad ...(9.3)$$

where n is the revolution per minute of shaft.

9.7 LOCKING ARRANGEMENTS

Screw threads have a helix angle that is much smaller than the effective angle of friction and are, therefore, self-locking. Threaded joints are also prevented from working loose by friction between the contact surfaces of nuts and screw heads. However, the tendency of nuts to loosen under vibration or changing forces acting thereon makes it necessary in a great many installations to secure the nut in position by some kind of locking device. A large number of devices are available. Some of these are described as follows:

Lock nuts: It is also known as Jamb nut or Check nut. It is the most common locking device. The lock nut also called the half nut is first tightened down on the bolt with ordinary force, then the upper main nut or full nut is tightened down upon it as shown in Fig. 9.9 (*a*). The upper nut is then held tightly while the lower lock nut is slackened back against it. Due to this action the threads in nuts get wedged tightly against those of the bolt. In slackening back the lock nut, a thin spanner is necessary, which is not always available in many shops. For this reason many times the lock nut is placed on the top. If the nuts are really tightened down as they should be, the upper nut carries whole of the tensile load on the bolt, where a lock nut cannot take that load. Therefore, the thick nut or main nut should be on top. To overcome both the difficulties, it is better practice, if saving in height is desired, to make both nuts of the same height, which may be about 3/4 times the bolt size as shown in Fig. 9.9 (*b*). This way of achieving the locking device is due to frictional action.

Spring lock washer: is shown in Fig. 9.9 (*c*). As the nut tightens the washer against the piece below, one edge of the washer is caused to dig itself into that piece, thus increasing the frictional resistance so that the nut will not loosen so easily. There are many different kinds of spring lock washers manufactured, some of which are fairly effective.

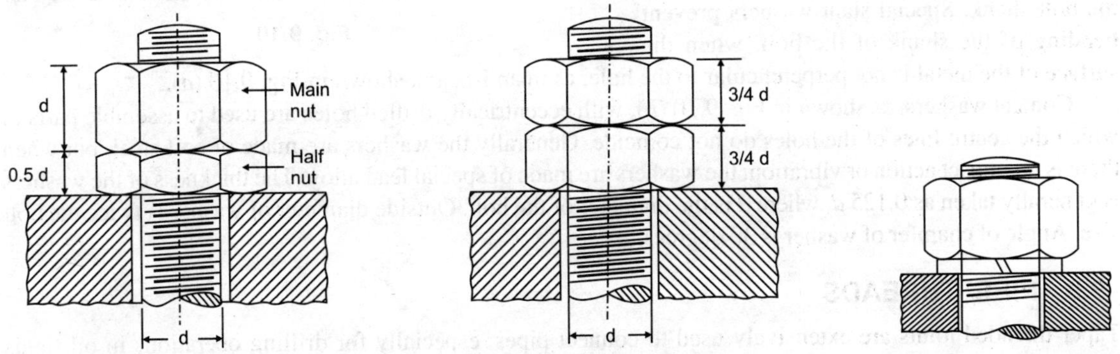

(a) Locking by nut *(b) Locking by nut* *(c) Spring lock washer*

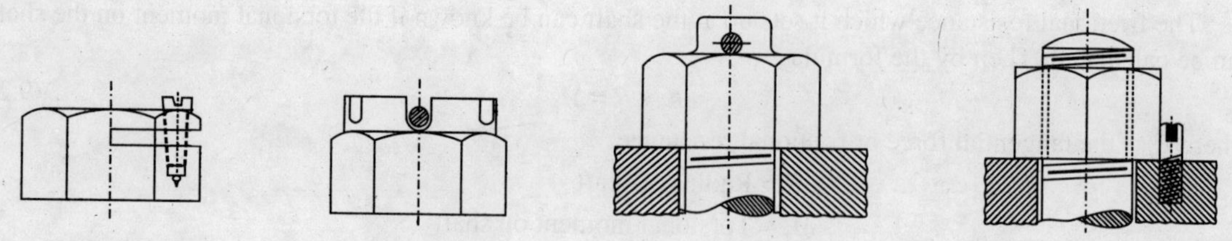

(d) Split nut (e) Castled nut (f) Locking by pin (g) Locking by threaded pin

Fig. 9.9

Split nut: or a slotted nut Fig. 9.9(*d*), is used successfully and has a cut on the sides. A small screw pulls the slot together and thus causes the nut to pinch the thread. This way of achieving the locking device is also due to friction.

Castled nut: Figure 9.9 (*e*). After this nut screwed tightly into position, a split pin is inserted through the groove of the nut and a hole through the bolt. This is a positive locking device and is used to a great extent in the automobile industry.

Locking by pin: The positive lock is the cottor pin Fig. 9.9 (*f*), which is inserted in a hole drilled through the bolt after the nut has been screwed home. Where the nut is to be removed frequently or pulled up in service, the use of castled nut with a finer thread is satisfactory solution. Insertion of the pin above the nut makes its removal easier but has the disadvantages that the nut can back up before it is stopped by the pin.

Locking by threaded pin: Figure 9.9 (*g*). This is used with large nuts, such as those hold pistons. This lock has two advantages, one is that the bolt is not weakened by having holes or grooves cut in it, and second the nut may be stopped in any position. It is a positive locking arrangement.

9.8 WASHERS

It is placed under a nut or the head of a bolt (screw) to transmit and distribute evenly the force exerted on the connected parts or to lock them. Furthermore, the spring action of washers maintains a high frictional resistance to loosening of the nut even when some loosening has taken place. They may also be used to prevent surface injury to parts or to serve as take-up where there is insufficient thread on the bolt shank. Special slant washers prevent bending of the shank of the bolt, when the surface of the metal is not perpendicular to the hole, as in an I-beam, shown in Fig. 9.10 (*a*).

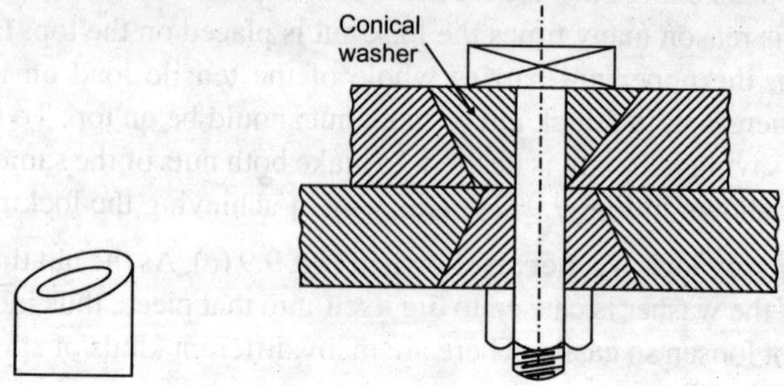

(a) Slant face washer (b) Use of conical washer

Fig. 9.10

Conical washers, as shown in Fig. 9.10 (*b*), with eccentrically drilled holes are used to assemble parts in which the centre lines of the holes do not coincide. Generally the washers are made of soft steel, but when there is no impact action or vibration, the washers are made of special lead alloy. The thickness of the washers is generally taken as 0.125 *d*, where *d* is the diameter of the bolt. Outside diameter of washer is twice the bolt size. Angle of chamfer of washer is 30 degrees.

9.9 PIPE THREADS

Taper threaded joints are extensively used to connect pipes, especially for drilling operations in oil fields where adequate tightness under high pressure and considerable load is the prime demand. Taper threads are formed with a taper 1 mm per 16 mm in order to make a tight joint. Taper threads are superior to cylindrical in that they:

1. ensure tightness of a joint without any sealing medium.
2. inaccuracies due to improper manufacture and wear of the thread working length cause the thread play which is also taken by taper thread by means of an axial shift.
3. spread the load more uniformly over the length of engagement.
4. save time in assembling and disassembling the joint.

9.10 SPECIAL SCREWS

Screw fastenings of special shapes are used freely in various design problems. Only a few typical examples will be given here.

T-head bolt or Anchor bolt: The usual proportions of a T-headed bolt are shown in Fig. 9.11 (*a*). The head of this bolt is like a T. It is used in machine tables. The T-head of the bolt can slide to the required position, through the T-slots cut on the table. A square neck is provided with the head. This bolt can be quickly removed.

Hook bolt: The shape and usual proportions of a hook bolt are shown in Fig. 9.11 (*b*). In this bolt the head projects only to one side of its shank. The shank of the bolt passes through a hole in one part and the other part to be joined comes under the bolt head and is supported by it. This bolt is also provided with a square neck to prevent its rotation in bolt hole.

Eye bolt: This bolt is used where frequently removal is required, when misplacing of the bolt must be prevented. An eye bolt, having a circular cross-section as head is shown in Fig. 9.12. A flat circular portion, integral with the head, is also provided. Below this circular portion, the shank of the bolt is of a cylindrical shape on which threads are cut. This type of bolt is used for lifting heavy machines. The bolt is screwed in threaded hole on the top of the machine directly above the centre of gravity, so that while lifting, the machine does not change from its usual working position. It can be used for lifting the machines such as turbine casings, heavy electric motors or generators, compressors, gear box, etc.

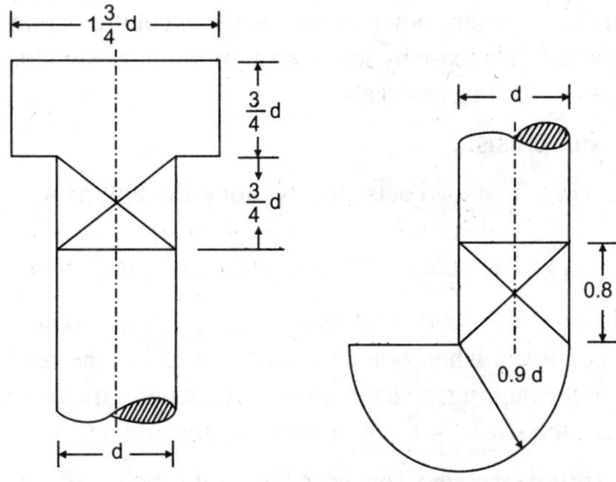

(a) T-bolt (b) Hook bolt

Fig. 9.11

9.11 BOLTS SUPPORTING TENSILE LOADS ONLY, WITHOUT AN INITIAL STRESS

When a bolt supports a load that acts along its axis and tends to stretch it as in Fig. 9.12. the bolt is under tension. In the case of bolts, the smallest section is at the root of the threads. this is a section (circular) with a diameter equal to the root diameter (d_c). Hence in designing a bolt, as loaded in figure, the section at the root of the threads is made large enough to resist the tensile load P. Other sections, such as those with the outside diameter (d) of the bolt will then be safe.

Thus $$P = \frac{\pi}{4} d_c^2 \times f_t$$

or $$d_c = \sqrt{\frac{4P}{\pi f_t}} \quad \text{...(9.4)}$$

where f_t is the allowable tensile stress of the material of bolt. Having found the root diameter the standard bolt can be selected from Tables 9.1 or 9.2. The one selected should always be the standard diameter that is just above the theoretical value.

9.12 CALCULATIONS OF SCREW FASTENINGS

Threaded joints are used in a wide variety of ways calling for a great number of designs. Given below are the basic types of bolted joints which differ according to assembly and loading conditions:

1. Calculations of Non-Tightened (without an initial stress) bolts: The joint shown in Fig. 9.12 is an example of such a case. Another example of such joints is the threaded end of a cargo lifting hook of a load-hoisting machine. The nut is freely screwed on the shank and secured by a pin. The bolt shank is loaded upon application of an external load and is subjected to tensile load only. Strength is the principal criterion of serviceability for threaded connections. In standard bolts, screws and studs, the tensile strength of the shank equals the shearing and bearing strengths of the thread. So, for simplicity the analysis and design are carried out for the shank (body) only, with the help of equ. (9.4). Such joints, that is those not subjected to preliminary tightening, loaded by an external force alone are used in mechanical engineering on comparatively rare occasions. However, many other examples of calculations may be reduced to this particular case by selecting appropriate design loads or allowable stresses (as discussed ahead).

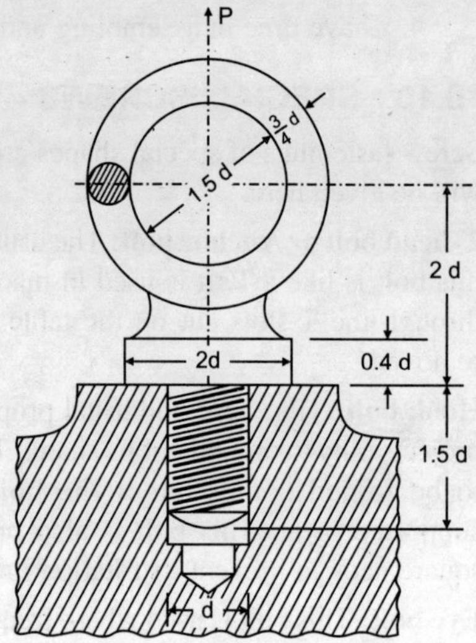

Fig. 9.12

Assumptions:

(*a*) The load acts strictly along the bolt axis.

(*b*) The stresses are uniformly distributed over the bolt section.

(*c*) The concentration of stresses is neglected.

Note: Bolts, studs and screws carry mainly axial loads, for which the design is handled as given above. Sometimes, when bolts are used to prevent the relative movement of two or more parts, such as bolts in a flange coupling, a shear stress is induced in the bolts. When the action of the load on a bolt is at angle to the axis, the stress in the bolt becomes tension and shear combined.

2. Initial stresses (Tensioned or Tightened, Carrying no external load): As a result of the turning of the nut, a bolt, screw, or stud is subjected to the following stresses:

(*a*) Tensile stress due to stretching of the bolt.

(*b*) Compression on the threads.

(*c*) Torsional stress caused by the frictional resistance of the thread during its tightening.

(*d*) Bending stress if surfaces under the head or nut are not perfectly normal to the bolt axis.

(*e*) Shear across the threads.

These stresses are induced in a majority of group joints used in mechanical engineering to fasten covers, flanges, base plates, etc. In some cases these joints should also be tight, e.g., the joint between the cover and cylinder of an internal combustion engine. In other cases the joint must not be allowed to yawn. Connecting rod and foundation bolts, etc., must comply with these conditions.

None of these stresses can be accurately determined, so bolts are designed on the basis of the direct tension stress with a comparatively high factor of safety to allow for the undetermined stresses. The following analysis will explain the procedure for determining the initial stresses.

According to experiments made at Sibley College the load in *N* produced by screwing a threaded member may be estimated as

$$P_i = 2840\, d\; N \qquad ...(9.5)$$

where *d* is the nominal diameter of the bolt in mm.

The above relation was confirmed by German experiments with screws having diameters of 12.5 mm, 18.5 mm and 25 mm. The above expression indicates that screws are subjected to an excessively high initial stress. Smaller screws are easily stretched above the elastic limit and even broken. For this reason, bolts that must carry external loads should not be under 16 mm in any case.

Relation (9.5) apply to cases where the nuts have metal to metal contact. If a bolt fastens two parts with a flexible gasket between them a smaller effort is applied in order not to crush the gasket. In such a case the initial tension calculated by relation (9.5) may be taken one half, thus

$$P_i = 1420\, d\, N \quad \text{...(9.6)}$$

The torsional stress f_s can be calculated by the relation

$$f_s = \frac{16 M_t}{\pi d_c^3} \quad \text{...(9.7)}$$

where M_t is the torsional moment applied which can be measured with the help of a torque wrench for assembly of bolted connections, and dc is the core diameter of the bolt.

The magnitude of the resultant stress can be found by combining the tensile stress produced by load P from relations (9.5) or (9.6) and the torsional stress from relation (9.7). This resultant stress can be found by using the appropriate theory of failure.

The normal stress in the lateral section of the threaded portion of the bolt, caused by the tightening force $P(P = P_i)$ is given by equ. (9.4) as:

$$f_t = \frac{P}{\frac{\pi}{4} \cdot d_c^2}$$

The maximum shear stress caused by the moment M_t twisting the bolt shank is given by equ. (9.7.), i.e., $f_s = 16M_t/\pi \cdot d_c^3$. The torsional moment can also be calculated by the relation:

$$M_t = P \cdot \frac{dp}{2} \cdot \tan(\alpha + \phi)$$

where P = Axial force of bolt tightening

d_p = pitch diameter of thread

α = helix angle of the thread (lead angle)

ϕ = reduced angle of friction between the surfaces of screw and the nut

$= \tan^{-1}\mu$ (μ, co-efficient of friction between the surfaces of the screw and the nut.)

So, when the bolt is subjected simultaneously to axial force and twisting moment by the tightening of the bolt, the equivalent stress will be given as,

$$f_{eq} = \sqrt{f_t^2 + 4 \cdot f_s^2}\,, \quad \text{Maximum shear stress theory}$$

$$= \sqrt{f_t^2 + 3 \cdot f_s^2}\,, \quad \text{Maximum distortion energy theory}$$

By considering the various factors of the joint (Magnitude of tightening force, parameters of the screw joint and the value of μ), it has been seen that for metric threads, the equivalent stress, on an average, is approximately 30% higher than stress f_t. Thus, the effect of shearing stresses can be allowed for by increasing the design tensile load (Initial tightening load) by 30%. Thus, the design equation will be,

$$f_t = \frac{1.3 \times P}{\frac{\pi}{4} \cdot d_c^2}$$

According to experiments conducted, it has been found that repeated unscrewing and screwing of steel nuts increases the torsional moment that must be applied, because there is gradual scoring of the threads. After screwing the nut 50 times, the torsional resistance is about double than the resistance at the first tightening, and after 200 times the increase is about 2.5 times.

Bending stresses in threaded Joints: If the outside surfaces of the parts that are bolted together are not parallel as shown in Fig. 9.13, the bolt will be subjected to bending. This bending stress is given by the relation.

$$f_b = \frac{aE}{2l} \quad ...(9.8)$$

where a is the difference in height between the extreme corners of the nut or head when the bolt is under bending, or out of parallelism of one face with respect to the other, Fig. 9.13. Here $2l$ is the total thicknesses of flanges.

E = modulus of elasticity of bolt material.

The analysis of the above relation reveals that greater the shank length, less is the bending stress induced in the bolt or stud. Experimental verification of the effect of bending stress due to misalignment has shown that at angles of misalignment $\alpha \leq 2$ degrees the static strength of bolts made from steels of low strength changes but little. This effect is considerable at small angles of misalignment only for parts made from high-strength steels, especially under loads producing varying stresses.

It follows from this that consideration should be given to proper working of contact surfaces. With structural surfaces, slant face washers, Fig. 9.10 (*a*) may be used.

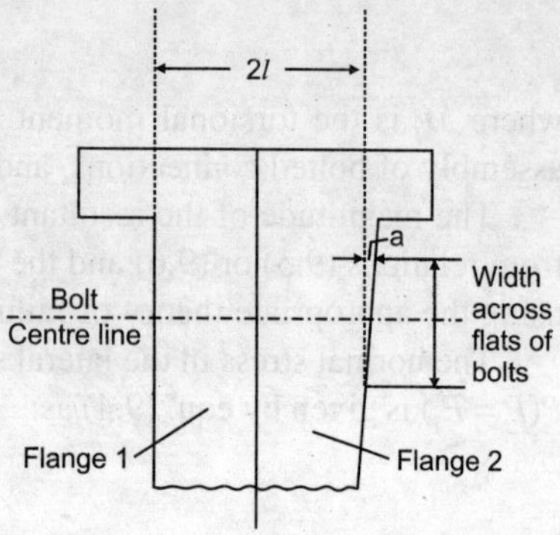

Fig. 9.13 *Non-parallel flanges*

The sum of this bending stress and the initial stress from tightening the nut can very easily exceed not only the elastic limit but even the ultimate strength of the bolt material. However, because of the ductility of steel the bolt will be bent before breaking, but this bending is not permissible in a machine part.

Thickness of Bolt Head: In a bolt subjected to tensile force only as in Fig. 9.14, which induces tension only, the size of the bolts is found from relation 9.4. The thickness of the head h can be found from its behaviour under shear, compression and bending forces. An analysis and the experience in use of bolts show that the thickness h calculated for bending stress satisfies the condition for shear and compression. The figure shows the beam under bending. Then

$$B.M. = M_b = P \times \frac{d}{2}$$

$$Z = \text{Modulus of section} = \frac{\pi d \times h^2}{6}$$

Thus the thickness h can be found from relation

$$f_b = \frac{M_b}{Z} = \frac{P\frac{d}{2}}{\pi d \frac{h^2}{6}} \quad ...(9.9)$$

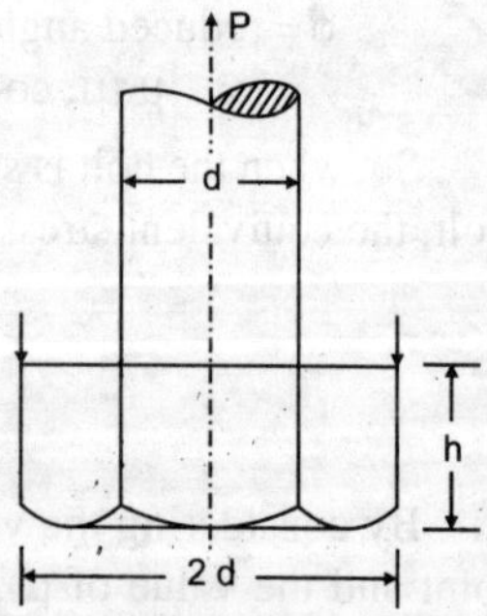

Fig. 9.14

Nut Thickness: The nut thickness can be found again from the strength condition of thread in shear, bending and compression.

(i) **Strength in shear:**

(a) For bolt (Screw):

$$f_s \leq \frac{P}{\pi \cdot d_1 \cdot h_n}; \; h_n = \text{height of nut}$$

where $h_n = z \times p$, z is the number of threads in engaged portion.

d_1 = core diameter of screw.

(b) For nut:

$$f_s \leq \frac{P}{\pi \cdot D \cdot h_n}$$

where D = Major diameter of nut.

(ii) **Crushing/Bearing**

$$P \leq \frac{\pi}{4}(d^2 - d_1^2) \cdot f_c \cdot z$$

From here, $z = ?$ $\therefore$ $h_n = p \times z$

(iii) **Bending:** The thread is regarded as cantilever beam having a restrained end. The load, P, is replaced by the concentrated load, P/z applied to the middle of the effective depth of thread.

Now lever arm for bending of thread,

$$l = \frac{D - D_2}{2} = \frac{d - d_2}{2}$$

$$\therefore \qquad \frac{P}{z} \cdot l = \frac{\pi d_1}{6} \cdot b \cdot f_b$$

where b = tooth width at root = p (see Fig. 9.2)

Bolted joints subjected to Transverse Load: When a bolt supports a load that acts perpendicular to the axis of the bolt as shown in Fig. 9.15, the bolt is under shear. The bolt can be under single or double shear. The minimum section under shear is at outside diameter.

Thus $$P = \frac{\pi}{4} d^2 \times f_s \qquad ...(9.10)$$

where f_s is the allowable stress in shear.

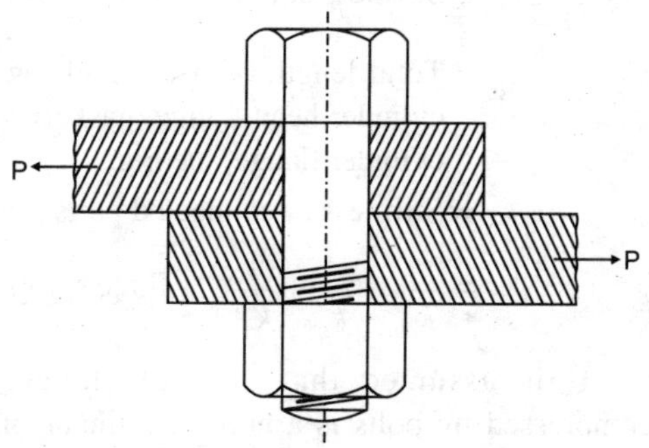

Fig. 9.15 *Bolt under shear*

The above relation is written for single shear. For a bolt under double shear, the area to be taken is double. Bolts should not be preferably subjected to shear.

When a combined stress *i.e.*, tension and shear are produced in a bolt, then the principal stresses are found, which help to evaluate the size of the bolt as discussed in Art. 9.12. Having found the root diameter, the standard size of the bolt can be selected from Tables 9.1 or 9.2. In case the standard tables for the metric screw threads are not available, the nominal diameter of the bolt is calculated by the following relation in case of coarse threads.

$$d_c = 0.84\, d \qquad ...(9.11)$$

(4) Stresses due to Combination of Forces: (Tensioned and carrying an external load). The resultant axial load on a bolt depends on the initial tension and on the external load. It also depends on the relative elastic yielding of the bolt and the connecting members. If the connecting members are very yielding as compared with the bolt, as shown in Fig. 9.16 (*a*), which shows a soft gasket, the resultant load on the bolt will closely approximate the sum of the initial tension and the external load. If the bolt, however, is very yielding as compared with the connecting members as shown in Fig. 9.16 (*b*), the resultant load will be either the initial or the external load, whichever is greater. The actual conditions will lie between the two extreme cases.

When a bolted assembly Fig. 9.16 (*a*) is tightened with an initial load P_i, the bolt will stretch elastically an amount δ_b, and the connected members will get compressed an amount δ_c. Applying Hooke's law, the load deformation diagram is shown in Fig. 9.17 (*a*) where the tensile deformation of bolt is measured from left to right and that of the compressed member from right to left on the abscissa of the diagram, being opposite in sign. Because of the difference in stiffness of the bolt and that of the connected members, the slopes of the deformation curves of the bolt and connected members will be different.

Now, when an external axial load P_e is applied to this initially tightened assembly, it will cause an additional extension of the bolt and hence an increased load on it and decrease in the compression in the connected members. Because the connected members are initially compressed, the increased load on the bolt, ΔP, will be only a fraction of the external load P_e. As long as the connected members are under compression, this increased load P on the bolt may be written as,

$$\Delta P = kP_e$$

where $k = \frac{a}{1+a}$, '*a*' being the ratio of the elasticity of connected parts to elasticity of bolts. '*k*' can also be determined as,

$$k = \frac{k_b}{k_b + k_c},$$

where k_b = Stiffness of bolt = $\frac{A_b \cdot E_b}{l_b}$, N/mm

l_b = Total length of parts held together by the bolts, i.e. cylinder head flange, gasket and cylinder flange, see Fig. 9.16, $= l_{ch} + l_g + l_f$

k_c = Stiffness of connected parts

$$= \frac{1}{k_{ch}} + \frac{1}{k_g} + \frac{1}{k_f} \quad \text{(Three springs in series).}$$

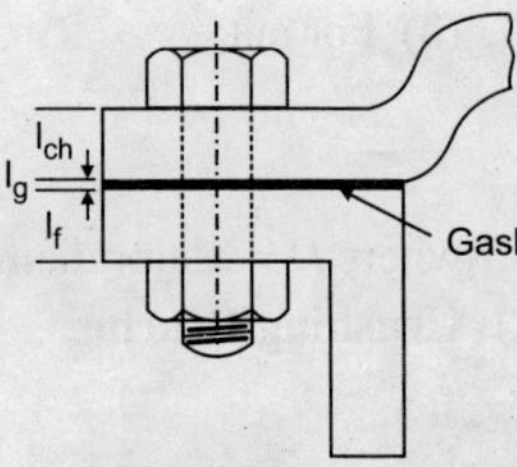

(a) Gasket joint

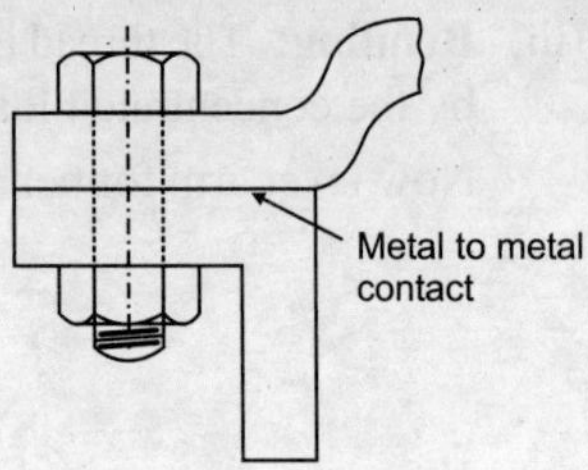

(b) Ground joint

Fig. 9.16

It is assumed that area of flanges compressed by bolts is a hollow cylinder of outer diameter '3*d*' and inner diameter '*d*'.

$$\therefore \quad A_c = \frac{\pi}{4}(9d^2 - d^2) = 2\pi d^2$$

$$\therefore \quad k_{ch} = \frac{2\pi d^2 \times E_c}{l_{ch}};$$

$$k_g = \frac{2\pi d^2 \times E_g}{l_g}; \quad \frac{2\pi d^2 \times E_f}{l_f}$$

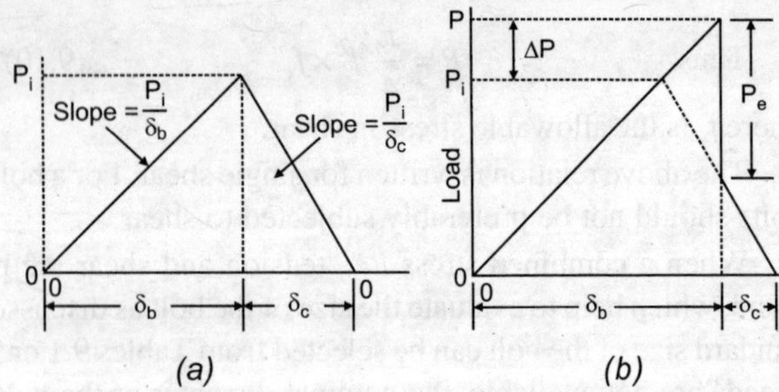

Fig. 9.17 *Elastic behaviour of a bolted joint*

The quantity *k* may have the values between zero and one, T 9.4. The resultant load is given as

$$P = P_i + k\,P_e \quad \text{...(9.12)}$$

Tables 9.4 Values of k

Type of joint	*k*
Metal to metal joint through bolts	0.00
Hard copper gasket with long through bolts	0.25
Soft copper gasket with long through bolts	0.50
Copper asbestos gasket	0.60
Soft packing with through bolts	0.75
Soft packing with studs	1.00
Narrow copper ring	0.01
Lead gasket with studs	0.10

The value of '*a*' may be estimated by the designer to obtain an approximate value of resultant load. The designer thus has control over the influence of the resultant load on a bolt by proportioning the sizes of the connected parts and bolts and by specifying initial tension in the bolt, *e.g.*, a connecting rod bolt may be tightened up with initial tension greater than the external load. The load on the bolt then will be static and the bolt may be designed on the basis of static failure. If the initial tension was not high enough, the resultant load on the bolt would be affected by the external load which is cyclic, and failure of the bolt would be on the basis of cyclic failure and governed by the endurance limit of the materials and stress concentration. This is the reason for practice in the assembly of automotive engines to tighten up connecting rod bolts so that the stress approaches the yield point.

Depending upon the resilient properties of the bolts and the connected parts, the resultant load actually acting on bolts will be within (see Fig. 9.16),

$$P = (1.1 \text{ to } 2) \times P_e$$

By virtue of above complexities, we can take

$$P = 2 \times P_e = P_e + P_e \qquad (\text{Here } P_i = P_e)$$

This relation ensures a pressure tight joint between the clamped parts.

However, the initial tightening of the bolts causes tangential stresses. Thus, from Art. 9.12, the design load on bolt will be

$$P = P_e + 1.3 \times P_e = 2.3\, P_e$$

∴ Design equation will be,

$$d_1 \geq \sqrt{\frac{4 \times 2.3 \times P_e}{\pi \times f_t}} \qquad \text{...(9.13)}$$

Thus, when initial tension is specified use eqn. 9.12, otherwise use eqn. 9.13.

9.13 DYNAMIC STRESSES IN SCREW FASTENINGS

There are many cases where bolts are subjected to suddenly applied loads, for example, the bolts holding the bearing caps on a connecting rod, in the case of a cylinder head bolt of an internal combustion engine, and tie rods used on bridges and on passenger and freight cars. It means the bolts subjected to dynamic loads should be designed not only for strength but also for their ability to absorb impact energy. Consider, for example, the rod shown in Fig. 3.13, which is subjected to a dynamic load due to the weight '*F*' falling through a distance *h*. Upon striking the plate, the rod yields and elongates a distance '*e*' in absorbing the impact energy. It is observed from the relation 3.53.

$$f_t' = \frac{F}{A} + \frac{F}{A}\sqrt{1 + \frac{2h}{e}}$$

that with all other factors remaining same, the stress f_t' becomes smaller the height the elongation '*e*' becomes larger. This observation suggests the use of stretchy material and the adoption of other expedients to increase the elongation, such as increasing the length of the rod, in order to increase the resistance to shock loads. The properties of resilience and toughness are important properties for materials under impulsive loading or shock loading.

When a bolt is subjected to shock loading the resilience of the bolt should be considered in order to prevent breakage at the thread. The resilience of a cm^3 of a material when it is stressed to the elastic limit is called the modulus of resilience and it is given by the equation (3.60) of Chapter 3,

$$U_p = \frac{f_{yp}^2}{2E} \times V$$

where $V = A \times l\ cm^3$.

If a bolt of the usual form having a full sized shank and threaded end is used to support a tensile load, the stress in the threaded part will be higher than that in the shank. If a tensile load is suddenly applied, i.e., impact loading is there, the energy absorbed by each unit volume of the bolt material will be proportional to the square of its stress at the same location (as indicated by the above equation), hence a large part of the energy will be absorbed at the region of the threaded part, making it vulnerable to failure.

In a usual bolt, the stress at the minor diameter must be kept within safe limits, it follows that the stress in the unthreaded shank will be unnecssarily low. Since this stress is less than it need be, the deformation '*e*' of the shank will not be as much as it could be. Now if the shank is reduced in size until it has the same area as the root area of the screw, the stress in the shank will be increased, the deformation of the shank will be increased, and consequently it follows that the capacity of the bolt to resist shock loading will be increased. By this the threaded portion of the bolt will be relieved of absorbing high energy. Thus the deformation of a member will be greatest (and hence the capacity to absorb impact energy will be greatest) when all its sections have the same maximum significant stress. Thus to best resist shock, a member must have the shape of uniform strength. Such a bolt is known as bolt of uniform strength. It is clear from equation 3.60 that the capacity of the bolt to resist the shock can also be increased by increasing the length of the bolt. Figure 9.18 shows a few methods making bolts of uniform strength.

Figure 9.18 (*b*) is turning down the diameter of the shank of the bolt to a diameter corresponding to that of the root diameter of the threads or sometimes even slightly less. Another way of having bolt of uniform strength is that the area of the shank is reduced to the root area by drilling a hole as shown in Fig. 9.18 (*a*), down to the threaded end.

Let d_h = diameter of the hole as shown in figure

Then,
$$\frac{\pi}{4} d_h^2 = \frac{\pi}{4} (d^2 - d_c^2)$$

Thus,
$$d_h = \sqrt{d^2 - d_c^2} \qquad \text{...(9.13)}$$

where d and d_c have the usual meanings.

The method of reducing the area in 9.18 (*b*) is cheaper, but the bolt is weaker in bending and twisting. The method of 9.18 (*a*) may be difficult to manufacture in long bolts, but there is very little weakening affect in bending or in twisting. It should be observed that the capacity of these uniform strength bolts for static tensile loads is the same as for regular bolts, so there is no reason to use these bolts except for live loads.

When a bolt is used to hold two or more members together, short lengths of original diameter must be left at each junction to serve as guide surface for the bolted parts and for shear, as shown in Fig. 9.18 (*c*).

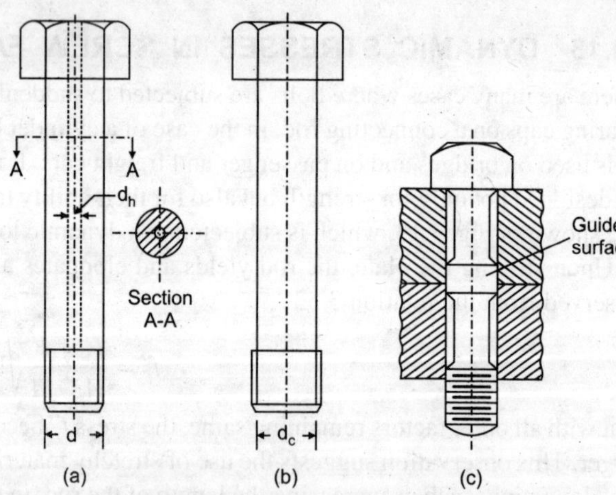

Fig. 9.18 *Uniform strength bolts*

For bolts subjected to dynamic loading, the initial load may be roughly taken as.

$$P_i = 2.5 \times \text{operating dynamic load in axial direction.}$$

9.14 BOLTS SUBJECTED TO FATIGUE LOADING

The elastic behaviour of a bolted joint subjected to fatigue loading is similar to that described in Art. 9.12, with difference that the external load P_e is varying and the increased load ΔP on the bolt will also be varying. The load deformation curve for this type of loading is shown in Fig. 9.19. The increased load ΔP varies from the value of the preload P_i to some maximum value in tension P_{max}", when the load on the joint is repeated and reversed. The mean load, P_m is given as,

$$P_m = \frac{P_{max} + P_{min}}{2} = P_i + \frac{\Delta P}{2}$$

The variable load,
$$P_v = \frac{P_{max} - P_{min}}{2} = \frac{\Delta P}{2}$$

Now writing p_m and p_v in terms of mean stress and variable stress and the cross sectional area of the bolt and using equation (4.46), the design equation can be written as,

$$A = \frac{P_m + (f_{yp} / f_e) K_f P_v}{f_{yp} / F_S} \quad \text{...(9.14)}$$

The area found by the above equation will be normally at the root of the thread, being the minimum cross-sectional area. However, because of the difference in severity of the stress concentration effects that may exist between the thread root and the fillet joining the head to the shank of the bolt, fatigue failures have been observed to occur at the bolt head. For example Peterson reports that about 15 percent of all bolt failures occur at the section in the bolt shank under the head and the rest occur at the thread root adjacent to the face of the nut in contact with the bolted member. The fatigue failure can be minimised by various means. One method is to tighten the bolted joint to produce a high pre-stress without breaking. For fatigue loaded bolted joints,

$$P_i = (2.5 \text{ to } 5) \times P_e \quad \text{...(9.15)}$$
$$\leq 0.7 \times f_{yp} \times A$$

The other method is by increasing the flexibility of the bolt so that it can be stretched enough to bring down the amplitude of the variable stress f_v. One way of increasing the flexibility of the bolt has already been discussed, that is, by using a reduced shank diameter bolt, Fig. 9.18 (*c*). Such a bolt will also be known as "anti-fatigue bolt", in which the entire bolt length between the head and the threads is reduced to the root diameter. The other ways of increasing the flexibility of the bolt and lowering the stress concentration factor K_f are given below:

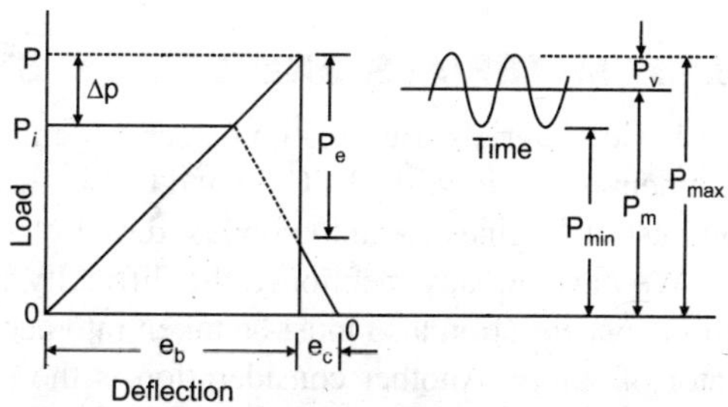

Fig. 9.19 *Load-deformation curve for a fatigue loaded bolted joint*

1. Provision of a considerable length of free thread beyond that required for engagement of the thread, Fig. 9.20 (*a*).
2. A radius between shank and head. This will help in reducing the stress concentration at this point, Fig. 9.20 (*a*).
3. The stress concentration is highest at the first threads. Therefore, the run-out angle for the thread should be small. Fig. 9.20 (*b*).
4. By providing a stress-relieving groove, Fig. 9.20 (*b*), where it is shown on the right end of the stud. The diameter of the groove should be equal to that of the root of the thread or slightly less. This groove should be merged with the shank by radii of considerable size. A large radius will cause less stress concentration in transferring the load from the larger to the smaller cross section. Also, if the stud gets bent due to misfits in assembly of parts, a smaller diameter will produce a smaller bending stress in the material.

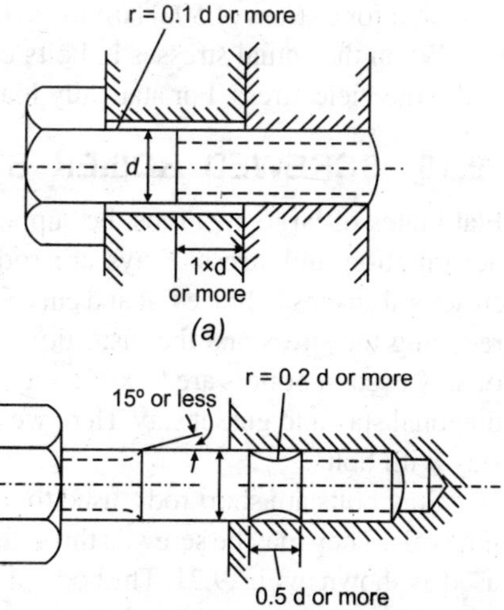

Fig. 9.20 *Methods for increasing the strength of the bolts*

As discussed in chapter 4 the endurance strength can be increased by increasing the surface finish. So, for an "anti-fatigue bolt" the threads are rolled and not cut. The usual manufacturing process is :

(*a*) Forging of bolt (*b*) Heat treatment
(*c*) Grinding to size (*d*) Thread rolling.

During thread rolling, the material is strained plastically and is workhardened, and is, therefore, stronger against both tension and fatigue, especially the latter. Increase in tensile strength is from 10 to 20% and that in fatigue strength, is from 10 to 75%.

Also, finer metric series of threads are used. Tests conducted give the following stress concentration factors. For carbon steels with metric threads K_f varies from 3 to 4.5, higher values refer to threads with $d \geq 24$ mm. With rolled threads, K_f gets reduced by 20 to 25%. For alloy steels, the respective values are 5.5 to 6.

The design of bolts subjected to fatigue-loading may also be done corresponding to static load condition like below:

For ordinary bolts,

$$P = 1.3 \text{ to } 1.6P_e$$

and
$$f_t = 0.3 f_{yp}$$

For anti-fatigue bolts,

$$P = (k + k_1)\, P_e$$

where $k_1 = 2.5$ to 5, and k is the same as discussed under Art. 9.12. The stress induced must be less than the elastic limit.

9.15 MATERIALS AND DESIGN STRESS

The basic materials used for bolts, screws and nuts are ordinary structural steels. If local or general heat treatment is required, steels (C-35 and C-45) are used. For strained joints of vital high speed parts, in internal combustion engines for automobiles, diesel-electric locomotives etc., alloy steel, are employed.

We have already mentioned the difficulty in determining the stress value caused by initial tightening. Stress concentration also must be taken into account, by either a stress concentration factor or an increased factor of safety. Another consideration is the possibility of impact loads and repeated stresses. All these considerations add up to a very large factor of safety and a very low allowable or working stress. Values of design stress for carbon steels should range from around 15.0 N/mm^2 for screws upto 12.5 mm in diameter and 30 N/mm^2 for screws upto 50 mm in diameter. For bolts more than 50 mm in diameter, bolts are generally designed for a stress of 50.0 N/mm^2 to 60.0 N/mm^2 for carbon steels and upto 100.0 N/mm^2 with alloy steels.

When the actual stresses in bolts can be fairly determined, the design stress of the bolt may be slightly under the yield stress. For statically loaded bolts, it may be taken as $f_t \leq (0.6 \text{ to } 0.8) \times f_{ypt}$.

9.16 SCREWED BOILER STAYS

Flat plates of large area must be supported by stays to reduce the deformation and stress. Stays are rods, plates, angles, or other structural shapes to brace flat and curved surfaces for the purpose of reducing the stress and the distortion. The different arrangements of stay used in boilers are flexible stay, angle brace, through stays, diagonal stay and gusset stay. Here we are concerned with through stay with bolts.

Stay bolts are short rods, used to join parallel plates at regular intervals. They may be screwed through the plates and nuts, may be used as shown in Fig. 9.21. The body of the stay (shank) is generally turned to a diameter smaller than the root diameter of the threads to reduce the local stresses in the threads. On account of unequal expansion of stayed surfaces and the relative motion between the

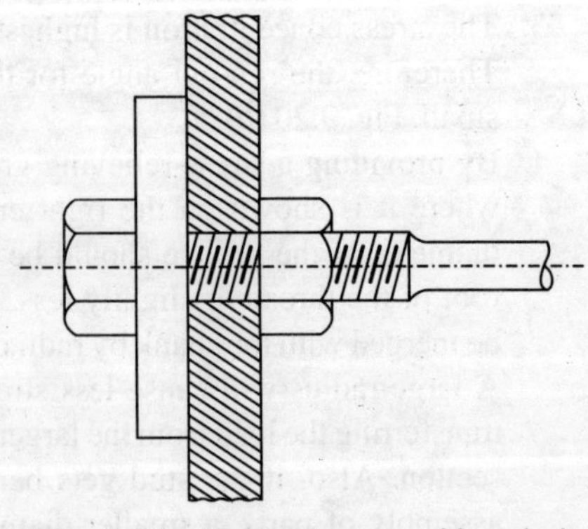

Fig. 9.21 *Through stay with a bolt and nut*

plates, stay bolts are subjected to unknown bending stresses in addition to the direct tensile stress. To reduce these bending stresses, flexible stays are used. Through stays joining surfaces at a considerable distance apart are generally spaced at least 0.4 m apart to allow a man's passing between them for inspection or for clearing the stayed vessel. Washers are used in the stays to distribute the load on a larger area of the stayed vessel.

The diameter of a stay bolt can be calculated by keeping in mind that the force on one stay bolt will be pressure acting in the stayed vessel multiplied by the area of the plate to be supported by the stay bolt under consideration.

Let d_c = core diameter of the stay bolt.
A = area of the plate supported by the stay bolt.
p = pressure of fluid acting in the stayed vessel.
f_t = design stress in tension of stay bolt.

Thus,

Force on the stay bolt supporting the plate = Tensile load on core

$$pA = \pi/4\, d_c^2 \times f_t$$

$$\therefore \quad d_c = 1.13\sqrt{\frac{pA}{f_t}} \qquad ...(9.16)$$

Once the core diameter of stay is known, the nominal diameter can be seen from Tables 9.1 or 9.2.

The following stresses for screwed stays are recommended by the board of trade.

35.0 N/mm^2 on welded iron stays.
50.0 N/mm^2 on iron stays.
65.0 N/mm^2 on steel stays.
42.0 N/mm^2 on iron stay tubes.
53.0 N/mm^2 on steel stay tubes.

SOLVED EXAMPLES

Example 9.1. What horse power can be transmitted safely by a 12.5 mm set screw employed in a pulley installation if the speed of the pulley at which it rotates is 850 rev/min and the diameter of the shaft is 37.5 mm.

Solution: From Table 9.3, the holding power or frictional resistance, F_t = 2250 N and from problem, R = 37.5/2 mm, n = 850 rev/min. Putting these values in relation (9.3), we get

$$\text{Power} = 2250 \times 2\pi \times \frac{37.5}{2\times 1000} \times \frac{850}{60} = 3753 \text{ W}$$

Thus with this screw 3753 W power can be transmitted.

We can check the diameter of the screw also by the relation 9.1.

$$d_0 = 0.125\, D + 8 \text{ mm}$$

$$0.125 \times 37.5 + 8 = 12.7 \text{ mm. Take the standard size 12.5 mm.}$$

Example 9.2. A dead load of 80 kN is to be lifted by an eye bolt. The bolt is of steel with an ultimate tensile strength of 400.0 N/mm^2. Use the metric standard coarse threads. Find the nominal diameter of the bolt.

Solution: Factor of safety for steel (supporting dead and steady load) on ultimate stress is 4.

Therefore, allowable stress in tension $= f_t = \dfrac{400.0}{4} = 100.0$ N/mm^2

Substituting the values in relation 9.4.

$$d_c = \sqrt{\frac{4\times 80000}{\pi \times 100.0}} = 32 \text{ mm}$$

Consulting Table 9.1, we find that the next higher size of core diameter is 34.093 mm. Hence accepting 34.093 mm as the standard core dia, the nominal or outside diameter

$$d = 39 \text{ mm.}$$

Example 9.3. Find the safe load which the following bolts will take in tension, assuming an allowable tensile stress of 50.0 N/mm².

M4, M8, M12, M16, M20, M24, M30 and M39

The class of tolerance used is coarse.

Solution: Safe load in *N* which the bolt can take up is given as cross-sectional area at bottom of the thread × allowable stress.

$$\therefore \quad P = 50 \times a_c$$

where a_c is the *X*-sectional area, this area can be seen from Table 9.1 for coarse series against the designation of bolt. Safe load taken by each bolt is shown below:

Designation	Cross-section area, *ac* mm²	$P = 50 \times a_c$ in N
M4	8.78	439
M8	36.6	1830
M12	84.3	4215
M16	157	7850
M20	245	12.25×10^3
M24	353	17.6×10^3
M30	561	28×10^3
M39	976	48.8×10^3

In case the bolt is not required to take any initial stress, then column third gives the force which can be applied to the bolt externally in tension.

Example 9.4. Two machine parts are fastened together tightly by means of a M24 *c* tap bolt as shown in Fig. 9.6 (*c*). If the load tending to separate these bolts is neglected, compute the stress that is set up in this bolt by the initial tightening.

Solution : M24 *c* means, a bolt of nominal diameter 24 mm having a coarse pitch of 3 mm.

From Table 9.1 see the core diameter corresponding to M24 *c*, which is 20.32 mm.

Now $d_c = 20.32$ mm and $d = 24$ mm

Initial tightening load $= P_i = 2840 \times 24 = 68 \times 10^3$ N

$$\text{Initial tightening stress} = \frac{68000}{\pi / 4 d_c^2} = \frac{68000}{\pi / 4 \times (20.32)^2} = 211.2 \text{ N/mm}^2.$$

Example 9.5. In a steam engine the maximum steam pressure is 1.0 N/mm² absolute and the back pressure is 0.014 N/mm² absolute. The cylinder diameter is 0.3 m. Determine the diameter of the screwed end of the piston rod when the allowable stress of mild steel is 45.0 N/mm² in tension.

Solution: The maximum tensile load on the piston

$$= \pi/4\, D^2 (p_1 - p_2) = \pi/4\, (300)^2\, (1.0 - 0.014) = 69.5 \text{ kN}$$

This load is to be resisted by the core area of the screwed end of piston, hence,

$$\pi/4\, d_c^2 \times f_t = 69500$$

or

$$\pi/4 \times d_c^2 \times 45.0 = 69500$$

gives,

$$d_c = 44.5 \text{ mm}$$

$$d = \frac{44.5}{0.84} = 53.5 \text{ mm}$$

Thus the nominal size of the screwed end is 53.5 mm.

Example 9.6. Two screws for a pipe hanger must hold a tensile load of 10 kN. Calculate the most suitable size. Use the metric fine series. The working stress under tension is 40.0 N/mm^2.

Solution: Load on each screw $= \dfrac{10000}{2} = 5000$ N

Minimum area at the bottom of the thread required $= \dfrac{5000}{40.0} = 125$ mm^2

From table 9.2 for fine series we adopt M14 × 1.5

Which has diameter of bolt as 14 mm and pitch as 1.5 mm.

Example 9.7. Find the diameter of the steel valve spindle. The area of the valve face may be taken as a rectangle 225 mm × 175 mm. Net pressure on the top of the valve is 0.8 N/mm^2, coefficient of friction between valve and seat is 0.2. The stress at the bottom of screw thread is not to exceed 55.0 N/mm^2 in tension. Assume fine threads.

Solution: Area of the valve face = 225 × 175 = 39.3 × 10^3 mm^2

P = Force acting normal to the valve face = 31.44 kN = 39300 × 0.8

Force of friction at right angles to the normal force = μP

$$F = 0.2 \times 31.44 \times 10^3 = 6288 \text{ N}$$

Therefore, the maximum force in the valve spindle = 6288 N

Let dc = core diameter of the screw on the spindle.

Equating the resistance of the valve spindle to the maximum force in the valve spindle,

$$\pi/4\, d_c^2 \times f_t = F$$

$$\pi/4\, d_c^2 \times 55.0 = 6288$$

gives

$$d_c = 12.1 \text{ mm}$$

From Table 9.2, the standard size of the valve spindle will be M14 × 1.5, i.e., the nominal diameter will be 14 mm.

Example 9.8. The **effective diameter of cylinder is 0.4 m. Highest pressure of steam acting on the cylinder head is 1.1. N/mm^2. Assume the allowable stress in tension of bolt material as 32 N/mm^2. Calculate the number and size of bolts required. Sketch the elevation for the arrangement. Initial tightening has been taken in the F.O.S.**

Solution: Let d_c = core diameter of bolts,

z = number of bolts used to connect the cylinder head.

P = Force acting on the cylinder head

$$= \frac{\pi}{4} D^2 \times p$$

where D is diameter of the cylinder head.

$$P = \pi/4 \times (400)^2 \times 1.1 = 148 \text{ kN}$$

Resisting force offered by the bolts = $\pi/4\, d_c^2 \times z \times f_t$

where f_t is the allowable stress in tension of bolt material.

From Table 9.1, take the standard bolt M24 × 3 c.

Core diameter, d_c = 20.32 mm

∴ Area under tension = $\pi/4\, d_c^2$ = 353 mm^2

Equating the resisting force to the force acting on the cylinder head

$$353 \times z \times 32.0 = 148000$$

$$z = 13.1$$

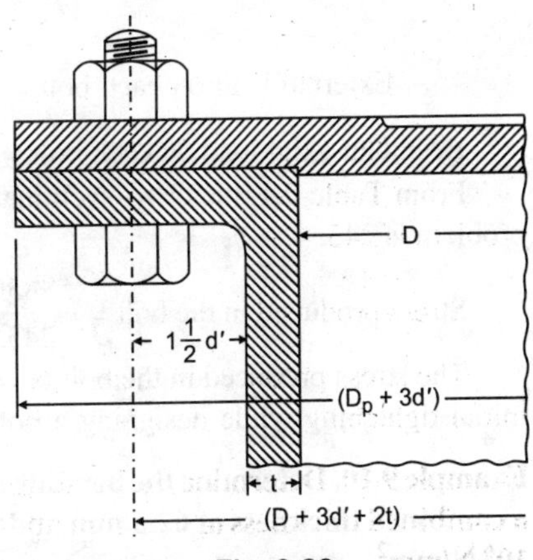

Fig. 9.22

Using next higher even number $z = 14$.

The edge distance is given as 1.5 d', where d' is the diameter of bolt hole, which is always bigger than the size of bolt. In this case take the diameter of the bolt hole as 25 mm.

∴ Edge distance = 1.5 × 25 mm = 37.5 mm

Now pitch circle diameter (P.C.D.), $D_p = D + 3d' + 2t$

where t is the thickness of cylinder. In this case assume the thickness of cylinder as 10 mm.

$$\therefore \quad D_p = 400 + 3 \times 25 + 2 \times 10 = 495 \text{ mm}$$

Now pitch of bolts on cylinder head

$$= \frac{\pi \times D_p}{z} = \frac{\pi \times 495}{14} = 111 \text{ mm.}$$

For the joint to be leak proof, the pitch of the bolts should lie between $20\sqrt{d'}$ to $30\sqrt{d'}$ when d' is in mm. The pitch is also given as ranging from 3.5 d' to 5 d'.

$$\therefore \quad 20\sqrt{d'} = 20 \times \sqrt{25} = 100 \text{ mm.}$$

$$30\sqrt{d'} = 30 \times \sqrt{25} = 150 \text{ mm.}$$

When in our case the pitch found lies between the two values, hence satisfactory.

Outer diameter of the cylinder cover

$$D_p + 3d' = 495 + 3 \times 25 = 570 \text{ mm}$$

Note: Bolts or studs on cylinder covers are screwed up tightly in order to make the joint leak proof and this tightening produces sufficient tensile stress in the bolts and sometimes breaking the bolts before the fluid pressure acts on the cylinder head. For this reason a bolt of less than 16 mm diameter in size should never be used.

Example 9.9. Pressure of steam acting on the head of cylinder is 0.7 N/mm², when the diameter of the cylinder is 0.3 m. The head is held in position by 12 bolts of size M20 × 2.5 c.

The soft copper gasket with long through bolts is used to make the joint.

Determine the stress induced in the bolt.

Solution: The resultant load on the bolt is given by relation (9.12),

$$P = P_i + kP_e$$

k from Table 9.4 for the kind of packing used = 0.5

Initial tension due to tightening of bolt is given by relation (9.5).

$$\therefore \quad P_i = 2840 \times d = 2840 \times 20 = 56.8 \text{ kN}$$

$$\text{External load on each bolt} = \frac{\pi/4 \times D^2 \times p}{z} = \frac{\pi/4 \times (300)^2 \times 0.7}{12} = 4100 \text{ N}$$

Resultant axial load on each bolt = 56800 + 0.5 × 4100 = 58850 N = 58.85 kN

From Table 9.1, the cross-sectional area at the bottom of the threads for the bolt size mentioned in the problem is 245.

$$\text{Stress produced in the bolt} = \frac{58850}{245} = 240.0 \text{ N/mm}^2.$$

The stress produced in the bolt is very high. The example shows that how important it is to consider the initial tightening while designing a bolt.

Example 9.10. Determine the bending stress induced in a M24 bolt that must hold two flanges which have a combined thickness of 62.5 mm and are so placed that the gap a in Fig. 9.13 is 0.2 mm. Take $E = 0.21 \times 10^6$ N/mm².

Solution: The bending stress induced in the bolt due to gap a is given by relation 9.8.

$$f_b = \frac{aE}{2l} = \frac{0.2 \times 0.21 \times 10^6}{62.5} = 672 \text{ N/mm}^2$$

We find that the stress produced is very high, inspite of the fact that the deviation of the surface is very small. The example shows, therefore, that how important it is to have the outside surface true.

Example 9.11. The cylinder of a stationary engine is 0.12 m in diameter and is held to the crank case by M12 × 1.75 *c*, nickel steel bolts. The maximum gas pressure in the cylinder is 3.5 N/mm². Assume the ultimate strength of this steel to be 800 N/mm² and the yield stress to be 600 N/mm². Determine the number of bolts required. Take the gasket factor as 0.75.

Solution: As the size of bolt is less than 20 mm, therefore, it is necessary to consider the initial tightening.

Initial tightening of the bolt is given by relation (9.5)

$$P_i = 2840\, d = 2840 \times 12 = 34.08 \text{ kN}$$

Stress produced due to initial tightening $= \dfrac{34080}{\pi/4\, d_c^2} = \dfrac{34080}{\pi/4\,(9.853)^2} = 500 \text{ N/mm}^2$

Available stress for the applied gas load = 600.0 – 500.0 = 100.0 N/mm².

Assuming a real F.O.S. = 2

Allowable stress to take the applied load $= \dfrac{100.0}{2} = 50.0 \text{ N/mm}^2$.

Let z = no. of bolts used to connect crank case to the cylinder.

Then load resisted by the bolts $= \pi/4\; d_c^2 \times z \times f_t$

Total load on all the bolts $= \pi/4\; (D^2) \times p_s$

where p_s is the steam pressure.

Therefore, equating the load on bolts to the load to be resisted by bolts.

$$\pi/4\; D^2 \times p_s = \pi/4\; d_c^2 \times z \times f_t$$

$$\pi/4 \times (120)^2 \times 3.5 = \frac{\pi}{4} \times (9.853)^2 \times z \times 50.0$$

gives $z = 10.4$ say 12.

Now the probable stress in each bolt

$$= 500.0 + \frac{0.75 \times \pi/4\, D^2 \times p_s}{\pi/4\, d_c^2 \times z} = 500.0 + \frac{0.75 \times (120)^2 \times 3.5}{(9.853)^2 \times 12}$$

$$= 500.0 + 33.0 = 533.0 \text{ N/mm}^2.$$

This stress is about 90 per cent of the yield stress and, therefore, satisfactory.

In order to maintain leak proof joint it is necessary to check the pitch of bolts as done in example 8.

Example 9.12. In the design of a new size reciprocating air compressor, it is proposed that eight bolts be used to fasten the head to the cylinder. The cylinder bore diameter is 75 mm, and the maximum working pressure will be 3.5 N/mm². We are asked to design the bolt for a F.O.S. of 3 under the following conditions:

(*a*) A compressible gasket seal will be used between the head and the cylinder. The gasket is made from a fibre sheet, and preliminary calculations indicated that an initial compression load of about 64 kN is required on the gasket for leak proof seal.

(*b*) An O-ring seal is used with metal to metal contact between the head and the cylinder, and a torque wrench is used to screw up the nut to give the desired initial tension load.

The stresses given for the bolt material are

$f_{yp} = 470.0$ N/mm², $f_e = 220.0$ N/mm², $f_u = 550.0$ N/mm². The material is machine finished. Stress concentration factor may be taken as $K_f = 2.0$.

Solution: Load per bolt due to maximum air pressure $= P_e$

$$P_e = \frac{\pi/4(75)^2 \times 3.5}{8} = 1920 \text{ N}$$

(*a*) The bolt is relatively rigid in compression with the joint, and the total load on the bolt will be considered to be the sum of the initial load and the external load due to the gas pressure. The unknown effects of surface finish and dimensional tolerances of the threads and the nut washer face give an unexact analysis, either analytical or experimental, of the relationship of the initial load to the wrenching torsional moment, it will be necessary to specify a value of torsional moment great enough to ensure that the gasket compression is sufficient to seal properly. The design procedure will be to design on the initial tension 30 per cent greater than that actually required.

Therefore, the initial tightening per bolt will be

$$P_i = \frac{1.3 \times 64000}{8} = 10.4 \times 10^3 \text{ N} \quad \Delta P = P_e = 1920 \text{ N}$$

The external load varies repeatedly from zero to 1920 N. Therefore, the design will be based on the equation (4.46)

$$A = \frac{P_m + (f_{yp} / f_e) K_f P_v}{f_{yp} / FS}$$

where A = stress area for bolt

$$P_m = \text{mean load} = (P_{max} + P_{min}) \frac{1}{2}$$

where $P_{max} = 10400 + 1920 = 12.32 \times 10^3$ N

$P_{min} = 10.4 \times 10^3$ N

$$\therefore \qquad P_m = (12320 + 10400) \frac{1}{2} = 11.36 \times 10^3 \text{ N}$$

$$P_v = (P_{max} - P_{min}) \frac{1}{2} = (12320 - 10400) \frac{1}{2} = 960 \text{ N}$$

Taking F.S. as 3.0,

$$A = \frac{11360 + (470.0/220.0) \times 2 \times 960}{470.0/3} = 98.5 \text{ mm}^2.$$

For coarse series, from table 9.1, the next higher stress area is 115 mm^2.

Therefore, the choice will be M14 × 2*c* standard bolt, using eight bolts to take necessary initial tension load of 10.4×10^3 N.

(*b*) **Metal to metal contact and initial tension:** In this case the bolt is relatively flexible, and the total load on the bolt, until the external air pressure load exceeds the initial tension load, will be considered to be the initial tension load. The bolt will be loaded statically and the relation becomes

$$A = \frac{FS}{f_{yp}} \times P$$

The value of torque required to give this initial load cannot be calculated exactly, it is necessary to specify a value of torque great enough to ensure that the initial tension at least three times the gas pressure load and then to provide an additional margin of safety beyond this to ensure that the bolt does not yield during tightening. The design procedure will be to specify the torsional moment required as 30 percent greater than that calculated for the desired initial tension load and the stress will be taken as 75 percent of the yield strength.

Therefore,
$$A = \frac{3 \times 1.3 \times 8000}{0.75 \times 470.0} = 21.2 \text{ mm}^2$$

From Table 9.1, the metric standard bolt will be M7 × 1*c*, this gives the nominal size of the bolt only 7 mm.

Example 9.13. A M30 bolt has an effective length of 0.65 m and is subjected to an impact load of 900 N acting through a distance of 6 mm. What is the stress induced in a standard bolt and in a bolt of uniform strength. Take $E = 0.21 \times 10^6$ N/mm².

What hole size is to be drilled to make a bolt of uniform strength?

Solution: From the knowledge of strength of materials, by equating the strain energy to the external work done, we get the relation for the stress induced in the shank of the bolt material as

$$f = \frac{2P}{A}\left[\frac{x}{e} + 1\right]$$

where P = Impact load on bolt.

A = cross-sectional area of the shank of bolt.

x = distance through which the impact load acts.

e = elongation of the bolt.

f = stress induced in the bolt material.

Now modulus of elasticity

$$E = \frac{\text{Stress}}{\text{Strain}} = \frac{f}{e/l}$$

where l = effective length of bolt.

$$\therefore \quad e = \frac{f \times l}{E} = \frac{f \times 650}{0.21 \times 10^6} = 310 f \times 10^{-6} \text{ mm.}$$

For the given size of the bolt, the cross-section of the shank of the bolt

$$= \pi/4\, d^2 = 0.7854 \times (30)^2 = 707 \text{ mm}^2.$$

On putting the values in the above relation,

$$f = \frac{2 \times 900}{707}\left[\frac{6}{310 f \times 10^{-6}} + 1\right]$$

gives $\quad f = 72.0 \text{ N/mm}^2.$

For the given size of the bolt, the cross-section of the bolt at the threaded part = 561 mm².

$$\text{Therefore, the stress at the threaded part will be} = \frac{72 \times 707}{561} = 90.8 \text{ N/mm}^2.$$

From this we can see that the bolt due to impact loading is stressed more at the threaded portion and stressed less at the bolt shank. To have bolt of uniform strength, the bolt shank area will be equal to root area of the threaded portion of the bolt. Thus the stress in the bolt will be

$$f = \frac{2 \times 900}{561}\left[\frac{6}{310 f \times 10^{-6}} + 1\right]$$

gives $\quad f = 81.0 \text{ N/mm}^2.$

This stress is more than the stress of 72 N/mm² for the standard bolt and less than the stress of 90.8 N/mm² which is at the root of the threads.

To provide the shank size of the bolt as the nominal diameter of the bolt, a hole can be drilled in the shank of the bolt as shown in Fig. 9.18 (*a*). This size of the bolt hole can be given by relation 9.13.

$$d_h = \sqrt{d^2 - d_c^2} = \sqrt{(30)^2 - (27.7)^2}$$

$$d_h = 11.6 \text{ mm.}$$

Example 9.14. The longitudinal bar stays of a short boiler are pitched at 0.375 m horizontally and vertically. The steam pressure is 1.0 N/mm². Find the size of bolts when the tensile stress of the material is 65.0 N/mm².

Solution: Area of plate supported by each stay = $375 \times 375 = 141 \times 10^3$ mm²

Load on the longitudinal stay = $1 \times 141 \times 10^3 = 141 \times 10^3$ N

Let d_c = core diameter of the stay.

Then,
$$\pi/4\, d_c^2 \times f_t = 141 \times 10^3$$
$$\pi/4\, d_c^2 \times 65.0 = 141 \times 10^3$$
$$d_c = 52.5 \text{ mm}$$

The nominal diameter of the bolt can be found by the relation 9.11.
$$d = \frac{52.5}{0.84} = 62.5 \text{ mm.}$$

Example 9.15. A steel stay for a marine boiler is 6 metre long and 80 mm in diameter. The steam pressure is 0.85 N/mm² and the area supported by the stay is square of 0.35 m. Find the stress induced in the stay taking its weight into account, when (a) the stay is fixed at the ends (b) supported at the ends. Take density of steel as 7900 kg/m³.

Solution : Core diameter of stay, $d_c = 0.84 \times 80 = 67.2$ mm.

Self weight of the stay, $F = \pi/4\, d^2 \times L \times$ density of the material $\times g = \pi/4\, (0.08)^2 \times 6 \times 7900 \times 9.81$
$= 2337.32$ N

(*a*) **Fixed at ends**

$$\text{B.M. due to weight of stay} = \frac{FL}{12} = \frac{2337.32 \times 6}{12} = 1168.66 \text{ Nm}$$

Bending stress induced due to the weight of the stay,
$$f_b = \frac{M_b}{Z} = \frac{M_b}{\pi/32\,(d_c)^3} = \frac{1168.66 \times 32}{\pi \times (0.0672)^3} = 39.23 \text{ MPa} = 39.23 \text{ N/mm}^2$$

Direct tensile stress on the stay
$$f_t = \frac{P}{A} = \frac{350 \times 350 \times 0.85}{\pi/4 \times (67.2)^2} = 29.30 \text{ N/mm}^2$$

Total stress = $39.23 + 29.30 = 68.53$ N/mm²

(*b*) **Supported at ends**

$$\text{B.M. due to the weight of stay} = \frac{FL}{8} = \frac{2337.32 \times 6}{8} = 1753 \text{ Nm}$$

Bending stress induced due to the weight of the stay,
$$f_b = \frac{M_b}{Z} = \frac{M_b}{\pi/32\,(d_c)^3} = \frac{1753 \times 10^3 \times 32}{\pi \times (67.2)^3} = 58.84 \text{ N/mm}^2$$

Direct tensile stress on the stay = 29.30 N/mm²

Total stress = $58.84 + 29.30 = 88.14$ N/mm²

Example 9.16. The diameter of the air cylinder of an air press is 200 mm and the cylinder assembly is held together by six bolts which are of the length of cylinder. The maximum operating air pressure in the cylinder is 0.9 N/mm². Experience with the previous designs indicates that the gasket must be preloaded with 20 kN forces to prevent air leakage. Suggest suitable diameter for the tension bolts which are made of steel having 540 N/mm² as the yield strength. Assume the factor of safety to be 2.7 and $E_c/E_b = a = 0.33$.

Solution: Initial load on bolts $P_i = 20000$ N

External load $P_e = \pi/4 \times D^2 \times p = \pi/4 \times (200)^2 \times 0.9 = 28520$ N

The resultant load on bolts = $P = P_i + kP_e$

where $k = \dfrac{a}{1+a} = \dfrac{0.33}{1+0.33} = 0.25$

Thus
$$P = 20000 + 0.25 \times 28520 = 27130 \text{ N}$$

$$\text{Load on each bolt} = \frac{27130}{6} = 4522 \text{ N}$$

$$\text{Allowable stress} = \frac{540}{2.7} = 200 \text{ N/mm}^2$$

$$\text{Core area} = \pi/4\ d_c^2 = \frac{4522}{200} = 22.61 \text{ mm}^2$$

From Table 9.1 the bolts selected are M7 of 1.0 mm pitch and 28.9 mm^2 stress area. The bolts are coarse series.

Example 9.17. A bolted joint is formed for a pressure vessel in which the bolt is tightened with an initial preload of 7000 N. The external load on a bolt varies from 0 to 9000 N. If $a = E_c/E_b = 0.5$, find out the size of the bolt. Take for bolt material yield stress $f_{yp} = 260$ N/mm^2, endurance limit in reversed axial loading of 130 N/mm^2. Factor of safety of 3 on Soderbergline and stress concentration factor of 2.8.

Solution: The resultant load on a bolt

$$P = P_i + kP_e$$

where
$$k = \frac{a}{1+a} = \frac{0.5}{1+0.5} = \frac{.5}{1.5} = 0.33$$

At zero external load

$$P_{min} = 7000 + 0.33 \times 0 = 7000 \text{ N}$$

At maximum value of external load

$$P_{max} = 7000 + 0.33 \times 9000 = 10000 \text{ N}$$

$$\therefore \quad f_{max} = \frac{P_{\max}}{\pi/4 d_c^2} = \frac{10000}{\pi/4 d_c^2}$$

$$f_{min} = \frac{P_{\min}}{\pi/4 d_c^2} = \frac{7000}{\pi/4 d_c^2}$$

$$\text{Mean stress on bolt} = f_m = \frac{f_{\max} + f_{\min}}{2} = \frac{8500}{\pi/4 d_c^2}$$

$$\text{Variable stress on bolt} = f_v = \frac{f_{\max} - f_{\min}}{2} = \frac{1500}{\pi/4 d_c^2}$$

Now applying Soderberg equation

$$\frac{1}{F_S} = \frac{f_m}{f_{yp}} + \frac{k_f f_v}{f_e}$$

$$\frac{1}{3} = \frac{\dfrac{8500}{\pi/4 d_c^2}}{260} + \frac{2.8 \times \dfrac{1500}{\pi/4 d_c^2}}{130}$$

$$\pi/4\ d_c^2 = 3\left[\frac{8500}{2600} + \frac{150 \times 2.8}{130}\right] = 195 \text{ mm}^2$$

From Table 9.1 (coarse series) the bolts selected are M20 of 2.5 mm pitch and 245 mm^2 stress area.

Example 9.18. The round cover of a gas tank is fastened by a group of bolts. Gas pressure, $p = 1.2 \times 10^6$ N/m². Tank diameter = 0.4 m, Bolt circle diameter, D_1 = 0.48 m. Find the diameter and the number of bolts.

Solution: This is a case of a strained pretensioned bolted joint, with subsequently imposed service load.

1. Total pressure applied to the gas tank cover, $P = \frac{\pi}{4} D^2 \times p$; D = Tank diameter
2. Effort applied to one bolt $Q = \frac{P}{z} = \frac{\pi D^2 p}{4z}$; z = Number of bolts
3. Let pitch of bolts (3.5 to 5d), $p = 4d$; d = major ϕ of bolt, $z = \frac{\pi D_1}{p} = \frac{\pi D_1}{4d}$; D_1 = Bolt circle diameter.
4. Required design load on the bolt to ensure a pressure tight joint, $F_b = 2.3Q = 2.3 \frac{\pi D^2 p \times 4d}{4\pi D_1} = 2.3 \frac{pdD^2}{D_1}$
5. Design equation is:

$$\frac{\pi d_c^2}{4} \cdot f_t = F_b = 2.3 \cdot \frac{pdD^2}{D_1}, \quad d_c = \text{minor } \phi$$

Let $\quad d \cong 1.2.d_c \; (d_c \cong 0.84d)$;

$$\frac{\pi d_c^2}{4} \cdot f_t \cong \frac{2.76\, pD^2 d_c}{D_1}$$

or

$$d_c = \frac{11.04 \times 1.2 \times 10^6 \times 0.4^2}{\pi \times 0.48 \times f_t} = \frac{1.429 \times 10^6}{f_t}$$

Let $\quad f_t = 80$ MPa

$$\therefore \quad d_c = \frac{1.429 \times 10^6}{80 \times 10^6} = 0.0178625 \text{ m}$$

This corresponds to 'd' lying in between (M20 × 2.5) and (M22 × 2.5).

Let us choose bolt diameter, d = (M22 × 2.5) for which d_c = 18.933 mm and f_t = 73 MPa which is < 80MPa, ∴ safe.

$$\therefore \quad z = \frac{\pi D_1}{4d} = 17$$

Use z = 16, as it is easier to divide the circumference into 16 parts than into 17.

EXERCISES

1. Discuss the advantages offered by the threaded joints.
2. In which cases the use of threaded joints is not recommended?
3. Explain the terms used in the case of threads for root, flank, crest, pitch, depth of thread, lead, major diameter, minor diameter, pitch diameter and thread angle.
4. How do you co-relate pitch with lead?
5. How do you co-relate pitch with number of threads?
6. Discuss the factors to be considered while selecting a thread profile.
7. What are the advantages and disadvantages of V-threads over square threads ? Discuss them in detail.
8. Why Acme threads are preferred over square threads?
9. In which case fine threads are preferred over coarse threads?
10. Distinguish among through bolt, carriage bolt, tap bolt, stud bolts and set screws.
11. How do you distinguish between tap bolt and cap screws?
12. When the necessity of a locking arrangement is felt?

13. Which are the different stresses set up in a bolt due to initial tightening? Explain these stresses in detail.
14. How do we take into account the initial tightening, when the external forces are acting on a bolt?
15. Discuss the bolt of uniform strength in detail.
16. An electric motor of 40 kN is to be lifted by an eye bolt. The allowable stress in tension of the bolt material is 55 N/mm^2. If the bolt is not initially stressed, design the eye bolt completely using coarse threads.
17. Calculate the load that can be taken by a tie rod, when the size of the rod is M 24 × 3 c. The allowable stress in tension is 45 N/mm^2.
18. Find the diameter of the screwed end of a piston rod. Diameter of the piston is 0.5 m, pressure of steam is 0.7 N/mm^2. The design stress for the material of the piston rod is 42 N/mm^2. Take the taper of the piston rod in the piston as 1 : 10. Sketch the arrangement.
19. Determine the size of bolt required for the head of 0.3 m cylinder containing steam at 1.4 N/mm^2. Assume a hard gasket used in making up the joint. Assume that the steam enters under the head of the bolt circle of diameter 0.35 m and number of bolts used is 12. Take the yield stress in the bolts as 600.0 N/mm^2.
20. In a steam engine the maximum steam pressure is 1.0 N/mm^2 absolute and the back pressure is 0.014 N/mm^2 absolute. The cylinder diameter is 0.3 m. The contact surfaces of the head and cylinder are ground together so that no packing is necessary, what must be the diameter of the bolt, so that the maximum stress in the bolt necessary to ensure a steam tight joint will not exceed 55.0 N/mm^2 ? Assume number of bolts used 8.
    If steel studs are used and rubber gasket 3 mm. thick be placed between the contact surfaces, what must be diameter of the studs?
21. Determine the safe useful load for a M 36 × 4 c bolt, 110 mm long, made of mild steel, for three conditions.
    (*a*) Static load.
    (*b*) Sudden change of load from zero to maximum.
    (*c*) Impact action resulting from a variable load when the nut is unscrewed one-sixteenth turn.
    The factor of safety may be taken as 1.5 on elastic stress. Stress concentration factor may be taken as 2.85.
22. Determine the diameter of the end of the piston rod of a double acting steam engine. The diameter of the cylinder bore is 0.45 m, the maximum steam pressure is 1.15 N/mm^2 and the steam is exhausted to the atmosphere. The diameter of the piston rod is 50 mm. The initial tightening may be taken as 25 percent of the tensile force due to steam pressure. Taking the loading as static and yield stress in the material of the piston rod as 350.0 N/mm^2. Factor of safety on yield stress 2.
23. What will be the efficient dimensions of bolts fastening the cover of a vessel under a working pressure of 0.5 N/mm^2? This working pressure varies from zero to 0.5 N/mm^2. The inside diameter of the cylinder bore is 1.0 m. Take E for steel as 0.21×10^6 N/mm^2 and the number of bolts as 20. The joint is made by a 4 mm gasket between the flanges.
24. Plates of a lap joint are fastened together by cap screws with fully threaded shanks. The joint must hold a tensile load of 120 kN. Calculate the number of M 20 × 1.5 *f* cap screws needed to ensure against a shearing failure of the screws. Use a design stress for shear in the screws of 17.5 N/mm^2.
25. D-slide valve of a non-condensing steam engine working at steam pressure of 1.0 N/mm^2 gauze has an effective area of 200 × 100 mm. Design the valve rod diameter which is threaded at the end to connect it to the valve. Give a sketch showing this end of the valve rod. The allowable design stress in the rod may be taken as 25 N/mm^2. The value of the coefficient of friction may be taken as 0.15.
26. Find the diameter of the copper screwed stays in a boiler. Each stay supports an area equal to square of 125 mm. Sides. Pressure of steam in the boiler is 1.25 N/mm^2. Safe tensile stress in the material of the stay is 35.0 N/mm^2.
27. In a boiler each bar stay supports an area 0.3 m × 0.3 m of the flat end plate. What pressure of steam within the boiler will produce a stress of 35 N/mm^2 on a wrought iron stay of M24 × 3*c*?
28. What horse power can be safely transmitted by two M16 × 2c set screws employed in a pulley installation if the rotative speed of the pulley is 250 rev/min and the diameter of the shaft is 62.5 mm?
29. A generator weighing 14 kN is furnished with an eye bolt in the housing for lifting purposes. Assume the bolt to be made of steel with metric coarse threads. Determine:
    (*a*) the required bolt diameter.
    (*b*) the depth to which the bolt should extend into the cast iron motor housing.
30. A 0.3125 m by 0.325 m by 325 rev/min diesel engine has a compression pressure of 2.25 N/mm^2 and a maximum explosion pressure 3.7 N/mm^2. The head is held in place by eight bolts M14 × 2c, arranged on a 0.4 m. bolt circle. A thin asbestos gasket is used.
    (*a*) Determine the stress due to the applied load.
    (*b*) Determine the stress due to tightening.

CHAPTER 10

Eccentric Loading of Riveted, Welded and Threaded Joints

10.1 ECCENTRIC LOADING OF RIVETED JOINTS

In general, the line of application of the load should pass through the centre of gravity of the rivet areas. In many cases the load acting on foundation plates fixed by means of rivets and brackets fixed to the walls or steel structures by means of rivets, is not always acting through the centre of gravity of the rivets system. This type of loading causes direct shear known as ''primary shear'' and in addition is supplemented by a ''secondary shear'' caused by the tendency of the force to twist the joint about the centre of gravity. A connection of eccentrically loaded joint is illustrated in Fig. 10.1 (*a*). The group of four rivets supports the load P with an eccentricity e. The actual forces resulting would be difficult to determine because of the influence of the stiffness of the bracket and the stiffness of the member to which the bracket is riveted. A readily solution is possible if following assumptions are made.

(a) Bracket and member (wall or stanchion) are assumed to be absolutely rigid.

(b) All deformations or relative movement of the bracket to the members takes place as a result of shear deformation of the rivets.

The first step in the solution of the problem is to locate the centre of gravity of the rivet group as shown in figure by 0. The next step is

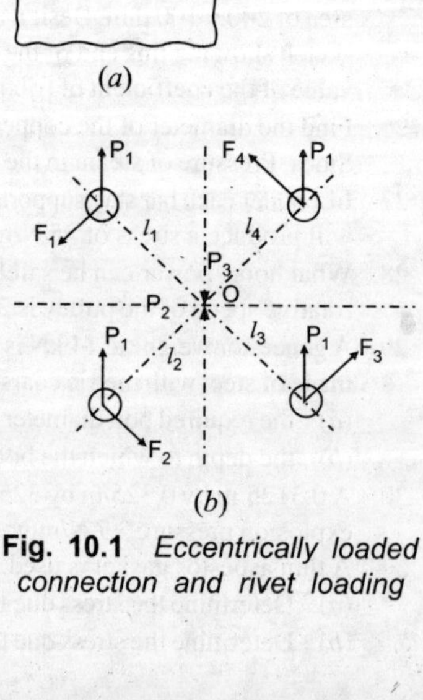

Fig. 10.1 *Eccentrically loaded connection and rivet loading*

to introduce at the centre of gravity of the rivet group two forces which are equal and opposite to P, as shown in Fig. 10.1 (*b*) P_2 and P_3. The result is a couple formed by P_2 and P, equal to P_e which tends to rotate the rivets about the centre of gravity of the group.

Now the primary shear on each rivet produced by $P_3 = P$, is the same for each rivet, and is

$$P_1 = \frac{P}{4} \qquad \text{...(10.1)}$$

The turning moment will induce loads on rivets 1, 2, 3 and 4 which are of different magnitudes and directions. The direction of the secondary shear on each rivet is at right angles to the direction of its moment arm l, as shown in Fig. 10.1 (*b*). Thus F_1, F_2, F_3 and F_4 are the secondary shear in rivets 1, 2, 3 and 4 respectively.

From the condition that the sum of the external turning moment and of the internal resisting moment equals zero, we have

$$Pe = F_1l_1 + F_2l_2 + F_3l_3 + F_4l_4$$

From the assumptions stated above, it follows that as a result of the turning of the bracket, deformation of each rivet will be proportional to its distance l from the centre of gravity of the group. Let l_1, l_2, l_3 and l_4 be the distances of rivets 1, 2, 3 and 4 from the centre of gravity respectively.

Then,

$$\frac{F_2}{F_1} = \frac{l_2}{l_1}, \frac{F_4}{F_1} = \frac{l_4}{l_1}, \frac{F_3}{F_1} = \frac{l_3}{l_1}$$

$$F_2 = F_1\frac{l_2}{l_1}, F_4 = F_1\frac{l_4}{l_1}, F_3 = F_1\frac{l_3}{l_1}$$

By putting the values of F_2, F_3 and F_4 in above relation, we get,

$$Pe = F_1l_1 + F1\frac{l_2}{l_1} \times l_2 + F_1\frac{l_3}{l_1} \times l_3 + F_1 \times \frac{l_4}{l_1} \times l_4$$

or

$$Pe = \frac{F_1}{l_1}\left[l_1^2 + l_2^2 + l_3^2 + l_4^2\right]$$

or

$$F_n = \frac{P_e \times l_n}{(l_1^1 + l_2^2 + l_3^2 + l_4^2)} \qquad \text{...(10.2)}$$

where the subscript n refers to any rivet.

The secondary shear loads are shown in Fig. 10.1 (*b*) as F_1, F_2, F_3 and F_4 along with primary shear load P_1 on each rivet.

The force tending to shear off each rivet is the resultant of the primary and secondary shear which can be found graphically as shown in Fig. 10.2. In the solution of the problem, the primary and secondary shear loads are laid off to scale and then the resultant load found out. The heavily loaded rivet will be one in which the included angle between the primary and secondary shear is the least.

The resultant shear can also be calculated by the relation

$$R = \sqrt{P_1^2 + F^2 + 2P_1 \times F \times \cos\theta} \qquad \text{...(10.3)}$$

where θ is the angle between primary and secondary shear.

Generally, the rivet having the maximum total shear load will be apparent by inspection, therefore, the value of the resultant load for that rivet may be calculated. The maximum loaded rivet becomes the critical one for determining the strength of an existing connection if all the rivets are to be of the same size, or for selecting the rivet size.

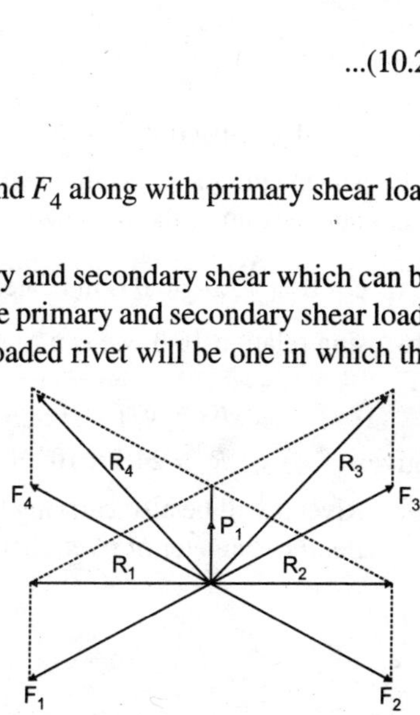

Fig. 10.2 *Resultant shear load*

SOLVED EXAMPLES

Example 10.1. Find the diameter of the rivet in joint shown in Fig. 10.3. Maximum shear stress in the rivet is limited to 60.0 N/mm². The maximum load carried by the bracket is 42 kN at an eccentricity of 200 mm. Also find the bearing stress if the thickness of the plate is 10 mm.

Solution: Direct shear on each rivet acting vertically downwards = $\dfrac{42000}{6} = P_1 = 7000$ N.

The centroid of the rivet group is at 0. The distance of rivets 3 and 6 from centre of gravity of rivets group is 50 mm. The distance of rivets 1, 2, 4 and 5 from centroid = $\sqrt{(50)^2 + (75)^2} = 90$ mm.

Due to eccentricity of 0.2 m, load has a tendency to turn the plate in clockwise direction about the centroid 0.

This turning moment = $42 \times 1{,}000 \times 200 = 8400.0$ Nm

The turning moment is resisted by six rivets as shown in Fig. 10.3. Let F_1, F_2, F_3, f_4, F_5 and F_6 be the secondary shear due to turning in rivets 1, 2, 3, 4, 5 and 6 placed at l_1, l_2, l_3, l_4, l_5 and l_6 respectively from the centroid.

Then $l_1 = l_2 = l_4 = l_5 = 90$ mm. $l_3 = l_6 = 50$ mm.

Now equating the resisting moment to turning moment

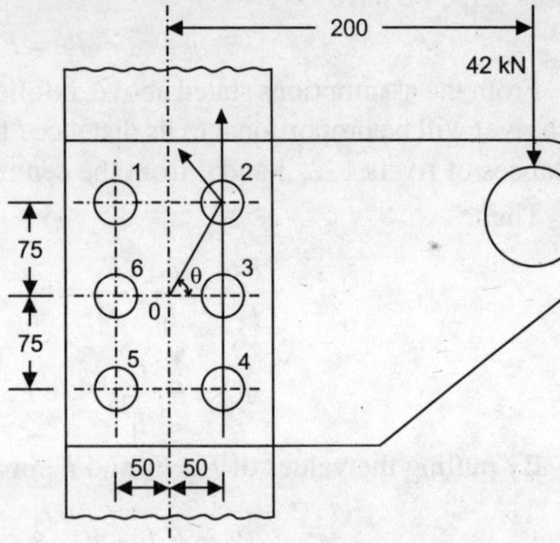

Fig. 10.3

$$\frac{F_1}{l_1}(l_1^2 + l_2^2 + l_3^2 + l_4^2 + l_5^2 + l_6^2) = P \times e$$

$$\frac{F_1}{90}(4 \times 90^2 + 2 \times 50^2) = 8400 \times 10^3$$

gives $\quad F_1 = 20 \times 10^3$ N

By inspection we can say that $F_1 = F_2 = F_4 = F_5 = 20 \times 10^3$ N

Since $\dfrac{F_1}{l_1} = \dfrac{F_3}{l_3} = \dfrac{F_6}{l_6}$, $\therefore F_3 = F_1 \times \dfrac{l_3}{l_1} = 20 \times 10^3 \times \dfrac{50}{90} = 11.1 \times 10^3$ N

Also by inspection $F_3 = F_6 = 11.1 \times 10^3$ N.

The magnitude of resultant will not be necessarily same for rivets 1, 2, 4 and 5, because the resultant load also depends upon the included angle.

$\cos\theta = \dfrac{5}{9}$, angle θ is indicated in the figure. The maximum resultant force will be on rivet 2. Therefore, by using relation 10.3, we can find R_2.

$$\therefore \quad R_2 = \sqrt{P_1^2 + F_2^2 + 2 \times P_1 \times F_2 \times \cos\theta} = \sqrt{(7000)^2 + (20000)^2 + 2 \times 7000 \times 20000 \times 5/9}$$

gives $\quad R_2 = 24.5 \times 10^3$ N

Rivet 5 will be also carrying the same resultant force.

Let d = diameter of rivet, then

$$\frac{\pi}{4} \times d^2 \times f_s = R$$

$$\frac{\pi}{4} \times d^2 \times 60.0 = 24.5 \times 10^3$$

gives $\quad d = 22.8$ mm

From Table 7.8, the standard size of the rivet will be 24 mm and size of rivet hole will be 25.5 mm. Bearing stress in the rivet.

$$f_b = \frac{F}{d \times t} = \frac{24500}{24 \times 10} = 102.0 \text{ N/mm}^2.$$

Example 10.2. Find the diameter of the rivet in arrangement of the joint shown in Fig. 10.4. Maximum shear stress in the rivet is limited to 65 N/mm². The maximum load carried by the joint is 109 kN. Assume that all the rivets are of the same size.

Solution: The direct, or primary shear force on each rivet = $\frac{109}{4}$ = 27.25 kN. This force will be acting in the direction of 109 kN. The centre of gravity of the rivet group will be at 0 as shown in the figure. The line of action of the load passes through a distance e from the centre of gravity of the rivet group. Therefore,

$$e = 112.5 \sin 30 = 56.25 \text{ mm.}$$

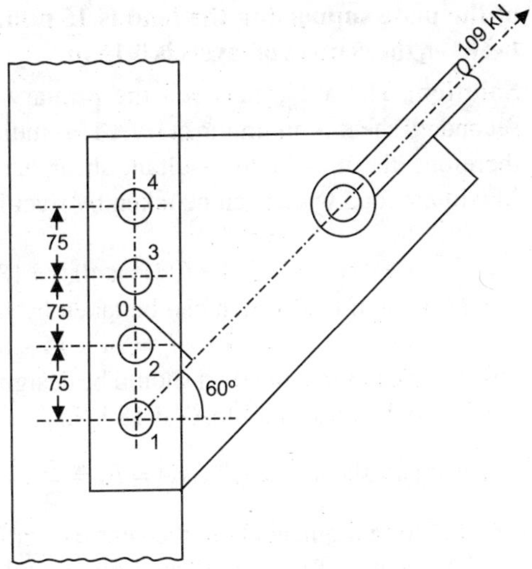

Fig. 10.4

Hence, the twisting moment about the centre of gravity of the rivet group will be

$$= 109 \times 10^3 \times 56.25 = 615 \text{ kN. cm.} = 6150 \text{ Nm.}$$

This moment will be resisted by the combined moments of the forces acting on the rivets 1, 2, 3 and 4 at distances l_1, l_2, l_3 and l_4 from the centroid.

$$6150 \times 10^3 = \frac{F_1}{l_1}(l_1^2 + l_2^2 + l_3^2 + l_4^2) = \frac{F_1}{112.5}[2 \times (112.5)^2 + 2 \times (37.5)^2]$$

gives $\quad F_1 = 24700 \text{ N} = 24.7 \text{ kN}$

The joint is assumed to be rigid, therefore the force F_1, F_2, F_3 and F_4 will be proportional to their distance from the centre of gravity.

$$\therefore \quad F_1 = F_4 = 24700 \text{ N}$$

$$F_2 = F_3 = \frac{l_2}{l_1} \times F_1 = \frac{37.5}{112.5} \times 24700 = 8230 \text{ N} = 8.23 \text{ kN}$$

Now the resultant load on rivets will be given by the relation 10.3.

$$R = \sqrt{P_1^2 + F^2 + 2P_1F\cos\theta}$$

in rivet 1, $\quad R_1 = \sqrt{(27.25)^2 + (24.70)^2 + 2 \times 27.25 \times 24.70 \times \cos 60} = 45.0 \text{ kN}$

in rivet 2, $\quad R_2 = \sqrt{(27.25)^2 + (8.23)^2 + 2 \times 27.25 \times 8.23 \times \cos 60} = 31.8 \text{ kN}$

in rivet 3, $\quad R_3 = \sqrt{(27.25)^2 + (8.23)^2 - 2 \times 27.25 \times 8.23 \times \cos 60} = 24.50 \text{ kN}$

in rivet 4, $\quad R_4 = \sqrt{(27.25)^2 + (24.70)^2 - 2 \times 27.25 \times 24.70 \times \cos 60} = 26.00 \text{ kN}$

So, the maximum resultant shear occurs in the rivet 1. Let d is the diameter of rivet, then

$$\pi/4\, d^2 \times f_s = R$$

$$\pi/4 \times d^2 \times 65.00 = 45000, \text{ gives } d = 30.9 \text{ mm.}$$

From Table 7.8, the standard size of the rivet will be 33 mm and the rivet hole size 35 mm.

Example 10.3. Find the maximum load that can be carried by the joint shown in Fig. 10.5, the load is put at an eccentricity of 200 mm. Maximum allowable shear stress in the rivet is 40 N/mm² and bearing stress in rivet is 65 N/mm². The diameter of the rivet is 24 mm and thickness of the plate supporting the load is 15 mm. The distance between the centres of rivets is 0.15 m.

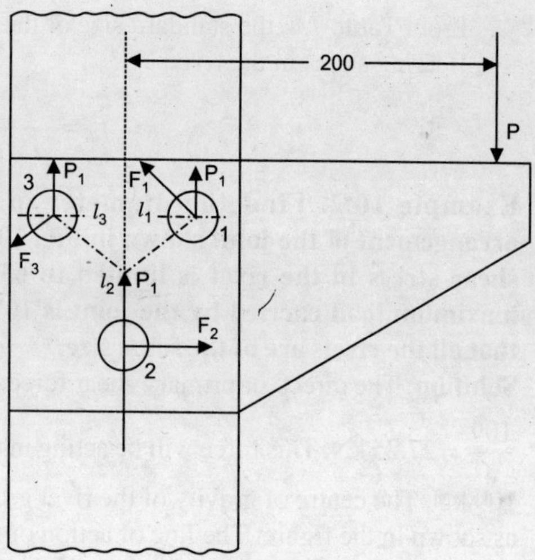

Fig. 10.5

Solution: The angle between the primary shear and the secondary shear is minimum at rivet 1, as indicated in figure, therefore, the maximum resultant shear occurs in rivet 1.

Maximum load which can be taken by rivet in shear

$$= \pi/4\, d^2 f_s$$
$$= \pi/4 \times (24)^2 \times 40.0 = 18 \text{ kN}$$

Maximum load which can be taken by rivet in bearing

$$= d \times t \times f_b = 24 \times 15 \times 65.0 = 23.3 \text{ kN}$$

which indicates that the rivet should be designed under shear to take up the maximum load of 18 kN.

Primary shear on each rivet $= P_1 = \dfrac{P}{3}$, where P is the maximum load that can be carried by the joint.

Twisting moment about the centre of gravity of the rivet group $= P \times 0.2$ Nm.

The centre of gravity of rivet group will be at 0 and its distance from each rivet will be $= l_1 = l_2 = l_3 = \dfrac{75}{\cos 30} = 86.5$ mm. Let F_1 be the force in rivet 1, due to twisting moment, then equating the twisting moment to the resisting moment of all rivets, we have

$$P \times 200 = \frac{F_1}{l_1}(l_1^2 + l_2^2 + l_3^2)$$

$$P \times 200 = \frac{F_1}{86.5}[3 \times (86.5)^2]$$

$$F_1 = \frac{P \times 200 \times 86.5}{3 \times (86.5)^2} = 0.77\text{ P}$$

Now the resultant load on the rivet will be given as

$$R_1 = \sqrt{P_1^2 + F_1^2 + 2P_1 \times F_1 \cos\theta}$$

$$18 \times 10^3 = \sqrt{\left(\frac{P}{3}\right)^2 + (0.77\,P)^2 + 2 \times \frac{P}{3} \times 0.77P \times \cos 30.}$$

gives, $\quad P = 16.9 \times 10^3$ N.

Example 10.4. A bracket is to be attached to a wall with the help of six rivets. The different arrangements in which the bracket can be attached to the wall with these rivets are shown in Fig. 10.6. The maximum allowable stress in shear is 60 N/mm². Determine the way in which the rivets should be arranged so that the design is economical. The bracket is required to support a load of 90 kN with an eccentricity of 200 mm.

Solution: As the number of rivets in each case is six, therefore the primary shear in each rivet, in every arrangement will be the same.

Therefore, $\quad P_1 = \dfrac{P}{6} = \dfrac{90 \times 10^3}{6} = 15 \times 10^3$ N

Now the secondary shear in each arrangement can be determined separately.

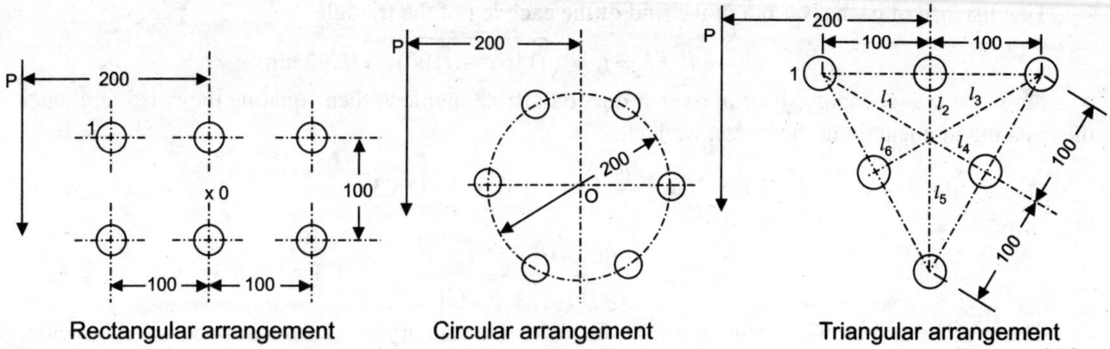

Fig. 10.6 *Arrangements*

Rectangular arrangement: The centroid of the rivet group is shown in figure by 0. The distance of corner rivets from centroid of rivet groups is,

$$= \sqrt{100^2 + 50^2} = 111.5 \text{ mm}$$

Now using the relation 10.2, we can find the secondary shear in a rivet which is farthest from centroid.

$$\therefore \qquad F_1 = \frac{90 \times 10^3 \times 200 \times 111.5}{4 \times (111.5)^2 + 2 \times (50)^2} = 39.6 \times 10^3 \text{ N}$$

The maximum load will occur in rivet in which the primary and secondary shear force include the minimum angle. Thus the maximum resultant force will occur in rivet 1, marked in figure.

Included angle between the two forces

$$\tan\theta = \frac{50}{100} = 0.5, \theta = 27°$$

Hence the resultant shear force in the rectangular arrangement will be given by relation 10.3.

$$R = \sqrt{P_1^2 + F_1^2 + 2P_1 \times F_1 \times \cos\theta} = \sqrt{(15000)^2 + (39.6 \times 10^3)^2 + 2 \times 15000 \times 39.6 \times 10^3 \times \cos 27}$$
$$= 53.3 \times 10^3 \text{ N.}$$

Circular arrangement: The centroid of the rivet group is shown in figure by 0. The distance of all the rivets from centroid of the rivet group is 100 mm. Using the relation 10.2, we can find the secondary shear, which will be equal in all the rivets in this arrangement.

$$\therefore \qquad F = \frac{90 \times 10^3 \times 200 \times 100}{6 \times (100)^2} = 30 \times 10^3 \text{ N.}$$

The maximum load will occur in a rivet in which the primary and secondary shear force include the minimum angle. Thus the maximum force will occur in rivet 1, marked in figure. For this the included angle is zero and direction of primary and secondary shear is same.

Therefore, the maximum resultant shear

$$= 15 \times 10^3 + 30 \times 10^3 = 45 \times 10^3 \text{ N}$$

Triangular arrangement: The centre of gravity of the rivet group will be at 0 as shown in figure and its distance form each rivet put at the corner will be

$$l_1 = l_3 = l_5 = \frac{100}{\cos 30} = 116 \text{ mm.}$$

The distance of each rivet put at the mid of the each leg of the triangle

$$= l_2 = l_4 = l_6 = \sqrt{(116)^2 - (100)^2} = 58.3 \text{ mm.}$$

Let F_1 be the secondary shear in rivet 1, due to twisting moment, then equating the twisting moment to the resisting moment of all the rivets, we have

$$90 \times 10^3 \times 200 = \frac{F_1}{l_1}[3 \times (116)^2 + 3 \times (58.3)^2]$$

$$F_1 = \frac{90 \times 10^3 \times 116 \times 200}{3 \times (116)^2 + 3 \times (58.3)^2} = 41.3 \times 10^3 \text{ N.}$$

The maximum load will occur in a rivet in which the primary and secondary shear force include the minimum angle. Thus the maximum force will occur in rivet 1, marked in figure, and this angle will be 30º. Therefore, the resultant load will be

$$R = \sqrt{P_1^2 + F_1^2 + 2P_1 \times F_1 \times \cos\theta} = \sqrt{(15 \times 10^3)^2 + (41300)^2 + 2 \times 15 \times 10^3 \times 41.3 \times 10^3 \times \cos 30}$$

$$= 54.7 \times 10^3 \text{ N.}$$

By comparing the loading in the above arrangement of rivets, we find that the placing of six rivets in circular arrangement is most economical, because the rivet size will be least in this case and triangular arrangement of putting the rivets is the worst. Hence the circular arrangement will be adopted.

Now, let d be the diameter of the rivet, then

$$\pi/4\, d^2 \times f_s = 45 \times 10^3$$

$$\pi/4\, d^2 \times 60.0 = 45000$$

gives, $d = 31$ mm.

From Table 7.8, the standard size of the rivet will be 33 mm and hole size corresponding to the rivet size will be 35 mm.

EXERCISES

1. Differentiate between 'primary shear' and 'secondary shear' in rivets.
2. Under what assumptions the design of eccentrically loaded riveted joints is done?
3. The operating arm of a brake shoe is attached to the brake shoe by six rivets arranged in two parallel rows. The distance between the rows is 75 mm, and the pitch in each row is 62.5 mm. A load of 2400 N is applied at a distance of 0.6 m from the centre of gravity of rivets. Determine the rivet diameter if the shear stress is not to exceed 50.0 N/mm^2. Also sketch the arrangement along with the brake shoe.

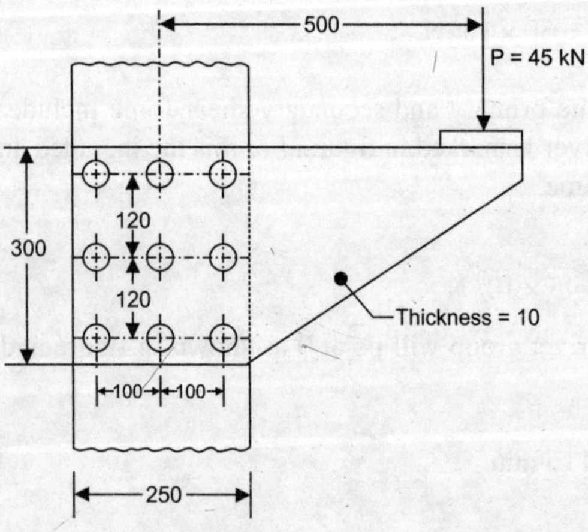

Fig. 10.7

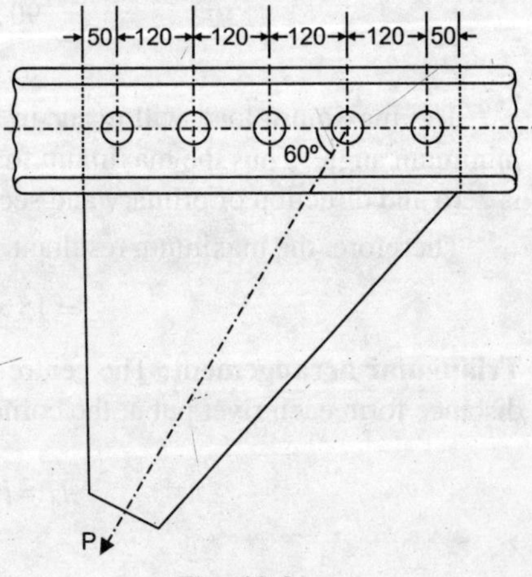

Fig. 10.8

4. The bracket in Fig. 10.7 is to carry a load of 45 kN at an eccentricity of 0.5 m from the centroid of the rivet group. If the rivets are 16 mm in diameter, what is the maximum shearing stress induced in the rivets? Also, find the bearing stress induced in the rivets, if the thickness of the plates is 10 mm.
5. Figure 10.8 shows the arrangement of a riveted joint. The rivets are 20 mm in diameter. What load may be applied at P if the allowable shear stress is not to exceed 70.0 N/mm^2?
6. Figure 10.9 shows riveted joint with an eccentric load of 15 kN. The rivets are 24 mm in diameter. Find the maximum shear stress induced in the rivet. If the thickness of the plate is 15 mm, determine the crushing stress on the rivet.

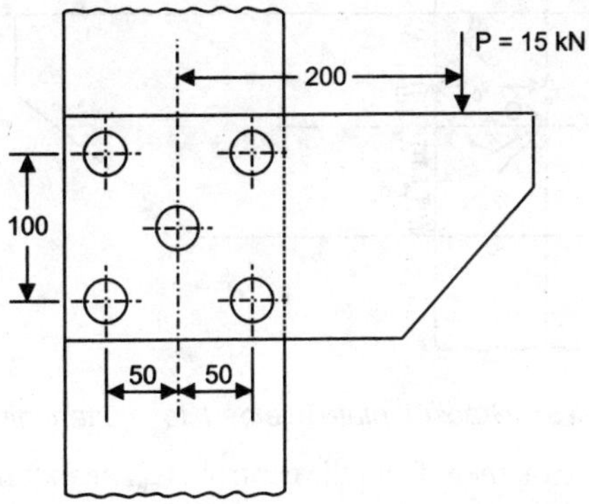

Fig. 10.9

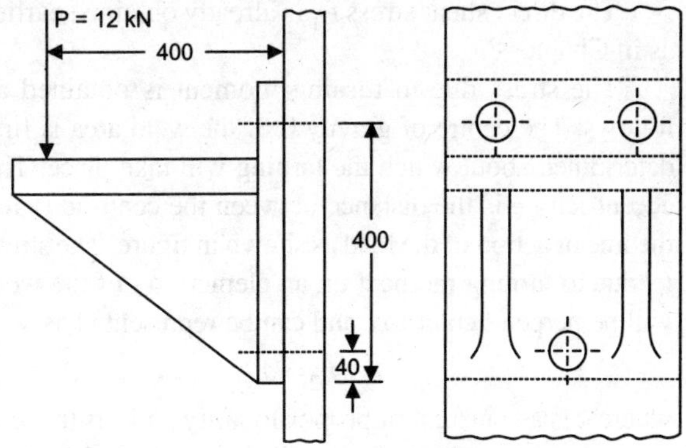

Fig. 10.10

7. A bracket is attached to a column by means of three rivets as shown in Fig. 10.10. Calculate the size of the rivets used if the permissible tensile and shear stress are 75 N/mm^2 and 55 N/mm^2 respectively. The eccentric load on the bracket is 12 kN at an eccentricity of 0.4 m. The distances from the centre of rivets to the edge of turning are mentioned in figure.
8. A bracket is attached to a column by means of five rivets as shown in Fig. 10.11. Find the diameter of the rivet for the riveted connection, if the maximum shear stress is not to exceed 90.0 N/mm^2 in the most heavily loaded rivet.

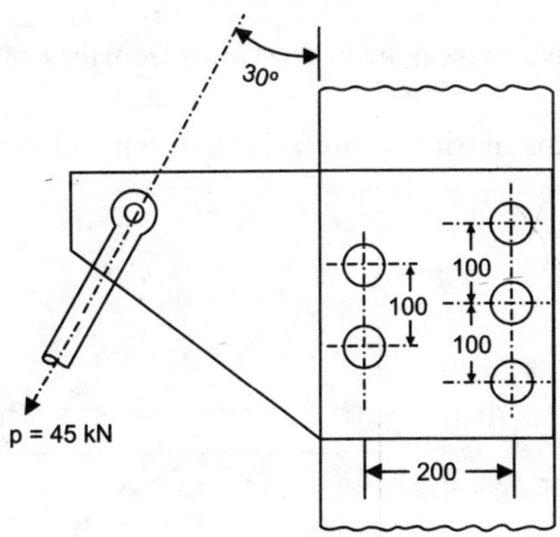

Fig. 10.11

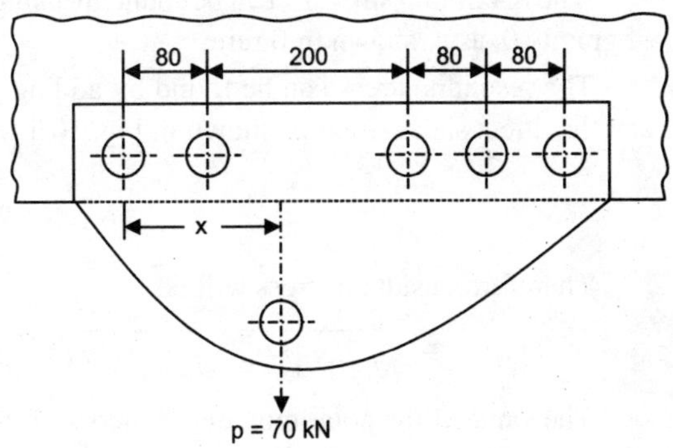

Fig. 10.12

9. A plate is attached to a girder by means of five rivets as shown in Fig. 10.12. Determine the distance x which permits the application of 70 kN load. The size of the rivets is 20 mm and the allowable shear stress in the rivets is 90.0 N/mm^2.

10.2 ECCENTRICALLY LOADING OF WELDED JOINTS

A general case of eccentric loading is shown in Fig. 10.13. In eccentric load of welded joints, two types of stresses are produced. One of these will be direct shear stress produced, which may be assumed to be uniformly distributed over the weld pattern and other is the shear stress caused by the turning moment. The state of stress in such a joint is complicated and in order to determine the value of the significant stress even approximately, it is necessary to assume that the torsional shear stress at any point is proportional to its distance from the centroid of all welds.

The direct shear stress f_{s1} is already obtained earlier as in Chapter 8.

The stress due to turning moment is obtained as follows: The centre of gravity 0 of the weld area is first determined about which the turning will take place. The eccentricity e is the distance between the centroid 0 and the line of action of the load as shown in figure. The stress f_{s2} due to turning moment on an element δA of the weld will be perpendicular to r and can be represented as

$$f_{s2} = kr$$

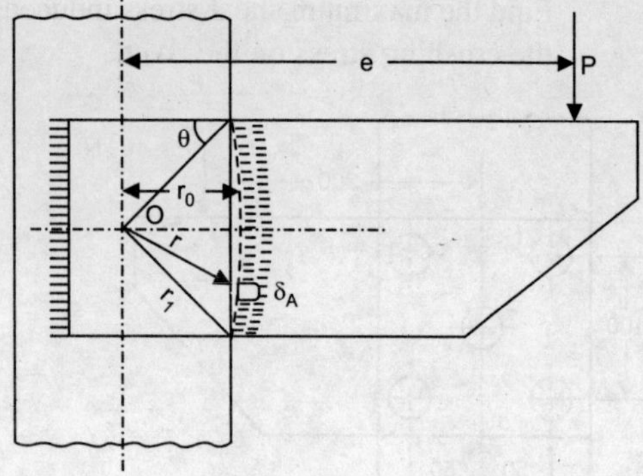

Fig. 10.13 *Bracket plate welded to a stanchion*

where k is a constant of proportionality and r is the distance of the element from the centroid 0. The external turning moment $M_t = P \times e$ will be equal to the torsional resistance of the element and the integral of the torsional resistance taken over the whole weld area in the joint. Thus,

$$M_t = P \times e = \int f_{s2} \delta A r$$

or
$$M_t = k \int r^2 \delta A = kJ \qquad \text{...(10.4)}$$

where J is the polar moment of inertia of the weld with respect to the centre of gravity 0.

The stress in any element can now be found by the relation

$$f_{s2} = \frac{M_t}{J} r \qquad \text{...(10.5)}$$

The maximum stress f_{s2} can be found by using for r the distance to the point farthest away from the centre of gravity 0, as r_1 shown in figure.

The resultant stress can be found by adding geometrically the maximum stress due to turning moment and the direct shear stress as shown in Fig. 10.14.

$$\cos \theta = \frac{r_0}{r_1}$$

Therefore, resultant stress will be

$$f_s = \sqrt{f_{s1}^2 + f_{s2}^2 + 2 f_{s1} f_{s2} \cos \theta} \qquad \text{...(10.6)}$$

The value of the polar moment of inertia of the weld about the centre of gravity of the weld system is given by parallel axis theorem, as,

$$J' = J_0 + A r_0^2 \qquad \text{...(10.7)}$$

where J_0 is the polar moment of inertia of the weld about the centre of gravity of the weld considered, and r_0 is the distance between the c.g. of the weld under consideration and the c.g. of the weld systems.

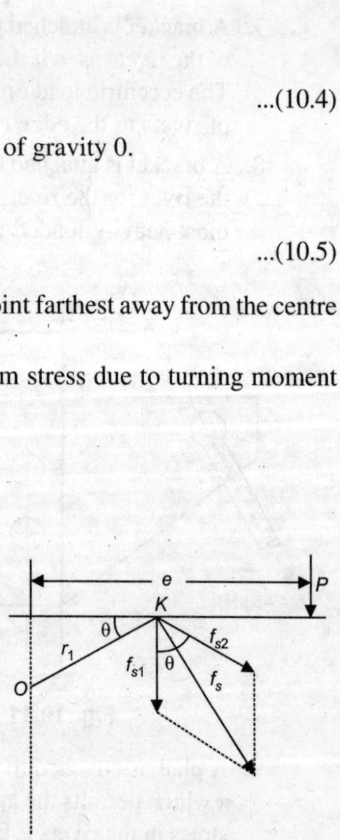

Fig. 10.14

In Fig. 10.13 the value of J_0 is given by

$$J_0 = \frac{Al^2}{12}$$

$$\therefore \quad J' = \frac{Al^2}{12} + Ar_0^2 = A\left[\frac{l^2}{12} + r_0^2\right] \qquad ...(10.8)$$

The sum of all these polar moment of inertia of the welds will give the numerical value of J to be used in relation 10.5.

Values of J for several combinations of eccentric loaded fillet welds are given in the Table 10.1.

Table 10.1 Values of J for fillet welds. Leg of weld = t

S. No.	Type of weld	Moment of inertia of weld J cm^4
1.	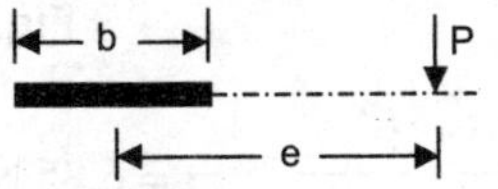	$\frac{tb^3}{12}$
2.	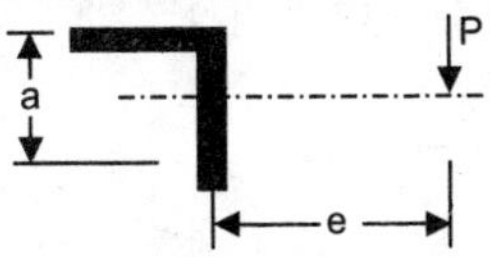	$\frac{ta^3}{12}$
3.	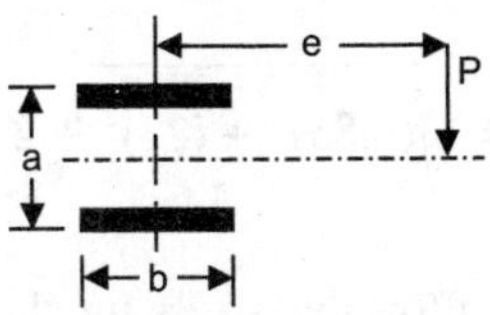	$\frac{tb(3a^2 + b^2)}{6}$
4.	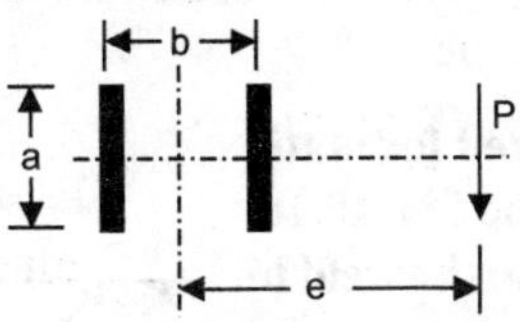	$\frac{ta(a^2 + 3b^2)}{6}$
5.	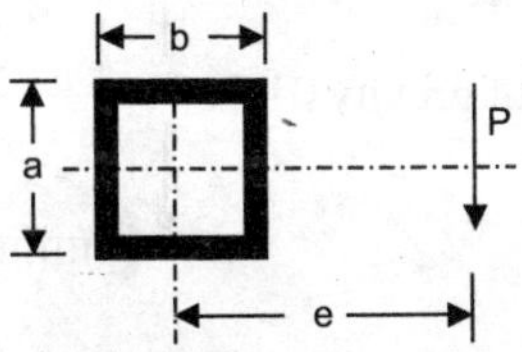	$\frac{t(a+b)^3}{6}$

SOLVED EXAMPLES

Example 10.5. Determine the maximum stress in the reinforced weld of the bracket plate shown in Fig. 10.15. Assume that the load varies from zero to the maximum value.

Solution: Effective length of each weld = 150 – 12.5 = 137.5 mm.

For each weld $r_0 = 50$ mm

Throat area of each weld = 6 × 137.5 × 1.2 × 0.7 = 730 mm^2

where 1.2 factor takes reinforcement of the weld.

$$\text{Polar moment of inertia of each weld} = A \times \left(\frac{l^2}{12} + r_0^2\right) = 730\left(\frac{(137.5)^2}{12} + (50)^2\right) = 297 \times 10^4 \text{ mm}^2$$

Total moment of inertia for both the welds $= 297 \times 10^4 \times 2 = 594 \times 10^4$ mm^4

Turning moment $= P \times e = 10 \times 10^3 \times 162.5 = 1625$ Nm

The distance r_1 to the point farthest away from the centre of gravity

$$r_1 = \sqrt{\left(\frac{137.5}{2}\right)^2 + (50)^2} = 85 \text{ mm}$$

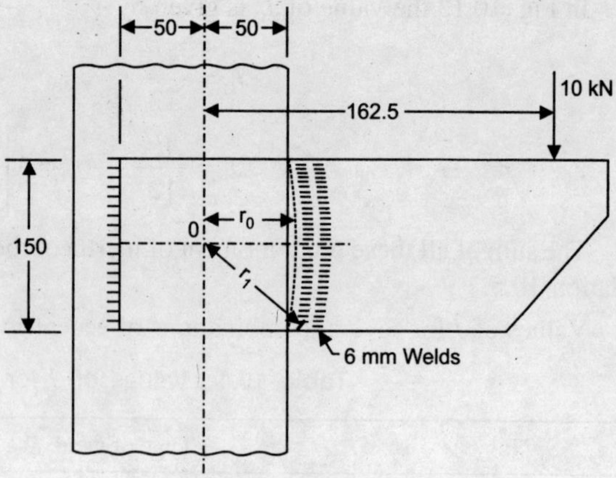

Fig. 10.15

Hence the maximum stress due to the turning moment, by relation 10.5

$$f_{s2} = \frac{1625 \times 10^3}{594 \times 10^4} \times 85 = 23.0 \text{ N/mm}^2$$

The direct stress is given by

$$f_{s1} = \frac{P}{2 \times 0.7t \times l \times 1.2} = \frac{10 \times 10^3}{2 \times 0.7 \times 6 \times 137.5 \times 1.2} = 6.85 \text{ N/mm}^2$$

$$\cos\theta = \frac{r_0}{r_1} = \frac{50}{85}$$

Therefore, the resultant stress will be

$$f_s = \sqrt{f_{s1}^2 + f_{s2}^2 + 2 f_{s1} f_{s2} \cos\theta} = \sqrt{(6.85)^2 + (23)^2 + 2 \times 6.85 \times 23.0 \times \frac{50}{85}}$$

$$= 27.5 \text{ N/mm}^2.$$

Which is with in the safe limits of 70.0 N/mm^2. The allowable stress for shear is mentioned in Table 8.3 of Chapter 8 for load varying from zero to maximum value.

Example 10.6. Determine the size of the weld required for static loading in the eccentrically loaded weld as shown in the Fig. 10.16. The effective length of the weld is 12.5 mm shorter than the weld to allow for starting and stopping of the weld.

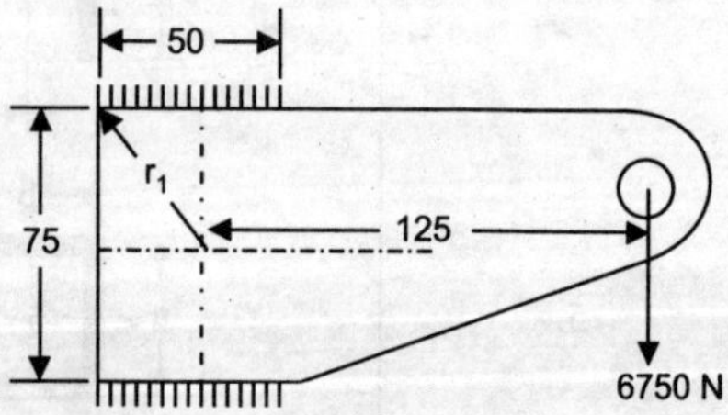

Fig. 10.16

Solution: The polar moment of inertia about the centre of gravity 0 is

$$J = \frac{bt(3a^2 + b^2)}{6}$$

where t is the leg of the weld to be determined.

$$J = \frac{50t[3 \times (75)^2 + (50)^2]}{6} = 161 \times 10^3 \times t \text{ mm}^4$$

The distance r_1 to the point farthest away from the centre of gravity

$$r_1 = \sqrt{(37.5)^2 + (25)^2} = 45 \text{ mm.}$$

We will determine the size of the weld excluding the reinforcement. The direct stress is given by

$$f_{s1} = \frac{P}{2 \times 0.7t \times l} = \frac{6750}{2 \times 0.7 \times 50 \times t} = \frac{96.4}{t} \text{ N/mm}^2$$

The max. stress due to the turning moment, by relation 10.2

$$f_{s2} = \frac{M_t r_1}{J}$$

The turning moment = $M_t = Pe = 6750 \times 125$

$$f_{s2} = \frac{6750 \times 125}{161 \times 10^3\, t} \times 45 = \frac{235}{t}\ \text{N/mm}^2.$$

The direct shear stress is always directed parallel to P. The secondary shear stress is maximum at the four corners of the weld and is in a direction perpendicular to its radius r_1 drawn from the centre of twist to the corner as indicated by point K in Fig. 10.14.

$$\cos\theta = \frac{r_0}{r_1} = \frac{37.5}{45} = 0.833$$

$$f_s = \sqrt{(f_{s1})^2 + (f_{s2})^2 + 2 f_{s1} f_{s2} \cos\theta}$$

using an allowable stress from Table 8.3 equal to 75 N/mm²

$$\therefore \qquad 75.0 = \sqrt{\left(\frac{96.4}{t}\right)^2 + \left(\frac{235}{t}\right)^2 + 2 \times \frac{96.4}{t} \times \frac{235}{t} \times 0.833}$$

gives $\qquad t = 4.25$ mm

Therefore, size of the weld is 4.25 mm.

Example 10.7. Calculate the size of the weld required for an eccentrically loaded weld as shown in Fig. 10.17. The allowable stress in the weld is 75.0 N/mm².

Solution: First of all the position of the point 0 (centre of gravity of the welds) will be determined by assuming the size of the weld as t.

Taking moments about the vertical left hand welded edge. Let a be the distance of centre of gravity from left edge. Then,

$$a = \bar{X} = \frac{A_1 x_1 + A_2 x_2 + \ldots}{A_1 + A_2 + \ldots}$$

$$a = \frac{50t \times 2 \times 25 + 75 \times t \times 0}{50t \times 2 + 75t} = 14.4 \text{ mm}$$

$\bar{y} = 37.5$ mm (by symmetry)

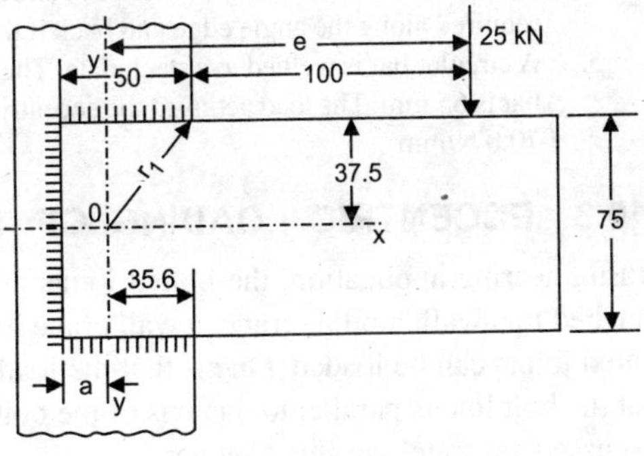

Fig. 10.17

Eccentricity = $e = 100 + 50 - 14.4 = 135.6$ mm.

Moment of inertia about x-axis = I_{xx}

$$I_{xx} = 2 \times 50 \times (37.5)^2 \times t + \frac{1}{12} t \times (75)^3$$

$$= 175 \times 10^3 . t \text{ mm}^4$$

Moment of inertia about y-axis = I_{yy}.

$$I_{yy} = 2 \times \frac{t \times (50)^3}{12} + 2 \times t \times 50 \times (25 - 14.4)^2 + 75 \times t \times (14.4)^2$$

$$= 47.5 \times 10^3\, t \text{ mm}^4$$

Polar moment of inertia = $J = I_{xx} + I_{yy} = 175 \times 10^3 t + 47.5 \times 10^3 t = 222.5 \times 10^3\, t$ mm⁴

Turning moment of the load = $25 \times 10^3 \times 135.6 = 3390$ Nm

The distance r_1 to the point farthest away from the centre of gravity

$$r_1 = \sqrt{(37.5)^2 + (35.6)^2} = 50.8 \text{ mm}$$

$$\cos\theta = \frac{35.6}{50.8} = 0.704$$

The direct shear stress is given by

$$f_{s1} = \frac{P}{0.7(2\times t\times 50 + t\times 75)} = \frac{25000}{0.7(100t + t\times 75)} = \frac{204}{t}\ \text{N/mm}^2$$

Maximum stress due to turning moment

$$f_{s2} = \frac{3390\times 10^3}{222.5\times 10^3\, t}\times 50.8 = \frac{774}{t}\ \text{N/mm}^2$$

Resultant stress is given by

$$f_s = \sqrt{(f_{s1})^2 + (f_{s2})^2 + 2 f_{s1}\times f_{s2}\cos\theta}$$

$$75 = \sqrt{\left(\frac{204}{t}\right)^2 + \left(\frac{774}{t}\right)^2 + 2\times\frac{204}{t}\times\frac{774}{t}\times .704}$$

gives $t = 12.4$ mm.

Therefore, the size of the weld will be 12.4 mm.

EXERCISES

1. What is the assumption made in the design of the welded joint of an eccentrically loaded welded joint?
2. Derive the relation to find out the resultant stress in an eccentrically loaded welded joint.
3. A 150 × 100 × 12.5 mm angle is welded to its support by two fillet welds. A load of 20 kN is applied normal to the gravity axis of the angle at a distance of 0.38 m from the centre of gravity of the welds. Determine the maximum shearing stress in the welds, assuming each weld to be 75 mm in length and parallel to the axis of the angle.
4. A 100 × 100 × 10 mm angle is used as a strut and is subjected to a load of 200 kN. Determine the length of the welds required along the angle edge and back if the maximum shear stress in the fillet welds is to be 30 N/mm^2.
5. A circular bar is welded to a steel plate. The bar acts as a cantilever of length 0.25 m. The diameter of the circular bar is 65 mm. The load acting at the eccentricity is 8000 N. Determine the size of the weld if the allowable stress is 70.0 N/mm^2.

10.3 ECCENTRIC LOADING OF THREADED JOINTS

In Engineering application, the bolted joints are many times subjected to eccentric loading, e.g., a bracket attached to a wall, a pillar crane, a wall crane joint, etc. In general there are two possible ways by which the bolted joints can be loaded. One is that the load line is perpendicular to the axis of the bolts and the other is that the bolt line is parallel to the axis of the bolts. The procedure for design of bolts in both the cases will be discussed separately in this Chapter.

Case (a) Loading parallel to the axis of bolt: The bracket of Fig. 10.18 is to be fastened to a rectangular base by means of four bolts as shown. Each bolt will be under a direct tensile load of $P_1 = P/z$, where z is the number of bolts, no matter where the line of action of P is situated.

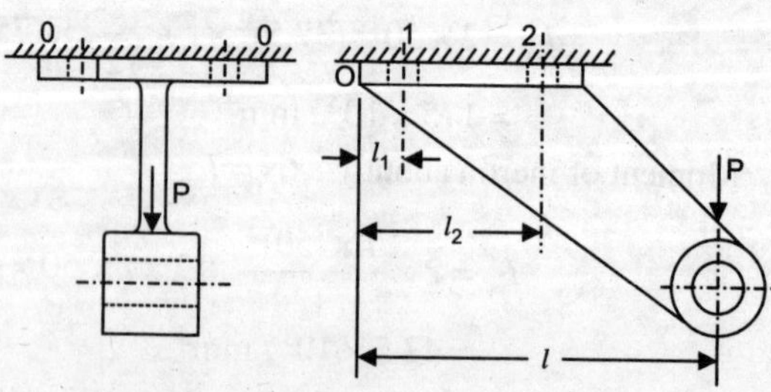

Fig. 10.18 *Bracket fastening*

The load P will tend to rotate bracket about the edge OO as shown in figure. Each bolt will be under a further tensile load due to this turning tendency of bracket. The elongation and hence the stress in a bolt will be proportional to its distance form OO, so that if all the bolts have equal cross-sectional areas, the load in a bolt will also be proportional to its distance from the turning edge OO.

When the flange is heavy, it may be considered to be a rigid body and all deformations or relative movement of the bracket to the members takes place as a result of deformation of bolts. The procedure for finding out the tensile force in the bolts due to turning will be as outlined below.

First of all fix the distance l_1 and l_2 from the turning edge OO to the centre of bolt line 1 and 2 respectively. Then turning moment due to load $(P) = P \times l$.

Now let w be the load due to turning in a bolt at unit distance from OO. Then in a bolt at distance l_1 from edge OO, the load in it will be wl_1, and the moment of this load about the edge OO will be $wl_1 \times l_1 = wl_1^2$. There are two bolts situated at bolt line 1, therefore, the resisting moment of these two bolts will be $2\,wl_1^2$. Similarly the resisting moment of the two bolts situated at the bolt line 2 will be $2wl_2^2$.

Now equating the moments of turning and resisting about the edge OO, we get

$$2wl_1^2 + 2wl_2^2 = P \times l \qquad \text{...(10.9)}$$

From this the load in a bolt at unit distance can be calculated.

The most heavily loaded bolts are those at the greatest distance from the edge of turning. In the bracket of Fig. 10.18, the bolts at line 2 are heavily loaded, due to eccentricity of the load P. Thus the load in a bolt at bolt line 2 will be wl_2 and at bolt line 1 will be wl_1.

Thus the total tensile load in the most heavily loaded bolt is

$$= P_1 + P_2 \qquad \text{... (10.10)}$$

where $P_2 = wl_2$, the load in a bolt at line 2.

Now for the load $(P_1 + P_2)$ the bolt may be designed.

Case (b) Loading perpendicular to the axis of the bolt: The bracket of Fig. 10.19 is to be fastened to a flange by four bolts as shown. There will be direct shear load on the bolts instead of direct tensile load as was in case (*a*). This shear load may be considered to be equally distributed among all the bolts if they are fitted in reamed holes and distributed between two bolts if rough bolts are used in clearance holes. Thus in the above example assuming that the bolts are fitted in reamed holes, the average shear load coming on a bolt will be given as

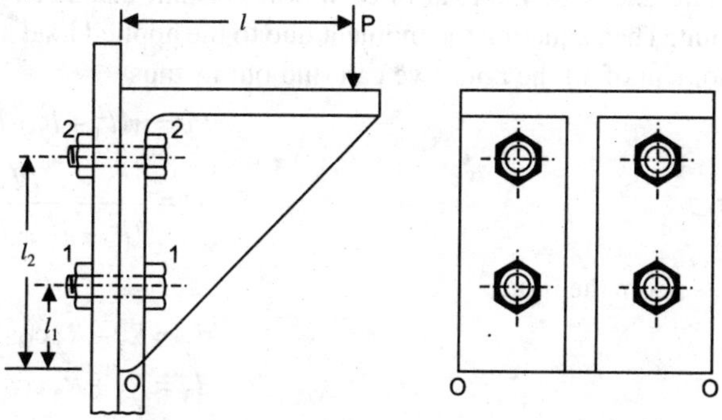

Fig. 10.19 *Bracket fastening to a flange*

$$F_s = \frac{P}{z} \qquad \text{...(10.11)}$$

Due to eccentricity the turning moment about the edge OO may be equated to the resisting moment of the bolts from turning edge. For this the relation 10.9 may be used which will help us to calculate the load (w) on a bolt per unit distance from the turning edge. From this we find that the bolts in line 2—2 are heavily loaded.

Therefore, the load on a bolt which is heavily loaded is

$$F_t = wl_2 \qquad \text{...(10.12)}$$

By knowing the shear and tensile loads on the bolt, the equivalent tensile and shear loads may be calculated.

$$F_{et} = \tfrac{1}{2}F_t + \tfrac{1}{2}\sqrt{(F_t)^2 + 4(F_s)^2} \qquad \text{...(10.13)}$$

where F_{et} is the equivalent tensile load.

and

$$F_{es} = \tfrac{1}{2}\sqrt{(F_t)^2 + 4(F_s)^2} \qquad \text{...(10.14)}$$

where F_{es} is the equivalent shear load.

Once we know the equivalent loads, then the size of bolt can be calculated with the help of relation 9.4 and check under shear load. For this the design stress may be seen from table for the bolt material.

In the bolts subjected to shear, the plane of shear should never be across the threaded portion of the shank. In the best designs, the diameter of the shank is made slightly larger than the threaded portion in order to avoid the injury to the threads by bearing pressure. When the bolt is subjected to both tension and shear loads, the root diameter can be approximated from the tension load, and the diameter of the shank may be approximated from the shear load. A diameter slightly larger than that required for either tension or shear can then be assumed, and the stresses due to combined tension and shear can be checked from relations (3.42) and (3.43).

In general, the bolts are not subjected to shear loads. Where it can not be avoided, the dowel pins are used to take the shear stresses as shown in Fig. 10.20 (*b*). Here the bolts should be designed to carry the tensile load only and the shear load will be taken by the dowel pins.

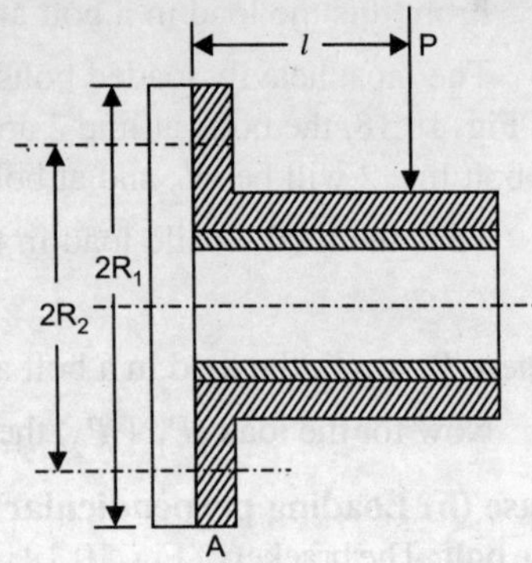

Case (c) Eccentric loading on the circular base: A machine member is frequently made with a circular base fastened by bolts located on a circle, e.g., a pillar crane, a round flange bearing of a machine tool, etc. The design procedure will be illustrated by a round flange bearing with four bolts as shown in Fig. 10.20. In the figure R_1 represents radius of the column flange and R_2 the radius of the bolt pitch circle. The numbering of the bolts has been done in the figure as 1, 2, 3 and 4. Let l_1, l_2, l_3 and l_4 be the distances of the bolt centres from the turning point A as shown in figure and w be the load in each bolt per unit distance from the turning point. Then equating the moment due to the applied load and the resisting moment of all the bolts, we can find out w, thus

$$Pl = w(l_1^2 + l_2^2 + l_3^2 + l_4^2)$$

or

$$w = \frac{Pl}{(l_1^2 + l_2^2 + l_3^2 + l_4^2)} \quad ...(10.15)$$

From the figure

$$l_1 = R_1 - R_2 \cos\theta, \quad l_2 = R_1 + R_2 \sin\theta$$

$$l_3 = R_1 + R_2 \cos\theta, \quad l_4 = R_1 - R_2 \sin\theta$$

On substituting these values in relation 10.15, we get

$$w = \frac{Pl}{4R_1^2 + 2R_2^2} \quad ...(10.16)$$

Load in bolt situated at $l = \dfrac{Pl(R_1 - R_2 \cos\theta)}{4R_1^2 + 2R_2^2}$

This force will be maximum when $\cos\theta$ is a minimum or when $\cos\theta = -1$, then $\theta = 180°$. It means the bolt placed at $\theta = 180°$ will be maximum stressed in tension.

$\therefore$ Maximum force $= \dfrac{Pl(R_1 + R_2)}{2(2R_1^2 + R_2^2)}$ ...(10.17)

If z is the number of bolts used, we can write the general expression for the load as

$$= \frac{2Pl(R_1 - R_2 \cos\theta)}{z(2R_1^2 + R_2^2)} \quad ...(10.18)$$

Maximum load in a bolt, when number of bolts used is z

$$F_{max} = \frac{2Pl(R_1 + R_2)}{z(2R_1^2 + R_2^2)} \quad ...(10.19)$$

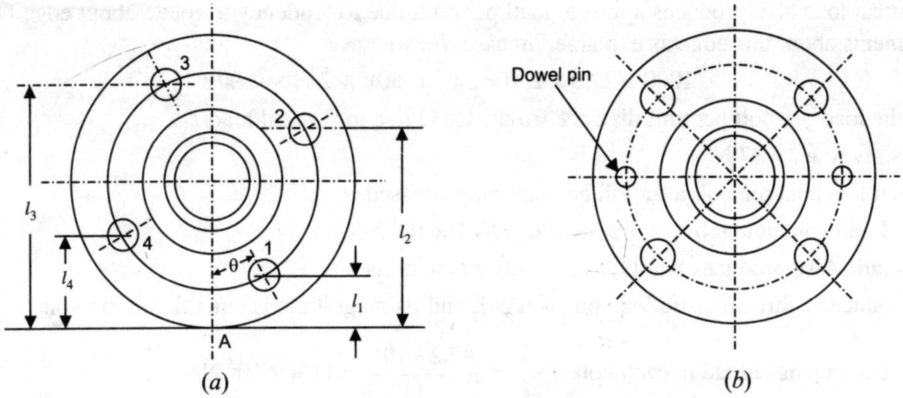

Fig. 10.20 *Fastening of a flangening bearing*

Relation 10.19 gives the absolute maximum value and should be used if the direction of the load P changes with relation to the bolts, as in the case of the base of a pillar crane. If the direction of the load P is fixed, the maximum load on the bolts can be reduced by locating the bolts so that two of them will be equally stressed, this is shown in Fig. 10.20 (*b*). In such a case, the relation 10.18 for maximum load will become

$$= \frac{2Pl\left[R_1 + R_2 \cos\left(\frac{180}{z}\right)\right]}{z(2R_1^2 + R_2^2)} \qquad ...(10.20)$$

Once the maximum load in a bolt is known, the size of the bolt can be calculated by the relation 9.4. The bolts should be relieved of shear stress by using dowel pins as shown in Fig. 10.20 (*b*).

Case (d) Eccentric loading acting in the plane containing the bolts: See Fig. 10.1. The method of analysis is exactly like that given under Art. 10.1.

Note: The methods discussed for the eccentric loading of riveted joints can also be used for bolted assemblies.

SOLVED EXAMPLES

Example 10.8. The hinged connection shown in Fig. 10.21 carries a load of 54.6 kN. All components are made from steel. Determine the size of the bolts. The allowable stresses in tension and shear are 75.0 N/mm² and 60.0 N/mm² respectively. Use the coarse threads.

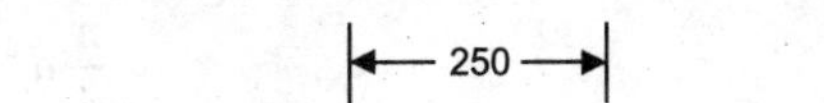

Fig. 10.21

Solution: The applied load can be resolved into two components.

Horizontal component $= T_h = 54.6 \cos 30$
$= 47.2$ kN

V ertical component $= T_v = 54.6 \sin 30$
$= 27.3$ kN

Four bolts are used to make the connection, two bolts in row A and two in row B.

Vertical load: Produces two types of loads on the bolts.

1. Shear load on each bolt

$$\frac{T_v}{4} = \frac{27.3 \times 10^3}{4} = 6.825 \times 10^3 \text{ N}$$

2. Vertical load also produces a tensile load per bolt, due to tendency to rotate about edge O. Taking moments about this edge as explained in case (*b*), we have

$$1000 \times 27.3 \times 250 = 2w_1 \times (50)^2 + 2w_1 \times (300)^2$$

where w_1 is the load per bolt per unit distance from edge O due to vertical load T_v.

This gives, $w_1 = 37$ N

Under this load the bolt at *A* will be maximum stressed.

$\therefore$ Load on a bolt at line $A = 37 \times 300 = 11.1 \times 10^3$ N

Horizontal load: Also produces two types of loads on the bolts.

1. It produces a direct tensile load on each bolt and its magnitude is equal in all the bolts.

$$\therefore \text{ Direct tensile load in each bolt} = \frac{T_h}{4} = \frac{47.2 \times 10^3}{4} = 11.8 \times 10^3 \text{ N}$$

2. Horizontal load also produces a tensile load per bolt, due to its tendency to rotate about edge 0. Taking moments about this edge as explained in case (*a*)

$$47.2 \times 1{,}000 \times 390 = 2w_2 \times (50)^2 + 2w_2 \times (300)^2$$

where w_2 is the load per bolt per unit distance from the edge 0 due to horizontal load T_h.

This gives, $w_2 = 99$ N.

The load due to turning will also produce the maximum load in the bolts at line A.

$\therefore$ Maximum load on a bolt at line $A = 99 \times 300 = 29.7 \times 10^3$ N.

Therefore, greatest tensile load on a bolt (bolts at line *A*) is,

$$F_t = 11.1 \times 10^3 + 11.8 \times 10^3 + 29.7 \times 10^3 = 52.6 \times 10^3 \text{ N}$$

Combine this load with the shear load 6.825×10^3 N to find out the equivalent tensile and shear loads. The equivalent loads can be found out by relations 10.13 and 10.14.

$$F_{et} = \frac{1}{2}F_t + \frac{1}{2}\sqrt{F_t^2 + 4F_s^2} = \frac{1}{2} \times 52.6 \times 10^3 + \frac{1}{2}\sqrt{(52.6 \times 10^3)^2 + 4(6.825 \times 10^3)^2} = 53.45 \times 10^3 \text{ N}$$

$$F_{es} = \frac{1}{2}\sqrt{F_t^2 + 4F_s^2} = \frac{1}{2}\sqrt{(52.6 \times 10^3)^2 + 4(6.825 \times 10^3)^2} = 27.15 \times 10^3 \text{ N}$$

The initial tightening has been already taken into account by taking a lower allowable stress.

Let *dc* = core diameter of each bolt, then

$$\frac{\pi}{4} d_c^2 \times f_t = F_{et}$$

$$\frac{\pi}{4} d_c^2 \times 75.0 = 53.45 \times 10^3$$

gives $d_c = 30.2$ mm.

From Table 9.1 for coarse series, the standard size of the bolt will be $M\,36 \times 4c$. For this bolt the stress area is 817 mm^2. Now we can check the bolt under shear stress when the load on the bolt in shear is 27.15×10^3 N.

$$\text{Shear stress produced} = \frac{27.15 \times 10^3}{817} = 33.0 \text{ N/mm}^2$$

which is much below the allowable shear stress of 60.0 N/mm^2, hence safe.

Example 10.9. A bracket is used to support a load *P* as shown in Fig. 10.18. What is the probable load which the bracket can support when $l_1 = 40$ mm, $l_2 = 300$ mm, and $l = 500$ mm? The maximum allowable stress in the bolt material is 75.0 N/mm^2 and the size of the bolt is $M\,27 \times 3c$.

Solution: Each bolt will be under a direct tensile load of $P_1 = P/z$

Therefore, $P_1 = P/z = 0.25\,P$ N.

With the help of relation (10.9), we can find the load in bolt per unit distance induced because of the turning of bracket.

$$2w \times 40^2 + 2w \times (300)^2 = P \times 500$$

gives, $w = 0.00273$ P N.

Now the bolts at the bolt line 2 will be maximum loaded due to turning.

∴ The load on a bolt at bolt line 2, $P_2 = wl_2 = 0.00273\, P \times 300 = 0.819P$ N.

Thus the maximum tensile load in the most heavily loaded bolt is $= P_1 + P_2 = 0.25P + 0.819P = 1.069$ PN.

From Table 9.1, we can see the core diameter of the bolt for the given standard size *M 27 × 3c*.

The core diameter is 23.32 mm.

Now $\pi/4\, d_c^2 \times f_t =$ load on the bolt

$$\pi/4 \times (23.32)^2 \times 75.0 = 1.069 \text{ P}$$

gives $P = 32.25 \times 10^3$ N.

Thus the maximum load which the bracket can support is 32.25 kN.

Example 10.10. A bracket is attached to the flange of a steel stanchion by means of four bolts as shown in fig. 10.22. Find the size of the bolts to be used for an allowable stress of 40.0 N/mm².

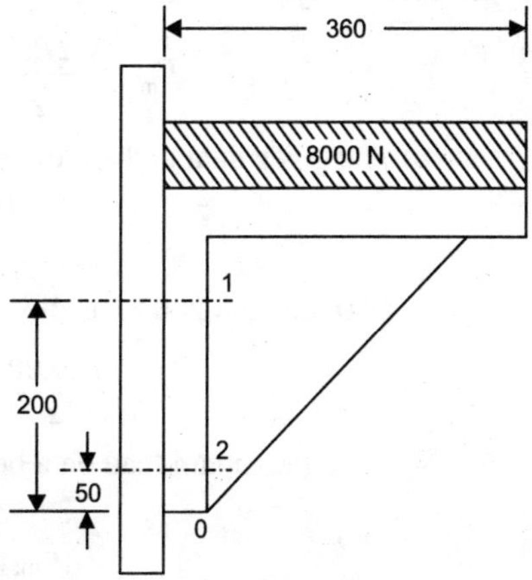

Fig. 10.22

Solution: The average shear load coming on a bolt will be given by relation (10.11).

$$F_s = \frac{P}{z} = \frac{8000}{4} = 2000 \text{ N}$$

The load of 8000 N acting at the centre of the bracket will tend to rotate the bracket about the edge 0. We can find the force per bolt per unit distance form the edge 0 due to the turning of bracket with the help of relation 10.9.

$$8000 \times 180 = 2w \times (50)^2 + 2w \times (200)^2$$

gives $w = 17$ N

The maximum load will occur in the bolts put at the bolt line 1.

Therefore, the maximum tensile load on a bolt line i

$$17 \times 200 = 3400 \text{ N}$$

$$F_t = 3400 \text{ N}.$$

By knowing the shear and tensile loads on a bolt, the equivalent tensile and shear loads may be calculated with the help of relations (10.13) and (10.14).

$$F_{et} = \frac{1}{2} \times 3400 + \frac{1}{2}\sqrt{(3400)^2 + 4(2000)^2} = 4340 \text{ N}$$

$$F_{es} = \frac{1}{2}\sqrt{(3400)^2 + 4(2000)^2} = 2640 \text{ N}$$

$$\pi/4\, d_c^2 \times f_t = F_{et}$$

$$\pi/4 \times d_c^2 \times 40.0 = 4340$$

gives $d_c = 11.75$ mm.

From Table 9.1 for coarse series, the standard bolt will be M 12 × 1.75 c. For this standard bolt the core diameter is 9.853 mm, and the stress area is 84.3 mm².

The shear stress induced in this bolt will be,

$$f_s = \frac{2640}{84.3} = 31.3 \text{ N/mm}^2, \text{ which is within safe limits.}$$

Example 10.11. A pillar crane shown in Fig. 10.23, is fastened to the foundation by 12 bolts spaced equally on a bolt circle. The bolt circle diameter is 2b and the pillar flange diameter is 2a. (a) Establish the formula for maximum load which may come on a bolt. (b) Determine the size of foundation bolts for a load P of 100 kN, if the radius l is 5.0 meters and the diameters of the pillar flange and the bolt circle are 1.8 meters and 1.6 metres respectively. for bolts use steel having an elastic stress of 280.00 N/mm^2 and a factor of safety 4.

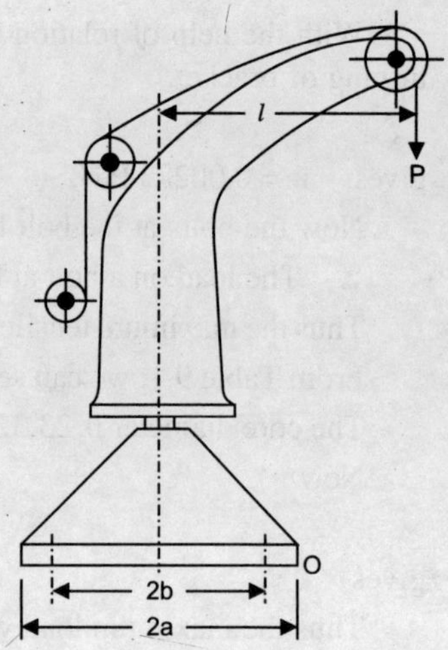

Fig. 10.23

Solution: (a) Pillar flange diameter = $2a$

Bolt circle diameter = $2b$

As the direction of the load P changes with relation to the bolts, therefore, the maximum load which may come on a bolt will be given by relation 10.19. The eccentricity of the load from the turning edge 0 will be $(l - a)$ according to given problem. Thus the relation for maximum load will be,

$$F_{max} = \frac{2P(l-a)(a+b)}{Z(2a^2+b^2)}$$

(b) The given data in the problem is,

$$P = 100 \times 10^3 \text{ N.}$$

$$(l - a) = 5 - 0.9 = 4.1 \text{ m}, z = 12$$

$a = 900$ mm, $b = 800$ mm, allowable tensile stress in bolt material

$$\frac{280.00}{4} = 70.00 \text{ N/mm}^2$$

Now the maximum load on a bolt can be calculated by the relation given above.

$$F_{max} = \frac{2 \times 100 \times 10^3 \times (4100)(900+800)}{12[2 \times (900)^2 + (800)^2]} = 51300 \text{ N.}$$

$$\pi/4 \, d_c^2 \times f_t = F_{Max}$$

$$\pi/4 \, d_c^2 \times 70 = 51300$$

gives, $d_c = 30.4$ mm.

From I.S. 1362-1959, i.e., form Table 9.1 coarse series, the standard size of the bolt is M 36 × 4 c.

Example 10.12. A pillar crane is fastened to the foundation of concrete base by means of four bolts of size M 33 × 3.5 c. The bolts are equally spaced on a bolt circle diameter of 0.5 m and the diameter of the flange of the crane is 0.6 m.

(a) Determine the distance of the load from the centre of the pillar along the axis x-x as shown in Figure 10.24. The load to be lifted by the pillar crane is 60 × 10^3 N. The allowable stress in the bolt material in tension is 60.0 N/mm^2.

(b) Figure 10.25 shows the bolts at an angle 45° to the x-axis. Determine the maximum stress in the bolts if the load is applied along a line y-y of the foundation. The distance of the load line from the centre of the pillar is same, as calculated in part (a).

Solution: The given data is,

$$z = 4, R_2 = 250 \text{ mm}, R_1 = 300 \text{ mm}$$

$$P = 60 \times 10^3 \text{ N}, f_t = 60.0 \text{ N/mm}^2$$

For the given size of the bolt, the area of the bolt which can take stress may be seen from table 9.1 and this area is 694 mm^2.

The maximum load which each bolt can take

$$= 694 \times 60.0 = 41.64 \text{ kN.}$$

The compressive load on four bolts is 60×10^3 N, while due to turning the tensile load is set up in the bolts.

Thus the maximum load that can be taken by each bolt at 1 and 2 due to eccentric load is,

$$= 41.64 \times 10^3 + \frac{60 \times 10^3}{4} = 56.64 \times 10^3 \text{ N}$$

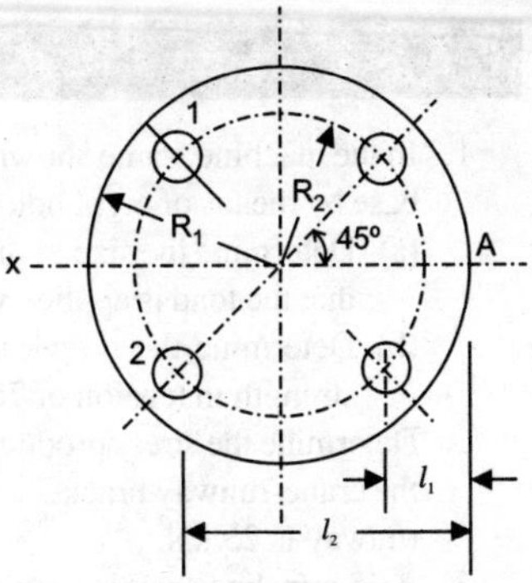

Fig. 10.24

Now, if w is the load in each bolt per unit distance from the turning point A,

then, $$wl_2 = 56.64 \times 10^3, w = \frac{56.64 \times 10^3}{l_2}$$

where $$l_2 = R_1 + R_2 \cos 45 = 300 + 250 \cos 45 = 477 \text{ mm.}$$

$\therefore$ $$w = \frac{56.64 \times 10^3}{477} = 117 \text{ N, also } l_1 = R_1 - R_2 \cos 45$$

$$= 300 - 250 \cos 45 = 123.5 \text{ mm}$$

Now resisting moment of all the bolts about the turning edge A

$$= 2wl_1^2 + 2wl_2^2 = 2 \times 117 \times (123.5)^2 + 2 \times 117 \times (447)^2$$

$$= 56.9 \times 10^6$$

$$= 56.9 \times 10^3 \text{ Nm}$$

If e is the eccentricity of the load from the turning point A, then the over-turning moment due to load P will be $= 60 \times 10^3 \times e$ Nm

Now equating the resisting moment to the overturning moment, we get,

$$56.9 \times 10^3 = 60 \times 10^3 \times e, \text{ gives } e = 0.948 \text{ m}$$

Thus, the distance of the load from the centre of pillar along the x-x axis will be $= 0.948 + 0.3 = 1.248$ m $= 1248$ mm.

(b) In this the load is on the y-axis. The turning due to this load will be on the edge B. This load acts at a distance of 0.948 m from the turning edge B.

Turning moment due to the load P will be

$$= 60 \times 10^3 \times 0.948 = 56.9 \times 10^3 \text{ Nm}$$

The resisting moment offered by all the bolts about the edge B

$$= w[(0.05)^2 + (0.3)^2 \times 2 + (0.55)^2] = 484.5\, w \text{ Nm}$$

Equating the turning moment to the resisting moment, we get

$$56.9 \times 10^3 = 484.5\, w, \text{ gives } w = 117.5 \text{ N.}$$

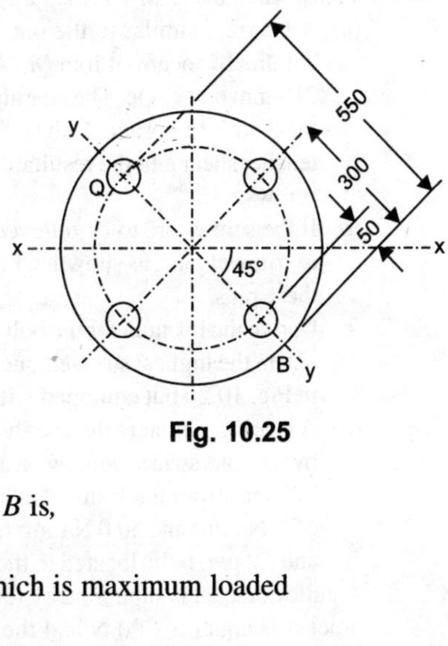

Fig. 10.25

Now the maximum load that can be taken by a bolt put at 'a',

$w \times l_3 = 117.5 \times 550 = 64.8$ kN.

The net force taken by bolt at 'a' due to overturning about edge B is,

$$= 64800 - 15000 = 49800 \text{ N}$$

Maximum stress induced due to overturning in the bolt at 'a' which is maximum loaded

$$= \frac{49800}{694} = 72 \text{ N/mm}^2.$$

EXERCISES

1. In the machine frame shown in Fig. 10.26, the guide post is bolted to the base by means of eight bolts on each side.
   (a) Determine the size of the bolts using ordinary bolt steel, considering that the load is applied with heavy shock.
   (b) Determine the proper size of bolts using steel having an ultimate strength in tension of 780 N/mm^2.

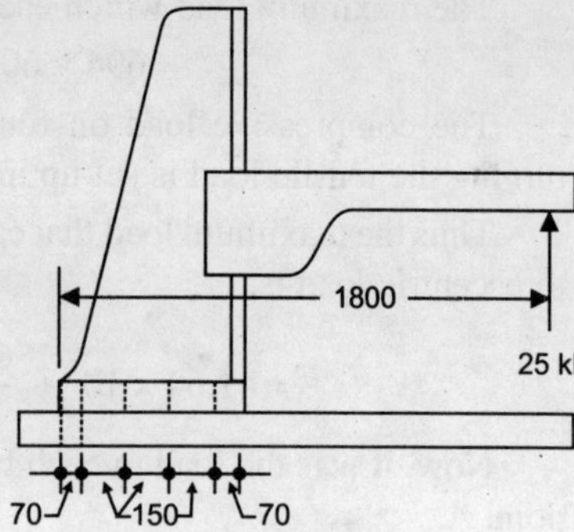

Fig. 10.26

2. Determine the stress produced in the M 30 × 2.5 c bolts used in fastening the crane-runway bracket as shown in Fig. 10.27. The load acting on the runway is 25 kN.
3. A 75 mm. line shaft is supported by a wall bracket so that the shaft centre is 180 mm from the wall. The bracket is supported by three M 20 × 2.5 c bolts. Two bolts are 140 mm below the shaft centre and the third bolt is 430 mm below the shaft centre. The lower edge of the bracket is 50 mm below the lower bolt. The shaft puts a vertical load of 20 kN on the bracket.
   (a) What is the shear stress induced on the bolts, assuming the shear to come on the body of the bolt?
   (b) What is the unit tensile stress in the bolt?
   (c) What is the maximum shearing and tensile stress on the bolts?

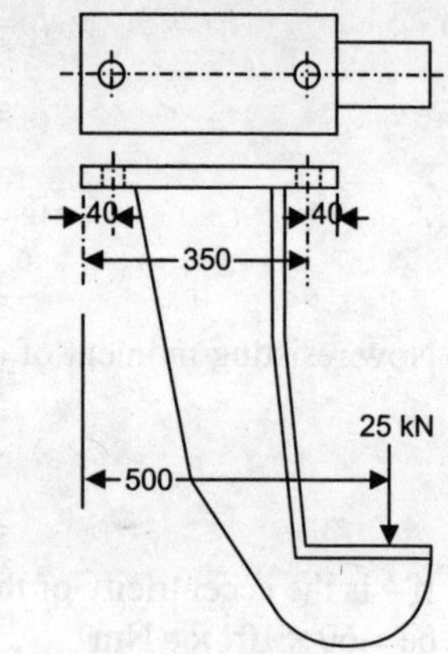

Fig. 10.27

4. The inside diameter of the stator of a 100 kW, 950 rev./min. motor is 950 mm. Starting torque may be assumed to be 200 percent of the running torque. The maximum belt pull on the motor shaft is 10 kN. The shaft centre is 600 mm above the base line. There are four foundation bolts spaced 600 mm on centres, axially, and 1050 mm on centres normal to the axis. The width of the base is 1200 mm. Bolt steel has an ultimate strength of 400.0 N/mm^2 with an elastic limit of 200.0 N/mm^2. Determine the load on each bolt and the proper diameter of these bolts.
5. A bracket, similar to the one shown in Fig. 10.19, carries a load P of 25 kN at a distance of 450 mm. It is carried by four bolts located so that $l_1 = 50$ mm and $l_2 = 550$ mm. Determine (*a*) the size of the bolts necessary, when the allowable stress is 60.0 N/mm^2 and (*b*) the tensile, shear and resultant stress, when using M 18 × 2.5 c bolts.
6. (a) A bearing, similar to the one shown in Fig. 10.20, is fastened to a column by means of four M 24 × 3 c cap screws spaced equally on a 210 mm bolt circle. The bearing flange diameter is $2R_1 = 260$ mm and the load P 15 kN acts with an eccentricity of 105 mm. Determine the tensile, shear and the resultant stresses coming upon each of the cap screws.
   (b) If the screws are to be relieved from shearing stresses by the use of two dowel pins, as shown in Fig. 10.20 (*b*), find the necessary size of these pins.

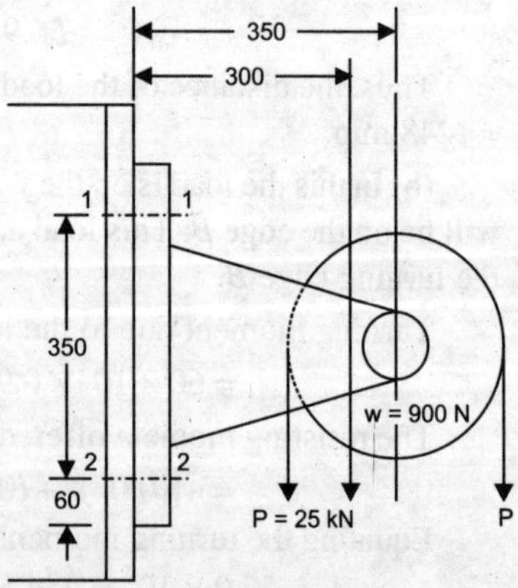

Fig. 10.28

7. (a) Derive the formula for the bolts in tension coming on (1) any bolt and (2) on the highest stressed one in a bearing similar to the one shown in Fig. 10.20 but equipped with six bolts.
   (b) A bearing similar to the one shown in Fig. 10.20 is fastened to a frame by six bolts spaced equally on a 265 mm bolt circle. The dimension $2R_1$ is 315 mm and the load of 35 kN is applied 375 mm from the frame. Determine the size of the bolts, when the allowable stresses in tension and shear are 65.0 N/mm^2 and 50.0 N/mm^2 respectively for (1) two bolts located in the vertical plane of symmetry of the bearing and (2) two bolts located in the horizontal plane of symmetry of the bearing.
8. A pulley bracket is supported by four bolts, two at 1 and two at 2, as shown in Fig. 10.28. The weight of the pulley and bracket is equal to 900 N and the load on the rope P is 25 kN. Taking an allowable tensile stress in the bolts as 40.0 N/mm^2 determine the size of the bolts required.

❑❑❑

CHAPTER 11 Cotter and Knuckle Joints

11.1 COTTERS

Cotters are temporary fasteners and are used to hold two parts together where the parts are subjected to axial forces only. A cotter is a wedge shaped piece of mild steel, often rectangular in section as shown in Fig. 11.1. A cotter is tapered on one side or both sides as shown in figure. It is usually driven in perpendicular to the axis of the connected parts. Taper on cotter is provided to facilitate its driving into position and withdrawal for lateral adjustments of connected parts. Usually the taper is 1 in 25. If the taper provided is in excess, say upto 1 in 8, there is a possibility of the cotter sliping back. To prevent sliping back of the cotter, some locking device is used. The locking device may be a taper pin or set screw used on the lower end of the cotter.

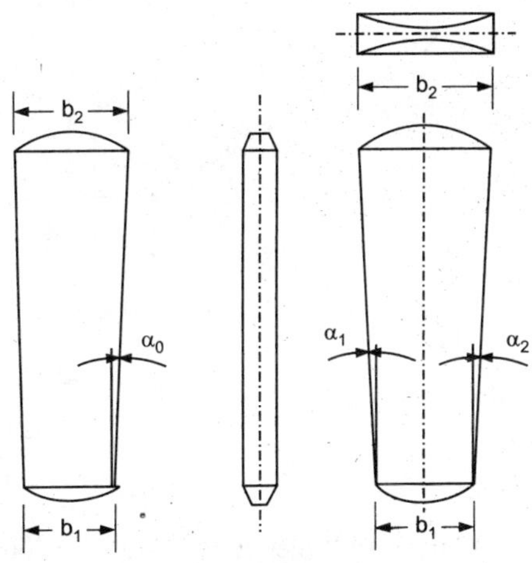

Fig. 11.1

11.2 COMPARISON BETWEEN KEYS AND COTTERS

Keys are usually driven parallel to the axis of the shafts which are subjected to torsional or twisting stress. Cotters are normally driven at right angles to the axes of the connected parts which are subjected to tensile or compressive stresses along their axes. A key resists shear over a longitudinal section, whereas a Cotter resists shear over two transverse sections.

11.3 COTTER JOINT

A joint formed by using a cotter is called a cottered joint. Cottered joints are often used to connect a rod which is subjected to axial loading, either to another rod or to some machine part such as a cross-head, base plate, connecting rod end of a steam engine, etc. In the cottered joint shown in Fig. 11.2, one rod end is formed into a socket and the other called spigot end is forged to a diameter a little larger than that of the rods. A hole to suit the dimensions of the cotter is provided in both the socket and spigot. The usual clearance, when the cotter is driven home varies from 1.5 mm to 3 mm, but 3 mm is usually the clearance used. When the cotter is driven in, it forces the spigot rod end inside the socket and makes the joint tight. Note that the spigot end is provided with a collar and a clearance is left between the end of the spigot and socket when the joint is made.

The dimensions of the rod end and socket end should be such that the area of each across the slot is not less than the area of the rod.

Cotter joints are advantageous in that they are simple in design and allow a convenient and rapid assembly and disassembly of the unit. The added feature of the joint is that the parts occupy exactly the same relative positions after assembly.

The main shortcomings of the cotter joint are production inadequacy of design, weakening of the main parts by the holes for cotters and the need for locking devices in important cases. These shortcomings prevent the use of a cotter joint on an extensive scale. However, they continue to be used where convenience and rapidity of assembly and disassembly are the prime considerations.

11.4 DESIGN OF COTTER JOINT

Figure 11.2 shows a cotter joint for two rods. If the initial stresses due to tightening up are ignored, and the rods are assumed of the same materials, equations by equating the resistances of the principal dimensions of the joint to failure in various possible ways may be obtained. Initial stresses are produced due to tightening of the cotter excessively which will bend in the plane of the side view, and this bending will aid in keeping the joint tight under the action of a variable load.

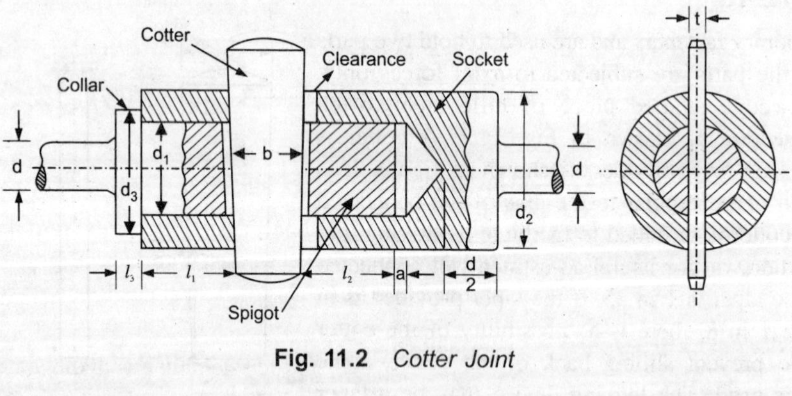

Fig. 11.2 *Cotter Joint*

Analysis of forces and stresses in a cotter joint shows that there are tensile, shearing and bearing stresses induced. Failure of the joint may occur in one of the following ways discussed below. Assume that the material of spigot, socket and cotter is same for which the tensile, shearing and bearing or compressive stresses are represented by f_t, f_s and f_c. Let F is the load acting on the rods in tension.

1. Tension failure of the rod: The load F is a tensile load on the rods, so that the diameter d can be found by equating the tensile resistance of the rod to the load F. Thus

$$F = \pi/4\, d^2 \times f_t \qquad ...(11.1)$$

Relation 11.1 will give the diameter d of the rod end.

2. Tension failure of the spigot at the location of cotter: At the section considered, the area in tension failure at the location of the cotter is shown in Fig. 11.3 (*a*).

The area resisting tearing of the spigot end = $(\pi/4\, d_1^2 - d_1 t)$

Resisting force in tension = $(\pi/4\, d_1^2 - d_1 t)\, f_t$

This tearing resistance must be equated to the tensile load on the joint.

$$F = (\pi/4\, d_1^2 - d_1 \times t)\, f_t \qquad ...(11.2)$$

From this relation, the diameter d_1 of the spigot end can be obtained when the thickness of the cotter is known.

Thickness of cotter t is generally = 0.4 d.

3. Tension failure of the socket at the location of cotter: At the section considered, the area in tension failure at the location of the cotter is shown in Fig. 11.3 (*b*).

The area resisting tearing of the socket end = $\pi/4 d_2^2 - \pi/4 d_1^2 - (d_2 - d_1)\, t$

Resisting force in tension = $[\pi/4 d_2^2 - \pi/4 d_1^2 - (d_2 - d_1) t]\, f_t$.

This tearing resistance must be equated to the tensile load on the joint.

$$\therefore \quad F = [\pi/4 d_2^2 - \pi/4 d_1^2 - (d_2 - d_1) t]\, f_t \qquad ...(11.3)$$

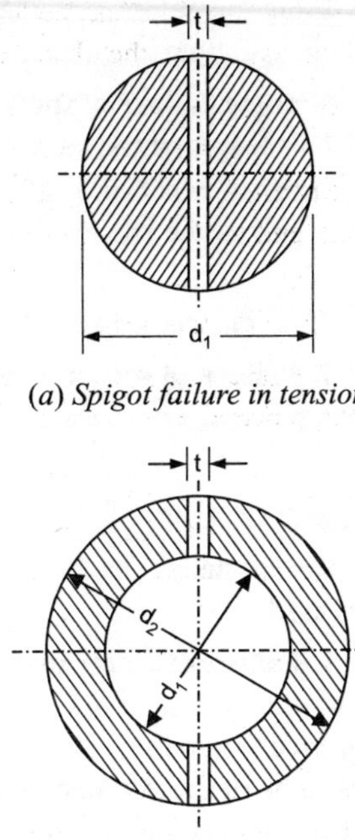

(*a*) *Spigot failure in tension*

(*b*) *Socket failure in tension*

Fig. 11.3

From this relation, the outside diameter d_2 of the socket end can be obtained. This dimension d_2 of the socket is satisfactory from a stand point of tension, it may be unsatisfactory from the stand point of crushing between the cotter and end of the socket. In other words, crushing may insist on a larger value of d_2 as compared to the value calculated by relation 11.3. Since d_2 and t are already known, it remains to check the socket end for crushing stress. Thus, crushing stress induced in socket end is given as

$$f_c = \frac{F}{(d_2 - d_1) \times t} \qquad ...(11.4)$$

Now, if the induced crushing stress is less than the allowable value, then the design is safe. If the result is more than the allowable crushing stress, then either d_2 or t must be increased. Mostly d_2 is increased.

4. Shear of the cotter: The cotter is in double shear as shown in Fig. 11.4(*a*). Since the cotter is tapered, the area in double shear will be based upon the mean width of the cotter. This mean width is shown as b in Fig. 11.4 (*a*).

The mean area of the cotter in double shear = $2bt$.

Resistance to shear of cotter = $2bt \times f_s$.

This shearing resistance of cotter must be equated to load causing shear of the cotter.

$$\therefore \quad F = 2bt \times f_s \qquad ...(11.5)$$

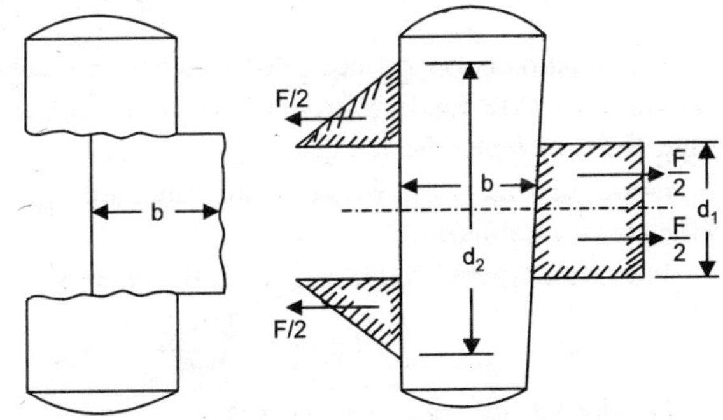

(*a*) *Shear of cotter* (*b*) *Bending of cotter.*

Fig. 11.4

From this relation, the mean width b of the cotter can be obtained.

5. Shear failure of the spigot end: Reference to Fig. 11.2, the length l_2 of the spigot end can be sheared off. This length of spigot end will be in double shear, whose area is $2l_2\, d_1$.

Shearing resistance offered by the spigot end = $2l_2\, d_1 \times f_s$.

Equating the shearing resistance of the spigot end to the shearing load.

$$F = 2l_2\, d_1 \times f_s \qquad ...(11.6)$$

From this relation, the length l_2 of the spigot end can be established. The dimension a in Fig. 11.2 may be kept as 2 times the clearance.

6. Shear failure of the Socket end: Reference to Fig. 11.2, the length l_1 of the socket end can be sheared off. This length of the socket end will be in double shear, whose area is $2l_1\,(d_2 - d_1)$.

Shearing resistance offered by the socket end = $2l_1\,(d_2 - d_1) \times f_s$. Equating the shearing resistance of the socket end to the shearing load.

$$F = 2l_1\,(d_2 - d_1) \times f_s \qquad ...(11.7)$$

This relation helps us to find the length l_1 of the socket end.

7. Failure of spigot or cotter under crushing: The spigot or cotter may fail by crushing over the area, $d_1 \times t$. Since d_1 and t have already been determined, it remains to check this area for the induced crushing stress.

$$\therefore \qquad f_c = \frac{F}{d_1 \times t} \qquad ...(11.8)$$

In case the induced crushing stress is more than the allowable crushing stress, then either d_1 or t may be increased.

8. Failure of cotter under bending: All the above relations assume that the load is uniformly distributed across the various cross-sections, which is not strictly true in case of cotter. Therefore, the cotter will be subjected to bending. The exact analysis of the bending of the cotter can be made. However, by assuming the load to be uniformly distributed on the cotter in the spigot end and varying uniformly on the cotter in the socket from zero at the outer diameter d_2 to a maximum at the inner diameter d_1. The loading of the cotter is shown in Fig. 11.4 (*b*). The maximum bending will occur at the centre of the cotter or spigot end.

Maximum bending moment

$$M_b = \left[\left(\frac{d_2 - d_1}{2} \times \frac{1}{3} + \frac{d_1}{2}\right) - \frac{d_1}{4}\right] \times \frac{F}{2}$$

$$M_b = \frac{F}{12}\left(d_2 + \frac{d_1}{2}\right)$$

Since $M_b = f_b \times Z$, where f_b is the bending stress in the cotter.

$$\frac{F}{12}\left(d_2 + \frac{d_1}{2}\right) = f_b \times \frac{t_b^2}{6} \qquad ...(11.9)$$

This relation helps in finding induced bending stress in the cotter. Induced bending stress should be less than the allowable bending stress of the cotter. In case induced bending stress is more than the allowable value, then t or b may be increased.

Once, the dimensions b and t of the cotter are adjusted from strength point of view, the taper of the cotter is then to be established.

For single tapered cotter, Fig. 11.1, the taper is

$$\tan \alpha_0 = \frac{b_2 - b_1}{l}, \text{ where } l \text{ is the length of the cotter.}$$

For double tapered cotter, taper is

$$\tan \alpha_1 + \tan \alpha_2 = \frac{b_2 - b_1}{l}$$

For small angles of taper $\tan \alpha_1 + \tan \alpha_2 \simeq \tan(\alpha_1 + \alpha_2)$, consequently if $\alpha = \alpha_1 + \alpha_2$, single and double tapered cotters have equal force transmitting features. Single tapered cotters are preferred as they are easy to manufacture.

Self-locking is considered adequate if taper angle α is less than the friction angle θ. When the taper is more than 1/25 some locking device is required so that the cotter does not slip out. If the joint is subjected to vibration the use of a locking device is advisable even with a smaller taper. Usually the locking used is a set screw.

In order to decrease the stress concentration in the slots of the spigot and of the socket, the edges of the cotter are rounded off or even sometimes the short sides are made semi circular.

When the cotter joint shown in Fig. 11.2 is required to transmit the compressive loads, then this rod will be under compressive force. For cotter joint to take up the compressive loads, the collar of the spigot end is to be designed.

9. Shear Failure of the Collar: The shearing failure of the collar will occur on diameter d_1 of spigot due to compressive load F.

Shearing area resisting the compressive load = $\pi d_1 l_3$

Shearing resistance of collar = $\pi d_1 l_3 \times f_s$

The shearing resistance of collar must be equated to the compressive load.

$$\therefore \qquad F = \pi d_1 l_3 \times f_s \qquad ...(11.10)$$

This relation helps us to find out the length l_3 of the collar.

10. Crushing failure of the Collar: The crushing failure of the collar is due to the compressive load F.

Crushing area resisting the compressive load = $\frac{\pi}{4}\left(d_3^2 - d_1^2\right)$

Crushing resistance of collar = $\frac{\pi}{4}\left(d_3^2 - d_1^2\right)f_c$

Crushing resistance of the collar must be equated to the compressive load.

$$\therefore \qquad F = \frac{\pi}{4}\left(d_3^2 - d_1^2\right)f_c \qquad ...(11.11)$$

Relation 11.11 helps us to find the diameter d_3 of the collar.

The stress concentration due to the change from d to d_3 with a fillet must be taken into account.

The dimensions of the cotter maybe calculated from the relations given above, or fixed empirically. For a cotter joint when all the components are made of steel, the empirical relations are given below.

$d_1 = 1.21\,d$	$b = 1.6\,d$	$a = 2 \times$ clearance
$d_2 = 1.75\,d$	$l_1 = l_2 = 0.75d$	$l = 4d$
$d_3 = 1.5\,d$	$l_3 = 0.45d$	
$t = 0.4\,d$	clearance 1.5 mm. to 3 mm.	
	Taper of cotter 1 : 25	

Sometimes the diameter of socket end at length l_1 is increased to take the crushing load. Empirically the increased diameter of the socket end is given as 2.4 times the rod end diameter.

Sometimes, instead of making the spigot and socket on each end of rod, a separate sleeve or muff fits over the end of each rod as shown in Fig. 11.5. In this joint the rod ends are circular in section, and are connected by using a sleeve or muff and then applying two cotters, one on each rod end. The two ends are made to meet at the centre of sleeve. When the cotters are driven in, they press the two rod ends against each other and make the joint tight.

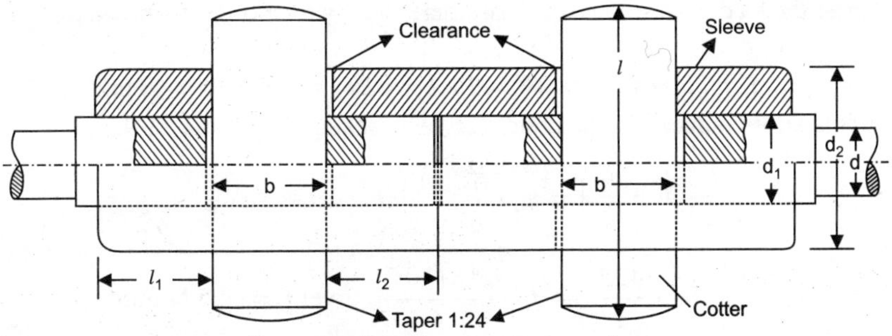

Fig. 11.5 *Sleeve and cotter joint*

The design procedure for sleeve and cotter joint is similar as discussed earlier for spigot and socket cotter joint. For a sleeve and cotter joint the empirical relations are given below in terms of rod diameter d.

$d_1 = 1.25\,d$ $l_1 = 1.25\,d$ $l = 3.5\,d$ Clearance = 3 mm.

$d_2 = 2.5\,d$ $l_2 = 2.5\,d$ $b = 1.25\,d$ Taper = 1 : 25

Overall length of sleeve = 8 d.

The change of diameter from d to d_1 must be very smooth to reduce the stress concentration.

Example 11.1. The cotter joint shown in Fig. 11.2 is commonly used for long pump rods and similar machine members. Design this joint to support a load varying from 28 kN in tension to 28 kN in compression. The material for all the components of the joint is steel with the allowable stresses in tension, compression and shear as 50.0 N/mm², 60 N/mm² and 35 N/mm² respectively.

Solution: The rod proper is in direct tension. Using the notations shown in Fig. 11.2, the diameter of the rod by relation 11.1 can be found out.

$$F = \pi/4d^2 \times f_t, \quad 28 \times 10^3 = \pi/4d^2 \times 50, \quad \text{gives } d = 26.7 \text{ mm, say } 27.5 \text{ mm.}$$

If the rod is very long between supported sections, it must be considered to be column and the diameter determined accordingly.

The maximum tension stress in the rod end is at the section through the cotter hole. The diameter d_1 of the spigot end can be found by relation 11.2. Thickness of cotter $t = 0.4\,d = 0.4 \times 27.5 = 11$ mm.

$$F = [(\pi/4d_1^2 - d_1 \times t)f_t],\ 28 \times 10^3 = (\pi/4d_1^2 - d_1 \times 11)\,50$$

gives, $d_1 = 35$ mm

The induced crushing stress can be checked by relation 11.8.

$$f_c = \frac{F}{d_1 \times t}, \quad f_c = \frac{28 \times 10^3}{35 \times 11} = 73 \text{ N/mm}^2$$

which indicates that spigot or cotter is unsafe under crushing. Therefore, increase the diameter $d_1 = 42.5$ mm which gives the induced crushing stress as 59.5 N/mm² hence safe.

In tension, the greatest stress in the socket is at the section through the cotter hole, thus using relation 11.3, we can find the diameter d_2

$$F = \left[\frac{\pi}{4}(d_2^2 - d_1^2) - (d_2 - d_1)t\right] f_t$$

$$28 \times 10^3 = [0.7854\,(d_2^2 - (42.5)^2) - (d_2 - 42.5)11]50 \qquad \text{gives,} \quad d_2 = 51.5 \text{ mm}$$

This diameter should be increased to 85 mm to take up crushing of cotter or socket. Induced crushing stress can be checked by relation 11.4.

$$f_c = \frac{F}{(d_2 - d_1) \times t} = \frac{28 \times 10^3}{(85 - 42.5)11} = 59.5 \text{N/mm}^2 \text{ hence safe.}$$

The cotter carries the load in shear on two areas, therefore, by relation 11.5 we can find the mean width of the cotter.

$$F = 2bt \times f_s,\ 28 \times 10^3 = 2 \times b \times 11 \times 35, \text{ gives } b = 40 \text{ mm.}$$

Now the bending stress induced in the cotter can be checked by relation 11.9.

$$\frac{F}{12}\left(d_2 + \frac{d_1}{2}\right) = f_b \times \frac{t_b^2}{6}$$

$$\frac{28 \times 10^3}{12}\left(85 + \frac{42.5}{2}\right) = f_b \times \frac{11 \times (40)^2}{6}, \text{ gives } f_b = 84.5 \text{ N/mm}^2$$

Hence, the induced bending stress is much more than the allowable value of 50 N/mm^2. To make the cotter safe under bending, increase the width of cotter to 52 mm, so that the induced bending stress is 50 N/mm^2. So take $b = 52$ mm.

The length l_z of the spigot end can be fixed by relation 11.6.

$$F = 2l_2 \times d_1 \times f_s, \qquad 28 \times 10^3 = 2l_2 \times 42.5 \times 35$$

gives, $l_2 \simeq 10$ mm.

Length l_1 of the socket end can be fixed by relation 11.8.

$$F = 2l_1 (d_2 - d_1) f_s, \qquad 28 \times 10^3 = 2l_1 (85 - 42.5)\, 35$$

gives, $l_1 \simeq 10$ mm.

But, in general l_1 and l_2 are usually made from 0.75 d.

Therefore, $\quad l_1 = l_2 = 0.75 \times 27.5 = 20.7$ mm say 21 mm.

The load varies from tension to compression. Therefore, the length of collar attached to spigot end can be found by relation 11.10

$$F = \pi d_1 l_3 \times f_s, \qquad 28 \times 10^3 = \pi \times l_3 \times 35 \times 42.5$$

gives, $l_3 = 6$ mm.

Usually l_3 is taken as 0.45 d.

Therefore, $l_3 = 0.45 \times 42.5 = 19$ mm.

The diameter d_3 of the collar can be found under consideration of crushing, i.e., by relation 11.11.

$$\text{F} = \pi/4(d_3^2 - d_1^2) f_c \qquad 28 \times 10^3 = 0.7854\,(d_3^2 - (42.5)^2) \times 60$$

gives $d_3 = 49$ mm.

But d_3 is generally taken as 1.5 d.

Therefore, $d_3 = 1.5 \times 42.5 = 64$ mm.

Length of the cotter will be taken as $4d$.

$$\therefore \qquad l = 4 \times d = 4 \times 27.5 = 110 \text{ mm.}$$

The taper in the cotter will be made 1 : 25, therefore,

$$\frac{1}{25} = \tan\alpha = \frac{b_2 - b_1}{l}, \quad \frac{1}{25} = \frac{b_2 - b_1}{110}$$

gives $b_2 - b_1 = 4.4$ mm, dimensions of the cotter will be $b_2 = 54.2$ mm, $b_1 = 49.8$ mm and $l = 110$ mm, mean width of cotter = 52 mm.

11.5 GIB AND ITS USE

It is a piece of mild steel having the same thickness as the cotter in combination with which it is to be used. Gib has the same taper as on the corresponding cotter. The two outer lines of the gib and cotter when joined are two parallel lines.

A gib is used in gib and cotter joint to achieve the following purposes:

(a) When the cotter without gib is driven in position, it tends to bend the lower part of the socket end away from the rod and this is prevented by using a gib as shown in Fig. 11.6.

(b) Gib helps to use parallel holes in the spigot and socket.

(c) To increase the bearing area of contact between the mating surfaces, a gib is inserted. When one gib is used the cotter is provided with taper only on one side and the gib is always on the outside and when two gibs are used the cotter is tapered on both sides, Fig. 11.6.

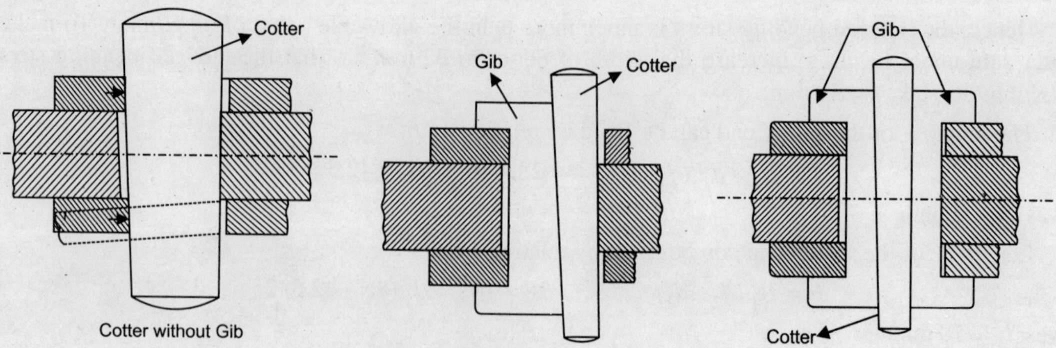

Fig. 11.6 *Use of gib*

11.6 GIB AND COTTER JOINT

This type of joint is mostly used for square rod ends as the case of big end of a connecting rod. One of the ends is formed into the shape of a fork or straps as shown in Fig. 11.7. Suitable hole of rectangular section, to accommodate the gib and cotter and necessary clearances, are made in rod ends. For making its assembly, the rod end is first pushed into the fork end till its movement is stopped by the solid part of the fork end. When the two ends are in this position, the gib is first inserted in the holes, which are to be in the same plane (offset by the clearance distance) and then after giving its proper position the cotter is driven in to make the joint tight. It should be noted here that the gib is to be placed on the open end side of the fork.

Sometimes to avoid the loosening of the cotter a set screw is used through the rod jamming against the cotter.

11.7 DESIGNING THE COTTER WITH GIB

The design procedure of the gib and cotter joint is similar to the one discussed in Article 11.4 for cotter joint. The rod end will be in tension failure. Each side of the rod end is assumed to be h. Let F to be the load in tension to be taken by the joint, then

$$F = h^2 \times f_t \qquad ...(11.12)$$

This relation will help us to find the dimension h of square end rod. The width of the strap is generally made equal to one side of the square end h. The thickness of the strap end is t_1 which can be calculated by considering the failure of strap under tearing at the hole.

The cotter and gib will be in double shear. For total width of gib and cotter to be B and thickness of each t, the shearing resistance offered by gib and cotter will be $2 \times B \times t \times f_s$. Thickness t is generally taken 0.3 times the width of the strap. Now equating resistance to the load F,

$$F = 2Bt \times f_s \qquad ...(11.13)$$

From this relation B can be calculated. This B is now proportioned between cotter and gib according to convenience. Generally,

width of gib = $0.55B$, width of cotter = $0.45B$.

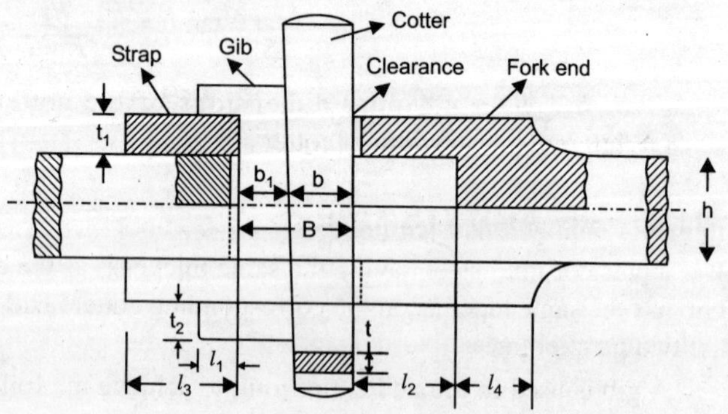

Fig. 11.7 *Gib and cotter joint for square rods*

In case two gibs are used with the cotter, then

width of each gib = $0.3B$, width of cotter = $0.4B$.

Thickness and dimension l_1 of the gib are taken equal to thickness of cotter. The dimension t_2 of the gib is taken equal to 1.3 times the thickness of the cotter. Clearance is usually kept 3 mm. Example 2 will explain the design procedure for a gib and cotter joint.

Example 11.2. Design a gib and cotter joint like that of Fig. 11.7 to resist safely a tensile load of 40 kN. The material of the gib, cotter and rods is same for which the allowable safe stresses are:

$$f_c = 60.0 \text{ N/mm}^2, f_t = 25.0 \text{ N/mm}^2 \text{ and } f_s = 20.0 \text{ N/mm}^2.$$

Solution: Let h be each side of the square section rod, then by relation 11.12, we have

$$F = h^2 \times f_t,\ 40 \times 10^3 = h^2 \times 25, \text{ gives } h = 40 \text{ mm}.$$

Width of strap $= h = 40$ mm. Thickness of cotter $= t = 0.3 \times h = 0.3 \times 40 = 12$ mm.

Now considering the shearing of the gib and cotter, we have

$$F = 2\,Bt \times f_s,\ 40 \times 10^3 = 2 \times B \times 12 \times 20, \text{ gives } B = 83 \text{ mm}$$

Single gib will be used to make this joint.

∴ Width of gib $= 0.55 \times B = 0.55 \times 83 = 45$ mm

Width of cotter $= 0.45B = 0.45 \times 83 = 38$ mm

Thickness of gib = Thickness of cotter $= t = 12$ mm

Dimension l_1 of the gib tongue $= t = 12$ mm

Height of the gib tongue $= t_2 = 1.3t = 1.3 \times 12 = 15.6$ mm say 16 mm.

The strap end will fail in tension at the location of the gib and cotter. Area resisting the tensile failure will be $= 2 \times (\text{width of strap} \times t_1 - t_1 \times t)$

Resisting tensile force = tensile force acting on the joint

$$2(40 \times t_1 - t_1 \times 12) \times 25 = 40 \times 10^3$$

gives, t_1 = thickness of strap = 28.5 mm

At the strap hole, either the strap or the gib can get crushed.

Area resisting the crushing action $= 2 \times t_1 \times t$

$$\text{Induced crushing stress} = \frac{F}{2 \times t_1 \times t} = \frac{40 \times 10^3}{2 \times 28.5 \times 12} = 58.5 \text{ N/mm}^2$$

Hence safe, as the induced stress is less than 60 N/mm^2.

Overall length of the gib $= h + 2 \times t_1 + 2 \times t_2 = 40 + 2 \times 28.5 + 2 \times 16 = 129$ mm

Length of the cotter is generally taken $= 4 \times h = 4 \times 40 = 160$ mm

Taper on one side of gib and cotter will be 1 : 25.

Length l_2 of the rod end can be established by considering the shearing.

This length of the spigot will be in double shear, whose area is $2l_2\,h$.

Shearing resistance offered by the rod end = tensile force

$$2l_2 \times h \times f_s = F,\ 2l_2 \times 40 \times 20 = 40 \times 10^3$$

gives, $l_2 = 25$ mm.

Length l_3 of the strap end can be fixed by considering the shearing.

This length of the strap end will be in double shear, whose area is $(2 \times l_3 \times t_1) \times 2$.

Shearing resistance offered by the strap end = tensile force

$$2 \times 2 \times l_3 \times t_1 \times f_s = F,\ 2 \times 2 \times l_3 \times 28.5 \times 20 = 40 \times 10^3$$

gives, $l_3 = 35$ mm.

Length l_4 of the strap end $= \frac{2}{3}h$, $= 2/3 \times 40 = 26.7$ mm say 27.5 mm.

This completes the design of the gib and cotter joint. All the corners are to be well rounded off to avoid stress concentration. The increase of size h to the strap size has to be very gradual as shown in Fig. 11.7.

Example 11.3. Design and sketch a cotter joint for fastening the piston rod to the cross-head of the engine having cylinder diameter 250 mm and steam pressure 1.05 N/mm². The thickness of the cotter is to be 0.3 times piston rod diamcter at the point where cotter is located. Allowable stresses in tension, shear and compression are 50 N/mm², 40 N/mm² and 84 N/mm² respectively.

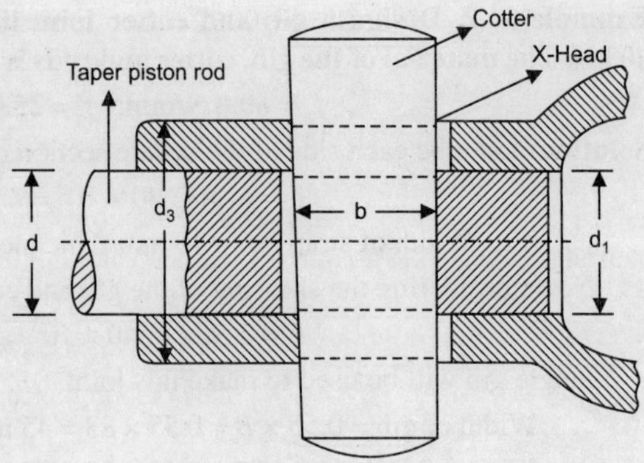

Fig. 11.8 *Connection of piston rod to X-head by means of a cotter*

Solution: The taper on the piston rod is taken as 1 : 20. The diameter d of the parallel part of the piston rod is calculated from column action.

The tapered length of the piston rod is generally kept as 2.2 times the diameter d_2.

Maximum load on the piston = $\pi/4 \times D^2 \times p$

$$F = \pi/4\,(250)^2 \times 1.05 = 51.5 \times 10^3 \text{ N}$$

Let d_2 be the diameter of the piston rod at the cotter.

Thickness of cotter = $0.3 \times d_2$.

Maximum load on the piston will cause the piston rod to fail in tension at cotter.

$$\therefore \quad (\pi/4 d_2^2 - d_2 \times t)\, ft = F$$

$$(\pi/4 d_2^2 - d_2 \times 0.3 d_2)\, 50 = 51.5 \times 10^3$$

gives, $d_2 = 46$ mm ; say 47.5 mm

Thickness of cotter = $0.3 \times 47.5 = 14.3$ mm say 15 mm.

Let b = width of cotter at the centre. The cotter will be in double shear, then

$$F = 2bt\,f_s,\ 51.5 \times 10^3 = 2 \times b \times 15 \times 40$$

gives $b = 43$ mm.

d_3 is the diameter of the socket through the cotter hole. This diameter can be obtained by considering the crushing of the cotter or socket. Then,

$$F = (d_3 - d_2)\, t \times f_c,\ 51.5 \times 10^3 = (d_3 - 47.5)\, 15 \times 84$$

gives, $d_3 = 87.5$ mm.

This diameter of the socket can be checked by considering the failure of the socket in tension at cotter point.

$$F = [\pi/4(d_3^2 - d_2^2) - (d_3 - d_2)\, t]\, f_t$$

$$51.5 \times 10^3 = [\pi/4(87.5^2 - 47.5)^2 - (87.5 - 47.5) \times 15]\, f_t$$

gives, $f_t = 13.7$ N/mm².

Which is much below the allowable tensile stress of 50 N/mm², hence safe. Length of the tapered portion of the piston rod = $2.2 \times d_2 = 2.2 \times 47.5 = 105$ mm, considering the taper of piston rod as 1:20, diameter d_1 of the piston rod will be

$$= d_2 - \frac{1}{20} \times \frac{105}{2} = 47.5 - 2.64 = 44.86 \text{ mm}$$

Diameter of the parallel part of the piston rod

$$= d = d_2 + \frac{1}{20} \times \frac{105}{2} = 50 \text{ mm}$$

Remaining dimensions of the cotter and socket can be fixed as discussed earlier.

Example 11.4. Design a cotter foundation bolt to support a tensile load of 100 kN. Assume the material of the cotter and the bolt to be same. Take the following values of the permissible stresses.

$$f_s = 40 \text{ N/mm}^2, f_t = 55 \text{ N/mm}^2$$
$$f_c = 105 \text{ N/mm}^2$$

Solution: Refer Fig. 11.9.

Assume

t = thickness of cotter
b = width of cotter
d = diameter of bolt.
d_1 = Enlarged diameter of the bolt.

The bolt will fail in tension.

$$\therefore \quad \frac{\pi}{4} d^2 \times f_t = F$$

$$\frac{\pi}{4} d^2 \times 55 = 100 \times 10^3$$

gives, $d = 48.2$ mm, say 50 mm.

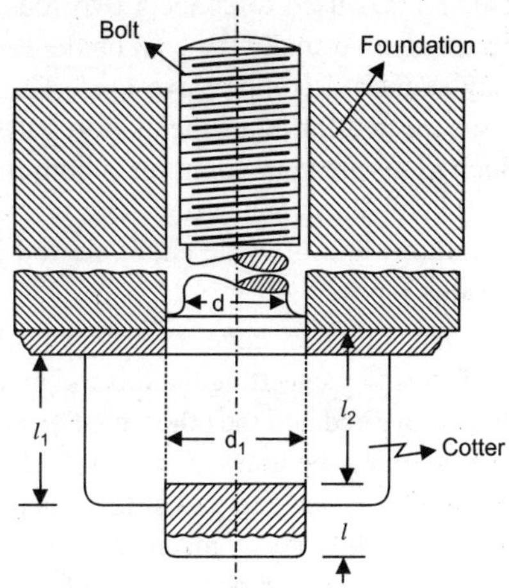

Fig. 11.9 *Foundation cotter bolt*

Now the enlarged end of the bolt will fail in tension at the cotter.

$$\therefore \quad \left(\frac{\pi}{4} d_1^2 - d_1 \times t\right) f_t = F$$

$$\left(\frac{\pi}{4} d_1^2 - d_1 \times \frac{d_1}{4}\right) 55 = 100 \times 10^3$$

Thickness of cotter t is taken as $\frac{d_1}{4}$

$$\therefore \quad d_1 = 58.4 \text{ mm. say } 60 \text{ mm}$$

$$\text{Thickness of cotter} = \frac{d_1}{4} = \frac{60}{4} = 15 \text{ mm}$$

The cotter will fail in double shear and the shearing area will be $2 \times b \times t$.

$$\therefore \quad 2 \times b \times t \times f_s = F,\ 2 \times b \times 15 \times 40 = 100 \times 10^3$$

gives, $b = 83$ mm, we adopt 85 mm.

Increase the thickness of the cotter to 17.5 mm, so that the induced crushing stress in the cotter is within safe limits.

$$\text{Induced crushing stress} = \frac{100 \times 10^3}{60 \times 17.5} = 96 \text{ N/mm}^2, \text{ which is safe.}$$

The length l of the enlarged portion of the bolt can be fixed from shear consideration.

$$2 \times l \times d_1 \times fs = F, \quad 2 \times l \times 60 \times 40 = 100 \times 10^3$$

gives, $l = 21$ mm.

This length may be increased to 40 mm, so that the induced shear stress is not more than 22.5 N/mm^2.

The length l_1 of the tongue of the cotter = b + thickness of cotter.

$$\therefore \quad l_1 = 85 + 17.5 = 102.5 \text{ mm}$$

Length l_2 will be kept more than l_1, therefore, let l_2 be 105 mm.

Practical considerations in cotter joint: Cotters are usually driven in to make a tight joint and the initial stresses are indeterminate, and as in the case of bolts, the applied loads may or may not increase these stresses. In many cases, therefore, it is necessary to alter the proportions as indicated by various strength relations to meet the conditions of service and the materials used. Empirical formulae mentioned in the text are based on experience, and prove to be of more practical use than the theoretical analysis.

11.8 KNUCKLE JOINT

This joint is used to connect two rods or bars when a small amount of flexibility or angular movement is necessary. Normally the load on the rods or bars to be connected is tensile, but if the joint is guided the rods may support a compressive load. The joint is used where it is desired to have a joint that can be readily disconnected for adjustments or repairs. The axis of the rods may be collinear or coplanar and intersecting. Such joints are very common in both machines and structures. Typical examples to which knuckle joints can be put are: Valve and eccentric rods, diagonal stays in boiler, lever and pump rod joint, tension links in bridge structures, link of cycle chain, tie rod joint for roof truss, link of suspension chain, and rod connections of various types.

The joint may have several forms, but a typical form is shown in Fig. 11.10. The ends of each of the rods are formed by upsetting the rod and machining. One of the rod ends is forged into the shape of fork or yoke or double eye end and the other into the shape of an eye. The eye is inserted between the jaws of the fork end and after aligning the holes in the fork end and eye end, the knuckle pin is inserted through them. The knuckle pin has a head at one end and at the other end a collar is secured to it by a taper pin. The pin is prevented from rotating in the fork by means of a small, stop, pin, peg or snug.

To get a better performance from the joint, the sides of the eye and fork end are machined, the pin is turned and the hole is accurately drilled. Material used for making the knuckle joint is usually steel or wrought iron.

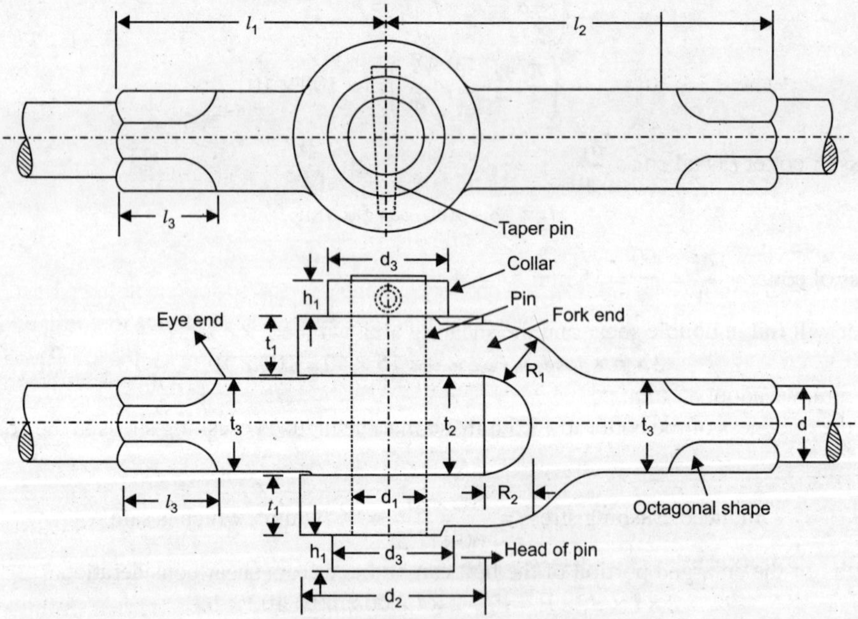

Fig. 11.10 *Knuckle joint*

The forged fork and eye ends are octagonal to a certain length and then square for some length. Octagonal shape is given to facilitate their adjustment while assembling the axis of the joint. The nature of the joint allows limited relative angular movement of the rods about the axis of the pin.

The methods of failure of the joint and the strength of the joint for each method of failure are discussed below. The assumptions are:

(a) There is no stress concentration.
(b) The load is uniformly distributed over each component.

With these assumptions, the strengths given below will be approximate, however they serve to indicate a well proportioned joint. Assume that a load F acts in tension along the axis of the joint, further, assume that the material of all the components is same having permissible tensile, compressive and shear stresses as f_t, f_c and f_s respectively.

(1) Failure of the solid rod: Due to the load F, the rod will fail in tension. If d is the diameter of the rod, then by equating the tensile resistance of the rod to the load F, we can find the rod diameter.

$$\therefore \quad F = \frac{\pi}{4} d^2 \times f_t \qquad ...(11.14)$$

(2) Failure of the pin: The pin is subjected to double shear. Equating the shear resistance of the pin to the load F, we can determine the pin diameter. If d_1 is the diameter of the pin then,

$$F = 2 \times \frac{\pi}{4} d_1^2 \times f_s \qquad ...(11.15)$$

From this relation, the diameter d_1 of the pin can be calculated. This assumes that, there is no clearance between the pin and the mating component, hence no bending of the pin will take place.

But, in practice the components are assembled with sufficient clearance to permit angular movement of one with respect to the other. Once, there is clearance between the components and the pin, the pin will be subjected to bending also.

Assuming the load to be uniformly distributed on the pin in the eye end and varying uniformly on the pin in the fork end from zero at the outside to a maximum at the inside. The loading of the pin is shown in Fig. 11.11. The maximum bending will occur at the mid section of the pin.

Maximum bending moment

$$= \frac{F}{2}\left[\left(\frac{t_1}{3} + \frac{t_2}{2}\right) - \frac{t_2}{4}\right] = \frac{F}{2}\left[\frac{t_2}{4} + \frac{t_1}{3}\right]$$

Now from relation

$$M_b = f_b \times z$$

$$\frac{F}{2}\left[\frac{t_2}{4} + \frac{t_1}{3}\right] = f_b \times \frac{\pi}{32} d_1^3 \qquad ...(11.16)$$

From relation 11.16, the diameter of the pin d_1 can be calculated, if the dimensions t_2 and t_1 are known. The dimension t_2 is generally taken as 1.25 d and t_1 as 0.75 d.

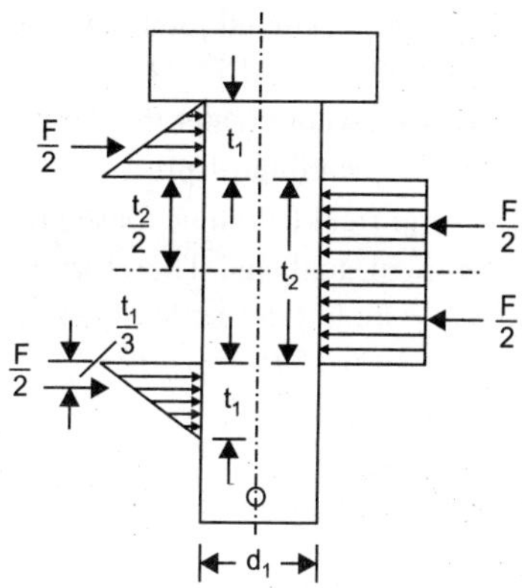

Fig. 11.11 *Bending of pin*

(3) Failure of eye end by shearing: The eye end will get sheared off due to the tensile force F.

$$F = (d_2 - d_1)\, t_2 \times f_s \qquad ...(11.17)$$

From this relation the diameter d_2 can be found out.

(4) Failure of eye end by tearing: The eye end will get teared off due to the tensile force F.

$$F = (d_2 - d_1)\, t_2 \times f_t \qquad ...(11.18)$$

This relation will help us to check the induced tensile stress in the eye end, if it comes out to be more than the allowable value, then the diameter d_2 can be increased.

(5) Failure of eye end in crushing: The pin or the eye end can get crushed under the tensile load F.

$$F = t_2 \times d_1 \times f_c \qquad ...(11.19)$$

This relation will help us to check the induced crushing stress in the eye end or pin, if it comes out to be more than the allowable value, then the thickness t_2 can be increased.

In the case of the valve rod and eccentric rod joint, the pin acts as a journal in the hole. As there is a relative motion between the eye end and the pin, a brass bush is used as a bearing. Thickness of the brass bush is taken as 2 mm to 3 mm. In such a case, in relation 11.18 in place of d_1 the outside diameter of bush will be taken. If p_b is the bearing pressure between the eye end and pin then,

$$F = t_2 d_1 \times p_b \qquad ...(11.20)$$

This relation will help us to check the bearing pressure produced. For slow speeds and gravity lubrication, the allowable bearing pressure is 7.5 N/mm^2 to 10 N/mm^2. Once the design is done from bearing point of view, the crushing limit will not arise.

(6) Failure of fork end by shearing : The fork end will get sheared off due to the tensile Force F.

$$F = (d_2 - d_1)\, 2 \times t_1 \times f_s \qquad ...(11.21)$$

From this relation the induced shearing stress can be checked. In case it comes out to be more than the allowable, then increase the thickness t_1.

(7) Failure of fork end in tension : The fork end will get teared off due to the tensile force F.

$$F = (d_2 - d_1)\, 2 \times t_1 \times f_t \qquad ...(11.22)$$

This relation will help us to check the induced tensile stress. In case a brass bush is used in the fork end also, then the diameter d_1 will be replaced by the outside diameter of brass bush.

(8) Failure of fork end in crushing: The pin or the fork end can get crushed under the tensile load F.

$$F = 2 \times t_1 \times d_1 \times f_c \qquad ...(11.23)$$

This relation will help us to check the induced crushing stress.

In case there is bearing in the fork and pin also, then a brass bush will be used here also. Therefore, it should be designed from the bearing point of view.

$$F = 2t_1 \times d_1 \times p_b \qquad ...(11.24)$$

From this relation the induced bearing pressure should be less than the allowable value.

A well designed joint is one that has equal strengths in failure for all the methods discussed above. From relations 11.17 and 11.21, it can be seen that the dimension t_1 should be half the t_2. However, t_1 is generally made somewhat greater than half of t_2 in order to prevent deflection or spreading of forks, which would cause excessive bending of pin.

Empirical formulae, based on wide experience in a particular service are of more practical use than the theoretical analysis. Therefore, the students are advised to see that the dimensions found by theoretical analysis should not be less than the empirical values given below. Assume d is the diameter of the rod.

$t_1 = 0.75d$ $d_1 = d$ $R_1 = 0.6d$

$t_2 = 1.25d$ $d_2 = 2d$ $R_2 = 0.8d$

$t_3 = 1.2d$ $d_3 = 1.5d$ $l_1 = 4d$

$l_2 = 4.5d$

$l_3 = t_3 = 1.2d$

mean diameter of taper pin = 0.25 d

$h_1 = 0.5d$

Figure 11.12 shows different types of pins used as fulcrums in knuckle joints.

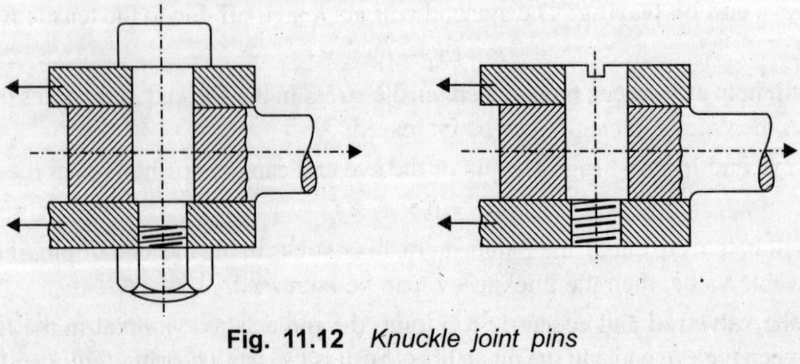

Fig. 11.12 *Knuckle joint pins*

Another form of knuckle is shown in Fig. 11.13. This is used to connect a hollow tube at one end or a rod, the one end of which is threaded to take the fork. The example of a threaded rod at one end is of hand pump in which the valve rod is connected to the lever. Instead of connecting the hollow tube with the threaded joint, it can be connected with the help of a pin to the fork also.

Figure 11.14 shows the use of knuckle joint for connecting three rods. In this joint two rods are forked and the third rod is plain having an eye formed at one end. A common pin passing through all the rods connects them. The dimensions of the joint can be obtained by the principles discussed in Article 11.8. The proportions given for connecting two rods require a little modification namely the pin being longer should be slightly larger in diameter. The diameter may be taken as 1.25 times the rod diameter or calculated.

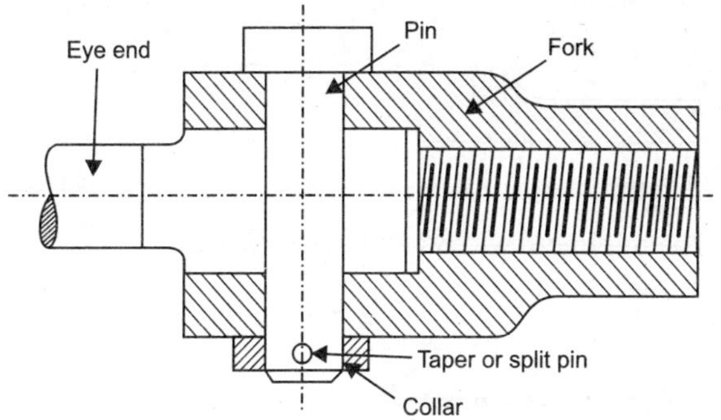

Fig. 11.13 *Knucle joint for hollow tube*

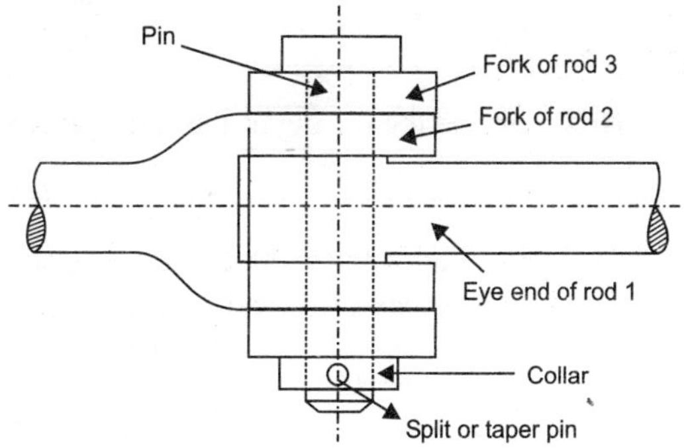

Fig. 11.14 *Knuckle joint connecting three rods*

11.9 SUSPENSION LINK JOINT

The ends of the links of a suspension bridge are usually enlarged. One method is shown in Fig. 11.15. This arrangement provides sufficient material to allow for a hole of a diameter d or the net sectional area of the enlarged end is the same as the section of the link.

Let a tensile load F, then the load on each side of the link will be $F/2$ as shown in figure . If b_1 denotes the width of the enlarged end, d the diameter of the pin and t the thickness of each side link then the values of b_1, d and t in terms of b may be found as follows.

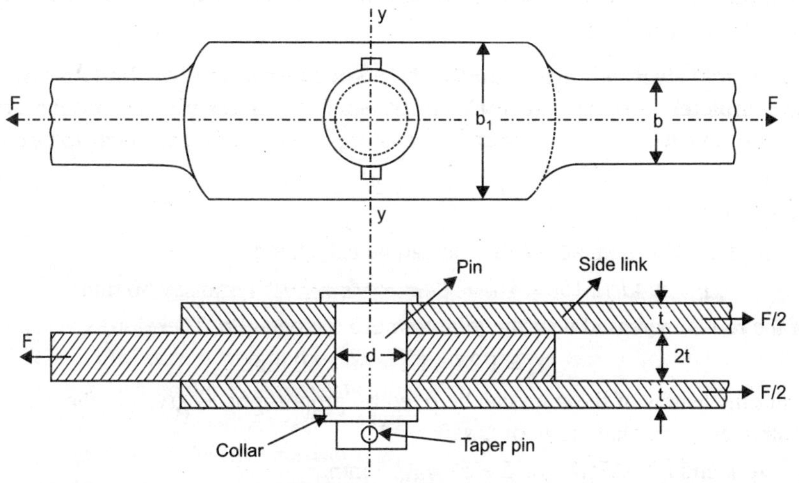

Fig. 11.15 *Suspension link joint*

Assuming permissible crushing, tensile and shearing stresses to be f_c, f_t, and f_s respectively.

The proportional unit usually adopted is b, the width of each link. Then, the tensile strength of each side link is given by the relation

$$f_t \times b \times t = \frac{F}{2} \quad \text{...(11.25)}$$

The relation will help us to calculate the thickness t.

Considering the crushing strength of pin or side link on projected area $d \times t$ is

$$f_c \times d \times t = \frac{F}{2} \quad \text{...(11.26)}$$

From this relation the diameter of pin can be calculated.

Taking $f_c = 1.5\, f_t$, and equating relation 11.25 to 11.26, we have

$$f_t \times b \times t = 1.5\, f_t \times d \times t$$

gives, $d = \frac{2}{3} b$.

Now, considering tearing of the side link at section $y - y$.

$$f_t (b_1 - d)\, t = \frac{F}{2} \quad \text{...(11.27)}$$

From this relation the width b_1 can be calculated. The minimum width will be $b_1 = (b + d)$.

Equating relations 11.25 and 11.27, we have

$$f_t\,(b_1 - d)\, t = f_t \times b \times t$$

Putting $d = \frac{2}{3} b$ in this, we get $b_1 = \frac{5}{3} b$.

Considering the shearing of the pin at the side link, the pin will be in single shear here.

$$f_s \times \pi/4 d^2 = \frac{F}{2} \quad \text{...(11.28)}$$

This relation will help us to check the induced shear stress in the pin. Equating relations 11.26 and 11.28

$$f_c \times d \times t = f_s \times \pi/4 d^2$$

Taking $f_c = 1.5 f_t$ and $f_s = 0.53\, f_t$, we get

$$t = 0.26\, d.$$

Thickness of the middle link will be taken as $2t$. The mean diameter of the taper pin will be taken as 0.20 times the pin diameter. Thickness of pin head and collar will be $0.3d$. Diameter of the pin head and collar will be 1.5 d where d is the diameter of pin.

Example 11.5. Two mild steel rods are connected by a knuckle joint to transmit an axial load of 150 kN. Design the joint completely. Assume the working stresses for both the pin and rod material as 80 N/mm^2 in tension, 68 N/mm^2 in shear and 160 N/mm^2 in crushing. Assume that the rods to be connected are not very long.

Solution: Refer to Fig. 11.10.

Using relation 11.14, the diameter of the rod can be calculated.

$$F = \pi/4 d^2 f_t,\ 150 \times 10^3 = \pi/4 d^2 \times 80,\ d = 48.8 \text{ mm say } 50 \text{ mm.}$$

The size of the rod is increased near the joint to make the octagon. Therefore,

$$t_3 = 1.2d = 1.2 \times 50 = 60 \text{ mm.}$$

The different dimensions of the joint can be fixed by proportions as given in the text and then checked for the induced stresses to see that the joint is safe.

Thickness of fork end $t_1 = 0.75d = 0.75 \times 50 = 37.5$ mm.

Thickness of eye end $t_2 = 1.25 \times d = 1.25 \times 50 = 62.5$ mm.

Thickness of head of pin or collar $h_1 = 0.5d = 0.5 \times 50 = 25$ mm.

Diameter of pin $d_1 = d = 50$ mm.

Diameter of fork end $d_2 = 2d = 2 \times 50 = 100$ mm.

Diameter of pin head or collar $d_3 = 1.5\,d = 1.5 \times 50 = 75$ mm.
Radius $R_1 = 0.6d = 0.6 \times 50 = 30$ mm.
Radius $R_2 = 0.8d = 0.8 \times 50 = 40$ mm.
$l_1 = 4d = 4 \times 50 = 200$ mm, $l_2 = 4.5d = 4.5 \times 50 = 225$ mm.
$l_3 = 1.2d = 1.2 \times 50 = 60$ mm.

Failure of pin: Pin will fail in double shear. The induced shearing stress can be checked by relation 11.15.

$$F = 2 \times \pi/4d_1^2 \times f_s$$

$$150 \times 10^3 = 2 \times \pi/4 \times (50)^2 f_s$$

gives, $$f_s = 38.5 \text{ N/mm}^2,$$

which is less than the allowable shear stress of 68 N/mm^2, hence safe. The pin can also fail under bending because of the clearance between pin and eye and fork ends. The induced bending stress can be checked by relation 11.16.

$$\frac{F}{2}\left(\frac{t_2}{4}+\frac{t_1}{3}\right) = f_b \times \frac{\pi}{32} d_1^3$$

$$\frac{150 \times 10^3}{2}\left(\frac{62.5}{4}+\frac{37.5}{3}\right) = f_b \times \frac{\pi}{32} \times (50)^3$$

gives, $$f_b = 172 \text{ N/mm}^2$$

which is much more than the allowable bending stress of 80 N/mm^2. Adopting the diameter of the pin as 65 mm gives the induced bending stress as 78.9 N/mm^2, hence safe.

Failure of the eye end can be checked under shearing by relation 11.17.

$$F = (d_2 - d_1)\, t_2 \times f_s,\ 150 \times 10^3 = (100 - 65)\ 62.5 f_s$$

gives, $f_s = 68$ N/mm^2, which is safe.

The eye end will get teared off, which can be checked by relation 11.18.

$$F = (d_2 - d_1)\, t_2 \times f_t,\ 150 \times 10^3 = (100 - 65)\ 62.5 f_t$$

gives, $f_s = 68$ N/mm^2, which is less then 80 N/mm^2, hence safe.

The pin or the eye end can get crushed under the tensile force F, this can be checked by relation 11.19.

$$F = t_2 \times d_1 \times f_c,\ 150 \times 10^3 = 62.5 \times 65 \times f_c$$

gives, $f_c = 37$ N/mm^2, which is much below the allowable value.

The fork end will get sheared off due to the tensile force F. This induced shearing stress can be checked by relation 11.21.

$$F = (d_2 - d_1) 2 \times t_1 \times f_s,\ 150 \times 10^3 = (100 - 65)\ 2 \times 37.5 \times f_s$$

gives, $f_s = 57.2$ N/mm^2, which is much safe.

The induced tensile stress in the fork end can be checked by relation 11.22.

$$F = (d_2 - d_1)\, 2 \times t_1 \times f_t,\ 150 \times 10^3 = (100 - 65)\ 2 \times 37.5 \times f_t$$

gives, $f_t = 30.8$ N/mm^2, which is much below 80 N/mm^2, hence safe.

The pin or the fork end can get crushed under the tensile load F. This induced crushing stress can be calculated by relation 11.23.

$$F = 2 \times t_1 \times d_1 \times f_c,\ 150 \times 10^3 = 2 \times 37.5 \times 65 \times f_c$$

gives, $f_c = 30.8$ N/mm^2, which is much below 160 N/mm^2, hence safe.

In this exercise the bearing pressure in the eye and pin is assumed to be high, because of low angular movement of the rod and this movement is intermittently.

Mean diameter of taper pin = $0.25d = 0.25 \times 50 = 12.5$ mm.

Hence, we see that the joint is safe as the stresses are within the allowable values.

Example 11.6. A tube outside diameter 75 mm is connected to the fork end by a pin as shown in Fig. 11.16. The tensile load on the tube is 20 kN. Calculate the diameter of the pin, the thickness of the tube, minimum thickness of the fork end and the dimension *l*. Material of the pin connecting the rod with fork is same. Working stresses in tension, shear and crushing are 80 N/mm^2, 62.5 N/mm^2 and 110 N/mm^2 respectively.

Solution: Assume d_p as the diameter of the pin. The pin will be under double shear.

$$2\frac{\pi}{4}d_p^2 \times f_s = F$$

$$2 \times \frac{\pi}{4}d_p^2 \times 62.5 = 20 \times 10^3$$

gives, d_p = 14.5 mm say 15 mm.

Outside diameter d_1 of the tube = 75 mm, d = inside diameter of tube.

The thickness of the tube can be calculated by considering the tearing of the tube at the pin section.

Tearing of the tube = $\pi/4(d_1^2 - d^2) - (d_1 - d)d_p$.

Equating the tearing resistance of the tube to the tensile force F.

$$[\pi/4(d_1^2 - d^2) - (d_1 - d)d_p] f_t = F$$

$$[\pi/4\{(75)^2 - d^2\} - (75 - d)\ 15]\ 80 = 20 \times 10^3$$

gives, d = 72.5 mm.

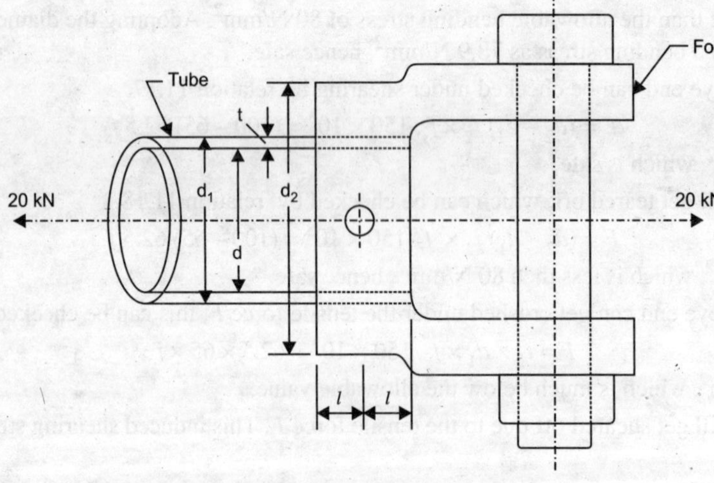

Fig. 11.16

The inside diameter of the tube can also be found by considering the crushing of the pin connecting the tube with the fork. Equating the crushing resistance of the pin to the tensile fore F.

$$(d_1 - d)\ d_p \times f_c = F,\ (75 - d)\ 15 \times 110 = 20 \times 10^3$$

gives, d = 63 mm.

The internal diameter of the tube will be kept as 63 mm to take up crushing of the pin.

Thickness of the tube, $t = \dfrac{d_1 - d}{2} = \dfrac{75 - 63}{2} = \dfrac{12}{2} = 6$ mm.

The length l of the tube can be established by considering the shearing of the tube which is in double shear.

$$2l(d_1 - d) f_s = F,\ 2l(75 - 63) \times 62.5 = 20 \times 10^3$$

gives, l = 13.3 mm.

The minimum thickness of the fork end can be calculated by considering the crushing of the pin in the fork. The diameter of the pin in the fork will be same as the diameter of pin connecting the tube with the fork.

Let d_2 be the outer dimension of the fork.

$$F = (d_2 - d_1)\, d_p \times f_c,\ 20 \times 10^3 = (d_2 - 75)\ 15 \times 110$$

gives, $d_2 = 87$ mm.

$$\text{Thickness of fork} = \frac{d_2 - d_1}{2} = \frac{87 - 75}{2} = 6 \text{ mm}.$$

Example 11.7. Design the suspension link of a structure which is subjected to a load of 160 kN. The allowable stresses in tension, shearing and crushing are 100 N/mm², 75 N/mm² and 150 N/mm² respectively. The width of each side link is 50 mm.

Solution: Refer Fig. 11.15, $F = 160$ kN.

The tensile strength of each side link is given by the relation 11.25.

$$f_t \times b \times t = \frac{F}{2},\ 100 \times 50 \times t = \frac{160 \times 10^3}{2}, \text{ gives } t = 16 \text{ mm.}$$

Considering the shearing of the pin at the side link, the pin diameter can be calculated by relation 11.28.

$$f_s \times \pi/4d^2 = \frac{F}{2},\ 75 \times \pi/4d^2 = \frac{160 \times 10^3}{2}, \text{ gives } d = 37 \text{ mm.}$$

The induced crushing stress in the pin can be checked by relation 11.26.

$$f_c \times d \times t = \frac{F}{2}, f_c \times 37 \times 16 = \frac{160 \times 10^3}{2}, \text{ gives } f_c = 135 \text{ N/mm}^2$$

which is within the safe limits of 150 N/mm².

By considering the tearing of the side link at the pin section, we can find the width of the enlarged portion.

$$f_t\,(b_1 - d)\, t = \frac{F}{2},\ 100 \times (b_1 - 37)\ 16 = \frac{160 \times 10^3}{2}, \text{ gives } b_1 = 87 \text{ mm.}$$

This width is not to be less than $b + d = 50 + 37 = 87$ mm, which is same as calculated.

The pin will be under push fit in the side link and the main link to avoid bending of the pin.

Thickness of the middle link = $2 \times t = 2 \times 16 = 32$ mm.

Mean diameter of the taper pin = $0.2 \times 37 = 7.4$ mm say 7.5 mm.

Thickness of pin head or collar = $0.3d = 0.3 \times 37 = 11$ mm.

Diameter of pin head or collar = $1.5d = 1.5 \times 37 = 55.5$ mm say 55 mm.

11.10 PIN JOINTS

Pins: Pins are examples of a large number of fastenings of a general type that is finding increasingly popular application in machinery, especially, in light machinery parts. Geometrically, pins may be divided into cylindrical pins, called 'straight pins', and conical pins, or 'taper pins'.

Cylindrical pins are held in the hole by frictional resistance which is due to an interference fit Fig. 11.17(*a*), caused by the opening of the pin ends, Fig. 11.17(*b*), or by other means. This pin is also used as a key and is considered to be most suitable for low power drives, but an adaption has been used in multiple form to anchor the disc of a steam turbine rotor to its shaft.

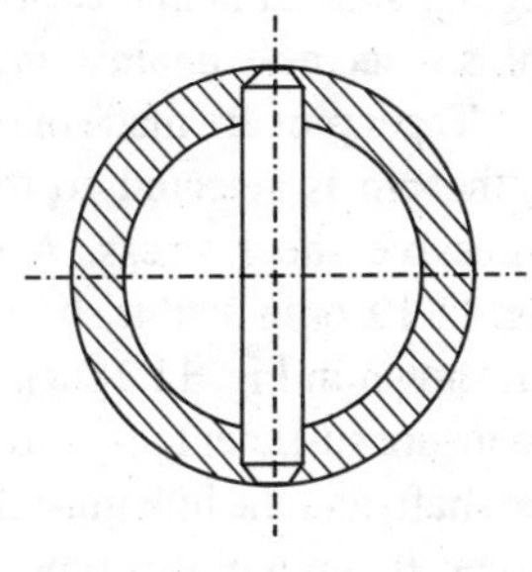

(*a*) Cylindrical pin

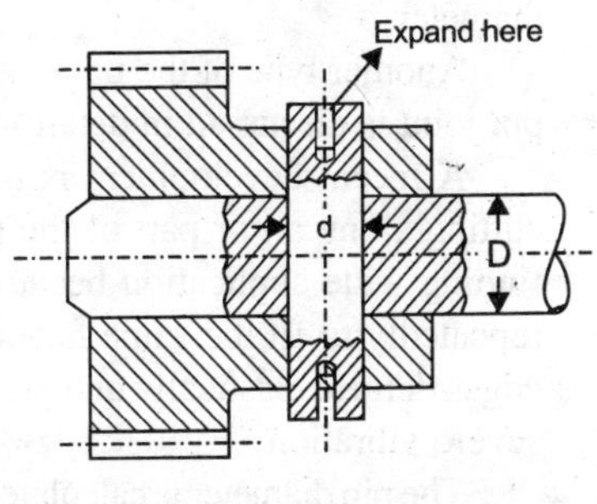

(*b*) Cylindrical with open end

Fig. 11.17

Standard taper pins may be used as fasteners for light work by placing them tangent to the shaft or on a diameter as shown in Fig. 11.18. Hollow shafts are connected frequently by means of a sleeve and taper pins as shown in Fig. 11.18 (*c*). The pin holes should be drilled and reamed with a taper reamer after the parts have been assembled. The diameter of the large end of the pin should be one fourth the shaft diameter and the taper is 1 : 48. To secure a part on a shaft in regard to both rotation and axial movement, the pin is inserted as shown in Fig. 11.18(*a*). However, if the torque transmitted to the hub is not heavy and it is desired to avoid unnecessary weakening of the shaft, the pin may be inserted as shown in Fig. 11.18(*b*).

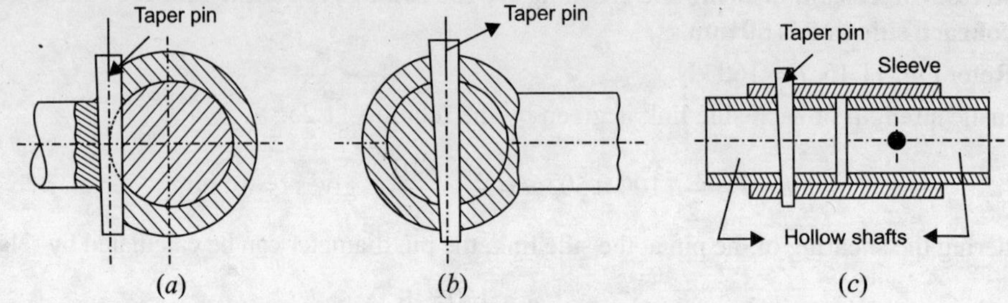

Fig. 11.18 *Use of taper pin*

When a transverse force F acts on the flange as shown in Fig. 11.19, the pin becomes a connecting pin which must be properly dimensioned to resist the force F. To remove the taper pin when there is no through hole, its large end may be threaded and a nut may be used as a puller.

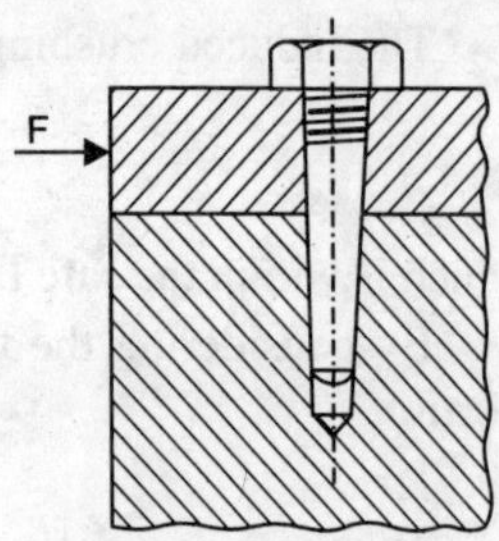

Fig. 11.19 *Taper pin*

Roll pin or split tubular spring pin is also a recent development to replace various tapered pins. This pin is made of steel or stainless steel, and is formed by rolling a plate into the shape of a cylinder with an axial gap. Both ends of the pin are chamfered, which permits it to be driven into a hole with a diameter slightly smaller than that of the pin. The spring action of the compressed pin holds it securely against any vibration of shock.

Taper pins are made of ordinary steel or of 0.10 carbon steel. The diameter of the pin is selected so that force acting on the pin does not induce an excessive shear stress. A pin may work in single shear like the one in Fig. 11.19, or in double shear, like the one shown in Fig. 11.18 (*b*). A tangential pin shown in Fig. 11.18(*a*), produces friction between the shaft and the hub. To insure sufficient pressure the hole is first reamed to a taper with the hub on the shaft, and the hub must then be taken off and the hole in it reamed slightly larger. In a pin of this type, satisfactory operation depends on the mechanic.

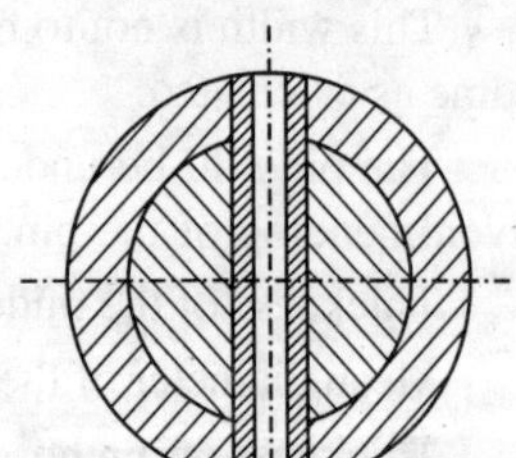

Fig. 11.20 *Split tubular spring pin*

The mean diameter of the pin is made $0.20d$ to $0.25d$, where d is the shaft diameter.

Another type of the pin is knuckle pin, which is shown in Figs. 11.10 and 11.12. The design of knuckle pin joint is discussed in detail in Article 11.8.

A recent development is the design of cylindrical and taper pins with grooves formed by coining or cutting along all or part of the pin length, such pins are called 'Grooved Pins', Fig. 11.21. These pins are finding wide application because they do not require reaming of the drilled holes. Such pins can be used repeatedly to fit the same holes. These pins have three tapered rolled grooves with protruding edges. The edges deform elastically and prevent loosening of the pin under the action of variable loading and even under severe vibration.

The pin diameter is calculated for shear. When the joint carries the shearing force F, the strength condition is

$$F = z \times \pi/4d^2 \times f_s \qquad \text{...(11.29)}$$

where z is the number of pins used. This relation is for single shear.

For the design shown in Fig 11.17 (*b*), which transmits torsional moment M_t over a shaft diameter D, the strength equation will be

$$\frac{2M_t}{D} = 2 \times \pi/4 d^2 \times f_s \qquad ...(11.30)$$

For taper pin, d is the mean diameter of the pin, f_s the allowable shear stress in the material of the pin.

Where the pin is loose in the joint, it should be designed from bearing, bending and crushing point of view also.

Pins are also used to locate the relative position of two parts when there is little or no force acting on the pin. These pins are called 'dowel pins' or simply 'dowels'. A connecting pin like dowel, may be used either as a permanent connection or as a fulcrum for a moveable joint. Fig. 11.22 (*a*) shows the use of a dowel pin. To insure accuracy of assembly the dowel is sometimes made tapered.

Pins are also used as a 'safety pin' which have the function of a shear pin which are designed to fail at a predetermined load and protect expensive parts from damage, Fig. 11.22 (*b*). Safety links, designed to fail in tension under the action of an overload, perform the same function as a safety pin.

Fig. 11.21 *Grooved pin*

Pins should be accurately fitted into the reamed holes and located so that bending of the pin will be avoided or reduced to a minimum. Provision should be made for pins to be removed, this may require a through hole or a threaded hole as shown in Fig. 11.19.

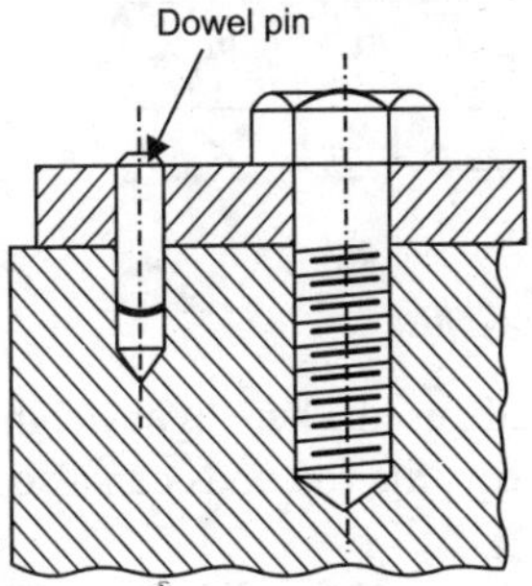

(*a*) *Dowel pin*

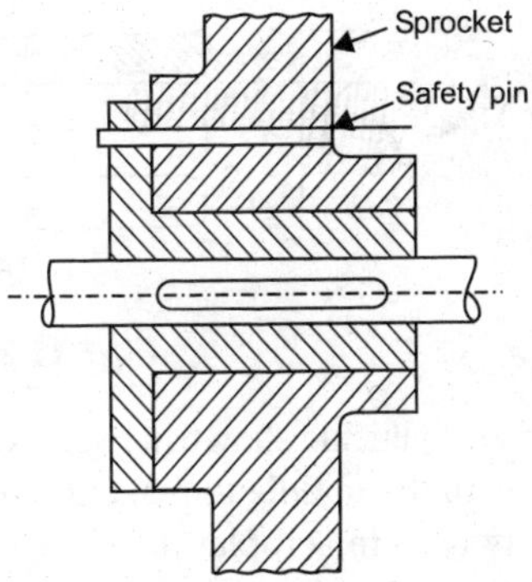

(*b*) *Safety pin*

Fig. 11.22

Designation of pins: (IS 2393 : 1980)

(a) **Cylindrical pins:** shall be designated by the nominal diameter, the tolerance on diameter and nominal length of pin.

Example. A cylindrical pin of nominal diameter 10 mm, tolerance h_8 and nominal length 20 mm shall be designated as:

Cylindrical pin $10h_8 \times 20$ IS : 2393

(b) **Solid taper pins:** shall be designated by the nominal diameter and nominal length, (IS : 6688-1981).

Example. A solid taper pin of 10 mm nominal diameter and 50 mm nominal length shall be designated as:

Solid taper pin A 10 × 50 IS : 6688.

(c) **Split taper pins :** shall be designated in the same way as solid taper pins.

Split taper pin C 10 × 50 IS : 6688.

Material: Cylindrical and taper pins conforming to this standard shall be manufactured from steel conforming to C 40.

For special applications, such as those required by the railways the taper pins may be manufactured from steel conforming to I.S. 226-1975 specification for structural steel (standard quality).

TURN BUCKLE:

11.11 ADJUSTABLE JOINTS

Joints which are designed for axial loading and in which subsequent tightening of joint is possible are adjustable joints. An example is provided by a turn buckle. An adjustable joint as given in Fig. 11.23 is very largely used in connecting round rods which are subjected to tensile loading. This type of joint is used where slight adjustment in length is required, e.g., for connection of tension members of steel structures, tie rod of a wall crane, telegraph wires, bridge structures, electric poles, etc. One end of each rod is threaded, one having right hand threads and the other having left hand threads. Two rods are connected by a coupler which has two threaded ends. The body of the coupler is made hexagonal or square, so that it can be turned easily with the help of a spanner. As the coupler is turned either both the rods move closer or away from each other depending upon the direction of rotation of the coupler. Sometimes, the coupler is turned with the help of a lever which is inserted in the hole shown dotted in the figure.

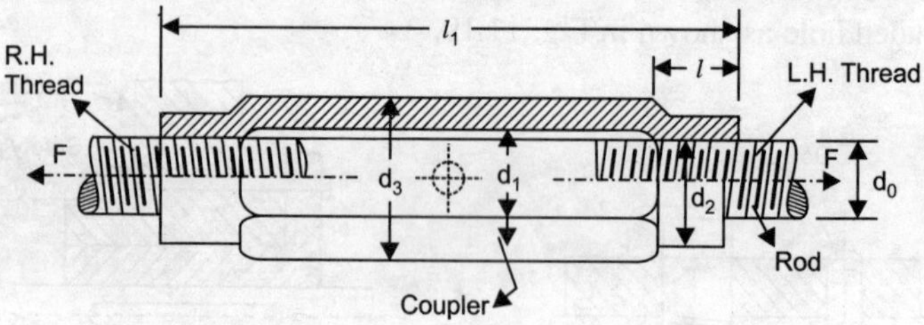

Fig. 11.23 *Turn buckle*

Another form of a turn buckle shown in Fig. 11.24 is generally used where light weight is the consideration. This coupler is hollow in the middle to make it lighter. Secondly, the rod ends can also be seen. Such a type of coupler is generally used in aeroplanes.

The material of the rod and coupler may be different or same, depending upon the tensile load on the joint. The methods of failure of the joint and the strength of the joint for each method of failure are discussed below.

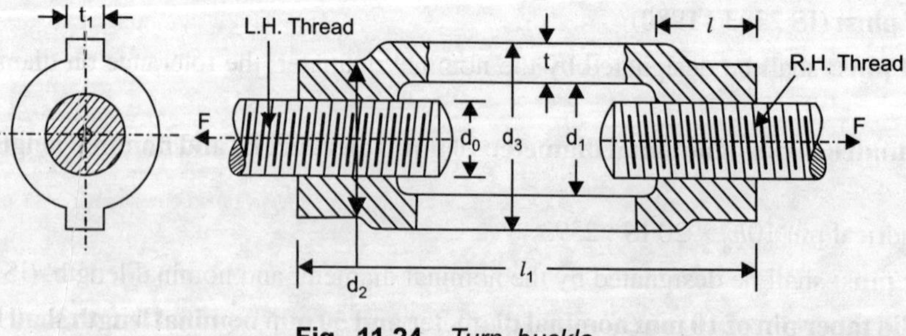

Fig. 11.24 *Turn buckle*

1. **Failure of solid rod :** In this case the bolt shank will carry tensile load F, in addition the thread on the rod end will carry torque which will be between the coupler end and the rod. The torsional moment M_t on the thread of rod and coupler end is given by theory of Mechanisms and machines.

$$M_t = F\frac{d_m}{2}\tan(\alpha+\phi) \quad ...(11.31)$$

where F = is the tensile load on the rod.

α = is the angle of friction = $\tan^{-1}\mu$

μ = is the coefficient of friction between threads of coupler and rod.

ϕ = inclination of the threads.

$= \tan^{-1}\dfrac{p}{\pi d_m}$ where p is the lead of threads.

d_m = is the mean diameter of threaded rod end.

$= (d_0 + d_c)\ 1/2.$

The case under consideration can be reduced, with an accuracy sufficient for engineering practice to the main task if we take $F_0 = 1.3$ F as the design load. If d_c is the core diameter of the rod, and the rod is likely to fail in tension,

Then
$$\pi/4 d_c^2 \times f_t = F_0 \quad ...(11.32)$$

where f_t is the allowable tensile stress in the material of the rod.

From above relation the core diameter of rod can be calculated. Generally coarse threads are used and by specifying the type of threads, the size of the rod can be given.

2. Failure of the threaded end of coupler: The length of the threaded portion of the coupler l can be established by considering the shearing of the threads on the coupler end.

$$\pi\, d_0 \times l \times f_s = F_0 \quad ...(11.33)$$

where f_s is the allowable shear stress in the coupler material.

The outside diameter d_2 of the coupler end can be calculated by considering the failure of the threaded end of coupler in tension. The area resisting the tension failure will be $\pi/4(d_2^2 - d_0^2)$.

$$\pi/4\ (d_2^2 - d_0^2) \times f_t = F, \quad ...(11.34)$$

where f_t is the allowable tensile stress in the coupler.

3. Failure of the coupler at the centre: The coupler can fail at the centre due to tensile failure. Equating the tensile resistance to the force F, we get

$$\pi/4(d_3^2 - d_1^2) \times f_t = F \quad ...(11.35)$$

The inside diameter d_1 of the coupler is generally taken as (d_0 + 5 mm). From this relation d_3 can be calculated if d_1 is known.

The length l_1 of the coupler is established, once the amount of adjustment is known. Usually this length is taken as 6 times the rod diameter. The threaded length l usually comes out to be very small as calculated by relation 11.33. This length has to be minimum equal to the diameter of the rod for steel coupler and 1.5 times the rod diameter for cast iron coupler.

The diameter d_2 is not to be less than 1.25 times the rod diameter.

The outside of the coupler has a hexagonal shape. This hexagonal shape is made on the diameter d_3. Sometimes instead of using a spanner, a lever is used to turn the coupler. The torsional moment to be supplied by the lever is given by relation 11.31. The length of the lever can be fixed by considering the average force applied by a man as 250 N and the lever can be designed under bending for a maximum bending moment equal to the torsional moment M_t.

For a turn buckle shown in Fig. 11.24, the different proportions are given below.

Length of coupler = $l_1 = 8d_0$ $\qquad t_1 = 0.75\ d_0$

Diameter of threaded end of coupler = 1.25 d_0 $\qquad t = 0.5\ d_0$

Length of threaded end = l = 1.25 d_0

Inside diameter of coupler = $d_1 = d_0 + 5$ mm.

Example 11.8. Two round tie rods of a wall crane are connected by means of a coupling joint as shown in Fig. 11.23. The maximum pull in the rod is 50 kN. Assuming that the allowable stress in tension, compression and shear as 70 N/mm^2, 80 N/mm^2 and 30 N/mm^2 respectively, design the joint.

Solution: Taking the same notations as shown on figure.

Design load for the threads $= F_0 = 1.3 \times F = 1.3 \times 50 \times 10^3 = 65 \times 10^3$N

By using the relation 11.32, we can find out the size of the rod.

$\pi/4 d_c^2 \times f_t = F_0$, $\quad \pi/4 d_c^2 \times 70 = 65 \times 10^3$

gives, $d_c = 34.5$ mm. $\quad d_0 = \dfrac{d_c}{0.84} = \dfrac{34.5}{0.84} = 41.8$ mm say 42.5 mm

Thus size of the rod = 42.5 mm $\quad d_c = 35.5$ mm.

The length of the threaded portion of the coupler l can be established by considering the shearing of the threads on the coupler end.

$\pi d_0 \, l \times f_s = F_0$, $\quad l \times \pi \times 42.5 \times 30 = 65 \times 10^3$, gives $l = 16.3$ mm.

In general practice l is not kept less than the size of the rod. Therefore, we keep the length of the screwed end as 42.5 mm.

The inside diameter d_1 of the coupler $= d_0 + 5$ mm.

$= 42.5 + 5 = 47.5$ mm.

The outside diameter d_2 of the coupler end can be fixed by relation 11.34.

$\pi/4(d_2^2 - d_0^2) f_t = F$, $\quad \pi/4(d_2^2 - (42.5)^2) \times 70 = 50 \times 10^3$

gives, $d_2 = 52.5$ mm.

But this diameter is not to be less than 1.25 d_0.

Therefore $d_2 = 1.25 \times 42.5 = 53.5$ mm. say 55 mm.

The diameter at the centre of the coupler can be fixed by relation 11.35.

$\pi/4(d_3^2 - d_1^2) \times f_t = F$, $\quad \pi/4(d_3^2 - 47.5)^2) \times 70 = 50 \times 10^3$

gives, $d_3 = 56.5$ mm.

To keep the thickness of the coupler uniform, adopt this diameter as 60 mm. On this diameter hexagonal shape will be made to put the spanner.

Length l_1 of the coupler $= 6\, d_0 = 6 \times 42.5 = 255$ mm.

If a joint shown in Fig. 11.24 is to be designed, the following procedure will be adopted.

The dimensions d_0, d_1, l_1 and d_2 will be designed in the same fashion as discussed above.

Thickness of coupler $t = 0.75 \times 42.5 = 0.75\, d_0 = 31.8$ mm.

Thickness of coupler $t_1 = 0.5\, d_0 = 0.5 \times 42.5 = 21.25$ mm.

The strength of the coupler can be checked under tearing when $d_3 = d_1 + 2t = 47.5 + 2 \times 31.8 = 47.5 + 63.6 =$ 111.1 mm. say 112.5 mm.

Now considering the tearing of the coupler

$F = 2 \times d_3 \times t_1 \times f_t$, $\quad 50 \times 10^3 = 2 \times 112.5 \times 21.25 \times f_t$

gives, $f_t = 10.5$ N/mm^2 which is much below the allowable tensile stress of 70 N/mm^2, hence safe.

The threaded length of the coupler end should be checked for crushing of threads. Assuming metric threads, number of threads per centimeter of length is 2.

The crushing area or projected area $= \pi/4(d_0^2 - d_c^2) \times$ no. of threads on length l

$= \pi/4\, [(42.5)^2 - (35.5)^2] \times 2 \times 4.25$

$= 3600$ mm^2

Crushing stress f_c induced $= \dfrac{F_o}{3600} = \dfrac{65 \times 10^3}{3600} = 18.1$ N/mm^2

which is much below 80 N/mm^2, hence safe.

EXERCISES

1. Define a cotter.
2. Differentiate between a cotter and a key.
3. Why the taper is provided in a cotter? A single taper cotter is preferred over double taper cotter, why?
4. Discuss the short-comings of a cotter joint. What are the advantages which make the use of cotter joint inspite of short-comings?
5. Design a cottered joint like that of Fig. 11.2 to safely resist a load of 70 kN which acts along the axis of the rods connected by the cotter. The material of the cotter and rods will permit the following safe stresses.

$$f_t = 55 \text{ N/mm}^2, f_c = 85 \text{ N/mm}^2, f_s = 43 \text{ N/mm}^2.$$

6. A tensile load is transmitted through a 45 mm. rod fitted with a flat cotter is 10 mm × 40 mm in cross-section. The edge of the opening for the cotter is 35 mm from the rod end. The opposite end of the rod is threaded, and the permissible stress in the threaded portion is 60 N/mm^2. Determine the weakest part of the joint, if the allowable stresses in all parts of the joint are as given in problem 5.
7. Make a neat sketch showing two views of a cotter joint and write equations showing the strength of the joint for the most probable methods of failure.
8. For the cottered connection shown in the sketch 11.25 determine the maximum load F that should be applied for the following allowable stresses:

$$f_t = 35 \text{ N/mm}^2, f_s = 17.5 \text{ N/mm}^2, \text{ and } f_c = 85 \text{ N/mm}^2. \text{ Neglect stress concentration.}$$

9. Determine the main dimensions of a cotter joint similar to that shown in Fig 11.5 in the text, assuming that a force of 150 kN tends to separate the two halves. The allowable stresses in tension, compression and shear are 100 N/mm^2, 160 N/mm^2 and 70 N/mm^2 respectively.

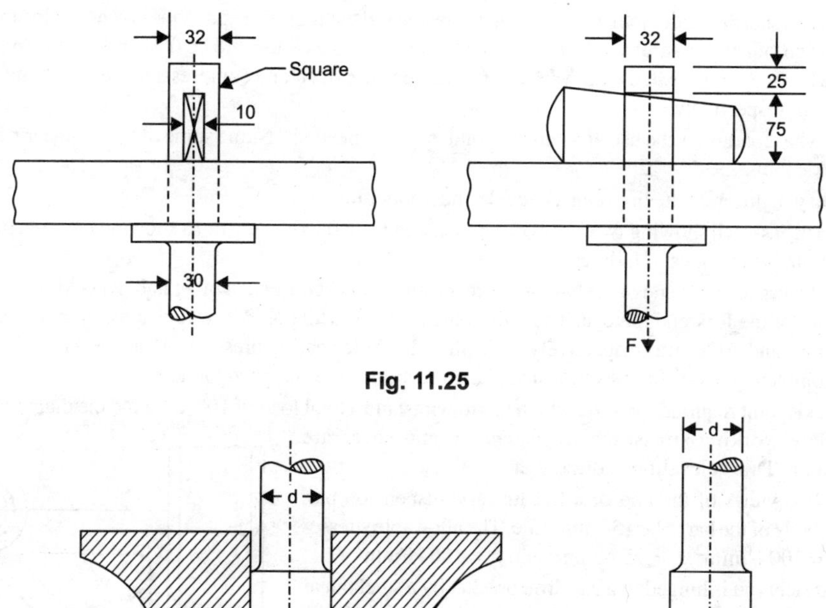

Fig. 11.25

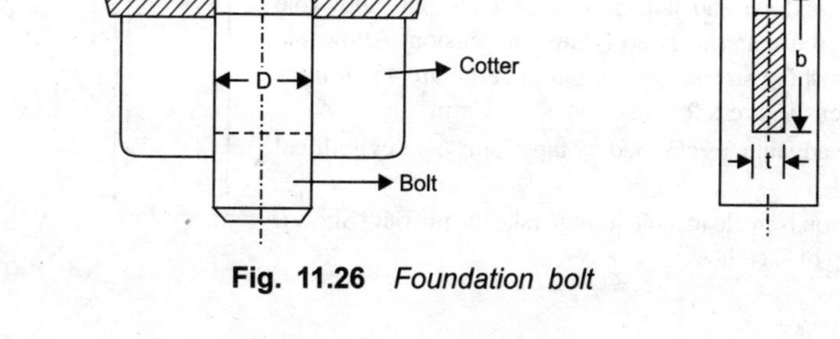

Fig. 11.26 *Foundation bolt*

10. Determine all the dimensions of cotter joint similar to that in Fig 11.2 in the text, for an external load of 10 kN, which changes continuously from tension to compression. The material of the cotter and rods will permit the following safe stresses.

$$f_t = 55 \text{ N/mm}^2, f_c = 85 \text{ N/mm}^2, f_s = 43 \text{ N/mm}^2.$$

Draw two views of the designed joint and dimension it completely.

11. A foundation bolt with a circular end is secured by means of cotter as shown in Fig. 11.26. Determine the dimensions D, b and t in terms of d, shearing stress on cotter to be 3/4, and bearing stress twice the tension stress in the bolt. Find the dimensions if $d = 63$ mm.
12. If a bolt and cotter are made of the same material, and if the depth of a cotter is equal to the bolt and the thickness 1/4 the diameter, prove that the cotter is weaker than the bolt.
13. Two round steel rods 45 mm. diameter are connected by a sleeve and cotter joint. Design and make detailed dimensioned drawing of the joint. Scale 1/2 full size. Assume tensile stress in rods 80 N/mm^2, shearing stress in cotter 70 N/mm^2. Thickness of cotter 0.25 d, where d is the diameter of the rod through which it passes. What is the crushing stress in the cotter ?
14. Discuss the use of a gib.
15. Sketch a cotter joint with double gibs for a connecting rod end.
16. How is the slipping of a cotter avoided, discuss any two methods?
17. The piston rod is keyed to a piston 0.6 m. in diameter by a cotter. Determine the two diameters of the piston rod, the dimensions of the cotter and the piston boss dimensions. Steam pressure is 1 N/mm^2 gauze. All the parts are of the same material, allowable stresses 25 N/mm^2 in tension and in shear and 75 N/mm^2 in crushing.
18. Design and draw a gib and cotter joint with two gibs to resist a safe tensile load of 50 kN. The material of the gib, cotter and rods is same for which the allowable safe stresses are.

$$f_c = 60 \text{ N/mm}^2, f_s = 20 \text{ N/mm}^2 \text{ and } f_t = 25 \text{ N/mm}^2.$$

19. Design and sketch a cotter joint for fastening the piston rod to the cross-head of the engine having cylinder diameter 0.24 m and steam pressure is 0.7 N/mm^2. The thickness of the cotter is to be 0.3 times piston rod diameter at the point where cotter is located. Allowable stresses in tension, shear and compression are 50 N/mm^2, 40 N/mm^2 and 80 N/mm^2 respectively.
20. Under what conditions the use of a knuckle joint is recommended ? Name some of the engineering applications of a knuckle joint.
21. How do you differentiate between a knuckle and a cotter joint?
22. Make a neat sketch showing two views of a knuckle joint, and write equations showing the strength of joint for the most probable methods of failure.
23. An eye is forged at one end of the half of the tie rod and a fork at the end of other half. Knuckle pin is passed through the holes of the fork and the eye. The pull in the tie rod is 150 kN. Allowable stresses in tension and shear are 65 N/mm^2 and 30 N/mm^2 respectively. Take the allowable bearing pressure on the pin as 10 N/mm^2. Design the joint completely. Draw the elevation and plan in section of the designed joint.
24. A knuckle joint required for a rod which has to withstand a total load of 100 kN. Find the diameters of the rod and pin. Safe working stresses both in tension and shear are 65 N/mm^2. Draw elevation and end view.
25. Draw two views of the end of a flat link for suspension bar, taking body of the bar to be 150 mm wide. The allowable stresses are, $f_t = 100$ N/mm^2, $f_s = 75$ N/mm^2 and $f_c = 150$ N/mm^2.
26. A mild steel rod is hinged to a cast iron bracket by a steel pin as shown in Fig. 11.27. The rod carries a tensile load of 45 kN. Design the tie rod, eye end of the rod, fork end and pin. Allowable stress in the rod material is 80 N/mm^2 in tension. Allowable shear and crushing stresses of the pin material are 45 N/mm^2. Allowable tensile stress for cast iron is 28 N/mm^2.
27. Mention the advantages claimed by taper pins over cylindrical pins.
28. What provision is made in a pin joint to take the pin out? Show it with the help of sketches.

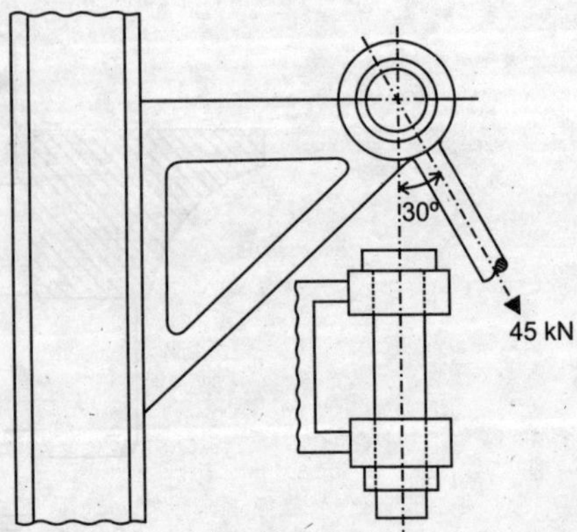

Fig. 11.27

29. Explain the advantage of roll pin over other types of pin.
30. How do you differentiate between a taper pin and grooved pin?
31. Differentiate between the use of a dowel and a safety pin with the help of sketches and explain.
32. (a) Determine all the dimensions of a knuckle pin which carries a load of 6000 N and has a rocking motion of 20°.
    (b) Find the stresses in the pin.
    (c) Give a sketch of the joint.
33. Make a neat sketch showing two views of an adjustable joint, and write equations showing the strength of the joint for the most probable methods of failure.
34. Two round tie rods of a wall crane are connected by means of a coupling joint as shown in Fig. 11.23. The maximum tensile load is 100 kN. Assuming that the allowable stresses in tension and shear are 70 N/mm^2 and 30 N/mm^2 respectively, design the joint.
35. Design the turn buckle shown in Fig. 11.24 for the same data as given in problem 34.
36. The pull in cast iron coupler of the stay rod of a telephone post is 15 kN. Assuming that the allowable stress in tension for rod material is 70 N/mm^2 and for cast iron coupler, the stresses in tension and shear are 35 N/mm^2, design the coupler completely. How much would be the rods drawn together for one turn of the coupler?
37. A turnbuckle is used to connect two round rods as shown in Fig. 11.28. The maximum tensile load is 40 kN. Assuming that the allowable stresses in tension, compression and shear are 70 N/mm^2, 80 N/mm^2 and 30 N/mm^2 respectively, design the turnbuckle completely. Draw and dimension the designed turnbuckle.

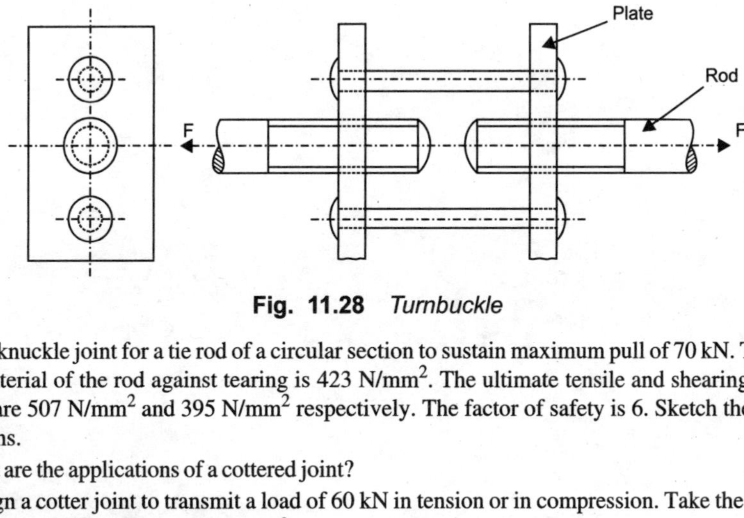

Fig. 11.28 *Turnbuckle*

38. Design a knuckle joint for a tie rod of a circular section to sustain maximum pull of 70 kN. The ultimate strength of the material of the rod against tearing is 423 N/mm^2. The ultimate tensile and shearing strengths of the pin material are 507 N/mm^2 and 395 N/mm^2 respectively. The factor of safety is 6. Sketch the designed joint with dimensions.
39. (a) What are the applications of a cottered joint?
    (b) Design a cotter joint to transmit a load of 60 kN in tension or in compression. Take the stresses as:
    Allowable tensile stress = 60 N/mm^2
    Allowable shear stress = 48 N/mm^2
    Allowable crushing stress = 110 N/mm^2.
    Sketch a sectional elevation of the joint to a suitable scale giving all the dimensions.
40. What are: spigot and socket in a cotter joint?
41. The two end rods of the knuckle joint are usually upset into octagonal shape. Why?
42. Design and sketch a gib and cotter joint. It is to take up a tensile load of 35 kN and is made of C 40 steel.
43. Design and sketch a knuckle joint. The tensile load is 60 kN.
44. A cotter joint is used to connect two rods which are in:
    (a) shear only (b) tension only
    (c) compression only (d) both tension and compression.
45. A Knuckle joint is used to connect two rods which lie in
    (a) same plane (b) different planes
    (c) parallel planes (d) perpendicular planes

46. A Knuckle joint may fail in:

(a) Bending mode (b) shear mode

(c) crushing mode (d) all the above modes

47. Explain the bending failure of cotter and the Knuckle pin.
48. What is uniform strength concept used in the design of cotter joint?
49. Why the body of the coupler in the turn-buckle is made hexagonal in section?
50. Why the body of the coupler in some designs of turn-buckle is made hollow?
51. Give the material of cylindrical and taper pin.
52. How the pins are designated as per BIS?
53. What are safety pins?
54. What are: eye end and Fork end of a Knuckle joint?
55. What is foundation cotter bolt? Sketch it.
56. Design a cotter joint for an axial load of 50 kN which slowly reverses in direction. Assume suitable material and factor of safety.

❑❑❑

CHAPTER 12 Keys and Couplings

12.1 KEYS

12.1.1 Introduction

The most common function of a key is to prevent relative rotation of a shaft and the member to which it is connected, such as the hub of a gear, pulley, disc, or crank. Keys give a temporary joint when torsional loads are to be transmitted from the shaft to the connecting members, or vice versa.

An extensive use of key joint is largely due to their simple and dependable design and construction, reliable operation, convenience of assembly and disassembly, low cost, etc.

The major disadvantages of this type of joint are:

1. The need for a keyway which not only makes the effective cross-section of the part smaller but also involves considerable stress concentration. Failures of shafts and axles are very often caused by high local stresses arising in the area of the keyway.
2. Keyed joints cannot be fitted closely enough to ensure accurate concentricity of parts and that one key cannot transmit large torques.

The greater accuracy required and complicated load conditions made the development of splines made integral with the shaft. These joints are superior to key joints because of the following reasons:

1. Greater strength of the splined shafts at varying and impact loads.
2. Larger contact area of splines which can transmit greater loads.
3. Better centering of the joined parts and better guidance of the hub along the axle.

Splined joints have the following main disadvantages:

1. Local stresses in the recentering corners of grooves, especially with straight-sided splines, which are, however, lower than in the case of keyways.
2. Inevitable uneven load distribution between the splines.
3. The need for special plant and cutting and measuring tools for the manufacture of splined shafts and hubs.

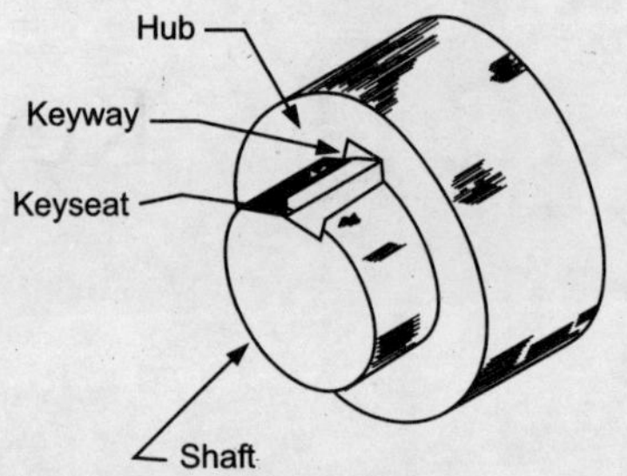

Fig. 12.1

As keys are subjected to considerable crushing and shearing loads, therefore, they are always made of steel. To accommodate the key, a groove called the 'keyseat' is cut in the shaft and a groove called ‘keyway’ is cut in the hub of the part to be mounted. Thus after assembly the key is partly in the shaft and partly in the hub of the mating part as shown in Fig. 12.1. This way the two parts are locked together so that one cannot rotate without the other.

A large number of types of keys are available and choice in any installation depends on a number of factors, such as power requirements, tightness of fit, stability of connection, and cost. For very light power requirements, a set screw may be tightened against the round shaft or against a flat spot on the shaft. For heavy power requirements a positive connection, such as that offered by a key, is necessary.

12.1.2 Type of Keys

The distinguishing features so overlap that a single, all embracing classification is impossible. Keys of various types will be described, but they will be classified only according to their duty.

Light duty and medium duty keys: Keys used for light duty and medium duty are shown in Fig. 12.2.

Square key: Figure 12.2 (*a*). The square key with the key sunk half in the shaft and half in the hub is the type most commonly used. This is used for light duty. Flat keys are used where the weakening of the shaft is serious. Although there is no universal standard, the square key has sides equal to one fourth the shaft diameter. It finds, its common use in industrial machinery.

Rectangular key: It is a modification of the square key. These are used where added stability of the connection is desired, as in the machine tools, Fig. 12.2 (*b*).

Square or rectangular key may be of uniform cross-section or they may be tapered. In tapered keys the width is uniform and the height of the key is tapered by 1 in 100. The bottom of the hub keyway has the same taper while the key seat in the shaft is not tapered. A taper key can be used for medium duty or a variable torsional moments.

Gib-head key: This is a key similar to the square or rectangular key but with a head projecting from one face, at the large end of the key, as shown in Fig. 12.2 (*c*). This head is used for driving the key when assembling and for drifting it out when dismantling. The latter feature is important if the small end of the key is not accessible for driving out. The taper of the Gib-head key is 1 : 100. The proportions of this key are shown in figure.

Wood ruff key: Figure 12.2 (*d*), it is used for small torsional moments to avoid troublesome fitting. The key seat in the shaft is milled with a special cutter. The key way in the hub is usually straight. The key will align itself in the key seat. It has one disadvantage of weakening the shaft more than by the use of a straight key. It is used extensively in the automotive industry and in machine tools, being particularly adaptable to gears. These keys are convenient during assembly and disassembly, simple to manufacture, but they tranmit comparatively low torsional moments.

Flat key: Figure 12.2 (*e*), is used for light duty when it is not desired to cut a key seat in the shaft. Usually the key has a taper of 1 : 100. A straight key can be used but it requires a set screw to hold it in place. It is a type of the saddle key. It consists of a square sided rectangular key slightly tapered and driven on a flat surface machined on the shaft. It is not suitable for heavy loads or sudden reversal of torsional moments. The disadvantage of this key is that it is liable to slip on the shaft.

Saddle key: Figure 12.2 (*f*), this key is similar to the flat saddle key and is suitable for the same class of service. It has concave bottom surface with a radius slightly less than that of the shaft. It should have a slight taper as it is entirely dependent on friction to prevent relative movement between shaft and hub. It requires no machining of the shaft. It is used for setting eccentrics or cams. If it is made straight, it requires a set screw. They are also used to hold parts during assembly until the permanent keyway can be located and machined. Figure 12.2 (g) shows a saddle key alone, where *l* is the length of the key.

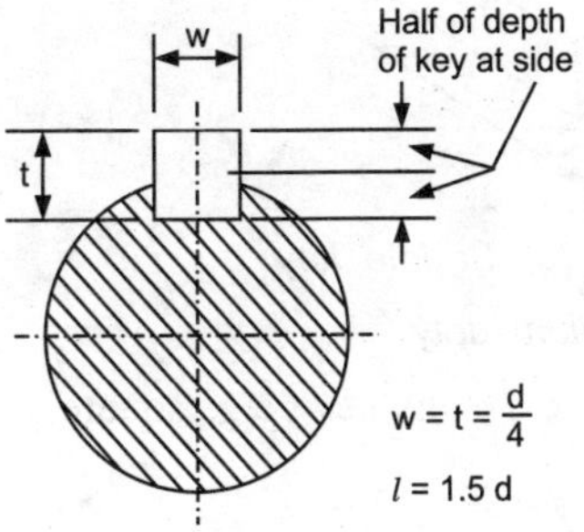

(a) Square key

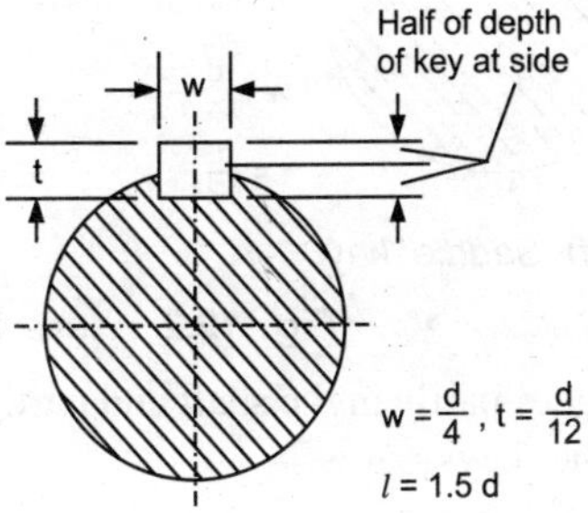

(b) Rectangular key

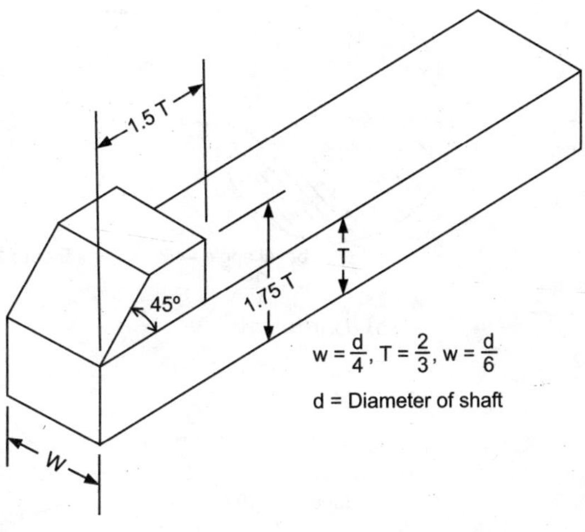

(c) Gib-head key

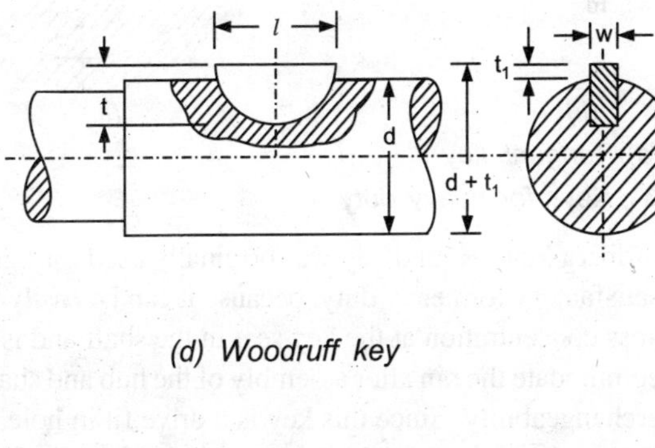

(d) Woodruff key

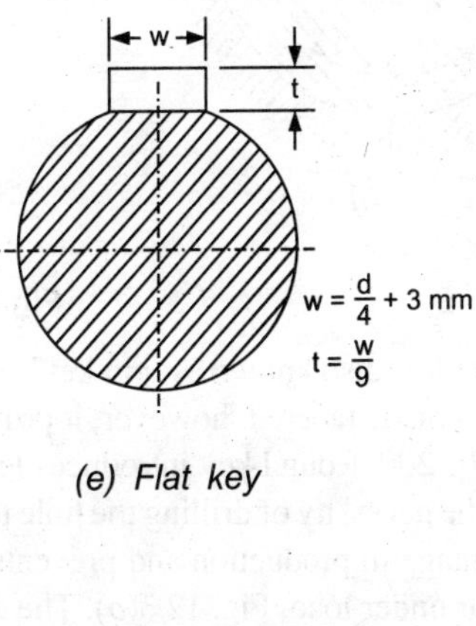

(e) Flat key

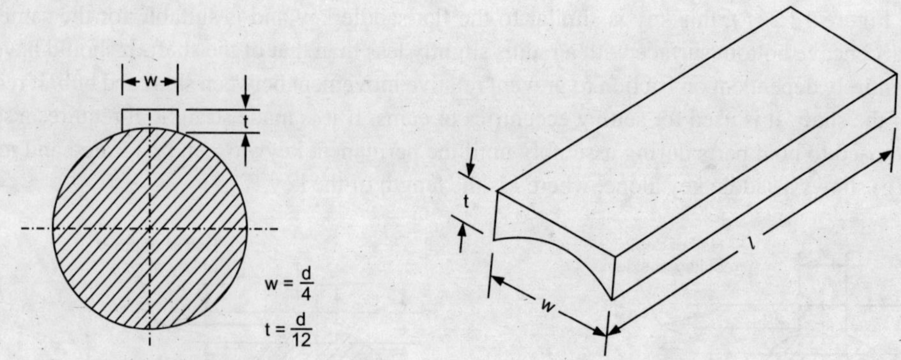

(f) Saddle key *(g) Saddle key*

Fig. 12.2 *Keys for light and medium duty*

A key which goes partly in the shaft and partly in the hub is called a 'sunk key', e.g., square, rectangular, woodruff and feather keys.

Heavy duty keys: Keys used for heavy duty are shown in Fig. 12.3.

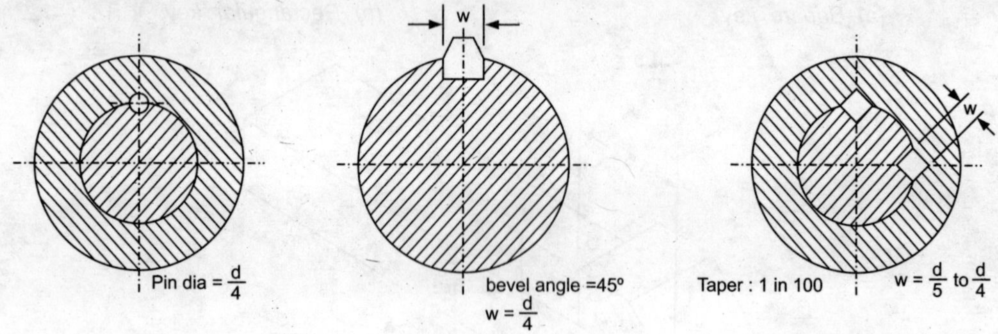

(a) Round key *(b) Barth key* *(c) Kennedy key*

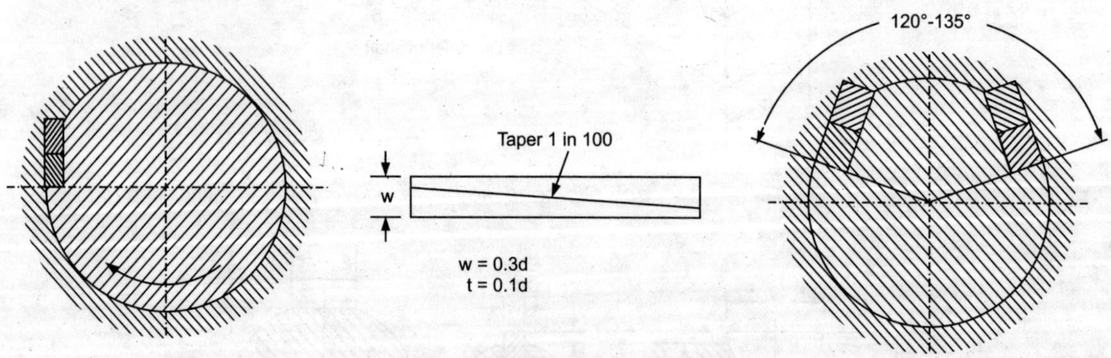

(i) *(ii)* *(iii)*

(d) Tangent key

Fig. 12.3 *Keys for heavy duty*

Round key: It is also known as 'pin key' or 'Nordberg key'. Round key was originally used for light and small work. When made tapered, however, it proved satisfactory for heavy duty, because it can be easily fitted. *It has a taper of 1 : 200*. Round key introduces less stress concentration at the key seat in the shaft and is satisfactory except for the necessity of drilling the hole to accommodate the pin after assembly of the hub and shaft. This may be disadvantage in production and prevents interchangeability. Since this key is a drive fit in hole, it is always in true shear under load, Fig. 12.3(*a*). The diameter at the centre of the pin is one sixth of the shaft diameter.

Barth key: Figure 12.3 (*b*) is a rectangular key with two ends beveled off. The double bevel ensures that the key will fit tightly against the top of the key way when the drive is in either direction, and lessens the tendency to twist. The key does not require a tight fit and the small clearance permits easy assembly and removal. This key is under compression rather than in being shear when under load. The beveled key is frequently used as a feather key. The portion of the key that fits into the shaft may be either square sided or dovetailed.

Kennedy key: Figure 12.3 (*c*), It is used for heavy duty and consists of two keys driven as shown in the figure. Shafts under 0.15 m in diameter use one key, larger shaft use two keys placed 90 or 120 degrees apart. To permit easy assembly, the hub is bored to fit the shaft and hub in concentric position. Like the barth key, they also work in compression. If the torsional moment is to be transmitted in either direction then for all diameters of shaft two keys are used, as in rolling mills.

Tangent key: A Tangent key or a Grip-spring is so placed that its broad side is directed along the tangent and one of its narrow sides is along the radius of the shaft, Fig. 12.3 *d* (*i*). It consists of a pair of wedges given a taper of 1 : 100 each, Fig. 12.3 *d* (*ii*). Since the tapered sides face each other, the keyway in the hub can be made parallel to the shaft axis. The wedges are driven home with a hammer, so that the joint finds itself in a stressed state. The interference fit between the shaft and the hub is in a tangential direction, hence the name — Tangent keys. The direction of rotation is very important for a tangential key joint, Fig. 12.3 *d* (*i*). A reversible unit requires two tangent keys usually located at an angle of 120° to 135°, Fig. 12.3 *d* (*iii*). These keys are used when the shaft diameter exceeds 60 mm and larger, suddenly applied torques are to be transmitted.

Feather keys: are used where it is necessary to slide a keyed gear or pulley along the shaft. Fig. 12.4 (*a*) and (*b*) show two types of feather keys. The key is made as tight fit and fastened in either the shaft or the hub and made a sliding fit in the mating member. At (a) the key is screwed to the shaft at (b) the key with gib-heads fits with the hub. Feather keys are parallel keys and permit relative axial movement of the pair. The pressure in bearing on feather keys should not be more than 7.5 N/mm^2, and if the members are to slide when under load, the pressure should be reduced below this value.

From above, it is clear that the keys can be grouped as follows :

(i) **Taper keys:** They produce strained joints. These include: Sunk or inserted, flat mounted and saddle keys. The disadvantage of these keys is the radial displacement of the axis of the mounted part relative to the shaft axis, which leads to an additional run out. So, such keys are preferred in low-speed transmission not calling for a high accuracy.

(ii) **Prismatic keys:** These keys have no taper, for example, feather keys. They are always of a sunk type. The ends of these keys may be rounded or flat.

(iii) In addition to the above types of keys, widely used types of keys are the woodruff keys, tangent keys and special designs of keys.

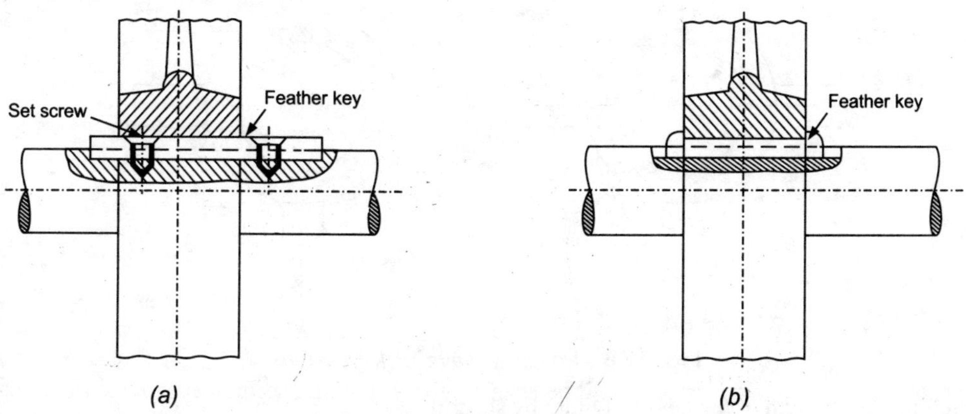

Fig. 12.4 *Feather keys*

12.1.3 Splines

Splines fittings were developed and are extensively used in the automobile industry. As already discussed a feather key is attached to a shaft by screws, but greater strength is obtained, if the feather instead of being attached separately, is actually solid part of the sliding member usually a shaft. Such a feather is called a spline. Splined shafts may have a number of key ways of equal size cut in their material as shown in Fig. 12.5. Splined shafts are used where in addition to transmitting heavy loads, axial sliding motion is also to be permitted. The hubs are made with a corresponding number of key seats. Increasing the number of keys, of course, increases the key strength proportionately. The splines are usually produced on shafts by means of hobs but splined holes are generally produced by broaching.

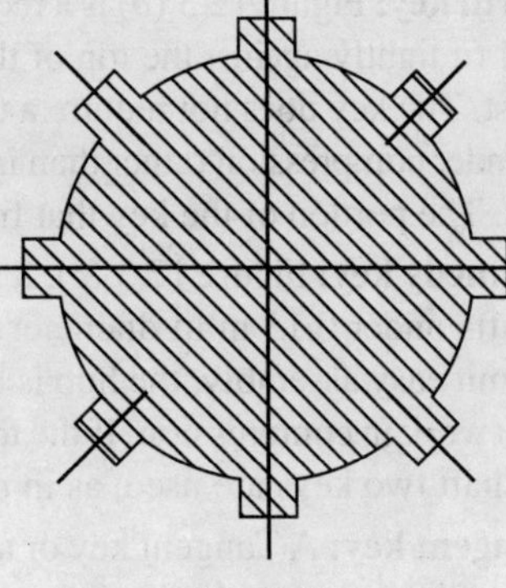

Fig. 12.5 *Splined shaft*

The designation of the splines shall consist of the width, height and length, followed by its Indian standard number (2610-1964). For example a straight side spline of width 6 mm, height 28 mm and length 34 mm shall be designated as :

Straight side spline 6 × 28 × 34 IS : 2610.

12.1.4 Strength of Parallel Rectangular and Square Key (Unstrained Joint)

Figure 12.6 shows two ways in which a key might fail if it were not strong enough to transmit the torsional moment between shaft and gear or pulley or coupling etc. Let M*t* be the torsional moment to be transmitted by the shaft. In each case the key was not able to the push from the shaft tending to drive the key in anticlockwise direction and the pulley or gear holding back this pushing force from the shaft.

These equal and opposite forces are assumed to be the same as the force at the shaft circumference and is given by

$$P = \frac{M_t}{d/2} = \frac{2M_t}{d}$$

In Fig. 12.6 (*a*) the portion of the key in the hub of pulley or gear has slid past the portion of the shaft. The shearing area is indicated in Fig. 12.7 and is equal to $w \times l$. This area was stressed beyond the ultimate shearing strength of the material.

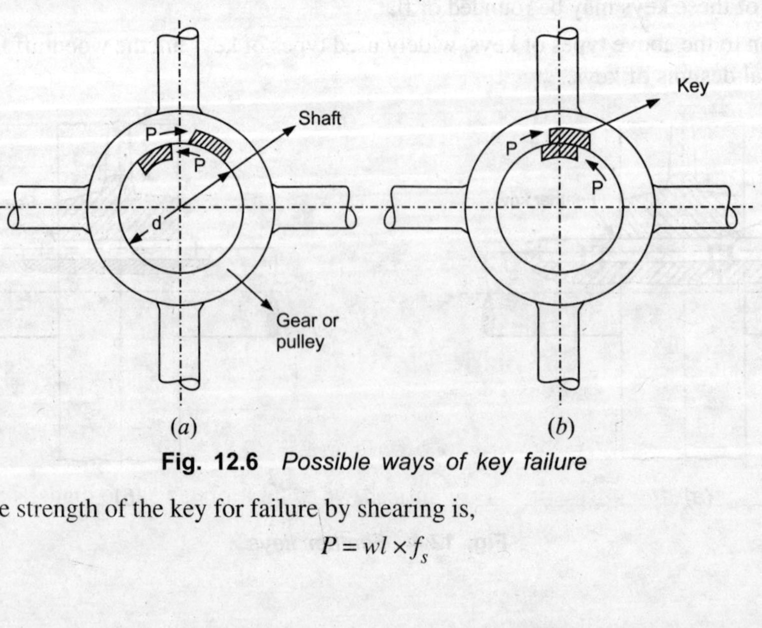

Fig. 12.6 *Possible ways of key failure*

Therefore, the strength of the key for failure by shearing is,

$$P = wl \times f_s$$

or $$f_s = \frac{2M_t}{dwl} \quad ...(12.1)$$

where, f_s is the allowable shear stress of the key material.

w is width of the key

l is the length of the key

and d is the diameter of shaft transmitting the torsional moment.

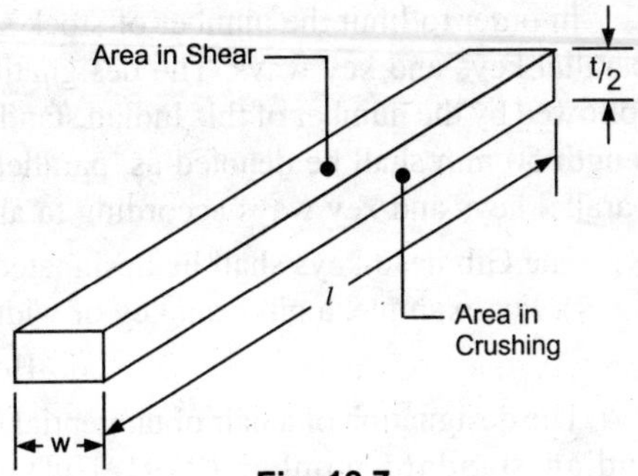

Fig. 12.7

The second method of failure of key is indicated in Fig. 12.6 (*b*).

In this case the key is too weak to resist successfully the push of the equal and opposite forces of the shaft and the hub, and thus the key was crushed between these forces. The crushing area is also shown in Fig. 12.7 and is equal to $l \times t/2$. This area was stressed beyond the ultimate crushing strength of the key material.

Therefore, the strength of the key for failure by crushing is,

$$P = l \times \frac{t}{2} \times f_c$$

or $$f_c = \frac{2M_t}{d} \times \frac{2}{l \times t} = \frac{4M_t}{d \times l \times t} \quad ...(12.2)$$

where f_c is the allowable crushing stress of the key material and t is the thickness of the key.

The force P in the shearing investigation is the same as P in the case of crushing, hence equating the two strengths.

$$wl\,f_s = l \times \frac{t}{2} \times f_c$$

or $$\frac{2w}{t} = \frac{f_c}{f_s} \quad ...(12.3)$$

When the key is perfectly fixed in the shaft and hub, the allowable crushing stress for usual key material (steel) is at least twice the allowable stress in shear. Taking f_c as twice the f_s, the relation (12.3) indicates that a square key is the proper size. In this case, it is necessary to check the key strength for shear only. When the key is not perfectly fixed in the shaft and the hub, the allowable crushing stress is about 70 percent more than the allowable shear stress and in this case the key must be checked for crushing also.

A square key whose sides are $d/4$ may be stronger or weaker than the shaft, depending on the length of the key. The length of the key required for equality strengths of key and the shaft may be determined by equating the strength of the key in shear to the strength of the shaft in shear. Therefore, neglecting all stress concentration.

$$wl\,f_s \frac{d}{2} = \frac{\pi d^3}{16} f_s \quad ...(12.4)$$

Substituting $d/4$ for w, we get

$$l = \frac{\pi d}{2} = 1.5\, d \text{ approximately.}$$

Thus for equal strengths of the key and shaft, the length of the key should be 1.5 times the shaft diameter. The above discussion is the basis for the proportions of a key that has wide industrial use. For gears, and for rotors of large diameter, stability may require a longer hub, and a key of corresponding length may be used. Also, shafts designed on the basis of stiffness are generally overly strong and hence short keys may be used.

It may be noted that a rectangular key is weaker in compression than in shear. The use of the rectangular key is generally limited to equipment with the shaft sizes that are designed for stiffness, which usually have an excess of strength, so that the standard size rectangular key has ample strength.

In order to limit the number of stock sizes of keys, I.S. 2048-1975 gives the standard specifications for parallel keys and key ways. The designation of parallel keys shall consist of the width, height and length, followed by the number of this Indian standard. For example, a parallel key of width 10 mm, height 8 mm and length 50 mm shall be denoted as, parallel key 10 × 8 × 50 IS : 2048. Table 12.1 gives the dimensions of parallel keys and key ways according to above mentioned standards.

The Gib head keys shall be designated by width, height and length followed by its I.S. number (2293-1974). For example, a gib head key of width 8 mm, height 7 mm and length 40 mm shall be designated as:

Gib-Head Key 8 × 7 × 40 I.S. : 2293

The designation of a pair of tangential keys shall consist of thickness, width and common length, and the Indian standard number (2291-1981). For example a pair of tangential keys with $t = 8$ mm, $b = 24$ mm and $l = 100$ mm, shall be designated as:

Tangential Keys 8 × 24 × 100 IS : 2291.

The woodruff keys shall be designated by their width, height and the number of its Indian Standard (2294-1980). For example, a Woodruff key of width 4 mm and height 5 mm shall be designated as :

Woodruff key 4 × 5 IS : 2294.

The taper key shall be designated by their type, width, height and length and the number of its Indian Standard 2292-1974. For example, a taper key of type A, width 10 mm, height 8 mm and length 80 mm shall be designated as:

Taper key A 10 × 8 × 80 IS : 2292.

All types of keys shall be manufactured from steel having a minimum tensile strength of 580 N/mm^2. Steel of designation C 40 or C 55 Mn 75 conform to this requirement.

The assumption that stress concentration may be neglected is no doubt justified, but the key is subject to a concentration of load and stress, also, the key way introduced stress concentration in the shaft so that the stress concentrations will affect both elements and can be neglected to obtain approximate relations of dimensions. In a steel shaft and key made of ductile material and subjected to a static load, stress concentration may not be serious. However, with a variable load, as in a shaft with torsional vibration, the stress concentration may be serious. For the effect of key way on the strength of shaft, see the chapter on design of shafts.

Generally, it is assumed that by cutting the key way the torsional strength of the shaft is reduced by 25 per cent, and then the relation 12.4 becomes,

$$wl\,f_s \times \frac{d}{2} = \frac{\pi d^3}{16} \times f_s \times 0.75 \qquad \text{...(12.5)}$$

Substituting $\frac{d}{4}$ for w, we get

$$l = \frac{\pi d}{2} \times 0.75 = 1.2\,d$$

Thus for equal strengths of the key and shaft, the length of the key should be 1.2 times the shaft diameter when the keyway effect on the torsional strength of the shaft has been taken into account.

Thus the design procedure will be, first to obtain the dimensions of the key by usual proportions as mentioned on Figs. 12.2 and 12.3 and then check for stresses.

Taper key (Strained Joint): The design equation is:

$$l \geq \frac{2.M_t}{d.w.\mu.f_c}$$ (Checking the surface that bears against the shaft, for bearing (crushing) strength.)

where μ = coefficient of friction between the shaft and the hub ≅ 0.25.

Note: Here, key section ($w \times t$) is the minimum.

Table 12.1

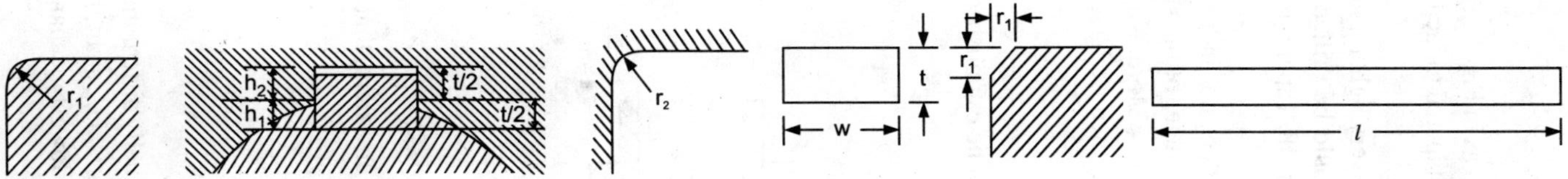

For Shaft	above	6	8	10	12	17	22	30	38	44	50	58	365	75
diameters	upto	8	10	12	17	22	30	38	44	50	58	65	75	85
key	w	2	3	4	5	6	8	10	12	14	16	18	20	22
cross-section	t	2	3	4	5	6	7	8	8	9	10	11	12	14
keyway	In shaft h_1	1.2	1.8	2.3	3.0	3.5	4.0	5.0	5.0	5.5	6.0	7.0	7.5	8.5
keyway	In shaft h_1	1	1.4	1.8	2.3	2.8	3.3	3.3	3.3	3.8	4.3	4.4	4.9	5.4
Length of	$1\frac{\text{Min.}}{\text{Max.}}$	6	6	8	10	14	18	22	28	36	45	50	56	63
Key		20	36	45	56	71	90	110	140	160	180	200	220	250
Chamfer or	$r\frac{\text{Min.}}{\text{Max.}}$		0.25			0.35				0.55				0.80
radius of key			0.16			0.25				0.40				0.60
Keyway radius	r_2 Max.		0.16			0.25				0.40				0.60

For Shaft	above	85	95	110	130	150	170	200	230	260	290	330	380	440
diameters	upto	95	110	130	150	170	200	230	260	290	330	380	440	500
key	w	25	28	32	36	40	45	50	56	63	70	80	90	100
cross-section	t	14	16	18	20	22	25	28	32	32	36	40	45	50
keyway	In shaft h_1	9	10	11	12	13	15	17	19	20	22	25	28	31
keyway	In shaft h_1	5.9	6.4	7.4	8.4	9.4	10.4	11.4	12.4	13.4	14.4	15.4	17.4	19.5
Length of	$1\frac{\text{Min.}}{\text{Max.}}$	71	80	90	100	110	125	140	160	180	200	220	250	280
Key		280	320	360	400	440	400	400	400	400	400	400	400	400
Chamfer or	$r\frac{\text{Min.}}{\text{Max.}}$					1.30			2.00				2.95	
radius of key						1.00			1.60				2.50	
Keyway radius	r_2 Max.					1.00			1.60				2.50	

Woodruff Keys: The projecting part of key can be checked for crushing.

$$\therefore \qquad l \geq \frac{2.M_t}{d.t_1.f_c} \quad \text{(key crushing)}$$

and

$$l \geq \frac{2.M_t}{d.w.f_s} \quad \text{(key shear).}$$

12.1.5 Pin Keys

A pin key can be used in two ways, as shown in Fig. 12.3 (a) and Fig. 12.8 (a). When used as in Fig. 12.3 (a), the pin is ananalogous to a square key and should be equally as strong. The pin is slightly tapered to facilitate the assembly. Hollow shafts are frequently connected by pins as shown in Fig. 12.8 (b).

The method of fitting the pin through the hub, perpendicular to the shaft is suitable to very light torsional moments only. It is used extensively for fastening collars on shafts. The pin holes should be drilled and reamed with a taper reamer after the parts have been assembled. The taper of the pins is l in 48. Taper pins are made of ordinary machine steel, or of 0.10 carbon steel.

The diameter of a taper pin can be found out so that the force acting upon the pin does not induce an excessive shear stress. A pin may work in single shear or in double shear. Let d is the diameter of the shaft, which is to transmit the torsional moment M_t. Then the force to be resisted by the pin will be $= 2M_t/d$. Now, if d_1 is the mean diameter of the pin, and the pin is in double shear as shown in Fig. 12.8 (a) then we have,

$$2 \times \pi/4\, d_1^2 \times f_s = \frac{2M_t}{d}$$

where f_s is the allowable shear strews in the pin material. From this the size of the pin can be established.

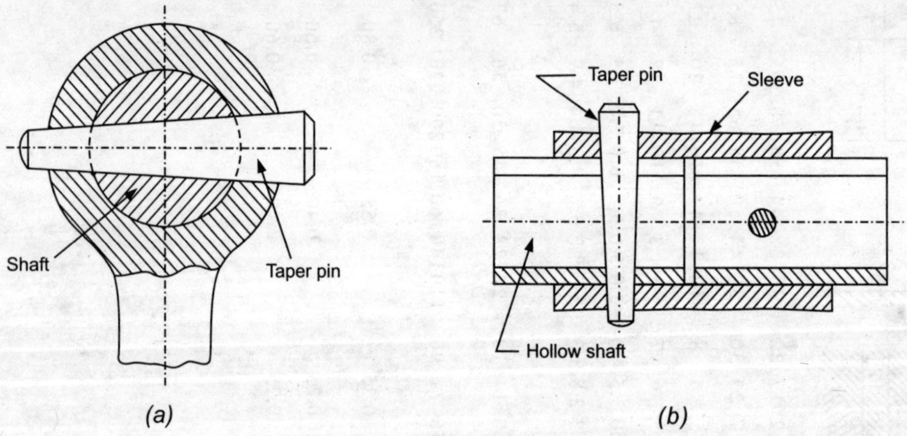

Fig. 12.8 *Use of pin keys*

Satisfactory operation of the taper pin depends on the skill of the mechanic to such an extent, that any calculations may prove to be futile, therefore, workmanship has to be perfect.

Empirically, the mean diameter of the pin is made,

$$d_1 = 0.20\, d \text{ to } 0.25\, d \qquad \text{...(12.7)}$$

12.1.6 Allowable Stresses in Key

The keys are usually made from cold drawn steel. The values of the crushing stress (f_c) are usually differentiated according to the type of the joint (fixed, sliding) and the nature of the load. If the joint parts, i.e., shaft, hub, and key are made from different materials, then the crushing stress for the material with the lowest mechanical properties is used. Table 12.2. gives the values of allowable stresses.

Table 12.2 Allowable stresses

Type of joint	*Material*	*Allowable-shear stress N/mm²*	*Allowable crushing stress N/mm² Nature of load*		
			Steady	*Intermittent*	*Impact*
Fixed	Steel Cast iron	120 87	150 80	100 53	50 27
Sliding	Steel	54	50	40	30

or allowable stresses can be taken as:

$$f_c = f_{yp}/F_S;\ f_s = f_{yp}/2F_S;\ F_S = 1.5 \text{ to } 2 \text{ for fixed key joints.}$$

In sliding key joints, the allowable stresses can be reduced to 1.5 to 2 times as compared to fixed joints.

12.1.7 Strength of Feather Key

One means of accomplishing the frequent engagement and disengagement of a gear is by the use of a feather key. This type of key fits loosely between shaft and hub, allowing the gear to be moved parallel to the shaft or away from the mating gear. In some types the key is constructed that it can be made fast to the hub (Fig. 12.4 (*b*)) and in other types it is fastened by counter sunk screws to the shaft as shown in Fig. 12.4. (*a*).

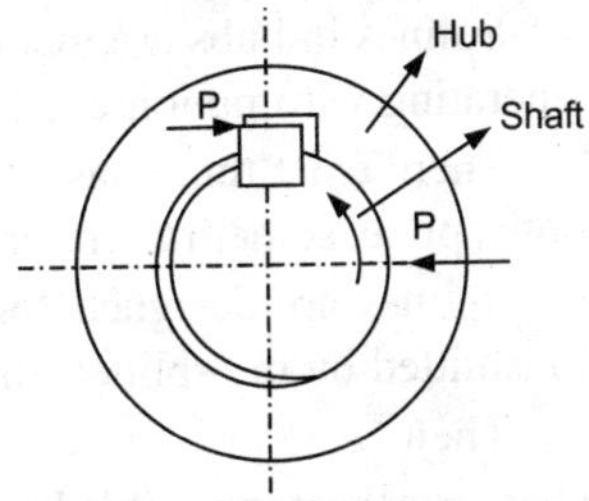

(a) One feather key

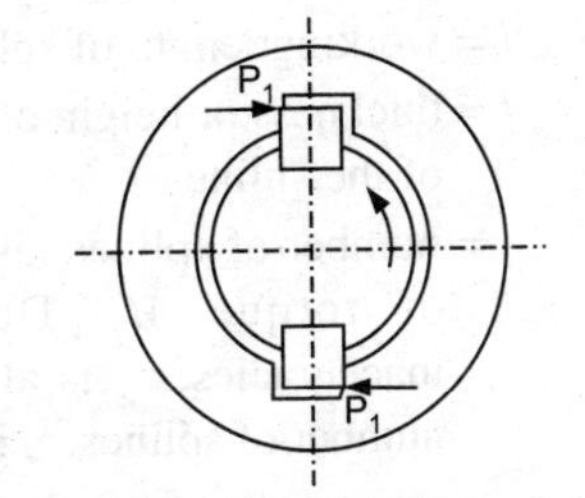

(b) Two feather keys

Fig. 12.9

It is desirable to provide for a minimum of axial force, when a pulley or gear must be moved axially along the shaft while power is being transmitted. The use of two feather keys equally spaced requires a smaller axial force than the use of one key. To show this, consider two cases as shown in Fig. 12.9. Figure 12.9(*a*) shows, when one feather key is used and Fig. 12.9(*b*) shows when two equally spaced feather keys are used. In both cases the keys are fixed in the shaft. The direction of rotation of the shaft is anticlockwise. Let *d* is the diameter of the shaft transmitting a torsional moment M*t*. Because of the clearance between the hub bore and shaft and between the key and the keyway of the hub, the shaft will take a position approximating to the one shown in the figure (a). Then the force *P* is given as,

$$P = \frac{M_t}{d/2} = \frac{2M_t}{d}$$

If μ is the coefficient of sliding friction between key and the hub, then the axial force required to move the hub axially will be,

$$\mu P + \mu P = 2\mu P = \frac{4\mu M_t}{d} \qquad ...(12.8)$$

where two feather keys are used, and will share the load equally, the shaft will take a position as shown in Fig. 12.9(*b*). For this case,

$$P_1 = \frac{M_t}{d}$$ which is half the force of single feather.

Now, the axial force required to move the hub axially will be,

$$P_1\mu + P_1\mu = 2P_1\mu = \frac{2\mu M_t}{d} \qquad ...(12.9)$$

By comparing the relations 12.8 and 12.9 for the axial force, we find that the axial force required for two feather keys is half that for one feather key. From this analysis the advantage of splined shaft can be seen, when the connecting component is to be moved axially. Thus more the number of splines on a shaft lesser the axial force required to move the hub axially.

Feathers are checked for crushing of the protruding narrow side,

$$l \geq \frac{2M_t}{d.t_1.f_c}; \quad t_1 = \text{protruding portion of the key}$$

Also, in shear, $l \geq 2M_t/d \,.\, w \,.\, f_s$.

12.1.8 Splined Joints

Splined joints are formed by projections on the shaft, which fit matching grooves in the hub. As compared with keyed joint, they offer higher load carrying capacity owing to an increased contact area, permit better alignment of the matching parts (which is a very useful property at high speeds), affect the fatigue strength of the shaft to a lesser extent (stress concentration is less at the roots of the teeth than in keyshafts), are easier to manufacture and can be made to a high degree of accuracy.

Splines in hubs are manufactured by broaching followed with grinding and those in shafts by hobbing, generating with pinion cutters, or broaching with subsequent grinding.

There exist three basic types of splines, namely parallel-sided splines, involute splines and triangular tooth spline, sometimes referred to as serrations. Fig. 12.10 (*a*).

Splines are designed for bearing strength of the side surfaces. Let M_t is the torsional moment to be transmitted by the splined shaft.

Then $$M_t = p \times A \times r_m$$

where p is the permissible bearing pressure on the splines.

A = Area of the splines under pressure = t/z_w

l = working length of splines = 1.5 d

t = thickness or height of side surface, in bearing, of the spline

z_w = number of splines taking part in transmission of torque M_t. Due to manufacturing inaccuracies, z_w is always less than the total number of splines, z. Let $z_w = 0.75\, z$.

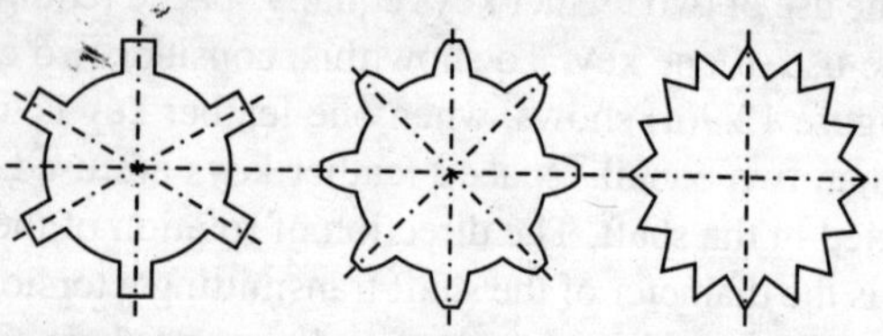

Fig. 12.10 *(a) Types of splines*

r_m = mean radius from the centre of the shaft to the centre of the spline.

$$r_m = \frac{d'+t}{2}$$

where d' is the root diameter of the splined shaft. Or, r_m, can be taken as,

$$r_m = \frac{D+d}{4},$$ where D and d are respectively, the major (outer) and minor (inner) diameters of the splines.

Also, $$t = \frac{1}{2}(D-d) \qquad \therefore \quad p \cong \frac{10.M_t}{(D^2-d^2)\times l \times z} \leq |p|_{allowable}$$

For an involute profile splined joint with D and d, the outer and P.C.D. diameters of the spline involute profile,

$$r_m = \frac{D+d}{4}$$

and bearing area $\cong 0.8\, m\, l\, 0.75\, z$, where m is the tooth module.

Table 12.3 gives the standard dimensions for splines on the shaft. The bearing pressure p depends on whether or not the shaft and hub are heat treated, on the type of the joint (fixed or sliding), etc.

Table 12.3 Standard splines

Type of loading	*4 Splines*	*6 Splines*	*10 Splines*
Permanent fit	$d' = 0.85d$ $w = 0.241d$ $t = 0.075d$	$d' = 0.90d$ $w = 0.25d$ $t = 0.05d$	$d' = 0.91d$ $w = 0.156d$ $t = 0.07d$
To slide when not under load	$d' = 0.75d$ $w = 0.241d$ $t = 0.125d$	$d' = 0.85d$ $w = 0.25d$ $t = 0.075d$	$d' = 0.86d$ $w = 0.156d$ $t = 0.07d$
To slide when under load	—	$d' = 0.80d$ $w = 0.25d$ $t = 0.10d$	$d' = 0.80d$ $w = 0.156d$ $t = 0.095d$

SOLVED EXAMPLES

Example 12.1. A solid machine shaft 50 mm in diameter, transmits a torsional moment of 1200 Nm. A square key is used whose one side is equal to 1/4 the shaft diameter and whose length is equal to 1.5 times the shaft diameter. Find the dimensions of key and check the key for its induced shearing and crushing stresses. Obtain the factor of safety of the key in shear and crushing when yield stress in shearing and crushing are 350 N/mm² and 425 N/mm² respectively.

Solution: Shaft diameter, $d = 50$ mm.

$\therefore$ Width of the key $= \dfrac{d}{4} = \dfrac{50}{4} = 12.5$ mm, length of key $= 1.5 \times 50 = 75$ mm

As the key is a square one, hence $w = t = 12.5$ mm.

Force P, responsible for shearing or crushing of the key $= \dfrac{2M_t}{d} = \dfrac{2\times1200\times10^3}{50} = 48000$ N.

The shearing stress induced in the key is obtained from relation 12.1.

$$f_s = \frac{2M_t}{dwl} = \frac{2\times1200\times10^3}{50\times12.5\times75} = 51 \text{ N/mm}^2$$

The crushing stress induced in the key is obtained from relation 12.2.

$$f_c = \frac{4M_t}{dlt} = \frac{4\times1200\times10^3}{50\times75\times12.5} = 102 \text{ N/mm}^2$$

F.O.S. in shear $= \dfrac{350}{51} = 6.86$; F.O.S. in crushing $= \dfrac{425}{102} = 4.17$.

Example 12.2. A medium steel shaft transmits a torsional moment of 1960 Nm. The diameter of the shaft is 58 mm. Select a standard key from Table 12.1 and check it for the induced unit shearing and crushing stresses. Obtain the factor of safety of the key in shear and crushing. Assume the material of the shaft and key to be same.

Solution: From Table 12.1, for the diameter of the shaft to be 58 mm, 16 × 10 × 90 I.S. 2048 parallel key is selected.

Shearing stress induced in this key from relation 12.1

$$f_s = \frac{2M_t}{dwl} = \frac{2\times1960\times10^3}{58\times16\times90} = 47 \text{ N/mm}^2$$

Crushing stress induced in the key is obtained from relation 12.2.

$$f_c = \frac{4M_t}{dlt} = \frac{4\times1960\times10^3}{58\times90\times10} = 150 \text{ N/mm}^2$$

The yield stresses in shear and crushing are 350 N/mm² and 425 N/mm² respectively.

$$\text{F.O.S. in shear} = \frac{350}{47} = 7.45$$

$$\text{F.O.S. in crushing} = \frac{425}{150} = 2.83.$$

Example 12.3. Determine the length and thickness of sunk key for a shaft of 0.09 m diameter. Assuming that the shearing resistance of the material of the key is the same as that of the shaft. Width of the key is 25 mm and $f_s = 0.4 \times f_c$.

Solution: Torsional resistance of the shaft $= \frac{\pi}{16} d^3 f_s = \frac{\pi}{16} \times (0.09)^3 \times f_s = \frac{143}{10^6} \cdot f_s$ Nm

Shearing resistance of the key $= wl \times f_s \times \frac{d}{2} = 0.025 \times l \times f_s \times \frac{.09}{2} = \frac{11.25}{10^6} \times l \times f_s$ Nm

As the shearing resistance of the key and shaft are equal,

$$\therefore \quad \frac{143}{10^6} \times f_s = \frac{11.25}{10^6} \times l \times f_s$$

gives, $l = 12.9$ cm $= 129$ mm.

Crushing resistance of the key

$$= l \times \frac{t}{2} \times f_c \times \frac{d}{2} = 12.9 \times \frac{t}{2} \times \frac{f_s}{0.4} \times \frac{9}{2} \times \frac{1}{10^6} = \frac{72.5}{10^6} \times t \times f_s \text{ Nm}$$

Equating this to shearing resistance of the shaft,

$$\frac{143}{10^6} f_s = \frac{72.5}{10^6} \times t \times f_s$$

gives, $t = 1.98$ cm say 20 mm.

Example 12.4. A 18.7 kW, 1400 rev/min motor has a steel shaft, the extension of the shaft is 75 mm. Diameter of the shaft is 45 mm. The motor pull out torsional moment is 3.4 times the average full load torsional moment of the motor and shock factor is 2.4. Assuming $f_s = 55$ N/mm² and $f_c = 110$ N/mm², find the size of the key. Check the shear strength of the key against the normal shear strength of the shaft.

Solution: Power $= \frac{2\pi n M_t}{60}$

$\therefore$ Average full load torsional moment of the motor $= \frac{18.7\times10^3\times60}{2\pi\times1400} = 130$ Nm

Motor pull out torsional moment = 3.4 × 130 = 442 Nm

$$\text{Allowable shear stress} = \frac{55}{2.4} = 23 \text{ N/mm}^2$$

$$\text{Allowable crushing stress} = \frac{110}{2.4} = 46 \text{ N/mm}^2$$

as $f_c = 2 \times f_s$, therefore, a square key will be adopted.

The extension of the shaft is 75 mm, thus the length of key will be 75 mm.

Now, to find the width of the key, the relation 12.1 can be used.

$$f_s = \frac{2M_t}{dwl},\ 23 = \frac{2 \times 442 \times 10^3}{45 \times w \times 75} \text{ gives } w = 10.9 \text{ mm.}$$

Thus cross-section of the key will be 10.9 × 10.9 mm.

The strength of a shaft with key way = k × strength of the shaft without key way,

$$= k \times f_s \times \frac{\pi}{16} d^3$$

where k, given by H.F. More $= 1 - 0.2\frac{w}{d} - 1.1.\frac{t/2}{d}$

$$\therefore \quad k = 1 - 0.2 \times \frac{10.9}{45} - 1.1 \times \frac{10.9}{2 \times 45} = 0.82$$

$$\frac{\text{Shear strength of the key}}{\text{Shear strength of the shaft}} = \frac{l \times w \times f_s \times d/2}{\pi/16 d^3 \times f_s \times k} = \frac{75 \times 10.9 \times 23 \times 45}{\pi/16 \times (45)^3 \times 23 \times 0.82 \times 2} = 1.30.$$

Example 12.5. A 100 mm shaft rotating at 100 rev/min transmits 224 kW. Power is taken off through a gear whose hub is 200 mm long. The key is made of steel having an ultimate shearing stress of 350 N/mm². Using a factor of safety of 5, determine the key desired.

Solution: Allowable stress in shear $= \frac{350}{5} = 70 \text{ N/mm}^2$

Torsional moment to be transmitted $= \frac{224 \times 60 \times 10^3}{2\pi \times 100} = 21.5 \times 10^3$ Nm

As the length of the hub is given, thus same is the length of the key.

Relation 12.1 will help to find the width of the key.

$$f_s = \frac{2M_t}{dwl},\ 70 = \frac{2 \times 21.5 \times 10^3 \times 10^3}{100 \times w \times 200}$$

gives, $w = 30.7$ mm.

Example 12.6. (a) A gear with a hub 75 mm long is fastened to a 50 mm commercial steel shaft by means of a 12.5 mm square feather key. What force is required to move the gear along the shaft when transmitting the full torsional moment capacity of the shaft? Assume the coefficient of friction to be 0.15.

(b) In case two keys located at 180 degrees from each other are used, then what is the force required to move the gear along the shaft?

Solution: Assuming an allowable stress in shear of the key material = 70 N/mm²

Torque, which can be resisted by the key will be found by the relation 12.1.

$$f_s = \frac{2M_t}{dwl},\ 70 = \frac{2 \times M_t}{50 \times 12.5 \times 75}, \text{ gives } M_t = 1640 \text{ Nm.}$$

(a) Axial force required to move the hub axially can be found by relation 12.8 and will be equal to

$$= \frac{4\mu \times M_t}{d} = \frac{4 \times 0.15 \times 1640 \times 10^3}{50} = 19.7 \text{ kN}$$

(b) Axial force required to move the hub axially, when two keys are used can be found by relation 12.9 and will be equal to

$$= \frac{2\mu M_t}{d} = \frac{2 \times 0.15 \times 1640 \times 10^3}{50} = 9850 \text{ N.}$$

Example 12.7. A 0.3 m lever is fixed to a 37.5 mm shaft by means of a taper pin passed through its hub perpendicular to the axis, the mean diameter of the pin being 10 mm. What pull on the end of this lever will cause a shearing stress on the pin of 65 N/mm², and what torsional stress will this produce on the shaft?

Solution: Mean diameter of pin = d_1 = 10 mm.

Shaft diameter = d = 37.5 mm.

By using relation 12.6, we can find the torsional moment on the shaft producing a shearing stress of 65 N/mm². Thus, assuming the pin to be in double shear, we have

$$2 \times \frac{\pi}{4} d_1^2 \times f_s = \frac{2M_t}{d}$$

$$2 \times \frac{\pi}{4} \times (10)^2 \times 65 = \frac{2M_t}{37.5}, \text{ gives } M_t = 192 \text{ Nm.}$$

As this torsional moment is to be applied by the lever, when the lever length is 300 mm. Let P be the force applied at the end of the lever then

$$P \times 300 = 192 \times 10^3, \text{ gives } P = 640 \text{ N.}$$

Let f_s be the stress produced in the shaft because of torque applied by the lever. Then,

$$M_t = \frac{\pi}{16} d^3 f_s,\ 192 \times 10^3 = \frac{\pi}{16} \times (37.5)^3 f_s$$

gives, f_s = 18.4 N/mm².

Example 12.8. An automobile engine has a capacity of 250 Nm. The clutch is attached to its driven shaft by means of six splines, that is to slide when not under load. Determine the dimensions of the shaft and spline. The material of the shaft and splines is steel and the load is applied with a moderate shock. The maximum pressure on the splines is 8.5 N/mm². Assume length of spline 1.5 times the shaft diameter.

Solution: From Table 12.3, the thickness of the spline when not under load is, $t = 0.125d$, number of splines $z = 6$, and $d' = 0.75d$.

Now the torsional moment to be transmitted by the splined shaft is given, as,

$$M_t = p \times A \times r_m$$

where p is the pressure on the splines and A is the area of the splines under pressure and is = tlz.

r_m = mean radius from the centre of the shaft to the centre of the splines.

$$= \frac{d' + t}{2} = \frac{0.75d + 0.125d}{2} = 0.4375d$$

Substituting different values in the above relation

$$M_t = p \times t \times l \times z \times r_m$$

$$250 \times 10^3 = 8.5 \times 0.125d \times 1.5d \times 6 \times 0.4375d$$

gives, d = 39.2 mm, say 40 mm.

∴ Diameter of shaft = 40 mm.

Length of spline = 1.5d = 1.5 × 40 = 60 mm.

Thickness of splines = 0.125d = 0.125 × 40 = 5 mm.

Root diameter of shaft = 0.75 × d = 0.75 × 40 = 30 mm.

Assuming the spline to be of square cross-section, gives width of key as 5 mm. This can be checked under shear for moderate shock taking a shear stress of 40 N/mm² from table 12.2 for steel when sliding.

EXERCISES

1. What is the function of a key?
2. What are the advantages and disadvantages of key joint?
3. Enlist the factors which determine the superiority of splines over the keys.
4. Mention the uses of square, rectangular, flat, Woodruff, Gibhead and saddle keys with respect to each other. Also explain, what is the advantage claimed by making a key tapered?
5. Explain in detail the uses of Nordberg, Barth, Kennedy, tangent and feather keys.
6. Describe in detail the method of determining the axial force required to move a hub over a shaft using two feather keys.
7. What is the relationship between the length of a standard key, whose width is one-fourth the shaft diameter if both the key and shaft are made from the same material and have the same strength under static torsional loads?
8. The breakdown torsional moment of a 11.2 kW, 1800 rev/min 3-phase induction motor is given as 200% of rated torsional moment. The motor shaft diameter is 40 mm, and the 112.5 mm long key is machined for a 9 × 9 mm square key. Assuming that the material of the key is steel, what will be the stresses induced in the key based upon the breakdown torsional moment of the motor ? What will be the factor of safety in shearing and crushing? What is the minimum length of the key you would recommend for use with this motor?
9. The shaft for a belt driven centrifugal blower transmits steadily a torsional moment of 250 Nm and the diameter required for the shaft at the pulley to transmit this torsional moment is 40 mm. If the key is made of steel and is fixed in the shaft and hub of pulley, what minimum length do you recommend if the key is (a) square (b) rectangular?
10. A belt pulley is fastened to a 75 mm shaft, running at 200 rev/min by means of a key 18 mm wide and 125 mm long. The permissible stresses in the key are 55 N/mm^2 in shear and 100 N/mm^2 in crushing.
    (a) Determine the horse power that can be transmitted by this key.
    (b) What width of the key is required?
11. A 1.2 meter, cast iron pulley is fastened to a 112.5 mm shaft by means of a 30 mm square key 180 mm long. Key and shaft are made of steel.
    (a) What force acting at the pulley rim will shear this key?
    (b) What force acting at the pulley rim will crush the cast iron keyway if the strength of cast iron is 170 N/mm^2 in tension and 680 N/mm^2 in compression?
    (c) What should be the maximum force applied at the rim of the pulley if the load is applied with moderate shock?
12. A 0.3 m gear transmitting 37 kW at 120 rev/min is to be fastened to a shaft of steel with a permanent fit by means of a six spline fitting. The hub is 1.5 times the shaft diameter. The ultimate shearing stress of the key material is 350 N/mm^2. Determine the spline dimensions, using a factor of safety of 5.
13. The transmission gears of an automobile are carried on a 58 mm, 10 spline when under load. The hub length of each gear is 40 mm. Determine the total horse power that can be transmitted at 3000 rev/min with 5.5 N/mm^2 permissible pressure on the splines.
14. A pin key of 10 mm average diameter is driven parallel to the axis of a 50 mm diameter shaft. Half of this protrudes into the shaft and half into the hub of a pulley. Calculate the maximum torsional moment that this pin can resist safely in shear and bearing per centimeter of length. Use a value of 70 N/mm^2 for the design stress in shear and 140 N/mm^2 for that in bearing. Hint : The area of hub key resisting bearing stress is considered to be equal to that of a square key.
15. A Propeller of an outboard motor has a 30 mm outside diameter hub which fits on a 15 mm diameter shaft. The hub and shaft are fastened by a brass shear pin. If an overload occurs at the propeller, the pin will shear, thus avoiding damage to the rest of the mechanism. Calculate the diameter of the shear pin which will fail at a torsional moment of 75 Nm with a shearing stress of 300 N/mm^2.

12.2 COUPLINGS

12.2.1 Introduction

Couplings are necessary to connect one shaft to another or to couple a drive shaft to a driven shaft. When the coupling can be arranged so as not to transmit torsional moment, it becomes a clutch. The essential difference between a clutch and a coupling is that a clutch can be disengaged or engaged at the will of the operator whereas a coupling is regarded as being fixed. Shaft couplings are used in machinery for several purposes, the most common of which are the following:

1. To provide for the connection of shafts of units that are manufactured separately, such as a motor and a generator, and to provide for disconnection for repairs or alterations.
2. Mild steel shafts are usually available in length of 6 meters to 7 meters. Sometimes it is in-convenient to use a shaft of large length due to space limitations or transport difficulties. Thus a long line shaft is composed of a number of small shaft pieces put together end to end by couplings and are supported by bearings.
3. To provide for misalignment of the shafts or to introduce mechanical flexibility.
4. To reduce the transmission of shock loads from one shaft to another.
5. To introduce protection against overloads.
6. To alter the vibration characteristics of rotating units.

Shaft couplings will be used under the following conditions:

1. With shafts having intersecting axes, that is axes in the same straight line. Rigid or flexible couplings of various forms are used here.
2. With shaft having intersecting axes, in which case a Universal coupling is employed.
3. With shafts whose axes are parallel and at a relatively small distance apart. Here the double - slider crank principle of mechanism is used.

There are two basic classes of coupling, namely (1) Rigid couplings and (2) Flexible couplings.

Rigid couplings do not accommodate misalignment and consequently should not be used in discriminately.

Flexible couplings allow slight angular deviations, also permit the axes of the shaft to float or run slightly out of alignment, and in some cases to move end wise, whilst still transmitting torsional moment from one shaft to another shaft. Further the use of a flexible coupling enables to compensate slight inaccuracy of workmanship in aligning the two connected machine elements. A typical example is a direct electric drive from an electric motor to a machine tool, where we have four bearings relatively close together.

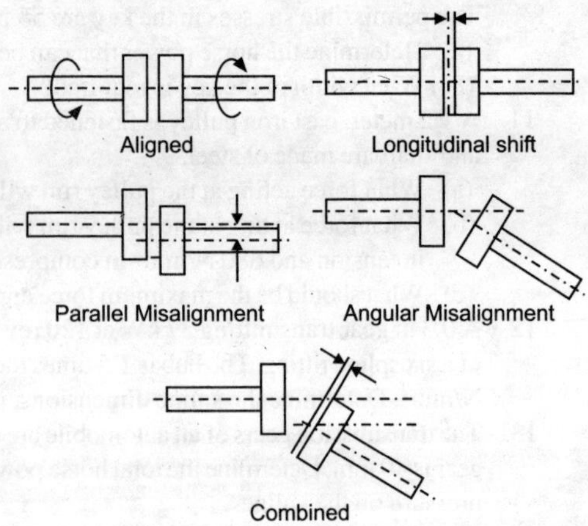

Fig. 12.10 *Alignment of shaft centre lines*

12.2.2 Rigid Couplings

This type of coupling has no flexibility or resilience, hence it is necessary for the shafts that are to be connected to be in good alignment, both laterally and angularly, in order to avoid excessive loads on the coupling, on the shafts, or on the shaft bearings. The various types of shaft misalignment are shown in Fig. 12.10. There is a large variety of Rigid couplings in use, but here some commonly used types of rigid couplings are described.

12.2.3 Sleeve or Muff Coupling

The coupling as shown in Fig. 12.11 is simply a hollow cylinder which is fitted over the ends of the two shafts to be connected and is then keyed there to by a sunk key. In this manner, the motion of one shaft is communicated to the sleeve and then to the other shaft, the three links acting as a single rigid member. Instead of securing the joint with the help of sunk keys, the joint can be secured by means of taper pins also as shown in Fig. 12.12. This type of coupling is simple in design and manufacture. It has no projecting parts as is evidenced by its perfect smooth exterior.

Among the disadvantages of the sleeve coupling are the difficulty in disassembly, that is, either the sleeve must be shifted along the shaft at least half its length or the shafts must be drawn apart to free the entire length of the coupling. The necessity for highly accurate axial alignment of the shafts since misalignment of shaft axes due to negligent assembly, or deformation under load, gives rise to forces that deflect the shafts.

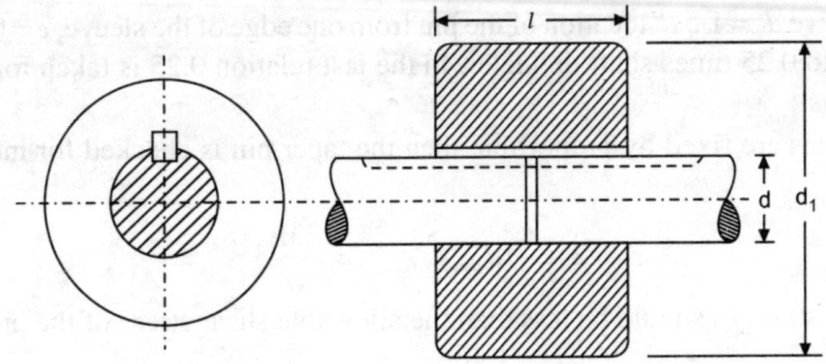

Fig. 12.11 *Sleeve or muff coupling with key*

The material of the sleeve is usually cast iron. It is evident that any coupling must be designed to transmit safely the full torsional moment of the shaft. Hence the sleeve and key of this coupling must safely resist the torsional moment, M*t*, of the shaft. The method of procedure in the design of the key has been considered in Section 12.1.4, where it was shown that the moments of the shearing and crushing resistance of the key must each be equal to the torsional moment of the shaft. The effective length of the key to be used in the cross-sectional area which provide the resistances of the key is only that part of the length of the coupling key that is in one of the two shafts connected. Often the coupling key is made in two parts, so that there are in reality two keys in line, one in each shaft. With the latter arrangement, the effective length of key is the length of one part, and with either arrangement becomes equal to approximately one half the length of the coupling.

In so for as the sleeve or muff is concerned, the sleeve is in reality acting as a section of a hollow shaft in transmitting the torsional moment through itself from one shaft to the other. Applying the notation of Fig. 12.11 to the polar section modulus for a hollow circular section, we have,

$$Z_p = \frac{\pi}{16}\left(\frac{d_1^4 - d^4}{d_1}\right)$$

From relation
$$\frac{M_t}{J} = \frac{f_s}{r}, \text{ or } M_t = Z_p \times f_s$$

Thus,
$$f_s = \frac{16Mtd_1}{\pi(d_1^4 - d^4)} \qquad \text{...(12.10)}$$

in which d_1 = outside diameter of the sleeve or muff.

d = Inside diameter of the sleeve or the diameter of the shaft.

f_s = induced shearing stress in the sleeve.

Mt = Torsional moment to be transmitted from one shaft to the other shaft.

To calculate the diameter of the sleeve from relation 12.10, it seems to be cumbersome. Therefore, first the dimensions are fixed by proportions and then the sleeve is checked for the induced shear stress so that it is safe.

The relationships between the proportion of sleeve couplings which satisfy the condition of equal strength of the elements are,

$d_1 = 2d + 12.5$ mm where d_1 is the outside diameter of the sleeve, length of the sleeve, $l = 3.5\ d$.

With the help of muff or sleeve the power from one shaft to the other shaft can also be transmitted with the help of taper pins as shown in Fig. 12.12. The usual proportions of sleeve coupling which satisfy the condition of equal strength of the elements are, length of the sleeve, $l = 3d$,

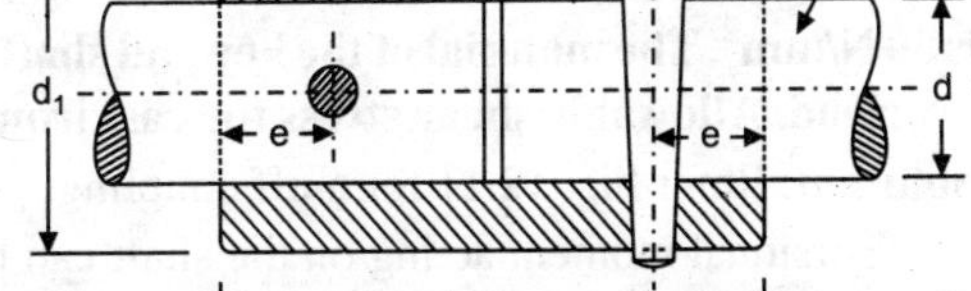

Fig. 12.12 *Sleeve or muff coupling with taper pin*

outside diameter of sleeve $d_1 = 1.5\,d$, location of the pin from one edge of the sleeve, $e = 0.75\,d$, mean diameter of the pin, $d_{pin} = 0.20$ to 0.25 times shaft diameter. In the last relation 0.25 is taken for small couplings and 0.20 for large couplings.

Once the dimensions are fixed by proportion, then the taper pin is checked for induced shear stress by relation 12.11.

$$2 \times \frac{\pi}{4} (d_{pin})^2 \times f_s = \frac{2M_t}{d} \qquad ...(12.11)$$

The induced shear stress f_s, should be less than the allowable shear stress of the pin material.

Example 12.9. A sleeve coupling is used to connect two 50 mm shafts whose safe shearing stress is 70 N/mm². Design the sleeve when it is made of cast iron with an ultimate shearing stress of 140 N/mm². Also determine the factor of safety for which the sleeve has been designed. The length of the key is half the length of sleeve and the section of the key is 12.5 mm × 12.5 mm, find the shearing and crushing stresses induced in the key used.

Solution: The relation, $M_t = \pi/16\, d^3 \times f_s$ will give the torsional moment which can be transmitted by this shaft.

$\therefore \quad M_t = \pi/16\,(50)^3 \times 70 = 1715$ Nm.

The torsional moment for which the sleeve is to be designed is the torsional moment to be transmitted by the shaft.

Now the proportions of the sleeve will be, outside diameter of sleeve, $d_1 = 2d + 12.5$ mm $= 2 \times 50 + 12.5 = 112.5$ mm, length of sleeve, $l = 3.5\,d = 3.5 \times 50 = 175$ mm.

Applying the relation 12.10, the shear stress induced in the sleeve can be checked.

$$f_s = \frac{16\,M_t d_1}{\pi(d_1^4 - d^4)} = \frac{16 \times 1715 \times 10^3 \times 112.5}{\pi\left[(112.5)^4 - (50)^4\right]} = 6.5 \text{ N/mm}^2.$$

$$\text{Factor of safety} = \frac{140}{6.5} = 21.5,$$

which indicates that the shear stress induced in the sleeve is much below the allowable shear stress. Hence the dimensions of the sleeve are

$$d_1 = 112.5 \text{ mm},\ l = 175 \text{ mm}$$

(b) Length of key = half the length of sleeve = 175/2 = 87.5 mm.

Width of key = 12.5 mm, thickness of key = 12.5 mm.

Then applying the relation 12.1, the shear stress induced in the key will be,

$$f_s = \frac{2M_t}{dwl} = \frac{2 \times 1715 \times 10^3}{50 \times 12.5 \times 87.5} = 62.5 \text{ N/mm}^2$$

Applying relation 12.2, the crushing stress induced in the key will be,

$$f_c = \frac{4M_t}{dlt} = \frac{4 \times 1715 \times 10^3}{50 \times 87.5 \times 12.5} = 125.0 \text{ N/mm}^2.$$

Example 12.10. Design a muff coupling for transmitting 45 kW at a speed of 350 rev/min. The muff is made of cast iron and is used to connect two shafts. The allowable shear stress in the material of the shaft is 50 N/mm². The material of the key and shaft is same and the coupling is required to transmit 25 percent overload. Allowable shear stress for cast iron is 16 N/mm².

Solution: Refer Fig. 12.11 for muff coupling.

Torsional moment acting on the shaft can be found from the relation,

$$\text{Power} = \frac{2\pi n M_t}{60},\quad M_t = \frac{60 \times 45 \times 10^3}{2\pi \times 350} = 1230 \text{ Nm}$$

Maximum torsional moment to be transmitted with 25 per cent over load,

$$M_{t\,max} = 1.25 \times 1230 = 1540 \text{ Nm}$$

If d is the diameter of the shaft, then

$$M_{t\,max} = \frac{\pi}{16} d^3 f_s, 1540 \times 10^3 = \frac{\pi}{16} d^3 \times 50$$

gives, $d = 54$ mm.

For material of the sleeve to be cast iron, the proportions of the sleeve will be

$$d_1 = 2d + 12.5 \text{ mm} = 2 \times 54 + 12.5 = 120.5 \text{ mm, say 120 mm.}$$

Length of sleeve = $3.5d = 3.5 \times 54 = 189$ mm, say 190 mm.

Now the muff can be checked for induced shear stress by relation 12.10.

$$f_s = \frac{16\, M_t d_1}{\pi(d_1^4 - d^4)} = \frac{16 \times 1540 \times 10^3 \times 120}{\pi\left[(120)^4 - (54)^4\right]} = 4.75 \text{ N/mm}^2.$$

This value is less than the allowable shear stress, hence safe. For a shaft diameter lying between 50 to 58 mm, the size of the key can be obtained from Table 12.1.

width of key, $w = 16$ mm

thickness of key, $t = 10$ mm

Length of key on each shaft = $\frac{1}{2}$ the muff length = $\frac{1}{2} \times 190 = 95$ mm

Shearing stress induced in the key can be calculated by relation 12.1, thus

$$f_s = \frac{2M_t}{dwl} \text{ where } l \text{ is the length of key in this relation,}$$

$$= \frac{2 \times 1540 \times 10^3}{54 \times 16 \times 95} = 38.2 \text{ N/mm}^2$$

which is well within the safe limits.

Crushing stress induced in the key,

$$f_c = \frac{4M_t}{dlt} = \frac{4 \times 1540 \times 10^3}{54 \times 95 \times 10} = 120 \text{ N/mm}^2\text{, safe.}$$

12.2.4 Clamp Coupling or Split Muff Coupling

It is also a rigid type of coupling. Clamp coupling is sometimes called a compression coupling or a ribbed coupling. Clamp coupling is a modification and improvement of the sleeve coupling. This coupling is made in two parts which are machined to fit the shaft and are finished off around the periphery and on both ends. The two halves of the coupling are clamped tightly against the surface of the shaft ends by through bolts and the entire torsional moment is transmitted entirely by friction. Small and medium size couplings transmit torsional moment through friction between the shafts and the coupling alone. In large couplings a key is incorporated between the shafts and coupling transmits the main torsional moment as shown in Fig. 12.13.

Its construction in two halves permits of its being readily assembled and removed which is distinct advantage. The bolt heads and nuts are protected by ribs to comply with safety laws. The sleeve or muff and clamp couplings we have examined are connected to the shafts after the latter have been assembled into bearing.

The different proportions of the clamp coupling are given below. The length of these coupling depends on the required number of bolts. But usually l = 5 to 3.5 times the shaft diameter.

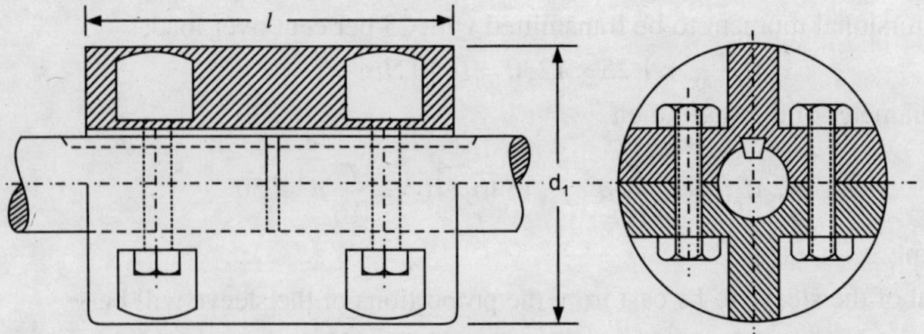

Fig. 12.13 *Clamp or compression coupling*

Outside diameter of coupling d_1 = 4 to 2 times the shaft diameter.

In these relations the greater values, i.e., 5 and 4 refer to couplings of small size, for shafts with size, d = 2.5 cm, while the smaller values 3.5 and 2 refer to large couplings, for shafts with size d = 30 cm.

Calculation of clamp couplings is now reduced to finding the proportions as given above and the number and size of bolts. It is done on the assumption that even with a key the torsional moment is transmitted only through friction between the shafts and coupling halves.

Let d_c = core diameter of the bolt used for clamping.

z = Total number of bolts used for clamping the coupling halves, the number of bolts used has to be a multiple of 4.

M_t = torsional moment to be transmitted by the coupling from one shaft to the other.

d = diameter of shaft.

μ = coefficient of friction between the shaft and sleeve material. An average value of μ is taken 0.3.

l = length of sleeve.

f_t = Allowable tensile stress in the bolt.

Load which can be taken by one bolt = $\pi/4\, d_c^2 \times f_t$

Clamping load on the shaft due to z number of bolts = $\dfrac{z}{2} \times \dfrac{\pi}{4} d_c^2 \times f_t$

where $\dfrac{z}{2}$ is the number of effective bolts for each shaft.

Clamping pressure intensity between the shaft and sleeve will be

$$p = \frac{\text{Force}}{\text{Projected Area}} = \frac{\frac{z}{2} \times \frac{\pi}{4} d_c^2 \times f_t}{d \times l/2}$$

Tangential force per unit area of shaft periphery will be

$$F = \mu \times \text{pressure} = \mu \times \frac{\pi}{4} \frac{d_c^2 \times f_t \times z}{dl}, \text{ where } z \text{ is the number of bolts.}$$

Now the torsional moment which can be transmitted by the coupling

$$= F \times \text{area} \times \frac{d}{2} = \mu \times \frac{\pi}{4} \frac{d_c^2 \times f_t \times z}{dl} \times \frac{\pi dl}{2} \times \frac{d}{2} = \frac{\pi^2}{16} \cdot d_c^2 \cdot f_t \cdot \mu \cdot z \cdot d$$

This torsional moment must be equal to the torsional moment M_t to be transmitted by the coupling, and then rearranging the terms, we get

$$M_t = \frac{\pi^2}{16} d_c^2 \times f_t \times \mu \times z \times d \qquad \text{...(12.12)}$$

Load which can be taken by each bolt, $P = \pi/4\, d_c^2 \times f_t$.

Substituting this value in relation 12.12, we get

$$M_t = P/4 \times \pi\mu\, zd$$

or $$P = \frac{4M_t}{\pi\mu zd} \quad ...(12.13)$$

By assuming the number of bolts, the relation 12.13 will help us to calculate the force P, from which the size of bolt can be calculated.

Example 12.11. Determine the main dimensions of a clamp coupling shown in Fig. 12.13, for a shaft diameter of 54 mm. The torsional moment to be transmitted by this coupling taking into account overloading is 1540 Nm. Assume the number of bolts used to connect two halves is 4. The design stress in the bolt material is limited to 70 N/mm^2. The coefficient of friction between the shaft and sleeve material is 0.3.

Solution: Effective number of bolts for each shaft = 2.

P is the load to be taken by each bolt, then from relation 12.13, we get

$$P = \frac{4M_t}{\pi\mu zd} = \frac{4 \times 1540 \times 10^3}{\pi \times 0.3 \times 4 \times 54} = 30.3 \text{ kN}$$

Load which can be taken by one bolt = $\pi/4\, d_c^2 \times f_t$

$\therefore$ $$P = \pi/4\, d_c^2 \times f_t,\ 30.3 \times 10^3 = \pi/4 \times d_c^2 \times 70.0$$

gives, $d_c = 23.6$ mm.

Outside diameter of bolt = $\frac{23.6}{0.84}$ = 28 mm.

As the size of the shaft is more than 25 mm, therefore, length of the sleeve = 3.5 × 54 = 190 mm.

Outside diameter of sleeve = $2d$ = 2 × 54 = 108 mm.

As the size of the shaft lies between 50 to 58 mm, the size of the key can be obtained from Table 12.1.

Width of key, w = 16 mm, thickness of key, t = 10 mm.

The size of the key can be checked for induced stresses as in example 2.

12.2.5 Flange Coupling

It is a rigid type of coupling. The flange coupling, Fig. 12.14, consists of two cast iron flanges. Keyed to the shaft ends and bolted together. The best practice is to press or shrink the flanges on the shafts, in which straight keys must be used. If the flanges are put on only with a push fit, taper keys must be used. For more accurate connection, the ends of the flange couplings and the centering surfaces are machined after assembly

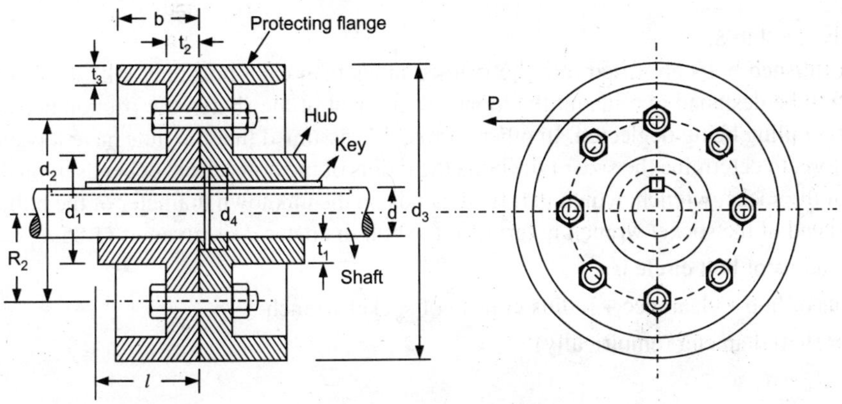

Fig. 12.14 *Flange coupling*

on the shaft. The hubs extend into flanges whose faces are brought and held together by a series of bolts arranged concentrically about the shaft so that their axes are parallel to the coincident or collinear axes of the shafts and coupling. These bolts must be carefully fitted in reamed holes so that each one will take its proportionate part of the load. An additional flange at the outside of the coupling is a safety flange to guard the projecting bolt heads and nuts and add to the rigidity of the coupling. The alignments of the shafts can also be obtained by extending one shaft through its flange and into the bore of the other. Flange type of coupling is adapted to heavy loads and hence is used almost exclusively on rather large shafting.

The presence of flanges on the shaft ends requires all parts (like toothed wheels, pulley, flywheels, bearing, etc.) fitted on the shaft to be of a split type. Inspite of this disadvantage, flange couplings ensure the most accurate, rigid and strong connection of shafts. For this reason, in important cases like the shafts of steam turbogenerators in three bearings, the shafts of vertical hydro-turbines, marine propeller shafts, etc, are usually connected by flange coupling.

The I.S. 6196-1971 has adopted various dimensions which are used by most manufacturers for the various sized couplings. These dimensions make the flange, the hub, and the bolts all capable of with-standing the amount of torsional moment so that one part is stronger than another. This latter feature should always be remembered when selecting a coupling although it might well be possible that one of a series of couplings should be selected so as to be weaker than the rest to serve as a safety device. In this regard, the bolts holding the flanges together are sometimes reduced in diameter so that they would readily break or shear off in the event of a damaging overload.

The design procedure followed for flange coupling will be as follows:

1. The hub will be designed by considering it to be a hollow shaft transmitting the torsional moment of the shaft, i.e., relation 12.10. The hubs of gears, pulley, clutches, and couplings are invariably proportioned empirically. The following empirical equation is frequently used.

$$d_1 = 1.75d + 6.5 \text{ mm} \qquad ...(12.14)$$

2. The length of the hub is fixed by the length of the key required. The width of the keys is very commonly made 1/4 the diameter of shaft. To obtain the length of the key it will be assumed that shaft and key are of same material and that the key is as strong in direct shear as the shaft is in torsion. The length found of the key fixes the length of the hub. Empirically the length of the hub is taken as 1.5 times the shaft diameter.
3. The thickness of the key is calculated as discussed in Article 12.14. A set screw is used to keep the key and hub from moving axially. The diameter of the set screw should be somewhat less than the width of key.
4. Empirically the number of bolts is to be fixed by the relation,

$$z = \frac{4}{150}d + 3; \quad d \text{ is in mm.} \qquad ...(12.15)$$

Usually, $z = 4$ to 8.

5. When finished bolts are fitted into the reamed holes then each bolt may take its share of load. The bolts will be designed to transmit the torsional moment of the shafts, the friction between the halves of the coupling being neglected. In other words, it is assumed that the nuts have loosened in service. Therefore, to determine the size of the bolts the radius of the bolts circle depends upon diameter of the head of the socket wrench, which in turn, depends on the unknown diameter of the bolt. The diameter of the head of the socket wrench in terms of the bolt diameter d_0 is given as ($1.85\, d_0 + 8$ mm).

Thus the radius of bolt circle is,

R_2 = radius of hub + clearance + radius of head of socket wrench.

= 1.5 × shaft diameter (empirically).

or $$R_2 = \frac{d_3 + d_4}{4}$$

where the clearance between the hub and the head of the socket wrench is taken as 5 mm. Once the radius of the bolt circle is known, then, shearing force P at the bolt circle of radius R_2 can be calculated when the shaft has to transmit a torsional moment M_t.

Thus, $$P = \frac{M_t}{R_2}$$

Let d_o be the outside diameter of the bolt, the area resisting the shearing load will be $\pi/4\ d_o^2$.

Shear load resisted by each bolt $= \pi/4\ d_o^2 \times f_s$

where f_s is the allowable shear stress in the bolt material.

If z is the number of bolts, calculated from step 4, then shear load resisted by z bolts

$$= \pi/4\ d_o^2 \times z \times f_s$$

Shear load resisted by bolt must be equal to the shearing force P at the bolt circle of radius R_2, thus

$$\frac{M_t}{R_2} = \frac{\pi}{4} d_o^2 \times z \times f_s\,,\ \ f_s \cong 0.25 f_{yp} \qquad \text{...(12.16)}$$

From this the diameters of the bolt can be calculated or by assuming the diameter of the bolt, it can checked for induced shear stress to be less than the allowable value.

Bolts with diameter 1.5 mm less than the diameters of the holes are used to allow for slight variations in the positions of the matching holes. This inaccuracy causes the flange of a flange coupling, when in use, to bear on some of the bolts only. For this reason, it is better practice to take the number of bolts acting to be half of the total number of bolts in relations 12.16 and 12.17.

6. The shearing force P also produces bearing stress in the flange which resists the push of the bolts. The area resisting this push will be $= z \times d_o \times t_2$, where t_2 is the thickness of the flange. Thus the total resisting force must be equal to the shearing force.

$$P = z \times d_o \times t_2 \times f_c \qquad \text{...(12.17)}$$

where f_c is the allowable compressive or bearing or crushing stress of the flange material.

From relation 12.17, thickness t_2 can be calculated. Empirically this thickness of the flange is given as,

$$t_2 = \frac{1}{2} t_1 + 6.5 \text{ mm} \qquad \text{...(12.18)}$$

where t_1 is the thickness of hub.

7. The thickness of the protecting flange should not be less than half that of the radial flange i.e., $t_3 \nless \frac{1}{2} t_2$. All the corners in the flange coupling should be well rounded to avoid stress concentration.

8. For safety in operation the halves of the flange coupling are to be provided with protecting flange as shown in Fig. 12.14 as guards for bolt heads and nuts. The width of the flange, $b \geq t_2 + 0.15\ d$.

9. The depth of the hub on the outer flange should not be less than 3 mm for the machinist to establish a fit without difficulty.

10. Transverse loads inducing bending stress in the shaft will also induce bending stress in the flange of the coupling. Thus stress reverse itself during complete rotation. If such stresses exceed the endurance limit of the material of the flange, fatigue failure will occur. But the flange of a standard coupling is of such thickness that safety from fatigue failure is ensured, except in very unusual cases.

11. Design of Bolts: (*a*) The bolts are designed in shear only, see equation 12.16, when they are accurately fitted in their holes, without any clearance. The holes are reamed and the fit is snug or with slight interference.

(*b*) More frequently, the bolts are installed with a certain clearance in their holes. In this case, they work under tension, because they have to be tightened sufficiently to create a force of friction capable of transmitting the required torque from one disc of the coupling to the other by friction alone.

The force of friction = $z \,.\, \mu \,.\, P_i$

where μ = Co-efficient of friction

and P_i = Force of tightening of one bolt

$$\therefore \quad z \,.\, \mu \,.\, P_i \geq P = \frac{M_t}{R_2}, \quad \text{where } P = \text{Turning force acting at the P.C.D. of bolts.}$$

Here the bolt circle radius R_2 (R_{bc}) is equal to the friction radius, R_f, which for uniformly distributed pressure is given as,

$$R_{bc} = R_f = \frac{2}{3} \cdot \frac{R_o^3 - R_i^3}{R_o^2 - R_i^2},$$

where R_o = Outer radius of the flange = $\dfrac{d_3}{2}$

R_i = Radius of recess = $\dfrac{d_4}{2}$

$$\therefore \quad P_i \geq \frac{P}{z.\mu}$$

During tightening of bolts, the bolt shank is also getting twisted which induces shear stress in it. Therefore, as per Art. 9.12, the design equation is,

$$\frac{\pi}{4} d_1^2 \times f_t \geq 1.3 \times P_i$$

$$\text{or} \quad d_1 = d_c \geq \sqrt{\frac{1.3 \times P_i \times 4}{\pi . f_t}} \quad \text{or} \geq \sqrt{\frac{4 \times 1.3 \times P}{\pi . f_t . z . \mu}} \qquad \text{...(12.19)}$$

Note: Bolts working in tension must be considerably larger in size than those working in shear.

12.2.6 Solid or Forged Flange Coupling

At one time the flanges of the rigid type of coupling were forged as an integral part of each end of the shaft, but this practice has been now almost entirely discontinued except in the case of very large shafts or when large torsional moments and forces are to be transmitted, such as those used for propeller shafts. When used for propeller shafts, the coupling is known as Marine type Flange Coupling as shown in Fig. 12.15. The alignment of the two shafts is ensured either by providing a ring on one flange and a similar recess on the other flange or by extending one shaft into flange of the other shaft. The machined faces are held together by means of tapered headless bolts numbering from four to twelve. The bolt holes are perfectly reamed so that the bolts are well fit in the holes, as a result each bolt will share equal load. The size of the bolt can be calculated similarly as discussed for flange coupling. Thickness of flanges can be established under the considerations of crushing. The different proportions of the solid Flanged Coupling are given below.

Diameter of the bolt circle = $d_2 = 1.5d$

where d is the shaft diameter.

Taper of bolt is 1 : 20 or 1 : 24.

Thickness of flange = $d/3 = t_2$

Outside diameter of flange = $d_3 = 2d$

Number of bolts will be calculated in the similar fashion as in Flange Coupling.

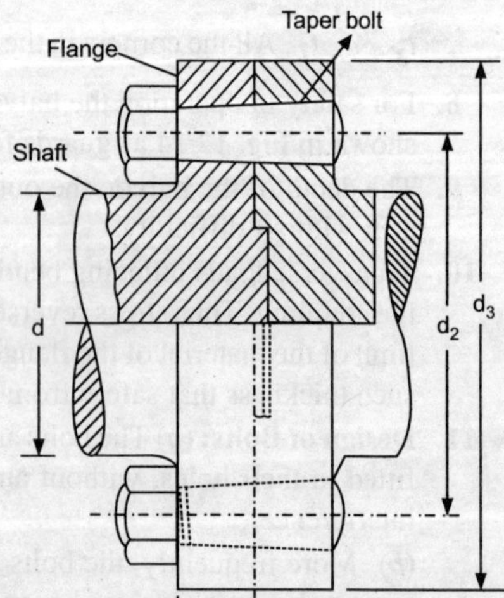

Fig. 12.15 *Solid flanged coupling*

Example 12.12. Two shafts 100 mm diameter are to be connected by means of two cast iron couplings. The allowable shearing stress of the bolt material is 45 N/mm². While that of the shaft material is 55 N/mm². Find the size of the bolts to be used. Check the flange for the induced crushing stress. Is it a safe stress?

Solution: From relation 12.15, the number of bolts used can be found out.

$$z = \frac{4}{150} \times 100 + 3 = 2.67 + 3 = 5.67 \text{ say } 6$$

Diameter of the hub of the flange from relation 12.14 is,

$$d_1 = 1.75d + 6.5 = 1.75 \times 100 + 6.5 = 181.5 \text{ mm say } 180 \text{ mm.}$$

Torsional moment to be transmitted by the shaft is,

$$M_t = \frac{\pi}{16} d^3 \times f_s = \frac{\pi}{16} \times (100)^3 \times 55.0 = 10.8 \times 10^3 \text{ Nm.}$$

Thus radius of bolt circle = radius of hub + clearance + radius of head of socket wrench.

$$R_2 = \frac{180}{2} + 5 + \frac{1.85 d_o + 8}{2} = 99 + 0.925\, d_o$$

Now the tangential or shear force P is given as,

$$P = \frac{M_t}{R_2} = \frac{10.8 \times 10^3 \times 10^3}{(99 + 0.925\, d_o)}$$

Now using the relation 12.16, the diameter of bolt can be calculated.

$$\frac{M_t}{R_2} = \frac{\pi}{4} d_o^2 \times z \times f_s, \frac{10.8 \times 10^3 \times 10^3}{(99 + 0.925\, d_o)} = \frac{\pi}{4} d_o^2 \times 6 \times 45$$

From this the value of d_o can be computed which is = 20.8 mm

Now the radius of bolt circle = 99 + 0.925 × 20.8 = 118.3 mm say 120 mm.

The thickness of the flange by relation 12.18 is,

$$t_2 = 0.5 \times t_1 + 6.5 = 0.5 \times \left(\frac{180 - 100}{2}\right) + 6.5 = 26.5 \text{ mm}$$

The induced crushing stress in the flange can be checked by relation 12.17.

$$P = \frac{M_t}{R_2} = z \times d_o \times t_2 \times f_c$$

$$\frac{10.8 \times 10^3 \times 10^3}{120} = 6 \times 20.8 \times 26.5 \times f_c$$

Thus, f_c = 27 N/mm². Ultimate crushing stress for cast iron = 550 N/mm², ultimate crushing stress for steel = 450 N/mm².

It is evident from these two values that there will be a greater tendency to crush the steel than to crush the cast iron. Hence we shall determine the factor of safety for steel.

Thus F.O.S. for crushing of steel = $\frac{450}{27}$ = 16.7, which is a proof that the induced stress is safe.

Example 12.13. It is required to design a coupling of flange type as shown in Fig. 12.14 for connecting the motor and centrifugal pump shafts. The details of the duty required from the coupling are:

Power to be transmitted ...18 kW.

Speed in rev/min ...1000

Find the diameter of the motor and pump shaft. The allowable stress in the shaft is limited to 50 N/mm² and the angle of twist is not to exceed 0.75 degrees in a length of 20 diameters. The allowable shear stress in the coupling bolts is 30 N/mm². Assume the torsional moment to be transmitted is 20% more than the mean torsional moment.

Solution: Maximum power to be transmitted by coupling = 18 × 1.2 = 21.6 kW

Maximum torsional moment to be transmitted by coupling $= \dfrac{21.6 \times 10^3 \times 60}{2\pi \times 1000} = 207$ Nm.

Let d be the diameter of the shaft, then

$$207 \times 10^3 = \frac{\pi}{16} d^3 \times f_s, \qquad 207 \times 10^3 = \frac{\pi}{16} \times d^3 \times 50, \text{ gives } d = 27.2 \text{ mm}$$

$$= 27.5 \text{ mm say.}$$

From rigidity point of view the induced stress in the shaft can be checked.

$\dfrac{f_s}{d/2} = \dfrac{G\theta}{l}$ where G is the modulus of rigidity in N/mm^2 and θ is the twist of shaft in radians.

$$f_s = \frac{0.084 \times 10^6 \times 27.5}{20 \times 2 \times 27.5} \times \frac{\pi}{180} 0.75 = 27.5 \text{ N/mm}^2$$

which indicates that the shaft diameter calculated for the allowable shear stress is safe to take up the twist of the shaft.

From Table 12.1 for shaft diameter of 27.5 mm the key dimensions can be taken.

Width of key $= w = 8$ mm, thickness of key $= t = 7$ mm.

Length of key $= 1.5\,d = 1.5 \times 27.5$ mm $= 42$ mm say 45 mm.

Shear stress induced in the key can be checked by relation 12.1.

$$f_s = \frac{2M_t}{dwl} = \frac{2 \times 207 \times 10^3}{27.5 \times 8 \times 45} = 41.7 \text{ N/mm}^2$$

which is less than the allowable shear stress of 50 N/mm^2.

Induced crushing stress in the key can be checked by relation 12.2.

$$f_c = \frac{4M_t}{dlt} = \frac{4 \times 207 \times 10^3}{27.5 \times 45 \times 7} = 95.5 \text{ N/mm}^2$$

which is again less than the allowable crushing stress of 100 N/mm^2.

The diameter of the hub $= d_1 = 1.75d + 6.5$ mm

$= 1.75 \times 27.5 + 6.5 = 55$ mm

Thickness of the hub $= t_1 = \dfrac{55 - 27.5}{2} = \dfrac{27.5}{2} = 13.75$ mm.

The induced stress in the hub can be checked by relation 12.10.

$$f_s = \frac{16 \times 207 \times 10^3 \times 55}{\pi[(55)^4 - (27.5)^4]} = \frac{16\, M_t d_1}{\pi(d_1^4 - d^4)} = 6.75 \text{ N/mm}^2$$

which is again less than the allowable shear stress of cast iron as flange material, hence safe.

Length of hub = length of the key $= l = 45$ mm.

Size of bolts: Bolt circle diameter = 3 times the shaft diameter

$= 3 \times d = 3 \times 27.5 = 82.5$ mm

Radius of bolt circle $= R_2 = \dfrac{82.5}{2} = 41.25$ mm.

Number of bolts is fixed by relation 12.15.

$$z = \frac{4}{150} d + 3 = \frac{4}{150} \times 27.5 + 3 = 0.735 + 3 = 3.735, \text{ say, } z = 4$$

The size of the bolt can be obtained by relation 12.16.

$$\frac{M_t}{R_2} = \frac{\pi}{4} d_o^2 \times z \times f_s$$

$$\frac{207 \times 10^3}{41.25} = \frac{\pi}{4} \times d_o^2 \times 4 \times 30, \text{ gives } d_o = 5.35 \text{ mm}$$

The accepted diameter of bolt will be 10 mm.

Thickness of the flange $= t_2 = \frac{1}{2} t_1 + 6.5 \text{ mm} = \frac{1}{2} \times 13.75 + 6.5 = 13.4$ mm.

The induced crushing stress in the flange of coupling can be checked by relation 12.17.

$$P = \frac{M_t}{R_2} = z \times d_o \times t_2 \times f_c$$

$$\frac{207 \times 10^3}{41.25} = 4 \times 10 \times 13.4 \times f_c, \text{ gives } f_c = 9.68 \text{ N/mm}^2$$

which is well within the limits.

The thickness of the protecting flange $= \frac{1}{2} t_2 = \frac{1}{2} \times 13.4 = 6.7$ mm

Outer diameter of flanges: Diameter of the head of the socket wrench $= 1.85\, d_o + 8$ mm
$= 1.85 \times 10 + 8 = 26.5$ mm

Outer diameter of flange $= d_3 = 2$ [radius of hub + clearance + diameter of head of socket wrench + clearance + thickness of protecting flange]
$= 2\,[27.5 + 5 + 26.5 + 5 + 6.7]$
$= 70.7$ mm, say 70 mm.

Example 12.14. A marine type flange coupling is required to transmit 2900 kW power at a speed of 100 rev/min. Flanges are connected by 8 taper bolts having an allowable shear stress of 60 N/mm². The material of shaft and bolts used is same. Design the flange coupling and determine the shaft diameter.

Solution: Torsional moment to be transmitted by shaft

$$M_t = \frac{60 \times 2900 \times 10^3}{2\pi \times 100} = 2.87 \times 10^5 \text{ Nm}$$

If d is the diameter of the shaft, then

$$M_t = \frac{\pi}{16} d^3 \times f_s,\ 2.87 \times 10^5 \times 10^3 = \frac{\pi}{16} d^3 \times 60, \text{ gives } d = 283 \text{ mm.}$$

we adopt $d = 285$ mm.

Diameter of bolt circle $= d_2 = 1.5d = 1.5 \times 285 = 427$ mm, say 430 mm.

Now using the relation 12.16, the mean diameter of the bolt can be established.

$$\frac{M_t}{R_2} = \frac{\pi}{4} d_o^2 \times z \times f_s, \text{ where } d_o \text{ is the mean diameter of bolt.}$$

R_2 is the bolt circle radius $= \frac{430}{2}$ mm.

$$\frac{2.87 \times 10^5 \times 10^3}{430/2} = \frac{\pi}{4} d_o^2 \times 8 \times 60, \text{ gives } d_o = 59.7 \text{ mm} = 60 \text{ mm say.}$$

Thickness of flange, $t_2 = d/3 = \frac{285}{3} = 95$ mm.

The flange can be checked under shear and crushing.

The flange under crushing can be checked by relation 12.17.

$$\frac{M_t}{R_2} = z \times d_o \times t_2 \times f_c, \quad \frac{2.87 \times 10^5 \times 10^3}{430/2} = 8 \times 60 \times 95 \times f_c$$

gives, $f_c = 29.3$ N/mm^2, Which is within safe limits.

The flange may get sheared off at the circumference of the shaft,

$$\therefore \qquad \pi d \times t_2 \times f_s = \frac{M_t}{d/2}, \quad \pi \times 285 \times 95 \times f_s = \frac{2.87 \times 10^5 \times 10^3}{(285)/2}$$

gives, $f_s = 23.7$ N/mm^2, which is well within limits.

12.2.7 Flexible Couplings

It is nearly impossible completely to eliminate all chances of misalignment between the shafts of separately built machines, and such misalignment of shafts always leads to eventual failure of bearings or fatigue failure of shafts. The relative position of the connected shafts, inaccurate at the outset due to inevitable errors of manufacture, is in the course of time aggravated by deformations caused by the working load, temperature fluctuations, the uneven sinking of foundations or supports, etc. In such cases ineffective rigid couplings are replaced by flexible couplings. Therefore, the purpose of a flexible coupling is to allow for imperfect alignment of two joining shafts, or to absorb impact from the fluctuation of torsional moment of angular speed. Hence, in most cases of joining shafts which transmit moderate to heavy power, it is necessary or at least desirable to use a flexible coupling of one type or another. The flexible couplings may be split into two categories from the stand point of design.

1. Couplings whose flexibility is obtained kinematically by the use of rigid members in which constraint is absent in certain directions. Such couplings should be lubricated. The couplings of this category are known as rigid slip coupling and examples are Oldham and Universal-joint couplings. This category of couplings can be used for any velocities and loads and cause comparatively small additional loads on shafts and bearings.
2. Couplings with incorporated flexible members. This type of coupling anticipates both misalignment and impact. Examples of such couplings are bushed pin type, leather pad type, etc. These couplings do not require any lubrication but bring to bear on the shaft and bearings an additional load, which depends on the rigidity of the coupling and relative shift of the connected shafts.

Flexible couplings are made in large variety of styles and principles of construction, some of which will be described briefly in this book.

12.2.8 Oldham Flexible Coupling

Falls in the category of rigid slip coupling. Oldham can be used for any velocities and loads and cause small additional loads on shafts and bearings. The special advantage of this coupling is that is can be used for shafts which are not in alignment but whose axes are parallel upto the extent of 0.05 times the shaft diameter. By this type of coupling also minor axial misalignment of shafts can be compensated. Oldham coupling consists of two flanges keyed or screwed to the respective shafts and the flanges have the slots on the face surface as shown in Fig. 12.16. A cylindrical piece called disc with a tongue running across each flat face is shown in figure by '*a*' and '*b*'. The tongues are at right angle to each other. The tongue '*a*' fits into the slot of the right handed flange and allows for to-and-fro relative motion of the shafts, while the tongue '*b*' fits into the slot of the left handed flange and allows for vertical relative motion of the parts. The resultant of these two components of motion will accommodate lateral misalignment of the shafts as they rotate. Because of the misalignment of the shafts, the tongues on the disc are continually sliding back and forth in the slots of the flanges as the shafts rotate. To keep the mechanism engaged, axial movement of both shafts must be prevented. The angular velocity of the driver and driven member is the same at any instant. The disc of this coupling is usually made of cast iron or steel, and if the shaft rotates at high speed with very much misalignment, its weight is apt to set

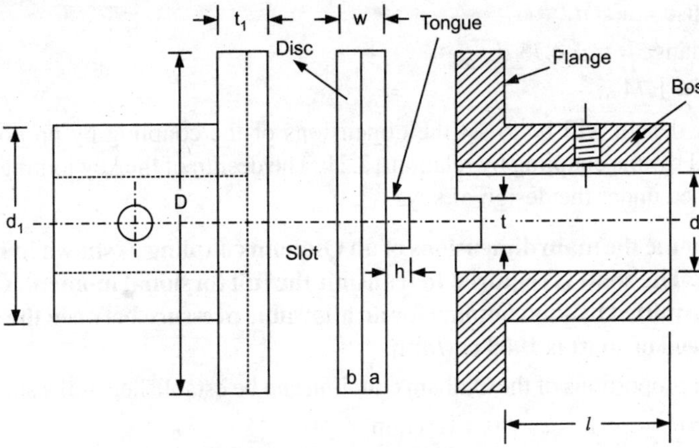

Fig. 12.16 *Oldham coupling in parts*

up undesirable vibration. For this reason speeds are usually restricted to a maximum of 1500 rev/min for this coupling. Fig. 12.17 shows assembled Oldham coupling.

The design of the Oldham coupling will be based on the allowable pressure between the faces of the slots and the tongues. Because of the elasticity of the material the pressure distribution for perfectly flat surfaces can be assumed to change on each side, as shown in Fig. 12.18, from a maximum value p at the periphery of the coupling to zero at the centre line.

Distance to the pressure area centroid from the centre line

$$= l = \frac{2}{3} \times \frac{1}{2} D = \frac{1}{3} D$$

where D is the diameter of the disc. Total pressure on each side $= F = 1/4 pDh$, where p is the maximum allowable pressure and h is depth of the slot or tongue.

Now torsional moment transmitted by both sides of the tongue will be

$$M_t = 2F \times l = 2 \times \frac{1}{4} p\, Dh \times \frac{1}{3} D = \frac{1}{6} p\, D^2 h \quad ...(12.20)$$

n = speed in r.p.m. of the shaft.

Then the power which can be transmitted by this coupling,

$$\text{Power} = \frac{2\pi n M_t}{60} = \frac{2\pi n}{60} \times \frac{1}{6} pD^2 h = \frac{pD^2 hn}{57} \quad ...(12.21)$$

The allowable pressure p must not exceed 8.5 N/mm^2 for normal misalignment. For a considerable misalignment, the maximum allowable pressure must not exceed 7 N/mm^2.

The different proportions of Oldham coupling are given below.

Shaft diameter $= d$

Diameter of boss $= d_1 = 2d$

Diameter of the flange or disc $= D = 3d$ to $4d$

Depth of slot or tongue $= h = t/2$

Width of slot or tongue $= t = D/6$ or $0.45d$

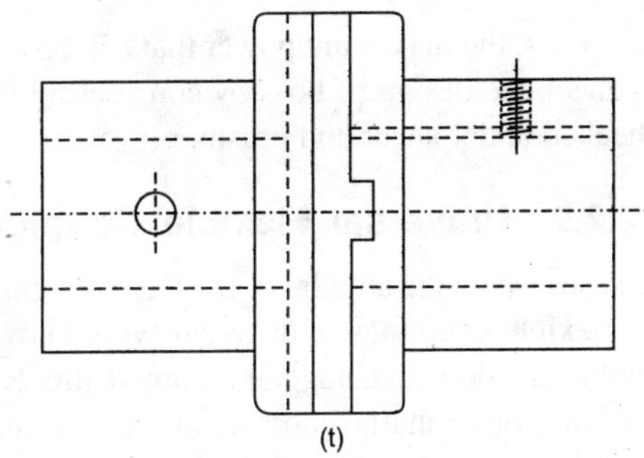

Fig. 12.17 *Assembled Oldham coupling*

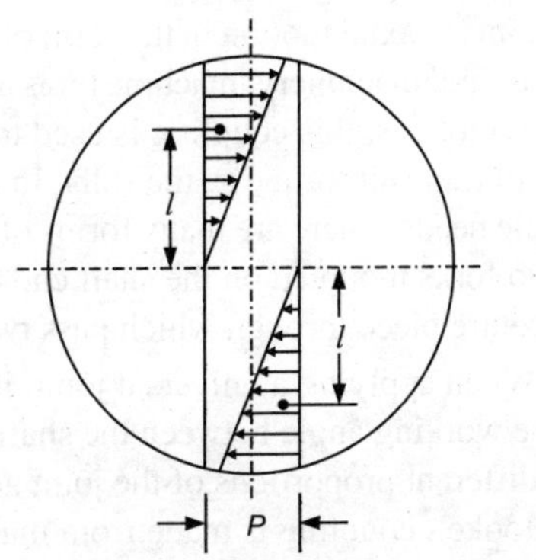

Fig. 12.18 *Pressure distribution in an Oldham coupling*

Thickness of the disc = $w = t/2$

Thickness of the flange = $t_1 = 0.75\ d$

Length of boss = $l = 1.74\ d$.

The procedure for design will be to set the dimensions of the coupling by proportions and check the power to be transmitted by this coupling by relation 12.21. The design of the key joining the flange with shaft will be done as discussed under the design of keys.

Example 12.15. Determine the main dimensions of an Oldham coupling as shown in Fig. 12.17, for a shaft diameter to be 90 mm. The shaft is required to transmit the full torsional moment. Check the maximum power that can be transmitted by this coupling for an allowable pressure between the slot and tongue to be 8.5 N/mm^2 and the speed of shaft is 1000 rev/min.

Solution: The different proportions of the Oldham coupling can be established as the shaft diameter is known.

Diameter of boss = $d_1 = 2 \times d = 2 \times 90 = 180$ mm.

Diameter of the disc or flange = $D = 3 \times d = 3 \times 90 = 270$ mm.

Width of slot or tongue = $t = 0.45d = 0.45 \times 90 = 40.5$ mm = 40 mm say.

Depth of slot or tongue = $h = t/2 = 40/2 = 20$ mm.

Thickness of the disc = $w = t/2 = 40/2 = 20$ mm.

Thickness of the flange = $t_1 = 0.75d = 0.75 \times 90 = 67.5$ mm.

Length of the boss = $l = 1.75 \times d = 1.75 \times 90 = 157.5$.

Now the power which can be transmitted by this coupling can be found by relation 12.21.

$$\text{Power} = \frac{pD^2hn}{57} = \frac{8.5 \times (270)^2}{57} \times \frac{20}{1000} \times 1000 = 217 \text{ kW}$$

Thus, the maximum power that can be transmitted by this coupling is 217 kW. The candidate is advised to check the design of boss by considering it to be a hollow cylinder and the design of the tongues may be checked under shear and crushing.

12.2.9 Universal Flexible Coupling or Articulated (or Joint) Coupling

In cases where the angular or off set misalignment of the shaft reaches values which cannot be accommodated by flexible couplings, a different type, known as Hooke's coupling or Universal Flexible coupling can be applied to good advantage. It is named after Robert Hooke, who developed the device. The universal coupling is used for two shafts whose centre lines form a large angle, as 5 to 15 degrees, or even 30 degrees. The joint itself has no torsional flexibility nor can it accommodate any parallel misalignment, i.e., shafts must intersect. If the off set changes during operation, as for the automobile propeller shaft, it is necessary to provide some freedom for axial motion in the form of a splined or some similar connection. This coupling is widely used in agricultural machinery, machine tools and automobiles, in instrument-making and other branches of industry. In automobiles, this coupling is used to connect shafts of the gearbox and the rear axle. In rolling mills, it is used to transmit torque to the rolls. In machine tools, it is used to transmit rotation to the spindles of multi-spindle heads. There are many forms of this coupling and one such coupling is shown in Fig. 12.19. It consists of two forks mounted on the shaft ends either with the help of keys/pins or screwed on the shaft ends. There is a centre piece through which pass two pins whose axes are at right-angles to each other.

When applying a universal joint, it is important to take into account the correction factor. This depends on the working angle between the shafts. Some typical correction factor values are given in the table below. The different proportions of the joint are shown in Fig. 12.19 in terms of the shaft diameter d. The choice of the Hooke's coupling is made from manufacture table for the power to be transmitted at the given speed.

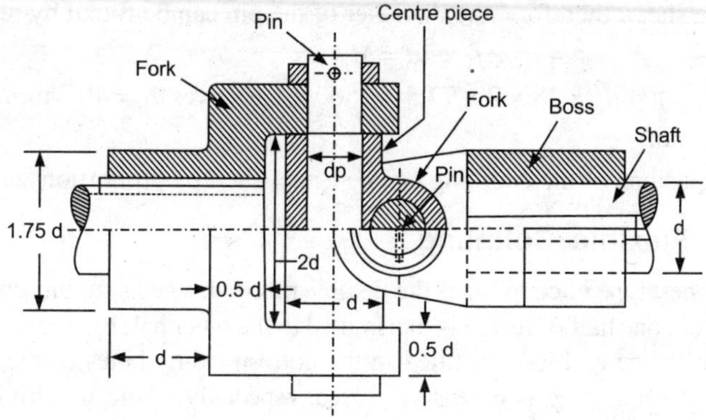

Fig. 12.19 *Hooke's coupling or universal flexible coupling.*

Correction Factors for Universal Joints

Working angle, degrees	0	5	10	20	30	40
Correction Factor	1.0	1.1	1.3	1.6	2.0	3.0

The pin is in double shear. If M_t is the torsional moment to be transmitted and d_p is the diameter of the pin, then diameter of the pin can be calculated by relation

$$\pi/4\, d_p^2 \times f_s \times 2d = M_t \qquad ...(12.22)$$

where f_s is the allowable shear stress in the material of the pin.

The induced stress in the boss of the fork can be checked by considering it to be a hollow cylinder. The safety of the fork can be checked under shear and crushing. Keys can be designed as discussed earlier. Rest of the dimensions can be set by proportions.

The central piece and the fork of the coupling are made from cast iron and the pins are made of steel. Universal joint should be lubricated and protected from dust and dirt.

Note: For shafts with diameter = 10 to 40 mm and for transmitted torques from 11.2 Nm to 1120 Nm, the standardised Universal joint couplings (Hooke's joint) are available.

Example 12.16. For a universal shaft to transmit 27.5 Nm torsional moment at 1000 rev/min. With an angle of 10 degrees between shafts a correction factor of 1.3 must be applied. Assume the usual factor of safety to be 6. Determine the power for which the universal coupling is to be designed.

Solution; The torsional moment to be transmitted by the coupling by considering the factor of safety and the correction factor = 27.5 × 6 × 1.3 = 214.5 Nm.

$$\text{Power involved} = \frac{214.5 \times 1000 \times 2\pi}{60} = 22.5 \text{ kW}$$

22.5 kW power at 1,000 rev/min would be the index for the choice of size in manufacturer's tables.

Example 12.17. In a universal coupling as shown in Fig. 12.19, shaft on the left hand side transmits 5500 Nm torsional moment to the other shaft. Find the diameters of the shafts and the pins. The allowable shearing stresses in the shaft and pin are 60 N/mm² and 28 N/mm² respectively. The shafts are assumed to be subjected to torsion only.

Solution: Torsional moment to be transmitted by coupling M_t = 5500 Nm

Let d be the diameter of the shaft, then

$$T = \pi/16 d^3 \times f_s,\ 5500 \times 10^3 = \pi/16 d^3 \times 60$$

gives, d = 77.5 mm.

The pin is in double shear, therefore, the diameter of the pin can be found by relation 12.22.

$$\pi/4\ d_p^2 \times f_s \times 2d = M_t$$

$$\pi/4\ d_p^2 \times 28 \times 2 \times 77.5 = 5500 \times 10^3, \text{ gives } d_p = 40.3 \text{ mm}$$

say diameter of pin = 40 mm.

The remaining dimensions of the coupling can be established by proportions as shown in Fig. 12.19.

12.2.10 Pin Type Flexible Coupling

The simplest and commonest type of coupling is the flexible rubber-bushed coupling shown is Fig. 12.20. In pin type couplings motion from one half of flange is transmitted to the other half by means of pins or bolts. The pins are rigidly bolted to one flange and loosely fitted in the corresponding holes of the other flange. In various designs this type of flexible coupling is extensively used, especially where the driving and driven units are mounted on a common base plate, where excessive misalignment is not likely, for example, a prime mover connected to generator, a compressor connected to an electric motor, an electric motor connected to a centrifugal blower, etc. This coupling makes up for the parallel misalignment of shaft upto 0.5 mm, angular misalignment upto 1.5 degrees and axial displacement. This coupling can drive in either direction, absorb impulses, shocks and vibrations. The pins are covered by rubber, leather or some other material suitable. Since the driving pins are easily removable, either shaft may be removed without having to be shifted endwise. These type of couplings are made with capacities 12000 kW at 2,500 rev/min. Rubber bushed couplings are standardised; their sizes should be selected to suit the torque transmitted. Such couplings are available for shaft diameters 9 to 160 mm and torques 6.3 to 16000 Nm. The coupling halves are mounted with axial clearance. The various considerations in the design of this type of flexible coupling are given below.

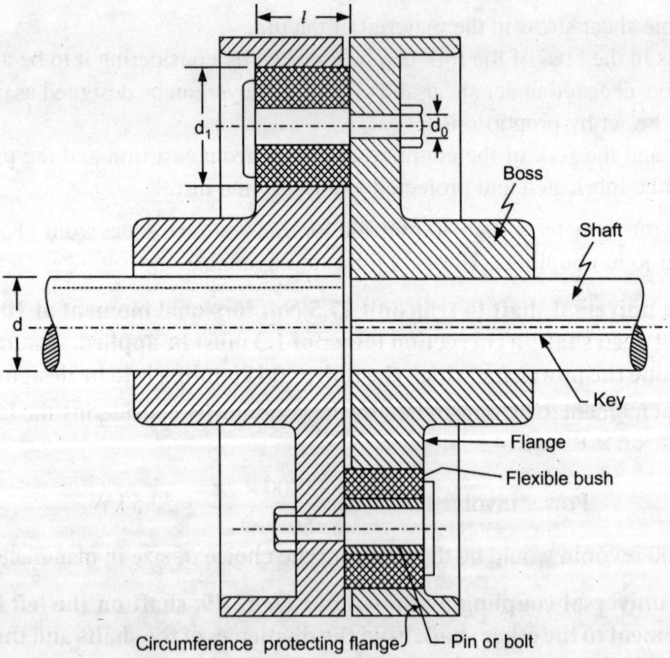

Fig. 12.20 *Pin type flexible coupling*

For the type of the coupling under consideration, it is necessary to modify the proportions from the rigid coupling, chiefly to reduce the bearing pressure on the rubber bushes which are the main feature of design. The different stresses acting over the pin will be:

1. Bearing Stress: Assume that M_t is the torsional moment to be transmitted by the coupling and R is the radius of the pitch circle. An increase in the pitch circle radius reduces the shear stress in the bolts necessary to transmit the given torsional moment M_t, but, of course requires an increase in the over-all diameter of the coupling. If z is the number of pins used, then force F applied at each pin to transmit the given torsional moment is,

$$F = \frac{M_t}{z \times R} \qquad \text{...(12.23)}$$

The number of pins to be used should be obtained by relation

$$z = \frac{d}{25} + 3 \qquad \text{...(12.24)}$$

where d is in mm.

The outer diameter d_1 of the rubber bush will be,

d_1 = diameter of pin + 2 × thickness of brass sleeve + 2 × thickness of rubber bush.

The rubber bushes are available in thickness of 6 mm and 7.5 mm for bores upto 25 mm and in thickness of 9 mm for large bores. The thickness of brass sleeve may be taken as 1.5 mm. Now if l is the length of the bush and p is the bearing pressure, then,

$$p \times \text{projected area} = \text{F}$$

$$p \times d_1 \times l = \text{F} \qquad \text{...(12.25)}$$

A suitable value of p, having regard to the life and durability of the rubber will be 0.35 N/mm^2, although higher values than this are sometimes specified. Relation 12.25 will help us to determine the length of the bush.

2. Shear stress: The threaded portion of the pin on which the nut is to be tightened should be as short as possible, so that shear can be taken by the unthreaded neck. Shear stress due to the transmitted torsional moment M_t on diameter d_0 will be,

$$f_s = \frac{F}{\text{area of neck}} = \frac{F}{\pi / 4\, d_0^2} \qquad \text{...(12.26)}$$

Check that this value is less than the allowable shear stress.

3. Bending stress: The pin is not rigidly held in the left hand flange, and the rubber is compressible, the force F at the enlarged end will exert a bending action on the pin. Effect of bending at the smaller diameter of the pin will be reduced by the longitudinal resistance of the shoulder of the pin, therefore, in estimating the bending stress the length under bending of pin will be 'l' and the longitudinal resistance to bending will be neglected. The type of the loading is indicated in Fig. 12.21. The clearance between the flanges is usually kept as 9 mm. The load F on the pin will be uniformly distributed on the length $(l - c)$, therefore the arm under bending moment upto the reduced area

$$= (l - c)\frac{1}{2} + c = l_1$$

Maximum bending moment = $F \times l_1$.

Induced bending stress will be

$$f_b = \frac{Fl_1}{\pi / 32\, d_0^3} \qquad \text{...(12.27)}$$

See that this value is less than the allowable bending stress.

Once the bending and shearing stresses have been calculated, the pin should be checked for the principal bending and shearing stresses.

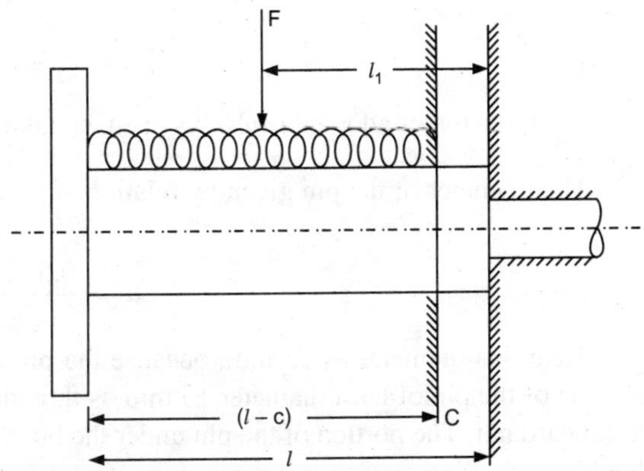

Fig. 12.21

4. Initial tightening stresses: Care should be taken during assembly that the nuts are not tightened down unnecessarily.

The remaining proportions of the coupling may now be taken as given below, where d is the shaft diameter, and t_1 is the boss thickness,

outer diameter of boss = $2 \times d$

length of the boss = $1.5\,d$

diameter of pin $d_0 = \dfrac{0.5\,d}{\sqrt{z}}$

diameter of the enlarged portion of pin, $d_p = 1.5\,d_0$

thickness of the right hand side flange = $t_2 = \dfrac{1}{2} \times t_1 + 6.5$ mm

t_1 = thickness of boss

thickness of the left hand side flange = length of bush – c

Width of the flanges = length of bush + 10 mm

To keep up the symmetry the width of both the flanges will be same. The thickness of the circumferential flange should not be less than half the thickness t_2.

The keys may be designed in the usual fashion. The length of the boss will also depend upon the length of the key.

As both the flanges are of different types, therefore, the patterns for both will be different. The coupling flanges are made from cast iron or steel forgings or steel castings. Pins are made from normalized steel.

Example 12.18. It is required to design a bushed pin type of flexible coupling for connecting the motor and centrifugal pump shafts. The details of the duty required from the pump are:

Power to be transmitted = 18.5 kW

Speed in rev/min = 1,000

The diameters of the motor and pump shafts are 50 mm and 45 mm respectively. Take the bearing pressure on the rubber bush as 0.35 N/mm² and the working shear stress in the material of the pins as 20 N/mm².

Solution: Torsional moment to be transmitted by the coupling is given as

$$\text{Power} = \frac{2\pi n M_t}{60},\quad M_t = \frac{18.5 \times 10^3 \times 60}{2\pi \times 1000}\ \text{Nm}$$

This gives $Mt = 179$ Nm

Number of pins to be used can be obtained by relation 12.24.

$$\therefore \quad z = \frac{d}{25} + 3 = \frac{50}{25} + 3 = 5$$

But the number adopted is always even, so take number of pins as 6.

The diameter of the pin given by relation, $d_0 = \dfrac{0.5d}{\sqrt{z}}$

$$\therefore \quad d_0 = \frac{0.5 \times 50}{\sqrt{6}} = 10.2\ \text{mm}$$

Take this diameter as 15 mm, because the pin is subjected to bearing and principal stresses. Now the portion of the pin of least diameter 15 mm is threaded and secured in the right hand flange with the help of a standard nut. The portion of the pin under the bush is enlarged in diameter, so that it can take up additional loads.

Let the enlarged diameter of the pin = 23 mm.

On the enlarged diameter a thin brass sleeve of thickness 1.5 mm is put. As the maximum diameter of the pin is less than 25 mm, therefore, the thickness of the bush will be 7.5 mm.

Now the outer diameter of the bush will be,

$$d_1 = 23 + 1.5 \times 2 + 7.5 \times 2 = 41 \text{ mm}$$

In the present design, a pitch circle diameter of 140 mm will be used to keep the dimensions of the rubber bush reasonably low. Now by relation 12.23, the force F acting on each pin can be calculated.

$$\therefore \qquad F = \frac{M_t}{zR} = \frac{179 \times 10^3}{6 \times 70} = 425 \text{ N}$$

The bearing pressure on the rubber bush is 0.35 N/mm^2, therefore, the length of rubber bush can be found by relation 12.25.

$$p \times d_1 \times l = \text{F}, \quad 0.35 \times 41 \times l = 425, \text{ gives } l = 29.7 \text{ mm.}$$

say the length of bush to be 30 mm.

The brass sleeve and rubber bush are slipped on the pin form the smaller end and fit against the head of the pin whose diameter may be taken as 30 mm. There is no rigid connection between the flanges and the drive takes place through the medium of the compressible rubber bushes. The head of the pin may be made thick.

Stresses in the pin: Shear stress on the pin by relation 12.26 is, $f_s = \dfrac{F}{\pi/4\, d_0^2} = \dfrac{425}{0.7854 \times (15)^2} = 2.42 \text{ N/mm}^2$.

The clearance between the two faces of the flanges is 9 mm, therefore, the length of pin under bending (Refer Fig. 12.21) = $(l - c)\ 1/2 + c = l_1$

$$l_1 = (30 - 9)\ \frac{1}{2} + 9 = 19.5 \text{ mm say 20 mm.}$$

Now the induced bending stress will be (by relation 12.27).

$$f_b = \frac{Fl_1}{\pi/32\, d_0^3} = \frac{425 \times 20}{(15)^3} \times \frac{32}{\pi} = 38.5 \text{ N/mm}^2$$

Now the pin can be checked for the principal stresses.

$$\text{Principal bending stress} = \frac{1}{2}\left(f_b + \sqrt{(f_b^2 + 4f_s^2}\right) = \frac{1}{2}\left(38.5 + \sqrt{(38.5)^2 + 4 \times (2.42)^2}\right) = 38.75 \text{ N/mm}^2$$

which is within safe limits of 40 N/mm^2

$$\text{Principal shear stress} = \frac{1}{2}\sqrt{f_b^2 + 4f_s^2} = \frac{1}{2}\sqrt{(38.5)^2 + 4(2.42)^2} = 19.5 \text{ N/m}^2$$

which is within safe limits of 20 N/mm^2.

Key design: Width of key = $\dfrac{d}{4} = \dfrac{50}{4} = 12.5$ mm

Adopting a square key of section 12.5 × 12.5 mm

Length of boss = 1.5d = 1.5 × 50 = 75 mm on motor side

Length of boss on pump side = 1.5 × 45 = 67.5 mm

Now the induced shear stress in the key can be checked by relation 12.1

$$f_s = \frac{2M_t}{dwl} = \frac{4 \times 179 \times 10^3}{50 \times 12.5 \times 75} = 7.65 \text{ N/mm}^2.$$

Induced crushing stress can be checked by relation 12.2.

$$f_c = \frac{4M_t}{dlt} = \frac{4 \times 179 \times 10^3}{50 \times 75 \times 12.5} = 15.3 \text{ N/mm}^2$$

The induced stresses are much below the allowable values.

diameter of boss on motor end = $2d = 2 \times 50 = 100$ mm

thickness of boss on motor end = $(100 - 50)\ \frac{1}{2} = 25$ mm

diameter of boss on pump end = $2d = 2 \times 45 = 90$ mm

thickness of boss on pump end = $(90 - 45)\ \frac{1}{2} = 22.5$ mm

thickness of flange on motor end = length of bush = 30 mm

thickness of flange on pump end = $\frac{1}{2} t_1 + 6.5$ mm = $\frac{1}{2} \times 22.5 + 6.5 = 17.8$ mm

width of flanges = length of bush + 10 mm = 30 + 10 = 40 mm

Thickness of the circumferential flange = $\frac{1}{2} \times 17.8 = 8.9$ mm

Overall diameter of flange = Pitch circle diameter = 2 × radius of rubber bush + 2 × thickness of circumferential flange + 2 × clearance.

$$= 140 + 2 \times \frac{41}{2} + 2 \times 8.9 + 2 \times 3 = 204.8 \text{ mm say } 210 \text{ mm.}$$

All the components of the coupling must be properly machined to ensure balance, particularly when required to run at high speeds.

12.2.11 Slip Coupling

The object of a slip coupling is to permit relative rotation, or slip, between the driving shaft and the driven shaft. A slip coupling is a safety device that prevents damage to rotating parts because of overloading. The slip coupling is adjusted so that it will begin to slip if the transmitted torsional moment exceeds a predetermined value. Usually the slip begins if the load exceeds by 10 to 20 percent the maximum load for which the shafts and other parts are designed. There are many variations in form of this type of coupling. A common form for medium and large capacities is shown in Fig. 12.22. The flanged hub *a* is keyed to the driving member. The driven member *b* is composed of a flanged hub with friction discs attached to both sides of the flanges. A number of small but heavy springs are loaded by screwing down the nuts, and thus create the necessary pressure on the friction surfaces. This ensures a constant maximum moment of friction transmitted by the coupling. In case of an overload, i.e., when the torsional moment on the shaft rises to a maximum friction moment of the coupling, the discs start to slip. Thus, further increase of the torsional moment on the shaft is safeguarded against failure.

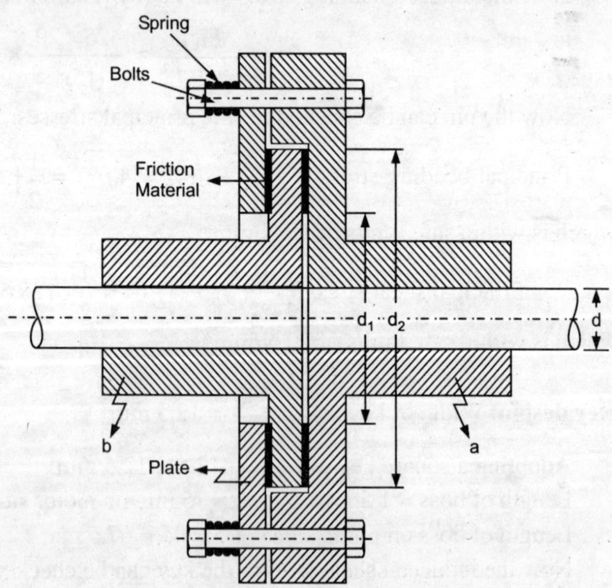

Fig. 12.22 *Slip coupling*

The design procedure for slip coupling will be as follows:

Slip in such a coupling occurs frequently, and no measurable wear takes place, therefore, it can be assumed that the unit pressure p between the discs is uniform and that the axial force is exerted by springs. The axial force Fa exerted by the spring is given by relation,

$$F_a = \pi/4\,(d_2^2 - d_1^2)\,p \quad ...(12.28)$$

Let μ is the coefficient of friction, then the tangential force for two pairs of friction surfaces, Fig. 12.22, is

$$F_t = 2\mu F_a = 2\mu\,\frac{\pi}{4}(d_2^2 - d_1^2)\,p = \frac{\mu\pi}{2}(d_2^2 - d_1^2)\,p \quad ...(12.29)$$

The mean radius at which the tangential force will be acting is given by relation,

$$r_m = \left(\frac{d_2 + d_1}{2\times 2}\right) \quad ...(12.30)$$

Now the frictional torsional moment transmitted by the coupling is given by relation,

$$M_t = F_t \times r_m = \frac{\mu\pi}{2}(d_2^2 - d_1^2)\,p \times \left(\frac{d_2 + d_1}{2\times 2}\right)$$

$$M_t = 0.3925\,(d_2^2 - d_1^2)\,(d_2 + d_1)\,p \times \mu \quad ...(12.31)$$

In the above relation, the torsional moment M_t can be equated to the torsional moment to be transmitted by the coupling. The unit pressure or bearing pressure p should be as low as possible to increase the life of the coupling. The desirable value for the bearing pressure is 0.4 to 0.5 N/mm^2 but in no case it should be higher than 0.7 N/mm^2. The diameter d_2 and d_1 may be selected so that, approximately, d_2/d_1 is 1.6. Larger ratios should be avoided as they unnecessarily increase the size of the coupling. Smaller ratios should not be used as they give appreciable different rates of wear and reduce the life of the bearing. The coefficient of friction for cast iron on cast iron may be taken as 0.15. The coefficient of friction can be increased by lining the coupling disc with friction materials, where the increase will depend upon the friction material used.

Thus relation 12.31 will help us to determine the values of d_2 and d_1.

The number of springs ranges from 6 for small couplings and for a shaft diameter of 55 mm, upto 16 for a shaft diameter of 160 mm.

The axial load (tensile load) coming on all the bolts can be calculated from relation 12.28. Thus, by fixing the number of bolts, the load coming on each bolt can be found out. The pitch of the bolts should be such that, it ensures the uniform distribution of pressure. The bolt should be designed for axial loading tension as discussed earlier.

The load coming on each spring will be in compression and magnitude of load will be the load on bolt in tension. The design method for springs is discussed in chapter 15. Remaining dimensions of the coupling can be fixed by proportions.

Example 12.19. It is required to design a slip coupling for connecting the motor and centrifugal pump shafts. The details of the duty required from the pump are:

Power = 18.75 kW, speed = 1,000 rev/min

The diameter of the shafts to be connected is 50 mm. The value of μ may be taken for cast iron wearing cast iron. The working stress in the bolts is 30 N/mm^2.

Solution: Torsional moment to be transmitted by the coupling can be calculated by the relation

$$\text{Power} = \frac{2\pi n M_t}{60},\; \text{M}_t = \frac{118.75\times 10^3 \times 60}{2\pi \times 1000}$$

$$\therefore \qquad Mt = 179 \text{ Nm}$$

Assuming the bearing pressure on the contact surfaces as 0.5 N/mm^2 and d_2/d_1 = 1.5. Then by relation 12.28, the axial force can be calculated.

$$F_a = \pi/4\,(d_2^2 - d_1^2)\,p,\; F_a = 0.7854\,[(1.5d_1)^2 - d_1^2]\,0.5$$

gives, $F_a = 0.49\,d_1^2$ N.

For cast iron surfaces wearing cast iron, $\mu = 0.15$.

Now by using relation 12.31, calculate the diameter d_1.

$$M_t = 0.3925\,(d_2^2 - d_1^2)\,(d_2 + d_1)\,p \times \mu$$

$$179 \times 10^3 = 0.3925\,[(1.5\,d_1)^2 - d_1^2)\,(1.5\,d_1 + d_1)]\,0.5 \times 0.15$$

gives, $d_1 = 125$ mm.

$$\therefore \quad d_2 = 1.5\,d_1 = 1.5 \times 125 \text{ mm} = 187.5 \text{ mm}$$

$$\text{Mean radius} = \frac{d_2 + d_1}{2 \times 2} = \frac{125 + 187.5}{4} = 78 \text{ mm}$$

Axial force, $F_a = 0.49 \times d_1^2 = 0.49 \times (125)^2 = 7600$ N

As the diameter of the shaft is 50 mm, therefore, the number of bolts used will be 6.

$$\text{Tensile load taken by each bolt} = \frac{7600}{6} = 1270 \text{ N}$$

If d_c is the core diameter of the bolt, then

$$\pi/4\; d_c^2 \times f_t = 1270,\; \pi/4\; d_c^2 \times 30 = 1270$$

gives, $d_c = 7.35$ mm

$$\text{Outside diameter of bolt} = \frac{7.35}{0.84} = \frac{d_c}{0.84} = 8.75 \text{ mm.}$$

To allow for extra tightening the diameter of the bolt may be taken as 10 mm.

Diameter of boss $= 2d = 2 \times 50 = 100$ mm.

Outside diameter of coupling $= d_2 + 2 \times$ diameter of bolt $\times 2 + 2 \times$ clearance.

$= 187.5 + 2 \times 10 \times 2 + 2 \times 10 = 247.5$ mm, say 250 mm.

The spring will be designed for a load of 1270 N in compression, with the condition, the inside coil diameter not to be less than 10 mm and if possible the outside coil diameter not be more than 20 mm.

EXERCISES

1. What is the function of a coupling?
2. How do you differentiate between a coupling and a clutch?
3. What are the purposes in machinery for which couplings are used?
4. Indicate what type of coupling is used under the following conditions:
   (a) With shafts having collinear axes.
   (b) Shafts having intersecting axes.
   (c) Shafts having parallel axes with a small distance apart.
5. How do you classify couplings ? Explain the function of each in detail.
6. What are the advantages and disadvantages of muff coupling?
7. A sleeve coupling is used to connect two 75 mm shafts whose safe shearing stress is 70 N/mm^2.
   (a) What torque is transmitted by the shafting?
   (b) To what torsional moment is the coupling subjected?
   (c) Find the diameter and the length of the coupling.
   (d) What will be the induced stress in the sleeve if the diameter of part (c) is used?
   (e) If cast iron sleeve with ultimate shearing stress of 140 N/mm^2 is used, what is the numerical value of the factor of safety?
8. Design a rigid sleeve coupling to connect two shafts, transmitting 18.75 kW at 1000 rev/min. The allowable shear stress in the material of the shaft is 55 N/mm^2. The material of the key and shaft is same and the coupling is required to transmit 20 percent overload. The material of sleeve is cast iron, the allowable shear stress for which is 16 N/mm^2. Make a neat sketch of the designed sleeve coupling showing side view and sectional elevation.
9. Do you recommend sleeve coupling with taper pin for example ? The material of the pin and shaft may be assumed to be same.

10. What are the advantages and disadvantages of flange coupling?
11. Outline the complete procedure for obtaining the force in the bolt of a clamp coupling shown in Fig. 12.13.
12. Design a clamp coupling completely for a shaft diameter of 50 mm. The torsional moment to be transmitted by this coupling taking into account overloading is 1400 Nm. The number of bolts used is 4 and the allowable stress in the material of the bolt is limited to 70 N/mm^2. The coefficient of friction between the shaft and muff material is 0.3. Draw the sectional elevation of the designed clamp coupling.
13. What are the advantages and disadvantages of flange coupling?
14. The bolt in the flange coupling should be made weaker than the other components of coupling, why?
15. (a) Make a neat sketch of flanged shaft coupling showing a face view and a cross-sectional view.
    (b) Derive expressions in terms of the torsional moment M_t in Nm for the following:
        (i) the shearing stress in the shaft.
        (ii) the shearing stress in the bolts.
        (iii) the bearing pressure on the bolts.

    The following notations may be used:

    d = shaft diameter. R = radius of bolt circle.
    z = number of bolts. d_1 = diameter of bolt.
    t = thickness of radial flange.
16. In a flange shafts coupling having a 37.5 mm. bore, it is desired that torsional stress in the shaft will not exceed 25 N/mm^2. The outside diameter of the coupling limited by space is 200 mm. There are three 15 mm bolts on a bolt circle diameter of 140 mm. The radial flange thickness is 18 mm. Determine the following:
    (a) the power that may be transmitted at 600 rev/min.
    (b) the shearing stress in the bolts.
    (c) the bearing pressure on the bolts.
17. Two 40 mm shafts are connected by a flange coupling. The flanges are fitted with six bolts of carbon steel on a 125 mm bolt circle. The shafts are run at 340 rev/min and transmit a torsional moment of 1000 Nm. Assume a factor of safety of 5.
    (a) What diameter bolts should be used?
    (b) How thick the flanges should be?
    (c) Determine the key dimensions.
    (d) Determine the hub length.
    (e) What power can be transmitted by this coupling?
18. A plane flange coupling for a 75 mm shaft has the following dimensions : bore of flange = 75 mm, hub diameter = 135 mm, hub length = 100 mm, flange diameter = 250 mm, flange thickness = 28 mm, bolt circle diameter = 200 mm, bolt diameter = 19 mm, number of bolts = 6, and key size is 18 mm square. All parts are made of carbon steel having a shearing strength of 400 N/mm^2 and a tensile strength of 650 N/mm^2. This coupling is rated at 37.3 kW power at 100 rev/min.
    (a) Determine the shearing, bearing and tensile stresses in all parts of the coupling.
    (b) What F.O.S. does this coupling have?
    (c) Is there a possibility of the flange shearing off at the hub ? Assume perfect shaft alignment.
19. When it is preferred to transmit torsional moment by friction between the contacting faces of the coupling?
20. What size of the bolt will be required if the torsional moment is transmitted from one half of the flanged shaft coupling in example 17 to the other by means of friction between the contact surfaces? The coefficient of friction between the contact surfaces may be taken as 0.1.
21. A 1.65 m diameter pulley transmits 164 kW power at a speed of 500 rev/min to a 0.55 m driven pulley on the shaft of which a flange coupling is to be attached. There are to be six bolts in a 150 mm bolt circle diameter.
    (a) Determine the required size of the bolts, using a design stress for bolt shear of 20 N/mm^2.
    (b) Calculate the hub length of each part of the coupling, if a 12.5 mm by 10 mm flat key is to be used. Assume design stress value of 55 N/mm^2 in shear and 140 N/mm^2 in bearing. The shaft is 50 mm in diameter.
22. Design a flange coupling completely to transmit 135 kW power at 120 rev/min from one shaft to another. Allowable shearing stress of the bolt material is 45 N/mm^2 while that of the shaft material is 55 N/mm^2. The flanges are made of cast iron for which the stresses in compression and shear are 75 N/mm^2 and 175 N/mm^2 respectively. The material of the shaft and key is same. The allowable stresses given take into account the service

factor and weakening effect of the key way in the shaft and flange. Draw the sectional elevation and the outside view of the designed coupling.

23. A flange coupling made for eight 10 mm diameter bolt on a 100 mm diameter bolt circle was selected for a shaft connection, but was found to be unobtainable. Make calculations to determine whether flange coupling with four bolts for 12.5 mm bolt on a 150 mm bolt circle diameter can be substituted with safety.
24. When a solid flanged coupling is preferred?
25. Design a solid flanged coupling completely for the data given in problem 22. Overload of 20% is to be transmitted by the coupling.
26. What are the advantages of flexible coupling?
27. How do we classify flexible coupling? Explain in detail.
28. What type of coupling is used to connect shafts whose axes are parallel?
29. Design an Oldham coupling completely for the data given in exercise 22. The allowable pressure between the tongue and the slot is 8.5 N/mm^2. Sketch the designed coupling in parts with dimensions.
30. Two shafts intersect at an angle of 30 degrees. What type of coupling can be used to connect these shafts?
31. A universal coupling as shown in Fig. 12.19 is required to transmit 6000 Nm torsional moment from one shaft to the other. The allowable shear stresses in the shaft and pin material are 55 N/mm^2 and 30 N/mm^2 respectively. Design the coupling completely assuming that the shafts are subjected to torsion only. The material of the shaft, fork and key is same. The allowable crushing stress in the material of the key is 90 N/mm^2. Draw the coupling designed and dimension it fully.
32. Under what conditions pin type Flexible Coupling is suggested?
33. It is required to design a bushed pin type flexible coupling for connecting two shafts of diameter 60 mm. The details of the duty required from the coupling are :

    Power to be transmitted by coupling = 135 kW

    Speed in rev/min = 120

    Take the bearing pressure on the bushes as 0.35 N/mm^2, and the allowable shear stress in the material of the pins as 25 N/mm^2. The allowable bending stress in the material of the pin is 50 N/mm^2.

    Draw to scale the sectional elevation and the side view of the designed coupling.
34. What are the factors to be investigated while selecting a coupling?

    When is the choice of a slip coupling made?
35. A slip coupling, Fig. 12.22, is assembled to slip at a torque corresponding to 20 percent overload on a 225 kW power motor running at 360 rev/min. The diameters are $d_2 = 0.8$ m and $d_1 = 0.5$ m, the coefficient of friction may be taken as 0.125, the number of bolts and springs as 12, and the spring scale is 450 N/mm. Determine the deflection with which each spring must be set up. Assume uniform pressure distribution.
36. (a) Determine the main dimensions of a slip coupling, Fig 12.22 to connect two 100 mm shafts. Assume a maximum shear stress in the shafts of 45 N/mm^2 and an overload of 20 percent when the coupling will slip. State reasons for all necessary assumptions and selections.

    (b) Design the springs and bolts to hold the torsional moment.

    The value of coefficient of friction may be taken as 0.15. The working stress in the bolt material is 30 N/mm^2. The allowable shear stress in the spring material is 30 N/mm^2.
37. A drive is to be designed from a 50 kW motor running at 2900 rpm, having 50 mm diameter shaft to connect an aircraft fuel pump testing Rig, through a flexible coupling in intermediate shaft having bearings in a housing at both the ends, and thence a splined shaft bush coupling to couple the motor shaft of the pump.

    Design a suitable flexible coupling to connect the motor and the intermediate shaft. The flexible coupling should be as compact as possible. Take the bearing pressure on the bushes as 0.35 N/mm^2 and the safe shear stress of the material of the pin as 25 N/mm^2. Safe bending stress of the pin material is 50 N/mm^2.

    Sketch the elevation and side view of the designed couplings.
38. (a) Give reasons for joining shafts with coupling.

    (b) Why are two universal joints often used when there is angular misalignment between two shafts?
39. A cast iron coupling is to be designed for power transmission of 1 kW at 150 rpm. The safe shear stress for the shaft and bolt material is 45 N/mm^2 and the bearing stress in the bolts is 90 N/mm^2. Determine the diameter of the shaft. Fix up the dimensions of the flange coupling, assuming shear stress in cast iron to be 60 N/mm^2. Use proportions wherever necessary.

40. Given: A splined joint between a shaft and a chain sprocket. Spline specifications: 8 × 36 × 42; pitch or mean diameter of splines = 39 mm.

Length: 40 mm; specific static moment of area, $A = 343$ mm^3 per mm. Dynamic load factor = 2; Load distribution factor in bearing strength analysis = 1.5. Analyse the joint for bearing strength, if torque at the shaft = 500 N.m.

Solution: Now specific total static moment of contact area about the shaft axis, $A = 0.5 \times d_p \times t \times z$, mm^3/mm.

where d_p = pitch or mean diameter of splines = $\dfrac{D+d}{2}$

t = working depth thickness of splines = $\dfrac{D-d}{2}$ (theoretically)

z = number of splines

D and d are outer and inner diameter of splines.

Now $D = 42$ mm, $d = 36$ mm, $d_p = 39$ mm, $z = 8$

Now Design Torque on the shaft. = 500 × 1.5 × 2 = 1500 N.m.

$$\therefore \quad \text{Mean bearing stress, } f_b \cong \frac{10 \times 10^3 \times M_t}{(D^2 - d^2) l \times z} \cong \frac{10 \times 10^3 \times 1500}{t \times 78 \times 40 \times 8}$$

$$\text{Now} \quad 343 = 0.5 \times \frac{D+d}{2} \times t \times 8$$

$$\therefore \quad t = \frac{343}{2 \times 78} = 2.2 \text{ mm.}$$

$$\therefore \quad f_b \cong \frac{10 \times 1000 \times 1500}{2.2 \times 78 \times 40 \times 8} \cong 273 \text{ MPa} < |f_b| = 400 \text{ MPa.}$$

∴ safe.

41. Discuss, with the help of neat sketches, the various type of misalignments, which can occur between two shafts.
42. What is the function of "register" in a flange coupling?
43. Which of the following couplings is a flexible coupling?
    (a) Muff (b) clamp
    (c) Flange (d) Bush type.
44. Write about the type of flexibility provided by the Oldham coupling.
45. How the bolts are designed under
    (a) shear (b) tension in flange coupling?
46. Can a universal Flexible coupling accommodate any parallel misalignment?

❑❑❑

CHAPTER 13

Pipe Joints

13.1 INTRODUCTION

In accordance with the process layout, the reservoirs and apparatus are interconnected by piping which conveys liquids or gaseous intermediate or end product (moving media). The media conveyed may be of different pressure and temperature. Various types, of pipes are used depending upon the nature of fluid, quantity to be transported, service conditions, cost of production and installation deterioration, liability of failure and results of such failure, its pressures and velocity.

The piping is comprised of individual pipes with other pipes apparatus and valves connected up by special fittings. Pipe valves are various devices designed to change the area through which the medium moves in the pipe. Valves can be installed not only on pipes but also on vessels, their purpose is to control the movement of the medium.

The pipe lines are made of cast iron, steel, wrought iron, red copper and brass pipes for the conveyance of steam, water, gas, oil and other fluids from one place to another. The sizes of pipes are standardized. A pipe is characterized by its outside diameter, wall thickness and the material it is made of. Normally several wall thicknesses for each outside diameter of the pipe are available. Thus, the inside diameter of the pipe varies with the wall thickness.

13.2 DESIGN OF PIPES

Cast iron pipe is extensively used for underground water lines, sewerage and gas mains. This material shows remarkable resistance to destruction. Being more durable than either wrought iron or steel, but is not suitable for lines subjected to high temperature stresses or severe vibration. Cast iron steam pipes can be used for pressures upto 0.7 N/mm^2. The code for pressure piping, limits its use to steam temperatures not exceeding 250ºC and for hot oil not over 180ºC. Although cast iron pipe in the first year of its installation becomes coated with rust the rate of penetration of corrosion decreases with time and is much lower after the pipe has been in the ground for a number of years than when first installed. This may also be used for process lines to convey corrosion liquids where its life may be just long enough to justify infrequent replacements.

Carbon steel pipes are used to convey water, oil, steam and fuels with different characteristics. Lengths from 10 meters to 15 meters are available. Stainless steel red copper and, red brass pipes are used for media causing considerable corrosion. Red brass and copper pipes are used for such services where their longer life expectancy compensates for the higher cost than that of ferrous pipe. Red brass and copper pipes resist many of the process solutions in the pulp and the paper and the chemical industries. Red brass which is 85 percent copper and 15 percent zinc alloy has a considerably greater allowable stress value does not permit the use of either of these materials at temperatures up to 180°C these become equal. The pressure piping ruling does not permit the use of either of these materials at temperatures over 225ºC. The pipes of these materials are available in lengths of 4 meters. Colours or the identification of the above standard pipe size are white for brass and black for copper. Red copper pipes are used for pressures lower than 12.5 N/mm^2 and rate of flow lower than 300 kg/min. For pressures greater than 43 N/mm^2 and temperatures more than 440ºC, thick walled seamless pipes made of heat-resistant chrome molybdenum alloy steels are used.

Concrete pressure pipe is extremely rugged but should not be used in the presence of acid seepage often encountered in the proximity of chemical plants. It is made in a variety of types each intended for its specific range of internal pressures. The choice of the wall thickness and type, being based not only on the internal pressure but also the external loading which the pipe will be called upon to withstand. Concrete pipes are used for pressures varying from 0.3 N/mm^2 to 2.5 N/mm^2. Care must be exercised in installing concrete pipe to insure that it conforms to line and grade as shown on plans. It should be installed upstream with the bell ends in the proper direction of flow and joined together so as to insure a maximum degree of straightness along the inside.

Continuous stave wood pipe is made in diameters as large as 6 meters. It is shipped in a knocked-down condition and erected on the site. In the western part of the United States and Canada where most of such pipe is in use, it is said to compare favourable on the basis of original cost, maintenance, and life expectancy with other piping materials.

The latest development in materials gives plastics as the materials of pipe. Plastics are highly resistant to corrosion. Therefore, they can be effectively used for conveying chemically active media. Plastics are lighter in weight also. In particular, Vinyl plastic pipes, polyethylene pipes are used.

When the material of the pipe is decided, then the main factors in the design of the pipe are thickness and the internal diameter of the pipe. The pipe installation is done to convey a certain quantity of fluid from one place to another, Diameter of the pipe required for the transmission of steam depends upon its velocity. The resistance to flow and consequent loss of pressure increases with the increase of velocity, this results to increase the size of the pipe to decrease the pressure loss. On the other hand, loss due to radiation and conduction increases with the increase in size of the pipe, how well covered the pipe may be. An increase in the size means a corresponding increase in thickness, thus making pipes more costly and less efficient. Thus a designer has to balance both the factors. The normal practice is to give the velocity by the relation:

$$V = \frac{21.66}{\sqrt{p}} \quad \text{...(13.1)}$$

where V is the velocity of steam in m/s

P is the pressure of steam in N/mm^2

In practice the mean velocity of steam through the piping is 30 to 40 meters/second.

Once the velocity of the fluid and quantity to be conveyed are known, the nominal bore of the pipe can be determined. Let Q be the quantity of the fluid to be conveyed in volume per unit time (m^3/s) by a pipe of internal diameter D_i and V be the velocity of fluid in the pipe per unit time (m/s). In case of steam flow in the pipe the specific volume can be seen from the steam tables. Now the relation for the flow can be written.

$$Q = \frac{\pi}{4} D_i^2 V \qquad \text{...(13.2)}$$

or

$$D_i = 1.13\sqrt{\frac{Q}{V}} \qquad \text{...(13.3)}$$

D_i = internal diameter in m.

On the other hand the internal diameter of the pipe can be given by the relation:

$$D_i = 18.8\sqrt{\frac{w}{V\rho}} \qquad \text{...(13.4)}$$

where D_i = internal diameter in mm

w = discharge or flow rate measured in kg/hour for gases and in tonne/hour for liquids.

V = flow velocity of the media in m/s.

ρ = density in kg/m^3 for gases and tonne/m^3 for liquid.

Once the minimum internal diameter or nominal bore of the pipe is known, the next higher size can be selected from the standard tables.

The next step is to calculate the thickness of the pipe. Pipes are generally designed on the basis of the thin cylinder theory. In thin cylinder theory it is assumed that the stresses across the section of the pipe are assumed to be uniform. This is only possible when the internal diameter D_i of the pipe exceeds 20 times the wall thickness t of the pipe, i.e., $D_i > 20t$. Another criterion used to check if the thin cyclinder theory can be applied is the ratio of the allowable stress of the pipe material to the pressure of the fluid inside the pipe. For pipes which are made of ductile materials, the allowable stress should be greater than or equal to six times pressure of the fluid,

i.e.,

$$\frac{f_t}{p} \geq 6.0.$$

Now by thin cyclinder theory for pipes subjected to internal pressure P, the wall thickness t is given by the relation:

$$t = \frac{PD_i}{2f_t} \qquad \text{...(13.5)}$$

If there is a longitudinal joint in the pipe, its efficiency has to be taken into account in determining the wall thickness of the pipe. If η is the efficiency of the longitudinal joint, then

$$t = \frac{PD_i}{2f_t \times \eta} \qquad \text{...(13.6)}$$

The thickness of the pipe calculated by above relation is to suit the pressure of the liquid or gas conveyed by the pipe line. However, there are other factors for which the pipe has to be designed. These factors are wear and corrosion which depend on the material of the pipe and on the type of the working media. Now the modified relation for the wall thickness of the pipe is

$$t = \frac{PD_i}{2f_t \times \eta} + C \qquad \text{...(13.7)}$$

where C is the allowance for threading, mechanical strength, and corrosion added to design thickness of wall. For a pipe made of carbon or alloy steel with thickness less than 6 mm, C is taken as 1 mm, and if wall thickness is greater than 6 mm, then C is taken as $0.18t$.

The thickness calculated by relation 13.7 in certain cases may not be safe from rigidity point. Particularly when the pipe is to be manufactured by casting, the thickness of the pipe can not be less than a certain, practicable minimum value. The minimum thickness will depend upon the size of the pipe and the material. From this consideration Weisback has recommended the following values for C.

Cast iron	9 mm	Copper	4 mm
Lead	5 mm	Mild steel	3 mm
Zinc	4 mm		

The efficiency of the longitudinal joint will depend upon the type of the joint. The allowable stress $f_t = f_u/n$, where f_u is the ultimate strength for pipe temperatures from 20°C to 250°C, n is the factor of safety, in this case $n = 4$, or it is yield strength for pipe temperature greater than 250°C then $n = 1.8$.

Where temperature is more than 350ºC, then the strength of the pipe has to be checked for creep limit, i.e., $f_t = f_{creep}$ for a given temperature, where f_{creep} is the creep limit. Creep stress is considerably lower than the yield point stress at elevated temperatures. The creep limit is the limit at which the creep rate is equal to a certain value determined by the specifications.

If the variation of stress across the thickness of the pipe is to be taken into account, Lame's Theory is used to obtain the thickness of the pipe. The Lame's formula is

$$t = \frac{D_i}{2}\left[\sqrt{\frac{f_t + p}{f_t - p}} - 1\right] \qquad ...(13.8)$$

where f_t is the allowable tensile or hoop stress in the pipe material.

The pipe joints are tested for hydraulic pressure. Therefore, this is the maximum pressure for which the strength of the joint should be checked. The test pressures for the different working pressures are given by Indian Standards. The following Table 13.1 gives various values of f_t to be used in above given relations.

Table 13.1

Pipe	N/mm^2
Cast iron steam or water pipes	140
Cast iron steam engine cyclinder	12.5
Solid drawn steel tubes	140
Lap welded wrought iron tubes	60
Copper steam pipes	25
Lead pipes	1.6

13.3 HYDRAULIC PIPE JOINTS

Pipe joints have to ensure the pressure tightness of the system under the effect of working pressures. In some cases pipe joints have to meet the additional requirement of easy disassembly. Pipe joints are made by one of the following types of joints: (1) Threaded joint, (2) Bell and spigot joint, (3) Brazed joint, (4) Welded joint, (5) Expansion joint, (6) Flanged joint.

1. Threaded Joint: Threaded pipe joints are probably the most widely used form of connection. When threads are clean and smoothly cut to the proper pitch, size and thread form by suitable dies and carefully assembled with the proper joint compound for service, this type of joint proves economical and Satisfactory for water as well as for steam, air, and process materials. Such a type of joint connecting straight pipes is shown in Fig. 13.1. Pipe threads for fastening purposes conforming to I.S. 2643–1975 shall be designated as:

(a) An external pipe thread for fastening purposes of size 2 with class B tolerance shall be designated as:
Ext – F P 2 B

(b) An internal pipe thread for fastening purposes of size 2 shall be designated as:
Int– F P 2

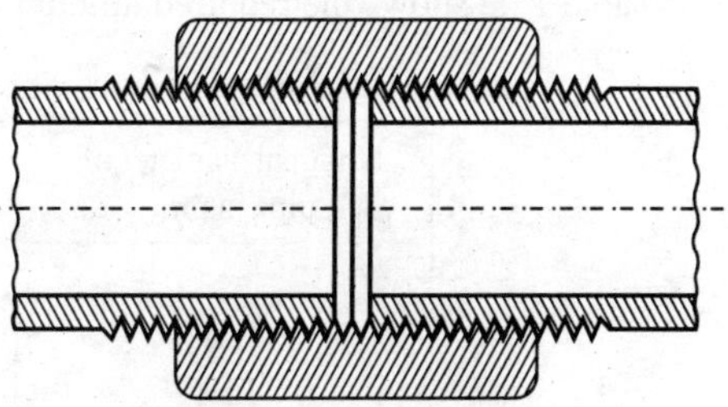

Fig. 13.1 *Threaded joint*

The various types of pipe fittings (sleeve joint, elbow joint and Tee-joint) are made of steel and malleable iron with cyclindrical and taper threads and are intended for nominal pressure of 2.5 to 3 N/mm^2.

In making up such joints, it is important that the threads be inspected thoroughly, cleaned and both the pipe and the fitting be lubricated before assembling. Pipe should be screwed into the fitting at a moderate speed so that binding due to friction does not prevent the pipe from being pressed home — a common cause for leakage joints.

Nipple and union joints are employed for pressures upto 10 to 15 N/mm^2 in hydraulic drive piping, etc. The parts of the nipple and union joints are standardized in the aircraft, automotive and machine tool industries. Screwed pipe joints are used where the overall smallness of size is most important.

2. Bell and Spigot Joint: These joints are used for low pressure where the overall smallness of size is most essential. They are mostly used for water under low pressure. The Bell and spigot joints (Figs. 13.2 and 13.3) are cheaper than flange joints, and are better for pipes laid in the ground. Since the connection is not rigid, the joint allows for any slight irregularities in the ground on which the pipes rest. Bell and spigot joints are made by partly filling the annular space with jute, rope or oakum which is lightly caulked. The remaining portion is then filled with molten lead by means of a runner which holds the lead inspace. After cooling, the lead in firmly caulked. Cold lead wool has in recent years been substituted for the poured lead methods. There are also a number of commercial poured joint compounds such as leadite, hydratite, and tegul used in place of lead. Neat porland cement is also successfully used to make up this type of joint. However, if a poured lead joint leaks, it may be recaulked, while if a cement joint leaks, it must be completely cut out and a new joint must be made. In the joint of Fig. 13.3 a strip of the socket marked S is made taper, and the joint is made by coating the tapered parts with red lead.

The proportions of a Bell and Spigot Joint may be as follows. All the dimensions are in mm,

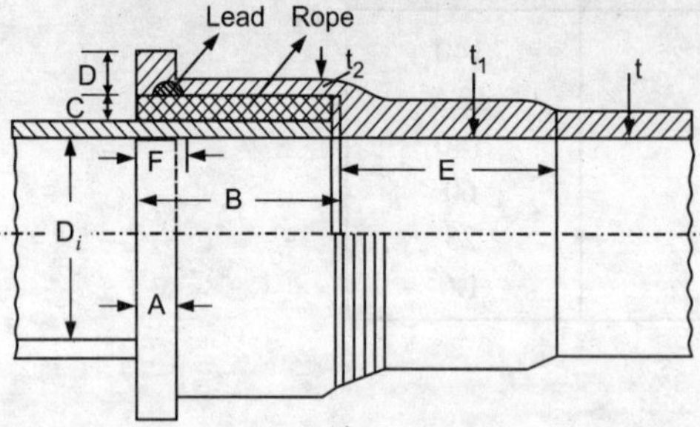

Fig. 13.2 *Bell and Spigot joint*

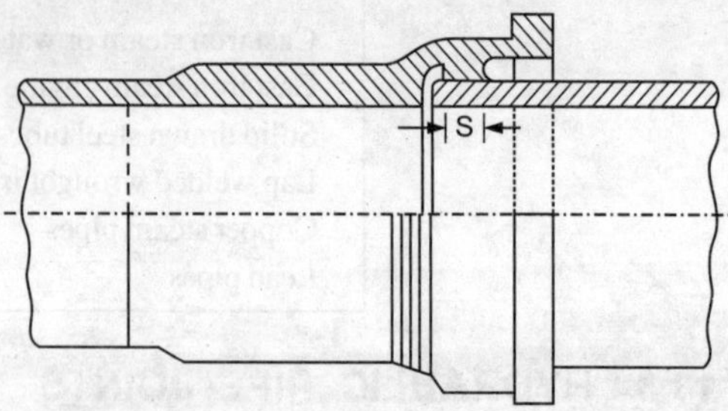

Fig. 13.3 *Bell and Spigot joint*

t = thickness of the pipe. D_i = Internal diameter of pipe

$t_1 = 1.1t + 1.75$ $A = 0.04\ D_i + 1.75$

$t_2 = 0.025\ D_i + 8.4$ $B = 0.1\ D_i + 68.8$

$C = 0.625$ to 12.5 $D = 0.04\ D_i + 20$

$E = 0.09\ D_i + 50$ $F = 0.06\ D_i + 25$

$S = 0.05\ D_i + 25$

Table 13.2 shows the required amount of lead and jute per joint.

Table 13.2 Required amount of lead and jute per joint

Nominal diameter of pipe in mm	Lead per Joint kg	Jute per Joint kg
75	2.73	0.082
100	3.40	0.096
150	4.65	0.141
200	6.00	0.200

Nominal diameter of pipe in mm	Lead per Joint kg	Jute per Joint kg
250	7.28	0.242
300	8.65	0.277
350	10.00	0.368
400	13.6	0.427
450	15.3	0.455
500	16.8	0.569
600	20.0	0.682
750	24.70	0.935
900	29.40	1.360

3. Brazed Joint: Malleable iron brazed fittings are finding an increasing field of application in the 'non-critical' pipe installation that does not warrant the cost of the welded joint. They are threadless fittings which if properly installed insure satisfactory leakproof connection. Their advantage lies in the elimination of the field cut pipe thread with its attended consumption of time and possible failure.

They are lighter in weight than the same size and kind of screwed fitting due to the elimination of the heavier wall required for threading. Their use is recommended by the manufacturer for meter and instrument lines, water, oil, and gas as well as process piping. To install this type of fitting, the pipe is cut off so that the end is square and free of burrs, full sized and not out of round if a tight connection is to be assured. The end of the pipe is sized for proper clearance, and after thoroughly cleaning and applying flux to both pipe and fitting, the pipe is inserted hard against the bottom of the cup. The assembly must be properly supported. The flame is applied until the flux bubbles and brazing alloy is then added which flows by capillary attraction into and clearance between pipe and cup. Enough alloy is added to fill the space and form a small fillet around the pipe. Copper and brass pipes are often connected by brazing. Examples of brazed joints are given in Fig. 13.4.

4. Welded Joints: Welded pipe joints are more reliable where pressure tightness and strength are concerned. However, this type of joint cannot be recommended when there is a need for frequent replacement of individual sections of pipes as is the case with pipes for chemically active media.

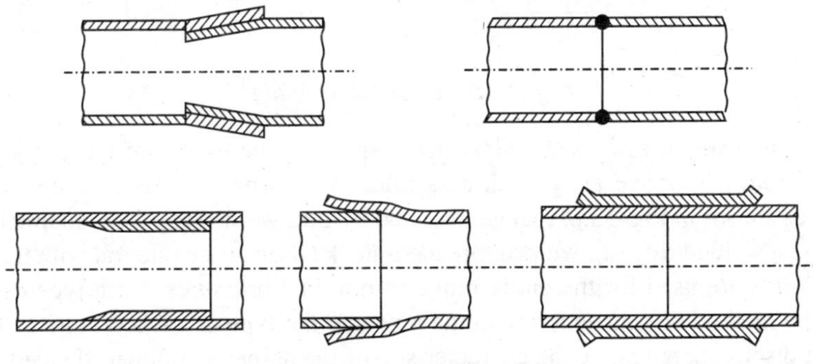

Fig. 13.4 *Brazed joints*

In case of longitudinal welded pipes, the edge of the plates of required thickness are trimmed by scarfed and then formed into cyclindrical shape and butt welded by the submerged arc process. The welded joint is equal to plate in strength.

Spiral seam fusion welded steel pipe is made of long strip of flat rolled steel spirally wound into a continuous tube and butt welded as it leaves the welding machine. Sizes from 100 to 900 mm nominal diameter are available.

Cold worked pipe is fabricated by the arc welding method and is then cold worked by being expanded by hydraulic pressure. This is said to increase the yield strength and the roundness of pipe and to test fully the strength of the welded seam.

Several methods of field welding are permitted: butt joint, single or double welded slip joint, single or double butt strap joint. Where enamel-lined steel pipe is used, 0.6 m is the smallest size which should be field welded, as this is the smallest pipe which can be entered in order to apply enamel as the welded joint.

A dresser coupling or welded slip joint should be placed every 120 or 150 meters when the temperature is lowest so as to put the pipe normally in compression rather than tension. The thrust, however, must be relieved adjacent to valves, pumps, etc., by means of expansion joints.

5. Expansion Joints: The variations in the temperatures of pipes caused by changes in temperatures of the ambient medium and of the working medium cause changes in the pipe length. To avoid unallowable temperatures stresses, deformations and even fracture of pipe lines with rigidity fastened pipes, special expansion joints are introduced. These joints are designed to allow for a considerable elongation or contraction of the pipeline due to temperature variations. Illustrated in Fig. 13.5 are expansion joints of various designs: (a) U-bend type, (b) circle bend joint, (c) offset and cross over pipe joints, and (d) Gland type slip joint.

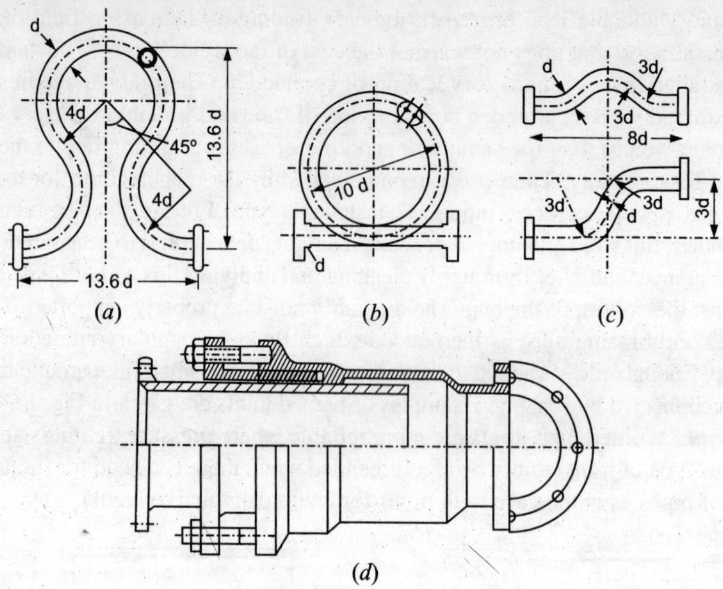

Fig. 13.5 *Expansion joints*

6. Flanged Joint: One of the most widely used types of pipe joints are flanged joints. Pipes with a nominal bore of 25 mm and above are joined, as a rule, by flanges. The flanges are available in a wide diversity of design as shown in Fig. 13.6. Flat weld (a and b), butt welded (c), free mounted on a beaded pipe (d), free mounted on a welded ring (e), welded threaded- neck (f) and case integral with the pipe (g).

Flanged pipe joints are used for the entire range of nominal pressures. Each type of the flanged joint (distinguished by the method of connection with the pipe and the type of gasket) has its field of application specified. We shall discuss here briefly the characteristic of the above mentioned flanged joints.

The flat welded flanges (Figs. 13.6 a, b) are simple, cheap and dependable for normal pressure to be less than 2.5 N/mm^2. The gasket for flat flanges is made to the longer diameter of the flanges and is compressed by the entire surface of the flanges. This makes it impossible to attain sufficiently high unit pressure and reliable pressure tightness. However, the seal can be improved by making a number of shallow concentric grooves on the flange faces and by reducing the area of the flange face surface which results in a higher unit pressure.

Steel threaded-neck flanges and steel butt-welded flanges are effective for the nominal pressures upto 10 N/mm^2. Steel flanges with lens-shaped circular gaskets are used for pressures upto 50 N/mm^2.

To save nonferrous metals, and simplify the assembly of pipes made of non ferrous metals, practical use has been made of steel flanges on the beaded pipes for pressures upto 0.6 N/mm^2. These flanges do not come

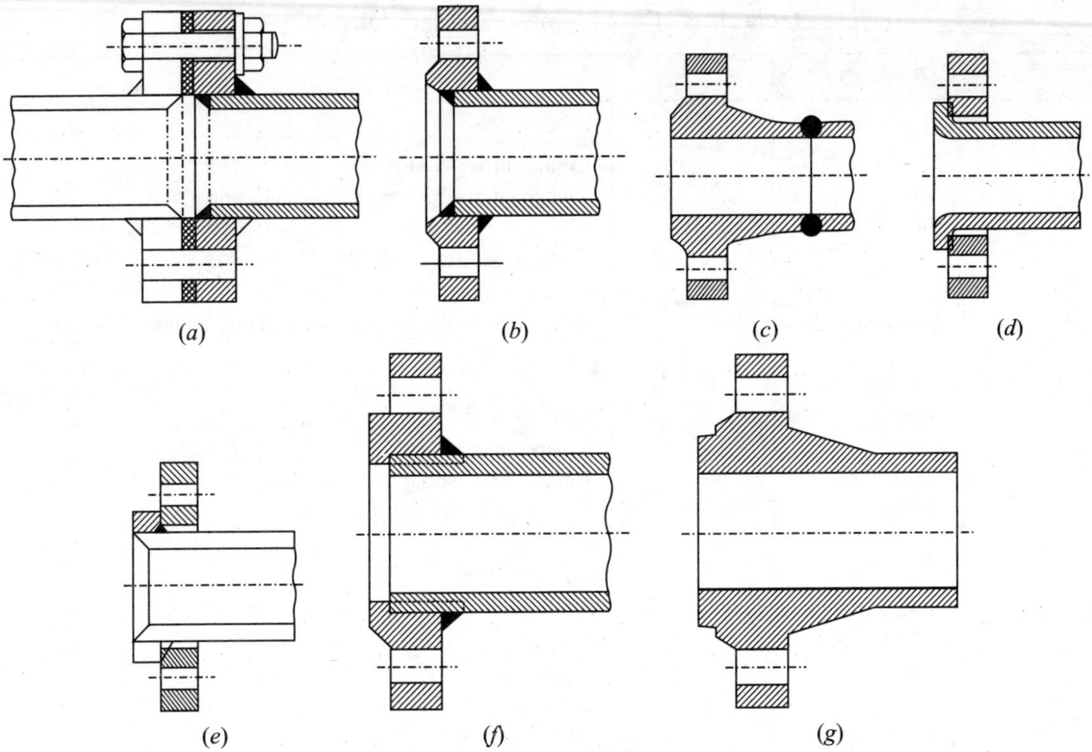

Fig. 13.6 *Flanges*

in contact with the working medium since their function is limited to holding together the beads of two pipes. Consequently, these flanges can be manufactured from carbon steel even if the working media are chemically active.

The flanges free-mounted on a welded ring simplify the assembly of pipes since they allow for a certain relative misalignment of the connected sections of pipes. This type of flange is used for pressure upto 2.6 N/mm^2.

Gaskets are used to maintain a tight seal between the faces of flanges. Such a seal could also be obtained without the use of gaskets if a perfect mating of the two surfaces were obtained by means of grinding and lapping. The cost of producing such an accurate finish is much more, therefore, the relatively inexpensive and easily replaceable gasket is employed. The seal is obtained by the effect of the gasket materials flowing into the imperfections of the jointing surfaces. Gasket materials employed may be either metallic or nonmetallic depending upon the condition of available bolt pressures, temperatures, and the corrosiveness of the substance conveyed within the pipe line. A tight seal is obtained only by reducing the clearance between the gasket and the flange surfaces to zero, which is brought about by the actual flow of the gasket material.

Temperature and the corrosive nature of the materials are a prime factor in the choice of gaskets. Non-metallic gaskets and those containing low melting point metals limited to temperatures below 400ºC. Metal clad asbestos may be used upto 460ºC, while for temperatures in excess of 460ºC only all metal gaskets of such material as monel should be employed. When high temperatures and pressures are to be encountered, it is wise to consult the gasket manufacturers for his recommendations. Table 13.3 gives recommended gasket materials for different service conditions.

Table 13.3 Recommended Gasket Materials.

	Service		Material
Steam			
	Exhaust	... ...	Composition asbestos
	Saturated	... ...	Composition asbestos or corrugated copper
	Superheated (over-340ºC)	... ...	Soft iron, steel, or monel metal.
Water			
	Boiler feed	... ...	Composition asbestos, corrugated monel, copper, or armco iron.
	Hydraulic	... ...	Fibre
	Cold water and brine	... ...	Red rubber
	Hot water	... ...	Composition asbestos
	Gas and Air	... ...	Composition asbestos
	Acids	... ...	Blue African asbestos, sheet lead, or alloy steel.
Ammonia			
	Hot	... ...	Thin asbestos
	Cold	... ...	Sheet lead
Oil			
	Heavy Cold	... ...	Composition asbestos or paper
	Light Gasoline or kerosene	... ...	Graphited asbestos
	Benzol	... ...	Composition asbestos.

Bolting materials for water as well as for steam pressures not exceeding 1.15 N/mm^2 and temperatures upto 250ºC may be carbon steel, unfinished square-headed machine bolts with cold punched heavy hexagonal nuts. While carbon steel machine bolts may be used for cold water service, they are not recommended for hot lines at that pressure. For steam pressures over 1.15 N/mm^2 and temperatures exceeding 250ºC, alloy steel heat-treated bolt studs with a hexagonal nut on each should be used.

13.4 HYDRAULIC PIPE JOINTS FOR HIGH PRESSURES

Pipes and pipe joints for high pressures fall into two main classifications.

(*a*) For hydraulic pressures upto 8.5 N/mm^2 and pipe bores from 50 to 175 mm, flanges and pipes are cast integrally from remelted cast iron, the flanges being elliptical and secured by two bolts. The proportions of these joints have been standardized for pipe diameters from 50 to 175 mm, the bore increasing by 25 mm from one size to the next.

(*b*) For pipes having fluid pressure above 8.5 N/mm^2 and with bores of 50 mm or below, the piping is of wrought steel, solid drawn (seamless), or rolled, The flanges may be of cast iron, steel mixture, or forged steel, screwed or welded on to the pipes and square in elevation, secured by four bolts. These joints are made for pipe bores from 12.5 to 50 mm rising in increments of 3 mm from 12 to 18 mm and 6 mm from 18 to 50 mm bore. The pipes and flanges in this division are suitable for hydraulic pressures upto 47.5 N/mm^2.

The joints in division (*a*) are further subdivided into two classes. Class A comprises pipes and flanges for fluid pressures from 5 N/mm^2 to 6.4 N/mm^2 and class B for pressures from, 6.4 N/mm^2 to 8.5 N/mm^2. The flanges in each class may be of two types. Type 1 is used for pipes from 50 to 100 mm bore in class A, and for 50 and 75 mm bore in class B. Flanges of type 2 are stronger than those of type 1 and are usually set well back on the pipe. In all the above classes, the joint is of the spigot and socket type made with a jointing ring of guttapercha.

Flanges used for connecting pipes are held together by means of bolts. The various types of flanges used for pipe joints are:

(*a*) Oval type flange, where two bolts are used.

(*b*) Square type flange, where four bolts are used.

(*c*) Round type flange, where four or more number of bolts are used but the number of bolts used will be even.

For underground water pipes oval type flanges are particularly preferred, because the flange may be laid horizontally and the bolts tightened or loosened without having to rotate the pipes, or digging too much of earth round the pipes.

Hydraulic pipe joints for high pressures differ from those used for low or medium pressures, in following ways.

1. For the joints in both divisions the jointing ring is of guttapercha, which is squeezed into the socket by the spigot on bolting up. The tightening up of bolts squeezes the ring into a triangular shape and makes a perfectly tight joint capable of withstanding pressures upto 47.5 N/mm^2.
2. For an oval or square flange integral with the pipe, the condition of bending is very clearly defined due to the flanges being set back on the pipe and the thickness of the flange can be accurately determined to withstand the bending action due to tightening of bolts.
3. Bolt holes are made square with sufficient clearance to accommodate square bolt heads and to allow for small movements due to the setting of the joint.
4. Oval and square flanges used for these joints are heavy. The number of bolts used is two or four which is a great advantage while assembling and disassembling the joint, especially these are useful in narrow space.
5. A short length of plain pipe beyond the threads is entered into the square flange, so that the threads are completely covered and reinforced by the latter. This is true in square flange pipe joint of Fig. 13.10.

(*a*) **Design of oval type Flanged Pipe Joint:** The oval type flanges are usually used for small pipes, say upto about 17.5 cm. in diameter. The flanges are generally cast integrally with the pipe ends.

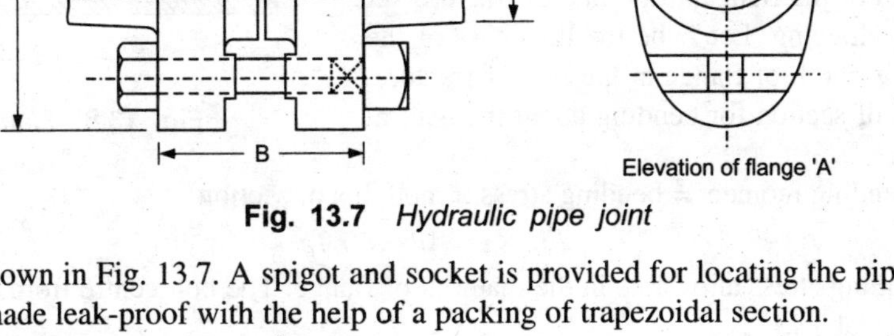

Fig. 13.7 *Hydraulic pipe joint*

The joint is shown in Fig. 13.7. A spigot and socket is provided for locating the pipe bore in a straight line. The joint is made leak-proof with the help of a packing of trapezoidal section.

The thickness of the pipes can be obtained by relation 13.8 and the internal diameter of the pipe can be obtained either by relation 13.3 or by 13.4.

Size of bolts: The force trying to separate the two flanges has to be resisted by the stress produced in the bolts. Consider a length of the pipe having its ends closed somewhere along its length, then the force tending to separate the two flanges due to fluid pressure *P* is given by

$$F_1 = \pi/4\, D_i^2 \times P$$

where D_i is the internal diameter of pipe.

The packing has also to be compressed to make the joint leak-proof. The intensity of pressure on packing should be greater than the pressure of the fluid inside the pipe, but for calculation purposes it can be assumed that the packing material is compressed to the same pressure as that inside the pipe. Therefore, the force F_2 tending to separate the flanges due to pressure in the packing will be given as

$$F_2 = \pi/4\,(D_1^2 - D_i^2) \times P$$

where D_1 is the outside diameter of packing of the socket.

The total force F, trying to separate the two flanges will be the sum of the two forces, i.e., F_1 and F_2

$$F = F_1 + F_2 = \pi/4\, D_i^2 \times P + \pi/4\,(D_1^2 - D_i^2) \times P$$

$$F = \pi/4\, D_1^2 \times P \quad \text{...(13.9)}$$

This relation could also be written by considering that the fluid pressure, tending to separate the flanges, is assumed to act over an area corresponding to the quarter diameter of the socket, since the contact load between the flanges must exceed the fluid load.

Since two bolts are used here to connect the flanges, the load on each bolt will be $F/2$.

Let d_c be the core diameter of the bolts.

$$F/2 = \pi/4\, d_c^2 \times f_t \quad \text{...(13.10)}$$

where f_t is the allowable stress in tension for the bolt material. The initial tightening of the bolts must be taken into account while selecting the allowable stress. It has been already mentioned that the allowable stresses in bolts increase with increase in size of bolts, due to decreasing effect of stresses from initial tightening.

By knowing the core diameter of bolts from the above relation, the standard size of the bolts can be seen from tables. Bolts of less than 12.5 mm diameter are never used for hydraulic pipes. In smaller bolts, initial tightening stresses of very heavy order are produced. The centre line of the bolt should be as near the centre of the pipe as possible to avoid the bending of the flange. But on the other hand sufficient clearance between the pipe surface and the bolt head should be provided for the tightening of the bolts without damaging the pipe material.

Thickness of Flange: In order to arrive at this dimension, the bending action of the bolt on the flange has to be considered, and the section at which fracture is most likely to occur is first found. First of all from standard tables the major and minor axes of the flange ellipse are taken for the given size of the pipe. Then the flange is laid to scale, and it is apparent from the drawing that the line of fracture is at x-x. Assume e as the eccentricity between the bolt line and the line of fracture x-x.

Then the bending moment at the section = $F/2 \times e$.

Now the width of the flange at the line of fracture x-x can be taken from drawing. Let t_1 be the thickness of the flange and b be the width of flange at the line of fracture. Then the modulus of section for bending about the neutral axis $AA = 1/6\, bt_1^2$.

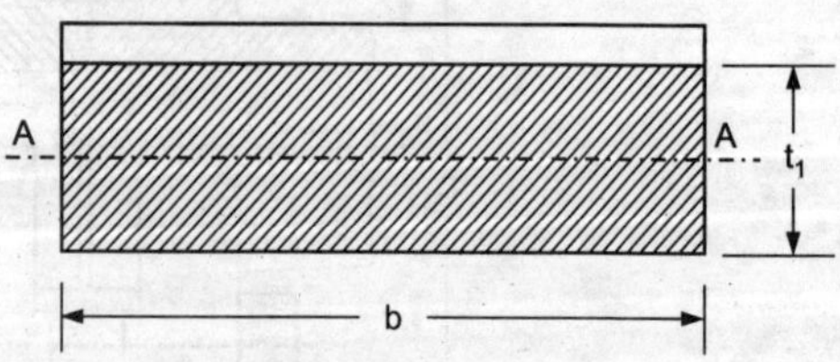

Fig. 13.8 *Flange section*

Then, since bending moment = bending stress × modulus of section.

$$F/2 \times e = 1/6\, bt_1^2 \times f_b \quad \text{...(13.11)}$$

where f_b is the bending stress allowable in the material of flange. The bolt centre from the pipe axis can be again seen from the tables.

$$e = l_1 - l_2 = l_1 - 1/2 \times \text{length of minor axis.}$$

A suitable value of the bending stress in cast iron, having regard to the weakness of the material in tension, will be 17.5 N/mm^2.

The bolts, provided with square neck and heads, will have ample clearance in the square holes cored in the flanges, thus permitting some degree of flexibility in the pipe line. Bolt used to hold the joint for testing at the test pressure, should be replaced when the joint is assembled for service, since there will be the possibility of overstraining at higher pressure. The standard length of piping is 2.75 meter for all bores above 50 mm, and for the latter the length is 1.8 meter. Reinforcement, as carried out in Fig. 13.7, consists in thickening the pipe in the neighbourhood of the flanges.

Hydraulic pipe joints with oval flanges are also known as Arms- strong's pipe joints.

The size of various parts of a hydraulic joint may be obtained from the following proportions.

$t_1 = 1.5t + 3$ mm number of bolts $= 0.0275D_i + 1.6$

$d = 0.75t + 10$ mm $D_2 = D_i + 7t + 37.5$ mm

$l_3 = 1.5t + 10$ mm

$l_4 = 2.3d$ $B = 4t + 17.5$ mm

The details of the packing in the oval type flange pipe joint of Fig. 13.7 are shown in Fig. 13.9.

(*b*) **Design of Square Type Flanged Pipe Joint:** The design of the square type flanged pipe joint is similar to the oval flanged type joint except that the load in this has to be divided into four bolts. In this case the flange will be made of mild steel as the pressure of fluid in the pipe is high. The bolts will be smaller than those required for the cast iron joint. The allowable stress in tension of the bolt material will be 57.5 N/mm^2. The axes of the bolts are arranged at the corners of a square of such size that the corners of the nuts clear the outside of the pipe.

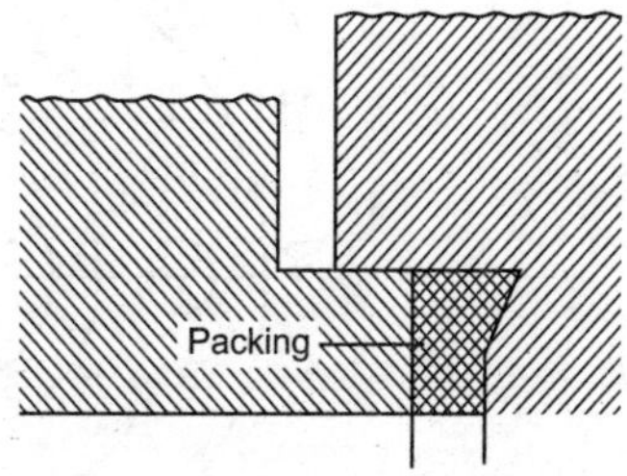

Fig. 13.9 *Details of packing*

Fracture of the flange due to bending action of the bolts and fluid pressure may take place at one of the sections AA, BB, or CC. It will be clear after considering the bending actions that these are most severe for the section AA, so the thickness of the flange will be determined for this section. In this investigation the stiffening and weakening effects of the spigot and socket, which are small, will be neglected. Fig. 13.10 indicates the square type flanged pipe joint.

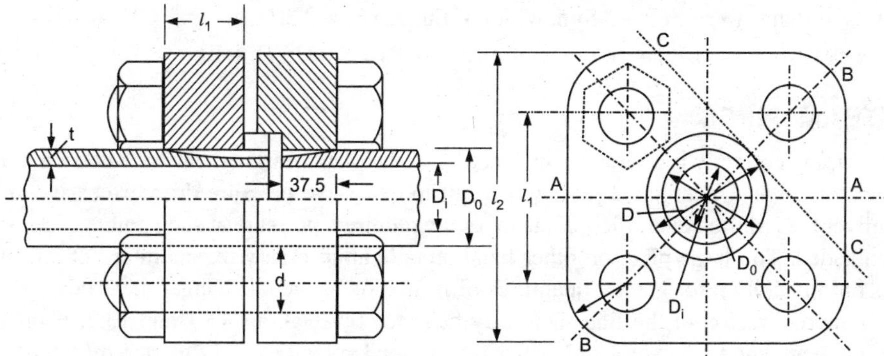

Fig. 13.10 *Hydraulic pipe joint*

The details of the packing in the square type flange pipe joint of Fig. 13.10 are shown in Fig. 13.11. All dimensions given are in mm.

The detailed design procedure for square type flange pipe joint is indicated in example number three of the chapter.

(*c*) **Design of Round type Flanged pipe Joint:** Cast iron, water and steam pipes are usually made in length of about 2.75 metre and these separate lengths may be connected by means of round type flanges. For the round type flanges it is generally assumed that the packing material extends upto the inside diameter of the bolts. The force in each bolt will be calculated in the similar way as Oval type flange pipe joint, except that the number of bolts in this will be even because of the symmetry of the section. To make the joint leakproof the circular pitch of the bolts should lie between $20\sqrt{d}$ to $30\sqrt{d}$, where d is the diameter of the bolt, in mm. Also to make the joint leakproof the diameter of the bolt should not be less than 16 mm. The thickness of the flange is obtained by considering a segment of the flange, as shown in Fig. 13.12. In this it is assumed that each of the bolt supports one segment of the round flange. The bending moment is taken about section xx. The effect of joining of these segments on the stresses induced is neglected. The diameter of the pipe and its thickness can be calculated as discussed earlier. The dimensions of the flanges may be obtained as follows.

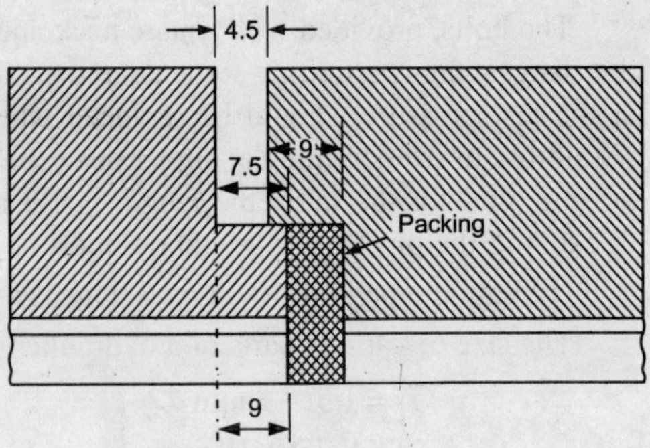

Fig. 13.11 *Details of packing*

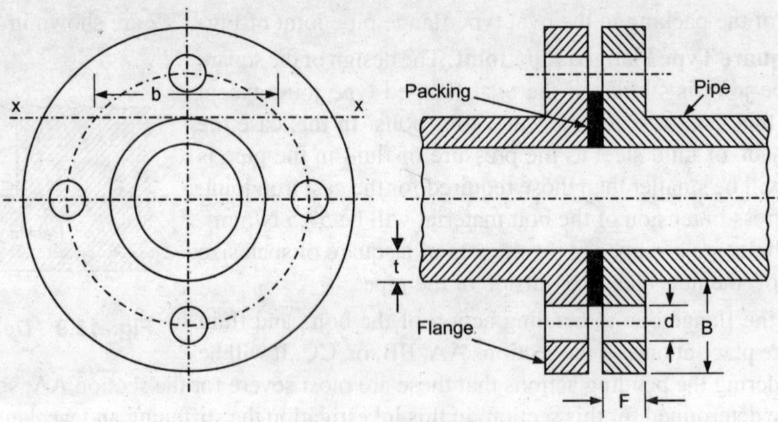

Fig. 13.12 *Round type flanged pipe joint*

Thickness of flange (F) = $3/2t + 3$ mm, width of flange (B) = $2.3d$

Diameter of bolts (d) = $0.75t + 10$ mm, number of bolts = $0.0275\, D_i + 1.6$

13.5 STEAM PIPES

Sizing steam piping correctly is an important function of the plant designer. The size chosen must be neither too small nor too large. Under-sized pipe will result in excessive pressure drop, excessive velocities, and failure to deliver the required quantities of steam of the required pressure at such points of the system where they are demanded. Too large pipe, on other hand, entails unnecessary expenditure not only for the more costly pipes but in addition for greater quantities of insulation as well as hanger materials.

A rough approximation of the line sizes may be made by relations 13.1 and 13.2, if the quantity and pressure of steam are known. Several authorities recommend velocities of 12 to 14.5 m/min per mm. of pipe diameter. Feed lines to pumps or reciprocating engines should have velocities one-third to one-quarter less than those given in Table 13.4.

Table 13.4 Steam velocities

Condition	Pressure N/mm^2	Use	Velocity m/min
Saturated	0–0.11	Heating	1200–1800.
Saturated	0.35–1.05	Miscellaneous	1800–3000
Superheated	1.4 and above	Boiler leads	3000–4500

It should be pointed out that the sizing of pipes is an art than a science. By this it meant that even a slight difference in any of the variables upon which a calculation is based, for example, the condition and variation of the inside diameter, would undoubtedly affect in a measurable quantity of the fluid delivered. The piping designer must, therefore, choose a pipe size which will cover demands of the installation both as to performance as well as economy.

Steel pipe is fabricated to conform to a number of specifications, the choice of which depends upon the required use. Such as working pressure and temperatures, costs and provisions of the I.B.R.

Welded or seamless steel pipes for ordinary uses such as plumbing and heating, or water, gas, and air lines may be either black or hot-dipped galvanized. It is not intended for use for high temperature service, close coiling or bending. No other than hydrostatic tests are required at the will. But welded pipe is usually limited to 100 mm and under. For larger sizes lap welding or electric welding is required. Length of 5 m to 7.0 m are commonly supplied. But welded pipe is not used for flanging. Unless otherwise specified, both ends of standard weight welded pipe are threaded.

Seamless carbon steel pipes are used for high temperature service and for bending, flanging, and other forming. For steam- service pressure of 2.8 N/mm^2 and over and high temperatures or where a superior grade of pipe is needed, additional tests are needed besides the chemical, mechanical and hydrostatic tests normally called for.

Welded wrought iron pipe finds its greatest use in condensate return lines where the pressure of air in the solution may cause a severe corrosion problem. There are two schools of thought on the wrought iron versus steel pipe question, some engineers maintaining it is worth the additional cost and others saying the contrary.

Cast iron pipes maintain the case of casting the flanges, accuracy of facing the flanges, sound joint and comparatively low cost. With the development in high pressures, the higher the weight of cast iron pipe is an apparent disadvantage. As the material is brittle, therefore, it is liable to fail from shock, particularly that due to water hammer. For this reason cast iron pipes are not used for pressures more than 0.7 N/mm^2.

Ductile iron pipe is made of a comparatively new material which has been developed since World War II. This material combines the mechanical properties of steel with the corrosion resistance of cast iron. It exhibits a remarkable degree of flexibility. Ductile iron has been found to be superior material to be used for flanges for stainless steel pipes.

13.6 STEAM PIPE FITTINGS

Pipe fittings employed on lines conveying steam may be classified by the method of making up a joint which may be *(a)* Flanged, *(b)* Welded, (c) Screwed, *(d)* Brazed.

(a) **Flanged pipe joints:** Are castings of either iron or steel. Flange joints in detail have been discussed in Article 13.3 of this chapter. Forged steel fittings are made in sizes upto 150 mm for working pressures of 4.25, 6.4 and 10 N/mm^2 at 500°C. Their dimensions coincide with those of similar cast steel fittings.

(b) **Welded Pipe Joints:** Forged steel welding fitting find their greatest application in steam and hot process lines of pressures 1 N/mm^2 and over. There are many definite advantages to this type of construction, and although the first cost may be higher than with screwed or flanged fittings, the lower cost of maintenance due to elimination of leaking gaskets and threaded connections probably equalizes the cost over a period of years. Welded type construction not only simplifies the installing and insulation but also cuts down on the weight by the elimination of flanges and bolts. A line may be of all welded construction, or flanges may be installed at suitable intervals to facilitate erection.

Forged fittings may be either of the welding or socket welding types. But welding fittings are used for long radius elbows, tees, return bends, and reducers in sizes 25 to 600 mm. Forged steel socket welding fittings are used for smaller size pipes rather than welding fittings. They have certain advantages, for example, the adjoining pipe need not be bevelled. It is, instead, slipped into the fitting and is supported without the use of special clamps. The joint is self aligning. This type of fitting may also be used in place of the forged steel screwed fittings, thereby eliminating the need for threaded connections which occasionally leak and may require back welding to insure a tight joint. Socket welding fittings for higher pressures and temperatures are made, on special order, of carbon molybdenum steel for use with carbon molybdenum pipe.

(*c*) Screwed fittings are discussed in Article 13.3.
(*d*) Brazed fittings are discussed in Article 13.3.

13.7 FLANGES

Flanges are used for the joining of lengths of pipe and also for connecting them to valves and equipment. The Indian Boiler Regulations (I.B.R.) 1950 (revised 1961) have standardized all dimensions of pipes and flanges based-upon steam pressure. They have been divided into five classes.

Class I. For steam pressure upto 0.35 N/mm^2 (water pressure upto 1.4 N/mm^2). This is not suitable for feed pipes or shocks.

Class II. For steam pressure over 0.35 N/mm^2, but not exceeding 0.7 N/mm^2.

Class III. For steam pressure over 0.7 N/mm^2, but not exceeding 1.05 N/mm^2.

Class IV. For steam pressure over 1.05 N/mm^2, but not exceeding 1.75 N/mm^2.

Class V. For steam pressure over 1.75 N/mm^2, but not exceeding 2.5 N/mm^2.

It is recommended by I.B.R. that for classes II, III, IV and V, it is desirable that the diameter of bolt circles, diameter of flanges and number of bolts should be identical and that difference should exist in variations of the flange thickness and diameter of bolts. It is also specified by I.B.R that the nuts should be chamfered on the side bearing on the flange and that the bearing surfaces of the flanges, heads and nuts should be true. The number of bolts in all cases should be multiple of four. I.B.R. recommends that the size of holes in the flanges should be 1.5 mm larger for bolts of 12.5 mm and 1.5 mm and 3.0 mm larger for bolts of higher sizes than 15 mm. For all the five classes mentioned above, the flanges having internal diameter of 12.5 mm to 600 mm are standardized for pipes, tees and bends. As soon as the size of the pipe is determined, the rest of the dimensions can be determined from the standards. The standards have been evolved on the basis of long experience, suitability and interchangeability.

The various dimensions may be calculated on the design principles as discussed in Article 13.4 for different flanges. The calculated dimensions do not agree with the standards, because design principles are not able to take practice considerations into account. It is of academic interest for the students to know how to find the thickness of the pipe, diameter of pipe, bolt size, number of bolts and the thickness of flanges from design principles. The rest of the dimensions may be obtained from standard table or by empirical relations.

13.8 OIL PIPING

Forged steel screw fittings are normally used in the oil industry for all installations requiring other than welded joints. Forged steel 14 N/mm^2 of the oil pipe fittings are designed for services beyond the temperature range of malleable iron fittings. They are used where a higher degree of safety is required, although operating temperature and pressure may not be high enough to require steel fittings. They are recommended for high pressure cold water, oil, or gasoline and for hot lines within the limits 50°C, 14 N/mm^2 and 500°C, 2.7 N/mm^2.

Forged steel 21 N/mm^2 fittings are most commonly used for high pressure hydraulic lines as well as high pressure and temperature services. They are forged from a special grade of steel. Couplings and half couplings are used for welding threaded connections to pipe threads and pressure vessels. They are available in size 3 to 100 mm in most of the common forms of screw fittings. Forged steel 42 N/mm^2 and 63 N/mm^2 fittings are similar to the 21 N/mm^2 fittings excepting for their greatly increased wall thickness.

Cast steel screw fittings are made in the 7 and 14 N/mm^2 classes. They are cast of either carbon steel or of carbon molybdenum or chrome molybdenum steels to withstand temperatures upto 500°C. These fittings are available in sizes from 62.5 mm to 200 mm.

Ring joint flanges are widely used in the oil and process industries for high temperatures and high ssure piping. Rings are made either oval or octagonal in shape. Grooves are also machined either with a ded or flat bottom. The octagonal ring is more specified by the oil refineries. Both shapes are highly t and will produce a tight seal at full bolt stress. The present tendency is toward the use of octagonal rings as they are more easily seated and provide a surface contact instead of line contact of oval shaped rings. For replacement, however, the oval ring is preferred as this can compensate for any warping and deformation of the groove resulting from operational stresses.

SOLVED EXAMPLES

Example 13.1. The internal pressure in a hydraulic pipe 75 mm bore is 5 N/mm². Find the thickness of the pipe if the stress must not exceed 14 N/mm².

Determine the suitable proportion of a spigot and socket joint to connect two lengths of pipes. Draw full sized sectional elevation.

Solution: Refer to Fig. 13.2. Thickness t of the pipe can be obtained by relation 13.8.

$$t = \frac{D_i}{2}\left[\sqrt{\frac{f_t+p}{f_t-p}}-1\right] = \frac{75}{2}\left[\sqrt{\frac{14+5}{14-5}}-1\right]$$

$$t = 17.1 \text{ mm say } 17.5, D_i = 75 \text{ mm.}$$

Outside diameter of pipe $D_o = D_i + 2t = 75 + 2 \times 17.5 = 110$ mm

$$t_1 = 1.1t + 1.75 = 19.25 + 1.75 = 21 \text{ mm}$$
$$A = 0.04D_i + 17.5 = 0.04 \times 75 + 17.5 = 20.5 \text{ mm}$$
$$t_2 = 0.025D_i + 8.4 = 0.025 \times 75 + 8.4 = 10.3 \text{ mm}$$
$$B = 0.1D_i + 68.8 = 0.1 \times 75 + 68.8 = 76.3 \text{ mm}$$
$$C = 7.5 \text{ mm}$$
$$D = 0.04D_i + 20 = 0.04 \times 75 + 20 = 23 \text{ mm}$$
$$E = 0.09D_i + 50 = 0.09 \times 75 + 50 = 56.75 \text{ mm}$$
$$F = 0.06D_i + 25 = 0.06 \times 75 + 25 = 29.5 \text{ mm}$$

The different notations are shown in the Fig. 13.2.

Example 13.2. Design completely an oval flanged pipe joint for pipes of internal diameter 150 mm subjected to a fluid pressure of 8.5 N/mm². The maximum tensile stress in the material is not to exceed 64 N/mm². The material of pipe is cast iron. The test pressure for the pipe joint is 23 N/mm².

Solution: It is usual to know the test pressure for pipes of this type, so that we can find the maximum allowable stress produced in the wall by the test pressure. The joint in question will belong to class *B* (reinforced) and will have flanges of Type 2. The joint is shown in Fig. 13.7. The thickness t of the pipes can be obtained by using the Lame's formula, i.e., relation 13.8.

$$t = \frac{D_i}{2}\left[\sqrt{\frac{f_t+p}{f_t-p}}-1\right] = \frac{150}{2}\left[\sqrt{\frac{64+23}{64-23}}-1\right]$$

In this *P* is the test pressure.

$$t = 34.5 \text{ mm, say } 35 \text{ mm.}$$

Outside diameter of the pipe = $D_i + 2t = 150 + 35 \times 2 = 220$ mm

From standard tables the outside diameter of the socket is 175 mm. Therefore, the total load *F* on the joint can be found out by relation 13.9.

$$F = \frac{\pi}{4} D_1^2 \times P$$

where D_1 is the diameter of the socket and *P* is the fluid pressure in the pipe.

$$F = \frac{\pi}{4}(175)^2 \times 8.5 = 204 \text{ kN}$$

Since there are two bolts, the load on each bolt

$$= \frac{204 \times 10^3}{2} = 102 \text{ kN}$$

Taking the initial tightening of the bolts into account for bolts of the size required here, a safe value of the working stress would be 70 N/mm^2. Now by using the relation 13.10, the diameter of the bolt can be calculated.

$$\frac{F}{2} = \frac{\pi}{4} d_c^2 \times f_t, \quad 102 \times 10^3 = \frac{\pi}{4} d_c^2 \times 70$$

gives, $d_c = 43$ mm

From this the standard size of the bolts can be obtained from tables. In the absence of the tables the outside diameter of the bolts will be

$$d = \frac{dc}{0.84} = \frac{43}{0.84} = 51.2 \text{ mm, say } 52.5 \text{ mm.}$$

The probable line of fracture will be at xx. The table of standards gives the major and minor axes of the flange ellipse as 475 mm and 265 mm respectively for the size of pipe as 150 mm. Again from standard tables, the bolt centre is at 175 mm from the axis of the plane of fracture from the bolt centre, then

$$e = l_1 - l_2 = 175 - \frac{1}{2} \times 265 = 42.5 \text{ mm}$$

The width b may now be measured by making the drawing of the elliptical flange, or calculated as follows.

If x and y coincide with the major and minor axes of the ellipse respectively, its equation is

$$\frac{x^2}{l^2} + \frac{y^2}{m^2} = 1$$

where l and m are semi axes, i.e., $l = 237.5$ mm and $m = 132.5$ mm. At the plane whose trace is xx, $x = 132.5$ mm and y can be obtained from the above relation.

$$\frac{(132.5)^2}{(237.5)^2} + \frac{y^2}{(132.5)^2} = 1, \quad \text{gives } y = 110 \text{ mm}$$

$\therefore$ width at the line of fracture $b = 2y = 220$ mm

Allowable stress in tension for cast iron $f_b = 17.5$ N/mm^2

Now by using the relation 13.11, we can find the thickness t_1 of the flange.

$$\frac{F}{2} \times e = \frac{1}{6} b t_1^2 \times f_b$$

$$\frac{204 \times 10^3}{2} \times 42.5 = \frac{1}{6} \times 220 \times t_1^2 \times 17.5, \text{ gives } t_1 = 83 \text{ mm.}$$

See that this is not less than the value given in standard tables.

The different proportions of the joint can be checked by empirical relations.

Thickness of flange $t_1 = 1.5t + 3 = 1.5 \times 35 + 3 = 55.5$ mm

diameter of bolt $d = 0.75t + 10 = 0.75 \times 35 + 10 = 36.25$ mm

$$l_3 = 1.5t + 10 = 1.5 \times 35 + 10 = 52.5 + 10 = 62.5 \text{ mm}$$

$$l_4 = 2.3d = 2.3 \times 36.25 = 83.4 \text{ mm}$$

number of bolts $= 0.0275\, D_i + 1.6 = 0.0275 \times 150 + 1.6 = 5.72$, say 6.0.

$$B = 4t + 17.5 = 4 \times 35 + 17.5 = 157.5 \text{ mm}$$

$$D_2 = D_i + 7t + 37.5 = 150 + 7 \times 35 + 37.5 = 432.5 \text{ mm}.$$

The number of bolts used are only two as in the problem it is required to design the oval type flange. The outer diameter D_2 will be slightly more than the empirical value.

$$D_2 = 175 \times 2 + 62.5 \times 2 = 475 \text{ mm}$$

Reinforcement has been done at section where the flange meets the pipe. The bolts provided with square necks and heads are given ample clearance in the square holes. The size of the square holes in the flanges is kept as $b_1 = 60$ mm. The dimension B calculated empirically is less, therefore, it will be kept as 250 mm.

Example 13.3. Design completely a square flanged pipe for pipes of internal diameter 50 mm subjected to a fluid pressure of 24 N/mm². The pipes used are seamless steel tubing. The flanges are of mild steel screwed on the pipes, and the joint is secured by four bolts.

Solution: Refer to Fig. 13.10 The yield stress for seamless steel tubing will lie between 260 N/mm² and 310 N/mm². So, let the test pressure should produce a stress not greater than about 240 N/mm². Assume the test pressure as twice the working value. i.e., 48 N/mm² then by using the Lame's formula the thickness of pipe can be calculated.

$$t = \frac{D_i}{2}\left[\sqrt{\frac{f_t + p}{f_t - p}} - 1\right] = \frac{50}{2}\left[\sqrt{\frac{240+48}{240-48}} - 1\right]$$

$$t = 5.625 \text{ mm, say } 6 \text{ mm}.$$

Outside diameter of steel tubing $D_o = D_i + 2t = 50 + 2 \times 6 = 62$ mm.

From standard tables, the outer diameter of socket is 80 mm for a pipe joint of 50 mm diameter. The total load F on the joint can be found out by relation 13.9.

$$F = \frac{\pi}{4} D_1^2 \times P = \frac{\pi}{4}(80)^2 \times 24 = 120 \text{ kN}$$

Since there are four bolts to make the pipe joint, therefore, the load per bolt

$$= \frac{120 \times 10^3}{4} = 30 \text{ kN}$$

Taking the allowable tension in the bolt as 57.5 N/mm², the diameter of the bolt can be found out by relation 13.10.

$$\frac{F}{4} = \frac{\pi}{4} d_c^2 \times f_t, 30 \times 10^3 = \frac{\pi}{4} d_c^2 \times 57.5$$

gives, $dc = 26$ mm.

In the absence of standard tables, the outside diameter of the bolt will be,

$$d = \frac{d_c}{0.84} = \frac{26}{0.84} = 31 \text{ mm}.$$

The axes of the bolts are arranged at the corners of a square of such size that the corner of the nuts clear the outside of the pipe.

The minimum length of a diagonal for this square = outside diameter of pipe + 2 × bolt diameter = $62 + 2 \times 31 = 62 + 62 = 124$ mm.

Therefore, the side of the square = 94 mm = l_1.

The sides of the flange must be of sufficient length to accommodate the nuts and bolt heads without overhang. Thus the length l_2 will be $= 94 + 2 \times 31 = 156$ mm.

Fracture of the flange due to the bending actions of the bolts and fluid pressure may take place at one of the sections AA. BB or CC.

Bending about section *AA*: Consider first the bending moments acting on the half flange and tending to produce rupture at the section AA. The half flange is in equilibrium under the following bending moment:

1. The bending action of the bolt about *AA*.
2. The bending action of the water pressure action over an area corresponding to the half socket, and concentrated along a semi circular arc at the mean radius of the pipe threads. The moment due to water pressure is in opposite direction to that of the bending action of the bolts.
3. The resisting moment of the flange at section *AA*. This is equal to the difference between (1) and (2). The sense of resisting moment will be same as that of (2), or we can say opposite to that of (1).

Bending about section *AA* will take place due to the force in two bolts. The central distance of the bolts from section *AA* is 47 mm. Therefore, the bending moment due to the bolts will be M_{b1}.

$$M_{b1} = 2 \times \frac{F}{4} \times \frac{l_1}{2} = 2 \times 30 \times 10^3 \times \frac{94}{2} = 2830 \text{ Nm.}$$

$$\text{Water pressure acting on half flange} = \frac{F}{2} = \frac{120 \times 10^3}{2} = 60 \text{ kN.}$$

The pipes are screwed with Metric threads of 0.44 threads per mm and the radii to the outside and to root respectively are 31 and (31 – 1.45) mm.

Therefore, outside diameter of pipe = 31 mm.

root diameter of pipe = 29.55 mm

So the mean of the arc over which the load due to fluid pressure, may be taken to be concentrated =

$$\frac{1}{2}(31 + 29.55) = 30.275 \text{ mm.}$$

The centroid of this arc = 0.6366 × radius of arc from the section *AA*

= 0.6366 × 30.275 = 19.2 mm

Therefore, the bending moment due to the water pressure will be Mb_2

$$M_{b2} = \frac{F}{2} \times 0.6366 \times 30.275 = 60 \times 10^3 \times 0.6366 \times 30.275 = 1160 \text{ Nm.}$$

The direction of bending moment M_{b1} and M_{b2} are opposite to each other, therefore, the residual bending moment to be resisted by the flange = 2830 – 1160 = $M_{b1} - M_{b2}$ = 1670 Nm.

The breadth of the flange at section *AA* taking bending

$$b = 156 - 62 = l_2 - D_o = 94 \text{ mm}$$

A moderate value for the bending stress will be assumed, since the flange , which is not supported at its outer edges, must be stiff enough to reduce the deflection there to a negligible amount. The allowable bending stress, therefore, will be taken as 47.5 N/mm². Let t_1 be the thickness of flange, then from the relation $M_b = f \times z$, we can calculate the thickness of flange.

$$M_b = M_{b1} - M_{b2} = f_b \times \frac{1}{6} b t_1^2$$

$$1670 \times 10^3 = \frac{1}{6} 94 \times t_1^2 \times 47.5$$

gives, t_1 = 48 mm, say 50 mm.

Bending about section *BB*: For bending about the axis *BB*, the bending will be because of force in one bolt placed diagonally.

$$\text{Length of the diagonal at which the bolts are placed} = \sqrt{(94)^2 + (94)^2} = 133 \text{ mm}$$

$$\text{eccentricity of bolt for bending about BB} = \frac{133}{2} = 66.5 \text{ mm.}$$

Bending moment due to force in one bolt = $\frac{F}{4} \times 66.5 = 30 \times 10^3 \times 66.5$

$$M_{b1} = 2000 \text{ Nm}$$

Bending moment due to water pressure is $= M_{b2} = 1160$ Nm.

The direction of bending moments M_{b1} and M_{b2} are opposite to each other, therefore the residual bending moment to be resisted by the flange = $M_b = M_{b1} - M_{b2} = 2000 - 1160 = 840$ Nm.

The breadth of the flange at section BB taking bending

b = length of diagonal – outside diameter of pipe – 2 × diameter of bolt hole.

Length of diagonal

$$= \sqrt{l_2^2 + l_2^2} = \sqrt{(156)^2 + (156)^2} = 220 \text{ mm}$$

$$\therefore \quad b = 220 - 62 - 2 \times 31 = 96 \text{ mm}$$

Using the determined value of 50 mm for t_1, the maximum bending stress can be found out.

$$M_b = f_b \times \frac{1}{6} b t_1^2, \; f_b = \frac{840 \times 10^3 \times 6}{96 \times (50)^2} = 21 \text{ N/mm}^2$$

which is less than the allowable stress in bending of 47.5 N/mm^2.

Bending about section CC: At section CC, the only applied bending moment will be that due to the bolt.

Eccentricity of bolt for bending about $CC = \frac{1}{2}(133 - 80) = 26.5$ mm.

Bending moment due to force in one bolt = $30 \times 10^3 \times 26.5 = 795$ Nm.

The breadth b at section CC can be measured by making the drawing for the flange which is 139 mm.

Using the determined value of 50 mm for t_1, the maximum bending stress induced can be found out.

$$M_b = f_b \times \frac{1}{6} b t_1^2, \; f_b = \frac{795 \times 10^3 \times 6}{139 \times (50)^2} = 13.85 \text{ N/mm}^2$$

which is again less than the allowable stress in bending of 47.5 N/mm^2. It is, therefore, clear that the section at which fracture is most likely to occur will be one such as AA, where there is bending action due to two bolts and due to the fluid pressure.

For a square flange, integral with pipe, the most severe bending action is that about section CC, which will be the probable section of fracture. The dimensions of packing are shown in Fig. 13.11. All the dimensions mentioned in Figs. 13.10 and 13.11 are in mm.

Example 13.4. In Fig. 13.13 is shown the elevation of a flange, for use with pipes carrying fluid with working pressures upto 0.5 N/mm^2. The thickness of the flange is 25 mm, determine the allowable stress in the flange. The internal diameter of the pipe is 200 mm. Eight bolts of diameter 16 mm are used to connect the flanges. The pitch circle diameter on which the bolts are put is 290 mm.

Solution: Thickness of the pipe can be found by

Lame's formula.

$$t = \frac{D_i}{2}\left[\sqrt{\frac{f_t + P}{f_t - P}} - 1\right]$$

Taking allowable tensile stress as 10 N/mm^2.

$$t = \frac{200}{2}\left[\sqrt{\frac{10 + 0.5}{10 - 0.5}} - 1\right] = 5.13$$

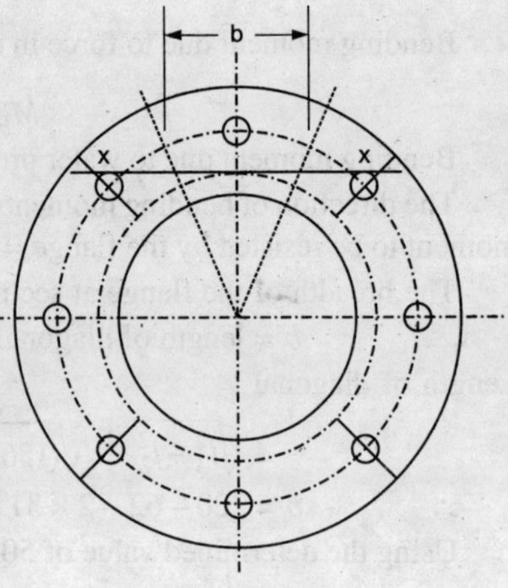

Fig. 13.13

Diameter on inside of the bolts

D_1 = pitch circle diameter – bolt diameter

$D_1 = 290 - 16 = 274$ mm

Total force acting on the bolts can be found out by relation 13.9.

$$F = \frac{\pi}{4} D_1^2 \times P = \frac{\pi}{4}(274)^2 \times 0.5 = 29.5 \text{ kN}$$

$$\text{Force to be resisted by each bolt} = \frac{F}{8} = \frac{29.5 \times 10^3}{8} = 3700 \text{ N}$$

The flange will fail along *x-x* due to bending. The force for bending is applied by one bolt. The width of the bending segment can be obtained by making the drawing which in this case is 92 mm.

$\therefore \quad b = 92$ mm.

The distance of the section *xx* from bolt centre (*e*)

$$= \frac{1}{2}[290 - 200 - 2 \times 5.13] = 40 \text{ mm}.$$

Then by applying the relation $M_b = f_b \times Z$, we can find the induced bending stress in the flange.

$$3700 \times 40 = f_b \times \frac{1}{6} \times 92 \times (25)^2, \quad \text{gives, } f_b = 13.8 \text{ N/mm}^2$$

Thus the allowable bending stress in the flange is 13.8 N/mm^2.

EXERCISES

1. Discuss the different relations for getting the thickness of a pipe.
2. Describe with sketches two methods by which the joints are made in connecting lengths of iron pipes.
3. Make neat sketches showing the construction of a Hydraulic pipe joint. Explain how the joint is made water tight? How are the bolts prevented from turning in the bolt holes when screwing up the nuts?
4. Sketch partly in section a union nut and joint for connecting two lengths of small piping.
5. Give sketches showing (*a*) The construction of 'Leaded' socket and spigot joint for large diameter water pipes, (*b*) of an expansion joint.
6. Make sketches of the following pipe connection. (*a*) Oval flange, (*b*) Square flange, (*c*) Round flange, (*d*) Screwed joint.
7. Show two methods of connecting the separate lengths of wrought iron or steel pipes.
8. Mention the specified advantages claimed by the following joints. (*a*) Threaded joint, (*b*) Bell and Spigot joint, (*c*) Brazed joint, (*d*) Welded joint, (*e*) Expansion joint and (*f*) Flanged joint. Give two practical applications in each case.
9. Discuss the advantages claimed by oval type flange pipe joint over the square or round type flange pipe joint.
10. What thickness of metal would be required for a cast iron water pipe 0.91 meter diameter under a head of 125 metres, safe working stress is 20 N/mm^2?
11. A marine boiler 4.5 metre diameter is required to carry a working pressure of 1.25 N/mm^2. The allowable stress in tension is not to exceed 75 N/mm^2, efficiency of rivetted joint is 80 percent. What thickness of the plates would be required?
12. In thin cylindrical shell subjected to internal fluid pressure, show that the tendency to burst length- wise is twice as great as at a transverse section.

    A boiler is 1.8 meter diameter, shell plates 20 mm thick, efficiency of the joint to tearing along a longitudinal seam is 72 percent. Calculate the safe working pressure if the tensile strength of plate is 450 N/mm^2, factor of safety is 6.

13. Determine the diameter and thickness of solid drawn steel tubes to carry steam from the boiler with a velocity of 30 m/sec. The rate of evaporation is 2000 m^3 of superheated steam per hour at a pressure of 0.8 N/mm^2 gauze. Assume the allowable tensile stress to be 35 N/mm^2.
14. Determine the suitable proportion of a spigot and socket joint to connect two lengths of pipes of diameter 100 mm. The internal pressure of fluid is 7.5 N/mm^2. Draw full sized sectional elevation.
15. Design a hydraulic pipe joint (Division 1, class *A*, flange type 1 for a fluid pressure of 6.4 N/mm^2 (test pressure 18 N/mm^2) and with a bore of 100 mm. The following data extracted from the tables of standards may be used. Flange length and width is 290 mm and 165 mm respectively, bolt centre 210 mm, diameter of facing *F* (Fig. 13.7) 165 mm, diameter of spigot 120 mm. Use a bending stress in the flange of 8.5 N/mm^2 and a maximum stress of 56 N/mm^2 in the pipe. Make a working drawing of the designed joint.
16. Design pipe joint with square, screwed flanges (Division 2) for a 25 mm bore piping, under a pressure of 14 N/mm^2. Assume pipe thickness 6 mm. screwed metric threads 0.435 mm, depth of thread 1.47 mm, bolt centre 60 mm, side of flange 100 mm, spigot diameter 50 mm, stress in bolts 10 N/mm^2, stress in flange 32 N/mm^2, depth of recess 7.5 mm.
    Prepare working drawing of the designed joint. For the designed pipe joint, calculate the bearing pressure on the threads of the pipe. Assume a length of thread of 37.5 mm.
17. A gun metal hydraulic fitting is connected by a 125 mm diameter branch to a container carrying a pressure of 7 N/mm^2. Design a suitable connection made by a flanged joint. The allowable stress of gun material is 20 N/mm^2 and 45 N/mm^2 for carbon steel studs. Assume the necessary data required.
    Make a working drawing of the designed joint.
18. A pipe carries the fluid at pressure of 0.75 N/mm^2. The length of the pipe extended by a round type flange pipe joint. The thickness of the flange is 30 mm, determine the allowable stress in the flange. The internal diameter of the pipe is 200 mm. Eight bolts of diameter 16 mm are used to connect the flanges. The pitch circle diameter on which the bolts are put is 300 mm.

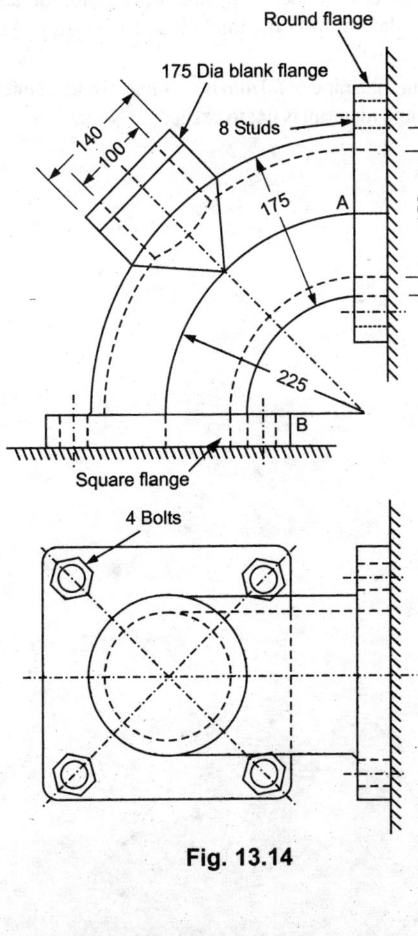

Fig. 13.14

19. Two cast iron pipes are joined together by means of round type flanges, bolted together. The inside diameter of the pipes is 100 mm and the pipes are subjected to internal fluid pressure of 8 N/mm^2. The permissible stresses are 32 N/mm^2 for bolt material and 16 N/mm^2 for pipe material. Determine.
    (*a*) Thickness of the pipe (*b*) Number of bolts (*c*) Diameter of bolts
    (*d*) Pitch circle diameter of bolts, and (*e*) Thickness of flange.
20. Determine the wall thickness required for a brass condenser tube of 25 mm inside diameter. The vacuum in the condenser is to be 6.50 mm of mercury and the water pressure is to be 0.25 N/mm^2 on the inside of the tube. Allow 1.5 mm on the wall thickness of corrosion.
21. A cast iron pipe is to deliver water at the rate of 2300 kg/s and a flow velocity of 0.5 m/s. The maximum pressure in the pipes is 0.85 N/mm^2. The permissible stress in the cast iron is 20 N/mm^2. Determine the diameter and the wall thickness.
22. Determine the pipe size for a demand of 7250 kg/hr of steam at 1.75 N/mm^2 and 260ºC.
23. It is said that the sizing of pipe is an art rather than a science, explain.
24. What are the advantages and disadvantages claimed by cast iron as the material of pipe carrying steam?
25. Design a cast iron flanged joint for a pipe of 100 mm internal diameter to carry steam at a pressure of 1.05 N/mm^2. Give a dimensioned sketch of the joint.
26. A cast steel right angled pipe bend for connecting a cylinder to a receiver is shown in Fig. 13.14. The top flange A is of circular form and is secured by 8 studs to cylinder face. The bottom flange B is of square form, as indicated and is held by 4 tap bolts. There is an inspection opening at 45 degrees on the bends as shown. The steam pressure is 1.75 N/mm^2 . Design the flange A and studs, presuming the diameter at the inside of the bolts to be 200 mm. Stress in studs is not to exceed 42 N/mm^2. The safe stress for the flange is 40 N/mm^2. Design also the square flange for the same size of the joint and the same stress in the tap bolt and flange.
    Determine the size of screw used for fixing the blank flange on the 100 mm diameter opening. The allowable stress in screws is 40 N/mm^2. The cover is made from mild steel plate, the stress for which is 70 N/mm^2. Use a concentration factor of 1.25 in determining the thickness of the cover. Assume the inside of the screws to be 140 mm.
27. Design an oval flanged pipe joint for a pipe of 50 mm bore subjected to an internal fluid pressure of 7.0 N/mm^2. The maximum tensile stress in the pipe material is not to exceed 20 N/mm^2 and in bolts 60 N/mm^2. Make a neat sketch of the designed joint.

CHAPTER 14 Levers

14.1 INTRODUCTION

A lever is a simple machine which is used to get mechanical advantage, roughly speaking it is a link or a rigid rod moving about some fixed point or axis. A lever is meant to raise a load with less effort, or sometimes simply to facilitate the application of force in a desired direction. It is customary to call a rotating link, a crank or handle, an oscillating link a lever or a beam, and the connecting rod. The fixed link is often enlarged and used as the supporting frame.

The lever may be straight or curved. The forces applied by the lever or on the lever may be parallel or may be inclined to one another. The principle of lever is the same as that of moments. Figure 14.1 explains the principle of moments for the case of parallel forces acting in the same plane. *O* indicates the fulcrum point, *A* the load point, and *B* the effort point. Taking moments about the fulcrum point

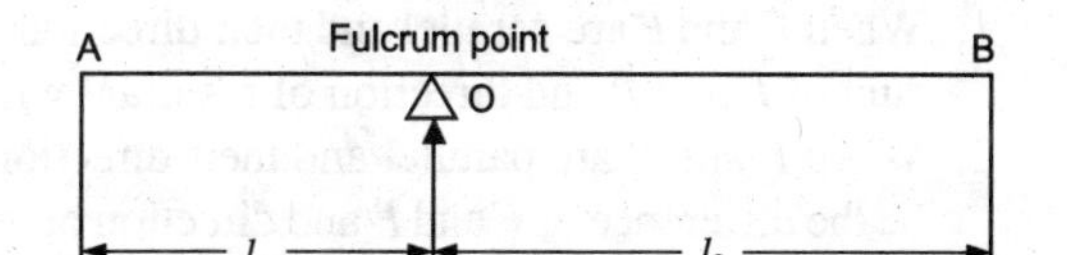

Fig. 14.1

$$P \times l_2 = F \times l_1 \qquad ...(14.1)$$

or

$$\frac{F}{P} = \frac{l_2}{l_1}$$

The advantage obtained by using a lever is called mechanical advantage and is the ratio of load to effort (*F/P*). or it is also equal to the ratio of effort arm to the load arm (l_2/l_1). The perpendicular distances of the fulcrum are called the arms of the lever. The ratio l_2/l_1 is also called the leverage. The displacement of effort to displacement of load is called displacement ratio. In the above figure F represent the load and *P* effort applied. Levers are of three types known as number 1, 2 and 3. In lever of type 1, the fulcrum is in between and the load and effort points are at the ends. In lever of type 2, the load point is in between and the fulcrum and effort points are at the ends. Whereas in lever of type 3, the effort point is in between and the load and fulcrum points are at the ends.

If a large load is to be removed by a small effort the arm of the effort must be very much greater than the arm of the load. It means the length of the effort arm required is very large which in many cases will not be advisable from space point of view. Therefore to remove larger loads in a limited space, the compound levers are used. The compound levers may be made up of straight pieces, which may be attached with one another by means of pin joints. Bell cranked levers may be used, instead of various levers together. The leverage of compound levers is the product of leverages of various levers.

Levers are used very commonly in engineering. Levers of type 1 are used in a bell crank lever, rocker arms in internal combustion engines, beam of a balance, handle of a hand pump, pincer and cutting pliers, scissors, etc. In this type of lever, the lever is only useful when the effort arm is longer than the load arm. Levers of type 2 are used in lever loaded safety valves, nut cracker, wheel barrow, etc. In this lever as the effort arm is always greater than the load arm, therefore mechanical advantage is always greater than one. In levers of type 3, the effort arm is always less than load arm, thus the mechanical advantage is always less than one. For this reason, the use of such levers is not recommended. However, they are being used for achieving some convenience in a mechanism as in pair of forceps or tongs, the treadle of a sewing machine, etc. In many applications in engineering, levers are used to change the direction of forces as in bell crank levers, in railways signals, rocker arms of I.C. engines, etc.

14.2 DESIGN PROCEDURE FOR LEVERS

Designing a lever means, determining physical dimensions of the lever when forces acting on the lever are specified. A lever is acted upon by two single forces or two sets of forces at effort and load points and by a reaction at the fulcrum. The forces acting on the effort and load point cause moments in the opposite directions. The effort arm and load arm are fixed either for requirements or from space considerations. Generally, load acting on one arm is known and the effort to be applied to resist this load can be calculated by the principle of moments. If the load and effort arms are equal then the effort to be applied is equal to the resistance to be overcome, If we neglect the friction at the different points of bearings.

The reaction at the fulcrum point *O* will depend upon that how F and P are acting. The different ways in which *F* and *P* can act are:

1. When *F* and *P* are parallel and their direction is same, then the resultant at the fulcrum point will be the sum of *F* and *P* and direction of resultant will be same as *F* and *P*.
2. When *F* and *P* are parallel and their direction is opposite, then the resultant at the fulcrum point will be the difference of *F* and *P* and direction of resultant will be the same as of *F* or *P* whichever is greater.
3. When forces *F* and *P* are inclined to each other, then, the resultant at the fulcrum point can be obtained by parallelogram of forces and its magnitude will depend upon *F* and *P* and the direction of the resultant will also depend upon the direction of *F* and *P*.
4. When two *F* and *P* act at right angles to two arms of lever, which include angle θ in between two arms, as in the case of exhaust valve mechanism, then the reaction at the fulcrum can be obtained by the relation.

$$R = \sqrt{F^2 + P^2 + 2FP\cos\theta} \qquad ...(14.2)$$

When the forces *F* and *P* act at right angles to each other as in the case of bell crank lever, then the reaction at the fulcrum will be

$$R = \sqrt{F^2 + P^2} \qquad ...(14.3)$$

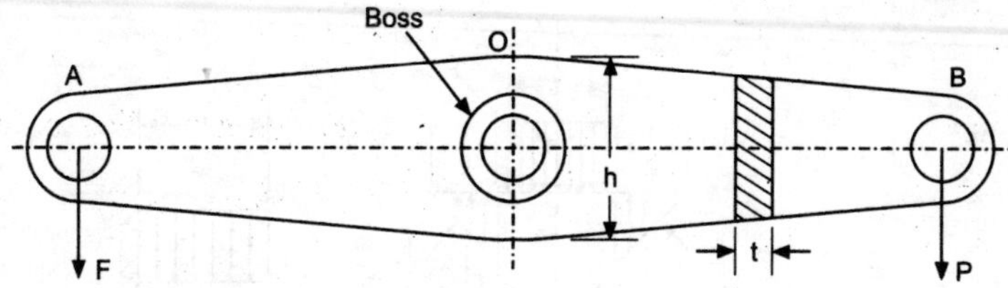

Fig. 14.2 *Lever*

Once the forces acting on the lever are known, the various dimensions of the lever can be obtained. In all types of levers when the load and effort are in a plane containing the lever arm, the section of the lever arm is mainly subjected to bending moment. Hence section of the arm is estimated from the bending moment considerations. Fig. 14.2 shows a lever fulcrum point in between the load and effort point. The figure indicates that the maximum bending moment occurs at the fulcrum and goes on reducing toward the ends. Thus theoretically the cross-section will be zero at the free end, but every object must have physical dimensions and hence the lever is suitably tapered and given a good looking shape.

For any kind of lever, the section at which maximum bonding moment occurs can be determined by the strength equation $M_b = f \times z$, where M_b is the maximum bending moment, f is the allowable stress of the material of lever and Z is the modulus section of lever. The section of the lever arm may be rectangular, elliptical, circular or of I-form. The modulus of section of the rectangular section is 1/6 th^2, where t is the thickness of the lever and h is the height of the lever. The height of the lever is generally 2 to 5 times the thickness of the lever. The modulus of section of an elliptical section is $\pi/32\ ab^2$ where a is the length of minor axis and b is the length of major axis of the ellipse. Usually b is 2 to 2.5 times a . The Modulus of section of circular section is $\pi/32\ d^3$, where d is diameter of the lever. For an I-section lever; it amy be assumed that the bending moment is resisted wholly by flanges. With this assumption the modulus of section for an I-section will be (flange area × depth of section).

The section of the lever arm is tapered from the fulcrum to the ends. The taper is given in height and not in thickness. The taper given depends upon the type of the free end. The end *A* and *B* should be checked for shear force. If the load at the end is applied by forked connections, then the dimensions of the lever at that end can be proportioned as a knuckle joint.

The fulcrum pin is designed for bearing and bending consideration with a proper check for shear and sometimes deflection also. The general procedure is to design the pin from bearing considerations and then to check it for shear. The permissible bearing pressure depends upon the amount of relative velocity between the pin and the lever and the frequency of motion. The sufficient length in bearing is usually obtained by providing bosses instead of choosing a thicker lever throughout. If the forces do not differ much, the same pin dimensions for pin at fulcrum, load and effort joint may be adopted to reduce spares.

The bending of the pin in the fork may take place in different ways. The different ways of force distribution on the pin are shown in Fig. 14.3. For the pin supported in fork, the stress distribution is assumed to be zero at the outside and maximum at the inside. The effect of clearance is neglected. The thickness of the lever is assumed to be t and the thickness of the fork is assumed to be half the thickness of lever.

For case (*a*) the maximum bending moment will occur at *y-y*.

$$\text{Maximum B.M. at } y\text{-}y = \frac{F}{2}\left[\frac{t}{2}+\frac{t}{6}\right] - \frac{F}{2}\times\frac{t}{4} = \frac{5}{24}Ft$$

$$\cong \frac{Ft}{4} \qquad \text{...(14.4)}$$

For case (*b*) the maximum bending moment will occur where the pin is supported.

$$\text{Maximum bending moment} = F \times \frac{t}{2} \qquad \text{...(14.5)}$$

Same will be the bending moment in case (*c*) also.

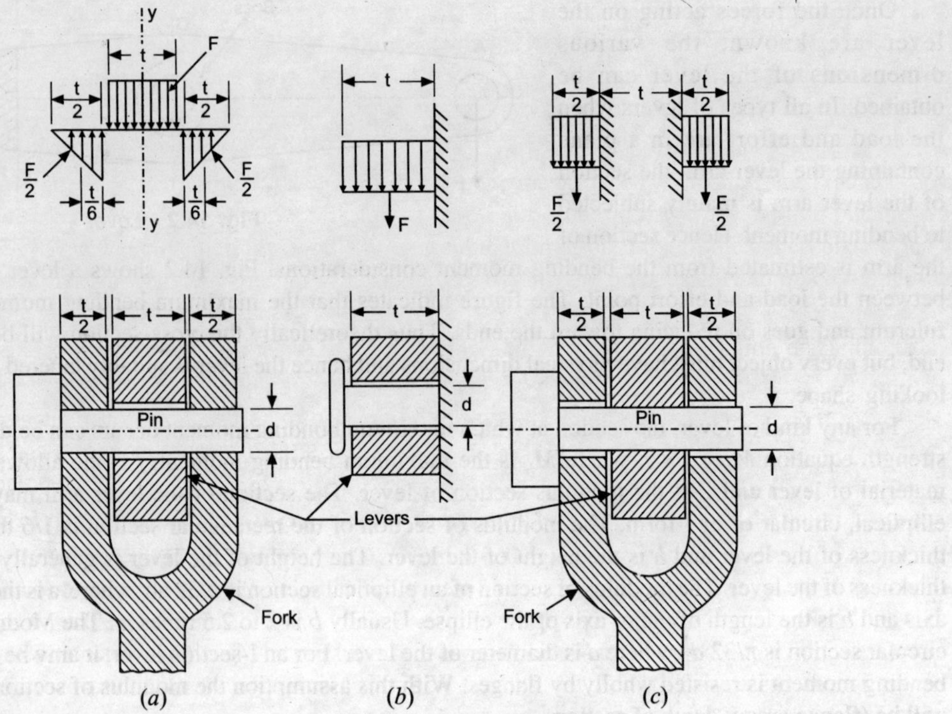

(a) Bearing in lever-pin simply supported in fork. (b) Bearing in lever-cantilever type pin. (c) Bearing in fork-cantilever type pin

Fig. 14.3 *Distribution of force on pin*

Bearing consideration of pin gives

$$F = d \times t \times p_b \qquad \text{...(14.6)}$$

where p_b is the allowable bearing pressure.

Co-relating (14.4) and (14.6) and equating bending moment to $f_b Z$, we have

$$d \times t \times p_b \times \frac{t}{4} = f_b \times \frac{\pi}{32} d^3$$

where f_b is the allowable bending stress and by solving we get

$$\frac{t}{d} \simeq 0.634 \sqrt{\frac{f_b}{p_b}} \qquad \text{...(14.7)}$$

Similarly for case (b) and (c), co-relating (14.5) and (14.6) gives

$$\frac{t}{d} = 0.488 \sqrt{\frac{f_b}{p_b}} \qquad \text{...(14.8)}$$

From the above relations the value of t and d can be calculated. Generally the diameter of the boos is taken twice the diameter of the pin and this may also taken as the height of the lever. It may further be noted that where relative velocity is small and bending is not so predominant, diameter of the pin may be obtained by shear consideration. But, normally, the diameter obtained by shear consideration is smaller, so getting diameter by shear consideration may not be encouraged, unless the clearance is very small.

For case *(a)* discussed above, $B.M. = \frac{Ft}{4} = f_b Z = f_b \times \frac{\pi}{32} d^3$

For this case the pin is in double shear.

∴ $F = 2 \times \pi/4\, d^2 \times f_s$ where f_s is the allowable shear stress. Dividing one by the other

$$\frac{(F \times t)/4}{F} = \frac{f_b \times (\pi/32)d^3}{2 \times (\pi/4)d^2 \times f_s}$$

On solving the relation, we get

$$\frac{f_s}{f_b} = \frac{d}{4t} \quad ...(14.9)$$

In the limiting case when the diameter of pin and thickness of lever are equal, then $f_s/f_b = 1/4$ which indicates that shear stress induced is 25 percent of the bending stress induced. This is permissible because for most of the material of the pin, the shear stress is 0.8 times the bending stress. If it was a case of cantilever, then the pin is in single shear? Again for thickness of lever to be equal to diameter of pin, we will get, $f_s = 0.25\, f_b$. Thus we find with this analysis, that, the checking of pin in shear is not necessary, except in the case when the pin is required to stop relative motion.

Normally, the suitable value of the ratio of length to diameter of pin is assumed and the dimensions of the pin can be obtained. The value of this ratio may be taken from 1 to 1.3. Therefore, the length of the boss will be 1 to 1.3 times the diameter of pin. Diameter of boss is twice the pin diameter. To increase the life of the lever and to reduce wear, it is customary to fit a phospher bronze bush of thickness 3 mm. To increase the life of the lever further, the dust proof lubrication arrangement should be provided. This helps in reducing the wear to a minimum, thus resulting in very less replacements of bush.

If the load or effort, are in plane containing lever arm but not normal to the axis of the lever then the lever arm is subjected to direct stress in addition to bending stress. It may be noted that the direct stress may be compressive or tensile, depending upon the direction of load. In certain cases the section of the lever may be under the action of bending moment and turning moment both, e.g., in an overhung crank.

SOLVED EXAMPLE

Example 14.1. The Fig. 14.4 shows the handle for turning the spindle of a large valve. The effort applied at each end of the handle is 400 N. The length of the handle from centre of spindle is 500 mm. The handle is attached to the spindle by means of a tapered round pin. Find the dimensions of the tapered pin and the handle, Take the spindle diameter as 50 mm and arm grip length 160 mm. Allowable stress in tension and shear are 100 N/.mm² and 50 N/mm².

Solution: The torsional moment to be resisted by the spindle of the valve

$$= 2 \times 500 \times 400 = 400 \text{ Nm}$$

This torsional moment is to resisted by the tapered pin. Let the diameter of the tapered pin be d_2 and this pin is in double shear, then

$$400 \times 10^3 = 2 \times \pi/4 d_2^2 \times f_s \times \frac{d}{2},$$

$$400 \times 10^3 = 2 \times \pi/4 d_2^2 \times 50 \times \frac{50}{2},$$

where d_2 is the mean diameter,

$d_2 = 4.5$ mm say 5 mm.

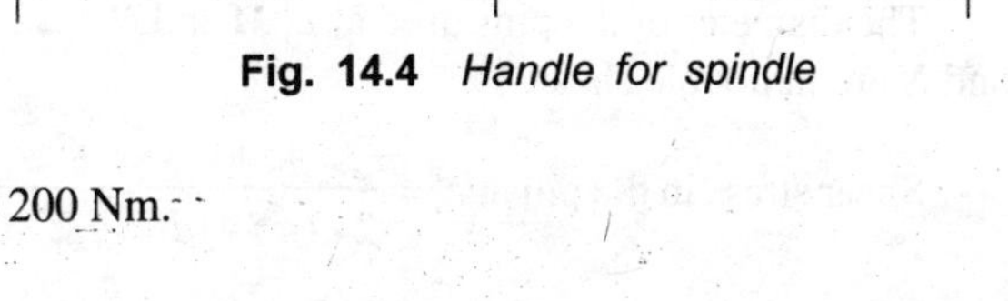

Fig. 14.4 *Handle for spindle*

The handle is also subjected to turning moment. Assuming that the handle extends upto the centre of the spindle.

Maximum bending moment, $M_b = 500 \times 400 = 200$ Nm.

The handle is also subjected to turning moment. Assuming that the effort of 400 N acts at the centre of the arm, then the turning moment will be

$$M_t = 400 \times 80 = 32 \text{ Nm}$$

The equivalent turning and bending moments can be now found out.

Mte = equivalent turning moment = $\sqrt{M_b^2 + M_t^2} = \sqrt{(200)^2 + (32)^2} = 226$ Nm

Mbe = equivalent bending moment = $\frac{1}{2}(M_b + \sqrt{M_b^2 + M_t^2}) = \frac{1}{2}(200 + \sqrt{(200)^2 + (32)^2} = 213$ Nm

Now the diameter of the handle can be obtained from shear considerations as the allowable stress in tension is very high.

$$\therefore \quad M_{te} = (\pi/16)d_1^3 \times f_s, 226 = \pi/16 d_1^3 \times 50$$

where d_1 is the diameter of the lever, $d_1 = 29$ mm say 30 mm.

For easy removal of the handle, the ends of the valve spindles are usually made square and by this the use of the tapered pin or key can also be eliminated. It may also be noted that if the force applied on each arm is same and the lever arms are equal, the spindle of the valve will not be subjected to bending moment. This also explains the advantage of hand- wheel. But as there are all chances of unequal force application or operation by one hand only, the spindle dimensions should be obtained taking into account bending moment.

The diameter of boss = 2 × diameter of spindle = 2 × 50 = 100 mm

Height of boss = 2 × diameter of handle = 2 × 30 = 60 mm.

Example 14.2. A horizontal lever LMN, 15 mm thick is attached by a fulcrum pin at *N* and to a vertical tie rod at *M* as shown in Fig. 14.5. An operating force of 900 *N* is applied vertically at *L*. Find the reactions at *N* and *M*. If the diameter of the tie rod at *M* is 12 mm, find the induced stress in it. Also find the induced shear stress in the pins at L, M and N, when the diameter of pins is 12 mm. What is the bearing stress on the lever at M?

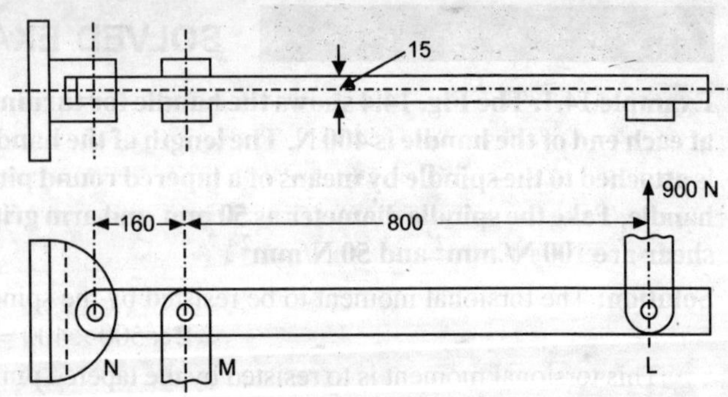

Fig. 14.5

Solution: Let R_1 be the force in the vertical tie rod at *M*. Then by taking moments about fulcrum point *N*, the force R_1 can be calculated. Thus,

$$R_1 \times 160 = 900 \times 960$$

$$R_1 = 5400 \text{ N}$$

As the applied effort at point *L* and the force in the tie rod are parallel and opposite in direction to each other, then the force at the fulcrum will be = 5400 – 900 = 4500 N. This force at the fulcrum will be acting vertically upwards.

Thus Reaction at *M* = 5400 N acting vertically downward,

Reaction at *N* = 4500 N acting vertically upwards.

Tensile stress induced in the tie rod at

$$M = \frac{R_1}{(\pi/4)d^2} = \frac{5400}{\pi/4(12)^2} = 48 \text{ N/mm}^2$$

The diameter of the pins used at *L*, *M* and *N* is 12 mm. The pin at *L* is in single shear while the pins at *M* and *N* are in double shear.

$$\text{Shear stress in the pin at } N = \frac{4500}{2\times(\pi/4)d^2} = \frac{4500}{2\times(\pi/4)(12)^2} = 20 \text{ N/mm}^2$$

Shear stress in the pin at $M = \frac{5400}{2\times(\pi/4)d^2} = \frac{5400}{2\times(\pi/4)(12)^2} = 24$ N/mm^2.

Shear stress induced in the pin at $L = \frac{900}{(\pi/4)d^2} = \frac{900}{(\pi/4)(12)^2} = 8$ N/mm^2.

Bearing area of the lever at the pin M = Thickness of lever × diameter of pin = 15 × 12 = 180 mm^2.

Bearing stress induced on the lever = $\frac{5400}{180} = 30$ N/mm^2.

14.3 HAND LEVER

In Fig. 14.6 is shown the ordinary form of the hand lever used in various machineries. P is the force manually applied at the handle and L is effective length of the lever from the centre of the shaft to the point where force P is applied. Let B and t be the height and thickness of the lever near the boss. The lever will fail in bending. Assuming that the lever will extend to the centre of the shaft, this assumption results to a stronger section of the lever.

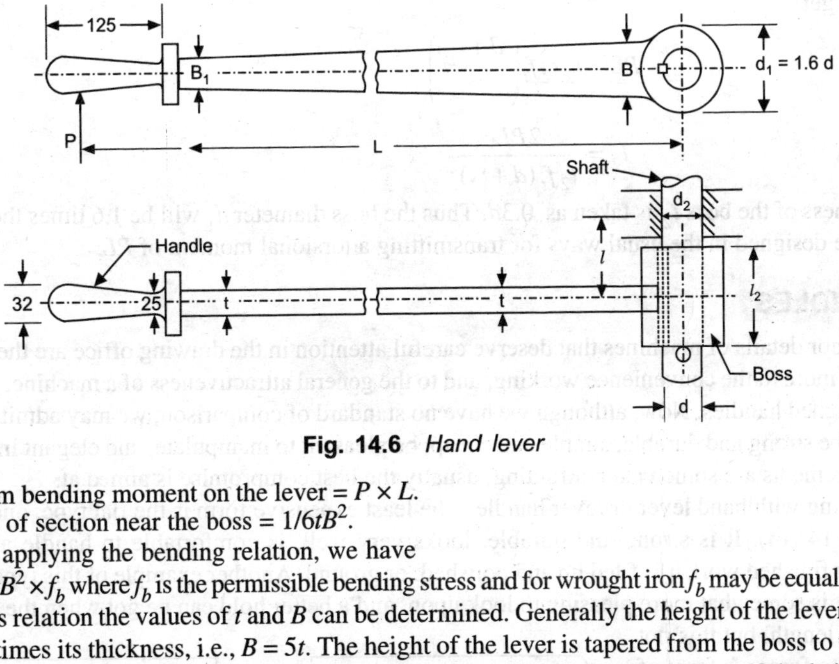

Fig. 14.6 *Hand lever*

Maximum bending moment on the lever = $P \times L$.
Modulus of section near the boss = $1/6tB^2$.
Then by applying the bending relation, we have
$PL = 1/6tB^2 \times f_b$ where f_b is the permissible bending stress and for wrought iron f_b may be equal to 70 N/mm^2.
From this relation the values of t and B can be determined. Generally the height of the lever near the boss is taken as 5 times its thickness, i.e., $B = 5t$. The height of the lever is tapered from the boss to the handle but thickness t is kept constant. The least depth B_1 near the grip of the handle is to be half the height B.

In manually operated levers, the design dimensions are usually obtained by considering maximum force a man can exert rather than the load actually necessary to be resisted. Approximate manual forces in N may be taken as given below:

Name of operation	Continuous	Intermitent	Instantaneous
Hand operation	70–100	200–250	350–400
Foot operation	150–200	400–500	700–800

The diameter d_2 is subjected to combined bending and twisting, therefore

$$\frac{\pi}{16}d_2^3 \times f_s = \sqrt{M_b^2 + M_t^2} = \sqrt{(Pl)^2 + (PL)^2} = P\sqrt{l^2 + L^2}$$

or
$$d_2 = \left[\frac{16P}{\pi f_s}\sqrt{(l^2+L^2)}\right]^{\frac{1}{3}} \quad ...(14.10)$$

The relation 14.10 determines the diameter d_2 according to the maximum shear stress theory.

The diameter d_2 can also be obtained by the Rankine's formula,

$$\frac{\pi}{16}d_2^3 \times f_b = P(l+\sqrt{l^2+L^2})$$

or
$$d_2 = \left[\frac{5.1P(l+\sqrt{l^2+L^2})}{f_b}\right]^{1/3} \quad ...(14.11)$$

where f_b and f_s are allowable bending and shear stresses. The values of f_b and f_s for wrought iron will be 70 and 60 N/mm^2.

The length of the boss l_2 from d to 1.25 d. It can be checked for a t_2 thickness by taking moments about the axis. Let t_2 be the thickness of the boss, then equating PL to the moment of resistance to tearing parallel to the axis, we get

$$PL = l_2 t_2 f_t \left(\frac{d+t_2}{2}\right)$$

or
$$l_2 = \frac{2PL}{t_2 f_t (d+t_2)} \quad ...(14.12)$$

The thickness of the boss t_2 is taken as 0.3d. Thus the boss diameter d_1 will be 1.6 times the diameter d. The key can be designed in the usual ways for transmitting a torsional moment of PL.

14.4 HANDLES

Among the minor details of machines that deserve careful attention in the drawing office are the handles, for few things add more to the convenience working, and to the general attractiveness of a machine, than suitable and neatly designed handles. Now, although we have no standard of comparison, we may admit that an ideal handle should be strong and durable, simple and cheap, be pleasant to manipulate, and elegant in appearance. As these requirements are somewhat conflicting, usually the best compromise is aimed at.

Commencing with hand lever or lever handles, the least expensive form is the flat type, one of which is shown in Fig. 14.7(a). It is strong and durable, looks very well, is comfortable to handle and needs no machining, For finished work it is filed up and polished, or ground. Another example of this type is shown in Fig. 14.7 (b). It is somewhat more pleasing to look upon, and a better hold can be got when the pull is in the direction of its length, but it is not quite so cheap to forge. A form of handle largely used is shown in Fig. 14.7 (c). It is easily made, but is clumsy in appearance, although comfortable to handle. Figure 14.7 (d) and 14.7 (e) show a type of lever handle which is more commonly used than any other, being suitable for any size work. They are easily made and cheaply turned and miled if for finished work, and they have a neat appearance and are strong.

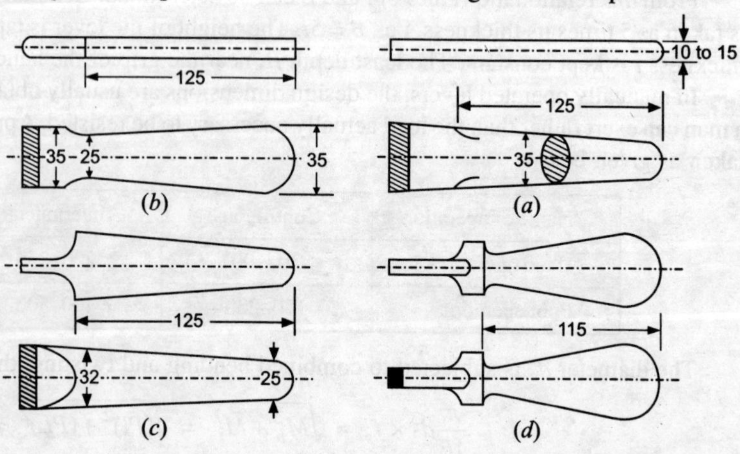

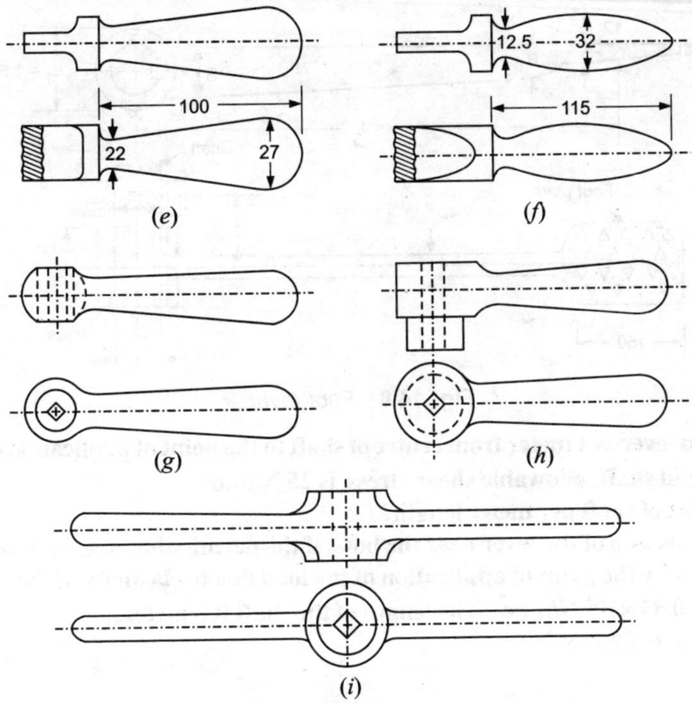

Fig. 14.7 *Machine handles*

Figure 14.7 (*f*) is an elegant form, designed to fit the palm of the hand, but it is not often met with now, except in small work, as it is more expensive to turn, requiring, if not finished by hand tools, a special former plate attachment to the slide rest of the lathe.

Two forms of Box-spanners are shown in Figs. 14.7 (*g*) and 14.7 (*h*). They both have taper handles. The boss of the latter is mainly on one side of the centre line, and it is turned small in diameter in relation to hole to suit a recess (such as we have in lathe dog- chucks), where it is desirable to keep down the size of this dimension.

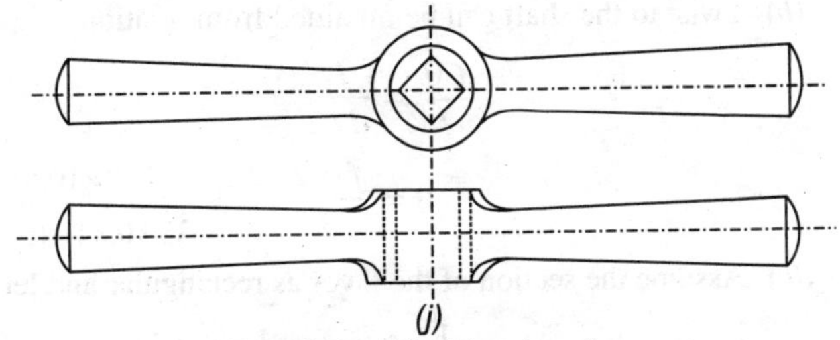

(*j*)

Fig. 14.7 *Machine handles*

Stop-valve handles are shown in Fig. 14.7 (*i*) and 14.7 (*j*), the latter with taper ends being the better looking and the more usual form. Tap wrenches also have this form.

In general the part grasped by hand may be 35 mm in greatest and 22 mm in smallest diameter and 125 mm long.

14.5 FOOT LEVER

Figure 14.8 show a foot lever. About 850 N is considered the maximum effort a man can exert in pushing a foot lever. Taking this value of P in Article 14.3, it can be dealt in the same way. The size of the foot plate is about 150 mm by 120 mm and 12 mm thick.

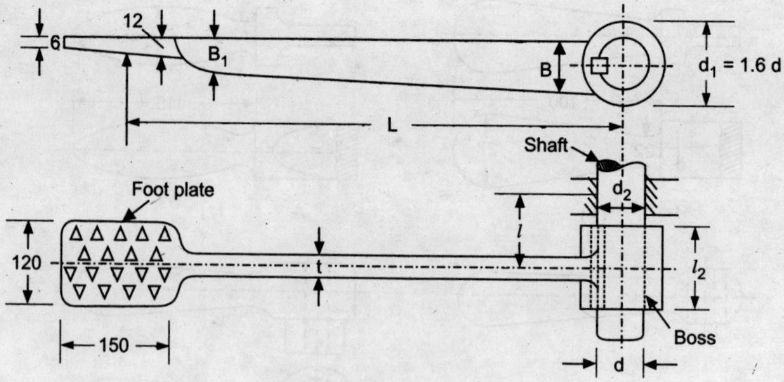

Fig. 14.8 *Foot lever*

Example 14.3. A hand lever is 1 meter from centre of shaft to the point of application of 200 N. Calculate

(*a*) Diameter of solid shaft, allowable shear stress is 25 N/mm^2.

(*b*) What is the twist of shaft per meter length?

(*c*) Determine the section of the lever near the boss, if the permissible stress is limited to 75 N/mm^2.

(*d*) Distance moved by the point of application of the load due to elasticity of the shaft.

Modulus of rigidity is 0.84×10^5 N/mm^2. The length of the shaft is 3 meter.

Solution: $L = 1000$ mm, $P = 200$ N

Refer Fig. 14.6.

(*a*) Torsional moment on shaft $M_t = 200 \times 1000 = 200$ Nm

Assuming pure torsion, resisting torsional moment $= \dfrac{\pi}{16} d^3 \times f_s$

where d is the diameter of shaft.

Then $$200 \times 10^3 = \frac{\pi}{16} d^3 \times 25, \text{ gives } d = 35 \text{ mm}$$

(*b*) Twist to the shaft can be obtained from relation

$$\frac{C\theta}{l} = \frac{f_s}{d/2}$$

$$\theta = \frac{f_s \times 2}{d} \times \frac{l}{c} = \frac{25 \times 2 \times 1000}{35 \times 0.84 \times 10^5} = 0.017 \text{ radian.}$$

(*c*) Assume the section of the lever as rectangular and let near the boss $B = 3t$.

Resisting moment $= \dfrac{1}{6} tB^2 \times f_b = \dfrac{1}{6} t(3t)^2 \times 75$

Equating the resisting moment to the maximum bending moment.

$$\frac{1}{6} t(3t)^2 \times 75 = 200 \times 10^3, \quad \text{gives } t = 13.35 \text{ mm say 15 mm}$$

Height $B = 3 \times t = 3 \times 15 = 45$ mm

Height of lever near handle $= B_1 = \dfrac{1}{2} B = \dfrac{45}{2} = 22.5$ mm

Length of boss $= l_2 = 1.25\, d = 1.25 \times 35 = 43.75$ mm say 45 mm

Diameter of boss $= d_1 = 1.6\, d = 1.6 \times 35 = 56$ mm.

(*d*) Distance moved by the point of application of load $= 0.017 \times 3000$ -

Example 14.4. A foot lever is 1000 mm from centre of shaft to point of application of load of 900 N. Calculate the diameter of shaft, if the permissible shear stress for shaft material is 75 N/mm². Determine the dimensions of key to secure lever to shaft, safe stress in shear for key material is 70 N/mm². Find the dimensions of rectangular arm of the foot lever at 100 mm from centre of shaft. Assume height of lever near boss as 3 times the thickness. Allowable stress is 70 N/mm².

Solution: Torsional moment on shaft $M_t = 900 \times 1000 = 900$ Nm

Refer Fig. 14.8.

Let d be the diameter of shaft.

$$900 \times 10^3 = \frac{\pi}{16} d^3 \times f_s,\ 900 \times 10^3 = \frac{\pi}{16} d^3 \times 75, \text{ gives } d = 39.2 \text{ mm say } 40 \text{ mm}$$

diameter of boss $d_1 = 1.6\, d = 1.6 \times 40 = 64$ mm

length of boss $l_2 = 1.2\, d = 1.2 \times 40 = 48$ mm

Considering combined bending and twisting on shaft. Taking l as 100 mm. Applying relation 14.10 we can determine diameter d_2.

$$d_2 = 3\sqrt{\frac{16P}{\pi f_s}\sqrt{l^2 + L^2}} = \left[\frac{16 \times 900}{\pi 75}\sqrt{(100)^2 + (1000)^2}\right]^{\frac{1}{3}}$$

$$d_2 = 39.7 \text{ mm say } 40 \text{ mm}$$

d_2 is the diameter of the shaft at the centre of the bearing and is put at 45 mm to allow for wear also.

KEY width of key $B = \frac{d}{4} + 3 = \frac{40}{4} + 3 = 13$ mm

depth of key $T = \frac{B}{2} = \frac{13}{2} = 6.5$ mm

Let l_3 be the length of key. Then considering shearing of key

$$l_3 \times B \times f_s = \frac{M_t}{d/2},\ l_3 \times 13 \times 70 = \frac{900 \times 10^3 \times 2}{40},\ l_3 = 50 \text{ mm}$$

which indicates that the length of boss will be 50 mm instead of 48 mm.

Section of lever: B.M. at 100 mm from centre of shaft = 900 × 900 = 810 Nm,

Now equating the bending moment to resisting moment,

$$810 \times 10^3 = f_b \times \frac{1}{6} tB^2,\ 810 \times 10^3 = 70 \times \frac{1}{6} t \times (3t)^2$$

gives $t = 19.6$ mm say 20 mm, $B = 3t = 3 \times 20 = 60$ mm.

Thickness of the lever will remain constant throughout. The depth of the lever will be tapered from B to $B_1 = 30$ mm. Other dimensions of the foot plate are shown in Fig. 14.8.

Example 14.5. Figure 14.9 shows a typical hand lever for operating a part of a machine tool. Assume the maximum force a man can apply is 400 N. Design the lever completely. The permissible stresses for pin material are 50 N/mm² in bending and 10 N/mm² bearing pressure. The allowable bending stress for lever is 70 N/mm².

Solution: Let F be the load moved by the lever.

Taking moments about the fulcrum point, we have

$$400 \times 280 = F \times 70, \text{ gives } F = 1600 \text{ N}$$

Direction of F and P both are vertically downwards. The reaction of the fulcrum will be

$$R = F + P = 1600 + 400 = 2000 \text{ N.}$$

Fulcrum Pin: Assuming the pin to be bearing in fork. Then by using the relation 14.8, we get

$$\frac{t}{d} = 0.488\sqrt{\frac{f_b}{p_b}} = 0.488\sqrt{\frac{50}{10}} = 1.0$$

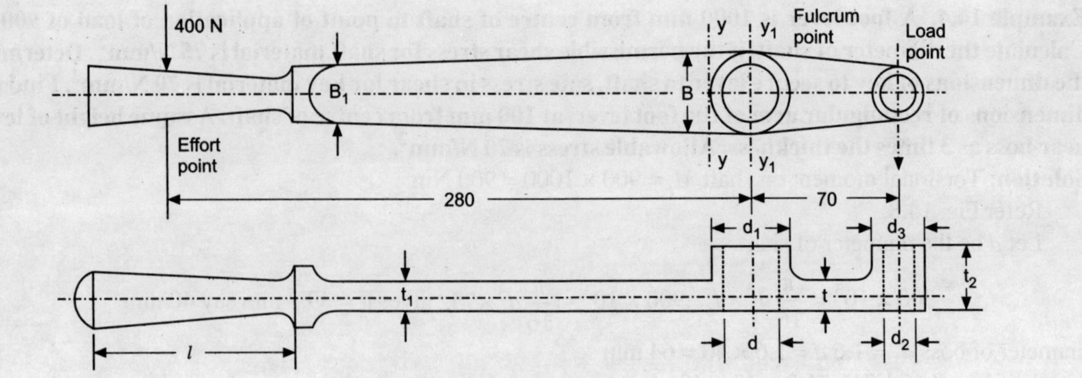

Fig. 14.9 *Hand lever*

Also considering bearing of the pin in the lever at fulcrum

$$t \times d \times p_b = 2000,\ t \times d \times 10 = 2000, \text{ gives } td = 200$$

From the two, we get d = 14.1 mm, say 15 mm

$$t = 14.1 \text{ mm, say } 15 \text{ mm}$$

Thus thickness of boss at fulcrum, t = 15 mm = length of boss.

Section of Lever: Outer diameter of boss, $d_1 = 2d = 2 \times 15 = 30$ mm. The maximum bending moment is at section $y_1 - y_1$ and is probably weakest, but for convenience we will design the section of lever at *yy* and check at section y_1y_1.

$$\text{Bending moment at section } y - y = P\left(280 - \frac{d_1}{2}\right) = 400\left(280 - \frac{30}{2}\right) = 106 \text{ Nm}$$

Let t_1 be the thickness of lever and B the height at section *yy*, then equating the bending moment to resisting moment.

$$M_b = f_b \times Z,\ 106 \times 10^3 = 70 \times 1/6\, t_1 B^2$$

Assuming thickness t_1 of the lever as 10 mm, we get B = 30 mm,

which is alright, as the minimum height of lever at section yy has to be equal to boss diameter.

Thus section of lever t_1 = 10 mm, B = 30 mm at section *yy*.

Now the lever can be checked at the fulcrum point. The figure shows the section at y_1y_1.

$$\text{Modulus of section } Z = \frac{\frac{1}{12}t\,d_1^2 - \frac{1}{12}td^3}{d_1/2} = \frac{\frac{1}{12}\times 15\times 30^3 - \frac{1}{12}\times 15\times(15)^3}{30/2} = 1.96\times 10^3 \text{mm}^3$$

Maximum bending moment at section $y_1\,y_1$ = 400 × 280 = 112 Nm

$$\text{Bending stress induced} = \frac{112\times 10^3}{1.96\times 10^3} = 57 \text{ N/mm}^2$$

which is safe, as it is less than the permissible stress of 70 N/mm^2.

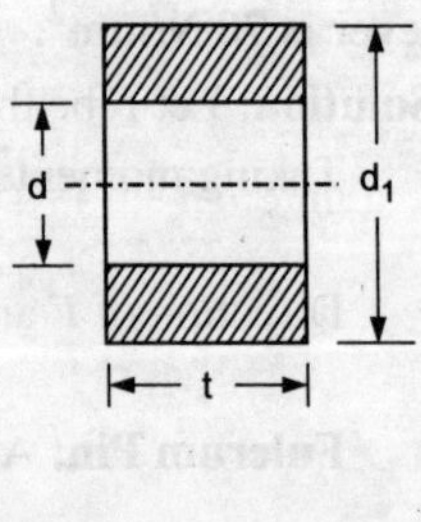

Similarly the pin at the load point may be designed for a load of 1600 N. The designed values obtained will be $d_2 = t_2$ = 13 mm. Thus the thickness t of the lever will be 10 mm and thickness of boss will be 6.5 mm. The diameter d_3 of the boss will be $2\,d_2$ = 26 mm.

The lever will be tapered towards the effort point. The height of the lever near the handle will be $B_1 = 1/2B = 30/2 = 15$ mm.

The handle shape will be chosen from Fig. 14.7. Let the handle be used having length l as 100 mm and minimum diameter 22 mm and maximum as 27 mm.

14.6 CRANKED LEVER FOR HAND OPERATED HOISTING WINCH

One of the commonly used hand levers for hoisting winches, etc. is shown in Fig. 14.10. This lever may be operated either by one person or two persons. The force exerted by one person for continuous operation may be taken as 150 to 200 N with a velocity of about 50 m/min. The maximum force for operation per person may be taken as 400 N. When two persons operate the lever the force will be doubled. The length l of the handle for one operator may be taken as 300 mm and for two operators it may be taken as 500 mm. The lever may be fitted to the shaft with the help of a key or many times the shaft is squared at the end to eliminate key and for easy removal of lever. The side of the square on the shaft end may vary from 30 to 40 mm. The length L is usually from 350 to 400 mm. The height of such levers from the ground to the centre of shaft is usually 1 m to 1.15 m.

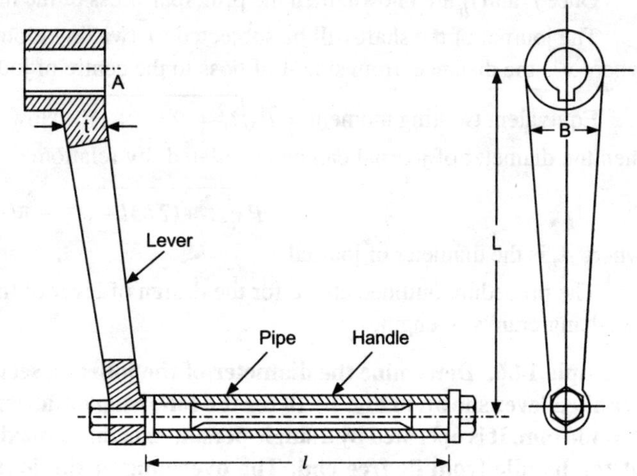

Fig. 14.10 *Cranked lever*

The section of the lever is generally rectangular with average width as 40 mm and thickness 25 mm. The handle of the lever is generally of circular section of diameter varying from 40 to 60 mm. The handle diameter is preferably raised at end portions and pipe is slipped over as shown in figure which prevent hand scoring. The inclination of the lever to the handle is to prevent projected end A to interfere with the operator's hand. This inclination may be made from 10ºC to 15º. The effort on the handle may be taken to act at 2/3 of its length instead 1/2 to be on safer side. Let P be the effort applied by the operator on the handle.

Then maximum bending moment = $(2/3)Pl$, equating the bending moment to the resisting moment

$$(2/3)\ Pl = (\pi/32)\ d^3 f_b \qquad ...(14.13)$$

where d is the diameter of the handle of the lever and f_b is the permissible bending stress of the handle material.

The section of the lever is rectangular and tapers from the boss to the handle. The lever is subjected to constant twisting moment (M_t) of the magnitude $(2/3)\ Pl$ and the varying bending moment which is maximum near the boss. Assuming the lever to be extending to the centre of the shaft, the maximum bending moment will be PL.

At present time, there is insufficient information on the subject of combined bending and twisting of rectangular sections to enable an equivalent bending or twisting moment to be calculated with sufficient accuracy. Therefore, indirect procedure will be adopted, We will design the lever section for 25% more bending moment, and determine the dimensions for width and thickness of lever and then check for the stresses induced. The induced shear stresses should not exceed the permissible values.

Now the relation for bending of lever will be

$$1.25\ PL = (1/6)\ tB^2 \times f_b \qquad ...(14.14)$$

where f_b is the bending stress allowable and t and B are thickness and width of lever near the boss. Once t and B are known, the induced bending stress can be found. The induced stress can be found in the section of lever near the boss by following relations.

For rectangular section of thickness t and width B, Refer Table (3.1)

$$M_t = (2/9)\ Bt^2 \times f_s \qquad ...(14.15)$$

For a square section of side t, Refer Table (3.1)

$$M_t = (2/9)\ t^3\ f_s \qquad ...(14.16)$$

For an elliptical section of major axis B and minor axis t, Refer Table (3.1)

$$M_t = \pi/16\, t^2 B f_s \quad \text{...(14.17)}$$

where $M_t = (2/3)\, Pl$ and f_s is the permissible shear stress.

Once f_s and f_b are known, then the principal stress or the maximum shear stress induced can be checked,

The journal of the shaft will be subjected to twisting moment PL and the bending moment $P\,(2/3\, l + x)$, where x is the distance from side A of boss to the centre of journal.

Equivalent twisting moment $= P\sqrt{L^2 + (2/3l + x)^2}$. Now if we know the value of allowable shear stress, then the diameter of journal can be calculated by relation

$$P\sqrt{L^2 + (2/3l + x)^2} = \pi/16\; d_2^3 \times f_s \quad \text{...(14.18)}$$

where d_2 is the diameter of journal.

The procedure outlined above for the design of lever of the cranked lever may be used for the design of overhang cranks in engines.

Example 14.6. Determine the diameter of the journal, section of lever and the diameter of handle of a cranked lever shown in Fig. 14.10 for the following data, length of handle $l = 300$ mm, length of the lever $L = 400$ mm. It is operated by a single person exerting a maximum force of 400 N at a distance of one-third of the handle from its free end. The overhang of the journal is 100 mm. The allowable stresses are $f_b = 50$ N/mm^2, $f_s = 40$ N/mm^2.

Solution: Let d_2 be the diameter of the journal which can be obtained by relation 14.18.

$$P = 400 \text{ N}, x = 100 \text{ mm}$$

$$P\sqrt{L^2 + (2/3l + x)^2} = \pi/16\; d_2^3 \times f_s$$

$$400\left[(400)^2 + \left(\frac{2}{3}\times 300 + 100\right)^2\right]^{\frac{1}{2}} = \frac{\pi}{16} d_2^3 \times 40$$

gives, $d_2 = 29.4$ mm say 30 mm

Let d be the diameter of handle which can be obtained by relation 14.13.

$$\frac{2}{3}Pl = \frac{\pi}{32}d^3 \times f_b,\ \frac{2}{3}\times 400 \times 300 = \frac{\pi}{32}\times d^3 \times 50$$

gives, $d = 25.3$ mm say 25 mm

Let t be the thickness of the lever and B its width. Assuming width as twice the thickness, i.e., $B = 2t$. Using relation 14.14.

$$1.25\, PL = \frac{1}{6}t B^2 \times f_b,\ 1.25 \times 400 \times 400 = \frac{1}{6}t\,(2t)^2 \times 50$$

gives, $t = 18.8$ say 20 mm

$$B = 2t = 2 \times 20 = 40 \text{ mm}$$

Now it can be checked for the induced bending and shearing stresses.

Turning moment (M_t) on lever section $= \frac{2}{3}\times 400 \times 300 = 80$ Nm

Bending moment (M_b) on lever section $= 400 \times 400 = 160$ Nm

Induced bending stress $f_b = \dfrac{M_b}{1/6\,t B^2} = \dfrac{160\times 10^3 \times 6}{20 \times (40)^2} = 30$ N/mm^2

Induced shearing stress can be checked by relation 14.15,

$$f_s = \frac{M_t}{(2/9)Bt^2} = \frac{80\times10^3\times9}{2\times40\times(20)^2} = 22.5 \text{ N/mm}^2$$

The induced stresses f_b and f_s are much below the allowable values, hence the principle stress will also be less. Hence the designed section above is correct.

14.7 LEVER OF LEVER LOADED SAFETY VALVE

Figure 14.11 shows a lever loaded safety valve. In this safety valve the controlling force for the valve is provided by the weights put at free end. The effect of the weights is transferred at the valve through the lever action.

The lever is pivoted at the fulcrum by a pin to the toggle. The force on the valve is through lever at the point A by putting the weights at the end of the lever. The weights put and the distance of weights from fulcrum is such, so that when the maximum allowable pressure is reached in the boiler the downward forces are overcome by the upward forces due to stream pressure and the valve opens. As a result of valve opening the excess steam escapes out, thus keeping the pressure in the boiler with in safe limits. The design procedure of the lever for the lever type safety valve is explained by an example.

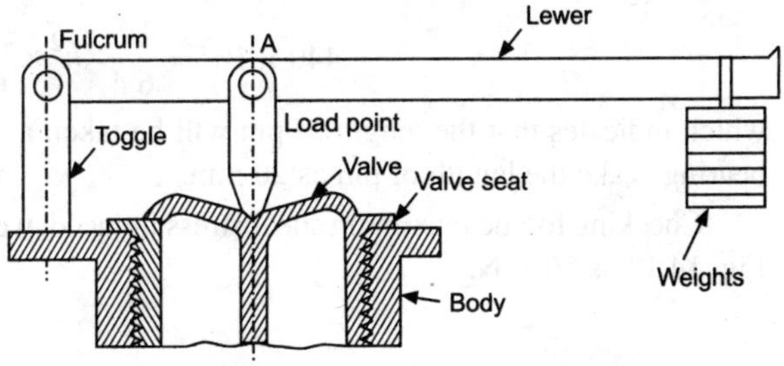

Fig. 14.11 *Lever loaded safety valve*

Example 14.7. Figure 14.12 shows the lever of a lever loaded safety valve. The diameter of the safety valve is 80 mm. It is required to blow off at 1.25 N/mm². Design the lever for the following permissible stresses. Stress in tension, shear and crushing are 35 N/mm², 20 N/mm² and 50 N/mm² respectively. The permissible bearing pressure for the pin may be taken as 20 N/mm². Give dimensioned sketch of the lever.

Solution: The maximum steam load at which the valve blows off

$F = (\pi/4)\, D^2 \times P = (\pi/4) \times (80)^2 \times 1.25 = 6300$ N

Taking moments about fulcrum A, $6300 \times 80 = P \times 800$

Gives $P = 630$ N. Thus the dead load to be kept at C in the form of effort is 630 N. The load F is acting upwards and P is acting downwards, therefore, force at A will be $R = F - P$.

$R = 6300 - 630 = 5670$ N. This reaction will be acting vertically downwards.

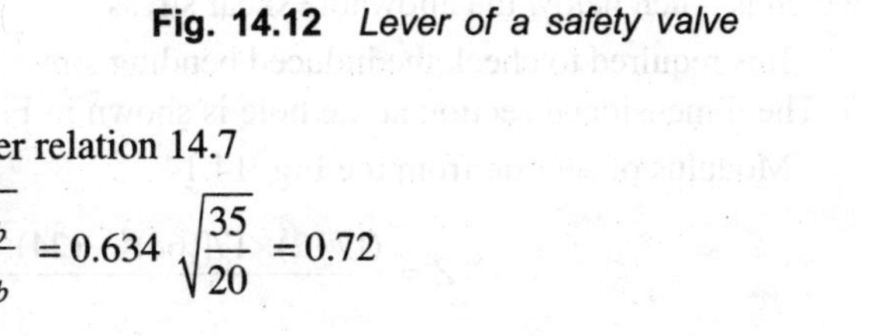

Fig. 14.12 *Lever of a safety valve*

Pins at A and B: Considering the pin at B, as per relation 14.7

$$\frac{t}{d} = 0.634\sqrt{\frac{f_b}{p_b}} = 0.634\sqrt{\frac{35}{20}} = 0.72$$

considering the bearing of the pin

$$d t\, p_b = F, \qquad d\times t\times 20 = 6300, \text{ gives } d\times t = 315$$

From the above two, $d = 21$ mm, $t = 15$ mm.

Thus, diameter of pin = 21 mm, length of pin = 15 mm.

As the load at A is not much different from that at B, take the pin A to be of the same size as that at B, to facilitate the inter-changeability of parts. To increase the life of the lever and to reduce the wear, a gun metal bush of 1.5 mm thickness will be used in the holes at A and B.

Hole diameter in the lever = 21 + 1.5 × 2 = 24 mm.

Diameter of boss = 2 × the hole size in lever = 2 × 24 = 48 mm.

Section of Lever: The lever is subjected to bending moment and shear force. The section is decided by considering bending at a section just near the boss at *B*.

Maximum bending moment at above section $= 630\left[720-\frac{48}{2}\right] = 440$ Nm

Equating the bending moment to the resisting moment.

$$440 \times 10^3 = \frac{1}{6}th^2 \times f_b \text{, taking } t = \frac{h}{4}$$

$$440 \times 10^3 = \frac{1}{6}\frac{h}{4} \times h^2 \times 35 \text{, gives } h = 68 \text{ mm, } t = 17 \text{ mm.}$$

which indicates that the length of pin will be taken a little more than 17 mm to provide adequate length in bearing. Take the length of pin as 20 mm.

Checking for the maximum shear stress induced, we have, maximum shear force from shear force diagram Fig. 14.13 as 5670 N.

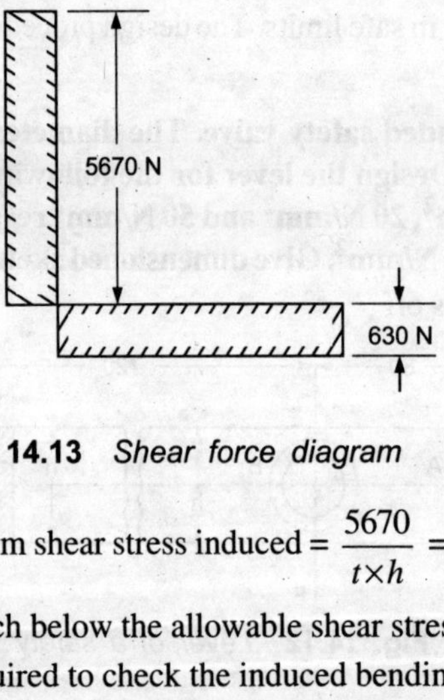

Fig. 14.13 *Shear force diagram*

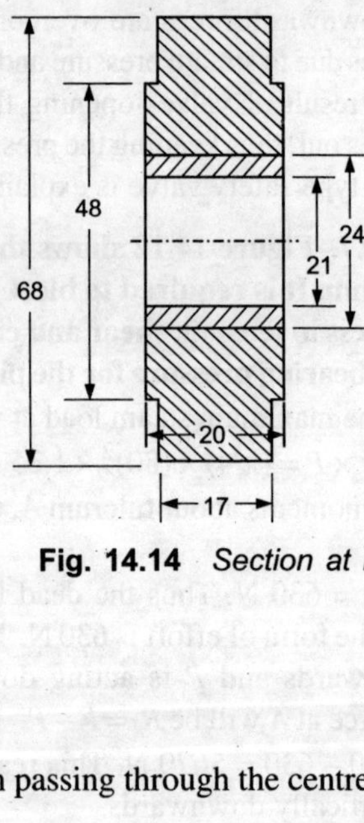

Fig. 14.14 *Section at B*

$$\text{Maximum shear stress induced} = \frac{5670}{t \times h} = \frac{5670}{68 \times 17} = 5 \text{ N/mm}^2$$

which is much below the allowable shear stress.

It is required to check the induced bending stress at the section passing through the centre of the hole at *B*. The dimensioned section at the hole is shown in Fig. 14.14.

Modulus of section from the Fig. 14.14.

$$Z = \frac{(1/12)\times 17[(68)^3 - (24)^3] + 2\times(1/12)\times 1.5\times[(48)^3 - (24)^3]}{68\times\frac{1}{2}}$$

$$Z = 13.2 \times 10^3 \text{ mm}^3.$$

$$\text{Hence the bending stress induced} = f_b = \frac{M_b}{Z} = \frac{630\times 720}{13.2\times 10^3} = 34.5 \text{ N/mm}^2$$

This stress is within limits.

14.8 ROCKER ARM FOR EXHAUST VALVE

The general arrangement of the exhaust valve gear is shown in Fig. 14.15. The maximum load on the valve will determine the bending moment, and hence the section of the arm, The rocker arm is generally of the I-section. For calculating the bending moment, it is assumed that the arm of the lever extends from the point of application of load to the centre of the pivot (fulcrum) of the rocker. This assumption results in a slightly stronger lever near the boss. The cam comes in contact with one end of the rocker arm and this end of the rocker arm has a forked end to receive the roller. The roller is put on a pin and is free to rotate to reduce the wear. No bush is inserted over this pin or roller because after sufficient service the roller is to be discarded due to wear at the profile of roller. The ratio of length to diameter of the fulcrum and roller pin is taken as 1.25. The allowable bearing pressure on these pins is 3 to 6 N/mm^2.

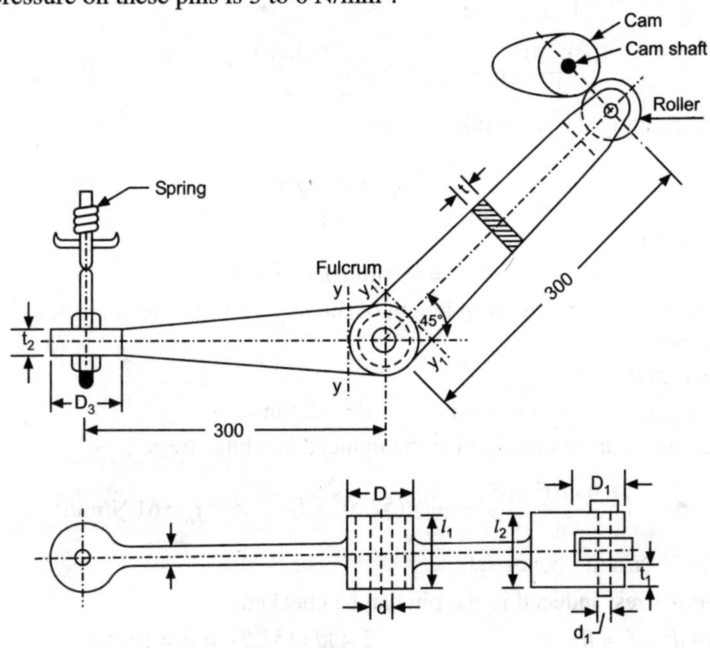

Fig. 14.15 *Exhaust valve gear rocker arm*

To take up the wear and to avoid the replacement of rocker arm, the boss at fulcrum is bushed by a 3 mm phospher bronze bush. The outside diameter of the boss is usually kept as twice the diameter of the pin at fulcrum. The outside diameter at the forked end is also kept twice the pin diameter. The roller diameter is kept slightly larger than the eye diameter of the forked end. The thickness of each fork end is kept as half the diameter of the pin. A slight clearance is kept between the roller and the forked eye so that the roller can move freely, for this reason the pin should be checked for bending as explained in Article 14.2.

The other end of the rocker arm is made circular to receive the tappet, which is a stud with a locknut. With this stud arrangement the tappet clearance can be adjusted. The outside diameter of the circular arm is twice the diameter of the stud and the depth of the section is also equal to twice the diameter of the stud.

Example 14.8. Design a cam lever as shown in Fig. 14.15 operating the exhaust valve of a diesel engine. Following data may be assumed.

Maximum load on the valve = 7000 N, Effective arm length = 300 mm

Roller diameter = 80 mm, Angle between the arms = 135°.

Bearing pressure between the pin and lever = 5 N/mm^2. The rocker arm is made of forged steel for which the allowable stress in tension is 80 N/mm^2. Assume the section of the rocker arm as rectangular.

Solution: The maximum effort P (vertical) to be exerted by the cam in order to open the valve can be calculated by taking moments about the central pin (fulcrum).

$$(P \cos 45°) \times 300 = 7000 \times 300, \text{ gives } P = 9900 \text{ N}.$$

Reaction at the fulcrum pin $R = 7000 + 9900 = 16.9$ kN

Considering the bearing of the fulcrum pin.

$$p_b \times l_1 \times d = R, 5 \times l_1 \times d = 16.9 \times 10^3$$

where l_1 and d are length and diameter of pin,

$$\therefore \quad l_1 \times d = 3380 \text{ mm}^2 \quad ...(i)$$

Considering the pin to be simply supported at the ends and having concentrated load of 16.9×10^3 N at the centre.

Bending moment $= M_b = \frac{RL}{4} = \frac{16.9 \times 10^3}{4} l_1$ equating the bending moment to the resisting moment

$$\frac{16.9 \times 10^3}{4} l_1 = \frac{\pi}{32} d^3 \times f_b, \frac{16.9 \times 10^3}{4} l_1 = \frac{\pi}{32} d^3 \times 60$$

where f_b is allowable bending stress for fulcrum pin.

$$\frac{l_1}{d^3} = 0.00165 \quad ...(ii)$$

From (i) and (ii) we get,

$$l_1 = 90 \text{ mm}, d = 38 \text{ mm}$$

The general design practice is to keep length of pin as 1.25 times the diameter of pin, so that the lever may not rock on the pin.

$\therefore$ taking $l_1 = 1.25\, d$

Then from (i) we get, $l_1 = 65$ mm $d = 52.5$ mm

Now the fulcrum pin can be checked for the induced bending stress.

$$\frac{16.9 \times 10^3 \times 65}{4} = \frac{\pi}{32} \times (52.5)^3 \times f_b, \quad \therefore \quad f_b = 61 \text{ N/mm}^2$$

which is very near to 60 N/mm^2, hence safe.

The average shear stress induced in the pin can be checked.

$$2 \times \pi/4\, d^2 \times f_s = R, \qquad 2 \times \pi/4\, (52.5)^2 \times f_s = 16.9 \times 10^3$$

$\therefore$ $f_s = 3.9$ N/mm^2, which is very much safe.

The external diameter of the boss will be $= 2d = D = 2 \times 52.5 = 105$ mm

Assuming 3 mm thick bush of brass, the internal diameter of the hole in the lever will be

$$= 52.5 + 3 \times 2 = 58.5 \text{ mm}.$$

The section of boss at the fulcrum can be checked for bending. The section of the boss is shown in Fig. 14.16. Maximum bending moment at this section

$$M_b = 7000 \times 300 = 2100 \text{ Nm}$$

The moment of resistance by bush can be neglected, then Z of the section.

$$Z = \frac{(1/12) \times 65[(105)^3 - (58.5)^3]}{\frac{105}{2}} = 99 \times 10^3 \text{ mm}^3$$

Induced bending stress $= \frac{M_b}{Z} = \frac{2100 \times 10^3}{99 \times 10^3} = 21.5$ N/mm^2

Hence safe.

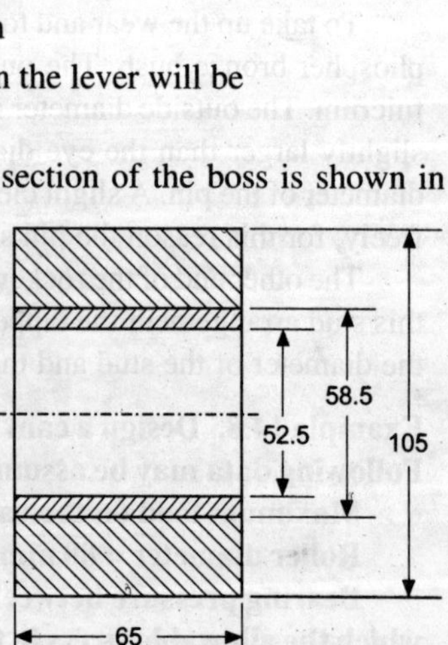

Fig. 14.16 *Section at fulcrum*

Roller Pin: The roller is carried by the fork end by means of pin, which is fixed in the eyes.

Considering the bearing of the pin in roller of length l'_2.

$P = p_b \times d_1 \times l'_2$ $\qquad$ $9900 = 5 \times d_1 \times l'_2$

assuming $l'_2 = 1.5\, d_1$, $\qquad$ $9900 = 5 \times d_1 \times 1.5\, d_1$

gives, $\quad d_1 = 36.5$ mm, $l'_2 = 55$ mm.

As this pin also is supported at both ends in the fork, then maximum bending moment

$$M_b = \frac{Pl'_2}{4} = \frac{9900 \times 55}{4} = 136 \text{ Nm}$$

Induced bending stress in the pin = $\dfrac{M_b}{Z} = \dfrac{136 \times 10^3}{\pi/32(36.5)^3} = 28.5$ N/mm^2 which is very much within the safe limits.

The pin can be checked for the induced shearing stress.

$$f_s = \frac{P}{2 \times (\pi/4)(d_1)^2} = \frac{9900 \times 4}{2 \times \pi \times (36.5)^2} = 4.8 \text{ N/mm}^2 \text{ which is also safe.}$$

The thickness t_1 of each eye will be = $\dfrac{1}{2} l'_2 = \dfrac{1}{2} \times 55 = 27.5$ mm.

Overall diameter of the eye will be $D_1 = 2d_1 = 2 \times 36.5 = 73$ mm.

The outer diameter of the roller should be as large as it is reasonably possible and should be at least 3 mm more than the outer diameter of the eye. In the present case it will be 80 mm.

The dimension $l_2 = 55 + 2 \times 27.5 + 2 \times 1.5 = 113$ mm.

The dimension 1.5 mm is kept as the clearance between the roller and the fork on either side of the roller.

Lever Section: The section just near the fulcrum boss on both the sides will be decided from the consideration of bending stress induced in it. Considering the bending of section at yy or y_1y_1

$$\text{B.M.} = M_b = 7000\left(300 - \frac{105}{2}\right) = 1732.5 \text{ Nm}$$

The lever arm will be of uniform thickness t throughout and depth will be h near the boss and varying towards the ends. The depth of the lever arm on the eye end will be the outside diameter of the eye, i.e., 73 mm. On the stud end side will depend upon the thickness t which is still to be calculated.

Z of the section at yy or $y_1y_1 = \dfrac{1}{6} th^2$

Assuming $h = 100$ mm.

Equating the bending moment to the resisting moment.

$$173.25 \times 10^3 = 80 \times \frac{1}{6} t(100)^2, \quad \text{gives } t = 13 \text{ mm say } 15 \text{ mm}$$

Adjusting Screw: The adjustable tappet screw has to carry 7000 N load in compression. Assume that the screw is made of mild steel for which the allowable crushing stress is 50 N/mm^2.

$$7000 = \frac{\pi}{4} \times d_c^2 \times f_c \text{ where } d_c \text{ is the core diameter of the stud.}$$

$$7000 = \frac{\pi}{4} \times d_c^2 \times 50, \text{ gives } d_c = 13.5 \text{ mm}$$

outside diameter of screw = $\dfrac{d_c}{0.84} = 16$ mm.

A standard stud can be chosen from the Indian standards.

Diameter D_3 at the screw end = $2 \times 16 = 32$ mm

depth at the screw end $t_2 = 2 \times 16 = 32$ mm

Therefore depth of the lever arm at the screw end will be 32 mm.

14.9 BELL CRANK LEVER

In Article 14.8 we have seen levers whose arms are inclined at a certain angle to each other and straight levers. When the angle between two arms is a right angle , then the lever is known as a bell crank lever. They are used in garrage jack; etc. The section of the arms of bell crank lever may be I-form, rectangular, or elliptical. The principal discussed in Article 14.2 may be applied for the design of bell crank levers, The design of such a lever is illustrated by an example.

Example 14.9. A bell crank lever to raise a vertical load is shown in Fig. 14.17. The vertical load to be lifted is 5000 N. The lever consists of steel forgings, turnings, on a pin at the fulcrum. The following data apply both to the pins and the lever.

Allowable stress in tension 75 N/mm²

Allowable stress in shear 65 N/mm²

Permissible bearing pressure for pin 9 N/mm².

Solution: The force at M can be obtained by taking moments about the fulcrum N. Let P be the force at M, then

$$P \times 120 = 5000 \times 480, \quad \therefore \quad P = 20 \text{ kN}$$

The reaction at the fulcrum will be $R = \sqrt{F^2 + P^2}$

$$R = \sqrt{(5000)^2 + (20 \times 10^3)^2} = 20.5 \text{ kN}$$

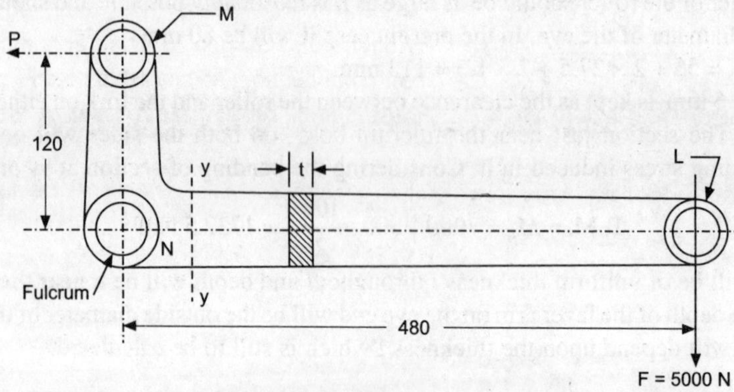

Fig. 14.17 *Bell crank lever*

The pin at fulcrum is in double shear and it should also be designed from bearing consideration and checked under bending. Assuming a ratio of length to diameter of the pin as 1.25. Let us design the pin from bearing consideration and then check it for shearing and bending:

$$R = l \times d \times p_b$$

where l and d are the length and diameter of pin.

$$20.5 \times 10^3 = 1.25\, d \times d \times 9$$

gives, $\quad d = 42.5 \text{ mm.}, \quad l = 53 \text{ mm}$

Shear stress induced in the pin

$$= \frac{20.5 \times 10^3}{2 \times (\pi / 4) \times (42.5)^2} = 7.1 \text{ N/mm}^2$$

which is much below the allowable value, hence safe.

Again considering the pin as a beam, supported at both ends, we can evaluate the induced bending stress. As the nature of the distribution of load cannot be exactly known, we will consider it to be between the uniformly distributed load and the concentrated load.

$\therefore$ Bending moment $= \dfrac{Rl}{6}$, average value between $\left(\dfrac{Rl}{8}\right)$ and $\left(\dfrac{Rl}{4}\right) = \dfrac{20.5\times10^3\times53}{6} = 181$ Nm.

Now equating the bending moment equal to the resisting moment.

$$181 \times 10^3 = \frac{\pi}{32}\times(42.5)^3 \times f_b, f_b = 24 \text{ N/mm}^2$$

which is much below value of 75 N/mm^2, hence safe.

It will be advisable to use brass bush pressed into the boss as a bearing so that renewal will be simple, when wear occurs. Taking the thickness of bush as 3 mm, the diameter of hole in the lever will be 48.5 mm. The section at the fulcrum pin can be checked for failure in bending.

Diameter of boss at fulcrum = 2 × 42.5 = 85 mm.

Modulus of section $Z = \dfrac{53}{12}[(85)^3 - (48.5)^3]\times\dfrac{2}{85} = 5.15 \times 10^3$ mm^3.

Maximum bending moment at fulcrum

$$M_b = 5000 \times 480 = 2400 \text{ Nm}$$

Hence, induced bending stress $= \dfrac{M_b}{Z} = \dfrac{2400\times10^3}{51.5\times10^3} = 46.5$ N/mm^2

which is less than 75 N/mm^2, hence safe.

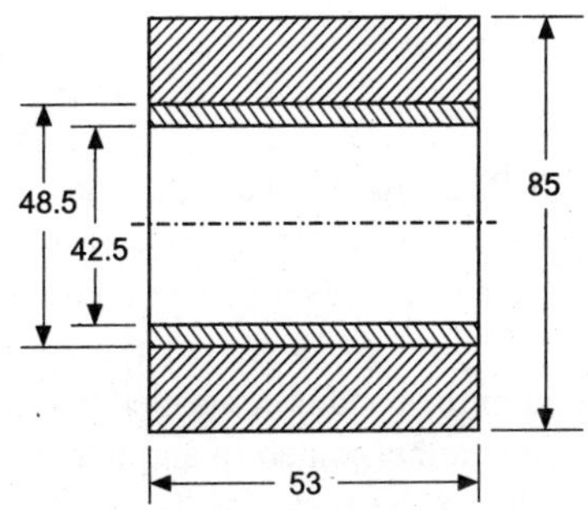

Section of lever: Assume that the lever extends upto the centre of the fulcrum from the point of application of load. This assumption is commonly made and results in a slightly stronger section. Assuming that the height h of the lever is thrice its thickness near the boss. The weakest section of failure will be yy, thus maximum bending moment = 5000 × 430 = 2150 Nm

Equating the bending moment to the resisting moment.

$$2150 \times 10^3 = \frac{1}{6}th^2 \times f_b, \quad 2150 \times 10^3 = \frac{1}{6}t\times(3t)^2\times75$$

gives $t = 40$ mm, $h = 3t = 3 \times 40 = 120$ mm.

Pin at M: The load carried by this pin in a single shear being 20 kN is not different from that carried by the pin at the fulcrum. Therefore the dimensions of pin at M may be taken to be same as that of fulcrum pin to reduce the spares. The dimension of the boss also will be same as those of the central boss.

Pin at L: The pin at end L is in double shear and it should also be designed from bearing consideration and checked for bending. Assuming the ratio of length to pin diameter as 1.25 and the same bearing pressure, we have

$$F = l_1 \times d_1 \times p_b, 5000 = 1.25\, d_1 \times d_1 \times 9$$

gives $d_1 = 21$ mm, $l_1 = 26.6$ mm

Shear stress induced in the pin $= \dfrac{5000}{2\pi/4(21)^2} = 7.25$ N/mm^2

which is much below the allowable value, hence safe.

As this is a fork end, each eye must be at least $0.5l_1$ wide in this case 13.3 mm. Again to provide for wear, chilled phosphor bronze bushes of 3 mm thickness may be provided in the eyes. Then the inner diameter of hole will be = 21 + 2 × 3 = 27 mm.

The pin can be checked under bending. The pin is neither simply supported nor rigidly fixed at its ends. Hence, the distribution of load will be as shown in Fig. 14.3(a). Maximum bending moment M_b for this pin can be obtained by relation 14.4.

$$M_b = \frac{F_t}{4} = \frac{Fl_1}{4} = \frac{5000 \times 26.6}{4} = 33 \text{ Nm}$$

Modulus of section $Z = \frac{\pi}{32}(21)^3 = 0.915 \times 10^3 \text{ mm}^3$

$\therefore$ Induced bending stress $= \frac{33 \times 10^3}{0.915 \times 10^3} = 36 \text{ N/mm}^2$

which is much below the allowable value, hence safe.

The diameter of the fork will be twice the pin diameter, in this case it will be 42 mm. The lever from the fulcrum to the load point and from fulcrum to the effort point will be tapered as shown in figure.

14.10 LEVER FOR THE BRAKE

The two types of brakes in common use are:

1. Block Brake.
2. Band Brake.

Figure 14.18, shows the arrangement of a Block Brake. P is the braking force which is transmitted to the block through the lever which operates with respect to the fulcrum. Assume that the pressure between block and wheel is F, then taking moments, about fulcrum

$$p \times b = F \times a$$

or $$F = P \times b/a \quad ...(14.19)$$

The length a and b of the lever will be fixed from convenience point of view. Once the forces acting on lever are known, the design of the lever can be done as explained earlier.

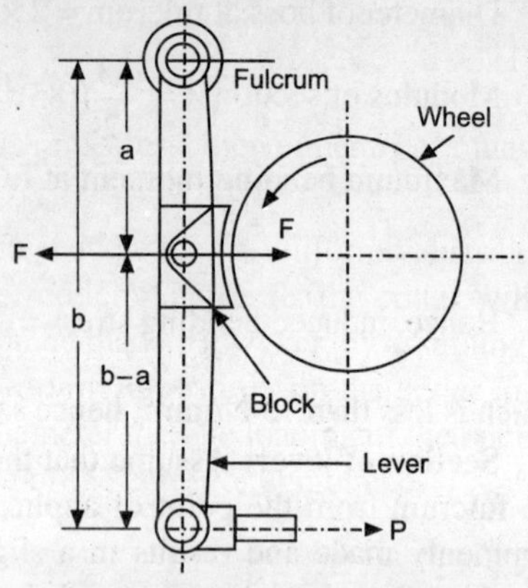

Fig. 14.18 *Block brake*

Figure 14.19, shows the arrangement of a band brake. One end of the band is connected to one arm of the lever and the other end is either joined to another arm of the same lever, or to a pin fixed on the frame of the machine. The balancing resistance is created by friction between the band and rim of the wheel. When the brake is in action there are tensions T_1 and T_2 in the straight part of the band RS and MD respectively, of which T_1 is greater. The difference of turning action of these forces on the lever is balanced by a force P applied to the lever at L, by the operator.

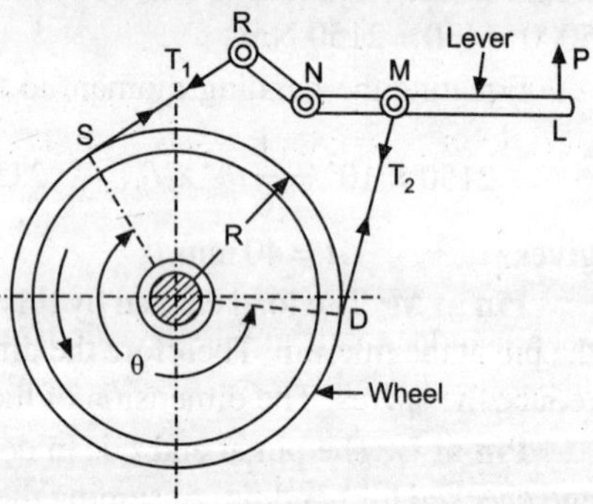

Fig. 14.19 *Band brake*

The resisting torsional moment due to friction of band wheel

$$= (T_1 - T_2)R \quad ...(14.20)$$

where R is the effective radius of wheel, i.e., the radius measured to the middle or the thickness of the band.

From belt tensions, $\frac{T_1}{T_2} = e^{\mu\theta}$ ...(14.21)

where θ is the angle of lap between belt and circumference of wheel.

From relations (14.20) and (14.21), the values of T_1 and T_2 can be calculated. Thus, once the forces acting on lever are known, the lever can be designed as explained earlier.

Figure 14.20 shows the various arrangements of levers for the band brake.

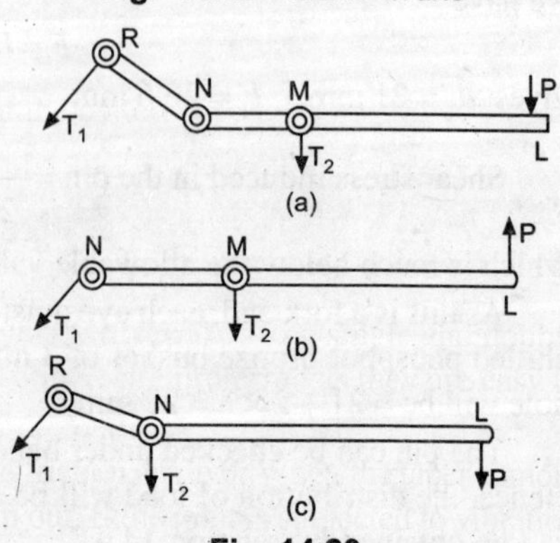

Fig. 14.20

Example 14.10. A drum. 600 mm diameter, of band brake sustains a torsional moment of 800 Nm. The coefficient of friction between drum and belt, when the brake is applied reaches upto 0.35 and the safe stresses for the materials of the brake lever and fulcrum pin in bending, shear and bearing are 60, 40 and 20 N/mm^2 respectively. Design the brake lever and pin and show the dimensions on a neat sketch. The data required of the lever is shown in Fig. 14.21.

Solution: When force P is applied at the end of lever, tensions T_1 and T_2 are induced in the belt. T_1 is the tension in tight side and T_2 in the slack of the belt. Then by relation 14.21, we get

$$\frac{T_1}{T_2} = e^{\mu\theta}$$

where $\theta = 240 \times \frac{\pi}{180} = \frac{4}{3}\pi$ radians

$\mu = 0.35$

$$\therefore \quad \frac{T_1}{T_2} = e^{0.35 \times 4/3\,\pi} = 4.365$$

or $T_1 = 4.365\, T_2$

Now, torsional moment

$$= (T_1 \times R - T_2 R) = (T_1 - T_2)\, R$$

$$800 \times 10^3 = (4.365\, T_2 - T_2)\, 300, \text{ gives } T_2 = 790 \text{ N}$$

and $T_1 = 4.365 \times 790 = 3460$ N

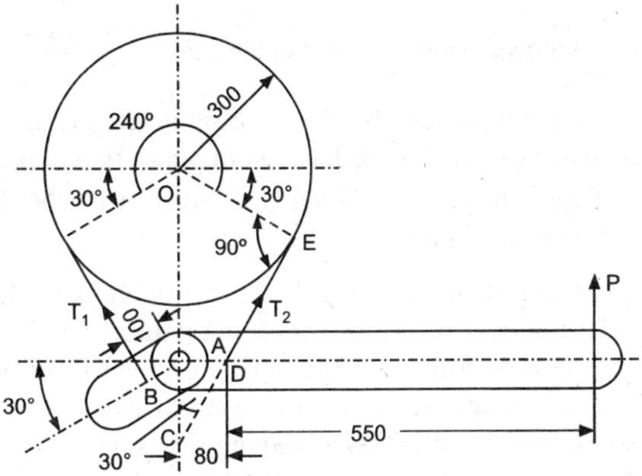

Fig. 14.21 *Band brake*

Now force P should be estimated. The different angles are shown in the figure as the fulcrum point is exactly below the centre of the drum, it is established from triangle OCE that, $\angle ACD = 30°$, $\angle CDA = 60°$. Since portion AB is set at 30° to the horizontal, it is obvious that tension T_1 acts perpendicular to AB. Taking moments about the fulcrum A, we can get the force P.

$$T_1 \times 100 = T_2 \cos 30° \times 80 + P \times (550 + 80)$$

or

$$3460 \times 100 = 790 \times 0.866 \times 80 + P \times 630$$

gives,

$$P = 470 \text{ N}$$

Lever section: Maximum bending moment = 3460 × 100 = 346 Nm. The bending moment has been calculated by assuming that the lever extends upto the centre of the fulcrum, which gives slightly a stronger lever. Assume the section of the lever to be rectangular with depth h three times its thickness t.

Equating the bending moment to the resisting moment,

$$346 \times 10^3 = \frac{1}{6} th^2 \times f_b\,,\ 346 \times 10^3 = \frac{1}{6} t(3t)^2 \times 60$$

gives, $t = 15.7$ mm say 16 mm, $h = 3t = 3 \times 16 = 48$ mm.

Fulcrum Pin: The resultant force acting on the pin is found out by resolving the three forces T_1, T_2, and P into their horizontal and vertical components.

$$\text{Horizontal component} = T_1 \cos 60° - T_2 \cos 60° = 3460 \times \frac{1}{2} - 790 \times \frac{1}{2} = 1335 \text{ N}$$

$$\text{Vertical component} = P + T_2 \cos 30° + T_1 \sin 60° = 470 + 790 \times 0.866 + 3460 \times 0.866 = 4150 \text{ N}$$

$$\text{Resultant force on pin} = \sqrt{(1335)^2 + (4150)^2} = 4360 \text{ N}$$

Considering the bearing of pin

$$p_b \times l \times d = 4360,\ 20 \times 16 \times d = 4360 \text{ gives, } d = 13.6 \text{ mm}$$

Therefore, the pin is 13.6 mm in diameter and 16 mm long in the lever. The pin can be checked in shear.

$$\text{Induced shear stress} = \frac{4360}{2 \times \pi/4\,(13.6)^2} = 15 \text{ N/mm}^2, \text{ hence safe.}$$

The thickness of the fork in which pin is put is half the pin length, i.e., 8 mm. Thus total length of pin

$$= 16 + 2 \times 8 + 1.5 \times 2 = 35 \text{ mm}$$

where 1.5 mm is the clearance on either side of the lever in fork. Diameter of fork is double the pin diameter, i.e., 27.2 mm. The pin can now be checked under bending. The maximum bending moment by relation 14.4 will be

$$M_b = \frac{F_t}{4} = \frac{4360 \times 16}{4} = 17.44 \text{ Nm}$$

$$\text{Induced bending stress in the pin} = \frac{M_b}{Z} = \frac{17.44 \times 10^3}{\pi / 32(13.6)^3} = 100 \text{ N/mm}^2$$

As the induced bending stress in the pin is more than the permissible value, therefore increase the diameter of pin. Taking diameter of pin as 16.5 mm, then the stress induced will be only 58 N/mm^2 hence safe.

Now the section of lever at the fulcrum can be checked for the induced bending stress, as explained in the previous examples.

Example 14.11. The Fig. 14.22 shows block brake with self energizing action. Design the fulcrum and suitable section of the lever. Assume the following particulars. For lever the tensile and crushing stresses are 75 N/mm^2, for pin the tensile stress is 75 N/mm^2, shear stress and the allowable bearing pressure is 20 N/mm^2. Maximum braking force is 5000 N.

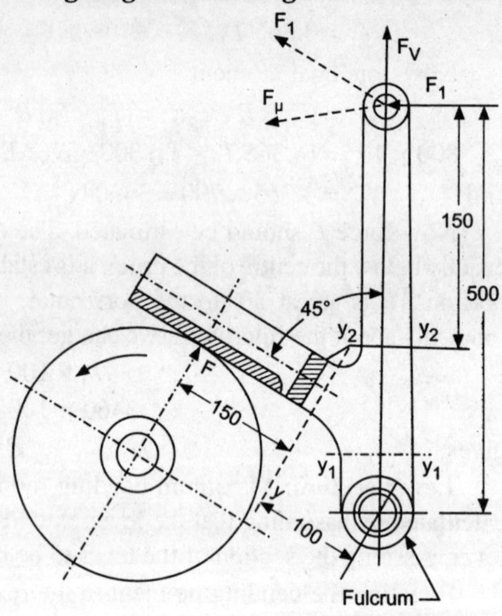

Fig. 14.22 Block brake

Solution: $F = 5000$ N

Let F_1 is the force applied at the lever end as shown in figure.

Taking moments about fulcrum point, we have

$$5000 \times 250 = F_1 \times 500$$

gives,

$$F_1 = 2500 \text{ N}$$

The resultant load on the fulcrum can be obtained by the relation

$$R = \sqrt{F^2 + F_1^2 + 2FF_1 \cos\theta}$$

$$= \sqrt{(5000)^2 + (2500)^2 + 2 \times 5000 \times 2500 \times \cos 45°}$$

$$= 7000 \text{ N}$$

Pin at Fulcrum: Now let l = length of pin,

d = diameter of pin.

Using the equation 14.7, we can find out the ratio of $\frac{l}{d}$,

$$\frac{l}{d} = 0.634\sqrt{\frac{f_b}{p_b}} = 0.634\sqrt{\frac{75}{20}} = 0.634 \times 1.94 = 1.24$$

Considering bearing of the pin, we get

$$l \times d \times p_b = R, l \times d \times 20 = 7000$$

or

$$1.24\, d \times d = 350,\ d = 16.85 \text{ mm say 20 mm.}$$

$$l = 26 \text{ mm}$$

The section of the lever can be checked under bending at *y-y*. Maximum bending moment = 5000 × 150 = 750 Nm.

A brass bush of 3 mm will be put in the fulcrum, so that wearing of the lever is avoided. Thus hole size in the lever = 20 + 3 × 2 = 26 mm. Boss diameter at fulcrum will be twice the size of the hole in the lever, in this it will be 52 mm. Same will be the height of the lever provided at section *y-y* and Y_1-Y_1. The bending moment will be maximum at *y-y*, we can find the thickness *t* of the lever.

$$750 \times 10^3 = \frac{1}{6}th^2 \times f_b, 750 \times 10^3 = \frac{1}{6}t \times (52)^2 \times 75$$

gives, $t = 22.2$ mm, say 23 mm,

Thus, provide the thickness of lever as 23 mm with length of boss 30 mm at fulcrum. The increased length of the boss is provided to get sufficient length in bearing. The section y_1-y_1 will not be stressed as much as section *y-y*. The bending moment at section of the fulcrum can be found and checked for the induced bending stress, this is left as an exercise for the student.

The bending moment at section y_2-y_2 is $2500 \times 150 = 375$ Nm for this bending moment the height of the lever can be found out, keep the thickness of the lever as 23 mm.

If the load F_1 is inclined to the lever arm as shown in the figure by the dotted line, then F_1 may be resolved into horizontal F_H and vertical F_V components. The effect of the horizontal component F_H will be similar as discussed for F_1 and for vertical component F_V is to cause direct tensile stress in the lever arm.

The lever in height will be tapered from the fulcrum to the end where effort F_1 is applied.

Example 14.12. A compound lever for braking car wheels is shown in Fig. 14.23. The compound lever is to be designed for the following data:

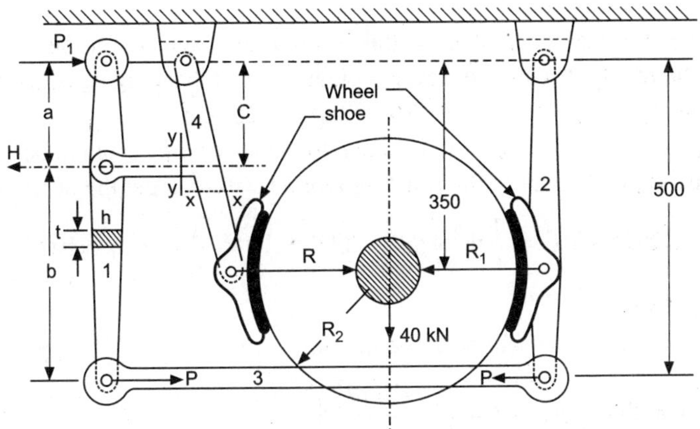

Fig. 14.23 *Lever arrangement for braking car wheels*

Wheel load $F = 40$ kN
Coefficient of friction between wheel and rail $\mu = 0.25$
Coefficient of friction between wheel and shoe $\mu' = 0.30$
Permissible stress in tension of pin = 100 N/mm².
Permissible bearing pressure between pin and lever = 40 N/mm².
Allowable bending stress in lever material = 100 N/mm².
Take $R = R_1$ and $a = b = c$.

Solution: R_2 = radius of car wheel,

Limiting braking force to be applied = $40 \times 10^3 \times 0.25 = \mu F = 10$ kN

Limiting torsional moment = $10 \times 10^3 \times R_2$, and this is to be equal to $2 R\mu' R_2$

$\therefore \quad 10 \times 10^3 \times R_2 = 2R \times 0.3 \times R_2$, gives $R = 16.67$ kN $= R_1$.

Let P be the pull in the lever 3. Then taking moments about the point, we have

$$P \times 500 = R_1 \times 350, P = 16.67 \times 10^3 \times \frac{35}{50} = 11.7 \text{ kN}$$

Since it is given that $a = b$, therefore $P = P_1$, and $H = 11.7 \times 10^3 + 11.7 \times 10^3 = 23.4$ kN

Again taking the moment about the above point, we have

$$H \times C = R \times 350, C = \frac{16.67 \times 10^3 \times 350}{23.4 \times 10^3} = 250 \text{ mm}$$

The pin at H of lever 1 can be designed for a load of 23.4 kN.

Using the relation 14.7, we can find the ratio of length of pin to diameter of pin.

$$\frac{l}{d} = 0.634 \sqrt{\frac{f_b}{p_b}} = 0.634 \sqrt{\frac{100}{40}} = 1.0$$

Again considering the bearing of the pin, we have

$$p_b \times d \times l = H, 40 \times d^2 = 23.4 \times 10^3, \text{ gives } d = 24.2 \text{ mm say } 25 \text{ mm}$$

$\therefore$ length of pin $l = 25$ mm, diameter of pin $d = 25$ mm

We know that $a = b = c = 250$ mm

The section of the pin can be checked under double shear.

$$\text{Induced shear stress} = \frac{23.4 \times 10^3}{2 \times \pi / 4 (25)^2} = 23.8 \text{ N/mm}^2, \text{ hence safe.}$$

The pin can also be checked under bending, this is left as an exercise for the student.

Lever Section: The required section of lever is obtained from bending consideration.

maximum bending moment $P \times b = 11.7 \times 10^3 \times 250 = 2925$ Nm.

Let t be the thickness of lever, which will be equal to the length of the pin in bearing, i.e., 25 mm..

By equating the bending moment to the resisting moment, we can get the depth of the lever.

$$2925 \times 10^3 = \frac{1}{6} th^2 \times fb', \quad 2925 \times 10^3 = \frac{1}{6} \times 25 \times h^2 \times 100$$

gives $h = 83.7$ mm say 85 mm.

The weakest section of the lever will be at the fulcrum.

$$\text{Modulus of section at the fulcrum } Z = \frac{1}{6}\left(\frac{th^3 - td^3}{h}\right)$$

$$Z = \frac{1}{6}\left[\frac{25 \times (85)^3 - 25 \times (25)^3}{85}\right] = 29.3 \times 10^3 \text{ mm}^3$$

$$\text{Induced bending stress} = \frac{M_b}{Z} = \frac{P \times b}{29.3 \times 10^3} = \frac{11.7 \times 10^3 \times 250}{29.3 \times 10^3} = 90 \text{ N/mm}^2$$

which is less than 100 N/mm^2, hence safe.

The pins at the ends of lever 1 can be designed for a load of 11.7 kN. Considering the pin under bearing,

$$d_1 \times t \times p_b = P, d_1 \times 25 \times 40 = 11.7 \times 10^3$$

gives $d = 11.7$ mm say 15 mm

The length of the pin will be same, because thickness of lever does not change.

Pins for reactions R, R_1 and H may be made 25 mm in diameter, because these forces are nearly equal and extra store is not to be kept. All other pins will be of diameter 15 mm. All links and lever will be made of the same thickness t, i.e. 25 mm. The weakest section for the lever 2 may be kept identical as that for lever 1.

Lever 3, can be designed from tension consideration. Let t is the thickness and h_1 be the depth of the lever 3, then

$$h_1 \times t \times f_t = P, h_1 = \frac{11.7\times10^3}{100\times25} = 4.7 \text{ mm say 5 mm}$$

The ends of the lever can be given the shape as shows in figure.

For lever 4, section at y-y may be fixed from tension consideration. Again let for this lever t be thickness and h_2 be the depth at this section, then

$$h_2 \times t \times f_t = H,\ h_2 = \frac{23.4\times10^3}{100\times25} = 9.4 \text{ mm say 10 mm}$$

Again for lever 4, section ar x-x may be fixed from bending consideration.

$$\text{Maximum bending moment} = R\left(350 - C - \frac{h_2}{2}\right) = 16.67 \times 10^3 \times (350 - 250 - 5) = 1580 \text{ Nm}$$

$$M_b = 16.67 \times 10^3 \times 95 = 1580 \text{ Nm}$$

By equating the bending moment to the resisting moment,

We can find the depth of the lever 4. Say the depth of the lever is h_3.

$$1580 \times 10^3 = (1/6)\, th_3^2 \times f_b. \quad 1580 \times 10^3 = 1/6 \times 25 \times h_3^2 \times 100$$

gives $h_3 = 33.6$ mm say 35 mm.

The other dimensions of the levers can be fixed while making the drawing.

Example 14.13. A band brake is required to balance a torsional moment of 300 Nm on the drum shaft. The drum is to be made 360 mm in diameter and keyed to the shaft. The band is to be lined with asbestos as fabric having coefficient of friction 0.3 and the arc of contact 270°. The brake lever and pins are to be made of mild steel. Show the calculations for the dimensions of the steel lever and pins. Permissible stresses for tension and crushing for steel used are 70 N/mm² and shear stress is 55N/mm².

Solution: Refer Fig. 14.24.

Torsional moment = 300 Nm

$$\theta = 270^\circ$$

$$\mu = 0.3$$

$$\theta = 270 \times \frac{\pi}{180} = \frac{3}{2}\pi \text{ radians}$$

$$\frac{T_1}{T_2} = e^{0.3 \times 3/2\, \pi} = 4.121$$

$$\text{Resisting torque} = (T_1 - T_2)R = 300 \times 10^3$$

$$(T_1 - T_2) \times 180 = 300 \times 10^3,$$

$$T_1 - T_2 = 1670 \text{ N}$$

From above $4.121\, T_2 - T_2 = 1670$

gives, $T_2 = 535$ N, $T_1 = 2200$ N

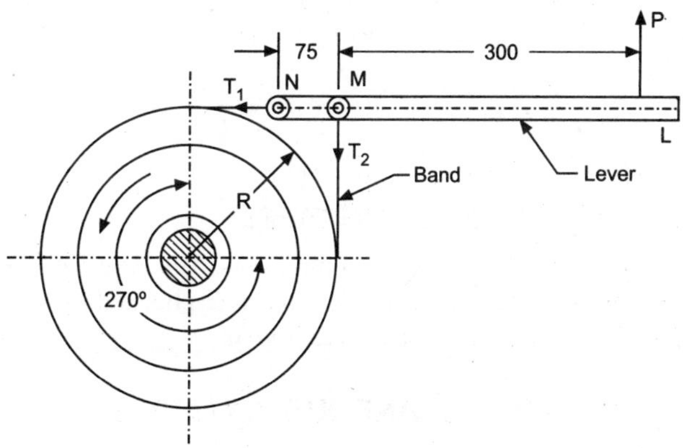

Fig. 14.24 *Band brake*

Design of pin: Both the pins at M and N will be designed for the force T_1 to avoid the store. Assuming that the bearing pressure between pin and lever to be 9 N/mm². Considering bearing of the pin and $l/d_1 = 1.25$, we have

$$p_b \times l \times d_1 = T_1, \quad 9 \times 1.25\, d_1 \times d_1 = 2200$$

gives, $d_1 = 14$ mm, $l = 17.5$ mm

Thus, length of pin $l = 17.5$ mm

diameter of pin $d_1 = 14$ mm

The lever provides the eye hole and connectors at band provide the fork. The thickness of each eye end will be half the length of pin in the fork, i.e., 9 mm. Providing a clearance of 1.5 mm on either side of the lever, the length of pin under bending will be

$$= 17.5 + 9 \times 2 + 2 \times 1.5 = 38.5 \text{ mm}$$

The pin will be in double shear. This pin can be checked under shear.

$$\text{Induced shear stress in the pin} = \frac{2200}{2 \times \pi / 4 (14)^2} = 7.2 \text{ N/mm}^2 \text{ hence safe.}$$

The pin can also be checked under bending.

$$\text{Bending moment } M_b \text{ by relation (14.4)} = \frac{Fl}{4} = \frac{2200 \times 38.5}{4} = 21 \text{ Nm}$$

$$\text{Induced bending stress} = \frac{M_b}{Z} = \frac{21 \times 10^3}{(\pi / 32)(14)^3} = 67 \text{ N/mm}^2$$

which is less than the allowable value, hence safe.

Lever section: By taking moments about fulcrum, we can find the effort P applied.

$$P \times 375 = 535 \times 75, \quad P = 107 \text{ N}$$

The lever will be considered as a cantilever supported at the fulcrum N. Neglecting the effect of T_2 on the lever for computing bending moment on MN, the error is on the safer side.

Maximum bending moment at N due to P at L.

$$107 \times 375 = 40.2 \text{ Nm}$$

Let h be the height of lever and t its thickness. Assuming the thickness of lever as half its height. Equating the bending moment to resisting moment, we get

$$40.2 \times 10^3 = \frac{1}{6} th^2 \times f_b, \quad 40.2 \times 10^3 = \frac{1}{6} t \times (2t)^2 \times 70$$

gives $t = 9.5$ mm say 10 mm

$h = 2t = 2 \times 10 = 20$ mm

The boss will be made at pin joints. The length of boss will be 17.5 mm and diameter of boss $2d_1$, i.e. 28 mm. Now the section at fulcrum can be checked for induced bending stress as done in previous examples. This is left as an exercise for the students.

14.11 MISCELLANEOUS EXAMPLES

Under this we will consider some examples, involving the principles of design of lever.

Example 14.14. Figure 14.25 shows a safety device. The door opens when the total load P_1 on it reaches 20 kN. Design the bell crank lever. If the permissible bending stress for the lever and pin is 75 N/mm² and the bearing pressure for the pin may be taken as 12.5 N/mm². The shorter side of the bell crank lever is 120 mm and longer side is 480 mm.

Also suggest the procedure for designing the bolts required to fasten the fulcrum bracket.

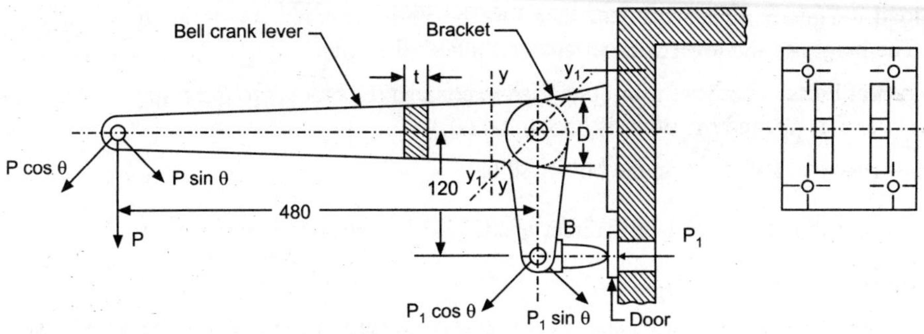

Fig. 14.25

Solution: Taking moments about the fulcrum, we have

$$20 \times 10^3 \times 120 = P \times 480, \qquad \therefore \quad P = 5000 \text{ N}$$

Resultant load on fulcrum can be obtained by relation 14.3.

$$R = \sqrt{P^2 + P_1^2} = \sqrt{(5000)^2 + (20 \times 10^3)^2} = 20.5 \text{ kN}$$

The ratio of the length of pin in bearing to diameter of pin can be obtained by relation 14.7.

$$\frac{l}{d} = 0.634\sqrt{\frac{f_b}{p_b}} = 0.634\sqrt{\frac{75}{12.5}} = 1.57$$

Considering bearing of the pin, we have

$$p_b \times l \times d = R,\ 12.5 \times l \times d = 20.5 \times 10^3,\ l \times d = 1640$$

$$\therefore \quad 1.57\, d \times d = 1640 \qquad d = 32.4 \text{ mm say } 32.5 \text{ mm}$$

$$l = 50 \text{ mm}$$

A brass bush of 3 mm thickness will be inserted in the lever to avoid replacement of lever. Thus hole size in lever will be.

$32.5 + 3 \times 2 = 38.5$ mm. Diameter of boss will be twice the diameter of pin, i.e., 65 mm.

Assuming height of lever section (tangential to boss) as the diameter of boss, i.e., $h = 65$ mm.

Assuming that the lever extends upto the centre of the fulcrum, the maximum bending moment will be

$$M_b = 20 \times 10^3 \times 120 = 2400 \text{ Nm}$$

By equating the bending moment to the resisting moment, we can get the thickness t of the lever.

$$2400 \times 10^3 = \frac{1}{6} t h^2 \times f_b, \qquad 2400 \times 10^3 = \frac{1}{6} \times t \times (65)^2 \times 75$$

gives, $\quad t = 46.8$ mm

Increase the value of h (tangential to boss) to 75 mm then

$t = 36$ mm. This gives 7 mm boss length on either side to make adequate bearing.

Checking the weakest section, say at 45°, we have

$$Z = \frac{1}{12} \frac{lD^3 - ld^3}{D/2} = \frac{1}{6} \times \frac{50}{65}[(65)^3 - (38.5)^3] = 27.7 \times 10^3 \text{ mm}^3$$

X-sectional area near the boss = 50 (75 – 36) = 1950 mm^2.

Now resolving the force of 20 kN into two components, parallel and normal to the section under consideration. 20 cos θ causes moment with lever arm (120 cos θ) and 20 sinθ causes direct stress. Here θ = 45°.

$$\therefore \quad \text{Induced stress} = \frac{20 \times 10^3 \times 120 \times \cos^2 45°}{27.7 \times 10^3} + \frac{20 \times 10^3 \sin 45°}{1950}$$

$$= 50.8 \text{ N/mm}^2, \text{ which is less than } 75/\text{mm}^2 \text{ hence safe.}$$

As the load at A and B are nearly equal, therefore the diameter of pin and its length in bearing will be same at A and B. The height of the lever will be tapered on both the arms.

Fulcrum bracket bolts: There are four bolts used to connect the bracket to the column. There are two forces acting, one is vertical (P) and the other is horizontal (P_1).

Horizontal Forces P_1 : Its effect on the bolts will be

(*a*) The bracket bolts are subjected to direct tensile load, Which is the same for each bolt, i.e., $\frac{P_1}{4}$ in this case.

Vertical Force P : Its effect on the bolt will be

(*a*) The bolts are subjected to tensile loading due to tilting of the bracket about the lower edge.

(*b*) Direct shear load which is the same for each bolt, i.e., $\frac{P}{4}$ in this case.

This is a case of eccentric loading of bolted joints. Once the permissible stresses are known and resultant load is calculated on the bolt heavily loaded, the size of the bolts can be determined. See (10.3) for eccentric loading of bolted joints.

Example 14.15. A cross lever for a double cylinder, double acting pump is shown in Fig, 14.26. The force of 5000 N acts downwards, while that of 7000 N acts upwards. Design the lever X- section and determine the size of pins at A, B, C and D. The permissible stress in tension and crushing for pins is 50 N/mm² and for lever is 200 N/mm². The permissible stress in shear for pins is 40 N/mm² and the bearing pressure is 20 N/mm².

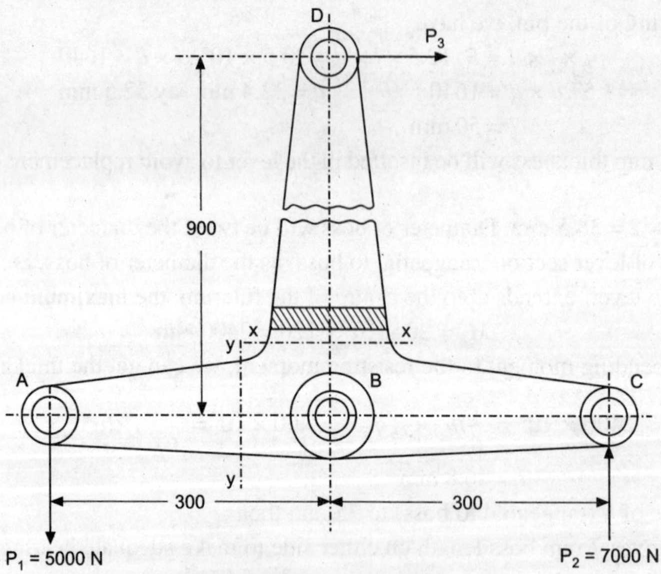

Fig. 14.26 *Cross levers*

Solution: Let P_3 be the effort applied at end D of lever,

Taking moments about the fulcrum B of the lever, we get

$$P_3 \times 900 = 5000 \times 300 + 7000 \times 300, \text{ gives } P_3 = 4000 \text{ N}$$

When one side of the pump does not work, the worst condition arises at that time, because the effort required increases at that time.

So under the worst condition, taking the moment about the fulcrum

$$P_3 \times 900 = 7000 \times 300, \quad \text{gives } P_3 = 2340 \text{ N}$$

In the worst condition, the reaction at the fulcrum will be = $\sqrt{(2340)^2 + (7000)^2} = 7400$ N

When both sides operate, the pin at B has a load of = 7000 – 5000 = 2000 N

Resultant force on pin at $B = \sqrt{(2000)^2 + (4000)^2} = 4500$ N

We find from above that the fulcrum pin is heavily loaded when one side of the pump does not work. Therefore the pin at A, B, C, and D are to be designed for the following forces.

Pins	Load (N)
A	5000
B	7400
C	7000
D	4000

From this we find that the pins at B and C can be designed for load of 7400 N and pins at A and D can be designed for a load of 5000 N. The bearing pressure for all pins is 20 N/mm^2.

Pins at B and C: By using the relation 14.7, we can find the ratio of length to diameter of pin.

$$\frac{l}{d} = 0.634\sqrt{\frac{f_b}{p_b}} = 0.634\sqrt{\frac{50}{20}} = 1.0$$

Now considering the bearing of the pin, we get

$$p_b \times l \times d = 7400, \qquad 20 \times d \times d = 7400$$

gives, $d = 19.3$ mm say 20 mm, $l = 20$ mm

A bush of 3 mm thickness will be inserted in the lever to increase its life. Thus the size of the hole in the lever will be 26 mm and the outside diameter of the boss will be twice the hole size, in this case 52 mm.

It should be remembered that the direction of the load will be reversed, consequently the loads will be changed, hence the pins at A and C must be identical. Thus, pins at A, B and C will be identical.

To avoid the extra store, the pin at D will also be of the same size. Thus, the size of the holes in the lever at all point will be identical.

Vertical Arm: Considering that the vertical arm extends upon the centre of the fulcrum, the maximum bending moment.

$$= 4000 \times 900 = 3600 \text{ Nm}$$

Equating the bending moment to the resisting moment at the section x-x, we get

$3600 \times 10^3 = (1/6)\, th^2 \times f_b$ where t is thickness and h is the depth of the lever. Assume the depth is three times the thickness.

$\therefore \quad 3600 \times 10^3 = (1/6) \times t \times (3t)^2 \times 200, \qquad$ gives $t = 23$ mm

depth of lever $h = 3 \times 23 = 69$ mm, say 70 mm.

The depth of the lever will be tapered towards the end D.

At D the maximum outside diameter is 52 mm. The thickness will remain constant.

Because of the three arms of the lever, there is no necessity of checking at fulcrum.

Horizontal Arms: Again considering that the horizontal arm extends upto the centre of the fulcrum, the maximum bending moment will be $7000 \times 300 = 2100$ Nm.

Equating the bending moment to the resisting moment at the section y-y, we get

$$2100 \times 10^3 = \frac{1}{6} t\, h_1^2 f_b$$

The thickness t of the horizontal arm will be same as of vertical, i.e., 23 mm, h_1 is the depth of horizontal arm at section y-y.

$\therefore \quad 2100 \times 10^3 = (1/6) \times 23 \times h_1^2 \times 200, \qquad$ gives $h_1 = 52$ mm.

This depth of the lever will also be tapered towards the ends A and C from bending consideration, but from wearing consideration we have already fixed the boss size, which incidentally are same, therefore in the horizontal arms no tapper will be done.

The designed pin can be checked under bending, and shear considerations, this is left as an exercise to the student.

Example 14.16. Fig. 14.27 Shows a bench shearing machine, calculate

(a) Forces on pins A , B, C and D.

(b) Cross-section of the lever.

(c) Weakest section of the lever.

Size of the section to be sheared off is 5 mm × 3 mm. Maximum operating force is 160 N. The materials to be sheared off is mild steel having an ultimate shearing strength of 400 N/mm^2. Permissible stress for pins, link and lever in tension is 80 N/mm^2 and the allowable bearing pressure is 20 N/mm^2.

Solution: Area to be sheared = 0.5 × 0.3 = 15 mm^2

Maximum force to shear the work = 15 × 400 = 6000 N.

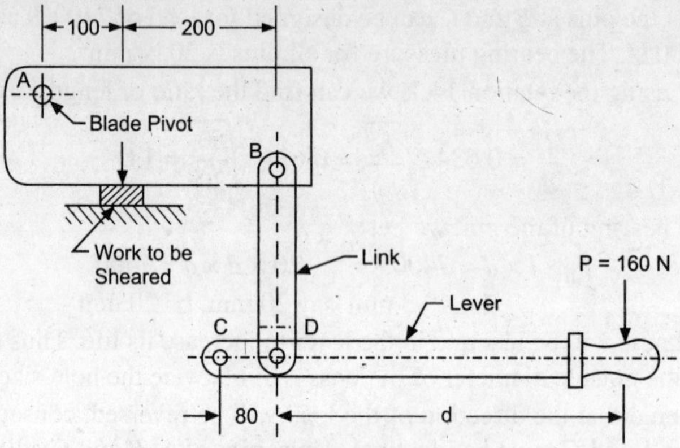

Fig. 14.27 *Bench shearing machine*

Assume that the force in the link is P_1, then taking moments about blade pivot,

$$P_1 \times 300 = 6000 \times 100, \text{ gives } \quad P_1 = 2000 \text{ N.}$$

Force in pin at Blade pivot (A) = 6000 – 2000 = 4000 N

Force in pin at B and D = 2000 N

Force in pin C = 2000 + 160 = 2160 N

By taking moments about the point D, we can find the distance d.

$$2160 \times 80 = d \times 160, \text{ gives } \quad d = \frac{2160 \times 80}{160} = 1080 \text{ mm.}$$

The force in the pins at B, C and D is nearly equal , therefore the size of the pin provided will be same. These pins will be designed for a load of 2160 N.

By using the relation 14.7, we can determine the ratio of length to diameter of pin.

$$\frac{l}{d} = 0.634 \sqrt{\frac{f_b}{p_b}} = 0.634 \sqrt{\frac{80}{20}} = 1.268$$

Considering bearing of the pin.

$$p_b \times l \times d = 2160, \qquad 20 \times 1.268\, d \times d = 2160$$

gives, $\quad d = 10$ mm, $\qquad l = 12.5$ mm

The pin can be checked for induced shear stress.

$$\text{Induced shear stress} = \frac{2160}{2 \times \pi / 4 (10)^2} = 13.7 \text{ N/mm}^2, \text{ hence safe.}$$

The lever can be bushed with brass of 2 mm thickness to increase the life of lever. Thus size of hole will be 14 mm. The diameter of the boss will be twice the hole size, in this case 28 mm.

Links Cross-section: There will be one link having fork at each end, otherwise the link BD will be circular in cross-section. Let d_1 be the diameter of the link and the allowable stress in tension is 80 N/mm^2, then

$$\frac{\pi}{4}d_1^2 \times f_t = 2160, \qquad \frac{\pi}{4}d_1^2 \times 80 = 2160,$$

gives, $d_1 = 5.9$ mm, adopt this diameter as 10 mm because forks are to be made at both ends.

Lever Section: The thickness of lever will be equal to the length in bearing of the pin, *i.e.*, 12.5 mm. Assume that h is the height of the lever.

Maximum bending moment = 2160 × 80 = 172.8 Nm.

equating the bending moment to the resisting moment.

$$172.8 \times 10^3 = \frac{1}{6}th^2 \times f_b, \qquad 172.8 \times 10^3 = \frac{1}{6} \times 12.5 \times h^2 \times 80,$$

gives, $h = 32.2$ mm.

This depth is more than the boss diameter , therefore the depth of the lever at D will be 32.5 mm and thickness 12.5 mm. The boss diameter at B and C will remain same, i.e., 28 mm, thus the lever from D to C can be tapered.

The section of the lever at D can be checked for induced tensile stress.

$$\text{Modulus of section} = \frac{1}{12} \times \frac{[12.5 \times (32.5)^3 - 12.5 \times (14)^3]}{32.5/2} = 2.23 \times 10^3 \text{ mm}^3$$

$$\text{Induced stress} = \frac{172.8 \times 10^3}{2.23 \times 10^3} = 77.5 \text{ N/mm}^2, \text{ hence safe}$$

The part to be grasped by hand will be made 35 mm in greatest and 25 mm in smallest diameter and 125 mm long.

Pin at A: This pin will be designed for a force of 4000 N. Let l_2 and d_2 be the length and diameter of the Pin at A respectively. The ratio l_2/d_2 is already calculated as 1.268.

Considering bearing of the pin,

$$p_b \times l_2 \times d_2 = 4000, \qquad 20 \times 1.268 \times d_2 \times d_2 = 4000$$

gives, $d_2 = 12.5$ mm , $l_2 = 16$ mm.

The pin can be checked under shear stress, supposing the pin is in double shear, then

$$\text{Induced shear stress} = \frac{4000}{2 \times \pi/4(12.5)^2} = 16.4 \text{ N/mm}^2, \text{ hence safe.}$$

By knowing the method of attachment of blade by the pin, the pin can be checked for induced bending stress.

Example 14.17. A pressure vessel as shown in Fig. 14.28 is used as a digester in a chemical process. It is designed to withstand a fluid pressure of 0.2 N/mm^2. The vessel and its cover are made of cast iron, all other links are of steel. The cover is held tightly against the vessel by means of a screw B which is turned down through the tapped hole in the beam A so that the end of the screw press firmly against the cover. The beam, has a rectangular section, in which depth is twice its thickness. The rectangular section is opened up at the centre to take the tapped hole as shown in figure. The beam is attached by pins C and D to the links G and H, which are secured by pins E and F, to the extensions cast on the vessel. If ft for cast iron = 15 N/mm^2, for steel = 45 N/mm^2, f_c for steel = 45 N/mm^2, f_s for steel = 40 N/mm^2, it is required to find.

(a) Solid rectangular section of the beam A. (b) section of beam at the tapped hole. (c) Diameters of links G and H. (d) Dimensions at the wall of the supports of pins E and F, assuming a rectangular section with $h_2 = 2t_2$.

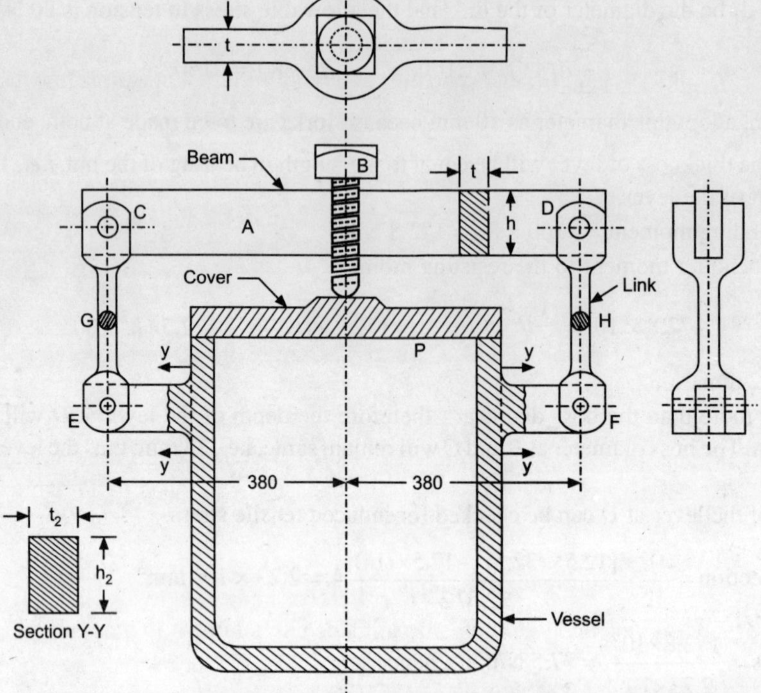

Fig. 14.28 *Pressure vessel*

Solution: (a) Total pressure P on the cover = $\pi/4D^2 \times P = \pi/4\,(600)^2 \times 0.2 = 56.5$ kN.

Assume that h is the depth and t the thickness of the beam A. $h = 2t$, is given. It is a simple beam with a single load at its centre. From the symmetry of loading, the reaction at C and D are equal, i.e., 28. 25 kN.

$$\therefore \quad R_c = R_d = \frac{P}{2} = \frac{56.5 \times 10^3}{2} = 28.25 \text{ kN}$$

The worst section for bending moment is a mid- point of the beam.

$$M_b = 28.25 \times 10^3 \times 380 = 10700 \text{ Nm}$$

Modulus of section $Z = (1/6)\, th^2 = (1/6)\, t\,(2t)^2 = (2/3)\, t^3$

Now $\quad M_b = f \times Z,\ 10700 \times 10^3 = 45 \times (2/3)\, t^3$

gives, $\quad t = 71$ mm, say 72.5 mm, $h = 2 \times t = 145$ mm.

(*b*) The axis of the tapped hole is parallel to the dimension t of the section, therefore the solid rectangular section of the beam can be split into two halves at the holes or opened up for the hole, so that the metallic area in a section through the hole is equal to area of the solid section obtained in part (d). The factor of safety will remain uncharged. The diameter of hole in the beam is 47.5 mm, this section will then have

$$(72.5 + 47.5) \times 145, \text{ i.e., } 120 \text{ mm} \times 145 \text{ mm}.$$

(*c*) The links G and H are in tension. Let d_1 be the diameter of the link. Equating the tensile force to the resisting tension in the link,

$$\frac{\pi}{4} d_1^2 \times f_t = 28.25 \times 10^3, \quad \frac{\pi}{4} d_1^2 \times 45 = 28.25 \times 10^3$$

gives, $\quad d_1 = 28.2$ mm say 30 mm.

(*d*) The extensions of brackets on the vessel are part of that casting and both act as cantilevers. They are of the same length and sustain equal end loads. Hence their sections at their juncture with the vessel become weakest sections, and their dimensions will be identical.

Length of cantilever from centre line of pin to the wall

$$= [380 - (300 + 6)] = 74 \text{ mm}$$

End load of cantilever = shearing load on pin $= 28.25 \times 10^3$ N.

$\therefore$ Bending moment $M_b = 28.25 \times 10^3 \times 74 = 2090$ Nm

$$\text{Modulus of section } Z = \frac{1}{6} t_2 h_2^2 = \frac{1}{6} t_2 \times (2t_2)^2 = (2/3)\, t_2^3$$

$$M_b = f \times Z,\ 2090 \times 10^3 = 15 \times (2/3)\, t_2^3$$

gives, $t_2 = 59.3$ mm. say 60 mm, $h_2 = 2 \times t_2 = 2 \times 60 = 120$ mm.

EXERCISES

1. Define the following terms: lever, mechanical advantage, arm of the lever, leverage, and the displacement ratio.
2. Explain the different types of levers with sketches. Indicate, why the lever of type 3 is not recommended in use?
3. Differentiate between simple and compound levers.
4. A heavy machine part is raised by a 1 m crow bar which is supported at 100 mm from the end where it takes the weight. The machine part weights 15 kN and half of this is taken by the crow bar, what effort must be applied at the end of crow bar in order to raise the machine part? Find the diameter of the crow-bar, if the allowable stress in the material is limited to 75 N/mm^2.
5. Figure 14.4 shows the handle for turning of a large valve. The effort applied at each end of the handle is 450 N. The length from the centre of the spindle is 750 mm. The handle is attached to the spindle by means of a tapered round pin. Find all the dimensions of the tapered pin and handle. Take the spindle diameter as 60 mm. Allowable stresses in tension and shear are 95 N/mm^2 and 50 N/mm^2 respectively.
6. A horizontal lever *LMN* is attached by a fulcrum pin at *N* and to a vertical tie rod at *M* as shown in Fig. 14.5. An operating force 1000 N is applied vertically at *L*. The distance of vertical rod *M* from the fulcrum point is 150 mm and the distance of effort point from fulcrum point is 1.15 m. Find the reactions at *N* and *M*.
   Determine the dimensions of the lever *LMN*, the pins and the tie rod at *M*. The allowable stresses for all the parts in tension, compression and shear are 60 N/mm^2, 60 N/mm^2 and 30 N/mm^2 respectively. The bearing pressure on pins may be taken as 23 N/mm^2. Make a dimensioned drawing of the designed lever.

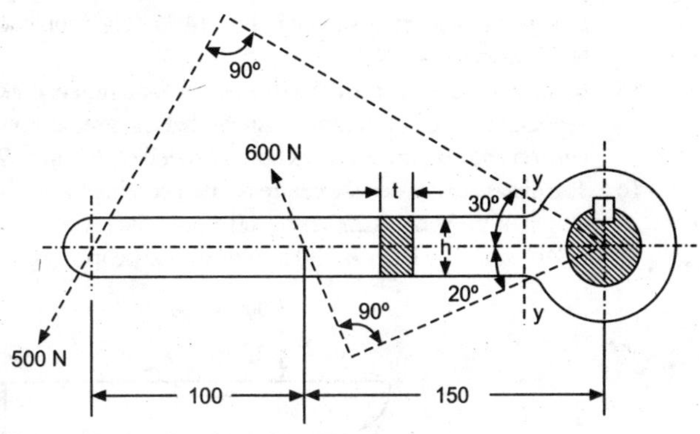

Fig. 14.29

7. A beam scale is designed to measure up a maximum of 5000 N. The distance between the two points of suspension of the scale pans is 2.0 metres. Assume that the beam of the scale is made of cast steel of rectangular section, depth being 4 times the thickness and that there is a 62.5 mm diameter hole in the centre of the beam to admit the central knife edge and its seat. calculate the dimensions of the beam. Draw a neat dimensioned sketch of the beam scale showng clearly the central suspension and the knife edges at the two ends to suspend the pans. The permissible stresses in tension and shear for cast steel are 85 N/mm^2 and 65 N/mm^2 respectively.
8. The Fig. 14.29 shows a link of the mechanism and is acted upon by the forces as indicated. Calculate:
   (a) Net torsional moment on the shaft, (*b*) shear stress induced by the torsional moment if the shaft is 20 mm in diameter, (*c*) section of the lever at *y-y*, if height of the lever is three times its thickness. Permissible stress in tension is 30 N/mm^2.
9. A hand lever with knurled handle as shown in Fig. 14.30 is used to overcome the resistance of 500 N. Design and prepare a dimensioned drawing of the lever if it is made of mild steel.

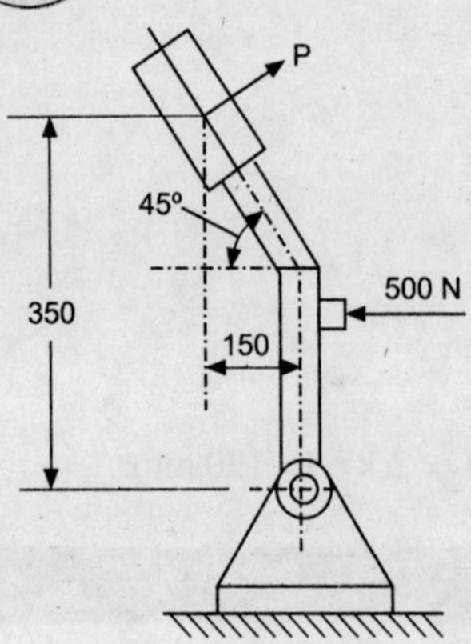

Fig. 14.30

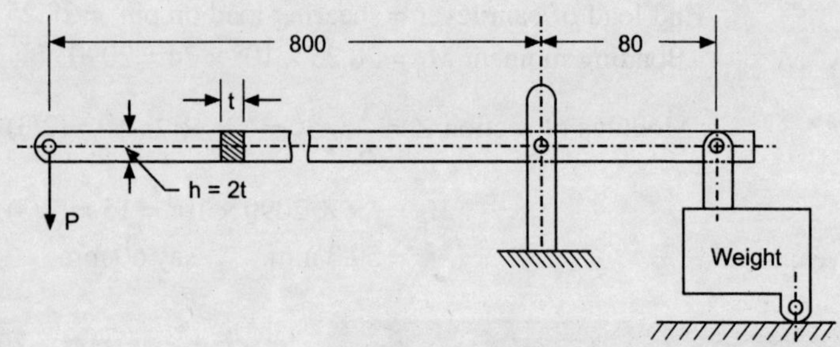

Fig. 14.31

10. A symmetrical body weighing 7500 N is lifted from one end by the arrangement shown in Fig. 14.31. If an effort *P* is applied to raise the body, then design the lever. The permissible stress in tension for the pin and lever material is 75 N/mm^2. Allowable bearing pressure for pins is 7.5 N/mm^2. Assume the thickness of the lever as half the depth of the lever.
11. A foot lever (Fig. 14.8) rectangular in section with height 3 times the thickness near the boss. The distance from the centre of the shaft to the effort point is 500 mm and the maximum foot load applied is 750 N. Calculate.
    (*a*) The shaft diameter for a permissible shear stress of 50 N/mm^2. The distance l may be assumed as 120 mm.
    (*b*) Dimensions of key. Allowable stress in key is 45 N/mm^2.
    (*c*) Dimensions of the lever. Take f_b = 100 N/mm^2 and f_c = 115 N/mm^2.
12. Design a foot lever as shown in Fig. 14.32. The foot load applied is 900 N. Take the bending and shear stresses as 60 N/mm^2 and 45 N/mm^2.
13. Make a designed sketch of a hand lever for a brake, making the length of lever 800 mm, effective overhang from nearest bearing is 120 mm. Assume that the greatest pull on the end is 350 N. The permissible stresses in shear, tension and compression are not to exceed 60 N/mm^2. What would be the diameter of the spindle?
14. The distance between the centres of two bearings which support a pedal lever brake shaft is 500 mm. The effective length of the pedal is 250 mm, and that of the lever which is in the same plane, but at the opposite side of the shaft is 100 mm. The latter is 330 mm from one bearing and the former 170 mm from the other bearing.

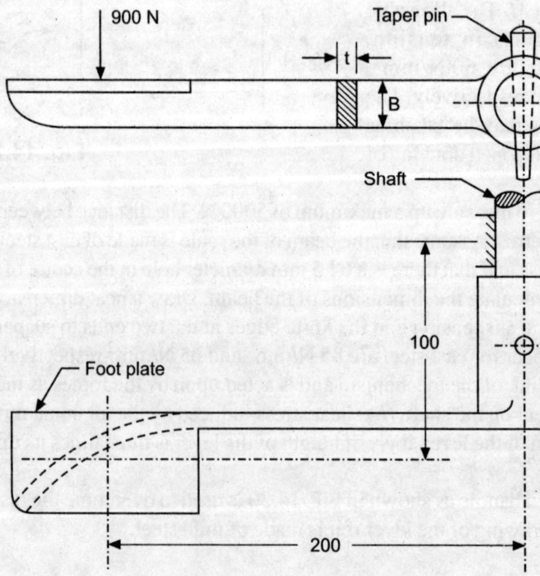

Fig. 14.32 *Foot lever*

Determine the diameter of the shaft if the permissible shear stress is not to exceed 50 N/mm^2. Determine the dimensions of the pedal lever at the boss if the width is 3 times the thickness. The value of the allowable stress in the material of the lever is 60 N/mm^2. The force acting on the pedal is 300 N.

15. The hand brake lever of a car is shown in Fig. 14.33. The cables attached to the points *L* and *M* are adjusted so that when a force is applied at the top of the lever the forces in the two cables are equal. Find
   (a) Dimensions of the fulcrum pin if allowable stress in tension is 25 N/mm^2 and bearing pressure is 10 N/mm^2.
   (b) The rectangular section of the lever if permissible bending stress for lever is 80 N/mm^2.

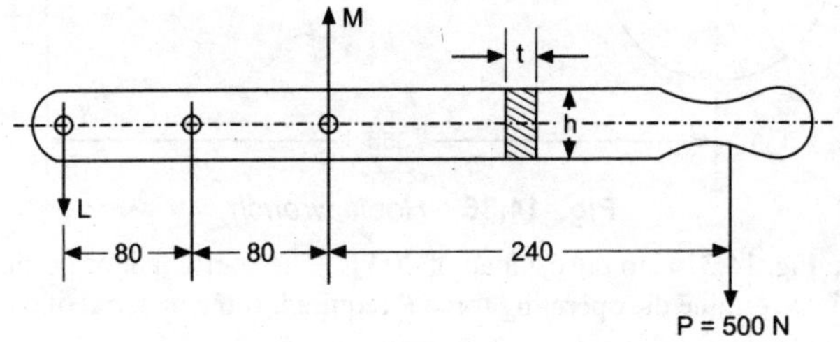

Fig. 14.33 *Hand brake lever of car*

16. Figure 14.34 shows the fitting of a shaft by the handle. Determine the diameter of the shaft if the allowable shear stress is 20 N/mm^2. Also find the diameter of the handle if the allowable tensile stress is 25 N/mm^2. Take the boss diameter as twice the shaft diameter.

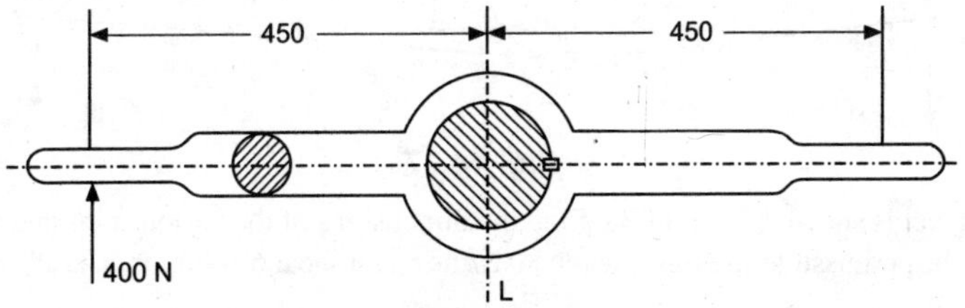

Fig. 14.34

17. A cranking lever for a diesel engine is shown in Fig, 14.35. *A* is a pin on which a force of 450 N is applied at a point 165 mm away from the centre line of the crank web. Crank radius is 280 mm. B is a boss, 30 mm long having a square hole to connect the lever with the square end of the crank shaft. Design the lever completely with its material as mild steel. Prepare a dimensioned drawing.

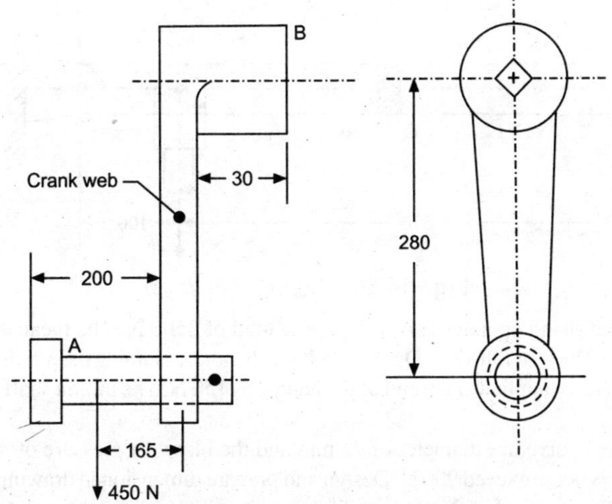

Fig. 14.35 *Cranking lever*

18. Figure 14.36 shows a mild steel hook wrench used for tightening circular locknuts. The force applied at the handle end is 275 N. Design the lever completely and prepare a dimensioned drawing.

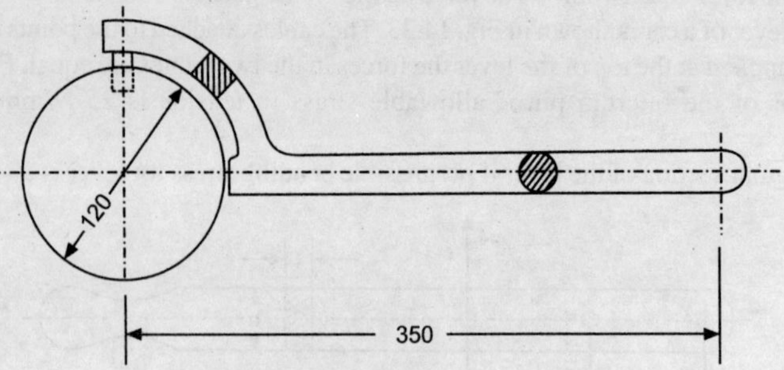

Fig. 14.36 *Hook wrench*

19. The lever shown in Fig. 14.37 is to raise a load of 30 kN. The coefficient of friction between the lever and the fulcrum pin is 0.06. Determine the operating force *P* required. If the material of the pins and lever is mild steel, design the lever completely.

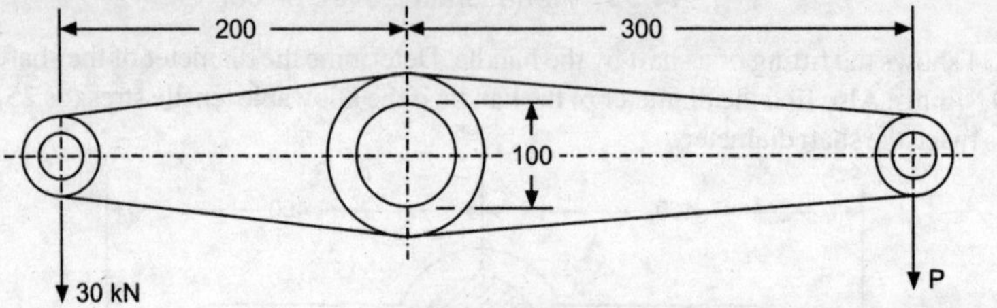

Fig. 14.37

20. A compound lever is shown in Fig. 14.38. Find the dimensions of the various pins and section of levers *AB*, *AC* and *CD*. Take the permissible stress in tension 80 N/mm^2 and shear 60 N/mm^2. The allowable bearing pressure is 20 N/mm^2.

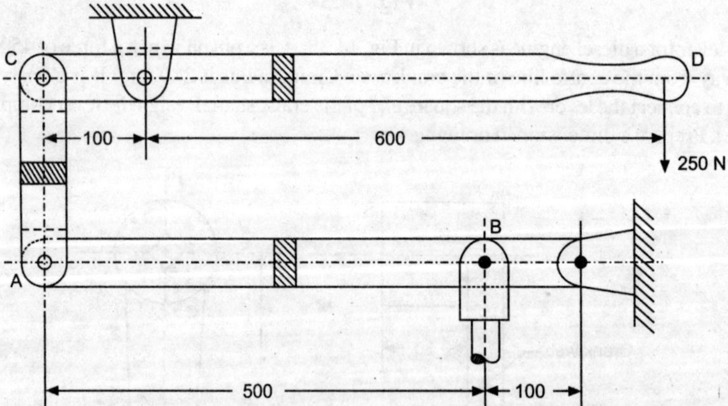

Fig. 14.38 *Compound lever*

21. A geared winch crab (hand operated) is used to lift a load of 2500 N. The mean drum diameter is 20 times the diameter of the rope. The gear ratio is 5. Design a suitable handle, assuming the efficiency of winch to be 70 percent. Take the maximum force applied at the end of the handle as 150 N. The pinion shaft is squared with 20 mm side to receive the handle.
22. A lever loaded safety valve has diameter of 75 mm and the blow off pressure of the boiler is 1.45 N/mm^2. The weight on the lever is not to exceed 900 N. Design and prepare dimensioned drawing of the lever, fulcrum, and the pivot of the lever. The material of the lever and fulcrum is mild steel and the weights are of cast iron. Choose the suitable stresses of the material.

23. Design the lever of a lever loaded safety valve for the maximum load of 5000 N on the valve when it blows off freely. The permissible stress may be taken as $f_t = 35$ N/mm^2, $f_s = 20$ N/mm^2, and $f_c = 50$ N/mm^2. The allowable bearing pressure is 20 N/mm^2. The length of the lever between the pivot and weight centre line is 750 mm and between pivot and load centre line is 75 mm.
Prepare neat dimensioned working drawing of the lever with the valve and valve seat arrangement shown in part.
24. Figure 14.39 shows a steel lever acted upon by horizontal and vertical forces. The lever is suspend from a fixed point x. Design the lever. The allowable stresses in bending and shear are 90 N/mm^2 and 40 N/mm^2 respectively.

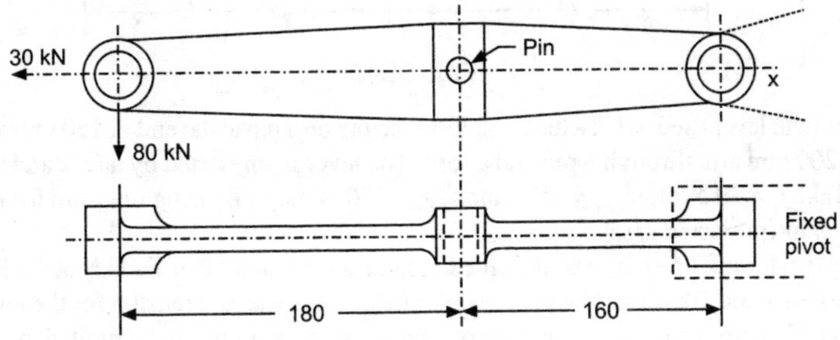

Fig. 14.39

25. Design a cam lever as shown in Fig. 14.15 operating the exhaust valve of a diesel engine. Following data may be assumed:

Maximum load on the valve	4500 N
Effective arms length	250 mm
Roller diameter	65 MM
Angle between the arms	125°
Bearing pressure between the pin and lever	5.25 N/mm^2.

Draw an elevation of the complete assembly adding the designed details. Draw an end view to the left of the elevation and a plan of the assembly. The section of the rocker may be assumed rectangular. Draw up a material list for the gear.
26. Figure 14.15 shows a rocker arm for an exhaust valve of a gas engine. Find out
    (*a*) dimensions of pin at roller end,
    (*b*) roller diameter and width,
    (*c*) section of lever, assuming the section to be I-from,
    (*d*) pin at fulcrum,
    (*e*) screw end of the rocker arm.

    Following data may be assumed. Maximum load on roller 4000 N. Effective arms length 175 mm. Angle between the arms 135°. The permissible stress in tension for pins is 60 N/mm^2 and the bearing pressure is 5 N/mm^2. The permissible stress in tension for section of lever is 100 N/mm^2 and in shear is 60 N/mm^2.
27. A push rod 375 mm long has forked end on one side and spherical one on the other side. It is connected to the tappet lever. Design the push rod and the tappet lever if the maximum resistance at the lever is 1000 N. Choose your own materials and suitable stresses.
28. Design a rocker arm for operating an exhaust valve of a gas engine. It has rocker arms of length 175 mm. Angle between the arms is 135°. The exhaust valve is 75 mm in diameter and the gas pressure, when the valve begins to open, is 0.37 N/mm^2. The greatest suction pressure is 0.02 N/mm^2 below atmospheric. The initial load may be taken as 0.06 N/mm^2 of valve area and the valve inertia and friction losses 200 N. The lever is to be of I-form, cast steel with ultimate strength 800 N/mm^2. The allowable bearing pressure is 6 N/mm^2 and the permissible tensile stress in pin material is 75 N/mm^2.
29. Design the rocker arm and the fulcrum pin of a diesel engine rocker arm shown in Fig. 14.40. The total load on the valve end is 1350 N and that at tappet is 2000 N. The rocker arm is to be I-section with an allowable stress in tension as 50 N/mm^2. The working bearing pressure at the pin is 5 N/mm^2. The diameter of the stud on the tappet end side is 16 mm. Give a neat and dimensioned sketch of the rocker arm designed.

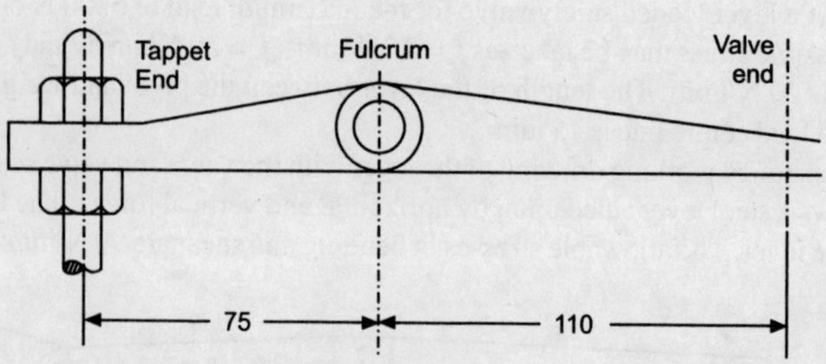

Fig. 14.40

30. Design a bell crank lever used to lift a load of 2750 N acting on a pin at the end of 150 mm arm. The effort is applied at the end of 200 mm arm through a pin and a fork. The lever is supported by a fork and a pin and the pin is fixed in the fork. Take $f_t = 70$ N/mm^2, $f_s = 55$ N/mm^2, $f_c = 100$ N/mm^2. Bearing pressure for the pins is not to exceed 17.5 N/mm^2 of the projected area.
31. A right angled bell crank lever is to be designed to raise a vertical load of 7000 N at the longer arm end. The arm lengths are 400 mm and 100 mm. The permissible stresses in shear and tension for the lever and pin materials are 70 N/mm^2 and 85 Nmm^2 respectively and the bearing pressure on pins is to be limited to 10 N/mm^2. Determine all the dimensions of the lever and make a dimensioned sketch.
32. A right angled bell crank lever has arms 1.05 m and 0.75 m and longer arm carries a load of 55 kN. The shaft transmits a torsional moment of 15.7 kNm through one bearing neck, the bearing being close to the lever. Adopt any suitable method of making the lever and comment on your design. Choose your own materials and stresses. Sketch two views of the lever and dimension it fully.
33. A spring loaded governor of the Hartnell type has arms of equal length. The weights rotate in a circle of 350 mm diameter when the sleeve is in mid position and the weight arms are vertical, the equilibrium speed for this position is 400 rev/min. The maximum sleeve movement is to be 40 mm and the maximum variation in speed allowed is 4 percent from the mean speed. The weight of the ball is 35 N. Design the spring and the bell crank lever for the governor. Assume your own materials and values of stress for the same.
34. Figure 14.41 shows an angular lever used on job shop printing press. The material of the lever is mild steel. A force *F* of 1000 N is applied at *P* in the direction shown. Design the lever and pins at *L*, *M* and *N*.
35. Design a bell crank lever having both arms of equal length. The stress in the materials of the lever is not to exceed 15 N/mm^2. The free ends of the arms terminate in a circular ring of 75 mm external diameter and 25 mm hole diameter to receive pins. The common end of the arms is formed into a boss of suitable diameter to receive a pin of 40 mm diameter.
A load of 1800 N is applied at one end of the free ends with an effective arm length of 400 mm between the ring and boss centres. The arms have an I-section between the boss and the ring, such that the depth of the section is about 2.25 times the flange width. Make the necessary calculations and draw a dimensioned sketch of the lever, giving sectional views where required.
36. A bell crank lever is shown in Fig. 14.42. The load on each end of the lever is 12.5 kN acting as indicated in figure. Find dimensions of the pin at fulcrum and the bronze pins at lever end, allowing a bearing pressure of 6 N/mm^2 and bending stress of 30 N/mm^2 in each case.
Fix suitable dimensions of the elliptical section of the lever at *AA*, and check the bending stress there. Hence specify a suitable material there.

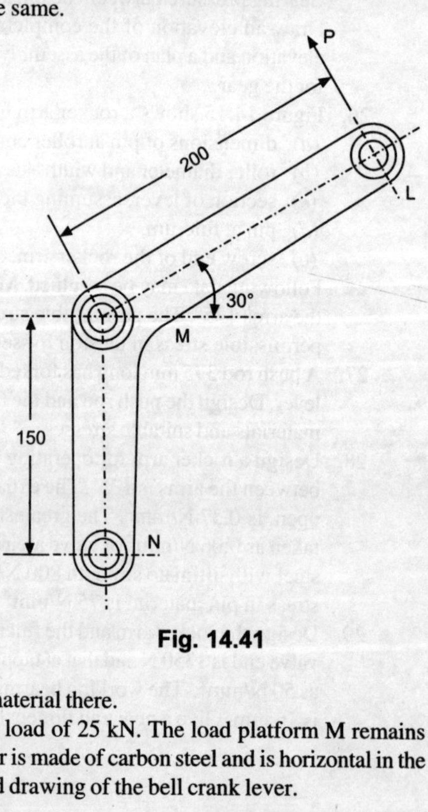

Fig. 14.41

37. Figure 14.43 shows a bell crank lever for a garrage jack to lift a load of 25 kN. The load platform M remains horizontal in all positions of the jack. The lift is 150 mm. If the lever is made of carbon steel and is horizontal in the lower most position of the jack, design and prepare a dimensioned drawing of the bell crank lever.

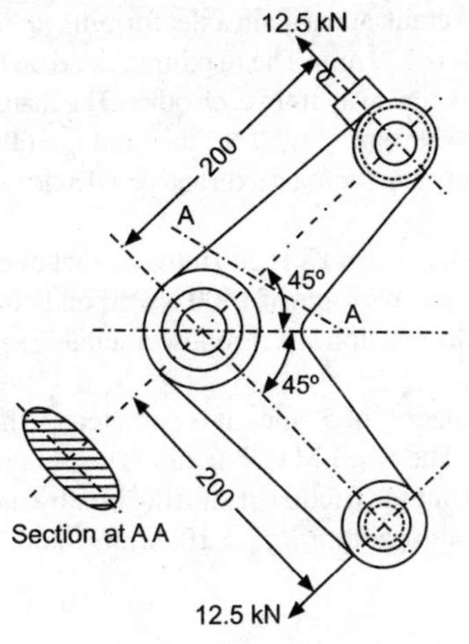

Fig. 14.42

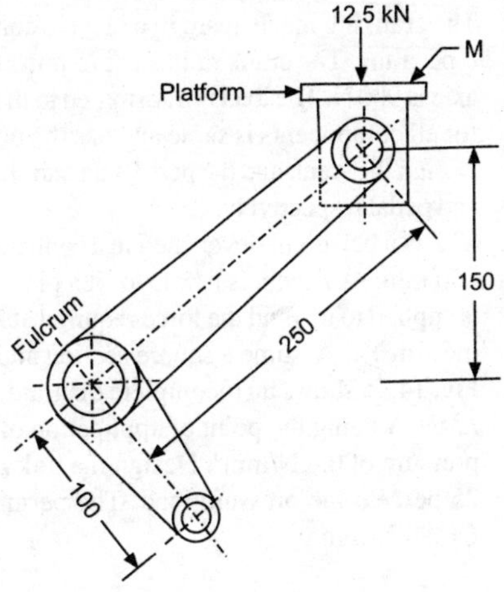

Fig. 14.43

38. A push rod 400 mm long has forked ends on one side and spherical one on other side. It is connected to a tappet lever, having an angle of 135° between the arms. If the maximum resistance at the valve is 1250 N, with valve arm 65 mm and push rod arm 100 mm, design the lever having I-section. Also find the size of the push rod. The permissible stress in tension is 75 N/mm^2 and the bearing pressure between the pin and lever is 9 N/mm^2.
39. A band brake for a crane is required to hold a load of 4500 N. This load is taken on a rope wound round the crane barrel which is 400 mm diameter. The brake drum, which is fixed to the barrel shaft is 600 mm diameter. The band embraces three quarter of the circumference of the drum and the coefficient of friction between band and drum is 0.3. The brake is to be applied by a hand lever above the drum and the operating force (vertically downwards) must not exceed 250 N. Design and sketch a suitable mild steel operating lever having one end of the band attached to its fulcrum, assuming a bending stress of 90 N/mm^2.
Devise and sketch a suitable catch for locking the brake in the "on" position.
Design the lever in both cases. The material of lever and pin is mild steel. Assume your own stresses and other data required.
40. The drum of a simple band brake is 460 mm diameter. The band is 4.5 mm thick and 62.5 mm wide. It embraces three fourth of the circumference of the drum. The hand lever is arranged as shown in Fig. 14.24. Take $NM = 75$ mm, $NL = 450$ mm. If the force at the end of the lever is 250 N and $\mu = 0.22$, what is the resisting torsional moment exerted on the brake drum? Draw a proportionate sketch of the brake arrangement. Show calculations for the dimensions of the shaft, key, lever and pin at fulcrum. Allowable working stresses for steel are as follows:
$$f_t = f_c = 70 \text{ N/mm}^2, f_s = 55 \text{ N/mm}^2.$$
41. Figure 14.44 shows the brake arrangement for a crane. The lever *ABCD* is on a fixed fulcrum *B* and is connected to the brake strap by pins at *A* and *C*. The pull on the tight side of the strap may be taken as three times the pull on the slack side. Find the force *P* required to support a load of 20 kN. Design the lever completely. Material of lever and pin is mild steel. Assume your own stresses and other data required.
42. (a) Sketch with detailed description of one type of beam scale used for weighing the heavy vehicles and give its design procedure.
(b) Draw a simple safety valve, giving all its details and relative dimensions. Which type of lever it should be?
(c) Sketch a tappet lever for engines. Which of its dimensions should be calculated with the help of the design formula?

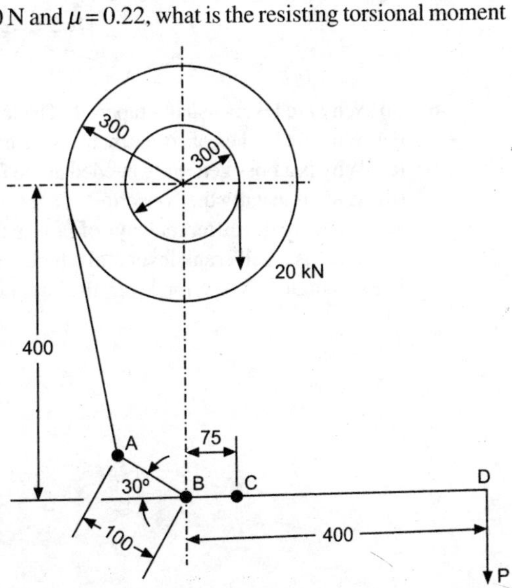

Fig. 14.44

43. The cranking mechanism in a bicycle consists of a pedal, an axle, a crank and a main axle, forming an overhung type crank. The crank radius is 175 mm. The length of the pedal axle is 120 mm. The maximum load on the pedal axle is 900 N. The load is transmitted to the pedal axle at two points 90 mm away from each other. The material used for all components is same and has the permissible stresses as $f_t = 90$ N/mm^2, $f_s = 50$ N/mm^2 and $f_b = 60$ N/mm^2. Design the crank and the pedal axle and show clearly the fixing arrangement for main axle and pedal axle with crank and pedal respectively.

44. *XYZ* is a bell crank lever used in a centrifugal governor. The length of the arm *YX* is 200 mm and that of arm *YZ* is 150 mm. At *Y* there is a hole to fit a pin for supporting it on a pivot. At *X* a force of 1500 N Perpendicular to *YX* is applied to it. Find the force required at *Z* perpendicular to *YZ* and the reaction at *X*. Propose a suitable section for the arm *YX*. Assume a square section and bending stress 15 N/mm^2.

45. Fig, 14.45 shows an incomplete diagram of a feed pump, whose plunger is of 50 mm. It is operated by the handle *LMN*, *N* being the point of application of the effort by an operator. The arm LM is 200 mm. The plunger exerts pressure of 0.8 N/mm^2. Design the link *LO* and handle *LMN*. Account for friction at stuffing box by increasing 25 percent the pressure load. The permissible stresses for pin, f_t = 60 N/mm^2, f_b = 10 N/mm^2 and for lever f_t = 80 N/mm^2.

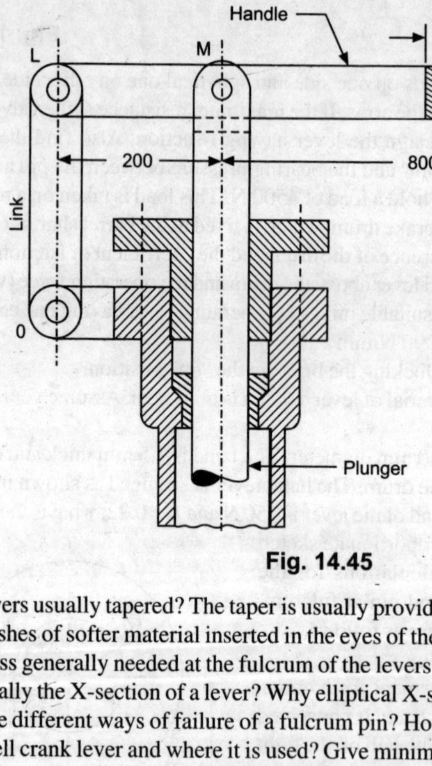

Fig. 14.45

46. (a) Why are levers usually tapered? The taper is usually provided in which dimension?
(b) Why are bushes of softer material inserted in the eyes of the levers?
(c) Why is a boss generally needed at the fulcrum of the levers?
(d) What is usually the X-section of a lever? Why elliptical X-section is preferred over round X-section?
(e) What are the different ways of failure of a fulcrum pin? How is usually the fulcrum pin designed?
(f) What is a bell crank lever and where it is used? Give minimum four examples.

47. In a rectangular X-section lever, the larger dimension is kept in the plane of motion of the lever. Why?

CHAPTER 15 Springs

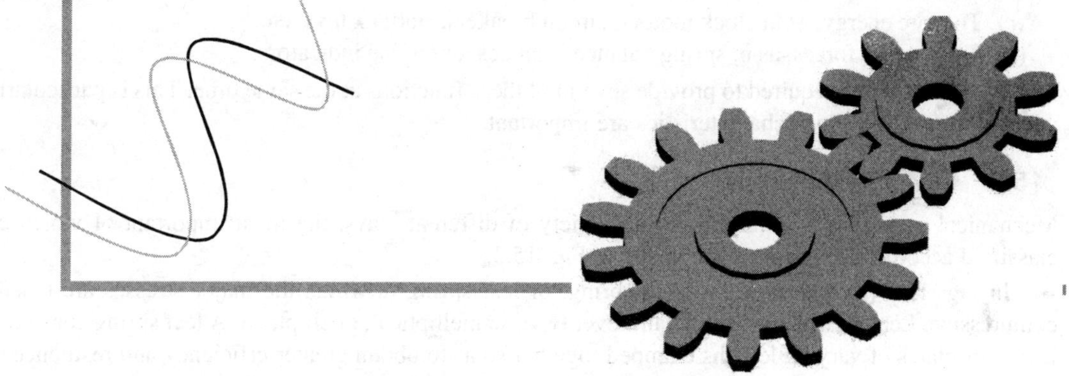

15.1 INTRODUCTION

A mechanical spring may be defined as an elastic body whose primary function is to deflect under load (or to absorb energy) and which recovers to its original shape when released after being distorted. Although most material bodies are elastic and will distort under load, they are not all considered springs. A mechanical spring should primarily deflect more and it attains there by greater elastic energy, e.g., a steel helical spring can be expanded to twice its length without loosing its elasticity whereas a bolt made from same grade of steel is overstrained when it deflects by 4/1000 of its original length.

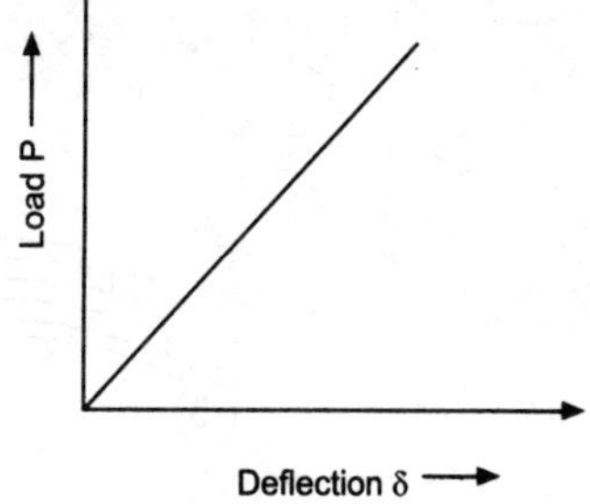

Fig. 15.1 *Load-deflection diagram*

Provided the material is not stressed beyond the elastic limit, the usual type of spring will have a straight line load deflection diagram as shown in Fig. 15.1. This means that deflection is proportional to the load, i.e., if the load is doubled, the deflection will be doubled. The relation will hold good even if the acting load is a torsional moment or moment, provided that linear deflection is replaced by angular deflection.

However, not all springs have linear load-deflection diagrams. In some cases load-deflection diagrams as shown in Fig. 15.2 may be found. Curve *A* may be obtained with a thin, flat circular plate or disc loaded to a large deflection. Curve *B* may result from an initially coned disc (or Belleville) spring. Another example of a non-linear characteristic is that of ordinary clock power spring when wound tightly and then allowed to unwind.

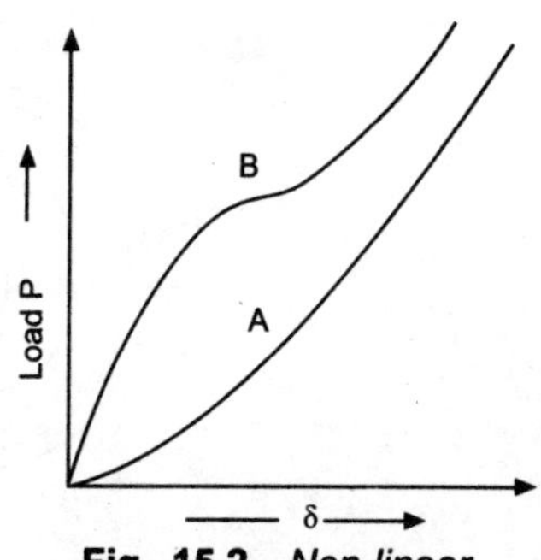

Fig. 15.2 *Non-linear deflection curve*

Among the primary functions of springs the following are perhaps the most important.

(*a*) To cushion, absorb, or control energy due either to shock, as in car springs, railway buffers, elevator buffer springs, in aircraft landing gears, or in artillery recoil system, or to vibration, as in spring supports or elastic motor drives, coil spring coupling and vibration dampers.

(*b*) To control motion by maintaining contact between two elements, such as a cam and its follower, by restoring a machine part to its normal position when the disturbing force is removed, in a governor or valve of an I.C. engine, or by producing the necessary pressure in a friction device, as in a brake or clutch.

(*c*) To store energy, as in clock motors, circuit breakers, starters, toys, etc.

(*d*) To measure forces, as in spring balances, gauges, or engine indicators.

Springs are often required to provide several of these functions at the same time. This is particularly true when vibration and shock characteristics are important.

15.2 CLASSIFICATION

Mechanical springs may come in a wide variety of different ways, the most important of which can be classified according to shape, as indicated in Fig. 15.3.

In Fig. 15.3 (*a*) is shown the plate spring, or leaf spring in which the major stresses are tensile and compressive. Leaf springs may be of cantilever type, semielliptical, or elliptical. A leaf spring consists of flat leaves or plates of varying lengths clamped together so as to obtain greater efficiency and resilience.

In Fig. 15.3 (*b*) is shown the cylindrical helical spring which may be in compression or tension. The major stresses produced in this are shear due to twisting. The load applied is parallel to the axis of the spring. The cross-section of the wire may be round, square or rectangular. These springs are wound in the form of a helix out of a wire.

In Fig. 15.3 (*c*) is shown a helical conical spring. The major stresses produced in this spring are also shear due to twisting. If the radius of the coils of a helical spring is constant, then it becomes a cylindrical helical spring.

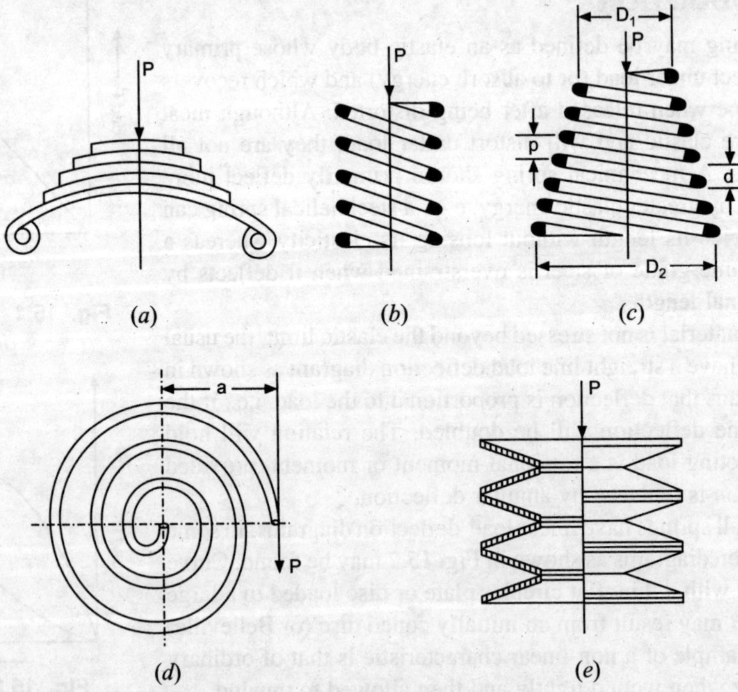

Fig. 15.3 *Shapes of springs*

If a helical spring works in torsion, i.e., the torsional moment is applied about the axis of the helix, then the spring obtained is helical torsion spring. Major stresses produced in this spring are tensile and compressive due to bending.

If the angle of the helix is zero, as in Fig. 15.3 (*d*), it is a spiral spring, consists of a flat strip wound in the form of a spiral and loaded in torsion as shown in figure. The major stresses produced in this are tensile and compressive due to bending by load *P*.

Special springs for large loads and deflections are made in the form of a series of flat tapered or dish-shaped discs as shown in Fig. 15.3 (*e*). In this spring the major stresses produced by load P are tensile and compressive.

Springs employing air or liquid are finding many applications these days. Fluids (air or liquid) can behave as a compression spring.

15.3 CYLINDRICAL HELICAL SPRING WITH AXIAL LOADING

The coil, or helical, spring consists of a wire or rod wound about a mandrel to form a helix and is primarily intended for axial direct compression or tension loads. A helical spring is considered close coiled if the plane containing each coil is nearly perpendicular to the axis of the helix. This will be the case when the helix angle is small and generally a spring is taken as close coiled if the angle be less than 10°.

In Fig. 15.4 (*a*) is shown a helical coil spring with ends adopted to support a compressive load. At (*b*) the spring has been deflected by the axial load *P*, which may be assumed to compress the spring as between two parallel plates. Compression springs may be made of round wire, or wire with rectangular section. The first part of this discussion will deal with round wire springs. The design of any spring must satisfy two requirements.

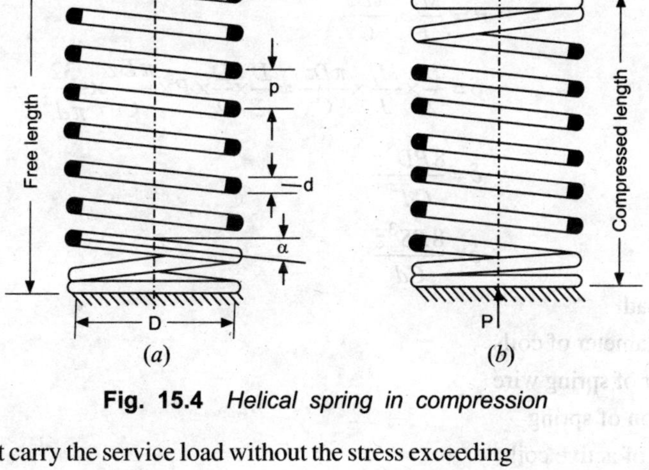

Fig. 15.4 *Helical spring in compression*

(*a*) The spring must carry the service load without the stress exceeding the safe value.

(*b*) The force deflection characteristics, i.e., spring rate must be satisfactory for the given application.

In Fig. 15.5, is shown part of a compression spring that supports a compressive load *P* and a section of the wire cut by axial plane. The action of the force *P* tends to rotate the wire, thereby causing torsional stresses in the wire. Also bending, direct compression, and direct shearing stresses, which are neglected in the conventional spring equations, are set up in the wire.

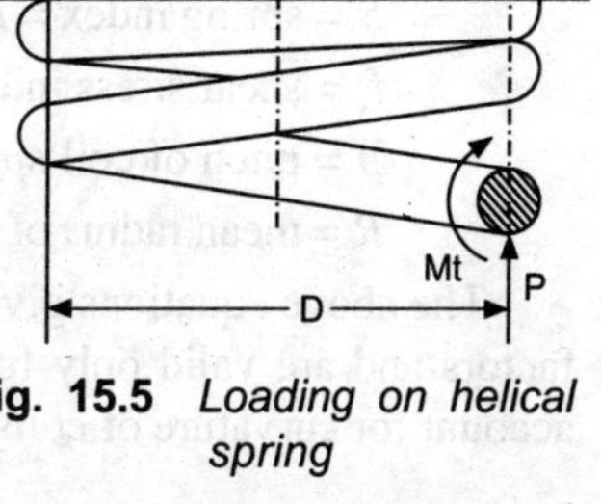

Fig. 15.5 *Loading on helical spring*

Considering only the torsion, the part of the spring shown in figure is in equilibrium under the action of the two forces P and the resisting torsional moment M_t, as shown in the figure. The torsional moment M_t is determined by the relations.

$$M_t = \frac{1}{2} PD \qquad ...(15.1)$$

The same torsional moment acts upon all sections or the loaded spring. The internal resisting moment will be

$$f_s \times Z_p = f_s \times \frac{\pi}{16} d^3 \qquad ...(15.2)$$

Equating the external moment to the internal one and solving for f_s, results in

$$f_s = \frac{8PD}{\pi d^3} \qquad ...(15.3)$$

and deflection of spring is given by relation

$$\delta = \frac{D}{2} \times \theta$$

Thus $$\frac{M_t}{J} = \frac{f_s}{r} = \frac{C\theta}{l}$$

or $$\theta = \frac{M_t}{J} \times \frac{l}{C}$$

where l equals the length of the twisted wire. The length of the wire in the helical spring equals $\pi Dz/\cos \alpha$, where α is the lead angle of the helix, which for close coiled springs is of the order of 5 degrees, hence cos α is approximately unity.

$\therefore$ $$\theta = \frac{M_t}{J} \times \frac{\pi Dz}{C}$$

hence $$\delta = \frac{D}{2} \times \frac{M_t}{J} \times \frac{\pi Dz}{C} = \frac{D}{2} \times \frac{D}{2} \times P \times \frac{\pi Dz}{C} \times \frac{32}{\pi d^4}$$

$\therefore$ $$\delta = \frac{8PD^3 z}{Cd^4} \qquad ...(15.4)$$

or $$\delta = \frac{8PS^3 z}{Cd} \qquad ...(15.5)$$

where P = Axial load

D = Mean diameter of coils

d = diameter of spring wire

δ = deflection of spring

z = number of active coils

C = modulus of rigidity

S = spring index = D/d for circular wires

f_s = shear stress induced in the wire of the spring

p = pitch of coil springs

R = mean radius of coils.

The above equations give satisfactory results for large spring index S, although equations neglect many factors and are valid only for static loads. For low values of spring index S, relation 15.3 is modified to account for curvature of coils and direct shear stress.

15.4 CURVATURE OF COIL

It should be noted that the fibre length on the inside of coiled wire is shorter than on the outside. This means that, for a given angular rotation of adjacent cross-sections of the wire, the inner-fibre has a

(*a*) Pure torsional stress.
(*b*) direct shear stress.
(*c*) Resultant of torsional shearing stress and direct shear stress.
(*d*) Resultant of direct, torsional and curvature shearing stress.

higher shearing stress. Second the direct shearing stress due to the load *P* is additive to the torsional shearing stress on the inner fibre and subtractive on the outer fibre. The effect of curvature and direct shear stress is shown by free body diagrams in Fig. 15.6.

If the effect of the curvature of the wire is neglected, the components and their sum will be as shown in Figs. 15.6 (*a*), (*b*) and (*c*).

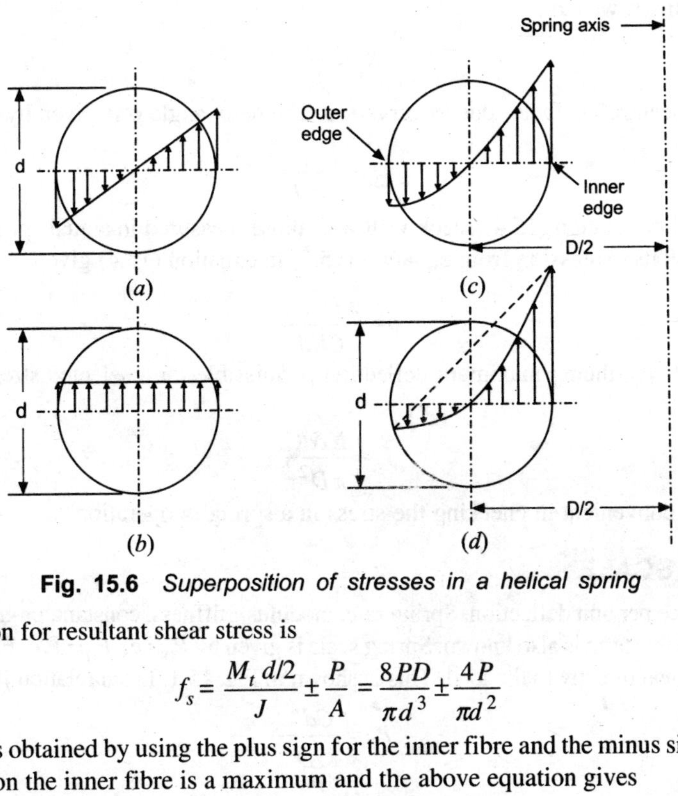

Fig. 15.6 *Superposition of stresses in a helical spring*

Then the equation for resultant shear stress is

$$f_s = \frac{M_t\, d/2}{J} \pm \frac{P}{A} = \frac{8PD}{\pi d^3} \pm \frac{4P}{\pi d^2}$$

In which stress is obtained by using the plus sign for the inner fibre and the minus sign for the outer fibre. Therefore the stress on the inner fibre is a maximum and the above equation gives

$$f_s = \frac{8PD}{\pi d^3}\left[1 + \frac{1}{2}\frac{d}{D}\right] = \frac{8PD}{\pi d^3}\left[1 + \frac{0.5}{S}\right]$$

This permits us to define the 'shear stress multiplication factor, i.e.,

$$K_s = \left(1 + \frac{0.5}{S}\right)$$

This accounts for the effect due to direct shear stress but not for curvature of the wire. The values of this factor have been calculated for various values of the spring index *S*. When the springs are subjected to static loads, the effect of curvature may be neglected, because yielding of the material will relieve the stresses. Thus,

$$f_s = K_s \frac{8PD}{\pi d^3} \quad \text{...(15.6)}$$

In order to include the effects of both direct shear and wire curvature, a stress factor K has been determined by the use of analytical method by A.M. Wahl which may be used with equation 15.3 to determine the maximum shearing stress in the wire as follows.

$$f_s = K\frac{8PD}{\pi d^3} = K\frac{8PS}{\pi d^2} \quad \text{...(15.7)}$$

where $K = \frac{4S-1}{4S-4} + \frac{0.615}{S}$

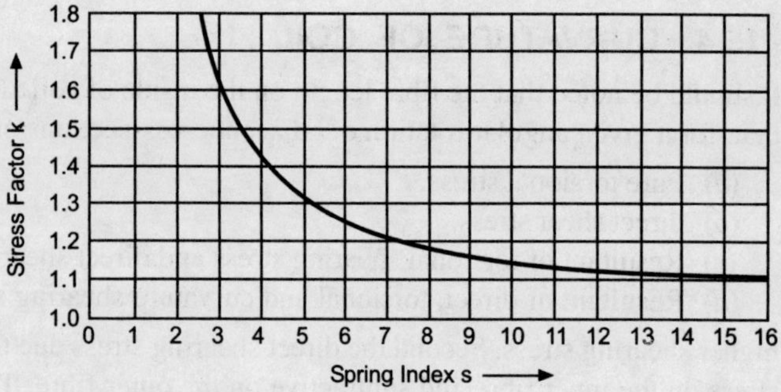

Fig. 15.7

Value of K may be read from Fig. 15.7.

This curve shows that the stress factor increases very rapidly as the spring index S decreases. It is advisable to have S minimum 3.

From above relations, we have

$$K_c = \frac{K}{K_s}$$

where K_c = stress concentration factor due to curvature. The helix angle α is given by relation

$$= \tan^{-1}\left(\frac{p}{\pi D}\right) \quad \text{...(15.8)}$$

values of δ calculated by relation (15.4) check well with those measured in actual springs.

Substituting for P its expression from equation (15.7) in equation (15.4) gives

$$\delta = \frac{\pi D^2 z f_s}{CKd} \quad \text{...(15.9)}$$

This equation allows finding maximum deflection permissible for a selected stress. Solving the above equation for f_s gives

$$f_s = \frac{K\delta dC}{\pi D^2 z} \quad \text{...(15.10)}$$

This equation is convenient in checking the stress in a spring in operation.

15.5 SPRING SCALE

It is defined as the force per unit deflection. Spring rate, modulus, stiffness, constant, or gradient are some of the other names by which this term is also known. Spring scale is given by P_0, i.e., $P_0 = P/\delta$. From relation (15.4) the deflection is proportional directly to the load. This is shown in Fig. 15.1. From relation (15.4) we get .

$$P_0 = \frac{Cd^4}{8zD^3} \quad \text{...(15.11)}$$

Also

$$\delta = \frac{P}{P_0} \quad \text{...(15.12)}$$

If the force P_1 exerted by a spring under certain conditions is given, and the force P_2 exerted with an additional compression δ' of the spring is also given, then

$$P_0 = \frac{P_2 - P_1}{\delta'}$$

Conversely, the total deflection δ by relation (15.12) is

$$\delta_2 = \frac{P_2}{P_0} = \frac{\delta' \times P_2}{P_2 - P_1} \quad \text{...(15.13)}$$

This relation is illustrated by Fig. 15.8. This relation will be useful to find the total deflection in a spring loaded safety valve.

15.6 RESILIENCE

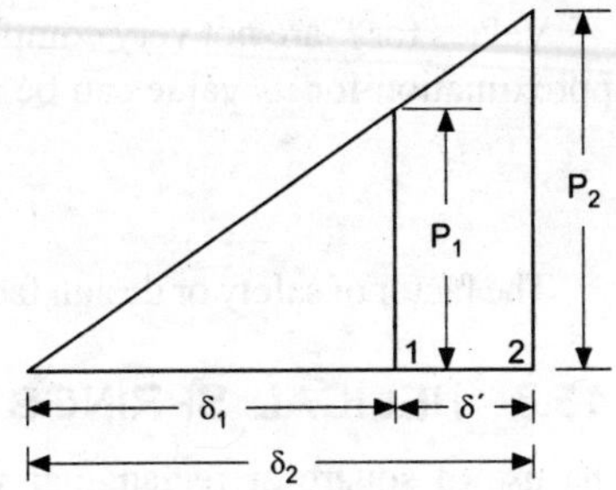

Fig. 15.8 *Relation between loads and deflection in a spring*

The resilience U of a spring is equal to the energy absorbed by it, which itself is equal to the work done on it by the external force. For gradually applied loads, the resilience of a spring is

$$U = \frac{P \times \delta}{2} = \frac{P}{2} \times \frac{8PD^3Z}{Cd^4} = \frac{4P^2D^3z}{Cd^4}$$

Putting for P its value from relation (15.7), we get

$$U = \frac{\pi^2 d^2 Dzf_s^2}{16K^2C}$$

Volume V of the spring is $= \pi Dz \times \frac{\pi}{4} d^2$

$$\therefore \qquad U = \frac{f_s^2}{4K^2C} \times V \qquad \text{...(15.14)}$$

Relation (15.14) shows that, in order to have a greater resilience and a better utilization of the material, K should be low. In turn the spring index S should be large. The checking of the resilience of a spring conveniently directly from its dimensions and deflection can be done by relation

$$U = \frac{\delta^2 d^4 C}{16zD^3} \qquad \text{...(15.15)}$$

15.7 IMPACT LOADING

It should be noted that in this case relations (15.4) and (15.7) cannot be used unless a fictitious force P is found which will produce the same deflection and stress as when the impact energy is absorbed by the spring. Suppose the load P falls through a distance h on the spring, then

$$U = P(h + \delta) = \frac{1}{2} Pe \times \delta \qquad \text{...(15.16)}$$

where δ is the deflection of the spring and P_e is the equivalent gradual load which shall give the same results as that of P which falls through a height h.

15.7.1 Helical Springs Subjected to Fatigue Loading

Many methods have been proposed for the design of helical springs subjected to fatigue loading. Here, we give one of the methods proposed by Wahl. It is based on the modified Soderberg diagram and the design equation is given as,

$$F_S = \frac{f_{yp}}{f_m - f_v + \dfrac{2f_v \times f_{yp}}{f_e}}$$

where f_{yp} and f_e are in shear as yield point stress and endurance limit, respectively.

The mean shear stress, f_m, is calculated by considering only the shear stress factor K_s, since experience suggests that the curvature correction factor is not needed for mean stress, so

$$f_m = \frac{8P_m D}{\pi d^3} K_s$$

$$f_v = \frac{8\,P_v\,D}{\pi d^3} K$$

where $\quad P_v = \text{Variable load} = \dfrac{P(\text{max.}) - P(\text{min.})}{2}.$

Values for f_e are not very complete, but for oil tempered carbon steel wire diameters upto 16 mm, a good approximation for its value can be obtained from the following equation,

$$f_e = \frac{475}{(d)^{0.2}} \text{ N/mm}^2, d \text{ is in mm}$$

The factor of safety or design factor may be taken as 1.8, more or less, depending upon working conditions.

15.8 HELICAL SPRINGS OF NON-CIRCULAR WIRE

The use of square or rectangular wire is not recommended for springs unless space limitations make it necessary. Wire sections are not made in large quantities, as are those of round wires. Square or rectangular wire is used to obtain the greatest load capacity in the smallest space, but this means that these springs are highly stressed. The non-circular section of the wire is also used to provide for predetermined altering of the spring rate by grinding off the outside of the coils. For grinding off the required calculations become very complicated for round wire springs but are relatively simple for rectangular wire. A spring with rectangular section of wire is shown in Fig. 15.9.

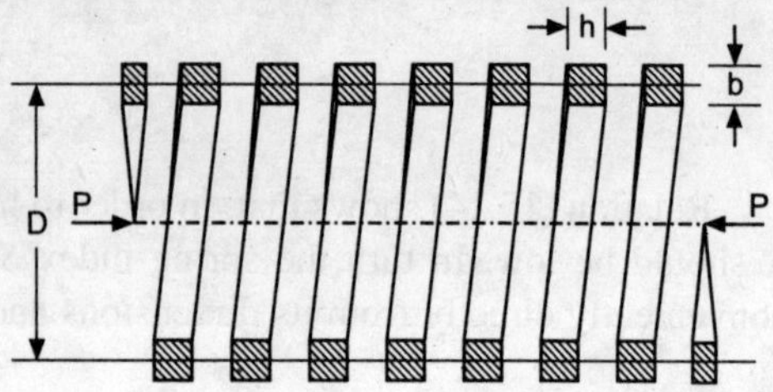

Fig. 15.9 *Spring with rectangular section*

An accurate determination of the stresses in a special section helical spring is difficult. The stress in a square wire spring, based on St. Venant's torsion theory for non-circular bars is

$$f_s = \frac{2.4\,PD}{b^3} K \qquad \text{...(15.17)}$$

where b is each side of square.

The deflection of a helical spring with square wire is

$$\delta = \frac{5.575\,PD^3 z}{Cb^4} \qquad \text{...(15.18)}$$

The stress in a rectangular wire section spring, if longer side is parallel to the axis of the spring.

$$f_s = \frac{PD(3h + 1.8b)}{2b^2h^2} K \qquad \text{...(15.19)}$$

For shorter side of the rectangular section parallel to the axis of the spring.

$$f_s = \frac{PD(3b + 1.8h)}{2b^2h^2} K \qquad \text{...(15.20)}$$

The deflection of a coil spring with rectangular wire is

$$\delta = \frac{2.45\,PD^3 z}{Cb^3(h - 0.56b)} \qquad \text{...(15.21)}$$

The expression for the stiffness of the spring of square and rectangular section wires can be obtained.

The resilience of a spring with a rectangular cross-section, generally speaking, is smaller than that of a round-wire spring. However, it increases as the ratio b/h decreases.

For square and rectangular section wire springs, the spring index S has the value D divided by the radial thickness of the wire, and K is

$$K = \frac{4S - 1}{4S - 4} + \frac{0.615}{S}$$

The main disadvantages of the non-circular wire section spring are:

1. Quality of the material used for springs is not so good.
2. The stress distribution is not as favourable as for circular wires, but this effect is negligible where loading is of static nature.
3. The shape of the wire does not remain square or rectangular while forming helix, this results in trapezoidal cross-sections. It reduces the energy absorbing capacity of the spring.

15.9 MATERIALS FOR COILED SPRINGS

The majority of coil springs are made of oil-tempered carbon steel wire containing 0.60 to 0.70 per cent carbon and 0.60 to 1.0 per cent manganese. Music wire is used for high grade springs using wire less than 3 mm diameter. Annealed wire containing 0.85 to 0.95 per cent carbon and 0.30 to 0.40 per cent manganese is used in the larger sizes that are coiled hot and hardened and drawn after coiling. Nickel steel, chromium steel (stainless steel), brass, phosphor bronze, monel metal and other metals that can be hard drawn are used in special cases to increase fatigue resistance, corrosion resistance, and temperature resistance.

A recent development in the method of producing high quality springs is from Nylons and plastics. Rubber is often used as spring materials. Spring operating in humid and chemical atmosphere may be made of bronze or brass.

The strength of a drawn wire increases with the reduction in size and hardness increases with a corresponding loss in ductility. The process of drawing if passing through a heated surface after which it is dipped in lead bath, subsequent operation makes the outer layer cold hardened and this improves the elastic property of material.

Springs are wound cold upto 12 mm wire size and for larger size are wound hot.

Since coil springs are in torsion, the design stress must be based on the torsional shear strength. Tests indicate that the yield stress in torsion is approximately 45 per cent of the ultimate strength in tension for oil-tempered wire, and 35 per cent for hard drawn wire. Safe working stress should not exceed 80 per cent of the yield stress in torsion, and if the spring is continuously subjected to rapid load fluctuations, the working stress must be reduced from 25 to 50 per cent. The safe working stress in shear and the modulus of rigidity for different materials are given in Table 15.1.

Table 15.1

Materials	Working stress in shear N/mm^2	Modulus of rigidity C, GN/m^2
Music wire	460	80 to 90
Stainless steel	325	80 to 90
Z-nickel spring wire	400	80 to 90
K-Monel	325	65
Beryllium Copper	225	55
Phosphor Bronze	225	48
Brass spring	160	40

Shear stress values for steel springs subjected to variable or reversing loads, are taken as low as 300 N/mm^2.

The value for other steels and beryllium copper may be taken as 0.8 times values of spring steel, where as 0.5 times values of spring steel are taken for bronze and monel metals, and 0.35 times values of spring steel for spring brass.

Springs are often grouped into classes, according to service conditions e.g., light service, average service, and severe service.

Light Service: Consists of operation under essentially static conditions and where the maximum load is carried in frequently, i.e., a total of less than 1,000. Examples are safety valves, slip coupling, and bolted joints.

Average Service: Includes springs subjected to the maximum load 1,000 to 100,000 times during service life. Examples are engine governor springs, automobile suspension, springs used in circuit breaker mechanisms.

Severe Service: Consists of a large number of cycles, i.e., greater than one million of varying load where the maximum load is one-half or less of the maximum load. The common example of severe service is that of automotive valve springs.

Tables of allowable stresses for different wire sizes and classes of service have been widely used for designing springs, e.g., Table 15.2 is for helical compression spring.

Table 15.2 Allowable shear stresses in N/mm^2

Wire diameter mm	Light Service	Average Service	Severe Service
Upto 2.12	650	525	420
2.24—4.5	600	490	400
4.75—8.00	520	420	340
8.5—13.2	460	370	300
14.0—22.4	400	320	250
23.6—37.5	350	280	225

15.10 SOLID STRESS

This is the stress induced in the spring wire when it is compressed to the extent that coils touch each other. Stresses at this stage are very high due to touching of coils and it is just possible that permanent set may take place. Allowable values are provided in Table 15.3. If the stress exceeds this given value, some method, like pre-stressing could be used to increase the strength of the spring. The beneficial results of pre-stressing will be decreased if the spring is subsequently heated above 300ºC and will be completely lost at 450ºC.

Table 15.3 Allowable Solid Stresses

Material	Maximum stress N/mm^2 × 10^{-3}
Hard drawn carbon steel $1.25 < d < 12.5$ mm	$0.75/d^{0.2}$
Oil tempered carbon steel $1.25 < d < 12.5$ mm	$1.026/d^{0.22}$
Oil tempered alloy steel $1.25 < d < 12.5$ mm	$1.03/d^{0.125}$
Music wire $0.75 < d < 3.75$ mm	$1.08/d^{0.17}$
Hard drawn stainless steel $1.25 < d < 12.5$ mm	$0.8/d^{0.22}$
Phosphor bronze $1.88 < d < 12.5$ mm	$0.67/d^{0.25}$
Spring brass $1.88 < d < 12.5$ mm	$0.50/d^{0.25}$

15.11 SOLID LENGTH AND FREE LENGTH

Compression springs are usually compressed solid when subjected to maximum permissible stress. The free length will be the closed length plus the maximum deflection. The closed length depends upon the number of effective coils and the type of the end provided. Different types of ends are illustrated in Fig. 15.10. Table 15.4 indicates the equations to determine the actual number of turns, solid length, and free length. Note that the actual number of turns is more than the effective number of turns. The pitch p of the coils is given by

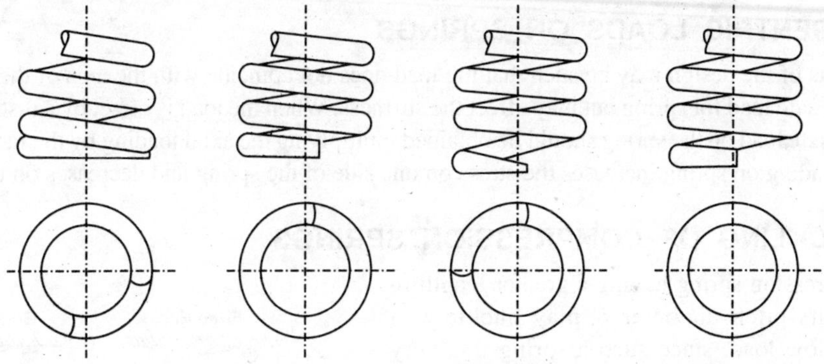

Fig. 15.10 *Different types of ends for compression springs*

Table 15.4

Type of end	Actual number of turns (z')	Solid length	Free length
(a)	z	$(z + 1)\,d$	$zp + d$
(b)	z	zd	zp
(c)	$z + 2$	$(z + 3)\,d$	$zp + 3d$
(d)	$z + 2$	$(z + 2)\,d$	$zp + 2d$

$$p = \frac{\text{free length} - \text{solid length}}{z'} + d \qquad \text{...(15.22)}$$

where z' is the actual number of turns.

Solid length of spring = $z' \times d$ ...(15.23)

Free length = Solid length + deflection under the load + clash allowance. ...(15.24)

Where z' is the actual number of turns and d is the diameter of the spring wire.

While choosing the pitch of the coils, the following points should be noted.

(a) The spring must not close up before the maximum service load is reached.

(b) If the spring is accidently or carelessly compressed, the pitch of the coils should be such that, the stress does not increase the yield point stress in torsion.

In general the active number of coils z does not equal the total number of turns z'. This is because of the use of different types of end construction. The plain ends are the most economical to produce. But, since the spring bows sideways, the loading, deviates considerably from the axial loading assumed in the derivation of the stress equations, and the stress on one side of the spring may be considerably higher than calculated. The closest approach to axial loading is given by squared and ground ends and the maximum deviation from axial loading is given by squared end but left ungrounded.

As the load increases the number of inactive coils also increases due to the seating of the end coils and the amount of increase varying from 0.5 to 1 turn at usual working loads.

Compression springs are seldom designed to close up under the maximum working load. An allowance, known as the clash allowance, of 15 per cent of the maximum working deflection is usually adequate to avoid clashing of the coils during service.

15.12 ECCENTRIC LOADS ON SPRINGS

The requirements of the design may be such that the load does not coincide with the axis of the coils. This not only reduces the safe load for spring but may affect the stiffness. When the load is offset by a distance e from the spring axis, the safe load on the spring should be obtained multiplying the axial loading by the factor $D/(2e + D)$. The eccentric loading on spring increases the stress on one side of the spring and decreases on the other side.

15.13 BUCKLING OF COMPRESSION SPRINGS

A helical compression spring having a greater length in proportion to its pitch diameter D may buckle at a comparatively low load, since such a spring is a very flexible column as shown in Fig. 15.11. It is found experimentally that a compression spring in which the free length is more than 4 times the pitch diameter D may fail by buckling. The critical axial load Pcr that can cause buckling may be found by the formula

$$P_{cr} = P_0 \times K_l \times l_0 \quad ...(15.25)$$

where P_0 is the spring scale in N/mm

l_0 is the free length of the spring in mm

K_l is a factor depending on the ratio l_0/D as shown in Fig. 15.12.

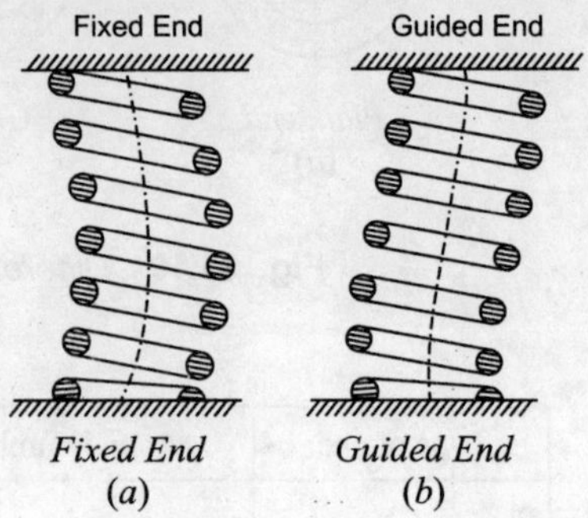

Fig. 15.11 *Buckling of compression spring*

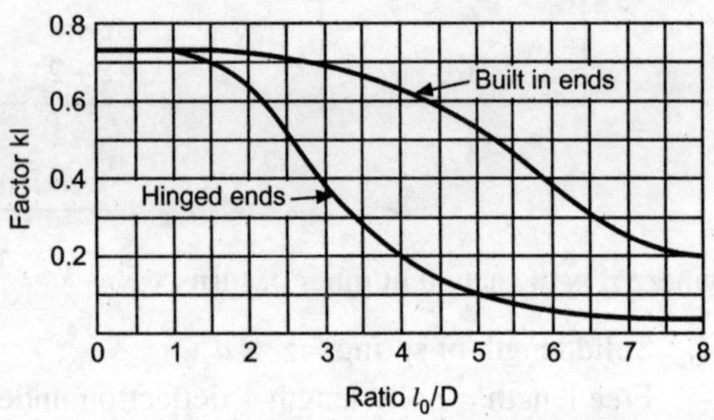

Fig. 15.12 *Buckling factor for helical compression springs*

In the selection of the value of K_l from Fig. 15.12, a hinged end spring may be considered as one supported on pivots at both ends as in case of plain ends, and a built in end as one in which a squared and ground-end spring is compressed between two rigid and parallel flat plates.

The buckling of the spring can be avoided by mounting it either on a central rod or by locating the spring on a tube. In the latter case, it is however desirable to keep the clearance between the tube walls and the spring as small as possible. But it must be sufficient to allow for increase in spring diameter during compression.

15.14 VIBRATION OF CYLINDRICAL SPRINGS

When a helical spring rests on a rigid support at one end and the other end is deflected by a suddenly applied load, the coils of the spring will not suddenly share the deflection equally, since time is required for the propagation of stress along the spring wire. An analysis of the progression of the deflection is that the end coil in contact with the applied load deflects and then this coil transmits a large part of its deflection to the adjacent coil. Thus, in progression, a wave of compression coil travels along the spring to the supported end, where it is reflected and travels back to the deflected end. This wave travels along the spring indefinitely. If the time required for the wave to travel from one end of the spring to the other and return coincides with the time interval between load applications, a condition of resonance is reached and very large deflections of the coils will be produced with correspondingly high stresses. Under this condition failure of the spring may occur. This phenomenon is termed as surge.

The speeds at which surging will occur correspond to the natural frequency of the spring, which depends upon the wire diameter, the coil diameter, the number of effective coils, and the elastic properties of the spring wire. The fundamental frequency is given by the expression,

$$f = \frac{d}{2\pi D^2 \cdot z} \cdot \sqrt{\frac{C \cdot g}{2w}}, \text{ cycles/sec.} \qquad ...(15.26)$$

For springs clamped between two plates. Here w = specific weight of the material.

For springs with one end free and the other clamped, the lowest natural frequency will be half of that if both the ends are clamped.

Experiments indicate that the actual frequency of the spring is from 10 to 15 per cent below that given by the equation 15.26. According to W.M. Griffith, the critical frequency of the spring should be at least twenty time the frequency of application of a periodic load, in order to avoid resonance with all harmonic frequencies upto twentieth order.

The surge in the springs can be eliminated by the following ways:

(*a*) Use of springs having high natural frequency.

(*b*) The presence of friction dampens on the centre coils, causes the wave propagation to die out.

(*c*) Use a modified cam contour.

(*d*) The pitch of the spring coils is made different at the ends than at the centre to have different natural frequencies when spring is compressed.

(*e*) In machine gun mechanism, the velocity of application of the load is high and the reflection of stress waves in the spring may give rise to resonant stresses. Springs made of stranded wire have shown improved performance in these applications. The springs are usually wound with wire made of three strands of performed wire in which the direction of winding the strands is opposite to the direction of winding the coils to form the spring, so that the deflection of the spring tends to wind the individual wires close together, and this introduces frictional damping and thus decreases the magnitude of the transient stresses. The fatigue life of standard wire springs has been increased as much as four times over life of conventional springs.

15.15 TENSION SPRINGS

Helical tension or extension springs are similar to helical compression springs. The equations already derived for compression springs apply as well to extension springs. The tension springs are usually wound with the coils close and under an initial tension, i.e. it is necessary to apply from 20 to 30 per cent of the maximum load before the coils begin to separate. The initial tension load can also be calculated by relation.

$$P_{\text{initial}} = \frac{fs \times d^3}{2.55D} N \qquad ...(15.27)$$

f_s is in N/mm^2 and d and D are in mm.

In the above equation, f_s is the torsional stress due to initial tension. Its values are given below:

S	3	4	5	6	7	8	9	10	11	12	13	14	15
f_s	168	157.5	140	126	113.4	101.5	91	81.2	74.2	67.9	61.6	55.3	49

Thus the total design load will be the sum of the initial load and the load due to the extension of the spring, or

$$P = P_i + P_0 \times \delta$$

where P = Total load applied

P_i = initial tensile load

P_0 = Spring scale

A few of the types of ends used on tension springs are shown in Fig. 15.13. When loops or hooks are provided the small radius where the hook joins the first coil is a region of high local stress. To provide for this local stress and for the greater possibility of excess deformation, the design stress for tension springs should not exceed 70 per cent of that used with compression springs (Tables 15.1 and 15.2). In tension springs to find the total number of active turns is a complicated affair. The total number of turns must be equal to the number of turns (z) between the points where the loops start plus the equivalent turns for the loops. It has been found experimentally that half turn should be added for each loop. Thus for a spring having loops on both ends, the total number of active turns will be $= z + 1$.

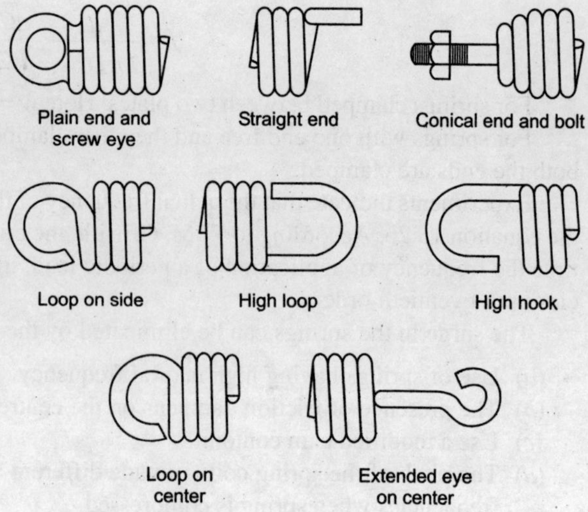

Fig. 15.13 *Extension spring ends*

Extension springs are used in machinery considerably less than compression springs. The main reasons for this are:

1. They are more expensive to manufacture.
2. There is more stress concentration at the end coils of a tension spring than at the loop or other attaching device of a compression spring.
3. They require more elaborate fastening of the ends.
4. Deflection of a compression spring to the solid condition acts as an automatic limit to the stress in the spring, whereas a tension spring is likely to be stressed beyond the elastic limit.
5. In case of failure, the kinematic constraint is lost.

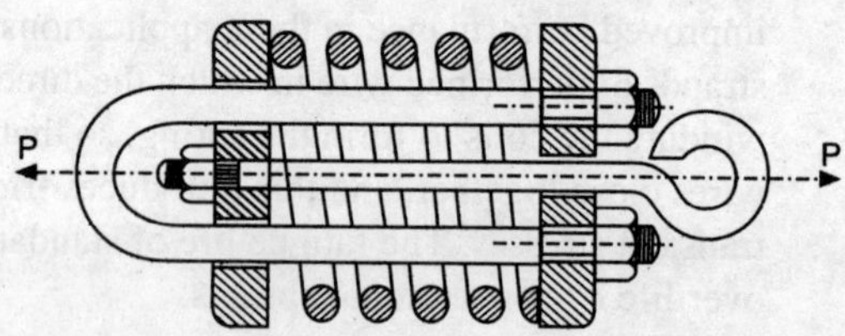

Fig. 15.14 *Compression spring for tension purposes*

To overcome the last two disadvantages listed, the arrangement shown in Fig. 15.14 may be used instead of an extension spring.

15.16 DESIGN PROCEDURE

The design of a coil spring involves a cut and try method, and the result should be checked by actual testing of the spring. Further a slight variation in the spring scale must be expected in springs that seem to have identical dimensions. The data usually known to design a helical spring are:

(*a*) Load on the spring, or the force which the spring must exert and the desired force increase.
(*b*) deflection.
(*c*) certain space limitations are given, such as the minimum inside diameter or the maximum outside diameter and the height of the spring under the working condition.
(*d*) Sometimes the energy to be absorbed by the spring is given and from this either deflection or load on the spring can be calculated.

The values to be found by the designer are the pitch diameter D, the size of the wire d, the number of turns z and z'. The free length l_0, the solid length, and the type of ends.

From the values known, the spring scale P_0 can be found.

The usual design procedure is as follows:

1. From space limitations or other conditions, the pitch diameter D is selected.
2. Select the material and see the permissible stress from Table 15.1. From this table the modulus of rigidity can also be seen.
3. By knowing this data, the wire diameter d can be found by relation 15.3.

4. The stress factor K can be seen from Fig. 15.7 or calculated by the relation given by A.M. Wahl. Then the more accurate value of d is found by relation 15.7. The more accurate value of the permissible stress can also be seen from table 15.2, corresponding to the wire diameter d calculated in step 3.
5. If the spring index S is directly assumed or known then $D = S \times d$. Experience has shown that values of S from 6 to 10 will result in a reasonably balanced design in relation to stability, size, and efficient use of material. If no other limitations need be considered, it is recommended that a spring index of 8 be used. For general industrial uses the spring index should be 8 to 10, for valve and cluch springs 5 is common, and 3 is a minimum value to be used in extreme cases. Finally, once S is known, then the value of K also be calculated or seen from Fig. 15.7. In this case the equation 15.7 involving S can be used for calculating the wire diameter.
6. The necessary minimum diameter d having been calculated, the actual diameter d is taken as the nearest size from I.S. 1137–1959. This standard gives the basic thickness of sheet and diameter of wire in millimeters, Table 15.5. A large size decreases the working stress and gives a slightly stiffer spring, while a smaller size increases the stress and gives a softer spring.
7. After a standard wire size has been determined, the number of active coils is calculated by use of the deflection equation (15.4) or (15.5). If deflection δ is not known, it must be determined from spring scale or equation 15.13. If z becomes too small, the spring is too soft. According to equation (15.4), the pitch diameter D must be decreased. This change will also slightly decrease d, and the calculations must be repeated. If z is too large, D must be increased.

Table 15.5

0.020	0.042	0.090	0.190	0.400	0.850	1.80	3.75	8.00	17.0
0.021	0.045	0.095	0.200	0.425	0.900	1.90	4.00	8.50	18.0
0.022	0.048	0.100	0.212	0.450	0.950	2.00	4.25	9.00	19.0
0.024	0.050	0.106	0.224	0.475	1.000	2.12	4.50	9.50	20.0
0.025	0.053	0.112	0.236	0.500	1.06	2.24	4.75	10.0	21.2
0.026	0.056	0.118	0.250	0.530	1.12	2.36	5.00	10.6	22.4
0.028	0.060	0.125	0.265	0.560	1.18	2.50	5.30	11.2	23.6
0.030	0.063	0.132	0.280	0.600	1.25	2.65	5.60	11.8	25.0
0.032	0.067	0.140	0.300	0.630	1.32	2.80	6.00	12.5	
0.034	0.071	0.150	0.315	0.670	1.40	3.00	6.30	13.2	
0.036	0.075	0.160	0.335	0.710	1.50	3.15	6.70	14.0	
0.038	0.080	0.170	0.355	0.750	1.60	3.35	7.10	15.0	
0.040	0.085	0.180	0.375	0.800	1.70	3.55	7.50	16.0	

8. When z is finally determined, the total number of coils z' can be determined from Table 15.4, after selecting the end type.
9. The free length of the spring should be calculated as given in table 15.4. To this should be added the clash allowance to prevent the spring from being compressed solid with the application of the maximum load. When the clash allowance is used, it is not necessary to check whether the stress exceeds the elastic limit (solid stress) because of mishandling, etc.
10. The solid length can also be calculated from Table 15.4.
11. The pitch of the coils can be calculated from relation (15.22) and the helix angle from relation (15.8).
12. Finally the designed spring can be checked for buckling, vibration of spring, etc.
13. The length of the wire for compression spring can be found from relation $\dfrac{\pi D \times z'}{\cos \alpha}$.
14. An additional factor that must often be considered is that the diameter of a helical spring increases under compression and decreases under tension. Adequate clearance should be provided to accommodate this motion without binding.

A similar analysis can be used to determine the dimension of a tension spring.

Example 15.1. A gas engine valve spring is to have a mean diameter of 37.5 mm. The maximum load it will have to sustain is 450 N with a corresponding deflection of 12.5 mm. The spring is to be made of tempered wire. Since the material is to be subjected to repeated loading and fatigue must be considered, a low working stress of 300 N/mm² will be used. Determine the size for the wire and the number of coils to be used.

Solution: Assuming the spring index $S = 6$

Then, $$K = \frac{4\times6-1}{4\times6-4} + \frac{0.615}{6} = 1.2525$$

Now by using the relation 15.7, the diameter d can be calculated,

$$f_s = K \times \frac{8PD}{\pi d^3}, \quad 300 = \frac{1.2525\times8\times450\times37.5}{\pi d^3}$$

gives, $$d = 5.65 \text{ mm.}$$

From Table 15.5, the standard wire size will be 5.6 mm.

Thus, actual value of $S = \frac{D}{d} = \frac{37.5}{5.6} = 6.7$

The assumption is quite close to 6.0. To be more accurate, the design procedure can be repeated for value of $S = 6.7$.

The number of turns can be calculated by relation 15.4.

$$\delta = \frac{8PD^3 z}{Cd^4}, 12.5 = \frac{8\times450\times(37.5)^3 z}{0.8\times10^5\times(5.6)^4}$$

gives, $z = 5.18$.

Since this is a compression spring a total of 1.82 inactive turns should be allowed, making the total number of turns = 5.18 + 1.82 = 7 turns.

Example 15.2. A spring is made from a wire of 1.25 mm diameter and 750 N/mm² as its yield strength. For a mean diameter of 12.5 mm and 14 active coils of the spring, determine (*a*) static load corresponding to the yield point of the material and deflection corresponding to that, (*b*) solid height assuming that the ends are squared and ground, (*c*) stiffness of spring, (*d*) pitch of the wire so that the solid stress will not exceed the yield point. Take $C = 0.85 \times 10^5$ N/mm².

Solution: Spring index $S = \frac{D}{d} = \frac{12.5}{1.25} = 10$

stress factor $$K = \frac{4\times10-1}{4\times10-4} + \frac{0.615}{10} = 1.145$$

The static load can be obtained by the use of relation 15.7,

$$f_s = K\frac{8PD}{\pi d^3}, 750 = \frac{1.145\times8\times P\times12.5}{\pi\times(1.25)^3}$$

gives, $P = 40$ N.

The deflection produced corresponding to this static loading can be calculated by relation 15.4.

$$\delta = \frac{8PD^3 z}{Cd^4}, \quad \delta = \frac{8\times40\times(12.5)^3\times14}{0.85\times10^5\times(1.25)^4}$$

gives, $\delta = 42$ mm.

Stiffness of spring $= \frac{P}{\delta} = \frac{40}{42} = 0.951$ N/mm

Solid length of spring $= (z+2)d = (14+2)\times1.25 = 20$ mm

Free length of spring = Solid length + deflection under load + clash allowance

= 20 + 42 + 15 per cent of the working deflection.

$$= 20 + 42 + \frac{15}{100} \times 42 = 68.3 \text{ mm}$$

pitch of the coil can be obtained by relation 15.22.

$$p = \frac{\text{free length} - \text{solid length}}{z'} + d = \frac{68.3 - 20}{16} + 1.25 = 4.27 \text{ mm.}$$

Example 15.3. A cylinder relief valve should blow at a pressure of 1.5 N/mm² and should lift by 6 mm for a 5 per cent increase in pressure. The diameter of the valve is 60 mm. Design the valve spring giving a suitable combination of (*a*) mean coil diameter, (*b*) wire diameter, (*c*) number of active and inactive coils, (*d*) pitch of coils, (*e*) type of ends, (*f*) solid length and free length of spring.

Permissible shear stress for the spring material may be taken as 500 N/mm² and $C = 0.85 \times 10^5$ N/mm².

Spring wire sizes are 2, 3, 4, 5, 6, 7.1, 8, 9, 10 and 11.2 mm. Also for S = 4, 5, 6, 7, 8, 9 and 10, the stress factor K is 1.4, 1.32, 1.2525, 1.21, 1.184, 1.187, and 1.15 respectively.

Solution: Deflection of spring = 6 mm

Load on valve before it lifts = $\pi/4 \times (60)^2 \times 1.5 = 1350\,\pi$ N.

Load on valve after it lifts = $\pi/4\,(60)^2 \times 1.5 \times 1.05 = 1417.5\,\pi$N

Load causing the deflection of 6 mm = $(1417.5 - 1350)\pi = 67.5\,\pi\text{N} = 213$ N

Assuming the spring index $S = 8$

Maximum load on spring = $1417.5\,\pi = 4750$ N

For spring index to be 8, the stress factor K will be 1.184. By using equation 15.7, we can find the wire diameter d.

$$f_s = K\frac{8PS}{\pi d^2},\; 500 = \frac{1.184 \times 4750 \times 8 \times 8}{\pi \times d^2}$$

gives, $d = 15.2$ mm.

The given sizes of wires are not suitable, therefore the standard size of wire from table 15.5 is 16 mm.

Adopt $d = 16$ mm., $D = 8 \times 16 = 128$ mm

The number of turns can be obtained from relation 15.5,

$$\delta = \frac{8PS^3z}{Cd} \quad \text{or} \quad \frac{P}{\delta} = \frac{Cd}{8S^3z}$$

$$\therefore \quad \frac{213}{6} = \frac{0.85 \times 10^5 \times 16}{8 \times (8)^3 \times z}, \text{ gives } z = 9.4$$

Assuming squared and ground end of the spring.

Total number of turns = $z' = z + 2 = 9.4 + 2 = 11.4$

Solid length of spring = $(z + 2)d = z' \times d = 11.4 \times 16 = 182.4$ mm

Free length of spring = Solid length + deflection under load + clash allowance

$$\text{deflection under load} = \frac{4750}{213} \times 6 = 133 \text{ mm}$$

∴ Free length = 182.4 + 133 + 0.15 × 133 = 335.4 mm

Pitch of the coils can be obtained by relation 15.22

$$p = (\text{Free length} - \text{Solid length})\frac{1}{z'} + d$$

$$p = \frac{335.4 - 182.4}{11.4} + 16 = 29.4 \text{ mm.}$$

Example 15.4. Calculate the dimensions of a helical spring for a spring loaded safety valve from the following data:

Valve diameter	**...65 mm**
Maximum pressure when the valve blows off freely	**0.73 N/mm²**
Valve lift when pressure rises from 0.7 to 0.73 N/mm²	**3.2 mm**
Maximum permissible stress	**500 N/mm²**
Spring index, S	**...6**
Modulus of rigidity	**...0.85 × 10⁵ N/mm²**

Give a neat sketch of the free spring showing the leading dimensions. Assume this to be a Ramsbottom safety valve.

Solution: As the spring is put in Ramsbottom safety valve, therefore it will be under tension

Tensile force on the spring before the valve lifts $= \frac{\pi}{4}(65)^2 \times 0.7 = 2300$ N

Tensile force on the spring after the valve lifts $= \frac{\pi}{4}(65)^2 \times 0.73 = 2430$ N

Force causing the deflection of 3.2 mm = 2430 – 2300 = 130 N

If may be noted that the maximum stress in the wire is produced at 2430 N of load.

$$\text{Stress factor } K = \frac{4S-1}{4S-4} + \frac{0.615}{S} = \frac{4\times6-1}{4\times6-4} + \frac{0.615}{6} = 1.2525$$

$$\text{Spring stiffness} = \frac{130}{3.2} = 40.7 \text{ N/mm}$$

Now the wire diameter can be found, by using relation 15.7.

$$f_s = K\frac{8PS}{\pi d^2},\ 500 = \frac{8\times2430\times6}{\pi d^2}\times1.2525$$

gives, $d = 9.75$ mm.

The standard size of wire from table 15.5 is 10 mm.

Mean coil diameter $D = d \times S = 6 \times 10 = 60$ mm.

Outside diameter of coil $D_o = D + d = 60 + 10 = 70$ mm.

Inside diameter of coil $D_i = D - d = 60 - 10 = 50$ mm.

Mean coil radius $R = 30$ mm.

In order to evaluate the number of coils required, we use the deflection equation 15.5.

$$\delta = \frac{8PS^3z}{Cd},\ \frac{P}{\delta} = \frac{Cd}{8S^3z}$$

$$40.7 = \frac{0.85\times10^5\times10}{8\times(6)^3\times z},\ \text{gives } z = 12.0$$

Thus for a spring having loops on both ends, the total number of active turns will be

$$z' = z + 1 = 12.0 + 1 = 13.0$$

The least gap between two coils when the spring is in free state will be 1 mm.

∴ Free length of spring $= zd + (z-1) \times 1$

$= 12 \times 10 + 11 \times 1 = 131$ mm.

$$\text{Pitch } (p) \text{ of the coil} = \frac{\text{free length}}{z-1} = \frac{131}{12-1} = 11.9 \text{ mm.}$$

$$\text{Total deflection of the spring} = \frac{P_2}{P_0} = \frac{2430}{40.7} = 60 \text{ mm}$$

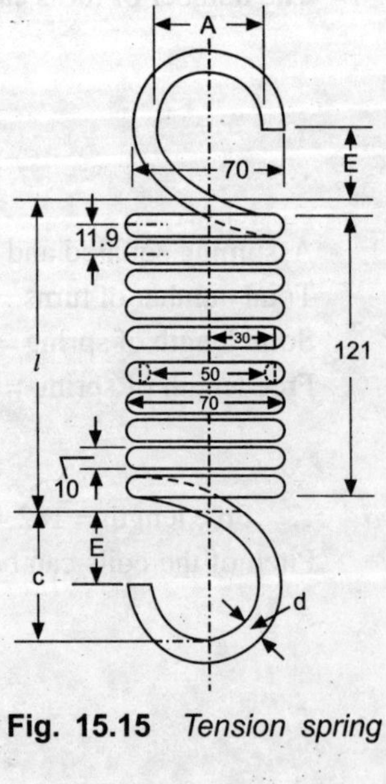

Fig. 15.15 *Tension spring*

Extended length of spring = free length + deflection

$= 131 + 60 = 191$ mm.

Tension spring with full loop over centre is shown in Fig. 15.15.

Dimensions A and C are approximately $= D_i = 50$ mm.

Dimension $E = \frac{A}{3} = \frac{50}{3} = 16.7$ mm.

Dimension $l = (z + 1)\, d + (z - 1) \times 1 = 13 \times 10 + 11 \times 1 = 141$ mm.

where z is the number of turns in dimension l between points where loops begin.

Thus overall free height of spring $= l + 2 \times c + 2 \times d = 141 + 2 \times 50 + 2 \times 10 = 261$ mm.

Example 15.5. The following particulars refer to the valve spring of a petrol engine.

Length of the spring when the valve is open	**...41 mm.**
Length of the spring when the valve is closed	**...49 mm.**
Spring load when the valve is open	**...360 N**
Spring load when the valve is closed	**...220 N**
Maximum inside diameter of spring	**25 mm.**

Design the spring for the maximum permissible shear stress of 400 N/mm² and modulus of rigidity 0..83 × 10⁵ N/mm². Also determine the dangerous engine speed.

Solution: Compression in the spring $= 49 - 41 = 8$ mm.

The compression of the spring is caused by a load of $= 360 - 220 = 140$ N

Maximum load for which the spring is to be designed is 360 N

The inside diameter of the coil is not to be more than 25 mm.

$\therefore$ Mean coil diameter $= 25 + d$ mm.

By using equation 15.3, we can find the wire diameter d.

$$f_s = \frac{8PD}{\pi d^3},\ 400 = \frac{8 \times 360 \times (25 + d)}{\pi d^3},\ \text{gives } d = 3.96 \text{ mm.}$$

$$\therefore \quad D = 25 + 3.96 = 28.96 \text{ mm., spring index } S = \frac{D}{d} = \frac{28.96}{3.96} = 7.3,$$

$$\text{Stress factor } K = \frac{4S - 1}{4S - 4} + \frac{0.615}{S} = \frac{4 \times 7.3 - 1}{4 \times 7.3 - 4} + \frac{0.615}{7.3} = 1.2$$

Again the wire diameter d can be calculated by relation 15.7, which takes the stress factor into account.

$$f_s = K\frac{8PS}{\pi d^2};\ 400 = \frac{1.2 \times 8 \times 360 \times 7.3}{\pi \times d^2},\ \text{gives } d = 4.5 \text{ mm.}$$

Again the procedure must be repeated to get the most accurate value of d, and it will be 4.25 mm.

Mean coil diameter $= 25 + 4.25 = 29.25$ mm.

Inside coil diameter $= 25$ mm.

Outside coil diameter $= D + d = 29.25 + 4.25 = 33.5$ mm.

Maximum deflection under 360 N load can be calculated by relation 15.13.

$$\delta_2 = \frac{\delta' \times P_2}{P_2 - P_1} = \frac{8 \times 360}{360 - 220} = 20.6 \text{ mm}$$

Now the number of turns can be calculated from relation

$$\delta = \frac{8PD^3 z}{Cd^4},\ 20.6 = \frac{8 \times 360 \times (29.25)^3 \times z}{0.83 \times 10^5 \times (4.25)^4}$$

gives $z = 7.65$. Assuming the square and ground end,

total number of turns will be $z' = z + 2 = 9.65$.

$$\text{solid length of spring} = z'd = 9.65 \times 4.25 = 41 \text{ mm.}$$
$$\text{free length of spring} = 41 + 20.6 + 0.15 \times 20.6 = 64.7 \text{ mm.}$$
$$\text{pitch of the coil} = \frac{64.7 - 41}{9.65} + 4.25 = 6.7 \text{ mm.}$$

The dangerous engine speed will be given by relation 15.25.

$$f = \frac{d}{2\pi D^2 z}\sqrt{\frac{Cg}{2w}}$$

For designed spring, d = 4.25 mm, D = 29.25 mm.

z = 9.65, $C = 0.83 \times 10^5$ N/mm², w = 0.0725 MN/m³

By putting these values

$$f = \frac{4.25}{2\pi(29.25)^2 \times 9.65}\sqrt{\frac{0.83 \times 10^5 \times 9810}{2 \times 0.0725 \times 10^6 \times 10^{-9}}}$$

$$f = 194 \text{ V.p.s.}$$
$$= 11640 \text{ V.p.m. or r.p.m.}$$

Since the cam shaft speed is half of the engine speed (for four stroke engine) the most dangerous engine speed would be 11640 × 2 = 23280. r.p.m. Other dangerous speeds would be $\frac{23280}{2}$ = 11640 r.p.m., $\frac{23280}{3}$ = 7760 r.p.m. and $\frac{23280}{4}$ = 5820 r.p.m. Lower engine speeds are not likely to set up surge action, because of the damping effect of the spring material between the impulses.

When one end of the spring is clamped in a plate and the other end is free, the fundamental frequency of the spring will be $\frac{194}{2}$ = 97 V.p.s.

Example 15.6. A loaded narrow gauge car weighing 16 kN and moving at a velocity of 1.2 m/s brought to rest by a bumper consisting of two helical steel springs of square section. The mean coil diameter of the springs is six times the side of the square. In bringing the car to rest the springs are compressed by 200 mm. The permissible shear stress is not to exceed 400 N/mm². Find the following: (a) Maximum load on each spring, (b) side of square section of wire, (c) mean diameter of coils, (d) number of active coils. Take C = 0.84 × 10⁵ N/mm².

Solution: The kinetic energy of the moving car is to be absorbed by two springs when each is compressed by 200 mm. kinetic energy of the moving car

$$= \frac{1}{2}\frac{W}{g}V^2 = \frac{1}{2} \times \frac{16 \times 10^3}{9.81} \times (1.2)^2 = 1175 \text{ Nm.}$$

Let maximum force on each spring = P.

If the springs are loaded from zero load to load P gradually, then

$$\text{Work done on each spring} = \frac{P}{2} \times \delta$$

$$\therefore \quad 2 \times \frac{P}{2} \times \delta = 1175, \; 2 \times \frac{P}{2} \times 0.2 = 1175$$

gives, P = 5870 N

$$\text{Spring stiffness} = \frac{5870}{200} = 29.35 \text{ N/mm.}$$

Spring index S = 6, then the stress factor K will be

$$K = \frac{4S - 1}{4S - 4} + \frac{0.615}{S} = \frac{4 \times 6 - 1}{4 \times 6 - 4} + \frac{0.615}{6} = 1.2525.$$

By using relation 15.17, the side of the square can be determined.

$$f_s = \frac{2.4PD}{b^3}K,\ 400 = \frac{2.4 \times 5870 \times 6b}{b^3} \times 1.2525$$

gives, $b = 16.25$ mm., we adopt 17 mm as the side of square.

Mean coil diameter $D = 6b = 6 \times 17 = 102$ mm.

The number of active coils can be obtained from relation 15.18

$$\delta = \frac{5.575PD^3z}{Cb^4},\ 200 = \frac{5.575 \times 5870 \times (102)^3 \times z}{0.84 \times 10^5 \times (17)^4}$$

gives, $z = 40.3$.

$\therefore$ active number of coils = 40.3.

Example 15.7. A motor vehicle single plate clutch is to have both sides of the plate effective. It is to transmit 22 kW power at a speed of 2800 rev./min. and a 20 percent overload is allowed. The pressure intensity on the friction surface is not to exceed 0.085 N/mm² and the surface speed at the mean radius must not be greater than 35 m/s. The coefficient of friction for the surface is 0.3. The outside diameter of the surfaces is to be 1.3 times the inside diameter. The axial thrust is to be provided by six springs of about 25 mm mean coil diameter. The maximum permissible stress in shear is limited to 400 N/mm². Take C = 0.84 × 10⁵ N/mm². Design the springs, when the spring wire sizes are 2.0, 2.12, 2.24, 2.36, 2.5, 2.65, 2.8, 3.0, 3.15, 3.35, 3.55, 3.75, 4.0, 4.25, 4.50, 4.75, 5.0, 5.3, 5.60 and 6.0 mm.

Solution: Torsional moment to be transmitted by the clutch $= \frac{60 \times W}{2\pi n}$ Nm

$$= \frac{60 \times 22 \times 10^3}{2\pi \times 2800} = 77 \text{ Nm.}$$

As 20 per cent over load is allowed, therefore design torsional moment for clutch $1.2 \times 77 = 92.4$ Nm.

Let R_1 be the inside radius of the friction lining, then its outside radius will be 1.3 R_1. Assuming the uniform wear of friction lining and the maximum intensity of wear occurs at the inner radius. From the mechanics of machine we have the relation.

$$p \times R = \text{constant} = C$$

where p is the pressure intensity which is not to exceed 0.085 N/mm². For the assumption of uniform wear. the torque transmitted $= \pi\mu C[R_2^2 - R_1^2]$

As both the sides of the friction plate are effective, therefore, the torsional moment to be transmitted by each surface $= \frac{92.4}{2} = 46.2$ Nm.

Equating, the two, we have

$46.2 \times 10^3 = \pi \times 0.3 \times 0.085\ R_1\ [(1.3\ R_1)^2 - R_1^2]$

gives, $R_1 = 95$ mm. say 100 mm, $R_2 = 1.3 \times R_1 = 130$ mm.

$$\text{Mean radius} = \frac{R_1 + R_2}{2} = \frac{100 + 130}{2} = 115 \text{ mm.}$$

$\therefore$ Surface speed at mean radius $= 2\pi \times \frac{115}{1000} \times 2800 = 2030$ m/min

which is less than the allowed value of 2100 m/min.

The axial load to be applied $= 2\pi C(R_2 - R_1) = 2\pi \times 0.085 \times 100(130 - 100) = 1600$ N

The axial load is to be provided by 6 springs.

$\therefore$ Maximum load on each spring $= \frac{1600}{6} = 267$ N

The mean coil diameter $D = 25$ mm.

The wire diameter can be obtained by relation 15.3,

$$f_s = \frac{8PD}{\pi d^3}, 400 = \frac{8\times267\times25}{\pi d^3}$$

gives, $d = 3.49$ mm, spring index $= D/d = S = \frac{25}{3.49} = 7.19$

$$\text{stress factor } K = \frac{4S-1}{4S-4}+\frac{0.615}{S} = \frac{4\times7.15-1}{4\times7.15-4}+\frac{0.615}{7.15} = 1.246.$$

Again the wire diameter can be calculated by relation 15.7, which takes the stress factor into account.

$$f_s = K\frac{8PD}{\pi d^3}, 400 = \frac{1.246\times8\times267\times25}{\pi d^3} \quad \text{gives, } d = 3.75 \text{ mm.}$$

The calculations can be again repeated for more correct size of wire. The standard size of wire will come out to be 3.75 mm.

Let us assume that 8 number of coils are there in the spring which are effective. To give the axial thrust of 267 N by each spring, the initial compression will be given by the relation 15.4

$$\delta = \frac{8PD^3z}{Cd^4}, \delta = \frac{8\times267\times(25)^3\times8}{0.84\times10^5\times(3.75)^4} \text{ gives, } \delta = 17.5 \text{ mm.}$$

Minimum free length of the spring $= z'd + \delta = 10\times3.75 + 17.5 = 55$ mm.

The ends of the spring have been assumed to be square and ground, giving two ineffective coils.

$$\text{Length of wire required will be } = \frac{\pi Dz'}{\cos\alpha} = \frac{\pi\times25\times10}{\cos5^\circ} = 790 \text{ mm.}$$

Example 15.8. In a free rolling conveyer shown in Fig. 15.16, a crate loaded with non-fragile material, 300 N tare, reaches station L at a velocity of 3 m/s and is normally unloaded at station M. In case it gets away from M, a pair of helical buffer springs at N are expected to arrest the motion of crate. It is recommended that the springs may be set up with an initial compression of about 50 mm and the compression of the spring due to the impact of load may be limited to 150 mm. Design the springs. Take modulus of rigidity as 0.8 × 10^5 N/mm^2 and the allowable shear stress of 800 N/mm^2.

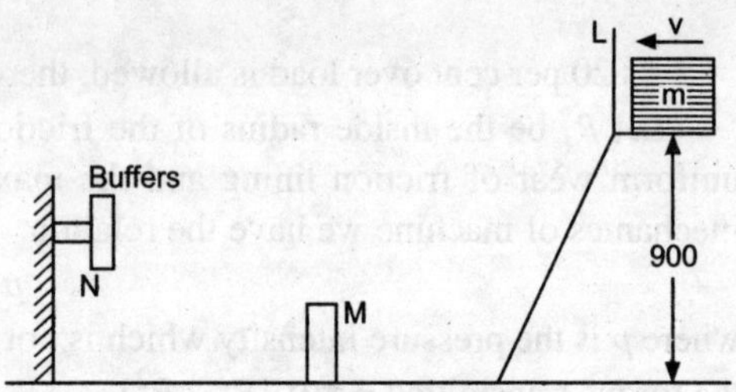

Fig. 15.16

Solution: The kinetic energy of the rolling conveyer $= \frac{1}{2}$ W/g V^2

$$= \frac{1}{2}\times\frac{300}{9.81}\times(3)^2 = 138 \text{ Nm.}$$

$$\text{Potential energy of mass} = Ph = 300\times\frac{900}{1000} = 270 \text{ Nm}$$

$$\text{Total energy, } E = 138 + 270 = 408 \text{ Nm}$$

Deflection of spring to absorb this energy = 150 mm.

Let P be the load on each spring, producing 150 mm deflection.

$$\text{Then stiffness of spring } P_0 = \frac{P}{150} \text{ N/mm}$$

$$\therefore \quad \frac{408\times10^3}{2} = \frac{1}{2}\left(\frac{P}{150}\times50+\frac{P}{150}\times200\right)\times150 \text{ gives, } P = 1632 \text{ N}$$

$$\text{spring stiffness } = P_0 = \frac{1632}{150} = 10.9 \text{ N/mm.}$$

Maximum load acting on spring = 200 × 10.9 = 2180 N

This load will produce a deflection of 200 mm.

Assume the spring index $S = 8$

$$\text{Stress factor } K = \frac{4\times8-1}{4\times8-4}+\frac{0.615}{8} = 1.187$$

Now the wire diameter can be calculated from relation

$$f_s = \frac{8PS}{\pi d^2}K,\ 800 = \frac{8\times2180\times8}{\pi d^2}\times1.187 \text{ gives, } d = 8.16 \text{ mm.}$$

Standard wire size from table 15.5 is 8.5 mm

Mean coil diameter = 8 × 8.5 = 68 mm

The number of turns can be calculated by relation

$$\delta = \frac{8PD^3z}{Cd^4},\ 200 = \frac{8\times2180\times(68)^3\times z}{0.8\times10^5\times(8.5)^4}, \text{ gives } z = 15.5$$

Assuming the squared end for construction of spring,

Actual number of turns, $z' = z + 2 = 15.5 + 2 = 17.5$

Solid length $= (z + 3)d = (15.5 + 3)\ 8.5 = 150.5$ mm.

Free length = 150.5 + 200 + 15/100 × 200 = 380.5 mm.

$$\text{Pitch of spring } = \frac{380.5-150.5}{17.5}+8.5 = 21.7 \text{ mm}$$

Check for buckling. Spring scale, $P_0 = 10.9$ N/mm

Free length of spring, $l_0 = 380.5$ mm

$$\text{Ratio, } \frac{l_0}{D} = \frac{380.5}{68} = 5.6$$

from Fig. 15.12, the factor K_l for built in ends will be = 0.49

Now the critical load that can cause buckling can be checked by relation 15.25.

$$P_{cr} = P_0 \times K_l \times l_0,\ P_{cr} = 10.9\times380.5\times0.49$$

$$P_{cr} = 2100 \text{ N}$$

which is very near to a load of 2180 N, hence safe.

Example 15.9. A spring of mean coil diameter 50 mm, wire diameter 10 mm, active number of coils, 11 and free length 180 mm, is available. It is to be used at a place where load varies from P to $3P$. Find the value of load this spring can take up. Given that f_{yp} = 700 N/mm², and design factor 1.3. Take $f_c = \frac{475}{d^{0.2}}$. N/mm², d is in mm. Also find the buckling load and critical frequency. Take $C = 0.8 \times 10^5$ N/mm², and $w = 0.08$ MN/m³.

Solution: f_c includes all the modifying factors $= \dfrac{475}{(10)^{0.2}} = 300$ N/mm²

$$\text{Mean load } = \frac{P+3P}{2} = 2P$$

$$\text{variable load } = \frac{3P-P}{2} = P$$

$$\text{Spring index, } S = D/d = \frac{5}{1} = 5.0$$

$$K = \frac{4S-1}{4S-4} + \frac{0.615}{S} = \frac{4\times5-1}{4\times5-4} + \frac{0.615}{5} = 1.313$$

$$\text{Shear stress multiplication factor} = K_s = 1 + \frac{0.5}{S} = 1 + \frac{0.5}{5} = 1.1$$

$$\text{Mean shear stress } f_m = \frac{8P_m D}{\pi d^3} K_s$$

$$f_m = \frac{8\times 2P\times 50}{\pi\times(10)^3}\times 1.1 = 0.281\text{ P N/mm}^2$$

$$\text{Variable shear stress } f_v = \frac{8P_v D}{\pi d^3} K$$

$$f_v = \frac{8\times P\times 50}{\pi\times(10)^3}\times 1.313 = 0.168\text{ P N/mm}^2$$

$$\text{Now, design factor} = \frac{f_{yp}}{\left(f_m - f_v + \dfrac{2f_v \times f_{yp}}{f_e}\right)}$$

$$1.3 = \frac{700}{\left(0.281P - 0.168P + \dfrac{2\times0.168P\times700}{300}\right)}$$

gives, $P = 600$ N, Maximum load $3P = 1800$ N.

The buckling load can be checked by relation 15.23.

$$P_{cr} = P_0 \times K_l \times l_0$$

Spring scale can be found from relation

$$\delta = \frac{8PD^3 z}{Cd^4},\ \frac{P}{\delta} = P_0 = \frac{Cd^4}{8D^3 z} = \frac{0.8\times10^5\times(10)^4}{8\times(50)^3\times11}$$

gives, $P_0 = 72.8$ N/mm.

Assuming the ends of the spring to be hinged and for $= \frac{l_o}{D} = \frac{180}{50} = 3.6$ we can see K_l from Fig. 15.12.

Thus $K_l = 0.28$, $l_o/D = 3.6$, $P_0 = 72.8$ N/mm

$$P_{cr} = 72.8\times0.28\times180 = 3650\text{ N}$$

which is more than the maximum load of 1800 N, therefore, the spring will not buckle.

The fundamental or critical frequency of the spring can be found by relation 15.26.

$$f = \frac{d}{2\pi D^2 z}\sqrt{\frac{Cg}{2w}} = \frac{10}{2\pi(50)^2\times11}\sqrt{\frac{0.8\times10^5\times9810}{2\times0.08\times10^6\times10^{-9}}}$$

$$f = 128\text{ V.p.s. or cycles/sec.}$$
$$= 7680\text{ r.p.m. or V.p.m.}$$

Example 15.10. A spring is subjected to a load varying from 400 N to 1000 N. It is to be made of oil tempered, cold wound wire. Design factor based on Wahl's line is 1.25. The spring index is to be 6.0. The free length of the spring should lie between 100 to 150 mm. The compression of spring under maximum load is 30 mm. Determine the diameter of the wire and the mean coil diameter of the spring. Take yield stress in shear $f_{yp} = 770$ N/mm² and endurance stress in shear $f_e = 350$ N/mm². Take $C = 0.8\times10^5$ N/mm².

Solution: For repeated load A.M. Wahl has proposed the relation as

$$\text{Design factor} = \frac{f_{yp}}{f_m - f_v + \dfrac{2f_v \times f_{yp}}{f_e}}$$

In the above equation

$$f_m = \text{mean shear stress} = \frac{8P_m D}{\pi d^3} K_s$$

$$f_v = \text{variable shear stress} = \frac{8P_v D}{\pi d^3} K$$

The reason for multiplying the mean shear stress by multiplication factor K_s is that, the experience suggests that curvature correction factor is not needed for mean stress.

$$K = \frac{4S-1}{4S-4} + \frac{0.615}{S} = \frac{4\times6-1}{4\times6-4} + \frac{0.615}{6} = 1.2525$$

$$K_s = 1 + \frac{0.5}{S} = 1 + \frac{0.5}{6} = 1.083$$

$$\text{Mean load} = P_m = \frac{400+1000}{2} = 700 \text{ N}$$

$$\text{Variable load} = P_v = \frac{1000-400}{2} = 300 \text{ N}$$

$$\text{Spring index} = S = D/d, \qquad \therefore \quad D = 6\times d$$

Assume that the wire diameter is d mm

Then,
$$f_m = \frac{8\times700\times6d}{\pi d^3}\times1.083 = \frac{11600}{d^2}\text{ N/mm}^2$$

$$f_v = \frac{8\times300\times6d\times1.2525}{\pi d^3} = \frac{5740}{d^2}\text{ N/mm}^2$$

Now putting the various values in the Wahl's relation

$$1.25 = \frac{770}{\dfrac{11600}{d^2} - \dfrac{5740}{d^2} + \dfrac{2\times(5740)/d^2\times770}{350}} \text{ gives, } d = 7 \text{ mm.}$$

Take the standard size from table 15.5, which is 7.1 mm.

Mean diameter of coil = 7.1 × 6 = 42.6 mm.

The number of turns can be obtained form relation

$$\delta = \frac{8PD^3 z}{Cd^4},\ 30 = \frac{8\times1000\times(42.6)^3\times z}{0.8\times10^5\times(7.1)^4} \text{ gives, } z = 9.8 \text{ turns.}$$

Assuming the ends of the spring to be squared and ground, total number of turns $z' = 9.8 + 2 = 11.8$.

$$\text{Solid height} = z'd = 11.8\times7.1 = 83.5 \text{ mm}$$

$$\text{Free height} = 83.5 + 30 + \frac{15}{100}\times30 = 118 \text{ mm say 120 mm.}$$

Example 15.11. A diesel engine weighs 800 kN and is mounted on 16 springs in order to protect the building from vibrations.

The section of the spring wire is rectangular with side ratio 1.8. One spring has four effective coils and the spring index is 6. Determine

(a) Section of the spring so that long side is parallel to the axis of the spring. (b) Deflection under load when engine is stationary. (c) Maximum coil diameter. (d) Shear stress induced if shorter side is parallel to the spring axis.

Permissible shear stress in spring wire material is 300 N//mm^2 and modulus of rigidity is 0.8×10^5 N/mm^2.

Solution: Load on each spring $= \dfrac{800 \times 10^3}{16} = 50$ kN

Let h be the longer side and b the shorter side of the wire cross-section, then $h = 1.8\ b$.

$$\text{Stress factor } K = \frac{4S-1}{4S-4} + \frac{0.615}{S} = \frac{4\times6-1}{4\times6-4} + \frac{0.615}{6} = 1.2525$$

(*a*) By using relation 15.19, we can find section of wire.

$$f_s = \frac{PD(3h+1.8b)}{2b^2h^2}K,\ 300 = \frac{6b\times50\times10^3(3\times1.8b+1.8b)}{2b^2\times(1.8b)^2}\times1.2525$$

gives, $b = 31.6$ mm say 32 mm, $h = 1.8 \times 32 = 57.6$ mm

(*b*) Deflection is given by relation 15.21.

$$\delta = \frac{2.45PD^3z}{Cb^3(h-0.56b)},\ \delta = \frac{2.45\times50\times10^3\times(6\times32)^3\times4}{0.8\times10^5(57.6-0.56\times32)\times32^3}$$

gives, $\delta = 31.85$ mm.

(*c*) Mean coil diameter $= D = 6b = 6 \times 32 = 192$ mm.

Outside coil diameter $= D_o = 192 + 32 = 224$ mm.

(*d*) Shear stress induced in the spring when shorter side is parallel to the spring axis can be found by relation 15.20.

$$f_s = \frac{PD(3b+1.8h)}{2b^2h^2}\times K, f_s = \frac{50\times10^3\times192\times1.2525(3\times32+1.8\times57.6)}{2\times(32)^2\times(57.6)^2}$$

gives, $f_s = 355$ N/mm^2, which indicates that the stress induced in the spring in more when the shorter side of the wire section is parallel to the spring axis.

Example 15.12. Design the springs for a mechanical sieve. The weight of the dropping and the material being shifted is 900 N, the height of drop is 12.5 mm, no steps except the springs themselves are used, 8 springs are taking the full impact and each time the frame is lifted by the mechanism they open up to their free length, the load may be assumed as distributed evenly among all eight springs. Under the impact each spring is deflected by 47.5 mm (approximately). The endurance limit for torsion is 340 N/mm^2. Also take the safety factor as 1.7, $C = 0.8 \times 10^5$ N/mm^2.

Solution: The energy which each spring must absorb is

$$U = \frac{900(12.5+47.5)}{8} = 6.75 \text{ Nm}$$

Select $S = D/d = 10$ then $D = 10\ d$ and from Fig. 15.7 the stress factor $K = 1.14$.

Design stress $f_s = \dfrac{340}{1.7} = 200$ N/mm^2

Assume the number of turns $z = 10$ free coils

Putting these values in relation 15.14, and solving it for d.

$$U = \frac{f_s^2 \times V}{4K^2C} = \frac{\pi^2 d^2 D z f_s^2}{16K^2C}$$

or $$6.75 \times 10^3 = \frac{\pi^2 d^2 \times 10d \times 10 \times (200)^2}{16 \times (1.14)^2 \times 0.8 \times 10^5} \text{ gives, } d = 5.7 \text{ mm}$$

The nearest standard size is 6 mm

The increase of d from 5.7 mm to 6 mm will lower the stress from 200 N/mm^2 to 190 N/mm^2. The corresponding actual deflection 'δ' found by equation 15.9, is

$$\delta = \frac{\pi D^2 z f_s}{CKd} = \frac{\pi \times 10 \times 190 \times (10 \times 6)^2}{0.8 \times 10^5 \times 1.14 \times 6} = 39.3 \text{ mm}$$

Therefore the energy to be absorbed will be a little smaller now.

It is, $$U = \frac{900(12.5 + 39.3)}{8} = 6.3 \text{ Nm}$$

A recheck of U by equation 15.15 gives

$$U = \frac{\delta^2 d^4 C}{16zD^3} = \frac{(39.3)^2 (6)^4 \times 0.8 \times 10^5}{16 \times 10 \times (10 \times 6)^3} = 5.8 \text{ Nm}$$

The two values are very close to each other.

The free length of the spring can be found by assuming the square and ground ends, i.e.,

$$z' = z + 2 = 10 + 2 = 12$$

$$\text{Free length} = 12 \times 6 + 39.3 + \frac{15}{100} \times 39.3 = 117.2 \text{ mm}$$

$$\text{Solid length} = 12 \times 6 = 72 \text{ mm}$$

The spring scale P_0 of the spring can be found from relation 15.4

$$\delta = \frac{8PD^3z}{Cd^4},\ P/\delta = P_0 = \frac{Cd^4}{8D^3z}$$

$\therefore$ $$P_0 = \frac{0.8 \times (10)^5 \times (6)^4}{8 \times (10 \times 6)^3 \times 10} = 60 \text{ N/mm}$$

Mean coil diameter $D = Sd = 10 \times 6 = 60$ mm.

Out side diameter of coil $= D + d = 60 + 6 = 66$ mm.

Example 15.13. Derive a relation for the strain energy stored in circular cross-section rod loaded by pure torsional moment. Hence, calculate the weight of the helical spring made of circular cross-section wire which is required to bring to rest weight of 250 N moving with a velocity of 2.75 m/s. The maximum allowable stress in the spring material is to be 550 N/mm^2 and $C = 0.84 \times 10^5$ N/mm^2.

If the spring is made from wire of diameter 7.1 mm with a mean coil diameter of 60 mm, calculate the stiffness of the spring and the number of coils required.

Solution: The strain energy of a circular bar loaded in torsion may be calculated from the diagram of torsion shown in Fig. 15.17. Within elastic limit the angle of twist (θ) is proportional to twisting moment (M_t). The small area shaded in figure represent the work done by the torsional moment during an increment $d\theta$ in the angle of twist θ.

Area OLM $= \dfrac{M_t\theta}{2}$. This area represents the total energy stored in bar during twist.

We know that $\theta = \dfrac{M_t l}{JC}$, then strain energy $= \dfrac{M_t^2 l}{2JC}$

By substitution of values

$$M_t = \frac{\pi}{16} d^3 \times f_s \text{ and } J = \frac{\pi}{32} d^4$$

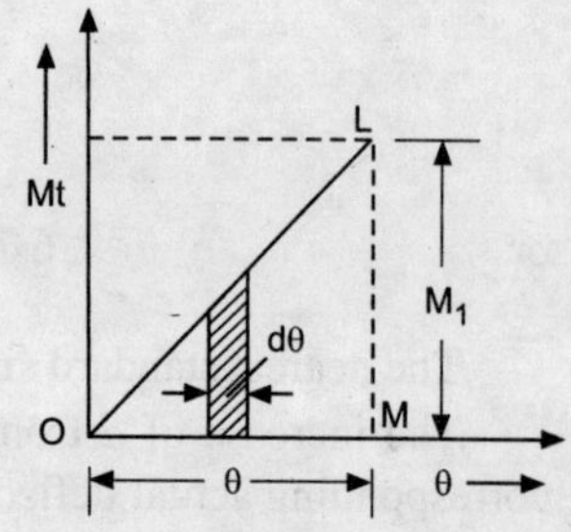

Fig. 15.17

We get, strain energy $= \frac{f_s^2}{4C} \times$ volume of the spring

K.E. to be stored in the spring $= \frac{1}{2}\frac{W}{g} v^2 = \frac{1}{2} \times \frac{250}{9.81} \times (2.75)^2 = 96$ Nm

Equating the strain energy to the K.E. to be stored by spring

$$\frac{(550)^2}{4 \times .84 \times 10^5} \times \text{volume of spring} = 96 \times 10^3$$

gives, volume of spring = 107×10^3 mm^3.

Let z be the number of effective coils of the spring, then

$$107 \times 10^3 = \pi D z \times \frac{\pi}{4} d^2.$$

$$107 \times 10^3 = \pi \times 60 \times z \times \pi/4\,(7.1)^2 \text{ gives, } z = 14.5.$$

$$\text{Spring index} = \frac{D}{d} = \frac{60}{7.1} = 8.45$$

$$\text{Stiffness of the spring } P_0 = \frac{Cd}{8S^3 z} = \frac{0.84 \times 10^5 \times 7.1}{8 \times (8.45)^3 \times 14.5}$$

$$P_0 = 8.55 \text{ N/mm}$$

Taking the density of spring material as 7250 kg/m^3.

$$\text{Weight of spring} = \frac{7250 \times 107 \times 10^3 \times 9.81}{10^9} = 7.61 \text{ N.}$$

15.17 CONCENTRIC SPRINGS

These springs are also known as composite, cluster, or parallel springs. Concentric springs are used for one of the following three purposes.

1. To obtain a greater spring force in a given space. i.e. when the outside coil diameter is limited, use is made of concentric springs.
2. To insure the operation of a mechanism in the event that one spring breaks.
3. To obtain a spring force which does not increase in a direct relation to the deflection, but increases faster.

To obtain the above three conditions, either a two-spring nest or a three spring nest may be used. Values of load division as suggested by the *SAE* committee are as follows. For a 2-spring nest, outer spring 2/3, inner spring 1/3 of the total load. For a 3-spring nest, outer spring 4/7, middle spring 2/7, and inner spring 1/7 of the total load.

In case (1) and case (2), two or more concentric springs have the same free length and are compressed equally. Such springs are used for automobile clutches and railway trucks. In case (3) springs are made of different lengths. The shorter springs to act only after the longer is compressed a certain amount. Such springs are used in governors of variable speed of the speed engines to take care of the variable centrifugal force, which is proportional to the square of the speed.

Figure 15.18 shows two coils and three coils springs. To prevent any tendency to bind, the adjacent coils are wound in opposite directions. If the material of both springs is same, it is desirable to have the same spring index in order to get same stress factor and hence same induced stress in the material. This will avoid overloading of one spring with respect to others.

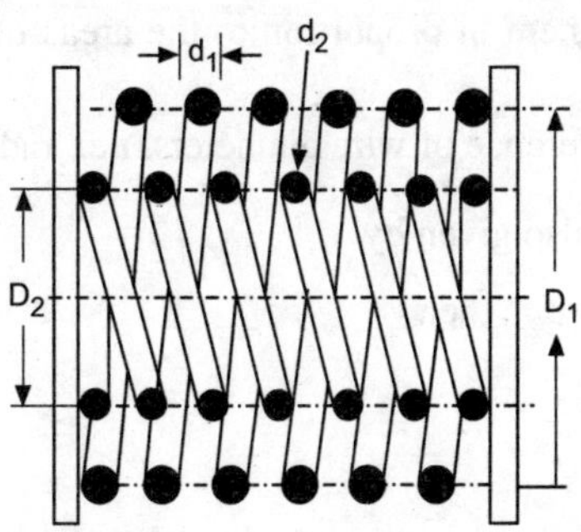

Fig. 15.18 *(a) Two springs arrangement*

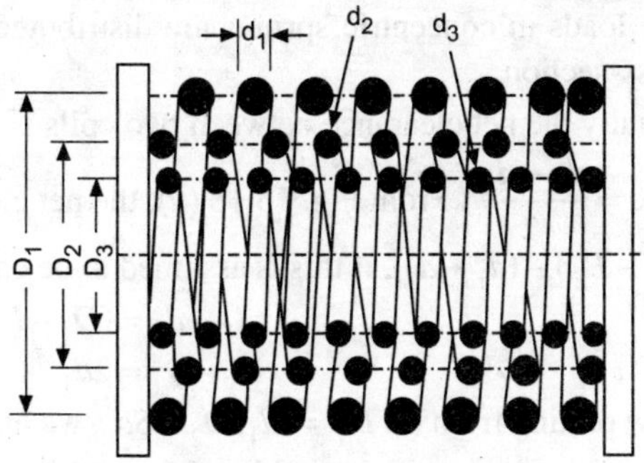

Fig. 15.18 *(b) Three springs arrangement*

The design procedure with following noteworthy points remains the same.

(*a*) Assuming same material for both springs, permissible shear stresses are practically same

i.e.,
$$f_s = \frac{8P_1D_1}{\pi d_1^3} = \frac{8P_2D_2}{\pi d_2^3}$$

or
$$\frac{P_1D_1}{d_1^3} = \frac{P_2D_2}{d_2^3} \quad ...(i)$$

(*b*) When both springs of same initial height deflect between parallel plates, their magnitude of deflection is same, i.e.,

$$\delta_1 = \delta_2$$

On the basis of this condition, we obtain

$$\frac{8P_1D_1^3z_1}{C_1d_1^4} = \frac{8P_2D_2^3z_2}{C_2d_2^4}$$

Since materials are same, $C_1 = C_2$

$$\therefore \quad \frac{P_1D_1^3z_1}{d_1^4} = \frac{P_2D_2^3z_2}{d_2^4} \quad ...(ii)$$

(*c*) Both springs, when compressed until the adjacent coils meet, have the same solid height, i.e.,

$$z_1d_1 = z_2d_2 \quad ...(iii)$$

(*d*) Dividing the equation (*ii*) by (*i*) we shall get,

$$\frac{D_1^2z_1}{d_1} = \frac{D_2^2z_2}{d_2} \quad ...(iv)$$

Now equations (*iii*) and (*iv*), we get

$$\frac{D_1}{d_1} = \frac{D_2}{d_2} = S \quad ...(v)$$

Thus, when designing the concentric springs, their proportions should be so chosen that their spring indexes are same.

(*e*) From equations (*i*) and (*v*), we get

$$\frac{P_1}{P_2} = \frac{d_1^2}{d_2^2} \quad ...(vi)$$

The loads in concentric springs are distributed between them in proportion to the areas of the coil cross-section.

(*f*) Usually the net clearance between two coils is kept as difference of wire diameters, i.e., radial clearance $= \dfrac{d_1 - d_2}{2}$. From Fig. 15.18 (*a*), the net clearance is also given by $(D_1 - D_2) - (d_1 + d_2)$, if this is assumed to be equal to $(d_1 - d_2)$, then

$$d_1 - d_2 = (D_1 - D_2) - (d_1 + d_2)$$

or

$$D_1 - D_2 = 2d_1$$

Now putting from (*v*) $D_1 = Sd_1, D_2 = Sd_2$, we have

$$Sd_1 - Sd_2 = 2d_1$$

$$\therefore \quad \frac{d_1}{d_2} = \frac{S}{S-2} \qquad ...(15.28)$$

15.18 STRANDED WIRE HELICAL SPRING

Stranded wire spring design requires that the helix of the strand should be opposite in direction to that of the coils so that they do not get unwound on deflection. These springs have an inherent tendency for damping as discussed in Article 15.14. Such types of springs are used in machine gun mechanism. The estimation of deflection on loads can be calculated by assuming stranded wire spring as many partial springs acting in parallel as there are number of wires in the strands.

$$\therefore \quad \frac{P}{\delta} = P_0 = \frac{Cd^4}{\pi D^2 z} K \times m \qquad ...(15.29)$$

and

$$f_s = \frac{Cd\delta}{\pi D^2 z} \qquad ...(15.30)$$

where $K = 1.05$ and m = number of wires in the strand.

Example 15.14. Figure 15.18 (a) shows two concentric springs made of same material, for a valve gear mechanism in an aircraft engine. The maximum load on the spring is 6500 N. The deflection under the load should not exceed 40 mm. The permissible shear stress for the springs material is 600 N/mm^2. Take the spring index for both springs as 6.0 and modulus of rigidity as 0.8×10^5 N/mm^2. Find main dimensions of the springs by assuming a standard clearance.

Solution: Let P_1 and P_2 be the load shared by outer and inner springs respectively, then, $P_1 + P_2 = 6500$ N.

From relation (15.28) we can find the wire diameter ratio.

$$\frac{d_1}{d_2} = \frac{S}{S-2} = \frac{6}{6-2} = 1.5$$

and

$$\frac{P_1}{P_2} = \left(\frac{d_1}{d_2}\right)^2 = (1.5)^2 = 2.25, P_1 + P_2 = 6500 \text{ N}$$

$$\therefore \quad 2.25\, P_2 + P_2 = 6500, \text{ gives } P_2 = 2000 \text{ N}, P_1 = 4500 \text{ N}$$

$$\text{Stress factor, } K = \frac{4S-1}{4S-4} + \frac{0.1.615}{S} = \frac{4\times6-1}{4\times6-4} + \frac{0.615}{6} = 1.2525$$

Now by using relating 15.7, the wire diameter can be calculated,

$$f_s = K\frac{8P_1D_1}{\pi d_1^3}, 600 = \frac{1.2525 \times 8 \times 4500 \times 6 \times d_1}{\pi d_1^3}$$

gives, $d_1 = 12$ mm. From Table (15.5) we find that this is the standard size.

Now, $d_2 = \frac{12}{1.5} = 8$ mm.

The mean coil diameter of outer spring $= 12 \times 6 = 72$ mm and of inner spring

$= 8 \times 6 = 48$ mm.

For finding number of active turns of spring, deflection may be considered for individual springs.

$$\therefore \quad \delta = \frac{8 P_1 D_1^3 z_1}{C d_1^4}, \; 40 = \frac{8 \times 4500 \times (72)^3 \times z_1}{0.8 \times 10^5 \times (12)^4}$$

gives, $z_1 = 5.68$.

Assuming square and ground ends for spring, the total number of turns $= 5.68 + 2 = 7.68$.

Since $\quad z_1 d_1 = z_2 d_2$

$$\therefore \quad \frac{z_2}{z_1} = \frac{d_1}{d_2} = 1.5$$

$$\therefore \quad z_2 = 1.5 \times 5.68 = 8.52$$

Solid height of outer spring $= z_1' d_1 = 7.68 \times 12 = 92.16$ mm.

Solid height of inner spring $= z_2' d_2 = 11.52 \times 8 = 92.16$ mm.

The number of coils in the inner spring will be taken as 11.52 to have the same solid height.

Free height of both springs $= 92.16 + 40 + \frac{15}{100} \times 40 = 138.16$ mm.

Outer coil diameter of outer spring $= 72 + 12 = 84$ mm.

Inner coil diameter of outer spring $= 72 - 12 = 60$ mm.

Outer coil diameter of inner spring $= 48 + 8 = 56$ mm.

Inner coil diameter of inner spring $= 48 - 8 = 40$ mm.

Example 15.15. The following table gives the particulars of a concentric helical spring. The spring is subjected to an axial load of 400 N. Determine for each spring.

(a) Change in length.

(b) Load shared by each spring.

(c) The shear stress induced in the wire.

Take the modulus of rigidity for both springs as 0.84×10^5 N/mm^2

	Mean coil diameter	Wire diameter	Number of turns	Free length
Inner Spring	**30 mm.**	**4.06 mm.**	**8**	**75 mm.**
Outer Spring	**40 mm.**	**4.87 mm.**	**10**	**90 mm.**

Solution: The inner spring does not take any load till the outer spring is compressed by 15 mm. After this both the springs are compressed together. Therefore, firstly we can calculate the load on the outer spring to deflect it by 15 mm.

$$\delta = \frac{8 P_1 D_1^3 z_1}{C d_1^4}, \; 15 = \frac{8 \times P_1 \times (40)^3 \times 10}{0.84 \times 10^5 \times (4.87)^4}$$

gives, $P_1 = 140$ N

Now remaining load to be taken together by both the springs $= 400 - 140 = 260$ N.

Assume x be further compression of outer spring and total deflection of inner spring.

Load taken by inner spring = Stiffness of inner spring × x

$$= \frac{Cd_2^4}{8D_2^3 z_2} \times x = \frac{0.84 \times 10^5 \times (4.06)^4}{8 \times (30)^3 \times 8} \times x = 13.2x \text{ N.}$$

Additional load taken by outer spring $= \frac{140}{15} x = 9.26\, x$ N

$\therefore$ $13.2x + 9.26\,x = 260$ gives, $x = 11.6$ mm.

(*a*) Change in length of outer spring = 15 + 11.6 = 26.6 mm.
Change in length of inner spring = 11.6 mm.

(*b*) Load shared by inner spring = 13.2 × 11.6 = 153 N
Load shared by outer spring = 140 + 9.26 × 11.6 = 247 N

(*c*) Shear stress induced in outer spring $= \frac{8P_1 D_1 K_1}{\pi d_1^3}$

Spring index of outer spring $= \frac{D_1}{d_1} = \frac{40}{4.87} = 8.2$

Spring index of inner spring $= \frac{D_2}{d_2} = \frac{30}{4.06} = 7.4$

Stress factor $K_1 = \frac{4 \times 8.2 - 1}{4 \times 8.2 - 4} + \frac{0.615}{8.2} = 1.175$

Stress factor $K_2 = \frac{4 \times 7.4 - 1}{4 \times 7.4 - 4} + \frac{0.615}{7.4} = 1.2$

Shear stress induced in the outer spring $f_{s1} = \frac{8P_1\, D_1\, K_1}{\pi d_1^3}$

$$f_{s1} = \frac{8 \times 247 \times 40 \times 1.175}{\pi (4.87)^3} = 250 \text{ N/mm}^2$$

Shear stress induced in inner spring, $f_{s2} = \frac{8\,P_2 D_2 K_2}{\pi d_2^3}$

$$f_{s2} = \frac{8 \times 153 \times 30 \times 1.2}{\pi \times (4.06)^3} = 210 \text{ N/mm}^2.$$

Example 15.16. A composite spring has closed coil helical springs. The outer spring is 10 mm larger than inner spring. The outer spring has 10 coils of mean diameter 25 mm and wire diameter 3.5 mm. Find the stiffness of the inner spring if an axial load of 260 N causes the outer one to compress by 20 mm. If the radial clearance between two springs is 1.7 mm, find the diameter of the inner spring wire for 8 active coils. Take $C = 0.84 \times 10^5$ N/mm².

Solution: Spring stiffness of the outer spring.

$$P_{01} = \frac{Cd_1^4}{8D_1^3 z_1} = \frac{0.84 \times 10^5 \times (3.5)^4}{8 \times (25)^3 \times 10} = 9.7 \text{ N/mm}$$

Load taken by outer spring = P_1 = 9.7 × 20 = 194 N

Load taken by inner spring = P_2 = 260 − 194 = 66 N

The inner spring gets compressed by 10 mm only.

$\therefore$ Stiffness of inner spring $= P_{02} = \frac{66}{10} = 6.6$ N/mm

Inside coil diameter of outer spring = 25 – 3.5 = 21.5 mm.
Outside coil diameter of inner spring = 21.5 – 2 × 1.7 = 18.1 mm.
Mean coil diameter of inner spring = (18.1 – d_2) mm.
Spring stiffness of the inner spring will be,

$$P_{02} = \frac{Cd_2^4}{8D_2^3 \times z_2}, 6.6 = \frac{0.84 \times 10^5 \times d_2^4}{8 \times (18.1 - d_2)^3 \times 8}$$

Solving by hit and trial, the wire diameter of inner spring will be 2.13 mm.

15.19 CONICAL SPRINGS

The springs so for described have linear load-deflection characteristics. A conical spring is used either where space limitations prohibit the use of a cylindrical spring or where a single spring with a variable stiffness is desired, This can be accomplished by winding the wire in a conical form so that the larger coils, which have the greater deflection rate, will successively drop out of action by seating on the next smaller coil. In order to save space, some springs are wound so that the coils telescope into each other. Springs of this type are made of round or rectangular wire, a spring of round wire is shown in Fig. 15.3 (*c*). The conical springs are sometimes used in vibration problems where springs are used to support a body that has a varying stress.

The conical spring has a draw back that it is difficult to manufacture and tilting of spring during deflection takes place. Another disadvantage is inefficient distribution of stress in the spring.

From Fig. 15.3 (*c*) and equation (15.7) it can be seen that the largest coil with the diameter D_2 is subjected to the highest stress, and that the stress reduces gradually toward the end with the smallest diameter D_1. Naturally, in designing the spring the highest stress must be taken into account. For a round stock this stress will be found by relation 15.7, i.e., $f_s = \frac{8PD}{\pi d^3} K$. For a square stock, it will be found by equation 15.18, with the substitution of D_2 for D.

If the spring is so designed that with an increase of the load the coils having the larger diameters will gradually compress solidly, the smallest coil must be designed for the maximum load.

The deflection of a conical spring of round wire is

$$\delta = \frac{2zP(D_2^3 + D_2^2 D_1 + D_2 D_1^2 + D_1^3)}{Cd^4} \quad ...(15.31)$$

By substitution for P its value from equation 15.7, the deflection can be expressed in terms of the maximum stress. Thus,

$$\delta = \frac{\pi z f_s (D_2^3 + D_2^2 D_1 + D_2 D_1^2 + D_1^3)}{4CdD_2 K} \quad ...(15.32)$$

The axial deflection of a conical spring made of rectangular stock with a radial thickness b and an axial dimension h, is

$$\delta = \frac{0.71zP(b^2 + h^2)(D_2^3 + D_2^2 D_1 + D_2 D_1^2 + D_1^3)}{b^3 h^3 C} \quad ...(15.33)$$

Example 15.17. A conical buffer spring is subjected to a maximum load of 30 kN, when maximum compression amounts to 100 mm. Find out the dimensions if the maximum and minimum radii of the coil are to be 80 mm and 25 mm respectively. Assume a round section and permissible shear stress as 800 N/mm² and modulus of rigidity as 0.8 × 10⁵ N/mm².

Solution: $D_1 = 50$ mm, $D_2 = 160$ mm, $P = 30$ kN, $\delta = 100$ mm, $f_s = 800$ N/mm²
$C = 0.8 \times 10^5$ N/mm².

To start with relation 15.7 stress factor can be assumed as one.

$$f_s = \frac{8PD_2}{\pi d^3},\ 800 = \frac{8\times 30\times 10^3 \times 160}{\pi d^3},\ \text{gives } d = 24.8 \text{ mm say } 25 \text{ mm}$$

Spring index $S = \frac{D_2}{d} = \frac{160}{25} = 6.4$

$$K = \frac{4S-1}{4S-4} + \frac{0.615}{S} = \frac{4\times 6.4-1}{4\times 6.4-4} + \frac{0.615}{6.4} = 1.236.$$

Again using the equation 15.7

$$800 = \frac{8\times 30\times 10^3 \times 160\times 1.236}{\pi d^3},\ \text{gives } d = 26.6 \text{ mm.}$$

Again the procedure can be repeated to find the most accurate size of 26 mm.

The upper diameter D_1 can be checked for induced stress.

$$S = \frac{D_1}{d} = \frac{50}{26} = 1.93$$

$$K_1 = \frac{4S-1}{4S-4} + \frac{0.615}{S} = \frac{4\times 1.93-1}{4\times 1.93-1} + \frac{0.615}{1.93} = 2.13$$

$$f_s = \frac{8PD_1K_1}{\pi d_1^3} = \frac{8\times 30\times 10^3 \times 50\times 2.13}{\pi\times (26)^3} = 462 \text{ N/mm}^2$$

hence safe.

To find out the number of turns, the relation (15.31) can be used,

$$\delta = \frac{2zP(D_2^3 + D_2^2 D_1 + D_2 D_1^2 + D_1^3)}{Cd^4}$$

$$100 = \frac{2\times z\times 30\times 10^3[(160)^3 + 50\times 160^2 + 160\times 50^2 + (50)^3]}{0.8\times 10^5 \times (26)^4}\ \text{gives, } z = 10.3.$$

As the spring is in compression, the total number of coils can be taken as 12.0.

Solid height of spring $z'd = 12\times 26 = 312$ mm.

Free length $= 312 + 100 + \frac{15}{100}\times 100 = 427$ mm.

Pitch $p = \frac{427-312}{12} + 26 = 35.6$ mm.

Example 15.18. In example 15.17, assume the wire section to be rectangular with side ratio of 20. Find out the dimensions of the spring.

Solution: $h = 20b$. The longer side is parallel to the spring axis, then making use of relation 15.19.

$$f_s = \frac{PD(3h+1.8b)}{2b^2h^2}K \qquad \text{(Take } K = 1\text{)}$$

$$800 = \frac{30\times 10^3 \times 160(3\times 20b + 1.8b)}{2\times b^2 \times (20b)^2},\ \text{gives } b = 7.8 \text{ mm}$$

adopt, $b = 8$ mm, $h = 20\times 8 = 160$ mm.

Spring index $S = \frac{D_2}{b} = \frac{160}{8} = 20$

Stress factor $K = \frac{4\times 20-1}{4\times 20-4} + \frac{0.615}{20} = 1.04$

With the same section of wire, we can find the induced stress in wire section

$$f_s = \frac{30\times10^3\times160(3\times160+1.8\times8)}{2\times(8)^2\times(160)^2}\times1.04 = 760\text{ N/mm}^2$$

hence safe.

The induced shear stress on the smaller coil diameter can also be checked.

$$\text{Spring index } S = \frac{D_1}{b} = \frac{50}{8} = 6.25$$

$$\text{Stress factor } K = \frac{4\times6.25-1}{4\times6.25-4}+\frac{0.615}{6.25} = 1.258$$

$$\text{Induced stress } f_s = \frac{PD_1(3h+1.8b)}{2b^2h^2}K$$

$$= \frac{30\times10^3\times50(3\times160+1.8\times8)}{2\times(8)^2\times(160)^2}\times1.258 = 282\text{ N/mm}^2.$$

hence safe,

To find out the number of turns, the relation (15.33) can be used.

$$\delta = \frac{0.71zP(b^2+h^2)(D_2^3+D_2^2D_1+D_2D_1^2+D_1^3)}{b^3h^3C}$$

$$100 = \frac{0.71\times z\times30\times10^3[(8)^2+(160)^2][(160)^3+(160)^2\times50+160\times(50)^2+(50)^3]}{(8)^3\times(160)^3\times0.8\times10^5}$$

gives, $z = 5.2$, as the spring is in compression, the total number of turns can be taken as 7.0.

Solid height of spring = $zh = 7\times160 = 1120$ mm.

$$\text{Free length} = 1120+100+\frac{15}{100}\times100 = 1235\text{ mm.}$$

$$\text{Pitch } p = \frac{1235-1120}{7}+h = \frac{115}{7}+160 = 176.4\text{ mm.}$$

15.20 HELICAL TORSION SPRINGS

The torsion springs illustrated in Fig. 15.19 are used in door hinges, levers, pawl ratcher, automobile starters, for brush holders in electric motors and, in fact for any application where torque is required. They are wound in the same manner as extension or compression springs but have the ends shaped to transmit torque. The primary stress in helical torsion springs is flextural in contrast to the helical compression or tension spring where the stresses are torsional shear stresses.

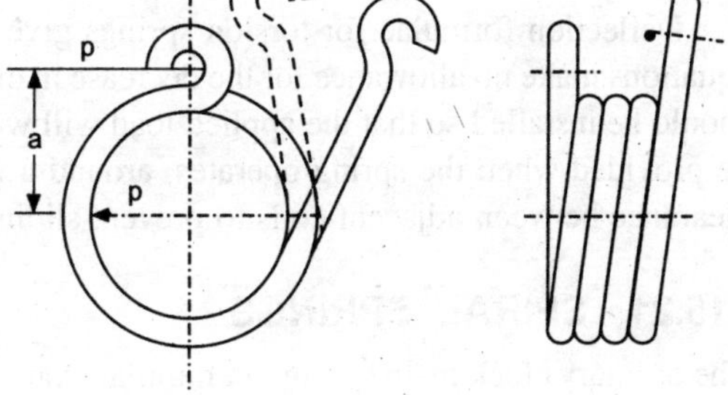

Fig. 15.19 *Torsion spring*

Normally such springs are made from wire up to 12 mm in diameter and wound not otherwise winding forces required are too high. Starting with the curved beam equations and making certain simplifications A.M. Wahl shows that the stress in a torsion spring of round wire is

$$f_b = \frac{32M_b}{\pi d^3}\left[\frac{4S^2-S-1}{4S(S-1)}\right] = K_1\frac{32M_b}{\pi d^3} = \frac{32Pa}{\pi d^3}K_1 \qquad \text{...(15.34)}$$

where M_b = bending moment = $P \times a$

K_1 = stress factor from Fig. 15.20

P = Load acting

a = moment arm of load or distance from load line to spring axis

d = diameter of wire

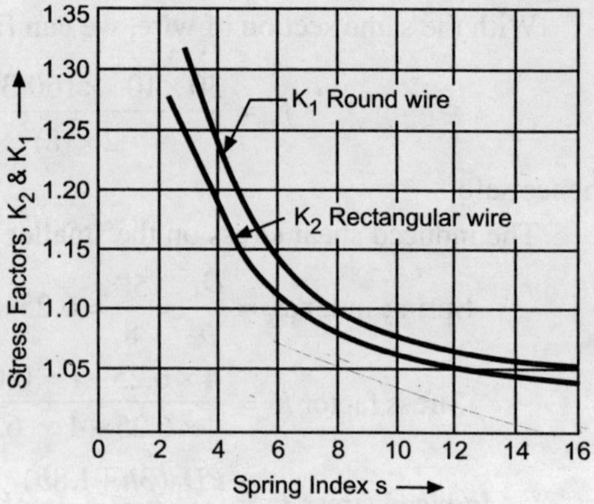

Fig. 15.20 *Stress factors for torsion spring*

For rectangular wire, the stress is

$$f_b = \frac{6M_b}{hb^2}\left[\frac{3S^2 - S - 0.8}{3S(S-1)}\right] = K_2\frac{6M_b}{hb^2}$$

$$= K_2\frac{6Pa}{hb^2} \qquad ...(15.35)$$

where h = wire dimension parallel to axis

b = wire dimension perpendicular to axis.

For square wire, the stress is

$$f_b = K_2\frac{6P \times a}{h^3} \qquad ...(15.36)$$

where h is each side of square.

The deflections of helical torsion springs are given by equations indicated below.

For round wire

$$\delta = \frac{64PDa^2z}{Ed^4} \quad \text{and} \quad \theta = \frac{64PDaz}{Ed^4} \qquad ...(15.37)$$

For rectangular wire

$$\delta = \frac{12\pi PDa^2z}{Ehb^3} \quad \text{and} \quad \theta = \frac{37.8PDaz}{Ehb^3} \qquad ...(15.38)$$

For square wire

$$\delta = \frac{12\pi PDa^2z}{Eh^4} \quad \text{and} \quad \theta = \frac{37.8PDaz}{Eh^4} \qquad ...(15.39)$$

where δ = Movement of point of load applications

θ = Total angle of twist in radians.

Deflection formulae for torsion springs givê values slightly less than the actual deflections, since the equations make no allowance for the decrease in diameter as the coils wind up under load. All torsion springs should be installed so that the applied load will wind up the wire, reducing the diameter and clearance must be provided when the spring operates, around a mandrel. The springs should also be wound with a small clearance between adjacent coils to prevent sliding friction.

15.21 SPIRAL SPRINGS

The ordinary clock main spring is a familiar example of this type of spring. This spring consists of a flat strip of rectangular section to form a spiral shape as shown in Fig. 15.3 (*d*). These springs are used to produce torsion on the axle to which one end is attached. They are usually made from flat ribbons of tempered steel. The load action on the spring is that of bending. If a force P, acting at a radius a, is applied to the axis of the spring, a twisting moment $P \times a$ is set up. This is balanced by an equal internal resisting couple in the spring material.

Assuming both ends on the spring to be clamped, we have

$$Pa = \frac{f_b \times bt^2}{12} \quad \text{or} \quad f_b = \frac{12 P \times a}{bt^2} \qquad ...(15.40)$$

where f_b = tensile or compressive stress in the spring
b = width of spring ribbon
t = thickness of spring ribbon.

The distance the force P would move through is given by the following expression:

$$\delta = \theta \times a = \frac{Pa^2 l}{EI} = \frac{f_b \times l \times a}{Et} \qquad ...(15.41)$$

where l = length of the ribbon forming the spring.
θ = angular deflection in radians.
E = modulus of elasticity of the spring material.

It will be noted that the deflection for a given stress is not affected by the width of ribbon.

Example 15.19. A torsion spring is wound from a round wire into coil with mean diameter 50 mm. The torsional moment applied on the spring is 7 Nm. Assume the spring index as 9. The permissible stress in the spring is 600 N/mm² and E = 2 × 10⁵ N/mm². What must be the diameter of wire and the corresponding deflection in degrees? Take the number of effective coils as 6.5.

Solution: $M_b = 7$ Nm, $S = 9$

From Fig. 15.20, for round wire $K_1 = 1.08$

Now by using relation (15.34), we can find wire diameter.

$$f_b = K_1 \frac{32 M_b}{\pi d^3}, \; 600 = \frac{1.08 \times 32 \times 7 \times 10^3}{\pi \times d^3} \text{ gives, } d = 5 \text{ mm.}$$

Deflection in radians can be calculated by relation (15.37)

$$\theta = \frac{64 (P \times a) Dz}{Ed^4} = \frac{64 \times 7 \times 10^3 \times 50 \times 6.5}{2 \times 10^5 \times (5)^4} = 1.16 \text{ radians}$$

$$= \frac{180}{\pi} \times 1.16 = 66.5°.$$

Example 15.20. A flat spiral spring with rectangular section 0.5 mm × 10 mm is 2000 mm long. One end of it is attached to a small spindle and other to a fixed point. If the spring is in a state of case of run down, find (a) what turning moment on spindle is required to give 4 complete turns? (b) Strain energy stored. (c) Maximum stress induced.

Take E = 2 × 10⁵ N/mm².

Solution: (a) The angular deflection = number of turns × 2π, $\theta = 4 \times 2\pi = 8\pi$ radians

Using relation 15.41, we have

$$\theta \times a = \frac{Pa^2 l}{EI}, \text{ or } \theta = \frac{Pal}{EI}$$

or
$$P \times a = \text{Turning moment on spindle} = M_b = \frac{\theta \times E \times I}{l}$$

or
$$M_b = \frac{8\pi \times 2 \times 10^5 \times 10 \times (0.5)^3}{2 \times 1000 \times 12} = 0.26 \text{ Nm.}$$

(b) Strain energy stored in spring $= \frac{1}{2} M_b \times \theta = \frac{1}{2} \times .26 \times 8\pi = 3.3$ Nm

(c) Maximum stress induced is given by relation (15.40).

$$f_b = \frac{12 Pa}{bt^2} = \frac{12 \times 0.26 \times 10^3}{10 \times (0.5)^2} = 1248 \text{ N/mm}^2.$$

Example 15.21. A flat spiral steel spring is to give a maximum torsional moment of 1 Nm for a maximum stress of 800 N/mm². What thickness and length are necessary to give three complete turns of motion when stress decreases from 800 to zero? Space consideration limit the width of the wire to 15 mm. Take $E = 2 \times 10^5$ N/mm².

Solution: Maximum stress is given by relation (15.40).

$$f_b = \frac{12\,Pa}{bt^2} = \frac{12 M_b}{bt^2},\ 800 = \frac{12 \times 1 \times 10^3}{15 \times t^2}$$

gives, $t = 1.0$ mm.

From Table 15.5, we find that 1.0 mm is the standard thickness of sheet available.

By using relation 15.41, we can find the length of wire.

$$\theta \times a = \frac{Pa^2 l}{EI},\ \text{or } \theta = \frac{(Pa)l}{EI}$$

or
$$l = \frac{\theta \times E \times I}{M_b} = \frac{3 \times 2\pi \times 2 \times 10^5 \times 15 \times (1)^3}{1 \times 10^3 \times 12} = 4710 \text{ mm.}$$

Example 15.22. A pivoted roller follower is held in contact with the cam by a torsion spring. The moment exerted by the spring varies from 2 Nm to 5 Nm as the follower oscillates through 60°. Design the suitable spring with a factor of safety 2 based on Wahl's line. Ultimate strength of steel is 1300 N/mm² and yield strength 1100 N/mm². Assume the stress concentration factor to be 1.1. Take the spring index as 12.5 and $E = 2.1 \times 10^5$ N/mm².

Solution: Mean torsional moment $= \dfrac{5+2}{2} = 3.5$ Nm

Variable torsional moment $= \dfrac{5-2}{2} = 1.5$ Nm

endurance limit of steel $= \dfrac{1300}{2} = 650$ N/mm²

For repeated loading A.M. Wahl has proposed the relation.

$$\text{Design factor} = \frac{f_{yp}}{f_m - f_v + \dfrac{2 f_v \times f_{yp}}{f_e}}$$

Now $f_{yp} = 1100$ N/mm², $f_e = 650$ N/mm²

Assume the wire diameter to be d, then

$$f_m = \frac{16 \times T_m}{\pi d^3} = \frac{16 \times 3.5 \times 10^3}{\pi d^3} = \frac{17760}{\pi d^3} \text{ N/mm}^2$$

$$f_v = \frac{16 \times T_v}{\pi d^3} K_c = \frac{16 \times 1.5 \times 10^3}{\pi d^3} \times 1.1 = \frac{7680}{\pi d^3} \times 1.1 \text{ N/mm}^2$$

Now substituting the various values in the above relation

$$2 = \frac{1100}{\dfrac{17760}{\pi d^3} - \dfrac{7680 \times 1.1}{\pi d^3} + \dfrac{2 \times 7680 \times 1.1}{\pi d^3} \times \dfrac{1100}{650}} \text{ gives, } d = 4 \text{ mm.}$$

From table 15.5 this is the standard size available.

The deflection of the helical torsion spring for round wire is given by relation 15.37.

$$\therefore\quad \theta = \frac{64\,PDaz}{Ed^4} = \frac{64 M_b Dz}{Ed^4}$$

$$D = \text{mean coil diameter} = Sd = 12.5 \times 4 = 50 \text{ mm}$$

and $$\theta = \frac{\pi}{3} \text{ radians}$$

By putting the values in above relation we can find z.

$$\frac{\pi}{3} = \frac{64 \times 5 \times 10^3 \times 50 \times z}{2.1 \times 10^5 \times (4)^4}$$

gives, $z = 3.5$.

Thus effective number of turns = 3.5

Outside diameter of coil = $D + d = 50 + 4 = 54$ mm

Inside diameter of coil = $D - d = 50 - 4 = 46$ mm

Effective length of wire out of
which spring is to be made = $\pi Dz = \pi \times 50 \times 3.5 = 550$ mm.

15.22 LEAF SPRING

The term flat spring is commonly applied to a wide variety of shapes made out of a flat strip. An advantage of the flat cantilever spring over the helical spring is that the end of the spring may be guided along a definite path as it deflects. Thus the spring may act as a structural member as well as an energy absorbing device. By making the spring in a particular shape it may be possible to combine several functions, thus simplifying the design. For example, an automobile leaf spring may be designed not only to absorb road shocks, but also to carry lateral loads, brake torque, driving torque, etc.

A flat spring may have the form of cantilever beam, as in Fig. 15.21(*a*), or it may have the form of a simple beam, as in Fig. 15.21(*b*) which can be considered as double cantilever beam. The main stresses are tension on one side of the neutral axis and compression on the other. The additional transverse shear may be neglected as far as strength is concerned.

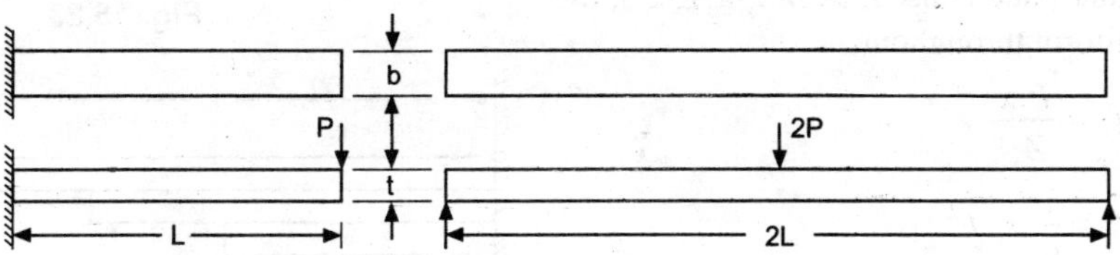

(*a*) Centilever (*b*) Beam

Fig. 15.21 *Flat springs*

The maximum stress in Fig. 15.21 is $f = \frac{M_b}{Z}$.

(*a*) Maximum bending moment M_b will be at the cantilever end of magnitude PL.

$$\therefore \quad f = \frac{M_b}{Z} = \frac{PL}{I/6bt^2} = \frac{6PL}{bt^2} \quad \text{...(15.42)}$$

where b is the width of plate and t its thickness. Similarly the maximum deflection is

$$\delta = \frac{Pl^3}{3EI} = \frac{1}{3}\frac{PL^3}{E1/12bt^3} = \frac{4PL^3}{Ebt^3} \quad \text{...(15.43)}$$

(*b*) In the case of a beam the maximum bending moment is at the centre of the beam and its magnitude will be

$$\frac{Wl}{4} = \frac{2P.2L}{4} = \text{P.L}$$

where W = load at the centre of the beam, and
l = span length of beam.

$$\therefore \qquad f = \frac{M_b}{Z} = \frac{PL}{1/6bt^2} = \frac{6PL}{bt^2}$$

where *b* and *t* for a beam and cantilever remain same. Similarly, the maximum deflection for a beam is

$$\delta = \frac{1}{48}\frac{Pl^3}{EI} = \frac{1}{48}\frac{(2P)\times(2L)^3}{EI/12bt^3} = \frac{4PL^3}{Ebt^3}$$

This analysis shows that an automobile spring having span length 2*L* loaded at the centre by a load 2*P* may be treated as a beam or a cantilever with load *P* at the end of span *L*. *The design analysis in both cases remains same. Therefore, in this book an automobile spring (semi-elliptical) will be treated as two cantilevers.*

If the plate of cantilever is cut into a series of *z* strips of width *b*, and these are placed as shown in Fig. 15.22, the equations 15.42 and 15.43 become,

$$f = \frac{6PL}{zbt^2} \qquad ...(15.44)$$

$$\delta = \frac{4PL^3}{Ezbt^3} \qquad ...(15.45)$$

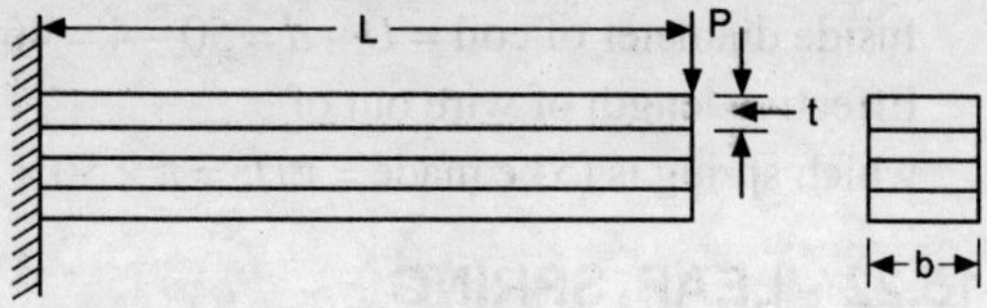

Fig. 15.22

Putting the value of *P* from (15.44) in (15.45), gives

$$\delta = \frac{2}{3}\frac{fL^2}{Et} \qquad ...(15.46)$$

This gives the stress and deflection in a leaf spring of uniform cross-section.

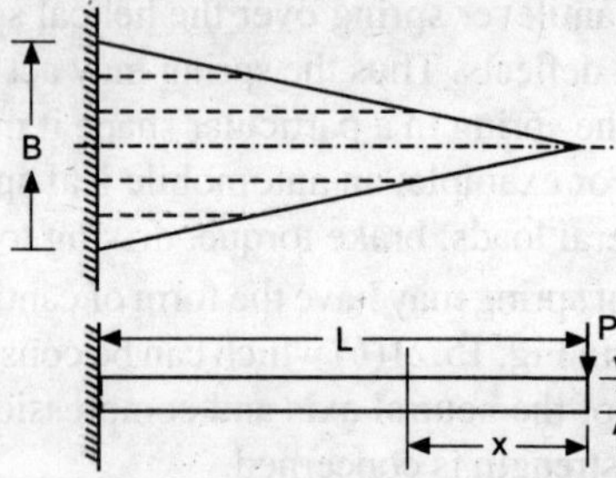

Fig. 15.23

The stress in such a spring is maximum at the support. If a triangular plate is used, as in Fig 15.23, the stress will be uniform throughout,

$$f_x = \frac{P.x}{Z_x}$$

$$Z_x = \frac{1}{6}.b_x.t^2$$

and, $$b_x = B.\frac{x}{L}$$

$$\therefore \qquad f_x = \frac{6P.x.L}{B.x.t^2} = \frac{6.P.L}{B.t^2}, \text{ which is constant.}$$

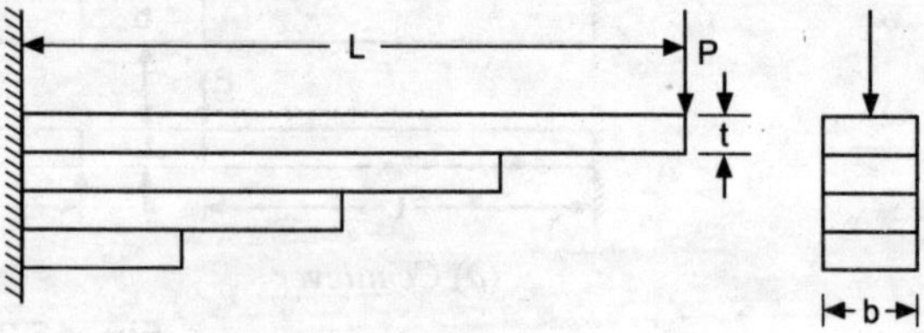

Fig. 15.24

∴ It is called as a beam of uniform strength.

If this triangular plate is cut into strips and placed as in Fig. 15.24 to form a graduated leaf spring, the stress in such a spring will be

$$f = \frac{6PL}{zbt^2} = \frac{6PL}{Bt^2} \qquad ...(15.47)$$

and deflection will be

$$\delta = \frac{1}{2}\frac{PL^3}{EI} = \frac{1}{2}\frac{PL^3}{EI/12bt^3z} = \frac{6PL^3}{Ezbt^3} = \frac{6PL^3}{EBt^3} \qquad ...(15.48)$$

Putting the value of *P* from (15.47) in relation (15.48), we get

$$\delta = \frac{fL^2}{Et} \qquad ...(15.49)$$

where *z* is the number of graduated plates.

In order to get compactness, flat spring is cut into z strips, the strips are piled one above the other and we get a familiar spring. The ends of the individual strips are usually modified by being made wider and thinner in order to make a neater appearance. The beam that is made by piling separate leaves, is just as strong as the beam with the leaves besides each other. In the former case the load is transmitted from one leaf to the other by bearing while in the latter by shear.

If bending stress alone is considered, the graduated spring may have zero width at the load end, but sufficient metal must be provided to support the shear. Hence it is necessary to have one or more leaves of uniform cross-section extending clear to the end.

The equation 15.46 is for z strips or leaves of uniform cross-section (full length) and equation 15.49 is for z leaves of graduated length. *The examination of these deflection equations indicates that, for the same deflection, the stress in the full length leaves (uniform cross-section leaves) is 50 percent greater than in the graduated leaves, assuming that each spring deflects according to its own elastic curve.*

Assume that the suffix f and g are used for full length and graduated leaves respectively, then

$$f_f = \frac{3 f_g}{2}$$

or

$$\frac{6.P_f.L}{z_f.bt^2} = \frac{3}{2}\left(\frac{6.P_g.L}{z_g.bt^2}\right)$$

where P_f is the load taken by full length leaves and P_g by graduated leaves and z_f is the number of full length leaves and z_g of the graduated leaves.

From the above relation

$$\frac{P_f}{P_g} = \frac{3}{2}.\frac{z_f}{z_g} \qquad ...(15.50)$$

Now adding 1 on both sides, we have

$$\frac{P_f}{P_g} + 1 = \frac{3}{2}.\frac{z_f}{z_g} + 1$$

$$\frac{P_f + P_g}{P_g} = \frac{3z_f + 2z_g}{2z_g}$$

gives,

$$P_g = \left(\frac{2z_g}{2z_g + 3z_f}\right) \times P \qquad ...(15.51)$$

Similarly we can write

$$P_f = \left(\frac{3z_f}{2z_g + 3z_f}\right) \times P \qquad ...(15.52)$$

Relations (15.51) and (15.52) give us the load shared by full length and graduated leaves in an automobile spring.

Now from relation (15.44), we have

$$f_f = \frac{6.P_f.L}{z_f.bt^2} = \frac{6.P.L}{z_f.bt^2}\left(\frac{3z_f}{2z_g + 3z_f}\right)$$

$$f_f = \frac{18PL}{bt^2(2z_g + 3z_f)} \qquad ...(15.53)$$

This is the relation between the maximum stress and the load applied at the end of the spring when all the leaves have the same thickness.

The stress in the graduated leaves will be

$$f_g = \frac{12\,PL}{bt^2\,(2z_g + 3z_f)} \quad ...(15.54)$$

The deflection in full length and graduated leaves can be obtained by putting the stress f_f from relation (15.53) for 'f' in relation (15.46)

$$\delta = \frac{2}{3}\frac{L^2}{Et} \times \frac{18PL}{bt^2\,(2z_g + 3z_f)} = \frac{12PL^3}{Ebt^3\,(2z_g + 3z_f)} \quad ...(15.55)$$

The value of deflection can also be obtained by putting the stress f_g from relation (15.54) for 'f' in relation (15.49).

15.23 LEAF SPRING CONSTRUCTION

The most common type of the leaf spring is the semi-elliptical leaf spring, which is shown in Fig. 15.25. For very heavy loads to be supported it can be elliptical leaf spring. The semi-elliptical spring may be considered as two cantilevers and elliptical spring as four cantilevers. In a full elliptical spring the stress induced by a given load is exactly the same as that induced in a semi-elliptical spring, but the deflection will be twice as great. Quarter-elliptical springs are also made use of. The unloaded spring is cambered, the magnitude of the camber being such that the spring is approximately straight under the full static load. Under the maximum dynamic load the camber will be negative. Normally the camber is taken half the full deflection, sometimes the camber is zero under no load condition.

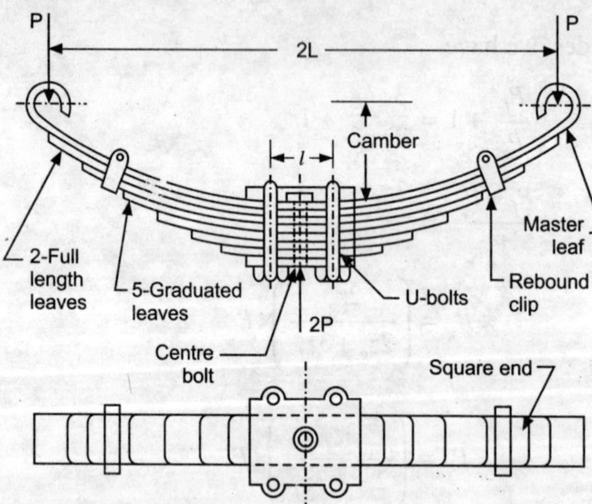

Fig. 15.25 *Leaf spring (semi-elliptical)*

The long leaf fastened to the supports is called the main leaf or master leaf. Its ends are bent to from an eye. If heavy loading is there, one or more full length leaves are provided below the master leaf and with this, spring no longer remains beam of uniform strength and correction must be applied for it.

The bundle of leaves is held together by either a **band** shrunk around them at the centre or by a **bolt** passing through the centre. In the first case the band exerts a stiffening and strengthening effect and this should be allowed for by subtracting the width of the band from the overall length of the spring to obtain the **effective length** for bending. In the second case the diameter of the bolt must be substracted from the width of the leaf when making the calculations for strength. The deflection will not be affected materially by the presence of the bolt hole. **Effective length** for bending of a spring with centre bolt should be measured to a point about one-third the distance from the U-bolt edge to the centre.

The **U-bolts** are used to hold the spring to the machine structure. The U-bolts create a stress concentration at the edge of the spring seat. A soft pad placed between the leaf and the seat will reduce the stress concentration.

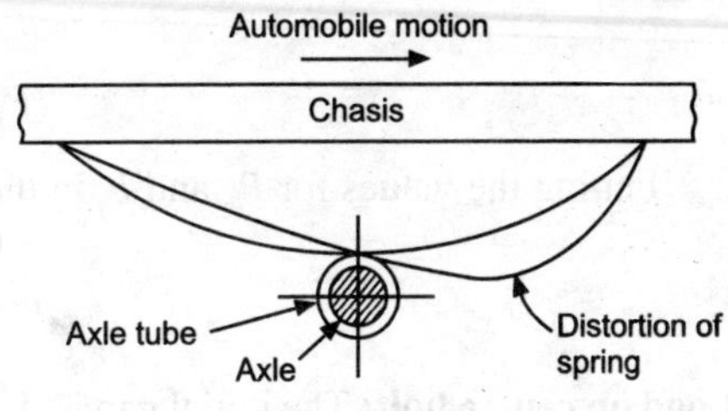

Fig. 15.26

The **rebound clip** helps to distribute to the graduated leaves some of the load of the rebound which otherwise would be taken by the master leaf alone.

Ends of the spring are fastened to the supports by means of hinges so that the length of the spring can increase while fully loaded. The spring is required to support the static and longitudinal loads and in addition driving and braking torsional moment cause additional stresses. For example, in automobiles, leaf springs connect the chasis to the axle. When brakes are applied, the axle is brought to rest, but the chasis and the automobile body due to its inertia keep to move ahead and thus additional torsional moment act on the spring.

15.24 EQUALISED STRESS IN SPRING LEAVES

In the spring discussed in Article 15.22, the stress in the full length leaves was found to be 50 per cent greater than the stress in the graduated leaves. For economical use of the material, we should design all the leaves to be stressed equally otherwise the master leaf will fail and is going to control the design. This condition can be obtained if the full-length leaves are given a greater radius of curvature than that used in the graduated leaves before the leaves are assembled to form a spring. As shown in Fig. 15.27, this will leave a gap or clearance space between the full length and graduated leaves. This initial gap is represented by C and is also known as **'Nip'**. When the centre bolt is drawn up tight, the full length leaf will bend back and have an initial stress opposite to that produced by the normal loading, and the graduated leaf will have an initial stress of the same nature as that produced by the normal loading. As the full load is applied the stress in the full length leaves is first relieved and then increases in the normal way.

The initial gap between the leaves can be adjusted so that under maximum load conditions the stress in all the leaves will be same, or, if desired the stress in the full length leaves may be less. In fact the full length leaf is designed for a lower stress, since it must carry additional loads due to swaying of the automobile, twisting, breaking, etc.

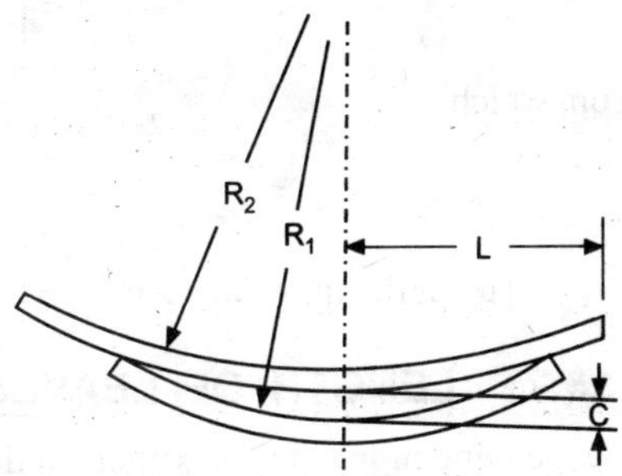

Fig. 15.27

Initial gap: Consider the case of equal stress in all the leaves at the maximum load condition. Now under the maximum load condition, the deflection of the graduated leaves will be more than the full length leaves by the amount approximately equal to initial gap C, Hence.

$$C = \delta_g - \delta_f$$

Assume P_g and P_f be the load shared by graduated and full length leaves respectively. Then from relations 15.48 and 15.45, we have

$$C = \frac{6P_g L^3}{E z_g b t^3} - \frac{4P_f L^3}{E z_f b t^3}$$

But we are designing the spring for equal stresses at maximum load, therefore

$$\frac{6P_g L}{z_g b t^2} = \frac{6P_f L}{z_f b t^2}$$

or $$P_g = \frac{z_g}{z} . P \text{ and } P_f = \frac{z_f}{z} . P$$

Putting the values for P_g and P_f in the above relation gives

$$C = \frac{6PL^3}{Ezbt^3} - \frac{4PL^3}{Ezbt^3} = \frac{2PL^3}{Ezbt^3} \quad ...(15.56)$$

Load on centre bolt: The initial gap is sum of initial deflections of the full length and graduated leaves. The initial deflection is caused by tightening the centre bolt so that the gap reduces to zero. Let *Pb* be the force produced in the bolt to reduce the gap to zero. Hence,

$$C = \frac{2PL^3}{Ezbt^3} = \left(\frac{4L^3}{Ez_f bt^3}\right)\frac{P_b}{2} + \left(\frac{6L^3}{Ez_g bt^3}\right)\frac{P_b}{2}$$

gives, $$P_b = \frac{2z_g z_f P}{z(2z_g + 3z_f)} \quad ...(15.57)$$

Final stress in leaves: The final stress in the spring leaves will be the stress in the graduated leaves due to the applied load plus the initial stress or the final stress in the spring leaves will be the stress in the full length leaves due to the applied load minus the initial stress. Solving it with the help of graduated leaves,

$$f = \left(\frac{6P_g L}{z_g bt^2}\right) + \left(\frac{6L}{z_g bt^2}\right) \times \frac{P_b}{2} = \left(\frac{6L}{z_g bt^2}\right)\left[P_g + \frac{P_b}{2}\right]$$

putting the value for *Pg* from equation 15.51 and *Pb* from 15.57.

$$f = \left(\frac{6L}{z_g bt^2}\right)\left[\frac{2z_g P}{(2z_g + 3z_f)} + \frac{z_g z_f P}{z(2z_g + 3z_f)}\right]$$

from which

$$f = \frac{6PL}{zbt^2} \quad ...(15.58)$$

The deflection caused by the applied load is the same as in the spring without initial stress.

15.25 LENGTH OF LEAVES

The bending moment is maximum at the centre and goes on decreasing towards the ends, therefore maximum resistance of moment is required at the centre and less moment of resistance towards the ends. So we make use of graduated leaves with their lengths decreasing in steps. The law of variation of bending moment is linear one. The procedure for obtaining the length of leaves is as given here.

Let $2L$ = Span of spring.

l = Width of the band, or if the U-bolts are used, the distance between the centre of U-bolts.

Effective length of spring = $2L - l$

(If band is used).

Effective length of spring = $2L - l + \frac{1}{3}\frac{l}{2} \times 2 = 2L - \frac{2}{3}l$

(If U-bolts are used)

The sum of the graduated leaves (z_g) and the full length leaves (z_f) will be the total number of leaves z. If $z_f = 1$, then number of leaves to be cut is z, if $z_f = 2$, the number of leaves to be cut is $(z - 1)$ and two of the full length leaves will be same length (this includes the master leaf). For $z_f = 2$.

$$\text{Smallest leaf length} = \frac{\text{effective length}}{(z-1)} + \text{Ineffective length of spring, } i.e., \ l \text{ or } \frac{2}{3}l.$$

$$\text{Next leaf length} = \frac{\text{effective length}}{(z-1)} \ 2 + \text{Ineffective length of spring.}$$

$$\text{Length of } (z-1)\text{th leaf} = \frac{\text{effective length}}{(z-1)} \ (z-1) + \text{Ineffective length of spring.}$$

The zth leaf or the top leaf will be of full length and will act as master leaf.

Length of master leaf = Span length of spring + Allowance for eyes.

$$= 2L + \pi\,(d + t)\,2$$

where d is the inside diameter of the eye at each end and t is the thickness of master leaf.

Similarly for $z_f = 3$, the number of leaves to be cut will be $(z-2)$ and three of the full length leaves will be of same length (including master leaf).

The relation between radius of curvature and the camber is given below:

$$R = \frac{l_1^2}{8\delta_0} \qquad \text{...(15.59)}$$

and exact relation is

$$\delta_0\,(2R - \delta_0) = \frac{l_1^2}{4} \qquad \text{...(15.60)}$$

where δ_0 = The camber

R = Radius of curvature.

l_1 = The span of the spring = $2L$.

15.26 MATERIAL

Plain carbon steel of 0.90 to 1.00 per cent carbon in annealed condition and heat treated after forming process is commonly used. Chrome-vanadium and silica-manganese steels are used for the better grade springs. The alloy steels do not have strengths greatly in excess of the carbon steels, but have greater toughness and higher endurance limit and are better suited to springs subjected to rapidly fluctuating loads. The working stress for a leaf spring should be based on the endurance limit and, of course should never exceed the yield stress. The endurance limit in reversed bending varies from 40 to 50 per cent of the ultimate strength in tension.

For static loads, as in governor springs, a safety factor of 1.5 is sufficient on endurance limit. In rail road service the values of safety factor can range from 2 to 2.25, in automobile design safety factor is taken as 2.25 to 2.5 for rear spring and from 3.25 to 3.5 for front springs based on endurance limit. If the design takes into account the shock action and is based on endurance limit, a factor of safety 1.5 or even 1.25 is sufficient.

For normally used spring steels, working bending stress is 1000 to 1300 N/mm^2 but the normal design stress does not account for sudden impact loads and other factors, so safe stresses used are 600 to 650 N/mm^2.

15.27 GENERAL CONSIDERATIONS IN DESIGN

The stress and deflection as determined by the usual spring equations are slightly altered by several items that must be considered. Before the final adoption of the design for quantity production, the spring performance should be checked under expected working conditions. The general points to be kept in mind while designing are:

1. In actual laminated spring the full length leaves must have square ends by which they are fastened to the supports. Sometimes, the graduated leaves are simply cut square at the ends. Heavy machinery, truck and railway springs are usually designed in this way. To make the transition from one leaf to the next gradual, it is better to taper the ends in width or thickness, or in both dimensions. Different end conditions are shown in Fig. 15.28. Automobile and similar springs are usually rounded and thinned at the ends. Thus change does not affect the stress found by equation 15.58, but it may influence the deflection found by equation 15.55.

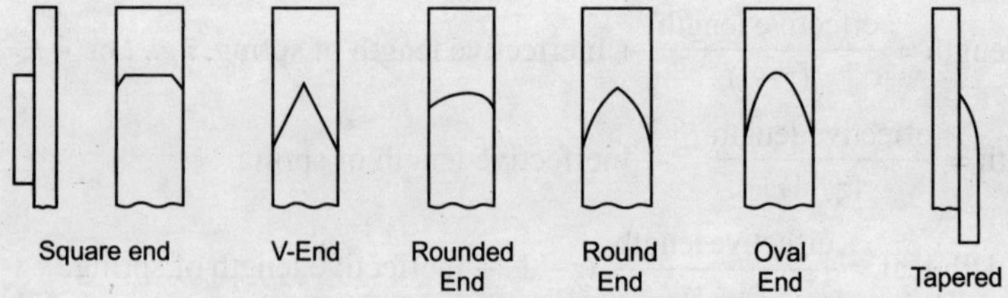

Fig. 15.28 *Types of leaf spring ends*

2. Some springs are made with master leaf thinner than the other. Equation 15.46 shows that as t is decreased the stress for the same deflection is decreased. Hence when the master leaf is subjected to twisting or compression loads, it is desirable to use a thin master leaf to increase the safety factor. Sometimes the master leaf is made of stronger material as compared to other leaves.

3. Laminated springs for heavy loads usually have under the master leaf one or more additional full length leaves, as in Fig. 15.25. With such an arrangement the spring ceases to be of uniform strength. The relation between the load P and the stress f remains the same. The deflection can be calculated by introducing a correction factor k_3. Thus,

$$\delta = \frac{12\,PL^3\,k_3}{Ebt^3\,(2z_g + 3z_f)} \quad \text{...(15.61)}$$

The factor k_3 may be determined by relation

$$k_3 = \frac{1 - 4r + 2r^2\,(1.5 - \log_e r)}{(1-r)^3} \quad \text{...(15.62)}$$

where $r = \dfrac{z_f}{z}$ with z_f, the number of full length leaves and z the number of leaves.

4. In automobile springs all leaves are nipped to certain extent. This helps to keep them in contact on the rebound and also keeps out the dirt.

5. Friction between the leaves tends to reduce the deflection and make the springs stiffer. To reduce wear and to obtain uniform spring action, some designers specifying lubricant oil in the spring assembly and providing spring coverings to retain the lubricant to keep out the dirt.

6. Centre bolt reduces the metal area at the centre and causes localized stresses near the hole. This is poor design and such springs usually break through the centre hole. Shrunk bands are superior construction but are used only on heavy springs.

7. To protect the spring from excessive stress during the rebound, short rebound leaves may be placed above the master leaves.

8. Loss of fatigue strength must be accounted in case of hot rolled leaf springs.

Laminated springs permit a saving in material and a greater deflection than a spring of constant depth, that is, they have greater resilience and shock absorbing ability.

15.28 DESIGN PROCEDURE

So many variable factors enter into the design calculations of a laminated spring that no definite rules can be given, cut and try methods have to be applied. The data usually known to design a laminated spring are.

(*a*) Load on the spring and the shocks it has to absorb.
(*b*) Span length, which will be decided by the designer from the space available for spring installation.
(*c*) Deflection.
(*d*) Load, span length and deflection will decide upon the type of the laminated spring to be used.

The usual design procedure is as follows:

1. The first step is to decide what material will be used. From chapter 2 the endurance limit stress can be determined. Having selected a suitable safety factor, the permissible stress is found.
2. With the help of table 15.5, assume the approximate thickness t of the leaves.
3. Decide about the centre to centre distance of U-bolt or width of band l. This will depend upon the diameter of axle-tube or the place where it will be mounted.
4. By knowing the step 3, the effective length for bending can be known.
5. The values of $z \times b$ can be found from relation $f = \dfrac{6PL}{zbt^2}$ by substituting the selected value of f, using the given load P, and by using a length L as determined in step 4.
6. The number of leaves z usually should be from 5 to 10, but in railroad cars it may be increased upto 14.
7. With the help of steps 5 and 6, the width b of the leaves can be determined. If this does not give a leaf width b within recommended limits of 45 mm and 100 mm, the thickness t must be changed and whole procedure repeated.
8. In a particular case other factors, such as nipping and variable leaf thickness, must be taken into consideration in finding both the maximum stress and deflection.
9. The curtailment of the plates can be done as explained in article 15.25. Generally in automobile leaf springs the number of full length plates used is two. At the eye end the induced shear stress can be checked due to load P.

$$f_s = \frac{P}{z_f \times b \times t} \quad \text{...(15.63)}$$

where z_f is the number of full length plates including the master leaf. If the induced shear stress happens to be more, then increase z_f.

10. The radius of curvature can be obtained by relation 15.60.
11. The centre bolt or the band will be designed for the load P_b which is the force produced in the bolt to reduce the gap to zero. The bolt or band will be under tension due to load P_b calculated by relation (15.57).
12. The master leaf can be attached at the ends to the chasis of an automobile in different ways. One of the methods is shown in Fig. 15.29. The width of eye of master leaf is b and on either side of it a slight clearance is left. Then length of pin in the plates will be b' $= b + 2 \times$ clearance. The clearance may be taken 1.5 mm on either side. Considering the bearing of the pin in the eye

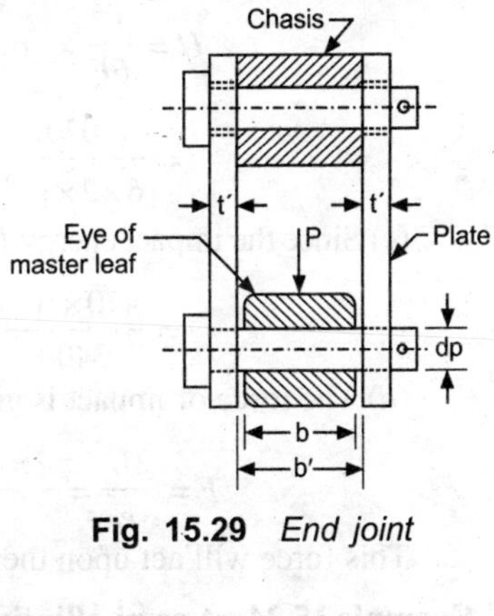

Fig. 15.29 *End joint*

$$d_p \times b \times p_b = \text{load on pin}$$

where d_p = diameter of pin

p_b = allowable bearing pressure

$$\text{Load on pin} = \frac{P}{\cos\alpha}$$

where α = inclination of plates, generally taken 45°, it can also be assumed as zero.

Then
$$d_p \times b \times p_b = \frac{P}{\cos\alpha} \quad \text{...(15.64)}$$

Check for Pin: The pin can be checked under bending,

$$M_b = \text{Maximum bending moment} = \frac{\text{Load on pin} \times b'}{4}$$

Load on the pin is assumed to be distributed uniformly over length b' and the pin is assumed to be fixed in the plates.

Now
$$M_b = f_b Z$$

$$\frac{\text{Load on pin} \times b'}{4} = f_b \times \frac{\pi}{32} d_p^3 \qquad ...(15.65)$$

The induced bending stress (f_b) should not be more than the permissible value.

The pin will be in double shear. Equating the shearing load to the resisting shear.

$$2 \times \pi/4\, d_p^2 \times f_s = \text{load on pin} \qquad ...(15.66)$$

The induced shear stress (f_s) will be less than the permissible value.

Same will be the dimensions of the pin required to attach the joint with the chasis.

Assume t' as the thickness of the plate and considering the crushing of the pin in the plate, we have

$$2 \times d_p \times t' \times f_c = \text{Load on pin} \qquad ...(15.67)$$

From this we can find the thickness of plate t' required.

The spring after design should be checked for performance under expected working conditions.

Example 15.23. A laminated leaf spring is to carry a load of 3400 N with a deflection of about 31 mm. The spring must be supported at ends, the distance between the supports being 650 mm and is loaded at the centre. Allow a maximum stress of 420 N/mm^2. Take E = 2 × 10^5 N/mm^2. Find (a) the stress which will be induced if the load comes down with a shock, deflecting the spring 75 mm. (b) The magnitude of impact energy which the spring will absorb in this case. (c) The height from which the load must drop, and (d) the corresponding impact force. Number of leaves is 6 of width 51 mm each. Thickness of each leaf is 5.46 mm.

Solution: (*a*) The stress under the increased deflection can be found by noticing that stress is proportional to deflection. Hence,

$$f = \frac{420 \times 75}{31} = 1020 \text{ N/mm}^2$$

(*b*) The impact is equal to the resilience of the spring, which is

$$U = \frac{f^2}{6E} \times \text{volume of spring material}$$

$$= \frac{(1020)^2}{6 \times 2 \times 10^5} \times \frac{650 \times 6 \times 51 \times 5.46}{2} = 470 \text{ Nm}$$

(*c*) Since the impact energy $U = P(h + e)$, where e is the deflection, or 75 mm, the height of drop h must be

$$= \frac{470 \times 10^3}{3400} - 75 = 63 \text{ mm}$$

(*d*) The force of impact is given by relation

$$F = \frac{2U}{e} = \frac{2 \times 470 \times 10^3}{75} = 12.6 \text{ kN}$$

This force will act upon the end supports of the spring.

Example 15.24. A semi-elliptical, laminated, automobile spring to carry a load of 2850 N is to consist of seven leaves 56.2 mm wide, two of the leaves extending the full of the spring. The spring is to be 1075 mm in length and is to be attached to axle by two U-bolts 75 mm. apart. These bolts clamp the central portion of the spring so rigidly that they may be considered equivalent to a band having width equal to the distance between the bolts. The leaves are to be made of silica-manganese steel. Assuming an allowable stress of 350 N/mm^2, determine thickness for the leaves and deflection.

Solution: Effective length = $2L - l = 1075 - 75 = 1000$ mm

$2P = 2850$ N $\quad P = 1425$ N $\quad b = 56.2$ mm

$z_f = 2, \quad z_g = 5 \quad f = 350$ N/mm^2

Assuming that the leaves are to have no initial tension. Relation 15.53, gives us the maximum stress, therefore

$$f_t = \frac{18\,PL}{bt^2\,(2z_g + 3z_f)}$$

$$350 = \frac{18 \times 1425 \times 500}{56.2 \times t^2\,(2 \times 5 + 3 \times 2)}, \text{ gives } t = 6.85 \text{ mm.}$$

From I.S. 1137–1959 (Table 15.5), we adopt 7.1 mm.

The deflection will be given by relation (15.61) as two of the leaves are of full length.

$$\delta = \frac{12\,PL^3 k_3}{E\,bt^3\,(2z_g + 3z_f)}$$

$$k_3 = \frac{1 - 4r + 2r^2\,(1.5 - \log_e r)}{(1-r)^3}, \; r = \frac{z_f}{z} = \frac{2}{7} = 0.286$$

$$k_3 = \frac{1 - 4\times 0.286 + 2\times(0.286)^2\,(1.5 - \log_e(0.286))}{(1\text{-}0.286)^3} = 0.435$$

$$\delta = \frac{12\times 1425\times(500)^3\times 0.435}{2.1\times 10^5\,(2\times 5 + 3\times 2)\times 56.2\times(7.1)^3} = 27.3 \text{ mm}$$

Example 15.25. A locomotive spring has an overall length of 1.1 m and sustains a load of 75 kN at its centre. The spring has 3 full length leaves and 15 graduated leaves with a central band of 100 mm wide. All leaves are to be stressed to 420 N/mm^2 when fully loaded. The ratio of the spring depth to width is to be approximately 2. Take E = 2.1 × 10^5 N/mm^2.

(a) Determine the width and thickness of the leaves.

(b) Determine the initial space that should be provided between the full length and graduated leaves before the band load is applied.

(c) What load is exerted on the band after the spring is assembled.

Solution: Overall length of spring = 1100 mm.

Effective length of spring = $2L$ = 1100–100=1000 mm.

$L = 500$ mm

Load at the centre = $2P$ = 75 kN

$P = 37.5$ kN

As all the leaves are stressed to the same stress when loaded, therefore the final stress is given by relation 15.58.

$$f = \frac{6PL}{zbt^2}, \; 420 = \frac{6 \times 37.5 \times 10^3 \times 500}{18bt^2} \text{ gives, } bt^2 = 15\times 10^3$$

$$\frac{\text{Spring depth}}{\text{Spring width}} = \frac{z\times t}{b} = 2, \; \frac{18\times t}{b} = 2$$

or

$$t = \frac{b}{9} \text{ or } b \times \frac{b^2}{81} = 15 \times 10^3$$

gives, $b = 105$ mm, and $t = 11.6$mm.

From Table 15.5, the standard thickness of plate available is 11.8 mm.

(*b*) The nip to be provided can be calculated by relation 15.56.

$$C = \frac{2PL^3}{E\,zbt^3} = \frac{2\times 37.5\times 10^3(500)^3}{2.1\times 10^5\times 18\times 105(11.8)^3} = 17.5 \text{ mm}$$

(*c*) The band load can be calculated by relation 15.57.

$$P_b = \frac{2zgzfP}{z(2zg+3zf)} = \frac{2\times 15\times 3\times 37.5\times 10^3}{18(2\times 15+3\times 3)} = 4800 \text{ N}$$

Example 15.26. A semi-elliptical laminated spring is to carry a load of 5000 N and consists 8 leaves 46 mm wide, two of the leaves being of full length. The spring is to be made 1000 mm between the eyes and is held at the centre by a 60 mm wide band. Assume that the spring is initially stressed so as to induce an equal stress of 500 N/mm² when fully loaded. Design the spring giving (a) thickness of leaves (b) eye diameter (c) length of leaves (d) Maximum deflection and camber. Assume E = 2.1 × 10⁵ N/mm².

Solution: Effective length of spring $= 2L = 1000 - 60 = 940$ mm

$$L = 470 \text{ mm}$$

Load at the centre $= 2P = 5000$ N

$$P = 2500 \text{ N}$$

(*a*) As the leaves are initially stressed, therefore the final stress will be given by relation 15.58.

$$f = \frac{6PL}{zbt^2},\ 500 = \frac{6\times 2500\times 470}{8\times 46\times t^2}$$

gives $t = 6.2$ mm, standard thickness is 6.3 mm.

(*b*) Assume d_p as the diameter of pin, then considering the bearing of pin

$$d_p \times b \times p_b = \text{load on pin}$$

taking $p_b = 10$ N/mm² and load on pin $= \dfrac{P}{\cos 45°}$

$$d_p \times 46 \times 10 = \frac{2500}{\cos 45°}, \text{ gives } d_p = 8.5 \text{ mm}$$

Length of pin under bending $b' = b + 2 \times 1.5 = 49$ mm.

$$\text{Maximum bending moment} = \frac{\text{Load on pin}\times b'}{4} = \frac{2500\times 49}{4\cos 45°}$$

$$M_b = 43.5 \text{ Nm}$$

By equating the bending moment to the resisting moment of the pin.

$$f_t \times \frac{\pi}{32} d_p^3 = M_b,\ 80 \times \frac{\pi}{32} d_p^3 = 43.5 \times 10^3$$

gives, $d_p = 17.4$ mm adopt 20 mm as the diameter of pin. The pin can be checked for induced shear stress.

$$f_s = \frac{\text{Load on pin}}{2\times\pi/4\,d_p^2} = \frac{2500}{2\times\pi/4(20)^2\times\cos 45°} = 5.7 \text{ N/mm}^2$$

hence safe. The pin can be checked for the induced principal bending and shear stresses.

$$(c)\ \text{Smallest leaf length} = \frac{\text{effective length}}{z-1} + \text{Ineffective length of spring}$$

$$= \frac{940}{8-1} + 60 = 134 + 60 = 194 \text{ mm}$$

$$\text{Second leaf length} = \frac{940}{7}\times 2 + 60 = 328 \text{ mm}$$

$$\text{Third leaf length} = \frac{940}{7} \times 3 + 60 = 462 \text{ mm}$$

$$\text{Fourth leaf length} = \frac{940}{7} \times 4 + 60 = 596 \text{ mm}$$

$$\text{Fifth leaf length} = \frac{940}{7} \times 5 + 60 = 730 \text{ mm}$$

$$\text{Sixth leaf length} = \frac{940}{7} \times 6 + 60 = 864 \text{ mm}$$

$$\text{Seventh leaf length} = \frac{940}{7} \times 7 + 60 = 1000 \text{ mm}$$

The seventh and eighth leaves will be of full length and top leaf will act as master leaf.

$$\text{Length of master leaf} = 1000 + \pi\,(20 + 6.3) \times 2 = 1165 \text{ mm}$$

(*d*) Deflection of spring will be given by relation

$$\delta = \frac{12PL^3}{E\,bt^3\,(2z_g + 3z_f)} = \frac{12 \times 2500 \times (470)^3}{2.1 \times 10^5 \times 46\,(6.3)^3\,(2 \times 6 + 3 \times 2)} = 75 \text{ mm}$$

Camber is usually taken as half of the deflection $= \frac{75}{2} = 37.5$ mm

The radius of curvature after the leaves are assembled will be given by relation 15.60.

$$\delta_0\,(2R - \delta_0) = \frac{l_1^2}{4},\ 37.5\,(2R - 37.5) = \frac{(1000)^2}{4}$$

gives, $R = 3354$ mm.

Example 15.27. A semi-elliptical spring has ten leaves in all, with the two full length leaves extending 625 mm. It is 62.5 mm wide and is made of strips 6 mm thick. Design a helical spring, with mean diameter of coil 100 mm., which will have approximately the same values of induced stress and deflection for any load.

Solution: Assume $2P$ be the load acting at the centre of the semi-elliptical or helical spring. The maximum stress induced in the leaf spring is given by

$$f = \frac{6PL}{zbt^2},\ f = \frac{6 \times P \times 625}{2 \times 10 \times 62.5(6)^2} = 0.0833\,P$$

The maximum shear stress induced in the helical spring is given by

$$f = \frac{8PD}{\pi d^3},\ f = \frac{8 \times (2P) \times 100}{\pi d^3} = \frac{500\,P}{d^3}$$

If the value of induced stress is to be the same, then

$$0.0833\,P = \frac{500\,P}{d^3},\ \text{gives } d = 18.5 \text{ mm}$$

From table 15.5, the standard wire size will be 19 mm

Outside coil diameter = 100 + 19 = 119 mm

Inside coil diameter = 100 – 19 = 81 mm

The deflection of a leaf spring will be

$$\delta = \frac{12PL^3}{Ebt^3\,(2z_g + 3z_f)} = \frac{12 \times P \times (312.5)^3}{E \times 62.5 \times (6)^3\,(2 \times 8 + 3 \times 2)}$$

$$\delta = \frac{1240\,P}{E}$$

The deflection of the helical spring will be

$$\delta = \frac{8PD^3z}{Cd^4}, \delta = \frac{8 \times (2P) \times (100)^3 \times z}{C \times (19)^4}$$

$$\delta = \frac{0.123\,Pz}{C} \times (10)^3$$

If the deflection is to be same, then

$$\frac{1240\,P}{E} = \frac{0.123\,Pz}{C} \times (10)^3, z = 10 \times \frac{C}{E}$$

The ratio of $\frac{E}{C}$ may be taken as 2.5.

Then, $z = \frac{10}{2.5} = 4.0$ coils.

The total number of coils may be taken as 6.0, to allow for two end coils.

Example 15.28. A car is provided with four leaf springs each having 9 leaves and supports are 1 m apart and section of the leaves is 40 × 5 mm. The full capacity load for springs amounts to 8000 N. The rear axle takes 60 per cent of load, breaking strength 1200 N/mm². Check the dimensions.

Solution: Load on rear axle = 0.6 × 8000 = 4800 N

Each spring on rear axle will take a load of $= 2P = \frac{4800}{2} = 2400$ N

$\therefore \quad P = 1200$ N

Load on front axle = 8000 – 4800 = 3200 N

Load on each spring at front axle = 1600 N

The stress induced in the spring at rear axle will be

$$f = \frac{6PL}{zbt^2} = \frac{6 \times 1200 \times 500}{9 \times 40 \times (5)^2} = 400 \text{ N/mm}^2$$

Stress induced in the spring at front axle will be $= 400 \times \frac{800}{1200} = 267$ N/mm².

Allowed value of stress = 0.35 f_{ut} = 0.35 × 1200 = 420 N/mm² which indicates that the stress induced is less than the permissible value, hence safe.

The deflection in a spring can be given by relation 15.48.

$$\delta = \frac{6PL^3}{Ezbt^3}, \delta = \frac{3(2P)L^3}{Ezbt^3}$$

$$\frac{(2P)}{\delta} = \text{spring stiffness} = \frac{Ezbt^3}{3L^3} = \frac{2 \times 10^5 \times 9 \times 40 \times (5)^3}{3 \times (500)^3}$$

Spring stiffness = 8 N/mm.

Since there are four springs, total stiffness = 8 × 4 = 32 N/mm.

Example 15.29. The rear spring (semi-elliptical) of an automobile has 9 leaves, each with an average thickness of 6 mm and a width of 50 mm. Span of the spring is 1.4 m and the total weight on the spring is 6000 N. Determine the rate of the spring, and maximum stress caused by the dead weight.

What approximate repeated maximum force (zero to P_{max}) would cause impending fatigue in 10^5 cycles, the number of applications of the maximum load expected during the ordinary life of a car? Ultimate strength for the material of spring is 1400 N/mm² and yield strength is 1200 N/mm². Endurance limit may be taken as half of the ultimate strength. For impending fatigue, the design factor will be one. The stress concentration factor may be taken as 1.4. Take E = 2 × 10⁵ N/mm².

Solution: Spring stiffness will be given by relation,

$$\text{Spring stiffness} = P_0 = \frac{2P}{\delta} = \frac{Ezbt^3}{3L^3} = \frac{2\times10^5\times9\times50\times(6)^3}{3\times(700)^3} = 19.8 \text{ N/mm}$$

Maximum stress caused by dead weight

$$= \frac{6PL}{zbt^2} = \frac{6\times3000\times700}{9\times50\times(6)^2} = 778 \text{ N/mm}^2$$

when the load varies from zero to P_{max}.

$$P_m = \text{mean load} = \frac{O + P_{max}}{2} = 0.5\,P_{max}$$

$$P_v = \text{variable load} = \frac{P_{max} - 0}{2} = 0.5\,P_{max}$$

$$\therefore \quad f_v = f_m = \frac{6PL}{zbt^2} = \frac{6\times0.5P_{max}\times700}{2\times9\times50\times(6)^2} = 0.065\,P_{max}\text{ N/mm}^2$$

Endurance limit = 0.5 × 1400 = 700 N/mm^2.

As we are designing for finite life of 10^5 cycles, the endurance limit will be

$$700\left[\frac{10^6}{10^5}\right]^{0.09} = 860 \text{ N/mm}^2$$

$$\text{design factor} = \frac{f_{yp}}{f_m - f_v + \dfrac{2f_v\times f_{yp}}{f_e}}$$

$$1 = \frac{1200}{0.065P_{max} - 0.065P_{max} + \dfrac{1.4\times2\times0.065P_{max}\times1200\times1.4}{860}}$$

gives, P_{max} = 5250 N.

15.29 DISC SPRINGS

Disc or "Belleville" springs are used where space limitations require high capacity units, i.e., applications requiring high spring rates and compact spring units. This is obtained at the expense of relatively non-uniform stress distribution across the section. Each spring consists of several annular discs that are dished to a conical shape, as in Fig. 15.30. They are stacked up on top of another, as in Fig. 15.3 (*e*), in order to increase the deflection. The unit may be held in alignment by a central bolt or a tube. Fig. 15.3(*e*) refers to springs placed in series and the deflection is proportional to the number of discs, Fig. 15.31 refers to springs placed in parallel and has load capacity proportional to the number of discs.

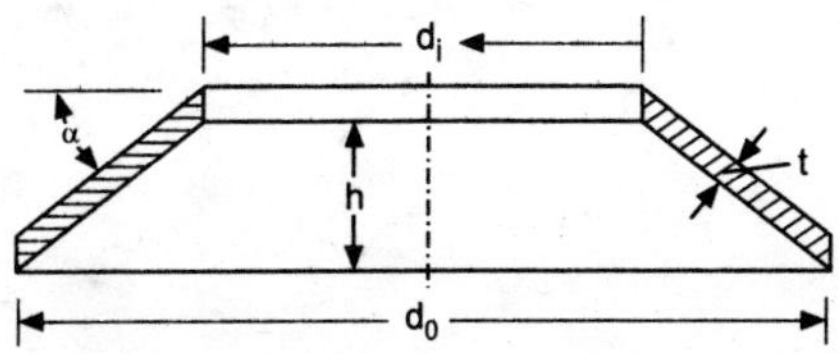

Fig. 15.30 *Disc spring*

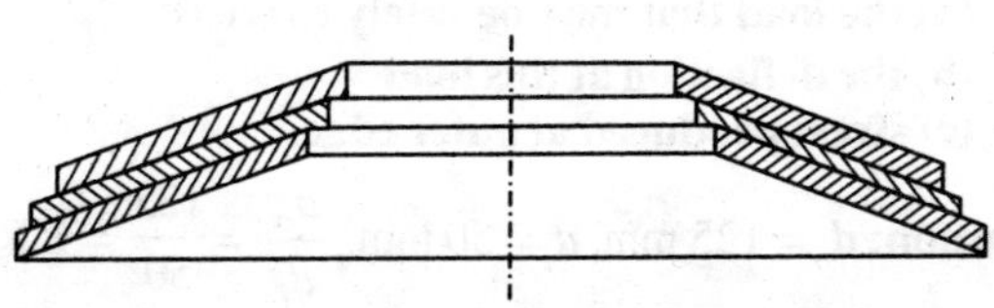

Fig 15.31 *Disc springs in parallel arrangement*

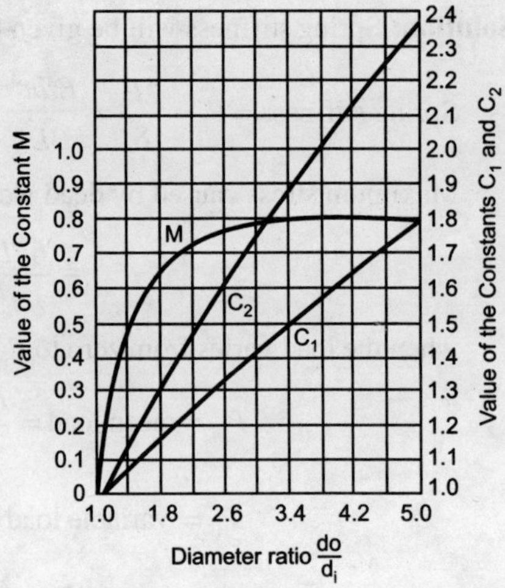

Fig. 15.32 *Disc spring constants*

These springs are cheap and permit different combinations, very easy to replace under failures. Disc type springs are used for buffers, impacts and suddenly applied loads as in punching, shearing, and drawing machines, for valves and supports of carriages, for machine foundations, and they are also used for frame structures as vibration insulators.

When the load is applied uniformly around the edge, the relation between the applied load P and the axial deflection is given by equation

$$P = \frac{4E\delta}{(1-\nu^2)Md_o^2}\,[(h-\delta)(h-\delta/2)\,t+t^3] \quad ...(15.68)$$

where δ = axial deflection of each disc.
t = Thickness of disc
d_o = Outside diameter of disc
ν = Poisson's ratio
h = free height of truncated cone
M = a constant depending upon ratio of outside to inside diameter, see Fig. 15.32.

The maximum stresses occurring at the edges are given by equations

$$f = \frac{4E\delta}{(1-\nu^2)Md_o^2}\,[C_1(h-\delta/2)+C_2t] \text{ at inner edge} \quad ...(15.69)$$

$$f = \frac{4E\delta}{(1-\nu^2)Md_o^2}\,[C_1(h-\delta/2)-C_2t] \text{ at outer edge} \quad ...(15.70)$$

where C_1 and C_2 are constants depending upon the ratio of the outside to inside diameter of disc. The constants can be taken from Fig. 15.32.

The actual stresses in these springs are unknown. The endurance life of a disc spring is considerably increased if the corners of the disc are rounded off. The load capacity of disc springs may be increased by moving the point of load application outward from the inner edge. The load capacity varies approximately as the ratio of the distances from the outer edge to the point of load application. The deflection varies approximately as this distance ratio. Initially flat springs have a maximum flexibility when the ratio of outside to inside diameters is approximately 2. In general the ratio d_o/d_i should be between 1.5 and 5.

Generally the material of the disc springs is alloy of chrome vanadium steel. The allowable stress for this material of disc spring is 1500 N/mm^2, with dynamic loading allowable stress (average) is 800 N/mm^2 with variation of $\pm$ 400 N/mm^2. The poisson's ration $\nu = 0.3$. The springs have $\alpha = 4°$ to $7°$, where the best value is 6.5°. The thickness t of the disc varies from 1 mm to 2 mm. Diameter at base is 28 to 300 mm and h/t = 0.4 to 0.75 for stiff springs and 0.75 to 1.42 for $\delta > 0.75\,h$ for greater elastic deflections. The ratio t/d_0 varies from 0.03 to 0.06.

Example 15.30. A disc spring is made of 3 mm sheet with an outside diameter of 125 mm and an inside diameter of 50 mm. The spring is dished 4.5 mm. The maximum stress is to be 560 N/mm^2. Determine:

(a) the load that may be safely carried

(b) the deflection at this load

(c) stress produced at outer edge.

Solution: $d_o = 125$ mm, $d_i = 50$ mm, $\dfrac{d_o}{d_i} = \dfrac{125}{50} = 2.5$

$$t = 3 \text{ mm}, f = 560 \text{ N/mm}^2, h = 4.5 \text{ mm}$$

From Fig. 15.32, we can see the constant M, C_1 and C_2 for diameter ratio of 2.5.

$$M = 0.75, C_1 = 1.325, \quad C_2 = 1.54$$

Assume the poisson's ratio = 0.3 and $E = 2.1 \times 10^5$ N/mm^2.

The maximum stress will be produced at the inner edge, therefore by using relation 15.69, we can find the deflection.

$$f = \frac{4E\delta}{(1-\nu^2)Md_o^2}\left[C_1\left(h-\frac{\delta}{2}\right)+C_2 t\right]$$

$$560 = \frac{4\times 2.1\times 10^5\times\delta}{[1-(0.3)^2]\,0.75\times(125)^2}\left[1.325\left(4.5-\frac{\delta}{2}\right)+1.54\times 3\right]$$

gives, $\delta = 0.75$ mm.

The load that may be safely carried can be calculated by relation (15.68).

$$P = \frac{4\times 2.1\times 10^5\times 0.75}{[1-(0.3)^2]\,0.75\times(125)^2}\left[(4.5\text{-}0.75)\left(4.5\text{-}\frac{0.75}{2}\right)3+(3)^3\right]$$

$$P = 43.3 \text{ kN}$$

The stress produced at outer edge can be calculated by relation 15.70.

$$f = \frac{4E\delta}{(1-\nu^2)Md_o^2}\left[C_1\left(h-\frac{\delta}{2}\right)-C_2 t\right]$$

$$f = \frac{4\times 2.1\times 10^5\times 0.75}{[1-(0.3)^2]\,0.75\times(125)^2}\left[1.325\left(4.5-\frac{0.75}{2}\right)-1.54\times 3\right]$$

$$f = 505 \text{ N/mm}^2.$$

EXERCISES

1. What is the function of a spring? In what type of springs the load-deflection curve is not straight line? Show it by a diagram.
2. Explain the uses of springs with minimum five examples in each case.
3. Generally how the springs are classified? Indicate the different types of springs by sketches and give minimum two practical applications of each.
4. When the use of the non-circular section wire springs is recommended? Why this section of the wire for springs is not generally used?
5. Differentiate between a close coiled and open coiled helical spring.
6. What are the requirements to be satisfied by a spring?
7. How the surge in the springs can be eliminated?
8. Extension springs are used in machinery considerable less than compression springs, why?
9. Give the design procedure for the design of compression spring of an I.C. Engine. Indicate the procedure for checking of this spring for buckling and surge.
10. A helical spring of round wire supports a static load of 1000 N. The inside diameter of the spring coil must not be smaller than 50 mm. The spring is to deflect 15 mm. The permissible shear stress in the spring material is 400 N/mm^2. Calculate the wire size, mean coil diameter, number of effective turns, number of actual turns, spring index, solid length, free length, and spring stiffness. Assume the plain ends of the spring, sketch the spring designed and give all the dimensions.
11. Design a spring for a Ramsbottom type spring loaded safety valve. The valve diameter is 60 mm. Maximum pressure when the valve blows of freely is 0.74 N/mm^2. Maximum lift of the valve is 4 mm when the pressure changes from 0.7 N/mm^2 to 0.75 N/mm^2. The permissible shear stress in spring wire material is 500 N/mm^2. Assume the spring index as 8. Take C = 0.8 × 10^5 N/mm^2. Sketch the spring indicating the types of ends.

12. Design a helical spring for a spring balance to weigh 50 N. Scale length is 50 mm and the spring index is to be 5. Choose your own material and the safe working stress for the same. Sketch the designed spring and its method of attachment.
13. Figure 15.14 shows compression spring arranged for tension rods. Calculate the size of the wire for spring, hook bolt diameter and U-bolt diameter. Also find the free length, and the number of coils of the spring. Indicate the type of the ends of the spring. Assume.
    Maximum tension in the rod = 25 kN
    Permissible stress in the rod = 70 N/mm^2
    Spring index = 7
    Allowable stress in the spring wire = 500 N/mm^2
    Modulus of rigidity = 0.84×10^5 N/mm^2
    Deflection of the spring when load changes from 22.5 kN to 25 kN is 10 mm. The material of hook bolt and U-bolt is same.
14. Design a coil spring for a spring loaded governor to give a deflection of 12.5 mm under an axial load of 115 N. The mean radius of the coil is 25 mm and the wire diameter is 6 mm. Take $C = 0.8 \times 10^5$ N/mm^2. Check the spring under buckling. Give a neat sketch of the spring in the uncompressed state.
15. The following particulars refer to the valve spring of petrol engine:
    length of spring when valve is open... 40 mm
    length of spring when valve is closed... 48 mm
    spring load when the valve is open... 400 N
    spring load when the valve is closed... 250 N
    Maximum inside diameter of spring... 26 mm
    Maximum allowable stress... 500 N/mm^2
    Modulus of rigidity... ... ...0.83×10^5 N/mm^2
    Calculate (*a*) size of the wire required (*b*) number of active coils required (*c*) spring length when completely closed assuming square and ground ends (*d*) pitch of the spring (*e*) dangerous engine speed.
16. Figure 15.33 shows a combined spring and weight loaded valve. The diameter of the valve is 30 mm and it weighs 10 N and weight of lever is 25 N. The free length of the spring is 100 mm and loaded length $h = 120$ mm. Stiffness of spring is 40 N/mm. What load P is needed for the valve to just open at 3.5 N/mm^2 pressure? A lift of 1/16 times the valve diameter is needed for escape. Assume spring index 6. Permissible shear stress for spring wire is 500 N/mm^2 and $C = 0.84 \times 10^5$ N/mm^2. Design the spring completely.

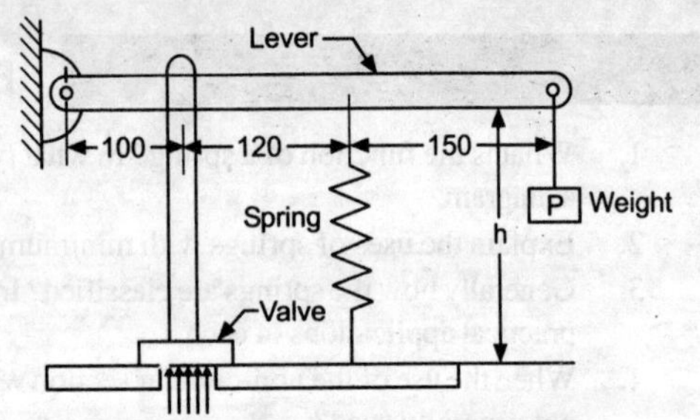

Fig. 15.33 *Spring and weight loaded valve*

17. A truck weighing 20 kN and moving at 25 km/hr has to be brought to rest by a buffer. Find how many springs each of 15 coils will be required to store the energy of motion during the compression of 150 mm. The available size of wire is 20 mm. The mean radius for the coils is 100 mm.
18. Compare the resilience for a cylindrical helical spring of (*a*) round section (*b*) rectangular section, if both springs are of same weight and material. Also state the effect on resilience of the side ratio for the rectangular section.
19. A helical spring whose mean diameter of coils is 8 times that of the wire is to absorb 500 Nm of energy. The initial compression of the spring is 50 mm and the additional compression is 100 mm while absorbing the shock. The maximum permissible stress is 450 N/mm^2 and modulus of rigidity is 0.83×10^5 N/mm^2. Design the spring completely.
20. Design a round wire helical compression spring for inside diameter of spring as 37 mm and ends are to be squared and ground. The maximum load is 500 N at 50 mm deflection. Spring is assembled by compressing it 15 mm. State the material used for spring. What is the margin of safety for each type of spring? Assume your own values for safe stress, etc.
21. A carbon steel spring is to be subjected to a load that varies from 2000 N to 5000 N. The outside diameter should be between 90 mm to 100 mm, the spring index between 5 and 10. The approximate stiffness of the spring is 90 N/mm. Choose your own carbon steel material, and for a design factor of 1.35 by the Wahl line, find the wire

diameter. Also determine the mean coil diameter, number of active turns, number of actual turns, solid length, free length for squared and ground ends, pitch of coils and inside diameter of coil.

22. A spring controlled governor of the Hartnell type has two balls, each weighing 25 N and each attached to the arms of a bell crank lever which pivots about a fixed fulcrum. The outer arms of the bell crank levers carry rollers which lift the sleeve against the pressure exerted by a spring surrounding the governor spindle. Two arms of the bell crank lever are of equal length and minimum and maximum radii of rotation of the governor balls are 75 mm and 90 mm. If the sleeve is to begin to lift at 250 rev/min and the increase of speed allowed is 5 per cent, find the initial load on the sleeve and required stiffness of the spring. Design the suitable spring for circular wire section. The permissible shear stress is not to exceed 250 N/mm^2 and $C = 0.84 \times 10^5$ N/mm^2. Spring index is 8. Clash allowance is 15 per cent of the deflection.

23. A coiled compression spring, of oil tempered steel wire has seven active coils of 10 mm wire diameter in a coil of 80 mm outside diameter. The spring is used to produce axial thrust on clutch. The free length of the spring is 190 mm with the clutch engaged, the length is 140 mm. Calculate (*a*) the stress in the wire (*b*) force exerted by the springs against the clutch plate (*c*) length of wire (*d*) power transmitted at 360 rev/min by clutch if friction plate radii are 153 mm and 200 mm and $\mu = 0.1$.

24. A rod of 3 mm diameter is coiled into a tension spring of 20 effective coils with a mean diameter of 50 mm. The modulus of rigidity is 0.84×10^5 N/mm^2
    (*a*) Determine the stress due to load of 13.5 N.
    (*b*) Determine the corresponding spring elongation.
    (*c*) If the initial tension in the spring was 4.5 N, what would be the stress and elongation when 13.5 N load is applied.

25. A gas engine valve spring is loaded to 340 N when the valve is closed and 520 N when the valve is open. The valve lift is .75 mm. The outside diameter is to be from 37.5 mm to 45 mm and the allowable stress is 400 N/mm^2.
    (*a*) Determine the wire diameter, outside diameter, inside diameter, number of effective coils and number of actual coils.
    (*b*) Determine the open or free length of the spring assuming the ends to be squared and ground. Also determine the pitch of coils.

26. Design a helical spring for the front wheel suspension of an automobile, using silicon manganese steel. The spring, must have a scale of 35.5 N/mm, an inside diameter of 105 mm, and a free length of 350 mm with squared and ground ends. The spring must carry a static load of 3650 N and when compressed solid it should not receive a permanent deformation.

27. A weight of 750 N located at a position 750 mm above the centre of a platen is to drop on the platen. The platen is to be supported on four helical springs. Assume the following:

Deflection of each spring	...75 mm
Spring index	..5.0
Maximum permissible shear stress	...400 N/mm^2

    Design the spring completely and make a dimensioned sketch.

28. A cluster spring is made up of two springs for a valve gear mechanism. The material of the spring wires is same. Using following particulars, design the springs completely.
    Maximum load = 9750 N, permissible shear stress = 60 N/mm^2.
    Maximum deflection = 45 mm, spring index = 6, modulus of rigidity = 0.84×10^5 N/mm^2.
    Adopt standard clearance between two coils and allow for the wire curvature.

29. Design a set of three concentric springs with gradually decreasing pitch diameter for a centrifugal variable speed governor. At a speed of 100 rev/min, loading on the springs due to centrifugal forces amounts to 200 N. The compression of the spring is 50 mm when the speed is 200 rev/min. Give a neat sketch of the springs designed by you indicating the specific advantage of this arrangement. Take the permissible shear stress as 600 N/mm^2, $C = 0.8 \times 10^5$ N/mm^2 and spring index as 6.

30. The movement of a low moving plunger is to be controlled by parallel springs arrangement, consisting of two concentric helical springs having common anchorage. Upward lift of the plunger is 20 mm. The spring force resisting upward movement on the plunger is 200 N. This is provided by the outer spring only. The spring force is 40 N/mm motion. The inner spring now makes contact with the plunger shoulder. Additional further lift of 10 mm, changes the spring load to 800 N. Design the spring, if outer diameter of outer spring is to be 30 to 40 mm

and mean diameter of the inner spring 20 mm. Take the permissible stress in shear as 300 N/mm^2 and $C = 0.8 \times 10^5$ N/mm^2 for the spring material.

31. One of the concentric springs in a pedestal type truck for a 500 kN capacity railroad freight car has free dimensions as given in table. The length of both the inner and outer spring is 165 mm, when carrying the rated load.
(*a*) What is the rated load for the concentric spring? What per cent of the rated load is carried by the outer spring? What by the inner spring?
(*b*) What is the design stress for the outer spring? For the inner spring?

	Number of coils	Wire diameter	Outer coil diameter	Free length
Outer spring	$5\frac{1}{8}$	39 mm	200 mm	197 mm
Inner spring	7	25 mm	115.6 mm	187.5 mm

32. A steel compression spring is to be manufactured in quantities of 100,000. The spring is subjected to a fluctuating force of 37.5 N maximum and 10 N minimum, at speed of 60 cycles/min. It is to deflect 9 mm under action of this force variation. There is no limitation on the length. The spring operates in a lubricated atmosphere whose temperature does not exceed 150°C. The service may be described as mild. Cost is a primary consideration. Specify the material treatment and the complete dimensions of the spring.

33. A clutch spring made of rectangular section 5 mm × 10 mm is wound in conical form with end outside diameters 150 mm and 100 mm. If there are 4.5 effective coils, calculate:
(*a*) Clutch pressure to produce a spring pressure of 300 N/mm^2.
(*b*) Free length of spring if operating length is 30 mm.
(*c*) Stress in spring section if additional compression is 5 mm while disengagement takes place.

34. (*a*) A conical spring with 10 effective coils and diameters 80 mm and 60 mm is loaded with 1000 N. Calculate the size of round wire if $f_s = 400$ N/mm^2 and deflection under load.
(*b*) For a rectangular section wire of a spring, find the deflection and size of wire for the data give in problem 34(*a*). Assume a rectangular section wire with side ratio 8.

35. A torsion spring is of 6.1 mm monal metal wire, wound with an inside diameter of 45 mm. Assuming a safety factor, find (*a*) the maximum force which can be applied with a 55 mm leverage, Fig. 15.19, and (*b*) the corresponding deflection in degrees, if spring has 4.5 coils.

36. (*a*) Design a torsion spring, Fig. 15.19, for a maximum torque of 2.5 Nm. The lever arm is 45 mm. Make the spring of round wire, and assume a sudden torque application.
(*b*) Using data of part (*a*) design a spring made of square wire.

37. (*a*) A flexible coupling transmits 15 kW power at 1,000 rev/min through a torsional helical spring made of 15 mm square wire. Find the pitch diameter of the coil, taking the permissible stress as 500 N/mm^2. Also find the number of active coils if the torsional deflection should not exceed 5°.
(*b*) Work part (*a*) using 15 mm round wire.

38. A window shade spring is made of steel wire 1.19 mm square. The outside diameter of coil is 20 mm. The spring must exert a pull on the shade of 11 N after it is wound 12 revolutions. The wooden roller is 32 mm in diameter. Determine:
(*a*) The number of coils to be used and (*b*) the maximum flexural stress.

39. A flat spiral spring for holding an electric motor brush against the commutator is to be made of brass ribbon of 12.5 mm wide. It is to exert pressure of 4.5 N against brush. The line of action of the thrust against the brush passes 75 mm from the axis of the spring. The length of the ribbon used is 450 mm. Assume the maximum allowable stress to be 100 N/mm^2 and the modulus of the brass to be 0.9×10^5 N/mm^2. Determine the thickness for the ribbon and the number of degrees through which the axis must be turned to produce the force required.

40. A laminated spring is made of six graduated and two full length leaves, each 3.35 mm thick and 18 mm wide. The effective length of spring is 750 mm, The permissible stress is 280 N/mm^2.
(*a*) Determine the central load that may be carried, assuming no initial stress in the spring.
(*b*) Determine the deflection.

41. A semi-elliptic laminated spring is made of 5 mm thick steel plate 50 mm wide. The length between the supports is 665 mm and the band is 65 mm wide. The spring has two full length and five graduated leaves. A central load of 1600 N is applied. Determine:
    (*a*) The maximum stress in each set of leaves for an initial condition of no stress in the leaves.
    (*b*) The maximum stress if initial stress is provided to cause equal stresses when loaded.
    (*c*) The deflection in part (*a*) and (*b*).
42. The spring of a small truck are to hold a maximum load of 4000 N each and to have a deflection rate 50 N/mm. The engine develops a maximum torsional moment of 300 Nm where the rear axle ratio is 3 : 1 and the 4-ply tyres are sized 170 × 760 mm. The coefficient of friction between the tyre and ground is 0.6. The span length of the spring may be taken as 1.5 times the tyre diameter. Design the leaf spring and make a neat sketch showing all necessary dimensions. Tabulate a satisfactory combination of:
    (*a*) thickness of leaf and width
    (*b*) number of leaves
    (*c*) camber in free position
    (*d*) radius of curvature
    (*e*) material used and heat treatment
    (*f*) length of each leaf, and
    (*g*) deflection.
43. A cantilever spring has an effective length of 550 mm and has two full length and eight graduated leaves, each 55 mm wide. The spring is to sustain a load of 400 N with a stress of 350 N/mm^2 in all leaves.
    (*a*) Determine the thickness of the spring leaves and the standard size from I.S. 1137 – 1959.
    (*b*) Determine the deflection at full load.
44. A spring of a truck has 10 leaves of graduated length. The spring supports are 1.06 m apart and the central band is 87.5 mm wide, The central load is to be 5000 N, with a permissible stress of 300 N/mm^2. Determine the width and thickness of the steel spring material and the deflection when loaded. The spring should have a ratio of total depth to width of about 2.5.
45. It is required to find out the number of leaves, width and thickness for a spring with following particulars.
    Effective span of spring between supports = 100 mm
    Total load = 5000 N
    Permissible stress in spring material = 480 N/mm^2
    Deflection = 75 mm
    At least two leaves must be of full length. Calculate the above factors for (*a*) leaves not stressed initially, (*b*) leaves stressed initially. Assume the width of leaf 20 times its thickness.
46. It is required to compare the helical spring and semi-elliptical spring for a given load of 15 kN and stiffness, 100 N/mm by way of designing the springs completely and then comparing the weights. Assume
    (*a*) Span of semi-elliptical spring as 800 mm. Use leaves 50 × 4.75 mm to be loaded centrally.
    (*b*) Outside diameter of helical spring is 125 mm.
    Density of steel is 7800 kg/m^3, $E = 2.0 \times 10^5$ N/mm^2.
    $C = 0.835 \times 10^5$ N/mm^2, safe working stress for leaf spring is 350 N/mm^2 and for helical spring is 250 N/mm^2.
47. (*a*) Under what conditions is the use of concentric springs recommended? Give two practical examples.
    (*b*) List important types of springs and comment on their applications.
    (*c*) A semi-elliptical laminated spring is to carry a load of 3000 N and consists of seven leaves 60 mm wide, two of the leaves extending the full length of the spring. The spring is to be 1.1 m in length and is to be attached to the axle by two U-bolts 75 mm apart. The leaves are to be made of silica-manganese steel. Assuming an allowable stress of 350 N/mm^2, determine the thickness for the leaves and the deflection.
    Draw to half or quarter full size the elevation and plan of the spring.
48. A car weighing 9000 N when loaded has a wheel base of 2.75 meters and its centre of gravity is 1.45 meter behind the front axle. The car has to be sprung on four similar longitudinal semi-elliptical springs 0.85 m between the shackle points. Design a suitable spring using a factor of safety of 1.5 on proof stress of 840 N/mm^2. The static loads should be multiplied by a load factor of 2.4 to allow for impact loads and the maximum deflection is not to exceed 75 mm. Strips are available in widths varying by 6 mm and in thickness varying by 0.8 mm. Assume the width of spring approximately 6 times thickness. Take $E = 2.1 \times 10^5$ N/mm^2.

Give a fully dimensioned and fully proportioned sketch of the spring and show the method of securing a rear spring to the axle of the car.

49. (*a*) Sketch the various arrangements by which a vehicle is supported on the axle of the wheel by means of carriage leaf spring.
    (*b*) What considerations are given in the design of such springs and what are usually left over? Why do they not make all the leaves of the same length?
    (*c*) Explain the utility of the centre bolt, U-clamps, rebound clips and camber.
50. Give in detail the design procedure for the design of a leaf spring of an automobile. Indicate, how the unequalised stresses produced in full length and graduated leaves at maximum load are counted in design?
51. A simple beam spring of uniform strength is required to store an impact energy of 200 Nm with a deflection of 50 mm. Taking working stress as 500 N/mm^2, find the length and maximum width of spring if the spring is made of uniform thickness 10 mm. Also suggest an equivalent laminated spring.
52. The top of shaking table is supported on four-flat spring leaves A, Fig. 15.34 with the ends rigidly fixed both in the floor and to the table. The free length is 650 mm and the spring leaf width is 40 mm. Find:
    (*a*) the force P necessary, the top 37.5 mm form its middle position and
    (*b*) maximum stress in the strip. Thickness of strip is 5 mm.

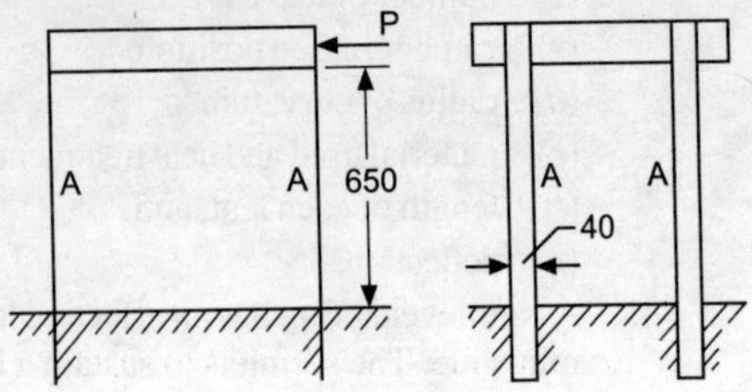

Fig. 15.34

53. Derive an equation for the energy absorbed by a cantilever spring of uniform strength. Express the answer in terms of metal volume and compare it with the energy absorbed by a beam of constant width and hence show that the laminated springs permit saving in material.
54. Find the maximum load carrying capacity of a plate spring with $d_o/d_i = 1.4$ and $d_i = 50$ mm. Take $t/d_o = 0.03$ to 0.06 and $v = 0.3$. take $h/t = 0.5$ for $\delta = 0.75h$.
    Also find the maximum stress induced, taking modulus of elasticity as 2.1×10^5 N/mm^2. What changes have to be made if the induced stress is more than the permissible value?
55. A disc spring is made of 3 mm sheet metal steel with an outside diameter of 125 mm and the diameter ratio as 2.5. The spring is dished 4.5 mm. The maximum stress to be produced at the outer edge is 560 N/mm^2. Determine.
    (*a*) The deflection.
    (*b*) The load that may be safely carried.
    (*c*) Check the stress produced at the inner edge.
56. Explain what do you understand by A.M. Wahl's factor and state its importance in the design of helical springs? A close coiled helical compression spring of 12 active coils has a spring stiffness of P_0. It is cut into two springs having 5 and 7 turns. Determine spring stiffnesses of resulting springs.
57. (*a*) Name the types of stresses that are produced when a compression helical spring is loaded.
    (*b*) Explain two methods of avoiding the tendency of a compression spring to buckle.
    (*c*) In what way can the fatigue resistance of the springs be increased?
    (*d*) In which type of springs the load versus deflection curve is not a straight line? Sketch the load deflection curves for various types of springs.
58. Explain the following terms of the spring:
    (*a*) free length, (*b*) solid length, (*c*) spring rate, (*d*) spring index, (*e*) active and inactive coils, (*f*) stress factor.
59. (*a*) Explain with the help of a neat sketch the superposition of stresses in a helical spring.
    (*b*) Classify the springs according to service conditions.
    (*c*) What is meant by solid stress in a spring wire?
60. (*a*) What are the general considerations to be kept in mind by the designer while designing a leaf spring?
    (*b*) Under what conditions a leaf spring is preferred over a coil spring?
61. Draw the distribution of stress produced in the wire of a close-coiled helical spring.
62. What is meant by buckling of springs?
63. What is meant by surging of springs?

64. How the values of "Wahl's factor" vary with the value of *S*?
65. Write the formula for calculating the shear stress induced in a close-coiled helical spring.
66. Concentric helical springs are wound:
    (a) In the same helix hand
    (b) in opposite helix hands
    (c) In any direction
    (d) depending on the load
67. The type of spring used in a clock is:
    (a) Leaf
    (b) helical
    (c) conical
    (d) spiral
68. How the deflection of a helical spring varies with wire diameter?
69. The stiffness of a conical spring is
    (a) constant
    (b) variable
    (c) unpredictable
70. To equalise the stress in leaf springs, the initial radius of curvature of full length leaves should be
    (a) Greater than the graduated leaves.
    (b) smaller than the graduated leaves.
    (c) equal to the graduated leaves.
    (d) equal and opposite to the graduated leaves.

❑❑❑

Section III

POWER TRANSMISSION

POWER TRANSMISSION

General: Power is transmitted from the prime mover to a machine, from one machine to another, or from one member of a machine to another, by means of intermediate mechanisms called 'drives'. The most common are the 'rotary motion drives' due to their numerous advantages over reciprocating motion. These intermediate mechanisms are necessary, instead of directly coupling the machine to the prime mover, due to the following reasons:

1. The optimal speeds of the prime movers or that of the standard motors may be different from the velocities required to operate the machines. The prime movers usually have higher angular velocities, which makes the designs more compact and light, while the operating members frequently require a large torque with relatively low velocities (the reverse case is rare).
2. The velocity of the driven machine may have to be frequently changed or regulated, whereas, the speed of the prime mover should be kept constant for its use to the full advantage.
3. In some cases, several machines may have to be operated from only one prime mover.
4. Certain periods of the machine operation may require torques far in excess of those available at motor shaft, e.g., punches, presses, etc.
5. Sometimes the machines are not coupled directly to the prime mover shaft due to the considerations of safety, convenience and maintenance.
6. The uniform rotary motion of the prime mover has to be converted into other kinds of motion : reciprocating, rocking, screw-type etc., of the driven members.

The various types of drives are: mechanical, hydraulic, pneumatic and electrical.. The intermediate mechanisms may be actuated by means of mechanical drives alone or by a combination of these with other transmissions. The choice of a proper drive for any case is arrived at through comparison of the technical and economical features of several designs. In this book, only the mechanical drives will be discussed.

Type of Mechanical Drives: The mechanical drives are classified in the following manner:

1. By the mode of Power Transmission:

(*a*) Transmission by friction. (*b*) Transmission by mesh.

Transmission by friction may be further classified as:

(*i*) With direct contact, e.g., friction drives.

(*ii*) With a flexible connection, e.g., Belt drives.

Similarly, the transmission by mesh may be further classified as:

(*i*) With direct contact, e.g., *toothed* and worm gears.

(*ii*) With a flexible connection e.g., chain drives.

2. By the change of velocity Ratio:

(*a*) Step by step change. (*b*) no change

(*c*) variable speed drives e.g. automotive vehicles.

3. By the position of shafts:

(*a*) Parallel shaft (*b*) Shafts at right angles

(*c*) Intersecting shafts.

Speed Ratio or Velocity Ratio:

The peripheral speed of the driving member can be found from the formula,

$$v_1 = \frac{\omega_1 D_1}{2}$$

or

$$v_1 = \frac{\pi D_1 n_1}{60}$$

The peripheral speed of the driven member

$$v_2 = \frac{\omega_2 D_2}{2}$$

or
$$v_2 = \frac{\pi D_2 n_2}{60}$$

In the absence of slipping, the peripheral speeds of both members should be equal, so

$$v_1 = v_2$$

i.e.,
$$\frac{\omega_1 D_1}{2} = \frac{\omega_2 D_2}{2} \text{ or } \frac{\pi D_1 n_1}{60} = \frac{\pi D_2 n_2}{60}$$

Whence
$$\frac{\omega_1}{\omega_2} = \frac{D_2}{D_1} = \frac{n_1}{n_2}$$

where D_1 and D_2 = diameter of the driving and driven members.

$\omega_1(n_1)$ and $\omega_2(n_2)$ = angular speeds of the driving and driven members.

The ratio of the angular speed of the driving member to that of the driven member.

$$(VR)_{1,2} = \frac{\omega_1}{\omega_2} = \frac{n_1}{n_2} = \frac{D_2}{D_1}$$

is referred to as speed ratio, or velocity ratio.

The sign of *VR* will depend on the directions of the angular velocities. If ω_1 and ω_2 are in same directions (open belt drive), *VR* has a positive sign. If ω_1 and ω_2 are in opposite directions (spur gear drive), *VR* has a minus sign.

If the power P_1 on the driving shaft is known, the power P_i on any intermediate shaft or last i.e. output shaft can be determined from the following relation:

$$P_i = P_1 \times \eta_{1i} \qquad \text{... (1)}$$

where η_{1i} = efficiency of the drive from the first to the *i*th, shaft. Power is derived from the formula (S.I. units).

$$P = \frac{M_t \times \omega}{1000} \text{ kW}$$

where M_t = torsional moment, newton metre, Nm

ω = angular speed, radians/s

$$\therefore \quad M_t = \frac{\text{kW} \times 1000}{\omega} = \frac{\text{kW} \times 1000 \times 60}{2\pi \text{ rev / min}} = \frac{9550 \times \text{kW}}{\text{rev/min}} \text{ Nm}$$

$\therefore$ From equation (1)

$$M_{ti} . \omega_i = M_{t1} . \omega_1 . \eta_{1i}$$

Hence,
$$M_{ti} = M_{t1} \cdot \frac{\omega_1}{\omega_i} \cdot \eta_{1i} = M_{ti} \cdot (VR)_{1,i} \cdot \eta_{1i}$$

$\therefore$ M_{t2} = torsional moment on the second shaft

$$= M_{t1} \cdot (V_R)_{1,2} . \eta_{1,2}$$

Thus rotary motion drives are used for transmitting energy and changing velocities and torques. In the multistage drive, where the speed ratios of the individual stages are $(V_R)_1$, $(V_R)_2$ ... $(V_R)_n$ the total speed ratio of the entire transmission is,

$$(V_R)_{1,n} = (V_R)_1 \times (V_R)_2 \times (V_R)_3 \ldots (V_7)_n$$

The total efficiency of the multistage drive is found from the formula:

$$\eta_{1,n} = \eta_1 \cdot \eta_2 . \eta_3 \ldots \eta_n .$$

Selection of a mechanical drive: The following data should be provided for the design of a mechanical drive:

1. Power to be transmitted from the driving shaft or the transmitted power on the driven shaft.
2. Angular velocities of the driving and driven shafts.
3. Distance between the shafts and their mutual location.
4. Overall dimensions.

For the above conditions, many drives of various types can be designed. The most advantageous design should be selected after the comparison of the following parameters of each design.

1. Efficiency
2. Weight
3. Size
4. Capital cost
5. Operational and maintenance cost.

The following factors will also be useful for selecting a proper drive:

1. **Velocity Ratio:** The desired velocity ratio can be obtained through a single stage of a drive of any type or several stages of drives of the same or different types.

   Toothed wheel gearing gives the highest speed ratio in a single stage. Worm gears give the highest value of velocity ratio, usually upto 100. After this the various drives in serial order are:

(i) Toothed gears	Velocity ratio	from 4 to 20
(ii) Roller chain drives	” ”	from 6 to 10
(iii) Silent chain drives	” ”	upto 15
(iv) V-belt drive	” ”	from 8 to 15
(v) Flat belt drive with idler	” ”	upto 10
(vi) Open flat belt drive	” ”	upto 5
(vii) Friction drives	” ”	from 5 to 10.

   The velocity ratio in toothed wheel gearing is limited only by the size of the drive and in belt drive, by the minimum allowable arc of contact on the smaller pulley. For large velocity ratios, it is better to employ multistage drives and these are accompanied by noise and vibrations. For overdrives, the velocity ratios are as below:

   (i) Toothed wheel gearing : Velocity ratio is 1 : 1.5 to 1 : 2
   (ii) Friction and belt drives : Velocity ratio is 1 : 3 to 1 : 5.

2. **Peripheral Velocity:** Other factors remaining the same, the transmitted power is directly proportional to the peripheral velocity which suggests the use of highest possible peripheral velocity to reduce the size of the drive. But the maximum allowable peripheral velocity is limited by many factors.

   The important factor in the belt drives is the centrifugal forces which increase the belt load and reduce its useful tension. For chain drives, the limiting factors are centrifugal forces and the knocking of the meshing links. In gear drives, the teeth should be cut with greater accuracy as the speed increases, otherwise the inaccuracies will cause added loads on the drives. In worm gears, the limiting factor is the wear resistance of the materials. The maximum allowable peripheral speeds for various drives are as below:

   (i) Flat belts, $v_{max} \leq 25$ m/s
   (ii) For special belts made from artificial fibre, $v_{max} \approx 50$ m/s
   (iii) For standard V belts, $v_{max} \approx 25$ to 30 m/s.
   (iv) Special quality V-belt with a steel core, $v_{max} \approx 40$ m/s
   (v) Steel band drives, $v_{max} \approx 80$ m/s
   (vi) Chain drives, $v_{max} \approx 25 - 30$ m/s
   (vii) In toothed gears, for spur gears with $v > 10$ m/s, and bevel gears at $v > 15$ m/s the teeth should be cut with precision. At present, the limit of v is upto between 120 to 150 m/s. In worm gears, v_{max} is ≤ 20 m/s.
   (viii) With friction drives, $v_{max} \ngtr 25 - 30$ m/s

3. **Transmitted Power:** The greatest power transmitted is by toothed gears. It can reach a limit of 36775 kW and more in gear reducers for marine turbines. The limit for worm gears is only upto 550 kW due to the great amount of heat given off. By reducing the amount of heat given off and by improving heat dissipation, the limit can be increased many times. In the case of V and chain drives, the power transmitted can be increased by increasing their size and number. However, the load distribution on individual belt or chain does not remain constant.

   For V-belts, the maximum is 735 to 1100 kW.
   For Flat belts, the maximum is 1835 kW.
   For chain drives, the maximum is 3670 kW.
   For friction drives, the limit is 150 to 225 kW.

4. **Losses in Transmission :** The losses in the drive consist of two parts : constant part which does not depend upon the load and a variable part which is proportional to the load. The constant part approximates the power of the idle run i.e. the power needed to rotate the drive without any power being transmitted. It depends on the velocity of rotation, weight of the drive parts, friction in the bearing and teeth. The losses in transmission for various drives are given below, for the sake of comparison:

   (*i*) Toothed gears : 1%
   (*ii*) Chain drive : 2%
   (*iii*) Flat belt drive : 2.5 %
   (*iv*) V-belt drive : 4%
   (*v*) Friction drive : 4%
   (*vi*) Worm gear : 10 to 25%

5. **Weight, Size and Cost of Drives:** This factor is also of great importance. The size of the mechanical drive may be designated in terms of centre distance and the width of the pulleys or wheels. For a given power and velocity ratio, the various drives may be written in the following order of decreasing centre distance:

   (*i*) Flat belt
   (*ii*) Flat belt with idler pulley
   (*iii*) V-belt
   (*iv*) Chain drive
   (*v*) Toothed gear
   (*vi*) Worm gear.

   On the basis of the width of the gear wheel or pulley, the various drives may be written as below in the decreasing order of the widths of the wheels or pulleys.

   (*i*) Chain
   (*ii*) Flat belt
   (*iii*) Flat belt with idler pulley
   (*iv*) Toothed gear
   (*v*) V-belt
   (*vi*) Worm gear.

   It can be inferred from the above analysis, that belt drives occupy more space whereas the toothed and worm gears have a more compact layout and can be easily fitted into machines.

   There is not much difference in the weight of the drives, of course, it will be least in the case of worm gear and largest in the case of toothed gear.

   Regarding the cost, if the cost of the V-belt drive is taken as unity, then for the other drives, it may be written approximately as follows:

   (*i*) Flat belt $\approx$ 1.06
   (*ii*) Flat belt with idler pulley = 1.25
   (*iii*) Worm gear = 1.25
   (*iv*) Chain drive = 1.40
   (*v*) Toothed gear = 1.65.

Some other characteristics can be compared for belt and gear drives, as follows:

Characteristics	Belt Drive	Gear Drive
1. Initial cost	Cheaper	Costly
2. Maintenance cost	Low	Considerable
3. Lubrication	None	Must
4. Shock absorption	Better	Not so good, depends upon gear material.
5. Life	1–5 years	Depends upon materials. Longest service life if properly lubricated, is generally more then 10 years.
6. Environment	Can not work under all types of conditions	If properly lubricated, can work under any condition and also under high temperature.
7. Noise	Low	Can be maintained low with proper lubrication and by adjusting shaft distances.
8. Force of Loads	Very heavy loads cannot be transmitted under normal conditions	Quite heavy loads can be transmitted as the gear size is practically unlimited.
9. Direction of input and output	Can be changed but not that easily and efficiently.	Can be changed easily.

The main characteristics of the various drives are given below:

Characteristics of Basic Types of Drives

Type of drive	Transmitted Power kW	Peripheral Speed m/s	Speed Ratio	Efficiency
1. Friction	upto 20 (100)	upto 20	upto 7 (25)	0.80–0.92
2. Flat belt	upto 100 (1500)	5–30 (100)	upto 4 (10)	0.92–0.98
3. V-bolt	upto 50 (1000)	5–30	upto 7 (15)	0.87–0.97
4. Straight tooth gear	upto 10000	25	upto 6 (10)	0.92–0.99
5. Helical tooth gear	upto 50000	25 (140)	upto 7 (20)	0.94–0.99
6. Worm gear	upto 100	upto 35	8 – 100 (1000)	0.75–0.90
7. Chain	upto 100 (5000)	upto 25	upto 15	0.94–0.98

Note: The highest known values are given in brackets.

CHAPTER 16

Shafts and Axles

16.1 GENERAL

Shafts and axles are the machine members, mostly cylindrical in cross-section, which support the revolving parts of a machine, such as pulleys, gears, flywheels, etc. An axle only supports a revolving part and may either be fixed relative to their support or rotate together with the parts mounted on them. Hence the load carried by the revolving parts is taken by the axle as a bending load and it is subjected to bending stress only. On the other hand, a shaft not only supports a revolving part but also transmits torsional moment. As a result, the shafts are subjected to bending as well as torsional stresses.

Depending upon the form of the geometrical axis of the shaft, it may be classified as: Straight Shaft and Crankshaft.

The design of the crankshafts is dealt with in the next chapters.

The straight shafts may be either Solid or Hollow. Shafts may also be flexible with changing form of the geometrical axis. Flexible shafts belong to a special group. However, these will also be discussed in this chapter.

The shafts can also be classified according to their use:

1. Prime Mover Shafts, e.g., turbine shafts, motor shafts, engine shafts, etc.

2. Machine shafts.

3. Power Transmission Shafts e.g. line shafts, countershafts, jack shafts, etc.

A line shaft is used in workshops and factories. It is a long and continuous shaft and is used for the distribution of power. The countershaft is employed in between a line shaft and a machine. The shaft directly connected to the power source is called as jackshaft. From this shaft, drive is given to other shafts.

16.2 CAUSES OF FAILURE IN SHAFTS AND AXLES AND STRESSES IN SHAFTS

The shafts and axles usually fail due to fatigue, which arises due to the following reasons:

1. The presence of cyclic over-loads.
2. Stress concentration. They may be due to production or operation causes, e.g., under cuts, machining traces, knotches, etc., or due to the presence of fillets, keyways, drilled holes, etc., which result in points of maximum stress concentration.
3. Wrong adjustment of bearings, insufficient clearances, etc.

A Shaft is subjected to torsion, bending or axial loads or a combination of these loads. Therefore, the shafts must be strong enough to resist these loads individually or in combination. In certain conditions, it is equally important to provide adequate bending stiffness of the shaft and to limit torsional deflection. Hence a shaft should be designed both for strength and rigidity or stiffness. Shafts may also fail due to transverse or torsional vibrations. The design of shafts for torsional vibrations is a special subject and it is out of the scope of this book.

16.3 MATERIALS FOR SHAFTS AND AXLES

The desirable properties for the materials for shafts and axles are:

1. Sufficient high strength.
2. A low sensitivity to stress concentration
3. Ability to withstand heat and case hardening treatment to lessen the effects of stress concentration and increase the wear resistance of the journals.
4. Good machinability.

To meet these requirements, the shafts are made from carbon steels of grades C-25, C-30 , C-35, C-40, C-45, C-50. Steel C-35 is most frequently used. It can be heat treated to impart high mechanical properties, i.e. with ultimate tensile strength of 800 MPa and elastic limit of 550 MPa. The heavily loaded shafts are made of alloy steels of various grades in order to minimise their diameter and to increase the wear resistance. These materials are treated by heat and case hardening methods to increase toughness, shock resistance and greater strength. Where surface wear resistance is the dominating factor, a carburizing grade steel can be used. Other case hardening methods such as nitriding, cyaniding, flame and induction hardening can also be used as a means of producing a wear resistant surface.

16.4 METHODS OF MANUFACTURING OF SHAFTS

The usual methods of shaft manufacture are:

1. Hot Rolling.
2. Cold drawing.
3. Turning or grinding from rough bars.

In the first method, hot rolled steel bars are passed through grooved finishing rolls until reduced to size by compression or elongation.

In the second method, the shafts are made to the required size by cold drawing the bars through dies of the required diameter. In both these methods, a hard surface layer may be formed on the surface of the shaft. When key way is cut on the shaft, this layer is penetrated, thereby, relieving the skin tension resulting in shaft warping. The required straightening adds to the cost, but can be beneficial if hammer-peening is used to straighten the shaft. This method produces a compressive surface stress that counter acts the effect of fatigue stress. Where warping is a serious consideration, and can not be easily rectified by straightening, an oil hardening alloy steel can be considered. Cold rolling and cold drawing methods are used for the manufacture of shafts of 76 to 89 mm diameter.

In the third method which is used for shafts larger than 89 mm diameter, the shafts are machined from hot rolled carbon steel. The machining must be deep enough to remove all of the decarburized scale caused by hot rolling. In this method, there is less tendency for the shaft to warp. Such shafts are more accurate for size and roundness and are more adaptable to machining and key seating. Shafts of sizes more than 15 cm are usually forged and then turned to size, examples being rail road axles, press cranks etc. Shafts of steel or nodular iron are also made by casting (for example, automotive engine crank shaft).

16.5 DESIGNING OF STRAIGHT SHAFTS

16.5.1 Design for Strength

(a) Pure Torsional Load: When the shaft is subjected to pure torsional load, the principal stress induced in the shaft will be the shear stress. The maximum shear stress induced in the shafts is given by relation:

$$f_s = \frac{16\ Mt}{\pi d^3}\ \text{N/m}^2 \qquad ...(16.1)$$

where M_t = Torsional moment, Nm
d = Diameter of shaft in m

The above relation is for a solid shaft. Sometimes, the shaft is made as hollow shaft. A hollow shaft has greater strength and stiffness than a solid shaft of equivalent weight. Moreover, the removal of core from the centre of the shafts increases their reliability. The shear stress relation for a hollow shaft is given as:

$$f_s = \frac{16\ M_t\ d_o}{\pi(d_o^4 - d_i^4)},\ \text{N/m}^2 \qquad ...(16.2)$$

$$= \frac{16\ M_t}{\pi d_o^3} \times \frac{1}{1 - R^4} \qquad ...(16.3)$$

where d_o = outer diameter of the shaft, m
d_i = inner diameter of the shaft, m
R = ratio of inside to outside diameter.

Shock and Fatigue Factor: This factor is taken into account to take care of the nature of load. This factor should be applied to the computed torsional moment. The factor which is designated as "K_s" has the following values:

For gradually applied or steady loads, $K_s = 1.0$
For suddenly applied loads, minor shock only $K_s = 1.0$ to 1.5
The suddenly applied loads, heavy shocks $K_s = 1.5$ to 3.0
∴ Design torsional moment = $K_s \cdot M_t$
The shear stress relations will be modified as:

$$K_s M_t = f_s \cdot \frac{\pi}{16} d^3,\ \text{solid shafts} \qquad ...(16.4)$$

$$K_s M_t = f_s \cdot \frac{\pi}{16} d_o^3\,(1 - R^4),\ \text{Hollow shafts} \qquad ...(16.5)$$

Design Stress: When the shaft is purchased according to carefully defined physical specifications, then the design stress may be taken according to the code for the design of Transmission shafts, given by ASME. It has been withdrawn since 1954, but, still extensively followed, the design stress is given as:

Design stress, f_s = 0.3 × tensile stress at the elastic limit
or = 0.18 × ultimate tensile stress
whichever is smaller.

For "commercial steel" shafting which has been purchased without specifying the physical properties.
Design stress, f_s = 55 MN/m².

Allowance for Keyway: The design stresses specified above do not take into account, the effect of keyway in the shaft. When a shaft has a keyway cut in it, it will have stress concentration at the keyway and allowance must be provided for this factor. This is done by modifying the design stress as below:

Design stress, $f_s = (0.75) \times$ design stress for shafts without keyway.

(b) Pure Bending Load: When a shaft is subjected to a pure bending load, the principal stresses induced in the shaft are tension and compression. The maximum stress induced is given by the relation:

$$f_t = \frac{32\,M_b}{\pi d^3}, \text{ for a solid shaft} \quad ...(16.6)$$

and

$$f_t = \frac{32\,M_b}{\pi d_0^3} \times \frac{1}{1 - R^4}, \text{ for a hollow shaft} \quad ...(16.7)$$

where M_b = Bending moment, Nm

In this case also, the effect of shock and fatigue loads must be taken into consideration. The factor K_b, for bending is given below:

Gradually applied, or steady load, $K_b = 1.5$
Suddenly applied load, minor shocks, $K_b = 1.5$ to 2.0
Suddenly applied load, heavy shocks, $K_b = 2.0$ to 3.0

$\therefore$ Design bending moment = $K_b \cdot M_b$

(c) Shaft Subjected to Combined Load.

***(i)* Shafts in Combined Torsion and Bending:** When a shaft is subjected to combined loading, the design is usually based on the maximum shear stress theory, since the shafts are usually made of ductile materials. Since under the combined loading, the shaft size will depend upon: (a) ratio of torsional moment to the bending moment, and (b) ratio of ultimate shear stress to ultimate tensile stress; it is suggested that the shaft size may be computed on the basis of both maximum shear stress theory and maximum normal stress theory and then, the larger size adopted.

Maximum shear stress is given as:

$$(f_s)_{max} = \frac{1}{2}\sqrt{f_t^2 + 4f_s^2} \quad ...(16.8)$$

and the maximum normal stress is given as:

$$(f_t)_{max} = \frac{1}{2}\left(f_t + \sqrt{f_t^2 + 4f_s^2}\right) \quad ...(16.9)$$

now

$$f_s = \frac{16}{\pi d_o^3} \cdot K_s \cdot M_t \frac{1}{1 - R^4}$$

and

$$f_t = \frac{32}{\pi d_0^3} \cdot K_b \cdot M_b \frac{1}{1 - R^4}$$

$\therefore$

$$(f_s)_{max} = \frac{16}{\pi d_0^3}\sqrt{(K_b \cdot M_b)^2 + (K_s \cdot M_t)^2} \cdot \frac{1}{1 - R^4} \quad ...(16.10)$$

and

$$(f_t)_{max} = \frac{16}{\pi d_0^3}\left[(K_b \cdot M_b) + \sqrt{(K_b \cdot M_b)^2 + (K_s \cdot M_t)^2}\right] \times \frac{1}{1 - R^4} \quad ...(16.11)$$

By putting $R = 0$, the above equations may be used for solid shafts also.

The term $\sqrt{(K_b \cdot M_b)^2 + (K_s \cdot M_t)^2}$ is termed as Equivalent Twisting moment, and

$\frac{1}{2}\left[(K_b \cdot M_b) + \sqrt{(K_b \cdot M_b)^2 + (K_s \cdot M_t)^2}\right]$ is referred as Equivalent Bending moment.

Design Stress: If the shaft is purchased according to physical specifications, then, the design stress may be taken as

Design stress, $f_s = (0.3) \times$ elastic limit in tension

or $f_s = (0.18) \times$ ultimate tensile strength

whichever is smaller,

and Design stress, $f_t = (0.6) \times$ elastic limit in tension

or $f_t = (0.36) \times$ ultimate tensile strength

whichever is smaller,

For commercial shafting,

Design stress, $f_s = 55$ MPa

and Design stress, $f_t = 110$ MPa

The above values are for shafts without keyways. For shafts with keyways:

Design stress = 0.75 × Design stress without keyway.

***(ii)* Shafts under Combined Torsional, Bending and Axial Loading:** If the shaft is subjected to an axial load in addition to torsional and bending loads as in a propeller shafts, then the average axial stress should be added to the bending stress and then the maximum shear stress or maximum normal stress induced in the shaft may be calculated. The direct stress due to axial load may be written as:

$$f_c = \alpha \cdot \frac{F_a}{A}$$

where F_a = axial load on the shaft

A = area of cross section of the shaft

α = constant depending upon the magnitude of column action of unsupported shaft length between the bearings.

If there is no column action of the shaft due to the axial load, $\alpha = 1$.

For column action, the value of α will depend upon L/k ratio.

If L/k is less than 115, the value of α is given by the straight line formula:

$$\alpha = \frac{1}{1 - 0.0044(L/k)}$$

where L = shaft length between the supporting bearings.

k = radius of gyration of the shaft.

If L/k, i.e., the slenderness ratio is above 115, α is given by Euler's formula

$$\alpha = \frac{f_{yp} \cdot L/k)^2}{c\pi^2 E}$$

where f_{yp} = yield stress in compression.

c = constant for the type of column end support.

= 1.0 for hinged ends.

= 2.25 for fixed ends

= 1.60 for ends partially restrained, as in bearings.

The maximum shear stress equation will become:

$$(f_s)_{\max} = \left[\left(\frac{K_s\, 16M_t}{\pi d^3}\right)^2 + \frac{1}{4}\left(\frac{K_b \cdot 32M_b}{\pi d^3} + \alpha \cdot \frac{4Fa}{\pi d^2}\right)^2\right]^{\frac{1}{2}} \quad \text{...(16.12)}$$

For hollow shafts,

$$(f_s)_{\max} = \frac{16}{\pi d_o^3 (1 - R^4)} \sqrt{\left[K_b \cdot M_b + \frac{\alpha \cdot F_a \cdot d_o \cdot (1 + R^2)}{8}\right]^2 + (K_s \cdot M_t)^2} \quad \text{...(16.12a)}$$

The torsional moment acting on the shaft can be determined, as already explained from

$$M_t = \frac{\text{kW} \times 1000 \times 60}{2\pi \text{ rev/min}} = \frac{9550 \times \text{kW}}{n} \text{ Nm}$$

For a belt drive, the torsional moment is found by

$$M_t = (F_1 - F_2) \times R_p \text{ Nm}$$

where F_1 = Tension in the tight side of belt, N

F_2 = Tension in the loose side of belt, N

R_p = Radius of the pulley, m

For a gear drive, the torsional moment is found from

$$M_t = F_t \times R_p \text{ Nm}$$

where F_t = Tangential force at the pitch radius, N

R_p = pitch radius, m.

16.5.2 Designing for Rigidity and Stiffness

Whereas, the shafts are mostly designed for strength, there are certain applications where the governing consideration is the stiffness or the rigidity of the shaft, e.g., long shaft of small diameter or where vibration is not permissible. It is generally more difficult to design for stiffness than for strength because of the lack of data to specify permissible deflections.

The deflection of the shaft for the various kinds of loads can be determined with the help of formulae given in Chapter 3. The deflection of the shaft can be minimized by mounting the supported parts near the bearings. A common rule for the transmission shafts is that the transverse deflection should be limited to 0.84 mm per metre of the shaft length. For machine shafts, the transverse deflection is given as

$$Y_{max} \le 0.75 \text{ mm/m length, or}$$

$$\le 0.06 \text{ L, mm}$$

where L = Distance between load and bearing, metres.

For shafts carrying gears, $Y_{max} \le 3/f$

where f = Gear face width, mm.

If the deflection of the shaft exceeds the above limits, there will be tendency for the shaft to bind in the bearings. However, this can be avoided by employing self adjusting bearings.

The torsional deflection may be found out by the relation:

$$\theta = \frac{M_t L}{JC}$$

where θ = angular deformation of shaft in length L, rad

J = Polar moment of inertia of cross-sectional area about axis of rotation, m^4

Here C = Modulus of rigidity in shear, N/m^2

It is a common practice to limit the angular deflection of the shaft to 1 degree in a length equivalent to 20 times the shaft diameter.

For machine tools $\theta \le 0.25$ °/m.

For camshafts of I.C. engines, $\theta \le 0.5°$.

16.5.3 Standard Available Sizes

The standard available sizes of circular shafts i.e., diameters in mm are: 5, 6, 7, 8, 9, 10, 11, 12, 14, 16, 18, 20, 22, 25, 28, 32, 36, 40, 45, 50, 56, 63, 71, 80, 90, 100, 110, 125, 140, 160, 180, 200.

16.5.4 Soderberg Criterion for Shaft Design

When a shaft is subjected to a combined fatigue loading (torsional moments and bending moment), it can be designed according to the procedure explained under Article 4.4.3.4, that is, by defining equivalent normal stress and equivalent shear stress. We shall write below the design equation based on 'Maximum Shear Stress Theory' and 'Distortion Energy Theory'.

(i) Maximum Shear Stress Theory: The design equation, according to this theory is equation (4.56), that is,

$$\frac{f_{yp}}{FS} = \left[\left(f_m + K_f f_v \cdot \frac{f_{yp}}{f_e}\right)^2 + 4\left(f_{ms} + K_f f_{vs} \cdot \frac{f_{yps}}{f_e}\right)^2\right]^{\frac{1}{2}} \qquad ...(16.13)$$

Writing the stresses in the above equation in terms of torsional moments and bending moments, we can write it as,

$$\frac{f_{yp}}{FS} = \left[\left(M_m + K_f M_v \cdot \frac{f_{yp}}{f_e}\right)^2 + \left(T_m + K_f T_v \cdot \frac{f_{yps}}{f_e}\right)^2\right]^{\frac{1}{2}} \cdot \frac{32}{\pi d^3}, \text{ for a solid shaft} \qquad ...(16.14)$$

and

$$\frac{f_{yp}}{FS} = \left[\left(M_m + K_f M_v \cdot \frac{f_{yp}}{f_e}\right)^2 + \left(T_m + K_f T_v \cdot \frac{f_{yps}}{f_e}\right)^2\right]^{\frac{1}{2}} \cdot \frac{32}{\pi d_o^3 [1-(d_i/d_o)^4]}, \text{ for a hollow shaft} \qquad ...(16.15)$$

where M_m = mean applied bending moment
M_v = variable applied bending moment
T_m = mean applied torsional moment
T_v = variable applied torsional moment.

(ii) Distortion Energy Theory: The design equations based on distortion energy theory and similar to equations (16.14 and 16.15), can be written from equation (4.60) as given below:

$$\frac{f_{yp}}{F_S} = \left[\left(M_m + k_f \cdot M_v \cdot \frac{f_{yp}}{f_e}\right)^2 + \frac{3}{4}\left(T_m + K_f \cdot T_v \cdot \frac{f_{yps}}{f_e}\right)^2\right]^{\frac{1}{2}} \times \frac{32}{\pi d^3}, \text{ for a solid shaft} \qquad ...(16.16)$$

$$\frac{f_{yp}}{F_S} = \left[\left(M_m + k_f \cdot M_v \cdot \frac{f_{yp}}{f_e}\right)^2 + \frac{3}{4}\left(T_m + K_f \cdot T_v \cdot \frac{f_{yps}}{f_e}\right)^2\right]^{\frac{1}{2}} \cdot \frac{32}{\pi d_o^3\left[1-(d_i/d_o)^4\right]}, \text{ for a hollow shaft} \qquad ...(16.17)$$

(a) Shaft Design Equations for Fatigue and Shock Loads: When shafts are subjected to shock loads alongwtih fatigue loads, such as punch presses, shafts of ball mills, and so on, design equations (16.14 to 16.17), should be modified to include the shock or impact factor, K_t, as discussed under Article 4.4.2. With this, the design equations (16.14 to 16.17), will get modified as,

(i) Maximum Shear Stress Theory:

$$\frac{f_{yp}}{F_S} = \left[\left\{K_i\left(M_m + K_f \cdot M_v \cdot \frac{f_{yp}}{f_e}\right)\right\}^2 + \left\{K_i\left(T_m + K_f \cdot T_v \cdot \frac{f_{yps}}{f_e}\right)\right\}^2\right]^{\frac{1}{2}} \frac{32}{\pi d^3},$$

for a solid shaft ...(16.18)

A similar equation can be written for a hollow shaft.

(ii) Distortion Energy Theory: From equation (16.16),

$$\frac{f_{yp}}{F_S} = \left[\left\{K_i\left(M_m + K_f \cdot M_v \cdot \frac{f_{yp}}{f_e}\right)\right\}^2 + \frac{3}{4}\left\{K_i\left(T_m + K_f \cdot T_v \cdot \frac{f_{yps}}{f_e}\right)\right\}^2\right]^{\frac{1}{2}} \cdot \frac{32}{\pi d^3},$$

for a solid shaft ...(16.19)

A similar equation can be written for a hollow shaft.

In the equations (16.18 and 16.19), the values of impact factor, K_i are as given under Article 4.4.2. These are being reproduced below:

K_i = 1.0 for gradually applied or steady loads.
= 1.0–1.5 for minor shocks.
= 1.5–2.0 for heavy shocks.

(b) Shaft Design for Steady Torsional Moment and Constant Bending Moment: The situation of a steady torsional moment and a constant bending moment acting on a shaft is very common and probably occurs more often than another type of loading. For a point on the surface of rotating shaft, a constant bending moment causes tensile stress which is completely reversed with each rotation, that is mean stress is zero. The effect of a steady torque is a shear stress on the surface that does not vary in time, that is, the variable component of shear stress is zero. Thus, for a rotating shaft transmitting power at a steady rate, we have

$$\therefore \; f_m = M_m = 0, \text{ and } f_{vs} = T_v = 0$$

∴ The design equations become

(i) Maximum Shear Stress Theory: From equation (16.18)

$$\frac{f_{yp}}{F_S} = \frac{32}{\pi d^3}\left[\left(K_i \cdot K_f \cdot M_v \cdot \frac{f_{yp}}{f_e}\right)^2 + (K_i \cdot T_m)^2\right]^{\frac{1}{2}}, \text{ for a solid shaft} \quad ...(16.20)$$

and

$$\frac{f_{yp}}{F_S} = \frac{32}{\pi d_o^3\left[1-(d_i/d_o)^4\right]}\left[\left(K_i \cdot K_f \cdot M_v \cdot \frac{f_{yp}}{f_e}\right)^2 + (K_i \cdot T_m)^2\right]^{\frac{1}{2}}$$

for a hollow shaft ...(16.21)

(ii) Distortion Energy Theory: From equation (16.19),

$$\frac{f_{yp}}{F_S} = \frac{32}{\pi d^3}\left[\left(K_i \cdot K_f \cdot M_v \cdot \frac{f_{yp}}{f_e}\right)^2 + \frac{3}{4}(K_i \cdot T_m)^2\right]^{\frac{1}{2}},$$

for a solid shaft ...(16.22)

and

$$\frac{f_{yp}}{F_S} = \frac{32}{\pi d_o^3\left[1-(d_i/d_o)^4\right]}\left[\left(K_i \cdot K_f \cdot M_v \cdot \frac{f_{yp}}{f_e}\right)^2 + \frac{3}{4}(K_i \cdot T_m)^2\right]^{\frac{1}{2}}$$

for a hollow shaft ...(16.23)

16.6 DESIGN OF AXLES

As mentioned earlier, the axles don't transmit any torque and are subjected to the effect of bending loads only, and so the axles are calculated for bending only. An axle may either be rotating along with the supported parts or be stationary relative to them. The design equation will be:

$$M_b = f_b \times Z$$

where f_b = allowable bending stress, N/m^2

$$Z = \text{Section modulus, } m^3 = \frac{\pi d^3}{32}$$

The bending moment should be modified to take into account the effect of shock and fatigue, the factors for this are given below:

Rotating Axles:

Gradually applied and steady loads, $K_b = 1.5$

Suddenly applied loads with minor shock, $K_b = 1.5 - 2.0$

Suddenly applied loads with heavy shock, $K_b = 2.0 - 3.0$

Stationary Axles:

Gradually applied and steady loads, $K_b = 1.5$

Suddenly applied loads with minor shock, $K_b = 1.5 - 2.0$

For commercial steel axles:

Design stress, f_b = 110 MPa without allowance for keyway.

= 82.50 MPa with allowance for keyway.

For steel purchased under definite specifications:

Design stress, $f_b = (0.6) \times$ elastic limit in tension

or $f_b = (0.36) \times$ ultimate strength in tension

whichever is smaller.

The axles are also checked sometimes for stiffness. The maximum deflection is given as:

$$Y_{max} = (0.001 - 0.003) \times L$$

where L = axle span, i.e., distance between bearings.

For extra stiffness of the axle:

$$Y_{max} = (0.0001 - 0.0025) \times L$$

16.7 DESIGN OF SHAFTS FOR CRITICAL SPEED

Sometimes, shafts fail as a result of transverse or torsional vibrations. Shafts should, therefore, always be checked for vibrational stability while measures should be taken to preclude dangerous transverse or Torsional vibrations. The calculations for the shaft for transverse vibrations are done to safeguard it against resonance when the vibration amplitude sharply increases and can reach value which will cause failure of the shaft. Resonance sets in when the shaft velocity reaches a critical point at which the frequency of variation of the external forces coincides with the frequency of vibration of the system comprising the shaft and parts fitted to it. Resonance can also occur when the frequency of vibration (natural frequency) of the external forces is a multiple of the frequency of the system vibrations.

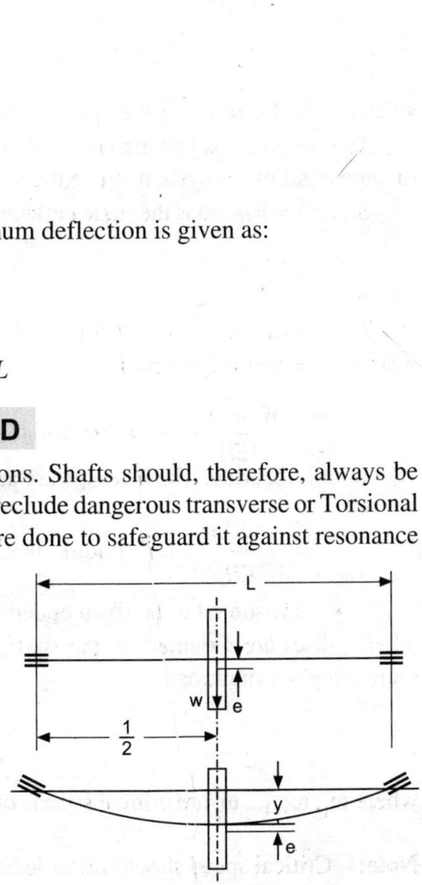

Fig. 16.1

Let a disc of weight W whose centre of gravity is displaced relative to the axis of rotation by the magnitude e be mounted on a shaft (Fig. 16.1). As the shaft and the disc rotate uniformly the centrifugal force C will cause the shaft to deflect.

The centrifugal force $C = m\omega^2(y + e)$. Neglecting the effect of the weight of the system the amount of deflection y for the accepted design (analogous to a simple beam) is equal to

$$y = \frac{CL^3}{48EI}$$

$$\therefore \qquad C = \frac{48EI}{L^3}y = ky,$$

where k is the force causing the shaft deflection equal to unity, for example 1 cm.

In this manner, $m(y + e)\omega^2 = ky$ and

$$y = \frac{e}{\dfrac{k}{m\omega^2} - 1} \qquad ...(16.24)$$

An increase in the angular velocity ω causes the value of the deflection y to grow and at $\omega = \sqrt{\frac{k}{m}}$, $y \to \infty$; this means that at such angular velocity the shaft must fail.

The angular velocity is approaching to which the deflection grows infinitely is called critical velocity.

Hence, $$\omega_{cr} = \sqrt{\frac{k}{m}},$$

Since $\omega_{cr} = \frac{\pi n_{cr}}{30}$ sec^{-1}, then the critical velocity of the shaft per minute is

$$n_{cr} = \frac{30}{\pi}\sqrt{\frac{k}{m}} = \frac{30}{\pi}\sqrt{\frac{k \times g}{W}} \approx 300\sqrt{\frac{k}{W}} \qquad ...(16.25)$$

where $g = 981$ cm/sec^2 is the gravity acceleration.

The conclusions just arrived at also hold for the case where account is taken of the shaft deflection due to the weight of the system, since the system rotates relative to a shaft flexed under it own weight.

Since $\Delta = W/k$ cm is the static deflection of the shaft from the weight (W) of the part fitted onto the shaft, then

$$n_{cr} \approx 300\sqrt{\frac{1}{\Delta}} \qquad ...(16.26)$$

In this way the critical velocity of shaft can be easily obtained from the static deflection Δ.

where Δ = Static deflection

$= \frac{W \cdot a^2b^2}{3EIL}$, when the disc is mounted at a distance 'a' from L.H.S. Bearing and at 'b' from R.H.S. bearing. The bearings are anti-friction bearings and the shaft is taken as simply supported beam.

$= \frac{W \cdot a^3b^3}{3EIL^3}$, for conditions same as above, but, bearings, are journal bearings and the shaft is assumed to be fixed ended.

when z discs are mounted on the shaft, the critical speed of the shaft is approximately determined by the Dunkerley's formula as,

$$\frac{1}{\omega_{cr}^2} = \frac{1}{\omega_1^2} + \frac{1}{\omega_2^2} + ... + \frac{1}{\omega_z^2} \qquad ...(16.27)$$

where $\omega_1, \omega_2, ..., \omega_z$ are critical speeds of the shafts for each disc respectively.

Note: Critical speed should be, at least, 20% away from the operating speed.

Excessive vibration of the shaft indicates that its velocity is close to the critical point. If this shaft continues to operate in these conditions it will inevitably fail.

Because of different magnitudes of resistance arising during vibration (internal friction, friction in the bearings, effect of environment) the shaft does not fail suddenly. Since at $\omega > \omega_{cr}$ the shaft deflection has the limit value, then if the shaft passes quickly through the zone of critical velocities its rotation becomes stable. Therefore, shafts can also operate at $n > n_{cr}$. Such shafts are called flexible shafts.

As $\omega \to \infty$, $y \to e$, i.e., the shaft self-centres. The transition through the zone of critical velocities should be affected as quickly as possible. Incorporation of vibration dampers into the system is also very useful. (See Art. 28.5)

Thus, resonance can be eliminated by employing rigid shafts with a high natural frequency which effectively resist bending strain or thin little-vibrating flexible shafts with a low natural frequency which bend freely under the action of centrifugal forces and take a new form of elastic equilibrium.

At high velocities, for example, in centrifuges with n = 20,000–40,000 rpm, flexible shafts make not only simple and cheap designs but are the only possible solution.

It follows from the above that for high-speed shafts the careful balancing of the shaft together with all parts fitted on it is of extreme importance.

16.8 FLEXIBLE SHAFTS

Flexible shafts are used under the following situations:

1. To transmit motion between parts whose axes of rotation are so arranged that it is impossible to form a rigid coupling between them.
2. To transmit motion when the motion position of the axes changes during the process of operation.

The above advantages of the flexible shafts are explained in Fig. 16.2.

These shafts are employed in the drives of concrete vibrators, devices for cleaning the hulls of ships, in mechanical picks, remote control instruments, etc.

There are three types of drive:

(*a*) Power drives.
(*b*) Control drives.
(*c*) Instrument drives.

In all the three types, the drive comprises the following three main elements:

(i) Flexible shaft.
(ii) Casing or protection.
(iii) Shaft end fittings.

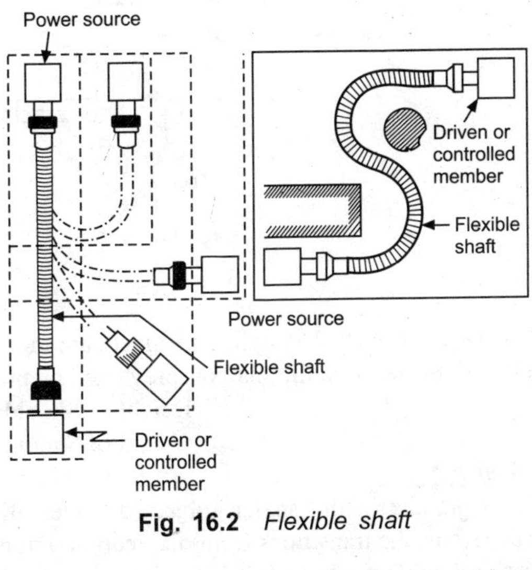

Fig. 16.2 *Flexible shaft*

A flexible shaft is made up of consecutively wound layers of carbon steel or bronze wire. The first layer is wound round a central core. This core can either be left inside or removed. Adjoining layers are wound in opposite directions. The direction of rotation of the shaft should be such that in transmitting torque, the outermost layer twists. The number of wires in each layer is 4 to 12 and the maximum number of layers is 8.

The selection of the power drive flexible shaft depends upon many factors—torque to be transmitted, curvature, operating and service conditions, direction of rotation and length of the application.

Casing: The functions of the casing are:

1. To take the forces transmitted to the shaft.
2. To retain the grease on the shaft surface.

3. To protect the shaft from dirt.
4. To safeguard personnel.
5. To prevent shaft failure.
6. To prevent formation of loops in operation.

The selection of a casing depends upon: Service conditions, degree of flexibility required, cost and appearance. The casing may be: metallic, made up of fabric, rubber covered or plastic covered.

End Fittings: These are employed to couple the flexible shaft to the motor shaft and the shaft of the service machine. The following rules apply for end fittings.

1. The shaft and casing are attached independent of each other since only the shaft rotates. The shaft end fittings are designed to engage the rotating parts of the connected members, while the casing end fittings are designed for rigid attachment to the stationary housing of the rotating parts.
2. One end of the shaft end fitting is designed to slide back and forth in the rotating part to accommodate any relative change in length between core and casing resulting during operation.

16.9 DESIGN OF STEPPED SHAFTS FOR ASSEMBLY

Figure 16.3 shows input Shaft of single-Stage Spur-Gear Reduction Unit. Power is transmitted to the input shaft from the electric motor by V-belts.

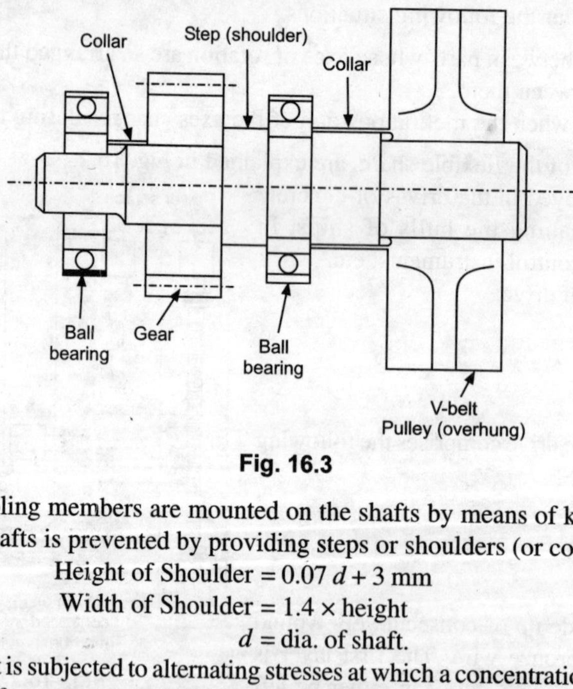

Fig. 16.3

Gears as well as coupling members are mounted on the shafts by means of key or splined joints. Axial play of the gears on the shafts is prevented by providing steps or shoulders (or collars) on the shaft.

$$\text{Height of Shoulder} = 0.07\,d + 3 \text{ mm}$$

$$\text{Width of Shoulder} = 1.4 \times \text{height}$$

where d = dia. of shaft.

Since the rotating shaft is subjected to alternating stresses at which a concentration of stresses is particularly dangerous, the transitions from one section of the shaft to another are smooth. The recommended fillet radii at the shoulders are, $r = 0.1\,d$.

The shaft diameter under the gear is made little larger than that under the L.H. Bearing, other wise, if the two diameters are equal, the gear would have been pulled over the seating surface for the bearing and impair the accuracy of fit.

The shaft diameter under the pulley is selected so that during assembly, the bearing can freely pass over this section of the shaft into which the keys have been pressed.

The necessity to have different diameter of the shaft facilitate assembly, the selected shaft diameter usually exceeds the required min. diameter.

Example 16.1. A solid shaft is subjected to a bending moment of 3.46 kNm and torsional moment of 11.5 kNm. The shaft is made of C- 45 steel, and factor of safety is 6. Determine the diameter of shaft.

Solution: For C-45 steel, hot rolled and normalised,

Ultimate tensile strength may be taken as about 630 – 710 MPa. Let us take it 690 MPa.

$$\therefore \quad f_t = \frac{690}{6} = 115 \text{ MPa}$$

Now ultimate shear stress is usually = 0.75 × ultimate tensile stress ≅ 517.5 MPa

$$\therefore \quad f_s = \frac{517.5}{6} \cong 86.0 \text{ MPa}$$

We shall design the shaft both for equivalent twisting moment and equivalent bending moment.

Equivalent twisting moment, $M_{te} = \sqrt{M_b^2 + M_t^2} = \sqrt{(3.46 \times 10^3)^2 + (11.5 \times 10^3)^2} = 12 \text{ kNm}$

$$\therefore \quad \frac{\pi}{16} d^3 \times f_s = 12 \times 1000$$

$\therefore \quad d = 0.0893$ m = 89.3 mm, choose 90 mm nearest standard size.

Equivalent bending moment, $M_{be} = \frac{1}{2}(M_b + \sqrt{(M_b^2 + M_t^2)}) = \frac{1}{2}(3.46 + 12) = 7.73 \text{ kNm}$

$$\therefore \quad \frac{\pi}{32} d^3 \times f_t = 7.73 \times 1000$$

$\therefore \quad d = 0.088$ m = 88 mm

Use 90 mm nearest standard size

$\therefore$ Diameter of the shaft is to be taken as 90 mm.

Example 16.2. A transmission shaft with keyways is subjected to a maximum torsional moment of 750 Nm and a maximum bending moment of 1200 Nm. The loads are suddenly applied, minor shocks are encountered, and the allowable shear stress is 42 MPa. Find the shaft diameter.

Solution: From equation (1610), for solid shaft

$$d^3 = \frac{16}{\pi \cdot f_s} \sqrt{(K_b M_b)^2 + (K_s M_t)^2}$$

For suddenly applied loads, with minor shocks,

$$K_b = 1.5 \text{ to } 2.0, \text{ take } 2.0$$

$$K_t = 1.0 \text{ to } 1.5, \text{ take } 1.5$$

and $\quad f_s = 42 \text{ MN/m}^2$

$$\therefore \quad d^3 = \frac{16}{\pi \times 42 \times 10^6} \sqrt{(2 \times 1200)^2 + (1.5 \times 750)^2}$$

$$\therefore \quad d = 0.0683 \text{ m}$$

Use 71 mm nearest standard size.

Example 16.3. Determine the diameter of a hollow shaft with a ratio of d_i/d_o of 0.8, capable of transmitting 300 kW at 225 rev/min, when subjected to a maximum bending moment of 5500 Nm. The load is suddenly applied with minor shocks for torsional moment, the bending moment is steady, and the allowable shearing stress is 56 MPa.

Solution: From equation (16.10),

$$d_o^3 = \frac{16}{\pi f s} \cdot \sqrt{(K_b \cdot M_b)^2 + (K_s \cdot M_t)^2} \times \frac{1}{1 - R^4}$$

$$f_s = 56 \text{ MPa}$$
$$R = d_i/d_o = 0.8$$
$$K_b = 1.5;\ K_s = 1.5$$
$$M_b = 5500 \text{ Nm}$$

$$M_t = \frac{9550 \times \text{kW}}{n} = \frac{9550 \times 300}{225} = 12.74 \text{ kNm}$$

$$\therefore \quad d_o^3 = \frac{16}{\pi \times 56 \times 10^6} \cdot \sqrt{(1.5 \times 5.5 \times 10^3)^2 + (1.5 \times 12.74 \times 10^3)^2} \times \frac{1}{1-(0.8)^4}$$

$$\therefore \quad d_o = 147.7 \text{ mm, say } 160 \text{ mm}$$
$$d_i = 160 \times 0.8 = 128 \text{ mm.}$$

Example 16.4. A machinery shaft is to transmit 61.5 kW at a speed of 1150 rev/min with mild shock. The shaft is subjected to a maximum bending moment of 900 Nm and an axial thrust of 70 kN. The shaft is supported at intervals of 2.5 metres. What should be its diameter when designed according to code?

Solution: According to code, shaft is made of a ductile material whose ultimate tensile strength is twice the ultimate shear strength, and the shaft diameter is controlled by the maximum shear theory regardless of the ratio of the twisting moment to the bending moment.

From equation 16.12,

$$f_s = \left[\left(\frac{K_s \cdot 16 M_t}{\pi d^3}\right)^2 + \frac{1}{4}\left(\frac{K_b \cdot 32 M_b}{\pi d^3} + \alpha \cdot \frac{4F_a}{\pi d^2}\right)^2\right]^{\frac{1}{2}}$$

$$f_s = 56 \text{ MPa}$$

Let $K_s = 1.5\ K_b = 1.5, F_a = 70 \text{ kN}; M_b = 900 \text{ Nm}$

$$M_t = \frac{9550 \times 61.5}{1150} = 510 \text{ Nm.}$$

The above equation cannot be solved, unless α is known, which can be found out only if d is known. Therefore, the problem should be solved by trial and error method. Neglecting F_a.

$$d^3 = \frac{16}{\pi f_s}\sqrt{(K_b \cdot M_b)^2 + (K_s \cdot M_t)^2}$$

$$\therefore \quad d^3 = \frac{16}{\pi \times 56 \times 10^6}\sqrt{(1.5 \times 900)^2 + (1.5 \times 510)^2}$$

From here, $d = 52$ mm.

Let us try $d = 60$ mm

$$\therefore \quad \text{Radius of gyration, } k = \frac{d}{4} = 15 \text{ mm}$$

$$\therefore \quad L/k = \frac{2.5 \times 1000}{15} = 166.7 > 115$$

$$\alpha = \frac{f_{yp} \cdot (L/k)^2}{c\pi^2 E}$$

Taking $f_{yp} = 300$ MPa for C–45 steel, annealed.

$c = 1.6$ for partially restrained ends

$$\alpha = \frac{300 \times 10^6 \times (166.7)^2}{1.6 \times \pi^2 \times 210 \times 10^9} = 2.51$$

$\therefore$ From equation (16.12)

$$f_s = \left[\left(\frac{1.5 \times 16 \times 510}{\pi \times 0.06^3}\right)^2 + \frac{1}{4}\left(\frac{1.5 \times 32 \times 900}{\pi \times 0.06^3} + 2.51 \times \frac{4 \times 70 \times 1000}{\pi \times 0.06^2}\right)^2\right]^{\frac{1}{2}}$$

= 65.5 MPa, which is more than the allowable stress.

$\therefore$ Unsafe.

If we try $d = 64$ mm and repeat the calculations, f_s will come out to be equal to 52.32 MPa, which is less than the allowable stress,

$\therefore$ Safe.

Example 16.5. The arrangement for a transmission shaft is shown in Fig. 16.4. 18.75 kW is to be transmitted through the shaft at 150 rev/min. The gear tooth reaction acts vertically upward as shown. The flywheel 'F' delivers the power vertically downward through a belt, the tension on the tight side F_1, being 2.5 times the tension on the slack side, F_2. The gear weighs 900 N, the flywheel 2700 N and both are keyed to the shaft. The shaft is made of 0.40 percent carbon steel that has an ultimate tensile strength of 490 MPa and an elastic limit of 280 MPa. Loads may be considered suddenly applied, but with minor shocks only. The weight of the shaft is assumed to be negligible in comparison to other loads. Determine the diameter of the shaft.

Solution: We should find out first of all, the maximum twisting moment and the maximum bending moment on the shaft.

Now $$M_t = \frac{9550 \times 18.75}{150} = 1193 \text{ Nm}$$

$\therefore$ $$\text{Gear tooth reaction} = \frac{M_t}{\text{gear radius}} = \frac{1193}{0.375} = 3185 \text{ N} \uparrow$$

Weight of gear = 900 N $\downarrow$

$\therefore$ $$\text{Net load at } G = 3185 - 900 = 2285 \text{ N} \uparrow$$

Again $(F_1 - F_2) \times$ radius of flywheel = M_t

$\therefore$ $$(F_1 - F_2) = \frac{1193}{0.625} = 1910 \text{ N}$$

But $$F_1/F_2 = 2.5$$

$\therefore$ $$F_1 = 3185 \text{ N}; F_2 = 1275 \text{ N}$$

$\therefore$ Total downward pull of the belt plus the weight of the flywheel at F

$$= F_1 + F_2 + 2700 = 3185 + 1275 + 2700 = 7160 \text{ N}\downarrow$$

Next step is to find reactions at bearings 1 and 2

$$R_1 \times 1.00 = 7160 \times 1.25 + 2285 \times 0.30$$

$\therefore$ $$R_1 = 7160 \times 1.25 + 2285 \times 0.30 = 9635.5 \text{ N} \uparrow$$

Similarly, $R_2 \times 1.00 = 7160 \times 0.25 + 2285 \times 1.30$

$$R_2 = 7160 \times 0.25 + 2285 \times 1.3 = 4760.5 \text{ N} \downarrow$$

As is clear from the S.F. diagram, the maximum bending moment acts at bearing 1.

$$M_b = 7160 \times 0.25 = 1790 \text{ Nm}$$

and $$d = \left(\frac{16}{\pi \cdot fs} \cdot \sqrt{(K_b \cdot M_b)^2 + (K_s \cdot M_t)^2}\right)^{\frac{1}{3}}$$

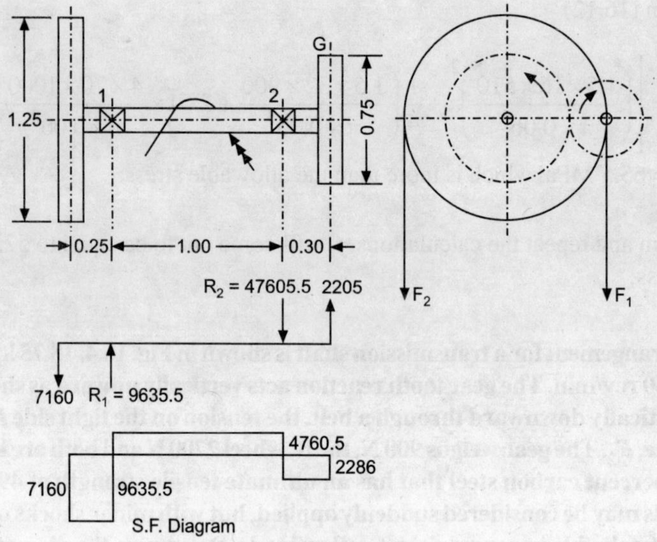

Fig. 16.4

Now $K_b = 2.0$ and $K_s = 1.5$

$$f_s = 0.3 \times 280 \times 0.75 = 63 \text{ MPa}$$

or $= 0.18 \times 490 \times 0.75 = 66.15$ MPa, whichever is smaller

$\therefore$ $f_s = 63$ MPa

$$\therefore \quad d = \left(\frac{16}{\pi \times 63 \times 10^6} \cdot \sqrt{(2 \times 1790)^2 + (1.5 \times 1193)^2} \right)^{\frac{1}{3}}$$

$$= 0.0686 \text{ m} = 68.6 \text{ mm}$$

Use 71 mm nearest standard size.

Example 16.6. A factory line shaft is 4.5 metre long and is to transmit 75 kW at 200 rev/min. The allowable fibre stress in shear is 49 MPa, and the maximum allowable twist is 1 deg. in a length of 20 diameters. Determine the required shaft diameter.

Solution: The twisting moment on the shaft is :

$$M_t = \frac{9550 \times 75}{200} = 3580 \text{ Nm}$$

Diameter of shaft on the strength basis :

$$M_t = f_s \times \frac{\pi}{16} d^3$$

$$\therefore \quad 3580 = 49 \times 10^6 \times \frac{\pi}{16} d^3$$

$$\therefore \quad d = 0.072 \text{ m}$$

Diameter of shaft on the basis of rigidity:

$$\theta = \frac{M_t L}{JC}$$

$$J = \frac{\pi d^4}{32}, \; C = 84.00 \text{ GPa}$$

$$L = 20\,d, \theta = 1^\circ = \frac{\pi}{180}\ \text{rad}$$

$$\frac{\pi}{180} = \frac{3580 \times 20\,d \times 32}{\pi d^4 \times 84 \times 10^9}$$

$$d = 0.0795\ \text{m}.$$

In this case, the shaft diameter required for strength is less than that required for torsional rigidity, and so, the shaft diameter should be made 0.0795 m, that is, 80 mm nearest standard size.

Example 16.7. A 11.25 kW, 1725 rev/min motor drives a centrifugal pump through a single set of 5 : 1 reduction gears. Determine the diameters of the shaft on the motor and the pump if ordinary steel shafting with standard keys is used. Load may be considered to be applied with minor shocks.

Solution:

(a) Motor Shaft: Twisting moment on the motor shaft

$$M_t = \frac{9550 \times 11.25}{1725} = 62.3\ \text{Nm}$$

Now shaft is of ordinary steel, therefore, allowing for keyways:

$$f_s = 56 \times 0.75 = 42\ \text{MPa}$$

now $$K_s \cdot M_t = \frac{\pi}{16} d^3 \times f_s$$

$$K_s = 1.5$$

$\therefore$ $$1.5 \times 62.3 = \frac{\pi}{16} d^3 \times 42.0 \times 10^6$$

$\therefore$ $$d = \text{Diameter of motor shaft} = 0.02246\ \text{m say } 25\ \text{mm}$$

(b) Pump Shaft: Speed of pump shaft $= \frac{1}{5} \times 1725 = 345$ rev/min

$\therefore$ Torsional moment on the pump shaft = 5 × 62.3 = 311.50 Nm

$\therefore$ Diameter of pump shaft $= 25 \times (5)^{\frac{1}{3}} = 42.75$ mm

Use 45 mm nearest standard size.

Example 16.8. A monel propeller shaft for a launch is to transmit 75 kW at 120 rev/min without a significant bending moment. The efficiency of the propeller is 70% at 10 Knots (1 Knot is 1.85 km per hr). If $L/k < 40$, what should be the diameter? Follow the "Code" procedure.

Solution: Equation (16.12) will be used, i.e.,

$$f_s = \left[\left(\frac{K_s \cdot 16 M_t}{\pi d^3}\right)^2 + \frac{1}{4}\left(\frac{K_b\, 32 M_b}{\pi d^3} + \alpha \cdot \frac{4F_a}{\pi d^2}\right)^2\right]^{\frac{1}{2}}$$

and $$M_b = 0$$

$\therefore$ $$f_s = \left[\left(\frac{K_s \cdot M_t \cdot 16}{\pi d^3}\right)^2 + \frac{1}{4}\left(\frac{4F_a}{\pi d^2} \cdot \alpha\right)^2\right]^{\frac{1}{2}}$$

Now $$M_t = \frac{9550 \times 75}{120} = 5970\ \text{Nm}$$

now $$F_a \times V = P \times \eta$$

$$V = \frac{10 \times 1.85 \times 1000}{3600} = 5.13 \ m/s$$

$$P = 75 \text{ kW}$$

$$\therefore \quad Fa = \frac{75 \times 1000 \times 0.7}{5.13} = 10.23 \text{ kN}$$

$K_s = 1.5$ for suddenly applied loads with minor shocks

$$\alpha = \frac{1}{1 - 0.0044(L/k)} = 1.213$$

For commercial shafting, $f_s = 56.00$ MPa

$$\therefore \quad 56 \times 10^6 = \left[\left(\frac{1.5 \times 16 \times 5970}{\pi d^3}\right)^2 + \frac{1}{4}\left(\frac{1.213 \times 4 \times 10.23 \times 10^3}{\pi d^2}\right)^2\right]^{\frac{1}{2}}$$

By trial, $d = 95$ mm.

Use 100 mm nearest standard size.

Example 16.9. A countershaft receives 26.25 kW from a motor through a coupling and transmits it via two belt drives to two machine tools each consuming 11.25 kW as shown as Fig. 16.5. Both the belt drives are arranged horizontally. The diameters of pulleys are 0.25 m and 0.6 m and their corresponding weights are 400 N and 750 N. Shaft speed is 30 rad/s. Find the diameter of the shaft.

Solution: To arrive at the diameter of the shaft, we must determine:

(*a*) Maximum twisting moment acting on the shaft.

(*b*) Maximum bending moment acting on the shaft.

The method employed to obtain the maximum bending moment is:

(*i*) To find the loads acting on the shaft due to the belt tensions and/or due to loads and due to the weights of the pulleys, flywheels and gear etc.

(*ii*) To resolve these loads in two convenient directions at right angles, usually vertically and horizontally.

(*iii*) To draw the bending moment diagrams or calculate the bending moments for each of these planes.

(*iv*) The bending moments in the two diagrams are combined to find the greatest bending on the shaft.

The forces of belt tensions will bend the shaft in a horizontal direction, whereas the weights of pulleys will bend it in a vertical plane.

1. Finding the belt tension :

The turning force on the pulley I

$F_{t1} = (F_1 - F_2)$ I

where F_1 = tension on the tight side, and F_2 = tension on slack side

$$\text{now} \qquad F_{t1} = \frac{(kW) \times 1000}{V_1}$$

$$V_1 = \frac{D_1 \times \omega}{2} = \frac{0.25 \times 30}{2} = 3.75 \text{m/s}$$

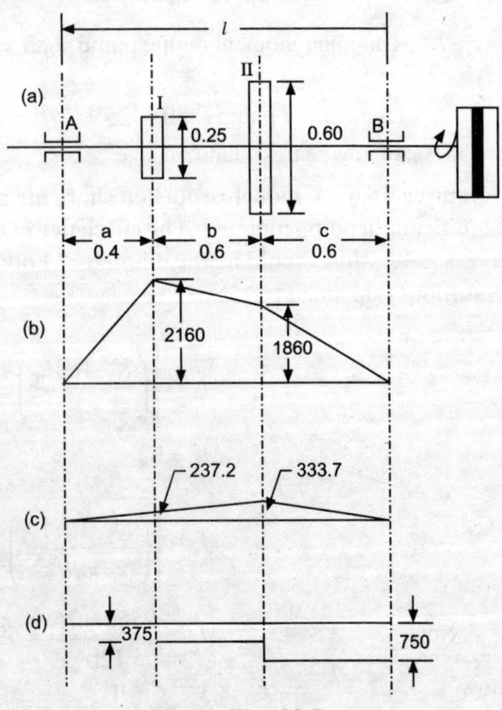

Fig. 16.5

$$\therefore \quad F_{t1} = \frac{11.25 \times 1000}{3.75} = 3000 \text{ N}$$

now $$V_2 = \frac{D_2 \times \omega}{2} = \frac{0.6 \times 30}{2} = 9 \text{ m/s}$$

$$\therefore \quad F_{t2} = (F_1 - F_2)\,\text{II} = \frac{11.25 \times 1000}{9} = 1250 \text{ N}$$

Assuming angle of contact of the belts as 180 degree and for an average value of 0.35 for the coefficient of friction.

$$F_1/F_2 = e^{\mu\theta} = 3$$

$\therefore$ For pulley I, $F_1 = 4500$ N and $F_2 = 1500$ N

For pulley II, $F_1 = 1875$ N and $F_2 = 625$ N

$\therefore$ Total belt pull for pulley I $= F_1 + F_2 = 6000$ N

and Total belt pull for pulley II = 2500 N

2. Find the bearing reactions in the horizontal plane.

$$R_A h \times l = (F_1 + F_2)_{\text{I}}(b + c) + (F_1 + F_2)_{\text{II}}\,\text{c}$$
$$= (6000) \times 1.1 + 2500 \times 0.6 = 8100$$

$$\therefore \quad R_A h = \frac{8100}{1.5} = 5400 \text{ N}$$

$$\therefore \quad R_B h = (6000 + 2500) - 5400 = 8500 - 5400 = 3100 \text{ N}$$

3. Construct the diagram of the bending moments in the horizontal plane Fig. 16.5 (b).
4. Find the bearing reactions in the vertical plane:

$$R_A v = \frac{W_1(b + c) + W_2 \times c}{l} = \frac{400 \times 1.1 + 750 \times 0.6}{1.5} = 593 \text{ N}$$

$$\therefore \quad R_B v = (W_1 + W_2) - R_A v$$
$$= 1150 - 593 = 557 \text{ N}$$

5. Draw the bending moment diagram in a vertical plane, Fig. 16.5 (c).
6. Determine the resultant bending moments for sections I and II.

$$M_{\text{I}} = \sqrt{(M_{hor.})^2 + (M_{ver.})^2} = \sqrt{2160^2 + 237.2^2} = 2172 \text{ Nm}$$

$$M_{\text{II}} = \sqrt{1860^2 + 333.7^2} = 1890 \text{ Nm}$$

7. Determine the Twisting moments:

$$T_{\text{I}} = \frac{1000 \times kW}{\omega} = \frac{1000 \times 11.25}{30} = 375 \text{ Nm}$$

$$T_{\text{II}} = \frac{11.25 \times 1000}{30} = 375 \text{ Nm}$$

Total torsional moment $M_t = T_{\text{I}} + T_{\text{II}} = 750$ Nm

This has been shown in Fig. 16.5 (d).

8. Equivalent twisting moment at section I

$$(T_e)_{\text{I}} = \sqrt{M_1^2 + T_1^2} = \sqrt{2172^2 + 375^2} = 2240 \text{ Nm}$$

Similarly $(T_e)_{\text{II}} = \sqrt{1890^2 + 750^2} = 2033$ Nm

$\therefore$ Section I is critical.

9. ∴ $$(T_e)_I = \frac{\pi}{16} d^3 \times f_s$$

f_s for commercial shafting, with keyways:

$$= 56 \times 0.75 = 42 \text{ MPa}$$

∴ $$2240 = \frac{\pi}{16} d^3 \times 42 \times 10^6$$

∴ $$d^3 = \frac{2240 \times 16}{\pi \times 42 \times 10^6} = \frac{27162}{10^6}$$

∴ $d = 0.0648$ m Use 71 mm nearest standard size.

Example 16.10. 0.60 m pulley B, keyed to the shaft, receives power from below at an angle of 45°, as shown in Fig. 16.6. A 0.45 m gear C delivers 40 percent of power horizontally to the right. A 0.30 m gear E delivers the remaining power downward toward the left at an angle of 30° below the horizontal. The gears and pulleys are keyed to the shaft. The power transmitted is 22.5 kW at 360 rev/min. a nickel steel with a yield point of 504 MPa and an ultimate stress of 700 MPa is specified for the shaft. The total load is gradually applied. Determine the diameter of the shaft.

Solution: First of all, the direction of the forces acting on the shafts are determined. The gear C delivers power to the right, therefore, the reaction of the shaft will be to the left. Again, the gear E delivers the power downward to the left, therefore, the reaction on the shaft will be upward and towards the right. This has been shown in the end view of Fig. 16.6.

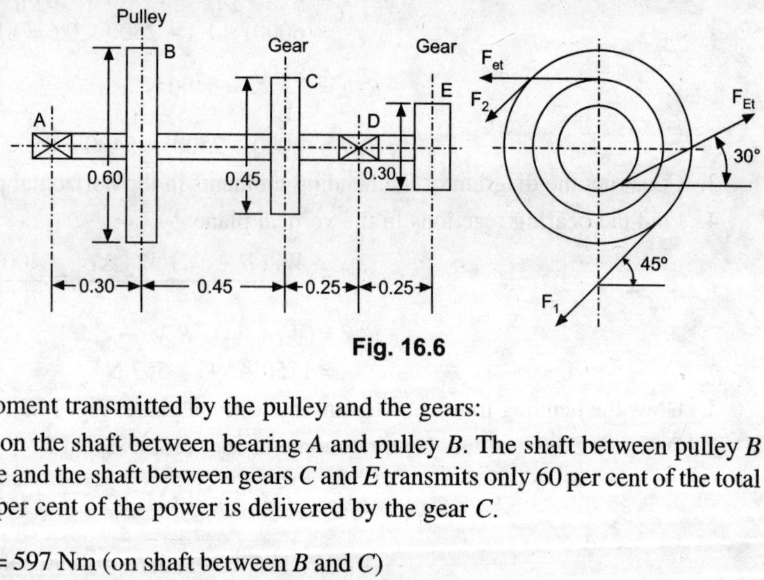

Fig. 16.6

1. Determine the torsional moment transmitted by the pulley and the gears:

There is no torsional moment on the shaft between bearing *A* and pulley *B*. The shaft between pulley *B* and gear *C* transmits the total torque and the shaft between gears *C* and *E* transmits only 60 per cent of the total torsional moment torque since 40 per cent of the power is delivered by the gear *C*.

$$T_B = \frac{9550 \times 22.5}{360} = 597 \text{ Nm (on shaft between } B \text{ and } C)$$

∴ $$T_C = 0.4 \times 597 = 238.8 \text{ Nm}$$

$$T_E = 358.2 \text{ Nm (on shaft between } C \text{ and } E)$$

2. Determine the bending forces:

(*i*) The net belt tension, $(F_1 - F_2) = \frac{T_B}{r_B} = \frac{597}{0.30} = 1990$ N

∴ Total belt pull, i.e., the bending force $F_B = F_1 + F_2 = 2\,(F_1 - F_2) = 3980$ N

assuming $\frac{F_1}{F_2} = 3.$

(*ii*) Tangential component of the tooth pressure in the gear *C*, is

$$F_{ct} = \frac{T_C}{r_C} = \frac{238.8}{0.225} = 1062 \text{ N}$$

$\therefore$ Normal tooth load, $F_{cn} = \dfrac{F_{ct}}{\cos\phi}$, where ϕ is the pressure angle now $\phi = 20°$.

$$\therefore \quad F_{cn} = \frac{1062}{0.93969} = 1107 \text{ N}$$

$\therefore$ Its components in the horizontal and vertical directions are (Fig. 16.7)

$$F_{cnh} = F_{cn} \times \cos\phi = F_{ct} = 1062 \text{ N} \leftarrow$$
$$F_{cnv} = F_{cn} \times \sin\phi = 1107 \times 0.342$$
$$= 378 \text{ N} \downarrow$$

(*iii*) Tangential component of the tooth pressure in the gear *E*, is

$$F_{Et} = \frac{T_E}{r_E} = \frac{358.2}{0.15} = 2388 \text{ N}$$

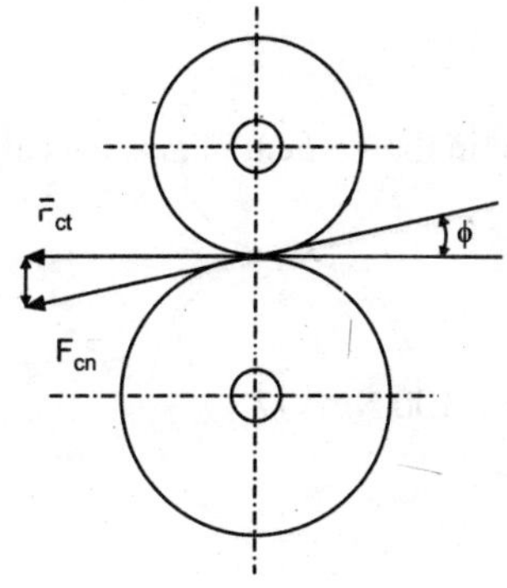

Fig. 16.7

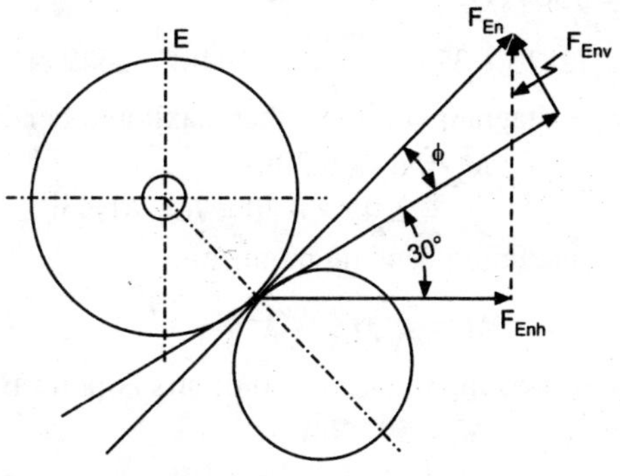

Fig. 16.8

$\therefore$ Normal tooth load, $F_{En} = \dfrac{2388}{0.93969} = 2540$ N

$\therefore$ From Fig. 16.8,

$$F_{E}nh = F_{E}n \cdot \cos(30 + \phi)$$
$$= 2540 \times \cos 50 = 2540 \times 0.643 = 1633 \text{ N} \rightarrow$$
$$F_{E}nv = 2540 \times \sin 50 = 2540 \times 0.766 = 1948 \text{ N} \uparrow$$

(iv) The horizontal and vertical components of the total belt tension are:

$$F_{Bh} = F_B \cos 45° = 3980 \times 0.707 = 2815 \text{ N} \leftarrow$$
$$F_{B}v = F_B \sin 45° = 3980 \times 0.707 = 2815 \text{ N} \downarrow$$

3. Find the maximum bending moment in the horizontal plane. The forces are shown in Fig. 16.9. To find the reactions at the bearing *A* and *D*, take moments about each in turn.

$$\therefore \quad \Sigma M_A = 0$$

$$\therefore \quad 2815 \times .30 + 1062 \times 0.75 + R_D \times 1.00 - 1633 \times 1.25 = 0$$

The plus sign for R_D assumes that it produces *cw* moment

$$\therefore \quad 844.5 + 796 + R_D \times 1.00 - 2045 = 0$$
$$\therefore \quad R_D \times 1.00 = 404.5$$
$$\therefore \quad R_D = 404.5 \text{ N} \leftarrow$$

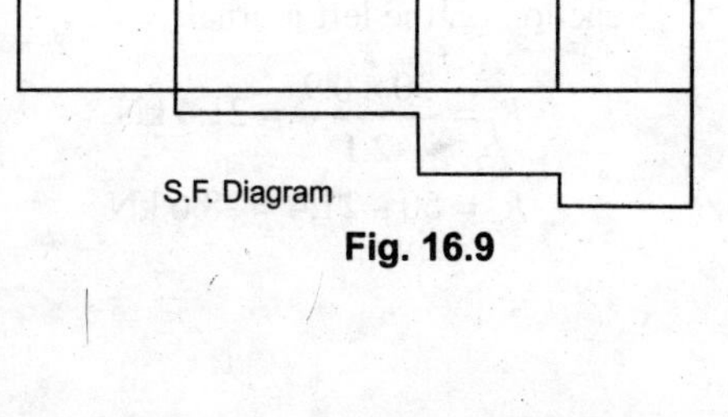

Fig. 16.9

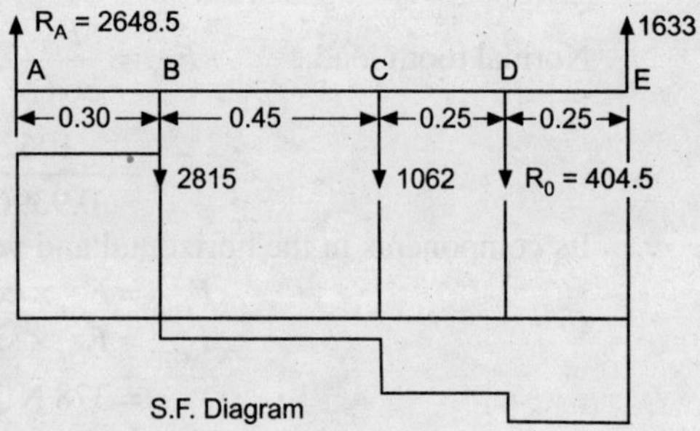

Fig. 16.10

$\therefore$ Our assumption is correct.

$\therefore R_A = (2815 + 1062 + 404.5) - (1633) = 2648.5 \text{ N} \rightarrow$

It is clear from the S.F. diagram, that maximum bending moment occurs at B.

$$\therefore \quad M_h = R_A \times 0.30 = 2648.5 \times 0.30 = 794.55 \text{ Nm}$$

4. Next, find the maximum bending moment in the vertical plane, the forces are shown in Fig. 16.10.

$$\Sigma M_A = 0$$

$\therefore 2815 \times 0.30 + 378 \times 0.75 - 1948 \times 1.25 + R_D \times 1.00 = 0$

R_D has been assumed to give *cw* moment.

$$\therefore \quad 844.5 + 283.5 - 2432 + R_D \times 1.00 = 0$$

$$\therefore \quad R_D = 1304 \text{ N} \downarrow$$

$$\therefore \quad R_A = (2815 + 378 + 1304) - (1948) = 2549 \text{ N} \uparrow$$

From the S.F. diagram, it is clear that maximum bending moment in the vertical planes acts also at *B*.

$$\therefore \quad M_v = R_A \times 0.30$$
$$= 2549 \times 0.30 = 764.70 \text{ Nm}$$

5. $\therefore$ Total maximum bending moment.

$$M_b = \sqrt{M_h^2 + M_v^2} = \sqrt{794.55^2 + 764.70^2} = 1100 \text{ Nm}$$

6. Maximum twisting moment also occurs at point B

$$M_t = 597 \text{ Nm}$$

7. $d^3 = \dfrac{16}{\pi f s}\sqrt{(K_b \cdot M_b)^2 + (K_s \cdot M_t)^2}$

$$K_s = 1; \quad K_b = 1.5$$

$f_s = 0.3 \times$ elastic limit, or

$= (0.18) \times$ ultimate tensile strength, whichever is smaller

$$= 0.3 \times 504 = 151.2 \text{ MPa}$$

or $= 0.18 \times 700.0 = 126 \text{ MPa}$

$\therefore$ Allowing for keyway,

$$f_s = 126 \times 0.75 = 94.5 \text{ MPa}$$

$$\therefore \quad d^3 = \frac{16}{\pi \times 94.5 \times 10^6}\sqrt{(1.5 \times 1100)^2 + (597)^2}$$

$$\therefore \quad d = 0.0450 \text{ m}$$

Use 45 mm diameter shaft.

Example 16.11. An axle is supported on two end journals and carries a load of 50 kN at a point 1.2 metre from one journal and 0.9 metre from the other. Determine the maximum diameter of the axle, if the stress allowed is 63.0 MPa.

Solution. Let the load be located 1.2 metre from the left journal.

$\therefore$ Reaction on the left journal.

$$R_l = \frac{50 \times 0.9}{2.1} = 21.4 \text{ kN}$$

$$\therefore \quad R_r = 50 - 21.4 = 28.6 \text{ kN}$$

Maximum bending moment will occur at the load point.

$$\therefore \quad M_b = R_l \times 1.20 = R_r \times 0.9$$
$$= 21.4 \times 1.2 = 25.68 \text{ kNm}$$

$$\therefore \quad M_b = \frac{\pi}{32} d^3 \times f_b$$

$$\therefore \quad d^3 = \frac{32 \times 25.68 \times 10^3}{\pi \times 63 \times 10^6} = 4.152 \times 10^{-3} \text{ m}^3$$

$$\therefore \quad d = 0.1607 \text{ m, say 161 mm.}$$

Example 16.12. In a railway wagon, the maximum load on a pair of wheels is 100 kN: one wheel take 70 kN, the other 30 kN. The distance between the rails is 1.45 m and between the centres of the axle boxes is 1.90 m. Find the diameter of the axle at the wheel. Safe stress is 77 MPa.

Solution. The location of the various loads is shown in Fig. 16.11.

R_1 and R_2 denote the loads on the axle boxes.

To find R_1 *Normal not italics* R_2 acts, therefore,

$$R_1 \times 1.90 = 70 \times 1.675 + 30 \times 0.225$$

$$\therefore \quad R_1 = \frac{124}{1.90} = 65.2 \text{ kN}$$

$$\therefore \quad R_2 = 100 - 65.2 = 34.8 \text{ kN}$$

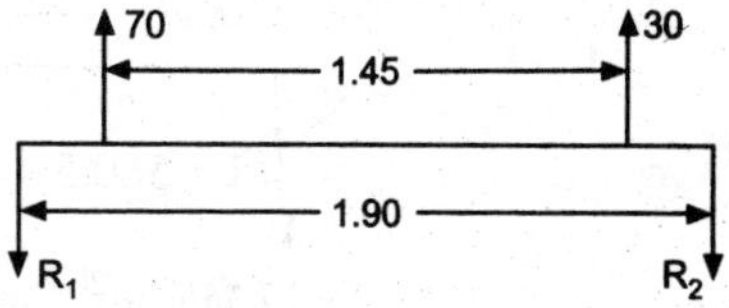

Fig. 16.11

Maximum bending moment on the wheel

$$M_b = 65.2 \times 0.225 = 14.68 \text{ kNm}$$

$$\therefore \quad M_b = \frac{\pi}{32} d^3 \times f_b$$

$$\therefore \quad d^3 = \frac{14.68 \times 1000 \times 32}{\pi \times 77 \times 10^6} = 1.94 \times 10^{-3} \text{ m}^3$$

$$\therefore \quad d = 0.01248 \text{ m, say 125 mm.}$$

Example 16.13. A hollow shaft 0.50 m outside diameter, and 0.30 m inside diameter is supported by two bearings 6 m apart, the shaft is driven by a flexible coupling at one end and drives a ship's propeller at 10.5 rad/s. The maximum thrust on the propeller is 540 kN when the shaft is transmitting 5885 kW. The shaft weight is 67.5 kN. Determine the maximum shear stress in the shaft considering the weight of the shaft producing the column effect. Assume $K_b = 1.5$ and $K_s = 1.0$.

Solution:

$$\text{Maximum bending moment, } M_b = \frac{FL}{8} = \frac{67.5 \times 1000 \times 6}{8} = 50.625 \text{ kNm}$$

$$\text{Maximum turning moment, } M_t = \frac{1000 \times \text{Power in kW}}{\omega} = \frac{1000 \times 5885}{10.5} = 560.5 \text{ kNm}$$

$$I = \frac{\pi}{64}(0.50^4 - 0.30^4) = 0.00267 \text{ m}^4$$

$$A = \frac{\pi}{4}(0.50^2 - 0.30^2) = 0.126 \text{ m}^2$$

$$\therefore \quad k = \sqrt{\frac{I}{A}} = \sqrt{\frac{0.00267}{0.126}} = 0.146 \text{ m}$$

$$\therefore \quad \frac{L}{k} = \frac{6}{0.146} = 41.1, \text{ which is less than 115.}$$

Therefore, $\alpha = \dfrac{1}{1 - 0.0044\left(\dfrac{6}{0.146}\right)} = 1.22 \qquad \left(\alpha = \dfrac{1}{1 - 0.0044(L/k)}\right)$

For hollow shafts, eqn. (16.12a) is

$$(f_s)_{max} = \frac{16}{\pi d_0^3 (1 - R^4)} \times \sqrt{(K_s \cdot M_t)^2 + \left[K_b \cdot M_b + \frac{\alpha \cdot Fa \cdot d_0 \cdot (1 + R^2)}{8}\right]^2}$$

Now $\quad d_o = 0.50 \text{ m}; d_i = 0.30 \text{ m}$

$\therefore \quad R = \dfrac{d_i}{d_o} = \dfrac{0.30}{0.50} = 0.6$

$$\therefore \quad (f_s)_{max} = \frac{16}{\pi (0.50)^3 \cdot (1 - 0.6^4)} \times \sqrt{(1 \times 560.5 \times 10^3)^2 + \left[1.5 \times 50.625 \times 10^3 + \frac{1.22 \times 540 \times 10^3 \times 0.50(1 + 0.6^2)}{8}\right]^2}$$

$= 27 \text{ MN/m}^2.$

Example 16.14. Determine the diameter below which the angle of twist of a shaft, and not the maximum stress, is the controlling factor in design of a solid shaft in torsion. The allowable shear stress is 56.00 MPa and the maximum allowable twist is 1/4 degree per metre.

$C = 84$ GPa

Solution: The torsional equation is,

$$\theta = \frac{Mt \cdot L}{J \cdot C}$$

Now $J = \dfrac{\pi d^4}{32}$

So, $\theta = \dfrac{584\, Mt \cdot L}{Cd^4}$ deg. for a solid circular shaft.

$= \dfrac{584\, M_t \cdot L}{C \cdot (d_o^4 - d_i^4)}$ deg. for a hollow circular shaft.

For solid shaft,

$$M_t = \frac{\theta . C . d^4}{584\, L}$$

Also $M_t = f_s \cdot \dfrac{\pi d^3}{16}$

$$\therefore \quad \frac{\theta . C . d^4}{584\, L} = f_s \cdot \frac{\pi d^3}{16}$$

$$\therefore \quad \frac{\left(\frac{1}{4}\right)(84 \times 10^9)(d^4)}{584 \times 1.00} = 56 \times 10^6 \times \frac{\pi \times d^3}{16}$$

$\therefore \quad d = 0.305$ m.

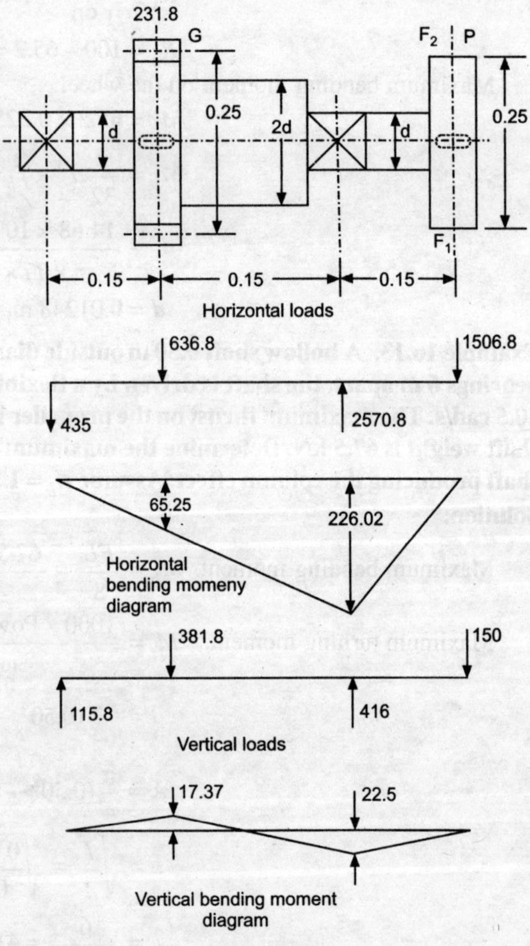

Fig. 16.12

Example 16.15. The shaft shown in Fig. 16.12 transmits 7.5 kW from the pulley A to the gear B at 900 rev/min. The shaft is machined from C-35 steel, hot rolled. The pulley diameter is 0.25 m and the pitch circle diameter of the gear is also 0.25 m. The weight of the pulley is 150 N and the weight of the gear is also 150 N. The ratio of the belt tension is 2.5. The gear pressure angle is 20°. Determine the size of the shaft using the Soderberg criterion. Assume steady load conditions. Take material properties as:
Ultimate tensile strength = 595 MPa
Yield point stress in tension = 385 MPa.

Solution. Since the torsional moment is constant, therefore, its variable component is zero.

(*i*) ∴ $$T_m = \frac{9550 \times 7.5}{900} = 79.6 \text{ Nm}$$

(*ii*) The belt tensions are found as:

$(F_1 - F_2) \times$ radius of pulley $= T_m$

$$\therefore \quad F_1 - F_2 = \frac{79.6}{0.125} = 636.8$$

Also $F_1/F_2 = 2.5$

$$F_1 = 1061.3 \text{ N}; \qquad F_2 = 424.5 \text{ N}$$

(*iii*) The gear tooth tangential force, $F_t = \dfrac{T_m}{R_g} = \dfrac{79.6}{0.125} = 636.8$ N.

(*iv*) The separating force is $F_r = F_t.\tan\phi = 636.8 \tan 20° = 231.8$ N.

(*v*) The belt tension and gear forces are assumed acting as shown. The loading and bending moment diagrams are drawn in figure.

The resultant bending moment at gear $= \sqrt{65.25^2 + 17.37^2} = 67.522$ Nm

The resultant bending moment at the right bearing $= \sqrt{226.02^2 + 22.5^2} = 227.137$ Nm.

(*a*) Let us consider the shaft section at the gear first. Of course, the bending moment here is not maximum, but the key-way might be an influencing factor.

The stress concentration factor, K_f due to the keyway in bending = 1.6.

The stress concentration factor in bending due to the fillet where the shaft portion of diameter '*d*' joins with shaft of diameter '2*d*' will depend upon the fillet radius. Neither the fillet radius is known, nor the shaft diameter is known, so, K_f will have to be found by trial and error.

Taking '*r/d*' = 0.083, with *r* = 1.6 mm (These values agree with final results), the theoretical stress concentration factor, K_t is 1.86 (Fig. 4.8). The notch sensitivity factor, q_f, from Fig. 4.10 is about 0.78.

$\therefore \quad K_f = 1 + q_f(K_t - 1) = 1.67$

∴ Fillet has adverse effect than the keyway.

Now in bending, mean stress is zero, since the bending stress undergoes complete reversal.

Again in torsion, the variable component is zero. Therefore, from equation (16.13), we have

$$\frac{f_{yp}}{FS} = \left[\left(K_f \cdot f_v \cdot \frac{f_{yp}}{f_e}\right)^2 + 4f_{ms}^2\right]^{\frac{1}{2}}$$

Now $$f_v = \frac{M_b}{Z} = \frac{67.522 \times 32}{\pi d^3} = \frac{687.773}{d^3}, \text{ N/m}^2$$

$$f_{ms} = \frac{16\,T_m}{\pi d^3} = \frac{16 \times 79.6}{\pi d^3} = \frac{405.4}{d^3}, \text{ N/m}^2$$

Now $f_{yp} = 385$ MPa, and let $FS = 1.5$.

Now $f_e = \frac{1}{2} \times 595 = 297.5$ MPa

$\therefore$ $f_e = C_F \cdot C_s \cdot C_L \cdot f_e^*$

$C_F = 0.87$ (Table 4.4), $C_s = 0.85$, say, $C_L = 1.0$

$\therefore$ $f_e = 0.87 \times 0.85 \times 1.0 \times 297.5 = 220.0$ MPa

$$\therefore \quad \frac{385 \times 10^6}{1.5} = \left[\left(1.67 \times \frac{687.773}{d^3} \times \frac{385 \times 10^6}{220 \times 10^6}\right)^2 + 4 \times \left(\frac{405.4}{d^3}\right)^2\right]^{\frac{1}{2}} = \frac{2167.383}{d^3}$$

$$\therefore \quad d^3 = \frac{8.444}{10^6}$$

$\therefore$ $d = 0.02036$ m.

(b) Next, we analyse the shaft at the right bearing. Here also, there is a sudden change in the shaft diameter, and stress concentration factor, K_t, cannot be determined directly.

Let $r/d = 0.22$, with $r = 6.4$

$\therefore$ $K_t = 1.4$ (Fig. 4.8), and $q_f = 0.95$, say since Fig. 4.10 is only for $r = 4.5$ mm

$\therefore$ $K_f = 1 + 0.95\,(1.4 - 1) = 1.38$

Hence, $$f_v = \frac{227.137 \times 32}{\pi d^3} = \frac{2313.5984}{d^3} \text{ N/m}^2$$

$$f_{ms} = \frac{405.4}{d^3} \text{ N/m}^2$$

$$\therefore \quad \frac{385 \times 10^6}{1.5} = \left[\left(1.38 \times \frac{2313.5984}{d^3} \times \frac{385 \times 10^6}{220 \times 10^6}\right)^2 + 4 \times \left(\frac{405.4}{d^3}\right)^2\right]^{\frac{1}{2}} = \frac{5645.8626}{d^3}$$

$$\therefore \quad d^3 = \frac{22}{10^6} \qquad \therefore \quad d = 0.028 \text{ m}$$

(c) So, final dimensions of the shaft are:

$d = 28$ mm; $2d = 56$ mm.

Example 16.16. A 0.60 m diameter pulley driven by a horizontal belt drive transmits power through a solid shaft to a 0.25 m diameter pinion which drives a mating gear. The weight of the pulley is 1500 N. The arrangement of elements, belt tensions and components of the gear reactions on the pinion are as shown in Fig. 16.13. Determine the shaft diameter. Take fatigue and shock factors as:

$K_b = 2$ and $K_s = 1.5$.

Solution: The bending moment diagrams have been drawn in Fig. 16.13. It is clear from the figure that maximum bending moments occur at the pinion. The maximum combined moment is,

$$M_b = \sqrt{1441.5625^2 + 1005.75^2} = 1676.70 \text{ Nm}$$

The twisting moment, M_t, is

$$M_t = (F_1 - F_2) \times R_p = (5445 - 1815) \times 0.30 = 1089 \text{ Nm}$$

From equation (16.10),

$$(f_s)_{\max} = \frac{16}{\pi d^3} \sqrt{(K_b \cdot M_b)^2 + (K_s \cdot M_t)^2}$$

$(f_s)_{\max}$ (allowable) $= 56$ MPa, without keyway

With keyway, f_s (allowable) $= 42$ MPa

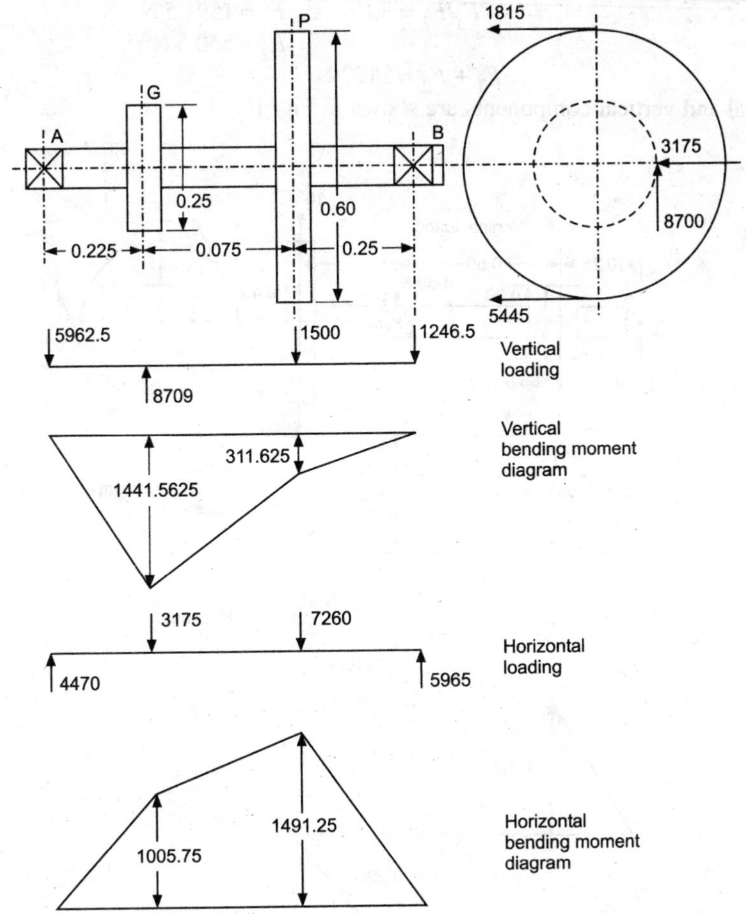

Fig. 16.13

$$42 \times 10^6 = \frac{16}{\pi d^3}\sqrt{(2\times 1676.7)^2 + (1.5\times 1089)^2} = \frac{16\times 3.73\times 10^3}{\pi d^3}$$

$\therefore \quad d^3 = 0.4523 \times 10^{-3}$

$\therefore \quad d = 0.0768$ m

Use 80 mm diameter.

Example 16.17. A machine shaft turning at 600 rev/min is supported on bearings 0.75 m apart (Fig. 16.14). 15 kW is supplied to the shaft through a 0.45 m pulley located 0.25 to the right of the right bearing. The power is transmitted from the shaft through a 0.20 m spur gear located 0.25 m to the right of the left bearing. The belt drive is at an angle of 60° above the horizontal. The pulley weighs 900 N. The ratio of belt tensions is 3 : 1. The gear has 20° pressure angle and meshes with another gear located directly above the shaft. The material properties are:
Ultimate strength = 490 MN/m^2
Yield point stress = 362 MN/m^2
Determine the shaft diameter. Take K_b = 1.5 and K_s = 1.0.

Solution: $M_t = \dfrac{9550\times 15}{600} = 238.75$ Nm

$\therefore \quad (F_1 - F_2)\times 0.225 = 238.75$

$\therefore \quad F_1 - F_2 = 1061$

Also $$F_1/F_2 = 3. \quad \therefore \; F_1 = 1591.5 \text{ N}$$
$$F_2 = 530.5 \text{ N}$$

So $$F_1 + F_2 = 2122 \text{ N}$$

Its horizontal and vertical components are shown in Fig. 16.14.

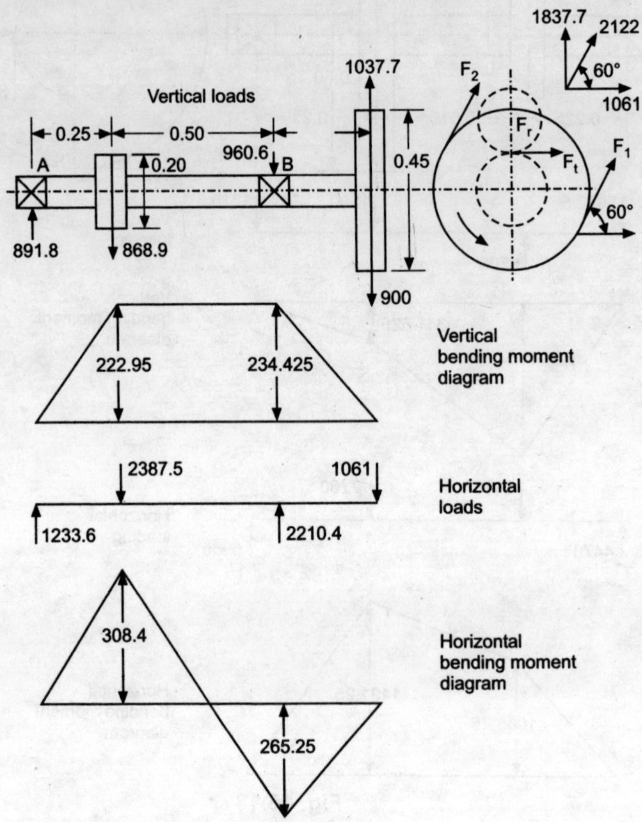

Fig. 16.14

Now Tangential force on gear, $F_t = \dfrac{M_t}{R_g} = 2387.5$ N

$\therefore$ Radial force $F_r = F_t . \tan 30° = 868.9$ N

From Fig. 16.14, it is clear that maximum bending moment occurs at gear, and it is equal to,

$$M_b = \sqrt{222.95^2 + 308.4^2} = 370.55 \text{ Nm}$$

Now design stress, $f_s = (0.3) \times$ elastic limit in tension

or $= (0.18) \times$ ultimate tensile strength

whichever is smaller.

Now 30% of 362 = 108.62 MPa

and 18% of 490.0 = 88.20 MPa

$\therefore$ f_s (allowable) = 88.20 MPa, without keyway and with keyway, f_s (allowable) = 75% × 88.2 = 66.15 MPa

Using relation (16.10)

$$66.15 \times 10^6 = \frac{16}{\pi d^3}\sqrt{(K_b \cdot M_b)^2 + (K_s \cdot M_t)^2} = \frac{16}{\pi d^3}\sqrt{(1.5 \times 370.55)^2 + (1.0 \times 238.75)^2} = \frac{16}{\pi d^3} \times 618.736$$

$$d = 0.036 \text{ m} = 36 \text{ mm}.$$

EXERCISES

1. What is the difference between a shaft and an axle?
2. Classify the shafts.
3. Enumerate the cases of shaft failures.
4. Name the various shaft materials.
5. What do you understand by cold rolled and cold drawn shafting?
6. What effect has a keyway on the strength of a shaft?
7. What effect has a keyway on the torsional rigidity of a shaft?
8. Is a hollow shaft stronger or weaker than a solid shaft of the same weight?
9. Has a hollow shaft more or less resistance to bending than a solid shaft of the same diameter?
10. A machinery shaft is subjected to torsion only. The bearingas are 2.40 metre apart. The shaft transmits 187.5 kW at 200 rev/min. Allow a shear stress of 42 MPa after an allowance for keyways.
    (*a*) Determine the shaft diameter for steady loading.
    (*b*) Determine the shaft diameter if the load is suddenly applied with minor shocks.
11. A cold rolled transmission shaft is to transmit a torsional moment of 3100 Nm with a maximum bending moment of 5000 Nm. Material is C-40 steel.
    (*a*) Determine the diameter for a mild shock load.
    (*b*) Determine the conventional factor of safety, based on yield strength.
12. 56.25 kW motor operating at 90 rev/min is to drive a lathe. The shaft is fitted with a standard key. Load may be taken as suddenly applied with minor shocks. Considering torsion only, determine the shaft diameter if the shaft material has a yield stress of 245 MPa in tension.
13. A machinery shaft supported on bearings 2.4 metre apart is to transmit 187.5 kW at 200 rev/min. It is subjected to a bending load of 5000 N located at a distance of 0.66 m from one bearing. Safe stress in shear is 42 MPa and in bending 84 MPa.
    (*a*) Determine the shaft diameter for steady loading.
    (*b*) Determine the shaft diameter if the transverse load is steady and the torsional load is suddenly applied.
14. An automobile engine delivers 56.25 kW at 3400 rev/min of the motor to the rear axle. The differential ratio is 4.7 to 1, and the transmission reduction in low gear is 2.8 to 1. If the axle diameter is 30 mm determine the maximum torsional stress in the rear axle for full load conditions:
    (*a*) In low gear.
    (*b*) In high gear.
    Assume a drive efficiency of 90 per cent in each case.
15. A 14 cm monel propeller shaft in a boat transmits 375 kW at 150 rev/min. The efficiency of the propeller is 60 per cent. A bending moment of 800 Nm is also applied to the shaft, and $L/K < 40$, for $K_s = 1.5$ and $K_b = 2$, determine the stress in the shaft.
16. A 13.5 kN load is raised by a cable working over a drum which is 0.90 m in diameter. Find the diameter of the shaft to which the drum is keyed and safe stress is 45 MPa.
17. A machine propeller shaft is 0.15 m in diameter and is subjected to a thrust of 68 kN and a torque of 22.5 kNm. Determine the maximum compressive and shear stress induced.
18. A steam engine having stroke of 0.30 m has an over hung crank. The maximum tangential force on the crank may be assumed as 35 kN. The distance between the centre of crank pin and centre of crankshaft bearing is 0.025 m. Assuming an allowable shear stress of 31 MPa, determine the diameter of the crank shaft.
19. A centrifugal circulating pump for a stream turbine condenser is to be direct connected to a motor through a short shaft and a pair of flexible couplings. The pump delivers 20 m^3 per minute at 865 rev/min against a dynamic head of 7.5 metres. The pump efficiency is 80 percent.
    (*a*) Determine the power of the motor required.
    (*b*) Assuming hot rolled C-40 steel, determine diameter of shaft.
20. A line shaft is to be driven from a motor under it by means of a belt running over 1.50 m pulley attached to the line shaft. The pull on the tight side of the belt is 5000 N and on the slack side 2000 N. The weight of the pulley is 2900 N. The pulley is overhung, the distance from the central plane of the pulley to the centre line of the supporting bearing is 0.30 m.

The ultimate tensile and shearing strengths of the shaft material may be taken as 440 MPa, 336 MPa respectively. Determine the diameter of shaft using a factor of safety of 6.

21. A steel shaft 50 mm in diameter is 1.8 m between supports. A vertical load of 2250 N is applied 0.45 m from the bearing and a load of 1800 N at 30 deg. with the vertical and perpendicular to the shaft is applied at a distance of 0.75 m from the same bearing. What is the maximum tensile stress in the shaft?

22. A shaft for a punch press is supported on two bearings, 0.60 m apart as shown in Fig. 16.15. It carries a flywheel *A*, gear *D*, and a pulley *C*. The weight of the flywheel is 6800 N and its radius of gyration is 0.15 m. The gear *D* delivers power horizontally to the right and is 0.20 m in diameter and is keyed to the shaft. The pulley *C* receives power from above at an angle of 45° to the right of vertical and is 1 m in diameter. If the maximum power transmitted is 18.75 kW at 250 rev/min with heavy shock, determine the diameter of the hot drawn C-40 steel shaft.

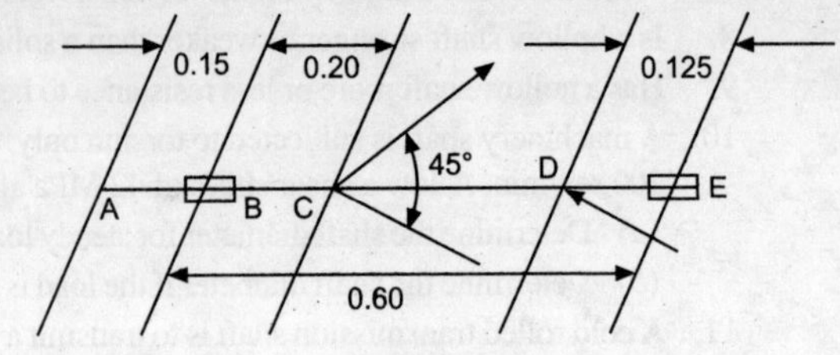

Fig. 16.15

23. A shaft is supported on two bearings 0.60 m apart. The shaft supports a 0.50 m pulley 0.25 m to the right of the left hand bearing, and the belt drives a pulley directly below. Another pulley 0.375 m in diameter is located 0.125 m to the right of the right hand bearing and the belt is driven from a pulley horizontally to the right. The coefficient of friction for the belt is 0.30 and the angle of contact 180 deg. The maximum tension in the belt on the smaller pulley is 3600 N. Find shaft diameter allowing $f_t = 56$ MPa, $f_s = 42$ MPa.

24. A 75 mm countershaft in a machine shop transmits 39 kW at 315 rev./min. It is made of cold rolled steel with ultimate tensile strength of 490.0 MPa and a yield strength of 310.0 MPa. It is supported on bearings 1.50 m apart. It has two pulleys mounted upon it. One is 0.45 m in diameter and is located 0.38 m to the right of left hand bearing and it receives power in a horizontal direction from a pulley to its right. The other pulley 0.635 m in diameter, is located 0.45 m to the left of the right hand bearing and delivers power in a vertical direction to a machine below it. For $K_b = 1.75$ and $K_s = 1.25$, find the stress in the shaft and compare it with the design stress as obtained from the code.

25. A mild carbon steel shaft transmitting 11.25 kW at 210 rev./min is supported on two bearings 0.58 m apart and has keyed to it two gears. An 18 tooth 20 deg. involute 10 mm module gear is located 0.125 m to the right of the right hand bearing and delivers power to a gear directly below the shaft. An 80 tooth, 6 mm module gear is located 0.15 m to the right of the left hand bearing and receives power from a gear directly over it. Calculate the diameter of the shaft assuming $f_t = 84$ MPa and $f_s = 70$ MPa.

26. A steel shaft 1.5 m long between bearings carries 1000 N pulley at its mid point. The pulley is keyed to the shaft and receives 20 kW at 200 rev./min. The belt drive is horizontal and the ratio of the belt tensions is 3 : 1. The diameter of the pulley is 600 mm. The load is applied with moderate shocks. Calculate the necessary diameter of the shaft.

27. A shaft is supported between two bearings placed 1.2 m apart. A 0.60 m diameter pulley is mounted at a distance of 0.30 m to the right of the left hand bearing. This drives a pulley directly below it with the help of a belt having a maximum tension of 2.5 kN. Another pulley 0.40 m in diameter is placed 0.20 m to the left of the right hand bearing and is driven with the help of an electric motor and belt which is placed horizontal to the right. The coefficient of friction for the belt drives is 0.24 and the angle of contact for both pulleys is 180°. Determine a suitable diameter of a commercial shafting. Take.

$$f_s = 62 \text{ MPa}, f_t = 102 \text{ MPa}, K_s = 1.0, K_b = 1.5$$

28. Three identical pulleys of 0.50 m diameter and weight 500 N are mounted on a line shaft supported on two bearings 4.0 m apart. The pulley '*A*' is mounted at 0.30 m to the right of the left hand bearing, and receives 30 kW at 200 rev./min from a pulley vertically below it. The pulley '*B*' is mounted 1.0 m to the right of the left hand bearing and delivers 6 kW to a pulley through a belt drive inclined backward at 45° to the vertical. The remaining power is taken out through another pulley '*C*' which is mounted 3.0 m to the right of the left hand bearing and drives a planing machine, the drive being 30° to the right and to the front. The angle of contact for all the pulleys can be taken as 180° and co-efficient of friction as 0.30. Determine the shaft diameter. For shaft material $f_s = 82.5$ MPa. Consider keyway effect.

29. A horizontal power transmission shaft is supported by two bearings 1.0 m apart. The shaft overhangs the right hand bearing and supports a 0.75 m pitch diameter straight tooth spur gear 0.30 m from the bearing. It supports a 1.25 m flat belt pulley 0.25 m to the left of the left hand bearing. The gear weighs 3 kN and the pulley weighs 10 kN. The gear is driven by a pinion such that the tangential turning force acts upward (ignore radial force). The pulley that delivers the

power vertically downwards has a belt tension ratio of 2.5 : 1. The shaft must transmit 20 kW at 150 rev./min. Assume transmission efficiency as 100%. Determine the shaft diameter based on maximum normal stress theory. Assume suitable other data.

30. What are the requirements of shaft material?
31. Give the different criteria of designing a shaft.
32. Name the shaft directly coupled to the shaft of the power source.
33. Name the shaft used for transmission of power.
34. What is meant by stiffness and rigidity of a shaft?
35. How the rigidity of a shaft is affected by its length?
36. What is "critical speed" of shafts?
37. What is "whirling" of a shaft?
38. Should the speed of rotation a shaft be above or below its critical speed?
39. The critical speed of a rotating shaft is dependent on:
    (a) Axial elongation (b) Transverse deflection
    (c) Torsional rotation (d) All of above
40. How the rigidity of a rotating shaft is affected by its diameter?
41. What are flexible shafts? When are they preferred?
42. How a flexible shaft is fabricated?
43. Write the functions of the casing of a flexible shaft.
44. Write on "End fittings" of a flexible shaft.
45. Write a short note on "Design of stepped" shaft for assembly.

❑❑❑

CHAPTER 17

Bearings

17.1 FUNCTION

A bearing is a machine part whose function is to support a moving element and to guide or confine its motion, while preventing motion in the direction of applied load. They take up the radial and axial loads imposed on the shaft or axle they carry, and transmit these to the casing or machine frame. Since there is a relative motion between the bearing and the moving element, there will be a power or energy loss due to friction and if their surfaces actually touch each other, wear may also take place. Hence a bearing should perform its function with a minimum loss of energy from friction and at the same time should control the rate of wear. This is achieved to a large extent by interposing layer of lubricant between the contact surface of the bearing and the moving element. In many applications, the bearings are located between the shafts and frame of a machine. They are also located between the contact members and the frame or between contact members and linkage members.

17.2 CLASSIFICATION

Bearings may be classified in the following many ways:

1. **Depending upon the direction of load to be supported:** On the basis of this criteria, the bearings are classified as:
   - ***(a)*** **Radial bearings:** These are also known as journal bearings. In these bearings the main load is perpendicular to the axis of rotation of the moving element.
   - ***(b)*** **Thrust bearings:** In these bearings, the load acts along the axis of rotation.
2. **Depending upon the nature of contact between the working surface:** Based on this criteria, the bearings are classified as:
   - ***(a)*** **Sliding contact bearings,** also known as plain bearings.
   - ***(b)*** **Rolling contact bearings** in which steel balls or steel rollers are interposed between the working surfaces.

3. **Depending upon the type of loading :** Based on this criteria the journal bearings may be classified as:
   (*a*) **Bearing with a steady load,** such as bearings of electric generators, and motors, centrifugal pumps, or various blowers or fans. Such bearings are called **"Power Bearings".**
   (*b*) **Bearings with a variable or fluctuating load** such as main, crankpin, wrist pin, or camshaft bearings in diesel engines.

17.3 SELECTION OF BEARING TYPE

There is no hard and fast rule or formula existing, for deciding between sliding and rolling bearings. Each type has its own advantages and disadvantages and there is no best bearing for all the applications. For a particular application, the designer must evaluate the advantages and disadvantages of each bearing and then reach the best compromise. For this, the performance of each bearing must be compared in relation to, load capacity, friction, space requirements, accuracy, noise, cost, etc.

A broad choice can be made with the help of Tables 17.1, 17.2 and 17.3.

Table 17.1 Bearing selection based on mechanical requirements

S.No.	Characteristics	Sliding	Rolling
1.	**Load:**		
	One way	Good	Excellent
	Both ways	Good	Excellent
	Starting	Poor	Excellent
	Unbalance	Good	Excellent
	Shock	Fair	Excellent
	Emergency	Fair	Fair
2.	**Speed:**		
	Limited by	Turbulance and temperature rises. Usual value is upto 75 m/s	Centrifugal loading and dynamic effects. Usual value is upto 50 m/s
3.	**Starting Friction**	Poor	Good
4.	**Space Requirements:** (Radial Bearings)		
	Radial dimension:	Small	Large
	Axial dimension	1/4 to 2 times the shaft diameter.	1/5 to 1/2 times the shaft diameter.
5.	**Type of failure**	Often permits limited emergency operation after failure	Limited operation may continue after fatigue failure but not after lubricant failure.
6.	**Damping of Vibrations**	Good	Poor
7.	**Type of Lubricant**	Oil or other fluid, grease, dry lubricants, air or gas,	Oil or grease
8.	**Quantity of Lubricant**	Large, except in low speed boundary lubrication types.	Very small, except where large amounts of heat must be removed.
9.	**Noise**	Quiet	May be noisy, depending upon quality of bearing and resonance of mounting.
10.	**Power Consumption**	Varies as $n^2 D^3 L$	Varies widely depending upon type of lubrication. Varies directly as speed. Usually lower than sliding bearings.

Table 17.2 Bearing selection based on environmental conditions

Characteristics	Sliding	Rolling
Low temp. Starting	Poor	Good
High temperature operation	Limited by lubricant	Limited by lubricant

Table 17.3 Bearing selection based upon economics

Characteristics	Sliding	Rolling
Life	Unlimited except for cyclic loading.	Limited by fatigue properties of bearing metal.
Maintenance	Clean lubricant is required.	Clean lubricant required only occasional attention grease.
Cost	Very small in mass production quantities or simple types.	Intermediate, but standardised, varies little with quantity.
Ease of replacement	Function of design and installation.	Function of type of installation.

Table 17.1 shows the variation of bearing load capacity with speed.

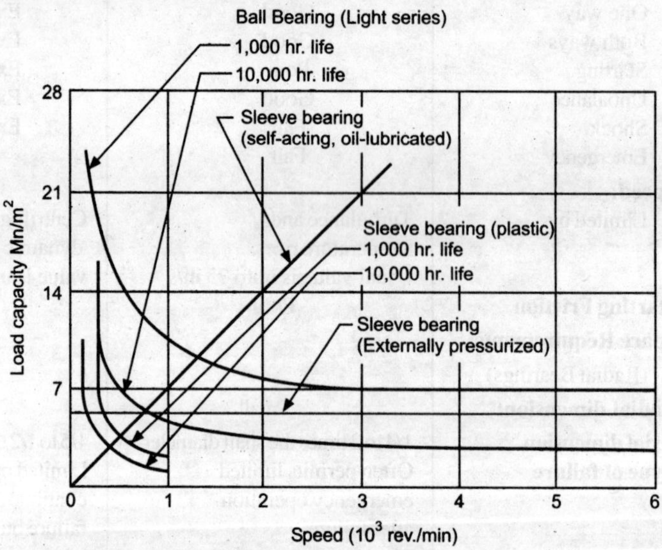

Fig. 17.1 *Speed, load capacity relationship*

17.4 SLIDING BEARINGS OR PLAIN BEARINGS

As already indicated, in sliding or plain bearings, the primary motion between the bearing and the moving element is sliding one. The sliding bearings can be classified into two groups depending upon the nature of motion of the moving element.

1. **Linear Bearings.** In these bearings, the sliding action is guided in a straight line e.g. engine cross heads or the motion between a planer table and its bed. Such bearings are also called as "Slipper or Guide Bearings".
2. **Sliding Bearings with motion of rotation.** In these bearings, the sliding motion between the bearing and the moving element, is the motion of rotation. The examples are "plain journal bearings" or "Sleeve bearings" in which the load acts at right angle to the axis of rotation, and the "Thrust Bearings" in which the loads act along the axis of the moving element. Thrust bearings will be discussed in next pages.

Plain Journal or Sleeve Bearings. These bearings are designed to carry radial loads. In these bearings, the bearing forms a sleeve around a shaft. The part of the shaft which rotates in the bearing is called "Journal". The sleeve bearings are classified as:

1. **Full Bearing.** Figure 17.2(a). In these bearings, the bearing completely envelops the journal.
2. **Partial Bearing.** Figure 17.2(b). The enveloping angle is not 360° but is 120°. The full bearing and partial bearings are also known as "clearance bearings" since the journal size is less than the bearing bore. The friction in a partial bearing will be less than in full bearing, but its applications are limited to only those situations where the load is always in one direction, e.g., rail road car axles.

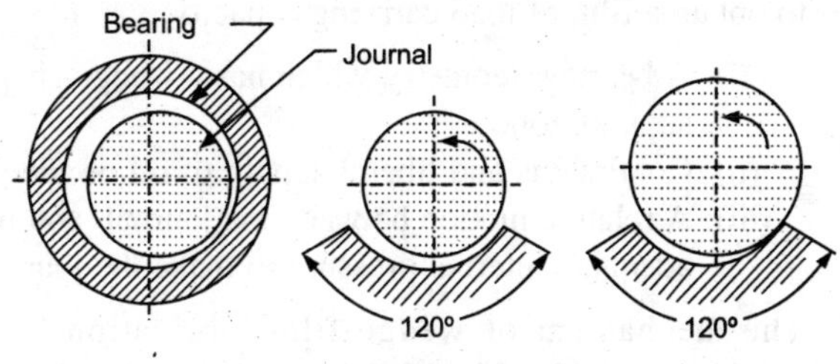

(*a*) *Full* (*b*) *Partial* (*c*) *Fitted*

Fig. 17.2 *Sleeve bearings*

3. **Fitted Bearing.** Figure 17.2(c). This is a special case of partial bearing in which the sizes of journal and bearing are equal and hence there is no clearance.

Sliding bearings can also be classified according to the types of friction present between the bearing and the moving element.

1. **Dry Friction Bearings:** No lubricant is supplied in between the rubbing surfaces and there is metal contact. The coefficient of friction may range from 0.1 to 0.25. These bearings may also be designated as "Self Lubricated Bearings" since no oil or grease are put in between the rubbing surfaces from an external source. They have their own lubricant as a part of the bearing material. Solid lubricants such as graphite, MoS_2 and WS_2 are applied to the bearing surfaces as pigments in organic or non-organic binders. Sometimes, oils, waxes, greases and solid lubricants are impregnated into porous materials to form self lubricating materials. Some materials, such as fluorocarbons and carbon-graphite are inherently self lubricating. Such bearings give satisfactory operation under low pressure and low speed conditions.
2. **Boundary Friction Bearings:** In these bearings, lubricant is supplied to the rubbing surfaces in scarce quantities. The lubricant is neither constant nor abundant. Such bearings are suitable only for low load and low speed conditions since at increased loads, the thin film of lubricant will break and the bearing surfaces approach each other, resulting in metal to metal contact and wear of the surfaces.
3. **Semi Fluid Friction:** This is the transition between boundary and fluid type of friction. Lubricant is supplied constantly but not abundantly and so the lubricant film is very thin and when it breaks, there is metal to metal contact. The coefficient of friction may range from 0.005 to 0.10.
4. **Bearings with Fluid Friction:** In these bearings, there is always a thick film of lubricant between the working surfaces and there will never be a metal to metal contact. The friction will be due to the internal friction between the lubricant particles and magnitude will depend upon the relative velocity of the bearing surfaces, thickness of lubricant layer, viscosity of lubricant.

   Plain bearings can also be classified according to the method of lubrication, as:

   (*a*) Unlubricated
   (*b*) Dry film lubricated, self lubricated.
   (*c*) Boundary lubricated.
   (*d*) Thick film lubricated or Hydrodynamic lubrication.
   (*e*) Hydrostatic or externally pressurized lubrication.

17.4.1 Hydro-dynamic Lubrication

In this type of lubrication, there is always a thick layer of lubricant between the working surfaces. The principle of this lubrication is based on the fact the viscous fluid resists being pushed around. Under proper conditions, this resistance to motion will develop a pressure distribution in the lubricant film that can support a useful load. In hydrodynamic bearings, this load supporting pressure develops due to two processes:

1. the flow of a viscous fluid in a converging channel, called as "Wedge Film" lubrication.
2. the resistance of the viscous fluid to being squeezed out from between the approaching surfaces, called "Squeeze film lubrication".

1. **Wedge Film Lubrication:** For wedge-film lubrication, the following conditions must be fulfilled so as to obtain a film of load carrying capacity.
   (i) A bearing geometry which has a wedge-shaped space or a converging flow channel in the load support zone.
   (ii) Continuous and liberal supply of oil of adequate viscosity for filling this space.
   (iii) A relative motion between the journal and bearing, sufficiently high to form an oil wedge with adequate internal pressure to make the journal float under load.

The mechanism of wedge-film lubrication is explained with the aid of Fig. 17.3, which shows a bearing and journal with clearance greatly exaggerated. The load is acting downward and when the journal is at rest, it will touch the bottom of the bearing, Fig. 17.3 (a). As the journal starts rotating, it rolls upto the left side of the bearing and a thin film of lubricant is formed between the contact surfaces. As the journal keeps rotating, it will draw more oil between the surfaces resulting in a thick film of lubricant and the journal will be lifted up, Fig. 17.3 (b).

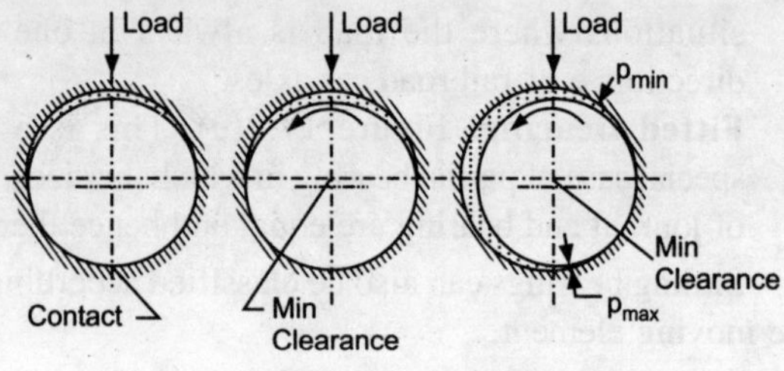

(a) (b) (c)

Fig. 17.3 *Wedge-film lubrication*

With the increase in the speed of rotation, the oil drawn under the journal builds up an internal pressure which raises the journal up and to the right until an equilibrium state is reached, Fig. 17.3 (c). The exact position of the minimum clearance between the bearing and the journal depends upon the initial clearance, speed of rotation, amount of load, and viscosity of oil. Increase of load, or decrease of speed or viscosity of oil will make the journal to settle lower in the bearing. The pressure distribution is shown in Fig. 17.4.

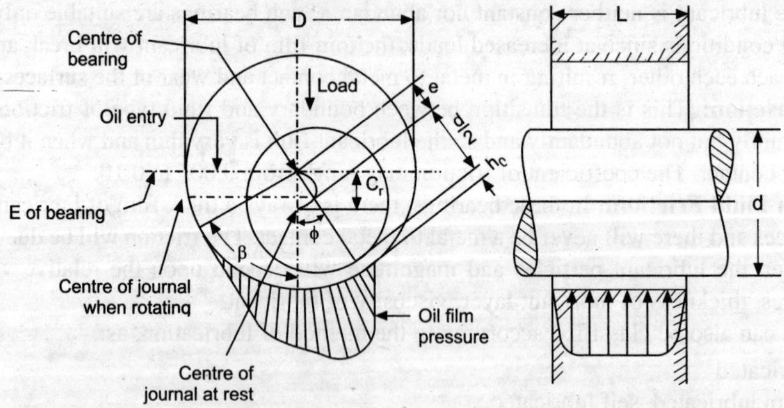

(a) (b)

Fig. 17.4 *Oil pressure distribution*

2. **Squeeze Film Lubrication:** The above analysis of wedge film lubrication is applicable, in general, to statically loaded bearings in which there is rotation of the journal relative to the bearing. In certain cases, e.g., wrist pin bearing of I.C. engines, the conditions are different and the theory of wedge film lubrication cannot be applied to such cases:
   (i) The wrist pin journal oscillates in its bearing never completing a full revolution, whereas in wedge film lubrication, the journal must revolve in the bearing to generate a load supporting film of lubricant.

(ii) Bearing based on thick film lubrication should carry relatively steady loads whereas in wrist pin bearing, there is wide variation of load during each revolution of engine.

(iii) The load on a wrist pin bearing varies in direction and for four-cycle engine reverses with each revolution of the engine.

Such cases in which the load reciprocates with no journal rotation are called, "Dynamically Loaded Bearings" and their design is based upon the "Squeeze film lubrication" since there is a tendency for the lubricant to be squeezed out of the clearance space under the relative motion of the journal and bearing.

17.4.2 Hydro-static Bearing

The bearings in which the load lifting capacity is produced by the rotation of the journal relative to the bearings, are called "Self-acting bearings" or "Self-contained bearings,". In hydrostatic bearings also called "Externally pressurised bearings", the load can be supported without any relative motion between the journal and the bearing. This is achieved by forcing externally pressurised lubricant between the journal and the bearing. The advantages of such bearings are that even at staring the friction is very less and these are capable of supporting a load on a fluid film in very slow moving or stationary bearings. Also there will be no wear of working surfaces. A hydro-static bearing system is shown in Fig. 17.5. It consists of an external pressure supply, flow restrictors, pressure pads, and clearance space. The pressurised lubricant is fed through the flow restrictors to load support pads which are generally provided in the stationary member of the bearing. The function of the flow restrictors is to automatically adjust the pressure in the pads according to load fluctuations, when the journal is centered in the bearing, the force exerted on the journal will be the same for every region. When the load is applied to the journal, it will move off the centre. This will result in decreased force on one side and increased force on the other. The load will be supported by the net force and the equilibrium position of the journal will depend upon the relative magnitudes of the load and the supply pressure.

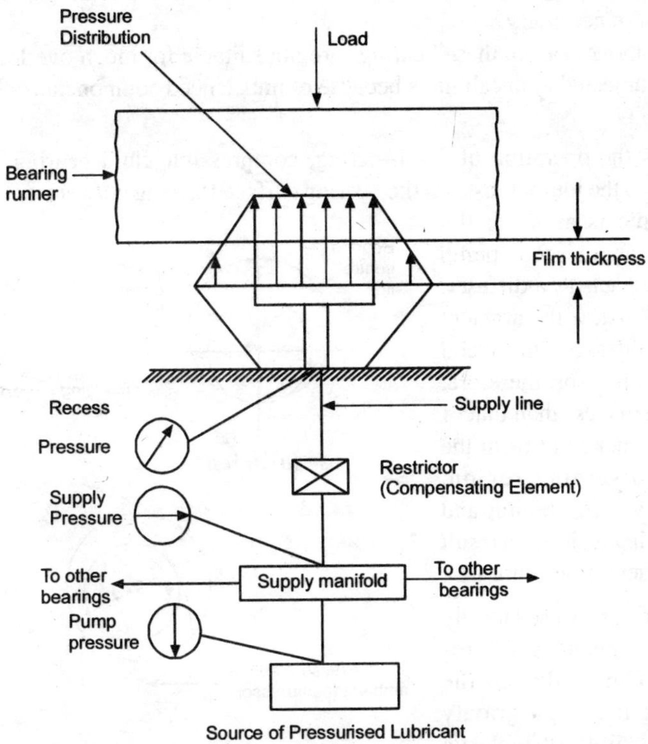

Fig. 17.5 *Hydro-static bearing system*

Hydro-static bearings are expensive and bulky auxiliary equipment is needed. But the saving in depreciation is compensating and also the machines using these bearings maintain their accuracy. These bearings are employed in machine tools, gyroscopes, large telescope azimuths, and precision spindles.

17.4.3 Gas Bearing

In these bearings, the lubricant is gas, usually air, instead of an oil. As for oil bearing, gas bearings are also of two types:

1. Externally pressurised gas bearing, or Hydrostatic gas bearing.
2. Self-acting gas bearing.

The advantages of gas bearings are:

(*a*) They can run at very high speeds of over 10^5 rev/min with little lubricant heating.
(*b*) They can operate at over 650ºC.
(*c*) Their relatively low constant torsional moment which is important in instruments and textile spindles.
(*d*) They can operate in a radioactive environment in which oil will break down.
(*e*) They are used where contamination by liquid lubricant is a problem.
(*f*) When necessary, these bearings can be used in a nonoxidizing environment by using an inert gas such as nitrogen instead of air.
(*g*) Resistance to thermal breakdown of the lubricant.
(*h*) High reliability and long life.
(*i*) Low and constant friction.
(*j*) Close position control and low vibration.

The limitations of gas bearings are:

(*i*) The clearance is very small, about one-tenth smaller than in oil bearings, so precise manufacturing techniques are necessary.
(*ii*) Due to the above reason, these bearings are not suitable for much overloads or vibration.
(*iii*) These are subjected to instabilities because of unbalanced components or due to the dynamics of the gas film.

Figure 17.6 shows the operation of a self-acting, compressible fluid bearing. At zero speed, in the presence of gravity force, the journal rests on the bearing surface forming converging and diverging passages i.e., wedges. As torque is applied, the converging wedge acts as a viscous pump and the diverging wedge acts as a diffuser. The pressure above and below the ambient are generated. Since the diffuser is in general less efficient than a compressor, therefore, (Δp) max, of the diffuser is less than that of the compressor. Gas is sucked in from the bearing ends in the diverging region, compressed in the converging region and flows out of the converging region as a result of end leakage and conservation of mass.

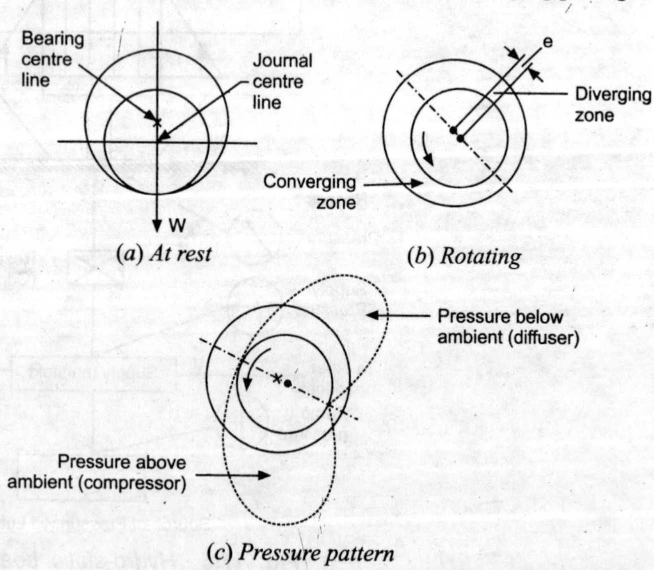

Fig. 17.6 *Working of a gas bearing*

Applications: Gas bearings are used mostly in instrument field, most commonly in gyros and accelerometers. This is due to the constant friction, long life, high gravity capability and the possibility of inelastic design of gas bearings. Other examples of large instruments are: altitude control simulator and a man carrying motion generator. Recently hydrostatic drums, tape guides and capstans have been used in strip mills for metal working as well as in computers and recorders for tape transport.

17.4.4 Spherical Bearing or Self-aligning Bearing

In certain applications, where certain misaligning factors can not be avoided, such as bending of the shaft, flexure of the structure etc., spherical or self aligning bearings are employed. These bearings are shown in Fig. 17.7. The ball is free to rotate in the socket of a plain, cylindrical, spherical bearing but is clamped tightly enough to relieve high misalignment torsional moments by tilting. The motion of the bearing is between the shaft and the cylindrical bearing. This motion is generally oscillatory in rod end spherical bearings and most of it takes place between the ball and its seat. Such bearings are employed in air frames, automobiles, and structural tie rods. Porous spherical bearings are used in electric motors.

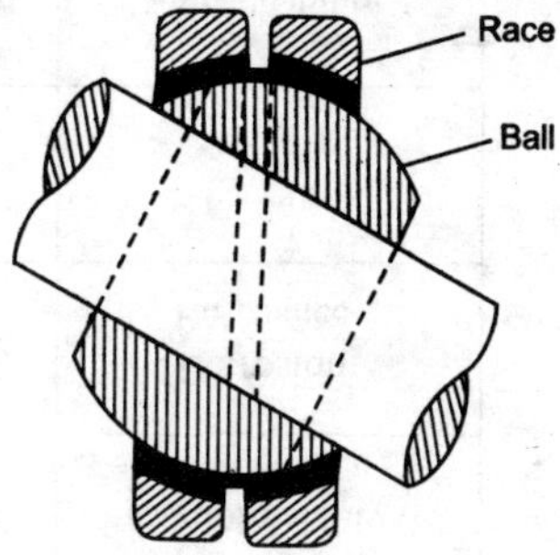

Fig. 17.7

17.4.5 Materials for Sliding Bearings

A bearing material is selected so as to meet the demands of its applications, since it has to work under different states of lubrication, and other conditions. Operating factors to be considered are:

(*i*) Load, (*ii*) Speed, (*iii*) Contamination, (*iv*) Life, (*v*) Temperature, (*vi*) Type and quantity of lubrication available.

No single bearing material can meet all the requirements and proper selection is a matter of judgement based on the following most essential factors:

1. Compatibility: It is a measure of the antiweld and antiscoring properties of a bearing material. Except externally pressurised bearings, there will be metal to metal contact between the journal and bearing, may be at starting. During this contact, the high spots on bearing and journal will rub against one-another resulting in localised welding due to the friction so created. This local welding will result in a damage resembling scoring. Hence a good bearing material should not weld easily to the journal material.

2. Conformability: It is the ability of the bearing material to compensate for the misalignment of the journal and to accommodate other geometric errors by plastic deformation or creep without excessive heating and wear. For this, the material should be soft with a low modulus of elasticity.

3. Embeddability: Inspite of the best efforts to keep the lubricant clean, it may be contaminated with dust, etc. This bearing material should be able to absorb these foreign particles to avoid scoring and wear.

4. Load Capacity or Compressive Strength: The material should be of high compressive strength to withstand the maximum hydro-dynamic pressure so as to avoid its extrusion or some other permanent deformation.

5. Fatigue Resistance: The material should be of sufficient fatigue strength so that it can withstand repeated loads without developing surface fatigue cracks. This is of particular importance in aircraft and automotive engines.

6. Corrosion Resistance: Corrosion resistance of bearing materials is important in certain applications such as internal combustion engines. The material can be made corrosion resistance by applying a thin layer of vanadium over cadmium or lead alloys, or by adding tin to lead babbitt.

7. Thermal Conductivity: The material should be of high thermal conductivity so as to allow the rapid removal of heat produced due to friction.

8. Bondability: For high capacity applications, the bearings are made by bonding one or more thin layers of a bearing material to a high strength steel shell. Hence, the material should possess the property of bondability.

Table 17.4 Performance characteristics of bearing materials

Bearing Material	B.H.N. Room Temp.	B.H.N. 150°C	Min Shaft B.H.N.	Load Capacity MN/m^2	Max. Operating Temp. °C	Compatibility	Conformability	Corrosion Resistance	Fatigue Strength	Embeddability	Thermal Conductivity
Tin base babbitt	20 — 30	6—12	130—170	5.6—10.5	150	E*	E	W	E	E	W
Lead base babbitt	15—20	6—12	130—170	5.6—8.5	150	E	E	W	F	E	W
Lead bronze	40—70	40—60	300	21—28	230	F	P	G	G	P	F
Tin bronze	60—80	60—70	300—400	28+	290+	W	W	E	G	W	W
Cadmium base	30—40	15	200—250	10.5—14	290	E	G	P	W	G	P
Copper lead	20—30	20—23	200	10.5—17.5	175	G	G	F	W	G	F to G
Aluminium alloy	45—50	40—45	200—300	35+	120	P	F	G	E	F	F
Silver overplate	25	25	300—400	35+	230	G	F	E	E	F	E

*E = Excellent; G = Good; F = Fair; P = Poor; W = worst.

9. Thermal Expansion: If the bearing is to operate at high temperatures or over a wide range of temperatures, then its coefficient of expansion and its effect on clearance should be considered.

10. Cost: The overall cost of a bearing is very frequently the deciding factor in the selection of the material.

Performance characteristics of the various bearing materials are listed in Table 17.4.

Babbitts are most widely used. In automobiles, these are used as thin layers 0.05 to 0.15 mm thick, bonded to an insert or steel shell. Tin Babbitt is composed of 80 to 90% tin, 3 to 8% copper, and 4 to 14% antimony. The composition of lead babbitts is 10 to 15% antimony; upto 10% tin and the remaining lead. To improve upon the strength and fatigue resistance of babbitt materials, a special type of bearing known as "Three-component bearings" are employed. They consist of a low carbon steel backing, an intermediate layer of copper, bronze aluminium, or silver and an overlayer of babbitt ranging from 0.013 to 0.13 mm thick. These bearings are frequently used for heavy duty service in automobiles, trucks, diesels, aircrafts and gear units.

Due to good fatigue strength, high load capacity, and good high temperature performance, the copper lead materials are extensively used for heavy duty main and connecting rod bearings for automobiles, trucks, diesels and aircrafts. For moderate loads and speeds, they are used to some extent in electric motors, turbines, and steam engines. The lead bronzes are generally a first choice for intermediate loads and speeds and are used in machine tools, home appliances, farm machinery, and pumps. Tin bronzes are used in high load, low speed applications, e.g., trunnion bearings, gear bushings for farm equipment, earth moving machinery, rolling mill bearings, and in internal combustion engines for connecting rod bearings, valve guides and starter bearings. Aluminium bronzes are best suited for heavy duty, low speed applications with plentiful lubrication.

The silver bearings are used almost exclusively in aircraft engines due to their excellent fatigue strength. For relatively light loads, cast iron and steel bearings may also be used.

Porous Metals: The bearings made of porous metals are employed for use with little or no attention of lubrication. Their low initial cost, simplicity of design have made them popular for home appliances, small motors, machine tools, aircraft automotive accessaries, business machines, instruments, and farm and construction equipment. The powder metal bearing materials include : lead bronzes, high and low density bronzes, alloys of iron copper, and iron copper-carbon; and aluminium. The porous metal bearings are made in the following manner:

1. The metals are powdered and blended into a uniform mixture.
2. The mixture is pressed under high pressure in hardened steel or tungsten-carbide dies.
3. It is then sintered below its melting point into a coherent mass.
4. The bore, outside diameter, and faces are sized in a burnishing operation.
5. The bearing which is porous is placed in a bath of lubricating oil to fill the voids.

Limiting conditions for operation of porous metal bearings are given in Table 17.5.

Table 17.5 Porous metal bearings

Material	pV kPa.m/s	p, MN/m^2 Static	p, MN/m^2 Dynamic	V, m/s
Bronze	1750	56	14	6
Lead bronze	2800	24.5	5.6	7.5
Copper iron	1225	140	28	1.125
Hardenable cu-Iron	2625	350	56	0.0175
Iron	1050	70	21	2
Bronze Iron	1225	73.5	17.5	4
Lead Iron	1750	28	7	4
Aluminium	1750	28	14	6

Non-metallic Bearings

The various non-metallic bearings are made of plastics, carbon-graphite, wood, rubber, and ceramics. Plastic bearings have freedom from corrosion, quiet operation, availability in moulded shapes, no need of lubrication. The various plastics used include: Phenolics, Nylon, polycarbonates, acetal, etc.

The other non-metallic bearings are self lubricating, high temperature resistant and low cost bearings. Operating limits for non-metallic bearings and their applications are listed in Table 17.6.

Table 17.6 Operating limits for non-metallic bearings

Material	Load Capacity MPa	Max. Temp. ºC	Max. Speed m/s	pV; kPa. m/s	Application
Phenolics	42.0	90	15.0	530	Pumps, Propellers
Nylon	7.0	90	5.0	105	Mill roll necks
Reinforced Teflon	17.5	260	5.0	350	Dairy, Textile, Food machinery
Poly carbonate	7.0	105	5.0	105	Optical devices, Electric shavers, egg beaters, door hinges
Acetal	7.0	80	5.0	105	Electric motors
Carbon Graphite	4.2	420	15	530	Meter, conveyors, Pumps, Turbines.
Rubber	0.35	65	20	—	Marine propellers
Wood	14	65	10	420	Conveyors, machinery

17.4.6 Lubricants

The main purposes of lubricants are:

1. To reduce friction losses and wear between rubbing surfaces.
2. To carry off heat into which frictional work is converted.
3. To provide protection against corrosion.

All lubricants can be grouped into three general types:

(*a*) Fluids
(*b*) Greases
(*c*) Solid film lubricants.

Both oils and greases are used to lubricate the sliding bearing and rolling bearings.

17.4.6.1 Oils

Both mineral oils and synthetic oils are used as bearing lubricants. Oils are slippery hydrocarbon liquids. Synthetic oils are better than mineral oils in some respects. Synthetic oils can operate over a wide temperature range with little change in viscosity. The most popular synthetic oils are polyalkylene glycols and the silicones. The other synthetic oils are : polymer oils, phosphate esters, organic chlorine compounds, and diesters.

Advantages of Oil

1. Oil is easy to drain and refill. It is easy to control the fill volume of oil in a housing or reservoir.
2. An oil lubricant for a bearing might also be usable at many other points in the machine.
3. The oils are indispensable for dissipation of large amount of heat from the bearing housing.
4. Oils are available for a greater range of operating speeds and temperature than grease.
5. It readily feeds into all areas of contact and can carry away dirt, water, and wear products.

Oil Selection: The important oil characteristics for its selection are:

1. Viscosity: This is the most important physical property of oils. It determines the oil's load carrying capacity, thickness of lubricating film maintained, operating temperature of bearing, and friction generated. A high viscosity oil can support heavier loads but it has also more internal friction. Viscosity decreases with an increase in oil temperature. Viscosity is a measure of the internal frictional resistance of a fluid. Newton's law of viscous flow states that the shear stress in a fluid varies directly with the velocity gradient, that is,

$$f_s \propto dV/dy$$

or

$$f_s = Z \times \frac{dV}{dy}$$

where the constant of proportionality '*Z*' is defined as the viscosity. This is also known as 'dynamic viscosity' or 'absolute viscosity'. In bearings, the film thickness is small and the velocity gradient is almost linear and constant. Therefore,

$$\frac{dV}{dy} = \frac{V}{y} = \frac{V}{h}, \; h \text{ is film thickness}$$

$$\therefore \quad f_s = Z\frac{V}{h}, \text{ N/m}^2$$

Units of *Z* are: $\left(\frac{\text{N}}{\text{m}^2}\right) \times (\text{m})/(\text{m/s}) = \text{Ns/m}^2$

It is conventional to express the units of 'Z' in mass units; therefore, these become kg/ms. The S.I. symbol of viscosity is P*as* ≡ *kg/ms.*

Viscosity is measured as the number of seconds for 60 ml of lubricant at a specified temperature to flow through a tube 17.6 mm in diameter 12.25 mm long and is expressed as 'seconds Saybolt'. If *S* is Saybolt reading in seconds then,

$$v = \left(0.22S - \frac{180}{S}\right), \text{ centistoke } (CSt)$$

is called the kinematic viscosity (stoke is cm^2/s and $CSt = \frac{1}{100}$ stoke).

In SI units, its units are m^2/s and are related to '*CSt*' as given below,

$$v\,(m^2/s) = v\,(CSt) \times 10^{-6}$$

Now dynamic viscosity and kinematic viscosity are related as,

$$Z = \rho \times v$$

where ρ = density of the lubricant, kg/m^3.

The specific gravity of most of the oils at 15.5°C ranges between 0.86 to 0.95, Table 17.7. The specific gravity at any other temperature '*t*' can be found out by

$$(\text{Specific gravity})_t = (\text{Specific gravity})_{15.5} - 0.000365(t - 15.5) \quad ...(17.1)$$

It is clear that for most of the oils an average value of 900 kg/m^3 can be taken for density.

The variation of absolute viscosity with temperature for commonly used lubricating oils is shown if Fig. 17.8. The specific gravities of the oils are listed in Table 17.7.

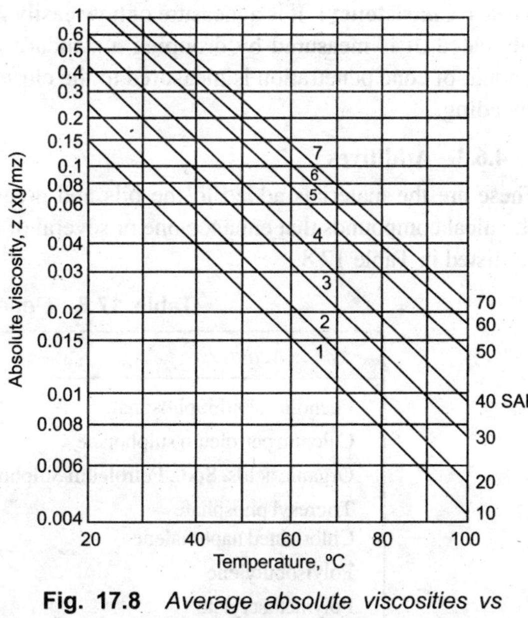

Fig. 17.8 *Average absolute viscosities vs temperature*

Table 17.7 Specific gravities of oils

Curve No.	Type of oil	Specific gravity at 15.5°C
1.	Turbine oil, ring oiled bearing.	0.8894
2.	Light transmission for gears, Automobile oil.	0.9254
3.	Automobile oil.	0.9263
4.	Medium transmission oils for large generators. Motors, Steam Turbines, Heavy Speed gears, Heavy motor oil, Automobile oil.	0.9275
5.	Light cylinder oil.	0.89
6.	Airplane oil.	0.8927
7.	Gear oil.	0.9153

2. **Viscosity Index:** It is the measurement of how much an oil's viscosity changes with temperature.
3. **Pour Point:** It is the lowest temperature at which the oil will flow.
4. **Flash Point:** It is the temperature at which an oil will give off ignitable vapours.
5. **Fire Point:** It is the temperature at which the oil will burn when ignited.

The flash point is not necessarily a safe limit for oil. Some decomposition will probably occur before the flash point is reached.

17.4.6.2 Greases

Grease is a semisolid, combining a fluid lubricant with a thickening agent, usually a soap. The most common soaps are, Calcium, Sodium and Lithium soaps.

Advantages of Grease

1. It can be more easily retained in a housing, since it does not flow as readily as oil.
2. Since grease is easily retained or contained, leak proof designs are unnecessary.
3. Maintenance is less. There is no oil level to be maintained and regreasing is less frequent.
4. Grease has better sealing abilities than oil. This helps to keep dirt and moisture out of the housing.

Grease Consistency: It is a measure of how easily grease may be squeezed out from between the parts being lubricated. It is measured by dropping a standard metal cone point downward into a grease sample. The amount of cone penetration is measured to the closest tenth of mm. It can be readily changed by working or kneading.

17.4.6.3 Additives

These are the materials added to the oils and greases to increase their lubricating qualities. They are the chemical compounds that enhance one or several of the lubricant's properties. The general types of additives are listed in Table 17.8.

Table 17.8 General types of additives

Additives	*Function*
Phenolics, Dithiophosphate	Oxygen inhibitor
Calcium petroleum sulphonate	Detergent
Organic acids, Sod., Petroleum Sulphonate	Rust inhibitor
Tricresyl phosphate	Wear preventive
Chlorinated naphthalene	Boundary lubrication
Polyisobutylene	Viscosity index improver
Polymethacrylate	Pour point depressant
Silicone oil	Defoaming agent

17.4.6.4 Solid and Bonded Film Lubricants

Whatever may be the smoothness of the rubbing surfaces, there will be microscopic peaks on the surface of journal and bearing. Under very heavy loads or very low speeds, these peaks will break through the lubricant layer and rub against one another causing wear and seizure. Under these circumstances, boundary lubrication is necessary. The boundary lubricant should resist the penetration of the peaks on the journal under extreme loads. Other requirements are load capacity, temperature resistance, etc. Solid lubricants include graphite, molybdenum disulphide, tungsten disulphide, and for low loads, fluorocarbons and nylon. These lubricants are applied as dry powders, mixtures with greases and oils, or mixture with binders which form dry films when applied and cured. The binders are usually resins. These dry film lubricants have the following advantages:

1. They provide low coefficients of friction (0.04 to 0.10) over a wide range of loads, speeds, and temperatures.
2. Provide life time lubrication.
3. Solve lubricant problems caused by abrasive contaminant.
4. Reduce fretting caused by vibration.

Many times, the solid lubricants are added to the bearing structure materials. For example, MoS_2 is added to phenolic compositions moulded into heavy machinery bearings. Graphite and MoS_2 are incorporated into sintered bronze or steel bearings when life time built in lubrication is needed.

Solid lubricants are recommended only when hydrodynamic lubrication is not adequate or can not be accommodated. Also these have the following limitations:

1. Wear life is less than oil lubricants.
2. They will not dissipate heat.

17.4.7 Lubricating Devices

To feed the lubricant to the bearing and to replenish it due to its loss from the bearing ends, the following devices are employed:

1. Drop Feed Oiler: Figure 17.9 (*a*). This is the simplest way for feeding the oil. Oil is put in the reservoir provided on the bearing cover and it flows down to the clearance space by gravity. Oil is distributed along the journal by means of an axial groove cut into the bearing. The drawback is that the flow is not uniform and it varies with the level in the reservoir and it requires constant attention and is suitable only for slow speeds and intermittent service.

2. Needle Oiler : Fig. 17.9 (*b*). It overcomes the drawback of the drop feed oiler. A needle passing through the oiler rests over the surface of journal. When it rotates, the needle is shaken by the irregularities of the journal surface and oil passes from the oil reservoir through the gap between the needle and the hole in the plug. When the journal is stationary, there is no oil feed to the bearing.

3. Wick Lubricator: Figure 17.9 (*c*). The operation of wick lubricator is based on the siphoning effect and it ensures a uniform supply of oil to the working surfaces. The drawback is that it will go on supplying the oil to the bearing even when the journal is at rest.

4. Ring Oilers: This method is commonly used in the lubrication of sliding bearings. The ring slipped on the journal is larger in diameter than the journal, Fig. 17.9 (*d*). As the shaft rotates, the ring will revolve alongwith the journal. The ring passes through an oil bath and the oil carried by it drips to the bearing thus lubricating it.

5. Grease Cups: Greases for lubrication of bearings are applied through grease cups. Grease is packed into the body and cap of the grease cup and the cap is screwed on the body. As the grease is used up, the cap is turned one or two revolutions to force the grease from the cup through the central hole towards the working surfaces.

6. Splash and pressure lubrication: Splash lubrication is a common method of lubrication in vertical internal-combustion engines, air compressors, etc. As the big end of the connecting rod dips into the oil put in the crank sump, the oil will be splashed up. The crank pin, piston pin and the cylinder walls will be lubricated by this method.

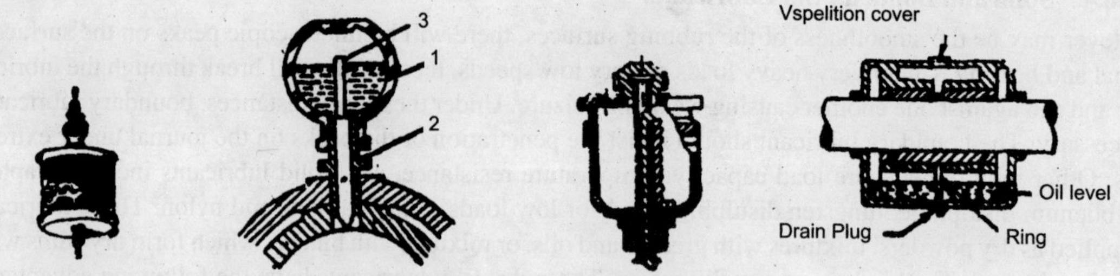

(a) *Drop oiler* (b) *Needle oiler* (c) *Wick lubricator* (d) *Ring oiler*

Fig. 17.9 Lubricating devices

For the continuous supply of lubricant and to avoid its wastage when the shafts are stationary, the bearings of high speed shafts operating continuously for a long time without intervals, pressure lubrication is employed. Oil is sucked by a pump from the sump of the engine, delivered under pressure through a filter into a distribution device and hence to the individual lubricating points. After lubricating the oil drops down into the oil sump.

17.4.7.1 Oil Grooves

The purpose of the oil grooves is to assist in the distribution of oil between the rubbing surfaces. The oil should be introduced in the region of low pressure. The grooves should be located along the generating line of the bearing shell, passing through the lubricant feed hole on the non-loaded side of the bearing. For horizontal bearings, the grooves must end about 12mm from each end of the bearing. The grooves should be as simple as possible and should be smoothly rounded. Some good oil grooves are shown in Fig. 17.10.

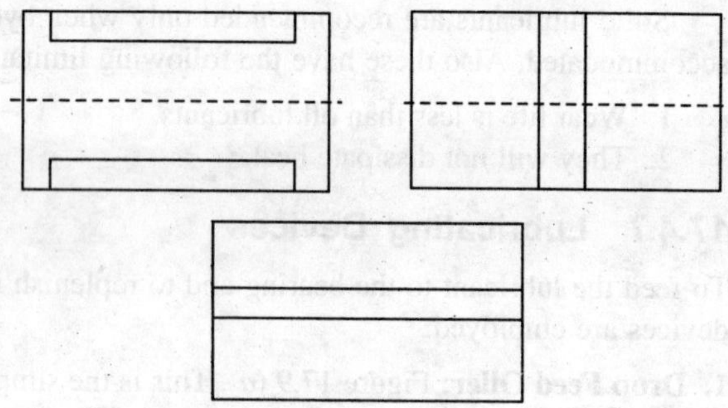

Fig. 17.10 *Oil grooves*

17.4.8 Wear in Sliding Bearings

Wear of plain bearings is strongly dependent upon the state of lubrication. There is practically no wear in hydrostatic bearings since journal and bearing never touch each other even when the shaft is stationary. However, if any defect builds up in flow restrictors due to plugging of wear debris in them, serious failure may result. In the case of gas lubricated, externally pressurised bearings occasional rubbing and wear of the working surfaces may occur due to impact of vibration load. In the case of hydro-dynamic bearings, wear takes place during start up since the speed is too low to build up the pressure to separate the bearing and the journal. Serious wear may also result during start up if hard wear debris get embedded in the bearing surface. These wear debris will scratch the journal surface, the phenomenon is known as "Wire Wooling". Boundary lubricated and self lubricated bearings wear much faster than fluid film bearings. Self lubricated plastic bearings wear at higher rates than boundary lubricated, metal alloy bearings.

17.4.9 Heat Generation in Sliding Bearings

In sliding members, the heat is generated either by shearing of oil film or by rubbing contact. In hydrodynamic and hydrostatic bearings, the heat generated at the running speed is due to the oil shear. Boundary lubricated and self lubricated bearings are more sensitive to sliding velocity than fluid form types because the coefficient of friction is quite high. Frictional heating is a function of bearing pressure, sliding velocity, and coefficient of friction.

In spite of the many advantages of sliding contact bearings over rolling contact bearings as listed in Tables 17.1, 17.2 and 17.3, the sliding contact bearings have the following major drawbacks: Comparatively high friction losses. It is not always possible to ensure fluid friction; excessive consumption of lubricants; great lengths of the bearing shells reaching $L = (2-3)\ D$; employment in a number of cases of scarce materials (bronze, babbitt) and the necessity for constant supervision. Due to these factor in modern mechanical engineering, sliding contact bearings are employed less frequently than rolling contact bearings. However, in many cases their use is preferred, for example with extra heavy shafts (for which there are no suitable rolling contact bearings), for applications involving impact or vibration loads, and in cases when split bearings are required (e.g., crankshafts etc.)

17.4.10 Journal Bearing Design

The design of a journal or sleeve bearing will depend upon the state of lubrication. Each case will be dealt separately.

1. Self-lubricated and Boundary Lubricated Bearings: For these bearings, the design variables are: material, lubricant, and bearing size. Lubricant and material determine the load limit. Size is based on keeping within the lubricant material limitation. In these bearings, the overheating is the most common cause of failure. If coefficient of friction is assumed to be constant, then the designer has control over two parameters: bearing pressure p, and sliding velocity V, which are defined as:

Bearing pressure, $$p = \frac{\text{Load}}{\text{Projected bearing area}}$$

or $$p = \frac{F}{LD}\ \text{N/m}^2 \quad \text{...(17.2)}$$

where L = length of bearing, m

D = Diameter of bearing, m

and, sliding velocity, $$V = \text{rev/s} \times \text{bearing circumference, m/s} \quad \text{...(17.3)}$$

Hence, the rough indicator for heat generation can be combined "pV" factor which can be used as a starting point for determining material and size. The limiting values for "pV" factor for porous metal bearings and non-metallic bearings are listed in Tables 17.5 and 17.6.

2. Hydro-dynamic Journal Bearings: A hydrodynamic journal bearing under load is shown in Fig. 17.4.

D = diameter of bearing.

d = diameter of journal

L = length of bearing or journal

The various terms connected with hydro-dynamic journal bearings are:

$$C_d = \text{diametral clearance} = D - d \quad \text{...(17.4)}$$

$$C_r = \text{radial clearance} = \frac{1}{2}Cd = \frac{1}{2}(D-d) \quad \text{...(17.5)}$$

$$\frac{C_d}{D} = \text{diametral clearance ratio} \quad \text{...(17.6)}$$

= 1.00 mm per m for plastic bearings, and 1.5 mm per m for hard bearings.

Square bearing $\frac{L}{D} = 1$, Long bearing $\frac{L}{D} > 1$, Short bearing $\frac{L}{D} =< 1$.

If L/D ratio is very less, the load carrying capacity of the bearing is impaired due to the effect of side leakages. Longer bearings are used to restrain the shaft from vibrations, as in line shafts, or to position the shaft accurately, as in machine tools. The angle ϕ = Attitude angle. It is the angle between the centre of the bearing and the line joining the centre of the bearing to the points on journal and bearing, pertaining to minimum oil film thickness.

Eccentricity, e = Radial distance between the centre of bearing and the displaced centre of the journal under load.

$\in$ = eccentricity ratio or attitude,

$$= \frac{\text{actual eccentricity } e \text{ under load conditions}}{\text{Maximum value of eccentricity when the journal touches the bearing at rest position}}$$

$$\therefore \quad \in = \frac{e}{C_r} = \frac{C_r - h_0}{C_r} = 1 - \frac{h_0}{C_r} \qquad \text{...(17.7)}$$

where h_0 = minimum oil film thickness.

It is clear from Fig. 17.4, that the minimum oil film thickness depends upon the position of the journal centre under running conditions. The centre position depends upon the unit pressure, p, the clearance ratio C_d/d, the oil viscosity Z, the speed of rotation n, and the angle β. This angle is between the load line and the beginning of the supporting oil wedge. The beginning of the oil wedge is the position where the oil pressure is equal to the atmospheric pressure. In many cases it is at an oil groove. The angle β may be taken as 60º for a complete 360º bearing, if no oil groove is located within 60º of the load line. The hydrodynamic theory of lubrication indicates that the journal centre moves along a semi circular arc of diameter C_r.

***(i)* Minimum Oil Film Thickness.** It determines the closest approach of the journal and the bearing surfaces with complete lubrication. Its value depends upon the degree of finish of these surfaces and the rigidity of the journal and bearing construction, and on the size and kind of the dirt particles present in the oil. If the bearing construction is quite rigid and if the oil is kept clean, then the minimum oil film thickness for hydrodynamic lubrication, must be equal to or greater than the sum of the average peak to valley roughness of the journal and bearing surfaces, i.e.,

$$h_0 \geq C_p(\text{rms}_B + \text{rms}_J),$$

where C_p = roughness peak factor, Table 17.9.

= ratio of peak to average roughness

rms_B, rms_J = bearing and journal roughness respectively, µm, Table 17.10.

To take into account, the dirt particles, shaft deflection, etc., a factor called "clearance factor" is incorporated to increase the specified minimum film thickness, so

$$h_0 \geq C_e \cdot C_p \cdot (\text{rms}_B + \text{rms}_J), \text{ where } \quad C_e = \text{clearance factor.}$$

For ordinary babbitt lined bearings the film thickness should not be less than 0.02 mm. On larger power machinery, the minimum film thickness may be limited to 0.025 to 0.15 mm. One rule of thumb is to limit the minimum film thickness to 0.00025D.

It has been found by investigators that, for bearings of certain geometry, the minimum oil film thickness is a function of Bearing Modulus or Bearing service characteristic number (defined ahead as equal to Zn/p), that is,

$$h_0 = \phi\,(Zn/p)$$

$$\therefore \quad h_0 \propto Z$$

$$\propto n$$

$$\propto 1/p$$

$\therefore$ For hydro-dynamic lubrication to occur.

1. Wedge-shaped gap should be formed between the sliding surfaces.
2. A lubricant of adequate viscosity should continually be introduced into the gap.
3. The relative speed of the journal should be enough to create the hydro-dynamic pressure in the fluid film necessary to balance the external load.

Table 17.9 Roughness peak factor C_p

Metallic Finish	Average value, C_p	Typical range, C_p
Ground	4.5	3.5–5.0
Hydrolapped	6.5	5.5–7.5
Loose abrasive lapped	10.0	7.5–13.0
Sand papered	7.0	5.5–9.0
Super finished	7.0	5.5–9.5
Turned	5.0	3.5–8.0

Table 17.10 Bearing surface roughness μm rms

Type of Surface	Surface Roughness
High Precision Journal	0.125–0.2
High Precision Bearing	0.125–0.5
Medium Precision Journal	0.4
Medium Precision Bearing	0.8
Commercial Grinding	0.4–1.6
Broaching, Reaming, Boring, Finish Turning	0.8–1.6

***(ii)* Bearing Modulus:** It is designated as "*C*" and is defined as;

$$\text{Bearing Modulus} = \frac{Zn}{p} \qquad \ldots(17.8)$$

The bearing modulus is of special interest in bearing design. Its variation with coefficient of friction is shown in Fig. 17.11. For any given bearing , there is a value for bearing modulus indicated by *C*, for which the coefficient of friction is minimum. The bearing should not be operated at this value of bearing modulus, since a slight decrease in speed or a slight increase in pressure will make the journal to operate in the partial lubrication state resulting in high friction, heating and wear.

To prevent this, the average value of Bearing modulus should be,

$$\frac{Zn}{p} \geq 3C$$

The values of *C* are given in Table 17.11.

p is in MPa

For large fluctuations and heavy impact loads,

$$\frac{Zn}{p} \cong 15C$$

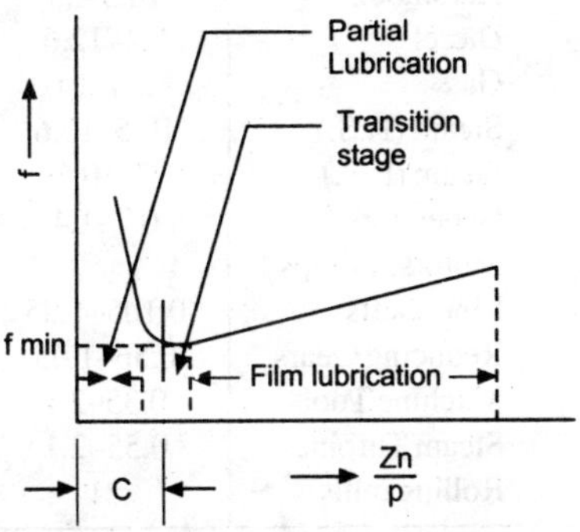

Fig. 17.11 *Variation of bearing modulus with coefficient of friction*

The design data for Journal bearings is listed in Table 17.12.

Table 17.11 Values of C

Shaft	Bearing	C
Hardened and ground steel	Babbitt	2.85
Machined soft steel	Babbitt	3.60
Hardened and ground steel	Plastic Bronze	4.30
Machined soft steel	Plastic Bronze	5.00
Hardened and ground steel	Rigid Bronze	5.70
Machined soft steel	Rigid Bronze	7.00

Table 17.12 Design Data for Journal Bearings

Bearing Types	p MN/m^2	Z kg/ms	$\frac{Zn}{p}$	$\frac{C_d}{D}$	$\frac{L}{D}$
Axles:					
Locomotive	3.85	0.600	4.5–7.35	0.001 max.	1.6–1.8
Railway Car	2.1–3.5	0.500	7.00–14.30	0.001 max.	1.8–2.0
Crankpins:					
Aircraft	5.25–35	0.008	1.5	—	0.5–1.5
Automobile	10.5–24.5	0.008	1.5	0.001 max.	0.5–1.4
Diesel	10.5–28	0.020–0.060	0.75–1.43	0.001	0.8–1.5
Gas	8.4–12.6	0.020–0.060	1.43–2.85	0.001	0.8–1.5
Steam (H.S.)	2.8–8.4	0.030	0.85–2.14	0.001 max.	1
Steam (L.S.)	5.6–10.5	0.080	0.85–1.14	0.001 max.	1–1.25
Shears, Punches	35–56	0.100	—	0.001	1–2
Crank shaft, main bearing:					
Aircraft	4.2–12.6	0.025–0.030	2.25–3.00	0.001 max.	0.6–1.5
Automobile	2.1–12.6	0.025–0.030	2.85–4.30	0.001 max.	0.8–1.5
Diesel	2.45–8.4	0.022–0.060	2.14–2.85	0.001 max.	0.8–1.5
Gas	3.5–7	0.020–0.060	2.85–4.30	0.001 max.	0.4–2.4
Steam (H.S.)	0.42–3.5	0.015–0.030	3.60–4.30	0.001 max.	1.3–1.7
Steam (L.S.)	0.56–2.8	0.070	2.85–4.30	0.001 max.	1.2–1.6
Shears, Punches	14.0–28.0	0.100	—	0.001	1–2
Wrist Pins:					
Aircraft	20–70	0.007–0.008	1.14–2.14	0.001 max.	0.6–1.5
Automobile	10.5–35	0.007–0.008	1.14–2.14	0.001 max.	0.8–1.5
Diesel	8.4–12.6	0.020–0.060	2.85–3.60	0.001 max.	0.8–1.4
Gas	8.4–14	0.020–0.060	0.75–1.43	0.001 max.	0.8–1.25
Steam (H.S.)	10.5–12.6	0.025	0.75	0.001	1.4–1.6
Steam (L.S.)	7–10.5	0.070	0.75	0.001	1.2–1.5
Generators	0.7–1.4	0.025	28.50	0.001	1–2
Motors, Pumps					
Line shafts	0.105–1.05	0.025–0.060	4.30–14.30	0.001	2.5–3
Reducing Gears	0.56–17.5	0.030–0.050	5.70–43.00	0.001	2–4
Machine Tools	0.35–2.1	0.040	0.3–1.43	0.001	1–3
Steam Turbines	0.55–2.1	0.010–0.020	14.30–28.50	0.001	1–2
Rolling mills	21	0.050	14.30	0.0015	1.1–1.5

***(iii)* Sommerfeld number or Bearing Characteristic Number:** This is also a very useful number in the design of sleeve bearings. It is defined as:

$$S = \frac{Zn'}{p}\left(\frac{D}{C_d}\right)^2 \quad \text{...(17.9) } n' \text{ is rev/s and } p \text{ is in N/m}^2$$

Figure 17.12 shows the graphs of the minimum film ratio h_0/C_r versus Sommerfeld number, S for bearings having various length diameter ratio L/D. These graphs can be easily used for the design of sleeve bearings. For example, corresponding to a known minimum permissible oil film thickness h_0 and an assumed radial clearance C_r a specific value of S can be obtained. Then if certain quantities in equation (17.9) are known, the remaining can be found out. Curves showing variation of coefficient of friction variable, oil flow variable, Temperature rise variable are also available.

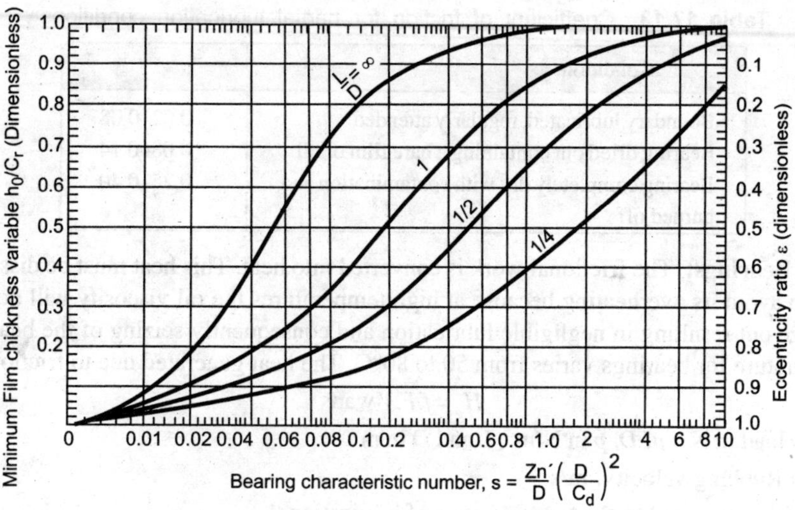

Fig. 17.12 *Minimum film ratio vs sommerfeld number*

***(iv)* Operating pressures:** The minimum operating pressure known as "critical pressure" is the pressure at which the oil film breaks down and metal to metal contact begins. It is given empirically as :

$$p = \frac{Zn}{4.75\times10^6}\cdot\left(\frac{D}{C_d}\right)^2\cdot\left(\frac{L}{D+L}\right), \text{MN/m}^2 \qquad ...(17.10)$$

In the above equation, it is assumed that $h_0 \cong \frac{C_d}{4}$.

***(v)* Coefficient of Friction:** The coefficient of friction is of great significance in bearing design. Its value can't be determined very accurately since the degree of lubrication is generally not known. For self-contained full bearings, for hydrodynamic lubrication, the coefficient of friction is given empirically as:

$$f = \frac{0\cdot326}{10^6}\cdot\left(\frac{Zn}{p}\right)\left(\frac{D}{C_d}\right) + k_f \qquad ...(17.11)$$

where k_f = factor to correct for end leakage. Its value varies with L/D ratio. Its value may be taken as 0.002 for $0.75 < L/D < 2.8$. Its value can also be found out from Fig. 17.13.

For pressure fed hydrodynamic bearings, and hydrostatic bearings, the correction factor is not considered, and, therefore, the coefficient of friction is given as:

$$f = \frac{0\cdot326}{10^6}\cdot\left(\frac{Zn}{p}\right)\left(\frac{D}{C_d}\right) \qquad ...(17.12)$$

The coefficient of friction for these conditions, is generally below than 0.001.

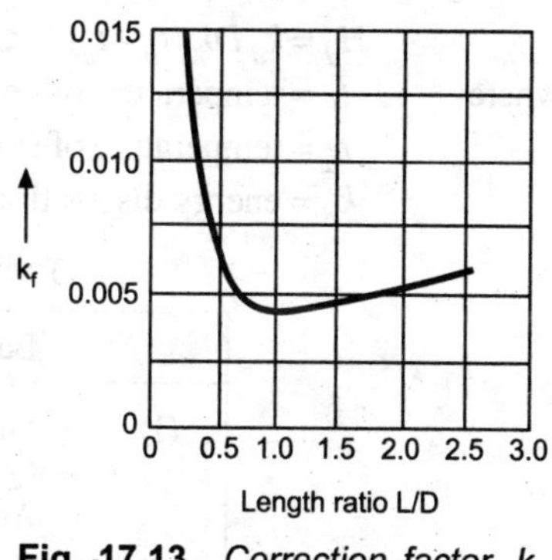

Fig. 17.13 *Correction factor, k_f*

For lightly loaded bearings, the coefficient of friction may be approximately calculated from Petroff's equation:

$$f = \frac{\pi^2}{30}\left(\frac{Zn}{p}\right)\left(\frac{D}{C_d}\right) \qquad ...(17.13)$$

here p is N/m^2.

When the bearings operate at boundary lubrication or partial lubrication conditions, the values of 'f' may be taken from Table 17.13, which is due to Fuller.

Table 17.13 Coefficient of friction for partial lubrication conditions

Condition	f
Boundary lubricated, regularly attended	0.02–0.08
Bearing dried out containing a bare film of oil	0.08–0.14
Bearing completely dry with contamination and burned off	0.25–0.40

(*vi*) Heating of Bearings: The frictional work is converted into heat. This heat must be dissipated from the bearing house to avoid its overheating because at high temperatures the oil viscosity will decrease, making the oil to squeeze out resulting in negligible lubrication and consequently seizing of the bearing. The usual operating temperature for bearings varies from 50 to 80°C. The heat generated due to friction is given as:

$$H_g = f.F.V \text{ watts}$$

where F = load in N = pLD, p in N/m^2, L and D in m

and V = Rubbing velocity, m/s

f will correspond to the type and state of lubrication.

$$V = \frac{\pi Dn}{60}$$

∴ Heat generated is:

$$H_g = f.F.\frac{\pi Dn}{60} \text{ watts} \qquad \text{...(17.14)}$$

After thermal equilibrium, the heat will be dissipated at the same rate at which it is generated. The heat dissipation will depend upon the method of lubrication:

(*a*) Self-contained bearings: In such bearings, the heat generated in the oil will be transmitted to the bearing body from where it must be dissipated into the surrounding area. The amount of heat dissipated will depend upon the temperature difference, size and mass of the radiating surface, and on the air flow around the bearing. The rate of heat dissipation is given as:

$$H_d = k_d DL(t_b - t_a) \qquad \text{...(17.15)}$$

where t_b = temperature of bearing °C

t_a = temperature of surrounding., °C

k_d = energy dissipation coefficient, W/m^2 °C, Table 17.14.

Table 17.14 Energy dissipation coefficient

Bearing Condition	k_d
(1) Simple Pillow-block bearing:	
Quiet Air	146.56 to 244.23
Air at 2.5 m/s	488.50 to 767.60
(2) Heavy duty self aligning ring-oil lubricated :	
Quiet Air	293.00 to 390.77
Air at 2.5 m/s	879.23 to 1228.13

The bearing temperature t_b is frequently assumed to be the mean between the average oil temperature t_o and the temperature of surroundings, i.e.,

$$t_b = t_a + \frac{1}{2}(t_o - t_a) \qquad \text{...(17.16)}$$

For usual industrial applications, the upper limit of permissible oil temperatures ranges from 70°C to 82°C. If the temperature is higher, bearing is cooled by water circulating through coils built in the bearing or using pressure feeding of the lubricant.

***(b)* Pressure Fed Bearings:** The amount of heat removed by the circulating oil is:

$$H_o = \rho \,.\, \phi \,.\, Q \,.\, (t_o - t_e), \text{ watts} \qquad ...(17.17)$$

where ρ = density of oil = 900 kg/m^3

ϕ = specific heat of oil = 1.84 to 2.05 kJ/kg °C

Q = quantity of oil, m^3/s

$t_o - t_e$ = (outlet temperature of oil – entrance temperature of oil) °C.

The average oil film temperature may be taken as:

$$t_a = t_e + \frac{1}{2}(t_o - t_e) \qquad ...(17.18)$$

The upper limit of the permissible oil temperature for usual industrial applications ranges from about 70° to 82° C. If this temperature is exceeded, either the oil feed pressure or oil viscosity must be changed. The oil is usually circulated through the bearings at pressures ranging from 0.07 to 0.7 N/mm^2. Another method to avoid overheating of oil is to circulate 3 to 5 times the amount of oil calculated from the equation (17.17).

***(vii)* Design Procedure:** There is no one design procedure for the design of journal bearings. There can be unlimited design procedures. Sommerfeld number, as discussed earlier, may be used for the bearing design. The design usually involves :

1. Optimizing clearance,
2. Optimizing bearing length, lubricating inlet slots, and
3. Calculation of energy balance.

Before we give the design procedure, the following design considerations should be kept in mind:

1. Lubricants: The lubricant to be used can be either liquid, solid or gas. The choice will depend upon such factors as type of machine, method of lubricant supply and load characteristics.

2. Bearing load: The load coming on a bearing is usually specified. However, its value per projected area can be determined by suitably selecting length and diameter of the bearing. The value of load per projected area will depend upon the bearing life desired. It is clear that smaller the load per projected area, the greater the bearing life and vice versa.

3. Length Diameter Ratio: This is a very important parameter in bearing design. Its value lies between 0.8 to 1.5, with a value of unity being most common. Long bearing (L/D ratio greater than 1) is used where misalignment must be avoided and a reduced load carrying capacity can be tolerated. In short bearings ($L/D < 1$), the danger of metal to metal contact due to large shaft deflections is greatly reduced. So, a rough rule for choosing a L/D ratio is:

(i) Use preferably, $L/D = 1.0$
(ii) Use $L/D < 1.0$, if deflections are expected to be severe
(iii) Use $L/D > 1.0$, if shaft alignment is important.

4. Clearance: The value of journal bearing clearance depends on: materials, manufacturing accuracy, load carrying capacity, minimum film clearance, oil flow, film temperature, etc. Large clearance will allow foreign material to pass-easily through the bearing, the increased oil flow will reduce film temperature and thus increase the bearing life. But a large clearance will result in a loose, noisy bearing and a resulting decrease in minimum film thickness. A reasonable clearance ratio (C_d/D) equal to 0.001 is most common.

When the bearing load, the diameter and speed of the shaft are known, the following design procedure may be followed:

1. Determine the bearing length by choosing a ratio of $\frac{L}{D}$ from Table 17.12.
2. Check the bearing pressure.

$p = \frac{F}{LD}$, from Table 17.12 for probable satisfactory value.

3. Assume a clearance ratio $\frac{C_d}{D}$ from Table 17.12.
4. Assume a lubricant oil from Table 17.7, and its operating temperature, the range of which has already been mentioned. From Fig. 17.8, find the value of Z corresponding to the operating temperature assumed for the selected oil.
5. Find $\frac{Zn}{p}$ and check this value with corresponding values in Table 17.12 to determine the possibility of maintaining fluid film lubrication.
6. Determine the coefficient of friction, from equations: (17.11), (17.12), (17.13) or Table 17.13.
7. Determine the heat generated by equation (17.14).
8. Determine the heat dissipated by equation (17.15) or equation (17.17).
9. Determine the thermal equilibrium to see that heat dissipated becomes atleast equal to the heat generated.

(*viii*) Mechanical Design: A sliding bearing consists of two basic elements:

(1) Housing (2) Liner or shell which is the working element of the bearing. The simplest one piece bearing or solid bearing will be with a hole bored directly in the boss of a bed or frame elementary machine. Such a construction is used only for slow speed machines. The next efficient bearings are those which are made separately and then bolted to the frame. The bearing may be without a liner. The liner is a bush made of C.I., bronze or some other antifriction material and is pressed into the hole. The liner may be coated with a thin layer of a antifriction alloy, e.g., babbitt, etc. The bearing housing is made of C.I. Fig. 17.14. Such a bearing is usually called " Bush Bearing". Such bearings are used only for slow speed shafts and small unit pressures. Also there is no provision for wear compensation and the shaft will have to be assembled into it axially. To overcome these difficulties, split bearings, Fig. 17.15 are used. It allows the shaft to be inserted perpendicular to the axis and the clearance to be adjusted as the bushing wears out. For this, thin shims are inserted between the bearing halves.

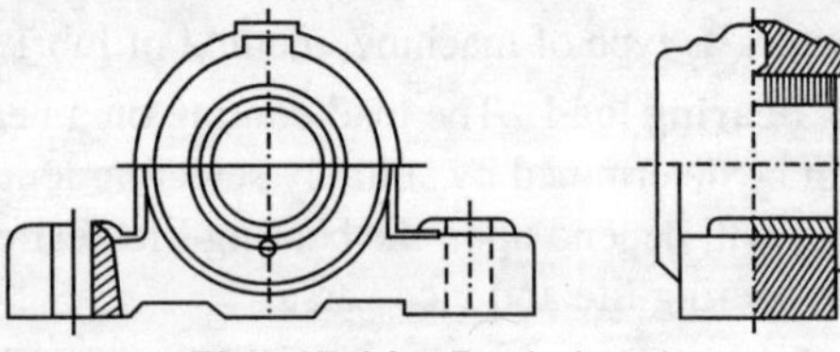

Fig. 17.14 *Bush bearing*

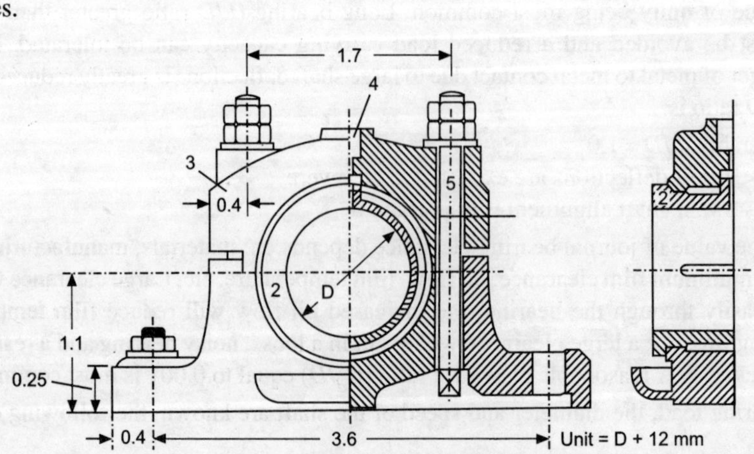

Fig. 17.15 *Split-bearing*

When the bottom wears out, it is turned true cylindrically and again the bearing is assembled by removing one or two shims and then moving down the upper half by tightening the bolts. The bushing is locked axially by shoulders and its rotation is prevented by providing a snug on its bottom which will go into a hole in the body or by giving square, hexagonal, or octagonal shapes to its bottom surface, i.e., outer surface. A sleeve bearing with a cap and other details is shown in Fig. 17.16.

The thickness of Bearing shells can be taken as given below :

(i) Small brass and bronze shells,
$t = 0.08D + 2.5$ mm

(ii) Medium size bearings (about 75 mm dia),
$t = 0.2\,D$ for shells of *C.I.*
$= 0.02\,D + 3.2$ mm for shells of babbitt

(iii) Engine main bearings (225 mm and above),
$t = 0.15\,D$ for shells of *C.S.*
$= 0.01\,D + 3.2$ mm for shells of babbitt.

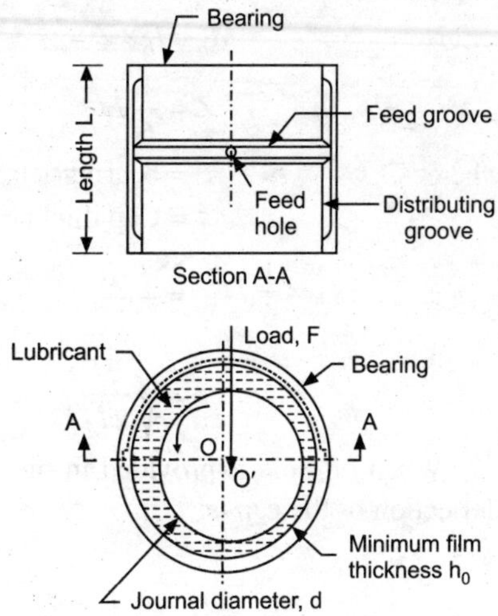

Fig. 17.16 *Sleeve bearing*

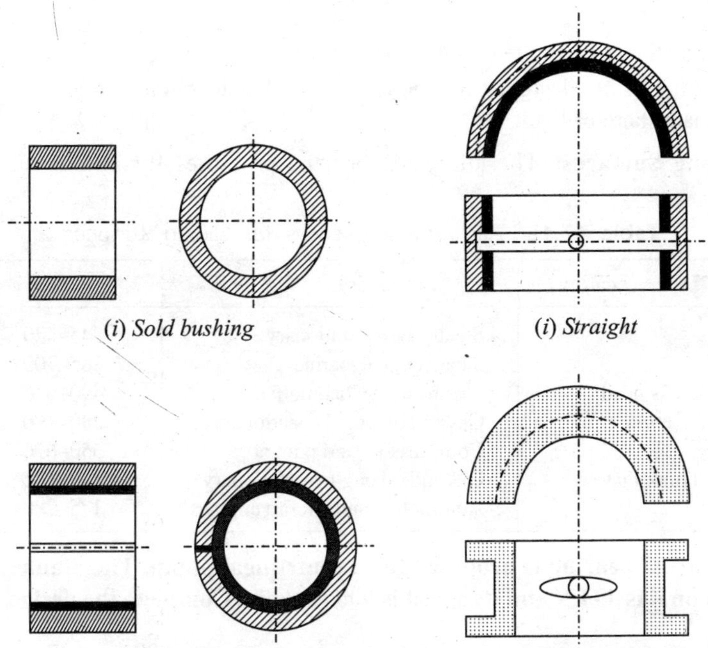

(*i*) *Sold bushing* (*i*) *Straight*

(*ii*) *Lined bushing* (*ii*) *Flanged*

Fig. 17.17 *Bearings*

(*ix*) **Bearing Cap:** Bearing cap is used when a split bearing is used. Ordinarily the load is directed against the bearing and not the cap. In some cases, the load must be carried by the cap, e.g., when split connecting rod ends are used in double acting engines, the full load comes on the cap. However, in all the cases, it is customary to design the cap and the holding down bolts on the basis of full load.

The cap is designed as beam loaded at the centre and supported at the bolt centres, Fig. 17.18. The maximum bending moment will be at the centre of cap, and is given as:

$$M_b = \frac{F \cdot a}{4}$$

where a = distance between bolt centres.

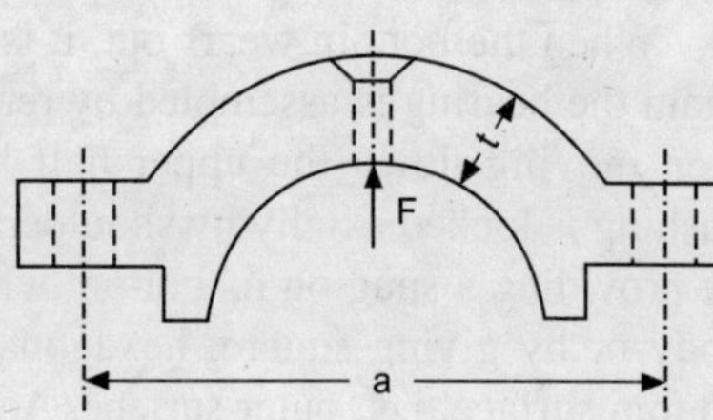

Fig. 17.18 *Bearing cap*

Now $M_b = f_b \times Z$

$$Z = \frac{1}{6} Lt^2$$

where L = bearing length

t = Cap thickness,

$$\therefore \quad f_b \times \frac{1}{6} Lt^2 = \frac{F \cdot a}{4}$$

$$\therefore \quad t = \sqrt{\frac{3Fa}{2Lf_b}} \qquad \text{...(17.19)}$$

When oil hole is provided in the cap, its diameter should be subtracted from the bearing length. The deflection of the cap is;

$$y = \frac{1}{48} \cdot \frac{Fa^3}{EI} = \frac{Fa^3}{4ELt^3}$$

or

$$t = 0 \cdot 63\, a \sqrt[3]{\frac{F}{ELy}} \qquad \text{...(17.20)}$$

The cap deflection should be limited to about 0.025 mm.

The holding down bolts are designed in simple tension. The load on each bolt is usually taken as 4/3 times their proportionate share of load.

***(x)* Design of Sliding Surfaces:** The allowable bearing pressures for sliding surfaces are given in Table 17.15.

Table 17.15 Allowable pressures for sliding surfaces

Type of bearing	Machinery	p, kPa
Cross head	Steam engine stationary	245–420
	Steam engine marine	385–700
	Steam engine locomotive	490–630
	Gas and oil engine, stationary	280–490
	Compressors and pumps	350–630
Trunk Pistons	Gas and oil engines, stationary	140–175
	Automotive and aircraft engines	175–280

Example 17.1. A journal bearing is proposed for a centrifugal pump. The diameter of the journal is 0.15 m and the load on it is 40 kN and its speed is 900 rev/min. Complete the design calculation for the bearing.

Solution:

1. Diameter of the journal, $D = 0.15$ m

From Table 17.12, $\frac{L}{D}$ ratio for pumps varies from 1 to 2.

Let us take $\frac{L}{D} = 1.6$

$\therefore$ Length of journal, $L = 1.6 \times 0.15 = 0.24$ m

2. Bearing Pressure, $p = \frac{F}{LD} = \frac{40 \times 1000}{0.24 \times 0.15} = 1.11$ MPa

From Table 17.12, the pressure is within the usual limits of 0.7 to 1.4 MPa for pumps.

3. Selecting turbine oil of ring oiled bearings from Table 17.7. Let us assume the operative temperature of oil film as 55°C. From Fig.17.8, the absolute viscosity of oil at this temperature is:

$$Z = 0.018 \text{ Ns/m}^2$$

$$\therefore \quad \frac{Zn}{p} = \frac{0.018 \times 900}{1.11} = 14.59$$

From Table 17.12, the operating value of $\frac{Zn}{p} = 28.50$.

As already indicated, the minimum value of bearing modulus at which the oil film will break is given as:

$$3C = \frac{Zn}{p} = 28.50$$

$\therefore$ C = Bearing modulus at the minimum point of friction
$= 9.50$

Since the calculated value of bearing modulus 14.59 is more than 9.50, it is clear from Fig. 17.11, that the bearing will operate under hydrodynamic conditions.

4. Now coefficient of friction is given as :

$$f = \frac{0.326}{10^6} \cdot \left(\frac{Zn}{p}\right)\left(\frac{D}{C_d}\right) + k_f, p \text{ is in MN/m}^2$$

Now from Table 17.12, $\frac{C_d}{D} = 0.001$.

Now from Fig. 17.13, for $\frac{L}{D} = 1.6$, $k_f \cong 0.0024$

$$\therefore \quad f = \frac{0.326}{10^6} \times 14.59 \times 10^3 + 0.0024 = 0.0072$$

5. Heat generated $= f.F.V.$ watts

$$V = \frac{\pi D n}{60} = \frac{\pi \times 0.15 \times 900}{60} = 7.07 \text{ m/s}$$

$\therefore$ $H_g = 0.0072 \times 40 \times 10^3 \times 7.07$
$= 2036.16$ W

6. Heat dissipated
$H_d = k_d LD\,(t_b - t_a)$ watts
Now from Table 17.14, let us take, $k_d = 1228.13$
Now from equation (17.16)

$$t_b - t_a = \frac{1}{2}(t_o - t_a) = \frac{1}{2}(55 - 15.5) = 19.75°\text{C}$$

$$\therefore \quad H_d = 1228.1 \times 0.24 \times 0.15 \times 19.75 = 873.20 \text{ W}$$

Since the heat dissipated is less than the heat generated, the bearing should be redesigned by making changes in the assumed values for oil temperature, viscosity and clearance ratio, or the bearing should be cooled artificially by water.

Let us assume $t_o = 65°\text{C}$
From Fig. 17.8, $Z = 0.0124 \text{ Ns/m}^2$

$$\therefore \quad \frac{Zn}{p} = \frac{0.0124 \times 900}{1.11} = 10.054 > 9.50 \therefore \text{O.K.}$$

Now let us take the correction factor $k_f = 0.002$, and $\frac{C_d}{D} = 0.0015$

$$\therefore \quad f = \frac{0.326}{10^6} \times 10.054 \times \frac{10^4}{15} + 0.002 = 0.00412$$

$$\therefore \quad H_g = 0.0042 \times 40 \times 10^3 \times 7.07 = 1165.14 \text{ W}$$

$$\text{Now} \quad t_b - t_a = \frac{1}{2}(65 - 15.5) = 24.75°\text{C}$$

$$\therefore \quad H_d = 1228.13 \times 0.24 \times 0.15 \times 24.75$$

$$= 1094.26 \text{ W, therefore, not much difference.}$$

Example 17.2. Design a bearing and journal to support a load of 4500 N at 600 rev/min using a hardened steel journal and a bronze backed babbitt bearing. The bearing is lubricated by oil rings. Take room temperature as 21°C and the oil temperature as 80°C.

Solution:

1. Let us assume the oil as light transmission oil corresponding to curve 2, in Fig. 17.8.

$\therefore$ $Z = 0.01$ kg/ms.

2. From Table 17.11, the minimum bearing modulus for the given journal and bearing is 2.85.

Therefore, operating $\frac{Zn}{p}$ is given as:

$$\frac{Zn}{p} = 3 \times 2.85 = 8.55$$

$\therefore$ $p = 0.702$ MPa

3. Required Projected area, $L \times D = \frac{F}{p} = \frac{4500}{0.702 \times 10^6} = 0.00641 \text{ m}^2$

Let us take $L = D$

$\therefore$ $D^2 = 0.00641$ $\therefore$ $D = 80$ mm

$\therefore$ $L = 80$ mm

Actual bearing pressure, $p = \frac{4500}{0.08 \times 0.08} = 0.703$ MPa

4. Now the permissible pressure for film lubrication is given as:

$$p = \frac{Zn}{4.75 \times 10^6} \cdot \left(\frac{D}{C_d}\right)^2 \left(\frac{L}{D+L}\right)$$

Let us take $\frac{C_d}{D} = 0.001$

$$\therefore \quad p = \frac{0.01 \times 600}{4.75 \times 10^6} \times 10^6 \times \frac{0.08}{0.08 + 0.08} = 0.6316 \text{ MPa}$$

which is less than the actual working pressure, therefore, the bearing will work under film lubrication.

5. Now $f = \frac{0.326}{10^6} \times \frac{Zn}{p} \times \frac{D}{C_d} + k_f = \frac{0.326}{10^6} \times \frac{0.01 \times 600}{0.703} \times 1000 + 0.002 = 0.004815$

6. Heat generated.

$H_g = fFV$ watts

$V = \pi \times \frac{80}{1000} \times \frac{600}{60} = 2.51$ m/s $\quad \therefore \quad H_g = 0.004815 \times 4500 \times 2.51 = 54.39$ W

7. Heat dissipated,

$$H_d = k_d.D.L\,(t_b - t_a)$$

Now $$t_b - t_a = \frac{1}{2}(80 - 21) = 29.5°C$$

Let us take $k_d = 390.77$ for still air

$\therefore$ $$H_d = 390.77 \times 0.08 \times 0.08 \times 29.5 = 73.78 \text{ W}$$

Since $H_d > H_g$ $\therefore$ Design is safe.

Example 17.3. A 49.917 mm diameter hardened and ground steel journal rotates uniformly at 1800 rev/min in a lathe turned stationary bronze journal bearing which is 50 mm in length. The effective temperature of the light transmission oil given by curve 2 in Fig. 17.8, is 65.5°C. Determine:

(*a*) Maximum radial load that the journal can carry.

(*b*) Power loss in the bearing.

Solution: (*a*) The radial load can be found out by using the Sommerfeld number graph.

1. Now diametral clearance, $C_d = D - d = 50.00 - 49.917 = 0.083$ mm

$\therefore$ $$\frac{C_d}{D} = \frac{0.083}{50} = 0.00166$$

2. Now minimum oil film thickness is given as :

$$h_0 \geq C_c.C_p({}^{rms}B + {}^{rms}J).$$

Now from Table 17.9,

C_p for Journal = 5.0, and for bearing = 4.5

From Table 17.10,

${}^{rms}B = 1.2\ \mu m$; and ${}^{rms}J = 0.15\ \mu m$

Let us assume that the oil is kept clean, therefore, $C_c = 1$

$\therefore$ $$h_o \geq (5 \times 0.15 + 4.5 \times 1.2)\ \mu m$$
$$\geq 6.15\ \mu m$$
$$\geq 6.15 \times 10^{-6} \times 10^3 \text{ mm}$$
$$\geq 0.00615 \text{ mm}$$

$\therefore$ $$h_o = 0.00615 \text{ mm}$$

$$\frac{h_o}{C_r} = \frac{0.00615}{0.0415} = 0.148$$

Now from Fig. 17.12.

$$\frac{h_o}{C_r} = 0.148$$

Sommerfeld number is 0.055

$\therefore$ $$\frac{Zn'}{p} \cdot \left(\frac{D}{C_d}\right)^2 = 0.055$$

Now from Fig. 17.8, $Z = 0.0175$ kg/ms

$\therefore$ $$\frac{0.0175 \times 30}{p} \cdot \left(\frac{1}{0.00166}\right)^2 = 0.055$$

$\therefore$ $$p = 3.464 \times 10^6 \text{ N/m}^2$$

$\therefore$ Maximum load on Journal, $F_{max} = pLD = 3.464 \times 10^6 \times 0.025 \times 0.05 = 4.33$ kN (*b*) Now $$f = \frac{0.326}{10^6}.\frac{Zn}{p}.\frac{D}{C_d} + k_f$$

Now from Fig. 17.13, for $\frac{L}{D} = 0.5, k_f = 0.005$

$$\therefore \quad f = \frac{0.326}{10^6}.\frac{0.0175 \times 1800}{3.464}.\frac{50}{0.083} + 0.005 = 0.00679$$

Now $$V = \pi \times \frac{50}{1000} \times \frac{1800}{60} = 4.712 \text{ m/s}$$

$$\therefore \quad \text{Power loss} = f.F.V = 0.00679 \times 4.33 \times 4.712 = 0.1385 \text{ kW}$$

Example 17.4. A full journal bearing appearing operates with the following data :

Length of bearing = 0.15 m

Diameter of bearing = 0.10 m

Radial load on bearing = 22.70 kN

Speed of Journal = 250 rev/min

Radial clearance = 0.045 mm

It is desired to limit the minimum oil film thickness to 0.025 mm. Determine a suitable value for the viscosity of the lubricant and the power loss due to friction.

The Sommerfeld equation and the relation for finding the coefficient of friction are given below:

$$\frac{Zn'}{p}.\left(\frac{R}{C_r}\right)^2 = \frac{0.0823(2 + \epsilon^2\sqrt{1-\epsilon^2})}{\pi^2 \epsilon}$$

and $$f = \frac{0.326}{10^6} \times \left(\frac{Zn}{p}\right) \times \left(\frac{D}{C_d}\right) + 0.002$$ **where ϵ = eccentricity ratio.**

Solution: (*a*) Now eccentricity ratio $\epsilon = 1 - \frac{h_o}{C_r}$

Now
$$h_o = 0.025 \text{ mm}$$
$$C_r = 0.045 \text{ mm}$$

$$\therefore \quad \epsilon = 1 - \frac{0.025}{0.045} = 1 - 0.55 = 0.45$$

$$R = 0.05 \text{ m}$$

$$p = \frac{F}{L \times D} = \frac{22.7 \times 1000}{0.15 \times 0.10} = 1.513 \text{ MP}a$$

Now L.H.S. of the Sommerfeld equation:

$$\frac{Zn'}{p}.\left(\frac{R}{C_r}\right)^2 = \frac{Z \times 250}{1.513 \times 10^6 \times 60}.\left(\frac{50}{0.045}\right)^2 = 3.4\,Z$$

and R.H.S. is:

$$\frac{0.0823\left(2 + 0.202\sqrt{1 - 0.202}\right)}{\pi^2 \times 0.45} = 0.03645$$

$$\therefore \quad 3.4\,Z = 0.03645$$

$$\therefore \quad Z = \frac{0.03645}{3.4} = 0.01072 \text{ kg/ms}$$

(*b*) Now
$$f = \frac{0.326}{10^6} \cdot \frac{Zn}{p} \cdot \left(\frac{D}{C_d}\right) + 0.002$$

$$= \frac{0.326}{10^6} \cdot \frac{0.01072 \times 250}{1.513} \cdot \left(\frac{100}{0.090}\right) + 0.002 = 0.00265$$

Now
$$V = \pi \times \frac{100}{1000} \times \frac{250}{60} = 1.308 \text{ m/s}$$

$\therefore$ Power loss due to friction = 0.00265 × 22.7 × 1000 × 1.308 = 78.68 W.

Example 17.5. A bearing cap is to withstand a load of 9000 N. Journal is 0.10 m in diameter by 0.25 m long. There are four holding bolts, two on each side of the shaft, placed on 0.20 m centres. Assume a factor of safety of 6 and that the cap is made of alloy cast iron with an ultimate strength of 292 MPa in tension. A 1 cm oil hole is provided. Determine the thickness of the bearing and the diameter of the bolts, assuming safe stress in bolts to be 70 MPa in tension. Check the deflection of cap taking E = 1.05 GN/m².

Solution: Now $F = 9000$ N

Cap: a = distance between bolt centres = 0.20 m

$\therefore$ Maximum bending moment on cap, $M_b = \frac{F \cdot a}{4} = \frac{9000 \times 0.20}{4} = 450$ Nm

Now $M_b = f_b \times Z$

$$f_b = \frac{292}{6} = 48.66 \text{ MPa}$$

$$Z = \frac{1}{6} Lt^2, \text{ now } L = \text{length of bearing} = 0.25 \text{ m}$$

$$\therefore \quad 48.66 \times 10^6 \frac{1}{6} \times 0.25 \times t^2 = 450$$

$$\therefore \quad t = 0.0149 \text{ m say } 0.015 \text{ m}$$

Bolts: Load on each bolt $= \frac{1.33F}{4} = \frac{1.33 \times 9000}{4} = 3000$ N

$$\therefore \quad \frac{\pi}{4} d_c^2 \times f_t = 3000; f_t = 70 \text{ N/mm}^2$$

$$\therefore \quad d_c = 7.4 \text{ mm}$$

$$\therefore \quad d = \frac{7.4 \text{ mm}}{0.8} = 9.25 \text{ mm say } 9.5 \text{ mm}$$

Deflection of Cap: It is given as:

$$y = \frac{1}{48} \cdot \frac{Fa^3}{EI} = \frac{Fa^3}{4ELt^3}$$

$$y = \frac{9000 \times (0.20)^3}{4 \times 1.05 \times 10^9 \times 0.25 \times (0.015)^3} = 0.203 \text{ mm.}$$

which is more than the limited value of 0.025 mm, let us take t = 30 mm

$$\therefore \quad y = \frac{9000 \times (0.20)^3}{4 \times 1.05 \times 10^9 \times (0.03)^3 \times 0.25} = 0.0254 \text{ mm which is } \textbf{O.K.}$$

Note: Do the calculations by considering the oil hole.

Example 17.6. A water cooled bearing having the following dimensions has operated at 600 rev/min under 9000 N load with automobile oil for a number of years without any trouble. The viscosity of this oil at the oil temperature of 50°C is 0.050 kg/ms.

Journal diameter = 50 mm

Bearing length = 80 mm

Radial clearance = 0.025 mm

It is proposed to use bearings of this design in several new machines which are under design.

(*a*) What changes do you recommend if a new machine operates at 300 rev/min ?

(*b*) What changes do you recommend if a new machine operates under 18 kN ?

(*c*) What changes should be made in oil if bearing is given radial clearance of 0.015 mm?

(*d*) If a new machine uses turbine oil whose viscosity at the oil operating temperature of 50°C is 0.0225 kg/ms, what changes do you recommend?

Solution:

(*a*) Bearing modulus under both the conditions should be the same, therefore,

$$\frac{Z_1 n_1}{p_1} = \frac{Z_2 n_2}{p_2}$$

Since oil is the same, $\therefore \quad Z_1 = Z_2$

$$\therefore \quad \frac{n_1}{p_1} = \frac{n_2}{p_2}$$

where $$p_1 = \frac{F}{LD} = \frac{9000}{80 \times 50} = 2.25 \text{ N/mm}^2$$

$$\therefore \quad \frac{600}{2.25} = \frac{300}{p_2}$$

$$\therefore \quad p_2 = \frac{300}{600} \times 2.25 = 1.125 \text{ N/mm}^2$$

(*b*) By the same above criteria,

$$\frac{n_1}{p_1} = \frac{n_2}{p_2}$$

now $$p_2 = \frac{F_2}{LD} = \frac{18 \times 1000}{80 \times 50} = 4.5 \text{ N/mm}^2$$

$$\therefore \quad \frac{600}{2{\cdot}25} = \frac{n_2}{4{\cdot}5}$$

$$\therefore \quad n_2 = \frac{4{\cdot}5}{2{\cdot}25} \times 600 = 1200 \text{ rev/min}$$

(*c*) Since radial clearance is changed, the Sommerfeld number should be compared:

$$\therefore \quad \frac{Z_1 n_1'}{p_1} \cdot \left(\frac{R}{C_{r_1}}\right)^2 = \frac{Z_2 n_2'}{p_2} \cdot \left(\frac{R}{C_{r_2}}\right)^2$$

since $$p_1 = p_2 \text{ and } n_1' = n_2'$$

$$\therefore \quad Z_1 \cdot \left(\frac{R}{C_{r_1}}\right)^2 = Z_2 \cdot \left(\frac{R}{C_{r_2}}\right)^2$$

$$C_{r1} = 0.025 \text{ mm} ; C_{r2} = 0.015 \text{ mm} ; Z_1 = 0.050 \text{ kg/ms}$$

$$\therefore \quad 0.050\cdot\left(\frac{25}{0\cdot025}\right)^2 = Z_2\cdot\left(\frac{25}{0.015}\right)^2$$

$$\therefore \quad Z_2 = 0.018 \text{ kg/ms}$$

(*d*) By the above criteria,

$$Z_1\cdot\left(\frac{R}{C_{r_1}}\right)^2 = Z_2\cdot\left(\frac{R}{C_{r_2}}\right)^2$$

now

$$Z_1 = 0.050 \text{ kg/ms}$$
$$C_{r1} = 0.025 \text{ mm}$$

$$\therefore \quad 0\cdot050\cdot\left(\frac{25}{0\cdot025}\right)^2 = 0\cdot0225\cdot\left(\frac{25}{C_{r_2}}\right)^2$$

$$\therefore \quad C_{r2} = 0.0168 \text{ mm}$$

Example 17.7. A journal bearing 75 mm long supports a load of 7500 N on a 50 mm diameter journal turning at 750 rev/min. The diameter clearance is 0.0693 mm. Determine the viscosity of the oil if the operating temperature of the bearing surface is to be limited to 77°C when in still air at 21°C.

Solution: By equation (17.11),

$$f = \frac{0\cdot326}{10^6}\left(\frac{Zn}{p}\right)\left(\frac{D}{C_d}\right) + 0\cdot002 = \frac{0\cdot326}{10^6}\left(\frac{Z\times750}{p}\right)\left(\frac{D}{C_d}\right) + 0\cdot002$$

now

$$p = \frac{F}{L\times D} = \frac{7500}{0\cdot075\times0\cdot05} = 2.0 \text{ MPa}$$

$$D = 50 \text{ mm};\ C_d = 0.0693 \text{ mm}$$

$$\therefore \quad f = \frac{0\cdot326}{10^6}\times\frac{Z\times750}{2\cdot0}\times\frac{50}{0\cdot0693} + 0\cdot002 = 0.\,0882\,Z + 0.002$$

Now heat generated, $H_g = f.F.V.$ watts

$$V = \text{rubbing velocity in m/s}$$

$$= \frac{\pi Dn}{60\times1000} = \frac{\pi 50\times750}{1000\times60} = 1.963 \text{ m/s}$$

$$\therefore \quad H_g = 7500\times1.963\,(0.0882\,Z + 0.002) \text{ watts}$$

Now heat dissipated is,

$$H_d = k_d DL\,(t_b - t_a)$$

From Table 17.14, let $k_d = 209.34$ W/m^2 °C

$$H_d = 209.34\times0.05\times0.075\times(77-21) = 43.96 \text{ W}$$

$$H_g = H_d$$

$$\therefore \quad 7500\times1.963\,(0.0882\,Z + 0.002) = 43.96$$

From here $\quad Z = 0.01118$ kg/ms.

Example 17.8. A journal bearing 0.15 m diameter and 0.225 m long is supporting a load of 9000 N at 1200 rev/min. If the radial clearance is 0.075 mm and the bearing wastes 1.5 kW in friction, what is the oil viscosity at the operating temperature?

Solution: Now $H_g = f.F.V.$ watts

$$F = 9000 \text{ N}$$

$$V = \frac{\pi \times 0{\cdot}15 \times 1200}{60} = 3\pi \text{ m/s}$$

$$f = \frac{0{\cdot}326}{10^6}\left(\frac{Zn}{p}\right)\left(\frac{D}{C_d}\right) + 0{\cdot}002$$

Now $$p = \frac{9000}{0{\cdot}15 \times 0{\cdot}225} = 0.267 \text{ MPa}$$

$$C_d = 2 \times 0.075 = 0.15 \text{ mm}$$

$$\therefore \quad f = \frac{0{\cdot}326}{10^6}\left(\frac{Z \times 1200}{0{\cdot}267}\right)\left(\frac{150}{0{\cdot}15}\right) + 0{\cdot}002 = 1.465\,Z + 0.002$$

$\therefore$ $(1.465\,Z + 0.002) \times 9000 \times 3\pi = 1.5 \times 1000$

From here,

$\therefore$ $Z = 0.0107$ kg/ms.

Example 17.9. A full journal bearing of 0.10 m diameter by 0.15 m long supports a radial load of 5500 N. The shaft speed is 500 rev/ min. The room temperature is 32°C and the surface of the bearing is to be limited to 63°C. Select a suitable oil to satisfy the above requirements, if the bearing is well ventilated and no artificial cooling is to be used. Assume $D/C_d = 1000$.

Solution: First of all the viscosity of the oil should be found out by equating heat generated to heat dissipated.

The coefficient of friction is

$$f = \frac{0{\cdot}326}{10^6}\left(\frac{Zn}{p}\right)\left(\frac{D}{C_d}\right) + 0{\cdot}002$$

Now $$p = \frac{5500}{0.10 \times 0.15} = 0.367 \text{ MN/m}^2$$

$$\therefore \quad f = \frac{0.326}{10^6}\left(\frac{Z \times 500}{0.367}\right)\left(\frac{1000}{1}\right) + 0.002 = 0.444\,Z + 0.002$$

$\therefore$ Heat generated, $H_g = f.F.V.$ watts

Now $$V = \frac{\pi \times 100 \times 500}{1000 \times 60} = 2.618 \text{ m/s}$$

$\therefore$ $H_g = (0.444Z + 0.002)\,(5500)\,(2.618)$ watts

Now heat dissipated is, $H_d = k_d DL\,(t_b - t_a)$ watts

From Table 17.14, let $k_d = 628$ W/m^2 °C for well ventilated bearing.

$\therefore$ $H_d = 628 \times 0.10 \times 0.15 \times (63 - 32) = 292.02$ watts

$H_g = H_d$

$\therefore$ $(0{\cdot}444Z + 0.002) \times 5500 \times 2{\cdot}618 = 292{\cdot}02$

$\therefore$ $Z = 0\cdot0412$ kg/ms

Now from equation 17.16.

$$t_b = t_a + \frac{1}{2}(t_o - t_a) \quad \therefore \quad \frac{1}{2}(t_o - t_a) = t_b - t_a = 63 - 32 = 31$$

$\therefore$ $(t_o - t_a) = 62$

$\therefore$ $t_o = 62 + 32 = 94$°C

$\therefore$ From Fig. 17.8, for Z = 0.0412 kg/ms oil temperature of 94°C, the suitable oil may be selected.

Example 17.10. A full journal bearing of diameter 0.075 m and a length of 0.125 m is to support a load of 20 × 10^3 N at the shaft speed of 1000 rev/min. The bearing temperature is to be limited to 77°C in a room temperature of 38°C. The viscosity of the oil used is 0.0098 kg/ms at 116°C. Find the amount of artificial cooling required, by means of an external oil cooler.

Solution : The heat generated and heat dissipated are to be found out.

Now, the coefficient of friction is,

$$f = \frac{0.326}{10^6}\left(\frac{Zn}{p}\right)\left(\frac{D}{C_d}\right) + 0.002$$

Now $\quad t_b - t_a = \frac{1}{2}(t_o - t_a)$

∴ $\quad t_o - t_a = 2\,(t_b - t_a) = 2 \times (77 - 38) = 78$

∴ $\quad t_o = 78 + 38 = 116°C$

∴ $\quad Z = 0.0098$ kg/ms

Now $\quad p = \frac{20 \times 10^3}{0.075 \times 0.125} = 2.13$ MN/m^2; let $\frac{D}{C_d} = 1000$

∴ $\quad f = \frac{0.326}{10^6} \times \frac{0.0098 \times 1000}{2.13} \times 1000 + 0.002$

$= 0.0015 + 0.002 = 0.0035$

∴ $\quad V = \frac{\pi \times 0.075 \times 1000}{60} = 3.927$ m/s

∴ $\quad H_g = 0.0035 \times 20 \times 10^3 \times 3.927 = 274.90$ watts

Now $\quad H_d = k_d \,\mathrm{DL}\,(t_b - t_a)$ watts

From table 17.14, let $k_d = 209.34$ W/m^2°C

$H_d = 209.34 \times 0.075 \times 0.125 \times 39 = 76.54$ watts

∴ Amount of artificial cooling needed $= H_g - H_d$

$= 274.90 - 76.54$

$= 198.36$ W.

Example 17.11. A bearing 0.05 m in diameter and 0.075 m in length supports a shaft running at 900 rev/min. The room temperature at 32°C and the bearing temperature is 82°C. The viscosity of the oil used is 0.0128 kg/ms at the operating temperature of 132°C. The diameteral clearance is 0.05 mm and the bearing is to operate in still air without any artificial cooling. Determine.

(*a*) The permissible load on the bearing.

(*b*) Power loss.

Solution: The coefficient of friction is,

$$f = \frac{0.326}{10^6}\left(\frac{Zn}{p}\right)\left(\frac{D}{C_d}\right) + 0.002$$

Now $\quad p = \frac{F}{L \times D} = \frac{F}{0.05 \times 0.075} = 266.67$ F N/m^2

∴ $\quad f = \frac{0.326}{10^6} \times \frac{0.0128 \times 900}{266{\cdot}67F \times 10^{-6}} \times \frac{50}{0.05} + 0.002 = \frac{14{\cdot}08}{F} + 0.002$

Now $\quad V = \frac{\pi \times 0.05 \times 900}{60} = 2.356$ m/s

$$\therefore \qquad H_g = \left(\frac{14.8}{F} + 0.002\right) \times F \times 2.356 \text{ watts}$$

From Table 17.14, Let $k_d = 209.34\ W/m^2°C$

$$\therefore \qquad H_d = 209.34 \times 0.05 \times 0.075 \times (82 - 32) = 39.25\text{W}$$

Equating $\qquad H_g = H_d$

$$F \times 2.356\left(\frac{14.08}{F} + 0.002\right) = 39.25$$

From here,

∴ (*a*) $\qquad F = 1289.69$ N

∴ (*b*) $\qquad$ Power loss $= f.\ F.\ V$ watts

Now $\qquad f = \dfrac{14\cdot08}{1289\cdot69} + 0.002 = 0.0109 + 0.002 = 0.0129$

∴ $\qquad$ Power loss = 0.0129 × 1289.69 × 2.356 = **39.25 W.**

(*xi*) Applications and limitations of Sliding Bearing: The use of sliding bearings in modern mechanical engineering is less frequent as compared to rolling contact bearings. However, their application is preferred with extra heavy shafts for which there are no suitable ball or rolling bearings, and in situations involving impact and vibration loads in cases when split bearings are required, e.g., *crankshaft*, etc. Sliding bearings have the following limitations:

1. Comparatively high friction losses.
2. Great length of bearing shells, reaching equal to 2 to 3 times the diameter.
3. Excessive consumption of lubricants.
4. Necessity of constant supervision.
5. Use in many cases of scarce materials, e.g., bronze, babbitt.

17.5 THRUST BEARINGS

As already indicated, thrust bearings are used to guide or support the shafts which are subjected to a load applied in the direction of the axis of the shaft, e.g., the thrust bearing carry the load in vertical steam turbines, water turbines, motors and pumps; and provide horizontal load reactions in machine tools and marine drive shafts, i.e., propeller shafts. Thrust bearings are classified as below:

1. **Step Bearing :** Such bearings are employed to support a loaded shaft in a vertical position.
2. **Collar Bearing :** These thrust bearings are for horizontal shafts. Thrust bearings are also classified according to the state of lubrication as:

(*a*) Marginally lubricated bearings.

(*b*) Hydrodynamically lubricated bearings i.e. externally pressurized bearings.

17.5.1 Marginally Lubricated Thrust Bearings

1. Step or Pivot Bearing: A simple form of step bearing is shown in Fig.17.19. A hardened steel disc supports the shaft. Now since the rubbing velocity is proportional to the radius of the shaft, the wear, which is proportional to the rubbing velocity, will be maximum at the outer radius and minimum at the centre. Due to this, the central portion will be left higher resulting in concentration of load there, which will lead to overheating and failure of the lubricant and the bearing.

The above trouble is overcome by:

(*a*) Counterboring the end of the shaft for a short distance to compensate on the end wear, Fig. 17.19.

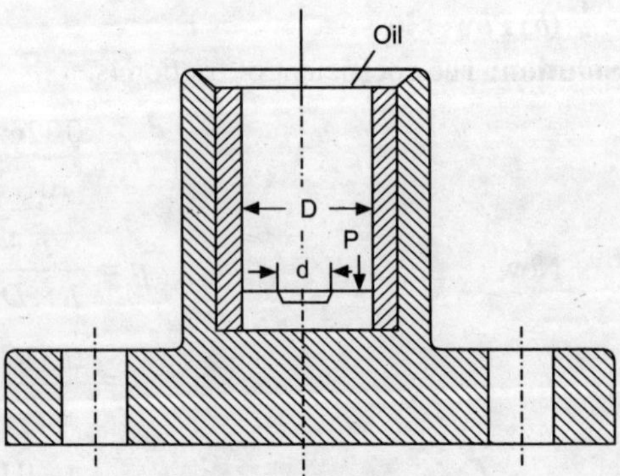

Fig. 17.19

(b) Supporting the shaft on a pile of disks instead of one disk. Each disk has a relative motion with respect to each other. This will reduce and distribute the wear. It is a usual practice to make alternate disks of unlike material, e.g., steel and bronze. Even if one pair of disks sticks due to improper lubrication, no harm will be done to the bearing as the other disks will be free to turn.

Materials: The bearing surface is usually made of C.I., bronze or steel with babbitt lining. It may either be fitted into position or pinned or keyed to prevent rotation. The lateral wear from the shaft is taken by bronze bushing.

Design Calculation. The total axial load on the shaft will be given as:

$$F_a = p \times A = p \times \frac{\pi}{4} D^2 \text{ for plain bearings}$$

$$= p \times \frac{\pi}{4}(D^2 - d^2) \text{ for bearings relieved at the centre.}$$

If uniform pressure distribution is assumed, the friction torsional moment referred to the mean diameter will be given as:

$$T_f = \frac{f.F_a.D_m}{2} = f.\,F_a.\,R_m$$

Now effective mean radius "R_m" at which the force acts, is given as:

$$R_m = \frac{2}{3}\left(\frac{R^3 - r^3}{R^2 - r^2}\right)$$

where $R = \frac{D}{2}$ and $r = \frac{d}{2}$

$$\therefore \quad T_f = f \cdot F_a \cdot \frac{2}{3}\left(\frac{R^3 - r^3}{R^2 - r^2}\right) \quad ...(17.21)$$

For plain bearing, $r = 0$

$$\therefore \quad T_f = f \cdot \frac{2}{3} F_a \cdot R \quad ...(17.22)$$

$\therefore$ Power lost in friction,

$$= T_f.\,\omega = T_f \cdot \frac{\pi n}{30} \text{ watts} \quad ...(17.23)$$

The value of 'f' for well lubricated step bearings may be taken as 0.015 to 0.020.

2. Collar Bearings: A simple multicollar thrust bearing for horizontal shafts is shown in Fig. 17.20. The bearing pressure will thus be distributed over a number of collars. The total axial load on the bearing will be given as:

$$F_a = p.z.\frac{\pi}{4}\left(D^2 - d^2\right)$$

where z = number of collars

D = outer diameter of collar

d = Inner diameter of collar

$\frac{D}{d}$ is usually from 1.4 to 1.8.

Width of collar is usually taken as $\frac{d}{6}$ and

Clearance between the collars as $\frac{d}{3}$.

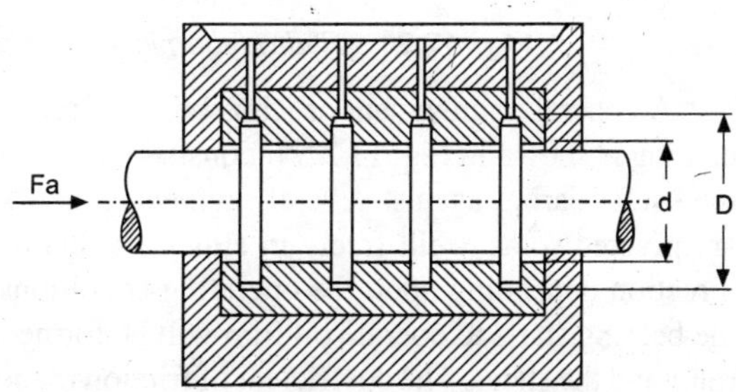

Fig. 17.20 *Collar bearing*

The collars may either be cut integral with shaft or held rigidity to it.

Frictional torsional moment will be given as

$$T_f = f \cdot F_a \cdot \frac{D_m}{2}$$

now
$$D_m = \frac{D + d}{2}$$

$$\therefore \quad T_f = \frac{f \cdot F_a \cdot (D + d)}{4} \quad \text{...(17.24)}$$

$$\therefore \quad \text{Power lost in friction} = T_f \cdot \frac{\pi n}{30} \text{ watts}$$

The coefficient of friction is taken as 0.03 to 0.05.

Allowable Bearing pressures: For rubbing velocities ranging from 0.25 to 1.0 m/s, the allowable bearing pressure may be taken as : $pV \le 700$, for step bearings, single collar bearings and water cooled multicollar thrust bearings. p is in kPa and V in m/s.

For step bearings, for speeds over 1 m/s the pressure may be as high as 0.7 MPa (Common range is 0.35 to 0.42 MN/m^2). However, for very low speeds; the pressure may be as high as 14 MPa and for intermittent service pressure upto 10.5 MPa be used. Common unit loadings for collar bearings range from 0.35 to 0.56 MPa.

17.5.2 Hydrodynamic Thrust Bearings

In these type of bearings, instead of marginal lubrication, the bearing surfaces are separated by a film lubricant. For this, two designs are in use:

1. Tapered Land Bearing. Figure 17.21(*a*). The bearing consists of fixed pads or sectors on a stationary flat plate. The pads are inclined planes sloping in one circumferential direction. A flat rotor slides against the pad plate. A short flat area is usually provided for the thrust rotor to seat when the bearing is stationary. In this construction, a wedge shaped hydrodynamic supporting oil film is formed at each tapered pad.

2. Tilting Pad Bearing: It is a tapered land bearing with a self adjusting taper angle for an optimum operation over a wide range of operating conditions. For shafts having rotation only in one direction, the ratio a/b is about 0.8 for optimum performance. For a shaft having rotation reversals, the pivot is placed at the centre of the pad.

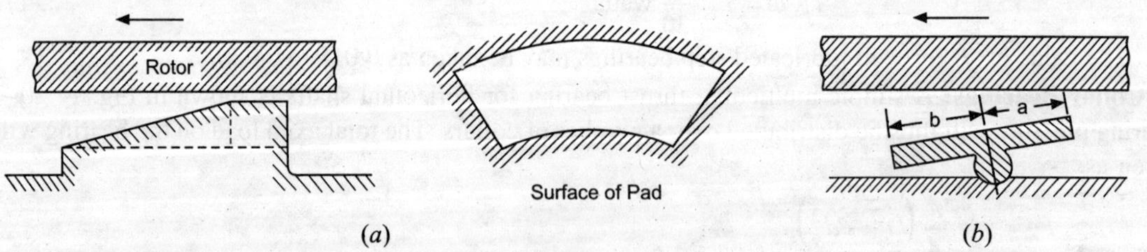

Fig. 17.21 *Hydrodynamic thrust bearings (a) Tapered land, (b) Tilting pad*

A general arrangement of Kingsbury hydrodynamics thrust bearing is shown in Fig. 17.22. It consists of a collar attached to the shaft bearing against several segmental shoes. These shoes are pivoted so as to tilt freely in either a radial or a tangential direction depending upon the operating conditions. As soon as the bearing starts to operate oil film will be formed between the rotor and the shoes. The coefficient of friction ranges from 0.001 to 0.003. Unit loadings upto 7.0 MPa may be used.

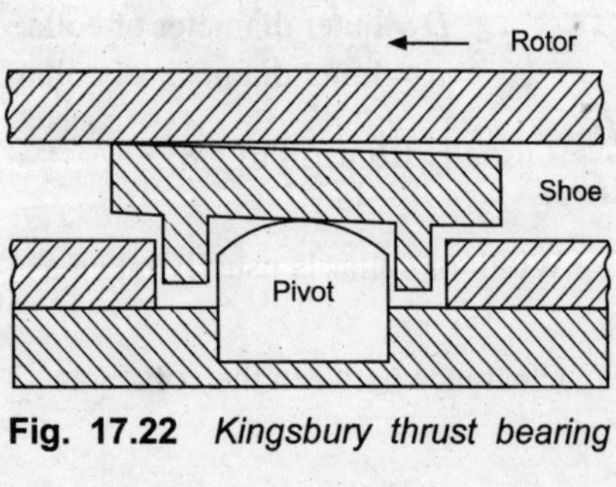

Fig. 17.22 *Kingsbury thrust bearing*

Example 17.12. Determine the number of thrust collars required to sustain a load of 36 kN on a shaft 0.15 m diameter if the allowable bearing pressure is 0.35 MPa. If the coefficient of friction is 0.03 and the shaft revolves at 200 rev/min, determine the power absorbed in friction work.

Solution: Shaft diameter $d = 0.15\text{m}$

Collar diameter, $D = (1.4 \text{ to } 1.8)\, d = 1.6\, d = 0.24 \text{ m}$

Now total axial load, $F_a = p.z.\frac{\pi}{4}(D^2 - d^2)$

and, $p = 0.35 \text{ MPa}$

$$\therefore \quad 36 \times 1000 = 0.35 \times 10^6 \times z \times \frac{\pi}{4}(0.24^2 - 0.15^2), \text{ gives } z = 3.75, \text{ say } 4$$

Now frictional torsional moment $T_f = \frac{f \cdot F_a \cdot (D+d)}{4} \text{ Nm} = 0.03 \times 36 \times 1000 \times \frac{0.39}{4} = 105.3 \text{ Nm}$

$\therefore$ Power absorbed in friction work $= 105.3 \times \frac{\pi \times 200}{30} = 2.2054 \text{ kW}.$

Example 17.13. Determine the main dimensions of a multicollar thrust bearing for a propeller shaft of a 450 kW power marine oil engine. The engine makes 220 rev/min, the shaft diameter is 0.15 m and the propeller has a pitch of 2.50 m. Assume a slip of 25%.

Solution: Pitch of the propeller is 2.50 m, it means that for each revolution of engine, the ship should move forward by 2.5 m but since the slip is 25%, therefore, the actual distance moved

$$= 2 \cdot 50 - \frac{25}{100} \times 2 \cdot 50 = 1.875 \text{ m for each revolution of engine.}$$

or

$$= 1.875 \times \frac{220}{60} = 6.875 \text{ m/s}$$

Now since total axial force × speed = work done

$$\therefore \quad F_a = \frac{450 \times 1000}{6 \cdot 875} = 65.50 \text{ kN}$$

Now $F_a = p.z.\frac{\pi}{4}(D^2 - d^2)$

$d = 0.15 \text{ m}$

Let us take $D = 1.6\, d = 0.24 \text{ m}$

Now mean rubbing velocity $V = \frac{\pi D_m . n}{60} \text{ m/s}$

Now, $D_m = \frac{D + d}{2} = \frac{0.24 + 0.15}{2} = 0.195 \text{ m}$

$$\therefore \quad V = \frac{\pi \times 0 \cdot 195 \times 220}{60} = 2.24 \text{ m/s}$$

Since it is more than 1 m/s, therefore, the bearing pressure should not exceed 0.7 MPa.

Let us take $p = 0.5 \text{ MPa}$

$$\therefore \quad 65.50 \times 10^3 = 0 \cdot 5 \times 10^6 \times z \times \frac{\pi}{4}\left(0 \cdot 24^2 - 0.15^2\right)$$

$$\therefore \quad z = 4.76, \text{ say 5 collars.}$$

Friction torsional moment $T_f = f \cdot F_a \cdot \frac{(D+d)}{4}$

Let $f = 0.03$

$\therefore$ $$T_f = 0.03 \times 65.50 \times 1000 \times \frac{0.39}{4} = 191.5 \text{ Nm}$$

$\therefore$ Loss of Power $= T_f \cdot w$ watts

$$= \frac{191.5 \times \pi \times 220}{30} = 4410 \text{ W}$$

$$= 4.41 \text{ kW}.$$

Example 17.14. Calculate the dimensions of a bearing disk of a step bearing. Load supported is 20 kN. Also find the horse power lost in friction, if the average speed of shaft is 750 rev/min. Take central hole diameter as one-fourth of the outer diameter : Assume p = 0.4 MN/m² and f = 0.015.

Solution: Refer Fig. 17.19,

$$d/D = 0.25$$

Now $$F_a = p \times \frac{\pi}{4}\left(D^2 - d^2\right)$$

$\therefore$ $$20 \times 1000 = 0.4 \times 10^6 \times \frac{\pi}{4}[D^2 - (0.25D)^2]$$

$\therefore$ $$D = 0.261 \text{ m}$$

$\therefore$ $$d = 0.06525 \text{ m}$$

Now $$T_f = f \cdot F_a \cdot \left(\frac{R^3 - r^3}{R^2 - r^2}\right) \cdot \frac{2}{3}$$

$$R = \frac{D}{2} = 0.1305 \text{ m}$$

$$r = \frac{d}{2} = 0.032625 \text{ m}$$

$\therefore$ $$T_f = 0.015 \times 20 \times 1000 \times \frac{2}{3} \times 0{\cdot}1364 = 27.28 \text{ Nm}$$

$\therefore$ Power lost in friction $= \frac{27{\cdot}28 \times 750 \times \pi}{30} = 2.142$ kW.

17.6 ROLLING CONTACT BEARINGS

In these bearings the contact between the bearing elements is rolling instead of sliding as in plain bearings. Since the rolling friction is very less as compared to the sliding friction, such bearings are also known as "Antifriction bearings". But this is misnomer because; although at starting, the friction in a plain bearing may be more than that in rolling contact bearings, but when the condition of full hydrodynamic lubrication is attained in a plain bearing, the running friction may be quite less as compared to rolling contact bearings. The variation of friction with speed of shaft for the sliding contact and the rolling contact bearings, is shown in Fig. 17.23.

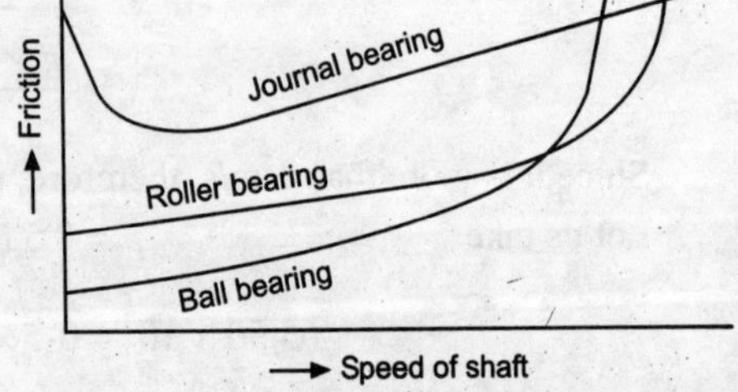

Fig. 17.23 *Comparison of friction of bearing*

Advantages:

1. Low starting and low running friction except at very high speeds.
2. Both radial as well as thrust loads can be carried by the bearings.
3. Momentary overloads can be carried without failure.
4. Shaft alignment is more accurate than in sliding bearings.
5. Easy replacement.
6. Dependability in service.
7. Simple lubrication.
8. Wear is practically nil.

Disadvantages:

1. More noisy especially at high speed.
2. Sensitive to dirt, etc.
3. Lower resistance to shock loading.
4. Initial cost is more.
5. Complicated design of bearing housing.

Classification: The rolling element bearings are classified into the following two main categories:

1. Ball Bearing: In these bearings, the rolling element is a spherical ball.

2. Roller Bearings: In these bearings, the rolling element is a roller, which might be cylindrical, conical, spherical, or concave.

Depending upon the load to be carried, the rolling contact bearings may be classified as :

1. Radial Bearings: In radial contact bearings, the load is primarily radial with only low magnitude thrust loads.

2. Angular Contact Bearings: These bearings are used where loads are combined radial and high thrust, and where accurate shaft location is required. These bearings are also called "Radial-thrust" bearings carrying both radial and axial loads.

3. Thrust Bearings: These bearings handle loads which are primarily thrust.

In accordance with their load-carrying capacity, bearings are classified as: light, 'Medium and 'Heavy' series, differing in dimensions, their inside diameter being the same.

From the viewpoint of the numbers of rows of the rolling elements, there are: 'Single-row' bearings, 'Double-row' bearings and 'Multi-row' bearings.

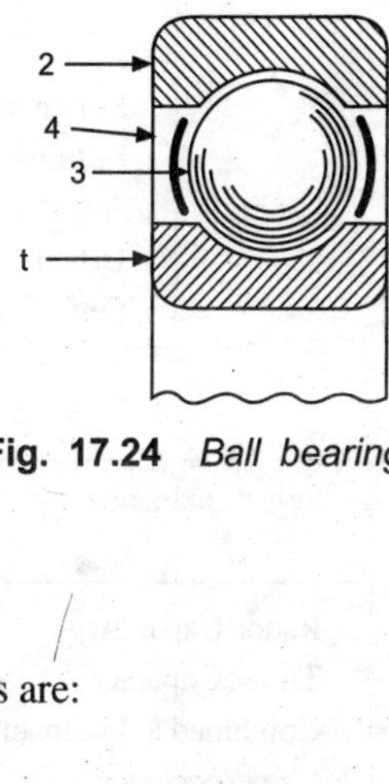

Fig. 17.24 *Ball bearings*

17.6.1 Ball Bearings

A ball bearing consists of four elements as shown in Fig. 17.24. The four elements are:

(*i*) Inner race 1, which is pressed or pushed or pressed on to the shaft.
(*ii*) Outer race 2, which is secured in the bearing housing.
(*iii*) Rolling element, i.e., balls, 3 which roll over the race tracks or raceways.
(*iv*) Separator, cage, or ball retainer, 4, for spacing or separating the balls so that they do not touch each other to reduce wear and noise.

Both the rolling elements and the races are made of high-quality hardened steel.

17.6.1.1 Radial Ball Bearings

The various types of radial ball bearings are shown in Fig. 17.25.

(*a*) Single row bearing: This bearing is usually made with deep groove. This is the most widely used type of ball bearing.

(*b*) Filling Notch Bearing: This is also known as **maximum capacity bearing.** Its geometry is identical to that of a deep groove bearing except for a filling notch. In this bearing, the ball complement is increased, thereby increasing the bearing load capacity.

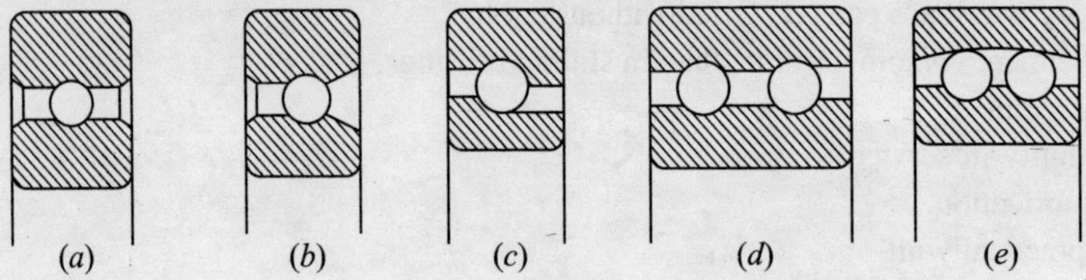

Fig. 17.25 *Radial ball bearing*

(*c*) Angular contact bearing: These bearings take an axial load in addition to the radial load.

(*d*) Double row bearing: By providing two rows of the balls, the load capacity of the bearing is increased. The bearings may also be multi-row bearings.

(*e*) Self-Aligning bearing: Such a bearing permits inclination of the inner race or shaft axis with relation to the axis of the outer race with 2–3°. There are two arrangements : internal and external. In internal arrangement, the inner surface of the outer race is ground as a spherical surface. In external arrangement, the outside diameter of the outer race is ground as a spherical surface which fits into a similar ground housing.

Since the single row deep groove bearing is the most popular type, its radial capacity is assigned a value of one. The relative capacity of other types of the same bore and outside diameter, is shown in Table 17.16.

A comparison guide for the basic ball bearing types is given in Table 17.17.

Table 17.16 Relative capacities of ball bearings

Type	Radial capacity	Thrust capacity
Single row deep groove	1.00	0.7 (2-direction)
Filling notch	1.2–1.4	0.2 (2-direction)
Angular contact	1.00–1.15	1.5–2.3 (1-direction)
Double row	1.5	1.5 (2-direction)
Self aligning	0.7	0.2 (1-direction)

Table 17.17 Comparison of various ball bearings

Selection Factor	Single row deep groove	Filling notch	Angular contact	Double Row	Self Aligning
Radial Capacity	G*	E	G	E	F
Thrust Capacity	F	F	E	G	P
Combined load capacity	G	F	E	G	G
Life expectancy	E	G	E	E	F
Operating Speeds	E	G	E	F	F
Compensation for misalignment	F	F	P	F	E
Grease Capacity	F	F	F	G	G
Sealing effectiveness	E	F	P	G	P
Shaft and Housing rigidity under moment loads	F	F	P	E	P

*E = excellent, G = good; F = fair, P = poor

Standard Dimensions and Designations of Ball Bearings: The various dimensions that have been standardised by International Standards are : Inside diameter, or bore, outside diameter, and width, as shown in Fig. 17.26. These standard dimensions are in mm. There is no standard for the number and size of steel balls. The standard ball bearings are available in the following five standard series, in accordance with their load carrying capacity:

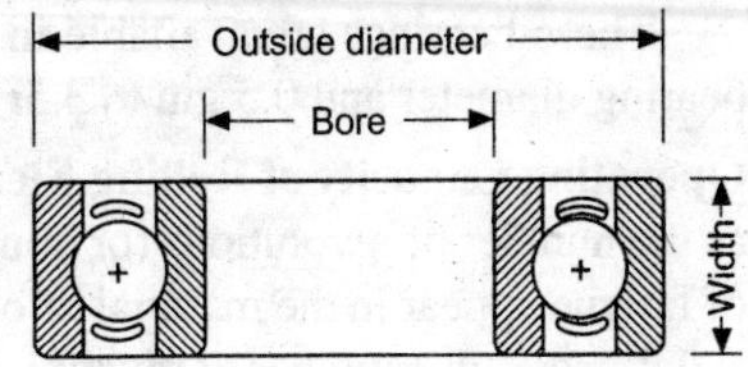

Fig. 17.26 *Standardised dimensions for ball bearings*

(*i*) Extremely light (1900), (*ii*) Extra light (100), (*iii*) Light (200), (*iv*) Medium (300), (*v*) Heavy (400).

Sometimes an extra heavy series (500) is also used. The bearings are designated by a number which consists of at least three digits. Additional digits are added to specify special features. The last three digits denote the series and the bore of the bearings. The last two digits from 04 onwards, when multiplied by 5, give the bore diameter in mm. Thus, if a bearing is designated by the No. 208, it is a bearing of light series with bore equal to 5 × 08, i.e., 40mm. The load capacity of a bearing of medium series is about 30 to 40 percent more than that of a light series bearing and the load capacity of heavy-series bearing is about 20 to 30 percent greater than those of the medium series bearings. The dimensions of some standard ball bearings are given in Table 17.18.

Table 17.18 Dimensions of Ball bearings

Bearing number	Bore	Outside diameter, mm			Width of Races, mm		
	mm	200 Series	300 Series	400 Series	200 Series	300 Series	400 Series
00	10	30	35		9	11	
01	12	32	37		10	12	
02	15	35	42		11	13	
03	17	40	47	62	12	14	
04	20	47	52	72	14	15	
05	25	52	62	80	15	17	21
06	30	62	72	90	16	19	23
07	35	72	80	100	17	21	25
08	40	80	90	110	18	23	27
09	45	85	100	120	19	25	29
10	50	90	110	130	20	27	31
11	55	100	120	140	21	29	33
12	60	110	130	150	22	31	35
13	65	120	140	160	23	33	37
14	70	125	150	180	24	35	42
15	75	130	160	190	25	37	45
16	80	140	170	200	26	39	48
17	85	150	180	210	28	41	52
18	90	160	190	225	30	43	54
19	95	170	200	240	32	45	
20	100	180	215	250	34	47	
21	105	190	225		36	49	
22	110	200	240		38	50	

These bearings are available in the range of 1 to 2600 mm in outside diameter, 0.35 to 203 mm in ball bearing diameter and 0.5 gm to 3.5t in mass.

Operating Capacity of Rolling Element Bearings. The life of an individual ball or roller bearing is defined as the number of revolutions (or hours at some given constant speed) which the bearing runs before the signs of fatigue appear in the material of one of the races or any of the rolling elements. The rating life of a bearing is defined as the number of revolutions (or hours at some given constant speed) during which not less than 90 per cent of a batch of bearings of the given type would operate without failure in similar working conditions. Even for the same batch of bearings operating under the same working conditions, the service life of the bearings varies within a wide range. This is due to different materials, heat treatment, surface finish and accuracy of the bearing elements. The rating life is commonly referred to as L_{10} life, B-10 life or minimum life.

Figure 17.27 shows a typical curve showing the number of bearings and the range of service life. This curve is obtained by testing a batch of apparently identical bearings under identical load, speed and operating conditions. It is clear from the figure that a wide scatter is obtained in the lives of the bearings. This is due to the factors discussed above. The graph shows the rating life also called the "90% survival life". Also shown is the 'average life' which is defined as the number of revolutions (or hours at some constant speed) that 50% of the group of ball bearings will complete or exceed and 50% will fail before the rated life is reached. It is clear from the graph that 'Average life' is appreciably greater than the 'rating life'. As presently determined, it is five times the 'rating life'.

It has been found experimentally that the life of a bearing is a function of the load. If one group of bearing of a definite type operates under constant load P_1 and have a life L_1, and another group, in similar conditions under the load P_2 has a life L_2 then the loads and lives are connected by the relation :

$$\frac{L_1}{L_2} = \left(\frac{P_2}{P_1}\right)^3, \text{ for ball bearings} \qquad ...(17.25)$$

i.e., life of bearing is inversely proportional to the third power of load.

If $L_2 = 1$ million revolutions of inner ring, P_2 can be written as equal to "C" which is known as "Basic load rating". It is defined as the radial load that a ball bearing can withstand for 1 million revolutions of the inner ring with only 10% failure. It is also known as "Specific dynamic capacity", "Basic dynamic capacity" or "Dynamic load rating". Hence the life of a bearing at any load 'P' can be written from equation 17.25 as

$$\text{Rating life, } L = \left(\frac{C}{P}\right)^3 \text{ millions of revolutions} = \left(\frac{C}{P}\right)^3 \times 10^6 \text{ revs.}$$

$$\therefore \quad P = C\left(\frac{1}{L}\right)^{\frac{1}{3}} \qquad ...(17.26)$$

If L = life in hours, and n = rev/min., then

$$L = \left(\frac{C}{P}\right)^3 \times \frac{10^6}{60n} \text{ hours}$$

$$P = C\left(\frac{10^6}{60nL}\right)^{\frac{1}{3}} \qquad ...(17.27)$$

The basic dynamic load rating for a bearing under variable load is given as:

$$C = \left(\frac{\Sigma L_i P_i^3}{10^6}\right)^{\frac{1}{3}} \qquad ...(17.28)$$

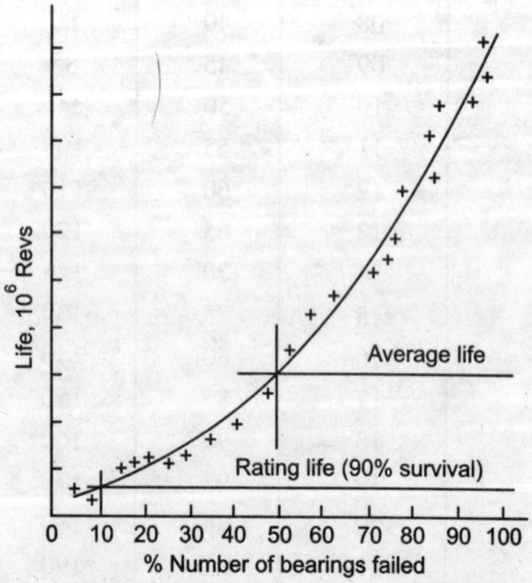

Fig. 17.27 *Service life of bearings*

The recommended bearing life for various types of machines is given in Table 17.19.

It is clear from the definition of Basic Load Rating that speed is not a consideration in its definition. The rating life of 1 million revolutions is a base value selected for each of calculations. The corresponding load rating is so high that the contacting surfaces would undergo plastic deformation, were it actually applied. Thus the "Basic Load Rating" is purely a reference figure, which would probably never be applied.

Calculations of Radial Ball Bearings

(*a*) Statically Loaded Bearings. The bearings which do not rotate or rotate at a very small speed ($n < 1$ rev/min) are called as statically loaded. An example of a 'stationary radial load', one that does not rotate, acts in a constant direction is the belt load. The limiting load of such bearings is determined by the magnitude of the permanent deformation of the contact surfaces, and not by the service life of the bearing parts. The earliest work on rolling element bearings was carried out by Hertz and Striebeck. The maximum load which would be carried by a ball is given by the formula:

$$P_0 = \frac{4.37F_r}{z} \quad ...(17.29)$$

where F_r = radial load on the bearings

z = number of balls in the bearing.

Table 17.19 Recommended bearing life

Type of Application	Life, hrs.
Instruments and apparatus for infrequent use.	Upto 500
Aircraft engines	500–2000
Machines for short or intermittent service where service interruption is of minor importance	4000–8000
Machines for intermittent service where reliable operation is of great importance	8000–14000
Machines for 8-hr. service which are not always fully utilized	14000–20000
Machines for continuous 24-hr. service	50000–60000
Machines of continuous 24-hr. service where reliability is of extreme importance	100000–200000

In this formula, it has been assumed that there is no radial clearance between balls and races. Radial clearance has a marked effect on the load distribution between balls and on the magnitude of P_0. For this reason, the above equation is modified as:

$$P_0 = \frac{5F_r}{z} \quad ...(17.30)$$

The basic static load rating "C_0" which corresponds to a total permanent deformation of ball and race, at the most heavily stressed contact point, equal to 0.0001 times the ball diameter, is given by the following formula:

$$C_0 = f_0 \cdot i \cdot z\,D^2 \cdot \cos\alpha \quad ...(17.31)$$

where i = number of rows of balls in any one bearing.

z = number of balls per row.

D = Ball diameter

α = Angle between the line of action of ball load and a plane normal to bearing axis, Fig. 17.27(a).

f_0 = a factor. In (N, m) units, its values are :

= 3.34×10^6 for self aligning ball bearings.

= 12.26×10^6 for radial contact and angular contact ball bearings.

If the bearing is under the combined radial and thrust loads, the static equivalent load P_0 shall be the greatest of those obtained from the following formulas:

$$P_0 = X_0 . F_r + Y_0 . F_a$$

and $P_0 = F_r$

where X_0 = a radial factor

Y_0 = a thrust factor

F_r = radial load

F_a = thrust load.

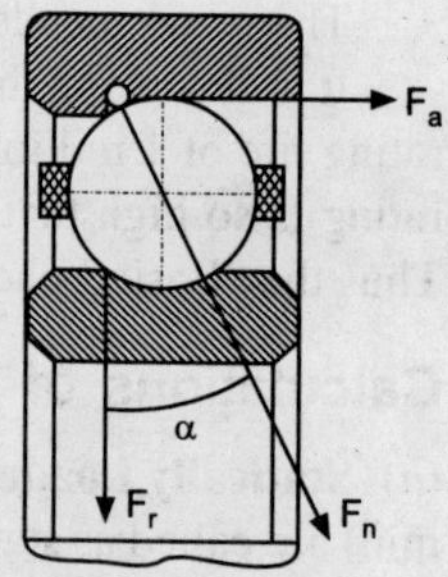

Fig. 17.27(a)

The values of X_0 and Y_0 are given in Table 17.20.

The equivalent load is defined as that hypothetical load which, if applied, will have the same effect on bearing life as the actual loads.

***(b)* Dynamically loaded Bearings.** The equivalent load for a dynamically loaded bearing under the combined action of radial and thrust loads is always referred as a radial load and is given by the formula:

$$P = XVF_r + YFa \qquad ...(17.32)$$

or $P = VF_r$, whichever is larger

Table 17.20 Value for X_0 and Y_0 [IS3823 (Part I) - 1966]

Bearing Type		Single row bearing		Double row bearing	
		X_0	Y_0	X_0	Y_0
Radial contact groove ball bearings		0.6	0.5	0.6	0.5
	α = 20°	0.5	0.42	1	0.84
Angular contact	= 25°	0.5	0.38	1	0.76
groove ball	= 30°	0.5	0.33	1	0.66
bearings	= 35°	0.5	0.29	1	0.58
	= 40°	0.5	0.26	1	0.52
Self aligning ball bearings		0.5	0.22 cot α	1	0.44 cot α

where V = Rotation factor

= 1.0 for all types of bearings if inner race is rotating

= 1.2 for all types of bearings except self aligning, if inner race is stationary.

= 1.0 for self aligning for rotation of either race

The values for the radial factor "X" and thrust factor "Y" for most of the bearings can be assumed as given below :

$$X = 1 : Y = 1.5$$

For angular contact bearings,

$$X = 0.50 ; Y = 1.0$$

The self aligning bearing :

$$X = 0.50 ; Y = 2.50.$$

***(c)* Selection of Radial Ball Bearings.** The design of a rolling element bearing concerns only the selection of the most suitable bearing from the catalogue of a manufacturer according to the procedure given in the catalogue.

The following procedure can be followed:

1. The equivalent load coming on the bearing is calculated by the help of equation (17.32). This equation does not account for any shock or impact forces and temperature conditions which a bearing may experience. Incorporating these two factors, the design load F, is calculated from the equivalent load, P, as follows:

$$F = P \times K_a \times K_T$$

where K_a = application factor or service factor, Table 17.21.

K_t = temperature factor, Table 17.22.

2. Now it is a usual practice with bearing manufacturers to specify the rated radial bearing load capacity to a certain speed in rev/min, and a certain life in hours (L_{10} or Average). For example, in Tables 17.23 and 17.27, the radial load ratings are at 10,000 hours of average life at 500 rev/min. The actual or design radial load, *F*, may have to be carried for a different life and at a different speed than given in the catalogues or Tables 17.23 and 17.27. It is, therefore, important that when comparing the rated capacity of bearings made by different manufacturers, capacities are all reduced to the same life expectancy and speed. We can find the required radial load rating, C_R, for these conditions with the help of equation (17.26) that is,

$$C_R = F \cdot \left(\frac{L_d}{L_c}\right)^{\frac{1}{3}} \left(\frac{n_d}{n_c}\right)^{\frac{1}{3}}$$

where L_d = Desired life of bearing, hrs.

L_c = Catalogue life of bearing, hrs.

n_d = Rotational speed of bearing, rev/min.

n_c = Catalogue rotational speed, rev/min.

or $C_R = F \times K_l \times K_s$

where $K_l = \text{life factor} = \left(\frac{L_d}{L_c}\right)^{\frac{1}{3}}$

$K_S = \text{Speed factor} = \left(\frac{n_d}{n_c}\right)^{\frac{1}{3}}$

$$\therefore \quad C_R = P \times K_a \times K_t \times K_l \times K_s \qquad \text{...(17.33)}$$

Table 17.21 Values for K_a

Type of Service	K_a	
	Ball Bearings	Roller Bearings
Uniform and Steady Load	1.0	1.0
Light shock load	1.5	1.0
Moderate shock load	2.0	1.3
Heavy shock load	2.5	1.7
Extreme shock load	3.0	2.0

Table 17.22 Values for K_t

$t°C$ =	125	150	175	200	225	250
K_t =	1.05	1.10	1.15	1.25	1.35	1.40

3. After determining the required radial load, the suitable bearing is selected from Table 17.23 so that the radial load capacity given in the table is more than the required radial load.

17.6.1.2 Thrust Ball Bearings

The three main types of thrust ball bearings are :

(*a*) Single direction thrust bearings with flat seats, Fig. 17.28 (*a*).
(*b*) One direction thrust bearings with grooved seats, Fig. 17.28 (*b*).
(*c*) Double direction thrust bearings with grooved seats, Fig. 17.28 (*c*).

(*a*) *One directional, flal race* (*b*) *One directional grooved race* (*c*) *Two directional grooved race*

Fig. 17.28 Thrust ball bearings

Assuming that only 80 percent of the balls take the axial load, the force acting on each ball will be given as:

$$P_0 = \frac{F_a}{0.8\,z} \qquad \text{...(17.34)}$$

The basic static load rating of the thrust ball bearing is given by the following formula :

$$C_{0a} = f_{0a} \cdot z\,D^2 \cdot \sin\alpha \qquad \text{...(17.35)}$$

where f_{0a} = constant, 49.05×10^6 for units of (N, m)
z = number of balls carrying thrust in one direction
D = Ball diameter
α = normal angle of contact.

The static equivalent load for thrust ball bearings with contact angle $\alpha \neq 90°$ under combined radial and thrust loads, is given by the following formula :

$$P_{0a} = F_a + 2.3\,F_r\,\tan\alpha \qquad \text{...(17.36)}$$

For a pure thrust ball bearing under dynamic loading, in equation (17.32), the equivalent load is simply equal to F_a. For the selection of a thrust ball bearing, the same procedure can be adopted as for radial ball bearing. The rated thrust capacity may be taken as 75 percent of the radial capacity as given in Table 17.23.

17.6.2 Roller Bearings

Roller bearings have higher load capacities than ball bearings for a given overall size. They are widely used in moderate speed, heavy duty applications. The four principal types of roller bearings are :

(*a*) Cylindrical, (*b*) Needle, (*c*) Tapered, (*d*) Spherical; Fig 17.28A

In cylindrical roller bearings, the rolling elements are cylinders with approximate length-diameter ratios ranging from 1 : 1 to 1 : 3. In needle bearings, the rollers have length-diameter ratios greater than 2.

These bearings occupy smaller space and consist of large number of very small cylindrical rollers. They have the highest load capacity for a given radial space of all rolling bearings but their size is limited upto about 250 mm bore diameter. These bearings are used in automotive transmissions, small gasoline engines, gear pumps, small appliance and tool motors, alternators, helicopter rotors, and aircraft controls. Tapered roller bearings have the best combination of high radial and thrust capacities at low to moderate speeds. The rollers and raceways are truncated cones. These bearings are widely used in rollneck application in rolling mills, transmissions, gear reducers, geared shafting (particularly worms and spiral bevels), steering mechanisms, and machine tool spindles where high load capacity and resistance to shock loading are required. Spherical roller bearings have the self aligning feature. This is achieved by grinding the outer or inner raceway spherical.

Table 17.23 Typical Radial capacity of ball bearings (based on average life L_c of 10000 hrs and n_c 500 rev/min.)

Bearing No.	Bore mm	Radial-load capacity, kN															
		Deep-groove type				Filling slot type			Angular type			Two row spherical type			Two row angular type		
		Series				Series			Series			Series			Series		
		100	200	300	400	200	300	400	200	300	400	200	300	400	200	300	400
00	10	0.743	0.846	1.469	—	—	—	—	1.179	1.469	—	0.912	1.38	—	—	—	—
01	12	0.823	0.979	1.691	—	—	—	—	1.517	1.825	—	1.113	1.602	—	1.491	—	—
02	15	0.890	1.157	2.092	—	—	—	—	1.847	2.292	—	1.291	1.914	—	2.092	—	—
03	17	0.952	1.469	2.492	4.050	—	—	—	2.025	2.715	4.561	1.602	2.247	3.427	2.292	4.139	7.521
04	20	1.629	1.780	2.893	6.141	—	4.317	—	2.737	3.271	6.898	2.203	2.848	4.673	3.427	4.539	9.256
05	25	1.758	2.225	4.050	7.165	—	5.34	—	2.937	4.560	8.455	2.715	3.427	5.073	3.738	6.809	11.615
06	30	2.420	2.715	5.296	8.655	4.183	6.497	—	4.539	5.942	10.191	3.382	4.673	6.097	5.162	8.989	13.751
07	35	3.013	3.783	6.675	10.191	5.652	8.099	—	5.652	7.491	11.971	3.916	5.518	7.743	6.900	11.303	17.845
08	40	3.190	4.450	8.188	11.704	6.942	9.79	—	7.454	9.292	14.618	4.45	6.497	8.722	7.387	13.884	22.161
09	45	4.094	4.940	9.879	13.662	7.387	12.816	—	7.899	9.990	16.087	5.385	8.322	10.502	8.100	16.643	25.338
10	50	4.325	5.429	11.080	15.620	8.188	14.910	—	8.678	12.927	19.002	5.785	9.434	12.722	9.434	19.803	33.331
11	55	5.874	6.853	13.083	17.800	10.235	17.133	—	10.013	15.264	21.672	7.076	11.259	13.795	11.303	26.478	39.027
12	60	6.186	7.298	13.573	20.070	—	—	—	11.214	17.8	24.899	8.233	12.905	15.126	13.84	28.124	42.943
13	65	6.484	8.411	17.750	22.475	—	—	—	13.061	20.448	27.345	8.767	13.840	17.444	14.329	34.043	48.95
14	70	8.188	9.212	19.080	27.635	—	—	—	14.819	23.318	33.598	10.013	16.243	21.36	16.198	38.715	55.18
15	75	8.566	10.146	21.36	30.349	—	—	—	16.465	24.965	36.957	10.502	17.444	22.606	17.711	45.835	66.305
16	80	10.146	11.081	22.562	33.242	—	—	—	18.023	27.50	40.406	11.437	19.269	23.808	19.714	46.715	71.2
17	85	11.058	13.083	25.276	36.179	—	—	—	20.248	31.551	44.033	12.816	21.271	29.815	22.918	55.18	73.87
18	90	12.238	13.973	27.59	39.272	—	—	—	22.873	33.642	47.793	14.863	24.386	35.511	26.255	57.405	90.335
19	95	12.816	16.109	30.705	45.740	—	—	—	26.277	38.938	55.670	16.643	27.145	37.825	29.815	71.645	101.015
20	100	13.350	17.444	34.176	52.620	—	—	—	29.993	41.341	64.036	18.245	31.595	47.17	31.862	77.875	—
21	105	15.041	18.334	37.38	—	—	—	—	31.662	46.636	—	19.847	33.153	—	37.825	84.105	—
22	110	16.732	20.025	41.83	—	—	—	—	33.308	49.707	—	21.36	35.467	—	42.275	—	—

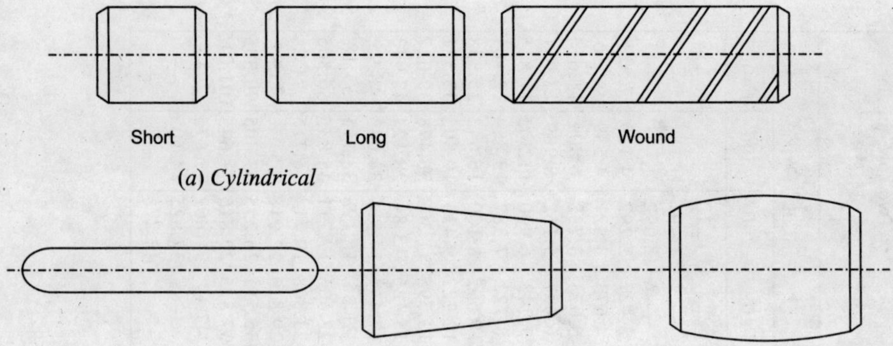

(b) *Needle* (c) *Tapered* (d) *Spherical (Barrel)*

Fig. 17.28 *(a) Roller bearings*

Bearings are made with either convex or concave rollers in both single and double row designs. These bearings are used where high load capacity, high tolerance to shock loads, and self aligning capability are needed. They are speed limited and are used in vibrators, shakers, conveyors, speed reducers, transmissions, and other heavy duty machinery. A relative comparison of these types of roller bearings is given in Table 17.24.

Table 17.24 Comparison of roller bearing types

Selection Factors	Cylindrical	Needle	Tapered	Spherical
Radial load capacity	E*	E	E	E
Thrust load capacity	P	P	G	F
Combined load capacity	P	P	E	F
Life expectancy	E	G	E	G
Operating speeds	G	F	G	F
Compensation for misalignment	F	P	F	E
Grease capacity	F	E	F	F
Sealing effectiveness	P	P	P	
Shaft and housing rigidity under moment loads	P	E	P	

*E = excellent; G = good; F = fair ; P = Poor

The approximate value of the rating life L for radial bearings is given by the following formula:

$$L = \left(\frac{C}{P}\right)^{\frac{10}{3}} \text{ million revolutions} \qquad ...(17.37)$$

The maximum static load which can be carried by a roller is given as:

$$P_0 = \frac{4.6 F_r}{z} \qquad ...(17.38)$$

The basic load rating for rolling bearings is given by the formula:

$$C_0 = f_0 \cdot i \cdot z\, l_{eff} D \cos \alpha \qquad ...(17.39)$$

where l_{eff} = effective length of contact between one roller and that ring where the contact is the shortest (overall roller length minus roller chamfers or minus grinding undercuts).

$$f_0 = 21.6 \times 10^6 \text{ for units of N, m)}$$

The static equivalent load P_0 for radial roller bearings under combined radial and thrust loads shall be the greatest magnitude of those obtained from the two following equations:

$$P_0 = X_0 . F_r + Y_0 . F_a \quad ...(17.40)$$

and

$$P_0 = F_r$$

The values of X_0 and Y_0 are given in Table 17.25.

Table 17.25 Values of factors, X_0 and Y_0 [IS 3823 (Part II)-1966]

Bearing Type	Single row bearing		Double row bearing	
	X_0	Y_0	X_0	Y_0
Self-aligning and tapered roller bearings. $\alpha \neq 0°$	0.5	0.22 cot α	1	0.44 cot α

The equivalent load P for self aligning and tapered roller bearing under combined constant radial and constant thrust loads, shall be given by the following formula:

$$P = XVF_r + YF_a \quad ...(17.41)$$

The values for V are the same as for radial ball bearings.

Values of factors X and Y are given in Table 17.26.

Table 17.26 Values of factors X and Y [IS 3824 (Part II)-1966]

Bearing Type	*X*				*Y*				*e*
	Single row Roller Bearing		*Double row Roller Bearing*		*Single row Roller Bearing*		*Double row Roller Bearing*		
	$\frac{F_a}{vF_r} \leq e$	$\frac{F_a}{vF_r} > e$	$\frac{F_a}{vF_r} \leq e$	$\frac{F_a}{vF_r} > e$	$\frac{F_a}{vF_r} \leq e$	$\frac{F_a}{vF_r} > e$	$\frac{F_a}{vF_r} \leq e$	$\frac{F_a}{vF_r} > e$	
Self aligning roller bearings and tapered roller bearings $\alpha \neq 0°$	1	0.4	1	0.67	0	0.4 cot α	0.45 cot α	0.67 cot α	1.5 tan α

For $\alpha = 0°$, $F_a = 0$ and X = 1.

In the Table 17.26 above, '*e*' is the axial load parameter. It is clear that axial force has no effect on the equivalent load '*P*', as long as the ratio F_a/VF_r does not exceed '*e*'.

Selection of Roller Bearings: The selection of a radial roller bearing is made in the same manner as for a radial ball bearing. The typical radial load capacity of straight cylindrical roller bearings is given in Table 17.27.

17.6.3 Comparison of Ball and Roller Bearings

The roller bearings have greater capacity for radial loads for a given overall size, due to the greater area of contact. Also they are more easily adopted to high speeds. On the other hand, they are costlier, more sensitive to misalignment and dirt. Also, they have considerably smaller axial thrust capacity.

17.6.4 Materials

The material for rolling element bearings is high quality steel, e.g., Special alloy steel, with Ni and either Cr or Mo. The material is given extraordinary and uniform hardness and toughness by heat treatment.

17.6.5 Lubrication

The functions of a lubricant in ball and roller bearings are:

1. To provide a separating film between rolling and sliding contacting surfaces and thus prevent wear.
2. To act as a coolant to maintain proper bearing temperature.
3. To prevent the bearing from being contaminated by dirt and other contaminants.
4. To prevent corrosion of bearing surfaces.

Table 17.27 Typical radial capacity of straight cylindrical roller bearings, based on average life of 10,000 hrs and speed 500 rev/min.

Bearing No.	Bore, mm	Radial Capacity, kN		
		200 Series	300 Series	400 Series
00	10	1.647	2.74	
01	12	1.847	3.54	
02	15	2.047	3.98	
03	17	2.648	4.85	9.078
04	20	3.315	5.30	11.926
05	25	3.983	7.832	15.00
06	30	5.874	10.90	18.468
07	35	6.853	11.837	22.25
08	40	8.188	14.33	26.70
09	45	9.568	18.02	31.462
10	50	10.903	22.16	36.668
11	55	11.615	23.99	42.186
12	60	12.282	26.70	48.105
13	65	14.329	31.15	54.290
14	70	16.109	33.375	60.743
15	75	18.023	42.765	67.507
16	80	20.248	45.835	74.627
17	85	24.920	56.115	82.014
18	90	26.300	60.120	89.801
19	95	30.038	64.169	97.722
20	100	33.375	70.844	106.088
21	105	36.668	80.946	
22	110	41.252	87.042	
24	120	54.112	99.46	
26	130	61.05	112.185	
28	140	71.734	132.61	

Vegetable oils should never be used for rolling element bearings, these tend to become acidic. Petroleum oils, greases are mostly used in these bearings. Because grease is less troublesome, it is most commonly used if the speed is not great. It does not leak from the housing as oils and also it provides a better seal for the exclusion of foreign matter. Too much grease should be avoided to prevent churning, resulting in increased bearing temperature. Oil is preferred at very high speeds. Oil should not be supplied in great quantities, because churning of oil generates excessive frictional heat resulting in hot bearings.

17.6.6 Manufacturing Processes

The balls are hot forged on hammers from steel rods. They are then heat treated, ground and polished. The races are also formed by forging and thus possess a fibre pattern. The races are then heat treated, ground and highly polished.

17.6.7 Mounting of Ball and Roller Bearings

The mounting of the rolling contact bearings on the shaft be done very accurately and with precision since an inaccurately mounted bearing may quickly fail. The type of mounting will depend upon the design of the shaft and bearing and the service conditions encountered. In majority of the cases, the shaft is the rotating member and the bearing housing is stationary. The type of the mounting will depend upon the following factors:

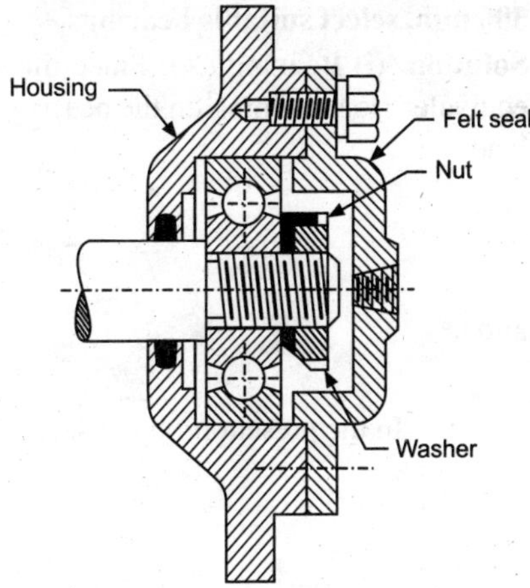

Fig. 17.29 *Bearing mounting*

1. Accuracy required, such as in precision spindle.
2. Speed, since fast speed requires rigid and accurate mounting.
3. The load requirements, since the greater the loading or shock, the better should be the mounting.
4. The desired life.
5. The economics of construction and the case of maintenance.

The usual and preferred method of mounting is shown in Fig. 17.29. The inner race is mounted on the rotating shaft with a press fit and it seats against a shoulder. It is secured with a nut threaded on the shaft with a washer in order to prevent the relative motion between the shaft and the bearing. The outer ring or race is mounted less tightly with a close push fit but is not loose in the housing.

The following rules should also be observed in the mounting of ball and rolling bearing:

(*i*) When the shaft is mounted on more than one bearing, then there should be only one locking bearing i.e., with no axial clearance between the outer race and the housing. In the case of the remaining bearings, the outer race is not secured axially and there is some axial clearance provided between the outer race and the housing. Such an arrangement is referred to as "floating support", Fig.17.30. This will allow the bearing to freely adjust itself against any expansion or contraction of the shaft due to temperature rise, inaccuracies in machining and shaft deflection.

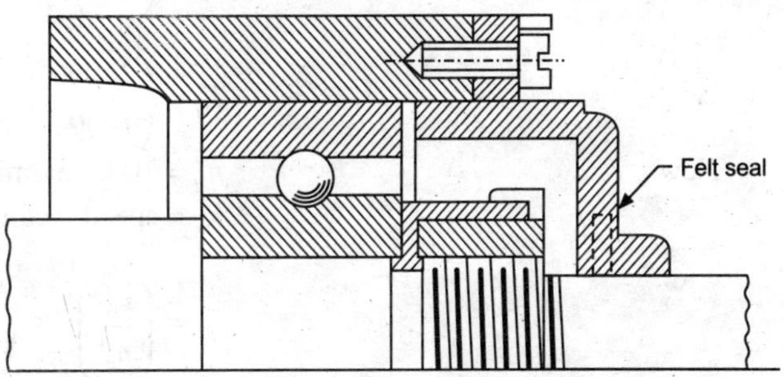

Fig. 17.30 *Floating ball bearing*

(*ii*) For the same reason, there should be only one thrust bearing of all the ball bearings mounted on the same shaft.

(*iii*) If the shaft is supported only on radial bearings, the thrust bearing should be the one that carries the minimum radial load.

(*iv*) In long line shafts, the bearing of one of the central sections is usually secured. In short transmissions, one of the most heavily loaded bearings is fixed.

(*v*) The floating supports are not used when the shaft is supported in radial thrust bearings, either ball bearings or tapered roller bearings.

(*vi*) The floating support can also be dispensed with short shaft supported on radial ball bearings.

(*vii*) When the bearing is housed in a split bearing, the nuts should be tightened carefully so that the housing cap does not deform the outer race in order to prevent the pinching of balls or rollers.

Example 17.15. The spindle of a wood working machine revolves at 1000 rev/min and it is to be mounted on two single row radial ball bearings. Bearing (X) is subjected to a radial load of 2250 N and a thrust load of 1900 N. Bearing (Y) is subjected to a radial load of 2250 N only. The machine is to be used approximately 8 hrs per day and an average service life of 10 yrs is desired. If the diameter of the spindle is not to exceed 300 mm, select suitable bearings.

Solution: (i) Bearing (X): Since this bearing is under the combined action of radial and thrust loads, the equivalent load coming on the bearing will have to be found out. A simplified formula to find the equivalent load, is:

$$P = \frac{F_r + 3F_a}{2} \text{ if } F_a > \frac{1}{3} F_r$$

and

$$P = \frac{F_r + 4F_a}{3} \text{ if } F_a > \frac{1}{2} F_r$$

∴ In the present problem since $F_a > \frac{1}{2} F_r$

$$\therefore \quad P = \frac{2250 + 4 \times 1900}{3} = 3283.30 \text{ N}$$

Now the desired radial load is given as :

$$C_R = P \times K_a \times K_l \times K_s \times K_t$$

Assuming moderate shock conditions, K_a will be equal to 2.0, Table 17.21.

Now desired life = 10 × 300 × 8 = 24000 hrs, assuming 300 working days per year.

$$\therefore \quad L_d = 24000 \text{ hrs.}$$

$$\text{Since} \quad L_c = 10000 \text{ hrs.}$$

$$\therefore \quad K_l = \text{life factor} = \left(\frac{L_d}{L_c}\right)^{\frac{1}{3}}$$

$$\therefore \quad K_l = \left(\frac{24000}{10000}\right)^{\frac{1}{3}} = 1.34$$

$$\text{Now} \quad n_d = 1000 \text{ rev/min and } n_c = 500 \text{ rev/min}$$

$$K_s = \text{speed factor} = \left(\frac{n_d}{n_c}\right)^{\frac{1}{3}} = \sqrt[3]{\frac{1000}{500}} = 1.26$$

Let K_t = temperature factor

= 1.0, if it is assumed that the temperature is within safe limits.

∴ Desired radial load = 3283.30 × 2 × 1.34 × 1.26 × 1 = **11.080 kN**

From Table 17.23. Bearing No. 310 deep groove type, may be selected with capacity of 11.080 kN

***(ii)* Bearing (Y):** For this bearing, the desired radial load will be

Desired radial load = 2250 × 2 × 1.34 × 1.26 × 1 = 7.58 *kN*

From Table 17.23. Bearing No. 308 deep groove type with radial capacity of 8.188 kN may be selected.

It is clear from above that the bore of bearing (X) will be 50 mm, and that of bearing (Y) 40 mm. For a well balanced design, the following bearings may be selected:

(X) Bearing No. 410, with radial capacity equal to 15.62 kN

(Y) Bearing No. 310, with radial capacity equal to 11.080 kN

Example 17.16. Determine the minimum value of the basic dynamic load rating for selecting ball bearings to give 5000 hrs of operations with not more than 10 percent failures. The radial load is 1800 N during 90 percent of the time and 7200 N during remaining 10 percent. The shaft is to rotate at 150 rev/min.

Solution: Since the bearing is under variable loads, the dynamic load rating is given by the following formula:

$$C = \left(\frac{\Sigma L_i \cdot P_i^3}{10^6}\right)^{\frac{1}{3}} = \left(\frac{L_1P_1^3 + L_2P_2^3}{10^6}\right)^{\frac{1}{3}}$$

Now $L_1 = 0.90 \times 150 \times 60 \times 5000 = 40.5 \times 10^6$ revolutions

$P_1 = 1800$ N

$L_2 = 0.10 \times 150 \times 60 \times 5000 = 4.5 \times 10^6$ revolutions

$P_2 = 7200$ N

$$C = \left[\frac{40.5\times10^6 \times 1800^3 + 4.5\times10^6 \times 7200^3}{10^6}\right]^{\frac{1}{3}} = 12.45 \text{ kN.}$$

Example 17.17. A bearing is required to carry 4500 N stationary radial load. The shaft rotates at 1000 rev/min and the life desired is 30000 hours. The running conditions are steady, no shock loading. Select a suitable bearing.

Solution :

(*i*) Ball bearing:

Desired Radial Load $C_R = P \times K_a \times K_l \times K_s \times K_t$

Now, $P = 4500$ N

K_a = application factor or service factor

= 1.0 from Table 17.21

$$K_l = \text{life factor} = \left(\frac{L_d}{L_c}\right)^{\frac{1}{3}}$$

Now L_d = Desired life of bearing = 30,000 hrs.

L_c = Catalogue life of bearing = 10,000 hrs.

∴ $K_l = \sqrt[3]{\dfrac{30000}{10000}} = 1.442$

K_s = Speed factor

$$= \sqrt[3]{\frac{n_d}{n_c}} = \sqrt[3]{\frac{1000}{500}} = 1.26$$

Let $K_t = 1.0$ for bearing running at normal temperature

∴ $C_R = 4500 \times 1.0 \times 1.442 \times 1.26 \times 1.0 = 8.176$ kN

From Table 17.23, A ball bearing of deep groove type with No. 308 is suitable whose radial load capacity is 8. 188 kN.

Example 17.18. A bearing is required for a 35 mm shaft. It is to operate for 8 hours per day, 5 days per week for 5 years and is to carry a stationary radial load of 2250 N at 1500 rev/min, inner ring rotating. There is a possibility of light shock. Select a suitable bearing.

Solution:

(*i*) Ball Bearing: Desired radial load $C_R = P \times K_a \times K_l \times K_s \times K_t$

P = equivalent load or radial load = 2250 N

$$K_a = 1.5, \text{ Table 17.21}$$

$$K_l = \sqrt[3]{\frac{L_d}{L_c}}$$

Now $$L_d = 8 \times 5 \times 52 \times 5 = 10400 \text{ hrs.}$$

∴ $$K_l = \sqrt[3]{\frac{10400}{10{,}000}} = 1.013$$

$$K_s = \left(\frac{n_d}{n_c}\right)^{\frac{1}{3}} = \sqrt[3]{\frac{1500}{500}} = 1.442$$

Let $K_t = 1.0$ for operation at normal temp.

∴ $$C_R = 2250 \times 1.5 \times 1.013 \times 1.442 \times 1.0 = 4970 \text{ N}$$

From Table 17.23. Bearing No. 307 (35 mm. bore) of deep groove type with radial load capacity of 6.675 kN and Bearing No. 207, of filling slot type with load capacity of 5.65 kN, are suitable. Since filling slot bearing is better than deep groove for radial loads, this should be preferred.

***(ii)* Roller Bearing:** From Table 17.27, Bearing No. 207 with load capacity of 6.853 kN will be quite suitable.

Example 17.19. Find the "rating life" of the 50 mm bore, light series, ball journal bearing under a 6800 N radial load at 600 rev/min. The load is out of balance and, therefore, rotates with the inner ring. There is no shock loading.

Solution: $$C_R = P \times K_a \times K_l \times K_s \times K_t$$

Now equivalent load P is given as,

$$P = XVF_r + YF_a$$

Here, $X = 1.0$, $F_a = 0$; V = rotational factor = 1.2

∴ $$P = 1.2 \times 6800 = 8160 \text{ N}$$

From Table 17.23 with bore of 50 mm light series bearing of filling slot type, i.e., bearing No. 210,

Radial load capacity = 8.188 kN

Now $K_a = 1.0$; $K_t = 1.0$

$$K_s = \sqrt[3]{\frac{600}{500}} = 1.062$$

∴ $$8.188 = 8.160 \times 1.0 \times K_l \times 1.062 \times 1.0$$

∴ $$K_l = \frac{8.188}{8.160 \times 1.062} = 0.962$$

Now $$K_l = \sqrt[3]{\frac{L_d}{L_c}}$$

∴ $$\frac{L_d}{L_c} = (0.962)^3 = 0.89$$

∴ $$L_d = 10{,}000 \times 0.89 = \mathbf{8900 \text{ hrs.}}$$

This is the average life at 500 rev/min. Therefore, rating life $L_{10} = \frac{8900}{5} = 1780$ hrs. at 500 rev/min.

Example 17.20. A radial ball bearing has a basic load rating of 50 kN. If the desired rating life of the bearing is 6000 hours, what equivalent radial load can the bearing carry at 500 rev/min.

Solution: Using the simple relation (17.27).

$$L = \frac{10^6}{60n}\left(\frac{C}{P}\right)^3$$

Here $n = 500$ rev/min, $L = 6000$ hours, $C = 50$ kN

$$\therefore \quad \left(\frac{C}{P}\right)^3 = \frac{6000 \times 60 \times 500}{10^6} = 180$$

$$\therefore \quad P = \frac{C}{(180)^{\frac{1}{3}}} = \frac{50}{5{\cdot}65} = \mathbf{8.85\ kN.}$$

Example 17.21. For a radial ball bearing, the desired rated life is 10000 hours for a speed of 600 rev/min and radial load of 5 kN. Find the basic load rating for the bearing.

Solution: Using the relation,

$$L = \frac{10^6}{60n}\left(\frac{C}{P}\right)^3$$

$n = 600$ rev/min, $L = 10000$ hours, $P = 5$ kN

$$\therefore \quad \left(\frac{C}{P}\right)^3 = \frac{10000 \times 60 \times 600}{10^6} = 360$$

$$\therefore \quad C = P{\cdot}(360)^{\frac{1}{3}} = 5 \times 7.12 = \mathbf{35.60\ kN.}$$

Example 17.22. A ball bearing is to withstand a radial load of 5 kN and have a rated life of 1500 hours at a speed of 600 rev/min. Determine the load rating by referring to Table 17.23.

Solution : Equation (17.33) can be written below (neglecting the service factor and temperature factor)

$$C_R = P{\cdot}\left[\left(\frac{L_d}{L_c}\right){\cdot}\left(\frac{n_d}{n_c}\right)\right]^{\frac{1}{3}}$$

Here L_d = desired life, 1500 hours, L_c = catalogue life, 10000 hours (Table 17.23)

n_d = 600 rev/min, n_c = catalogue rated speed, 500 rev/min (Table 17.23)

$$\therefore \quad C_R = 5{\cdot}\left[\frac{3}{20} \times 1.2\right]^{\frac{1}{3}} = 5 \times 0.565 = \mathbf{2.825\ kN.}$$

∴ From Table 17.23 deep groove type No. 107 will be suitable.

17.6.8 Selection of Rolling Contact Bearings on the Basis of Basic Load Rating

In this method, the life factor and the speed factors are not involved. Selection is made as per the following steps:

I. Determine the equivalent Radial load by using eqn. 17.32 for Ball bearings and eqnuation (17.41) for Roller bearing), that is,

$$P = (XVF_r + YF_a) \,.\, K_a \,.\, K_t$$

II. Determine the Basic Load Capacity by using eqn. 17.26, if rating life is in millions of revolutions and equation (17.27), if the rating life is in hours, for Ball bearings and for Roller bearings use equation (17.37).

III. Select the bearing by referring to Tables (Appendix II).

Note: For additional reading on Rolling Contact bearings, the readers can refer to the following Indian Codes:

1. Radial Ball bearings	:	IS : 3824 (Part I) 1966
2. Radial Roller bearings	:	IS : 3824 (Part II) 1966
3. Thrust Ball bearings	:	IS : 3824 (Part III) 1966
4. Thrust Roller bearings	:	IS : 3824 (Part IV) 1966

Example 17.23. Select a suitable ball bearing to carry a radial load of 10,000 N and an axial load of 4000 N. The shaft rotates at 1000 rpm. Average life is 5000 hours. Inner race rotates. Take mild shock.

Solution: Now $P = (XVF_r + YF_a) \times K_a$

Now $V = 1$ and for most of the bearings except angular contact bearings and self aligning bearings, let $X = 1$ and $Y = 1.5$

From Table 17.21, $K_a = 1.5$

$\therefore$ Equivalent Radial load, $P = (1 \times 1 \times 10000 + 1.5 \times 4000) \times 1.5 = 24$ kN.

Now Average life = 5000 hours

$$\therefore \quad \text{Rated life} = \frac{5000}{5} = 1000 \text{ hours}$$

Rated life in cycles = $1000 \times 60 \times 1000 = 60 \times 10^6$ rev.

$$\text{Now Rating life, } L = \left(\frac{C}{P}\right)^3 \times 10^6 \text{ rev.}$$

$$\therefore \quad 60 \times 10^6 = \left(\frac{C}{P}\right)^3 \times 10^6$$

$$\therefore \quad \frac{C}{P} = (60)^{\frac{1}{3}} = 3.91$$

$\therefore$ Dynamic load capacity = $24 \times 3.91 = 93.83$ kN.

$\therefore$ Appendix II, Table II.2, Deep groove bearing No. 6316 will be suitable.

Example 17.24. A ball bearing supporting a bevel gear shaft is subjected to a radial load of 1600 N and a thrust load of 550 N. Select the proper bearing for a speed of 150 rev/min. Life expectancy is 50×10^3 cycles.

Solution: Equivalent radial load is,

$$P = XVF_r + YF_a$$

Now $\quad F_r = 1600$ N, $F_a = 550$ N.

If it is assumed that inner race is rotating, then $V = 1$. For most of the ball bearings, except angular contact bearings and self-aligning bearings,

let $\quad X = 1$ and $Y = 1.5$

$$P = 1 \times 1 \times 1600 + 1.5 \times 550 = 1600 + 825 = 2425 \text{ N}$$

Rated life, $\quad L = 50 \times 10^3$ cycles

$$\therefore \quad 50 \times 10^3 = \left(\frac{C}{P}\right)^3 \times 10^6$$

$$\therefore \quad \frac{C}{P} = \left(\frac{50}{1000}\right)^{0.333} = 0.3688$$

$$\therefore \quad C = 894.27 \text{ N.}$$

Appendix II, Table II. 1, Bearing (Deep groove ball) No. 6200 is more than suitable. Even a bearing of extra-light series will be appropriate.

EXERCISES

1. What is a bearing?
2. How will you classify the bearings?
3. On what factors the selection of a bearing for a particular application depends?
4. Describe the various stages of friction in the case of plain bearings.
5. Differentiate between "Hydro-dynamic lubrication"; "Wedge film lubrication" and "Squeeze film lubrication".
6. Distinguish between "Hydro-dynamic bearings" and "Hydro-static bearings".
7. What are the factors to be considered while selecting the material for plain bearings?
8. Name the various bearings metals.
9. What is the function of lubrication in bearings?
10. Give the advantages and disadvantages of oils and greases.
11. Describe the various lubrication devices.
12. What is the importance of oil grooves in bearings?
13. Define "Bearing number".
14. Define "Sommerfeld number".
15. What is meant by "Self contained bearings"?
16. Give the applications and limitations of sliding bearings.
17. Enumerate the advantages and disadvantages of rolling contact bearings.
18. How will you classify the rolling contact bearings?
19. What are the four main parts of a ball bearing?
20. Compare ball and roller bearings.
21. What is the material for rolling contact bearings?
22. How the rolling contact bearings are manufactured?
23. What bearings are called "Needle bearings"? Give their application.
24. A 150 mm diameter Diesel engine main bearing turns at 450 rev/min. It is desirable to limit the bearing temperature to 55°C. Determine the allowable bearing pressure per square mm of projected area, if the journal is hardened and ground and a hard bronze bearing liner is used. Determine also the load that the bearing can carry.
25. A sleeve bearing 130 mm long supports a load of 2.75 kN from a 75 mm shaft rotating at 100 rev/min. Select a suitable oil and clearance, and check for satisfactory operation.
26. A steam turbine bearing 100 mm in diameter and 150 mm long supports a load of 45 kN. The operating speed is 1800 rev/min. (*a*) Select on oil to give a (*Zn/p*) value of about 14.00. (*b*) What heat is generated under these conditions? (*c*) To limit the temperature to 80°C is artificial cooling necessary for a room temperature of 38°C ?
27. A diesel engine crankshaft operates at 240 rev/min. A flywheel weighing 27 kN overhangs the bearing by 0.25 m. The maximum torque on the shaft is 3200 Nm. Average load on the bearing is about 40 kN and shear stress for the shaft may be taken as 28 MN/m^2.
    (*a*) Determine the proportions of the shaft.
    (*b*) Check the (pV) value.
    (*c*) Select an oil and clearance.
    (*d*) Calculate the heat generated.
28. Design a journal bearing to support a load of 7500 N while the shaft runs at 700 rev/min using a hardened steel journal and bronze backed babbitt bearing. The oil used has a viscosity of 0.0087 kg/ms at 82°C. Clearance may be assumed as 0.0025 cm per cm diameter and ambient temperature 29°C. If the heat radiating capacity of the bearing is 150 W per sq meter of projected area of bearing per degree C, is artificial cooling necessary?
29. Determine the dimensions of a bearing and journal to support a load of 5500 N at 800 rev/min using hardened steel journal and bronze-backed bearing. An abundance of oil is provided which has a specific gravity of 0.9 at 15.5°C and a viscosity of 10.2 centistokes at 82°C that may be taken to the limiting temperature for oil. Assume a clearance of 0.0025 cm per cm of diameter. Calculate also the rate of heat generated in the bearing.
30. A 75 mm bearing using oil with an absolute viscosity of 0.07 kg/ms running at 300 rev/min gives satisfactory operation with a bearing pressure of 1.4 MPa. The bearing clearance is 0.125 mm.
    (*a*) Determine the unit pressure at which this bearing should operate if the speed is changed to 400 rev/min.
    (*b*) If the bearing clearance is 0.075 mm, what change should be made in the oil?

31. Design a sleeve for the following data:
    Journal diameter = 50 mm
    Journal speed = 300 rev/min
    Journal finish = 0.40 μm
    Bearing finish = 0.80 μm
    Finish method = Equivalent to lap
    Clearance factor, $C_r = 2$
    Clearance ratio = 0.0015–0.002 mm per mm
    Oil viscosity = Oil No. 30 at 60°C.
32. A stationary journal bearing 47.5 mm in diameter and 38 mm long is bored and lathe turned from a bronze casting. The 1600 rev/min journal is 47.3 mm in diameter and is of hardened and ground steel. The No. 20 oil has an operating temperature of 77°C. The radial load on the bearing is 51.20 kN. Does the bearing operate under hydrodynamic or marginal lubrication conditions ? What is the probable power loss in the bearing due to shear of the oil film?
33. The No. 20 oil at an oil film temperature 49°C is employed to lubricate a 50.7 mm diameter cylindrically ground steel journal rotating at 1200 rev/min. The lathe turned bronze bearing is stationary, has a 50.8 mm bore and has a length of 25.4 mm. Under constant loading conditions, determine the probable maximum radial load the bearing can carry and still maintain hydrodynamic lubrication.
34. The following data apply to a lubricated bearing:
    Load on bearing = 3600 N
    Speed of Journal = 360 rev/min
    Diameter of Journal = 0.075 m
    Length of Journal = 0.15 m.

    Assume a clearance of 0.001 cm per cm of diameter, Oil No. 10, oil temperature 60°C, and ambient temperature of 15.5°C.

    Determine:

    (a) Actual value of *Zn*/*p* *(b)* Coefficient of friction. *(c)* Heat generated in watts. *(d)* Heat dissipated in watts.
35. The main bearing of an engine generator set is 0.20 m in diameter by 0.30 m long and supports a load of 54 kN. The bearing is ring oiled, and the coefficient of friction may be taken as 0.0125. The engine rotates at 85 rev/min in a room where the temperature is 27°C.

    (a) Determine the probable operating oil temperature. *(b)* Determine the power loss in the bearing.
36. A ship travelling 30 km per hr requires 1750 kW at the propeller, which turns at 80 rev/min. The thrust collars on the propeller shaft are 0.45 m in outside diameter and 0.30 m in inside diameter. Assume the coefficient of friction to be 0.08 and the permissible pressure on the bearing surfaces to be 0.35 MPa.

    (a) Determine the number of thrust collars required. *(b)* Determine the power lost in the bearing.
37. A rotary pump has shaft diameter of 60 mm and it is required to take a load of 4000 N. If the maximum speed of the pump is 1000 rev/min, calculate the collar bearing dimensions and power lost in friction. Take coefficient of friction as 0.04.
38. Design a horizontal shaft multiple collar bearing to carry a 90 kN thrust if the inside diameter of each thrust area is 0.15 m, the ratio of the outside to inside diameter of each thrust area is approximately 1.75, the coefficient of friction is 0.08; and the minimum allowable unit loading is 3.5 MPa.
39. The oil engine of a motor boat develops 750 kW at 200 rev/min. The speed of the boat is 14 knots (1 knot = 1.85 km/hr) and the propeller efficiency is 27 percent. Find the main dimensions of a multi-collar thrust bearing.
40. A ball bearing supporting a bevel gear shaft is subjected to a radial load of 1600 N and a thrust load of 550 N. Select the proper bearing for a speed of 150 rev/ min, under favourable conditions. Life expectancy is 50×10^3 hours.
41. The bearing of a cantilever gear carries a radial load of 4000 N and an axial load of 1000 N. Angular speed of gear is 60 rad/s; required bearing life = 10×10^3 hours. Working temperature of the bearing is less than 373 K. Diameter of gear shaft extension is 65 mm; Inner race revolves; load with light jerks. Select the type and find the dimensions of the bearing.
42. The spindle of a drilling machine journal diameter of 50 mm carries an axial load of 8000 N at an angular speed of 30 rad/s. Select the roller contact bearing for the required bearing life of 10×10^3 hours.

43. Select a roller bearing for the low speed shaft of a gear reduction unit operating at 65 rev/min with an applied radial load of 13 kN. The shaft size has been determined and is 60 mm.
44. What stationary radial load can be carried by an 85 mm bore medium series, roller journal bearing running at 1200 rev/min to give a rating life of 10000 hrs? The inner ring rotates and the service factor may be taken as 2.0.
45. Select a bearing for a 40 mm shaft to carry 1800 N stationary radial load and 2700 N axial load at a shaft speed of 500 rev/min. The rating life required is 6000 hours. There is no shock loading.
46. What will be the rating life of 50 mm bore, medium series ball bearing running at 1000 rev/min under 3600 N stationary radial load and 4500 N axial load?
47. A 0.10 m diameter shaft operating at 2000 rev/min is supported by means of a 0.15 m long full journal bearing which is supporting radial load of 44 kN. The oil temperature is to be limited to 60°C, the surrounding air temperature 38°C.

    Assume $Zn/p = 28.50$. Determine:

    (*a*) Coefficient of friction, (*b*) bearing pressure, (*c*) heat generated,
    (*d*) heat dissipated, (*e*) oil to be used, (*f*) Is artificial cooling required?
48. A 0.06 m journal bearing 0.10 m long supports a load of 2250 N at 600 rev/min. If the room temperature is 24°C, what should be the viscosity of the oil used to limit the bearing surface temperature to 60°C?
49. A 5500 N rotor is supported midway between two journal bearings on a 0.05 m shaft which rotates at 800 rev/min. The oil selected has viscosity of 0.0072 kg/ms at the operating temperature of 107°C. If the bearing is to operate in an atmosphere 32°C without artificial cooling, find the length of the bearing.
50. A journal bearing 0.075 m long on a 0.0025 m shaft supports a load of 2700 N. The bearing has clearance of 0.05 mm and the viscosity of the oil used is 0.027 kg/ms at the operating temperature. If the bearing is capable of dissipating 150 W/m^2 °C, determine the maximum speed of rotation.
51. A journal bearing 0.15 m in diameter and 0.225 m long carries a radial load of 900 N at 1200 rev/min. The diametral clearance is 0.075 mm. Find the viscosity of the oil being used at the operating temperature if 1.5 kW is being wasted in friction.

❑❑❑

18 CHAPTER

Clutches and Brakes

18.1 CLUTCHES

The basic function of the clutch is to control the flow of mechanical power within a machine. It must be capable of transmitting specified amount of torsional moment. The clutch is also called upon to convert potential or kinetic energy into internal energy and to dissipate it in the form of heat, transferred from the clutch to the atmosphere. The other main functions of a clutch are:

1. To start and stop a machine or a rotating element without starting and stopping the prime mover.
2. To maintain constant speed, torsional moment and power.
3. To limit torsional moment.
4. For automatic disconnect, quick start and stop, gradual starts and non-reversing and over running functions.
5. To connect and disconnect the shafts at will.
6. To reduce shocks transmitted between machine shafts.

One very important requirement to be satisfied by the shafts and other parts connected by clutches is that they should be strictly collinear. Misalignment greatly deteriorates the performance of a clutch and results in rapid failure.

18.2 CLUTCH-LOAD CHARACTERISTICS

The two basic things required for the design of a clutch are:

1. The torque it must be able to transmit.
2. The energy it must be able to dissipate during the clutch slip period.

The clutch divides a kinematic transmission into two parts: the driving or the input part and the driven of the follower part (Fig. 18.1). The clutch torsional moment capacity must be sufficient to overcome the

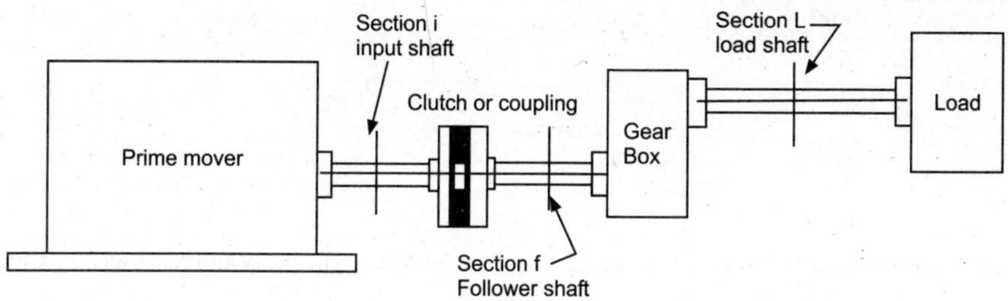

Fig. 18.1 *Kinematic transmission*

torsional moment of the load and the inertia of all the parts connected to the follower side of the clutch. When the clutch has been engaged, it should transmit the torsional moment from the driving shaft to the driven shaft without slipping under normal load.

Let us consider three sections on the kinematic transmission : Section *i* on the input shaft, section *f* on the follower shaft, and section *l* on the load shaft. The moments of inertia, torsional moment and angular velocity at the input shaft, follower shaft and load shaft are represented respectively by the symbols I_i, T_i, w_i ; I_f, T_f, w_f; and I_l, T_l, w_l.

Now when the clutch is under the process of engagement, the clutch torsional moment T_c during the period when the follower shaft is being accelerated, is given as:

$$T_c = T_f + I_f \cdot \frac{dw_f}{dt} \qquad \text{...(18.1)}$$

now $T_f = \dfrac{T_l}{m_g}$; and $I_f = I'_f + \dfrac{I_l}{m_g^2}$

where, I_f = moment of inertia of follower shaft and load referred to it.

I'_f = moment of inertia of follower shaft and all parts joined to it.

I_l = moment of inertia of load shaft and all parts joined to it.

m_g = gear ratio = $\dfrac{w_f}{w_l}$

If Torsional moment capacity of the clutch T_c is $\geq T_i$, and assuming that T_i and T_f remain constant during clutch slip period then solving eqn. 18.1 and putting $w_f = 0$ when, $t = 0$, we have

$$w_f = \frac{T_i - T_f}{I_f} \cdot t \qquad \text{...(18.2)}$$

Now the energy of friction work done during the clutch slip period is given as

$$dE = T_i d\theta \qquad \text{...(18.3)}$$

and $d\theta = (w_i - w_f)\, df$ ...(18.4)

where θ = Angle of clutch slip, rad.

Due to the friction work done, the internal energy of the clutch parts gets increased, thereby, raising their temperature,

$$\therefore \quad dE = T_i \cdot (w_i - w_f)\, dt$$

$$= T_i \cdot \left(w_i - \frac{T_i - T_f}{I_f} \cdot t \right) dt = w_i\, T_i \cdot dt - \frac{T_i^2}{I_f} t \cdot dt + \frac{T_i \cdot T_f}{I_f} \cdot t\; dt.$$

$$\therefore \quad E = w_i T_i \Delta t - \frac{T_i^2 \cdot \Delta t^2}{2I_f} + \frac{T_i \cdot T_f}{2I_f} \cdot \Delta t^2 \qquad \text{...(18.5)}$$

where Δt = clutch slip period, sec.

Since at the end of clutch slip period. $w_f = w_i$

$$\therefore \quad w_f = w_i = \frac{T_i - T_f}{I_f} \cdot \Delta t$$

and

$$E = T_i \cdot \frac{T_i - T_f}{2I_f} \Delta t^2 \qquad \text{...(18.6)}$$

The temperature rise, ΔT, of the clutch assembly can be approximately determined by the equation:

$$\Delta T = \frac{E}{Cm_c}, \text{°C} \qquad \text{...(18.6a)}$$

where C = specific heat, J/kg °C

and m_c = mass of clutch assembly, kg. Take the value for C as 500 J/kgC° for steel or cast iron.

To avoid overheating of the clutch, this heat must be dissipated to the atmosphere.

18.3 TYPES OF CLUTCHES

The clutches may be classified in two ways:

1. Based on the means adopted to couple the two halves of the clutch.
2. Based on the means used to actuate them.

Mode to couple the two clutch halves: Based on this principle, there are three major types of clutches:

1. Mechanical clutches.
2. Electro-magnetic fluid and power clutches.
3. Fluid clutches.

In this chapter, only mechanical clutches will be discussed.

18.3.1 Mechanical Clutches

This type may be further classified as:

(*i*) Positive contact clutches. (*ii*) Friction clutches.

(*i*) Positive Contact Clutches: These clutches operate by machine jaws or teeth of the driving member with corresponding elements of the driven member. Such a clutch is used when a positive drive is required and when a gradual engagement between the driving and the driven members is not a must. The fixed half of the clutch is mounted tightly on its shaft with the help of a key, whereas the other half which is the movable is free to slide axially on its shaft but it is prevented from turning relatively to its shaft by means of splines or one or more feather keys. The positive contact clutches have the following advantages:

(a) They will not slip to the point of destruction.

(b) For comparable torque capacity, this clutch is smaller, lighter and less costly than a friction clutch.

(c) It does not develope heat appreciably during use and so it may be cycled frequently.

(d) It is simpler and does not require adjustment for wear.

There are two constructions of the positive clutches:

(1) Jaw clutch (2) Tooth clutch.

(1) Jaw clutch: There are two types of the Jaw clutch:

(a) Square Jaw Clutch: It is also known as "claw" clutch and is shown in Fig. 18.2. It is of simple construction, compact and low in cost. It produces no heat during operation. It is not in wide use since it can't be safely engaged during operation. Also it picks up the load at the instant of engagement, and if the load inertia is high, destructive shock may be transmitted to the drive. Although it is best to stop the unit before engaging this type of clutch, running engagement is possible at

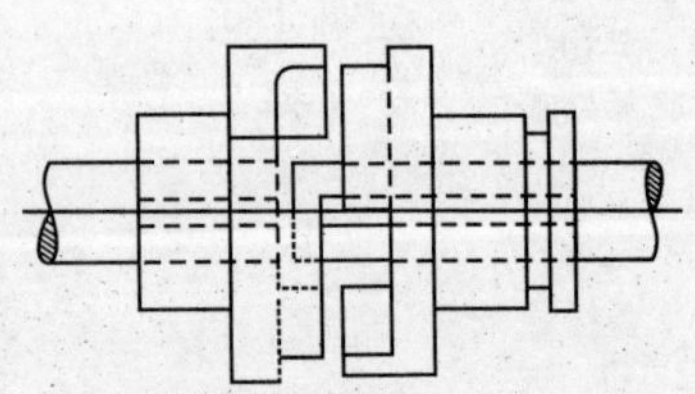

Fig. 18.2 *Square jaw clutch*

2 to 10 rev./min. Machined square jaw clutches are available generally in sizes from 25 to 150 mm in capacity ranging from 1 to 200 kW at 100 rev/min. It can transmit the power in either direction.

(b) Spiral Jaw Clutch: Figure 18.3. This type has eliminated many of the undesirable features of the square jaw clutch. These clutches are usually used as cast. Where the load is to be disengaged while running, finished jaws should be employed and lubrication provided. This clutch can be engaged at higher speeds than the square jaw clutch, but will shock load the driver at engagement, if load inertia is high. It drives in one direction only and has a tendency to free-wheel. Running engagement is possible at 2 to 150 rev./min. They are available in the same size and capacity range as the square jaw clutch.

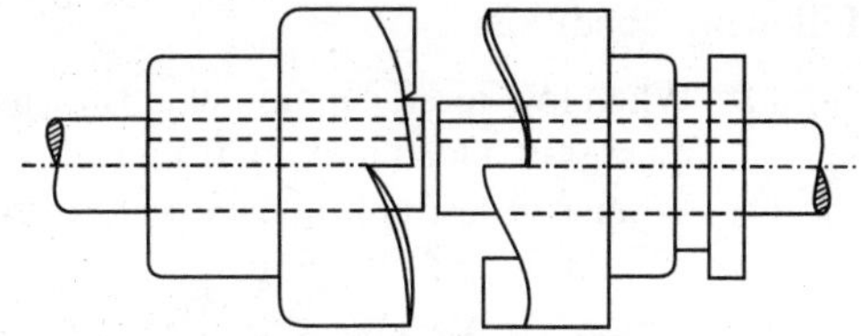

Fig. 18.3 *Spiral jaw*

(2) Tooth Clutch: This clutch has multiple teeth modified in various ways to provide strength or high speed capability. Sizes generally range from 25 to 162.5 mm bore with capacities from 0.5 to 225 kW per 100 rev/min depending upon the type of clutch. Running engagement is possible from 65 to 300 rev/min depending upon type.

(i) Design Procedure: Figure 18.4 shows half a four-jaw clutch. The torque of the shaft must be transmitted by the projecting jaws. Two types of stresses are induced in the jaws. (1) Shear stress, (2) Bearing stress and the corresponding areas are shown as *S* and *B* respectively in the figure. Considering the shearing of the jaws:

$$\text{Torque to be transmitted, } M_t = \frac{\pi}{16} d^3 \times f_s = z \,.\, A \,.\, f'_s \,.\, r$$

where d = diameter of the shaft
f_s = safe shear stress of the shaft material
z = number of Jaws.
A = shear area of one Jaw
f'_s = safe shear stress of the jaw material.
r = mean radius of the jaw.

Considering the crushing of the jaw:

$$\frac{\pi d^3}{16} \cdot f_s = (r_1 - r_2) \,.\, h.z.f_c.r.$$

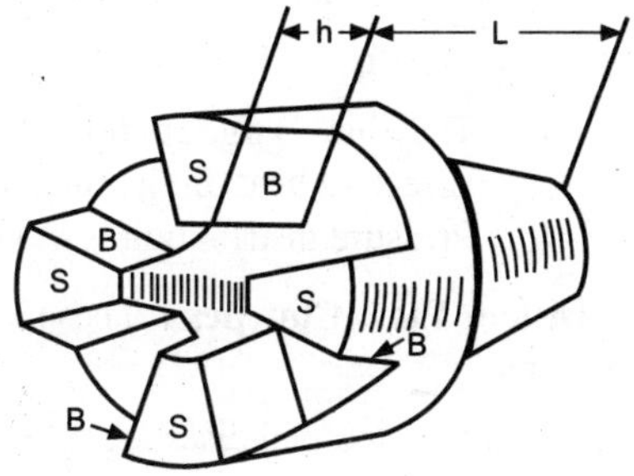

Fig. 18.4 *Jaw clutch*

where r_1 = outside radius of Jaw.
r_2 = inside radius of Jaw.
h = length of Jaw.
f_c = safe bearing stress of the jaw material.

In the above analysis, it is assumed that the effective moment arm is the mean radius of the jaw.

The distance *L* may be defined by the length of the key necessary. There is no rule for the number of jaws. If the number is small, it takes longer for the engagement to take place.

***(ii)* Friction Clutches:** The design and operation of friction clutches is based on the forces developed by surfaces in direct contact. Friction clutches have generally two or more rotating concentric surfaces which can be forced firmly into contact with one another so that the tangential friction force transmits torque from the input to the output shaft. When engaged the clutch members tend to rotate as a single unit, but they are free to slip as long as they are not in equilibrium. As a rule, at least one surface is lined with a material having high coefficient of friction. Friction clutches have the following advantages:

1. They can slip during engagement which enables the drive to pick up and accelerate the load with minimum shock.
2. Since they do not use teeth or jaws, they can be used at high engagement speeds.
3. They need not be unloaded before engagement.
4. They will slip momentarily under shock loads, thus providing cushioning.

The clutch slip creates problems of heat dissipation and due to this the clutch performance is affected. Of all various types of clutches, the friction clutches are most widely used. Friction clutches find use under the following conditions:

(*a*) When low-torsional moment prime movers such as internal combustion engines and certain motors, must start a load moving from rest.
(*b*) When it is desirable to have one source of power drive two or more shafts which have different cycles of starting and stopping.

In automobiles, tractors, power shovels, winches and some machine tools, friction clutches are the main mechanisms on which unfailing operation of the machine depends.

The friction clutches are classified according to the manner of applying the contact pressure as:

1. Axial friction clutches: In these clutches, the contact pressure is applied by shift movement along the shaft. This type includes the following clutches :

(a) Disk clutches: Which may be single disk clutch or multi-disk clutch.
(b) Cone clutch: Which may be single cone or double cone clutch.

2. Radial friction clutches : These are also known as "Rim or Drum clutches". In these clutches the contact pressure is applied normal to the shaft axis against a rim or drum. This type includes the following clutches :

(*a*) Tyre-Pneumatic clutch.
(*b*) Centrifugal clutch.
(*c*) Over running or free wheeling clutch.

The axial type clutches have the following advantages:

(*a*) They have large friction surfaces for a given size and weight.
(*b*) Effective heat dissipating surfaces are large.
(*c*) Pressure distribution is more uniform over the friction lining area.

I. Disk or Plate Clutches : These are the most common variety of the friction clutches due to the following reasons :

1. Solid or strapped discs are easier to manufacture than friction elements of other forms.
2. Clutches for transmitting various torsional moments can be manufactured by varying the number of discs of identical size. This facilitates the standardisation and centralisation of their production.
3. They require less space than other types.
4. Their worn parts can more conveniently be replaced.

Due to the above advantages, disk clutches are extensively used in automobiles, tractors, machine tools, etc. Single disk clutch is employed in all the modern cars. Figure 18.5 illustrates the operating principle of such a clutch. There are three main elements of the clutch:

(*a*) In the centre is the "clutch plate" which is a steel disc which is surfaced on both sides with friction material. It is mounted in such a manner that it is free to slide axially on the splines and through which it transmits the torsional moment to the output shaft or the follower shaft.
(*b*) Pressure plate which is spring loaded and which pushes the clutch plate firmly against the machined rear face of the flywheel. Both the pressure plate and the flywheel rotate with the engine crankshaft or the input shaft. When the clutch pedal is pressed down,

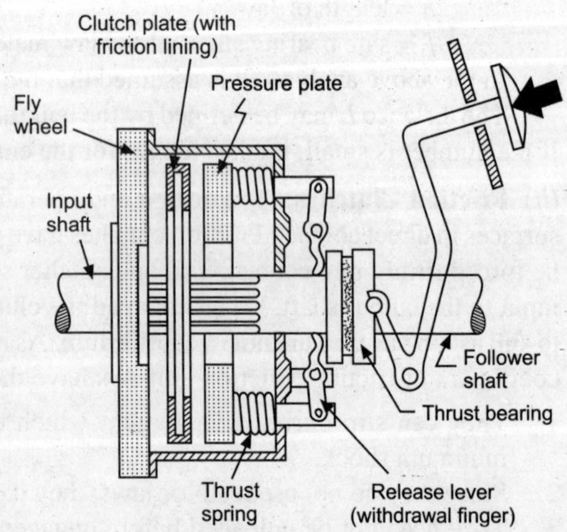

Fig. 18.5 *Single disk clutch*

the pressure plate moves back slightly and relieves pressure on the clutch plate. The clutch plate slips back from the flywheel face and the follower shaft becomes stationary. The movement of the clutch linkage is transferred to the pressure plate by a thrust bearing. When the foot is taken off from the clutch pedal, the thrust bearing moves back by three levers known as "Release levers". Due to spring action the pressure plate pushes ahead and forces the clutch plate against the flywheel face. To help the smooth take up of the loads the clutch plate is usually made with a flexible centre i.e. the outer portion is mounted upon a separate hub, with a spring or rubber interposed to allow a limited amount of angular movement which absorb torsional vibration.

Design of Disk Clutches: As an approach to the design problems of single or multi-disk clutches, it is necessary to establish the relation between the dimensions of the disk, the pressure applied to the friction surfaces, the coefficient of friction and the torsional moment developed. Now the torsional moment transmitted by a disk clutch is a function of:

(i) Normal force on the clutch (ii) the coefficient of friction (iii) the effective or mean radius of the friction forces and (iv) number of planes of friction contacts.

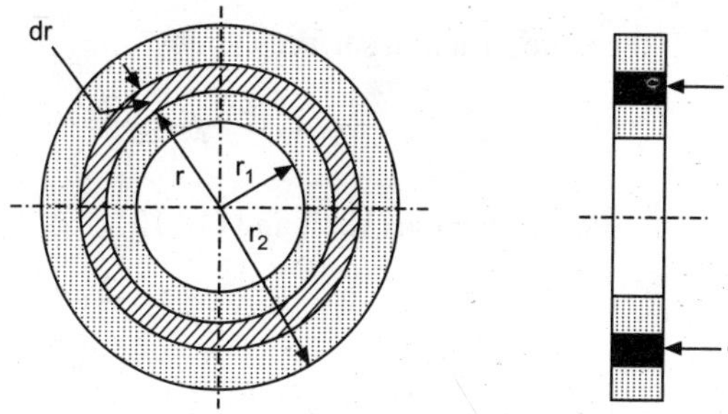

Fig. 18.6 *Forces on clutch disk face*

i.e., $M_t = P.\mu.r_m.z$ ...(18.7)

where M_t = clutch torsional moment, Nm

P = normal force on clutch disk, N i.e., the axial force.

r_m = effective radius of friction forces, m

z = number of friction planes, i.e., number of pair of surfaces in contact.

The value of r_m will depend upon the assumption made in design. The design of plate clutches is based upon one of two premises:

1. Uniform Pressure.
2. Uniform Axial wear.

Consider a single friction surface, Fig. 18.6. Consider an elemental area of radius r and radial width dr. The tangential friction force acting on this area will be given as:

$$dF = \mu . dP$$

now $dP = p.\, 2\pi r\, .\, dr$

where p = pressure on the elemental area N/m^2

∴ $dF = \mu . p.\, 2\pi r.\, dr$

∴ Total tangential friction force, $F = 2\pi\mu \int_{r_1}^{r_2} p \cdot r \cdot dr$ ...(18.8)

where r_2 = outer radius of contact surfaces

r_1 = inner radius of contact surfaces.

Now the friction torsional moment for the elemental area is given as:

$$dM_t = r\, .\, dF = \mu\, p 2\pi\, r^2\, dr$$

∴ Total friction torsional moment, $M_t = 2\pi\mu \int_{r_1}^{r_2} p \cdot r^2 \cdot dr$ ...(18.9)

Uniform Pressure: When the friction surface is new and unworn and is rigidly mounted, it is usually assumed that the normal force P is uniformly distributed over the entire area, or

$$p = \frac{P}{A} = \frac{P}{\pi\left(r_2^2 - r_1^2\right)} \quad ...(18.10)$$

Integrating (18.9), $M_t = \frac{2}{3}\pi\mu \cdot p\left(r_2^3 - r_1^3\right)$

Putting the value of p in this,

$$M_t = \frac{2P\mu}{3}\left(\frac{r_2^3 - r_1^3}{r_2^2 - r_1^2}\right) \quad ...(18.11)$$

If there are z friction surfaces, then

$$M_t = \frac{2P\mu z}{3}\left(\frac{r_2^3 - r_1^3}{r_2^2 - r_1^2}\right) \quad ...(18.12)$$

∴ From equation (18.7) and (18.12),

$$r_m = \frac{2}{3}\left(\frac{r_2^3 - r_1^3}{r_2^2 - r_1^2}\right)$$

Note: The assumption of uniform pressure shall be more appropriate where the plates are flexible to permit deflection when wear occurs.

Uniform Axial Wear: The clutch is subjected to wear due to sliding friction, therefore, normal wear is a work of friction.

$$\alpha\, p\, V$$

where V = sliding velocity

when the friction surface is new and there is uniform pressure distribution, wear will be greatest where the sliding velocity is maximum and due to this pressure between the surfaces will be reduced. This wearing in process will continue until the product pV is a constant over the entire area. After this, wear will be uniform as shown in Fig. 18.7. After the running in wear has ended

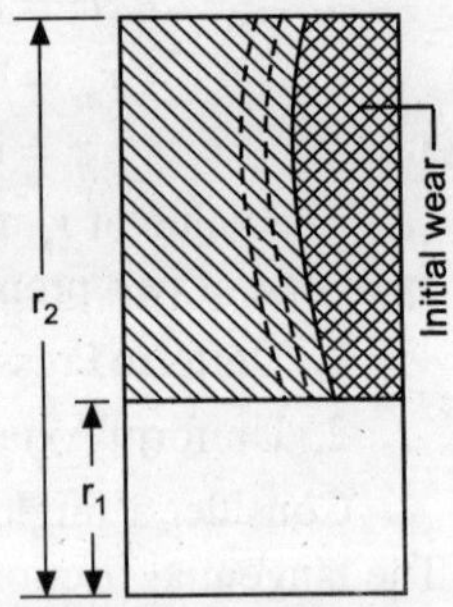

Fig. 18.7 *Uniform wear*

$$p = \frac{C}{V} = \frac{C'}{r}$$

here, C and C' are constants.

Now, the total normal force or the operating force is,

$$P = \int p\,dA = 2\pi\int_{r_1}^{r_2} p r\,dr = 2\pi\int_{r_1}^{r_2} C'\,dr = 2\pi \,.\, C'(r_2 - r_1) \quad ...(18.13)$$

$$\therefore \quad C' = \frac{P}{2\pi\,(r_2 - r_1)}$$

now put the value of p in equation (18.9)

$$\therefore \quad M_t = 2\pi\mu\int_{r_1}^{r_2}\frac{C'}{r}\cdot r^2\,dr = \pi\mu C'\left(r_2^2 - r_1^2\right)$$

Putting the value of C' in this equation,

$$M_t = \pi\mu\cdot\frac{P}{2\pi\cdot(r_2 - r_1)}\cdot\left(r_2^2 - r_1^2\right)$$

$$\therefore \quad M_t = \mu\cdot P\cdot\left(\frac{r_2 + r_1}{2}\right)$$

∴ For z friction faces

$$M_t = \mu \cdot P \cdot z\left(\frac{r_2 + r_1}{2}\right) \quad ...(18.14)$$

∴ Comparing equations (18.7) and (18.14)

$$r_m = \frac{r_2 + r_1}{2} \quad ...(18.15)$$

Now torsional moment to be transmitted,

$$M_t = \frac{\text{Power to be transmitted (kW)} \times 1000}{\omega \text{ (rad/s)}}$$

∴ Design torsional moment, $M_d = \beta \times M_t$.

where β is the engagement factor or the friction margin factor (1.25 to 1.50). It prevents the slipping of clutch discs when the assigned torque is transmitted.

If M_{fr} is the frictional moment developed by the clutch, then the condition for clutch serviceability is,

$$M_{fr} \geq M_d$$

Clutch Disk Radii: A basic consideration in the design of a clutch unit to meet specific limiting dimensions is how to obtain the best balance between maximum torsional moment capacity and effective operating life. For a particular application, the following, information will be usually available:

1. Torsional moment to be transmitted; 2. Limiting dimensions, or the maximum outside diameter of the clutch unit; 3. Pressure with which the clutch plates may be engaged.

Another point which may be predetermined is whether single disc construction or multi-disk mechanism is to be used. The torsional moment capacity can be increased by:

1. Using materials of high coefficient of friction. 2. Using increased plate pressure.

Friction materials are expensive and since the depth of wear is proportional to the normal force between two surfaces in sliding motion, the increased plate pressure will result in increased wear of friction surfaces. Torsional moment capacity may also be increased by increasing the number of disks. This factor will be discussed later. Without altering the above factors, torsional moment capacity can still be increased by increasing the inside diameter of the friction surfaces as is clear from the torsional moment equations. But increasing the inside diameter will reduce the amount of wearable material and the life and durability of the friction disks will suffer, since the area of contact of the clutch material is decreased thereby increasing the contact pressure which will result in more wear. So wear is the limiting consideration for increasing the value of the ratio r_1/r_2.

If equal importance is given to the durability and torsional moment output, then it has been found that,

$$r_1/r_2 \cong 0.47936$$

In practice, the ratio r_1/r_2 is seldom less than 0.5 and is usually between 0.6 and 0.7.

r_1 is usually = (2 to 5) times the shaft diameter.

Other design factors also influence the ratio r_1/r_2. For example, the steering clutch of a heavy tractor needs a high torsional moment capacity and the wear problem is not serious, the ratio is taken as 0.748. Where wear problem is more important, the ratio may be less than the optimum value of about 0.479.

It is general practice in design to make the clutch as large as possible in outside diameter and as narrow as possible across, the face. Thus the torsional moment will be relatively high, the pressure distribution will be good, the temperature difference between inside diameter and outside diameter will be minimized and stresses and distortion will be correspondingly reduced.

Normal Pressure. As outlined above, the wear of the friction surfaces is directly proportional to the normal pressure between them. Current design pressures range from about 0.105 N/mm^2 to about 1.225 N/mm^2.

However, pressure considerably higher than 1.225 N/mm^2 are used sometimes and for certain applications the pressure may be as low as 0.0005 N/mm^2.

Number of Clutch Disks. Although the torsional moment capacity may be increased and the engaging force can be cut down appreciably by increasing the number of discs (called the multi-plate clutch) the increase in the number of disks is limited due to two reasons:

1. Space limitations 2. Mechanical considerations.

In the multi-disk clutch, when the pressure is applied to engage it, the pressure is first felt by the disk nearest the pressure plate. Some torsional moment is developed by this disc and the friction with the spline in the containing drum impedes further motion. Thus, some small resistance must be over come by the applied pressure before the second disc is reached. When pressure is first applied against the second disc, it will also develope some torsional moment which will be accompanied by further resistance to application of pressure to the succeeding disks and so on. The pressure on the last plate is not as great as the applied pressure, and the torsional moment transmitted is not as great as calculated on the basis of applied pressure. Hence due to the frictional resistance to axial motion at the spline connections between the hub and clutch disks and between the drum and driving discs, the performance of multi-disk clutches is generally limited. In multi-disk clutch, there will be difference in wear on the first and on the last clutch plate. Therefore, the recommended number of clutch discs for applications requiring the starting of the load from rest, is limited to four. However, when the static torsional moment is to be transmitted, e.g., earth moving tractor steering clutch or when two shafts are to be coupled with zero relative rotational motion during clutch actuation, the number of clutch disks may be 10 to 20.

Figure 18.7(a) shows a multi-disc clutch. It has two groups of discs, inner discs (driven) and outer discs (driving). The outer plates are splined to the drum (mounted on the driving shaft) and the inner plates are splined to the driven shaft. The clutch has several friction planes and is, therefore less bulky than a single disc clutch. Such a clutch is small in size and requires little effort for engagement. When the levers are pressed, z pairs of friction surfaces are created on the discs.

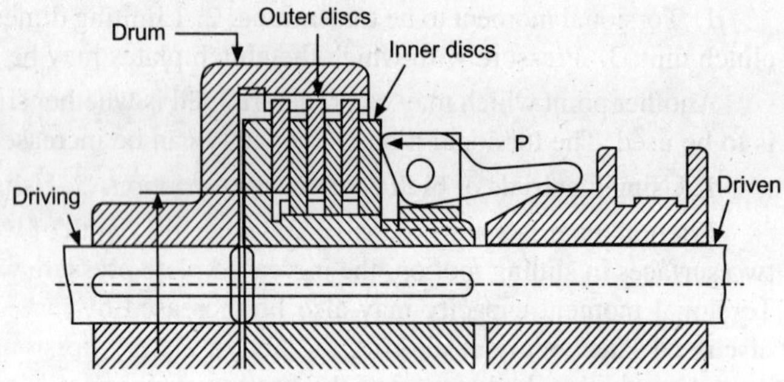

Fig. 18.7 *(a) Multi-disc clutch*

The friction force is shared by all the friction planes. When the clutch is disengaged, a clearance of about 0.2 to 1 mm is formed between the adjacent mating surfaces, which depends upon the facing material. All the discs in a clutch must be perfectly parallel and co-axial to prevent local wear and overheating.

If z = Number of pairs of contact surfaces, then

$$\text{Number of plates (outer) on driving shaft} = \frac{z}{2}.$$

$$\text{Number of inner plates (driven)} = \frac{z}{2} + 1.$$

Many authors take the number of plates on the driven shaft as $z/2$ and that on the driving shaft as $z/2 + 1$.

Multidisk clutches are usually used in wet clutch design. They are intended to transmit small and medium torques, such as in lathes.

Friction Materials: The three factors for selecting a material for facing of a friction clutch disc are: Coefficient of friction, design pressure of contact, and maximum operating temperature. A good friction material should have the following qualities:

1. A high coefficient of friction capable of remaining uniform over of wide range of surface velocities, temperatures and loads.
2. Adequate mechanical and thermal strength.
3. Little wear and no scoring.
4. High heat conductivity for rapid transfer of heat from the friction surfaces.

Some of the most widely used friction materials are given in Table 18.1.

Table 18.1 Friction materials

Material of friction Surfaces		Operating conditions	Coefficient of friction	Unit pressure N/mm^2	Maximum operating temperature °C
Hardened steel	Hardened Steel	In oil	0.08	0.6–0.8	250
Cast Iron	Cast/ Iron or Steel	In oil	0.06	0.6–0.8	250–300
Cast Iron	do	Dry	0.15	0.25–0.4	250–300
Bronze	do	In oil	0.05	0.4	150
Pressed asbestos	do	Dry	0.3	0.2–0.3	150–250
Powder-metal	do	Dry	0.4	0.3	550
Powder metal	do	In oil	0.1	0.8	550

Dry or Wet Clutch Design: A dry clutch has the following advantages as compared to the wet clutch:

1. High coefficient of friction. 2. High torsional moment transmission capacity.

However, if a clutch has been designed to operate in dry environment, it must always be kept dry even in the proximity of otherwise copiously lubricated bearing, gears and shafts. This requirement leads to difficult sealing problems. Hence under oily conditions, a wet clutch may be better choice. The advantages offered by wet friction clutch are:

1. Smoothness of operation and engagement.
2. Greater operating life.
3. Ease of cooling.
4. Simplification of sealing problem.

However, the wet friction is about 1/3 that of dry friction and hence the torsional moment capacity of a wet clutch is less than a comparable dry clutch. It may be compensated by increasing the diameter or the number of discs.

If the clutch has been designed for wet operation, it is essential to maintain the oil film, otherwise the torsional moment developed may become several times the design capacity and cause damage through high wear rate and harsh engagement.

Thickness of clutch Plates: Having found out the suitable dimensions and number of disks, the next step is to find the thickness of clutch discs or plates. Under this, two dimensions are necessary:

1. Thickness of the steel core. 2. Thickness of the friction material face.

Thickness of the steel core is dependent upon the diameter of the disks and the typical values are below, given in Table 18.2.

Table 18.2 Thickness of core plates

Range of clutch diameter, m	0–0.15	0.15–0.30	0.30–0.45	0.45–0.60
Core thickness, mm	0.75	1.5	2.25	3.0

The thickness of the friction faces may be very small and in wet clutch design, it may be still further reduced because the wear rate is small. The usual thickness of wet friction sintered metal friction face is 0.375 to 1 mm. In clutches and brakes using a small number of discs, the friction faces may be thicker and where single disks are used, it may be as large as 6 mm. For example, a typical design for a single disc clutch as found in automobile clutches, has a 3 mm thick rubber-asbestos facing on a 4 mm thick core plate.

Typical Disk Design: The face of the clutch disk may or may not be provided with radial slots, radial grooves or spiral grooves. Radial grooves and/or spiral grooves are not needed in dry clutch design but these are usually provided in the wet clutch design for the proper circulation of oil through the clutch plate since the energy released as heat in wet clutch design is carried away by oil. The oil is usually supplied by pumping outward from the inner hub of the clutch.

Radial slots are also provided sometimes to prevent coning and warping of the friction disks due to their expansion under temperature. The radial grooves perform the following functions:

1. To provide a channel for circulation of cooling oil through the clutch pack.
2. To collect and throw out loose particles.
3. To help maintain the minimum oil film.

The radial slots also perform the same function as the radial grooves. The function of the spiral grooves is:

1. To prevent the forming of a hydrodynamic oil film which would reduce friction.
2. It provides a place for the surface oil to be squeezed into during engagement.
3. It helps to circulate this oil.
4. It provides a place into which the products of wear may be swept.
5. It helps to maintain the desirable minimum oil film by continuously sweeping the oil over the surface.

A typical design of driving disk and clutch disk is shown in Fig. 18.8.

It is general to make pitch of spiral grooves about 1.875 to 2.5 mm and the width of groove about 0.5 to 0.75 mm. Depth is usually less than the width, or about 0.375 to 0.5 mm, reaching nearly but not fully, to the depth of the friction face.

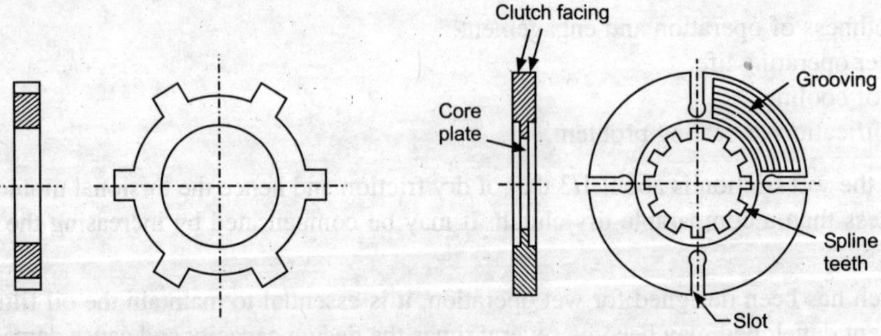

(a) Driving disk (b) Clutch disk

Fig. 18.8

Table 18.3 gives a typical grooving practice for wet clutch friction disks.

Table 18.3 Grooving practice for wet clutch

Grooves of Slots	Disk in Diameter, mm
No grooves	0–230
Spiral grooves	125–500
Radial slots	200–450
Radial grooves	230–450
Radial grooves and slots (alternating)	350–500

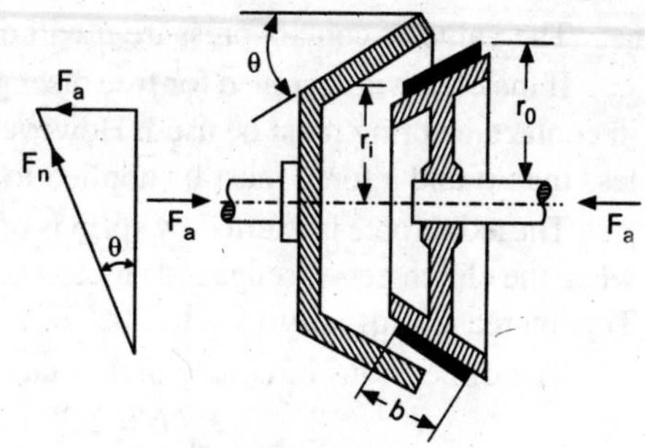

Fig. 18.9 Cone clutch

II. Cone Clutch: The principle of a cone clutch is explained in Fig. 18.9. Cone clutch is simple in construction and is comparatively powerful for a given outside diameter and axial energizing force. This was in extensive use in automobiles and other machines but in recent years it has been completely displaced by disk clutch. Some advantages of the cone clutch are:

1. The force is concentrated at a large diameter.
2. Sufficient friction surface width is provided for durability.
3. A full 360° of lining material is available.

The chief objection to the cone clutch is the difficulty of disengagement which develops after operation from exposure to dust and dirt. Score marks and ridges on the drum and lining surfaces may impede sliding. By increasing the cone angle, the ability to release is improved but the mechanical advantage of the clutch assembly is lowered.

The only difference between the analysis of disk clutch and cone clutch is that in the later the pressure must be measured normal to the cone surface. After engagement, if the normal force between the contact surfaces is F_n, then the axial force F_a needed to produce this will be:

$$F_a = F_n \sin\theta$$

$$\therefore \quad F_n = \frac{F_a}{\sin\theta}$$

∴ Frictional force, $F = \mu \,.\, F_n = \dfrac{\mu \cdot F_a}{\sin\theta}$

∴ Torsional moment produced by the cone clutch $= F \,.\, r_m = \dfrac{\mu \cdot F_a \cdot r_m}{\sin\theta}$

now

$$r_m = \frac{r_i + r_o}{2}$$

$$\therefore \quad M_t = \frac{\mu \cdot F_a}{2\sin\theta}(r_i + r_o) \qquad ...(18.16)$$

If the clutch is engaged when one member is stationary and other rotating, then the cone faces will tend to slide on each other in the direction of an element of the cone. This will resist the engagement, and therefore,

$$F_a = F_n(\sin\theta + \mu\cos\theta)$$

It has been found that the term $\mu \cos\theta$ is only 25% effective, therefore,

$$F_a = F_n\left(\sin\theta + \frac{\mu\cos\theta}{4}\right) \qquad ...(18.17)$$

It is clear that the torsional moment developed by cone clutch is a function of inner and outer radii of the friction surface, the cone angle, the force of engagement and the coefficient of friction at the contact surfaces. As cone angle θ approaches 0, the torsional moment developed approaches infinity. At θ = 0, the cone becomes a drum and at θ = 90°, the cone becomes a single face disk clutch. The disadvantage of too small an angle is that the clutch will seize and the disengaging force will become very large. The value of θ generally varies from 7½ deg. to 30 deg. depending upon the clutch facing material and the type of disengagement required. The most common angle is 12½ deg. The face of the cloth is generally covered with an asbestos composition. Sometimes, there may be metal to metal contact in which case proper lubrication should be maintained. The width of the cone face may be determined as below:

$$b = \frac{F_n}{p \cdot 2\pi \cdot r_m} \qquad ...(18.18)$$

The value of contact pressure p will depend upon the severity of service and the surface materials.

If the clutch is designed for free disengagement, tan θ must be greater than μ and to keep the cone surface in contact, a spring must be used. However, if the clutch is not to disengage of its own accord, tan θ must be less than μ and a force must be applied to disengage the clutch.

The axial force is exerted by springs or by other means. If springs are used, these are designed for the force when the clutch gets disengaged since at the instant, the spring gets more compressed and load on it increases. This increase is usually 15% to 20% of the operating load.

The dimensions of cone clutches can be taken empirically as:

$$r_m / b = 2.25 \text{ to } 4$$

$$r_m = (2.5 \text{ to } 5) \times \text{shaft diameter}$$

Peripheral velocity = 10 to 25 m/s for leather faced clutches.

Speed = 5 to 17 rev/s for metal to metal contact.

III. Tyre-Pneumatic Clutch: This clutch incorporates a flexible rubber tube. When compressed air is forced into this tube, the friction shoes which are attached to one side of the tube will either contract or expand to engage a drum fitted on the driven shaft. Thus clutch my be:

1. Expanding shoe type. 2. Contracting shoe type.

In the expanding type, air pressure inflates a rubber tube which forces the friction shoes outward into engagement with the drum. In the contracting shoe type, the air expanding rubber tube forces the friction shoes inward.

A radial tube-pneumatic clutch of the contracting shoe type is shown in Fig. 18.10. It consists of a rim 1 connected to the driving shaft and tyre 2 with shoes on the side facing rim 3 of the drum mounted on the driven shaft. The tyre is oval in cross-section. The external tread of the tyre is bonded to the rim 1. The friction shoes are attached to the tyre by means of smooth pins. A friction strap is fastened to the shoe side which faces the driven shaft. When compressed air is forced into the tyre, it expands and forces the shoes to the rim. Due to the friction between the shoes and rim, the driving shaft gets engaged with the driven shaft.

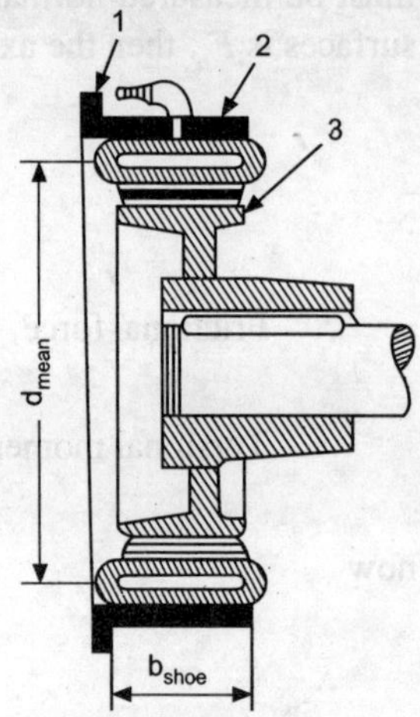

Fig. 18.10 *Tube pneumatic clutch*

Such a clutch has the following advantages:

1. Contact is made on the full width of the 360° friction surface.
2. Applied pressure is uniform over the friction shoes so that the loading is distributed evenly.
3. Torsional moment is at the maximum diameter, providing a constant torsional moment arm.
4. Constant velocity contact assures even wear of friction surfaces.
5 No lubrication or adjustment is necessary.
6. The resilience of the tube cushions vibrations, permits some parallel and angular shaft misalignment, dampness, torsional shocks and automatically compensates for the lining wear.

The effect of centrifugal force on the shoes should be allowed for. In the clutch shown in Fig. 18.10, the centrifugal force will decrease the effort of pressing the shoes against the rim. At the same time, this effect will help in early separation of the friction surfaces on release.

The total force developed by the tyre is determined by the pressure of the compressed air and the size of the tube:

$$F = (p - \Delta P) \,.\, A, N \qquad \text{...(18.19)}$$

where p = pressure of compressed air in the tube in N/mm^2

Δp = pressure expanded on the deformation of the tyre sides

A = active area of the tube, mm^2

p is usually 0.6 to 0.8 N/mm^2 and Δp is taken as about 0.05 N/mm^2.

Now $A = \pi d_m b_m$ mm^2

where d_m = mean diameter of the tube.

b_m = mean width of the tube.

The tyre-pneumatic clutch can be made to operate in the axial direction also. The tube in this case will serve to press the discs together.

IV. Centrifugal Clutch: This type of clutch is used to connect two parts of a transmission after the driving part has attained a definite velocity of rotation. This type is particularly useful with prime movers which can not be started under load, e.g., internal combustion engines or the machines of considerable flywheel torsional moment which are to be driven from induction motors, e.g., centrifuges, drums of washing machines, belt conveyors, etc. This is due to the fact that the ordinary induction motors possess a comparatively low starting torsional moment smaller than the maximum torsional moment developed during rotation; whereas the machines with large masses require considerable starting torsional moment to set them into motion. A centrifugal clutch will allow the motor to pick up the speed and then, after a definite speed has been attained, to start the machine. Thus the centrifugal clutch has the following advantages:

1. It enables a motor or prime mover to start up under no load conditions.
2. The load is picked up gradually as the speed increases.
3. Very useful when the power unit has a low starting torsional moment.
4. The power unit will not stall if it is subjected to an over load, because the clutch will slip after the speed drops below that required to transmit the torsional moment needed by the load.

The outward movement of the shoes is controlled by flat or helical spring, so that a predetermined speed has to be reached before the engagement takes place. The spring control may be "light" or "full". Manufacturers specify 25% of the running speed before engagement and 75% for full spring control.

The principle of centrifugal clutch is explained in Fig. 18.11. Two weights are attached to the spider which is mounted on the input shaft. The outer faces of the weights are lined with friction material. When the input shaft starts rotating, the weights move radially out due to centrifugal force and press against the rim which is joined to the output shaft. The torsional moment is transmitted from the input shaft to the output shaft by means of clutch facing material. The radial outward movement of the weights is controlled by means of one or more tension springs connected to the weights.

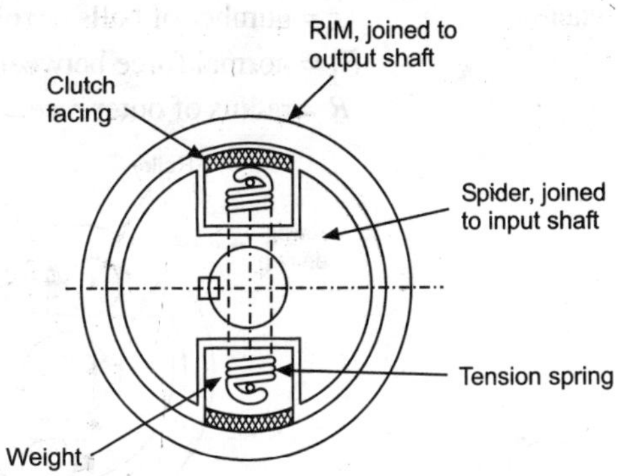

Fig. 18.11 *Centrifugal clutch*

If the angle subtended by the arc of contact of the weight with the rim is small, then the contact of the weight with the rim may be assumed as point contact. So the net radial force required to produce the required torsional moment is given as:

$$F = \frac{M_t}{\mu \cdot R_m \cdot z} \quad \text{...(18.20)}$$

where R_m = radius of rim

z = number of weights.

The net radial force may be found out as below:

$$F = F_c - F_s = \frac{W_r \cdot r \cdot \omega^2}{g} - F_s \quad \text{...(18.21)}$$

where W_r = weight of the revolving mass.

r = radius of gyration of the mass.

ω = angular velocity of the rotating shaft.

F_c = centrifugal force.

F_s = spring force.

The spring force may be calculated from equation (18.21) by putting $F = 0$, when ω = angular engagement velocity of the clutch.

The above analysis is valid if angle of contact is upto 40 deg. The error due to this assumption will be very small.

V. Over-running or Free-wheeling Clutches: These clutches automatically connect and disconnect the shafts depending on the ratio of their angular velocities. When the speeds of the driving shaft exceeds the speed of the driven shaft, the clutch connects the two. However, when this ratio is reversed, the clutch is released, thus the torsional moment is transmitted only in one direction.

These clutches are widely used in various starting devices, bicycles, motor-cycles, metal cutting machines tools, and automobile transmissions.

The principle of working of these clutches is either wedging or wrapping. In either case, the driving and the driven members are coupled by the elastic tension of an intermediate member. Depending upon the design of intermediate member, two major types of these clutches are:

1. Roller type. 2. Sprag type.

Roller Type Over-running Clutch : The clutch is shown in Fig. 18.12. The members are wedged together by the action of concentric balls or rollers riding on a clamming race. In this case shown, the torsional moment is being transmitted from the input shaft rotating clockwise of the outer race. If the outer race starts revolving clockwise relative to the inner race, the balls will freewheel and the clutch will get released. The wedging action is due to the small angle α between the inner cam flat and the tangent at the point of contact of the rollers on the outer race. Assuming that the rollers do not slip, the torsional moment transmitted by the clutch is given as:

$$M_t = z \cdot F_1 \tan \frac{\alpha}{2} \qquad \text{...(18.22)}$$

where z = number of balls or rollers.

F_1 = normal force between the roller and the race.

R = radius of outer race.

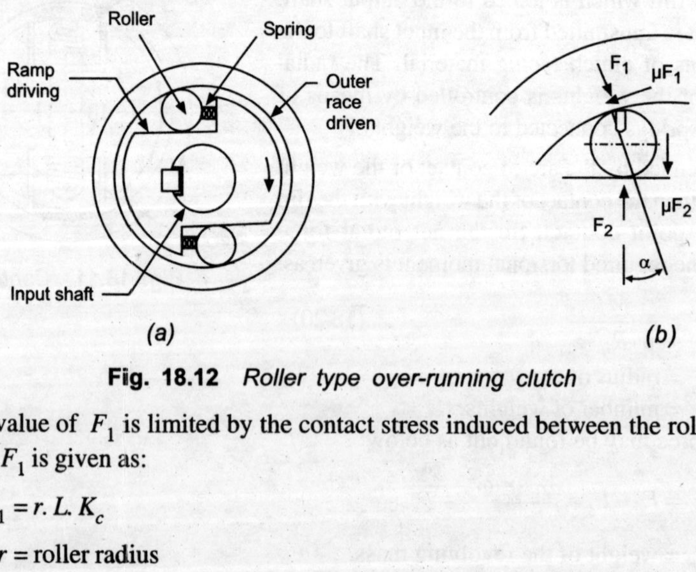

Fig. 18.12 *Roller type over-running clutch*

The maximum value of F_1 is limited by the contact stress induced between the roller and the ramp. For a flat ramp, F_1 is given as:

$$F_1 = r \cdot L \cdot K_c \qquad \text{...(18.23)}$$

where r = roller radius

L = length of roller, is usually less than 2½ times roller diameter.

K_c = contact load-stress factor N/mm^2

Typical values of K_c are:

$= 105$ to $168\ N/mm^2$ for carburized carbon steel rollers and mating surfaces.
$= 63$ to $98\ N/mm^2$ for medium carbon steel rollers and mating surfaces.
$= 7$ to $38.3\ N/mm^2$ for C.I. rollers with hardened steel mating surfaces.

The usual values of α are 3 to 6 deg. or to avoid slipping of the rollers,

$$\alpha \le 2 \tan^{-1} \mu$$

where μ = coefficient of friction between roller and ramp or between roller and race.

Sprag clutch: In this clutch sprags are interposed between the inner race and the outer race and the torsional moment is transmitted due to the wedging action of the sprags in between the two races. This type of clutch has the following advantages over the roller type clutch:

1. Large torsional moment is transmitted due to the greater number of sprags between the races.
2. Since the radius of contact of the sprag is greater than that of the roller in an equivalent clutch, the allowable normal force between the sprag and race will be greater than between roller and race.

A sprag clutch is shown in Fig. 18.13. The shape of the sprag is such that for one direction of relative rotational motion between the two races, it becomes more upright and so becomes tightly wedged between the two races. If the relative motion is in the opposite direction, the sprags roll down and the clutch is released.

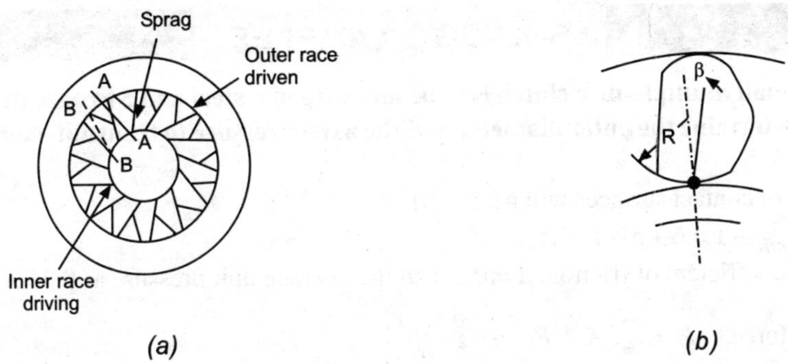

Fig. 18.13 *Sprag clutch*

For this, the dimension along *A-A* is made larger than along *B-B*,
The torsional moment capacity of sprag clutch is given as :

$$M_t = \frac{r_1 \cdot L \cdot z \cdot K_c \cdot \tan\beta}{\dfrac{1}{r_1} + \dfrac{1}{R}} \qquad ...(18.24)$$

where r_1 = radius of inner race.
L = length of sprag.
z = number of sprags.
R = sprag radius at contact of inner race.
β = sprag gripping angle, and is usually 2° to 7°.

Mode of Control of Engagement of the Clutch: Basically there are two modes of controlling the engagement and disengagement of clutches:

1. Mechanical Controls; 2. Automatic Controls.

Mechanical Controls: These controls incorporate levers and linkages, overcentre toggles, balls or rollers working over cams, and other forms of mechanisms which have been developed to obtain mechanical advantage at maximum efficiency. These mechanical linkages or shifts can be operated:

(i) Manually

(ii) or by external source of power.

In manually operated clutches, the clutches are operated by hand or foot either by direct levers for small clutches or through compound linkages or hand driven gearing for the really big ones.

The common external sources of power which are employed to actuate the mechanical linkages are:

1. Solenoid
2. Air Cylinder
3. Hydraulic Ram.

The sources of power are generally employed to compress the friction surfaces through linkages.

The clutches actuated by solenoid are termed as "electromagnetic clutches". In these clutches, the friction surfaces are compressed by the force of attraction of a solenoid i.e., electromagnet forming an integral part of the clutch. Only the disc type of electromagnetic clutches have been widely used. These clutches are spring released. This makes them fail-safe, because the clutch is automatically disengaged if there is power failure.

Automatic Controls: The clutches with automatic controls are self actuating and there is no external control what so ever. The examples are:

Centrifugal clutches, over running or free wheeling clutches, etc.

SOLVED EXAMPLES

Example 18.1. A small multiple-disc clutch is to be made up of 6 steel and 5 bronze disks, with an inner radius of 25 mm. Determine the outer diameter and the axial pressure to transmit a torsional moment of 17.5 Nm.

Solution: Now, no. of contact surfaces will be, $z = 10$

i.e., $z = z_{steel} + z_{bronze} - 1 = 6 + 5 - 1 = 10$

Let us assume, coefficient of friction, $\mu = 0.1$, and the average unit pressure as 0.35 N/mm^2

Now, the axial force, $P = p_m \cdot A = p_m \cdot \pi (r_2^2 - r_1^2)$

$= 0{\cdot}35\pi (r_2^2 - r_1^2)\, N$, where p_m = mean or average unit pressure

Also, the torsional moment transmitted is given as:

$$M_t = \mu \cdot z \cdot P \cdot r_m$$

now, assuming uniform rate of wear,

$$r_m = \frac{r_1 + r_2}{2}$$

$$\therefore \quad 17.5 \times 1000 = 0{\cdot}1 \times 10 \times 0{\cdot}35\pi \times (r_2^2 - r_1^2) \times \frac{r_1 + r_2}{2}$$

$$\therefore \quad (r_2^2 - r_1^2)(r_2 + r_1) = 31831$$

now $r_1 = 25$ mm

$$\therefore \quad (r_2^2 - 625)(r_2 + 25) = 31831$$

If we take $r_2 = 35$ mm

then $(r_2^2 - 625)(r_2 + 25)$ comes out to be 36000, which is safe.

Taking $r_2 = 34$ mm

$(r_2^2 - 625)(r_2 + 25)$ comes out to be 31329 which is less.

So, $r_2 = 35$ mm is O.K.

$\therefore$ Axial Force $= 0.35\pi (35^2 - 25^2) = 659.73$ N

Example 18.2. A hydraulically operated clutch is to be designed for an automatically operated lathe. Determine the number of plates and the operating force required for a clutch which is to transmit a torsional moment of 35 Nm under normal operating conditions. The clutch is to be designed to slip under 300 per cent of rated torsional moment to protect the gears and other parts of the drive. The limits for the diameters of the friction surface due to space limitations are 100 mm and 62.5 mm. The clutch is to operate in an oily atmosphere.

Solution: For almost all practical purposes, it is both satisfactory and desirable to design a clutch on the basis of uniform axial wear. The design equation is:

$$M_t = \mu \cdot P \cdot z \cdot \frac{r_2 + r_1}{2}$$

Hardened steel plates will be used for the clutch

Now $$M_t = 3 \times 35 = 105 \text{ Nm}$$

$$\mu = 0.08 \text{ , from Table 18.1}$$

$$r_2 = \frac{100}{2} = 50 \text{ mm, maximum for space limitation}$$

$$r_1 = \frac{62{\cdot}5}{2} = 31.25 \text{ mm maximum for space limitation}$$

$$P = \text{axial force} = \frac{\pi}{4}(D_2^2 - D_1^2) \cdot p_m$$

The mean or average pressure p_m may be taken as 0.7 N/mm^2 for long wear life and frequent use.

$$\therefore \quad P = \frac{\pi}{4}(100^2 - 62{\cdot}5^2) \times 0{\cdot}7 = 3350.2 \text{ N}$$

$$\therefore \quad 105 \times 1000 = 0{\cdot}08 \times 3350{\cdot}2 \times z \times \frac{50 + 31{\cdot}25}{2}$$

$$\therefore \quad z = \frac{105 \times 1000 \times 2}{0{\cdot}08 \times 3350{\cdot}2 \times 81{\cdot}25} = 9.64$$

The next larger even whole number of friction planes is 10. Therefore, 10 friction planes and 11 plates will be specified.

There are two choices now, either to reduce the outer diameter or to reduce the operating force and thus the pressure on the plates, by the ratio 9.64/10. For the first choice, the pressure between the plates will be kept at 0.7 N/mm^2. In the present problem, space is an important consideration. Therefore, the outer diameter will be determined by keeping:

$$z = 10, \text{ and } p = 0.7 \text{ N/mm}^2$$

$$\therefore \quad P = \frac{\pi}{4}(D_2^2 - D_1^2) \times p_m$$

∴ From the torsional moment equation,

$$M_t = \mu \cdot z \cdot p_m \frac{\pi}{4}(D_2^2 - D_1^2)\frac{D_2 + D_1}{4}$$

$$\therefore \quad 105 \times 1000 = 0{\cdot}08 \times 10 \times 0{\cdot}7 \times \pi \times \frac{D_2^2 - 62{\cdot}5^2}{4} \times \frac{D_2 + 62{\cdot}5}{4}$$

Solving this cubic equation, $D_2 = 99$ mm

$$\therefore \quad P = \frac{\pi}{4}(99^2 - 62.5^2) \times 0{\cdot}7 = 3245 \text{ N}$$

∴ **Specifications:**

Plates: 11, hardened steel, outer diameter of friction surface = 99 mm inner diameter of friction surface = 62.5 mm, operating force = 3245 N.

Example 18.3. Design a single-plate automobile clutch for a six cylinder engine which has a maximum torsional moment of 245 Nm at 1900 rev/min. The fly-wheel diameter is large enough to accommodate a clutch 230 mm outside diameter. The clutch is engaged at 50 km per hr. Determine:

(a) Number of revolution of clutch slip during engagement.

(b) Heat to be dissipated by the clutch for each engagement.

The following data is given:

Engine speed at beginning of clutch engagement = 1900 rev/min

Engine torsional moment, during engagement = 99 Nm

Automobile weight, loaded = 14200 N

Automobile wheel diameter = 0.71 m

Moment of inertia of combined engine rotating parts, flywheel and input side of the clutch = 1.025 kgm²

Gear reduction at differential = 4.11

Torsional moment at rear wheels available for accelerating automobile = 173 Nm

Coefficient of friction of clutch material = 0.3

Design pressure of clutch material = 0.245 N/mm²

Solution: The torsional moment capacity of the clutch should be sufficient to exceed the maximum engine torsional moment. The maximum engine torsional moment is:

$$M_{tE} = 245 \text{ Nm at } 1900 \text{ rev/min}$$

Assuming uniform wear conditions at the clutch facing, the clutch torsional moment is

$$M_t = \mu . z P. r_m,$$

now

$$\mu = 0.3$$

$$z = 2$$

$$P = p_m \cdot \pi (r_2^2 - r_1^2) \text{, N}$$

$$= 0.245\pi (115^2 - r_1^2) = 0.77 (13225 - r_1^2) \text{ N}$$

$$r_m = \frac{r_2 + r_1}{2} = \frac{115 + r_1}{2}$$

$$\therefore \quad M_t = 0.3 \times 2 \times 0.77 (13225 - r_1^2) \cdot \frac{115 + r_1}{2}$$

$$= 0.231 (13225 - r_1^2) \cdot (115 + r_1) \times 10^{-3} \text{ Nm}$$

Let us assume the ratio r_1/r_2 ranging from 0.5 to 0.7 and calculate P and M_t:

r_1/r_2	r_1, mm	P, N	M_t, Nm
0.5	57.5	7640	395.24
0.6	69	6520	360
0.7	80.5	5190	304.5

We can choose either $r_1 = 69$ mm or $r_1 = 80.5$ mm. Both will give sufficient margin over the maximum engine torsional moment.

For $\quad r_1 = 69 \text{ mm, margin is} = \dfrac{360}{245} - 1 = 47.0\%$

$$r_1 = 80.5 \text{ mm, margin is} = \frac{304.5}{245} - 1 = 24.5\%$$

Both the margins are sufficient to prevent the clutch from slipping at any time after the engagement period is complete. To be sufficiently safe, let us take $r_1 = 69$ mm.

Now the contact pressure on the clutch facing varies from

$$p = \frac{P}{2\pi (r_2 - r_1) r} = \frac{6520}{2\pi (115 - 69) \times 69} = 0.327 \text{ N/mm}^2, \text{ at } r = 69 \text{ mm}$$

to
$$p = \frac{6520}{2\pi (115 - 69) \times 115} = 0.196 \text{ N/mm}^2, \text{ at } r = 115 \text{ mm}$$

(*a*) At 50 km per hr automobile speed, the angular velocity of the clutch follower shaft is:

$$\omega_f = \frac{m_g \cdot V_a}{R_t}$$

where $V_a = 50$ km/hr
R_t = tyre radius = 0.355 m
m_g = gear ratio = 4.11

$$\therefore \quad \omega_f = \frac{4.11 \times 50 \times 1000}{0.355 \times 3600} = 161 \text{ rad/s}$$

The angular velocity of engine is equal to the angular velocity of the clutch input shaft.

$$\therefore \quad \omega_i = 1900 \times \frac{2\pi}{60} = 199 \text{ rad/s}$$

Now at the end of engagement period,

$$\therefore \quad \omega_i + \alpha_i \,.\, \Delta t = \omega_f + \alpha_f \,.\, \Delta t$$

where Δt = clutch slip period in sec.

$$\therefore \quad \Delta t = \frac{\omega_i - \omega_f}{\alpha_f - \alpha_i}$$

Now, the angular acceleration of the engine during the clutch slip period of the clutch is,

$$\alpha_i = \frac{M_{ti} - M_t}{I_E}$$

Now $M_{ti} = 99$ Nm

$$\therefore \quad \alpha_i = \frac{99 - 360}{1.025} = -254.5 \text{ rad/s}^2$$

The angular acceleration of the clutch output is

$$\alpha_f = \frac{m_g \,.\, a_t}{R_t}$$

where a_t = acceleration of the tyre = $\dfrac{F_a}{F/g}$

where F_a = acceleration force on automobile.
F = automobile weight.

Now
$$F_a = \frac{M_{ta}}{R_t} = \frac{173}{0.355} = 487 \text{ N}$$

$$\therefore \quad a_t = \frac{487 \times 9.81}{14200} = 0.336 \text{ m/s}^2$$

$$\therefore \quad \alpha_f = \frac{4.11 \times 0.336}{0.355} = 3.88 \text{ rad/s}^2$$

∴ Clutch slip period, $\Delta t = \dfrac{199 - 161}{3.88 + (254.4)} = \mathbf{0.147\ secs}$

This angle through which the input side of the clutch rotates during engagement time Δt is

$$\theta_i = \omega_i \cdot \Delta t + \frac{1}{2}\alpha_i \cdot \Delta t^2$$

$$= 199 \times 0.147 + \frac{1}{2} \times (-254.5)\,(0.147)^2 = 29.5 - 2.75 = 26.75 \text{ rad}$$

For the output side,

$$\theta_f = 161 \times 0.147 + \frac{1}{2} \times 3.88 \times (0.147)^2 = 23.6915 \text{ rad}$$

∴ The angle of slip between clutch input and output during engagement is, therefore,

$$\Delta\theta = \theta_i - \theta_f = 26.75 - 23.69 = 3.06 \text{ rad}$$
$$= 0.487 \text{ rev}$$

(*b*) Energy of friction work to be dissipated by the clutch for each engagement is

$$E = M_t \,.\, \Delta\theta = 360 \times 3.06 = 1101.6 \text{ J}$$

Example 18.4. The following data is given for a dry single plate clutch:

Power = 18.65 kW

Speed = 1500 rev/min

number of springs = 6

Ratio $\dfrac{\textbf{mean radius of friction faces}}{\textbf{radial width of friction faces}} = \mathbf{4.5}$

Find : (a) Mean radius and radial width of the friction faces.

(b) Dimensions of the clutch plate.

(c) Dimensions of the spring.

Solution:

(a) (i) Mean radius: In this type of clutch, in order to ensure reasonable uniform wear, the width of the friction faces is kept small. The following approximate method will also give accurate results:

Area of friction faces = $2\pi\, w r_m$

where w = radial width of friction faces = $r_2 - r_1$

∴ $$P = p_m\, 2\pi\, w\, r_m = p_m \times 2\pi \frac{r_m^2}{4.5}$$

Torsional moment to be transmitted,

$$M_t = \frac{9550 \times kW}{n} = \frac{9550 \times 18.65}{1500} = 118.74 \text{ Nm}$$

now $M_t = \mu P . z r_m$

$z = 2$

Let $p_m = 0.07 \text{ N/mm}^2$ for Ferodo fabric

and $\mu = 0.25$

∴ $$118.74 \times 1000 = 0{\cdot}25 \times 0{\cdot}07 \times 2\pi \times r_m^2 \times \frac{1}{4.5} \times 2 \times r_m$$

∴ $r_m = 135$ mm

(ii) Radial width:

$$w = \frac{135}{4{\cdot}5} = 30 \text{ mm}$$

(b) Clutch Plate: The clutch plate is made of high carbon steel 3 mm thick. It is provided with radial slots to prevent distortion when the plate becomes hot through slipping.

(c) Spring Design:

Now $$M_t = \mu \cdot P \cdot z \cdot r_m$$

∴ Total axial force, $$P = \frac{118 \cdot 74 \times 1000}{0 \cdot 25 \times 2 \times 135} = 1759 \text{ N}$$

∴ $$\text{Load per spring} = \frac{1759}{6} \cong 293 \text{ N}$$

To allow for adjustment and for maximum engine torsional moment, this value is increased by 25 per cent, therefore,

Design load per spring = 293 × 1.25 = 366.25 N

The helical coil spring can now be designed according to the procedure outlined in Chapter 15.

Example 18.5. A cone clutch with leather to iron contact is to transmit power of 22 kW at 94.25 rad/s. The mean diameter is limited to 0.30 m and the slope of the face is to be 12 deg. Assuming coefficient of friction as 0.25.

Determine:

(a) Axial force required to engage the clutch.

(b) Width of the face.

Solution: (*a*) Torsional moment to be transmitted.

$$M_t = \frac{1000 \times (kW)}{\omega} \text{ Nm}, = \frac{1000 \times 22}{94 \cdot 25} = 233.42 \text{ Nm}$$

Now $$M_t = \mu F n r_m$$

∴ $$233.42 = 0.25 \times Fn \times 0.15$$

∴ Normal force, $$F_n = \frac{233.42}{0.25 \times 0.15} = 6224.5 \text{ N}$$

Now the axial force is given as :

$$F_a = F_n \times \left(\sin\theta + \frac{\mu \cos\theta}{4} \right)$$

Now $$\theta = 12 \text{ deg.}$$

∴ $$F_a = 6224 \cdot 5 \left(0 \cdot 2079 + \frac{0 \cdot 25 \times 0 \cdot 9781}{4} \right) = \mathbf{1675.8 \text{ N}}$$

(*b*) Width of the cone face is given as ;

$$b = \frac{F_n}{p \times 2\pi r_m}$$

Assuming an allowable contact pressure of 0.07 N/mm^2

$$b = \frac{6224 \cdot 5}{0 \cdot 07 \times 2\pi \times 150} = 94.3 \text{ mm say } 95 \text{ mm}$$

Example 18.6. (a) Determine the main dimensions of a cone clutch. It is to be faced with leather and is to transmit 30 kW at 750 rev/min from an electric motor to an air compressor. (*b*) Find the axial force that must be produced by the spring.

Solution: Now, the torsional moment to be transmitted is:

$$M_t = \frac{9550 \times 30}{750} = 382 \text{ Nm}$$

Now, to find the shaft diameter,

$$K_t M_t = \frac{\pi}{16} d^3 \times f_s$$

Let $K_t = 1.5$ and $f_s = 42$ N/mm^2, allowing for keyway,

$$\therefore \quad 382 \times 1000 \times 1.5 = \frac{\pi}{16} d^3 \times 42$$

$$\therefore \quad d = 41.11 \text{ mm, say } 41.50 \text{ mm}$$

Now $M_t = \mu \cdot F_n \cdot r_m$

Now $F_n = p \times 2\pi r_m \times b$

$$\therefore \quad M_t = \mu \times 2\pi r_m \times b \times r_m \times p$$

now $\frac{r_m}{b}$ varies from 2.25 to 4

Let us take this as 3

$$\therefore \quad M_t = \mu \times 2\pi\, r_m \times \frac{r_m}{3} \times r_m \times p = \mu \cdot p \cdot \frac{2\pi}{3} \cdot r_m^3$$

$$\therefore \quad r_m = \sqrt[3]{\frac{3}{2\pi} \times \frac{M_t}{\mu p}}$$

Let us assume, $\mu = 0.20$ and $p = 0.105$ N/mm^2

$$\therefore \quad r_m = \sqrt[3]{\frac{3 \times 382 \times 1000 \times 1{\cdot}5}{2\pi \times 0{\cdot}105 \times 0{\cdot}20}} = 235 \text{ mm}$$

Now in commercial clutches, r_m varies from 2.5 d to 5d

Now $\frac{r_m}{d} = \frac{235}{41{\cdot}5} = 5.66$, which is just O.K.

Now peripheral velocity,

$$V = \frac{2\pi \times 235 \times 750}{1000 \times 60} = 18.46 \text{ m/s}$$

For high speed, leather faced clutches, V should be between 10 to 25 m/s, therefore, the value achieved in the design is O.K.

Let us take a trial value of b, as

$$b = \frac{r_m}{3} = \frac{235}{3} = 78.3 \text{ mm}$$

Taking, $\theta = 12.5°$

Small cone diameter $= 2r_m - b \sin\theta = 470 - 78.3 \times 0.2164 = 470 - 16.9 = 453$ mm

Outside cone diameter $= 470 + 16.9 = 487$ mm

Corresponding face width is:

$$b = \frac{487 - 453}{2\sin\theta} = \frac{34}{2 \times 0{\cdot}2164} = 78.6 \text{ mm}$$

(*b*) Now $F_a = F_n \cdot \sin\theta = p \times 2\pi r_m \times b \times \sin\theta$

$= 0.105 \times 2\pi \times 235 \times 78.6 \times 0.2164 =$ **2640 N**

Example 18.7. Design a centrifugal clutch for a belt drive from a shunt wound d.c. motor. The clutch is to be incorporated in the motor pulley, and full spring control is required. The following data are specified:

Power (25 per cent overload to be allowed) = 15 kW
Speed = 720 rev/min
engagement to begin at 75 per cent of the running speed
Number of shoes = 4
Outside diameter of pulley = 35 cm
Inside diameter of pulley rim = 32.5 cm
Width of pulley = 25 cm
Spring type = flat, rectangular section
Initial clearance between friction surfaces, when linings are new = 1.6 mm.

Solution: (a) Weight of Shoes. Let ω_2 = running speed = 720 rev/min

$$\omega_1 = \text{engagement speed} = 0.75 \times 720 = 540 \text{ rev/min}$$

∴ $$\omega_2 = \frac{2\pi \times 720}{60} = 75.3 \text{ rad/s}$$

and $$\omega_1 = 0.75 \times 75.3 = 56.5 \text{ rad/s}$$

Centrifugal force at speed $\omega_2 = F_2c = \dfrac{W_s}{g}\omega_2^2\, r$

and centrifugal force at speed $\omega_1 = F_1c = \dfrac{W_s}{g}\omega_1^2 \cdot r$

where W_s = weight of the shoe
r = radius of gyration of the shoe.

The value of r will be greater at ω_2 by 1.6 mm, but the difference is small as compared to r.

The force F_1c is opposed by an equal and opposite force provided by the spring since the radial movement of the shoe commences only when the speed reaches ω_1.

∴ Net outward radial force on the inside rim of the pulley at speed ω_1 is:

$$F = F_2c - F_1c = \frac{W_s}{g} \cdot r\,(\omega_2^2 - \omega_1^2)$$

∴ Torsional moment to be transmitted,

$$M_t = \mu . R_m . z . F$$

Now $$M_t = \frac{9550 \times 15 \times 1.25}{720} = 249 \text{ Nm}$$

$$R_m = \text{rim radius} = 162.5 \text{ mm}$$

r will be equal to R_m minus the half radial depth of the friction lining.

Let us take r = 150 mm

∴ $$249 \times 1000 = \mu \times 162{\cdot}5 \times 4 \times \frac{W_s}{g} \cdot 150\,(\omega_2^2 - \omega_1^2)$$

Now $$\omega_1 = 0.75\,\omega_2$$

∴ and let us assume μ = 0.25 for Ferodo lining,

∴ $$249 \times 1000 = 0{\cdot}25 \times 162{\cdot}5 \times 4 \times \frac{W_s}{9810} \times 150 \times \frac{7}{16}\,\omega_2^2$$

$$= 0{\cdot}25 \times 162{\cdot}5 \times 4 \times \frac{W_s}{9810} \times 150 \times \frac{7}{16} \times 5670$$

∴ $$W_s = 40.4 \text{ N}$$

If the shoe is made up of C.I., for which density is 7260 kg/m^3

$$\text{Volume of the shoe} = \frac{40\cdot4}{7260} = 0.556\times10^{-3}\ \text{m}^3$$

The dimensions of the shoe can be suitably proportioned in the space available.

(b) Spring: The maximum load on the spring is the centrifugal force at the engagement speed, i.e.,

$$F_s = \text{load on spring} = F_1c = \frac{W_s}{g}\cdot\omega_1^2\cdot r = \frac{40\cdot4}{9.81}\times(56\cdot5)^2\times0\cdot15 = \mathbf{1972\ N}$$

The spring can be designed accordingly.

Example 18.8. In a plate clutch, the axial force is 4500 N. The inner radius of contact is 50 mm and the outside radius is 100 mm. Assuming uniform wear, determine:

(a) Maximum pressure.

(b) Minimum pressure.

(c) Average pressure.

Solution: We know that for uniform wear, the pressure variation is given by:

$$p = \frac{c'}{r},\ \text{where } c' \text{ is a constant given as,}$$

$$c' = \frac{P}{2\pi(r_2 - r_1)}$$

$$\therefore\quad p = \frac{P}{2\pi(r_2 - r_1)\cdot r}\qquad \text{where } P = 4500\ \text{N};\ r_2 = 100\ \text{mm};\ r_1 = 50\ \text{mm}$$

(*a*) Maximum pressure occurs at $r = r_1 = 50$ mm

$$\therefore\quad p_{max} = \frac{4500}{2\pi(100-50)\times50} = \frac{4500}{5000\pi} = 0.2865\ \text{N/mm}^2$$

(*b*) Minimum pressure occurs at $r = r_2 = 100$ mm

$$\therefore\quad p_{min} = \frac{4500}{2\pi(100-50)\times100} = \frac{4500}{10000\pi} = 0.1433\ \text{N/mm}^2$$

(*c*) Average or mean pressure is given as,

$$\therefore\quad p_m = \frac{P}{\text{area of contact}} = \frac{4500}{\pi(r_2^2 - r_1^2)} = \frac{4500}{\pi(100^2 - 50^2)} = 0.191\ \text{N/mm}^2$$

Example 18.9. A multiple disk clutch, steel on bronze, is to transmit 3.68 kW at 750 rev/min. The inner radius of contacts is 38 mm and the outer radius of contact is 70 mm. The clutch operates in oil with an expected coefficient of friction 0.10. The average pressure is 0.35 N/mm² maximum. Determine:

(a) total disks of steel and bronze required.

(b) axial force required.

Solution: (*a*) Let us find the torsional moment capacity of one pair of surfaces in contact.

Assuming uniform wear,

$$M_t = \mu.P.r_m = \mu\cdot P\cdot\left(\frac{r_2 + r_1}{2}\right)$$

$$\text{where}\quad P = p_m\cdot\pi(r_2^2 - r_1^2) = 0.35\,\pi(70^2 - 38^2) = 3800\ \text{N}$$

$$\therefore\quad M_t = 0\cdot10\times3800\times\frac{70+38}{2} = \frac{20520}{1000} = 20.52\ \text{Nm}$$

The total torsional moment applied is $= \dfrac{9550 \times \text{kW}}{\text{rev/min}} = \dfrac{9550 \times 3{\cdot}68}{750} = 46.86$ Nm

$\therefore$ Number of pair of surface $= \dfrac{46{\cdot}86}{20{\cdot}52} = 2.28.$

Now the number must be even. Therefore use 4 pairs of surfaces with $z_{steel} = 3$ and $z_{bronze} = 2$

(*b*) Now, total torsional moment is given as,

$$M_t = \mu \cdot P \frac{r_2 + r_1}{2} \cdot z$$

$$\therefore \quad 46.86 \times 1000 = 0.10 \times P \times \frac{70 + 38}{2} \times 4$$

$$\therefore \quad P = \frac{46{\cdot}86 \times 1000}{0{\cdot}10 \times 4 \times 54} = \mathbf{2169.4\ N}$$

Example 18.10. A multiple disk clutch consists of 5 steel and 4 bronze disks. The torsional moment to be transmitted is 18.5 Nm. The inner diameter is restricted to 50 mm. Determine:

(a) the necessary outer diameter of the disks.

(b) the necessary axial force.

Coefficient of friction may be taken as 0.10. The average pressure is not to exceed 0.35 N/mm^2. Assume uniform wear.

Solution: Number of pair of contact surface.

$$z = z_{steel + bronze} - 1 = 5 + 4 - 1 = 8$$

$$\therefore \quad M_t = \mu . P . r_m . z = \mu \cdot p_m \cdot \pi (r_2^2 - r_1^2) \cdot \frac{r_2 + r_1}{2} \cdot z$$

$$\therefore \quad 18.5 \times 1000 = 0{\cdot}10 \times 0{\cdot}35 \times \pi (r_2^2 - 25^2) \times \frac{r_2 + 25}{2} \times 8$$

From here, by trial and error

$$r_2 = 36.5 \text{ mm}$$

$$\therefore \quad M_t = \mu . P . r_m . z$$

$$\therefore \quad 18.5 \times 1000 = 0{\cdot}10 \times P \times \frac{36{\cdot}5 + 25}{2} \times 8$$

$$\therefore \quad P = \frac{18500}{0{\cdot}1 \times 61{\cdot}5 \times 4} = 752 \text{ N}$$

$$\therefore \quad p_m = \frac{752}{\pi (36.5^2 - 25^2)} = \frac{752}{\pi \times 707} = 0.3385 \text{ N/mm}^2 < 0.35 \text{ N/mm}^2$$

$\therefore$ **O.K.**

Example 18.11. A soft surface cone clutch is to handle 215 Nm of torsional moment at 1250 rev/min. The large diameter of the clutch is 35 cm, the cone pitch angle is 6¼°, the face width *b* is 6.25 cm and the coefficient of friction is 0.20. Determine:

(a) the axial force required to transmit the torsional moment.

(b) the axial force required to engage the clutch, engagement taking place when the clutch is not rotating.

(c) the average normal pressure on the contact surfaces when the maximum torsional moment is being transmitted.

(d) the maximum normal pressure assuming uniform wear.

Solution : Cone angle $\theta = 6¼°$

$$2 \times r_o = 35 \text{ cm};$$

$$\therefore \quad r_o = 17.5 \text{ cm}$$

$$\therefore \quad r_i = r_o - b \sin\theta = 17.5 - 6.25 \times \sin 6\frac{1°}{4} = \mathbf{16.82 \text{ cm}}$$

$$\therefore \quad r_m = \frac{r_o + r_i}{2} = \frac{17{\cdot}5 + 16{\cdot}82}{2} = \mathbf{17.16 \text{ cm}}$$

Now $$M_t = \frac{\mu \cdot F_a}{\sin\theta} \cdot r_m$$

$$\therefore \quad 215 \times 100 = \frac{0{\cdot}20 \times F_a \times 17{\cdot}16}{0{\cdot}1088}$$

(*a*) $$\therefore \quad F_a = \frac{215 \times 100 \times 0{\cdot}1088}{0{\cdot}20 \times 17{\cdot}16} = 682 \text{ N}$$

(*b*) F_a = axial force to engage the clutch

$$= F_n \,.\, (\sin\theta + \mu\cos\theta) = \frac{M_t}{\mu \cdot r_m}(\sin\theta + \mu\cos\theta) = \frac{215 \times 100}{0{\cdot}20 \times 17{\cdot}16}(0{\cdot}1088 + 0{\cdot}20 \times 0{\cdot}994)$$

$$\therefore \quad F_a = \frac{215 \times 100 \times 0{\cdot}3076}{0{\cdot}20 \times 17{\cdot}16} = \mathbf{1930 \text{ N}}$$

(*c*) $$p_{ave} = \frac{F_a}{\pi\,(r_o^2 - r_i^2)} = \frac{682}{\pi(17.5^2 - 16{\cdot}82^2)} = \mathbf{0.0944 \text{ N/mm}^2}$$

(*d*) $$p_{max} = \frac{F_a}{2\pi\,(r_o - r_i) \times r_i} = \frac{682}{2\pi\,(175 - 168{\cdot}2) \times 168{\cdot}2} = \mathbf{0.0947 \text{ N/mm}^2}$$

18.4 BRAKES

The brake is a frictional device whose primary function is to control the motion of a machine or a machine member. In doing so, it is called upon to bring to rest a body which is in motion or to slow it down or to hold it in state of rest or of uniform motion against the action of external forces or couples. Other uses of brakes are to release a load and control its speed or to prevent an unwanted reversal of the direction of rotation. In the process of performing this primary function, it is required to absorb kinetic energy of the moving parts or the potential energy of the objects being lowered by hoists, elevators, etc., and convert these energies into internal energy of the brake components and dissipate it in the form of heat. This heat is dissipated by means of convection heat transfer to the surrounding air or by means of water which is circulated through the passages in the brake drum. The difference between a clutch and a brake is that whereas in the former both the members to be engaged are in motion, a brake connects a moving member to a stationary member.

Two factors are of paramount importance in the design of a brake:

1. Determination of the torsional moment it must provide.
2. Amount of energy which the brake must dissipate repeatedly.

18.5 ENERGY TO BE ABSORBED BY THE BRAKE

The type of energy to be absorbed by the brake will depend upon the type of motion of the moving body. The motion can be: (1) Pure translation (2) Pure Rotation (3) A combination of translation and rotation, e.g. a rolling wheel. The energy corresponding to these motions is kinetic energy. Sometimes the brake will have to absorb potential energy also in addition to Kinetic energy, e.g. a vertical or an inclined bucket elevator. The Kinetic energy of a body with a pure translatory motion is:

$$K.E. = \frac{W_b V^2}{2g} \text{ Nm} \qquad ...(18.25)$$

where W_b = weight of the body, N, and V = velocity in m/s

If the brake does not stop a moving body but only reduces its speed , then the energy to be absorbed by the brake is the difference of Kinetic energies at the two speeds.

The Kinetic energy of a rotating body about its centre of rotation is:

$$K. E. = \frac{I\omega^2}{2} \qquad ...(18.26)$$

where I = moment of inertia of the mass of body about its axis of rotation.

ω = angular velocity of the rotating body.

For the rolling body, the total Kinetic energy will be the sum of the Kinetic energy of translation and that of rotation.

If F is the frictional force and the body is to be brought to rest in a distance 's', then Frictional loss is given as:

$$\text{Frictional loss} = F \times s$$

Now $s = \pi Dn$

where D = diameter of brake drum, and n = number of turns of the brake wheel during which the body is to be brought to rest.

Since the Kinetic energy dissipated must be equal to the frictional loss,

Therefore, total K.E. absorbed = $F \times \pi Dn$

Since $F \times \frac{D}{2}$ = Frictional torsional moment of the brake, T_f

$$\therefore \quad T_f = \frac{K.E.}{2\pi n} \qquad ...(18.27)$$

In the case of bucket elevator, if 's' is the vertical distance required to stop the load, then

Total Kinetic energy plus potential energy to be absorbed by the brake will be given as:

= K. E. due to translation + K. E. due to rotation + Potential energy

$$= \frac{W_b V^2}{2g} + \frac{I\omega^2}{2} + W_b \cdot s.$$

18.5.1 Dissipation of Heat

For the useful service by a brake, it should not be allowed to overheat. Heat dissipating capacity of the brake depends upon the braking surface and the unit pressure. Increased braking surface and reduced unit pressure results in longer life for the brake.

Thus, heat generated is,

$$H_g = \mu.Fn.V, \text{ watts}$$

$$= \mu.p.A_c.V, \text{ watts} \qquad ...(18.27a)$$

where μ = coefficient of friction, p = average contact pressure, N/m^2

A_c = contact area, m^2, V = rubbing velocity, m/s

The heat generated can also be found out by considering the amount of kinetic energy and/or potential energy that is being absorbed. Thus,

$$H_g = (E_k + E_p) \text{ watts}$$

Now, the heat energy dissipated may be estimated by,

$$H_d = C\,\Delta T\,.\,A_r \text{ watts}$$

where C = coefficient of heat transfer, W/m^{2}K

ΔT = temperature difference between the exposed radiating surface and surrounding air

A_r = area of radiating surface, m^2

The value of C may be about 29.5 for a ΔT of 40°C and 44 for a ΔT of 200°C.

Another simple criterion of brake capacity is,

$$= \frac{kW}{wD}, \text{ which is limited to 350 W/m}^2$$

where w = width of band or shoe, m

D = diameter of drum, m.

Or the temperature rise of the brake assembly can be found out by equation (18.6*a*).

All the above relations are approximate due to a large number of factors involved.

Now in a unit time, the heat generated will be equal to the frictional force times the rubbing velocity (equation. 18.27*a*). Now if the coefficient of friction is fixed, then

Heat generated in a unit time = Normal force × rubbing velocity. For a given braking surface,

Heat generated in a unit time = Unit pressure × rubbing velocity = $p \times V$.

Thus the product pV is the common criterion for the design of brakes. The upper limits of pV that will give a reasonable wear life for the brake are given in Table 18.4, p is in N/m^2 and V in m/s.

Table 18.4 Operating limits for brakes

Operating Condition	Upper limit of pV
1. Continuous application of load and poor dissipation of energy.	10.5×10^5
2. Intermittent application of load, comparatively longer periods of rest, poor dissipation.	12×10^5
3. Continuous application of load and good dissipation of heat, as in an oil bath.	29.4×10^5

Some typical values of the maximum temperature are given below:

(i) Leather, fibre and wood facing = 65–70°C
(ii) Asbestos = 90–105°C
(iii) Asbestos block lining for automotives = 180–225°C.

Some values for coefficients of friction and allowable pressure are given in Table 18.5.

Table 18.5 Coefficient of friction and allowable pressures

Material in contact	Coefficient of friction			Allowable pressure N/mm^2
	Dry	Greasy	Oily	
Cast Iron on Cast Iron	0.2–0.15	0.10–0.06	0.10–0.05	1.05–1.75
Bronze on Cast Iron	—	0.10–0.05	0.10–0.05	0.56–0.84
Steel on Cast Iron	0.35–0.25	0.12–0.07	0.10–0.06	0.84–1.40
Wood on Cast Iron	0.35–0.20	0.12–0.08	—	0.42–0.63

Material in contact	Coefficient of frictions			Allowable pressure N/mm^2
	Dry	Greasy	Oily	
Fibre on metal	—	0.20–0.10	—	0.07–0.28
Cork on metal	0.35	0.30–0.25	0.25–0.22	0.055–0.105
Leather on metal	0.5-0.3	0.20–0.15	0.15–0.12	0.07–0.28
Wired asbestos on metal	0.5–0.35	0.30–0.25	0.25–0.20	0.21–0.56
Asbestos in rubber compound on metal	0.3–0.4	—	—	0.525–0.700
Asbestos in resin binder, on metal	0.3–0.4	—	0.10	0.525–0.700 4.2 for oily
Powdered metal on metal	0.2–0.4	—	0.05–0.08	2.8

18.6 TYPES OF BRAKES

The brakes are usually classified according to the means for transforming energy by the braking element. The common types of brakes are:

1. **Hydraulic Brakes:**
   (*a*) Pump or Hydrodynamic; (*b*) Agitator
2. **Electric Brakes :**
   (*a*) Generator; (*b*) Eddy Current.
3. **Mechanical Brakes:**
   (*a*) Block; (*b*) External and internal shoe; (*c*) Band; (*d*) Disk; (*e*) Slipper.

Both hydraulic and electric brakes are mostly used where large amounts of energy are to be transformed while retarding the load e.g. laboratory dynamometer, highway trucks and electric locomotive. They also find application in retarding or controlling the speed of a vehicle on long downgrades. Their use for the retarding of moving loads is also limited. The most important consideration in the use of these brakes is that these should not be dependent upon to hold a load stationary since the braking torsional moment is a function of the speed of rotation of the brake rotor.

In this chapter, only the mechanical brakes shall be discussed. The mechanical brakes may be divided into two groups according to the direction of the acting force: (*a*) Radial Brakes; (*b*) Axial Brakes.

Radial brakes may be subdivided as "External Brakes" and "Internal Brakes". Depending upon the shape of the friction elements, these brakes may be named as "Block or Shoe Brakes" and "Band Brakes".

Axial brakes may be further classified as "Disk Brakes" and "Cone Brakes".

18.6.1 Block or Shoe Brakes

In these brakes, relatively rigid blocks or shoes press against the inside or outside surfaces of the brake drum. The brake drum is keyed on to the shaft. A single block brake is shown in Fig. 18.14. P is the force which is applied at the end of the brake arm or lever to press the block against the brake drum. F_n is the normal reaction between the block and the drum. The brake arm is pivoted about the pin A.

In the Fig. the equilibrium of the lever and the shoe has been considered. Therefore, F_n, is the normal reaction exerted by the brake wheel on the block. This is in opposite direction to the force applied on the drum by the shoe, by virtue of the application of the brake actuating force, P. Similarly, for counter clockwise motion of the drum, the frictional force, F, applied by the drum on the block is in the L.H.S. horizontal direction. Here, the shoe is considered to be a short shoe so that the pressure distribution along the shoe is uniform. Hence the normal force F_n and frictional force F may be shown acting at the midpoint of contact of the shoe.

Braking Force $= \mu . F_n$

$\therefore$ Brake torsional moment $= \mu . F_n . r$

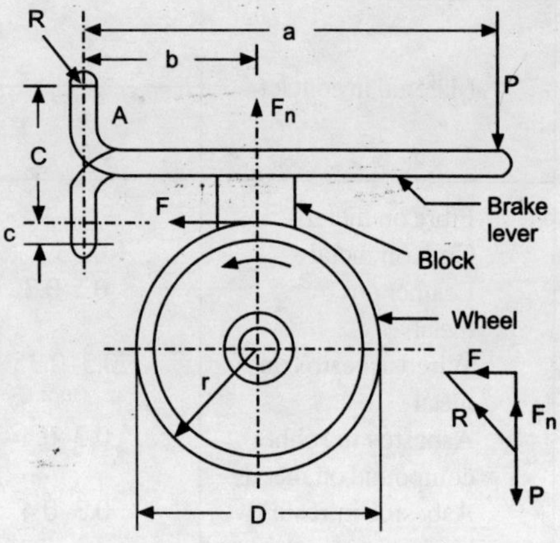

Fig. 18.14 *Block brake*

The brake arm may be considered as a "Free body" under the action of the forces, which are:

(*i*) The brake actuating force, P

(*ii*) The braking force $\mu . F_n$

(*iii*) The normal reaction F_n between the block and the drum.

(*iv*) The pin reaction R, which is unknown in magnitude and direction. Since the arm is in equilibrium, therefore, the moments about any point must be zero. Taking moments about the point A,

$$\Sigma M_A = 0$$

$$\therefore \quad M_p + M_n + M_\mu = 0 \quad \text{...(18.28)}$$

where M_p = moment due to the actuating force P.

M_n = moment due to the normal reaction F_n.

M_μ = moment due to the friction force or braking force.

$$\therefore \quad P_a - F_n . b + F.c = 0$$

or
$$P = \frac{F_n \cdot b - F \cdot c}{a}$$

now
$$F = \mu . F_n$$

$$\therefore \quad P = \frac{F}{\mu a}(b - \mu c) \quad \text{...(18.29)}$$

This is the force which should be applied to produce a friction force F.

When the friction acts so as to assist in applying the brake, the brake is said to be "Self-actuating" or "Self-energizing" For this M_μ acts in the same direction as M_p, or when M_μ and M_n have opposite signs. It is clear from the above equation (18.28), that when

$$M_\mu = M_n, \; M_p = 0 \text{ or } P = 0$$

i.e., the extent of self energizing is so much that no external force is required to apply the brake. Such a brake is called "Self locking", i.e., when the block touches the drum, the brake gets locked instantaneously. Hence the conditions for self locking are:

1. brake is self energizing.
2. $M_\mu \geq M_n$

Self-locking is needed only for back-stop applications. For other conditions, it is desirable to avoid the self locking or to prevent the brake from "grabbing". Hence the friction moments should be less than the total required braking. This margin should be enough to take into account the change in coefficient of friction.

From equation (18.29), it is clear that the brake shall be self-locking, if

$$\mu . c \geq b$$

Taking an average value of 0.3 for μ, for self-locking property,

$$b = 0.3\, c$$

Now braking torsional moment,. $T_f = F.\, r = \mu\, .\, F_n\, .\, r.$

$\therefore$ from equation (18.29), multiplying both sides by r

$$T_f = \frac{P.r\cdot\mu\cdot a}{(b-\mu\cdot c)} \quad ...(18.30)$$

Knowing the various forces acting on the brake arm, it can be designed. Again the value and direction of the pin reaction R may be determined by drawing the force diagram as shown. Knowing the magnitude of R, the pivot pin and the pin bearing may be designed.

However, the horizontal and vertical components of the pin reaction at A can be found by summing the horizontal and vertical forces:

$$R_x = F = \mu\,.\,F_n$$

$$R_y = F_n - P$$

The brake shown in the Fig. 18.14 is self energizing because the frictional force helps the actuating force. If the pivot point A is located below the line of action of the frictional force (dotted line), the moment due to frictional force will oppose that of the actuating force, and the brake would not be self-energizing. However, if the direction of rotation is reversed, the situation discussed above will be reversed.

(a) Double-block Brakes: Although the single-block brake is simple and reliable, the major objection is the unbalanced normal force *Fn* which exerts a heavy pressure on the shaft bearings and produces a bending of the shaft. Due to this reason the block brakes are usually made with two opposed shoes or blocks so as to eliminate or reduce the unbalanced force. Such a brake is frequently used in hoisting machinery, e.g., electric cranes and block brakes with either one or two shoes are universally used as Railway brakes. A double shoes brake is shown in Fig. 18.15. The setting of the brake is done by means of the spring '*s*' which pulls the upper ends of the brake arms together, and thus the brake is applied. The force *P* at the end of the lever *b, is* applied with the help of a solenoid. When the solenoid is energised, it exert a force on the lever sufficient to overcome the spring force and thus brake gets released and the drum starts rotating. When the current goes off, the force *P* will become zero and the brake will get engaged automatically due to the spring force and hence there will be no down movement of the load. Each link of the brake may be considered separately as a free body for force analysis.

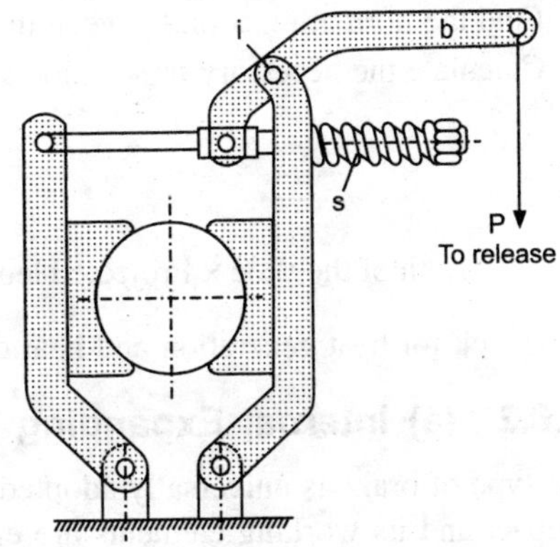

Fig. 18.15 *Double shoe brake*

(b) Symmetrically Pivoted Shoe Brakes: In the case of block or shoe brakes discussed earlier, it is assumed that the pressure distribution between the shoe and the drum is uniform and that there is point contact between the shoe and the drum. These brakes are usually used for low value of braking load. A pivoted shoe brake is shown in Fig. 18.16. It is assumed that the braking force acts through the point B which is coincident with the brake shoe pivot.

The normal reaction will also pass through this point and hence no moment will act on the shoe. This will result in more uniform pressure distribution and uniform wear of the lining. Therefore, these brakes have more life and can provide a greater braking torsional moment. The distance '*h*' of the pivot from the drum centre is given by relation :

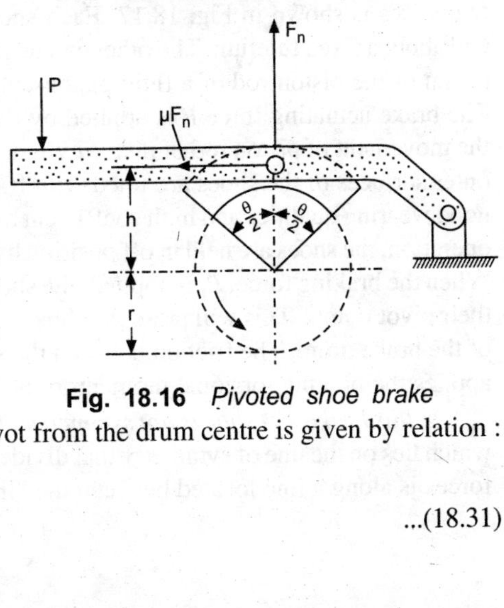

Fig. 18.16 *Pivoted shoe brake*

$$h = \frac{4\sin\dfrac{\theta}{2}}{\theta + \sin\theta} \quad ...(18.31)$$

where θ = shoe angle, rad.

It has been found that if θ is less than about 60°, there is no significant difference between h and r. So, the design of a simple block brake and pivoted shoe brake will differ only if θ is more than 60°. The braking torsional moment of a pivoted shoe brake will be given as:

$$T_f = \mu \,.\, F_n \,.\, h \quad ...(18.32)$$

$$= \mu \cdot F_n \cdot r \cdot \frac{4 \sin \frac{\theta}{2}}{\theta + \sin \theta} \quad ...(18.33)$$

The drawbacks of the pivoted shoe brake are:

1. Physical problem in locating the pin so near to the drum surface.
2. Maintaining the correct position of the pivot as the lining wears out.

(c) Design Procedure of Block Brakes:

1. Select the brake drum diameter in accordance with the given load, given speed and other data.

$$= \frac{M_t}{(0{\cdot}4335)^{1/3}} < D < \frac{M_t}{(0{\cdot}3472)^{1/3}}\,;\ M_t = \text{Design braking torque}$$

2. Determine the friction force.
3. Calculate the torsional moment capacity of the brake.
4. Select the friction material and the corresponding coefficient of friction.
5. Find the normal reaction between the drum and the block or shoe.
6. Calculate the necessary brake shoe area by the relation

$$A = \frac{F_n}{p}$$

A = width of the shoe × Projected length of shoe = $w \times 2r \sin \theta/2$ 'w' is taken as $\left(\frac{1}{4} \text{ to } \frac{1}{3}\right) \times D$

7. Check for heat generation and heat dissipation.

18.6.2 (a) Internal Expanding Shoe Brake

This type of brake is universally adopted for automobiles. It is compact and its working elements are enclosed and protected from dirt, etc., by the rotating drum. This brake with one of the two shoes is shown in Fig. 18.17. Each shoe is pivoted at one end about a fixed fulcrum. The other end rests against the face of a cam or the piston rod of a fluid piston-cylinder arrangement. The brake actuating force P is applied by the cam operation or the movement of piston rod of piston cylinder arrangement. The outer surfaces of the shoes are lined with friction material with good wearing qualities and high coefficient of friction. In normal operation, the shoes are held in off position by means of a spring. When the braking force, P, is applied, the shoes will rotate about their pivot points. This will press the shoes against the inner rim of the brake drum. The friction between the shoes and the drum applies the braking torsional moment to the drum.

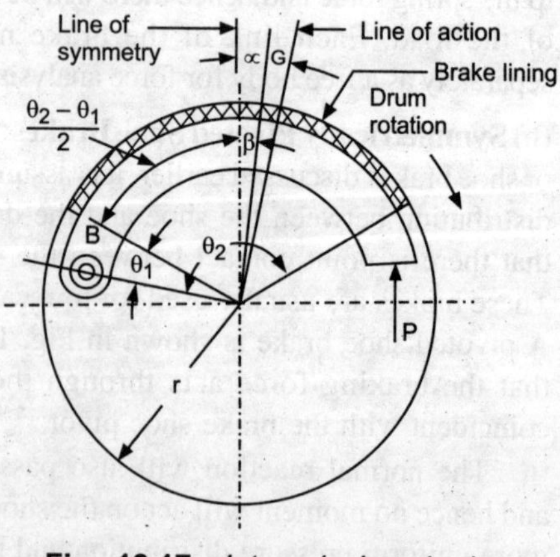

Fig. 18.17 *Internal shoe brake*

In this brake, the shoe is not symmetrically pivoted, hence the friction forces will not pass through a point which lies on the line of symmetry that divides the brake lining angle $(\theta_2 - \theta_1)$. Instead, the resultant of friction forces is along a line located between the "line of symmetry" and "line of action" as shown in Fig. 18.17.

The angle β between the line of symmetry and the line on which the point G lies, is given by the relation:

$$\tan\beta = \tan\alpha \cdot \frac{(\theta_2-\theta_1)-\sin(\theta_2-\theta_1)}{(\theta_2-\theta_1)+\sin(\theta_2-\theta_1)} \quad ...(18.34)$$

where α = angle between the line of symmetry and line of action.

Force analysis of Internal expanding Shoe brake: The complete internal expanding brake with two identical shoes is shown in Fig. 18.18. The shoes are hinged at pivot points O_1 and O_2. The brake drum rotation is C.C.W. For this case, the left shoe is called "Leading shoe" and the right shoe, the "Trailing shoe". In such brakes, at least one shoe per wheel is made self-energizing. This results in great braking power. Hence, much pedal pressure need not be applied.

Let us consider a small element of brake lining at an angle 'θ' from OO_1 (line joining pivot point to the drum centre) and subtending an angle $\delta\theta$ at the drum centre. We are considering the leading shoe.

θ_1 = Angle between the line OO_1 and the start of the brake lining,

θ_2 = Angle between the line OO_1 and the end of the brake lining,

$\therefore \quad \theta_2-\theta_1$ = Brake lining included angle.

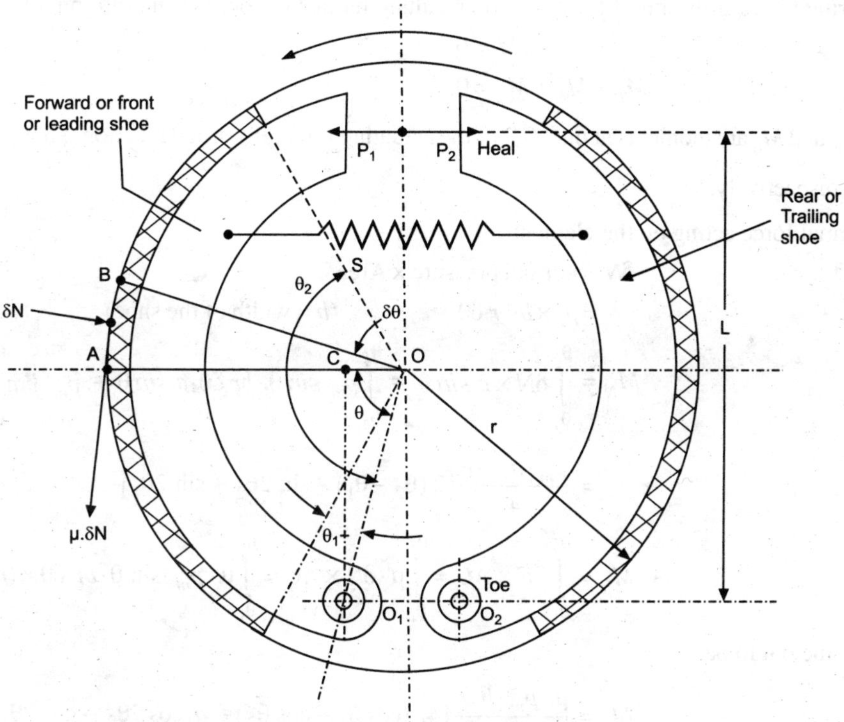

Fig. 18.18 *Force analysis of internal expanding shoe brake*

Assumptions: 1. Forces applied to the shoes P_1 and P_2 are equal in magnitude ($P_1 = P_2 = P$) and act in one line.

2. Shoe is rigid as compared to brake lining. The brake lining when compressed obey's Hooke's law. That is, the compression pressure on the lining and hence the normal reaction at any point on the contact surface is proportional to its distance away from the pivot point, i.e., perpendicular distance from O_1 on to the line OA, which is a sin θ ($OO_1 = a$).

$$\therefore \quad p_n \propto a \sin\theta$$

So, the normal pressure distribution is not uniform in such a long shoe.

$$\therefore \quad p_n = K.a.\sin\theta, \text{ where}$$

$$K = \text{Brake constant, N/m}^3.$$

Let p_m = Maximum normal pressure on the shoe and θ_m is the angle with OO_1 at which it occurs, i.e.,

$$p_m \propto a \,.\, \sin \theta_m$$

or $$p_m = K \,.\, a \,.\, \sin \theta_m$$

$\therefore$ $$\frac{p_m}{\sin \theta_m} = \frac{p_n}{\sin \theta}$$

or $$p_n = p_m \cdot \frac{\sin \theta}{\sin \theta_m}$$

It is clear that p_m occurs at $\theta_m = 90°$.

$\therefore$ $$p_n = p_m \cdot \sin \theta.$$

Note: Even though, the actuating force, P, is assumed to be equal in magnitude and act in one line, but K_L is > K_T, that is, the leading shoe contributes greater braking torque (more than 50%) on the drum. So, p_{mL} will be > p_{mT}.

Considering the equilibrium of the shoe, the resultant moment about the fulcrum pin O_1 is zero, i.e.,

$$\Sigma M = 0$$

or $$M_P + M_N + M_\mu = 0,$$

where M_P, M_N and M_μ are moments about 'O_1' of the actuating force, P, normal reaction $\int \delta N$ and the friction force $\mu \int \delta N$ respectively.

Now normal force acting on the element,

$$\delta N = \text{normal pressure} \times \text{Area}$$

$$= p_n \times b \,.\, r\, \delta\theta \qquad (b = \text{width of the shoe})$$

$\therefore$ $$M_N = \int_{\theta_1}^{\theta_2} \delta N \times a \sin \theta = \int_{\theta_1}^{\theta_2} p_m \sin \theta \cdot br\, \delta\theta \cdot a \sin \theta = p_m \cdot bra \int_{\theta_1}^{\theta_2} \sin^2 \theta \, d\theta$$

$$= \frac{p_m \cdot b \cdot r \cdot a}{4} [2(\theta_2 - \theta_1) - \sin 2\theta_2 + \sin 2\theta_1]$$

$$M_\mu = \int_{\theta_1}^{\theta_2} \delta F \times AC = \int_{\theta_1}^{\theta_2} \mu \cdot \delta N \times AC = \int_{\theta_1}^{\theta_2} \mu \cdot p_m \cdot \sin \theta \cdot br\, \delta\theta \times (r - a \cos \theta)$$

It will come out to be

$$M_\mu = \frac{\mu \cdot p_m \cdot b \cdot r}{4} [4r (\cos \theta_1 - \cos \theta_2) + a(\cos 2\theta_2 - \cos 2\theta_1)]$$

Now $$M_P = P \times L$$

$\therefore$ $$M_P + M_\mu - M_N = 0$$

M_μ supports M_P, therefore, the brake is self-energizing. It becomes self-locking if $M_\mu = M_P$

$\therefore$ $$P = \frac{M_N - M_\mu}{L}$$

Now braking torque on the element

$$= \delta F \times r$$

$\therefore$ Brake torque of the leading shoe $= \int_{\theta_1}^{\theta_2} \delta F \times r$

$= \mu \cdot p_{mL} \cdot br^2 (\cos\theta_1 - \cos\theta_2)$ (p_{mL} = maximum pressure on the leading shoe).

$\therefore$ Total braking torque of the complete brake,

$$T_b = \mu \cdot br^2 \cdot (\cos\theta_1 - \cos\theta_2)(p_{mL} + p_{mT})$$

where p_{mT} = maximum pressure on the trailing shoe.

Hinge Pin Reaction:

$$R_X = p_m \cdot br \left[\int_{\theta_1}^{\theta_2} \sin\theta \cos\theta \, d\theta - \mu \int_{\theta_1}^{\theta_2} \sin^2\theta \, d\theta \right] - P_X$$

That is,

$$= \int \delta N \cdot \cos\theta - \int \mu \, \delta N \sin\theta - P_X$$

$$R_Y = p_m \cdot br \left[\int_{\theta_1}^{\theta_2} \sin^2\theta \, d\theta + \mu \int_{\theta_1}^{\theta_2} \sin\theta \cos\theta \, d\theta \right] - P_Y$$

Note: Here, a general case is taken when P is inclined to the heel end of the shoe.

For Clock-wise Rotation

$$P = \frac{M_N + M_\mu}{L}$$

$$R_X = p_m \cdot br \left[\int_{\theta_1}^{\theta_2} \sin\theta \cos\theta \, d\theta + \mu \int_{\theta_1}^{\theta_2} \sin^2\theta \, d\theta \right] - P_X$$

$$R_Y = p_m \cdot br \left[\int_{\theta_1}^{\theta_2} \sin^2\theta \, d\theta - \mu \int_{\theta_1}^{\theta_2} \sin\theta \cos\theta \, d\theta \right] - P_Y$$

(b) External Contracting Shoe Brake: Refer Fig. 18.19. The drum rotation is C.W. Therefore, the shoe shown is "Leading shoe".

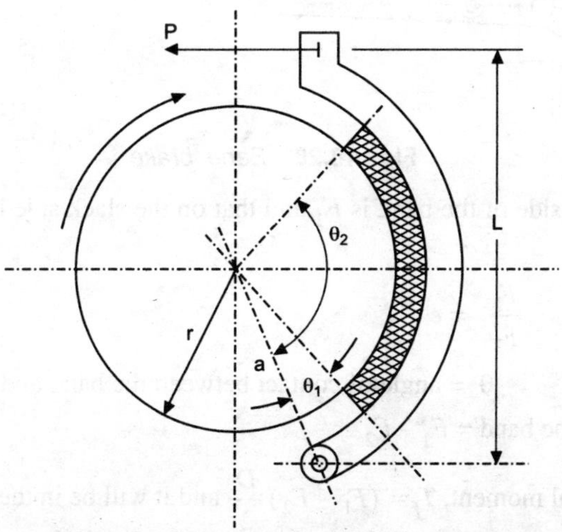

Fig. 18.19 *External contracting shoe brake*

(*i*) **For Clock-wise Rotation:**

$$M_N = p_m \cdot br \cdot a \int_{\theta_1}^{\theta_2} \sin^2 \theta \, d\theta$$

$$M_\mu = \mu \cdot p_m \cdot br \cdot a \int_{\theta_1}^{\theta_2} \sin \theta \, (r - a \cos \theta) \, d\theta$$

$$P = \frac{M_N + M_\mu}{L}$$

(*ii*) **For C.C.W. Rotation:**

$$R_X = p_m \cdot br \left[\int_{\theta_1}^{\theta_2} \sin \theta \cos \theta \, d\theta + \mu \int_{\theta_1}^{\theta_2} \sin^2 \theta \, d\theta \right] - P_X$$

$$R_Y = p_m \cdot br \left[\mu \int_{\theta_1}^{\theta_2} \sin \theta \cos \theta \, d\theta - \int_{\theta_1}^{\theta_2} \sin^2 \theta \, d\theta \right] - P_Y$$

$$P = \frac{M_N + M_F}{L}$$

18.6.3 Band Brakes

The general arrangement of a band brake is shown in Fig. 18.20. It consists of a steel band or strap lined on one side with leather, wooden blocks or an asbestos composition material. The actuating force is applied so that it tightens the band around the brake drum.

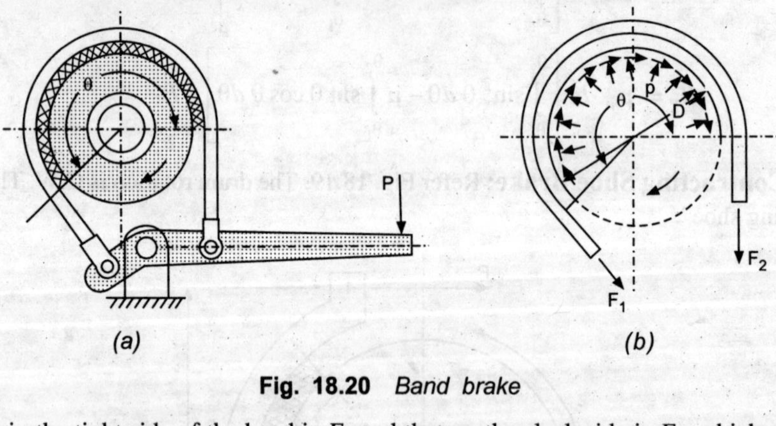

Fig. 18.20 *Band brake*

The tension in the tight side of the band is F_1 and that on the slack side is F_2, which are related by the formula:

$$\frac{F_1}{F_2} = e^{\mu \theta} \qquad \text{...(18.35)}$$

where θ = angle of contact between the band and the brake drum.

Total friction force on the band = $F_1 - F_2$

∴ The braking torsional moment, $T_f = (F_1 - F_2)\dfrac{D}{2}$ and it will be in the direction of F_1.

A free body diagram of the brake lever is shown in Fig. 18.21.

The sum of the moments about the fulcrum should be zero. Therefore,

$$\Sigma M_0 = 0$$

i.e.,
$$P \cdot a + F_1 \cdot c - F_2 \cdot b = 0 \quad \text{...(18.36)}$$

As discussed in the case of block brakes, a brake will be self energizing if the friction force assists the braking effort. Since the friction force is $(F_1 - F_2)$ and it acts in the direction of F_1, therefore, a band brake will be self-energizing when F_1 acts to apply the brake as in Fig. 18.20.

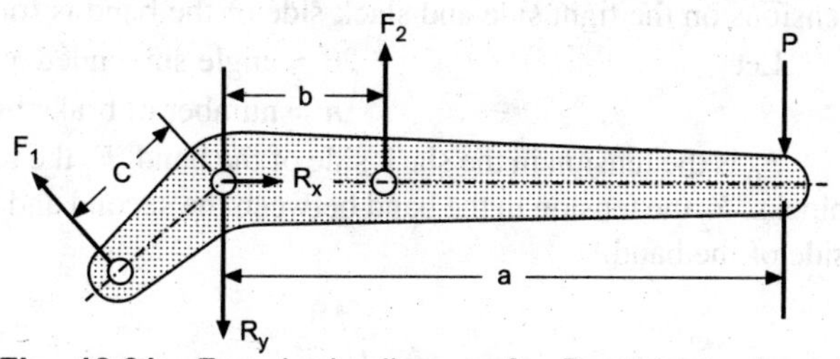

Fig. 18.21 *Free body diagram for Band brake lever*

From equation (18.36), if $F_1 \cdot c$ is $\geq F_2 \cdot b$, the actuating force will be negative or equal to zero as the case may be. This means that no external force is needed to apply the brake. Such a brake is called as self-locking.

If the direction of rotation of the drum is reversed, the forces, F_1 and F_2 will exchange their places since F_1 must always act to resist the rotation of the drum. In this case a greater value of P will be required to apply the brake which is not a desirable feature.

From equation (18.36)

$$P.a = F_2 b - F_1 c = F_2(b - c.\, e^{\mu\,\theta})$$

$\therefore$ For the brake to be self-locking,

$$b \leq c.e^{\mu\theta}$$

or
$$\frac{b}{c} \leq e^{\mu\theta}$$

The band brake shown in Fig. 18. 20 is called " Differential band brake". If one of the ends of the band is attached to the fixed fulcrum, then the band brake is called as a "Simple band brake".

The average pressure on the band will be given as:

$$p = \frac{F_1 + F_2}{D \cdot w} \quad \text{...(18.37)}$$

where w = width of the band.

Since the tension in the band is not constant, the pressure is the maximum at the tight side and minimum at the slack end side, so.

$$p_{max} = \frac{2F_1}{D \cdot w} \text{ and } p_{min} = \frac{2F_2}{D \cdot w}$$

The maximum bearing pressure between steel band and friction lining on C.I. drum varies from 0.5 to 0.7 MPa.

The band thickness is given empirically as,

$$h = 0.005\, D.$$

now
$$w.h = \frac{F_1}{f_t}$$

where f_t = safe stress in tension for the band material which is usually mild steel.

Since the application of load may be sudden, a factor of safety of 4 should be used based on the yield strength to find out the value of f_t which is usually 50 to 80 N/mm^2.

18.6.4 Band and Block Brake

In this type of brake, wooden blocks placed between the steel band and the brake drum act as friction material. These are easily replaceable when these get worn out. Instead of a steel band, holes may be drilled in the blocks and ropes passed through them. A band and block brake is shown in Fig. 18.22. The ratio of the tensions on the tight side and slack side of the band is found as follows:

Let 2θ = angle subtended at the drum centre by each block.

n = number of brake blocks.

F_0 is the tension in the slack side of the band, F_1 the tension on the band between the first and the second blocks. F_2 the tension in the band between the second and the third blocks and so on. F_n is the tension in tight side of the band.

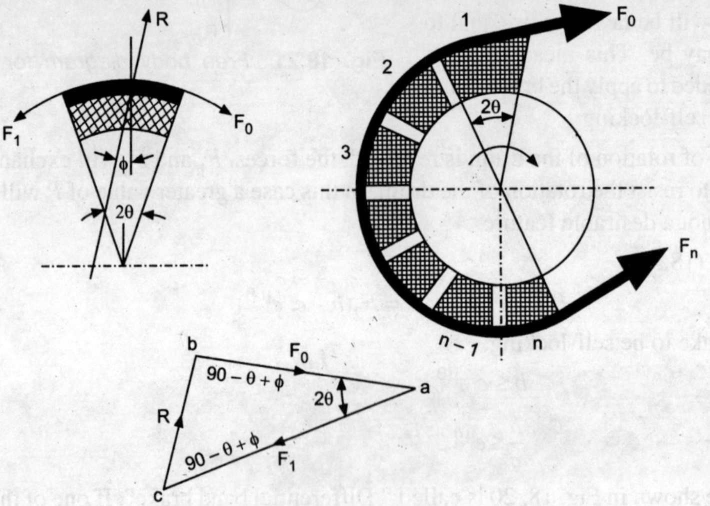

Fig. 18.22 *Band and block brake*

Considering the first block. It is in equilibrium under the action of three forces:

(i) Tension F_0; (ii) Tension F_1; (iii) Reaction R of the drum on the block.

From the triangle of forces :

$$\frac{F_1}{F_0} = \frac{\sin abc}{\sin acb} = \frac{\sin(90-\theta+\phi)}{\sin(90-\theta-\phi)} = \frac{\cos(\theta-\phi)}{\cos(\theta+\phi)}$$

$$= \frac{\cos\theta\cos\phi + \sin\theta\sin\phi}{\cos\theta\cos\phi - \sin\theta\sin\phi} = \frac{1+\tan\theta\tan\phi}{1-\tan\theta\tan\phi} = \frac{1+\mu\tan\theta}{1-\mu\tan\theta}$$

The triangle of forces of each block is similar and therefore the ratio of the tensions in the band for each block is same.

Therefore,
$$\frac{F_n}{F_0} = \frac{F_n}{F_{n-1}} \cdots \frac{F_3}{F_2}\cdot\frac{F_2}{F_1}\cdot\frac{F_1}{F_0} = \left(\frac{F_1}{F_0}\right)^n = \left(\frac{1+\mu\tan\theta}{1-\mu\tan\theta}\right)^n \qquad \text{...(3)}$$

Procedure: From Power

$$P(\text{kW}) = \frac{2\pi NT}{60\times 1000}$$

$$T = (F_n - F_0)\times r \qquad \text{...(1)}$$

where r = Radius of drum.

Equation of Lever: relation between F_0 and F_n. ...(2)

From equation (1) and (2). Find F_0 and F_n. Then from equation (3), find n.

18.6.5 Disk Brakes

The analysis of disk brakes is similar to the one for disk clutches.

Example 18.12. The brake shown in Fig. 18.23, is fitted with a cast iron brake shoe. The coefficient of friction is 0.30. The braking torsional moment is to be 346 Nm.

(a) Determine the force, P, for counter clockwise rotation.

(b) Determine the force, P, with clockwise rotation.

(c) Where must the pivot be placed to make the brake self-locking with the counter clockwise direction.

Solution: The clockwise direction is shown by full line and the counter clockwise direction by the dotted line.

(*a*) Taking moment about the fulcrum point:

$$P \times 0.80 + F \times 0.05 - F_n . 0.20 = 0$$

now $\quad F \times 0.18 = 346$

$$\therefore \quad F = \frac{346}{0 \cdot 18} = 1922 \text{ N}$$

Also $\quad F = \mu . F_n$

$$\therefore \quad F_n = \frac{1922}{0 \cdot 3} = 6410 \text{ N}$$

$$\therefore \quad P \times 0.80 + 1922 \times 0.05 - 6410 \times 0.20 = 0$$

$$\therefore \quad P = 1482.4 \text{ N}$$

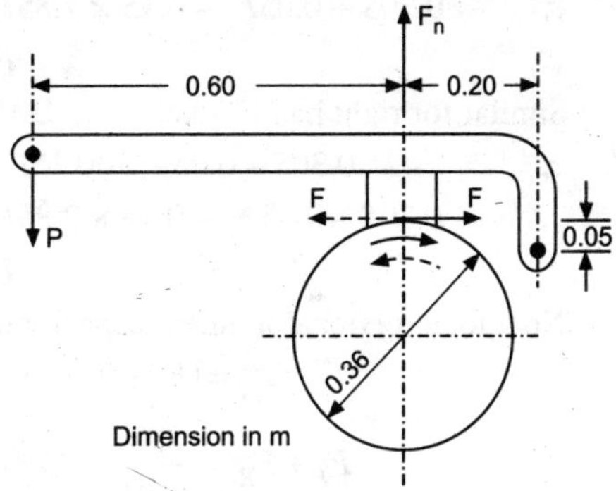

Fig. 18.23 *Block brake*

(*b*) Again taking moments about the pivot.

$$P \times 0.80 - 1922 \times 0.05 - 6410 \times 0.20 = 0$$

$$\therefore \quad P = 1722.5 \text{ N}$$

(*c*) Let the distance of the line of action of *F* from pivot be, *x*

$\therefore$ The moment equation is :

$$P \times 0.80 + Fx - F_n . 0.20 = 0$$

Now for self-locking, *F* acts to produce the braking effect and *F.x* must be at least equal to F_n.0.20, so that *P* is zero.

$$\therefore \quad F . x = F_n . 0.20$$

$$\mu . Fn\, x = F_n . 0.20$$

$$x = \frac{0 \cdot 20}{0 \cdot 30} = \mathbf{0.667 \text{ m}}$$

Example 18.13. The layout of a double block brake is shown in Fig. 18.24. The brake is rated at 240 Nm at 600 rev/min. The angle of contact for each shoe is 120°. Assuming coefficient of friction to be 0.3 and for conditions of service a *pV* value of 17.5 × 10[5] may be assumed. Determine:

(a) Spring force S required to set the brake.

(b) Width of shoes.

Solution: From equation (18.33), the relation between the normal force and the friction force may be written as:

$$F = \frac{4\mu \cdot F_n \sin \frac{\theta}{2}}{\theta + \sin \theta}$$

$$\therefore \quad \frac{F}{F_n} = \frac{4\mu \cdot \sin \frac{\theta}{2}}{\theta + \sin \theta}$$

$$\mu = 0.3, \text{ and } \theta = 120°$$

$$\therefore \quad \frac{F}{F_n} = \frac{4 \times 0{\cdot}3 \times 0{\cdot}866}{2{\cdot}09 + 0{\cdot}866} = 0.351$$

$$\therefore \quad F_n = \frac{F}{0{\cdot}351} = 2.85\,F.$$

(*a*) **Spring Force:**

Considering the left hand Yoke, $\Sigma M_L = 0$

$$\therefore \quad S \times 0.30 + F_L \times 0.05 - F_{nL} \times 0.15 = 0$$

$$\therefore \quad 0.30S + 0.05F_L - 0.15 \times 2.85F_L = 0$$

$$\therefore \quad F_L = 0.795\,S$$

Similar for right hand Yoke, $\Sigma M_R = 0$

$$\therefore \quad 0.30S - 0.05F_R - 0.15F_{nR} = 0$$

$$\therefore \quad 0.30S - 0.05F_R - 0.15 \times 2.85F_R = 0$$

$$\therefore \quad F_R = 0.628S$$

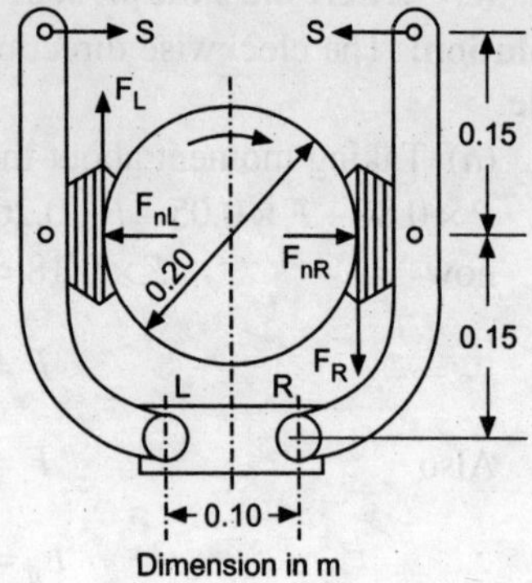

Fig. 18.24 *Double block brake*

Now total torsional moment capacity of the brake is:

$$T_f = (F_L + F_R) \times r$$

$$\therefore \quad F_L + F_R = \frac{240}{0{\cdot}10} = 2400 \text{ N}$$

$$\therefore \quad S\,(0.795 + 0.628) = 2400$$

$$\therefore \quad S = 1685 \text{ N}$$

When releasing the brake, the spring is further compressed, therefore, the spring release will be greater than *S* equal to 1685 N. This is usually 10 to 15 per cent greater.

(*b*) **Shoe Width :** Now $F_{nL} = 2.85F_L$

and $F_{nR} = 2.85F_R$

$$F_L = 0.795S = 0.795 \times 1685 = 1340 \text{ N}$$

$$F_R = 0.628S = 0.628 \times 1685 = 1060 \text{ N}$$

$\therefore$ Maximum normal force is $F_{nL} = 2.85 \times 1340 = 3840$ N

If w = width of the shoe, m

Then projected bearing area for one shoe is = $w \times 2r \sin\frac{\theta}{2} = w \times 2 \times 0.10 \times \sin 60 = 0.1732\,w$ m²

Now rubbing velocity $V = \frac{\pi Dn}{60} = \frac{\pi \times 0{\cdot}20 \times 600}{60} = 6.28$ m/s,

Now unit pressure on the shoe, $p = \frac{F_{nL}}{0.1732w} = \frac{3840}{0.1732w}$ N/m²

Now $pV = 17.5 \times 10^5$

$$\therefore \quad \frac{3840}{0.1732w} \times 6{\cdot}28 = 17.5 \times 10^5$$

$$\therefore \quad w = 0.079 \text{ m, say 80 mm}$$

Now $\frac{w}{D} = \frac{80}{200} = 0.4.$

This ratio is usually held between 1/4 and 1/3 in well-proportioned brakes, therefore, the design is satisfactory.

Brake's shoe may be made of wood and are attached to the Yoke by means of bolts. For mechanical drive, the shoes are made of C.I. and provided with special brake linings. The linings may be secured with rivets or with countersunk screws.

Brake lining: Brake lining should satisfy the following requirements:

1. Have a high coefficient of friction.
2. Retain braking capacity at temperatures up to 300° C.
3. Effectively resist wear at the highest speeds, unit pressures, and temperatures.
4. Easily yield to treatment.
5. Have a low cost.

Brake wheel: For hand operated hoisting machinery, the wheels are usually made of Cast Iron, whereas, for power drives, the wheels are made of Cast Steel. Brake wheels should be dynamically balanced. The width of the wheel should be about 5 to 10 mm more than the shoe width. For better neat dissipation, the wheels are always finned.

For normal operating conditions, the size of the wheels may be selected from Table 18.6.

Table 18.6 Wheel size

Brake wheel		Power of motor at rev/min, kW		
Diameter, mm	Width, mm	570	730	960
160	50	6.75	7.5	8.25
200	65	9.75	11.25	12
250	80	14.25	15.75	18
320	100	23.25	26.25	30
400	125	38.25	44.25	50.25
500	160	63.75	72	84
630	200	101.25	117	—
800	250	165	195	—

Brake levers: Brake levers are manufactured from die-forged or Cast steel. The safe bending stress of levers taking into account braking shocks is taken from 40 to 80 MPa depending upon the size of the brake.

Example 18.14. A simple band brake is shown in Fig. 18.25. The brake drum is 0.75 m in diameter and is fitted with a steel band 4 mm thick lined with brake lining having a coefficient of friction of 0.25 when not sliding. The arc of contact is 245°. This brake drum is attached to a 0.60 m hoisting drum that sustains a rope load of 8200 N. The operating force has a moment arm of 1.5 m and the band is attached 0.125 m from the pivot point.

(a) Determine the force required to just support the load.
(b) What force will be required if the direction of rotation is reversed.
(c) What width of steel band is required if the tensile stress is limited of 52.5 N/mm^2.

Solution :

Given ; $D = 0.75$ m; $\theta = 245°$; $\mu = 0.25$

now $$\frac{F_1}{F_2} = e^{\mu\theta} = e^{0.25 \times \frac{\pi \times 245}{180}} = 2.914 \quad ...(1)$$

Now rope load = 8200 N ; and hoisting drum diameter = 0.60 m

$\therefore$ Torsional moment on the shaft $= 8200 \times \frac{0.60}{2} = 2460$ Nm

Also, torsional moment $= (F_1 - F_2)\times\frac{D}{2}$

$$\therefore \quad (F_1 - F_2)\times\frac{0.75}{2} = 2460,\ F_1 - F_2 = 6560\text{ N} \quad ...(2)$$

$$\therefore \quad \text{From (1) and (2)}; \quad F_2 = \frac{6560}{2.914} = 3425\text{ N}$$

and $$F_1 = 2.914\times 3425 = 9980\text{ N app.}$$

(*a*) Taking moments about fulcrum point A,

$$P\times 1.50 = F_2\times 0.125,\ P = \frac{3425\times 0.125}{1.50} = 285.4\text{ N}$$

(*b*) When the direction of rotation is reversed, F_1 and F_2 interchange their places. Again taking moments about point *A*,

$$P\times 1.50 = F_1\times 0.125,\ P = \frac{9980\times 0.125}{1.50} = 831.6\text{ N}$$

(*c*) Now $h = 4$ mm

Also, $$w.h = \frac{F_1}{f_t},\ f_t = 52.5\text{ N/mm}^2$$

$$\therefore \quad w = \frac{9980}{52.5\times 4} = 47.5\text{ mm, say } \mathbf{50\ mm.}$$

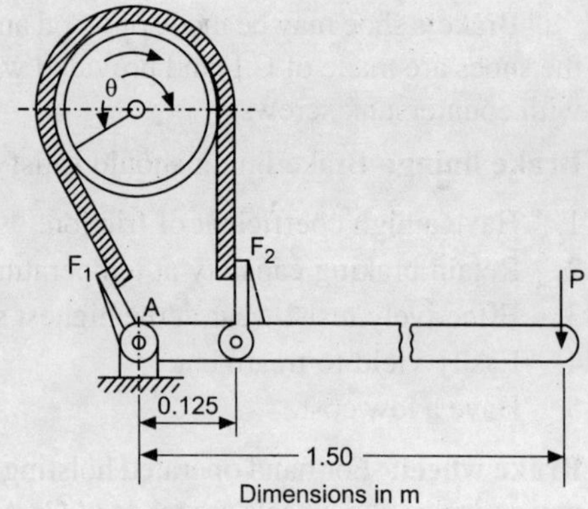

Fig. 18.25 *Simple band brake*

Example 18.15. A differential band brake is shown in Fig. 18.26. The angle of contact is 280 deg. The brake is to sustain a torque of 400 Nm. The band has a compressed woven lining and presses against Cast Iron drum of 0.375 m diameter. Assuming coefficient of friction as 0.3 :

(*a*) Determine the necessary operating force, P.

(*b*) Find the width and thickness of the steel band, assuming safe stress in tension as 55 N/mm².

(*c*) Find the section of the brake lever.

(*d*) Design the pins and show the attachment arrangements of the band with the pins.

Solution:

Now $F_1 / F_2 = e^{\mu\theta}$

$$\mu = 0.3,\ \theta = 280\text{ deg.} = 280\times\frac{\pi}{180} = 4.88\text{ radians}$$

$$\therefore \quad \frac{F_1}{F_2} = e^{0.3\times 4.88} = 4.33 \quad ...(1)$$

Also $$(F_1 - F_2)\times\frac{D}{2} = 400,$$

$$D = 0.375\text{ m}$$

$$\therefore \quad F_1 - F_2 = \frac{2\times 400}{0.375} = 2130\text{ N} \quad ...(2)$$

From equations (1) and (2),

$$F_2 = \frac{2130}{3.33} = 640\text{ N},\ F_1 = 2770\text{ N}$$

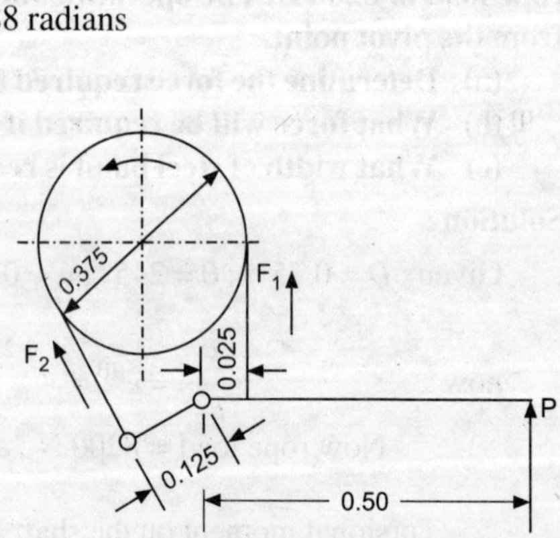

Fig. 18.26 *Band brake*

(*a*) **Operating force, P:**

We know that $\Sigma M_0 = 0$

$\therefore \quad P \times 0.50 + F_1 \times 0.025 - F_2 \times 0.125 = 0$

$\therefore \quad 0.50\,P + 2770 \times 0.025 - 640 \times 0.125 = 0$

$\therefore \quad P = 21.5 \text{ N}$

(*b*) **Dimensions of band:** To ensure a tight contact with the brake drum, the width of the band should not exceed 150 mm for drum diameters of $D > 1000$ mm and 100 mm for $D < 1000$ mm. For brakes of hand operated winches, the band can be of ordinary plain steel of the following sizes :

Width, *w* mm	25–30	40–60	80	100	140–200
Thickness *h*, mm	3	3–4	4–6	4–7	6–10

The steel bands are lined with friction material for power driven mechanisms. Wooden shoes are attached to the band by two or four wooden screws. A gap of 4–6 mm should be left between the blocks. The brake lining is fastened to the strap with counter sunk copper rivets or screws.

The band thickness is also given empirically:

$$h = 0.005\,D = 0.005 \times 375 = 1.875 \text{ mm, say 2 mm}$$

The brake band is under tensile rupture due to maximum tension, F_1.

$$\therefore \quad w.h = \frac{F_1}{f_t}$$

f_t ranges from 50 to 80 N/mm^2. Given value is 55 N/mm^2.

$$\therefore \quad w = \frac{2770}{2 \times 55} = 25.18 \text{ mm, say 26 mm}$$

(*c*) **Brake lever:** Maximum bending moment close to the fulcrum pin is:

$$M_b = F_2 \times 0.125 = 640 \times 0.125 = 80 \text{ Nm}$$

Assuming rectangular section of the lever with thickness = $\frac{1}{3} \times$ height

i.e. $\quad t = \frac{1}{3}h \quad$ or $h = 3t$

$\therefore \quad M_b = f_b \times Z$

Now $\quad Z = 1/6\,th^2 = 1/6 \times t \times (3t)^2 = 1.5\,t^3$

The safe tensile stress for brake lever material is 40 to 80 N/mm^2.

Let $\quad f_b = 60 \text{ N/mm}^2$

$\therefore \quad 80 \times 1000 = 60 \times 15\,t^3$

$\therefore \quad t = 10 \text{ mm}, h = 30 \text{ mm}$

(*d*) (*i*) **Fulcrum Pin:** The resultant reaction at this pin can be found out by calculating the horizontal and vertical components due to the forces F_1, F_2 and P. Then the pin may be designed in double shear and bearing and can be checked in bending.

(*ii*) **Left Hand and Right Hand Pins:** The attachment of the band to the left hand and right hand end pins is shown in Fig. 18.27.

The load on the left pin = F_2 = 640 N

Considering the bearing of the pin:

Load = projected area of pin × bearing pressure

$\therefore \quad 640 = l \times d \times p_b$

Let $\quad p_b = 10 \text{ N/mm}^2, \ \frac{l}{d} = 2$

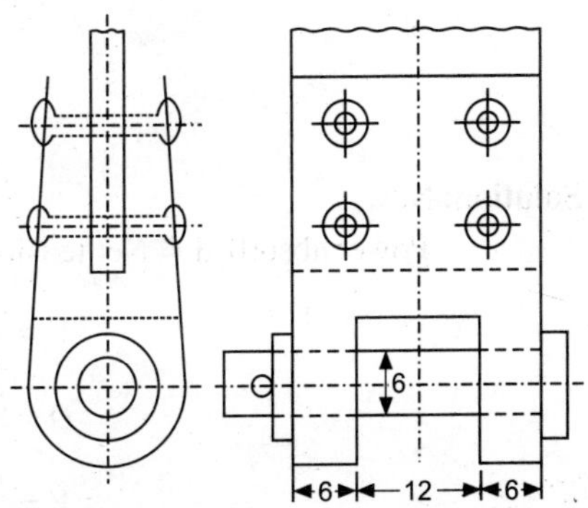

Fig. 18.27 *Band attachment*

$\therefore$ $$640 = 2d^2 \times 10$$

$\therefore$ $$d \cong 5.7 \text{ mm, say 6 mm}$$

$\therefore$ Length in bearing = 12 mm

The other dimension of connection are shown in the figure.

Checking in shear :

$$640 = 2 \times \frac{\pi}{4} d^2 \times fs$$

$$f_s = \frac{640 \times 2}{\pi \times 36} = 11.3 \text{ N/mm}^2 \text{ which is quite low.}$$

The pin can also be checked in bending.

The right hand pin can be designed in the above manner.

Example 18.16. In the band and block brake shown below, the brake is lined with 12 blocks, each of which subtends an angle of 15° at the centre of the braking drum. Thickness of block: 75 mm, and diameter of drum: 850 mm. If, when the brake is in action, the greatest and least tensions in the brake strip are T_n and T_0, show that, $T_n/T_0 = \left[\frac{1 + \mu \tan 7\frac{1}{2}°}{1 - \mu \tan 7\frac{1}{2}°}\right]^{12}$, where, μ = Coefficient of friction of the blocks.

(b) With the lever arrangement as shown, find the least force at C for the blocks to absorb 225 kW at 240 rev./min. Take μ = 0.4 **(PTU)**

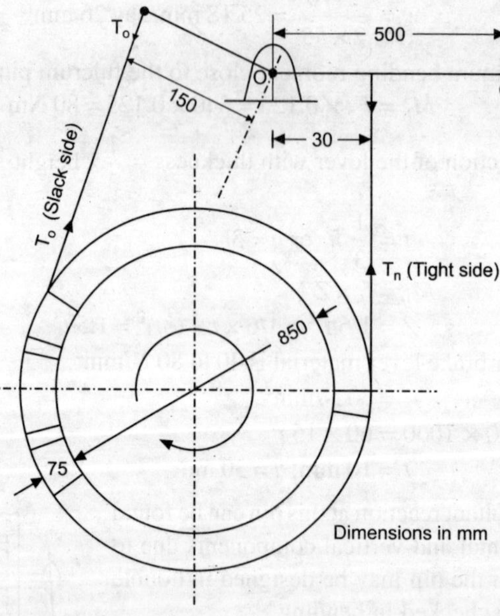

Solution: Now,

Power absorbed = Net tension × Peripheral velocity

$$V = \frac{\pi D n}{60 \times 1000}, \text{ m/s}$$

$$D = 850 + 2 \times 75 = 1000 \text{ mm}$$

$\therefore$ $$V = \frac{\pi \times 1000 \times 240}{60 \times 1000} = 4\pi \text{ m/s}$$

$\therefore$ Net tension $= (T_n - T_o) = \dfrac{225 \times 1000}{4\pi} = 17905$ N ...(1)

Now $\dfrac{T_n}{T_o} = \left[\dfrac{1 + 0.4 \times 0.13165}{1 - 0.4 \times 0.13165}\right]^{12} = 3.543$ ...(2)

Note: For the derivation of the relation, See Art. 18.6.4. Here, $\theta = 7\frac{1}{2}°$, and $n = 12$.

From equations (1) and (2)

$$T_o = 7041 \text{ N}, \quad \text{and} \quad T_n = 24946 \text{ N}.$$

(*b*) Free body diagram of the lever:

$$P \times 500 + T_n \times 30 - T_o \times 150 = 0$$

$$\therefore \quad P \times 500 + 24946 \times 30 - 7041 \times 150 = 0$$

From here, $P = 615.54$ N.

Example 18.17. Refer to Fig. 18.18. The following data is given:
Drum rotation: Clock-wise; $\theta_1 = 15°$, $\theta_2 = 105°$
Drum radius: 125 mm ; $\mu = 0.25$; $b = 65$ mm ; $p_m = 0.6$ N/mm^2
Vertical distance from the drum centre to the fulcrum pin = 90 mm
Vertical distance from the drum centre to the line of actuating force = 90 mm
Horizontal distance of fulcrum pins from centre of drum = 40 mm
Determine: (a) Actuating force; (b) Braking torque of the shoe brake.

Solution: It is clear that $L = 90 + 90 = 180$ mm

and $OO_1 = a = \sqrt{90^2 + 40^2} = 98.5 \text{ mm} = OO_2$

Now it is clear that the R.H.S. is the leading shoe and the left hand shoe is trailing shoe. Therefore, the maximum pressure given is

$$p_{mR} = p_{mL} \text{ (leading shoe)} = 0.6 \text{ N/mm}^2.$$

$$\therefore \quad M_N = \frac{p_m \cdot b \cdot r \cdot a}{4}[2(\theta_2 - \theta_1) - \sin 2\theta_2 + \sin 2\theta_1]$$

$$= \frac{0.6 \times 65 \times 125 \times 98.5}{4}\left[2 \times \frac{\pi}{2} - \sin 210° + \sin 30°\right]$$

$$= 497.23 \text{ N.m}$$

$$M_\mu = \frac{\mu \cdot p_m \cdot b \cdot r}{4}[4r(\cos\theta_1 - \cos\theta_2) + a(\cos 2\theta_2 - \cos 2\theta_1)]$$

$$= \frac{0.25 \times 0.6 \times 125 \times 65}{4}[4 \times 125(\cos 15° - \cos 105°) + 98.5(\cos 210° - \cos 30°)]$$

$$= 134.64 \text{ N.m}$$

$$\therefore \text{ Actuating force, } P = \frac{M_N - M_\mu}{L} = \frac{497.23 - 134.64}{180} \times 1000 = 2014 \text{ N}$$

Let p_{mT} = Maximum pressure on the trailing (L.H.S.) shoe,

$$\therefore \quad M_N = \frac{497.23}{0.6} \times p_{mT} = 828.72 \times p_{mT} \text{ N.m}$$

$$M_\mu = \frac{134.64}{0.6} \times p_{mT} = 224.4 \times p_{mT} \text{ N.m}$$

Now for the trailing shoe (L.H.S.)

$$P = \frac{M_N + M_\mu}{L}$$

$$\therefore \qquad p_{mT} = \frac{2014 \times 180}{1053.12 \times 1000} = 0.344 \text{ N/mm}^2$$

(*b*) Braking capacity of the brake,

$$= (p_{mL} + p_{mT}) \times \mu \times br^2 (\cos\theta_1 - \cos\theta_2)$$
$$= (0.6 + 0.344) \times 0.25 \times 65 \times 125^2 (\cos 15° - \cos 105°)$$
$$= 0.944 \times 0.25 \times 65 \times 125 \times 125 (0.966 + 0.259)$$
$$= 293.62 \text{ N.m.}$$

18.7 MATERIALS FOR BRAKE DRUMS AND FOR BRAKE LININGS

The most common material for brake drums is cast iron with some alloying elements added. When good heat conduction is needed, stainless steel, aluminium and model can also be used. Brake linings are usually made of some form of asbestos and binders. Most of the time, the asbestos based linings are moulded. These linings are used upto temperatures of about 200°C. For temperature range of 200°C to 400°C, sintered metal linings are usually employed. Sintered metal linings with ceramic particles can be used for temperature range of 400°C to 540°C. By increasing the ceramic content, temperatures as high as 1000°C can be tolerated. Ceramic linings have high thermal conductivity, long life and very stable friction properties. However, these are expensive, have lower coefficient of friction and must be made to shape because of their rigidity. Linings are attached to the drum usually by riveting or bonding. Riveting is cheaper, and simple, but its drawback is that the usable lining thickness depends upon the depth to which the rivet head can be counter sunk. Bonded linings have greater effective thickness and more friction area but more expensive.

18.8 DESIGN CONSIDERATIONS

Clutches and brakes are usually analysed for the torsional moment capacity required and the energy that must be absorbed and dissipated. Other factors to be considered are — required lining area, method of actuation, available space, effect of actuating force on the structure, the characteristics of the friction material and the environment in which the machine is working. Economic product is a compromise of all these factors.

EXERCISES

1. What is the function of a clutch?
2. What are the basic design requirements of a clutch?
3. Enumerate the various types of clutches.
4. Which are the most commonly used types of clutches?
5. Enumerate the advantages of friction clutches.
6. How can you increase the capacity of a disk clutch?
7. What is the common radii ratio for a disk clutch?
8. What are the limitations on the number of disks of a disk clutch?
9. Enumerate the qualities of the friction material used in clutches.
10. Compare dry and wet clutch design.
11. What is the included angle of a cone clutch?
12. Enumerate the advantages of a tyre pneumatic clutch.
13. Name a few automatic clutches.
14. What are the various means for controlling the engagement of a clutch?
15. What is the function of a brake? How its functions differ from that of a clutch?
16. What are the important factors in brake design?
17. What are the various types of brakes?
18. What is a self-actuating or self-energizing brake? When a brake becomes self-locking?

19. Is self-locking desirable? If so, under what conditions?
20. What is the drawback of a single shoe brake? How do you over come this?
21. A disk clutch consists of two steel disks in contact with one asbestos fabric-faced disk having an outside diameter of 250 mm and an inside diameter of 200 mm. Assuming coefficient of friction as 0.35 and if the disks are pressed together by an axial force of 9000 N, determine the power that can be transmitted at 1000 rev/min.
22. A Jaw clutch for a 105 mm shaft has three jaws with radial faces. The dimensions are : inside diameter of jaws, 11.2 cm, outside diameter 36 cm; axial height of jaws 5 cm; and key 2.5 by 2.5 by 16.2 cm. Assume a clearance of 3.2 mm between the jaws and a working stress in the shaft of 42 MPa.
    (*a*) What power can be transmitted at 100 rev/min?
    (*b*) Determine the shearing and bearing stresses in the key and the shearing stress on the jaw faces.
23. A square jaw clutch is required for a shaft 50 mm diameter and the maximum torsional moment carried by the engaged clutch is to be 865 Nm. If the force is assumed to act at the mean radius of the jaws. Determine:
    (*a*) minimum outside diameter of jaws. (*b*) minimum axial length of the jaws.
24. A multiple disk clutch is to be used on machine tools. There are 8 driven disks having an outside diameter of 75 mm and an inside diameter of 56 mm. The disks are of metal and run in oily environment. The coefficient of friction may be assumed to be 0.2 and the permissible unit pressure 0.7 N/mm^2.
    (*a*) Determine the axial pressure required. (*b*) Determine the power that can be transmitted at 600 rev/min.
25. A six cylinder engine is rated at 45 kW at 2000 rev/min. The maximum torsional moment is developed at 1200 rev/min and 30 kW. The multiple disc clutch consists of fabricated disks of 0.216 m outside diameter and 0.159 m inside diameter in contact with five driven disks. The coefficient of friction is 0.20. Determine the axial pressure to engage this clutch.
26. A multiple disk clutch with 6 driven disks faced with asbestos clutch fabric with coefficient of friction as 0.25, is used with a tractor. The outside and inside diameters of the disks are 0.30 m and 0.20 m respectively. Five springs are used with the clutch. Length of the spring when the clutch is engaged is 70 mm and the spring diameter is 38 mm. The length of spring when the clutch is disengaged is 63.5 mm and the pressure is 45 per cent higher than when engaged. The maximum stress in the springs is not to exceed 490 MPa. Determine the power that can be transmitted at 1200 rev/min allowing 0.007 N/mm^2 pressure on the disks.
27. Hardened steel clutch plates with inner and outer friction surface diameters of 50 mm and 75 mm respectively are available in your plant. A clutch is to be designed utilising these plates. The clutch is to transmit 4 kW at 825 rev/min in an oil atmosphere. Determine:
    (*a*) the number of plates required. (*b*) the magnitude of axial force required during engagement.
28. The maximum capacity of a simple plate clutch is 55 kW at 1800 rev/min. The clutch facing has a coefficient of friction of 0.40 and a permissible pressure of 0.21 N/mm^2. The clutch is engaged through 12 springs of 32 mm mean diameter. The springs compress 1.6 mm for disengagement with an increase in pressure of 10 per cent. Determine the diameters of the clutch facing if the inner diameter is 0.7 of the outer diameter.
29. A cone clutch has a face angle of 15 deg. with a maximum diameter of 0.60 m with a face width of 80 mm. The coefficient of friction is 0.20 and the permissible pressure is 0.085 N/mm^2.
    (*a*) What torsional moment may be transmitted? (*b*) What power may be transmitted at 800 rev/min?
    (*c*) What axial pressure must be exerted at this power ?
30. An engine developing 30 kW at 1250 rev/min is fitted with a cone clutch built into the flywheel. The cone has a face angle of 12½° and a maximum diameter of 0.35 m. The coefficient of friction is 0.20. The normal pressure on the clutch face is not to exceed 0.085 N/mm^2.
    (*a*) Determine the face width required.
    (*b*) Determine the spring pressure required to engage this clutch.
31. A cone clutch faced with leather is to transmit 15 kW at 1000 rev/ min. The cone angle is 25 deg; the mean radius is 0.165 m; the coefficient of friction is 0.20, and the friction surface is 75 mm wide. Determine :
    (*a*) the axial force required in the spring. (*b*) the normal pressure at the friction surfaces.
32. A friction clutch engaging with the rim of a pulley is to be designed with 4 pairs of fabric lined shoes, half the shoes in contact with the inner surface of the rim and half in contact with the outer surface. The mean rim diameter of pulley is 0.50 m and 15 kW is to be transmitted at 200 rem/min. If the coefficient of friction is assumed to be 0.25 and the permissible bearing pressure as 0.21 N/mm^2; determine the area required for each shoe.
33. Calculate the friction work for the brake on a hoist which in lowering 27 kN load through 60 metre at the rate of 1.5 m/s. The load must be stopped in the last 6 metre of descent. The rope drum diameter is 1.5 meter and brake drum, which is keyed to the rope drum shaft, has a diameter of 1.2 metre. The moment of inertia of all the rotating parts is 1660 kgm^2.

34. The block brake shown in Fig. 18.28 is to balance a torsional moment of 575 Nm on the drum shaft. Assuming the coefficient of friction between brake shoe and drum to be 0.25 determine the operating force, P, required.

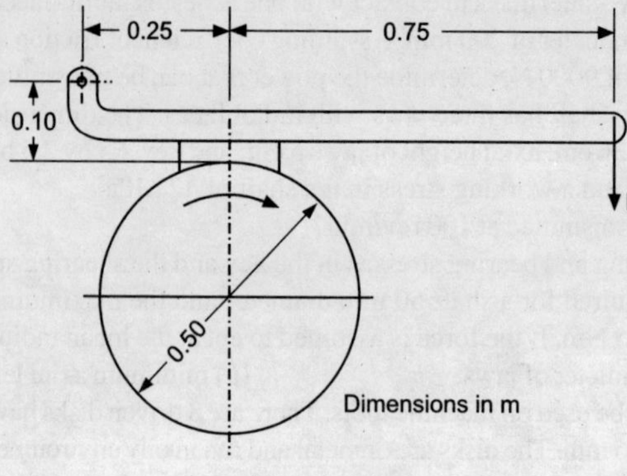

Fig. 18.28

35. The block brake shown in Fig. 18.29 is to balance a torsional moment of 1725 Nm acting on the shaft. The coefficient of friction between the brake shoe and drum is 0.2.
   (*a*) Determine the operating force, P, required.
   (*b*) Can this brake be made self acting? If so, state the change in dimensions,

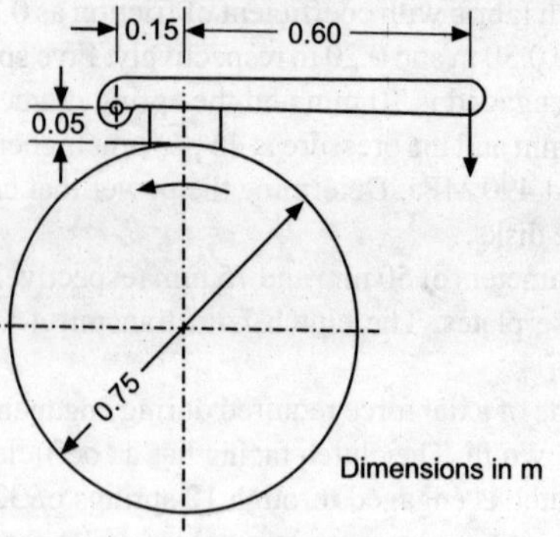

Fig. 18.29

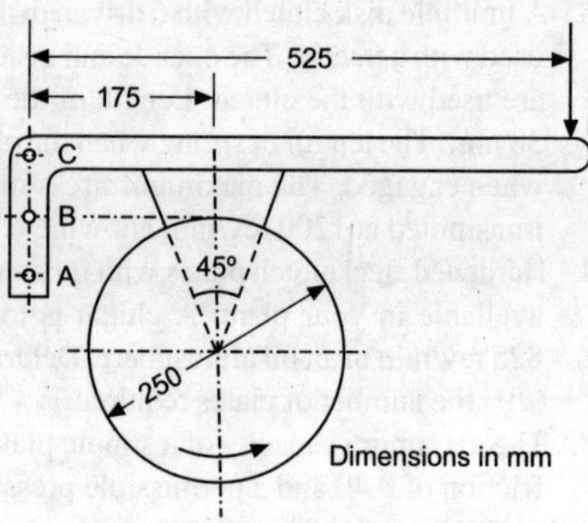

Fig. 18.30

36. Where along the vertical line through A-B-C must the lever pivot be located for the brake in Fig. 18.30 to be self-locking when $\mu = 0.35$?

37. The brake in Fig. 18.31 is to be self-energizing and is to have a braking torsional moment of 22 Nm. Assuming the coefficient of friction to be 0.25:

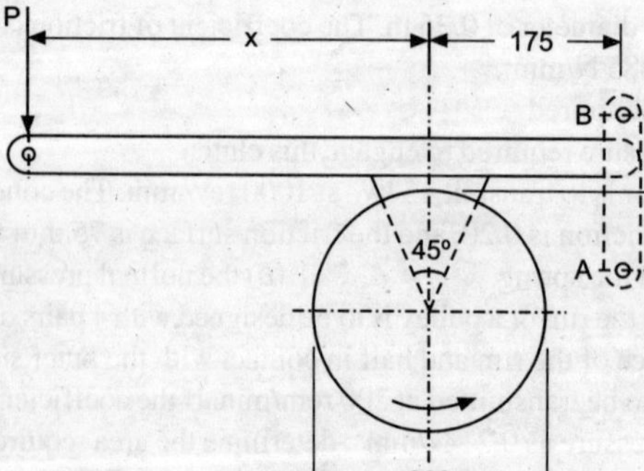

Fig. 18.31

(*a*) Determine, which point, *A* or *B*, should be the brake lever pivot point.

(*b*) Determine the required value of *x*.

(*c*) Determine the power absorbed if the drum is rotating at 105 rev/min.

38. The dimensions of a typical double shoe brake in m are shown in Fig. 18.32. The spring pressure is 13.65 kN when the brake is set. Assuming coefficient of friction as 0.35, determine:

(*a*) Total braking capacity of the brake.

(*b*) Force that must be applied to the operating arm by the magnet to release brake.

39. An elevator brake is shown in Fig. 18.33. Each brake shoe is 125 mm long. The coefficient of friction is 0.30; and the permissible pressure 0.35 N/mm^2 average. The operating force is 450 N.

(*a*) Determine the braking torsional moment for clockwise rotation.

(*b*) Determine the braking torsional moment for CCW direction.

(*c*) Determine the required width of the brake shoes.

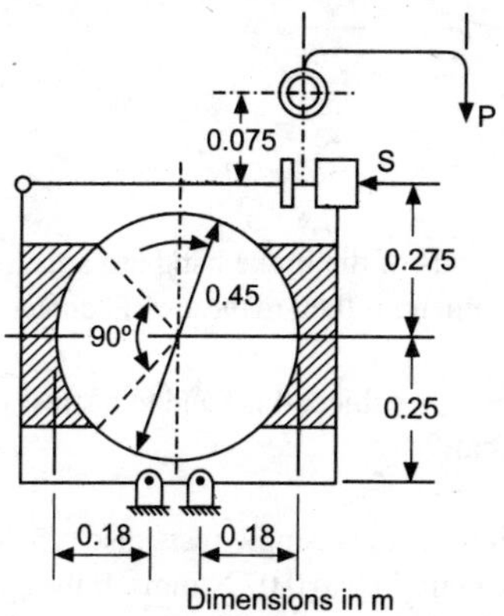

Fig. 18.32

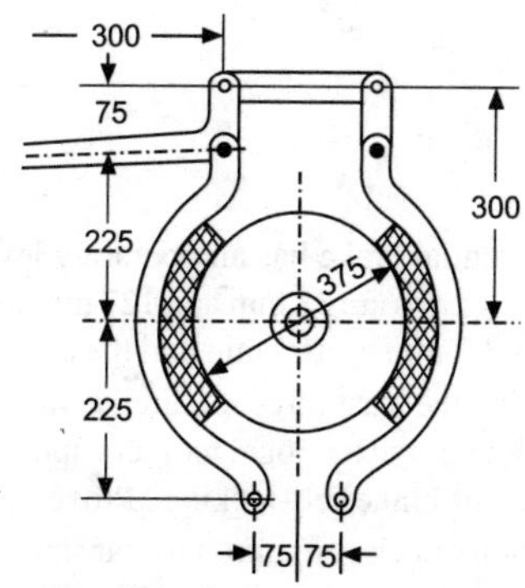

Fig. 18.33

40. The rope drum of an elevator hoist is 0.60 m in diameter and the speed of the elevator is 3 m/s. This drum is fitted with a brake drum 0.90 m in diameter having four C.I. brake shoes each subtending an arc of 45 deg on the brake drum. This elevator weighs 18 kN loaded, and the brake is to have sufficient capacity to stop the elevator in 3.6 metres. The coefficient of C.I. may be taken as 0.20.

(*a*) Determine the radial pressure on each brake shoe.

(*b*) Determine the width of shoes if the allowable pressure on the brake shoe is 0.35 N/mm^2.

(*c*) How much heat is generated in stopping this elevator?

41. The spring loaded brake is shown in Fig. 18.34. It is to be used intermittently to extra a braking torsional moment of 1385 Nm on a drum rotating at 200 rev/min. The shoes are lined with asbestos friction material:

(*a*) Determine the required spring force.

(*b*) If the shoes are identical, what should be their minimum width?

(*c*) For the direction of rotation shown, which shoe will have the greater rate of wear and what will be the ratio of the rates of wear of the two shoes?

42. A simple band brake operates on a drum 0.60 m in diameter that is running at 200 rev/min. The coefficient of friction is 0.25. The brake band has a contact of 270 deg., and one end is fastened to a fixed pin and the other end to the brake arm 0.125 m from the fixed pin. The straight brake arm is 0.75 m long and is placed perpendicular to the diameter that bisects the angle of contact.

(*a*) What is the minimum pull necessary on the end of the brake arm to stop the wheel if 30 kW is being absorbed? What is the direction of rotation for this minimum pull?

(*b*) If the maximum tensile stress is not to exceed 56 MPa. Find the width of the 2.4 mm thick steel band.

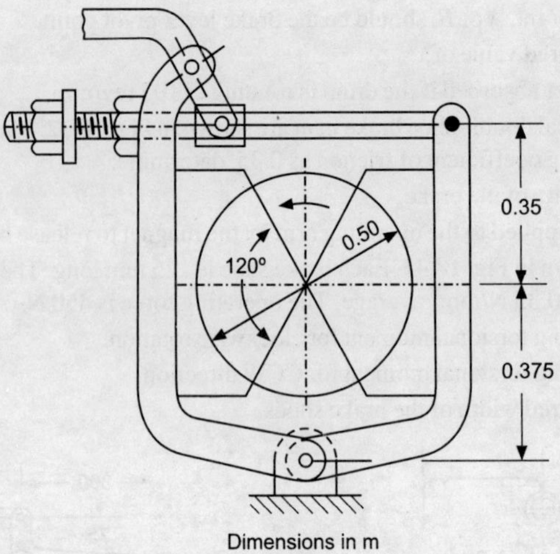

Fig. 18.34

43. A differential brake has an operating lever 0.225 m long. The ends of the brake band are attached so that their operating arms are 38 mm and 127 mm long. The brake drum diameter is 0.60 m the arc of contact 300 deg. Brake band is 3.2 mm by 100 mm and the coefficient of friction is 0.22.
    (*a*) Find the least force required at the end of the operating lever to subject this band to a stress of 55 MPa.
    (*b*) What is the torsional moment applied to the brake-drum shaft?
    (*c*) Is this brake self-locking? Prove your answer.
44. A friction clutch is required to transmit 33 kW at 2000 rev/min. It is to be single plate disc type with both sides effective. The pressure is exerted axially by means of springs and limited to 0.07 N/mm^2. If the outer diameter of the plate is to be 0.30 m., find the required inner diameter of the clutch ring and the total force exerted by the springs. Assume the wear to be uniform and coefficient of friction 0.3. Draw a neat dimensioned sketch of the clutch showing main dimensions.
45. A cone clutch transmits more power than a disk clutch. How?
46. The number of effective surfaces of a mutli-plate clutch with 5 steel plates and 4 brass plates is:
    (a) 4 (b) 5
    (c) 8 (d) 9.
47. The type of clutch used on prime movers with low starting torque is:
    (a) single plate clutch (b) multi-plate clutch
    (c) cone clutch (d) centrifugal clutch.
48. Self-braking (locking) is not possible with the following type of brakes:
    (a) simple band (b) simple block
    (c) differetial band (d) internal expanding shoe brake.
49. Short shoe brakes have angle of contact less than:
    (a) 10° (b) 20°
    (c) 45° (d) 60°
50. in block brakes, the asumption of normal reaction and frictional force passing through the mid point of the block is valid only if the angle if contact is:
    (a) less than 45° (b) greater than 45°
    (c) less than 60° (d) greater than 60°
51. The ratio of inner radius to outer radius of a plate type clutch varies as
    (a) 0.1 to 0.25 (b) 0.5 to 0.7
    (c) 1 to 3 (d) 5 to 7.

❑❑❑

CHAPTER 19

Belt Drives

19.1 GENERAL

Belt drives are being most widely used in industry for the transmission of power from one shaft to another over considerable distances. It is successfully competing with individual drives in which the machine is directly coupled to the shaft of its own motor. In belt drives, there are two distinct arrangements:

1. Group drive in which the belts are driven from short line shafts and several small motors.
2. Belt drives using line shafts driven by a single large motor.

In its simplest form, the belt drive consists of an endless belt which is wrapped tightly over two pulleys known as driving and driven pulleys which in turn are mounted on the driving and driven shafts respectively. The motion from the driving pulley is transmitted to the driven pulley by frictional resistance between the belt and the surface of the pulleys. Since there will always be some slip between the belt and the surface of pulleys, the velocity ratio in belt drives is never absolutely constant.

There are basically four types of belts used for power transmission:

1. Flat, including link.
2. V-section including round.
3. Ribbed.
4. Toothed, or timing.

These four types of belts are shown in Fig. 19.1.

Selection of a belt drive depends upon:

1. Power to be transmitted.
2. Speed of driver and driven shafts.

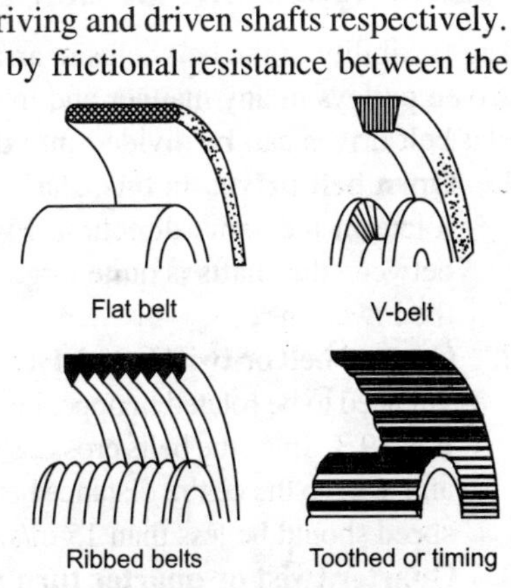

Fig. 19.1 *Types of belts*

3. Reduction ratio.
4. Centre distance.
5. Shaft relationship.
6. Positive drive requirements.
7. Service conditions.
8. Space available.

The two most widely used types of belts are "Flat" and "V-belt". On the basis of peripheral speed, the belt drives are classified as below:

(a) **Light Drives:** These are used for the transmission of small powers at belt speeds upto 12 m/s, e.g., agricultural machines, fitter machines, and small machine tools.
(b) **Medium Drives:** These are used for medium powers at 12 to 24 m/s. Good quality belts should be used.
(c) **Heavy Drives:** These are for large powers and speeds above 24 m/s, e.g. generators, compressors and main drives.

19.2 FLAT BELT DRIVES

The advantages of flat belt drives are:

1. Flexibility, shock absorption and efficiency at high speeds.
2. Resistance to abrasive and other harmful environments.
3. Simplicity, low cost, smoothness of operations, low maintenance, long life, and noiselessness as compared with gear transmission.
4. It allows long distances between shafts and the belts can be spliced or connected for endless operations.
5. Excellent performance in serpentine drives.

At the same time, the flat belt drives have got the following disadvantages:

(*a*) Velocity ratio is not constant.
(*b*) Comparative large size.
(*c*) Sometimes noisier than other belt drives.
(*d*) Slip and have comparatively low efficiencies and moderate speeds.
(*e*) Tension in the belt should be kept high to limit the amount of slip. These high tensions result in high bearing loads and belt stresses.
(*f*) Stretching of the belt calls for resewing if the centre distance is constant or a tensioning device is required.

19.2.1 Types of Flat Belt Drives

The flexibility of the belt helps in arranging the shifts of the driving and driven pulleys in any manner and in using as many pulleys as possible. Flat belt drives can be divided into the following:

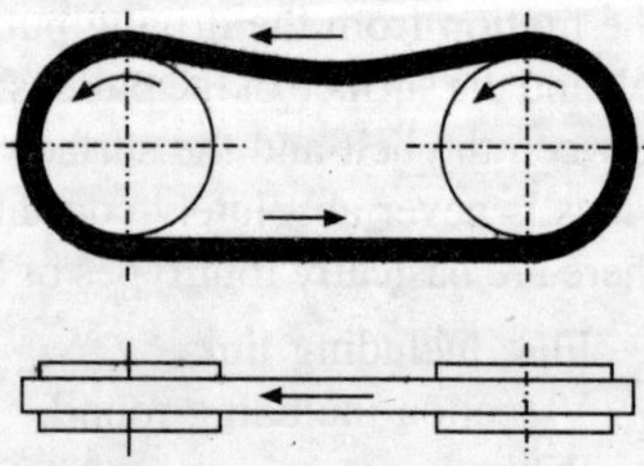

Fig. 19.2 *Open belt drive*

1. **Open belt drive:** In this, shafts are arranged in parallel and these rotate in the same direction, Fig. 19.2. When the centre distance between the shafts is quite large, the tight side of the belt should be the lower one.
2. **Crossed belt or twist belt drive:** If the shafts are arranged in parallel but need to be rotated in opposite directions, this type of drive is used, Fig. 19.3. Since the belts cross each other, there will be too much wear and tear, so the centre distance between the shafts must be more then 20 *b*, where *b* is the belt width, and the speed should be less than 15 m/s.
3. **Quarter twist or quarter turn drive :** In this, the shafts are at right angles and rotate in some definite direction. The face of the pulley should be sufficiently wide ≥ 1.4 *b* so that the belt does not leave the pulleys Fig. 19.4.

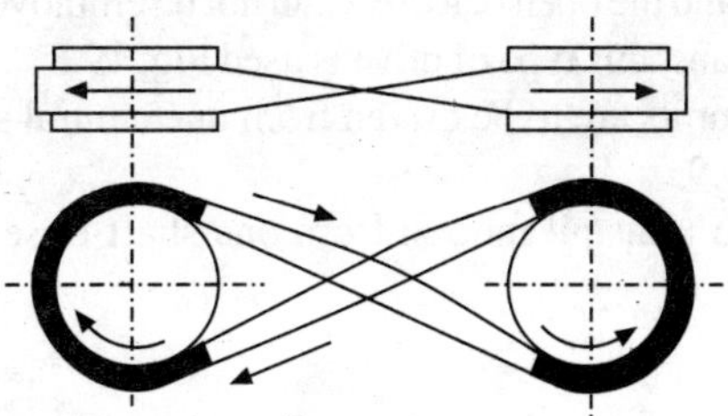

Fig. 19.3 *Crossed belt drive or Twist belt drive*

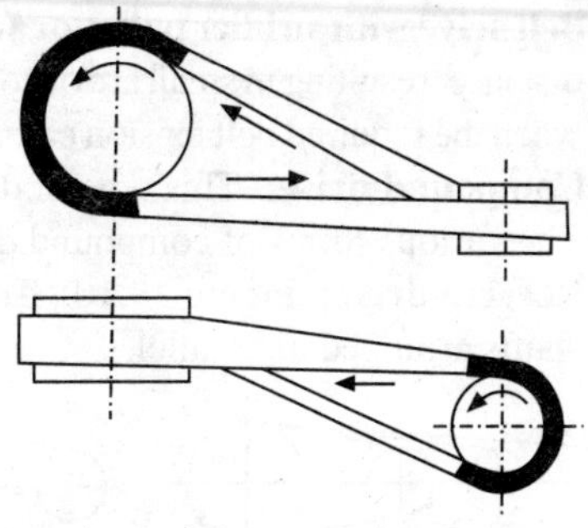

Fig. 19.4 *Quarter twist or quarter turn drive*

4. **Right angle drive:** Used with shafts at right angles when the pulleys can not be arranged as in Fig. 19.4. or when reversible motion is required, Fig. 19.5.
5. **Stepped pulley drive:** This drive, Fig 19.6, is used for stepped changing of angular speed of the driven shaft when the angular speed of the driving shaft is constant.

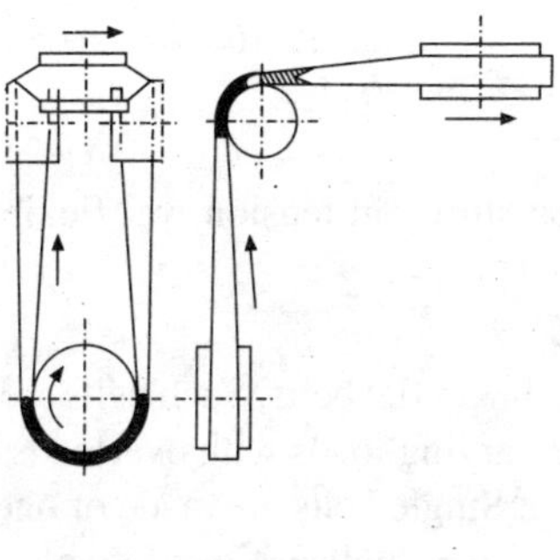

Fig. 19.5 *Right angle drive*

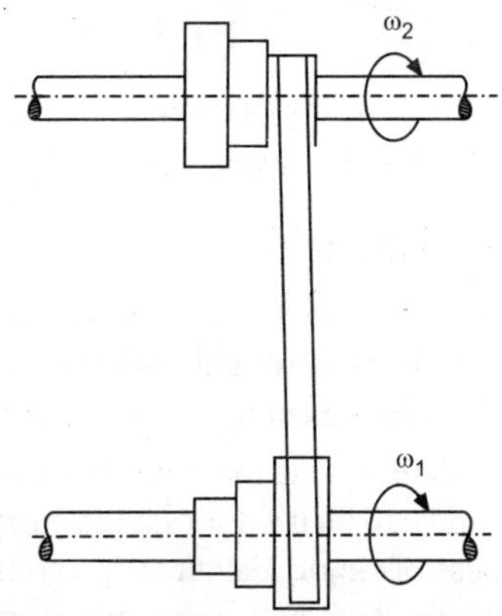

Fig. 19.6 *Stepped pulley drive*

6. **Fast and loose pulley drive:** When the driven shaft is to be rotated and stopped too often, two pulleys are provided on this shaft. One, which is keyed on to the shaft is called working or fast pulley and the other one known as idler or loose pulley rotates freely over the shaft. When the driven shaft is required to be stopped, the belt will be pushed on to the loose pulley from the fast pulley by means of a lever. The driving shaft keeps on running at uniform speed. By this drive the starting and stopping of the prime mover too often is avoided, Fig. 19.7.

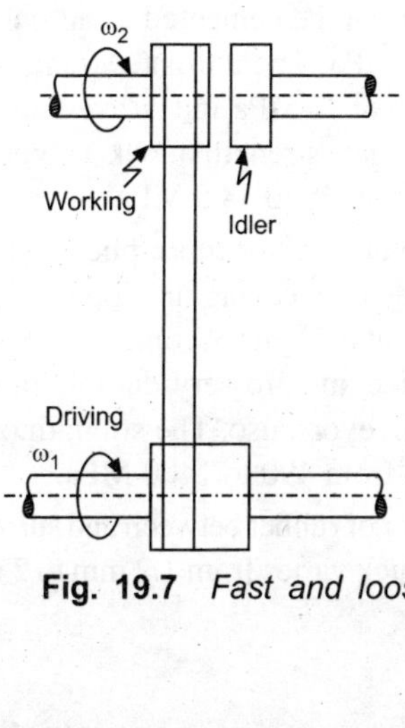

Fig. 19.7 *Fast and loose pulley drive*

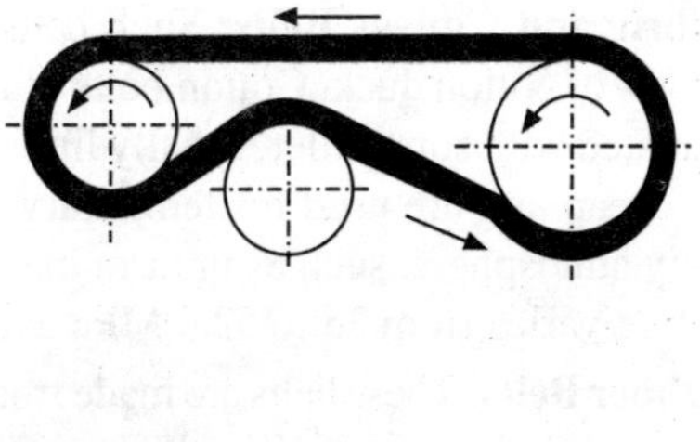

Fig. 19.8 *Jockey pulley drive*

7. **Belt drive with an idler pulley or Jockey pulley drive:** When the high velocity ratio is desired in short centre distance, resulting in small arc of contact on the smaller pulley and the open belt drive can not be employed, or when the required belt tension cannot be obtained by other means, this type of drive is used Fig. 19.8.
8. **Compound drive:** This type of drive is used when several units are to be driven from one central shaft. The various forms of compound drives are shown in Fig. 19.9.
9. **Reverse drive:** Figure 19.10., This type of drive is used to transmit motion from one shaft to several shafts arranged in parallel.

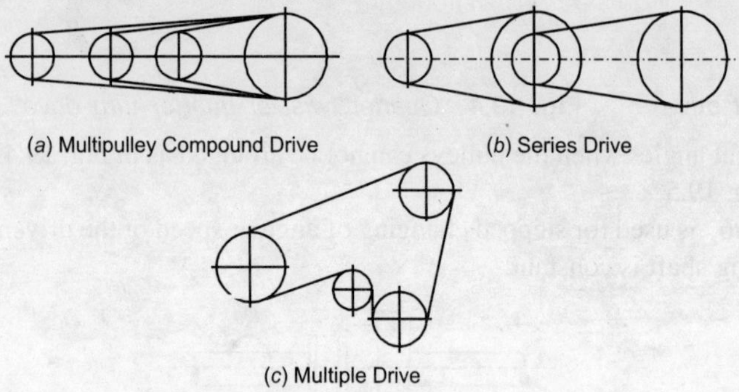

Fig. 19.9 *Compound drive*

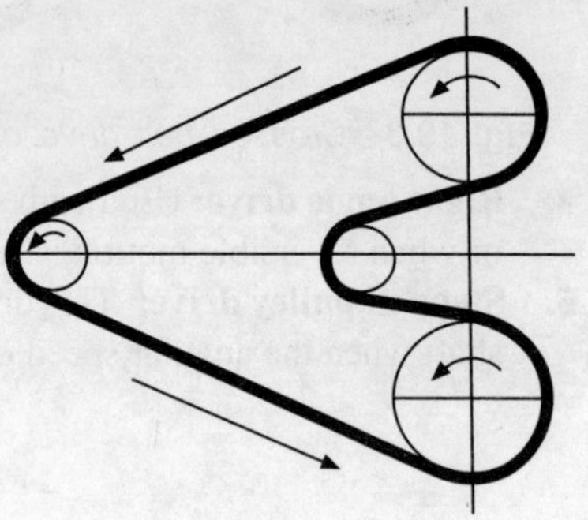
Fig. 19.10 *Reverse drive*

19.2.2 Belt Materials

The materials for the belts to be used for transmission must be strong in tension, yet flexible and pliable, durable, relatively light in weight, and have a good frictional surface.

The main materials used for flat belts are:

1. **Leather:** Leather is the material most commonly associated with flat belts. Such belts are noted for high flexibility and can be used for drives carrying considerable varying loads with overloads reaching 40%. The thickness of single leather strip varies from 3 to 5.5 mm. Single belts are made of one thickness and double belts or triple belts are made from two thickness or three thickness respectively. Such belts are called as "Single Ply", "Double Ply" and "Triple Ply" belts. The separate thicknesses are joined together by cementing. In the same manner, the belts of any length or width can be obtained by cementing together the individual strips lengthwise or widthwise respectively. It has been found that the capacity of two-ply belt is reduced to 85% and that of the triple ply belt 75% of the capacity calculated on the basis of a single ply of the same thickness. Leathers may be tanned with oak, chestnut bark, vegetable compounds, alum salts, and chrome salts. Oak tanned leather is fairly stiff, whereas chrome leather is soft and pliable. Oak tanned leather does not have as good friction as the chrome tanned, nor it is as resistant to water. To combine their qualities, a ply of chrome tanned leather can be cemented to an oak tanned leather. The modulus of elasticity of leather varies from 140 to 315 MPa. The ultimate strength of best leather is about 28 MPa. The first grade belts have a minimum strength of 21 MPa and an average of about 26.5 MPa for a single ply and about 24.5 MPa for double ply. The ultimate strength of oak tanned leather ranges from 21 to 31.5 MPa and of chrome tanned leather it varies from 28 to 38.5 MPa.
2. **Fabric and Canvas Belts:** Such belts are made by stitching together three or more plies or layers of canvas of cotton duck. Cotton belts of any thickness can also be made by weaving in a loom. The fabric is treated with some filler, usually linseed oil, to make it water proof but the belts become stiff. Such belts are cheap and are used for temporary, intermittent, and tough service and are very durable in dry and dusty atmosphere, such as in farm machinery and sometimes for conveyors also. The strength of cotton belting varies from 35 to 52.5 MPa and modulus of elasticity varies from 490 to 1400 MPa.
3. **Rubber Belts:** These belts are made from folded canvas duck with layer of rubber between and surrounding the whole belt structure. The thickness of one ply or layer of canvas duck varies from 1.2 mm to 2 mm and

the belt is usually from three to ten plies in thickness. Of course, the number of plies can be up to 15. The belt is vulcanized under heat and pressure. These belts are used preferably for outdoor service, as these are resistant to moisture, acids, alkalies, oils, etc., e.g., in saw-mills, oil fields, and paper mills. But these are quickly damaged by light, heat and oil. The ultimate strength of these belts range from about 6.5 to 10.5 MPa and the modulus of elasticity varies from about 910 to 1470 MPa.

4. **Balata Belts:** Balata is gum similar to rubber. Balata belts are made in the same manner as the rubber belts except that balata is substituted for rubber. Such belts are acid proof and water proof and are about 25% more strong than rubber belts. The operating temperature should not exceed about 40 to 50°C.
5. **Camel's Hair Belts:** The belts made from camel's hair are very suitable in an atmosphere charged with moisture, steam or mineral oil vapour. Effective tension in the belt varies between about 180 to 450 N for about 5 mm to 15 mm thickness.

19.2.3 Belt Joints and Fasteners

Unless the belts are endless, some type of fastener is required to join the two ends of the belt length. The various types of joints in use are:

(*a*) Cementing, (*b*) Lacing, Laces are made of raw-hide, tanned lace leather, or wire. (*c*) Metal hooks. These joints are shown in Fig. 19.11.

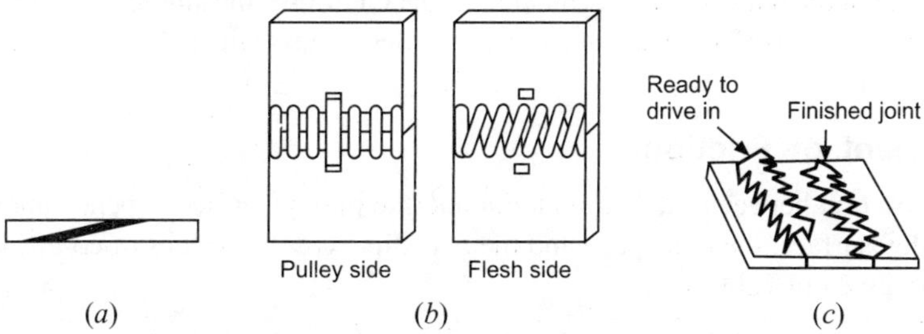

Fig. 19.11 *Flat belt joints*

The efficiencies of these joints are given below:

S. No.	Joint	Efficiency
1.	Cemented by belt maker	1.00
2.	Cemented	0.98
3.	Wire laced by machine	0.90
4.	Wire laced by hand	0.82
5.	Raw-hide laced	0.60
6.	Metal hooks	0.35

19.2.4 Working Stresses

Factor of safety for leather belts may be taken equal to 8 to 10. Factor of safety for rubber and balata belting, because of their uniform structure, may be taken from 6 to 8. The working stresses so obtained must be multiplied by the efficiency of the joint to find out the safe allowable stress.

19.2.5 Density of Belt Materials

$$\text{Leather} = 1000 \text{ kg/m}^3$$
$$\text{Rubber} = 1400 \text{ kg/m}^3$$
$$\text{Balata} = 1110 \text{ kg/m}^3$$
$$\text{Canvas} = 1220 \text{ kg/m}^3$$

19.2.6 Belt Speeds

The belt speed should not be very high, since, as the belt speed increases, centrifugal force begins to pull the belt away from the pulley, thus reducing its effectiveness. For maximum power and economy, belt speeds of 25 to 30 m/s may be used, but for efficiency and longer belt life, speeds of 15-20 m/s should be used.

19.2.7 Belt Slip and Creep

Some amount of slip is always present between the belt and pulley surface when power is being transmitted. Amount of slip increases if the belt is overloaded. It has been found that a slip of upto about 3 per cent increases the coefficient friction. Under normal conditions, a slip of about 1½ to 2 per cent is present. Due to slip, the belt speed is less than the peripheral speed of the driving pulley and slightly more than the peripheral speed of the driven pulley.

Creep is a phenomenon caused by the elasticity of the material. The belt is stretched more on the tight side than on the slack side. When a unit length of belt reaches the driving pulley and leaves it after moving around it, its length gets decreased. It means that the driving pulley is receiving more length of belt and delivering a smaller length. The reverse happens on the driven pulley. The difference in the length being received and being delivered by a pulley is called "creep". Due to creep, the average velocity of the belt is slightly lower than that of the driving pulley surface. On the driven pulley, the average velocity of belt is slightly more than that of the pulley surface. Under normal conditions, the amount of creep is about 1 per cent.

In practice, the combined effect of slip and creep is called simply slip and the combined effect should not exceed 3 per cent.

19.2.8 Coefficient of Friction

The coefficient of friction between the belt material and the pulley surface depends upon the belt material, material of the pulley surface, the belt speed and the belt slip. Average values of co-efficients of friction for design purposes are given in Table 19.1.

Table 19.1 Coefficient of friction for belts

Belt Material	Pulley Material						
	Cast Iron, Steel			Wood	Compressed Paper	Leather face	Rubber face
	Dry	Wet	Greasy				
Leather, oak tanned	0.25	0.20	0.15	0.30	0.33	0.38	0.40
Leather, chrome tanned	0.35	0.32	0.22	0.40	0.45	0.48	0.50
Canvas, stitched	0.20	0.15	0.12	0.23	0.25	0.27	0.30
Cotton, woven	0.22	0.15	0.12	0.25	0.28	0.27	0.30
Camel hair, woven	0.35	0.25	0.20	0.40	0.45	0.45	0.45
Rubber	0.30	0.18	—	0.32	0.35	0.40	0.42
Balata	0.32	0.20	—	0.35	0.38	0.40	0.42

19.2.9 Centre Distance

The life of a belt is a function of the centre distance between the driver and driven shafts. The shorter the belt, the more often it will be subjected to additional bending stresses while running around the pulleys at a given speed, and quicker it will be destroyed due to fatigue. So the centre distance is chosen so as to increase the life of the belt. Under shock loading, a longer belt will last more than a short one. In flat belt drives, the centre distance may reach 15 m and, in exceptional cases, even more. Of course, the good centre distance is 6 m to 7.5 m. The centre distance can be selected from the following empirical relations:

Centre distance, $C \geq (0.07 \text{ to } 0.1) \times V$

or $C \geq (1.5-2)(D_1 + D_2)$

or $C \geq 3.5\, D_2$

where V = belt speed in m/s

D_1 = Diameter of smaller pulley.

D_2 = Diameter of larger pulley.

If the maximum limit of centre distance is exceeded, there will be too much flapping of the belt and trouble will be encountered in keeping the belt tight.

19.2.10 Losses in Transmission

The losses in a belts drive are due to:

1. Slip and creep of the belt on the pulleys.
2. Windage or air resistance to the movement of belt and pulleys.
3. Bending of the belt over the pulley.
4. Friction in the bearings of pulley.

The losses due to windage and bending of belt are usually negligible. As already indicated, the total loss due to slip and creep should not exceed 3 per cent. Average loss due to bearing friction is 1 per cent and it should not exceed 2 per cent. Therefore, the over all efficiency of the drive is about 95% to 96%.

19.2.11 Flat-Belt Drive Formulas

Open belt drive: Refer Fig. 19.12.

Angle of contact on smaller pulley, $\theta_1 = 180 - 2\sin^{-1}\dfrac{(D_2 - D_1)}{2C}$ degrees

Angle of contact on larger pulley, $\theta_2 = 180 + 2\sin^{-1}\dfrac{(D_2 - D_1)}{2C}$ degrees.

Length of belt, $$L = 2C + \frac{(D_2 - D_1)^2}{4C} + 1{\cdot}57\,(D_2 + D_1)$$

If the length of the belt is known, the centre distance can be found out as below:

$$C = p + \sqrt{p^2 - q}$$

where $$p = \frac{L}{4} - \frac{\pi}{8}(D_2 + D_1)$$

and $$q = \frac{(D_2 - D_1)^2}{8}$$

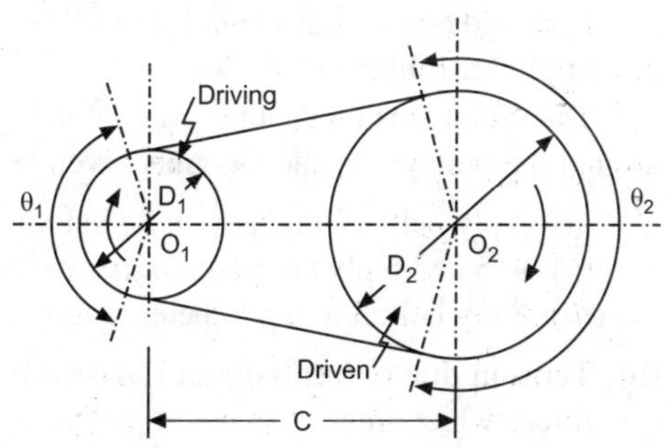

Fig. 19.12 *Open belt Drive*

Note: The value of 'θ_1' should never be less than 2.5 radians. If so, the value of centre distance must be increased.

Cross-belt drive:

$$\theta_1 = \theta_2 = 180 + 2\sin^{-1}\frac{(D_2 - D_1)}{2C} \text{ degrees.}$$

$$L = 2C + 1.57\,(D_2 - D_1) + (D_2 + D_1)^2 / 4C.$$

Quarter-turn drive:

$$L = 1.57\,(D_2 + D_1) + \sqrt{C^2 + D_2^2} + \sqrt{C^2 + D_1^2}$$

As already indicated, the peripheral speed of belt on the driving pulley, neglecting slip, will be

$$V_1 = \frac{\pi D_1 n_1}{60} \text{ m/s}$$

Peripheral speed of belt on driven pulley, $V_2 = \dfrac{\pi D_2 n_2}{60}$ m/s

If there is no slip, $V_1 = V_2$

$\therefore$ $D_1 n_1 = D_2 n_2$

Hence, velocity ratio, $(V_R) = \dfrac{D_2}{D_1} = \dfrac{n_1}{n_2} = \dfrac{\omega_1}{\omega_2}$

If slip is taken into account, then $V_2 = (1 - s)\, V_1$

Where s = slip factor (0.01 – 0.03)

$$\therefore \quad V_R = \frac{\omega_1}{\omega_2} = \frac{n_1}{n_2} = \frac{D_2}{D_1\,(1-s)} = \frac{D_2}{D_1}(1+s),$$

If thickness of belt is also taken into account, then $V_R = \dfrac{\omega_1}{\omega_2} = \dfrac{n_1}{n_2} = \dfrac{D_2 + t}{D_1 + t}(1+s)$

where t = belt thickness.

19.2.12 Stresses in the Belt

The various stresses in the belt will be due to the following forces:

(a) Initial Tension : When the belt is put around the pulleys, then whether the belt is stationary or running idly without transmitting any power, the tension induced in the belt material is known as Initial Tension, F_i. The initial tension is given to develop adequate traction. As the drive is loaded, the tension on the driving side of the belt i.e., the tight side rises to F_1 and that of the driven side i.e., the slack side, reduces to F_2. The three tensions are connected by the following empirical relations, which are not very accurate:

$$F_1 + F_2 = 2F_i$$

or $$F_i = \frac{1}{4}\left[\left(\sqrt{F_1} + \sqrt{F_2}\right)\right]^2$$

or $$\sqrt{F_1} + \sqrt{F_2} = 2\sqrt{F_i}$$

F_i should range from about 1.4 to 1.7 N/mm^2 for leather belts and from about 0.275 to 0.335 N/mm width of belt/ply for rubber belts.

The initial tension is obtained by making the belt length shorter than the actual steel tape measurement around the pulleys, by the amounts given below:

(*i*) 3 ply belt: 15 mm per metre length
(*ii*) 4, 5 and 6 ply belt: 10 mm per metre length
(*iii*) 8 ply belt: 5 mm per meter length.

(b) Tension due to Centrifugal Forces : In operation, the curved portions of the drive develop centrifugal forces which impose an additional load on the belts. This force is given as,

$$F_c = \rho A V^2, N$$

where ρ is the density of the belt material, kg/m^3

A = cross-sectional area of belt, m^2

V = belt speed, m/s.

For flat belts,

$$A = \text{width of belt} \times \text{thickness of belt}$$
$$= b \times t$$

The centrifugal force gives rise to a tensile stress in the belt, which is given by,

$$f_c = \frac{F_c}{A} = \rho V^2, \text{N/m}^2$$

For V-belt drive, total centrifugal force is given as,

$$F_c = \rho A V^2, \text{N}$$

where $A = A_1 z$, z is the number of belts in the drive

and A_1 = cross-sectional area of one belt.

(c) Tension due to Useful Load: When a useful power is transmitted by the drive, then as written above, the total tension in the belt will rise to F_1 due to frictional forces on the tight (driving) side and fall to F_2 on the slack (driven) side.

We know that, useful power, P (watts) is given by,

$$P = F_t \times V$$

where F_t is the tangential force arising in the drive, that is the net belt pull, and is given as,

$$F_t = (F_1 - F_2), \text{N}$$

$$\therefore \quad P = (F_1 - F_2) \,.\, V, \text{watts}$$

Assuming that the load is transmitted over the entire arc of contact and neglecting the centrifugal forces, we can relate F_1 and F_2 as,

$$F_1 = F_2 . e^{\mu\theta} = F_2 . k$$

where θ is the angle of contact and μ is the coefficient of friction between the belt and the pulley surface.

It can be easily shown, that,

$$F_1 = F_t \cdot \frac{k}{k-1}$$

and

$$F_2 = F_t \cdot \frac{1}{k-1}$$

If centrifugal forces are taken into account, then the total tension in the belt on the tight and slack sides may be found as,

$$F_1 = F_t . \frac{k}{k-1} + F_c \qquad \text{...(19.1)}$$

$$F_2 = F_t . \frac{1}{k-1} + F_c$$

The tight side carries a heavier load than does the slack side in operation. The associated tensile stresses are given as,

$$f_t = \frac{F_1}{A} = \left(\frac{F_t}{A}\right) . \frac{k}{k-1} + \frac{F_c}{A} = \frac{k}{k-1} . f_e + f_c$$

where f_e = effective stress, or specific tangential force, F_t/A.

With allowances for centrifugal forces, the relationship between initial tension F_i and the tension in the loaded belt may be defined as,

$$F_i = 0.5\,(F_1 + F_2) + q \,.\, F_c \qquad \text{...(19.2)}$$

where q accounts for the compliance of the belt and the shaft bearings. For standard drives utilizing fixed shafts, it may be taken as 0.1 to 0.3. If belt tension is maintained automatically, $q = 1$.

(d) Bending Stresses: When it goes around the pulley, the belt experiences bending stresses, given as,

$$f_b = E \,.\, \frac{t}{D}, \text{N/m}^2$$

where E = modulus of elasticity of the belt material.

D = diameter of pulley.

Maximum Stress: The stress is a maximum where the belt wraps around the smaller pulley, and is given as,

$$f_{max} = f_t + f_b = \frac{k}{k-1} \cdot f_e + f_c + f_b$$

19.2.13 Design of Belt Section

In practice the belt drive is designed for the power to be transmitted and not for the strength. The power transmitted by a belt depends upon:

(*i*) Net belt tension, that is, difference in belt tensions on the tight and slack side.
(*ii*) Coefficient of friction between belt material and pulley face.
(*iii*) Angle of contact.
(*iv*) Centre distance.

(*a*) The first is to calculate the design power, which is given as,

$$\text{Design power} = \text{Power to be transmitted} \times \text{Service factor}$$

The service factor is given below:

S. No.	Service conditions	Service factor
1.	Oily, wet or dusty environments	1.35
2.	Jerky loads	1.20
3.	Shock and reversed loads	1.40

(*b*) Now the belt section will be based on the maximum tension, that is tension on the tight side of the belt.

As discussed in the previous article,

$$\text{Power} = F_t \times V \qquad ...(19.3)$$

where $F_t = (F_1 - F_2)$

Now F_1 and F_2 are total tensions on the tight side and slack side of the belt.

(*c*) It is usual to neglect the effect of centrifugal stress if the belt speed is upto 10 m/s. Then the tensions F_1 and F_2 are connected by the relation:

$$\frac{F_1}{F_2} = e^{\mu\theta} \qquad ...(19.4)$$

θ = angle of contact on the smaller pulley, if both the pulleys are of same material and if of different materials, the design will refer to the pulley with smaller value of "$\mu\theta$".

So, from equations (19.3) and (19.4), F_1 can be calculated.

Then belt section is,

$$b \times t = \frac{F_1}{f_t \times C_1}, \text{ For } V \text{ upto 10 m/s} \qquad ...(19.5)$$

where f_t = allowable stress in belt, N/m^2.

C_1 = a correction factor, depending on the angle of centre line of drive with horizontal, Table 19.2.

(*d*) If the belt speed is more than 10 m/s, the effect of centrifugal force must be taken into account.

So, from equation (19.1),

$$\frac{F_1 - F_c}{F_2 - F_c} = k = e^{\mu\theta}$$

$$\therefore \quad (F_1 - F_2) = (F_1 - F_c)\left(\frac{e^{\mu\theta} - 1}{e^{\mu\theta}}\right) \qquad ...(19.6)$$

The effect of centrifugal force is considered indirectly. We proceed on the same lines as for case when the centrifugal force is neglected, but while calculating belt section, the allowable stress is reduced by the centrifugal stress. Thus,

Belt section, $$b \times t = \frac{F_1}{(f_t - \rho V^2) \times C_1}, \text{ for } V > 10 \text{ m/s} \quad ...(19.7)$$

From equation (19.1),

$$F_1 - F_2 = F_t$$

Hence, it is clear that the centrifugal force F_c does not figure in power transmission relationships.

The equation (19.6) can be written as,

$$(F_1 - F_2) = bt\,(f_t - \rho V^2)\left(\frac{e^{\mu\theta} - 1}{e^{\mu\theta}}\right) \quad ...(19.8)$$

This equation is for horizontal drives and is known as general belt equation.

In equations (19.5) and (19.7), the belt thickness 't' is taken so that the ratio D_1/t, is not smaller than the values given below:

Leather : $\frac{D_1}{t} \geq 30$ to 35

Rubberized fabric : $\frac{D_1}{t} \geq 30$ to 40

Cotton fabric : $\frac{D_1}{t} \geq 25$ to 30

Then the belt width is determined. The relations of width and thickness of belts are given in Tables 19.3 and 19.4. Finally, the standard belt width and thickness are taken as mentioned ahead.

Note: Some authors determine the belt section on the basis of net belt pull, that is, F_t. However, design of belt section on the basis of tension on tight side, F_1, will be still safer.

Table 19.2 Factor C_1

Type of Drive	Inclination to Horizontal		
	0–60°	60–80°	80–90°
Open belt drive	1.0	0.9	0.8
Cross belt drive	0.9	0.8	0.7
Quarter turn drive	0.8	0.7	0.6

Note: In the above analysis, the tight side of the belt is the lower one and the slack side, the upper one and it should be so for horizontal or nearly so drives to prevent the pulling away of the belt from the pulley surfaces due to the weight of the belt and to increase the arc of contact due to the sag of the slack side.

Table 19.3 Thickness and width for leather belts

Grade of Belt	Approximate thickness mm			Width, mm
	Single	Double	Triple	
1. Light	3	6	—	upto 200 mm for single ply belt
2. Medium	4	8	12.5	upto 300 mm for double ply belt
3. Heavy	5.5	11	16.5	upto 600 mm for triple ply belt

Table 19.4 Width and thickness for rubber and balata belts

Thickness, mm	Number of plies	Width, mm
1.2 mm to 2 mm per ply	3	25 to 200
	4	25 to 600
	5	40 to 900
	6	50 to 1500
	7	100 to 1500
	8–15	150 to 1800

For rubber belts, the number of plies may be found out empirically as:

Number of plies = 0.0 13 D, for Low Speed

and = 0.01 D, for High Speed

Standard belt width and thickness are given below ;

(i) **Width:** Standard belt width is given in R_{10} series from 25 mm to 63 mm width and in R_{20} series from 71 mm to 600 mm width. Referring to Table 5.1, the standard widths will be : 25, 32, 40, 50, 63, 71, 80, 90, 100, 112, 125, 140, 160, 180, 200, 224, 250, 280, 315, 355, 400, 450, 500, 560 and 600 mm.

(ii) **Thickness:**

5 mm for belt width of 25 to 63 mm
6.5 mm for belt width of 50 to 140 mm
8.0 mm for belt width of 90 to 224 mm
10.0 mm for belt width of 125 to 400 mm
12.0 mm for belt width of 250 to 600 mm

Condition for the transmission of maximum power: We know that power transmitted is given as

$$P = (F_1 - F_2) \times V \text{, kW}$$

now
$$F_2 = \frac{F_1}{e^{\mu\theta}}$$

$\therefore$
$$P = F_1\left(1 - \frac{1}{e^{\mu\theta}}\right) \times V = F_1 \,.\, V.l$$

where the constant
$$l = 1 - \frac{1}{e^{\mu\theta}}$$

If centrifugal tension is considered, then

Tension on the tight side of belt $F_1 + F_c$

and Tension on the slack side of belt = $F_2 + F_c$.

$\therefore$ Maximum tension to which the belt can be subjected is,

$$F = F_1 + F_c$$

$\therefore$
$$F_1 = F - F_c$$

$\therefore$
$$P = (F - F_c) \times V \times l$$

now
$$F_c = f_c \times \text{belt section}$$
$$= \rho V^2 \times b \times t = mV^2 \text{, N}$$

where m = mass density per unit length of belt, kg/metre

$\therefore$
$$P = (F \times V - mV^3) \times l$$

For maximum power,
$$\frac{dP}{dV} = 0$$

$$\therefore \qquad F - 3m\,V^2 = 0$$

$$F - 3F_c = 0$$

$$\therefore \qquad F = 3F_c \text{ or } F_c = 1/3F$$

$\therefore$ For maximum power transmitted, one third of the maximum tension in the belt is absorbed as centrifugal tension.

Belt velocity for maximum power transmitted,

$$V = \sqrt{\frac{F}{3m}} = \sqrt{\frac{F.g}{3w}},$$

where w = Weight density of belt per unit length.

19.2.14 (1) Rock Wood Drive

In this drive, the motor base is pivoted for tension adjustment. This is also a short centre drive superior to an idler drive because it eliminates the bending of the belt in opposite directions. Such a drive is shown in Fig. 19.13.

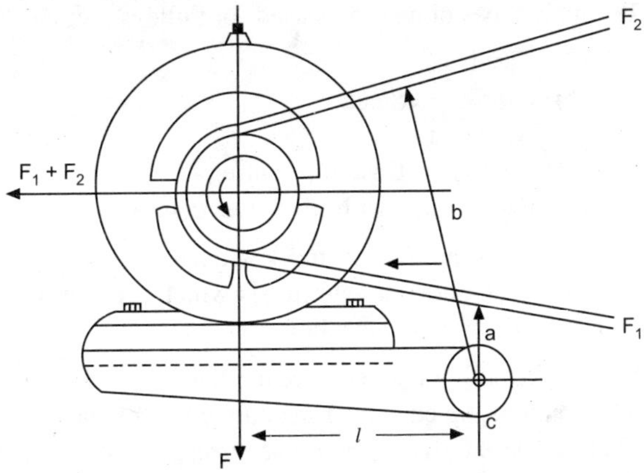

Fig. 19.13 *Rock wood drive*

The base of the motor is pivoted on the pin or shaft support '*C*' and the weight of the motor provides the necessary belt tension. The lever arm '*l*' is made adjustable so that the motors of different weights may be balanced. The arm length can be found out by taking the moments about the pivot *C*.

$$\therefore \qquad F_1 \times a + F_2 \times b = F \times l$$

or

$$l = \frac{F_1 a + F_2 b}{F}$$

where F = weight of motor.

If the direction of rotation is reversed to that shown in Fig. 19.13, the belt tensions F_1 and F_2 get interchanged, then

$$l = \frac{F_1.b + F_2.a}{F}$$

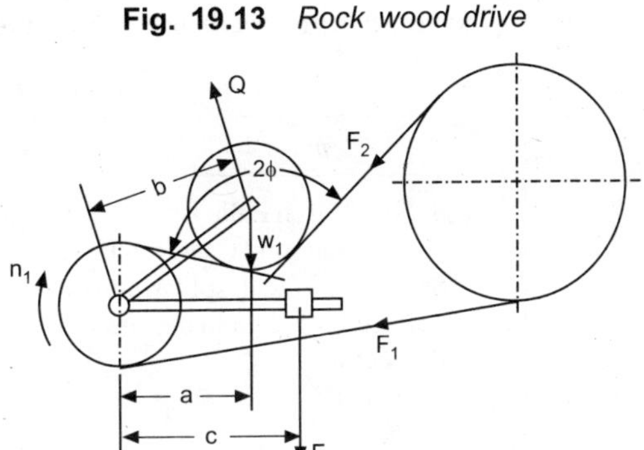

Fig. 19.14 *Short centre drive with idler roller*

2. **Short Centre Drive with Idler Roller:** This drive is shown in Fig. 19.14. The idler acts as belt tightener and it increases the angle of contact on the small pulley and it ensures a constant tension of the belt irrespective of its stretch. The idler is usually put on the slack side of the belt. The disadvantages of this drive are:

   (*a*) It is non-reversible.

   (*b*) Too much bends of the belt reduces the belt life.

   The weight of the load *F* is determined by taking moments about the hinge of the arm, so considering the equilibrium of the lever.

$$\therefore \qquad Q_1.b - W_1.a - F.c = 0$$

$$\therefore \qquad F = \frac{Q_1.b - W_1.a}{c}$$

Force Q is a resultant of tensions F_2 in the slack side of the belt, i.e.,

$$Q = 2F_2.\cos\phi.$$

19.2.15 Flat Belt Pulleys

The pulleys for the transmission of power should have the following qualities:

1. Ability to absorb shock.
2. Ability to conduct heat.
3. Ability to resist corrosive environment.
4. Smooth surface for low belt wear.
5. Coefficient of friction of face.
6. Strength combined with low weight.
7. Minimum air resistance.

The main types of materials used for pulleys are:

(*a*) Cast Iron.
(*b*) Fabricated Steel.
(*c*) All wood.
(*d*) Wood rim, Cast iron centre.
(*e*) Paper rim, Cast Iron or steel centre.

Cast iron pulleys are most widely used in practice. They are most suitable in plants where there is excessive dampness or acid fumes which prevent the use of other pulley materials, e.g., in chemical plants, tanneries, paper mills, dye houses.

Split Pulleys: For convenience in mounting pulleys on shafts, it is a common practice to cast them in halves, which can be joined at the rim and the hub by means of bolts or links. Such pulleys can be very conveniently mounted on the shaft at any place irrespective of the position of bearings.

Rim Speeds: The normal rim speeds for various pulleys are given in Table 19.5.

Table 19.5 Rim speeds

Type	*Speed, m/s*
Solid cast Iron Pulleys	25
Good grade solid Cast Iron Pulleys	27.5
Split pulleys, with split at arms	22.5
Split pulleys, Split in between arms	20
Pressed steel	20–25
All wood	25–30
Wood rim	40–50
Paper rim	40–50

Pulley Size: The minimum pulley size is limited by the elongation of the outer fibres of the belt as it bends or wraps around the pulley circumference. This elongation of the belt fibres due to bending over the pulley surface is one cause of failure of belts. The trouble increases as the size of the pulley decreases. According to Spooner the diameter of pulley should not be made less, than,

$$D = 10^4\, t^2$$

where D = diameter of pulley in m

t = thickness of belt in m

According to another empirical relation, the minimum diameter of pulley for leather belt is given as :

$$D \geq 50\, t.$$

For rubber belts, a good practical rule is to have about 0.075 m of pulley diameter per ply of belt at lower speeds and at higher speeds to have about 0.10 m of pulley diameter per ply.

According to another empirical relation,

$$D_{min} = (1.10 \text{ to } 1.33) \sqrt[3]{\frac{P}{n_{max}}}, \text{ m} \quad \text{or} \quad D_{min} = (0.525 \text{ to } 0.630) \sqrt[3]{\frac{P}{\omega_{max}}}, \text{ m}$$

or $D_{min} = D_1 = (53 \text{ to } 65)\,(M_{t1})^{\frac{1}{3}}$, mm, where

M_{t1} = Torque at driving shaft, Nm

where the power P is in kW.

Since excessive flexing of a belt will result in a shortened life, therefore, for reasonable life, a minimum ratio of the diameter of a pulley to the thickness of the belt is about 50.

A cast iron pulley as per IS 1691–1980 is shown in Fig. 19.15. The width of the pulley will depend on the width of belt. It is given by the relation,

$$B = [1.1\,b + (0.01 \text{ to } 0.015)] \text{ m}$$

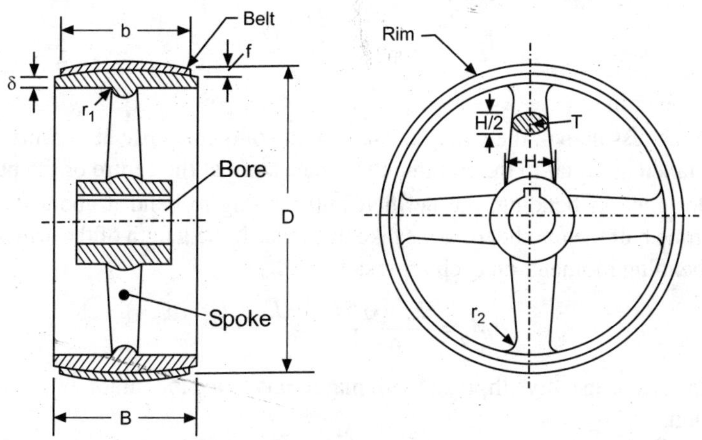

Fig. 19.15 *Cast iron pulley*

According to ISI, the pulley widths are given below in Table 19.6

Table 19.6 Pulley width (IS 2122 (Part I)-1973)

Belt Width mm	Pulleys to be wider than belt width by
Upto including 125	13
125–250	25
250–375	38
375–500	50

The face of the pulley rim is crowned to keep the belt on the pulley, since inaccurate alignment of the pulleys causes the belt to run off side ways. The crown will force the belt to return to the centre of pulley.

The usual value of crown '*f*' is about 15 mm per metre of face width. Or to obtain the crown height, a curve with centre on the vertical axis of pulley and radius of 3*B* to 5 *B*, can be drawn passing through the two points of the pulley face width.

The thickness of the rim edge of C.I. pulley is given as:

$$\delta = (0.005D + 0.003) \text{ m}$$

The rim thickness is also given as :

$$\delta = \frac{D}{200} + 3 \text{ mm for single belt}$$

$$= \frac{D}{200} + 6 \text{ mm for double belt}$$

The inside of the section should taper slightly so that the pattern may leave the mould easily.

Radius of the cross-section of arms, $r = \frac{3}{4} H$

Radius $r_1 = r_2 = \frac{H}{2}$ (near rim).

Arms: The number of arms do not depend on a fixed criteria. Small pulleys are provided with solid web or a web with holes for handling convenience instead of arms or spokes.

For D upto 200 mm, use web.

For D, above 200 mm and upto 450 mm number of arms = 4.

For D, above 450 mm number of arms = 6.

The number of arms is also given by the following relation :

$$z = (9 \text{ to } 10) \frac{\sqrt{D}}{2}$$

where D = pulley diameter in m

If z comes out to be less than 4, the pulley is made with solid disc instead of arms. The thickness of the solid web is usually taken as equal to the rim thickness measured at the centre of the pulley face.

The arms are designed in bending, the net belt pull tending to bend it about the centre of hub. It is assumed that at one time half the number of arms take this load. If the length of the arm is considered upto the centre of hub, then bending moment, on each arm is:

$$M_b = \frac{F_t \times 0.5D}{0.5\, z} = \frac{F_t \times D}{z}$$

The section of the arm is usually elliptical, with major axis twice the minor axis, the major axis is placed in the place of rotation.

Now section modulus for this section is given as

$$Z = \frac{\pi}{32} hH^2$$

where H = major axis, h = minor axis

$$\therefore \quad Z = \frac{\pi}{32} \frac{H}{2} . H^2 \approx 0.05 \text{ H}^3$$

$$\therefore \quad f_b \times Z = M_b$$

Since the arms are subjected to complete reversal of stress, a factor of safety of at least 8 should be used. Therefore the allowable stress f_b for C.I. arms will be 14 to 21 MPa, since the ultimate strength of C.I. in tension ranges from 126 to 210 MPa.

At the rim, the arm dimensions are :

$H' = 0.8H$ and $h' = 0.8\, h$, or a taper of 4 mm per 100 mm is given to the arm from boss to rim.

Along its length, the arm may be curved or straight. Curved arms are easy to cast but are heavy. With modern advanced foundry technology there is no justification for curved arms.

If the face width of pulley is more than 30 cm, two rows of arms are provided, instead of one row, as shown in Fig. 19.16. This is done to avoid too heavy arms in one row.

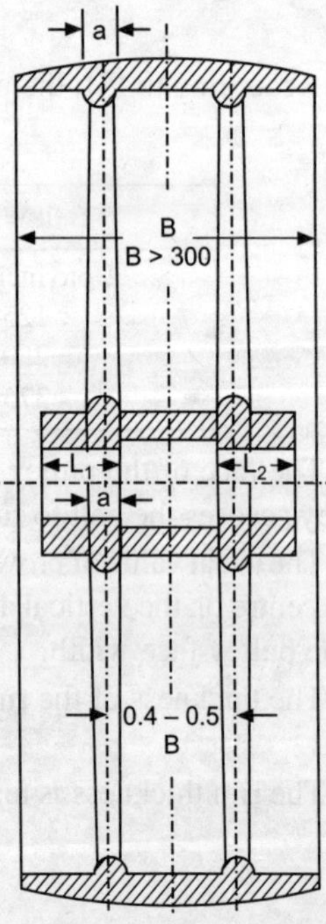

Fig. 19.16 *Cast iron pulley with two rows of arms*

Shaft. The shaft will be under the following forces:

1. Torsional moment due to transmission of power.
2. Bending due to total belt pull.
3. Bending due to the weight of the pulley.

Torsional moment on shaft $M_t = F_t \times D/2$.

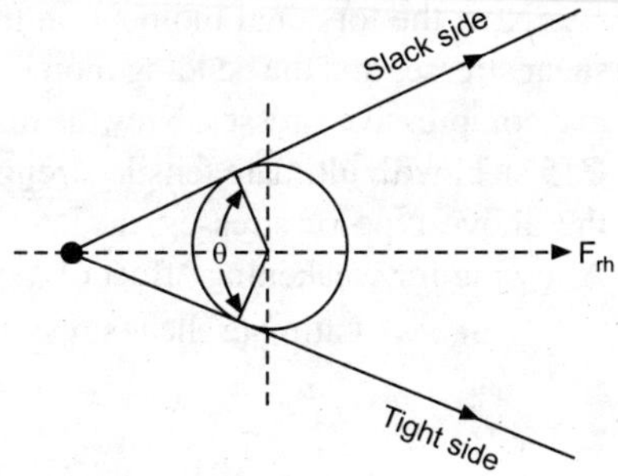

Fig. 19.17 *(a) Pressure on shaft*

Pressure on the shaft due to belt tension : As already discussed, the total pull in the belt is given as,

Total pull on tight side = F_1 for slow drives ($V \leq 10$ m/s)

Total pull of slack side = F_2

and for fast drives ($V \geq 10$ m/s)

Total pull on tight side = $F_1 + F_c$

and total pull on slack side = $F_2 + F_c$

The pressure on shaft imposed due to belt pulls will be given vectorially as Fig. 19.17(*a*).

$$\overrightarrow{F_{sh}} = \vec{F_1} + \vec{F_2} \text{ for slow drives}$$

and
$$\overrightarrow{F_{sh}} = (F_1 \vec{+} F_c) + (F_2 \vec{+} F_c) \text{ for fast drives}$$

If θ is taken as π radians, then

$$F_{sh} = F_1 + F_2 \text{ for slow drives.}$$

and
$$= F_1 + F_2 + 2F_c \text{ for fast drives}$$

If coefficient of friction is taken to be $\mu = 0.22$ (average value for leather belts), then it will be seen that,

$$F_{sh} = 3\,F_t \text{ for slow drives}$$

and
$$= 3F_t + 2F_c \text{ for fast drives}$$

To create enough frictional force between the belt and the pulley, the pulley must be laid with a pre-tension. This tension is mostly determined by feel and is usually greater than necessary. This results in greater loads on shafts. So, from the safety point of view, the shaft pressure is taken as,

$$F_{sh} = (4 \text{ to } 6)\,F_t$$

The shaft pressure is of great significance since it determines the bearing loads for the design of bearings.

The pressure imposed on the shaft by the tension force in the sides of the belt can be determined with a sufficient degree of accuracy by the formula,

$$F_{sh} = 2F_i \sin\frac{\theta}{2} \quad \text{...(19.9)}$$

where F_i = initial tension = $f_i \times bt$.

Here f_i = initial tension stress, whose average value may be taken as 1.8 MPa.

Now the force F_{sh} acts almost horizontally, and the weight of pulley, F, acts vertically downward, therefore, the resultant is,

$$R = \sqrt{F_{sh}^2 + F^2}$$

If the pulley is fitted in between the bearings, the bending moment due to R can be found out very easily. If the pulley is over hanging as shown in Fig. 19.17, then

$$M_b = R \times \text{overhang.}$$

∴ Equivalent torsional moment $M_{te} = \sqrt{M_t^2 + M_b^2}$

∴
$$M_{te} = \frac{\pi}{16} d^3 \times fs$$

where d = shaft diameter

f_s = allowable stress in shear.

Now the torsional moment on the shaft will produce fluctuating shear stresses and the bending moment will produce alternating tensile and compressive stresses. Now the materials for shaft is usually C45 to C55 steel with ultimate tensile strength equal to 450 to 550 MPa. For the above type of stresses, a factor of safety of 8 should be used. Allowing for weakening effect of key also,

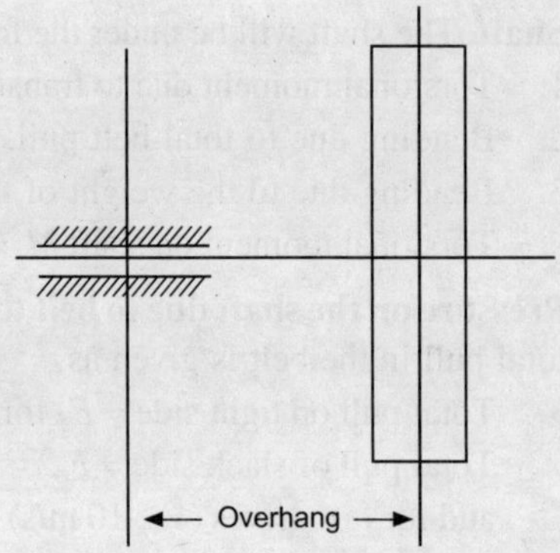

Fig. 19.17 *Over hanging pulley*

$$\text{Ultimate shear stress} = 0.75\ (450 \text{ to } 550)$$
$$= 340 \text{ to } 412.5 \text{ MPa}$$
$$\therefore \quad f_s = 42.5 \text{ to } 50 \text{ MPa}$$
$$\text{Now hub diameter} = (1.7 \text{ to } 2)\ d$$
$$\text{hub length} = (1.5 \text{ to } 2)\ d$$
$$\text{or} \quad = \frac{2}{3}\ B \text{ to } B.$$

Key: Using standard square sunk key,

$$\text{width of key} = \text{thickness} = \frac{d}{4}$$

Minimum length of key can be found out by equating its moment of resistance to the shaft torsional moment.

$$\therefore \quad M_{te} = l \times w \times f_s \times \frac{d}{2}$$

where l = length of key, w = width of key.

f_s may be taken higher than for shaft. Length of key should be more than hub length.

19.2.16 Steel Pulleys [IS : 1691–1980]

A mild steel pulley is shown in Fig. 19.18. The rim is generally made in two halves, the hub and arms are pressed separately. Then, the rim sections, arms and the hub are riveted and bolted together resulting in a split pulley. It is then mounted on the shaft without the help of a key since the pulley will be held to the shaft by the clamping action of the hub. However, under most severe service conditions, a key is used. Generally, interchangeable bushings are provided with steel pulleys to permit their use with shafts of different sizes.

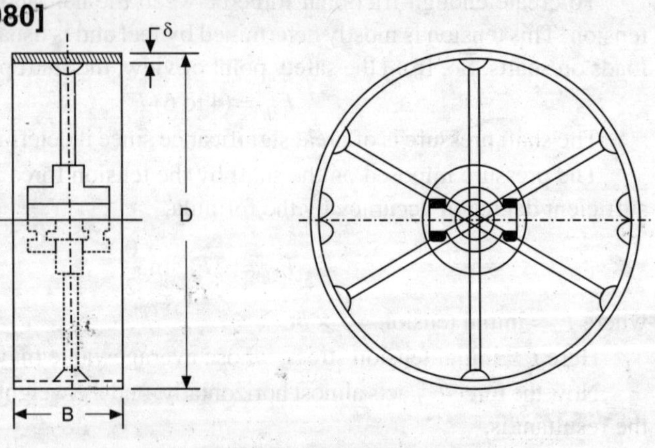

Fig. 19.18 *Mild steel pulley*

Steel pulley have the following advantages:

1. They possess great strength and durability as these are fabricated from pressed steel sheets.
2. They are lighter as compared to C.I. pulleys of the same size.
3. Coefficient of friction obtained with leather belts is the same as for C.I. pulleys.
4. Since the arms are very narrow, the loss due to air fanning is minimum.

The standard number of spokes and their sizes are given below:

Minimum length of boss $= 1/2 \times$ face width

$\nless$ 100 mm for pulley with 19 mm spokes.

$\nless$ 138 mm for pulley with 22 mm spokes.

The thickness of the rim of all mild steel pulleys is usually 5 mm.

The nominal diameters of C.I. and Mild steel pulleys are : 40, 45, 50, 56, 63, 71, 80, 90, 100, 112, 125, 140, 160, 180, 200, 224, 250, 280, 315, 355, 400, 450, 500, 560, 630, 710, 800, 900, 1000, 1120, 1250, 1400, 1600, 1800, 2000, 2240, 2500, 2800, 3150, 3550, 4000, 4500, and 5000 mm.

Pulley Diameter, mm	No. of Spokes	Diameter of spoke, mm
280 to 500	6	19
560 to 710	8	19
860 to 1000	10	22
1120	12	22
1250	14	22
1400	16	22
1600	18	22
1800	20	22
2000	22	22

The nominal widths of flat cast iron and mild steel pulleys are : 20, 25, 32, 40, 50, 63, 71, 80, 90, 100, 112, 125, 140, 160, 180, 200, 224, 250, 280, 315, 355, 400, 450, 500, 560, 630 mm.

The standard material for the pulleys should be taken as below:

(a) C.I. Pulley : Grade 20 of IS : 210–1970. **(b) MS Pulley :** IS : 226–1669.

Example 19.1. Design completely belt drive to drive a winch from an electric motor of 11 kW power. Speed of motor shaft is 750 rev/min. Speed ratio is 4. Belt position is horizontal and there is considerable variation of load.

Solution: 1. Find the diameter of the motor shaft pulley by the relation:

$$D_1 = (1.10 \text{ to } 1.33)\sqrt[3]{\frac{P}{n_{max}}}$$

The power P, in this empirical relation is in kW.

Since there is considerable variation in load, let us take the service factor as 1.3. Therefore, design power is,

$$P = \text{Power to be transmitted} \times \text{Service factor} = 11 \times 1.3 = 14.3 \text{ kW}$$

Now n_{max} will naturally be of the motor shaft,

$$\therefore \quad n_{max} = 750 \text{ rev/min}$$

$$\therefore \quad D_1 = (1.10 \text{ to } 1.33)\sqrt[3]{\frac{14.3}{750}}$$

$$\text{say} = 1.10 \times \sqrt[3]{\frac{14.3}{750}} = \frac{1.10}{3.74} = 0.295 \text{ m, say } 0.30 \text{ m}$$

2. Calculate the diameter of the driven pulley, by the relation:

$$V_R = \frac{D_2}{D_1(1-s)}$$

now $V_R = 4$; and let us assume, $s = 0.02$

$$\therefore \quad D_2 = 4 \times 0.30 \times 0.98 = 1.177 \text{ m, say } 1.2 \text{ m}$$

$\therefore$ Corrected velocity ratio is

$$V_R = \frac{1.20}{0.30 \times 0.98} = 4.08$$

Deviation from the given velocity ratio = $\frac{4.08-4}{4} = 2\%$

which is quite allowable as for general purpose drives, the deviation can be ± (3 – 5)%.

3. Determine the belt speed,

$$V = \frac{\pi D_1 n_1}{60} = \frac{\pi \times 0.3 \times 750}{60} = 11.85 \text{ m/s}$$

4. Find the centre distance,

$C \geq (1.5-2)(D_1 + D_2)$, say 2 (0.30 + 1.20) = 3 metres

Also, $C \geq 3.5\, D_2$, i.e., 4.2 metres

let us take centre distance = 4 metres.

5. Determine the length of the belt,

$$L = 2C + \frac{(D_2 - D_1)^2}{4C} + 1.57\,(D_2 + D_1)$$

$$= 2 \times 4 + \frac{(0.9)^2}{4 \times 4} + 1.57\,(1.2 + 0.3) = 10.40 \text{ m}$$

6. Determine the angle of contact on the smaller pulley,

$$\theta_1 = 180 - 2\sin^{-1}\frac{D_2 - D_1}{2C} = 180 - 2\sin^{-1}\frac{1.2 - 0.3}{2 \times 4} = 167° = 2.91 \text{ radians}$$

which is more than 2.5 radians, therefore, O.K.

7. **Belt Section:** Since V is > 10 m/s, therefore the centrifugal force will be considered,

now power $P = \frac{(F_1 - F_2) \times V}{1000}$, kW

$\therefore$ $F_1 - F_2 = \frac{14.3 \times 1000}{11.85} = 1206.75$ N ...(1)

Now, $\frac{F_1}{F_2} = e^{\mu\theta}$

The coefficient of friction for leather belt on C.I. pulley may be taken as 0.35.

$\theta = 2.91$ radians

$\therefore$ $\frac{F_1}{F_2} = e^{0.35 \times 2.91} = 2.77$

$\therefore$ From = n (1),

$$F_2\,(2.77 - 1) = 1206.75, \therefore F_2 = \frac{1206.75}{1.77} = 681.78 \text{ N}$$

$\therefore$ $F_1 = 1206.75 + 681.78 = 1888.53$ N

$\therefore$ Belt section, $b \times t = \frac{F_1}{(f_t - \rho V^2) \times C_1}$

Now the working stress for oak tanned leather is $\frac{21}{10}$ to $\frac{31.5}{10}$, i.e., 2.1 to 3.15 MPa say 3.0 MPa.

If the belt is completed by a wire laced joint by machine, the efficiency of the joint is 0.90.

$\therefore$ $f_t = 0.9 \times 3.0 = 2.7$ MPa

Now for a horizontal drive, $C_1 = 1$

Centrifugal stress, $\rho V^2 = 1000 \times (11.85)^2 = 1.40 \times 10^5$ N/m^2 = 0.14 MPa

$$\therefore \quad b \times t = \frac{1888.53}{(2.7 - 0.14) \times 10^6} = 0.00074 \text{ m}^2$$

Taking, $t = 0.03\, D_{min} = 9$ mm

$$\therefore \quad b = \frac{0.00074 \times 10^6}{9} = 82.2 \text{ mm, say } 83 \text{ mm}$$

8. Pulley on the winch shaft:

Diameter = 1.2 m

Width of the pulley face, $B = [1.1b + (0.01 \text{ to } 0.015)]$ m = $[1.1 \times 0.083 + 0.010]$ m
= 0.1013 m, say 102 mm

Crown height = 15 mm per metre of face width

i.e., about **1.5 mm.**

Rim thickness, $\delta = (0.005D_2 + 0.003)$m
= $(0.005 \times 1.2 + 0.003) = 0.009$ m, i.e., 9 mm

Arms: For pulley diameter of 1.2 m let us take the number of arms as 6.

Now B.M. per arm $= \dfrac{F_t \times D_2}{z} = \dfrac{1206.75 \times 1.2}{6} = 241.35$ Nm

$\therefore \quad M_b = f_b \times Z$

$\therefore \quad 241.35 = 14 \times 10^6 \times 0.05\, H^3$

$\therefore \quad H$ = major axis at the hub = 0.07 m

$\therefore \quad h$ = minor axis at the hub = $\dfrac{H}{2}$ = 0.035 m

$\therefore$ At the rim, $H' = 0.8H = 0.056$ m
$h' = 0.8\, h = 0.028$ m

Shaft: Torsional moment on the shaft $M_t = F_t \times \dfrac{D_2}{2} = \dfrac{1206.75 \times 1.2}{2} = 724.05$ Nm

Now total belt pull, $F_{sh} = F_1 + F_2 + 2F_c$

Now $F_c = \rho\, V^2 \times bt = 0.14 \times 10^6 \times 0.00074 = 103.6$ N

$\therefore \quad F_{sh} = 1888.53 + 681.78 + 207.2 = 2777.51$ N

Now, from the equation (19.9),

$F_{sh} = 2f_i \times bt \times \sin \theta/2$
$\cong 2 \times 1.8 \times 83 \times 9 \times 1$ (taking $\theta = \pi$ radians) $\cong$ 2689.2 N

Now the weight of the pulley = weight of hub + weight of arms + weight of rim. Assuming the weight of hub as 10% of rim weight. Mass of hub + mass of rim = $1.1 \times \pi[D_2 + (f + \delta)] \times (f + \delta) \times B \times \rho$ kg

where f = Crown = 0.15 cm, δ = rim thickness = 0.9 cm
ρ = density of C.I. = 0.0072 kg/cm^3

$\therefore$ Mass of hub + mass of rim = $1.1\, \pi[120 + 1.05] \times 1.05 \times 10.2 \times 0.0072 = 28.46$ kg

$\therefore$ Weight of (hub + rim) = $28.46 \times 9.81 = 279.22$N

Now area of cross-section of elliptical section = $\dfrac{\pi}{4} . h . H$

Now mean $H = \dfrac{7 + 5.6}{2} = 6.3$ cm

And mean $h = \dfrac{3.5 + 2.8}{2} = 3.15$ cm

Taking the length of the arm as $\frac{D_2}{2}$, i.e., 0.6 m or 60 cm

$$\text{Mass of arms} = \frac{\pi}{4} \times h \times H \times 60 \times \rho \times 6 = \frac{\pi}{4} \times 3.15 \times 6.3 \times 60 \times 0.0072 \times 6 = 40.4 \text{ kg}$$

$\therefore$ Weight of arms $= 40.4 \times 9.81 = 396.32$ N

$\therefore$ Total weight of pulley, $F = 279.22 + 396.32 = 675.54$ N

$$\therefore \quad R = \sqrt{F_{sh}^2 + F^2} = \sqrt{2777.51^2 + 675.54^2} = 2858.5 \text{ N}$$

Referring to Fig. 19.17 let us take overhang = 10 cm

$$\therefore \quad M_b = 2858.5 \times 0.10 = 285.85 \text{ Nm}$$

$$\therefore \quad M_{te} = \sqrt{M_b^2 + M_t^2} = \sqrt{285.85^2 + 724.05^2} = 778.4 \text{ Nm}$$

$$\therefore \quad M_{te} = \frac{\pi}{16} f_s d^3 \quad \therefore 778.4 = \frac{\pi}{16} \times 42.5 \times 10^6 \times d^3$$

$$\therefore \quad d = 0.04535 \text{ m, say 46 mm}$$

$\therefore$ Hub diameter $= 2d = 92$ mm, hub length $= 2d = 92$ mm

Key: Width = thickness $= \frac{d}{4} = \frac{46}{4} = 11.50$ mm

$$\therefore \quad M_{te} = l \times w \times f_s \times \frac{d}{2}$$

$$\therefore \quad 778.4 \times 1000 = l \times 11.50 \times 50 \times 23.0 \qquad (\text{Here } f_s = 50 \text{ N/mm}^2)$$

$\therefore$ Length of key $= l = 60.2$ mm

But it should be at least equal to hub length, so let us take $l = 92$ mm.

9. Motor Pulley: $D_1 = 300$ mm, $B = 102$ mm, $f = 1.5$ mm

$\delta = (0.005\, D_1 + 0.003)$ m, $= 0.0045$ m, i.e., 4.5 mm

Arms: Let the number of arms be 4,

$$\therefore \quad \text{B.M. per arm} = \frac{F_t \times D_1}{z} = \frac{1206.75 \times 0.3}{4} = 90.5 \text{ Nm}$$

$$\therefore \quad 0.05H^3 \times 14 = 90.5 \times 1000,$$

$\therefore$ H = major axis of oval section of arm at hub = 51 mm

At rim, $H' = 0.8 \times 51 = 40.8$ mm

and $h' = 20.4$ mm

Shaft: $M_t = F_t \times \frac{D_1}{2} = \frac{1206.75 \times 0.3}{2} = 181$ Nm

Since the pulley is small, if we neglect its weight, no significant error will be caused.

$$\therefore \quad M_b = (F_1 + F_2 + 2F_c) \times \text{overhang} = 2777.51 \times 0.1 = 277.75 \text{ Nm}$$

$$\therefore \quad M_{te} = \sqrt{M_b^2 + M_t^2} = \sqrt{277.75^2 + 181^2} = 331.52 \text{ Nm}$$

$$\therefore \quad M_{te} = \frac{\pi}{16} d^3 \times f_s,\ 331.52 \times 1000 = \frac{\pi}{16} d^3 \times 42.5,\ d = 34.12, \text{ say 35 mm}$$

$(f_s = 42.5 \text{ N/mm}^2)$

$\therefore$ Hub diameter = 70 mm, hub length = 70 mm

Check in bending. $M_{be} = \frac{1}{2}\left[M_b + \sqrt{M_b^2 + M_t^2}\right] = \frac{1}{2}[277.75 + 331.52] = 304.635$ Nm

$$\therefore \quad M_{be} = \frac{\pi}{32} d^3 \times f_b$$

$$f_b = 304.635 \times 1000 \times \frac{32}{\pi} \times \frac{1}{35^3} = 72.37 \text{ N/mm}^2$$

Safe stress in bending should be 90 to 110 N/mm^2, $\therefore$ **Safe.**

Key: width = thickness = $\frac{d}{4}$ = 8.75 mm square sunk key.

length = 70 mm

Example 19.2. Find out the diameter of cast iron pulleys and the thickness and width of a leather belt to transmit 128.7 kW power from a shaft that is directly connected to a steam engine running at 31.4 rad/s, to a centrifugal pump with a speed ratio of 1 : 3.5.

Solution: To proceed with the solution of the problem, let us assume the belt speed as 20 m/s

$$\therefore \quad V = \frac{D_1 w_1}{2} = \frac{D_1 \times 31.4}{2}$$

$\therefore \quad D_1$ = diameter of pulley on the engine shaft *i.e.,* driving pulley

$$= \frac{20 \times 2}{31.4} = 1.275 \text{ m}$$

Assuming a slip of 2%,

$$V_R = \frac{D_2}{D_1(1-s)} \quad \therefore \quad \frac{1}{3.5} = \frac{D_2}{1.275 \times 0.98}$$

$\therefore \quad D_2$ = diameter of the driven pulley = 0.357 m, say 360 mm

Assuming a service factor of 1.2.

Design Power = 128.7 × 1.2 = 154.45 kW

$$\therefore \quad F_1 - F_2 = \frac{\text{Power in kW} \times 1000}{V} = \frac{154.45 \times 1000}{20} = 7722.5 \text{ N}$$

Now angle of contact on smaller pulley

$$\theta = \left(\pi - \frac{D_1 - D_2}{C}\right) \text{ radians}$$

Now centre distance, $C \geq 3.5 \times$ larger diameter

or $\quad C = 3.5 \times 1.275 = 4.46$ m, say 5 m

$$\therefore \quad \theta = \pi - \frac{1.275 - 0.36}{5} = 2.96 \text{ radians}$$

Let us assume, coefficient of friction between leather and C.I. pulley as 0.35.

$$\therefore \quad \frac{F_1}{F_2} = e^{\mu\theta} = e^{0.35 \times 2.96} = 2.82$$

$$\therefore \quad F_2 = \frac{7722.5}{1.82} = 4243.15 \text{ N}, \; F_1 = 7722.5 + 4243.15 = 11965.6 \text{ N}.$$

Now assuming a working stress of $\frac{31.5}{9}$ = 3.5 N/mm^2 and the belt joined as wire laced by machine for which the efficiency is 0.9

$$\therefore \quad f_t = 3.5 \times 0.9 = 3.15 \text{ N/mm}^2$$

Since V is > 10 m/s, therefore, effect of centrifugal force must be considered.

$\therefore$ Centrifugal stress, $f_c = \rho V^2 = 1000 \times 400 = 4 \times 10^5 \text{ N/m}^2 = 0.4 \text{ N/mm}^2$

$$\therefore \quad b \times t = \frac{F_1}{(f_t - f_c)} = \frac{11965.6}{3.15 - 0.400} = 4362.2 \text{ mm}^2$$

Now belt thickness, t should be $\leq 0.02 \times$ smaller pulley diameter

or $$t \geq 0.02 \times 360 \text{ or } 7.2 \text{ mm}$$

$\therefore$ Taking light double ply belt with $t = 6$ mm

$$b = \frac{4362.2}{6} \cong \mathbf{730 \text{ mm}}$$

Example 19.3. The diameters of the driving and driven pulleys are 900 mm and 1200 mm respectively and the centre distance is 3 metres. The output of the driven shaft is 110 kW. Assuming: Belt Speed = 21 m/s, Coefficient of friction = 0.30

Slip = 1.5 per cent at each pulley, friction loss at each shaft = 5 per cent. Determine:

(a) The rev/min of each shaft.

(b) Difference in belt tensions.

(c) Size of machine wire laced leather belt required.

(d) The required shaft sizes, assuming pure torsion and an allowable stress of 56 MPa.

(e) The overall efficiency of this transmission.

Solution: Let n_1 and D_1 be the rev/min and diameter of the driving pulley and n_2 and D_2 the corresponding values for the driven pulley.

(*a*) Now belt speed = (peripheral velocity of driving pulley) × (1– *s*)

where $$s = 0.015, D_1 = 0.9 \text{ m } D_2 = 1.2 \text{ m}$$

$$\therefore \quad 21 = \frac{\pi D_1 n_1}{60} \times 0.985 \text{ , } n_1 = 452 \text{ rev/min}$$

Again peripheral velocity of the driven pulley = (belt speed) × (1 – *s*)

$$\therefore \quad \frac{\pi D_2 n_2}{60} = 21 \times 0.985, \eta_2 = 329 \text{ rev/min}$$

(*b*) Now the output at the driven shaft = 110 kW

$\therefore$ Torsional moment acting on the driven shaft $M_t = \dfrac{1000 \times 110 \times 30}{\pi \times 329} = 3192.36$ Nm

$\therefore$ Torsional moment transmitted by the belt = 1.05 × 3192.36 = 3352 Nm

$\therefore$ Difference of belt tensions, $(F_1 - F_2) = \dfrac{3352}{0.60} = \mathbf{5586.63 \text{ N}}$

(*c*) Now Angle of contact, $\theta = \left(180 - 2\sin^{-1} \dfrac{D_2 - D_1}{2C} \text{ degrees}\right)$

$$= \left[180 - 2\sin^{-1}\left(\frac{120 - 90}{600}\right)\right] \text{ degrees} = 169.5 \text{ degrees} = 2.96 \text{ radians}$$

$$\frac{F_1}{F_2} = e^{\mu\theta} = e^{0.3 \times 2.96} = e^{0.888} = 2.43$$

$$\therefore \quad F_1 = 2.43\, F_2, F_1 = F_2 + 5586.63 \text{ N}$$

$$\therefore \quad F_1 = \frac{5586.63 \times 2.43}{1.43} = 9493.36 \text{ N}$$

Since the belt velocity is greater than 10 m/s, therefore, centrifugal stress will be taken into account. It is given as:

$$f_c = \rho V^2 \text{, For leather } \rho = 1000 \text{ kg/m}^3$$

$$\therefore \quad f_c = 1000 \times 21^2 = 4.41 \times 10^5, f_c = 0.441 \text{ MPa}$$

Let f_t, for leather = 3.0 MPa
and efficiency of the joint = 90 per cent.

$\therefore$ Net allowable tensile stress, $f_t = 3.0 \times 0.9 = 2.7$ MPa

Assuming the drive to be horizontal one,

$$b \times t = \frac{9493.36}{(2.7 - 0.441)} = 4202.46 \text{ mm}^2$$

Now empirically, $D \geq 50t$

$\therefore$ Thickness of belt, $t \leq \frac{D}{50} \leq \frac{900}{50}$, i.e., ≤ 18 mm

Let $t = 18$ mm

$$\therefore \quad b = \frac{4202.46}{18} = 233.5 \text{ mm}$$

(*d*) Now $M_t = \frac{\pi}{16} d^3 \times f_s$

For driven shaft,

$$3352 = \frac{\pi}{16} d^3 \times 56 \times 10^6$$

$$d = 0.0673 \text{ m} = 67.30 \text{ mm}$$

Now Torsional moment at the driving pulley = $5586.63 \times \frac{0.90}{2} = 2514$ Nm

$\therefore$ Torsional moment input to the driving shaft = 2514 × 1.05 = 2639.7 Nm

$$\therefore \quad d^3 = \frac{2639.7}{\pi \times 3.5 \times 10^6} = 0.00024 \text{ m}^3, d = 0.0062 \text{ m} = 6.20 \text{ mm}$$

(*e*) Now Total losses = 2 (1.5 + 5) = 13 per cent.

$\therefore$ Overall efficiency of the transmission

= 100 – 13 = 87 per cent.

Example 19.4. A 7.355 kW, 1200 rev/min motor is arranged for a rock-wood drive as shown in Fig. 19.13. The motor weighs 1600 N and pulley diameter is 0.20 m. The distance "b" is 0.40 m and the distance *l* is 0.45 m. The starting torsional moment on the motor is 250 per cent of rating. Determine:

(a) The effective tension required for starting the drive and the values of F_1 and F_2.
(b) The distance '*a*' from the hinge point to the motor centre line.
(c) The tension F_1 and F_2 in the belt at rated load.
(d) The tension at Zero load.
(e) Compare the results of (a), (c), and (d) with the tensions required with a simple open belt drive.

Solution: (*a*) Rated Torsional moment $M_t = \frac{1000 \times \text{Power in kW}}{\omega}$

Now $w = \frac{\pi n}{30} = \frac{\pi \times 1200}{30} = 125.66$ rad/s

$$\therefore \quad M_t = \frac{1000 \times 7.355}{125.66} = 58.53 \text{ Nm}$$

$\therefore$ Starting Torsional moment = 58.53 × 2.5 = 146.325 Nm

$\therefore$ Effective tension, $F_1 - F_2 = \frac{146.325 \times 2}{0.20} = 1463.25$ N

Let $\mu = 0.3$ and $\theta = \pi$ radians

$$\therefore \quad \frac{F_1}{F_2} = e^{0.3 \times \pi} = e^{0.942} = 2.57$$

$$\therefore \quad F_2 = \frac{1463.25}{1.57} = 932 \text{ N and } F_1 = 2.57 \times 932 = 2395.25 \text{ N}$$

(*b*) Now,

$$\therefore \quad l = \frac{F_1 \times a + F_2 \times b}{F}, \; 0.45 = \frac{2395.25 \times a + 932 \times 0.40}{1600}$$

$$\therefore \quad a = \frac{347.2}{2395.25} = \mathbf{0.145 \text{ m}}$$

(*c*) At rated load,

$$M_t = 58.53 \text{ Nm}, \; F_1 - F_2 = 585.3 \text{ N}$$

Again, $F_1 \times a + F_2 \times b = F \times l$

$\therefore$ $0.145F_1 + 0.40F_2 = 720$

$\therefore$ $0.145F_1 + 0.40(F_1 - 585.3) = 720$, gives $F_1 = 1750.68$ N

$\therefore$ $F_2 = 1750.68 - 585.3 = 1165.38$ N

(*d*) At no load, $F_2 = F_1$

$\therefore$ $F_1(0.145 + 0.40) = 720$

$$\therefore \quad F_1 = F_2 = \frac{720}{0.545} = 1321.10 \text{ N}$$

(*e*) Now the total load on the motor bearing will be,

At maximum load = $F_1 + F_2 = 2395.25 + 932 = 3327.25$ N

At normal rated load = $1750.68 + 1165.38 = 2916.06$ N

At no load = 2642.20 N.

In the case of a simple open belt drive with fixed centres, the belt tensions will remain practically constant at all loads. Hence the bearing load will always be 3327.25 N. In the case of Rockwood drive, this value is reached only when drive is operating at maximum load.

EXERCISES

1. What are the various types of belts?
2. On what factors, the selection of a belt drive depends?
3. Classify the belt drives. Enumerate their fields of application.
4. Enumerate the advantages and disadvantages of a flat belt drive.
5. Enumerate the various types of flat belt drives.
6. What materials are used for flat belts?
7. What are the safe speeds for belt drives?
8. What is belt slip and creep? Clearly make a difference between the two.
9. What is the effect of centre distance and diameter of pulley on the life of a belt? What are the limits of safe centre distance for flat belt drives?
10. What are the various losses in the power transmission by belts?
11. What is the safe velocity ratio for flat belt drive?
12. What are the various stresses set up in a belt?
13. On what factors, the design of a belt drive depends?
14. Name the short centre drives.
15. Name the materials for pulleys.
16. What are split pulleys and what are their advantages?
17. Why the face of the pulley is crowned?

18. Why the section of the arms is usually elliptical?
19. In which plane the major axis of the arm section is placed and why?
20. What are the advantages and disadvantages of straight and curved arms?
21. Design completely a flat belt drive for the following data :
An A.C. motor drives an air compressor, power of motor = 15 kW, rev/min of motor = 720, Diameter of motor pulley = 0.30 m, rev/min of compressor = 180. Centre distance = 1.50 m, slip 2%.
22. Design an overhanging pulley for the following specifications :

Power	= 18 kW
Speed	= 200 rev/min
Angle of contact	= 165°
Coefficient of friction	= 0.25
Overhanging length	= 0.30 m
Belt thickness	= 10 mm
Safe stress for arms	= 14 MPa
Safe shear stress for shaft	= 40 MPa
Safe stress for belt	= 2.5 MPa
Safe stress for rim	= 4.0 MPa

23. The following particulars refer to a belt drive from a C.I. pulley mounted on a shaft:

Power transmitted	= 18 kW
rev/min of shaft	= 250
Ratio of tensions	= 3
Velocity of leather belt	= 12.5 m/s
Allowable belt tension	= 4200 N/m of belt width
Number of arms with oval section	= 6

Design : (*a*) Pulley; (*b*) Shaft; (*c*) Key.
Given, safe shear stress for shaft and key = 56 MPa
Safe tensile stress for C.I. = 14 MPa
24. A 75 kW high speed engine running at 250 rev/min is belted to an electric generator. The band wheel on the engine is 2 metre in diameter. Assuming an arc of contact on the generator pulley to be 150°, determine the necessary width for a double ply oak tanned leather belt 9.5 mm thick.
25. Determine the dimensions of a cast iron pulley which is 1.20 m in diameter and transmits 75 kW at 200 rev/min. It is to be used with a heavy double belt of leather.
26. A 3.6 kW electric motor running at 1500 rev/min is arranged for pivoted drive with the tight side of the belt horizontal and 0.25 m above the pivot. The diameter of the motor pulley is 0.20 m and motor weighs 450 N. The driven pulley is required to run at 300 rev/min. The distance between the pulley centres is 0.90 m. Determine the leverage for full load operation of the drive.
27. A set of 3- step cone pulleys are required for a leather belt drive connecting two parallel shafts 3 m apart and running in opposite directions. The driving shaft has a speed of 150 rev/min and the driven shaft is required to run at 140, 200 and 260 rev/min. Maximum power to be transmitted is 4.5 kW, ratio of tensions 3 and belt tension per mm width is not to exceed 12.5 N. Design the stepped pulleys.
28. The following data relate to an overhung flat belt drive:

Power transmitted	= 22 kW
rev/min	= 300
Angle of contact	= 165°
	μ = 0.3

Distance of centre of pulley from centre of nearest bearing = 0.25 m
Take into account centrifugal tension, but neglecting weight of C.I. pulley, find:
(*a*) Diameter of pulley.
(*b*) Width of 8 mm thick leather belt.
(*c*) Diameter of shaft (*d*) Size of key.
(*e*) Dimensions of four elliptical arms having major axis = 0.6 minor axis, permissible stresses :

Shaft and key,	f_s = 42 MPa
Arms,	f_t = 14 MPa
Pulley rim,	f_t = 4.9 MPa
Belt	= 2.1 MPa

29. The following particulars refer to a belt drive:

    Power transmitted = 11 kW
    Speed = 600 rev/min
    Angle of contact = 210°
    Coefficient of friction between belt and pulley = 0.22
    Pulley material: C.I., having its density = 7200 kg/m^3
    Safe tensile stress of pulley rim = 4.2 MPa

    Number of arms in the pulley = 4 of rectangular section having sides in the ratio 2 : 1.
    Safe tensile stress of pulley arms = 14 MPa.
    Determine: (*i*) The mean diameter of pulley.
    (*ii*) Width of belt 6.5 mm thick and having a safe tensile stress of 2.1 MPa.
    (*iii*) Shaft diameter. Neglect the bending in the shaft due to loads if any, and take the maximum shear stress in the shaft as 42 MPa.
    (*iv*) Size of pulley boss and arms.

19.3 V-BELT DRIVES

19.3.1 Drives Using V-belts Have Got the Following Advantages Over Flat Belt Frives

1. Higher velocity ratio, upto 7 and 8, the maximum being 10.
2. Provide long life, 3 to 5 years.
3. Easily installed and removed.
4. As the V-belts are made endless, there is no joint trouble, so the drive is smooth.
5. Possibility of using with small centre distances.
6. The small centre distances give compactness as required in many machine layouts.
7. More positive and reliable drive and less slippage than with flat belts, and these can be used in any position including vertical, and even with vertical shafts.
8. Quiet and low in maintenance.
9. Provide shock absorption between driver and driven shafts.
10. Possibility of powering a number of driven shafts from a single drive without the use of belt tighteners.
11. V-belts are able to transmit higher torsional moment at less width and tensions than flat belts.

V-belt drives have also got the following limitations as compared to flat belts:

1. These cannot be used with large centre distances.
2. V-belts are subjected to a certain amount of creep, so these should not be used when synchronous speeds are required, such as in timing devices.
3. V-belt are not so durable as flat ones.
4. The design of pulleys for V-belt drives is more complicated than the design for flat belt drives.
5. Improper belt tensioning and mismatching of belt lengths can reduce service life.
6. Belt life in temperatures above about 82° C and below about – 15° C is significantly shortened.
7. Centrifugal force prevents the use of V-belts at speeds above 50 m/s and at speeds below 5 m/s, they are usually not economical.
8. Relatively low efficiency due to considerable losses in belt deformation and likely dimensional instability along the belt.

V-belts are made in trapezoidal section. They are made of cotton fabric, cords and rubber and are vulcanised in moulds of any desired cross-section. A profile of the V-belt is shown in Fig. 19.19. Basically, the V-belt consists of following five component sections:

1. Flexible section consisting of several layers of rubberized fabric carrying the tension due to bending when the belt runs around the pulley.
2. One layer of load carrying cords of laminated or multilayer cord construction. Load carrying tensile cords are made of rayon, nylon, dacron, steel and glass fibre.

3. The bottom rubber of compression section subjected to compression during bending. It is located below the pitch line and it serves two functions. It is a platform for supporting the tensile section and it transmits through its side walls the wedging pressure from the tensile section to the sheave groove.
4. The cushion section serves several purposes. It keeps the cords bonded between the top tensile and bottom compression sections and it absorbs some of the shock of the varying load conditions.
5. The cover is made of rubberized fabric.

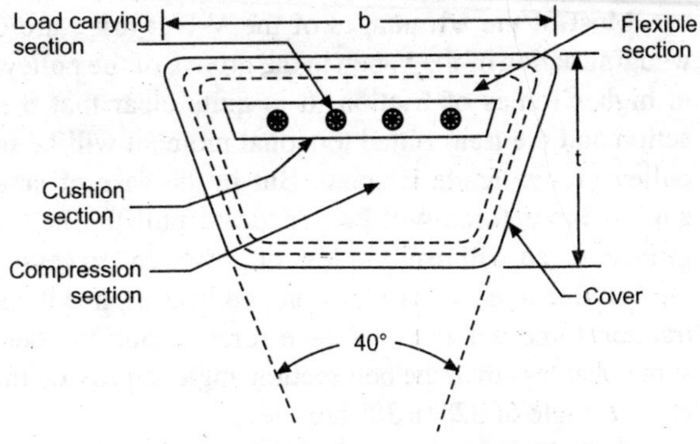

Fig. 19.19 *Basic V-belt construction*

The included angle of the belt section is 40°. V-belt are available in five standard sections, the properties of which are given in Table 19.7.

Table 19.7 Properties of V-belt (IS 2494 : 1974)

Belt Type	Cross-section Dimensions			Minimum recommended grooved pulley diameter, mm	Correction for pitch diameter, mm	Recommended power ranges, kW	Recommended number of sheave grooves
	b_{mm}	t_{mm}	Area of section cm^2				
A	13	8	0.81	76	9.5	0.5–3.75	6
B	17	11	1.4	137	12.7	1.5–15	9
C	22	14	2.3	229	19	11–75	14
D	32	19	4.75	330	22.2	37.5–150	14
E	38	23	6.95	533	28.6	75–260	20

Note: Correction for pitch diameter should be subtracted from the outside diameter of the sheave to get pitch diameter.

Standard Pulley Diameter: The standard pulley diameter for V-belt drives are: 20, 22, 25, 28, 32, 36 mm and all other diameters used for flat pulleys given under Art. 19.2.

The standard pitch lengths of the V-belts are given in Table 19.8.

Table 19.8 Pitch lengths of belts (IS 2494 ; 1974)

Belt Type	Pitch lengths of belts, mm
A	645, 696, 747, 823, 848, 925, 950, 1001, 1026, 1051, 1102, 1128, 1204, 1255, 1331, 1433, 1458, 1509, 1560, 1636, 1661, 1687, 1763, 1814, 1941, 2017, 2068, 2093, 2195, 2322, 2474, 2703, 2880, 3084, 3287, 3693.
B	932, 1008, 1059, 1110, 1212, 1262, 1339, 1415, 1440, 1466, 1567, 1694, 1770, 1821, 1948, 2024, 2101, 2202, 2329, 2507, 2583, 2710, 2888, 3091, 3294, 3701, 4056, 4158, 4437, 4615, 4996, 5377.
C	1351, 1580, 1783, 1961, 2113, 2215, 2342, 2494, 2723, 2901, 3104, 3205, 3307, 3459, 3713, 4069, 4171, 4450, 4628, 5009, 5390, 6101, 6863, 7625, 8387, 9149.
D	3127,3330, 3736, 4092, 4194, 4473, 4651, 5032, 5413, 6124, 6886, 7648, 8410, 9172, 9934, 10696, 12220, 13744, 15268, 16792.
E	5426, 6173, 6899, 7661, 8423, 9185, 9947, 10709, 12233, 13757, 15281, 16805.

Most of the advantages of the V-belt drive are due to the wedging action of the V-belt in the groove of the pulley, resulting in higher forces of friction. It is quite clear that the wedging action and the transmitted torsional moment will be more if the pulley groove angle is small. But in the case of small groove angles, more force will be needed to pull the belt out of the groove which will result in loss of power and excessive wear of the belt due to excessive friction and heat. Hence the selected groove angle is a compromise to secure large traction force without excessive force to pull the belt from the groove. The sheave groove angle is made somewhat less than the belt section angle to provide the wedging action of the belt in the groove. Usually the groove angle of 32° to 38° are used.

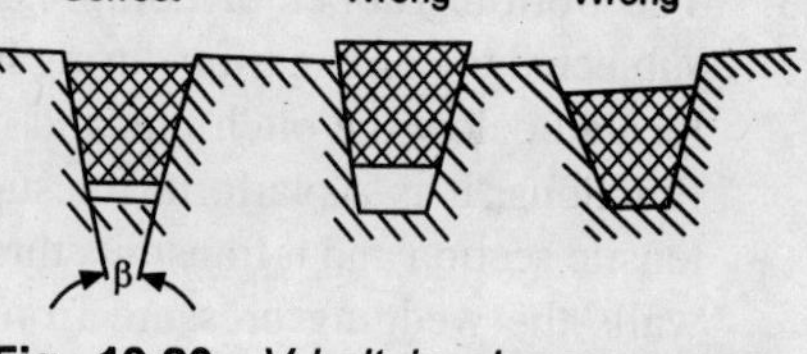

Fig. 19.20 *V-belt in sheave groove*

To facilitate the installation and removal of the endless V-belt and to adjust its tension as it stretches out, the mounting of one of the pulleys, usually the motor pulley has provisions for movement square to the shaft axis. A properly installed V-belt should not touch the bottom of the groove nor it should project out of groove, Fig. 19.20.

V-belt drives are used for prime mover power capacities upto about 200 kW power at peripheral speeds upto 25 to 30 m/s. For best results, the belt speed should be 20 m/s. For belt velocities in excess of 25 m/s, and less then 5 m/s the drive becomes troublesome and uneconomical respectively.

The tensions relationship for V-belts is

$$\frac{F_1}{F_2} = e^{\mu\,\theta/\sin\beta/2}$$

where θ = arc of contact on smaller pulley.

β = pulley groove angle.

To prevent the sticking of the belt in the groove, the groove angle should be larger than the double the angle of friction, i.e.,

$$\beta > 2\tan^{-1}\mu$$

Since $\mu = 0.30$ for rubber fabric on C.I. pulley.

$\therefore$ $\beta > 2\tan^{-1} 0.3 \cong 34°$

As already indicated, β varies from 32 to 38°.

Within these limits, the mean value of the coefficient of friction is

$$\mu_e = \frac{\mu}{\sin 17.5°} = \frac{\mu}{0.3} \cong 3.3\mu$$

Thus all other conditions being equal, an element of arc of V-belt is capable of transmitting a peripheral effort nearly three times greater than in a flat belt drive.

19.3.2 Formulas for V-belt Drive

The formulas for the V-belt drive are the same as for flat belt drive excepting that in design calculations, the pitch diameters of the pulleys instead of the outer diameters are used. Again, the design belt length pertains to the pitch point which refers to the neutral section of the belt.

If L_d = design belt length given by the neutral section.

and L_{in} = inner length of the belt.

then $L_d - L_{in} = 2\pi\,(t - t_0)$

where t = height of the belt section.

t_0 = the distance between the neutral section and its larger base $\cong \frac{1}{3}t$.

The V-belt shall be designated by the cross-section symbol, nominal inside length, a number to denote the nominal length in inches and the number of the standard, e.g., a V-belt of cross-section 'C' and of nominal inside length of 3048 mm shall be designated as : C 3048/120, IS : 2494.

19.3.3 Centre Distance

V-belt drives are not recommended for large centre distance, because the excessive vibrations of the slack side will shorten the belt life. In general, the following ranges for the centre distances are recommended:

$$C \ngtr 3\,(D_1 + D_2)$$

$$C \nless D_2$$

or

$$C_{min} = 0.5\,(D_1 + D_2) + 3t$$

Unless it is governed by the drive layout, the centre distance can conveniently be found out by using the relationship,

$$C = \frac{1.5\,D_2}{(VR)^{1/3}}$$

19.3.4 Service Factor

The power to be transmitted or the rated power must be multiplied by the service factor to get the design power. The service factor takes into account the severity of the load transmitted which depends upon the characteristics of the driving and driven units. The service factors are given in Table 19.9.

Table 19.9 V-belt service factor

	Driven machine	Service factor
1.	Agitators, paddle propellers	1.0–1.2
2.	Bakery machinery	1.0–1.2
3.	Flour, feed and cereal mills	1.0–1.4
4.	Conveyors	1.0–1.5
5.	Generators and exciters, Printing machinery	1.2
6.	Laundry machines, Paper machinery	1.2
7.	Compressors, screens	1.2–1.4
8.	Brick and clay machinery, oil field machinery	1.2–1.6
9.	Textile machinery	1.2–1.8
10.	Fans, blowers, pumps	1.2–2.0
11.	Mills, Crushing machinery	1.4–1.6
12.	Line shafts, Rubber plant machinery	1.4–2.0

The lower values are to be used when shock and peak loads are light and for severe loads, use the higher values.

19.3.5 V-belt Pulleys

The materials used for V-belt pulleys are:

1. Cast Iron.
2. Formed and pressed steel.
3. Moulded or machined plastic.
4. Diecast aluminium.

Cast Iron is economical, stable and long lived and has excellent friction characteristics on V-belts. Pressed steel is lighter and cheaper but it may give rise to excessive belt slip, wear, and noise. Formed steel sheaves are used primarily in automotive and agricultural purposes. For special applications, the sheaves may be made of hard-anodized aluminium alloy. In each case, the pulley

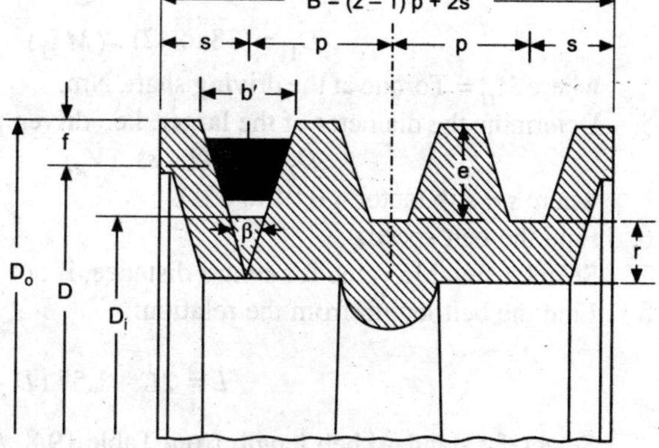

Fig. 19.21 *V-belt pulley*

surfaces contacting the belts should be finely machined to reduce elastic creep losses. Cast iron sheaves are generally limited to 32.5 m/s rim speeds. For speeds upto 50 m/s aluminium, steel, and ductile iron are used.

A section of the rim of the pulley for V-belt drives is shown in Fig. 19.21.

The various dimensions of the pulley rim are given in Table 19.10.

Table 19.10 Dimensions of pulley grooves for V-belt drives

Type of Belt	Various profile dimensions, mm							
	f	*e*	*p*	*s*	b′ for groove angles of			*r*
					34°	36°	38°	
A	3.5	12.5	16	10	13.1	13.3	13.4	6.35
B	5	16	20	12.5	17.0	17.2	17.4	8.00
C	6	21	26	17	22.7	22.9	23.1	8.75
D	8.5	28.5	37.5	24	—	32.5	32.8	11.10
E	10	34	44.5	29	—	38.5	38.9	14.30

19.3.6 Design Procedure for V-belt Drives

The design of the V-belt drive concerns primarily with the selection of a belt of standard profile and length and the determination of the number of belts required for the transmission of the given power. The calculations for a V-belt drive should be made in the following sequence:

1. Find the design power by multiplying the service factor given in Table 19.9 by the power to be transmitted or by the name-plate power of the motor.
2. Select the belt profile to suit the given power capacity from Table 19.7. Two or even three profiles might be suitable for this capacity. The adopted profile should ensure the smallest size and the largest service of life.
3. Determine the speed ratio,
$$V_R = \frac{w_1}{w_2} = \frac{n_1}{n_2}$$
4. Select the pitch diameter of the smaller pulley. Usually a diameter giving 20 to 25 m/s belt speed results in the most economical drive. The selected diameter should be more than the minimum recommended in Table 19.7, i.e.,
$D_1 \geq D_{min}$
D_1 can also be found from,
$$D_1 = (38 \text{ to } 42) \cdot (M_{t1})^{\frac{1}{3}}, \text{ mm}$$
where M_{t1} = Torque at the driving shaft, Nm.
5. Determine the diameter of the larger, i.e., driven pulley by the formula :
$$D_2 = D_1 (1 - s) \cdot (V_R)$$
where s = slip factor = 0.01 to 0.03.
6. Find the belt speed.
7. Select a suitable value for centre distance, if not given.
8. Find the belt length from the relation:
$$L = 2C + 1.57\,(D_2 + D_1) + \frac{(D_2 - D_1)^2}{4C}$$
Select the standard belt length from Table 19.8. After this, the exact centre distance is calculated by the relation:

$$C = p + \sqrt{p^2 - q}$$

where $p = \frac{L}{4} - \frac{\pi}{8}(D_2 + D_1)$ and $q = \frac{(D_2 - D_1)^2}{8}$.

The selected length should be checked for endurance (longevity of the belt) by the number of turns per second,

$$x = \frac{V}{L_0} \le 10/s.$$

where L_0 = standard selected belt length.

If the above condition is not satisfied, the value of 'C' (and therefore of L) should be increased.

9. Determine, the angle of contact of the smaller pulley, from the relation:

$$\theta = \pi - 2\sin^{-1}\frac{D_2 - D_1}{2C}, \theta \ge 2.1 \text{ radians}$$

10. Find the maximum belt tension from the relations:

$$\text{Design Power} = \frac{(F_1 - F_2)\,V}{1000}, \text{kW}$$

and

$$\frac{F_1}{F_2} = e^{\mu_e \theta}$$

where $\mu_e = \frac{\mu}{\sin \beta/2}$, where μ = Coefficient of friction of belt on the pulley surface.

11. Determine the number of belts by the relations,

$$z \ge \frac{F_1}{ft\,A.C_1.C_2}, \text{ if } V < 10 \text{ m/s}$$

and

$$z \ge \frac{F_1}{(ft - \rho V^2)\,A.C_1.C_2}, \text{ if } V > 10 \text{ m/s}$$

where A = area of cross-section of the belt profile chosen, Table 19.7

C_1 = Arc of contact factor on the smaller pulley, Table 19.11.

C_2 = belt length correction factor, Table 19.12.

Arc of contact factor is taken into account because the power transmitted may be limited by slipping of the belt of the smaller pulley.

Table 19.11 Angle of contact factor, C_1

θ, degrees	C_1	θ, degrees	C_1
180	1.0	130	0.87
170	0.98	120	0.83
160	0.96	110	0.79
150	0.93	100	0.74
140	0.90	90	0.68

For a given sheave speed, longer belt has more life than a short one, because the short belt is subjected to the action of load a greater number of times. For this reason, a correction factor known as belt length correction factor is used.

Table 19.12 Belt length correction factor, C_2 (IS : 2494-1974)

Nominal Inside Length, mm	Belt A		Belt B		Belt C		Belt D		Belt E	
	Pitch Length, mm	C_2	Pitch Length, mm	C_2	Pitch Length, mm	C_2	Pitch Length, mm	C_2	Pitch Length, mm	C_2
610	645	0.80	—	—	—	—	—	—	—	—
660	696	0.81	—	—	—	—	—	—	—	—
711	747	0.82	—	—	—	—	—	—	—	—
787	828	0.84	—	—	—	—	—	—	—	—
813	848	0.85	—	—	—	—	—	—	—	—
889	925	0.87	9.32	0.81	—	—	—	—	—	—
914	950	0.87	—	—	—	—	—	—	—	—
965	1001	0.88	1008	0.83	—	—	—	—	—	—
991	1026	0.88	—	—	—	—	—	—	—	—
1016	1051	0.89	1059	0.84	—	—	—	—	—	—
1067	1102	0.90	1110	0.85	—	—	—	—	—	—
1092	1128	0.90	—	—	—	—	—	—	—	—
1168	1204	0.92	1212	0.87	—	—	—	—	—	—
1219	1255	0.93	1262	0.88	—	—	—	—	—	—
1295	1331	0.94	1339	0.89	1351	0.80	—	—	—	—
1372	—	—	1415	0.90	—	—	—	—	—	—
1397	1433	0.96	1440	0.90	—	—	—	—	—	—
1422	1458	0.96	1466	0.90	—	—	—	—	—	—
1473	1509	0.97	—	—	—	—	—	—	—	—
1524	1560	0.98	1567	0.92	1580	0.82	—	—	—	—
1600	1637	0.99	—	—	—	—	—	—	—	—
1626	1661	0.99	—	—	—	—	—	—	—	—
1651	1687	1.00	1694	0.94	—	—	—	—	—	—
1727	1763	1.00	1770	0.95	1783	0.85	—	—	—	—
1778	1814	1.01	1824	0.95	—	—	—	—	—	—
1905	1941	1.02	1948	0.97	1961	0.87	—	—	—	—
1981	2017	1.03	2024	0.98	—	—	—	—	—	—
2032	2068	1.04	—	—	—	—	—	—	—	—
2057	2093	1.04	2101	0.98	2113	0.98	—	—	—	—
2159	2195	1.05	2202	0.99	2215	0.90	—	—	—	—
2286	2322	1.06	2329	1.00	2342	0.91	—	—	—	—
2438	2474	1.08	—	—	2494	0.92	—	—	—	—
2464	—	—	2507	1.02	—	—	—	—	—	—
2540	—	—	2583	1.03	—	—	—	—	—	—
2667	2703	1.10	2710	1.04	2723	0.94	—	—	—	—
2845	2880	1.11	2888	1.05	2901	0.95	—	—	—	—
3048	3084	1.13	3091	1.07	3104	0.97	3127	0.86	—	—
3150	—	—	—	—	3205	0.97	—	—	—	—
3251	3287	1.14	3294	1.08	3307	0.98	3330	0.87	—	—
3404	—	—	—	—	3459	0.99	—	—	—	—
3658	3693	—	3701	1.11	3713	1.00	3736	0.90	—	—

...(Contd)

Nominal Inside Length, mm	Belt A		Belt B		Belt C		Belt D		Belt E	
	Pitch Length, mm	C_2	Pitch Length, mm	C_2	Pitch Length, mm	C_2	Pitch Length, mm	C_2	Pitch Length, mm	C_2
4013	—	—	4056	1.13	4069	1.02	4092	0.92	—	—
4115	—	—	4158	1.14	4171	1.03	4194	0.92	—	—
4394	—	—	4437	1.15	4450	1.04	4473	0.93	—	—
4572	—	—	4615	1.16	4628	1.05	4651	0.94	—	—
4953	—	—	4996	1.18	5009	1.07	5032	0.96	—	—
5334	—	—	5377	1.19	5390	1.08	5413	0.96	5426	0.94
6045	—	—	—	—	6101	1.11	6124	1.00	6137	0.96
6807	—	—	—	—	6863	1.14	6886	1.03	6899	0.99
7569	—	—	—	—	7625	1.16	7648	1.05	7661	1.01
8331	—	—	—	—	8387	1.19	8410	1.07	8423	1.03
9093	—	—	—	—	9143	1.21	9172	1.09	9185	1.05
9855	—	—	—	—	—	1.23	9934	1.11	9947	1.07
10617	—	—	—	—	—	1.24	10696	1.12	10709	1.09
12141	—	—	—	—	—	—	12220	1.16	12233	1.12
13665	—	—	—	—	—	—	13744	1.18	13757	1.14
15189	—	—	—	—	—	—	15268	1.20	15281	1.17
16713	—	—	—	—	—	—	16792	1.23	16805	1.19

The number of belts should not exceed 8 to 12, otherwise the next larger section should be used, because with large number of belts the distribution of load is difficult to maintain due to variations in the actual lengths of belts and in the dimensions of the individual pulley grooves. Due to all this, the belts will have different service lives.

12. **Design of Pulley and Shaft:** The dimensions of the rim of the V-belt pulley can be calculated with the help of Fig. 19.21 and Table 19.10. The design of arms of the pulleys (number and size of section) will be done on the same lines as explained for flat belt pulleys.

For the design of shafts, the pressure on shaft due to belt tension may be taken with a sufficient degree of accuracy by the formula,

$$F_{sh} = 2.f_i.A.z.\sin\theta/2 \qquad \text{...(19.10)}$$

where f_i = initial tension stress, usually taken = 1.5 MPa.

Example 19.5. A V-belt drive is to transmit 14.7 kW to a compressor. The motor speed is 1150 rev/min and the compressor pulley runs at 400 rev/min. Determine the size and number of belts required.

Solution: The problem can be solved by following the procedure as below :

1. Form Table 19.9, the service factor is 1.2–1.4, say 1.3

   ∴ Design Power = Power to be transmitted × service factor = 14.7 × 1.3 = 19.11 kW

2. From Table 19.7, the belt section 'C' is selected.

3. Determine the speed ratio, $V_R = \dfrac{n_1}{n_2} = \dfrac{1150}{400} = \dfrac{23}{8} = 2\dfrac{5}{8}$

4. Select the pitch diameter of the smaller pulley. From Table 19.7, the minimum pulley diameter is given as 229 mm and correction factor for this is 19 mm.

   Therefore the pitch diameter, $D_1 \geq (229 - 19)$ mm

   or $D_1 \geq 210$ mm

Now $V_1 = \frac{\pi D_1 n_1}{60}$, For most economical drive, let $V = 20$ m/s

$\therefore \quad D_1 = \frac{20 \times 60}{\pi \times 1150} = 332$ mm

or $\quad D_1 = (38 \text{ to } 42) . (M_{t1})^{\frac{1}{3}}$, mm

Now $\quad M_{t1} = \frac{1000 \times kW}{w_1}$

and $\quad w_1 = \frac{\pi \ n_1}{30}$

$\therefore \quad M_{t1} = \frac{1000 \times 19.11 \times 30}{\pi \times 1150} = 158.68$ Nm

$\therefore \quad D_1 = 40 \times (158.68)^{\frac{1}{3}} = 40 \times 5.41 = 21655$ mm

Let $\quad D_1 = 340$ mm

5. Determine the diameter of the larger pulley:

$$D_2 = D_1 (1 - s) . (V_R)$$

Let $\quad s = 0.02$

$\therefore \quad D_2 = 340 \times 0.98 \times \frac{23}{8} = 960$ mm

6. Find the belt speed $V = \frac{\pi D_1 n_1}{60} = \frac{\pi \times 0.34 \times 1150}{60} = 20.5$ m/s

7. Select a suitable centre distance.

Now $\quad C \not> 3 (D_1 + D_2)$

$\quad \not< D_2$

i.e., C should be between 960 mm and 3 (340 + 960) mm

i.e., between 960 and 3900

Also $\quad C_{min} = 0.5 (D_1 + D_2) + 3t$

where $\quad t$ = height of belt section = 14 mm

or $\quad C_{min} = 0.5 (340 + 960) + 3 \times 14 = 650 + 42 \cong 700$ mm

let us take $\quad C = 3000$ mm

8. Find the belt length, referring to the neutral axis of belt section, i.e.,

$$L = 2C + 1.57(D_1 + D_2) + \frac{(D_2 - D_1)^2}{4C}$$

$$= 2 \times 3.0 + 1.57(0.34 + 0.96) + \frac{(0.96 - 0.34)^2}{4 \times 3.0} = 8.072 \text{ m}$$

Now from Table 19.8, the nearest standard belt length for 'C' belt is 8387 mm.

Now the exact centre distance should be found out by the relation:

$$C = p + \sqrt{p^2 - q}$$

Now $\quad p = \frac{L}{4} - \frac{\pi}{8} \times (D_2 + D_1) = \frac{8.387}{4} - \frac{\pi}{8} \times (0.96 + 0.34)$

or $\quad p = \frac{1}{4} [8.387 - 1.57 \times 1.30] = 1.587$

$$q = \frac{(D_2 - D_1)^2}{8} = \frac{(0.96 - 0.34)^2}{8} = 0.048$$

$$\therefore \quad C = 1.587 + \sqrt{(1.587)^2 - 0.048} = 1.587 + 1.572 = 3.16 \text{ m}$$

The minimum centre distance needed to fit the belt is,

$$C_{\min} = C - 2t = 3160 - 2 \times 14 = 3132 \text{ mm.}$$

The maximum centre distance needed to compensate for stretch is

$$C_{\max} \cong (1.05 \text{ to } 1.10) \times C \cong 1.05 \times 3160 = 3318 \text{ mm}$$

Now the number of runs of the belt per second,

$$x = \frac{V}{L_0} = \frac{20.5}{8.387} = 2.44 < 10/\text{s}$$

$\therefore$ The condition for longevity of belt is met.

9. Determine the angle of contact on the smaller pulley,

$$\theta = \pi - 2\sin^{-1}\frac{D_2 - D_1}{2C} = 180° - 2\sin^{-1}\frac{96 - 34}{2 \times 316} = 168.7°$$

$$= 2.94 \text{ radians} > 2.1 \text{ rad}, \therefore \text{ O.K.}$$

10. Find the maximum belt tension:

$$\text{Design Power} = \frac{(F_1 - F_2)V}{1000}, \text{ kW}$$

$$\therefore \quad F_1 - F_2 = \frac{19.11 \times 1000}{20.5} = 932.2 \text{ N}$$

Now $$\frac{F_1}{F_2} = e^{\mu_e \theta}$$

Now $$\mu_e = \frac{\mu}{\sin \beta/2}$$

Let $\mu = 0.3$ for rubber fabric on C.I. pulley.

and β = groove angle = 38°

$$\therefore \quad \mu_e = \frac{0.3}{\sin 19°} = \frac{0.3}{0.3255} = 0.922$$

$$\therefore \quad \frac{F_1}{F_2} = e^{0.922 \times 2.94} = e^{2.71} = 15$$

$$\therefore \quad 14 F_2 = 932.2$$

$$\therefore \quad F_2 = 66.586 \text{ N}$$

$$F_1 = 932.2 + 66.586 = 998.79 \text{ N}$$

11. Find the number of belts by the relation,

$$z \geq \frac{F_1}{(f_t - \rho V^2).A.C_1.C_2}$$

Now $$f_t = \frac{1}{6} \text{ to } \frac{1}{8} \text{ of } (6.5 \text{ to } 10.5) \text{ MPa} = 1.5 \text{ MPa} = 1.5 \text{ N/mm}^2$$

$$\rho V^2 = 1400 \times \mathbf{(20.5)^2} = \mathbf{0.588 \text{ MPa}} = 0.588 \text{ N/mm}^2$$

Now From Table 19.7, the area of section of "C" belt = 2.3 cm² = 230 mm²

C_1 = arc of contact factor, Table 19.11 $\cong$ 0.98

Now From Table 19.12, the inside length = 8331 mm

$\therefore$ C_2 = belt length correction factor = 1.19

$$\therefore \quad z \geq \frac{998.79}{(1.5 - 0.588) \times 230 \times 0.98 \times 1.19} \geq 4.08 \text{, say 4 belts}$$

12. **Size of drive pulleys:** The size of the drives pulleys according to Fig. 19.21 and Table 19.10, in mm:

$$f = 6, e = 21, p = 26, s = 17, r = 8.75, b' = 23.1 \text{ (for } \beta = 38°)$$

The outside diameters of pulleys are:

$$D_{out1} = D_1 + 2f = 340 + 12 = 352 \text{ mm.}$$
$$D_{out2} = D_2 + 2f = 960 + 12 = 972 \text{ mm}$$

The inside diameter of the pulleys are:

$$D_{in1} = D_{out1} - 2e = 352 - 2 \times 21 = 310 \text{ mm}$$
$$D_{in2} = D_{out2} - 2e = 972 - 2 \times 21 = 930 \text{ mm.}$$

The nearest standard diameters (outer) are 355 mm and 1000 mm.

The width of pulleys, $B_1 = B_2 = (z - 1)\, p + 2s = (4 - 1) \times 26 + 2 \times 17 = 78 + 34 = 112$ mm.

Pressure exerted on the shaft due to belt tension is,

$$F_{sh} = 2\, f_i.A.z. \sin\frac{\theta}{2} \quad \text{(eqn. 19.10)}$$
$$= 2 \times 1.5 \times 230 \times 4 \times \sin\frac{168.7°}{2} = 2746.6 \text{ N}$$

After this, the design of pulley arms and shaft can be completed as in the case of flat belt drive.

EXERCISES

1. Calculate the V-belt drive from an electric motor of a gang saw. Motor power rating = 55 kW. Pulley diameter on gang saw, D_2 = 900 mm. Motor angular speed 75 rad/s; driven pulley angular speed, 30 rad/s.
2. Select a V-belt drive for a 7.5 kW, 1160 rev/min induction motor to drive a ventilating fan at approximately 400 rev/min. The minimum centre distance between the sheaves is 1000 mm.
3. Design a V-belt drive from an electric motor to the flywheel of a forging press from the following data; motor power = 80 kW, during operation the motor is overloaded, the overload factor being 1.5. Rated speed of motor is 730 rev/min and of the press flywheel 300 rev/min; the flywheel with diameter of 1500 mm is preferably used as a driven pulley. The distance between the axles of the flywheel and the motor pulley can be about 1650 mm.
4. A 7.5 kW motor running at 1750 rev/min is to be used to drive a rotary pump. The pump should run at approximately 1175 rev/min. The centre distance should not exceed 1100 mm. Space limits the diameter of the driven sheave to 290 mm. Determine the sheave diameters, the belt size and the number of belts.
5. A single acting reciprocating pump making 60 strokes a minute is driven from a geared speed reducer having a 10 to 1 ratio. The reducer is to be driven through a V-belt drive by a motor turning at 1750 rev/min. The diameter of the smaller sheave is 225 mm, 30 kW power is required. Determine the size and number of belts required.
6. The drive from a motor to a centrifugal pump consists of three '*B*' V-belts. The motor pulley has 110 mm pitch diameter, and the pump pulley has a pitch diameter of 390 mm. The motor runs at 1200 rev/min. What power can be transmitted if the centre distance is 920 mm.
7. An oil field pumping jack is fitted with a gear reduction unit delivering 3000 Nm torsional moment at 25 rev/min. The total reduction in the gears is 40 : 1. The gear unit is fitted with a 330 mm pitch diameter V-belt pulley and is driven by a 1200 rev/min motor. The centre distance is 1220 mm. Determine the number of size "*C*" belts required.
8. Assuming the density of the leather to be 1000 kg/m^3, find the speed at which maximum power will be transmitted if the stress in the belt is to be 2.8 MN/m^2.
9. A 1000 mm pulley transmits power to a 500 mm pulley by means of a medium double leather belt which is 500 mm wide. The centre distance is 4.2 meters. Assume the coefficient of friction as 0.3 (*a*) What will be the speed of the 1000 mm pulley in order that the belt be stressed to 2.0 MPa with zero power transmitted ? (*b*) What maximum power can be transmitted if the indicated stress in the belt is not to exceed 2.1 MPa? What is the speed of the belt when this power is transmitted ?

10. A motor is driving a compressor through a 150 mm heavy single ply leather belt. The motor pulley turning at 1750 rev/min is 200 mm in diameter, the compressor pulley is 300 mm in diameter, and the centre distance is 1.5 m. If the net belt pull $(F_1 - F_2)$ is 7000 N/m of width and $F_1/F_2 = 3$ (*a*). What is the stress in the belt? (*b*) What is the coefficient of friction? (*c*) What is the length of the belt?
11. The drive from a motor to a machine consists of three V-belts of size 'B'. The motor pulley has a pitch diameter of 150 mm and the machine pulley diameter is 450 mm. If the centre distance is 1.5 m, suggest a suitable A.C. motor to transmit maximum power.
12. A press is to punch holes in parts formed from steel plate. The press is to make 40 punches per minute, with each punch requiring 7000 Nm of energy. The Bull - Gear and pinion provide 8 : 1 ratio. A 1000 rev/min A.C. Induction motor is to drive the press through V-belts. Establish a satisfactory combination of: (*a*) Sheave diameter; (*b*) Belt size and number of belts, (*c*) Approximate centre distance of sheaves.
13. Design a V-belt drive actuated by an electric motor and driving a belt conveyor. The power supplied by the electric motor is 4 kW and the angular speed at the input shaft is 150 rad/s. The speed ratio of the drive is to approximately 2.8. The drive will be used for two shifts daily and subjected to a moderately varying load with a start-up overload of 150%.

Narrow V-belts: These V-belts are manufactured with the ratio of width (at the neutral plane) to the thickness of the belt equal to 1.050 to 1.1. There are four standard sizes: $(b \times t) = 10 \times 8$; 13×10; 17×13 and 22×18 mm section.

These belts find applications in machine tools, automobiles and tractors. They can handle greater tensions than normal section V-belts and, the cross-sectional area being the same, can transmit twice as much power. The operating velocity can be upto 50 m/s.

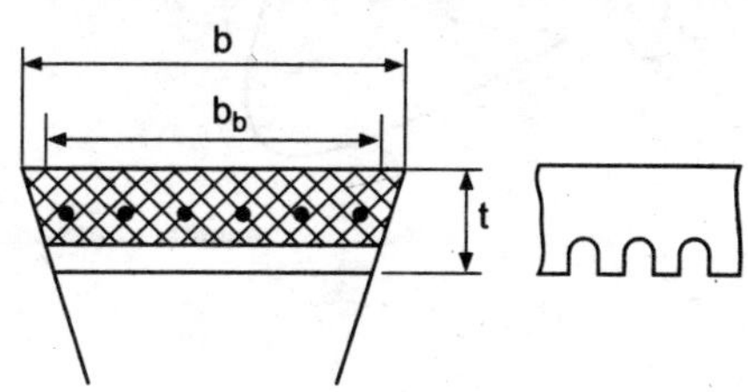

Fig. 19.22

Wide V-Bolt: In these V-belts, the ratio of the width (at the neutral plane) to the thickness of the belt lies between 2 to 4. These belts are used in continuous variable-speed drives. To extend the range of speeds, the belt section included angle is reduced to 28° in place of 40° as in conventional V-belts. The sheaves used are of the minimum allowable diameters. These belts are mostly reinforced with continuous filament cords and have cross-slots on the undersigned for flexibility, Fig. 19.22.

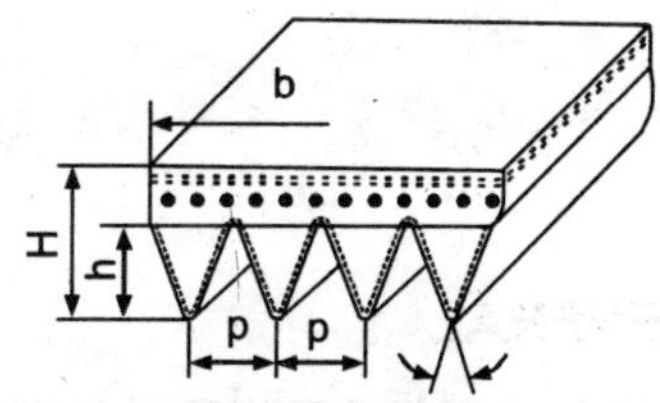

Fig. 19.23

Ribbed V-belts: These belts combine the advantages of conventional V-belts (improved grip) with those of flat belts (appreciable flexibility). The traction is provided principally by friction between the ribs and the sheave grooves, rather than the wedging action as in conventional V-belts. This makes them a good choice for sheaves of small diameter, and also where vibration is not permitted. The operational speed can be upto 50 m/s and the velocity ratio upto 12. These belts are impregnated with rubber and have a polyester-filament cord in the thin flat section and triangle-shaped ribs on the underside, Fig. 19.23. These belts can transmit large amounts of power. The width of a ribbed V-belt is 50 to 60% of that of a multiple V-belt composed of a number of normal section V-belts.

Endless (continuous loop) ribbed V-belts are commercially available in three sizes, Table 19.23. For ribbed V-belts sheaves have the same contact surfaces as those for conventional V-belts. The belt section included angle is taken as equal to 40°, irrespective of the sheave diameter.

Table 19.23 Specifications of ribbed V-belts

Type of Belt	Dimensions, mm			Area of Cross section mm^2	Maximum Design Length, mm	No. of Ribs		Mass per metre length kg
	h	*H*	*p*			Recommended	Allowable	
I	2.35	4	2.4	54	400–2000	2–36	36	0.09
II	4.85	9.5	4.8	291	1250–4000	4–20	50	0.45
III	10.35	16.7	9.5	954	2000–4000	4–20	50	1.60

❏❏❏

CHAPTER 20

Rope Drives

GENERAL

The method of transmitting of power by means of ropes has been superseded of late to a large extent by other methods of power transmitting, e.g., electrical transmission of power. But rope drives still find wide application for power transmission under certain conditions, e.g., when large amount of power are to be transmitted over a considerable distance. The flat belt drive cannot be used under these conditions because for this, the economical centre distance is limited to about 8 m and if large amounts of power are to be transmitted, it would result in excessive belt cross-sections which are not desirable due to many reasons. On the other hand, the rope drives will operate successfully over considerable longer distances, fibre ropes upto a centre distance of about 60 m and wire ropes even for a longer centre distance of upto 150 m. The rope drives use two types of ropes:

1. Fibre Ropes.
2. Wire Ropes.

20.1 FIBRE ROPES

The ropes made from fibrous material are used both for power transmission and for hoisting and lifting of loads. They are also widely used in printing machines, household appliances, clocks and various devices. The various fibrous materials used for the manufacture of ropes are:

(i) Hemp and Manila: The hemp and manila fibres are rough and the ropes made from these fibres are not very flexible and possess poor mechanical properties. Hemp ropes have less strength as compared to manila ropes. These ropes are subjected to rapid abrasion, rapid damage from sharp objects and atmospheric effects, etc. When hemp and manila ropes are bent over the sheaves, there is some sliding of the rough fibres of which these are composed, causing the rope to wear and chafe internally. To minimise this trouble, the ropes are impregnated with a lubricant such as tar, tallow or graphite. This will also make the ropes rot proof and moisture proof. But the ropes become heavier and are not sufficiently flexible. Moreover, the strength of a tarred rope is about 20% less than that of a white rope. Hemp ropes are suitable only for hand operate hoisting machinery and are used predominately as tie ropes for lifting tackle, hooks, etc.

(ii) Cotton: The cotton fibres are very soft and smooth. Due to this, there will be no internal damage or deterioration in the cotton ropes. Therefore, the lubrication is not necessary for cotton ropes. But if it is done, it will reduce the external wear between the rope and the grooves of its sheaves. However, Manila ropes are more durable and are stronger than cotton ropes. Cotton ropes are costlier than manila ropes.

Other rope materials can be: Leather, nylon and rubberized fabric.

Classification of Ropes: The ropes are classified according to the mode of manufacture and the number of strands in each rope, as below:

(a) Hawser laid or plain laid: This is a rope in which 3 strands are twisted together without a central core. The direction of twisting is known as "lay". Each strand is made from separate yarns by twisting the yarns together left handed. The yarns are spun from the fibres with a right handed twist. The rope is formed by twisting 3 strands together right handed. This lay for the rope is designated as '*Z*' lay in which the strands run from left to right across the top of the rope as in right hand screw thread. A plain laid rope is shown in Fig. 20.1(*a*).

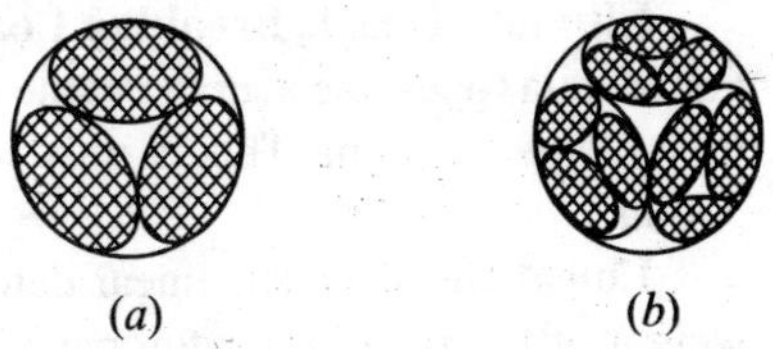

Fig. 20.1 *(a) Plain laid rope (b) Cable laid rope*

If in the rope, the strands coil in the same direction as the left-hand screw thread, then the rope is said to be of left-hand or "S" lay. If the fibres in the strand coil in the same direction as the strand, then the rope is of Lang lay or Parallel lay, and if they coil in the opposite direction to that of the strand, then the rope is in Regular or Cross or Ordinary lay.

(b) Shroud laid: The number of strands in this rope is 4 and the lay is '*Z*'.

(c) Cable laid: Figure 20.1(*b*). The number of strands is 9 and the lay is '*Z*' for primary ropes and '*S*' for final ropes. The rope is made by twisting together 3 primary ropes without a central core. Each primary rope consists of 3 strands and is plain laid.

20.1.1 Rope Drives for Power Transmission

The fibrous rope drives for power transmission have got the following advantages:

1. High mechanical efficiency
2. Low cost and economy of operation.
3. The shafts may be out of strict alignment.
4. Smooth, steady and quiet operation regardless of the amount of power and the distance over which it is transmitted.
5. Ease of operation with both long and short shaft centres and saving of space compared to belt drives.
6. The ropes are little affected by outdoor conditions.
7. Power may be taken off in any direction and in fractional parts of the whole amount.

Ropes made of cotton and manila are widely used for power transmission. Where small powers are to be transmitted or where small pulleys are employed, cotton ropes will be preferred because of their greater flexibility.

Systems of Rope Driving: The two systems employed in rope drives are:

1. Multiple system.
2. Continuous system.

1. Multiple System: In this system, one or more independent ropes run side-by-side in the grooves of the sheaves. Each rope is supposed to transmit the same amount of power. For this, the initial tension in all the ropes should be the same, but this condition is difficult to achieve and is therefore a disadvantage of this system. However a clear advantage is that if one rope breaks, the drive will not stop but it will go on operating even though at reduced load. In this system, it is possible to drive different machines from the same driving sheave. Multiple rope drives are seldom used.

2. Continuous System: In this system, only one rope is wound around the driving and the driven pulley several times. Its disadvantage is that if the rope breaks, the whole transmission comes to a stand still.

Note : The ropes are made endless by connecting their ends by splicing. To preserve a uniformity in the rope, the splicing should be done carefully and the length of the splice should be at least 60 times the diameter of the rope.

Ropes for rope drives: The diameter of the manila and cotton ropes used for power transmission usually ranges from 38 mm to 50 mm. The size of the rope is usually designated by its circumference or "girth". The net section of the rope is about 90 per cent of the circumscribing circle.

Ultimate Tensile Breaking Load: The ultimate tensile breaking load of the fibrous ropes varies greatly. For Manila ropes, the average value may be taken as, Ultimate Tensile Breaking Load = 5000 d^2N where d is rope diameter in cm. This is only an average value. For cotton ropes, the average value may be taken as 3500d^2N.

Linear density: The linear density of the ropes is designated in 'tex' or in 'kilotex'. Tex indicates the weight of the rope in newton per kilometre of length. Kilotex indicates the weight of the ropes in kN per kilometre of rope.

One kilotex = 1000 tex

The average value for the weight of rope may be taken as 0.70 d^2 newton per metre where d is in cm. The standardised requirements of cotton ropes of Power Transmission are given in Table 20.1. These are the minimum values for the linear density and ultimate tensile breaking load for the particular size of rope.

Table 20.1 Requirements for cotton ropes (IS : 3143–1965)

Size of Nominal Diameter, mm	Linear weight density, N/m	Ultimate Tensile Breaking Load, kN
19	2.90	10.34
22	3.70	15.21
25	4.60	19.64
28	6.05	24.63
32	7.30	32.17
38	10.80	45.17
44	14.20	60.83
50	17.35	78.55

Factor of Safety: To ensure durability, a high factor of safety should be used. Factor of safety for ropes is defined as:

$$F_S = \frac{\text{Ultimate tensile breaking strength}}{\text{Normal working load on the rope}}$$

For cotton ropes, the minimum value for FS should be at least 30. The factor of safety for manila ropes is somewhat higher due to its inferior properties as compared to cotton. The value may be taken as 35 or 36.

Velocity of Fibrous Ropes: The velocity of ropes varies between 15 to 30 m/s. The most economical speed for ropes is from 20 to 25 m/s.

Sheaves for Fibrous Ropes: As with all the flexible connectors, bending stresses are set up in these as they bend around the pulleys. This bending stress decreases as the sheave size increases. The usual sheave size is 40 d and the minimum should be at least 36d, where d is the rope diameter in cm.

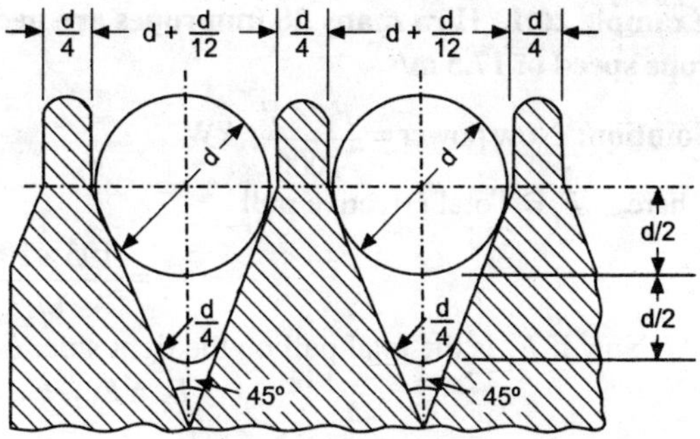

Fig. 20.2 *Typical rope pulley grooves*

The sheaves for rope curves are given special attention. The form of the grooves and the effective diameters should be the same for all the grooves of the same sheave. The surfaces should be accurately finished and well polished, as any roughness or unevenness seriously affects the life of the rope. The best groove angle which provides greatest adhesion without slipping, prevents undue wedging and offers least resistance to the rope while leaving the groove is 45°. The groove should be deep enough to prevent the rope reaching the bottom. The recommended dimensions for pulley grooves are shown in Fig. 20.2. (IS 3143–1965). Since the wedging action of the grooves is harmful for the ropes, the sheave when used only as an idler or tension sheave is grooved as in Fig. 20.3. The rim thickness in Fig. 20.2 is 0.5 *d*.

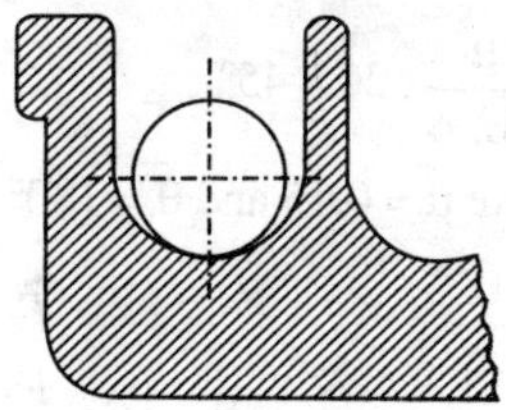
Fig. 20.3 *Rope idler sheave*

The maximum number of grooves should be 24.

Designations: The ropes shall be designated by their name, nominal diameter and the number of the standard. For example a rope of 25 mm diameter shall be designated as:

Cotton rope, 25, IS : 3143

Design Relation: The design relations and design procedure is on the same lines as the V-belt drives. The length of one rope is given by the formula:

$$L = \frac{\pi}{2}(D_1 + D_2) + (D_2 - D_1)\sin^{-1}\frac{D_2 - D_1}{2C} + 2C\cos\phi$$

where D_1 = Diameter of smaller pulley
D_2 = Diameter of larger pulley
C = Centre distance $\quad 2\phi$ = groove angle

The minimum centre distance of rope drives is given empirically as,

$$C_{\min} = (D_2 - D_1) + \frac{3}{2}D_2$$

The relation between the tension on the tight side and slack side is

$$\frac{F_1}{F_2} = e^{\mu_e \cdot \theta} \text{ where } \mu_e = \frac{\mu}{\sin\phi}$$

The average value for 'μ' with ropes on C.I. pulleys may be taken as 0.12 to 0.135. Some authors give the value for 'μ' as between 0.15 and 0.33.

If the centrifugal stress is taken into consideration, then

$$\frac{F_1 - F_c}{F_2 - F_c} = e^{\mu_e \cdot \theta}$$

$$\text{Total power transmitted} = \frac{(F_1 - F_2)\,V \times z}{1000}, \text{ kW}$$

where z = number of ropes
V = velocity of rope in m/s

Note: In the above relations, the sheave diameters are taken upto the centre of the rope section.

Example 20.1. How many 38 mm ropes are necessary for a rope drive that is to transmit 195 kW at a rope speed of 17.5 m/s.

Solution: Now power $= \frac{F_t \times V}{1000}$, kW

where F_t = Total effective pull

$$\therefore \quad F_t = \frac{195 \times 1000}{17.5} = 11.14 \text{ kN}$$

Now if F_1 is the total pull on the tight side, F_2 is the total pull on the slack side, then

$$F_t = F_1 - F_2$$

Again
$$\frac{F_1}{F_2} = e^{\mu_e . \theta}$$

where $\mu_e = \frac{\mu}{\sin \phi}$, $2\phi = 45°$

Let us take $\mu = 0.15$ and $\theta = 180°$

$$\therefore \quad \frac{F_1}{F_2} = e^{\frac{0.15 \times \pi}{\sin 22.5}} = e^{1.22} = 3.42.$$

$$\therefore \quad F_1 - \frac{F_1}{3.42} = 11.14 \times 1000$$

$$\therefore \quad F_1 = 11.14 \times 1000 \times \frac{3.42}{2.42} = 15.75 \text{ kN}$$

Now maximum allowable tensile load per rope

$$= \frac{\text{Maximum tensile breaking load}}{F_S} = \frac{5000\, d^2}{35} \cong 140\, d^2 \text{N}$$

$$= 140 \times (3.8)^2 = 2020 \text{ N}$$

Now the centrifugal effect $= \frac{wV^2}{g}$ N, V is in m/s

where w = weight of rope per metre length
$= 0.70\, d^2$ N/m $= 0.70\, (3.8)^2 = 10.1$ N/m

$$\therefore \quad \text{Centrifugal effect} = \frac{10.1 \times 17.5^2}{9.81} = 314 \text{ N}$$

$\therefore$ Net effective tensile load per rope = 2020 – 314 = 1706 N

$$\therefore \quad \text{Number of Ropes} = \frac{F_1}{1706} = \frac{15.75 \times 1000}{1706} = 9.2 \text{ say } z = 10$$

Example 20.2. How many cotton ropes 32 mm diameter will be required to transmit 375 kW over pulleys 12 m apart at a speed of 30 m/s.

Solution: Now from Table 20.1, the minimum ultimate tensile breaking load for a rope of 32 mm diameter 32.17 kN

Taking $FS = 30$

Minimum load which the rope must carry $= \frac{32.17 \times 1000}{30} = 1072$ N

Now centrifugal effect $= \frac{wV^2}{g}$ N

$$w = 0.70\, d^2 \text{ N/metre} = 0.70 \times (3.2)^2 = 7.14 \text{ N/m}$$

From the Table 20.1, $w = 7.3$ N/m

$$\therefore \quad \text{Centrifugal effect} = \frac{7.3 \times (30)^2}{9.81} = 670 \text{ N}$$

$\therefore$ Net effective load per rope = 1072 – 670 = 402 N

$$\therefore \quad \text{Now power} = \frac{F_t \times V}{1000}, \text{kW}$$

$$\therefore \quad F_t = \frac{375 \times 1000}{30} = 12.5 \text{ kN}$$

Now assuming values as for example 1.

$$\frac{F_1}{F_2} = 3.42$$

$$F_1 = 12.5 \times \frac{3.42}{2.42} = 17.58 \text{ kN}$$

$$\text{Maximum number of ropes} = \frac{17.58 \times 1000}{402} = 44$$

This is too high a number.

If the maximum tensile breaking load is taken as 3500 d^2N

$$\therefore \quad \text{Safe tensile load per rope} = \frac{3500}{30} d^2\text{N} = 117\, d^2 = 117 \times (3.2)^2 = 1195 \text{ N}$$

Effective tensile load per rope = 1195 – 670 = 525 N

$$\therefore \quad z = \frac{17.58 \times 10^3}{525} = 33.5, \text{ say } 34$$

As already indicated, the maximum number of grooves should be 24 as otherwise the load per rope may not be uniform. So, taking the diameter of rope, i.e., 38 mm

$$\text{Effective load per rope} = 117\,(3.8)^2 - \frac{10.8}{9.81} \times 30^2 = 1685 - 990 = 695 \text{ N}$$

$$\therefore \quad z = \frac{17.58 \times 10^3}{695} = 25.3$$

We can take z = 24 and the ropes made of fine quality cotton for which the maximum tensile breaking load will be higher.

Example 20.3. It is desired to transmit 225 kW by 44 mm diameter ropes of cotton running at 20 m/s. Show how many ropes are required if the speed ratio of 4 : 1 is to be produced from the source to its appliance. Design a suitable drive.

Solution: Now minimum diameter of sheave = 40d = 40 × 4.4 = 176 cm

$\therefore$ Diameter of Driving Pulley, D_1 = 176 cm

and Diameter of Driven Pulley, D_2 = 176 × 4 = 704 cm

$$\text{Now} \quad C_{min} = (D_2 - D_1) + \frac{3}{2} D_2 = (704 - 176) + \frac{3}{2} \times 704$$

$$= 528 + 1056 = 1584 \text{ cm, say 16 m}$$

$$\therefore \quad \theta = \pi - 2\sin^{-1}\frac{D_2 - D_1}{2C} = 180° - 2\sin^{-1}\frac{704 - 176}{3200} = 161° = 2.81 \text{ radians}$$

Now weight of rope per metre = 14.20 N from Table 20.1.

$$\therefore \quad \text{Centrifugal tension in rope,} \quad F_c = \frac{wV^2}{g} \text{ N}$$

$$V = 20 \text{ m/s}$$

$$\therefore \quad F_c = \frac{14.2 \times 20^2}{9.81} = 580 \text{ N}$$

Now $\quad e^{\mu_e \cdot \theta} = e^{\frac{\mu \cdot \theta}{\sin \phi}}$

Assuming $\quad \mu = 0.2, \phi = 22.5°$

$$\therefore \quad e^{\mu_e \cdot \theta} = e^{\frac{0.2}{\sin 22.5} \times 2.81} = e^{1.465} = 4.33$$

Now $\quad \dfrac{F_1 - F_c}{F_2 - F_c} = e^{\mu_e \cdot \theta}$

$$\therefore \quad \frac{F_1 - 580}{F_2 - 580} = 4.33$$

Now F_1 may be taken as $\dfrac{3500\, d^2}{F_S} = \dfrac{3500 \times 4.4^2}{30} = 2250 \text{ N}$

$$\therefore \quad \frac{2250 - 580}{F_2 - 580} = 4.33$$

$$\therefore \quad F_2 = \frac{1670}{4.33} + 580 = 966 \text{ N}$$

$$\therefore \quad (F_1 - F_2) \text{ per rope} = 2250 - 966 = 1284 \text{ N}$$

Now $\quad \text{power} = \dfrac{(F_1 - F_2) \times z \times V}{1000}$

$$\therefore \quad z = \text{number of ropes} = \frac{225 \times 1000}{20 \times 1284} = 8.76, \text{ say 9 Ropes.}$$

The design of the drive is on the same lines as for V-belts.

20.1.2 Fibrous Ropes Used in Hoisting Tackle

The ropes used in hoisting tackle are:

1. **Cotton Ropes:** It is available in two constructions:
   (a) Hawser laid or plain laid
   (b) Cable laid.
2. **Hemp Ropes:** Hemp ropes for lifting purposes are made by twisting three strands together in Z-direction. The hemp ropes are available in two grades:
   (a) **Grade 1.** This rope is spun from good quality hemp rope and is intended for lifting weights and hauling loads.
   (b) **Grade 2.** This rope is intended for various tying purposes and is made of average quality fibre.

(3) **Manila Ropes :** The ropes used for lifting purposes are made in three types:
   (a) Hawser laid or plain laid.
   (b) Shroud laid.
   (c) Cable laid.

Manila ropes are made in three grades: Grade 1, Grade 2, Grade 3.

For details about fibrous ropes for hoisting, the reader may refer to the following Indian Standards: IS 2542-1963;IS 2543-1963; IS 5176-1969; IS 1084-1969.

Factor of Safety: Velocity of rope upto 0.50 m/s, $F_S = 7$
Velocity of rope upto 1.50 m/s, $F_S = 18$
Velocity of rope upto 4.0 m/s, $F_S = 36$

Maximum Working load per rope:

Maximum permissible working load, N

$= 140d^2$, V from 2 to 4 m/s

$= 280\ d^2$, V from 0.75 to 1.5 m/s

$= 700\ d^2$, $V < 0.5$ m/s

Here d is in cm.

Sheaves : The diameter of the sheaves for fibrous ropes used for lifting purposes, should be at least "10d". The pulleys are manufactured in two designs:

1. **Fixed Pulley:** These pulleys are with fixed axles. These are also called guiding pulleys because they change the direction of the rope.
2. **Movable Pulleys:** These pulleys have movable axles to which either a load or force (effort) is applied.

Pulley System: A pulley system is a combination of several movable and fixed pulleys or sheaves. The system can be for a gain in force or for a gain in speed. Hoisting devices employ pulleys for a gain in force predominantly.

Pulley systems for a gain in force are designed:

1. With the rope running off a fixed pulley, Fig. 20.4 (*a*).
2. With the rope running off a moveable pulley, Fig. 20.4 (*b*).

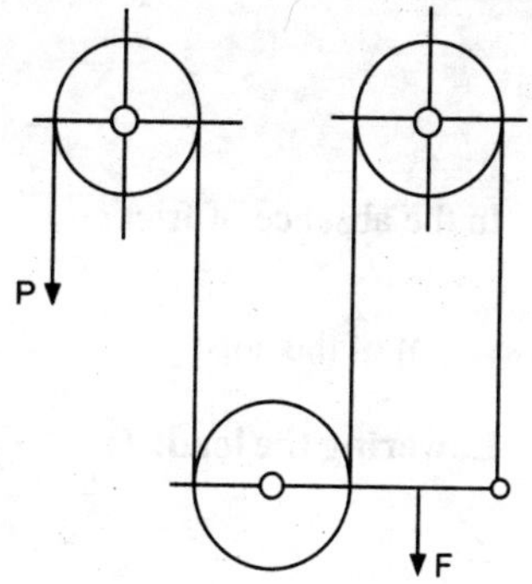

(*a*)

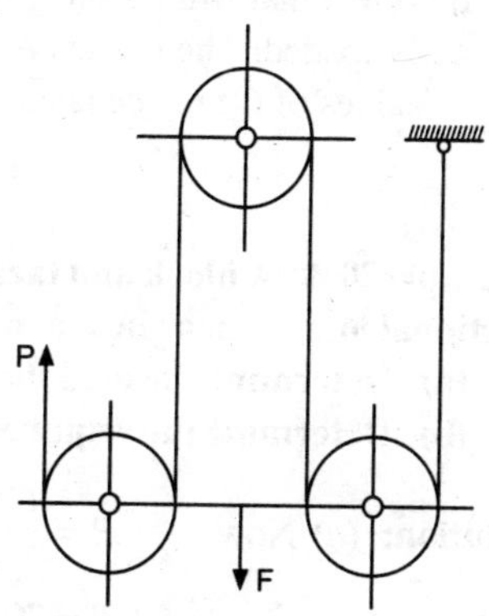

(*b*)

Fig. 20.4 *Pulley systems*

Design of Pulley System: When a rope passes over a pulley, and to its one end is attached a weight 'F' and an effort 'P' is applied at the other end, then theoretically,

$$P = F \text{ but in actual fact, } P > F, \text{ and}$$
$$P = C.F$$

where $C = P/F$, is called pulley factor of resistance. Its values depends upon:

1. Size of rope.
2. On the relative size of pulley and pin.
3. Coefficient of friction.
4. Type of bearings and the lubricants used for pulley axles.

The average value of 'C' for manila ropes may be taken as 1.14.

$$\eta = \text{efficiency of pulley} = 1/C$$

Raising the load: Figure 20.5. The $T_1, T_2, T_3, \ldots T_5$ be the tension in the various ropes.

Fig. 20.5 *Loads on a hoist*

Now $$T_2 = CT_1$$
$$T_3 = CT_2 = C \times CT_1 = C^2T_1$$

Similarly $$T_4 = CT_3 = C^3T_1$$
$$T_5 = CT_4 = C^4T_1$$

and $$P = CT_5 = C^5T_1$$

Now, from the equilibrium of forces on the lower block,

$$F = T_1 + T_2 + T_3 + T_4 + T_5 = T_1\,(1 + C + C^2 + C^3 + C^4)$$

$$= T_1\frac{C^5 - 1}{C - 1} = P\cdot\frac{C^5 - 1}{C^5(C - 1)} \quad \text{or} \quad P = F\cdot\frac{C^5\,(C - 1)}{C^5 - 1}$$

Therefore if z = number of times the ropes passes over the pulleys.

$$P = F \cdot \frac{C^z (C-1)}{C^z - 1}$$

In the absence of friction, $P_i = \dfrac{F}{z}$

$\therefore$ η of the hoist $\dfrac{F}{z} = \dfrac{C^z - 1}{zC^z (C-1)}$

Lowering the load: $T_1 = CT_2; T_2 = CT_3; T_3 = CT_4; T_4 = CT_5; T_5 = C.P$

$$\therefore \quad P = \frac{C-1}{C(C^5 - 1)} \cdot F \quad \text{or} \quad P = \frac{C-1}{C(C^z - 1)} \cdot F$$

It is clear that while raising the load, the rope '*P*' is most heavily loaded and while lowering, the rope T_1 is heavily loaded. The capacity of the hoist depends upon the most heavily loaded rope.

Values of *C* may be taken as = 1.15 for poorly lubricated bronze bearings.
= 1.08 for well lubricated bronze bearings.
= 1.05 for ball and roller bearings.

Example 20.4. A block and tackle consists of three-sheave block and one four-sheave block. Assume the frictional loss at each sheave to be 3 per cent of the rope tension on the right of the sheave.

(a) Determine the load that can be lifted when the applied rope pull is 700 N.

(b) Determine the required size of Manila Rope.

Solution: (*a*) Now $P = \dfrac{C^z (C-1)}{C^z - 1} \cdot F$

$\therefore$ $P = 700$ N, $z = 7$, efficiency of each sheave $= 1 - 0.03 = 0.97$

$$\therefore \quad C = \frac{1}{0.97} = 1.03$$

$$\therefore \quad 700 = \frac{1.03^7 (1.03 - 1)}{1.03^7 - 1} \cdot F = \frac{1.23 (1.03 - 1)}{1.23 - 1} \cdot F$$

$$\therefore \quad F = \frac{700 \times 0.23}{1.23 \times 0.03} = 4360 \text{ N}$$

(*b*) The maximum load on any rope will be *P*, i.e., 700 N

Now since the speed of rope is not given, let us arbitrarily choose *FS* equal to 30.

$\therefore$ Maximum Breaking Load for the rope = 700 × 30 = 21 kN

Taking an average of 5000 d^2 N as the breaking load for Manila Ropes.

$$\text{Safe load on the rope} = \frac{5000}{30} d^2 = \frac{500}{3} d^2 \text{ N}$$

$$\therefore \quad \frac{500}{3} d^2 = 700$$

$\therefore$ $d = 2.05$ cm, say 20 mm, which will give a slightly less *FS* than 30.

Example 20.5. A 16 mm Manila hoisting rope is used in a block and tackle having one block with two 10 cm sheaves and one single sheave block. The rope passes from the single sheave through two sheave block, through the single sheave block, back to and through the two sheave block to the point of application of the operating force. Assume 90 per cent efficiency for each sheave.

(a) Determine the maximum safe load that can be handled.

(b) Determine the operating force required to move this load.

(c) Determine the overall efficiency of this block and tackle.

Solution: (*a*) It has been mentioned that the maximum permissible working load for rope is,

$$140d^2 \text{ N if } 2.0 < V < 4 \text{ m/s}$$
$$280d^2 \text{ N if } 0.75 < V < 1.5 \text{ m/s}$$
and
$$700d^2 \text{ N if } V < 0.5 \text{ m/s}$$

∴ Maximum safe load that can be handled = $140\, d^2 = 140\,(1.6)^2 = 358$ N

(*b*) Now
$$P = \frac{C^z\,(C-1)}{C^z - 1} \cdot F$$

Now
$$C = \frac{1}{0.9} = 1.11, z = 3$$

∴
$$P = \frac{1.11^3\,(1.11-1)}{1.11^3 - 1} \times 358 = \frac{1.368 \times 0.11}{0.368} \times 358 = 146.5 \text{ N}$$

(*c*)
$$\eta = \frac{C^z - 1}{zC^z\,(C-1)} = \frac{1.368 - 1}{3 \times 1.368\,(1.11-1)} = 81.5\ \%$$

Rope Drums: Drums for fibrous ropes are usually of the plain type with high flanges so that the rope could be wound around it in several layers. This enables to have less length of the drum. The drum diameter should be ≥ 10*d*.

20.2 WIRE ROPES

Wire ropes were preferred in the past for long range power transmission but at present this use has become obsolete because of the other modes of power transmission *e.g.*, electric transmission. Wire ropes at present are employed for the following purposes:

1. Elevators
2. Oil well drilling
3. Mine hoist
4. Haulage devices
5. Cranes
6. Aerial conveyors
7. Tramways
8. Suspension bridges

The wire ropes have the following advantages as compared to chains, in hoisting machinery:

1. Lighter weight.
2. Less danger for damage due to jerks
3. More reliable in operation
4. Silent operation even at high working speeds

20.2.1 Materials

The materials used for wire ropes in the order of increasing strength are : Wrought iron, Cast steel, Extra strong Cast Steel, Alloy steel. For certain purposes, the ropes may also be made of copper, bronze, aluminium alloys and stainless steels.

20.2.2 Wire Rope Construction

Wire ropes are manufactured from steel wire with an ultimate strength ranging from 1100 to 2250 N/mm^2. During manufacturing, a special heat treatment is given to the wire which combined with cold drawing imparts high mechanical properties to the wire.

Steel wire ropes are made by special machines: first separate wires are twisted into strands and then the strands are twisted to form a rope. The strands are laid on a core made of hemp, jute, asbestos or a wire of softer steel. The core is saturated with a lubricant. Asbestos or wire core is used when the rope is subjected to radiant heat, e.g., cranes operating near furnaces. But the wire core reduces the flexibility of the rope, therefore, such ropes are used only where they are subjected to high compression, e.g., when being wound on a drum in several layers.

Cranes operating in dry environments use ropes made from bright uncoated wire. However, if the rope is to be used in damp conditions, these are coated with Zinc (galvanised). Corrosion of the ropes due to damp conditions can also be overcome by using suitable lubricants. For extreme corrosive conditions, ropes made of bronze, copper and stainless steels are used. Load lifting capacity of a Galvanised rope is reduced by 10% due to the tempering effect of hot zinc coating. Ropes are classified in the following manner:

1. Depending upon the direction of twist of the individual wires and that of the strands, relative to each other.
2. Depending upon the number of strands in the rope and the number of wires in each strand.

Depending upon the first criteria, the ropes are called as:

1. **Cross or Regular lay Ropes:** In such ropes, the strands are twisted into a rope in the opposite direction to that of the wires in the strands. Such ropes find the greatest application.
2. **Parallel or Lang lay Ropes:** In these ropes, the direction of twist of the wires in the strand is the same as that of the strands in the rope. These ropes are more flexible and resist wear more effectively.

   But these ropes have the tendency to spin. Such ropes are used in lifts and other hoists with guideways and also as haulage ropes.
3. **Composite or Reverse laid Ropes:** In such ropes, the wires in two adjacent strands are twisted in the opposite direction.

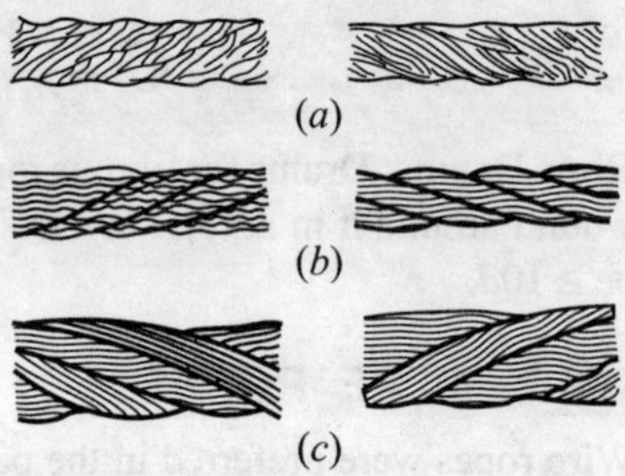

Fig. 20.6 *Lays of steel wire ropes (a) Regular lay (b) Lang lay (c) Composite lay*

The above types of lay are explained in Fig. 20.6.

The direction of the lay of the ropes can be either right hand or left hand accordingly, the ropes are known as **"Right lay Ropes"** and **"Left lay Ropes"**. Right-hand lay ropes are more frequently used.

Depending upon the second criteria, the ropes are standardised as: 6 × 7 rope, 6 × 37 rope and so on. A 6 × 7 rope means a rope made from six strands with seven wires in each strand. A 6 × 7 rope is shown in Fig. 20.7.

Fig. 20.7 *6 × 7 Rope*

The following standard constructions cover most of the applications and most services:

6 × 7	:	Standard coarse laid rope
6 × 19	:	Standard hoisting rope.
6 × 37	:	Extra flexible hoisting rope.
8 × 19	:	Extra flexible hoisting rope.

20.2.3 Application of Ropes

6 × 7 Rope: It is used as haulage rope in mine haulage, tramways, power transmission, factory yard haulage.

6 × 19 Rope: Hoisting rope in mines, quarries, ore docks, cargo hoists, cranes, derricks, Dredges, Elevators, Tramways, well drilling.

8 × 19 Rope: Hoisting Rope.

6 × 37 Rope: Hoisting rope in steel mill laddles, cranes, high speed elevators.

20.2.4 Properties of Various Types of Ropes

The breaking strength, the linear density etc., of a rope depends upon the purpose for which it is to be used. The properties of the various types of ropes are given below:

Table 20.2 Steel wire ropes for haulage purposes in mines

Type of Rope	Nominal diameter mm	Average weight N/m	Tensile Strength kN		
			Tensile strength of wire 1600 N/mm^2	Tensile strength of wire 1800 N/mm^2	
6 × 7	8, 9, 10, 11, 12, 13, 14, 16, 18, 19, 20, 21, 22, 24, 25, 26, 27, 28, 29, 31, 35	3.47 d^2	53 d^2	59.75 d^2	*d* in cm
6 × 19	13, 14, 16, 18, 19, 20, 21, 22, 24, 25, 26, 28, 29, 32, 35, 36, 38.	3.63 d^2	53 d^2	59.50 d^2	

Table 20.3 Suspension ropes for lifts and hoists

Type of Rope	Nominal diameter mm	Average weight N/m	Tensile Strength kN		
			Tensile strength of wire1000-1250 N/mm^2	Tensile strength of wire1250-1400 N/mm^2	
6 × 19	6, 8, 10, 12, 14, 16, 18, 20, 22, 25.	3.83 d^2	38.50 d^2	43.50 d^2	*d* in cm
8 × 19	8, 10, 12, 14, 16, 18, 20, 22, 25, 35, 36, 38	3.4 d^2	35.5 d^2	44.50 d^2	

Table 20.4 Steel wire rope for oil wells and oil well drilling

Type of Rope	Nominal diameter mm	Approxi-mate weight N/m	Ultimate Tensile Strength kN			
			Tensile strength of wire 1600–1800 N/mm^2	Tensile strength of wire 1800–2000 N/mm^2	Tensile strength of wire 2000–2250 N/mm^2	
6 × 7	10, 11, 13, 14, 16, 19, 22, 25	3.7 d^2	55.0 d^2	61.0 d^2	—	*d* in cm
6 × 19	13, 14, 16, 19, 22, 25, 29, 32, 35, 38	3.7 d^2	51.0 d^2	57.0 d^2	63.0 d^2	
6 × 37	13, 14, 16, 19, 22, 25, 29, 32, 35, 38	3.7 d^2	49.0 d^2	54.0 d^2	60.0 d^2	
8 × 19	13, 14, 16, 19, 22, 25, 29	3.38 d^2	—	53.0 d^2	—	
18 × 7	13, 14, 16, 19, 22, 25	4.0 d^2	—	57.50 d^2	—	

Table 20.5 Galvanised steel wire ropes for shipping purpose

Type of Rope	Nominal diameter mm	Average weight N/m	Tensile Strength kN		
			Tensile strength of wire 1400–1600 N/mm^2		
6 × 7	8, 10, 12, 14, 16, 18, 20, 22, 24 26, 28, 32, 36, 40	3.45 d^2	49.00 d^2	d in cm	
6 × 19	8, 9, 10, 11, 12, 14, 16, 18, 20, 22, 24, 26, 28, 32, 36, 40, 44 48, 52, 56	3.45 d^2	48.0 d^2	d in cm	
6 × 37	8, 9, 10, 11, 12, 14, 16, 18, 20, 22, 24, 26, 28, 32, 36, 40, 48 52, 56.	3.45 d^2	51.00 d^2	d in cm	

Table 20.6 Steel wire ropes for general engineering purposes, e.g. cranes, excavators, etc.

Type of Rope	Nominal diameter mm	Average weight N/m	Tensile Strength kN		
			Tensile strength of wire 1600–1750 N/mm^2	Tensile strength of wire 1750–1900 N/mm^2	
6 × 19	8, 9, 10, 11, 12, 13, 14, 16, 18, 20, 22, 24, 26, 28, 32, 36, 38, 40	3.75 d^2	54.00 d^2	59.00 d^2	d in cm
6 × 37	8, 9, 10, 11, 12, 13, 14, 16, 18, 20, 22, 24, 26, 28, 32, 36, 40, 44, 48, 52, 56.	3.80 d^2	51.00 d^2	55.00 d^2	

Note: Other ropes used for this purpose are 6 × 24; 34 × 7; 17 × 7; 18 × 7.

20.2.5 Approximate Wire Diameters and Effective Cross-section of Ropes

The diameter of the rope and the diameter of wire are connected by the formula:

$$d = 1.5\, dw \sqrt{z}$$

where dw = diameter of one wire

z = number of wires in the rope.

Approximate wire diameters and effective cross-section of ropes are given in Table 20.7, where d is in cm.

Table 20.7 Approximate wire diameters and areas

Rope	dw cm	A, cm^2
6 × 7	0.106 d	0.38 d^2
6 × 19	0.0663 d	0.38 d^2
6 × 37	0.045 d	0.38 d^2
8 × 19	0.050 d	0.35 d^2

20.2.6 Fibre Cores for Steel Wire Ropes

The fibre materials used as cores for steel wire ropes are of two types:

1. **Soft Fibres:** Italian or any other suitable hemp, and jute.
2. **Hard Fibres:** Manila.

The requirements of these cores for various ropes are given in Table 20.8.

Table 20.8 Fibre cores (IS: 1804—1976)

S. No.	Wire Ropes for	Hard Fibre Diameter Range, mm	Soft Fibre Diameter Range, mm
1.	Cranes	5 to 46	1 to 46
2.	Lifts and Hoists	-do-	-do-
3.	Shipping	-do-	-do-
4.	Oil wells drilling Cable	-do-	Not permissible
5.	Winding	-do-	-do-
6.	Man-riding and Rope ways	-do-	-do-
7.	Haulage, Mines and other purposes	-do-	1 to 46
8.	General Engineering Purposes	-do-	-do-

20.2.7 Working Loads

The working loads for the ropes for various applications are chosen so that the rope selected on the basis of strength has a reasonable life and it should not fail as a result of wear and fatigue. To find the safe working load, the factor of safety based on the ultimate strength are given in Table 20.9.

Table 20.9 Factor of safety for wire ropes

Service		F_S	Rope Speed m/s
(i)	Elevators	8–12	1.5–6.25
(ii)	Mine Hoist:		
	Depth To 150 m	8	6–8
	300–600 m	7	10–12.5
	600–900 m	6	12.5–15.0
	over–900 m	5	15.0
(iii)	Miscellaneous hoists	5	1.0
(iv)	Haulage Ropes	6	10.0
(v)	Cranes :		
	Motor driven	4–6	1.0
	Hand power	3–5	0.50
(vi)	Derricks	3–5	10.00
(vii)	Small electric and Air Hoists	7	1.0
(viii)	Hot ladle cranes	8	1.0
(ix)	Slings	8	1.0
(x)	Track-Cables, Guys	3.5	1.0

20.2.8 Friction and Efficiency

When a wire rope winds round a pulley, the friction between the wire fibres resists its bending. This resistance must be overcome by the pull applied to the running off side of the rope. The value of factor 'C' for wire ropes are given below in Table 20.10.

Table 20.10 Pulley factor of resistance "C"

Rope diameter mm	C
9	1.09
11	1.083
13	1.075
14	1.070
16	1.064
19	1.054
22	1.046
25	1.040

20.2.9 Wire Rope Sheaves and Drums

In order to reduce the bending stresses in the ropes as they bend around the sheaves or pulleys, the sheave diameters should be as large as possible. The sheave diameters for various types of ropes are given in Table 20.11, where space allows, larger sheave diameters than given in Table 20.11 should be used for better and economical service.

Table 20.11 Wire rope sheave diameters

Type of Rope	Recommended Sheave Diameters	
	Preferred	Minimum
6 × 7	72 *d*	42 *d*
6 × 19	45 *d*	30 *d*
6 × 37	27 *d*	18 *d*
8 × 19	31 *d*	21 *d*

The sheave groove has a great influence on the life and service of rope. The rim of a rope sheave is shown in Fig. 20.8.

Rope sheaves are usually made in the form of a casting (grey iron or steel) or weldment. Small sheaves are usually cast in one piece without ribbing. Large sheaves are provided with ribs and holes or with cross-shaped spokes. The sheaves are usually mounted on fixed axles on antifriction bearings or bronze bushings.

The bushings of the sheaves can be checked for unit pressure by the formula

$$p_b = \frac{F}{l \times d_a}$$

where l = length of bushing or sheave axle

d_a = diameter of sheave axle

F = working load.

l/d_a is usually taken as 1.5 to 1.8.

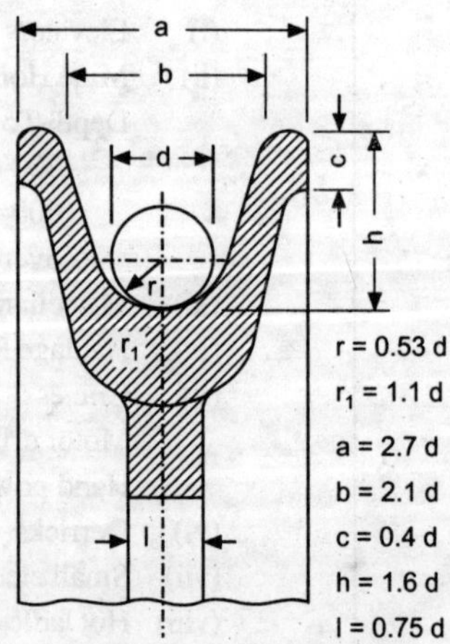

Fig. 20.8 *Wire rope sheave rim*

The unit bearing pressures should not exceed the values given below:

Peripheral velocity, m/s

0.1	0.2	0.3	0.4	0.5	0.6	0.7	0.8	0.9	1.0	1.1	1.2	1.3	1.4	1.5

p_b N/mm^2

7.5	7.0	6.6	6.2	6.0	5.7	5.5	5.4	5.3	5.2	5.1	5.0	4.9	4.8	4.7

A standard grooved drum for wire ropes is shown in Fig. 20.9.

The wall thickness of C.I. drum can be found out approximately by the following empirical formula.

$$h = 0.02\,D + (6 \text{ to } 10) \text{ mm}$$

where D = drum diameter, mm

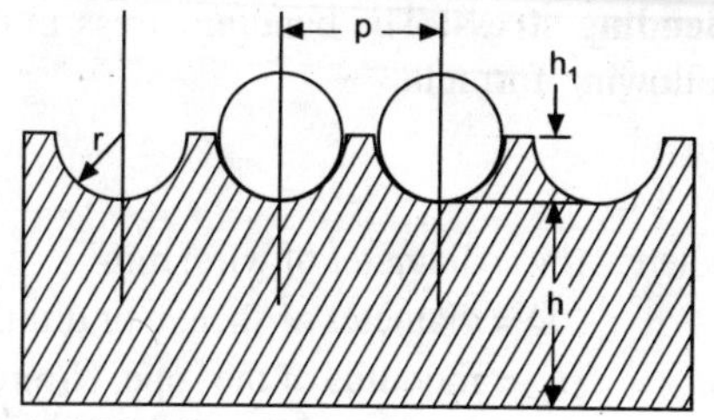

$p = 1.15\,d,\ h_1 = 0.25\,d,\ r = 0.53\,d,\ h = 1.1\,d$

Fig. 20.9 *Grooved rope drum*

20.2.10 Rope Fasteners

Spliced wire ropes should be avoided as far as possible as the strength of the spliced wire rope is reduced by 25 per cent. The various rope fasteners are shown in Fig. 20.10. The efficiencies of these fasteners are listed in Table 20.12.

Table 20.12 Efficiencies of rope fasteners

Type of Fastening	η
(*i*) Wire rope socket with zinc.	1.00
(*ii*) Thimble with four or five wire tucks.	0.90
(*iii*) Special offset thimble with clips	0.90
(*iv*) Regular thimble with clips.	0.85
(*v*) Three bolt wire clamps.	0.75

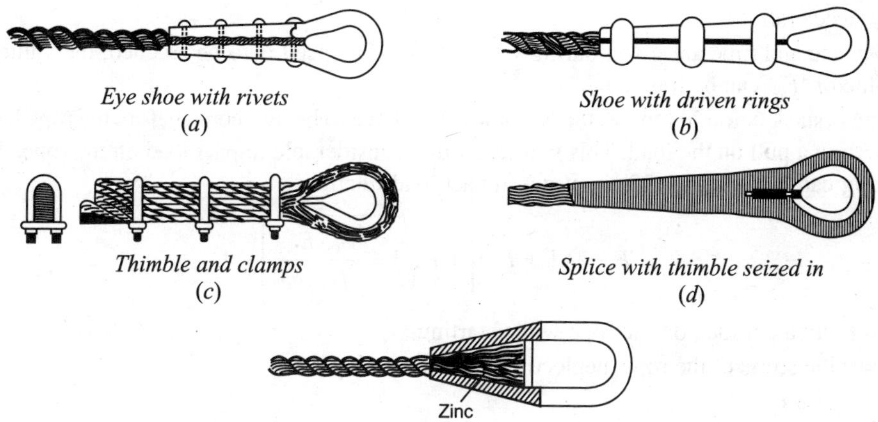

Eye shoe with rivets (*a*)

Shoe with driven rings (*b*)

Thimble and clamps (*c*)

Splice with thimble seized in (*d*)

Rope socket (*e*)

Fig. 20.10 *Wire rope, fasteners*

20.2.11 Selection of Wire Rope

The stresses set up in a wire rope are very complex. These stresses in a bent loaded rope consist of tension, bending and twisting stresses combined with mutual compression and rubbing of wire and strands. When the rope bends around a sheave or a drum, its wires on the outside are stretched while those on the inside get shortened. This action produces additional tension in the outer fibres. Due to the complexity of the stresses their complete determination with accuracy is not possible. The various types of stresses induced in a hoisting rope under load may be listed as below:

1. Direct stress due to the weight hoisted and the weight of the rope.
2. Bending stress when the rope winds around the sheave or drum.
3. Starting stresses.
4. Stress due to change of rope speed.

Bending stress: The bending stress in the outer wires of the rope may be calculated with the help of the following formula:

$$f_b = \frac{E_r \cdot dw}{D}$$

where dw = diameter of wire, mm

D = diameter of sheave or drum, mm

E_r = modulus of the rope, N/mm^2

= 0.84×10^5 N/mm^2 for steel ropes of ordinary construction. The equivalent bending load on the rope will be,

$$F_b = \frac{E_r \cdot dw}{D} \cdot A, N \quad \text{... (20.1)}$$

where A = effective cross-sectional area of the rope, mm^2

Stresses during Starting: When the rope starts moving such as in mine hoists and elevators, force is needed to accelerate the load supported by the rope and also the weight of the rope. This force induces an additional load in the rope. This additional load can be found out by the relation:

$$F_a = (F + F_r)\frac{a}{g} \quad \text{... (20.2)}$$

where F = weight of load

F_r = weight of the rope

a = acceleration.

If the distance and time are given during which a specific speed is to be reached, the value of 'a' and hence the value of 'F_a' can be found out.

If the rope is slack before it moves, this slackness will have to be overcome before the rope becomes taut and starts exerting a pull on the load. This will result in a considerable impact load on the rope. This impact load on starting can be found out by the usual impact load relation:

$$F_{st} = (F + F_r)\left(1 + \sqrt{1 + \frac{v_s^2 \cdot E_r}{f \cdot l \cdot g}}\right) \quad \text{... (20.3)}$$

where F_{st} = Equivalent load on the rope while starting

f = static stress of the rope, neglecting the effect of core.

$$= \frac{(F + F_r)}{\frac{\pi}{4} dw^2 \times z}$$

l = length of the rope

v_s = velocity of rope at the instant when the rope is taut.

$= \sqrt{2ah}$

where h = rope slackness

a = acceleration to be imparted to the drum and rope in order to take up the slack.

If there is no slackness, $h = 0$, $v_s = 0$

$$\therefore \quad F_{st} = 2\,(F + F_r) \quad \text{...(20.4)}$$

Stress due to Change in Speed: The equivalent load on the rope due to a change in speed can be found out with the help of equation 20.3. The acceleration may be found by the equation.

$$a = \frac{(V_2 - V_1)}{t} \quad \text{...(20.5)}$$

When the hoist drum is suddenly stopped while lowering the load, severe stresses are set up in the rope, since kinetic energy of the moving mass is suddenly made zero. This kinetic energy is absorbed by the rope and to find the stress induced in the rope due to this kinetic energy should be equated to the resilience of the rope. If during stopping, the load moves a certain distance, the potential energy due to this should also be added to the kinetic energy.

20.2.12 Design Procedure

The total load coming on the rope during operation is calculated and factor of safety on the basis of the ultimate tensile strength of rope found out. This factor of safety should be within the limits given in Table 20.9.

Effective load on the rope $= F + F_r + F_b$

Load on the rope during starting $= F_{st} + F_b$

Load on the rope during acceleration $= F + F_r + F_b + F_a$.

Example 20.6. Determine the size of a wire rope necessary for a mine hoist carrying a load or 69.50 kN to be lifted from a depth of 225 metres. A rope speed of 7.9 m/s is to be attained in 10 second.

Solution: 1. The first step is to select the proper rope for the purpose. From Table 20.2, it is clear that ropes 6 × 7; and 6 × 19 are used for mines. Let us try 6 × 19 rope.

2. The second step is to choose the diameter of the rope. From Table 20.9, the factor of safety for the problem should be about 7. Also we know that at starting the stress is doubled and again the bending stress is to be considered. Therefore, let us take tentatively the factor of safety as 15.

∴ Minimum breaking strength of rope = 15 × 69.50 = 1042.50 kN

Now from Table 20.2, the tensile strength of 6 × 19 rope made of wire with tensile strength of 1800 N/mm^2 is 59.50 d^2 kN.

$$\therefore \quad 59.50\, d^2 = 1042.50$$

$$\therefore \quad d = 4.12 \text{ cm say } 42 \text{ mm}$$

3. Find the various loads on the rope:

$$F_b = \frac{E_r \cdot dw}{D} \cdot A$$

Now $dw = 0.063\, d = 0.063 \times 4.2 = 0.265$ cm

D = diameter of sheave = 60 to 100 times the rope diameter for mines.

$= 100 \times 4.2 = 420$ cm

$A = 0.38\, d^2 = 0.38 \times 4.2^2 = 6.68 \text{ cm}^2 = 668 \text{ mm}^2$

$E_r = 0.84 \times 10^5 \text{ N/mm}^2$

$$\therefore \quad F_b = \frac{0.84 \times 10^5 \times 2.65}{4200} \times 6.68 = 35.45 \text{ kN}$$

Now weight of rope, $F_r = 3.63 \times (4.2)^2 \times 225 = 14.40$ kN

The starting load without slack, from equation 20.4.

$$F_{st} = 2\,(F + F_r) = 2\,(69.50 + 14.40) = 167.80 \text{ kN}$$

Now acceleration, from equation

$$a = \frac{V}{t} = \frac{7.9}{10} = 0.79 \text{ m/sec}^2$$

$$\therefore \quad F_a = \text{mass} \times a = \frac{69.50 + 14.40}{9.81} \times 0.79 = 6.75 \text{ kN}$$

4. Find F_S:

(*i*) Effective load during starting $= F_b + F_{st} = 35.45 + 167.80 = 203.25$ kN

$$\therefore \quad FS = \frac{69.50 \times 4.2^2}{203.25} = 5.11$$

(*ii*) Total load during first 10 seconds after starting

$$= F + F_r + F_b + F_a = 69.50 + 14.40 + 35.45 + 6.75 = 126.10 \text{ kN}$$

$$\therefore \quad FS = \frac{69.50 \times 4.2^2}{126.10} = 8.32$$

(*iii*) Total load during the uniform lifting or lowering of the load

$$= F + F_r + F_b = 69.50 + 14.40 + 35.45 = 119.35 \text{ kN}$$

$$\therefore \quad FS = \frac{69.50 \times 4.2^2}{119.35} = 8.79$$

∴ 42 mm, 6 × 19 rope is satisfactory.

Example 20.7. Select a wire rope for a vertical mine hoist to lift 12.40 MN of ore in each 8 hr shift from a depth of 720 m. Assume a two compartment shaft with the hoisting skips in balance.

Solution: 1. Assume: (*i*) maximum rope velocity of 12.5 m/s Table 20.9.

(*ii*) Acceleration and retardation periods of 12 seconds each.

(*iii*) Rest period of 10 secs for loading and discharging.

2. Acceleration, $a = \frac{V}{t} = \frac{12 \cdot 5}{12} = 1.04 \text{ m/sec}^2$

The distance covered during acceleration $= \frac{Vt}{2} = \frac{12 \cdot 5 \times 12}{2} = 75$ m

Total distance travelled at full speed = 720 – 2 × 75 = 570 m

Time needed for this distance $= \frac{570}{12 \cdot 5} = 45.6$ secs

3. Time for one trip:

Acceleration = 12 secs
Full speed = 45.6 secs
Deceleration = 12 secs
Rest period = 10 secs
Total = 79.6 secs

$$\therefore \quad \text{Total trips per hour} = \frac{3600}{79 \cdot 6} = 45$$

$$\therefore \quad \text{Capacity of skip} = \frac{12 \cdot 40 \times 10^6}{8 \times 45} = 34.44 \text{ kN, say } 35 \text{ kN}$$

Weight of skip may be taken equal to 60% of its load capacity,

i.e., $= 0.6 \times 35 \simeq 20$ kN

∴ Total load to be lifted by the ropes = 35 + 20 = 55 kN

4. Select 6 × 19 Rope. The factor of safety desired is about 6, Table 20.9. To take into account the complex stresses induced in the rope, the rope should be designed for a load = 55 × 14 = 770 kN

$$\therefore \quad 59.50\, d^2 = 770$$

∴ $d = 3.6$ cm, select standard 36 mm diameter rope.

∴ $dw = 0.063\, d = 0.063 \times 3.6 = 0.227$ cm

D = sheave diameter = 60 to 100 d, say 300 cm

$A = 0.38\, d^2 = 0.38 \times 3.6^2 = 4.94 \text{ cm}^2$

weight per meter $= 3.63\, d^2 = 3.63 \times 3.6^2 = 47.2$ N/m

5. Various Loads are:

(*i*) Weight of ore + weight of skip, $F = 55$ kN

(*ii*) Weight of rope, $F_r = 47.2 \times 720 = 34.05$ kN

(*iii*) Equivalent load due to acceleration of load and rope.

$$F_a = \text{mass} \times a = \frac{55 + 34.05}{9.81} \times 1.04 = 9.43 \text{ kN}$$

(*iv*) Bending load, $F_b = \dfrac{E_r \cdot dw}{D} \cdot A = \dfrac{0 \cdot 84 \times 10^5 \times 2 \cdot 27}{3000} \times 494 = 31.10$ kN

Total load $= 55 + 34.05 + 9.43 + 31.1 = 129.58$ kN

$$\therefore \qquad F_S = \frac{59.50 \times 3.6^2}{129.58} = 5.95$$

$\therefore$ Rope selected is O.K.

Example 20.8. Select a wire rope for the elevator in a building where the total lift is 180 m. The rope velocity is 4.5 m/s and the full speed is to be attained in 12 m. The lifting sheaves are of the traction type. The weight of the elevator cage is 11.35 kN and the weight of passengers is 9.10 kN.

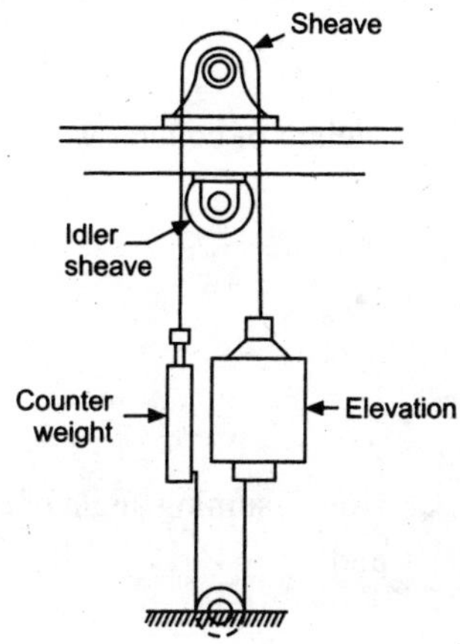

Fig. 20.11 *Elevator*

Solution: Refer Fig. 20.11.

1. From Table 20.9, the minimum factor of safety for elevators = 8.

$\therefore$ Total load for which the ropes should be designed $= 2 \times 8 \times (11.35 + 9.10) = 327.20$ kN

Note that the factor '2' is taken to account for weight of rope, acceleration and bending load.

Usually 4 ropes are provided in elevators, therefore,

$$\text{load per rope} = \frac{327.20}{4} = 81.80 \text{ kN}$$

2. From Table 20.3, select 16 mm diameter 8 × 19 rope.

$$\text{Weight per meter} = 3.4\, d^2 = 8.7 \text{ N}$$

$$dw = 0.05\, d = 0.05 \times 16 = 0.08 \text{ mm}$$

$$A = 0.35\, d^2 = 0.35 \times 2.56 = 0.896 \text{ cm}^2$$

$$D = 31\, d = 31 \times 1.6 = 50 \text{ cm}$$

3. Acceleration of rope: $\quad a = \dfrac{V^2}{2S} = \dfrac{4.5^2}{2 \times 12} = 0.844 \text{ m/sec}^2$

4. Rope loads: (*i*) Elevator and passengers $= \dfrac{11.35 + 9.10}{4} = 5.11$ kN

(*ii*) Weight of rope $= 8.7 \times 180 = 1.565$ kN

(*iii*) Acceleration load $= (5.11 + 1.565) \times \dfrac{0.844}{9.81} = 0.57$ kN

(*iv*) Equivalent bending load $= \dfrac{0.84 \times 10^5 \times 0.80}{500} \times 89.6 = 12.042$ kN

$$\text{Total load} = 5.11 + 1.565 + 0.57 + 12.042 = 19.287 \text{ kN}$$

$$\text{Now Tensile strength of rope} = 44.50\, d^2 = 113.92 \text{ kN}$$

$$\therefore \quad F_S = \frac{113.92}{19.287} = 5.9, \text{ which is very less.}$$

We can increase the diameter of sheave to reduce the bending load.

Let $D = 130$ cm

$$\therefore \quad \text{Equivalent bending load} = \frac{0.84 \times 10^5 \times 0.8}{1300} \times 89.6 = 4.63 \text{ kN}$$

$$\therefore \quad \text{Total load} = 5.11 + 1.565 + 4.63 = 11.875 \text{ kN}$$

$$\therefore \quad F_S = \frac{113.92}{11.875} = 9.6, \text{ which is O.K.} > 8$$

5. Number of wraps of rope around the driving sheaves :

When the elevator is at the bottom, the tensions in the rope are

$$F_1 = 5.11 + 1.565 + 0.57 = 7.245 \text{ kN}$$

$$\text{Weight of counter weight may be taken as} = \text{weight of elevator} + \frac{1}{3} \times \text{live weight}$$

$$= 11.35 + \frac{1}{3} \times 9.10 = 14.38 \text{ kN}$$

$$\therefore \quad F_2 = \frac{14.38}{4}\left(1 - \frac{0.844}{9.81}\right) = 3.285 \text{ kN}$$

$$\therefore \quad \text{Total } F_1 - F_2 = 7.245 - 3.285 = 3.96 \text{ kN}$$

Now assuming angle of contact = π radians

and $\mu = 0.18$ to 0.23, say 0.18

$$\mu.\theta = 0.18 \times \pi$$

$$\therefore \quad \frac{F_1}{F_2} = e^{\mu\theta} = e^{0.18 \times \pi} = 1.74$$

$$\therefore \quad F_2 = \frac{F_1}{1.74} = \frac{7.245}{1.74} = 4.11 \text{ kN}$$

$$\therefore \quad F_1 - F_2 = 7.245 - 4.11 = 3.145 \text{ kN for one wrap around sheave}$$

$$\therefore \quad \text{Number of wraps needed} = \frac{3.96}{3.145} = 1.263, \text{ say 2 wraps.}$$

EXERCISES

1. A rope drive of sixteen 38 mm cotton ropes travels 20 m/s. Determine the power that is being transmitted.
2. How many 25 mm Manila ropes are necessary for a rope drive that is to transmit 110 kW at a rope speed of 20 m/s?
3. A 38 mm, 6 × 19 steel rope is used to lift the cage of a mine hoist. The mine is 180 m deep, the rope sheave is 3.6 m in diameter and the cage acceleration is 1.5 m/sec^2.
   (*a*) Determine the equivalent bending load.
   (*b*) Determine the load due to rope.
   (*c*) Determine the useful load with a factor of safety of 5.

4. In an office building the elevator rises 350 m with an operation speed of 5 m/s and reaches full speed in 10 m. The loaded elevator weighs 20 kN. Determine the number of 16 mm, 6 × 19 steel ropes required if the factor of safety is 6. The sheave diameter is 0.90 m.
5. Select a wire rope for a vertical mine hoist to lift 15000 kN of ore in each 8 hr shift from a depth of 900 m. Assume a two compartment shaft with the hoisting skips in balance. Use a minimum velocity of 12.5 m/s with acceleration and deceleration of 15 s each and a rest period of 10 s for discharging and loading the skips. A hoisting skip weighs approximately 0.6 of its load capacity. The factor of safety should be about 5.
6. The pulley shown in Fig 20.12 is used to raise a load F of 27 kN with a maximum acceleration of 2.4 m/sec^2. Assume 6 × 19 steel rope, sheaves 42 rope diameters in diameter, and loss of 5 per cent with each pulley.
   (*a*) Determine the diameter of the rope required for a factor of safety of 5.
   (*b*) Determine the effort P.
7. The block and tackle in Fig. 20.13 is to raise a load of 18 kN with a maximum acceleration of 2.25 m/sec^2. Use a sheave diameter as 30 times the diameter of rope. Using a factor of safety of 6, determine:
   (*a*) The size of a steel rope.
   (*b*) The effort needed to raise the load.

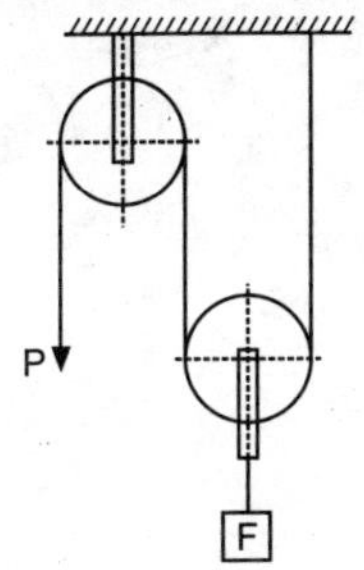

Fig. 20.12

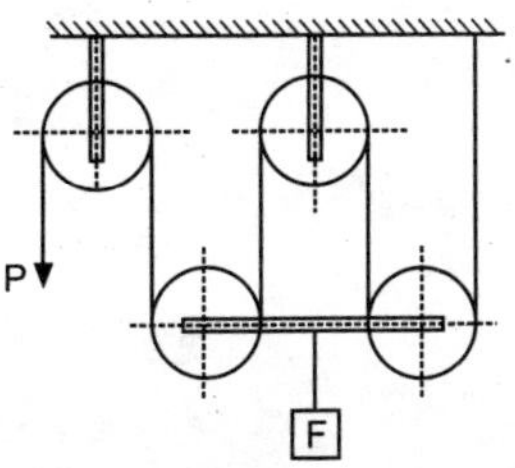

Fig. 20.13

8. A hoist in a copper mine lifts ore to a maximum of 600 m. The weight of the car and cage is 27 kN and 18 kN of ore are lifted per trip. The maximum rope speed is 10 m/s and the time of acceleration to this speed is 6 s. The diameter of the drum is 1.8 m.
   Determine the diameter of the 6 × 19 steel rope required for this purpose.

❑❑❑

CHAPTER 21

Chain Drives

21.1 GENERAL

In its simple form, a chain drive consists of an endless chain running over two sprockets – driver and driven. A sprocket is a wheel with teeth of a special profile. A chain drive operates without slipping like a gear drive and thus provides a positive drive, but where precise timing is required these can't be used. The alignments of the shafts must be more accurate than for belt drives.

The principal advantages of a chain drive are as given below:

1. It can be employed both for relatively long or short centre distances. The centre distance may be as large as 8 metres.
2. A small and compact size as compared to a flat belt drive.
3. Less loads on shafts compared to belt drives.
4. It can have efficiency as high as 98 to 99 percent when operating under ideal conditions.
5. It is possible to transmit power or motion to several shafts with the help of one chain only.

At the same time, a chain drive has got the following disadvantages.

1. Due to wear of chain joints, the chain gets stretched that increases the chain pitch resulting in velocity fluctuations and necessitating the use of take up drives.
2. Relatively high production costs.
3. Noisy operation.
4. A chain drive requires more amount of servicing, maintenance and repairs as compared to belt drives.
5. More complicated design than for a belt drive.

Chain drives find wide applications in agricultural machinery, bicycles, motor cycles, metal and wood working machines, lathes, conveyors, coal cutters, the transmission of auxiliary mechanism in rolling mills and in transport mechanism, etc.

Chain drives can either be open or closed like a gear drive.

Modern chain drives are used for velocity ratios of less than 8 with chain velocities upto 25 m/s and power ratings upto 110 kW or even more.

21.2 TYPES OF CHAIN DRIVES

From the functioning point of view, the chain drives may be classified as follows:

1. Driving or Power Transmission chain.
2. Crane Chains.
3. Tractive Chains or Pulling Chains.

Only driving chains will be dealt with in this chapter. There are six main types of chains used for power transmission, Table 21.1, but the three most commonly used chains are :

(a) Roller chains.
(b) Bush chains.
(c) Silent chains or inverted tooth chains.

For these chain drives, the shafts should be parallel. In general, the roller chains and bush chains are recommended for low and medium speed service, whereas silent chain is employed for high speed ratios due to its more smooth and quiet action.

These two types of chains are compared in Table 21.2.

Table 21.1 Power transmission chains

S. No.	Type of Chain	Pitch, p mm	Ultimate Strength kN	Application
1.	Bush	15–9100	12.5–500	General mechanical engineering
		8–9.525	5–12	Drives in motor-cycles
2.	Bush-Roller	12.7–15.875	7.5–20	Drives in motor cycles and bicycles
		15.875–41.3	17.5–50	Agricultural machinery
		8–50	2–160	General mechanical engineering
		20–65	38–430	Oil-field machinery
3.	Bush-Roller with bent plates	40–100	80–500	Drives operating at heavy duty
4.	Silent	12.7–31.75	19–690	Main drives in various machines
5.	Hooked link-belt chain	30–38	6.0–9.0	Agricultural machinery
6.	Stud link-belt chain	42	25	Drives operating under impact load

Table 21.2 Chains compared

Design Factor	Type of Chain	
	Roller	Silent
Speed, m/s	25	32.5
Velocity Ratio	8 : 1	12 : 1
Load Equivalent width	1.6	1.9
Life, hrs.	15000	15000
Noise (relative)	2	1

21.3 CONSTRUCTION OF CHAINS

Any chain is designated by three dimensions: Pitch, Width and Diameter.

Pitch: It is the distance from centre to centre of adjacent pins or rivets.

Width: It is the nominal width of the link or the length of the pin or the rivet measured within the side plates, that is, within the roller link plates.

Diameter: It is used primarily in roller chain and it refers to the actual outside diameter of the roller.

21.3.1 Roller Chains

The construction details of a roller chain are shown in Fig. 21.1. It consists of two rows of outer and inner plates. The outer row of plates is known as pin link or coupling link whereas the inner row of plates is called roller link. The pins are press fitted in the pin links and these pass through the bushings which are press fit in the roller links and join these. The chain rollers are mounted on the bushings and they roll over the sprocket teeth. They are designed to turn freely on the outside of the bushings. The pins are usually free to oscillate in side the bushings. The various components of the roller chains are made from alloy steel, case hardened and ground. Other materials such as brass or bronze are sometimes used in special applications *e.g.* food processing machine or where corrosion is a problem. The load transmitting capacity of a roller chain can be increased by having a longer pitch and larger sprocket diameter. To avoid longer sprocket diameter, multi-row-roller chains or chains with multiple strand width are used. Theoretically, the power capacity of multistrand chain is equal to the capacity of the single chain multiplied by the number of strands, but actually this quantity is usually reduced by 10 per cent. Or more accurately, the power capacity of a duplex, triplex and quadruplex chain may be taken as 1.7, 2.5 and 3.3 times respectively, the power capacity of single chain. A single row chain is called as simple chain. A two row chain is termed as Duplex Chain and a three row chain is called as Triplex Chain and so on. The multiple strands can be upto octoplex by roller chains with strands upto quadruplex are more usual.

The bushed chains are similar in design to other chains, except that they do not have any rollers.

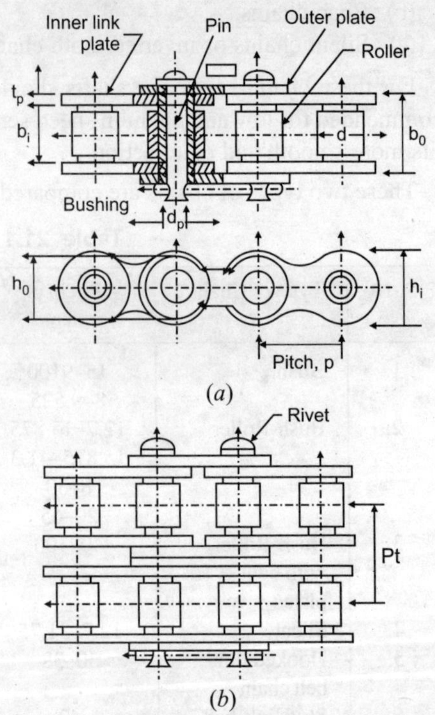

Fig. 21.1 *Construction of bush roller chain*

21.3.2 Silent Chains

These chains are not exactly silent but these are much smoother and quieter in action than a roller chain. The construction detail of a silent chain is shown in Fig. 21.2. The chain capacity is proportional to the chain width which is usually taken as 1½ to 12 times the pitch. In design, the silent chains are more complex than bush roller types, more expensive and require more careful maintenance.

The dimensions of the various parts of the chain are usually given in terms of the pitch, the approximate values being given below:

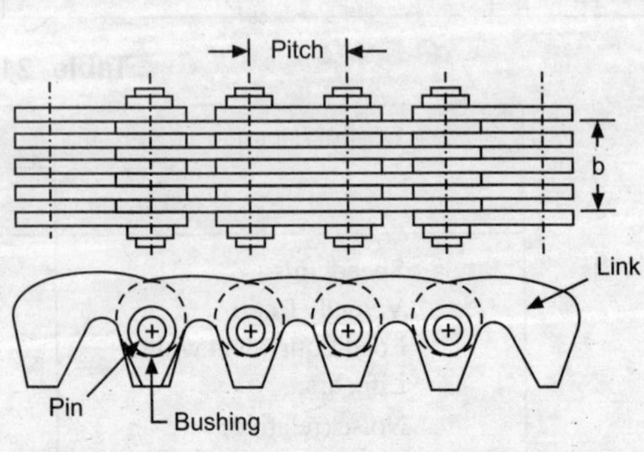

Fig. 21.2 *Construction of a silent chain*

(i) Roller diameter, $d = \frac{5}{8} \times$ pitch

(ii) Pin diameter, $d_p = \frac{5}{16} \times$ pitch

(iii) Chain width, i.e., the distance between the roller link plates,

$$b_i = \frac{5}{8} \times \text{ pitch}$$

(iv) Thickness of link plates, $t_p = \frac{1}{8} \times$ pitch

(v) Width between outer plates, $b_o = b_i + 2t_p$

(vi) Max. height of pin link plates or Coupling link plates, i.e., the outer plates,

$$h_o = 0.82 \times \text{pitch}$$

(vii) Max. height of roller link plates, i.e., the inner plates, $h_i = 0.95 \times$ pitch

(viii) Length of Roller, $l = 0.90\, b_i - 0.15$

21.4 SELECTION OF A CHAIN

For a given application, more than one chain, either roller or silent chain can be selected. Best drive will be determined by the factors: life expectance, space, speed and cost. Choice of a drive will depend upon: pitch, number of chains and sprocket size. The following factors should be analysed while selecting a satisfactory chain drive.

1. Type of Chain: As already mentioned, for low and medium speeds roller chain is recommended, for high speeds the silent chain is preferred.

Chain Velocity and Drive Ratio: Drive ratios greater than 7 : 1 are not generally recommended for roller chains. For greater speed ratios, it is desirable and economical to compound two or more drives. For silent chain a drive ratio of maximum 12 : 1 is recommended. However, it might be more economical to consider compound drives when the ratio is 8 : 1 or larger.

The chain velocity for roller chains can be as high as 20 m/s. However, unless the chain is enclosed and properly lubricated, the normal chain velocities range from 2.5 to 7.5 m/s. Silent chains can be operated upto 40 m/s but the normal chain velocity is of the order of 15 m/s.

2. Basic Parameters and Kinematics: The main parameter of any chain drive is the pitch. Pitch values have been appropriately standardised. The load carrying capacity, that is, the power rating of a chain drive increases with increasing pitch which, however, should be kept within reasonable limits.

The links of a chain running onto sprockets fall into place instantaneously, thus suffering dynamic loads and impacts which speed up their wear. The impacts, the non-uniformity of rotation of sprockets (due to chordal action discussed ahead), and the vibration of chain become more pronounced with increasing pitch (so decreasing number of teeth on the driving sprocket) and rotational speed. It is clear from above, that the pitch, number of teeth on sprockets and the rotational speed are interconnected. So, appropriate limiting values of maximum angular speed for smaller sprocket have been established (both theoretically and experimentally) for various pitch values and number of teeth on the sprocket, as will be explained ahead.

(a) Number of teeth in the sprockets: As explained above, if the number of teeth in the sprocket is very small, that is the pitch is large, it affects badly the chain service life and leads to excessive noise and heavier impact. The maximum number of teeth is limited by the amount of allowable chain stretch.

(i) Minimum number of teeth: Where the mass and size of chain drives are critical, the minimum number of teeth on the smaller sprocket is taken as $z_{\min} = 13$. Most commonly $z_{\min}$ in roller-chain drives, is found by the following relationship:

$$z_{1\min} = 29 - 2 \times V_R$$

It may be increased to z_{1min} = 13 to 15 for $V < 2$ m/s and to z_{1min} = 19 for $V > 2$ m/s. For drives subjected to impact loads in service, the recommended value is z_{1min} = 23. For the sprockets of silent-chain drives, the above values should be increased by 20 to 30%. The number of teeth for the smaller sprocket can also be taken from Table 21.3 depending upon the velocity ratio. As a rule, the selected number of teeth is 1.5 to 2 times the minimum recommended.

Table 21.3 Number of teeth in smaller sprocket

Type of Chain	Number of teeth at velocity ratio					
	1	2	3	4	5	6
Bush-Roller	31	27	25	23	21	17
Silent	40	35	31	27	23	19

(ii) Maximum number of teeth: The design number of teeth on the larger sprocket is dependent upon the velocity ratio,

$$V_R = \frac{w_1}{w_2} = \frac{z_2}{z_1}$$

Note: It should be noted that the speed ratio of chain drives is not expressed in terms of the sprocket pitch diameters:

$$V_R \neq \frac{D_2}{D_1}$$

So, once the desired V_R and z_1 have been determined,

$$z_2 = z_1 \times V_R$$

However, z_2 should not exceed a maximum value of

$$z_{2max} = 100 \text{ to } 120 \text{ for roller-chain drives}$$
$$= 120 \text{ to } 140 \text{ for silent chain drives.}$$

This limit is fixed because otherwise the stretched chain may run off the sprocket or break, as discussed below:

It can be seen that if the chain pitch gets increased by Δp, the displacement of the stretched chain from the centre of the sprocket as compared to the position occupied by a new chain, will be:

$$\Delta D = \frac{\Delta p}{\sin \frac{180}{z}} \quad ...(1)$$

where D = diameter of sprocket

z = number or teeth in the sprocket.

The allowable wear rate, which is defined as : $\frac{\Delta p}{p}$ is limited to 1.25 percent. The maximum allowable value of $\frac{\Delta p}{p}$ can be found experimentally.

Therefore, $$\frac{\Delta D}{p} = \frac{\Delta p}{p} \cdot \frac{1}{\sin \frac{180°}{z}} \leq \text{experimentally found value.}$$

Based on the above analysis, the maximum value of z has been found out to be:

$$z_{max} = 120 \text{ for roller chains}$$

and $$z_{max} = 140 \text{ for silent chains.}$$

The preferred range of number of teeth are:

17, 19, 21, 23, 25, 38, 57, 76, 95, and 114.

(iii) Chordal (Polygonal) Action: When a chain passes over a sprocket, it moves as a series of chords instead of a continuous arc as in the case of a belt drive. Thus the centre line of a chain is not at a uniform radius. This chordal action of the chain is shown in Fig. 21.3. When the driving sprocket moves at a constant speed, the driven sprocket rotates at a varying speed due to the continually varying radius of chain line. This variation in speed ranges from

$$V_{min} = 2\pi n_1 \ R \cdot \cos\frac{180}{z_1}$$

to

$$V_{max} = 2\pi n_1 R$$

where R = pitch radius of the sprocket

n_1 = speed of driving sprocket, rev/min

z_1 = number of teeth on driving sprocket.

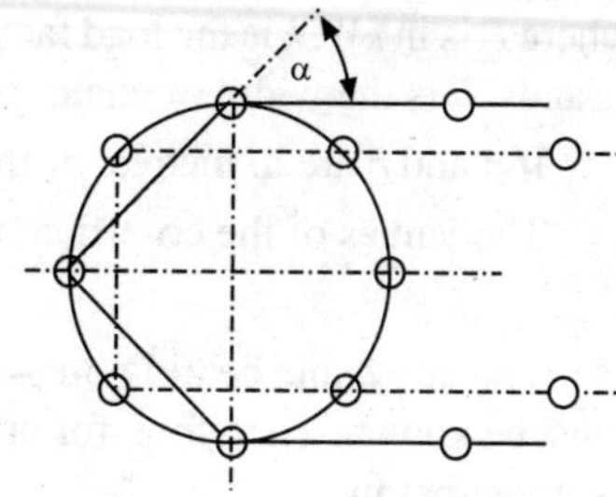

(a) Four tooth sprocket

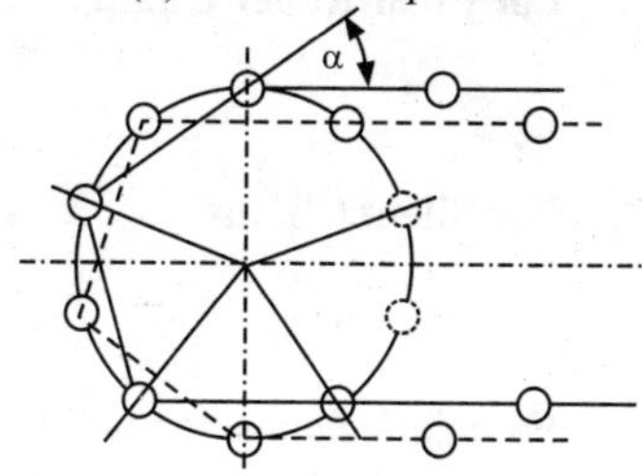

(b) Five tooth sprocket

Fig. 21.3 *Chordal action of chain drive*

It is clear from above that for the same pitch, the variation decreases, if the number of teeth in sprocket is increased. It is also clear from the figure that the angle α, known as angle of articulation decreases as the number of sprocket teeth are increased. Since the rate of wear of the chain pins and bushings is proportional to the angle of articulation, the greater number of sprocket teeth will result in less wear and longer life of the chain. It is recommended that for smoother operation, and longer chain life, the pinion should have 21 or more teeth. It has been found that if the number of teeth is 25 or more, the chordal action becomes negligible. The average speed of the chain sprocket is given as:

$$V = \frac{pnz}{60 \times 1000} \text{ m/s} \qquad ...(2)$$

where p = pitch of chain, mm

Velocity ratio is given: $$V_R = \frac{z_2}{z_1} = \frac{n_1}{n_2} = \frac{w_1}{w_2} \qquad ...(3)$$

n_1, z_1 = speed and number of teeth on smaller sprocket, and

n_2, z_2 = speed and number of teeth on larger sprocket.

The preferred speed ratios or the transmission ratios are given by R_{20} series as : 1, 1.12, 1.25, 1.4, 1.6, 1.8, 2, 2.25, 3.15, 4, 4.5, 5, 5.6, 6.3, 7.1, 8, 9, 10.

(b) Chain Pitch: It is the basic dimension of the chain. The selection of the pitch depends on : power to be transmitted, chain speed, number of revolutions of the smaller chain wheel. Its value is given by the following formulas:

When the centre distance is known, then

$$p = \frac{C}{30 \text{ to } 60}, \text{ for slow speed drives} \qquad ...(4)$$

The pitch is also given as, $p \le 10\,(3640/n_1)^{2/3}$ ...(5)

Pitch can also be determined by the following relations:

For bushed and Roller Chains:

$$p = C_1 \cdot \sqrt[3]{\frac{P \times k}{z_1\omega_1 \cdot p_b \cdot x}} \qquad ...(6)$$

For silent Chains:

$$p = C_2 \cdot \sqrt[3]{\frac{P \times k}{z_1\omega_1 \cdot p_b \cdot b}} \qquad ...(7)$$

where P is in kW, k is the load factor, ω_1 is the angular velocity of the driving sprocket, x is the number of chain strands, b is the width of chain, p_b is the allowable bearing pressure.

If p and b are in metres, p_b in N/m², then

The values of the co-efficients C_1 and C_2 are:

$$C_1 = 28, C_2 = 60$$

The allowable bearing pressure, p_b, is given in Table 21.5. But to find its value from this table, the pitch must be known. Therefore, for preliminary calculations, the allowable value of unit pressure, p_b, can be taken as given below:

For Bush-Roller Chains

$$p_b = 35 \text{ N/mm}^2 \text{ at } n_1 \leq 50 \text{ rev/min. to}$$
$$= 13.7 \text{ N/mm}^2 \text{ at } n_1 = 2800 \text{ rev/min.}$$

For Silent Chains

$$p_b = 20 \text{ N/mm}^2 \text{ at } n_1 \leq 50 \text{ rev/min. to}$$
$$= 7.8 \text{ N/mm}^2 \text{ at } n_1 = 2800 \text{ rev/min.}$$

For silent chains, the width of chain,

$$b = C_3 \times p$$

The value of C_3 ranges from 1.5 to 8.

For preliminary calculations, the recommended value for C_3 = 3 to 4.

The load factor k is given as, $k = k_1 \times k_2$.

k_1, takes into account the nature of load and service conditions, Table 21.4.

Table 21.4 Factor k_1

Daily service of driven Unit	Service factor k_1, for	
	Electric Motor or Steam Turbine	Diesel or Gasoline engine
Even load, 8-hour day	1.0	1.2
Uneven load, 8-hour day, or Even load 16 to 24-hour day	1.2	1.4
Heavy shock load, 8-hour day or Uneven load 16 to 24 hour day.	1.4	1.7

k_2 takes into account the method of lubrication,

= 1.0 for continuous lubrication

= 1.3 for periodic lubrication

After finding approximate values for 'p' from equation (6) or (7) the value of p_b is taken from Table 21.5 and equation (6) or (7) again used to find the exact value for p.

The value obtained by the above formulas or from equations (4) to (5) is rounded off to the standard value.

The standard pitches used are:

6.25, 8, 9.525, 12.7, 15.875, 19.05, 25.4, 31.75, 38.1, 44.45, 50.8, 63.5, and 76.2 mm.

The tentative values of z_1, z_2 and p are adopted as final, if the angular speed n_1 does not exceed the maximum permissible value n_{1max} given in Table 21.6.

Table 21.5 Allowable bearing pressure in chain joints

Speed of smaller sprocket rev/min	p_b, MN/m^2 with chain pitch, p, mm						
	12 to 15.875		19.05 to 25.4		30 to 38.1		40 to 50.8
	a	b	a	b	a	b	a
50	34.3	19.6	34.3	19.6	34.3	19.6	34.3
200	30.9	17.6	29.4	16.7	28.1	16.1	25.7
400	28.1	16.1	25.7	14.7	23.7	13.7	20.6
800	23.7	13.7	20.6	11.8	18.1	10.3	14.7
1200	20.6	11.8	17.2	9.81	14.7	8.43	—

Note: a — for roller and bush chains (z_1 = 15 to 30)
b — for silent chains (z_1 = 17 to 35)

Table 21.6 Maximum speed of smaller sprocket, $n_{1\,max}$

Type of chain	Number of teeth in smaller Sprocket, z_1	Chain pitch, p, mm				
		12	15	20	25	30
Roller Chain	15	2300	1900	1350	1150	1000
	19	2400	2000	1450	1200	1050
	23	2500	2100	1500	1250	1100
	27	2550	2150	1550	1300	1100
	30	2600	2200	1550	1300	1100
Silent Chain	17 to 35	12.7	15.875	19.05	25.4	31.75
		3300	2650	2200	1650	1300

For roller-chains, the maximum chain speed is given as below in Table 21.7.

Table 21.7 Chain speed for roller chains

S. No.	Pitch, mm	Maximum chain Speed, m/s	Maximum n_1
1	9.525	13.7	3500
2	12.7	13.4	2500
3	15.875	12.0	2300
4	19.05	10.9	1800
5	25.4	9.7	1500
6	31.75	8.1	1000
7	38.1	7.3	800
8	44.45	6.5	600
9	50.8	6.0	400
10	63.5	4.7	300
11	76.2	4.0	200

3. Wheel Centres and Length of Chain: The minimum centre distance is selected so as to obtain an adequate angle of wrap on the smaller sprocket for a given velocity ratio. A large angle of wrap or arc of contact assures a more uniform distribution of load on the sprocket teeth and better conditions of engagement. An angle of 90° is regarded as an absolute necessity and 120° is considered desirable.

It is evident that to avoid interference,

The minimum centre distance is given as:

$$C_{min} = C' + (30 \text{ to } 50 \text{ mm}) \text{ for } V_R \text{ upto } 3$$
$$= 1.2\, C', \text{ for } V_R = 3 \text{ to } 4$$
$$= 1.3\, C', \text{ for } V_R = 4 \text{ to } 5$$
$$= 1.4\, C', \text{ for } V_R = 5 \text{ to } 6$$
$$= 1.5\, C', \text{ for } V_R = 6 \text{ to } 7$$

where $C' = \dfrac{D_{a_1} + D_{a_2}}{2}$

D_{a_1} = Addendum, outer or tip diameter of sprocket pinion

D_{a_2} = Addendum, outer or tip diameter of sprocket wheel

Tip diameter may be taken as,

Tip diameter = Pitch diameter + 0.8 × Roller diameter

= Pitch diameter + 0.6 × pitch

From experience, the best results are obtained with

$$C = 30 \text{ to } 50\, p, \text{ with normal value of } 40\, p$$

For highly pulsating loads, $C = 20$ to $30\, p$

When C is more, the chain length becomes excessive and it rapidly stretches and causes jerky operation, therefore, it is recommended that $C \leq 80\, p$, so as to avoid excessive tension due to the self weight of the chain.

The chain length is given in terms of number of links,

$$L = m \times p$$

where m = number of chain links = $\dfrac{2C}{p} + \dfrac{z_2 + z_1}{2} + \dfrac{p(z_1 - z_2)^2}{4\pi^2 C}$

The number of chain links so obtained are approximated to the nearest even number.

Then, the accurate centre distance will be given as:

$$C = \frac{p}{4}\left[m - \frac{z_1 + z_2}{2} + \sqrt{\left(m - \frac{z_1 + z_2}{2}\right)^2 - 8\left(\frac{z_2 - z_1}{2\pi}\right)^2}\,\right] \quad \text{... (8)}$$

For fixed centre distance drives and drives with the centre line inclined at an angle equal to 50 to 70° to the horizontal, the slack (following) side of the chain should have a slight sag, for which purpose the design centre distance given by equation (8) is reduced by 0.2 to 0.4%, or by 2 to 5 mm. However, if the angle of inclination is within 70 to 90°, the value found by equation (8) is left unchanged.

For horizontal drives, the chain sag can be taken as

$$s = 2\% \text{ of } C = 0.02\, C$$

For the transmission inclined at on angle over 20°, $s = (0.01 \text{ to } 0.015)\, C$

For vertical transmission, no slack in the Chain is allowed.

Decrease in the centre distance for an initial sage can be taken as,

$$\Delta C \cong 0.5\, s, \quad \text{or} \quad \cong 0.002\, C \text{ to } 0.004\, C$$

∴ Exact centre distance = $C - \Delta C$

One method of controlling chain tension in assembly and in service can be: Remove one link at $s = 0.71\sqrt{Cp}$. This is for light loads with V upto 5.0 m/s, C up to 30 p and angle of inclination of the drive upto 30°.

Note: With odd z_1 and z_2 and even number of chain links, the sprockets and chain wear out more uniformly.

4. Sprockets: The operation of a chain drive is largely dependent on the quality of the sprockets. Therefore, special attention should be given to : material, manufacture, heat treatment and surface finish of the teeth. The sprockets are made of good grade C.I. and carburised or hardened steel. The shape of the sprocket teeth depends upon the type of chain, the main object is that the teeth may easily disengage from the chain links during operation.

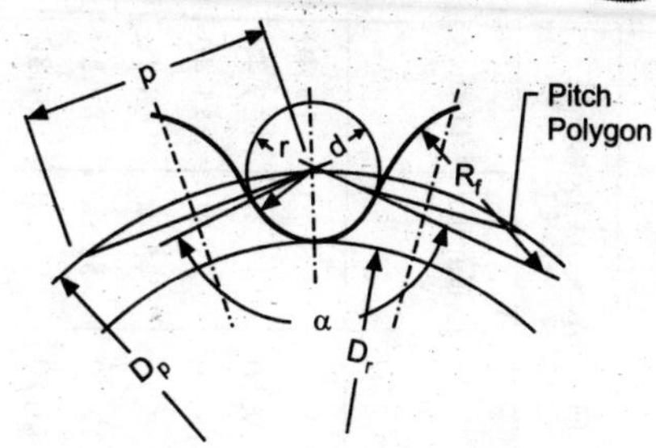

Fig. 21.4 *Form of wheel tooth gap*

The profile of teeth of the sprockets for bush and Roller Chains is: Convex, Concave, rectilinear, and combined rectilinear and convex. For high velocities, the concave profile is preferred due to its greater wear resistance owing to the greater length of the working section of the tooth profile.

The form of the wheel tooth gap is shown in Fig. 21.4 (IS 2403–1975).

Pitch diameter of the sprocket is given as,

$$D_p = \frac{p}{\sin \pi / z} \quad \ldots (9)$$

The other dimensions are given as:

Roller seating radius, $r = 0.505\, d$

Roller seating angle, $\alpha = \left(140^\circ - \frac{90^\circ}{z}\right) \text{to} \left(120^\circ - \frac{90^\circ}{z}\right)$

Outer diameter, $D_o = D_p + d\left(0.6 - \frac{2.56}{z}\right) \text{ to } D_p + d.$

Root diameter, $D_r = D_p - 2r$

Tooth flank radius, $R_f = 0.12\, d\,(z + 2) \text{ to } 0.008\, d\,(z^2 + 180).$

The profiles of the wheel rim are shown in Fig. 21.5 (IS 2403 — 1975).

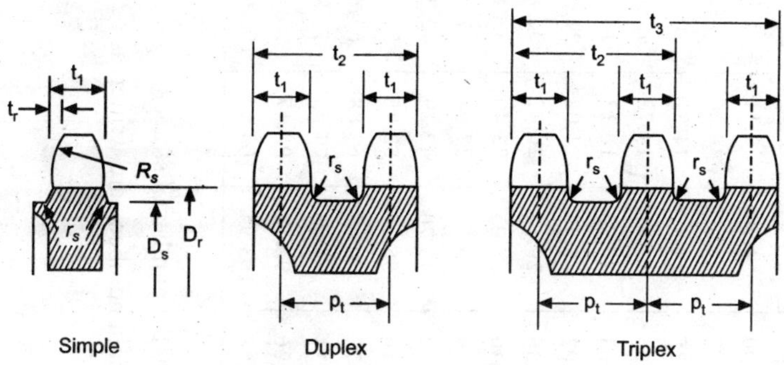

Fig. 21.5 *Wheel rim profiles*

The various dimensions are given below:

Shroud diameter,

$$D_s = p \cot \frac{180^\circ}{z} - 13h_i$$

Min. tooth side radius, $R_s = p$

Tooth width t_1:

(i) For $p \le 12.7$ mm

$t_1 = 0.93\, b_i$ for simple chain wheels.

$= 0.91\, b_i$ for duplex and triplex chain wheels

$= 0.88\, b_i$ for quadruplex chain wheels and above.

(ii) For $p > 12.7$ mm the corresponding values of t_1 are:

$0.95\, b_i$, $0.93\, b_i$ and $0.93\, b_i$

Tooth side relief, $t_r = 0.1\, p$ to $0.15\, p$

Shroud radius, $r_s = 0.76$ mm max

The widths t_2 and t_3 are given as: = (Number of chain strands – 1) × p_t + t_1

where p_t = Transverse pitch of strands or strand spacing.

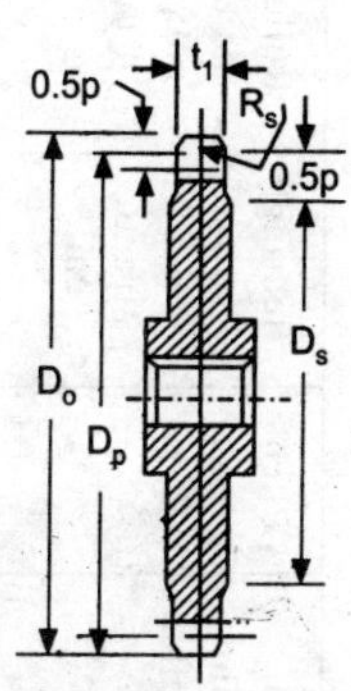

Fig. 21.6 *Sprocket*

Table 21.8 Characteristics of Roller Chains (IS–2403–1975)

ISO Chain Number	Pitch p, mm	Bearing* area per strand, cm^2	Roller dia d max. mm	Pin dia, d_p max. mm	Width between inner plates b_i, mm max.	Width between outer plates b_0, mm. min.	Height of inner plates h_i, mm max.	Height of outer plates h_o, mm max.	Average Weight, N/m			Min. Breaking Load kN		
									Simple	Duplex	Triplex	Simple	Duplex	Triplex
05 B	8.00	0.11	5.00	2.31	3.00	4.90	7.11	7.11	1.8	3.3	5.2	4.60	8.00	11.40
06 B	9.525	0.28	6.35	3.28	5.72	8.66	8.26	8.26	4.1	7.7	10.9	9.10	17.30	25.40
08 B	12.70	0.50	8.51	4.45	7.75	11.43	11.81	10.92	7.1	13.6	20.2	18.20	31.80	45.40
10 B	15.875	0.67	10.16	5.08	9.65	13.41	14.73	13.72	9.1	18.2	27.3	22.70	45.40	68.10
12 B	19.05	0.89	12.07	5.72	11.63	15.75	16.13	16.13	11.7	23.6	35.4	29.50	59.00	88.50
16 B	25.40	2.10	15.88	8.28	17.02	25.58	21.08	21.08	27.0	53.0	78.5	43.10	86.20	129.30
20 B	31.75	2.96	19.05	10.19	19.56	29.14	26.42	26.42	36.5	72.0	119.0	65.80	131.60	197.40
24 B	38.10	5.55	25.40	14.63	25.40	38.05	33.40	33.40	68.5	135.0	200.0	99.80	199.60	299.40
28 B	44.45	7.41	27.94	15.90	30.99	46.71	37.08	37.08	83.0	166.0	249.0	131.60	263.20	394.80
32 B	50.80	8.47	29.21	17.81	30.99	45.70	42.29	42.29	105.0	210.0	315.0	172.40	344.80	517.20
40 B	63.50	12.76	39.37	22.89	38.10	55.88	52.96	52.96	160.0	320.0	480.0	267.70	535.40	803.10
48 B	76.20	20.63	48.26	29.24	45.72	70.69	63.88	63.88	250.0	500.0	750.0	408.30	816.50	1224.70
56 B	88.90	27.91	53.98	34.32	53.34	81.46	77.85	77.85	—	—	—	553.40	1106.80	—
64 B	101.60	36.30	63.50	39.40	60.96	92.15	90.17	90.17	—	—	—	725.80	1451.50	—
72 B	114.30	46.17	72.39	44.48	68.58	103.94	103.63	103.63	—	—	—	916.30	1832.60	—

* For multi-strand chains, multiply this area by the number of strands.

Note: Depending upon the number of strands, the chain numbers are written as : 24B–1, 20 B–2, 16 B–3 and so on.

Its values are:

p 8.00	9.525	12.70	15.875	19.05	25.4	31.75	38.10	44.45	50.80	63.50	76.20
p_t 5.64	10.24	13.92	16.59	19.46	31.88	36.45	48.36	59.56	58.55	72.29	91.27

The sectioned view of a simple sprocket is shown in Fig. 21.6.

5. Chain Designation and Specifications: Chains are designated according to ISO numbers as given in Table 21.8. The chain number is followed by a suffix (1 for simple, 2 for duplex and 3 for triplex chain and so on) and the number of the standard. For example a chain of ISO number 24 B and of triplex construction shall be designated as

Chain 24 B–3IS : 2403–1975

The characteristics of Roller and Bush Chains are given in Table 21.8 and Table 21.9.

Table 21.9 Characteristics of bush chains

Chain number (ROLON)	Pitch, p mm	Width between inner plates b_i, min. mm	Pin dia d_p max. mm	Bush dia max mm	Plate depth max mm	Transverse Pitch p_t mm	Bearing Area cm^2	Average Weight N/m	Min. Breaking load, kN
		SINGLE STRAND							
B35	9.525	4.77	3.59	5.08	8.64	—	0.27	4.00	10.30
B975	9.525	7.55	3.59	5.08	9.19	—	0.39	5.30	13.00
B995	9.525	9.55	4.45	6.00	9.74	—	0.58	6.20	12.00
B995H	9.525	8.00	4.45	6.00	9.74	—	0.58	7.60	15.50
B1511	15.875	11.40	7.00	10.07	16.74	—	1.48	20.80	47.00
		DOUBLE STRAND							
DB35	9.525	4.77	3.59	5.08	8.64	10.15	0.54	7.10	20.50

Table 21.10 gives the characteristics of silent chains.

Table 21.10 Characteristics of silent chains

Pitch, p, mm	Width, b, mm	Breaking Load, kN	Pitch, p, mm	Width, b, mm	Breaking Load, kN
12.7	20.5	20.5	19.5	33.0	49.0
	26.5	26.5		39.0	58.0
	33.0	33.0		45.5	68.0
	39.0	39.0		51.5	77.0
	45.6	45.6		63.0	94.5
	51.5	51.5		78.5	117.0
	63.0	63.0		91.0	136
	78.5	78.5		104.5	156.0
	91.0	91.0		123.0	184.0
	104.5	104.5		135.5	203.0
15.87	26.5	31.8		148.0	222.0
	33.0	39.6		173.0	259.0
	39.0	46.8		198.0	297.0
	45.6	54.6	25.4	51.5	87.0
	51.5	61.8		76.5	130.0
	63.0	75.6		93.0	158.0
	78.5	94.2		101.0	171.0
	91.0	109.2		121.0	205.0
	104.5	125.4		154.5	262.0
	123.0	147.6		179.5	305.0
				204.5	347.0
				253.5	430.0

6. Lubrication: The service life of a roler chain depends essentially on a correct and careful lubrication, which is based upon chain speed as shown in Table 21.11.

In principle, a chain is lubricated most efficiently by oil. In case of very slowly running chain drives and if for constructional reasons, oil lubrication is impossible, a grease lubrication may be chosen.

Table 21.11 Chain lubrication

Chain Speed m/s	Lubrication	
	Favourable	Admissible
to 4	Light dropwise lubrication	Grease lubrication
to 7	Submersion lubrication in oil both	Dropwise lubrication
to 12	Forced lubrication	Oil bath
> 12	Spray lubrication (forced feed)	Forced feed lubrication.

21.5 DESIGN OF THE CHAIN DRIVE

The important parameters in chain drive design are : wear, vibrations and impact. Design calculations are carried out to find the optimal proportions of chain and sprocket at which the drive will be able to operate under the assigned duty without risk of damage. The design is carried out in the following steps:

1. Select the type of chain.
2. Find the number of teeth of the smaller sprocket from Table 21.3.
3. The chain pitch is selected from formulas (4) (5) (6) or (7). With selected values of z_1 and p, the value of $n_{1\,max}$ is to be satisfied as given in Table 21.6. If n_1 is given, then pitch can be selected from Table 21.6, after z_1 has been selected.
4. Total load on the driving side of the chain is determined and factor of safety found out. Total force is:

$$F_0 = F + F_c + F_f$$

where F = driving force = Tension on tight side, tension on slack side is considered zero.

$$= \frac{\text{Power in kW} \times 1000}{V}, \text{N}$$

where V = Chain velocity in m/s and F_c = Chain tension due to centrifugal force of inertia

$$= \frac{wV^2}{g}$$

where w = weight per metre of chain length, Tables 21.8 and 21.9.

F_f = Tensioning due to sagging

$= k_f \cdot w \cdot C$

where k_f = Coefficient to take into account the arrangement of chain drive.

k_f = 1.0, for vertical position.

= 2.0, for centre line inclined at an angle upto 45°.

= 4.0, for Horizontal drives.

Then $$F_S = \frac{F_u}{F_0}$$

where F_u = Breaking strength of chains.

$= 106\, p^2 N$, p in mm for roller chains

$= 106\, pN$ per mm width of chain for silent chains, p in mm

or from Tables 21.8 and 21.9.

The values of factor of safety are given in Table 21.12.

Table 21.12 Factor of safety for chain drives

Type of Chain	Speed of Pinion, rev/min								
	upto 50	200	400	600	800	1000	1200	1600	2000
Bush Roller Chains:									
p = 12.7 – 15.875 mm	7	7.8	8.55	9.35	10.2	11.0	11.7	13.2	14.8
= 19.05 – 25.4 mm	7	8.2	9.35	10.3	11.7	12.9	14.0	16.3	—
= 31.75 – 38.1 mm	7	8.55	10.2	13.2	14.8	16.3	19.5	—	—
Silent Chain:									
p = 12.7 – 15.875 mm	20	22.2	24.4	28.7	29.0	31.0	33.0	37.8	42.0
= 19.5 – 25.4 mm	20	23.4	26.7	30.0	33.4	36.8	40.0	46.5	53.5
= 31.75 mm	20	25	32	36.5	41	46	51	—	—

5. Then the chain is checked for wear by unit pressure in the joints given as:

$$p_b = \frac{F \cdot k}{A} \text{ N/mm}^2$$

where k = load factor; A = Projection of the joint bearing surface

For roller chains, $A = (d_p \times l_b) \times$ number of chain strands, or refer Table 21.8.

For silent chains, $A = 0.76\, d_p b$

where d_p = diameter of joint pivot; b = chain width; l_b = bush length = $(b_i + 2t_p)$

Check the value of p_b with values given in Table 21.5.

6. The force acting on the drive shaft is,

$$F_{sh} = F + 2k_f \cdot w \cdot C \qquad \text{...(10)}$$

21.5.1 Good Design Practice

For maximum operating life and efficiency, the following points should be considered.

(i) Small sprocket should have at least 23 teeth.

(ii) Smoothest, quiestet and economical operation will result with the smallest pitch chain, which can transmit the power required.

(iii) There should not be more than 100 teeth in large sprocket.

(iv) Desirable speed ratio range is not over 6 : 1 and in no case a ratio of 9 : 1 be exceeded.

(v) Unless smallest possible chain size is desired, a single strand chain drive is usually preferred to a drive of multiple strands.

(vi) Proper lubrication will assure maximum chain life.

Note: A chain is considered to have completed its service life when its permanent extension reaches about 3% of its original length.

Example 21.1. A compressor is to be actuated from a 10 kW electric motor. The speed of motor shaft is 970 rev/min and that of the compressor is to be 330. The compressor operates in two shifts. The minimum centre distance should be 550 mm. Calculate a suitable chain drive.

Solution:

1. Since the speed of the driver is not very high, we can presume that chain velocity will be low to average and so a roller chain is selected, being simpler and cheaper.

2. Velocity ratio is:

$$V_R = \frac{n_1}{n_2} = \frac{970}{330} \cong 2.94$$

From Table 21.3, for velocity ratio of 3, the number of teeth in the pinion is 25.

$\therefore \quad z_1 = 25$

$\therefore \quad z_2 = 25 \times \dfrac{970}{330} \cong 74$

which is well within the maximum limit for roller chains.

3. The chain pitch can be determined by the formula (4), as

$$p = \frac{C}{(30 \text{ to } 60)}$$

Now $\quad C = 550$ mm

$\therefore \quad p = \dfrac{550}{(30 \text{ to } 60)}$, i.e., 18.33 to 9.17 mm

Let us take the standard pitch as $p = 12.7$ mm

with $z_1 = 25$mm and $p = 12.7$ mm, $n_{1\,max} \cong 2500$ rev./min. (Table 21.6). Given $n_1 = 970$ rev/min. So, selected values of z_1 and p are O.K.

$\therefore$ From Table 21.8 let us select Roller chain with number 08 B.

$\therefore$ Chain velocity is,

$$V = \frac{p \cdot n_1 \cdot z_1}{60 \times 1000}, \text{ m/s} = \frac{12.7 \times 970 \times 25}{60 \times 1000} = 5.12 \text{ m/s}$$

4. Total load on the driving side of chain is,

$$F_0 = F + F_c + F_f$$

Now $\quad F = \dfrac{\text{Power in kW} \times 1000}{V}, N = \dfrac{10 \times 1000}{5.12} = 1960 \text{ N}$

$$F_c = \frac{wV^2}{g}$$

From Table 21.8, $w = 13.6$ N/m for a Duplex chain. A duplex chain is selected because from Table 21.12, the factor of safety for Roller chain with $p = 12.7$ mm and Pinion rev/min of 970 is about 11. Therefore, even by considering F only, the breaking load should be

$$F_u = 11 \times 1960 = 21.56 \text{ kN}$$

Therefore, a single chain will not be suitable for which

$$F_u = 18.20 \text{ kN}$$

$\therefore \quad F_c = \dfrac{13.6 \times (5.12)^2}{9.81} = 36.3 \text{ N}$

Now $\quad F_f = k_f\, w\, C$

If horizontal drive is assumed $k_f = 4.0$

$\therefore \quad F_f = 4 \times 13.6 \times 0.55 = 29 \text{ N}$

$\therefore \quad F_0 = 1960 + 36.3 + 29 = 2025.3 \text{ N}$

From Table 21.8, $F_u = 31.80$ kN

$\therefore \quad F_S = \dfrac{31.80 \times 1000}{2025.3} = 15.7 > 11. \quad \therefore$ O.K.

$\therefore$ Chain selected is : 08 B2 – IS : 2403.

5. **Check for Wear:**

$$pb = \frac{F \times k}{A} \text{ N/mm}^2$$

$$k = k_1 \times k_2$$

From Table 21.4, $k_1 = 1.2$.

For continuous lubrication, $k_2 = 1.0$

Now $$A = (d_p \times l_b) \times 2$$

Now $$d_p = \frac{5}{10} \times p = \frac{5}{16} \times 12.7 \cong 4$$

$$l_b = b_i \times 2t_p$$

Now $$b_i = \frac{5}{8} \times p \cong 8 \text{ mm}$$

$\therefore$ $$l_b = 8 + 3.2 = 11.2 \text{ mm}$$

$\therefore$ $$A = 4 \times 11.2 \times 2 = 89.6 \text{ mm}^2$$

$$P_b = \frac{1960 \times 1.2}{89.6} = 26.2 \text{ N/m}^2 \text{ which is quite O.K. (See Table 21.5)}$$

If we take the standard dimensions from Table 21.8,

$$d_p = 4.45\text{mm},\ b_i = 7.75\text{mm}$$

$$A \text{ per stand} = 0.50 \text{ cm}^2$$

$\therefore$ $$p_b = \frac{1960 \times 1.2}{0.50 \times 2} = 23.52 \text{ N/mm}^2$$

6. Now length of chain

$$L = m \times p$$

$$m = \frac{2C}{p} + \frac{z_2 + z_1}{2} + \frac{p(z_2 - z_1)^2}{4\pi^2 \times C} = \frac{2 \times 550}{12.7} + \frac{74 + 25}{2} + \frac{12.7(74 - 25)^2}{4\pi^2 \times 550} = 137.3 \text{ pitches}$$

The nearest even number is 138

$\therefore$ $L = 138 \times 12.7 = 1750$ mm

$\therefore$ Exact centre distance is

$$C = \frac{p}{4}\left[m - \frac{z_1 + z_2}{2} + \sqrt{\left(m - \frac{z_1 + z_2}{2}\right)^2 - 8\left(\frac{z_2 - z_1}{2\pi}\right)^2}\right]$$

$$= \frac{12.7}{4}\left[138 - \frac{25 + 74}{2} + \sqrt{\left(138 - \frac{25 + 74}{2}\right)^2 - 8\left(\frac{74 - 25}{2\pi}\right)^2}\right] = 553 \text{ mm}$$

To allow for chain sag, let

$C = 550$ mm.

7. Dimensions of the Chain Components:

(i) Roller diameter, $d = \frac{5}{8} p \cong 8$ mm

(ii) Pin diameter, $d_p = 4$ mm

(iii) Chain width, $b_i = 8$ mm

(iv) Thickness of link plates, $t_p = 1.6$ mm

(v) Max. height of the outer plates, $h_o = 0.82 \times p = 0.82 \times 12.7 = 10.4$ mm

(vi) Max. height of the inner plates, $h_i = 0.95 \times p = 0.95 \times 12.7 \cong 12$ mm

(vii) Length of Roller, $l = 0.9\, b_i - 0.15 = 7.2 - 0.15 = 7.05$ mm

These are approximate dimensions. For exact dimensions, refer Table 21.8.

With the help of these dimensions, the chain can be drawn.

8. Form of wheel tooth gap : Refer Fig. 21.4.

(i) Pitch circle diameter of sprocket pinion

$$D_p = \frac{p}{\sin \frac{\pi}{z_1}} = \frac{12.7}{\sin \frac{180}{25}} \cong 101 \text{ mm}$$

(ii) P.C.D. of sprocket wheel = 297 mm

(iii) Roller seating radius, $r = 0.505\, d = 0.505 \times 8 = 4.04$ mm

(iv) Let outer diameter of sprocket pinion, $D_0 = D_p + 0.8\, d = 107.4$ mm

(v) Outer diameter of sprocket wheel = 297 + 6.4 = 303.4 mm

(vi) Root diameter of sprocket pinion = $D_r = D_p - 2r = 101 - 8.08 = 92.92$ mm.

(vii) Root diameter of sprocket wheel = 288.92 mm

(viii) Roller seating angle of pinion, $\alpha = \left(140^\circ - \frac{90^\circ}{25}\right) \text{to} \left(120^\circ - \frac{90^\circ}{25}\right)$

say = 125°

(ix) Roller seating angle of wheel, $= \left(140^\circ - \frac{90^\circ}{74}\right) \text{to} \left(120^\circ - \frac{90^\circ}{74}\right)$

say = 125°

(x) Let tooth flank radius of pinion, $R_f = 0.12\, d\, (z + 2) = 0.12 \times 8 \times 27 = 25.9$ mm

(xi) Tooth flank radius of sprocket wheel = 73.8 mm

With the help of above dimensions, the form of wheel tooth gap can be drawn.

9. Wheel Rim Profiles : Refer Fig. 21.5.

(i) Shroud diameter for pinion, $D_s = p \cot \frac{180^\circ}{z_1} - 1.3\, h_i$

$$= 12.7 \times \cot \frac{180^\circ}{25} - 1.3 \times 12 = 84.6 \text{ mm}$$

(ii) Shroud diameter of wheel = 287.4 mm

(iii) Min. tooth side radius, $R_s = p = 12.7$ mm

(iv) Tooth width, $t_1 = 0.91 \times b_i = 0.91 \times 8 = 7.28$ mm

(v) Tooth side relief $t_r = 0.1 \times p$ to $0.15\, p$, say 1.5 mm

(vi) Shroud radius, $r_s = 0.76$ mm

Hence, the rim profile can be drawn.

EXERCISES

1. Enumerate the advantages and disadvantages of chain drive.
2. Write the applications of chain drive.
3. Classify the chains.
4. What are the minimum and maximum number of sprocket teeth for roller chains and silent chains?
5. What is the normal drive ratio for roller chain drive and silent chain drive?
6. What are the normal and maximum chain velocities for roller and silent chain drives?
7. What is the normal centre distance for a chain drive?
8. Explain the chordal action of a chain drive.
9. A roller chain is to be used in a power plant to transmit 180 kW from an engine to a generator. The generator speed is 525 rev/min, and the engine speed is approximately 75 rev/min. Determine the size of the sprockets, the pitch and number of strands required.

10. Design a roller chain drive to transmit power from a 15 kW motor to a reciprocating pump. The pump is to operate continuously 24 hr per day. The speed of the motor is 570 rev/min and that of the pump is 200 rev/min. Decide upon the number of teeth on each sprocket. Find the pitch and width of the chain.
11. It is desired to transmit 45 kW with a roller chain at a sprocket speed of 720 rev/min. The pitch diameter of the smaller sprocket must not exceed 168 mm. Determine the pitch and width of chain to transmit the power continuously under pulsating loads.
12. A roller chain operating under steady load conditions, transmits 4 kW from a shaft rotating at 600 rev/min to one operating at 750 rev/min.
    (*a*) Determine the chain required using at least 15 teeth in the sprockets.
    (*b*) Determine the sprocket pitch diameters.
    (*c*) Determine the shortest advisable centre distance.
    (*d*) Determine the number of links of chain required.
13. A 7.5 kW, 1200 rev/min motor drives a line shaft at 250 rev/min. The shaft centre distance is to be approximately 0.60 m. The motor shaft has a diameter of 38 mm. The starting torsional moment of the motor is from 1.75 to 2.00 times the running torsional moment. The load applied is with moderate shock.
    (*a*) Select a roller chain for this drive.
    (*b*) Determine the sprocket pitch diameters.
    (*c*) How many chain links are required?
    (*d*) What is the exact centre distance?
14. A silent chain operating under good service conditions transmits 75 kW from a 600 rev/min motor to a shaft running at 167 rev/min.
    (*a*) Determine the suitable number of teeth for each sprocket.
    (*b*) Determine the pitch and width of chain.
15. A truck equipped with a 37.5 kW engine uses a roller chain as the final drive to the rear axle. The driving sprocket runs at 225 rev/min and the driven sprocket at 100 rev/min with a centre distance of approximately 0.92 m. The chain speed is to be approximately 2.5 m/s. The transmission efficiency between the engine and the driving sprocket is 85 percent.
    (*a*) Determine the pitch and type of chain to be used.
    (*b*) Determine the number of teeth in each sprocket and the pitch diameters.
16. A bucket elevator is to be driven by a gear motor and roller chain drive. Gear motor = 4 kW; Speed of Gear motor = 100 rev/min; Bucket elevator speed = 55 rev/min; Load = even, centre distance = 1220 mm. Elevator will operate 16-24 hrs per day. Select the chain.

Timing-Belt Drives: These drives which are also known as positive belt drives, synchronous belt drives or toothed belt drives transmit motion without slipping. A timing belt drive, Fig. 21.7 combines the advantages of belt drives with those of chains and gears, but it is more compact than ordinary belt drives and is smoother and less noisy in operation and requires no maintenance and lubrication as compared with chain drives.

These drives are used to transmit a power of upto 100 kW. Velocities can be from 5 to 50 m/s, *VR* upto 12 and maximum efficiency of 0.98.

A typical timing belt consists of a load carrying element 1, Fig. 21.7, composed of metal cables embedded into a matrix of oil-resistant rubber or plastic 2, and can, therefore, work in a pool of oil. The belt has axial cogs or teeth moulded on the underside which fit into grooves on the pulley. The cogs prevent slippage and make accurate timing possible. For better wear resistance, the cogs are covered with nylon or similar fabric.

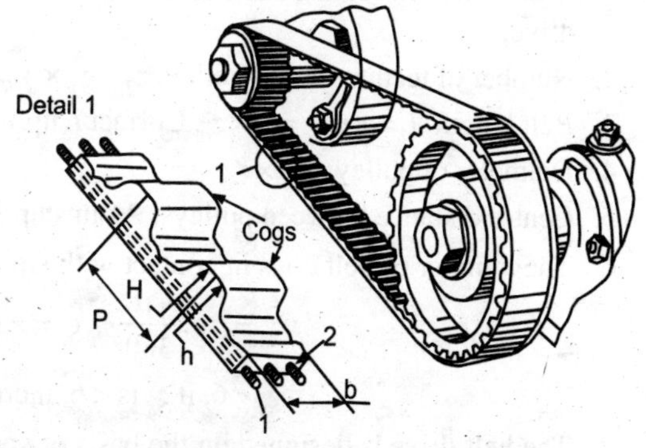

Fig. 21.7 *Timing belt drive*

The main parameter of a timing-belt drive is the tooth module, *m*, (pitch diameter of the pulley divided by the number of teeth) or the pitch of the cogs. All other parameters are expressed in terms of *m*, Table 21.13. Cog tooth included angles is 40°.

Table 21.13 Parameters of Timing Belts

Parameters	Module m, mm		
	3	5	7
Torque at smaller pulley, M_{t_1}, Nm	2.4	49	190
Specific tangential force, P_0, N/mm	10	35	45
Rotational speed of input shaft, n_1, min^{-1}	1000–8000	1000–5000	1000–4000
No. of teeth on smaller pulley, z_1	14–20	18–24	22–36
No. of belt cogs, z_c	36 – 160	45–140	45–125
Belt width, b, mm	12.5 – 25	25–50	50–80
Belt thickness H, mm	4	6	8
Cog pitch, p, mm	9.42	15.71	21.99
Tooth depth h, mm	1.8	3	4.2

As per BIS (IS: 9884–1981), the proportions of timing belts are as follows:

Table 21.14 Specifications of a timing belt

Pitch Code	Pitch, mm	Standard Width, mm
XL	5.080	6.4, 7.9, 9.5
L	9.525	12.7, 19.1, 25.4
H	12.700	19.1, 25.4, 38.1, 50.8, 76.2
XH	22.225	50.8, 76.2, 101.6
XXL	31.750	50.8, 76.2, 101.6, 127.0

Pulleys: Timing belt pulleys are similar to uncrowned belt pulleys, except that the grooves for the belt's teeth are cut in the pulley's face parallel to the axis. One of the pulleys in a drive (usually the smaller one) has flanges 1.5 to 4 mm deep to prevent the belt from running off the rim. The material of the pulleys can be: C.I., steel, light alloys or plastic. The teeth are shaped by generating, or, these may also be precision cast needing no further machining.

Design Analysis:

Steps:

1. Power to be transmitting will be given. So determining M_{t_1}, the module can be selected from Table 21.13. In case of a more than one choice, go in for smaller value of m, to get overall smaller dimensions for the drive,
2. Number of teeth or larger pulley, $z_2 = z_1 \times V_R$
3. Belt length, $L = \pi m z_c = p \times z_c$. Correct L upto 0.1 mm
4. Diameter of pulley $= m \times z$
5. Centre distance between pulleys: As in step 8, Art. 19.3.6.
6. The number of belt cogs in contact with smaller pulley,

$$z_s = z_1 \cdot \frac{\alpha}{360°}, \quad \alpha = \text{Angle of contact at the smaller pulley.}$$

$z_s \not< 6$, if z_s is < 6, increase the centre distance.

7. The belt drive is designed on the basis of specific tangential force, Table 21.13

8. Under given service conditions, the allowable specific tangential force is given as:

$$[P] = P_0 \times C_s \times C_o \times C_z$$

where C_s is the service factor, given as:

Table 21.15 Service factor, C_s

Load conditions	Overloal in start up %	C_s
Steady operation	upto 120	1.0
Moderate vibration	upto 150	0.9
Considerable vibration	upto 200	0.8
Variable impact load	upto 300	0.7

C_0 accounts for the effect of belt tighteners
= 0.9 for one belt tightener
= 0.8 for two belt tighteners

C_z = 1, 0.8 and 0.6 for z_s = 6, 5, 4, respectively.

The necessary belt width is determined by calculations.

❑❑❑

CHAPTER 22

Gear Drives

22.1 GENERAL

Gears are the most common means of transmitting power in the modern mechanical engineering world. They vary from a tiny size used in watches to the large gears used in marine speed reducers, bridge lifting mechanisms and railroad turn table drives. They form vital elements of main and ancillary mechanisms in many machines such as automobiles, tractors, metal cutting machine tools, rolling mills, hoisting and transmitting machinery, marine engines, etc. Toothed gears are used to change the speed and power ratio as well as direction between an input and output shaft.

The advantages of gear drives over other transmission means are:

1. They give positive drives and constancy of speed ratio without any slippage.
2. The drive is very compact due to short centre distances used in such drives.
3. High efficiency, reliable service and simple operation.
4. They are capable of driving loads subjected to shock at speeds upto 20 m/s.
5. Maintenance is inexpensive and if properly lubricated and operated, gear drives have the longest service life as compared to other mechanical drives.
6. Gear drives can be used where precise timing is desired.
7. Gear drives can drive much heavier loads than other drives due to their unlimited sizes.
8. Gear drives can be used for a wide range of transmitted power, i.e., from one-tenths of a kW to tens of thousands of kW.

On the other hand, the gears have got the following disadvantages:

1. The manufacture is complex. Special tools and equipment are needed to manufacture them.
2. Due to errors and inaccuracy in their manufacture, the drive may become noisy accompanied by vibrations at high speeds.
3. They are not suitable for large centre distances because the drive will become bulky.

22.2 CLASSIFICATION OF GEARS

The gears can be classified in the following ways:

1. Depending on the relative position of the geometrical axes of the driving and driven shaft:

(a) **Parallel Shaft:** The following types of gears are used between parallel shafts:

(i) **Spur gears:** In these gears the teeth are straight and parallel to the shafts axis. Spur gears are the most common type and their advantages are simplicity in design, economy of manufacture and maintenance and absence of end thrust on the bearing. They impose only radial loads on the bearings. These are usually known as slow speed gears but if noise is not a serious design problem, these can be used at almost any speed which can be handled by other types of gears. The most usual arrangement of spur gears is an external gear and pinion combination, but if centre distance is to be reduced, then internal gear and external pinion combination is used.

(ii) **Helical gear:** In these gears, the teeth are inclined to the axis of the shafts and are in the form of helix. These gears are usually thought of as high speed gears and they can take higher loads as compared to equal sized spur gears. The motion is smoother and quieter than the spur gears. Single helical gears impose both radial loads and thrust loads on their bearings and so require the use of thrust bearings. The helix angle must be the same on both gear and pinion, but the hand of helix on the pinion is opposite to that on the gear, i.e., a right hand pinion meshes with a left hand gear and a left hand pinion with a right hand gear.

(iii) **Double helical gears:** These gears have two sets of opposed helical teeth. With this arrangement, the thrusts are counterbalanced and there is no thrust loading on the bearings.

(iv) **Herringbone gears:** These gears are essentially the same as the double helical gears but in these gears there is no space separating the two opposed sets of helical teeth.

(b) **Intersecting Shafts:** The following types of gears are used between intersecting shafts:

(i) **Straight Bevel gears:** Such gear can be used on shafts at any angle, but 90° is the most usual. The blanks of these gears are conical in shape. The teeth are tapered in both thickness and tooth height. At one end, the tooth is small while at the other end it is large. The dimensions of the tooth are usually specified at the large end. Bevel gears are designed in pairs and hence are not interchangeable.

(ii) **Spiral Bevel gears:** In these bevel gears, the teeth are curved and oblique. Such gears run more quietly and can take up more load as compared to straight bevel gears, but the thrust loads are greater.

(iii) **Zerol Bevel gears:** These gears are similar to straight bevel gears but the teeth are curved in the lengthwise direction. These curved teeth are arranged in such a manner that the effective spiral angle is zero.

(iv) **Face gears:** In these gears, the teeth are cut on the flat face of the blank. They mesh at right angles with spur or helical pinions. They are called on centre face gears when the shafts intersect.

(c) **Non-parallel, Non-intersecting Shafts:** The axes of such shafts cross in space. The shafts which lie in parallel planes may be skewed at any angle between zero and 90°. The following types of gears are used between such shafts:

(i) **Spiral gears:** These gears are also known as crossed helical gears. These gears are mounted on shafts that do not intersect and that are at an angle to each other. The angle is usually 90°.

(ii) **Hypoid gears:** These gears are similar to spiral bevel gears with the difference that the axes of the shafts do not intersect. The distance between a hypoid pinion axis and the axis of the hypoid gear is called the "offset". If the offset is zero, hypoid gears become spiral bevel gears. Shaft angle is usually 90° but other angles are also possible. Because the shafts do not intersect, bearings can be provided on both sides of the gear which will result in extra rigidity. High reduction hypoid gears permit ratios between 10 : 1 and 120 : 1 and even as high as 360 : 1 in fine pitches.

(iii) Worm gears: In such gears, one gear has screw threads. Due to this factor, they are quiet, vibration free and give a smooth output. Worm and worm gear shafts are almost invariable at right angles.

Other gears used between these shafts are : Convex worm gearing, cone drive gearing, face gears (offset), spiroid gears and Helicon gears, Beveloid gears.

The comparison of single mesh gears is given in Table 22.1.

Table 22.1 Comparison of single mesh gears

Type of Gearing	Gear-Ratio Range	Efficiency at rated power %	Maximum Pitch line velocity m/s
(i) Spur	1 to 10	98	10
(ii) Helical and herring bone	1 to 15	98	25
(iii) Helical and Double helical high speed	1 to 15	98	150
(iv) Crossed helical	1 to 10	98	20
(v) Straight Bevel	1 to 6	98	5
(vi) Spiral Bevel	1 to 9	98	40
(vii) Zerol	1 to 9	98	20
(viii) Hypoid	1 to 9	98	20
(ix) High reduction Hypoid	10 to 90	80	
(x) Worm	1½ to 90	50 to 90	30
(xi) Double enveloping worm	3½ to 90	50 to 98	20
(xii) Face	3 to 8	95 to 99	20
(xiii) Spiroid	10 to 100	50 to 97	30
(xiv) Helicon	3 to 100	50 to 98	30
(xv) Beveloid	1 to 100	50 to 95	20

2. Depending upon the number of steps:

(i) One step gears; (ii) Two step gears; (iii) Three step gears and so on.

3. Depending upon the housing design:

(i) **Open drives:** The gear drive is without a casing and is subjected to the action of dust and dirt.

(ii) **Closed drives:** The gear drives are enclosed in special casings and are protected against dirt and dust and are properly lubricated.

4. Depending upon the peripheral velocity:

(i) **Low velocity:** V is < 3 m/s

(ii) **Medium velocity:** V = 3 to 15 m/s

(iii) **High velocity:** V is >15 m/s

5. Depending upon the accuracy of manufacture: According to International standards, there are 12 degrees of accuracy of manufacture of gears, 1 being the highest.

22.3 BASIC RACK

It is defined as a rack whose teeth are proportioned according to a standard tooth form. The proportions and form of the teeth of wheels when they are cut, is determined by the basic rack. It is very clear that toothed pairs made from different wheels must be identical with the form of the basic rack. The profile and proportions of the basic rack in terms of module have been standardised (IS 2535–1978). Figure 22.1 shows the basic rack for straight tooth spur gears. The line *a–a*, which is a datum line for constructing the profile and where the thickness of the tooth is equal to the space width between the teeth, is known as "Pitch Line". The distance between the corresponding sides of two adjacent teeth of a gear measured on the "pitch line" is called the "circular pitch", p_c.

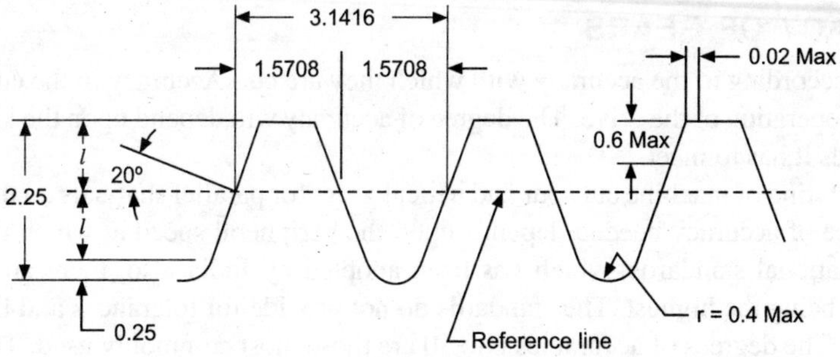

Fig. 22.1 *Basic rack for spur gears*

$$p_c = \pi m$$

where, *m* is module

a = addendum = distance from the pitch line to the tip of the tooth = m

d = dedendum = distance from the pitch line to the root of the tooth = 1.25 m.

Thickness of tooth = Space width = $\frac{P_c}{2}$ = 1.57 m

It is clear from above that dedendum is always greater than addendum, so that when the tooth of the mating gear goes into the space width of the other gear, its tip should not touch the root of the other gear and there is always some clearance.

Clearance = 0.25 m

∴ Working depth = $2a = 2m$

Now '*m*' is defined as the 'module' and it is given as the ratio of the pitch diameter of the wheel to the number of teeth on the wheel. Tne "pitch diameter" is the diameter of the wheel corresponding to the pitch line.

Half the angle between the sides of the basic rack tooth is called the "Profile angle".

The proportions of the basic rack for Straight Bevel gears are :

$$p_c = \pi m\text{; thickness of tooth = Space width} = \frac{p_c}{2} = \frac{\pi}{2} m$$

$$a = m;\ d = 1.2\ m : \text{clearance} = 0.2\ m$$

Working depth = 2*m*.

In actual practice, for normal functioning of the mating gears, and to allow for thermal and centrifugal expansion, the tooth is made somewhat thinner, to provide easier engagement of the corresponding tooth space. Hence, the tooth thickness is less than the width of space, the difference is known as "Backlash".

Sometimes the tip and the root of the tooth profile are relieved. This will ensure sliding into and out of mesh without knocking and also will take care of any misalignment.

The modules have been standardised and the Indian Standards give the following modules—Table 22.2 (IS 2535-1978).

Table 22.2 Recommended series of modules, mm

Preferred	Choice 2	Choice 3
1, 1.25, 1.5, 2, 2.5, 3, 4, 5, 6, 8, 10, 12, 16. 20, 25, 32, 40, 50.	1.125, 1.375, 1.75, 2.25, 2.75, 3.5, 4.5, 5.5, 7, 9, 11, 14, 18, 22, 28, 36, 45.	3.25, 3.75, 6.5

22.4 ACCURACY OF GEARS

The gears are rated according to the accuracy with which they are cut. Accuracy in the cutting of the gears is essential for normal operation of the drive. The degree of accuracy will depend upon the kind of service of the gear and the demands it has to meet.

The method for rating of machine cut spur and helical gears for parallel shafts, is on the basis of degree of accuracy. The degree of accuracy needed depends upon the peripheral speed at which the gear is to rotate. According to international standards, which has been adopted by India also, there are twelve degrees of accuracy, 1 to 12, 1 being the highest. The standards do not provide for tolerances and limits in the degrees of accuracy 1, 2, 12. The degrees of accuracies 6 to 10 are those most commonly used. Three types of norms have been established for each degree of accuracy:

(a) the kinematic accuracy of the wheel.
(b) its smooth operation (c) contact between the teeth.

Table 22.3 gives the relationship between the peripheral speeds of gears and degrees of accuracy. Table 22.4 gives the values of backlash for gears.

Table 22.3 Peripheral speed of gears

Degrees of Accuracy	Peripheral speed of Gears, m/s
1–6	> 15
6–7	7.5–15
7–8	5–7.5
8–9	2.5–5
10–12	Below 2.5

The standards also provide for 12 degrees of accuracy for Bevel gears. The most frequently used degrees are 7, 8, 9, 10 and 11. The standards don't provide for limits and tolerances for degrees 1–4 and 12.

Table 22.4 Backlash for gears, mm

Peripheral Speed, < 8 m/s			Peripheral Speed >8 m/s	
	BACKLASH, mm			
Module, mm	Minimum	Maximum	Module, mm	Backlash, mm
20	0.75	1.25	8	0.40
16	0.50	0.85	7	0.38
12	0.35	0.60	6	0.36
10	0.30	0.51	5	0.28
8	0.22	0.40	4	0.23
6	0.20	0.33	3.5	0.22
5	0.15	0.25	3	0.21
4	0.13	0.20	2.75	0.20
3	0.10	0.15	2.5	0.19
2.5	0.08	0.13	2	0.18
2	0.08	0.13		
1.5 and finer	0.00	0.10		

To select a correct degree of accuracy for gears for various machines, the guidelines given in Table 22.5 may be followed.

Table 22.5 Degrees of accuracy for various machines

Type of Machine	Degree of Accuracy
(1) Reduction gears in turbines and turbo machines	3–6
(2) Metal cutting machine tools	3–8
(3) Automobiles	5–8
(4) Trucks	7–9
(5) Tractors	8–10
(6) General purpose reduction gears	6–9
(7) Gear wheels of rolling mills	6–10
(8) Mine winders	8–10
(9) Crane mechanisms	7–10
(10) Farm Machinery	8–11

22.5 MULTI-MESH GEARS

Gear drives of higher speed ratios and greater versatility can be obtained by the combination of two or more gear pairs on parallel and non-parallel, and on intersecting and non-intersecting shafts. The speed ratio which is defined as the ratio of the number of revolutions of the driver to the number of revolutions of the driven gear, can be written in the following manner for a single mesh gear drive :

$$V_R = \frac{\omega_1}{\omega_2} = \frac{n_1}{n_2} = \frac{D_2}{D_1}$$

Now from the definition of "pitch" of gear:

$$\text{pitch} = \frac{\pi D_1}{z_1} = \frac{\pi D_2}{z_2}$$

where D_1 = pitch circle diameter of the driving gear.
D_2 = pitch circle diameter of the driven gear.
z_1, z_2 = number of teeth on the driving and driven gears.

$$\therefore \quad V_R = \frac{\omega_1}{\omega_2} = \frac{n_1}{n_2} = \frac{D_2}{D_1} = \frac{z_2}{z_1}$$

Note: The angular velocity, peripheral velocity and other dimensions of the gear always refer to the pitch diameter of the gear.

Now if the centre distance between the driving and driven shafts is quite large, and if these are connected by a single mesh gears, (Fig. 22.2 dotted line), the drive will become very bulky. To avoid this, the same velocity ratio can be transmitted by means of a large number of gears mounted on parallel shafts.

The intermediate gears between the driving gear 1 and driven gear 4, are called idlers or idler gears. The speed ratio of first pair is:

$$(V_R)_1 = \frac{\omega_1}{\omega_2} = \frac{d_2}{d_1}$$

$$\text{second pair, } (V_R)_2 = \frac{\omega_2}{\omega_3} = \frac{d_3}{d_2}$$

$$\text{third pair, } (V_R)_3 = \frac{\omega_3}{\omega_4} = \frac{d_4}{d_3}$$

$$\therefore \quad (V_R)_1 \times (V_R)_2 \times (V_R)_3 = \frac{\omega_1}{\omega_2} \times \frac{\omega_2}{\omega_3} \times \frac{\omega_3}{\omega_4} = \frac{d_2}{d_1} \times \frac{d_3}{d_2} \times \frac{d_4}{d_3} = \frac{d_4}{d_1} = \frac{z_4}{z_1}$$

$\therefore$ Desired overall speed ratio, is

$$V_R = (V_R)_1 \times (V_R)_2 \times (V_R)_3 = \frac{\omega_1}{\omega_4} = \frac{z_4}{z_1}$$

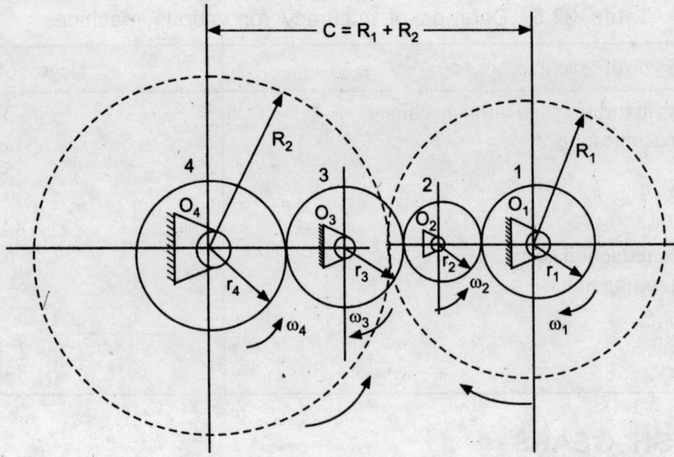

Fig. 22.2 *Gear drive with idlers*

Thus the employment of a large number of gears does not change the overall velocity ratio but helps in reducing the weight and dimensions of the drive.

Again a high speed ratio cannot be transmitted in a proper way by a single mesh or a single pair of gears (usually $V_{R\,max} < 10$). High speed ratios are obtained by the use of "multistage drives", Fig. 22.3. The speed ratio of such a drive is:

$$(V_R)_{\text{I-IV}} = \frac{\omega_{\text{I}}}{\omega_{\text{IV}}}$$

Now

$$(V_R)_1 = (V_R)_{\text{I-II}} = \frac{\omega_{\text{I}}}{\omega_{\text{II}}} = \frac{z_2}{z_1}$$

$$(V_R)_2 = (V_R)_{\text{II-III}} = \frac{\omega_{\text{II}}}{\omega_{\text{III}}} = \frac{z_4}{z_3}$$

$$(V_R)_3 = (V_R)_{\text{III-IV}} = \frac{\omega_{\text{III}}}{\omega_{\text{IV}}} = \frac{z_6}{z_5}$$

$$\therefore \quad (V_R)_1 \times (V_R)_2 \times (V_R)_3 = \frac{\omega_{\text{I}}}{\omega_{\text{IV}}} = (V_R)_{\text{I–IV}} = \frac{z_2 \times z_4 \times z_6}{z_1 \times z_3 \times z_5}$$

Thus the overall speed ratio of a multistage drive is equal to the product of the speed ratios of the individual gear pairs included in the drive.

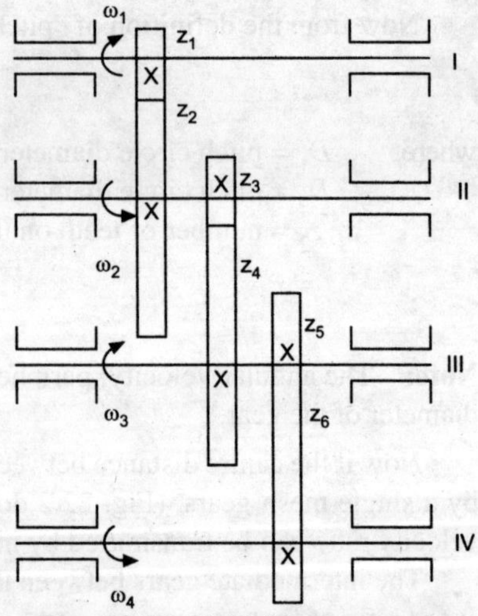

Fig. 22.3 *Multistage gear drive*

22.6 GEAR NOMENCLATURE

The different terms in connection with toothed gears are explained below with the help of Fig. 22.4.

1. **Pinion and Gear:** Whenever two toothed wheels are in mesh with each other, the large wheel will be termed as gear and the smaller one as the pinion no matter which wheel is the driver or the driven.
2. **Pitch Surfaces:** These are imaginary planes, cylinders, or cones that roll together without slipping.
3. **Pitch Circle:** The distance between the corresponding sides of the adjacent teeth of a gear is measured on the arc of a circle. The circle on which this distance becomes equal to the rack tooth pitch is called the pitch circle.
4. **Pitch circle diameter:** D. It is the diameter of the pitch circle.

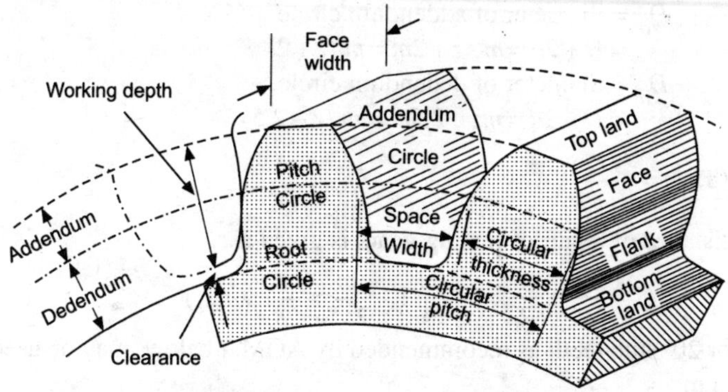

Fig. 22.4 *Gear nomenclature*

5. **Circular Pitch:** It is the distance between the corresponding sides of two adjacent teeth of a gear measured on the pitch circle. It is written as

$$p_c = \frac{\pi D}{z}, \; z = \text{number of teeth on the gear or pinion} \qquad \text{... (22.1)}$$

6. **Diametral pitch, p_d:** This is a definition mostly used in F.P.S system. It is defined as the ratio of the number of teeth per unit pitch diameter, i.e.,

$$p_d = \frac{z}{D} \qquad \text{... (22.2)}$$

From equations (1) and (2),

$$p_c \cdot p_d = \pi \qquad \text{... (22.3)}$$

7. **Module, *m*:** This is the standard term used in S.I. units. It is the reverse of the diametral pitch. It can be defined as the length on the pitch circle diameter per tooth, i.e.,

$$m = \frac{D}{z} \qquad \text{... (22.4)}$$

8. **Addendum, *a*:** It is defined as the radial distance between the pitch circle and the circle coinciding with the tops of the teeth in a cross-section. This circle is known as "Addendum Circle". As already indicated, the standard value of '*a*' for Spur gears is, $a = m$. ... (22.5)
9. **Root circle:** This circle is also known as "dedendum circle or "clearance circle". It can be defined as the circle described around the gear centre and passing through the bottom of the tooth spaces.
10. **Dedendum:** It is the radial distance between the pitch circle and the dedendum circle. As already indicated, its value for spur gears is:

$$d = 1.25\, m \qquad \text{... (22.6)}$$

11. **Height of tooth:** Or total tooth depth = $a + d$.
12. **Working depth:** It is the working depth of engagement of two mating gears. It is twice the addendum.
13. **Clearance:** It is the difference between the dedendum of one gear and addendum of the mating gear. Therefore, *c* is given as:

$$c = d - a = 1.25\, m - m = 0.25\, m \qquad \text{... (22.7)}$$

14. **Face, Flank of the tooth:** Face of the tooth is the portion of the tooth between the pitch circle and the addendum circle. Flank of the tooth is the portion of the tooth between the pitch circle and the root circle.
15. **Thickness of tooth:** It is the thickness of the tooth measured along the pitch circle.
16. **Tooth Space:** It is the width of the empty space in between the two consecutive teeth, measured on the pitch circle.
17. **Back lash:** It is the difference between the tooth space and thickness of tooth, as measured on the pitch circle.

Now, D_a = diameter of addendum circle

$$= D + 2a = m \cdot z + 2m = m(z+2) \quad \text{... (22.8)}$$

D_d = diameter of dedendum circle.

$$= D - 2d = mz - 2.5m = m(z - 2.5) \quad \text{... (22.9)}$$

$$\therefore \text{ From equation (8), } m = \frac{D_a}{z+2} \quad \text{... (22.10)}$$

Again the centre distance of a pair of mating gears is given as:

$$C = \frac{D_1}{2} + \frac{D_2}{2} = \frac{z_1 \cdot m}{2} + \frac{z_2 \cdot m}{2} = \frac{m}{2}(z_1 + z_2) \quad \text{... (22.11)}$$

The proportions for 20º stub teeth as recommended by AGMA below, may be used:

Addendum, $a = 0.8$ m

Minimum dedendum, $d = 1.0$ m

Minimum total depth = 1.8 m

Minimum clearance = 0.2 m

Tooth thickness = 1.571 m

22.7 TOOTH PROFILES

The condition to be satisfied by the tooth profile so that the teeth of two mating gears will transmit uniform angular velocity, is that the common normal to the mating tooth surfaces at their point of contact must always pass through the same pitch point, i.e., the point where the line of centres intersects the pitch circles. Two tooth profiles satisfying this condition and in common use are

(1) Cycloidal, (2) Involute.

A cycloidal tooth has different surfaces for face and flank, *i.e.*, convex for face and concave for flank or vice versa. It is clear that if two gears mesh with each other, then convex surface on the tooth on one gear will be in mesh with concave surface of the tooth on the second gear. This will result in very little wear of the tooth surfaces. But the difficulty encountered in producing exact profile has forced this system into obsolescence.

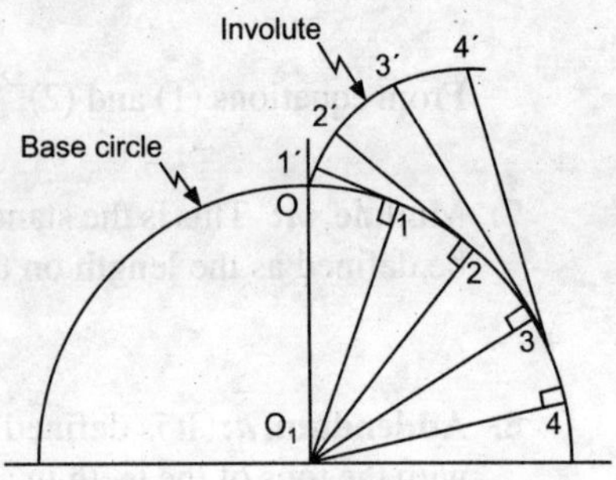

Fig. 22.5 *Involute profile*

The surfaces of the involute system are either straight or curved (convex). An involute is the curve traced by a point on a line as the line rolls on another curve, Fig. 22.5.

The circle on which the line roles is called the "Base Circle".

The universal employment of Involute Profile is due to the following reasons:

1. Simplicity of manufacture.
2. Interchangeability of gears in case of changes in the speed ratio.
3. Possibility of a certain increase in the centre distance without affecting the velocity ratio or the accuracy of engagement.

However, the involute profile has the drawback of limited loading capacity.

In Fig. 22.5, it should be noted that 01 = 11′; 02 = 22′, etc. An apparent property of the involute is that any line tangent to the base circle will be perpendicular to the involute curve at its point of intersection. It is clear, therefore, that if any surface is in contact with involute curve and a normal is drawn common to both the curves, it will be tangent to base circles. Thus, if two involutes are in contact, Fig. 22.6 the common normal is the straight line tangent to both base circles. This line is called the "transmission line" or "line of action" and the contact between two

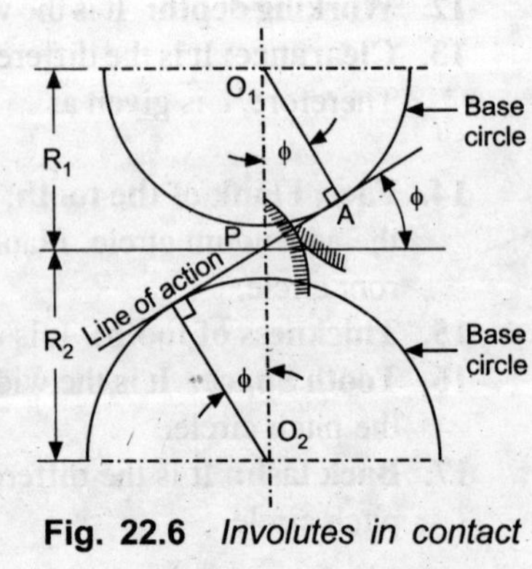

Fig. 22.6 *Involutes in contact*

involutes will always lie on this line. This line will always intersect the line of centres at the same point, the pitch point P. Hence, the involute profile satisfies the basic law of gear tooth action.

"Arc of action" is the path covered by the tooth profile along the rolling circle within the period of actual contact. For the continuity of tooth contact, the arc of action should be greater than the circular pitch of the tooth. If this condition is not fulfilled, one pair of teeth will disengage before the next pair comes into contact. This will result in intermittent operation of the gear drive and tooth shocks. The ratio of the arc of action to the circular pitch of tooth is known as "Contact Ratio" ε. Hence, for smooth drive

$$\varepsilon = \frac{\text{Arc of action}}{p_c} = \frac{S}{p_c} > 1$$

For spur gear drives : $\varepsilon \geq 1.1$.

Pressure Angle, ϕ: It is also known as "angle of obliquity". It is the angle between the line of action and the perpendicular to the line of centres. This angle is important in relation to the strength and wear of the tooth and in relation to the forces acting on the gear shaft and the bearings. It contributes to the forces which tend to separate the gears. If the pressure angle is large, the base circles will be smaller resulting in smaller contact ratio. But it will result in a higher capacity tooth and also gears and pinions with fewer teeth can be used without undercutting. Also the separating force which is undesirable will be greater with a higher value of pressure angle. The two pressure angles most commonly used are 14½° and 20°. Indian Standards provide mostly 20° pressure angle.

It is apparent that the involute profile exists outside the base circle, only. Hence only the portion of tooth which lies outside the base circle is of involute profile and will mesh with another involute of the same module. The portion of the tooth inside the base circle does not come into contact with the mating tooth. Hence this portion can be given any shape. Formerly it used to be a radial profile but now a days it is automatically undercut by the gear machining methods.

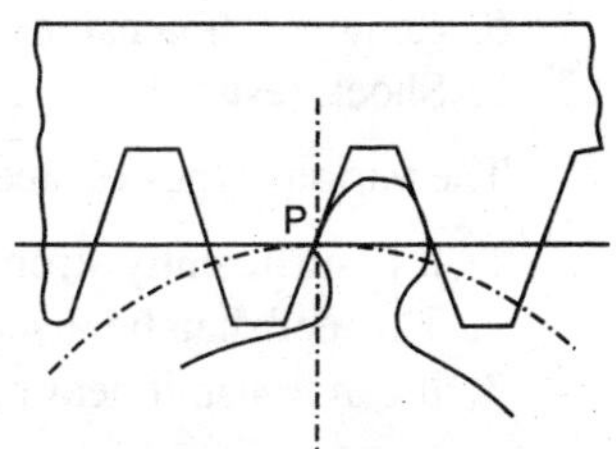

Fig. 22.7 *Interference*

Interference in Involute Gears: It is mentioned above that tooth portion inside the base circle is just useless. Now, if for a given pressure angle, the number of teeth on pinion is decreased below a certain limit, the useless tooth portion below the base circle of the pinion will interfere with the addendum of the large gear. To avoid this, one method is to undercut the flank of the pinion, Fig. 22.7. But this would impair the strength of the tooth significantly. For this, the minimum number of teeth on pinion are specified, which will mesh with any gear, even rack, without interference. Thus, the minimum number of teeth on a pinion, for

14½° Full depth system, is 31
20°, Full depth system, is 17
20°, Stub system, is 14
14½°, Composite system, is 12.

For gear drives used at low speeds and for manual work, e.g., winches, the minimum number of teeth for 20°. Involute system can be reduced to 10.

22.8 SYSTEMS OF GEAR TEETH

The following systems of gear teeth are used in practice

1. 14½° Full depth system
2. 14½° composite system
3. 20° Full depth system
4. 20° Stub system.

Composite system has been developed to avoid interference without undercutting. The tooth form is true 14½°, involute near the pitch line, but the outer part of the face and the inner part of the flank are made cycloidal.

In the full depth system, the full height of the tooth is full addendum plus full dedendum *i.e.*, $m + 1.25m = 2.25m$ for Spur gears and 2.20 m for Bevel gears. The employment of 20°, pressure angle helps in reducing the minimum number of teeth on the pinion as compared to 14½° pressure angle. Moreover, the tooth becomes stronger as its breadth at the root increases.

In the Stub tooth system, the tooth has less working depth, usually 20% less than full depth system. The addendum is made shorter. Stub system has the following advantages.

1. Due to shorter addendum; the interfering portion of the tooth is removed and hence there will be less interference in stub system.
2. The tooth becomes stronger as the lever arm of the bending moment on tooth is reduced.
3. Machining cost is reduced as compared to full depth system.

Stub system is advantageous mostly in gears with a small number of teeth. With a large number of teeth, the full depth systems perform better.

22.9 MATERIALS

The various materials used for gears include a wide variety of steels, cast irons, bronzes and non-metallic materials such as texolite, wood laminate, pressed wood, capron, etc. The selection of the gear material depends upon:

1. Type of service
2. Peripheral speed
3. Degree of accuracy desired
4. Method of manufacture
5. Required dimensions and weight of the drive.
6. Allowable stress
7. Shock resistance
8. Wear resistance.

The various types of steels meet most of the above demands. Steel has the following advantages:

1. It is sufficiently strong and highly resistant to wear by abrasion.
2. The teeth can be machined to the required degree of accuracy.
3. It can transmit heavy loads.

Steels used for gears may range from plain carbon steels to highly alloyed steels and from low to high carbon steels. Choice will depend upon many factors including, size, service and design.

To combine properly the toughness and tooth hardness, gears made of steels are heat treated. The various heat treatment processes employed are case hardening, through hardening, nitriding, precipitation hardening. To achieve this, the methods employed are furnace hardening, carburizing, induction hardening, flame hardening. High hardness heat treated steels give smaller gears for a given load and also the service life is increased by about ten times without increasing the size or weight of gears.

Cast iron is popular due to its good wearing properties, excellent machinability, and ease of producing complicated shapes by the casting method.

Bronzes are very popular in worm gear drives because of their ability to withstand heavy sliding loads and ability to "wear in" to fit hardened steel worms. They are also very useful for corrosive conditions. Like C.I., they can also be easily cast into complex shapes.

Gears made of non-metallic materials give noiseless operation at high peripheral speeds, but their load carrying capacity is low.

Laminated gears produced from phenolic resins used in various combinations, have good load bearing capacities inspite of the low physical strength of the material.

New materials such as nylon, titanium and sintered iron have also become important in recent years, in gear work.

Note: The material for pinion should always be better than for gear as it takes part in larger number of cycles than the teeth of gear and this factor may cause quicker wear.

The properties of the various materials used for the gears are given in Table 22.6.

Table 22.6 Gear materials

S. No.	Material	Condition (Finished Gear)	Minimum Tensile Strength, N/mm^2	Brinell Hardness number
1.	**Malleable C.I.**			
	(i) White heart Castings			
	Grade B	—	280	217 max
	(ii) Black heart Castings			
	Grade B	—	320	149 max
2.	**Cast Iron**			
	(Iron castings for gears and gear blanks).			
	(i) Grade 20	As cast	200	179 min.
	(ii) Grade 25	As cast	250	197 min.
	(iii) Grade 35	As cast	350	207 min.
	(iv) Grade 35	Heat treated	350	300 min.
3.	**Phosphor Bronze**			
	(Castings for gear blanks)	Sand Cast	160	60 min.
		Chill Cast	240	70 min.
		Centrifugal Cast	267.7	90 min.
4.	**Cast Steel**			
	Grade 1	—	550	145 min.
5.	**Forged Steel**			
	(a) Carbon Steels			
	C 30 Steel	Normalised	500	143
	C 30 Steel	Hardened and Tempered	600	152
	C 40 Steel	Normalised	580	152
	C 40 Steel	Hardened and Tempered	600	179
	C 55 Mn 75 Steel	Normalised	720	201
	C 55 Mn 75 Steel	Hardened and Tempered	700	223
	C 55 Mn 75 steel	—	800	248
	(b) Carbon Chromium Steels			
	55 Cr 70 steel	Hardened and Tempered	900	225 min.
	55 Cr 70 steel	Hardened and Tempered	1000	285 min.
	40 Cr 1 steel	Hardened and Tempered	800	229 min.
	(c) Carbon Manganese Steel			
	27 Mn 2 Steel	Normalised	550	—
	27 Mn 2 Steel	Hardened and Tempered	600	170 min.
	37 Mn 2 Steel	Hardened and Tempered	700	201 min.
	37 Mn 2 Steel	—	700	201 min.
	37 Mn 2 Steel	—	800	229 min.

...*(Contd)*

S. No.	Material	Condition (Finished Gear)	Minimum Tensile Strength, N/mm^2	Brinell Hardness number
	(d) Manganese Molybdenum Steel			
	45 Mn 2 Mo 28 Steel	Hardened and Tempered	700	201 min.
	35 Mn 2 Mo 28 Steel	Hardened and Tempered	800	229 min.
	35 Mn 2 Mo 45 Steel	Hardened and Tempered	800	229 min.
	35 Mn 2 Mo 45 Steel	Hardened and Tempered	900	225 min.
	(e) Chromium Molybdenum Steel			
	40 Cr 1 Mo 28 Steel	Hardened and Tempered	700	201 min.
	40 Cr 1 Mo 28 Steel	Hardened and Tempered	800	229 min.
	40 Cr 1 Mo 28 Steel	Hardened and Tempered	900	225 min.
	40 Cr 1 Mo 28 Steel	Hardened and Tempered	1000	285 min.
	40 Cr 1 Mo 60 Steel	Hardened and Tempered	900	248 min.
	40 Cr 1 Mo 60 Steel	Hardened and Tempered	1000	293 min.
	15 Cr 1 Mo 55 Steel	Hardened and Tempered	900	255 min.
	15 Cr 1 Mo 55 Steel	Hardened and Tempered	1550	144 min.
	(f) Nickel Steel			
	40 Ni 3 Steel	Hardened and Tempered	800	229 min.
	40 Ni 3 Steel	Hardened and Tempered	900	225 min.
	(g) Nickel Chromium Steel			
	30 Ni 4 Cr 1 Steel	Hardened and Tempered	1540	444 min.
	(h) Nickel Chromium Molybdenum Steel			
	40 Ni 2 Cr 1 Mo 28 Steel	Hardened and Tempered	900	225 min.
	40 Ni 2 Cr 1 Mo 28 Steel	Hardened and Tempered	1550	444 min.
	31 Ni 3 Cr 65 Mo 55 Steel	Hardened and Tempered	900	255 min.
	31 Ni 3 Cr 65 Mo 55 Steel	Hardened and Tempered	1550	444 min.
	31 Ni 3 Cr 65 Mo 55 Steel	Hardened and Tempered	1100	311 min.
	31 Ni 3 Cr 65 Mo 55 Steel	Hardened and Tempered	1000	285 min.
	31 Ni 3 Cr 65 Mo 55 Steel	Hardened and Tempered	1550	444 min.
	31 Ni 3 Cr 65 Mo 55 Steel	Hardened and Tempered	1200	341 min.
6.	**Surface Hardened Steel**			
	(a) Carbon Steel			
	C 40 steel	—	551.2	145 Core 460 Case
	C 55 steel	—	708.6	200 Core 520 Case

...*(Contd)*

S. No.	Material	Condition (Finished Gear)	Minimum Tensile Strength, N/mm^2	Brinell Hardness number
	(b) C-Cr Steel			
	C 55 Cr steel	—	866.1	250 Core 500 Case
	1% Cr steel	—	708.6	500 Case
	(c) Ni-Steel			
	3½% Ni steel	—	708.6	200 Core 300 Case
	(d) Ni-Cr Steel			
	3½% Ni steel	—	866.1	250 Core 500 Case
	1½% Ni, 1% Cr steel	—	866.1	250 Core
7.	**Case Hardened Steel**			
	(a) C-Steel			
	0.12 to 0.22 % C steel	—	503.9	650 Case
	0.15% C steel	—	503.9	140 Core
	0.20% C steel	—	503.9	140 Core 640 Case
	(b) Ni-Steel			
	3%, Ni steel	—	708.6	200 Case
	3½% Ni steel	—	708.6	200 Core 620 Case
	5% Ni steel	—	866.1	250 Core 600 Case

22.10 MANUFACTURING METHODS

The various manufacturing methods used for gears can be listed under the following headings:

1. Metal Removal
2. Casting
3. Forming.

22.10.1 Metal Removal Methods

In these methods, the metal is removed from the gear blanks to produce the gears. Under this, the main gear cutting methods are:

(*a*) Profiling
(*b*) Generation.

In the profiling method, the spaces between the teeth are cut by means of a disc cutter or an end milling cutter. The cutting portions of the cutter have the ground gear tooth contour. Almost any tooth form can be obtained by milling. But generally, only spur, helical and straight bevel gear are milled. Surface finish can be held to about 3.20 micrometers.

In the generating process, teeth are formed in a series of passes by a generating tool. If it is threaded and gashed, it is called a "hob" and the process is known as "hobbing". When the cutting tool is shaped like a pinion or a section of a rack, the method si called "Shaping". Surface finishes as fine as about 1.60 micro-m can be obtained. In another metal removal method, a tool or a series of tools are used which wrap around the gear and cut all teeth at the same time. Methods of this type are "broaching", "punching" and "shear cutting".

22.10.2 Gear Casting

The casting process is most often used for making blanks for gears on which the teeth will be cut. However, this method is also used to make toothed gears in one operation. The toothed gears made by sand casting are not very accurate and they find limited applications in farm machinery and in some hand operated devices. Large quantities of small gears are made by die casting. The materials used are low melting ones such as alloys of zinc, aluminium, or copper. The disadvantage is that the gears made out of these materials are not suitable for high load carrying capacity. Light gears of thermoplastics are made by injection moulding. Gears can also be cast by investment casting process.

22.10.3 Forming Methods

Methods included under this are cold drawing and "Roll forming". The roll forming method is greatly and rapidly being developed. Roll formed gears have better physical properties than cut teeth and in large quantities are less expensive to produce. Presently, only helical gears are being rolled, but spur gears should also be soon available.

In the "Plunge" method of rolling, the only method currently used, the gear blanks is mounted on a shaft between two dies. As the blank meshes with the dies, a complete involute profile is formed.

22.10.4 Gear Finishing Operations

After the teeth are cut, these are finally finished to produce high accuracy and surface smoothness. It is done by : Grinding, Shaving, Lapping, or Honing. Shaving removes only a small amount of surface metal. A very hard mating gear with many small cutting edges is run with the gear to be shaved. Surface finish can be as fine as about 0.80 micro-meters. Grinding is the most accurate methods of gear finishing. Surface finishes of about 0.50 micro-metres can be obtained.

The lapping process only corrects minute heat treatment distortion errors in hardened gears. The gear to be finished is run in mesh with a gear shaped lapping tool or another mating gear. An abrasive lapping compound is used in between them. Only the tooth contact is improved by this method and not the accuracy. Finish is of the order of about 0.80 micro-metres.

Honing is employed mainly to remove burrs and nicks and thus for improving the surface finish. Surface finish of the order to about 0.40 micro-metres can be obtained.

Note: The finishing operations are usually done for gears not under the 8th degree of accuracy.

22.11 TYPES OF GEAR TOOTH FAILURES

The failure of a gear tooth can occur due to the following causes:

1. **Tooth Fracture:** When two gears mesh with each other to transmit a load, the teeth of each gear are under bending action. The bending stress will be maximum at the root of the tooth. Due to the periodical effect of the load, fatigue cracks may occur near the tooth base resulting in ultimate failure of the tooth.
2. **Abrasion and run-in wear:** When some foreign material, e.g., metal particles, dust or dirt gets deposited in between the mating teeth, there will be wear of tooth surfaces. This wear is known as abrasion wear. It is larger in open drives than in closed drives. Wear due to insufficient smooth surface is called run-in wear. It occurs during the early operation of the drive.

3. **Pitting:** Pitting takes place due to the large contact pressures occurring between the tooth surfaces. Fatigue cracks are set up due to these repeated contact stresses. These cracks get developed into pits resulting in ultimate failure of the tooth.
4. **Seizure:** Due to the high loads applied to the tooth surfaces the lubrication oil film is forced out and the mating tooth surfaces mesh snugly. Due to the relative motion between them, the material gets scratched from the surface of the softer material resulting in scores and scratches on its surface.

22.12 LUBRICATION OF GEARS

To obtain the proper service and maximum life, the gears must be generously lubricated with a proper lubricant. The lubricant must possess the following properties:

1. Enough oiliness to reduce the friction between the rubbing tooth surfaces.
2. Enough viscosity to develop a suitable film between the tooth surfaces. The film should be able to withstand the high contact pressures and the temperature in the contact region.
3. It should remove the heat developed by the gears fast enough.
4. It should flush away the wear products and corrosion formed on the tooth surfaces.
5. No dirt, sand metal particles, sludge or acids should mix with the lubricant.
6. It should wet all the tooth surfaces before they go into mesh.

22.13 DESIGN CONSIDERATIONS

For the design of a gear drive, the following data is usually given :

1. Power to be transmitted.
2. Speed of the driving gear.
3. Speed of the driven gear, or the velocity ratio.
4. Centre distance.

While attempting the design of a gear drive, the try must be made to meet the following requirements:

1. The teeth must be sufficiently strong so as to withstand the static loading as well as dynamic loading on the drive.
2. The teeth should have good wear resisting properties for their long life.
3. Economy of space and material.
4. The gears must operate without tooth interference, with a proper length of contact and without undue noise.
5. The alignment of gears and deflection of shaft must be taken into consideration since these affect the gear performance.
6. The teeth must be able to resist tooth deflection and accelerations and stress concentration.
7. Proper lubrication of gear.

22.14 DESIGN CALCULATIONS OF STRAIGHT TOOTH SPUR GEARS

The design of any gear drive is extremely involved in view of the many complex requirements (listed above) to be met by it. Until about 1932, gear design was done by the help of Lewis equation which was based on the static strength of the tooth in bending, by considering it as a cantilever beam. Stress concentration at the tooth root was not considered since stress concentration factors were not in use when Lewis had first proposed his equation (towards the end of 19th century). At that time, the working of gear drives used to be intermittent and at low pitch line velocities. Then, as the pitch line velocity increased, the resulting dynamic effects were considered by incorporating a velocity factor or dynamic factor (first given by Barth) in the Lewis equation. Then, after 1932, Buckingham proposed a new method of gear design based on dynamic load, endurance limit of the material and the wear load. For calculating dynamic load, he proposed equation in which machining errors and tooth stiffness were considered. Greater the machining errors, greater the dynamic load. May be at that time, the methods of gear cutting were not very accurate and precise. The combined method of gear design, that is, first finding the preliminary dimension by Lewis equation and then checking these and modifying if need be, by the Buckingham method has been in use for long. The stress concentration effects were considered indirectly by comparing the dynamic load with flexural endurance limit of the gear materials.

Afterwards, AGMA (American Gear Manufacturers Association) came up with a modified Lewis equation in which many factors, like, velocity factor, stress concentration factor etc., were incorporated. Since, the present gear cutting and finishing methods can produce very precise gear teeth, resulting in appreciably reduced dynamic effects, AGMA has proposed relations for evaluating dynamic factor which give higher values as compared to earlier relations. The number of various factors to be considered in gear design depends upon the sophistication sought and also whether the exact design conditions are accurately known for which the various factors may be considered. The aim of any design course is to teach the students the basic and fundamental methods of design. So in the present chapter, the authors have considered the design factors about which an accurate guess can usually be made in the common class room design problems. For example, stress concentration factor is considered while deriving Lewis equation for beam strength of teeth, or while calculating the safe design stress. Dynamic effects are considered by incorporating the velocity factor. It may be mentioned here, that for non-metallic gears (also called silent gears) and worm gears, the dynamic load was used to be calculated by the use of velocity factor, even though for cut metallic gears, Buchingham's equation was preferred. The service conditions are considered by incorporating overload factor. The design is finally checked for wear by the equation given by Buckingham. To the authors, this approach results in more straight forward and convenient gear design procedure which seems to be quite adequate for normal design. However, the design by Buckingham's equation has also been explained.

22.14.1 Strength of Spur Gears

Each tooth may be considered as a cantilever beam loaded as shown in Fig. 22.8 (a). The force F_n between the tooth surfaces is normal to the surface acting at the tip of the tooth. This normal force F_n is resolved into two components, tangential F_t, and radial F_r, acting at a point A, the inter-section of the line of action of the normal tooth load and the centre of the tooth. F_t will induce a bending stress in the tooth and F_r will induce a uniform direct compressive stress, over the tooth section. The maximum value of bending stress is

$$f_b = K_f \frac{M.c}{I} = K_f \frac{F_t h \cdot t/2}{ft^3/12} = K_f \frac{6F_t \cdot h}{ft^2} \qquad ...(22.12)$$

where f = face width of tooth

K_f = stress concentration factor, where the tooth joins the bottom land. The maximum value of the compressive stress is:

$$f_c = K_f \frac{F_r}{f.t}$$

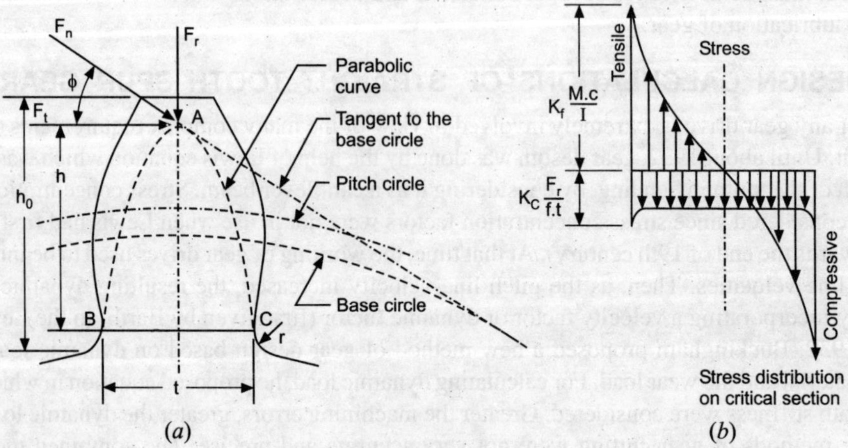

Fig. 22.8

Figure 22.8 (b) gives stress distribution at the base of the tooth. It is clear that the direct compressive stress is quite small as compared to the bending stress. It is also clear that it will increase the bending stress on the compressive side of the tooth and decrease the resultant stress on the tensile side. Now, since the gear teeth are under fatigue loading, fatigue failure will start on the tension side of the tooth, the direct compressive stress will strengthen the tooth by reducing the resultant tensile stress. Hence, direct compressive stress is ignored in determining the strength of the tooth. From equation (22.12),

$$F_t = \frac{f_b \cdot f \cdot t^2}{6h.K_f} \qquad \text{...(22.12a)}$$

The gear tooth is being designed as a cantilever beam and in a well designed beam, the stress should be uniform in every section of the beam. Thus in a gear tooth of uniform strength, the stress is a constant and since F_t and f are also constants, we write equation (22.12a), as

$$h = \frac{f_b \cdot f \cdot t^2}{6K_f F_t} = \text{constant} \times t^2$$

which is the equation of a parabola. This parabola can be inscribed in the tooth through point A and through points B and C at which the parabola is tangent to the tooth profile. Now, the tooth is larger than the parabola at every section except BC. Thus BC is the critical section in a gear tooth and maximum stress will occur there. Hence, equation (22.12) was derived for the section having the maximum stress.

Even if the contact ratio is 1.0, a single tooth theoretically never carries the full load at the tip of the tooth. But even a minute error in tooth spacing may prevent the sharing of the load. Therefore, it will be a conservative design if it is assumed that the total load is carried by a single pair of teeth through the tips of the teeth and the load is taken as uniformly distributed on the tooth face width. Therefore, substituting the maximum height 'h_0' and the corresponding root thickness 't_0' for 'h' and 't' in equation (22.12).

$$f_b = K_f \frac{6F_t \cdot h_0}{f \cdot t_0^2}$$

$$\therefore \quad F_t = \frac{f_b \cdot f \cdot t_0^2}{6K_f \cdot h_0} \qquad \text{...(22.13)}$$

Multiplying the numerator and denominator on the right hand side by p_c.

$$\therefore \quad F_t = \frac{f_b \cdot f \cdot p_c}{K_f} \cdot \frac{t_0^2}{6h_0 \cdot p_c} \qquad \text{...(22.14)}$$

The term $\dfrac{t_0^2}{6h_0 \cdot p_c}$ is non-dimensional and depends only upon the size of the tooth and its profiles and is called "Lewis form factor" and is denoted as y. Therefore, equation (22.14) becomes:

$$F_t = \frac{f_b \cdot f \cdot p_c \cdot y}{K_f}, N \qquad \text{...(22.15)}$$

We know that $p_c = \pi \cdot m$

$$\therefore \quad F_t = \frac{f_b \cdot f \cdot \pi \cdot m \cdot y}{K_f}, N \qquad \text{...(22.16a)}$$

This equation is known as "Lewis Equation" written also as

$$F_t = \frac{f_b \cdot f \cdot mY}{K_f}, N \qquad \text{...(22.16b)}$$

Now $$y = 0.154 - \frac{0.912}{z} \text{ for } 20^\circ \text{ pressure angle, full depth system}$$

$$= 0.175 - \frac{0.841}{z} \text{ for } 20^\circ \text{ pressure angle, stub tooth system} \quad ...(22.17)$$

$$= 0.124 - \frac{0.684}{z} \text{ for } 14\frac{1}{2}^\circ \text{ pressure angle}$$

Y is also form factor and is equal to $= \pi \times y$.

If the face width is very long in relation to pitch, the load may not remain uniformly distributed which will be against the assumption made. On the other hand, a narrow face width will give less smooth action and poor wearing qualities. The usual value is taken as $f = 3$ to $4\ p_c$ or 9.5 m to 12.5 m.

Under certain conditions, due to space limitations as in automobile change gears, f may be as low as 6 m and for turboreduction gears, it may be as high as 25 m. Lower values are for a lower manufacturing accuracy of gears and also in the case of low stiffness of shafts and their bearings.

The ratio f/m is sometimes called "face width factor" or "proportion factor".

The stress concentration factor K_f is a function of the ratio of fillet radius thickness and the ratio of tooth thickness to tooth height. For most design purposes, its value may be taken as,

$$K_f = 1.6 \text{ for full depth } 14\frac{1}{2}^\circ \text{ teeth}$$

$$= 1.5 \text{ for } 20^\circ \text{ full depth and stub teeth.}$$

Design Stress: The value of the bending stress 'f_b' in equation (22.20) may be taken as equal to the elastic limit of the material in bending devided by the factor of safety which may be taken as 1.5 to account for cyclic one way bending.

Many authors recommend that the design stress in the Lewis equation may be taken as one-third of the ultimate strength. The safe design stress can also be taken directly from Table 22.7.

Table 22.7 Safe working stress for gear materials

Material	f_b, MPa	Bhn
Cast iron, ordinary	56	180
Cast iron, good grade	70	180
Cast iron, High grade	105	180
Cast steel, untreated	140	180
Cast steel, heat treated	196	250
Forged steel, C–30	175	150
Steel C–30, heat treated	224	300
Steel C–40, untreated	210	150
Alloy steel, case hardened	350	650
Cr-Ni steel, heat treated	455	475
Cr-Va steel heat treated	525	450
Manganese Bronze	140	100
Gear Bronze	70	80
Aluminium Bronze	154	180
Phosphor Bronze	84	100
Rawhide, Fabroil, Bakelite, Micarta	56	

To base the design bending stress f_b, on elastic limit or yield strength, is valid only when fatigue is not a problem, that is when the gear drive is run not frequently. In many cases, fatigue is a problem and then 'f_b' must be based on the endurance limit of the material. For steel gears, it may be taken as,

$$f_b = 0.9 \times \text{endurance limit}$$

Thus relatively high value can be used, because, the teeth are usually subjected to repeated and not reverse loads. For steel gears, the endurance limit may be taken as,

$$\text{Endurance limit} = 1.75 \times \text{BHN, MPa}$$

Dynamic Effects: The tangential load used in the above analysis may be taken as the actual load on the tooth only for very slow speed (practically stationary) gears. But in actual practice, the actual load coming on the tooth is always higher. The increased load is due to the fact that in most cases, the gears are operating at appreciable speeds and also there is impact between the mating teeth. This impact action is due to the following reasons:

1. Manufacturing in accuracies resulting in errors in spacing and profiles of the teeth. This will cause non-uniform rotation of the driver gear and induce additional dynamic loads.
2. Deflection of the teeth under load relative to the teeth next coming into mesh.

The magnitude of the additional load depend upon the class of accuracy and peripheral speed of the gear. Thus the simplest approach to consider the effect of the above mentioned factors is by incorporating a factor known as "Velocity factor" or dynamic factor" C_v into the Lewis equation.

Many authors use this factor to find the design load or dynamic load to be used on the left side of the Lewis equation, so that, Dynamic load or Design load, $F_d = F_t/C_v$.

Others use this factor to reduce the allowable stress to be used in the Lewis equation. The safe working stress given in Table 22.7 is for almost stationary drives, that is, the static stress. According to these authors, the actual stress to be used in the Lewis equation (22.16a or 22.16b) will be:

Safe Stress = Safe Static Stress (Table 22.7) × C_v.

In whatever way the factor C_v is considered, the Lewis equation will ultimately have the form :

$$F_t = \frac{f_b \cdot f \cdot mY \cdot C_v}{K_f}, N \qquad \text{...(22.17a)}$$

The values for "C_v" are specified in terms of the quality of gear and pitch velocity.

$C_v = \dfrac{3}{3+V}$ for ordinary cut gears operating at V upto 7.5 m/s. This is known as Barth's equation.

$C_v = \dfrac{4.5}{4.5+V}$ for carefully cut gears, with V upto 12.5 m/s.

$C_v = \dfrac{6}{6+V}$ for very accurately cut and ground metallic gears with V upto 20 m/s. ...(22.18)

$C_v = \dfrac{5.6}{5.6+\sqrt{V}}$ for hardened steel, ground and lapped in precision gears for V upto 20 m/s.

$C_v = \dfrac{0.75}{1.0+V} + 0.25$ for non-metallic gears.

AGMA has proposed the following dynamic factors which are intended to account for, the effect of tooth spacing and profile errors, effect of pitch line velocity and angular velocity, inertia and stiffness of all rotating elements, transmitted load per cm of face width and tooth space.

$C_v = \dfrac{50}{50+\sqrt{200\,V}}$ for gears whose teeth are finished by hobbing or shaping.

$= \left[\dfrac{78}{78+\sqrt{200\,V}}\right]^{1/2}$ for gears with high precision shaped or ground teeth and if dynamic (appreciable) load is developed.

= 1, for gears with high precision shaped or ground teeth and there is no appreciable dynamic load.

To decide whether an appreciable dynamic load exists or not, the designer should examine the driving and driven machinery. If the drive is between a motor and a fan, it is certain that not much dynamic load will be developed. However, if the drive involves a single cylinder engine and the blade of a chain saw, a considerable dynamic load will be experienced.

The tangential load, F_t, to be used in the Lewis equation (22.16) is the load for which the drive is to be designed and includes the effect of service conditions of the drive. It may be obtained from the power equation,

$$F_t = \frac{1000 \times P \times C_0}{V}, \text{N} \qquad ...(22.19)$$

where P = power to be transmitted, or rated power of the prime mover, kW.

C_0 = service factor or overload factor, Table 22.8.

V = Pitch line velocity, m/s.

Table 22.8 Overload factors. Co * [8 hr/day operation]

Power Source	Load on driven Machine		
	Uniform	Moderate Shock	Heavy Shock
Uniform [a]	1.00	1.25	1.75
Light Shock[b]	1.25	1.50	2.00
Medium Shock[c]	1.50	1.75	2.25

a = Electric motor; *b* = multicylinder engine.
c = Single cylinder engine.
* = Alter the factors as : multiply by 0.8 for 3 hr/day operation and by 1.25 for 24 hr/day operation.

The stress concentration factor, K_f can also be taken care of while calculating the design stress instead of incorporating it in the Lewis equation (original Lewis equation did not consider the effect of stress concentration). Thus,

$$\text{Design stress} = \frac{\text{elastic limit of material}}{R \times K_f}$$

where R = Reliability factor, that is, factor of safety.

So, Lewis equation will be written as:

$$F_t = f_b \cdot f \cdot m \cdot Y \cdot Cv, N \qquad ...(22.20)$$

22.14.2 Buckingham's Equation for Dynamic Load

As already mentioned, Buckingham based his design on the dynamic load (instead of transmitted load) and the endurance limit of the material. Although one way of considering dynamic effects (due to reason already discussed) is the introduction of velocity factor in the Lewis equation, the method of calculating the dynamic load as given by Buckingham is as follows:

The maximum instantaneous load on the tooth, or the dynamic load, is

$$F_d = F_t + F_i = F_t + \frac{21\,V(c \cdot f \cdot + F_t)}{21\,V + \sqrt{c \cdot f + F_t}}, \text{N} \qquad ...(22.21)$$

where, c = a factor depending upon machining errors, and is given as,

$$c = \frac{k \cdot e}{\frac{1}{E_p} + \frac{1}{E_g}}, \text{N/m}$$

where k = a factor depending upon the form of the tooth

$= 0.107$ for $14\frac{1}{2}^{\circ}$ system

$= 0.111$ for 20° full depth system

$= 0.115$ for 20° stub tooth system

E_p and E_g are the Young's modulus for pinion and gear, N/m^2 and e = tooth action error, m, Table 22.9.

The class of gear cutting required depends upon the speed of operation, and noise is a good measure of the accuracy required. The maximum errors that will permit reasonably quiet operation at different speeds is given in Table 22.10. Knowing the class of gear cutting required, the probable error can be taken from Table 22.9.

Table 22.9 Values of 'e' mm

Module mm	Class 1 industrial	Class 2 accurate	Class 3 precision
Upto 4	0.05	0.025	0.0125
5	0.056	0.028	0.015
6	0.066	0.033	0.018
8	0.080	0.040	0.020
12	0.100	0.050	0.025
25	0.122	0.061	0.030

Table 22.10 Maximum permissible error in gear at various speeds

V m/s	error mm	V m/s	error mm	V m/s	error mm
1.25	0.094	8.75	0.043	16.25	0.020
2.50	0.081	10.00	0.038	17.50	0.018
3.75	0.071	11.25	0.033	20.00	0.015
5.00	0.061	12.50	0.030	22.50	0.015
6.25	0.053	13.75	0.025	25 and above	0.013
7.50	0.048	15.00	0.023		

Stress Concentration: There is a concentration of stress where the tooth joins the bottom land. This causes the actual stress to be from 1.2 to 2 times the stress as obtained by substituting F_d in the Lewis equation. The exact value, however, will depend upon the material, fillet radius, pressure angle, thickness of the tooth at root and the load position on the tooth. The values of K_f have already been mentioned under Art. 22.14.1.

Buckingham obtained the dynamic stress by putting the value of F_b in the Lewis equation and compared it with the endurance limit of the material. The dynamic stress will be,

$$f_d = \frac{K_f \cdot F_d}{m \cdot f \cdot Y}, \text{ (Eqn. 22.16 without } C_v) \qquad ...(22.22)$$

The dynamic stress should be less than the flexural endurance limit and give a reasonable margin of safety, depending upon the type of service. For steel, the flexural endurance limit, that is, in bending is given as :

$$f_{ef} = 1.75 \times \text{BHN, MPa}$$

For BHN = 400 and above, f_{ef} = 700 MPa

For other materials,

f_{ef} = 84 MPa for gray C.I. with BHN 160
= 126 MPa for Semi-steel with BHN 200
= 168 MPa for Phosphor bronze with BHN 100

Another way of considering the effect of dynamic load is to compare it with the beam strength or endurance strength of the tooth. The beam strength will be given as,

$$F_b = F_{en} = f_{ef} \cdot f \cdot m \cdot Y \quad \text{...(22.23)}$$

For safety against breakage, the beam strength of the tooth should be greater than the dynamic load that is,

For steady load, $F_b \geq 1.25\, F_d$
For Pulsating load, $F_b \geq 1.35\, F_d$ ...(22.24)
For Shock load, $F_b \geq 1.50\, F_d$

See, how the stress concentration factor appears in equation (22.24), indirectly.

Note: Apply the check to weaker gear, that is, the gear with lower value of '$f_b \cdot Y$'.

22.14.3 Check for Surface Durability for Wear

After designing the greater tooth for beam strength, the tooth must be checked for wear. The cause of tooth failure by wear have already been explained. According to Buckingham the limiting load for wear is given as:

$$F_w = D_p f \cdot K_w\, Q, \text{N} \quad \text{...(22.25)}$$

where D_p = pitch diameter of pinion, m
f = face width, m
K_w = load stress factor for fatigue, N/m²
Q = Ratio factor.

K_w is given as:

$$K_w = \frac{f_{es}^2 \sin \phi}{1.4} \left[\frac{1}{E_p} + \frac{1}{E_g} \right]$$

and

$$Q = \frac{2z_g}{z_g + z_p} \quad \text{for external gears}$$

$$= \frac{2z_g}{z_g - z_p} \quad \text{for internal gears.}$$

The subscripts 'g' and 'p' refer to gear and pinion,

ϕ = pressure angle
f_{es} = surface endurance limit of a gear pair, N/m²

It must be ensured that $F_w \geq F_d$

If F_w comes out to be less than F_d, it means that the gears are not hard enough. Their hardness will have to be increased by heat treatment.

It is a general rule that less wear occurs if different materials are in contact. Exceptions are : Cast iron on cast iron and heat treated steel tempered to a very hard surface, against equally heat treated hard steel. If the materials are different, the harder material will mechanically work harden the softer material raising its compressive endurance limit. The pinion should always be the harder to allow for work hardening of the gear, to preserve the involute profile, to allow for greater abrasive wear on the pinion, and to decrease the possibility of seizing.

The values for 'f_{es}' for various combinations of materials of gear and pinion are given in Table 22.11.

Table 22.11 Value for f_{es}

Pinion Material	BHN	Gear Material	BHN	f_{es}, MPa
Steel	150	Steel	150	350
Steel	200	Steel	150	420
Steel	250	Steel	150	490
Steel	200	Steel	200	490
Steel	250	Steel	200	560
Steel	300	Steel	200	630
Steel	250	Steel	250	630
Steel	300	Steel	250	700
Steel	350	Steel	250	770
Steel	300	Steel	300	770
Steel	350	Steel	300	840
Steel	400	Steel	300	875
Steel	350	Steel	350	910
Steel	400	Steel	350	980
Steel	400	Steel	400	1050
Steel	150	C.I.	—	350
Steel	200	C.I.	—	490
Steel	250	C.I.	—	630
Steel	150	Phosphor Bronze	—	350
Steel	200	Phosphor Bronze	—	490
Steel	250	Phosphor Bronze	—	595
C.I.	—	C.I.	—	630

Note: If the material of both pinion and gear is steel, then the surface endurance limit may be estimated from the simple relation,

$$f_{es} = (2.8 \times \text{BHN} - 70)\ \text{MPa}$$

where BHN is the average Brinell hardness of the pinion and gear pair.

22.14.4 Design of Cast Teeth

Low speed drives are sometimes made with cast teeth. They are mostly employed for rough outdoor service where noise is not objectionable. For such drives, the peripheral speed should not exceed about 2.5 m/s. The dimensions for the cast teeth can be calculated from the modified beam strength equation:

$$F_b = f_b f \pi m y, \text{N}$$

f_b, may be taken as = 56 MN/m^2 for ordinary C.I.

= 84 MN/m^2 for high grade or Nickel C.I.

= 140 MN/m^2 for cast steel

$f = 2$ to $2.5\ p_c$

= 6.25 m to 8 m

22.14.5 Design Procedure

Basically, the design procedure for finding the tooth proportions consists of the following steps:

1. If the peripheral speeds are not given, these should be selected. Its value may be taken as 3 to 5 m/s for a low rotatory speed of the pinion and a low to moderate power transmission. For higher values of pinion speed and higher power, greater values for peripheral speed may be assumed.

2. If the materials for pinion and gear are also not specified, these should also be selected. For a low velocity, C.I. and low grade steel may be used but for a higher velocity, better grade materials should be selected.
3. Design transmitted load, F_t is found out, by equation (22.19).
4. To solve the Lewis equation (22.16), the preliminary values for 'f' and Y are assumed:

$$f = 9.5 \text{ m to } 12.5 \text{ m}$$

and $Y = 0.25$ for 14½° tooth form.
$= 0.29$ for 20° full depth tooth form.
$= 0.35$ for 20° stub tooth.

The Lewis equation is then solved for m and the nearest standard value is selected. It is better to select a lower value of 'm' to have an overall small size of the drive.

If both the gears are made of the same material, only the weaker pinion is to be considered for using Lewis equation to determine the preliminary value of 'm'.

Note: A weaker gear is one with the smaller value of $f_b \cdot Y$.

5. The approximate pinion diameter is determined from the selected value of V and the number of teeth z_g and z_p are found out.
6. Exact values for Y are calculated both for pinion and gear.
7. The Lewis equation is solved both for pinion and gear and for each the value for 'f' determined. Greater value is chosen.
8. The gears are checked and modified if need be by Buckingham's equation for dynamic load and for wear to ensure that $F_w \geq F_d$ and $F_b \geq F_d$.

Note: Sometimes, for guidance sake, the following empirical relations may be used:

$$m = (0.01 - 0.02)\ C,\ C = \text{Centre distance}$$

The lower value are taken for more precise and rigid drives.

and $\dfrac{f}{C} = 0.12$ to 0.40, the higher values are for more precise and rigid drives.

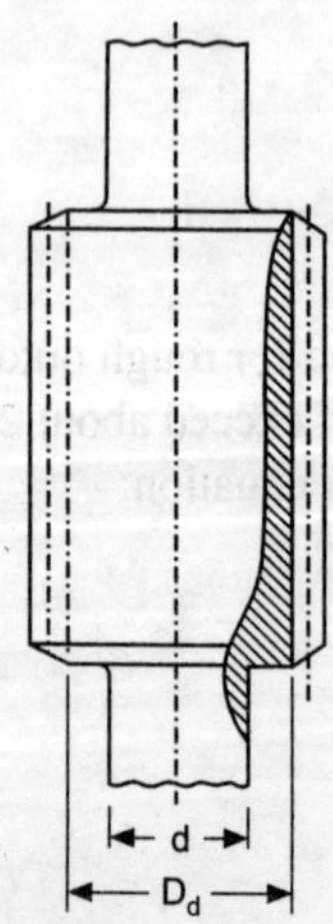

Fig. 22.9 *Pinion integral with shaft*

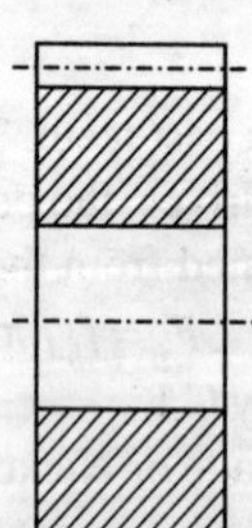

Fig. 22.10 *Solid pinion*

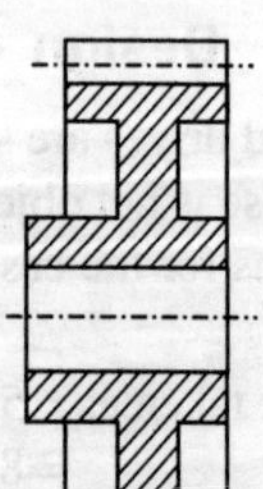

Fig. 22.11 *Pinion with web*

Note: In the above design procedure, the design parameters are to be adjusted so that the beam and wear strengths are equal and both equal or exceed the dynamic load. In many instances, however, it is neither practical nor desirable to go to the limits that might be required to equalise the beam and wear strengths of the gears. In such a situation, the lower of the two strengths must be greater than the dynamic load.

22.14.6 Gear Proportions

The gear construction may have different designs depending upon size and application. If the dedendum circle diameter D_d is just greater than the shaft diameter by an amount less than half the shaft diameter, the pinion is made integral with the shaft, Fig. 22.9. If the diameter of the pinion is:

$$D \geq 14.75\ \text{m} + 60\ \text{mm}$$

then the pinion is made solid with uniform thickness equal to the face width *f*, Fig. 22.10, The limit for a pinion or gear with a web is given by the equation,

$$D \geq 23.5\ \text{m} + 85\ \text{mm}$$

The web thickness should be from 1.6 m to 1.9 m. The web may be solid, Fig. 22.11, or with recesses to reduce its weight.

Large gears have arms joining the rim to the hub, the number of arms depends on the size of the gear. The number of arms may be decided as below :

4 or 5 arms for diameter upto 0.50 m
6 arms for diameter from 0.50 to 1.50 m
8 arms for diameter from 1.50 to 2.00 m
10 arms for diameter greater than 2.00 m

Design of Arms : The arms are designed by assuming them as cantilevers fixed at the hub end and loaded at the pitch line, with the load equally distributed to all arms. The load to be considered is the stalling load, *i.e.*, the load when maximum stress will be induced in the arms and in the teeth. This will happen at zero velocity when the drive is just to start operating. This stalling load may be taken as the design transmitted load divided by the velocity factor C_v.

$$\therefore \qquad \text{Stalling load, } F_0 = \frac{F_t}{C_v}$$

Thus, the bending moment on each arm will be :

$$M_b = \frac{F_0 \times r_g}{z_a}$$

where r_g = pitch radius of the gear wheel, z_a = number of arms.

$\therefore$ Section modulus of arm, $Z = \dfrac{M_b}{f_b}$, where f_b = allowable tensile stress in the arm material.

Arms of different cross-sections are used in gears. For small and medium sized gears, the usual section used is elliptical with major axis twice the minor axis. For this

$$Z = \frac{\pi}{64} b^3, \text{ where } b = \text{major axis}$$

$$M_b = f_b \times Z = f_b \times \frac{\pi}{64} b^3.$$

From this equation, the value of '*b*' at the outside of the hub can be determined. The arms are usually tapered towards the rim, about 65 mm per metre length of the arm. A gear with elliptical arms is shown in Fig. 22.12.

The other arm sections used on large and wide gears are shown in Fig. 22.13.

In addition to the maximum transmitted load coming on the arms, the arms are also under centrifugal force on the rim which produces a combination of bending and tension. It is very difficult to determine very accurately this effect. There are also stress concentration effects where the arm joins the rim. The arms are designed on the assumption that they are to have the same theoretical strength as the teeth. However, while designing the teeth, the stress concentration factor was incorporated in the Lewis equation. So, while determining the safe bending stress for arms, stress concentration factor should be considered. The stress concentration in the case of arms is not so severe as in tooth root. But, since centrifugal force on the rim is not being considered in the design of arms, therefore, a value of stress concentration factor as given below, will compensate for this,

$$K_f = 1.5 \text{ for ductile materials}$$
$$= 1.75 \text{ for heat treated materials}$$

so,
$$f_b = \frac{\text{elastic limit}}{R \times K_f}$$

where $R = 1.5$ as already discussed.

Or on the basis of ultimate strength, the factor of safety may be taken as $k_f \times 3$ (used for gear tooth design), that is, 4.5 for ductile materials and 5.25 for heat treated materials.

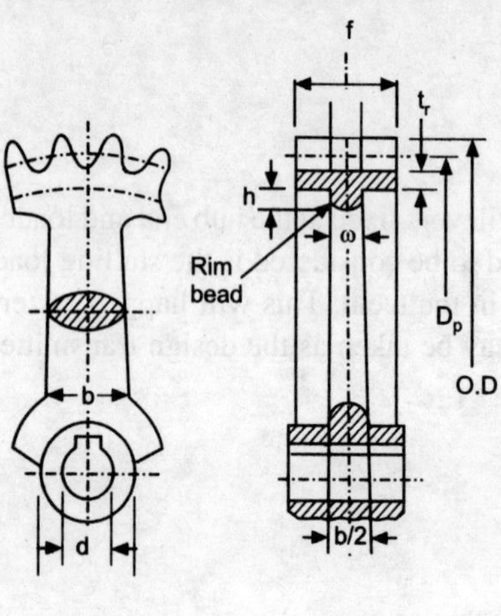

Fig. 22.12 *Gear proportions*

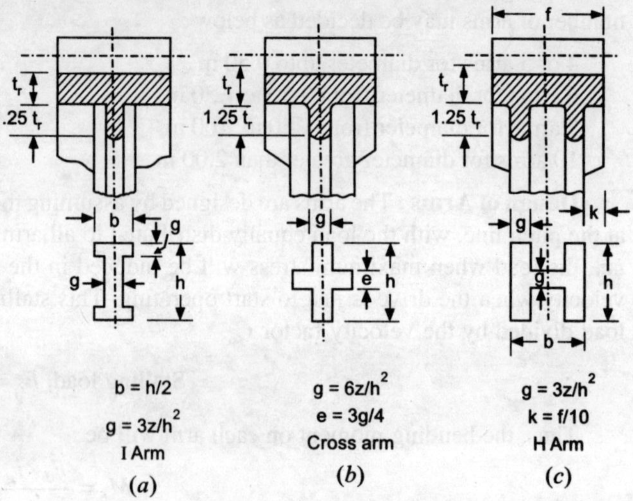

Fig. 22.13 *Gear arm sections*

Rim : Figure 22.12, the minimum thickness of rim below the root circle is usually given empirically as 1.6 m to 1.9 m. It can be determined by the following relation also:

$$t_r = m\left(\frac{z}{2z_a}\right)^{\frac{1}{3}}$$

where z = number of teeth on gear, z_a = number of arms.

If by this formula, t_r comes out to be less than 1.6 m then make t_r = 1.6 m. The rim is stiffened by providing a circumferential rib, with depth,

$h = t_r$ and width w = thickness of arm at this end.

Shaft and Hub: The diameter of the shaft can be found out by designing it under the combined effect of torsional moment transmitted by it and its bending. The bending moment acting on the shaft will be due to the weight of the gear or pinion and due to the normal force acting between the tooth surfaces. Figure 22.14 shows the various forces acting on a pair of gears. The normal force F_n can be found as:

$F_n = F_t/\cos\phi$, A thrust parallel and equal to F_n will act at gear centre.

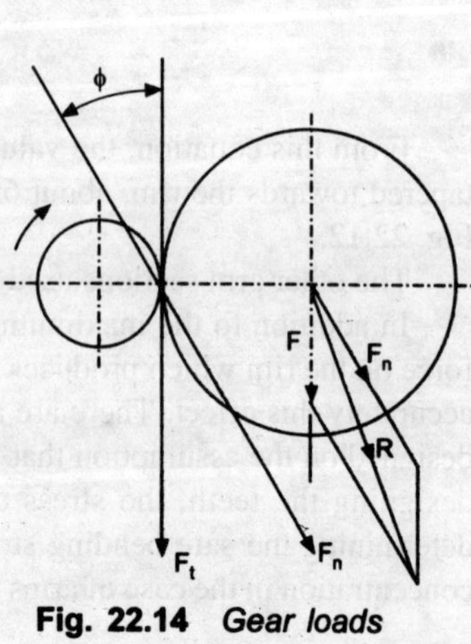

Fig. 22.14 *Gear loads*

If F is the weight of gear, then the resultant of F_n and F will be:

$$R = \sqrt{F_n^2 + F^2 + 2F_n F \cos\phi}$$

the weight of the gears may be taken approximately as,

$$F = kzf\, p_c^2,\ N\ (f \text{ and } p_c \text{ are in cms})$$

$$k = 0.17 \text{ for spur gears} = 0.098 \text{ for bevel gears}$$

Now the bending moment due to this force will depend upon the manner of gear mounting. If the gear is overhanging, then bending moment on the shaft will be

$$M_b = R \times \text{overhang}$$

Torsional moment $M_t = F_t \times \dfrac{D_g}{2}$

Note: The maximum torque on the shaft will be when the drive is just going to start operating. At that moment, the tooth load F_t will be equal to the stalling load F_o. Therefore, in the above analysis, do calculations after making $F_t = F_o$.

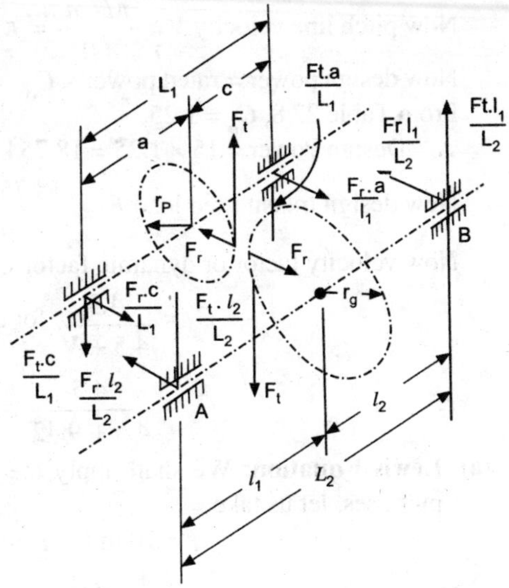

Fig. 22.15 *Spur gear bearing loads*

$\therefore$ Equivalent torsional moment $M_{te} = \sqrt{M_b^2 + M_t^2}$

$$= \frac{\pi}{16} d^3 \times f_s$$

where d = diameter of shaft

f_s = safe shear stress for the shaft material.

$\therefore$ $d = ?$

To find f_s take factor of safety equal to 3 (same as taken for gear tooth design) or determine its value according to the procedure given in Chapter 16.

If the gear is mounted in between bearings, then the bending moment can again be found out if the position of the gear in relation to the bearings is known. After this, the shaft diameter can be determined as before. The bearing loads in that case are shown in Fig. 22.15.

Note: Torsional moment on pinion shaft = $\dfrac{\text{Torsional moment on gear shaft}}{\text{Gear ratio}}$

Hub : The diameter of hub = 1.8 *d* for steel gear = 2*d* for C.I. gear

= 1.65 *d* for forged steel gear.

Minimum Hub Length = 1.25 *d*, for light service and ≮ *f*.

Example 22.1. A compressor running at 300 rev/min is driven by a 15 kW, 1200 rev/min motor through a 14½° full depth gears. The centre distance is 0.375 m. The motor pinion is to be of C-30 forged steel hardened and tempered, and the driven gear is to be of cast steel. Assuming medium shock condition:

(a) Determine the module, the face width, and the number of teeth on each gear.

(b) Check the gears for dynamic load and wear.

(c) Design the drive completely.

Solution: Now centre distance = $\dfrac{D_p}{2} + \dfrac{D_g}{2}$

$$\therefore \quad \frac{D_p}{2} + \frac{D_g}{2} = 0.375$$

Hence $D_p = 0.15$ m and $D_g = 0.60$ m since the speed ratio is

$$V_R = \frac{\omega_1}{\omega_2} = \frac{n_1}{n_2} = \frac{1200}{300} = 4$$

Now pitch line velocity $V = \dfrac{\pi D \cdot n}{60} = \pi \times \dfrac{0.15}{60} \times 1200 = 9.42$ m/s

Now design power = rated power × C_o

From Table 22.8, $C_o = 1.25$

∴ Design Power = 15 × 1.25 = 18.75 kW

Now design transmitted load $F_t = \dfrac{18.75 \times 1000}{9.42} = 1990$ N

Now velocity factor or dynamic factor C_v is given as

$$C_v = \frac{4.5}{4.5 + V} \quad \text{for carefully cut gears with } V \text{ upto 12.5 m/s}$$

$$\therefore \quad C_v = \frac{4.5}{4.5 + 9.42} = \frac{4.5}{13.92} = 0.323$$

(a) Lewis Equation: We shall apply the Lewis equation to gear for preliminary calculations. For trial purposes, let us take

$f = 10$ m for $Y = 0.25$ as listed in the design procedure.

Now $$F_t = \frac{f_b \cdot fmY}{K_f} \cdot C_v, N$$

From Table 22.7,

f_b, design stress is 140 MPa, for Cast Steel

Now K_f = stress concentration factor

= 1.6 for full depth $14\frac{1}{2}$° teeth

$$\therefore \quad 1990 = \frac{140 \times 10^6 \times 10m \times m \times 0.25}{1.6} \times 0.323$$

$$\therefore \quad m = 0.0053 \text{ m}$$

Now from Table 22.2 let us take the standard module of 5 mm

∴ z_p = number of teeth on pinion

$$= \frac{D_p}{m} = \frac{0.15}{0.005} = 30$$

z_g = number of teeth on gear

$$= \frac{D_g}{m} = \frac{0.60}{0.005} = 120$$

Now the exact value of Lewis form factor Y can be determined.

For pinion, $$Y = \pi\left(0.124 - \frac{0.684}{z_p}\right) = \pi\left(0.124 - \frac{0.684}{30}\right) = 0.318$$

Fro gear, $$Y = \pi\left(0.124 - \frac{0.684}{120}\right) = 0.375$$

Now applying the Lewis strength equation for pinion and gear to find the tooth face width.

For pinion:

$$F_t = \frac{f_b \times f \times m \times Y}{K_f} C_v$$

$$\therefore \quad f = \frac{1990 \times 1.6}{224 \times 10^6 \times 0.005 \times 0.318 \times 0.323} = 0.0277 \text{ m}, \; [(f_b = 224 \text{ MPa (Table 22.7)}].$$

For Gear:

$$\therefore \quad f = \frac{1990 \times 1.6}{140 \times 10^6 \times 0.005 \times 0.375 \times 0.323} = 0.376 \text{ m}$$

Let us take the value 'f' as 45 mm, both for pinion and gear, since it should be between 9.5 m to 12.5 m.

Gear Tooth Proportions:

Addendum, $a = m = 5$ mm

Dedundum, $d = 1.25\ m = 6.25$ mm

Clearance $= 0.25\ m = 1.25$ mm

Total depth $= a + d = 11.25$ mm

Working depth $= 11.25 - 1.25 = 10$ mm

Circular pitch, $p_c = \pi m = 15.71$ mm

Now from Table 22.4, for $m = 5$ mm and $V > 8$ m/s, backlash $= 0.28$ mm

Now tooth thickness + width of space = circular pitch

$\therefore$ (Width of space-backlash) + width of space = circular pitch

$$\text{Width of space} = \frac{15.71 + 0.28}{2} = 7.995 \text{ mm}$$

Tooth thickness $= 7.995 - 0.28 = 7.715$ mm

Diameter of Addendum circle for pinion $= D_p + 2a = 0.15 + 2 \times 0.005 = 0.16$ m

Diameter of Addendum circle for gear $= D_g + 2a = 0.60 + 2 \times 0.005 = 0.61$ m

Diameter of Dedendum circle for pinion $= D_p - 2d = 0.15 - 2 \times 0.00625 = 0.1375$ m

Diameter of Dedendum circle for gear $= D_g - 2d = 0.60 - 2 \times 0.00625 = 0.5875$m

(b) Check for dynamic Load: By Buckingham's equation,

$$F_d = F_t + F_i$$

$$F_i = \frac{21V(c \cdot f + F_t)}{21V + \sqrt{c \cdot f + F_t}}$$

Now for $V = 9.42$ m/s, the permissible error from Table 22.10 is approximately 0.040 mm. For module equal to 5 mm, a class 2 gear would have an error of 0.028 mm and class 1 gear of 0.056 mm. Since the permissible error is 0.040 mm, therefore, a class 2 gear must be used with error of 0.028 mm.

$$\text{Now} \quad c = \frac{k \cdot e}{\frac{1}{E_p} + \frac{1}{E_g}}$$

$k = 0.107$ for 14½° system

$E_p = E_g = 210$ GPa

$$\therefore \quad c = \frac{0.107 \times 0.028 \times 210 \times 10^9}{2 \times 10^3} = 314.58 \text{ kN/m}$$

$$\therefore \quad F_i = \frac{21 \times 9.42(314.58 \times 10^3 \times 0.045 + 1990)}{21 \times 9.42 + \sqrt{314.58 \times 10^3 \times 0.045 + 1990}} = 9831.4 \text{ N}$$

$$\therefore \quad F_d = 1990 + 9831.4 = 11821.4 \text{ N}$$

Now beam strength for Gear, (gear is weaker than pinion, compare $f_b \cdot Y$)

$$F_b = f_{ef} \cdot f \cdot m \cdot Y, \text{N}$$

Now from Table 22.6, $f_{ef} = 225$ MPa with BHN = 160.

$$\therefore \quad F_b = 225 \times 10^6 \times 0.045 \times 0.005 \times 0.375 = 18984.3 \text{ N}$$

which is $> 1.5\ F_d$ $\therefore$ safe.

Now for pinion, let us take $F_b = 1.5\, F_d = 17.732$ kN

$$\therefore \quad f_{ef} = \frac{17.732 \times 10^3}{0.045 \times 0.005 \times 0.318} = 247.83 \text{ MPa}$$

$$\therefore \quad \text{BHN of pinion material} = \frac{247.83}{1.75} = 141.6 \text{ say } 150$$

Check for Wear: The limiting wear load is,

$$F_\omega = D_p f K_w Q, N$$

Now $D_p = 0.15$ m, $f = 0.045$ m

$$K_w = \frac{f_{es}^2 \sin\phi}{1.4}\left(\frac{1}{E_p} + \frac{1}{E_g}\right)$$

f_{es} = surface endurance limit = 2.8 × BHN – 70 MPa

Now average value of BHN for gear pair = 155

Now $E_p = E_g = 210$ GPa

$\therefore \quad f_{es} = 2.8 \times 155 - 70 = 364$ MPa

$\sin\phi = \sin 14\tfrac{1}{2}° = 0.25$

$$\therefore \quad K_w = \frac{(364 \times 10^6)^2 \times 0.25}{1.4}\left(\frac{2}{210 \times 10^9}\right) = 225.9 \times 10^3, \text{N/m}^2$$

Q = ratio factor

$$= \frac{2z_g}{z_g + z_p} = \frac{2 \times 120}{120 + 30} = \frac{240}{150} = 1.6$$

$\therefore \quad F_w = 0.15 \times 0.045 \times 225.9 \times 10^3 \times 1.6 = 2439.74$ N, which is quite less than F_d.

∴ The gear pair will fail in wear. Therefore, it would be necessary to heat treat the gear pair to an average BHN of 315.

$\therefore \quad f_{es} = 2.8 \times 315 - 70 = 812$ MPa

$$\therefore \quad K_w = \text{load stress factor} = \frac{(812 \times 10^6)^2 \times 0.25}{1.4}\left(\frac{2}{210 \times 10^9}\right) = 1124 \times 10^3, \text{N/m}^2$$

$\therefore \quad F_w = 0.15 \times 0.045 \times 1124 \times 10^3 \times 1.6 = 121.4 \text{ kN} > F_d$, hence, O.K., also $F_b \geq F_w$, ∴ O.K.

(c) Design of Drive:

(i) Pinion: Now since the diameter of the pinion is 0.15 m and the limit of diameter for the provision of web, is

$$D \leq 23.5\, m + 85$$

$$\leq 23.5 \times 5 + 85, \leq 207.5 \text{ mm or } 0.2075 \text{ m}$$

∴ The pinion will be provided with a web and not arms.

web thickness = 1.6 m to 1.9 m

i.e., 8 mm to 9.5 mm, let us take 9 mm

Rim thickness: $t_r = 1.6$ m i.e., 8 mm

Shaft : The shaft should be designed for the maximum torsional moment to be transmitted and not for the useful torsional moment. Maximum torsional moment will be induced in the shaft when the driver is started from stationary position. The tooth load acting at the pitch line at that moment will be, corresponding to the stalling load.

Maximum Tangential load $= F_0 = \dfrac{F_t}{C_v} = \dfrac{1990}{0.323} = 6161$ N

$\therefore$ Torsional moment on the pinion shaft $= 6161 \times \dfrac{0.15}{2} = 462.07$ Nm

$$\text{Now } F_n = \frac{6161}{\cos\phi} = \frac{6161}{\cos 14\frac{1}{2}^\circ} = \frac{6161}{0.968} = 6364.67 \text{ N}$$

Neglecting the weight of the pinion, if we consider the pinion to be over hanging.
Bending moment, $M_b = F_n \times$ over hang.
Let us take overhang, distance from the centre of pinion to centre of bearing 0.10 m.

$\therefore$ $M_b = 6364.67 \times 0.10 = 636.47$ Nm

$\therefore$ Equivalent torsional moment, $M_{te} = \sqrt{M_b^2 + M_t^2} = \sqrt{636.47^2 + 462.07^2} = 786.52$ Nm

$$M_{te} = \frac{\pi}{16} d^3 \times f_s$$

Now the shaft material is usually C–45 and C–55 steel with ultimate tensile strength of 450 MPa and 550MPa. Let us choose C–55 steel.

Now shear stress (ultimate) = 60% of 550 = 330 MPa.

$\therefore$ $f_s = \dfrac{330}{3} = 110$ MPa

Allowing for keyway, $f_s = 110 \times 0.75 = 82.5$ MPa
Now as per procedure given in Chapter 16, $f_s = 550 \times 0.18 \times 0.75 = 74.25$ MPa
Let us take, $f_s = 82.5$ MPa

$\therefore$ $786.52 = \dfrac{\pi}{16} d^3 \times 82.5 \times 10^6$

$\therefore$ $d = 0.0365$ m, say 40 mm
Hub diameter $= 40 \times 1.65 = 66$ mm
Hub length $= 1.25 \times 40 = 50$ mm

(ii) Gear: Since the diameter of gear is 0.60 m, six arms will be provided.
Arms Stalling Load $F_0 = 6161$ N

$\therefore$ Bending moment, per arm,

$$M_b = \frac{6161 \times r_g}{z_a} = \frac{6161 \times 0.30}{3} = 308.05 \text{ Nm}$$

Taking elliptical section for the arm. Let us take F.O.S. as 5 as already discussed.

Design stress in bending $f_b = \dfrac{550}{5} = 110$ MPa

Now section modulus, $Z = \dfrac{\pi}{64} b^3$, b = major axis at hub

$\therefore$ $308.05 = \dfrac{\pi}{64} b^3 \times 110 \times 10^6$

$\therefore$ $b = 0.385$ m that is 40 mm

$\therefore$ minor axis at hub = 20.0 mm

Now major axis at rim end $= b -$ taper

$$= 40 - \frac{65}{1000} \times 300 = 20.5 \text{ mm}$$

$\therefore$ Minor axis at the rim end = 10.25 mm

Rim: The rim thickness is given as

$$t_r = m \cdot \left(\frac{z}{2z_a} \right)^{\frac{1}{3}}$$

Now $$z = 120, \text{ and } z_a = 6$$

$$t_r = 5 \cdot \left(\frac{120}{2 \times 6} \right)^{\frac{1}{3}} = 5 \times 2.16 = 10.8 \text{ mm, say 11 mm}$$

Depth of circumferential rib, $h = t_r = 11$ mm

Thickness of rib, w = thickness of arm at this end = 10.25 mm

Shaft: Torsional moment on the shaft may be considered as the stalling torsional moment to be transmitted. Therefore,

$$M_t = F_0 \times r_g = 6161 \times 0.30 = 1848.3 \text{ Nm}$$

Now $$F_n = \frac{F_0}{\cos \phi} = 6364.67 \text{ N}$$

Neglecting the bending moment due to the weight of gear, bending moment on the shaft,

$$M_b = F_n \times \text{overhang} = 6364.67 \times 0.10 = 636.47 \text{ Nm}$$

$$\therefore \quad M_{te} = \sqrt{M_t^2 + M_b^2} = \sqrt{1848.3^2 + 636.47^2} = 1954.8 \text{ Nm}$$

$$\frac{\pi}{16} d^3 \times 82.5 \times 10^6 = 1954.8$$

$$\therefore \quad d = 0.0494 \text{ m, say 50 mm}$$

Hub diameter = 1.8 d = 1.8 × 50 = 90 mm

Length of Hub = 1.25 d = 1.25 × 50 = 62.5 mm.

Example 22.2. Determine the proper pitch, module, face width, number of teeth, and outside diameters of a pair of 20° involute full depth spur gears to transmit 112.5 kW, from a pinion running at 750 rev/min to a gear running at 140 rev/min. The service is intermittent with light shocks.

Solution: Since nothing is mentioned about diameters, centre distance, materials and pitch line velocity, suitable assumptions will have to be made for these and then the drive calculated and if need be suitable changes made afterwards.

Pitch line velocity : Since the speed of pinion and of gear is not very high, a moderate value of pitch line velocity may be selected. As the pitch line velocity for medium velocity drives ranges from 3 to 15 m/s let us choose a velocity of 8 m/s.

Materials: Suitable materials for pinion and gear may be taken as : Forged C-30 steel with ultimate tensile strength of 500 MPa and Cast Iron, grade 35, heat treated, with ultimate tensile strength of 350 MPa respectively, from Table 22.6.

Now the approximate pitch diameter of pinion.

$$D_p = \frac{60 \times V}{\pi n_p} = \frac{60 \times 8}{\pi \times 750} = 0.204 \text{ metre}$$

(1) Design load :

Now useful transmitted load, $$F_t = \frac{1000 \times 112.5}{8} = 14.0625 \text{ kN}$$

Now, $$C_v = \frac{4.5}{4.5 + V} = \frac{4.5}{4.5 + 8} = \frac{4.5}{12.5} = 0.360.$$

If the power source is taken as electric motor, then since the service is intermittent, service factor C_0 from Table 22.8.

$$= 1.25 \times 0.8 = 1.00$$

$\therefore$ Design tangential load, $F_t = 1.00 \times 14.0625 = 14.0625$ kN

Gear Strength : The Lewis equation is :

$$F_t = \frac{f_b f\, m Y C_v}{K_f}, \text{N}$$

Let us take $Y = 0.29$ and $f = 10\,m$, $K_f = 1.5$

Let us apply the gear strength equation for the pinion,

$$FS = 3$$

$$\therefore \quad \text{Design stress } f_b = \frac{500}{3} = 166.7 \text{ MPa}$$

$$\therefore \quad 14.0625 \times 10^3 = \frac{166.7 \times 10^6 \times 10m \times m \times 0.29 \times 0.36}{1.5}$$

$$m = 0.011 \text{ m, that is 11 mm}$$

Let us choose a standard value of 'm' as 10 mm (First choice, Table 22.2)

Now number of teeth on pinion $= \dfrac{D_P}{m} = \dfrac{0.204 \times 1000}{10} = 20.4 > 17$, minimum needed $\therefore$ O.K.

Let us make $z_p = 20$

$$\therefore \quad \text{Exact, } D_p = 20 \times 10 = 200 \text{ mm} = 0.20 \text{ m}$$

$$V = \pi \times 0.20 \times \frac{750}{60} = 7.85 \text{ m/s, and } C_v = \frac{4.5}{12.35} = 0.364$$

$$\therefore \quad F_t = \frac{1000 \times 112.5}{7.85} = 14.2 \text{ kN}$$

$\therefore$ Exact value of Y, for pinion

$$= \pi\left(0.154 - \frac{0.912}{z_p}\right) = \pi\left(0.154 - \frac{0.912}{20}\right) = 0.341$$

$$\therefore \quad m^2 = \frac{14.2 \times 10^3 \times 1.5}{166.7 \times 10^6 \times 10 \times 0.341 \times 0.364} = 1.0294 \times 10^{-4} \text{ m}^2$$

$$\therefore \quad m = 0.010 \text{ m, that is, 10 mm}$$

$\therefore$ Chosen standard value of $m = 10$ mm is O.K.

Now from the Lewis equation for pinion,

$$f = \frac{14.2 \times 10^3 \times 1.5}{166.7 \times 10^6 \times 0.01 \times 0.341 \times 0.364} = 0.103 \text{ m} \doteq 103 \text{ mm}$$

For Gear :

$$z_g = 20 \times \frac{750}{140} = 107$$

Making $z_g = 107$, will give $n_g = 749$ rev/min, i.e., an error of only 0.133% which is negligible.

$$Y = \pi\left(0.154 - \frac{0.912}{107}\right) = 0.457$$

$$f_b = \frac{350}{3} = 117 \text{ MPa}$$

$$\therefore \quad f = \frac{14.2\times10^3\times1.5}{117\times10^6\times0.457\times0.01\times0.364} = 0.1094 \text{ m} = 109.4 \text{ mm}$$

Now since f lies between 9.5 m to 12.5 m, $\therefore$ Let us take $f = 110$ mm

Outside diameters :

$$(D_a)\,p = m\,(z_p + 2) = 10\,(20 + 2) = 220 \text{ mm}$$
$$(D_a)\,g = m\,(z_g + 2) = 10\,(107 + 2) = 1090 \text{ mm}$$

Check by Buckingham's equation for dynamic Load :

Now $F_i = \dfrac{21\,V\,(c.f. + F_t)}{21V + \sqrt{c.f. + F_t}}$

Now for $V = 7.83$ m/s, the maximum permissible error from Table 22.10 is approximately 0.047 mm. For module equal to 10 mm, a class 2 gear would have an error of about 0.045 mm. Therefore, a class 2 gear must be used, with $e = 0.045$ mm.

Now $c = \dfrac{k.e}{\dfrac{1}{E_p} + \dfrac{1}{E_g}}$

For 20º full depth profile, $k = 0.111$

$$E_p = 210 \text{ GPa}, E_g = 105 \text{ GPa}$$

$$\therefore \quad c = \frac{0.111\times0.045\times210\times10^9}{3\times10^3} = 349.65, \text{ kN/m}$$

$$\therefore \quad F_i = \frac{21\times7.85(349.65\times10^3\times0.11 + 14.2\times10^3)}{21\times7.85 + \sqrt{349.65\times10^3\times0.11 + 14.2\times10^3}} = 22 \text{ kN}$$

$$\therefore \quad F_d = F_t + F_i = 36.2 \text{ kN}$$

Now strength factor for pinion,

$$f_b \cdot Y = 116.7\times10^6\times0.341 = 56.845\times10^6$$

and for gear, $f_b \cdot Y = 117\times10^6\times0.457 = 53.47\times10^6$

$\therefore$ Gear is weaker.

Now Beam strength of gear$= f_{ef} f.\, m\,.\, Y, N$

Now f_{ef} for C.I. with BHN 160= 84 MPa

$$\therefore \quad F_b = 84\times10^6\times0.11\times0.01\times0.457 = 42.27 \text{ kN}$$

Now, for light shocks, $F_b \geq 1.25\,F_d$, that is ≥ 45.25 kN

Now from Table 22.6, iron grade 35 heat treated has BHN = 300

$\therefore$ The material would have much higher value of endurance limit (in bending) and hence beam strength. So, the design is safe for dynamic load.

Now f_{ef} for steel pinion $= \dfrac{45.25\times10^3}{0.11\times0.01\times0.341} = 120.63\times10^6$

$\therefore$ BHN for steel pinion (required) $= \dfrac{120.63}{1.75} = 68.93$

Check for Wear

Now Limiting Wear Load, $F_w = D_p K_w fQ, N$

$$D_p = 0.20 \text{ m}, f = 110 \text{ mm}$$

$$Q = \frac{2z_g}{z_g + z_p} = \frac{2 \times 107}{107 + 20} = 1.685$$

Now $$K_w = \frac{f_{es}^2 \times \sin\phi}{1.4}\left[\frac{1}{E_p} + \frac{1}{E_g}\right]$$

$$\sin\phi = \sin 20^\circ = 0.342$$

$$E_p = 210 \text{ GPa}, E_g = 105 \text{ GPa}$$

Now from table 22.6, BHN value for the forged C-30 steel with $f_{ut} = 500$ MPa, is 143. Now from Table 22.11 for steel pinion with BHN = 150 and C.I. gear, $f_{es} = 350$ MPa.

$$\therefore \quad K_w = (350 \times 10^6)^2 \times 0.342 \frac{1}{1.4} \times \frac{3}{210 \times 10^9} = 427.5 \times 10^3, \text{N/m}^2$$

$$\therefore \quad F_w = 0.20 \times 427.5 \times 10^3 \times 0.11 \times 1.685 = 15.80 \text{ kN}$$

which is less than F_d, $\therefore$ The design is not safe for wear.

Let us make $F_w \geq F_d = 39.5$ kN, say

$$\therefore \quad K_w = \frac{39.5 \times 10^3}{0.20 \times 0.11 \times 1.685} = 1065.55 \times 10^3, \text{N/m}^2$$

$$\therefore \quad f_{es}^2 = \frac{1065.55 \times 1.4 \times 210 \times 10^9 \times 10^3}{0.342 \times 3} = 305333 \times 10^{12}$$

$$\therefore \quad f_{es} = 552.57 \text{ MPa}$$

Now from Table 22.11 the required BHN for steel pinion will be about 225.

Example 22.3. A rawhide pinion is to transmit 30 kW at 1150 rev/min. Select a standard module for 20° full depth involute teeth, assuming medium shock condition and the power source as multicylinder engine.

Solution: Design Load: Design transmitted load $F_t = \dfrac{kW \times 1000 \times C_0}{V}$, N

From Table 22.8, overload factor $C_0 = 1.5$

Now $$V = \frac{\pi D n}{60} \text{ m/s}$$

Now $D = mz$, where z = number of teeth in the pinion

$$\therefore \quad F_t = \frac{kW \times 60 \times 1000 \times 1.5}{\pi\, m\, n\, z}$$

Now for a 20° full depth involute teeth, the minimum number of teeth is 17. So, let us take $z = 17$ in the above equation.

$$\therefore \quad F_t = \frac{30 \times 60 \times 1000 \times 1.5}{\pi m \times 17 \times 1150} = \frac{43.96}{m} \text{ N} \qquad \text{...(i)}$$

Now $$C_v = \frac{0.75}{1.00 + V} + 0.25$$

Now $$V = \frac{\pi m z n}{60} = \frac{\pi m \times 17 \times 1150}{60} = 1023.64\, m \text{ m/s}$$

$$\therefore \quad C_v = \left(\frac{0.75}{1.00 + 1023.64\, m} + 0.25\right)$$

Now Lewis equation is :

$$\therefore \quad F_t = \frac{f_b fm\, YC_v}{K_f}, \text{N}$$

$$\text{Now} \quad Y = \pi\left(0.54 - \frac{0.912}{17}\right) = 0.3142$$

$$\text{Let} \quad K_f = 1.5 \text{ and } f = 10\,m$$

$$\therefore \quad F_t = \frac{f_b \times 10m^2 \times 0.3142 \times C_v}{1.5} = 2.09\, f_b\, m^2 \times C_v \qquad \text{...(ii)}$$

∴ From equations (i and ii)

$$2.09 f_b m^2 = \frac{43.96}{m\left[\dfrac{0.75}{1.00 + 1023.64\,m} + 0.25\right]}$$

Let us assume, $m = 10$ mm

$$\therefore \quad \text{Stress induced, } f_b = \frac{43.96}{2.09 \times 0.01^3 \times 0.3157} = 66.41 \text{ MPa}$$

which is higher than the allowable stress of 56 MPa, Table 22.7.
Let us take $m = 11$ mm $= 0.011$ m

$$\therefore \quad \text{Stress induced } f_b = \frac{43.96}{0.011^3 \times 2.09 \times 0.3112} = 50.78 \text{ MPa}$$

which is OK.

∴ Pinion diameter $= 17\,m = 17 \times 11 = 187$ mm

face width $= 10\,m = 10 \times 11 = 110$ mm.

Example 22.4. Two shafts 0.30 m apart transmitting 18.75 kW are to be connected by a steel pinion meshing with a cast iron gear. The velocity ratio is to be 3 to 1 and the smaller gear is to run at 600 rev/min. The ultimate strength of the material for gear is 168 MPa and the factor of safety is 4. Design the arms for the gear and find the diameter of the gear shaft.

Solution: Now if D_1 = diameter of pinion
and D_2 = diameter of gear,

$$\text{then} \quad \frac{D_1}{D_2} = \frac{200}{600} \text{ (velocity ratio is 3)}$$

Also $D_1 + D_2 = 0.60$ m

∴ $D_1 = 0.15$ m and $D_2 = 0.45$ m

$$\text{Now velocity at pitch } V = \pi \times 0.15 \times \frac{600}{60} = 4.71 \text{ m/s}$$

$$\text{Now} \quad C_v = \frac{3}{3+V} = \frac{3}{3+4.71} = \frac{3}{7.71} = 0.389$$

$$\text{Now} \quad F_t = \frac{kW \times 1000}{V} = \frac{18.75 \times 1000}{4.71} = 3.98 \text{ kN}$$

Now since nothing is mentioned about service factor, so let us neglect it.

$$\therefore \quad \text{Stalling load} = \frac{F_t}{C_v} = \frac{3.98}{0.389} = 10.23 \text{ kN}$$

Let the number of arms = 5

$\therefore$ Bending moment per arm $= 10.23 \times 1000 \times \dfrac{0.45}{2} \times \dfrac{1}{5} = 460.35$ Nm

Now allowable stress in arm $= \dfrac{168}{4} = 42$ MPa

$\therefore$ Section modulus of arm, $Z = \dfrac{460.35}{42 \times 10^6} = 10.96 \times 10^{-6}$ m^3.

$\therefore \quad \dfrac{\pi}{64} b^3 = 10.96 \times 10^{-6} \quad \therefore \quad b = 0.0607$ m, say 0.061 m

$\therefore$ Major axis of arm at centre of wheel = 0.061 m = 61 mm

and minor axis of arm at centre of wheel $= \dfrac{0.061}{2} = 0.0305$ m = 30.5 mm

Major axis at pitch line $= 61 - \dfrac{65}{1000} \times 225 = 61 - 14.6 = 46.4$ mm

$\therefore$ Minor axis at pitch line $= \dfrac{46.4}{2} = 23.2$ mm

Shaft: Neglecting bending moment,

$$\frac{\pi}{16} d^3 f_s = M_t, \text{ let } f_s = 110 \text{ MPa}$$

$$\therefore \quad \frac{\pi}{16} d^3 \times 110 \times 10^6 = 10.23 \times 10^3 \times \frac{0.45}{2}$$

$$\therefore \quad d = 0.047 \text{ m, say 47.50 mm.}$$

Example 22.5. In an automotive type of gear box, the second speed gear shaft is to be driven from the main shaft with a velocity ratio of 1.5 to 1. The main shaft transmits 45 kW at 3000 rev/min. The shaft centre distance must not exceed 90 mm. Find the diameter of the gear shaft if the gear is mounted on it in the manner shown in Fig. 22.15. The distances are:
$L_2 = 0.25$ m, $L_1 = 0.15$ m, $l_2 = 0.10$ m. Take the allowable stress in shear for the shaft material as 55 MPa. Assume medium shock conditions.

Solution : Now $\dfrac{D_g + D_p}{2} = 90$

$\therefore \quad D_g + D_p = 180$ mm

Again $\quad \dfrac{D_g}{D_p} = 1.5 \quad \therefore \quad D_g = 1.5\, D_p$

$\therefore \quad D_p = \dfrac{180}{2.5} = 72$ mm and $D_g = 108$ mm

Now useful transmitted load, $F_t = \dfrac{kW \times 1000}{V}$, N

Now $\quad V = \pi \times \dfrac{72}{1000} \times \dfrac{3000}{60} = 11.31$ m/s

$\therefore \quad F_t = \dfrac{45 \times 1000}{11.31} = 3978.78$ N

Service factor for medium shock conditions from Table 22.8, is 1.5

∴ Actual transmitted load, $F_t = 3978.78 \times 1.5 = 5968.17$ N

Now Radial component, $F_r = F_t \tan \phi$

Let pressure angle, $\phi = 20°$

∴ $F_r = 5968.17 \times \tan 20° = 5968.17 \times 0.364 = 2172.41$ N

Vertical reaction at bearing B due to $F_t = \dfrac{F_t \cdot l_1}{L_2}$, Bearing B, the more heavily loaded

$$= \frac{5968.17 \times 0.15}{0.25} = 3580.9 \text{ N}$$

Horizontal reaction at bearing B due to $F_r = \dfrac{F_r \cdot l_1}{L_2} = \dfrac{2172.41 \times 0.15}{0.25} = 1303.45$ N

Neglecting the weight of the gear,

Vertical moment on shaft, $M_v = 3580.9 \times 0.10 = 358.1$ Nm

Horizontal moment on shaft, $M_h = 1303.45 \times 0.10 = 130.345$ Nm

Refer Table 3.4, Beam case (d). Gear position will be the dangerous section.

∴ Resultant B.M., $M_b = \sqrt{M_v^2 + M_h^2} = 381.08$

Twisting moment on the shaft, $M_t = F_t \times r_g = 5968.17 \times 0.054 = 322.28$ Nm

∴ $$M_{te} = \sqrt{M_b^2 + M_t^2} = \sqrt{381.08^2 + 322.28^2} = 499.09 \text{ Nm}$$

∴ $$\frac{\pi}{16} d^3 \times f_s = 499.09; f_s = 55 \text{ MPa}$$

∴ $$d = 0.0359 \text{ m, say 36 mm.}$$

22.15 DESIGN CALCULATIONS OF HELICAL SPUR GEARS

Helical gears are also known as spiral gears. They are used to connect non-intersecting shafts at any angle in parallel planes. Their difference from the straight tooth spur gears is that the teeth are not parallel to the axis of the shaft but are cut in the spiral form around the pitch cylinder. Helical gears were developed to overview the following disadvantages of spur gears:

1. There is instantaneous contact along the entire length of a tooth on one gear with a tooth on the mating gear. Due to this, there must be a sudden deformation of teeth to compensate for the deflection under load of the previously engaged teeth.
2. Manufacturing inaccuracies will result in shocks at the beginning of the engagement. The intensity of the shocks increases with the increase of speed. Hence the straight tooth spur gears are not suited for very high speeds unless they are very accurately cut.
3. The operation of the drive becomes noisy at high speeds.

The above difficulties are overcome by using helical gears in which the contact between the mating teeth does not occur at once but gradually, from one end of the tooth to the other. Due to this the contact line at any instant will be along a diagonal across the teeth. Due to the gradual and smooth engagement of the teeth the operation of the drive becomes quiet and there is very slight wear of the mating tooth surfaces and hence the efficiency of the helical gear drive is very high. Again in straight tooth spur gear the lever arm for the bending of the tooth is its full height since at some time during each contact period, the tooth load is concentrated at the upper edge of the tooth. But in helical gears, since the contact is distributed over the entire working surface of the tooth, the lever arm for the bending action will be the mean of the tooth height, i.e., about one half the tooth height. Due to this fact the helical teeth are much stronger than the straight tooth spur gears.

Moreover, the tooth deflection is very much reduced. Hence the helical gears are employed for drives required to transmit very heavy loads or to run at high speeds. They can satisfactorily operate at pitch line velocities ranging from 20 to 40 m/s, although speeds of upto 600 m/s have been successfully employed. Helical gears are also employed where quiet operation is necessary. Typical application of the helical gearing include the following drives:

1. Rolling mills
2. Steam turbine reducing gears.
3. Automobile timing gears.

Helical gears require more accurate machining and better lubrication facilities for the progressive contacting of mating teeth and increased sliding action which are needed for the quiet operation of the drive. The disadvantage of the helical gear drive is that an axial thrust is set up along the axis of the shaft. This will require the provision of a thrust bearing which may result in a more costly type of construction and an increase in the space requirements.

22.15.1 Definitions

Figure 22.16 shows a portion of a helical gear. The teeth are inclined to the axis of rotation and they form true helices. The constant angle made by the helices with the axis of rotation is known as "Helix Angle Ψ".

Hand of gear: This is an important term in helical gears. The gear shown in Fig. 22.16 is a left handed gear. It is apparent that on parallel shafts the helical gears have the same helix angle but are of the opposite hand.

Pitch and module: In helical-gears, the circular pitch and module of the teeth are measured in two planes. One in the diametral plane or the plane of rotation and the other in the normal plane which is the plane perpendicular to the teeth. The circular pitch in the plane of rotation is designated as "p_c" as in spur gears and that in the normal plane as "p_{cn}", the relation between them is

$$p_{cn} = p_c \cos \Psi \qquad \text{...(22.26)}$$

In the same manner the module in the plane of rotation is designated as 'm' as in spur gears and that in the normal plane as "m_n", the relation between them is:

$$m_n = m \cos \Psi \qquad \text{...(22.27)}$$

If the gears are cut by standard hobs, the module and pressure angle of the hob will apply in the normal plane and the standard value for module must be specified. Hence the value of module in the plane of rotation, i.e., "m" will contain a fraction and so also the pitch diameter. However, if the gears are cut by the Fellows gear shaper method, the module and pressure angle will apply to the plane of rotation. If the gears are cut by the hobbing method, the cutter should be selected on the basis of the normal module and for the number of teeth in a circumference corresponding to the helix. This number is known as "virtual or formative number of teeth" and is given as:

$$zf = \frac{z}{\cos^3 \Psi} \qquad \text{...(22.28)}$$

where z = number of teeth on the gears.

The minimum number of teeth z' for the helical gears which is adopted to avoid interference is given by,

$$z' = z \cos^3 \Psi$$

where z = minimum number of teeth in straight tooth spur gears to avoid interference.

Helix Angle and face width: In order to secure the continuity and smoothness of action, the tooth displacement 'd' or the overlap, Fig. 22.16 or the advancement of one end of tooth over the other end, should be at least given as:

$$d = p_c = f \tan \Psi \qquad \text{...(22.29)}$$

It is clear that as the helix angle increases, the tooth overlap increases but at the same time, the axial thrust, F_a, Fig. 22.16, which is given as:

$$F_a = F_t \tan \Psi$$

(Also, see Fig. 22.18 for Force Analysis)

increases, which is undersirable. Therefore, there is no need of increasing the helix angle beyond a value necessary to obtain the helical tooth action. In single helical gears, the value used in practice ranges from 20 to 35°. Unless otherwise specified, the helix angle refers to the plane of rotation. Also, as the face width of the tooth decreases, a greater value of helix angle is needed to achieve the helical tooth action. It is usually recommended that the overlap should be about 15 percent of the circular pitch, or

$$d = f \tan \Psi = 1.15\, p_c$$

$$f = 1.15 \frac{p_c}{\tan \Psi} \qquad \text{...(22.30)}$$

or

$$f = \frac{1.15 \pi m}{\tan \Psi} \qquad \text{...(22.31)}$$

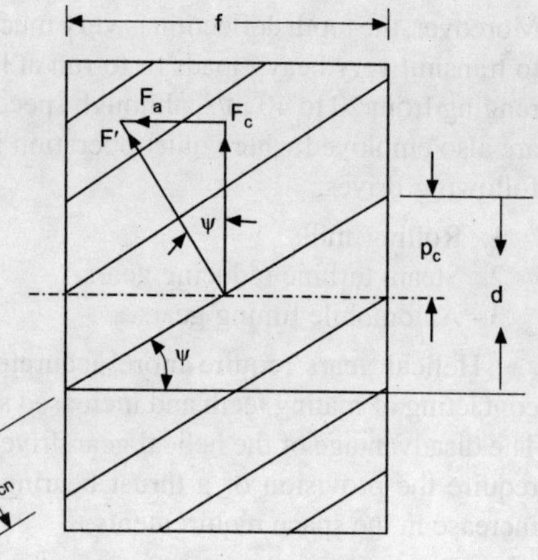

Fig. 22.16 *Helical gear*

The maximum face width is given as 12.5 m to 20 m. In terms of pinion diameter, it should be from 1.5 to 2 times the pinion diameter, although 2.5 times has been used. The face width is also given as:

$$f = \frac{4.5 \text{ m}}{\tan \Psi} \text{ to } \frac{20 \text{ m}}{\tan \Psi}$$

Note: Unless otherwise specified, the pressure angle ϕ is referred to the plane of rotation. The relation between the pressure angle in the plane of rotation ϕ and that in the normal plane ϕ_n is given as:

$$\tan \phi_n = \tan \phi \cos \psi \qquad \text{...(22.32)}$$

The standard pressure angle in the plane of rotation is 20°. However, AGMA, gives the limits for ϕ of 15° – 23′ and 25°.

22.15.2 Double Helical or Herring Bone Gears

The main disadvantage of single helical gears is the side thrust on the shaft bearings. This side thrust can be eliminated by using the helical gears in pairs, one having the right helix and the other having the left hand helix, and of equal magnitude. These gears are mounted on the shaft side by side and the end thrust produced by one gear is neutralised by the other. If both the gears are formed on one blank, the resulting gear is called herring bone gear or double helical gear, Fig. 22.17. The double helical gear teeth may be continuous, Fig. 22.17(a), or separated by a groove, Fig. 22.17(b). The first one are cut by the profiling method with end mills or with special shaping cutters, by generation. A gear with a groove is cut by hobbing (groove provides space for the hob cutter).

Herring bone gears are expensive to manufacture, so, their use is warranted only for heavy duty, enclosed mechanisms.

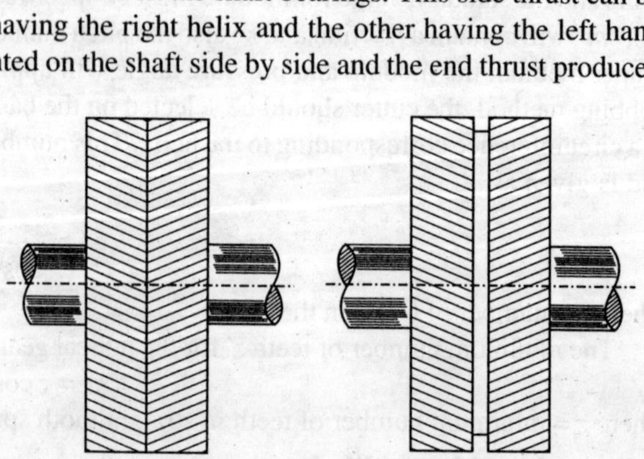

(a) (b)

Fig. 22.17 *Herring bone gear or double helical gear*

Gears with groove can be cut to a high degree of accuracy and are, therefore, employed in powerful high speed drives, such as, in turbine gear units driving screws of sea vessels. Gears with continuous teeth can also transmit high power when the velocity is not high, such as in rolling mills. The equation for the minimum face width for herring bone gear is:

$$f \geq 2.3 \frac{p_c}{\tan \Psi}$$

or
$$f \geq 2.3 \frac{m\pi}{\tan \Psi} \quad \text{...(22.33)}$$

The maximum value for 'f' ranges from 20 m to 30 m. Since there is no axial thrust, the helix angle can be made upto 45°.

22.15.3 Gear Tooth Proportions

There are no standard proportions for helical gears. They are usually made as stub teeth. Helical gears are not interchangeable. The following proportions are given for helical and herring bone gears, as recommended by AGMA :

Addendum, a, maximum = 0.8 m
Dedendum d, minimum = 1.0 m
Minimum total depth = 1.8 m
Minimum clearance = 0.2 m
Thickness of tooth = 1.571 m

22.15.4 Other Relations

The addendum and dedendum circle diameters:

$$D_a = D + 2m_n = m \times z + 2m_n = \frac{m_n}{\cos \Psi} z + 2m_n$$

∴
$$D_a = m_n \left(2 + \frac{z}{\cos \Psi} \right) \quad \text{...(22.34)}$$

Similarly,
$$D_d = m_z - 2.5\, m_n = m_n \left(\frac{z}{\cos \Psi} - 2.5 \right) \quad \text{...(22.35)}$$

Taking full depth tooth.
The centre distance,

$$C = \frac{m}{2}(z_1 + z_2) = m_n \times \frac{z_1 + z_2}{2 \cos \Psi}$$

22.15.5 Design Calculations

The following considerations apply both to helical and herring bone gears.

Strength of Helical Gears: In Fig. 22.16, F_t is the load tangent to the pitch cylinder in the plane of rotation (similar to case of spur gears). The load F_n equal to $F_t / \cos \Psi$ is the load in the same tangent plane normal to the pitch element of the helical tooth. As already explained, the mating teeth of accurate helical spur gears make contact along a diagonal straight line. The intensity of pressure along the line of contact will, therefore, vary because of the varying flexibility of tooth sections. It is estimated that the intensity of pressure near the middle of the tooth is 1.33 times the average intensity when the width of the face is such that the diagonal line of contact extends from the root to the tip of the tooth. Hence, if uniform intensity is assumed, the total safe load can be taken about 0.75 that to be transmitted. Therefore, the Lewis equation for an helical spur gear tooth may be expressed in the form given below. Load F_t is not normal to the pitch element of the helical tooth, but load F' is. This is the load causing the bending stress and is normal to the tooth surface in the normal plane. Now, the length of tooth surface in normal plane, $f_n = f/\cos \Psi$.

$$\therefore \qquad F' = 0.75\, f_b \cdot \frac{f}{\cos \Psi} \cdot m_n Y_f \cdot C_v$$

or

$$\frac{F_t}{\cos \Psi} = 0.75\, f_b \cdot \frac{f}{\cos \Psi} \cdot m_n Y_f \cdot C_v$$

or

$$F_t = 0.75 f_b . f . m_n . Y_f C_v \qquad \text{...(22.36)}$$

where m_n = module in the normal plane

Y_f = Lewis form factor corresponding to z_f teeth.

The values for the velocity factor C_v may be determined in the following manner:

$$C_v = \frac{4.5}{4.5 + V} \text{ for low angle helical gears with } V < 5 \text{ m/s}$$

$$= \frac{6}{6 + V}, \text{ for } 5 < V < 10, = \frac{15}{15 + V}, \text{ for } V = 10 \text{ to } 20 \text{ m/s}$$

$$= \frac{5.6}{5.6 + \sqrt{V}}, \text{ for } V > 20 \text{ m/s}$$

and for non-metallic gears

$$C_v = \frac{1.0}{1.0 + V} + 0.25$$

AGMA has given the following formula for helical gear,

$$C_v = \left[\frac{78}{78 + \sqrt{200\, V}}\right]^{\frac{1}{2}}$$

Design Producer: It is similar to that for straight tooth spur gears. Firstly, the preliminary design will be based on the Lewis equation (22.36). Then it will be checked by Buckingham's equation for dynamic load. For helical gears, the equation is:

$$F_d = F_t + \frac{21V(cf \cos^2 \Psi + F_t) \cos \Psi}{21V + \sqrt{cf \cos^2 \Psi + F_t}} \qquad \text{...(22.37)}$$

This equation will give less dynamic load than for a pair of straight spur gears, because of progressive contact and decreased deflection of teeth.

Beam Strength equation:

$$F_b = 0.75 \times f_{ef} \cdot f \cdot m_n \cdot Y_f$$

Strength for Wear: Finally the teeth are checked for wear. The limiting load for wear for helical gears is given by the modified Buckingham equation as,

$$F_w = \frac{D_p \cdot f \cdot K_w \cdot Q}{\cos^2 \Psi} \qquad \text{...(22.38)}$$

The symbols in the above equation have the same meaning as for spur gears except that K_w is determined by using the normal pressure angle ϕ_n.

Note: The stress concentration factor, K_f does not appear in Lewis equation (22.36). It is taken into account for finding the design stress as:

$$f_b = \frac{\text{Yield point stress}}{RK_f}$$

or $\quad f_b$ can be taken directly from Table 22.7.

Note: Many authors recommend the use of the following modified form of Lewis equation for helical gears:

$$F_t = \frac{f_b \cdot f \cdot m_n \, Y_f \cdot C_v}{C_w} \quad \text{...(22.39)}$$

where C_w = wear and lubrication factor.
= 1.15, for enclosed gears continuously lubricated with oil of proper viscosity and character.
= 1.25, for scant lubrication but regular and frequent inspection
= 1.35, for indifferent lubrication.

22.15.6 Bearing Loads

The loads transmitted to the bearings of straight spur gears are due to the tangential load and the radial load between the tooth surface. With helical gears, there is an additional load caused by the end thrust at the pitch line of the gear. Fig. 22.18, shows the various forces acting on the teeth of a helical gear. The normal force F_n, i.e., the force of tooth interaction directed along the common normal to the contacting profiles, is resolved into two components F' and F_r. The force F' is further resolved into two forces F_t and F_a. The force F_t is directed tangentially to the pitch circle and known as "Tangential load" or "Turning force". The force F_a is parallel to the gear axis and acting at the pitch line of gear is called "Axial Force" or "End Thrust". The forces F_n, F' and F_r lie in plane B–B perpendicular to the tooth direction. The force F_r which tends to move the gears apart in the radial direction is called the "radial force". As for spur gears:

$$F_r = F_t \tan\phi = F_t \frac{\tan\phi_n}{\cos\Psi} \quad \text{...(22.40)}$$

$$F_a = F_t \tan\Psi$$

The direction of the axial force depends upon the direction of rotation and the hand of teeth. Fig. 22.19 illustrates the bearing loads due to all the forces acting on a helical gear tooth. The bearing loads due to F_t and F_r can be found out in the same manner as for spur gears. The reactions at the bearings due to the axial force will be resisting the couple due to F_a. These reactions are given as:

$$R_E a = R_F a = \frac{F_a \times D_g}{2(l_1 + l_2)} = \frac{F_a \times r_g}{L_2} \text{ for gear, Fig. 22.19.}$$

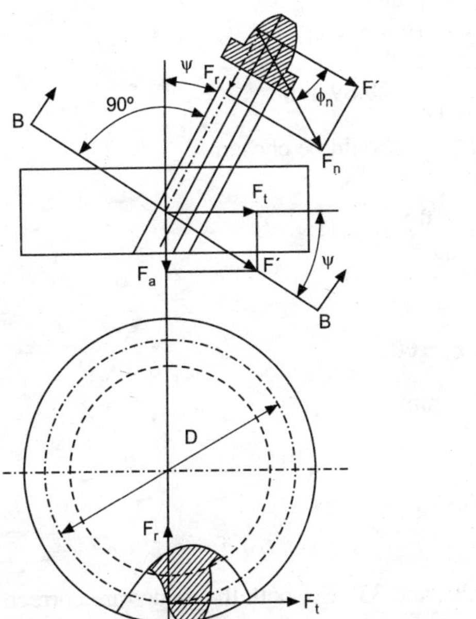

Fig. 22.18 *Forces on a helical gear tooth*

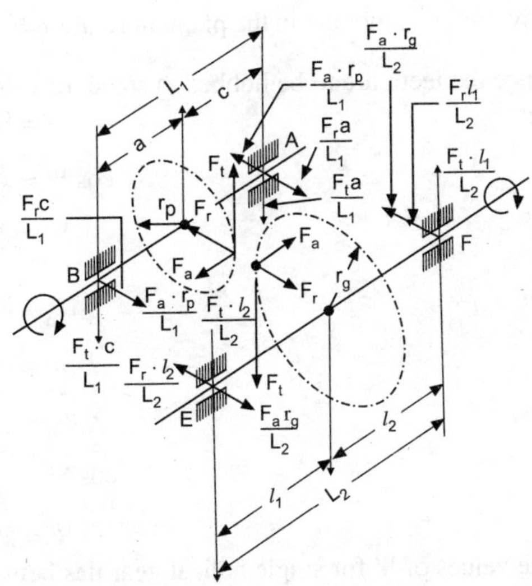

Fig. 22.19 *Bearing loads of helical gears*

Note that these reactions on the bearings are in opposite directions and the loads carried by the bearings will be different. The bearings should be so designed that the axial force F_a is taken by the bearing carrying the smaller radial load.

Example 22.6. Two helical gears are used in a speed reducer that is to be driven by an internal combustion engine. The rated power of the speed reducer is 75 kW at a pinion speed of 1200 rev/min. The speed ratio is 3 to 1. Assume medium shock conditions and 24 hr operation.

(a) Determine the module, face, number of teeth in each gear, and the material and heat treatment if the teeth are 20° full depth in the normal plane.

(b) Design the arms and shaft for the gear.

Solution: The power to be transmitted is 75 kW at a pinion speed of 1200 rev./min, both the values are neither low nor high, so let us assume a peripheral pitch line velocity of 10 m/s. Therefore, the pitch diameter of the pinion is:

$$D_p = \frac{V \times 60}{\pi n} = \frac{10 \times 60}{1200 \times \pi} = 0.159 \text{ m, say } 0.16 \text{ m}$$

$\therefore$
$$\text{Exact } V = \pi \times 0.16 \times \frac{1200}{60} = 10.05 \text{ m/s}$$

Tangential Design Load: $F_t = \dfrac{kW \times 1000 \times C_0}{V}$

Now service factor, from Table 22.8 $C_0 = 1.50 \times 1.25 = 1.875$

$\therefore$
$$F_t = \frac{75 \times 1000 \times 1.875}{10.05} = 14 \text{ kN}$$

Now
$$C_v = \frac{15}{15 + V} = \frac{15}{15 + 10.05} = 0.599$$

Lewis Equation:

$$F_t = 0.75 f_b \cdot f \cdot m_n \cdot Y_f \cdot C_v$$

Let the number of teeth in the pinion be 18.

$\therefore$
$$z_p = 18 \text{ and } z_g = 18 \times 3 = 54$$

Now 'm', i.e., module in the plane of rotation $= \dfrac{D}{z} = \dfrac{160}{18} = 8.89$ mm

Since the teeth are to be hobbed, a standard value of "m_n" should be chosen.

Let
$$m_n = 8 \text{ mm}$$

$\therefore$
$$\cos \Psi = \frac{m_n}{m} = \frac{8.0}{8.89} = 0.899$$

$\therefore$
$$\Psi = 25.85°$$

or

Let
$$z_p = 20, \therefore z_g = 60$$

$\therefore$
$$m = \frac{160}{20} = 8 \text{ mm}$$

Let
$$m_n = 7 \text{ mm},$$

$\therefore$
$$\cos \Psi = \frac{7}{8} = 0.875$$

$\therefore$
$$\Psi = 29°$$

The values of Ψ for single helical gear lies between 20° and 35°. So, both the values are correct. Let us take $\Psi = 29°$.

Now $\tan \phi_n = \tan \phi \cos \Psi$

$$\therefore \quad \tan \phi = \frac{\tan 20^\circ}{0.875} = \frac{0.364}{0.875} = 0.416.$$

∴ Pressure angle, in the plane of rotation $\phi = 22^\circ - 35'$, which is within the limits of 15°–23′ and 25°.
Now formative number of teeth in the pinion,

$$z_f = \frac{z_p}{\cos^3 \Psi} = \frac{20}{0.875^3} = 30$$

$$\therefore \quad Y_f = \pi\left(0.154 - \frac{0.912}{z_f}\right) = \pi\left(0.154 - \frac{0.912}{30}\right) = 0.388$$

Material for the pinion may be taken as C-30 forged steel, from table 22.7, f_b = 175 MPa, BHN = 150
∴ From the Lewis equation, the active width of the pinion,

$$\therefore \quad f = \frac{F_t}{0.75 f_b \cdot m_n \cdot Y_f \cdot C_v}$$

$$\therefore \quad f = \frac{14 \times 1000}{0.75 \times 175 \times 10^6 \times 0.007 \times 0.388 \times 0.599} = 0.0656 \text{ m, say 66 mm}$$

Now minimum face width is given as;

$$f \geq \frac{1.15 \pi m}{\tan \Psi} \geq \frac{1.15 \pi \times 8}{0.554} \geq 52.2 \text{ mm}$$

$$\therefore \quad f = \text{say 66 mm is O.K.}$$

Check by Buckingham's equation for dynamic load:

$$\therefore \quad F_i = \frac{21\, V\,(cf \cos^2 \Psi + F_t) \cos \Psi}{21V + \sqrt{cf \cos^2 \Psi + F_t}}$$

Now for V = 10.05 m/s, the maximum permissible tooth action error from Table 22.10 is 0.038 mm. From Table 22.9, for module equal to 8 mm, the probable error for class 2 gear is 0.04 mm and for class 3 gears it is 0.020 mm. Since the maximum permissible error is 0.038 mm, therefore, use class 3 gears with probable error of 0.02 mm.

$$c = \frac{k.e}{\frac{1}{E_p} + \frac{1}{E_g}}$$

$$k = 0.111$$

Let the material for gear be also steel, its exact composition will be determined later. Let it be cast steel with f_b = 140 MPa (Table 22.7), with BHN = 180

$$\therefore \quad E_p = E_g = 210 \text{ GPa}$$

Now for gear, $z_f = \dfrac{60}{0.875^3} = 90$

$$\therefore \quad Y_f = \pi\left(0.154 - \frac{0.912}{90}\right) = 0.452$$

Now strength factor for pinion, $f_b \cdot Y_f = 175 \times 0.388 = 67.9$ and strength factor for gear,

$$f_b \cdot Y_f = 140 \times 0.452 = 63.28$$

Therefore, gear is weaker,

Now $$c = \frac{0.111 \times 0.02 \times 210 \times 10^9}{10^3 \times 2} = 233.1 \text{ kN/m}$$

$$\therefore \quad F_i = \frac{21 \times 10.05(233.1 \times 10^3 \times 0.066 \times 0.766 + 14 \times 10^3) \times 0.875}{21 \times 10.05 + \sqrt{233.1 \times 10^3 \times 0.066 \times 0.766 + 14 \times 10^3}} = 12.81 \text{ kN}$$

$$\therefore \quad F_d = 14 + 12.81 = 26.81 \text{ kN}$$

Now Beam strength of gear,

$$F_b = 0.75 f_{ef} \cdot f \cdot m_n \cdot Y_f$$

Now it is clear from Table 22.7 that the BHN of gear material = 180

$$\therefore \quad f_{ef} = 1.75 \times 180 = 315 \text{ MPa}$$

$$\therefore \quad F_b = 0.75 \times 315 \times 10^6 \times 0.066 \times 0.007 \times 0.452 = 48.75 \text{ kN}$$

Now Beam strength is greater that $1.5 \times F_d$, therefore, the design is safe for dynamic loading.

Check for Wear:

$$F_w = \frac{D_p f K_w Q}{\cos^2 \Psi}$$

$$Q = \frac{2z_g}{z_p + z_g} = \frac{2 \times 60}{20 + 60} = 1.5$$

$$K_w = \frac{f_{es}^2 \times \sin \phi_n}{1.4}\left(\frac{1}{E_p} + \frac{1}{E_g}\right)$$

The average BHN of gear pair $= \dfrac{180 + 150}{2} = 165$

$\therefore$ f_{es} of the gear pair $= 2.8 \times 165 - 70 = 392$ MPa

$$\sin \phi_n = \sin 20^\circ = 0.342$$

$$\therefore \quad K_w = \frac{(392 \times 10^6)^2 \times 0.342}{1.4} \times \frac{2}{210 \times 10^9} = 358 \times 10^3, \text{N/m}^2$$

$$\therefore \quad F_w = \frac{0.16 \times 0.066 \times 358 \times 10^3 \times 1.5}{(0.875)^2} = 7.48 \text{ kN}$$

which is quite less than F_d, therefore, the teeth will fail in wear.

Heat treat the gear pair to an average BHN value of 300.

$$\therefore \quad f_{es} = 2.8 \times 300 - 70 = 770 \text{ MPa}$$

K_w will come out to be $= 1381.3$, kN/m^2

and $\quad F_w = 28.86$ kN which is $> F_d \quad \therefore$ safe.

Final values are :

$$m = 8 \text{ mm}; m_n = 7 \text{ mm}; f = 66 \text{ mm}; z_p = 20; z_g = 60.$$

(b) Arms: Gear diameter will be $= 0.16 \times 3 = 0.48$ m

Let us take number of arms = 4

Now stalling load, $F_0 = \dfrac{F_t}{C_v} = \dfrac{14}{0.599} = 23.25$ kN

$\therefore$ Bending moment per arm $= \dfrac{23.25 \times 1000 \times 0.24}{4} = 1395$ Nm

Taking elliptical section, with major axis = 2 minor axis

$$\therefore \quad \frac{\pi}{64} b^3 \times f_b = 1395$$

Let $$f_b = \frac{900}{5} = 180 \text{ MPa}$$

$\therefore$ b = major axis at the hub centre = 0.054 m = 54 mm

$\therefore$ a = minor axis at the hub centre = 27 mm

Taper of arm = 65 mm per metre length = 15.6 mm

$\therefore$ Major axis at the rim section = 54 – 15.6 = 38.4 mm

$\therefore$ Minor axis at the rim section = 19.2 mm

Gear Shaft: Maximum useful turning force,

$$F_o = 23.25 \text{ kN}$$

Since the bearing positions are not given, let us assume that gear is over hanging.

Axial force, $F_a = 23.25 \times \tan 29° = 23.25 \times 0.554 = 12.88$ kN

B.M. due to $F_o = 23.25 \times$ overhang (let overhang = 0.12m)

$= 23.25 \times 0.12 = 2.79$ kNm

B.M. due to $F_a = F_a \dfrac{D_g}{2} = 12.88 \times \dfrac{0.48}{2} = 3.09$ kNm

Since these act at right angles to each other, therefore, resultant, B.M. is:

$$M_b = \sqrt{2.79^2 + 3.09^2} = 4.164 \text{ kNm}$$

Torsional moment on gear shaft, $M_t = F_0 \times \dfrac{D_g}{2} = 23.25 \times \dfrac{0.48}{2} = 5.58$ kNm

$$\therefore \quad M_{te} = \sqrt{4.164^2 + 5.58^2} = 6.9625 \text{ kNm}$$

$$\therefore \quad \frac{\pi}{16} d^3 \times f_s = 6.9625 \times 100, \text{ let } f_s = 82.5 \text{ MPa}$$

$$\therefore \quad d = 0.0755 \text{ m, say } 76 \text{ mm}$$

Direct compressive or tensile stress on the shaft due to F_a

$$= \frac{12.88 \times 1000}{0.7854 \times 0.076^2} = 2.84 \text{ MPa}$$

Now Actual $$f_s = \frac{6.9625 \times 1000 \times 16}{\pi \times 0.076^3} = 80.778 \text{ MPa}$$

$\therefore$ Principal shear stress $= \dfrac{1}{2}\sqrt{2.84^2 + 4 \times 80.778^2} = 80.8 \text{ MPa} < 82.5 \text{ MPa}$ $\therefore$ safe.

Example 22.7. Determine the power capacity of a pair of helical turning gears having a transmission ratio of 10 : 1. The teeth are 20 full depth involute, 6 mm module. The pinion has 25 teeth and rotates at 5000 rev/min. The active face width is 76 mm and the material is C-40 steel untreated.

Solution: Nothing is mentioned about the service conditions.

Many authors recommend the following form of the Lewis equation for helical gears:

$$F_t = \frac{f_b \cdot f \cdot m_n \cdot Y_f \cdot C_v}{C_w} \quad \text{...(22.41)}$$

Now $$f = 76 \text{ mm}, m = 6 \text{ mm}$$

Now $$V = \frac{\pi D}{1000} \times \frac{n}{60}$$

Now $$D_p = m \times z_p = 6 \times 25 = 150 \text{ mm}$$

$\therefore$ $$V = \pi \times \frac{150}{1000} \times \frac{5000}{60} = 39.27 \text{ m/s}$$

$$C_v = \frac{5.6}{5.6 + \sqrt{39.27}} = 0.472$$

Let helix angle be = 20º

$\therefore$ Virtual number of teeth in the pinion $(z_p)f = \dfrac{25}{(\cos 20^\circ)^3} = 30.12$

$$m_n = m \cos \Psi = 5.64 \text{ mm}$$

$\therefore$ $$Y_f = \pi\left(0.154 - \frac{0.912}{30.12}\right) = 0.389$$

Let $$C_w = 1.25$$

Now $$f_b = 210 \text{ MPa (Table 22.7)}$$

$\therefore$ $$F_t = \frac{210 \times 10^6 \times 0.076 \times 0.00564 \times 0.389 \times 0.472}{1.25} = 16.68 \text{ kN}$$

$\therefore$ $$\text{Power} = 16.68 \times 39.27 = 655 \text{ kW.}$$

Example 22.8. Design a pair of equal diameter, 20 deg. stub tooth helical gears to transmit 37.5 kW with moderate shock at 1200 rev/min. The two shafts are parallel and 0.45 m apart. Each gear is to be of steel. Find the module and face width of the teeth.

Solution: It is clear that,

$$D_p = D_g = 0.45 \text{ m}$$

$\therefore$ $$V = \frac{\pi \times 0.45 \times 1200}{60} = 28.27 \text{ m/s}$$

(i) Now, $C_0 = 1.75$ from Table 22.8

(ii) Tooth load, $F_t = \dfrac{37.5 \times 1000 \times 1.75}{28.27} = 2.32 \text{ kN}$

$$C_v = \frac{5.6}{5.6 + \sqrt{26.27}} = 0.513$$

The minimum number of teeth for 20 deg-stub system is 14.

Let $$z_p = z_g = 15$$

$\therefore$ Module, $m = \dfrac{D_p}{z_p} = \dfrac{450}{15} = 30 \text{ mm}$

The nearest standard module is either 25 mm or 32 mm. It will be better to take $m = 25$ mm, so as to get safer number of teeth than the minimum.

$\therefore$ $$z_p = z_g = \frac{450}{25} = 18$$

Now $$m_n = m \cos \Psi$$

The helix angle Ψ is selected so as to get a standard value for m_n.

Let $$m_n = 20 \text{ mm}$$

$\therefore \qquad \cos\Psi = \frac{20}{25} = 0.8$

$\Psi = 36°, 48'$, which is a higher value.

Let $\qquad m_n = 22$ mm (second choice)

$\therefore \qquad \cos\Psi = \frac{22}{25} = 0.88$

$\Psi = 28.3°$, which is within the recommended limits,

$$z_f = \frac{18}{(\cos 28.3°)^3} = 26.37$$

Now, the tooth form factor Y_f for 20° stub system is given as

$$\therefore \qquad Y_f = \pi\left(0.175 - \frac{0.841}{z_f}\right) = \pi\left(0.175 - \frac{0.841}{26.37}\right) = 0.45$$

The material is steel. Let it be cast steel untreated with $f_b = 140$ MPa, Table 22.7.

Let $C_w = 1.25$, since at 28.27 m/s, the gears cannot run in an oil bath. Therefore, from Lewis equation (36a)

$$2.32 \times 10^3 = \frac{140\times10^6 \times f \times 0.022 \times 0.45 \times 0.513}{1.25}, \text{ gives } f = 0.004, m = 4 \text{ mm}$$

The minimum width should be

$$f = \frac{1.15\,\pi\, m}{\tan\Psi} = \frac{1.15\times\pi 25}{0.538} = 167.88 \text{ mm}$$

In general practice, f varies between $\frac{4.5\, m}{\tan\Psi}$ to $\frac{20\, m}{\tan\Psi}$

i.e., between 209 mm to 928 mm

Let us take, $f = 400$ mm.

Example 22.9. Determine the safe power which can be transmitted by a pair of helical gears, 20-deg full depth, 25-deg helix, having a normal module of 5 mm. Both the gears are made of forged C-30 steel and have a face width of 76.2 mm. The pinion speed is 2000 rev/min and it has 20 teeth. The velocity ratio is to be 5 to 1.

Solution: The beam strength is,

$$F_b = \frac{f_b \cdot f \cdot m_n \cdot Y_f}{C_w \cdot K_f}, \text{ from equation (36a) and using } f_b \text{ instead of } f_{ef}$$

Normal module is used because the tooth load causing the bending stress is normal to the tooth surface in the normal plane. The stress concentration factor has been included here (as was done with spur gears) instead of considering it while determining the design stress.

$$\text{Now } z_f = \frac{z_p}{\cos^3\Psi} = \frac{20}{(\text{Cos } 25)^3} = 27$$

$$\therefore \qquad Y_f = \pi\left[0.154 - \frac{0.912}{z_f}\right] = \pi\left[0.154 - \frac{0.912}{27}\right] = 0.377$$

Let $\qquad C_w = 1.15$ and $K_f = 1.5$

From Table 22.7, $f_b = 175$ MPa

$$\therefore \quad F_b = \frac{175\times10^6 \times 0.0762\times 0.005\times 0.377}{1.15\times1.5} = 14.572 \text{ kN}$$

Now $$m = \frac{m_n}{\cos\Psi} = \frac{5}{0.906} = 5.52 \text{ mm}$$

$$\therefore \quad D_p = z_p \cdot m = 20 \times 5.52 = 110.4 \text{ mm}$$

$$z_g = 5 \times 20 = 100$$

Now wear load is, $$F_w = \frac{D_p \cdot f \cdot K_w \cdot Q}{\cos^2 \Psi}$$

$$Q = \frac{2z_g}{z_p + z_g} = \frac{200}{120} = 1.67$$

$$K_w = \frac{f_{es}^2 \sin\phi_n}{1.4}\left[\frac{1}{E_p} + \frac{1}{E_g}\right]$$

Since wear load is also a normal load to the tooth surface, it is more appropriate to use normal pressure angle in the above equations.

Now from Table 22.7, BHN = 150

$$\therefore \quad f_{es} = 2.8 \text{ BHN} - 70 = 350 \text{ MPa}$$

Now $$\tan\phi_n = \tan\phi \cos\Psi = \tan 20 \times \cos 25 = 0.364 \times 0.906 = 0.33$$

$$\therefore \quad \phi_n = 18.3^\circ$$

$$\therefore \quad K_w = \frac{(350\times10^6)^2 \times 0.314}{1.4}\left[\frac{2}{210\times10^9}\right] = 261.67 \times 10^3, \text{ N/m}^2$$

$$\therefore \quad F_w = \frac{0.1104\times0.0762\times261.67\times10^3\times1.67}{(\cos 25)^2} = 4480 \text{ N} = 4.48 \text{ kN}$$

Since $F_b > F_w$, the design is safe for wear.

Now dynamic load can be approximated as, (see Art. 22.14.1)

$$F_d = F_t / C_v$$

Now $$F_b \geq F_d \text{ and } F_b \geq F_w$$

Since we are to determine the safe power which can be transmitted so let us make

$$F_d = F_w = 4.48 \text{ kN}$$

$$V = \frac{\pi\times110.4\times2000}{1000\times60} = 11.56 \text{ m/s}$$

$$\therefore \quad C_v = \frac{15}{15+V} = 0.565$$

$$\therefore \quad F_t = 4.48 \times 0.565 = 2.53 \text{ kN}$$

$$\therefore \quad \text{Power} = 2.53 \times 11.56, \text{kW} = 29.25 \text{ kW}.$$

22.16 DESIGN CALCULATIONS OF BEVEL GEARS

22.16.1 Uses of Bevel Gears

As already indicated, bevel gears are used to transmit power between two shafts whose geometrical axes intersect at a certain angle. The difference between spur gears and bevel gears is that in spur gears, the pitch surfaces are cylindrical while in bevel gears, these are conical. Two types of bevel gears are used in general practice:

1. **Straight tooth Bevel Gears:** Fig. 22.20. In these gears, the teeth of the gear converge to a common point *O*, along straight lines. The point is called the "apex" and is also the point of intersection of the gear axes.
2. **Spiral Bevel Gears:** Fig. 22.21. In spiral gears, the teeth are curved instead of straight. These gears have the same advantages over straight tooth bevel gears as the helical gears have over spur gears, i.e.
   (*a*) Due to the curvature of the teeth, there is an overlapping of contact and the teeth mesh with each other gradually resulting in smooth and quiet operation.
   (*b*) There is gradual transfer of load from one tooth to another eliminating sudden impacts and changes in tooth pressure as will occur in straight tooth bevel gears, especially at high pitch line velocities where the tooth engagement is sudden along the full line contact resulting in noise and vibrations.

Almost universal application of spiral bevel gears is for the rear axle drive of the automobiles and other high grade machinery.

Another form of spiral bevel gears is the "Hypoid Gears". It is similar to spiral gears with the difference that the pinion is offset from the centre of the gear axis Fig. 22.22. The shafts don't intersect at one point but cross each other. These types of gears are also finding application in rear axle drives of automobiles. These have the following advantages:

(*i*) Possibility of extending the shafts in either direction beyond the limits of gearing.
(*ii*) Gears can be mounted in between bearing instead of being over hanging and so are more robust.
(*iii*) In automobiles, the drive permits the use of a lower body construction without sacrificing road clearance.
(*iv*) Greater quietness and greater pinion diameter and pinion shaft can be accommodated.

However, due to sliding action between the teeth, lubrication is more difficult and there will be increased end thrust of the pinion on forward drive.

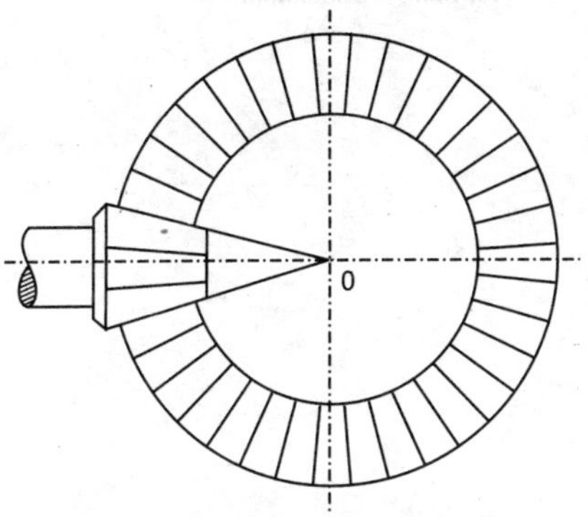

Fig. 22.20 *Straight tooth bevel gears*

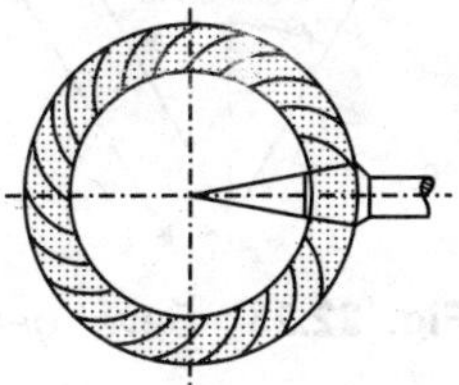

Fig. 22.21 *Spiral bevel gears*

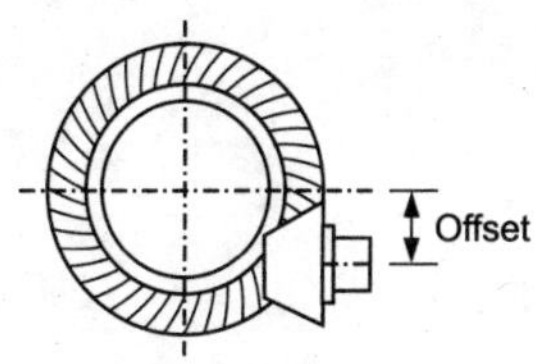

Fig. 22.22 *Hypoid gears*

22.16.2 Definitions

The various definitions concerning the bevel gears are explained below with the help of the Fig. 22.23.

Pitch Cone: As already indicated, the pitch surfaces of the bevel gears are conical. Pitch cone is the cone containing the pitch elements of the teeth.

Cone Centre: The apex of the pitch cone, *O*, or the point where the axes of the two mating gears interest each other is called the Cone Centre.

Pitch distance L: It is the length of the pitch cone element. It is also called pitch cone radius.

Pitch angle: It is angle made by the pitch line of a gear with the gear axis.

Face Angle: It will be equal to Pitch angle + addendum angle = $\gamma + \alpha$.

Root Angle: It will be equal to Pitch angle — Dedendum angle = $\gamma - \beta$.

Back Cone: It is an imaginary cone whose elements are perpendicular to the pitch cone elements at the large end of the tooth. The length of the back cone elements is called the "Back Cone Radius".

The distance *C* is called "Crown Height", the distance *B* is called "Backing" and the distance *M* is called the "Mounting Height".

Note: *(i)* The shaft angle $\theta = \gamma_1 + \gamma_2$ may be an acute angle, equal to 90º or an obtuse angle. A pair of bevel gears of equal size and having the same number of teeth, i.e., of unity velocity ratio and with $\theta = 90º$, are called "Miter Gears".

(ii) Unless otherwise stated, a pair of bevel gears is assumed to be straight toothed and to have their axes intersect at right angles.

(iii) The pitch diameters of gears, the addendum and dedendum, the module, the pitch line velocities are all measured at the larger end of the tooth.

(iv) In a pair of bevel gears, if $\theta > 90º$ and $\gamma_g = 90º$, the gear is called "Crown Gear".

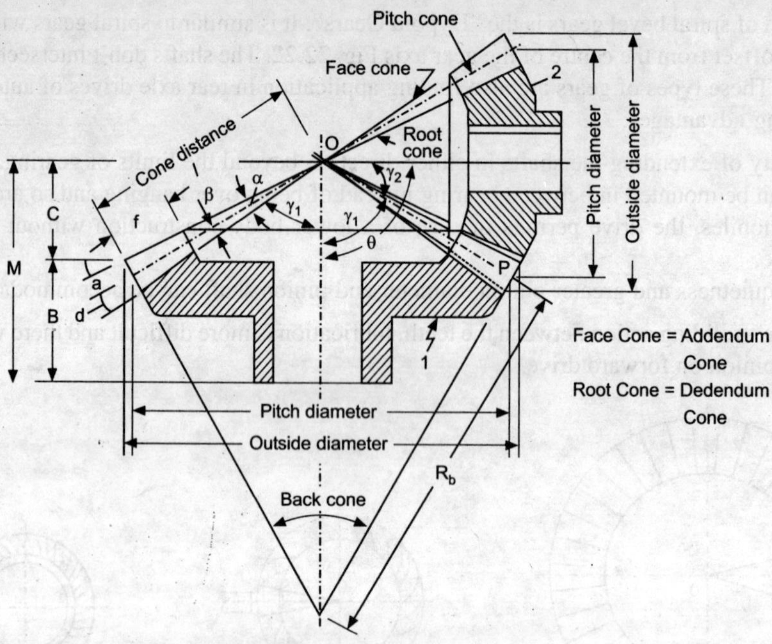

Fig. 22.23 *Bevel gear definitions*

Outside diameter: Outside diameter = Pitch diameter + $2a \cos \gamma$.

$$\therefore \qquad D_a = D + 2a \cos \gamma.$$

This is also called "Addendum cone diameter", *Da*.

Inside diameter or Dedendum cone diameter:

$$D_d = D - 2d \cos \gamma.$$

22.16.3 Formative Number of Teeth

The tooth profile at the larger end of the tooth can theoretically be laid out on a sphere. But it is not possible to represent this profile on a plane surface. To approximate the bevel gear tooth profiles as accurately as possible, the profile can be assumed to be laid out on the surface of the back cone and then the cone can be developed into a plane surface and spur gear teeth corresponding to the pressure angle and module of the bevel gear and radius of the developed cone can be drawn. The number of teeth on this imaginary spur gear is called the virtual number or formative number of teeth. From Fig. 22.23, it is clear that the radius of the back cone is given as:

$$R_b = \frac{R}{\cos\gamma}$$, where, R = pitch radius of pinion or gear as the case may be.

Now z_e = equivalent number of teeth or formative number of teeth

$$\frac{2R_b}{m} = \frac{2R}{m\cos\gamma}$$ (The shape of the teeth of a straight bevel gear when projected on the back cone, is the same as in a spur gear having a radius = Back cone radius)

$$\therefore \quad z_e = z_f = z_v = \frac{z}{\cos\gamma} \qquad \text{... (22.42)}$$

where z = actual number of teeth on the gear.

Note: (i) The virtual number of teeth and not the actual number of teeth must be used for selecting the cutters and in all calculations for the strength of bevel gears.

(ii) The smallest number of teeth in the bevel gear consistent with the absence of undercutting (z_b) min. is determined from the relation:

$$(z_b)\text{ min.} \geq (z_s)\text{ min} \times \cos\gamma$$

where (z_s) min. = the smallest number of teeth in a straight spur gear.

22.16.4 Proportions of Bevel Gear

As already indicated, the proportions of bevel gears are given as:

$$a = m \qquad p_c = \pi m$$

$$d = 1.2\text{ m, thickness of tooth = space width} = \frac{p_c}{2} = 1.57\text{ m}$$

working depth = 2 m

clearance = 0.2 m

Note: Bevel gears are not interchangeable and are designed in pairs.

22.16.5 Design Calculations

(i) Strength of Gears: Since the tooth thickness varies along its length, a modified Lewis Equation has been developed for Bevel gears as below:

$$F_t = f_b f m Y\left(1 - \frac{f}{L}\right) \cdot C_v \cdot N \qquad \text{... (22.43)}$$

where F_t = Design Tangential load at large end, N

m = module at large end

Y = Form factor for the virtual number of teeth.

The cone distance L is given as:

$$L = \sqrt{R_g^2 + R_p^2}$$

where R_g = pitch radius of gear

R_p = pitch radius of pinion

If $L < 30$ m; $f = 6$ m to 7 m, If $L > 30$ m; $f = 7$ m to 10 m

Also $$\frac{L}{4} \leq f \leq \frac{L}{3}$$ As before, $$f_b = \frac{f_{yp}}{RK_f}$$

For a right angle drive :

$$\tan\gamma_p = \frac{z_p}{z_g} \text{ and } \tan\gamma_g = \frac{z_g}{z_p}$$

The pitch line speed at the large end is about 6 to 6.5 m/s for ordinary cut teeth and about 10 m/s for generated teeth.

The velocity factor is determined as below:

$$C_v = \frac{3}{3+V} \text{ for teeth cut by form cutters}$$

and $$C_v = \frac{6}{6+V} \text{ for teeth generated with precision machines.}$$

V is at the large end of the tooth.

(ii) Check by Buckingham's equation for dynamic load: The dynamic load is calculated with the help of equation (22.21) in which the velocity 'V' at the large pitch circle diameter is used and the tangential load F_t is also based on this velocity. The remaining procedure is the same as for spur gears.

The Beam strength equation for Bevel gears is written as,

$$F_b = f \cdot m \cdot Y \cdot f_{ef}\left(1 - \frac{f}{L}\right), N;$$

The check is the same as for spur gears.

(iii) Check for Wear: The limiting load for wear is given as:

$$F_w = \frac{D_p f K_w Q}{\cos \gamma_p}, \text{N}$$

The value of ratio factor Q is determined as:

$$Q = \frac{2z_{fg}}{z_{fp} + z_{fg}}, \text{ that is on the formative number of teeth.}$$

22.16.6 Forces Acting in Bevel Gear

A gear pinion pair in mesh and the forces acting are shown in Fig. 22.24. As before, the normal force between the teeth F_n can be resolved into two components: The turning force F_t on the mean diameter of the gear and the radial force F_r also acting on the mean diameter. These will be given as:

$$F_t = F_n \cos \phi, \text{ line CD}$$
$$F_r = F_n \sin \phi, = F_t \tan \phi, \text{ line CB}$$

Now this radial force on the mean diameter can be further resolved in two directions parallel to the wheel axes. It is clear that the radial force, EB, on the pinion shaft is:

$$F_{rp} = F_t \tan \phi \cdot \cos \gamma_p$$

The axial force on the pinion shaft will be component CE.

$$F_{ap} = F_t \tan \phi \cdot \sin \gamma_p$$

It is clear that the magnitude of the radial force on the pinion shaft equals the axial force on the gear shaft, but their directions are opposite:

$$F_{rg} = -F_{ap}$$

and similarly $$F_{ag} = -F_{rp}$$

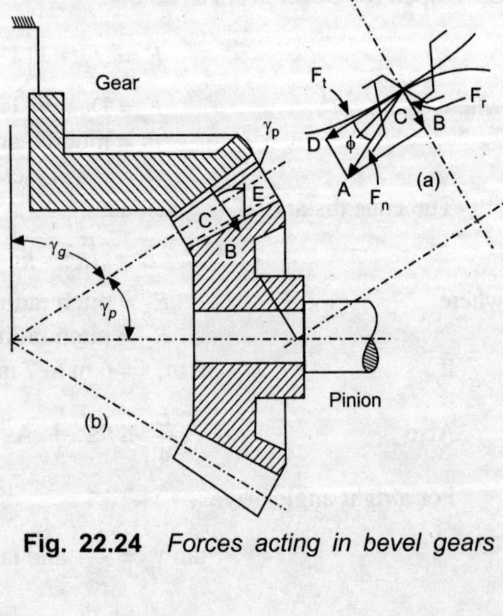

Fig. 22.24 *Forces acting in bevel gears*

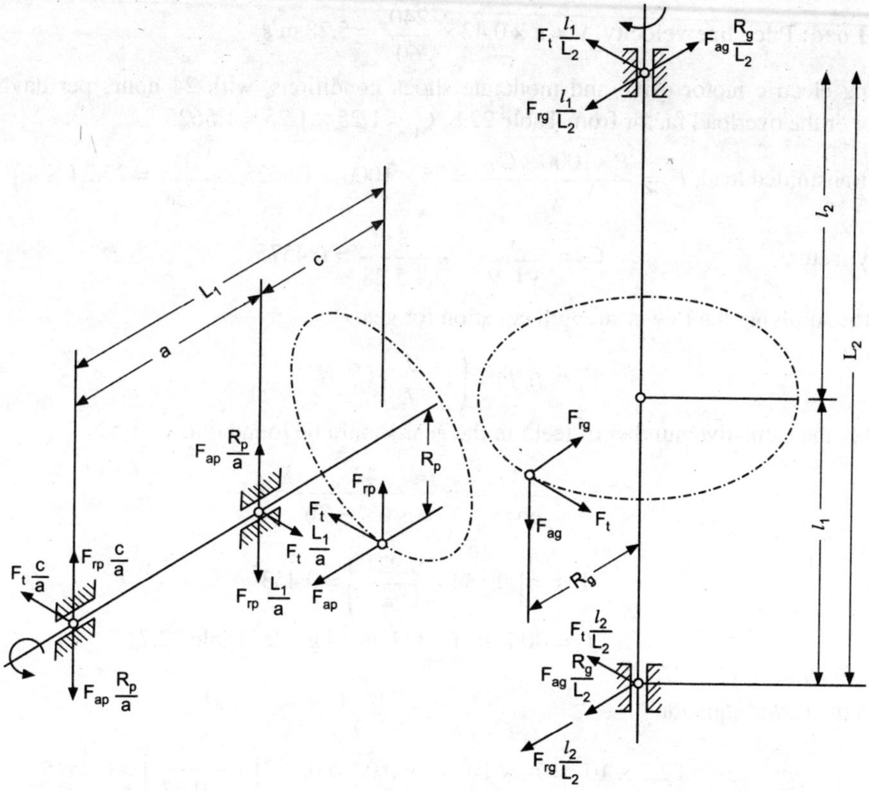

Fig. 22.24(a)

The forces F_t, F_r and F_a can be used to calculate the bearing pressures and then for the design of the shafts and bearings. This is illustrated in Fig. 22.24(a). Here the gear is mounted in between the bearings, whereas, the pinion is overhanging. The pinion should be mounted as close to the bearings as possible, to minimise the bending moments. The distance '*a*' between the bearings of the cantilever shaft should be equal (2.5 to 3.0) of the distance '*c*' from the middle of the pinion to the middle of the first bearing.

Example 22.10. Design a bevel gear drive between two shafts whose axes are at right angles. Speed of pinion shaft is 240 rev/min and that of the gear shaft is 120 rev/min. Pinion is to have 21 teeth of involute profile with module of 20 mm and a pressure angle of 20° and is to be of suitable material. Gear is of cast iron. Power at gear shaft = 75 kW.

Solution: Now $D_p = m \times z_p = 20 \times 21 = 420 \text{ mm} = 0.42 \text{ m}$

Since velocity ratio is 2 : 1.

$$\therefore \quad D_g = 2 \times 0.42 = 0.84 \text{ m, and } z_g = 21 \times 2 = 42$$

Now pitch cone angles may be determined as below:

$$\tan \gamma_p = \frac{z_p}{z_g} = \frac{21}{42} = 0.5$$

$$\therefore \quad \gamma_p = 26° - 34'$$

$$\therefore \quad \gamma_g = 90 - (26° - 34') = 63° - 26'$$

Now cone distance, $$L = \sqrt{R_g^2 + R_p^2} = \sqrt{0.42^2 + 0.21^2} = 0.47 \text{ m}$$

Design Load: Pitch line velocity, $V = \pi \times 0.42 \times \dfrac{240}{60} = 5.28$ m/s

Assuming electric motor drive and moderate shock conditions, with 24 hours per day running, the service factor or the overload factor from Table 22.8, $C_0 = 1.25 \times 1.25 = 1.5625$

Design transmitted load, $F_t = \dfrac{P \times 1000 \times C_0}{V} = 75 \times 1000 \times 1.5625 \times \dfrac{1}{5.28} = 22.2$ kN

Velocity factor, $$C_v = \frac{3}{3+V} = \frac{3}{3+5.28} = 0.4575$$

Strength: Applying the Lewis strength equation for gear

$$F_t = f_b\, fmY\left(1 - \frac{f}{L}\right) \cdot C_v \cdot N$$

To find Y, the formative number of teeth in the gear should be found out.

$$z_{fg} = \frac{z_g}{\cos\gamma_g} = \frac{42}{\cos 63° - 26'} = 94$$

$$\therefore \quad Y = \pi\left(0.154 - \frac{0.921}{94}\right) = 0.453$$

$f_b = 70$ MPa for C.I. good grade, Table 22.7

Now $L = 0.47$ m

∴ From the Lewis equation,

$$22.2 \times 10^3 = 70 \times 10^6 \times f \times 0.02 \times 0.453\left(1 - \frac{f}{0.47}\right) \times 0.4575$$

From here, $f = 0.086$ m

Now f should lie between $\dfrac{L}{4}$ and $\dfrac{L}{3}$, i.e., about 0.12 m to 0.16 m.

∴ Let us take, $f = 0.12$ m, i.e.. equal to $\dfrac{L}{4}$

Check for Dynamic Load:

$$F_d = F_t + \frac{21V(cf + F_t)}{21V + \sqrt{cf + F_t}}$$

Now $(f_b \times Y)$ for gear $= 70 \times 0.453 = 31.71$

Now $$z_{fp} = \frac{z_p}{\cos\gamma_p} = \frac{21}{\cos(26° - 34)} = \frac{21}{0.8944} = 23.5$$

$$\therefore \quad Y \text{ for pinion} = \pi\left(0.154 - \frac{0.921}{23.5}\right) = 0.36$$

Let the material for pinion be cast steel, heat treated with $f_b = 196$ MPa and BHN = 250, Table 22.7

∴ $(f_b \times Y)$ for pinion $= 196 \times 0.36 = 70.56$

∴ Check for dynamic load will be applied to gear.

Now $k = 0.111$

$$c = \frac{k \times e}{\dfrac{1}{E_p} + \dfrac{1}{E_g}}$$

Now at V = 5.28 m/s, the maximum permissible error from Table 22.10 is about 0.059 mm. From Table 22.9 with module of 20 mm the error for class 2 gear is about 0.0568 mm. Since the maximum permissible error is 0.059 mm, use gears of class 2.

$$\therefore \quad e = 0.0568 \text{ mm}$$

$$E_p = 210 \text{ GPa}, E_g = 105 \text{ GPa}$$

$$\therefore \quad c = \frac{0.111 \times 0.0568 \times 210 \times 10^9}{10^3 \times 3} = 441.34 \text{ kN/m}$$

$$\therefore \quad F_i = \frac{21 \times 5.28(441.34 \times 10^3 \times 0.12 + 22.2 \times 10^3)}{21 \times 5.28 + \sqrt{441.34 \times 10^3 \times 0.12 + 22.2 \times 10^3}} = 21.65 \text{ kN}$$

$$\therefore \quad F_d = 22.2 + 21.65 = 43.85 \text{ kN}$$

Now Beam strength of gear,

$$\therefore \quad F_b = f \cdot m \cdot Y \cdot f_{ef} \cdot \left(1 - \frac{f}{L}\right)$$

f_{ef} for C.I. = 84 MPa

$$\therefore \quad F_b = 0.12 \times 0.02 \times 0.453 \times 84 \times 10^6 \left(1 - \frac{0.12}{0.47}\right) = 68.00 \text{ kN}$$

Now F_b/F_d = 1.55, therefore, design is very much safe for dynamic load.

Check for Wear:

$$F_w = \frac{D_p f k_w Q}{\cos \gamma_p}$$

Now $\quad z_{fg} = 94, \quad z_{fp} = 23.5$

$$\therefore \quad Q = \frac{2z_{fg}}{z_{fp} + z_{fg}} = \frac{2 \times 94}{23.5 + 94} = 1.6$$

$$K_w = \frac{f_{es}^2 \sin \phi}{1.4}\left(\frac{1}{E_p} + \frac{1}{E_g}\right)$$

Now from Table 22.11, for steel pinion and C.I. gear, with BHN of 250 for Steel, f_{es} = 630 MPa

$$\therefore \quad K_w = \frac{(630 \times 10^6)^2 \times 0.342 \times 3}{1.4 \times 210 \times 10^9} = 1385 \text{ kPa}$$

$$\therefore \quad F_w = \frac{0.42 \times 0.12 \times 1385 \times 10^3 \times 1.6}{0.894} = 125 \text{ kN}$$

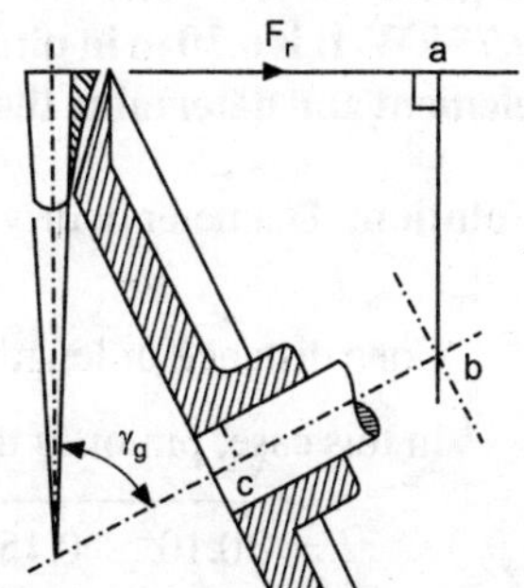

Fig. 22.25

Since $F_w > F_d$, $\quad \therefore$ Design is safe for wear also.

Tooth Proportions:

Addendum = m = 20 mm
Dedendum = 1.2 m = 24 mm
Working depth = 2.0 m = 40 mm
Clearance = 0.2 m = 4 mm
$p_c = \pi m = 20\,\pi$ mm

Tooth thickness = space width = $\frac{p_c}{2}$ = 10 π mm

For the gear, the blank diameter:

$$D_{og} = D_g + 2a \cos \gamma_g = 0.84 + 2 \times 0.02 \times 0.447 = 0.84 + 0.0179 = 0.858 \text{ m}$$

For the pinion, $D_{op} = D_p + 2a \cos \gamma_p = 0.42 + 2 \times 0.02 \times 0.894 = 0.456$ m

Gear shaft: Turning force, $F_t = 22.2$ kN

Radial force at the pitch diameter of gear, $F_r = F_t \tan \phi = 22.2 \times \tan 20° = 8.075$ kN

If in Fig. 22.25, the point '*b*' is the centre of the gear shaft bearing, then $bc = 0.23$ m

∴ Bending moment due to $F_t = F_t \times bc = 22.2 \times 0.23 = 5.10$ kNm

The bending moment on shaft, due to $F_r = F_r \times ab = 8.075 \times 0.275 = 2.22$ kNm

Since the B.M. due to F_t and F_r act in planes at right angles. ∴ Their resultant is

$$M_b = \sqrt{5.10^2 + 2.22^2} = 5.56 \text{ kNm}$$

Torsional moment on the shaft is, $M_t = F_t \times \frac{D_g}{2} = 22.2 \times 0.42 = 9.32$ kNm

$$\therefore \quad M_{te} = \sqrt{M_b^2 + M_t^2} = 10.86 \text{ kNm}$$

The shaft is subjected to direct stress, shear stress and alternating bending stress.

In the tooth design, we have taken *FS* as 1.5 on elastic limit or about 3 on ultimate tensile strength. Now, while calculating equivalent torque above, we have not considered the design load which includes dynamic factor but only the design transmitted load. Therefore, for shaft design, shock factor of about 1.5 (as discussed in Chapter 4) can be taken. Thus, the *FS* may be taken as 4.5, say 5 to account for direct stress also. Therefore, as before, f_s may be taken as,

$$f_s = \frac{412.5}{5} = 82.5 \text{ MPa}$$

or $f_s = 61.9$ MPa, after considering keyway effect.

$$\therefore \quad \frac{\pi}{16} d^3 \times 61.9 = 10.86 \times 1000$$

$$\therefore \quad d = 0.0961 \text{ m, say } 100 \text{ mm}$$

Hub diameter = 1.8 *d* = 1.8 × 100 = 180 mm

Hub length = 1.25 *d* = 1.25 × 100 = 125 mm

Pinion Shaft: Pinion shaft can also be designed in the above manner.

Example 22.11. A pair of high grade cast Iron bevel gears having shaft at right angles are to have an angular velocity ratio of driver to driven of 2 to 3. The driver is to rotate at 180 rev/min and is to transmit 9.75 kW. It is 0.30 m in pitch diameter. Take the width of face as about one third of the length of the pitch element and determine the pitch of the gears. Assume 24 hour/day operation.

Solution: Diameter of driven gear = $0.30 \times \frac{2}{3} = 0.20$ m

Cone distance or length of the pitch element, $L = \sqrt{R_p^2 + R_g^2}$

In this case, pinion is the driven and gear is the driver.

$$\therefore \quad L = \sqrt{0.10^2 + 0.15^2} = 0.18 \text{ m}$$

∴ Width of face, $f \cong L/3$, say 60 mm

Pitch line velocity, $V = \pi \times \frac{0.30}{60} \times 180 = 2.827$ m/s

Assuming electric motor drive with moderate shock conditions, $C_0 = 1.25 \times 1.25 = 1.5625$

$\therefore$ Design transmitted load, $F_t = \frac{9.75 \times 1000 \times 1.5625}{2.827} = 5390.6$ N

Velocity factor, $C_v = \frac{3}{3 + 2.827} = \frac{3}{5.827} = 0.518$

Since the material for both pinion and gear is C.I., the pinion becomes the basis of design.

Beam Strength equation:

$$\therefore \qquad F_b = f_{ef} \cdot f \cdot m \cdot Y\left(1 - \frac{f}{L}\right), \text{N}$$

Now $$z_{fp} = \frac{z_p}{\cos \gamma_p}$$

Now $$\tan \gamma_p = \frac{z_p}{z_g}$$

Let $$m = \frac{f}{7} \text{ to } \frac{f}{6}, \text{ i.e., 8.55 to 10 mm say 10 mm}$$

$$\therefore \qquad z_p = \frac{D_p}{m} = \frac{200}{10} = 20$$

and $$z_g = \frac{D_g}{m} = \frac{300}{10} = 30$$

$$\therefore \qquad \tan \gamma_p = \frac{200}{300} = 0.667 \qquad \therefore \gamma_p = 33° - 42'$$

$$\therefore \qquad z_{fp} = \frac{20}{0.832} = 24$$

$$\therefore \qquad Y \text{ for pinion} = \pi\left(0.154 - \frac{0.921}{24}\right) = 0.361$$

For high grade C.I. $f_{ef} = 84$ MPa

$$\therefore \qquad F_b = 84 \times 10^6 \times 0.06 \times 0.01 \times 0.361\left(1 - \frac{0.06}{0.18}\right) = 12.13 \text{ kN}$$

Now dynamic load can be found out as follows also:

$$F_d = F_t / C_v = 5390/0.518 = 10.04 \text{ kN}$$

Now since $F_b > F_d$, $\therefore$ Design is safe for dynamic loading

Check for Wear: $$F_w = \frac{D_p f K_w Q}{\cos \gamma_p}$$

Now $$zf_g = \frac{z_g}{\cos \gamma_g}; \text{ now } \gamma_g = 90° - (33° - 42') = 56° - 18'$$

$$\therefore \qquad zf_g = \frac{30}{0.555} = 54$$

$$\therefore \qquad Q = \frac{2zf_g}{zf_p + zf_g} = \frac{2 \times 54}{24 + 54} = \frac{108}{78} = 1.385$$

Now
$$K_w = \frac{f_{es}^2 \times \sin\phi}{1.4}\left(\frac{1}{E_p} + \frac{1}{E_g}\right)$$

Now $\quad f_{es}$ for C.I. = 630 MPa

Now $\quad E_p = E_g = 105$ GPa

$$\therefore \qquad K_w = \frac{(630 \times 10^6)^2 \times 0.0342}{1.4} \times \frac{2}{105 \times 10^9} = 1846.8 \times 10^3 \text{ N/m}^2$$

$$\therefore \qquad F_w = \frac{0.20 \times 0.06 \times 1846.8 \times 10^3 \times 1.385}{0.832} = 36.89 \text{ kN}$$

which is quite large as compared to F_d ∴ safe.

Example 22.12. A cast iron bevel gear has a module of 2.5 mm and its pitch diameter is 0.60 m. The pitch angle is 30° and the teeth are 20° full depth. Determine the permissible endurance load.

Solution: The permissible endurance load is the beam strength of the tooth, which is given as,

$$F_b = f \cdot m \cdot Y \cdot f_{ef}\left(1 - \frac{f}{L}\right), \text{N}$$

Now number of teeth on the gear, $z_g = \dfrac{D}{m} = \dfrac{600}{2.5} = 240$

∴ Formative Number of teeth, $z_f = \dfrac{z}{\cos\gamma} = \dfrac{240}{\cos 30^\circ} = 277$

$$\therefore \qquad Y = \pi\left(0.154 - \frac{0.921}{z_f}\right) = \pi\left(0.154 - \frac{0.921}{277}\right) = 0.473$$

The cone distance, $\qquad L = \dfrac{D}{2\sin\gamma} = \dfrac{0.60}{2 \times \sin 30} = 0.60$ m

Let, the face width, $\qquad f = \dfrac{L}{3} = \dfrac{0.60}{3} = 0.20$ m

Now f_{ef} = flexural endurance limit for C.I. is 84 MPa

$$\therefore \qquad F_b = 0.20 \times 0.0025 \times 0.473 \times 84 \times 10^6 \times \left(1 - \frac{0.20}{0.60}\right) = 13.244 \text{ kN.}$$

Example 22.13. Two steel bevel gears, both having a Brinell hardness of 250, connect shafts at 90°. The teeth are 14½° full depth and the module is 5 mm. The number of teeth on the pinion and gear are 30 and 48. The face width is 38 mm. Determine the wear load.

Solution: The wear load is given as,

$$F_w = \frac{D_p f K_w Q}{\cos\gamma_p}$$

$$D_p = z_p \times m = 30 \times 5 = 150 \text{ mm} = 0.15 \text{ m}$$

$$D_g = z_g \times m = 48 \times 5 = 240 \text{ mm} = 0.24 \text{ m}$$

$$f = 38 \text{ mm},$$

$$L = \sqrt{R_g^2 + R_p^2}$$

$\therefore$ $$L = \sqrt{0.12^2 + 0.075^2} = 0.1415 \text{ m}$$

Now $$\cos\gamma_p = \frac{R_g}{L} = \frac{0.12}{0.1415} = 0.848$$

$$\cos\gamma_g = \frac{R_p}{L} = \frac{0.075}{0.1415} = 0.53$$

$$z_{fg} = \frac{z_p}{\cos\gamma_g} = \frac{48}{0.53} = 90.5$$

$$z_{fp} = \frac{z_p}{\cos\gamma_p} = \frac{30}{0.848} = 35.4$$

$\therefore$ $$Q = \frac{2z_{fg}}{z_{fp} + z_{fg}} = \frac{2 \times 90.5}{35.4 + 90.5} = 1.44$$

$$K_w = \frac{f_{es}^2 \times \sin\phi}{1.4}\left(\frac{1}{E_p} + \frac{1}{E_g}\right)$$

Now $$E_p = E_g = 210 \text{ GPa}$$

$$f_{es} = 2.8 \times \text{B.H.N.} - 70 = 2.8 \times 250 - 70 = 630 \text{ MPa}$$

$\therefore$ $$K_w = \frac{(630 \times 10^6)^2 \times (\sin 14\tfrac{1}{2})^\circ}{1.4}\left(\frac{2}{210 \times 10^9}\right) = 676.026 \text{ kPa}$$

$\therefore$ $$F_w = \frac{0.15 \times 0.038 \times 676.026 \times 1.44}{0.848} = 6.54 \text{ kN.}$$

Example 22.14. Two cast iron bevel gears have pitch diameters of 75 mm and 100 mm respectively and are to transmit 2.25 kW at 1100 rev/min of the pinion. The tooth profiles are 14½° system.

(a) Determine the face width and the required module from the standpoint of strength using Lewis equation.

(b) Check the design for dynamic load and wear.

Solution: Since both the gears are of the same material, pinion is the weaker out of the two.

(a) The Lewis equation is

$$F_t = f_b\, f\, m\, \pi y\, C_v\left(1 - \frac{f}{L}\right)$$

When designing for strength, it is better to use modified form of the Lewis equation as,

$$\frac{1}{my} = \frac{f_b f \pi C_v}{F_t}\left(1 - \frac{f}{L}\right) = \text{an allowed value.}$$

Now $$V = \frac{\pi \times 0.075 \times 1100}{60} = 4.32 \text{ m/s}$$

$\therefore$ $$C_v = \frac{3}{3 + 4.32} = 0.414$$

Let $$f_b = 84 \text{ MPa, for C.I. good grade.}$$

Cone distance, $$L = \sqrt{R_p^2 + R_g^2}$$

$\therefore$ $$L = \sqrt{0.0375^2 + 0.05^2} = 0.0625 \text{ m}$$

Let $$f = \frac{L}{3} = \frac{62.5}{3} = 20.83 \text{ mm, say } 20 \text{ mm}$$

Now $$F_t = \frac{P \times 1000}{V} = \frac{2.25 \times 1000}{4.32} = 520.83 \text{ N}$$

Allowed value of $$\frac{1}{my} = \frac{84 \times 10^6 \times 0.02 \times \pi \times 0.414}{520.83}\left(1 - \frac{0.02}{0.0625}\right) = 2.853 \times 10^3$$

Let $$y = 0.1, \frac{1}{0.1\, m} = 2.853 \times 10^3, m = 3.505 \times 10^{-3} \text{ m} = 3.505 \text{ mm}$$

Let us try with $m = 3$ mm

This value of module will not give the number of teeth as whole number.

Let us take $m = 5$ mm

$$\therefore \quad z_p = \frac{D_p}{m} = \frac{75}{5} = 15, z_g = \frac{100}{5} = 20$$

$$\therefore \quad z_{fp} = \frac{z_p}{\cos \gamma_p}, \cos \gamma_p = \frac{R_g}{L} = \frac{50}{62.5}$$

$$\therefore \quad z_{fp} = \frac{15 \times 62.5}{50} = 18.75$$

$$\therefore \quad y = 0.124 - \frac{0.684}{18.75} = 0.124 - 0.0365 = 0.0875$$

$$\therefore \quad \frac{1}{my} = \frac{1}{0.005 \times 0.0875} = 2.286 \times 10^3 < 2.853 \times 10^3 \quad \therefore \text{ O.K.}$$

(b) Check for dynamic load:

$$F_d = F_t + \frac{21V(cf + F_t)}{21V + \sqrt{cf + F_t}}$$

$$c = \frac{k \times e}{\frac{1}{E_p} + \frac{1}{E_g}}$$

$k = 0.107$ for 14½° system.

From Table 22.10, at $V = 4.32$ m/s, the maximum permissible error is about 0.066 mm. From Table 22.9 let us use class 2 accurate gears with $e = 0.028$ mm.

$$E_p = E_g = 105 \text{ GPa}$$

$$\therefore \quad c = \frac{0.107 \times 0.028 \times 105 \times 10^9}{2 \times 1000} = 157.3 \text{ kN}$$

$$\therefore \quad F_d = 520.83 + \frac{21 \times 4.32(157.3 \times 10^3 \times 0.02 + 520.83)}{21 \times 4.32 + \sqrt{157.3 \times 10^3 \times 0.02 + 520.83}} = 2719.5 \text{ N}$$

Now beam strength is

$$F_b = f \cdot m \cdot Y \cdot f_{ef} \cdot \left(1 - \frac{f}{L}\right), \text{N}$$

For C.I., $$f_{ef} = 84 \text{ MPa}$$

$$\therefore \quad F_b = 0.20 \times 0.005 \times 84 \times 10^6 \times \pi \times 0.0875 \left(1 - \frac{0.02}{0.0625}\right) = 1573 \text{ N}$$

Since F_b is < F_d, therefore, better material should be specified for pinion so that its beam strength or endurance load is increased to atleast 2720 N.

Check for wear:

$$F_w = \frac{D_p\, f\, K_w\, Q}{\cos \gamma_p}, \; Q = \frac{2z_{fg}}{z_{fp} + z_{fg}}$$

Now $$z_{fg} = \frac{z_g}{\cos \gamma_g} = \frac{20}{3.75} \times 6.25 = 33.3$$

$$\therefore \quad Q = \frac{2 \times 33.3}{18.75 + 33.30} = \frac{66.6}{52.05} = 1.28$$

$$K_w = \frac{f_{es}^2 \times \sin \phi}{1.4}\left(\frac{1}{E_p} + \frac{1}{E_g}\right)$$

Now f_{es} = 630 MPa for Gray C.I. Gear and pinion, Table 22.11

$$\therefore \quad K_w = \frac{(630 \times 10^6)^2 \times 0.25}{1.4}\left(\frac{2}{105 \times 10^9}\right) = 1350 \times 10^3, \text{ N/m}^2$$

$$\therefore \quad F_w = \frac{0.075 \times 0.02 \times 1350 \times 10^3 \times 1.28}{0.8} = 3240 \text{ N} > F_d$$

Design is safe for wear.

Example 22.15. A pair of bevel gears is to be used to transmit 9 kW. Determine the required module and gear diameters for the following specification:

		Pinion	**Gear**
Number of teeth	:	**21**	**60**
Material	:	**Semi-Steel**	**Cast Iron**
f_b	:	**84 MPa**	**56 MPa**
BHN	:	**200**	**160**
Speed	:	**1200 rev/min**	**420 rev/min**
Tooth Profile	:	**14° system**	**same.**

Solution: Since the diameters are not known, the Lewis equation can be written as,

$$f_b C_v = \frac{F_t}{f\, m\, \pi\, y}\left(\frac{L}{L - f}\right)$$

Now $$F_t = \frac{M_t}{D/2} = \frac{2M_t}{mz}$$

$$\therefore \quad f_b C_v = \frac{2M_t}{fm^2 \pi yz}\left(\frac{L}{L - f}\right)$$

where z = actual number of teeth of the gear for which the equation is to be used.

Now let $$f = \frac{L}{3} \quad \therefore \quad \frac{L}{L - f} = \frac{3}{2}$$

Now $$L = \sqrt{R_p^2 + R_g^2} = R_p\sqrt{1+R^2}$$

where R = ratio of angular velocity of the pinion to the angular velocity of the gear.

$$\therefore \quad f = \frac{2R_p}{6}\sqrt{1+R^2} = \frac{D_p}{6}\sqrt{1+R^2} = \frac{m \times z_p}{6}\sqrt{1+R^2}$$

It can be seen, that,

$$\tan\gamma_p = \frac{z_p}{z_g} \text{ and } \tan\gamma_g = \frac{z_g}{z_p}$$

$$\therefore \quad \tan\gamma_p = \frac{21}{60} = 0.35 \qquad \therefore \gamma_p = 19^\circ - 18'$$

$$\therefore \quad \tan\gamma_g = \frac{60}{21} = 2.86 \qquad \therefore \gamma_g = 70^\circ - 42'$$

$$\therefore \quad z_{fp} = \frac{z_p}{\cos\gamma_p} = \frac{21}{0.9438} = 22.2$$

$$z_{fg} = \frac{60}{\cos\gamma_g} = \frac{60}{0.3305} = 182$$

$$\therefore \quad y_p = 0.124 - \frac{0.684}{22.2} = 0.124 - 0.0308 = 0.0932$$

$$y_g = 0.124 - \frac{0.684}{182} = 0.12024$$

$$\therefore \quad (f_b \times y) \text{ pinion} = 84 \times 10^6 \times 0.0932 = 7.83 \times 10^6$$

$$(f_b \times y) \text{ gear} = 56 \times 10^6 \times 0.12024 = 76.73 \times 10^3$$

∴ Gear is weaker and so the design will be based on gear. The modified Lewis equation will be used.

$$(f_b \times C_v)\frac{2M_t}{f \times m^2 \pi \times y \times z}\left(\frac{L}{L-f}\right) < \text{allowed stress}$$

Now $$M_t = \frac{9 \times 1000 \times 60}{2\pi \times 420} = 204.63 \text{ Nm}$$

$$y = y_g = 0.120; \; z = z_g = 60$$

$$L = \frac{m \times z_p}{6}\sqrt{1+R^2} = \frac{m \times 21}{6} \times \sqrt{1+\left(\frac{20}{7}\right)^2} = 31.8\, m$$

$$f = \frac{L}{3} = 10.6\, m, \text{ and } \frac{L}{L-f} = \frac{3}{2}$$

$$\therefore \quad (f_b \times C_v) = \frac{2 \times 204.63}{10.6m \times m^2 \times \pi \times 0.12 \times 60} \times \frac{3}{2} = \frac{2.56}{m^3}$$

Let velocity factor $C_v = 0.5$

$$\therefore \quad (f_b \times C_v) = 56 \times 0.5 = 28 \text{ MPa}$$

$$\therefore \quad \frac{2.56}{m^3} \le 28 \times 10^6 \text{ or } m \ge 0.0045 \text{ m}$$

Let $m = 5$ mm

$$\therefore \quad D_p = 21 \times 5 = 105 \text{ mm}$$

$$D_g = 60 \times 5 = 300 \text{ mm}$$

$$\therefore \quad V = \frac{\pi \times 300 \times 420}{1000 \times 60} = 6.59 \text{ m/s}$$

$$F_t = \frac{1000 \times 9}{6.59} = 1366 \text{ N}$$

$$\therefore \quad C_v = \frac{6}{6 + 6.59} = 0.481$$

$$\therefore \quad (f_b \times C_v) = 56 \times 0.481 = 27 \leq 28, \ \therefore \text{ O.K.}$$

Therefore, module = 5 mm

$$\therefore \quad f = 10.6 \times 5 = 53 \text{ mm}, L = 159 \text{ mm}.$$

(b) Check for Wear and Dynamic load:

Now beam strength of the gear tooth is,

$$\therefore \quad F_b = f \cdot m\pi \cdot y\, f_{ef} \left(1 - \frac{f}{L}\right)$$

For Grey C.I. with B.H.N. = 160 ; f_{ef} = 84 MPa

$$\therefore \quad F_b = 0.053 \times 0.005 \times \pi \times 0.12 \times 84 \times 10^6 \left(1 - \frac{1}{3}\right) = 5594.53 \text{ N}$$

Now $$F_w = \frac{D_p \times f \times K_w \times Q}{\cos \gamma_p}$$

$$Q = \frac{2z_{fg}}{z_{fp} + z_{fg}} = \frac{2 \times 182}{22.2 + 182} = \frac{364}{204.2} = 1.78$$

$$K_w = \frac{(f_{es})^2 \times \sin\phi}{1.4} \left(\frac{1}{E_p} + \frac{1}{E_g}\right)$$

$$= \frac{(630 \times 10^6)^2 \times 0.25}{1.4} \left(\frac{1}{210 \times 10^9} + \frac{1}{84 \times 10^9}\right) = 1180 \times 10^3, \text{ N/m}^2$$

$$\therefore \quad F_w = \frac{0.105 \times 0.053 \times 1180 \times 10^3 \times 1.78}{0.9438} = 12.38 \text{ kN}$$

Now $$F_d = F_t + \frac{21V\,(cf + F_t)}{21V + \sqrt{cf + F_t}}$$

$$c = \frac{k \times e}{\dfrac{1}{E_p} + \dfrac{1}{E_g}}$$

From Table 22.10, for V = 6.59 m/s, the maximum permissible error is about 0.0516 mm. From Table 22.9 with module equal to 5 mm, class 2 gears will have to be used with e = 0.028

$$k = 0.107$$

$$\therefore \quad c = \frac{0.107 \times 0.028 \times 210 \times 84 \times 10^9}{294 \times 1000} = 179.8 \text{ kN/m}$$

$$\therefore \qquad F_d = 1366 + \frac{21 \times 6.59(179.8 \times 10^3 \times 0.053 + 1366)}{21 \times 6.59 + \sqrt{179.8 \times 10^3 \times 0.053 + 1366}} = 7576.9 \text{ N}$$

Now $F_w > F_d$, therefore, design is safe for wear.

But $F_b < F_d$, therefore, F_b will have to be increased to equal at least F_d. This can be achieved by using a bigger module, increasing the face width or by using better materials.

22.17 WORM GEAR DRIVES

If motion is to be transmitted between shafts whose axes are non-coincident, non parallel, and non-intersecting, the following gear drives may be employed:

1. Worm and Worm gear drive
2. Crossed helical gears
3. Hypoid gear drive.

The last two drives are suitable only at comparatively low speed ratios, (upto 7) and low power ratings. But worm and worm gear drive can be employed for large speed ratios ranging from 10 : 1 to 100 : 1 and ratios as high as 500 : 1 have been used. This drive is used principally to transfer motion from the input shaft to the follower shaft for shaft angles which cross at 90º. Occasionally, this angle may be different from 90º. Worm and worm gear drive is mostly used as speed reducer with worm as the driver and worm gear or wheel as the driven member.

22.17.1 Worm

Two types of worms are in use:

1. Cylindrical or straight worm, Fig. 22.26(a).
2. Cone, hourglass or double enveloping worm, Fig. 22.26(b).

The Cylindrical form is most commonly used. It is a cylinder on which one or more threads are cut to form a helix. The shape of the thread is involute helicoid of pressure angle of 20º. Worm threads in the axial section have the shape of a rack with an included profile angle of $2\phi = 40º$, where ϕ is the pressure angle. The cone worm is used to some extent but it requires extremely accurate alignment. The thread of the worm may be single start, double start and so on. The worm gear drive can be classed as **"Screw-and-gear drive"**, i.e., a drive combining the features of the gear and screw drives. According to the location of the worm, they are distinguished as: Overheads (Fig. 22.29), Underslung (Fig. 22.31) and side mounted. Worms can be made either integral with the shaft or as a separate part mounted on it (This design being seldom used).

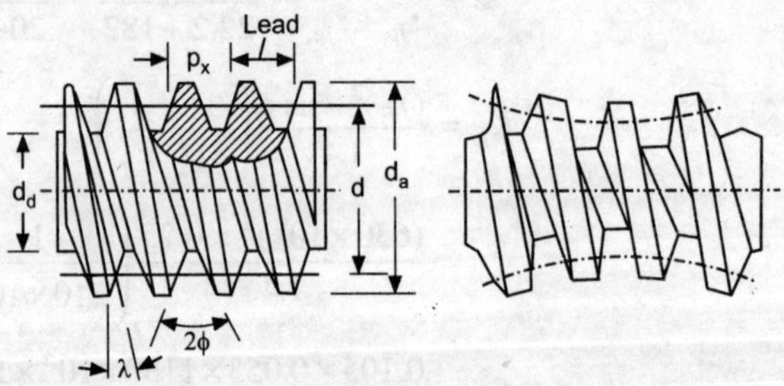

(a) Cylindrical *(b) Cone*

Fig. 22.26 *Worms*

Definitions:

(i) Axial pitch or linear pitch, p_x: It is the distance from any point on one thread of worm to the corresponding point on the adjacent thread, measured parallel to the worm axis.

(ii) Lead L: It is the distance travelled by a thread when one complete revolution is given to the worm.

$$L = Z_1 p_x,$$

where Z_1 = number of worm starts.

So, for single start worm, lead is equal to axial pitch and for double start worm thread, lead is twice the axial pitch and so on. It is clear that the axial pitch of the worm will be equal to the circular pitch of the worm gear.

(iii) Lead angle, λ: It is the angle between the tangent to the thread helix on the pitch cylinder and the plane normal to the axis of the worm. On the other hand, the helix angle of the thread is the angle between the tangent to the thread helix on the pitch cylinder and the axis of worm. Thus,

$$\text{Helix angle} = 90° - \lambda°$$

(iv) Velocity Ratio: The linear velocity of the worm is

$$V_w = \frac{L \times n_w}{60}, \text{ where } n_w = \text{rev/min. of worm.}$$

Linear velocity of worm gear is

$$V_g = \frac{\pi D \times n_g}{60}$$

where D = pitch diameter of worm gear or wheel

and n_g = rev/min of the wheel

Since $V_w = V_g \quad \therefore \frac{L \times n_w}{60} = \frac{\pi \times D \times n_g}{60}$

$$\therefore \quad Ln_w = \pi D n_g$$

$$\therefore \quad \text{Velocity ratio, } (V_R) = \frac{n_w}{n_g} = \frac{\pi D}{L}$$

Now $\pi D = z_g \times p_c$

where z_g = number of teeth in worm gear

and p_c = circular pitch of gear

and $L = Z_1 \times p_x$

$$\therefore \quad (V_R) = \frac{z_g \times p_c}{Z_1 \times p_x} = \frac{z_g}{Z_1} \quad (\text{since } p_c = p_x)$$

$$\therefore \quad (V_R) = \frac{\text{number of teeth in gear}}{\text{number of worm starts}}$$

$$\therefore \quad V_R = \frac{w_w}{w_g} = \frac{n_w}{n_g} = \frac{z_g}{Z_1} = \frac{\pi D}{L}$$

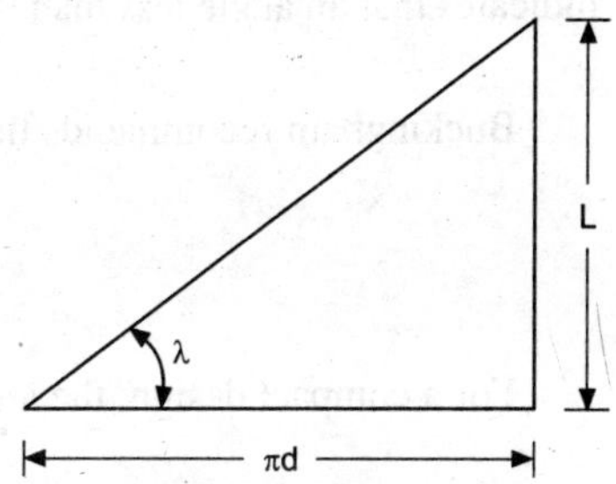

Fig. 22.27 *Lead angle*

(v) Other Relations: $p_x = \pi m = \dfrac{L}{Z_1}$

Here m = module

Now q = diametral quotient = $\dfrac{d}{m}$

where d = pitch diameter of worm.

Values of standard modules are : 1, 1.25, 1.5, 2, 2.5, 3, 4, 5, 6, 8, 10, 12, 16, 20 mm.

Values of 'q' range from 9 to 18, higher values usually used for low modules.

From experience, there are established values for 'Z' with various speed ratios, as given in Table 22.12. The number of worm starts usually range from 1 to 4 of course, it may go upto 6 to 8. $z_g > 29$ normally.

Table 22.12 Worm data

Velocity Ratio	Number of Threads
36 and over	One
12 to 36	Double
8 to 12	Triple
6–12	Quadruple
4–10	Sextuple

Addendum circle diameter of worm is:

$$d_a = d + 2m$$

Dedendum circle diameter of worm is:

$$d_d = d - 2.4\,m$$

or $d_d = d - 2.5\,m$ (assuming radial clearance to be 0.25 m)

The lead angle λ is given as, Fig. 22.27.

$$\lambda = \tan^{-1}\frac{L}{\pi d}$$

Now $L = Z_1 \times p_x$

$$\therefore \quad \lambda = \tan^{-1}\frac{Z_1 \times p_x}{\pi d} = \tan^{-1}\frac{Z_1}{q}$$

Now lead angle λ may vary from 9° to 45°. It is usually taken as, λ = 6° per worm thread. Experience indicates that an angle less than 9° results in rapid wear. A safe value for λ is given as:

$$\lambda \geq 12½°$$

Buckingham recommends the following limiting values for lead angles.

$$\phi = 14\frac{1}{2}^{\circ} \quad 20° \quad 25° \quad 30°$$

$$\lambda = 16° \quad 25° \quad 35° \quad 45°$$

For a compact design, the angle λ may be selected approximately from the relation:

$$\tan\lambda = \left(\frac{n_g}{n_w}\right)^{\frac{1}{3}}$$

The length of the threaded portion of the worm is taken from the empirical relations:

With $Z_1 = 1 - 2\,;\, l \geq (11 + 0.06\,z_g)\,m$, mm

$Z_1 = 3 - 4\,;\, l \geq (12.5 + 0.09\,z_g)\,m$, mm

For a worm with ground side surfaces of the thread profile, the calculated value of length should be increased depending upon module, by 25 to 30 mm, to exclude feed marks produced by the vibrating grinding wheel as it leaves the thread root.

The minimum pitch diameter of worm is empirically given as:

$d = 2.35\,p_x + 10$ mm for worm cut on shaft

$= 2.40\,p_x + 28$ mm for worms bored to fit over the shaft.

22.17.2 Worm Wheel

There are three types of worm wheel rim construction as shown in Fig. 22.28.

1. Figure 22.28(a), straight teeth are cut with a form cutter. This construction is not efficient and is employed for light and intermittent service and for small power transmissions.

2. Figure 22.28(b). The ends of the gear teeth are cut parallel to the gear axis.
3. Figure 22.28(c). The teeth are cut radially towards the worm axis. This construction is used for worms with a small lead angle. The included angle is called "face angle" and its value is $2\gamma = 60^\circ$ to 75° (may be 90° to 110°). It characterizes the degree of worm envelopment by the wheel rim. The recommended value of face angle is:

$$\tan \gamma \leq \frac{\tan \phi}{\tan \lambda}$$

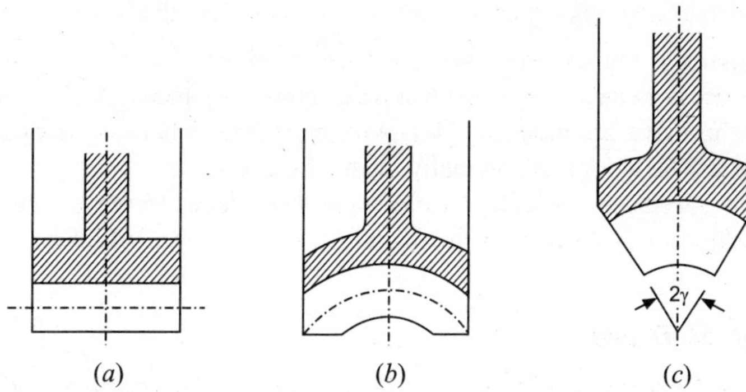

Fig. 22.28 *Worm wheel rim constructions*

Proportions: Refer Fig. 22.29. The diameter of the worm gear or wheel is:

$$D = z_g m$$

Diameter of addendum circle is

$$D_a = D + 2m$$

Diameter of dedendum circle is

$$D_d = D - 2.4\, m$$

The outside diameter D_o of the wheel is larger than the addendum circle diameter since the wheel rim envelops the worm. It is given as:

$$D_a < D_o < D_a + (1 - 2)\, m$$

The width of the wheel rim B is given as:

$$B \leq 0.75\, d_a \text{ with } Z_1 = 1 - 3$$

and

$$B \leq 0.67\, d_a \text{ with } Z_1 = 4$$

The design face width of the tooth along the wheel circle of diameter $d_a - 2m$, i.e., along the pitch circle of the worm, will be:

$$f = (d_a - 2m) \times \frac{\gamma^\circ \times 3.14}{180}$$

Its value is also given empirically as:

$$f = 2.38\, p_c + 6.5 \text{ mm, with } Z_1 = 1 - 2$$
$$= 2.15\, p_c + 5.0 \text{ mm, with } Z_1 = 3 - 4$$

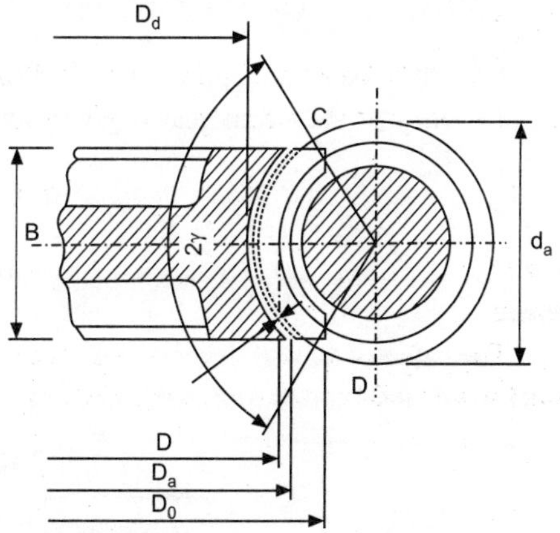

Fig. 22.29 *Worm gear proportions*

22.17.3 Designation of Worm Gear Drive

A worm gear drive is designated as:

$$Z_1 / z_g / q/m$$

Supposing we write a drive as R6/31/9/8, 160 mm, then it means that a right hand worm of starts 6, and of diameter quotient 9, meshes with a worm wheel of 31 teeth and module 8 mm, at a centre distance of 160 mm.

22.17.4 Materials

Worm is always made of a better grade material. Since in worm gear drive the predominant relative motion is sliding, unlike metals give better results. The material is decided as below:

(a) **Light loads, low speed:** Worm is of steel and worm gear of C.I.
(b) **Medium service conditions:** Worm is of case hardened steel of BHN 250 and worm gear is of phosphor bronze.
(c) **High speeds, Heavy loads with shock conditions:** Worm is of hardened molybdenum steel or chrome vanadium steel and gear is of phosphor bronze chilled to increase the hardness.

Note: The common practice is to make only the rim of the worm wheel of high-quality material, while the middle of the rim or its centre is made of C.I. The rim and the centre are joined by bolts or lock screws. For low-speed drives, the worm wheel rims are made of C.I. But for high speed and heavy duty drives, they are made of various grades of bronze. The worms are normally made of steel.

The worms of the low-speed drives usually receive no heat treatment. However, the worms of high speed drives are heat treated to $Rc \geq 45$. Also, a worm mating with a tinless bronze wheel rim should be heat treated to $Rc \geq 45$.

22.17.5 Efficiency of Drive

The losses in a worm gear drive consist of : losses in the gearing in the worm and worm wheel bearings, and those due to the churning and splashing of lubricating oil. Thus, the efficiency of the drive can be written as,

$$\eta = \eta_g \cdot \eta_b \cdot \eta_{ch}, \text{ where}$$

η_g = efficiency accounting for the losses due to gearing
η_b = efficiency accounting for the losses due in bearings
η_{ch} = efficiency accounting for the losses due to churning and splashing of oil.

The major losses are due to the gearing, which may be approximated as those in a screw-nut pair of same materials and lead angle as the worm-gear drive. Thus,

$$\eta_g = \frac{\tan \lambda}{\tan(\lambda + \phi_1)}$$

The other losses are approximately 4 to 5%. Therefore,
the efficiency of the worm gear is given approximately as:

$$\eta = (0.95 - 0.96)\frac{\tan \lambda}{\tan(\lambda + \phi_1)}$$

where ϕ_1 = reduced friction angle and given as $\tan^{-1} \mu_1$
where μ_1 = reduced coefficient of friction

The value of ϕ_1 will depend upon the materials of worm and gear, lubrication and sliding velocity. For a steel worm and bronze gear, the values of ϕ_1 are given is Table 22.13.

Table 22.13 Values of ϕ_1

Sliding Velocity, m/s	ϕ_1
0.1	3°–40′ to 4°–30′
0.5	3°–10′ to 3°–40′
1.0	2°–30′ to 3°–10′
2.0	2° to 2°–30′
4.0	1°–20′ to 1°–40′
10.0	0°–50′ to 1°–20′

The sliding velocity or the rubbing velocity is given as :

$$V = \frac{\pi d n_w}{60 \cos \lambda} \text{ m/s}$$

For ordinary industrial worm:

$$V \not> 6 \text{ m/s}$$

and for well designed, precision machined, hardened and ground worms, the usual range of V is, $6.67 < V < 16.67$

μ_1 is also given empirically as:

$$\mu_1 = 0.025 + \frac{V}{300}, \text{ for } V > 2.75 \text{ m/s}$$

and for V ranging from 0.2 to 2.75 m/s

$$\mu_1 = \frac{0.042}{V^{0.28}}$$

For low speeds, μ_1 may be taken as equal to 0.15.

In the tentative determination of efficiency, when the sliding velocity is still unknown, the value of ϕ_1, can be taken 3° for steel worm and a bronze wheel and as 6°–30′ for a steel worm and a C.I. wheel. Tentatively, the total efficiency of a worm gear drive taken for preliminary calculations

For single start worm = 0.7 to 0.75

For double start worm = 0.75 to 0.82

For quadruple start worm = 0.85 to 0.90

22.17.6 Heat Dissipation

The work of friction is converted into heat and this heat must be radiated out to avoid overheating of the lubricating oil. The heat dissipation will depend upon the size and surface of the housing and on the velocity of air surrounding the housing. The dissipating capacity will be given as:

$Q = A(t_2 - t_1)k$, where k is heat dissipating capacity, W/m² °C

where A = Heat dissipating area of worm and worm gear, m²

A = Projected area of worm (A_w) + Area of gear (A_g)

$$= l \cdot d + \frac{\pi}{4} D^2$$

$(t_2 - t_1)$ = Difference of gear temperature and room temperature, Δt

$\not>$ 27 to 38° C, t_1 is usually taken as 20°C.

The average value of k may be taken as 387 W/m² °C.

The heat dissipating capacity can be increased by:

1. Walls of the housing are made ribbed, so as to increase the cooling surface area.
2. Forced or Artificial cooling by a built in (mounted on worm shaft) or a separate fan.
3. By cooling the oil with water circulated through an immersed coil.
4. By using a circulatory lubrication system with an oil cooler.

The heat that must be dissipated = Power (kW) × 1000 × (1 – η) watts

The heat liberated in the gearing should be rejected to the ambient air through the walls of the reduction unit housing. Thus

$$t_{oil} = t_{air} + \Delta t \qquad [\text{here } t_{air} = t_1]$$

Now toil $\not>$ 60 to 70°C

22.17.7 Design of Worm Gearing

The quantities to be determined in the worm gear design are Centre distance, lead angle, the lead, and the number of threads on the worm.

To find a satisfactory combination of lead angle and centre distance, the following procedure may be adopted:

1. The centre distance between the worm and the gear is used as the starting point.

$$C = \frac{d + D}{2} = \frac{L}{2\pi}(\cot \lambda + V_R)$$

In terms of the normal lead L_n, the above equation can be written as:

$$\frac{C}{L_n} = \frac{1}{2\pi}\left(\frac{1}{\sin \lambda} + \frac{V_R}{\cos \lambda}\right); L_n = L \cos \lambda \quad ...(22.44)$$

Since velocity ratio is usually given, there are three unknown quantities, C, L_n, λ. If at a particular velocity ratio, values for $\frac{C}{L_n}$ are calculated for various values of lead angles, and a curve is plotted between $\frac{C}{L_n}$ and λ, it will be of the shape shown in Fig. 22.30. The lowest point on the curve gives the lead angle for minimum value of $\frac{C}{L_n}$. This value of $\frac{C}{L_n}$ represents the minimum centre distance that can be used with a given lead or inversely the maximum lead that can be used for a given centre distance. After this, with the help of Table 22.12 and standard modules, a suitable combination of lead angle, lead and centre distance for given design data, can be determined.

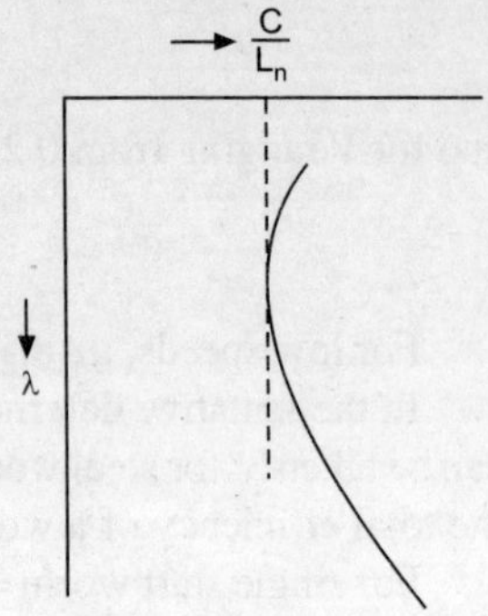

Fig. 22.30 *Worm Design Curve*

The lowest point on the curve can be found out mathematically also by differentiating equation (22.44) with respect to λ and equating it to zero. So,

$$\frac{(V_R)\sin^3 \lambda - \cos^3 \lambda}{\sin^2 \lambda \cos^2 \lambda} = 0$$

$$\therefore \quad (V_R) = \cot^3 \lambda \quad ...(22.45)$$

Again, $$C = \frac{d + D}{2} = \frac{m}{2}(q + z_g)$$

or $$m = \frac{2C}{(q + z_g)} \quad \begin{pmatrix} q = 9 \text{ to } 18 \\ z_g \not< 29 \end{pmatrix}$$

2. **Lewis equation for Strength:** The strength equation can be written as :

$$F_t = f_b f m Y C_v, \text{N} = \frac{f_b f p_c Y C_v}{\pi}, \text{N}$$

When the thread angle is 14½°, $Y = 0.314$
when thread angle is 20°, $Y = 0.393$, $Y = 0.470$ for $\phi = 25°$, $Y = 0.55$ for $\phi = 30°$.
or $Y = 0.314 + 0.015 (\phi - 14.5°)$

The velocity factor is given:

$$C_v = \frac{6}{6 + V_g}, \text{ For metal gears,}$$

and $$C_v = \frac{1 + 0.25 V_g}{1 + V_g}, \text{ For non-metallic gears,}$$

V_g = pitch line velocity of the worm gear, m/s

3. **Check for dynamic load:** The dynamic load for worm gear may be estimated by the relation :

$$F_d = F_t / C_v$$

Beam Strength : The beam strength or the endurance limit may be found as:

$$F_b = f_{ef}\, f.m.Y, \text{N}$$

as before, $F_b \geq F_d$

Since the material of the worm possesses higher mechanical properties than those of the worm wheel, the calculations (Steps 2 and 3) are done to determine the strength of the worm wheel teeth.

4. Check for Wear: The limiting load, F_w, for wear may be determined from the relation:

$$F_w = D \cdot f \cdot K_l, \text{N}$$

where D = pitch diameter of gear

K_l = load stress factor

The values of factor K_l are given in Table 22.14.

Table 22.14 Load stress factor K_l

Worm Material	Gear Material	K_l, kN/m^2
250–BHN Steel	Phosphor Bronze	415
Hardened Steel	Phosphor Bronze	550
Hardened Steel	Chilled Phosphor Bronze	830
Hardened Steel	Antimony Bronze	830
Hardened Steel	C.I.	345
Cast Iron	Phosphor Bronze	1035

5. At the end, the heat dissipating capacity of the drive housing must be checked.

The values given in Table 22.14 are suitable for lead angles upto 10°. For lead angles between 10° and 25°, increase K_l by 25%. For lead angles greater than 25°, increase K_l by 50%.

For estimating the approximate proper proportions of the gear unit, the following empirical relations recommended by AGMA may be used:

The pitch diameter of the worm should be selected so as to fall into the range:

$$\frac{C^{0.875}}{3.0} \leq d \leq \frac{C^{0.875}}{1.7} = 3p_c$$

Here, C and d are in cm.

Face width of gear, $f = 0.73 \times$ worm pitch diameter

Axial length of worm, $l = p_c\left(4.5 + \frac{z_g}{50}\right)$

The limiting power input, $P(kW) = \frac{3650 \times C^{1.7}}{V_R + 5}$; ($C$ in m)

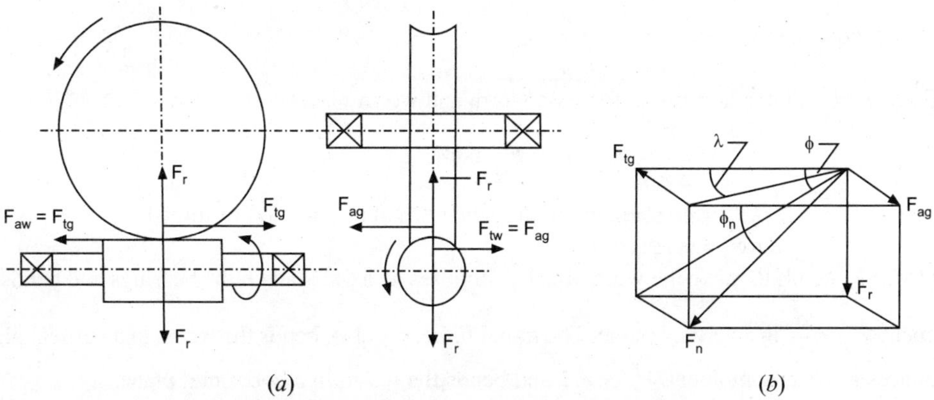

Fig. 22.31 *(a, b) Forces on Worm Teeth*

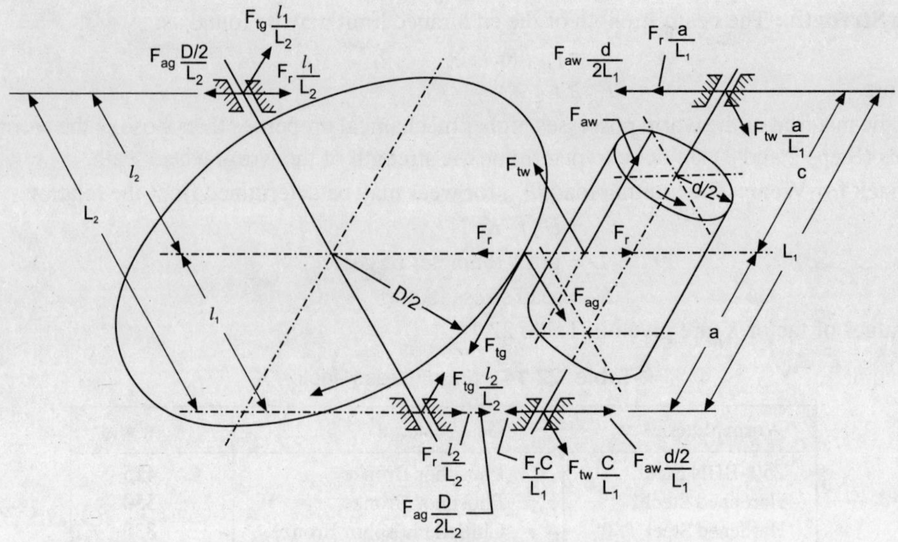

Fig. 22.31 *(c) Bearing loads*

22.17.8 Forces on Worm and Worm Wheel

The forces acting in the worm and worm wheel pair are shown in Fig. 22.31 *(a, b)*. The normal force, F_n can be resolved into three components:

1. Force F_{tg} = tangential force on wheel $= \dfrac{2 \times \text{Torsional moment on wheel}}{D} = \dfrac{2 \times M_{tg}}{D}$

$$F_{aw}, \text{ Axial force on worm} = F_{tg} \qquad \text{...(22.46)}$$

2. Force F_{ag} = Axial force on the worm wheel

$$= \text{Tangential force on the worm, } F_{tw} = F_{tg} \tan \lambda \qquad \text{...(22.47)}$$

In writing this relation, the friction forces between the worm threads and wheel teeth are neglected. We can also write,

$$F_{tw} = F_{ag} = \frac{2 \times \text{Torsional moment on worm}}{d} = \frac{2 \times M_{tw}}{d} \qquad \text{...(22.48)}$$

The torsional moments on the worm shaft (M_{tw}) and wheel shaft (M_{tg}) are related as:

$$M_{tg} = M_{tw} \times (V_R) \times \eta \qquad \text{...(22.49)}$$

Considering friction forces, we can write,

$$M_{tw} = F_{tg} \times \frac{d}{2} \tan(\lambda + \phi_1) \qquad \text{...(22.50)}$$

∴ From equations (48 and 50),

$$F_{ag} = F_{tg} \tan(\lambda + \phi_1) \qquad \text{...(22.51)}$$

3. Force F_r = radial or separating force on worm and worm wheel

$$= \frac{F_{tg}}{\cos \lambda} \cdot \tan \phi_n = F_{tg} \cdot \tan \phi,$$

where ϕ_n = pressure angle in plane normal to direction of thread.

$\tan \phi_n = \tan \phi . \cos \lambda$

The force F_{tg} tends to move the worm axially, imposes an axial load on the bearings and bends the worm with moment $F_{tg} \times \dfrac{d}{2}$ in a vertical plane. The radial force, F_r, also, bends the worm in a vertical plane. Force (F_{tw}) produces a twisting moment, $F_{tw} \times d/2$ and bends the worm in a horizontal plane.

Figure 22.31 (c) shows the bearing loads. With these, the bearings and the worm and worm wheel shafts can be designed.

In addition to the strength calculations, the worm must be checked for stiffness by limiting its deflection, that is, $y \leq 0.01\text{m}$, where

$$y = \frac{L_1^3}{48\, EI}\left(\sqrt{F_r^2 + F_{tw}^2}\right), \text{ here } I = \frac{\pi}{64} d_d^4 .$$

If the stiffness of the worm shaft is in sufficient, it is the usual practice to in crease worm thickness.

Example 22.16. Find the 20 degree involute worm and gear which will transmit 11.25 kW between shafts that are 0.25 m apart, if the speed reduction is to be 10.5 to 1, and the driving shaft is turning at 1200 rev/min.

Solution: The centre distance equation is:

$$\frac{C}{L_n} = \frac{1}{2\pi}\left(\frac{1}{\sin\lambda} + \frac{V_R}{\cos\lambda}\right)$$

As already indicated, the low point on the curve corresponds to

$$\cot^3\lambda = V_R = 10.5, \therefore \lambda = 24° - 32'$$

$$\therefore \quad \frac{0.25}{L_n} = \frac{1}{2\pi}\left(\frac{1}{0.4152} + \frac{10.5}{0.9097}\right), \text{ gives } L_n = 0.1122 \text{ m}$$

$$\therefore \text{ Axial lead } L = \frac{L_n}{\cos\lambda} = \frac{0.1122}{0.9097} = 0.1235 \text{ m}$$

From Table 22.12 let us choose, the number of worm thread = 4

$$\therefore \text{ Axial pitch of the threads, } p_x = \frac{0.1235}{4} = 30.875 \text{ mm}$$

Now $\quad p_x = \pi m$

$$\therefore \quad \text{Module } m = \frac{p_x}{\pi} = \frac{30.875}{\pi} = 9.828 \text{ mm}$$

The nearest standard value of module is 10 mm.

$$\therefore \quad L = Z_1 \times p_x = 4 \times \pi \times 10 = 125.6 \text{ mm}$$

$$\therefore \quad L_n = L\cos\lambda = 125.6 \times 0.9097 = 114.2 \text{ mm}$$

$$\therefore \quad C = \frac{5.27 \times 114.2}{2\pi \times 0.3777} = 254.5 \text{ mm}$$

Number of teeth in gear, $\quad z_g = 10.5 \times 4 = 42$

$$\therefore \quad \text{Pitch Diameter of worm, } \quad d = \frac{L}{\pi\tan\lambda} = \frac{125.6}{\pi \times 0.456} = 91.5 \text{ mm}$$

$\therefore$ Addendum circle diameter $d_a = d + 2\,m = 91.5 + 2 \times 10 = 111.5$ mm

Dedendum circle diameter $d_d = d - 2.4\,m = 91.5 - 2.4 \times 10 = 67.5$ mm

Length of thread portion, $l \geq (12.5 + 0.09\, z_g)\, m$, mm

$$\geq (12.5 + 0.09 \times 42)\, 10 \geq 162.8 \text{ mm.}$$

If the worm is to be ground, the length l should be increased by 25 to 30 mm to exclude the feed marks produced by the vibrating grinding wheel as it leaves the thread root.

$\therefore \quad$ Let us take, $l = 190$ mm

Proportions of Worm Wheel :

Number of teeth, $z_g = 42$

Pitch diameter of worm wheel $= z_g \cdot m = 42 \times 10 = 420$ mm

Addendum circle diameter, $D_a = D + 2m = 420 + 2 \times 10 = 440$ mm

Dedendum circle diameter, $D_d = D - 2.4\,m = 420 - 2.4 \times 10 = 396$ mm

Outside diameter, $D_o = D_a + 1.5\,m = 440 + 1.5 \times 10 = 455$ mm

Rim width, $B \le 0.67\,d_a \le 0.67 \times 111.5 \le 74.6$ mm

$\therefore$ Let us take, $B = 74$ mm

Let face angle, $2\gamma = 90°$

$$\therefore \quad f = (d_a - 2m) \times \frac{\gamma° \times 3.14}{180}$$

$$\therefore \quad f = (111.5 - 2 \times 10) \times \frac{45 \times 3.14}{180} = 71.8 \text{ mm}$$

By the relation, $f = 2.15\,p_c + 5.00$ mm (for $Z_1 = 3$ to 4) $= 2.15 \times 30.875 + 5.00 =$ **71.38 mm.**

Design tangential load: The pitch line velocity of gear $V_g = \dfrac{\pi D\, n_g}{60} = \dfrac{\pi \times 0.42 \times n_g}{60}$

Now $\quad n_w = 1200$ rev/min

$$\therefore \quad n_g = \frac{1200}{10.5} = \frac{800}{7} \text{ rev/min}$$

$$\therefore \quad V_g = \pi \times 0.42 \times \frac{800}{7 \times 60} = 2.51 \text{ m/s}$$

$\therefore$ Useful transmitted torsional moment load, $F_t = \dfrac{11.25 \times 1000}{2.51} = 4482$ N

Now $$C_v = \frac{6}{6 + V_g} = \frac{6}{6 + 2.51} = 0.709$$

Taking overload factor $C_o = 1.25$

Design Tangential Load, $F_t = C_o \times F_t = 1.25 \times 4482 = 5602.5$ N

Check for dynamic load: The dynamic load is, $F_d = F_t / C_v = 5602.5/0.709 = 7902$ N

Now beam strength is given as, $F_b = F_{ef} \cdot f \cdot m \cdot Y$

Let $\quad F_b = 1.5\,F_d = 11853$ N $= 11.853$ kN

$$\therefore \quad f_{ef} = \frac{11.853 \times 10^3}{0.0718 \times 0.01 \times 0.393} = 42 \text{ MPa}$$

Now if worm is of hardened steel of BHN = 250 and gear material Phosphor bronze, then,

$$f_{ef} = \frac{250}{1.75} = 143 \text{ MPa for worm material and}$$

$$f_{ef} = 168 \text{ MPa for gear material with BHN} = 100$$

$\therefore$ Both worm and gear are safe for dynamic loading.

Check for Wear. $\quad F_w = D f K_l$, N

Now from Table 22.14, $K_l = 830$ kN/m^2, if worm is to be of hardened steel.

$\therefore \quad F_w = 0.42 \times 0.0718 \times 830 = 25.03$ kN

which is quite large as compared to F_d, $\quad \therefore$ safe.

Heat Dissipation : The rubbing velocity is

$$V = \frac{\pi d\, n_w}{60 \cos \lambda} = \frac{\pi \times 0.0915 \times 1200}{60 \times 0.9097} = 6.32 \text{ m/s}$$

$\therefore \quad \mu_1 = 0.025 + \dfrac{6.32}{300} = 0.025 + 0.021 = 0.046 \quad \therefore \quad \phi_1 = 2° - 40$

$\therefore \quad \eta = (0.95 - 0.96) \times \dfrac{\tan\lambda}{\tan(\lambda + \phi_1)} = 0.955 \times \dfrac{\tan 24° - 32'}{\tan 27° - 12'} = 84.8\%$

∴ The heat that must be dissipated is

$$Q = \text{kW} \times 1000 \times (1 - \eta), \text{ watts}$$

Now $\quad \text{Power} = C_o \times 11.25 = 1.25 \times 11.25 = 14.0625 \text{ kW}$

$\therefore \quad Q = 14.0625 \times 1000 \times 0.152 = 2137.5 \text{ W}$

Now $\quad$ Area of gear, $A_g = \dfrac{\pi}{4} \times D^2 = \dfrac{\pi}{4} \times 0.42^2 = 0.1385 \text{ m}^2$

Projected area of worm, $\quad A_w = l \times d = 0.190 \times 0.0915 = 0.017385 \text{ m}^2$

∴ Total heat dissipating area $A = A_g + A_w = 0.1385 + 0.017385 = 0.155885 \text{ m}^2$

$\therefore \quad Q = A\,(t_2 - t_1) \times k$

$\therefore \quad t_2 - t_1 = \dfrac{2137.5}{0.155885 \times 378} = 36.275°\text{C}$

The permissible value for $(t_2 - t_1)$ is 38°C, ∴ safe

Worm Shaft : Now torsional moment acting on the gear shaft.

$$M_{tg} = \frac{9550 \times 14.0625}{800/7} = 1175.1 \text{ Nm}$$

∴ Moment on the worm shaft, $M_{tw} = \dfrac{M_{tg}}{(VR) \times \eta} = \dfrac{1175.1}{10.5 \times 0.848} = 132 \text{ Nm}$

∴ Turning force on the worm = Axial force on worm wheel

$\therefore \quad F_{tw} = \dfrac{2M_{tw}}{d} = \dfrac{2 \times 132 \times 1000}{91.5} = 2885 \text{ N}$

∴ Axial force on worm = torsional moment force on wheel or gear

$$F_{aw} = \frac{2 \times M_{tg}}{D} = \frac{2 \times 1175.1 \times 1000}{420} = 5600 \text{ N}$$

Radial force on worm = $F_{aw} \tan\phi = 5600 \times \tan 20°$

$\therefore \quad F_r = 5600 \times 0.354 = 2040 \text{ N}$

If the distance between the bearings of the worm shaft is taken as $\approx D$

∴ $L_1 = 0.42$ m, where L_1 is the distance between the bearings of worm shaft.

B.M. due to F_r in the vertical plane $= \dfrac{F_r \times L_1}{4} = \dfrac{2040 \times 0.42}{4} = 214 \text{ Nm}$

B.M. due to F_{aw} in the vertical plane $= F_{aw} \times \dfrac{d}{4} = 5600 \times \dfrac{0.0915}{4} = 128.2 \text{ Nm}$

∴ Total B.M. in the vertical plane = 214.0 + 128.2 = 342.2 Nm

B.M. due to F_{tw} in the horizontal plane $= F_{tw} \times \dfrac{L_1}{4} = 2885 \times \dfrac{0.42}{4} = 303 \text{ Nm}$

∴ Resultant B.M. on the worm shaft, $M_b = \sqrt{342.2^2 + 303.0^2} = 458.0 \text{ Nm}$

∴ Equivalent Twisting moment on the worm shaft

$$M_{te} = \sqrt{M_{tw}^2 + M_b^2} = \sqrt{132.0^2 + 458.0^2} = 476.0 \text{ Nm}$$

$\therefore \quad \frac{\pi}{16} dw^3 \times f_s = 476.0$

Taking $\quad f_s = 55$ MPa

$\therefore \quad dw = 0.0354$ m, say 36 mm

$\therefore$ Actual $\quad f_s = \frac{476.0 \times 16}{\pi \times 0.036^3} = 51.96$ MPa

Now direct compressive stress on the shaft, due to axial force

$$= \frac{F_{aw}}{\frac{\pi}{4} dw^2} = \frac{5600 \times 4}{\pi \times 0.036^2} = 5.50 \text{ MPa}$$

$\therefore \quad$ Resultant stress $= \frac{1}{2}\sqrt{5.5^2 + 4 \times 51.96^2} = \frac{1}{2} \times 104 = 52$ MPa < 55 MPa $\quad \therefore$ O.K.

If the worm is to be cut directly on the shaft, then the shaft diameter will be equal to dedendum circle diameter of the worm, i.e., equal to 67.5 mm. In that case the shaft will be very safe for the loads acting on it.

Gear Shaft : Axial force on worm wheel = 2885 N

Turning force on worm wheel = 5600 N

Radial force on worm wheel = 2040 N

Bending moment due to axial force will be in the vertical plane and

$$= \text{Axial force} \times \frac{D}{4} = 2885 \times \frac{0.42}{4} = 303 \text{ Nm}$$

B.M. due to Radial force will also be in the vertical plane and

$$= \text{Radial force} \times \frac{L_2}{4}$$

where L_2 = distance between the bearing centres of worm wheel shaft.

Let $\quad L_2 = 0.25$ m

$\therefore$ B.M. due to Radial force $= 2040 \times \frac{0.25}{4} = 127.5$ Nm

$\therefore$ Total B.M. in the vertical plane = 303 + 127.5 = 430.5 Nm

B.M. due to turning force will be in the horizontal plane and

$$= \text{Turning force} \times \frac{L_2}{4} = 5600 \times \frac{0.25}{4} = 350 \text{ Nm}$$

$\therefore$ Resultant B.M. on wheel shaft $= \sqrt{430.5^2 \times 350.0^2} = 555$ Nm

Now twisting moment on wheel shaft = 1175.1 Nm

$$M_{te} = \sqrt{1175.1^2 + 555.0^2} = 1300 \text{ Nm}$$

$\therefore \quad \frac{\pi}{16} d_g^3 \times f_s = 1300, d_g = 0.459$ m, say 50 mm.

$\therefore$ Actual $\quad f_s = 1300 \times \frac{16}{\pi} \times \frac{1}{0.05^3} = 52.97$ MPa

Direct compression stress $= \frac{\text{Axial force}}{\frac{\pi}{4} \times (0.05)^2} = \frac{2885 \times 4}{\pi \times 0.05^2} = 1.47$ MPa

$\therefore \quad$ Resultant stress $= \frac{1}{2}\sqrt{1.47^2 + 4 \times 52.97^2} = 52.975$ MPa < 55 MPa $\therefore$ safe.

Example 22.17. A speed reducer unit is to be designed for an input of 0.75 kW with a transmission ratio of 27. The speed of the hardened steel worm is 1750 rev/min. The worm wheel is to be made of phosphor bronze. The tooth form is to be 20° involute.

Solution: The centre distance will have to be assumed to start solving the problem. This is usually based on past experience and let us assume that for this size unit, $C = 100$ mm.

$$\text{Now } d \text{ lies between } \frac{C^{0.875}}{1.7} \text{ and } \frac{C^{0.875}}{3.0}$$

i.e., between 4.41 cm and 2.5 cm

Let us take $d = 40$ mm

$\therefore$ D = pitch diameter of worm gear = 200 – 40 = 160 mm

Now z_g must be some multiple of 27

We know that,

$$\pi D = z_g p_c$$

$$\therefore \quad \frac{\pi D}{p_c} = 27, 54, 81, \text{etc.}$$

Also $\quad d = 3p_c \quad \therefore p_c = \frac{d}{3} = 13.3$ mm

If $\quad z_g = 27, p_c = \frac{\pi \times 160}{27} = 18.62$ mm

If $\quad z_g = 54, p_c = 9.31$ mm

Let us use, $p_c = 9.3$ mm and $z_g = 54$

$$\text{Now (VR)} = \frac{\text{number of teeth in gear}}{\text{number of worm starts}}$$

$$\therefore \quad Z_1 = \text{Number of worm starts} = \frac{54}{27} = 2$$

now $\quad D = \frac{z_g \times p_c}{\pi} = \frac{54 \times 9.3}{\pi} = 160$ mm

$\therefore \quad d = 200 - 160 = 40$ mm

Now face width of gear, $f = 0.73 \times 40 = 29.2$ mm, say 29 mm.

Checks: Now load to be transmitted, $F_t = \frac{0.75 \times 1000}{0.509} = 1473.48$ N

Dynamic load: $\quad F_d = \frac{F_t}{C_v}$, C_o being not given

$$C_v = \frac{6}{6 + V_g}$$

$$V_g = \text{pitch line velocity of gear} = \frac{\pi D n_g}{1000 \times 60} = \frac{\pi \times 160 \times n_g}{1000 \times 60}$$

now $\quad n_g = \frac{1750}{27} = 64.8$ rev/min

$$\therefore \quad V_g = \frac{\pi \times 160 \times 64.8}{1000 \times 60} = 0.509 \text{ m/s}$$

$$\therefore \quad C_v = \frac{6}{6 + 0.509} = 0.922$$

$\therefore \quad F_d = \dfrac{1473.48}{0.922} = 1598.13$ N

Beam Strength = The load which the design can transmit is

$$F_b = \frac{f_{ef} \cdot f \cdot p_c Y}{\pi}, \text{N}$$

$$Y = 0.393$$

Let $\quad F_b = 1.5\, F_d$ (for extreme conditions) = 2397.2 N

$$\therefore \quad f_{ef} = \frac{2397.2 \times \pi}{0.029 \times 0.0093 \times 0.393} = 28 \text{ MPa}$$

Now f_{ef} for hardened steel is 143 MPa and for Phosphor bronze it is 168 MPa.
Therefore, the design is safe for dynamic load.

(ii) Wear: $\quad F_w = DfK_l$

From Table 22.14, $K_l = 550 \times 10^3$

$\therefore \quad F_w = 0.15 \times 0.0365 \times 550 \times 10^3 = 3011.25 \text{ N} > F_d$. $\quad \therefore$ safe.

Example 22.18. Design a worm gear set to transmit 11.25 kW from a shaft rotating at 1500 rev/min to another at 75 rev/min.

Solution: Assume normal pressure angle, $\phi = 20°$

Therefore, as given earlier, lead angle, $\lambda = 25°$

Now lead angle is usually 6° per worm thread. Therefore, the worm should be quadruple. Let us try a triple threaded worm.

Now, velocity ratio, $V_R = \dfrac{n_w}{n_g} = \dfrac{1500}{75} = 20$

V_R is also given as $= \dfrac{n_g}{Z_1}$

where $\quad Z_1$ = number of worm starts = 3

$\therefore$ Number of gear teeth, $z_g = 20 \times 3 = 60$

Since, most of the design data, for example, lead, worm diameter, centre distance are not given, we will have to make suitable assumptions and then check the validity of assumptions. It is more convenient to assume centre distance which is based upon space limitations and judgement of the designer. Now, let us assume, $C = 0.25$ m = 25 cm

Now worm diameter is given as,

$$\frac{C^{0.875}}{3.0} \le d \le \frac{C^{0.875}}{1.7}$$

i.e., $\quad 5.573 \le d \le 9.834$

Let us try $\quad d = 9.6$ cm

Now, $\quad d = 3p_c$

$\therefore \quad$ Circular pitch of gear, $p_c = \dfrac{1}{3} d$, Try $p_c = 3$ cm.

$$\therefore \quad \frac{\pi D}{z_g} = 3$$

$\therefore \quad D$ = diameter of gear $= \dfrac{3 \times 60}{\pi} = 57.3$ cm

$\therefore$ Actual $\quad C = \dfrac{d + D}{2} = 33.45$ cm

Check *d*: $$d = \frac{C^{0.875}}{2.35} = \text{(average)} = 9.81 \text{ cm.}$$

$\therefore$ $d = 9.6$ cm is satisfactory

Now lead, $L = Z_1\, p_x = 3 \times 3 = 9 \text{ cm } (p_x = p_c)$

$$\therefore \quad \tan\lambda = \frac{L}{\pi d} = \frac{9}{\pi \times 9.6} = 0.2984$$

$$\therefore \quad \lambda = 16.616 \text{ deg}$$

$$V_g = \frac{\pi D n_g}{1000 \times 60} = \frac{\pi \times 57.3 \times 75}{1000 \times 60} = 2.25 \text{ m/s}$$

$$\therefore \quad F_t = \frac{11.25 \times 1000}{2.25} = 5000 \text{ N}$$

Now Dynamic load or Design load,

$$F_d = \frac{F_t}{C_v} = \left(\frac{6 + V_g}{6}\right) \cdot F_t = \frac{6 + 2.25}{6} \times 5000 = 6843.3 \text{ N}$$

Now in a worm gear set, the teeth on the worm are always stronger than worm gear teeth. Now, beam strength of worm gear is given as,

$$F_b = \frac{f_{ef} \cdot f \cdot p_{cn} \cdot Y}{\pi}$$

For $\phi = 20°, Y = 0.393$

$$p_{cn} = p_c \cos\lambda = 3 \times 0.958 = 2.875 \text{ cm}$$

Now $f = 0.73 \times d = 0.73 \times 9.6 = 7.008$ cm

Face width of gear is also given empirically as,

$$f = 2.15\, p_c + 0.50 \text{ cm} = 2.15 \times 3 + 0.50 = 6.95 \text{ cm}$$

Let us take $f = 7$ cm

Now f_{ef} for gear material is 168 MPa

$$\therefore \quad F_b = \frac{168 \times 10^6 \times 0.07 \times 0.02875 \times 0.393}{\pi} = 42.3 \text{ kN}$$

Since $F_b \geq 1.5\, F_d$, $\therefore$ Design is safe for F_d.

Or the design can be iterated by altering the assumptions, so that the design is not too safe as above.

Check for Wear: Let the material for worm be hardened steel.

$$F_w = D \cdot f \cdot K_l$$

Now since $\lambda = 16.616$ deg.

$$K_l = 550 \times 1.25 = 687.5 \text{ kN/m}^2 \text{ (Table 22.14)}$$

$$F_w = 0.573 \times 0.07 \times 687.5 \times 10^3 = 27.57 \text{ kN}$$

Since $F_w > F_d$, therefore, the gear set is satisfactory for wear.

Axial length of worm, $l \geq (12.5 + 0.09\, z_g)\, m$, mm

now $$m = \frac{p_x}{\pi} = \frac{30}{\pi}$$

$$\therefore \quad l \geq (12.5 + 0.09 \times 60) \times \frac{30}{\pi} \geq 170.93 \text{ mm}$$

According to AGMA, $l = p_c\left(4.5 + \frac{z_g}{50}\right) = 30 \times 5.7 = 171$ mm. Take $l = 171$ mm

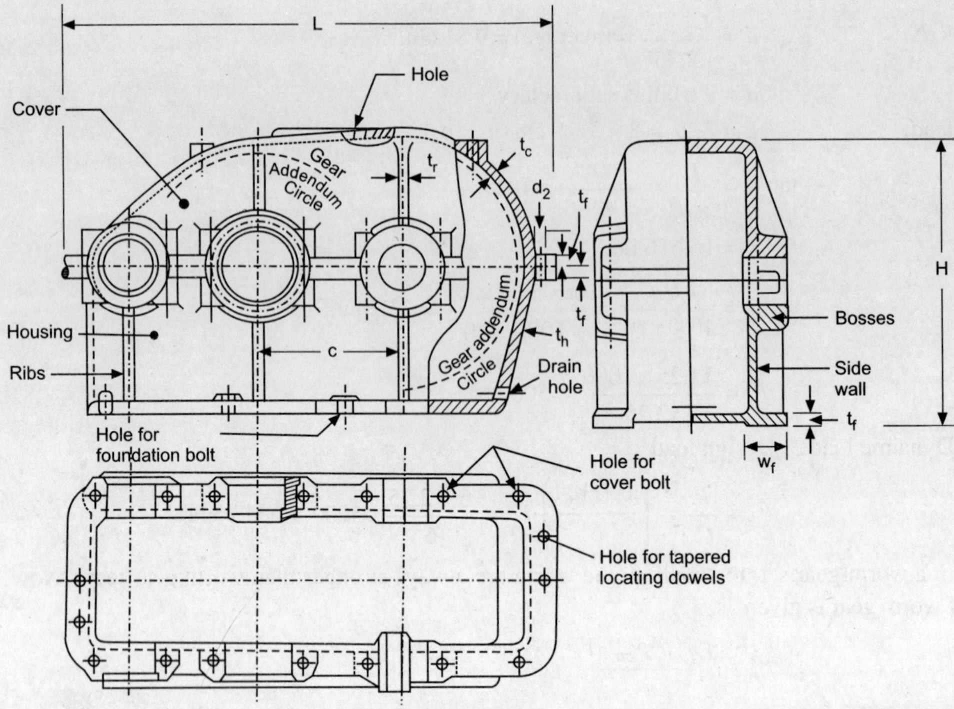

Fig. 22.32 *A gear housing*

Check for Heat Dissipation: Now sliding velocity V is

$$\therefore \quad V = \frac{\pi\, d\, n_w}{60 \cos \lambda}$$

$$V = \frac{\pi \times 9.6 \times 1500}{60 \times 100 \times 0.9582} = 8 \text{ m/s}$$

$\therefore$ From table 22.13, ϕ_1 is about 1 deg,

$$\therefore \quad \text{Efficiency of drive, } \eta = 0.955 \times \frac{\tan 16.616}{\tan 17.616} = 89.75\%$$

$$\therefore \quad \text{Heat to be dissipated} = 11.25 \times 1000 \times 0.1025 = 1153.125 \text{ watts}$$

$$\text{Total heat dissipating area, } A = A_g + A_w = \frac{\pi}{4} D^2 + l \times d = \frac{\pi}{4}(0.573)^2 + 0.171 \times 0.096 = 0.2741542 \text{ m}^2$$

Now
$$Q = A\,(t_2 - t_1) \times k$$

$$\therefore \quad t_2 - t_1 = \frac{1153.125}{378 \times 0.2741542} = 11.13^\circ \text{C} \;(< 27^\circ \text{C})$$

$\therefore$ Design is safe for heat dissipation also.

22.17.9 Advantages and Disadvantages of Worm Gear Drives

Advantages

1. Quiet and noiseless operation.
2. High speed ratios with comparatively small overall dimensions of the drive.
3. Small size.
4. Irreversibility or the self locking ability.

Disadvantages

1. Comparatively low efficiency.
2. Intensive heating of the drive due to conversion of friction losses into thermal energy.
3. Low power capacity, usually not over 75 kW.

22.18 GEAR HOUSINGS

When toothed gears are used as independent units to reduce the velocity of driven shafts, the combined assembly of gears, shafts and bearings is enclosed in separate oil-tight and dust proof rigid housings. The housings serve as a base for mounting the gears and ensure proper relative alignment of their axes, support for the bearings and bearing seals and as an oil bath for the gears. These gears are called 'reduction gears'. A set of gears used to increase the velocity of driven shafts is called 'multiplying gears'.

A gear housing results in the following advantages as compared to open drives:

1. The location of the bearings of the reduction unit shafts in a common rigid housing guarantees a constant relative position of the shaft axes, and ensures correct alignment of shafts.
2. The above factor favours the use of wide gears with a small module. The use of small modules results in the following benefits:

   (*i*) increased accuracy (*ii*) reduced operating noise of a drive (*iii*) reducted cost of manufacture.
3. Abundant lubrication minimizes wear and contributes to high efficiency of the reduction unit.
4. The housing promotes operations safety.

The gear housing parts should be as simple as possible in form with a minimum number of projections, ribs, beads, etc., to ensure stiffness. The housings are usually cast from grey C.I. Sometimes, they may be steel castings. If individually manufactured, the housing can be made welded of steel plates.

Figure 22.32. illustrates the housing and cover of a three shaft two- stage reduction unit. External ribs are provided to promote the stiffness of the unit. The side walls of the housing have cylindrical bosses whose sockets receive the bearing units with preassembled shafts. These sockets are bored with a high degree of accuracy. Both halves of the housing are bored together; they are preassembled, bolted and accurately matched by means of tapered pins. To avoid misalignment of the shafts, strict co-axiality of the opposite sockets is necessary. The outer diameter of the bearings is determined by the diameters of these covers. Bolts securing the cover to the body are preferably arranged as close to the bearings as possible, the nuts being provided with bosses.

Accurate positioning of the cover on the housing is ensured by two locating taper pins installed diagonally and passing through the cover flange and the upper flange of the housing. The top of the cover has a special hole covered with a cap or a threaded plug. This hole is provided for lubrication and inspection of the reduction unit. Another hole with a plug is there at the bottom of the housing. Through this, the lubricant is drained and a device for checking the lubricant level is installed. Eye bolts screwed into the cover are provided for lifting the cover.

Gears are mounted and secured on the input and output shafts by means of key or splined joints. Axial play of the gears on the shafts is prevented by providing steps or shoulders (collars) on the shafts.

Most reduction units are built with rolling contact bearings. The mounting of bearings has already been discussed in chapter 17.

The housing dimensions can be determined from the following empirical relations:

Thickness of housing wall, $t_h = 0.025\ C + 5$ mm

Thickness of housing cover wall, $t_c = 0.02\ C + 5$ mm

Thickness of housing upper flange and cover flange $t_f = 1.5\ t_h$

Thickness of housing lower flange, $t_{fl} = (2 \text{ to } 2.4)\ t_h$

Thickness of ribs, $t_r = 0.85\ t_h$

Diameter of foundation bolts, $d_f = 0.036\ C + 12$ mm

Diameter of bolts near bearings, $d_1 = 0.75\ d_f$

Diameter of bolts connecting the housing base and cover $d_2 = (0.5 \text{ to } 0.6)\ d_f$

Diameter of bearing cover bolts, $d_3 = (0.4 \text{ to } 0.5)\ d_f$

Width of lower flange, $w_f = 2.2\ d_f + (7 \text{ to } 10)$ mm

Width of upper flange = $2.2\ d_2 + (7 \text{ to } 10)$ mm

Some typical overall dimensions of a gear housing are given below, for a two shaft reduction unit:

C, mm	*L*, mm	*H*, mm	Width, mm
133.5	343.0	292.00	216.00
101.5	317.5	235.00	197.00

EXERCISES

1. Select a pitch of a 14½° involute tooth spur gear to transmit 11.25 kW at 320 rev/min. The load may be considered as steady. The gear is to be cast iron with cut teeth having a face width 3 and 3½ times the circular pitch.
2. In an automotive type of gear box, the second speed gear shaft is to be driven from the main shaft with a velocity ratio of 1.5 to 1. The main shaft transmits 45 kW at 3000 rev/min. The shaft centre must not exceed 90 mm. Determine a satisfactory module and the number of teeth for 20° stub-tooth gears of heat treated steel.
3. Design a pair of 20 deg. gears with a velocity ratio of 5 to 1 for the second reduction in a crane which has a capacity of a 50 kW rope pull on the drum. The first gear reduction at the drum is 4 to 1; the drum diameter is 0.66m and has a circumferential speed of 0.6 m/s. To conserve space, the shaft centre distance should be kept as small as possible.
4. A 20 deg. full depth pinion drives a gear and transmits 12.25 kW at 1200 rev/min. The observed data are : $m - 5$ mm; $z_p = 24$; $z_x = 54$; gear faces = 50 mm. The gears are made of the same steel with B.H.N. of 300. Check the gears for strength and wear.
5. A 64 tooth full depth 20 deg. involute gear rotates at 100 rev/min and drives a second gear.
   (a) What is the highest speed at which gear can run?
   (b) How many teeth are on the second gear?
6. A pair of carefully cut gears with 20 deg. stub teeth is to transmit 22.5 kW at 300 rev/min of the gear with a speed reduction of 5 : 1. The 76 mm pinion made of forged C-30 steel with a hardness of 250 BHN and drives a cast iron gear. Determine the module, the face width required if the load is steady.
7. A pair of 20 deg. full depth spur gears must transmit 18.75 kW at a speed of 5000 rev/min of the pinion. The velocity ratio is 10 and the diameter of the pinion is 76 mm. There should be between 15 and 24 teeth in the pinion. The pinion is made of C-30 forged steel with a B.H.N. of 250 and the gear is made of ordinary cast iron. Determine the module, the number of teeth, and the face width of these gears.
8. Determine the module, face, and number of teeth for a pair of gears if they are to transmit 3.75 kW at 1800 rev/min. of the pinion, for intermittent service. The material is to be good grade cast iron and the pinion is to have 18 teeth 20° full depth, $V_R = 2.5$.
9. Design the teeth for a pair of cut-teeth spur gears with 20° full depth teeth to be made of nickeled chromium steel. The gears are to transmit 45 kW at 3000 rev/min. of the pinion. The pinion has 21 teeth and the gear 25 teeth. Find the module, face, and diameters for continuous service. Will it be necessary to surface harden the teeth?
10. Design the 20° full depth teeth of a pair of gears to transmit a smooth continuous load of 30 kW at 1200 rev/min of the 20-tooth pinion. The velocity ratio is 3 and the material is to be steel. The surface of the teeth may be heat treated as needed for wear. Determine the module and face, and specify the grade of steel and its heat treatment.
11. What should be the module, face and number of teeth on each gear of a pair of spur gears, a texolite pinion driving a cast iron gear, if they are to transmit 18.75 kW at 700 rev/min of the 0.18 m pinion in continuous service? The velocity ratio is 3, the teeth are 20° full depth and the load is smooth. Determine the outside and root diameters.
12. A bakelite pinion is used to transmit power at a speed of 400 rev/min. The module is 10 mm the pitch diameter is 0.25 m and the face is 0.127 m. The teeth are standard 20° involute. (a) Determine the number of teeth, circular pitch, and outside diameter of pinion. (b) Determine the power the pinion should transmit for smooth intermittent service. (c) Determine the power for continuous service.

13. A pair of cast steel annealed spur gears with cast teeth transmits 18.75 kW and 60 rev/min of the pinion. The pitch diameter of the pinion is approximately 0.23 m and the velocity ratio is 2.5. Determine the module and face of the gears; also the correct pitch diameters and tooth numbers.
14. Design the teeth for a pair of cast iron spur gears with cast teeth to transmit 26 kW. The pinion runs at 50 rev/min and the velocity ratio is to be 2.5. Decide upon a suitable grade of cast iron and find the module, face, and diameters, and centre distance for the gears.
15. It is required to determine the proportions of the spur gear drive to transmit 7.5 kW from a shaft rotating at 1170 rev/min to a low speed shaft with a speed reduction of 3 to 1. Assume that the teeth are 20-deg. full depth system with 24 teeth on the pinion. The pinion is to be C-40 forged steel and the gear of cast steel. Assume that the torsional moment at starting is 150 per cent torsional moment at rating. Determine the B.H.N. for the pinion and gear to check for wear.
16. A 0.61 m cast iron gear transmits 22.5 kW at 240 rev/min. The pitch of the 20° full depth teeth is 0.10 m and the face is 63.5 mm long. The shaft on which the gear is mounted is 57 mm in diameter : (a) Determine the dimensions for hub diameter, rim thickness, and rib thickness. (b) Assuming ellipitcal shaped arms, determine the dimensions of the arm at the hub and the pitch circle. (c) Determine the dimensions of a cross-shaped arm. (d) Determine the dimensions of an I-shaped arm.
17. A pair of helical gears is to transmit 37.5 kW at 1750 rev/min of the pinion. The velocity ratio is to be 4.25 and the helix angle is to be 15°. The gears are subjected to a heavy shock load 24 hr per day. The minimum pitch diameter of the pinion is 0.116 m. Determine the module, face, material heat treatment, if the teeth are 20° full depth in the normal plane.
18. A steel pinion having diameter of about 0.20 m meshes with a steel gear which turns at one third the angular speed of the pinion. The pinion rotates at 1150 rev/min and carries a torsional moment of 2060 Nm. The angie of cut of the teeth is 20°. Determine the normal module of the teeth, the number of teeth required, and the hardness of the gear and pinion.
19. Design the teeth for two herringbone gears for a single reduction speed reducer whose velocity ratio is 3.80. The speed reducer is to transmit 27 kW when the pinion turns 3000 rev/min. The helix angle should be 30° and the teeth are to be 20° stub in the plane of rotation. The length of the face of the pinion should not exceed twice the pitch diameter. The material of the gears is a high carbon steel, heat treated to have a yield point of approximately 420 MPa and a B.H.N. of 450.
20. A 20 tooth helical gear has a pitch diameter of 0.25 m. The helix angle is 23 deg and the pressure angle measured in a plane perpendicular to the axis of rotation is 20°. (a) Determine the module. (b) Find the circular pitch in a plane normal to the teeth. (c) Find the pressure angle in a plane normal to the teeth.
21. A 56.25 kW motor, running at 450 rev/min is geared to a pump by a double helical gearing. The C-30 forged steel pinion on the motor shaft is 0.20 m in diameter and drives the good grade cast iron gear on the pump shaft at 120 rev/min. Determine the module and the face width.
22. Design a herring bone drive from a 2.25 kW steam turbine, running at 30000 rev/min to a speed reducer that should run at 2500 rev/min.
23. A steel helical pinion having a B.H.N. of 250 has 60, 20 deg. stub teeth, and rotates at 1000 rev/min. The normal module is 2.5 mm and the width of the face is five times the normal pitch. The angle of cut is 30°. The mating gear has 200 teeth, and a hardness of 250 Brinell. What power can the pair transmit quietly.
24. A continuous tooth herring bone cast steel gear with 30 deg. helix angle is to transmit 27 kW at 1200 rev/min. If the gear has 24 teeth, determine the necessary module, pitch diameter, and width of face for 20 deg. full depth teeth.
25. A pair of helical gears connecting parallel shafts is to transmit 415 Nm torsional moment at 2800 rev/min of the pinion. The teeth are to be 20° stub of heat treated alloy steel. The width of the face is limited to 38 mm. The driven gear rotates at 1800 rev/min. For a heiix angle of 30 deg. select the necessary module and check for wear. What is the axial thrust developed.
26. A helical gear 0.40 m in diameter is carried on a shaft between two bearings, 0.25 m from one end and 0.38 m from the other. The driving and driven shafts are in the same horizontal plane and transmit 460 Nm torsional moment. The helix angle is 45 deg. and in such a direction as to put the thrust on the inside of the far bearing. The pressure angle is 20 deg. in the plane of rotation. (a) Determine the end thrust and reaction on the bearings. (b) Determine the necessary diameter of commercial steel shafting for steady load.
27. Design the full depth 20 deg. teeth of a pair of cast iron straight bevel gears to transmit intermittently 7.5 kW at 625 rev/min of the pinion. The velocity ratio should be about 3.5. The pinion is to have not less than 22 teeth. Find the module, face width and diameters of the gears, strength alone considered.

28. Design the teeth of a pair of bevel gears to transmit 18.75 kW at 600 rev/min of the pinion. The velocity ratio should be about 3 and the pinion should have about 20 teeth which are full depth 20° involute. Specify the material and heat treatment, which seems best to you; also the module, face width and diameters of the gears.
29. Two cast iron bevel gears with interchangeable type teeth transmit 18 kW at 500 rev/min of the pinion. The velocity ratio is 4, module is 8 mm and the face width is 50 mm. The pinion has 21 full depth, 20° involute teeth. (a) Compute the stress in the teeth of each gear. (b) What grade of cast iron should be used for each gear?
30. Decide upon the module, face and number of teeth of a pair of straight bevel gears to transmit continuously 3.75 kW at 1150 rev/min of the pinion. Driving and driven loads are uniform. The velocity ratio is to be about 1¾. Both the gears are to be made of cast iron and the pinion overhangs its bearings. Check the strength.
31. A 30 tooth pinion drives a 90 tooth gear. The module is 32 mm and face width 63.5 mm. (a) Determine the diameter of gear shaft. (b) Determine the hub diameter and length and the thickness of rim for the gear (c) Determine the dimensions of the T-shaped arm.
32. 18.75 kW 900 rev/min electric motor is required to drive a centrifugal pump 8 hr./day at 300 rev/min by means of a straight bevel gears connected at 90° shaft angle. Starting torsional moment is 230 Nm. Design the gears : (a) Gear and pinion made from cast iron. (b) Gear and pinion made from bronze. (c) Gear and pinion made from case hardened steel.
33. A pair of straight bevel gears must transmit 15 kW at 1250 rev/min of the 18 tooth pinion. Speed reduction ratio is 3.5 : 1. Use 14½ full depth teeth. Select the materials to obtain a compact design. Determine the module, the gear face, the pitch diameters and pitch cone angle for both gears.
34. A pair of straight bevel gears consists of a pinion with 16 teeth and gear with 48 teeth. The module is 10 mm, the face is 63.5 mm, the material is nickel cast iron, and the service is continuous with light shocks. Find what power can be transmitted if the pinion speed is 1200 rev/min.
35. A pair of high grade cast iron bevel gears having shafts at right angles are to have an angular velocity ratio of driver to driven of 2 to 3. The driver is to rotate at 180 rev/min. and is to transmit 10 kW. It is 0.30 m in pitch diameter. Take the width of face as about one third the length of the pitch element and determine the module of gears.
36. A vertical drill spindle of a drilling machine is to be driven by means of a pair of straight bevel gears with the pinion mounted on a horizontal drive shaft. The approximate speed ratio is 2.25 to 1 and drill must have a torsional moment of 400 Nm at 414 rev/min. Determine the proportions of gears.
37. Determine the pitch diameters, number of threads and axial pitch for a worm drive having a velocity ratio of 14 : 1 and centre distance of approximately 0.15 m.
38. Design a high efficiency worm gear speed reducer to transmit continuously the rated output of a 15 kW motor running at 1750 rev/min. The steel worm with a BHN of 250 is to be integral with the motor shaft. $VR = 11$. The phosphor bronze gear should not have less than 40 teeth. (a) Determine the pressure angle, pitch, face and diameter of the gears.
39. Design a worm and worm gear for a speed reducer to deliver 32 kW at a speed of 580 rev/min of the worm. The velocity ratio is to be 26. The material of the gear is phosphor bronze and that of the worm is hardened steel.
40. The input to 0.10 m double threaded worm is 24 kW at 1750 rev/min of the worm. The efficiency is 85%. The 0.50 m gear has 52 teeth. What is the driving force on the gear? What is the force opposing rotation of the worm?
41. Design a worm drive for a speed reducer to transmit 30 kW at a worm speed of 600 rev/min. The desired velocity ratio is 25 : 1, and an efficiency of at least 87 per cent is desired. Assume that the worm is made of hardened steel, and select the material of the gear.
42. A three thread worm rotating at 100 rev/min drives a 31-tooth worm gear and transmits 11.25 kW. The worm has 20-deg. teeth with 6 mm. module, 50 mm pitch diameter and an included face angle of 60°. The coefficient of friction is 0.05. (a) Determine the helix angle of the worm. (b) Determine the speed ratio. (c) Determine the centre distance. (d) Find the efficiency of the worm gear set.
43. A triple thread cast iron worm running at 22.5 rev/min receives 5.6 kW through its shafts. The total efficiency of the worm and its bearings is 92 per cent. The velocity ratio is to be 10 : 1 and the distance between the shafts is to be 0.20 m. The worm is to have 14½° pressure angle and a lead angle greater than 15°. Supporting bearings of both worm and worm gear placed on 0.15 m centres. (a) Determine the worm shaft diameter with a permissible stress of 45.5 MPa in torsion. (b) using a standard module determine the number of teeth and pitch diameters of the worm and worm gear. (c) Determine the heat that must be dissipated per hour.

44. Calculate an enclosed irreversible worm gear drive, Power at gear shaft = 4.5 kW; $V_R = 35$; $\omega_w = 100$ rad/s

Solution: 1. From Table 22.12, for $VR = 35$, let us assume number of thread starts of worm, $Z_1 = 1$. This will result in a drive which is not too bulky. The worm is made of steel, case hardened to $R_C > 45$, and ground. The worm wheel (gear) rim material is bronze.

2. The centre distance, C, amy be determined by the relation.

$$C = 30{,}500\left(\frac{z_g}{q}+1\right)\sqrt[3]{\frac{P_{2d}}{\left(\frac{z_g}{q}\cdot f_c\right)^2\cdot\omega_g}},\ m \qquad \text{...(1)}$$

Now, $z_g = Z_1 \times V_R = 1 \times 35 = 35$

q = diametral quotient (8 to 13), let it be = 10.

P_{2d} = Design power at the drive shaft. Let service factor be = 1.3

$\therefore$ $P_{2d} = 4.5 \times 1.3 = 5.85$ kW.

$$\omega_g = \text{angular speed of gear} = \frac{\omega_w}{V_R} = \frac{100}{35} = 2.85 \text{ rad./s}$$

f_c = allowable normal contact stress for wheel material, N/m^2.

Its value can be taken as below:

(*i*) In case of rim made of C.I.,

$f_c = (188 - 115\,V)$, MPa, with $V \le 0.9$ m/s

$= (188 - 58\,V)$, MPa, with $V \le 2$ m/s

C.I. wheel rims should not be used, with $V > 2$ m/s.

(ii) For tin less bronze rim material, fc may be taken as:

V, m/s	0.5	1.0	2.0	3.0	5.0
f_c, MPa	67	65	63	61	57

(*iii*) For tin-bronze rim material:

Material	Casting Method of Material	f_c, MPa	
		$R_C < 45$	$R_C \ge 45$
Tin Bronze	Loam casting Chill casting Centrifugal casting	145 200 230	145 243 278
Tin Bronze	Loam Chill Centrifugal	127 145 180	153 175 210

From this table, let $f_c = 243$ MPa for Tin bronze, chill casting, with $R_c \ge 45$.

Solving the equation (1), $C = 0.196\ m = 196$ mm

3. Determine the module,

$$C = \frac{d+D}{2} = \frac{m}{2}(q + z_g)$$

$$\therefore \quad m = \frac{2 \times 196}{10 + 35} = 8.72 \text{ mm, standardise it and take } m = 10 \text{ mm}$$

The remaining steps are similar, as detailed in Art. 22.17.

❑❑❑

CHAPTER 23

Power Screws

23.1 FUNCTION

A power screw is employed to convert rotary motion into rectilinear motion. Power screws should produce uniform, slow and powerful motion. Depending on the application of the drive, the screw of the screw drives are classified into:

1. **Translating screws:** Such drives are used as load drives to transmit large forces at comparatively low speed, e.g., jack screws, screw presses, valves and pressing screw of rolling mills.
2. **Lead screws:** Serving to accurately transmit motion in lathes, machine tools, measuring instruments and machines such as screw operated tensile testing machine, etc.

The efficiency of the power screws should be highest possible so that the power lost in friction is as small as possible. A power screw should possess high resistance to wear and retain its precision for a long time.

Advantages:

1. Simplicity of design.
2. Simplicity of obtaining slow translation motion and a great gain in power.
3. Smoothness, noiselessness.
4. Ability to carry heavy loads.
5. High velocity ratio.
6. Self locking.
7. High accuracy of motions.

Disadvantages:

1. High friction losses resulting in rapid wear.
2. Low efficiency.

23.2 TYPES OF POWER SCREWS

The following thread forms are used for power screws:

(*i*) **Square threads:** The form and the proportions of the square thread are shown in Fig. 23.1(a).

Such threads have the highest efficiency but their cutting is difficult. The various proportions of standard square threads are given in Table 23.1 (IS 4694-1968). The various proportions of the thread are given below:

$$h_1 = 0.5\,p\,;\, e = 0.5\,p$$
$$H = 0.5\,p + a\,;\, h_2 = 0.5\,p - b.$$

From the Table 23.1, the various diameters for the nut and the bolt can be found out in the following manner:

Major diameter of bolt = nominal diameter

Major diameter of nut = nominal diameter + $2a$

Core diameter of bolt and nut = major diameter of bolt – p.

The values of 'a' and 'b' are given below:

Fine Series

a = 0.25 mm for nominal diameter upto 400 mm
= 0.50 mm for nominal diameter from 420 mm to 640 mm

b = 0.25 mm for nominal diameter upto 110 mm
= 0.50 mm for nominal diameter from 115 to 400 mm
= 1.00 mm for nominal diameter from 420 to 640 mm

Normal Series :

a = 0.25 mm for nominal diameter upto 110 mm
= 0.50 mm for nominal diameter from 115 to 300 mm

b = 0.50 mm for nominal diameter upto 110 mm
= 1.00 mm for nominal diameter from 115 to 300 mm

Table 23.1 Proportion of square threads

FINE SERIES			NORMAL SERIES			COARSE SERIES		
Nominal Diameter, mm	Common Difference mm	Pitch mm	Nominal Diameter mm	Common Difference mm	Pitch mm	Nominal Diameter mm	Common Difference mm	Pitch mm
10–22	2	2	22–28	2	5	22–28	2	8
22–62	2	3	30–36	2	6	30–38	2	10
65–110	5	4	38–44	2	7	40–52	2	12
115–175	5	6	46–52	2	8	55–58	3	14
180–200	5	8	55–58	3	9	58–62	2	14
200–240	10	8	58–62	2	9	65–80	5	16
250–300	10	12	65–80	5	10	80–82	2	16
300–400	20	12	80–82	2	10	85–95	5	18
420–500	20	18	85–110	5	12	95–98	3	18
520–640	20	24	115–145	5	14	100–110	5	20
			150–175	5	16	115–130	5	22
			180–200	5	18	135–155	5	24
			210–230	10	20	160–180	5	28
			240–260	10	22	185–200	5	32
			270–290	10	24	210–240	10	36
			300	—	26	250–280	10	40
						290–300	10	44
						300–340	20	44
						360–400	20	48

Coarse Series:

a = 0.25 mm for nominal diameter upto 52 mm
= 0.50 mm for nominal diameter from 55 to 400 mm
b = 0.50 mm for nominal diameter upto 52 mm
= 1.00 mm for nominal diameter from 55 to 400 mm

Designation: A square thread shall be designated by the letter SQ followed by nominal diameter and pitch, the two being separated by the sign (×) and the number of the standard. For example, a square thread of nominal diameter 30 mm and pitch 6 mm shall be designated as:

SQ 30 × 6 — IS 4694 — 1968.

(*ii*) **Acme or Trapezoidal thread:** Such threads are employed when loss of motion is objectionable such as in lead screws of lathes and similar uses. This thread is shown in Fig. 23.1(b). For Acme threads, the included angle $2\beta = 29°$, and for Trapezoidal threads, $2\beta = 30°$.

Table 23.2 Proportions of trapezoidal threads, mm

Nominal diameter, d	Common difference	Pitch	Other Pitches
8	—	1.5	—
9–11	1	2	1.5, 3
12–14	2	3	2
16–20	2	4	2
22–28	2	5	3, 8
30–36	2	6	3, 10
38–44	2	7	3, 10
46–52	2	8	3, 12
55–60	5	9	3, 14
65–80	5	10	4, 16
85–95	5	12	4, 18
100–110	5	12	4, 20
115–130	5	14	6, 22
135–145	5	14	6, 24
150–155	5	16	6, 24
160–170	5	16	6, 28
175	—	16	8, 28
180	—	18	8, 28
185–200	5	18	8, 32
210–230	10	20	8, 36
240	—	22	8, 36
250–260	10	22	12, 40
270–280	10	24	12, 40
290–300	10	24	12, 44

In Fig. 23.1(b)

$$a = \frac{p}{2.7} \text{ and } h = \frac{p}{2} + 0.25 \text{ mm}$$

Figure 23.1(c) shows the profile of ISO metric Trapezoidal screw threads. The various proportions of the thread are given below:

$$H_1 = 0.5\,p\,;\, z = 0.25\,p + H_1/2$$
$$H_3 = h_3 = H_1 + a_c = 0.5\,p + a_c$$

The values of the nominal clearance 'a_c' are:

a_c = 0.15 mm for p = 1.50 mm
= 0.25 mm for p = 2 to 5 mm

$= 0.50$ mm for $p = 6$ to 12 mm
$= 1.00$ mm for $p = 14$ to 44 mm

In the Fig. , d = nominal major diameter of the external thread
d_2 = nominal pitch dia. of the external thread = $d - 2z$
d_3 = nominal minor diameter of the external threads = $d - 2h_3$
D_1 = nominal minor diameter for internal threads = $d - 2H_1 = d - p$
D_2 = nominal pitch diameter for internal threads = d_2
D_4 = nominal major diameter for internal threads = $d + 2a_c$

The standard dimensions for the Trapezoidal screw threads are given in Table 23.2 (IS 7008 (Part II and III) – 1973).

Designation: 1. Single start ISO metric Trapezoidal screw threads shall be designated by the letters 'T_r' followed by the values of the nominal diameter and of the pitch expressed in mm and separated by the sign 'x', *e.g.*, $T_r 40 \times 7$.

2. Multiple start screw threads shall be designated by the letters 'T_r' followed by the values of the nominal diameter and lead for the multiple start threads, separated by sign 'x' and in brackets the letter 'p' and the value of the pitch all in mm.

Ex: $T_r\, 40 \times 14\ (p\ 7)$

(Number of starts = $\dfrac{\text{Lead}}{\text{Pitch}} = \dfrac{14}{7}$ defines a thread of 40 diameter with 2 starts).

3. To designate left hand ISO metric trapezoidal screw threads, the letter 'LH' shall be added to the thread designation:

Ex: $T_r\, 40 \times 14\ (p\ 7)$ LH

***(iii)* Buttress thread:** Such threads are also used sometimes for power screws but they can transmit the power only in one direction, e.g., screw jack, vices, etc. The flat side acts as the driving side and so has the high efficiency of square thread. The sloping side permits the use of a split nut for the adjustment of wear. Their cost of manufacture is also slightly smaller as compared to square threads. Buttress thread is shown in Fig. 23.1(d).

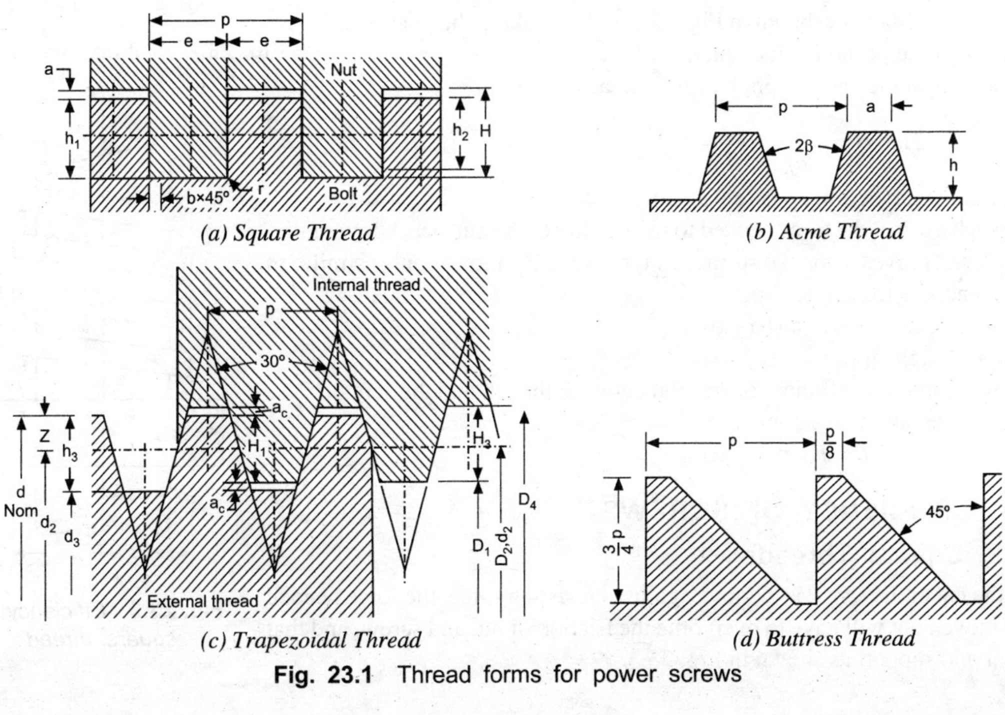

(a) Square Thread *(b) Acme Thread* *(c) Trapezoidal Thread* *(d) Buttress Thread*

Fig. 23.1 Thread forms for power screws

23.2.1 Multiple Threads

If a large transitory movement of the screw drive per revolution is needed, then screws with two or more parallel threads are used. Such threads are known as multiple threads and are designated as double, triple and so on, start threads. Like this, they allow the use of fine threads resulting in high efficiency because of the increase of helix angle.

23.2.2 Comparison of Square and Trapezoidal Threads

1. Square thread has the highest efficiency as compared to other power screws.
2. Square thread is more difficult to cut than trapezoidal and so more costly. It can be cut only on a turning lathe, whereas, trapezoidal thread can be cut on a thread milling machine which is more efficient.
3. Square thread is less sturdy than trapezoidal. For the same pitch, the thickness at the root of the thread is more in the case of trapezoidal threads.
4. The adjustment for the resulting wear is very difficult in the case of square threads but it can be overcome in the case of trapezoidal threads by using a split nut and clamping the two halves together. Hence such threads are used for drives where there must be little or no back lash, such as feed screws and lead screws of machine tools.

23.3 POWER SCREW DRIVE

In any power screw drive a screw and nut pair is employed, which have relative motion with respect to each other. This relative motion may be any one of the following types:

1. Screw rotates in its bearings, while the nut moves axially against the resisting force.
2. Nut is stationary while the screw rotates and moves axially against the resisting force.
3. Nut rotates, while the screw moves axially with no rotation.

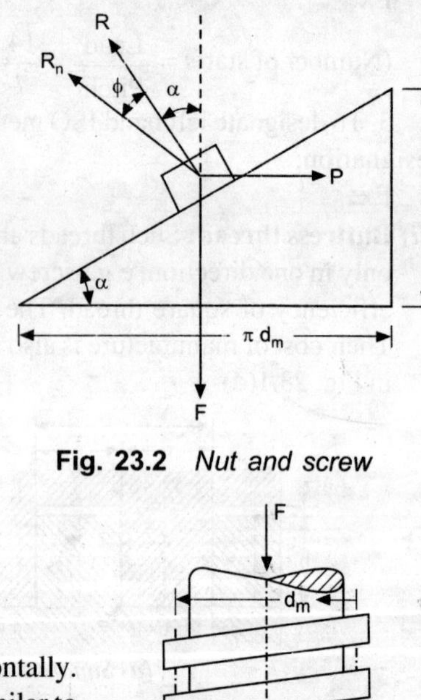

Fig. 23.2 *Nut and screw*

The movement of the nut on the screw is similar to the movement of a weight on an inclined plane, since when one thread is developed, it is an inclined plane as shown in Fig. 23.2 and nut taking the place of weight. The base of the inclined plane will be $= \pi d_m$

α = helix angle and p = pitch of the threads so,

$$\tan \alpha = \frac{p}{\pi d_m}$$

For multistart screws, p is replaced by lead 'l'.

In such a drive, the force needed to raise or lower the nut, acts horizontally. When the nut moves up on the surface of the screw, the movement is similar to up the plane, the force P will be:

$$P = F \tan(\alpha + \phi)$$

where ϕ = angle of friction

so $\mu = \tan \phi$ = coefficient of friction between the nut and the screw.

When the nut moves down,

$$P = F \tan(\phi - \alpha)$$

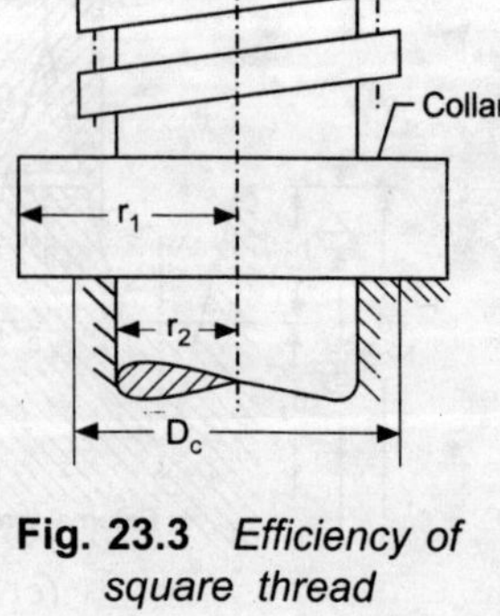

Fig. 23.3 *Efficiency of square thread*

23.4 EFFICIENCY OF SCREWS

23.4.1 Square Threads

When the nut or the screw is to move against a resisting force, the force needed for the movement will have to overcome the friction of nut and screw, and that of collar and support as shown in Fig. 23.3.

Force needed to overcome the nut and screw friction,

$$P = F.\tan(\alpha + \phi)$$

Force needed to overcome the friction of collar on support = $\mu_c F$

where F = weight of the load.

μ_c = coefficient of friction between collar and support.

P = force acting at the mean radius of thread and normal to F.

∴ Total torsional moment needed to raise the load = thread torsional moment + collar friction torsional moment, $M_t = M_{t1} + M_{t2}$

$$P.\frac{d_m}{2} + \mu_c.F.\frac{D_c}{2},$$

where D_c = friction diameter of the collar

$= \frac{4}{3}.\frac{r_1^3 - r_2^3}{r_1^2 - r_2^2}$, considering uniform pressure and when power lost is friction is to be found out.

$= (r_1 + r_2)$, considering uniform rate of wear and when power transmitted is to be determined.

∴ Total torsional moment, needed to raise the load is,

$$M_t = F.\frac{d_m}{2}.\tan(\alpha + \phi) + \mu_c.F.\frac{D_c}{2} \quad \text{...(23. 1)}$$

∴ Work done in raising the load, when one complete turn is given to the screw is,

$$2\pi M_t = F.\frac{d_m}{2}.\tan(\alpha + \phi).2\pi + \mu_c.F.\frac{D_c}{2}.2\pi$$

∴ Work input, $2\pi M_t = F.\pi d_m.\tan(\alpha + \phi) + \mu_c.\pi F.D_c$

Now Efficiency of a screw thread is given as :

$$\eta = \frac{\text{useful work}}{\text{work input}}.$$

Now when the screw moves through one revolution, the load moves up by a distance equal to pitch for a single start thread.

$$\therefore \quad \eta = \frac{F.p}{F.\pi d_m.\tan(\alpha + \phi) + \mu_c.\pi F.D_c}$$

Now $\frac{p}{\pi d_m} = \tan\alpha, \therefore p = \pi d_m.\tan\alpha$

$$\therefore \quad \eta = \frac{F.\pi d_m.\tan\alpha}{F.\pi d_m.\tan(\alpha + \phi) + \mu_c.\pi F.D_c} = \frac{\tan\alpha}{\tan(\alpha + \phi) + \mu_c.\frac{D_c}{d_m}} \quad \text{...(23.2)}$$

$$= \frac{\tan\alpha}{\frac{\tan\alpha + \tan\phi}{1 - \tan\alpha\tan\phi} + \mu_c.\frac{D_c}{d_m}} = \frac{\tan\alpha(1 - \tan\alpha\tan\phi)}{\tan\alpha + \tan\phi + \mu_c\frac{D_c}{d_m}.(1 - \tan\alpha\tan\phi)}$$

Now $\tan\phi = \mu$

$$\therefore \quad \eta = \frac{\tan\alpha(1 - \mu\tan\alpha)}{\tan\alpha + \mu + \mu_c.\frac{D_c}{d_m}.(1 - \mu\tan\alpha)}$$

Neglecting collar friction, from equation (23.2)

$$\eta = \frac{\tan\alpha}{\tan(\alpha + \phi)} \quad \text{...(23.3)}$$

Efficiency in lowering the load, $\eta = \dfrac{\tan\alpha}{\tan(\phi - \alpha)}$ ...(23.4)

For maximum efficiency, $\dfrac{d_\eta}{d_\alpha} = 0$, and it will be seen that $\alpha = 45°$

From equation (23.3) $\eta_{max} = \dfrac{1-\tan\phi}{1+\tan\phi}$ ...(23.5)

If for ordinary values of μ, a curve is plotted between the efficiency and the helix angle, then efficiency increases rapidly upto helix angle of 20°, and then the increase is slow and maximum is reached for helix angle of 40 to 45 degrees. There is not much difference in efficiency between range of helix angle of 30 and 50 degrees. As the helix angle increases, it becomes difficult to cut the threads and the mechanical advantage decreases. In practice, helix angles more than 30° are seldom used.

If an external force "*T*" is applied at a distance, *l*, from the axis of the screw,

Then wrench torsional moment = Screw torsional moment + Collar friction torsional moment

$$\therefore \quad T.l = F.\frac{d_m}{2}\tan(\alpha+\phi) + \mu_c.F.\frac{D_c}{2} \quad \text{...(23.6)}$$

Self locking Screws and Overhauling of Screws: If the coefficient of friction between the nut and screw is small or the lead is large, the axial load may be sufficient to turn the nut, and the load will start moving downward without the application of any torsional moment. Such a condition is known as overhauling of screws. In the limiting condition, the friction forces just balance the axial load and no torsional moment is required to lower the load. As we know and if the load moves down, the friction forces are reversed, therefore from equation, (23.1), $M_t = 0$,

$$\therefore \quad F.\frac{d_m}{2}\tan(\phi-\alpha) + \mu_c.F.\frac{D_c}{2} = 0$$

$$\therefore \quad d_m \times \frac{\tan\phi - \tan\alpha}{1+\tan\phi\tan\alpha} + \mu_c.D_c = 0$$

$$\therefore \quad d_m(\mu - \tan\alpha) + \mu_c.D_c.(1+\mu\tan\alpha) = 0$$

$$\therefore \quad \tan\alpha(d_m - \mu.\mu_c.D_c) - (\mu.d_m + \mu_c.D_c) = 0$$

or
$$\tan\alpha = \frac{\mu.d_m + \mu_c D_c}{d_m - \mu.\mu_c.D_c}$$

$$\therefore \quad \text{For overhauling, } \tan\alpha \geq \frac{\mu.d_m + \mu_c D_c}{d_m - \mu.\mu_c.D_c}$$

If a ball or thrust bearing is used to support the screw, μ_c becomes practically zero,

$$\therefore \quad \tan\alpha \geq \mu$$

or
$$\alpha \geq \phi.$$

If $\phi \geq \alpha$, the screw will not overhaul, i.e., the load will not move downward without the application of a torsional moment directed downward. Such a condition is known as "Self-locking" or "Self-braking".

Now
$$\eta = \frac{\tan\alpha}{\tan(\alpha+\phi)}$$

Now for self-locking, $\alpha \leq \phi$

$$\therefore \quad \eta \leq \frac{\tan\phi}{\tan(\phi+\phi)} \leq \frac{\tan\phi}{\tan 2\phi} \leq \frac{\tan\phi(1-\tan^2\phi)}{2\tan\phi} \leq \frac{1-\tan^2\phi}{2}$$

∴ For self-locking,

$$\eta \le \frac{1-\tan^2\phi}{2}$$

Hence it is clear that for self locking, the efficiency can not be more than 50%. It is also clear from here, that if the efficiency is more than 50%, the screw will not remain self locking and it will overhaul. This can be proved in the following manner also:

Supposing a load F is to be lifted up through a distance h

∴ useful work $= F.h.$

∴ work input $= \dfrac{F.h}{\eta}$

∴ work needed to overcome friction $= \dfrac{F.h}{\eta} - Fh = Fh\left(\dfrac{1}{\eta}-1\right)$

If the same load is to move downward through the same distance h, the potential energy of the load is $= F.h.$

For overhauling, the potential energy should be more than the frictional resistance, therefore,

$$F.h \ge F.h\left(\frac{1}{\eta}-1\right)$$

or $\eta \ge 1-\eta$

or $2\eta \ge 1$ or $\eta \ge 50\%$

Coefficient of friction: The coefficient of friction for the threads of power screws depends upon : materials for the nut and screw, workmanship in cutting the threads, lubrication, and degree of 'running in' of the threads, and the unit pressure on the threads. If the unit pressure is less than about 100 N/mm^2 and the rubbing velocity is less than about 0.25 m/s the values for coefficient may be taken from Table 23.3. If screw is operated very slowly, then the coefficient may be taken as 30% higher. The coefficient of friction for plain thrust collars are given in Table 23.4. Most power screws are made of steel with nuts of C.I. or bronze.

Table 23.3 Values for m

Screw	Nut			
	Steel	Brass	Bronze	C.I.
Steel, dry	0.15–0.25	0.15–0.23	0.15–0.19	0.15–0.25
Steel, Machine oil	0.11–0.17	0.10–0.16	0.103–0.15	0.11–0.17
Bronze	0.08–0.12	0.04–0.06	—	0.06–0.09

Table 23.4 Values for m_c

Material	Running μ_c	Starting μ_c
Soft steel on C.I.	0.121	0.170
Hardened steel on C.I.	0.092	0.147
Soft steel on bronze	0.084	0.101
Hardened steel on bronze	0.063	0.081

23.4.2 Trapezoidal Threads

Figure 23.4 shows a section of an Acme or Trapezoidal thread taken at angle of 90° to its elements. R_n is the normal reaction that an axial load F would induce in a square thread having the same helix angle α, as the acme or trapezoidal thread. Because of the sloping sides of the angular thread, the normal reaction will be Q, normal to the thread face, where

$$Q = \frac{R_n}{\cos \beta}$$

where $2\beta = 29°$ for Acme threads

and $= 30°$ for trapezoidal threads.

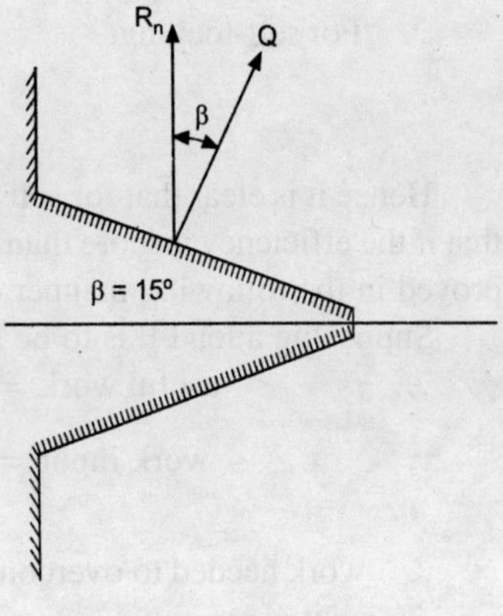

Fig. 23.4 *Section of a trapezoidal thread*

Since the thread friction varies as the normal pressure, the equations for trapezoidal or acme threads can be derived from those for square threads, by dividing the friction factors by cos β. Only the friction for nut and screw are to be divided by cos β.

$\therefore$ $P = F . \tan(\alpha + \phi)$, for square threads

$$= F.\frac{\tan\alpha + \tan\phi}{1 - \tan\alpha\tan\phi} = F.\frac{\tan\alpha + \mu}{1 - \mu\tan\alpha}$$

$\therefore$ For trapezoidal threads,

$$P = F.\frac{\tan\alpha + \mu/\cos\beta}{1 - \mu\tan\alpha/\cos\beta} = F.\frac{\tan\alpha + \mu.\sec\beta}{1 - \mu\sec\beta\tan\alpha}$$

$\therefore$ From equation (23.6),

$$\therefore \quad T.l = F.\frac{\tan\alpha + \mu.\sec\beta}{1 - \mu\sec\beta\tan\alpha}.\frac{d_m}{2} + \mu_c.F.\frac{D_c}{2} \quad ...(23.7)$$

Neglecting collar friction, the force P, when friction between nut and screw is zero, $P_i = F.\tan\alpha$

$$\therefore \quad \eta = \frac{P_i}{P} = \frac{\tan\alpha\,(1 - \mu.\sec\beta\tan\alpha)}{\tan\alpha + \mu\sec\beta} \quad ...(23.8)$$

Example 23.1. What force will be required at a radius of 80 mm to raise and lower a 11 kN cross bar of a planer ? The bar is raised and lowered by two 38 mm square thread screws, having a pitch of 7 mm. The screw is of steel and nut is of bronze 38 mm thick. The collar is of steel and it has an outside diameter of 76 mm and inside diameter of 38 mm. Assume the coefficient of friction at the threads as 0.11, at the collar as 0.13.

Solution: Now major diameter of thread = 38 mm, pitch = 7 mm

$\therefore$ Mean diameter of thread $= 38 - \frac{7}{2} = 34.5$ mm

$$\therefore \quad \tan\alpha = \frac{p}{\pi d_m} = \frac{7}{\pi \times 34.5} = 0.0646$$

$$\therefore \quad \alpha = 3°\text{–}42'.$$

Now $\tan\phi = 0.11, \therefore \phi = 6°\text{–}18'$.

Now total torsional moment needed to raise the cross bar,

$$M_t = F.\tan(\alpha + \phi).\frac{d_m}{2} + \mu_c.F.\frac{D_c}{2}$$

Now $$\frac{D_c}{2} = \frac{2}{3}\left(\frac{r_1^3 - r_2^3}{r_1^2 - r_2^2}\right)$$

where $r_1 = 38$ mm, and $r_2 = 19$ mm

$$\therefore \quad \frac{D_c}{2} = \frac{2}{3}\left(\frac{38^3 - 19^3}{38^2 - 19^2}\right) = \frac{2}{3}\left(\frac{47700}{1080}\right) = 29.5 \text{ mm}$$

Now load on each screw $= \dfrac{11 \times 1000}{2} = 5500$ N

$$\therefore \quad M_t = 5500.\tan(10°).\frac{34.5}{2} + 0.13 \times 5500 \times 29.5 = 37.83 \text{ Nm}$$

$\therefore$ From equation (23.5),

$$T \times l = 37.83$$

$$T \times 80 = 37.83 \times 1000, \text{ gives } T = 473 \text{ N}$$

(*ii*) Lowering of the bar:

$$M_t = F.\frac{d_m}{2}\tan(\phi - \alpha) + \mu_c.F.\frac{D_c}{2}$$

$$= 5500 \times \frac{34.5}{2}.\tan(2°\text{–}36') + 21092.5 = 25.4 \text{ Nm}$$

$$\therefore \quad T = \frac{25.4 \times 1000}{80} = 317.5 \text{ N.}$$

Example 23.2. In a large gate valve used in a high pressure water line, the gate weighs 5 kN and the friction, due to water pressure and causing resistance to opening is 2.5 kN. The valve stem is 38 mm in diameter. The valve stem is non-rotating and is raised by a rotating wheel with internal threads acting as a rotating nut on the valve stem which is fitted with square threads of 7 mm pitch. The wheel presses against a supporting collar of 38 mm inside diameter and 76 mm outside diameter. Assume the coefficient for threads to be 0.15 and for the collar 0.25.

(*a*) Determine the efficiency of the screw and collar.

(*b*) Determine the torsional moment that must be applied to the wheel to raise the load.

Solution: Outside diameter of screw = 38 mm

$\therefore$ Core diameter $= 38 - 7 = 31$ mm

$\therefore$ Mean diameter $= 34.5$ mm

$$\therefore \quad \tan\alpha = \frac{p}{\pi d_m} = \frac{7}{\pi \times 34.5} = 0.0646, \therefore \alpha = 3° - 42'$$

Now $\tan\phi = 0.15$,

$\therefore$ $\phi = 8° - 33'$

Now equivalent radius of collar,

$$\frac{D_c}{2} = \frac{2}{3}\left(\frac{r_1^3 - r_2^3}{r_1^2 - r_2^2}\right)$$

$$r_1 = 38 \text{ mm}, r_2 = 19 \text{ mm}, \frac{D_c}{2} = 29.5 \text{ mm}$$

now $\tan(\alpha + \phi) = \tan 12° - 15' = 0.217$

(*a*)

$$\eta = \frac{\tan\alpha}{\tan(\alpha + \phi) + \mu_c.\dfrac{D_c}{d_m}}$$

$$= \frac{0.0646}{0.217 + 0.25 \times \dfrac{59}{34.5}} = \frac{0.0646}{0.616} \times 100 = 10.5\%$$

(*b*)

$$M_t = F.\frac{d_m}{2}.\tan(\alpha + \phi) + \mu_c.F.\frac{D_c}{2}, \text{ for lifting the gate.}$$

Now $F = 5.0 + 2.5 = 7.5$ kN since friction acts downward.

$$M_t = 7.5 \times 1000 \times \frac{34.5}{2} \times 0.217 + 0.25 \times 7.5 \times 1000 \times 29.5$$

$$= 83.5 \times 10^3 \text{ Nmm} = 83.5 \text{ Nm}$$

Note: For lowering the gate, the friction will act upwards,

Therefore, $F = 5.0 - 2.5 = 2.5$ kN.

Example 23.3. The lead screw of a lathe has a 50 mm Acme thread, one thread per cm. To drive the tool carriage, this screw must exert an axial pressure of 3000 N. The thrust is carried on a collar 110 mm outside diameter and 55 mm inside diameter. The lead screw revolves at 3.14 rad/s.

(*a*) Determine the efficiency of the screw and collar assuming a coefficient of friction of 0.15 for threads and 0.12 for collar.

(*b*) Determine the power required to drive the screw.

Solution: Axial load, $F = 3000$ N

Outside diameter = 50 mm

Mean diameter = $50 - p/2$

Now $p = 10$ mm

$\therefore$ $d_m = 45$ mm

$$\tan\alpha = \frac{p}{\pi d_m} = \frac{10}{\pi \times 45} = 0.0707$$

$\therefore$ $\alpha = 4° - 3'$

Now $2\beta = 29°$

$\therefore$ $\text{Sec}\,\beta = \text{Sec}\,14.5° = 1.0329$

$\mu_c = 0.12 \quad \mu = 0.15$

Collar friction diameter,

$$D_c = \frac{4}{3}\left(\frac{r_1^3 - r_2^3}{r_1^2 - r_2^2}\right)$$

$$r_1 = \frac{110}{2} = 55 \text{ mm}; \; r_2 = \frac{55}{2} = 27.5 \text{ mm}$$

$$\therefore \quad D_c = \frac{4}{3}\left(\frac{55^3 - 27.5^3}{55^2 - 27.5^2}\right) = 86.5 \text{ mm}$$

(*a*)

$$\eta = \frac{\tan\alpha\,(1 - \mu\tan\alpha\sec\beta)}{(\tan\alpha + \mu\sec\beta) + \mu_c.\frac{D_c}{d_m}(1 - \mu\tan\alpha\sec\beta)}$$

$$= \frac{0.0707\,(1 - 0.15 \times 0.0707 \times 1.0329)}{(0.0707 + 0.15 \times 1.0329) + 0.12 \times \frac{86.5}{45}(1 - 0.15 \times 0.0707 \times 1.0329)} = 15.45\%$$

(*b*) Now

$$M_t = F.\frac{\tan\alpha + \mu\sec\beta}{1 - \mu\tan\alpha\sec\beta}.\frac{d_m}{2} + \mu_c.F.\frac{D_c}{2}$$

$$= 3000.\frac{0.0707 + 0.15 \times 1.0329}{1 - 0.15 \times 0.0707 \times 1.0329} \times \frac{45}{2000} + 0.12 \times 3000 \times \frac{86.5}{2 \times 1000} = 31 \text{ Nm}$$

$$\therefore \quad \text{Power} = \frac{31 \times w}{1000} \text{ kW} = \frac{31 \times 3.14}{1000} = 0.0974 \text{ kW}.$$

Example 23.4. In a hand vice the screw has a double start square threads of 26 mm diameter. If the lever is 0.25 m and maximum force that can be applied at the end of the lever is 300 N, determine the force with which the job is held in the jaws of vice. Take coefficient of friction as 0.14.

Solution: Now torsional moment applied at the lever = 300 × 0.25 = 75 Nm

$$\therefore \quad M_t = F.\frac{d_m}{2}.\tan(\alpha + \phi)$$

F = axial load or the force with which the job is held in the jaws

Now outside diameter of screw = 26 mm

Assuming normal series of square threads, $p = 5$ mm

$$\therefore \quad d_m = 26 - \frac{5}{2} = 23.5 \text{ mm}$$

Now lead $\quad 2p = 10$ mm

$$\therefore \quad \tan\alpha = \frac{\text{lead}}{\pi d_m} = \frac{10}{\pi \times 23.5} = 0.1353,$$

$$\therefore \quad \alpha = 7.7°$$

Again $\quad \tan\phi = 0.14, \phi = 8°$

$$\therefore \quad M_t = F \times \frac{23.5}{2} \times \tan 15.7°$$

$$\therefore \quad 75 \times 1000 = F \times \frac{23.5}{2} \times 0.281, \text{ gives } F = 22.70 \text{ kN.}$$

23.5 STRESSES IN SCREWS

A power screw is subjected to an axial load and to a turning moment. Due to these the following types of stresses are set up in the screws:

1. **Direct tension or compression due to axial load.** The method of mounting the screw will determine whether the stress is tensile or compressive. If possible the thrust collar should be provided so that the screw is in tension. The direct stress in the body of the screw can be found out by dividing the axial load by the minimum cross-sectional area of the screw. If the screw is under compression and its unsupported length exceeds six to eight times the core diameter, it must be checked for stability against buckling.
2. **Shear stress due to axial load.** This stress is usually not dangerous. Due to the axial load, the threads of screw will shear at the core or root diameter and the threads of nut will shear at the major diameter which is nothing but the outside diameter of the screw. Therefore:

$$f_{s\,(\text{screw})} = \frac{F}{z.\pi d_c.t} \text{ N/mm}^2$$

$$f_{s\,(\text{nut})} = \frac{F}{z.\pi d_0.t} \text{ N/mm}^2$$

where F = axial load on the screw, N

z = number of threads in engagement

d_c = core diameter of screw mm

d_o = outer diameter of screw, mm

t = thickness of the thread or width of the thread, mm

3. **Compression between the thread surfaces, due to axial load.** This would result in a bearing pressure between the thread surfaces. This bearing pressure is not uniform since under the load, the nut and the screw are under different kinds of stresses, i.e., if there is tension in the screw there will be compression in the nut and vice versa. However, the assumption of uniform bearing pressure simplifies the design.
4. **Torsional shear stress in the screw due to the external turning moment.**

23.6 DESIGN CALCULATIONS

Design calculations of the screw drive include:

1. Calculation of the wear of the thread working surfaces.
2. Strength of the screw and nut.
3. Buckling stability of the screw under compression.

Calculation for Wear: Wear is the most important factor in the power screws. It occurs between the sliding surfaces of the nut and screw threads when these rotate under load. Many failures of the screw drive occur due to this wear. To ensure a normal working life of the drive, the bearing pressures on the thread surfaces should be within limits. It is given by the relation:

$$F = \frac{\pi}{4}(d_o^2 - d_c^2) \times p_b \times z,$$

where p_b = bearing pressure

z = number of threads in the nut.

$$\therefore \quad p_b = \frac{4F}{\pi(d_o^2 - d_c^2) \times z} \leq \text{safe value.}$$

The allowable values for bearing pressure are given in Table 23.5. ...(23.9)

Table 23.5 Safe bearing pressures in power screws

Service	Material		p_b, N/mm^2	Remarks
	Screw	Nut		
1. Hand Press	Steel	Bronze	17.5–24.5	Low speed, well lubricated
		C.I.	12.5–17.5	Low speed ≤ 0.04 m/s
2. Screw Jack	Steel	Bronze	11.0–17.5	Low speed ≤ 0.05 m/s
		C.I.	4.0–7.0	Medium speed, 0.1 to 0.2 m/s
3. Hoisting Screw	Steel	Bronze	5.5–10.0	Medium speed, 0.1 to 0.2 m/s
4. Lead Screw	Steel	Bronze	1.05–1.70	High speed > 0.25 m/s

Calculation for Strength:

***(i)* Screw:** The body of the screw is subjected to joint action of compression (tension) and the twisting moment:

$$f_t = \frac{4F}{\pi d_c^2}$$

$$f_s = \frac{16 M_t}{\pi d_c^3}$$

$$\therefore \quad \text{Equivalent shear stress,} \quad f = \frac{1}{2}\sqrt{f_t^2 + 4f_s^2}$$

$$\text{Equivalent normal stress} = \frac{1}{2}\left[f_t + \sqrt{\left(f_t^2 + 4f_s^2\right)}\right]$$

The twisting moment is the only which is acting on screw and nut. The collar twisting moment is not to be considered here. The above analysis can not give the value of 'd_c' because the twisting moment can be found out only if the thread profile is known. Therefore, initially, the core diameter is calculated on the basis of compression only, and taking load as:

$$F_{des} = \beta F,$$

where β = 1.25 to 1.30, to account for the effect of twisting moment

$$\therefore \qquad d_c = \sqrt{\frac{4\beta F}{\pi . f_c}}$$

Then from the tables, the standard size of the thread is selected. Twisting moment is calculated and the screw is checked for principal stress.

After this, the screw is checked for self locking.

(ii) **Nut:** Nut is shown in Fig. 23.5. From equation (23.9), the number of threads can be determined,

i.e., $$z \geq \frac{4F}{\pi (d_o^2 - d_c^2) \times p_b}$$

$\therefore$ Height of the nut, $H = p . z.$

The nut is also checked for tension in the circular section,

i.e., $$f_t = \frac{4\beta . F}{\pi (D_1^2 - d_o^2)}$$

The factor β is considered to account for torsion of the nut body by the twisting moment. As the collar of the nut will have to bear the whole load when the load rests on it, it might get crushed or sheared.

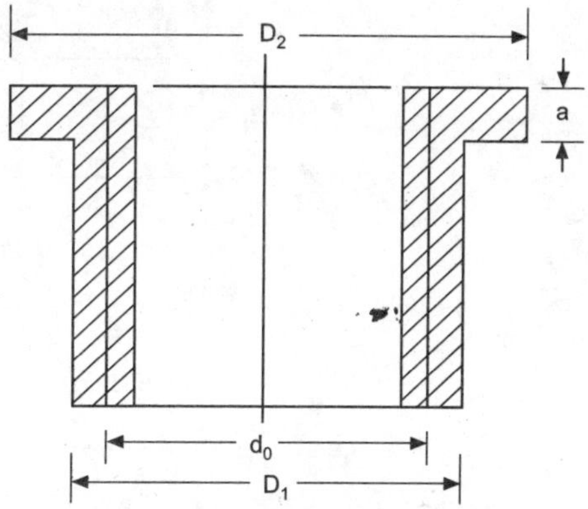

Fig. 23.5 *Nut*

$\therefore$ Considering crushing of the collar of the nut,

$$F = \frac{\pi}{4}(D_2^2 - D_1^2) \times f_c$$

Considering shearing action,

$$F = \pi D_1 \times a \times f_s.$$

Check for Stability of the Screw: The screw can be checked for stability against buckling by the use of Rankine Gordon formula or the Johnson's formula given below:

$$F_{cr} = A.f_{yp}\left[1 - \frac{f_{yp}}{4c\pi^2 E} . \left(\frac{L}{k}\right)^2\right]$$

$$\therefore \qquad f_c = \frac{F}{A}\left[\frac{1}{1 - \frac{f_{yp}}{4c\pi^2 E} . \left(\frac{L}{k}\right)^2}\right], f_c = \text{stress induced due to load } F.$$

where F_{cr} = critical load causing failure

f_{yp} = yield stress

k = least radius of gyration of the section,

c = end fixity, coefficient, L = length of screw.

23.7 DESIGN PROCEDURE

Let us consider the case of a Screw Jack, Fig. 23.6.

1. Find the core diameter of the screw on the basis of compression or tension only.
2. Select the standard outer diameter, core diameter and pitch of the threads.
3. Find the torsional moment to be transmitted by the nut and screw.
4. Find the direct stress and the torsional shear stress and check the screw for combined principal stress.
5. Check the screw for stability.
6. Calculate the thread for wear and determine the height of the nut. Also find its other dimensions.

7. Determine the dimensions of the jack lever. For this find the turning moment required to lift the load and overcome the friction of the bearing surface of the jack cup.

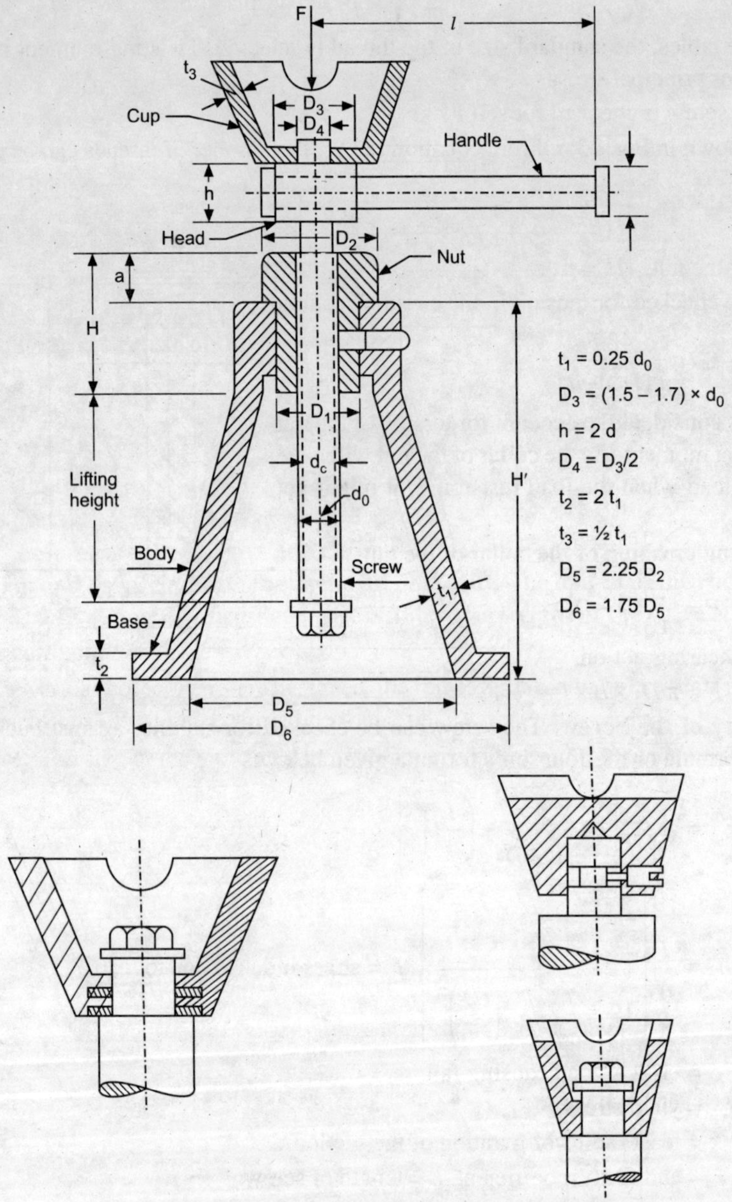

Fig. 23.6 *Screw Jack*

Then $M_t = T \times l,$

where T = force applied at the end of the lever.

l = lever length.

The effect of a normal man ranges between 200 to 400 N or even smaller in the case of continuous work.

The diameter of the lever can be found out by calculating it for bending. The total torsional moment provided by the lever will act as bending moment on the lever. Therefore,

$$B.M. = M_t = \frac{\pi}{32} d^3 \times f_b, \;\therefore\; d = ?$$

Example 23.5. Calculate the screw drive of Jack for load capacity of 100 kN. Lifting height is 0.5 m.

Solution: Refer Fig. 23.6. The drive can be calculated according to the procedure given in Art. 23.7.

1. **Screw:** The material for the screw may be selected as C-25 steel with ultimate strength in tension ranging from 470 to 570 N/mm^2.

   Let us take the ultimate strength in tension = 525 N/mm^2

   Yield stress in tension and compression = 294 N/mm^2

   Yield stress in shear = 182 N/mm^2

   Now Axial load, F = 100 kN

   ∴ Design load = $\beta \cdot F$ = 1.3 × 100 = 130 kN

   Assuming a factor of safety of 2.25 on yield stress,

$$f_c = \frac{294}{2.25} = 130 \text{ N/mm}^2$$

$$\therefore \quad d_c = \sqrt{\frac{4\beta F}{\pi f_c}} = \sqrt{\frac{4 \times 130 \times 1000}{\pi \times 130}} = 35.7 \text{ mm}$$

Choosing square threads, normal series, Table 23.1, let us select a thread for which outside diameter = 46 mm, p = 8 mm

$$\therefore \quad d_c = 38 \text{ mm}, d_m = 42 \text{ mm}$$

$$\therefore \quad \tan\alpha = \frac{p}{\pi d_m} = \frac{8}{\pi \times 42} = 0.0606, \alpha = 3° - 33'$$

Let us assume a value for μ, i.e., coefficient of friction for nut and screw to be 0.125.

$$\therefore \quad \tan\phi = 0.125,$$

$$\therefore \quad \phi = 7° - 9'$$

$$\therefore \quad \text{Turning moment } M_{t1} = F.\frac{d_m}{2}.\tan(\alpha + \phi) = 100 \times 10^3 \times \frac{0.042}{2} \times \tan 10° - 42' = 396.8 \text{ Nm}$$

Check for Principal Stress:

$$\text{Direct compression stress, } f_c = \frac{4F}{\pi d_c^2} = \frac{4 \times 100 \times 1000}{\pi \times (38)^2} = 88.4 \text{ N/mm}^2$$

$$\text{Torsional shear stress, } f_s = \frac{16M_t}{\pi d_c^3} = \frac{16 \times 396.8 \times 1000}{\pi \times (38)^3} = 37 \text{ N/mm}^2$$

∴ Maximum shear stress

$$f_{s\,max} = \frac{1}{2}\sqrt{f_c^2 + 4f_s^2} = \frac{1}{2}\sqrt{(88.4)^2 + 4 \times 37.0^2} = 57.6 \text{ N/mm}^2$$

Maximum normal stress,

$$f_{s\,max} = \frac{1}{2}\left[f_c + \sqrt{f_c^2 + 4f_s^2}\right] = \frac{1}{2}[88.4 + 115.2] = 101.8 \text{ N/mm}^2$$

Now we know that the factor of safety for a service like this is 1.5 to 2 on the basis of yield stress. Both the above stresses give higher factor of safety and are, therefore, O.K.

Check for Stability: Check for stability can be applied only if the height of the nut is known.

2. **Nut:** Let us take the material of the nut as phosphor bronze.

   Yield stress in tension = 126 N/mm^2

   Yield stress in compression = 112 N/mm^2 and in the shear = 105 N/mm^2

Now applying the equation for wear.

$$F = \frac{\pi}{4}(d_0^2 - d_c^2) \times p_b \times z$$

Let $p_b = 15\ \text{N/mm}^2$, Table 23.5

$$\therefore \quad 100 \times 1000 = \frac{\pi}{4}(46^2 - 38^2) \times 15 \times z, \therefore z = 11.9, \text{ say } 12$$

$\therefore$ Height of nut, $H = p \times z = 8 \times 12 = 96$ mm

Refer Fig. 23.5,

$$f_t = \frac{4\beta.F}{\pi(D_1^2 - d_0^2)}$$

Now f_t for phosphor bronze $= \dfrac{126.0}{2} = 63\ \text{N/mm}^2$

$$\therefore \quad \pi(D_1^2 - 64^2) = \frac{4 \times 1.3 \times 100 \times 1000}{63}, \text{ gives } D_1 = 70 \text{ mm}$$

Considering the crushing of the collar of the nut,

$$F = \frac{\pi}{4}(D_2^2 - D_1^2) \times f_c$$

Now

$$f_c = \frac{112}{2} = 56\ \text{N/ mm}^2$$

$$\therefore \quad 100 \times 1000 = 0.785(D_2^2 - 70^2) \times 56$$

$$\therefore \quad D_2 = 85 \text{ mm}$$

Considering shear of the collar of nut,

$$F = \pi D_1 \times a \times f_s$$

Now

$$f_s = \frac{105}{2} = 52.5\ \text{N/mm}^2$$

$$\therefore \quad 100 \times 1000 = \pi \times 70 \times a \times 52.5 \qquad a = 8.66 \text{ mm, say } 9 \text{ mm}$$

Check for Stability of Screw:

Unsupported length $=$ lifting height $+ \dfrac{H}{2} = 0.50 + 0.048 = 0.548$ m

(Now crippling load),

$$F_{cr} = A.\, f_{yp}\left[1 - \frac{f_{yp}}{4c\pi^2 E}.\left(\frac{L}{k}\right)^2\right]$$

$$A = \frac{\pi}{4} d_c^2 = 0.785 \times (0.038)^2 = 0.00113 \text{ m}^2$$

$$L = 548 \text{ mm}, k = 0.25\, d_c = 9.5 \text{ mm}$$

$$\therefore \quad \frac{L}{k} = \frac{548}{9.5} = 57.6$$

$$f_{yp} = 294 \text{ MN/m}^2, E = 210 \text{ GN/m}^2$$

$c = 0.25$, assuming one end of the column as fixed and the other free.

$$F_{cr} = 0.00113 \times 294 \times 10^6 \left[1 - \frac{294 \times 10^6}{4 \times 0.25 \times \pi^2 \times 210 \times 10^9} \times (57.6)^2\right]$$

$$= 176 \times 10^3 \text{ N} > 100 \times 10^3 \text{ N}, \therefore \text{ safe.}$$

Now $$F_S = \frac{F_{cr}}{F} = 1.76$$

The usual value for steady load is 2, but if the conditions of loading and quality of material are accurately known, a value as low as 1.25 may be used. Therefore, the design is quite safe if not very safe.

3. Lever: Total torsional moment to be provided by the lever,

$$M_t = F.\frac{d_m}{2}.\tan(\alpha + \phi) + \mu_c.F.\frac{D_c}{2}$$

Let $\mu_c = 0.147$, Table 23.4.

Now $$\frac{D_c}{2} = \frac{2}{3}\left(\frac{r_1^3 - r_2^3}{r_1^2 - r_2^2}\right), \text{ Refer Fig. 23.6.}$$

Now $D_3 = (1.5 - 1.7)\, d_o = 69$ to 78.2 mm, say 78 mm

$$D_4 \cong \frac{D_3}{2} = 39 \text{ mm}$$

$$\therefore \quad \frac{D_c}{2} = \frac{2}{3} \times \frac{39^3 - 19.5^3}{39^2 - 19.5^2} = 30.2 \text{ mm}$$

$$\therefore \quad M_t = 396.8 + 0.147 \times 100 \times 10^3 \times 0.0302 = 839.8 \text{ Nm}$$

Now $T \times l = M_t$

Now one man can apply a force of 400 N maximum

$\therefore$ Lever length, $$l = \frac{839.8}{400} = 2.1 \text{ m}$$

To give a grip for the hands, the length may be made as 2.25 m.

Now to find the diameter of lever, its bending can be considered.

$$\therefore \quad \frac{\pi d^3}{32} \times f_b = 839.8 \times 1000$$

Now $$f_b = \frac{294}{2} = 147 \text{ N/mm}^2$$

$$\therefore \quad \frac{\pi}{32} d^3 \times 147 = 839.8 \times 1000, \text{ gives } d = 38.9 \text{ mm, say 40 mm}$$

$\therefore$ From Fig. 23.6, $h \cong 2d = 80$ mm

Other dimensions: Thickness of the body, $t_1 \cong 0.25\, d_o = \frac{1}{4} \times 46$, say 12 mm

Diameter of body at top $\cong 1.5\, D_4 = 1.5 \times 85 = 127.5$ mm, say 130 mm

For stability, the body tapers outward towards the base.

Let $D_5 \cong 2.25\, D_2 = 190$ mm, $D_6 \cong 1.75\, D_5$, say 330 mm

$$t_2 = 2t_1 = 24 \text{ mm} \quad t_3 = \frac{t_1}{2}, \text{ say 7 mm}$$

Example 23.6. A screw press is to exert a force of 60 kN with an applied torsional moment of 550 Nm. The unsupported length of the screw is 0.45 m and a thrust bearing of hardened steel on C.I. is provided at the power end. The screw is to be made of steel having an ultimate strength of 530 MN/m^2 and a yield stress of 265 MN/m^2. The design stresses are to be 87.5 MN/m^2 in tension and compression, 52.5 MN/^{2} in shear and 14 MN/m^2 in thread bearing. The nut is of C.I. and the permissible shear is 21 MN/m^2. Determine the dimension of the screw and nut.

Solution: Axial load $F = 60$ kN

Design load, $\beta F = 1.25 \times 60 = 75$ kN

$$\therefore \quad d_c = \sqrt{\frac{4\beta F}{\pi . fc}} = \sqrt{\frac{4 \times 75 \times 1000}{\pi \times 87.5}} = 33 \text{ mm}$$

Using Normal series of square threads and selecting a thread with outside diameter, $d_o = 46$ mm, $p = 8$ mm

$$\therefore \quad d_c = 38 \text{ mm}, d_m = 42 \text{ mm}$$

Now $$\tan \alpha = \frac{p}{\pi d_m} = \frac{8}{\pi \times 42} = 0.606, \therefore \alpha = 3° - 33'$$

Assuming $\mu = 0.15$

$$\therefore \quad \tan \phi = 0.15 \therefore \phi = 8° - 33'$$

Now $$M_t = F.\frac{d_m}{2}. \tan (\alpha + \phi) + \mu_c.F.\frac{D_c}{2}$$

Now μ_c may be taken as 0.147

$$\therefore \quad 550 \times 1000 = 60 \times 1000 \times 21 \tan (12° - 6') + 0.147 \times 60 \times 1000 \times \frac{D_c}{2}, \text{ gives}$$

$$D_c = 63.5 \text{ mm}$$

Check for maximum Stress:

$$f_s = \frac{16\, M_{t_1}}{\pi d_c^3} = \frac{16 \times 270}{\pi \times (0.038)^3} = 25.2 \text{ MN/m}^2 \text{ [Here } M_{t1} = F.\frac{d_m}{2}. \tan (\alpha + \phi)]$$

The stress due to column action is:

$$f_c = \frac{F}{A}\left[\frac{1}{1 - \left(\frac{L}{k}\right)^2 . \frac{f_{yp}}{4\pi^2\, cE}}\right]$$

Assuming one end to be fixed and the other free, $c = 0.25$

$$L = 450 \text{ mm}, k = 0.25 \times d_c = 9.5 \text{ mm}$$

$$\therefore \quad \frac{L}{k} = \frac{450}{9.5} = 47.4, A = \frac{\pi}{4}\, d_c^{\,2} = \frac{\pi}{4} \times (0.038)^2 = 0.00113 \text{ m}^2$$

$$E = 210 \text{ GN/m}^2, f_{yp} = 265 \text{ MN/m}^2$$

$$\therefore \quad f_c = \frac{60 \times 1000}{0.00113}\left[\frac{1}{1 - \frac{(47.4)^2 \times 265 \times 10^6}{4 \times 0.25 \times \pi^2 \times 210 \times 10^9}}\right] = 73.4 \text{ MN/m}^2$$

Check for combined stress:

$$f_{s\,max} = \frac{1}{2}\sqrt{f_c^2 + 4f_s^2} = \frac{1}{2}\sqrt{73.4^2 + 4 \times 25.2^2} = 44.8 \text{ MN/m}^2 < 52.5 \text{ MN/m}^2, \therefore \text{ O.K.}$$

$$\therefore \quad f_{c\,max} = \frac{1}{2}\left[f_c + \sqrt{f_c^2 + 4f_s^2}\right] = \frac{1}{2}[73.4 + 89.6] = 81.5 \text{ MN/m}^2 < 87.5 \text{ MN/m}^2, \therefore \text{O.K.}$$

Nut:

$$F = \frac{\pi}{4}(d_o^2 - d_c^2) \times p_b \times z$$

$$\therefore \quad 60 \times 1000 = \frac{\pi}{4}(46^2 - 38^2) \times 14 \times z, \; z = 8.15, \text{ say 9 threads.}$$

$\therefore$ Length of nut $= p \times z = 8 \times 9 = 72$ mm.

Check for shear:

$$W = \pi \,.\, d_0 .\, t\, z f_s$$

$$\therefore \quad f_s = \frac{60 \times 1000}{\pi \times 46 \times 4 \times 9} = 11.5 \text{ N/mm}^2 < 21 \text{ N/mm}^2 \text{ for C.I.}; \left(t = \frac{p}{2}\right)$$

Example 23.7. The force required to open or close a certain submerged sliding water gate is estimated at 27 kN. It is required to design a single threaded steel screw such that one man exerting a pull of 360 N at the periphery of a hand wheel 1.0 m in diameter, attached directly to the screw, can operate the gate. The greatest unsupported length of the screw is found to be 1.5 m. Coefficient of friction for screw and collar is 0.15. The crushing strength of the material is 210 MN/m², the maximum working tensile or compressive stress is 70 MN/m². E = 210 GN/m².

Solution: The screw will be in compression in closing the gate, it must be, therefore, designed as a column fixed at both the ends.

$$\therefore \quad d_c = \sqrt{\frac{4F}{\pi \cdot f_c}} = \sqrt{\frac{4 \times 27 \times 1000}{\pi \times 70}} = 22.35 \text{ mm } (f_c = 70 \text{ N/mm}^2)$$

Since the long column is also under torsion, let d_c be taken as 30 mm.

Now from the Johnson's formula:

$$\frac{F}{A} = f_c\left[1 - \frac{f_{yp}}{4c\pi^2 E} \cdot \left(\frac{L}{k}\right)^2\right]$$

Now $$A = \frac{\pi}{4} d_c^2 = 7.06 \times 10^{-4} \text{ m}^2, \; k = 0.25.\, d_c = 7.5 \times 10^{-3} \text{ m}$$

$$\therefore \quad \frac{L}{k} = \frac{1.5}{7.5 \times 10^{-3}} = 200, \text{ and } c = 4 \text{ for both ends fixed.}$$

$$\therefore \quad \frac{27 \times 1000}{7.06 \times 10^{-4}} = f_c\left[1 - \frac{210 \times 10^6}{4 \times 4 \times \pi^2 \times 210 \times 10^9} \times 200^2\right]$$

$$\therefore \quad \text{Stress induced in the screw, } f_c = \frac{270 \times 10^6}{7.06 \times 0.746} = 51.2 \text{ MN/m}^2$$

Let us assume the efficiency of screw and collar = 15 per cent.

Energy input by the operator in one revolution of handwheel = $\pi \times 1.0 \times 360 = 1130$ Nm

$\therefore$ Useful work = 1130 Nm × 0.15 = 169.5 Nm, $\therefore F \times p = 169.5$

$$\therefore \quad p = \frac{169.5 \times 1000}{27 \times 1000} = 6.28 \text{ mm, say 6 mm Normal Series}$$

$$\therefore \quad d_o = 30 + 6 = 36 \text{ mm}, \quad d_m = 33 \text{ mm}$$

$$\therefore \quad \tan \alpha = \frac{p}{\pi d_m} = \frac{6}{\pi \times 33} = 0.0579, \quad \therefore \alpha = 3^\circ - 19'$$

Now,
$$\eta = \frac{\tan \alpha (1 - \mu \tan \alpha)}{\tan \alpha + \mu + \mu_c \cdot \frac{D_c}{d_m}(1 - \mu \tan \alpha)}$$

Now
$$\mu = \mu_c = 0.15$$

Since the wheel is attached directly to screw, r_2 of the collar $= \frac{d_c}{2} = 15$ mm

Let diameter of boss of the wheel $= 2d_c = 60$ mm

$$\therefore \quad r_1 = 30 \text{ mm}$$

$$\therefore \quad \frac{D_c}{2} = \frac{2}{3} \cdot \frac{r_1^3 - r_2^3}{r_1^2 - r_2^2} = \frac{2}{3} \times \frac{30^3 - 15^3}{30^2 - 15^2} = \frac{2}{3} \times 35 = 23.3 \text{ mm}$$

or
$$r_c = \frac{1}{2}(r_1 + r_2) = 22.5 \text{ mm}$$

$$\therefore \quad \eta = \frac{0.0579(1 - 0.15 \times 0.0579)}{0.0579 + 0.15 + 0.15 \times \frac{45}{33}(1 - 0.15 \times 0.0579)} = \frac{0.0574}{0.41} = 14\%$$

Assumed value was 15%, not much difference. Also, the value of D_c may not be accurate.

Now turning moment applied to the wheel = 360 × 0.50 = 180 Nm

Frictional moment at the collar,

$$M_{t2} = \mu_c \cdot F \cdot \frac{D_c}{2}$$

$$= 0.15 \times 27 \times 1000 \times 0.0225 = 91 \text{ Nm}$$

Hence torsional moment on the screw,

$$M_{t1} = 180.0 - 91.0 = 89 \text{ Nm}$$

$$\therefore \quad f_s = \frac{16 M_{t1}}{\pi d_c^3} = \frac{16 \times 89 \times 1000}{\pi \times (30)^3} = 16.8 \text{ N/mm}^2$$

$$\therefore \quad f_{c\max} = \frac{1}{2}\left[f_c + \sqrt{f_c^2 + 4f_s^2}\right]$$

$$= \frac{1}{2}\left[51.2 + \sqrt{51.2^2 + 4 \times 16.8^2}\right]$$

$$= \frac{1}{2}(51.2 + 61.2) = 56.2 \text{ N/mm}^2 < 70 \text{ N/mm}^2, \quad \therefore \quad \text{Safe.}$$

23.8 OTHER TYPES OF SCREWS

23.8.1 Differential and Compound Screws

In some applications, a very slow motion of the screw is desired while in others a rapid movement is desired. Slow motion can be obtained with a single screw by adopting a small pitch and so a small helix angle. This would result in a weak thread. In the same manner, a rapid movement can be achieved by employing a single screw with a large pitch and so a large helix angle. This will result in difficulty of thread cutting. Both these problems can be overcome by using a combination of two screws in place of a single screw. Such an arrangement is shown in Fig. 23.7. The hub of the handwheel '*a*' has got coarse pitch outside threads and a finer pitch inside threads. The hands of the threads are same but these have different pitches. The pitch 'p_2' of the outer thread is more than the pitch 'p_1' of the inner thread. Therefore, when a complete revolution is given to the handwheel '*a*' the screw '*b*' will move a distance equal to the difference of pitches. Therefore, the distance moved is given as;

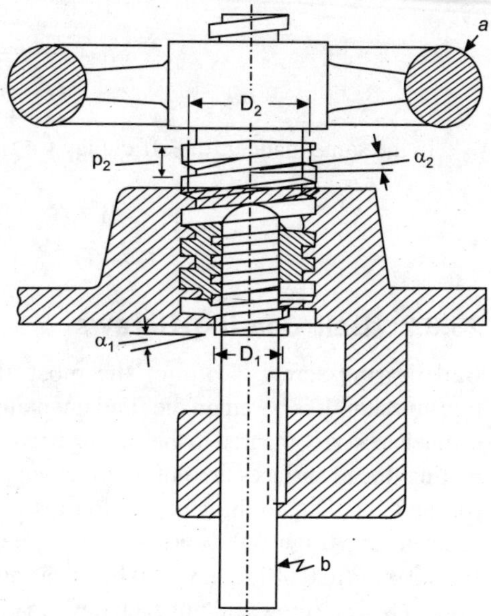

Fig. 23.7 *Differential screws*

$$h = p_2 - p_1 = \frac{1}{z_2} - \frac{1}{z_1}$$

where z_2 = number of threads of the coarser pitch
and z_1 = number of threads of the finer pitch.

Such a screw is known as "Differential Screw". If the threads are of different hands, the screw will move equal to the sum of the two pitches. Such a screw is known as "Compound Screw".

Efficiency of Differential Screws: When the coarse thread raises the load, it is lowered by the finer thread. Therefore, force applied at the coarse thread to raise a load F is $P_2 = F.\tan(\alpha_2 + \phi)$.

In the same manner, the force applied at the finer thread to lower the load F is $P_1 = F \tan \cdot (\phi - \alpha_1)$

where α_2 = helix angle of the coarse thread.
α_1 = helix angle of the finer thread.

$\therefore$ Torsional moment applied at the coarse thread, $M_{t2} = F \cdot \dfrac{D_2}{2} \tan(\alpha_2 + \phi)$

and Torsional moment applied at the finer thread, $M_{t1} = F \cdot \dfrac{D_1}{2} \tan(\phi - \alpha_1)$

where D_2 = mean diameter of the coarse thread.
and D_1 = mean diameter of the finer thread.

$\therefore$ Total torsional moment applied for the differential screw is,

$$M_t = M_{t2} + M_{t1} = F \cdot \frac{D_2}{2} \tan(\alpha_2 + \phi) + F \cdot \frac{D_1}{2} \tan(\phi - \alpha_1)$$

$$= F \cdot \frac{D_2}{2} \cdot \frac{\tan \alpha_2 + \mu}{1 - \mu \tan \alpha_2} + F \cdot \frac{D_1}{2} \cdot \frac{\mu - \tan \alpha_1}{1 + \mu \tan \alpha_2}$$

Now the torsional moment needed when friction is zero, *i.e.*, $\mu = 0$

$$M_{ti} = F \cdot \frac{D_2}{2} \cdot \tan \alpha_2 - F \cdot \frac{D_1}{2} \cdot \tan \alpha_1$$

$$\therefore \quad \eta = \frac{M_{ti}}{M_t} = \frac{D_2 \tan \alpha_2 - D_1 \tan \alpha_1}{D_2 \cdot \dfrac{\tan \alpha_2 + \mu}{1 - \mu \tan \alpha_2} - D_1 \cdot \dfrac{\tan \alpha_1 - \mu}{1 + \mu \tan \alpha_1}}$$

Note: The coefficient of friction in the coarse threads and finer threads is assumed to be the same. If they are different, then

$$\therefore \qquad \eta = \frac{D_2 \tan \alpha_2 - D_1 \tan \alpha_1}{D_2 \cdot \dfrac{\tan \alpha_2 + \mu_2}{1 - \mu_2 \tan \alpha_2} - D_1 \cdot \dfrac{\tan \alpha_1 - \mu_1}{1 + \mu_1 \tan \alpha_1}}$$

In the same manner, the efficiency for the compound screw will be:

$$\eta = \frac{D_2 \tan \alpha_2 - D_1 \tan \alpha_1}{D_2 \cdot \dfrac{\tan \alpha_2 + \mu_2}{1 - \mu_2 \tan \alpha_2} + D_1 \cdot \dfrac{\tan \alpha_1 + \mu_1}{1 - \mu_1 \tan \alpha_1}}$$

28.8.2 Ball Bearing Screws

Ball bearing screw is one of the most efficient methods of power transmission. It is based on the fundamental principle that rolling friction is much less as compared to the sliding friction. In ball bearing screws, the mating thread surfaces are not in direct engagement but are separated by rolling balls. Due to this, the friction losses are very less and operating efficiencies as high as 90% are achieved. The coefficient of friction ranges from 0.0003 to 0.0015. A typical ball bearing screw assembly is shown in Fig. 23.8. The screw and nut each have mating helical ball races. Within the nut, the length of this race is filled with polished steel balls. A tubular ball guide on the nut interrupts the path of the balls, deflects them from the helical race and guides them diagonally across the outside of the nut and back into the helical race. This construction results in a closed circuit through which the rolling balls, as the medium of force transfer, recirculate continually as the screw and nut are rotated relative to each other. Such an arrangement was first developed for steering gears of automobiles, but now its applications include a wide range of mechanical functions either as a power transmitting unit or as a linear or angular positioning device. Typical applications include power actuators, landing gear of airplanes.

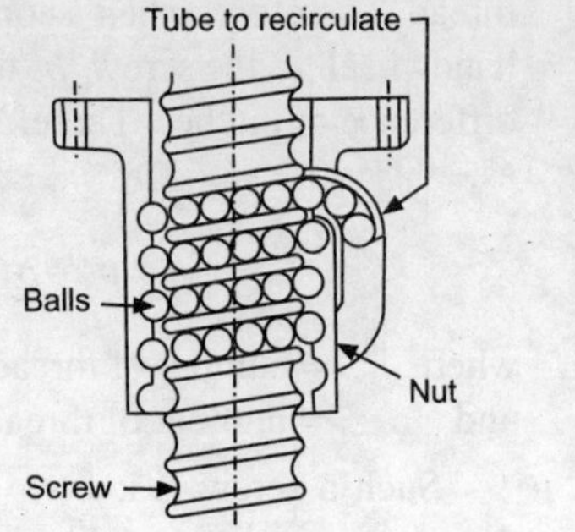

Fig. 23.8 *Ball bearing screw assembly*

The conventional function of the ball bearing screw is to convert rotary to linear motion. The nut generally serves as the output member and is driven along the longitudinal axis of the screw as the screw rotates. Other operating arrangements, i.e., screw acting as the output member are also possible. Figure 23.9 shows a portion of the assembly press, illustrating the member of the screw mechanisms, applied to the power train of the cross head of a press. An important feature of the ball bearing screw is its "reversibility" characteristic which permits adaptation of the device for linear to rotary motion conversion. The efficiency of such a system will be less than that of the other system.

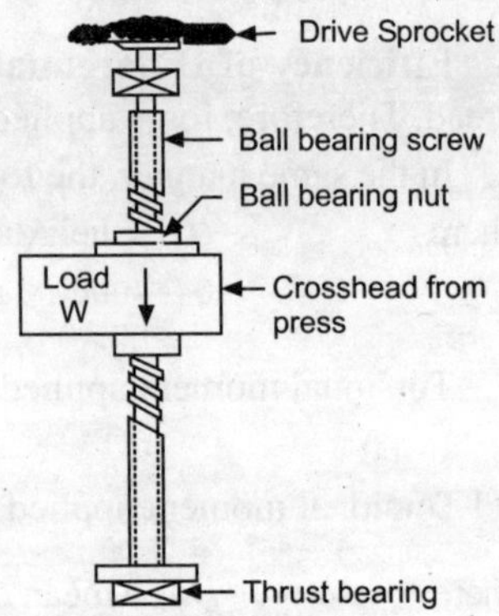

Fig. 23.9 *Ball bearing screw machinism*

An important term in ball bearing screw is "Ball circle diameter". It is defined as the diameter of a circle whose radius is the distance from the screw axis to the centre of one of the active bearing balls. This dimension is always slightly larger than the outside diameter of the screw thread. Screw threads are generally produced by two methods:

1. Machine Grinding
2. Rolling.

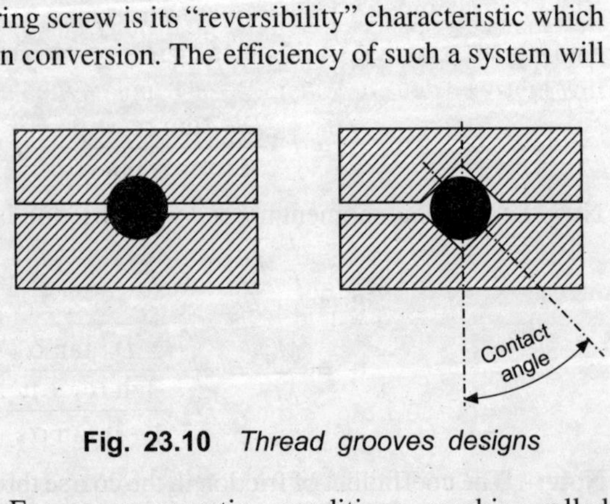

Fig. 23.10 *Thread grooves designs*

Rolled threads are suitable for most general purpose applications where the working conditions are normal. For severe operating conditions, machine rolled threads are preferred. Standard sizes of the rolled threads are given in Table 23.6.

Two thread groove designs have been developed for ball bearing screws (Fig. 23.10). (1) circular arc design, (2) gothic arc design.

Gothic arc design is better because if any foreign matter enters the effective ball groove, it will be pushed aside by the balls into the space provided by the divergence arcs.

Ball nuts are essentially of two types:

1. Single Circuit.
2. Multi Circuit.

Table 23.6 Standard sizes of rolled threads

Ball circle diameter mm	Ball diameter mm	Thread lead mm
9.525	1.588	3.175
16.027	3.175	5.08
25.40	3.97	6.350
29.74	7.144	10.49
38.10	8.73	12.03
57.15	9.525	12.7
76.20	12.7	16.76

To increase the load of the ball bearing screw, either the ball diameter, screw diameter or number of bearing balls should be increased. The last method is usually adopted. One way to increase the number of balls is to increase the number of turns in the single circuit design. This may lead to a jamming tendency. Due to this, more circuits are provided, the total load being divided equally. There may be 1½, 2½, or 3½ active turns of ball.

Materials: Chrome steel is the standard ball material, and the nut and screw are usually heat treated alloy steels. Stainless steels are also commonly employed. Also a number of special ferrous and non-ferrous alloys and heat treatments have been used to meet severe or unusual operating requirements. For providing high wear resistance, stellite balls have been used. To reduce noise, moulded nylon balls can be employed in combination with a steel screw and a nut. However, the load capacity will be reduced.

All screws 12.5 mm ball circle diameter are supplied with solid centre; however, the larger sizes can be obtained with centre bores. The centre bore is usually 0.50 to 0.75 times the ball circle diameter.

Comparison with Acme threads: The ball bearing screws have the following advantages over Acme threads:

1. At least 90% efficiency, as compared with 15% to 20% efficiency of Acme threads.
2. Require less than 1/3 as much torsional moment as Acme screws for same amount of linear output.
3. Save substantially on cost, size and weight of motors, gearboxes and auxiliary equipment.
4. Far less wear, lesser maintenance, longer life.
5. Precision positioning, free play can be virtually eliminated where necessary.
6. Operate dependably with or without lubrication at temperatures from about –25° C to +80° C.

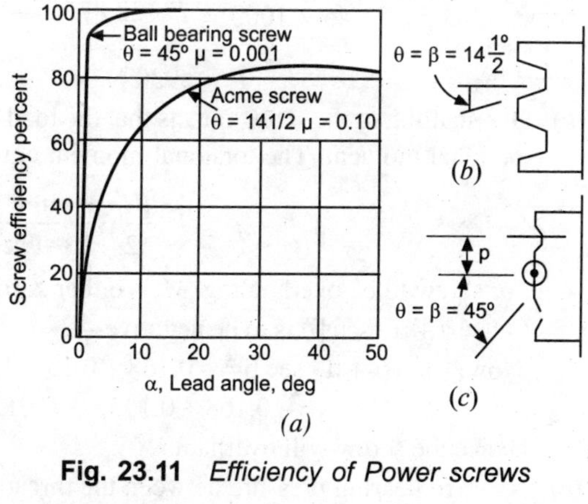

Fig. 23.11 *Efficiency of Power screws*

A comparison of efficiencies of Acme thread and Ball bearing screw is shown in Fig. 23.11.

The same formulas on torsional moment and efficiency can be applied for ball bearing screws as with Acme or Trapezoidal thread, the only difference being the included angle,

$$2\beta = 90° \text{ for ball bearing screws.}$$

The torsional moment equation is,

$$M_t = F \cdot \frac{d_m}{2} \cdot \frac{\cos\theta \sin\alpha + \mu\cos\alpha}{\cos\theta\cos\alpha - \mu\sin\alpha}, \text{ neglecting collar friction,}$$

where d_m = diameter of mean helix

θ = Angle between the normal force on the ball and plane tangent to pitch cylinder of the screw.

$\cong \beta$.

Efficiency is given as :

$$\eta = \tan\alpha \cdot \frac{\cos\theta\cos\alpha - \mu\sin\alpha}{\cos\theta\sin\alpha + \mu\cos\alpha}$$

Example 23.8. The screw shown in Fig. 23.12 is operated by a torsional moment applied at the lower end. The nut is loaded and prevented from turning by guides. The outside diameter of the screw is 50 mm and the thread is Acme triple start. The coefficient of friction of the threads is 0.15. Assume the friction in the ball bearing as negligible. If the torsional moment M_t is 45 Nm,

(*a*) Determine the load which could be raised.

(*b*) Would the screw be overhauling?

(*c*) Determine the average bearing pressure between the screw and nut thread surfaces.

Solution: We know that,

$$M_t = F \cdot \frac{d_m}{2} \cdot \frac{\tan\alpha + \mu\sec\beta}{1 - \mu\sec\beta\tan\alpha} + \mu_c \cdot F \cdot \frac{D_c}{2}, \text{ for Acme threads.}$$

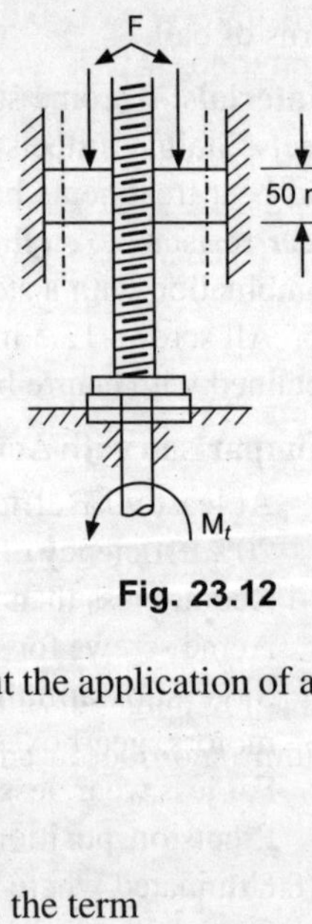

Fig. 23.12

From Table 23.2, pitch of the threads = 8 mm

$\therefore$ Thread depth $= \frac{p}{2} + 0.25 \text{ mm} = 4 + 0.25 = 4.25$ mm

$$\therefore \quad r_m = 25 - \frac{4.25}{2} = 25 - 2.125 = 22.875 \text{ mm}$$

$$\text{Now } \tan\alpha = \frac{\text{lead}}{2\pi r_m} = \frac{3\times 8}{2\pi \times 22.875} = 0.168$$

$$\beta = 14.5°, \mu = 0.15, \mu_c = 0$$

$$\therefore \quad 45 \times 1000 = F \times 22.875 \times \frac{0.168 + 0.15 \times 1.0329}{1 - (0.15)(1.0329)(0.168)}$$

$$\therefore \quad F = 5930 \text{ N}$$

(*b*) Overhauling of the screw means that the load will start moving downward without the application of any torsional moment. The torsional moment equation will become

$$M_t = F \cdot \frac{d_m}{2} \cdot \frac{-\tan\alpha + \mu\sec\beta}{1 + \mu\sec\beta\tan\alpha}$$

For screw to be overhauling. M_t is either zero or negative. For M_t to be negative, the term $(-\tan\alpha + \mu \cdot \sec\beta)$ is to be negative.

Now, $-\tan\alpha + \mu \cdot \sec\beta = -0.168 + 0.15 \times 1.0329$

$= -0.168 + 0.155 = -0.013$

Hence the screw will overhaul.

(*c*) Average bearing pressure between the thread surfaces is given as:

$$p_b = \frac{4F}{\pi(d_o^2 - d_c^2) \times z} = \frac{4F}{\pi(d_o - d_c)(d_o + d_c) \times z} = \frac{F}{2\pi z r_m h},$$

where h = depth = 4.25 mm

z = number of threads in contact

$$= \frac{\text{nut length}}{\text{pitch}} = \frac{50}{8.0} = 6.25$$

$$p_b = \frac{5930}{2\pi \times 6.25 \times 22.875 \times 4.25} = 1.552 \text{ N/mm}^2$$

Example 23.9. Estimate the maximum Wrench torsional moment which can be applied in tightening a 19 mm bolt if the shear stress in the body of the bolt is not to exceed 140 N/mm².

Outside diameter of the bolt = 19 mm

Root diameter = 15.75 mm

Included angle of thread section = 60°

Threads per cm = 4

Effective friction radius under nut, $\frac{D_c}{2}$ = 12.5 mm

Thread and collar friction coefficient = 0.10.

Solution: The maximum shear stress in the body of the bolt is given as:

$$(f_s)_{max} = \sqrt{\left(\frac{1}{2} f_t\right)^2 + f_s^2}$$

Now, $$f_t = \text{direct tensile stress} = \frac{F}{\pi r_0^2} = \frac{F}{\pi \times (7.875)^2} = 0.00513 \text{ F, N/mm}^2$$

$$f_s = \frac{M_{t1} \cdot r_c}{J} = \frac{M_{t1} \cdot r_c}{\frac{1}{2} \pi r_c^4} = \frac{2M_{t1}}{\pi (7.875)^3} = 0.013\, M_{t1} \text{ N/mm}^2$$

where M_{t1} = screw torsional moment = wrench torsional moment – collar friction torsional moment

But M_t = Wrench torsional moment = Screw torsional moment + collar friction torsional moment.

$$= F\left[r_m\left(\frac{\tan\alpha + \mu \sec\beta}{1 - \mu \sec\beta \tan\alpha}\right) + \frac{D_c}{2} \cdot \mu_c\right]$$

Now $$\beta = 30°,\ r_m = \frac{19.00 + 15.75}{4} = 8.6875 \text{ mm}$$

and $$\tan\alpha = \frac{\text{pitch}}{2\pi r_m} = \frac{2.5}{2\pi \times 8.6875} = 0.0468$$

$$\mu = \mu_c = 0.10;\ \frac{D_c}{2} = 12.5 \text{ mm}$$

$$\therefore \quad M_t = F\left[8.6875\left(\frac{0.0468 + 0.10 \times 1.15}{1 - (0.10)(1.15)(0.0468)}\right) + 12.5 \times 0.10\right]$$

$$M_t = 1.42 \text{ F} + 1.25 \text{ F} = 2.67 \text{ F}$$

$$\therefore \quad M_{t1} = 1.42 \text{ F}$$

$$\therefore \quad f_s = 0.013 \times 1.42 \text{ F} = 0.01846 \text{ F N/mm}^2$$

$$\therefore \quad 140 = \sqrt{(\tfrac{1}{2}\times 0.00513F)^2 + (0.01846F)^2} = 0.00316\text{ F}$$

$$\therefore \quad F = \frac{140}{0.00316} = 44.30 \text{ kN}$$

$\therefore$ Wrench torsional moment = 2.67 × 44.30 = 118.20 kN mm

= 118.20 Nm

Example 23.10. The following data apply to the C-clamp shown in Fig. 23.13;

Trapezoidal threads

5 threads per centimeter (single-threaded)

Outside diameter = 12 mm

Coefficient of thread friction = 0.12

Coefficient of collar friction = 0.25

Mean collar radius = 6.4 mm

Load = 4500 N

Force exerted by the operator at the end of lever = 90 N

Determine:

(*a*) length of the handle, *L*

(*b*) maximum shear stress in the body of the screw and where does this exist?

(*c*) bearing pressure on the threads.

Solution:

$$M_t = F\cdot\left[r_m\left(\frac{\tan\alpha + \mu.\sec\beta}{1-\mu\sec\beta\tan\alpha}\right) + \mu_c\cdot\frac{D_c}{2}\right]$$

Now, d_o = 12 mm, pitch = 2 mm

$$\therefore \quad h = \text{depth} = \frac{p}{2} + 0.25 \text{ mm} = 1.25 \text{ mm}$$

$$\therefore \quad r_m = 6 - \frac{1}{2}\times 1.25 = 5.375 \text{ mm}$$

$$\tan\alpha = \frac{p}{2\pi r_m} = \frac{2}{2\pi\times 5.375} = 0.0592$$

$$\therefore \quad M_t = \left[5.375\left(\frac{0.0592 + 0.12\times 1.0353}{1-(0.12)\,(1.0353)\,(0.0592)}\right) + 0.25\times 6.4\right]\times 4500$$

$= 4500\,(0.993 + 1.60) =$ **11668.5 Nmm = 11.6685 Nm**

$$\therefore \quad L = \frac{11.6685}{90} = \mathbf{0.13\ m}$$

(*b*) Consider two sections A – A and B – B. Section A – A is subjected to torsional moment and bending and section B – B, which is just below the nut is subjected to torsional moment and direct compressive load.

Section A – A:

Shear stress, $f_s = \dfrac{M_t\cdot r}{J}$

$$r = r_c = \frac{1}{2}(12 - 2 - 0.5) = \frac{1}{2}\times 9.5 = 4.75 \text{ mm}$$

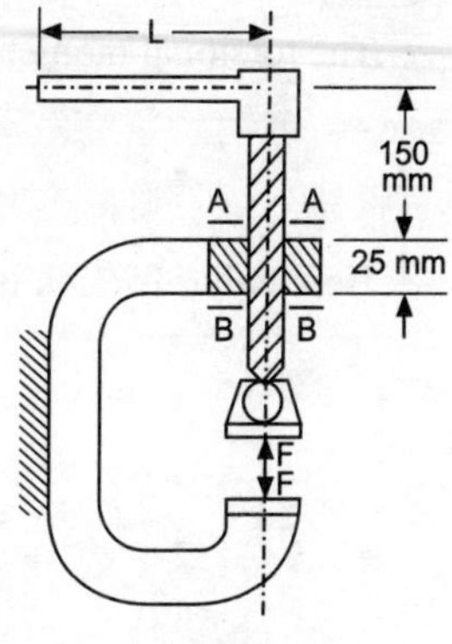

Fig. 23.13

$$J = \frac{1}{2}\pi r_c^4 = \frac{1}{2}\pi(4.75)^4 = 797 \text{ mm}^4$$

$$\therefore \quad f_s = \frac{11.67 \times 1000 \times 4.75}{797} = 69.4 \text{ N/mm}^2$$

Bending stress, $f_t = \dfrac{M_b \cdot c}{l}$

$$M_b = 90 \times 0.15 = 13.50 \text{ Nm}$$

$$c = r_c = 4.75$$

$$l = \frac{1}{4}\pi r_c^4 = 398.5 \text{ mm}^4$$

$$\therefore \quad f_t = \frac{13.50 \times 1000 \times 4.75}{398.5} = 160.8 \text{ N/mm}^2$$

$$\therefore \quad (f_s)_{max} = \sqrt{(\tfrac{1}{2}f_t)^2 + f_s^2} = \sqrt{(80.4)^2 + (69.4)^2} = \mathbf{106.58 \text{ N/mm}^2}$$

Solution B – B: This section is subjected to the collar friction torsional moment.

$$\therefore \text{ Shear stress, } f_s = \frac{\left(F \cdot \mu_c \cdot \frac{D_c}{2}\right) \cdot r_c}{J} = \frac{4500 \times 1.6 \times 4.75}{797} = 42.5 \text{ N/mm}^2$$

$$\text{Direct compressive stress, } f_c = \frac{F}{A} = \frac{4500}{\pi \times (4.75)^2} = 63.5 \text{ N/mm}^2$$

$$\therefore \quad (f_s)_{max} = \sqrt{(31.75)^2 + (42.5)^2} = 53.05 \text{ N/mm}^2$$

$\therefore$ Maximum shear stress occurs at section $A – A$.

(*c*)

$$p_b = \frac{F}{2\pi z r_m h}$$

$$z = \frac{2.5}{0.2} = 12.5$$

$$\therefore \quad p_b = \frac{4500}{2\pi \times 12.5 \times 1.25 \times 5.375} = \mathbf{8.52 \text{ N/mm}^2}.$$

Example 23.11. A screw jack is to be made as shown in Fig. 23.14. Neither screw rotates. Outside screw diameter is 50 mm. Thread is square, single thread, and coefficient of thread friction is 0.14. Determine

(*a*) Efficiency of the screw jack.

(*b*) The load that can be lifted if the shear stress in the bodies of the screws in limited to 28 N/mm².

Solution: The arrangement of the screw involves the differential screw action. In one rotation of the nut, the lift of the load equals the difference of the pitches of the two screws.

(*a*) Now efficiency of the screw jack is,

$$\eta = \frac{\text{Output work per turn}}{\text{Input work per turn}}$$

Output work per turn $= F \times$ (lead or pitch of upper screw – pitch or lead of the lower screw)

$$= F\,(16 - 12) = 4\,F \text{ N mm}$$

The torsional moment required to turn the nut relative to the upper screw is,

$$M_{t1} = F\left[r_m\left(\frac{\tan\alpha + \mu}{1 - \mu\tan\alpha}\right)\right]$$

Here collar friction is neglected since the screw does not rotate to which the load is applied.

$$r_m = r_o - \frac{\text{thread depth}}{2} = 25 - 4 = 21 \text{ mm}$$

$$\tan\alpha = \frac{p}{2\pi r_m} = \frac{16}{2\pi\times 21} = 0.121$$

$$\therefore \quad M_{t1} = F\left[21\left(\frac{0.121 + 0.15}{1 - 0.15\times 0.121}\right)\right] = 7.72 \text{ F Nmm}$$

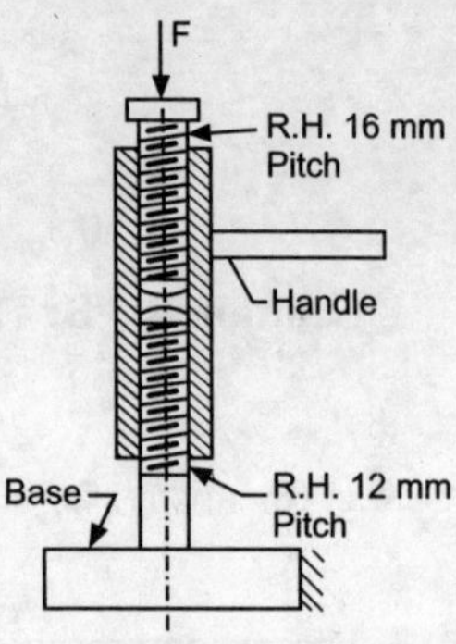

Fig. 23.14

For the lower screw,

$$M_{t2} = F\left[r_m\left(\frac{-\tan\alpha + \mu}{1 + \mu\tan\alpha}\right)\right]$$

$$r_m = 25 - 3 = 22 \text{ mm}$$

$$\tan\alpha = \frac{12}{2\pi\times 22} = 0.0968$$

$$\therefore \quad M_{t2} = F\left[22\left(\frac{-0.0968 + 0.15}{1 + 0.15\times 0.0968}\right)\right] = 3.37 \text{ F Nmm}$$

$\therefore$ Total input torsional moment, $M_t = M_{t1} + M_{t2} = 7.72\text{ F} + 3.37\text{F} = 11.09\text{ F}$

$$\therefore \quad \eta = \frac{4F}{2\pi\times 11.09F} = 5.76\%$$

(*b*) The upper screw is critical

Direct compressive stress, $f_c = \dfrac{F}{\pi\times r_c^2}$

Now $\quad r_c = r_o - \text{thread depth} = 25 - 8 = 17 \text{ mm}$

$$\therefore \quad f_c = \frac{F}{\pi\times(17)^2} = 0.0011 \text{ F N/mm}^2$$

$$f_s = \frac{M_t\cdot r}{J}, r = r_c = 17 \text{ mm}$$

$$J = \frac{1}{2}\pi r_c^4 = 131.2\times 10^3 \text{ mm}^4; M_t = M_{t1} = 7.72 \text{ F}$$

$$\therefore \quad f_s = \frac{7.72\ F\times 17}{131.2\times 10^3} = 0.001 \text{ F N/mm}^2$$

$$\therefore \quad (f_s)_{\text{max.}} = \sqrt{(\tfrac{1}{2} f_c)^2 + f_s^2} = 28 \text{ N/mm}^2$$

$\therefore \quad 0.0014 \text{ F} = 28, \quad \therefore F = 24.56 \text{ kN}.$

EXERCISES

1. A 52 mm square thread steel screw is used in a common screw jack. There are 1.25 threads per cm. The body of the jack is made of C.I. The coefficient of friction on the threads is 0.15. The thrust collar at the top has an inside diameter of 38 mm and an outside diameter of 76 mm. The coefficient of friction for the collar is 0.25. The load to be lifted by the jack is 20 kN.
   (*a*) Determine the efficiency of screw alone.
   (*b*) Determine the efficiency of the screw and the thrust collar.
   (*c*) Determine the pull necessary on the end of a 0.60 m lever to operate this screw Jack.
2. A sluice gate weighing 600 kN is raised and lowered by means of two 70 mm square thread screws. The screws are operated by an electric motor running at 600 rev/min. A ball thrust bearing is used, reducing the apparent friction coefficient to 0.003 on a 50 mm radius. Bronze nuts and fair lubrication are used. It the gate must be raised at the rate of 10 mm/s. Determine:
   (*a*) The number of revolutions per minute of the screws.
   (*b*) The power of the motor required to raise the load, assuming a mechanical efficiency of 0.85 for the speed reduction mechanism.
   (*c*) The power requirement to lower the gate.
3. A 70 mm screw with square threads is used in a press. It has a maximum unsupported length of 0.45 m. The screw is of C-25 steel and nut is of phosphor bronze, determine:
   (*a*) The safe capacity of the press.
   (*b*) The proper length of the nut.
   (*c*) The necessary torsional moment, assuming 0.13 as the friction coefficient in the threads and 0.15 at the thrust collar, which has an outside diameter of 90 mm and an inside diameter of 26 mm.
4. A screw having a double square thread is to raise a load of 100 kN at a speed of 1/60 m/s. The following data apply:

Outside diameter of thread	: 75 mm
Pitch	: 25 mm
Mean collar diameter	: 127 mm
Coefficient of thread friction	: 0.10
Coefficient of collar friction	: 0.12

   Determine:
   (*a*) The power required to drive the screw.
   (*b*) The efficiency of the screw.
5. (*a*) Show that if the power screw is to be self locking, its efficiency can never be more than 50%.
   (*b*) A power screw is capable of exerting a force of 15 kN and is required to have a linear movement of 0.30 m. Establish a satisfactory combination of:
      (*i*) Screw diameter and number of threads per cm.
      (*ii*) Number of start on the screw.
      (*iii*) Thread form
      (*iv*) Length and outside diameter of the nut.
   Assume suitable materials for the screw and nut and their permissible values of stresses.
6. The two screws that support the moving head of a 275 kN hydraulic testing machine have 58 mm trapezoidal threads. Each one exerts half the load in tension. The nuts on these screws are used only for adjustment and do not turn under load (the force is supplied by oil pressure).
   (*a*) Determine the tensile stresses in these screws under the maximum load of 275 kN.
   (*b*) Determine the number of threads required in the nuts if the permissible bearing pressure is 14 N/mm^2.
7. A square thread screw, 50 mm in diameter is used to exert a force of 110 kN in a shaft straightening press. The maximum unsupported length of the screw is 0.40 m and the material is C-25 steel annealed.
   (*a*) What is the equivalent compressive stress in the screw? Is this a satisfactory value?
   (*b*) What torsional moment is necessary to turn a screw against the load for $\mu = 0.15$?
   (*c*) What is the efficiency of the screw?
   (*d*) What torsional moment is necessary to lower the load?

8. Select and calculate a power screw for the travel of a radial drilling machine. The maximum force acting on the screw = 36.40 kN arises as the arm is lifted; it is composed of the arm weight and the frictional resistance when the arm travels along the column. The thread pitch is 8 mm and the thread is single start. During the handling of the machine with a sling strapped round the arm, the arm lifting screw is loaded for compression with the force of 30 kN by the weight of column and the foundation plate. The maximum length of screw from the thrust bearing to the nut is 1500 mm.
9. The capstan and square thread combination, shown in Fig. 23.15 raises and lowers a gate weighing 20 kN against a maximum frictional resistance between the gate and its guides of 4.5 kN induced by water pressure against the

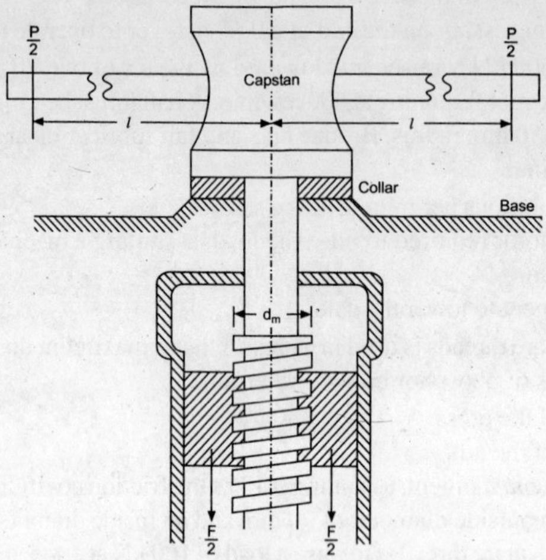

Fig. 23.15 *Capstan and Screw*

gate when in its lowest position. Determine the maximum force to be exerted at the end of the capstan bar for raising the load. Check the combination for self locking quality, and determine the force and its direction necessary to lower the gate. Also determine the number of threads required in the nut for an allowable bearing pressure in the threads of 7 N/mm^2 of the projected area. The following data are available:
Length of leverage, l = 1.2 m
Material of screw, steel : of nut, brass : of washer, steel
Diameter of washer (collar): inside, 50 mm, : outside 200 mm
Screw : single, square thread : pitch 8 mm; shank diameter, 50 mm
Assumed coefficient of friction : at washer, 0.15; at thread, 0.13.
10. Design a screw jack for lifting a load of 50 kN through a height of 0.25 m. The screw is made of steel for which Rankine's constant are 330 N/mm^2 and $\dfrac{1}{7500}$ for column pin joined at both ends. Use a factor of safety of 4. The nut is made of gun metal for which f_s = 25 N/mm^2. Bearing pressure on the threads is limited to 10 N/mm^2. The nut is fixed in C.I. housing and the screw is turned in the nut by using a tommy bar inserted in suitable holes in the head of the screw. The coefficient of friction of the threads and collar may be taken as 0.12.
11. Design a simple screw jack for lifting a load of 10 kN and having a maximum lift of 0.25 m. The elastic limit of the screw material may be taken as 240 N/mm^2 in tension and compression and 150 N/mm^2 in shear. The material for nut is to be phosphor bronze for which the elastic strengths may be taken as 128 N/mm^2 in tension and 115 N/mm^2 in compression and 105 N/mm^2 in shear. The bearing pressure between the threads of screw and nut is 17.5 N/mm^2. The coefficient of friction between nut and screw is 0.13.
12. A screw jack suitable to lift a maximum load of 50 kN through a height of 200 mm with a ground clearance of 300 mm, is to be designed with the following data:
Allowable stress for screw,

Tensile stress	=	90 N/mm^2
Compressive stress	=	110 N/mm^2
Shearing stress	=	60 N/mm^2

Allowable stress for nut,

Tensile stress = 60 N/mm^2

Compressive stress = 50 N/mm^2

Shearing stress = 45 N/mm^2.

Allowable bearing pressure between nut and screw is 16 N/mm^2, coefficient of friction of threads and collar may be taken as 0.14. Allowable bending stress for M.S. handle is 100 N/mm^2.

Design : (*i*) Screw, (*ii*) Nut, (*iii*) Handle.

13. Design the screw of a jack. Thread type : self locking, buttress. Load carrying capacity (axial load) is 60 kN. Screw length : 500 mm, screw material : 0.35 C steel. Nut material : anti-friction grey cast iron. The screw rests on a ball thrust bearing.
14. What is the function of a power screw?
15. What are "Translating" and "Leading" screws?
16. What are the requirements of a power screw?
17. Give the advantages and limitations of power screws.
18. Write on screws used for power screws.
19. Compare square and trapezoidal threads.
20. What are multiple threads?
21. What is "self-locking" of power screws?
22. How is it related to helix angle of the screws?
23. What is the usual value of helix angle for power screws?
24. Why is the nut of a power screw made of soft material?
25. In power screws, Acme threads are preferred over square threads, why?
26. What is the value of helix angle for maximum efficiency?
27. Write the expression for maximum efficiency of a screw in terms of angle of friction
28. The condition for self locking of a power screw is:

    (a) $\alpha > \phi$ (b) $\alpha < \phi$

    (c) $\alpha = \phi$ (d) None of above.

    α = helix angle; ϕ = angle of friction.
29. What are differential and compound screws?
30. Write on "Ball bearing screws".
31. The efficiency of a screw is given as, for lifting the load

    (a) $\dfrac{\tan\phi}{\tan\alpha}$ (b) $\dfrac{\tan\alpha}{\tan(\phi-\alpha)}$

    (c) $\dfrac{\tan\alpha}{\tan(\alpha+\phi)}$ (d) $\dfrac{\tan(\phi-\alpha)}{\tan\alpha}$
32. The efficiency of square threads in power screws is

    (a) Minimum (b) Maximum

    (c) Unpredictable (d) Less than the acme thread
33. In power screws, the buttress thread can take up load in:

    (a) one direction (b) two directions

    (c) any direction (d) not sure.

Section IV

PRESSURE VESSELS

PRESSURE VESSELS

General: Pressure vessels are those vessels which operate under pressure. Such vessels find wide applications in chemical industry and also in the water, steam, gas and air supply systems of industrial enterprises. The term "Pressure Vessels" refers to those reservoirs and apparatus which work under internal or external pressures greater than 0.7 atm. gauge.

Classification of Pressure Vessels: The pressure vessels may be classified into different categories as follows:

(1) **According to the thickness of the Wall.** According to this criteria, the pressure vessels are classified as:

(*a*) Thin-walled pressure vessels (*b*) Thick-walled pressure vessels.

In the thin-walled pressure vessels, the ratio of the wall thickness to the inner diameter of the vessel is less than 0.07 and in thick walled pressure vessels this ratio is greater than 0.07. Thin walled vessels are used in boiler shells, pressure tanks and pipes, pressure cylinders in small presses and in low pressure processing equipment. Thick walled vessels find use in high pressure cylinders, tanks, gun barrels, etc.

(2) **According to the geometric shape.** Based on this criteria the pressure vessels are classified as:

(*a*) cylindrical, Fig. IV-1*a*, *b*
(*b*) conical, Fig. IV-1*c*
(*c*) cylindrical vessels with on or two cones, Fig. IV-1*d* and 1*e*.

Type shown in Fig. IV-1*a*, finds applications in boiler drums, chemical vessels and gas tanks and that shown in fig. IV-1*b*, is used for heat exchangers and chemical vessels. Type IV-1*c* is used for chemical vessels and digestors, i.e. cooking boilers and types IV-1*d* and 1*e* find applications in chemical vessels, cyclone dust separators and water purification vessels.

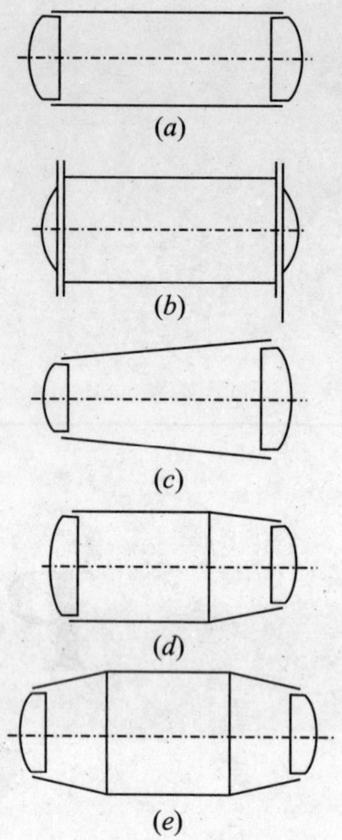

(*a*) (*b*) (*c*) (*d*) (*e*)

Fig. IV-1 *Shapes of pressure vessels.*

(3) **According to the end construction:** Depending upon the end construction, the pressure vessels may be:

(*a*) Open ended, e.g., a piston cylinder arrangement of a Press.
(*b*) Closed ended, e.g., a tank.

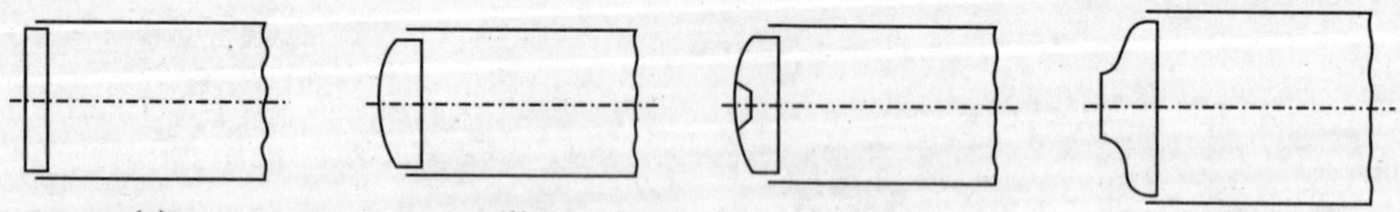

(*a*) (*b*) (*c*) (*d*)

Fig. IV-2 *End head shapes*

According to the end head shape, the vessels can be classified as:

(*i*) Flat head, Fig. IV-2*a*.
(*ii*) Convex head, Fig. IV-2*b*
(*iii*) Convex head with a man hole, Fig. IV-2*c*.

The heads are fastened to the shell either permanently, i.e., by riveting or welding; or temporarily by bolted joints.

(4) **According to the direction of force acting on the walls of the vessels.** Based on this criteria, the pressure vessels may be classified as:

(*a*) those subjected to internal pressures.
(*b*) those subjected to external pressures.

(5) **According to the material.** The material may be brittle or ductile and the criteria of design and failure of the pressure vessels will be chosen accordingly. The vessels are made of sheet steel by riveting and welding, of cast iron by casting and of non-ferrous alloys by riveting or by casting. The vessels may be constructed of stainless steel or a corrosion resisting lining-may be provided for vessels used in chemical apparatus or refinement apparatus.

(6) **According to the type of service.** According to this, the vessels are divided depending upon the pressure, temperature and environment conditions. Choice of the material and design stress will be influenced according to whether the temperature is low or high. The range of temperature will also decide the basis for design. If the temperature is normal, the elastic state of the material will be design basis, whereas it will be creep if the temperature is more than about 340°C

CHAPTER 24

Cylinders, Pipes and Tubes

24.1 CYLINDERS

24.1.1 Cylinders Subjected to Internal Pressure

1. **Thin Cylinders.** When a cylinder is subjected to an internal pressure, two types of tensile stresses are set up in the wall thickness of the cylinder:

(*a*) Circumferential or hoop stress.
(*b*) Longitudinal stress.

The relationship between these stresses, the internal pressure, the internal diameter of the cylinder and the wall thickness is arrived at by making the following assumptions:

(*i*) The stress induced in the cylinder wall is uniform throughout its thickness since the thickness is very small as compared to the cylinder diameter.
(*ii*) Effect of the curvature of the cylinder wall is neglected.
(*iii*) Restraining action of the heads at the ends of the cylinder is neglected.

The relationship for the hoop stress is:

$$f_1 = \frac{pd}{2t} \text{ N/mm}^2 \qquad \text{... (24.1)}$$

and for longitudinal stress:

$$f_2 = \frac{pd}{4t} \text{ N/mm}^2 \qquad \text{... (24.2)}$$

where f_1 = hoop stress
f_2 = longitudinal stress
p = internal pressure, N/mm^2
d = internal diameter of the cylinder, mm
t = cylinder wall thickness, mm.

It is clear from the above equations, that the stress in the cylinder wall parallel to its axis is half the circumferential, hoop or the tangential stress. Hence, the cylinder is twice strong in this direction as compared to the other direction. Therefore, the longitudinal joint in the boiler shell is made stronger than that in a ring joint.

When there is a joint or seam in the cylinder, the efficiency of the joint should be taken into account, therefore,

$$f_1 = \frac{pd}{2t \cdot \eta} \qquad \text{... (24.3)}$$

and

$$f_2 = \frac{pd}{4t \cdot \eta} \qquad \text{... (24.4)}$$

where η = efficiency of the joint.

For ductile material f_1 may be taken as equal to $0.8 f_{ypt}$ and for brittle materials, f_1 may be taken as equal to $1/8 f_{ut}$.

For the design of the boilers, the minimum thickness of boiler plate should conform to that given by the 'Boiler Code', as given in Table 24.1. If the calculated thickness comes out to be less than that given in the table, then the table value is to be adopted. On the other hand, if the computed value is greater, then this is to be taken.

Table 24.1 Minimum plate thickness

Boiler Diameter, m	Plate Thickness mm
≤ 0.90	6.35
0.94 to 1.37	8.00
1.4 to 1.80	9.525
> 1.80	12.70

According to the code, a factor of safety of 5 should be used and the steel of the plates should have minimum the following ultimate stresses:

In tension, = 385 N/mm^2
In compression = 665 N/mm^2
In shear = 308 N/mm^2.

Engine cylinders are usually made of C.I. due to ease in casting. Their thickness is given as below:

$$t = \frac{pd}{2f_1} + 6 \text{ to } 12 \text{ mm for reboring}$$

f_1 is taken as 8.75 N/mm^2.

For thin spherical shells, subjected to internal pressure, the stress induced in the wall is the same as the stress in a cylinder parallel to its axis. So, for a spherical shell,

$$t = \frac{pd}{4f_2}$$

2. **Thick Cylinders.** In the case of thick cylinders, the stress across the wall thickness can't be assumed to be uniform as in the case of thin cylinders. The stress will be maximum at the inner radius and minimum at the outer radius.

The analysis of thick cylinders is based on 'Lame's equations'. These equations for tangential and radial stresses are:

$$f_t = a + \frac{b}{r^2}$$

and
$$f_r = a - \frac{b}{r^2}$$

where
$$a = \frac{p_i d_i^2 - p_o d_o^2}{d_o^2 - d_i^2}$$

$$b = \frac{d_i^2\, d_o^2}{4}\left(\frac{p_i - p_o}{d_o^2 - d_i^2}\right)$$

where p_i = internal pressure
p_o = external pressure
d_i = inner diameter
d_o = outer diameter
r = any radius.

Since we are considering internal pressures only, $\therefore p_o = 0$

$$\therefore \quad f_t = \frac{p_i d_i^2}{d_o^2 - d_i^2} + \frac{d_i^2\, d_0^2 \cdot p_i}{4r^2\left(d_o^2 - d_i^2\right)} = \frac{p_i d_i^2}{4r^2}\left(\frac{4r^2 + d_o^2}{d_o^2 - d_i^2}\right)$$

and
$$f_r = \frac{p_i d_i^2}{4r^2}\left(\frac{4r^2 - d_o^2}{d_o^2 - d_i^2}\right)$$

These stresses shall be maximum at $r = \frac{d_i}{2}$

$$\therefore \quad f_{t\,\max} = p_i\left(\frac{d_i^2 + d_o^2}{d_o^2 - d_i^2}\right) \quad \text{... (24.5)}$$

and $f_{r\,\max} = -p_i$, i.e., the compressive stress.

Brittle Materials: If the cylinder is made of brittle materials such as C.I., hard steel, and cast *Al*, the design criteria is maximum normal stress theory of failure. Accordingly, the stress induced in the cylinder wall:

$$f_t = f_{t\,\max} = p_i\left(\frac{d_i^2 + d_o^2}{d_o^2 - d_i^2}\right)$$

Now $d_o = d_i + 2t$, the above equation will become

$$t = \frac{d_i}{2}\left[\sqrt{\frac{f_t + p_i}{f_t - p_i}} - 1\right] \quad \text{... (24.6)}$$

f_t may be taken as $= \frac{1}{8} f_{ut}$.

This equation is applicable to both open and close ended cylinders.

Ductile Materials:

(a) Cylinder with closed ends: The thickness is calculated with the help of Clavarino's equation based on the maximum strain energy theory:

$$t = d_i \left[\sqrt{\frac{f_t + (1 - 2/m)\, p_i}{f_t - (1 + 1/m)\, p_i}} - 1 \right] \qquad \text{... (24.7)}$$

where $\frac{1}{m}$ = Poisson's ratio; $f_t = 0.8\, f_{ypt.}$

(b) Cylinder with open ends: In such cylinders, the seal is provided by a piston. The wall thickness is obtained with the help of Birnie's equation:

$$t = d_i \left[\sqrt{\frac{f_t + (1 - 1/m)\, p_i}{f_t - (1 + 1/m)\, p_i}} - 1 \right] \qquad \text{... (24.8)}$$

where $f_t = 0.8\, f_{y\,pt.}$

Barlow's equation:

$$f_t = \frac{p_i d_0}{2t} \qquad \text{... (24.9)}$$

It is derived from Lame's equation when t is small compared to d_i so that t^2 can be neglected. It is similar to the thin cylinder equation except 'd_o' has been written for 'd_i'. This equation is usually applied for calculating the wall thickness for high pressure oil and gas pipes. For ductile materials, $f_t = 0.8\, f_{y\,pt.}$ and for brittle materials, $f_t = 1/8\, f_{ut}$.

24.1.2 Vessels Subjected to External Pressure

The vessels subjected to external pressure are designed for strength and stiffness, i.e., they should have the ability to withstand the external load and retaining their shape.

24.2 PIPES AND TUBES

The pipes are extensively used for conveying all types of fluids ; liquid, gases, and, vapours from one place to another. The usual materials for pipes are : C.I., Steel, Copper and Brass. The selection of the material and the method of manufacture will depend upon the type and characteristics of working fluid.

C.I. pipes are used seldom and only for non-pressure pipe lines. They are manufactured by vertical casting or by centrifugal casting. The pipes made by the latter method are called "Spun pipes".

Steel pipes are made from plates, formed while hot. The longitudinal joint is either lap welded or butt welded. Seamless tubes or pipes have no longitudinal joint and are formed by drawing.

The iron and steel pipes may be galvanized to avoid corrosion. Copper and brass pipes are made either by drawing or from sheets by brazing or welding the seams.

Carbon steel pipes are used to convey water, oils and fuels and fuel gases. Stainless steel, copper and brass pipes are employed for fluids causing corrosion.

Thick walled seamless pipes made of heat resistance chrome-molybdenum alloy steels are used for high pressures and temperatures, i.e., greater than 4 N/mm^2 pressure and 450° C temperature. Copper and thin walled seamless steel pipes are used for pipes of the low power hydraulic drives, i.e., pressure less than 12 N/mm^2 and rate of flow less than 0.30 m^3/min. Recently, the increased use is also being made of plastic pipes.

Design of Pipes: The sizes of the pipes and tubes have been standardised. A pipe is designated by : its diameter, its wall thicknesses and its material. For each outside diameter, several wall thickness are specified. Thus, the inside diameter of the pipe varies with the wall thickness. Pipe is also characterised by the "nominal bore" which does not coincide with its inside diameter. As a rule, they differ somewhat. The nominal bores range from 1 to 4000 mm.

The thickness of the pipe wall is selected to suit the pressure of the liquid or gas conveyed by the pipe line. The thickness obtained by the thin cylinder formulae is too small to be of any practical value for the pipes. The modified relation for pipes is:

$$t = \frac{pd}{2f_t} + C,$$

where C is an allowance for corrosion and for wear under service conditions. It depends upon the pipe material and on the type of working media. It is given as :

$$C = 0.18\,t \text{ for } t > 6 \text{ mm}$$
$$= 1 \text{ mm for } t < 6 \text{ mm}$$

The following values for '*C*' have been recommended by Weisbach:

C.I. = 9 mm; Copper = 4 mm; Steel = 3 mm; Zinc = 4 mm.

Example 24.1. Find the necessary thickness for the shell plates of a Lancashire boiler 2.25 metre in diameter, boiler pressure 0.85 N/mm². The material is mild steel, tensile strength 470 N/mm². Assume the efficiency of the longitudinal joint to be 70 per cent. Factor of safety 5. Find the tensile stress in the unperforated plate and in the plate at the joint.

Solution:

(*i*) $$t = \frac{pd}{2f_t \times \eta}; f_t = \frac{470}{5} = 94 \text{ N/mm}^2$$

$$= \frac{0.85 \times 2250}{2 \times 94.0 \times 0.7} = 14.53 \text{ mm, say 15 mm}$$

(*ii*) $$f_t = \frac{pd}{2t} = \frac{0.85 \times 2250}{2 \times 15} = 63.75 \text{ N/mm}^2$$

(*iii*) $$f_t = \frac{pd}{2t \times \eta} = \frac{63.75}{0.7} = 91 \text{ N/mm}^2$$

Example 24.2. If the pressure in cast iron pipe 100 mm bore is 0.35 N/mm² find the thickness of the pipe. Safe stress is 14 N/mm².

Solution: $$t = \frac{p \times d}{2ft} + C$$

C, for C.I. = 9 mm

$\therefore$ $$t = \frac{0.35 \times 100}{2 \times 14} + 9$$

= 10.25 mm, say 11 mm or the next standard value is selected.

Example 24.3. A cast iron hydraulic cylinder has a bore of 250 mm and is subjected to an internal pressure of 8.5 N/mm². Determine the wall thickness if the tensile stress is not to exceed 35 N/mm².

Solution: Since the material is brittle, equation (24.6) will be used:

$$t = \frac{d_i}{2}\left[\sqrt{\frac{f_t + p_i}{f_t - p_i}} - 1\right] = \frac{250}{2}\left[\sqrt{\frac{35.0 + 8.5}{35.0 - 8.5}} - 1\right], \text{ gives } t = 35 \text{ mm}$$

According to Barlow's equation:

$$t = \frac{p_i\, d_o}{2f_t} = \frac{8.5 \times d_o}{2 \times 35} = 0.1214\, d_o$$

Now $$t = \frac{d_o - d_i}{2}$$

$\therefore$ $$0.2428\, d_o = d_o - d_i$$

$\therefore$ $$0.7572\, d_o = 250,\ d_o = 330 \text{ mm}$$

$\therefore$ $$t = \frac{330 - 250}{2} = 40 \text{ mm.}$$

Example 24.4. A steel tank for shipping gas is to have an inside diameter of 200 mm and a length of 0.10 m. The gas pressure is 10.5 N/mm². The permissible stress is to be 56 N/mm².

(a) Determine the required wall thickness, using the thin cylinder equation.

(b) Determine the thickness using Clavarino's equation.

Solution:

(*a*) $$t = \frac{pd}{2f_t} = \frac{10.5 \times 200}{2 \times 56} = 18.75 \text{ mm, say } 19 \text{ mm}$$

(*b*) $$t = \frac{d_i}{2}\left[\sqrt{\frac{f_t + p_i}{f_t - p_i}} - 1\right]$$

$$= \frac{200}{2}\left[\sqrt{\frac{56.0 + 10.5}{56.0 - 10.5}} - 1\right] = 20 \text{ mm.}$$

Example 24.5. A cast iron steam engine cylinder is 0.30 m in diameter. The maximum steam pressure in the cylinder is 1.05 N/mm². Find the thickness of the cylinder wall.

Solution:

$$t = \frac{pd}{2f_t} + (6 \text{ to } 12 \text{ mm})$$

If the C.I. is an ordinary one, $f_t = 8.75$ N/mm²

$$\therefore \quad t = \frac{1.05 \times 300}{2 \times 8.75} + 9 \text{ mm} = 27 \text{ mm.}$$

Example 24.6. A cast iron pipe is to deliver water at the rate of 125 m³ per minute and the velocity is 0.5 m/s. The maximum pressure in the pipe is 0.875 N/mm². The permissible stress in the C.I. is 21 N/mm². Determine the pipe diameter and the wall thickness.

Solution: (*a*) The flow equation is:

$$Q = A \times V$$

$$Q = 125 \text{ m}^3/\text{min.} = \frac{125}{60} \text{ m}^3/\text{s}, \ V = 0.5 \text{ m/s}$$

$$\therefore \quad \frac{125}{60} = A \times 0.5$$

$$\therefore \quad A = \text{cross-sectional area of pipe}$$

$$= \frac{125}{60 \times 0.5} = \frac{125}{30} \text{ m}^2$$

$$\therefore \quad \frac{\pi}{4} d^2 = \frac{125}{30} \text{ m}^2 \quad \therefore d = 2.30 \text{ m}$$

(*b*) $$t = \frac{pd}{2f_t} + C$$

C, for C.I. = 9 mm

$$\therefore \quad t = \frac{0.875 \times 2300}{2 \times 21} + 9 = 47.8 + 9 = 56.8 \text{ mm, say } 57 \text{ mm.}$$

Example 24.7. A hydraulic control for a straight line motion utilizes a spherical pressure tank A (Fig. 24.1) that is connected to a work cylinder B. A pump maintains pressure in the tank equal to 2.8 N/mm².

(a) Determine the thickness of the plate of the tank, if its diameter is 0.80 m. The tank is welded with joints having strength equal to that of the plate. The allowable tensile strength of the steel plates used is 52.5 N/mm².

(b) Assuming a pressure drop of 0.21 N/mm² between the tank and the cylinder, determine the piston diameter required to produce an operating force *F* of 22.50 kN. Assume an allowance for friction in the cylinder and packing equal to 10 per cent of *F*.

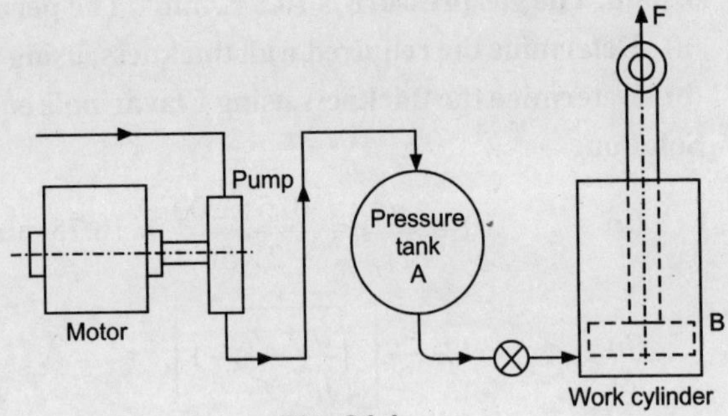

Fig. 24.1

(c) Determine the thickness of C.I. cylinder if the safe stress is 28 N/mm².

(d) Determine the power output of the cylinder during a working stroke, if the piston stroke is 0.45 m and that 5 s are needed for each stroke.

(e) If the work cycle of the piston rod occurs once very 30 s and the overall efficiency of the hydraulic control is 80 per cent and that of the pump is 60 per cent, find the motor power for continuous pump operation.

Solution:

(*a*) $$t = \frac{pd}{4 f_t \eta}, \text{ for spherical shell}$$

$$\eta = 100\%$$

∴ Thickness of plate for pressure tank :

$$t = \frac{2.8 \times 0.80 \times 1000}{4 \times 52.5} = 10.68 \text{ mm, say 11 mm}$$

(*b*) Total force to be produced by the piston:

$$= F + 10\% \text{ of } F = 1.1\,F = 1.1 \times 22.50 = 24.75 \text{ kN}$$

Now F = Area of piston × pressure in the cylinder

Pressure in the cylinder = 2.8 – 0.21 = 2.59 N/mm²

$$\therefore \quad \frac{\pi}{4} d_p^2 \times 2.59 = 24.75 \times 1000$$

∴ Diameter of piston, d_p = 111 mm

(*c*) Thickness of the cylinder:

$$t = \frac{pd}{2f_t} = \frac{2.59 \times 111}{2 \times 28.0} = 5.13 \text{ mm, say 6 mm}$$

(*d*) Net operating force produced by the piston = 22.50 kN

∴ W.D. per sec. = Force × distance travelled per sec.

$$= 22.50 \times \frac{450}{5} = 2025 \text{ kN mm} = 2.025 \text{ kNm}$$

∴ Power output of cylinder = 2.025 kW

(*e*) Power of the motor:

The motor is to provide a force of 24.75 kN to the cylinder.

$$\therefore \quad \text{Power of motor} = 24.75 \times \frac{0.45}{5} \times \frac{1}{0.8 \times 0.6} \times \frac{5}{30}$$

Since the Power which is to be produced by cylinder in 5 s is to be provided by the motor in 30 s.

$\therefore$ Power of motor = 0.773 kW.

Example 24.8. A pressure vessel is shown in Fig. 14.28.

(a) Determine the thickness of the cylinder wall, for the data given in Ex. 17 (Chapter 14)

Solution : (*a*)

$$t = \frac{pd}{2f_t}$$

$$p = 0.2\ \text{N/mm}^2;\ d = 600\ \text{mm};\ f_t = 17.5\ \text{N/mm}^2$$

$$\therefore \quad t = \frac{0.2 \times 600}{2 \times 17.5} = 3.43\ \text{mm}$$

But the minimum thickness of C.I. casting is 6 mm

So let us take $t = 6$ mm.

Example 24.9. An ordinary or a simple hydraulic accumulator is shown in Fig. 24.2. It is required to store 0.12 m^3 of water at a pressure of 30 N/mm^2. Assume the length of the stroke to be 3 metres, and compute the dimensions of the remaining parts.

Solution: The function of the hydraulic accumulator is to store the liquid under pressure delivered by the pump when it is not needed by a machine and to supply it to the machine later as and when needed. The accumulator shown in Fig. 24.2 consists of a stationary cylinder in which moves a weighted ram. The ram is weighted so as to get the required pressure. The load may either consist of a sufficient number of C.I. discs or a tank filled with heavy scrap metal. Water is forced into the cylinder from the pump through the inlet pipe. It raises the ram against the resistance offered by the weights. When the *ram* reaches its upper limit of stroke, some device is provided to shut off the pump.

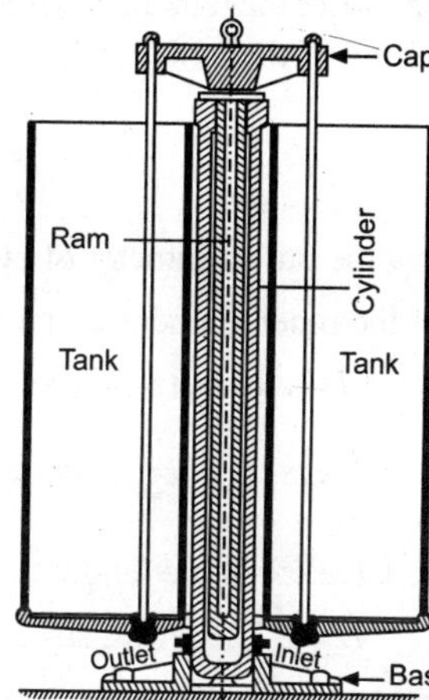

Fig. 24.2 *Hydraulic accumulator*

1. Ram:

$\therefore$ Volume of water to be stored = $0.12 \times (1000)^3$ mm^3

$$\therefore \quad \frac{\pi}{4} d_r^2 \times \text{stroke} = \text{volume}$$

$$\therefore \quad \frac{\pi}{4} d_r^2 \times 3 \times 1000 = 0.12 \times (1000)^3$$

$$\therefore \quad d_r = \text{diameter of } ram \text{ (outer)} = 225.4\ \text{mm, say } 226\ \text{mm}$$

$$\text{Now the load required} = \frac{\pi}{4} d_r^2 \times p = \frac{\pi}{4} \times (226)^2 \times 30 = 1200\ \text{kN}$$

The *ram* may be made of C.I. and it will act as a strut. If *d* is the internal diameter of the *ram*, then

$$\frac{\pi}{4}(226^2 - d^2) \times f_c = 1200 \times 1000$$

Taking $f_c = 56\text{ N/mm}^2$

$$\therefore \quad \frac{\pi}{4}(226^2 - d^2)\times 56 = 1200 \times 1000, \quad \therefore d = 154.5\text{ mm.}$$

$$\therefore \quad \text{Thickness of metal} = \frac{(226 - 154.5)}{2} = \frac{71.75}{2} = 35.75\text{ mm.}$$

2. **Cylinder:** We may use cast steel for cylinder for such pressures. The test pressure is usually double the working pressure. The test pressure should produce a stress which should not exceed 63 N/mm^2 for C.I. and 175 N/mm^2 for C.S.

The inside of the cylinder is usually unmachined. So, it should be rather larger than the ram to allow for irregularities. Let us take, inner diameter of cylinder, $d_i = 240$ mm.

The thickness of cylinder is found from Lame's equation:

$$t = \frac{d_i}{2}\left[\sqrt{\frac{f_t + p_i}{f_t - p_i}} - 1\right]$$

Let us take $f_t = 140\text{ N/mm}^2$, now $p_i = 2 \times 30 = 60.0\text{ N/mm}^2$

$$\therefore \quad t = \frac{240}{2}\left[\sqrt{\frac{140.0 + 60.0}{140.0 - 60.0}} - 1\right] = 69.6\text{ mm, say 70 mm.}$$

3. **Tank:** Let the tank be loaded with scrap iron. Taking the density of scrap Cast Iron as 7210 kg/m^3, total volume of the tank needed is

$$= \frac{\text{mass}}{\text{density}} = \frac{1200 \times 1000}{9.81 \times 7210} = 16.966\text{ m}^3$$

Now the outer diameter of cylinder = 240 + 140 = 380 mm

Let the inner diameter of tank = 450 mm

$\therefore$ If D is the outer diameter of tank, then

$$\frac{\pi}{4}(D^2 - 0.45^2)\times 3 = 16.966.$$

We have taken the length of the tank as 3 meters

$\therefore$ $D = 2.72$ metres, say 2.75 meters.

4. **Cap:** A suitable form of cap is shown in Fig. 24.3, where the dimensions are in cm. Bending moment across the section CD, $M_b = F \times l$,

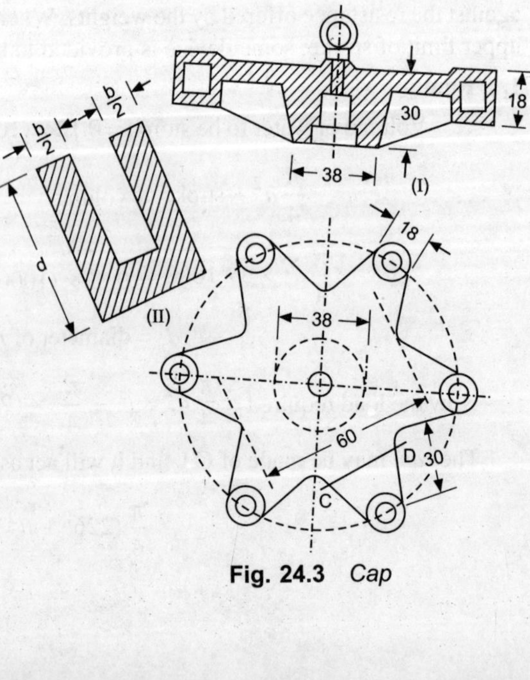

Fig. 24.3 *Cap*

where $F = \frac{1200}{6} = 200\text{ kN}$

and $l = 0.30$ m

$$\therefore \quad M_b = 200 \times 0.30 = 60\text{ kNm}$$

A section of the arm is shown.

Let $b = 5\text{ cm} = 50\text{ mm}$

$$\therefore \quad M_b = f_b \times Z$$

Let $f_b = 126\text{ N/mm}^2$ for C.S.

$$Z = \frac{b\,d^2}{6}, \text{ approximately, considering the section to consist of two rectangles only.}$$

$$\therefore \quad 60 \times 1000 \times 1000 = 126 \times \frac{50d^2}{6},$$

$$d = 239.05 \text{ mm, say 24 cm.}$$

5. **Bolts:** There are six bolts supporting the load. Therefore, load per bolt = $\frac{1200}{6}$ = 200 kN

$$\therefore \quad \frac{\pi}{4} d_c^2 \times f_t = 200 \times 1000$$

Let $f_t = 63 \text{ N/mm}^2$

$\therefore$ $d_c = 63.56$ mm.

The next standard core diameter = 67.73 mm, with outside diameter 80 mm.

6. **Foot or Base Plate:** One of several forms of base plates is shown in Fig. 24.4. The total weight on the bed plate will be the load required *i.e.,* the weight of tank, weight of ram and weight of cylinder. The weight of the ram and the cylinder can be Very easily calculated.

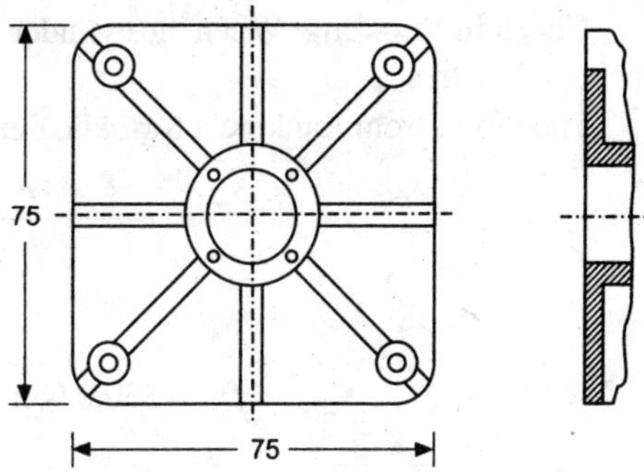

Fig. 24.4 *Base plate*

Example 24.10. Design a differential accumulator for the following details:
Water pressure = 15.68 MPa
Stroke = 1.8 meters
Losses at glands, etc = 4 per cent
Capacity: To deliver power corresponding to 7.5 kW as the cylinder descends uniformly in 3/4 minutes, the supply pump being stopped.

Solution: In this type of accumulator, Fig. 24.5, the ram is stationary and has two different diameters. The water from the pump enters the accumulator through the inlet pipe and is delivered to the cylinder through the duct D at a point which pertains to the maximum lift of the accumulator cylinder which is moving. The cylinder is weighted to give the necessary pressure. The capacity of the differential accumulator is smaller than that of the ordinary one and is dependent upon the difference in the diameters of the two portions of ram.

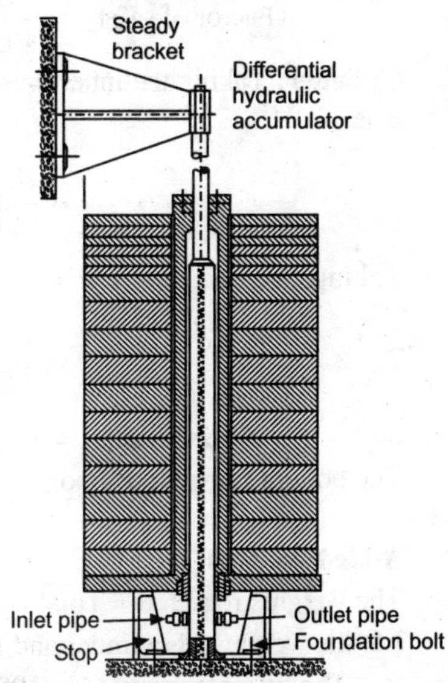

Fig. 24.5 *Differential accumulator*

This type has got the following advantages:

1. Large pressure can be obtained with a relatively smaller machine.
2. It is suitable when small quantities of water at high pressure are needed.

1. **Ram:** Work done by ram

= load on the ram × distance travelled.

Now load on ram = water pressure × diff. in the area of the two portions of the

$$\text{ram} = 15.68 \times \text{A, N}$$

$$\therefore \quad W.D. = 15.68 \times 10^6 \times \text{A} \times 1.80 \text{ Nm}$$

Also $\quad W.D$ = Equivalent of 7.5 kW in $\frac{3}{4}$ minutes = 7.5 × 1000 × 45 = 33750 kNm

∴ $15.68 \times 10^6 \times A \times 1.80 = 337.50 \times 1000$ gives $A = 0.01196$ m^2.

It can be seen from the diagram that the total height of the ram will be about 4 meters, therefore, the smaller section should be stiff enough to avoid the buckling of the ram. Let the smaller diameter be 0.09 m.

$$\therefore \quad \frac{\pi}{4}(d_o^2 - 0.09^2) = 0.01196, \; d_o = 0.1527 \text{ m, say 153 mm.}$$

The ram may be casted from C.I. or C.S. or it may be forged from M.S. In this problem, a forged ram will be used.

Check in Buckling: When the cylinder is at the top of its stroke, the length of the ram supporting the load is about 2 metres.

Therefore, from Rankine's formula, the buckling load on the ram is:

$$P = \frac{f_{yp} \times A}{1 + a\left(\frac{L}{k}\right)^2}$$

$$f_{yp} = 330.0 \text{ N/mm}^2 \text{ for M.S. ; } a = \frac{1}{7500}$$

$$L = 2 \text{ metres for both ends free, } k = \frac{153}{2} = 76.5 \text{ mm}$$

$$\therefore \quad P = \frac{330.0 \times \frac{\pi}{4} \times 153^2}{1 + \frac{1}{7500} \times \left(\frac{2000}{76.5}\right)^2} = 5561 \text{ kN}$$

$$\text{Now the actual load on the ram} = \frac{p \times A}{\eta} = \frac{15.68 \times 10^6 \times 0.01196}{0.96} = 195.35 \text{ kN}$$

$$\therefore \quad \text{Factor of safety} = \frac{5561}{195.35} = 28.47, \text{ which is very safe.}$$

2. **Cylinder:** Taking the inner diameter of cylinder = 170 mm. The cylinder is made of close grained cast iron,

$$t = \frac{d_i}{2}\left[\sqrt{\frac{f_t + p_i}{f_t - p_i}} - 1\right]$$

Taking $\quad f_t = 35 \text{ N/mm}^2 ; p_i = 15.68 \text{ N/mm}^2$

$$\therefore \quad t = \frac{170}{2}\left[\sqrt{\frac{35.0 + 15.68}{35.0 - 15.68}} - 1\right] = 49.8 \text{ mm, say 50 mm.}$$

$$\therefore \quad \text{Outer diameter} = 170 + 2 \times 50 = 270 \text{ mm.}$$

The bottom flange to support the added weights is cast integral with the cylinder and it will be designed later.

3. **Added Weights:**

The weights required = Total load-weight of cylinder-weight of the bottom flange.

Let the weights of cylinder and flange be 15 kN

∴ The weights required = 195.35 – 15 = 180.35 kN.

Inner diameter of weights = outer diameter of cylinder + some clearance

$= 270 + 10 = 280$ mm

The height of the weight may be taken as equal to stroke, i.e., 1.8 m. If D is the diameter of outer portion of the weights, then

$$\frac{\pi}{4}(D^2 - 0.28^2) \times 1.80 \times 7210 = \frac{180.35 \times 1000}{9.81}$$

$$\therefore \quad D = 1.36 \text{ m} = 136 \text{ cm}.$$

4. **Bottom Flange:** It is shown in Fig. 24.6. Assuming that the fracture occurs across a diameter XX and neglecting the strengthening effect of the fillet and the cylinder.

Centroid of a semicircular area = 0.424 × radius along the middle radius from the centre.

To find the centroid of the annulus, take moments about XX,

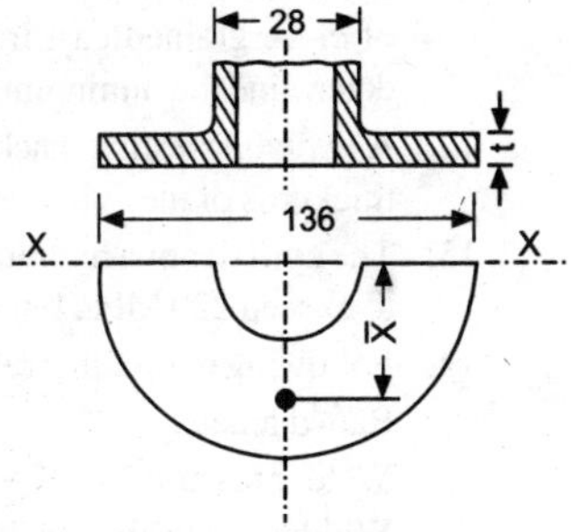

Fig. 24.6 *Bottom Flange*

$$\therefore \quad \frac{1}{2}\pi \times 68^2 \times 0.424 \times 68$$

$$= \frac{1}{2}\pi(68^2 - 14)^2 \times \overline{X}$$

$$+ \frac{1}{2}\pi \times 14^2 \times 0.424 \times 14$$

$$\therefore \quad \overline{X} = 29.8 \text{ cm}$$

B.M. about $XX = \frac{1}{2}\text{load} \times \overline{X} = \frac{1}{2} \times 180.35 \times 0.298 = 26.872$ kNm.

$$\therefore \quad M_b = f_t \times Z$$

$$Z = 2 \times \frac{1}{6}bt^2 = 2 \times \frac{1}{6} \times \frac{1.36 - 0.28}{2} \times t^2 = 0.18\,t^2$$

Taking $f_t = 32$ MPa

$\therefore \quad 26.872 \times 1000 = 0.18\,t^2 \times 32 \times 10^6$, $t = 0.0683$ m, say 7 cm.

EXERCISES

1. Define a "Pressure Vessel".
2. How would you classify the Pressure Vessels?
3. What is the difference between thin and thick cylinders?
4. A cylinder 0.25 m diameter is to be filled with fluid at a pressure of 21 MPa. The cylinder is made of steel with an ultimate strength of 420 MPa. Determine the outside diameter of cylinder using an apparent factor of safety of 5.
5. A closed end C.I. cylinder of 0.20 m inside diameter is to carry an internal pressure of 14 MPa with a permissible stress of 21 MPa.
   (*a*) Determine the wall thickness by means of Lame's and Clavarino's equations.
   (*b*) Which result would you use ? Give reasons for your conclusions.
6. A C.I. pipe is 0.25 m in inside diameter and the metal is 9.5 mm thick. The pipe contains water under a head of 75 metres.
   (*a*) What is the apparent factor of safety considering water pressure only?
   (*b*) Determine the stress caused by bending if the pipe is full of water, horizontal, 7 meter long and simply supported at the ends.
   (*c*) What is the maximum combined tensile stress?

7. The drum of a steam generating unit is 5.5 meter long and 1.20 m in diameter. The steam pressure is 3.5 MPa gauge. Assuming an allowable tensile stress of 70 MPa and that the efficiencies of the circumferential and longitudinal joints are 65 and 90 per cent , respectively, recommend the thickness of the shell.
8. The cylinder of a hydraulic accumulator has an internal diameter of 0.40 m and is 50 mm thick. The internal pressure at its maximum is to be 5 MPa. Determine the maximum and minimum tangential stresses induced.
9. The pressure within the cylinder of a hydraulic press is 7 MPa. The diameter of the cylinder is 0.38 m. Determine the thickness of the cylinder, allowing a tensile stress of 17.5 MPa.
10 A cast iron cylinder of close grained iron is 0.10 m in diameter and is subjected to an internal pressure of 18 MPa. The ultimate tensile stress of the material is 280 MPa. Using a factor of safety of 8, determine the wall thickness required.
11. A 0.25 m diameter Diesel engine cylinder is subjected to an explosion pressure of 3.15 MPa. The casting is made of close grained cast iron having an ultimate tensile strength of 245 MPa. Use a factor of safety of 10, and determine the minimum thickness of the bored cylinder.
12. A hydraulic testing machine has a maximum capacity of 1000 kN. The piston diameter is 0.25 m. Calculate the wall thickness of the cylinder.
13. The cylinder of a hydraulic intensifier is subjected to an internal pressure of 125 MPa. The maximum stress is not to exceed 220 MPa. Internal diameter is 0.10 m. Find the thickness and the hoop stress at the outer circumference.
14. Following particulars relate to a Hydraulic-Accumulator:

    Ram diameter = 0.20 m

    Water pressure = 23.5 MPa

    Width of U-leather packing =19 mm

    Design the gland and prepare a dimensioned sketch.

❑❑❑

CHAPTER 25

Cylinder Heads and Cover Plates

25.1 FLAT HEADS

The heads of pistons, cylinders, boilers and the sides of the rectangular of square tanks usually consist of flat plates or slightly dished plates. These plates, can either be cast integral with the supporting members or fixed to them by means of bolts, welds or rivets. Depending upon the type of connection between the head and the supporting members, the flat plates forming the heads can be divided into two types:

(a) Simply supported at the edges or centre.
(b) Rigidly fixed to supporting members.

The two usual forms of flat plates commonly used are: Circular, and Rectangular. Their design will depend upon two factors:

(i) Type of connection with the supporting members.
(ii) Nature of the load, i.e. uniformly distributed or concentrated.

The analysis of the flat plates is complicated and the exact solution is not available. The formulas used for the design of flat plates are based upon the work of Grash and of Back.

25.1.1 Circular Flat Plates

(a) **Plates with Uniformly Distributed Load:** The thickness 't_c' of the plate of diameter 'd' supported at the circumference and subjected to a pressure 'p' uniformly distributed over the area is given as:

$$t_c = k_1 d \sqrt{\frac{p}{f_t}} \qquad ...(25.1)$$

where f_t = allowable design stress.

k_1 is a coefficient, whose values are given below. It depends upon the material and method of holding the edges.

C.I.: Freely supported , k_1= 0.54, the maximum stress occurs at the centre of head.
Fixed k_1= 0.44, the maximum stress occurs at the edge.

MS: Freely supported, k_1= 0.42, the maximum stress occurs at the centre of head.
Fixed k_1 = 0.35, the maximum stress occurs at the edge.

The following equations are also used to find the thickness of the circular plates:

$$t_c = \sqrt{\frac{3F\left(3+\frac{1}{m}\right)}{8\pi.f_t}} \text{ for plates with edges freely supported,} \quad ...(25.2)$$

and

$$t_c = \sqrt{\frac{3F}{4\pi.f_t}} \text{ for plates with edges rigidly fixed,} \quad ...(25.3)$$

where F = total load on the plate = $p.\pi r^2$
$1/m$ = Poisson's ratio.

(b) Plates Loaded Centrally: If the plate is subjected to a load, F, distributed uniformly over an area $1/4\pi d_0^2$, the thickness of the C.I. flat circular plate may be determined as given below:

$$t_c = 3.0\sqrt{\left(1-\frac{0.67\,d_o}{d}\right)\cdot\frac{F}{f_t}} \text{ for freely supported plate.} \quad ... (25.4)$$

$$= 1.65\sqrt{\frac{F}{f_t}\cdot\log_e\left(\frac{d}{d_o}\right)}. \text{ for fixed plate} \quad ... (25.5)$$

25.1.2 Rectangular Flat Plates

(a) Uniform Load: The thickness of the plate is given as:

$$t_c = a\,b\,k_3\sqrt{\frac{p}{f_t(a^2+b^2)}} \quad ... (25.6)$$

where a = larger side of the rectangle,
b = smaller side of the rectangle
k_3 = a coefficient.

For C.I. k_3 = 0.75 for freely supported plate, the maximum stress occurs at the centre and k_3 = 0.62 for fixed plate, the maximum stress occurs at the centre of longer edge.

For M.S. k_3 = 0.60 for freely supported plate
= 0.49 for fixed plate.

The other commonly used formulas are:

$$t_c = ab\sqrt{\frac{p}{2f_t(a^2+b^2)}} \text{ for freely supported edges,} \quad ... (25.7)$$

and

$$t_c = ab\sqrt{\frac{p}{3f_t(a^2+b^2)}}, \text{ for fixed plates.}$$

(b) Concentrated Load: If the load is concentrated at the intersection of the diagonals,

$$t_c = k_4\sqrt{\frac{abF}{f_t\,(a^2+b^2)}}, \text{ for freely supported edges,} \quad ... (25.8)$$

where k_4 is a coefficient, whose values are given as:

C.I.:	$k_4 = 4.3$,	for freely supported edges.
	$= 4.0$	for fixed plates
M.S.	$k_4 = 3.45$	for freely supported edges.
	$= 2.8$ to 3.2	for fixed edges.

25.1.3 Elliptical Covers

$$t_c = k_5\, a\, b \sqrt{\frac{p}{f_t\,(a^2 + b^2)}}$$

The values of k_5 are as given below:

C.I.: $k_5 = 1.2$, fixed edges; $= 1.5$, free edges.

M.S.: $k_5 = 0.9$, fixed edges; $= 1.2$, free edges.

a and b are the major axis and minor axis respectively.

25.2 DISHED HEAD

It was noticed in the last chapter that a spherical shell is stronger than a cylindrical shell. Therefore, to utilize this strength benefit, many cylinders are provided with semispherical heads, as shown in Fig. 25.1. The thickness of such a head that is riveted or welded to a cylindrical shell, is given as:

$$t_c = \frac{8.33\, pR}{2 f_u}, \qquad \text{... (25.9)}$$

where R = inside radius of curvature of the head

f_u = ultimate strength of the material.

R should not be greater than d, the shell diameter.

If $R < 0.8\,d$, R should be taken equal to $0.8\,d$ in the above equation.

If there is an opening in the head, 150 mm or larger, the thickness of the head should be increased by 15 per cent. This increase should not be less than about 3.2 mm.

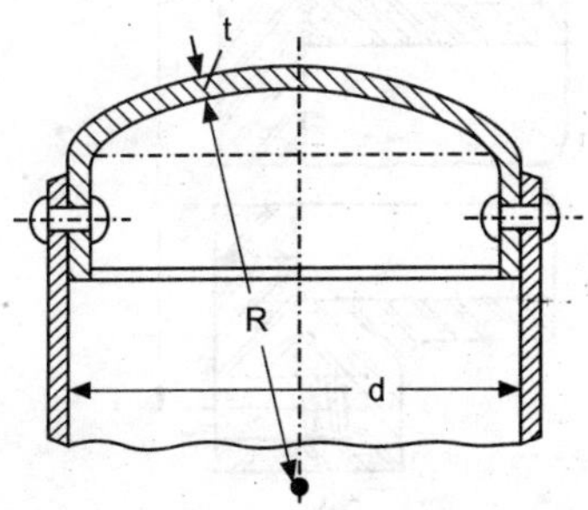

Fig. 25.1 *Dished head*

The thickness of the cover of a dished head (integral or welded) may also be calculated by the following relation:

$$t_c = \frac{pR\cdot k}{2 f_t \eta}, \qquad \text{... (25.10)}$$

where the constant k is given as;

d/R :	0	0.4	0.8	1.2	1.6	2.0
k :	1.0	1.3	1.05	1.08	1.035	?

25.3 STANDARD ELLIPSOIDAL COVER (INTEGRAL OR WELDED) FIG. 25.2.

$$t_c = \frac{p\cdot d}{4 f_t \eta} \qquad \text{...(25.11)}$$

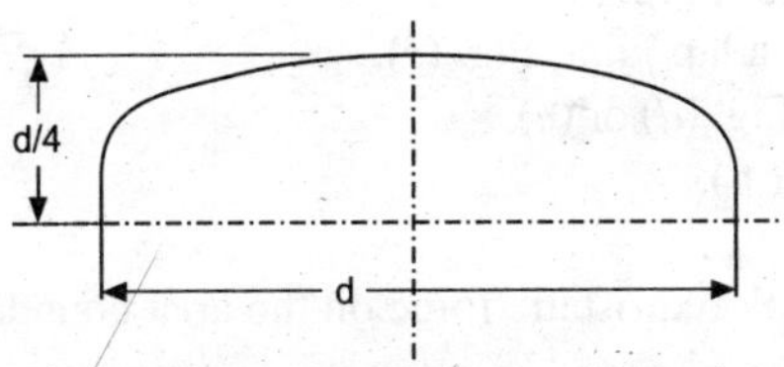

Fig. 25.2

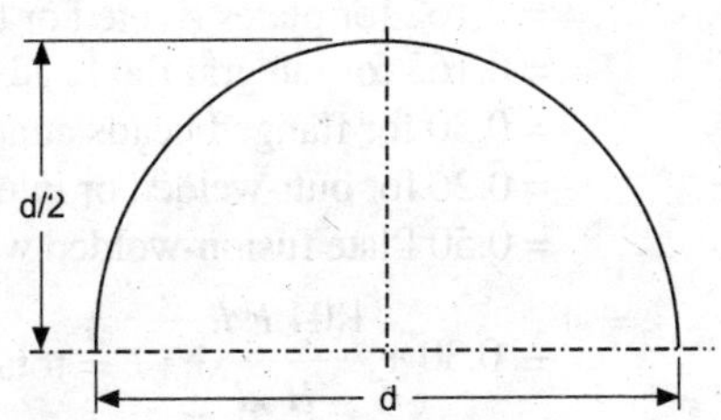

Fig. 25.3

25.4 HEMISPHERICAL COVER (INTEGRAL OR WELDED) FIG. 25.3.

$$t_c = \frac{p \cdot d}{4 f_t \eta} \qquad \text{...(25.12)}$$

25.5 CONICAL COVER (INTEGRAL OR WELDED) FIG. 25.4.

$$t_c = \frac{p \cdot d}{2 f_t \eta (\cos \theta)} \qquad \text{... (25.13)}$$

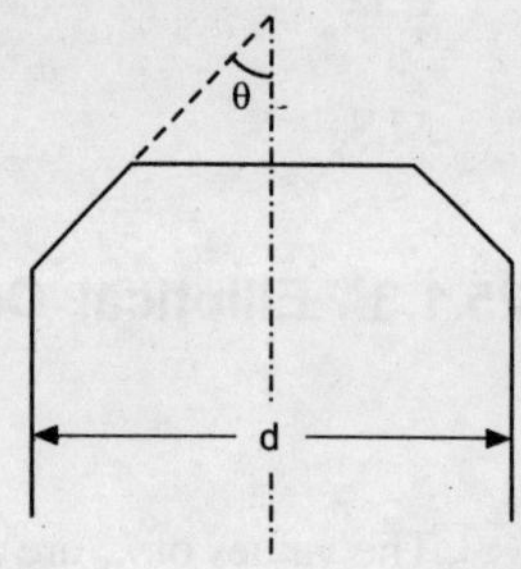

Fig. 25.4

25.6 BOLTED AND WELDED HEADS

Figure 25.5 shows the construction of the various types of unstayed steel heads, cover plates, blind flanges, etc. The minimum thickness of such flat plates is given by the relation:

$$t_c = d\sqrt{\frac{k\,p}{f_t}} \qquad \text{... (25.14)}$$

k is an empirical coefficient.

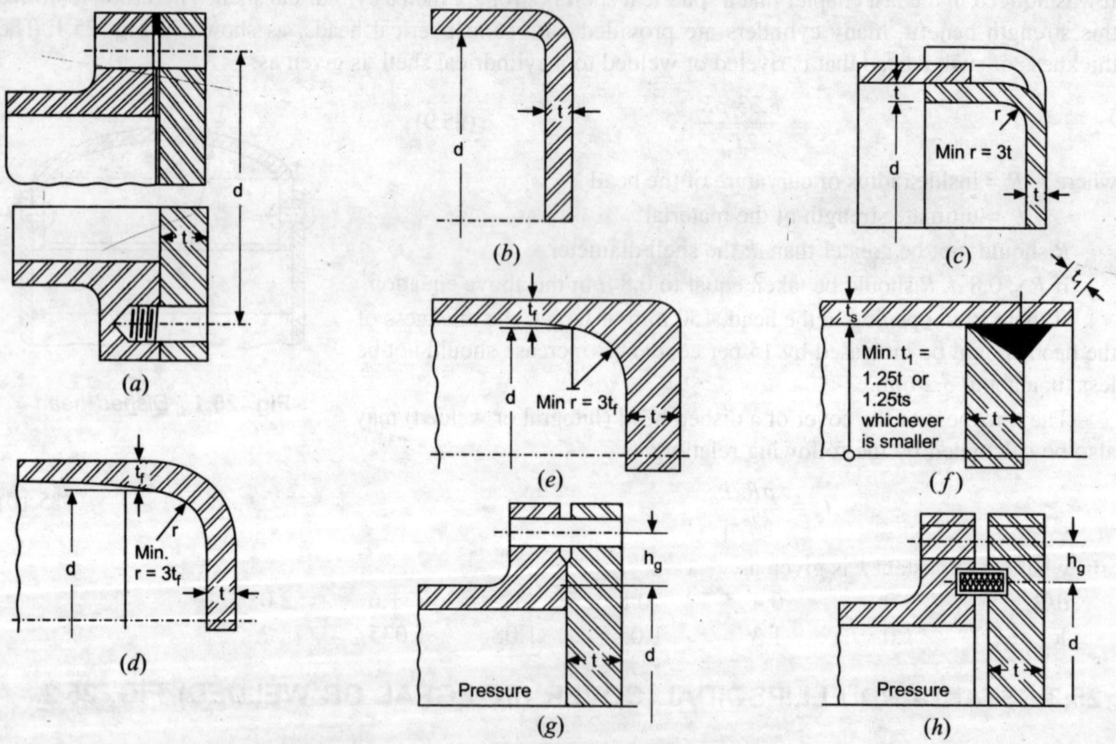

Fig. 25.5 *Types of cylinder heads and covers*

= 0.162 for plates riveted or bolted rigidly to flanges, Fig.(*a*).

= 0.162 for integral flat heads; $d \le 60$ cm; $t \ge 0.05d$, Fig. (*b*).

= 0.30 for flanged heads attached to the shell by a lap joint, Fig. (*c*).

= 0.26 for butt-welded or integral forged heads, Fig. (*d*) or (*e*).

= 0.50 Plate fusion-welded with fillet weld, Fig. (*f*).

$= 0.30 + \dfrac{1.04\, F\, h_g}{H \cdot d}$; F = total bolt load; H = total hydrostatic force on the area bounded by the outside diameter of the gasket. Fig. (*g*) or (*h*).

25.7 FLANGE AND BOLT OR STUD ARRANGEMENT

(i) Cover with bolts:

Thickness of flange, $t_f = (1.1 \text{ to } 1.25) \times t = (1.25 \text{ to } 1.5) \times d_b$

The distance, $l = t + d_b + 5$

$= t + 0.5d_b + 5$, when nut is spot faced, so as to keep the arm of the bending moment 'l' small.

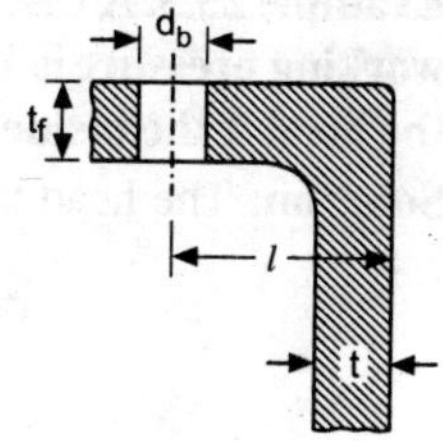

Fig. 25.6

(ii) Cover with studs:

$t_f = (1.25 \text{ to } 1.50) \times d_b$

$h = (1.50 \text{ to } 1.75) \times d_b$, C.I. $= 1.25 \times d_b$, steel casting

$= (1.75 \text{ to } 2.00) \times d_b$, *Al* and other nonferrous materials.

$l = (1.25 \text{ to } 1.50) \times d_b$

(iii) Pitch of the bolts and studs is given as

$p = 7d_b$ for pressures from 0 to 0.35 N/mm^2

$= 3.5\, d_b$ for pressures from 1.225 to 1.4 N/mm^2

and is $\nless 3d_b$, in order to leave room for the wrench when tightening nuts.

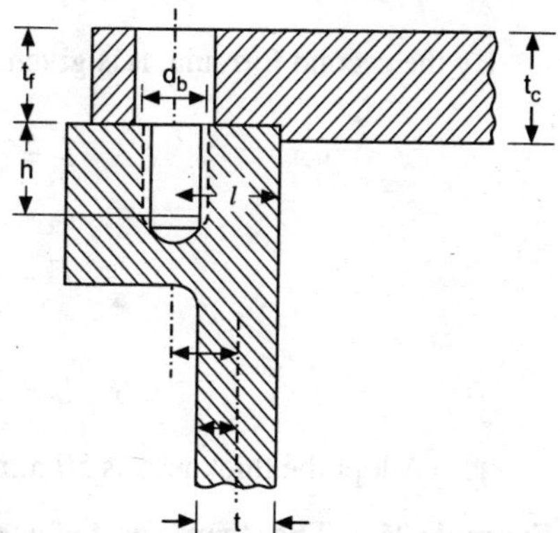

Fig. 25.7

Example 25.1. A flat circular plate of 0.30 m diameter is supported around the edge and is subjected to a uniform pressure of 1.05 N/mm². The stress is to be limited to 70 N/mm². Determine the thickness of the steel plate required, if 1/m = 0.3.

Solution: The load is uniformly distributed and the plate is supported around the edge, therefore, the thickness will be given by the relation:

$$t_c = \sqrt{\frac{3F\left(3+\frac{1}{m}\right)}{8\pi f_t}}, \quad F = p\,\pi r^2$$

$$\therefore \quad t_c = \sqrt{\frac{3 \times 1.05 \times \pi \times 225 \times 100 (3 + 0.3)}{8\pi \times 70}} = 20.4 \text{ mm, say 21 mm.}$$

The thickness is also given as:

$$t_c = k_1 d \sqrt{\frac{p}{f_t}}, \quad k_1 = 0.42$$

$$\therefore \quad t_c = 0.42 \times 300 \times \sqrt{\frac{1.05}{70}} = 15.35 \text{ mm, say 16 mm.}$$

$\therefore$ Adopt a thickness of 21 mm.

Example 25.2. The piston of a gas engine is 115 mm is diameter and is made of C.I. The maximum pressure acting on it is 2.45 N/mm². Since the skirt of the piston is quite thin, and, therefore, flexible, the face will be considered as a flat plate merely supported at the edges. Determine the thickness of the face if the maximum tensile stress is to be 28 N/mm².

Solution : The thickness shall be given as:

$$t_c = k_1\, d \sqrt{\frac{p}{f_t}}, \quad k_1 = 0.54$$

$$\therefore \quad t_c = 0.54 \times 115 \sqrt{\frac{2.45}{28.0}} = 18.38 \text{ mm, say 19 mm.}$$

Example 25.3. A cast iron cylinder is to have a flat head cast integral with the cylinder walls. The internal working pressure is 14 N/mm^2 and the permissible working stress 24.5 N/mm^2. Determine the thickness of the head if the inside diameter is 150 mm.

Solution: The head thickness is given by the relation:

$$t_c = k_1\, d\sqrt{\frac{p}{f_t}},\ k_1 = 0.44$$

$$\therefore \quad t_c = 0.44 \times 150\sqrt{\frac{14}{24.5}} = 0.44 \times 150 \times 0.756 = 50 \text{ mm.}$$

By the second formula, it is given as:

$$t_c = \sqrt{\frac{3F}{4\pi f_t}}$$

$$F = 14.0 \times \frac{\pi}{4} \times (150)^2 = 247.50 \text{ kN}$$

$$\therefore \quad t_c = \sqrt{\frac{3 \times 247.50 \times 1000}{4\pi \times 24.5}} = 49.11 \text{ mm}$$

$\therefore$ Adopt the thickness as 50 mm.

Example 25.4. The steam chest of a steam engine is covered by a flat rectangular plate that is 250 mm × 380 mm. The plate is made of C.I. and is subjected to a steam pressure of 1.225 N/mm^2. The plate is to be assumed to be uniformly loaded and supported at the edges. Determine the thickness of plate for a design factor of 6.

Solution: The thickness is given by the relation:

$$t_c = a\, b\sqrt{\frac{p}{2f_t(a^2 + b^2)}}$$

$a = 380$ mm; $b = 250$ mm, $p = 1.225$ N/mm^2

$f_u = 210$ N/mm^2 for good grade grey C.I.

$$\therefore \quad f_t = \frac{210.0}{6} = 35.0 \text{ N/mm}^2$$

$$\therefore \quad t_c = 380 \times 250\sqrt{\frac{1.225}{2 \times 35.0\,(380^2 + 250^2)}} = 28 \text{ mm.}$$

Example 25.5. A tank for holding hydrogen gas under a pressure of 7 N/mm^2 is 250 mm in diameter. One end is closed by a dished head with the inner side convex. The radius of curvature of the head is 212.5 mm. What should the head thickness be?

Solution: The head thickness is given as:

$$t_c = \frac{8.33\, pR}{2f_u}$$

$$p = 7 \text{ N/mm}^2,\ R = 212.5 \text{ mm}$$

Let the material of the tank be good grade grey C.I. with $f_u = 210$ N/mm^2

$$t_c = \frac{8.33 \times 7 \times 212.5}{2 \times 210} = 29.5 \text{ mm, say 30 mm}$$

The cover thickness can also be calculated by the relation (25.10):

$$t_c = \frac{p\,R\,k}{2 f_t \eta}$$

Now $\quad d/R = \dfrac{25}{21.25} = 1.2, \; \therefore\; k = 1.08$

Let $\quad f_t = 35$ N/mm^2; and $\eta = 1$ for integral head.

$\therefore \quad t_c = \dfrac{7 \times 212.5 \times 1.08}{2 \times 35 \times 1} = 23$ mm

Hence $\quad t_c = 30$ mm should be taken.

Example 25.6. A hydraulic press is required to press cotton bales of about 300 mm cube. Available fluid pressure is 14 N/mm^2. The maximum pressure to be applied is 60 kN. Give a sectional sketch of the press showing : the ram, cylinder, base plate, top plate, sliding platform, and the four supporting columns. Take the following safe stresses:

f_t for C.S. = 105 N/mm^2
f_t for M.S. = 56 N/mm^2
f_t for cylinder material = 28 N/mm^2
Friction factor for gland = 1.25

Solution:

1. Ram: Maximum pressure to be applied by ram = 60 kN

$\therefore \quad \dfrac{\pi}{4} d_r^2 \times p = 60 \times 1000, p = 14$ N/mm^2

$\therefore \quad \dfrac{\pi}{4} d_r^2 \times 14.0 = 60 \times 1000, d_r = 74$ mm, say 7.5 cm.

If in order to reduce the weight of the ram, it is to be made hollow, it can be designed as a thick cylinder subjected to external pressure.

From the Lame's equation, by putting $p_i = 0$

Max. tangential stress, $f_{t\,max} = -\,p_o \cdot \dfrac{d_o^2 + d_i^2}{d_o^2 - d_i^2}$, i.e., compressive.

Max. radial stress, $f_{r\,\max} = -\,p_o$, i.e., compressive.

Now according to maximum shear stress theory for ductile materials,

$$\text{Max. shear stress}, f_{s\,\max} = \frac{f_{t\,\max} - f_{r\,\max}}{2} = -\frac{p_o\, d_i^2}{d_o^2 - d_i^2}$$

The max. shear stress is half the maximum principal stress, which in this case is compressive in nature,

$$\therefore \quad f_c = 2 f_{s\,\max} = \frac{2 p_o\, d_i^2}{d_o^2 - d_i^2}$$

Let $\quad \dfrac{d_o}{d_i} = k$

$$\therefore \quad f_c = \frac{2 p_o}{k^2 - 1}, \text{ or } k^2 - 1 = \frac{2 p_o}{f_c}$$

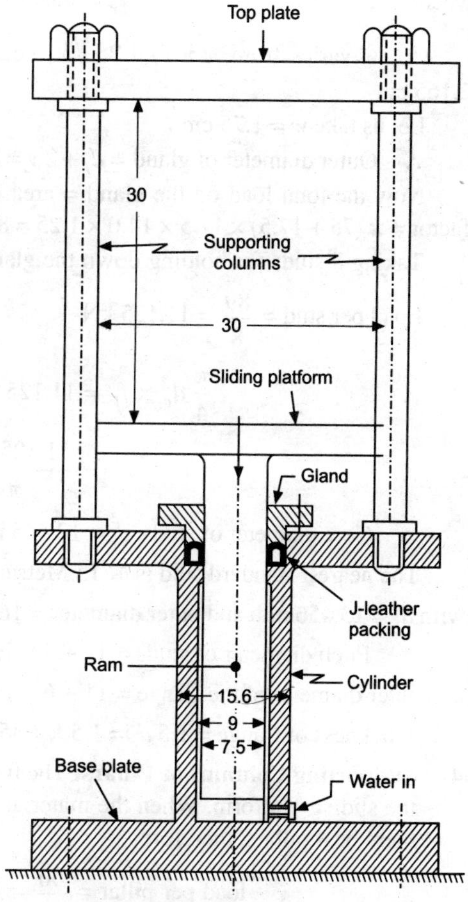

Fig. 25.8 *Hydraulic press*

Now $p_o = 14\ \text{N/mm}^2$ and $f_c = 56\ \text{N/mm}^2$

$$\therefore \quad k^2 - 1 = \frac{2 \times 14}{56.0},\ k = 1.225$$

$$d_i = \frac{7.5}{1.225} = 6.12\ \text{cm, say } \mathbf{6.15\ cm.}$$

2. **Cylinder:** The material of the cylinder is C.I. It will be designed as a thick cylinder. Since the inside of the cylinder is unmachined, therefore, let the inner diameter of cylinder, $d_i = 7.5 + 1.5 = 9$ cm.

From Lame's equation:

$$t = \frac{d_i}{2}\left[\sqrt{\frac{f_t + p_i}{f_t - p_i}} - 1\right] = \frac{90}{2}\left[\sqrt{\frac{28+14}{28-14}} - 1\right] = 33\ \text{mm.}$$

$\therefore$ Outside diameter of cylinder = 9 + 2 × 3.3 = **15.6 cm.**

3. **Gland:** The width of the U-leather packing for a ram of 7.5 cm diameter is about 1.6 cm which fixes the width 'w' of the gland Fig. 25.9. The width of the gland is also given empirically as:

$$w = \sqrt{\frac{d_r}{2.5}} \text{ to } 1.25\sqrt{\frac{d_r}{2.5}}$$

where d_r = dia. of ram.

$\therefore$ w varies from $\sqrt{3}$ to $1.25\sqrt{3}$, i.e., 1.732 cm to 2.165 cm

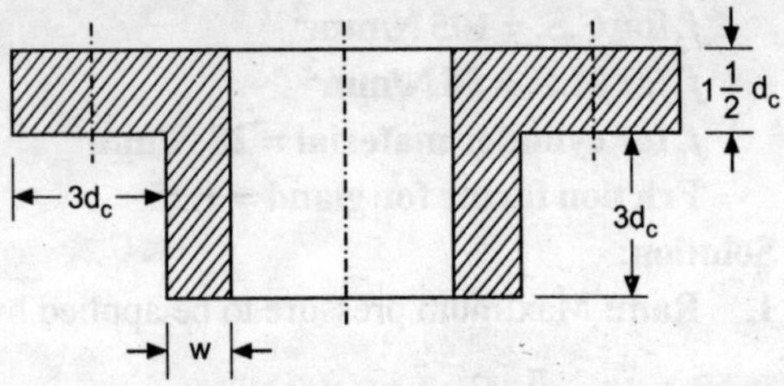

Fig. 25.9 *Gland*

Let us take $w = 1.75$ cm

$\therefore$ Outer diameter of gland = $d_r + 2w = 7.5 + 2 \times 1.75 = 11$ cm.

Now the total load on the gland = area of gland exposed to fluid pressure × fluid pressure × friction factor = π (75 + 17.5) × 17.5 × 14.0 × 1.25 = 89 kN

Taking 8 studs for holding down the gland,

$$\text{Load per stud} = \frac{89}{8} = 11.125\ \text{kN}$$

$$\therefore \quad \frac{\pi}{4} d_c^2 \times f_t = 11.125 \times 1000, \text{ taking } f_t = 90\ \text{N/mm}^2$$

$$\therefore \quad d_c^2 = \frac{11.125 \times 1000 \times 4}{\pi \times 90.0} = 157.386\ \text{mm}^2$$

$\therefore$ Core diameter of studs, $d_c = 12.545$ mm.

The nearest standard stud is M 15 Metric Threads Series, Coarse with $d_c = 13.456$ mm and outer diameter = 16 mm

$\therefore$ Pitch diameter of studs = 11 + 3 × 1.3546 = 15.0638 cm

and outer diameter of the flange = 11 + 6 × 1.3546 = 19.1276 cm, say 19.2 cm

Thickness of flange = 1.5 d_c = 1.5 × 1.3546 = 2.0319 cm, say 21 mm.

4. **Supporting Columns or Pillars:** The function of the pillars is to support the top plate. It may also guide the sliding platform. When the material is being pressed, these pillars will be under direct tension. Let there be 4 pillars, therefore,

$$\text{load per pillar} = \frac{60}{4} = 15\ \text{kN}$$

$\therefore \quad \frac{\pi}{4} d_c^2 \times f_t = 15 \times 1000$, taking $f_t = 56$ N/mm^2

$\therefore \quad d_c = 18.5$ mm

For the rigidity and stability of the columns, let us select M 24 × 2 fine series thread, with $d_c = 21.545$ mm and outer diameter 24 mm

5. **Upper Cylinder Flange:** The centre to centre distance of the pillars is equal to=30+2.4=32.4cm

$\therefore$ Outer diameter of the cylinder flange = 32.4 + 3 × 2.4 = 32.4 + 7.2 = 39.6 cm, say 40 cm.

The top view of the flange is shown in Fig. 25.10.

Maximum B.M. at section XX = $\frac{60 \times 1000}{4} \times \frac{8}{100} = 1200$ Nm

Width of the section is about 0.20 m

$$\therefore \quad M_b = f_t \times Z = f_t \times \frac{1}{6} bt^2$$

$$f_t = 28 \text{ MPa}$$

$$\therefore \quad 1200 = 28 \times 10^6 \times \frac{1}{6} \times 0.20 \times t^2$$

$\therefore$ Thickness of flange = 0.036 m = 36 mm.

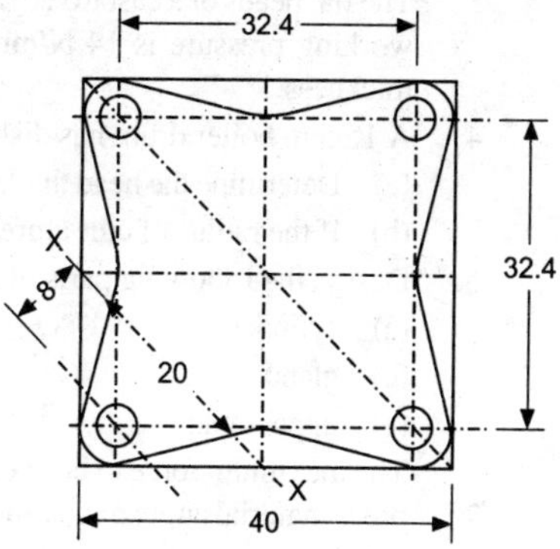

Fig. 25.10 *Upper cylinder flange*

6. **Top Plate:** It can be designed as a rectangular flat plate, uniformly loaded and supported at the four corners. The thickness of the plate whose material is taken as C.S. is given as:

$$t_c = a \cdot b \sqrt{\frac{p}{2 f_t (a^2 + b^2)}}$$

$$a = b = 300 \text{ mm}, f_t = 105 \text{ N/mm}^2$$

$$p = \text{intensity of pressure} = \frac{60 \times 1000}{300 \times 300} = 0.67 \text{ N/mm}^2$$

$$\therefore \quad t_c = 300 \times 300 \sqrt{\frac{0.67}{2 \times 105 \times 18 \times 10^4}} = 37.9 \text{ mm, say 40 mm.}$$

7. **Sliding Platform:** This is nothing but a square plate, loaded uniformly on the whole face and supported at the centre on a circle of diameter of the ram. The analysis of this plate can be approximated by considering it as a circular plate. The relation used to find the thickness is due to "Grashof", i.e.,

$$t_c = \frac{d}{2} \times \sqrt{\frac{p}{f_t}} \left[2 \log_e \frac{d}{d_o} + 2.5 \right] \text{mm},$$

where $\quad d$ = diameter of plate = 300 mm

d_o = diameter of support = 75 mm

f_t for M.S. = 56 N/mm^2

$$p = \text{intensity of pressure} = \frac{60 \times 1000}{\frac{\pi}{4} \times 300^2} = 0.848 \text{ N/mm}^2$$

$$\therefore \quad t_c = \frac{300}{2} \sqrt{\frac{0.848}{56.0}} \left[2 \log_e \frac{300}{75} + 2.5 \right] = 56 \text{ mm.}$$

EXERCISES

1. A circular plate 0.60 m in diameter and supported but not fixed at the edges, is subjected to a uniformly distributed load of 0.875 N/mm^2. The material is C-45 steel rolled. Take factor of safety of 2.5 based on the yield strength. Determine the thickness of the plate.
2. The cylinder head of a steam engine is 0.30 m in diameter and is subjected to a steam pressure of 1.75 N/mm^2. Assuming that the head is a circular C.I. plate with ultimate tensile strength of 140 N/mm^2, supported but not fixed at the edges, find the thickness of the plate for a design factor of 6.
3. The flat heads of a cast iron cylinder are cast integral with the cylinder walls. The inside diameter is 0.15 m and the working pressure is 14 N/mm^2 and the permissible stress is 24.5 N/mm^2. Determine the head and the wall thickness.
4. A 1.35 m boiler drum has dished heads of 1.20 m radius of curvature. The boiler pressure is 1.725 N/mm^2.
   (a) Determine the head thickness.
   (b) If the radius of curvature is 0.90 m, what thickness is required?
5. Design the following parts of a hydraulic press, for the given data:
   (a) cylinder (b) pillars
   (c) gland (d) lower movable plate
   (e) upper table (f) plunger.
   The maximum force to be exerted is 800 kN and the normal working pressure of water is 20 MPa. It is required to press materials upto a maximum size of 0.9 × 1.00 m high and stroke length is 0.10 m. The allowable stress may be taken as:
   Compressive stress for cast steel = 100 MPa
   Compressive stress for cast iron = 20 MPa
   Compressive stress for mild steel = 80 MPa
   Tensile stress for M.S. = 80 MPa (P.U. 1971, 74)
6. A hydraulic testing machine has a maximum capacity of 100 kN. The piston diameter is 0.25 m. Calculate the wall thickness of cylinder. Also design the necessary gland, gland bolts and hydraulic seal for the cylinder. Give a neat sketch of the cylinder gland in position.

COMPOUND CYLINDERS

In case of thick cylinders subjected to very high internal pressures, $p_i \geq f_t$, (examples: high pressure cylinders, gun barrels and so on), no thickness of single cylinder will avoid its failure, as is clear from equations 24.6, 24.7 and 24.8. In equation 24.6, the denominator will become zero or negative and in equations 24.7 and 24.8, it will become negative. This problem is overcome by prestressing the cylinder before it is put into service, thereby, inducing residual compressive stresses in the material. When such a cylinder is put into use, firstly the residual compressive stresses will be relieved and then the tensile stresses will be induced. Thus, the internal pressure capacity of the cylinder gets enhanced. This process of inducing residual compressive stresses in the cylinder is called: "Autofrettage" or "Self-hooping". One way of doing this is the use of compound cylinders. A compound cylinder consists of two cylinders with an interference fit. The inner diameter of the outer cylinder (also called Jacket) is slightly smaller than the outer diameter of the inner cylinder. On heating the outer cylinder, it expands and passes over the inner cylinder. On cooling, the jacket gets shrunk on the inner cylinder, producing an interference fit. This produces contact pressure, p, at the interface or junction of the jacket and the inner cylinder, resulting in compressive tangential stresses being induced in the inner cylinder and tensile tangential stresses in the jacket.

If d_1 = inner diameter of the cylinder.
d = outer diameter of the cylinder, or
= inner diameter of the jacket, also known as common/interface/junction/rated diameter of mating surfaces.
d_2 = outer diameter of the jacket.

Then $d = \sqrt{d_1 \times d_2}$, for maximum pressure.

The unit contact pressure at the interface is give as,

$$p = \frac{\delta}{d\left(\frac{a}{E_1} + \frac{b}{E_2}\right)}, \text{N/mm}^2.$$

where δ = Shrinkage allowance, sum of increase in inner diameter of the jacket and decrease in the outside diameter of the cylinder.

E_1 and E_2 = Young's moduli of the materials of cylinder and jacket.

a and b = The coefficients determined by Lame's formula

For thick walled cylinders, deformed symmetrically relative to their axes

$$a = \left(\frac{d^2 + d_1^2}{d^2 - d_1^2} - \mu_1\right); \quad b = \left(\frac{d_2^2 + d^2}{d_2^2 - d^2} + \mu_2\right)$$

where μ_1 and μ_2 = values of Poisson's ratio

$\cong 0.30$ for steel; 0.25 for C.I. etc.

If both the cylinders are of the same material, then $E_1 = E_2$ and p will be given as,

$$p = \frac{E\delta}{d}\left[\frac{(d^2 - d_1^2)(d_2^2 - d^2)}{2d^2(d_2^2 - d_1^2)}\right].$$

If the size of the members and the mechanical properties of their materials are known, then the value of, p, can be determined from the above equation, provided, the design interference 'δ' is known. 'p' will act as external pressure on cylinder and internal pressure on the jacket.

Stress-analysis due to contact pressure: For compound cylinders made of ductile material, and open at ends, Birnie's equations will be used. Since the ends are open, there will be no longitudinal stress in the material. Out of the other two principal stresses, the tangential (hoop) stress, f_t, is of major interest as compared to the radial stress, f_r, which will always be compressive and it vanishes at the outer surface.

Tangential stress at the inner surface of the cylinder

$$= -\frac{2pd^2}{d^2 - d_1^2} \text{ (Compressive)}$$

Tangential stress at the outer surface of the cylinder

$$= -p\left[\frac{d^2 + d_1^2}{d^2 - d_1^2} - \mu\right] \text{ (Compressive)} \qquad \text{...(A)}$$

Tangential stress at the inner surface of the jacket,

$$= p\left[\frac{d_2^2 + d^2}{d_2^2 - d^2} + \mu\right] \text{ (Tensile)}$$

Tangential stress at the outer surface of the jacket,

$$= \frac{2pd^2}{d_2^2 - d^2} \text{ (Tensile)}$$

Now, when this compound cylinder is subjected to internal pressure, p_i, then, the tangential stress at any radius is given as,

$$f_t = (1-\mu)\frac{p_i\, d_1^2}{(d_2^2 - d_1^2)} + (1+\mu)\frac{p_i\, d_1^2\, d_2^2}{4r^2\,(d_2^2 - d_1^2)}\text{; tensile}$$

(*i*) At the inner surface of the cylinder, i.e., $r = \frac{1}{2}\, d_1$

$$f_t = (1-\mu)\frac{p_i\, d_1^2}{d_2^2 - d_1^2} + (1+\mu)\frac{p_i\, d_2^2}{d_2^2 - d_1^2},$$

(*ii*) At $r = \frac{1}{2}\, d$, that is, at the outer surface of the cylinder or the inner surface of the jacket

$$f_t = (1-\mu)\cdot\frac{p_i\, d_1^2}{d_2^2 - d_1^2} + (1+\mu)\frac{p_i\, d_1^2\, d_2^2}{d^2\,(d_2^2 - d_1^2)}\text{, Tensile}$$

(*iii*) At $r = \frac{1}{2} d_2$, that is, at the outer surface of the jacket

$$f_t = (1-\mu)\frac{p_i\, d_1^2}{d_2^2 - d_1^2} + (1+\mu)\frac{p_i\, d_1^2}{d_2^2 - d_1^2} \quad \text{...(B)}$$

$$= \frac{2p_i\, d_1^2}{d_2^2 - d_1^2}\text{, Tensile}$$

Resultant Stresses at different surfaces of the compound cylinder can be obtained by combining stresses given by (A) for contact pressure and those given by (B) for internal pressure.

Note: If the material is brittle, uses Lame's equation.

Other Methods of Auto-frettage: 1. Closely winding wire under tension, around the cylinder, thereby, inducing residual compressive stresses in the material.

2. By overloading the cylinder before it is put into service. For this, the overloading pressure is such that in a portion of the cylinder near its inner surface, the stresses induced are in the plastic range and thereby undergoes permanent deformation. When the overloading pressure is released the outer portion of the cylinder will induce residual compressive stresses in the inner surface.

Problem: A compound cylinder is formed by a shrinkage process. The dimensions are: d_2 = 60 cm; d_1 = 20 cm; d = 40 mm. The maximum principal stress induced is 100 MPa. Determine the contact pressure and the original dimensions of the inner cylinder and the jacket. Take E = 210 GPa, for steel and μ = 0.3.

Solution: The maximum principal tangential stress (tensile) will occur at the inner surface of the jacket and is given as:

$$f_t = p\left[\frac{d_2^2 + d^2}{d_2^2 - d^2} + \mu\right], \quad \text{see eqn. (A)}$$

from here $p \cong 34.50$ MPa. $\cong 34.50$ N/mm^2.

$$\therefore \quad \delta = \frac{p\cdot d}{E}\left[\frac{2d^2\,(d_2^2 + d_1^2)}{(d^2 - d_1^2)(d_2^2 - d^2)}\right] \cong 0.0285 \text{ mm.}$$

$\therefore$ Outside diameter of inner cylinder = 40 mm, and

Inner diameter of jacket = 40 – 0.0285 = 39.9715 mm

Problem: In the above problem, if the compound cylinder is subjected to internal pressure of 300 MPa, determine the stress distribution in the material.

Solution: (a) **Stresses due to contact pressure:**

Tangential stress at the inner surface of the cylinder $= -\dfrac{2pd^2}{d^2 - d_1^2} = -92$ MPa.

Tangential stress at the outer surface of the cylinder $= -p\left(\dfrac{d_1^2 + d^2}{d^2 - d_1^2} - \mu\right) = -48.30$ MPa.

Tangential stress at the inner surface of the jacket $= p\left[\dfrac{d_2^2 + d^2}{d_2^2 - d^2} + \mu\right] = 100$ MPa.

Tangential stress at the outer surface of the jacket, $= \dfrac{2pd^2}{d_2^2 - d^2} = 55.2$ MPa.

(b) **Stresses due to Internal Pressure:** $pi = 300$ MPa.

(i) **Inner surface of cylinder:**

$$f_t = (1-\mu)\frac{p_i d_1^2}{d_2^2 - d_1^2} + (1+\mu)\frac{p_i d_2^2}{d_2^2 - d_1^2} = 465 \text{ MPa.}$$

(ii) Outer surface of cylinder or inner surface of jacket

$$f_t = (1-\mu)\frac{p_i d_1^2}{d_2^2 - d_1^2} + (1+\mu)\,p_i \frac{d_1^2 \cdot d_2^2}{d^2(d_2^2 - d_1^2)} = 135.9375 \text{ MPa.}$$

(iii) Outer surface of the jacket

$$f_t = \frac{2p_i d_1^2}{d_2^2 - d_1^2} = 75 \text{ MPa.}$$

(c) **Resultant Stresses:**

(*i*) At $r = \dfrac{d_1}{2}$ $= -92 + 465 = 373$ MPa (Tensile).

(*ii*) At $r = \dfrac{d}{2}$; outer surface of cylinder $= -48.30 + 135.9375 = 87.6375$ MPa

Inner surface of Jacket $= 100 + 135.9375 = 235.9375$ MPa

(*iii*) At $r = \dfrac{d_2}{2}$ $= 55.2 + 75 = 130.2$ MPa.

❑❑❑

Section V

I.C. ENGINE PARTS

CHAPTER 26

General Design

26.1 SELECTION OF TYPE

The selection of the type of I.C. Engine will depend upon the power output needed and the type of service required. The following factors may be considered while arriving at the type of the engine.

(a) **Type of Process:** The two processes on which I.C. Engine works are "Otto Process" and "Diesel Process". At present, the Diesel process is usually used for very large engines in marine and stationary service and also for heavy duty automobiles, whereas the Otto process is used in very large gas engines. For small engines, either of the process may be used, but first cost always favours Otto process. The type of process will also be decided by the fuel cost.

(b) **Type of Cycle:** Both two stroke and four stroke cycles have their advantages and disadvantages and the selection of one depends upon personal preference and past experience. Where weight and space count, two stroke cycle will be preferred, and for low fuel consumption and long runs at full load, the four stroke cycle has the advantage. Small oil engines for which simplicity is more important than fuel economy are built on two stroke cycle. Similarly large oil engines for marine service work on two stroke cycle due to weight considerations. Light weight high speed petrol engines usually work on four stroke cycle. Medium size oil and petrol engines are built to work on either cycle.

(c) **Number of Cylinders:** For smoother torsional moment on crank shaft, lower engine weight and higher efficiency, it is better to use a greater number of small cylinders instead of less number of large cylinders. For this, the small cylinders must run at high engine speed. Large cylinders are used as in low crankshaft speeds are needed, as in marine propulsion. Therefore, the desirable crankshaft speed is an important factor in the selection of size and number of cylinders. The power of the engine can be increased by a greater number of cylinders instead of increasing its dimensions. For horizontal engines, the number of cylinders may be one, two or four. These may be arranged either all in parallel or in tandem and double tandem arrangement for larger sizes. For vertical engines, the number of cylinders can be from one to sixteen.

(d) **Arrangement of cylinders:** The arrangement of cylinders should be such as to give balance and degree of compactness to the engine depending on the service. Low speed engines and for which floor space is no consideration, are built with horizontal cylinders. For higher speeds, vertical-in-line arrangement is preferred. For compact, shorter and lighter constructions, V-type and radial engines are best.

(e) **Single and Double acting:** Due to the complexity of the construction and operation in the case of double acting engines, the great majority of I.C. Engines are built single acting. Double acting engines are used only for very large power outputs with large cylinders.

(f) **Engine Speed:**

1. **Rotational Speed:** The rotational speed is usually fixed by the use to which the engine is to be put or by the speed of the machinery to which it is to be connected. There are no set rules for its selection. The usual range of rotational speed is from 100 to 5000 rev/min, with a tendency to use higher speeds to lower the weight and cost of engine.
2. **Piston Speed:** As for piston speed, the eventual use of the engine has considerable bearing on its selection. For present engines, it varies from about 3 to 18 m/s with a tendency to use higher values to lower the weight and cost of the engine. As piston speeds have direct effect on gas velocities, the four stroke engine of lower stroke-bore ratio (below 1.25) can run at higher piston speeds than the long strokes types. This is due to the fact, that in lower ratio engines, larger valve ports can be had in the cylinder head which will keep the gas velocities within limits. On the other hand, higher piston speeds will raise the inertia stresses but these can be reduced by using aluminium-alloy pistons.
3. **Speed Factor:** The engines may be classified as "low speed", "medium speed" and "high speed". Neither the rotational speed "n" nor the 'piston Speed' 'C' are helpful for this purpose because rotational speed does not take into account the size of the engine and piston speed depends upon the stroke of engine. Due to this a term "speed factor, C_s" is used, which is given as,

$$C_s = \frac{n \cdot C}{10^5}, \text{ where } C \text{ is in m/min.}$$

Now $$C = 2nL$$

where L is the piston stroke in metres.

$$\therefore \quad C_s = \frac{2n^2 L}{10^5}$$

The engine is classified as,

(i) Low speed if C_s is less than 1.
(ii) Medium speed if C_s is between 1 and 3.
(iii) High speed if C_s is between 3 and 9.
(iv) Super speed if C_s is > 9.

The speed is limited by internal stresses, the inability of certain parts to function properly and the rapid decrease in volumetric efficiency at high speeds. Practice indicates that the maximum desirable mean piston speeds range from 12.5 to 17.5 m/s for aircraft and automobile engines. For heavy duty stationary engine including high speed diesel engines, the maximum mean piston speeds range from about 7.5 to 12.5 m/s. The value for large gas diesel engines is 5 to 7.5 m/s. High piston speeds are used in small stroke engines, whereas low piston speeds are used in large stroke engines.

(g) **Stroke Bore Ratio:** The performance and design of the engine is indirectly affected by *L/D* ratio. If *L/D* is about unity, then the valve sizes can be increased resulting in reduced gas velocities even if the piston speed is high. But there might be certain difficulty in the design of combustion chamber. A certain clearance is always there between the piston head and the cylinder and the air trapped in this space is useless for combustion purpose and this space varies as the square of the bore. Therefore, from the consideration of combustion chamber design, it has been found that *L/D* = 1.5 gives best results for four strokes. In the case of two stroke engine, a slightly higher value may be used. The dimensions of the engines can be reduced to about one third, i.e., three times the reduction in piston stroke. But to obtain the same piston speed and output, crankshaft speed will have to be increased. Again for high surface volume ratios, the *L/D* ratio should be about unity. In practice *L/D* ratio, varies from 1 to 2, the more common value being towards unity. Fair average values for *L/D* ratios is 1.0 for air cooled air craft engines and 1.2 to 1.35 for all other engines.

26.2 PRINCIPLE OF SIMILITUDE

According to the principle of similitude, similar engines are constructed of the same material and their dimensions are proportional to same dimension, usually the cylinder bore. The ratio of the lengths, widths and thicknesses will be similar and so, the similar engines are scale reproductions of one another. The similarity must be geometrical, kinetical and dynamical. Hence the similar engines have:

1. The same *L/D* ratio.
2. The same compression ratio.
3. The same mean effective pressure.
4. The same volumetric efficiency.
5. The same speed factor.

Therefore, the relationship between the cylinder bore and the various parameters of the engine is:

(a) RPM $\alpha\, D^{-1}$
(b) Piston displacement $\propto D^3$.
(c) Weight of any part $\propto D^3$.
(d) Bearing areas $\propto D^2$.

(e) Weight/Piston displacement $\propto \dfrac{D^3}{D^3}$ = const.

26.3 GENERAL DESIGN CONSIDERATIONS

The size of the engine as a whole is determined on the basis of mean cylinder pressure and the speed, but the design of the individual parts depends on the maximum cylinder pressure and the acceleration of the moving parts. The stresses in the various parts of the engine come from the following types of loads.

1. Due to gas pressure in the cylinder.
2. Due to inertia and centrifugal forces.
3. Due to frictional forces.
4. Due to torsional moment reaction and weight.
5. Due to vibrations, especially of the crank shaft.
6. Thermal loads.

The gas pressure should be regarded as the shock load while designing the running gear.

The inertia load is proportional to the square of the speed. The inertia load due to the reciprocating masses is in opposition to the gas load at the TDC and so it reduces the load on the connecting rod and the bearings. On TDC of the exhaust stroke in four stroke engines, the inertia load will induce a tensile load in the connecting rod. Hence there will be some reversal of stress in the connecting rod.

The load on the valve gear is entirely due to inertia forces. For high speed engines, the inertia load will be large due to high accelerations of the valve gear. This would necessitate heavy valve springs.

The centrifugal forces due to the unbalanced rotating masses induce loads on the bearings when the crankshaft rotates. These are in the opposite direction to that of the gas load, but due to their unbalanced nature, induce stresses in other parts of the engine, e.g., the engine holding down bolts.

There will be transverse bending of the connecting rod due to the centrifugal forces set up by the rotating mass of the connecting rod. Fortunately this load is maximum when the gas load is very much reduced and vice versa, otherwise a complex type of compound fluctuating stresses would occur in the connecting rod. However, this is a reversible load on the connecting rod.

Centrifugal forces due to balanced mass will induce hoop stresses in the material of the parts and this should be kept within limits.

If friction forces are large, failure of some parts may occur due to overheating and seizure. When the crankshaft rotates, the load due to the turning moment comes on the crankshaft coupling and the foundation bolts. The bolts are usually made quite strong to resist the forces due to unbalanced couples, however, turning moment should also not be neglected. In some cases, there might be end thrust on the crankshaft and the bearings should be designed accordingly.

The stresses due to vibrations are very important, especially, in the case of crankshaft. To reduce their effects, the designer should observe the following points:

1. The dimensions should be suitably selected to avoid resonance.
2. The parts should be made quite stiff to reduce the amplitude of vibrations.
3. The material and the shape of the parts should be such as to avoid failure by fatigue.
4. Damping in the shaft should be increased by the use of suitable materials.

There will be thermal loads also on the engine parts since about one half of the unused heat leaves through the water jacket. There would be temperature gradients in the metal and due to this stresses will be set up because of unequal expansion. The stresses due to thermal loads will be less if the section is thinner. While designing the various parts of the internal combustion engine, the following temperature effects should be considered:

1. Expansion of metals.
2. Stresses induced in metals due to temperature differences.
3. Reduction in strength of metals at high temperatures.
4. Deterioration of the metal surfaces which are subjected to high and varying temperatures.
5. Growth of Cast Iron.

EXERCISES

1. What is the effect of number of cylinders on the performance of an I.C. Engine?
2. What is the effect of stroke bore ratio on the design of an I.C. Engine?
3. What is the importance of Principle of Similitude in the design of an I.C. Engine?
4. Determine the piston speed, speed factor, and speed classification of 0.25 m × 0.30 m engine running at 515 rev/min.
5. Discuss the various design considerations for an I.C. Engine.

□□□

CHAPTER 27

Design of Principal Parts

27.1 CYLINDER, CYLINDER HEAD

27.1.1 Design of Cylinder

The cylinder of an internal combustion engine performs a number of duties. Its primary function is to contain the working fluid and the secondary function is to guide the trunk piston.

(a) Construction: In very small engines, the cylinder, the water jacket and the frame may be manufactured as one piece, while in smaller engines, the cylinder and the jacket can be cast as one piece. But all the larger engines and most of the high speed engines have separate cylinder liners or sleeves. Separate liners are more economical as these can be easily replaced after wear and tear. Moreover better grade, wear resistant cast iron can be used for liner and a cheaper grade for jacket. Also this construction takes care of the longitudinal expansion.

The cylinder liners are made in two types: "Wet Liner" and "Dry Liner". In the case of wet liner, water in jacket is in direct contact with the outer wall of the liner, whereas the dry liner is pressed into the cylinder proper. The two types are shown in Figs. 27.1 and 27.2. In the engines over about 13 cm bore, usually the wet type of liner is used.

The advantages of dry liner are:

1. It is simpler to replace a dry liner.
2. There is no danger of water leakage either into the crankcase or the combustion chamber.
3. Due to the absence of a heavy flange at the top of the liner, cylinder centres can be reduced.
4. There is better cooling of the upper part of the liner.

Its disadvantages are:

1. Complicated casting.
2. Decreased heat flow through the composite wall.

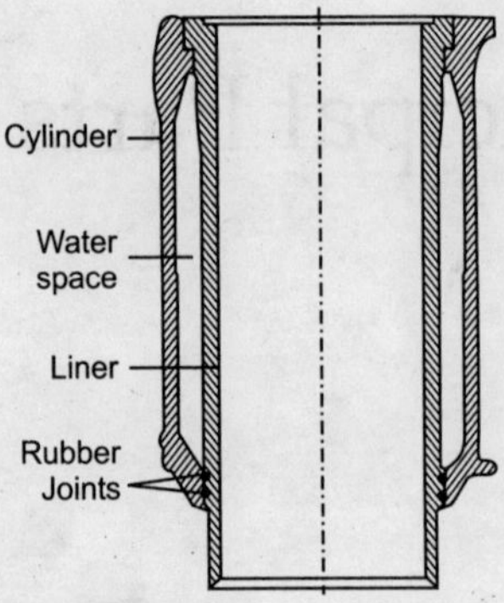

Fig. 27.1 *Wet liner*

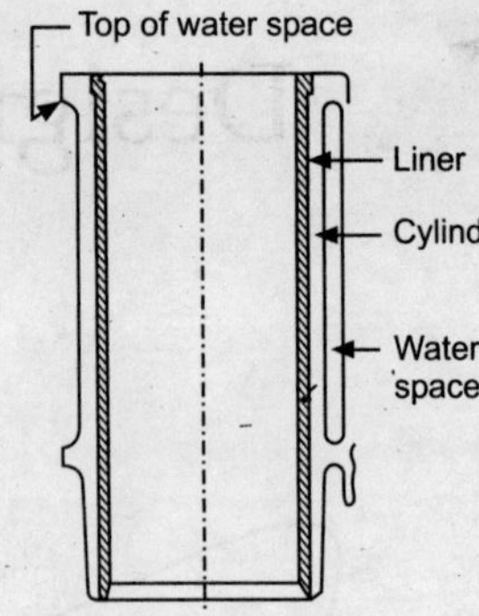

Fig. 27.2 *Dry liner*

The advantages of wet liner are:

1. The foundry problem is considerably eased, since the large internal cores of the cylinder block can be properly supported.
2. The cylinder block is relieved of the stresses due to longitudinal expansion of the liner.

Its disadvantages are:

1. Difficult replacement.
2. Danger of water leakage into the crankcase and the combustion space if the casting is defective.

The cylinders of the multicylinder engines are cast either separately or in blocks, even in larger sizes.

(b) **Qualities of a good liner:** A good cylinder liner must possess the following qualities:

1. Strength to resist the gas pressure.
2. Sufficiently hard to resist wear.
3. Strength to resist the thermal stresses due to the heat flow through the liner wall.
4. Corrosion resisting.
5. Capable of taking a good bearing surface.
6. It should be symmetrical in shape to avoid unequal deflection due to gas load and unequal expansion due to thermal load.
7. No distortion of the inner surface due to restraining fixings.

(c) **Liner materials:** The liner material should be strong, hard and corrosion resistant and produce a good bearing surface. The liner materials in order of preference are:

1. A good grade grey cast iron with homogeneous and close grained structure, *i.e.*, Perlitic and similar cast irons.
2. Nickle cast iron and Nickle chromium cast iron.
3. Nickle-chromium cast steel with molybdenum in some cases.

The cast iron liners can be centrifugally cast. In large engines, usually the cast steel cylinders are used, but may be with cast iron liners. For aircraft engines, the liners are often turned of alloy steel forgings to get very light cylinders.

(d) **Surface treatment of cylinder liner:** As already has been noted that the inner surface of the cylinder liners should be quite hard to reduce their wear. The following methods are adopted for this purpose:

1. In the case of dry liner, after it has been initially machined, it is heated in an oven to about 840°C. The material of the liner is cast iron with 2.00 percent nickle and 0.65 percent chromium, After heating the liner is quenched in oil and then tempered for about two hours at about 180°C. After this the liner is finished by a honing operation by which the bore is held within close limits of the desired roundness and straightness. The hardness would be about Rock-well C 50.
2. In another method, the cast iron liners are heated from inside by a high frequency induction current. After this these are water quenched giving a surface hardness of Rock well C 52 to 55. The liners are then tempered to lower the hardness. After this, the bore is honed to the final finish.

3. In some cases, the cylinder bore is chromium electroplated. The special procedure adopted gives a very hard, and porous surface. A good oil film is formed and retained in this surface due to which the surface wear is appreciably reduced.

Due to hardened liner surface, the wear of piston surface and the piston rings is also very much reduced.

(e) **Total liner stress:** Two types of stresses are induced in the cylinder liner–thermal stresses and pressure stresses. Compressive stress is induced in the inner fibres of the wall and tensile stress is induced in the outer fibres of the wall due to thermal stresses, while the pressure stress will be throughout tensile. So, on the inside of the wall, compressive stress due to temperature will be relieved by the pressure stress, while on the outside of the wall, the tensile stress due to temperature will be increased by the pressure stress. The temperature stress is proportional to the temperature difference, to the wall thickness and to the modulus of elasticity of the wall material. The pressure stress is also a function of the wall thickness and decreases with its increase.

The temperature stress can be reduced by two methods:

1. By making the wall thickness thinner by the use of a material of better tensile strength.
2. By lowering the temperature difference between the inside and outside of wall by increasing the velocity of water flow through the jacket.

It has been determined by 'Pye' and others that about 60 per cent of the heat loss to cooling water flows through the cylinder liner proper. This includes the heat flowing from the piston head to the liner via the piston rings. The liner failure due to pressure and temperature stresses are almost unknown. The two main reasons of liner failure are distortion and wear.

(f) **Liner Distortion:** Cylinders, cylinder blocks and liners should be designed so as to reduce the distortion to a minimum. The liner distortion can be produced due to the following reasons:

1. Due to temperature stresses: As indicated above, a compressive stress is induced on the inside and a tensile stress is induced on the outside of the wall. Any defect in the casting on the inner wall will change the stress distribution and the liner may go out of round. Similarly any ribs or bosses on the outer wall of the liner impose local restraint on the liner wall, they being under tensile stresses. A circumferential rib will try to distort the liner out of parallel while an axial rib will distort it out of round. Therefore, ribs and bosses on the outer wall of the liner should be avoided as far as possible.
2. Distortion may also occur due to the axial expansion of the liner. Due to this reason, one end should be left free.
3. Distortion is also caused by the gas pressure.
4. Distortion may also be caused by bolting on the cylinder head or any other part.
5. Distortion due to fixing, especially at the lower water joint. Here rubber rings are used and as it is not compressible, if it is pushed out of shape, it must have room to move.

(g) **Cylinder Lubrication:** The importance of cylinder lubrication may be seen from the fact that even with careful arrangements for cylinder lubrication, about 3 to 6% of the indicated power of the engine is lost due to friction between the cylinder liners and the piston rings to maintain a satisfactory gas seal. Just sufficient lubrication of the liner is needed as excess of oil will do more harm than good. When the engine is running and attains its operating temperature, a splash of lubricating oil is thrown up from the big end and the main bearings of the crank shaft. Due to this splash, the cylinder walls are very well lubricated. But in the beginning, when the engine starts from cold, the oil will just ooze from the bearings, being quite thick. Due to this reason, for initial running period of the engine, the cylinder walls might be oil starved, and there might be high wear of the liner, For this reason, a separate arrangement should be provided for cylinder lubrication. Such an arrangement is shown in Fig. 27.3.

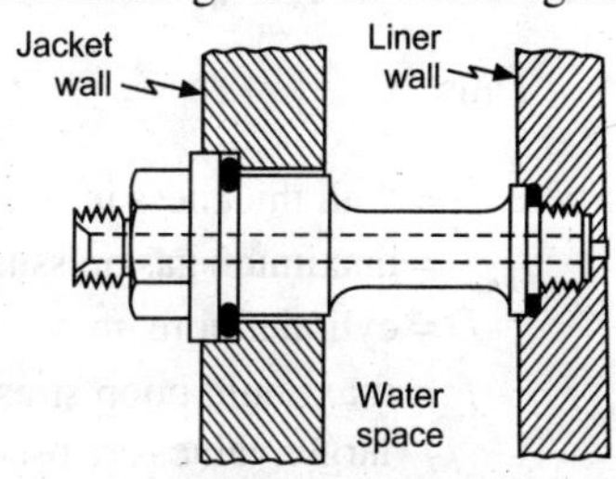

Fig. 27.3 *Liner lubrication*

(h) Liner Joints: The upper joint should not touch at the corners. The stress concentration should be avoided. The joint may be metal to metal or a copper joint may be used. Some typical joints are shown in fig. 27.4. At the lower or bottom free end of the liner, expansion joints are provided to prevent water leakage. Some typical lower liner water joints are shown in Fig. 27.5.

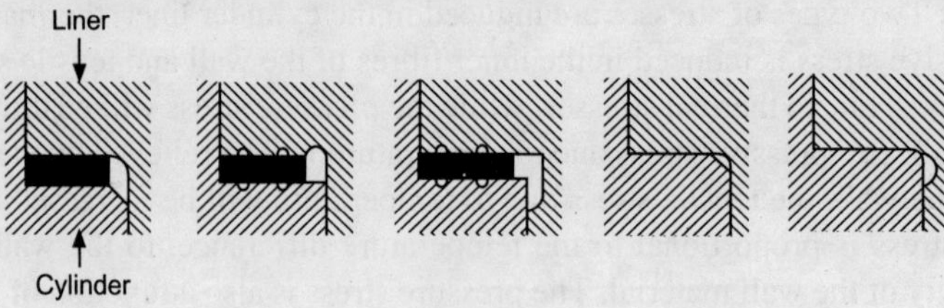

Fig. 27.4 *Upper liner water joints*

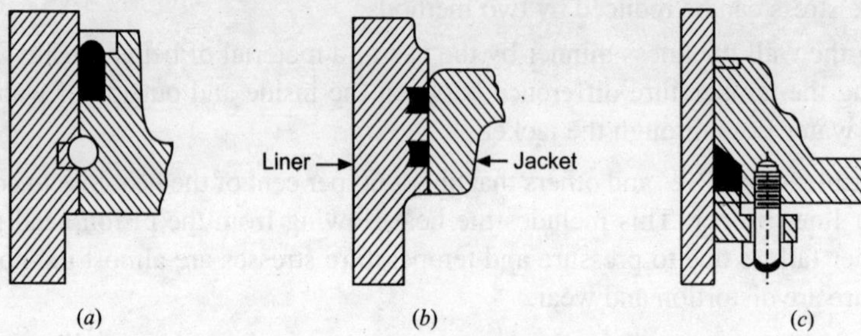

Fig. 27.5 *Lower liner water joints*

(i) Wall Thickness: The cylinder wall thickness is stressed due to gas pressure and the piston side thrust. There may be bending of the wall due to the piston side thrust. But at the point of maximum side thrust, the gas pressure has decreased sufficiently, so the stress in the wall due to side thrust will be very small and can be neglected. Due to gas pressure, two types of stresses will be set up in the cylinder wall, longitudinal and circumferential. These act at right angle to each other and so the net stress in each direction is reduced. The longitudinal stress is usually small and can be neglected.

Net $$f_l = f_l - \frac{f_c}{m}$$

and Net $$f_c = f_c - \frac{f_l}{m},$$

where f_l = longitudinal stress.

f_c = circumferential stress.

$\frac{1}{m}$ = Poisson's ratio.

The wall thickness is usually calculated by applying the formula for a thin cylinder.

Thus $$t = \frac{p_{max.} \times \text{D}}{2f_c} + k$$

where t = wall thickness in mm

$p_{max.}$ = maximum gas pressure, N/mm^2

D = cylinder bore, mm

f_c = maximum hoop stress and is equal to 35.0 to 105.0 N/mm^2 depending on the size and material, larger values are used for smaller bores.

k = reboring factor, given in Table 27.1.

Table 27.1

Cylinder bore in mm	75	100	150	200	250	300	350	400	450	500
k, in mm	1.5	2.3	4.0	6.0	7.5	9.5	10.5	12.5	12.5	12.5

The thickness of the cylinder walls usually varies from about 4.5 mm to more than 25 mm, depending upon the cylinder size. According to an empirical relation, for liners of oil engines,

$$t \geq \frac{D}{15}$$

near the top portion and through 20% of the stroke. For dry liners, the total thickness 't' is the thickness of the liner and that of cylinder wall. The thickness of the dry liner is given as

$$t' = 0.030\,D \text{ to } 0.035\,D$$

The thickness of the inner walls of the auto-mobile engine cylinders is usually given empirically as

$$t = 0.045\,D + 1.6 \text{ mm}$$

The thickness of jacket wall is given as

$$= \frac{1}{3} \text{ to } \frac{3}{4} t, \text{ larger ratio for smaller cylinder.}$$

or $$= 0.032\,D + 1.6 \text{ mm}$$

The water space between the outer cylinder wall and the inner jacket wall is

= 10 mm for a 75 mm cylinder to about 75 mm for a 750 mm cylinder.

or $$= 0.08\,D + 6.5 \text{ mm}$$

Example 27.1. Determine the thickness of a cast iron cylinder wall and the stresses for a 300 mm Petrol, engine, with a maximum gas pressure of 3.5 N/mm².

Solution: For cylinder diameter of 300 mm the reboring factor is 9.5 mm

∴ Wall thickness $$t = \frac{p_{max} \times D}{2 f_c} + k$$

$$p_{max.} = 3.5 \text{ N/mm}^2;\ D = 300 \text{ mm}$$

f_c may be taken as 45 N/mm²

∴ $$t = \frac{3.5 \times 300}{2 \times 45.0} + 9.5 = 21.5 \text{ mm}$$

Now apparent, longitudinal stress, $$f_l = \frac{\text{Force}}{\text{Area}} = \frac{\frac{\pi}{4} D^2 \times p_{max.}}{\frac{\pi}{4}(D_o^2 - D^2)} = \frac{p_{max.} \times D^2}{D_o^2 - D^2}$$

Now $$D_o = D + 2t = 300 + 43 = 343 \text{ mm}$$

∴ Apparent $$f_l = \frac{3.5 \times 300^2}{343^2 - 300^2} = 11.45 \text{ N/mm}^2$$

Again, apparent $$f_c = \frac{\text{Force}}{\text{Area}} = \frac{p_{max} \times D}{2t} = \frac{3.5 \times 300}{2 \times 21.5} = 24.4 \text{ N/mm}^2$$

∴ Net $$f_l = f_l - \frac{f_c}{m}$$

Now $\frac{1}{m}$ is usually equal to $\frac{1}{4}$

$\therefore$ Net $f_l = 11.45 - \frac{24.4}{4} = 11.45 - 6.1 = 5.35 \text{ N/mm}^2$

Net $f_c = 24.4 - \frac{11.45}{4} = 22.4 - 2.86 = 21.54 \text{ N/mm}^2$

(j) Cylinder Diameter and Length:

The output of a given cylinder can be written as

$$\text{Power} = \frac{pLAn}{60}, \text{W}$$

where L = stroke in m

A = Piston area, mm^2

n = Number of working strokes per minute.

p = indicated *m.e.p.* if power is indicated and is *B.mep.* if output is in Brake power, N/mm^2.

Now η_m (mechanical efficiency) $= \frac{\text{Brake power}}{\text{Indicated power}} = \frac{\text{B.P.}}{\text{I.P.}}$

Note: As a guide, the maximum gas pressure can be taken as 9 to 10 times the brake mean effective pressure.

With the help of above equation, the diameter and length of the cylinder can be calculated very easily.

Example 27.2. A vertical four stroke compression ignition engine has the following specifications:

Brake power = 4.5 kW

Speed = 1200 rev/min

I. *m e p*. = 0.35 N/mm^2

η_m = 0.80

Determine the dimensions of the cylinder.

Solution: Since $\eta_m = \frac{\text{Brake power}}{\text{Indicated power}}$

$\therefore$ Indicated Power $= \frac{4.5}{0.8} = 5.625 \text{ kW}$

$\therefore$ Indicated Power $= \frac{p\,\text{LA}n}{60}$

Now $n = \frac{1200}{2} = 600$

$\therefore$ $5.625 \times 10^3 = \frac{0.35 \text{ LA } 600}{60}$

$\therefore$ $LA = \frac{5.625 \times 60 \times 10^3}{0.35 \times 600} = 1.608 \times 10^3$

$\therefore$ $L \times \frac{\pi}{4} \times D^2 = 1.608 \times 10^3$

Now $\frac{L}{D}$ can be taken as 1.35, L being in mm

$\therefore$ $\frac{\pi}{4} \times \frac{1.35\,D}{1000} \times D^2 = 1.608 \times 10^3$, gives $D = 115$ mm

∴ Diameter of cylinder = 115 mm

Stroke = 115 × 1.35 = 155 mm

Now Length of cylinder = stroke + clearance on both sides.

Now clearance on both sides may be taken as 10 to 15% of the stroke.

∴ Length of cylinder = 155 + 155 × 0.15 = 178.5 mm

(k) Cylinder Flanges and Studs: The cylinder is either cast integral with the upper half of the crank case or attached to it with the help of flanges, studs and nuts.

The cylinder flange is made thicker than the wall of the cylinder.

Flange thickness should not be less than 1.1 to 1.25*t*.

Common value for flange thickness = 1.2 to 1.4 *t*.

or = 1.25 to 1.50 *d*,

where *d* = bolt diameter, nominal.

The distance of the end of the flange from the centre of the stud or bolt should not be less than $d + 6$ mm, and not more than 1.5 *d*.

The use of studs decreases the bending stress at the flange root since the moment arm can be made very small.

The material of the studs or bolts is usually nickle steel with a yield point of 630 to 945 MPa.

The diameter of the bolt or stud is calculated by equating of the gas load to the area of all the studs at the root of the threads multiplied by the allowable fibre stress.

Therefore, $$\frac{\pi}{4} D^2 \times p_{max} = z \times \frac{\pi}{4} d_c^2 \times f_t$$

∴ $$p_{max} \times D^2 = z d_c^2 f_t.$$

∴ $$d_c = D\sqrt{\frac{p_{max}}{z f_t}},$$

where f_t = allowable fibre stress, 35 to 70 N/mm²

d_c = core diameter.

Low value of f_t is taken since there is already high stress in the studs due to tightening of the nuts.

The number of studs '*z*' may be taken as

$$\left(\frac{D}{100} + 4\right) \text{ to } \left(\frac{D}{50} + 4\right), D \text{ in mm}$$

or the pitch of the bolts may be taken as $19\sqrt{d}$ to $28.5\sqrt{d}$, where *d* is in mm.

In practice, *d* generally varies from $\left(\frac{3}{4} \text{ to } 1\right)$ times the thickness of flange. In no case, *d* should be less than 16 mm.

(l) Cylinder Head: Usually a separate cylinder cover or head is provided with all but the smallest engines. A box type section is employed of considerable depth to accommodate parts. The general design of the cover is governed by the following factors along with the strength considerations:

1. Air and gas passages.
2. Accommodation of valves and their gear.
3. Accommodation of the atomiser at the centre of the cover in the case of diesel engines.

Cylinder head is the most difficult part to be designed and manufactured. The cylinder heads are usually made of close grained cast iron or alloy cast iron containing nickle, chromium and molybdenum, for small and medium sized engines, while for large engines, the material is low C-steel.

The thickness of the cylinder wall ranges from about 6.5 mm for small engines to proportionately larger values for large engines. The thickness depends on the shape of the head. If the cylinder head is approximately a flat circular plate, the thickness can be determined by the relation:

$$t = D\sqrt{\frac{Cp_{\max}}{f_t}},$$

where C = constant, in this case equal to 0.1.

f_t = allowable stress, taken to be 35 to 56 N/mm^2.

A low value of 'f_t' is taken because both pressure and temperature stresses are induced in the cylinder head and the above equation is based upon only the cylinder pressure. The heat transfer through the head is about 5 to 13 times as much heat per unit area as the cylinder walls, depending on the design and amount of cooling.

The proportions of a cylinder cover for four stroke engine covers from about 0.10 m to 0.45 m diameter of bore, is shown in Fig. 27.6.

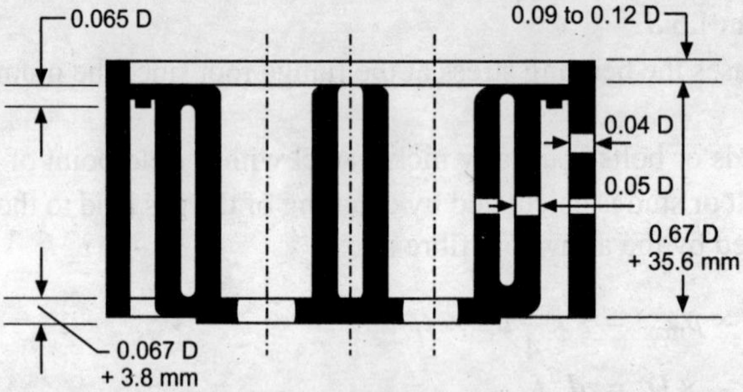

Fig. 27.6 *Cylinder cover*

The typical methods of securing a tight joint between the cylinder and the cylinder cover are shown in Fig. 27.7.

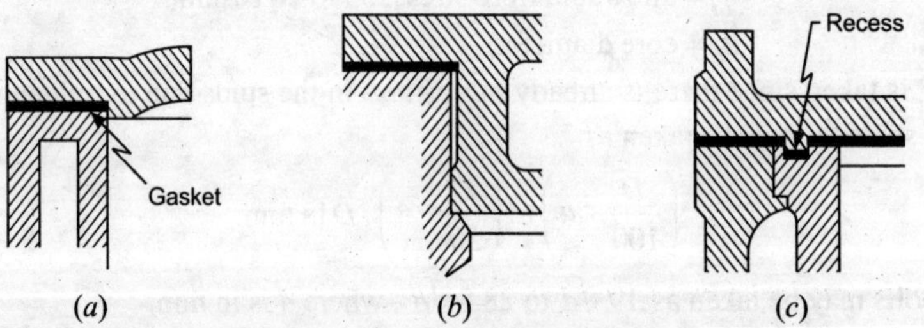

Fig. 27.7 *Cylinder head joints*

The design of cylinder head bolts or studs is similar to as treated above. However, in this case, the area of the cylinder head subjected to gas pressure may be larger than the piston area depending upon the type of joint.

Example 27.3. Determine the thickness of a plain cylinder head for 0.30 m cylinder. The maximum gas pressure is 3.2 N/mm^2. Design the studs also for the cylinder cover.

Solution: The thickness of the cylinder cover is given as

$$t = D\sqrt{\frac{Cp_{\max}}{f_t}}$$

where $D = 300$ mm, $p_{max} = 3.2$ N/mm^2

Take $f_t = 42$ N/mm^2, for good grade cast iron, $C = 0.1$

$$t = 300 \times \sqrt{\frac{0.1 \times 3.2}{42.0}} = 26.2 \text{ mm, say } 26.5 \text{ mm}$$

Studs: The gas pressure will actually act upon the pitch circle diameter of the studs, but as the stud diameter is not known initially, the pressure may be assumed to be acting on the cylinder diameter. Or, it is a common practice that the centre of the stud should be 1.25 *d* to 1.50 *d* from the inner wall of the cylinder.

$\therefore$ Pitch circle diameter $D_p = D + 3d = 300 + 3d$ mm

$$\therefore \quad \text{Load on the stud} = \frac{\pi}{4} D_p^2 \times \text{maximum gas pressure}$$

$$= \frac{\pi}{4} \times (300 + 3d)^2 \times 3.2$$

But, $\text{load} = z \times \frac{\pi}{4} d_c^{\,2} \times f_t$

Let us take core diameter = 0.8 × nominal diameter

$$\therefore \quad d_c = 0.8\, d$$

$$\therefore \quad \frac{\pi}{4} \times (300 + 3d)^2 \times 3.2 = z \times \frac{\pi}{4} (0.8\, d)^2 \times f_t$$

Now
$$z = \left(\frac{D}{100} + 4\right) \text{ to } \left(\frac{D}{50} + 4\right) = \left(\frac{300}{100} + 4\right) \text{ to } \left(\frac{300}{50} + 4\right)$$

= 7 to 10, let us take 8.

Take $f_t = 63 \text{ N/mm}^2$

$$\therefore \quad (300 + 3\, d)^2 \times 3.2 = (0.8\, d)^2 \times 63 \times 8$$

gives $d = 43$ mm, $D_p = 300 + 3 \times 43 = 249$

$$\therefore \quad \text{Pitch of the studs} = \frac{\pi D_p}{z} = \frac{\pi \times 429}{8} = 168.5 \text{ mm}$$

Now minimum pitch should be = $3d$ = 129 mm

and maximum pitch lies between $19\sqrt{d}$ to $28.5\sqrt{d}$, i.e., 124.5 mm to 187 mm.

Therefore, both the conditions are satisfied.

EXERCISES

1. What are the advantages and disadvantages of "Wet Liner" and "Dry Liner"?
2. Enumerate the qualities of a good cylinder liner.
3. List the various liner materials.
4. How would you reduce the distortion of the liners?
5. What is the use of surface treatment of cylinder liners? Describe the various methods.
6. Determine the wall thickness of a 0.30 m gray cast iron gas-engine cylinder; the maximum pressure being 2.1 N/mm^2.
7. Determine the wall thickness of a 0.225 m gas engine cylinder made of nickle cast iron, the maximum gas pressure being 2.2 N/mm^2.
8. Determine the wall thickness of a wet cast iron cylinder liner with a 5 mm thick dry cast iron inside a 0.21 m bore compression ignition oil engine with a maximum cylinder pressure of 5.25 N/mm^2.
9. Determine the thickness of the cylinder head and the number and size of the hold down bolts for the cylinder in problem 7.

27.2 PISTON

(a) **Function:** The piston used in internal combustion engines is usually of Trunk type. Such pistons are open at one end. The main function of the piston is to receive the impulse from the expanding gas and transmit the energy to the crank shaft through the connecting rod. At the same time, the piston must also disperse a large amount of heat from the combustion chamber to the cylinder walls.

Trunk pistons consist of:

1. Head or crown to carry the cylinder pressure.
2. Skirt to act as a bearing for connecting rod side thrust.
3. A piston pin to connect it to the connecting rod.
4. Piston rings to seal the cylinder.

The profile of the piston head or crown will depend upon the design of the combustion chamber. The thermal stresses will depend upon the rating and the efficiency of the engine. Amount of side thrust on the cylinder wall will be a function of the connecting rod/stroke ratio and the gas pressure. Many factors including the bottom end design, i.e., the balancing weights on the crank, etc. will influence the number and type of piston rings.

(b) **Design Objects:** The modern trend is to develop internal combustion engines of increased "power capacity". This has been made possible by improved engine design including increased compression ratio and engine speed. Due to these reasons, the role of the piston which is one of the fastest moving parts of the engine, has become of vital importance. The piston speed may reach 15 m/s or even more and at such high speeds, heavy reciprocating pistons develop high inertia forces, which are undesirable. Therefore, for high speed engines, weight of the piston is of primary importance along with other design factors. In summing up, in the design of internal combustion engine piston, the following objects must be sought:

1. The piston should have enormous strength and heat resistance properties to withstand high gas pressures and inertia forces.
2. Minimum weight to minimise the inertia forces.
3. Good and quick dissipation of heat from the crown to the rings and bearing area and then to cylinder walls.
4. Form an effective gas and oil seal.
5. Sufficient rigid construction to withstand thermal and mechanical distortion.
6. Sufficient bearing area to prevent undue wear.
7. Symmetrical design for even expansion under thermal loads, as free as possible from discontinuities.
8. High speed reciprocation without noise.
9. Minimum work of friction.
10. Material of the piston must possess good wearing qualities, so that the piston is able to maintain the surface hardness upto the operating temperatures.
11. Have little or no tendency towards corrosion or "picking up".

(c) **Materials:** Commonly used materials for the pistons of internal combustion engines are: Cast Iron, Cast Aluminium, Forged Aluminium, Cast Steel, and Forged Steel.

Cast iron is the ideal material for the rubbing surfaces of piston. Cast iron and steel have the high strength required, good wearing qualities at the required temperature, and low thermal expansion. Their disadvantages are that they give heavier piston than aluminium and have low thermal conductivity. However, with proper design, the weight difference can be reduced. Also, if the cast iron pistons are tin plated, they give less cylinder liner wear than the aluminium pistons. A close grained perlitic material is almost universally used for iron pistons. Addition of various alloys depends upon the various manufactures.

Aluminium alloys give light pistons and for better heat dissipation, aluminium alloys are the ideal materials due to their very high thermal conductivity. Aluminium is 3 times lighter than cast iron. Its strength is good at low temperatures but it looses about 50% of its strength at temperatures above about 320°C. Its expansion is about 2½ times that of cast iron and the resistance to abrasion is low at high temperatures. However, these disadvantageous properties of aluminium have now been overcome by alloying it with other materials and by developing advanced designs of pistons. The split skirt, T-slotted as well as cam ground, oval sectioned pistons made from aluminium alloys are mostly used which can be tightly fitted into the cylinder bore to eliminate "Piston Slap". A coating of aluminium oxide or tin on aluminium alloys pistons has been found to be protective against "Scuffing" or "partial siezure" during running in period after overhaul.

The strength of the cast aluminium alloy is approximately the same as that of cast iron. It gives lighter pistons and better heat dissipation but is costlier than cast iron.

The strength of the forged aluminium pistons is greater by about 40% than the aluminium piston made by die casting process and is still lighter. Such pistons are chiefly used for heavy duty, high speed compression ignition oil engines and for aircraft engines.

Alloy cast steel pistons are used in some automotive engines and forged steel pistons are used in some aircraft engines. But both require liners of high surface hardness.

From the above analysis, it may be concluded that for moderately rated engines with piston speeds below 6 m/s, cast iron pistons are employed and for more highly rated engines and those running at higher piston speeds, aluminium alloy pistons are used.

Pistons over 300 mm diameter in four stroke engines or 200 mm in two stroke engines should be oil cooled or water cooled. For very heavy ratings, oil cooling is carried down to as low as 115 mm diameter.

In some cases, with aluminium alloy pistons, there is wear of the top piston ring groove or grooves. For this, C.I. inserts carry the top pair of rings.

A typical cast iron piston of a diesel engine of medium size and speed is shown in Fig. 27.8.

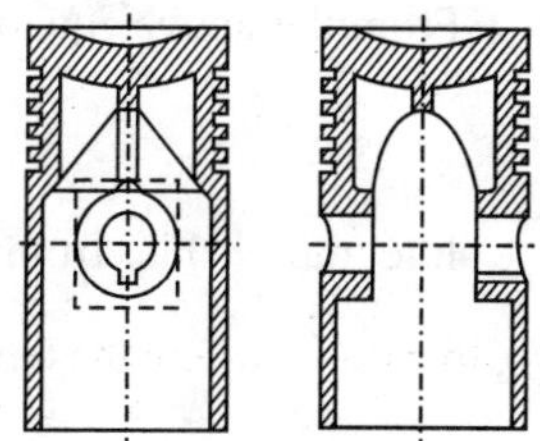

Fig. 27.8 *Trunk piston*

(d) Piston Head: The thickness of the piston head or crown is controlled by two requirements:

(i) Strength
(ii) Heat dissipation.

To provide a large area for heat flow, the crown thickness is made greater than is necessary for strength purpose. Part of the gas load is taken by ribs and piston pin. They also help to dissipate the heat absorbed from the gases.

Sometimes, the head of the piston is made of different material than that of the lower part and joined to it with the help of studs. Its purpose is to use a material of high strength, e.g., cast steel for the head, in large oil engines. Such a construction will help in replacing the head only if it is damaged due to high pressure and temperature. In such constructions, it should be noted that the upper part should never touch the cylinder walls even at the highest temperature.

(i) **Strength:** The head thickness on the basis of strength can be calculated by assuming the head to be a flat plate of uniform thickness and fixed at the edges. The gas load is considered as a uniform load. The following equation can then be applied.

$$t_h = \sqrt{\frac{3 p_{max} D^2}{16 f_t}} = 0.433 D \sqrt{\frac{p_{max}}{f_t}}$$

where t_h = cylinder head thickness, mm
D = cylinder bore, mm
p_{max} = max gas pressure, N/mm^2
f_t = allowable stress in bending, N/mm^2

For gray cast iron, with ultimate strength in tension about 140 N/mm^2, the allowable stress may be taken as 35.0 to 37.5 N/mm^2, using a factor of safety of about 4. For nickle cast iron, semi steel or special aluminium alloy (normalised) with ultimate strength of about 250.0 N/mm^2, f_t may be taken as approximately 60.0 N/mm^2. If the material is forged steel with ultimate strength of about 370.0 N/mm^2, f_t may be taken as approximately 85.0 to 90.0 N/mm^2.

(ii) **Heat Dissipation:** The formula for calculating the head thickness on the basis of heat dissipation can be derived in the following manner, assuming the crown to be flat and of constant thickness:

Heat absorbed by the piston from the gases = Heat flow through the head from centre to the cylinder walls.

$$\therefore \quad \pi r^2 \times q = \frac{dT}{dr} \times k\ 2\pi r \times t_h,$$

where r = Piston radius

q = heat flow from gases, W/m^2

T = temperature at the circumference with radius r, °C,

$\frac{dT}{dr}$ = temperature gradient,

k = Heat conductivity factor, W/m °C

From the above equation,

$$dT = \frac{q}{2k\, t_h} r\, dr$$

Integrating this, from r = 0, i.e., the middle of piston where the temperature may be designated as T_c, to $r = \frac{D}{2}$, where the temperature may be designated as T_e

$$\therefore \quad T_c - T_e = \frac{D^2 q}{16\, k\, t_h}$$

$$\therefore \quad t_h = \frac{D^2 q}{16\, k\, (T_c - T_e)}$$

If H = Heat flow from gases in watts

then $$q = \frac{H}{\frac{\pi}{4} D^2} \quad \therefore \quad qD^2 = \frac{H}{\frac{\pi}{4}}$$

$$\therefore \text{ Head thickness,} \quad t_h = \frac{H}{16\, k (T_c - T_e) \times \frac{\pi}{4}}$$

$$\therefore \quad \boxed{t_h = \frac{H}{12.56\, k (T_c - T_e)}}$$

The heat flowing through the head may be estimated by the formula:

$$H = Cw \times HCV \times BP \text{ kW},$$

where w = mass of fuel used, per kW per second

HCV = Higher calorific value of the fuel, kJ/kg

BP = brake power of engine per cylinder, kW

C = constant representing that portion of the heat supplied to the engine which is absorbed by the piston, approximately 0.05.

BP can be calculated with the help of the following formula:

$$BP = \frac{p_m \, AL \cdot n}{60}, \text{ watts}$$

where p_m = brake mean effective pressure, N/mm^2

A = Piston area, mm^2

L = Stroke, m

n = Number of working strokes per minute

The value of 'k' can be taken as:

k = 46.6 for gray cast iron.

= 174.75 for aluminium alloys; and = 51.25 for steel.

For C.I., $T_c - T_e = 220°C$ with T_c about 425°C.

For Al, $T_c - T_e = 75°C$ with T_c about 260°C.

The average consumption of fuel oil for an average diesel engine is from 0.24 to 0.30 kg per kW per hour. Higher calorific value for diesel fuel may be taken as 44 × 10^3 kJ/kg.

It will be noted that the head thickness will come out to be more from the heat analysis than from the strength analysis. However, if the piston is cooled by oil or water, it can be made equal to the thickness calculated on the strength basis.

If the piston is convex or concave instead of flat, it will be able to withstand higher temperatures, being less rigid than the flat one.

The thickness of head, as found in practice, for various engines is given below in Table 27.2.

Table 27.2

Type of Engine	Piston	Four Stroke	Two Stroke
Compression ignition, oil engines	C.I.	0.11 D to 0.13 D	0.16 D to 0.18 D
Compression ignition, oil engines	Al	0.13 D to 0.16 D	0.17 D to 0.20 D
Spark ignition gas engines	C.I.	0.12 D to 0.14 D	0.20 D to 0.23 D

(e) Ribs: A number of ribs are provided in the piston for various purposes. Side thrust from the connecting rod is transmitted to the skirt through the piston pin boss. Therefore, to distribute this load properly to avoid the distortion of the skirt, a stiffening rib should be provided at the centre line of the boss and extending around the skirt. Triangular ribs are provided just below the ring section and extending to the pin bosses to carry the gas load. Sometimes, these ribs also extend across the head. These ribs will carry the gas load and it will help in carrying a large part of the heat from the piston head to the piston rings and then to the cylinder wall, thus reducing the temperature difference. These ribs make the head rigid, capable of withstanding gas and thermal loads. But at the same time, the head expansion may cause the skirt out of round and according to some designers, it is a bad practice to provide ribs on a surface subjected to expansion, and in designs without ribs, the piston pin boss is extended right upto the crown. In general, four to six ribs are provided radially in the head. The thickness of the ribs may be taken as equal to 1/3 to 1/2th.

(f) Piston Rings: The main function of the piston rings is to impart the necessary radial pressure to maintain the seal between the piston and the cylinder bore. The piston rings can be put into two groups:

1. Compression rings or pressure rings.
2. Oil scraper or oil control rings.

The compression rings are inserted at the top portion of the piston. Their function is:

1. to seal the space between the piston and the cylinder liner and thus prevent the high pressure gases or air charge from leaking down the liner.

2. to transfer the heat from the piston head to the cylinder walls.
3. to absorb some part of the piston fluctuations due to side thrust.

There is no strict rule regarding the number of piston rings. Automobile and air craft pistons have three or sometimes four rings. Stationary compression ignition engines have five to seven piston rings. Since the heat transfer is better from rings to the liner than from the piston wall to the rings, it is better to use a number of narrow rings than a few wide shallow rings.

The function of oil-scraper rings is to scrape the lubricating oil from the surface of the liner, so as to minimise the flow of oil into the combustion chamber. Also during the upward stroke, these should allow sufficient oil to go upward for the proper lubrication of the liner. In small engines, there may be one oil ring and in large engines, it may be two or three. The oil rings are provided below the compression rings. One ring may be fitted below the gudgen pin and the remaining above the gudgen pin. In some engines, the last piston compression ring is made as oil ring by bevelling off the lower edge of the groove and drilling small holes towards the inside of the piston so that the excess oil scraped by the ring flows through these holes into the piston and falls into the sump, Otherwise, the oil rings are slotted all around its periphery to provide an escape for the oil that the slot edges cut from the cylinder wall. Various types of piston rings and suitable grooves are shown in Fig. 27.9.

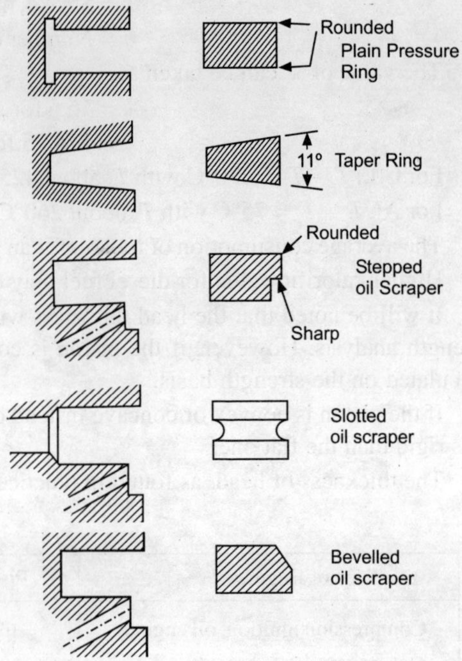

Fig. 27.9 *Types of pston rings and grooves*

Compression rings are machined from gray-iron castings. To have a gas tight seal, high stresses are used in piston ring design to obtain high wall pressure, espectally for high engine speed. The high ring tension increases the metal to metal friction. For this reason, composite rings have been developed. In the outer surface, soft metal inserts, usually of bronze are provided (Fig. 27.10(*a*)). To reduce friction and the resulting wear, the outer surface of the rings can be treated. The usual method is to chromeplate the outer surface. The hard and smooth chrome surface so obtained has great affinity for lubricating oil and liner wear and ring wear is sufficiently reduced. In some engines, bimetallic rings are used to increase the strength and to reduce probability of ring fracture. The steel ring is brazed to the inner surface of the cast iron ring (Fig. 27.10(*b*)).

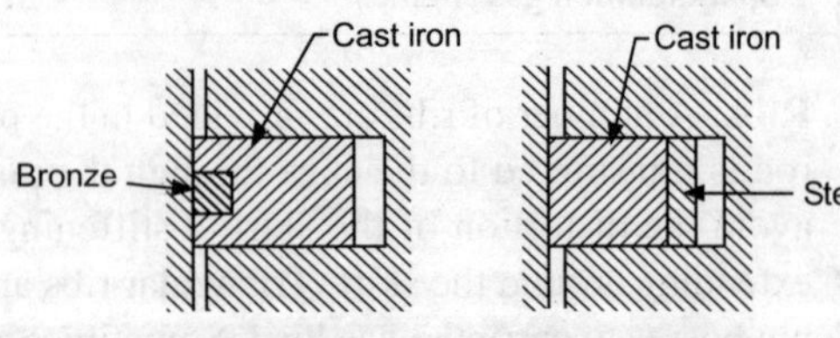

Fig. 27.10 *Compression rings*

The rectangular cross section is the most widely used type of compression rings. The diameter of the ring is slightly larger than the cylinder bore, To permit it to go into the cylinder, a part of the ring is cut off. Due to this diameter difference, pressure is produced against the liner wall. Also, the gases leak behind the surface of the upper rings and this pressure is increased. The various methods of cutting the ends are shown in Fig. 27.11. The gap between the ends should be sufficiently large when the ring is put cold in the cylinder so that even at the highest temperature, the ends will not touch with each other when the ring expands, otherwise there might be the buckling of the ring. The square cut ends are most commonly used. In the case of two stroke engines, thin rings are avoided to reduce the possibility of their fracture when riding over the ports. Also, the ring ends must never come over the ports due to the rotation of the piston rings in the grooves. For this, the piston rings are pinned or anchored to avoid their rotation, so that the ring gap is always confined to the plain part of the liner. The ring joints should be spaced equally round the piston and should not come one below the

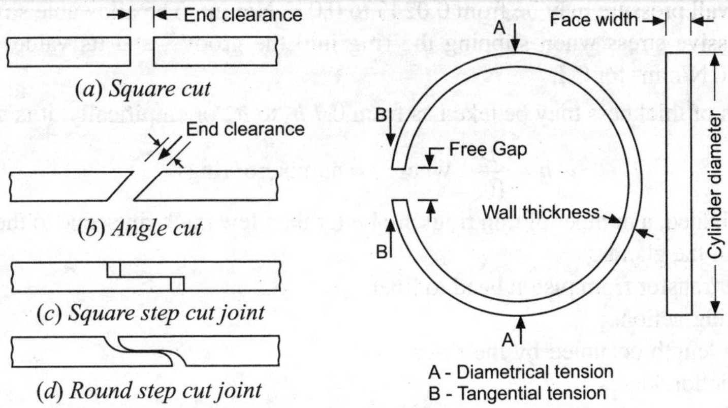

Fig. 27.11 *Piston rings*

other. In the case of horizontal engines also, the rings must be located by pins, so that all gaps do not come in one line due to the rotation of the rings. If this thing happens, then the gases might leak from the combustion chamber to the crankcase and the phenomenon is called "Blow-by". Blow-by can also occur if there is too much lubricating oil on the cylinder walls. The gas pressure might reach in greater measure on the face of the rings than behind the rings, resulting in partial loss of seal. Blow-by is indicated by increased crankcase pressure and ultimately piston siezure may occur if this defect is not diagnosed and corrected.

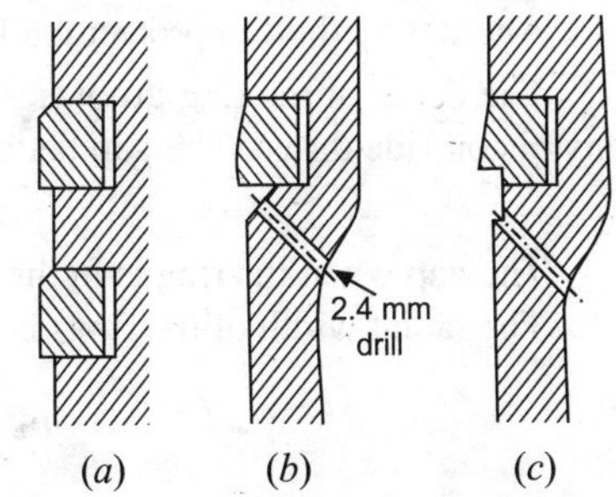

Fig. 27.12 *Methods of preventing oil pumping*

The different ways of preventing oil pumping are shown in Fig. 27.12 The upper edges of the piston rings are rounded off (Fig. 27.12(a)). For the last piston ring, the lower groove edge is bevelled off and small holes are drilled towards the inside of the piston to assist in removal of excess oil (Fig. 27.12(*b*)). Figure 27.12(*c*) shows a more effective scraper ring with an under cut, for the cases where oil consumption is excessive.

Because the rings will expand under temperature therefore a ring gap clearance of about 0.005 mm/mm of cylinder bore is provided. Also enough clearance must be provided between the ring and the sides of the ring groove for the unrestricted movement of the ring in the ring groove. Axial clearance in the groove varies from 0.05 mm for small engines to 0.125 mm in large pistons. Clearance behind the ring should be from 0.05 mm to 0.10 mm for pressure rings and from 0.125 to 0.55 mm for oil rings.

Piston ring materials: The material for the piston rings is usually cast iron and alloy cast iron due to their good wearing qualities and also they retain the spring characteristics even at high temperatures. The ultimate tensile strength varies from 175.0 to 350.0 N/mm^2 depending upon the composition. In high output air craft engine, the material is heat treated high carbon cast iron alloy containing nickle, chromium and molybdenum and with tensile strength of 425 to 700 N/mm^2. As already indicated, to avoid the wear of the liner and the rings, the rings are chrome plated.

Design Calculations: The radial thickness of the ring depends upon the wall pressure desired. It can be found out by the following equation:

$$b_r = D\sqrt{\frac{3p_w}{f_t}}, \; D \text{ in mm}$$

where b_r = radial width or thickness, mm
p_w = wall pressure, N/mm^2
f_t = allowable stress in bending, N/mm^2

The desirable wall pressure may be from 0.0245 to 0.042 N/mm^2. The allowable stress 'f_t' is taken quite high to avoid excessive stress when slipping the ring into the groove, and its value may be taken from 84.0 N/mm^2 to 112.0 N/mm^2 for C.I.

The axial width or thickness may be taken as from 0.7 b_r to b_r, or empirically, it is gives as

$$h = \frac{D}{10z}, \text{ where } z = \text{number of rings.}$$

As already explained, a number of thin rings are better than few thick rings due to the following reasons:

1. less wear of the glands.
2. better heat transfer from piston head to liner.
3. better sealing action.
4. less piston length occupied by the rings.
5. reduced friction losses.

The thickness between the ring grooves i.e. the land may be taken as equal to or slightly less than the axial thickness of the ring 'h' or

$$\text{width of ring lands} = 0.75\,h \text{ to } h.$$

The width of the top land is made larger than the other ring lands to protect the top ring from the high temperature conditions existing at the top of the piston.

$$\therefore \quad \text{width of top ring land} = t_h \text{ to } 1.2\,t_h; \text{ or } = 0.2 \text{ to } 0.3D.$$

Another method to protect the lands and rings, is to have an unsplit "fire ring" above the other rings.

The depth of the ring grooves should be more than the ring depth so that the ring does not have to take any piston side thrust. The gap between the free ring ends is given as

$$G = 3.5\,b_r \text{ to } 4\,b_r.$$

The gap when the ring is in the cylinder should be 0.002 D to 0.004 D.

The radial width of the rings is given empirically also as below:

$$b_r = \frac{D}{25} \text{ for concentric rings.}$$

$$= \frac{D}{27.5} \text{ opposite the joint of concentric rings.}$$

$$= \frac{D}{55} \text{ at the joint of eccentric rings.}$$

Concentric rings are uniform in size around their periphery, while the eccentric rings are thicker opposite the slot and are no longer in use.

(g) Piston Barrel: For ideal flow of heat, the thickness of the piston barrel should be equal to the head thickness at the top, tapering down to zero at the bottom-end, But for strength considerations, the wall thickness is never zero at bottom. The thickness at the ring section must be modified to have the equivalent areas across the ring grooves. For the same amount of heat flow, the area under the head should be equal to the edge area of the piston head.

Therefore,

$$\pi D t_h = \pi (D_1 - t_1).\, t_1,$$

where D_1 = diameter of the bottom of the ring groove.

t_1 = thickness under the ring groove.

$$\therefore \quad t_1 = \left(\frac{1}{2} D_1 \pm \sqrt{(D_1^2 - 4Dt_h)}\right)$$

allowing for the ring clearances,

$$D_1 = D - (2b_r + 0.006D + 0.5 \text{ mm}) \text{ at compression rings.}$$

and

$$D_1 = D - (2b_r + 0.006D + 1.50 \text{ mm}) \text{ at oil grooves.}$$

or, empirically the maximum thickness of the barrel, is given as,

$$t_1 = 0.03\,D + 4.5\text{ mm} + b$$

where b = depth of ring groove, about 0.4 mm larger than b_r.

At the open end, the wall thickness is made as,

$$t_2 = 0.25\,t_1 \text{ to } 0.35\,t_1.$$

Piston Skirt: The portion of the piston barrel below the ring section upto the open end is known as "skirt". It is to take the side thrust of the connecting rod. Its length should be such that the side thrust pressure should not exceed 0.21 to 0.28 N/mm^2 for low speed engines and this may go up to 0.49 N/mm^2 for pistons of the high speed engines. The normal pressure values are given below:

Normal side thrust R = (0.03 to 0.10) × F,

where F = maximum gas force = $\frac{\pi}{4} D^2 \times p_{max}$

∴ $R = l \times D \times p_b$

where l = skirt length, mm p_b = side pressure, N/mm^2

∴ Length of piston, $L = l$ + length of ring section + top land.

Empirically, $L = D$ to $1.5\,D$.

The length of the piston determines the side thrust on the liner. A longer piston runs more quietly and has less "piston slap". For high speed engines, the pistons are made of shorter length to reduce the reciprocating weight. This will result in greater side thrust and wear of the liner but the friction of the big end bearings and crank shaft bearing will be reduced.

(h) Piston pin or Gudgeon pin: Piston pin is the only one most important link between the piston and the connecting rod. Whole of the gas load developed in the cylinder passes through the piston pin, gudgeon pin called wrist pin also. The connection between the piston pin and the connecting rod small end can be of the following types:

1. **Full floating:** In this type, the pin works both in the piston boss and the small end bush of the connecting rod. The pin is secured by circlips against end movements, (Fig. 27.13(*a*)), so that the ends of the pin do not touch and score the liner.
2. **Semi-floating:** This type of pin is free to move either in the piston boss but clamped in the small end of the connecting rod (Fig. 27.13(*b*)) or is free to move in the small end of the connecting rod and secured to the piston bosses by means of a screw. (Fig. 27.8)
3. **Fixed :** If the pin is locked in the piston at both the ends, it is called fixed pin. Its disadvantage is that whole of the swinging motion of the connecting rod is confined to the small end bearing and its wear will not be uniform.

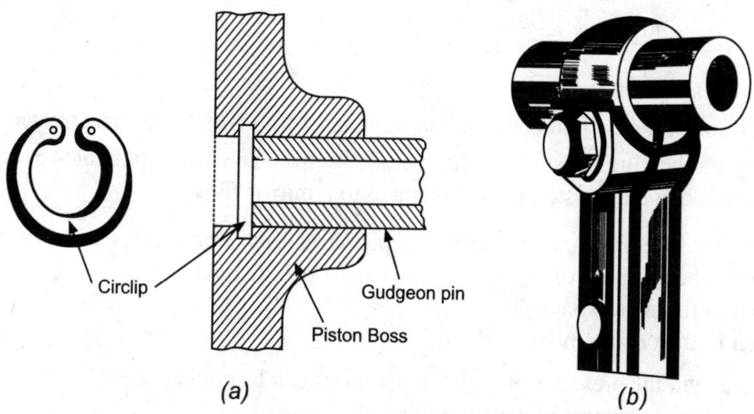

Fig. 27.13 *Piston pin connections*

To offset the turning effect of the friction and to obtain uniform distribution of pressure between the piston and the liner, the piston pin should be placed in the centre of the skirt. It should be somewhat above the centre line of skirt, say 0.02 to 0.04 D, but never below it for smooth running of the piston. In some engines, to shorten the height of the engine; the pin is moved towards the top of the piston.

Piston pin materials: The usual material is nitrogen hardened or case hardened steel alloy containing nickle, chromium, molybdenum or vandadium, having tensile strength of from 700.0 to 910.0 N/mm^2.

Design calculations: The piston pin should be designed for the maximum gas load or the inertia force of piston, whichever is larger. The piston pin bearing area should be roughly equally divided between the piston pin boss and the connecting rod bearing. So, the length of the pin in the small end bearing of the connecting rod should be about 0.45 D, allowing for end clearances of pin, etc. Bearing pressure is the design criteria, and the diameter of the pin is determined by equating the gas load and the bearing load.

$$\therefore \quad l_1 \times d \times p_b = \frac{\pi}{4} D^2 \times p_{max},$$

where l_1 = length of piston pin in the connecting rod bearing.

d = diameter of the pin, p_b = bearing pressure.

Due to the space limitations, a high value of bearing pressure is taken, usually from about 15.0 N/mm^2 for large pistons to about 31.5 N/mm^2 for smaller pistons.

The piston pin is checked in bending with the gas load taken as uniformly distributed load over the length l_1, and the pin supported at the centre of the bosses at two ends, (Fig. 27.14). The length of the beam between the supports, is,

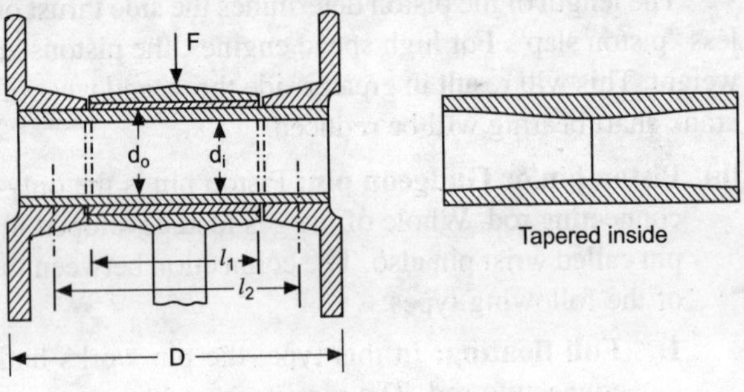

Fig. 27.14 *Piston pin*

$$l_2 = l_1 + \frac{D - l_1}{2} = \frac{l_1 + D}{2}$$

The maximum bending moment will be at the centre of the pin, or

$$M_b = \frac{F}{2}\left(\frac{l_2}{2} - \frac{l_1}{4}\right) = \frac{FD}{8} \quad \therefore \quad M_b = \frac{\pi}{32} d^3 \times f_b$$

To reduce the weight of the pin, it is usually made hollow, with outside diameter d_o and inside diameter d_i

$$M_b = \frac{\pi}{32} \times \frac{d_o^4 - d_i^4}{d_o} \times f_b.$$

where f_b = allowable stress in bending,

= 84.0 N/mm^2 for case hardened carbon steel and 140.0 N/mm^2 for heat treated alloy steel.

The connecting rod bearing is usually of high grade bronze. The piston pin holes in all aluminium alloy pistons are superfinished and all piston pins are lapped to a mirror finish.

(i) **Piston Clearance:** The clearance between the piston and the liner varies with the engine design and piston diameter. The problem of piston clearance is of two distinct types:

1. Those due to thermal expansion.
2. Those due to distortion under load.

The expansion due to the presence of high temperatures can be counteracted by making the piston more or less taper. The amount of taper will depend upon:

1. Operating temperature of the piston.
2. The relative expansion between piston and liner.
3. The running clearance adopted.

If excessive clearance is provided it will lead to an objectionable and damaging phenomenon known as "Piston Slap". If the clearance is too less, piston seizure may occur. Figure 27.15 shows the allowances for piston expansions for cast iron pistons of oil engines.

For Alalloy pistons, the clearance is about twice that for C.I. However, the following clearances are recommended:

1. Tractor: 0.075 to 0.1875 mm
2. Stationary oil engines: 0.0625 to 0.125 mm
3. Scooters, Tempos, Motorcycles and Three wheelers: 0.0375 to 0.075 mm
4. Automobiles: 0.0375 to 0.15 mm.

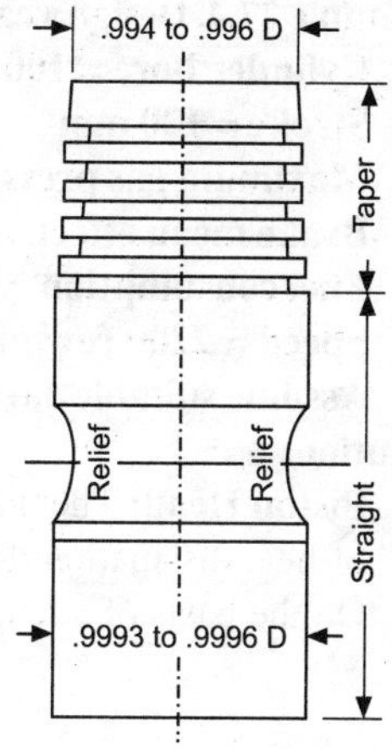

Fig. 27.15 *Allowances for piston expansion*

The above values are for A*l* alloy pistons. The skirt of A*l* alloy pistons is usually spilt to counteract the high expansion of these pistons. In airplane engines, still greater clearances are provided.

Distortion due to load is caused by the flexure of the piston pin bosses and pistons. Also, due to the local heating of the pin because of the friction at the small end bearing, the pin can expand along its length. In order to avoid excessive friction and seizure, the sides of the piston near the piston pin are relieved. Or the piston can be made oval.

The above clearances are for uncooled pistons. These can be reduced if cooling of the piston is adopted.

Pistons are usually finished by grinding for C.I. and by diamond turning for aluminium alloy.

(j) Piston Cooling: As already indicated, to cope with the heat transfer to the piston head, the piston should be cooled. The usual medium is oil. In small engines, the oil of the piston crown is led through a hole in the connecting rod. The oil will travel from the crank shaft bearings to the big end bearing of the connecting rod and then to the piston pin through the hole in the connecting rod. This will also lubricate the gudgeon pin and from here the oil will be splashed upwards the underneath of piston crown. After that, the oil will drop down the sump. Another method is to convey the oil into the underside of the piston head by means of telescopic tubes.

(k) Piston Troubles: The following are some of the troubles encountered in pistons:

1. Seizure of piston.
2. Cracking of piston crowns and side walls.
3. Burning of piston crowns.
4. Excessive lubricating oil consumption.
5. Sticking of piston rings.
6. Breaking of piston rings.
7. Scuffing of piston rings and cylinder liners. Scuffing is a form of surface damage caused by the adhesion and welding of high spots. When these high spots fracture, particles so created will cause scuffing by abrasive action. It usually occurs at around the top dead centre position where the temperatures are highest and the thickness of oil film is thinnest. It can be avoided by better lubrication and using oil of better grade, and of higher viscosity capable of maintaining an oil film under the more strenuous conditions.

Example 27.4. Design a cast iron piston for a single acting four stroke engine for the following specifications:
Cylinder bore = 100 mm
Stroke = 120 mm
Maximum gas pressure = 5 N/mm²
Brake mean effective pressure = 0.65 N/mm²
Fuel consumption = 0.227 kg/kW/hr
Speed = 2200 rev/min
Assume suitable data.

Solution:

1. **Piston Head:** The thickness of the piston head can be found out on the basis of strength and on the basis of heat dissipation through the piston head.

On the basis of strength, the head thickness is given as:

$$t_h = 0.433D\sqrt{\frac{p_{max}}{f_t}}$$

where $D = 100$ mm; $p_{max} = 5$ N/mm²

f_t for C.I. may be taken as 37.5 N/mm²

$$\therefore \quad t_h = 0.433 \times 100\sqrt{\frac{5.0}{37.5}} = 15.81 \text{ mm, say 16 mm}$$

On the basis of heat dissipation, the head thickness is given as:

$$t_h = \frac{H}{12.56\,k\,(T_c - T_e)}$$

where H = Heat flow through the head, watts = $C \times w \times HCV \times BP$

Now $$BP = \frac{p_m \times L \times A \times n}{60} = \frac{0.65 \times 0.12 \times 0.785 \times (100)^2 \dfrac{2200}{2}}{60 \times 1000} = 11.225 \text{ kW}$$

Now, $C = 0.05$; and $w = \dfrac{0.227}{3600}$ kg/kW/s

HCV of the fuel may be taken as 41870 kJ/kg

$$\therefore \quad H = \frac{0.05 \times 0.227 \times 41870 \times 11.225}{3600} = 1.48 \text{ kW}$$

∴ Now k, for CI = 46.6

and $T_c - T_e$ can be taken as 220°C.

$$\therefore \quad t_h = \frac{1.48 \times 1000}{12.56 \times 46.6 \times 220} = 0.0115 \text{ m, i.e., 11.5 mm}$$

Let us take t_h = 16 mm.

2. **Radial Ribs:** Four to six ribs may be provided. The thickness of ribs

$$= \frac{1}{3} \text{ to } \frac{1}{2} t_h = 5.33 \text{ to 8 mm; say 7 mm.}$$

3. **Piston Rings:** Let us take three compression rings and one oil ring. The radial thickness of the rings is given as:

$$b_r = D\sqrt{\frac{3p_w}{f_t}}$$

$$p_w = 0.0245 \text{ to } 0.042 \text{ N/mm}^2, \text{ take } p_w = 0.035 \text{ N/mm}^2$$

Taking f_t as 90.0 N/mm²

$$b_r = 100 \times \sqrt{\frac{3 \times 0.035}{90.0}} = 3.416 \text{ mm, say 3.5 mm}$$

Axial width = (0.7 to 1) × b_r *i.e.* 2.45 to 3.5 mm, say 3 mm

∴ Section of ring is 3.5 mm × 3 mm

Empirically, the minimum axial width, $h = \frac{D}{10z} = \frac{100}{10 \times 4} = 2.5$ mm, hence O.K.

Width of top land = t_h to 1.2 t_h = 16 mm to 19.2 mm, say 18 mm

Width of other ring lands = 0.75 h to h = 2.25 to 3 mm, say 2.5 mm

The gap between the free ring ends, G = 3.5 b_r to 4 b_r = 12.25 to 15 mm, say 13 mm

The gap when the ring is in the cylinder = 0.002D to 0.004D = 0.20 to 0.40 mm, say 0.3 mm

4. **Piston Barrel:** The thickness of barrel is,

$$t_1 = b + 0.03D + 4.5 \text{ mm} = (3.5 + 0.4) + 0.03 \times 100 + 4.5 = 11.4 \text{ mm}$$

At the open end, the wall thickness is, $t_2 = 0.25\, t_1$ to $0.35\, t_1$ = 2.85 mm to 3.99 mm, say 3.5 mm

Skirt Length:

Maximum gas load, $$F = \frac{\pi}{4} D^2 \times p_{max} = \frac{\pi}{4} \times (100)^2 \times 5 = 39.25 \text{ kN}$$

Now side thrust on the liner = (0.03 to 0.10) × F

∴ Let $R = l \times D \times p_b$

Let p_b, the bearing pressure = 0.45 N/mm², the engine being high speed.

∴ 39.25 × 1000 = l × 100 × 0.45, gives l = 87 mm, say 90 mm

∴ Piston length = skirt length + ring section + top land.

$$= 90 + 4 \times 3 + 3 \times 2.5 + 18 = 127.5 \text{ mm}$$

5. **Piston Pin:** The piston is designed in bearing for maximum gas load, therefore, $F = l_1 \times d \times p_b$

l_1 = length of pin in bearing may be taken as 0.45D. i.e. 45 mm

Let p_b = 25 N/mm² for a high quality bronze bushing

∴ 39.25 × 1000 = 45 × d × 25.0

∴ d, the diameter of piston pin = $\frac{39.25 \times 1000}{45 \times 25.0}$ = 34.9 mm, say 35 mm

Let the material of pin be heat treated alloy steel.

To check the pin in bending, the maximum bending moment on the piston pin,

$$M_b = \frac{F \times D}{8} = \frac{39.25 \times 1000 \times 100}{8 \times 1000} = 490.625 \text{ Nm}$$

$$\therefore \quad 490.625 \times 10^3 = \frac{\pi}{32} \times d^3 \times f_b$$

$$\therefore \quad f_b = \frac{490.625 \times 10^3 \times 32}{\pi \times 35^3} = 116.56 \text{ N/mm}^2,$$ which is well within the safe limit of 140.0 N/mm² for the pin of heat treated alloy steel.

To make the pin lighter, it is usually made hollow, so

Let d_o = 35 mm; d_i = 20 mm;

$$\therefore \quad M_b = \frac{\pi}{32} \times \frac{d_o^4 - d_i^4}{d_o} \times f_b$$

$$\therefore \quad 490.625 \times 1000 = \frac{\pi}{32} \times \frac{35^4 - 20^4}{35} \times f_b$$

$$\therefore \quad f_b = 130.46 \text{ N/mm}^2,$$ which is safe.

EXERCISES

1. What is the difference between the barrel piston and trunk piston?
2. List the various parts of a Trunk piston.
3. What are the functions of a Trunk piston?
4. List the various materials used for pistons.
5. Enumerate the advantages and disadvantages of aluminium pistons.
6. How are the pistons, finished on the outer surface?
7. Explain the terms "Piston Seizure", "Piston Slap", "Blow By", "Scuffing".
8. What are the advantages of composite pistons?
9. What is the use of ribs?
10. What are the functions of piston rings?
11. What is the purpose of oil scraper rings?
12. What is the material of the piston rings?
13. What forces press the piston rings against the cylinder wall?
14. How is oil pumping prevented?
15. What is the function of piston skirt?
16. When the cooling of the pistons in necessary? How it is done?
17. What is the function of piston pin?
18. What are the materials used for piston pins?
19. How the piston pins are finished?
20. Why the piston pin is located at or above the middle of the skirt length?
21. Why piston clearances are necessary? What is its usual value?
22. Enumerate the various piston troubles.
23. How the wear of the piston rings is prevented?
24. Find the thickness of the head of a cast iron oil engine trunk piston 0.30 m diameter, the maximum pressure is 3.85 MPa.
25. Find the thickness of the head of a heavy duty oil engine if the diameter is 0.20 m, the maximum pressure is 8.55 MPa and if made of: (a) cast iron, (b) cast aluminium alloy, (c) an aluminium forging.
26. Determine the thicknesses of the barrel at the head, just below the piston rings, and at the bottom. Find also the thickness and width of the piston rings. Find the length of piston and design the wrist pin for both bearing pressure and bending stresses for the piston in problem 24.
27. Design and sketch a cast iron piston for a single acting four stroke I.C. Engine with a 0.14 m cylinder bore 0.19 m stroke and 0.375 m connecting rod length. The maximum gas pressure is 3.5 N/mm^2 and the engine speed is 600 rev/min. It develops a brake mean effective pressure of 0.7 N/mm^2 and uses 4.65 kJ per kW per second, check the piston for heat flow.
28. Design an aluminium alloy piston for a single acting four stroke engine for the following specifications:

Cylinder bore	=	0.30 m
Stroke	=	0.375 m
Maximum gas pressure	=	8 MPa
Brake mean effective pressure	=	1.15 MPa
Fuel consumption	=	0.22 kg/kW/hr
Speed	=	500 rev/min.

29. The piston of an automobile engine is held to the connecting rod by a wrist pin having an external diameter of 20 mm and internal diameter of 11 mm. The length of the bearing in the rod end is 25 mm and the bearing in each boss of the piston is 15 mm. The maximum load transmitted is 14.50 kN.
    (a) Determine the bearing pressure between the rod end and the wrist pin.
    (b) Compare the maximum bending stress in the pin, assuming that it is a simple beam with a uniform load.
    (c) Compute the maximum deflection of the pin.

30. Design a suitable aluminium alloy piston with two compression rings and one oil ring for a petrol engine of following particulars:

Cylinder diameter	=	0.10 m
Peak gas pressure	=	3.2 MPa
Mean effective pressure	=	0.8 MPa
Average side thrust	=	2400 N
Skirt bearing pressure	=	0.22 MPa
Bending stress in piston crown	=	36 MPa
Crown temperature difference	=	70°C
Heat dissipated through crown	=	157 kJ/m^{2}s
Allowable radial pressure	=	0.04 MPa
Bending stress in rings	=	90 MPa
Heat conductivity, k	=	160 W/m°C

Draw a full scale dimensioned drawing and indicate the method of reducing the thermal expansion in the skirt of designed piston. (P.U. 1973)

27.3 CONNECTING ROD

(a) Function: The main function of the connecting rod is to transmit the push and pull from the piston pin to the crank pin. In many cases, its secondary function is to convey the lubricating oil from the bottom end to the top and *i.e.* from the crank pin to the piston pin and then for splash of jet cooling of piston crown. The usual form of the connecting rod used in internal combustion engines and automobiles is shown in Fig. 27.16. It has an eye at the small end for the piston pin bearing, a long shank, and a big end opening which is usually spilt to take the crankpin bearing shells.

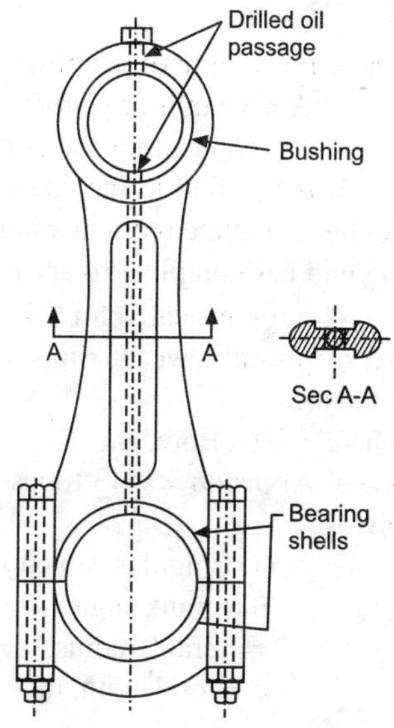

Fig. 27.16 *Connecting rod*

(b) Materials: The connecting rods of internal combustion engines are mostly manufactured by drop forging. The connecting rod should have adequate strength and stiffness with minimum weight. The materials for connecting rods range from mild or medium carbon steels to alloy steels, In industrial engines, carbon steel with ultimate tensile strength ranging from 550 to 670 MPa is used. In transport engines, alloy steel having a strength of about 780 to 940 MPa is used, e.g., manganese steel is commonly used. In aero engines, nickel chrome steel having ultimate tensile strength of about 940.0 to 1350.0 MPa is most commonly used. For connecting rods of low speed horizontal gas engines, the material may be sometimes steel castings. For high speed engines, the connecting rods may also may be made of duralumin and aluminium alloys. After drop forging, the outer surfaces are kept unfinished but in high duty aero engines, the outer surfaces are highly finished.

(c) Shape of Connecting Rod: The usual shapes of the connecting rod are: rectangular, circular, tubular, I-section, and H-section. In low speed large engines, the section is usually circular with flattened sides, or rectangular, the larger dimension being in the plane of rotation. In high speed engines, lightness of the connecting rod is a major factor. Therefore, tubular, I-section or H-section rods are used, the last two being most common.

(d) Length: The length of the connecting rod depends upon the ratio of connecting rod length and stroke, i.e. l/r ratio; on l/r ratio depends the angularity of the connecting rod with respect to the cylinder centre line. The shorter the length of the connecting rod l in respect to the crank radius 'r', the smaller the ratio l/r, and the greater the angularity. This angularity also produces a side thrust of the piston against the liner. This side thrust and the resulting wear of the linear decreases with a decrease in the angularity, i.e., with an increase of l/r ratio increases. However, an increase of l/r ratio increases the over all height of the engine. Due to these two factors, the common values of l/r ratio are 4 to 5.

(e) Stresses in the Connecting Rod: The stresses in the connecting rod are set up by a combination of forces. The various forces acting on the connecting rod are:

1. The combined effect of gas pressure on the piston and the inertia of the reciprocating parts.
2. Friction of the piston rings and of the piston.
3. Inertia of the connecting rod.
4. The friction of the two end bearings, i.e., of the piston pin bearing and the crank pin bearing.

1. Load due to gas pressure and Piston Inertia: The direct load on the piston due to the gas pressure can be calculated with the help of indicator diagram for the engine. From the indicator diagram, the gas pressure at various crank angles can be noted, and multiplying these gas pressures by the piston area, the gas loads can be calculated at various crank angles. Then a curve may be drawn between the gas load and the crank angle, covering two complete revolutions for four stroke and one revolution for two stroke engine. The load due to the piston inertia

= Weight of the reciprocating masses × acceleration

$$\therefore \quad F_i = \frac{F}{g}\omega^2 r\left(\cos\theta + \frac{r\cos 2\theta}{l}\right),$$

where F = weight of reciprocating masses,

N = weight of piston including that of rings + weight of piston pin + weight of one third portion of connecting rod (small end portion).

It is justified to include the weight of the eye end of the connecting rod, since the connecting rod has neither complete reciprocating nor complete rotary motion. Small end has complete reciprocating while the big end has complete rotary motion.

Now empirically, for C.I.

wt. of piston + wt. of rings + wt. of piston pin $= 0.0415D^3, N$

where D = Bore, cm

wt. of Al piston = 0.43 times the weight of C.I. piston,

ω = angular velocity of crank, rad/s

θ = crank angle from T.D.C.

r = crank radius, m

l = rod length, m

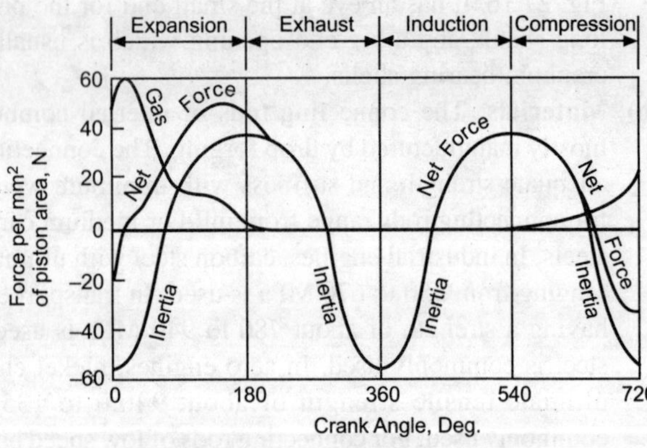

Fig. 27.17 *Net force diagram*

With the help of the above relation a curve can be plotted between the inertia load F_i and the crank angle θ, preferably on the same ordinates as the gas load and then a net load curve is plotted, Fig. 27.17. The effect of the inertia force is to reduce the maximum compressive stress at the end of the exhaust stroke.

The net vertical load on the connecting rod will be the algebraic sum of the gas load and the inertia load. The actual axial load on the connecting rod will be more than this net load due to the angularity of the rod. It is clear from Fig. 27.18, that the load on the connecting rod, Q will be maximum when the crank angle $\theta = 90°$. But at this position, the gas pressure would have decreased considerably. Therefore, it is usual to take the load on the connecting rod

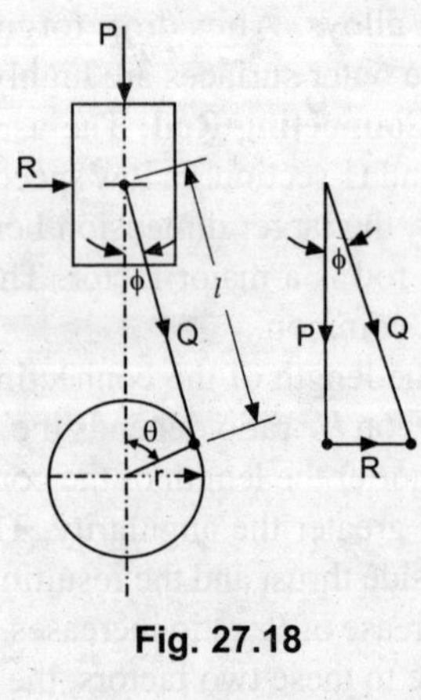

Fig. 27.18

corresponding to the maximum gas load. However, if a complete analysis is to be made, the following expression may be used.

$$Q = \frac{P}{\cos\phi} = \frac{P}{\sqrt{1-\sin^2\phi}} \qquad \left[\because \sin\theta = \frac{l}{r}\sin\phi\right]$$

$$= \frac{P}{\frac{r}{l}\sqrt{\left(\frac{l}{r}\right)^2 - \sin^2\theta}}$$

2. Force due to the friction of piston rings and of piston:

The friction force due to the piston rings may be determined by the following relation:

$$P_f = h\,\pi D z p_r\,\mu$$

where h = axial width of the rings

D = cylinder bore.

z = number of rings

p_r = pressure of rings, 0.0245 to 0.042 N/mm^2

μ = coefficient of friction, about 0.1.

The friction of the piston is produced by the normal component of the piston pressure. The average value of normal component has been already mentioned and it varies from 3 to 10 % of the piston pressure. If the coefficient of friction is about 0.06, the friction force will be at the most 0.6% of the piston pressure, which will be very low. Therefore, in the design calculations, the effect of friction of piston rings and of the piston can be neglected.

3. Inertia of connecting rod: The inertia of the connecting rod will have two components: along the rod, i.e., the longitudinal component, and normal to the rod, i.e., the transverse component. The longitudinal component is taken into account by considering about one third portion of the connecting rod on the small end side as reciprocating and the remaining two third as rotating with the crank.

Due to the transverse component, a centrifugal force will act on every part of the rod, This force tends to bend the rod outwards, away from the line. The stress in the rod will be alternating one. For investigation purpose, the rod may be assumed to be of uniform section throughout its length. The bending force will be zero at the piston pin and maximum at the crank pin. The variation can be assumed to be triangular. It is clear that bending stress will be maximum when the crank and the connecting rod are at right angles, Fig. 27.19.

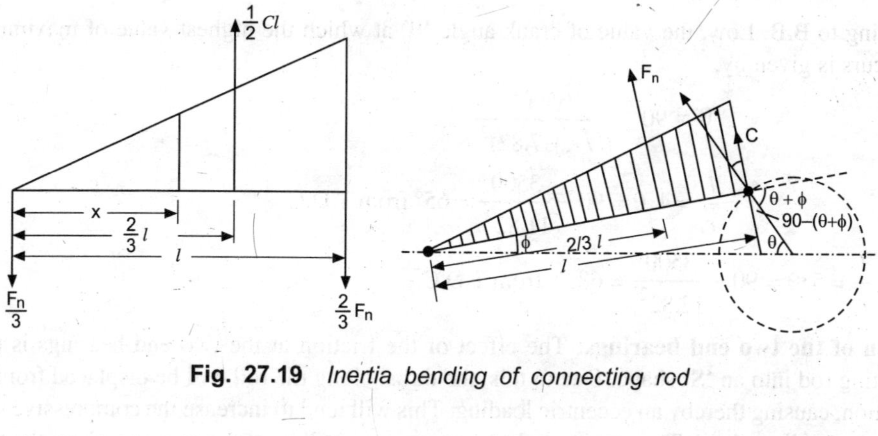

Fig. 27.19 *Inertia bending of connecting rod*

If 'C' is the centripetal force acting on a unit length at the crank-pin, then

$$C = \rho A\, \omega^2 r \sin(\theta + \phi)$$

where ρ = density of the material, kg/m^3

A = cross-sectional area of rod, m^2

ω = angular speed, rad/s

r = crank radius, m

Now C is maximum, when the crank and the connecting rod are at right angles, *i.e.*, $\theta + \phi = 90°$.

$\therefore$ Maximum value of $C = \rho A\, \omega^2 r \sin 90° = \rho A\, \omega^2 r$

The resultant normal force acting on the connecting rod, will be, $F_n = 1/2\ Cl$ and it will act at a distance $2/3\ l$ form the piston pin.

Now, since it has been assumed that one third weight of the connecting rod is concentrated at the piston pin and two third at the crank pin, the reactions at the two bearings due to F_n will also be in the same ratio, i.e., $2/3\ F_n$ at crank pin and $1/3\ F_n$ at the piston pin.

Now bending moment on the connecting rod, at a distance 'x' from the piston pin,

$$M_x = \frac{F_n}{3} \times x - \frac{1}{2}\frac{Cx}{l} \times x \times \frac{x}{3} = \frac{F_n \times x}{3} - \frac{Cx^3}{6l} = \frac{F_n \times x}{3} - \frac{F_n \times x^3}{3\,l^2} = \frac{F_n \cdot x}{3\,l^2}(l^2 - x^2)$$

For maximum bending moment,

$$\frac{d\,M_x}{dx} = 0$$

$$\therefore \quad \frac{F_n}{3} - \frac{3F_n \times x^2}{3l^2} = 0$$

$$\therefore \quad \frac{F_n}{3} = \frac{3F_n \times x^2}{3\,l^2} \quad \therefore \quad 3x^2 = l^2 \quad \therefore \quad x = \frac{1}{\sqrt{3}}$$

i.e. maximum bending moment occurs at a distance $\frac{1}{\sqrt{3}}$ form the piston pin.

$$\therefore \quad M_{\max} = \frac{F_n}{3l^2} \times \frac{1}{\sqrt{3}}\left(l^2 - \frac{l^2}{3}\right) = \frac{2F_n \times l}{9\sqrt{3}} = 0.128\,F_n\,l$$

$$\therefore \quad \text{Maximum bending stress} = \frac{M_{\max}}{Z}$$

According to B.B. Low, the value of crank angle 'θ' at which the highest value of maximum bending moment occurs is given by,

$$\theta = 90 - \frac{3500}{(l/r + 7.82)^2}$$

For $\frac{l}{r} = 4, \theta = 90 - \frac{3500}{11.82^2} = 65°$ from T.D.C.

and for $\frac{l}{r} = 5, \theta = 90 - \frac{3500}{12.82^2} = 68.7°$ from T.D.C.

4. **Friction of the two end bearings:** The effect of the friction at the two end bearings is to bend the connecting rod into an 'S' shape. Due to this, the thrust along the rod will be displaced from the axis of the section, causing thereby an eccentric loading. This will tend to increase the compressive stress on the rod due to the direct load. The maximum bending moment will be at the sections where the shank of the connecting rod joins the piston pin and the crank pin ends.

(f) Design Procedure: It is clear from the above stress analysis, that the maximum compressive load on the rod will be algebraic sum of the gas load and the inertia load due to the reciprocating masses and the effect of the inertia load is to decrease the maximum compressive gas load. This load is further reduced slightly by the piston rings and piston friction. Since the gas pressure is maximum very near the dead centre, it is not considered necessary to resolve the piston load along the connecting rod since the angularity of the rod will be very small.

Again there is bending stress on the connecting rod due to its inertia, the total stress in the rod is the sum of the direct compressive stress due to net piston load and the transverse bending stress also known as whipping stress. The whipping stress is maximum at about 65° to 68° form T.D.C., where the gas load has decreased considerably as it is maximum at T.D.C. Therefore, the maximum direct compressive and the maximum whipping stress do not occur at one crank angle. If the direct compressive stress due to net piston load and the whipping stress are plotted against Q, and then the maximum stress due to the combined effect is found out, it usually occurs at about $\theta = 90°$, and its value is quite low as compared to the direct stress due to maximum gas load. Therefore, the usual procedure to find the section of the connecting rod is to use the maximum gas load instead of the net piston load, with a reasonable factor of safety. Like this, a robust section will be obtained which can withstand the maximum whipping stress as well as the bending stress due to bearing friction.

Due to maximum gas load, the rod will be subjected to alternating direct compression and tensile stresses. Now, since the direct compression stress corresponds to compression and explosion strokes, it will be numerically much larger than the tensile stress. Therefore, the connecting rod is designed as a strut. Due to the comparatively large diameter or radius of gyration in the case of I-section, with respect to the rod length, Rankine Gordon formula is used.

$$\text{Buckling load} = \frac{f_{cu} \times A}{1 + a\left(\frac{l}{k}\right)^2}, \text{ N}$$

where f_{cu} = ultimate crushing stress, N/m^2

A = section area, m^2

l = equivalent length, m

k = radius of gyration about the axis of buckling, m

a = constant.

Now $$a = \frac{1}{7500} \text{ for M.S.} = \frac{1}{9000} \text{ for W.I.} = \frac{1}{1600} \text{ for C.I.}$$

$$\text{Buckling load} = \text{maximum gas load} \times \text{factor of safety} = \frac{\pi D^2}{4} p_{max} \times \text{F.O.S.}$$

Now F.O.S. for columns:

Steady load: 3.375 to 5.

Light shock: 4.2 to 7.5.

Heavy shock: 6.75 to 15.

If, instead of f_{cu} the failure stress considered is the yield point stress, f_{yp}, then the F.O.S. will be correspondingly reduced. Then the Rankine Gordon formula will be written as,

$$\text{Buckling load} = \frac{f_{yp} \times A}{1 + a\left(\frac{l}{k}\right)^2}, \text{ N}$$

The Rankine Gordon formula can also be written as,

$$\text{Maximum gas load} = \frac{f_c \times A}{1 + a\left(\frac{l}{k}\right)^2}, \text{N}$$

where f_c = safe stress in compression = $\frac{f_{cu}}{\text{F.O.S.}}$ or $\frac{f_{yp}}{\text{F.O.S.'}}$ as the case may be.

Example 27.5. Design a connecting rod for four stroke Petrol engine, with the following data:

Piston diameter = 0.10 m Stroke = 0.14 m
Length of connecting rod, centre to centre = 0.315 m
Weight of reciprocating parts = 18.2 N
Speed = 1500 rev/min with possible overspeed of 2500.
Compression ratio = 4 : 1
Probable maximum explosion pressure = 2.45 MPa.

Solution: Section of Rod: In I.C. engines, the most widely used section of rod is I due to its lightness as it will keep the inertia forces small and it can withstand the high gas pressures also.

In the plane of motion, the ends of the rod are direction free and so freely hinged at the piston pin and crank pin. Hence, for buckling about the neutral axis *xx*, the strut is freely hinged, Fig. 27.20. In the plane, perpendicular to the plane of motion, for buckling about the axis *yy*, the strut is fixed ended due to the constraining effect of the bearings at piston and crankpins, Therefore, for buckling about axis *yy*, the rod is four times as strong as for buckling about axis *xx*. But the rod should be equally strong in both the planes,

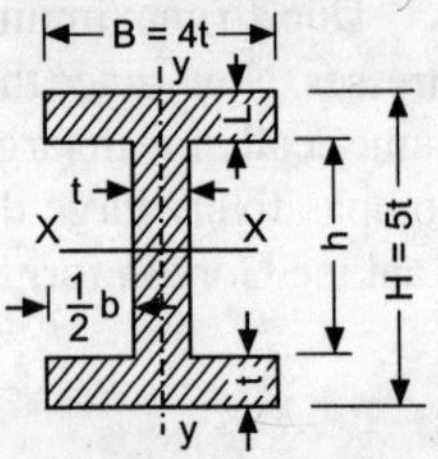

Fig. 27.20 *I-section*

$$4I_{yy} = I_{xx}$$

$$\therefore \quad k_{yy}^2 = \frac{1}{4} k_{xx}^2$$

Let the flange and web thickness of section = t

Depth of section = $5t$

Width of section = $4t$.

$\therefore$ Area of the section, A = $11t^2$

$$I_{xx} = \frac{1}{12}(BH^3 - bh^3) = \frac{1}{12}[4t \times (5t)^3 - 3t \times (3t)^3] = \frac{419}{12}t^4$$

$$I_{yy} = \frac{1}{12}(2tB^3 + ht^3) = \frac{1}{12}[2t \times (4t)^3 + 3t \times t^3] = \frac{131}{12}t^4$$

$$\therefore \quad k_{xx}^2 = \frac{I_{xx}}{A} = \frac{419t^4}{12 \times 11t^2} = 3.18\,t^2$$

$$k_{yy}^2 = \frac{I_{yy}}{A} = \frac{131t^4}{12 \times 11t^2} = 0.995\,t^2$$

$$\therefore \quad \frac{k_{yy}^2}{k_{xx}^2} = \frac{0.995t^2}{3.18t^2} = \frac{1}{3.2}$$

which is slightly higher than 1/4, therefore, the proportions of the section are satisfactory.

Now section of the rod can be determined by considering the buckling about axis, xx by the Ranking formula:

$$\text{Buckling load} = \frac{f_{cu} \times A}{1 + a\left(\frac{l}{k}\right)^2}$$

Buckling load = maximum explosion load × F.O.S.

Let us take F.O.S. as 5

$$\therefore \quad \text{Buckling load} = \frac{\pi}{4} \times (0.10)^2 \times 2.45 \times 10^6 \times 5 = 96 \text{ kN}$$

Now f_{cu} for low carbon steel = 330.0 MPa

$$A = 11t^2; \; a = \frac{1}{7500}$$

$$l = 0.315 \text{ m, for hinged ends. } k = k_{xx}$$

$$\therefore \quad 96 \times 1000 = \frac{330 \times 10^6 \times 11t^4}{1 + \frac{1}{7500} \times \frac{0.315^2}{3.18t^2}}$$

$$\therefore \; 3630 \times 10^3 \times t^4 - 96t^2 - 0.0004 = 0$$

Let $T = t^2$

$$\therefore \quad 3630 \times 10^3 \times T^2 - 96T - 0.0004 = 0 \qquad \text{From here } T = 2.645 \times 10^{-5}$$

$$\therefore \quad t = 5.14 \text{ mm say } 5.5 \text{ mm}$$

$$\therefore \quad \text{Depth} = 5t = \mathbf{27.5\ mm}, \quad \text{Width} = 4t = \mathbf{22\ mm.}$$

Since the connecting rod is to be manufactured by forging, the sharp corners are rounded off for the easy withdrawal of the section from the dies. Therefore, the modified section is shown in Fig. 27.21.

Fig. 27.21 *Modified I-section*

The section obtained above is at the middle of rod. The width is kept constant throughout the length, but the depth varies.

Depth at crank end = 1.1 to 1.25 × depth at mid point.

Depth at piston end = 0.75 to 0.9 × depth at mid point.

Section at crank end = 34 mm × 22 mm

Section at small end = 24 mm × 22 mm.

Dimensions of Big End: The greatest load to be carried by the big end bearing is the maximum gas load. The crank pin is designed in bearing. The allowable bearing pressure depends upon the bearing material, method of lubrication and space limitations. In most of the cases, the big end has removable precision bearing shells. The shells may be of bronze or steel, or brass and a thin lining of bearing metal, which is generally 1.0 mm or less in thickness, is given on the inner surface of shells. The bearing metal may be tin or lead base babbit, copper lead, cadmium, silver, etc.

Due to space limitations, a high bearing pressure is used. With babbit and centrifugal force lubrication from a ring fastened to the crank shaft (usually horizontal engines) the bearing pressure should not exceed 7 N/mm^2. With pressure lubrication it can be 8.4 to 10.5 N/mm^2. With copper lead bearing and pressure lubrication, it can go up to 12.5 N/mm^2.

$$\therefore \quad l_c \times d_c \times p_b = P$$

where l_c = length of crank pin, mm; d_c = diameter of crank pin, mm

p_b = allowable bearing pressure, N/mm^2

$$P = \text{max. gas load, N} = \frac{\pi}{4} D^2 \times p_{max} = 19.20 \text{ kN}$$

Now $\frac{l_c}{d_c}$ is usually 1.25 to 1.5

$$\therefore \quad 1.4 d_c \times d_c \times 10 = 19.20 \times 1000$$

$$\therefore \quad d_c = \mathbf{37\ mm}$$

$$\therefore \quad l_c = 37 \times 1.4 \text{ say } \mathbf{52\ mm}$$

Empirically d_c should be = (0.50 to 0.65) × cyl. bore = 50 to 65 mm

l_c should be = (0.40 to 0.60) × cyl. bore = 40 to 60 mm.

Design of small end: The design of the small end, *i.e.*, the piston pin or the gudgeon pin has already been discussed under the design of piston. Piston pin bearing is usually a high quality bronze bushing, but it may also be made A*l* alloy. Its thickness can be 0.04 to 0.05 times the cylinder bore. When the cylinder bore exceeds 0.125 to 1.15 m, it is usual to lock the bush against rotation by means of a set screw. In some heavy duty engines, the antifriction needle bearing is used. It consists of an inner and outer face between which small diameter rollers are located. The piston pin is used as inner race.

The bearing pressure is usually from 15 to 31.5 N/mm^2

$$\therefore \quad l_p \times d_p \times p_b = P$$

$$\frac{l_p}{d_p} = 1.5 \text{ to } 2$$

$$\therefore \quad 1.5\, d_p^2 \times 15 = 19.20 \times 1000$$

$$\therefore \quad d_p = 29.2 \text{ mm, say } \mathbf{30\ mm}$$

$$\therefore \quad l_p = 30 \times 1.5 = \mathbf{45\ mm}$$

Bolts for Big end Cap: The maximum load on the bolts and the big end cap will be of inertia force at T.D.C. on the exaust stroke. The compression and explosion loads will be compressive and will not effect the parts. Also the inertia load at B.D.C. is compressive. The number of bolts for fastening the crank pin bearing to the rod may be two or four bolts.

The material for connecting rod bolt should be nickel alloy steel having a tensile strength of 700 to 1050 MPa. The bolts are under repeated but not alternating stresses, therefore, the factor of safety may be taken as 6. Therefore, allowable stress can be about 120 to 175 MPa.

Now inertia force,
$$F_i = \frac{F}{g}\omega^2 r\left(\cos\theta + \frac{r}{l}\cos 2\theta\right)$$

Now
$$F = 18.2 \text{ N}, \quad \frac{l}{r} = \frac{31.5}{7} = 4.5, \quad r = 0.07 \text{ m}$$

$$\omega = \frac{2\pi \times 2500}{60} = 260.2 \text{ rad/s}, \quad \theta = 0°$$

$$\therefore \quad F_i = \frac{18.2}{9.81} \times (260.2)^2 \times 0.07\left(1 + \frac{1}{4.5}\right) = 10.88 \text{ kN}$$

If two bolts are used, and the core diameter is d_c

then
$$2 \times \frac{\pi}{4} d_c^2 \times f_t = 10.88 \times 1000$$

$$\therefore \quad 2 \times \frac{\pi}{4} d_c^2 \times 120 = 10.88 \times 1000; \text{ giving } d_c = 7.6 \text{ mm, say } 8 \text{ mm}$$

$$\therefore \quad \text{Outer diameter} = \frac{d_c}{0.8} = \mathbf{10\ mm}$$

Connecting rod big end bearing cap : The bearing cap should be strong and as rigid as possible. The cap is designed as a beam freely supported at the bolt centres. The cap may be assumed to be loaded in between uniformly distributed load and centrally concentrated load. Therefore, the maximum bending moment may be taken as $\frac{F_i\, s}{6}$

where F_i = load on the cap, i.e. the inertial load.

s = distance between bolt centres.

= diameter of bearing + twice the thickness of bearing liner + diameter of bolt + some clearance.

Thickness of bearing liner = thickness of shell + thickness of bearing metal.

Thickness of steel bearing shell may be taken empirically as = 0.05 × cylinder bore = 0.05 × 100 = 5 mm

Thickness of bearing metal may be taken as 1 mm

$\therefore \quad s = 37.0 + 2(5 + 1) + 10 + \text{say } 1.5 \text{ mm} = 37.0 + 12 + 10 + 1.5 = 60.5 \text{ mm, say } \mathbf{60\ mm}$

$$\therefore \quad \text{Bending moment} = \frac{F_i\, s}{6} = \frac{10.88 \times 1000 \times 60}{6 \times 1000} = 108.8 \text{ Nm}$$

Now $\quad M_b = f_b \times Z$

f_b may be taken as 120 N/mm², $Z = \frac{bc^2}{6}$

where b = width of cap = length of bearing = 52 mm

c = thickness of cap, mm

$$\therefore \quad 108.8 \times 1000 = 120 \times \frac{52c^2}{6}$$

$\therefore \quad c = 10.35$ mm say **11 mm**

Transverse Inertia Bending Stress:

Centripetal force per unit of connecting rod length, acting at the crank pin,

$$C = \rho A \omega^2 r, \text{ newton/meter}$$

Now ρ for steel = 7830 kg/m³

$$A = 11t^2 = 11 \times (0.0055)^2 = 3.33 \times 10^{-4} \text{ m}^2, r = 0.07 \text{ m}$$

$$\omega = \frac{2\pi n}{60} = \frac{2\pi \times 2500}{60} = \frac{500\pi}{6} \text{rad/s}$$

$$\therefore \quad C = 7830 \times 3.33 \times 10^{-4} \times \left(\frac{500\pi}{6}\right)^2 \times 0.07 = 12509.54 \text{ N/m}$$

Now maximum B.M. = 0.128 $F_n\, l$

Now $\quad F_n = \frac{1}{2} Cl$

$$\therefore \quad \text{Maximum B.M.} = \frac{0.128\, Cl^2}{2} = \frac{0.128 \times 12509.54 \times (0.315)^2}{2} = 79.44 \text{ Nm.}$$

$$\therefore \quad f_b = \frac{M_{max}}{Z}$$

Now $\quad Z = \frac{I_{xx}}{\text{Depth}/2}$

Now $I_{xx} = \frac{419}{12} t^4 = \frac{419}{12} (0.0055)^4 = 3.195 \times 10^{-8}$ m^4

$\therefore \quad Z = \frac{3.195 \times 10^{-8}}{0.01375} = 232.4 \times 10^{-8}$ m^3

$\therefore \quad f_b = \frac{79.44}{232.4 \times 10^{-8}} = \mathbf{34.18\ MPa.}$

Example 27.6. Design a connecting rod for a single cylinder four stroke diesel engine with following specifications:

Power = 7.5 kW
Mechanical efficiency = 80%
Weight of reciprocating parts = 20 N
Length of connecting rod = 0.30 m
Speed = 1500 rev/min with a possible over speed to 2500 rev/min
Assume suitable missing data.

Solution: The diesel cycle is shown in Fig. 27.22.

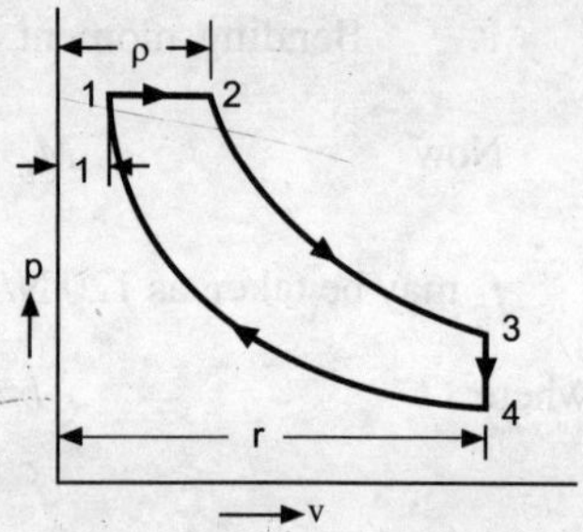

Fig. 27.22 *Diesel cycle*

Let compression ratio = 15

$\therefore \quad \frac{V_4}{V_1} = 15$

Let us take or assume, $V_1 = 1$

$\therefore \quad V_4 = 15$

$\therefore \quad$ Stroke volume $= V_4 - V_1 = 14$

Let us assume that the cut off is 6% of stroke volume.

$\therefore \quad V_2 - V_1 = 0.06 \times 14 = 0.84$

$\therefore \quad V_2 = 1 + 0.84 = 1.84$

$\therefore \quad$ Constant $\rho = \frac{V_2}{V_1} = \frac{1.84}{1} = 1.84$

Let p_4, the exaust pressure = atmospheric pressure = 0.103 MPa
Now mean effective pressure (indicated) is given as

$$\text{I.M.E.P} = \frac{p_4 \times r^{\gamma} \times [\gamma(\rho - 1) - r^{1-\gamma}(\rho^r - 1)]}{(\gamma - 1)(r - 1)}$$

where $\gamma = 1.4$, and $r = 15$

$$\therefore \quad \text{I.M.E.P.} = \frac{0.103 \times 15^{1.4}\,[1.4 \times 0.84 - 15^{-0.4}\,(1.84 - 1)]}{0.4 \times 14} = 0.585 \text{ MPa}$$

Now $\quad \text{I.H.P.} = \frac{pLAn}{60}$, W

$$\therefore \quad \frac{7.5 \times 1000}{0.8} = \frac{0.585 \times 10^6 \times L \times A \times 1500}{60 \times 2}$$

From here $\quad LA = 1282.05 \times 10^{-6}$, Let $\frac{L}{D} = 1.5$

$\therefore \quad 1.5D \times \frac{\pi}{4} D^2 = 1282.05 \times 10^{-6}$

$\therefore \quad D = 0.10294$ m, say 105 mm

$\therefore \quad$ Stroke, $L = 105 \times 1.5 = 157.5$ mm

Now maximum explosion pressure may be taken as about 7 to 9 times the mean effective pressure. Therefore, maximum explosion pressure = 8 × 0.585 = 4.68 MPa.

(g) Lubrication: The lubrication of the two end bearings of the connecting rod is very important. There are two methods for the lubrication of connection rod and bearings:

1. Splash lubrication. 2. Pressure feed lubrication.

In splash lubrication, a dipper or spout is attached to the big end of the connecting rod. During the downward motion of the connecting rod, the spout will dip into the lubricating oil in the sump. Through this spout, the oil will flow into the big end bearing. Now as the connecting rod moves up, a splash of oil will be produced by the spout. This splashed up lubricant will find its way into the small end bearing through many widely chamfered holes which are provided on the upper surface of the small end.

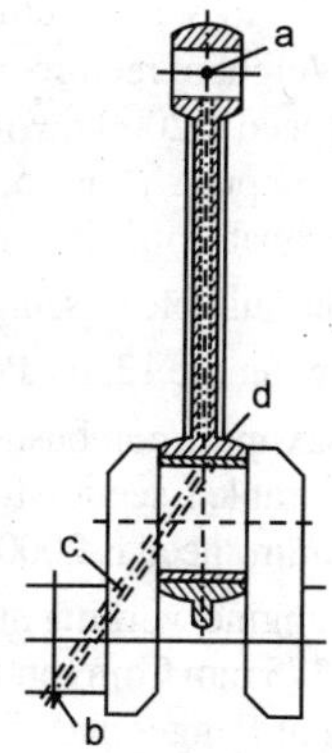

Fig. 27.23 *Pressure lubrication of crank pin and piston pin*

In the pressure feed system oil is fed under pressure to the crank pin bearing through holes drilled through the crank shaft, crank webs and crank pin. From the crank pin bearing, the oil is fed to the small end bearing through the hole rifle drilled in the connecting rod shank, This arrangement is shown in Fig. 27.23.

In some cases, the oil scraped form the walls of the liner by the scraper rings, will flow to the piston pin bearing holes through drilled in the walls of the piston.

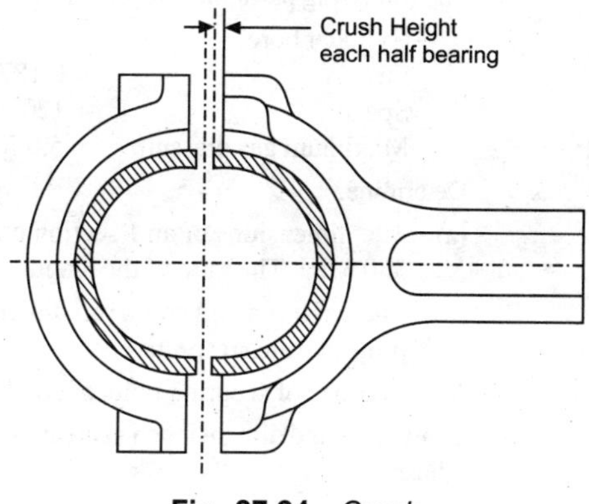

Fig. 27.24 *Crush*

(h) Crush: Good seating of the bearings in their housings is accomplished by making each half of the bearings shell slightly lower than the semi-circular housing in which it is clamped. The difference between 'the bearing arc length and the housing arc length is called 'Crush". (fig. 27.24). The amount of crush will depend upon the bearing material, the housing material, the temperature conditions and the magnitude of load. Too much crush will force the housing out of round, distorting the desired clearance between the journal and bearing. Too little crush may cause the bearing lining to loosen.

Note: The wear of the big end bearing is allowed for by thin metallic strips known as "Shims" between the cap and the fixed half. As the wear takes place, one or more strips are removed and the bearing is trued up.

EXERCISES

1. What is the function of connecting rod of I.C. Engines?
2. Enumerate the materials for the connecting rods.
3. Why 'I' section is chosen for high speed internal combustion engines?
4. Enumerate the various stresses induced in the connecting rod.
5. What is the effect of the angularity of the connecting rod upon the side pressure of the piston on the cylinder liner?
6. What is the usual ratio of the connecting rod length and the stroke of piston? How it is decided upon?
7. Under what force, the big end bolts and the cap are designed?
8. What are the materials of the piston pin bearing and the crank pin bearing?

9. How the small end bearing and the crank pin bearing of the connecting rod are lubricated?
10. What are "Shims" and what is their use?
11. What is "Crush"?
12. Design the connecting rod of an Internal Combustion Engine, for the following data:

Piston diameter	=	0.1525 m
Stroke	=	0.216 m
Length of connecting rod, centre to centre	=	0.495 m
Weight of reciprocating masses	=	27.5 N

Speed = 2000 rev/min with possible over speed of 3000 rev/min.

Compression ratio	=	5 : 1
Probable maximum explosion pressure	=	3.5 MPa

Assume suitable missing data.
13. Solve example 12, for Petrol Engine.
14. The maximum combustion pressure in an automotive Diesel Engine is 7 MPa. Cylinder bore is 0.127 m. Determine the desirable I-section for the middle of the connecting rod 0.3175 m long. If the engine stroke is 0.127 m and the maximum speed is 2000 rev/min, then determine the whipping stress in the connecting rod.
15. A gas engine working on Otto cycle has a piston of 0.1 m diameter. The explosion pressure is 3 MPa. Connecting rod is 373 mm from centre and to be a mild steel forging of I-section having same thickness for the flanges and web. Width of flanges and overall depth of the section are respectively 4 times and 5 times the web thickness. Design economical cross-section for the connecting rod assuming suitable stress and a factor of safety. Find also the dimensions for the small end assuming a bearing pressure of 14.0 MPa.
16. The following particulars refer to four stroke cycle diesel engine cylinder:

Cylinder bore	= 0.15 m
Stroke	= 0.1875 m
Speed	= 1200 rev/min
Maximum gas pressure	= 5.6 MPa

Determine:

(a) The dimensions of an I-section connecting rod of forged steel with an elastic limit compressive stress of 350 MPa. The ratio of the length of connecting rod to the length of crank is 4 and the factor of safety is 5.

(b) The wrist pin and crank pin dimensions on the basis of bearing pressure of 10.5 MPa and 6.5 MPa of the projected area respectively.

(c) The size of securing bolts of crank pin end if the allowable stress is not to exceed 35 MPa.

Draw freehand dimensioned sketches of connecting rod showing provision if any made for lubrication and water adjustment.

17. Design a mild steel connecting rod with an I-section for a single cylinder spark ignition engine from the following data:

Diameter of the piston	= 0.104 m
Weight of reciprocating parts	= 18.2 N
Length of connecting rod from centre to centre	= 0.314 m
Stroke length	= 0.14 m
Speed of the engine	= 1500 rev/min
Maximum explosion pressure	= 2.28 MPa gauge.

Assume that the maximum thrust takes place at T.D.C. during the explosion stroke.

18. A four stroke diesel engine has the following particulars:

Length of connecting rod	=	550 mm
Cylinder bore	=	150 mm
Stroke	=	240 mm
Maximum combustion pressure	=	4.7 N/mm^2

Determine the diameter of connecting rod of circular section. Take F.O.S. = 3 on f_{yp} = 330 MPa. Determine whipping stress if the engine runs at 1000 rev/min. Specific weight of material in 7800 kg/m^3.

27.4 CRANK SHAFTS

(a) Function: The function of a crank shaft is to transform reciprocating motion into rotary one or vice-versa. Crank shaft consists of the shaft parts which revolve in the main bearings, the crank pins to which the big ends of the connecting rod are connected, the crank webs or cheeks which connect the crank pins and the shafts parts.

(b) Types: The crank shafts can be divided into two types:

1. Crank shaft with a side crank or overhung crank, Fig. 27.25(*a*).
2. Crank shaft with a centre crank, Fig. 27.25(*b*).

A crank shaft can be made with two side cranks on each end or with two or more centre cranks. A crank shaft with only one side crank is called a *single throw crank shaft* and the one with two side cranks or two centre cranks as a *multithrow crank shaft.*

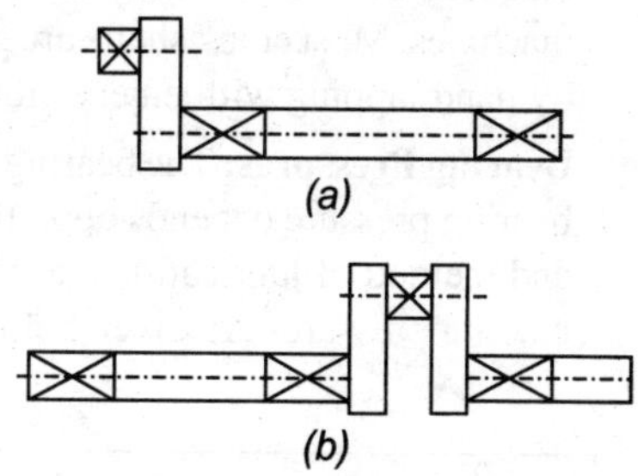

Fig. 27.25 *Types of crank shafts*

The overhung crank shaft is used for medium size and large horizontal engines. Its main advantage is that only two bearings are needed, in either the single crank or two crank, crank shafts. Misalignment causes most crank shaft failures and this danger is less in shafts with two bearings than with three or more supports. Hence, the number of bearings is a very important factor in design. Centre crank shafts of large engines generally have a bearing between each pair of cranks and a bearing on each end shaft. To make the engine lighter and shorter, the number of bearings in automobiles should be reduced.

(c) Materials: For the proper functioning, the crank shaft should fulfill the following conditions:

1. Enough strength to withstand the forces to which it is subjected, i.e., the bending and twisting moments.
2. Enough rigidity to keep the distortion to a minimum.
3. Stiffness to minimise, and strength to resist, the stress due to torsional vibrations of the shaft.
4. Sufficient mass properly distributed to see that it does not vibrate critically at the speeds at which it is operated.
5. Sufficient projected areas of crank pins and journals to keep down the bearing pressure to a value dependent on the lubrication available.
6. Minimum weight, especially in aero engines.

The crank shafts are made much heavier and stronger than necessary from the strength point of view so as to meet the requirements of rigidity and vibrations. Therefore, the weight cannot be reduced appreciably by using a material with a very high strength. The material to be selected will also depend upon the method of manufacture, i.e., cast, forged, or built up. Built up crank shafts are sometimes used in aero engines where light weight is very important.

In industrial engines, 0.35 carbon steel of ultimate tensile strength 500 to 525 MPa and 0.45 carbon steel of ultimate tensile strength of about 627 to 780 MPa are commonly used.

In transport engines, alloy steel, e.g., manganese steel having ultimate strength of about 784 to 940 MPa is generally used.

In aero engines, nickle chromium steel having ultimate tensile strength of about 940 to 1100 MP*a* is generally used. Heavy duty cast iron is being successfully used for crank shafts, especially for industrial engines of comparatively low speeds and it can replace easily the plain carbon steels. Cast iron usually used for crank shafts is nickle cast iron with ultimate tensile strength ranging form 350 to 525.0 MPa. Cast steel is also used as material for crank shafts, its ultimate tensile strength ranging from 560.0 to 600.0 MPa.

(d) Manufacturing: Great care must be observed in the manufacture of crank shafts since it is the most important part of the engine. Small crank shafts are drop forged. Larger shafts are forged and machined to shape. Casting of the crank shafts allows a theoretically desirable but complicated shape with a minimum of machining and at the smallest cost. These are cast in permanent moulds for maximum accuracy and a minimum of machining. While machining, the shaft must be properly supported between centres and special precautions should be taken to avoid springing. The journals and the crank pins are ground to exact size after turning. After this, the crank shaft is balanced. Large shafts of low speed engines are balanced statically, crank shafts of high speed engines are balanced dynamically on special balancing machines. Most crank shafts are ground at the journals and crank pins. In some cases, grinding is followed by hand lapping with emery cloth.

(e) **Bearing Pressures:** The bearing pressures are very important in the design of crank shafts. The allowable bearing pressure depends upon the journal velocity, change of direction of the bearing pressure, amount and method of lubrication and the maximum gas pressures and space limitations. Maximum allowable bearing pressures are given below in Table 27.3.

Table 27.3

Type of Engine	Kind of bearing	Highest allowable bearing pressure p MPa	Minimum Zn/p	Clearance c/d	l/d ratio
Automotive and aircraft engines	Main	5.62–11.95	2.13	0.001	0.6–1.8
	Crank pin	10.5–24.5	1.42	0.001	0.6–1.4
	Piston pin	26.0–35.0	1.14	0.001	1.5–2.0
Gas and oil engines, four stroke	Main	5.0–8.5	2.85	0.001	0.6–2.0
	Crank pin	10.0–14.0	1.42	< 0.001	0.6–1.5
	Piston pin	12.5–17.0	0.86	< 0.001	1.5–2.0
Gas and oil engines, Two stroke	Main	3.5–5.5	3.60	0.001	0.6–2.0
	Crank pin	7.0–10.5	1.72	< 0.001	0.6–1.5
	Piston pin	8.5–12.5	1.14	< 0.001	1.5–2.0

(f) **Stresses:** The stresses induced in the crank shafts are bending and also shear stresses due to torsional moment on the shaft. Most crank shafts fail due to progressive fracture due to repeated bending or reversed torsional stresses. Thus the type of loading on the crank shafts is fatigue loading, therefore, the design should be based on endurance limit.

To avoid stress concentration and fatigue, abrupt changes in the section of crank shaft should be avoided. Two different cross sections must be blended with a large filled 'r', if possible, r should not be less than 0.2 d. If there is no space for fillet, the crank web should be undercut to obtain the fillet. This will make the web weak and to compensate for it, the width is increased, Fig. 27.26.

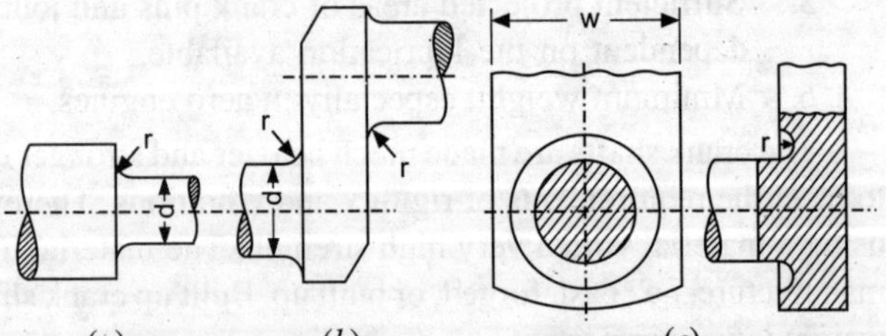

Fig. 27.26 *Use of fillets*

Since failure of the crank shaft is serious for the engine, and also because of the inaccuracy in determining all the forces and stresses, a high factor of safety based on endurance limit, from 3 to 4 should be used. To be on the safe side, the endurance limits for complete reversal of bending and torsional stresses are taken.

For chrome nickle and other alloy steels, the endurance limit is about 525.0 MPa in bending and about 290.0 MPa in shear.

For carbon steel and cast steel, the endurance limit is about 225 MPa in bending and about 124 MPa in shear. For alloy cast iron, the endurance limit is about 140MPa in bending and in shear.

Therefore, the allowable stress is:

For Carbon Steel:

$$\text{Bending} = 56.0 \text{ to } 75.0 \text{ MPa}$$
$$\text{Shear} = 31.0 \text{ to } 42.0 \text{ MPa}$$
$$\text{Combined Stress} = \frac{1}{2} \times \text{elastic limit in tension..}$$

For alloy Cast Iron:

$$\text{Bending} = \text{Shear} = 35.0 \text{ to } 47.0 \text{ MPa}$$
$$\text{Combined Stress} = \text{elastic limit in pure tension.}$$

For Chrome nickle and other alloy steels:

$$\text{Bending} = 130.0 \text{ to } 175.0 \text{ MPa}$$
$$\text{Shrear} = 72.5 \text{ to } 97.0 \text{ MPa}$$
$$\text{Combined Stress} = \frac{1}{2} \times \text{elastic limit in pure tension}.$$

(g) Design Procedure:

1. Determine the magnitudes of the various loads acting on the crank shaft.
2. Determine the distances between supports. The distances will depend upon the lengths of the bearing. The lengths and diameters of the bearing are determined on the basis of maximum permissible bearing pressure, *l/d* ratios and the acting loads, Table 27.3.
3. For the sake of simplicity and safety, the shaft is considered to be supported at the centres of the bearings.
4. Now calculate the distances between supports.
5. The thickness of the crank webs is assumed, about 0.5 *d* to 0.6 *d*. where *d* is the shaft diameter, or from 0.22*D* to 0.32*D*, where *D* is the cylinder bore.
6. Assume allowable bending and shearing stresses.
7. Compute the necessary dimensions of the crank shafts.

Note: All the forces and reactions are assumed to be acting at the centres of the bearings.

(h) Design Calculations: In the design of the crank shafts, it is assumed that the crank shaft is a beam with two or more supports. Every crank shaft must be designed or checked at least for two crank positions, one when the bending moment is maximum, and other when the twisting moment is maximum. In addition, the additional moment due to the flywheel weight, belt tension and other forces must be considered.

To make the calculations simpler, without loosing accuracy, it is assumed that the effect of the bending forces does not extend two bearings between which a force is applied.

(i) Analysis of Centre Crank Shaft

1. **Crank on Dead centre:** When the crank is on dead centre, Fig. 27.27, maximum bending moment will act in the crank shaft. The thrust in the connecting rod will be equal to the piston gas load 'P'. F is the weight of the flywheel acting downward and $F_1 + F_2$, is the belt pull acting horizontally.

Now
$$P = \frac{\pi}{4} D^2 \times p_{max}$$

Due to this, there will be two horizontal reactions, H_1 at bearing 1, and H_2 at bearing 2, so that

$$H_1 = H_2 = \frac{P}{2} \text{ (assuming } a = a').$$

otherwise,
$$H_1 = \frac{Pa'}{f} \text{ and } H_2 = \frac{P \times a}{f}$$

The weight of the flywheel '*F*' will give two vertical reactions. V_2 and V_3 at bearings 2 and 3 respectively.

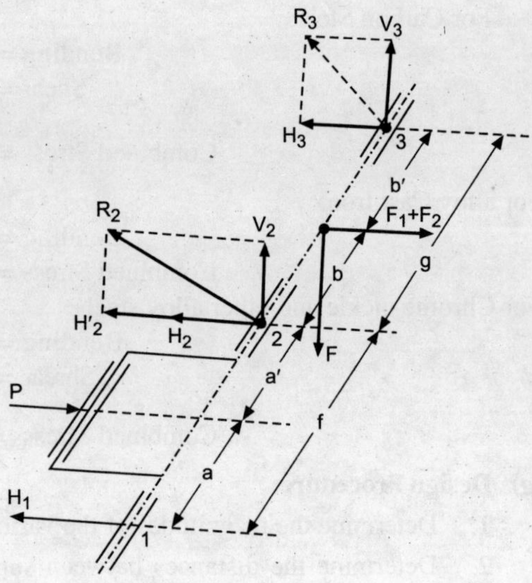

Fig. 27.27 *Crank at dead centre*

$$V_2 = V_3 = \frac{F}{2}, \text{ if } b = b' = \frac{g}{2}$$

otherwise, $$V_2 = \frac{F}{g} \times b', \text{ and } V_3 = \frac{F}{g} \times b$$

The belt pull '$F_1 + F_2$' will give two horizontal reactions at bearings 2 and 3.

$$H_2' = H_3 = \frac{F_1 + F_2}{2}, \text{ if } b = b' = \frac{g}{2}$$

otherwise, $$H_2' = \frac{F_1 + F_2}{g} \times b',$$

and $$H_3 = \frac{F_1 + F_2}{g} \times b$$

In this position of the crank, there will be no twisting moment, and the various parts will be designed for bending only.

Crank pin: The bending moment at the centre of the crank pin,

$$M_b = H_1 \times \frac{f}{2}$$

and $$M_b = \frac{\pi}{32} d^3 \times f_b,$$

where d = diameter of pin, m

f_b = allowable stress in bending, N/m^2

∴ d can be calculated.

Allowable bearing pressure can be assumed and the length of the pin can be determined from the relation,

$$P = l \times d \times p.$$

$$\therefore \quad l = ?$$

Left Hand Crank Web: The crank web be designed for eccentric loading. There will be two stresses on it, one direct compressive stress and the other bending stress due to the gas load P.

The thickness 't' of the crank web may be assumed to be equal to,

$$t = 0.22D \text{ to } 0.32D$$

or $$= 0.5d \text{ to } 0.6d$$

or $$= 0.65d + 6.35 \text{ mm}$$

The width 'w' may be assumed to be as follows:

$$w = \frac{9}{8} d + 12.7 \text{ mm}$$

now bending arm for reaction $H_1 = \left(\frac{f}{2} - \frac{l}{2} - \frac{t}{2}\right)$

∴ Total stress on the web = direct stress + bending stress

$$= \frac{H_1}{wt} + \frac{H_1\left(\frac{f}{2} - \frac{f}{2} - \frac{t}{2}\right)}{\frac{1}{6} wt^2}$$

The total stress must not exceed the allowable stress in bending.

Right Hand Crank Web: Since the bearings 1 and 2 are usually of the same length and symmetrical to the cylinder centre line, therefore, $H_1 = H_2$ and hence the dimensions of the right hand web will be same as the left hand web.

Shaft Under the flywheel: Bending moment due to flywheel weight is

$$M_f = V_3 \times \frac{g}{2} = \frac{F}{2} \times \frac{g}{2} = \frac{Fg}{4}$$

Bending moment due to the belt pull, is

$$M_b = H_3 \times \frac{g}{2} = (F_1 + F_2)\frac{g}{4}$$

Since these act at right angles to each other, the combined bending moment is given as

$$M_{to} = \sqrt{M_f^2 + M_b^2} = \frac{\pi}{32} dw^3 \times f_b,$$

where dw = diameter of the shaft under flywheel.

f_b = allowable stress in bending. Its value should be taken low to take care of reversal of stresses in each revolution and to ensure necessary rigidity.

2. Crank at angle of Maximum Twisting moment: The twisting moment on the crank shaft will be maximum when the tangential force 'P_t', is maximum and this can be calculated graphically by taking pressures from the net-effort diagram for different crank angles. The angle usually lies between 25° to 35° from the dead centre for a constant volume combustion engine and between 30 to 40° for constant pressure combustion engines. At this angle, the gas pressure will not be maximum. If P' is the gas load along the cylinder centre line, then the thrust Q along the connecting rod is given by, (Fig. 27.28)

$$Q = \frac{P'}{\cos\varphi}; \left[\sin\phi = \frac{\sin\theta}{L/R}\right]$$

Here, L = length of connecting rod

and R = crank radius

Q will have two components, the tangential force 'P_t' and the radial force 'P_r' along the crank, given by

$$P_t = Q\sin(\theta + \phi)$$

$$P_r = Q\cos(\theta + \phi)$$

F_t will have two reactions H_{t1} and H_{t2} at bearings 1 and 2 respectively.

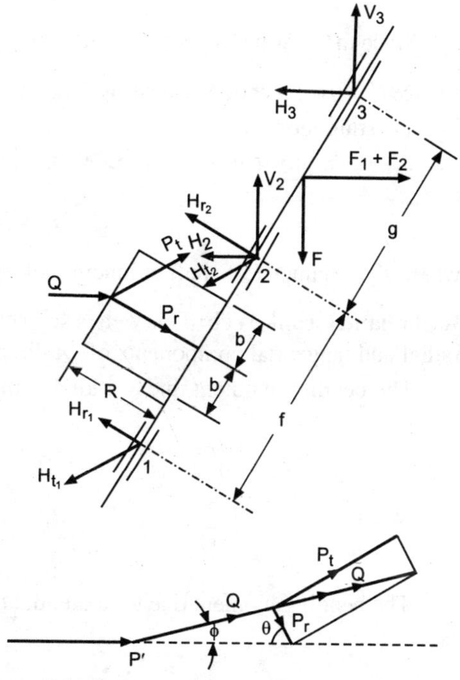

Fig. 27.28 *Crank on angle of maximum twisting movement*

F_r will have two reactions H_{r1} and H_{r2} at bearings 1 and 2 respectively.

The reactions at the bearings, due to F and $F_1 + F_2$ will be same as before.

In this position of the crank shaft, the different sections will be subjected to both bending and torsional moments and these must be checked for combined stress. It may be noted at this point that the usual criterion of failure for crank shafts is the shear stress.

Crank pin: Bending moment at the centre of the crank pin is,

$$M_b = H_{r1} \times \frac{f}{2}$$

Twisting moment is, $M_t = H_{t1} \times R$.

Equivalent twisting moment, $M_{te} = \sqrt{M_b^2 + M_t^2} = \frac{\pi}{16} d^3 \times f_s$. $\therefore$ $d = ?$

where f_s = allowable shear stress, N/m^2.

Shaft under the Flywheel: The combined bending moment due to flywheel and the belt pull will be the same as before.

The twisting moment, $M_t = P_t \times R$

$\therefore$ Equivalent twisting moment,

$$M_{te} = \sqrt{M_t^2 + M_{to}^2} = \frac{\pi}{16} d\omega^3 \times f_s$$

$$d\omega = ?$$

Shaft at Juncture of Right hand Crank Arm:

Bending moment, $M_b = H_1'\left(\frac{f}{2}+\frac{l}{2}+\frac{t}{2}\right) - Q\left(\frac{l}{2}+\frac{t}{2}\right)$

Since the bearings are symmetrical, $H_1' = \frac{Q}{2}$

where H_1' = total reaction at bearing 1 due to Q.

Twisting moment, $M_t = P_t \times R$

$\therefore$ Equivalent twisting moment

$$M_{te} = \sqrt{M_t^2 + M_b^2} = \frac{\pi}{16} {d_2}^3 \times f_s \quad \therefore \quad d_2 = ?$$

where d_2 = diameter of shaft at junction of right hand crank.

Right hand Crank Web: This web is subjected to bending stresses in two planes normal to each other, due to radial and tangential components of Q; direct compression; and to torsion.

The bending moment due to radial component is

$$M_{br} = H_{r_2}\left(\frac{f}{2} - \frac{l}{2} - \frac{t}{2}\right)$$

Now $M_{br} = f_{br} \times Z, Z = \frac{1}{6} wt^2$

$\therefore$ $f_{br} = ?$

The bending moment due to tangential component is maximum at the juncture of the crank and shaft. It is

$$M_{bt} = P_t\left(R - \frac{d_2}{2}\right)$$

where d_2 = Shaft diameter at bearing 2

Now $M_{bt} = f_{bt} \times Z, Z = \frac{1}{6} t\omega^2$

$\therefore$ $f_{bt} = ?$

The stress due to direct compression, $f_d = \frac{P_r}{2 \times \omega t}$

The total compressive stress will be maximum at the upper left corner to the cross-section of crank, and will be

$$f_c = f_{br} + f_{bt} + f_d.$$

Now the twisting moment, on the arm is

$$M_t = H_{t_1}\left(\frac{f}{2}+\frac{l}{2}\right) - P_t \times \frac{l}{2}$$

or

$$= H_{t_2}\left(\frac{f}{2}-\frac{l}{2}\right)$$

$\therefore$

$$f_s = \frac{M_t}{Z_p}$$

where Z_p = Polar section modulus $= \dfrac{\omega t^2}{4.5}$

$\therefore$ Total combined stress is,

$$f_{max} = \frac{f_c}{2} + \sqrt{\left(\frac{f_c}{2}\right)^2 + f_s^2}$$

f_{max} must be within limits. If it exceeds the safe limit, w can be increased since if does not affect any other dimension.

Left hand Crank: This crank web is less severely stressed than the right hand crank since it is not to transmit a power while the right hand crank transmit the power to the flywheel and to the power take off. Hence, there is no need to check the left hand crank and its dimensions may be taken as that of the right hand crank.

Crank shaft bearings: The distance between bearing 1 and 2 may be assumed to be equal to twice the cylinder diameter. From the length of the crank pin and the thickness of the arm, the lengths of the bearing can be found out. Bearing 2 is the most heavily loaded, therefore, only this bearing may be checked for the safe bearing pressure. The total pressure on bearing 2 will be equal to

$$= \frac{P}{2} + \frac{F}{2} + \frac{(F_1 + F_2)}{2}$$

(j) Analysis of the Side Crank Shaft: The analysis of the side crank shaft is on the same lines as for centre crank shaft. Before the crank shaft is checked for the positions of maximum bending moment and that of maximum twisting moment, the approximate dimensions for the crank pin and the journals may be found.

Crank Pin: As already indicated, the dimensions of crank pin are determined from the bearing and lubrication requirement and then checked for strength and deflection. The crank pins of the side crank engines are cantilever beams. Therefore, if the piston gas load is 'P' and to take into account the possibility that the load may not be evenly distributed over the crank pin length, the bending moment may be taken as $0.75Pl$ instead of $0.5Pl$. The maximum bending stress will be at the junction of the pin and the crank web.

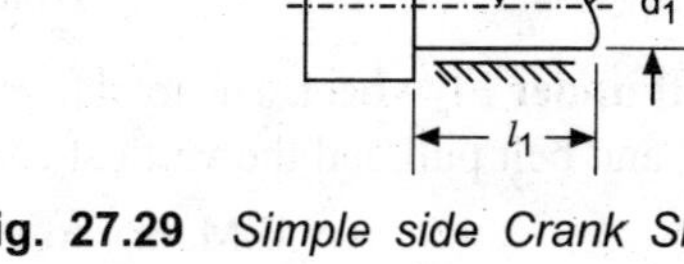

Fig. 27.29 *Simple side Crank Shaft*

$\therefore$

$$0.75Pl = \frac{\pi}{32} d^3 \times f_b$$

Assuming l/d ratio as from 0.6 to 1.4, the diameter of the pin 'd' and its length 'l' can be determined, 'f_b' is the allowable working stress in bending.

After this, the bearing pressure, $p_b = P/ld$ should be checked.

Journal: The bending moment at the middle of the journal is

$$M_b = P\,(0.75\,l + t + 0.5\,l_1),$$

where t = crank web thickness = 0.45 to 0.75 d.

l_1 can be assumed to be equal to from 1.5 d to $2d$.

$$\therefore \quad M_b = \frac{\pi}{32}\,d_1^3 \times f_b \quad \therefore d_1 = ?$$

1. Crank on Dead Centre. Figure 27.30 shows a side crank shaft at dead centre with its loads and distances of their application.

The gas load P gives two horizontal reactions at bearings 1 and 2

$$H_1 = P \cdot \frac{(a+b)}{b} \text{ and } H_2 = -\frac{Pa}{b}$$

The weight of the flywheel 'F' gives two vertical reactions,

$$V_1 = \frac{F_c'}{b} \text{ and } V_2 = \frac{F_c}{b}$$

The belt pull '$F_1 + F_2$' gives two horizontal reactions,

$$H_1' = (F_1 + F_2) \times \frac{c'}{b}$$

and $$H_2' = (F_1 + F_1) \times \frac{c}{b}.$$

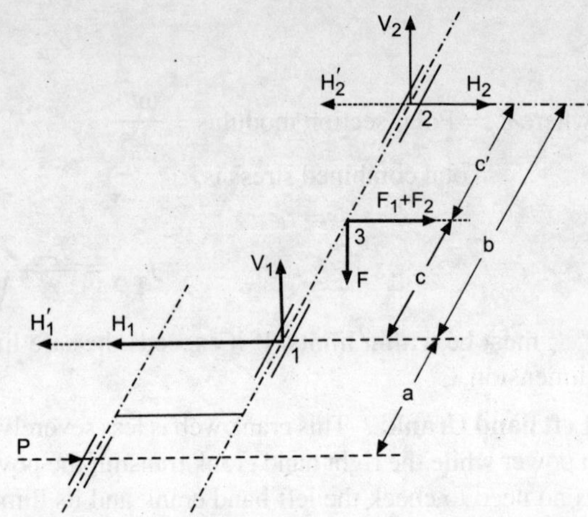

Fig. 27.30 *Side crank shaft on dead centre*

Crank Web: Under this crank position, the crank arm is subjected to a bending moment and to a direct compression.

$$\text{Bending moment, } M_b = P\left(0.75\,l + \frac{t}{2}\right)$$

$$\therefore \quad f_b = \frac{M_b}{Z},\ Z = \tfrac{1}{6}\,wt^2$$

$$\text{Direct stress,} \quad f_d = \frac{P}{wt}$$

Total stress $= f_b + f_d$, should not exceed the safe limit.

Shaft under Flywheel. The total bending moment will be resultant of horizontal bending moment due to gas load and belt pull and the vertical bending moment due to the flywheel weight.

B.M. due to gas load $= P\,(a+c) - (H_1 + H_1')\,(c)$.

B.M. due to belt pull $= (F_1 + F_2) \times b/4$

Total horizontal B.M., $M_h = M_g + M_b$

$$\text{Vertical B.M.,} \quad M_v = \frac{F \times b}{4}$$

$$\therefore \text{ Net B.M.,} \quad M_e = \sqrt{M_h^2 + M_v^2} = \frac{\pi}{32}\,d_w^3 \times f_b \ \therefore d_w = ?$$

2. Crank at Angle of Maximum Torsional Moment. As in the case of centre crank shaft, the tangential component 'P_t' and the radial component 'P_r' of the connecting rod thrust Q, can be determined. The forces acting on the shaft at this position are shown in Fig. 27.31.

The radial components, i.e., reaction due to the radial force 'P_r' at bearings 1 and 2 are:

$$H_{r_1} = P_r \times \frac{a+b}{b} \text{ and } H_{r_2} = -P_r \times \frac{a}{b}$$

Tangential reaction due to tangential force P_t are:

$$H_{r_1} = P_t \times \frac{a+b}{b} \text{ and } H_{t_2} = -P_t \times \frac{a}{b}.$$

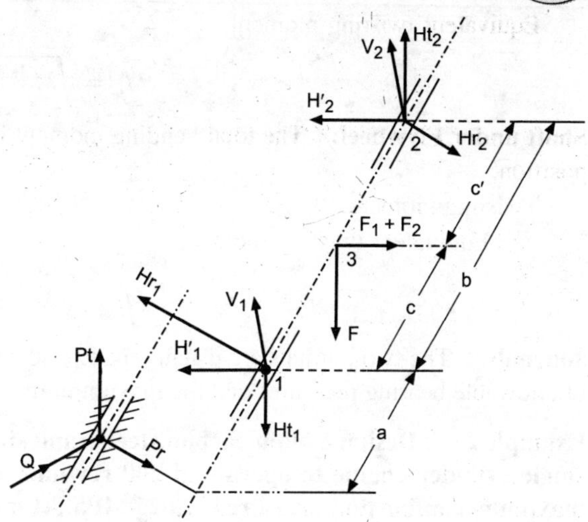

Fig. 27.31 *Crank at angle of maximum torsional moment*

Crank Web. The dangerous section is where the web joins the shaft. This section is subjected to bending due to P_t, bending due to P_r, direct compression due to P_r, and to torsion.

$$\text{B.M. due to } P_t = P_t\left(R - \frac{d_1}{2}\right) = M_{bt}$$

Now $$M_{bt} = f_{bt} \times Z,\ Z = \frac{1}{6} t\, w^2$$

$\therefore$ $$f_{bt} = ?$$

$$\text{B.M. due to } P_r, M_{br} = P_r\left(0.75\, l + \frac{t}{2}\right) = f_{br} \times Z,\ Z = \tfrac{1}{6} wt^2$$

$\therefore$ $$f_{br} = ?$$

Direct compression stress, $$f_d = \frac{P_r}{\omega t}$$

Total compression stress, $$f_c = f_{bt} + f_{br} + f_d.$$

The torsional moment, $$M_t = P_t\left(0.75\, l + \frac{t}{2}\right)$$

$\therefore$ $$\text{Shear stress } f_s = \frac{M_t}{Z_p},$$

where Z_p = polar section modulus = $\dfrac{t^2\omega}{4.5}$

$$\text{Total stress}, f_{max} = \frac{f_c}{2} + \sqrt{f_s^2 + \left(\frac{f_c}{2}\right)^2}$$

f_{max} should not exceed the safe limit.

Shaft at Junction of Crank:

Bending moment, $$M_b = Q\,(0.75\, l + t)$$

Torsional moment, $$M_t = P_t \times R$$

Equivalent twisting moment,

$$M_{te} = \sqrt{M_t^2 + M_b^2} = \frac{\pi}{16} d_1^3 \times f_s$$

Shaft under Flywheel: The total bending moment is found in the same manner as before for dead centre position.

Twisting moment, $M_t = P_t \times R$

$\therefore$ Equivalent twisting moment

$$M_{te} = \sqrt{M_t^2 + M_e^2} = \frac{\pi}{16} d_\omega^3 \times f_s.$$

Journals: The two journals are usually made of the same diameter. The lengths are found out on the basis of allowable bearing pressures and the maximum reactions at the journals.

Example 27.7. Design a plain carbon steel crank shaft for a 0.40 m by 0.60 m single acting four stroke single cylinder engine to operate at 200 rev/min. The mean effective pressure is 0.49 MPa, and the maximum combustion pressure is 2.625 MPa. At maximum torsional moment, when the crank angle is 36°, the gas pressure is 0.975 MPa. The ratio of the connecting rod length to the crank radius is 4.8. The flywheel is used as a pulley. The weight of the flywheel is 54.50 kN and the total belt pull is 6.75 kN. Assume suitable values for the missing data.

Solution: 1. **Crank at Dead Centre.** Refer Fig. 27.27.

Piston gas load, $P = \frac{\pi}{4} D^2 \times p_{\max} = \frac{\pi}{4} \times (0.40)^2 \times 2.625 \times 10^6 = 330 \text{ kN}$

Let us assume, $f = 2 \times \text{cylinder diameter} = 2 \times 0.40 = 0.80 \text{ m}$

Again taking $a = a' = \frac{f}{2} = 0.40 \text{ m}$

$$H_1 = H_2 = \frac{P}{2} = 165 \text{ kN}$$

In such shafts, the lengths of the main bearings are taken to be equal.

$$V_2 = V_3 = \frac{F}{2} = \frac{54.50}{2} = 27.25 \text{ kN} \qquad [b = b']$$

$$H_2' = H_3 = \frac{F_1 + F_2}{2} = \frac{6.75}{2} = 3.375 \text{ kN}$$

Crank pin: The bending moment at the centre of pin is

$$M_b = H_1 \times \frac{f}{2} = 165 \times 1000 \times 0.40 = 66 \text{ kNm}$$

$$\therefore \quad \frac{\pi}{32} d^3 \times f_b = 66 \times 1000$$

Taking $f_b = 75 \text{ MPa}; \ \frac{\pi}{32} d^3 \times 75 \times 10^6 = 66 \times 1000$

$\therefore$ $d = 0.208 \text{ m, say } 0.21 \text{ m}$

Allowable bearing pressure may be taken as 10 MPa

$\therefore$ $P = l d_p$

$\therefore$ $l = \text{length of crank pin} = \frac{330 \times 1000}{0.21 \times 10 \times 10^6} = 0.157 \text{ m, say } 0.16 \text{ m}$

Left Hand Crank: Let the thickness of web, $t = 0.65\,d + 6.35$ mm

$$= 0.65 \times 0.21 + 0.00635 = 0.143 \text{ m}$$

and width 'w' $= \frac{9}{8} \times d + 12.7$ mm

$$= \frac{9}{8} \times 0.21 + 0.0127 = 0.249 \text{ m}$$

Now bending moment $= H_1 \left(\frac{f}{2} - \frac{l}{2} - \frac{t}{2} \right) = 165\,(0.40 - 0.08 - 0.0715) = 41$ kNm

Now section modulus of arm, $Z = \frac{1}{6}\,\omega t^2 = \frac{1}{6} \times 0.249 \times 0.143^2 = 8.5 \times 10^{-4}\ \text{m}^3$

$\therefore$ Bending stress, $f_b = \frac{M_b}{Z} = \frac{41 \times 1000}{8.5 \times 10^{-4}} = 48.2$ MPa

Direct compressive stress, $f_d = \frac{H_1}{\omega t} = \frac{165 \times 1000}{0.249 \times 0.143} = 4.62$ MPa

Total stress $= f_b + f_d = 48.2 + 4.62 = 52.82$ MPa, which is well below the limit.

Right hand Crank: Since it has been assumed that the main bearings are of the same length and $H_1 = H_2$, therefore, the analysis of the right hand crank arm will be same as for left hand crank arm.

Shaft under the flywheel: As already indicated, in such shafts, the lengths of the bearings (main) are taken to be equal.

$$\therefore \quad l_1 = l_2 = l_3 = \left(\frac{f}{2} - \frac{l}{2} - t \right) \times 2 = 2(0.40 - 0.08 - 0.143) = 0.354 \text{ m}$$

Assuming width of the flywheel equal to 0.30 m

$$g = 0.354 + 0.30 = 0.654, \text{ say } 0.66 \text{ m}$$

Allowing space for gearing and clearance, let us take 'g' = 0.80 m

$\therefore$ Bending moment due to the flywheel weight

$$M_f = \frac{Fg}{4} = \frac{54.50 \times 0.80}{4} = 10.9 \text{ kNm}$$

Bending moment due to the belt pull,

$$M_b = (F_1 + F_2) \times \frac{g}{4} = \frac{6.75 \times 0.80}{4} = 1.35 \text{ kNm}$$

$\therefore$ Total bending moment,

$$M_{to} = \sqrt{M_f^2 + M_b^2} = \sqrt{10.9^2 + 1.35^2} = 10.95 \text{ kNm}$$

$$\therefore \quad \frac{\pi}{32}\,d\omega^3 \times f_b = 10.95 \times 1000$$

Take a low value of f_b, say 42 MPa

$$\therefore \quad d\omega = 0.139 \text{ m, say } 0.14 \text{ m}$$

2. **Crank at angle of Maximum Torsional moment:**

Refer Fig. 27.28.

Piston gas load, $P' = \frac{\pi}{4} \times 0.40^2 \times 0.975 \times 10^6 = 122.52$ kN

$\therefore$ Connecting rod thrust, $Q = \dfrac{P'}{\cos\phi}$

Now $\sin\phi = \dfrac{\sin\theta}{L/R}$

and $\theta = 36°$, and $L/R = 4.8$

$\therefore$ $\sin\phi = \dfrac{\sin 36°}{4.8} = \dfrac{0.58778}{4.8} = 0.1225$

$\therefore$ $\phi \cong 7°$

Now $P_t = Q\sin(\theta+\phi)$ and $P_r = Q\cos(\theta+\phi)$

$$Q = \frac{122.52}{\cos 7°} = \frac{122.52}{0.99255} = 123.44 \text{ kN}$$

$\therefore$ $P_t = 123.44 \times \sin 43° = 84.40$ kN, $P_r = 123.4 \text{ kN} \times \cos 43° = 90.28$ kN

$\therefore$ $H_{t_1} = H_{t_2} = \dfrac{P_t}{2} = 42.20$ kN

and $H_{r_1} = H_{r_2} = \dfrac{P_r}{2} = 45.14$ kN

Crank pin: Bending moment at the centre of crank pin,

$$M_b = H_{r_1} \times \frac{f}{2} = 45.14 \times 0.40 = 18.056 \text{ kNm}$$

Twisting moment, $M_t = H_{t_1} \times R = 42.20 \times 0.30 = 12.66$ kNm

Equivalent twisting moment $= \text{M}_{te} = \sqrt{\text{M}_t^2 + \text{M}_b^2} = \sqrt{18.056^2 + 12.66^2}$

$= 22.052$ kNm

$\therefore$ $\dfrac{\pi}{16} d^3 \times f_s = 22.052 \times 1000$

Taking $f_s = 35$ MPa $\therefore$ $d = 0.146$ m,

which is small as compared to the diameter of pin already calculated. Instead of revising the dimensions and the calculations, let us keep d as 0.21 m.

Shaft under the fly wheel: Total bending moment will be the same as before,

$\therefore$ $M_{to} = 10.95$ kNm

Twisting moment, $M_t = P_t \times R = 84.40 \times 0.30 = 25.257$ kNm

$\therefore$ $M_{te} = \sqrt{M_t^2 + M_{to}^2} = \sqrt{25.257^2 + 10.95^2} = 27.529$ kNm

$\therefore$ $\dfrac{\pi}{16} d\omega^3 \times f_s = 27.529 \times 1000$

$\therefore$ $f_s = 27.529 \times 1000 \times \dfrac{16}{\pi} \times \dfrac{1}{0.14^3} = 51.095$ MPa

which is more than the safe limit of 31.0 to 42.0 MPa

Let us take $f_s = 35.0$ MPa

$\therefore$ $d\omega = 0.159$ m, say 0.16 m

Shaft at juncture of Right hand crank arm:

Bending moment, $M_b = H_1'\left(\dfrac{f}{2} + \dfrac{l}{2} + \dfrac{t}{2}\right) - Q\left(\dfrac{l}{2} + \dfrac{t}{2}\right)$

$\therefore \qquad M_b = 61.72\ (0.40 + 0.08 + 0.0715) - 123.44\ (0.08 + 0.0715) = 15.4\ \text{kNm}$

Twisting moment, $\qquad M_t = P_t \times R = 25.257\ \text{kNm}.$

$$\therefore \qquad M_{te} = \sqrt{25.257^2 + 15.4^2} = 29.582\ \text{kNm}$$

$$\therefore \qquad \frac{\pi}{16} d_2^3 \times f_s = 29.582 \times 1000$$

taking $\qquad f_s = 42\ MPa \therefore d_2 = 0.152$ m, say 0.155 m

Right hand Crank Web: Bending moment due to radial component,

$$M_{br} = H_{r2}\left(\frac{f}{2} - \frac{l}{2} - \frac{t}{2}\right) = 45.14\ (0.40 - 0.08 - 0.0175) = 11.12\ \text{kNm}$$

$$f_{br} = \frac{M_{br}}{\frac{1}{6}\omega t^2} = \frac{11.12 \times 1000 \times 6}{0.249 \times 0.143^2} = 13.22\ \text{MPa}$$

Bending moment due to tangential component,

$$M_{bt} = P_t\left(R - \frac{d_2}{2}\right) = 84.40\ (0.30 - 0.0775) = 18.73\ \text{kNm}$$

$$\therefore \qquad f_{bt} = \frac{M_{bt}}{\frac{1}{6}t\omega^2} = \frac{18.73 \times 1000 \times 6}{0.143 \times 0.249^2} = 12.68\ \text{MPa}$$

$$\text{Direct compressive stress, } f_d = \frac{P_r}{2 \times \omega t} = \frac{90.28 \times 1000}{2 \times 0.249 \times 0.143} = 1.27\ \text{MPa}$$

$\therefore$ Total compressive stress,

$$f_c = f_{br} + f_{bt} + f_d = 13.22 + 12.68 + 1.27 = 27.17\ \text{MPa}$$

The twisting moment on the arm is

$$M_t = H_{t2}\left(\frac{f}{2} - \frac{l}{2}\right) = 42.20\ (0.40 - 0.08) = 13.5\ \text{kNm}$$

$$\therefore \qquad f_s = \frac{M_t}{Z_p} = \frac{13.5 \times 4.5}{0.249 \times 0.143^2} = 11.93\ \text{MPa}$$

$\therefore$ Combined stress,

$$f_{c\,max} = \frac{f_c}{2} + \sqrt{f_s^2 + \left(\frac{f_c}{2}\right)^2} = 13.585 + \sqrt{11.93^2 + 13.585^2}$$

$$= 13.585 + 18.08 = 31.665\ \text{MPa}.$$

which is quite below the safe limit.

Journals: As already indicated, bearing 2 is the most heavily loaded, therefore, only this bearing need be checked.

$$\text{Total reaction at this bearing} = \frac{P}{2} + \frac{F}{2} + \frac{F_1 + F_2}{2} = \frac{330}{2} + \frac{54.50}{2} + \frac{6.75}{2} = 195.625\ \text{kN}$$

$$\therefore \qquad \text{Bearing pressure} = \frac{195.625 \times 1000}{l_2 \times d_2} = \frac{195.625 \times 1000}{0.354 \times 0.155} = 3.57\ \text{MPa},$$

which is quite low as compared to the safe limit of 5.0 to 8.5 MPa.

Example 27.8. Design an overhung crank shaft with two main bearings and a flywheel in between them for an I.C. engine, single cylinder 0.25 m × 0.30 m. The flywheel weighs 27 kN. The maximum pressure is 2.1 MPa. The torsional moment is maximum when the crank is at 35° form the I.D.C, while the pressure is 1.05 MPa.

Assume missing data.

Solution:

1. **Crank on Dead Centre:** Refer Fig. 27.30.

Connecting rod thrust = Piston gas load

$$\therefore \quad P = \frac{\pi}{4} D^2 \times p_{max} = \frac{\pi}{4} \times 0.25^2 \times 2.1 \times 10^6 = 103 \text{ kN}$$

Crank pin:

Bending moment on pin, $M_b = 0.75\ Pl$

$$\therefore \quad \frac{\pi}{32} d^3 \times f_b = 0.75\ Pl$$

$\frac{l}{d}$ varies form 0.6 to 1.4, take $\frac{l}{d} = 0.8$

Take $f_b = 56.0$ MPa

$$\therefore \quad \frac{\pi}{32} d^3 \times 56 \times 10^6 = 0.75 \times 103 \times 1000 \times 0.8\ d$$

$\therefore$ d = diameter of pin = 0.106 m, say 0.11 m

$\therefore$ l = length of pin = 0.800 × 11 = 0.088 m, say 0.09 m

Checking the bearing pressure,

$$p = \frac{P}{l \times d} = \frac{103 \times 1000}{0.09 \times 0.11} = 10.4 \text{ MPa}$$

which is safe, being below 14 MPa.

Journal 1. The bending moment at the middle of the journal is

$$M_b = P \times a,\ a = 0.75\ l + t + 0.5 l_1$$

where t = thickness of web = 0.45 d to 0.75 d

or t = 0.0495 to 0.0825 m, say 0.06 m

l_1 = length of shaft in journal 1

= 1.5 d to 2 d = 0.165 m to 0.22 m, say 0.18 m

$$\therefore \quad M_b = 103\ [0.75 \times 0.09 + 0.06 + 0.5 \times 0.18] = 22.4 \text{ kNm}$$

$$\therefore \quad \frac{\pi}{32} d_1^3 \times f_b = 22.4 \times 1000$$

Take $f_b = 56.0$ MPa, $\therefore\ d_1 = 0.16$ m

Assuming the width of the flywheel as 0.25 m, and allowing for certain clearances and taking $l_1 = l_2 = 0.18$ m, the distance $b = 0.48$ m.

$\therefore$ Components of the gas load at bearings 1 and 2.

$$H_1 = P\frac{(a+b)}{b} = \frac{103\ (0.2175 + 0.48)}{0.48} = 150 \text{ kN}$$

$$H_2 = -P \cdot \frac{a}{b} = -103 \times \frac{0.2175}{0.48} = -46.6 \text{ kN}$$

Vertical reactions due to flywheel weight,

$$V_1 = V_2 = \frac{F}{2} = \frac{27}{2} = 13.50 \text{ kN}$$

There is no belt pull.

Crank web: Bending moment on the crank web,

$$M_b = P\left(0.75\,l + \frac{t}{2}\right) = 103\left(0.75 \times 0.09 + \frac{0.06}{2}\right) = 10.04 \text{ kN}$$

$$\therefore \qquad f_b = \frac{M_b}{Z} = \frac{10.04 \times 10^3 \times 6}{w \times 0.06^2} = \frac{16.75}{w} \text{ MPa}$$

$$\text{Direct compressive stress, } f_d = \frac{P}{wt} = \frac{103 \times 10^3}{w \times 0.06} = \frac{1.72}{w} \text{ MPa}$$

$$\therefore \qquad \text{Total stress} = f_b + f_d = \frac{18.47}{w} \text{ MPa}$$

$$\not> 75 \text{ MPa}$$

$$\therefore \qquad \frac{18.47}{w} = 75, \text{ gives } w = 0.25 \text{ m}$$

Shaft under flywheel:

B.M. due to gas load, $M_h = P(a + c) - H_1 \times c$

$$c = \frac{0.48}{2} = 0.24 \text{ m}$$

$$\therefore \qquad M_h = 103\,(0.2175 + 0.24) - 150 \times 0.24 = 11.2 \text{ kNm}$$

Vertical B.M. due to flywheel weight

$$M_v = \frac{F \times b}{4} = \frac{27 \times 0.48}{4} = 3.24 \text{ kNm}$$

$$\therefore \quad \text{Net bending moment, } M_e = \sqrt{M_b^2 + M_v^2}$$

$$= \sqrt{11.2^2 + 3.24^2} = 11.64 \text{ kNm}$$

$$\therefore \qquad \frac{\pi}{32} dw^3 \times f_b = 11.64 \times 10^3$$

Take $f_b = 42.0$ MPa, then $d_w = 0.1415$ m

It is less than the shaft diameter, but usually, it is greater than the shaft diameter.

2. Crank at position of maximum torsional moment:

Piston gas load, $P' = \frac{\pi}{4} \times 0.25^2 \times 1.05 \times 10^6 = 51.50$ kN

$$\therefore \quad \text{Connecting rod thrust, } Q = \frac{P'}{\cos\phi}$$

Now $\sin\phi = \frac{\sin\theta}{L/R}$

Let L/R ratio be 4.5

$$\therefore \qquad \sin\phi = \frac{\sin 35^\circ}{4.5} = 0.1275. \ \therefore\ \phi = 7.32^\circ$$

$$\therefore \qquad Q = \frac{51.50}{0.992} = 52 \text{ kN}$$

$$\therefore \quad P_t = Q \sin(\theta + \phi) = 52 \times \sin 42.32° = 35 \text{ kN}$$

$$P_r = Q \cos(\theta + \phi) = 52 \cos 42.32° = 38.40 \text{ kN}$$

$$\therefore \quad H_{r_1} = Pr \times \frac{a+b}{b} = 38.40 \times \frac{0.6975}{0.48} = 55.80 \text{ kN}$$

$$H_{r_2} = -P_r \times \frac{a}{b} = -38.40 \times \frac{0.2175}{0.48} = -17.40 \text{ kN}$$

$$H_{t_1} = P_t \times \frac{a+b}{b} = 35 \times \frac{0.6975}{0.48} = 50.8 \text{ kN}$$

$$H_{t_2} = -P_t \times \frac{a}{b} = -35 \times \frac{0.2175}{0.48} = -15.75 \text{ kN}$$

Crank web:

Bending moment due to $P_t = M_{bt} = P_t\left(R - \frac{d_1}{2}\right) = 35\,(0.15 - 0.08) = 2.45 \text{ kNm}$

$$\therefore \quad f_{bt} = \frac{M_{bt}}{Z} = \frac{2.45 \times 1000 \times 6}{0.06 \times 0.25^2} = 3.92 \text{ MPa}$$

Bending moment due to $P_r = M_{br} = P_r\left(0.75\,l + \frac{t}{2}\right) = 38.40\,(0.75 \times 0.09 + 0.03) = 3.744 \text{ kNm}$

$$\therefore \quad f_{br} = \frac{M_{br}}{Z} = \frac{M_{br}}{\frac{1}{6}wt^2} = \frac{3.744 \times 1000 \times 6}{0.25 \times 0.06^2} = 24.96 \text{ MPa}$$

Direct compressive stress, $f_d = \frac{P_r}{wt} = \frac{38.4 \times 1000}{0.25 \times 0.06} = 2.56 \text{ MPa}$

∴ Total compressive stress at the upper left corner.

$$f_c = f_{bt} + f_{br} + f_d = 3.92 + 24.96 + 2.56 = 31.44 \text{ MPa}$$

Now twisting moment on the web,

$$M_t = P_t\left(0.75\,l + \frac{t}{2}\right) = 35 \times 0.0975 = 3.41 \text{ kNm}$$

∴ Shear stress, $f_s = \frac{M_t}{Z_p} = \frac{M_t}{\frac{1}{4.5}wt^2} = \frac{3.41 \times 1000 \times 4.5}{0.25 \times 0.06^2} = 17 \text{ MPa}$

∴ Total stress, $f_{max} = \frac{f_c}{2} + \sqrt{f_s^2 + \left(\frac{f_c}{2}\right)^2} = 15.72 + \sqrt{17^2 + 15.72^2} = 38.87 \text{ MPa}$

which is well below the safe limit of 56 to 75 MPa.

Shaft at junction of Crank:

Bending moment, $M_b = Q\,(0.75\,l + \text{t}) = 52\,(0.0675 + 0.06) = 6.63 \text{ kNm}$

Twisting moment, $M_t = Pt \times R = 35 \times 0.15 = 5.25 \text{ kNm}$

$$\therefore \quad M_{te} = \sqrt{M_b^2 + M_t^2} = 8.45 \text{ kNm}$$

$$\therefore \quad \frac{\pi}{16} d_1^3 \times f_s = 8.45 \times 1000$$

$$\therefore \quad f_s = \frac{8.45 \times 1000 \times 16}{\pi \times 0.16^3} = 10.5 \text{ MPa}$$

Shaft under flywheel:

B.M. due to gas load, $M_h = Q\,(a + c) - \sqrt{H_{r1}^2 + H_{t1}^2} \times c = 52 \times 0.4575 - 75.60 \times 0.24 = 5.65$ kNm

B.M. due to flywheel weight = $M_w = 3.24$ kNm

$$\text{Net B.M.} = M_e = \sqrt{M_h^{\,2} + M_w^{\,2}} = 6.51 \text{ kNm}$$

Now Twisting moment, $M_t = P_t \times R = 5.25$ kNm

$$\therefore \quad M_{te} = \sqrt{M_e^{\,2} + M_t^{\,2}} = 8.35 \text{ kNm}$$

$$\therefore \quad \frac{\pi}{16}\, dw^3 \times f_s = 8.35 \times 1000$$

Due to complexity of forces, let us take f_s to be equal to 28 to 35 MPa, say 30 MPa.

$$\therefore \quad dw = 0.12 \text{ m}$$

Actually dw should be taken as more than shaft diameter:

$\therefore$ Let us take $dw = 0.18$ m

Journals: As already indicated, both the bearings are made of equal size. The maximum reaction will be at dead centre position of crank.

Horizontal reaction, $H_1 = 150$ kN

Vertical reaction, $V_2 = 13.5$ kN

$$\therefore \quad \text{Net reaction} = \sqrt{H_1^2 + V_1^2} = 150.6 \text{ kN}$$

$$\therefore \quad \text{Bearing pressure,} \quad p = \frac{150.6 \times 1000}{0.18 \times 0.16} = 5.25 \text{ MPa}$$

which is well within the range of 5.0 to 8.5 MPa.

EXERCISES

1. What are the main requirements of an internal combustion engine crank shaft?
2. What is the influence of the number of bearings on the shaft characteristic?
3. Compare centre and side crank, crank shafts.
4. What determines the arrangements of crank shaft throws in internal combustion engines?
5. What are the methods and materials used in manufacturing crank shafts?
6. What are the advantages of making a crank shaft from cast iron alloys?
7. What are the two most usual causes of failure of the crank shafts?
8. What may cause heavy bending stresses in crank shafts?
9. Explain why torsional vibrations are dangerous.
10. What is the function of counterweight in crank shafts?
11. Design the overhung crank shaft for a 0.25 m × 0.40 m horizontal gas engine, explosion pressure 2.38 MPa, weight of flywheel 16 kN, total belt pull 3 kN. When the crank is at 30°, the torque on the crank shaft is maximum and the gas pressure at this position is 1.015 MPa. Length of the connecting rod is 0.95 m.
12. Design the crank pin and the crank for a side crank shaft for a 0.50 m × 0.60 m gas engine, the weight of the flywheel is 84 kN, the explosion pressure is 2.45 MPa. At maximum torque when the crank angle is 36° the gas pressure is 0.945 MPa, the connecting rod length is 4.75 times crank radius.
13. Solve problem 11, for centre crank shaft.
14. Solve problem 12, for centre crank shaft.

27.5 VALVE GEAR MECHANISM

(a) **Definition:** The term valve gear mechanism designates the combination of those parts which operate or actuate the various intake, exaust, fuel and air starter valves, open and close them at the proper time with respect to the position of the piston and crank pin, and keep them open during the required time.

The valve gears of internal combustion engines vary considerably in their construction, depending upon type, speed, and size of the engines. Figure 27.32 shows a simple arrangement of a valve gear for vertical, overhead valve, four stroke engine, and Fig. 27.33 shows the arrangement for a horizontal engine. For vertical engines, the rotating cam shaft with the cam pushes the cam follower and push rod upward and thus transmits the cam action to the rocker arm. The rocker arm moves on a fulcrum pin and changes the above motion into a downward motion of the valve and thus the valve opens. When the cam follower moves over the circular portion of cam, the pushing action of the rocker arm on the valve is released and the valve returns to its seat and closes it by the action of the valve spring. In some vertical engines, the cam shaft is located at the cylinder head level. In this arrangement the push rod is eliminated and the roller type cam follower is made part of the rocker arm. Similar is the arrangement for horizontal engines as shown in Fig. 27.33.

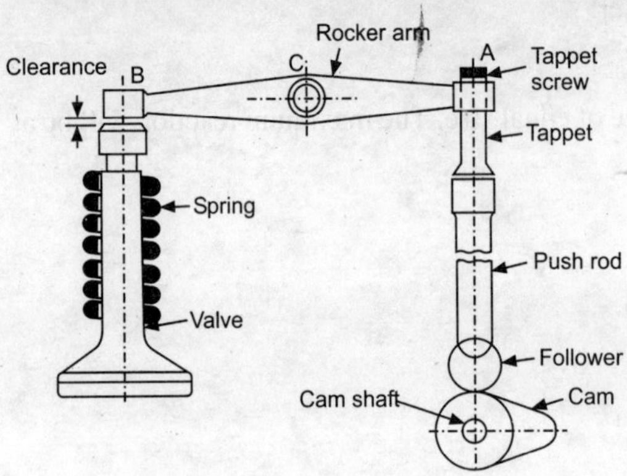

Fig. 27.32 *Valve gear arrangement for vertical engines*

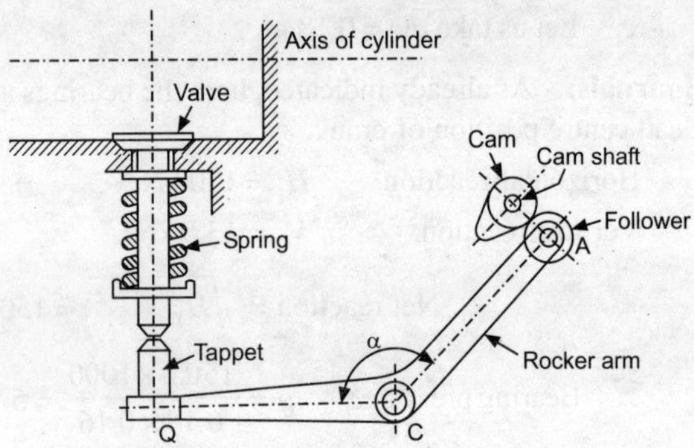

Fig. 27.33 *Valve gear arrangement for horizontal engines*

(b) **Valves:** The valves used in internal combustion engines are of two types: Poppet or Mushroom valve and Sleeve valves. Sleeve valves are used only in a few automobile engines. The poppet type is shown in Fig. 27.34. It consists of a disk or head at the end of a rod or stem. The margin is provided to avoid a thin edge.

The requirements of a good valve are:

1. The valve must open and close at definite crank angles and with a minimum of noise and wear.
2. The opening and closing of the intake and exaust valves must be properly timed to secure the maximum power and efficiency of the engine.
3. Valves must seat tightly and prevent the escape of gases during the working process.
4. Valve openings should be sufficient to obtain ample breathing capacity.
5. Material of the valves should be such so as to withstand the various temperature, erosive and corrosive conditions.
6. The valve should be as light as possible to reduce the inertia forces.
7. The valves must also be sturdy to withstand the continuous pounding action due to the continued opening and closing.

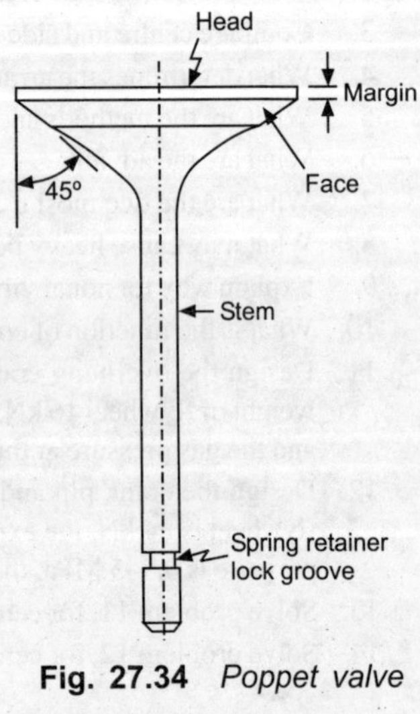

Fig. 27.34 *Poppet valve*

Size of Valve ports: The size of valve ports depends upon the cylinder diameter and the cylinder centre distances. If the cylinder bore is large, large valve ports can be provided. Large ports give a better breathing capacity for a given displacement in the case of large bore small stroke engines. However, the size of the valve opening can be determined with the help of the following relation:

$$V_p \times a_p = V_P \times a_P,$$

where a_P = piston area, V_P = mean piston speed

a_p = port area, V_p = mean gas velocity through the port.

The gas velocity should not be very high, since greater the gas velocity, the greater the pressure difference to maintain the flow. Good practice is to obtain the following gas velocities:

35 to 50 m/s for stationary engines, and 50 to 75 m/s for automotive and aircraft engines.

Better cylinder charging and better cooling of the exaust valves are required for higher engine output. For this purpose, the inlet ports are made 20 to 40 percent larger than the exaust valves. Moreover, if the cylinder pressure decreases by 0.007 MPa due to restricted air intake system, the output decreases by about 7 per cent whereas for 0.007 MPa increase of back pressure due to restricted gas outlet system, the decrease in output is only about 1.5 percent. To attain maximum flow with minimum pressure drop, the following points must be given weightage:

1. Use a small seat angle and obtain maximum opening with a given lift.
2. Round all the entering and leaving edges of the valves and seats.
3. Use large and smooth passage ways to and from the valves.
4. The pressure drop should never exceed the critical value, i.e., 0.53 times the manifold pressure.
5. The flow area between the valves and the combustion chamber walls should not restrict the flow.

Valve-seat angles, Lifts, valve seats and inserts: The seating surface of the valves i.e. the head may either be flat or beveled. The beveled valves have conically shaped heads which give them a self centering action and so these are universally used. The valve head may be beveled at an angle of 30° or 45°, Fig. 27.35. To obtain a greater opening area for a limited lift, the inlet valves may be made with α = 30°, but with exaust valves, α is never less than 45°, otherwise the head will become thin to withstand the high gas temperatures.

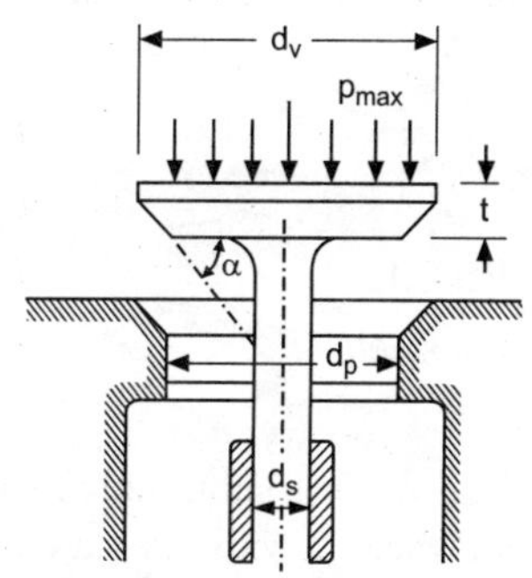

Fig. 27.35 *Conical poppet valve in the port*

The valve lift can be obtained by equating the area across the valve seat to the port area. In the case of flat headed valves, the opening area is directly proportional to the valve lift. Therefore,

$$\pi d_p\, l = \frac{\pi}{4} dp^2, \text{ neglecting the valve stem}$$

$$\therefore \quad \text{Lift,} \quad l = \frac{d_p}{4}$$

For conical valves, the ring shaped valve opening depends upon the perpendicular distance between the seating surfaces. Thus,

$$\frac{\pi d_p^{\,2}}{4} = \pi\, d_p\, l \cos \alpha \text{ (approximately).}$$

$$\therefore \quad l = \frac{d_p}{4 \cos \alpha}, \text{ approximately.}$$

Valve seats usually have the same angle as the valve seating surface. But it is a good practice to make the angle of valve seat as 1/2 to 1 degree larger than the valve angle as shown in Fig. 27.36.

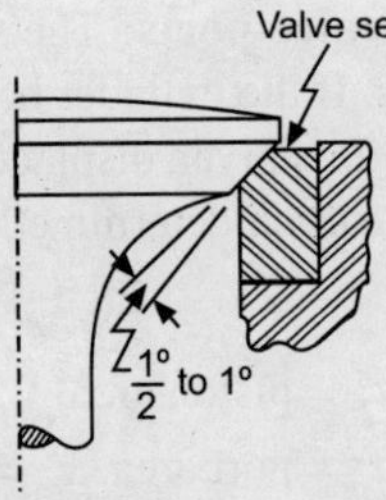

Fig. 27.36 *Valve interference angle and valve seat insert*

This interference angle results in more effective seal. In the past, the valves used to seat directly on the surface of the cast iron cylinder head. But due to the continuous pounding of the valves under high temperature conditions, the seats are eroded. To overcome this trouble, separate replaceable valve seats of hard and heat resisting alloys, are provided as shown in Fig. 27.36. The material of the valve seats inserts may be cast iron, high speed steels, monel metal, and various alloys and bronzes. Stellite is used for the face of an insert made of steel.

Design of Valves: Inlet valves run cooler than exaust valves. So the material for inlet valves may be carbon steel, nickle steel, chrome nickle steel and chrome molybdenum alloys, which may be hardened to withstand the repeated high stresses. Material for exaust valves must be able to maintain their strength at high temperatures and offer resistance to air hardening, deformation, oxidation, wear, pitting and etching. The austenitic types of chrome nickle steels give best service. Of course, the standard chrome nickle steel, cobalt nickle steel, high speed steel and stainless steel are also used.

The thickness 't' of the valve disk or head can be determined by considering it a flat plate subjected to maximum gas load taken as uniformly distributed. Thus,

$$t = k \times d_p \times \sqrt{\frac{p_{max}}{f_t}}$$

where d_p = port diameter, mm

p_{max} = maximum gas pressure, N/mm^2

k = constant

f_t = allowable stress in bending, N/mm^2.

For C.I, $k = 0.54$ and $f_t = 28$ N/mm^2

For steel, $k = 0.42$ and $f_t = 56.0$ N/mm^2 for carbon steel

= 105.0 N/mm^2 for high grade alloy steel.

The valve is subjected to valve spring force also which is taken as concentrated load at the centre. This occurs when the valve is seated. This stress is given as:

$$f_t = \frac{1.4F}{t^2}\left(1 - \frac{2d_s}{3d_p}\right)$$

where F = spring force, N

$$d_s = \text{valve stem diameter}, = \frac{d_p}{8} + 6.35 \text{ mm} \quad \text{to} \quad \frac{d_p}{8} + 11 \text{ mm}$$

Heat flows into both valves from the hot gases. The temperature of the inlet valve head seldom exceeds 250 to 275°C but that of the exaust valves, it is generally about 700 to 760°C. The valves, therefore, should be designed so as to absorb least heat and dissipate most. In the case of heavy duty engines, e.g., aircraft engines, the exaust valve is usually hollow and filled with sodium. Due to the movement of the valve, the melted sodium will move up and down and thus helps to transfer the heat to the outside and prevents the valve from over heating, Fig. 27.37.

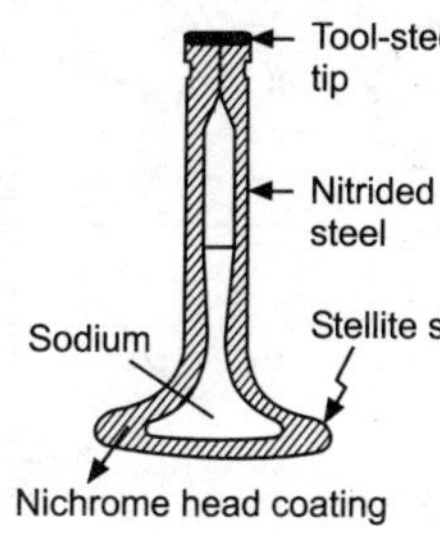

Fig. 27.37 *Sodium cooled air craft engine exaust valve*

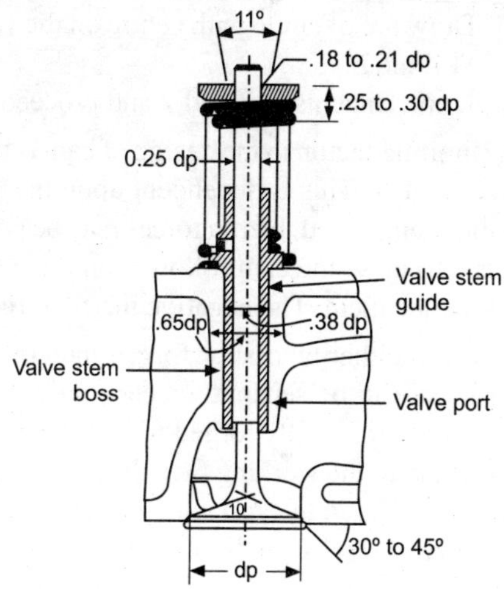

Fig. 27.38 *Proportions of valve*

The valve cages are made of hard, close grained cast iron. Figure 27.38. gives proportions of valves. Sometimes the valves are rotated during operation to prevent overheating and failure of the seating surface due to poor seating at some spot. Also, the heat picked up by the valves is dissipated through the valve guide. Therefore, a long guide may be advantageous. The material used for valve guides is iron.

(c) Cams: A cam is an eccentric projection on a revolving disk used for the opening and closing of a valve through the intermediate parts. The cams are of the following four types:

1. Tangent cams.
2. Concave flank cams.
3. Convex flank cams.
4. Proportional cams.

The tangential cam is constructed of a circle, a nose circle, and two flanks tangent to both the circles. In the case of round flank cams, the flanks are arcs of circles tangent to both base circle and nose circle. A proportional cam with a roller follower is designed to give simple harmonic motion to the valve so that at all lifts, the accelerating forces are proportional to the spring forces. In all the cases, the approach ramp should be provided to take up the tappet clearance gradually. When the valve is closed, a slight clearance is necessary in the straight poppet valve between the valve lifter and the valve stem and in the overhead poppet valve, between the rocker arm and the valve stem. This clearance allows for expansion of the valve stem as the engine becomes heated. This clearance is also known as "Valve Lash". Its value may be taken as 0.10 to 0.15 mm.

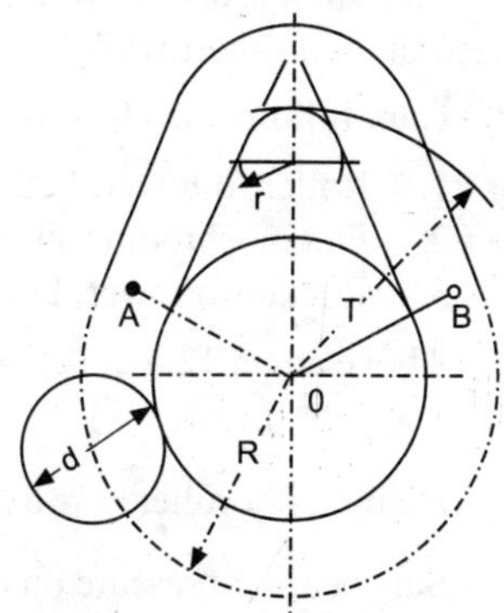

Fig. 27.39 *Tangent cam without approach ramp*

Drawing of Tangent Cam:

(i) Without provision for approach ramp:

1. Draw the base circle required.
2. Draw *OA* and *OB*, so that angle *AOB* is equal to the necessary cam period in degrees.
3. Draw radius '*T*' = base circle radius + cam lift required.
4. Draw tangents at *A* and *B*.
5. Fill in the nose radius to touch the tangents and the lift circle.

(ii) With approach ramp:

1. Set off angle *AOB* as before.
2. Set up *AX* = *BY* = valve lash.

3. Draw arc of circle with centre on the vertical line to pass through X, Y and Z.
4. Draw tangents at X and Y and proceed as before.

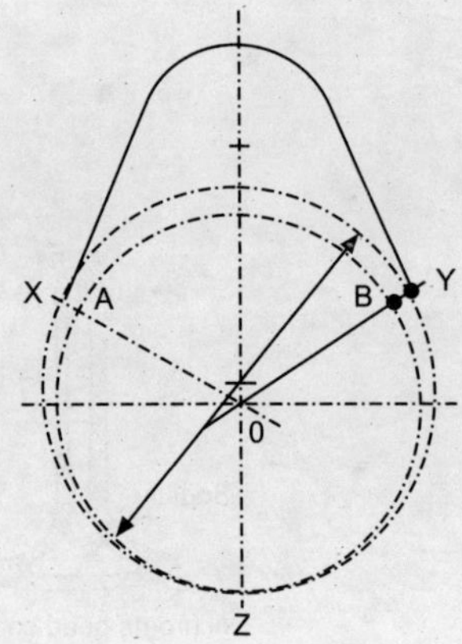

Fig. 27.40 *Tangent cam with approach ramp*

The limiting factor in the design of cam is the inertia loading imposed on the valve gear. This is dependent upon the weight of the gear and the accelerations employed. Inertia forces may be positive or negative. Positive forces tend to press the cam follower on the cam and the negative forces tend to separate them. The negative inertia forces are opposed by springs.

Cams can either be cast or forged integral with the cam shaft as one piece or these can be separate pieces, machined and keyed to the shaft. Separate cams are usually forged (drop), case hardened, and ground both in the bore and in the working face. In most modern engines, the cams are usually cast or forged integral with the cam shaft. The material for the cam shafts is forged steel usually of nickle chromium alloy steel. These are heat treated and are carried in plain bearings. Larger cam shafts are often bored hollow, The base circle of the cam should be kept as small as possible to avoid excessive accelerations and the roller or follower radius as large as possible.

The stiffness of cam shaft is essential, therefore, its diameter will depend upon its length. But the average practice is:

Diameter of cam shaft = 0.16 × cylinder bore + 12.7 mm, when cams are solid with shaft.
= 0.175 × cylinder bore + 21.6 mm, when cams are keyed on.

Base circle diameter of the cam is about 3 mm greater than the cam shaft diameter for solid cams but for keyed on cams:

Base circle diameter = 1.5 × shaft diameter + 20 mm

The width of the cam depends upon the load to be carried. The following figures may be used for well lubricated roller followers:

215 to 250 N per mm width if cam is face hardened.

110 to 140 N per mm width if the cam is of plain cast iron or steel.

For flat mushroom followers, the figure is upto 125 N per mm width if both cam and follower are hard.

With rocker followers, the load should be kept down to about 55 to 70 N per mm of cam width.

For roller followers, the width of cam is given empirically as:

Cam width = 0.09 × cylinder bore + 6 mm to 0.11 cylinder bore + 12 mm

Cam shaft is driven from crank shaft by means of gears of chain. In four stroke engine, its speed is half the speed of crank shaft while for two stroke engines it is equal to the crank shaft.

(d) Cam followers: The most commonly used forms of cam followers are:

1. Roller follower, Fig. 27.41(*a*).
2. Flat mushroom follower, Fig. 27.41(*b*).
3. Rocker follower, Fig. 27.41(*c*).

Above about 12.5 cm cylinder bore, roller followers appear to be universal, but below this size, either the mushroom or the rocker follower may be used.

Diameter of roller = $\frac{1}{2}$ to $\frac{2}{3}$ × cam base circle diameter.

Safe bearing pressure on roller pin = 7 N/mm^2

Shear stress in roller pin = 14.0 to 21.0 N/mm^2

The pin should be force lubricated.

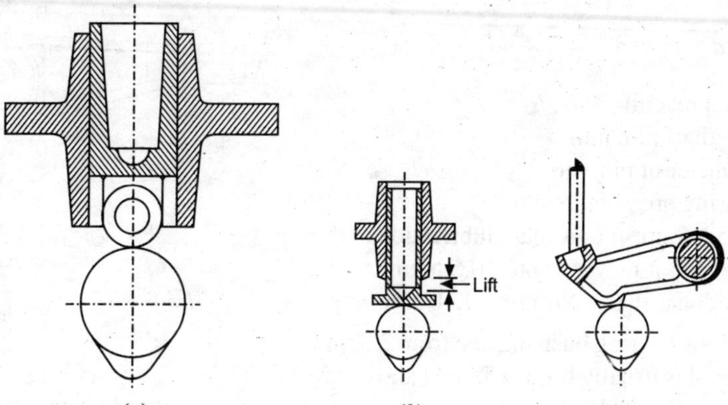

Fig. 27.41 *Cam followers*

(e) Rocker Arm: In four stroke engines, the rocker arm for exaust valve is the most heavily loaded. Although the force required to operate the inlet valve is relatively small, it is usual to make the inlet valve lever of the same dimension as the exaust valve lever for the sake of uniformity of appearance. Apart from the weight of the valve itself, the following forces tend to open it when it should be shut, i.e., the exaust valve.

1. Vacuum on the suction stroke.
2. Inertia of the valve.

The above two problems are counteracted by fitting a spring, the normal load of which is equivalent to a pressure of about 0.07 N/mm^2 of valve area in the case of slow speed engines and it may go upto 0.14 N/mm^2 of valve area or more for high speed engines.

Therefore, referring to Fig. 27.32 the force at *B* during periods of positive acceleration = gas load + spring force + force due to valve acceleration.

Now when the exaust valve opens, the cylinder pressure ranges from about 0.35 to 0.42 N/mm^2.

$\therefore$ Gas load = area of valve × cylinder pressure when the exhaust valve opens

Spring force = area of valve × negative pressure at suction stroke.

$$\text{Force due to valve acceleration} = \frac{W_v}{g} \times f,$$

where W_v = weight of valve, f = valve acceleration.

Now force at *A* = force at *B* + force due to rocker acceleration.

Force at *C* = Force at *A* + Force at *B*.

The rocker arm may be treated as a simply supported beam loaded at a point corresponding to the fulcrum *C*. Section of the arm adjacent to the boss *C* may be found out by considering the bending. Therefore,

$$M_b = f_b \times Z$$

f_b = safe stress in bending, N/mm^2

The material for the arm may be C.I., C.S., or mild steel stamping. Since the arm will be under a varying stress of some kind but with some shock, the factor of safety may be taken as about 10. To reduce the interia of rocker arm its section for high speed engines can be I-section but for low speed engines, it may be rectangular one. I-section rocker arm is shown in Fig. 27.42. The fulcrum pin can be designed in bearing and checked in double shear. If each rocker pin has its own support, then its bending may be neglected, but if a number of rockers are mounted on the same spindle, then the bending stress must be calculated. The load will be the total reaction at *C*.

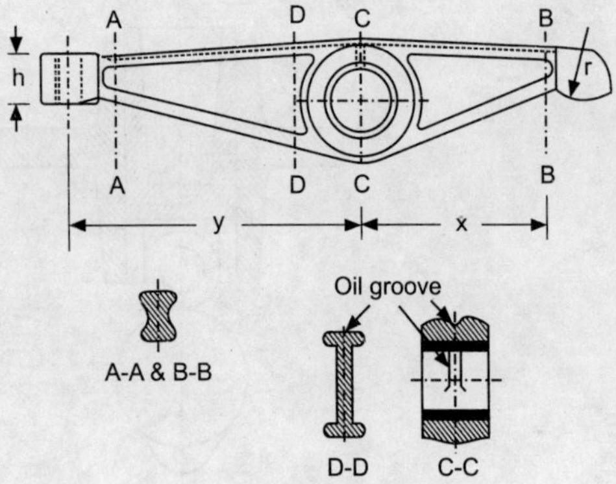

Fig. 27.42 *Rocker arm*

$$\therefore \quad P = l \times d \times p_b \text{ and } P = 2 \times \frac{\pi}{4} d^2 \times f_s ,$$

where P = Total reaction at C, N

l = length of pin, mm

d = diameter of pin, mm

p_b = bearing pressure, N/mm^2

= 4.55 N/mm^2 for ordinary lubrication of the pin and it may go upto 10.5 N/mm^2

f_s = safe shear stress, N/mm^2 = 42.0 N/mm^2

The thickness of the bearing bush may be from 2.3 mm to 4.5 mm. The material is usually bronze, but Al alloy may also be used. To find the outside diameter of the boss, the bending at section CC may be considered.

For equal strength, the rocker arm is tapered towards the ends both in width and depth but keeping the thickness of flange and web as constant. Or, the depth of the section may be determined to carry a tappet adjusting screw.

h = 1.0 to 1.4 × nominal diameter of tappet screw. The radius 'r' should be such that the contact between the rocker and valve or push rod is always fully tangential.

The lever ratio y/x is often fixed by space considerations. For moderate and low speed engine, the aim is to make $y = x$. But for high speed engines, y/x should be decided carefully so as to minimise the inertia forces. The desirable value of y/x is about $\frac{1}{1.5}$ to $\frac{1}{1.6}$, $\frac{1}{1.3}$ is common figure. The distance between the fulcrum pin centre and the top end of the valve in closed position affects the side thrust on the valve guide. In practice, this distance is taken as equal to one third of valve lift. Again, the distance x should not be less than five times the valve lift.

(f) **Valve Springs:** The purpose of the valve spring is to provide sufficient force during the valve lifting process to overcome the inertia of the valve gear and keep it in contact with the cam without bouncing. This is obtained by putting the spring under compression while installing it. When the valve is opened, this force is increased by the additional compression of the spring. The total load for which the spring should be designed is given as, Total load = Initial load + load at full lift.

Now initial load = Force on the valve tending to draw it into the cylinder on the suction stroke. and load at full lift = full lift × stiffness of spring.

To ensure a small difference of tension between the position, where the valve is closed and fully open, a spring with a greater number of coils, 10 to 14 is selected.

If the force to be exerted by the spring is very large, the wire diameter and the mean diameter of coils may be large. To make the spring fit better in available space, two concentric springs will be used in such cases. Also in air craft engines, concentric springs are provided for the sake of safety so that if one spring fails, the other will continue to operate the valve.

(g) **Push Rods:** Push rods are subjected to buckling and may be designed by the use of Euler's formula or Rankine Gordon formula. If the slenderness ratio is not greater than 80 for cast iron columns and 100 for steel and other ductile material the columns are considered as short columns and Rankine Gordon formula is used. If the slenderness ratio is more than 100 for ductile materials and greater than 80 for cast iron, then the columns are considered as long columns and Euler's formula is used. The body of the rod may be solid or tubular but it is usually tubular to make it as light as possible. Great care should be taken in their manufacture to ensure absolute straightness. If the push rods are short, they may safely be made of solid centreless ground stock.

The end conditions of the push rods will depend on the general design of the gear. If the push rods are guided, then the ends may simply consist of flat ended plugs pressed in and brazed or welded. If the ball and socket arrangement is used, the socket should be sufficiently deep to prevent the rod from jumping out in the event of valve sticking. These arrangements are shown in fig. 27.43. In the ball and socket arrangement, the ends will not be as a rule guided.

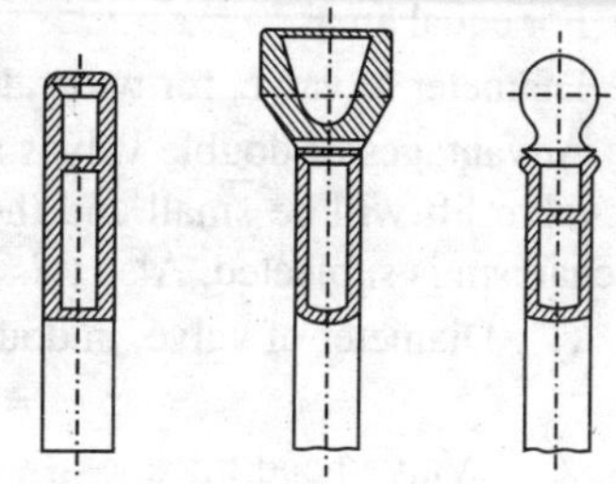

(a) (b)

Fig. 27.43 *(a) Flat ended plates, (b) Ball and socket arrangement*

The material employed for push rods is usually bright drawn mild steel with 0.4% C content. The material may be in the form of tubes or solid bars. The yield stress for the material is about 420 N/mm^2. On this a factor of safety of not less than 3 or 4 should be used as the rod is under fatigue loading. If the safe stress in buckling is based on endurance limit, a factor of safety of about 2 to 3 may be used with light shock conditions. For the design of push rods, a handy table is given below giving the buckling stresses for various *l/d* ratios of the push rods. On these values a factor of safety of about 3 to 4 should be used to determine the safe stress and it is desirable that the safe stress should not exceed about 70.0 N/mm^2.

Table 27.4

l/d ratio	Buckling Stress, N/mm^2	
	Tubular rod	Solid
10	420	400
20	294	231
30	203	147
40	140	91
50	98	63
60	73.5	49
70	57.5	36.5
80	45.5	28.0
90	38.5	21.0
100	31.5	17.5

In the tubular push rods, the inside diameter is usually from 0.6 to 0.8 times the outside diameter.

Example 27.9. Determine thickness of a single 50 mm valve keeping the stress at 42.0 N/mm^2. Gas pressure is 3.5 N/mm^2.

Solution: The valve thickness is given as,

$$t = k \times d_p \times \sqrt{\frac{p_{max}}{f_t}}$$

Now $d_p = 50$ mm $p_{max} = 3.5$ N/mm^2 $f_t = 42.0$ N/mm^2 and for steel, $k = 0.42$

$$\therefore \qquad t = 0.42 \times 50 \times \sqrt{\frac{3.5}{42.0}} = 6.05 \text{ mm.}$$

Example 27.10. Determine thickness of the two valves to replace a single valve of Ex. 1. Assume the valve have a 45° seat.

Solution: When two valves are used instead of one, the combined area of the two should at least equal the area of opening of the one valve. If the valve seat angles are equal and the lift diameter ratio is also same, then from the equation

$$A = \pi d_p\, l,$$

For equal areas,

Diameter of valve, for two valves = 0.707 × diameter of valve, for single valve.

Advantages of double valves are that greater opening area and lower gas velocities are obtained. Also, the valve lift will be small and the valve weight will be less which reduces the forces to which the valve mechanism is subjected. Also this arrangement results in more valve seat area with better valve cooling.

∴ Diameter of valve, in double valve system,

$$= 0.707 \times 50 = 35.35 \text{ mm}$$

∴ Valve head thickness,

$$t = \frac{0.42 \times 35.35}{3.464} = 4.28 \text{ mm, say } 4.3 \text{ mm.}$$

Example 27.11. The arrangement of the exhaust valve gear for a horizontal diesel engine is similar to shown in fig. 27.33. The particulars of the engine are:

Power = 8.25 kW

Bore = 0.143 m

Stroke = 0.267 m

Speed = 475 rev/ min

Maximum gas pressure = 3.5 N/mm²

The valve opens 33° before outer dead entre and closes 1° after inner dead centre. It opens and closes with constant acceleration and deacceleration for each half of the lift. The length of the rocker arm on either side of the fulcrum is 0.15 m and the included angle is 160°. The weight of the valve is 3 N.

Design the various components of the valve gear.

Solution: Before, the various components are designed, the forces acting at various points must be determined and for this, the dimensions of the valve should be found out.

Design of Valve: The size of the valve head may be determined from the relation.

$$a_p \times V_p = a_P \times V_P$$

Now a_P = area of piston $= \frac{\pi}{4} D^2 = \frac{\pi}{4} \times 0.143^2$

V_P = mean piston velocity = 2 *LN*

$$= 2 \times \frac{0.267}{60} \times 475 = 4.23 \text{ m/s}$$

V_p = mean velocity of gas through the port, may be taken as 35 m/s

∴ a_p = area of port $= \dfrac{a_P \times V_P}{V_p}$

∴ $$\frac{\pi}{4} d_p^{\,2} = \frac{\pi}{4} \times \frac{0.143^2 \times 4.23}{35}$$

∴ d_p = 0.0498 m, say 0.05 m = 50 mm

For seat angle of 45° the projected width of the valve seat may be taken empirically as = 0.05 d_p to 0.07 d_p = say 3 mm.

∴ Diameter of valve head = $d_p + 2 \times 3 = 56$ mm

Now the lift of the valve is given approximately as,

$$l = \frac{d_p}{4 \cos \alpha} = \frac{50}{4 \times \cos 45°} = 17.8 \text{ mm, say } 18 \text{ mm}$$

The thickness of the valve head is given as,

$$t = k \times d_p \sqrt{\frac{p_{max}}{f_t}} = 0.42 \times 50 \sqrt{\frac{3.5}{56}} = 5.25 \text{ mm}$$

Forces at Various Points: As already indicated, the maximum load on the valve, *i.e.* at point *B*, is equal to gas load when it opens + initial spring force tending to hold on its seat against suction on the suction stroke. + Inertia of the valve.

As already written, when the valve opens, the back pressure in the cylinder varies from 0.35 to 0.42 N/mm^2. Let us take it as 0.4 N/mm^2.

$$\therefore \quad \text{Gas load} = \text{area of valve} \times \text{back pressure} = \frac{\pi}{4} \times 56^2 \times 0.4 = 985 \text{ N}$$

Considering the weight of valve also.

Actual gas load = 985 + 3.0 = 988 N

Initial spring force = area of valve × vacuum pressure – weight of valve.

Now this vacuum pressure may reach 0.05 to 0.056 N/mm^2

Since the engine is a slow speed engine, let us take the vacuum pressure as 0.025 N/mm^2

$$\therefore \quad \text{Initial spring force} = \left(\frac{\pi}{4} \times 56^2 \times 0.025 \right) - 3 = 58.5 \text{ N}$$

Again inertia of valve = mass × acceleration = $\dfrac{W_v}{g} \times f$

Now, total angle for which the valve is open = 33 + 1 + 180 = 214°

$$\therefore \quad \text{Cam shaft angle, if the engine is four stroke} = \frac{214}{2} = 107°$$

Now when the cam shaft turns through $\dfrac{107}{2} = 53.5°$, the valve lifts by distance of 18 mm. Now half of this period is occupied by constant acceleration and half by constant deceleration. The same precess occurs when the valve closes. Therefore, period for constant acceleration is equal to cam shaft rotation of $\dfrac{53.5}{2} = 26.75°$, while the valve lifts through a distance of 9 mm.

$$\therefore \quad \text{Time of constant acceleration} = \frac{26.75}{360} \times \frac{2}{475} \times 60 = 0.01877 \text{ secs}$$

$$\therefore \quad \text{From the equation,} \quad S = \frac{1}{2} f \cdot t^2 ; = \frac{9}{1000} = \frac{1}{2} \times f \times 0.01877^2$$

$$\therefore \quad f = \frac{9 \times 2}{1000 \times 0.01877^2} = 5.14 \text{ m/s}^2$$

$$\therefore \quad \text{Inertia force of valve} = \frac{3}{9.81} \times 51.4 = 15.75 \text{ N}$$

Considering the weight of valve also.

Accelerating force = 15.75 + 3 = 18.75, say 20 N

∴ Total load at point *B* = 988 + 58.5 + 20 = 1066.5 N

Since the lever ratio of the rocker arm is unity, the total load at point *A* will be = 1066.5 N.

$$\therefore \quad \text{Total load at fulcrum point, } C = \sqrt{1066.5^2 + 1066.5^2 + 2 \times 1066.5 \times 1066.5 \times \cos 20°}$$

$$\therefore \quad P = 2100 \text{ N}$$

It is clear that force at A acts along the line passing through the axis of cam shaft and that of roller.

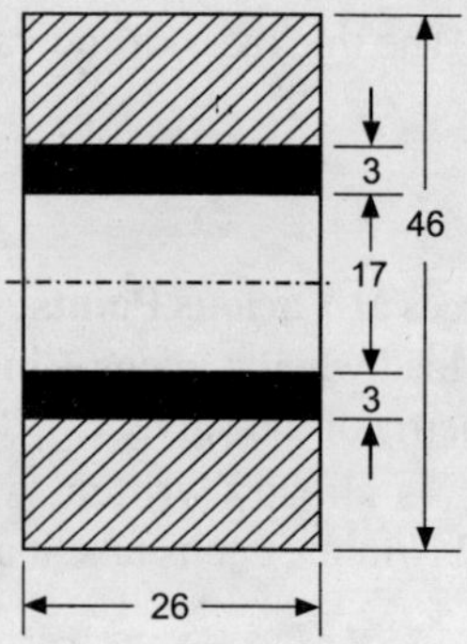

Fig. 27.44 *Section of boss*

Fulcrum Pin:

For bearing $$P = l \times d \times p_b$$

and for shear $$P = 2 \times \frac{\pi}{4} d^2 \times f_s$$

$$\therefore \quad 2100 = 2 \times \frac{\pi}{4} d^2 \times 42$$

$$\therefore \quad d = 5.66 \text{ mm, say } 6 \text{ mm}$$

$\therefore$ From the bearing equation,

$$2100 = l \times 6 \times p_b$$

p_b may be taken as 5 N/mm^2 for ordinary lubrication,

$$\therefore \quad 2100 = l \times 6 \times 5 \quad \therefore \quad l = 70 \text{ mm}$$

In practice l/d ranges from 1.25 to 2

Taking $l/d = 1.5$, from the bearing equation,

$$2100 = 1.5\, d^2 \times 5$$

$$\therefore \quad d = 16.75 \text{ mm, say } 17 \text{ mm}$$

$$\therefore \quad l = \text{length of pin} = 17 \times 1.5 = 25.5 \text{ mm, say } 26 \text{ mm.}$$

If now the pin is checked in double shear, it is very safe. The thickness of the bronze bearing bush may be taken as 3 mm. The diameter of the pin boss may be taken as equal to $2 \times (17 + 6)$, i.e., 46 mm, or a moment may be taken at the centre of the fulcrum pin axis.

$$\therefore \quad \text{B.M.} = 1066.5 \times 0.15 = 160 \text{ Nm}$$

Now section of the boss is shown in Fig. 27.44.

Moment of inertia,

$$\text{M.O.I.} = \frac{26 \times 46^3}{12} - \frac{26 \times 17^3}{12} = 200 \times 10^3 \text{ mm}^4$$

$$\therefore \quad Z = \frac{200 \times 10^3}{46/2} = 8.69 \times 10^3 \text{ mm}^3$$

$$\therefore \quad f_b = \frac{160 \times 1000}{8.69 \times 1000} = 18.4 \text{ N/mm}^2\text{, which is quite safe.}$$

Rocker Arm: As the engine is slow speed engine, rectangular section may be selected for the arm.

Let t = thickness, near the pin boss.

and h = depth, near the pin boss.

Considering the bending of the arm,

bending arm = 0.15 – 0.02 = 0.13 m

$$\therefore \quad \text{B.M.} = 1066.5 \times 0.13 = 138.5 \text{ Nm}$$

Now 'h' can be taken equal to or more than the boss dimension, say 46 mm

$$\therefore \quad M_b = f_b \times Z$$

If the material is taken as cast steel for which ultimate strength in tension is about 420 N/mm^2

$$\therefore \quad f_b = \frac{420}{F_S} = \frac{420}{10} = 42 \text{ N/mm}^2$$

$$\therefore \quad 138.5 \times 1000 = \tfrac{1}{6} \times t \times 46^2, \; t = 10 \text{ mm}$$

Adjusting Screw: This screw is under compression, the load to be carried by it is 1066.5 N. The permissible working stress in bolts is considerably smaller than the working tensile stress, when it is used for other purposes. Assuming a safe stress of 50 N/mm^2,

$$\therefore \quad 1066.5 = \frac{\pi}{4} d_c^2 \times 50, \quad \therefore \quad d_c = 5.2 \text{ mm}$$

$$\therefore \quad \text{Nominal diameter} = \frac{d_c}{0.8} = \frac{5.2}{0.8} = 6.5, \text{ say } 7 \text{ mm}$$

Suitable dimensions for the circular end of the rocker arm to carry this stud will be,

$$\text{Diameter} = 2 \times 7 = 14 \text{ mm}$$

$$\text{Depth} = 2 \times 7 = 14 \text{ mm}$$

Forked end of Rocker Arm: The roller is carried by the forked end by means of a pin. To make the wear uniform, it should revolve in the eyes. The load on the pin is 1066.5 N. Assuming a safe bearing pressure of 7 N/mm^2.

$$1066.5 = l_p \times d_p \times 7$$

Taking $l_p/d_p = 1.25$

$$\therefore \quad 1066.5 = 1.25\, d_p^2 \times 7$$

$$\therefore \quad d_p = \text{diameter of pin} = 11.1 \text{ mm, say } 12 \text{ mm}$$

$$\therefore \quad l_p = \text{length of pin} = 1.25 \times 12 = 15 \text{ mm}$$

This length is in between the two eyes. Thickness of each eye should be atleast $1/2 l_p$, let us take it as 8 mm. To provide for wear, phosphor bronze bushes 3 mm thick may 'be provided in the eyes.

Radial thickness of eye = 2 × 12 + 2 × 3 = 24 + 6 = 30 mm

The diameter of the roller should be as large as possible and should be at least 2 d_p, i.e., 24 mm. Let us take a value of 32 mm, so that the roller projects out of the eyes.

A clearance is necessary between the roller and the inside of fork, therefore, there will be some bending on the pin. Leaving a clearance of 3 mm on each side of roller, the dimension of the inside of fork will be equal to 15 + 2 × 3 = 21 mm.

Now the portion of the pin in the eyes may be taken as fixed and that in between the fork as guided with load acting on it as uniformly distributed, the bending on the pin may be written approximately as,

$$\therefore \quad \text{B.M.} = \frac{1066.5}{2}\left(\frac{0.021}{2} + \frac{0.008}{3}\right) = 4.22 \text{ Nm}$$

$$\therefore \quad \frac{\pi}{32} d_p^3 \times f_b = 4.22 \times 1000$$

$$\therefore \quad f_b = \frac{4.22 \times 1000 \times 32}{\pi \times 12^3} = 24.88 \text{ N/mm}^2$$

which is quite low.

A detailed sketch of the rocker arm is shown in Fig. 27.45.

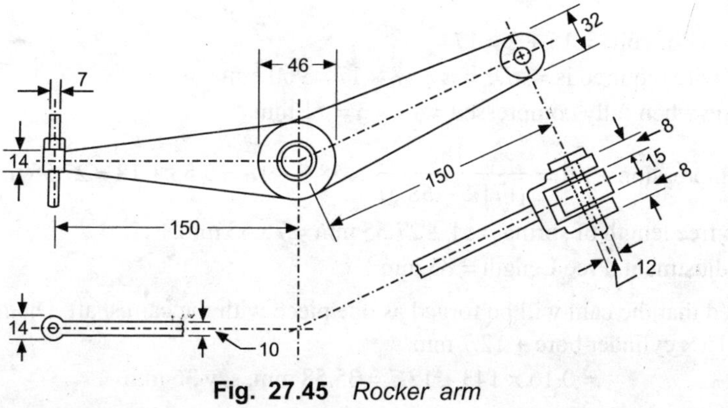

Fig. 27.45 *Rocker arm*

Springs: The maximum load on the spring consists of two parts:

Initial load and load at full lift. Initial load has been already calculated as equal to 58.5 N. The load at full lift will be equal to lift × stiffness of spring.

If the spring stiffness is not given in the problem, then as already indicated, the total load on the spring may be taken as equivalent to a pressure of 0.07 N/mm² of valve area in the case of slow speed engines.

$$\therefore \quad \text{Total load on spring} = \frac{\pi}{4} \times 56^2 \times 0.07 = 171.5 \text{ N}$$

Now the shear stress induced in the spring is given as $f_s = \frac{8FD}{\pi d^3} \times K$

where F = spring load, 171.5 N

D = coil diameter, mm

d = wire diameter, mm

K = stress increase coefficient $= \frac{4S-1}{4S-4} + \frac{0.615}{S}$,

where S = spring index $= \frac{D}{d}$ and it should be ≥ 5.

Let us take $S = 6 \quad \therefore \quad D = 6d$

$$\therefore \quad K = \frac{24-1}{24-4} + \frac{0.615}{6} = 1.2525$$

f_s for the usual spring material may be taken as 420 N/mm²

$$\therefore \quad 420 = \frac{8 \times 171.5 \times 6d}{\pi d^3} \times 1.2525$$

$$\therefore \quad d = 2.8 \text{ mm, say 3 mm, } D = 6 \times 3 = 18 \text{ mm}$$

Now deflection of the spring is given as, $y = \frac{8FS^3z}{Cd}$

where z = number of effective coils

and C = modulus of rigidity, say 0.8×10^5 N/mm²

Now $y = 18$ mm

$$\therefore \quad 18 = \frac{8 \times 171.5 \times 216 \times z}{0.8 \times 10^5 \times 3}, \ z = 14.6, \text{ say } 15$$

Since the spring is to be used as a compression spring, two more coils should be provided to act as bearings at the ends.

∴ Total number of coils = 15 + 2 = 17

∴ Length of wire required is $= \pi Dz = \pi \times 18 \times 17 = 960$ mm

Length of spring when fully compressed = 17 × 3 = 51 mm

$$\text{Now initial compression + lift} = \frac{18}{(171.5 - 58.5)} \times 58.5 + 18 = 9.55 + 18 = 27.55 \text{ mm}$$

∴ Minimum free length of spring = 51 + 27.55 mm = 78.55 mm.

To allow for adjustment, Free Length = 85 mm

Cams: It is assumed that the cam will be forged as one piece with the cam shaft. Diameter of shaft is given empirically as = 0.16 × cylinder bore + 12.7 mm.

$$= 0.16 \times 143 + 12.7 = 35.58 \text{ mm, say 36 mm}$$

Base circle diameter is about 3 mm greater, therefore, base circle diameter = 40 mm
Width of the cam will be equal to the width of the roller, i.e., 15 mm
or it is given empirically also, as

$$\text{Cam width} = 0.09 \times \text{cylinder bore} + 6 \text{ mm}$$
$$= 0.09 \times 143 + 6 = 18.87 \text{ mm, say } 19 \text{ mm}$$

Let us take a cam width of 18 mm.
Now the cam can be drawn according to the following procedure:
Refer Fig. 27.46.

1. Draw a line *OA* to any scale, Fig. 27.46(*a*).
2. Draw a vertical line *OB* equal to valve lift *i.e.* 18 mm and complete the rectangle *OBCA*.
3. Divide the line *OA* into say eight equal parts.
4. From its mid point draw a vertical line, i.e., 4*Y*.
5. This line is also divided into eight equal parts.
6. Points 1, 2, 3, and 4 on this vertical line are joined to point *O*, and points 4, 5, 6, 7 and 8 on this line are connected to point *C*.
7. Now vertical ordinates are drawn through points 1, 2, 3,.., 8 on the line *OA*. Where these ordinates meet the lines drawn in step 6, points are marked.
8. A smooth curve is drawn passing through all these points. Then *OL* gives the curve of constant acceleration and *LC* gives the curve of constant deacceleration.
9. The ordinates of this curve from the line *OA* give the displacement of the follower from its lowest point during the lift.
10. Now to draw the cam, base circle is drawn with P as centre and diameter equal to 40 mm, Fig. 27.46(*b*).
11. Then angle ∝, i.e., equal to 53.5° is measured and it is divided into the same equal number of parts, i.e., eight.
12. Then lines P_a...*PQ* are drawn through *P*.
13. Then a chain arc *QZ* are drawn with *P* as centre and radius equal to base circle radius plus the roller radius.

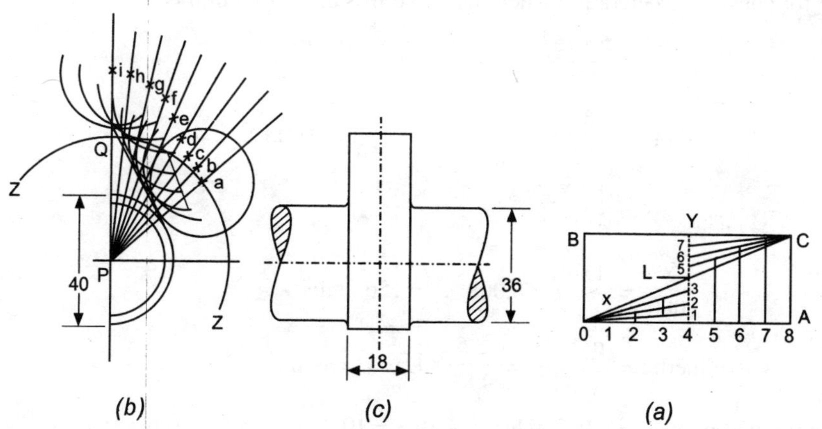

Fig. 27.46 *Cam profile*

14. Now the distances equal to 1*x*, 2*y*, etc. are taken outwards from the arc *QZ* along the corresponding lines P_a, P_b, etc. These points represent the successive positions of the roller centre.
15. With these points as centres and radius equal to the roller radius, arcs are drawn.
16. A curve is then drawn tangential to these arcs and that gives the profile of the cam.

Example 27.12. Design the rocker arm and push rod for the exhaust valve gear of a four stroke vertical, overhead valve, diesel engine with piston having a diameter of 0.14 m and a stroke of 0.18 m. The operating speed of the engine is 1200 rev/min. The diameter of the exhaust valve is 54 mm and it is to start to open 39° before the crank reaches its bottom dead centre position and it is to close 8° after the crank has passed its top dead centre position. The gas pressure in cylinder when the valve starts to open is 0.49 N/mm² absolute and the pressure on the top side of the valve may be taken as 0.105 N/mm² absolute. The weight of the valve is 2.5 N and lift is 10 mm. Draw the tangent cam also. Take clearance at the cam equal to 0.17 mm and deflection at the cam as equal to 0.12 mm.

Solution. As we know, that the projected width of the valve seat angle of 45° may be taken as $0.05\,d_p$ to $0.07\,d_p$

$$\therefore \quad d_p + 2 \times 0.06\,d_p = 54$$

$$\therefore \quad d_p = \text{diameter of port} = \frac{54}{1.12} = 48.2 \text{ mm; say } 49 \text{ mm}$$

As already indicated the total load at the exhaust valve, i.e., point B, Fig. 27.32 is equal to gas load when it opens + spring load + inertia of valve. Now when the valve opens, the cylinder pressure acting on its face, i.e., on 5 mm diameter is 0.49 N/mm², while the pressure on its top side, i.e., on 49 mm diameter is 0.105 N/mm².

$$\text{Therefore, net gas load} = \frac{\pi}{4} \times 54^2 \times 0.49 - \frac{\pi}{4} \times 49^2 \times 0.105$$

$$= 1122.2 - 198 = 924.2 \text{ N}$$

Spring load may be taken as 0.05 N/mm² of valve area, i.e.,

$$= \frac{\pi}{4} \times 54^2 \times 0.05 = 114.5 \text{ N}$$

Now the total angle through which valve remains open = 39 + 8 + 180 = 227°

$$\therefore \quad \text{Cam shaft angle} = \frac{227}{2} = 113.5°$$

$$\therefore \quad \text{Angle for valve lift} = \frac{113.5}{2} = 56.75°$$

∴ Angle for constant acceleration when the valve lifts equal to 5 mm is

$$= \frac{56.75°}{2}$$

$$\therefore \quad \text{Time} = \frac{56.75}{2} \times \frac{2}{1200} \times 50 \times \frac{1}{360} = 0.008 \text{ s}$$

$$\therefore \quad S = \frac{1}{2} \times f \times t^2$$

$$\therefore \quad \frac{5}{1000} = \frac{1}{2} \times f \times 0.008^2, f = 156.2 \text{ m/s}^2$$

$$\therefore \quad \text{Valve inertia} = \frac{W_v}{g} \times f = \frac{2.5}{9.81} \times 156.2 = 39.8 \text{ N}$$

∴ Total load at point B = 924.2 + 114.5 + 39.8 = 1078.5 N, say 1080 N

The load at point A will depend upon the lever ratio $\frac{y}{x}$, let us take $\frac{y}{x} = \frac{1}{1.3}$. Now x should be at least = 5 × lift of valve = 5 × 10 = 50 mm

$$\therefore \quad y = \frac{50}{1.3} \text{ say, } 40 \text{ mm}$$

∴ Refer to Fig. 27.42.

Load at point $A = 1080 \times \frac{50}{40} = 1350$ N

∴ Load at point $C = 1080 + 1350 = 2430$ N

Rocker Arm: Now to find the dimensions of the rocker arm adjacent to the boss, the lever arm from point $B = 50 - 13 = 37$ mm

∴ $M_b = 1080 \times 0.037 = 40$ Nm

If the material of the rocker arm is C.S., $f_b = 42$ N/mm^2

∴ $40 \times 1000 = f_b \times Z = 42 \times Z, Z = \frac{20}{21} \times 10^3$ mm^3

Now the depth of *I* section can be taken as 26 mm

Now the usual *I*-section employed is shown in Fig. 27.47.

Depth = 6 *t*, width= 2.5 *t*

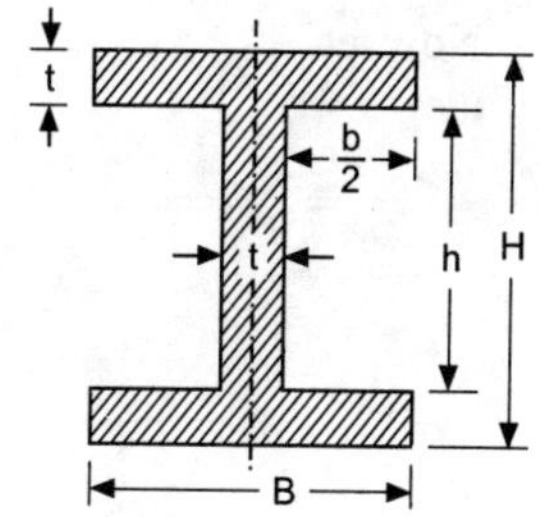

Fig. 27.47 *I-section*

∴ $6t = 26 \therefore t = \frac{26}{6} = 4.33$ mm, say 4.5 mm

or $I_{xx} = \frac{1}{12}(BH^3 - bh^3) = \frac{1}{12}(2.5 \times 6^3 \times t^4 - 1.5 \times 4^3 \times t^4) = \frac{444}{12} t^4$

∴ $Z_{xx} = \frac{I_{xx}}{y} = \frac{444\, t^4}{12} \times \frac{1}{3t} = \frac{111}{9} t^3 = 12.33\, t^3$

∴ $12.33\, t^3 = \frac{20}{21} \times 1000$

∴ $t = 4.27$ mm, say 4.5 mm, therefore, it is O.K.

∴ Depth = 27 mm, width = $2.5 \times 4.5 = 11.25$ mm, say 12 mm.

To make the arm uniform in strength, it is tapered in depth as well as width towards the ends.

Fulcrum Pin: Reaction at fulcrum pin $C = 2430$ N

∴ $2430 = l \times d \times p_b$ in bearing

Also $2430 = 2 \times \frac{\pi}{4} d^2 \times f_s$ in shear

∴ $2430 = 2 \times \frac{\pi}{4} d^2 \times 42$, gives $d = 6.1$ mm, say 7 mm

Now in the bearing equation, p_b may be taken as 10 N/mm^2 for forced lubrication, therefore,

$2430 = l \times 7 \times 10$, gives $l = 35$ mm

Now in practice l/d varies form 1.25 to 2, taking $l/d = 1.5$,

∴ from bearing equation, $2430 = 1.5\, d^2 \times 10$

∴ $d = 12.75$ mm, say 13 mm

∴ l = length of pin = $13 \times 1.5 = 19.5$ mm, say 20 mm

A bush 3 mm thick may be provided in the boss.

∴ Inside diameter of boss = $13 + (2 \times 3) = 19$ mm

∴ Outside boss diameter = 32 mm, which will accommodate the depth of the rocker arm section adjacent to the pin.

Push Rod: The load on the push rod = 1350 N

The length of the push rod will depend upon the length, i.e., height of the engine and other design, let us assume the length as 500 mm.

Rankine Gordon formula, is

$$P = \frac{f_c A}{1 + a\left(\frac{l}{k}\right)^2}, P = 1350 \text{ N}$$

f_c = safe buckling stress = 70 N/mm²

Now let d_o = outer diameter and d_i = inner diameter of rod.

$$\therefore \quad A = \frac{\pi}{4}(d_o^2 - d_i^2), a = \frac{1}{7500} \text{ for steel}$$

l = length of rod = 500 mm, taking the ends of rod as guided.

$$k^2 = \frac{I}{A} = \frac{\frac{\pi}{64}(d_o^4 - d_i^4)}{\frac{\pi}{4}(d_o^2 - d_i^2)}$$

$$\therefore \quad k^2 = \frac{d_o^2 + d_i^2}{16}$$

Now $\frac{d_o}{d_i}$ is usually $\frac{1}{0.6}$ to $\frac{1}{0.8}$

$\therefore$ let $d_i = 0.6\, d_o$

$$\therefore \quad A = \frac{\pi}{4}(d_o^2 - 0.36\, d_o^2) = 0.16\, \pi\, d_o^2$$

and

$$k^2 = \frac{d_o^2 + 0.36\, d_o^2}{16} = \frac{1.36}{16} d_o^2 = 0.085\, d_o^2$$

$$\therefore \quad 1350 = \frac{70 \times 0.16\, \pi\, d_o^2}{1 + \frac{1}{7500} \times \frac{(500)^2}{0.085\, d_o^2}}$$

$\therefore$ $0.8976\, d_o^4 - 34.425\, d_o^2 - 1350 = 0$

From here, $d_o \approx 12$ mm

$$\therefore \quad \frac{l}{d_o} \text{ ratio} = \frac{500}{12} = 41.7$$

Now for this ratio, the safe stress from Table 27.4 is about 40 N/mm²

$\therefore$ Safe stress will have to be taken less than 70 N/mm²

Let us take f_c = 35 N/mm²

$\therefore$ From Rankine Gordon Formula, $0.45\, d_o^4 - 34.425\, d_o^2 - 13500 = 0$

From here $d_o \approx 15$ mm

$$\therefore \quad \frac{l}{d_o} \text{ ratio} = \frac{500}{15} = 33.3$$

For l/d_o ratio equal to 35, the safe stress is about 50 N/mm²

$\therefore$ d_o = 15 mm is O.K., $\therefore$ d_i = 9 mm

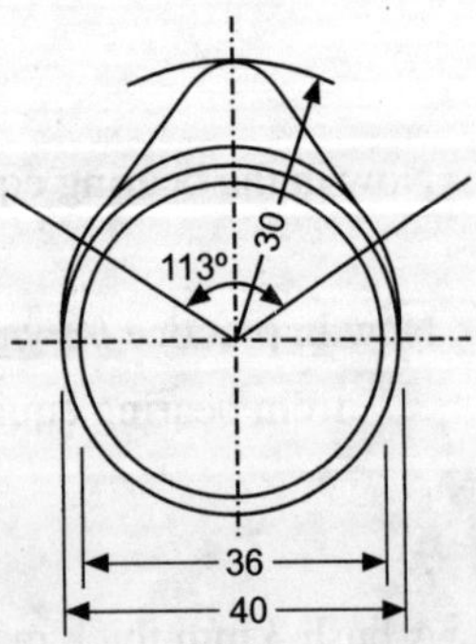

Fig. 27.48 *Tangent cam*

Cam: It is drawn according to the procedure already explained in fig. 27.46. Cam will be taken as forged as one piece with cam shaft.

Cam shaft diameter = 0.16 cylinder bore + 12.7 mm
= 0.16 × 140 + 12.7 = 35.1 mm
= say 36 mm

Base circle diameter = say 40 mm

Cam width = 0.09 × cylinder bore + 6 mm

= 0.09 × 140 + 6 = 18.6 mm say 19 mm

Cam profile is drawn in Fig. 27.48.

Example 27.13. The exhaust valve gear of a 4-stroke diesel engine running at 350 rev/min is made of forged steel. The valve is to open against an exhaust pressure of 0.315 MPa and an initial spring pressure of 566.5 N. The valve has weight of 7 N, a diameter of 0.075 m and a lift of 0.02 m and is to open and close with simple harmonic motion. The valve begins to open 50° before outer dead centre and is completely closed 10° after inner dead centre.

Assume:

Weight of the lever = 23 N

Radius of gyration of the lever about the rocker pin = 0.05 m

Length BC = Length CA = 0.125 m (Refer Fig. 27.32).

Design:

(a) Section of the lever near the pin boss at C.

(b) Lever boss, to contain a bronze bush under a bearing pressure suitable for lubrication.

(c) Check the boss for bending and rocker pin for shear.

Solution: Lever is to be designed under bending. So, the forces at points A, B and C should be known.

Total force at B = Force to open valve against gas pressure + Initial spring load + Force to accelerate valve – Weight of the valve

Now force against gas pressure = Valve area × pressure

$$= \frac{\pi}{4} \times (0.075)^2 \times 0.315 \times 10^6 = 1391.6 \text{ N}$$

Initial spring load = 566.5 N

Since valve opens at 50° before outer dead centre and closes at 10° after inner dead centre, the valve is open during a crank angle of

$$= 50 + 180 + 10 = 240°$$

Now the angle turned through per second by crank shaft

$$= \frac{350 \times 360}{60} = 2100°$$

$$\therefore \quad \text{Time for valve to open and close} = \frac{240}{2100} = 0.1143 \text{ s}$$

Now maximum acceleration of valve = $\omega^2 r$

where $\omega = \frac{2\pi}{t} = \frac{2\pi}{0.1143} = 54.97$ rad/s

and r = half valve lift = 0.01 m

∴ Maximum acceleration of valve = $(54.97)^2 \times 0.01 = 30.22$ m/s^2

∴ Force to accelerate the valve = mass of valve × acceleration

$$= \frac{7}{9.81} \times 30.22 = 21.6 \text{ N}$$

∴ Total force at B = 1391.6 + 566.5 + 21.6 – 7 = 1972.7 N

Now force at A = force at B + force required at A to accelerate rocker angularly

Angular acceleration of rocker, α = Liner acceleration of end B/CB

$$= \frac{30.22}{0.125} = 241.76 \text{ rad/s}^2$$

Since torsional moment $= I\alpha = mk^2\alpha$

$$\therefore \text{ (Force at A to accelerate rocker)} \times CA = \frac{23}{9.81} \times (0.05)^2 \times 241.76 = 1.42 \text{ Nm}$$

$$\therefore \quad \text{Accelerating force at } A = \frac{1.42}{0.125} = 11.36 \text{ N}$$

$$\therefore \quad \text{Total force at } A = 1972.7 + 11.36 = 1984.06 \text{ N}$$

$$\therefore \quad \text{Reaction at } C = 1972.7 + 1984.06 = 3956.76 \text{ N}$$

The remaining design can be completed as in the previous example.

EXERCISES

1. Which of the opening is more critical, the intake or the exhaust valve? Explain why.
2. Which of the valves should be and actually sometimes is made larger, the intake or exhaust valve?
3. Why do intake and exhaust valves have conical heads and seats?
4. What is interference angle between valve seat and valve seating surface. Why it is provided?
5. What are the materials used for intake and exhaust valves of internal combustion engines?
6. Why separate seat insert is provided? What is its material?
7. What is the duty of a valve spring?
8. What is 'valve flash' and why it is provided?
9. Name the various types of cam followers.
10. Determine the stress in a flat valve head having a diameter of 50 mm and a thickness of 6.5 mm. The maximum pressure in the combustion chamber is 3.5 N/mm^2. Determine the total stress if the stem diameter is 12.5 mm and the spring force is 680 N.
11. Exhaust valve of an oil engine is operated by a rocker arm. Maximum force on the roller is 1850 N. The effective length of each arm is 0.15 m and they include and angle of 135 degree. The load and effort are applied at right angles to the lever arm. Design
    (a) The lever of rectangular cross section whose thickness is half its depth.
    (b) Fulcrum pin.
    Assume suitable materials and safe stresses.
12. The cam shaft of an overhead valve petrol engine drives the rocker arm and valves through a push rod. If the push rod is 0.30 m long, the equivalent weight of the moving parts is 3N and the maximum acceleration is 12000 m/s^2, design the push rod and give a neat dimensioned sketch. Neglect the weight of the push rod.
13. Find the main dimensions of the exhaust valve lever for a 0.50 m stroke cylinder of a four stroke vertical diesel engine. The diameter of the exhaust valve is 0.165 m. Assume inertia load to be equal to 180 N. Make suitable assumptions for the missing data.
14. Design the complete valve gear arrangement for the exhaust valve of a four stroke, overhead valve type, vertical diesel engine. The diameter of the valve head is 40 mm and its lift is 9.65 mm. It weighs 2.5 N. The exhaust valve opens 46° before B.D.C. and closes 14° after T.D.C. The cylinder diameter is 100 mm and stroke is 120 mm and the engine runs at 2200 rev/min.
15. Design an exhaust valve gear for a horizontal gas engine with the following data:
    Diameter of valve is 75 mm, its weight is 5N, and its lift is 25 mm. The valve opens 33° before O.D.C and closes 1° after I.D.C. and it is to open with constant acceleration and deceleration for each half of the lift. The greatest back pressure when the exhaust valve opens is 0.4 N/mm^2 and the greatest suction pressure is 0.035 N/mm^2 below atmospheric. The engine runs at 300 rev/min. The effective length of each arm of the rocker lever is 0.175 m and the included angle is 140 degree. Take cam shaft diameter as 50 mm and the minimum cam radius equal to 50 mm.

□□□

Section VI

MISCELLANEOUS TOPICS

CHAPTER 28

Miscellaneous Topics

28.1 FLYWHEELS

28.1.1 Function

The primary function of a flywheel is to act as "energy accumulator". It is a device which acts as energy redistributor and is interposed between the source of energy and the place where it is needed, when the supply of energy does not vary in the same manner as does the demand for energy. It will absorb energy when demand is less than the supply of energy and will give out the energy when the demand is more than the energy being supplied. We know that the power generated in the cylinders of steam engines and internal combustion engines is variable but the power needed at the crank shaft is at uniform rate. Therefore, when the flywheel is attached to the shafts of these prime movers, it will help in the withdrawal of power at a constant rate equal to average rate of production. For example, in the case of gas or diesel engine, the energy developed during the power stroke is much more than the engine load. This excess energy will be absorbed by the flywheel and will be returned to the engine crank shaft during the periods when no energy is being developed in the cylinders, i.e., during suction, compression stroke in two stroke engines. In this manner, it helps to keep the crank shaft rotating at a uniform speed. It should be clear that the flywheel helps to smooth out only the cyclic speed fluctuations caused by the non uniform flow of power from the pistons. It is apparent that when the flywheel absorbs energy, its velocity increases and when it gives up the energy, its speed decreases. The permissible fluctuation of speed depends upon the purpose for which the engine is used.

In machines like punches, shears, etc., the energy is consumed at a highly variable rate and that only during a small portion of the working cycle. If such a machine is driven directly by the motor or with the help of belt drive, the later must be powerful so as to supply large amount of energy to the machine during a short period. For the remaining period, the energy source will run idle. To avoid this, a flywheel is put in between the machine and the energy source so that it goes on storing energy from the power source during the greater portion of the operating cycle and gives it up during a small period of the cycle. This will help in transmitting the energy from the power source to the machine at a practically constant rate throughout the cycle. This will result in a great reduction in the size and cost of the power source.

28.1.2 Weight of Flywheel

The energy stored in a flywheel is in the form of kinetic energy. The kinetic energy of a revolving mass is given by the expression,

$$E = \frac{W_f}{2g} v^2 = \frac{W_f}{2g} \cdot \omega^2 k^2 = \frac{1}{2} I\omega^2$$

where W_f = weight of revolving mass, that is, of flywheel, N

ω = angular velocity, rad/s

k = radius of gyration, m

I = mass moment of inertia = $\frac{W_f k^2}{g}$, kgm^2

Now if the flywheel rim is rectangular with inner radius R_2 and outer radius R_1.

Then $$k = \sqrt{\frac{1}{2}(R_1^2 + R_2^2)}$$

If the thickness of rim is small as compared to the radius of rim, the radius of gyration may be taken to be equal to the mean radius of rim, without much error. Therefore, the kinetic energy of revolving rim of weight W_r is:

$$E_r = \frac{W_r}{2g} \omega^2 R^2, \text{ where } R = \text{mean radius of rim}$$

It is a usual practice to neglect the effect of flywheel hub and arms and so 'E_r' will be taken as the kinetic energy of the flywheel. Where great accuracy is needed the effect of hub and arms may be considered as follows:

For usual flywheel proportions, the radius of gyration of hub and arms is about 0.57 the mean radius of rim. The weight of hub and arms is usually about 0.25 to 0.56 times the total weight of the wheel for high speed engines and from 0.32 to 0.41 for low speed engines. It is usual to take the weight of hub and arms as about 0.40 the weight of rim for both low and high speed engines. So the kinetic energy of hub and arm is:

$$E_h = \frac{W_a}{2g} \cdot \omega^2 ka^2 = \frac{0.40\, W_r}{2g} \cdot \omega^2 \cdot (0.57R)^2 = 0.13 \frac{W_r}{2g} \cdot \omega^2 R^2$$

where W_a = weight of hub and arms, and k_a = radius of gyration of hub and arms.

It is clear, therefore, that the inertia effect of hub and arms is equivalent to about 1/8th, the mass of the rim at the mean radius of rim. Therefore, the kinetic energy of the revolving flywheel is:

$$E = E_r + E_h \cong 1.125 \frac{W_r}{2g} \cdot \omega^2 R^2 = 1.125 \frac{W_r}{2g} v^2$$

where v = wheel speed at mean radius of the rim, m/s.

Now $W_f R^2$ is known as "flywheel effect".

The energy stored or given up during a change in speed from v_1 to v_2 becomes:

$$\Delta E = 1.125 \frac{W_r}{2g} (v_1^2 - v_2^2)$$

where v_1 = maximum velocity

v_2 = minimum velocity

∴ Fluctuation of energy in the flywheel is

$$\Delta E = 1.125\frac{W_r}{2g}(v_1 - v_2)(v_1 + v_2) = \frac{1.125 \times C_s \cdot W_r \cdot v^2}{g} \quad ...(28.1)$$

where C_s = Coefficient of fluctuation of speed

$$= \frac{\text{Change in speed}}{\text{mean speed}} = \frac{v_1 - v_2}{v}, v = \frac{v_1 + v_2}{2}$$

The reciprocal of the coefficient of fluctuation or co-efficient of uniformity is termed as "co-efficient of steadiness". The mass of the flywheel rim is thus given as:

$$m_r = \frac{W_r}{g} = \frac{1}{1.125} \cdot \frac{\Delta E}{C_s v^2} = \frac{k\Delta E}{C_s v^2} \quad ...(28.2)$$

where k = a constant to account for the effect of the hub, arms shafts etc. It can be taken as 0.90 unless otherwise specified.

From the above equation it is clear that the weight and design of the fly wheel for any given limit of fluctuation of speed depends upon the maximum excess or deficiency of the available energy with respect to the load during the energy cycle. As already indicated, the coefficient of fluctuation of speed depends upon the type of service. The usual range for this coefficient is given in Table 28.1.

Table 28.1

Type of machine or class of service	C_s
Crushers, hammers	0.200
Punches and shears	0.1 to 0.15
Pumps	0.03 to 0.15
Flour-milling machinery	0.02
Machine tools	0.200
Paper making machinery	0.025
Transmission by belt	0.030
Transmission by gear train	0.020
Spinning machines	0.030 - 0.050
Weaving machines	0.025
Electric generators	0.003 to 0.006

The fluctuation of energy can be determined in the following manner:

28.1.3 Engine Flywheels

The fluctuation of energy may be found out from the net turning moment diagram of the engine for one complete energy cycle. The load curve will be a straight line parallel to the base line and the effort curve will be an irregular one. The fluctuation of energy will be represented by the largest area above or below the load line or the mean torsional moment line. Typical turning moment diagrams for steam engine and internal combustion engine are shown in Figs. 28.1 and 28.2 respectively.

For most purposes, the fluctuation of energy can be found out with the help of coefficient of fluctuation of energy, which is given as:

$$C_e = \frac{\Delta E}{E_a}$$

where E_a = Average energy delivered per cycle, by the engine shaft.

Now $$E_a = 1000\,P \times \frac{60}{n}, \text{ for steam engines and 2-stroke I.C. engines}$$

and $$= 1000\,P \times \frac{60}{n/2}, \text{ for 4-stroke I.C. engines}$$

where P = Power in kW.

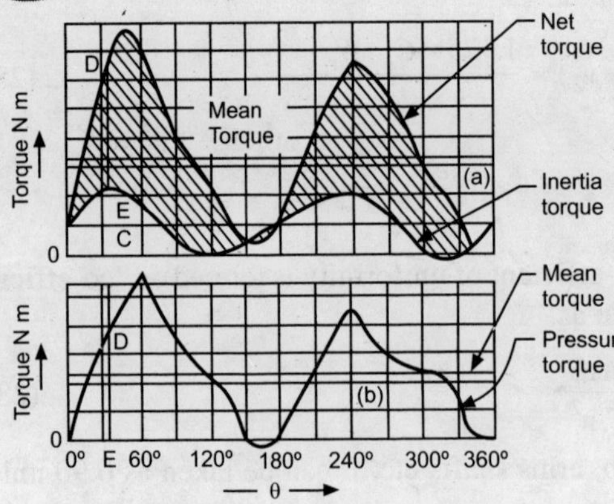

Fig. 28.1 *Net turning moment diagram for steam engine.*

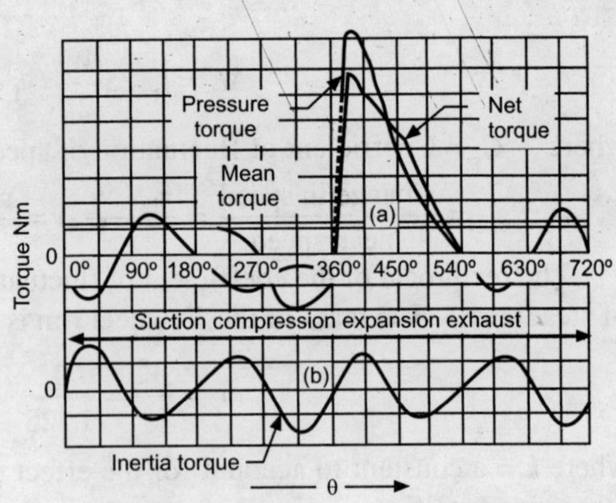

Fig. 28.2 *Net turning moment diagram for Internal combustion engine*

$\therefore$ $$\Delta E = C_e \times E_a \qquad ...(28.2a)$$

$\therefore$ From equations 28.1 and 28.2*a*

$$C_e \times E_a = \frac{1.125 \times C_s \cdot W_r \cdot v^2}{g}$$

$\therefore$ $$W_r = \frac{C_e}{C_s}\,\frac{E_a}{v^2}\cdot\frac{g}{1.125} \qquad ...(28.3)$$

or $$W_r = \frac{C_e k E_a}{C_s v^2} \qquad ...(28.3a)$$

The values of 'C_e' for steam engines and internal combustion engines are given in Tables 28.2 and 28.3.

Table 28.2 Values of C_e for steam engines

% Cut off	Single Cylinder	Two Cylinder Crank at 90°	Three Cylinders Cranks at 120°
0.10	0.35	0.088	0.040
0.20	0.33	0.082	0.037
0.40	0.31	0.078	0.034
0.60	0.29	0.072	0.032
0.80	0.28	0.070	0.031
1.00	0.27	0.068	0.030

Table 28.3 Values of C_e for Internal combustion engines

Types of Engine	Number of cylinders	Angles between cranks	C_e	
			4-stroke cycle	2-stroke cycle
Single acting :	1	360	2.35–2.40	0.95–1.00
Twin	2	360	0.92–1.04	0.75–0.85
Opposed	2	180	0.92–1.04	0.75–0.85
Tandem	2	0	1.50–1.60	0.75–0.85
Twin	2	180	0.92–1.04	0.20–0.25
Opposed	2	360	1.50–1.60	0.20–0.25
Vertical	3	120	0.60–0.75	0.15–0.18
Vertical	4	180 and 90	0.15–0.20	0.075–0.10
Vertical	6	180 and 60	0.10–0.12	0.016–0.02
Double acting:	1	—	1.50–1.60	0.20–0.20
Twin or tandem	2	—	0.18–0.20	0.18–0.20
Twin-tandem	4	90	0.08–0.09	0.07–0.08

28.1.4 Flywheels for Punches

As already indicated, in the case of punches, presses, shears, etc., the energy is consumed at a highly variable rate but the energy is supplied at a comparatively constant rate by the motor or by the belt. About 85 per cent of the energy supplied during an energy cycle is consumed in about 1/5th of the cycle. The great excess of demand over that which can be supplied by the motor or the belt is supplied by the flywheel and causes the flywheel to decrease in speed.

The force required to punch a hole in a plate is:

$$F = \pi \cdot d' \cdot t \times f_{us}$$

where d' = diameter of punch or hole

t = thickness of plate

f_{us} = ultimate shear strength of the plate material.

To find the size of the flywheel, the work done in punching a hole and the energy to be supplied during an energy cycle must be known. It is assumed that the resistance to the passage of punch through the hole varies in the triangular form.

∴ Work done in punching a hole = average force × plate thickness = $(1/2)F \times t$. Now the energy to be supplied by the flywheel can be found out and by equating it to equation 28.3, the weight of flywheel can be determined. It is clear that the capacity of the flywheel can be increased by increasing the weight, diameter or the velocity. To keep the diameter and weight of the wheel within reasonable limits, the flywheel should be mounted on a high speed auxiliary shaft.

28.1.5 Rim Dimensions

After finding out the mass of the rim, its dimensions can be found out as:

$$m_r = \text{Volume of rim} \times \text{density of material} = \pi D \times w \times h \times \rho \text{ kg}$$

where D = mean diameter of the wheel, m

w = width of the rim, m

h = depth or thickness of rim, m

ρ = mass density of wheel material, kg/m^3

= 7200 for C.I. and 7830 for steel, cast or forged.

The section of the rim is usually rectangular. The rim thickness h is made equal to or somewhat greater that w, or w/h may be selected between 0.65 and 2. If the flywheel is also used as a pulley, the rim width will be about 20 to 40 mm wider than the belt.

Note: The diameter of the flywheel for steam engines is usually from three to five times the stroke of the piston.

28.1.6 Stresses in Flywheel Rims

The stresses in the flywheel rims are set up due to the following reasons:

1. Centrifugal stresses
2. Stresses due to change of speed and load
3. Shrinkage stresses while casting.

Centrifugal stresses can be estimated with considerable accuracy and the stresses due to change of speed and load can be calculated only with fair accuracy while the shrinkage stresses, which might be quite large, cannot be computed with any degree of accuracy. It is taken into account by careful proportioning and arrangement of flywheel parts and by using a large factor of safety.

The calculation of the stresses in the rim due to centrifugal stresses will depend upon two distinct possibilities:

(*i*) The arms are absolutely flexible and they do not exert a restraining effect on the rim.
(*ii*) The arms are rigid and these completely restrain the rim at the places of attachment.

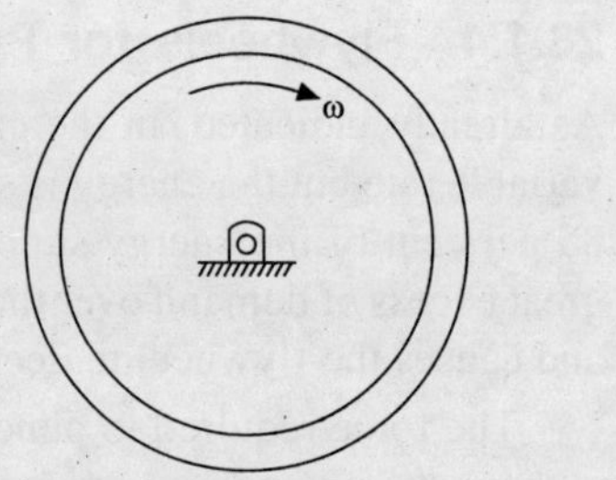

Fig. 28.3

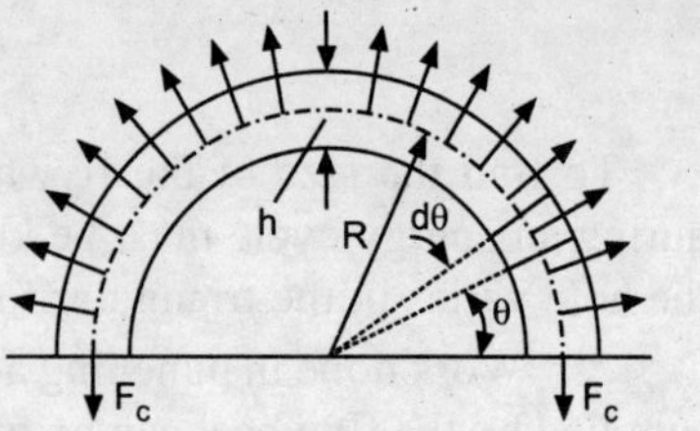

Fig. 28.3 *(a)*

In the first case, the wheel rim will act as a free rotating ring subjected to internal pressure (Fig. 28.3). The stress induced in the rim will be tangential stress and it may be estimated in the following manner: Consider half the ring as shown in Fig. 28.3(a).

Consider a small section of rim, so that $\Delta\theta$ (radians) is very small.

Now mass of section $dm = w \cdot h \cdot R\Delta\theta \cdot \rho$ kg

$\therefore$ Centrifugal force or inertia load on the small section

$$= \text{mass} \times \text{acceleration} = (w \cdot h \cdot R\Delta\theta \cdot \rho)(R\omega^2)$$

Its horizontal components will balance. Its vertical component is

$$= (w \cdot h \cdot R\Delta\theta \cdot \rho)(R\omega^2)\sin\theta$$

$\therefore$ Vertical component of the inertia load of half the rim

$$= \int_0^{\pi} (w \cdot h \cdot R\Delta\theta \cdot \rho)(\omega^2 R)\sin\theta = 2R^2 wh\rho\omega^2$$

This is balanced by the tensile forces at the cut section

$$\therefore \quad 2F_c = 2R^2 wh\rho\omega^2$$

Assuming that the tensile stress is uniformly distributed across a section

$$2F_c = 2(whf_t)$$

$$\therefore \quad f_l = \rho\omega^2 R^2 = \rho V^2 \text{ N/m}^2$$

In the second case, when the arms are rigid, the portion of the rim between two arms will act as a fixed beam and uniformly loaded, as shown in Fig. 28.4.

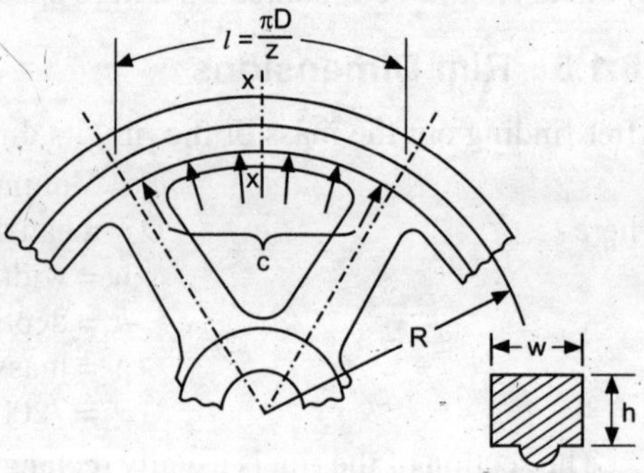

Fig. 28.4

The uniformly distributed load,

$$C = \frac{\rho \cdot w \cdot h \cdot v^2}{R} \text{ N/m of arc of rim.}$$

Now the maximum bending moment due to this will act at the arms and is approximately equal to

$$= \frac{Cl^2}{12}$$

where l = distance between the arms along the rim arc.

$$= \frac{2\pi R}{z}, \quad \text{where } z = \text{number of arms.}$$

$$\therefore \quad \text{Max. B.M.} = \frac{\rho \cdot w \cdot h \cdot v^2}{R} \cdot \left(\frac{2\pi R}{z}\right)^2 \cdot \frac{1}{12} = f_b \times Z, \text{ where } f_b = \text{bending stress, } Z = \frac{1}{6} wh^2$$

$$\therefore \quad f_b = \frac{2\pi^2 \rho v^2 R^2}{z^2 Rh} = \frac{2\pi^2 \rho v^2 R}{z^2 h} \text{ N/m}^2$$

Now it has been seen experimentally that the arms stretch approximately three fourth the amount required if the rim expanded freely under the action of the centrifugal force. Therefore, the total stress in the rim at junctions with the arms will be,

$$f = \frac{3}{4} f_t + \frac{1}{4} f_b$$

Now $f_t \not> 7$ MN/m^2 for C.I. wheels

$\not> 28$ MN/m^2 for C.S. wheels.

28.1.7 Rim Velocity

It is clear from the above analysis that the stress in the rim depends upon the rim speed. It has been experimentally determined that the "bursting speed" for two piece cast iron wheels with thick, heavy rims was about 90 m/s and about 67.5 m/s for two piece wheels with wide thin rims divided along a diameter through diametrically opposite arms. Using a factor of 3 with respect to the bursting speed, which is equivalent to a factor of safety of 9 with respect to the stress, the safe speed will be about 30 m/s for the first type of wheel and 22.5 m/s for the second type of wheel. The safe rim speeds are given below:

- 25 m/s for C.I. wheels of engines under 75 kW
- 30 m/s for C.I. wheels of larger machines
- 35 m/s with special design of rim which eliminates the danger of blowholes.
- 50 m/s with a special arm design, this and the above values are for C.I.
- 60 m/s with cast steel wheels.
- 75 m/s with C.S. arms and laminated steel arms and flywheels of all welded construction with forged steel rims.
- 50 m/s with C.I. flywheels machined all over in automotive engines.
- 75 m/s with semisteel flywheels of automotive engines.
- 60 m/s with steel flywheels of automotive engines.

28.1.8 Stresses in Arms

The arms of the flywheel are subjected to following types of stresses:

1. Direct pull due to the expansion or enlargement of the rim under centrifugal action.
2. Bending action due to the transmission of torsional moment from the rim to the shaft or from the shaft to the rim.
3. Initial stresses due to unequal cooling and shrinkage.
4. Acceleration forces transmitted to the rim.

Initial stress due to shrinkage and unequal cooling and the accelerating forces are difficult to be ascertained correctly. For the later case, the worst happens when the piston or bearings size and the engine comes to a halt. The flywheel seldom fails due to this and it is assumed that if the flywheel is strong enough to withstand the direct pull and bending action due to (1) and (2) respectively, it will be able to resist the acceleration forces. The shrinkage and cooling stresses will be higher for one-piece wheel than for two-piece wheel. In proportioning flywheel arms low allowable stresses are used to cover the uncertainties regarding the initial casting stresses and bending action on the arms. Moreover, the arms are subjected to complete reversal of stresses, therefore, the minimum factor of safety for C.I. wheels should be 10 and for steel wheels 8. It is quite common to take the allowable stress of 1/12 the ultimate strength of the material. When the flywheel is used on punches and other machines subjected to severe shock, the factor of safety should be as high as 15 for C.I. and 13 for W.I. and steel.

The direct stress due to the expansion or enlargement of the rim under centrifugal action is determined on the assumption that the stretch of the arms is equal to that corresponding to free, unrestrained radial expansion of the rim. As already indicated this will be equal to three-fourths full radial expansion.

Direct stress, $$f_d = \frac{3}{4} f_t = \frac{3}{4} \rho v^2 \text{ N/m}^2 \qquad \text{...(28.4)}$$

The arms will be under bending action due to transmission of torsional moment from the rim to the shaft or from the shaft to the rim, therefore, they will support maximum full torsional moment capacity of the machine. This bending stress will depend upon two distinct possibilities:

(*i*) For wheels with thick rims, the arms act as cantilevers fixed at the hub but free and guided at the rim joint.

(ii) For wheels with wide, thin rims, the arms act more like cantilevers fixed at the hub, but free and unguided at the outer ends.

The arm will bend as shown in Fig. 28.5 for the second case and for the first case, as shown in Fig. 28.6.

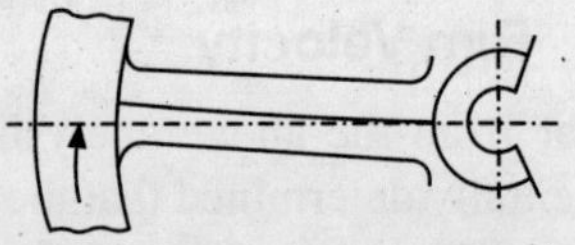

Fig. 28.5 *Arm as a free cantilever*

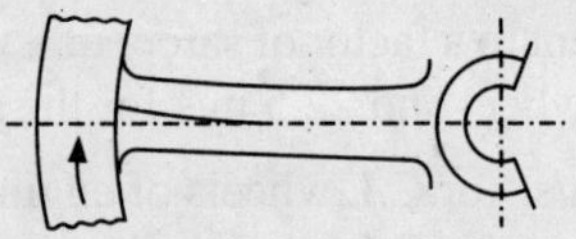

Fig. 28.6 *Arm as a guided cantilever*

To calculate the bending moment on the arm, the usual assumption is that load applied at the rim is sufficient to produce a torsional moment at the shaft equal to the maximum torsional moment. Therefore, force at the mean radius of the rim, is $F_r = \frac{M_t}{R}$, where M_t = maximum driving torsional moment on the shaft.

$$\therefore \quad \text{Load per arm} = \frac{F_r}{z} = \frac{M_l}{R \cdot z}$$

∴ Bending moment on the arm at the hub, for free cantilever,

$$M_1 = \frac{F_r}{z} \times \frac{(D - d_h)}{2} = \frac{M_t}{R \cdot z} \cdot \frac{(D - d_h)}{2} = \frac{M_t}{z} \cdot \frac{D - d_h}{D} \qquad \text{...(28.5)}$$

where d_h = hub diameter.

If the arm is taken as a guided cantilever, the maximum bending moment will be one-half that for the free cantilever, and is given as

$$M_2 = \frac{M_t}{2z} \cdot \frac{D - d_h}{D} \qquad \text{...(28.6)}$$

The sum of the direct stress and bending stress should not exceed the safe allowable stress. The bending moment M_2 will occur at both the ends of the arm.

28.1.9 Construction of Flywheels

Whether the flywheel is to be provided with web, arms, or it is to be a split flywheel, it all depends upon the diameter of the flywheel and the diameter of the shaft. To select the particular type of construction, the following criteria may be used as a guide:

(i) If D_o (outer diameter) is $< (7.5\,d + 80)$ mm, then the flywheel is provided with a solid web, where d is the diameter of the shaft. The web thickness may be taken as equal to $\frac{w}{20}$.

The flywheel with a solid web is shown in Fig. 28.7. Sometimes, holes are made in the web for handling facilities.

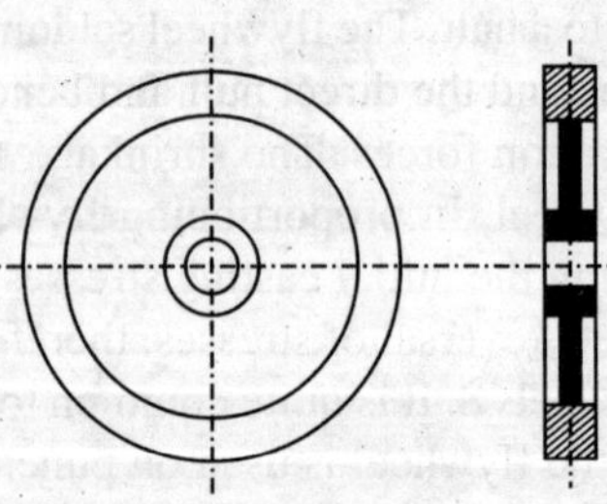

Fig. 28.7

(ii) When D_o is $> (7.5\,d + 80)$ mm, the flywheel is provided with arms. The number of arms is selected as:

$$z = 4; \quad D < 400 \text{ mm}$$
$$z = 6; \quad 400 < D < 3000 \text{ mm}$$
$$z = 8; \quad D > 3000 \text{ mm}$$

Sometimes, even upto 12 arms may be provided.

In smaller sized flywheels of this type, the rim, arms and hub are cast in one piece as shown in Fig. 28.8.

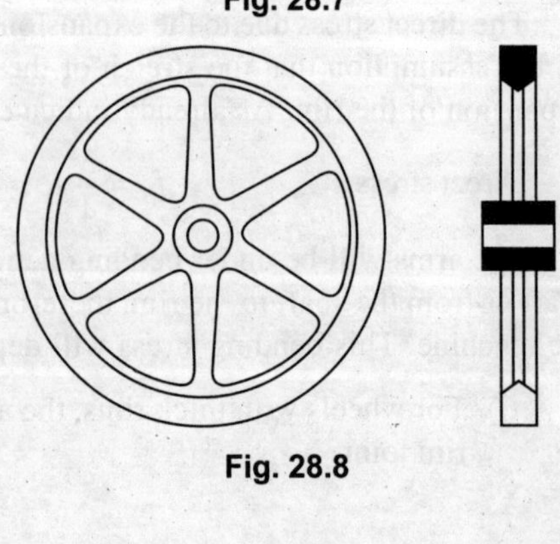

Fig. 28.8

But the wheels with diameter larger than about 2.5 metres, the flywheels are cast in two pieces, may be due to foundry limitations, or transporting difficulties. Also as the size increases, casting stresses are set up due to the unequal rates of cooling of hub, rim and arms. To relieve these stresses, sometimes the hub may only be split in large single piece flywheels, as shown in Fig. 28.9.

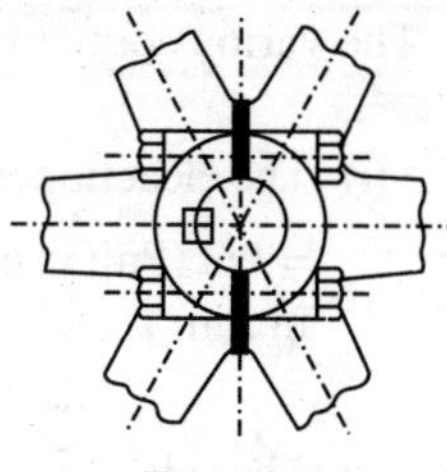

Fig. 28.9

A flywheel made in two halves should be parted at the arms instead of in between the arms due to better strength of the joint. The two halves are joined at the rim by means of:

1. Shrunk on steel cleats.
2. Dowel and cotter joint. Fig. 28.16.
3. Tie-bolts.

At the hub, the joint is made by means of:

1. Shrunk on steel cleats.
2. Tie-bolts

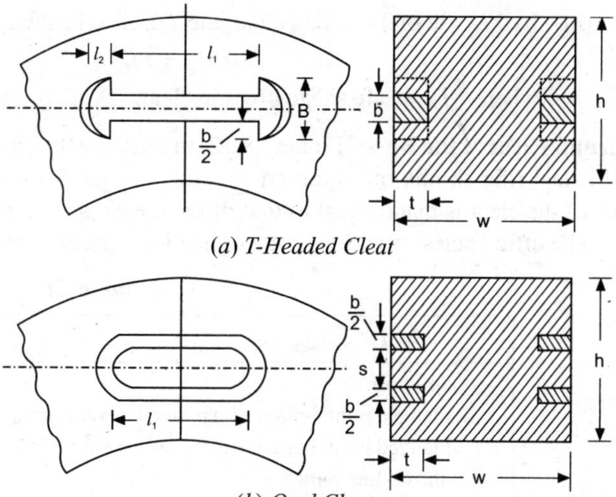

(a) *T-Headed Cleat*

(b) *Oval Cleat*

Fig. 28.10 *Links*

The material of the bolts, cleats and dowels and cotters is steel. While determining the size of the cleats or dowels and cotters at the rim, the holding power of the bolts at the hub is neglected and vice versa. The joint at the rim is subjected to the full centrifugal load. To allow for the reduction in the relative strength of the joint, the factor of safety for links is taken as equal to 0.75 times that for the rim. The two forms of steel cleats which are in common use are shown in Fig. 28.10. The T-headed cleat is also shown separately in Fig. 28.11.

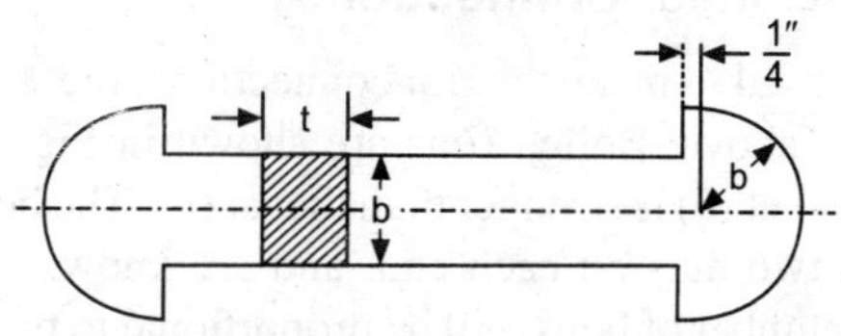

Fig. 28.11 *T-headed cleat*

The width or depth for the cleat '*b*' is taken as 1.25 to 1.35 the thickness, i.e., $b = (1.25 - 1.35)\, t$. The length of the slot in the rim into which the cleat will be put is made (3.5 to 4.5) $\sqrt{a}$, where *a* is the cross-sectional area of a link in square mm. The effective length of the link is made smaller than the length of the slot so that when the link is heated and put into the slot, the stress induced in the link by shrinkage due to its cooling does not exceed the yield point of the material. To calculate the shrinkage allowance, it is clear that the rim section is stressed, where the link bears against it. According to Prof. C.D. Albert, the average compressive stress in the part of rim affected is:

$$f_c = \frac{F_s}{\sqrt{AA_b}}$$

where F_s = Total pull in the link = shrinkage × stress cross sectional area of links

A = Bearing area at plane of division of the rim.

A_b = Bearing area under two link heads.

As already indicated, the total force at the rim connection is

$$P = f_t \times A$$

∴ If there are two cleats, one on either side of the rim face,

$$2 \times b \times t \times f_1 = P$$

where f_1 = Safe tensile stress in cleat.

The various dimensions of the cleat can be checked as follows:

(*i*) The cleat imposes a shear stress in the plug or shoulder $\frac{1}{2} l_1 \cdot s \cdot f_s$ $= P/2$ (Pull in each cleat), fix l_1 and s, where f_s = safe shear stress in rim.

$$B \approx 2b, \quad \therefore \quad s \approx b$$

(*ii*) There is a crushing action where the cleat bears on the rim.

$$\therefore \quad (B - b) \cdot t \cdot f_c = P/2$$

(*iii*) In the case of T-headed cleat, the *T* is also in shear.

$$\therefore \quad 2l_2 t \cdot f_{s1} = P/2$$

where f_{s1} = Safe shear stress in cleat.

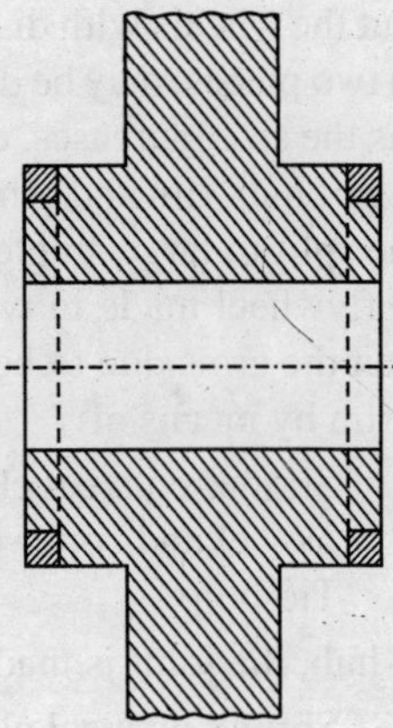

Fig. 28.12 *Shrunk cleat at the boss*

Shrunk Cleat at the Boss: Figure 28.12. In this case, the cleat is simply a mild steel ring shrunk on to turned shoulders or the boss. The sectional area of the cleat is made equal to that of the cleats at the rim.

The efficiencies of various rim connections are given below in Table 28.4.

Table 28.4

Type of Construction	Efficiency of Joint
Solid rim	1.00
Flanged joint, bolted, rim parted between arms	0.25
Flanged joint, bolted, rim parted on arms	0.50
Shrink-link joint	0.60
Dowel cotter joint	0.70

28.1.10 Bolted Connections

If bolts are used to make the rim connections, these can be designed as above. Bolted rims are shown in Fig. 28.13. Arrangement at (*b*) is stronger than that at (*a*). The bolts are made with two nuts on each end, and are known as stud bolts. The number of bolts will be proportional to the width of the rim. The distance '*d*' must be such that the bolt can be entered into the fitted holes. Cleats are sometimes used (at the rim) in conjunction with bolts. The arrangement of the hub bolts is shown in Fig. 28.14. The bolt area is made equal to that at rim. Or, it can be assumed that the centrifugal force tending to separate the wheel is taken up by bolts and to use a factor of safety of 6 to find the size of the bolts. In all cases, the joint face is relieved to concentrate the load on to a reduced surface.

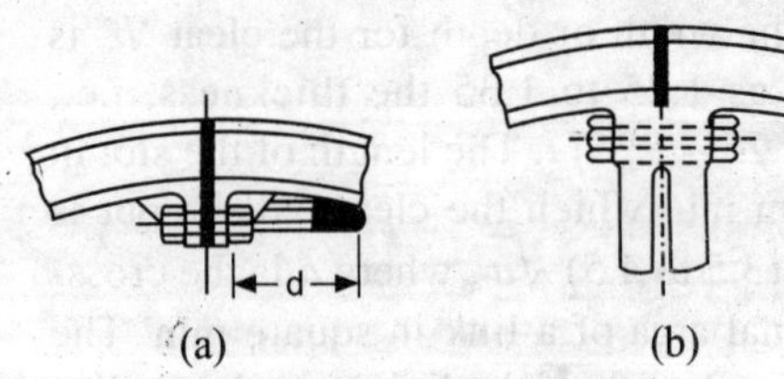

Fig. 28.13 *Bolted rims*

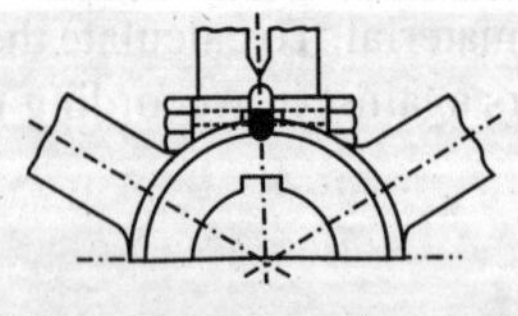

Fig. 28.14 *Bolted hub*

28.1.11 Design of Arms

The dimensions of the arms will be based on the direct stress and the bending stress. As already indicated to cover up the accelerating stresses and the shrinkage stresses, a low allowable stress is used in design. Therefore, the sum of the direct tensile stress due to the radial expansion of the rim and that due to bending should not exceed the safe allowable stress. From this allowable stress, the direct tensile stress is subtracted and the difference is taken as the allowable stress for bending and the arms are designed for bending only

Major axis is placed in the plane of rotation. This will reduce the windage loss and also the arms are used to maximum advantage for strength and stiffness.

The dimensions of the arm at the hub are calculated for the maximum bending moment M_1.

$\therefore \quad M_1 = f_b \times Z, Z = 0.05\ H^3, \quad \text{where } H = \text{major axis.}$

The dimensions at the rim are based on the bending moment M_2 which can act there. Or the arms are tapered from the hub to the rim, the taper being about 20 mm/metre length of arm for the major axis; for the minor axis, the taper is half this amount.

28.1.12 Shaft, Hub and Key

The diameter of the flywheel shaft may be tentatively based on the maximum torsional moment it is to transmit. The maximum torsional moment may be taken as twice the mean torsional moment.

Then, Diameter of hub, $d_h = 2 \times$ shaft diameter.

Length of hub, = 2 to 2.5 times the shaft diameter.

The length of the standard sunk key can be found out on the basis of maximum torsional moment. The arms are never attached directly on the hub and the rim but a circumferential rib is provided both at the hub and the rim. This is shown in fig. 28.15 which is a section taken on the hub.

Width of rib, $w' = 0.67$ H and height of rib, $h' = 0.67\ w'$.

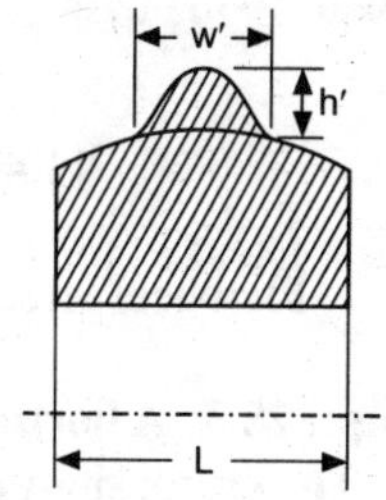

Fig. 28.15 *Section of hub*

Example 28.1. A steam engine having a piston of 0.50 m in diameter runs at 100 rev./min. From the tangential effort diagram, the area representing the maximum excess or deficiency of energy is found to be 9.65 sq.cm. The diagram is drawn to a force scale of 1 cm = 14 N/sq. cm of piston area and a space scale of 1 cm = 0.03 m. Assuming a mean diameter of 3 metres for the flywheel, find the weight of the rim required if the engine is to run a direct current generator.

Solution: Energy scale of diagram = 14 × 0.03 = 0.42 Nm/sq. cm of piston area

Now $\quad$ Area of piston $= \frac{\pi}{4} \times 50^2 = 1960$ sq. cm

$\therefore$ Maximum amount of energy to be handled by the flywheel at one time,

i.e., $\quad \Delta E = 0.42 \times 1960 \times 9.65 = 7940$ Nm

Now $\quad$ Value of $C_s = 0.006$, from Table 28.1

$$\text{Mean velocity of rim} = \frac{\pi \times 3 \times 100}{60} = 15.7 \text{ m/s}$$

Now $\quad$ $\text{Weight of the rim, } W_r = \frac{g \cdot k \cdot \Delta E}{C_s v^2} \text{ N} = \frac{9.81 \times 0.9 \times 7940}{0.006 \times 15.7^2} = 47.40 \text{ kN}$

Example 28.2. A single cylinder, single acting, four stroke oil engine develops 18.75 kW indicated power at 300 rev./min. The work done by the gases during the expansion stroke is 2.3 times the work done on the gases during the compression stroke and work done during the suction and exhaust strokes is negligible. The speed is to be maintained within 1% of the mean speed, find the flywheel effect.

Solution: The indicated work done per revolution $= 1000\,P \times \frac{60}{n} = \frac{1000 \times 18.75 \times 60}{300} = 3750$ Nm

$\therefore$ Indicated work done per cycle = 2 × 3750 = 7500 Nm

$\therefore$ Indicated work done per stroke $= \frac{7500}{4} = 1875$ Nm

Let $\quad W_e$ = Work done by the gases during expansion stroke.

Then $\quad W_e/2.3$ = work done on the gases during compression stroke.

$\therefore$ Work done per cycle $= W_e - \dfrac{W_e}{2.3} = \left(1 - \dfrac{1}{2.3}\right) W_e = \dfrac{1.3}{2.3} W_e$

$\therefore$ $\dfrac{1.3}{2.3} W_e = 7500$ $\therefore$ $W_e = 13.275$ k Nm

$\therefore$ Energy to be absorbed by the flywheel during the expansion stroke 13.275 – 1.875 = 11.4 kNm

Now $C_s = 0.02$

and $v = \dfrac{\pi Dn}{60} = \dfrac{\pi \times 2R \times 300}{60} = 10\pi$ R m/s

Now $\Delta E = \dfrac{C_s \cdot W_r \cdot v^2}{g}$, neglecting the effect of hub and arms.

$\therefore$ $11.4 \times 1000 = \dfrac{0.02 \cdot W_r\, 100\pi^2 \cdot R^2}{9.81}$

$\therefore$ Flywheel effect, $W_r R^2 = \dfrac{11.4 \times 1000 \times 9.81}{2\pi^2} = 5660 \text{ Nm}^2$.

Example 28.3. A horizontal press is driven by means of a belt. The pulley and the flywheel are on the same shaft. The crank shaft of the press is driven from the flywheel shaft with the help of two gears, the speed reduction being 6.5 to 1. The actual work done during each revolution of crank shaft is equal to 9350 Nm and it is done during four-tenths of a revolution. If the mechanical efficiency of the machine is 82 percent, find the power needed to drive the press and find the flywheel effect. The belt pulley makes 250 rev./min.

Solution: The crank shaft speed is $= \dfrac{250}{6.5} = 38.45$ rev./min

$\therefore$ Power needed to drive the press is

$$P = \frac{9350 \times 38.45}{1000 \times 0.82 \times 60} = 7.3 \text{ kW}$$

Now total energy to be supplied to the press $= \dfrac{9350}{0.82} = 11.4$ kNm

Now this energy is needed during four-tenths of a revolution. It means that during the remaining six-tenths of a revolution, the flywheel is storing the energy.

$\therefore$ Fluctuation of energy, $\Delta E = 11.4 \times \dfrac{6}{10} = 6.84$ kNm

$\therefore$ $6840 = \dfrac{C_s \times W_r \cdot v^2}{g}$, neglecting the effect of hub and arms.

Now $v = \dfrac{\pi \times D \times 250}{60} = \dfrac{\pi \times 2R \times 250}{60} = \dfrac{25\pi}{3}$ R m/s

$\therefore$ $6840 = \dfrac{C_s\, W_r \cdot 625\pi^2 R^2}{9 \times 9.81}$

Now from the table, $C_s = 0.15$

$\therefore$ $W_r R^2 = \dfrac{6840 \times 9 \times 9.81}{625\pi^2 \times 0.15} = 652 \text{ Nm}^2$.

Example 28.4. A single cylinder double acting steam engine delivers 187.5 kW at 100 rev./min. The maximum fluctuation of energy/revolution is 15 percent. The speed variation is limited to 1 percent either way from the mean. The mean diameter of the rim is 2.4 m. Design a cast iron flywheel for the engine.

Solution: Average energy delivered/revolution,

$$E_a = 1000\,P \times \frac{60}{n} = 1000 \times 187.5 \times \frac{60}{100} = 112.5 \text{ kNm}$$

$\therefore \qquad \Delta E = 112.5 \times \frac{15}{100} = 16.875, C_s = 0.02$

Now $\qquad m_r = \frac{k \cdot \Delta E}{C_s \cdot v^2}$

$$v = \frac{\pi Dn}{60} = \frac{\pi \times 2.4 \times 100}{60} = 12.55 \text{ m/s}$$

$\therefore \qquad m_r = \frac{0.9 \times 16.875 \times 1000}{0.02 \times 12.55^2} = 4821.35 \text{ kg}$

Rim Dimensions: Now $m_r = \pi D \times w \times h \times \rho$

$$4821.35 = \pi \times 2.4 \times w \times h \times 7200$$

$\therefore \qquad w \times h = 0.0888 \text{ m}^2$

Now $w/h = 0.65$ to 2, let us take $w/h = 1$

$\therefore \qquad w = h = \sqrt{0.0888} = 0.298$ m, say 0.30 m

Check for Stresses in the rim:

Now $\qquad f_t = \rho v^2 = \frac{7200 \times 12.55^2}{10^6} = 1.134 \text{ MN/m}^2$

Let the number of arms be 6.

Therefore, the uniformly distributed load on the rim,

$$C = \frac{\rho \cdot w \cdot h \cdot v^2}{R} \text{ N/m of arc of rim.}$$

$$= \frac{7200 \times 0.3 \times 0.3 \times 12.55^2}{1.20} = 85.05 \text{ kN/metre of rim arc}$$

Now B.M. on the rim due to this U.D.L. will be maximum at the arms and is given as

$$M_b = \frac{Cl^2}{12}$$

$$l = \frac{2\pi R}{z} = \frac{2\pi \times 1.20}{6} = 1.256 \text{ m}$$

$\therefore \qquad M_b = \frac{85.05 \times (1.256)^2}{12} \text{ kNm} = f_b \times \frac{1}{6} wh^2$

$\therefore \qquad f_b$ = bending stress

$$= \frac{6M_b}{wh^2} = \frac{6 \times 85.05 \times 1000 (1.256)^2}{12} \times \frac{1}{0.30 \times 0.09} = 2.485 \text{ MN/m}^2$$

Now the total stress will be maximum at the junction with the arm and is equal to

$$= \frac{3}{4} f_t + \frac{1}{4} f_b = \frac{3}{4} \times 1.134 + \frac{1}{4} \times 2.485 = 1.47 \text{ MN/m}^2$$

As already indicated, the factor of safety should be 8 or 9.

So, if the ultimate strength of C.I. is taken as 126 MN/m^2, the allowable stress = $\frac{126}{9}$ = 14 MN/m^2, hence the stress begin induced in the rim is quite low.

Shaft. The mean torsional moment on the shaft

$$T_m = \frac{9550 \times 187.5}{100} = 17.9 \text{ kNm}$$

Now let us take maximum torsional moment on the shaft twice the mean torsional moment,

$\therefore \qquad T_{max} = 2 \times 17.9 = 35.8$ kNm

Now the ultimate shear stress for mild steel ranges from 225 to 300 MN/m². Taking a factor of safety of 5, the safe stress will be 45 to 60 MN/m². To allow for bending of the shaft, let us take stress in shear as 40 MN/m².

$$\therefore \quad T_{max} = \frac{\pi}{16} d^3 \times f_s$$

$$\therefore \quad 35.8 \times 1000 = \frac{\pi}{16} \times d^3 \times 40 \times 10^6, d = 0.166 \text{ m}$$

Allowing for weakening effect of the key, let us take the diameter of shaft as 0.17 m

$\therefore$ Diameter of hub, $d_h = 2d = 0.34$ m

Length of hub = 2 to 2.5 d, say 0.40 m

Key: Using a square sunk key,

$$\text{Width} = \text{Thickness of key} = \frac{d}{4} = \frac{0.17}{4}, \text{ say } 0.0425 \text{ m}$$

Length of the key can be determined by considering the shearing action of the key, therefore,

$$T_{max} = \frac{1}{2} \times l \times w_i \times f_s \times d$$

where w_i = width of key = 0.0425 m

$$\therefore \quad 35.8 \times 1000 = \frac{1}{2} \times l \times 0.0425 \times 40 \times 10^6 \times 0.17$$

$$\therefore \quad l = 0.288 \text{ m}$$

But it should be equal to at least hub length $\therefore$ let $l = 0.40$ m.

Arms: Using 6 arms of elliptical section with major axis twice the minor axis.

Now maximum bending moment on the arm will be near the boss and is given as,

$$M_b = \frac{T_{max}}{z} \cdot \frac{D - d_h}{D} = \frac{35.8 \times 1000}{6} \times \frac{2.4 - 0.34}{2.4} = 5121.4 \text{ Nm}$$

$$\therefore \quad M_b = f_b \times Z$$

$$Z = 0.05 \text{ H}^3, \text{ where } H \text{ is major axis.}$$

Now safe stress for arms $= \frac{126}{9} = 14$ MN/m². To find the allowable stress in bending, the direct stress due to expansion of rim is to be subtracted from this value.

$$\therefore \quad f_b = 14 - \frac{3}{4} f_t = 14 - \frac{3}{4} \times 1.134 = 13.15 \text{ MN/m}^2$$

$$\therefore \quad 5121.4 = 13.15 \times 10^6 \times 0.05 \text{ H}^3$$

$$\therefore \quad H = \text{major axis} = 0.20 \text{ m, therefore, minor axis} = 0.10 \text{ m}$$

Now Length of arm = 1.20 – 0.15 – 0.17 = 0.88 m

$\therefore$ Taper of arm towards the rim is about 20 mm/metre length of the arm for the major axis.

$\therefore$ Major axis at the section where the arms join the rim = 0.20 – 0.02 = 0.18 m

$\therefore$ Minor axis = 0.09 m.

Example 28.5. Design completely the flywheel, the flywheel shaft and the key for securing the flywheel to the shaft, for a punching machine having a capacity for producing 30 holes of 19 mm diameter/minute in steel plate 16 mm thick, the material of the plate has an ultimate shear stress of 350 MPa. The actual punching operation is estimated to last for 1/5th of the angular rotation of the crank-shaft. The crank shaft is powered by a flywheel shaft through a reduction gearing having ratio of 1 : 10.

Assume:

(*a*) Mechanical efficiency of punching machine is 0.75.

(*b*) During the punching operation, the flywheel speed is reduced by a maximum of 10 per cent of the mean speed.

(*c*) The space limitations restrict the flywheel diameter to 0.92 m. ***(P.U. 1960)***

Solution: Maximum force required to punch the hole in the plate is,

$$F = \pi \cdot d' \cdot t \cdot f_{us} = \pi \times 0.019 \times 0.016 \times 350 \times 10^6 = 334.27 \text{ kN}$$

∴ Work done in punching a hole = average force × plate thickness

$$= \frac{1}{2} Ft = \frac{1}{2} \times 334.27 \times 1000 \times 0.016 = 2674 \text{ Nm}$$

If one hole is punched in every crank shaft rotation, therefore, the crank shaft speed is = 30 rev./min

∴ Speed of flywheel shaft = 30 × 10 = 300 rev./min

Power needed to drive the press is $= \frac{2674}{1000} \times \frac{30}{0.75 \times 60} = 1.78$ kW

Now, total energy to be supplied to the press, that is, energy at the flywheel shaft $= \frac{2674}{0.75} = 3565.3$ Nm.

Now the work done in punching a hole takes place in 1/5 rotation of the crank shaft. It means, during the remaining 4/5 rotation, the flywheel is absorbing the energy which it gives up during the punching operation.

∴ Maximum fluctuation of flywheel energy $= \frac{4}{5} \times 3565.3 = 2852.24$ Nm

Now $$m_r = \frac{k\Delta E}{C_s v^2}$$

Now $C_s = 0.1$. Taking mean diameter as 0.86 m,

$$v = \text{mean speed of flywheel shaft} = \frac{\pi \times 0.86 \times 300}{60} = 13.5 \text{ m/s}$$

∴ Mass of the rim is $m_r = \frac{0.90 \times 2852.24}{0.1 \times 13.5^2} = 140.85$ kg

∴ $\pi D \times w \times h \times \rho = 140.85,\ h = 0.06$ m

∴ $\pi \times 0.86 \times w \times 0.06 \times 7200 = 140.85$ gives $w = 0.12$ m

Now w/h varies from 0.65 to 2; ∴ w does lie in this range.

∴ Rim section is = 120 mm × 60 mm

Now the rim speed is 13.5 m/s. Now the safe limit for one piece cast iron flywheel is 30 m/s, therefore, the flywheel is quite safe for centrifugal stress. Or the total stress can be checked:

$$f_t = \rho v^2 = \frac{7200 \times 13.5^2}{10^6} = 1.31 \text{ MN/m}^2$$

Now $$f_b = \frac{2\pi^2 \rho v^2 R}{z^2 h}, \text{ let } z = 6$$

∴ $$f_b = \frac{2\pi^2 \times 7200 \times (13.50)^2 \times 0.43}{36 \times 0.06} = 5.16 \text{ MN/m}^2$$

∴ $$\text{Total stress} = \frac{3}{4} f_t + \frac{1}{4} f_b$$

$= 0.983 + 1.29 = 2.273$ MN/m^2, which is quite low and hence safe.

Shaft: Now the maximum torsional moment on the shaft occurs when the punching takes place. Now, the punching operation is completed in 1/5 rotation of crank-shaft, i.e., in 72 degrees of rotation of crank shaft.

∴ Rotation of flywheel shaft = 72 × 10 = 720° = 4π

∴ Work done $= T_{max} \times \theta$

∴ $3565.3 = T_{max} \times 4\pi$

$$\therefore \quad T_{max} = \frac{3565.3}{4\pi} = 283.6 \text{ Nm}$$

Now the torsional and bending moments on the shaft vary from practically zero to maximum values. Therefore, the shaft is under fatigue loading with shock conditions. Now ultimate shear stress for the shaft material, which is usually 0.45 percent carbon steel, ranges from 330 to 400 MPa. Using a factor of safety of 9 and allowing for key, the allowable stress ranges from $\frac{0.8 \times 330}{9}$ to $\frac{0.8 \times 400}{9}$, i.e., about 30 MPa to 35 MPa (0.8 is the weakening effect due to key way). Let us take a value of 32 MPa. Since the flywheel is light, the bending moment may be neglected, otherwise, if the bearing conditions are given, e.g., if the flywheel is overhanging, then bending moment due to flywheel weight = weight × overhang.

Overhang = half the rim width + half the bearing length + some clearance

$$\therefore \quad T_{max} = \frac{\pi}{16} d^3 \times f_s$$

$$\therefore \quad 283.6 = \frac{\pi}{16} \times d^3 \times 32 \times 10^6$$

∴ d will come out to be very low.

Let the overhang = 0.12 m

$$\therefore \quad \text{B.M.} = 140.85 \times 9.81 \times 0.12 = 165.8 \text{ Nm}$$

$$\therefore \quad T_e = \sqrt{(283.6)^2 + (165.8)^2} = 328.5 \text{ Nm}$$

$$\therefore \quad d^3 = \frac{328.5}{2 \times 10^6 \pi} = \frac{0.05227}{10^3} \text{ m}^3$$

$$d = 0.0374 \text{ m, say 38 mm}$$

Using a square sunk key,

$$\text{width} = \text{thickness} = \frac{d}{4}, \text{ say 10 mm}$$

$$\therefore \quad T_{max} = \frac{1}{2} \times l \times w_i \times f_s \times d$$

where w_i = width of key

$$\therefore \quad 283.6 = \frac{1}{2} \times l \times 0.01 \times 0.038 \times 32 \times 10^6$$

$$l = 0.0466 \text{ m, say} = 50 \text{ mm}$$

Now l should be at least 2 d, i.e., 76 mm, let us take l = 80 mm.

Example 28.6. Two steel links having a width of 0.0475 m and a thickness of 0.038 m are to be shrunk into a slot 0.1775 m long. The shrinkage stress to be induced is to be 210 MPa. The bearing area under the link heads is 0.0036 m², and that of the contact surface of the parts joined is 0.0257 m². The modulus of elasticity of the parts to be joined is 105 GPa. Determine the lengths of the links.

Solution: Cross-sectional area of link shank = 0.0475 × 0.038 = 0.001805 m²

∴ Total pull in two links due to shrinkage = 2 × 0.001805 × 210 × 10⁶ = 758.1 kN

Now average compressive stress between link heads,

$$f_c = \frac{F_s}{\sqrt{AA_b}}$$

Now $A = 0.0257 \text{ m}^2$ and $A_b = 0.0036 \text{ m}^2$

$$\therefore \quad f_c = \frac{758.1 \times 1000}{\sqrt{0.0257 \times 0.0036}} = 78.8 \text{ MN/m}^2$$

Shortening of slot due to pull of links, $\delta l_s = \frac{f_c}{E} \times l_s = \frac{78.8 \times 10^6}{105 \times 10^9} \times 0.1775 = 0.133$ mm

$\therefore$ Final length of slots, $(l_s - \delta l_s) = 177.5 - 0.133 = 177.367$ mm

$= \text{Final length of link} = l_f$

Now Unit elongation of link $= \frac{210 \times 10^6}{210 \times 10^9} = \frac{l_f - l_o}{l_o} = \frac{f_c}{E}$

$\therefore$ Original length of line $l_o = \frac{177.367}{1.001} = 177.189$ mm.

Example 28.7. The cast iron flywheel of a steam engine is 3 metre in diameter and has six arms. The rim width is 0.15 m and depth 0.20 m. The speed of the engine is 100 rev./min. The flywheel is to be built in two pieces held together at the rim by four steel links shrunk into place. The area of the links is to be such that they will carry the centrifugal force of half the rim with a factor of safety three-fourths as large as that in the remainder of the rim. The stress due to shrinkage in place is to be 210 MPa. Assume the cast iron between the link heads to be incompressible, the length of the slot between the bearing surfaces to be 0.175 m.

Solution: Centrifugal force of half the rim $= 2f_t \cdot A$

Now $f_t = \rho v^2$ N/m^2

Now $v = \frac{\pi D n}{60} = \frac{\pi \times 3 \times 100}{60} = 15.7$ m/s

$\therefore$ $f_t = \frac{7200 \times 15.7^2}{10^6} = 1.8$ MPa

$\therefore$ Total centrifugal force of half the rim $= 2 \times 1.8 \times 10^6 \times 0.15 \times 0.20 = 108$ kN

Now f_b = bending stress in the rim near the arm $= \frac{2\pi^2 \rho v^2 R}{z^2 h}$ N/m^2

Let z = number of arms = 6

$\therefore$ $f_b = \frac{2\pi^2 \times 7200 \times (15.7)^2 \times 1.5}{36 \times 0.20 \times 10^6} = 7.4$ MPa

$\therefore$ Total stress in the rim $= \frac{3}{4} f_t + \frac{1}{4} f_b = \frac{3}{4} \times 1.8 + \frac{1}{4} \times 7.4 = 3.2$ MPa

The ultimate strength of C.I. may be taken as 126.0 MN/m^2, therefore, the factor of safety for the rim

$= \frac{126}{3.2} = 39.4$

$\therefore$ Factor of safety for the links $= \frac{3}{4} \times 39.4 = 29.6$

Taking the ultimate strength of the link material as 490 MPa the allowable stress

$= \frac{490}{29.6} = 16.55$ MPa

$\therefore$ Area of shank of one link $= \frac{108 \times 10^3}{4 \times 16.55 \times 10^6} = 0.00163$ m^2

Now width of the link $= (1.25 \text{ to } 1.35) \times t$, say $1.30t$

$\therefore$ $1.3t^2 = 0.00163$

$\therefore$ **t = thickness of link $= \sqrt{\frac{0.00163}{1.3}} = 0.0354$ m = 35.4 mm**

$\therefore$ b = width of link = 35.4 × 1.3 = 46 mm

Now unit elongation due to shrinkage in place

$$= \frac{f_c}{E} = \frac{210 \times 10^6}{210 \times 10^9} = 0.001 = \frac{l_f - l_o}{l_o} = \frac{175 - l_o}{l_o}$$

$$\therefore \quad l_o = \text{Original length of link} = \frac{175}{1.001} = 174.825 \text{ mm.}$$

Example 28.8. Design the dowel and cotter joint for the rim and the bolts for the hub, if the flywheel is to be made in two pieces. The following data is provided:

Speed = 180 rev/min, Peripheral speed = 30 m/s.

Flywheel effect = 117 kNm^2. Maximum twisting moment due to engine = 52 kNm.

Solution: Assuming that the data refers to the mean diameter of flywheel, mean radius of wheel

$$= \frac{30 \times 60}{2\pi \times 180} = \text{about } 1.6 \text{ m}$$

$$\therefore \quad W_f R^2 = 117 \times 1000$$

$$\therefore \quad W_f \times (1.60)^2 = 117 \times 1000$$

$$\therefore \quad W_f = \text{weight of flywheel, neglecting the effect of hub and arms}$$

$$= \frac{117 \times 1000}{2.56} = 45.7 \text{ kN}$$

$$\therefore \quad m_f = \frac{45.7 \times 1000}{9.81} = 4658.5 \text{ kg}$$

$$\therefore \quad 2\pi R \times w \times h \times \rho = 4658.5$$

$$\therefore \quad w \times h = \frac{4658.5}{2\pi \times 1.6 \times 7200} = 0.0644 \text{ m}^2$$

Taking width of rim = height of rim, $w = h = \sqrt{0.0644} = 0.254$ m

Dowel and Cotter Joint: Refer Fig. 28.16

Total tension at each joint of rim $= f_t \cdot A$

Now $$f_t = \rho v^2 = \frac{7200 \times 30^2}{10^6} = 6.48 \text{ MPa}$$

$\therefore$ Total tension at each joint of rim = 6.48 × 10^6 × 0.0644 = 417.3 kN

Now the dowel and cotter joint is to be designed for this pull.

Let the allowable stress in the dowel be = 50 MPa

$$\therefore \quad \text{Effective area of dowel section} = \frac{417.3 \times 1000}{50 \times 10^6} = 0.00835 \text{ m}^2$$

Let the dowel be of rectangular section, with width b and thickness $= \dfrac{b}{2}$

Let thickness of cotter be, $t = \dfrac{b}{3}$.

$$\therefore \quad \text{Net area of dowel} = \left(b - \frac{b}{3}\right) \times \frac{b}{2} = \frac{b^2}{3}$$

$$\therefore \quad \frac{b^2}{3} = 0.00835$$

$$\therefore \quad b = 0.158 \text{ m}$$

$$\therefore \quad \text{Thickness of dowel} = 0.079 \text{ m, and}$$

$$\text{Thickness of cotter, } t = 0.053 \text{ m}$$

Now, Bearing pressure of cotter on dowel, Fig. 28.16

$$= \frac{P}{\text{thickness of dowel} \times \text{thickness of cotter}}$$

$$= \frac{417.3 \times 10^3}{0.079 \times 0.053} = 99.66 \text{ MPa}$$

which is safe (being less than the normal value of 100 MPa), therefore, the dimensions of dowel and cotter are O.K.

The hole of the dowel is usually made about 6 mm wider than the dowel itself, therefore, the bearing length of the cotter on the rim will be

$$= 0.254 - (0.079 + 0.006) = 0.169 \text{ m}$$

$\therefore$ Bearing pressure of cotter on rim $= \dfrac{417.3 \times 10^3}{0.053 \times 0.169} = 46.6$ MPa

which is very safe for C.I. rim.

The cotter is in double shear. Taking safe shear stress as 40 MPa, therefore, width of cotter at centre,

$$b_c = \frac{417.3 \times 10^3}{2 \times 0.053 \times 40 \times 10^6} = 0.10 \text{ m}$$

Now the distance between the inside edge of the cotter hole and the rim joint, l_2, must be sufficient to avoid tearing out of the intervening metal due to double shear. Let shear stress of rim be 7 MPa, therefore,

$$l_2 = \frac{417.3 \times 10^3}{2 \times 0.169 \times 7 \times 10^6} = 0.176 \text{ m}$$

The distance, l_1 can be determined from,

$$2 \times \frac{b}{2} \times l_1 \times f_s = P$$

$$\therefore \quad l_1 = \frac{417.3 \times 10^3}{2 \times 0.079 \times 40 \times 10^6} = 0.066 \text{ m}$$

Hub-bolts: The hub bolts are sometimes made as strong as the rim joint. Thus the core area of two bolts will be the same as the net effective area ofone dowel.

$$\therefore \quad 2 \times \frac{\pi}{4} d_c^2 = 0.00835,$$

$$d_c = 0.073 \text{ m}$$

$$\therefore \text{ Nominal Diameter} = \frac{d_c}{0.8} = \frac{0.073}{0.8} = 0.091 \text{ m}$$

Or the bolts can be designed to take up the centrifugal force, using a factor of safety as 6. Taking the ultimate strength as 600 MPa, safe stress $= \dfrac{600}{6} = 100$ MPa.

$$\therefore \; 2 \times \frac{\pi}{4} d_c^2 \times 100 \times 10^6 = 417.3 \times 10^3$$

$$\therefore \quad d_c = 0.052 \text{ m}$$

Therefore, nominal diameter of 0.091 m is O.K.

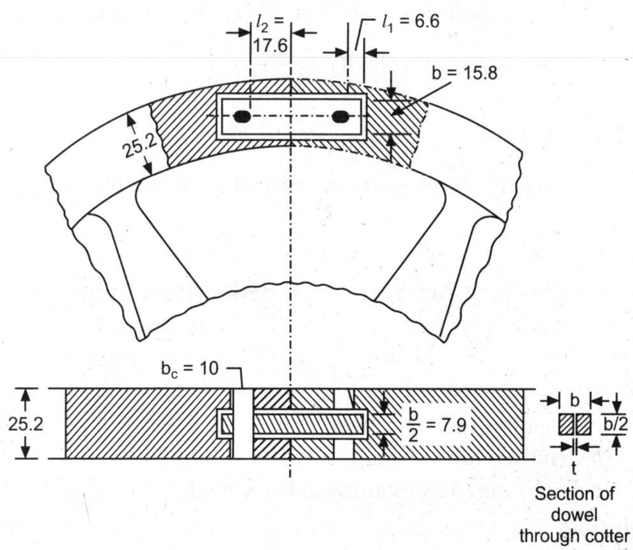

Fig. 28.16 *Dimension in cm*

Example 28.9. A small shear is to cut 25 mm square steel bars. The work required/cut is 2875 Nm. The cutting operation is to occupy one tenth of a revolution and there are to be 15 cuts/minute. The flywheel has a mean diameter of 0.90 m and is mounted on a shaft running at 120 rev/min. If the arms and hub supply 5 percent of the flywheel energy, determine:

(*a*) Weight of the flywheel
(*b*) Section of the rim.
(*c*) Average force exerted by the flywheel in supplying the energy.

Solution: (*a*) Let there be a cut at every stroke of the shear.

$$\text{Now cutting time} = \frac{1}{15} \times \frac{1}{10} = \frac{1}{150} \text{ minute} = \frac{2}{5} \text{ s}$$

The useful work supplied by the drive during this time is $= \frac{1}{10} \times 2875 = 287.5$ Nm

∴ Energy to be suppled by the flywheel = 2875 – 287.5 = 2587.5 Nm

$$\text{Now rim speed} = \frac{\pi \times 0.90 \times 120}{60} = 5.66 \text{ m/s}$$

Let us assume that the flywheel is slowed down 10 percent while delivering its energy. Therefore, the speed would drop to 5.094 m/s.

$$\text{Now} \qquad \Delta E = \frac{m_f (v_1^2 - v_2^2)}{2}$$

$$\therefore \qquad m_f = \frac{2587.5 \times 2}{(5.66^2 - 5.094^2)} = 849.75 \text{ kg}$$

(*b*) Mass of rim of flywheel = 0.95 × 849.75 = 807.26 kg

$$\therefore \qquad \pi D \times w \times h \times \rho = 807.26$$

Let the flywheel be made of C.I., so

$$\rho = 7200 \text{ kg/m}^3$$

$$\text{Taking} \qquad h/w = 1.5 \quad \therefore \quad \pi \times 0.90 \times w \times 1.5w \times 7200 = 807.26$$

$$\therefore \qquad w = 0.163 \text{ m}$$

$$\therefore \qquad h = 0.244 \text{ m}$$

(*c*) Now the flywheel is to deliver 2587.5 Nm of energy in 2/5 s

$$\text{Average mean rim velocity} = \frac{5.66 + 5.094}{2} = 5.377 \text{ m/s}$$

$$\therefore \quad \text{Distance travelled by rim in 2/5 sec} = 5.377 \times \frac{2}{5} = 2.1508 \text{ m}$$

∴ Average force exerted by flywheel to supply the energy

$$= \frac{2587.5}{2.1508} = 1203 \text{ N.}$$

Example 28.10. A flywheel is to be designed for a punching machine. The largest hole to be punched is 25 mm in diameter in a steel plate 12.5 mm thick. The maximum speed of the flywheel is 200 rev/min and because of the location of the shaft, the wheel may not be larger than 1.05 m in diameter. The ultimate shear stress of the material is 350 MPa. Neglecting the effect of hub and arms, determine:

(*a*) Weight of the flywheel
(*b*) Rim section
(*c*) Inner and outer radius of the wheel.

Solution: (*a*) It is assumed that the force required at the punch is 25 percent larger than that found by the ordinary formula.

$$\therefore \qquad F = \pi \cdot d' \cdot t f_{us} \times 1.25 = \pi \times 0.025 \times 0.0125 \times 350 \times 10^6 \times 1.25 = 428 \text{ kN}$$

Maximum punching effort will be at the time when the punch begins to penetrate the metal. Based on experience, it has been shown that the slug is free when the penetration is 1/3rd of the plate thickness. Therefore, the work done by the punch is,

$$= F \times \frac{1}{3} t = 428 \times 10^3 \times \frac{1}{3} \times 0.0125 = 1783.3 \text{ Nm}$$

Let the efficiency of the machine be 85 percent. Therefore, the energy at the flywheel shaft,

$$E = \frac{1783.3}{0.85} = 1995 \text{ Nm}$$

Let the mean diameter of wheel = 0.90 m

$$\therefore \quad v_1 = \frac{\pi \times 0.90 \times 200}{60} = 9.42 \text{ m/s}$$

If the greatest permissible slowing up of the flywheel shaft is assumed to be 15 percent, Table 28.1,

$$v_2 = 0.85 \times 9.42 = 8 \text{ m/s}$$

$$\therefore \quad \frac{m_f(9.42^2 - 8^2)}{2} = 1995$$

$$\therefore \quad m_f = 158 \text{ kg}$$

(*b*) $\pi D \cdot w \cdot h\rho = 158$

For C.I. Wheel, $\rho = 7200 \text{ kg/m}^3$

$\therefore \quad \pi \times 0.90 \times w \times h \times 7200 = 158$

$$\therefore \quad w \times h = 7.76 \times 10^{-3} \text{ m}^2$$

Let $w = h$

$$\therefore \quad w = h = 0.088 \text{ m}$$

(c) If the mean radius is taken as the sum of the inside and outside radii divided by 2:

$$\therefore \quad \text{Outer radius} = 0.45 + \frac{0.088}{2} = 0.494 \text{ m}$$

$$\text{Inner radius} = 0.45 - \frac{0.088}{2} = 0.405 \text{ m}$$

Example 28.11. Design a Cast Iron flywheel to store 675 kNm at 150 rev./min. Radius of gyration is 1.20 m. Determine:

- **(*a*) Weight and thickness of rim if width is 0.35 m (assuming whole energy in rim).**
- **(*b*) Diameter of shaft if 262.5 kW is transmitted.**
- **(*c*) Dimension of six arms, cross section to be elliptical, minor axis 0.65 major axis, safe stress 14 MN/m².**
- **(*d*) Dimension of key; safe stress for shaft and key, 42 MN/m².**
- **(*e*) Dimension of boss of wheel.**

Show in dimensioned sketches, sectional elevation and end view.

Solution: (*a*) Now

$$E = \frac{1}{2} \cdot \frac{W_f k^2}{g} \cdot \omega^2$$

k = radius of gyration = 1.20 m

$$\omega = \frac{2\pi \times 150}{60} = 15.7 \text{ rad/s}$$

$$\therefore \quad 675 \times 1000 = \frac{W_f \times (1.2)^2}{9.81} \cdot (15.7)^2$$

$\therefore$ **Weight of flywheel,** $W_f = 37.30$ kN

$$\text{Now weight of rim} = \text{weight of flywheel} = 37.30 \text{ kN}$$

$$\therefore \quad 37.3 \times 1000 = \pi \times 2 \times 1.20 \times w \times h \times 7200 \times 9.81$$

Now $w = 0.35$ m

$\therefore$ $h = 0.20$ m

(*b*) Since nothing is mentioned about maximum torsional moment, the shaft will be designed on the basis of mean torsional moment of the shaft.

$$\therefore \quad T_m = \frac{9550 \times 262.5}{150} = 16.7125 \text{ kNm}$$

$$\therefore \quad \frac{\pi}{16} d^3 \times 42 \times 10^6 = 16.7125 \times 1000$$

$$\therefore \quad d = 0.142 \text{ m}$$

(*c*) Torsional moment on shaft = B.M. on the arms

$$\therefore \quad \text{B.M. per arm} = \frac{16.7125}{6} = 2.7854 \text{ kNm}$$

$$\therefore \quad \frac{\pi}{32} \times H^2 \times 0.65 \times 14 \times 10^6 = 2785.4$$

$\therefore$ H = major axis of arm section at the hub = 0.146 m, say 0.15 m

$\therefore$ Minor axis at hub = 0.65 × 0.15 = 0.975 m

Taper for major axis = 20 mm/metre length of arm = 24 mm

Taper for minor axis = 10 mm/metre = 12 mm

$\therefore$ Major axis at the rim = 0.15 – 0.024 = 0.126 m

Minor axis at the rim = 0.0975 – 0.012 = 0.0855 m

(*d*) Diameter of hub = 2 × d = 0.284 m

Length of hub = 2.5 × d = 0.355 m

(*e*) Let the key be square key.

$$\text{Width} = \text{thickness} = \frac{d}{4} = \frac{0.142}{4} = 0.0355 \text{ m} = 35.5 \text{ mm}$$

$$\text{Now} \quad T_m = \frac{1}{2} l \times w_i \times f_s \times d$$

$$\therefore \quad \text{Length of key, } l = \frac{16.7125 \times 1000 \times 2}{0.0355 \times 42 \times 10^6 \times 0.142} = 0.158 \text{ m.}$$

But the length of the key cannot be less than the length of boss, therefore, let l = 0.36 m.

Example 28.12. Figure 28.17 shows the turning moment curve for one revolution of a multi-cylinder engine. The intercepted areas above and below the line of resistance torsional moment are:

– 0.32, + 4.06, – 2.71, + 3.29, – 3.16, + 2.32, – 3.74, + 2.71, – 2.45 cm^2.

The vertical and horizontal scales are: 5400 Nm and 24°/cm respectively. The engine has a mean speed of 250 rev/min and the fluctuation of speed is not to exceed 1.5% of the mean speed.

Determine the suitable diameter and cross-section of flywheel rim if the centrifugal stress is not to exceed 5.6 MPa. The density of the material is 7200 kg/m^3 and the width of rim is to be two times the thickness. The arms and hub of the flywheel provide for 5 percent of the energy.

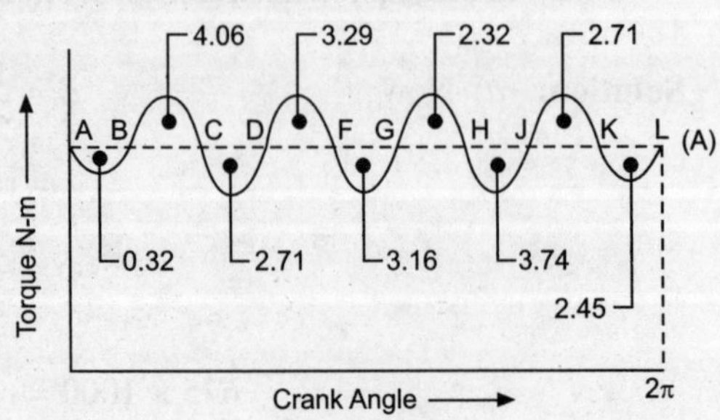

Fig. 28.17

Solution: Let the total energy at the beginning of the cycle, i.e., point A be E. Therefore, the energies at the points B, C, D, F, G, H, J, K and L will be : E – 0.32, E + 3.74, E + 1.03, E + 4.32, E + 1.16, E + 3.48, E – 0.26, E + 2.45 and E respectively.

$\therefore$ Max. Energy $= E + 4.32$, i.e., at point F.

Min. Energy $= E - 0.32$, i.e., at point B.

$\therefore$ Maximum fluctuation of energy,

$$\Delta E = (E + 4.32) - (E - 0.32) = 4.64 \text{ cm}^2$$

Now 1 cm^2 of turning moment diagram $= 5400 \times \dfrac{\pi \times 24}{180} = 2260$ Nm

$$\therefore \quad \Delta E = 2260 \times 4.64 = 10.5 \text{ kNm}$$

Now Mean speed, $\omega = \dfrac{2\pi \times 250}{60} = 26.17$ rad/s

$$\text{Fluctuation of speed} = \omega_1 - \omega_2 = \frac{1.5}{100} \times 26.17 = 0.3926 \text{ rad/s}$$

Now
$$\Delta E = \frac{1}{2} I(\omega_2^2 - \omega_1^2) = I\omega(\omega_2 - \omega_1)$$

$$\therefore \quad I = \frac{10.5 \times 1000}{26.17 \times 0.3926} = 1020 \text{ kgm}^2$$

Now the max. stress in the rim = 5.6 MN/m^2

$$\rho \times V_{max}^2 = 5.6 \times 10^6$$

$$V_{max} = \omega_2 \times R_{max}$$

Now
$$\omega_2 = 1.0075 \times \omega = 26.36 \text{ rad/s}$$

$$\therefore \quad 7200 \times (26.36)^2 \times R_{max}^2 = 5.6 \times 10^6$$

$$\therefore \quad \text{Max. radius of flywheel, } R_{max} = 1.058 \text{ m}$$

Let the mean radius of wheel be 0.90 m

$$\therefore \quad \frac{W_r}{9.81} \times (0.90)^2 = 1020 \times 0.95 = I \times 0.95$$

$$\therefore \quad \text{Weight of rim, } W_r = 11.736 \text{ kN}$$

$$\therefore \quad 2\pi \times 0.90 \times w \times h \times 7200 = \frac{11.736 \times 1000}{9.81}$$

Now
$$w = 2h$$

$$\therefore \quad 2\pi \times 0.90 \times 2h^2 \times 7200 = \frac{11.736 \times 1000}{9.81}$$

$$\therefore \quad h = 0.12 \text{ m}$$

$$\therefore \quad w = 0.24 \text{ m.}$$

Example 28.13. Figure 28.18 represents the crank effort diagram for a double acting steam engine. The two peaks *A* and *B* represent crank torsional moments of 2500 and 2300 Nm, respectively. The mean speed is 500 rev/min and the fluctuation of speed is to be limited to ±2% of the mean. The radius of gyration of the flywheel is 0.30 m. Determine:

(*a*) Mean torsional moment.
(*b*) Maximum fluctuation of energy.
(*c*) Weight of flywheel neglecting the effect of the hub and arms.
(*d*) Dimensions of the rim if the density of cast iron is 7250 kg/m^3 and the rim width is twice its thickness.

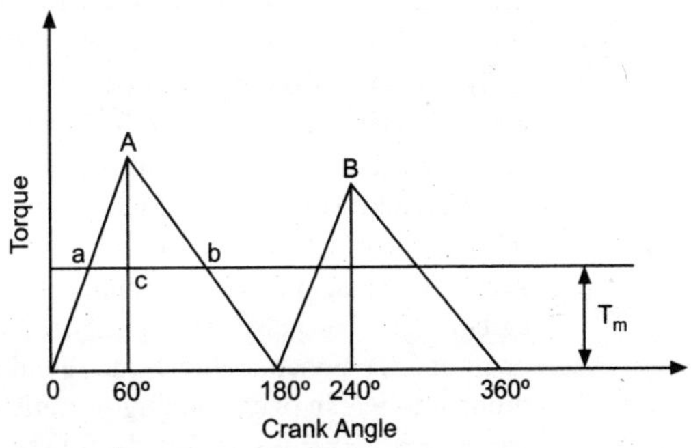

Fig. 28.18

Solution: Energy required per cycle = Area under torsional moment – crank angle curve

$$= \frac{1}{2} \times \pi \times 2500 + \frac{1}{2} \times \pi \times 2300 = 2400\,\pi \text{ Nm}$$

(*a*) Mean torsional moment $= \dfrac{2400\pi}{2\pi} = 1200$ Nm

(*b*) Maximum fluctuation of energy = Area of the triangle $aAb = \dfrac{1}{2} ab \times Ac$

Now $Ac = 2500 - 1200 = 1300$ Nm

$$ab = \frac{\pi}{2500} \times 1300 = \frac{13\pi}{25} \text{ radians}$$

$\therefore$ $\Delta E = \dfrac{1}{2} \times \dfrac{13\pi}{25} \times 1300 = 1060$ Nm

(*c*) Now $\Delta E = I \cdot \omega \cdot \Delta\omega$

$$\omega = \text{mean speed} = \frac{500 \times 2\pi}{60} = 52.5 \text{ rad/s}$$

$$\Delta\omega = \frac{4}{100} \times 52.5 = 2.10 \text{ rad/s}$$

$\therefore$ $1060 = \dfrac{W_f k^2}{9.81} \times 52.5 \times 2.10 = \dfrac{W_f (0.3)^2}{9.81} \times 52.5 \times 2.10$

$W_f = 1048$ N

(*d*) Now $W_f = 2\pi R \times w \times h \times \rho \times g = 2\pi \times 0.30 \times 2h \times h \times 7250 \times 9.81$

$\therefore$ $h^2 = 0.003906$

$\therefore$ $h = 0.0625$ m

$\therefore$ $w = 0.125$ m.

EXERCISES

1. A flywheel running at 300 rev/min is to absorb 70 kNm of energy. The maximum velocity at the mean radius is to be 25 m/s and is to vary not more than 5 percent. Find the weight, width and depth of the rim.
2. The speed of a flywheel changes from 300 rev/min to 290 rev/min in 0.05 s. The rim weighs 1150 N and has a mean diameter of 1.5 metres.
   (*a*) Determine the energy given up during the change in speed.
   (*b*) Determine the average torsional moment required to cause this change in speed.
3. A cast iron flywheel running at 40 rev/min is to furnish 105 kNm of energy to a punch during 0.15 revolution with a 15% change in speed. The maximum velocity at the mean radius is to be 1000 m/min.
   (*a*) Determine the dimensions of the rim.
   (*b*) Determine the power of the motor required to drive this punch, if the overall efficiency of the machine is 80%.
4. A flywheel for a four cylinder petrol engine develops 22.5 kW at 1000 rev/min. The fluctuation of energy is 25% of that of one revolution, fluctuation of speed 2% and the radius of gyration 0.18 m. Design:
   (*a*) The shaft assuming a safe stress of 35 MN/m^2.
   (*b*) Standard key for securing the flywheel to the shaft.
   (*c*) Steel flywheel taking the density of steel as 7830 kg/m^3. *(P.U. 1959)*
5. A punching machine has maximum capacity of punching 32 mm diameter holes in 22 mm thick plate of 0.4% carbon steel. The punching capacity is 24 holes/minute and the speed ratio of the driving shaft to the eccentric operating the punch is 9 to 1 and the mechanical efficiency of the press and drives is 72%. If the punch is brought to rest in one punching operation after the power is shut off and the diameter of flywheel cannot be larger than 1.2 metres, design the flywheel. *(P.U. 1961)*

6. A small shear is to have a capacity to cut 6.5×10^{-4} m^2 steel bar. The work required/cut is 2880 Nm. The cutting operation is to occupy one tenths of a revolution. There are to be 15 cuts/minute. The flywheel has a mean diameter of 1.2 metres and is mounted on a shaft running at 120 rev/min. Design the flywheel if it is to be of C.I.
7. A flywheel 3.6 metres in diameter rotates at 160 rev/min. The rim has a rectangular section with a face of 0.40 m and a depth of 0.15 m and is cast in two parts held together by two shrink links at one joint. Determine the main dimensions, the cross-section and the lengths of the links.
8. (*a*) A press is designed to punch 20 mm holes in a 20 mm thick plate. The material of the plate has an ultimate shear stress of 360 MN/m^2. The press is operated at 30 rev./min, through gears having a ratio of 8 : 1. The punching operation takes place over 75 degrees of the eccentric travel and the speed fluctuation is to be limited to 15%. The overall mechanical efficiency is 80%.
   (*i*) For a radius of 0.50 m determine the necessary cross-section of the flywheel rim.
   (*ii*) What is the total weight of the flywheel?
   (*iii*) What kW motor must be supplied to the press if
      (*a*) the flywheel is fitted?
      (*b*) the flywheel is not there?

   (*b*) What type of stresses can you expect in the flywheel rim? (P.U. 1973)
9. A cast iron flywheel with a mean diameter of 0.90 m changes speed from 400 rev/min to 380 rev/min while it gives up 9500 Nm of energy? What is the coefficient of fluctuation, the weight and the sectional area of the rim?
10. A 1.2 m flywheel, with a rim 0.10 m thick and 0.075 m wide rotates at 400 rev/min. If there are six arms, what is the approximate stress in the rim? Is this a safe stress? At what maximum speed should this flywheel rotate?
11. The energy required to shear a 25 mm round steel bar is approximately 1180 Nm. In use, the shearing machine is expected to make a maximum of 40 cutting strokes a minute. The frictional losses should not exceed 15% of the useful work. The shaft carrying the flywheel is to average 150 rev/min (a) what motor power is required? (b) Assuming a size of flywheel and choosing C_s find the weight and sectional dimensions of the rim of a C.I. flywheel. The width of the rim is to equal the depth and is not to exceed 0.09 m. It would be safe to assume that all the work of shearing is supplied by the kinetic energy given up by the flywheel.
12. Design a flywheel for a motor four cylinder engine to develop 22.5 kW at 1200 rev/min. Fluctuation of energy 30% of one rev; of speed 20%; radius of gyration 0.175 m. Calculate:
    (*a*) Energy to be absorbed/revolution.
    (*b*) Weight of steel rim.
    (*c*) Thickness of rim width is 75 mm.
    (*d*) Inner and outer radii of rim.
    (*e*) Diameter of shaft, safe stress 35 MN/m^2.
    (*f*) Diameter of boss.
    (*g*) Length of key; width 9.5 mm.

    Sketch with dimensions a sectional elevation and end view.
13. The turning moment diagram for an engine is drawn to a base of crank angle. The areas above and below the mean torsional moment line are:

$$+ 34.8, - 9.03, + 10.3 \text{ and } - 36.07 \text{ cm}^2$$

the scales being: 1 cm = 11 Nm; and 1 cm = 12° of crank angle. Determine the weight of the flywheel to keep the speed between 297 and 303 rev/min if its radius of gyration is 0.54 m. Also determine the suitable dimensions of the cast iron flywheel rim whose breadth is twice its radial thickness. The density of cast iron is 7250 kg/m^3 and assume that 90% of the flywheel effect is provided by the rim.

28.2 CURVED BEAMS

Curved beams find applications in many machine members such as *C* clamps; Crane hooks; frames of presses, riveters, nibblers, shears, punches, planers, boring machines, etc. The difference in the analysis for the straight beam and a curved beam is due to the fact that while in the case of a straight beam, the neutral axis of the cross-section passes through the centroid of the section, it does not do so in the case of a curved beam but is shifted towards the centre of curvature of the beam. While in the case of a straight beam under bending, the variation of the bending stress is linear, magnitude being proportional to the distance of a fibre from the neutral axis; the distribution of the stress in the case of a curved beam is non-linear because of the gravity axis being initially curved.

In all the machine members mentioned above, the line of action of the load does not pass through the centroid of the section over which the stress is to be found out. Due to this, the total stress over any section of the member will be the algebraic sum of the direct stress and the flexural or bending stress, i.e.,

$$f = f_d \pm f_b$$

In the case of straight beam or a beam whose axis is bent to a large radius, the total stress is given by the relation

$$f = \frac{P}{A} \pm \frac{M_b \cdot c}{I} \quad \text{...(28.7)}$$

where P = Applied load,

A = Area of the section over which the stress is to be obtained.

M_b = Bending moment about the centroid of the section.

c = Distance from the centroid to the extreme fibre.

I = Moment of inertia of the cross-section.

Now, $M_b = P \times a$

where a = eccentricity of the load with respect to the centroid of the cross-section,

$$f = \frac{P}{A} \pm \frac{P \cdot a \cdot c}{I} \quad \text{...(28.8)}$$

Maximum tensile stress on the section will be given as,

$$f = \frac{P}{A} + \frac{P \cdot a \cdot c_1}{I} \quad \text{...(28.9)}$$

and the maximum compressive stress on section will be,

$$f = \frac{P}{A} - \frac{P \cdot a \cdot c_2}{I} \quad \text{...(28.10)}$$

where c_1 = distance from the centroid to the tensile fibre and

c_2 = distance from the centroid to the compressive fibre.

If the beam is originally curved as shown in Fig. 28.19, the application of the above formulas will not give accurate results. The action of the load is to straighten out the beam, which will produce tensile stress in the fibres at x, and compressive stress in the fibres at y. The total stress at any point of the cross-section will be the sum of a unit direct stress and a unit bending stress due to the eccentricity of the load.

The maximum tension at x is given by the relation:

$$f_t = \frac{P}{A}\left[1 + \frac{a}{R - c_1}\left(\frac{c_1}{R} \times \frac{A'}{A' - A} - 1\right)\right] \quad \text{...(28.11)}$$

and maximum compressive stress at y, as

$$f_c = \frac{P \cdot a}{A(R + c_2)}\left(\frac{c_2}{R} \times \frac{A'}{A' - A} - 1\right) - \frac{P}{A} \quad \text{...(28.12)}$$

where R = radius of curvature, passing through the gravity axis O and having a centre at C

$$A' = \text{derived area} = R\int \frac{dA}{R - y} \quad \text{...(28.13)}$$

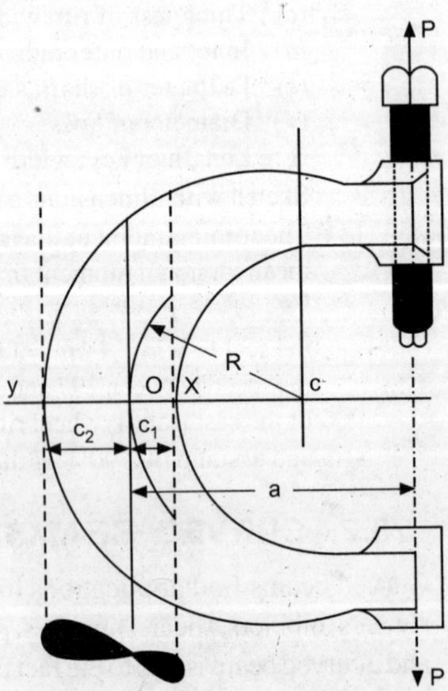

Fig. 28.19 *Curved beam*

y is positive if measured from the gravity axis towards the centre of curvature and is negative if measured away from the centre of curvature.

Another relation for the fibre stresses is given below:

$$f_o(\text{stress on outer fibres}) = \frac{M_b c_o}{Ae\,(R_n + c_o)} \text{ (Compressive)}$$

$$f_i(\text{stress on inner fibres}) = \frac{M_b c_i}{Ae\,(R_n - c_i)} \text{ (tensile)}$$

$\therefore$
$$\text{Total stress on inner fibres} = \frac{P}{A} + \frac{M_b c_i}{Ae\,(R_n - c_i)} \quad ...(28.14)$$

and
$$\text{Total stress on outer fibres} = \frac{P}{A} - \frac{M_b c_o}{Ae\,(R_n + c_o)} \quad ...(28.15)$$

where M_b = applied bending moment.
c_i = distance from the neutral axis to inner radius
c_o = distance from the neutral axis to outer radius
e = distance from the neutral axis to gravity axis of the section
$= R_g - R_n$

where R_g = Radius of curvature of the gravity axis.
and R_n = Radius of curvature of the neutral axis

Now $c_i = c_1 - e$ and $c_o = c_2 + e$

The value of R_n depends upon the shape of the beam. Its values for some common beam sections are given in Table 28.5.

Table 28.5 Values of R_n for curved beams

Shape of the section	R_n
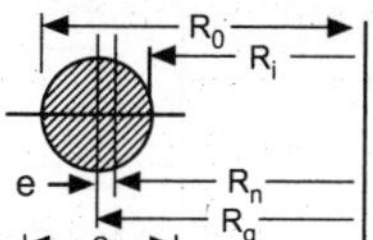	$R_n = \dfrac{(\sqrt{R_o} + \sqrt{R_i})^2}{4}$
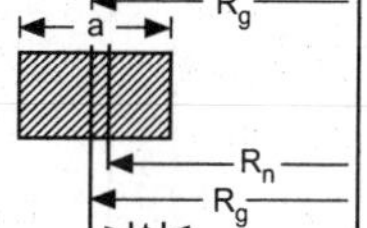	$R_n = \dfrac{a}{\log_e R_o / R_i}$
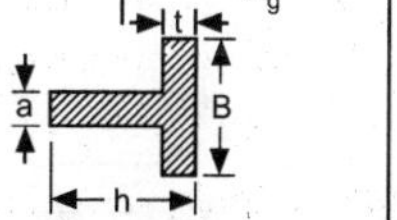	$R_n = \dfrac{(B-a)t + ah}{(B-a)\log_e \dfrac{R_i + t}{R_i} + a \log_e R_o / R_i}$
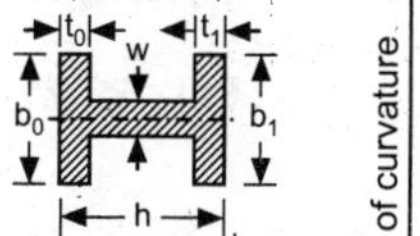	$R_n = \dfrac{(b_i - w)t_i + (b_o - w)t_o + w \cdot h}{b_i \log_e \dfrac{R_i + t_i}{R_i} + w \log_e \dfrac{R_o - t_o}{R_i + t_i} + b_o \log_e \dfrac{R_o}{R_o - t_o}}$
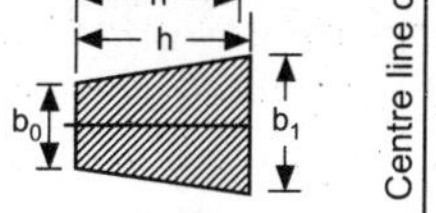	$R_n = \dfrac{(1/2)h(b_i + b_o)}{\left(\dfrac{b_i R_o - b_o R_i}{h}\right)\log_e \dfrac{R_o}{R_i} - (b_i - b_o)}$
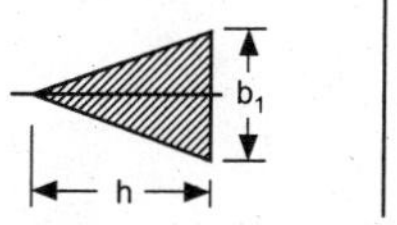	$R_n = \dfrac{(1/2)b_i \cdot h}{\dfrac{b_i r_o}{r_o - r_i} \log_e \dfrac{r_o}{r_i} - b_i}$

In the above analysis, it is assumed that planes normal to the gravity axis before bending remain planes normal to the gravity axis after bending.

Example 28.14. Determine the maximum stress created in the offset link shown in Fig. 28.20 by a force P = 4500 N. The cross-section of the link is: its height h = 50 mm, its width b = 12.5 mm. The eccentricity 'e' of the force is 12.5 mm.

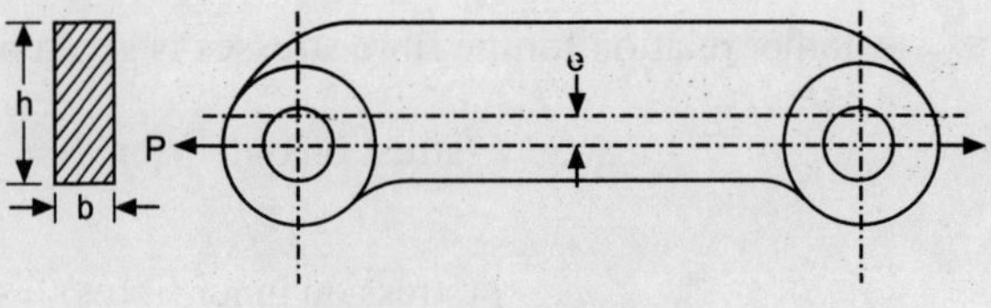

Fig. 28.20 *Offset link*

Solution: Maximum stress in the link is given as : $f = f_d + f_b$

Now $$f_d = \text{direct stress} = \frac{P}{A} = \frac{4500}{b \times h} = \frac{4500}{50 \times 12.5} = 7.2 \text{ N/mm}^2 \text{ (tensile)}$$

f_b = bending stress (tensile) at the inner edge and compressive at the outer edge.

$$= \frac{M_b}{Z}$$

$$M_b = P \times e = 4500 \times 0.0125 = 56.25 \text{ Nm}$$

$$Z = \frac{1}{6} bh^2 = \frac{1}{6} \times 12.5 \times 2500 = \frac{31.25}{6} \times 10^3 \text{ mm}^3$$

$$\therefore \quad f_b = \frac{56.25 \times 10^3}{31.25 \times 10^3} \times 6 = 10.8 \text{ N/mm}^2$$

$\therefore$ Maximum stress in the fibres on the face near to the line of action of the force,

$$P = f_d + f_b = 7.2 + 10.8 = 18.0 \text{ (tensile)}$$

Stress in the fibres at the outer edge

$$f = f_d - f_b = 7.2 - 10.8 = -3.6 \text{ N/mm}^2 \text{ (Compressive).}$$

Example 28.15. A C-clamp is shown in Fig. 28.21. The load P on the clamp is 22.50 kN. Clamp is made of steel casting, with h = 2b, and e = 0.15 m and there is an allowable stress of 98.0 N/mm², determine the recommended dimensions b and h.

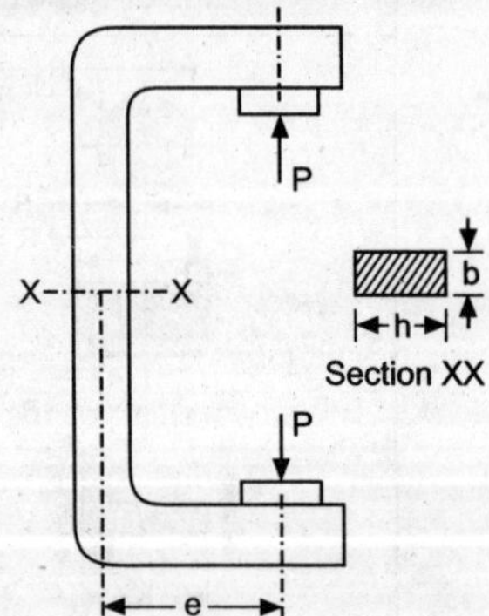

Fig. 28.21 *C-clamp*

Solution: The maximum stress will be on the inner edge and it will be tensile.

$$f = f_d + f_b$$

$$f_d = \frac{P}{A} = \frac{22.50 \times 10^3}{b \times 2b} = \frac{11.25 \times 10^3}{b^2} \text{ N/mm}^2$$

$$M_b = P \times e = 22.50 \times 0.15 = 3.375 \text{ kNm}$$

$$Z = \frac{1}{6} bh^2 = \frac{1}{6} \times b \times (2b)^2 = \frac{2}{3} b^3 \text{ mm}^3$$

$$\therefore \quad f_b = \frac{M_b}{Z} = \frac{3.375 \times 10^3 \times 10^3 \times 3}{2b^3} = \frac{5062.5 \times 10^3}{b^3} \text{ N/mm}^2$$

Now $$f = 98.0 \text{ N/mm}^2$$

$$\therefore \quad \frac{11.25 \times 10^3}{b^2} + \frac{5062.5 \times 10^3}{b^3} = 98.0$$

By hit and trial, b = 38.2 mm; say 38.5 mm

$$\therefore \quad h = 2b = 77 \text{ mm.}$$

Example 28.16. The frame of a screw clamp is shown in Fig. 28.22. The distance between the centre of screw to the inside of the frame is 15 cm. The maximum axial thrust obtainable along the centre line of the screw is 50 kN. Design a suitable frame for the clamp.

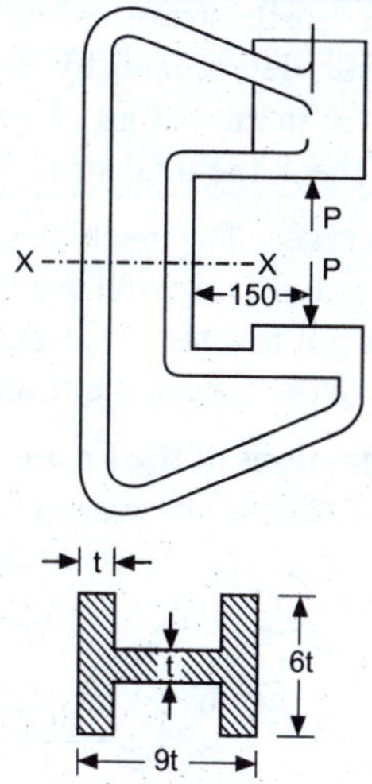

Fig. 28.22 *Frame of a screw clamp*

Solution: The frame is to be made of cast or drop forged steel. Since, steel possesses almost equal resistance to tension and compression, a symmetrical section as shown in the Fig. 28.22 is selected. The total stress over the section X-X will be the combination of direct stress (tensile) and the bending stress which will be tensile on the inner edge and compressive on the outer edge.

Area of the section $= 2 \times 6t \times t + 7t \times t = 19t^2$ mm^2

Section modulus, $Z = \frac{1}{12}[\{6t \times (9t)^3\} - (5t \times (7t)^3)] \times \frac{1}{4.5t} = 49.3\, t^3$ mm^3

Direct stress on the section, $f_d = \frac{\text{Load}}{\text{Area}} = \frac{50 \times 1000}{19t^2}$ N/mm^2 (tensile)

Bending stress, tensile at the inner edge,

$$f_b = \frac{M_b}{Z}$$

Now $M_b = 50\,(150 + 4.5t) \times 10^{-3}$ kNm

$$\therefore \quad f_b = \frac{50 \times 1000(150 + 4.5t)}{49.3t^3}$$

∴ Maximum tensile stress at the inner edge, will be

$$f_t = f_d + f_b = \frac{50 \times 1000}{19t^2} + \frac{50 \times 1000(150 + 4.5t)}{49.3t^3}$$

Let the suitable safe value for f_t be 125 N/mm^2, since it is only a case of one-directional stress.

$$\therefore \quad \frac{50 \times 1000}{19t^2} + \frac{50 \times 1000(150 + 4.5t)}{49.3t^3} = 125$$

By trial and error, $t = 12.5$ mm

∴ Overall dimensions of the frame section will be $9t \times 6t$, i.e., 112.5 mm × 75 mm

28.2.1 Machine Frames

The function of a machine frame is to support, guide and hold in accurate alignment all the moving members of an operating machine.

A machine frame is required to resist:

1. The forces transmitted by the machine in doing its work.
2. The acceleration effects of the moving parts.

The two basic requirements of a machine frame are:

(*i*) Strength to resist the induced forces or stresses.
(*ii*) Stiffness or rigidity to maintain accurate alignment.

A typical frame for a punching machine is shown in Fig. 28.23. The stresses induced at any section of the frame are of a complex nature and can be determined accurately by mathematical analysis only for simple forms of frames. As already discussed, the total stress over any section will be the algebraic sum of the direct stress and the bending stress due to eccentricity of load. Due to the bending moments, the fibres on the inner edge will be

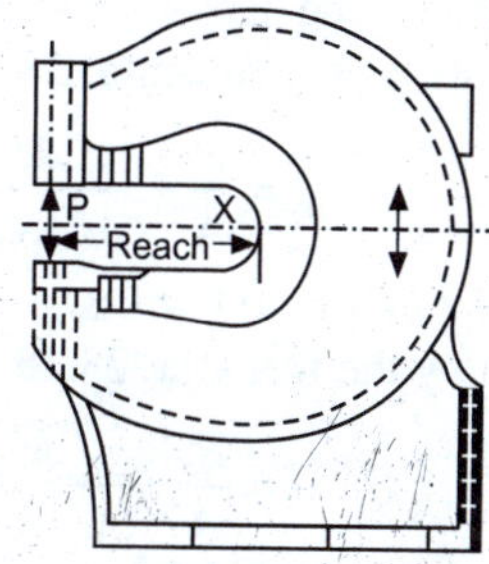

Fig. 28.23 *Punch*

under tensile stress and those on the outer edge under compressive stress. The total stress over any fibre shall be determined by means of formulas (28.11, 28.12) or (28.14 and 28.15). The usual procedure adopted in the design of machine frames is to assume some suitable dimensions for the frame section and then check the total stress induced in any section of the frame.

Materials: The machine frames are generally of intricate shapes and so, they are usually cast. Cast iron is used extensively because it is cheap, easily machined and finished and it offers good wear resistance. But it has got low tensile strength and is brittle. Therefore, for heavy duty machines, frames are made of cast steel. Now a days, the frames are also being made from simple rolled steel shapes assembled by welding.

Proportions of the Frame: The most commonly used sections for the machine frames are: "Box", "I" and "T" sections. These are shown in Fig. 28.24 with their common dimensional proportions. For box section,

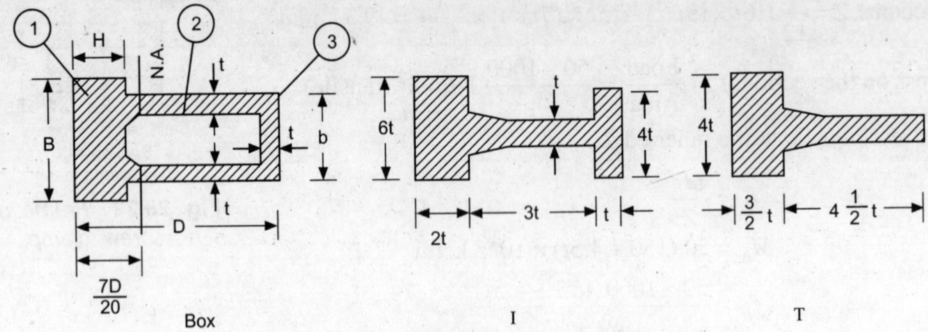

Fig. 28.24 *Section for machine frames*

If D = depth of the section, then

$B = 0.5\,D$ to D with an average value of $0.75\,D$

$b = 0.6\,B$ to B, $H = 0.2\,D$, $t = 0.06\,D$

If the frames are made of cast iron, for which the tensile strength is much less than its compressive strength, more metal is provided on the tension side as shown in Fig. 28.23. This will result in great saving of material and also in making the tensile and compressive stresses more in proportional to the strength of the material. If the frames are made of steel, a symmetrical section or a modified I-section may be adopted. When the frame is subjected to flexure combined with torsion, as in a shear frame, box section is superior. Such a frame is more pleasing and facilitates the mounting of brackets and other attachments. When the predominating load is flexure, and there are few attachments to be provided, I-section will be lighter and cheaper as compared to a box section frame.

Example 28.17. The frame of a 100 kN punch press is shown in Fig. 28.25. Determine the maximum stress in the frame.

Solution: The maximum stress in the inner fibres will be given by the formula, 28.14, i.e.,

$$f = \frac{P}{A} + \frac{M_b c_i}{Ae(R_n - c_i)}$$

and the combined stress in the outer fibres will be given by the formula, 28.15, i.e.,

$$f = \frac{P}{A} - \frac{M_b c_o}{Ae(R_n + c_o)}$$

Now R_n for the section is given as:

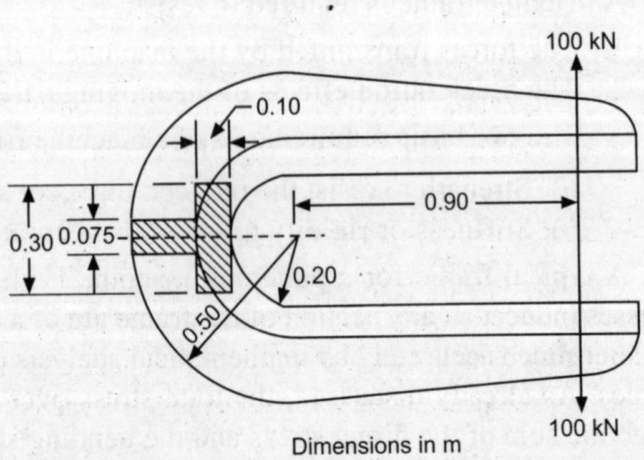

Fig. 28.25 *Frame of a punch press*

$$R_n = \frac{(B-a)t + a \times h}{(B-a)\log_e \frac{R_i + t}{R_i} + a \log_e \frac{R_o}{R_i}} \quad \text{(Table 28.5)}$$

Now $B = 0.30$ m, $a = 0.075$ m, $h = 0.30$ m, $R_i = 0.20$ m, $R_o = 0.50$ m, $t = 0.10$ m

$\therefore \quad R_n = 0.28125$ m

The distance from the centroid axis to the inner fibre can be found out by applying the relation given in Table 3.2.

$$c_1 = \frac{0.075 \times 0.30^2 + 0.225 \times 0.10^2}{2(0.075 \times 0.30 + 0.225 \times 0.10)} = 0.10 \text{ m}$$

$\therefore \quad c_2 = 0.30 - 0.10 = 0.20$ m

$\therefore \quad e = R_g - R_n = (0.20 + 0.10) - 0.28125 = 0.01875$ m

Now $\quad c_i = c_1 - e = 0.10 - 0.01875 = 0.08125$ m

and $\quad c_0 = c_2 + e = 0.20 + 0.01875 = 0.21875$ m

The bending moment referred to the neutral axis is,

$$M_b = P(l + R_n) = 100\,(0.90 + 0.28125) = 118.125 \text{ kNm}$$

$\therefore$ Compressive stress at the outer fibre is

$$f_o = \frac{M_b \times c_o}{A \times e(R_n + c_o)}$$

Now $\quad A = 0.30 \times 0.10 + 0.20 \times 0.075 = 0.045 \text{ m}^2$

$$\therefore \quad f_o = \frac{118.125 \times 1000 \times 0.21875}{0.045 \times 0.01875 \times 0.50} = 61.25 \text{ MPa (Compressive)}$$

Tensile stress at the inner fibres is,

$$f_i = +\frac{M_b \times c_i}{A \times e(R_n - c_i)} = +\frac{118.125 \times 1000 \times 0.08125}{0.045 \times 0.01875 \times 0.20} = +\,56.8 \text{ MPa (Tensile)}$$

Now combined stress in the inner fibres $= f_d + f_i$

$$\text{Now} \quad f_d = \frac{P}{A} = \frac{100 \times 1000}{0.045} = 2.22 \text{ MPa (Tensile)}$$

$\therefore$ Combined stress in the inner fibres = 2.22 + 56.8 = 59.02 MPa (Tensile)

Combined stress in the outer fibres = 2.22 – 61.25 = – 59.03 MPa (Compressive).

Example 28.18. The frame of a punching and shearing machine is to be designed. The reach or thrust is 0.60 m. The depth of the frame may be assumed to be as 0.75 m and the centre of curvature is 0.125 m from the inner edge. The punching capacity is a hole 25 mm in a diameter, in metal 19 mm thick. The shearing capacity is 150 by 22 mm flats, and 44 mm rounds. The frame is to be cast of close grained iron having ultimate tensile and compressive strengths of 175 and 700 MPa respectively. The shearing strength of the steel is taken as 336 MPa. The maximum unit tensile stress in the frame will be limited to 35 MPa and the maximum unit compressive stress is 140 MPa.

Solution: The force needed to punch a hole in a plate is given by the relation $\pi d t f_{su}$.

Let us design a box section for the frame. The depth of the frame is given as equal to 0.75 m. Referring to Fig. 28.24.

So, $\quad D = 0.75$ m

The other dimensions can be suitably assumed

$B = 0.5\,D$ to D, say $0.8\,D = 0.60$ m

$b = 0.6\,B$ to B, say $0.8\,B = 0.48$ m

$H = 0.2\,D = 0.15$ m

$t = 0.06\,D = 0.06 \times 0.75 = 0.045$ m say 0.05 m

Now the position of the centroid may be determined by taking the moments about the *B* edge and neglecting the fillets. So the distance of the centroid from the *B* edge

$$= \frac{(0.60 \times 0.15) \times 0.075 + 2 \times (0.55 \times 0.05) \times 0.025 + 0.48 \times 0.05 \times 0.725}{0.09 + 0.055 + 0.024} = 0.2812 \text{ m}$$

$\therefore$ a = eccentricity of the load from the centroid = 0.60 + 0.2812 = 0.8812 m

$$P = \pi \cdot d \cdot t \cdot f_{su}$$

In practice, the force is taken as 25% more than calculated above as this depends upon the clearance between the punch and the die.

$$\therefore \quad P = 1.25\, \pi \cdot d \cdot t \cdot f_{su}$$

Now $d = 0.025$ m, $t = 0.019$ m, $f_{su} = 336$ MPa

$$\therefore \quad P = 1.25\, \pi \times 0.025 \times 0.019 \times 336 \times 10^6 = 626 \text{ kN}$$

To reduce the force required for shearing flat plates, the shear blades are bevelled. This is done to reduce the force on the blade. The force required for shearing is given by the relation:

$$\boldsymbol{P} = \frac{0.225 \cdot f_{su} \cdot t^2}{\tan \alpha}$$

Now α varies from 5 to **10** degree, let us take $\alpha = 8$ degree, $t = 0.022$ m

$$\therefore \quad \boldsymbol{P} = \frac{0.225 \times 336 \times 10^6 \times (0.022)^2}{\tan 8^\circ} = 259.5 \text{ kN}$$

Shearing force for the circular bar, $P = \frac{\pi}{4} \times (0.044)^2 \times 336 \times 10^6 = 510$ kN

$\therefore$ Maximum force is due to punching operation,

So, $P = 626$ kN for design purposes.

Next, the factor A' should be determined,

For rectangle (1), $A' = R \int \frac{dA}{R - y}$

Now $R = 0.125 + 0.2812 = 0.4062$ m

$$\therefore \quad A' = 0.4062 \int_{0.2812}^{0.1312} \frac{0.60 d_y}{0.4062 - y}$$

y is taken as positive since it is measured towards center of curvature.

$$\therefore \quad A' = 0.4062 \times 0.60 \log_e \frac{0.4062 - 0.1312}{0.4062 - 0.2812} = 0.24372 \log_e \frac{0.27}{0.12} = 0.1975 \text{ m}^2$$

For rectangle (2), $A' = 2 \times 0.4062 \int_{0.1312}^{-0.4188} \frac{0.05 d_y}{0.4062 - y}$

$$= 0.04062 \log_e \frac{0.4062 + 0.4188}{0.4062 - 0.1312} = 0.04062 \log_e \frac{0.825}{0.275} = 0.0446 \text{ m}^2$$

For rectangle (3), $A' = 0.4062 \int_{-0.4188}^{-0.4688} \frac{0.48 d_y}{0.4062 - y} = 0.4062 \times 0.48 \log_e \frac{0.4062 + 0.4688}{0.4062 + 0.4188}$

$$= 0.4062 \times 0.48 \log_e \frac{0.875}{0.825} = 0.01135 \text{ m}^2$$

$$\therefore \quad \text{Total } A' = 0.1975 + 0.0446 + 0.01135 = 0.25345 \text{ m}^2$$

Now maximum tension on the inner side is

$$f_t = \frac{P}{A}\left[1 + \frac{a}{R - c_1}\left(\frac{c_1}{R} \times \frac{A'}{A' - A} - 1\right)\right]$$

$$= \frac{626 \times 1000}{0.1691}\left[1 + \frac{0.8812}{0.4062 - 0.2812}\left(\frac{0.2812}{0.4062} \times \frac{0.25345}{0.25345 - 0.1690}\right) - 1\right]$$

Maximum compressive stress at the outer fibres is

$$f_c = \frac{P}{A}\left[1 + \frac{a}{R + c_2}\left(\frac{c_2}{R} \times \frac{A'}{A' - A} - 1\right) - 1\right]$$

Now $c_2 = 0.75 - 0.2812 = 0.4688$ m

$$\therefore \quad f_c = \frac{626 \times 1000}{0.1690}\left[\frac{0.8812}{0.4062 + 0.4688}\left(\frac{0.4688}{0.4062} \times \frac{0.25345}{0.25345 - 0.1690} - 1\right) - 1\right] = 9.04 \text{ MPa}$$

Both the values are within the limits, therefore, the design is O.K.

Example 28.19. The section of a C-clamp is shown in the Fig. 28.25 (a). Determine the force *P* which can be exerted by the screw if the maximum tensile stress in the clamp is not to exceed 140 N/mm². Dimensions in the Fig. are in mm.

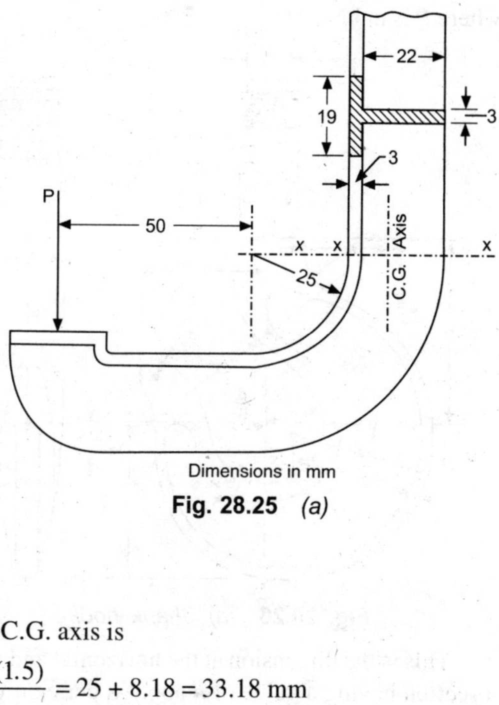

Fig. 28.25 *(a)*

Solution:

(*i*) It is clear from the Fig. 28.25 (a) that the maximum tensile stress will occur at point *x* on the section X-X.

(*ii*) The maximum tensile stress is given by the formula 28.14, as equal to

$$= \frac{P}{A} + \frac{M_b c_i}{Ae(R_n - c_i)}$$

(*iii*) $$R_n = \frac{(B - a)\cdot t + a \cdot h}{(B - a)\log_e \dfrac{R_i + t}{R_i} + a \log_e R_o / R_i}$$

Now $B = 19$, $a = 3$; $t = 3$; $h = 25$; $R_i = 25$; $R_o = 50$

$\therefore \quad R_n = 31.55$ mm

(*iv*) Now R_g = distance from centre of curvature of the C.G. axis is

$$= 25 + \frac{(22)(3)(11 + 3) + (19)(3)(1.5)}{(22)(3) + (19)(3)} = 25 + 8.18 = 33.18 \text{ mm}$$

(*v*) $\therefore \quad e = R_g - R_n = 33.18 - 31.55 = 1.63$ mm

(*vi*) $c_i = 31.55 - 25 = 6.55$ mm

(*vii*) $A = 22 \times 3 + 19 \times 3 = 123 \text{ mm}^2$

(*viii*) M_b = B.M. about C.G. $= P(50 + 33.18) = 83.18P$

(*ix*) $$\therefore \quad 140 = \frac{P}{123} + \frac{83.18P \times 6.55}{123 \times 1.63 \times (31.55 - 6.55)}$$

$$= 0.00813P + 0.1087P$$

$$P = 1200 \text{ N.}$$

28.2.2 Crane Hook

Two forms of crane hooks used in practice are shown in Fig. 28.26. The bed diameter C and the throat width E must be sufficient to accommodate the necessary slings. However, these dimensions should not be more than which are necessary, otherwise the hook will look clumsier. The bed diameter C is given by the relation:

$$C = x\sqrt{P} \text{ mm} \qquad ...(28.16)$$

where P = load in kN and x = constant varying 12 to 24.

The value of x should be taken as small as possible for the economy of the material.

The most economical section for the crane hook is modified triangular section or trapezoidal section as shown in Fig. 28.27. The depth of the section 'd' is given by the relation:

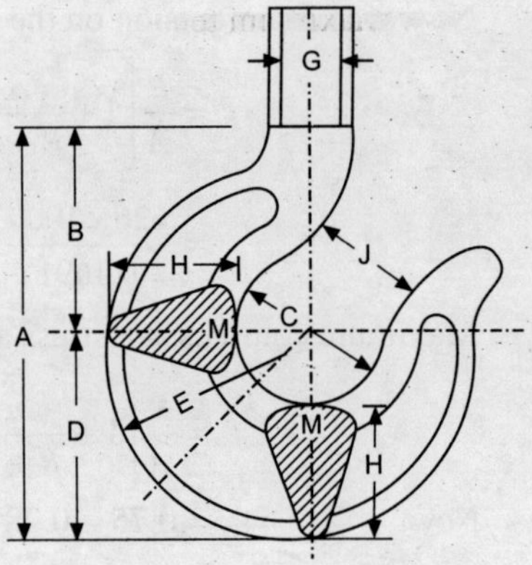

Fig. 28.26 *Crane hook*

$$d = \left(10\sqrt{P} + \frac{C}{10}\right) \text{ mm} \qquad ...(28.17)$$

where P is in kN.

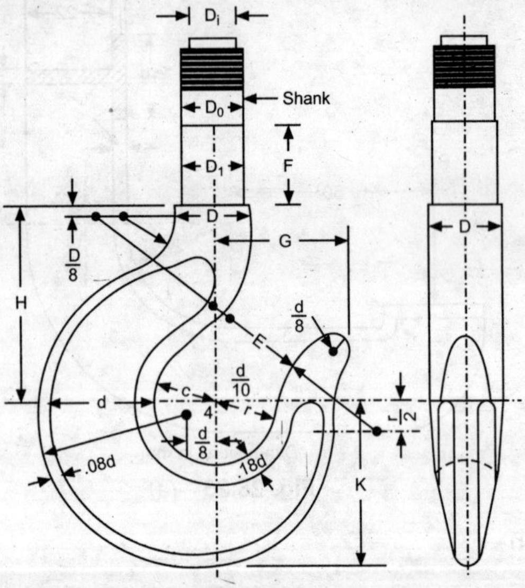

Fig. 28.26 *(a) Shank hook*

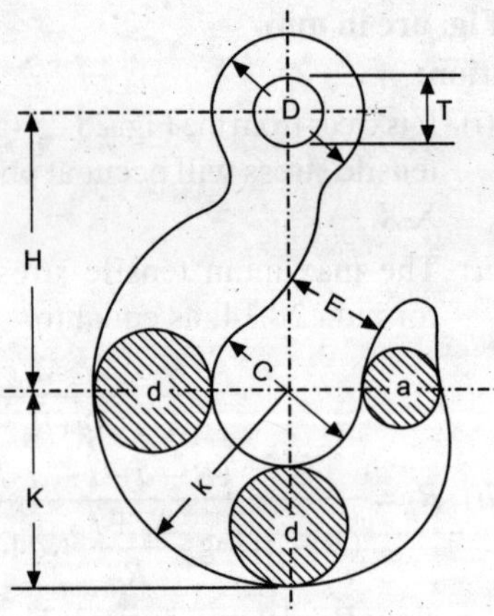

Fig. 28.26 *(b) Eye hook*

This is the dimension at the horizontal and vertical lines of the hook. At a plane midway between these a section having a value of d about 8 per cent greater is used.

The other dimensions of the crane hook are given empirically as below:

Shank Hook: E = Throat diameter = 0.75 C

$$G = r + \frac{2}{3}d; \quad H = r + 0.65d + 0.70\,D$$

K = 1.36 C for trapezoidal section, = 1.32 C for circular section.

Eye Hook: E = 0.75 C; H = 2 C; T = 0.5 C; D = 1.3 C; K = 1.32 C

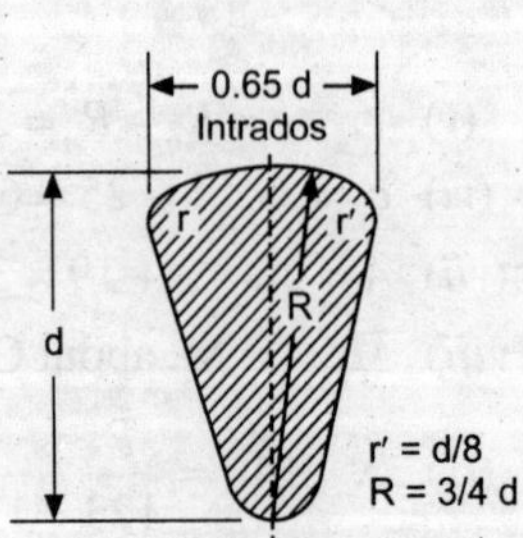

Fig. 28.27 *Modified triangular section*

Crane hooks are generally forged, either from high grade wrought iron or low carbon open hearth steel. The design of the crane hooks is based on the relations for curved beams. The maximum permissible stress for wrought iron should be about 140 N/mm^2 and for steel about 167.5 N/mm^2.

The shank of the hook in the case of hook shank is checked for the tensile stress in the threaded portion. The unit tensile stress is given as:

$$f_t = \frac{P}{\pi/4\, D_i^2} \qquad \text{...(28.18)}$$

The safe stress f_t should not exceed 50 N/mm^2.

The minimum height of the hook nut is determined from the permissible compressive stress in the threads given as:

$$h = \frac{4 \cdot P \cdot p}{\pi(D_o^2 - D_i^2) \times f_c} \qquad \text{...(28.19)}$$

where p = thread pitch

f_c = safe compressive stress = 30.0 to 35.0 N/mm^2 (steel on steel).

The cylindrical portion of the hook shank entering the hole in the cross piece is subjected to tensile stress. But due to accidental misalignment, this portion may also be subjected to bending also. Therefore, the safe stress is reduced correspondingly, the stress being given as,

$$f_t = \frac{P}{(\pi/4)D_1^2}$$

f_t is $\leq$ 50.0 N/mm^2

The height of the straight portion is given empirically:

$$F = 9.525\sqrt{P} \text{ mm}$$

where P is the load in kN.

Example 28.20. Design a crane hook with the useful load lifting capacity of the crane as 50 kN. The weight of the hook with grabbing tongs is 10 kN.

Solution: Total load, $P = 50 + 10 = 60$ kN

Bed diameter, $C = (12 \text{ to } 24)\sqrt{P}$ mm say $16\sqrt{P} = 16\sqrt{60} = 123.9$ mm, say 125 mm

The depth of the modified triangular section is,

$$d = 10\sqrt{P} + \frac{C}{10} = 10\sqrt{60} + \frac{125}{10} = 90 \text{ mm}$$

$\therefore$ Width of the section at the intrados = $0.65d = 0.65 \times 90 = 58.5$ mm.

(*a*) Checking the total stress in the section:

Radius of the intrados $= \frac{C}{2} = 62.5$ mm

Radius of extrados $= 62.5 + d = 62.5 + 90 = 152.5$ mm

$\therefore$ R_g = radius of curvature at the centroid

$$= 62.5 + \frac{1}{3}d = 62.5 + \frac{1}{3} \times 90 = 62.5 + 30 = 92.5 \text{ mm}$$

Now
$$R_n = \frac{(1/2)b_i h}{\frac{b_i r_o}{r_o - r_i}\log_e \frac{r_o}{r_i} - b_i}$$

Now $b_i = 58.5$ mm; $h = 90$ mm

$r_i = 62.5$ mm $r_o = 152.5$ mm

$\therefore$ $R_n = 88$ mm

$\therefore$ $e = R_g - R_n = 92.5 - 88 = 4.5$ mm

Stress on the inner fibre, $f_i = \frac{M_b c_i}{A.e.(R_n - c_i)}$

M_b = Bending moment referred to neutral axis = $60 \times 88 = 5280$ Nm

$$A = 25.8 \times 10^2 \text{ mm}^2$$

$$c_i = c_1 - e = \frac{1}{3} \times 90 - 4.5 = 30 - 4.5 = 25.5 \text{ mm}$$

$$\therefore \quad f_i = \frac{5280 \times 1000 \times 25.5}{25.8 \times 10^2 \times 4.5 \times (88 - 25.5)} = 177 \text{ N/mm}^2 \text{ (Tensile)}$$

$$\text{Direct stress}, f_d = \frac{60 \times 1000}{25.8 \times 100} = 23.25 \text{ N/mm}^2 \text{ (Tensile)}$$

$$\text{Total stress the intrados} = 177 + 23.25 = 200.25 \text{ N/mm}^2$$

which is somewhat higher than the safe allowable stress for open hearth steel hooks. Due to this, a better quality of C-steel will have to be employed. In the above analysis, the load line passes through the centre of curvature.

(*b*) To find the diameter of the threaded portion of the shank:

$$f_t = \frac{P}{(\pi/4)D_i^2}$$

Let $f_t = 50 \text{ N/mm}^2$

$$\therefore \quad D_i^2 = \frac{60 \times 1000}{(\pi/4) \times 50} = 15.28 \times 10^2 \text{ mm}^2$$

$$\therefore \quad D_i = 39.1 \text{ mm}$$

A metric Vee thread with major diameter of 48 mm is selected, i.e., *M* 48 thread with pitch of 5 mm.

(*c*) Height of the hook nut is given as:

$$h = \frac{4 \cdot P \cdot p}{\pi(D_o^2 - D_i^2) \times f_c}$$

Let $f_c = 30 \text{ N/mm}^2$

$$\therefore \quad h = \frac{4 \times 60 \times 1000 \times 5}{\pi(48^2 - 41^2) \times 30} = 20.5 \text{ mm, say 21 mm}$$

(*d*) The diameter of the cylindrical portion of the hook shank is given as:

$$f_t = \frac{P}{(\pi/4)D_1^2}$$

Now $f_t \le 50 \text{ N/mm}^2$

Let $f_t = 40 \text{ N/mm}^2$

$$\therefore \quad D_1^2 = \frac{60 \times 1000 \times 4}{\pi \times 40} = 19.1 \times 10^2 \text{ mm}^2$$

$$\therefore \quad D_1 = 43.6 \text{ mm}$$

Let us make it more than the dimension D_o, so, let $D_1 = 50$ mm.

EXERCISES

1. A C-clamp frame is shown in Fig. 28.28. The section of the frame is rectangular, 10 by 50 mm. The centre line of the screws is 75 mm from the neutral axis of the frame. Find the load to be applied by the screws to produce a tensile stress of 85 N/mm^2. Also, calculate the compressive load stress at this load.
2. A steel link has a cross-section 12.5 mm square. To provide clearance, it is to be replaced by an offset link. The inner edge of the offset link is 25 mm from the line of action of the load. The depth of this link is 50 mm. Determine the width required so that the strength of the offset link is equal to that of the straight link.

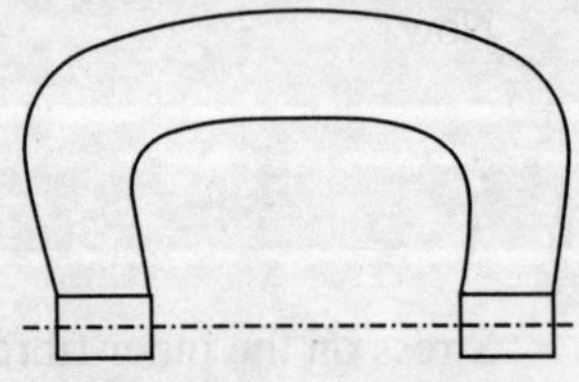

Fig. 28.28 *C-clamp*

3. A small punch has a capacity sufficient to punch a 16 mm hole in a 6.4 mm steel plate whose ultimate strength in tension is 630 N/mm^2. The frame is of the shape of a C-clamp with a T cross-section. The dimensions (Table 28.5) are $B = 150$ mm; $t = 30$ mm; $a = 25$ mm. The centre line of the punch is 160 mm from the inner surface of the flange. Determine the depth h of the section if the stress is not to exceed 98 N/mm^2.
4. Determine the maximum stress induced in the horizontal section of a punch frame similar to shown in Fig. 28.25. The section of the frame is 'T' section, with dimensions (Table 28.5) as $B = 100$ mm; $t = 25$ mm; $a = 25$ mm; $h = 175$ mm. The centre of curvature of the inner edge is 50 mm. The load is 40 kN.
5. Determine the maximum tensile stress induced in the clamp shown in Fig. 28.29. The effect of the curvature is to be considered. The load P is 60 kN. The section of the frame is 90 mm by 90 mm.

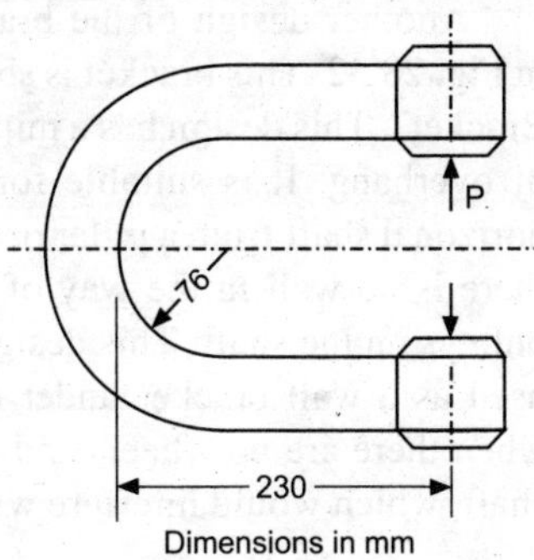

Fig. 28.29 *Clamp*

6. The bed diameter of a crane hook is 95 mm. The section of the hook is trapezoidal with depth equal to 190 mm. The width of the section at the larger end is 125 mm and at the smaller end is 89 mm. The load on the hook is 135 kN. Determine the maximum unit stresses in tension and compression.
7. Design a crane hook for a lifting capacity of 30 kN with 50 percent over load.
8. A crane hook has a round cross-section with diameter 95 mm. The bed diameter is 125 mm. (a) Determine the load which will produce a maximum stress of 122.5 N/mm^2 in the inner fibres; (b) determine the load that will produce the corresponding stress in the outer fibres.

28.3 WALL BRACKETS, HANGERS AND WALL BOXES

When the shafts in a workshop or a factory run parallel to and near a wall, then to hold the bearings which support them, wall brackets are employed. There are many forms of wall brackets but their design principles are mostly similar. They consist of a horizontal plate which is casted with a vertical plate and a web. The web supports the horizontal plate. The horizontal plate supports the bearing whereas the vertical plate is attached to the wall or a pillar by means of bolts. The shape of the brackets usually resembles a cantilever. The common material for the wall brackets is Cast Iron.

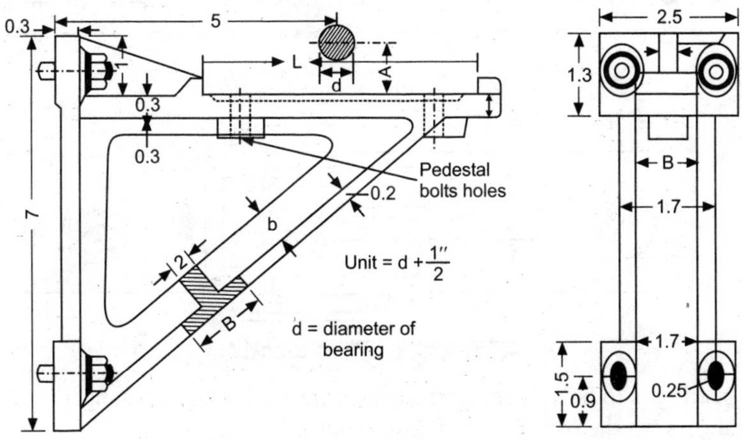

Fig. 28.30 *Wall bracket for pedestal bearing*

Wall brackets can be made in two designs. In Fig. 28.30 which shows the wall bracket for a pedestal bearing, the bearing is bolted to the horizontal plate of the bracket. The dimensions 'B' and 'L' are made to suit the dimensions of the pedestal bearing which the bracket is to carry and so also the distance between the holes for bolts which secure the bearing to the bracket.

In Fig. 28.31 the bearing is cast integral with the bracket.

Another design of the bracket is shown in Fig. 28.32. This bracket is shown as "Pillar Bracket". This design has a minimum amount of overhang. It is suitable for supporting a horizonal shaft from a pillar or column where there is no wall in the way of the wheels or pulleys on the shaft. This design may also be used as a wall bracket under the conditions when there are no wheels and pulleys on the shaft, which would interfere with wall.

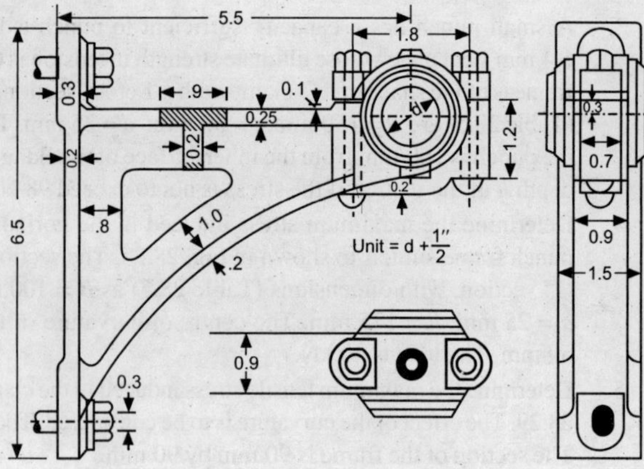

Fig. 28.31 *Wall bracket and bearing*

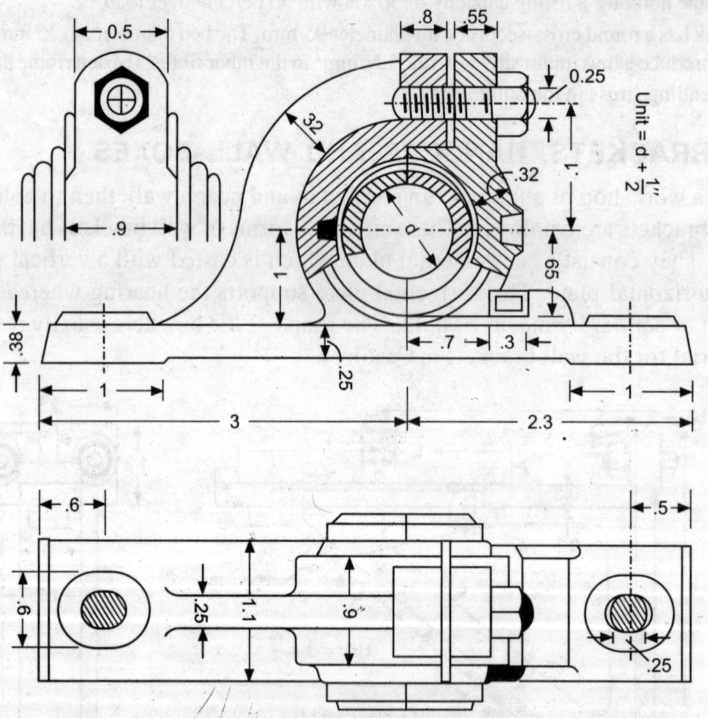

Fig. 28.32 *Pillar bracket*

When a shaft bearing is to be suspended from a ceiling or a beam, a "hanger" is used. There may be various forms of hangers but the two most common forms are:

'J-hanger' and 'U-hanger'. In the case of a 'U-hanger' Fig. 28.33, the bearing is supported between two arms. The dimensions K, L and A depend on the particular pedestal bearing to be carried by the hanger. Usually, the dimension 'H' is from four to eight times the unit. The average is about six. Usually 'T' section is used for this hanger. 'J' hanger is shown in Fig. 28.34. The bearing is secured by a bolt and a key.

When a shaft runs from one room to the other, then to support the length of the shaft which runs between the two rooms, the bearing is fitted in a structure called as "Wall Box". The box is built into

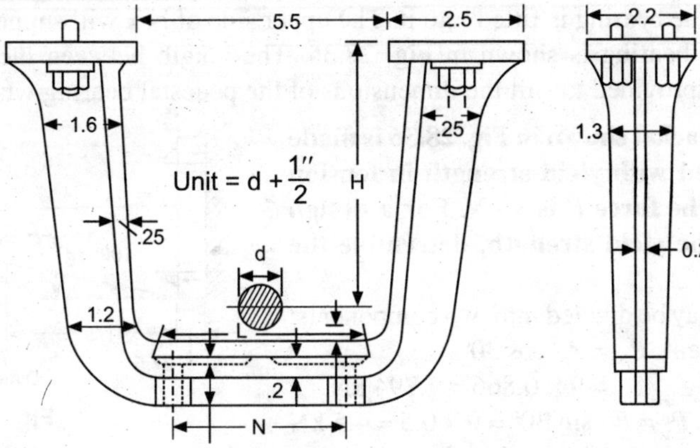

Fig. 28.33 *U-hanger*

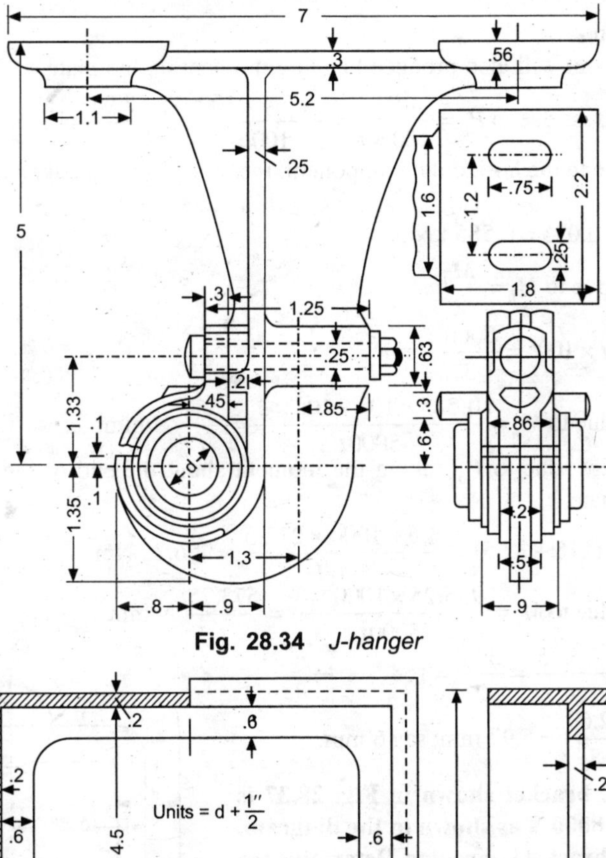

Fig. 28.34 *J-hanger*

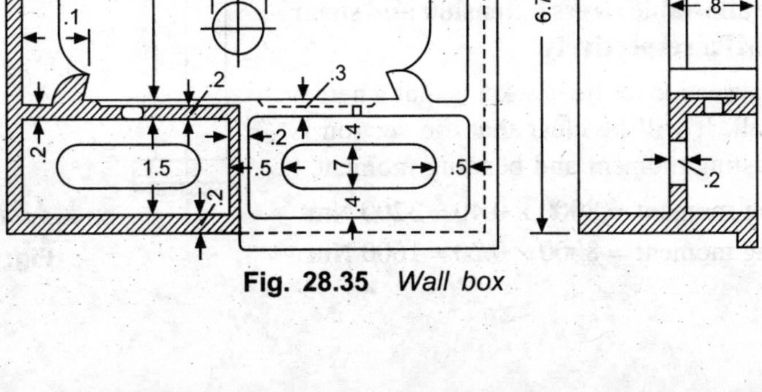

Fig. 28.35 *Wall box*

the separating wall and the bearing is fitted into it. The upper side of box will support the wall above it. A wall box for a pedestal bearing is shown in Fig. 28.35. The length between the joggles and the wall dimensions 'B' must be provided to suit the dimensions of the pedestal bearing which is to be supported.

Example 28.21. The bracket shown in Fig. 28.36 is made of carbon steel as rolled with yield strength in tension equal to 336 N/mm². The force *P* is 9 kN. For a design factor of 2 based on the yield strength, determine the thickness '*t*' of the arm.

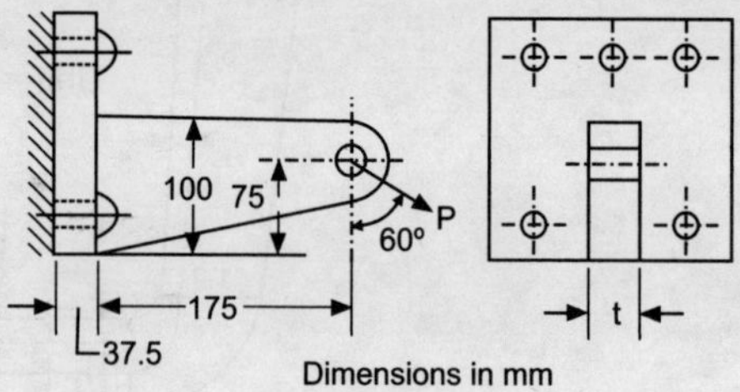

Fig. 28.36

Solution: The force *P* may be divided into two components:

Horizontal component, $P_h = P \cdot \cos 30°$

$= 9 \times 0.866 = 7.794$ kN

Vertical component, $P_v = P \cdot \sin 30° = 9 \times 0.5 = 4.5$ kN

The horizontal component will produce on the plate:

(*i*) Direct tension

(*ii*) Bending tension.

The vertical component will also produce bending moment on the plate.

Now direct tensile stress due to $P_h = \dfrac{P_h}{100 \times t} = \dfrac{7.794 \times 1000}{100 \times t} = \dfrac{77.94}{t}$ N/mm²

Bending moment due to the horizontal component, the moment being taken about the bottom edge of the plate, is

$$M_h = P_h \times 0.075 = 0.585 \text{ kNm}$$

$\therefore$ Bending stress due to $M_h = \dfrac{M_h}{Z}$

Now $\quad Z = \dfrac{1}{6} \times t \times 100^2 = \dfrac{5000}{3} t \text{ mm}^3$

$\therefore$ Bending stress due to $M_h = \dfrac{0.585 \times 10^3 \times 10^3 \times 3}{5000t} = \dfrac{351}{t}$ N/mm²

Now bending moment due to P_v, taking the moment about the end of the rivet plate for stronger section for the bracket plate,

$$M_v = P_v \times (175 + 37.5) = \frac{4.5 \times 1000 \times 212.5}{1000} = 956.25 \text{ Nm}$$

$\therefore$ Bending stress due to $M_v = \dfrac{956.25 \times 1000 \times 3}{5000t} = \dfrac{573.75}{t}$ N/mm²

$$\therefore \quad \frac{77.94}{t} + \frac{351}{t} + \frac{573.75}{t} = \frac{336}{2} = 168$$

$$\therefore \quad t = \frac{1002.69}{168} = 5.97 \text{ mm; say 6 mm.}$$

Example 28.22. A.M.S. bracket shown in Fig. 28.37 is subjected to a pull, *P* = 8000 N as shown in the diagram. The cross-section of the bracket is circular. Determine the dimensions '*d*' if the allowable stress in tension and shear are 60 MPa and 40 MPa respectively.

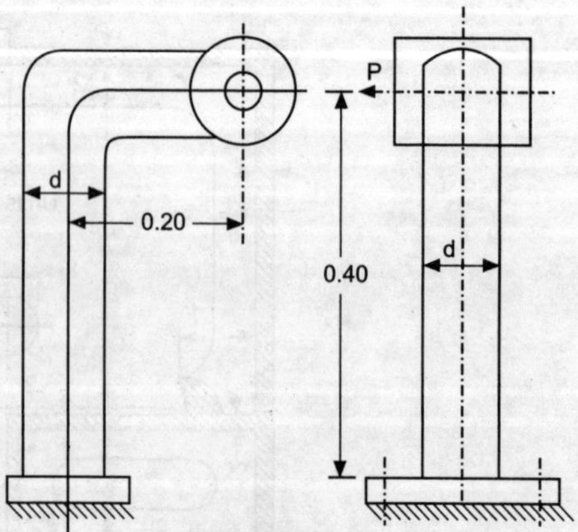

Fig. 28.37

Solution: The critical section of the bracket is that where it is attached to the wall. It will be clear that the section is subjected to both twisting moment and bending moment.

Bending moment $= 8000 \times 0.40 = 3200$ Nm

Twisting moment $= 8000 \times 0.20 = 1600$ Nm

Now, equivalent twisting moment,

$$T_e = \sqrt{3200^2 + 1600^2} = 3577.7 \text{ Nm}$$

Equivalent bending moment,

$$M_e = \frac{1}{2}\left[3200 + \sqrt{3200^2 + 1600^2}\right] = \frac{1}{2}(3200 + 3577.7) = 3388.85 \text{ Nm}$$

$$\therefore \quad T_e = \frac{\pi}{16} d^3 \times f_s$$

$$\therefore \quad d^3 = \frac{3577.7 \times 16}{\pi \times 40 \times 10^6} = 0.456 \times 10^{-3} \text{ m}^3$$

Given $\quad d = 0.078$ m

Also $\quad M_e = \frac{\pi}{32} \times d^3 \times f_t$

$$\therefore \quad d^3 = \frac{3388.85 \times 32}{\pi \times 60 \times 10^6} = 0.575 \times 10^{-3} \text{ m}^3\text{, gives } d = 0.084 \text{ m}$$

So, let us take $d = 0.084$ m, that is 84 mm.

PROBLEMS

1. Determine the thickness of the plate for the bracket shown in Fig. 28.38 where $P = 36$ kN.
2. A pulley bracket is shown in Fig. 28.39. The weight of the pulley and the bracket is $F = 800$ N and the load on the belt is $P = 27$ kN. Take allowable stress in tension 35 MPa.

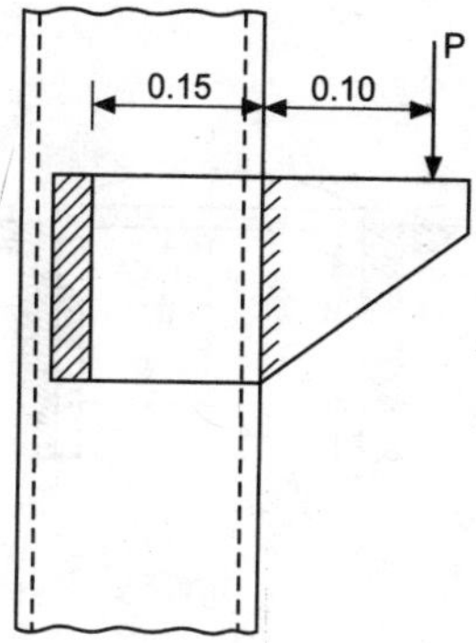

Fig. 28.38

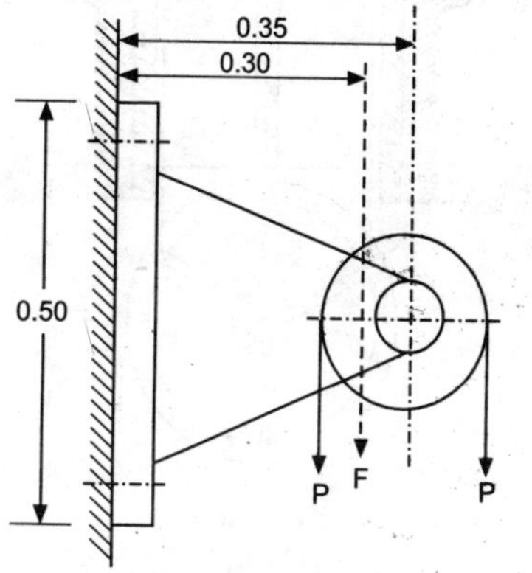

Fig. 28.39

28.4 SOME BOILER-MOUNTINGS

28.4.1 Safety Valves

Steam boilers are designed to sustain a definite internal pressure per sq. mm of internal surface. This pressure is known as "test pressure". But in actual practice, for the sake of safety a pressure lower than test pressure is allowed to be built up inside the boiler. This pressure is termed as "Working pressure". Hence a device should be fitted to the steam chest to prevent an undue rise in the pressure since too great an internal pressure might.

(*i*) cause explosions.
(*ii*) damage the plant itself.
(*iii*) spoil the material.
(*iv*) hurt the operators.

Such a device is known as "Safety valve". In principle it consists of an exit tube on top of which is a valve which is so loaded by means of weights or springs that it confines the steam and prevents its waste, but at the same time does not allow a dangerous built up of steam inside the boiler. Hence, the safety valve should be entirely automatic in its action, i.e., it should operate without the help of any human agency.

The safety valves may be grouped into the following categories:

1. Lever and weight type
2. Spring-balance type
3. Dead weight type
4. Spring loaded type.

Spring balance type has long gone out of favour. At present, the most commonly used type is the spring loaded safety valve.

Valve: A valve is a circular disc of brass, gun metal, bronze or steel, bevelled off at an angle of 45° on its lower edges. It is provided with light wings, feathers or guides on its under side, which fit into the cylindrical parts of their seats. These help to keep the valve central to its seat when raised above it. Some forms of the valves are shown in Fig. 28.40. In Fig. 28.40 (*a*) the valve has three wings or feathers while in Fig. 28.40(*b*) it has four wings. In Fig. 28.40(*c*) the feathers are of spiral or screw form. The upward fluid pressure will act on the curved surfaces and this will give a rotary motion to the valve as it rises. This will prevent the parts of the valve face from always beating on the same parts of the seat and hence there will be more uniform wear.

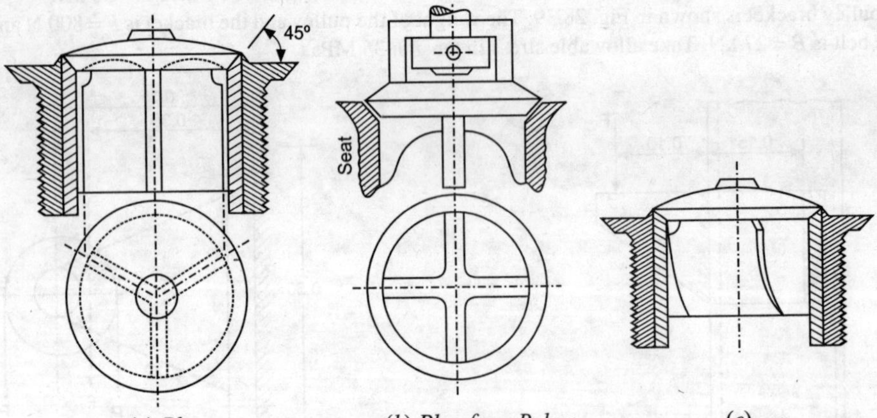

(*a*) *Plan* (*b*) *Plan from Below* (*c*)

Fig. 28.40 *Valves*

Safety Valve Seatings: The seating is the heart of any valve and so, is the starting point in any valve design. For better performance of the safety valve and to avoid its "chattering" and tilting, the following fundamental requirement should be observed. "The thrust of the spindle to be imparted to the valve member should be arranged at a point situated below the level of the seat contacting face". This fact is explained in Fig. 28.41. The valve seats are generally made of brass.

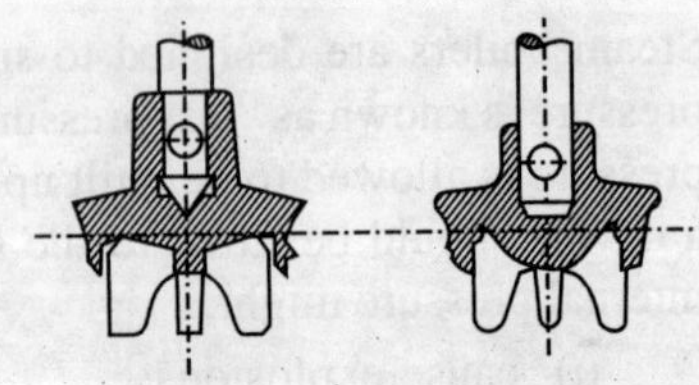

Fig. 28.41 *Methods of arranging the point of thrust in safety valve members*

Lift of Valve: The valve lift is the height to which it should be lifted to permit a free escape of fluid equal to the area of the underside of the valve, i.e, if l = lift of the valve, then

$$\frac{\pi D^2}{4} = \pi D \times l \qquad \text{...(28.20)}$$

where D = diameter of the valve opening.

In actual practice, when the valve is controlled automatically by the action of the fluid, the lift is generally much less than this. When the pressure on the valve is quite high, the lifting should not exceed about 6 mm.

Loading of Safety Valves: The total downward force on a safety valve is given as:

Total downward force = downward force opposing the upward force due to fluid pressure + downward force to produce a clamping pressure between the seating faces so as to prevent the leakage across the faces.

$$= a \cdot p + k \qquad ...(28.21)$$

where a = cross-sectional area of the inside of the seating and p = fluid pressure.

The clamping pressure is usually taken as 1.5 to 2 times the pressure of the fluid being controlled, and it will act on the valve seating. Therefore, the total downward force shall be given as:

$$P = \frac{\pi}{4} D^2 \times p + (1.5 \text{ to } 2) \times p \times (\pi D w) \qquad ...(28.22)$$

where w = width of the valve seating.

In the above equation the lower factor, i.e., 1.5 may be adopted for accurately lapped surfaces.

It has been found in practice that only few steam safety valves will settle down at the set blowing off pressure but continue to blow until the pressure has dropped appreciably. Most boiler safety valves are set to blow off at a pressure 5 percent in excess of the normal working pressure. According to Boiler Regulations, the valve shall shut down at a pressure not more than 5 percent and not less than $2\frac{1}{2}$ percent below the blowing-off pressure. For valves less than 30 mm bore, the limits shall be increased to 10 percent and 5 percent respectively. It is quite clear, therefore, that no safety valve can be expected to open and shut promptly at the set pressure.

The width of the seating edge is given as:

w = 0.02 × effective diameter of the seating for valves less than 30 mm diameter.

= 0.01 × effective diameter of the seating for valves 30 mm diameter and above.

In actual practice somewhat wider seating faces are adopted to avoid their failure due to "wire drawing" and distortion.

Minimum Aggregate Area: As already indicated a steam safety valve should not permit a dangerous increase in pressure within the boiler. For this, it should have sufficient area to discharge the steam as quickly as it is generated when the supply of steam from the boiler is stopped due to certain reasons. According to the Boiler Codes, the safety valves have been classified as under:

1. **Ordinary-lift Safety Valves:** In these valves, the valve member lifts automatically at least $D/24$.
2. **High-lift Safety Valves:** Wherein the lift is at least $D/12$.
3. **Full-lift Safety Valves:** In these valves, the valve member lifts automatically a distance such that the discharge area around the edge of the seating is equal to the area through the valve orifice, making due allowance for the loss of area due to guides or other obstructions.

In the Boiler Regulations, the following rules are given for determining the minimum aggregate areas of safety valves:

(*a*) **Saturated Steam:** The minimum aggregate area of the orifices through the seating of the safety valves on each boiler is given by the formula:

$$A = \frac{E}{C \cdot p} \qquad ...(28.23)$$

where A = The aggregate area in mm^2 of the orifices through the seatings of the valves for ordinary and high-lift valves

= For full-lift safety valves, the net area in mm^2 through the seats after deducting the area of other obstructions when the valves are fully lifted

E = Total peak load evaporation in kg/hr for which the boiler is specified.

p = Higher pressure to which any safety valve is to be set to lift in bar absolute (1 bar = 10^5 N/m^2)

C = A constant, whose values are given below:

Type of Valve	C
Ordinary lift	0.05
High-lift	0.10
Full-lift	0.24

Where two valves are loaded by a single spring, the areas shall be increased by 50 percent.

(*b*) **Super-heated Steam:** When the valves have to pass superheated steam, the area shall be increased as given below:

$$A_s = A \cdot \sqrt{1 + \frac{2.7T}{1000}} \qquad \text{...(28.24)}$$

where A_s = Minimum aggregate area in mm^2 for superheated steam.

A = Minimum aggregate area in mm^2 for saturated steam.

T = Degree of superheated in °C.

According to a clause in Boiler Regulations: The valve shall be designed so that the maximum peak load evaporation for which the boiler is specified will be completely discharged with a rise in pressure of not more than 10 percent of the safety valve blow-off pressure. This permissible rise is known as "Accumulation".

The minimum aggregate area of locked up safety valves of ordinary spring loaded type can also be obtained from the following formula which is based on the heating surface:

$$A = \frac{HK}{p} \qquad \text{...(28.25)}$$

where A = Aggregate area of locked-up safety valves in m^2

H = Total heating surface of boiler in m^2.

p = Working pressure of boiler in N/m^2, aboslute.

K = A constant

= 62.5 for coal fired cylindrical boilers situated in open stroke-holds.

= 75 for coal-fired cylindrical boilers with closed strokeholds and forced drought.

= 75 for oil-fired cylindrical boilers.

= 55 for coal fired water tube boilers, with natural drought.

= 62.5 for coal fired water-tube boilers, forced drought.

= 62.5 for oil-fired water tube boilers.

28.4.2 Lever Safety Valves

The lever safety valve is one of the simplest of all safety valves. The loading is attained by a pendulous dead weight or a cheese weight which in conjunction with a lever mechanism produces the desired axial thrust on the spindle and hence the valve member, as shown in Fig. 28.42.

To determine the weight to be suspended at the end of the lever, consider the Fig. 28.43. The axial force, P due to the set or blowing-off pressure acting on the underside of the valve member is found out in the manner as already described, i.e.,

$$P = \text{effective area of seating} \times \text{blowing-off pressure} + \text{clamping force}$$

$$= \frac{\pi}{4} D^2 \times p + 1.5\, p \cdot \pi D w$$

To find 'F', moments are taken about the fulcrum pin (FP). The frictional resistance of the pins and other rubbing surfaces is neglected but the free movement at these points is very important. Pins should be made of non-ferrous material. However, if steel pins are preferred, bronze bushes should be provided in the lever. Neglecting friction, and for the time being, neglecting the weight of the lever, spindle, valve member, etc. Since the whole system is static equilibrium, the algebraic sum of moments about the fulcrum pin is zero, therefore,

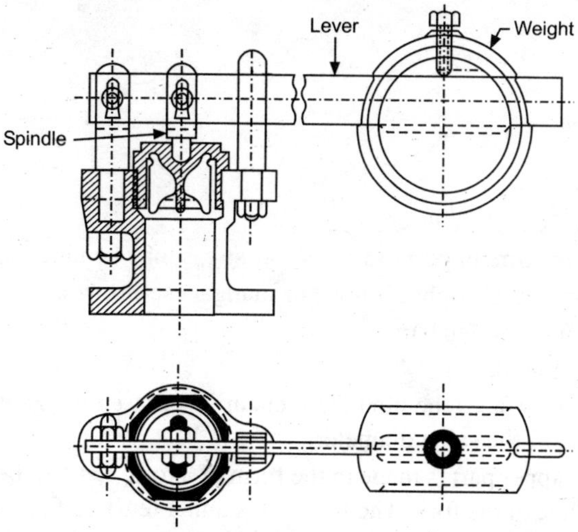

Fig. 28.42 *Lever safety valve*

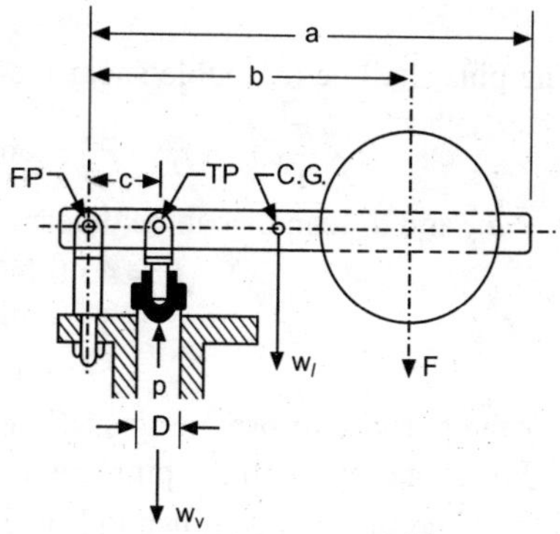

Fig. 28.43

$$P \times c - F \times b = 0 \quad \text{or} \quad F = \frac{P \times c}{b} \qquad \text{...(28.26)}$$

Now, if $\quad W_l$ = Weight of the lever,

W_v = Weight of the valve member and spindle,

then
$$P \times c - F \times b - W_l \times \frac{a}{2} - W_v \cdot c = 0$$

$\therefore$
$$F = \frac{P \times c - W_l \times \dfrac{a}{2} - W_v \times c}{b} \qquad \text{...(28.27)}$$

The weight of the lever 'W_l' can be determined by knowing the over all length of the lever and assuming its cross-section. The lever should be subsequently checked for strength. The fulcrum distance 'c' is dependent upon the outside diameter of the valve seating or the valve member. It should be kept as small as possible to obtain the lightest possible weight. The weight should be attached to the lever in such a way that it cannot be moved inadvertently.

Spindle: The spindle is subjected to a direct compressive force. The maximum compressive force on the spindle being the axial thrust on the spindle, i.e., 'P'.

$\therefore$
$$P = \frac{\pi}{4} d_1^2 \times f_c$$

where $\quad d_1$ = diameter of spindle at its lower extremity

f_c = allowable compressive stress.

In actual practice, the diameter is taken larger than 'd_1' to take into account:

(*i*) torsional stress if the valve member rotates on its seating.

(*ii*) weakening effect of the transverse cotter hole.

(*iii*) tapering of the spindle nose.

Hence, the spindle should be made as robust as related components and space will limit.

Pins: Both the fulcrum pin and the thrust pin securing the spindle to the lever will be subjected to shearing and bending forces. The shearing forces are generally greater than bending forces.

Taking moments about the thrust pin, the force on it will be given as:

$$P_{fp}\uparrow = \frac{F(b-c)}{c} \quad ...(28.28)$$

Similarly,

$$P_{tp}\downarrow = \frac{F \times b}{c} \quad ...(28.29)$$

The pins shall be in double shear therefore,

$$2 \times \frac{\pi}{4} d_{fp}^2 \times f_s = P_{fp} \quad \text{and} \quad 2 \times \frac{\pi}{4} d_{tp}^2 \times f_s = P_{tp}$$

∴ d_{fp} and d_{tp} can be can be determined. The maximum permissible shear stress may be taken as:

f_e = 45.0 N/mm² for rolled phosphor bronze or manganese-bronze.
= 31.5 N/mm² for brass, rolled bars
= 56.0 for "Monel" metal.

For the purpose of providing sufficient wearing surfaces the pin diameters are increased by about 6 mm. For economy, both the pins are usually made of the same diameter.

The linked fulcrum is shown in Fig. 28.44. The upper part is made in the form of a fork. Let 'w_f' be the width of the fork and 't_f' be the thickness of each blade of the fork. The fulcrum is under tension, therefore,

$$P_{fp} = 2 \times t_f \times (w_f - d_{fp}) \times f_t \quad ...(28.30)$$

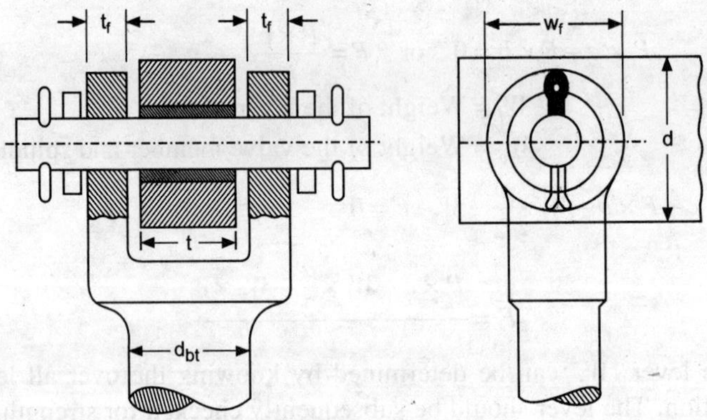

Fig. 28.44 *Linked fulcrum*

The safe stress in tension is given as:

f_t = 35.0 N/mm² for gun metal
= 70.0 N/mm² for mild steel.

Now, if we consider the bearing of the pin in the fulcrum,

$$P_{fp} = 2 \cdot t_f \cdot d_{fp} \cdot f_b \quad ...(28.31)$$

The safe bearing stress is given as:

f_b = 14.0 to 17.5 N/mm² for gun metal or phosphor bronze.

With the help of equations (28.30) and (28.31), 'w_f' and 't_f' can be determined.

The fulcrum bolt is under pure tension, therefore, its diameter at the root will be given as:

$$d_{bt} = \sqrt{\frac{4 \times P_{fp}}{\pi f_t}} + 6 \text{ mm} \quad ...(28.32)$$

where $f_t = 42.0$ N/mm^2 for phosphor bronze (cast)
$= 31.5$ N/mm^2 for gun met or bronze
$= 70.0$ N/mm^2 for mild steel.

Lever: For the sake of economy in machining, the lever is usually cut from a length of mild steel bar of rectangular section. The only machining needed will be of drilling the pin holes and trimming off the ends. The maximum bending moment in the lever will occur at the section over the valve spindle. The section of the lever found at this place is maintained throughout its length for the sake of economy.

Bending moment, $M_b = F(b - c)$

$\therefore \quad M_b = f_b \times Z$

where $$Z = \frac{1}{6} \cdot t \cdot (d^2 - d_h^2)$$

where t = breadth of lever
d = total depth of lever
d_h = diameter of hole = pin diameter if no bush is used.
f_b = maximum safe stress in bending, say 70.0 N/mm^2 for mild steel.

Valve Chest: No exact treatment is possible to find the thickness of the valve chest due to its complex shape. Lame's formula may be used to find the thickness of the valve chest:

$$t = \frac{d_i}{2}\left[\sqrt{\frac{f_t + p}{f_t - p}} - 1\right]$$

Under normal temperature conditions, (say upto 180°C)
$f_t = 21.0$ N/mm^2 for ordinary cast iron.
$= 39.0$ N/mm^2 for ordinary gun metal castings.
$= 42.0$ N/mm^2 for good gun metal castings.
$= 87.5$ N/mm^2 for carbon steel castings.
$= 112.0$ N/mm^2 for carbon steel forgings.

According to Indian Boiler Regulations, for valve chests of circular cross-section:

$$t = \frac{pd_i}{C} + X \qquad ...(28.33)$$

where t = thickness of wall in 0.80 mm. The constants C and X are given in Table 28.6. For values of 'C' in Table 28.6, p is in MPa and t and d_i are in mm.

Table 28.6

Casting Material	C	X
Cast Steel, 440-550 N/mm^2 U.T.S.	70	8
Cast Iron, at least 140 N/mm^2 U.T.S.	28	6
Bronze, at least 220 N/mm^2 U.T.S.	30	4

The foregoing formula gives results which comply with the practical needs. The constant X has been included to provide for a minimum wall thickness compatible with moulding requirements.

Thus, when the pressure is negligible, the minimum wall thickness will be at least equal to X, that is, $8 \times 0.8 = 6.4$ mm for cast steel, 4.8 mm for cast iron and 3.2 mm for bronze even at zero pressure.

The Lame's relation can be further modified by taking into account the constant, X therefore,

$$t = \frac{d_i}{2}\left[\sqrt{\frac{f_t + p}{f_t - p}} - 1\right] + X \qquad ...(28.34)$$

The values of X will be as given below:

$$X = 6.4 \text{ mm for cast steel}$$
$$= 4.8 \text{ mm for cast iron and } 3.2 \text{ mm for bronze.}$$

For steam pressures upto 1.05 N/mm^2 saturated, the chest is usually made in Cast Iron. In the event of any doubt regarding the choice of the material, cast steel should be adopted.

28.4.2.1. Spring loaded Safety Valve: Figure 28.45 shows a spring loaded safety valve which is often favoured on boilers of the Economic type. Two valves are loaded by a single spring which is in tension. The discharge of the steam is direct to atmosphere. The valve members are in no way connected to the loading lever. If the spring breaks, they are prevented from being blown away by the slotted restraining shackles as shown in the Fig. In the case of direct spring loaded safety valve, the spring will be under compression.

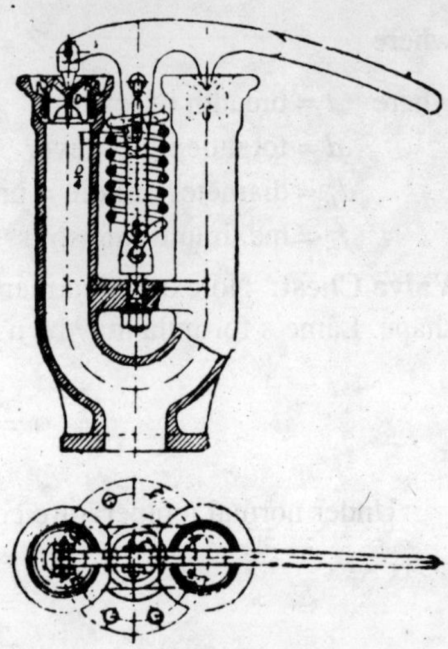

Fig. 28.45

Spring design: The design of the compression or the tension spring can be arrived at as already discussed in chapter 15. However, the following rules should be followed for the springs of safety valves:

(*a*) Initial compression to give the desired load is one-quarter the diameter of the valve seating but the Admiralty specify double this amount.

(*b*) For ordinary-lift valves, the total compression or extension of the springs required to load the valves to set pressure shall not be less than one-quarter of the valve diameter.

(*c*) The proportion of unloaded length to external diameter of the spring shall not exceed 4 to 1. These requirements do not apply to springs for full-lift safety valves.

(*d*) The maximum shear stress in the spring shall not exceed 700 N/mm^2 for compression springs and 560 N/mm^2 for extension springs.

Example 28.23. Design a simple lever safety valve for a boiler of 0.7 N/mm^2 (gauge) pressure. The valve should be 50 mm in size, fitted over a seat of proper material. The fulcrum pins should be of hard steel and cheese weight of cast iron. Lever should be of sufficient strength to ensure rigidity and move in guiding slot freely. Weight at the end of the lever should not exceed 300 N.

Solution: The total axial downward force on the valve is

$$P = \frac{\pi}{4} D^2 \times p + 1.5 \times p \times \pi D w$$

Now $\quad w = 0.01 \times D = 0.01 \times 50 = 0.50$ mm

$$\therefore \quad P = \frac{\pi}{4} \times 2500 \times 0.8 + 1.5 \times 0.8 \times \pi \times 50 \times 0.50 = 1665 \text{ N}$$

Referring to Fig. 28.43, let the distance c = 60 mm, since it depends upon the outside dimensions of the valve member.

Now neglecting the weight of the valve and the lever,

$$P \times c = F \times b$$

$$\therefore \quad \text{Distance, } b = \frac{1665 \times 60}{300} = 333 \text{ mm, say } 335 \text{ mm}$$

$$\therefore \quad \text{Exact value of, } F = \frac{1665 \times 60}{335} = 298 \text{ N}$$

Spindle. The spindle is under direct compressive force,

$$P = \frac{\pi}{4} d_1^2 \times f_c$$

The material of the spindle is M.S., so let $f_c = 60$ N/mm^2

$$\therefore \quad d_1^2 = \frac{1665 \times 4}{\pi \times 60} = 35.3 \text{ mm}^2$$

$$\therefore \quad d_1 \approx 6 \text{ mm, i.e., at the bottom end.}$$

$\therefore$ To take into account the effect of other factors, let $d_1 = 8$ mm

Let the spindle diameter above the bottom end = 10 mm.

Pins: It is clear that the thrust pin is subject to a greater shearing force, this force is given as:

$$P_{tp} = \frac{F \cdot b}{c} = \frac{298 \times 335}{60} = 1665 \text{ N}$$

$$\therefore \quad 2 \times \frac{\pi}{4} \times d_{tp}^2 \times f_s = 1665$$

Let $f_s = 45$ N/mm^2 for steel pins

$$\therefore \quad d_{tp} = 4.85 \text{ mm}$$

To provide sufficient wearing surfaces, $d_{tp} = 10$ mm

For the sake of economy, let $d_{fp} = d_{tp} = 10$ mm

Fulcrum. The fulcrum is under tension, Referring to Fig. 28.44

$$P_{fp} = 2 \times t_f \times (w_f - d_{fp}) \times f_t$$

Now $$P_{fp} = \frac{F(b-c)}{c} = \frac{298 \times (335 - 60)}{60} = 1367 \text{ N}$$

Also, considering the bearing of the fulcrum pin,

$$P_{fp} = 2t_f \cdot d_{fp} f_b$$

Let $$f_b = 20 \text{ N/mm}^2$$

$$\therefore \quad t_f = \text{thickness of the fulcrum}$$

$$= \frac{1367}{2 \times 10 \times 20} = 3.42 \text{ mm; say 4 mm}$$

$$\therefore \quad 1367 = 2 \times 4 \times (w_f - 10) \times f_t$$

Let $$f_t = 70 \text{ N/mm}^2$$

$$\therefore \quad w_f = 10 + \frac{1367}{2 \times 4 \times 70} = 12.44 \text{ mm, say 12.5 mm}$$

Fulcrum-bolt: The fulcrum bolt is under tension, its diameter is given as:

$$d_{bt} = \sqrt{\frac{4 \times P_{fp}}{\pi f_t}} + 6 \text{ mm} = \sqrt{\frac{4 \times 1367}{\pi \times 70}} + 6 \text{ mm} \approx 11 \text{ mm}$$

Lever: Maximum bending moment on the lever,

$$M_b = F(b-c) = \frac{298(335 - 60)}{1000} = 82 \text{ Nm}$$

Now $$Z = \frac{1}{6} \cdot t \cdot (d^2 - d_h^2)$$

Now $$d_h = \text{diameter of hole}$$

Since the pin is of steel, it will have to be bushed, therefore, the diameter of hole will be greater than the diameter of pin, so let $d_h = 12$ mm

Let $$t = \text{breadth of lever} = 2t_f = 8 \text{ mm}$$

$$\therefore \quad M_b = Z \times f_b$$

$$82 \times 1000 = \frac{1}{6} \times 8 \times (d^2 - 12^2) \times 70$$

$\therefore$ d = depth of lever = 32 mm

Valve Chest: The thickness of the valve chest is given as

$$t = \frac{d_i}{2}\left[\sqrt{\frac{f_t + p}{f_t - p}} - 1\right] + X$$

Now $d_i = 50$ mm; $f_t = 21$ N/mm^2 for C.I.

$p = 0.8$ N/mm^2, $X = 4.8$ mm for C.I.

$\therefore$ $$t = \frac{50}{2}\left[\sqrt{\frac{21.8}{20.2}} - 1\right] + 4.8 = 5.71 \text{ mm, say 6 mm.}$$

Example 28.24. Design a closed helical compression spring for a spring loaded valve from the following data:

Valve diameter = 75 mm

Blow off pressure = 0.95 N/mm^2 (absolute)

Lift of the valve when pressure rises from 0.875 to 0.95 N/mm^2 (absolute) is 3.5 mm. Assume spring index as 5. Take allowable shear stress for the spring material as 595 MPa and the value of modulus of rigidity for the spring material as 84 GPa.

Solution: The axial load on the valve is given as:

$$P = \frac{\pi}{4} D^2 \times p + 1.5 \times p \times (\pi D w)$$

Now $D = 75$ mm; $p = 0.95$ N/mm^2

Let $w = 0.02\, D = 0.02 \times 75 = 1.5$ mm

$\therefore$ $$P = \frac{\pi}{4} \times (75)^2 \times 0.95 + 1.5 \times 0.95 \times (\pi \times 75 \times 1.5) = 4197 + 503.6 = 4700 \text{ N}$$

Now according to Boiler Regulations, the initial compression of the springs is not to be less than one-quarter of the valve diameter, but the Admiralty specify double this amount.

So, let the initial compression of the springs be = 30 mm

$\therefore$ Stiffness of the spring $= \frac{4700}{30} \approx 156.7$ N/mm

Now the total compression of the spring when the valve lifts, is

$= 30 + 3.5 = 33.5$ mm

$\therefore$ Maximum load on the spring = $156.7 \times 33.5 = 5250$ N

Now Spring index, $S = \frac{\text{Mean diameter of coil}}{\text{Diameter of the spring wire}}$

Let 'd' be the diameter of the spring wire,

$\therefore$ Mean diameter of coil = $5d$

Now torsional moment on the spring $= 5250 \times \frac{5d}{2} = 13.125d$ kNm

Now $$M_t = \frac{\pi}{16} \times d^3 \times f_s \times \frac{1}{K}$$

where K = stress concentration factor

$$= \frac{4S - 1}{4S - 4} + \frac{0.615}{S} = \frac{19}{16} + \frac{0.615}{5} = 1.1875 + 0.123 = 1.3105$$

$$\therefore \quad 13.125 \times 1000\, d = \frac{\pi}{16} \times d^3 \times 595 \times 10^6 \times \frac{1}{1.3105}$$

$$\therefore \quad d = 0.0121 \text{ m; say } 12.5 \text{ mm}$$

$\therefore$ Mean diameter of coil $= 5d = 62.5$ mm

Now z = number of effective coils

$$= \frac{C \cdot d^4 \cdot \delta}{64\, FR^3} = \frac{C \cdot d \cdot \delta}{8 FS^3}$$

Now δ = deflection of spring = 3.5 mm

Now the load to produce this deflection, $F = 156.7 \times 3.5 = 548.45$ N

$$\therefore \quad z = \frac{84 \times 10^9 \times 0.0125 \times 0.0035}{8 \times 548.45 \times 125} = 6.7 \text{ turns}$$

Now about 1½ to 2 turns are needed at the end of the spring to support it. These are known as dead coils and are not relied upon to take any load.

$\therefore$ Total number of coils, $z' = 6.7 + 1.5 = 8.2$, say 8 coils

$\therefore$ Free length of spring $= (z' - 1) \times \text{gap} + z' \times d + \delta \max$

Now gap between coils is about 0.8 mm to 1.6 mm; say 1 mm

$\therefore$ Free length of spring $= 7 \times 1 + 8 \times 12.5 + 33.5 = 7 + 100 + 33.5 = 140.5$ mm

Now external diameter of coil = 62.5 + 12.5 = 75 mm

$$\therefore \quad \frac{\text{Free length of spring}}{\text{External diameter of coil}} = \frac{140.5}{75}$$

is < 4, as desired by Boiler Regulation and hence the design is O.K.

Example 28.25. Design the spring of a safety valve for a steam boiler working at a pressure of 1.575 MN/m² abs, and having an evaporative capacity of 1363 kg/hr.

Solution: Let us assume that the steam is saturated and the valve is of high-lift safety valve.

Now set or blow off pressure = 5 percent in excess of normal pressure

$$= 1.05 \times 1.575 = 1.655 \text{ MN/m}^2 = 16.55 \text{ bar}$$

$\therefore$ Total area of the valve,

$$A = \frac{E}{C \cdot p} \quad \text{and} \quad C = 0.10$$

$$\therefore \quad A = \frac{1360}{0.10 \times 16.55 \times 10^6} = 821.75 \text{ mm}^2$$

If the valve is of Ramsbottom type, there will be two valves. Therefore, area of each valve

$$= \frac{1}{2} \times 821.75 = 410.875 \text{ mm}^2$$

$\therefore$ $\frac{\pi}{4} D^2 = 410.875 \times 1.5$, since the two valves are loaded by the same spring, the area is to be increased by 50 per cent.

$$\therefore \quad D^2 = 1.5 \times 410.875 \times \frac{4}{\pi} = 784.61$$

$$\therefore \quad D = 28 \text{ mm}$$

Now lift of the valve is to be at least $\frac{D}{12}$; say 2.5 mm.

Now, the maximum load on the spring can be calculated as before and its design can be completed.

28.4.3 Steam Stop Valve

Steam Stop Valve is usually a screw down stop valve. This is a valve in which the disc is lifted from and lowered on to the body seat through the agency of a screw-and-nut mechanism. Screw down stop valves are generally of two types:

1. **Globe Valve:** In this valve, the body ends of the valve are in line with each other and the disc is lifted from and lowered on the body seat by a spindle whose axis is at right angles to that of the body ends. This is inserted in a straight pipe line.
2. **Junction Valve or Angle Valve:** In this valve, the body ends are at right angles with each other, and in which the disc is lifted from and lowered on the body seat by a spindle whose axis is in line with that of one body end. In this valve, there is a right-angled change in the direction of flow. This type is generally employed on the steam pipe of a boiler to serve as the main stop valve. The screw-down stop valves can also be classified as:

(*a*) **Outside Screw type:** In this type, the screwed portion of the actuating spindle is outside the pressure space of the valve. This is a very good feature and is particularly suitable for steam stop valve.

(*b*) **Inside Screw type:** In this type, the actuating thread of the spindle is engaged within the bonnet or the cover of the valve body. This design is not recommended for steam stop valves.

A common steam stop valve of outside screw type is shown in Fig. 28.46. These valves can be grouped into three categories according to their pressure and temperature ratings:

(*i*) **Class 1 Valves:** These valves are suitable for non-shock cold working pressure upto 1.2 N/m^2 (cold service means a temperature not exceeding 45ºC).

(*ii*) **Class 2 Valves:** These valves are suitable for maximum non-shock cold working pressures upto 1.4 N/mm^2.

(*iii*) **Class 3 Valves:** These valves are suitable for use on steam pipe lines for maximum non-shock cold working pressures upto 2.8 N/mm^2.

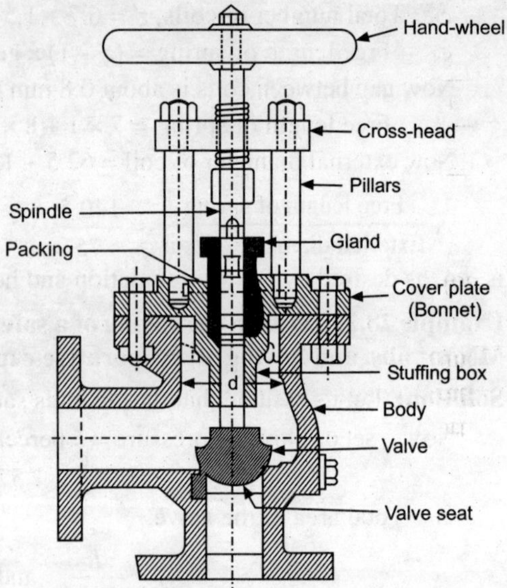

Fig. 28.46 *Steam stop valve*

The pressure and temperature ratings for these valves are listed in Table 28.7.

Table 28.7 Pressure and temperature rating of valves

Service Temp. ºC	Maximum Non-Shock service pressure, N/mm^2		
	Class 1	Class 2	Class 3
45	1.2	1.4	2.8
66	0.7	1.4	2.8
93	0.4	1.3	2.6
121	0.2	1.2	2.4
150	—	1.1	2.1
170	—	0.8	1.8
178	—	0.7	1.7
186	—	—	1.6
198	—	—	1.5
208	—	—	1.4

Valve Seatings: As already mentioned under safety valves, the valve is the most important item of any valve. Hence, careful thought should be paid to the design of the seatings and to the materials employed. The contact face of the valve member may be:

1. flat
2. mitred
3. radiused.

These three types are shown in Fig. 28.47. The radiused type of seating is more difficult to machine and lap than a mitred or flat one. Also it is expensive as more material is used in this type.

The total axial force to close down the valve effectively is given as:

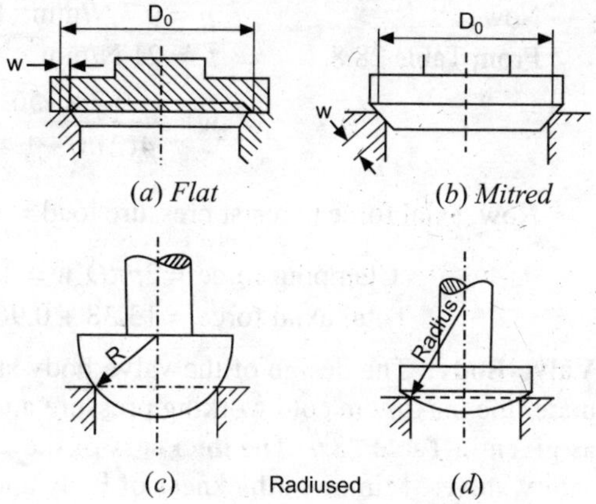

Fig. 28.47 *Screw-down stop Valve Seatings*

P = axial force to resist pressure load + axial clamping force at the seating.

As already mentioned under safety valves, the clamping pressure should be at least 1½ to 2 times the line pressure. Therefore,

$$P = \text{effective area} \times \text{fluid pressure} + 2 \times \text{line pressure} \times \text{area of seating edge}$$

$$= \frac{\pi}{4} D_o^2 \times p + 2p \times \pi D_o \times w \qquad \text{...(28.35)}$$

or

$$P = \pi D_o p\left(\frac{D_o}{4} + 2w\right)$$

$\therefore$ Load per unit area of the seating is given as:

f_s = maximum unit surface stress

$$= \frac{\text{load}}{\text{area}} = \frac{\pi D_o p\left(\frac{D_o}{4} + 2w\right)}{\pi D_o w} = \frac{pD_o}{4w} + 2p$$

or

$$w = \text{width of seating} = \frac{pD_o}{4(f_s - 2p)} \qquad \text{...(28.36)}$$

The maximum permissible unit surface stresses for valve seatings are given in Table 28.8.

Table 28.8 Values of unit surface stress

Seating Material	Max. B.H.N. (30 kN)	f_s N/mm^2
Cast Iron	180	7.0
Fibre composition	45	3.5
Gun-Metal	75	14.0
Monel-Metal	270	63.0
Nickle bronze	240	42.0
Nickle Chrome Steel	440	84.0
Phosphor Bronze	100	21.0
Stellite	600	105.0

Example 28.26. Determine the total axial force required to effect positive closure of a 150 mm screw down stop valve having flat-faced phosphor bronze seatings. The steady steam pressure is 0.7 N/mm². Assume D_0 = 156 mm. neglect gland friction.

Solution: By equation 28.36, the width of the seating can be found out:

$$w = pD_0/4(f_s - 2p)$$

Now $p = 0.7\ \text{N/mm}^2, D_o = 156$ mm

From Table 28.8, $f_s = 21\ \text{N/mm}^2$

$$\therefore \quad w = \frac{0.7 \times 156}{4(21.0 - 1.4)} = \frac{0.7 \times 156}{4 \times 19.6} = 1.392 \text{ mm, say } 1.4 \text{ mm}$$

Now, axial force to resist pressure load $= \frac{\pi}{4} D_o^2 \times p = \frac{\pi}{4} \times (156)^2 \times 0.7 = 13.38$ kN

$$\text{Clamping force} = 2p\pi D_o w = 2 \times 0.7 \times \pi \times 156 \times 1.40 = 960 \text{ N}$$

$$\text{Total axial force} = 13.38 + 0.96 = 14.34 \text{ kN}$$

Valve-Body: The design of the valve body shall be such so as to provide ample resistance to distortion under the maximum cold working pressure and under the specified pressure at the maximum temperature as given in Table 28.7. The thickness of the valve body can be found out by the formula as given under safety valves. Minimum thickness of body and cover is given in Table 28.9.

Valve Actuating Screws: The threads on the spindle shall be of the square, acme form or truncated V form. The design of these valves have already been dealt with in chapter 23.

Table 28.9 Minimum body and cover thickness, (dimensions in mm)

Class of valve	Nominal Size of Valve										
	8	10	15	20	25	32	40	50	65	80	100
1	1.7	1.8	1.9	2.0	2.0	2.3	2.5	2.7	2.8	3.2	3.8
2	1.9	2.0	2.2	2.3	2.6	2.7	2.8	3.2	3.6	3.8	4.5
3	2.3	2.4	2.7	2.9	3.4	3.7	3.8	4.4	5.0	5.5	6.8

Example 28.27. Determine the net tangential effort to be applied to the rim of a hand wheel of a 0.275 m screw down stop valve for a working steam pressure 0.7 N/mm². The spindle is of gun metal and operates in a mild steel nut. Its outside diameter is 50 mm. The thread is single start, square thread, lubricated. The valve seating is of nickel bronze. Assume the diameter of the handwheel as 0.5 m.

Solution: Assuming normal series for the threads, from Table 23.1, the pitch of the threads = 8 mm

$\therefore$ Mean diameter of spindle, $d_m = 50 - 4 = 46$ mm

$$\therefore \quad \tan\alpha = \frac{p}{\pi d_m} = \frac{8}{\pi \times 46} = 0.0553$$

$\therefore$ Slope of the thread, $\alpha = 3° - 10'$

Assuming coefficient of friction as 0.10

$$\therefore \quad \tan\phi = 0.10 \quad \therefore \quad \phi = 5° - 43'$$

Now, efficiency of the thread, $\eta = \frac{\tan\alpha}{\tan(\alpha + \phi)} = \frac{\tan(3° - 10')}{\tan(8° - 53')} = \frac{0.0553}{0.1563} = 35.4\%$

Now the total axial force to be applied to the valve,

$$P = \pi D_o p \left(\frac{D_o}{4} + 2w \right)$$

Now $$w = \frac{pD_o}{4(f_s - 2p)}$$

Now, $p = 0.7\ \text{N/mm}^2$ f_s from Table 28.8 $= 42\ \text{N/mm}^2$

Let $D_o = 181$ mm

$$\therefore \quad w = \frac{0.7 \times 181}{4(42.0 - 1.4)} = \frac{0.7 \times 181}{4 \times 40.6} = 0.78 \text{ mm}$$

$$\therefore \quad P = \pi \times 181 \times 0.7\left(\frac{181}{4} + 2 \times 0.78\right) = 18.7 \text{ kN}$$

$\therefore$ Tangential effort at the end wheel, $F = \dfrac{P \times p}{2\pi r \eta}$

Now $r = 250$ mm

$$\therefore \quad F = \frac{18.70 \times 10^3 \times 8}{2\pi \times 250 \times 0.354} = 269 \text{ N}$$

Note: Other things being equal, the effort required may be reduced by:

1. increasing the diameter of the handwheel and/or.
2. increasing the diameter of the actuating spindle.
3. increasing the number of threads per cm of the actuating spindle.
4. reducing the frictional resistance of the screw and nut mechanism by providing means of lubrication.

Actuating Spindle: The connection between the spindle and the valve member should be such that the two are able to rotate independent of other. If it is rigid, there will be on the spindle, torsional as well as compressive forces during the closing stages, whereas if the connection is flexible, there will be only compressive forces on the spindle. Again, if the connection is rigid, the valve member shall not have self aligning tendency, which is not at all desired.

Therefore, $$P = \frac{\pi}{4} d_s^2 \times f_c$$

where d_s = spindle diameter

f_c = allowable compressive stress for spindle material.

The values for 'f_c' are given below:

f_c = 10.0 to 12.0 N/mm^2, for brass (cast)

= 49.0 to 70.0 N/mm^2, for gunmetal

= 98.0 to 105.0 N/mm^2, for Manganese bronze (rolled)

= 91.0 to 120.0 N/mm^2, for Mild Steel

= 84.0 to 112.0 N/mm^2, for Monel metal

= 98.0 to 140.0 N/mm^2, for Nickle-chrome steel

In actual practice, the diameter of the spindle as found above is increased as below:

$$d_s = \sqrt{\frac{4P}{\pi f_c}} + (2.5 \text{ to } 12 \text{ mm})$$, as the valve size increases, this factor increases.

The minimum diameter of spindles 'as per ISI' are given in Table 28.10. The upper end of the spindle shall be provided with a tapered square for mounting the handwheel. The handwheel shall be held in place by a nut screwed on to the spindle or by a set screw and washer.

Table 28.10 Minimum Spindle diameters, mm

Class of valve	Nominal Valve Size, mm										
	8	10	15	20	25	32	40	50	65	80	100
1,2	6.0	6.0	8.0	9.0	9.5	10	12	12	14	16	20
3	7	8	9	9.5	11	12	14	16	18	20	22

The length of spindle thread engagement can be found out as in chapter 23, considering the wear of the threads. It will be given as:

$$l_t = \frac{2P}{\pi d_m \cdot p_b}$$

where p_b = maximum allowable bearing pressure. It will depend upon the material and the amount of lubrication provided.

= 1.4 to 8.4 N/mm^2 for gunmetal spindle and forged steel cross head. It is given empirically that,

$$l_t \nless 2d_m + 6 \text{ mm}$$

As per ISI: The minimum length of the actuating thread in engagement at the open or closed position of the valve shall not be less than two thirds of the external diameter of the thread. Also, the total length of the spindle shall be such that the handwheel may be freely gripped by hand when the valve is closed and also repacking is possible without taking off the handwheel when the valve is fully open.

Pillars and Cover Bolts: The design of the pillars and the cover bolts will depend upon the type of the joint between the cover plate and the valve body and whether the pillars are simply screwed in the flange of the bonnet or it passes through the flanges of both cover plate and valve body and secured by nut.

(a) **Pillars simply screwed into the top plate and joint is made with the help of gasket:** In this case, the pillars will be under direct tension due to the total axial force required to close down the valve. Usually there are two pillars. Therefore, the load per pillar can be determined and its diameter calculated.

The total clamping force to be exerted by the cover bolts will be:

F = Force due to pressure tending to separate the two flanges
+ force required to maintain the fluid tightness of the gasket
+ upward thrust of the spindle in the process of valve closure, which will be transmitted to the cover plate via the pillars.

In the worst case, the steam can leak upto the inside of the bolts. So, if

D = Distance between the inside of bolts, and, therefore,

Force due to pressure tending to separate the two halves

$$= p \times \frac{\pi}{4} D^2$$

To maintain the fluid tightness of the gasket, a unit clamping pressure of 0.3 p shall be sufficient. Hence force required to maintain the fluid tightness of the gasket is equal to

$$= 0.3p \times \frac{\pi}{4}(D^2 - d^2)$$

where d = diameter corresponding to the inner edge of the cover plate

$$\therefore \qquad F = p \times \frac{\pi}{4} D^2 + 0.3p \times \frac{\pi}{4}(D^2 - d^2) + P$$

Therefore, if there are z bolts.

The load per bolt = $\frac{F}{z}$ and hence the bolts can be designed. The flanges of the bonnet and valve body should be sufficiently thick to provide a leak prof joint. Their thickness should never be less than the nominal diameter of the bolts used connecting these flanges. The unit stress on the area of the minimum diameter of any bolt shall not exceed 50 N/mm^2 corresponding to the force on diameter "D". Bolts or studs less than 16 mm diameter shall not be used for making body to bonnet connection for valves of size 25 mm and above.

(b) **Pillar ends continuous through both flanges and nutted and the joint made with the help of gasket:** In this case, the axial force due to the valve closure shall not be transmitted to the cover plate via the pillars.

∴ Total force on the bolts shall be given as:

$$F = p \times \frac{\pi}{4} D^2 + 0.3p \times \frac{\pi}{4}(D^2 - d^2)$$

∴ Force to be exerted by each bolt and pillar $= \frac{F}{z}$, where z = number of bolts, including pillars. But, the maximum force to be sustained by each pillar shall be equal to

$$= \frac{F}{z} + \frac{P}{2}$$

(c) Pillar screwed in cover plate and cover joint with metal to metal joint (Spigoted) on inside of cover plates: For steam pressure above about 1.75 N/mm^2, usually a metal to metal spigot and socket joint is used instead of employing a gasket. In this case,

$$F = p \times \frac{\pi}{4} D_g^2 + 2p \times \pi \cdot s \cdot (D_g + s) + P$$

Here 2 p is taken as unit joint to joint clamping pressure,

where D_g = diameter of the inside spigot joint.

s = width of spigot, usually taken as $\frac{D_g}{20} + 2.5$ mm

$$\therefore \quad \text{Load per bolt} = \frac{F}{z}$$

$$\text{Load on each pillar} = \frac{P}{2}.$$

Example 28.28. Determine the maximum load in the cover bolts and the pillars of a 150 mm screw down stop valve, working pressure 1.75 N/mm^2 for the three cases. The following data is given:

Distance between inside of bolts = 300 mm
Diameter of inside of the cover plate = 230 mm
Axial force to effectively close the valve = 45 kN
Number of bolts including pillars = 10
Design the bolts and pillars for case (*a*).

Solution:

$$F = p \times \frac{\pi}{4} D^2 + 0.3p \times \frac{\pi}{4}(D^2 - d^2) + P$$

$$= 1.75 \times \frac{\pi}{4} \times 300^2 + 0.3 \times 1.75 \times \frac{\pi}{4}(300^2 - 230^2) + 45 \times 10^3$$

$$= 124.8 + 15.28 + 45 = 185.08 \text{ kN}$$

Number of cover bolts $= 10 - 2 = 8$

$$\therefore \quad \text{Load per bolt} = \frac{185.08}{8} = 23.135 \text{ kN}$$

Assuming M 22 metric coarse threads, stress area = 3.03 cm^2 (Table 9.1)

∴ Stress in the bolt due to the steam load corresponding to '*D*',

$$f_t = \frac{124.8 \times 1000}{8 \times 3.03 \times 100} = 51.5 \text{ N/mm}^2, \text{ which is} > 50.0 \text{ N/mm}^2$$

∴ Take M 24 threads, stress area = 3.53 cm^2; f_t will come out to be < 50.0 N/mm^2

$$\text{Load on each pillar} = \frac{45}{2} = 22.5 \text{ kN}$$

$$\therefore \quad \frac{\pi}{4} d_c^2 \times f_t = 22.5 \times 1000$$

Let $f_t = 50$ N/mm^2

$$\therefore \quad \text{Stress area} = \frac{22.5 \times 1000}{50} = 450 \text{ mm}^2 = 4.5 \text{ cm}^2$$

From Table 9.1, select M-27 coarse thread with stress area = 4.59 cm^2

(*b*) Total force, $F = p\times\frac{\pi}{4}D^2 + 0.3p\times\frac{\pi}{4}(D^2-d^2) = 124.8 + 15.28 = 140.08$ kN

$\therefore$ Clamping force on each bolt and pillar

$= \frac{F}{z}$, *z* in this case will be 10

$= 14.008$ kN

Total force on each pillar $= 14.008 + \frac{45}{2} = 36.508$ kN

(*c*) Total clamping force, $F = p\times\frac{\pi}{4}\times D_g^2 + 2p\times\pi\times s\times(D_g+s) + P$

Let, the mean diameter of spigot,

$$D_g = \frac{D+d}{2} = \frac{300+230}{2} = 265 \text{ mm}$$

$$\therefore \quad s = \frac{D_g}{20} + 2.5 = \frac{265}{20} + 2.5 = 15.75 \text{ mm}$$

$$\therefore \quad F = 1.75\times\frac{\pi}{4}\times(265)^2 + 2\times1.75\times\pi\times15.75\times(265+15.75) + 45\times1000$$

$$= 190.2 \text{ kN}$$

$\therefore$ Load on each bolt $= \frac{190.2}{8} = 23.775$ kN

Load on each pillar $= \frac{45}{2} = 22.50$ kN

Cross Head: This member is subjected to the full axial force of the spindle in the process of closing the valve and by the pressure acting on the valve member when the valve is closed. The following assumptions are made in its design:

(i) It acts as a beam loaded at the centre of its span.

(ii) It is simply supported at the pillar centres.

The bending moment will be maximum at the centre of span, and is

$$M_b = \frac{P\times L}{4}$$

where L = the span (distance between pillar centres)

Also, $M_b = f_b\times Z$

Now $Z = \frac{h^2}{6}(b-d_m)$, refer Fig. 28.48

$$\frac{h^2}{6}(b-d_m)\times f_b = \frac{PL}{4}$$

$$\therefore \quad b = \frac{1.5PL}{f_b\cdot h^2} + d_m \quad ...(28.37)$$

Now $h \nless l_t$

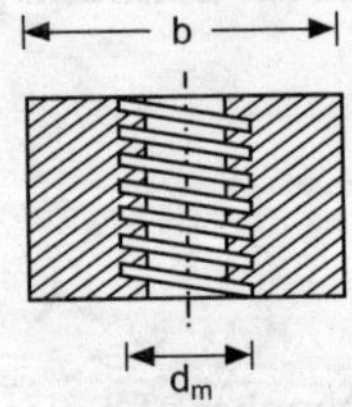

Fig. 28.48 *Transverse section of cross head*

The cross head, Fig. 28.49 is usually carbon steel forgings or stampings, or cut from mild steel bar stock

f_b = allowable bending stress

= 140.0 N/mm² for steel forging

= 70.0 N/mm² for mild steel.

For small valves, sometimes sand cast gun metal cross heads are used, and $f_b = 42$ N/mm².

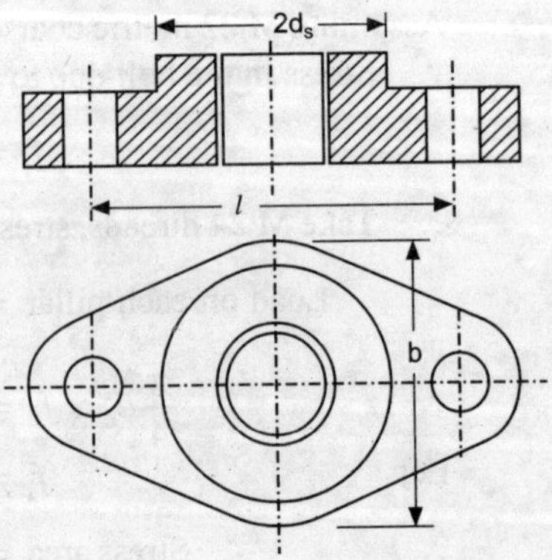

Fig. 28.49 *Cross head*

Hand-wheel: The hand wheel will be designed in the same manner as explained in Chapter 23, in the case of Power Screw 76 cm diameter handwheel is regarded as the largest that can be easily operated by one man unaided.

Stuffing Box and Packing: The stuffing box is usually made integral with the bonnet. The depth of the stuffing box cavity depends upon the number of turns of the packing needed to give effective sealing, and this in turn depends upon the pressure and temperature conditions. A good packing is good quality square section graphited asbestos, preferably reinforced by antifriction wire. The depth shall not be less than the diameter of the spindle for class 1 and 2 valves and for class 3 valve, it shall not be less than $1\frac{1}{4}$ times the spindle diameter.

The arrangement of stuffing box and gland is shown in Fig. 28.50.

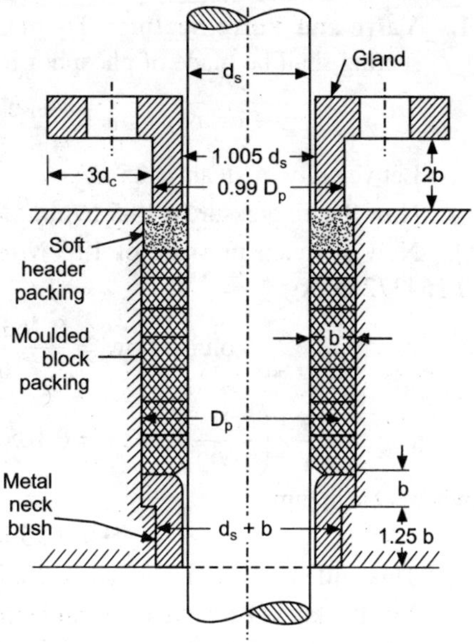

Fig. 28.50 *Stuffing box and gland*

The outer diameter of packing is given as,

$$D_p = 1.25\, d_s + 10 \text{ mm}$$

Width of the packing, $b = \frac{1}{2}(D_p - d_s) = 0.125 d_s + 5$ mm

According to "ISI" the stuffing box dimensions are given Table 28.11.

Table 28.11 Stuffing box dimensions, mm

Spindle diameter	Stuffing Box Diameter Min.	Packing Size of Space Min.
6.0	10	2
10	15	2.5
12	17	2.5
16	23	3.5
20	28	4.0
22	30	4.0
25	35	5.0

Gland. The dimensions of the gland are shown in Fig. 28.50. The gland may be a smooth casting. The studs to connect the gland to the bonnet should be provided with coarse threads which will fit them for service over a long period without stripping since renewal is not always easy. The steam load acting underneath the land can be calculated and the diameter of the studs determined accordingly. In no case, the diameter of the stud should be less than 12.5 mm except in the case of small gun metal valves. Regarding the number of studs, no hard and fast rule can be laid down. For valves upto 100 mm bore to operate for normal pressures, two studs will suffice. For larger size, three or four studs are recommended. The thickness of the flange of the gland may be taken as 1½ times the core diameter of the studs.

Steam Velocities. The average values of the steam velocities are given below:

1. Main smooth pipes : 22.5 to 27 m/s for Saturated Steam
   : 30 to 42 m/s for Superheated Steam
2. Stop valves: 20 to 30 m/s.

Example 28.29. Design the various parts of a steam stop valve of a boiler working at a pressure of 1.75 N/mm^2 gauge, and temperature of 370°C and supplying 2300 kg of steam per hour.

Solution: It is clear, from Table 28.7 that for steam pressure of 1.75 N/mm^2 gauge and temperature of 370°C, valve of class 3 will be used.

1. Valve and Valve Seating: To guard against rusting, as already mentioned, both the valve and valve seating shall be made of phosphor bronze. To find the diameter of the valve of the inlet steam pipe,

$$\text{Volume flow of steam} = \frac{\pi}{4} D^2 \times \text{Velocity of steam}$$

Let velocity of steam = 25 m/s

Now Steam pressure = 1.85 N/mm^2 absolute

Now at steam pressure of 1.85 N/mm^2 and 370ºC (Superheated Steam), specific volume of steam = 0.164372 m^3/kg.

$$\therefore \quad \text{Volume flow} = \frac{0.164372 \times 2300}{60 \times 60} = 0.105 \text{ m}^3/\text{s}$$

$$\therefore \quad \frac{\pi}{4} \times \left(\frac{D}{1000}\right)^2 \times 25 = 0.105$$

where D is in mm

$$\therefore \quad D = 72.5 \text{ mm, let us take standard valve of } D \text{ as 80 mm.}$$

This will give steam velocity of 20.5 m/s which is also O.K.

The thickness of valve disc can be found out by considering it a flat plate uniformly loaded and simply supported at the edges. The formula is, as given in Chapter 25.

$$t = 0.54 D \sqrt{\frac{p}{f_t}}$$

Let f_t = 35.0 N/mm^2 for Phosphor bronze

$$\therefore \quad t = 0.54 \times 80 \sqrt{\frac{1.85}{35.0}} = 9.9 \text{ mm, say 12 mm}$$

Width of the seating is given as:

$$w = \frac{pD_o}{4(f_s - 2p)}, \text{ let } D_o = 85 \text{ mm}$$

Now from Table 28.8, $f_s = 21$ N/mm^2

$$\therefore \quad w = \frac{1.85 \times 85}{4 \times 17.3} = 2.27 \text{ mm; say 3 mm}$$

2. Valve Body: The thickness of the valve body can be found out by the modified Lame's equation 28.34.

$$t = \frac{d_i}{2}\left[\sqrt{\frac{f_t + p}{f_t - p}} - 1\right] + X$$

The material of the valve body is taken as cast steel, therefore,

$$f_t = \frac{\text{Ultimate strength}}{\text{FS}}$$

Now from Table 28.6, U.T.S. of Cast Steel for valve bodies is 440 to 550 N/mm^2. FS should be taken as 10, therefore,

$$f_t = 44.0 \text{ to } 55.0 \text{ N/mm}^2\text{; let us take } f_t = 50.0 \text{ N/mm}^2$$

Now d_i = 80 mm; p = 1.85 N/mm^2, X = 6.4 mm for C.S.

$$\therefore \quad t = \frac{80}{2}\left[\sqrt{\frac{51.85}{48.15}} - 1\right] + 6.4 = 1.2 + 6.4 = 7.6 \text{ mm; say 8 mm}$$

Now from Table 28.9, the minimum thickness for class 3 valve of nominal size of 80 mm is 5.5 mm, therefore, our result is on safer side. The thickness of the bonnet or the cover plate may also be taken as 8 mm.

3. **Spindle:** The spindle shall be designed under direct compressive force.

Compressive force on the valve = Total force to close down the valve

$$= \frac{\pi}{4} D_o^2 \times p + 2p \times \pi D_o \times w \quad \text{(Eqn. 28.35)}$$

$$\therefore \quad P = \frac{\pi}{4} \times (85)^2 \times 1.85 + 2 \times 1.85 \times \pi \times 85 \times 3 = 13.46 \text{ kN}$$

$$\therefore \quad P = \frac{\pi}{4} d_s^2 \times f_c$$

Let the material for spindle be gun-metal, for which

$$f_c = 49.0 - 70.0 \text{ N/mm}^2\text{, let it be } 60 \text{ N/mm}^2$$

$$\therefore \quad d_s = \sqrt{\frac{4 \times 13.46 \times 1000}{\pi \times 60.0}} = 17 \text{ mm}$$

In actual practice, the spindle diameter is increased by 2.5 to 12 mm.

∴ Let the core diameter of the screwed parts of spindle be 20 mm

If square threads of coarse series are selected with pitch = 8 mm

∴ Nominal diameter of spindle = 28 mm

Now from Table 28.10, for valve of class 3 and size 80 mm, the minimum nominal spindle diameter should be 20 mm, therefore, our design is safe.

Now the length of the screwed portion of the spindle in engagement is:

$$l_t = \frac{2P}{\pi \cdot d_m \cdot p_b}, \text{ let } p_b = 5 \text{ N/mm}^2$$

$$\therefore \quad l_t = \frac{2 \times 13.46 \times 1000}{\pi \times 24 \times 5} = \text{about } 70 \text{ mm}$$

Empirically, $l_t \nless 2d_m + 6$ mm, i.e., 54 mm, therefore, this condition is also satisfied.

4. **Pillars and Cover Bolts:** Let the pillars be of screwed type, and the joint is made with the help of a gasket.

Load on two pillars = Total axial force required to close down the valve = 13.46 kN

∴ Load on each pillar = 6.73 kN

$$\therefore \quad \frac{\pi}{4} d_c^2 \times f_t = 6730; \text{ let } f_t = 50 \text{ N/mm}^2$$

$$\therefore \quad \frac{\pi}{4} \times d_c^2 = \frac{6730}{50} = 134.6 \text{ mm}^2 = 1.346 \text{ cm}^2$$

Metric coarse threads M 16 with stress area of 1.57 cm^2 will be selected.

Now the total clamping force to be exerted by the cover bolts,

$$F = p \times \frac{\pi}{4} D^2 + 0.3p \times \frac{\pi}{4}(D^2 - d^2) + P$$

where D = distance between the inside of bolts.

and d = diameter corresponding to inner edge of cover plate.

As nothing is mentioned about these two dimensions, and as the valve size is 80 mm; let us take $d = 120$ mm and $D = 160$ mm

$$\therefore \quad F = 1.85 \times \frac{\pi}{4} \times 160^2 + 0.3 \times 1.85 \times \frac{\pi}{4}(160^2 - 120^2) + 13.46 \times 1000$$

$$= 54.66 \text{ kN}$$

Let $f_t = 50$ N/mm^2

$\therefore$ Total stress area $= \dfrac{54.66 \times 1000}{50} = 1093.2 \text{ mm}^2 = 10.932 \text{ cm}^2$

Now bolts less than 16 mm diameter are not to be used.

Let us select M 18 × 1.5 metric fine threads, with stress area = 2.16 cm^2

$\therefore$ Number of studs $= \dfrac{10.932}{2.16} = 5.06$; say 6 studs.

5. **Cross-Head:** The cross-head is designed as a beam loaded at the centre and simply supported at the pillar centers.

Span L = Diameter of spindle boss + 3 × diameter of pillar + some clearance between the spindle boss and pillar nut.

$= 2 \times 28 + 3 \times 16 + 10 = 114$ mm;

say 120 mm, because gland is also to be accommodated within the pillars.

$\therefore$ From equation 28.37, and referring to Fig. 28.48.

$$b = \frac{1.5PL}{f_b \cdot h^2} + d_m$$

Let $f_b = 70$ N/mm^2 for mild steel cross-head

Now $h \nless l_r$, so let $h = 70$ mm

$$\therefore \quad b = \frac{1.5 \times 13.46 \times 1000 \times 120}{70 \times 4900} + 24 = 31 \text{ mm}$$

But it cannot be less than diameter of spindle boss, therefore, let $b = 60$ mm.

6. **Hand-Wheel:** The hand wheel is made of Cast Iron. When the force is applied at its periphery to raise or lower the stop valve, torsional moment is transmitted to the spindle. This torsional moment should be sufficient to overcome the frictional resistance at the bridge between the spindle and the nut and also at the collar on the valve.

The frictional torsional moment between the screw and nut is given as:

$$T_1 = P \times \frac{d_m}{2} \times \tan(\alpha + \phi)$$

where P is the load on the spindle = 13.46 kN

No $\quad \tan\alpha = \dfrac{p}{\pi d_m} = \dfrac{8}{\pi \times 24} = 0.01 \quad \therefore \quad \alpha = 5° - 47'$

If the coefficient of friction of the screw and nut is assumed as 0.1, then

$$\phi = 0.1 \quad \therefore \quad \phi = 5° - 43'$$

$$\therefore \quad T_1 = 13.46 \times 10^3 \times \frac{0.024}{2} \times \tan(11° - 30') = 34.8 \text{ Nm}$$

The frictional torsional moment at the collar may be written as:

$$T_2 = \frac{2}{3} \mu c \cdot P \cdot \frac{d_m}{2}$$

where μc is the coefficient of friction at the collar, let it also be 0.1.

$$\therefore \quad T_2 = \frac{2}{3} \times 0.1 \times 13.46 \times 10^3 \times 0.012 = 10.768 \text{ Nm}$$

$\therefore$ Total torsional moment to be provided by the hand-wheel

$$= M_t = T_1 + T_2 = 45.568 \text{ Nm}$$

Now a single man can apply a maximum force of 400 N and the hand-wheel should not be more than 0.75 m in diameter.

Let the mean radius of the wheel be taken as 0.30 m, therefore, force to be applied by a man

$$= \frac{45.568}{0.30} = 151.9 \text{ N, which is very nominal.}$$

Let the wheel be provided with four arms, therefore, bending moment per arm,

$$M_b = \frac{45.568}{4} = 11.392 \text{ Nm, at the centre of spindle.}$$

Let the section of the arm be elliptical, with major axis = 2 × minor axis.

Now $M_b = f_b \times Z$

where $Z = \frac{\pi}{32} H^2 \times h$ where H = major axis, h = minor axis

or $Z \approx 0.05 \text{ H}^3$

Let $f_b = 21 \text{ N/mm}^2$ for C.I.

$\therefore \quad 0.05H^3 \times 21 = 11.392 \times 1000$

or $H = 22.1$ mm; say 24 mm $\therefore\ h = 12$ mm

Now the taper of the arm may be taken as about 20 mm/metre

$\therefore$ Major axis at the rim $= 24 - \frac{20}{100} \times 30 = 24 - 6 = 18$ mm

$\therefore$ Minor axis at the rim = 9 mm

A hand wheel is shown in Fig. 28.51. The inside and outside of the rim are notched to provide maximum hand grip.

Rim thickness may be taken as 28 mm and the rim width may be taken as 25 mm.

Check for total stress in Spindle: The spindle is subjected to direct compressive stress in the threaded portion due to the torsional moment applied to open or close the valve.

$$f_b = \frac{4P}{\pi d_s^2} = \frac{4 \times 13.46 \times 1000}{\pi \times 400} = 42.8 \text{ N/mm}^2$$

Here d_s = Spindle core diameter of threaded portion.

$$f_s = \frac{16 M_t}{\pi d_s^3} = \frac{16 \times 45.568 \times 1000}{\pi \times 8000} = 28.0 \text{ N/mm}^2$$

$$\therefore \quad f_{max} = \frac{1}{2}\left[f_b + \sqrt{f_b^2 + 4f_s^2}\right] = \frac{1}{2}[42.8 + 70.4]$$

$= 56.6 \text{ N/mm}^2$, which is well within the limits.

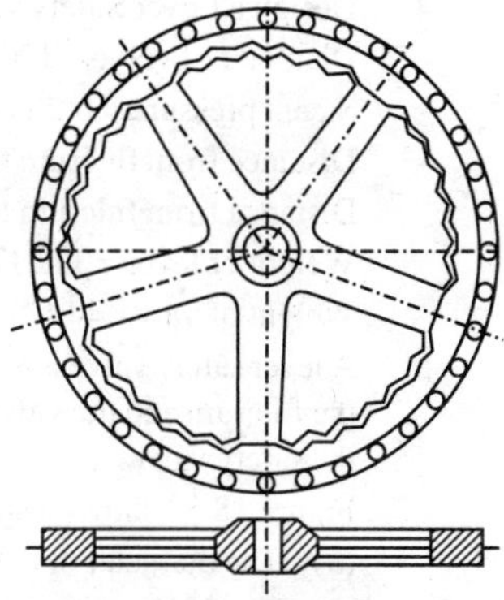

Fig. 28.51 *Hand-wheel*

7. **Stuffing Box and Packing:** Referring Fig. 28.50.

$D_p = 1.25\, d_s + 10$ mm, here d_s = outer diameter of spindle

$= 1.25 \times 28 + 10 = 45$ mm

$\therefore$ Width of packing, $b = \frac{45 - 28}{2} = 8.5$ mm

Outer diameter of neck bush body $= d_s + b = 28 + 8.5 = 36.5$ mm

Thickness of neck bush flange $= b = 8.5$ mm

For class 3 valve, depth of stuffing box is $> 1\frac{1}{4} \times d_s$, i.e., > 35 mm, say 40 mm

8. **Gland:** Outer diameter of the gland body $= 0.99 \times D_p = 0.99 \times 45 = 44.6$ mm

Inner diameter of gland body $= 1.005 \times d_s = 28.14$ mm

$\therefore$ Thickness of the gland body $= \frac{44.6 - 28.14}{2} = 8.23$ mm

$$\text{Depth of gland} = 2b = 17 \text{ mm}$$

The gland is attached to the cover plate with two studs, the total load on the studs

$$= p \times \frac{\pi}{4}(45^2 - 28^2) \times k$$

where k = friction factor, 1.25 to 3.0

$$\therefore \quad \text{Total load on studs} = 1.85 \times \frac{\pi}{4}(45^2 - 28^2) \times 1.5 = 2700 \text{ N}$$

$$\therefore \quad \text{Load per stud} = 1350 \text{ N}$$

$$\therefore \quad \text{Stress area} = \frac{1350}{35} = 38.55 \text{ mm}^2$$

where 35 N/mm^2 is taken as the safe stress.

Using coarse, series, M-10 threads will be O.K. but the diameter of the studs is not to be less than 12.5 mm. Therefore, let us select M 14 threads with pitch 2 mm.

Thickness of flange of gland = 1.5 × stud diameter; say 20 mm

The other dimensions of the gland are shown in Fig. 28.50.

EXERCISES

1. What is the function of a safety valve?
2. Name the various types of safety valves.
3. How would you classify the safety valves based on the valve lift?
4. Design a Lever Safety valve for the following given data:
   Diameter of valve = 100 mm
   Steam pressure = 0.2 N/mm^2
   Distance from fulcrum to the valve = 100 mm
   Distance from fulcrum to the weight = 720 mm
   Weight of lever = 112 N
   Weight of valve = 14 N
5. A lever safety valve is 80 mm in diameter and blows off at pressure of 0.85 N/mm^2. The weight is hung 1.2 m from the fulcrum and the valve is 95 mm from the fulcrum. Neglect the weight of the lever and of the valve and design the safety valve.
6. Figure 28.52 shows a spindle of a 200 mm steam stop valve subjected to a pressure of 1.4 N/mm^2 gauge. Find:
   (*a*) The diameter of the spindle.
   (*b*) The pitch of the screw whose outside diameter is equal to that of the spindle, if the torsional moment on the hand wheel is 135 Nm.
   (*c*) The height of the gun metal nut, whose allowable bearing pressure is 14 N/mm^2.
   (*d*) Size of eight studs on 300 mm pitch diameter cover, assuming allowable stess in these 35 N/mm^2.
   (*e*) Size of four arms of rectangular section of 380 mm diameter hand-wheel, if the stress in these is limited to 14 N/mm^2.

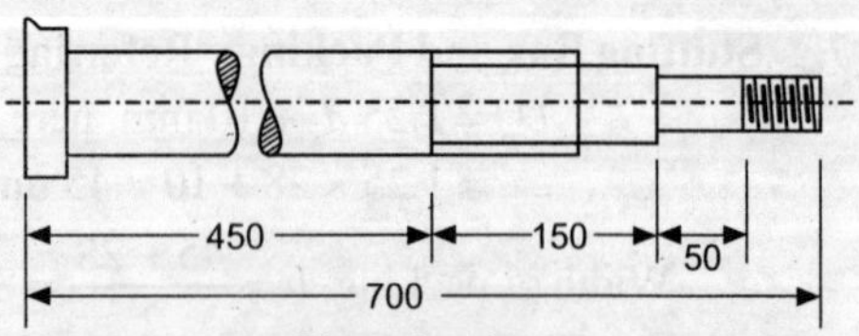

Fig. 28.52

   The spindle is made of mild steel and assume the coefficient of friction between screw and nut as 0.08.
7. Design a spring loaded safety valve for a boiler operating at a pressure of 3.15 N/mm^2 and generating 27500 kg of steam per hour. Give full calculations of the valve and spring and prepare a neat sketch.

28.5 GOVERNORS

The function of the governor is to maintain, as closely as possible, a constant mean speed of rotation of the crank shaft over long periods during which the load on the engine may vary. Its function differs from that of a flywheel in that it has no control over the cyclic fluctuations of speed which arise due to the fluctuations of turning moment on the crank shaft. The governors are generally of two types: (a) Centrifugal, (b) Inertia.

In the centrifugal type, two or more masses known as balls revolve about the axis of a shaft which is driven from the engine crank shaft through a suitable drive. Each ball is acted upon by a radially inward acting force which is provided by a dead weight or a spring or both. This force is termed as controlling force and is equal to the outward centrifugal force which acts on the balls as these revolve around the axis of the shaft, at a uniform speed. When the load on the engine decreases, its speed will increase and the governor balls will fly outwards until the centrifugal force is balanced by the controlling force. The reverse will happen when the load on the engine increases. The movement of the balls is transmitted by the governor mechanism to a valve which controls the amount of energy supplied to the engine. Figure 28.53 shows a well known spring loaded "Hartnell" governor.

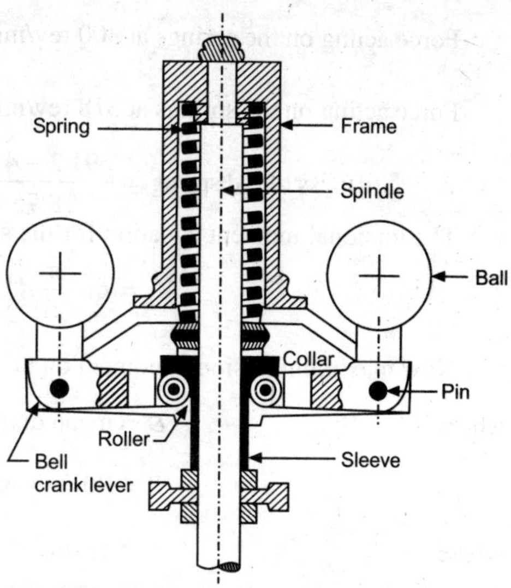

Fig. 28.53 *Spring loaded governor*

Example 28.30. In a Hartnell governor of the type shown in Fig. 28.53 the following data is given:

Weight of each ball = 13.6 N

Length of vertical arm of the bell crank lever = 100 mm

Length of horizontal arm of the bell crank lever = 50 mm

Distance of the fulcrum pin of each bell crank lever from the axis of the rotation = 87.5 mm.

Maximum radius of rotation of the governor balls = 112.5 mm

Minimum radius of rotation of the governor balls = 75 mm

Minimum equilibrium speed = 300 rev/min

Maximum equilibrium speed = 318 rev/min

Design the various components of the governor.

Solution:

1. **Spring:** When the governor balls fly outward due to the centrifugal force, the rollers move in the vertical direction. The deflection of the spring is the distance moved by the rollers as the speed changes from 300 rev/min to 318 rev/min.

As the speed varies, the movement of the ball centres = 112.5 – 75 = 37.5 mm

If the movement of the rollers is assumed to be straight line in the vertical direction, then deflection of the spring $= 37.5 \times \dfrac{50}{100} = 18.75$ mm

Now the centrifugal force acting on the balls at the speed of 300 rev/min is,

$$F_1 = \frac{W_b}{g} \cdot \omega_1^2 \cdot r_1$$

Now $\quad r_1 = 75$ mm; $\omega_1 = \dfrac{2\pi \times 300}{60} = 10\pi$ rad/s

$$\therefore \quad F_1 = \frac{13.6}{9.81} \times (10\pi)^2 \times \frac{75}{1000} = 102.5 \text{ N}$$

$\therefore$ F_2 = centrifugal force acting on the balls at the speed of 318 rev/min

$$= F_1 \cdot \left(\frac{n_2}{n_1}\right)^2 \cdot \frac{r_2}{r_1} = 102.5 \times (1.06)^2 \times \frac{112.5}{75} = 172.8 \text{ N}$$

Force acting on the springs at 300 rev/min $= 2 \times F_1 \times \frac{100}{50} = 410$ N

Force acting on the springs at 318 rev/min $= 2 \times F_2 \times \frac{100}{50} = 691.2$ N

$\therefore$ Stiffness of the spring $= \frac{691.2 - 410}{18.75} = 15$ N/mm

The torsional moment equation for the spring is:

$$M_t = f_s \cdot \frac{\pi}{16} d^3 \cdot \frac{1}{K}$$

Now maximum torsional moment on the spring $= 691.2 \times \frac{D}{2}$ Nm

where D = mean diameter of coil.

K = Stress concentration factor $= \frac{4S-1}{4S-4} + \frac{0.615}{S}$

where S = Spring index, say 6, i.e., $D/d = 6$; d = wire diameter

$$\therefore \quad K = \frac{23}{20} + \frac{0.615}{6} = 1.15 + 0.1025 = 1.2525$$

$$\therefore \quad M_t = 691.2 \times 3d = 2073.6\, d \text{ Nm}$$

Let $f_s = 450$ N/mm^2

$$\therefore \quad 2073.6\,d = 450 \times \frac{\pi}{16} d^3 \times \frac{1}{1.2525}$$

$$\therefore \quad d = 5.4 \text{ mm; say 6 mm}$$

$\therefore$ Diameter of coil, $D = 6 \times 6 = 36$ mm

Now, the deflection equation is $\delta = \frac{8FS^3 z}{Cd}$

$\therefore$ z = number of effective coils $= \frac{\delta \cdot Cd}{8FS^3} = \frac{18.75 \times 84 \times 10^3 \times 6}{8 \times (691.2 - 410) \times 216} = 19.45$

Total number of coils, $z' = z + 1.5 = 19.45 + 1.5 = 20.95$, say 21 turns

$\therefore$ Free length of spring $= (z' - 1) \times \text{gap} + z \times d + \delta_{max}$

Let the gap between two consecutive coils when the spring is fully compressed be 1 mm. (it varies from 0.8 to 1.6 mm).

Now $\delta_{max} = 18.75 \times \frac{691.2}{281.2} = 46.1$ mm

$\therefore$ Free length of spring $= 20 \times 1.0 + 21 \times 6 + 46.1 = 192.1$ mm

2. Lever: (*i*) **Fulcrum-Pin:** The maximum load on the pin will be given as:

$$P = \sqrt{(172.8)^2 + (345.6)^2} = 385.5 \text{ N}$$

The pin shall be designed in bearing,

$$\therefore \quad P = l \times d \times p_b$$

Let $l = d$ and let $p_b = 75$ N/mm^2

$$\therefore \quad 385.5 = 75\, d^2$$

$\therefore \qquad d = 7.17$ mm say 8 mm

$\therefore \qquad l = 8$ mm

Let $\qquad$ boss diameter $= 2d = 16$ mm

(*ii*) **Roller Pin:** Load on each pin $= \dfrac{691.2}{2} = 345.6$ N

$\therefore \qquad P = l \times d \times p_b$

Let $\qquad l = d$; and let $p_b = 75$ N/mm^2

For the sake of the interchangeability, let diameter and lengths of the fulcrum pin and roller pin be same.

This end of the lever will be formed into a fork. The thickness of the eye of the fork can be taken as $\dfrac{d}{2}$, i.e., 4 mm.

The pin shall be in double shear, therefore,

$$f_s = \frac{345.6}{2 \times (\pi/4) \times (8)^2} = 3.44 \text{ N/mm}^2\text{, which is quite low.}$$

(*iii*) **Lever Arm:** Let the material of the lever arm be cast steel.

The maximum bending moment will be just adjacent to the boss diameter of the fulcrum pin, therefore,

$$M_b = 172.8\left(100 - \frac{1}{2} \times 16\right) = 172.8 \times \frac{92}{1000} = 15.9 \text{ Nm}$$

If t = thickness and h = depth of the lever, then

$$M_b = \frac{1}{6} \times t \times h^2 \times f_b$$

Let $h = 3t$ and $f_b = 70$ N/mm^2

$$\therefore \qquad 15.9 \times 1000 = \frac{1}{6} \times \frac{h^3}{3} \times 70,\ h = 16 \text{ mm}$$

$$\therefore \qquad t = 5.33 \text{ mm, say } 5.5 \text{ mm}$$

(*iv*) **Screwed end of lever:** The end of the vertical arm of the bell crank lever is screwed into the governor balls. The maximum bending moment on the screwed end will be given as:

$$M_b = 172.8 \times \text{radius of ball}$$

If the balls are of C.I., then

$$13.6 = \frac{4}{3}\pi r^3 \times 7220 \times 9.81$$

$\therefore \qquad$ Radius of ball, $r = 35.75$ mm; say 36 mm

$$\therefore \qquad M_b = 172.8 \times 0.036 = 6.22 \text{ Nm}$$

$\therefore$ If stress in bending is taken as 70 N/mm^2 and d_c is the core diameter of the screwed end of the lever, then

$$\frac{\pi}{32} d_c^3 \times 70 = 6.22 \times 1000$$

$$\therefore \qquad d_c = 9.68 \text{ mm}$$

$\therefore$ Metric thread M 14 × 1.5 fine series may be selected.

3. Spindle: The spindle is under direct tensile load and a twisting moment.

Maximum tensile load = 691.2 N

Let the weight of the lever, cap etc. be taken as equivalent to 20 N at the ball centre, therefore, twisting moment,

$$M_t = (13.6 + 20) \times (2 \times 112.5) = 33.6 \times \frac{225}{1000} = 7.56 \text{ Nm}$$

$$\therefore \quad f_t = \frac{691.2}{(\pi/4)\times d_c^2} = \frac{880}{d_c^2}\ \text{N/mm}^2$$

$$\therefore \quad f_s = \frac{7.56\times 1000}{(\pi/16)\times d_c^3} = \frac{385\times 100}{d_c^3}\ \text{N/mm}^2$$

Let the safe stress in the spindle be 70 N/mm^2. Considering only the direct tensile load, stress area

$$= \frac{691.2}{70} = 9.874\ \text{mm}^2$$

Let us select M 6 Metric coarse threads, therefore,

$$d_c = 0.84\times 6 = 5.04\ \text{mm}$$

$$\therefore \quad f_t = \frac{880}{(5.04)^2} = 34.6\ \text{N/mm}^2$$

and $$f_s = \frac{385\times 100}{(5.04)^3} = 300\ \text{N/mm}^2\text{, which is a very high value.}$$

Let us try M 12 threads

$$\therefore \quad d_c = 12\times 0.84 = 10.08\ \text{mm}$$

$$\therefore \quad f_t = \frac{880}{(10.08)^2} = 8.65\ \text{N/mm}^2$$

$$f_s = \frac{385\times 100}{(10.08)^3} = 37.25\ \text{N/mm}^2$$

$$\therefore \quad \text{Maximum principal stress} = \frac{f_t}{2} + \sqrt{\left(\frac{f_t}{2}\right)^2 + (f_s)^2} = 4.325 + \sqrt{(4.325)^2 + (37.25)^2}$$

$$= 4.325 + 37.665 = 42\ \text{N/mm}^2\text{, which is O.K.}$$

28.6 VIBRATIONS AND VIBRATION CONTROL

Vibration is the energy which is diverted from a useful purpose to a destructive end. It is a periodic motion which takes place when elastic system is displaced from its equilibrium position and released. Vibration should not be confused with shock which is a disturbance produced by a suddenly applied force. The oscillations due to work die out before the next impulse whereas vibration is a continuous steady state condition. For the proper understanding of vibrations and vibration control, the following terms should be understood;

(*i*) **Free or natural vibration:** When after the initial displacement of the body from its equilibrium position, no external forces act on the body and the motion of the body is maintained by the internal elastic forces, the vibration is known as free natural vibration.

(*ii*) **Damped vibration:** In the case of free vibration, the internal elastic energy will be used up in due course of time in overcoming the internal and external resistances to motion and the body will come to rest to its original equilibrium position. Such a vibration is termed as damped vibration.

(*iii*) **Forced vibration:** When a periodic disturbing force is applied to the body, vibration thus generated is termed as forced vibration.

(*iv*) **Amplitude:** It is the extent of vibration in a body. Full amplitude, (Fig. 28.54) measured from peak to peak in thousandths of a mm, helps in evaluating machine's condition. Half amplitude is the distance moved by a body from its neutral position and it is used to calculate vibration energy.

(*v*) **Period:** It is the interval of time after which the motion of a body i.e. vibration, is repeated.

(*vi*) **Frequency:** The frequency of vibration is the number of complete cycles per sec or per minute.

(*vii*) **Natural frequency:** Every part of a structure, and about everything in nature, has a natural frequency of vibration. It is the frequency at which it will vibrate on receiving a shock. Natural frequency of a mounted machine depends upon the static deflection or resilient supports independent of the mass.

(*viii*) **Forced frequency:** In the case of forced vibration, the body will vibrate at the same frequency as the frequency of the periodic applied force. This frequency is termed as forced frequency or disturbing frequency.

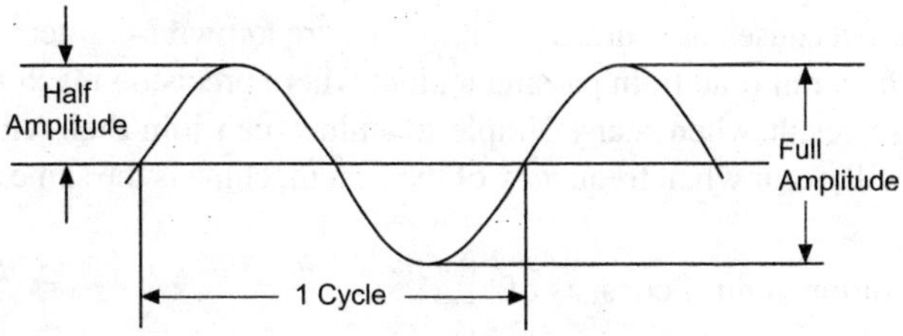

Fig. 28.54 *Frequency waveform*

(*ix*) **Beat frequency:** This frequency occurs when two vibrations of almost equal frequency are present.

(*x*) **Resonance:** When the frequency of the external force applied to a body is the same as its natural frequency, resonance occurs. This results in the body vibrating with larger amplitudes.

(*xi*) **Critical Speed:** Resonance in a rotating shaft will occur when it passes through a speed corresponding to its natural frequency. This speed of the shaft is termed as critical speed and it causes the shaft to bow. The shaft will straighten out again as the critical speed is exceeded. The machine must be designed so that the shaft does not rotate at this speed.

(*xii*) **Steady state condition:** This condition exists when the same vibration cycle repeats itself continuously.

(*xiii*) **Transient:** This is an interruption of vibration pattern over a short period.

A vibration is generally of two types:

(*a*) Rectilinear

(*b*) Torsional.

1. Longitudinal vibration, in which the compression and extension of bars and wires takes place along the axis, e.g., the compression and extension of coil springs.
2. Transverse vibration in which the motion takes place at right angle to the axis of bars, etc.

Torsional vibration is in the form of twisting or rotation such as in shaft. Its amplitude is measured in radians or degrees whereas the magnitude of rectilinear vibration is measured in cms.

The term "degrees of freedom" designates the number of co-ordinates needed to specify the position of an elastic system.

The major cause of vibration in machinery is unbalanced. So, it can be said that vibration is the symptom of unbalance. The basic cause of unbalance is the non-symmetrical weight distribution in the various machine components. The various causes of unbalance can be named as below:

1. Non-symmetry in cast parts.
2. Non-homogeneity of materials, since it is a characteristic of all the materials, regardless of how they are produced.
3. Distribution in the rotating part at operating speed.
4. Poor distribution of mass in a completely machined component.
5. Variation in mass distribution.
6. *Eccentricity:* Even if it is held to a few ten thousandths of a mm, is also a mark of unbalance in rotating equipment, e.g., the inner race of a ball bearing.
7. Bent shafts.
8. Shaft flexing.

9. Shifting of parts that are either pressed on to a shaft or held by other methods can cause a condition similar to flexing of non-symmetrical shafts.
10. Hydraulic vibrations may occur from cavitations in pumps, vibrations excited by fluid flows and other conditions.
11. Temperature may cause changes in structure which increase vibrations by bringing critical speed closer to operating speed. Thus, a machine that is smooth when cold may be rough when hot.

All the above mentioned causes or sources of vibrations are known as "Internal". The vibratory sources can also be "external", e.g., a rail road train passing a plant where precision machining is done. The external vibration sources can also result when many simple machines can join their vibratory forces to set up a resonant condition. This will occur when frequency of the first machine is the same as, or a multiple of natural frequency of the second.

Vibration Control: Vibration control consists of:

(*i*) removing or controlling or balancing the disturbing forces.

(*ii*) or changing the natural frequency so that it is far removed from the operating frequency so as to avoid resonance to occur.

All the vibratory sources mentioned above can be corrected to a degree by dynamic balancing and static balancing. The causes which are not possible to be corrected can be isolated. The natural frequency of a machine can be altered by mounting it on flexible supports. In the case of torsional vibrations, the natural frequency of a shaft can be decreased by : decreasing its diameter, increasing its length or by employing a flexible coupling. The natural frequency of the shaft can be increased by making it more rigid i.e. by increasing its diameter.

Vibration Isolation. If perfect balancing of the machine were possible, there would be no need of isolation of vibrations. But perfection is far off. Hence isolating steps in where balance leaves off. The various steps in isolating the machine vibration are shown in Fig. 28.55.

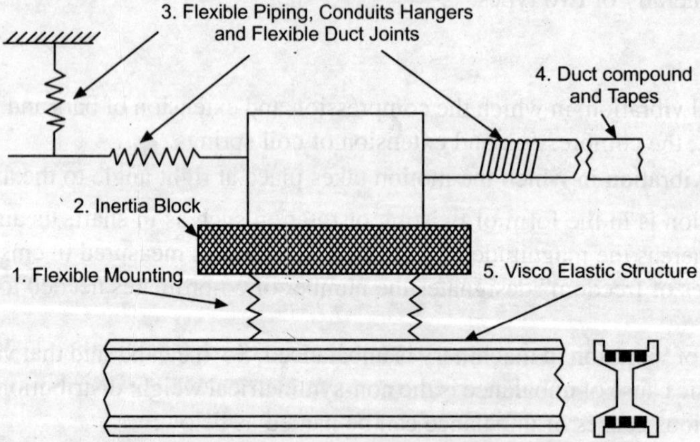

Fig. 28.55 *Vibration isolation*

These steps are:

1. In many cases, it will be satisfactory to place the isolator directly below the machine.
2. If the above is not possible, then the practicality of increasing the machine mass can be investigated by knowing the isolation efficiency required. Isolation efficiency is the ability of any mounting to isolate vibration. It can be expressed as:

$$\eta = \left[1 - \frac{1}{(f_d / f_n)^2 - 1}\right]$$

where f_d = disturbing or forced frequency, and f_n = natural frequency.

3. Once the machine is isolated from floor, then the ways to avoid vibration from transmitting through connected pipes and ducts should be looked into. For this flexible connectors, spring hangers and duct wrappings may be employed.
4. Finally, the building structure can be considered.

Selecting the isolator: Static deflection is the most important single factor in selecting material of the isolator. Another important factor is the damping characteristic of the material. The various isolator materials are:

(*a*) **Springs:** Springs take high static deflection and so steel mountings are widely used. Springs may be made of hard drawn wire, oil tempered wire or a high tensile material such as chrome vanadium.

(*b*) **Rubber:** Where deflection requirements are fairly low, rubber, mountings are widely used. It is practically non-compressible since it is a visco-elastic substance like water.

(*c*) **Felt:** Felt is a fabric built up by the inner locking of fibres through mechanical work, chemical action, moisture and heat, without spinning, knitting or weaving.

(*d*) **Glass Fibre:** Glass fibre blankets are used as the resilient element in "floating floor" construction, because of their valuable dampening properties, chemical inertness and resistance to moisture.

(*e*) **Lead Asbestos:** Pads of lead asbestos find use in building construction and under heavy machinery to dampen vibration.

(*f*) **Cork:** Cork is a good material for isolating vibration, shock and sound. The ground cork particles are compressed under high pressure and then baked with super-heated steam to form the slabs.

(*g*) **Fabric:** Fabric can take big loads. It reduces heavy impact shock and high frequency vibration and noise. It is made of layers of tightly twisted, closely woven light weight cotton duck. Each layer is impregnated with rubber compound. Layers are built up to the needed thickness and vulcanized together.

(*h*) **Plastic:** Synthetic-rubber like plastic has qualities of rubber but retains resiliency over long periods and does not oxide or harden.

28.7 SELECTION OF ELECTRIC MOTOR

The first step in selecting a motor for a particular drive is to obtain the following data:

1. Requirement of the driven machine
2. Electric supply.

A properly selected drive results in lowest cost and optimum performance, hence the drive system should be selected so that its characteristic match, as far as possible, the inherent characteristics of the driven machine or load, which may be listed as below:

(*i*) Power needed
(*ii*) Torsional moment range
(*iii*) Operating cycle - frequency of starts and stops
(*iv*) Speed
(*v*) Operating position - horizontal, vertical or tilted
(*vi*) Direction of rotation
(*vii*) End play and thrust
(*viii*) Ambient temperature
(*ix*) Surrounding conditions - water, gas, corrosion, dust, outdoor, etc.

The information needed about electric supply is

(*i*) Voltage of power system
(*ii*) Number of phases
(*iii*) Frequency

(*iv*) Limitations on starting current

(*iv*) Effect of demand, energy on power rates.

Motors generally operate at best power factor and efficiency when fully loaded.

(*a*) **Torque:** The starting torque needed by the driven machine must be less than the required starting torque of the proposed motor. Motor torque must always be more than the torque needed by the load in going from standstill to full speed.

In some machines, there may be wide fluctuations in their torque requirements. Even if the average torque is low, the many torque peaks may be more than the full load torque. If the load torque impulses are repeated quite frequently as in air compressor, it is best to use a high slip motor with a flywheel. But if the load is generally steady at full load, a more efficient low slip motor may be used. Also the maximum motor torque (breakdown torque) must be higher than load peak torque.

(*b*) **Enclosures:** The environment in which a motor will run is a very important factor to be considered when selecting the motor as on this factor depends the type of insulation and motor enclosure used. The important environmental factors which should be considered are:

(*i*) **Unusual Temperature:**

1. Operation below 10ºC
2. Operation in ambient temperatures above 40ºC
3. Operation in poorly ventilated rooms
4. Operation in places of restricted air circulation.

(*ii*) **Surrounding Air Contaminants:**

1. Exposure to chemical fumes
2. Operation to damp places or exposed to steam
3. Exposure to dust that is combustible, gritty or conductive
4. Exposure to salt water.

(*iii*) **Altitude:** Motor operation at any altitude upto about 1000 m above sea level is usually considered normal. At greater heights, there will be less effective cooling due to reduced density of air. Therefore, the sea level temperature rise of the motor must be less than the standard or the motor should be de-rated when used at high altitudes.

(*iv*) **Outdoor Installation:** When the motor is to be installed outdoor, special considerations should be observed. If special housings are not used, antirust treatment of all the motor parts is necessary even in temperature conditions. Antifungus treatment is also needed in the more troupial climate.

The more enclosed a motor, the more it costs and the hotter it tends to run. Totally enclosed motors may require a large frame size for a given up than open or protected motors.

(*c*) **Insulation:** The type of the insulation to be used is dependent upon the surrounding atmosphere and operating temperature. The life of the motor winding is adversely affected by the temperature. Long insulation life and higher than rated temperature are incompatible. All motors are rated by temperature rise in a standard ambient. The standard ambient is at 40ºC. For special ambient above 40º C, a standard motor can be used at reduced load or reduced life, or a higher temperature insulation class may be used.

(*d*) **Variable Cycle:** When the load variation is according to some regular cycle, it will be economical to select a motor that matches the peak load. Instead, calculate the power needed on the root mean square basis.

28.7.1 Motor Types and Their Characteristics

The types of motors which are in common use are: Induction Motors; Synchronous Motors; and D.C. Motors.

Induction Motors: Because of its ruggedness and simplicity, the induction motor is used more than any other type. The induction motors can be further classified as:

1. Squirrel cage motors
2. Wound rotor motors.

The squirrel cage induction motor is the most common type of motor for industrial purposes. This motor is available in four types:

(*i*) General purpose motor
(*ii*) High Torque motor
(*iii*) High Slip motor
(*iv*) Low Torque motor.

Table 28.12 gives the summary of motor characteristics and applications of Induction Motors which can act as a guide for the selection of an Induction Motor.

Synchronous Motors: These motors are literally "in time" with the power supply. Their average speed is directly proportional to line frequency and inversely proportional to the number of poles in the stator and correspondingly in the rotor. The efficiency is 1 to 3% higher than that of same size and speed induction or dc motor. Also, these can be operated at power factors from 1.0 down to 0.2 leading for plant power factor correction.

Synchronous motors are made in all sizes, ranging from sub-fractional non-excited units to large kW dc excited motors used for industrial drives. Their electric input may vary from a few watts as in the case of electric clocks and timing motors, to thousands of kilo watts for large industrial motors driving compressors and pumps. Small high speed synchronous motors cost more than comparable induction motors, but with low speed and large high speed motor, costs favour the synchronous motors. Hence synchronous motors find applications for the constant speed service, direct connection to slow speed machines and where power factor correction is required.

D.C. Motors: Fundamentally, the d.c. motor is an ac motor with a commutating device. These motors are inherently adjustable speed machines. With DC motors, wide and economical ranges of speed control and starting torques are possible. But, for constant speed service, ac motors are generally preferred because they are more rugged and have lowest first cost. DC motors are excellently suited to drives where:

(*a*) Wide speed range, small incremental speed changes, or speed matching is required.
(*b*) Smoothness of acceleration or deceleration is important.
(*c*) Controlled rate of acceleration or deceleration is needed.
(*d*) Torque for tension needs to be limited or controlled.

For most industrial uses, dc motors are of integral kW ratings, but fractional kW sizes can also be made available.

DC motors are classified as below according to the type of field winding.

1. Shunt Wound.
2. Series Wound.
3. Compound Wound.

In Table 28.13 are given the characteristics and applications of the various types of DC motors.

Table 28.12 Characteristics and applications of induction motors

Type of Motor	Speed Regulation	Speed Control	Starting Torque	Maximum Torque	Slip	Starting Current	Applications
1. General purpose Squirrel Cage	Drops about 3% for large to 5% for smaller sizes.	None, except multispeed types.	Normal	Design 1: High Design 2: Normal	Low	Design 1: Normal Design 2: Low	Constant speed service where starting torque is not high, e.g., fans, blowers; rotary; compressors, centrifugal pumps, Machine tools, generators.
2. High Torque Squirrel Cage	Drops about 3% for large to 6% for small sizes	None, except multispeed types	High	Normal	Low	Low	Constant speed service where fairly high starting torque is required at infrequent intervals, e.g., reciprocating pumps and compressors, crushers, etc.
3. High slip Squirrel Cage	Drops about 10% to 15% from no load to full load.	None, except multispeed types.	Very high	Very high	High	Low	Constant speed service and high starting torque, if starting is not too frequent, and for taking high peak loads with or without flywheels, e.g., punch presses, shears, elevators etc.
4. Low-Torque Squirrel Cage	Drops about 3% for large to 5 % for small sizes.	None, except multispeed types.	Low	Low	...	Low	Constant speed service where starting duty is light, e.g., fans, blowers, centrifugal pumps, etc.
5. Wound-Rotor	With rotor rings short circuited drops about 3% for large to 5% for small sizes.	Speed can be reduced to 5% by rotor resistance to obtain stable operation. Speed varies inversely as load.	High	Normal	Low	Low	Where high starting torque with low starting current or where limited speed control is required, e.g., fans, centrifugal and plunger pumps, compressors, conveyors, hoists, cranes.

Table 28.13 Characteristics and applications of D.C. motors

Type of Motor	Speed Regulations	Speed Control	Starting Torque	Break down Torque	Applications
1. Series	Varies inversely as the load, Races on light load and full voltage.	Zero to maximum depending on control and load.	High, varies as square of voltage limited by commutation, heating and line capacity.	High limited by commutation heating and line capacity.	Where high starting torque is required and speed can be regulated, e.g., traction, hoists, bridges, gates, car retarders etc.
2. Shunt	Drops 3 to 5% from no load to full load.	Any desired range, depending upon motor design and type of system.	Good, with constant field, varies directly as voltage applied to armature.	-do-	Where constant or adjustable speed is required and starting conditions are not severe, e.g., fans, blowers, centrifugal pumps, conveyors, wood working machines, metal working machines, elevators.
3. Compound	Drops 7 to 20% from no load to full load depending upon amount of compounding.	Any desired range, depending on motor design and type of control.	Higher than for shunt, depending upon amount of compounding.	-do-	Where high starting torque, combined with fairly constant speed is required, e.g., plunger pumps, shears, punch presses, bending rolls, hoists, elevators, geared elevators.

❑❑❑

APPENDIX I

SI UNITS

The "System International d′ Unite's"— The International System of Units abbreviated as SI was formally recognised by the General Conference on Weights and Measures in 1960. SI base units, supplementary units and derived units are given below in Table I.1

Table I.1 SI units

S. No.	Physical Quantity	Unit	Symbol
	Base Units		
1.	Mass	kilogram	kg
2.	Length	metre	m
3.	Time	second	s
4.	Temperature	degree kelvin	K
5.	Current	ampere	A
6.	Luminous intensity	candela	cd
7.	Amount of substance	mole	mol

Supplementary Units

S. No.	Physical Quantity	Unit	Symbol
1.	Plane angle	radian	rad
2.	Solid angle	steradian	sr

Derived Units

S. No.	Physical Quantity	Unit	Symbol
1.	Force	newton	N, kgm/s^2
2.	Energy, work, heat	joule	J = Nm, kg m^2/s^2
3.	Power	watt	$W = \frac{J}{s} = \frac{\text{Nm}}{\text{s}}$, kgm^2/s^3
4.	Pressure, stress	pascal	$Pa = N/m^2$, kg/ms^2
5.	Frequency	hertz	$Hz = s^{-1}$
6.	Potential	volt	$V = W/A$
7.	Resistance	ohm	$\Omega = V/A = W/A^2$
8.	Charge	coulomb	$C = As$
9.	Capacitance	farad	$F = As/V = A^2s/W$
10.	Magnetic flux	weber	$Wb = As/V = A^2s/W$
11.	Inductance	henry	$H = Vs/A = Ws/A^2$
12.	Conductance	siemen	$S = V/A = W/A^2$
13.	Magnetic induction	tesla	$T = Wb/m^2 = Vs/m^2 = Ws/Am^2$
14.	Luminous flux	lumen	$lm = cd\ sr$
15.	Illuminance	lux	$lm/m^2 = cd\ sr/m^2$

S.I. units of common physical quantities used in mechanical engineering are given in Table I.2, below:

Table I.2 SI Units of common physical quantities

S. No.	Physical Quantity	Unit	Unit Symbol
1.	Acceleration	metre/second^2	m/s^2
2.	Angular acceleration	radian/second^2	rad/s^2
3.	Angular displacement	radian	rad
4.	Angular momentum	kilogram metre^2/second	kgm^2/s
5.	Angular velocity	radian/second	rad/s
6.	Area	square metre	m^2
7.	Calorific value	joule/metre^3	J/m^3
8.	Density	kilogram/metre^3	kg/m^3
9.	Discharge	metre^3/second	m^3/s
10.	Displacement	metre	m
11.	Energy, Enthalpy	joule	J
12.	Entropy	joule/kilogram kelvin	J/kg K
13.	Force	newton	N
14.	Frequency	per second	Hz, s^{-1}
15.	Heat	joule	J, Nm
16.	Length	metre	m
17.	Mass	kilogram	kg
18.	Moduli of elasticity	newton/metre^2	Pa, N/m^2
19.	Moment	newton metre	Nm
20.	Momentum	kilogram metre/second	kgm/s
21.	Moment of Inertia	kilogram metre^2	kgm^2
22.	Plane angle	radian	rad
23.	Power, Radiant flux	watt	W, Nm/s
24.	Pressure, stress	newton/metre^2	Pa, N/m^2
25.	Solid angle	steradian	sr
26.	Specific energy and Specific enthalpy	joule/kilogram	J/kg
27.	Specific volume	metre^3/kilogram	m^3/kg
28.	Specific heat	joule/kilogram K	J/kg K
29.	Speed	metre/second	m/s
30.	Surface tension	netwon/metre	N/m
31.	Temperature	kelvin	K
32.	Time	second	s
33.	Thermal capacity	joule/kilogram K	J/kg K
34.	Torque	newton metre	Nm
35.	Velocity	metre/second	m/s
36.	Viscosity (kinematic)	metre^2/second	m^2/s
37.	Vicosity (dynamic)	newton second/metre^2 or kilogram/metre second	Ns/m^2 or kg/ms
38.	Volume	metre^3	m^3
39.	weight	newton	N
40.	Work	joule	J, Nm

In S.I. units, multiples and submultiples find great use. These are given in Table I.3

Table I.3 Multiples and submultiples

exa	E	10^{18}	deci	d	10^{-1}
peta	P	10^{15}	centi	c	10^{-2}
tera	T	10^{12}	milli	m	10^{-3}
giga	G	10^{9}	micro	μ	10^{-6}
mega	M	10^{6}	nano	n	10^{-9}
kilo	k	10^{3}	pico	p	10^{-12}
hecto	h	10^{2}	femto	f	10^{-15}
deca	da	10^{1}	atto	a	10^{-18}

Note: In SI units, the following units are acceptable for the various physical quantities:

(i) Length : m, km, cm, mm
(ii) Area : m^2, cm^2, mm^2
(iii) Volume : m^3, cm^3, mm^3
Angle : rad, sr, degrees
(iv) Mass : kg, Mg, g, mg
(v) Velocity and Acceleration : m/s, km/h, Hz, rad/s, m/s^2, rad/s^2
(vi) Density and Specific volume : kg/m^3, g/cm^3, kg/cm^3, m^3/kg, cm^3/kg
(vii) Discharge : m^3/s, *l*/s, kg/s, cm^3/s
(viii) Force and Weight : N, kN, MN, μN, mN
(ix) Pressure, Stress and Elastic Moduli : Pa, kPa, MPa, GPa, bar, mPa, μPa
(x) Energy, Work, Heat : J, kJ, MJ, mJ, kWh
(xi) Power : W, kW, MW, GW, mW, μW
(xii) Torque and Moment : Nm, kNm, MNm, μNm
(xiii) Momentum and Angular Momentum : kgm/s. kgm^2/s
(xiv) Moment of Inertia : kgm^2
(xv) Concentration : kg/m^2, g/m^2
(xvi) Viscosity : Ns/m^2, kg/ms, Pas

Conversion factors for converting common MKS units into S.I. units are given in Table I.4.

Table I.4 Conversion tables

S. No.	To Convert the Following	Into SI	Multiply By	Conversely Multiply By
1.	Litres (*l*)	m^3	10^{-3}	10^3
2.	degrees (Angles)	rad	0.017453	57.2958
3.	minutes (Angles)	rad	0.000291	3.436×10^3
4.	second (Angles)	rad	0.0000048	0.2083×10^6
5.	slug	kg	9.807	0.1020
6.	r.p.m.	rad/s	0.1047	9.55
7.	c.p.s.	Hz	1	1
8.	kgf	N	9.807	0.1020
9.	kgf/metre^2	Pa, N/m^2	9.807	0.1020
10.	kgf/cm^2	kPa	98.07	0.0102
11.	bar	kPa	100	0.01

S. No.	To Convert the Following	Into SI	Multiply By	Conversely Multiply By
12.	atm (atmospheres)	kPa	101.325	9.862×10^{-3}
13.	mm of mercury	kPa	0.1333	7.501
14.	kgf.metre	J	9.807	0.1020
15.	ergs	J	10^{-7}	10^{7}
16.	kilowatt hours	MJ	3.6	0.2778
17.	calories	J	4.187	0.2388
18.	horse power hours	MJ	2.685	0.3724
19.	ergs/second	W	10^{-7}	10^{7}
20.	horse power	kW	0.7355	1.36
21.	kilo calories/minute	W	69.77	0.0143
22.	k cal/m hour°C	W/m°C	1.163	0.86
23.	k cal/kg	kJ/kg	4.187	0.2388
24.	k cal/kg°C	kJ/kg K	4.187	0.2388
25.	k cal/m^2 hour °C	W/m^2 °C	1.163	0.86
26.	poise, P, (gram/cm second)	kg/ms (Ns/m^2)	0.1	10
27.	centipoise, cP	Ns/m^2	0.001	1000
28.	stokes, St (centimeter^2/second)	m^2/s	10^{-4}	10^{4}
29.	centistokes, cSt	m^2/s	10^{-6}	10^{6}

Some suggestions for the use of SI units:

1. The names of the units after the scientists are not written in capital letters. For example, the unit of force is written as newton and not as Newton after whom the unit is named.
2. The symbols of the units named after the scientists are written in capital letters. For example, N for newton and J for joule.
3. Symbols for units do not have plural forms. For example, 40 metres will be written as 40 m and not as 40 ms which would mean milli seconds, because, 's' is the symbol for seconds. Similarly, it will be wrong to write 'cms' and 'kgs' for centimetres and kilograms respectively.
4. No full stop, no dots, no dashes are put between the symbols for units, except at the end of sentences. For example, torsional moment is Nm and not N.m or N-m.
5. Decimal fractions always begin with 0. For example, half metre will be written as 0.5 m and not as .5 m.
6. Since 'm' is the symbol of both 'metre' and 'milli', avoid writing the symbol for metre (m) before the symbol of the name of the unit. For example, for torque, it will be Nm and not mN which would mean milli newton.
7. The multiple and submultiple prefixes in steps of 1000 only are recommended. For example, length can be expressed in mm, m, or km but not in cm unless a valid reason exists.
8. When SI units are raised to a power, the prefixes are also raised to the same power. Thus, $km^2 = (1000\ m)^2 = (1000)^2\ (m^2) = 10^6\ m^2$.
9. Except for kilogram which is a base unit, prefixes should not be used in the denominators of derived units. Thus meganewton per square metre (MN/m^2) is satisfactory, but newton per square millimetre N/mm^2 is not recommended. However, N/mm^2 has gained international acceptance and seems to be favoured over MN/m^2 for stress, although they are equal.
10. Double prefixes should not be used. For example, we do not write Mkm for 10^6 km. Here '*k*' is a prefix for 10^3 and 'Mkm' has a double prefix 'Mk' with m. This is not allowed. Similarly, instead of millimillimetres mmm, use micrometres, μm.

11. No prefix is to be used with 'kg'. First convert the mass in kilogram to that in gram and then use the prefix. Thus 10^{-6} kg is not to be written as μkg, but,
10^{-6} kg = 10^{-3} g = mg
12. No degree mark (°) is used with kelvin, the unit of temperature. A temperature interval can also be expressed in 'C'.
13. The expression 'per' in symbols of derived units is always indicated by a fraction line as m/s or m/s. The word 'per' is never used.
14. No RPM or RPS, but rev/min or rev/s are allowed.
15. Numbers having four or more digits are placed in groups of three and separated by a space instead of a comma. For example,
192 423. 618 50 but not 192, 423 61850

However, the space may be omitted for the special case of numbers having four digits.

APPENDIX II
BASIC LOAD RATINGS OF ROLLING CONTACT BEARINGS

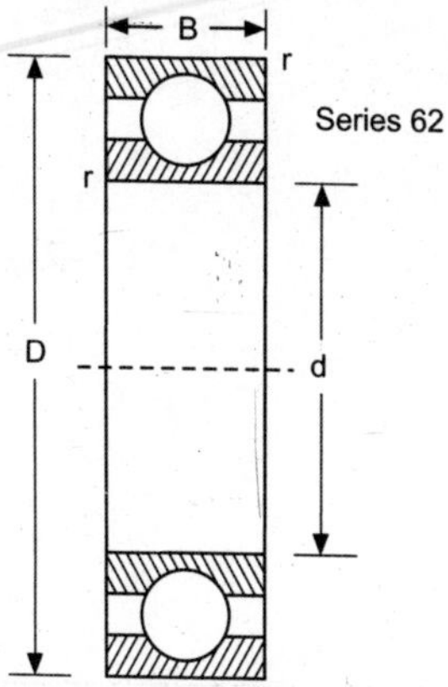

Table II.1 Deep Groove Ball Bearings

ISI NO	Bearing of basic design No. (SKF)	*d* mm	*D* mm	*B* mm	*r* mm	Basic capacity, *N*		Max. permissible speed rev/min
						Static C_o	Dynamic C	
(1)	(2)	(3)	(4)	(5)	(6)	(7)	(8)	(9)
10BC02	6200	10	30	9	1	2160	3925	20000
12BC02	01	12	32	10	1	2940	5250	20000
15BC02	02	15	35	11	1	3430	5980	16000
17BC02	6203	17	40	12	1	4315	7355	16000
20BC02	04	20	47	14	1.5	6375	9805	16000
25BC02	05	25	52	15	1.5	6965	10690	13000
30BC02	6206	30	62	16	1.5	9805	14710	13000
35BC02	07	35	72	17	2	13535	19615	10000
40BC02	08	40	80	18	2	15495	22165	10000
45BC02	6209	45	85	19	2	17750	24910	8000
50BC02	10	50	90	20	2	20595	27070	8000
55BC02	11	55	100	21	2.5	25300	33340	8000
60BC02	6212	60	110	22	2.5	31580	40210	6000
65BC02	13	65	120	23	2.5	35715	42905	6000
70BC02	14	70	125	24	2.5	38440	47070	5000
75BC02	6215	75	130	25	2.5	41385	50600	5000
80BC02	16	80	140	26	3	44130	55505	5000
85BC02	17	85	150	28	3	53450	63745	4000
90BC02	6218	90	160	30	3	60800	74040	4000
95BC02	19	95	170	32	3.5	71100	82870	4000
100BC02	20	100	180	34	3.5	79925	97145	3000
105BC02	6221	105	190	36	3.5	90710	101500	3000
110BC02	2	110	200	38	3.5	101010	108850	3000
120BC02	24	120	215	40	3.5	101010	110815	3000
	622	130	230	40	4	112780	120130	2500
	28	140	250	42	4	127680	126510	2500
	30	160	270	45	4	139745	135330	2500
	32	170	31[illegible]	48	4	151020	140235	2000
	6234	180	320	52	5	184365	159850	2000
	36	190	340	5[illegible]	5	200060	169165	1600
	38	200	360	58	5	235360	196130	1600
	40					259880	205940	1600

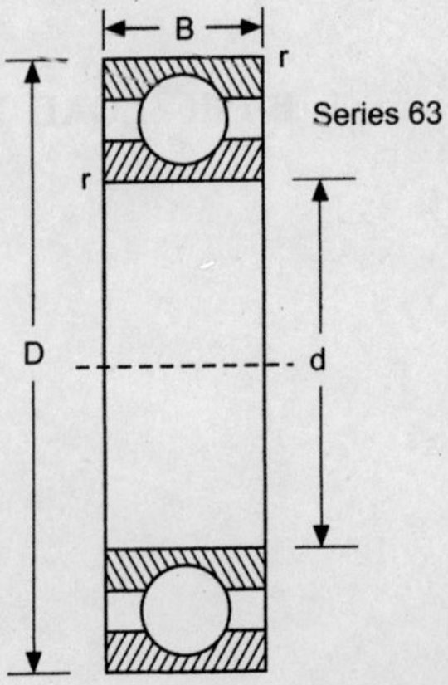

Table II.2 Deep groove ball bearings

ISI NO	Bearing of basic design No. (SKF)	*d* mm	*D* mm	*B* mm	*r* mm	Basic capacity, *N* Static C_o	Dynamic C	Max. permissible speed rev/min
(1)	(2)	(3)	(4)	(5)	(6)	(7)	(8)	(9)
10BC03	6300	10	35	11	1	3570	6080	16000
12BC03	01	12	37	12	1.5	4220	7550	16000
15BC03	02	15	42	13	1.5	5100	8580	16000
17BC03	6303	17	47	14	1.5	6080	10300	13000
20BC03	04	20	52	15	2	7550	12260	13000
25BC03	05	25	62	17	2	10150	15985	10000
30BC03	6306	30	72	19	2	14220	20990	10000
35BC03	07	35	80	21	2.5	16970	25300	8000
40BC03	08	40	90	23	2.5	20990	31380	8000
45BC03	6309	45	100	25	2.5	29225	40700	8000
50BC03	10	50	110	27	3	34720	47070	6000
55BC03	11	55	120	29	3	41190	53940	6000
60BC03	6312	60	130	31	3.5	47070	62270	5000
65BC03	13	65	140	33	3.5	53450	71100	5000
70BC03	14	70	150	35	3.5	60800	79925	5000
75BC03	6315	75	160	37	3.5	71100	89280	4000
80BC03	16	80	170	39	3.5	78450	94140	4000
85BC03	17	85	180	41	4	85810	101500	4000
90BC03	6318	90	190	43	4	96110	107870	3000
95BC03	19	95	200	45	4	107870	117680	3000
100BC03	20	100	215	47	4	129450	1352[illegible]	3000
105BC03	6321	105	225	49	4	140240	140235	2500
110BC03	22	110	240	50	4	16[illegible]2790	152980	2500
120BC03	24	120	260	55	4	191230	157890	2500
	6326	130	280	58	5	215750	173580	2500
	28	140	300	[illegible]5	5	249090	196130	2000
	30	150	320	[illegible]	[illegible]	[illegible]	210840	2000

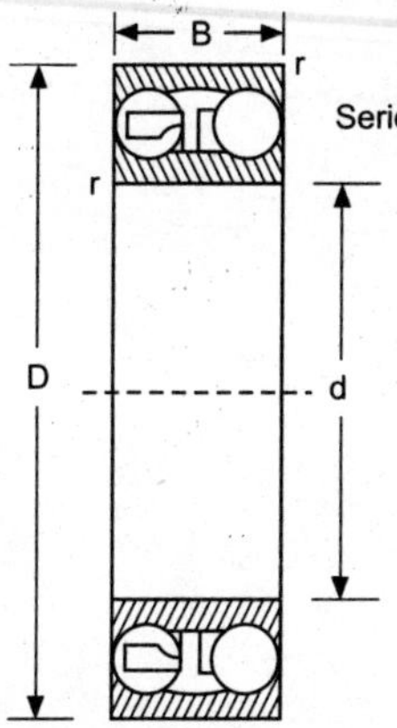

Series 22

Table II.3 Self-aligning ball bearings

Bearing with cylindrical bore No. (SKF)	Bearing with taper bore No. (SKF)	*d* mm	*D* mm	*B* mm	*r* mm	Basic capacity, *N*		Max. permissible speed rev/min
						Static C_o	Dynamic C	
(1)	(2)	(3)	(4)	(5)	(6)	(7)	(8)	(9)
2200		10	30	14	1	1670	5540	20000
01		12	32	14	1	1960	5640	20000
02		15	35	14	1	2110	5740	16000
2203		17	40	16	1	2745	7500	16000
04	2204 K	20	47	18	1.5	3825	9610	16000
05	05 K	25	52	18	1.5	4120	9610	13000
2206	2206 K	30	62	20	1.5	5390	11770	13000
07	07 K	35	72	23	2	7845	16180	10000
08	08 K	40	80	23	2	8825	16920	10000
2209	2209 K	45	85	23	2	9810	17410	8000
10	10 K	50	90	23	2	10490	17410	8000
11	11 K	55	100	25	2.5	12455	20200	8000
2212	2212 K	60	110	28	2.5	15300	25810	6000
13	13 K	65	120	31	2.5	19610	33340	6000
14	—	70	125	31	2.5	21080	34080	5000
2215	2215 K	75	130	31	2.5	21575	34080	5000
16	16 K	80	140	33	3	24520	37850	5000
17	17 K	85	150	36	3	29030	44620	4000
2218	2218 K	90	160	40	3	35550	53450	4000
19	19 K	95	170	43	3.5	42170	63740	4000
20	[illegible]	100	180	46	3.5	49030	76000	3000
2221	—	105	190	50	3.5	53940	82870	3000
22	2222 K	110	200	53	3.5	62270	96600	3000

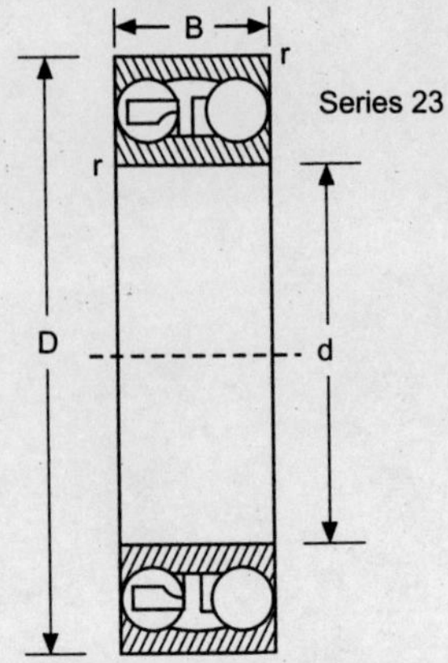

Table II.4 Self-aligning ball bearings (Contd...)

Bearing with cylindrical bore No. (SKF)	Bearing with taper bore No. (SKF)	d mm	D mm	B mm	r mm	Basic capacity, N Static C_o	Dynamic C	Max. permissible speed rev/min
(1)	(2)	(3)	(4)	(5)	(6)	(7)	(8)	(9)
2301		12	37	17	1.5	2940	8925	16000
02		15	42	18	1.5	3285	9070	13000
03		17	47	19	1.5	4070	10790	13000
2304	2304 K	20	52	21	2	5345	13730	10000
05	05 K	25	62	24	2	7600	18490	10000
06	06 K	30	72	27	2	9810	24030	8000
2307	2307 K	35	80	31	2.5	12945	29810	8000
08	08 K	40	90	33	2.5	15450	34720	6000
09	09 K	45	100	36	2.5	19220	40940	6000
2310	2310 K	50	110	40	3	23630	50750	5000
11	11 K	55	120	43	3	28050	56390	5000
12	12 K	60	130	46	3.5	32750	66693	4000
2313	2313 K	65	140	48	3.5	38490	73550	4000
14	—	70	150	51	3.5	43640	82375	4000
15	15 K	75	160	55	3.5	50750	92670	3000
16	16 K	80	170	58	3.5	56390	102970	3000
2317	2317 K	85	180	60	4	59580	106890	2500
18	18 K	90	190	64	4	68160	115720	2500

Table II.5 Single row angular contact ball bearings

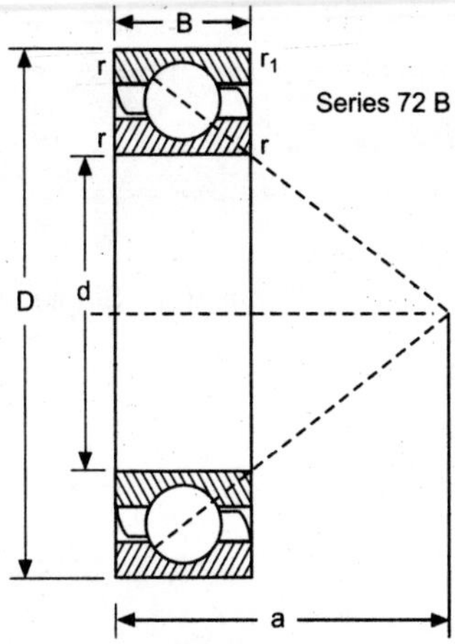

ISI No.	Bearing No. (SKF)	d mm	D mm	B mm	r mm	r_1 mm	a mm	Basic capacity, N Static C_o	Dynamic C	Max. permissible speed rev/min
(1)	(2)	(3)	(4)	(5)	(6)	(7)	(8)	(9)	(10)	(11)
15BA02	7202 B	15	35	11	1	0.5	16	3680	6080	13000
17BA02	03 B	17	40	12	1	0.8	18	4460	7700	13000
20BA02	04 B	20	47	14	1.5	0.8	21	6370	10150	10000
25BA02	7205 B	25	52	15	1.5	0.8	24	7700	11280	10000
30BA02	06 B	30	62	16	1.5	0.8	27	10790	15400	10000
35BA02	07 B	35	72	17	2	1	31	14710	20590	8000
40BA02	7208 B	40	80	18	2	1	34	18490	24520	8000
45BA02	09 B	45	85	19	2	1	37	21180	27655	6000
50BA02	10 B	50	90	20	2	1	39	23140	28440	6000
55BA02	7211 B	55	100	21	2.5	1.2	43	29175	36285	6000
60BA02	12 B	60	110	22	2.5	1.2	47	36285	42900	5000
65BA02	13 B	65	120	23	2.5	1.2	50	42410	49030	5000
70BA02	7214 B	70	125	24	2.5	1.2	53	44620	52470	5000
75BA02	15 B	75	130	25	2.5	1.2	56	49030	54430	4000
80BA02	16 B	80	140	26	3	1.5	59	55650	60800	4000
85BA02	7217 B	85	150	28	3	1.5	64	63740	69630	4000
90BA02	18 B	90	160	30	3	1.5	67	75760	81400	4000
95BA02	19 B	95	170	32	3.5	2	71	85810	92670	3000
100BA02	7220 B	100	180	34	3.5	2	76	90710	98070	3000
105BA02	21 B	105	190	36	3.5	2	80	101500	106890	2500
110BA02	22 B	110	220	38	3.5	2	84	112780	117680	2500

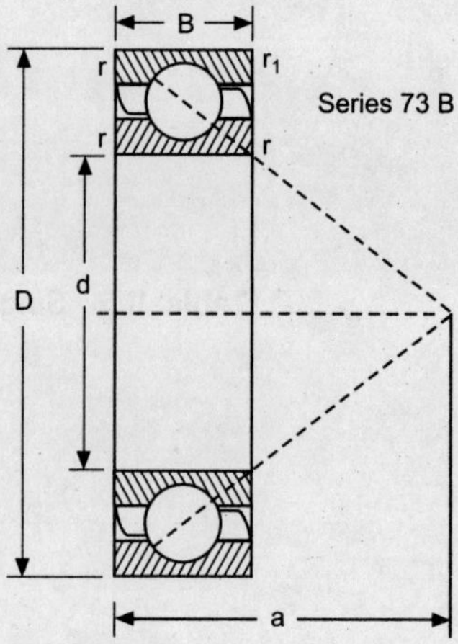

Table II.6 Single row angular contact ball bearings (Contd...)

ISI No.	Bearing No. (SKF)	d mm	D mm	B mm	r mm	r_1 mm	a mm	Basic capacity, N Static C_o	Dynamic C	Max. permissible speed rev/min
(1)	(2)	(3)	(4)	(5)	(6)	(7)	(8)	(9)	(10)	(11)
17BA03	7303 B	17	47	14	1.5	0.8	21	7110	11283	10000
20BA03	04 B	20	52	15	2	1	23	8140	13580	10000
25BA03	05 B	25	62	17	2	1	27	12260	18930	10000
30BA03	7306 B	30	72	19	2	1	31	15840	24025	8000
35BA03	07 B	35	80	21	2.5	1.2	35	20005	28050	8000
40BA03	08 B	40	90	23	2.5	1.2	39	24910	34715	6000
45BA03	7309 B	45	100	25	2.5	1.2	43	33340	44620	6000
50BA03	10 B	50	110	27	3	1.5	47	40210	51485	6000
55BA03	11 B	55	120	29	3	1.5	52	46580	59820	5000
60BA03	7312 B	60	130	31	3.5	2	55	53450	69630	5000
65BA03	13 B	65	140	33	3.5	2	60	61290	78450	5000
70BA03	14 B	70	150	35	3.5	2	64	72570	87280	4000
75BA03	7315 B	75	160	37	3.5	2	68	80415	96600	4000
80BA03	16 B	80	170	39	3.5	2	72	89240	102970	4000
85BA03	17 B	85	180	41	4	2	76	98070	111550	3000
90BA03	7318 B	90	190	43	4	2	80	111500	120130	3000
95BA03	19 B	95	200	45	4	2	84	122580	129250	2500
100BA03	20 B	100	215	47	4	2	90	149060	144650	2500
105BA03	7321 B	105	225	49	4	2	94	160090	153230	2500
110BA03	22 B	110	240	50	4	2	99	188780	168920	2500

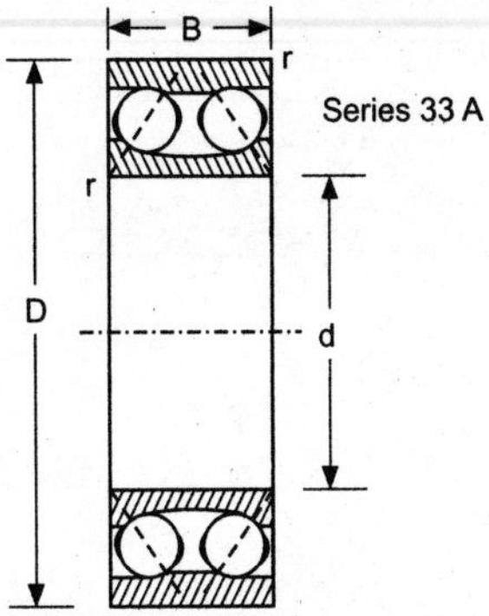

Table II.7 Double row angular contact ball bearings

Bearing No. (SKF)	d mm	D mm	B mm	r mm	Basic capacity, N Static C_o	Dynamic C	Max. permissible speed rev/min
(1)	(2)	(3)	(4)	(5)	(6)	(7)	(8)
3302 A	15	42	19.0	1.5	9070	13730	10000
03 A	17	47	22.2	1.5	12650	18930	8000
04 A	20	52	22.2	2	13730	18930	8000
3305 A	25	62	25.4	2	19615	26085	6000
06 A	30	72	30.2	2	27165	35300	6000
07 A	35	80	34.9	2.5	35600	43640	5000
3308 A	40	90	36.5	2.5	44620	53450	5000
09 A	45	110	39.7	2.5	54430	62270	4000
10 A	50	110	44.4	3	72570	80170	4000
3311 A	55	120	49.2	3	78450	85910	4000
12 A	60	130	54.0	3.5	94630	98070	3000
13 A	65	140	58.7	3.5	108950	115720	3000
3314 A	70	150	63.5	3.5	126510	135820	3000
15 A	75	160	68.3	3.5	138080	140235	2500
16 A	80	170	86.3	3.5	153965	157890	2500
3317 A	85	180	73.0	4	173580	173580	2500
18 A	90	190	73.0	4	205940	200150	2500

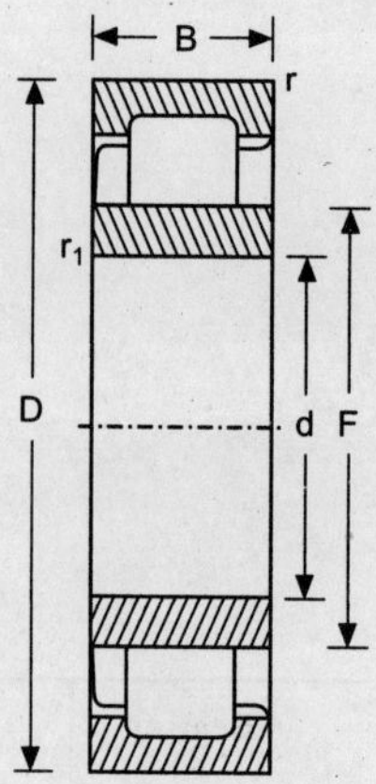

Table II.8 Cylindrical roller bearings, series NU 22

Bearing No. (SKF)	D mm	D mm	B mm	R mm	R_1 mm	F mm	Basic capacity, N Static C_o	Dynamic C	Max. permissible speed rev/min
(1)	(2)	(3)	(4)	(5)	(6)	(7)	(8)	(9)	(10)
NU 2205	25	52	18	1.5	1	32	11030	15790	13000
2206	30	62	20	1.5	1	38.5	16970	23140	13000
2207	35	72	23	2	1	43.8	27655	35600	10000
NU 2208	40	80	23	2	2	50	32750	40700	10000
2209	45	85	23	2	2	55	35600	43640	8000
2210	50	90	23	2	2	60.4	38540	45500	8000
NU 2211	55	100	25	2.5	2	66.5	45500	52560	8000
2212	60	110	28	2.5	2.5	73.5	59820	69630	6000
2213	65	120	31	2.5	2.5	79.6	74040	81640	6000
NU 2214	70	125	31	2.5	2.5	84.5	78450	85910	5000
2215	75	130	31	2.5	2.5	88.5	84730	96350	5000
2216	80	140	33	3	3	95.3	98070	109340	5000
NU 2217	85	150	36	3	3	101.8	117680	127000	4000
2218	90	160	40	3	3	107	135820	140235	4000
2219	95	170	43	3.5	3.5	113.5	160340	173580	4000
NU 2220	100	180	46	3.5	3.5	120	184855	196130	3000
2222	110	200	53	3.5	3.5	132.5	226435	254170	3000
2224	120	215	58	3.5	3.5	143.5	271645	280470	3000
NU 2226	130	230	64	4	4	156	306700	296650	2500
2228	140	250	68	4	4	169	378540	362850	2500
2230	150	270	73	4	4	182	436400	423650	2500

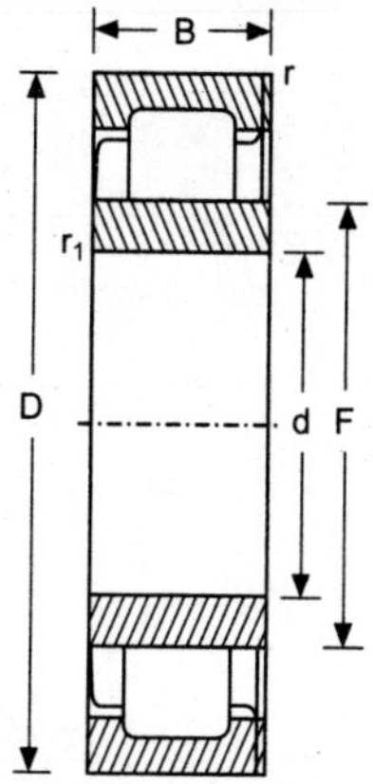

Table II.9 Cylindrical roller bearings, series NU 23

Bearing No. (SKF)	d mm	D mm	B mm	r mm	r_1 mm	F mm	Basic capacity, N Static C_o	Dynamic C	Max. permissible speed rev/min
(1)	(2)	(3)	(4)	(5)	(6)	(7)	(8)	(9)	(10)
NU 2205	25	62	24	2	2	32	21280	30205	10000
2306	30	72	27	2	2	42	26430	35305	10000
2307	35	80	31	2.5	2	46.2	30695	41780	8000
NU 2308	40	90	33	2.5	2.5	53.5	47270	58840	8000
2309	45	100	36	2.5	2.5	58.5	55700	74040	8000
2310	50	110	40	3	3	65	74040	90960	6000
NU 2311	55	120	43	3	3	70.5	82965	106890	6000
2312	60	130	46	3.5	3.5	77	103460	129250	5000
2313	65	140	48	3.5	3.5	83.5	120130	142690	5000
NU 2314	70	150	51	3.5	3.5	90	138270	155925	5000
2315	75	160	55	3.5	3.5	95.5	166710	200150	4000
2316	80	170	58	3.5	3.5	103	184855	213785	4000
NU 2317	85	180	60	4	4	108	196130	227415	3000
2318	90	190	64	4	4	115	222610	254000	3000
2319	95	200	67	4	4	121.5	260760	292140	3000
NU 2320	100	215	73	4	4	129.5	309890	341170	3000
2322	110	240	80	4	4	143	423650	455030	2500
2324	120	260	86	4	4	154	499160	557510	2500
NU 2326	130	280	93	5	5	167	637430	609970	2500
2328	140	300	102	5	5	180	726670	736480	2000
2330	150	320	108	5	5	193	815910	815910	2000

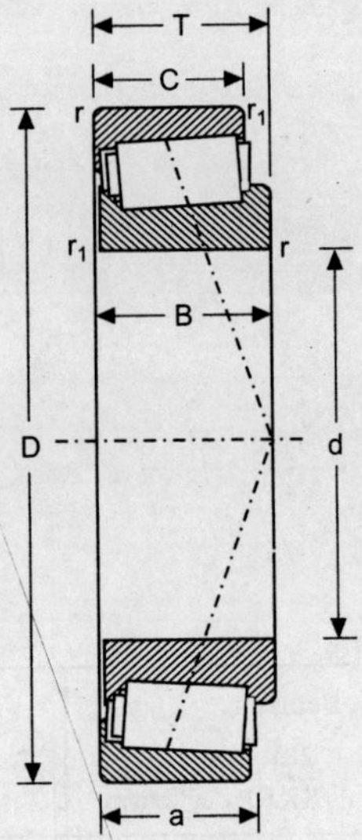

Table II.10 Taper roller bearings, series 322

Bearing No. (SKF)	d mm	D mm	B mm	T mm	C mm	r mm	r_1 mm	a mm	Basic capacity, N Static C_o	Dynamic C	Max. permissible speed rev/min
(1)	(2)	(3)	(4)	(5)	(6)	(7)	(8)	(9)	(10)	(11)	(12)
32206	30	62	20	21.25	17	1.5	0.5	15	27165	31675	6000
07	35	72	23	24.25	19	2	0.8	18	36285	41430	6000
08	40	80	23	24.75	19	2	0.8	19	40210	45500	6000
32209	45	85	23	24.75	19	2	0.8	20	45500	49915	5000
10	50	90	23	24.75	19	2	0.8	21	47270	51580	5000
11	55	100	25	26.75	21	2.5	0.8	22	61050	65115	4000
32212	60	110	28	29.75	24	2.5	0.8	24	75710	78450	4000
13	65	120	31	32.75	27	2.5	0.8	26	90810	96105	4000
14	70	125	31	33.25	27	2.5	0.8	28	90810	96105	3000
32215	75	130	31	33.25	27	2.5	0.8	29	99930	101450	3000
16	80	140	33	35.25	28	3	1	30	113760	117680	3000
17	85	150	36	38.5	30	3	1	33	134840	135820	2500
32218	90	160	40	42.5	34	3	1	36	160730	158080	2500
19	95	170	43	45.5	37	3.5	1.2	38	180440	180440	2500
20	100	180	46	49	39	3.5	1.2	41	207705	202510	2500
32221	105	190	50	53	43	3.5	1.2	44	241240	236340	2000
22	110	200	53	56	46	3.5	1.2	46	271645	254000	2000
24	120	213	53	61.5	50	3.5	1.2	52	334410	298120	2000
32226	130	230	64	67.75	54	4	1.5	56	407960	369710	1600
28	140	250	68	71.75	58	4	1.5	60	481510	422670	1600
30	150	270	73	77	60	4	1.5	64	534460	480525	1600

Table II.11 Single thrust ball bearings

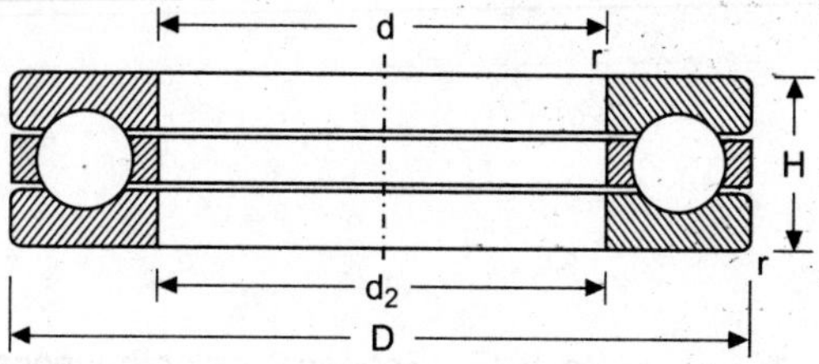

ISI No.	Bearing No. (SKF)	d mm	D mm	H mm	d_2 mm	r mm	Basic capacity, kg Static C_0	Dynamic C	Max. permissible speed rev
(1)	(2)	(3)	(4)	(5)	(6)	(7)	(8)	(9)	(10)
10TA12	51200	10	26	11	10.2	1	1400	1000	10000
12TA12	01	12	28	11	12.2	1	1560	1040	8000
15TA12	02	15	32	12	15.2	1	2040	1220	8000
17TA12	51203	17	35	12	17.2	1	2200	1270	8000
20TA12	04	20	40	14	20.2	1	3100	1730	6000
25TA12	05	25	47	15	25.2	1	4150	2160	6000
30TA12	51206	30	52	16	30.2	1	4800	2280	6000
35TA12	07	35	62	18	35.2	1.5	6400	3050	5000
40TA12	08	40	68	19	40.2	1.5	7650	3450	5000
45TA12	51209	45	73	20	45.2	1.5	8650	3650	4000
50TA12	10	50	78	22	50.2	1.5	9150	3750	4000
55TA12	11	55	90	25	55.2	1.5	13200	5500	3000
60TA12	51212	60	95	26	60.2	1.5	14600	5700	3000
65TA12	13	65	100	27	65.2	1.5	15600	5850	2500
70TA12	14	70	105	27	70.2	1.5	16300	6000	2500
75TA12	51215	75	110	27	75.2	1.5	17300	6100	2500
80TA12	16	80	115	28	80.2	1.5	18000	6200	2000
85TA12	17	85	125	31	85.2	1.5	22000	7500	2000
90TA12	51218	90	135	35	90.2	2	27000	9150	2000
100TA12	20	100	150	38	100.2	2	34000	11400	1600
110TA12	22	110	160	38	110.2	2	37500	12000	1600
120TA12	51224	120	170	39	120.2	2	39000	12000	1300
130TA12	26	130	190	45	130.3	2.5	51000	16000	1300
140TA12	28	140	200	46	140.3	2.5	54000	16300	1000
150TA12	51230	150	215	50	150.3	2.5	60000	17600	1000
160TA12	32	160	225	51	160.3	2.5	62000	18000	1000
170TA12	34	170	240	55	170.3	2.5	73000	21200	800
180TA12	51236	180	250	56	180.3	2.5	76500	21600	800
190TA12	38	190	270	62	190.3	3	91500	25000	800
200TA12	40	200	280	62	200.3	3	96500	25500	600
	51244	220	300	63	220.3	3	1040000	26000	600
	48	240	340	78	240.0	3.5	143000	33500	500
	51252	260	660	79	263.3	3.5	156000	34500	500
	56	280	380	80	280.3	3.5	163000	35500	400
	60	300	420	95	300.3	4	204000	45500	400
	51264	320	440	95	320.4	4	236000	46500	400
	68	340	460	96	340.4	4	245000	47500	300
	72	360	500	100	360.4	5	315000	58500	300

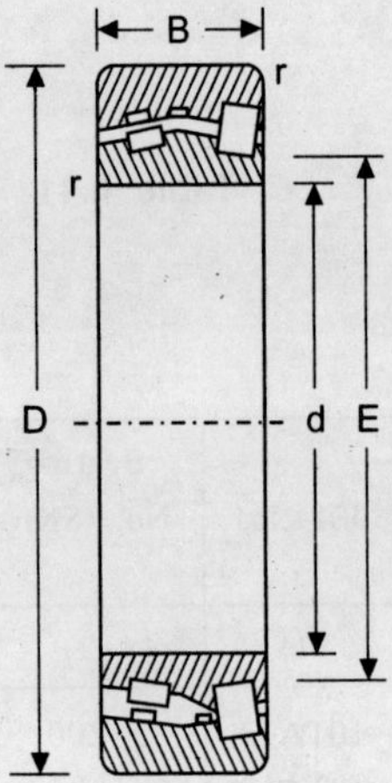

Table II.12 Spherical roller bearings

Bearing with Cylindrical bore	d mm	D mm	H mm	d_2 mm	r mm	Basic capacity, kg Static C_0	Basic capacity, kg Dynamic C	Max. permissible speed rem.
(1)	(2)	(3)	(4)	(5)	(6)	(7)	(8)	(9)
*22205 C	25	52	18	32	1.5	2000	3000	8000
*60 C	30	62	20	38	1.5	3100	4500	6000
70 C	35	72	23	44	2	4050	6200	5000
22208 C	40	80	23	50	2	4800	6200	5000
09 C	45	85	23	55	2	5200	6400	5000
10 C	50	90	23	60	2	5500	6700	400
22211 C	55	100	25	66	2.5	6800	8300	400
12 C	60	110	28	73	2.5	8500	10000	4000
13 C	65	120	31	79	2.5	10200	11800	3000
22214 C	70	125	31	84	2.5	10800	12200	3000
15 C	75	130	31	90	2.5	11200	12700	3000
16 C	80	140	33	95	3	14000	15300	2500
22217 C	85	150	36	100	3	16300	18000	2500
18 C	90	160	40	107	3	19600	20800	2500
19 C	95	170	43	113	3.5	22800	24500	2000
22220 C	100	180	46	120	3.5	25500	27000	2000
22 C	110	200	53	132	3.5	34500	34500	2000
24 C	120	215	58	143	3.5	40000	40000	1600
22226 C	130	230	64	153	4	48000	46500	1600
28 C	140	250	68	167	4	54000	53000	1600
30 C	150	270	73	179	4	65500	64000	1300
22232 C	160	290	80	191	4	78000	73500	1300
34 C	170	310	86	204	5	90000	83000	1000
36 C	180	320	86	213	5	95000	88000	1000
22238 C	190	340	92	226	5	102000	950000	1000
40 C	200	360	98	238	5	116000	108000	800
44 C	220	400	108	264	5	143000	129000	800

Note: For bearing taper bore add K-Example 22208 CK. Table 1 : 12 on diameter.
*Note available with tapper bore.

Table II.13 Double thrust ball bearings flat housing washers—Series 522

Series 522

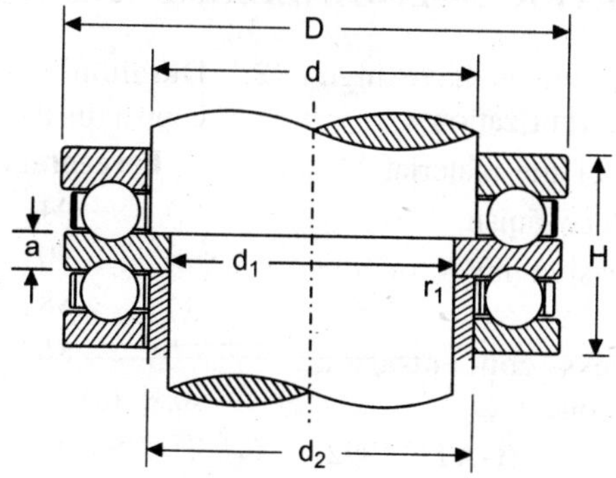

Bearing No. (SKF)	d mm	1 mm	D mm	H mm	d_2 mm. mm	a mm	r ≈ mm	r_1 ≈ mm	Basic capacity, N		Max. permissible speed rev/min
									Static C_o	Dynamic C	
(1)	(2)	(3)	(4)	(5)	(6)	(7)	(8)	(9)	(10)	(11)	(12)
52202	15	10	32	22	15.2	5	1	0.5	20	1220	8000
04	20	15	40	26	20.2	6	1	0.5	3100	1730	6000
05	25	20	47	28	25.2	7	1	0.5	4150	2160	6000
52206	30	25	52	29	30.2	7	1	0.5	4800	2280	6000
07	35	30	62	34	35.2	8	1.5	0.5	6400	3050	5000
08	40	30	68	36	40.2	9	1.5	1	7650	3450	5000
09	45	35	73	37	45.2	9	1.5	1	8650	3650	4000
10	50	40	78	39	50.2	9	1.5	1	9150	3750	4000
52211	55	45	90	45	52.2	10	1.5	1	13200	5500	3000
12	60	50	95	46	60.2	10	1.5	1	14600	5700	3000
13	65	55	100	47	65.2	10	1.5	1	15600	5850	2500
14	70	55	105	47	70.2	10	1.5	1.5	16300	6000	2500
15	75	60	110	47	75.2	10	1.5	1.5	17300	6100	2500
52216	80	65	115	48	80.2	10	1.5	1.5	18000	6200	2000
17	85	70	125	55	85.2	12	1.5	1.5	22000	7500	2000
18	90	75	135	62	90.2	14	2	1.5	27000	9150	2000
2040	100	85	150	67	100.2	15	2	1.5	34000	11400	1600

APPENDIX III

PROBLEMS FROM COMPETITIVE EXAMINATIONS (GATE, IES, IAS)

CHAPTER 2—ENGINEERING MATERIALS

1. Carburized machine components have high endurance limit because carburization

(*a*) raises the yield point of the material.

(*b*) produces a better surface finish.

(*c*) introduces a compressive layer on the surface.

(*d*) suppresses any stress concentration produced in the component.

Ans: (*c*) **(GATE 1992)**

2. The alloying element mainly used to improve the endurance strength of steel materials is

(*a*) nickel. (*b*) Vanadium.

(*c*) molybdenum. (*d*) tungsten.

Ans: (*b*) **(GATE 1997)**

3. Match list I with list II and select the correct answer using the codes given below the lists:

List I (Steel type)	**List II (Product)**
A Mild steel	1. Screw driver
B Tool steel	2. Commercial beams
C Medium carbon steel	3. Crane hooks
D High carbon steel	4. Blanking dies

Codes:	A	B	C	D
(*a*)	1	4	3	2
(*b*)	2	4	1	3
(*c*)	1	3	4	2
(*d*)	2	4	3	1

Ans: (*d*) **(IES 1993)**

4. Which of the following statements are true of annealing of steels?

1. Steels are heated to 500° to 700°C.
2. Cooling is done slowly and steadily.
3. Internal stresses are relieved.
4. Ductility of steel is reduced.

Codes:

(*a*) 2, 3 and 4 (*b*) 1, 3 and 4

(*c*) 1, 2 and 4 (*d*) 1, 2 and 3

Ans: (*d*) **(IES 1993)**

5. Duralumin alloy contains aluminium and copper in the ratio of

	Percentage Al	**Percentage Cu**
(*a*)	94	4
(*b*)	90	8
(*c*)	88	10
(*d*)	86	12

Ans: (*a*) **(IES 1993)**

6. The blade of power saw is made of

(*a*) Boron steel. (*b*) High speed steel.

(*c*) stainless steel. (*d*) malleable cast iron.

Ans: (*b*) **(IES 1993)**

7. Quartz is a

(*a*) ferroelectric material.

(*b*) ferromagnetic material.

(*c*) piezoelectric material.

(*d*) diamagnetic material.

Ans: (*c*) **(IES 1993)**

8. Match list I with list II and select the correct answer using the codes given below the lists:

List I (material properties)	**List II** (tests to determine material properties)
A Ductility	1. Impact test
B Toughness	2. Fatigue test
C Endurance limit	3. Tension test
D Resistance to penetration	4. Hardness test

Codes:	A	B	C	D
(*a*)	3	2	1	4
(*b*)	4	2	1	3
(*c*)	3	1	2	4
(*d*)	4	1	2	3

Ans: (*c*) **(IES 1993)**

9. For a linearly elastic, isotropic and homogeneous material, the number of elastic constants required to relate stress and strain is:

(*a*) two. (*b*) Three.

(*c*) four. (*d*) six.

Ans: (*c*) **(IAS 1994)**

10. The unit of elastic modulus is same as those of

(*a*) stress, shear modulus and pressure.

(*b*) strain, shear modulus and force.

(*c*) shear modulus, stress and force.

(*d*) stress, strain and pressure.

Ans: (*a*) **(IAS 1994)**

11. Assertion (A) : Endurance limits for all materials are always less than the ultimate strength of the corresponding materials.

Reason (R) : Stress concentraton in a machine part due to any dislocation is very damaging when the part is subjected to variable loading.

Ans: Both A and R are true. **(IAS 1994)**

12. Which one of the following materials is highly elastic?

(*a*) Rubber (*b*) Brass

(*c*) Steel (*d*) Glass

Ans: (*c*) **(IAS 1995)**

13. Match list I with list II and select the correct answer using the codes given below the lists:

List I (Property)	List II (Testing Machine)
A Tensile strength	1. Rotating bending machine
B Impact strength	2. Three point loading machine
C Bending strength	3. Universal testing machine
D Fatigue strength	4. Izod testing machine

Codes:	A	B	C	D
(*a*)	4	3	2	1
(*b*)	3	2	1	4
(*c*)	2	1	4	3
(*d*)	3	4	2	1

Ans: (*d*) **(IAS 1995)**

14 Match list I with list II and select the correct answer using the codes given below the lists:

List I (Material/part)	List II (Techniques)
A Ductile iron	1. Inoculation
B Malleable iron	2. Chilled
C Rail steel joints	3. Annealing
D White cast iron	4. Thermit welding
	5. Isothermal annealing

Codes:	A	B	C	D
(*a*)	1	3	4	2
(*b*)	5	3	2	1
(*c*)	2	1	4	5
(*d*)	1	4	2	3

Ans: (*a*) **(IAS 1995)**

15. Assertion (*A*) : In high speed steels, alloying elements tungsten, chromium, and Vanadium are added to make them suitable to work at higher cutting speeds than tool steel or low alloy steels.

Reason (R) : Vanadium adds to the property of red hardness and tungsten and chromium adds to high wear resistance.

Ans: Both A and R are true. **(IAS 1995)**

16. Consider the following treatments:

1. Normalising 2. Hardening
3. Martempering 4. Cold working

Hardness and tensile strength in austenitic stainless steel can be increased by

(*a*) 1, 2 and 3. (*b*) 1 and 3.

(*c*) 2 and 4. (*d*) 4 alone.

Ans: (*d*) **(IES 1994)**

17. Killed steels:

(*a*) have minimum impurity level.

(*b*) have almost zero percentage of phosphorous and sulphur.

(*c*) are produced by L.D. process.

(*d*) are free from oxygen.

Ans: (*d*) **(IES 1994)**

18. Which of the following pairs are correctly matched?

1. Silicon steels-transformer stampings
2. Duralumin-Cooking utensils
3. Gun metal-bearings

Codes:

(*a*) 1, 2 and 3 (*b*) 1 and 2

(*c*) 1 and 3 (*d*) 2 and 3

Ans: (*a*) **(IES 1994)**

19. Which of the following are correctly matched?

1. Lead screw nut-phosphor bronze
2. Piston - Cast iron
3. Cam - EN - 31 steel
4. Lead screw - wrought iron

Codes:

(*a*) 2, 3 and 4 (*b*) 1, 3 and 4

(*c*) 1, 2 and 4 (*d*) 1, 2 and 3

Ans: (*d*) **(IES 1994)**

20. Assertion (A) : Non-ferrous materials are best machined with diamond tools.

Reason (R) : Diamond tools are suitable for high speed machining.

Ans: A is false and R is true **(IES 1995)**

21. Assertion (*A*) : An aluminium alloy with 11 percent silicon is used for making engine pistons by die casting technique.

Reason (R) : Aluminium has low density and addition of silicon improves its fluidity and, therefore, its castability

Ans: A and R, both are true. **(IES 1995)**

22. Match list I with list II and select the correct answer using the codes given below the list:

List I (Heat treatment)	**List II (Effect on the properties)**
A Annealing	1. Refined grain structure
B Nitriding	2. Improves the hardness of the whole mass
C Martempering	3. Increases surface hardness
D Normalising	4. Improves ductility

Codes:	A	B	C	D
(*a*)	4	3	2	1
(*b*)	1	3	4	2
(*c*)	4	2	1	3
(*d*)	2	1	3	4

Ans: (*a*) **(IES 1995)**

23. Consider the following work materials

1. Titanium 2. mild steel
3. stainless steel 4. Grey Cast iron

The correct sequence of these materials in terms of increasing order of difficulty in machining is

(*a*) 4, 2, 3, 1. (*b*) 4, 2, 1, 3.

(*c*) 2, 4, 3, 1. (*d*) 2, 4, 1, 3.

Ans: (*a*) **(IES 1995)**

24. Addition of magnesium to cast iron increases its

(*a*) hardness

(*b*) ductility and strength in tension

(*c*) corrosion resistance

(*d*) creep strength

Ans: (*b*) **(IES 1995)**

25. Small amounts of which one of the following elements/pair of elements is added to steel to increase its machinability?

(*a*) Nickel

(*b*) Sulphur and phosphorous

(*c*) Silicon

(*d*) Manganese and copper

Ans: (*b*) **(IES 1996)**

26. Which of the following pairs regarding the effect of alloying elements in steel are correctly matched?

1. Molybdenum: Forms abrasion resisting particles
2. Phosphorous: Improves machinability in free cutting steels
3. Cobalt: Contributes to red hardness by hardening ferrite
4. Silicon: Reduces oxidation resistance

Codes:

(*a*) 2, 3 and 4 (*b*) 1, 3 and 4

(*c*) 1, 2 and 4 (*d*) 1, 2 and 3

Ans: (*a*) **(IES 1996)**

27. 18/8 stainless steel contains

(*a*) 18% steel, 8% chromium

(*b*) 18% chromium, 8% nickel

(*c*) 18% tungsten, 8% nickel

(*d*) 18% tungsten, 8% chromium

Ans: (*b*) **(IES 1996)**

28. Tin base white metals are used where the bearings are subjected to

(*a*) Large surface wear.

(*b*) elevated temperature.

(*c*) light load and pressure.

(*d*) high pressure and load.

Ans: (*d*) **(IES 1996)**

29. Alloy steel which is work hardenable and which is used to make the blades of bulldozers, bucket wheel excavators and other earth moving equipment contain iron, carbon and

(*a*) Chromium. (*b*) silicon.

(*c*) manganese. (*d*) magnesium.

Ans: (*c*) **(IES 1996)**

30. The highest stress that a material can withstand for a specified length of time without excessive deformation is called

(*a*) Fatigue strength.

(*b*) endurance strength.

(*c*) creep strength.

(*d*) creep rupture strength.

Ans: (*c*) **(IES 1997)**

31. Match list I with list II and select the correct answer using the codes given below the list:

List I	List II
A Toughness method	1. Moment area method
B Endurance strength	2. Hardness
C Resistance to abrasion	3. Energy absorbed before fracture in a tension test
D Deflection in a beam	4. Fatigue loading

Codes:	A	B	C	D
(*a*)	4	3	1	2
(*b*)	4	3	2	1
(*c*)	3	4	2	1
(*d*)	3	4	1	2

Ans: (*c*) **(IES 1997)**

32. Which one of the following materials is used for car tyres as a standard material?

(*a*) Styrene-butadiene rubber (SBR)

(*b*) butyl rubber

(*c*) Nitrile rubber

(*d*) Any of the above depending on need

Ans: (*a*) **(IES 1997)**

33. Assertion (A) : The notch sensitivity of cast iron component is zero.

Reason (R) : Cast iron does not have a yield point.

Ans: Both A and R are correct. **(IES 1997)**

34. Match list I (alloying element in steel) with list II (property conferred on steel by the element) and select the correct answer using the codes given below the list.

List I	List II
A Nickel	1. Corrosion resistance
B Chromium	2. Magnetic permeability
C Tungsten	3. Heat resistance
D Silicon	4. Hardenability

Codes:	A	B	C	D
(*a*)	4	1	3	2
(*b*)	4	1	2	3
(*c*)	1	4	3	2
(*d*)	1	4	2	3

Ans: (*a*) **(IES 1998)**

35. Match list I (alloys) with list II (applications) and select the correct answer using the codes given below the list:

List I	List II
A Chromel	1. Journal bearing
B Babbit alloy	2. Milling cutter
C Nimonic alloy	3. Thermocouple wire
D High speed steel	4. Gas turbine blades

Codes:	A	B	C	D
(*a*)	3	1	4	2
(*b*)	3	4	1	2
(*c*)	2	4	1	3
(*d*)	2	1	4	3

Ans: (*a*) **(IES 1998)**

36. Assertion (A) : Cast iron is generally hard, brittle and wear resistant.

Reason (R) : Cast iron contains more than 2.0 percent Carbon and as such the percentage of cementite in it is higher.

Ans: Both A and R are correct. **(IES 1998)**

37. Match list I and list II and select the corrrect answer using the codes given below

List (Materials)	**List II (Applications)**
A Tungsten carbide	1. Abrasive wheels
B Silicon Nitride	2. Heating elements
C Aluminium oxide	3. Pipes for carrying liquid metals
D Silicon Carbide	4. Drawing dies

Codes:	A	B	C	D
(*a*)	3	4	1	2
(*b*)	4	3	2	1
(*c*)	3	4	2	1
(*d*)	4	3	1	2

Ans: (*d*) **(IES 1999)**

38. Heating the steels to 30ºC above the critical temperature line, soaking at that temperature and then cooling slowly to room temperature to form a pearlite and ferrite structure is known as

(*a*) hardening. (*b*) normalising.
(*c*) tempering. (*d*) annealing.

Ans: (*d*) **(IES 1999)**

39. Cast iron is used for machine beds because of its high

(*a*) tensile strength.
(*b*) endurance strength.
(*c*) damping capacity.
(*d*) compressive strength.

Ans: (*c*) **(IES 1999)**

40. The total area under the stress strain curve of a mild steel specimen tested upto failure under tension is a measure of

(*a*) ductility. (*b*) ultimate strength.
(*c*) stiffness. (*d*) Toughness.

Ans: (*d*) **(GATE 2002)**

CHAPTER 4—DESIGNING FOR STRENGTH

1. Fatigue strength of a rod subjected to cyclic axial force is less than that of a rotating beam of the same dimensions subjected to steady lateral force because

(*a*) axial stiffness is less than bending stiffness.
(*b*) of absence of centrifugal effects in rod.
(*c*) the number of discontinuities vulnerable to fatigue are more in the rod.
(*d*) at a particular time, the rod has only one type of stress whereas the beam has both the tensile and compressive stresses.

Ans: (*d*) **(GATE 1992)**

2. A 4.5 cm diameter steel shaft is subjected to a static axial load of 125 kN (tensile). A completely reversed bending moment of 358 Nm is superposed on the shaft. Due to the presence of notches, the stress concentration factor was found to be 1.5. Find the factor of safety if the following material data is given: Ultimate tensile strength = 600 MPa, Endurance limit = 250 MPa.

Ans: (4) **(GATE 1992)**

3. The yield strength of a steel shaft is twice its endurance limit.Which of the following torque fluctuations represent the most critical situation according to Soderberg criterion?

(*a*) – T to + T (*b*) – T/2 to + T
(*c*) 0 to + T (*d*) + T/2 to + T.

(GATE 1993)

Solution:

Hint: Soderberg criterion is

$$\frac{1}{FS} = \frac{f_m}{f_{yp}} + \frac{f_v}{f_e}$$

Now $f_{yp} = 2f_e$

$$\therefore \quad \frac{1}{FS} = \frac{f_m}{2f_e} + \frac{f_u}{f_e} \quad \text{or} \quad \frac{f_e}{FS} = \frac{f_m}{2} + f_v$$

Find FS for all the four cases. It will be seen that it comes out to be least for case (*a*). Thus, it is the most critical situation.

4. A two dimensional state of stress at a point is given by $\sigma_x = 60$ MN/m^2, $\sigma_y = 20$ MN/m^2 and τ_{xy}.
   (*a*) What is the maximum permissible magnitude of the shear stress τ_{xy} if the larger principal stress is not to exceed 75 MN/m^2?
   (*b*) What is the magnitude of the other principal stress and the maximum shear stress?

   Ans: (52.20 MN/m^2, – 31 MN/m^2, 51.20 MN/m^2)

   (GATE 1994)

5. A test specimen is stressed slightly beyond the yield point and then unloaded. Its yield strength will
   (*a*) decrease.
   (*b*) increase.
   (*c*) remain same.
   (*d*) become equal to ultimate tensile strength.

   Ans: (*b*) **(GATE 1995)**

6. A thin cylinder of 100 mm internal diameter and 5 mm thickness is subjected to an internal pressure of 10 MPa and a torque of 2000 Nm. Calculate the magnitude of principal stresses.

   Ans: (100.13 MPa, 49.87 MPa) **(GATE 1996)**

7. A cube of 5 mm side is loaded as shown
   (*a*) Determine the principal stresses.
   (*b*) Will the cube yield if the yield strength of the material is 70 MPa. Use Von Mise's theory. **(GATE 1996)**

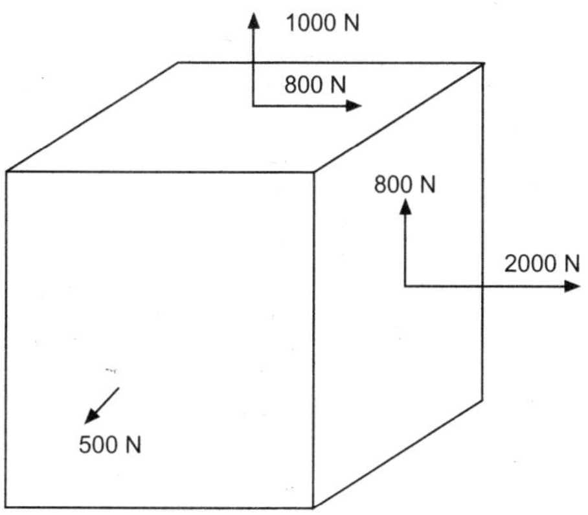

8. A component in an aircraft flap actuator can be adequately modelled as a cylindrical bar subjected to an axial force of 8.0 kN, a bending moment of 55 Nm and a torsional moment of 30 Nm. A 20 mm solid bar of 7075-*t*6 aluminium having Sut = 591MPa, S_{yp} = 542 MPa and yield point stress in shear = 271 MPa, is recommended for its use. Determine the factor of safety available as per stress strain theory.

   Ans: (3.985 on the basis of yield stress in Shear). **(GATE 1997)**

9. Decreasing grain size in a polycrystalline material
   (*a*) increases yield strength and corrosion resistance.
   (*b*) decreases yield strength and corrosion resistance.
   (*c*) decreases yield strength but increases corrosion resistance.
   (*d*) Increases yield strength but decreases corrosion resistance.

   Ans: (*a*) **(GATE 1998)**

10. Which theory of failure will you use for aluminium components under steady loading:
    (*a*) Principal stress theory
    (*b*) Principal strain theory
    (*c*) Strain energy theory
    (*d*) Maximum shear stress theory.

    Ans: (*d*) **(GATE 1999)**

11. Under repeated loading, a material has the stress strain curve as shown. Which of the following statements is true?

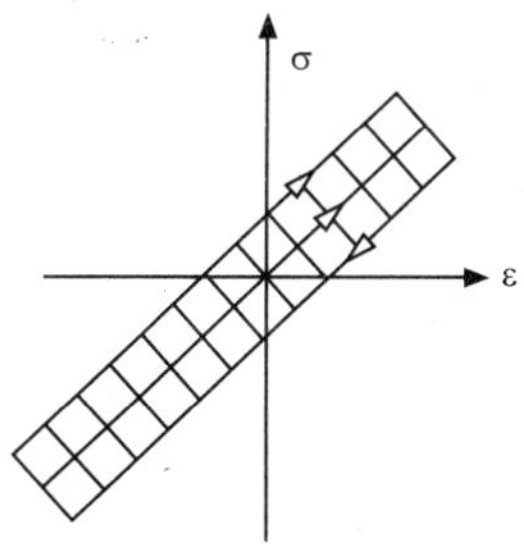

    (*a*) The smaller the shaded area, the better the material damping.
    (*b*) The larger the shaded area, the better the material damping.
    (*c*) Material damping is an independent material property and does not depend on this curve.
    (*d*) None of the above. **(GATE 1999)**

    Ans: (*a*)

12. The peak bending stress at critical section of a component varies between 100 MN/m^2 and 300 MN/m^2. The ultimate tensile strength of the material is 700 MN/m^2, yield point stress is 500 MN/m^2, endurance limit for reversed bending is 350 MN/m^2. Find the factor of safety.

Ans: (Soderberg criterion 1.46; Goodman criterion 1.75) **(GATE 1999)**

13. The maximum principal stress for the stress state shown in the figure is

(*a*) σ. (*b*) 2σ.

(*c*) 3σ. (*d*) 1.5σ.

Ans: (*b*) **(GATE 2001)**

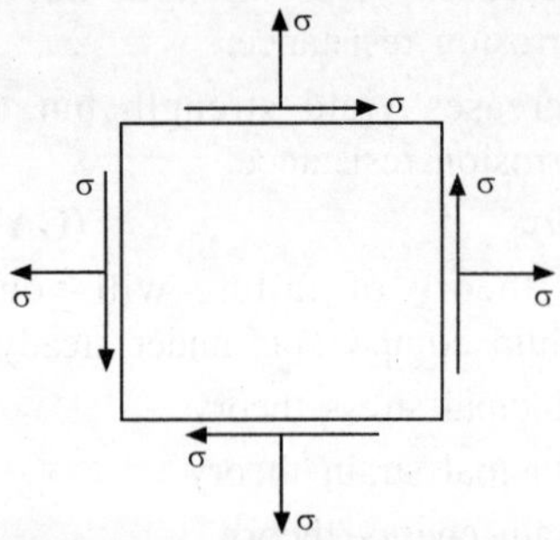

14 If principal stresses corresponding to a two dimensional state of stress are σ_1 and σ_2 and σ_1 is greater than σ_2 and both are tensile, then which one of the following would be the correct criterion for failure by yielding, according to the maximum shear stress criterion?

(*a*) $\dfrac{(\sigma_1 - \sigma_2)}{2} = \pm\dfrac{\sigma_{yp}}{2}$

(*b*) $\dfrac{\sigma_1}{2} = \pm\dfrac{\sigma_{yp}}{2}$

(*c*) $\dfrac{\sigma_2}{2} = \pm\dfrac{\sigma_{yp}}{2}$

(*d*) $\sigma_1 = \pm 2\sigma_{yp}$

Ans: (*b*) **(IES 1993)**

15. A solid circular shaft is subjected to a maximum shear stress of 140 MPa. The magnitude of the maximum normal stress developed in the shaft is

(*a*) 140 MPa. (*b*) 80 MPa.

(*c*) 70 MPa. (*d*) 60 MPa.

Ans: (*a*) **(IAS 1995)**

16. If σ_y is the yield strength of a particular material, then the distortion energy theory is expressed as

(*a*) $(\sigma_1 - \sigma_2)^2 + (\sigma_2 - \sigma_3)^2 + (\sigma_3 - \sigma_1)^2 = 2\sigma_y^2$.

(*b*) $(\sigma_1^2 - \sigma_2^2 + \sigma_3^2) - 2v(\sigma_1\sigma_2 + \sigma_2\sigma_3 + \sigma_3\sigma_1) = \sigma_y^2$

(*c*) $(\sigma_1 - \sigma_2)^2 + (\sigma_2 - \sigma_3)^2 + (\sigma_3 - \sigma_1)^2 = 3\sigma_y^2$.

(*d*) $(1 - 2v)(\sigma_1 + \sigma_2 + \sigma_3)^2 = 2(1 + v)\,\sigma_y^2$.

Ans: (*a*) **(IES 1994)**

17. The design calculations for members subjected to fluctuating loads with the same factor of safety yield the most conservative estimates when using

(*a*) Gerber relation. (*b*) Soberberg relation.

(*c*) Goodman relation.(*d*) None of the above.

Ans: (*b*) **(IES 1995)**

18. If a shaft made from ductile material is subjected to combined bending and twisting moments, calculations based on which one of the following failure theories would give the most conservative value?

(*a*) Maximum principal stress theory.

(*b*) Maximum shear stress theory.

(*c*) Maximum strain energy theory.

(*d*) Maximum distortion energy theory.

Ans: (*b*) **(IES 1996)**

19. If failure in shear along 45° planes is to be avoided, then a material subjected to uniaxial tension should have its shear strength equal to at least

(*a*) tensile strength.

(*b*) compressive strength.

(*c*) Half the difference beteen the tensile and compressive strengths.

(*d*) half the tensile strength.

Ans: (*c*) **(IES 1994)**

20. A block of steel is loaded by a tangential force on its top surface while the bottom surface is held rigidly. The deformation of block is due to

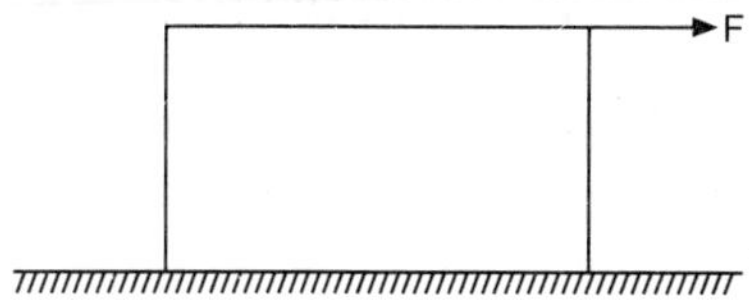

(*a*) Shear only. (*b*) Bending only.

(*c*) Shear and bending.

(*d*) Torsion.

Ans: (*c*) **(GATE 1992)**

21. A large uniform plate containing a rivet hole is subjected to uniform axial tension of 95 MPa. The maximum stress in the plate is

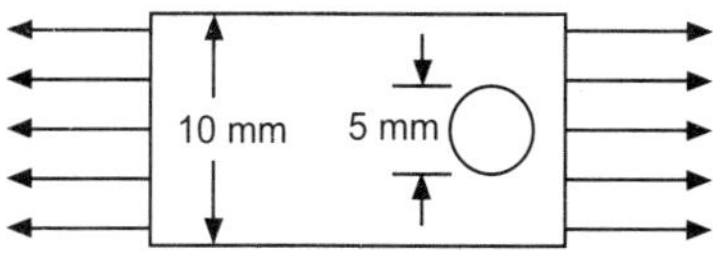

(*a*) 100 MPa. (*b*) 285 MPa.

(*c*) 190 MPa. (*d*) Indetermine.

Ans: (*a*) **(GATE 1992)**

22. The normal stresses at a point are: f_x = 10 MPa, f_y = 2 MPa. The shear stress at the point is 4 MPa. The maximum principal stress at this point is

(*a*) 16 MPa. (*b*) 14 MPa.

(*c*) 11 MPa. (*d*) 20 MPa.

Ans: (*c*) **(GATE 1998)**

23. The most suitable theory of failure for brittle materials is

(*a*) Maximum normal stress theory

(*b*) Maximum strain energy theory

(*c*) Maximum distortion energy theory

(*d*) Maximum shear stress theory.

Ans: (*a*)

24. The endurance limit of a material can be improved by

(*a*) Polishing. (*b*) Coating.

(*c*) Heat treatment. (*d*) Shot peening.

Ans: (*a*), (*c*), (*d*)

25. Infinite life of machine members is generally taken as

(*a*) 10^4 cycles. (*b*) 10^5 cycles.

(*c*) 10^7 cycles. (*d*) 10^8. cycles

Ans: (*c*)

26. If the area of cross-section of a wire is circular and if the radius of the circle decreases to half its original value due to the stretch of the wire by a load, then the modulus of elasticity of the wire will be

(*a*) One-fourth of its original value.

(*b*) Halved.

(*c*) Doubled. (*d*) Unaffected.

Ans: (*d*) **(IES 1993)**

27. Which of the following pairs are correctly matched?

1. Resilience—Resistance to deformation
2. Malleability—Shape change
3. Creep—Progressive deformation
4. Plasticity—Permanent deformation

Select the correct answer using the codes

(*a*) 2, 3, 4. (*b*) 1, 2, 3.

(*c*) 1, 2, 4. (*c*) 1, 3, 4.

Ans: (*d*) **(IES 1994)**

28. During a tensile testing of a specimen using a UTM, the parameters actually measured include

(*a*) True stress and True strain.

(*b*) Poisson's ratio and Young's modulus.

(*c*) Engineering stress and engineering strain.

(*d*) Load and elongation.

Ans: (*d*) **(IES 1996)**

29. A static load is mounted at the centre of a shaft rotating at uniform angular velocity. This shaft will be designed for

(*a*) the maximum compressive stress (static).

(*b*) the maximum tensile stress (static).

(*c*) the maximum bending moment (static).

(*d*) Fatigue loading.

Ans: (*d*) **(GATE 2002)**

CHAPTER 5 — MANUFACTURING CONSIDERATIONS IN DESIGN

1. Interchangeability can be achieved by
(*a*) Standardisation.
(*b*) better process planning.
(*c*) Simplification.
(*d*) Better product planning.
Ans: (*a*) **(IES 1993)**

2. In low carbon steels, presence of small quantities of sulphur improves
(*a*) Weldability. (*b*) Formability.
(*c*) Machinability. (*d*) Hardenability.
Ans: (*c*) **(IES 1995)**

3. The correct sequence of the given materials in ascending order of their weldability is
(*a*) M.S., Copper, C.I., Al.
(*b*) C.I., M.S., Al, Copper.
(*c*) Copper, C.I., M.S., Al.
(*d*) Al, Copper, C.I., M.S.
Ans: (*d*) **(IES 1999)**

4. Which one of the following is the correct temperature range for hot extrusion of aluminium?
(*a*) 300 – 340ºC. (*b*) 350 – 400ºC.
(*c*) 430 – 480ºC. (*d*) 550 – 650ºC.
Ans: (*b*) **(IES 1999)**

CHAPTER 7—RIVETED JOINTS

1. If the tearing efficiency of a riveted joint is 75 percent, then the ratio of rivet hole diameter to the pitch is equal to
(*a*) 0.2. (*b*) 0.25.
(*c*) 0.50. (*d*) 0.60.
Ans: (*b*) **(GATE 1996)**

2. A manufacturer of rivets claims that the failure load in shear of his product is 500 ± 25 N. This specification implies that
(*a*) no rivet is weaker than 475 N and stronger than 525 N.
(*b*) the standard rivet deviation of random sample of rivets is 25 N.
(*c*) there is an equal probability of failure strength to be either 475 N or 525 N.
(*d*) there is approximately two to one chance that the strength of a rivet lies between 475 N to 525 N.
Ans: (*a*) **(GATE 1992)**

3. The most efficient riveted joint possible is one which would be as strong in tension, shear and bearing as the original plates to be joined. But this can never be achieved because.
(*a*) rivets can be made with the same material.
(*b*) rivets are weak in compression.
(*c*) there should be atleast one hole in the plate reducing its strength.
(*d*) clearnace is present between the plate and the rivet.
Ans: (*c*) **(IES 1993)**

4. Which one of the following structural joints with 10 rivets and same size of plate and material will be the most efficient?

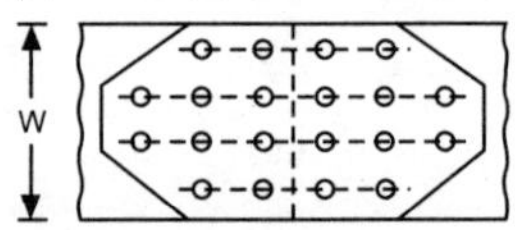

(a)

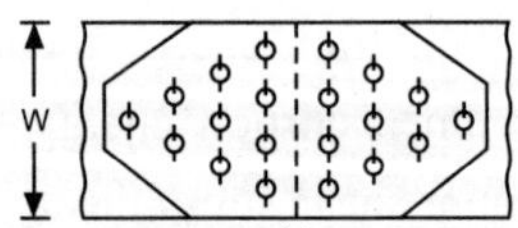

(b)

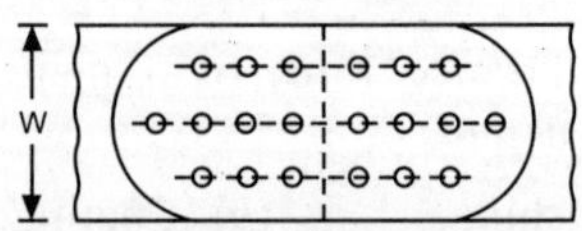

(c)

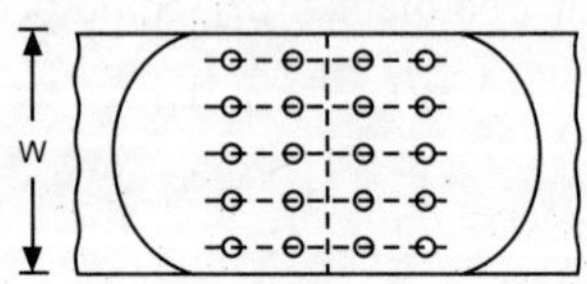

(d)

Ans: (*b*) **(IES 1994)**

5. Match list I with list II and select the correct answer

List I (Type of joints)	List II (an element of the joint)
A Riveted joint	1. Pin
B Welded joint	2. Strap
C Bolted joint	3. Lock washer
D Knuckle joint	4. Fillet

Codes:	A	B	C	D
(a)	4	3	2	1
(b)	2	3	4	1
(c)	2	4	3	1
(d)	2	4	1	3

Ans: (c) **(IES 1994)**

CHAPTER 8—WELDED JOINTS

1. Which of the following methods is not used for increasing the fatigue strength of a welded joint
   (a) hammer peening. (b) heat treatment.
   (c) coating. (d) grinding.
   Ans: (b) **(IES 1993)**
2. For resistance spot welding of 1.5 mm thick steel sheets, the current required is of the order of
   (a) 10 A. (b) 100 A.
   (c) 1000 A. (d) 10,000 A.
   Ans: (d) **(GATE 1992)**
3. The following two figures show welded joints (xxxxx indicates welds), for the same load and same dimensions of plate and weld, the joint shown in

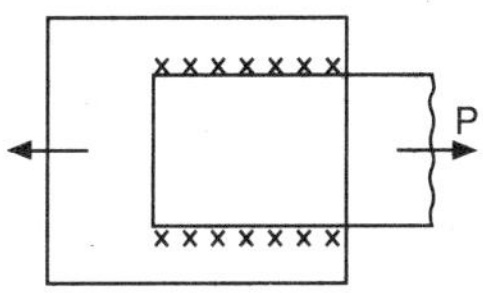

Fig. I

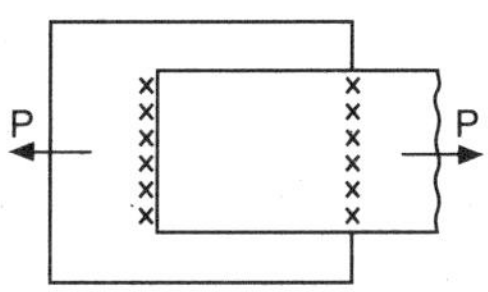

Fig. II

   (a) Figure I is better because the weld is in shear and the principal stress in the weld is not in line with P.
   (b) Figure I is better because the load transfer from the tie bar to the plate is not direct.
   (c) Figure II is better because the weld is in tension and safe stress of weld in tension is greater than that in shear.
   (d) Figure II is better because it has less stress concentration.
   Ans: (c) **(IES 1994)**
4. The permissible stress in a fillet weld is 100 N/mm^2. The fillet weld has equal lengths of 15 mm each. The allowable shearing load on weldment per cm. length of the weld is
   (a) 22.5 kN. (b) 15.0 kN.
   (c) 10.6 kN. (d) 7.5 kN.
   Ans: (c) **(IES 1995)**
5. A double fillet welded joint with parallel weld of length L and leg B is subjected to a tensile force P. Assuming uniform stress distribtion, the shear stress in the weld is given by
   (a) $\frac{\sqrt{2}P}{BL}$. (b) $\frac{P}{2BL}$.
   (c) $\frac{P}{\sqrt{2}BL}$. (d) $\frac{2P}{BL}$.
   Ans: (c) **(IES 1996)**
6. In a butt welded joint, subjected to tensile force P as shwon in figure, h = throat of butt weld H = Total height of weld. The average tensile stress in the weld is given by

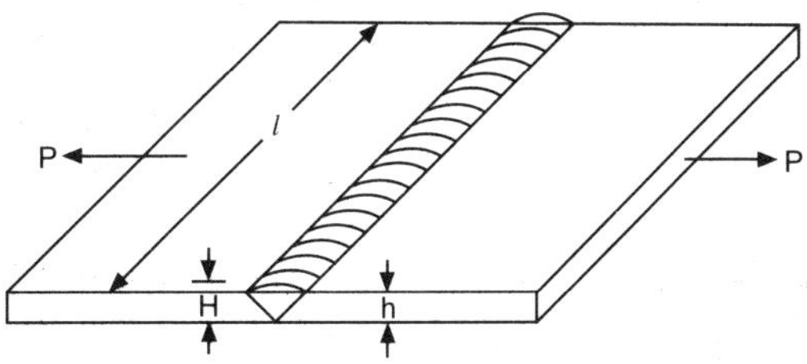

   (a) $\frac{P}{Hl}$. (b) $\frac{P}{hl}$.
   (c) $\frac{P}{2hl}$. (d) $\frac{2P}{Hl}$.
   Ans: (b) **(IES 1997)**

CHAPTER 9 — BOLTED JOINTS

1. Bolts in the flanged end of pressure vessel are usually pre-tensioned. Indicate which of the following statements is not true.

(*a*) Pre-tensioning helps to seal the pressure vessel.

(*b*) Pre-tensioning increases the fatigue life of bolts

(*c*) Pre-tensioning reduces the maximum tensile stress in the bolts

(*d*) Pre-tensioning helps to reduce the effect of pressure pulsations in the pressure vessel.

Ans: (*c*) **(GATE 1999)**

2. Which of the following stresses are associated with the tightening of a nut on a stud?

1. Tensile stress due to stretching of a stud
2. Bending stress of stud
3. Transverse shear stress across threads
4. Torsional shear stresses in threads due to frictional resistance.

Codes:

(*a*) 1, 2 and 3 (*b*) 1, 2 and 4

(*c*) 2, 3 and 4 (*d*) 1, 3 and 4

Ans: (*a*) **(IES 1993)**

3. Assertion (A) : Uniform strength bolts are used for resisting impact loads.

Reason (R) : The area of cross-section of the threaded and unthreaded parts are made equal.

Ans: Both A and R are true. **(IES 1994, 1995)**

4. Which of the following stresses are associated with the tightening of a nut on a bolt?

1. Tensile stress due to stretching of a bolt
2. Bending stress due to bending of a bolt
3. Crushing and shear stresses in threads
4. Torsional shear stress due to frictional resistance between the nut and bolt.

Select the correct answer using the codes given below:

(*a*) 1, 2 and 4 (*b*) 1, 2 and 3

(*c*) 2, 3 and 4 (*d*) 1, 3 and 4

Ans: (*d*) **(IES 1998)**

CHAPTER 11 — COTTER AND KNUCKLE JOINTS

1. Match list I with list II and select the correct answer using the codes given below the lists

List I (Type of joint)	**List II (Modes of joining members)**
A Cotter joint	1. Connects two rods or bars permitting small amount of flexibility.
B Knuckle joint	2. Rigidly connects two members.
C Turn buckle	3. Connects two rods having threaded ends.
D Riveted joint	4. Permanent fluid tight joint between two flat pieces.
	5. Connects two shafts and transmits torque.

Codes:	A	B	C	D
(*a*)	5	1	3	2
(*b*)	2	1	3	4
(*c*)	5	3	2	4
(*d*)	2	3	1	4

Ans: (*b*) **(IES 1993)**

2. Match list I with list II and select the correct answer using the codes given below the list:

List I (Types of joint)	**List II (Use of these joints)**
A Cotter joint	1. Tie rod of a wall crane.
B Knuckle joint	2. Suspension bridges
C Suspension link joint	3. Diagonal stays in boiler.
D Turn buckle	4. Cross-head of a steam engine.

Codes:	A	B	C	D
(*a*)	4	2	3	1
(*b*)	4	3	2	1
(*c*)	3	2	1	4
(*d*)	2	1	4	3

Ans: (*b*) **(IES 1995)**

3. Consider the following design considerations
1. Tensile failure
2. creep failure
3. bearing failure
4. shearing failure
5. bending failure.

The design of the pin of a rocker arm of an I.C. engine is based on

(*a*) 1, 2 and 4 (*b*) 1, 3 and 4
(*c*) 2, 3 and 5 (*d*) 3, 4 and 5

Ans: (*d*) **(IES 1995)**

4. Assertion (A) : When the coupler of a turn buckle is turned in one direction both the connecting rods either move closer or move away from each other depending upon the direction of rotation of the coupler.

Reason (R) : A turn buckle is used to connect two round rods subjected to tensile loading and required subsequent adjustment for tightening and loosening.

Ans: Both A and R are true. **(IES 1996)**

5. Match list I (machine element) with list II (casue of failure) and select the correct answer using the codes given below the list:

List I	List II
A Axle	1. Shear stress
B Cotter	2. Tensile/compressive stress
C Connecting Rod	3. Wear
D Journal bearing	4. Bending stress

Codes:	A	B	C	D
(*a*)	1	4	2	3
(*b*)	4	1	2	3
(*c*)	4	1	3	2
(*d*)	1	4	3	2

Ans: (*b*) **(IES 1998)**

CHAPTER 12—KEYS AND COUPLINGS

1. A key connecting a flange coupling to a shaft is likely to fail in

(*a*) Shear. (*b*) tension.
(*c*) torsion. (*d*) bending.

Ans: (*a*) **(GATE 1995)**

2. The bolts in a rigid flange coupling connecting two shafts transmitting power are subjected to

(*a*) shear force and bending moment.
(*b*) axial force.
(*c*) torsion.
(*d*) torsion and bending.

Ans: (*a*) **(GATE 1996)**

3. In the assembly of pulley, key and shaft

(*a*) pulley is made the weakest.
(*b*) key is made the weakest.
(*c*) key is made the strongest.
(*d*) all the three are designed for equal strength.

Ans: (*b*) **(IES 1993)**

4. Assertion (A) : The effect of keyways on a shaft is to reduce its load carrying capacity and to increase its torsional rigidity.

Reason (R) : Highly localised stresses occur at or near the corners of keyways.

Ans: Both A and R are true. **(IES 1994)**

5. Assertion (A) : Oldham coupling is used to transmit power between two parallel shafts which are slightly offset.

Reason (R) : There is no sliding member to reduce power in Oldham coupling.

Ans: A is true and R is false. **(IES 1994)**

6. Match list I with list II and select the correct answer using the codes given below the list:

List I (Name)	List II (Type)
A Oldham coupling	1. Joins collinear shafts and is of rigid type.

B Flange coupling	2. Joins non-linear shafts and is adjustable.
C Universal coupling	3. Joins collinear shafts and engages and disengages them during motion.
D Friction coupling	4. Compensates peripheral shafts, longitudinal and angular shifts of shafts.

Codes:	A	B	C	D
(*a*)	2	1	4	3
(*b*)	3	2	1	4
(*c*)	1	4	2	3
(*d*)	2	4	1	3

Ans: (*a*) **(IES 1995)**

7. Match list I with list II and select the correct answer using the codes given below the list:

List I	**List II**
A Parallel shafts with slight offset	1. Hooke's joint
B Parallel shafts at a reasonable distance	2. Worn and worm wheel
C Perpendicular shafts	3. Oldham's coupling
D Intersecting shafts	4. Belt and pulley

Codes:	A	B	C	D
(*a*)	4	3	2	1
(*b*)	4	3	1	2
(*c*)	3	4	1	2
(*d*)	3	4	2	1

Ans: (*d*) **(IES 1997)**

8. Match list I with list II and select the correct answer using the codes given below:

List I (Type of key)	**List II (Characteristic)**
A Woodruff key	1. Loose fitting, light duty
B Kennedy Key	2. Heavy duty
C Feather key	3. Self aligning
D Flat key	4. Normal industrial use.

Codes:	A	B	C	D
(*a*)	2	3	1	4
(*b*)	3	2	1	4
(*c*)	2	3	4	1
(*d*)	3	2	4	1

Ans: (*d*) **(IES 1997)**

9. The shearing area of a key of length L, breadth B and depth H is equal to

(*a*) B × H. (*b*) L × H.

(*c*) L × B. (*d*) $L \times \frac{H}{2}$.

Ans: (*c*) **(IES 1998)**

10. In the assembly design of shaft, pulley and key, the weakest member is

(*a*) pulley. (*b*) key.

(*c*) shaft. (*d*) none.

Ans: (*b*) **(IES 1998)**

11. When two shafts are neither parallel nor intersecting, power can be transmitted by using

(*a*) a pair of spur gears.

(*b*) a pair of helical gears.

(*c*) an oldham's coupling.

(*d*) a pair of spiral gears.

Ans: (*c*) **(IES 1998)**

12. Which of the following stresses are associated with the design of pins in bushed pin type flexible coupling?

1. Bearing stress 2. Bending stress
3. Axial tensile stress
4. Transverse shear stress.

Codes:

(*a*) 1, 3 and 4 (*b*) 2, 3 and 4

(*c*) 1, 2 and 3 (*d*) 1, 2 and 4

Ans: (*d*) **(IES 1998)**

13. The coupling used to connect two shafts with large angular misalignment is

(*a*) a flange coupling.

(*b*) an Oldham's coupling.

(*c*) a flexible bush coupling.

(*d*) a Hook's joint.

Ans: (*d*) **(GATE 2002)**

CHAPTER 13 — PIPE JOINTS

1. A metal pipe of 1.0 m diameter contains a fluid having a pressure of 10 kg/cm^2. If the permissible tensile stress in the metal is 200 kg/cm^2, then the thickness of the metal required for making the pipe would be

(*a*) 5 mm. (*b*) 10 mm.

(*c*) 20 mm. (*d*) 25 mm.

Ans: (*d*) **(IES 1993)**

2. Which of the followign joining processes are best suited for manufacturing pipes to carry gas products?

1. Riveting 2. Welding
3. Bolts and nuts

Select the correct answer using the codes given below:

(*a*) 1 and 2 (*b*) 1 and 3

(*c*) 2 alone (*d*) 1, 2 and 3

Ans: (*c*) **(IES 1998)**

3. For tight leakage joints, the following type of thread is best suited

(*a*) metric (*b*) buttress

(*c*) square

(*d*) NPT (National Pipe Threads)

Ans: (*d*)

4. Which of the following pipe joints would be suitable for pipes carrying steam

(*a*) flanged (*b*) threaded

(*c*) ball and spigot (*d*) expansion

(*e*) compression

Ans: (*d*)

5. Pressure vessels that are to contain poisonous or toxic substances are classified as

(*a*) class I (*b*) class II

(*c*) class III (*d*) class IV

Ans: (*a*)

CHAPTER 14—LEVERS

1. Which statement is wrong in respect of levers of class one?

(*a*) The fulcrum is in between the load and effort.

(*b*) The effort arm is longer than the load arm.

(*c*) The mechanical advantage is less than one.

(*d*) The leverage is the ratio of effort arm to load arm.

Ans: (*c*)

2. Which statement is wrong in respect of levers of class two?

(*a*) The load point is in between the fulcrum and effort point.

(*b*) The effort arm is always smaller than the load arm.

(*c*) The mechanical advantage is always greater than one.

(*d*) The leverage is the ratio of effort arm to load arm.

Ans: (*b*)

3. The application of the lever of class three is found in

(*a*) Rocker arm in I.C. engine

(*b*) Wheel barrows

(*c*) pair of forceps or tongs.

(*d*) lever of lever loaded safety valve.

Ans: (*c*)

4. Which amongst the following is not the application of lever of class one?

(*a*) Handle of a hand pump.

(*b*) Rocker arm in an I.C. engine.

(*c*) Beam of a beam balance.

(*d*) Wheel barrow.

Ans: (*d*)

5. Which amongst the following is not the application of lever of class two

(*a*) handle of a hand pump.

(*b*) wheel barrow.

(*c*) Lever of a lever loaded safety valve.

(*d*) Nut cracker.

Ans: (*a*)

6. Spot the odd one out.

(*a*) bell crank lever

(*b*) pincer or cutting pliers

(*c*) Rocker arm in an I.C. engine

(*d*) treadle of a sewing machine

Ans: (*d*)

7. Which is the incorrect statement about a bell crank lever.
   (*a*) It is a lever of class one.
   (*b*) The effort arm and load are perpendicular to each other.
   (*c*) Used in railway signalling.
   (*d*) None of the above.

   Ans: (*d*)

8. In lever of class three, the mechanical advantage is always
   (*a*) equal to one. (*b*) less than one.
   (*c*) more than one. (*d*) more than two.

   Ans: (*b*)

9. In lever of class two, the mechanical advantage is always
   (*a*) equal to one. (*b*) less than one.
   (*c*) more than one. (*d*) more than two.

   Ans: (*c*)

10. Which of the following statement is wrong in the case of a lever.
    (*a*) Mostly a lever is designed under bending.
    (*b*) The section of the lever generally can be round, rectangular, elliptical or I-form.
    (*c*) The lever is tapered.
    (*d*) The lever isalso designed under torsion.

    Ans: (*d*)

CHAPTER 15 — SPRINGS

1. A length of 10 mm diameter steel wire is coiled to a close coiled helical spring having 8 coils of 75 mm mean diameter, and the spring has a stiffness K. If the same length of wire is coiled to 10 coils of 60 mm mean diameter, then the spring stiffness will be
   (*a*) K. (*b*) 1.25 K.
   (*c*) 1.56 K. (*d*) 1.95 K.

   Ans: (*c*) **(IES 1993)**

2. The ends of the leaves of a semi-elliptical leaf spring are made triangular in plan in order to
   (*a*) obtain variable I in each leaf.
   (*b*) permit each leaf to act as a ovehanging beam.
   (*c*) have variable bending moment in each leaf.
   (*d*) Make $\frac{M}{I}$ constant throughout the length of the leaf.

   Ans: (*d*) **(IAS 1994)**

3. An open coiled helical spring is subjected to an axial force, the wire of the spring is subjected to
   (*a*) direct shear only.
   (*b*) combined shear and bending only.
   (*c*) combined shear, bending and twisting.
   (*d*) combined shear and twisting only.

   Ans: (*d*) **(IAS 1995)**

4. Cracks in helical springs used in Railway carriages usually start on the inner side of the coil because of the fact that
   (*a*) it is subjected to higher stress than the outer side.
   (*b*) it is subjected to a higher cyclic loading than the outer side.
   (*c*) it is more stretched than the outer side during the manufacturing process.
   (*d*) it has a lower curvature than the outer side.

   Ans: (*d*) **(IES 1994)**

5. Assertion (A) : Two concentric helical springs used to provide greater spring force are wound in opposite directions.

   Reason (R) : The winding in opposite directions in the case of helical springs prevents buckling.

   Ans: Both A and R are correct. **(IES 1995)**

6. In the calculation of induced shear stress in helical springs, the Wahl's correction factor is used to take care of
   (*a*) combined effect of transverse shear stress and bending stress in the wire.
   (*b*) combined effect of bending stress and curvature of the wire.
   (*c*) combined effect of transverse shear stress and curvature of the wire.
   (*d*) combined effect of torsional shear stress and transverse shear stress in the wire.

   Ans: (*c*) **(IES 1995, 1997)**

7. A long helical spring having a spring stiffness of 12 kN/m and number of turns 20, breaks into two parts with number of turns 10 each in both the parts. If the two parts are connected in series, then the stiffness of the resultant spring will be

(*a*) 6 kN/m. (*b*) 12 kN/m.
(*c*) 24 kN/m. (*d*) 30 kN/m.

Ans: (*b*) **(IES 1996)**

8. The maximum bending moment in a simply suppored beam of length L loaded by a concentrated load W at the midpoint is given by

(*a*) WL. (*b*) $\frac{WL}{2}$.
(*c*) $\frac{WL}{4}$. (*d*) $\frac{WL}{8}$.

Ans. (*c*) **(IES 1996)**

9. Given that:

d = diameter of spring wire
R = mean radius of coils
n = number of coils
G = modulus of rigidity

the stiffness of the closely coiled helical spring subject to an axial load W is equal to

(*a*) $\frac{Gd^4}{64R^3n}$. (*b*) $\frac{Gd^3}{64R^3n}$.
(*c*) $\frac{Gd^4}{32R^3n}$. (*d*) $\frac{Gd^4}{64R^3n}$.

Ans: (*a*) **(IES 1996)**

10. When a closely coiled helical spring is subjected to a couple about its axis, the stress induced in the wire material of the spring is

(*a*) bending stress only.
(*b*) direct stress only.
(*c*) a combination of torsional shear stress and bending stress.
(*d*) a combination of bending stress and direct shear stress.

Ans: (*a*) **(IES 1996, 1998)**

11. A simply supported beam of constant flexural rigidity and length 2L carries a concentrated load P at its mid span and the deflection under the load is δ. If a cantilever beam of the same flexural rigidity and length L is subjected to load P at its free end, then the deflection at the free end will be

(*a*) $\frac{1}{2}\delta$. (*b*) δ.
(*c*) 2δ. (*d*) 4δ.

Ans: (*c*) **(IES 1998)**

12. Two closely coiled springs are made from the same small diameter wire, one wound on 2.5 cm diameter core and the other on 1.25 cm diameter core. If each spring had n coils, then the ratio of their spring constants would be

(*a*) $\frac{1}{16}$. (*b*) $\frac{1}{8}$.
(*c*) $\frac{1}{4}$. (*d*) $\frac{1}{2}$.

Ans: (*b*) **(IES 1998)**

13. A cantilever beam of rectangular cross-section is subjected to a load W at its free end. If the depth of the beam is doubled and load is halved, the deflection of the free end as compared to original deflection will be

(*a*) half. (*b*) one-eighth.
(*c*) one-sixteenth. (*d*) double.

Ans: (*c*) **(IES 1999)**

14 The maximum shear stress occurs on the outermost fibers of a circular shaft under torsion. In a closely coiled helical spring, the maximum shear stress occurs on the

(*a*) outermost fibres.
(*b*) fibres at mean diameter.
(*c*) innermost fibres.
(*d*) end coils.

Ans: (*c*) **(IES 1999)**

15. A helical spring has N turns of coil of diameter D, and a second spring, made of same wire diameter and of same material has $\frac{N}{2}$ turns of coil diameter $2D$. If the stiffness of the first spring is K, then the stiffness of second spring will be

(*a*) $\frac{K}{4}$. (*b*) $\frac{K}{2}$.
(*c*) 2 K. (*d*) 4 K.

Ans: (*a*) **(IES 1999)**

16. If a compression coil spring is cut into two equal parts and the parts are then used in parallel, the ratio of the spring rate to its initial value will be

(*a*) 1. (*b*) 2.

(*c*) 4.

(*d*) indeterminate for want of sufficient data.

Ans: (*c*) **(IES 1999)**

17. A spring of stiffness 1000 N/m is stretched initially by 10 cm from the undeformed position. The work required to stretch it another 10 cm is

(*a*) 5 Nm. (*b*) 7 Nm.

(*c*) 10 Nm. (*d*) 15 Nm.

Ans: (*d*) **(IAS 1995)**

18. A truck weighing 150 kN and travelling at 2 m/s impacts with a buffer spring which compresses 1.25 cm per 10 kN. The maximum compression of the spring is

(*a*) 20.00 cm. (*b*) 22.85 cm.

(*c*) 27.66 cm. (*d*) 30.00 cm.

Ans: (*c*) **(IAS 1995)**

19. Two helical springs of same material and of equal circular cross-section and length and number of turns, but having radii 20 mm and 40 mm, kept concentrically (smaller radius spring within the layer of larger radius spring), are compressed between two planes with a load P. The inner spring will carry a load equal to

(*a*) $\frac{P}{2}$. (*b*) $\frac{2}{3}P$.

(*c*) $\frac{P}{9}$. (*d*) $\frac{8}{9}P$.

Ans: (*d*) **(IES 1994)**

CHAPTER 16—SHAFTS

1. The fig. shows a shaft from a hand-operated machine. The frictional torque in the journal bearings at A and B is 15 Nm each. Find the diameter of the shaft step (on which the pulley is mounted) using maximum distortion energy criterion. The shaft material is C40 steel (yield stres in tension = 380 MPa) and the factor of safety is 1.5. **(GATE 1992)**

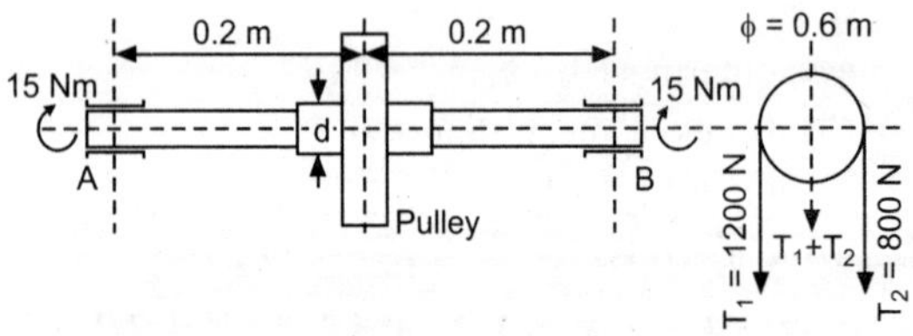

Solution: Total torque on shaft

= Torque transmitted by pulley

+ Frictional torque at the two bearings

$= (T_1 - T_2)r + 2 \times 15$

$= 400 \times 0.3 + 30 = 150$ Nm

$$\text{Maximum B.M.} = \frac{(T_1 + T_2) \times 0.4}{4} = 200 \text{ Nm}$$

Using distortion energy theory,

$d = 22$ mm

2. The outside diameter of a hollow shaft is twice its inside diameter. The ratio of the torque carrying capacity to that of a solid shaft of the same material and of the same outside diameter is

(*a*) 15/16. (*b*) 3/4.

(*c*) 1/2. (*d*) 1/16.

Ans: (*a*) **(GATE 1993)**

3. Two shafts A and B are made of the same material. The diameter of shaft B is twice that of shaft A. The ratio of power which can be transmitted by shaft A to that of shaft B is

(*a*) $\frac{1}{2}$. (*b*) $\frac{1}{4}$.

(*c*) $\frac{1}{8}$. (*d*) $\frac{1}{16}$.

Ans: (*c*) **(GATE 1994)**

4. Find the maximum principal stress developed in a cylindrical shaft, 8 cm in diameter and subjected to a bending moment of 2.5 kN-m and a twisting moment of 4.2 kN-m. If the yield stress of the material is 300 MN/m^2, determine the factor of safety of the shaft according to the maximum shear stress theory of failure.

Ans: 2.66 on the basis of yield stress in shear

(GATE 1994)

5. A solid shaft can resist a bending moment of 3.0 kN-m and a twisting moment of 4.0 kN-m together, then the maximum torque that can be applied is

(*a*) 7.0 kN-m. (*b*) 3.5 kN-m.

(*c*) 4.5 kN-m. (*d*) 5.0 kN-m.

Ans: (*d*) **(GATE 1996)**

6. Design of shafts made of brittle materials is based on

(*a*) Guest's theory.

(*b*) Rankine's theory.

(*c*) St. Venant's theory.

(*d*) Von Mises theory.

Ans: (*b*) **(IES 1993)**

7. A shaft is subjected to a maximum bending stress of 80 N/mm^2 and maximum shear stress equal to 30 N/mm^2 at a particular section. If the yield point in tension of the material is 280 N/mm^2, and the maximum shear stress theory of failure is used. Then the factor of safety obtained will be

(*a*) 2.5. (*b*) 2.8.

(*c*) 3.0. (*d*) 3.5.

Ans: (*b*) **(IES 1994)**

8. A shaft is subjected to fluctuating loads for which the normal torque (T) and bending moment (M) are 1000 N-m and 500 N-m respectively. If the combined shock and fatigue factor for bending is 1.5 and combined shock and fatigue factor for torsion is 2, then the equivalent twisting moment for the shaft is

(*a*) 200 N-m. (*b*) 2050 N-m.

(*c*) 2100 N-m. (*d*) 2136 N-m.

Ans: (*d*) **(IES 1994)**

9. Match list I with list II and select the correct answer using the codes given below the lists:

	List I		**List II**
A	Crankshaft	1.	Supports the revolving parts and transmits torque.
B	Wire shaft	2.	Transmits motion between shafts where it is not possible to effect a rigid coupling between them.
C	Axle	3.	Converts linear motion into rotary motion.
D	Plain shaft	4.	Supports only the revolving parts.

Codes:

	A	B	C	D		A	B	C	D
(*a*)	3	2	1	4	(*b*)	4	2	3	1
(*c*)	3	2	4	1	(*d*)	1	4	2	3

Ans: (*c*) **(IES 1995)**

10. A transmission shaft subjected to bending loads must be designed on the basis of

(*a*) maximum normal stress theory.

(*b*) maximum shear stress theory.

(*c*) maximum normal stress and maximum shear stress theories.

(*d*) fatigue strength.

Ans. (*c*) **(IES 1995)**

11. The equivalent bending moment under combined action of bending moment M and torque T is

(*a*) $\sqrt{M^2 + T^2}$. (*b*) $\frac{1}{2}\sqrt{M^2 + T^2}$.

(*c*) $M + \sqrt{M^2 + T^2}$.

(*d*) $\frac{1}{2}\left[M + \sqrt{M^2 + T^2}\right]$.

Ans: (*d*) **(IES 1996)**

12. Match List I (Positioning of two shafts) with List II (Possible connection) and select the correct answer using the codes given below the Lists:

	List I		**List II**
A	Parallel shafts with slight offset	1.	Hooke's joint
B	Parallel shafts at reasonable distance	2.	Worm and worm wheel
C	Perpendicular shafts	3.	Oldham coupling
D	Intersecting shafts	4.	Belt and Pulley

Codes:

	A	B	C	D		A	B	C	D
(*a*)	4	3	2	1	(*b*)	4	3	1	2
(*c*)	3	4	1	2	(*d*)	3	4	2	1

Ans: (*c*) **(IES 1997)**

CHAPTER 17 — BEARINGS

1. Spherical roller bearings are normally used
   (*a*) for increased radial load.
   (*b*) for increased thrust load.
   (*c*) when there is less radial space.
   (*d*) to compensate for angular misalignment.
   Ans: (*a, d*) **(GATE 1992)**

2. The life of a ball-bearing is inversely proportional to
   (*a*) $(\text{Load})^{1/3}$. (*b*) $(\text{Load})^3$.
   (*c*) $(\text{Load})^{3.3}$. (*d*) $(\text{Load})^2$.
   Ans: (*b*) **(GATE 1993)**

3. The dynamic load capacity of a 6306 bearing is 22 kN. The maximum radial load it can sustain to operate at 600 rev/min, for 2000 hours is
   (*a*) 4.16 kN. (*b*) 3.60 kN.
   (*c*) 6.25 kN. (*d*) 5.29 kN.
   Ans: (*d*) **(GATE 1997)**

   Hint: $L = \dfrac{10^6}{60n}\left(\dfrac{C}{P}\right)^3$

   L = 2000 h, n = 600, C = 22 kN, P = ?

4. A full journal bearing, 200 mm in diameter and 200 mm long, is to support a radial load of 45 kN at an operating speed of 960 rev/min. If it is to operate at a Sommerfeld number of 0.08 when the radial clearance is 0.2 mm, determine the viscosity of the oil to be used. Estimate also the power loss in the bearing if the resulting coefficient of friction is 0.00315.
   (GATE 1997)

Solution: Sommer field number is given as,

$$S = \frac{Zn'}{p}\left(\frac{D}{C_d}\right)^2 \quad \text{...(i)}$$

D = 200 mm, Cd = 2 × 0.2 = 0.4 mm,
n′ = 16 rev/s

$$p = \frac{45 \times 10^3}{0.2 \times 0.2} = 1.125 \text{ MPa}$$

Put all these values in equation (i), then

$Z = 0.0225 \text{ Ns/m}^2$

$$\text{Now } V = \frac{\pi \times 200 \times 960}{1000 \times 60} = 10 \text{ m/s}$$

Power loss in heating,

Hg = f FV

$= 0.00314 \times 45 \times 10^3 \times 10 =$ **1425 watts.**

5. The basic load rating of a ball bearing is
   (*a*) the maximum static radial load that can be applied without causing any plastic deformation of bearing components.
   (*b*) the radial load at which 90% of the group of apparently identical bearings run for 1 million revolutions before the first evidence of failure.
   (*c*) the maximum radial load that can be applied during operation without any plastic deformation of bearing components.
   (*d*) a combination of radial and axial loads that can be applied without any plastic deformation.
   Ans: (*b*) **(GATE 1998)**

6. Select one of the following two bearings and justify why did you reject the other:
   (*a*) **Deep groove ball bearings:**
   Dynamic load rating = 47,000 N
   Radial factor (X) = 0.56
   Axial load factor (Y) = 1.83
   (*b*) **Cylindrical Roller bearing:**
   Dynamic load rating = 28,000 N
   Radial load factor = 1.0
   Axial load factor = 0.0
   Life expectancy = 10,000 h
   Rotation factor = 1
   Reliability factor = 0.62

Load conditions:
Radial load = 3,000 N
Axial load = 1,600 N
RPM = 1000 **(GATE 1998)**

Solution: Dynamic load is given as

P = XVFr + YFa

(*a*) X = 0.56, Y = 1.83, V = 1

∴ P = 0.56 × 1 × 3000 + 1.83 × 1000
= 4608 N

Now C = 47,000 N

$$\therefore \quad C/P = \frac{47000}{4608} = 10.2$$

(*b*) $P = 1 \times 1 \times 3000 + 0 \times 1000$

$= 3000$ N

$\therefore \quad C/P = \dfrac{28000}{3000} = 9.33$

$\therefore$ Bearing (*a*) is to be selected.

7. The life of a ball bearing at a load of 10 kN is 8000 hrs. Its life in hrs. if the load is increased to 20 kN, keeping all other conditions same, is

(*a*) 4000. (*b*) 2000.

(*c*) 1000. (*d*) 500.

Ans: (*c*) **(GATE 2000)**

Hint: $L_1 \cdot P_1^3 = L_2 \cdot P_2^3$.

8. A full journal bearing with a journal of 75 mm diameter and a bearing of 75 mm length is subjected to a radial load of 2500 N at 400 rev/min. The lubricant is SAE 50 at 75ºC having a viscosity of 16.5×10^{-3} kg/ms. Radial clearance is 0.03 mm. Eccentricity of the bearing is 0.27. Find the Sommerfeld number and the minimum film thickness.

(GATE 2000)

Solution: $S = \dfrac{Zn'}{p}\left(\dfrac{D}{C_d}\right)^2$

$$p = \frac{2500}{0.075 \times 0.075} = 4.44 \text{ MPa};$$

$$n' = \frac{400}{60} \text{ rev/s}$$

$$\therefore \quad S = \frac{16.5 \times 10^{-3} \times 400}{4.44 \times 10^6 \times 60} \times \left(\frac{75}{0.06}\right)^2$$

$$= 0.0387$$

$$\varepsilon = 1 - \frac{h_0}{C_r}$$

$$\therefore \quad \frac{h_0}{C_r} = 1 - \varepsilon = 1 - 0.27 = 0.73$$

$\therefore \quad h_0$ = minimum film thickness

$= 0.73 \times 0.03 =$ **0.0219 mm.**

9. In a flat collar pivot bearing, the moment due to friction is proportional to (r_1 and r_2 are the outer and inner radii respectively)

(*a*) $\dfrac{r_1^2 - r_2^2}{r_1 - r_2}$. (*b*) $\dfrac{r_1^2 - r_2^2}{r_1 + r_2}$.

(*c*) $\dfrac{r_1^3 - r_2^3}{r_1^2 - r_2^2}$. (*d*) $\dfrac{r_1^3 - r_2^3}{r_1 - r_2}$.

Ans: (*c*) **(IES 1993)**

10. A friction circle is drawn when a journal rotates in a bearing. Its radius depends on the coefficient of friction and the

(*a*) magnitudes of the forces on the journal.

(*b*) angular velocity of the journal.

(*c*) clearance between the journal and the bearing.

(*d*) radius of the journal.

Ans: (*d*) **(IES 1993)**

11. Consider the following statements about anti-friction bearings:

1. They have low starting and low running friction at moderate speeds.
2. They have high resistance to shock loading.
3. They can carry both radial and thrust loads.
4. Their initial cost is high.
5. They can accommodate some amount of misalignment of shaft.

Of these statements:

(*a*) 1, 2, 3 and 4 are correct.

(*b*) 1, 3 and 4 are correct.

(*c*) 1, 4 and 5 are correct.

(*d*) 1, 2, 3 and 5 are correct.

Ans: (*a*) **(IES 1994)**

12. The most suitable bearing for carrying very heavy loads with slow speed is

(*a*) hydro-dynamic bearing.

(*b*) ball bearing.

(*c*) roller bearing.

(*d*) hydrostatic bearing.

Ans: (*d*) **(IAS 1994)**

13. Thrust bearings of sliding type are often provided with multiple sector shaped bearing pads of the tilting type instead of a continuous annular bearing surface in order to (*a*) distribute the thrust load more non-uniformly.

(*b*) provide limited adjustments to shaft misalignments.

(*c*) enable the formation of a wedge-shaped oil film.

(*d*) enabling lubricating oil to come into contact with the total bearing area.

Ans: (*c*) **(IAS 1994)**

14 Deep groove ball bearings are used for

(*a*) heavy thrust load only.

(*b*) small angular displacements of shafts.

(*c*) radial load at high speed.

(*d*) combined thrust and radial loads at high speed.

Ans: (*d*) **(IAS 1995)**

15. In an oil lubricated journal bearing, coefficient of friction between the journal and the bearing,

(*a*) remains constant at all speeds.

(*b*) is minimum at zero speed and decreases monotonically with increase in speed.

(*c*) is maximum at zero speed and decreases monotonically with increase in speed.

(*d*) becomes minimum at an optimum speed and then increases with further increase in speed.

Ans: (*d*) **(IES 1995)**

16. Removal of metal particles from the raceway of a rolling contact bearing is a kind of failure of bearing known as

(*a*) pitting. (*b*) wearing.

(*c*) spalling. (*d*) scuffing.

Ans: (*a*) **(IES 1995)**

17. Match list I with List II and select the correct answer using the codes given below the lists:

List I (Requirement)	**List II (Type)**
A High temperature service:	1. Teflon bearing
B High load	2. Carbon bearing
C No lubrication	3. Hydrodynamic bearing
D Bushings	4. Sleeve bearing

Codes:

	A	B	C	D		A	B	C	D
(*a*)	1	2	3	4	(*b*)	4	1	2	3
(*c*)	2	1	3	4	(*d*)	2	3	1	4

Ans: (*d*) **(IES 1995)**

18. Tapered roller bearings can take

(*a*) radial load only.

(*b*) axial load only.

(*c*) both radial and axial loads and the ratio of these loads being less than unity.

(*d*) both radial and axial loads and the ratio of these loads being greater than unity.

Ans: (*c*) **(IES 1996)**

19. The bearing characteristic number in a hydrodynamic bearing depends on

(*a*) length, width and load.

(*b*) length, width and speed.

(*c*) viscosity, speed and load.

(*d*) viscosity, speed and bearing pressure.

Ans: (*d*) **(IES 1996)**

20. In journal bearings, the radius of the friction circle increases with the increase in

(*a*) load.

(*b*) radius of the journal.

(*c*) speed of the journal.

(*d*) viscosity of the lubricant.

Ans: (*b*) **(IES 1997)**

21. Match List I with List II and select the correct answer using the codes given below the Lists:

List I (Bearing)	**List II (Purpose)**
A Ball bearing	1. Heavy loads with oscillatory motion.
B Tapered Roller bearings.	2. Light loads.
C Spherical roller bearing.	3. Carrying both radial and thrust loads.
D Needle Roller bearings.	4. Self-aligning property.

Codes:

	A	B	C	D		A	B	C	D
(*a*)	4	1	3	2	(*b*)	2	1	4	3
(*c*)	2	3	1	4	(*d*)	2	3	4	1

Ans: (*d*) **(IES 1997)**

22. Which one of the following statements is NOT true of rolling contact bearings?

(*a*) The bearing characteristic number is given by ZN/p, where Z is the absolute viscosity of lubricant, N is the shaft speed and p is the bearing pressure.

(*b*) Inner race of a radial ball bearing has an interference fit with the shaft and rotates along with it.

(*c*) Outer race of a bearing has an interference fit with bearing housing and does not rotate.

(*d*) In some cases, the inner race is stationary and outer race rotates.

Ans: (*a*) **(IES 1997)**

23. Match List-I with List-II and select the correct answer using the codes given below the lists:

List I (Bearing materials)	List II (Properties)
A Babbits	1. Porous
B Bronze	2. Good embeddability
C C.I.	3. Suitable for high loads and low speeds.
D Sintered powder	4. Runs with C.I. journals metal

Codes:

	A	B	C	D		A	B	C	D
(*a*)	2	3	4	1	(*b*)	3	2	1	4
(*c*)	2	3	1	4	(*d*)	3	2	4	1

Ans: (*a*) **(IES 1997)**

24. Match List I with List II and select the correct answer using the codes given below the lists:

List I	List II
A End thrust	1. Plain bearing
B No Cage	2. Ball bearing
C More accurate centering	3. Needle bearing
D Can be overloaded	4. Tapered roller bearing

Codes:

	A	B	C	D		A	B	C	D
(*a*)	3	4	2	1	(*b*)	4	3	1	2
(*c*)	[illegible]	4	1	2	(*d*)	4	3	2	1

Ans: (*d*) **(IES 1998)**

25. In a single row deep groove ball bearing, cages are needed to

(*a*) separate the two races.

(*b*) separate the balls from the inner race.

(*c*) separate the outer race fromthe balls.

(*d*) ensure that the balls do not cluster at one point and maintain proper relative angular positions.

Ans: (*d*) **(IES 1999)**

26. Which one of the following types of bearings is employed in shafts of gearboxes of automobiles?

(*a*) Hydrodynamic journal bearing.

(*b*) Multi-lobed journal bearings.

(*c*) Antifriction bearings.

(*d*) Hybrid journal bearings.

Ans: (*c*) **(IES 1999)**

27. Starting friction is low in

(*a*) Hydro-static lubrication

(*b*) Hydro-dynamic lubrication

(*c*) Mixed (or Semi-fluid) lubrication

(*d*) Boundary lubrication.

Ans: (*a*) **(GATE 1992)**

28. In thick film hydrodynamic journal bearings, the coefficient of friction

(*a*) increases with increase of load.

(*b*) is independent of load.

(*c*) decreases with increase of load.

(*d*) may increase or decrease with increase in load.

Ans: (*c*) **(GATE 1996)**

29. To restore stable operating condition in a hydrodynamic journal bearing, when it encounters higher magnitude loads,

(*a*) oil viscosity is to be decreased.

(*b*) oil viscosity is to be increased.

(*c*) oil viscosity index is to be increased.

(*d*) oil viscosity index is to be decreased.

Ans: (*b*) **(GATE 1997)**

[**Hint:** If load increases, p increases,

$$\therefore \frac{Zn}{p} \text{ decreases,}$$

$\therefore$ to restore stable operating condition, Z should increase].

30. The SI unit of kinematic viscosity (ν) is

(*a*) m^2/s. (*b*) kg/(m – sec).

(*c*) m/sec^2. (*d*) m^3/sec^2.

Ans: (*a*) **(GATE 2001)**

31. Match correct pairs in List I and List II:

List I	List II
(*a*) Automobile wheel mounting on axle	1. Magneto bearing
(*b*) High speed grinding wheel	2. Angular contact bearing
(*c*) I.C. Engine connecting rod	3. Taper roller bearing
(*d*) Leaf spring eye mounting	4. Hydrodynamic journal bearing
	5. Sintered metal bearing
	6. Teflon/Nylon bush

Ans: a-3, b-2, c-4, d-6 **(GATE 1997)**

32. Babbit lining is used on brass/bronze bearings to

(*a*) increase bearing resistance.

(*b*) increase compressive resistance.

(*c*) provide antifriction properties.

(*d*) increase wear resistance.

Ans: (*d*) **(IES 1995)**

CHAPTER 18—CLUTCHES AND BRAKES

1. In designing a plate clutch, assuming of uniform wear conditions is made because

(*a*) it is closer to real life situation.

(*b*) it leads to a safer design.

(*c*) it leads to cost effective design.

(*d*) no other assumption is possible.

Ans: (*a*) **(IES 1996)**

2. In the multiple disc clutch, if there are 6 discs on the driving shaft and 5 discs on the driven shaft, then the number of pairs of contact surfaces will be equal to

(*a*) 11. (*b*) 12.

(*c*) 10. (*d*) 22.

Ans: (*c*) **(IES 1997)**

3. Match List I with List II and select the correct answer using the codes given below the lists:

List I	List II
A Single plate friction clutch	1. Scooters
B Multi-plate friction clutch	2. Rolling mills
C Centrifugal clutch	3. Trucks
D Jaw Clutch	4. Mopeds.

Codes:

	A	B	C	D
(*a*)	1	3	4	2
(*b*)	1	3	2	4
(*c*)	3	1	2	4
(*d*)	3	1	4	2

Ans: (*d*) **(IES 1998)**

4. A multiple disc friction clutch is installed on a shaft whose angular velocity is 100 rad/s and which transmits a power of 5 kW. Check whether the number of friction surface pairs $z = 3$ is sufficient if the inside and outside radii of the friction surfaces are 50 mm and 75 mm respectively. Take $\mu = 0.08$, $p = 0.49$ MPa.

Solution: There are two ways to solve the problem. We can check whether the number of discs is sufficient or not, and whether the unit pressure is within the limits or not.

I. Frictional moment capacity of clutch is

$$M_{fr} = \frac{2}{3}\pi \cdot \mu \cdot p \cdot \left(r_2^3 - r_1^3\right) \cdot z$$

Now $M_{fr} \le \beta \cdot Mt$

where β is known as friction margin factor or the grip safety factor (1.25 to 1.50). This factor guarantees against slipping of the clutch discs during the transmission of the given torque, and M_t is the torque to be transmitted,

$$M_t = \frac{1000 \times 5}{100} = 50 \text{ Nm} \qquad \left(M_t = \frac{1000P}{\omega}\right)$$

$$\therefore \quad 1.25 \times 50 = \frac{2}{3}\pi \times 0.08 \times 0.49 \times 10^6 \times (0.075^3 - 0.05^3) \times z$$

From here $z = 2.05$

∴ Required z is less than the actual z (3),

∴ O.K.

II. $p = \dfrac{1.25 \times 50 \times 3}{2 \times \pi \times 0.08 \times (0.075^3 - 0.05^3) \times 3}$

$= 0.41 \times 10^6 = 0.41$ MPa

which is less than 0.49 MPa, so O.K.

5. A friction clutch is required to transmit 34.5 kW at 2000 rev/min. It is to be single plate disc type clutch with both sides of the plate effective, the pressure being applied axially by means of springs and limited to 70 kN/m^2 on the plate. If the outer diameter of the friction lining is 1.5 times the internal diameter, find the required dimensions d_1 and d_2 of the clutch ring and total force exerted by the springs. Assume the wear to be uniform, and coefficient of friction = 0.3.

(GATE 1994)

Solution: Refer Fig. 18.7

$$\frac{r_2}{r_1} = 1.5$$

For uniform wear,

$p_r = C$, with (p) max = 70 kN/m^2, and it will be at r_1.

Now Torque to be transmitted,

$$M_t = \frac{\text{Power}}{2\pi n} = \frac{34.5 \times 1000 \times 60}{2\pi \times 2000} = 164.75 \text{ Nm}$$

Now normal axial force,

$$P = 2\pi C(r_2 - r_1)$$
$$= 2\pi \times p \cdot r_1(1.5r_1 - r_1)$$
$$= 2\pi \times 70 \times 10^3 \times r_1 \times 0.5r_1$$
$$= 219911.50\ r_1^2, \text{ newtons}$$

$$\text{Friction torque} = \mu \cdot P \cdot z\left(\frac{r_1 + r_2}{2}\right)$$

$\therefore$ $0.3 \times 219911.50\ r_1^2 \times 2 \times 1.25\ r_1 = 164.75$

$\therefore$ From here, $r_1 = 100$ mm, $\therefore$ $d_1 = 200$ mm

$\therefore$ $r_2 = 150$ mm, $\therefore$ $d_2 = 300$ mm.

and $P = 219911.50 \cdot (100)^2 \cdot \dfrac{1}{10^6} = 2200$ N.

6. A clutch has disc plates with outer radius 120 mm. An axial force of 4 kN is acting on them. The disc plates are new and have a coefficient of friction 0.7. Assuming uniform pressure and uniform wear, find the torque transmitted.

(*a*) When the disc plates are solid.

(*b*) When the disc plates are hollow with outer radius 120 mm and inner radius 60 mm.

(GATE 1995)

Solution: (*a*) **Solid disc plates:**

(i) **Uniform pressure:**

$$M_t = \frac{2P\mu z}{3}\left(\frac{r_2^3 - r_1^3}{r_2^2 - r_1^2}\right)$$

$r_1 = 0$, $\quad r_2 = 120$ mm

$\mu = 0.7$, $\quad P = 4$ kN

Assume single plate disc clutch, $\therefore$ $z = 2$

$\therefore$ $M_t = 448$ Nm.

(ii) **Uniform Wear:**

$$M_t = \mu P z\left(\frac{r_1 + r_2}{2}\right)$$ **Ans.** 336 Nm.

(*b*) **Hollow disc plates:**

$r_2 = 120$ mm and $r_1 = 60$ mm.

$\therefore$ Solve the two equations for uniform pressure and uniform wear.

7. A multiple disc clutch is to transmit 4 kW at 750 rev/min. Available steel and bronze discs of 40 mm inner radius and 70 mm outer radius are to be assembled alternately in appropriate numbers. The clutch is to operate in oil with expected co-efficient of friction of 0.1 and maximum allowable pressure is not to exceed 350 kPa. Assume uniform wear conditions to prevail. Specify the number of steel (driving) and bronze (driven) discs required. Also, determine what axial force is to be applied to develop the full torque. **(GATE 1997)**

Solution: For uniform wear condition,

$$M_t = \mu \cdot P \cdot z \cdot \left(\frac{r_2 + r_1}{2}\right)$$

$r_1 = 0.04$ m, $r_2 = 0.07$ m, $\mu = 0.1$,

$P = p \cdot \pi\left(r_2^2 - r_1^2\right)$,

$p = 350 \times 10^3$ N/m^2

With these values, solve the twisting moment equation to find z. It will come out to be 2.55. The next larger even number is 4

$\therefore$ No. of pairs of contact surfaces = 4

$\therefore \quad 4 = z_{steel} + z_{bronze} - 1$

$\therefore \quad z_{steel} + z_{bronze} = 5$

From here, $z_{steel} = 3$ and $z_{bronze} = 2$

8. A plate clutch with maximum diameter 60 mm has maximum lining pressure of 0.35 MPa. The power to be transmitted at 400 rpm is 135 kW and $\mu = 0.3$. Find inside diameter and spring force required to engage the clutch. **(PTU)**

9. A cone clutch is to be designed to transmit a torque of 1000 Nm. Space restrictions limit the outside and inside radii to the values of 75 mm and 45 mm respectively and the semi cone angle of the value of 15º. A moulded friction lining (operating under dry conditions) is to be used for which the average value of the coefficient of friction is 0.35. Using uniform wear theory, find the required clamping force. The clamping force is supplied by one spring of the spring constant = 1500 N/mm. If the friction lining wears out by 0.4 mm, what is the reduction in the torque capacity of the clutch?

Ans: (12.325 kN, 190 Nm) **(GATE 1992)**

10. A simple band brake is to absorb 4.5 kW at 150 rpm. The coefficient of friction is 0.12 and the maximum pressure between the lining and the drum is to be 700 kPa. Determine the width of the band if the drum diameter is 300 mm and the angle of wrap is 215º.

Ans: (50.2 mm) **(GATE 1993)**

Solution:

$$\text{Brake torque} = \frac{9550 \times P}{n}$$

$$= \frac{9550 \times 4.5}{150} = 286.5 \text{ N-m}$$

$\therefore \quad (F_1 - F_2) \times r = 286.5$

Also $\quad F_1/F_2 = e^{\mu\theta}$

From these two equations,

$F_1 = 5272.7$ N

and $\quad F_2 = 3362.7$ N

$$\text{Now} \quad p_{max} = \frac{2F_1}{D \cdot w}$$

$$\therefore \quad 700 \times 10^3 = \frac{2 \times 5272.7}{0.3 \times w}$$

From here, $w = 50.2$ mm.

CHAPTER 19, 20 AND 21—BELT, ROPE AND CHAIN DRIVES

1. In a belt drive, if the pulley diameter is doubled, keeping the tension and belt width constant, then it will be necessary to

(*a*) increase the key length.

(*b*) increase the key depth.

(*c*) increase the key width.

(*d*) decrease the key length.

Ans: (*c*) **(IES 1993)**

2. In a multiple V-belt drive, when a single belt is damaged, it is preferable to change the complete set to

(*a*) reduce vibration.

(*b*) reduce slip.

(*c*) ensure uniform loading.

(*d*) ensure proper alignment.

Ans: (*c*) **(IES 1993)**

3. The following data refers to an open belt drive:

	Pulley A	Pulley B
Purpose ...	Driving	Driven
Diameter ...	450 mm	750 mm
Angle of contact ...	$\theta_A = 150º$	$\theta_B = 120º$
Coefficient of friction between belt and pulley	$f_A = 0.36$	$f_B = 0.22$

The ratio of tensions may be calculated using the relation $(T_1/T_2) = \exp(z)$ where z is

(*a*) $f_A\theta_A$ (*b*) $f_B\theta_B$

(*c*) $(f_A + f_B)(\theta_A + \theta_B)/4$

(*d*) $(f_A\theta_A + f_B\theta_B)/2$.

Ans: (*b*) **(IES 1993)**

4. When compared to a rod of same diameter and material, a wire rope

(*a*) is less flexible.
(*b*) has a much smaller load carrying capacity.
(*c*) does not provide much warning before failure.
(*d*) provides much greater time for remedial action before failure.

Ans: (*d*) **(IES 1994)**

5. A 50 kW motor using six V-belts is used in a pulp mill. If one of the belts breaks after a month of continuous running, then
(*a*) the broken belt is to be replaced by a similar belt.
(*b*) all the belts are to be replaced.
(*c*) the broken belt and two adjacent belts are to be replaced.
(*d*) the broken belt and one adjacent belt is to be replaced.

Ans: (*b*) **(IAS 1994)**

6. Which one of the following pairs is not correctly matched?
(*a*) Positive drive ... Belt drive.
(*b*) High velocity ratio ... Worm gearing.
(*c*) To connect non-parallel and non-intersecting shafts ... spiral gearing.
(*d*) Diminished noise and smooth operation ... Helicalgears.

Ans: (*a*) **(IES 1995)**

7. Sources of power loss in a chain drive are given below:
1. Friction between chain and sprocket teeth.
2. Overcoming the chain stiffness.
3. Overcoming the friction in shaft bearing.
4. Frictional resistance to the motion of chain in air or lubricant.

The correct sequence of descending order of power loss due to these sources is
(*a*) 1, 2, 3, 4 (*b*) 1, 2, 4, 3
(*c*) 2, 1, 3, 4 (*d*) 2, 1, 4, 3

Ans: (*a*) **(IES 1995)**

8. In a flat belt drive, if the slip between the driver and the belt is 1%, that between belt and follower is 3% and driver and follower pulley diameters are equal, then the velocity ratio of the drive will be:
(*a*) 0.99 (*b*) 0.98
(*c*) 0.97 (*d*) 0.96

Ans: (*d*) **(IES 1996)**

9. Effective stress in wire ropes during normal working is equal to the stress due to
(*a*) axial load plus stress due to bending
(*b*) acceleration/retardation of masses plus stress due to bending
(*c*) axial load plus stress due to acceleration/retardation.
(*d*) bending plus stress due to acceleration/retardation.

Ans: (*a*) **(IES 1996)**

10. When a belt drive is transmitting maximum power
(*a*) effective tension is equal to centrifugal tension.
(*b*) effective tension is half of centrifugal tension.
(*c*) driving tension on slack side is equal to the centrifugal tension.
(*d*) driving tension on tight side is twice the centrifugal tension.

Ans: (*d*) **(IES 1996)**

11. If μ is the actual coefficient of friction in a belt moving in grooved pulley, the groove angle being 2α, the virtual coefficient of friction will be :
(*a*) $\mu/\sin\alpha$. (*b*) $\mu/\cos\alpha$.
(*c*) $\mu\sin\alpha$. (*d*) $\mu\cos\alpha$.

Ans: (*a*) **(IES 1997)**

12. In the case of a vertical belt pulley drive with T_c as centrifugal tension and T_0 as the initial tension, the belt would tend to hang clear of the lower pulley when
(*a*) $T_c < T_0$. (*b*) $T_c = T_0/3$.
(*c*) $T_c > T_0$. (*d*) $T_c = T_0/2$.

Ans: (*c*) **(IES 1997)**

13. Given that T_1 and T_2 are the tensions on the tight and slack sides of the belt respectively, the initial tension of the belt taking into account centrifugal tension T_c, is equal to

(*a*) $\dfrac{T_1+T_2+T_c}{3}$. (*b*) $\dfrac{T_1+T_2+2T_c}{2}$.

(*c*) $\dfrac{T_1+T_2+3T_c}{3}$. (*d*) $\dfrac{T_1+T_2+3T_c}{3}$.

Ans: (*b*) **(IES 1997)**

14 Given that W = weight of the load handled, Wr = weight of the rope and f = acceleration, the additional load in ropes of a hoist during starting is given by

(*a*) $F_a = \left(\dfrac{W - W_r}{g}\right) f.$

(*b*) $F_a = \left(\dfrac{W + W_r}{g}\right) f.$

(*c*) $F_a = \dfrac{W}{g} \cdot f.$ (*d*) $F_a = \dfrac{W_r}{g} \cdot f.$

Ans: (*b*) **(IES 1997)**

15. In the assembly design of shaft, pulley and key, the weakest member is

(*a*) pulley. (*b*) key.
(*c*) shaft. (*d*) none.

Ans: (*b*) **(IES 1998)**

16. Which one of the following belts should not be used above 40ºC?

(*a*) Balata belt (*b*) Rubber belt
(*c*) Fabric belt (*d*) Synthetic belt.

Ans: (*b*) **(IES 1999)**

17. The difference between tensions on the tight side and slack side of a belt drive is 3000 N. If the belt speed is 15 m/s, the transmitted power in kW is

(*a*) 45. (*b*) 22.5.
(*c*) 90. (*d*) 100.

Ans: (*a*) **(IES 1998)**

18. With regard to belt drives with given pulley diameters, centre distance and coefficient of friction between the pulley and the belt materials, which of the statements below are FALSE?

(*a*) A crossed flat belt configuration can transmit more power than an open flat belt configuration.
(*b*) A "V" belt has greater power transmission capacity than an open flat belt.
(*c*) Power transmission is greater when belt tension is higher due to centrifugal affects than the same belt drive when centrifugal affects are neglected.
(*d*) Power transmission is the greatest just before the point of slipping is reached.

Ans: (*c*) **(GATE 1999)**

19. A flat belt open drive transmits 1.5 kW of power. The coefficient of friction between the belt and the drive pulley is 0.25 and the angle of lap is 159º. The drive pulley is rotating in the clockwise direction with the linear belt velocity of 3.75 m/s. Determine the X and Y components of the reaction force on the drive pulley shaft.

(GATE 1999)

20. The ratio of tension on the tight side to that on the slack side in a flat belt drive is

(*a*) Proportional to the product of coefficient of friction and lap angle.
(*b*) an exponential function of the product of coefficient of friction and lap angle.
(*c*) proportional to the lap angle.
(*d*) proportional to the coefficient of friction.

Ans: (*b*) **(GATE 2000)**

21. A flat belt transmits 112 kW at a belt speed of 24 m/s between two pulleys of diameters 250 mm and 400 mm, having a pulley centre distance of 1000 mm. The allowable belt stress is 8.6 MN/m^2, and the belts are available having thickness-to-width ratio of 0.1 and a material density of 1100 kg/m^3. If coefficient of friction between the belt and pulleys is 0.3, determine the minimum required width of the belt. **(GATE 1994)**

Solution: Given data:

Transmitted power = 112 kW

$V = 24$ m/s; $D_1 = 250$ mm, $D_2 = 400$ mm,

$C = 1000$ mm, $f_t = 8.6$ MN/m^2, $\rho = 1100$ kg/m^3,

$\mu = 0.3$, $t/b = 0.1$

$$\text{Now } P = \frac{(F_1 - F_2)V}{1000} \text{ kW}$$

$$\therefore \quad F_1 - F_2 = \frac{122 \times 1000}{24} = 4666.7 \text{ N} \quad ...(1)$$

Angle of wrap,

$$\theta = 180° - 2\sin^{-1}\frac{D_2 - D_1}{2C}$$

$$= 180° - 2\sin^{-1}\frac{400 - 250}{2 \times 1000}$$

$$= 180° - 2 \times 4.3 = 171.4°$$

$$\therefore \quad \frac{F_1}{F_2} = e^{\mu\theta} = e^{0.3 \times 2.99} = 2.452 \qquad ...(2)$$

From (1) and (2)

$$F_1 = 7880.7 \text{ N}, F_2 = 3214 \text{ N}$$

$$\text{Now } b \times t = \frac{F_1}{ft - \rho v^2}$$

$$\therefore \quad 0.1b^2 = \frac{7880.7}{8.6 \times 10^6 - 1100 \times 24^2}$$

$$b^2 = 9.75768 \times 10^{-3} \text{ m}^2$$

$$\therefore \quad b = 0.09878 \text{ m} \cong 9.90 \text{ cm.}$$

22. A belt drive has an angle of wrap of 160° on the smaller pulley. Adding an idler increases the wrap angle to 200°. The slack side tension is the same in both the cases and the centrifugal force is negligible. By what percentage is the torque capacity of the belt drive increased by adding the idler. Use coefficient of friction $\mu = 0.3$

Ans: (35.4%) **(GATE 2001)**

23. Total slip will occur in a belt drive, when

(*a*) Angle of rest is zero.

(*b*) Angle of creep is zero.

(*c*) Angle of rest is greater than angle of creep.

(*d*) Angle of creep is greater than angle of rest.

Ans: (*a*) **(GATE 1997)**

24. The percentage improvement in power capacity of a flat belt drive, when the wrap angle at the driving pulley is increased from 150° to 210° by an idler arrangement for a friction coefficient of 0.3, is

(*a*) 25.21. (*b*) 33.92.

(*c*) 40.17. (*d*) 67.85.

Ans: (*b*) **(GATE 1997)**

CHAPTER 22—GEAR DRIVES

1. In spur gears having involute teeth, the product of circular pitch and diametral pitch is .... .

Ans: (π) **(GATE 1994)**

2. Interference in a pair of gears is avoided, if the addendum circles of both the gears common tangent to the base circles within the points of tangency. True or False?

Ans: (True) **(GATE 1995)**

3. In spur gears, the circle on which the involute is generated is called the

(*a*) Pitch circle. (*b*) Clearance circle.

(*c*) Base circle. (*d*) Addendum circle.

Ans: (*c*) **(GATE 1996)**

4. A 1.5 kW motor is running at 1440 rev/min. It is to be connected to a stirrer at 36 rev/min. The gearing arrangement suitable for this application is

(*a*) Differential gear. (*b*) Helical gear.

(*c*) Spur gear. (*d*) Worm gear.

Ans: (*d*) **(GATE 2000)**

5. In gears, interference takes place when

(*a*) the tip of a tooth of a mating gear digs into the portion between base and root circles.

(*b*) gears do not move smoothly in the absence of lubrication.

(*c*) pitch of the gear is not same.

(*d*) gear teeth are undercut.

Ans: (*a*) **(IES 1993)**

6. Match list I with list II and select the correct answer using codes given below the lists:

List I (Standard tooth forms)	**List II (Advantages or disadvantages)**
A 20° and 25° systems	1. Result in lower loads on bearing.
B 14½° stub-tooth system	2. Broadest at the base and strongest in bending
C 25° Full depth system	3. Obsolete
D 20° Full depth system	4. Standards for new applications

Codes:	A	B	C	D
(*a*)	4	3	2	1
(*b*)	3	1	2	4
(*c*)	3	2	1	4
(*d*)	4	2	3	1

Ans: (*a*) **(IES 1993)**

7. In involute gears, the pressure angle is
(*a*) dependent on the size of teeth.
(*b*) dependent on the size of gears.
(*c*) always constant.
(*d*) always variable.
Ans: (*c*) **(IES 1993)**

8. The lead angle of a worm is **22.5º**. Its helix angle will be
(*a*) 22.5º. (*b*) 45º.
(*c*) 67.5º. (*d*) 90º.
Ans: (*c*) **(IES 1994)**

9. For an involute gear, the ratio, pitch circle radius/base circle radius is (ϕ is the pressure angle)
(*a*) sin ϕ. (*b*) cos ϕ.
(*c*) sec ϕ. (*d*) cosec ϕ.
Ans: (*b*) **(IAS 1994)**

10. Consider the following expressions for spur gears:
1. Addendum = $1/p_d$
2. Clearance = $p_c/20$
3. Centre distance for internal gears = $(T_1 + T_2)/2p_d$
4. Dedendum = $1.157\ p_c/\pi$

Of these expressions
(*a*) 1, 2, 3 and 4 are correct.
(*b*) 1 and 2 are correct.
(*c*) 1, 2, and 3 are correct.
(*d*) 1, 2 and 4 are correct.
Ans: (*a*) **(IES 1995)**

11. Which one of the following is true for involute gears?
(*a*) Interference is inherently absent.
(*b*) Variation in centre distance of shafts increases radial force.
(*c*) A convex flank is always in contact with concave flank.
(*d*) Pressure angle is constant throughout the teeth engagement.
Ans: (*d*) **(IES 1995)**

12. The gear train usually employed in clocks is a
(*a*) reverted gear train
(*b*) simple gear train
(*c*) sun and planet gear
(*d*) differential gear.
Ans: (*a*) **(IES 1995)**

13. Match List I with List II and select the correct answer using the codes given below the lists:

List I	List II
A Interference	1. Arc of approach, arc of recess, circular pitch.
B Dynamic load on tooth	2. Lewis equaton
C Static load	3. Minimum number of teeth of pinion
D Contact ratio	4. Inaccuracies in tooth profile.

Codes:	A	B	C	D
(*a*)	3	4	1	2
(*b*)	1	2	3	4
(*c*)	4	3	2	1
(*d*)	3	4	2	1

Ans: (*d*) **(IES 1995)**

14 An involute pinion and gear are in mesh. If both have the same size of addendum, then there will be an interference between the
(*a*) tip of the geartooth and flank of pinion
(*b*) tip of the pinion and flank of gear.
(*c*) flanks of both gear and pinion.
(*d*) tips of both gear and pinion.
Ans: (*a*) **(IES 1996)**

15. Match List I with List II and select the correct answer

List I	List II
A Helical gears	1. Non-interchangeable
B Herringbone gears	2. Zero axial thrust
C Worm gears	3. Quiet motion
D Hypoid gears	4. Extreme speed reduction

Codes:

	A	B	C	D		A	B	C	D
(*a*)	1	2	3	4	(*b*)	3	2	1	4
(*c*)	3	1	4	2	(*d*)	3	2	4	1

Ans: (*d*) **(IES 1996)**

16. Match List I (Positioning of two shafts) with List II (Possible connection) and select the correct answer using the codes given below the lists:

List I	List II
A Parallel shafts with slight offset	1. Hooke's joint
B Parallel shafts at a reasonable distance.	2. Worm and wheel
C Perpendicular shafts	3. Oldham coupling
D Intersecting shafts	4. Belt and Pulley

Codes:

	A	B	C	D		A	B	C	D
(*a*)	4	3	2	1	(*b*)	4	3	1	2
(*c*)	3	4	1	2	(*d*)	3	4	2	1

Ans: (*d*) **(IES 1997)**

17. When a shaft transmits power through gears, the shaft experiences

(*a*) torsional stresses alone.

(*b*) bending stresses alone.

(*c*) constant bending and varying torsional stresses.

(*d*) varying bending and constant torsional stresses.

Ans: (*d*) **(IES 1997)**

18. The limiting wear load of spur gear is proportional to (where E_p = Young's modulus of pinion material, E_g = Young's modulus of gear material)

(*a*) $(E_p + E_g)^{-1}$. (*b*) $\left(\dfrac{E_p + E_g}{E_p E_g}\right)$.

(*c*) $\left[1 + \dfrac{E_p}{E_g}\right]$. (*d*) $\left[1 + \dfrac{E_g}{E_p}\right]$.

Ans: (*b*) **(IES 1997)**

19. The working surface above the pitch surface of the gear tooth is termed as

(*a*) addendum. (*b*) dedendum.

(*c*) flank. (*d*) face.

Ans: (*d*) **(IES 1998)**

20. Consider the following characteristics:

1. Small interference
2. Strong tooth
3. Low production cost
4. Gear with small number of teeth

Those characteristics which are applicable to Stub 20° involute system would include

(*a*) 1 alone. (*b*) 2, 3 and 4.

(*c*) 1, 2 and 3. (*d*) 1, 2, 3 and 4.

Ans: (*d*) **(IES 1998)**

21. When two shafts are neither parallel nor intersecting, power can be transmitted by using

(*a*) a pair of spur gears.

(*b*) a pair of helical gears.

(*c*) an Oldham's coupling.

(*d*) a pair of spiral gears.

Ans: (*c*) **(IES 1998)**

22. In the formulation of Lewis equation for toothed gearing, it is assumed that tangential load F_t, acts on the

(*a*) pitch point. (*b*) tip of the tooth.

(*c*) root of the tooth.

(*d*) whole face of the tooth.

Ans: (*b*) **(IES 1998)**

23. A certain minimum number of teeth is to be kept for a gear wheel

(*a*) so that the gear is of a good size.

(*b*) for better durability.

(*c*) to avoid interference and undercutting.

(*d*) for better strength.

Ans: (*c*) **(IES 1999)**

24. A single stage spur gear reducer is found to have a pinion with 15 teeth and addendum (tip) diameter of 85 mm. The mating gear has 33 teeth and 175 mm tip diameter. If the gears are supposed to have standard 20° pressure angle, full depth involute teeth, what should be the module and the centre distance between them.

Ans: 5 mm, 120 mm. **(GATE 1997)**

25. Which of the following statements are correct?

1. For constant velocity ratio transmission between two gears, the common normal at the point of contact must always pass through a fixed point on the line joining the centres of rotation of the gears.

2. For involute gears, the pressure angle changes with change in centre distance between gears.
3. The velocity ratio of compound gear pairs depends upon the number of teeth of the input and output gears only.
4. Epicyclic gear trains involve rotation of atleast one gear axis about some other gear axis.

(*a*) 1, 2, 3. (*b*) 1, 3 and 4.
(*c*) 1, 2 and 4. (*d*) 2, 3 and 4.

Ans: (*c*) **(IES 1996)**

26. Consider the following statements:
The form factor of a spur gear tooth depends upon the
1. number of teeth.
2. pressure angle.
3. addendum modification coefficient.
4. circular pitch.

Of these statements
(*a*) 1 and 3 are correct.
(*b*) 2 and 4 are correct.
(*c*) 1, 2 and 3 are correct.
(*d*) 1 and 4 are correct.

Ans: (*c*)

27. The minimum number of teeth on the pinion to operate without interference in standard full height involute teeth gear mechanism with 20º pressure angle is

(*a*) 14. (*b*) 12.
(*c*) 18. (*d*) 32.

Ans: (*c*) **(GATE 2002)**

28. Large speed reduction (greater than 20) in one stage of a gear train are possible through:

(*a*) Spur gearing. (*b*) Worm gearing.
(*c*) Bevel gearing. (*d*) Helical gearing.

Ans: (*b*) **(GATE 2002)**

CHAPTER 23—POWER SECREWS

1. For a self-locking screw, the efficiency of screw must be ... .

Ans: 50%. **(GATE 1994)**

2. Self-locking in a power screw is better achieved by decreasing the helix angle and increasing the coefficient of friction: True/False

Ans: True. **(GATE 1995)**

3. A power screw moves a nut against a force of 10 kN at a speed of 6 mm/s. The screw has a single square thread of 6 mm pitch on a major diameter of 45 mm. Calculate the power of the driving motor. The coefficient of friction at the screw thread is 0.15.

Ans: 0.0118 kW. **(GATE 1996)**

4. The efficiency of a power screw is maximum when the lead angle is

(*a*) $\frac{\pi}{2} - \frac{\phi}{2}$. (*b*) $\frac{\pi}{2} - \phi$.
(*c*) $\pi - \frac{\phi}{2}$. (*d*) $\pi - \phi$.

Where ϕ is the friction angle.

Ans: (*b*) **(GATE 1997)**

5. Consider the following statements regarding power screws:
1. The efficiency of a self locking screw can not be more than 50%.
2. If the friction angle is less than the helix angle of the screw, then the efficiency will be more than 50%.
3. The efficiency of ACME (trapezoidal) thread is less than that of a square thread.

Of these statements
(*a*) 1, 2 and 3 are correct.
(*b*) 2 and 3 are correct.
(*c*) 1 and 3 are correct.
(*d*) 1 and 2 are correct.

Ans: (*a*) **(IES 1994)**

6. The load cup of a screw jack is made separate from the head of the spindle to
(*a*) enhance the load carrying capacity of the jack.
(*b*) reduce the effort needed for lifting the work load.
(*c*) reduce the value of frictional torque required to be countered for lifting the load.

(*d*) prevent the rotation of load being lifted.

Ans: (*d*) **(IES 1995)**

7. While designing a screw in screw jack against buckling failure, the end conditions for the screw are taken as

(*a*) both ends fixed.

(*b*) both ends hinged.

(*c*) one end fixed and other end hinged.

(*d*) one end fixed and the other end free.

Ans: (*d*) **(IES 1995)**

8. To ensure self locking in a screw jack, it is essential that helix angle is

(*a*) larger than friction angle.

(*b*) smaller than friction angle.

(*c*) equal to friction angle.

(*d*) such as to give maximum efficiency in lifting.

Ans: (*b*) **(IES 1996)**

CHAPTER 28—FLYWHEELS

1. If the rotating mass of a rim type flywheel is distributed on another rim type flywheel whose mean radius is half mean radius of the former, the energy stored in the latter at the same speed will be

(*a*) four times the first one.

(*b*) same as the first one.

(*c*) one-fourth of the first one.

(*d*) one and half times the first one.

Ans: (*c*) **(IES 1993)**

2. A flywheel is fitted to the crankshaft of an engine having 'E' amount of indicated work per revolution and permissible limits of coefficients of fluctuation of energy and speed as K_e and K_s respectively. The kinetic energy of the flywheel is then given by

(*a*) $\frac{2K_eE}{K_s}$. (*b*) $\frac{K_eE}{2K_s}$.

(*c*) $\frac{K_eE}{K_s}$. (*d*) $\frac{K_s \cdot E}{2K_e}$.

Ans: (*b*) **(IES 1993)**

3. In the case of a flywheel, the maximum fluctuation of energy is the

(*a*) sum of maximum and minimum energies.

(*b*) difference between the maximum and minimum energies.

(*c*) ratio of maximum and minimum energy.

(*d*) ratio of minimum and maximum energy.

Ans: (*b*) **(IES 1998)**

4. A four-stroke single cylinder, I.C. engine developes 80 kW at 300 rpm. The fluctuation of energy can be assumed to be 0.9 times the energy developed per cycle. If the coefficient of fluctuation of speed is not to exceed 2 per cent and the maximum centrifugal stress in the rim of the flywheel is limited to 6 MN/m^2, estimate the mean diameter of the rim and the moment of inertia of the flywheel. The density of the flywheel material is 7500 kg/m^3.

(GATE 1994).

Solution: Engine speed = 300 rev/min,

$\therefore$ Number of cycles/min = 150

Energy produced = 80 × 1000 × 60 J/min.

$\therefore$ Energy produced per cycle

$$= \frac{80 \times 1000 \times 60}{150} = 32000 \text{ J}$$

$\therefore$ Fluctuation of energy per cycle

$= \Delta E = 0.9 \times 32000 = 28800$ J

Now, $\rho V_1^2 = 6 \times 10^6$

$$\therefore \quad V_1 = \sqrt{\frac{6 \times 10^6}{7500}} = 28.28 \text{ m/s}$$

Now C_s = coefficient of fluctuation of speed

$$\therefore \quad 0.02 = \frac{2(V_1 - V_2)}{V_1 + V_2}$$

$$\therefore \quad V_2 = \frac{99}{100} V_1$$

$\therefore$ Average velocity of the flywheel

$$= \frac{V_2 + V_1}{2} = \frac{199}{200} \cdot V_1 = 28.14 \text{ m/s}$$

$$\therefore \quad \frac{\pi DN}{60} = 28.14$$

∴ Mean diameter of flywheel rim,

$$D = \frac{28.14 \times 60}{\pi \times 300} = 1.79 \text{ m}$$

$$\text{Now } \Delta E = I\omega(\omega_1 - \omega_2) = I\omega^2\left(\frac{\omega_1 - \omega_2}{\omega}\right)$$

$$\text{Now } \omega = \frac{2\pi n}{60} = \frac{2\pi \times 300}{60} = 31.42 \text{ rad./s}$$

$$\therefore \quad 28800 = I \cdot (31.42)^2 \cdot \frac{2}{100}$$

$$\therefore \quad I = \frac{28800 \times 50}{31.42 \times 31.42} = 1458.65 \text{ kgm}^2$$

5. A machine punching 40 mm diameter holes in 30 mm thick plate requires 7 Nm of energy per square mm of shear area. The punch has a stroke of 100 mm and it takes 10 s to complete one cycle. The mean speed of the flywheel is 25 m/s, and the fluctuation of speed should not exceed 3% of the mean speed. Assuming that the motor supplies energy to the machine at a uniform rate, determine the power of the motor and the mass of the flywheel required.

(GATE 1995)

Solution: Energy needed to punch one hole

$= 7 \times \pi \times 40 \times 30 = 26389.38$ Nm

$$\text{Holes punched per minute} = \frac{60}{10} = 6$$

∴ Energy needed per minute

$= 26389.38 \times 6$ Nm

$$\therefore \quad \text{Power of motor} = \frac{26389.38 \times 6}{60}, \text{ J/s}$$

$= 2638.94$ watt

Now stroke of press = 100 mm

Time neded to complete one stroke

$$= \frac{10}{2} = 5\text{s}$$

∴ Time needed to punch hole in 30 mm thick plate,

$$= \frac{5}{100} \times 30 = 1.5 \text{ s}$$

∴ Energy supplied by motor to punch one hole

$= 2638.94 \times 1.5 = 3958.41$ Nm

∴ Energy supplied by flywheel

$= 26389.38 - 3958.41 = 22430.97$ Nm

∴ $\Delta E = 22430.97$

Now $\Delta E = C_s \cdot m \cdot v^2$ ...(eqn. 28.1)

$C_s = 0.03, \quad v = 25$ m/s

$$\therefore \quad m = \frac{22430.97}{0.03 \times 25^2} = 1196.32 \text{ kg.}$$

6. A punch press, fitted with a flywheel having a radius of gyration of 0.5 m, runs at 260 rev./min. The press is capable of punching 600 holes per hour. Each punching operation takes 1.5 s and requires a work of 10,000 Nm. The rating of the electric motor is 2 kW. Determine the mass of flywheel, if the speed of the flywheel is not to drop below 240 rev./min.

(GATE 2000)

Solution: Rating of motor = 2 kW

∴ Energy supplied by motor in 1.5 s

$= 2000 \times 1.5 = 3000$ Nm

∴ ΔE = Energy supplied by flywheel

$= 10000 - 3000 = 7000$ Nm

$$\Delta E = \frac{1}{2} \cdot m \cdot k^2 \cdot (\omega_1^2 - \omega_2^2)$$

$$= m \cdot k^2 \cdot \omega^2 \left(\frac{\omega_1 - \omega_2}{\omega}\right)$$

$$\omega = \frac{2\pi \times 250}{60} = 26.18 \text{ rad/s}$$

$$\frac{\omega_1 - \omega_2}{\omega} = \frac{260 - 240}{250} = 0.08$$

$$\therefore \quad 7000 = m \cdot (0{\cdot}5)^2 \cdot (26.18)^2 \times 0.08$$

$$\therefore \quad m = 510.66 \text{ kg}$$

7. A flywheel of moment of inertia 9.8 kgm^2 fluctuates by 30 rpm for a fluctuation of energy of 19360 J. The mean speed of the flywheel is (rpm)

(*a*) 600. (*b*) 900.

(*c*) 968. (*d*) 2940.

Ans: (*a*) **(GATE 1998)**

Hint: $\Delta E = I \cdot \omega(\omega_1 - \omega_2)$.

8. Which of the following statements is corrrect?

(*a*) Flywheel reduces speed fluctuations during a cycle for a constant load, but flywheel does not control the mean speed of the engine if the load changes.

(*b*) Flywheel does not reduce speed fluctuations during a cycle for a constant load, but flywheel does control the mean speed of the engine if the load changes.

(*c*) Governor controls speed fluctuations during a cycle for a constant load, but governor does not control the mean speed for the engine if the load changes.

(*d*) Governor controls speed fluctuations during a cycle for constant load, and governor also controls the mean speed of the engine if the load changes.

Ans: (*a*) **(GATE 2001)**

IES 2002

1. The dynamic load on a gear is due to
1. inaccuracies of tooth spacing.
2. irregularity in tooth profile.
3. deflection of teeth under load.
4. type of service (intermittent, one shift per day, continuous per day).

Which of the above statements are correct?
(*a*) 1, 2 and 3 (*b*) 2, 3 and 4
(*c*) 1, 3 and 4 (*d*) 1, 2 and 4
Ans: (*a*)

2. Consider the following statements:
A 20° stub tooth system is generally preferred in spur gears as it results in
1. Stronger teeth.
2. lesser number of teeth on the pinion.
3. lesser chances of surface fatigue failure.
4. reduction of interference.

Which of the above statements are correct?
(*a*) 1, 2 and 4 (*b*) 3 and 4
(*c*) 1 and 3 (*d*) 1, 2, 3 and 4.
Ans: (*a*)

3. If the rotating mass of a rim type flywheel is distributed on another rim type flywheel whose mean radius is half the mean radius of the former, then the energy stored in the latter at the same speed will be
(*a*) four times the first one
(*b*) same as the first one
(*c*) one-fourth of the first one
(*d*) two times the first one. *Ans:* (*c*)

4. In a thrust collar bearing, the number of collars having been doubled while maintaining coefficient of friction and axial thrust same. It will result in
(*a*) same friction torque and same bearing pressure.
(*b*) double friction torque and half bearing pressure.
(*c*) double friction torque and same bearing pressure.
(*d*) same friction torque and half bearing pressure. *Ans:* (*c*)

5. The radius of friction circle in a journal bearing is dependent on coefficient of friction and the
(*a*) angular velocity of the journal.
(*b*) radius of the journal.
(*c*) magnitudes of the forces on the journal.
(*d*) journal and bearing clearance. *Ans:* (*b*)

6. A screw jack is said to be self locking, if its efficiency is
(*a*) less than 50%. (*b*) equal to 50%.
(*c*) more than 50%. (*d*) 100%. [*Ans:* (*b*)]

7. If 'P' is the pitch of a square thread, then the depth of thread 'd' is given by
(*a*) 0.5 P. (*b*) P.
(*c*) 1.5 P. (*d*) 2.0 P. [*Ans:* (*a*)]

8. The arm of a radial drilling machine is being raised at a speed of 3.9 m/min by single start square threads of 6 mm pitch and 30 mm diameter. The speed of screw
(*a*) is 650 rpm. (*b*) is 180 rpm.
(*c*) is 130 rpm.
(*d*) can not be determined as the data is insufficient. [Ans: (*a*)]

9. A cotter joint is used when no relative motion is permitted between the rods joined by the cotter. It is capable of transmitting
(*a*) twisting moment.
(*b*) an axial tensile as well as compressive load.
(*c*) the bending moment.
(*d*) only compressive axial load. [*Ans:* (*b*)]

10. In a fillet welded joint, the weakest area of the weld is
(*a*) toe. (*b*) root.
(*c*) throat. (*d*) face. [*Ans:* (*c*)]

11. The length of the belt in the case of cross-belt drive is given in terms of centre distance between pulleys (C), diameters of the pulleys D and d as

(*a*) $2C + \frac{\pi}{2}(D+d) + \frac{(D+d)^2}{4C}$.

(*b*) $2C + \frac{\pi}{2}(D-d) + \frac{(D+d)^2}{4C}$.

(*c*) $2C + \frac{\pi}{2}(D+d) + \frac{(D-d)^2}{4C}$.

(*d*) $2C + \frac{\pi}{2}(D-d) + \frac{(D-d)^2}{4C}$. [*Ans:* (*b*)]

12. The power transmitted by a belt is dependent on the centrifugal effect in the belt. The maximum power can be transmitted when the centrifugal tension is

(*a*) $\frac{1}{3}$ of tension (T_1) on the tight side.

(*b*) $\frac{1}{3}$ of total tension (T_t) on the right side.

(*c*) $\frac{1}{3}$ of the tension (T_2) on the slack side.

(*d*) $\frac{1}{3}$ of sum of tensions T_1 and T_2, i.e., $\frac{1}{3}(T_1 + T_2)$ [*Ans:* (*b*)]

13. A shaft can safely transmit 90 kW while rotating at a given speed. If this shaft is replaced by a shaft of diameter double of the previous one and rotated at half the speed of the previous, the power that can be transmitted by the new shaft is

(*a*) 90 kW. (*b*) 180 kW.
(*c*) 360 kW. (*d*) 720 kW. [*Ans:* (*a*)]

14. A cold rolled steel shaft is designed on the basis of maximum shear stress theory. The principal stresses induced at its critical section are 60 MPa and – 60 MPa respectively. If the yield stress for the shaft material is 360 MPa, the factor of safety of the design is

(*a*) 2. (*b*) 3.
(*c*) 4. (*d*) 6. [Ans: (*b*)]

15. The bolts in a rigid flanged coupling connecting two shafts transmitting power are subjected to

(*a*) shear force and bending moment.
(*b*) axial force.
(*c*) torsion and bending moment.
(*d*) torsion. [*Ans:* (*a*)]

16. An eccentrically loaded riveted joint is shown with 4 rivets at P, Q, R and S. Which of the rivets are the most loaded?

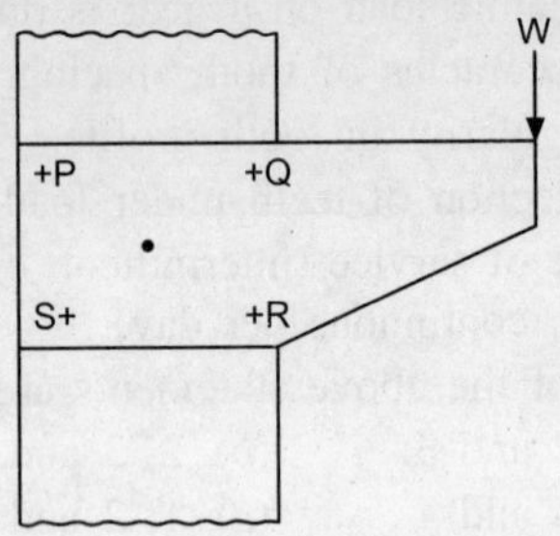

(*a*) P and Q. (*b*) Q and R.
(*c*) R and S. (*d*) S and P. [*Ans:* (*b*)]

17. When a helical compression spring is cut into two equal halves, the stiffness of each of the resulting springs will be

(*a*) unaltered. (*b*) double.
(*c*) one-half. (*d*) one-fourth.
[*Ans:* (*b*)]

18. While calculating the stress induced in a closed coil helical spring, Wahl's factor must be considered to account for

(*a*) the curvature and stress concentration effect.
(*b*) shock loading.
(*c*) poor service conditions.
(*d*) fatigue loading. [*Ans:* (*a*)]

19. Strain energy stored in a body of volume V subjected to uniform stress s is

(*a*) $\frac{sE}{V}$. (*b*) $\frac{sE^2}{V}$.

(*c*) $\frac{sV^2}{E}$. (*d*) $\frac{s^2V}{2E}$. [*Ans:* (*d*)]

20. Match List I (Bearings) with List II (Load type) and select the correct answer using the codes given below the lists:

List I (Bearings)	List II (Load type)
A Hydro-dynamic journal bearings	1. High radial and thrust load combined
B Rectangular hydro-static bearings	2. Radial load only
C Taper roller bearing	3. Thrust load only
D Angular contact ball bearing	4. Medium to Low radial and thrust load combined

Codes:	A	B	C	D
(*a*)	2	3	1	4
(*b*)	4	1	3	2
(*c*)	2	1	3	4
(*d*)	4	3	1	2

[*Ans:* (*a*)]

21. Why are bolts subjected to impact, made longer?

22. Why are antifriction bearings not used at very high speeds?

23. Differentiate clearly between 'shaft' and 'axle'.

24. If two rectangular beams of identical dimensions but made of mild steel and gray cast iron are subjected to same point load at the free end, within elastic limit, which one will deflect more and why?

GATE 2003

1. A wire rope is designated as 6 × 19 standard hoisting. The numbers 6 × 19 represent
(*a*) diameter in millimeter × length in meter.
(*b*) diameter in centimeter × length in meter.
(*c*) number of strands × number of wires in each strand.
(*d*) number of wires in each strand × number of strands. [*Ans:* (*c*)]

2. Maximum shear stress developed on the surface of a solid circular shaft under pure torsion is 240 MPa. If the shaft diameter is doubled, then the maximum shear stress developed corresponding to the same torque is
(*a*) 120 MPa. (*b*) 60 MPa.
(*c*) 30 MPa. (*d*) 15 MPa. [*Ans:* (*c*)]

3. For a certain engine having an average speed of 1200 rpm, a flywheel approximated as a solid disc, is required for keeping the fluctuation of speed within 2% about the average speed. The fluctuation of kinetic energy per cycle is found to be 2 kJ. What is the least possible mass of the flywheel if its diameter is not to exceed 1 m? [*Ans:* 12.665 kg]

4. In a band brake, the ratio of tight side band tension to the tension on slack side is 3. If the angle of overlap of the band on the drum is 180°, the coefficient of friction required between drum and the band is
(*a*) 0.20. (*b*) 0.25.
(*c*) 0.30. (*d*) 0.35. [*Ans:* (*d*)]

IES 2004

1. Which one of the three materials listed here has the highest ultimate tensile strength? What is the approximate Carbon content in each material?
(*i*) Mild steel (*ii*) Cast iron
(*iii*) Spring steel
Ans (*a*) : (*iii*)
(*b*): 0.3%; > 2%; 0.65%

2. State the Von-Mise's theory. Also give the mathematical expression.

3. Why are shafts not made out of Cast iron?

4. When are I-sections preferred in engineering applications? Elaborate your answer.

5. For the three velocity ratios specified below, which one of the following gear drive would be preferable for transmitting power with reductions of 4, 20 and 70?
The drives are: single stage spur gear reduction, single start worm drive and triple start worm drive.

6. In a typical power transmission from an induction motor running at 1440 rev./min. to an output speed of 1 rev./min. which order of reduction of the three drives listed below is preferred? Why? Worm gear reduction, belt reduction and spur gear reduction.

7. An anti friction bearing used in a gear box has a specification 6108. What do the number 6, 1 and 08 signify?

8. Mention three different methods used to prevent an anti-friction bearing from sliding axially in the shaft at the inner race and in a bore at outer race.

9. Which one of the three bearings listed below will you choose to take a heavy axial load on the bearing:

(*i*) self aligning ball bearing
(*ii*) deep groove ball bearing
(*iii*) Thrust bearing

Give reasons for your selection.

10. Consider the following statements regarding improvement of efficiency of worm gear drive:

1. Efficiency can be improved by increasing the spiral angle of worm thread to 45° or more.
2. Efficiency can be improved by adopting proper lubrication.
3. Efficiency can be improved by adopting worm diameter as small as practicable to reduce sliding between worm-threads and wheel teeth.
4. Efficiency can be improved by adopting Convex tooth profile both for worm and wheel

Which of the statements given above are correct?

(*a*) 1, 2 and 3 (*b*) 1, 2 and 4
(*c*) 2, 3 and 4 (*d*) 1, 3 and 4 *Ans:* (*a*)

11. Match list-I (Items in joints) with list-II (Type of failure) and select the correct answer using the codes given below the lists:

List-I (*Items in joints*)	*List-II* (*Type of failure*)
A. Bolts in bolted joints of engine cylinder cover plate.	1. Double transverse shear
B. Cotters in cotter joint	2. Torsional shear
C. Rivets in lap joint	3. Single transverse shear
D. Bolts holding two flanges in a flange coupling	4. Tension

Codes:

	A	*B*	*C*	*D*		*A*	*B*	*C*	*D*
(*a*)	4	1	3	2	(*c*)	3	1	4	2
(*b*)	4	2	3	1	(*d*)	3	2	4	1

Ans: (*a*)

12. Which one of the following statements is correct? While designing a parallel sunk key it is assumed that the distribution of force along the length of the key

(*a*) varies linearly
(*b*) is uniform throughout
(*c*) varies exponentially, being more at the torque input end.
(*d*) varies exponentially, being less at the torque output end.

13. Match list-I (Different systems) with list-II (Associated terminology) and select the correct answer using codes given below the lists:

List-I (*Different systems*)	*List-II* (*Associated terminology*)
A. Riveted joints	1. Nipping
B. Welded joints	2. Angular movement
C. Leaf springs	3. Fullering
D. Knuckle joints	4. Fusion

Codes:

	A	*B*	*C*	*D*		*A*	*B*	*C*	*D*
(*a*)	3	2	1	4	(*c*)	1	4	3	2
(*b*)	1	2	3	4	(*d*)	3	4	1	2

Ans: (*d*)

14. In a cotter joint, the width of the cotter at the centre is 50 mm and its thickness is 12 mm. The load acting on the cotter is 60 kN. What is the shearing stress developed in the cotter?

(*a*) 120 N/mm^2 (*b*) 100 N/mm^2
(*c*) 75 N/mm^2 (*d*) 50 N/mm^2

15. Match list-I (Coupling) with list-II (Purpose) and select the correct answer using the codes given below the lists:

List-I (*Coupling*)	*List-II* (*Purpose*)
A. Muff coupling	1. To transmit power between two parallel shafts
B. Flange coupling	2. To transmit power between two intersecting shafts with flexibility
C. Old ham's coupling	3. For rigid connection between two aligned shafts for power transmission
D. Hooke's joint	4. For flexible connection between two shafts with some misalignment for transmitting power.

Codes:

	A	B	C	D		A	B	C	D
(a)	1	4	3	2	(c)	3	2	1	4
(b)	3	4	2	1	(d)	1	2	3	4

Ans: (*b*)

16. Which one of the following is the correct expression for the torque transmitted by a conical clutch of outer radius R, inner radius r and semi-cone angle α assuming uniform pressure? (where W = total axial load and μ = Coefficient of friction).

(*a*) $\dfrac{\mu W(R+r)}{2\sin\alpha}$ (*b*) $\dfrac{\mu W(R+r)}{3\sin\alpha}$

(*c*) $\dfrac{2\mu W(R^3-r^3)}{3\sin\alpha(R^2-r^2)}$ (*d*) $\dfrac{3\mu W(R^3-r^3)}{4\sin\alpha(R^2-r^2)}$

Ans: (*c*)

17. Which one of the following is the value of helix angle for maximum efficiency of a square threaded screw? ($\phi = \tan^{-1}\mu$)

(*a*) $45° + \phi$ (*b*) $45° - \phi$

(*c*) $45° - \dfrac{\phi}{2}$ (*d*) $45° + \dfrac{\phi}{2}$

18. Which type of gears are used for shaft axes having an offset?

(*a*) Mitre gears (*b*) Spiral bevel gears
(*c*) Hypoid gears (*d*) Zerol gears

Ans: (*c*)

19. Which one of the following statements is correct? Certain minimum number of teeth on the involute pinion are necessary in order to

(*a*) provide an economical design
(*b*) avoid interference
(*c*) reduce noise in operation
(*d*) overcome fatigue failure of the teeth.

Ans: (*b*)

20. Match List -I (Types of gear failure) with List-II (Reasons) and select the correct answer using the codes given below the lists:

List-I **(*Types of gear failure*)**	***List-II*** **(*Reasons*)**
A. Scoring	1. Oil film breakage
B. Pitting	2. Yielding of surface under heavy loads
C. Scuffing	3. Cyclic loads causing high surface stress
D. Plastic flow	4. Insufficient lubrication

Codes:

	A	B	C	D		A	B	C	D
(a)	2	1	3	4	(c)	4	3	1	2
(b)	2	3	1	4	(d)	4	1	3	2

Ans.: (*d*)

21. A helical gear has the active face width equal to b, pitch p and helix angle α. What should be the minimum value of b in order that the contact is maintained across the entire active face of the gear?

(*a*) $p\cos\alpha$ (*b*) $p\sec\alpha$
(*c*) $p\tan\alpha$ (*d*) $p\cot\alpha$ *Ans:* (*d*)

22. Which one of the following statements is correct? Shafts used in heavy duty speed reducers are generally subjected to

(*a*) bending stress only
(*b*) shearing stress only
(*c*) combined bending and shearing stresses
(*d*) bending, shearing and axial thrust simultaneously. *Ans:* (*c*)

23. The diameter of a shaft is increased from 30 mm to 60 mm, all other conditions remaining unchanged. How many times is its torque carrying capacity increased?

(*a*) 2 times (*b*) 4 times
(*c*) 8 times (*d*) 16 times *Ans:* (*c*)

24. It is seen from the curve that there is a minimum value of the co-efficient of friction (μ) for a particular value of the Bearing Characteristic Number denoted by α. What is this value of the Bearing Characteristic Number called?

(*a*) Mc kee Number (*b*) Reynolds Number
(*c*) Bearing Modulus (*d*) Sommerfeld Number

Ans: (*c*)

25. Which one of the following statements is correct? Antifriction bearings are:

(*a*) Sleeve bearings
(*b*) gas lubricated bearings
(*c*) ball and roller bearings
(*d*) journal bearings *Ans:* (*c*)

26. The modulus of elasticity for a material is 200 GN/m^2 and Poisson's ratio is 0.25. What is the modulus of rigidity?

(*a*) 80 GN/m^2 (*b*) 125 GN/m^2
(*c*) 250 GN/m^2 (*c*) 320 GN/m^2

Ans: (*a*)

❑❑❑

INDEX

W

❑❑❑

Win Prizes!

ATTENTION READERS

We request you for your frank assessment regarding some of the aspects of the book given as under:

MACHINE DESIGN

By

Dr. P.C. Sharma **Dr. D.K. Aggarwal**

Please fill up the given space in neat capital letters.

(i) What topics of your syllabus that are important from you examination point of view are not covered in the book ?

....

....

....

....

(ii) What are the chapters wherein the treatment of the subject matter is not systematic or organised or updated ?

....

....

....

....

(iii) Have you come across misprints/mistakes/factual inaccuracies in the book? Please specify the chapters and the page numbers.

....

....

....

....

(iv) What other books on the same subject have you found/heard better that the present book? Please specify in terms of price and quality.

....

....

....

....

(v) Further suggestions and comments for the improvement of the book:

....

....

....

....

(PS: If need be please attach a separate sheet to write your views.)

Other Details :

(i) Who recommended you the book ? (Please tick in the box near the option relevant to you.)

☐ Teachers ☐ Friends ☐ Bookseller

(ii) Name of the recommending teacher, his designation and address.

....

....

....

....

(iii) Name and address of the bookseller you purchased the book form :

....

....

....

....

(iv) Name and address of your institution (Please mention the University or Board, as the case may be)

....

....

....

....

(v) Your name and complete postal address :

....

....

....

....

(vi) Course you are enrolled in :

....

....

....

....

The best assessment will be awarded each month. The award will be in the form of our publications as decided by the Editorial Board.

Please mail the filled up coupon at your earliest to:

S.K. Kataria & Sons®
4885/109, Prakash Mahal, Dr. Subhash Bhargav Lane,
Opposite Delhi Medical Association, Daryaganj, New Delhi-110002
Phone: +91-11-23243489, 43551243; Telefax: +91-11-23243489
e-mail: katariabooks@yahoo.com; katariabook@gmail.com
Website: www.skkatariaandsons.com